ins in flow-through sequence can produce water of very high quality and are available commercially.

Acid Solutions

Prepare the following reagents by cautiously adding required amount of concentrated acid, with mixing, to designated volume of proper type of distilled water. Dilute to 1,000 mL and mix thoroughly.

See Table A for preparation of HCl, H_2SO_4, and HNO_3 solutions.

Alkaline Solutions

a. Stock sodium hydroxide, NaOH, 15*N* (for preparing 6*N*, 1*N*, and 0.1*N* solutions): Cautiously dissolve 625 g solid NaOH in 800 mL distilled water to form 1 L of solution. Remove sodium carbonate precipitate by keeping solution at the boiling point for a few hours in a hot water bath or by letting particles settle for at least 48 hr in an alkali-resistant container (wax-lined or polyethylene) protected from atmospheric CO_2 with a soda lime tube. Use the supernate for preparing dilute solutions listed in Table B.

TABLE B: PREPARATION OF UNIFORM SODIUM HYDROXIDE SOLUTIONS

Normality of NaOH Solution	Required Weight of NaOH to Prepare 1,000 mL of Solution *g*	Required Volume of 15*N* NaOH to Prepare 1,000 mL of Solution *mL*
6	240	400
1	40	67
0.1	4	6.7

Alternatively, prepare dilute solutions by dissolving weight of solid NaOH indicated in Table B in CO_2-free distilled water and diluting to 1,000 mL.

Store NaOH solutions in polyethylene (rigid, heavy-type) bottles with polyethylene screw caps, paraffin-coated bottles with rubber or neoprene stoppers, or borosilicate glass bottles with rubber or neoprene stoppers. Check solutions periodically. Protect them by attaching a tube of CO_2-absorbing granular material such as soda lime, Ascarite,* Caroxite,† or equivalent. Use at least 70 cm of rubber tubing to minimize vapor diffusion from bottle. Replace absorption tube before it becomes exhausted. Withdraw solution by a siphon to avoid opening bottle.

b. Ammonium hydroxide solutions, NH_4OH: Prepare 5*N*, 3*N*, and 0.2*N* NH_4OH solutions by diluting 333 mL, 200 mL, and 13 mL, respectively, of the concentrated reagent (sp gr 0.90, 29.0%, 15*N*) to 1,000 mL with distilled water.

Indicator Solutions

a. Phenolphthalein indicator solution: Use either the aqueous (1) or alcoholic (2) solution.

1) Dissolve 5 g phenolphthalein disodium salt in distilled water and dilute to 1 L.

2) Dissolve 5 g phenolphthalein in 500 mL 95% ethyl or isopropyl alcohol and add 500 mL distilled water.

If necessary, add 0.02*N* NaOH dropwise until a faint pink color appears in solution 1) or 2).

b. Methyl orange indicator solution: Dissolve 500 mg methyl orange powder in distilled water and dilute to 1 L.

STANDARD METHODS

For the Examination of Water and Wastewater

STANDARD METHODS

For the Examination of Water and Wastewater

FIFTEENTH EDITION

Prepared and published jointly by:

American Public Health Association
American Water Works Association
Water Pollution Control Federation

Publication Office:

American Public Health Association
1015 Fifteenth Street NW
Washington, DC 20005

30M2/81
The Library of Congress has cataloged this work as follows:
American Public Health Association.
Standard methods for examination of water and wastewater.
Library of Congress Catalog Number: 55-1979
International Standard Book Number: 0-87553-091-5

Printed and bound in the United States of America
Typography: Byrd PrePress, Springfield, VA
Set in: Times New Roman
Text and Binding: R. R. Donnelley & Sons Company

Cover Design: Donya Melanson Assoc., Boston, MA

PREFACE TO THE FIFTEENTH EDITION

The Fourteenth and Earlier Editions

The first edition of *Standard Methods* was published in 1905. Each subsequent edition presented significant improvements of methodology and enlarged its scope to include technics suitable for examination of many types of samples encountered in the assessment and control of water quality and water pollution.

A brief history of *Standard Methods* is of interest because of its contemporary relevance. A movement for "securing the adoption of more uniform and efficient methods of water analysis" led in the 1880's to the organization of a special committee of the Chemical Section of the American Association for the Advancement of Science. A report of this committee, published in 1889, was entitled: A Method, in Part, for the Sanitary Examination of Water, and for the Statement of Results, Offered for General Adoption.* Five topics were covered: (1) "free" and "albuminoid" ammonia; (2) oxygen-consuming capacity; (3) total nitrogen as nitrates and nitrites; (4) nitrogen as nitrites; and (5) statement of results.

In 1895, members of the American Public Health Association, recognizing the need for standard methods in the bacteriological examination of water, sponsored a convention of bacteriologists to discuss the problem. As a result, an APHA committee was appointed "to draw up procedures for the study of bacteria in a uniform manner and with special references to the differentiation of species." Submitted in 1897,† the procedures found wide acceptance.

In 1899, APHA appointed a Committee on Standard Methods of Water Analysis, charged with the extension of standard procedures to all methods involved in the analysis of water. The committee report, published in 1905, constituted the first edition of *Standard Methods* (then entitled *Standard Methods of Water Analysis*). Physical, chemical, microscopic, and bacteriological methods of water examination were included. In its letter of transmittal, the Committee stated:

> The methods of analysis presented in this report as "Standard Methods" are believed to represent the best current practice of American water analysts, and to be generally applicable in connection with the ordinary problems of water purification, sewage disposal and sanitary investigations. Analysts working on widely different problems manifestly cannot use methods which are identical, and special problems obviously require the methods best adapted to them; but, while recognizing these facts, it yet remains true that sound progress in analytical work will advance in proportion to the general adoption of methods which are reliable, uniform and adequate.
>
> It is said by some that standard methods within the field of applied science tend to stifle investigations and that they retard true progress. If such standards are used in the proper spirit, this ought not to be so. The Committee strongly desires that every effort shall be continued to improve the techniques of water analysis and especially to compare current methods with those herein recommended, where different, so that the results obtained may be still more accurate and reliable than they are at present.

Revised and enlarged editions were published by APHA under the title *Standard Methods of Water Analysis* in 1912 (Second Edition), 1917 (Third), 1920 (Fourth),

---

**J. Anal. Chem.* 3:398 (1889).

†*Proc. Amer. Pub. Health Ass.* 23:56 (1897).

and 1923 (Fifth). In 1925, the American Water Works Association joined APHA in publishing the Sixth Edition, which had the broader title, *Standard Methods of the Examination of Water and Sewage*. Joint publication was continued in the Seventh Edition, dated 1933.

In 1935, the Water Pollution Control Federation (then the Federation of Sewage Works Associations) issued a committee report, "Standard Methods of Sewage Analysis."‡ With minor modifications, these methods were incorporated into the Eighth Edition (1936) of *Standard Methods*, which was thus the first to provide methods for the examination of "sewages, effluents, industrial wastes, grossly polluted waters, sludges, and muds." The Ninth Edition, appearing in 1946, likewise contained these methods, and in the following year the Federation became a full-fledged publishing partner. Since 1947, the work of the *Standard Methods* committees of the three associations—APHA, AWWA, and WPCF—has been coordinated by a Joint Editorial Board, on which all three are represented.

The Tenth Edition (1955) included methods specific for the examination of industrial wastewaters; this was reflected by a new title: *Standard Methods for the Examination of Water, Sewage and Industrial Wastes*. In order to describe more accurately and concisely the contents of the Eleventh Edition (1960), the title was shortened to *Standard Methods for the Examination of Water and Wastewater*. It remained unchanged for the Twelfth Edition (1965), the Thirteenth Edition (1971), and the Fourteenth Edition (1976).

In the Fourteenth Edition, the separation of test methods for water from those for wastewater was discontinued. All methods for a given component or characteristic appeared under a single heading. The coordination of methods was reflected in the revised numbering system. The major divisions of the Fourteenth Edition were as follows:

Part 100—General Introduction
Part 200—Physical Examination
Part 300—Determination of Metals
Part 400—Determination of Inorganic Nonmetallic Constituents
Part 500—Determination of Organic Constituents
Part 600—Automated Laboratory Analyses
Part 700—Examination of Water and Wastewater for Radioactivity
Part 800—Bioassay Methods for Aquatic Organisms
Part 900—Microbiological Examination of Water
Part 1000—Biological Examination of Water

The Fifteenth Edition

With minor differences, the organization of the Fourteenth Edition has been retained. Numerous changes, revisions, and improvements in methods have been made and the most noteworthy are mentioned in this preface. Two major policy

‡*Sewage Works J.* 7:444 (1935).

decisions have been made by the Joint Editorial Board and these were implemented as far as possible. First, the International System of Units (SI) has been adopted. Except where prevailing field systems or practices require English units, these have been replaced by SI units. The most obvious changes consistent with this decision are the use of L as the abbreviation of liter (instead of l) and the use of pascal (Pa) or kilopascal (kPa) for pressure. For the Sixteenth Edition it is planned that this conversion will be completed. Second, it was decided to eliminate the use of trade names or proprietary materials, thereby avoiding possible claims regarding restraint of trade or commercial favoritism. Wherever generic substance names were available, these have been used. Terms such as borosilicate glass, polytetrafluorethylene (TFE), etc., have been substituted for trademarks or copyrighted names. While this usage may be unfamiliar or awkward, the Joint Editorial Board hopes that users of this book will adjust to the changes without difficulty. Again, the conversion will be completed in the Sixteenth Edition.

The Fifteenth Edition retains the General Introduction (Part 100), containing important information on proper execution of procedures. Every user of this manual must study both the General Introduction and the introductions to all other parts. Each introduction discusses vital matters of general application within the specific subject area to minimize repetition in the succeeding text. Successful analysis rests on close adherence to the introductory recommendations and cautions. Before undertaking an analysis, read and understand the complete discussion of each procedure, including method selection, sampling and sample storage, and interferences.

Two subjects have received increased attention. The first of these is laboratory safety, which is stressed under the individual tests. New or proposed OSHA (Occupational Safety and Health Administration) regulations probably will require still greater concern because laboratories are in fact places of employment where hazardous or toxic materials may be used. A brief section (108) summarizes some of these concerns. For the Sixteenth Edition it is planned to include an integrated, more comprehensive section on laboratory safety. The other area of concern is quality assurance, for which regulatory and legal actions have created special needs. Sections 104 (for chemical analyses), 701 (for radiological analyses), 801 (for bioassays), and a totally new Section 902 (for bacteriological analyses) deal with these problems. They should be studied carefully to insure that laboratory results correctly and reliably reflect sample composition.

Part 200 (physical examination) includes a new section on floatables with tentative procedures for particulate floatables and floatable oil and grease. Extensive modifications have been made in the sections dealing with color and conductivity.

Part 300 (metals) has been revised to emphasize atomic absorption spectrophotometry as the method of choice in the analysis of most metals. The determination of arsenic has been moved into this part to indicate correctly the metallic nature of the element. The borohydride reduction method for arsenic and selenium has been added. A general discussion of the determination of microquantities of metals by electrothermal atomic absorption spectrophotometry (flameless) now is included. The section on polarographic methods has been deleted; however, it is planned to include updated material, emphasizing anodic stripping voltammetry, in the Sixteenth Edition. Hardness has been redefined as the calcium carbonate equivalent of only calcium and magnesium, thus making the definition consonant with the analytical method. Both the gravimetric procedure for sodium and the colorimetric method for potassium have been eliminated, leaving atomic absorption and flame photometry methods for these metals. In the zincon method for zinc, chloral hydrate, a scheduled drug that is difficult to obtain, has been replaced by cyclohexanone.

In Part 400 (inorganic nonmetals) a DPD method for chlorine dioxide has been added and the section on pH has been rewritten completely with the definition modified to conform to national and international practice. Consistent with the goal of presenting together all the analytical methods for each constituent, the automated methods of Part 600 of the Fourteenth Edition have been redistributed under the individual constituents. Part 600 (automated methods) has been reduced to general information on the approach and use of automated analytical equipment.

Extensive changes have been made in the determination of some organic constituents (Part 500) and an entirely new section on halogenated methane and ethane derivatives, relating to the current concerns about trihalomethanes and other synthetic organics in drinking water, has been added. The section on total organic carbon has been expanded to include instruments other than those based on combustion-infrared analysis. In the BOD determination the immediate dissolved oxygen demand has been eliminated, changes have been made in the preparation and evaluation of dilution water, the reference data on glucose-glutamic acid have been updated, changes have been made in permissible sample storage time, and a procedure for nitrification control has been added. Procedures for pesticides have been modified by deleting the cholinesterase inhibition test for organophosphates and carbamates.

Part 700 (radioactivity) has been reedited and new sections for radium 228 and radioactive cesium and iodine have been added. In standardizing counters for alpha activity use of plutonium 239 or americium 241 has been added while for beta activity cesium 137 is the sole standard.

Part 800 (bioassays) has been severely edited to eliminate repetition and improve clarity but there have been no fundamental changes in the methodologies.

In Part 900 (microbiology) numerous minor changes have been made, including deletion of lactose broth medium in the presumptive test for coliform bacteria, addition of A-1 medium in a single-step test for fecal coliforms, and substitution of PSE agar for ethyl violet broth as the confirmatory medium for fecal streptococci. The sections on pathogenic organisms have been reworked and a procedure for *Giardia* has been added. The section on nematodes has been expanded by inclusion of a detailed taxonomic key. Two entirely new sections have been added; these deal with special procedures for the recovery of environmentally stressed organisms and rapid microbiological technics that attempt to obtain results more closely related to real-time information. As already mentioned, a completely rewritten and considerably expanded section on quality assurance has been included.

Part 1000 (biological examinations) has been revised and updated with considerable expansion of the section on aquatic plants.

Selection and Approval of Methods

For each new edition, both the technical criteria for selection of methods and the formal procedures for their approval and inclusion are reviewed critically. In regard to the approval procedures, it is considered particularly important to assure that the methods presented have been reviewed and are supported by the largest number of qualified persons, so that they may represent a true consensus of expert opinion.

For the Fourteenth Edition a Joint Task Group was established for each test. This scheme has been continued for the Fifteenth Edition. Appointment of an individual to a Joint Task Group generally was based on the expressed interest or recognized expertise of the individual. The effort in every case was to assemble a group having maximum available expertise in the test methods of concern.

Each Joint Task Group was charged with reviewing the pertinent methods in the Fourteenth Edition along with other methods from the literature, recommending the methods to be included in the Fifteenth Edition, and presenting those methods in the form of a proposed section manuscript. Subsequently, each section manuscript was ratified by vote of the membership of the pertinent committees of the societies. Every negative vote and every comment submitted in the balloting was reviewed by the Joint Editorial Board. All relevant suggestions were referred to the appropriate Joint Task Groups for resolution. When negative votes on the first ballot could not be resolved by the Joint Task Group, or the Joint Editorial Board, the section was reballoted among all who voted (affirmatively or negatively) on the original ballot. Only a few positions could not be resolved in this manner and the Joint Editorial Board made the final decision.

The methods presented here, as in previous editions, are believed to be the best available and generally accepted procedures for the analysis of water, wastewaters, and related materials. They represent the recommendations of specialists, ratified by a large number of analysts and others of more general expertise, and as such are truly consensus standards, offering a valid and recognized basis for control and evaluation.

The technical criteria for selection of methods were applied by the Joint Task Groups and by the individuals reviewing their recommendations, with the Joint Editorial Board providing only general guidelines. In addition to the classical concepts of precision, accuracy, and minimum detectable concentration, selection of a method also must recognize such considerations as the time required to obtain a result, needs for specialized equipment and for special training of the analyst, and other factors related to the cost of the analysis and the feasibility of its widespread use.

Status of Methods

All methods in the Fifteenth Edition are "standard" unless designated "tentative". No other categories are used. Methods with "standard" status have been studied extensively and accepted as applicable within the limits of sensitivity, precision, and accuracy given. "Tentative" methods are those still under investigation that have not yet been evaluated fully or are not considered sufficiently tested at present to be designated "standard".*

Technical progress makes advisable the establishment of a program to keep *Standard Methods* abreast of advances in research and general practice. The Joint Editorial Board has developed the following procedure for effecting interim changes in methods between editions:

1. Any method given "tentative" status in the current edition may be elevated to "standard" by action of the Joint Editorial Board, on the basis of adequate published data supporting such a change as submitted to the Board by the appropriate Joint Task Group. Notification of such a change in status shall be accomplished by publication in the official journals of the three associations sponsoring *Standard Methods*.

2. No method having "standard" status may be abandoned or reduced to "tentative" status during the interval between editions.

---

*The Committee on Laboratory Standards and Practices (CLaSP) of APHA adopted a somewhat different methods classification scheme, which is presented here for consideration and comment by the reader.

Class O—a method or procedure that has been subjected to a thorough evaluation, has been widely used, and through wide use has demonstrated its utility by extensive application, but has not been formally, collaboratively tested. This classification will include methods that are referred to as standard methods in the current APHA publications; essentially it is a grandfather clause.

Class A—a method or procedure that has been subjected to a thorough evaluation, has demonstrated its applicability for a specific purpose on the basis of extensive use, and has been successfully, collaboratively tested.

Class B—a method that has been used successfully in research or other disciplines, has been devised or modified explicitly for routine examination of specimens, has had limited evaluation, and has not been tested collaboratively.

Class C—(1) a new unproved or suggested method not previously used but one that has been proposed by recognized laboratory workers as useful or gives promise of being suitable; (2) a method that previously has been placed in Classes, O, A, or B but which, through technological advances or significant change in numerical level of acceptable exposure or other circumstances, has been rendered not suitable for its intended purpose and presumably has been superceded by a method of a higher classification. In essence C-1 includes proposed new methods and C-2 includes methods no longer recommended.

Except for Class O, the scheme allows for a progression from Class C to Class A, thereby permitting a new unproven method or procedure to be made available pending further evaluation (Class C). As the procedure is tested and evaluated it progresses to Class B and, after thorough evaluation and a successful collaborative test, it becomes a Class A method.

The scheme is most readily applied to manuals of methods that are periodically reissued and the additions, deletions and changes are a challenge to the user. The scheme could foreseeably be applied to a wider range of publications other than manuals.

Note, however, that "standard" methods as defined herein are comparable to CLaSP's Class O or A while "tentative" methods are comparable to Class B or (possibly) C.

3. A new method may be adopted as "tentative" or "standard" by the Joint Editorial Board between editions, such action being based on adequate published data as submitted by the Joint Task Group concerned. Upon adoption, the details of the method, together with a resume of the supporting data, must be published in the official journal of any one of the three sponsoring associations, and reprints shall be made available at a nominal charge. Notice of such publication and of the availability of reprints shall appear in the official journals of the other two sponsors.

Even more important to maintaining the current status of these standards is the intention of the sponsors and the Joint Editorial Board that subsequent editions will appear regularly at reasonably short intervals. Reader comments and questions concerning this manual should be addressed to: STANDARD METHODS, American Water Works Assoc., 6666 West Quincy Avenue, Denver CO 80235.

Acknowledgments

For the major portion of the work in preparing and revising the methods in the Fourteenth Edition, the Joint Editorial Board gives full credit to the Standard Methods Committees of the American Water Works Association and of the Water Pollution Control Federation, and to the Subcommittee on Standard Methods for the Examination of Water and Wastewater and the Committee on Laboratory Standards and Practices of the American Public Health Association. Members of these committees chair and serve as members of the Joint Task Groups. They were assisted often by advisors, not formally members of the committees, and in many cases not members of the sponsoring societies. To the advisors, special gratitude is extended in recognition of their efforts. A list of the committee members and advisors follows these pages.

The Joint Editorial Board expresses its appreciation to William H. McBeath, M.D., Executive Director, American Public Health Association, to Eric F. Johnson and his successor, David B. Preston, Executive Director, American Water Works Association, and to Robert A. Canham, Executive Director, Water Pollution Control Federation, for their continuous cooperation and helpful advice. Paul A. Schulte, Deputy Executive Director, American Water Works Association, acted as secretary to the Joint Editorial Board for this edition and provided an endless variety of helpful services as well as useful advice. In the final stages of compiling the manuscript, obtaining approval of every section, and readying it for printing, Mr. Schulte and the Joint Editorial Board were assisted most ably by Jon De Boer, Water Quality Engineer, American Water Works Association. Special recognition for her valuable services is due to Mary Ann H. Franson, Managing Editor of the Fifteenth Edition, who has discharged most efficiently the extensive and detailed responsibilities on which a complete volume depends.

Joint Editorial Board
Arnold E. Greenberg, American Public Health Association (Chairman)
Joseph J. Connors, American Water Works Association
David Jenkins, Water Pollution Control Federation

JOINT EDITORIAL BOARD

COMMITTEES FOR THE FIFTEENTH EDITION

Joint Task Group Chairmen

The following served as Chairmen for the Joint Task Group(s) developing the section(s) following their names.

RODGER B. BAIRD, *428*
ROBERT A. BAKER, *207, 211*
ROBERT B. BAKER, *312*
ERVIN BELLACK, *411*
THOMAS BELLAR, *514*
I. BOB BLUMENTHAL, *216*
ROBERT L. BOOTH, *212, 600*
ROBERT H. BORDNER, *902*
FRANCIS T. BREZEUSKI, *910*
PATRICK L. BREZONIK, *415, 417*
ROBERT L. BUNCH, *512*
SHIH L. CHANG, *917*
KENNETH A. CHEN, *301, 304, 308, 311*
LARRY H. K. CHENG, *419*
NORMAN A. CLARKE, *913*
NICHOLAS N. CLESCERI, *422*
WILLIAM B. COOKE, *915*
SAMUAL D. FAUST, *510*
PETER E. GAFFNEY, *508*
EDWIN E. GELDREICH, *908, 909, 911, 912, 919, 920*
WALTER GINSBURG, *907*
MARVIN L. GRANSTROM, *409*
PIERCE HAYWARD, *326, 328*
JOHN D. HEM, *325*
DON E. HENLEY, *802, 803*
THOMAS B. HOOVER, *402, 403, 404*
DAVID L. HOWARD, *317*
BERNARD HUTCHINS, *324*
BILLY G. ISOM, *1005*
ROBERT A. JUNG, *405, 424*
O. D. KEATON, *327*
MICHAEL A. KESSICK, *324*
RILEY N. KINMAN, *415*
YEN C. KWAN, *421*
LESLIE E. LANCY, *410*
JAMES J. LICHTENBERG, *509*
FREDERIC J. LINNIG, *513*
WARREN LITSKY, *914*
LINDA W. LITTLE, *204*
LESTER J. LOUDEN, *315, 322*
ANTHONY G. MACEJUNAS, *325*
G. N. MCDERMOTT, *507*
DANIEL A. MCLEAN, *504*
JAMES W. MULLINS, *710*
REMO NAVONE, *321, 324, 328*
RICHARD J. NOGAJ, *208*
JOHN A. OSBORNE, *1004*
DENNIS H. PATTERSON, *328*
HAROLD E. PEARSON, *211*
WESLEY O. PIPES, JR., *918*
FREDERICK G. POHLAND, *502, 511*
RICHARD D. POMEROY, *425*
DONALD B. PORCELLA, *800, 804, 805, 806, 807, 808, 809, 810*
T. B. S. PRAKASAM, *415, 418*
HUGH D. PUTNAM, *1003*
OLSEN J. ROGERS, *307*
HARVEY M. ROSEN, *420*
JOHN R. ROSSUM, *205, 502*
ROBERT S. SAFFERMAN, *916*
F. MICHAEL SAUNDERS, *211*
JAN W. SCHERFIG, *206*
EDWARD J. SHERVIN, *327*
KENNETH J. SHULL, *316*

LOUIS SNYDER, *326*
FRANK W. SOLLO, JR., *203*
ROBERT G. SPICHER, *325*
ALAN A. STEVENS, *506*
TERRY SURLES, *326, 327*
CORNELIUS I. WEBER, *Part 1000*
RICHARD L. WEITZEL, *1002*
GEORGE P. WHITTLE, *407, 409, 413*
ROBERT T. WILLIAMS, *505*

Standard Methods Committee Joint Task Group Members

The following served as Group Members of the Joint Task Group(s) developing the section(s) following their names.

JOHN C. ADAMS
V. DEAN ADAMS, *424, 506, 801*
FRANKLIN J. AGARDY
DONALD G. AHEARN, *915*
RICHARD ALEXANDER, *508*
HERBERT E. ALLEN, *203, 407*
MARTIN ALLEN, *920*
MORRIS L. ALLEN, *908, 911, 920*
OSMAN ALY, *510*
CHARLES W. AMAN, *207, 211, 510*
CHARLES W. AMELOTTI, *312, 411*
BERTIL G. ANDERSON, *801*
JIM ANDERSON, *316*
JOHN ANDERSON, *316*
LARS W. J. ANDERSON, *1004*
NORVAL E. ANDERSON, *208*
RICHARD L. ANDERSON, *901*
BRIAN ARMITAGE, *1003*
DAVID E. ARMSTRONG, *424*
JOHN A. ARRINGTON, II., *214*
J. W. ARTHUR, *800*
ROBERT M. ARTHUR, *213, 507*
DONALD B. AULENBACH, *418, 428*
DAVID S. BAILEY, *801*
RODGER B. BAIRD, *316, 323, 428, 509*
ROBERT A. BAKER, *207, 211, 504, 510, 512*
ROBERT B. BAKER, *404*
ROBERT J. BAKER, *410*
EDMOND J. BARATTA, *701–710*
JOHN BATES, *1005*
GERALD BERG, *412*
WILLIAM L. BERK, *208*
KENNETH E. BIESINGER, *806, 807, 808, 809*
FRANCIS B. BIRKNER, *320*
RICHARD L. BIXBY, *214, 424*
CURTIS BLAIR, *314, 413, 908, 909*
R. O. BLOSSER
I. BOB BLUMENTHAL, *216*
FRED H. BOATWRIGHT
ROBERT L. BOOTH, *214, 412, 502, 600*
ROBERT H. BORDNER, *901, 902, 903, 904, 905, 906, 920*
DENNIS L. BORTON
WILLIAM H. BOUMA
WILLIAM C. BOYLE
LLOYD W. BRACEWELL, *206*
WESLEY L. BRADFORD, *418, 419, 420, 421, 425*
ROBERT X. BRAZEAU, *205, 317, 325, 326*
FRANCIS T. BREZENSKI, *901, 903, 904, 905, 906, 910, 914*
FRANK J. BAUMANN, *427*
ROBERT J. BECKER, *409, 419*
ERVIN BELLACK, *415*
THOMAS BELLAR, *514*
DANIEL F. BENDER, *426*
LARRY D. BENEFIELD, *213, 511*
FRED BENFIELD, *1005*
RICK BENNETT, *319*

Gary W. Holcombe, *800*
Albert C. Holler, *315, 324*
Wayne C. Holley
Joseph E. Holliday, Jr., *212, 319, 325*
Thomas B. Hoover, *402, 403, 407*
Stanley O. Hopkins, *415, 424*
Donald L. Hoven, *212, 311, 319*
David L. Howard, *307, 323, 509*
Lyman H. Howe, *418*
Chin Pao Huang
Duane E. Humble, *911, 913*
John S. Hunter, III, *208*
Joseph V. Hunter, *502, 507, 508*
Emanuel F. Hurwitz, *208, 502, 507, 508*
Bernard Hutchins, *324*
Joseph A. Hutchinson, *710*
Robert Hutton, *909*
Norman J. Hyatt, *212, 325*
A. Y. Hyndshaw, *207, 211, 505*
Billy G. Isom, *1005*
Herbert W. Jackson, *801*
Joseph F. Jadlocki, Jr., *801*
Walter Jakubowski, *913*
Victor J. Janzer, *701–710*
Stephen R. Jenkins, *402, 403, 407*
Everett A. Jenne
Robert R. Jennings, *418, 419, 420, 421, 425, 507, 801*
B. G. Johnson
B. T. Johnson, *800*
J. Donald Johnson, *409, 415, 423*
Patricia G. Johnson, *317, 319*
Robert W. Johnson, *413, 426, 509*
Waynon W. Johnson, *801*
P. Jokiel, *800*
Bernard R. Jones, *800*
P. B. Joshipura, *315, 316*
Robert A. Jung, *209, 408, 428*
Lawrence J. Kamphake, *419, 420, 423*
Larry Kanipe, *701–710*
Ray Kaplan, *314*
E. N. Karabaic, *413*
Irwin Katz, *910*
Fred W. Kawahara, *207, 211, 506*
O. D. Keaton, *323, 324, 327*
Floyd D. Kefford, *600*
Michael A. Keirn, *1003*
John J. Kelly, *404*
Harriet Kennedy, *914*
Michael A. Kessick, *319*
Stephen R. Kin, *413*
Paul H. King, *204, 509*
Troy E. King, *323, 404, 505*
Riley N. Kinman, *414, 415*
William H. Kirchhoff
William L. Klein, *810, 909*
Donald Klemn, *1005*
Max Kloster, *212, 427*
Kenneth F. Knowlton
Shigeru Kobayashi, *424*
Frederick C. Kopfler, *509*
F. R. Kopperdahl, *800*
R. E. Kreider, *801*
Peter A. Krenkel
Ed Kreusch, *428*
Robert C. Kroner, *507, 508, 510*
Yen C. Kwan, *424*
Aubrey Lafargue, *204, 317, 505*
Leslie E. Lancy, *413*
Melvin O. Lamble, *1005*
Russell W. Lane, *203, 205*
Mary C. Lantieri, *205, 316, 323*
Guy R. Lanza, *801*
Alexander Lapteff, *409*
T. E. Larson, *405*
David R. S. Lean, *424*
Hubert Lechevalier, *916*
Jean Lecocoq, *505*
G. Fred Lee, *417, 418, 419, 420, 424*
Harry Leibee, *504*
Armond E. Lemke, *800*
Ronald Lewis, *916, 918*
James J. Lichtenberg, *506, 509*
Maxim Lieber, *312*
Robert Lieberman, *701–710*
Christopher Lind, *306*

Frederic J. Linnig, *513*
K. D. Linstedt
Warren Litsky, *901, 903, 904, 905, 906, 910, 914, 920*
Linda W. Little, *204, 800*
Larry Lobring, *412*
Haines B. Lockhart, Jr., *801*
R. Addis Lockwood
J. J. Long
Maxine Long, *907*
Karl E. Longley, *213, 909, 911, 912*
Lester L. Louden, *315, 316, 317, 318, 319, 322, 324, 408, 428*
James C. S. Lu, *300*
Richard G. Luthy, *412, 502*
Anthony G. Macejunas, *320*
Claude Z. Maehler, *509*
Gerald L. Mahon, *415*
Thomas E. Maloney, *801, 802, 810*
Sursit S. Mar, *424*
L. L. Marking, *800*
John R. Marsden
Theodore Martin, *320*
Charles Mashni, *317*
G. Maske, *908*
Bill Mason, *1005*
Paul J. Mason, *204, 320, 507*
Willy J. Masschelein, *422*
Owen B. Mathre, *413*
Vincent R. Mattson, *800*
Foster L. Mayer, *810*
Tilden F. McCommas, *205, 216*
J. Howard McCormick, *801*
Ross F. McCurdy, *600*
G. N. McDermott, *502, 507*
Earl F. McFarren, *506, 509, 600*
G. A. McFeters, *909, 914, 920*
Marvin J. McGarity, *205, 212, 423*
James J. McKeown, *204, 205, 212, 422*
J. M. McKim, *800*
Daniel A. McLean, *418, 419, 420, 421, 425, 504*
D. J. McLeay, *800*
Lilia McMillan, *916, 918*
Dale McMurtrey, *506, 509*
Joseph L. Melnick, *913*
Basil W. Mercer, Jr., *419, 420*
Theodore Metcalf, *913*
J. Middlebrooks, *803*
Donald Miller, *402, 403, 407, 424*
James A. Miller, *908*
William E. Miller, *600, 801*
Jimmie G. Minchey, *327*
Glenn E. Moore, *317*
Robert L. Morris, *307, 506, 509*
Sumner Morrison, *907*
James Morrow, *409*
Patricia W. Mueller, *320, 321, 324, 326, 324*
James W. Mullins, *700*
J. W. Murphrey, *319, 320*
R. Bruce Murray, *800*
Alan I. Mytelka, *502, 507*
Carl Nabel, *423*
J. Nagano
S. Nagano
Walter Nahulak, *204, 325*
Harry D. Nash
Remo Navone, *315, 316, 318, 319*
Alan V. Nebeker, *800*
N. Newton, *216*
Richard J. Nogaj, *208*
John T. O'Connor
William A. O'Connor, *420*
Viola K. Ohr, *316, 317, 319, 418, 505, 507*
Donovan W. Oldridge, *409, 418, 419*
Betty H. Olson, *919*
Donald T. Olson, *800*
John A. Osborne
A. T. Palin, *408, 410, 413*
J. M. Pappenhagen
Thomas R. Parr, *909, 911*
R. Parrish, *800*
Robert A. Paterson, *915*

Dennis H. Patterson, *323, 413, 514*
Maria Pavlova, *910*
Wayne L. Paulson, *208*
Harry M. Pawlowski, *207, 211, 212, 418, 419, 420, 421*
David A. Payne, *409, 418*
Stephen Pearlman, *505*
Harold E. Pearson, *209, 506*
R. K. Peddicord, *800*
Richard Peoples, *205, 310, 312*
Arthur H. Perler, *214, 306, 505*
Spencer A. Peterson, *1004*
Frederick K. Pfaender, *502*
Bernhard H. Pfeil, *209, 315*
Quentin R. Pickering, *800*
Wesley O. Pipes, Jr., *213, 901, 903, 904, 905, 906, 908, 911, 918*
Laurence R. Pittwell
John T. Pivinski, *415, 424*
Marvin D. Piwoni, *418, 419, 420, 421*
Russell H. Plumb, Jr., *801*
Franklin Pogge
Frederick G. Pohland, *402, 403, 407, 502, 504, 511*
Robert B. Pojasek, *413, 507*
Richard D. Pomeroy, *206, 208, 421, 428, 502*
Romola Popper, *320, 322, 404, 902, 907, 911, 912, 916, 919*
Donald B. Porcella, *800*
Robert D. Potts, *415, 506*
T. B. S. Prakasam, *418, 419, 420, 421, 504*
Chandan Prasad
Stanley J. Prazer, *214, 325*
William B. Prescott, *204*
Richard Prober, *413*
Donald B. Procella, *800*
Donald Ptak, *911*
Frank A. Puglisi, *800*
Hugh D. Putnam
Clifford W. Randall, *209, 213, 507*
Ronald L. Raschke, *1003*
D. J. Reasoner, *912, 919*
John J. Redys, *914*
Donald J. Reish, *801*
David J. Rexing, *203*
George G. Richards, *203*
E. Rider, *800*
Lloyd R. Robinson, Jr., *315, 505*
John A. Roeber, *208, 213, 422*
I. H. Rogers, *801, 810*
J. L. Rogers, *412, 417*
Olsen J. Rogers, *405*
Antonio A. Romano, *916*
Aaron A. Rosen, *207, 211, 502, 506*
Harvey M. Rosen, *423*
James E. Rosencrance, *324*
John R. Rossum, *205, 214, 323, 324, 428, 502*
Emanuel Rudolph, *915*
Robert A. Ryder, *423*
Sham L. Sachdev, *310, 405, 505*
Robert S. Safferman, *901, 903, 904, 905,906, 913, 916*
Arnold C. Salinger
Gerry Salpeter, *413, 600*
Ernest Sanchez, *700–710*
Thomas J. Sargent
Michael F. Saunders, *213*
Bernard Saunier, *409, 411*
Jan. W. Scherfig, *206*
John A. Schmitt, *915*
William B. Schworm
I. Seidenberg, *910*
Robert E. Selleck, *206*
Eugene A. Setzkorn, *512*
Roland L. Seymour, *915*
Joseph Shapiro, *424*
Gerry L. Shell, *208*
B. F. Shema, *800*
Joseph H. Sherrard, *507, 508, 801, 918*
Edward J. Shervin, *322*
Peter J. Shuba, *800*
Kadhim M. Shuker

Kenneth E. Shull, *306*
John E. Shultz, *800*
John Sickafoose, *308, 423*
Philip C. Singer
Chanchal Singh, *323, 424*
J. Edward Singley, *203, 204*
E. W. Skepnek
Vernon Snoeyink, *505*
Louis Snyder, *316*
Mark D. Sobsey, *913*
Frank W. Sollo, Jr., *203*
Paul C. Soltow, Jr., *801*
Charles A. Sorber, *403, 403, 409, 410, 412*
Anne Spacie, *810*
Richard E. Sparks, *801*
David T. Specht, *802*
Robert L. Spehar, *800*
P. A. M. Sperring
Robert C. Spicher, *315, 325, 505, 506*
Stuart Spiegel, *909*
Neal C. Spivey, *323, 324, 327*
John B. Sprague, *801*
Otis J. Sproul
Jack R. Stabley, Jr., *504*
John F. Stafford, *311, 314, 505*
Francis S. Stay, Jr.
Charles E. Stephan, *801*
Myron Stephenson, *504*
Alan A. Stevens, *506*
Nelson E. Steward
Raymond M. Stewart, *214, 311, 314, 318, 428, 504, 912, 913*
R. F. Stone, *415*
James A. Stoner, *506, 510*
Charles E. Stringer, *307, 323, 418*
Terry Surles, *317, 322, 325, 326, 509*
Sidney Sussman, *426*
Ogden F. Swan
Richard Swartz, *800*
Robert D. Swisher, *512*
Gladys H. Swope
Albert Sylvia, *506*
A. F. Tabri, *506, 509*
Armen C. Tarjan, *917*
Clarence M. Tarzwell, *801*
Howard Taylor
Ray Taylor, *907*
Richard B. Taylor, *319*
John I. Teasley, *410, 412*
Lee Tebo, *1005*
Nelson A. Thomas, *800*
Brian Thompson, *509*
Cliff G. Thompson, *216, 311, 314, 318*
James C. Thurrott, *312, 424*
R. V. Thurston, *800*
Edwin C. Tifft, Jr., *411*
Michael Trehy, *214, 306, 316, 330*
Richard L. Trombly, *321, 328*
R. Rhodes Trussell, *203*
James H. Tucker, *800*
Pamela J. Tucker, *404, 405, 423*
Lloyd Tyler, *321, 328*
George S. Uyesugi, *701–710*
Michael J. Vandaveer, *504*
Man Mohan Varma, *422, 507*
Peter E. Ventura
F. John Vernberg, *801*
Hugo T. Victorian, *907, 918*
Craig O. Vinson
Charles T. Walbridge, *800*
Richard D. Walker, *512*
William T. Waller, *801*
Gerald E. Walsh
Lawrence K. Wang, *209, 213, 512*
Timothy Ward, *800*
Tak Watanabe, *404*
B. L. Weand, *316, 317, 319*
Cornelius I. Weber, *917, 1000*
Richard L. Weitzel, *1002, 1003*
Sally D. Wenris, *1005*
Clark Wescott, *424*
Foymae K. West
Arnold F. Westerhold, *509*
Warren C. Westgarth, *209, 213*

ROBERT E. WHITE, *509, 801*
GEORGE P. WHITTLE, *409, 410, 412, 414, 415*
BRANNON H. WILDER, *319*
BENJAMIN F. WILLEY
ROBERT WILLIAMS, *600*
ROBERT T. WILLIAMS, *505*
THEODORE J. WILLIAMS, *320, 408, 409, 512*
A. L. WILSON
RON WILLSON, *1005*
RONALD WINBORNE, *407, 426*
E. WINDLE-TAYLOR, *908, 909*
JOHN W. WINTER, *902*
DONALD WITHCOMB, *307, 323*
C. E. WOELKE, *800*
JOHN D. WOLSZON, *402, 403, 407*
LEON A. WOODS, *509*
CHARLES C. WRIGHT, *424*
RODGER A. YORTON, *404, 415, 424, 505*
DAVID YOUNG
J. C. YOUNG, *213, 507*
NICHOLAS S. ZALEIKO, *209, 214, 507, 508, 600, 801*
RAY ZEHNPFENNIG, *323, 325, 326, 415*
JOHN ZINK, *307, 404, 505*

FOREWORD

With this, its 15th edition, *Standard Methods for the Examination of Water and Wastewater* achieves its 75th anniversary of contribution to the health and safety of water users throughout the world, particularly in the United States.

Its first edition articulated early principles of water sanitation based upon the then-developing germ theory of disease. The American Public Health Association, American Water Works Association, and Water Pollution Control Federation are proud and privileged to provide you with this current text that embodies developments in water methodology to the present time. We do so with firm dedication to the principles of accuracy, precision, sensitivity, and reproducibility reflected in this text.

Approximately 60,000 copies of the 14th edition are in circulation, a fact that underscores the high level of regard and credibility that *Standard Methods* has achieved in the scientific community.

Our Societies have made a very substantial and broad organizational commitment to maintenance of this reputation. Manpower and financial resources devoted to assurance of consensus among the 400 participants in editorial development of each edition of *Standard Methods* exceed many times over the costs of production and distribution of the manual. The highest councils of our organizations are deeply involved in decisions affecting management of the work. Publication of *Standard Methods* and maintenance of its credibility are central to the educational and professional purposes of our Societies.

We wish to express the sincere gratitude of our organizations and membership to the Joint Editorial Board, Arnold E. Greenberg (*Chairman*), Joseph J. Connors, and David Jenkins, for their tireless dedication and professional contribution to *Standard Methods*. To each participant in the many subcommittees, committees, and Joint Task Groups that provide consensus on the text of *Standard Methods* we extend our thanks.

To the users of the manual, who make the final determination of value of our work, we offer an additional benefit with the publication of the 15th Edition. We have researched the full range of U.S. Federal Regulations for water analysis, and we are providing to all users of this manual a supplement of those governmental methods that are cited by the Code of Federal Regulations for regulatory use, but that have not been subjected to the rigors of the consensus process that underlies inclusion of techniques in *Standard Methods*.

William H. McBeath
David B. Preston
Robert A. Canham

TABLE C. INTERNATIONAL RELATIVE ATOMIC WEIGHTS, 1977

Scaled to the relative atomic mass, $A_r(^{12}C) = 12$

The atomic weights of many elements are not invariant but depend on the origin and treatment of the material. The footnotes to this table elaborate the types of variation to be expected for individual elements. The values of $A_r(E)$ given here apply to elements as they exist naturally on earth and to certain artificial elements. When used with due regard to the footnotes they are considered reliable to ±1 in the last digit or ±3 when followed by an asterisk *. Values in parentheses are used for certain radioactive elements whose atomic weights cannot be quoted precisely without knowledge of origin; the value given is the atomic mass number of the isotope of that element of longest known half life.

Name	Symbol	Atomic number	Atomic weight	Footnotes	Name	Symbol	Atomic number	Atomic weight	Footnotes
Actinium	Ac	89	227.0278	d	Molybdenum	Mo	42	95.94*	
Aluminium	Al	13	26.98154		Neodymium	Nd	60	144.24*	b
Americium	Am	95	(243)		Neon	Ne	10	20.179*	c
Antimony	Sb	51	121.75*		Neptunium	Np	93	237.0482	d
Argon	Ar	18	39.948*	a, b	Nickel	Ni	28	58.70	
Arsenic	As	33	74.9216		Niobium	Nb	41	92.9064	
Astatine	At	85	(210)		Nitrogen	N	7	14.0067	
Barium	Ba	56	137.33*	b	Nobelium	No	102	(259)	
Berkelium	Bk	97	(247)		Osmium	Os	76	190.2	b
Beryllium	Be	4	9.01218		Oxygen	O	8	15.9994*	a
Bismuth	Bi	83	208.9804		Palladium	Pd	46	106.4	b
Boron	B	5	10.81	a, c	Phosphorus	P	15	30.97376	
Bromine	Br	35	79.904		Platinum	Pt	78	195.09*	
Cadmium	Cd	48	112.41	b	Plutonium	Pu	94	(244)	
Cesium	Cs	55	132.9054		Polonium	Po	84	(209)	
Calcium	Ca	20	40.08	b	Potassium	K	19	39.0983*	
Californium	Cf	98	(251)		Praseodymium	Pr	59	140.9077	
Carbon	C	6	12.011	a	Promethium	Pm	61	(145)	
Cerium	Ce	58	140.12	b	Protactinium	Pa	91	231.0359	d
Chlorine	Cl	17	35.453		Radium	Ra	88	226.0254	b, d
Chromium	Cr	24	51.996		Radon	Rn	86	(222)	
Cobalt	Co	27	58.9332		Rhenium	Re	75	186.207	
Copper	Cu	29	63.546*	a	Rhodium	Rh	45	102.9055	
Curium	Cm	96	(247)		Rubidium	Rb	37	85.4678*	b
Dyprosium	Dy	66	162.50*		Ruthenium	Ru	44	101.07*	b
Einsteinium	Es	99	(252)		Samarium	Sm	62	150.4	b
[illegible]	Er	68	167.26*		Scandium	Sc	21	44.9559	

Europium	Eu	63	151.96	b
Fermium	Fm	100	(257)	
Fluorine	F	9	18.998403	
Francium	Fr	87	(223)	
Gadolinium	Gd	64	157.25*	b
Gallium	Ga	31	69.72	
Germanium	Ge	32	72.59*	
Gold	Au	79	196.9665	
Hafnium	Hf	72	178.49*	
Helium	He	2	4.00260	b
Holmium	Ho	67	164.9304	
Hydrogen	H	1	1.0079	a
Indium	In	49	114.82	b
Iodine	I	53	126.9045	
Iridium	Ir	77	192.22*	
Iron	Fe	26	55.847*	
Krypton	Kr	36	83.80	b, c
Lanthanum	La	57	138.9055*	b
Lawrencium	Lr	103	(260)	
Lead	Pb	82	207.2	a, b
Lithium	Li	3	6.941*	a, b, c
Lutetium	Lu	71	174.967*	
Magnesium	Mg	12	24.305	b
Manganese	Mn	25	54.9380	
Mendelevium	Md	101	(258)	
Mercury	Hg	80	200.59*	
Selenium	Se	34	78.96*	
Silicon	Si	14	28.0855*	
Silver	Ag	47	107.868	b
Sodium	Na	11	22.98977	
Strontium	Sr	38	87.62	b
Sulfur	S	16	32.06	a
Tantalum	Ta	73	180.9479*	
Technetium	Tc	43	(98)	
Tellurium	Te	52	127.60*	b
Terbium	Tb	65	158.9254	
Thallium	Tl	81	204.37*	
Thorium	Th	90	232.0381	b, d
Thulium	Tm	69	168.9342	
Tin	Sn	50	118.69*	
Titanium	Ti	22	47.90*	
Tungsten	W	74	183.85*	
Unnilhexium	Unh	106	(263)	
Unnilpentium	Unp	105	(262)	
Unnilquadium	Unq	104	(261)	
Uranium	U	92	238.029	b, c
Vanadium	V	23	50.9415*	
Xenon	Xe	54	131.30	b, c
Ytterbium	Yb	70	173.04*	
Yttrium	Y	39	88.9059	
Zinc	Zn	30	65.38	
Zirconium	Zr	40	91.22	b

a. Element for which known variations in isotopic composition in normal terrestrial material prevent a more precise atomic weight being given. A_r (E) values should be applicable to any 'normal' material.
b. Element for which geological specimens are known in which the element has an anomalous isotopic composition, such that the difference between the atomic weight of the element in such specimens and that given in the table may exceed considerably the implied uncertainty.
c. Element for which substantial variations in A_r from the value given can occur in commercially available material because of inadvertent or undisclosed change of isotopic composition.
d. Element for which the value of A_r is that of the radioisotope of longest half-life.

SOURCE: International Union of Pure and Applied Chemistry.

TABLE OF CONTENTS

TABLES

PAGE

FIGURES

PLATES

Black and White plates of aquatic organisms

Color plates of blue-green algae (unpaginated special color section)

PART 100

GENERAL

INTRODUCTION

101 APPLICATIONS

The procedures described in these standards are intended for the examination of waters of a wide range of quality, including water suitable for domestic or industrial supplies, surface water, groundwater, cooling or circulating water, boiler water, boiler feed water, and treated and untreated municipal or industrial wastewater. The unity of the fields of water supply, receiving water quality, and wastewater treatment and disposal is recognized by presenting methods of analysis for each constituent in a single section for all types of waters.

An effort has been made to present methods that apply as generally as possible, and where alternative methods are necessary for samples of different composition, to present as clearly as possible the basis for selecting the most appropriate method. However, samples with extreme concentrations or otherwise unusual compositions may present difficulties that preclude the direct use of these methods. Hence, some modification of a procedure may be necessary in specific instances. Whenever a procedure is modified, state plainly the nature of modification in the report of results.

Certain procedures are intended for use with sludges and sediments. Here again, the effort has been to present methods of the widest possible application, but when chemical sludges or slurries or other samples of highly unusual composition are encountered, the methods of this manual may require modification or may be inappropriate.

The analysis of bulk chemicals received for water treatment is not included herein. A committee of the American Water Works Association prepares and issues standards for water treatment chemicals.

102 LABORATORY APPARATUS, REAGENTS, AND TECHNICS

1. Containers

For general laboratory use, the most suitable material for containers is resistant borosilicate glass.* Special glassware is available with characteristics such as high resistance to alkali attack, low boron content, or exclusion of light. Choose stoppers, caps, and plugs to resist the attack of material contained in the vessel. Cork stoppers wrapped with a relatively inert metal foil are suitable for many samples. Metal screw caps are a poor choice for samples that will cause them to corrode readily. Glass stoppers are unsatisfactory for strongly alkaline liquids because of their tendency to stick fast. Rubber stoppers are excellent for alkaline liquids but unacceptable for organic solvents, in which they swell or disintegrate. Use

*Pyrex, manufactured by Corning Glass Works; Kimax, Kimble Glass Co., Division of Owens-Illinois; or equivalent.

polytetrafluoroethylene (TFE)† or silver plugs for burets that contain strongly alkaline liquids. When appropriate, use other materials such as porcelain, nickel, iron, platinum, stainless steel, and high-silica glass.‡

Collect and store samples in bottles made of borosilicate glass, hard rubber, plastic, or other inert material.

For relatively short storage periods, or for constituents that are not affected by storage in soft glass, such as calcium, magnesium, sulfate, chloride, and perhaps others, the 2.5-L acid-bottle "bell closure" is satisfactory. This closure holds a glass or polyethylene disk against the ground-glass surface or a polyethylene insert on the bottle lip and insures adequate protection. If part of the sample is to be analyzed later for silica, sodium, or other substances that would be affected by prolonged storage in soft glass, transfer it to a small plastic bottle, while leaving the remainder of the sample in the soft-glass bottle.

Carefully clean sample bottles before each use. Rinse glass bottles, except those to be used for chromium or manganese analyses, either with a cleaning mixture made by adding 1 L conc H_2SO_4 slowly, with stirring, to 35 mL saturated sodium dichromate solution, or with 2% $KMnO_4$ in 5% KOH solution followed by an oxalic acid solution. Rinse with other concentrated acids to remove inorganic matter. Detergents are excellent cleansers for many purposes; use either detergents or conc HCl for cleaning hard-rubber and plastic bottles. After the bottles have been cleaned, rinse thoroughly with tap water and then with distilled water.

For shipment, pack bottles in wooden, metal, plastic, or heavy fiberboard cases, with a separate compartment for each bottle. Line boxes with corrugated fiber paper, felt, or other resilient material, or provide with spring-loaded corner strips, to prevent breakage. Alternatively use lined wicker baskets. Samples stored in plastic bottles need no protection against breakage by impact or through freezing.

†Teflon or equivalent.

‡Vycor, manufactured by Corning Glass Works, or equivalent.

2. Distilled Water

Some tests described in this manual are sensitive enough to detect even the minute traces of impurities present in ordinary distilled water. In such cases, use double- or triple-distilled water. The material of which the still is constructed may contribute impurities to the distillate. Most commercial stills, for example, are constructed in part of copper, and distilled water from them frequently contains 10 to 50 μg Cu/L. For special purposes, distill water from an all-borosilicate-glass apparatus or from an apparatus in which the condenser is made of glass, fused quartz, silver, or block tin.

Ordinary distillation of water will not remove ammonia (NH_3) or carbon dioxide (CO_2); in fact, distilled water often is supersaturated with CO_2 because of the decomposition of raw-water bicarbonates to carbonates in the boiler. Remove NH_3 by distilling from acid solution or by passing the water through a column of mixed anionic and cationic resins. Remove CO_2 by distilling from a solution containing an excess of alkali hydroxide, by boiling for a few minutes, by vigorously aerating the water with a stream of inert gas (nitrogen, for example) for a sufficient period, or by passing the water through a column of strong anion-exchange resin in the hydroxide form. Distilled water kept in glass containers slowly leaches the more soluble materials from the glass and the concentration of total dissolved solids increases.

Demineralized water from a mixed-bed ion exchanger is satisfactory for many applications in this manual. However, because ion exchange fails to remove such nonelectrolytes and colloids as plankton, nonionic organic materials, and dissolved air, it is not suitable for determinations where such constituents interfere. Some ion-exchange resins also release traces of organic matter, making the demineralized water unsuitable for use in certain tests.

Produce a very high-purity water, with a conductivity of less than 0.1 μmhos/cm, by passing ordinary distilled water through a mixed-bed exchanger and discarding the effluent until the desired quality is obtained. Water prepared in this way often is satisfactory for use in the determination of trace cations and anions.

Three types of special distilled water are specified for various methods in this manual. For easy reference, the preparation of these waters is described on the inside front cover.

3. Reagents

Use only the best quality chemical reagents even though this instruction is not repeated in the description of a particular method. Order chemicals for which the American Chemical Society has published specifications in the "ACS grade." Order other chemicals as "analytical reagent grade" or "spectral grade organic solvents." Methods of checking purity of suspect reagents are found in books of reagent specifications listed in the bibliography under laboratory reagents.

Unfortunately, many commercial dyes for which the ACS grade has not been established fail to meet exacting analytical requirements because of variations in the color response of different lots. In such cases, use dyes certified by the Biological Stain Commission.

Where neither an ACS grade nor a certified Biological Stain Commission dye is available, purify the solid dye through recrystallization.

The following standard substances, each bottle of which is accompanied by a certificate of analysis, are issued by the National Bureau of Standards, Department of Commerce, Washington, D.C., for the purpose of standardizing analytical solutions:

Acidimetric:

84h—Acid potassium phthalate
350—Benzoic acid

Oxidimetric:

40h—Sodium oxalate
83c—Arsenic trioxide
136c—Potassium dichromate

Buffer:

185e—Acid potassium phthalate
186Ic—Potassium dihydrogen phosphate
186IIc—Disodium hydrogen phosphate
187b—Borax
188—Potassium hydrogen tartrate
189—Potassium tetroxalate
191—Sodium bicarbonate
192—Sodium carbonate

Many hundreds of other standards issued by NBS are described in its Special Publication 260.

A successful dithizone test demands reagents of the highest purity. Chloroform and carbon tetrachloride are available in a grade declared to be suitable for dithizone methods. Select reagents of this quality for the dithizone methods described in this manual.

Water-soluble sodium salts of the common indicators usually are recommended for indicator preparation in this manual because of their general availability and reasonable cost.

When alcohol or ethyl alcohol is specified for preparing such solutions as phenolphthalein indicator, use 95% ethyl alcohol. Alternatively, use a similar grade of isopropyl alcohol.

Certain organic reagents are somewhat unstable upon exposure to the atmosphere. If the stability of a chemical is limited or unknown, purchase small lots at frequent intervals.

Dry all anhydrous reagent chemicals required for preparation of standard calibration solutions and titrants in an oven at 105 to 110 C for at least 1 to 2 hr and preferably overnight. After cooling to room temperature in an efficient desiccator, promptly weigh the proper amount for dissolution. Should a different drying temperature be necessary, this is specified for the particular chemical. For hydrated salts, substitute milder drying in an efficient desiccator for oven-drying.

4. Common Acid and Alkali Solutions

a. Concentration units used: Reagent concentrations are expressed in this manual in terms of normality, molarity, and additive volumes.

A *normal solution (N)* contains one gram equivalent weight of solute per liter of solution.

A *molar solution (M)* contains one gram molecular weight of solute per liter of solution.

In additive volumes $(a+b)$, the first number, a, refers to the volume of concentrated reagent; the second number, b, refers to the volume of distilled water required for dilution. Thus, "1 + 9 HCl" denotes that 1 volume of concentrated HCl is to be diluted with 9 volumes of distilled water.

To make a solution of exact normality from a chemical that cannot be measured as a primary standard, prepare a relatively concentrated stock solution and then make an exact dilution to the desired strength. Alternatively, make a solution of slightly higher concentration than that desired, standardize, and make suitable adjustments in concentration by dilution; or, use the solution as first standardized and modify the calculation factor. This last procedure is useful especially for solutions that slowly change strength and must be restandardized frequently—for example, sodium thiosulfate solution. Adjustment to exact normality specified is desirable when a laboratory makes a large number of determinations with one standard solution.

Determinations are in accord with the instructions in this manual as long as normality of a standard solution does not result in a titration volume so small as to preclude accurate measurement or so large as to cause abnormal dilution of the reaction mixture, and as long as the solution is standardized properly and the calculations are made properly.

b. Preparation and dilution of solutions: If a solution of exact normality is to be prepared by dissolving a weighed amount of a primary standard or by diluting a stronger solution, bring it up to exact volume in a volumetric flask.

Accurately prepare stock and standard solutions prescribed for colorimetric determinations in volumetric flasks. Where concentration does not need to be exact, mix the concentrated solution or solid with measured amounts of water, using graduated cylinders for these measurements. There is usually a significant change of volume when strong solutions are mixed, so that the total volume is less than the sum of volumes used. For approximate dilutions, volume changes are negligible when concentrations of 6*N* or less are diluted.

Mix thoroughly and completely when making dilutions. One of the most common sources of error in analyses using standard solutions diluted in volumetric flasks is failure to attain complete mixing.

c. Storage of solutions: Some standardized solutions alter slowly because of chemical or biological changes. The practical life, required frequency of standardizations, or storage precautions are indicated for such standards. Others, such as dilute HCl, are nonreactive. Yet their strength, too, may change by evaporation that is not prevented by a glass stopper. Changes in temperature cause a bottle to "breathe", and allow some evaporation.

Do not consider a standard valid for

more than one year unless it is restandardized. It is valid for that length of time only if conditions minimize evaporation. If the bottle is opened often or if it is much less than half full, significant evaporation occurs in a few months.

Use glass bottles of chemically resistant glass. For standard solutions that do not react with rubber or neoprene, use stoppers of these materials, because they can, if properly fitted, prevent evaporation as long as the bottle is closed. Screw-cap bottles also are effective. If the cap has a gasket of a reasonably resistant material, permissible usage will be about the same as for rubber stoppers.

d. Hydrochloric and sulfuric acid as alternatives: Dilute standardized H_2SO_4 and HCl are called for in various procedures. Often these solutions are interchangeable. Where one is mentioned, the other may be used if it is known to make no difference.

e. Preparation: Although instructions usually describe preparation of 1 L of solution, prepare smaller or larger volumes as needed. Instructions calling for the preparation of 100 mL usually involve either short-life reagents or those used in small amounts.

A safe general rule is to add more concentrated acid or alkali to water, with stirring, in a vessel that can withstand thermal shock, and then to dilute to final volume after cooling to room temperature.

f. Uniform reagent concentrations: An attempt has been made to establish a number of uniform common acid and base concentrations that will serve for adjustment of pH of samples before color development or final titration. The following acid concentrations are recommended for general laboratory use: the concentrated reagent of commerce, 6*N*, 1*N*, 0.1*N*, and 0.02*N*. See the inside front cover for directions for preparing these acid concentrations, as well as the required 15*N*, 6*N*, and 1*N* NaOH solutions, and 5*N*, 3*N*, and 0.2*N* NH_4OH solutions.

5. Volumetric Glassware

Calibrate volumetric glassware or obtain a certificate of accuracy from a competent laboratory. Volumetric glassware is calibrated either "to contain" (TC) or "to deliver" (TD). Glassware designated "to deliver" will do so with accuracy only when the inner surface is so scrupulously clean that water wets it immediately and forms a uniform film upon emptying. Whenever possible, use borosilicate glassware.

Carefully measure weights and volumes in preparing standard solutions and calibration curves. Observe similar precautions in measuring sample volumes. Use volumetric pipets or burets where the volume is designated to two decimal places (X.00 mL) in the text. Use volumetric flasks where specified and where the volume is given as 1,000 mL rather than 1 L.

6. Nessler Tubes

Unless otherwise indicated use "tall"-form nessler tubes made of resistant glass and selected from uniformly drawn tubing. The glass should be clear and colorless and the tube bottoms should be plane-parallel. When the tubes are filled with liquid and viewed from the top with a light source beneath, there should be no dark spots nor any lenslike distortion of the transmitted light. The best quality tube is manufactured by fusion-sealing a separately prepared, ground, and polished circle of glass to the tube to form its bottom. Less expensive tubes are manufactured with integral bottoms that cannot be made perfectly flat, but may appear satisfactory. The tops of the tubes should be flat, preferably fire-polished, and smooth enough to permit cover slips to be cemented on for sealing. Nessler tubes with standard-taper clear glass tops are available commercially. Graduation marks should completely encircle the tubes.

The 100-mL tubes should have a total length of approximately 375 mm. Their inside diameter should approximate 20 mm and the outside diameter 24 mm. The graduation mark should be as near as possible to 300 mm above the inside bottom. Tubes sold in sets should be of such uniformity that this distance does not vary more than 6 mm. (Sets are available commercially in which the maximum difference between tubes is not more than 2 mm.) A graduation mark at 50 mL is permissible.

The 50-mL tubes should have a total length of about 300 mm. Their inside diameter should approximate 17 mm and the outside diameter 21 mm. The graduation mark on the tube should be as near as possible to 225 mm above the inside bottom. Tubes sold in sets should be of such uniformity that this distance does not vary more than 6 mm. (Sets are available commercially in which the maximum difference between tubes is not more than 1.5 mm.) A graduation mark at 25 mL is permissible.

Tubes for Jackson candle turbidimeters, in addition to conforming precisely to the measurements given in Section 214, Turbidity, should conform to all requirements of quality, glass color, and workmanship pertaining to nessler tubes.

7. Colorimetric Equipment and Technic

a. General: Many procedures depend on matching colors, either by eye or with a photometric instrument. To obtain the best possible results, understand the principles and limitations of these methods, especially because the choice of instrument and of technic is discretionary. Both visual and photometric methods are included.

Tall-form nessler tubes provide a 30-cm light path. This is highly desirable when very faint colors are to be compared. Nessler tubes are inexpensive, their use does not require much training, they are not subject to mechanical or electrical failure, and in general they are entirely satisfactory for most routine work. Because they are portable and do not require a source of electric light, they can be used in the field.

Photometric instruments are more versatile than nessler tubes. They generally are capable of superior accuracy if used properly and they do not depend on external lighting conditions or on the analyst's eyesight. Therefore, results obtained with photometric instruments are less subject to personal bias and are more reproducible. Their use often allows corrections to be made for interfering color or turbidity. It is not necessary to prepare a complete set of standards for each determination if a photometric instrument is used, whereas it is necessary to prepare such a set, or to maintain permanent standards, if nessler tubes are used for visual comparison.

Photometric methods are not free from specific limitations. While an analyst will recognize that something has gone wrong on seeing an unusual color or turbidity when making a visual comparison, such a discrepancy easily may escape detection during a photometric reading, for the instrument always will yield a reading, whether meaningful or not. Frequently check sensitivity and accuracy by testing standard solutions to detect electrical, mechanical, or optical problems in the instrument and its accessories. Testing, maintaining, and repairing such instruments call for specialized skills.

A photometer is not uniformly accurate over its entire scale. At very low transmittances the scale is crowded in terms of concentration, so that a considerable change in relative concentration will cause only a slight change in position of the indicator dial or needle. At very high transmittances, slight differences between optical cells, the presence of condensed moisture, dust, bubbles, fingerprints, or a slight lack of reproducibility in positioning the cells can cause as great a change in readings as would a considerable change in concentration. The difficulties are minimized if readings are made to fall between 0.1 and 1.0 absorbance by diluting or concentrating the sample or varying the light path by selecting cells of appropriate size.

Some suggestions for suitable ranges and light paths are offered under individual methods in this manual, but much reliance necessarily must be placed on the knowledge and judgment of the analyst. Most photometers are capable of their best performance when readings fall in the range of approximately 1 to 0.1 absorbance with respect to a blank adjusted to read 0 absorbance. The closer the readings approach 0 or 3.0 absorbance, the less accurate they become. If it is impractical to use an optical cell with a sufficiently long light path—as in some commercial instruments or to concentrate the sample or select a more sensitive color test, then it may be more accurate to compare very faint colors in nessler tubes than to attempt photometric readings close to 100% transmittance.

If it is impractical to use an optical cell with a sufficiently long light path—as in some commercial instruments—or to concentrate the sample or select a more sensitive color test, then it may be more accurate to compare very faint colors in nessler tubes than to attempt photometric readings close to 100% transmittance.

In general, the best wavelength or filter to select is that which produces the largest spread of readings between a standard and a blank. This usually corresponds to a visual color for the light beam that is complementary to that of the solution—for example, a green filter for a red solution, a violet filter for a yellow solution.

Absorptivities are useful in comparing method sensitivities and in estimating the concentration of absorbing solutions such as dithizone. The absorptivity may be computed from:

$$a = \frac{A}{b\,c}$$

where:

a = absorptivity, L/(g·cm),
A = absorbance of a solution, dimensionless,
b = concentration, g/L, and
c = cell path length, cm.

The molar absorptivity, E, is the absorptivity multiplied by the molecular weight of a substance, L/(mole·cm).

Use of a photoelectric instrument makes unnecessary the preparation of a complete set of standards for each set of samples to be analyzed. However, prepare a reagent blank and at least one standard in the upper end of the optimum concentration range, with every group of samples, to verify the constancy of the calibration curve. This precaution will reveal any unsuspected changes in the reagents, the instrument, or the technic. At regular intervals, or if at any time results fall under suspicion, prepare a complete set of standards—at least five or six spaced to cover the optimum concentration range—to check the calibration curve. Also valuable in this regard is the absorptivity information given in this manual for a number of photometric methods.

Use the utmost care with calibration curves supplied by the instrument manufacturer or in the use of commercial permanent standards of colored liquids or glasses. Verify frequently the accuracy of the curves or permanent standards by comparing with standards prepared in the laboratory, using the same set of reagents, the same instrument, and the same procedures as those used for analyzing samples. Even if permanent calibration curves or artificial standards have been prepared accurately by the manufacturer, they may not be valid under conditions of use. Permanent standards may be subject to fading or color alteration. Their validity may depend also on certain arbitrary lighting conditions. Standards and calibration curves may be incorrect because of slight differences in reagents, instruments, or technics between the manufacturer's and the analyst's laboratories.

If a photometer provides readings in terms of absorbance, plot calibration curves on arithmetic coordinates; if readings are in terms of percentage transmittance, convert to absorbance before plotting. Usually, such graphs will give straight, or nearly straight, lines.

Photometric determinations measure concentration. The volume of solution used for the actual photometric measurement varies and contains only a fraction of the total weight of constituent present in the final volume of solution. If the final volumes of the standards and the samples are the same then the concentration, expressed as micrograms per final volume, is numerically, but not dimensionally, equal to the number of micrograms in the final volume. Thus it is possible, but not good practice, to plot absorbance, A, against weight of constituent, as well as the more proper plot of A against concentration. If the plot is A against weight, be aware that the results need correcting when the final volume of sample and standard are not the same.

Use photometric compensation to correct for interference caused by color or turbidity and also for impurities in the chemicals and distilled water used in the reagent blank. Do not use it to compensate for interfering substances that react with the color-developing reagents to produce a color (i.e., positive interferences). The principle involved is the additivity of absorbances.

If there is a significant reagent blank but no sample color or turbidity, make the necessary correction by adding the color-developing reagents to distilled water and adjusting the photometer to the null point with the resulting solution.

If there is color or turbidity or both in the sample, but a negligible reagent blank, correct by carrying an additional sample portion through the procedure, but either omit one of the essential color-developing reagents or, preferably, bleach out the color after it has been produced, but in such a way that the interfering color or turbidity is not bleached. Use this special blank for nulling the photometer. Take into account any significant change in volume produced by the addition or omission of reagents.

If color or turbidity or both are present in the sample, and if, in addition, the reagent blank is significant, a slightly more complicated procedure is needed to correct for both interferences: Prepare the calibration curve by setting the photometer to zero absorbance with plain distilled water and read all the standards, including a zero standard or reagent blank, against the distilled water. If the graph is plotted in the recommended manner, and Beer's law holds, a straight line will be obtained; but if there is a measurable reagent blank, this line will not pass through the point of origin.

For each sample, prepare a special blank by either omitting a reagent or bleaching out the color as described above. Place each special blank in the photometer in turn, adjust the instrument each time to read zero absorbance, and read each regularly developed sample against its corresponding blank. Interpret the observed absorbances from the calibration graph. As before, consider any significant increase or decrease in volume caused by addition or omission of reagents in the calculations.

In visual color comparison with some instruments, compensate for color and turbidity by the Walpole technic. View the treated sample, after color development, through distilled water, but view the color standard through an untreated sample. It is inconvenient to use the Walpole technic when viewing tall-form nessler tubes axially, because of their clumsy length.

Sometimes none of the cited expedients will apply. In such an event, several approaches are available to separate turbidity from a sample. The nature of the sample, the size of suspended particles, and the reasons for conducting the analysis will all combine to dictate the method for turbidity removal. Turbidity may be coagulated by adding zinc sulfate and an alkali, as is done in the direct nesslerization method for ammonia nitrogen. For samples of relatively coarse turbidity, centrifuging may suffice. In some instances, glass fiber filters, filter paper, or sintered-glass filters of fine porosity will serve the purpose. For very small particle sizes, membrane filters may provide the required retentiveness. Used with discretion, each of these methods will yield satisfactory results in a suitable situation. However, it must be emphasized that no single universally ideal method of turbidity removal is available. Moreover, be perpetually alert to adsorption losses possible with any flocculating or filtering procedure.

b. Dithizone solutions: Several colorimetric methods for metals (Cd, Pb, Hg, Ag, Zn) use dithizone (diphenylthiocarbazone) as an extractable, colored, metal-complexing agent. The methods presented later in this text have been based on three stock dithizone solutions, the preparation of which is described below. The dithizone concentration in the stock dithizone solutions is based on having a 100% pure dithizone reagent. Some commercial grades of dithizone are contaminated with the oxidation product diphenylthiocarbodiazone or with metals. Purify dithizone as directed below. For dithizone solutions not stronger than 0.001% (*w/v*), calculate the exact concentration by dividing the absorbance of the solution in a 1.00-cm cell at 606 μm by 40.6×10^3, the molar absorptivity.

Adjust dilutions of stock dithizone solutions to produce working dithizone solutions of the indicated strength based on the measured stock dithizone solution concentration.

1) *Stock dithizone solution I,* 100 mg dithizone/1,000 mL $CHCl_3$: Dissolve 100 mg dithizone in 50 mL $CHCl_3$ in a 150-mL beaker and filter through a 7-cm-diam paper§. Receive filtrate in a 500-mL separatory funnel or in a 125-mL erlenmeyer flask under slight vacuum; use a filtering device designed to handle the $CHCl_3$ vapor. Wash beaker with two 5-mL portions $CHCl_3$ and filter. Wash the paper with three 5-mL portions $CHCl_3$, adding final portion dropwise to edge of paper. If filtrate is in flask, transfer with $CHCl_3$ to a 500-mL separatory funnel.

Add 100 mL 1 + 99 NH_4OH to separatory funnel and shake moderately for 1 min; excessive agitation produces slowly breaking emulsions. Let layers separate, swirling funnel gently to submerge $CHCl_3$ droplets held on surface of aqueous layer. Transfer $CHCl_3$ layer to 250-mL separatory funnel, retaining the orange-red aqueous layer in the 500-mL funnel. Repeat extraction, receiving $CHCl_3$ layer in another 250-mL separatory funnel and transferring aqueous layer, using 1 + 99 NH_4OH, to the 500-mL funnel holding the first extract. Repeat extraction, transferring the aqueous layer to 500-mL funnel. Discard $CHCl_3$ layer.

To combined extracts in the 500-mL separatory funnel add 1 + 1 HCl in 2-mL portions, mixing after each addition, until dithizone precipitates and solution is no longer orange-red. Extract precipitated dithizone with three 25-mL portions $CHCl_3$. Dilute combined extracts to 1,000 mL with $CHCl_3$; 1.00 mL = 100 μg dithizone.

2) *Stock dithizone solution II,* 250 mg dithizone/250 mL $CHCl_3$: Dissolve 250 mg dithizone in 50 mL $CHCl_3$ in a 150-mL

§Whatman No. 42 or equivalent.

beaker and filter through a 7-cm-diam paper‖. Receive filtrate in a 1,000-mL separatory funnel or in a 125-mL erlenmeyer flask under slight vacuum; use a filtering device designed to handle $CHCl_3$ vapor. Wash beaker with two 5-mL portions $CHCl_3$ and filter. Wash paper with three 5-mL portions $CHCl_3$, adding the last portion dropwise to edge of paper. If filtrate is in flask, transfer with $CHCl_3$ to 1,000-mL separatory funnel.

Add 200 mL 1 + 99 NH_4OH to separatory funnel and shake moderately for 1 min; excessive agitation produces slowly breaking emulsions. Let layers separate, swirling funnel gently to submerge $CHCl_3$ droplets held on surface of aqueous layer. Transfer $CHCl_3$ layer to 500-mL separatory funnel, retaining the orange-red aqueous layer in the 1,000-mL funnel. Repeat extraction of $CHCl_3$ layer with 200 mL 1 + 99 NH_4OH, transferring $CHCl_3$ layer to another 500-mL separatory funnel. Transfer aqueous layer to 1,000-mL funnel holding the first extract. Repeat extraction with third 200-mL portion 1 + 99 NH_4OH. Discard $CHCl_3$ layer, transfer aqueous layer to 1,000-mL funnel. To the combined extracts add 1 + 1 HCl in 4-mL portions, mixing after each addition, until dithizone precipitates and solution is no longer orange-red. Extract precipitated dithizone with four 25-mL portions $CHCl_3$. Dilute combined extracts to 250 mL with $CHCl_3$; 1.00 mL = 1,000 μg dithizone.

3) *Stock dithizone solution III,* 125 mg dithizone/500 mL CCl_4: Dissolve 125 mg dithizone in 50 mL $CHCl_3$ and proceed as for dithizone solution II, but extract precipitated dithizone with 25-mL portions CCl_4. Dilute CCl_4 extracts to 500 mL; 1.00 mL = 250 μg dithizone. (CAUTION: *CCl_4 is toxic—avoid inhalation, ingestion, and contact with the skin.*)

‖Whatman No. 42 or equivalent.

8. Other Methods of Analysis

The use of an instrumental method of analysis not specifically described in procedures in this manual is permissible provided that the results so obtained are checked periodically, either against a standard method described herein or against a standard sample of undisputed composition. Identify any such instrumental method used in the laboratory report along with the analytical results.

a. Atomic absorption spectrophotometry: Atomic absorption spectrophotometry has been applied to the determination of metals in water without the need for prior concentration or extensive sample pretreatment. The use of organic solvents coupled with oxyacetylene, oxyhydrogen, or nitrous oxide-acetylene flames enables the determination of metals that form refractory oxides. These standards include atomic absorption methods for many metals including certain flameless and electrothermal (heated graphite) technics.

b. Flame photometry: Flame photometry is used for the determination of sodium, potassium, lithium, and strontium.

c. Emission spectroscopy: Arc-spark emission spectroscopy is an important analytical tool and is proving valuable both for trace metal analysis and for certain determinations not easily made by any other method. Considerable specialized training and experience with this technic are required to obtain satisfactory results, and frequently it is practical to obtain only semiquantitative results from such methods. An arc-spark emission spectrograph is relatively expensive when used exclusively for routine water testing, but its purchase can be justified if it is used as a general laboratory analytical instrument.

d. Polarography and related analytical systems: Polarography is suggested for scanning industrial wastes for various metal ions, especially where the possible interferences in the colorimetric procedures are unknown. Pulse polarography has enabled the determination of seven or more metals at the low microgram-per-liter level when a single 100-mL sample has been ashed with HNO_3. Differential pulse voltammetry and differential pulse anodic stripping voltammetry also have gained acceptance for the determination of heavy metals and their speciation, in water and wastewater.

A method closely allied to polarography is amperometric titration, which is suitable for determining residual chlorine, chlorine dioxide, and iodine, and in other iodometric methods.

e. Potentiometric titration: Many titrimetric methods can be performed potentiometrically, by using a millivoltmeter or pH meter with suitable electrodes.

f. Selective ion electrodes: Selective ion electrodes are available for rapid estimation of certain constituents in water. These electrodes function best in conjunction with an expanded-scale pH meter or a suitable millivoltmeter. For the most part, the electrodes operate on the ion-exchange principle. Selective ion electrodes now available are designed for measuring ammonia, cadmium, calcium, divalent copper, hardness, lead, potassium, silver, sodium, total monovalent and total divalent cations, and bromide, chloride, cyanide, fluoride, iodide, nitrate, perchlorate, and sulfide anions, among others.

These devices are subject to varying degrees of interference from other ions in the sample and many still must receive thorough study to warrant adoption as tentative and standard methods. Nonetheless, their value for monitoring activities is readily apparent. To remove all doubt of variations in reliability, check each electrode in the presence of interferences as well as the ion for which it is intended. This manual details the electrode method for fluoride (Section 413) and ammonia nitrogen (Section 417).

The commercial dissolved oxygen (DO) probes vary considerably in their dependability and maintenance requirements. Despite these shortcomings, they have been applied to the monitoring of DO in a variety of waters and wastewaters. Most probes embody an electrode covered by a thin layer of electrolyte held in place by an oxygen-permeable membrane. The DO in solution diffuses through the membrane and electrolyte layer to react at the electrode, inducing a current proportional to the activity and hence, in solution of low or constant ionic strength, essentially proportional to concentration of DO. Satisfactory DO electrodes also are available without a membrane. In either case, keep the face of the DO sensor well agitated and provide temperature compensation to insure acceptable results.

g. Gas chromatography: Considerable work is under way in the development of gas chromatographic methods suitable for water and wastewater analysis. Such methods appear in this manual for determining chlorinated hydrocarbon pesticides, components in sludge digester gas, phenols, and halogenated methanes and ethanes.

h. Automated analytical instrumentation: Automated analyses are discussed in Part 600 in general terms. Specific automated technics are presented for chloride (Section 407), fluoride (Section 413), nitrogen (ammonia) (Section 417), nitrogen (nitrate) (Section 418), phosphate (Section 424), silica (Section 425), and sulfate (Section 426).

i. Other methods of analysis: Instrumentation and new methods of analysis always are under development. The analyst will find it advantageous to keep abreast of current progress. Reviews of each branch of analytical chemistry are published regularly in *Analytical Chemistry* and the annual literature review in *Journal Water Pollution Control Federation*.

9. Interferences

Many analytical procedures are subject to interference from substances present in the sample. The more common and obvious interferences are known and information about them has been given in the details of individual procedures. It is inevitable that unknown or unexpected interferences may be encountered. Such occurrences are unavoidable because of the diverse nature of waters and particularly of wastewaters. Therefore, be alert to hitherto untested analates, new treatment compounds—especially complexing agents—and new industrial wastes and their potential threat to the accuracy of chemical analyses.

Any sudden change in the apparent composition of water that has been rather constant, any abnormal color observed in a colorimetric test or during a titration, any unexpected turbidity, odor, or other laboratory finding is cause for suspicion. Such change may be due to normal variation in the relative concentrations of the usual constituents or it may be caused by the introduction of an unforeseen interfering substance.

A few substances—such as chlorine, chlorine dioxide, alum, iron salts, silicates, copper sulfate, ammonium sulfate, and polyphosphates—are so widely used that they deserve special mention as possible causes of interference. Of these, chlorine is probably the worst offender, in that it bleaches or alters the colors of many sensitive organic reagents that serve as titration indicators and as color developers for photometric methods. Among the methods that have proved effective in removing chlorine residuals are the addition of minimal amounts of sulfite, thiosulfate, or arsenite; exposure to sunlight or an artificial ultraviolet source; and prolonged storage.

Whenever interference is encountered or suspected and no specific recommendations are given for overcoming it, determine what technic, if any, will eliminate the interference without adversely affecting the analysis itself. If two or more choices of procedure are offered, often one procedure will be less affected than another by the presence of the interfering substance. If different procedures yield considerably different results, it is likely that interference is present. Some interferences become less severe upon dilution or upon use of smaller samples; any tendency of the results to increase or decrease in a consistent manner with dilution indicates the likelihood of interference effects.

a. Types of interference: Interference may cause analytical results to be either too high or too low as a consequence of one of the following:

1) An interfering substance may react as though it were the substance sought and thus produce a high result—for example, bromide will respond to titration as though it were chloride.

2) An interfering substance may react with the substance sought and thus produce a low result.

3) An interfering substance may combine with the analytical reagent and prevent it from reacting with the substance sought—for example, chlorine will destroy many indicators and color-developing reagents.

Nearly every interference will fit one of these classes. For example, in a photometric method, turbidity may be considered as a "substance" that acts like the one being determined—that is, it reduces light transmission. Occasionally, two or more interfering substances, if present simultaneously, may interact in a nonadditive fashion, either canceling or enhancing one another's effects.

b. Counteracting interference: The best way to minimize interference is to remove the interfering substance or to render it innocuous by one of the following methods:

1) Physically remove either the substance sought or the interfering substance. For example, distill off fluoride and ammonia, leaving interferences behind. The interferences also may be absorbed on an ion-exchange resin, as described more fully in Section 106.

2) Adjust the pH so that only the substance sought will react. For example, adjust the pH to 2 so that volatile acids will distill from a solution.

3) Oxidize (digest) or reduce the sample to convert the interfering substance to a harmless form—for example, reduce chlorine to chloride by adding thiosulfate; digest samples for analysis by atomic absorption spectrometry with one of a variety of digestion reagents to destroy organic matter.

4) Add a suitable agent to complex the interfering substance so that it is innocuous although still present: For example, complex iron with pyrophosphate to prevent it from interfering with the copper determination; complex copper with cyanide or sulfide to prevent interference with the titrimetric hardness determination.

5) A combination of the first four technics may be used: For example, distill phenols from an acid solution to prevent amines from distilling; use thiosulfate in the dithizone method for zinc to prevent most of the interfering metals from passing into the CCl_4 layer.

6) Color and turbidity sometimes may be destroyed by wet or dry ashing or may be removed by using a flocculating agent. Some types of turbidity may be removed by filtration. These procedures, however, introduce the danger that the desired constituent also will be removed.

c. Compensation for interference: If none of these technics is practical, several methods of compensation can be used:

1) If color or turbidity initially present interferes in a photometric determination, it may be possible to use photometric compensation. The technic is described in Section 102.7*a* preceding.

2) Determine the concentration of interfering substances and add identical amounts to the calibration standards. This involves much labor.

3) If the interference does not continue to increase as the concentration of interfering substance increases, but tends to level off, add a large excess of interfering substance to all samples and to all standards. This is called "swamping."

4) The presence in the chemical reagents of the substance sought may be accounted for by carrying out a blank determination.

10. Special Requirements

Many of the methods described in the main body of this manual call for apparatus, reagents, and technics that are more or less specific to the particular determination being performed. Such requirements are described under the methods to which they apply. In addition observe the general considerations presented in this section. The requirements of radiological, bacteriological, biological, and bioassay methods tend to differ in many respects from those of chemical and physical tests. Special attention is directed to the descriptions of apparatus and procedures in the sections dealing with those methods.

103 EXPRESSION OF RESULTS

1. Units

In this text, chemical and physical results are expressed in milligrams per liter (mg/L). Record only the significant figures. If concentrations generally are less than 1 mg/L, it may be more convenient to express results in micrograms per liter (μg/L). Use μg/L when concentrations are less than 0.1 mg/L.

Express concentrations greater than 10,000 mg/L in percent, 1% being equal to 10,000 mg/L when the specific gravity is 1.00. In solid samples and liquid wastes of high specific gravity, make a correction if the results are expressed as parts per million (ppm) or percent by weight:

$$\text{ppm by weight} = \frac{\text{mg/L}}{\text{sp gr}}$$

$$\text{\% by weight} = \frac{\text{mg/L}}{10{,}000 \times \text{sp gr}}$$

In such cases, if the result is given as milligrams per liter, state specific gravity.

The unit grains per gallon (gpg) is encountered occasionally. (1 gpg = 17.1 mg/L.) The use of this unit is decreasing, and it is not encouraged.

The unit equivalents per million (epm), or the identical and less ambiguous term milligram-equivalents per liter, or milliequivalents per liter (me/L), can be valuable for making water treatment calculations and checking analyses by anion-cation balance.

Table 103:I presents factors for converting concentrations of common ions from milligrams per liter to milliequivalents per liter, and vice versa. The term milliequivalent used in this table represents 0.001 of an equivalent weight. The equivalent weight, in turn, is defined as the weight of the ion (sum of the atomic weights of the atoms making up the ion) divided by the number of charges normally associated with the particular ion. The factors for converting results from milligrams per liter to milliequivalents per liter were computed by dividing the ion charge by the weight of the ion. Conversely, factors for converting results from milliequivalents per liter to milligrams per liter were calculated by dividing the weight of the ion by the ion charge.

2. Significant Figures

To avoid ambiguity in reporting results or in presenting directions for a procedure, it is the custom to use "significant figures." All digits in a reported result are expected to be known definitely, except for the last digit, which may be in doubt. Such a number is said to contain only significant figures. If more than a single doubtful digit is carried, the extra digit or digits are not significant. If an analytical result is reported as "75.6 mg/L," the analyst should be quite certain of the "75," but may be uncertain as to whether the ".6" should be .5 or .7, or even .4 or .8, because of unavoidable uncertainty in the analytical procedure. If the standard deviation were known from previous work to be ±2 mg/L, the analyst would have, or should have, rounded off the result to "76 mg/L" before reporting it. On the other hand, if the method were so good that a result of "75.61 mg/L" could have been conscientiously reported, then the analyst should not have rounded it off to 75.6.

Report only such figures as are justified by the accuracy of the work. Do not follow the all-too-common practice of requiring that quantities listed in a column have the same number of figures to the right of the decimal point.

a. Rounding off: Round off by dropping digits that are not significant. If the digit 6, 7, 8, or 9 is dropped, increase preceding digit by one unit; if the digit 0, 1, 2, 3, or 4 is dropped, do not alter preceding digit. If

TABLE 103:I. CONVERSION FACTORS*
(Milligrams per Liter—Milliequivalents per Liter)

Ion (Cation)	me/L = mg/L×	mg/L = me/L×	Ion (Anion)	me/L = mg/L×	mg/L = me/L×
Al^{3+}	0.1112	8.994	BO_2^-	0.02336	42.81
B^{3+}	0.2775	3.603	Br^-	0.01252	79.90
Ba^{2+}	0.01456	68.67	Cl^-	0.02821	35.45
Ca^{2+}	0.04990	20.04	CO_3^{2-}	0.03333	30.00
Cr^{3+}	0.05770	17.33	CrO_4^{2-}	0.01724	58.00
			F^-	0.05264	19.00
Cu^{2+}	0.03147	31.77	HCO_3^-	0.01639	61.02
Fe^{2+}	0.03581	27.92	HPO_4^{2-}	0.02084	47.99
Fe^{3+}	0.05372	18.62	$H_2PO_4^-$	0.01031	96.99
H^+	0.9922	1.008	HS^-	0.03024	33.07
K^+	0.02558	39.10	HSO_3^-	0.01234	81.07
			HSO_4^-	0.01030	97.07
Li^+	0.1441	6.941	I^-	0.007880	126.9
Mg^{2+}	0.08229	12.15	NO_2^-	0.02174	46.01
Mn^{2+}	0.03640	27.47	NO_3^-	0.01613	62.00
Mn^{4+}	0.07281	13.73	OH^-	0.05880	17.01
Na^+	0.04350	22.99	PO_4^{3-}	0.03159	31.66
NH_4^+	0.05544	18.04	S^{2-}	0.06238	16.03
Pb^{2+}	0.009653	103.6	SiO_3^{2-}	0.02629	38.04
Sr^{2+}	0.02283	43.81	SO_3^{2-}	0.02498	40.03
Zn^{2+}	0.03059	32.69	SO_4^{2-}	0.02082	48.03

* Factors are based on ion charge and not on redox reactions that may be possible for certain of these ions. Cations and anions are listed separately in alphabetical order.

the digit 5 is dropped, round off preceding digit to the nearest even number: thus 2.25 becomes 2.2 and 2.35 becomes 2.4.

b. Ambiguous zeros: The digit 0 may record a measured value of zero or it may serve merely as a spacer to locate the decimal point. If the result of a sulfate determination is reported as 420 mg/L, the report recipient may be in doubt whether the zero is significant or not, because the zero cannot be deleted. If an analyst calculates a total residue of 1,146 mg/L, but realizes that the 4 is somewhat doubtful and that therefore the 6 has no significance, the answer should be rounded off to 1,150 mg/L and so reported but here, too, the report recipient will not know whether the zero is significant. Although the number could be expressed as a power of 10 (e.g., 11.5 × 10^2 or 1.15 × 10^3), this form is not used generally because it would not be consistent with the normal expression of results and might be confusing. In most other cases, there will be no doubt as to the sense in which the digit 0 is used. It is obvious that the zeros are significant in such numbers as 104 and 40.08. In a number written as 5.000, it is understood that all the zeros are significant, or else the number could have been rounded off to 5.00, 5.0, or 5, whichever was appropriate. Whenever the zero is ambiguous, it is advisable to accompany the result with an estimate of its uncertainty.

Sometimes, significant zeros are dropped without good cause. If a buret is read as "23.60 mL," it should be so recorded, and not as "23.6 mL." The first number indicates that the analyst took the trouble to estimate the second decimal place; "23.6 mL" would indicate a rather careless reading of the buret.

c. The plus-or-minus (±) notation: If a calculation yields as a result "1,476 mg/L" with a standard deviation estimated as ±40 mg/L, report it as 1,480 ± 40 mg/L. However, if the standard deviation is estimated as ±100 mg/L round off the answer still further and report as 1,500 ± 100 mg/L. By this device, ambiguity is avoided and the report recipient can tell that the zeros are only spacers. Even if the problem of ambiguous zeros is not present, showing the standard deviation is helpful in that it provides an estimate of reliability.

d. Calculations: As a practical operating rule, round off the result of a calculation in which several numbers are multiplied or divided to as few significant figures as are present in the factor with the fewest significant figures. Suppose that the following calculation must be made to obtain the result of an analysis:

$$\frac{56 \times 0.003462 \times 43.22}{1.684}$$

A ten-place calculator yields an answer of "4.975740998". Round off this number to "5.0" because one of the measurements that entered into the calculation, 56, has only two significant figures. It was unnecessary to measure the other three factors to four significant figures because the "56" is "the weakest link in the chain" and limits accuracy of the answer. If the other factors were measured to only three, instead of four, significant figures, the answer would not suffer and the labor might be less.

When numbers are added or subtracted, the number that has the fewest decimal places, not necessarily the fewest significant figures, puts the limit on the number of places that justifiably may be carried in the sum or difference. Thus the sum

$$\begin{array}{r} 0.0072 \\ 12.02 \\ 4.0078 \\ 25.9 \\ 4{,}886 \\ \hline 4{,}927.9350 \end{array}$$

must be rounded off to "4,928," no decimals, because of the addends, 4,886, has no decimal places. Notice that another addend, 25.9, has only three significant figures and yet it does not set a limit to the number of significant figures in the answer.

The preceding discussion is necessarily oversimplified. The reader is referred to the bibliography for more detailed sources.

104 PRECISION, ACCURACY, AND CORRECTNESS OF ANALYSES

104 A. Precision and Accuracy

A clear distinction should be made between the terms "precision" and "accuracy" when they are applied to methods of analysis. *Precision* refers to the reproducibility of a method when it is repeated on a homogeneous sample under controlled conditions, regardless of whether or not the observed values are widely displaced from the true value as a result of systematic or constant errors present throughout the measurements. Precision can be expressed by the standard deviation. *Accuracy* refers to the agreement between the amount of a component measured by the test method and the amount actually present. *Relative error* expresses the difference between the measured and the actual amounts, as a percentage of the actual amount. A method may have very high precision but recover only a part of the constituent being determined; or an analysis, although precise, may be in error because of poorly standardized solutions, inaccurate dilution technics, inaccurate balance weights, or improperly calibrated equipment. On the other hand, a method may be accurate but lack precision because of low instrument sensitivity, variable rate of biological activity, or other factors beyond the analyst's control.

It is possible usually to determine both precision and accuracy of a test method by analyzing samples to which known quantities of standard substances have been added. It is possible to determine precision, but not accuracy, of such methods as those for suspended solids, BOD, and numerous physical characteristics because of the unavailability of standard substances that can be added in known quantities on which percentage recovery can be based. For further information on recovery, see 104C.4.

Precision and accuracy data presented in this volume are explained in Section 104A.2, below.

1. Statistical Approach

a. Standard deviation (σ): Experience has shown that if a determination is repeated a large number of times under essentially the same conditions, the observed values, x, will be distributed at random about an average as a result of uncontrollable or experimental errors. If there is an infinite number of observations from a common universe of causes, a plot of relative frequency against magnitude will produce a symmetrical bell-shaped curve known as the Gaussian or normal curve (Figure 104:1). The shape of this curve is defined by two statistical parameters: (1) the mean or average, $\bar{x}$, of n observations; and (2) the standard deviation,

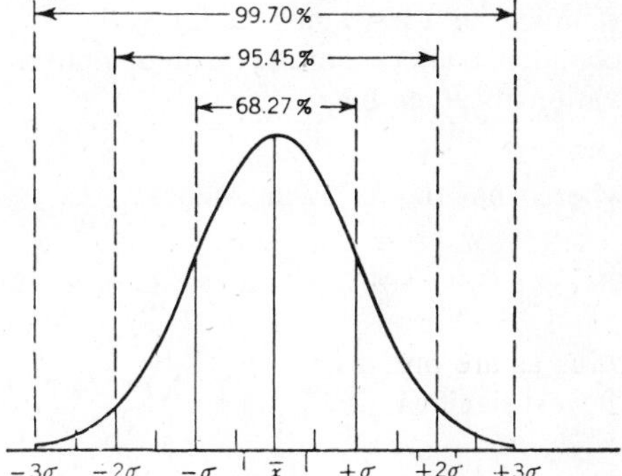

Figure 104:1. Gaussian or normal curve of frequencies.

σ, which fixes the width or spread of the curve on each side of the mean. The formula is:

$$\sigma = \sqrt{\frac{\Sigma (x - \bar{x})^2}{n - 1}}$$

The proportion of the total observations lying within any given range about the mean is related to the standard deviation. For example, 68.27% of the observations lie between $\bar{x} \pm 1\ \sigma$, 95.45% between $\bar{x} \pm 2\ \sigma$, and 99.70% between $\bar{x} \pm 3\ \sigma$. These limits do not apply exactly for any finite sample from a normal population; the agreement with them may be expected to be better as the number of observations, n, increases.

b. Application of standard deviation: If the standard deviation, σ, for a particular analytical procedure has been determined from a large number of samples, and a set of n replicates on a sample gives a mean result $\bar{x}$, there is a 95% chance that the true value of the mean for this sample lies within the values $\bar{x} \pm 1.96\ \sigma/\sqrt{n}$. This range is known as the 95% confidence interval. It provides an estimate of reliability of the mean and may be used to forecast the number of replicates needed to secure suitable precision.

If the standard deviation is not known and is estimated from a single small sample,* or a few small samples, the 95% confidence interval of the mean of n observations is given by:

$$\bar{x} \pm t\ \ \sigma/\sqrt{n},$$

where t has the following values:

n	t
2	12.71
3	4.30
4	3.18
5	2.78
10	2.26
∞	1.96

The use of t compensates for the tendency of small samples to underestimate the variability.

c. Range (R): The difference between the smallest and largest of n observations also is closely related to the standard deviation. When the distribution of errors is normal in form, the range, R, of n observations exceeds the standard deviation times a factor d_n only in 5% of the cases. Values for factor d_n are:

n	d_n
2	2.77
3	3.32
4	3.63
5	3.86
6	4.03

As it is rather general practice to run replicate analyses, use of these limits is very convenient for detecting faulty technic, large sampling errors, or other assignable causes of variation.

d. Rejection of experimental data: Quite often in a series of observations, one or more of the results deviate greatly from the mean whereas the other values are in close agreement with the mean. At this point, decide whether to reject disagreeing values. Theoretically, no results should be rejected, because the presence of disagreeing results shows faulty technics and therefore casts doubt on all results. Reject the result of any test in which a known error has occurred.

For methods for rejection of other experimental data, consult standard texts on analytical chemistry or statistical measurement.

2. Evaluation of Methods

Data on precision of methods first appeared in the 10th Edition. In preparation of the 11th Edition, a concerted effort was made to offer an idea of the precision and accuracy with which selected methods can be applied on a broad geographic basis in examination of the relatively simpler water samples. The manner of best expressing the resulting data is a matter of continuing study. The 11th and 12th Editions presented both precision and accuracy in terms of milligrams per liter. This practice is retained where such data continue to be cited in this edition. However, more recent experience suggests that data can be presented more briefly and clearly in the form of a percentage. By this system, the standard deviation is expressed as a percentage of the mean and is termed the relative standard deviation or coefficient of variation. It measures the precision or reproducibility of a method, independent of the known concentration of sample constituent. Similarly, the relative error gives the difference between the mean of a series of test results and the true value, expressed as a percentage of the true value. Thus, the relative error repre-

*A "small sample" in statistical discussions means a small number of replicate determinations, n, and does not refer to the quantity used for a determination.

sents the measure of accuracy of a method. The relative standard deviation and relative error are preferred in quoting precision and accuracy of a method.

The information regarding precision and accuracy of the methods presented in this volume has been obtained from a number of sources and at various times.

a. Water Pollution Control Federation: Some of the oldest data are presented in connection with certain wastewater methods. These were collected on the initiative of the Standard Methods Committee of the Water Pollution Control Federation and some of the data appeared first in the 10th edition of this manual. For many methods, results were obtained from 10 replicate determinations on 10 different days, or when necessary, from 5 replicate samples on 20 days.

Most methods studied were found to be statistically reliable and the standard deviation given may be used with some confidence in statistical prediction. If a method has been found statistically unreliable, this is indicated in the statements on precision under the method. The standard deviations of unreliable methods cannot be used safely for statistical prediction but may be of some value for indicating roughly the variation that may be expected.

In expressing evaluation data on each test, the number of analysts and determinations is given in shorthand form; for example, "n = 5; 56 × 10," means that 5 different analysts ran 56 separate sets of 10 determinations each, making a total of 560 determinations. Usually precision is expressed as the standard deviation in original units of measurement—i.e., milligrams or milliliters. In a few instances, precision is expressed as the coefficient of variation C_v (ratio of standard deviation to the average), expressed as percentage:

$$C_v = \frac{100\sigma}{\bar{x}}$$

The standard deviation given with each method is based on careful laboratory examination. No attempt was made to obtain the standard deviation under research conditions or with the use of specially calibrated apparatus or glassware. The values given are to be regarded as provisional and subject to change on further study. In general, the standard deviations given may be regarded as being too high rather than too low.

b. Analytical Reference Service: For a number of methods applicable to relatively clean water samples, precision and accuracy data are based on results of studies by the Analytical Reference Service, which was conducted formerly by the U.S. Public Health Service.

This activity was devoted to the collaborative study of water chemistry methodology. Nearly 300 agencies, including public, private, and university laboratories, were involved in collaborative testing. The results provide an evaluation of selected analytical methods and supply a factual basis for judgment of the reliability that may be expected in the practical application of these methods.

c. Environmental Protection Agency: Currently, the United States Environmental Protection Agency (EPA) is involved in an extensive program for evaluating both analytical methods and the performance of its own laboratories. This activity is directed by the Environmental Monitoring and Support Laboratory (EMSL) of EPA. Data from these studies have been made available for inclusion in this volume and for a number of methods they have replaced older results.

3. Graphical Representation of Data

Graphical representation of data is one of the simplest methods for showing the influence of one variable on another. Graphs frequently are desirable and advantageous in colorimetric analysis because they show any variation of one variable with respect to the other within specified limits.

a. General: Ordinary rectangular-coordinate paper is satisfactory for most purposes. For some graphs, semi-logarithmic paper is preferable.

The five rules listed by Worthing and Geffner (see Section 109) for choosing coordinate scales are useful. Although these rules are not inflexible, they are satisfactory. When doubt arises, use common sense. The rules are:

1) Plot the independent and dependent variables on abscissa and ordinate in a manner that can be comprehended easily.

2) Choose the scales so that the value of either coordinate can be found quickly and easily.

3) Plot the curve to cover as much of the graph paper as possible.

4) Choose the scales so that the slope of the curve approaches unity as nearly as possible.

5) Other things being equal, choose the variables to give a plot that will be as nearly a straight line as possible.

Entitle a graph to describe adequately what the plot is intended to show. Present legends on the graph to clarify possible ambiguities. Include in the legend complete information about the conditions under which the data were obtained.

b. Method of least squares: If sufficient points are available and the functional relationship between the two variables is well defined, a smooth curve can be drawn through the points. If the function is not well defined, as is frequently the case when experimental data are used, use the method of least squares to fit a straight line to the pattern.

Any straight line can be represented by the equation $x = my + b$. The slope of the line is represented by the constant m and the slope intercept (on the x axis) is represented by the constant b. The method of least squares has the advantage of giving a set of values for these constants not dependent upon the judgment of the investigator. Two equations in addition to the one for a straight line are involved in these calculations:

$$m = \frac{n\,\Sigma xy - \Sigma x\,\Sigma y}{n\,\Sigma y^2 - (\Sigma y)^2}$$

$$b = \frac{\Sigma y^2\,\Sigma x - \Sigma y\,\Sigma xy}{n\,\Sigma y^2 - (\Sigma y)^2}$$

n being the number of observations (sets of x and y values) to be summed. To compute the constants by this method, first calculate Σx, Σy, Σy^2, and Σxy. Carry out these operations to more places than the number of significant figures in the experimental data because the experimental values are assumed to be exact for the purposes of the calculations.

Example: Given the following data to be graphed, find the best line to fit the points:

Absorbance	Solute Concentration *mg/L*
0.10	29.8
0.20	32.6
0.30	38.1
0.40	39.2
0.50	41.3
0.60	44.1
0.70	48.7

Let y equal the absorbance values that are subject to error and x the accurately known concentration of solute. First find the summations (Σ) of x, y, y^2, and xy:

x	y	y^2	xy
29.8	0.10	0.01	2.98
32.6	0.20	0.04	6.52
38.1	0.30	0.09	11.43
39.2	0.40	0.16	15.68
41.3	0.50	0.25	20.65
44.1	0.60	0.36	26.46
48.7	0.70	0.49	34.09
Σ=273.8	2.80	1.40	117.81

Next substitute the summations in the

equations for m and b; $n = 7$ because there are seven sets of x and y values:

$$m = \frac{7\ (117.81) - 2.80\ (273.8)}{7\ (1.40) - (2.80)^2} = 29.6$$

$$b = \frac{1.4\ (273.8) - 2.80\ (117.81)}{7\ (1.40) - (2.80)^2} = 27.27$$

To plot the line, select three convenient values of y—say, 0, 0.20, 0.60—and calculate the corresponding values of x:

$$x_0 = 29.6\ (0) + 27.27 = 27.27$$
$$x_1 = 29.6\ (0.20) + 27.27 = 33.19$$
$$x_2 = 29.6\ (0.60) + 27.27 = 45.04$$

When the points representing these values are plotted on the graph, they will lie in a straight line (unless an error in calculation has been made) that is the line of best fit for the given data. These data points are plotted in Figure 104:2.

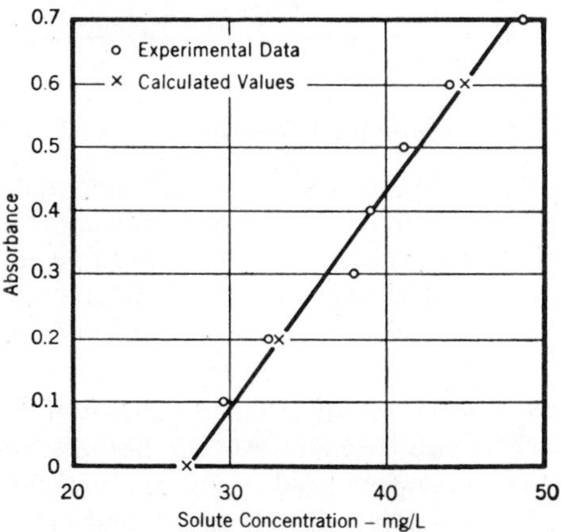

Figure 104:2. Example of least-squares method.

4. Self-Evaluation (Desirable Philosophy for the Analyst)

A good analyst tempers confidence with doubt. Such doubt stimulates a search for new and different methods of confirmation for reassurance. Frequent self-appraisals should embrace every step—from collecting samples to reporting results.

The analyst's first critical scrutiny should be directed at the entire sample collection process to guarantee a representative sample for analysis and to avoid any possible losses or contamination during collection. Attention also should be given to the type of container and to the manner of transport and storage, as discussed elsewhere in this volume.

A periodic reassessment should be made of available analytical methods, with an eye to applicability for the purpose and the situation. In addition, each method selected must be evaluated by the analyst for sensitivity, precision, and accuracy, because only in this way can it be determined whether the analyst's technic is satisfactory and whether directions have been interpreted properly. Self-evaluation on these points can give the analyst confidence in the value and significance of reported results.

The benefits of less rigid intralaboratory as well as interlaboratory evaluations deserve serious consideration. The analyst regularly can check standard or unknown concentrations with and without interfering elements and compare results on the same sample with results obtained by others. Such programs can uncover weaknesses in the analytical chain and permit improvements to be instituted without delay. The results can disclose whether the trouble stems from faulty sample treatment, improper elimination of interference, poor calibration practices, sloppy experimental technic, impure or incorrectly standardized reagents, defective instrumentation, or even inadvertent mistakes in arithmetic.

Other checks of an analysis are described in Section 104C and involve anion-cation balance, conductivity, ion exchange, and recovery of added substance in the sample.

All these approaches are designed to appraise and upgrade the level of laboratory performance and thus inspire greater faith in the final reported results.

104 B. Quality Control in Chemical Analysis*

1. Introduction

Quality assurance in the laboratory has come to mean many things and to some is merely equated with good laboratory operations such as:

1. Adequately trained and experienced personnel,
2. Good physical facilities and equipment,
3. Certified reagents and standards,
4. Frequent servicing and calibration of instruments, and
5. A knowledgeable and understanding management.

While all of these are important, none in itself assures reliability of laboratory data. A good analytical quality control program consists of three factors: (1) using only methods that have been studied collaboratively and found acceptable (this generally implies "Standard Methods"),[1,2] (2) routinely analyzing a control sample at least once each day[3] on which unknown samples are being analyzed, and (3) confirming the ability of a laboratory to produce acceptable results by requiring analysis of a few reference samples once or twice a year. The second factor may be designated internal or statistical quality control, while the third is external quality control, proficiency testing, or laboratory evaluation. In the following discussion, internal quality control will be emphasized. It is based on a system developed for the control of production processes and product quality, but the same concepts are adapted readily to laboratory operations.

2. Internal Quality Control

a. Control charts: The applicability of control chart technics is based on the assumption that laboratory data approximate a normal distribution like that shown in Figure 104:1. The data from such a system, however, can be presented in a different graphic way by plotting on the vertical scale the units of the test results and on the horizontal scale the order or sequence in which the results were obtained. The mean and limits of dispersion in terms of the standard deviation are calculated and plotted as in the control chart of Figure 104:3.

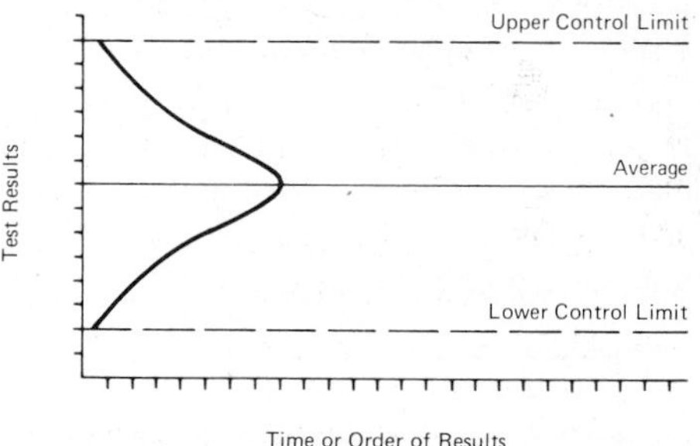

Figure 104:3. Control chart.

For best results make at least 20 determinations on a sample before calculating the standard deviation or plotting the control chart. All results need not be obtained on the same day; optimally accumulate them as part of a day-to-day operation. A result may be obtained on a control sample each time an unknown or group of unknowns is analyzed.

For example, consider a collection of results, in milligrams per liter, obtained by analysis of a water sample for copper, as follows:

1. 0.251	11. 0.229
2. 0.250	12. 0.250
3. 0.250	13. 0.283
4. 0.263	14. 0.300
5. 0.235	15. 0.262
6. 0.240	16. 0.270
7. 0.260	17. 0.225
8. 0.290	18. 0.250
9. 0.262	19. 0.256
10. 0.234	20. 0.250

*For discussion of quality control in bacteriological testing see Section 902.

The mean of this series is 0.256 mg/L, the standard deviation is 0.020 mg/L, and the resulting control chart is that of Figure 104:4.

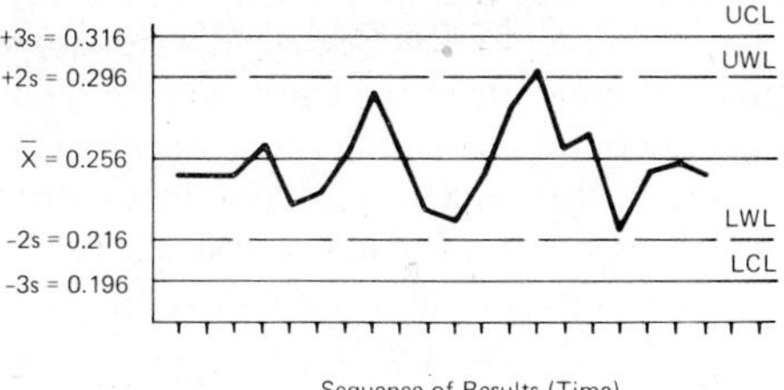

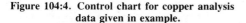

Figure 104:4. Control chart for copper analysis data given in example.

The upper and lower control limits (UCL and LCL) are at +3 and −3 standard deviations from the mean, respectively, and the upper and lower warning limits (UWL and LWL) at +2 and −2 standard deviations.

This is a control chart for individuals, and is used most frequently in chemical analysis. In quality control on a production line, every fifth or tenth item can be measured or tested and returned. Laboratory control requires adding known or standard samples to the unknown samples, thus increasing both the cost and time of analysis.

If a result falls outside the control limits on an individual control chart, the analysis is said to be "out of control"; take immediate action to determine the cause of the outlying result. Consider unreliable any analytical results obtained for samples on the same day as the erroneous result occurred on the known sample and repeat the analyses after corrective action has been taken and the procedure is back in control. It sometimes is desirable to take milder action when the results exceed the warning limits, because an unduly high percentage of results exceeding these limits is a warning that laboratory precision may not be as good as expected or that distribution of results is not normal.

b. Multiple sample control chart: Preferably use several control samples that span a range of concentrations. This avoids unintentional bias on the part of an analyst who becomes familiar with the assay value of a single control. When using several control samples, modify the control chart so that instead of the analytical results, the deviation of the results from the mean concentration of the corresponding control sample is plotted (Figure 104:5). In general, use such a control chart only for control samples that do not differ widely from each other in concentration. In this case, it can be assumed that the standard deviation is constant over the range of concentrations spanned by the control samples. Set the upper and lower control limits at ±3 standard deviations and the upper and lower warning limits at ±2 standard deviations.

c. The $\bar{x}$-R control chart: The $\bar{x}$-R control chart (really two charts but usually treated as one) is probably the most widely used in industrial applications where large numbers of a product are being sampled because the $\bar{x}$ chart is more sensitive to change in the mean (which is sometimes referred to as a shift or trend), and the R chart detects changes in the dispersion or variability (standard deviation).

Figure 104:5. Multiple sample control chart.

To construct an $\bar{x}$-R control chart, accumulate at least 20 pairs of duplicate determinations on the control sample. Triplicate, quadruplicate, or even more replicate determinations can be used. Although improved results are obtained, the time and cost involved become self-limiting, so that, if the $\bar{x}$-R chart is used at all, it is limited to the analysis of duplicates.

After collecting 20 pairs of duplicate determinations, compute the averages ($\bar{x}_i$) of each pair of results and the range (R) of each pair. Then calculate the grand mean ($\bar{\bar{x}}$):

$$\bar{\bar{x}} = \Sigma \bar{x}_i / n$$

and the mean range ($\bar{R}$):

$$\bar{R} = \Sigma R / n$$

Calculate upper and lower control limits using Table 104:I, which gives control chart constants (essentially 90% confidence limits for various sample sizes), and the following equations:

Control limits for averages:

$$UCL = \bar{\bar{x}} + A_2\bar{R}$$
$$LCL = \bar{\bar{x}} - A_2\bar{R}$$
$$UWL = \bar{\bar{x}} + 2/3(A_2\bar{R})$$
$$LWL = \bar{\bar{x}} - 2/3(A_2\bar{R})$$

Control limits for ranges:

$$UCL = D_4\bar{R}$$
$$LCL = D_3\bar{R}$$
$$UWL = \bar{R} + 2/3(D_4\bar{R} - \bar{R})$$

where A_2, D_3, and D_4 are constants depending on subgroup size (Table 104:I):

TABLE 104:I. FACTORS FOR COMPUTING CONTROL CHART LINES[3]

Observations in Subgroup (n)	Factor A_2	Factor D_3	Factor D_4
2	1.88	0	3.27
3	1.02	0	2.58
4	0.73	0	2.28
5	0.58	0	2.12
6	0.48	0	2.00
7	0.42	0.076	1.92
8	0.37	0.136	1.86

As an example, Table 104:II gives an array of data and Figure 104:6 is an $\bar{x}$-R chart for these data.

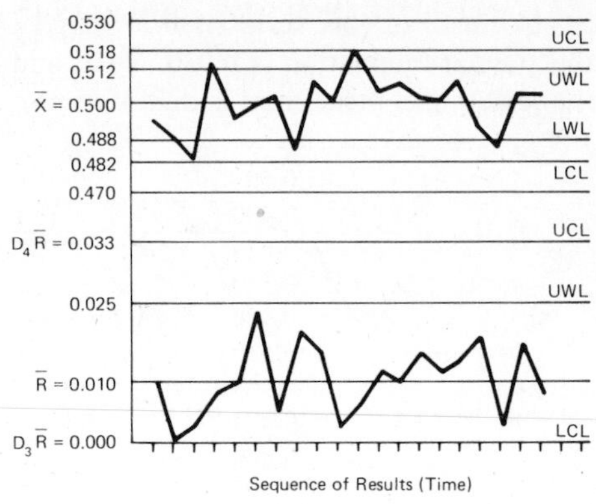

Figure 104:6. $\bar{x}$ − R chart.

After constructing the $\bar{x}$-R chart use it as you would the individual control chart in day-to-day operation, but obtain duplicate determinations (if the control chart was constructed using duplicates; otherwise whatever number was used) on the control sample every time an unknown or group of unknowns is analyzed. Calculate average and range of replicate determinations on control sample and plot them on the $\bar{x}$-R chart. If either falls outside the control limits, take corrective action.

d. Moving averages and ranges: A study of the control chart constants (Table 104:I) shows that the $\bar{x}$-R chart is more efficient for detecting modest changes in the process as the subgroup size increases. A reasonable compromise between individual values and larger subgroups is to use moving averages. Such a set of data is given in Table 104:III. The moving average for n samples is the mean of the results for those samples, $\Sigma x/n$, and the moving range is the difference between the extreme results in the group of n samples. The moving average smooths out variations in results. The moving range also can be plotted to serve as a measure of dispersion. Thus, it is possible to obtain nearly the same information from individual determinations on a control sample while avoiding the additional cost and time in-

TABLE 104:II. DATA AND COMPUTATIONS FOR CONSTRUCTION OF EXAMPLE $\bar{x}$-R CHART

Sequence of Results	Results x_i	Results x'_i	Average $\bar{x}_i$	Range R
1	0.501	0.491	0.496	0.010
2	0.490	0.490	0.490	0.000
3	0.479	0.482	0.480	0.003
4	0.520	0.512	0.516	0.008
5	0.500	0.490	0.495	0.010
6	0.510	0.488	0.499	0.022
7	0.505	0.500	0.502	0.005
8	0.475	0.493	0.484	0.018
9	0.500	0.515	0.508	0.015
10	0.498	0.501	0.500	0.003
11	0.523	0.516	0.520	0.007
12	0.500	0.512	0.506	0.012
13	0.513	0.503	0.508	0.010
14	0.512	0.497	0.504	0.015
15	0.502	0.500	0.501	0.002
16	0.506	0.510	0.508	0.004
17	0.485	0.503	0.494	0.018
18	0.484	0.487	0.486	0.003
19	0.512	0.495	0.504	0.017
20	0.509	0.500	0.504	0.009
			$\Sigma\bar{x}_i = 10.005$	$\Sigma R = 0.191$

$\bar{\bar{x}} = \frac{10.005}{20} = 0.500$ $\bar{R} = \frac{0.191}{20} = 0.0096 \simeq 0.010$

volved in making duplicate determinations.

e. Accuracy: Accuracy in analytical chemistry is a measure of the difference between the mean of the results and the true value. Most frequently, accuracy is expressed in terms of the mean error:

$$\text{mean error} = \bar{x} - T.V.$$

where:

$\bar{x}$ = mean
$T.V.$ = true value

Accuracy can be determined only if the control sample is prepared in the laboratory from known amounts of pure reagents (see Section 104B.4). The accuracy of the determination is the difference between the mean of the 20 determinations (collected for the construction of the initial control chart) and the true value or known amount of chemical added.

The true value ($T.V.$) also can be plotted on the control chart. Usually it falls only slightly above or below the mean. If the true value is more than half way between the mean and the upper or lower warning limit (i.e., the mean error is greater than one standard deviation), check the method, reagent, glassware, technic, or instruments for bias and take corrective action.

3. External Quality Control

The ability of a laboratory to produce acceptable results can be confirmed by requiring analysis once or twice a year of a few reference samples. These reference samples may be no different from the control samples that the laboratory has been

TABLE 104:III. MOVING AVERAGE AND RANGE TABLE (n = 2)

Sample No.	Assay Value	Sample Nos. Included	Moving Average	Moving Range
1	17.09	—	—	—
2	17.35	1–2	17.22	+0.26
3	17.40	2–3	17.38	+0.05
4	17.23	3–4	17.32	−0.17
5	17.09	4–5	17.16	−0.14
6	16.94	5–6	17.02	−0.15
7	16.68	6–7	16.81	−0.26
8	17.11	7–8	16.90	+0.43
9	18.47	8–9	17.79	+1.36
10	17.08	9–10	17.78	−1.39
11	17.08	10–11	17.08	0.00
12	16.92	11–12	17.00	−0.16
13	18.03	12–13	17.45	+1.11
14	16.81	13–14	17.42	−1.22
15	17.15	14–15	16.98	+0.34
16	17.34	15–16	17.25	+0.19
17	16.71	16–17	17.03	−0.73
18	17.28	17–18	17.00	+0.57
19	16.54	18–19	16.91	−0.74
20	17.30	19–20	16.92	+0.76

preparing for its own use, with the exception that the amount of each substance present is unknown to the analysts.

A check by one laboratory of the proficiency or ability of another laboratory to obtain acceptable results (i.e., an external quality control program) is a check on whether the laboratory being tested has an acceptable internal quality control program. The analysis of an occasional reference sample may, however, serve another useful purpose in that it may reveal an error in the preparation of standards or control samples or the use of poor-quality distilled water, reagents, etc., that may be causing extreme variability of results.

In evaluating proficiency testing results, the laboratory issuing the reference samples may use control charts. Control limits are generally larger than those for internal quality control charts because variation between laboratories is always larger than variation within one laboratory as a result of differences in instruments, glassware, etc.

If data for use in constructing an external or interlaboratory quality control chart are not available from the results of analysis of the reference sample by many laboratories, use data from collaborative studies of the method to establish control limits.

4. Preparation of Control Samples

Control samples can be natural water samples. In fact, with data obtained by recycling about 20% of the normal stream of samples to obtain duplicate determinations, one can construct a control chart. However, such a chart will provide information only about precision of the determination.

To obtain information on the accuracy of the determination, prepare a synthetic sample by adding known amounts of a pure chemical or chemicals to distilled water. Such samples may contain substances known to interfere with the determination, thus simulating natural samples. The true value of the constituents of these synthetic samples is the amount added.

Control samples also may be natural water samples or samples with standard additions (natural water with one or more chemicals added) that have been assayed by several "referee" laboratories, preferably by several different methods providing good agreement, so that a "known" value can be assigned to the sample.

One of the most important factors in a quality control program is the availability of an adequate supply of a stable known control. Ideally, prepare enough of the control so that it can be used for as long as 6 months; this will eliminate the need for preparing new quality control charts at more frequent intervals. Preserve samples to prevent changes due to bacterial or mold growths, precipitation, or plating out on container walls.

Concentrates often are more stable than dilute solutions that resemble natural waters. Therefore, prepare concentrates that are diluted before use.

In Section 4*b*, below, the compositions of five synthetic concentrated samples are presented. They are most useful in controlling analysis of potable water. The concentrates are quite stable and if kept in a cool place out of direct sunlight they may be used for at least 6 months. Prepare new dilutions (the working control samples) weekly and preferably each day of use.

a. Preparation: Dilute all concentrates (Solutions A, B, C, D, and E) 5 mL to 1,000 mL with good-quality distilled water just before use. If several control samples are wanted, prepare by diluting 4 mL or 6 mL (or some other volume) of the concentrate to 1,000 mL with distilled water. Calculate the concentrations in these other dilutions.

b. Synthetic concentrated samples:

Solution A

Metal	Salt*	Stock Solutions *g/L*	mL stock/L concentrate	mg/L concentrate†	mg/L diluted 5 mL/L
Zinc	Zn metal in HNO_3	1.0000	10	10	0.05
Cadmium	$Cd(NO_3)_2 \cdot 4H_2O$	2.7442	4	4	0.02
Lead	$Pb(NO_3)_2$	1.5984	10	10	0.05
Iron	$Fe(NO_3)_3 \cdot 9H_2O$	7.2359	30	30	0.15
Manganese	Mn metal in HNO_3	1.0000	20	20	0.10
Chromium	$K_2Cr_2O_7$	2.8281	10	10	0.05
Silver	$AgNO_3$	1.5748	10	10	0.05
Copper	Cu metal in HNO_3	1.0000	5	5	0.025
Cobalt	$Co(NO_3)_2 \cdot 6H_2O$	4.9383	8	8	0.04
Barium	$Ba(NO_3)_2$	1.9029	30	30	0.15
Mercury	$Hg(NO_3)_2$	1.6184	1.0	1.0	0.005

* Use nitrate salts because sulfates precipitate barium and chlorides precipitate silver.
† If necessary, add HNO_3 until pH of concentrate is about 2, before making up to a final volume of 1L.

Solution B

Dry KH_2PO_4 in a desiccator, weight 54.436 g, and transfer into a 1-L volumetric flask. Add 180.2 mL 1.0000 *N* NaOH. Make solution up to volume with distilled water. When diluted 5 mL to 1,000 mL with distilled water this will produce:

Total dissolved solids	293 mg/L
Conductivity	303 μmhos/cm
pH	6.87

Solution C

Anion	Salt	Stock* Solution *g/L*	mL stock/L concentrate	mg/L in concentrate	mg/L diluted 5 mL/L
NO_3	KNO_3	8.1525	undiluted	5,000.0	25.0
SO_4	Na_2SO_4	5.9144	undiluted	4,000.0	20.0
LAS	Standard†		as required	30.0	0.15
Cl	KCl	23.1335	undiluted	11,000.0	55.0
F	NaF	2.2104	240	240.0	1.2

* Preserve with 1 mL/L of $HgCl_2$ (27.1 g/L).
† From Environmental Monitoring and Support Laboratory, U.S. EPA, Cincinnati, Ohio 45268.

Solution D

Standardize a solution of sodium or potassium cyanide containing approximately 2.0 mg CN/mL by titrating with standard silver nitrate that has been standardized the same day with standard sodium chloride. If this stock standard solution is found to contain, for example, 1.89 mg CN/mL take 42.3 mL of this solution, make up to about 3.5 L, adjust to pH 11.5 with NaOH, and dilute to 4,000 mL. This concentrate contains 20.0 mg CN/L, and is quite stable. Dilute 5 mL to 1,000 mL just before use to obtain 0.10 mg CN/L.

5. References

1. Evaluation of Laboratory Methods for the Analysis of Inorganics in Water. 1968. Advan. Chem. Ser. No. 73, p. 253, American Chemical Soc.
2. McFarren, E.F., R.J. Lishka & J.H. Parker. 1970. Criterion for judging and acceptability of analytical methods. *Anal. Chem.* 42:358.
3. Bennett, C.A. & N.L. Franklin. 1954. Statistical Analysis in Chemistry and the Chemical Industry. John Wiley and Sons, Inc., New York, N.Y.

Solution E

Metal	Salt	Stock Solution *g/L*	mL stock/L concentrate	mg/L concentrate	mg/L diluted 5 mL/L
Arsenic	$Na_2HAsO_4 \cdot 7H_2O$	4.1653	8.0	8.0	0.04
Selenium	SeO_2	1.4052	1.0	1.0	0.005

104 C. Checking Correctness of Analyses

The following procedures for checking the correctness of analyses are applicable specifically to water samples for which relatively complete mineral analyses are made.

1. Anion-Cation Balance

Theoretically, the sum of the anions, expressed in milliequivalents per liter (me/L), must equal exactly the sum of the cations, similarly expressed, in any sample. In practice, the sums seldom are equal because of unavoidable variations in analysis. This inequality increases as the ionic concentration increases. A control chart can be constructed so that it will be evident immediately if the difference between the sums of the anions and cations falls between acceptable limits, which have been taken as ±1 standard deviation.

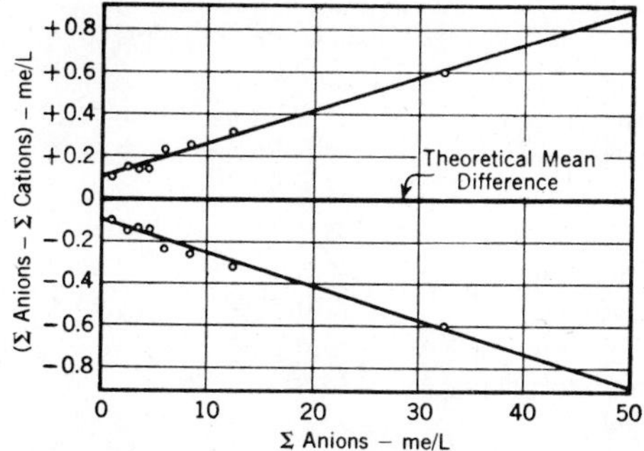

Figure 104:7. Control chart for anion-cation balances.

If the difference is plotted against the sum of the anions, lines showing ±1 standard deviation, that is, the acceptable limits, are given by the equation:

$$\Sigma \text{ anions} - \Sigma \text{ cations} = \pm (0.1065 + 0.0155\ \Sigma \text{ anions})$$

This is shown in Figures 104:7 and 104:8, which represent modified control charts. Values of differences of the sums falling outside of the limits set by the equations indicate that at least one of the determinations should be rechecked.

A fortuitous combination of erroneous analyses resulting in the balancing of errors (compensating errors) may produce agreement between the sum of the anions and the sum of the cations even though two or more individual analytical results are seriously incorrect. The additional methods of checking that follow are useful for detecting such discrepancies.

2. Conductivity

In using conductivity check analyses, two methods of calculation may be used.

a. Rough calculation: In most natural waters it has been found that when conductivity (in micromhos per centimeter at 25 C) is multiplied by a factor ordinarily in the range of 0.55 to 0.7, the product is equal to milligrams per liter total filtrable residue. For waters containing appreciable concentrations of free acid or caustic alkalinity, the factor may be much lower than 0.55 and for highly saline waters it may be much higher than 0.7. An approximate check will reveal gross mistakes in analysis.

b. More refined calculations: To obtain better results based on electrical conductance dilute sample so that its conductivity falls within a narrow range and take into account the contribution of each separate ion to total measured conductivity. Use distilled water, boiled and cooled, to dilute sample in a known ratio until the conductivity lies between 90 and 120 μmhos/cm. Some trial may be neces-

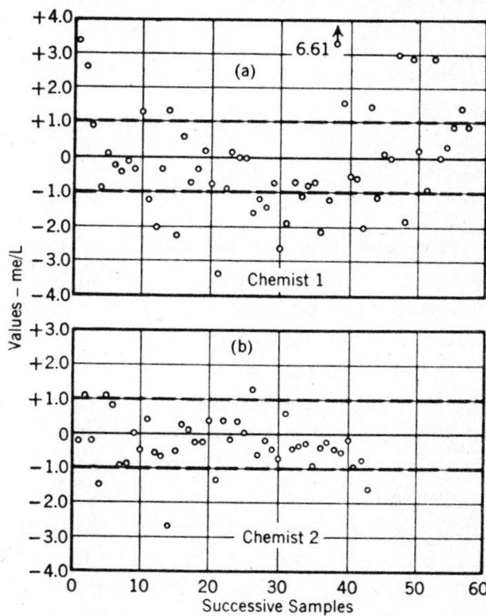

Figure 104:8. Control chart with transformed limits. The vertical scale is plotted:

$$\frac{\Sigma\ anions - \Sigma\ cations}{0.1065 + 0.0155\ \Sigma\ anions}$$

sary to achieve the proper dilution because conductivity does not vary in exact ratio to dilution; if it did, dilution would be unnecessary. The exact dilution ratio, D, must be known:

$$D = \frac{V_s + V_w}{V_s}$$

where:
V_s = volume of sample and
V_w = volume of distilled water.

Determine conductivity of distilled water, which should be less than 2 μmhos/cm. Determine conductivity in the usual way and calculate "diluted conductivity", K_d, from the equation:

$$K_d = \frac{AD \times 10^6}{R_d} - (D - 1)\, K_w$$

where:
A = cell constant,
R_d = measured resistance of the diluted sample, ohms, and
K_w = distilled-water conductivity.

Next, compute the diluted conductivity from the chemical analysis by multiplying the concentration found (as either milliequivalents per liter or milligrams per liter) by the appropriate factor in Table 104:IV and summing the products. If the computed diluted conductivity is more than 1.5% greater or more than 2% lower than the measured value of the diluted conductivity, the chemical analysis probably is in error and should be rechecked.

The diluted-conductivity method of checking is not applicable to samples having conductivities lower than 90 μmhos/cm or pH values less than 6 or greater than 9, or to samples that contain significant quantities of ions not listed in Table 104:IV. The conductivities due to hydrogen ion and hydroxyl ion are much greater than those due to other ions and will cause the method to be invalid for samples outside the pH 6 to 9 range.

TABLE 104:IV. CONDUCTIVITY FACTORS OF IONS COMMONLY FOUND IN WATER

Ion	Conductivity (25 C) *μmhos/cm*	
	Per me/L	Per mg/L
Bicarbonate	43.6	0.715
Calcium	52.0	2.60
Carbonate	84.6	2.82
Chloride	75.9	2.14
Magnesium	46.6	3.82
Nitrate	71.0	1.15
Potassium	72.0	1.84
Sodium	48.9	2.13
Sulfate	73.9	1.54

3. Ion Exchange

Accuracy of the chemical analysis also can be checked when all major ionic constituents of a simple natural water are determined quantitatively by means of an ion-exchange method. A serious discrepancy between the cations, milliequivalents per liter, obtained by titration of the sample after ion exchange, and the total cation concentration found by summing the determined constituents can uncover a gross error in analysis.

The ion-exchange method is based on replacement of cations in the original sample with hydrogen ions supplied by a strongly acidic cation-exchange resin. The acid produced by mixing the sample with the resin is titrated with standard sodium hydroxide, NaOH. Total alkalinity of the original sample must be measured to complete the calculation.

The apparatus required for the ion-exchange method consists of a pH meter (line- or battery-operated) or methyl orange indicator, a magnetic or mechanical strirrer provided with a speed control, a 10-mL buret graduated in 0.05-mL steps, and glass wool. The reagents are 0.02*N* standard acid, 0.02*N* standard NaOH, a strongly acidic cation-exchange resin* of analytical grade, and distilled water.

Pipet a sample containing 0.1 to 0.2 me cations into a 250-mL erlenmeyer flask or beaker and add enough distilled water to bring volume to 100 mL. Add 2.0 g cation-exchange resin and stir at moderate speed for 15 min. Remove resin by filtering through a plug of glass wool placed in neck of a 10-cm borosilicate glass funnel and wash with two 15-mL portions distilled water. Titrate combined filtrate and washings to pH 4.5 with 0.02*N* NaOH, using a pH meter or methyl orange as end-point indicator and stirring during titration.

The cation exchange also can be done by the column method described in Section 106.

Determine alkalinity in the sample, in milliequivalents per liter, by titrating a portion to a pH of 4.5 with 0.02*N* standard acid, using a pH meter or methyl orange as end-point indicator.

Compute total cations in sample by the equation:

$$E = \frac{AB \times 1{,}000}{C} + D \qquad (1)$$

where:

A = volume of standard NaOH used in titration, mL,
B = normality of NaOH,
C = volume of sample taken, mL,
D = alkalinity, me/L, and
E = total cations in sample, me/L.

The column method of ion exchange is preferred for checking results of gravimetric determinations for filtrable residues because less attention to total milliequivalents of cations in the sample is required.

Determine filtrable and fixed filtrable residues by multiplying the quantity ($E - D$) from Equation 1 by a factor—usually between 70 and 90, depending on the water tested and the resin used in the column—and adding 50*D*. For more accurate determinations, derive factors for type of water tested routinely by subtracting 50*D* from the filtrable residues and dividing results by the quantity ($E - D$), thus:

$$F_1 = \frac{R_{fg} - 50D}{E - D} \qquad (2)$$

$$F_2 = \frac{R_{ffg} - 50D}{E - D} \qquad (3)$$

where:

R_{fg}, R_{ffg} = filtrable and fixed filtrable residues determined gravimetrically and
F_1, F_2 = resulting factors.

The equations for calculating the filtrable residues are then:

$$R_{fx} = (E - D)\,F_1 + 50D \qquad (4)$$

$$R_{ffx} = (E - D)\,F_2 + 50D \qquad (5)$$

where:

R_{fx}, R_{ffx} = filtrable and fixed filtrable residues determined by ion exchange.

Once these factors are derived for a given water, filtrable-residue values can be determined by ion exchange and gravimetric values obtained less frequently, or only when there is a major change in the water being checked.

For approximation purposes, use a factor of 80 in either Equation 4 or Equation 5. If the water tested has an unusually high sulfate or bicarbonate content compared to chloride (in milligrams per liter), or has unusually high organic content, the factor in Equation 4 approaches 90. If the water tested has an unusually high chloride content compared to sulfate (in milligrams per liter), the factor more nearly approximates 70.

4. Recovery

A qualitative estimate of the presence or absence of interfering substances in a par-

*Amberlite IR 120 (II), Rohm and Haas Company, Philadelphia, Pa., or equivalent.

ticular determination may be made by means of a recovery procedure. Although this does not enable the analyst to apply any correction factor to the results of an analysis, it does give some basis for judging applicability of a particular method of analysis to a particular sample. Furthermore, it enables the analyst to obtain this information without an extensive investigation to determine exactly which substances interfere. It also eliminates the necessity of making separate determinations on the sample for the interfering substances themselves.

Test for recovery at the same time as sample analysis. Recoveries are not run routinely with samples with a known general composition or when using a method of well-established applicability. Recovery methods are tools to remove doubt about the applicability of a method to a sample. In brief, the recovery procedure involves applying the analytical method to a reagent blank; to a series of known standards covering the expected range of concentration of the sample; to the sample itself, in at least a duplicate run; and to recovery samples, prepared by adding known quantities of the substance sought to separate portions of the sample, each portion equal to the size of sample taken for the run. Add the substance sought in sufficient quantity to overcome the limits of error of the analytical method, but not to cause the total in the sample to exceed the range of the known standards used.

Correct results by subtracting reagent blank from each other determined value. Graphically represent the resulting known standards. From this graph, determine amount of sought substance in sample. Subtract this value from each of the analyses of sample plus known added substance. The resulting amount of substance divided by the known amount added, multiplied by 100, gives percentage recovery.

The procedure outlined above may be applied to colorimetric or instrumental methods. It also may be applied in a more simple form to titrimetric, gravimetric, and other types of analyses.

Rigid rules concerning percentage recoveries required for acceptance of results of analyses for a given sample and method cannot be stipulated. Recoveries in the range of the sensitivity of the method may be very high or very low and approach a value nearer to 100% as the error of the method becomes small with respect to the amount of substance added. In general, intricate and exacting procedures for trace substances that have inherent errors due to their complexity may give recoveries that would be considered very poor and yet, from the practical viewpoint of usefulness of the result, may be quite acceptable. Poor results may reflect either interferences present in the sample or real inadequacy of the method of analysis in the range in which it is being used.

Judicious use of recovery methods for evaluating analytical procedures and their applicability to particular samples is an invaluable aid to the analyst, but analytical skill is required to assure the validity of the methods. Recovery using colorimetric methods can be particularly deceptive, depending on the nature of the sample, its pretreatment, and concentration of the constituent being measured. Most analytical methods have satisfactory precision and accuracy over a limited range, with the lower range limit being controlling in trace analysis. A known addition to a sample may bring the concentration into the range where the method is reliable; if the addition is too large the apparent precision and accuracy apply at the higher concentration and not at the concentration originally present. Natural waters frequently contain complexing materials that combine with metals to the extent that the complexed metals will not react in some colorimetric methods. If nearly all of a metal is complexed, a known addition may be recovered completely but the method will still not recover the metal originally present in a complexed state. In such cases pretreatment to destroy the complex is necessary. The analyst must account for such pitfalls in designing recovery procedures.

Special methods for the treatment of data from radiological, biological, bacteriological, and bioassay methods are presented in the corresponding sections of this manual.

105 COLLECTION AND PRESERVATION OF SAMPLES

It is an old axiom that the result of any test procedure can be no better than the sample on which it is performed. It is not possible to specify detailed procedures for the collection of all samples because of the varied purposes and analytical procedures. More detailed information appears in connection with specific methods. This section presents general considerations.

The objective of sampling is to collect a portion of material small enough in volume to be transported conveniently and handled in the laboratory while still accurately representing the material being sampled. This implies that the relative proportions or concentrations of all pertinent components will be the same in the samples as in the material being sampled, and that the sample will be handled in such a way that no significant changes in composition occur before the tests are made.

A sample may be presented to the laboratory for specific analyses with the collector taking responsibility for its validity. Often, in water and wastewater work, the laboratory conducts or prescribes the sampling program, which is determined in consultation with the user of the test results. Such consultation is essential to insure selecting samples and analytical methods that provide a true basis for answering the questions that prompted the sampling.

The sampling program defines the portion of the whole to which the test results apply. Account must be taken of the variability of the whole with respect to time, area, depth, and in some cases, rate of flow.

1. General Precautions

Obtain a sample that meets the requirements of the sampling program and handle it in such a way that it does not deteriorate or become contaminated before it reaches the laboratory. Before filling, rinse sample bottle out two or three times with the water being collected. Representative samples of some sources can be obtained only by making composites of samples collected over a period of time or at many different sampling points. The details of collection vary so much with local conditions that no specific recommendations would be universally applicable. Sometimes it is more informative to analyze numerous separate samples instead of one composite.

Sample carefully to insure that analytical results represent the actual sample composition. Important factors affecting results are the presence of suspended matter or turbidity, the method chosen for its removal, and the physical and chemical changes brought about by storage or aeration. Particular care is required when processing (grinding, blending, sieving) samples to be analyzed for trace constituents, especially metals. Some determinations, particularly of lead, can be invalidated by contamination from such processing. Treat each sample individually with regard to the substances to be determined, the amount and nature of turbidity present, and other conditions that may influence the results.

It is impossible to give directions covering all conditions, and the choice of technic must be left to the analyst's judgment. In general, separate any significant amount of suspended matter by decantation, centrifugation, or an appropriate filtration procedure. Often a slight turbidity can be tolerated if experience shows that it will cause no interference in gravimetric or volumetric tests and that it can be corrected for in colorimetric tests, where it has potentially the greatest interfering effect. When relevant, state whether or not the sample has been filtered.

Make a record of every sample collected and identify every bottle, preferably by attaching an appropriately inscribed tag or label. Record sufficient information to provide positive sample identification at a later date, as well as the name of the sample collector, the date, hour, and exact location, the water temperature, and any other data that may be needed for correlation, such as weather conditions, water level, stream flow, etc. Provide space on the label for the initials of those assuming sample custody and for the time and date of transfer. Fix sampling points by detailed description, by maps, or with the aid of stakes, buoys, or landmarks in a manner that will permit their identification by other persons without reliance on memory or personal guidance.

Cool hot samples collected under pressure while still under pressure (see Section 107.3 and Figure 107:1).

Before collecting samples from distribution systems, flush lines sufficiently to insure that the sample is representative of the supply, taking into account the diameter and length of the pipe to be flushed and the velocity of flow.

Collect samples from wells only after the well has been pumped sufficiently to insure that the sample represents the groundwater source. Sometimes it will be necessary to pump at a specified rate to achieve a characteristic drawdown, if this determines the zones from which the well is supplied. Record pumping rate and drawdown.

When samples are collected from a river or stream, analytical values may vary with depth, stream flow, and distance from shore and from one shore to the other. If equipment is available, take an "integrated" sample from top to bottom in the middle of the stream in such a way that the sample is composited according to flow. If only a grab or catch sample can be collected, take it in the middle of the stream and at mid-depth.

Lakes and reservoirs are subject to considerable variations from normal causes such as seasonal stratification, rainfall, runoff, and wind. Choose location, depth, and frequency of sampling depending on local conditions and the purpose of the investigation.

These general directions do not provide enough information for collecting samples in which dissolved gases are to be determined. For specific instructions see the sections that describe these determinations.

Use only representative samples (or those conforming to a sampling program) for examination. The great variety of conditions under which collections must be made makes it impossible to prescribe a fixed procedure. In general, take account both of the tests or analyses to be made and the purpose for which the results are needed.

2. Types of Samples

a. Grab or catch samples: Strictly speaking, a sample collected at a particular time and place can represent only the composition of the source at that time and place. However, when a source is known to be fairly constant in composition over a considerable period of time or over substantial distances in all directions, then the sample may be said to represent a longer time period or a larger volume, or both, than the specific point at which it was collected. In such circumstances, some sources may be quite well represented by single grab samples. Examples are some water supplies, some surface waters, and rarely, some wastewater streams.

When a source is known to vary with time, grab samples collected at suitable intervals and analyzed separately can document the extent, frequency, and duration of these variations. Choose sampling intervals on the basis of the frequency with which changes may be expected, which may vary from as little as 5 min to as long as 1 hr or more.

When the source composition varies in space rather than time, collect samples from appropriate locations.

Use great care in sampling wastewater sludges, sludge banks, and muds. No definite procedure can be given, but take every possible precaution to obtain a representative sample or one conforming to a sampling program.

b. Composite samples: In most cases, the term composite sample refers to a mixture of grab samples collected at the same sampling point at different times. Sometimes the term time-composite is used to distinguish this type of sample from others. Time-composite samples are most useful for observing average concentrations that will be used, for example, in calculating the loading or the efficiency of a wastewater treatment plant. As an alternative to the separate analysis of a large number of samples, followed by computation of average and total results, composite samples represent a substantial saving in laboratory effort and expense. For these purposes, a composite sample representing a 24-hr period is considered standard for most determinations. Under certain circumstances, however, a composite sample representing one shift, or a shorter time period, or a complete cycle of a periodic operation, may be preferable. To evaluate the effects of special, variable, or irregular discharges and operations, collect composite samples representing the period during which such discharges occur.

For determining components or characteristics subject to significant and unavoidable changes on storage, do not use composite samples. Make such determinations on individual samples as soon as possible after collection and preferably at the sampling point. Analyses for all dissolved gases, residual chlorine, soluble sulfide, temperature, and pH are examples of this type of determination. Changes in such components as dissolved oxygen or carbon dioxide, pH, or temperature may produce secondary changes in certain inorganic constituents such as iron, manganese, alkalinity, or hardness. Use time-composite samples only for determining components that can be demonstrated to remain unchanged under the conditions of sample collection and preservation.

Take individual portions in a wide-mouth bottle having a diameter of at least 35 mm at the mouth and a capacity of at least 120 mL. Collect these portions every hour—in some cases every half hour or even every 5 min—and mix at the end of the sampling period or combine in a single bottle as collected. If preservatives are used, add them to the sample bottle initially so that all portions of the composite are preserved as soon as collected. Analysis of individual samples sometimes may be necessary.

It is desirable, and often essential, to combine individual samples in volumes proportional to flow. A final sample volume of 2 to 3 L is sufficient for sewage, effluents, and wastes.

Automatic sampling devices are available; however, do not use them unless the sample is preserved as described below.

Clean sampling devices, including bottles, daily to eliminate biological growths and other deposits.

c. Integrated samples: For certain purposes, the information needed is provided best by analyzing mixtures of grab samples collected from different points simultaneously, or as nearly so as possible. Such mixtures sometimes are called integrated samples. An example of the need for such sampling occurs in a river or stream that varies in composition across its width and depth. To evaluate average composition or total loading, use a mixture of samples representing various points in the cross-section, in proportion to their relative flows. The need for integrated samples also may exist if combined treatment is proposed for several separate wastewater streams, the interaction of which may have a significant effect on treatability or even on composition. Mathematical prediction of the interactions may be inaccurate or impossible and testing a suitable integrated sample may provide more useful information.

Both natural and artificial lakes often show variations of composition with both depth and horizontal location. However, under most conditions, neither total nor average figures are especially significant, and local variations are more important. In such cases, examine samples separately rather than integrate them.

Preparation of integrated samples usually requires special equipment to collect a sample from a known depth without contaminating it with overlying water. Knowledge of the volume, movement, and composition of the various parts of the water being sampled usually is required. Therefore, collecting integrated samples is a complicated and specialized process that cannot be described in detail.

3. Quantity

Collect a 2-L sample for most physical and chemical analyses. For certain determinations, larger samples may be necessary. Table 105:I shows the volumes ordinarily required for analyses.

Do not use the same sample for chemical, bacteriological, and microscopic examinations because methods of collecting and handling are different.

4. Preservation

Complete and unequivocal preservation of samples, whether domestic wastewater, industrial wastes, or natural waters, is a practical impossibility. Regardless of the sample nature, complete stability for every constituent never can be achieved. At best, preservation technics only retard chemical and biological changes that inevitably continue after sample collection. Changes that take place in a sample are either chemical or biological.

Some determinations are more likely than others to be affected by sample storage before analysis. Certain cations are subject to loss by adsorption on, or ion exchange with, the walls of glass containers. These include aluminum, cadmium, chromium, copper, iron, lead, manganese, silver, and zinc, which are best collected in a separate clean bottle and acidified with nitric acid to a pH below 2.0 to minimize precipitation and adsorption on container walls.

Temperature changes quickly; pH may change significantly in a matter of minutes; dissolved gases may be lost (oxygen, carbon dioxide). Determine temperature, pH, and dissolved gases in the field. With changes in the pH-alkalinity-carbon dioxide balance; calcium carbonate may precipitate and cause a decrease in the values for calcium and for total hardness.

Iron and manganese are readily soluble in their lower oxidation states but relatively insoluble in their higher oxidation states; therefore, these cations may precipitate out or they may dissolve from a sediment, depending upon the redox potential of the sample. Microbiological activity may be responsible for changes in the nitrate-nitrite-ammonia content, for decreases in phenol concentration and in BOD, or for reducing sulfate to sulfide. Residual chlorine is reduced to chloride. Sulfide, sulfite, ferrous iron, iodide, and cyanide may be lost through oxidation. Color, odor, and turbidity may increase, decrease, or change in quality. Sodium, silica, and boron may be leached out of the glass container. Hexavalent chromium may be reduced to chromic ion.

Biological changes taking place in a sample may change the oxidation state. Soluble constituents may be converted to organically bound materials in cell structures, or cell lysis may result in release of cellular material into solution. The well-known nitrogen and phosphorus cycles are examples of biological influences on sample composition.

The foregoing discussion is by no means inclusive. Clearly it is impossible to prescribe absolute rules for preventing all possible changes. Additional advice will be found in the discussions under individual determinations, but to a large degree the dependability of water analyses rests on the experience and good judgment of the analyst.

a. Time interval between collection and analysis: In general, the shorter the time that elapses between collection of a sample and its analysis, the more reliable will be the analytical results. For certain constituents and physical values, immediate analysis in the field is required.

It is impossible to state exactly how much elapsed time may be allowed between sample collection and its analysis; this depends on the character of the sample, the analyses to be made, and the conditions of storage. Changes caused by growth of microorganisms are greatly retarded by keeping the sample in the dark and at a low temperature. When the interval between sample collection and analysis is long enough to produce changes in either the concentration or the physical state of the constituent to be measured, follow the preservation practices given in Table 105:I. Record time elapsed between sampling and analysis, and which preservative, if any, was added.

b. Preservation methods: Sample preservation is difficult because almost all preservatives interfere with some of the tests. Immediate analysis is ideal. Storage at low temperature (4 C) is perhaps the best way to preserve most samples until the next day. Use chemical preservatives only when they are shown not to interfere with the analysis being made. When they are used, add them to the sample bottle initially so that all sample portions are preserved as soon as collected. No single method of preservation is entirely satisfactory; choose the preservative with due regard to the determinations to be made. All methods of preservation may be inadequate when applied to suspended matter. Because formaldehyde affects so many analyses, do not use it.

Methods of preservation are relatively limited and are intended generally to retard biological action, retard hydrolysis of chemical compounds and complexes, and reduce volatility of constituents.

Preservation methods are limited to pH control, chemical addition, refrigeration, and freezing. Table 105:I lists preservation methods by constituent.

Clean sampling devices, including bottles, daily to eliminate biological growths and other deposits.

c. Integrated samples: For certain purposes, the information needed is provided best by analyzing mixtures of grab samples collected from different points simultaneously, or as nearly so as possible. Such mixtures sometimes are called integrated samples. An example of the need for such sampling occurs in a river or stream that varies in composition across its width and depth. To evaluate average composition or total loading, use a mixture of samples representing various points in the cross-section, in proportion to their relative flows. The need for integrated samples also may exist if combined treatment is proposed for several separate wastewater streams, the interaction of which may have a significant effect on treatability or even on composition. Mathematical prediction of the interactions may be inaccurate or impossible and testing a suitable integrated sample may provide more useful information.

Both natural and artificial lakes often show variations of composition with both depth and horizontal location. However, under most conditions, neither total nor average figures are especially significant, and local variations are more important. In such cases, examine samples separately rather than integrate them.

Preparation of integrated samples usually requires special equipment to collect a sample from a known depth without contaminating it with overlying water. Knowledge of the volume, movement, and composition of the various parts of the water being sampled usually is required. Therefore, collecting integrated samples is a complicated and specialized process that cannot be described in detail.

3. Quantity

Collect a 2-L sample for most physical and chemical analyses. For certain determinations, larger samples may be necessary. Table 105:I shows the volumes ordinarily required for analyses.

Do not use the same sample for chemical, bacteriological, and microscopic examinations because methods of collecting and handling are different.

4. Preservation

Complete and unequivocal preservation of samples, whether domestic wastewater, industrial wastes, or natural waters, is a practical impossibility. Regardless of the sample nature, complete stability for every constituent never can be achieved. At best, preservation technics only retard chemical and biological changes that inevitably continue after sample collection. Changes that take place in a sample are either chemical or biological.

Some determinations are more likely than others to be affected by sample storage before analysis. Certain cations are subject to loss by adsorption on, or ion exchange with, the walls of glass containers. These include aluminum, cadmium, chromium, copper, iron, lead, manganese, silver, and zinc, which are best collected in a separate clean bottle and acidified with nitric acid to a pH below 2.0 to minimize precipitation and adsorption on container walls.

Temperature changes quickly; pH may change significantly in a matter of minutes; dissolved gases may be lost (oxygen, carbon dioxide). Determine temperature, pH, and dissolved gases in the field. With changes in the pH-alkalinity-carbon dioxide balance; calcium carbonate may precipitate and cause a decrease in the values for calcium and for total hardness.

Iron and manganese are readily soluble in their lower oxidation states but relatively insoluble in their higher oxidation states; therefore, these cations may precipitate out or they may dissolve from a sediment, depending upon the redox potential of the sample. Microbiological activity may be responsible for changes in the nitrate-nitrite-ammonia content, for decreases in phenol concentration and in BOD, or for reducing sulfate to sulfide. Residual chlorine is reduced to chloride. Sulfide, sulfite, ferrous iron, iodide, and cyanide may be lost through oxidation. Color, odor, and turbidity may increase, decrease, or change in quality. Sodium, silica, and boron may be leached out of the glass container. Hexavalent chromium may be reduced to chromic ion.

Biological changes taking place in a sample may change the oxidation state. Soluble constituents may be converted to organically bound materials in cell structures, or cell lysis may result in release of cellular material into solution. The well-known nitrogen and phosphorus cycles are examples of biological influences on sample composition.

The foregoing discussion is by no means inclusive. Clearly it is impossible to prescribe absolute rules for preventing all possible changes. Additional advice will be found in the discussions under individual determinations, but to a large degree the dependability of water analyses rests on the experience and good judgment of the analyst.

a. Time interval between collection and analysis: In general, the shorter the time that elapses between collection of a sample and its analysis, the more reliable will be the analytical results. For certain constituents and physical values, immediate analysis in the field is required.

It is impossible to state exactly how much elapsed time may be allowed between sample collection and its analysis; this depends on the character of the sample, the analyses to be made, and the conditions of storage. Changes caused by growth of microorganisms are greatly retarded by keeping the sample in the dark and at a low temperature. When the interval between sample collection and analysis is long enough to produce changes in either the concentration or the physical state of the constituent to be measured, follow the preservation practices given in Table 105:I. Record time elapsed between sampling and analysis, and which preservative, if any, was added.

b. Preservation methods: Sample preservation is difficult because almost all preservatives interfere with some of the tests. Immediate analysis is ideal. Storage at low temperature (4 C) is perhaps the best way to preserve most samples until the next day. Use chemical preservatives only when they are shown not to interfere with the analysis being made. When they are used, add them to the sample bottle initially so that all sample portions are preserved as soon as collected. No single method of preservation is entirely satisfactory; choose the preservative with due regard to the determinations to be made. All methods of preservation may be inadequate when applied to suspended matter. Because formaldehyde affects so many analyses, do not use it.

Methods of preservation are relatively limited and are intended generally to retard biological action, retard hydrolysis of chemical compounds and complexes, and reduce volatility of constituents.

Preservation methods are limited to pH control, chemical addition, refrigeration, and freezing. Table 105:I lists preservation methods by constituent.

106 ION-EXCHANGE RESINS

Ion-exchange resins are useful and flexible tools. Used in a series of mixed beds of both cation and anion exchangers and of organic absorbents, they produce reagent water of high quality. Used separately in atmospheric- or high-pressure columns they effect analytical separations of both inorganic and organic ions. However, high-pressure applications have not yet reached the status of standard methods for water and wastewater analysis.

Ion-exchange resins commonly are in the form of sodium or hydrogen counter-ions (attached to the matrix) for cation exchangers and of chloride, formate, acetate, and hydroxide counter-ions for anion exchangers. The user may substitute other counter-ions by passing regenerating solutions through a resin column as recommended by the manufacturer. The form to be used in a specific case will depend not only on the relative affinity of the resin for counter-ions and sample ions, but also on the ions that can be tolerated if the concentrated sample ions are to be eluted for analysis. Sequential elution of organics usually is done with carefully selected buffer solutions.

In water analysis, ion exchangers can be applied to: (a) remove interfering ions, (b) determine total ion content, (c) indicate the approximate volume of sample for certain gravimetric determinations, (d) concentrate trace quantities of cations, and (e) separate anions from cations. This manual recommends the use of ion-exchange resins for the removal of interference in the sulfate determination and for the determination of total ion content (see 104C.3 above). Inasmuch as the ion-exchange process can be applied in other determinations, a brief description of typical operations will be given here for convenient use where supplementary applications are warranted.

1. Selection of Method

The batch method of ion exchange is satisfactory for sample volumes of less than 100 mL while the column method can be used for any sample volume. In the batch method, the resin is agitated with the sample for a given time, after which the resin is removed by filtration. The column method is more efficient in that it provides continuous contact between the sample and the resin, thereby enabling the exchange reaction to go to completion. In this modification, the solution passes slowly through the resin bed and ions are removed quantitatively from the sample. Elution of the resin permits recovery of the exchanged substances.

2. Procedure

Use resins specifically manufactured for analytical applications. Prepare the ion exchanger by rinsing the resin with several volumes of ion-free water (good-quality distilled water) to remove any fines or coloring matter and other leachable material that might interfere with subsequent colorimetric procedures.

a. Batch method for cation removal: Pipet a sample portion containing 0.1 to 0.2 me of cations into a 250-mL erlenmeyer flask or beaker and add enough distilled water to bring final volume to 75 mL. Add 2.0 g strongly acidic cation-exchange resin and stir at moderate speed for 15 min. Filter through a plug of glass wool placed in the neck of a 10-cm (4-in.) borosilicate glass funnel. When filtration is complete, wash resin with two 10-mL portions of distilled water and make up to 100 mL total volume with distilled water.

Regeneration and storage of resin: Transfer spent resin from the batch procedure to a flask containing 500 mL 3*N* nitric acid (HNO_3). When sufficient resin has accumulated, wash into a column (Figure 106:1) and regenerate by passing 3*N* HNO_3 through the column at a rate of 0.1 to 0.2 mL of acid/mL resin/min. Use about 20 mL 3*N* HNO_3/mL resin in the column. Finally, wash resin with sufficient distilled water until the effluent pH is 5 to 7, using the same rate of flow as in the regeneration step. Remove

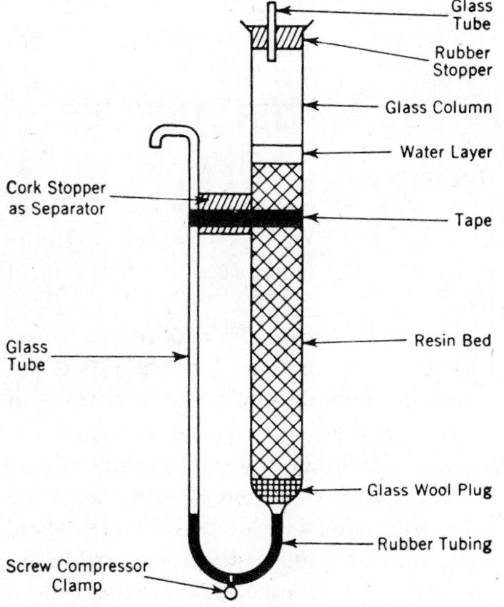

Figure 106:1. Ion-exchange column.

resin from column and store under distilled water in a wide-mouth container. Should the water become colored during storage, decant and replace with fresh distilled water. Before use, filter resin through a plug of glass wool placed in the neck of a funnel, wash with distilled water, and let drain. The resin is then ready for use.

b. Column method for cation removal: Prepare column as depicted in Figure 106:1 (length of resin bed, 21.5 cm; diameter of column, 1.3 cm; representing approximately 21 mL, or 20 g of resin). Other ion-exchange columns can be used equally well. One of the simplest consists of a buret containing a plug of glass wool immediately above the stopcock.

To make an effluent tube for a buret, *slowly,* bend TFE tubing, 1/8 in. ID × 1/4 in. OD, into an elongated S-shape; bend to 45-deg angle to the straight portion, secure with rubber bands or adhesive tape, and set aside for at least 48 hr to relieve stress. Repeat until desired conformation is achieved. Retain curved ends in place with plastic strips approximately 2 cm × 10 cm × 2 mm with suitably placed 3/16 in. drilled holes.

(Whatever type of column is adopted, never let liquid level in column fall below upper resin surface because trapped air causes uneven flow rates and poor efficiency of ion exchange. Adjust sample and column to the same temperature.)

Charge column by stirring resin in a beaker with distilled water and then carefully wash the suspension through a funnel into the column already filled with distilled water. Backwash column if necessary by introducing distilled water at the bottom and passing it upward through the column until all air bubbles and channels are removed. Connect a separatory funnel to the column top or use an inverted volumetric flask to feed sample, regenerating, or rinsing solutions; make certain diameter of flask neck is large enought to permit automatic feed. More efficiently, use a small controllable peristaltic pump to apply solutions to column. Let sample flow through column at the rate of 0.2 mL solution/mL resin/min. After sample has passed through column, wash resin with distilled water until effluent pH is 5 to 7. Use a pH meter or pH test papers to determine when column has been washed free of acid. For convenience, when rinsing column or absorbing cations from a sample of one or more liters, start this operation before the close of a workday and let exchange process proceed overnight. The column will not dry because of the curved outlet.

Column elution: After distilled-water wash, elute adsorbed cations by passing 100 mL 3*N* HNO_3 through the column at a rate of 0.2 mL acid/mL resin/min. Because a volume of 100 mL 3*N* HNO_3 quantitatively removes 3 me of cations, use additional increments of 100 mL 3*N* HNO_3 for quantities of adsorbed cations in excess of 3 me. After elution, rinse column free of acid with enough distilled water to produce an effluent pH of 5 to 7. Wash at same flow rate for acid elution. The acid elution and washing regenerate the column for future use. The combined acid eluates contain the cations orginally present in the sample.

107 EXAMINATION OF INDUSTRIAL WATER SUPPLIES

The following discussion summarizes the reasons for conducting an examination of an industrial water supply. Because this section is limited in scope, the reader is urged to refer to comprehensive reference books on industrial water treatment[1–4] for further information on this subject.

1. Industrial Needs

Industrial water is water used directly or indirectly in an industrial process. Volumetrically, the most important industrial usage of water is for cooling, either on a once-through basis or with cooling towers. Of almost equal importance is industrial water destined for steam generation.

Many industries, particularly those producing foods and beverages, use water as a raw material. Still others, such as the dye industry, take advantage of the solvent power. Water is used in the nuclear industries, for radiation shielding, reactor modulation, and cooling. Other industries use water to carry matter, as in hydraulic classifiers, or to carry energy, as in high-pressure sprays for debarking timber or descaling steel.

2. Industrial Water Treatment

Natural waters seldom are suitable for industrial use without some treatment. Water treatment can be divided into two phases. Primary, external, or pretreatment is treatment applied to water before it reaches the point of use. Secondary, internal, or post-treatment takes place after the water reaches the point of use.

Although the number of basic water treatment processes and operations is not large, the combinations and variations of these steps are almost infinite. For example, treatment for a boiler feedwater may consist of sedimentation, chlorination, softening with lime and soda ash or with sodium phosphate (hot or cold), filtration, ion exchange, acidizing, aeration, and degassing. A discussion of these treatment processes is beyond the scope of this book.

3. Analysis of Industrial Water

The chemical examination of water is extremely important in selecting a proper supply for a specific industrial application or a variety of uses.[5–7]

Diagnosis of existing or potential water problems requires a maximum of information about the particular system involved. Thorough and accurate analyses of makeup water and of water from operating systems, analyses and examination of deposits and sections removed from operating systems, plus full information on the size, design characteristics, and operating conditions of the water system or water-using equipment are desirable in helping the wa-

TABLE 107:I. ROUTINE AND SPECIAL DETERMINATIONS ON INDUSTRIAL WATER SAMPLES

Determination	Applications*
Acidity	B, C, P
Alkalinity:	
Hydroxyl (OH)	B, P
Phenolphthalein (P)	B, C, P
Total, methyl orange or mixed indicator (M)	B, C, P
Ammonia	B, P
Boron	C, P
Calcium	B, C, P
Carbon dioxide	B, P
Chloride	B, C, P
Chlorine, residual	C, P
Chromium, hexavalent	B, C, P
Color	P
Conductivity	B, C, P
Copper	B, C, P
Fluoride	C, P
Hardness	B, C, P
Hydrazine	B
Iron	B, C, P
Lead	P
Magnesium	B, C, P
Manganese	C, P
Morpholine	B
Nickel	B, P
Nitrate	B, C, P
Nitrite	B, C, P
Octadecylamine	B
Oil and grease	B, C, P
Oxygen, dissolved	B, P
pH	B, C, P
Phosphate:	
Ortho	B, C, P
Poly	B, C, P
Residue, total:	
Filtrable	B, C, P
Nonfiltrable	B, C, P
Silica	B, C, P
Sodium	B, C, P
Sulfate	B, C, P
Sulfide	C, P
Sulfite	B, C, P
Tannin and lignin	B, C, P
Turbidity	P
Zinc	B, C, P

* Key: B—boiler water, feedwater, or condensate; C—cooling water, recirculating (open and closed systems) or once-through; P—industrial process applications.

ter treatment specialist make a decision about the materials that must be removed from or added to the water to make it suitable for the intended process.

Once the most suitable raw water supply has been chosen and a treatment system has been established, start a testing program to insure adequate treatment rather than expensive over-treatment.[2–4] A control testing schedule can vary from a simple color comparison for chromate in a closed cooling system to a determination of a dozen or more ions in the microgram-per-liter range for supercritical boiler operations. Tailor the control program to the individual case. For example, analysis for only chloride and phosphate concentrations may be necessary for treatment control in one type of system, whereas for a very high-pressure boiler, alkalinity, chloride, copper, hydrazine or sulfite, iron, morpholine, pH, phosphate, sulfate, and other analyses may be required.

Table 107:I lists determinations most frequently made to control the quality of water destined for steam generation, heating, cooling, and other manufacturing processes. The method of analysis for each constituent is set forth subsequently in these standards. Procedures for chromate,[8] hydrazine,[9] morpholine,[9] nickel,[9] and octadecylamine[9] are described in the references.

As already indicated, an analysis and the conclusions drawn from it can be no better than the sample on which it is made. Use proper sampling technics to obtain the specificity or representativeness required (Section 105). Give proper attention to flushing sample lines and preventing contamination; make sure the sample is representative.

Obtain samples not collected under pressure in the usual manner. Cool boiler waters collected under pressure to approximately 20 C while still under pressure. Figure 107:1 illustrates typical sampling-cooling installations. Certain determinations demand specialized sampling equipment and procedures and/or analy-

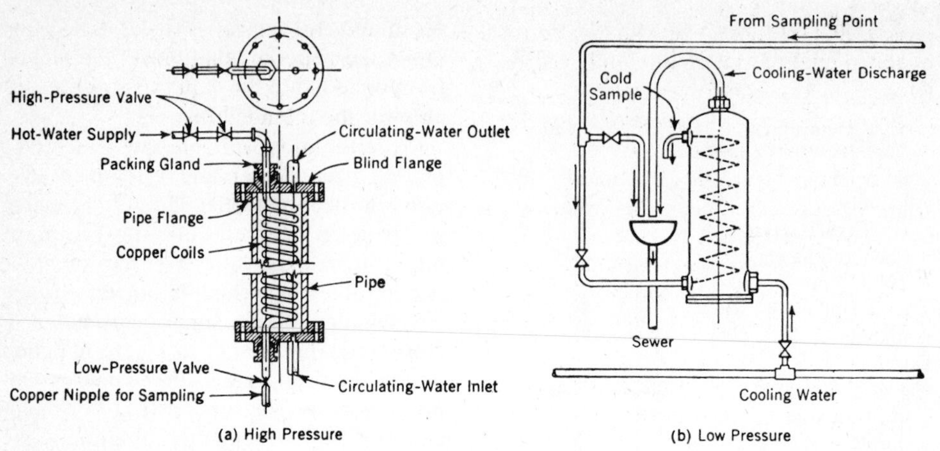

Figure 107:1. Cooling coils for boiler water sampling. These illustrations do not represent standardized equipment, but exemplify the characteristics of devices that should be used.

ses conducted immediately after sample collection. Specific examples include carbon dioxide (CO_2), dissolved oxygen (DO), dissolved and total iron, hydrogen sulfide, or octadecylamine in condensate. For such determinations use the sampling procedures specified as part of the analytical method.

4. Scale

Scale or sludge formation in industrial water systems may induce equipment failures, such as boiler tube ruptures or plugged heat-exchanger tubing. Such occurrences frequently result from increased temperature, evaporation, or aeration, which cause insolubility of some ionic combinations present. Common scale deposits may consist of calcium carbonate ($CaCO_3$), phosphate, silicate, or sulfate, or of magnesium hydroxide, phosphate, and silicate. Others may be from oxides, silica, or related substances.

In distribution systems with temperatures of 0 to 93 C (32 to 200 F), estimate the tendency of water to form $CaCO_3$ scale by calculating the Langelier pH of saturation (pH_S) from the alkalinity, calcium, dissolved solids, and temperature of the water (see Section 203, Calcium Carbonate Saturation) and subtracting pH_S from the measured pH of the water. A positive value indicates scale-forming tendency; a negative value indicates scale-dissolving (or possibly corrosive) tendency. A stable water usually has a zero or slightly positive index. Ryznar[10], using Langelier's method for determining pH_s, calculated the Stability Index, ($2pH_s - pH$). For natural waters the Langelier Index can be positive or negative; the Ryznar Index will be positive. In either case the correlation of the indices with corrosion must be made by impartial determinations of the accompanying corrosion. The literature also offers methods for calculating indices or solubilities of calcium phosphate,[11] calcium sulfate,[12] and magnesium hydroxide.[13]

Scale formation is prevented in boilers[2,14] by ion exchange or other hardness-reducing or solubilizing methods applied to the feedwater. Residual hardness is precipitated inside the boiler by addition of phosphate or carbonate and organic sludge-conditioning chemicals, to produce a nonadhering sludge rather than a scale; or is reacted with chelating agents to form soluble complexes.

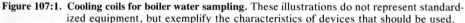

For sludge removal and maintenance of

scale-free conditions, and for the control of steam quality, maximum limits are designated for total dissolved solids. Control is accomplished by blowdown or sidestream treatment as necessary.

In cooling towers, acid is added for alkalinity reduction and scale prevention; however, presoftening of makeup water is practiced in some cases. Polyphosphates and other chemicals also are added to control $CaCO_3$, calcium phosphate, and silica deposit formation.[15,16] Proper and adequate bleedoff is an important part of treatment of cooling water.

5. Corrosion

Deterioration of metals caused by corrosion may produce dissolved metal ions that may redeposit as oxides. Certain constituents or contaminants, including DO, CO_2, anions (such as chloride and sulfate), and acidity and low pH increase the corrosive tendency. Microorganisms such as sulfate-reducing or iron bacteria also can be involved in corrosion and/or deposit formation.

Development of a thin protective layer of $CaCO_3$ on metal surfaces often is effective in preventing corrosion in the distribution system. Use the Stability Index or Ryznar Stability Index[10] to indicate the tendency to dissolve $CaCO_3$ or the degree of stability or balance.

Another procedure[17] for estimating corrosive tendency is calculation of the ratio

$$\frac{\text{me } (Cl^- + SO_4^{2-})/L}{\text{me alkalinity as } CaCO_3/L}$$

In the neutral pH range (7 to 8) and in the presence of DO, ratios equal to or below about 0.1 are said to indicate general freedom from corrosion, whereas increasingly higher ratios generally indicate more aggressive waters.

Corrosion in boilers is controlled by removal of most of the dissolved oxygen from feedwater by deaeration and by the reaction of residual oxygen with sodium sulfite or hydrazine. Corrosion of iron also is reduced effectively by application of alkaline chemicals such as caustic soda and sodium phosphates to provide boiler water pH in the range of 10 to 12. At high boiler pressures (about 1,000 kPa and above) it may be preferable to use a "low solids" type of treatment (ammonia, amines, hydrazine) to avoid localized corrosion associated with high pH. Because of the lack of buffering and softening agents in this treatment, high-quality demineralized water very low in silica is necessary and only minimum condenser leakage can be tolerated.

Carbonates and bicarbonates decompose in boilers to release CO_2 in the steam and subsequently form carbonic acid in the steam condensate.[18,19] Acid corrosion may result. The presence of DO accelerates the carbonic acid attack.

Steam condensate corrosion can be minimized by maintaining low carbonate content in the boiler water and by using low-bicarbonate or low-carbonate makeup waters. Volatile neutralizing amines such as morpholine and cyclohexylamine reduce corrosion in condensate systems by raising and controlling condensate pH. Filming amines such as octadecylamine,[20] which forms a monomolecular film on walls of condensate piping to prevent contact with an acidic environment, also is effective in reducing corrosion.

Inhibition of corrosion in cooling towers is provided by addition of corrosion inhibitors such as chromates, phosphates, and silicates, singly or in various combinations. Open recirculating cooling systems[14,15] in which water is continuously saturated with DO pose more difficult corrosion-control problems than do closed recirculating systems. The generally low makeup water requirements of closed systems account for the minimum scale and corrosion problems. Where excessive makeup water is used because of poor design or leakage, scaling or corrosion may occur unless corrective treatment is used.

6. Special Problems

Fouling is deposition of a material, normally in suspension, on a surface in contact with water. Deposition may cause reduced flow, inefficient heat transfer, and localized corrosion. Fouling also may result from reaction of dissolved minerals and water-conditioning chemicals and from growth of microorganisms.

Sanitation practices to retard growth of undesirable organisms may include destruction of pathogenic organisms by heat, as used in the food industries, or the addition of selective biocides to kill slime-forming or sulfate-reducing organisms growing on heat-transfer surfaces. Also included is destruction of algae growing in cooling towers, water plants growing in reservoirs, and mollusks growing in cooling-water pipes of industries using seawater for once-through cooling water. Because many industries rely on biological methods to treat wastewater, biocides must not inhibit growth of desirable organisms in the waste treatment system and in its effluent.

Taste and odor result from organic or inorganic matter. Taste- and odor-forming substances may be removed by adsorption on a number of materials such as activated carbon, or they may be oxidized by chlorine, chlorine dioxide, oxygen, ozone, or permanganate. In some instances they may be rendered nonoffensive by masking.

A problem of disposal may arise when some industries reuse water to such a degree that high concentrations of solids or impurities develop. Wastewater injected into a disposal well must be chemically compatible with the aquifer; otherwise aquifers will clog and shorten the life of the well. Wastewater effluents discharged to surface waters must be chemically, biologically, and thermally compatible with the receiving stream or lake to avert serious upset of the ecological balance.

7. References

1. AMERICAN SOCIETY FOR TESTING AND MATERIALS. 1969. Manual on Water and Industrial Waste Water, 3rd ed. ASTM Spec. Tech. Publ. 442. Philadelphia, Pa.
2. Boiler water chemistry symposium. 1954. *Ind. Eng. Chem.* 46:953.
3. NORDELL, E. 1961. Water Treatment for Industrial and Other Uses, 2nd ed. Reinhold Publishing Corp., New York, N.Y.
4. HAMER, P., J. JACKSON & E.F. THURSTON. 1961. Industrial Water Treatment Practice. Butterworths, London.
5. U.S. ENVIRONMENTAL PROTECTION AGENCY. 1976. National Interim Primary Drinking Water Regulations. EPA-570/9-76-003, Washington, D.C.
6. AMERICAN WATER WORKS ASSOCIATION. 1971. Water Quality and Treatment, 3rd ed. McGraw Hill, Inc., New York, N.Y.
7. CALIFORNIA STATE WATER POLLUTION CONTROL BOARD. 1963. Water Quality Criteria, 2nd ed. Calif. State Water Pollution Control Board, Sacramento.
8. FURMAN, N.H., ed. 1962. Standard Methods of Chemical Analysis, reprinted 1975. D. Van Nostrand Co., Princeton, N.J., Vol. 1, pp. 354, 360.
9. AMERICAN SOCIETY FOR TESTING AND MATERIALS. 1979. Book of ASTM Standards, Part 31. Water. ASTM, Philadelphia, Pa.
10. RYZNAR, J.W. 1944. A new index for determining amount of calcium carbonate scale formed by a water. *J. Amer. Water Works Ass.* 36:472.
11. GREEN, J. & J.A. HOLMES. 1947. Calculation of the pH of saturation of tricalcium phosphate. *J. Amer. Water Works Ass.* 39:1090.
12. DENMAN, W.L. 1961. Maximum re-use of cooling water. *Ind. Eng. Chem.* 53:817.
13. LARSON, T.E., R. W. LANE & C. H. NESS. 1959. Stabilization of magnesium hydroxides in the solids-contact process. *J. Amer. Water Works Ass.* 51:1551.
14. APPLEBAUM, S.B. & R.J. ZUMBRUNNEN. 1956. Selecting water-treating processes for medium-pressure boilers. Amer. Soc. Mechan. Eng. Paper No. 56-A-191.

15. LANE, R.W. & T.E. LARSON. 1963. Role of water treatment in the economic operation of cooling towers. *Amer. Power Conf.* XXV: 687.
16. APPLEBAUM, S.B. 1950. Treatment of cooling water. *Combustion* 22:5 (Nov.), 41.
17. LARSON, T.E. & R.V. SKOLD. 1958. Laboratory studies relating mineral quality of water to corrosion of steel and cast iron. *Corrosion* 14:6, 285t.
18. COLLINS, L.F. 1943. More information concerning corrosion in steam heating systems. 4th Water Conf., Eng. Soc. Western Pennsylvania, 33.
19. BERK, A.A. & J. NIGON. 1948. Amine volatility and alkalinity in relation to corrosion control in steam heating systems. U.S. Bur. Mines Tech. Pap. 714.
20. WILKES, J.F. et al. 1955. Filming amines—Use and misuse in power plant water-steam cycles. *Amer. Power Conf.* XVII: 527.

108 SAFETY

1. General Discussion

Safety is a subject of concern for all. In the specific context of laboratories it is especially important because of the presence and use of hazardous chemicals or equipment, microorganisms that may produce human disease, and radioactive substances. Laboratory management has a vital role to play in insuring a safe working environment. This role is dictated by federal or state regulations as well as by the appropriate interests of management in the well-being of its employees and the overall productivity of the laboratory. In the final analysis, however, safety is the responsibility of the individual laboratory worker. While management must deal with the work place, the provision of safe procedures, and the training of employees in safe practices, it is the analyst who ultimately makes the laboratory safe or unsafe.

An institutional employer has legal responsibilities to its employees and individual managers may have personal liabilities. OSHA, the Occupational Safety and Health Administration (federal or state), may have established general industry safety and health standards[1] applicable to laboratories. As an example, OSHA has published a list of chemical carcinogens and has defined highly restrictive conditions for their use. This list may be amended as additional chemicals become known as carcinogens. OSHA regulations provide details for some laboratory operations and are the basis for many laboratory controls. Laboratory management must be thoroughly familiar with such pertinent regulations.[2]

2. Laboratory Hazards

Laboratory hazards fall into two major categories: Those of universal significance, such as fire, electrical, and mechanical hazards, and those specific to a laboratory, such as chemical or biological hazards. Because it is not intended to compile a manual of safe practices here, general safety problems will not be discussed. However, the use of combustible or explosive reagents and the widespread use of electrical equipment may pose special problems in the laboratory. Mechanical hazards include broken glass and other sharp objects that frequently are present in the laboratory.

From the standpoint of chemical and biological safety it must be recognized that many reagents require handling with the utmost care, either in their original state or in solution, or both. This care is exercised either to protect the health and safety of the analyst-user or because of environmental hazard or damage, particularly in connection with laboratory waste disposal. A number of reagents specified in this manual may bear commercial labels with the words POISON, DANGER, CAUTION, FLAMMABLE, or comparable warnings. Handle these with special care. The safest way to deal with hazardous materials is to avoid their use, and efforts to eliminate such reagents have been made in selecting analytical methods for inclusion in this manual. For example, the recognition that orthotolidine is a carcinogen has been one of the reasons for abandoning those procedures for measuring chlorine residual based on its use. When avoidance is impossible, hazardous materials have been identified and special precautions against exposure by inhalation, by ingestion, or through the skin have been given. Simple safety practices include using mechanical pipettors instead of mouth pipets and the use of safety clothing and fume hoods. Safety glasses or, still better, full face masks offer important protection against explosions or implosions.

Important information on safety and health is available from the American Water Works Association,[3] the American Public Health Association,[4] the Manufacturing Chemists' Association,[5] or the general literature.[6]

3. Organizing for Safety

Depending on the size and nature of the laboratory there should be a safety officer and/or a safety committee. This does not mean that laboratory safety is the exclusive responsibility of the safety officer because awareness of safety and the use of safe practices is required of all laboratory personnel. However, the safety officer should be well informed and serve as the leader in hazards recognition and the development of procedures to minimize or eliminate them. Laboratories using radioisotopes under state or federal license are required to have a safety officer and an active safety program. This requirement has been useful in maintaining the radiological laboratory as a safe working place; comparable attention to all aspects of laboratory safety is equally rewarding.

It is essential that management provide a structurally safe housing for the laboratory in which adequate provision for fire escape is provided. Additionally, properly located safety equipment is required. This equipment should include such items as fire extinguishers, fire blankets, safety showers, eye wash fountains, safety glasses, face shields, and first aid and spill control kits. Employees should know the location of these safety devices and be familiar with their use. The safety officer should be responsible for insuring adequate supplies and providing training in safety procedures.

4. Hazardous Materials in *Standard Methods*

Special attention is called to a number of hazardous materials specified in this manual. The discussion below is not exhaustive but it should serve to emphasize some safety concerns.

a. Acids and alkalies: Concentrated acids and bases may cause chemical burns and are especially hazardous if spilled or splashed into the eyes. Always handle with extreme care to avoid contact. In diluting concentrated acids, always add the acid to water to prevent local overheating and possible injury; never add water to the acid.

b. Arsenic: Inorganic arsenic compounds are used to prepare standards and may be present in samples. Arsenic is highly toxic and may cause lung cancer; avoid inhalation, ingestion, and skin exposure.

c. Azides: Sodium azide is used in a number of procedures including the test for dissolved oxygen. It is toxic and reacts with acid to produce the still more toxic hydrazoic acid. When discharged to a drain it may react with, and accumulate on, copper or lead plumbing fixtures. The metal azides are explosive and detonate readily. Avoid inhalation, ingestion, and skin exposure. Destroy azides by adding a concentrated solution of sodium nitrite, $NaNO_2$ (1.5 g $NaNO_2$/g sodium azide). To remove accumulated metal azides from drainpipes and traps, treat overnight with a 10% solution of sodium hydroxide.

d. Beryllium and its salts: A beryllium salt is used to prepare beryllium standards and beryllium may be present in waste samples. The element is highly toxic and may yield volatile products. Always handle beryllium-containing materials in a hood and avoid ingestion.

e. Biohazards: Samples may contain pathogenic microorganisms. Exposure to these organisms may be incidental to chemical or biological examination or occur in the specific examination for certain disease-producing organisms. In either case, avoid ingestion, particularly in culturing pathogens. Use aseptic technics and sterilize all discarded cultures.[4,7]

f. Compressed gases: Compressed gases are used widely in most laboratories, especially if an atomic absorption spectrophotometer or a gas-liquid chromatograph is used. The gases may be flammable or explosive and require careful handling. Protect the cylinders themselves from freezing, overheating, and mechanical damage. Chain, lock, or otherwise prevent the cylinders from moving or falling over. Use the appropriate pressure-reducing valve for each type of gas cylinder.

g. Cyanides: Cyanides are used as reagents or may be present in samples. Most cyanides are toxic; avoid ingestion. Handle such solutions in a fume hood and avoid inhalation. In acid solution the toxic gas, hydrogen cyanide, may be produced; therefore, do not acidify cyanide solutions.

h. Mercury: Mercury and its compounds are used to prepare standards, displace gases, serve as indicator liquid in thermometers, and preserve samples. Liquid mercury is a toxic volatile element. Handle spills expeditiously to prevent inhalation. Keep powdered sulfur on hand to spread immediately on mercury spills to minimize volatilization before cleanup. Disposal of samples containing mercury may be environmentally damaging.

i. Perchloric acid: Perchloric acid is used in digesting organic matter. It can react explosively with organic matter and must be handled with care; predigest samples containing organic matter with nitric acid before adding perchloric acid and do not add perchloric acid to a hot solution. Like azides in a drain, perchlorates may accumulate in a hood or air exhaust system. Accumulated perchlorates may react explosively with organic matter; use special perchloric acid fume hoods and ducting if perchloric acid digestion is done frequently.

j. Toxic or carcinogenic organic compounds: Organic solvents and solid organic reagents are used in many determinations. These may be flammable or explosive, and as such require special handling and storage, or they may be toxic or carcinogenic. Handle solvents such as chloroform or carbon tetrachloride in a fume hood and avoid inhalation, skin contact, and ingestion. Diaminobenzidine (used in the determination of selenium) and dimethylphenylenediamine oxalate (used in the sulfide determination) are suspect carcinogens; handle with extreme care and avoid ingestion and skin contact.

5. References

1. OCCUPATIONAL SAFETY AND HEALTH ADMINISTRATION. 1976. General Industry Safety and Health Standards. OSHA 2206 (29 CFR 1910), U.S. Dept. of Labor, Washington, D.C.
2. SANDERS, H.J. 1976. Chemical lab safety

and the impact of OSHA. *Chem. Eng. News* 54:22, 15.
3. American Water Works Association. 1958. Safety Practices for Water Utilities. Manual M6, Amer. Water Works Ass., New York, N.Y.
4. Inhorn, S.L., ed. 1978. Quality Assurance Practices for Health Laboratories. Amer. Pub. Health Ass., Washington, D.C.
5. Manufacturing Chemists' Association, General Safety Committee. 1972. Guide for Safety in the Chemical Laboratory, 2nd ed. D. Van Nostrand Co., New York, N.Y.
6. Steere, N.V., ed. 1971. Handbook of Laboratory Safety, 2nd ed. Chemical Rubber Company, Cleveland, Ohio.
7. National Institutes of Health. 1974. Biohazards Safety Guide. Environmental Services Branch, National Inst. Health, U.S. Public Health Service.

109 BIBLIOGRAPHY

General

Hauck, C.F. 1949. Gaging and sampling waterborne industrial wastes. Amer. Soc. Testing & Materials Bull. (Dec.)

Black, H.H. 1952. Procedures for sampling and measuring industrial wastes. *Sewage Ind. Wastes* 24:45.

Welch, P.S. 1952. Limnology, 2nd ed. McGraw-Hill Book Co., New York, N.Y.

Taylor, E.W. 1958. Examination of Waters and Water Supplies, 7th ed. Little, Brown & Co., Boston, Mass.

Hem, J.D. 1959. Study and interpretation of the chemical characteristics of natural water. U.S. Geol. Surv. Water Supply Pap. No. 1473.

Klein, L. 1959. River Pollution. I. Chemical Analysis. Academic Press, New York, N.Y.

Rainwater, F.H. & L.L. Thatcher. 1960. Methods for Collection and Analysis of Water Samples. U.S. Geol. Surv. Water Supply Pap. No. 1454.

U.S. Public Health Service. 1962. Public Health Service Drinking Water Standards, 1962. PHS Publ. No 956.

Taylor, F.B. 1963. Significance of trace elements in public finished water supplies. *J. Amer. Water Works Ass.* 55:619.

Camp, T.R. 1963. Water and Its Impurities. Reinhold Publishing Corp., New York, N.Y.

California State Water Pollution Control Board. 1963. Water Quality Criteria, 2nd ed. Calif. State Water Pollution Control Board, Sacramento.

Sawyer, C.N. & P.L. McCarty. 1967. Chemistry for Sanitary Engineers, 2nd ed. McGraw-Hill Book Co., New York, N.Y.

American Society for Testing and Materials. 1969. Manual on Industrial Water and Industrial Waste Water, 3rd ed. Spec. Tech. Publ. 148-I, ASTM, Philadelphia, Pa.

American Water Works Association. 1971. Water Quality and Treatment: a Handbook of Public Water Supplies, 3rd ed. McGraw-Hill, New York, N.Y.

World Health Organization. 1971. International Standards for Drinking Water, 3rd ed. WHO, Geneva.

Department of the Environment. 1972. Analysis of Raw, Potable and Waste Waters. Her Majesty's Stationery Office, London.

Wetzel, R.G. & G.E. Likens. 1979. Limnological Analyses. W.B. Saunders Co., Philadelphia, Pa.

Water Supply Data

Lohr, E.W. & S.K. Love. 1954. The industrial utility of public water supplies in the United States. 1952. Parts 1 and 2. U.S. Geol. Surv. Water Supply Pap. No. 1299 and 1300.

Laboratory Reagents

Rosin, J. 1967. Reagent Chemicals and Standards, 5th ed. D. Van Nostrand Co., Princeton, N.J.

American Chemical Society. 1974. Reagent Chemicals—American Chemical Society Specifications, 5th ed. ACS, Washington, D.C.

The United States Pharmacopeia. 1975.

19th rev. U.S. Pharmacopeial Convention Inc., Rockville, Md.

NATIONAL BUREAU OF STANDARDS. 1979. Catalog of NBS Standard Reference Materials. NBS Spec. Pub. 260, 1979–80 ed.

General Analytical Technics

FOULK, C.W., H.V. MOYER & W.M. MACNEVIN. 1952. Quantitative Chemical Analysis. McGraw-Hill Book Co., New York, N.Y.

WILLARD, H.H., N.H., FURMAN & C.E. BRICKER. 1956. Elements of Quantitative Analysis, 4th ed. D. Van Nostrand Co., Princeton, N.J.

HUGHES, J.C. 1959. Testing of glass volumetric apparatus. Nat. Bur. Standards Circ. No. 602.

WILSON, C.L. & D.W. WILSON, eds. 1959, 1960, 1962. Comprehensive Analytical Chemistry, Vol. 1A, Vol. 1B, Vol. 1C. Elsevier Publishing Co., New York, N.Y.

WELCHER, F.J., ed. 1963, 1966. Standard Methods of Chemical Analysis. 6th ed. Vol. IIA, & Vol. IIIA. D. Van Nostrand Co., Princeton, N.J.

MEITES, L., ed. 1963. Handbook of Analytical Chemistry. McGraw-Hill Book Co., New York, N.Y.

KOLTHOFF, I.M., E.J. MEEHAN, E.B. SANDELL & S. BRUCKENSTEIN. 1969. Quantitative Chemical Analysis, 4th ed. Macmillan Co., New York, N.Y.

PECSOK, R. et al. 1976. Modern Methods of Chemical Analysis, 2nd ed. John Wiley & Sons, New York, N.Y.

VOGEL, A.I. 1978. Textbook of Quantitative Inorganic Analysis. Including Elementary Instrumental Analysis. 4th ed. Revised by J. Basset. Longman, New York, N.Y.

Colorimetric Technics

MELLON, M.G. 1947. Colorimetry and photometry in water analysis. *J. Amer. Water Works Ass.* 39:341.

NATIONAL BUREAU OF STANDARDS. 1947. Terminology and Symbols for Use in Ultraviolet, Visible, and Infrared Absorptiometry. NBS Letter Circ. LC-857 (May 19).

GIBSON, K.S. & M. BALCOM. 1947. Transmission measurements with the Beckman quartz spectrophotometer. *J. Res. Nat. Bur. Standards* 38:601.

SNELL, F.D. & C.T. SNELL. 1948. Colorimetric Methods of Analysis, 3rd ed. D. Van Nostrand Co., Princeton, N.J., Vol. 1.

MELLON, M.G., ed. 1950. Analytical Absorption Spectroscopy. John Wiley & Sons, New York, N.Y.

DISKANT, E.M. 1952. Photometric methods in water analysis. *J. Amer. Water Works Ass.* 44:625.

SANDELL, E.B. 1959. Colorimetric Determination of Traces of Metals, 3rd ed. Interscience Publishers, New York, N.Y.

BOLTZ, D.F., ed. 1978. Colorimetric Determination of Nonmetals, 2nd ed. John Wiley & Sons, New York, N.Y.

Other Methods of Analysis

LEDERER, E. & M. LEDERER. 1957. Chromatography, 2nd ed. Elsevier Press, Houston, Tex.

LINGANE, J.J. 1958. Electroanalytical Chemistry, 2nd ed. Interscience Publishers, New York, N.Y.

SAMUELSON, O. 1963. Ion Exchangers in Analytical Chemistry. John Wiley & Sons, New York, N.Y.

HARLEY, J.H. & S.E. WIBERLEY. 1967. Instrumental Analysis, 2nd ed. John Wiley & Sons, New York, N.Y.

EWING, G.W. 1975. Instrumental Methods of Chemical Analysis, 4th ed. McGraw-Hill Book Co., New York, N.Y.

HEFTMANN, E., ed. 1975. Chromatography, 3rd ed. Reinhold Publishing Corp., New York, N.Y.

General Analytical Reviews and Bibliographies

WEIL, B.H. et al. 1948. Bibliography on Water and Sewage Analysis. State Engineering Experiment Station, Georgia Institute of Technology, Atlanta.

Annual reviews of analytical chemistry. 1953–1979. *Anal. Chem.* 25:2; 26:2; 27:574; 28:559; 29:589; 30:553; 31:776; 32:3R; 33:3R; 34:3R; 35:3R; 36:3R; 37:1R; 38:1R; 39:1R; 40:1R; 41:1R; 42:1R; 43:1R; 44:1R; 45:1R; 46:1R; 47; 48; 49; 50; 51.

WATER POLLUTION CONTROL FEDERATION RESEARCH COMMITTEE. 1960–1979. Annual literature review, nature and analysis of chemical species. *J. Water Pollut. Control Fed.* 32:443; 33:445; 34:419; 35:553; 36:535; 37:735; 38:869; 39:867; 40:897; 41:873; 42:863; 43:933; 44:903; 45:979; 46:1031; 47:1118; 48:998; 49:901; 50:1000; 51:1093.

Statistics

WORTHING, A.G. & J. GEFFNER. 1943. Treatment of Experimental Data. John Wiley & Sons, New York, N.Y.

AMERICAN SOCIETY FOR TESTING AND MATERIALS. 1950. Symposium on application of statistics. ASTM Spec. Tech. Publ. 103.

AMERICAN SOCIETY FOR TESTING AND MATE-

RIALS. 1951. Manual on quality control of materials. ASTM Spec. Tech. Publ. 15C, revised 1960.

DEAN, R.B. & W.J. DIXON. 1951. Simplified statistics for small numbers of observations. *Anal. Chem.* 23:636.

YOUDEN, W.J. 1951. Statistical Methods for Chemists. John Wiley & Sons, New York, N.Y., reprinted 1977.

GORE, W.L. 1952. Statistical Methods for Chemical Experimentation. Interscience Publishers, New York, N.Y.

Guide for measures of precision and accuracy. 1962. *Anal. Chem.* 34:364R.

OSTLE, B. 1963. Statistics in Research. Iowa State Univ. Press, Ames.

YOUDEN, W.J. 1967. Statistical Techniques for Collaborative Tests. Ass. Official Analytical Chemists, Washington, D.C.

DIXON, W.J. & F.J. MASSEY, JR. 1969. Introduction to Statistical Analysis, 3rd ed. McGraw-Hill Book Co., New York, N.Y.

GREENBERG, A.E., N. MOSKOWITZ, B.R. TAMPLIN & J. THOMAS. 1969. Chemical reference samples in water laboratories. *J. Amer. Water Works Ass.* 61:599.

KOLTHOFF, I.M. & E.B. SANDELL. 1969. Quantitative Inorganic Analysis, 4th ed. Macmillan Co., New York, N.Y. Chapter 15.

HOEL, P.G. 1971. Introduction to Mathematical Statistics, 4th ed. John Wiley & Sons, New York, N.Y.

DUNCAN, A.J. 1974. Quality Control and Industrial Statistics, 4th ed. Richard R. Irwin, Inc., Homewood, Ill.

Evaluation of Methods

KRAMER, H.P. & R.C. KRONER. 1959. Cooperative studies on laboratory methodology. *J. Amer. Water Works Ass.* 51:607.

KRONER, R.C., D.G. BALLINGER & H.P. KRAMER. 1960. Evaluation of laboratory methods for analysis of heavy metals in water. *J. Amer. Water Works Ass.* 52:117.

MULLINS, J.W. et al. 1961. Evaluation of methods for counting gross radioactivity in water. *J. Amer. Water Works Ass.* 53:1466.

LISHKA, R.J., F.S. KELSO & H.P. KRAMER. 1963. Evaluation of methods for determination of minerals in water. *J. Amer. Water Works Ass.* 55:647.

Checking Analyses

ROSSUM, J.R. 1949. Conductance method for checking accuracy of water analyses. *Anal. Chem.* 21:631.

ROBERTSON, R.S. & M.F. NIELSEN. 1951. Quick test determines dissolved solids. *Power* 97:87 (Feb.).

NAVONE, R. 1954. Sodium determination with ion-exchange resin. *J. Amer. Water Works Ass.* 46:479.

GREENBERG, A.E. & R. NAVONE. 1958. Use of the control chart in checking anion-cation balances in water. *J. Amer. Water Works Ass.* 50:1365.

Collection of Samples

AMERICAN SOCIETY FOR TESTING AND MATERIALS. 1974. Water, Practices for Sampling. ASTM Publ. D3370-74T, Philadelphia, Pa.

PART 200
PHYSICAL
EXAMINATION

201 INTRODUCTION

This section deals primarily with measurement of the physical properties of a sample, as distinguished from the concentrations of chemical or biological components. Many of the determinations included here, such as color, electrical conductivity, and turbidity, fit this category unequivocally. However, physical properties cannot be divorced entirely from chemical composition, and some of the technics of this section measure collective properties resulting from the presence of a number of constituents. Others, for example, calcium carbonate saturation and oxygen transfer, are related to, or depend on, chemical tests. Also included here are tests for appearance, odor, and taste, which have been classified traditionally among physical properties, although the point could be argued. Finally, Section 213, Tests on Sludges, includes certain biochemical tests. However, for convenience they are grouped with the other tests used for sludge.

With these minor exceptions, the contents of this section have been kept reasonably faithful to its name. Most of the methods included are either inherently or at least traditionally physical, as distinguished from the explicitly chemical, radiological, biological, or bacteriological methods of other sections.

202 APPEARANCE

To record the general physical appearance of a sample, use any terms that briefly describe its visible characteristics. These terms may state the presence of color, turbidity, suspended solids, crustacea, larvae, worms, sediment, floating material, and similar particulate matter detectable by the unaided eye. Use numerical values when they are available, as for color, turbidity, and suspended solids.

203 CALCIUM CARBONATE SATURATION

The pH at which a water is just saturated with calcium carbonate ($CaCO_3$) is known as the pH of saturation, or pH_S. The Langelier saturation index[1] (SI) is defined as the actual pH minus pH_S. A negative index indicates a tendency to dissolve $CaCO_3$ and a positive index indicates a tendency to deposit $CaCO_3$. This index is not related directly to corrosion, but deposition of a thin, coherent carbonate scale may be protective. Thus a slight positive index frequently is associated with noncorrosive conditions, whereas a negative index indicates the possibility of corrosion.

To calculate the saturation index, determine methyl orange alkalinity, calcium ion concentration, pH, temperature, and total filtrable residue.

1. Calculation

The pH_S value can be calculated, with good precision, by using equilibrium expressions for the solution of $CaCO_3$ and the second hydrolysis of carbonic acid (H_2CO_3):

$$CaCO_3(s) \leftrightarrows Ca^{2+} + CO_3^{2-}$$

$$K_S = (Ca^{2+})(CO_3^{2-}) \quad (1)$$

$$HCO_3^- \leftrightarrows H^+ + CO_3^{2-}$$

$$K_2 = \frac{(H^+)(CO_3^{2-})}{(HCO_3^-)} \quad (2)$$

Dividing Equation 1 by Equation 2 and rearranging gives:

$$(H^+) = \frac{K_2}{K_S}(Ca^{2+})(HCO_3^-) \quad (3a)$$

or, taking negative logarithms, and expressing (Ca^{2+}) and (HCO_3^-) as $CaCO_3$,

$$pH_S = p[Ca^{2+}] + p[HCO_3^-] + p[K'_2/K'_S] \quad (3b)$$

Obtain values of conditional equilibrium constants from the literature.[2–5]

Accounting for the effect of temperature and ionic strength on equilibrium constants, Larson[5] formulated the expression for the pH at $CaCO_3$ saturation, pH_S, as:

$$pH_S = A + B - \log[Ca^{2+}] - \log[\text{alkalinity}] \quad (4a)$$

where calcium ion concentration and alkalinity are expressed as milligrams $CaCO_3$ per liter. Values for constants and logarithms in Equation 4a are given in Tables 203:I through III. This method of calculation is adequate for waters having a pH_S of 9.3 or less. Above pH_S 9.3, use bicarbonate alkalinity rather than total alkalinity. This can be determined with a nomograph such as that reported by Dye.[6]

For example, for a water having a calcium ion concentration of 200 mg as $CaCO_3$/L, an alkalinity of 60 mg as $CaCO_3$/L, a temperature of 16 C, and a total filtrable residue of 650 mg/L, this equation is solved:

$$pH_S = 2.20 + 9.88 - 2.30 - 1.78 = 8.00 \quad (4b)$$

TABLE 203:I. CONSTANT *A* AS FUNCTION OF WATER TEMPERATURE

Water Temperature *C*	*A**
0	2.60
4	2.50
8	2.40
12	2.30
16	2.20
20	2.10
25	2.00
30	1.90
40	1.70
50	1.55
60	1.40
70	1.25
80	1.15

*Calculated from K_2 as reported by Harned and Scholes[7] and K_s as reported by Larson and Buswell.[3] Values above 40 C involve extrapolation.

TABLE 203:II. CONSTANT *B* AS FUNCTION OF TOTAL FILTRABLE RESIDUE

Total Filtrable Residue *mg/L*	*B*
0	9.70
100	9.77
200	9.83
400	9.86
800	9.89
1,000	9.90

TABLE 203:III. LOGARITHMS OF CALCIUM ION AND ALKALINITY CONCENTRATIONS

Ca^{2+} or Alkalinity *mg $CaCO_3$/L*	log
10	1.00
20	1.30
30	1.48
40	1.60
50	1.70
60	1.78
70	1.84
80	1.90
100	2.00
200	2.30
300	2.48
400	2.60
500	2.70
600	2.78
700	2.84
800	2.90
900	2.95
1,000	3.00

If the measured pH of this water is 9.0, the saturation index is 9.0 − 8.0 or +1.0, and the water is supersaturated with respect to $CaCO_3$.

The pH_S value of 8.0 means that the water will neither dissolve nor precipitate $CaCO_3$ at that pH. While it does not indicate specifically anything about corrosion of any metal in contact with the water, it is widely assumed that maintaining the water pH above pH_S will result in deposition of a protective coating of $CaCO_3$ on distribution system piping. Frequently, this does not occur because of nonuniform deposition or sloughing of materials from pipe walls. The formation of protective $CaCO_3$ coatings may be inhibited further by application of polyphosphates to finished water as sequestering agents.

Additional aids for calculating SI are available.[8–13] Particularly useful is the Caldwell-Lawrence diagram,[13] which facilitates estimation of chemical dosages for softening as well as equilibrium conditions. The saturation pH is obtained readily for most waters from a convenient nomogram.*

The pH_S with respect to $CaCO_3$ frequently has been estimated experimentally by equilibrating $CaCO_3$ chips with a given water (marble test). This test suffers because equilibrium may not be attained and the partial pressure of carbon dioxide in the atmosphere over the sample may influence the results adversely.

Another useful measure of $CaCO_3$ saturation is the driving force index, or DFI, defined by McCauley[14] as:

$$DFI = \frac{[Ca^{2+}]\,[CO_3^{\,2-}]}{10^{10}K_s}$$

Both calcium and carbonate concentrations are expressed as milligrams $CaCO_3$ per liter.

The DFI is calculated easily from the SI as follows:

$$DFI = 10^{(SI)}$$

This value is not a logarithm but a simple ratio of the existing ion product to that which would exist at equilibrium. A value of 1.0 indicates equilibrium, values above 1.0 indicate the possibility of $CaCO_3$ deposition, and values below 1.0 indicate a tendency to dissolve $CaCO_3$ and possible corrosion.

2. References

1. LANGELIER, W.F. 1936. The analytical control of anticorrosion water treatment. *J. Amer. Water Works Ass.* 28:1500.
2. MOORE, E.W. 1938. Calculation of chem-

*Graph and Nomogram for Determination of pH Saturation by Langelier's Formula, originally prepared by C.P. Hoover and M.L. Riehl, obtainable from American Water Works Association, 6666 West Quincy Ave., Denver, Colorado 80235, for $1.00, Order No. 50005.

ical dosages required for the prevention of corrosion. *J. New England Water Works Ass.* 52:311.
3. LARSON, T.E. & A.M. BUSWELL. 1942. Calcium carbonate saturation index and alkalinity interpretations. *J. Amer. Water Works Ass.* 34:1667.
4. LANGELIER, W.F. 1946. Effect of temperature on the pH of natural waters. *J. Amer. Water Works Ass.* 38:179.
5. LARSON, T.E. 1951. The ideal lime-softened water. *J. Amer. Water Works Ass.* 43:649.
6. DYE, J.F. 1952. Calculation of the effect of temperature on pH, free carbon dioxide, and the three forms of alkalinity. *J. Amer. Water Works Ass.* 44:356.
7. HARNED, H.S. & S.R. SCHOLES, JR. 1941. The ionization constant of HCO_3^- from 0 to 50°. *J. Amer. Chem. Soc.* 63:1706.
8. LANGELIER, W.F. 1946. Chemical equilibria in water treatment. *J. Amer. Water Works Ass.* 38:169.
9. HOOVER, C.P. 1938. Practical application of the Langelier method. *J. Amer. Water Works Ass.* 30:1802.
10. BLACK, A.P. 1948. The chemistry of water treatment. *Water Sewage Works* 95:369.
11. HIRSCH, A.A. 1942. A special slide rule for calcium carbonate equilibrium problems. *Ind. Eng. Chem.*, Anal. Ed. 14:178.
12. HIRSCH, A.A. 1942. A slide rule for carbonate equilibrium and alkalinity in water supplies. *Ind. Eng. Chem.*, Anal. Ed. 14:943.
13. CALDWELL, D.H. & W.B. LAWRENCE. 1953. Water softening and conditioning problems. *Ind. Eng. Chem.* 45:535.
14. MCCAULEY, R.F. 1960. Use of polyphosphates for developing protective calcite coatings. *J. Amer. Water Works Ass.* 52:721.

204 COLOR

Color in water may result from the presence of natural metallic ions (iron and manganese), humus and peat materials, plankton, weeds, and industrial wastes. Color is removed to make a water suitable for general and industrial applications. Colored industrial wastewaters may require color removal before discharge into watercourses.

1. Definitions

The term "color" is used here to mean true color, that is, the color of water from which turbidity has been removed. The term "apparent color" includes not only color due to substances in solution, but also that due to suspended matter. Apparent color is determined on the original sample without filtration or centrifugation. In some highly colored industrial wastewaters color is contributed principally by colloidal or suspended material. In such cases both true color and apparent color should be determined.

2. Selection of Pretreatment for Removal of Turbidity

To determine color by currently accepted methods, turbidity must be removed before analysis. The optimal method for removing turbidity without removing color has not been found yet. Filtration yields results that are reproducible from day to day and among laboratories. However, some filtration procedures also may remove some true color. Centrifugation avoids interaction of color with filter materials, but results vary with the sample nature and size and speed of the centrifuge. When sample dilution is necessary, whether it precedes or follows turbidity removal, it can alter the measured color if large color-bodies are present.

Acceptable pretreatment procedures

are included with each method. State the pretreatment method when reporting results.

3. Selection of Method

The visual comparison method is applicable to nearly all samples of potable water. Pollution by certain industrial wastes may produce unusual colors that cannot be matched. In this case use an instrumental method. A recent modification of the tristimulus and the spectrophotometric methods allows calculation of a single color value representing uniform chromaticity differences even when the sample exhibits color significantly different from that of platinum cobalt standards; it is included as a tentative method. For comparison of color values among laboratories, calibrate the visual method by the instrumental procedures.

204 A. Visual Comparison Method

1. General Discussion

a. Principle: Color is determined by visual comparison of the sample with known concentrations of colored solutions. Comparison also may be made with special, properly calibrated glass color disks. The platinum-cobalt method of measuring color is the standard method, the unit of color being that produced by 1 mg platinum/L in the form of the chloroplatinate ion. The ratio of cobalt to platinum may be varied to match the hue in special cases; the proportion given below is usually satisfactory to match the color of natural waters.

b. Application: The platinum-cobalt method is useful for measuring color of potable water and of water in which color is due to naturally occurring materials. It is not applicable to most highly colored industrial wastewaters.

c. Interference: Even a slight turbidity causes the apparent color to be noticeably higher than the true color; therefore remove turbidity before approximating true color by differential reading with different color filters[1] or by differential scattering measurements.[2] Neither technic, however, has reached the status of a standard method. Remove turbidity by centrifugation or by the filtration procedure described under Method B. Centrifuge for 1 hr unless it has been demonstrated that centrifugation under other conditions accomplishes satisfactory turbidity removal.

The color value of water is extremely pH-dependent and invariably increases as the pH of the water is raised. When reporting a color value, specify the pH at which color is determined. For research purposes or when color values are to be compared among laboratories, determine the color response of a given water over a wide range of pH values.[3]

d. Field method: Because the platinum-cobalt standard method is not convenient for field use, compare water color with that of glass disks held at the end of metallic tubes containing glass comparator tubes filled with sample and colorless distilled water. Match sample color with the color of the tube of clear water plus the calibrated colored glass when viewed by looking toward a white surface. Calibrate each disk to correspond with the colors on the platinum-cobalt scale. The glass disks give results in substantial agreement with those obtained by the platinum-cobalt method and their use is recognized as a standard field procedure.

e. Nonstandard laboratory methods: Using glass disks or liquids other than water as standards for laboratory work is permissible only if these have been individually calibrated against platinum-cobalt standards. Waters of highly unusual color, such as those that may occur by mixture with certain industrial wastes, may have hues so far removed from those of the platinum-cobalt standards that comparison by the standard method is difficult or impossible. For such waters, use the methods in Sections 204B and C. However, results so obtained are not directly comparable to those obtained with platinum-cobalt standards.

f. Sampling: Collect representative samples in clean glassware. Make the color determination within a reasonable period because biological or physical changes occurring in storage may affect color. With naturally colored waters these changes invariably lead to poor results.

2. Apparatus

a. Nessler tubes, matched, 50-mL, tall form.

b. pH meter, for determining sample pH (see Section 423).

3. Preparation of Standards

a. If a reliable supply of potassium chloroplatinate cannot be purchased, use chloroplatinic acid prepared from metallic platinum. Do not use commercial chloroplatinic acid because it is very hygroscopic and may vary in platinum content. Potassium chloroplatinate is not hygroscopic.

b. Dissolve 1.246 g potassium chloroplatinate, K_2PtCl_6 (equivalent to 500 mg metallic Pt) and 1.00 g crystallized cobaltous chloride, $CoCl_2 \cdot 6H_2O$ (equivalent to about 250 mg metallic Co) in distilled water with 100 mL conc HCl and dilute to 1,000 mL with distilled water. This stock standard has a color of 500 units.

c. If K_2PtCl_6 is not available, dissolve 500 mg pure metallic Pt in aqua regia with the aid of heat; remove HNO_3 by repeated evaporation with fresh portions of conc HCl. Dissolve this product, together with 1.00 g crystallized $CoCl_2 \cdot 6H_2O$, as directed above.

d. Prepare standards having colors of 5, 10, 15, 20, 25, 30, 35, 40, 45, 50, 60, and 70 by diluting 0.5, 1.0, 1.5, 2.0, 2.5, 3.0, 3.5, 4.0, 4.5, 5.0, 6.0, and 7.0 mL stock color standard with distilled water to 50 mL in nessler tubes. Protect these standards against evaporation and contamination when not in use.

4. Procedure

a. Estimation of intact sample: Observe sample color by filling a matched nessler tube to the 50-mL mark with sample and comparing it with standards. Look vertically downward through tubes toward a white or specular surface placed at such an angle that light is reflected upward through the columns of liquid. If turbidity is present and has not been removed, report as "apparent color." If the color exceeds 70 units, dilute sample with distilled water in known proportions until the color is within the range of the standards.

b. Measure pH of each sample.

5. Calculation

a. Calculate color units by the following equation:

$$\text{Color units} = \frac{A \times 50}{B}$$

where:

A = estimated color of a diluted sample and

B = mL sample taken for dilution.

b. Report color results in whole numbers and record as follows:

Color Units	Record to Nearest
1–50	1
51–100	5
101–250	10
251–500	20

c. Report sample pH.

6. References

1. KNIGHT, A.G. 1951. The photometric estimation of color in turbid waters. *J. Inst. Water Eng.* 5:623.
2. JULLANDER, I. & K. BRUNE. 1950. Light absorption measurements on turbid solutions. *Acta Chem. Scand.* 4:870.
3. BLACK, A.P. & R.F. CHRISTMAN. 1963. Characteristics of colored surface waters. *J. Amer. Water Works Ass.* 55:753.

204 B. Spectrophotometric Method

1. General Discussion

a. Principle: The color of a filtered sample is expressed in terms that describe the sensation realized when viewing the sample. The hue (red, green, yellow, etc.) is designated by the term "dominant wavelength", the degree of brightness by "luminance", and the saturation (pale, pastel, etc.) by "purity". These values are best determined from the light transmission characteristics of the filtered sample by means of a spectrophotometer.

b. Application: This method is applicable to potable and surface waters and to wastewaters, both domestic and industrial.

c. Interference: Turbidity interferes. Remove by the filtration method described below.

2. Apparatus

a. Spectrophotometer, having 10-mm absorption cells, a narrow (10-nm or less) spectral band, and an effective operating range from 400 to 700 nm.

b. Filtration system, consisting of the following (see Figure 204:1):

1) *Filtration flasks,* 250-mL, with side tubes.

2) *Walter crucible holder.*

3) *Micrometallic filter crucible,* average pore size 40 μm.

4) *Calcined filter aid.**

5) *Vacuum system.*

*Celite No. 505 (Johns Manville Corp.) or equivalent.

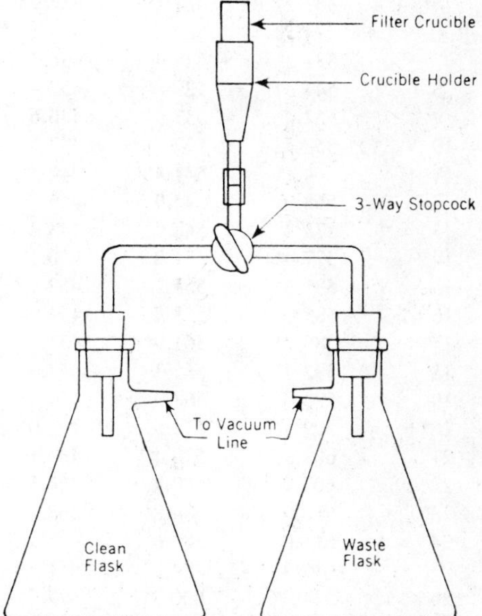

Figure 204:1. Filtration system for color determinations.

3. Procedure

a. Preparation of sample: Bring two 50-mL samples to room temperature. Use one sample at the original pH; adjust pH of the other to 7.6 by using sulfuric acid (H_2SO_4) and sodium hydroxide (NaOH) of such concentrations that the resulting volume change does not exceed 3%. A standard pH is necessary because of the variation of color with pH. Remove excessive quantities of suspended materials by centrifuging. Treat each sample separately, as follows:

TABLE 204:I. SELECTED ORDINATES FOR SPECTROPHOTOMETRIC COLOR DETERMINATIONS†

Ordinate No.	X	Y	Z
	Wavelength *nm*		
1	424.4	465.9	414.1
2*	435.5*	489.5*	422.2*
3	443.9	500.4	426.3
4	452.1	508.7	429.4
5*	461.2*	515.2*	432.0*
6	474.0	520.6	434.3
7	531.2	525.4	436.5
8*	544.3*	529.8*	438.6*
9	552.4	533.9	440.6
10	558.7	537.7	442.5
11*	564.1*	541.4*	444.4*
12	568.9	544.9	446.3
13	573.2	548.4	448.2
14*	577.4*	551.8*	450.1*
15	581.3	555.1	452.1
16	585.0	558.5	454.0
17*	588.7*	561.9*	455.9*
18	592.4	565.3	457.9
19	596.0	568.9	459.9
20*	599.6*	572.5*	462.0*
21	603.3	576.4	464.1
22	607.0	580.4	466.3
23*	610.9*	584.8*	468.7*
24	615.0	589.6	471.4
25	619.4	594.8	474.3
26*	624.2*	600.8*	477.7*
27	629.8	607.7	481.8
28	636.6	616.1	487.2
29*	645.9*	627.3*	495.2*
30	663.0	647.4	511.2
Factors When 30 Ordinates Used			
	0.03269	0.03333	0.03938
Factors When 10 Ordinates Used			
	0.09806	0.10000	0.11814

†Insert in each column the transmittance value (%) corresponding to the wavelength shown. Where limited accuracy is sufficient, use only the ordinates marked with an asterisk.

Thoroughly mix 0.1 g filter aid in a 10-mL portion of centrifuged sample and filter to form a precoat in the filter crucible. Direct filtrate to waste flask as indicated in Figure 204:1. Mix 40 mg filter aid in a 35-mL portion of centrifuged sample. With vacuum still on, filter through the precoat and pass filtrate to waste flask until clear; then direct clear-filtrate flow to clean flask by means of the three-way stopcock and collect 25 mL for the transmittance determination.

b. Determination of light transmission characteristics: Thoroughly clean 1-cm absorption cells with detergent and rinse with distilled water. Rinse twice with filtered sample, clean external surfaces with lens paper, and fill cell with filtered sample.

Determine transmittance values (in percent) at each visible wavelength value presented in Table 204:I, using the 10 ordinates marked with an asterisk for fairly accurate work and all 30 ordinates for increased accuracy. Set instrument to read 100% transmittance on the distilled water blank and make all determinations with a narrow spectral band.

4. Calculation

a. Tabulate transmittance values corresponding to wavelengths shown in Columns X, Y, and Z in Table 204:I. Total each transmittance column and multiply totals by the appropriate factors (for 10 or 30 ordinates) shown at the bottom of the table, to obtain tristimulus values X, Y, and Z. The tristimulus value Y is *percent luminance*.

b. Calculate the trichromatic coefficients x and y from the tristimulus values X, Y, and Z by the following equations:

$$x = \frac{X}{X + Y + Z}$$

$$y = \frac{Y}{X + Y + Z}$$

Locate point (x, y) on one of the chro-

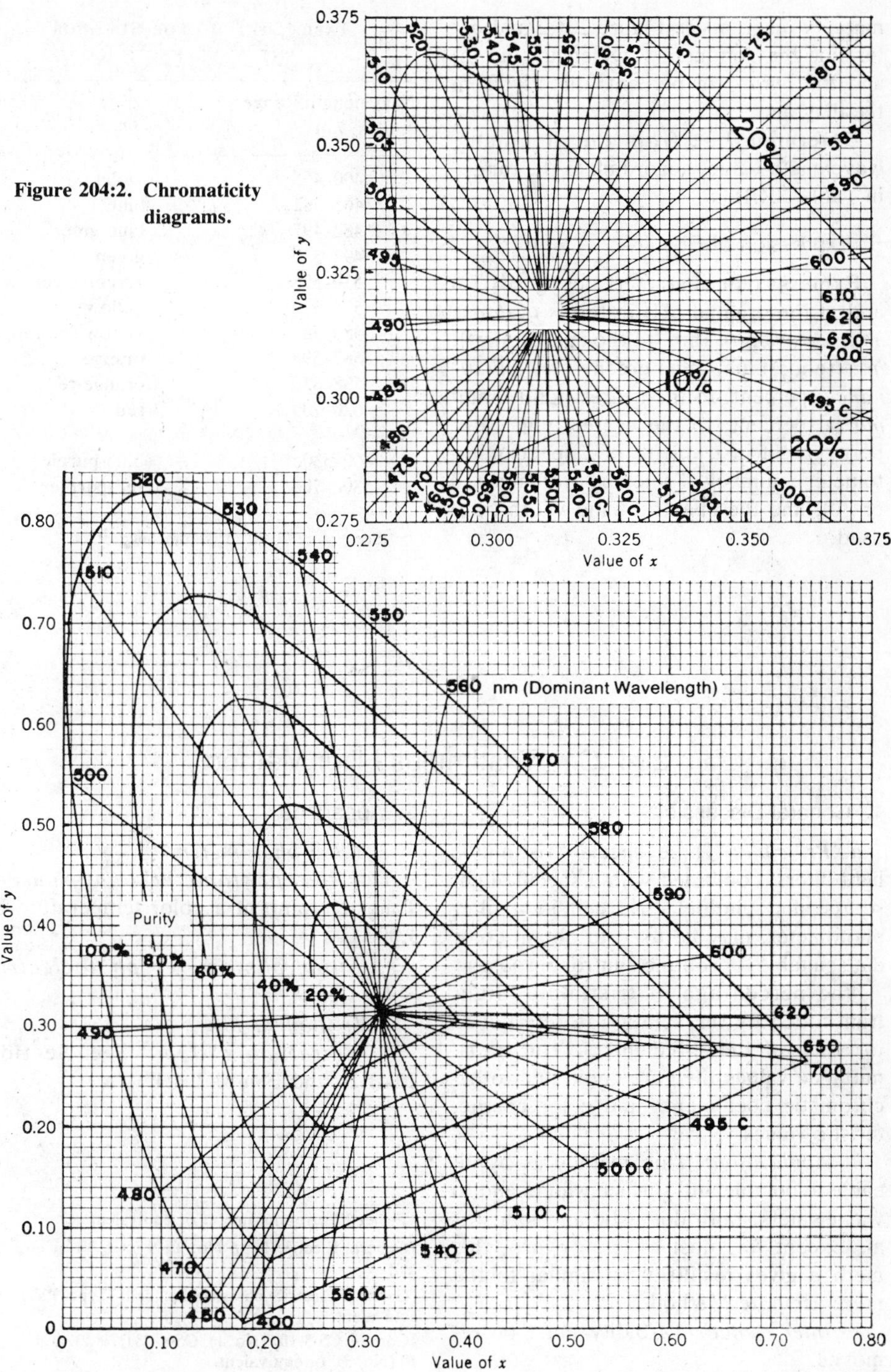

Figure 204:2. Chromaticity diagrams.

maticity diagrams in Figure 204:2 and determine the dominant wavelength (in nm) and the purity (in percent) directly from the diagram.

Determine hue from the dominant-wavelength value, according to the ranges in Table 204:II.

5. Expression of Results

Express color characteristics (at pH 7.6 and at the original pH) in terms of *dominant wavelength* (nm, to the nearest unit), *hue* (e.g., blue, blue-green, etc.), *luminance* (percent, to the nearest tenth), and *purity* (percent, to the nearest unit). Report type of instrument (i.e., spectrophotometer), number of selected ordinates (10 or 30), and the spectral band width (nm) used.

TABLE 204:II. COLOR HUES FOR DOMINANT WAVELENGTH RANGES

Wavelength Range *nm*	Hue
400–465	violet
465–482	blue
482–497	blue-green
497–530	green
530–575	greenish yellow
575–580	yellow
580–587	yellowish orange
587–598	orange
598–620	orange-red
620–700	red
400–530c*	blue-purple
530c–700*	red-purple

* See Figure 204.2 for significance of "c".

204 C. Tristimulus Filter Method

1. General Discussion

a. Principle: Three special tristimulus light filters, combined with a specific light source and photoelectric cell in a filter photometer, may be used to obtain color data suitable for routine control purposes.

The percentage of tristimulus light transmitted by the solution is determined for each of the three filters. The transmittance values then are converted to trichromatic coefficients and color characteristic values.

b. Application: This method is applicable to potable and surface waters and to wastewaters, both domestic and industrial. Except for most exacting work, this method gives results very similar to the more accurate Method B.

c. Interference: Turbidity must be removed.

2. Apparatus

*a. Filter photometer.**

b. Filter photometer light source: Tungsten lamp at a color temperature of 3,000 C.†

c. Filter photometer photoelectric cells, 1 cm.‡

d. Tristimulus filters.§

e. Filtration system: See Section 204B.2*b* and Figure 204:1.

*Fisher Electrophotometer or equivalent.

†General Electric lamp No. 1719 (at 6 V) or equivalent.

‡General Electric photovoltaic cell, Type PV-1, or equivalent.

§Corning CS-3-107 (No. 1), CS-4-98 (No. 2), and CS-5-70 (No. 3), or equivalent.

3. Procedure

a. Preparation of sample: See Section 204B.3*a*.

b. Determination of light transmission characteristics: Thoroughly clean (with detergent) and rinse 1-cm absorption cells with distilled water. Rinse each absorption cell twice with filtered sample, clean external surfaces with lens paper, and fill cell with filtered sample.

Place a distilled water blank in another cell and use it to set the instrument at 100% transmittance. Determine percentage of light transmission through sample for each of the three tristimulus light filters, with the filter photometer lamp intensity switch in a position equivalent to 4 V on the lamp.

4. Calculation

a. Determine luminance value directly as the percentage transmittance value obtained with the No. 2 tristimulus filter.

b. Calculate tristimulus values X, Y, and Z from the percentage transmittance (T_1, T_2, T_3) for filters No. 1, 2, 3, as follows:

$$X = T_3 \times 0.06 + T_1 \times 0.25$$
$$Y = T_2 \times 0.316$$
$$Z = T_3 \times 0.374$$

Calculate and determine trichromatic coefficients x and y, dominant wavelength, hue, and purity as in Section 204B.4*b* above.

5. Expression of Results

Express results as prescribed in Section 204B.5.

204 D. ADMI Tristimulus Filter Method (TENTATIVE)

1. General Discussion

a. Principle: This method is an extension of Tristimulus Method 204C. By this method a measure of the sample color, independent of hue, may be obtained. It is based on use of the Adams-Nickerson chromatic value formula for calculating single number color difference values, i.e., uniform color differences. For example, if two colors, A and B, are judged visually to differ from colorless to the same degree, their ADMI color values will be the same. The modification was developed by members of the American Dye Manufacturers Institute (ADMI).[1]

b. Application: This method is applicable to colored waters and wastewaters having color characteristics significantly different from platinum-cobalt standards, as well as to waters and wastewaters similar in hue to the standards.

c. Interference: Turbidity must be removed.

2. Apparatus

*a. Filter photometer** equipped with CIE tristimulus filters (see 204C.2*d*).

b. Filter photometer light source: Tungsten lamp at a color temperature of 3,000 C. (see 204C.2*b*).

c. Absorption cells and appropriate cell holders: For color values less than 250 ADMI units, use cells with a 5.0-cm light path; for color values greater than 250, use cells with 1.0-cm light path.

*Fisher Electrocolorimeter, Model 181, or equivalent.

d. Filtration system: See Section 204B.2*b* and Figure 204:1; or a centrifuge capable of achieving 1,000 times the acceleration of gravity. (See Section 204A.)

3. Procedure

a. Instrument calibration: Establish curves for each photometer; calibration data for one instrument cannot be applied to another one. Prepare a separate calibration curve for each absorption cell path length.

1) Prepare standards as described in 204A.3. For a 5-cm cell length prepare standards having color values of 25, 50, 100, 200, and 250 by diluting 5.0, 10.0, 20.0, 30.0, 40.0, and 50.0 mL stock color standard with distilled water to 100 mL in volumetric flasks. For the shorter path-length, prepare appropriate standards with higher color values.

2) Determine light transmittance (see 204D.3*c*, below) for each standard with each filter.

3) Using the calculations described in 204D.3*d* below, calculate the tristimulus values (X_s, Y_s, Z_s) for each standard, determine the Munsell values, and calculate the intermediate value (*DE*).

4) Using the *DE* values for each standard, calculate a calibration factor F_n for each standard from the following equation:

$$F_n = \frac{(APHA)_n\ (b)}{(DE)_n}$$

where:

$(APHA)_n$ = APHA color value for standard *n*,

$(DE)_n$ = intermediate value calculated for standard *n*, and

b = cell light path, cm.

Placing $(DE)_n$ on the X axis and F_n on the Y axis, plot a curve for the standard solutions. Use calibration curve to derive the *F* value from *DE* values obtained with samples.

b. Sample preparation: Prepare two 100-mL sample portions (one at the original pH, one at pH 7.6) as described in Section 204B.3*a*, or by centrifugation. (NOTE: Centrifugation is acceptable only if turbidity removal equivalent to filtration is achieved.)

c. Determination of light transmission characteristics: Thoroughly clean absorption cells with detergent and rinse with distilled water. Rinse each absorption cell twice with filtered sample. Clean external surfaces with lens paper and fill cell with sample. Determine sample light transmittance with the three filters to obtain the transmittance values: T_1 from Filter 1, T_2 from Filter 2, and T_3 from Filter 3. Standardize the instrument with each filter at 100% transmittance with distilled water.

d. Calculation of color values: Tristimulus values for samples are X_s, Y_s, and Z_s; for standards X_r, Y_r, and Z_r; and for distilled water X_c, Y_c, and Z_c. Munsell values for samples are V_{xs}, V_{ys}, and V_{zs}; for standards V_{xr}, V_{yr}, and V_{zr}; and for distilled water V_{xc}, Y_{yc}, and V_{zc}.

For each standard or sample calculate the tristimulus values from the following equations:

$$X = (T_3 \times 0.1899) + (T_1 \times 0.791)$$
$$Y = T_2$$
$$Z = T_3 \times 1.1835$$

Tristimulus values for the distilled water blank used to standardize the instrument are always:

$$X_c = 98.09$$
$$Y_c = 100.0$$
$$Z_c = 118.35$$

Convert the six tristimulus values (X_s, Y_s, Z_s, X_c, Y_c, Z_c) to the corresponding Munsell values using published tables† 2,

†By permission of Instrumental Colour Systems, Newbury, Berkshire, England, copies of the tables for this conversion may be obtained from the American Water Works Association, 6666 West Quincy Avenue, Denver, Colo. 80235, at nominal cost.

3, 4 or by the equation given by Bridgeman.[5]

Calculate the intermediate value of *DE* from the equation:

$$DE = \{(0.23\ \Delta V_y)^2 + [\Delta(V_x - V_y)]^2 + [0.4\ \Delta(V_y - V_z)]^2\}^{\frac{1}{2}}$$

where:

$$V_y = V_{ys} - V_{yc}$$
$$\Delta(V_x - V_y) = (V_{xs} - V_{ys}) - (V_{xc} - V_{yc})$$
$$\Delta(V_y - V_z) = (V_{ys} - V_{zs}) - (V_{yc} - V_{zc})$$

when the sample is compared to distilled water.

With the standard calibration curve, use the *DE* value to determine the calibration factor *F*.

Calculate the final ADMI color value as follows:

$$\text{ADMI value} = \frac{(F)\ (DE)}{b}$$

where:

b = absorption cell light path, cm.

Report ADMI color values at pH 7.6 and at the original pH.

4. Alternate Method

The ADMI color value also may be determined spectrophotometrically, using a spectrophotometer with a narrow (10-nm or less) spectral band and an effective operating range of 400 to 700 nm. This method is an extension of 204B. Tristimulus values may be calculated from transmittance measurements, preferably by using the weighted ordinate method or by the selected ordinate method. The method has been described by Allen et al.,[1] who include work sheets and worked examples.

5. References

1. ALLEN, W., W.B. PRESCOTT, R.E. DERBY, C.E. GARLAND, J.M. PERET & M. SALTZMAN. 1973. Determination of color of water and wastewater by means of ADMI color values. *Proc. 28th Ind. Waste Conf.*, Purdue Univ., Eng. Ext. Ser. No. 142:661.
2. MCLAREN, K. 1970. The Adams-Nickerson colour-difference formula. *J. Soc. Dyers Colorists* 86:354.
3. WYSZECKI, G. & W.S. STILES. 1967. Color Science. John Wiley & Sons, New York, N.Y. (See Tables 6.4, A,B,C, pp. 462–467.)
4. JUDD, D.B. & G. WYSZEKI. 1963. Color in Business, Science, and Industry, 2nd ed. John Wiley & Sons, New York, N.Y. (See Tables A,B, and C in Appendix.)
5. BRIDGEMAN, T. 1963. Inversion of the Munsell value equation. *J. Opt. Soc. Amer.* 53:499.

204 E. Bibliography

HAZEN, A. 1892. A new color standard for natural waters. *Amer. Chem. J.* 14:300.

HAZEN, A. 1896. The measurement of the colors of natural waters. *J. Amer. Chem. Soc.* 18:264.

Measurement of Color and Turbidity in Water. 1902. U.S.Geol. Surv., Div. Hydrog. Circ. 8, Washington, D.C.

HARDY, A.C. 1936. Handbook of Colorimetry. Technology Press, Boston, Mass.

OPTICAL SOCIETY OF AMERICA. 1943. Committee Report. The concept of color. *J. Opt. Soc. Amer.* 33:544.

RUDOLFS, W. & W.D. HANLON. 1951. Color in industrial wastes. *Sewage Ind. Wastes* 23:1125.

JONES, H. et al. 1952. The Science of Color. Thomas Y. Crowell Co., New York, N.Y.
PALIN, A.T. 1955. Photometric determination of the colour and turbidity of water. *Water Water Eng.* 59:341.
CHRISTMAN, R.F. & M. GHASSEMI. 1966. Chemical nature of organic color in water. *J. Amer. Water Works Ass.* 58:723.
GHASSEMI, M. & R.F. CHRISTMAN. 1968. Properties of the yellow organic acids of natural waters. *Limnol. Oceanogr.* 13:583.

205 CONDUCTIVITY

1. General Discussion

Conductivity is a numerical expression of the ability of an aqueous solution to carry an electric current. This ability depends on the presence of ions, their total concentration, mobility, valence, and relative concentrations, and on the temperature of measurement. Solutions of most inorganic acids, bases, and salts are relatively good conductors. Conversely, molecules of organic compounds that do not dissociate in aqueous solution conduct a current very poorly, if at all.

The physical measurement made in a laboratory determination of conductivity is usually of resistance, measured in ohms or megaohms. The resistance of a conductor is inversely proportional to its cross section and directly proportional to its length. The magnitude of the resistance measured in an aqueous solution therefore depends on the characteristics of the conductivity cell used, and is not meaningful without knowledge of these characteristics. Specific resistance is the resistance of a cube 1 cm on an edge. In aqueous solutions such a measurement is rare because of the difficulties of electrode fabrication. Practical electrodes measure a given fraction of the specific resistance, the fraction being the cell constant, C:

$$C = \frac{\text{Measured resistance, } R_m}{\text{Specific resistance, } R_s}$$

The reciprocal of resistance is conductance. It measures the ability to conduct a current and is expressed in reciprocal ohms or mhos. A more convenient unit in water analysis is micromhos. When the cell constant is known and applied, the measured conductance is converted to the specific conductance or conductivity, K_s, the reciprocal of the specific resistance:

$$K_s = \frac{1}{R_s} = \frac{C}{R_m}$$

The term "conductivity" is preferred and customarily is reported in micromhos per centimeter (μmhos/cm). In the International System of Units (SI) the reciprocal of the ohm is the siemens (S) and conductivity is reported as millisiemens per meter (mS/m); 1 mS/m = 10 μmhos/cm. To report results in SI units divide μmhos/cm by 10.

Freshly distilled water has a conductivity of 0.5 to 2 μmhos/cm, increasing after a few weeks of storage to 2 to 4 μmhos/cm. This increase is caused mainly by absorption of atmospheric carbon dioxide, and, to a lesser extent, ammonia.

The conductivity of potable waters in the United States ranges generally from 50 to 1,500 μmhos/cm. The conductivity of domestic wastewaters may be near that of the local water supply, although some industrial wastes have conductivities above 10,000 μmhos/cm. Conductivity instruments are used in pipelines, channels, flowing streams, and lakes and can be incorporated in multiple parameter monitoring stations using recorders.

Laboratory measurement of conductivity is relatively accurate but less accurate means of determining conductivity find numerous applications such as signalling exhaustion of ion-exchange resins and rapid determination of large changes in inorganic content of waters and wastewaters. Monitoring devices can give continuous, unattended records of conductivity if they are properly installed and maintained. Most problems in obtaining good records with monitoring equipment are related to electrode fouling and to inadequate sample circulation.

Laboratory conductivity measurements are used to:

a. Establish degree of mineralization to assess the effect of the total concentration of ions on chemical equilibria, physiological effect on plants or animals, corrosion rates, etc.

b. Assess degree of mineralization of distilled and deionized water.

c. Evaluate variations in dissolved mineral concentration of raw water or wastewater. Minor seasonal variations found in reservoir waters contrast sharply with the daily fluctuations in some polluted river waters. Wastewater containing significant trade wastes also may show a considerable daily variation.

d. Estimate sample size to be used for common chemical determinations and to check results of a chemical analysis.

e. Determine amount of ionic reagent needed in certain precipitation and neutralization reactions, the end point being denoted by a change in slope of the curve resulting from plotting conductivity against buret readings.

f. Estimate total filtrable residue in a sample by multiplying conductivity (in micromhos per centimeter) by an empirical factor. This factor may vary from 0.55 to 0.9, depending on the soluble components of the water and on the temperature of measurement. Relatively high factors may be required for saline or boiler waters, whereas lower factors may apply where considerable hydroxide or free acid is present. Even though sample evaporation results in the change of bicarbonate to carbonate the empirical factor is derived for a comparatively constant water supply by dividing dissolved residue by conductivity. Approximate milliequivalents per liter of either cations or anions in some waters by multiplying conductivity (in micromhos per centimeter) by 0.01.

Use a conductivity cell and a Wheatstone bridge for measuring the sample electrical resistance or measure conductivity as the ratio of electrical current through the cell to the applied voltage.

Electrolytic conductivity (unlike metallic conductivity) increases with temperature at a rate of approximately 1.9%/degree C. Significant errors can result from inaccurate temperature measurement. Potassium chloride (KCl) solutions have a lower temperature coefficient of conductivity than typical potable water. Sodium chloride (NaCl), on the other hand, has a temperature coefficient that closely approximates that found in most waters from wells and surface sources. Note that each ion has a different temperature coefficient; thus, for precise work, determine conductivity at 25.0 C. The significance of the temperature correction, one part in 500 for 25 ± 0.1 C, depends on available equipment and precision desired.

2. Apparatus

a. Self-contained conductivity instruments: Use an instrument consisting of a source of alternating current, a Wheatstone bridge, a null indicator, and a conductivity cell or other instrument measuring the ratio of alternating current through the cell to voltage across it. The latter has the advantage of a linear reading of conductivity. Choose an instrument capable of measuring conductivity with an error not exceeding 1% or 1 μmho/cm, whichever is greater.

b. Thermometer, capable of being read to the nearest 0.1 C and covering the range 23 C to 27 C. An electrical thermometer having a small thermistor sensing element is convenient because of its rapid response.

c. Conductivity cell:

1) Platinum-electrode type—Conductivity cells containing platinized electrodes are available in either pipet or immersion form. Cell choice depends on expected range of conductivity and resistance range of the instrument. Experimentally check range for complete instrument assembly by comparing instrumental results with the true conductivities of the KCl solutions listed in Table 205:I. Clean new cells with chromic-sulfuric acid cleaning mixture and platinize the electrodes before use. Subsequently, clean and replatinize them whenever the readings become erratic, when a sharp end point cannot be obtained, or when inspection shows that any platinum black has flaked off. To platinize, prepare a solution of 1 g chloroplatinic acid, $H_2PtCl_6 \cdot 6H_2O$, and 12 mg lead acetate in 100 mL distilled water. A stronger solution reduces the time required to platinize electrodes and may be used when time is a factor, e.g., when the cell constant is 1.0/cm or more. Immerse the electrodes in this solution and connect both to the negative terminal of a 1.5-V dry cell battery. Connect positive side of battery to a piece of platinum wire and dip the wire into the solution. Use a current such that only a small quantity of gas is evolved. Continue electrolysis until both cell electrodes are coated with platinum black. Save platinizing solution for subsequent use. Rinse electrodes thoroughly and when not in use keep immersed in distilled water.

2) Nonplatinum-electrode type—Use conductivity cells containing electrodes constructed from durable common metals (stainless steel among others) for continuous monitoring and field studies. Calibrate such cells by comparing sample conductivity with results obtained with a laboratory instrument. Use properly designed and mated cell and instrument to minimize errors in cell constant.

TABLE 205:I. CONDUCTIVITY OF POTASSIUM CHLORIDE SOLUTIONS AT 25 C*

Concentration *N*	Equivalent Conductivity *mho/cm/equiv.*	Conductivity *μmhos/cm*
0	149.85	
0.0001	149.43	14.94†
0.0005	147.81	73.90
0.001	146.95	147.0
0.005	143.55	717.8
0.01	141.27	1.413
0.02	138.34	2.767
0.05	133.37	6.668
0.1	128.96	12.900
0.2	124.08	24.820
0.5	117.27	58.640
1	111.87	111.900

*Data drawn from Robinson & Stokes.[1]
†Computed from equation given in Lind et al.[2]

3. Reagents

a. Conductivity water: Pass distilled water through a mixed-bed deionizer and discard first 1,000 mL. Conductivity should be less than 1 μmho/cm.

b. Standard potassium chloride solution, KCl, 0.0100*N*; Dissolve 745.6 mg anhydrous KCl in conductivity water and dilute to 1,000 mL at 25 C. This is the standard reference solution, which at 25 C has a conductivity of 1,413 μmhos/cm. It is satisfactory for most samples when the cell has a constant between 1 and 2. For other cell constants, use stronger or weaker KCl solutions listed in Table 205:I. Store in a glass-stoppered borosilicate glass bottle.

4. Procedure

a. Determination of cell constant: Rinse conductivity cell with at least three portions of 0.01*N* KCl solution. Adjust temperature of a fourth portion to 25.0 ± 0.1 C. Measure resistance of this portion and note temperature. Compute cell constant, *C*:

$$C = (0.001413)\,(R_{KCl})\,[1 + 0.0191\,(t - 25)]$$

where:

R_{KCl} = measured resistance, ohms, and
t = observed temperature, degrees C.

b. Conductivity measurement: Rinse cell with one or more portions of sample. Adjust temperature of a final portion to 25.0 ± 0.1 C. Measure sample resistance or conductivity and note temperature.

5. Calculation

The temperature coefficient of most waters is only approximately the same as that of standard KCl solution; the more the temperature of measurement deviates from 25.0 C, the greater the uncertainty in applying the temperature correction. Report all conductivities at 25.0 C.

a. When sample resistance is measured, conductivity at 25 C is:

$$K = \frac{(1{,}000{,}000)\,(C)}{R_m\,[1 + 0.0191\,(t - 25)]}$$

where:

K = conductivity, μmhos/cm,
C = cell constant, cm^{-1},
R_m = measured resistance of sample, ohms, and
t = temperature of measurement.

b. When sample conductivity is measured, conductivity at 25 C is:

$$K = \frac{(K_m)\,(1{,}000{,}000)\,(C)}{1 + 0.0191\,(t - 25)}$$

where:

K_m = measured conductivity, mhos at t C, and other units are defined as above.

Note: If conductivity readout is in micromhos per centimeter, delete the factor 1,000,000 in the numerator.

6. Precision and Accuracy

Three synthetic samples were tested with the following results:

Conductivity *μmhos/cm*	No. of Results	Relative Standard Deviation %	Relative Error %
147.0	117	8.6	9.4
303.0	120	7.8	1.9
228.0	120	8.4	3.0

7. References

1. ROBINSON, R.A. & R.H. STOKES. 1959. Electrolyte Solutions, 2nd ed. Academic Press, New York, p. 466.
2. LIND, J.E., J.J. ZWOLENIK & R.M. FUOSS. 1959. Calibration of conductance cells at 25 C with aqueous solutions of potassium chloride. *J. Amer. Chem. Soc.* 81:1557.

8. Bibliography

JONES, G. & B.C. BRADSHAW. 1933. The measurement of the conductance of electrolytes. V. A redetermination of the conductance of standard potassium chloride solutions in absolute units. *J. Amer. Chem. Soc.* 55:1780.

206 FLOATABLES

One important criterion for evaluating the possible effect of waste disposal into surface waters is the amount of floatable material in the waste. Two general types of floating matter are found: particulate matter that includes "grease balls," and liquid components capable of spreading as a thin, highly visible film over large areas. Floatable material in wastewaters is important because it accumulates on the surface, is highly visible, is subject to wind-induced transport, and may contain pathogenic bacteria and/or viruses associated with individual particles. Colloidally-dispersed oil and grease behave like other dispersed organic matter and are included in the material measured by the COD, BOD, and TOC tests. The floatable oil test indicates the readily separable fraction. The results are useful in designing oil and grease separators, in ascertaining the efficiency of operating separators, and in monitoring raw and treated wastewater streams. Many cities and districts have specified floatable oil limits for wastewater discharged to sewers.

206 A. Particulate Floatables (TENTATIVE)

1. Discussion

a. Principle: This method depends on the gravity separation of particles having densities less than that of the surrounding water. Particles that collect on the surface and can be filtered out and dried at 103 to 105 C are defined by this test as floatable particles.

b. Application: This method is applicable to raw wastes, treated secondary effluents, and industrial wastes.

c. Precautions: Even slight differences in sampling and handling during and after collection can give large differences in the measured amount of floatable material. For a reproducible analysis treat all samples uniformly, preferably by mixing them in a standard manner, before flotation. Because the procedure relies on the difference in specific gravity between the liquid and the floating particles, temperature variations may affect the results. Conduct the test at a constant temperature of 20 C.

d. Minimum detectable concentration: The minimum detectable concentration is approximately 1 mg/L.

2. Apparatus

a. Floatables sampler with mixer: Use a metal container of at least 5 L capacity equipped with a propeller mixer on a separate stand (Figure 206:1), and with a 20-mm-ID bottom outlet cocked at an angle of 45° to the container wall against the direction of fluid movement. The 45° angle assures that even large particles will flow

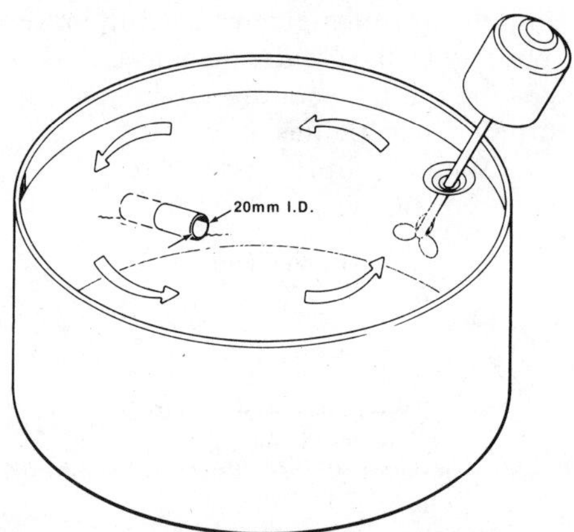

Figure 206:1. Floatables sampler with mixer.

from the container into the flotation funnels when the sample is withdrawn. Fit exterior of bottom outlet with a short piece of tubing and a pinch clamp to allow unrestricted flow through the outlet. Coat the inside of the container with TFE* to pre-

*Teflon or equivalent.

vent oil and grease from sticking to the surface.

b. Flotation funnel: Use an Imhoff cone provided with a TFE stopcock at the bottom and extended at the top to a total volume of 3.5 L (Figure 206:2). TFE-coat the

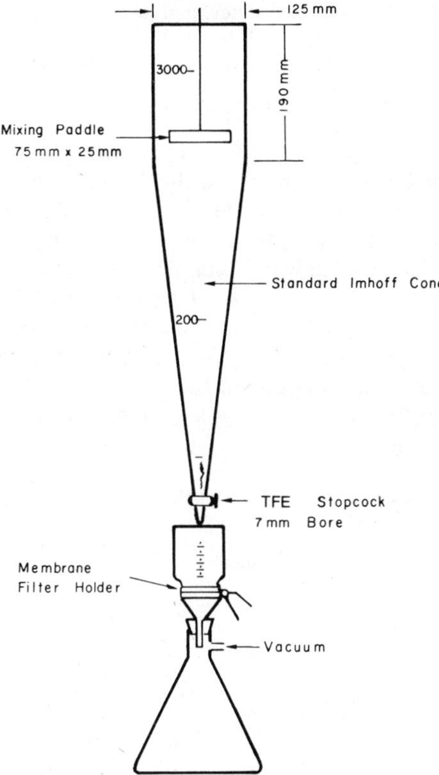

Figure 206:2. Floatables flotation funnel and filter holder.

inside of the flotation funnel to prevent floatable grease particles sticking to the sides. Mount flotation funnels as shown in Figure 206:3 with a light behind the bottom of the funnels to aid in reading levels.

c. Filter holder: TFE-coat the inside of the top of a standard 500-mL membrane filter holder.

d. Filters, glass fiber, fine porosity.†

e. Vacuum flask, 500 mL.

f. TFE coating: Follow instructions that accompany commercially available coating kits. Alternatively, have necessary glassware coated commercially.

3. Procedure

a. Preparation of glass fiber filters: See Section 209D.3*a*.

b. Sample collection and treatment: Collect sample in the floatables sampler at a point of complete mixing, transport to the laboratory, and place 3.0 L in the flotation funnel. While the flotation funnel is being filled, mix sampler contents with a small propeller mixer. Adjust mixing speed to provide uniform distribution of floating particles throughout the liquid but avoid extensive air entrapment through formation of a large vortex.

c. Flotation: Mix flotation funnel contents at 40 rpm for 15 min using a paddle mixer (Figure 206:3). Let settle for 5 min,

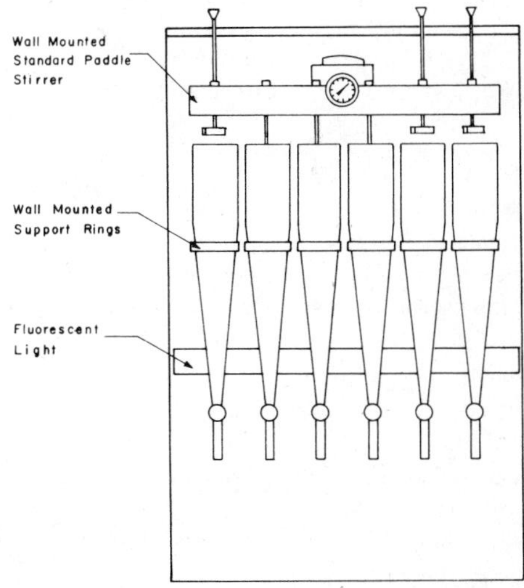

Figure 206:3. Flotation funnels and mixing unit.

mix at 100 rpm for 1 min, and let settle for 30 min. Discharge 2.8 L through bottom stopcock at a rate of 500 mL/min. With

†Whatman GF/C or equivalent.

distilled water from a wash bottle, wash down any floatable material sticking to the sides of the stirring paddle and funnel. Let remaining 200 mL settle for 15 min and discharge settled solids and liquid down to the 40-mL mark on the Imhoff cone. Let settle again for 10 min and discharge until only 10 mL liquid and the floating particles remain in funnel. Add 500 mL distilled water and stir to separate entrapped settleable particles from the floatable particles. Let settle for 15 min, then discharge to the 40-mL mark. Let settle for 10 min, then discharge dropwise to the 10-mL mark. Filter remaining 10 mL and floating particles through a preweighed glass fiber filter. Wash sides of flotation funnel with distilled water to transfer all floatable material to filter.

d. Weighing: Dry and weigh glass fiber filter at 103 to 105 C for exactly 2 hr (see Section 209D.3*b*).

4. Calculation

$$\text{mg particulate floatables/L} = \frac{(A - B)}{C}$$

where:

A = weight of filter + floatables, mg,
B = weight of filter, mg, and
C = sample volume, L.

5. Precision and Accuracy

Precision varies with the concentration of suspended matter in the sample. There is no completely satisfactory procedure for determining the accuracy of the method for wastewater samples but approximate recovery can be determined by running a second test for floatables on all water discharged throughout the procedure, with the exception of the last 10 mL. Precision and accuracy are summarized in Table 206:I.

TABLE 206:I. COEFFICIENT OF VARIATION AND RECOVERY FOR PARTICULATE FLOATABLES TEST

Type of Wastewater	Average Floatables Concentration *mg/L*	No. of Samples	Coefficient of Variation *%*	Recovery *%*
Raw*	49	5	5.7	96
Raw	1.0	5	20	92
Primary effluent	2.7	5	15	91

*Additional floatable material added from skimmings of a primary sedimentation basin.

206 B. Trichlorotrifluoroethane-Soluble Floatable Oil and Grease (TENTATIVE)

1. Discussion

The floatable oil and grease test does not measure a precise class of substances; rather, the results are determined by the conditions of the test. The fraction measured includes oil and grease, both floating and adhering to the sides of the test vessel. The adhering and the floating portions are of similar practical significance because it is assumed that most of the adhering portion would otherwise float under stream conditions. The results have been found to represent well the amount of oil removed in separators having overflow rates equivalent to the test conditions.

2. Apparatus

a. Floatable oil tube (Figure 206:4): Before use, carefully clean tube by brushing with a mild scouring powder or immersing in chromic acid cleaning solution for several hours. Water must form a smooth film on inside of cleaned glass. Do not use lubricant on stopcock.

b. Conical flask: 300 mL.

3. Reagents

a. 1,1,2 trichloro-1,2,2 trifluoroethane‡: See Oil and Grease, Partition-Gravimetric Method (Section 503A).

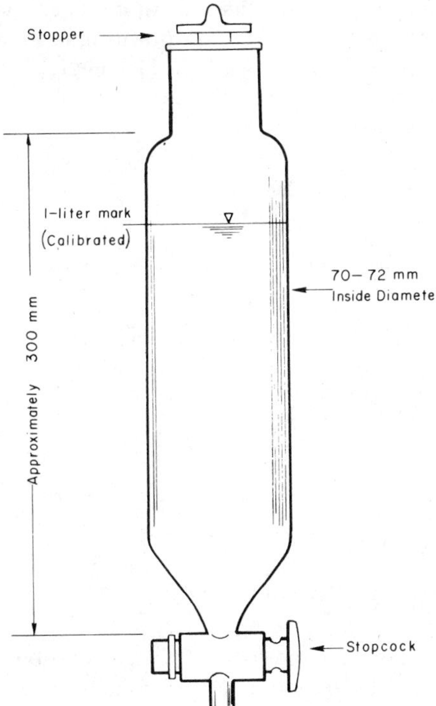

Figure 206:4. One-liter capacity floatable oil tube.

‡Freon or equivalent.

b. Hydrochloric acid, HCl, 6*N*.

c. Filter paper.§

4. Procedure

a. Sampling: Collect samples at a place where there is strong turbulence in the water and where floating material is not trapped at the surface. Fill floatable oil tube to mark by dipping into water. *Do not use samples taken to the laboratory in a bottle, because oil and grease cannot be redispersed to their original condition.*

b. Flotation: Support tube in a vertical position. Start flotation period at sampling site immediately after filling tube. The standard flotation time is 30 min. If a different time is used, state this in reporting results. At end of flotation period, discharge the first 900 mL of water carefully through bottom stopcock, stopping before any surface oil or other floating material escapes. Rotate tube slightly back and forth about its vertical axis to dislodge sludge from sides, and let settle for 5 min. Completely discharge sludge that has settled to the bottom or that comes down from the sides with the liquid. Scum on top of the liquid may mix with the water as it moves down the tube. If this occurs, stop drawing off water before any floatables have been lost. Let settle for 5 min before withdrawing remainder of water. After removing water, return tube to laboratory to complete test.

c. Extraction: Acidify to pH 2 or lower with a few drops of 6 *N* HCl, add 50 to 100 mL trichlorotrifluoroethane to the tube, and shake vigorously. Let settle and draw off solvent into a clean dry beaker. Filter solvent through a dry filter paper into a tared 300-mL conical flask, taking care not to let any water get on filter paper. Add a second 50-mL portion of trichlorotrifluoroethane to the tube and repeat extraction, settling, and filtration into the same 300-

§Whatman No. 40 or equivalent.

mL flask. A third extraction may be needed if the amount of floatables in sample exceeds 5 mL/L. Wash filter paper carefully with fresh solvent discharged from a wash bottle with a fine tip. Evaporate solvent from flask as for determination of Oil and Grease (Section 503A). For each solvent batch, determine weight of residue left after evaporation from the same volume as used in the analysis.

5. Calculations

Report results as "soluble floatable oil and grease, 30 min (or other specified) settling time, mg/L."

Trichlorotrifluoroethane-soluble floatable oil and grease, 30 min settling time, mg/L

$$= \frac{(A - B) \times 1{,}000}{\text{mL sample}}$$

where:

A = weight of flask + soluble floatable oil and grease, g, and

B = weight of flask + residue from evaporation of the same volume of solvent as that used in the test.

6. Precision and Accuracy

There is no standard against which accuracy of this test can be determined. Variability of replicates is influenced by sample heterogeneity. If large grease particles are present, the element of chance in sampling may be a major factor. One municipal wastewater stream and two wastewater streams from meat-packing plants, both containing noticeable particles of grease, were analyzed in triplicate. Averages for the three wastewaters were 48, 57, and 25 mg/L; standard deviations averaged 11%. An oil refinery made duplicate determinations of its separator effluent on 15 consecutive days, obtaining results ranging from 5.1 mg/L to 11.2 mg/L. The average difference between pairs of samples was 0.37 mg/L.

206 C. Bibliography

SCHERFIG, J. & H.F. LUDWIG. 1967. Determination of floatables and hexane extractables in sewage. *In* Advances in Water Pollution Research, Vol. 3, p. 217, Water Pollution Control Federation, Washington, D.C.

POMEROY, R.D. 1953. Floatability of oil and grease in wastewaters. *Sewage Ind. Wastes* 25:1304.

207 ODOR

Odor, like taste, depends on contact of a stimulating substance with the appropriate human receptor cell. The stimuli are chemical in nature and the term "chemical senses" often is applied to odor and taste. Water is a neutral medium, always present on or at the membranes that perceive sensory response. In its pure form, water cannot produce odor or taste sensations. No satisfactory theory of olfaction ever has been devised, although many have been formulated. Man and animals can avoid many potentially toxic foods and waters because of adverse sensory response. Without this form of primitive sensory protection many species would not have survived. Today, these same senses often continue to provide the first warning of potential hazards in the environment.

Odor is recognized[1] as a quality factor affecting acceptability of drinking water (and foods prepared with it), tainting of fish and other aquatic organisms, and aesthetics of recreational waters. Most organic and some inorganic chemicals contribute taste or odor. These chemicals may originate from municipal and industrial waste discharges, from natural sources (such as decomposition of vegetable matter), or from associated microbial activity.

Technological expansion in varieties and quantities of waste materials, demands for water disposal of former air pollutants, and continuous population growth with consequently increased reuse of available water supplies increase the potential for impairment of sensory water quality. Domestic consumers and process industries such as food, beverage, and pharmaceutical manufacturers require water essentially free of tastes and odors.

Some substances, such as certain inorganic salts, produce taste without odor and are evaluated by a Taste Test (Section 211). Many other sensations ascribed to the sense of taste actually are odors, even though the sensation is not noticed until the material is taken into the mouth. Despite rapid strides in relating sensory qualities to chemical analyses, [2] most odors are too complex and are detectable at concentrations too low to permit their definition by isolating and determining the odor-producing chemicals. The ultimate odor-testing device is the human nose. Odor tests are performed to provide qualitative descriptions and approximate quantitative measurements of odor intensity. The method for intensity measurement presented here is the *threshold odor* test, based on a method of limits.[2] *Suprathreshold* methods are not included.

Sensory tests are useful as a check on the quality of raw and finished water and for control of odor through the treatment process. They can assess the effectiveness of different kinds of treatment and provide a means of tracing the source of contamination.

1. General Discussion

a. Principle: Determine the threshold odor by diluting a sample with odor-free water until the least definitely perceptible odor is achieved. There is no absolute threshold odor concentration because of inherent variation in individual olfactory capability. A given individual varies in sensitivity over time. Day-to-day and within-day differences occur. Furthermore, individuals vary in response to the characteristic, as well as concentration of odorant. The number of persons selected to measure threshold odor will depend on the objective of the tests, economics, and available personnel. Larger-sized panels are needed for sensory testing when the results must represent the population as a whole or great precision is desired. Under such circumstances, panels of not less than five persons, and preferably ten or more, are recommended.[2] Measurement of threshold levels by a single individual is often a necessity at water treatment plants. Interpretation of the single tester result requires knowledge of the relative acuity of that person. Some investigators have used specific odorants, such as *m*-cresol or *n*-butanol, to calibrate an individual's response.[3]

b. Application: This threshold method is applicable to samples ranging from nearly odorless natural waters to industrial wastes with threshold numbers in the thousands. There are no intrinsic difficulties with the highly odorous samples because they are reduced in concentration proportionately before being presented to the test observers.

c. Qualitative descriptions: A satisfactory system for characterizing odor has not been developed despite efforts over more than a century. Previous editions of this book contained a table of odor descriptions proposed as a guide in expressing odor quality. The reader may continue to encounter the obsolete standard abbreviations of that table. The 12th Edition presents an explanation of such terms.

d. Sampling and storage: Collect samples for odor testing in glass bottles with glass or TFE-lined closures. Complete tests as soon as possible after sample collection. If storage is necessary, collect at least 500 mL of sample in a bottle filled to the top; refrigerate, making sure that no extraneous odors can be drawn into the sample as it cools. Do not use plastic containers.

e. Dechlorination: Most tap waters and some wastewaters are chlorinated. Often it is desirable to determine the odor of the chlorinated sample as well as that of the same sample after dechlorination. Dechlorinate with arsenite or thiosulfate in exact stoichiometric quantity as described under Nitrogen (Ammonia), Section 417. CAUTION—*Do not use arsenic compounds as dechlorinating agents on samples to be tasted.*

f. Temperature: Threshold odor values vary with temperature. For most tap waters and raw water sources, a sample temperature of 60 C will permit detection of odors that might otherwise be missed; 60 C is the standard temperature for hot threshold tests. For some purposes—because the odor is too fleeting or there is excessive heat sensation—the hot odor test may not be applicable; where experience shows that a lower temperature is needed, use a standard test temperature of 40 C. For special purposes, other temperatures may be used. *Report temperature at which observations are made.*

2. Apparatus

To assure reliable threshold measurements, use odor-free glassware. Clean glassware shortly before use with nonodorous soap and acid cleaning solution and rinse with odor-free water. Reserve this glassware exclusively for threshold testing. Do not use rubber, cork, or plastic stoppers. Do not use narrow-mouth vessels.

a. Sample bottles, glass-stoppered or with TFE-lined closures.

b. Constant-temperature bath: A water bath or electric hot plate capable of temperature control of ±1 C for odor tests at elevated temperatures. The bath must not contribute any odor to the odor flasks.

c. Odor flasks: Glass-stoppered, 500-mL (ST 32) erlenmeyer flasks, to hold sample dilutions during testing.

d. Pipets:

1) *Transfer and volumetric pipets or graduated cylinders:* 200-, 100-, 50-, and 25-mL.

2) *Measuring pipets:* 10-mL, graduated in tenths.

e. Thermometer: Zero to 110 C, chemical or metal-stem dial type.

3. Odor-Free Water

a. Sources: Prepare odor-free dilution water as needed by filtration through a bed of activated carbon. Most tap waters are suitable for the preparation of odor-free water except that it is necessary to check water for residual chlorine, unusual salt concentrations, or unusually high or low pH. All these may affect some odorous samples. Where supplies are adequate use distilled water to prepare odor-free water. A convenient odor-free-water generator is shown in Figures 207:1 and 207:2.

*b. Odor-free-water generator:**

1) *Borosilicate glass pipe,* 3-in. diam, 18-in. length.

2) *Asbestos inserts* (two), for 3-in. pipe.

*For approximate metric dimensions in centimeters multiply dimensions in inches by 2.54.

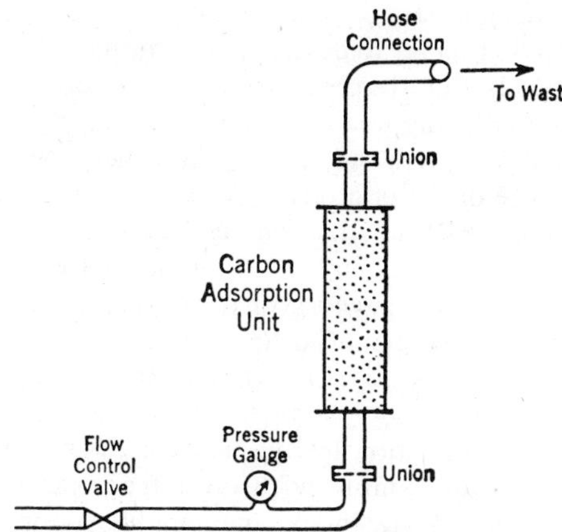

Figure 207:1. Odor-free-water generator.

3) *Flange sets* (two), for 3-in. pipe.

4) *Gaskets*† (two), 1/4 in. thickness, with 3-in. hole slotted to 3/8 in. depth to take screen. Drill 3 holes, 5/16-in. diam, to match flange.

5) *Stainless-steel screens*† (two), 40-mesh, 3-3/4-in. diam.

6) *Brass plates* (two), 3/16-in. thickness × 6-1/4-in. diam. Tap hole in center for 3/4-in. nipple. Score a circular groove (1/16-in. depth × 1/16-in. width and 3-3/8-in. diam) into the plate to prevent leakage. Drill 3 holes, 5/16-in. diam, to coincide with the flange.

7) *Galvanized nipples* (two), 3/4-in. × 3 in. Thread nipple into brass plate and weld in place.

8) *Aluminum bolts and nuts* (six), 5/16-in. × 2 in., for holding assembly together.

9) *Activated carbon* of approximately 12 to 40 mesh grain size.‡

Attach end fittings of adsorption unit to the glass pipe. Draw bolts up evenly, holding brass plate to glass pipe to get a good seal on gasket. Fill unit with carbon. Tap the cylinder gently but do not tamp the carbon. Attach end fittings on adsorption unit and connect to water source as shown in Figure 207:1.

Avoid organic contaminants in making pipe joints or other plumbing. Use TFE-type tape or a paste made by mixing red lead powder and water. Clean all new fittings with kerosene and follow with a detergent wash. Rinse thoroughly with clean water.

c. Generator operation: Pass tap or distilled water through odor-free-water generator at a rate of 100 mL/min. When generator is started, flush to remove carbon fines and discard product.

Check quality of water obtained from the odor-free-water generator daily at

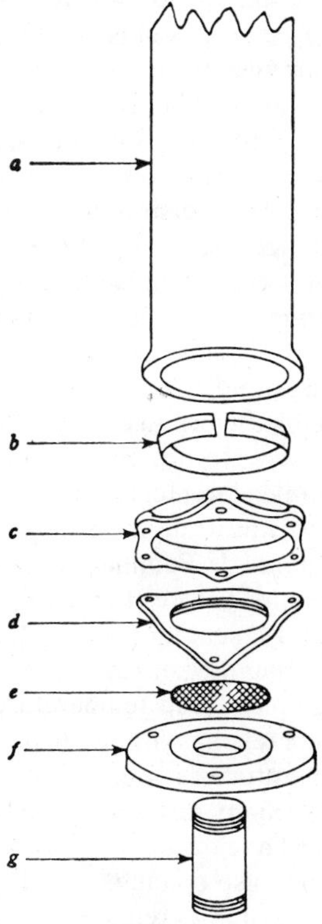

Figure 207:2. End assembly of odor-free-water generator.

†Neoprene, such as can be obtained from Netherland Rubber Co., Cincinnati, Ohio.

‡Nuchar WV-G, Westvaco, Covington, Va.; Filtrasorb 200, Calgon Corp., Pittsburgh, Pa.; or equivalent.

40 C and 60 C before use. The life of the carbon will vary with the condition and amount of water filtered. Subtle odors of biological origin often are found if moist carbon filters stand idle between test periods. Detection of odor in the water coming through the carbon indicates that a change of carbon is needed.

4. Procedure

a. Precautions: Carefully select by preliminary tests the persons to make taste or odor tests. Although extreme sensitivity is not required, exclude insensitive persons and concentrate on observers who have a sincere interest in the test. Avoid extraneous odor stimuli such as those caused by smoking and eating before the test or those contributed by scented soaps, perfumes, and shaving lotions. Insure that the tester is free from colds or allergies that affect odor response. Limit frequency of tests to a number below the fatigue level by frequent rests in an odor-free atmosphere. Keep room in which tests are conducted free from distractions, drafts, and odor.[3] If necessary, set aside a special odor-free room ventilated by air that is filtered through activated carbon and maintained at a constant comfortable temperature and humidity.[4]

For precise work use a panel of five or more testers. Do not allow persons making odor measurements to prepare samples or to know dilution concentrations being evaluated. Familiarize testers with the procedure before they participate in a panel test. Present most dilute sample first to avoid tiring the senses with the concentrated sample. Keep temperature of samples during testing within 1 C of the specified temperature.

Because many raw and waste waters are colored or have decided turbidity that may bias results, use opaque or darkly colored odor flasks, such as red actinic erlenmeyer flasks.

b. Characterization: As part of the threshold test or as a separate test, direct each observer to describe in his or her own words the characteristic sample odor. Compile the consensus that may appear among testers and that affords a clue to the origin of the odorous pollutant. The value of the characterization test increases as observers become more experienced with a particular category of odor, such as algae, chlorophenol, or mustiness.

c. Threshold measurement:|| The "threshold odor number," designated by the abbreviation T.O.N., is the greatest dilution of sample with odor-free water yielding a definitely perceptible odor. Bring total volume of sample and odor-free water to 200 mL in each test. Follow dilutions and record corresponding T.O.N. presented in Table 207:I. These numbers have been computed thus:

TABLE 207:I. THRESHOLD ODOR NUMBERS CORRESPONDING TO VARIOUS DILUTIONS

Sample Volume Diluted to 200 mL *mL*	Threshold Odor No.	Sample Volume Diluted to 200 mL *mL*	Threshold Odor No.
200	1	12	17
140	1.4	8.3	24
100	2	5.7	35
70	3	4	50
50	4	2.8	70
35	6	2	100
25	8	1.4	140
17	12	1.0	200

||There are numerous methods of arranging and presenting samples for odor determinations. The methods offered here are practical and economical of time and personnel and generally are adequate. If extensive tests are planned and statistical analysis of data is required, become familiar with the triangle test and the methods that have been used extensively by flavor and allied industries.[5]

$$\text{T.O.N.} = \frac{A + B}{A}$$

where:

A = mL sample and
B = mL odor-free water.

1) Place proper volume of odor-free water in the flask first, add sample to the water (avoiding contact of pipet or sample with lip or neck of flask), mix by swirling, and proceed as follows:

Determine approximate range of the threshold number by adding 200 mL, 50 mL, 12 mL, and 2.8 mL of sample to separate 500-mL glass-stoppered erlenmeyer flasks containing odor-free water to make a total volume of 200 mL. Use a separate flask containing only odor-free water as reference for comparison. Heat dilutions and reference to desired test temperature.

2) Shake flask containing odor-free water, remove stopper, and sniff vapors. Test sample containing least amount of odor-bearing water in the same way. If odor can be detected in this dilution, prepare more dilute samples as described in ¶5) below. If odor cannot be detected in the first dilution, repeat above procedure using sample containing the next higher concentration of odor-bearing water, and continue this process until odor is detected clearly.

3) Based on results obtained in the preliminary test, prepare a set of dilutions using Table 207:II as a guide. Prepare the five dilutions shown in the appropriate column and the three next most concentrated in the next column to the right in Table 207:II. For example, if odor was first noted in the flask containing 50 mL sample in the preliminary test, prepare flasks containing 50, 35, 25, 17, 12, 8.3, 5.7, and 4.0 mL sample, each diluted to 200 mL with odor-free water. This array is necessary to challenge the range of sensitivities of the entire panel of testers.

Insert two or more blanks in the series near the expected threshold, but avoid any repeated pattern. Do not let tester know which dilutions are odorous and which are blanks. Instruct tester to smell each flask in sequence, beginning with the least concentrated sample, until odor is detected with certainty.

TABLE 207:II. DILUTIONS FOR VARIOUS ODOR INTENSITIES

Sample Volume in Which Odor First Noted			
200 mL	50 mL	12 mL	2.8 mL
Volume of Sample to be Diluted to 200 mL *mL*			
200	50	12	(Inter-
140	35	8.3	mediate
100	25	5.7	dilution)
70	17	4.0	
50	12	2.8	

4) Record observations by indicating whether odor is noted in each test flask. For example:

mL Sample Diluted to 200 mL	12	0	17	25	0	35	50
Response	−	−	−	+	−	+	+

5) If the sample being tested requires more dilution than is provided by Table 207:II, prepare an intermediate dilution consisting of 20 mL sample diluted to 200 mL with odor-free water. Use this dilution for the threshold determination. Multiply T.O.N. obtained by 10 to correct for the intermediate dilution. In rare cases more than one tenfold intermediate dilution step may be required.

5. Calculation

The threshold odor number is the dilution ratio at which taste or odor is just detectable. In the example above, ¶4*c*4), the first detectable odor occurred when 25 mL sample was diluted to 200 mL. Thus, the threshold is 200 divided by 25, or 8. Table 207:I lists the threshold numbers corresponding to common dilutions.

The smallest T.O.N. that can be observed is 1, as in the case where the odor flask contains 200 mL undiluted sample. If no odor is detected at this concentration, report "No Odor Observed" instead of a threshold number. (In special applications, fractional threshold numbers have been calculated.[6])

Anomalous responses sometimes occur; a low concentration may be called positive and a higher concentration in the series may be called negative. In such a case, designate the threshold as that point after which no further anomalies occur. For instance:

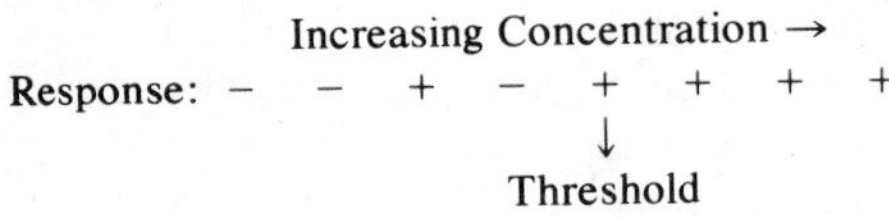

where:

− signifies negative response and
+ signifies positive response.

Occasionally a flask contains residual odor or is contaminated inadvertently. For precise testing repeat entire threshold odor test to determine if the last flask marked "−" was actually a mislabelled blank of odor-free water or if the previous "+" was a contaminated sample.

Use appropriate statistical methods to calculate the most probable average threshold from large numbers of panel results. For most purposes, express the threshold of a group as the geometric mean of individual thresholds.

6. Interpretation of Results

A threshold number is not a precise value. In the case of the single observer it represents a judgment at the time of testing. Panel results are more meaningful because individual differences have less influence on the result. One or two observers can develop useful data if comparison with larger panels has been made to check their sensitivity. Do not make comparisons of data from time to time or place to place unless all test conditions have been standardized carefully and there is some basis for comparison of observed intensities.

7. References

1. U.S. ENVIRONMENTAL PROTECTION AGENCY. 1973. Proposed Criteria for Water Quality. Vol. 1, Washington, D.C.
2. AMERICAN SOCIETY FOR TESTING AND MATERIALS COMMITTEE E-18. 1968. STP 433, Basic principles of sensory evaluation; STP 434, Manual on sensory testing methods; STP 440, Correlation of subjective-objective methods in the study of odors and taste. ASTM, Philadelphia, Pa.
3. BAKER, R.A. 1962. Critical evaluation of olfactory measurement. *J. Water Pollut. Control Fed.* 34:582.
4. BAKER, R.A. 1963. Odor testing laboratory. *J. Water Pollut. Control Fed.* 35:1396.
5. Flavor Research and Food Acceptance. 1958. Reinhold Publishing Corp., New York, N.Y.
6. ROSEN, A.A., J.B. PETER & F.M. MIDDLETON. 1962. Odor thresholds of mixed organic chemicals. *J. Water Pollut. Control Fed.* 34:7.

8. Bibliography

HULBERT, R. & D. FEBEN. 1941. Studies on accuracy of threshold odor value. *J. Amer. Water Works Ass.* 33:1945.

SPAULDING, C.H. 1942. Accuracy and application of threshold odor test. *J. Amer. Water Works Ass.* 34:877.

THOMAS, H.A., JR. 1943. Calculation of threshold odor. *J. Amer. Water Works Ass.* 35:751.

MONCRIEFF, R.W. 1946. The Chemical Senses. John Wiley & Sons, New York, N.Y.

CARTWRIGHT, L.C., C.T. SNELL & P.H. KELLY. 1952. Organoleptic panel testing as a research tool. *Anal. Chem.* 24:503.

LAUGHLIN, H.F. 1954. Palatable level with the threshold odor test. *Taste Odor Control J.* 20:No. 8 (Aug.).

SECHENOV, I.M. 1956 and 1958. Problem of hygenic standards for waters simultaneously polluted with harmful substances [in Russian]. *Gig. Sanit.* Nos. 10 and 8.

SHELLENBERGER, R.D. 1958. Procedures for determining threshold odor concentrations in aqueous solutions. *Taste Odor Control J.* 24:No. 5 (May).

Taste and Odor Control in Water Purification, 2nd ed. 1959. West Virginia Pulp & Paper Co. Industrial Chemical Sales Division, New York. [Contains 1,063 classified references.]

BAKER, R.A. 1961. Problems of tastes and odors. *J. Water Pollut. Control Fed.* 33:1099.

LAUGHLIN, H.F. 1962. Influence of temperature in threshold odor evaluation. *Taste Odor Control J.* 28:No. 10 (Oct.).

BAKER, R.A. 1963. Odor effects of aqueous mixtures of organic chemicals. *J. Water Pollut. Control Fed.* 35:728.

ROSEN, A.A., R.T. SKEEL & M.B. ETTINGER. 1963. Relationship of river water odor to specific organic contaminants. *J. Water Pollut. Control Fed.*35:777.

STAFF REPORT. 1963. The threshold odor test. *Taste Odor Control J.* 29:Nos. 6, 7, 8 (June, July, Aug.).

WRIGHT, R.H. 1964. The Science of Smell. Basic Books, New York, N.Y.

AMERINE, M.A., R.M. PANGBORN & E.B. ROESSLER. 1965. Principles of Sensory Evaluation of Food. Academic Press, New York, N.Y.

ROSEN, A.A. 1970. Report of research committee on tastes and odors. *J. Amer. Water Works Ass.* 62:59.

SUFFET, I.H. & S. SEGALL. 1971. Detecting taste and odor in drinking water. *J. Amer. Water Works Ass.* 63:605.

GELDARD, F.A. 1972. The Human Senses. John Wiley & Sons, New York, N.Y.

STAHL, W.H., ed. 1973. Compilation of Odor and Taste Threshold Values Data. Amer. Soc. Testing & Materials Data Ser. DS 48, Philadelphia, Pa.

AMERICAN SOCIETY FOR TESTING AND MATERIALS. 1973. Annual Book of ASTM Standards. Part 23, D-1292-65, ASTM, Philadelphia, Pa.

208 OXYGEN TRANSFER

In several unit processes, air is introduced into wastewater to oxidize objectionable materials. The rate at which supplied oxygen is dissolved in liquid is called oxygen transfer. Several methods for measuring oxygen transfer by steady- and non-steady-state technics have been reported. Presented here are the common technics for the most reliable measurements of oxygen transfer. Transfer rates still are subject to variations because the measurements are affected by variables such as type of aeration equipment, aeration tank geometry, power input per unit volume, temperature, barometric pressure, and various liquid characteristics. Because this is a field engineering technic, units used are English rather than SI.

208 A. Oxygen Transfer in Clean Water (TENTATIVE)

1. General Discussion

Oxygen transfer in clean water is defined by the equation:

$$\frac{dc}{dt} = K_La\,(C_s - C_t) \qquad (1)$$

where:

$\frac{dc}{dt}$ = oxygen transfer rate, mg/L/hr,

K_La = mass transfer coefficient, hr^{-1},

C_s = dissolved oxygen (DO) saturation concentration at test temperature and

pressure, mg/L, and
C_t = DO at time t, mg/L.

The rate at which oxygen enters solution is proportional to the DO deficit, $(C_s - C_t)$, and to the area of the air-water interface per unit volume of water, a. The coefficient K_La depends on the hydrodynamics of the interfacial area and increases with increased turbulence at the interface. The standard oxygen transfer rate is defined for clean water, at 20 C, with an initial DO of zero, and an atmospheric pressure of 760 torr.

The mass transfer coefficient (K_La) is a composite of the liquid film coefficient K_L and the interfacial area available for mass transfer.

$$K_La = \frac{\ln\left[\frac{C_s - C_1}{C_s - C_2}\right]}{t_2 - t_1} \quad (2)$$

where:

C_1 = DO, mg/L at time t_1, hr,
C_2 = DO, mg/L at time, t_2, hr,
t = time, hr, and
C_s = DO saturation concentration, mg/L.

C_s varies with oxygen partial pressure in contact with the water and water composition and temperature. C_s values for clean tap water at temperatures between 0 and 50 C, in contact with wet air at a total atmospheric pressure of 760 torr, are given in Table 421:I. For other atmospheric pressures, calculate the saturation values:

$$C_s = C_{s760}\frac{P_b - P_w}{760 - P_w} \quad (3)$$

where:

C_{s760} = DO saturation at 760 torr total pressure, mg/L,
C_s = DO saturation concentration during test, mg/L,
P_b = barometric pressure during test, torr, and
P_w = vapor pressure of water, torr.

For surface aerators use Equation 3 to calculate C_s. For diffused-air systems, the air-bubble pressure is greater than atmospheric and C_s commonly is taken as the DO saturation concentration corresponding to the average oxygen partial pressure in the bubbles at mid-depth.

The value of K_La at any temperature can be calculated from the equation:

$$(K_La)_T = (K_La)_{20}\ \theta^{T-20} \quad (4)$$

where:

$(K_La)_{20}$ = K_La at 20 C,
$(K_La)_T$ = K_La at T,
θ = temperature correction coefficient, 1.016 to 1.047, with an accepted value of 1.024, and
T = temperature, C.

K_La is affected by type of aeration equipment and tank geometry. While some observers have discovered empirical relationships between K_La and depth, width, and length of tank, there are no universally-accepted factors that can be applied to all aerators and test conditions. As a guide, the power applied per unit volume is used widely for comparative purposes. An acceptable range is 0.01 to 0.04 kW/m^3 (0.05 to 0.20 hp/1,000 gal).

From K_La and the volume of the tank, the rate of oxygen transfer in pounds per hour at the standard condition of zero DO and 20 C can be calculated.

2. Apparatus

a. Aeration equipment: Use an aeration device capable of operating at a constant power input. Hold volume of water under aeration constant with no change in relative setting of aeration device.

b. Sampling pumps: For basins with more than 200 m^3 (50,000 gal) volume locate six submersible sample pumps at various depths and at least at two places that divide the basin into approximately equal-volume sampling zones. Pump at a rate that displaces the volume of each sample bottle at least 3 and, preferably 10 times, between samples. Limit detention time between pump and sample outlet to 5 to 10 sec. Use sample pumps and lines of equal size and length to minimize time differences among samples taken at a given time. Measure interval between samples with a stopwatch.

c. Dissolved oxygen meters with membrane-covered electrodes: Use for DO determination in removed samples and in the basin itself to indicate proper time to start test and withdraw samples. Calibrate and check meter for linearity against titrated DO values for distilled water or against the oxygen partial pressure in air at the same temperature and barometric pressure using the procedure outlined in Section 421F.3*a*. Perform calibrations on water from test basin. Make calibration measurements at 100%, 70 to 80%, 45 to 55%, and 20 to 30% of saturation to check linearity of electrode response. As a minimum, calibrate electrode before and after each aerator test run and check electrode linearity at the beginning and end of each testing day.

d. Mixing tank, for pre-mixing chemicals before discharge into the aeration tank.

e. Electrical instruments: Ammeters and kilowatt meters to measure drawn power.

f. Sample bottles: 300-mL BOD bottles with gastight caps. Fifty to 60 are sufficient for one test. Use bottle holders or trays so bottle overflow is returned to test tank.

3. Reagents

a. Reagents required for Dissolved Oxygen (Section 421).

b. Sodium sulfite, Na_2SO_3, technical grade.

c. Cobaltous chloride, $CoCl_2 \cdot 6H_2O$.

4. Procedure

a. Thoroughly clean aeration basin and fill with clean tap water.

b. Maintain test water temperature as close to 20 C as possible. Air temperature during test should be not more than 10 C different from water temperature.

c. Dissolve $CoCl_2$ catalyst in warm water and add to tank contents at the rate of 8.05 g/m³ (6.7×10^{-5} lb/gal). Run aerator for at least 30 min to insure complete distribution. Add catalyst only once.

d. Run aeration device and release Na_2SO_3, preferably in solution, into tank. For a liquid with an initial DO of 10 mg/L, use approximately 117 g Na_2SO_3/m³ (9.7×10^{-4} lb/gal) per test.

e. Collect samples in 300-mL BOD bottles at specific time intervals and seal until analyzed. Handle sample lines carefully to avoid entraining air bubbles. Begin sampling when the DO just has begun to rise from zero as indicated by a DO electrode located near the tank bottom and outer wall. As DO increases, draw samples at 1- to 3-min intervals or one at approximately every 1.0 mg/L increase in DO. Determine DO on at least six sets of samples between 10 and 80% saturation. A convenient equation for determining the required interval is:

$$t = \frac{138}{OC} \cdot 10^6 \cdot W\, 1.024^{(20 - T)} \qquad (5)$$

where:

t = sampling interval, min,
W = test liquid weight, lb,
OC = expected oxygenation capacity, lb/hr,
138 = constant for 80% saturation termination, and
T = temperature, C.

f. If individual DO values from the various sampling points in the tank are within 0.25 mg/L of the average DO, make only a single determination of K_La by a semilog plot of average DO deficit vs. time. If individual DO values deviate by more than 0.25 mg/L from the average DO, prepare individual semilog plots for each sampling point and determine K_La for each sampling point. If the plots of individual sampling points have parallel slopes, the deviation results from differences in flow time in sampling pumps and tubing and the test is valid. Nonparallel slopes indicate poor mixing. If any K_La values vary more than 7.5% from the average, discard results because of inadequate mixing.

5. Calculation

Obtain a DO saturation value from Table 421:I for the appropriate test temperature and correct for barometric pressure using Equation 3. Compare corrected table value with measured saturation value. If the corrected table value does not agree with the measured saturation value, construct separate semilog plots of DO deficit vs. time and draw the straight line of best fit through the plotted points, selecting the C_s value that the line appears to approach asymptotically.

Construct a semilog plot of DO deficit against time and draw the straight line of best fit through the plotted points. Determine K_La from the slope of this plot using:

t_1 = time, hr, at which $C_1 = 0.2\ C_s$, and
t_2 = time, hr, at which $C_2 = 0.80\ C_s$.

Correct K_La to 20 C using Equation 4. Calculate weight of oxygen dissolved at standard conditions (20 C, zero DO, and 760 torr):

$$N = K_La_{20}\ C_{s20} W \tag{6}$$

where:
N = lb O_2 dissolved/hr,
C_{s20} = DO saturation value of water at 20 C, mg/L, and
W = test liquid weight, lb.

Determine efficiency of aeration device as:

$$E = \frac{N}{P}$$

where:
E = efficiency, lb O_2/hp · hr and
P = power input, hp.

6. Precision and Accuracy

There is no standard against which accuracy of the oxygen transfer test can be measured. The precision of the DO determination on which the oxygen transfer test is based is given in Section 421. The application of statistical analysis to typical aeration data shows that for controlled test facilities a single K_La value is reproducible within 15% of the mean for tests involving multiple aerators in a single basin and 8% for tests involving a single aerator in a single basin.

208 B. Nonsteady-State Oxygen Transfer in Activated Sludge (TENTATIVE)

1. General Discussion

The oxygen transfer rate under process conditions can be approximated if certain wastewater characteristics are known. Under process conditions, the rate of change of DO is given by the equation:

$$\frac{dc}{dt} = K_La(C_s - C_t) - r \tag{1}$$

where:
r = rate of oxygen uptake of the wastewater and the other terms are as defined in Section 208A. At steady state (with a constant DO) $dc/dt = 0$, and Equation 1 becomes:

$$K_La = \frac{r}{C_s - C_t} \tag{2}$$

In wastewater, both the values of K_La and C_s are different than in clean tap water. The ratio of K_La for wastewater to that for clean tap water is designated α and the ratio of C_s for wastewater to that for clean tap water is designated β.

In an activated sludge system, the mixed liquor solids exert a DO demand and the basic equation for the rate of change of DO is modified to:

$$\frac{dc}{dt} = \alpha\, K_La(\beta\, C_s - C_t) - r \quad (3)$$

where:

r = DO uptake rate of wastewater, mg/L/hr,
α = ratio of K_La of mixed liquor to K_La at standard conditions,
β = ratio of C_s in mixed liquor to C_s at standard conditions, and
C_t = DO at time t, mg/L.

K_La depends on temperature; determine its value at any temperature using Equation 4, 208A.1.

2. Apparatus

a. Model aerator suitable for bench-scale tests: Use mechanical or diffused-air aeration devices. With a diffused-air system use a rotameter for flow control; with a mechanical aerator, control mixing intensity with a variable-speed motor and rheostat.

b. Dissolved oxygen analyser and membrane electrode (self-stirring if desired): See Section 208A.2.

c. Magnetic stirrer.

d. BOD bottles, 300-mL capacity.

3. Reagents

a. Reagents listed in Section 208A.3.

b. Nitrogen gas, technical grade.

4. Procedure

a. Determination of uptake rate, r: Remove a wastewater sample containing at least 0.5 to 1.0 mg DO/L from aeration tank and transfer to a BOD bottle. Using a DO membrane electrode and with constant mixing, measure change in DO with time. The uptake rate during measurement should be linear with time for the results to be valid.

b. Determination of alpha, α: Fill the model aeration tank with a measured volume of tap water and record temperature. Deoxygenate by stripping with nitrogen. Reaerate at the desired rate and record DO at regular intervals. Plot DO deficit on semilog paper and determine K_La from the slope. Repeat for different mixing intensities and temperatures.

Repeat above procedure using the same volume of wastewater. Because α is a function of the degree of biological stabilization, use a system similar in characteristics to the mixed liquor anticipated in the prototype aeration basin.

Alternatively, determine K_La for the wastewater as follows: Aerate wastewater to increase its DO by several milligrams per liter and monitor the increase of DO with time. The rate of DO increase is given by the equation:

$$\frac{dc}{dt} = (K_La\, C_s - r) - K_La\, C_t \quad (5)$$

Plot dc/dt versus C_t to obtain a straight line with a slope of K_La. Calculate ratio of K_La values for mixed liquor to K_La values for tap water for each temperature and mixing rate. Extrapolate the plot to a C_t value of zero and determine $(K_La\, C_s - r)$ from the value of dc/dt. This procedure assumes that the uptake rate, r, does not change appreciably during the test.

c. Determination of beta, β: Saturate a settled wastewater sample with DO by violent hand mixing for several minutes in a half-full 1-L jar. Record temperature and withdraw a sample for DO measurement, preferably by the membrane electrode method. Divide the measured DO value by the handbook saturation value for clean tap water at the same temperature (after it has been corrected for atmospheric pressure) to determine β, which is seldom less than 0.8.

5. Calculations

Calculate the actual oxygen transfer rate in wastewater either as:

$$\text{Actual oxygen transfer rate } (R) = rW$$

where:

R = weight of oxygen dissolved/hr, lb/hr, and

W = weight of liquid aerated, 10^6 lb.

or as:

Actual oxygen transfer rate:

$$R = K_L a W(C_s - C_t)$$

These two calculations should give comparable results.

6. Precision and Accuracy

The determination of oxygen transfer rate under process conditions gives only approximate results.

208 C. Bibliography

NOGAJ, R.J. & E. HURWITZ. 1963. Determination of aerator efficiency under process conditions. *Proc. 18th Ind. Waste Conf.*, Purdue Univ., p. 674.

GLOPPEN, R.C. & J.A. ROEBER. 1965. Rating and application of surface aerators. *TAPPI* 49: No. 12.

ECKENFELDER, W.W., JR. & D.L. FORD. 1970. Water Pollution Control. Pemberton Press, New York, N.Y.

WATER POLLUTION CONTROL FEDERATION. 1970. Aeration in Wastewater Treatment. Manual of Practice No. 5, Water Pollut. Control Fed., Washington, D.C.

209 RESIDUE

The term "residue" refers to solid matter suspended or dissolved in water or wastewater. Residue may affect water or effluent quality adversely in a number of ways. Waters with high residue generally are of inferior palatability and may induce an unfavorable physiological reaction in the transient consumer. Highly mineralized waters also are unsuitable for many industrial applications. For these reasons, a limit of 500 mg residue/L is desirable for drinking waters. Waters with very high levels of nonfiltrable residues may be esthetically unsatisfactory for such purposes as bathing.

1. Definitions

"Total residue" is the term applied to the material left in the vessel after evaporation of a sample and its subsequent drying in an oven at a defined temperature. Total residue includes "nonfiltrable residue," that is, the portion of total residue retained by a filter, and "filtrable residue," the portion of total residue that passes through the filter.

The earlier-used terms "suspended" and "dissolved" (residue) correspond to nonfiltrable and filtrable residue, respectively. The chemical and physical nature of the material in suspension, the pore size of the filter, the area and thickness of the filter mat, and the amount and physical state of the materials deposited on it are the principal factors affecting separation of nonfiltrable from filtrable residue. A method designed to control all variables affecting filtration would be too cumbersome for practical use. It must be recognized, therefore, that residue determinations are not subject to the usual criteria of accuracy. The types of residue are defined arbitrarily by the methods used for their determination, and these in turn represent practical approaches to what otherwise would be exceedingly complex operations.

2. Sources of Error and Variability

Analyses performed for some special purposes may demand deviation from the stated procedures to include an unusual constituent with the measured residue. Whenever such variations of technic are introduced, record and present them with the results.

In interpreting results, recognize the following sources of error: Results for total, volatile, and fixed residues are subject to considerable error because of losses of volatile compounds during evaporation and of carbon dioxide (CO_2) and volatile minerals during ignition; results for residues high in oil or grease content may be questionable because of the difficulty of drying to constant weight in a reasonable time.

The temperature at which the residue is dried has an important bearing on results, because weight losses due to volatilization of organic matter, mechanically occluded water, water of crystallization, and gases from heat-induced chemical decomposition, as well as weight gains due to oxidation, depend on temperature and time of heating. A choice of two drying temperatures is provided and the analyst should be familiar with the probable effects of each.

"Fixed residue"—the residue remaining after ignition for 1 hr at 550 ± 50 C—does not distinguish precisely between organic and inorganic residue because the loss on ignition is not confined to organic matter. It includes losses due to decomposition or volatilization of certain mineral salts. A better characterization of the organic matter in water can be made by methods such as total organic carbon, BOD, or COD, described in Sections 505, 507, and 508, respectively.

Conductivity measurements are approximately proportional to the filtrable residue and may be used in selecting proper sample size for residue determinations. However, close correlation of results of the two tests is not obtained always.

An additional possibility for checking fixed filtrable residue is by use of ion-exchange procedures described in the Introduction, Section 106.

Selection of drying temperature: The methods described are gravimetric and permit a choice of drying temperature.

Residues dried at 103 to 105 C may retain not only water of crystallization but also some mechanically occluded water. Loss of CO_2 will result in conversion of bicarbonate to carbonate. Loss of organic matter by volatilization usually will be very slight at this temperature. Because removal of occluded water is marginal at 105 C, attainment of constant weight is very slow.

Residues dried at 180 ± 2 C will lose almost all mechanically occluded water. Some water of crystallization may remain, especially if sulfates are present. Organic matter is lost by volatilization but is not completely destroyed. Bicarbonates are converted to carbonates and carbonates may be decomposed partially to oxides or basic salts. Some chloride and nitrate salts may be lost. In general, evaporating and drying water samples at 180 C yields values for total residue closer to those obtained through summation of individually determined mineral species than the values for total residue secured through drying at a lower temperature.

Select drying temperature best suited to the sample. Examine waters low in organic matter and total mineral content and intended for human consumption at either temperature, but dry waters containing considerable mineral salts or those with pH over 9.0 at the higher temperature. In any case, report drying temperature.

3. Sample Handling and Preservation

Begin analysis as soon as possible because of the impracticality of preserving the sample. Exclude large floating particles or submerged agglomerates of nonhomogeneous materials from the sample in Methods A, D, and E.

Water has considerable solvent action on glass. Use resistant-glass bottles or plastic bottles provided that the material in suspension does not adhere to container walls. Analyze samples likely to contain iron or manganese promptly to minimize the possibility of chemical or physical change during storage.

4. Selection of Method

Methods A through F are suitable for the determination of residue in potable, surface, and saline waters, as well as domestic and industrial wastewaters in the range up to 20,000 mg/L.

Historically, Method C, determining total filtrable residue dried at 103 to 105 C has been used by most laboratories. Because of problems discussed above, Method B, specifying that the residue be dried at 180 C, is preferable for drinking waters, waters low in organic matter, and waters with high mineral content.

Method G is applicable to determining volatile and fixed fractions in sediments, suspended matter, and solid and semisolid materials produced during water and wastewater treatment.

The amount and type of suspended matter, the purpose of the analysis, and the relative ease of making the determination will dictate whether the nonfiltrable residue is obtained directly or by calculation of the difference between total and filtrable residues.

209 A. Total Residue Dried at 103-105 C

1. General Discussion

a. Principle: A well-mixed sample is evaporated in a weighed dish and dried to constant weight in an oven at 103 to 105 C. The increase in weight over that of the empty dish represents the total residue. Although the results may not represent the weight of actual dissolved and suspended solids in wastewater samples, the determination is useful for plant control. In some instances, correlation may be improved by adding 1*N* sodium hydroxide (NaOH) to wastewater samples with a pH below 4.3 and maintaining the pH of 4.3 during evaporation. Correct final calculation for added sodium.

b. Interferences: Exclude large, floating particles or submerged agglomerates of nonhomogeneous materials from the sample. Disperse visible floating oil and grease with a blender before withdrawing a sample portion for analysis.

2. Apparatus

a. Evaporating dishes: Dishes of 100-mL capacity made of the following materials:

1) Porcelain, 90-mm diam.

2) Platinum—Generally satisfactory for all purposes.

3) High-silica glass.*

*Vycor, product of Corning Glass Works, Corning, N.Y., or equivalent.

b. Muffle furnace for operation at 550 ± 50 C.

c. Steam bath.

d. Drying oven, for operation at 103 to 105 C.

e. Desiccator, provided with a desiccant containing a color indicator of moisture concentration.

f. Analytical balance, 200-g capacity, capable of weighing to 0.1 mg.

3. Procedure

a. Ignite clean evaporating dish at 550 ± 50 C for 1 hr in a muffle furnace.

b. Cool, desiccate, weigh, and store dish in desiccator until ready for use.

c. Transfer a measured volume of sample to preweighed dish and evaporate to dryness on a steam bath or in a drying oven. Choose a sample volume that will yield a residue between 2.5 mg and 200 mg. Volume required may be estimated from conductivity. If necessary, add successive sample portions to the same dish. When evaporating in a drying oven, lower temperature to approximately 2 C below boiling to prevent splattering.

d. Dry evaporated sample for at least 1 hr at 103 to 105 C.

e. Cool dish in desiccator to balance temperature and weigh.

f. Repeat cycle of drying at 103 to 105 C, cooling, desiccating, and weighing until a constant weight is obtained, or until weight loss is less than 4% of previous weight.

4. Calculation

$$\text{mg total residue/L} = \frac{(A - B) \times 1{,}000}{\text{sample volume, mL}}$$

where:

A = weight of sample + dish, mg, and
B = weight of dish, mg.

5. Precision and Accuracy

Precision is about ±4 mg or ±5%. When the residue from a 50- to 100-mL sample of raw sewage was weighed, the standard deviation of the weighing was 1.9 mg (n = 3; 60 × 10), but the data are considered statistically unreliable because of sampling errors. On settled effluents a statistically reliable standard deviation of 0.9 mg (n = 1; 5 × 20) was found.

209 B. Total Filtrable Residue Dried at 180 C

1. General Discussion

Filtrable residue is material that passes through a standard glass fiber filter and remains after evaporation and drying to constant weight at 180 C.[1] The determined values may not check with the theoretical value for solids calculated from chemical analysis of water. Approximate methods for correlating chemical analysis with residue are available.[2]

The filtrate from the total nonfiltrable residue (Section 209D) may be used for determination of total filtrable residue.

Interferences: Highly mineralized waters with a considerable calcium, magnesium, chloride, and/or sulfate content may be hygroscopic and require prolonged drying, proper desiccation, and rapid weighing. Samples high in bicarbonate require careful and possibly prolonged drying at 180 C to insure complete conversion of bicarbonate to carbonate.

2. Apparatus

All of the apparatus listed in Section 209A.2 is required and in addition:

*a. Glass-fiber filters**, circular, without organic binder.

b. Filtration apparatus suitable for filter selected:

1) *Filter holder:* Gooch crucible adapter or membrane filter funnel.

2) *Gooch crucible,* 25-mL to 40-mL capacity, suitable for filter size selected.

c. Suction flask, 500-mL capacity.

3. Procedure

a. Preparation of glass-fiber filter: Place filter either on membrane filter apparatus or bottom of a suitable Gooch crucible. Apply vacuum and wash filter with three successive 20-mL volumes of distilled water. Continue suction to remove all traces of water. Discard washings.

b. Preparation of evaporating dish: Ignite cleaned evaporating dish at 550 ± 50 C for 1 hr in a muffle furnace. Cool and store in desiccator until needed. Weigh immediately before use.

c. Sample analysis: Because excessive residue in the evaporating dish may form a water-entrapping crust, use a sample yielding between 2.5 mg and 200 mg total filtrable residue. If sample contains less than 10 mg filtrable residue/L, use 250 mL. Under vacuum, filter well-mixed sample through glass-fiber filter, wash with three successive 10-mL volumes of distilled water, and continue suction for about 3 min after filtration is complete. Transfer filtrate to a weighed evaporating dish and evaporate to dryness on a steam bath. Dry for at least 1 hr in an oven at 180 ± 2 C, cool in a desiccator to balance temperature, and weigh. Repeat drying cycle until a constant weight is obtained or until weight loss is less than 4% of previous weight or 0.5 mg, whichever is less. Base calculation on original sample volume because all filtrate is evaporated.

*Whatman grade 934AH and 984H; Gelman type A/E; Millipore type AP40; or equivalent. Available in diameters of 2.2-cm to 4.7-cm.

4. Calculation

$$\text{mg total filtrable residue at 180 C/L} = \frac{(A - B) \times 1{,}000}{\text{sample volume, mL}}$$

where:

A = weight of dried residue + dish, mg, and
B = weight of dish, mg.

209 C. Total Filtrable Residue Dried at 103-105 C

Follow procedure described in Section 209B. Dry filtrate at 103 to 105 C instead of 180 C.

Precision and accuracy: In 18 laboratories, a synthetic sample containing 134 mg filtrable residue/L was analyzed at a drying temperature of 103 to 105 C with a standard deviation of 13 mg/L.

209 D. Total Nonfiltrable Residue Dried at 103-105 C (Total Suspended Matter)

1. General Discussion

Total nonfiltrable residue is the retained material on a standard glass-fiber filter after filtration of a well-mixed sample. The residue is dried at 103 to 105 C. If the suspended material clogs the filter and prolongs filtration, the difference between the total residue and the total filtrable residue provides an estimate of the total nonfiltrable residue.

Volatile nonfiltrable residue and fixed nonfiltrable residue can be determined on the material retained on the glass-fiber filters in the Gooch crucibles on completion of the drying at 103 to 105 C.

2. Apparatus

Apparatus listed in Sections 209A.2 and 209B.2 is required.

3. Procedure

a. Preparation of glass-fiber filter: Place filter either on membrane filter apparatus or the bottom of a suitable Gooch crucible. Apply vacuum and wash filter with three successive 20-mL portions of distilled water. Continue suction to remove all traces of water, and discard washings. Remove filter from membrane filter apparatus and transfer to an aluminum or stainless steel planchet as a support. Remove crucible and filter combination if a Gooch crucible is used. Dry in an oven at 103 to 105 C for 1 hr. Store in desiccator until needed. Weigh immediately before use.

b. Sample treatment: Because excessive residue on the filter may entrap water and extend drying time, take for analysis a sample volume that will yield between 2.5 mg and 200 mg total nonfiltrable residue. As a practical limit, filter 100 mL of well-mixed sample under vacuum. Wash filter with three successive 10-mL portions of distilled water. Carefully remove filter from membrane filter funnel assembly and transfer to an aluminum or stainless steel planchet as a support. Alternatively remove crucible and filter combination from crucible adapter if a Gooch crucible is used. Dry for at least 1 hr at 103 to 105 C, cool in a desiccator to balance temperature, and weigh. Repeat drying cycle until a constant weight is attained or until weight loss is less than 4% of previous weight, or 0.5 mg, whichever is less.

c. The dried residue in the Gooch crucible may be used for determining volatile and fixed matter at 550 C in Section 209G.3*b*4).

4. Calculation

$$\text{mg total nonfiltrable residue/L} = \frac{(A - B) \times 1{,}000}{\text{sample volume, mL}}$$

where:

A = weight of filter + residue, mg, and
B = weight of filter, mg.

5. Precision and Accuracy

The precision of the determination varies directly with the concentration of suspended matter. The standard deviation was 5.2 mg/L (coefficient of variation 33%) at 15 mg/L, 24 mg/L (10%) at 242 mg/L, and 13 mg/L (0.76%) at 1,707 mg/L (n = 2; 4 × 10). There is no satisfactory procedure for obtaining the accuracy of the method on wastewater samples because the true concentration of suspended matter is unknown. See Section 209A.5 for other comments.

209 E. Total Volatile and Fixed Residue at 550 C

1. General Discussion

The volatile and fixed components in the total residue of Method A may be determined by igniting the sample at 550 ± 50 C. The determination is useful in control of wastewater treatment plant operation because it offers a rough approximation of the amount of organic matter present in the solid fraction of wastewater, activated sludge, and industrial wastes.

2. Apparatus

See Sections 209A.2 and 209B.2.

3. Procedure

Ignite residue produced by Method A to constant weight in a muffle furnace at a temperature of 550 ± 50 C. Constant weight has been reached when two successive weighings do not differ by more than 4%. Have furnace up to temperature before inserting sample. Usually, 15 to 20 min ignition are required. Let dish cool partially in air until most of the heat has been dissipated. Transfer to a desiccator for final cooling in a dry atmosphere. Do not overload desiccator. Weigh dish as soon as it has cooled completely. Report loss of weight on ignition as total volatile residue and weighed residue as total fixed residue.

4. Calculation

$$\text{mg volatile residue/L} = \frac{(A - B) \times 1{,}000}{\text{sample volume, mL}}$$

$$\text{mg fixed residue/L} = \frac{(B - C) \times 1{,}000}{\text{sample volume, mL}}$$

where:

A = weight of residue + dish before ignition, mg,

B = weight of residue + dish after ignition, mg, and

C = weight of dish, mg.

5. Precision and Accuracy

Three laboratories examined four samples by means of 10 replicates with a standard deviation of 11 mg/L at 170 mg/L volatile residue concentration.

209 F. Settleable Matter

1. General Discussion

Settleable matter in surface and saline waters as well as domestic and industrial wastes may be determined and reported on either a volume (milliliters per liter) or a weight (milligrams per liter) basis.

2. Apparatus

The apparatus listed under Sections 209A.2 and 209B.2, and an Imhoff cone, are required for a gravimetric test. The volumetric test requires only an Imhoff cone.

3. Procedure

a. By volume: Fill an Imhoff cone to the 1-L mark with a thoroughly mixed sample. Settle for 45 min, gently stir sides of cone with a rod or by spinning, settle 15 min longer, and record volume of settleable matter in the cone as milliliters per liter. If the settled matter contains pockets of liquid between large settled particles, estimate volume of these and subtract from volume of settled matter. The practical lower limit of measurement is about 1 mL/L. Where a separation of settleable and floating materials occurs, do not estimate the floating material as settleable matter.

b. By weight:

1) Determine total nonfiltrable residue of well-mixed sample (Section 209D).

2) Pour a well-mixed sample into a glass vessel of not less than 9 cm diam. Use a sample of not less than 1 L and sufficient to give a depth of 20 cm. Alternatively use a glass vessel of greater diameter and a

larger volume of sample. Let stand quiescent for 1 hr and, without disturbing the settled or floating material, siphon 250 mL from center of container at a point halfway between the surface of the settled sludge and the liquid surface. Determine nonfiltrable residue (milligrams per liter) of this supernatant liquor (Section 209D). This is the nonsettling matter.

4. Calculation

mg settleable matter/L
= mg suspended matter/L
− mg nonsettleable matter/L

209 G. Volatile and Fixed Matter in Nonfiltrable Residue and in Solid and Semisolid Samples

1. General Discussion

This method is applicable to the determination of total residue on evaporation and its fixed and volatile fractions in such solid and semisolid samples as river and lake sediments, sludges separated from water and wastewater treatment processes, and sludge cakes from vacuum filtration, centrifugation, or other sludge dewatering processes.

The determination of both total and volatile residue in these materials is subject to negative error due to loss of ammonium carbonate [$(NH_4)_2CO_3$] and volatile organic matter while drying. Although this is true also for wastewater, the effect tends to be more pronounced with sediments, and especially with sludges and sludge cakes.

The mass of organic matter recovered from sludge and sediment requires a longer ignition time than that specified for residue from wastewaters, effluents, or polluted waters. Carefully observe specified ignition time and temperature to control losses of volatile inorganic salts.

Make all weighings quickly because wet samples tend to lose weight by evaporation. After drying or ignition, residues often are very hygroscopic and rapidly absorb moisture from the air.

2. Apparatus

See Sections 209A.2 and 209B.2.

3. Procedure

a. Solid and semisolid samples:

1) Total residue and moisture—

a) Preparation of evaporating dish—Ignite a clean evaporating dish at 550 ± 50 C for 1 hr in a muffle furnace. Cool in a desiccator, weigh, and store in a desiccator until ready for use.

b) Fluid samples—If the sample contains enough moisture to flow more or less readily, stir to homogenize, place 25 to 50 g in a prepared evaporating dish, and weigh to the nearest 10 mg. Evaporate to dryness on a water bath, dry at 103 C for 1 hr, cool in an individual desiccator containing fresh desiccant, and weigh.

c) Solid samples—If the sample consists of discrete pieces of solid material (dewatered sludge, for example), take cores from each piece with a No. 7 cork borer or pulverize the entire sample coarsely on a clean surface by hand, using rubber gloves. Place 25 to 50 g in a prepared evaporating dish and weigh to the nearest 10 mg. Place in an oven at 103 C overnight. Cool in an individual desiccator containing fresh desiccant and weigh. Prolonged heating may result in a loss of volatile organic matter and $(NH_4)_2CO_3$, but it usually is necessary to dry samples thoroughly.

2) Volatile residue—Determine volatile residue, including organic matter and volatile inorganic salts, on the total residue

obtained in 1) above. Avoid loss of solids by decrepitation by placing dish in a cool muffle furnace, heating furnace to 550 C, and igniting for 60 min. (First ignite samples containing large amounts of organic matter over a gas burner and under an exhaust hood in the presence of adequate air to lessen losses due to reducing conditions and to avoid odors in the laboratory.) Cool in a desiccator and reweigh. Report results as fixed residue (percent ash) and volatile residue.

b. Nonfiltrable residue (suspended matter):

1) Preparation of glass-fiber filter—Place a glass-fiber filter in a membrane filter holder, Hirsch funnel, or Buchner funnel, with wrinkled surface of filter facing upward. Apply vacuum to the assembled apparatus to seat filter. With vacuum applied, wash filter with three successive 20-mL portions of distilled water. After the water has filtered through, disconnect vacuum, remove filter, transfer to an aluminum or stainless steel planchet as a support, and dry in an oven at 103 C for 1 hr (30 min in a mechanical convection oven). If volatile matter is not to be determined, cool filter in a desiccator to balance temperature and weigh. If volatile matter is to be determined, transfer filter to a muffle furnace and ignite at 550 C for 15 min. Remove filter from furnace, place in a desiccator until cooled to balance temperature, and weigh.

2) Treatment of sample—Except for samples that contain high concentrations of filtrable matter, or that filter very slowly, select a sample volume $\geq$14 mL/cm^2 filter area.

Place prepared filter in membrane filter holder, Hirsch funnel, or Buchner funnel, with wrinkled surface upward. With vacuum applied, wet filter with distilled water to seat it against holder or funnel. Measure well-mixed sample with a wide-tip pipet or graduated cylinder. Filter sample through filter using suction. Leaving suction on, wash apparatus three times with 10-mL portions of distilled water, allowing complete drainage between washings. Discontinue suction, remove filter and dry to constant weight (see 209B.3*c*) at 103 C for 1 hr in an oven (30 min in a mechanical convection oven). After drying, cool filter in a desiccator to balance temperature and weigh.

3) Filtration with Gooch crucibles—Alternatively, use glass-fiber filters of 2.2 or 2.4 cm diam with Gooch crucibles and follow the procedure in Section 209D.3*b*.

4) Ignition—Ignite filter with its nonfiltrable residue (total suspended matter) for 15 min at 550 ± 50 C, transfer to a desiccator, cool to balance temperature, and weigh.

4. Calculation

a. Solid and semisolid samples:

$$\%\ \text{total residue} = \frac{A \times 100}{B}$$

$$\%\ \text{volatile residue} = \frac{(A - C) \times 100}{A}$$

$$\%\ \text{fixed residue} = \frac{C \times 100}{A}$$

b. Nonfiltrable residue (suspended matter):

$$\text{mg nonfiltrable volatile residue/L} = \frac{(D - E) \times 1{,}000}{\text{sample volume, mL}}$$

$$\text{mg nonfiltrable fixed residue/L} = \frac{C \times 1{,}000}{\text{sample volume, mL}}$$

where:

A = weight of dried solids, mg,
B = weight of wet sample, mg,
C = weight of ash, mg,
D = weight of residue before ignition, mg, and
E = weight of residue after ignition, mg.

5. Precision and Accuracy

See Section 209D.5.

209 H. References

1. Methods for Chemical Analysis of Water and Wastes. 1974. U.S. EPA, Technology Transfer, 625-/6-74-003, pp. 266–267.
2. SOKOLOFF, V.P. 1933. Water of crystallization in total solids of water analysis. *Ind. Eng. Chem.*, Anal. Ed. 5:336.

209 I. Bibliography

THERIAULT, E.J. & H.H. WAGENHALS. 1923. Studies of representative sewage plants. *Pub. Health Bull.* No. 132.

HOWARD, C.S. 1933. Determination of total dissolved solids in water analysis. *Ind. Eng. Chem.*, Anal. Ed. 5:4.

SYMONS, G.E. & B. MOREY. 1941. The effect of drying time on the determination of solids in sewage and sewage sludges. *Sewage Works J.* 13:936.

FISCHER, A.J. & G.E. SYMONS. 1944. The determination of settleable sewage solids by weight. *Water Works Sewage* 91:37.

DEGEN, J. & F.E. NUSSBERGER. 1956. Notes on the determination of suspended solids. *Sewage Ind. Wastes* 28:237.

CHANIN, G., E.H. CHOW, R.B. ALEXANDER & J. POWERS. 1958. Use of glass fiber filter medium in the suspended solids determination. *Sewage Ind. Wastes* 30:1062.

NUSBAUM, I. 1958. New method for determination of suspended solids. *Sewage Ind. Wastes* 30:1066.

SMITH, A.L. & A.E. GREENBERG. 1963. Evaluation of methods for determining suspended solids in wastewater. *J. Water Pollut. Control Fed.* 35:940.

GOODMAN, B.L. 1964. Processing thickened sludge with chemical conditioners. Pages 78 et seq *in* Sludge Concentration, Filtration and Incineration. Univ. Michigan Continued Education Ser. No. 113, Ann Arbor.

WYCKOFF, B.M. 1964. Rapid solids determination using glass fiber filters. *Water Sewage Works* 111:277.

210 SALINITY

Salinity is an important measurement in the analysis of certain industrial wastes and seawater. It is defined as the total solids in water after all carbonates have been converted to oxides, all bromide and iodide have been replaced by chloride, and all organic matter has been oxidized. It is numerically smaller than the filtrable residue and usually is reported as grams per kilogram or parts per thousand (‰).

Associated terms are chlorinity, which includes chloride, bromide, and iodide, all reported as chloride, and chlorosity, which is the chlorinity multiplied by the water density at 20 C. An empirical relationship[1] between salinity and chlorinity often is used:

$$\text{Salinity, ‰} = 0.03 + 1.805\ (\text{chlorinity, ‰})$$

Selection of method: Three procedures are presented. The electrical conductivity (A) and hydrometric (B) methods are suited for field use along a shoreline or in a small boat. For laboratory or field analysis of estuarine or coastal inlet waters the argentometric method (C) is recommended.

210 A. Electrical Conductivity Method

See Conductivity, Section 205. Because of the relatively high concentration of ions in seawater and the effect of temperature on conductivity, standardize each instrument against seawater samples of known salinity (as determined by the argentometric method below).

210 B. Hydrometric Method

1. General Discussion

Principle: Salinity is determined by measuring specific gravity with a hydrometer, correcting for temperature, and converting specific gravity to salinity at 15 C by means of density salinity tables.

2. Apparatus

a. Hydrometer jar: Use a special jar 400 mm high with 45 mm ID, a rubber-stoppered transparent plastic tube with the same dimensions, or a 500-mL graduated cylinder.

b. Thermometer, graduated in 0.2 C divisions.

c. Hydrometer, seawater*: Use a set of three, with specific gravity ranges of 0.966 to 1.011, 1.010 to 1.021, and 1.010 to 1.031. Hydrometer divisions should be 0.002. Have a set calibrated by the National Bureau of Standards for specific gravity of NaCl solutions at 15/4 C.

*Such as that available from Kahl Scientific Instrument Corp., P.O. Box 1166, El Cajon, Calif. 92022.

3. Procedure

a. Fill hydrometer jar 2/3 full of sample.

b. While holding jar vertically, place thermometer and hydrometer in jar.

c. Read and record temperature.

d. Read and record specific gravity. Estimate the fourth decimal place.

e. Make temperature corrections for specific gravity reading from factors listed in Table 210:I.

4. Calculation

Determine salinity from Table 210:II. Locate corrected density and read salinity from opposite column. Report salinity as parts per thousand ($^0/_{00}$).

TABLE 210:I. VALUES FOR CONVERTING HYDROMETER READINGS AT CERTAIN TEMPERATURES TO DENSITY AT 15 C*

Observed Reading	Temperature of Water in Jar, *C*												
	−2.0	−1.0	0.0	1.0	2.0	3.0	4.0	5.0	6.0	7.0	8.0	9.0	10.0
0.9960													
0.9970													
0.9980													
0.9990	−1	−2	−3	−4	−5	−5	−6	−6	−6	−6	−6	−5	−5
1.0000	−2	−3	−4	−5	−5	−6	−6	−6	−6	−6	−6	−5	−5
1.0010	−3	−4	−4	−5	−6	−6	−6	−7	−7	−6	−6	−6	−5
1.0020	−3	−4	−5	−6	−6	−7	−7	−7	−7	−7	−6	−6	−5
1.0030	−4	−5	−6	−6	−7	−7	−7	−7	−7	−7	−6	−6	−5
1.0040	−4	−5	−6	−7	−7	−7	−8	−8	−7	−7	−7	−6	−6
1.0050	−5	−6	−6	−7	−8	−8	−8	−8	−8	−7	−7	−6	−6
1.0060	−6	−6	−7	−8	−8	−8	−8	−8	−8	−8	−7	−6	−6
1.0070	−6	−7	−8	−8	−8	−8	−8	−8	−8	−8	−7	−7	−6
1.0080	−7	−8	−8	−9	−9	−9	−9	−9	−8	−8	−7	−7	−6
1.0090	−7	−8	−9	−9	−9	−9	−9	−9	−9	−8	−8	−7	−6
1.0100	−8	−9	−9	−10	−10	−10	−10	−9	−9	−8	−8	−7	−6
1.0110	−9	−9	−10	−10	−10	−10	−10	−10	−9	−9	−8	−7	−6
1.0120	−9	−10	−10	−10	−10	−10	−10	−10	−10	−9	−8	−7	−7
1.0130	−10	−10	−11	−11	−11	−11	−11	−10	−10	−9	−8	−8	−7
1.0140	−10	−11	−11	−11	−11	−11	−11	−11	−10	−10	−9	−8	−7
1.0150	−11	−11	−12	−12	−12	−12	−11	−11	−10	−10	−9	−8	−7
1.0160	−12	−12	−12	−12	−12	−12	−12	−11	−11	−10	−9	−8	−7
1.0170	−12	−12	−12	−13	−13	−12	−12	−12	−11	−10	−9	−8	−7
1.0180	−13	−13	−13	−13	−13	−13	−12	−12	−11	−10	−9	−8	−7
1.0190	−13	−13	−14	−14	−13	−13	−13	−12	−12	−11	−10	−9	−8
1.0200	−14	−14	−14	−14	−14	−13	−13	−12	−12	−11	−10	−9	−8
1.0210	−14	−14	−14	−14	−14	−14	−13	−13	−12	−11	−10	−9	−8
1.0220	−15	−15	−15	−15	−15	−14	−14	−13	−12	−11	−10	−9	−8
1.0230	−15	−15	−15	−15	−15	−15	−14	−13	−12	−12	−10	−9	−8
1.0240	−16	−16	−16	−16	−15	−15	−14	−14	−13	−12	−11	−10	−8
1.0250	−16	−16	−16	−16	−16	−15	−15	−14	−13	−12	−11	−10	−8
1.0260	−17	−17	−17	−16	−16	−16	−15	−14	−13	−12	−11	−10	−8
1.0270	−18	−17	−17	−17	−17	−16	−15	−14	−14	−12	−11	−10	−9
1.0280	−18	−18	−18	−17	−17	−16	−16	−15	−14	−13	−11	−10	−9
1.0290	−19	−18	−18	−18	−17	−17	−16	−15	−14	−13	−12	−10	−9
1.0300	−19	−19	−19	−18	−18	−17	−16	−15	−14	−13	−12	−10	−9
1.0310	−20	−19	−19	−19	−18	−17	−16	−16	−15	−13	−12	−10	−9

*Add tabular values to the last decimal of observed reading. For example, an observed reading of 1.0000 at 10.0 C is converted to 1.0000 + (−0.0005) or 0.9995 at 15 C.

TABLE 210:I, CONT.

Observed Reading	Temperature of Water in Jar, C											
	11.0	12.0	13.0	14.0	15.0	16.0	17.0	18.0	18.5	19.0	19.5	20.0
0.9960												
0.9970												
0.9980							3	4	5	6	7	8
0.9990	−4	−3	−2	−1	0	1	3	4	5	6	7	8
1.0000	−4	−3	−2	−1	0	1	3	4	5	6	7	8
1.0010	−4	−3	−2	−1	0	1	3	4	5	6	7	8
1.0020	−4	−3	−2	−1	0	1	3	4	5	6	7	8
1.0030	−4	−3	−2	−1	0	1	3	4	5	6	7	8
1.0040	−5	−4	−3	−1	0	2	3	5	6	6	7	8
1.0050	−5	−4	−3	−1	0	2	3	5	6	7	8	9
1.0060	−5	−4	−3	−1	0	2	3	5	6	7	8	9
1.0070	−5	−4	−3	−2	0	2	3	5	6	7	8	9
1.0080	−5	−4	−3	−2	0	2	3	5	6	7	8	9
1.0090	−5	−4	−3	−2	0	2	3	5	6	7	8	9
1.0100	−5	−4	−3	−2	0	2	3	5	6	7	8	9
1.0110	−5	−4	−3	−2	0	2	3	5	6	7	8	9
1.0120	−6	−4	−3	−2	0	2	3	5	6	7	8	9
1.0130	−6	−4	−3	−2	0	2	4	5	6	7	8	10
1.0140	−6	−4	−3	−2	0	2	4	5	6	8	9	10
1.0150	−6	−4	−3	−2	0	2	4	5	6	8	9	10
1.0160	−6	−5	−3	−2	0	2	4	6	7	8	9	10
1.0170	−6	−5	−3	−2	0	2	4	6	7	8	9	10
1.0180	−6	−5	−3	−2	0	2	4	6	7	8	9	10
1.0190	−6	−5	−3	−2	0	2	4	6	7	8	9	10
1.0200	−6	−5	−3	−2	0	2	4	6	7	8	9	10
1.0210	−6	−5	−3	−2	0	2	4	6	7	8	9	10
1.0220	−7	−5	−3	−2	0	2	4	6	7	8	9	11
1.0230	−7	−5	−4	−2	0	2	4	6	7	8	9	11
1.0240	−7	−5	−4	−2	0	2	4	6	7	8	10	11
1.0250	−7	−5	−4	−2	0	2	4	6	7	8	10	11
1.0260	−7	−5	−4	−2	0	2	4	6	7	9	10	11
1.0270	−7	−5	−4	−2	0	2	4	6	7	9	10	11
1.0280	−7	−6	−4	−2	0	2	4	6	8	9	10	11
1.0290	−7	−6	−4	−2	0	2	4	6	8	9	10	11
1.0300	−7	−6	−4	−2	0	2	4	6	8	9	10	12
1.0310	−8	−6	−4	−2	0	2	4					

TABLE 210:I, CONT.

Observed Reading	Temperature of Water in Jar, *C*												
	20.5	21.0	21.5	22.0	22.5	23.0	23.5	24.0	24.5	25.0	25.5	26.0	26.5
0.9960											19	20	21
0.9970			10	11	12	14	15	16	17	18	19	20	22
0.9980	9	10	11	12	13	14	15	16	17	18	19	21	22
0.9990	9	10	11	12	13	14	15	16	17	18	20	21	22
1.0000	9	10	11	12	13	14	15	16	17	19	20	21	22
1.0010	9	10	11	12	13	14	15	17	18	19	20	21	23
1.0020	9	10	11	12	13	14	16	17	18	19	20	22	23
1.0030	9	10	11	12	13	15	16	17	18	19	21	22	23
1.0040	9	10	11	12	14	15	16	17	18	20	21	22	23
1.0050	10	11	12	13	14	15	16	17	19	20	21	22	24
1.0060	10	11	12	13	14	15	16	18	19	20	21	23	24
1.0070	10	11	12	13	14	15	17	18	19	20	21	23	24
1.0080	10	11	12	13	14	16	17	18	19	20	22	23	24
1.0090	10	11	12	13	15	16	17	18	19	21	22	23	25
1.0100	10	11	12	14	15	16	17	18	20	21	22	24	25
1.0110	10	12	13	14	15	16	17	19	20	21	22	24	25
1.0120	10	12	13	14	15	16	18	19	20	21	23	24	25
1.0130	11	12	13	14	15	16	18	19	20	22	23	24	26
1.0140	11	12	13	14	15	17	18	19	20	22	23	24	26
1.0150	11	12	13	14	16	17	18	20	21	22	23	25	26
1.0160	11	12	13	14	16	17	18	20	21	22	24	25	26
1.0170	11	12	13	15	16	17	18	20	21	22	24	25	27
1.0180	11	12	14	15	16	17	19	20	21	23	24	25	27
1.0190	11	12	14	15	16	18	19	20	21	23	24	26	27
1.0200	11	13	14	15	16	18	19	20	22	23	24	26	27
1.0210	12	13	14	15	17	18	19	21	22	23	25	26	27
1.0220	12	13	14	15	17	18	19	21	22	23	25	26	28
1.0230	12	13	14	16	17	18	20	21	22	24	25	26	28
1.0240	12	13	14	16	17	18	20	21	22	24	25	27	28
1.0250	12	13	15	16	17	18	20	21	23	24	25	27	28
1.0260	12	13	15	16	17	19	20	22	23	24	26	27	29
1.0270	12	14	15	16	17	19	20	22	23	24	26	27	29
1.0280	12	14	15	16	18	19	20	22	23	25	26	28	29
1.0290	13	14	15	16	18	19	21	22	23				
1.0300	13	14	15	16	18								
1.0310													

TABLE 210:I, CONT.

Observed Reading	Temperature of Water in Jar, C												
	27.0	27.5	28.0	28.5	29.0	29.5	30.0	30.5	31.0	31.5	32.0	32.5	33.0
0.9960	23	24	25	27	28	29	31	32	34	35	37	38	40
0.9970	23	24	26	27	28	30	31	33	34	36	37	39	40
0.9980	23	25	26	27	29	30	31	33	34	36	38	39	41
0.9990	24	25	26	28	29	30	32	33	35	36	38	39	41
1.0000	24	25	26	28	29	31	32	34	35	37	38	40	41
1.0010	24	25	27	28	30	31	32	34	35	37	39	40	42
1.0020	24	26	27	28	30	31	33	34	36	37	39	41	42
1.0030	25	26	27	29	30	32	33	35	36	38	39	41	42
1.0040	25	26	28	29	30	32	33	35	36	38	40	41	43
1.0050	25	26	28	29	31	32	34	35	37	38	40	42	43
1.0060	25	27	28	30	31	32	34	36	37	39	40	42	44
1.0070	26	27	28	30	31	33	34	36	38	39	41	42	44
1.0080	26	27	29	30	32	33	35	36	38	39	41	43	44
1.0090	26	28	29	30	32	33	35	36	38	40	41	43	45
1.0100	26	28	29	31	32	34	35	37	38	40	42	43	45
1.0110	27	28	30	31	32	34	36	37	39	40	42	44	45
1.0120	27	28	30	31	33	34	36	37	39	41	42	44	46
1.0130	27	29	30	32	33	35	36	38	39	41	43	44	46
1.0140	27	29	30	32	33	35	36	38	40	41	43	45	46
1.0150	28	29	31	32	34	35	37	38	40	42	43	45	47
1.0160	28	29	31	32	34	35	37	39	40	42	44	45	47
1.0170	28	30	31	33	34	36	37	39	40	42	44	46	47
1.0180	28	30	31	33	34	36	38	39	41	42	44	46	48
1.0190	29	30	32	33	35	36	38	39	41	43	44	46	48
1.0200	29	30	32	33	35	37	38	40	41	43	45	47	48
1.0210	29	31	32	34	35	37	38	40	42	43	45	47	49
1.0220	29	31	32	34	36	37	39	40	42	44	45	47	49
1.0230	30	31	33	34	36	37	39	41	42	44	46	47	49
1.0240	30	31	33	34	36	37	39	41	42	44	46	48	49
1.0250	30	31	33	35	36	38	39	41	43	44	46	48	50
1.0260	30	32	33	35	37	38	40	41	43	45	46	48	50
1.0270	30	32	34	35	37	38	40						
1.0280	31	32											
1.0290													
1.0300													
1.0310													

SOURCE: ZERBE, W. B. and C. B. TAYLOR. 1953. Sea Water Temperature and Density Reduction Tables. U.S. Dept. Commerce Spec. Publ. No. 298, Washington, D.C.

TABLE 210:II. CORRESPONDING DENSITIES AND SALINITIES*

Density	Salinity	Density	Salinity	Density	Salinity	Density	Salinity
0.9991	0.0	1.0036	5.8	1.0081	11.6	1.0126	17.5
0.9992	0.0	1.0037	5.9	1.0082	11.8	1.0127	17.7
0.9993	0.2	1.0038	6.0	1.0083	11.9	1.0128	17.8
0.9994	0.3	1.0039	6.2	1.0084	12.0	1.0129	17.9
0.9995	0.4	1.0040	6.3	1.0085	12.2	1.0130	18.0
0.9996	0.6	1.0041	6.4	1.0086	12.3	1.0131	18.2
0.9997	0.7	1.0042	6.6	1.0087	12.4	1.0132	18.3
0.9998	0.8	1.0043	6.7	1.0088	12.6	1.0133	18.4
0.9999	0.9	1.0044	6.8	1.0089	12.7	1.0134	18.6
1.0000	1.1	1.0045	6.9	1.0090	12.8	1.0135	18.7
1.0001	1.2	1.0046	7.1	1.0091	12.9	1.0136	18.8
1.0002	1.3	1.0047	7.2	1.0092	13.1	1.0137	19.0
1.0003	1.5	1.0048	7.3	1.0093	13.2	1.0138	19.1
1.0004	1.6	1.0049	7.5	1.0094	13.3	1.0139	19.2
1.0005	1.7	1.0050	7.6	1.0095	13.5	1.0140	19.3
1.0006	1.9	1.0051	7.7	1.0096	13.6	1.0141	19.5
1.0007	2.0	1.0052	7.9	1.0097	13.7	1.0142	19.6
1.0008	2.1	1.0053	8.0	1.0098	13.9	1.0143	19.7
1.0009	2.2	1.0054	8.1	1.0099	14.0	1.0144	19.9
1.0010	2.4	1.0055	8.2	1.0100	14.1	1.0145	20.0
1.0011	2.5	1.0056	8.4	1.0101	14.2	1.0146	20.1
1.0012	2.6	1.0057	8.5	1.0102	14.4	1.0147	20.3
1.0013	2.8	1.0058	8.6	1.0103	14.5	1.0148	20.4
1.0014	2.9	1.0059	8.8	1.0104	14.6	1.0149	20.5
1.0015	3.0	1.0060	8.9	1.0105	14.8	1.0150	20.6
1.0016	3.2	1.0061	9.0	1.0106	14.9	1.0151	20.8
1.0017	3.3	1.0062	9.2	1.0107	15.0	1.0152	20.9
1.0018	3.4	1.0063	9.3	1.0108	15.2	1.0153	21.0
1.0019	3.5	1.0064	9.4	1.0109	15.3	1.0154	21.2
1.0020	3.7	1.0065	9.6	1.0110	15.4	1.0155	21.3
1.0021	3.8	1.0066	9.7	1.0111	15.6	1.0156	21.4
1.0022	3.9	1.0067	9.8	1.0112	15.7	1.0157	21.6
1.0023	4.1	1.0068	9.9	1.0113	15.8	1.0158	21.7
1.0024	4.2	1.0069	10.1	1.0114	16.0	1.0159	21.8
1.0025	4.3	1.0070	10.2	1.0115	16.1	1.0160	22.0
1.0026	4.5	1.0071	10.3	1.0116	16.2	1.0161	22.1
1.0027	4.6	1.0072	10.5	1.0117	16.3	1.0162	22.2
1.0028	4.7	1.0073	10.6	1.0118	16.5	1.0163	22.4
1.0029	4.8	1.0074	10.7	1.0119	16.6	1.0164	22.5
1.0030	5.0	1.0075	10.8	1.0120	16.7	1.0165	22.6
1.0031	5.1	1.0076	11.0	1.0121	16.9	1.0166	22.7
1.0032	5.2	1.0077	11.1	1.0122	17.0	1.0167	22.9
1.0033	5.4	1.0078	11.2	1.0123	17.1	1.0168	23.0
1.0034	5.5	1.0079	11.4	1.0124	17.3	1.0169	23.1
1.0035	5.6	1.0080	11.5	1.0125	17.4	1.0170	23.3

* Density at 15 C. Salinity o/oo or g/kg.

TABLE 210:II, CONT.

Density	Salinity	Density	Salinity	Density	Salinity	Density	Salinity
1.0171	23.4	1.0211	28.6	1.0251	33.8	1.0291	39.0
1.0172	23.5	1.0212	28.8	1.0252	34.0	1.0292	39.2
1.0173	23.7	1.0213	28.9	1.0253	34.1	1.0293	39.3
1.0174	23.8	1.0214	29.0	1.0254	34.2	1.0294	39.4
1.0175	23.9	1.0215	29.1	1.0255	34.4	1.0295	39.6
1.0176	24.1	1.0216	29.3	1.0256	34.5	1.0296	39.7
1.0177	24.2	1.0217	29.4	1.0257	34.6	1.0297	39.8
1.0178	24.3	1.0218	29.5	1.0258	34.8	1.0298	39.9
1.0179	24.4	1.0219	29.7	1.0259	34.9	1.0299	40.1
1.0180	24.6	1.0220	29.8	1.0260	35.0	1.0300	40.2
1.0181	24.7	1.0221	29.9	1.0261	35.1	1.0301	40.3
1.0182	24.8	1.0222	30.1	1.0262	35.3	1.0302	40.4
1.0183	25.0	1.0223	30.2	1.0263	35.4	1.0303	40.6
1.0184	25.1	1.0224	30.3	1.0264	35.5	1.0304	40.7
1.0185	25.2	1.0225	30.4	1.0265	35.7	1.0305	40.8
1.0186	25.4	1.0226	30.6	1.0266	35.8	1.0306	41.0
1.0187	25.5	1.0227	30.7	1.0267	35.9	1.0307	41.1
1.0188	25.6	1.0228	30.8	1.0268	36.0	1.0308	41.2
1.0189	25.8	1.0229	31.0	1.0269	36.2	1.0309	41.4
1.0190	25.9	1.0230	31.1	1.0270	36.3	1.0310	41.5
1.0191	26.0	1.0231	31.2	1.0271	36.4	1.0311	41.6
1.0192	26.1	1.0232	31.4	1.0272	36.6	1.0312	41.7
1.0193	26.3	1.0233	31.5	1.0273	36.7	1.0313	41.9
1.0194	26.4	1.0234	31.6	1.0274	36.8	1.0314	42.0
1.0195	26.5	1.0235	31.8	1.0275	37.0	1.0315	42.1
1.0196	26.7	1.0236	31.9	1.0276	37.1	1.0316	42.3
1.0197	26.8	1.0237	32.0	1.0277	37.2	1.0317	42.4
1.0198	26.9	1.0238	32.1	1.0278	37.3	1.0318	42.5
1.0199	27.1	1.0239	32.3	1.0279	37.5	1.0319	42.7
1.0200	27.2	1.0240	32.4	1.0280	37.6	1.0320	42.8
1.0201	27.3	1.0241	32.5	1.0281	37.7		
1.0202	27.5	1.0242	32.7	1.0282	37.9		
1.0203	27.6	1.0243	32.8	1.0283	38.0		
1.0204	27.7	1.0244	32.9	1.0284	38.1		
1.0205	27.8	1.0245	33.1	1.0285	38.2		
1.0206	28.0	1.0246	33.2	1.0286	38.4		
1.0207	28.1	1.0247	33.3	1.0287	38.5		
1.0208	28.2	1.0248	33.5	1.0288	38.6		
1.0209	28.4	1.0249	33.6	1.0289	38.8		
1.0210	28.5	1.0250	33.7	1.0290	38.9		

210 C. Argentometric Method

1. General Discussion

This procedure is similar to that specified for chloride, Section 407A. It is not as precise as the longer Knudsen method[1] but is less time-consuming.

2. Sample Handling

Collect a sample in a 240-mL (8-oz) glass bottle with a No. 6 cork stopper. Pretreat stopper by soaking in melted paraffin wax for 30 to 40 sec, draining, and drying. Remove excess wax. To collect sample, rinse bottle three times with water being sampled and fill bottle to shoulder. Seal bottle by forcing waxed cork below level of neck. If samples are not examined within 2 days, dip neck of bottle in melted wax. The sealed sample is stable indefinitely. Titrate unsealed samples within a few minutes of collection; do not hold unsealed samples more than 1 hr before analysis.

3. Apparatus

Automatic zero-adjusting 35-mL buret: Lubricate if necessary with paraffin stopcock grease, never silicone.

4. Reagents

a. Standard seawater: Standard seawater of known chlorinity (''Eau de Mer Normale'') is available from the Depot d'Eau Normale, Laboratory Hydrographique, Charlottenlund Slot, Copenhagen, Denmark.* A secondary standard may be prepared by filtering seawater (chlorinity about 18 ‰) collected from the open ocean at a depth of at least 50 m. Stabilize with a few crystals of thymol and seal in sample bottles. Use the mean of 10 or more sample titrations as the chlorosity (20 C) of this secondary standard.

b. Silver nitrate solution, approximately 0.28*N*: Dissolve 48.5 g $AgNO_3$ in 500 mL distilled water and dilute to 1,000 mL. Store in glass-stoppered brown glass bottle at room temperature.

c. Potassium chromate indicator solution: Dissolve 63 g K_2CrO_4 in 100 mL distilled water. Add a few drops of 0.28*N* $AgNO_3$ until a definite red precipitate persists. Let stand to settle, filter, and store in glass dropping bottle.

d. Standard sodium chloride: Dry about 35 g NaCl to constant weight. Cool and weigh out 29.674 g. Dissolve in distilled water and dilute to 1,000 mL. Check this standard against Copenhagen water and periodically against the secondary seawater standard.

Standardization: Place 25.0 mL standard NaCl solution in a 150-mL erlenmeyer flask. Add 6 drops of chromate indicator and titrate with $AgNO_3$ solution in yellow light until a red precipitate just forms. Stopper flask with rubber stopper and shake vigorously to break curds of AgCl. Wash down stopper and continue titration to brown end point. Be consistent in endpoint recognition.

$$\text{Normality} = \frac{12.69}{\text{mL } AgNO_3}$$

5. Procedure

Let sample and $AgNO_3$ titrant come to same temperature. Use a 25.0-mL sample and titrate as directed above.

6. Calculation

a. Calculate the chlorosity equivalent of 1 mL $AgNO_3$ solution:

$$ClEq = N \times 0.0355$$

*Domestic sources are listed in AMERICAN SOCIETY OF LIMNOLOGY AND OCEANOGRAPHY. COMMITTEE ON APPARATUS AND SUPPLIES. 1964. Sources of Limnological and Oceanographic Apparatus and Supplies. Spec. Publ. No. 1, 3rd rev. *Limnol. Oceanogr.* 9 (suppl. Apr. 1964).

where:

$ClEq$ = chlorosity equivalent and
N = normality of $AgNO_3$.

b. Calculate the chlorosity

$$Cl_0 = d \times ClEq \times \frac{1{,}000}{25}$$

where:

d = mL titrant used and
$ClEq$ = chlorosity equivalent $AgNO_3$.

c. Convert chlorosity to salinity by using Table 210:IV. Record salinity, S, as $^0/_{00}$. For example, if 47.23 mL titrant (0.2859N) are used for a 25.0 mL sample, chlorosity = 47.23 × 0.2859 × 0.0355 × 1,000/25 = 19.17 $^0/_{00}$. From Table 210:IV, salinity = 33.82 $^0/_{00}$.

d. Alternatively, convert chlorosity to chlorinity by subtracting the appropriate factor given in Table 210:III. Record chlorinity as Cl $^0/_{00}$.

7. Precision and Accuracy

This procedure is suitable for salinities ranging from 4 to 40 $^0/_{00}$. It is accurate to between 0.05 and 0.1 $^0/_{00}$ salinity.

TABLE 210:III. CONVERSION OF CHLOROSITY, Cl_0, AT 20 C TO CHLORINITY, Cl $^0/_{00}$.

Calculated Chlorosity	Subtract for Chlorinity
9.95–10.35	−0.12
10.36–10.75	−0.13
10.76–11.15	−0.14
11.16–11.46	−0.15
11.47–11.76	−0.16
11.77–12.06	−0.17
12.07–12.46	−0.18
12.47–12.86	−0.19
12.87–13.07	−0.20
13.08–13.37	−0.21
13.38–13.67	−0.22
13.68–14.02	−0.23
14.03–14.27	−0.24
14.28–14.52	−0.25
14.53–14.82	−0.26
14.83–15.09	−0.27
15.10–15.37	−0.28
15.38–15.68	−0.29
15.69–15.87	−0.30
15.88–16.17	−0.31
16.18–16.32	−0.32
16.33–16.62	−0.33
16.63–16.82	−0.34
16.83–17.11	−0.35
17.12–17.32	−0.36
17.33–17.57	−0.37
17.58–17.82	−0.38
17.83–18.02	−0.39
18.03–18.27	−0.40
18.28–18.47	−0.41
18.48–18.67	−0.42
18.68–18.97	−0.43
18.98–19.17	−0.44
19.18–19.32	−0.45
19.33–19.52	−0.46
19.53–19.77	−0.47
19.78–19.97	−0.48

TABLE 210:IV. CONVERSION OF CHLOROSITY TO SALINITY

Conversion of 20 C chlorosity, $Cl/\text{liter}_{(20)}$, to salinity, $S‰$, from the expression $S‰ = 0.03 + [1.8050 \times Cl/\text{liter}_{(20)} \times 1/\rho_{(20)}]$ where $\rho_{(20)}$ is the density of seawater at chlorosity $Cl/\text{liter}_{(20)}$.

$Cl/\text{liter}_{(20)}$	$S‰$	$Cl/\text{liter}_{(20)}$	$S‰$	$Cl/\text{liter}_{(20)}$	$S‰$	$Cl/\text{liter}_{(20)}$	$S‰$
2.00	3.64	2.40	4.36	2.80	5.07	3.20	5.79
.01	.66	.41	.37	.81	.09	.21	.81
.02	.68	.42	.39	.82	.11	.22	.82
.03	.69	.43	.41	.83	.13	.23	.84
.04	.71	.44	.43	.84	.14	.24	.86
.05	.73	.45	.45	.85	.16	.25	.88
.06	.75	.46	.46	.86	.18	.26	.90
.07	.77	.47	.48	.87	.20	.27	.91
.08	.78	.48	.50	.88	.22	.28	.93
.09	.80	.49	.52	.89	.24	.29	.95
2.10	3.82	2.50	4.54	2.90	5.25	3.30	5.97
.11	.84	.51	.55	.91	.27	.31	5.99
.12	.86	.52	.57	.92	.29	.32	6.00
.13	.87	.53	.59	.93	.31	.33	.02
.14	.89	.54	.61	.94	.32	.34	.04
.15	.91	.55	.63	.95	.34	.35	.06
.16	.93	.56	.64	.96	.36	.36	.08
.17	.95	.57	.66	.97	.38	.37	.09
.18	.96	.58	.68	.98	.40	.38	.11
.19	3.98	.59	.70	.99	.41	.39	.13
2.20	4.00	2.60	4.71	3.00	5.43	3.40	6.15
.21	.02	.61	.73	.01	.45	.41	.16
.22	.03	.62	.75	.02	.47	.42	.18
.23	.05	.63	.77	.03	.48	.43	.20
.24	.07	.64	.79	.04	.50	.44	.22
.25	.09	.65	.80	.05	.52	.45	.24
.26	.11	.66	.82	.06	.54	.46	.25
.27	.12	.67	.84	.07	.56	.47	.27
.28	.14	.68	.86	.08	.57	.48	.29
.29	.16	.69	.88	.09	.59	.49	.31
2.30	4.18	2.70	4.89	3.10	5.61	3.50	6.33
.31	.20	.71	.91	.11	.63	.51	.34
.32	.21	.72	.93	.12	.65	.52	.36
.33	.23	.73	.95	.13	.66	.53	.38
.34	.25	.74	.97	.14	.68	.54	.40
.35	.27	.75	4.98	.15	.70	.55	.42
.36	.29	.76	5.00	.16	.72	.56	.43
.37	.30	.77	.02	.17	.74	.57	.45
.38	.32	.78	.04	.18	.75	.58	.47
.39	.34	.79	.06	.19	.77	.59	.49

TABLE 210:IV, CONT.

Cl/liter$_{(20)}$	$S^0/_{00}$	Cl/liter$_{(20)}$	$S^0/_{00}$	Cl/liter$_{(20)}$	$S^0/_{00}$	Cl/liter$_{(20)}$	$S^0/_{00}$
3.60	6.50	4.05	7.31	4.50	8.11	4.95	8.92
.61	.52	.06	.33	.51	.13	.96	.94
.62	.54	.07	.35	.52	.15	.97	.95
.63	.56	.08	.36	.53	.17	.98	.97
.64	.58	.09	.38	.54	.18	.99	.99
.65	.59	4.10	7.40	.55	.20	5.00	9.01
.66	.61	.11	.42	.56	.22	.01	.02
.67	.63	.12	.43	.57	.24	.02	.04
.68	.65	.13	.45	.58	.26	.03	.06
.69	.67	.14	.47	.59	.27	.04	.08
3.70	6.68	.15	.49	4.60	8.29	.05	.10
.71	.70	.16	.51	.61	.31	.06	.11
.72	.72	.17	.52	.62	.33	.07	.13
.73	.74	.18	.54	.63	.35	.08	.15
.74	.76	.19	.56	.64	.36	.09	.17
.75	.77	4.20	7.58	.65	.38	5.10	9.18
.76	.79	.21	.60	.66	.40	.11	.20
.77	.81	.22	.61	.67	.42	.12	.22
.78	.83	.23	.63	.68	.44	.13	.24
.79	.84	.24	.65	.69	.45	.14	.26
3.80	6.86	.25	.67	4.70	8.47	.15	.27
.81	.88	.26	.68	.71	.49	.16	.29
.82	.90	.27	.70	.72	.51	.17	.31
.83	.92	.28	.72	.73	.52	.18	.33
.84	.93	.29	.74	.74	.54	.19	.34
.85	.95	4.30	7.76	.75	.56	5.20	9.36
.86	.97	.31	.77	.76	.58	.21	.38
.87	6.98	.32	.79	.77	.60	.22	.40
.88	7.01	.33	.81	.78	.61	.23	.42
.89	.02	.34	.83	.79	.63	.24	.43
3.90	7.04	.35	.85	4.80	8.65	.25	.45
.91	.06	.36	.86	.81	.67	.26	.47
.92	.08	.37	.88	.82	.69	.27	.49
.93	.10	.38	.90	.83	.70	.28	.50
.94	.11	.39	.92	.84	.72	.29	.52
.95	.13	4.40	7.93	.85	.74	5.30	9.54
.96	.15	.41	.95	.86	.76	.31	.56
.97	.17	.42	.97	.87	.77	.32	.58
.98	.18	.43	7.99	.88	.79	.33	.59
.99	.20	.44	8.01	.89	.81	.34	.61
4.00	7.22	.45	.02	4.90	8.83	.35	.63
.01	.24	.46	.04	.91	.85	.36	.65
.02	.26	.47	.06	.92	.86	.37	.67
.03	.27	.48	.08	.93	.88	.38	.68
.04	.29	.49	.10	.94	.90	.39	.70

TABLE 210:IV, CONT.

Cl/liter(20)	*S*‰	*Cl*/liter(20)	*S*‰	*Cl*/liter(20)	*S*‰	*Cl*/liter(20)	*S*‰
5.40	9.72	5.85	10.52	6.30	11.32	6.75	12.12
.41	.74	.86	.54	.31	.34	.76	.14
.42	.75	.87	.56	.32	.36	.77	.16
.43	.77	.88	.57	.33	.37	.78	.17
.44	.79	.89	.59	.34	.39	.79	.19
.45	.81	5.90	10.61	.35	.41	6.80	12.21
.46	.83	.91	.63	.36	.43	.81	.23
.47	.84	.92	.64	.37	.44	.82	.24
.48	.86	.93	.66	.38	.46	.83	.26
.49	.88	.94	.68	.39	.48	.84	.28
5.50	9.90	.95	.70	6.40	11.50	.85	.30
.51	.91	.96	.72	.41	.52	.86	.31
.52	.93	.97	.73	.42	.53	.87	.33
.53	.95	.98	.75	.43	.55	.88	.35
.54	.97	.99	.77	.44	.57	.89	.37
.55	9.99	6.00	10.79	.45	.59	6.90	12.39
.56	10.00	.01	.81	.46	.60	.91	.40
.57	.02	.02	.82	.47	.62	.92	.42
.58	.04	.03	.84	.48	.64	.93	.44
.59	.06	.04	.86	.49	.66	.94	.46
5.60	10.07	.05	.88	6.50	11.68	.95	.47
.61	.09	.06	.89	.51	.69	.96	.49
.62	.11	.07	.91	.52	.71	.97	.51
.63	.13	.08	.93	.53	.73	.98	.53
.64	.15	.09	.95	.54	.75	.99	.55
.65	.16	6.10	10.97	.55	.76	7.00	12.56
.66	.18	.11	10.98	.56	.78	.01	.58
.67	.20	.12	11.00	.57	.80	.02	.60
.68	.22	.13	.02	.58	.82	.03	.62
.69	.24	.14	.04	.59	.84	.04	.63
5.70	10.25	.15	.05	6.60	11.85	.05	.65
.71	.27	.16	.07	.61	.87	.06	.67
.72	.29	.17	.09	.62	.89	.07	.69
.73	.31	.18	.11	.63	.91	.08	.71
.74	.32	.19	.12	.64	.92	.09	.72
.75	.34	6.20	11.14	.65	.94	7.10	12.74
.76	.36	.21	.16	.66	.96	.11	.76
.77	.38	.22	.18	.67	11.98	.12	.78
.78	.40	.23	.20	.68	12.00	.13	.79
.79	.41	.24	.21	.69	.01	.14	.81
5.80	10.43	.25	.23	6.70	12.03	.15	.83
.81	.45	.26	.25	.71	.05	.16	.85
.82	.47	.27	.27	.72	.07	.17	.86
.83	.48	.28	.28	.73	.08	.18	.88
.84	.50	.29	.30	.74	.10	.19	.90

TABLE 210:IV, CONT.

Cl/liter$_{(20)}$	$S^0/_{00}$	Cl/liter$_{(20)}$	$S^0/_{00}$	Cl/liter$_{(20)}$	$S^0/_{00}$	Cl/liter$_{(20)}$	$S^0/_{00}$
7.20	12.92	7.65	13.72	8.10	14.51	8.55	15.31
.21	.94	.66	.73	.11	.53	.56	.33
.22	.95	.67	.75	.12	.55	.57	.34
.23	.97	.68	.77	.13	.57	.58	.36
.24	12.99	.69	.79	.14	.58	.59	.38
.25	13.01	7.70	13.80	.15	.60	8.60	15.40
.26	.02	.71	.82	.16	.62	.61	.41
.27	.04	.72	.84	.17	.64	.62	.43
.28	.06	.73	.86	.18	.65	.63	.45
.29	.08	.74	.88	.19	.67	.64	.47
7.30	13.10	.75	.89	8.20	14.69	.65	.48
.31	.11	.76	.91	.21	.71	.66	.50
.32	.13	.77	.93	.22	.72	.67	.52
.33	.15	.78	.95	.23	.74	.68	.54
.34	.17	.79	.96	.24	.76	.69	.56
.35	.18	7.80	13.98	.25	.78	8.70	15.57
.36	.20	.81	14.00	.26	.80	.71	.59
.37	.22	.82	.02	.27	.81	.72	.61
.38	.24	.83	.03	.28	.83	.73	.63
.39	.25	.84	.05	.29	.85	.74	.64
7.40	13.27	.85	.07	8.30	14.87	.75	.66
.41	.29	.86	.09	.31	.88	.76	.68
.42	.31	.87	.11	.32	.90	.77	.70
.43	.33	.88	.12	.33	.92	.78	.71
.44	.34	.89	.14	.34	.94	.79	.73
.45	.36	7.90	14.16	.35	.95	8.80	15.75
.46	.38	.91	.18	.36	.97	.81	.77
.47	.40	.92	.19	.37	14.99	.82	.79
.48	.41	.93	.21	.38	15.01	.83	.80
.49	.43	.94	.23	.39	.03	.84	.82
7.50	13.45	.95	.25	8.40	15.04	.85	.84
.51	.47	.96	.27	.41	.06	.86	.86
.52	.49	.97	.28	.42	.08	.87	.87
.53	.50	.98	.30	.43	.10	.88	.89
.54	.52	.99	.32	.44	.11	.89	.91
.55	.54	8.00	14.34	.45	.13	8.90	15.93
.56	.56	.01	.35	.46	.15	.91	.94
.57	.57	.02	.37	.47	.17	.92	.96
.58	.59	.03	.39	.48	.18	.93	15.98
.59	.61	.04	.41	.49	.20	.94	16.00
7.60	13.63	.05	.42	8.50	15.22	.95	.01
.61	.65	.06	.44	.51	.24	.96	.03
.62	.66	.07	.46	.52	.25	.97	.05
.63	.68	.08	.48	.53	.27	.98	.07
.64	.70	.09	.50	.54	.29	.99	.09

TABLE 210:IV, CONT.

Cl/liter$_{(20)}$	$S^0/_{00}$	*Cl*/liter$_{(20)}$	$S^0/_{00}$	*Cl*/liter$_{(20)}$	$S^0/_{00}$	*Cl*/liter$_{(20)}$	$S^0/_{00}$
9.00	16.10	9.45	16.89	9.90	17.69	10.35	18.48
.01	.12	.46	.91	.91	.70	.36	.50
.02	.14	.47	.93	.92	.72	.37	.52
.03	.16	.48	.95	.93	.74	.38	.53
.04	.17	.49	.96	.94	.76	.39	.55
.05	.19	9.50	16.98	.95	.77	10.40	18.57
.06	.21	.51	17.00	.96	.79	.41	.59
.07	.23	.52	.02	.97	.81	.42	.60
.08	.24	.53	.03	.98	.83	.43	.62
.09	.26	.54	.05	.99	.85	.44	.64
9.10	16.28	.55	.07	10.00	17.87	.45	.66
.11	.30	.56	.09	.01	.88	.46	.67
.12	.31	.57	.11	.02	.90	.47	.69
.13	.33	.58	.12	.03	.92	.48	.71
.14	.35	.59	.14	.04	.94	.49	.73
.15	.37	9.60	17.16	.05	.95	10.50	18.74
.16	.38	.61	.18	.06	.97	.51	.76
.17	.40	.62	.19	.07	17.99	.52	.78
.18	.42	.63	.21	.08	18.01	.53	.80
.19	.44	.64	.23	.09	.02	.54	.81
9.20	16.45	.65	.25	10.10	18.04	.55	.83
.21	.47	.66	.26	.11	.06	.56	.85
.22	.49	.67	.28	.12	.08	.57	.87
.23	.51	.68	.30	.13	.09	.58	.88
.24	.53	.69	.32	.14	.11	.59	.90
.25	.54	9.70	17.33	.15	.13	10.60	18.92
.26	.56	.71	.35	.16	.15	.61	.94
.27	.58	.72	.37	.17	.16	.62	.96
.28	.60	.73	.39	.18	.18	.63	.97
.29	.61	.74	.40	.19	.20	.64	18.99
9.30	16.63	.75	.42	10.20	18.22	.65	19.01
.31	.65	.76	.44	.21	.23	.66	.03
.32	.67	.77	.46	.22	.25	.67	.04
.33	.68	.78	.47	.23	.27	.68	.06
.34	.70	.79	.49	.24	.29	.69	.08
.35	.72	9.80	17.51	.25	.30	10.70	19.10
.36	.74	.81	.53	.26	.32	.71	.11
.37	.75	.82	.54	.27	.34	.72	.13
.38	.77	.83	.56	.28	.36	.73	.15
.39	.79	.84	.58	.29	.38	.74	.17
9.40	16.81	.85	.60	10.30	18.39	.75	.18
.41	.82	.86	.62	.31	.41	.76	.20
.42	.84	.87	.63	.32	.43	.77	.22
.43	.86	.88	.65	.33	.45	.78	.24
.44	.88	.89	.67	.34	.46	.79	.25

TABLE 210:IV, CONT.

Cl/liter$_{(20)}$	$S^0/_{00}$	Cl/liter$_{(20)}$	$S^0/_{00}$	Cl/liter$_{(20)}$	$S^0/_{00}$	Cl/liter$_{(20)}$	$S^0/_{00}$
10.80	19.27	11.25	20.06	11.70	20.85	12.15	21.64
.81	.29	.26	.08	.71	.87	.16	.66
.82	.31	.27	.10	.72	.89	.17	.68
.83	.32	.28	.11	.73	.90	.18	.69
.84	.34	.29	.13	.74	.92	.19	.71
.85	.36	11.30	20.15	.75	.94	12.20	21.73
.86	.38	.31	.17	.76	.96	.21	.75
.87	.39	.32	.18	.77	.97	.22	.76
.88	.41	.33	.20	.78	20.99	.23	.78
.89	.43	.34	.22	.79	21.01	.24	.80
10.90	19.45	.35	.24	11.80	21.03	.25	.82
.91	.47	.36	.26	.81	.04	.26	.83
.92	.48	.37	.27	.82	.06	.27	.85
.93	.50	.38	.29	.83	.08	.28	.87
.94	.52	.39	.31	.84	.10	.29	.89
.95	.54	11.40	20.33	.85	.11	12.30	21.90
.96	.55	.41	.34	.86	.13	.31	.92
.97	.57	.42	.36	.87	.15	.32	.94
.98	.59	.43	.38	.88	.17	.33	.96
.99	.61	.44	.40	.89	.18	.34	.97
11.00	19.62	.45	.41	11.90	21.20	.35	21.99
.01	.64	.46	.43	.91	.22	.36	22.01
.02	.66	.47	.45	.92	.24	.37	.03
.03	.68	.48	.47	.93	.26	.38	.04
.04	.69	.49	.48	.94	.27	.39	.06
.05	.71	11.50	20.50	.95	.29	12.40	22.08
.06	.73	.51	.52	.96	.31	.41	.09
.07	.75	.52	.54	.97	.33	.42	.11
.08	.76	.53	.55	.98	.34	.43	.13
.09	.78	.54	.57	.99	.36	.44	.15
11.10	19.80	.55	.59	12.00	21.38	.45	.16
.11	.82	.56	.61	.01	.40	.46	.18
.12	.83	.57	.62	.02	.41	.47	.20
.13	.85	.58	.64	.03	.43	.48	.22
.14	.87	.59	.66	.04	.45	.49	.23
.15	.89	11.60	20.68	.05	.47	12.50	22.25
.16	.90	.61	.69	.06	.48	.51	.27
.17	.92	.62	.71	.07	.50	.52	.29
.18	.94	.63	.73	.08	.52	.53	.30
.19	.96	.64	.75	.09	.54	.54	.32
11.20	19.97	.65	.76	12.10	21.55	.55	.34
.21	19.99	.66	.78	.11	.57	.56	.36
.22	20.01	.67	.80	.12	.59	.57	.37
.23	.03	.68	.82	.13	.61	.58	.39
.24	.04	.69	.83	.14	.62	.59	.41

TABLE 210:IV, CONT.

Cl/liter$_{(20)}$	$S^0/_{00}$	*Cl*/liter$_{(20)}$	$S^0/_{00}$	*Cl*/liter$_{(20)}$	$S^0/_{00}$	*Cl*/liter$_{(20)}$	$S^0/_{00}$
12.60	22.43	13.05	23.21	13.50	24.00	13.95	24.79
.61	.44	.06	.23	.51	.02	.96	.80
.62	.46	.07	.25	.52	.03	.97	.82
.63	.48	.08	.27	.53	.05	.98	.84
.64	.50	.09	.28	.54	.07	.99	.85
.65	.51	13.10	23.30	.55	.09	14.00	24.87
.66	.53	.11	.32	.56	.10	.01	.89
.67	.55	.12	.34	.57	.12	.02	.91
.68	.57	.13	.35	.58	.14	.03	.92
.69	.58	.14	.37	.59	.16	.04	.94
12.70	22.60	.15	.39	13.60	24.17	.05	.96
.71	.62	.16	.41	.61	.19	.06	.98
.72	.64	.17	.42	.62	.21	.07	24.99
.73	.65	.18	.44	.63	.23	.08	25.01
.74	.67	.19	.46	.64	.24	.09	.03
.75	.69	13.20	23.48	.65	.26	14.10	25.05
.76	.71	.21	.49	.66	.28	.11	.06
.77	.72	.22	.51	.67	.30	.12	.08
.78	.74	.23	.53	.68	.31	.13	.10
.79	.76	.24	.55	.69	.33	.14	.12
12.80	22.78	.25	.56	13.70	24.35	.15	.13
.81	.79	.26	.58	.71	.37	.16	.15
.82	.81	.27	.60	.72	.38	.17	.17
.83	.83	.28	.62	.73	.40	.18	.19
.84	.85	.29	.63	.74	.42	.19	.20
.85	.86	13.30	23.65	.75	.44	14.20	25.22
.86	.88	.31	.67	.76	.45	.21	.24
.87	.90	.32	.69	.77	.47	.22	.26
.88	.92	.33	.70	.78	.49	.23	.27
.89	.93	.34	.72	.79	.51	.24	.29
12.90	22.95	.35	.74	13.80	24.52	.25	.31
.91	.97	.36	.76	.81	.54	.26	.32
.92	22.99	.37	.77	.82	.56	.27	.34
.93	23.00	.38	.79	.83	.58	.28	.36
.94	.02	.39	.81	.84	.59	.29	.38
.95	.04	13.40	23.83	.85	.61	14.30	25.39
.96	.06	.41	.84	.86	.63	.31	.41
.97	.07	.42	.86	.87	.65	.32	.43
.98	.09	.43	.88	.88	.66	.33	.45
.99	.11	.44	.89	.89	.68	.34	.46
13.00	23.13	.45	.91	13.90	24.70	.35	.48
.01	.14	.46	.93	.91	.72	.36	.50
.02	.16	.47	.95	.92	.73	.37	.52
.03	.18	.48	.96	.93	.75	.38	.53
.04	.20	.49	.98	.94	.77	.39	.55

TABLE 210:IV, CONT.

Cl/liter$_{(20)}$	$S^0/_{00}$	Cl/liter$_{(20)}$	$S^0/_{00}$	Cl/liter$_{(20)}$	$S^0/_{00}$	Cl/liter$_{(20)}$	$S^0/_{00}$
14.40	25.57	14.85	26.35	15.30	27.13	15.75	27.91
.41	.59	.86	.37	.31	.15	.76	.93
.42	.60	.87	.39	.32	.17	.77	.95
.43	.62	.88	.40	.33	.18	.78	.97
.44	.64	.89	.42	.34	.20	.79	.98
.45	.66	14.90	26.44	.35	.22	15.80	28.00
.46	.67	.91	.46	.36	.24	.81	.02
.47	.69	.92	.47	.37	.25	.82	.03
.48	.71	.93	.49	.38	.27	.83	.05
.49	.72	.94	.51	.39	.29	.84	.07
14.50	25.74	.95	.53	15.40	27.31	.85	.09
.51	.76	.96	.54	.41	.32	.86	.10
.52	.78	.97	.56	.42	.34	.87	.12
.53	.79	.98	.58	.43	.36	.88	.14
.54	.81	.99	.59	.44	.38	.89	.16
.55	.83	15.00	26.61	.45	.39	15.90	28.17
.56	.85	.01	.63	.46	.41	.91	.19
.57	.86	.02	.65	.47	.43	.92	.21
.58	.88	.03	.66	.48	.44	.93	.23
.59	.90	.04	.68	.49	.46	.94	.24
14.60	25.92	.05	.70	15.50	27.48	.95	.26
.61	.93	.06	.72	.51	.50	.96	.28
.62	.95	.07	.73	.52	.51	.97	.29
.63	.97	.08	.75	.53	.53	.98	.31
.64	25.99	.09	.77	.54	.55	.99	.33
.65	26.00	15.10	26.79	.55	.57	16.00	28.35
.66	.02	.11	.80	.56	.58	.01	.36
.67	.04	.12	.82	.57	.60	.02	.38
.68	.06	.13	.84	.58	.62	.03	.40
.69	.07	.14	.86	.59	.64	.04	.42
14.70	26.09	.15	.87	15.60	27.65	.05	.43
.71	.11	.16	.89	.61	.67	.06	.45
.72	.13	.17	.91	.62	.69	.07	.47
.73	.14	.18	.92	.63	.71	.08	.49
.74	.16	.19	.94	.64	.72	.09	.50
.75	.18	15.20	26.96	.65	.74	16.10	28.52
.76	.19	.21	.98	.66	.76	.11	.54
.77	.21	.22	26.99	.67	.77	.12	.55
.78	.23	.23	27.01	.68	.79	.13	.57
.79	.25	.24	.03	.69	.81	.14	.59
14.80	26.26	.25	.05	15.70	27.83	.15	.61
.18	.28	.26	.06	.71	.84	.16	.62
.82	.30	.27	.08	.72	.86	.17	.64
.83	.32	.28	.10	.73	.88	.18	.66
.84	.33	.29	.12	.74	.90	.19	.68

TABLE 210:IV, CONT.

Cl/liter$_{(20)}$	$S^0/_{00}$	Cl/liter$_{(20)}$	$S^0/_{00}$	Cl/liter$_{(20)}$	$S^0/_{00}$	Cl/liter$_{(20)}$	$S^0/_{00}$
16.20	28.69	16.65	29.47	17.10	30.25	17.55	31.03
.21	.71	.66	.49	.11	.27	.56	.04
.22	.73	.67	.51	.12	.28	.57	.06
.23	.75	.68	.52	.13	.30	.58	.08
.24	.76	.69	.54	.14	.32	.59	.10
.25	.78	16.70	29.56	.15	.34	17.60	31.11
.26	.80	.71	.58	.16	.35	.61	.13
.27	.82	.72	.59	.17	.37	.62	.15
.28	.83	.73	.61	.18	.39	.63	.17
.29	.85	.74	.63	.19	.41	.64	.18
16.30	28.87	.75	.65	17.20	30.42	.65	.20
.31	.88	.76	.66	.21	.44	.66	.22
.32	.90	.77	.68	.22	.46	.67	.23
.33	.92	.78	.70	.23	.47	.68	.25
.34	.94	.79	.71	.24	.49	.69	.27
.35	.95	16.80	29.73	.25	.51	17.70	31.29
.36	.97	.81	.75	.26	.53	.71	.30
.37	28.99	.82	.77	.27	.54	.72	.32
.38	29.00	.83	.78	.28	.56	.73	.34
.39	.02	.84	.80	.29	.58	.74	.36
16.40	29.04	.85	.82	17.30	30.60	.75	.37
.41	.06	.86	.84	.31	.61	.76	.39
.42	.07	.87	.85	.32	.63	.77	.41
.43	.09	.88	.87	.33	.65	.78	.42
.44	.11	.89	.89	.34	.66	.79	.44
.45	.13	16.90	29.90	.35	.68	17.80	31.46
.46	.14	.91	.92	.36	.70	.81	.48
.47	.16	.92	.94	.37	.72	.82	.49
.48	.18	.93	.96	.38	.73	.83	.51
.49	.20	.94	.97	.39	.75	.84	.53
16.50	29.21	.95	29.99	17.40	30.77	.85	.55
.51	.23	.96	30.01	.41	.79	.86	.56
.52	.25	.97	.03	.42	.80	.87	.58
.53	.26	.98	.04	.43	.82	.88	.60
.54	.28	.99	.06	.44	.84	.89	.61
.55	.30	17.00	30.08	.45	.85	17.90	31.63
.56	.32	.01	.09	.46	.87	.91	.65
.57	.33	.02	.11	.47	.89	.92	.67
.58	.35	.03	.13	.48	.91	.93	.68
.59	.37	.04	.15	.49	.92	.94	.70
16.60	29.39	.05	.16	17.50	30.94	.95	.72
.61	.40	.06	.18	.51	.96	.96	.74
.62	.42	.07	.20	.52	.98	.97	.75
.63	.44	.08	.22	.53	30.99	.98	.77
.64	.45	.09	.23	.54	31.01	.99	.79

TABLE 210:IV, CONT.

Cl/liter$_{(20)}$	*S*°/oo	*Cl*/liter$_{(20)}$	*S*°/oo	*Cl*/liter$_{(20)}$	*S*°/oo	*Cl*/liter$_{(20)}$	*S*°/oo
18.00	31.80	18.45	32.58	18.90	33.36	19.35	34.13
.01	.82	.46	.60	.91	.37	.36	.15
.02	.84	.47	.61	.92	.39	.37	.16
.03	.86	.48	.63	.93	.41	.38	.18
.04	.87	.49	.65	.94	.42	.39	.20
.05	.89	18.50	32.67	.95	.44	19.40	34.22
.06	.91	.51	.68	.96	.46	.41	.23
.07	.92	.52	.70	.97	.48	.42	.25
.08	.94	.53	.72	.98	.49	.43	.27
.09	.96	.54	.73	.99	.51	.44	.28
18.10	31.98	.55	.75	19.00	33.53	.45	.30
.11	31.99	.56	.77	.01	.54	.46	.32
.12	32.01	.57	.79	.02	.56	.47	.34
.13	.03	.58	.80	.03	.58	.48	.35
.14	.05	.59	.82	.04	.60	.49	.37
.15	.06	18.60	32.84	.05	.61	19.50	34.39
.16	.08	.61	.86	.06	.63	.51	.40
.17	.10	.62	.87	.07	.65	.52	.42
.18	.11	.63	.89	.08	.67	.53	.44
.19	.13	.64	.91	.09	.68	.54	.46
18.20	32.15	.65	.92	19.10	33.70	.55	.47
.21	.17	.66	.94	.11	.72	.56	.49
.22	.18	.67	.96	.12	.73	.57	.51
.23	.20	.68	.98	.13	.75	.58	.52
.24	.22	.69	32.99	.14	.77	.59	.54
.25	.23	18.70	33.01	.15	.79	19.60	34.56
.26	.25	.71	.03	.16	.80	.61	.58
.27	.27	.72	.05	.17	.82	.62	.59
.28	.29	.73	.06	.18	.84	.63	.61
.29	.30	.74	.08	.19	.85	.64	.63
18.30	32.32	.75	.10	19.20	33.87	.65	.64
.31	.34	.76	.11	.21	.89	.66	.66
.32	.36	.77	.13	.22	.91	.67	.68
.33	.37	.78	.15	.23	.92	.68	.70
.34	.39	.79	.17	.24	.94	.69	.71
.35	.41	18.80	33.18	.25	.96	19.70	34.73
.36	.42	.81	.20	.26	.97	.71	.75
.37	.44	.82	.22	.27	33.99	.72	.77
.38	.46	.83	.23	.28	34.01	.73	.78
.39	.48	.84	.25	.29	.03	.74	.80
18.40	32.49	.85	.27	19.30	34.04	.75	.82
.41	.51	.86	.29	.31	.06	.76	.83
.42	.53	.87	.30	.32	.08	.77	.85
.43	.55	.88	.32	.33	.09	.78	.87
.44	.56	.89	.34	.34	.11	.79	.89

TABLE 210:IV, CONT.

Cl/liter$_{(20)}$	S ‰	Cl/liter$_{(20)}$	S ‰	Cl/liter$_{(20)}$	S ‰	Cl/liter$_{(20)}$	S ‰
19.80	34.90	20.25	35.68	20.70	36.45	21.15	37.22
.81	.92	.26	.70	.71	.47	.16	.24
.82	.94	.27	.71	.72	.48	.17	.25
.83	.95	.28	.73	.73	.50	.18	.27
.84	.97	.29	.74	.74	.52	.19	.29
.85	34.99	20.30	35.76	.75	.53	21.20	37.30
.86	35.01	.31	.78	.76	.55	.21	.32
.87	.02	.32	.80	.77	.57	.22	.34
.88	.04	.33	.82	.78	.59	.23	.36
.89	.06	.34	.83	.79	.60	.24	.37
19.90	35.07	.35	.85	20.80	36.62	.25	.39
.91	.09	.36	.87	.81	.64	.26	.40
.92	.11	.37	.88	.82	.65	.27	.42
.93	.13	.38	.90	.83	.67	.28	.44
.94	.14	.39	.92	.84	.69	.29	.46
.95	.16	20.40	35.93	.85	.71	21.30	37.47
.96	.18	.41	.95	.86	.72	.31	.49
.97	.19	.42	.97	.87	.74	.32	.51
.98	.21	.43	35.99	.88	.76	.33	.53
.99	.23	.44	36.00	.89	.77	.34	.54
20.00	35.25	.45	.02	20.90	36.79	.35	.56
.01	.27	.46	.04	.91	.81	.36	.58
.02	.28	.47	.06	.92	.83	.37	.59
.03	.30	.48	.07	.93	.84	.38	.61
.04	.32	.49	.09	.94	.86	.39	.63
.05	.34	20.50	36.11	.95	.88	21.40	37.65
.06	.35	.51	.12	.96	.89	.41	.66
.07	.37	.52	.14	.97	.91	.42	.68
.08	.39	.53	.16	.98	.93	.43	.70
.09	.40	.54	.18	.99	.94	.44	.71
20.10	35.42	.55	.19	21.00	36.96	.45	.73
.11	.44	.56	.21	.01	36.98	.46	.75
.12	.46	.57	.23	.02	37.00	.47	.77
.13	.47	.58	.24	.03	.01	.48	.78
.14	.50	.59	.26	.04	.03	.49	.80
.15	.51	20.60	36.28	.05	.05	21.50	37.82
.16	.52	.61	.30	.06	.06	.51	.83
.17	.54	.62	.31	.07	.08	.52	.85
.18	.56	.63	.33	.08	.10	.53	.87
.19	.58	.64	.35	.09	.12	.54	.89
20.20	35.59	.65	.36	21.10	37.13	.55	.90
.21	.61	.66	.38	.11	.15	.56	.92
.22	.63	.67	.40	.12	.17	.57	.94
.23	.64	.68	.41	.13	.18	.58	.95
.24	.66	.69	.43	.14	.20	.59	.97

TABLE 210:IV, CONT.

Cl/liter(20)	S‰	Cl/liter(20)	S‰	Cl/liter(20)	S‰	Cl/liter(20)	S‰
21.60	37.99	21.70	38.16	21.80	38.33	21.90	38.50
.61	38.00	.71	.17	.81	.34	.91	.51
.62	.02	.72	.19	.82	.36	.92	.53
.63	.04	.73	.21	.83	.38	.93	.55
.64	.06	.74	.23	.84	.40	.94	.57
.65	.07	.75	.24	.85	.41	.95	.58
.66	.09	.76	.26	.86	.43	.96	.60
.67	.11	.77	.28	.87	.45	.97	.62
.68	.12	.78	.29	.88	.46	.98	.63
.69	.14	.79	.31	.89	.48	.99	.65
						22.00	38.67

SOURCE: A Manual of Sea Water Analysis. Department of Public Printing and Stationery, Ottawa, Canada.

210 D. Reference

1. FORD, W.L. & E.S. DEEVEY, JR. 1946. The Determination of Chlorinity by the Knudsen Method. Woods Hole Oceanographic Inst., Woods Hole, Mass.

210 E. Bibliography

THOMPSON, T.G. 1928. Standardization of silver nitrate solutions used in chemical studies of sea water. *J. Amer. Chem. Soc.* 50:618.

SVERDRUP, H.V., M.W. JOHNSON & R.H. FLEMING. 1942. The Oceans. Prentice-Hall, Inc., Englewood Cliffs, N.J.

ZERBE, W.B. & C.B. TAYLOR. 1953. Sea Water Temperature and Density Reduction Tables. Coast & Geodetic Survey, U.S. Dept. Commerce, Spec. Publ. No. 298. U.S. Govt. Print. Off., Washington, D.C.

VAN ARX, W.S. 1962. Introduction to Physical Oceanography. Addison-Wesley Publ. Co., Reading, Mass.

STRICKLAND, J.D.H. & T.R. PARSONS. 1968. A Practical Handbook of Seawater Analysis. Fisheries Research Board, Ottawa, Canada.

211 TASTE

Taste, like odor, is one of the chemical senses. Most of the general principles of sensory methods described in Section 207 (Odor) apply equally to the taste determination and should be reviewed as background to this section.

The differences between the two sense modes are reflected in their corresponding measurement methods. Taste and odor differ in the nature and location of the receptor nerve sites: high in the nasal cavity for odor, and primarily on the tongue for taste.[1] The odor sensation is stimulated by vapors without physical contact with a water sample, while taste requires contact of the taste buds with the water. Taste is simpler than odor—there may be only four true taste sensations: sour, sweet, salty, and bitter. Dissolved inorganic salts of copper, iron, manganese, potassium, sodium, and zinc[2] can be detected by taste. The taste sense is moderately sensitive. Concentrations producing taste range from a few tenths to several hundred milligrams per liter. The complex sensation experienced in the mouth during the act of tasting is a combination of taste, odor, temperature, and feel; this combination often is called flavor. Taste tests usually have to deal with this complex combination. If a sample contains no detectable odor and is presented at near body temperature, the resulting sensation is predominantly true taste.

It may not be assumed that a tasteless water is most desirable; it has become almost axiomatic that distilled water is less pleasant to drink than certain high-quality waters. Accordingly, there are two distinct purposes of taste tests. The first is to measure taste intensity by the so-called threshold test. The test results are used to assess treatment or pollution abatement required to convert a water source into a high-quality drinking water supply or to measure the taste impact of specific contaminants.[3] The second purpose is to evaluate the quality of a drinking water on the basis of the consumer's judgement. This involves a panel evaluation of undiluted samples presented as ordinarily consumed.[3] A mean acceptance rating is determined, based on a specified rating scale.[4]

Values representing mean thresholds or quality ratings for a laboratory panel are only estimates of these values for the entire consuming population.[5]

Make taste tests only on samples known to be safe for ingestion. Do not use samples that may be contaminated with bacteria, viruses, parasites, or toxic chemicals such as arsenic dechlorinating agents, or that are derived from an unesthetic source. Do not make taste tests on wastewaters or similar untreated effluents. Observe all sanitary and esthetic precautions with regard to apparatus and containers contacting the sample. Practice hospital-level sanitation of these items. Make analyses in a laboratory free from interfering background odors.[6] If possible provide carbon-filtered air at constant temperature and humidity,[7] because without such precautions the test measures flavor, not taste.

Use the procedures described in Section 207 with respect to purity of taste and odor-free water and use of panels of observers.

211 A. Taste Threshold Test

1. General Discussion

The threshold test is used when the purpose is quantitative measurement of detectable taste. When odor is the predominant sensation, as in the case of chlorophenols, the threshold odor test of Section 207 takes priority.

2. Apparatus

a. Preparation of dilutions: Use the dilution system described for odor tests in preparing taste samples.

b. For tasting: Present each dilution and blank to the observer in a clean 50-mL beaker filled to the 30-mL level or use ordinary restaurant drinking glasses to preclude multiple use and contamination. Do not use glassware used in sensory testing for other analyses. An automatic dishwasher supplied with water at not less than 60 C is convenient for sanitizing these containers between tests.

c. Temperature control: Maintain sample presentation temperature of 40 ± 1 C by use of a water bath apparatus.

3. Procedure

Prepare a dilution series (including random blanks) as described in Section 207 and bring to test temperature in the water bath. Present the series of samples to each tester. Pair each sample with a known blank; each should contain 30 mL water. Have the tester taste the sample by taking into the mouth whatever volume is comfortable, holding it for several seconds, and discharging it without swallowing the water. Have the tester compare sample with blank and record whether a taste or aftertaste is detectable in the sample. Submit samples in an increasing order of concentration until the tester's taste threshold has been passed.

Calculate individual threshold and threshold of a panel in the manner described for threshold odor tests.

211 B. Taste Rating Test

1. General Discussion

When the purpose of the test is to estimate taste acceptability, follow the taste rating procedure described below. This procedure has been used with water samples from public sources in laboratory research and consumer surveys to recommend standards governing mineral content in drinking water.[8] Each tester is presented with a list of nine statements about the water ranging on a scale from very favorable to very unfavorable. The tester's task is to select the statement that best expresses his opinion. The scored rating is the scale number of the statement selected. The panel rating is the arithmetic mean of the scale numbers of all testers.

2. Apparatus

a. Preparation of samples: Samples for this test usually represent finished water ready for human consumption; however, experimentally treated water may be used *if the sanitary requirements given above are met fully*. Use taste- and odor-free water and a 2,000-mg/L solution of NaCl prepared with taste- and odor-free water as reference samples.

b. For tasting: Follow procedure given in ¶A.2*b* above.

c. Temperature control: Present samples at a temperature that the testers will find pleasant for drinking water; maintain this temperature by a water bath apparatus. A temperature of 15 C is recommended, but in any case, do not let the test temperature exceed tap water temperatures that are customary at the time of the test. Specify test temperature in reporting results.

3. Procedure

For test efficiency, a single rating session may contain up to 10 samples, including reference samples noted above. Give testers thorough instructions and trial or orientation sessions followed by questions and discussion of procedures. In testing samples testers work alone. Select panel members on the basis of performance in trial sessions. Rating involves the following steps: *a)* initial tasting of about half the sample by taking water into the mouth, holding it for several seconds, and discharging it without swallowing; *b)* forming an initial judgment on the rating scale; *c)* a second tasting made in the same manner as the first; *d)* a final rating made for the sample and the result recorded on the appropriate data form; *e)* rinsing mouth with taste- and odor-free water; and *f)* resting 1 min before repeating Steps *a* through *e* on the next sample. Independently randomize sample order for each tester. Allow at least 30 min rest between repeated rating sessions. Testers should not know the composition or source of specific samples. Use the following scale for rating and record ratings as integers ranging from one to nine, with one given the highest quality rating.

Calculate mean and standard deviation of all ratings given each sample.

4. Rating Scale

Action tendency scale:

1) I would be very happy to accept this water as my everyday drinking water.
2) I would be happy to accept this water as my everyday drinking water.
3) I am sure that I could accept this water as my everyday drinking water.
4) I could accept this water as my everyday drinking water.
5) Maybe I could accept this water as my everyday drinking water.
6) I don't think I could accept this water as my everyday drinking water.
7) I could not accept this water as my everyday drinking water.
8) I could never drink this water.
9) I can't stand this water in my mouth and I could never drink it.

211 C. References

1. GELDARD, F.A. 1972. The Human Senses. John Wiley & Sons, New York, N.Y.
2. COHEN, J.M. L.J. KAMPHAKE, E.K. HARRIS & R.L. WOODWARD. 1960. Taste threshold concentrations of metals in drinking water. *J. Amer. Water Works Ass.* 52:660.
3. BRUVOLD, W.H., H.J. ONGERTH & R.C. DILLEHAY. 1967. Consumer attitudes toward mineral taste in domestic water. *J. Amer. Water Works Ass.* 59:547.
4. BRUVOLD, W.H. 1968. Scales for rating the taste of water. *J. Appl. Psychol.* 52:245.
5. BRUVOLD, W.H. 1970. Laboratory panel estimation of consumer assessments of taste and flavor. *J. Appl. Psychol.* 54:326.
6. BAKER, R.A. 1962. Critical evaluation of olfactory measurement. *J. Water Pollut. Control Fed.* 34:582.
7. BAKER, R.A. 1963. Odor testing laboratory. *J. Water Pollut. Control Fed.* 35:1396.
8. BRUVOLD, W.H., H.J. ONGERTH & R.C. DILLEHAY. 1969. Consumer assessment of mineral taste in domestic water. *J. Amer. Water Works Ass.* 61:575.

211 D. Bibliography

COX, G.J. & J.W. NATHAUS. 1952. A study of the taste of fluoridated water. *J. Amer. Water Works Ass.* 44:940.

LOCKHART, E.E., C.L. TUCKER & M.C. MERRITT. 1955. The effect of water impurities on the flavor of brewed coffee. *Food Res.* 20:598.

CAMPBELL, C.L., R.K. DAWES, S. DEOLALKAR

& M.C. MERRITT. 1958. Effect of certain chemicals in water on the flavor of brewed coffee. *Food Res.* 23:575.

COHEN, J.M. 1963. Taste and odor of ABS in water. *J. Amer. Water Works Ass.* 55:587.

BRUVOLD, W.H. & R.M. PANGBORN. 1966. Rated acceptability of mineral taste in water. *J. Appl. Psychol.* 50:22.

BRUVOLD, W.H. & W. R. GAFFEY. 1969. Rated acceptability of mineral taste in water. II. Combinatorial effects of ions on quality and action tendency ratings. *J. Appl. Psychol.* 53:317.

BRUVOLD, W.H. & H.J. ONGERTH. 1969. Taste quality of mineralized water. *J. Amer. Water Works Ass.* 61:170.

BRYAN, P.E., L.N. KUZMINSKI, F.M. SAWYER & T.H. FENG. 1973. Taste thresholds of halogens in water. *J. Amer. Water Works Ass.* 65:363.

212 TEMPERATURE

1. General Discussion

Temperature readings are used in the calculation of various forms of alkalinity, in studies of saturation and stability with respect to calcium carbonate, in the calculation of salinity, and in general laboratory operations. In limnological studies, water temperatures as a function of depth often are required. Elevated temperatures resulting from heated water discharges may have significant ecological impact. Identification of source of water supply, such as deep wells, often is possible by temperature measurements alone. Industrial plants often require data on water temperature for process use or heat-transmission calculations.

Normally, temperature measurements may be made with any good mercury-filled Celsius thermometer. As a minimum, the thermometer should have a scale marked for every 0.1 C, with markings etched on the capillary glass. The thermometer should have a minimal thermal capacity to permit rapid equilibration. Periodically check the thermometer against a precision thermometer certified by the National Bureau of Standards* that is used with its certificate and correction chart. For field operations use a thermometer having a metal case to prevent breakage.

Depth temperature required for limnological studies may be measured with a reversing thermometer, thermophone, or thermistor. The thermistor is most convenient and accurate; however, higher cost may preclude its use. Calibrate any temperature measurement devices with a National Bureau of Standards certified thermometer before field use. Make readings with the thermometer or device immersed in water long enough to permit complete equilibration. Report results to the nearest 0.1 or 1.0 C, depending on need.

2. Reversing Thermometer

The thermometer commonly used for depth measurements is of the reversing type. It is often mounted on the sample collection apparatus so that a water sample may be obtained simultaneously. Correct readings of reversing thermometers for changes due to differences between temperature at reversal and temperature at time of reading. Calculate as follows:

$$\Delta T = \left[\frac{(T^1 - t)\,(T^1 + V_0)}{K}\right] \times \left[1 + \frac{(T^1 - t)\,(T^1 + V_0)}{K}\right] + L$$

*Some commercial thermometers may be as much as 3 C in error.

where:

ΔT = correction to be added algebraically to uncorrected reading,
T^1 = uncorrected reading at reversal,
t = temperature at which thermometer is read,
V_0 = volume of small bulb end of capillary up to 0 C graduation,
K = constant depending on relative thermal expansion of mercury and glass (usual value of K = 6,100), and
L = calibration correction of thermometer depending on T^1.

If series observations are made it is convenient to prepare graphs for a thermometer to obtain ΔT from any values of T^1 and t.

3. Bibliography

WARREN, H.F. & G.C. WHIPPLE. 1895. The thermophone—A new instrument for determining temperatures. *Mass. Inst. Technol. Quart.* 8:125.

SVERDRUP, H.V., M.W. JOHNSON & R.H. FLEMING. 1942. The Oceans. Prentice-Hall, Inc., Englewood Cliffs, N.J.

AMERICAN SOCIETY FOR TESTING AND MATERIALS. 1949. Standard Specifications for ASTM Thermometers. No. E1-58, ASTM, Philadelphia, Pa.

REE, W.R. 1953. Thermistors for depth thermometry. *J. Amer. Water Works Ass.* 45:259.

213 TESTS ON SLUDGES

This section presents a series of tests uniquely applicable to sludges or slurries. The test data are useful in designing facilities for solids separation and concentration and for assessing operational behavior, especially of the activated sludge process.

213 A. Oxygen-Consumption Rate

1. General Discussion

This test is used to determine the oxygen-consumption rate of a sample of a biological suspension such as activated sludge. It is useful in laboratory and pilot-plant studies as well as in the operation of full-scale treatment plants. When used as a routine plant operation test, it often will indicate changes in operating conditions at an early stage.

2. Apparatus

a. Oxygen-consumption rate device: Either:

1) A probe with an oxygen-sensitive electrode (polarographic or galvanic), or

2) A manometric or respirometric device with appropriate readout and sample capacity of at least 300 mL. The device should have an oxygen supply capacity greater than the oxygen consumption rate of the biological suspension.

b. Stopwatch or other suitable timing device.

c. Thermometer to read to ±0.5 C.

3. Procedure

a. Calibration of oxygen-consumption rate device: Either:

1) Calibrate the oxygen probe and meter according to the method given in Oxy-

gen (Dissolved), Membrane Electrode Method, Section 421, or

2) Calibrate the manometric or respirometric device according to manufacturer's instructions.

b. Volatile nonfiltrable residue determination: See Residue, Section 209.

c. Preparation of sample: Increase DO concentration of sample by shaking it in a partially filled bottle or by bubbling air or oxygen through it.

d. Measurement of oxygen consumption rate:

1) Fill sample container with an appropriate volume of biological suspension. Adjust temperature of suspension to that of the basin from which it was collected or to required evaluation temperature, and maintain constant during analysis. Record temperature.

2) If an oxygen-sensitive probe is used, insert it into a BOD bottle containing a magnetic stirring bar and the biological suspension. Displace enough suspension with probe to fill flared top of bottle and isolate its contents from the atmosphere. Activate probe stirring mechanism and magnetic stirrer. (NOTE: Adequate mixing is essential. For suspensions with high concentrations of nonfiltrable residue (e.g. >5,000 mg/L), more vigorous mixing than that provided by the probe stirring mechanism and magnetic stirrer may be required.) If a manometric or respirometric device is used, follow manufacturer's instructions for startup.

3) After meter reading has stabilized, record initial DO and manometric or respirometric reading, and start timing device. Record appropriate DO, manometric, or respirometric data at time intervals of less than 1 min, depending on rate of consumption. Record data over a 15-min period or until DO becomes limiting, whichever occurs first. The oxygen probe may not be accurate below 1 mg DO/L. If a manometric or respirometric device is used, refer to manufacturer's instructions for lower limiting DO value. Low DO may limit oxygen uptake by the biological suspension and will be indicated by a decreasing rate of oxygen consumption as the test progresses. Reject such data as being unrepresentative of suspension oxygen consumption rate and repeat test using higher initial DO levels.

4. Calculations

a. If an oxygen probe is used, plot observed readings (DO, milligrams per liter) versus time (minutes) on arithmetic graph paper and determine the slope of the line of best fit. The slope is the oxygen consumption rate in milligrams per liter per minute.

b. If a manometric or respirometric device is used, refer to manufacturer's instructions for calculating the oxygen consumption rate.

c. Calculate specific oxygen consumption rate in milligrams per gram per hour as follows:

$$\text{Specific oxygen consumption rate} = \frac{\text{oxygen consumption rate, (mg/L)/min}}{\text{volatile nonfiltrable residue, g/L}} \times \frac{60\ \text{min}}{\text{hr}}$$

5. Precision and Accuracy

This determination is quite sensitive to temperature and has poor precision unless replicate determinations are made at the same temperature. When oxygen consumption is used as a plant control test, run periodic (at least monthly) replicate determinations to establish the precision of the technic.

213 B. Settled Sludge Volume

1. General Discussion

The settled sludge volume of a biological suspension is useful in routine monitoring of biological processes. For activated sludge plant control, a 30-min settled sludge volume or the ratio of the 15-min to the 30-min settled sludge volume has been used to determine the returned-sludge flow rate and when to waste sludge. The 30-min settled sludge volume also is used to determine sludge volume index[1] (Section 213C).

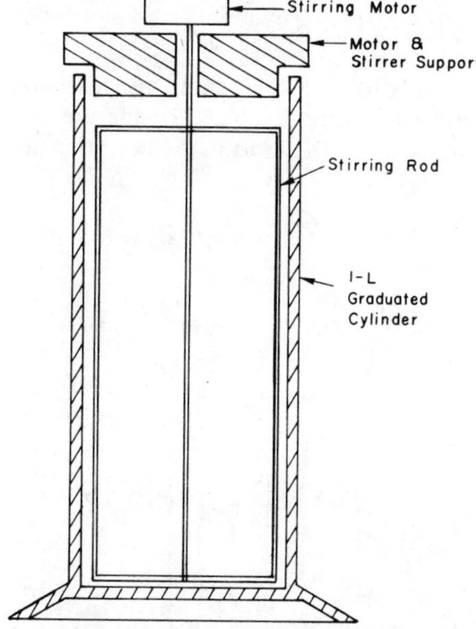

Figure 213:1. Schematic diagram of settling vessel for settled sludge volume test.

2. Apparatus

a. Settling column: Use 1-L graduated cylinder equipped with a stirring mechanism consisting of one or more thin rods extending the length of the column and positioned near the cylinder wall. Provide a stirrer able to rotate the stirring rods at no greater than 4 rpm (peripheral tip speed of approximately 1.3 cm/sec). (See Figure 213:1).

b. Stopwatch.

c. Thermometer.

3. Procedure

a. Place 1.0 L of sample in settling column, activate stirring mechanism, and let suspension settle. Continue stirring throughout test. Maintain suspension temperature during test at that in the basin from which the sample was taken.

b. Determine volume occupied by suspension at measured time intervals, e.g., 5, 10, 15, 20, 30, 45, and 60 min.

c. Report settled sludge volume of the suspension in milliliters for an indicated time interval.

4. Precision and Accuracy

Variations in suspension temperature, sampling and agitation methods, diameter of settling column, and time between sampling and start of the determination significantly affect results. Hence, follow exactly the same procedure each time the test is performed.

213 C. Sludge Volume Index

1. General Discussion

The sludge volume index (SVI) is the volume in milliliters occupied by 1 g of a suspension after 30 min settling. SVI typically is used to monitor settling characteristics of activated sludge and other biological suspensions.[2] Although SVI is not supported theoretically,[1] experience has shown it to be valuable in routine process control.

2. Procedure

a. Determine the nonfiltrable residue of a well-mixed sample of the suspension (See Section 209D).

b. Determine the 30 min settled sludge volume (See Section 213B).

3. Calculations

$$SVI = \frac{\text{settled sludge volume (mL/L)} \times 1{,}000}{\text{nonfiltrable residue (mg/L)}}$$

4. Precision

Precision is determined by the precision achieved in the nonfiltrable residue measurement, the settling characteristics of the suspension, and variables associated with the measurement of the settled sludge volume.

213 D. Zone Settling Rate

1. General Discussion

At high concentrations of nonfiltrable residue, suspensions settle in the zone-settling regime. This type of settling takes place under quiescent conditions and is characterized by a distinct interface between the supernatant liquor and the sludge zone. This distinct sludge interface settles as a blanket, the height of which is measured with time. Zone settling data for suspensions that undergo zone settling, e.g., activated sludge and metal hydroxide suspensions, can be used in the design, operation, and evaluation of sedimentation basins.[3–5]

2. Apparatus

a. Settling vessel: Use a transparent cylinder at least 1 m high and 10 cm in diameter. To improve the similarity between laboratory results and full-scale thickener results use larger diameters and taller cylinders.[1,3,5] Attach a calibrated millimeter tape to outside of cylinder. Equip cylinder with a stirring mechanism, e.g., one or more thin rods positioned near internal wall of settling vessel. Stir suspension near vessel wall over the entire depth of suspension at a peripheral speed no greater than 1 cm/sec. Greater speeds may interfere with the thickening process and yield inaccurate results.[6] Provide the settling vessel with a port in the bottom plate

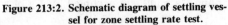

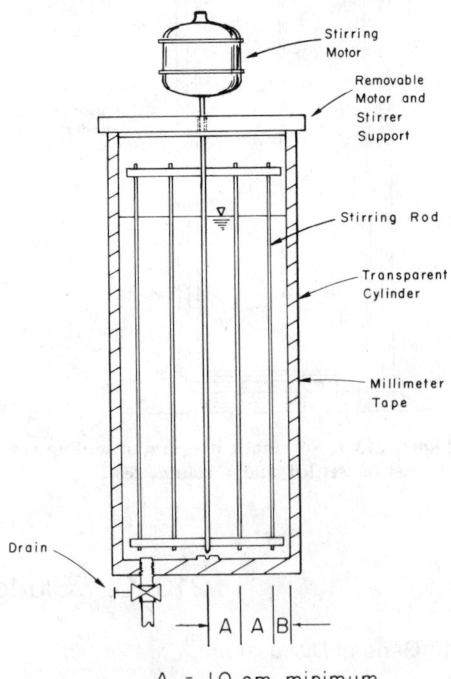

Figure 213:2. Schematic diagram of settling vessel for zone settling rate test.

for filling and draining (See Figure 213:2).

b. Stopwatch.

c. Thermometer.

3. Procedure

a. Maintain suspension in a reservoir in a uniformly mixed condition. Adjust temperature of suspension to that of the basin from which it was collected or to required evaluation temperature. Record temperature. Remove a well-mixed sample from reservoir and measure nonfiltrable residue concentration (Section 209D).

b. Activate stirring mechanism.

c. Fill settling vessel to a fixed height by pumping suspension from reservoir or by gravity flow. Fill at a rate sufficient to maintain a uniform nonfiltrable residue concentration throughout settling vessel at end of filling.

d. The suspension should agglomerate, i.e., form a coarse structure with visible fluid channels, within a few minutes. If suspension does not agglomerate, test is invalid and should be repeated.

e. Record height of solids-liquid interface at intervals of about 1 min. Collect data for sufficient time to assure that suspension is exhibiting a constant zone-settling velocity and that any initial reflocculation period, characterized by an accelerating interfacial settling velocity, has been passed.

4. Calculations

Plot interface height in centimeters vs. time in minutes.[3,5] Draw straight line through data points, ignoring initial shoulder or reflocculation period. Calculate interfacial settling rate as slope of line in centimeters per minute.

5. Precision and Accuracy

Zone settling rate is a function of concentration of nonfiltrable residue and suspension height as well as laboratory artifacts.[5,7] Using the filling method described above and a sufficiently large cylinder, these artifacts should be minimized. However, even with careful testing suspensions often may behave erratically. This unpredictable behavior increases as solids concentrations increase, cylinder diameters decrease, and settling characteristics become poorer.

213 E. Specific Gravity

1. General Discussion

The specific gravity of a sludge is the ratio of the masses of equal volumes of a sludge and distilled water. It is determined by comparing the mass of a known volume of a homogeneous sludge sample at a specific temperature to the mass of the same volume of distilled water at 4 C.

2. Apparatus

Container: A marked flask or bottle to hold a known sludge volume during weighing.

3. Procedure

a. Procedure A: Record sample temperature, *T*. Weigh empty container and record weight, *W*. Fill empty container to mark with sample, weigh, and record weight, *S*. Fill empty container to mark with water, weigh, and record weight, *R*. Measure all masses to the nearest 10 mg.

b. Procedure B: If sample does not flow readily, add as much of it to container as possible without exerting pressure, record volume, weigh, and record mass, *P*. Fill container to mark with distilled water, taking care that air bubbles are not trapped in the sludge or container. Weigh and record mass, Q. Measure all masses to nearest 10 mg.

4. Calculation

a. Procedure A:

sp. gr. T C/4 C

$$= \frac{\text{weight of sample}}{\text{weight of equal volume of water at 4 C}}$$

$$= \frac{S - W}{R - W} \times F$$

The values of the temperature correction factor F are given in Table 213:I.

b. Procedure B:

sp. gr. T C/4 C

$$= \frac{\text{weight of sample}}{\text{weight of equal volume of water at 4 C}}$$

$$= \frac{(P - W)}{(R - W) - (Q - P)} \times F$$

TABLE 213:I. TEMPERATURE CORRECTION FACTOR

Temperature C	Temperature Correction Factor
15	0.9991
20	0.9982
25	0.9975
30	0.9957
35	0.9941
40	0.9922
45	0.9903

213 F. References

1. DICK, R.I. & P.A. VESILIND. 1969. The SVI—what is it? *J. Water Pollut. Control Fed.* 41:1285.
2. FINCH, J. & H. IVES. 1950. Settleability indexes for activated sludge. *Sewage Ind. Wastes* 22:833.
3. DICK, R.I. 1972. Sludge treatment. *In* W.J. Weber, ed., Physicochemical Processes for Water Quality Control. Wiley-Interscience, New York, N.Y.
4. DICK, R.I. & K.W. YOUNG. 1972. Analysis of thickening performance of final settling tanks. *Proc. 27th Ind. Waste Conf.*, Purdue Univ., Eng. Ext. Ser. No. 141, 33.
5. VESILIND, P.A. 1975. Treatment and Disposal of Wastewater Sludges. Ann Arbor Science Publishing Co., Ann Arbor, Mich.
6. VESILIND, P.A. 1968. Discussion of Evaluation of activated sludge thickening theories. *J. San. Eng. Div., Proc. Amer. Soc. Civil Eng.* 94: SA1, 185.

213 G. Bibliography

DONALDSON, W. 1932. Some notes on the operation of sewage treatment works. *Sewage Works J.* 4:48.

MOHLMAN, F.W. 1934. The sludge index. *Sewage Works J.* 6:119.

RUDOLFS, W. & I.O. LACY. 1934. Settling and compacting of activated sludge. *Sewage Works J.* 6:647.

UMBREIT, W.W., R.H. BURRIS & J.F. STAUFFER. 1964. Manometric Techniques. Burgess Publishing Co. Minneapolis, Minn.

DICK, R.I. & R.B. EWING. 1967. Evaluation of activated sludge thickening theories. *J. San. Eng. Div., Proc. Amer. Soc. Civil Eng.* 93:SA4, 9.

DICK, R.I. 1969. Fundamental aspects of sedimentation I & II. *Water Wastes Eng.* 6:2 & 3:47 and 45.

DICK, R.I. 1970. Role of activated sludge final settling tanks. *J. San. Eng. Div., Proc. Amer. Soc. Civil Eng.* 96:SA2, 423.

214 TURBIDITY

Clarity of water is important in producing products destined for human consumption and in many manufacturing uses. Beverage producers, food processors, and treatment plants drawing on a surface water supply commonly rely on coagulation, settling, and filtration to insure an acceptable product. The clarity of a natural body of water is a major determinant of the condition and productivity of that system.

Turbidity in water is caused by suspended matter, such as clay, silt, finely divided organic and inorganic matter, soluble colored organic compounds, and plankton and other microscopic organisms. Turbidity is an expression of the optical property that causes light to be scattered and absorbed rather than transmitted in straight lines through the sample. Correlation of turbidity with the weight concentration of suspended matter is difficult because the size, shape, and refractive index of the particulates also affect the light-scattering properties of the suspension.

The standard method for determination of turbidity has been based on the Jackson candle turbidimeter; however, the lowest turbidity value that can be measured directly on this instrument is 25 units. Because turbidities of treated water generally fall within the range of 0 to 1 unit, indirect secondary methods also are required to estimate turbidity. Unfortunately, no instrument yet devised will duplicate the results obtained on the Jackson candle turbidimeter for all samples. Because of fundamental differences in optical systems, the results obtained with different types of secondary instruments frequently will not check closely with one another, even though the instruments are precalibrated against the candle turbidimeter.

Most commercial turbidimeters available for measuring low turbidities give comparatively good indications of the intensity of light scattered in one particular direction, predominantly at right angles to the incident light. These nephelometers are unaffected relatively by small changes in design parameters and therefore are specified as the standard instrument for measurement of low turbidities. Nonstandard turbidimeters, such as forward-scattering devices, are more sensitive than nephelometers to the presence of larger particles and are useful for process monitoring.

A further cause of discrepancies in turbidity analysis is the use of suspensions of different types of particulate matter for the preparation of instrumental calibration curves. Like water samples, prepared suspensions have different optical properties depending on the particle size distributions, shapes, and refractive indices. A standard reference suspension having reproducible light-scattering properties is specified for nephelometer calibration.

Because there is no direct relationship between the intensity of light scattered at 90 deg and Jackson candle turbidity, there is no valid basis for the practice of calibrating a nephelometer in terms of candle units. To distinguish between turbidities derived from nephelometric and visual methods, report the results from the former as nephelometric turbidity units (NTU) and from the latter as Jackson turbidity units (JTU).

1. Selection of Method

Its greater precision, sensitivity, and applicability over a wide turbidity range make the nephelometric method preferable to visual methods. The candle turbidimeter, with a lower limit of 25 turbidity units, has its principal usefulness in examining highly turbid waters. The bottle standards offer a practical means for checking raw and conditioned water at various stages of the treatment process.

2. Storage of Sample

Determine turbidity on the day the sample is taken. If longer storage is unavoidable, store samples in the dark for up to 24 hr. Do not store for long periods because irreversible changes in turbidity may occur. Vigorously shake all samples before examination.

214 A. Nephelometric Method—Nephelometric Turbidity Units

1. General Discussion

a. Principle: This method is based on a comparison of the intensity of light scattered by the sample under defined conditions with the intensity of light scattered by a standard reference suspension under the same conditions. The higher the intensity of scattered light, the higher the turbidity. Formazin polymer is used as the reference turbidity standard suspension. It is easy to prepare and is more reproducible in its light-scattering properties than clay or turbid natural water. The turbidity of a specified concentration of formazin suspension is defined as 40 nephelometric units. This suspension has an approximate turbidity of 40 Jackson units when measured on the candle turbidimeter; therefore, nephelometric turbidity units based on the formazin preparation will approximate units derived from the candle turbidimeter but will not be identical to them.

b. Interference: Turbidity can be determined for any water sample that is free of debris and rapidly settling coarse sediments. Dirty glassware, the presence of air bubbles, and the effects of vibrations that disturb the surface visibility of the sample will give false results. "True color," that is, water color due to dissolved substances that absorb light, causes measured turbidities to be low. This effect generally is not significant in the case of treated water.

2. Apparatus

a. Turbidimeter consisting of a nephelometer with a light source for illuminating the sample and one or more photoelectric detectors with a readout device to indicate intensity of light scattered at 90 deg to the path of incident light. Use a turbidimeter designed so that little stray light reaches the detector in the absence of turbidity and free from significant drift after a short warmup period. The sensitivity of the instrument should permit detecting turbidity differences of 0.02 NTU or less in waters having turbidity of less than 1 NTU with a range from 0 to 40 NTU. Several ranges are necessary to obtain both adequate coverage and sufficient sensitivity for low turbidities.

Differences in turbidimeter design will cause differences in measured values for turbidity even though the same suspension is used for calibration. To minimize such differences, observe the following design criteria:

1) Light source—Tungsten-filament lamp operated at a color temperature between 2,200 and 3,000 K.

2) Distance traversed by incident light and scattered light within the sample tube—Total not to exceed 10 cm.

3) Angle of light acceptance by detector—Centered at 90 deg to the incident light path and not to exceed ±30 deg from 90 deg. The detector, and filter system if used, shall have a spectral peak response between 400 and 600 nm.

b. Sample tubes, clear colorless glass. Keep tubes scrupulously clean, both inside and out, and discard when they become scratched or etched. Never handle them where the light strikes them. Use tubes with sufficient extra length, or with a protective case, so that they may be handled properly. Fill tubes with samples and standards that have been agitated thoroughly and allow sufficient time for bubbles to escape.

3. Reagents

a. Turbidity-free water: Turbidity-free water is difficult to obtain. The following method is satisfactory for measuring turbidity as low as 0.02 NTU.

Pass distilled water through a membrane filter having precision-sized holes of 0.2 μm*; the usual membrane filter used for bacteriological examinations is not satisfactory. Rinse collecting flask at least twice with filtered water and discard the next 200 mL.

Some commercial bottled demineralized waters are nearly particle-free. These may be used when their turbidity is lower than can be achieved in the laboratory. Dilute samples to a turbidity not less than 1 with distilled water.

b. Stock turbidity suspension:

1) Solution I—Dissolve 1.000 g hydrazine sulfate, $(NH_2)_2 \cdot H_2SO_4$, in distilled water and dilute to 100 mL in a volumetric flask.

2) Solution II—Dissolve 10.00 g hexamethylenetetramine, $(CH_2)_6N_4$, in distilled water and dilute to 100 mL in a volumetric flask.

3) In a 100-mL volumetric flask, mix 5.0 mL Solution I and 5.0 mL Solution II. Let stand 24 hr at 25 ± 3 C, dilute to mark, and mix. The turbidity of this suspension is 400 NTU.

4) Prepare solutions and suspensions monthly.

c. Standard turbidity suspension: Dilute 10.00 mL stock turbidity suspension to 100 mL with turbidity-free water. Prepare weekly. The turbidity of this suspension is defined as 40 NTU.

d. Dilute turbidity standards: Dilute portions of standard turbidity suspension with turbidity-free water as required. Prepare weekly.

4. Procedure

a. Turbidimeter calibration: Follow the manufacturer's operating instructions. In the absence of a precalibrated scale, prepare calibration curves for each range of the instrument. Check accuracy of any supplied calibration scales on a precalibrated instrument by using appropriate standards. Run at least one standard in each instrument range to be used. Make certain that turbidimeter gives stable readings in all sensitivity ranges used. High turbidities determined by direct measurement are likely to differ appreciably from those determined by the dilution technic, ¶ 4*c*.

b. Measurement of turbidities less than 40 NTU: Thoroughly shake sample. Wait until air bubbles disappear and pour sample into turbidimeter tube. When possible, pour shaken sample into turbidimeter tube and immerse it in an ultrasonic bath for 1 to 2 sec, causing complete bubble release. Read turbidity directly from instrument scale or from appropriate calibration curve.

c. Measurement of turbidities above 40 NTU: Dilute sample with one or more volumes of turbidity-free water until turbidity falls between 30 and 40 NTU. Compute turbidity of original sample from turbidity of diluted sample and the dilution factor. For example, if five volumes of turbidity-free water were added to one volume of sample and the diluted sample showed a turbidity of 30 NTU, then the turbidity of the original sample was 180 NTU.

d. Calibrate continuous turbidity monitors for low turbidities by determining turbidity of the water entering or leaving

*Nucleopore Corporation, 7035 Commerce Circle, Pleasanton, Calif., or equivalent.

them, using a laboratory-model turbidimeter. When this is not possible, use an appropriate dilute turbidity standard, ¶ 3*d*. For turbidities above 40 NTU use undiluted stock solution.

5. Calculation

$$\text{Nephelometric turbidity units (NTU)} = \frac{A \times (B + C)}{C}$$

where:

A = NTU found in diluted sample,
B = volume of dilution water, mL, and
C = sample volume taken for dilution, mL.

6. Interpretation of Results

a. Report turbidity readings as follows:

Turbidity Range *NTU*	Report to the Nearest *NTU*
0–1.0	0.05
1–10	0.1
10–40	1
40–100	5
100–400	10
400–1,000	50
>1,000	100

b. For comparison of water treatment efficiencies estimate turbidity more closely than is specified above. Uncertainties and discrepancies in turbidity measurements make it unlikely that two or more laboratories will duplicate results on the same sample more closely than specified.

214 B. Visual Methods—Jackson Turbidity Units

1. General Discussion

a. Principle: Turbidity measurements by the candle turbidimeter are based on the light path through a suspension that just causes the image of the flame of a standard candle to disappear—that is, to become indistinguishable against the general background illumination—when the flame is viewed through the suspension. The longer the light path, the lower the turbidity.

b. Interference: Turbidity can be determined for any water sample that is free of rapidly settling debris and coarse sediments. Dirty glassware, the presence of air bubbles, and the effects of vibrations that disturb the surface visibility of the sample give false results.

2. Apparatus

a. Candle turbidimeter consisting of a glass tube calibrated according to Table 214:I, a standard candle, and a support that aligns candle and tube. The glass tube and candle are supported in a vertical position so that the center line of the tube passes through the center line of the candle. The candle is supported by a spring-loaded cylinder designed to keep the top of the candle pressed against the top of the support as the candle gradually burns away. The top of the support for the candle is 7.6 cm below the bottom of the glass tube. The glass tube has a flat, polished optical-glass bottom and conforms to specifications for nessler tubes given in Section 102.6. It is graduated to read directly in JTU. Keep tube clean and free from scratches. Keep most of glass tube enclosed within a metal tube when observations are being made, both to protect against breakage and to exclude extraneous light.

Use a candle made of beeswax and spermaceti, designed to burn within the limits of 114 to 126 grains/hr. To insure uniform results, keep flame as near constant size and constant distance from bottom of glass tube as possible by frequently trimming charred portion of the wick and making sure that the candle is pushed to the top of its support. Eliminate all drafts during measurements to prevent flame from flickering. Do not burn candle for more than a few minutes at a time because the flame tends to increase in size. Before lighting candle each time, remove any portions of charred wick that can be broken off easily when manipulated with the fingers.

b. Bottles for visual comparison: A matched set of 1-L-capacity, glass-stoppered bottles made of borosilicate or other resistant glass.

TABLE 214:I. GRADUATION OF CANDLE TURBIDIMETER

Light Path* *cm*	Jackson Turbidity Units *JTU*	Light Path* *cm*	Jackson Turbidity Units *JTU*
2.3	1,000	11.4	190
2.6	900	12.0	180
2.9	800	12.7	170
3.2	700	13.5	160
3.5	650	14.4	150
3.8	600	15.4	140
4.1	550	16.6	130
4.5	500	18.0	120
4.9	450	19.6	110
5.5	400	21.5	100
5.6	390	22.6	95
5.8	380	23.8	90
5.9	370	25.1	85
6.1	360	26.5	80
6.3	350	28.1	75
6.4	340	29.8	70
6.6	330	31.8	65
6.8	320	34.1	60
7.0	310	36.7	55
7.3	300	39.8	50
7.5	290	43.5	45
7.8	280	48.1	40
8.1	270	54.0	35
8.4	260	61.8	30
8.7	250	72.9	25
9.1	240		
9.5	230		
9.9	220		
10.3	210		
10.8	200		

* Measured from inside bottom of glass tube.

3. Preparation of Standard Suspensions

a. Turbidity-free water: See Section 214A.3*a*.

b. Visual comparison standards: Prepare from natural turbid water or kaolin.

1) Natural water—For best results, prepare from natural turbid water from the same source as that to be tested. Determine turbidity with candle turbidimeter, then dilute portions of suspension to turbidity values desired.

Weekly prepare suspensions of turbidities below 25 units by diluting a freshly checked, more concentrated suspension.

2) Kaolin—Add approximately 5 g kaolin to 1 L distilled water, thoroughly agitate, and let stand for 24 hr. Withdraw supernatant without disturbing sediment. Determine turbidity with candle turbidimeter. Dilute to turbidity values desired. Preserve standard suspensions by adding 1 g $HgCl_2$/L suspension. Shake suspensions vigorously before each reading and check monthly with candle turbidimeter.

4. Procedure

a. Estimation with candle turbidimeter:

1) Turbidities between 25 and 1,000 JTU—Light trimmed candle, pour a minimum of 10 mL well-shaken, bubble-free sample into cool Jackson tube, and place in turbidimeter. Slowly add sample in small increments until flame image just disappears from view. Make certain that a uniformly illuminated field with no bright spots materializes. After the image has been made to disappear, remove enough sample, up to 20%, to make flame image visible again. Then add small increments of removed sample to approach the end point carefully. Keep glass tube clean on both inside and outside and avoid scratching the glass. The tube must be free of soot and condensed moisture. Remove tube immediately after final reading is obtained. Never place cold water in tube hot to the touch nor put an empty tube in turbidimeter.

2) Turbidities exceeding 1,000 JTU—Dilute sample with one or more volumes of turbidity-free water until turbidity falls below 1,000 JTU. Compute turbidity of original sample from turbidity of diluted sample and dilution factor. For example, if five volumes of turbidity-free water were added to one volume of sample and the diluted sample showed a turbidity of 500 JTU, the turbidity of the original sample was 3,000 JTU.

b. Estimation with bottle standards: In the range of 5 to 100 JTU, compare shaken samples with standard suspensions made by diluting concentrated standard suspensions with turbidity-free water in known ratios. Place sample and standards in bottles of the same size, shape, and type; leave enough empty space at top of each bottle to permit shaking. Compare sample and standards through sides of bottles by looking through them at the same object and noting the distinctness with which such objects as ruled lines or newsprint can be seen. Arrange artificial lighting above or below bottles so that no direct light reaches the eye. Record sample turbidity as that of the standard producing the visual effect most closely approximating that of the sample.

5. Calculation

See Section 214A.5. Report results from visual methods as Jackson turbidity units, JTU.

6. Interpretation of Results

Record turbidity readings in the following manner:

Turbidity Range *JTU*	Record to Nearest *JTU*
25–40	1
40–100	5
100–400	10
400–700	50
700 or more	100

Identify the visual method, i.e., candle turbidimeter or bottle standards.

214 C. Bibliography

WHIPPLE, G.C. & D.D. JACKSON. 1900. A comparative study of the methods used for the measurement of turbidity of water. *Mass. Inst. Technol. Quart.* 13:274.

AMERICAN PUBLIC HEALTH ASSOCIATION. 1901. Report of Committee on Standard Methods of Water Analysis. *Pub. Health Papers & Rep.* 27:377.

WELLS, P.V. 1922. Turbidimetry of water. *J. Amer. Water Works Ass.* 9:488.

BAYLIS, J.R. 1926. Turbidimeter for accurate measurement of low turbidities. *Ind. Eng. Chem.* 18:311.

WELLS, P.V. 1927. The present status of turbidity measurements. *Chem. Rev.* 3:331.

BAYLIS, J.R. 1933. Turbidity determinations. *Water Works Sewage* 80:125.

ROSE, H.E. & H.B. LLOYD. 1946. On the measurement of the size characteristics of powders by photo-extinction methods. *J. Soc. Chem. Ind.* (London) 65:52 (Feb.); 65:55 (Mar.).

ROSE, H.E. & C.C.J. FRENCH. 1948. On the extinction coefficient: Particle size relationship for fine mineral powders. *J. Soc. Chem. Ind.* (London) 67:283.

GILLETT, T.R., P.F. MEADS & A.L. HOLVEN. 1949. Measuring color and turbidity of white sugar solutions. *Anal. Chem.* 21:1228.

JULLANDER, I. 1949. A simple method for the measurement of turbidity. *Acta Chem. Scand.* 3:1309.

ROSE, H.E. 1950. Powder-size measurement by a combination of the methods of nephelometry and photo-extinction. *J. Soc. Chem. Ind.* (London) 69:266.

ROSE, H.E. 1950. The design and use of photoextinction sedimentometers. *Engineering* 169:350, 405.

BRICE, B.A., M. HALWER & R. SPEISER. 1950. Photoelectric light-scattering photometer for determining high molecular weights. *J. Opt. Soc. Amer.* 40:768.

KNIGHT, A.G. 1950. The measurement of turbidity in water. *J. Inst. Water Eng.* 4:449.

HANYA, T. 1950. Study of suspended matter in water. *Bull. Chem. Soc. Jap.* 23:216.

JULLANDER, I. 1950. Turbidimetric investigations on viscose. *Svensk Papperstidn.* 22:1.

ROSE, H.E. 1951. A reproducible standard for the calibration of turbidimeters. *J. Inst. Water Eng.* 5:310.

AITKEN, R.W. & D. MERCER. 1951. Comment on "The measurement of turbidity in water." *J. Inst. Water Eng.* 5:328.

ROSE, H.E. 1951. The analysis of water by the assessment of turbidity. *J. Inst. Water Eng.* 5:521.

KNIGHT, A.G. 1951. The measurement of turbidity in water: A reply. *J. Inst. Water Eng.* 5:633.

STAATS, F.C. 1952. Measurement of color, turbidity, hardness and silica in industrial waters. Preprint 156, Amer. Soc. Testing & Materials, Philadelphia, Pa.

PALIN, A.T. 1955. Photometric determination of the colour and turbidity of water. *Water Water Eng.* 59:341.

SLOAN, C.K. 1955. Angular dependence light scattering studies of the aging of precipitates. *J. Phys. Chem.* 59:834.

CONLEY, W.R. & R.W. PITMAN. 1957. Microphotometer turbidity analysis. *J. Amer. Water Works Ass.* 49:63.

PACKHAM, R.F. 1962. The preparation of turbidity standards. *Proc. Soc. Water Treat. Exam.* 11:64.

BAALSRUD, K. & A. HENRIKSEN. 1964. Measurement of suspended matter in stream water. *J. Amer. Water Works Ass.* 56:1194.

HOATHER, R.C. 1964. Comparison of different methods for measurement of turbidity. *Proc. Soc. Water Treat. Exam.* 13:89.

EDEN, G.E. 1965. The measurement of turbidity in water. A progress report on the work of the analytical panel. *Proc. Soc. Water Treat. Exam.* 14:27.

BLACK, A.P. & S.A. HANNAH. 1965. Measurement of low turbidities. *J. Amer. Water Works Ass.* 57:901.

HANNAH, S.A., J.M. COHEN & G.G. ROBECK. 1967. Control techniques for coagulation-filtration. *J. Amer. Water Works Ass.* 59:1149.

REBHUN, M. & H.S. SPERBER. 1967. Optical properties of diluted clay suspensions. *J. Colloid Interface Sci.* 24:131.

DANIELS, S.L. 1969. The utility of optical parameters in evaluation of processes of flocculation and sedimentation. *Chem. Eng. Progr. Symp. Ser.* No. 97, 65:171.

LIVESEY, P.J. & F.W. BILLMEYER, JR. 1969. Particle-size determination by low-angle light scattering: new instrumentation and a rapid method of interpreting data. *J. Colloid. Interface Sci.* 30:447.

OSTENDORF, R.G. & J.F. BYRD. 1969. Modern monitoring of a treated industrial effluent. *J. Water Pollut. Control Fed.* 41:89.

EICHNER, D.W. & C.C. HACH. 1971. How clear is clear water? *Water Sewage Works* 118:299.

HACH, C.C. 1972. Understanding turbidity

measurement. *Ind. Water Eng.* 9:18, No. 2.

SIMMS, R.J. 1972. Industrial turbidity measurement. *ISA Tran.* 11:146, No. 2.

TALLEY, D.G., J.A. JOHNSON & J.E. PILZER. 1972. Continuous turbidity monitoring. *J. Amer. Water Works Ass.* 64:184.

PART 300

DETERMINATION OF METALS

301 INTRODUCTION

The effects of metals in water and wastewater range from beneficial through troublesome to dangerously toxic. Some metals are essential, others may adversely affect water consumers, wastewater treatment systems, and receiving waters. Some metals may be either beneficial or toxic, depending on their concentrations.

Metals may be determined satisfactorily by instrumental or colorimetric methods. The instrumental methods are preferred because they are rapid and matrix effects often are controllable without extensive separations. Colorimetric methods are applicable when interferences are known to be within the capacity of the particular method. Preliminary treatment of samples often is required. Appropriate pretreatment methods are described for each type of analysis.

1. Definition of Terms

a. Filtrable (dissolved) metals: Those constituents (metals) of an unacidified sample that pass through a 0.45-μm membrane filter.

b. Nonfiltrable (suspended) metals: Those constituents (metals) of an unacidified sample that are retained by a 0.45-μm membrane filter.

c. Total metals: The concentration of metals determined on an unfiltered sample after vigorous digestion, or the sum of the concentrations of metals in both filtrable and nonfiltrable fractions.

d. Acid-extractable metals: The concentration of metals in solution after treatment of an unfiltered sample with hot dilute mineral acid.

To determine filtrable and nonfiltrable metals, filter immediately after sample collection. Do not preserve with acid until after filtration.

2. Sampling and Sample Preservation

Before collecting a sample, decide on the metal fraction to be analyzed (filtrable, nonfiltrable, total, or acid-extractable). This decision will determine in part whether the sample is acidified with or without filtration and the type of digestion required.

Serious errors may be introduced during sampling and storage because of (*a*) contamination from sampling device, (*b*) failure to remove residues of previous samples from sample container, and (*c*) loss of metals by adsorption on and/or precipitation in sample container by failure to acidify the sample properly where required. The best sample containers are made of quartz or TFE. Because these containers are expensive, the preferred sample container is made of linear polyethylene with a polyethylene cap. Borosilicate glass containers also may be used, but avoid soft glass containers for samples containing metals in the microgram-per-liter range. Store silver samples in a light-absorbing container.

Preserve samples immediately after sampling by acidifying with concentrated nitric acid (HNO_3) to pH <2. Usually 1.5 mL conc HNO_3/L sample (or 3 mL 1 + 1 HNO_3/L sample) is sufficient for short-term preservation. For samples with high buffer capacity, increase amount of acid (5 mL may be required for some alkaline or highly buffered samples). Use commercially available high-purity acid* or prepare high-purity acid by sub-boiling distillation of acid.

*Ultrex, J.T. Baker, or equivalent.

After acidifying sample, store it in a refrigerator at approximately 4 C to prevent change in volume due to evaporation. Under these conditions, samples with metal concentrations of several milligrams per liter are stable for up to 6 months (except mercury, for which the limit is 38 days in glass and 14 days in plastic). For microgram-per-liter metal levels, analyze as soon as possible after sample collection.

Alternatively, preserve samples for mercury analysis by adding 2 mL/L of 20% (w/v) $K_2Cr_2O_7$ solution (prepared in 1 + 1 HNO_3). Store in a refrigerator not contaminated with mercury (CAUTION: Mercury may increase in samples stored in plastic bottles in mercury-contaminated laboratories).

3. General Precautions

Avoid introducing contaminating metals from containers and distilled water. Thoroughly clean sample containers with a detergent solution, rinse with tap water, soak in acid, and then rinse with metal-free water. For quartz, TFE, or glass materials, use 1 + 1 HNO_3, 1 + 1 HCl, or aqua regia (conc HCl + conc HNO_3) for soaking. For plastic material, use 1 + 1 HNO_3 or 1 + 1 HCl. Reliable soaking conditions are 24 hr at 70 C. Chromic acid may be used to remove organic deposits from containers, but rinse containers thoroughly with water to remove traces of chromium. Do not use chromic acid for plastic containers or if analysis for chromium is to be made. Always use metal-free water in analysis and reagent preparation (see 303A.2*c*). In these methods, the word "water" means metal-free water.

For analysis of microgram-per-liter concentrations of metals, airborne contaminants in the form of volatile compounds, dust, soot, and aerosols present in laboratory air may become significant. To avoid contamination use "clean laboratory" facilities such as commercially available laminar-flow clean-air benches or custom-designed work stations.

4. Bibliography

ROBERTSON, D.E. 1965. The absorption of trace elements in sea water on various container surfaces. *Anal. Chim. Acta* 42:533.

ROBERTSON, D.E. 1968. Role of contamination in trace metal analysis of seawater. *Anal. Chem.* 40:1067.

MITCHELL, J.W. 1973. Ultrapurity in trace analysis. *Anal. Chem.* 45:492A.

STRUEMPLER, A.W. 1973. Adsorption characteristics of silver, lead, calcium, zinc and nickel on borosilicate glass, polyethylene and polypropylene container surfaces. *Anal. Chem.* 45:2251.

FELDMAN, C. 1974. Preservation of dilute mercury solutions. *Anal. Chem.* 46:99.

KING, W.G., J.M. RODRIGUEZ & C.M. WAI. 1974. Losses of trace concentrations of cadmium from aqueous solution during storage in glass containers. *Anal. Chem.* 46:771.

BATLEY, G.E. & D. GARDNER. 1977. Sampling and storage of natural waters for trace metal analysis. *Water Res.* 11:745.

SUBRAMANIAN, K.S., C.L. CHAKRABARTI, J.E. SUETIAS & I.S. MAINES. 1978. Preservation of some trace metals in samples of natural waters. *Anal. Chem.* 50:444.

302 PRELIMINARY TREATMENT OF SAMPLES

Samples containing particulates or organic material require pretreatment before analysis. This section describes general pretreatment for samples in which metals are to be determined according to Sections 303 through 328 with several exceptions. The special digestion technics for mercury are given in Sections 303F.3*b* and *c*, and those for arsenic and selenium in Section 303E.I.4*b*.

302 A. Preliminary Filtration

If filtrable (dissolved) metals are to be determined, filter sample through a 0.45-μm-pore-diam membrane filter on collection, acidify filtrate to pH ≤ 2 with conc nitric acid (HNO_3), and analyze directly. If a precipitate forms on acidification, digest acidified filtrate before analysis. Soak membrane filters in approximately 0.5*N* hydrochloric acid (HCl) and rinse with water before use.

302 B. Preliminary Treatment for Acid-Extractable Metals

Because extractable metals are lightly adsorbed on particulate material, use rigidly controlled conditions to obtain meaningful and reproducible results. At collection, acidify entire sample with 5 mL conc HNO_3/L sample. To prepare sample, mix well, transfer 100 mL to a beaker or flask, and add 5 mL 1 + 1 high-purity HCl. Heat 15 min on a steam bath. Filter through a 0.45-μm-pore-diam membrane filter, adjust filtrate volume to 100 mL with water, and analyze. Results represent extractable metals.

302 C. Preliminary Digestion for Metals

"Total metals" includes all metals, inorganically and organically bound, and both dissolved and particulate. To analyze for total metals, digest sample without preliminary filtration. To analyze for filtrable metals, digest filtrate or analyze it directly. To determine nonfiltrable (suspended) metals, digest filter and material on it. Run a filter blank to obtain a blank correction.

Convert metals associated with organic matter and particulates to a form (usually the free metal) that can be determined by atomic absorption spectroscopy. Use one of the digestion technics presented below. Optimally, determine recovery of specific metals by each of the digestion technics described below for the specific type of sample being tested using a standard addition technic. Use the least complicated or least rigorous digestion method compatible with the analytical method and the metal being analyzed. For flame atomic absorption analysis, chlorides or perchlorates are the preferred matrix; nitrate is acceptable; avoid sulfates whenever possible. For electrothermal atomic absorption (304) nitrates are the preferred matrix; generally avoid chlorides, perchlorates, and sulfates for most metals. For Ag and Th, omit chloride, while for Sb, Ru, and Sn its addition is preferred. In the analysis of Ti, add H_2SO_4 after HNO_3 digestion to dissolve titanium oxide. For total As and Se determinations use the special digestion technic given in 303E.I.4*b*. For total Hg use the digestion technic in 303F.3*b*. For Ag analysis, if adsorption to container walls or formation of AgCl is suspected, make sample basic with conc NH_4OH and add 1 mL cyanogen iodide (CNI) solution per 100 mL sample (see 303A.2*k*). Mix sample and let stand for 1 hr before proceeding with analysis.

Where compatible with method of analysis and metal being analyzed, use HNO_3

digestion for analysis of alkali and alkaline earth metals in relatively clean water or highly treated wastewaters; use HNO_3-H_2SO_4 digestion for samples containing readily-oxidized organic matter; use HNO_3-$HClO_4$ digestion for samples containing difficult-to-oxidize organic matter. Ash samples containing large amounts of organic matter. For sludges with high or refractory organic matter contents use digestion technic 302H. Acid digestion eliminates interferences from cyanide, nitrite, sulfide, sulfite, thiosulfate, and thiocyanate.

Report digestion technic used.

Most reagent-grade acids contain trace amounts of metals. Use acids with total iron and heavy-metal impurities less than 0.0001%. If the metallic impurities exceed 0.0001%, the blank values will be too large and variable for precise determination of metals. With specified amounts of acids, 0.0001% of heavy-metal impurity will add about 0.03 mg heavy metals to a sample. For acid digestion, use of purified-grade acid* may be necessary. Check certification label of the acid for content of metal(s) of interest to determine need for high-purity acid.

Always prepare acid blanks. Experience indicates that a blank made with the same acids and subjected to the same digestion procedure as the sample can correct for impurities present in acids, reagent water, or potentially present on glassware.

*Utrex, J.T. Baker, or equivalent.

302 D. Nitric Acid Digestion

1. Apparatus

Hot plate: A 30- × 50-cm heating surface is adequate.

2. Reagents

Nitric acid, HNO_3, conc.

3. Procedure

Mix sample and transfer a suitable volume (50 to 100 mL) to a beaker. Add 5 mL conc HNO_3. Bring to a slow boil and evaporate on a hot plate to the lowest volume possible (about 15 to 20 mL) before precipitation or salting-out occurs. Add 5 mL conc HNO_3, cover with a watch glass, and heat to obtain a gentle refluxing action. Continue heating and adding conc HNO_3 as necessary until digestion is complete as shown by a light-colored, clear solution.

Add 1 to 2 mL conc HNO_3 and warm slightly to dissolve any remaining residue. Wash down beaker walls and watch glass with water and then filter. Transfer filtrate to a 100-mL volumetric flask with two 5-mL portions of water, adding these rinsings to the volumetric flask. Cool, dilute to mark, and mix thoroughly. Take portions of this solution for required metal determinations.

302 E. Nitric Acid-Sulfuric Acid Digestion

1. Apparatus

a. Hot plate: A 30- × 50-cm heating surface is adequate.

b. Sintered-glass filter crucibles, fine porosity, with holder. Gooch crucibles with glass fiber filters may be used.

c. Conical flasks, 125-mL, acid-washed and rinsed with water.

d. Evaporating dishes or casseroles, glass or TFE.

2. Reagents

a. Methyl orange indicator solution.

b. Nitric acid, HNO_3, conc.

c. Hydrogen peroxide, H_2O_2, 30%.

d. Sulfuric acid, H_2SO_4, conc.

3. Procedure

Mix sample and pipet a suitable volume (as determined from the tabulation below) into an evaporating dish or casserole. (If the volume required exceeds 250 mL, add portions as the sample evaporates.) Acidify to methyl orange with conc H_2SO_4 and add 5 mL conc HNO_3 and 2 mL 30% H_2O_2. Evaporate on a steam bath or hot plate to 15 to 20 mL. Cover with a watch glass if necessary to avoid sample loss by spattering. An infrared lamp placed over the sample speeds evaporation.

Concentration *mg/L*	Volume *mL*
1	1,000
1–10	100
10–100	10
100–1,000	1

Transfer concentrate and any precipitate to a 125-mL conical flask using 5 mL conc HNO_3. Add 10 mL conc H_2SO_4 and a few boiling chips, glass beads, or Hengar granules (not suitable for selenium analysis). Evaporate on a hot plate in a hood until dense white fumes of SO_3 just appear. If solution does not clear, add 10 mL conc HNO_3 and repeat evaporation to fumes of SO_3. Remove all HNO_3 before continuing treatment. All HNO_3 will be removed when the solution is clear and no brownish fumes are evident.

Cool and dilute to about 50 mL with water. Heat to almost boiling to dissolve slowly soluble salts. Filter, then complete procedure as directed in Section 302D.3 beginning with, "Transfer filtrate . . ."

302 F. Nitric Acid-Perchloric Acid Digestion

1. Apparatus

a. Hot plate: A 30- × 50-cm heating surface is adequate.

b. Safety shield.

c. Safety goggles.

d. Conical flasks, 125 mL, acid-washed and rinsed with water.

e. Sintered-glass filter crucibles, fine porosity, with holder. Gooch crucibles with glass fiber filters may be used.

2. Reagents

a. Nitric acid, HNO_3, conc.

b. Perchloric acid, $HClO_4 \cdot 2H_2O$, purchased as 70 to 72% $HClO_4$, reagent grade.

c. Sodium hydroxide, NaOH, 6*N*.

d. Methyl orange indicator aqueous solution.

e. Ammonium acetate solution: Dissolve 500 g $NH_4C_2H_3O_2$ in 600 mL water.

3. Procedure

CAUTION—*Heated mixtures of $HClO_4$ and organic matter may explode violently. Avoid this hazard by taking the following precautions: (a) do not add $HClO_4$ to a hot solution containing organic matter; (b) always pretreat samples containing organic matter with HNO_3 before adding $HClO_4$; (c) use a mixture of HNO_3 and $HClO_4$ to start digestion even if organic matter is absent; and (d) avoid repeated fuming* *with $HClO_4$ in ordinary hoods. For routine operations, use a water pump attached to a glass fume eradicator.* * *Stainless steel fume hoods with adequate water wash-down facilities are available commercially and are acceptable for use with $HClO_4$; and (e) never let samples being digested with $HClO_4$ evaporate to dryness.*

Measure sample into a 125-mL conical flask. Acidify to methyl orange with conc HNO_3, add 5 mL conc HNO_3 more, and evaporate on a steam bath to 15 to 20 mL. Add 10 mL each of conc HNO_3 and $HClO_4$, cooling flask between additions. Add a few boiling chips, heat on a hot plate, and evaporate gently until dense white fumes of $HClO_4$ just appear. If solution is not clear, cover flask neck with a watch glass and keep solution just boiling until it clears. Never let $HClO_4$ digestate go to dryness. If necessary, add 10 mL conc HNO_3 to complete digestion. Cool, dilute to about 50 mL with water, and boil to expel any chlorine or oxides of nitrogen. Filter through a sintered-glass filter crucible or a glass-fiber filter into a clean filter flask. Rinse sample flask with two 5-mL portions of water, passing them through the crucible to wash any residue on the filter. (A filtering device that allows direct filtrate collection in a 100-mL volumetric flask is desirable.)

Proceed as directed in 303D.3 beginning with, "Transfer filtrate. . ."

If lead is to be determined in the presence of high amounts of sulfate (e.g., analysis of Pb in power plant fly ash samples), dissolve $PbSO_4$ on the filter as follows: add 50 mL ammonium acetate solution to conical flask in which digestion was carried out and heat to incipient boiling. Rotate flask occasionally to wet all interior surfaces and dissolve any deposited residue. Reconnect filter and slowly draw solution through it. Transfer filtrate to a 100-mL volumetric flask, cool, dilute to mark, mix thoroughly, and set aside for determination of lead.

*Such as those obtainable from G.F. Smith Chemical Co., Columbus, Ohio.

302 G. Dry Ashing

1. Apparatus

See Sections 209A.2 and 209B.2.

2. Procedure

Evaporate a sample of suitable size to dryness on a steam bath in a platinum or high-silica glass* evaporating dish. Transfer dish to a muffle furnace and heat sample to a white ash. If volatile elements are to be determined, keep temperature at 400 to 450 C. If sodium only is to be determined, ash sample at a temperature up to 600 C. Dissolve ash in a minimum quantity of conc HNO_3 and warm water. Filter diluted sample and adjust to a known volume, preferably so that the final HNO_3 concentration is about 1%.

Take portions of this solution for metals determination.

*Vycor, a product of Corning Glass Works, Corning, N.Y., or equivalent.

302 H. Digestion of Sludge with High or Refractory Organic Content

1. Apparatus

a. Hot plate: A 30- × 50-cm heating surface is adequate.

b. Conical flasks, 250 mL, acid-washed and rinsed with water.

c. Evaporating dishes or casseroles, glass or TFE.

2. Reagents

a. Nitric acid, HNO_3, conc and 1 + 1.

b. Perchloric acid, $HClO_4 \cdot 2H_2O$, 70 to 72% $HClO_4$.

c. Hydrofluoric acid, HF, 48 to 51%, reagent grade.

3. Procedure

Transfer a sample of suitable volume (see 302E.3) to an evaporating dish or casserole. Evaporate on a steam bath to 15 to 20 mL. Add 12 mL conc HNO_3 and evaporate on a hot plate to near dryness following the precautions given in 302D.3. Repeat HNO_3 addition and evaporation.

Using 25 mL 1 + 1 HNO_3, transfer residue to a 250-mL conical flask (use 10 mL to transfer residue and 15 mL to rinse dish). Add 20 mL $HClO_4$ and boil until nearly dry or until solution is clear and white fumes of $HClO_4$ have appeared. (See precautions for using $HClO_4$ in 302F.) Cool, add about 50 mL water, filter, and proceed as directed in 302D beginning with, "Transfer filtrate . . ." NOTE: *Use of a few drops of HF aids digestion of siliceous materials. Handle HF with extreme care and provide adequate safety measures.*

303 METALS BY FLAME ATOMIC ABSORPTION SPECTROPHOTOMETRY

Because requirements for determining metals by atomic absorption spectrophotometry vary with metal and/or concentration to be determined, the method is divided into six sections, as follows:

A. Determination of antimony, bismuth, cadmium, calcium, cesium, chromium, cobalt, copper, gold, iridium, iron, lead, lithium, magnesium, manganese, nickel, platinum, potassium, rhodium, ruthenium, silver, sodium, strontium, thallium, tin, and zinc by direct aspiration into an air-acetylene flame.

B. Determination of low concentrations of cadmium, chromium, cobalt, copper, iron, lead, manganese, nickel, silver, and zinc by chelation with ammonium pyrrolidine dithiocarbamate (APDC), extraction into methyl isobutyl ketone (MIBK), and aspiration into an air-acetylene flame.

C. Determination of aluminum, barium, beryllium, molybdenum, osmium, silicon, thorium, titanium, and vanadium by direct aspiration into a nitrous oxide-acetylene flame.

D. Determination of low concentrations of aluminum and beryllium by chelation with 8-hydroxyquinoline, extraction into MIBK, and aspiration into a nitrous oxide-acetylene flame.

E. Determination of arsenic and selenium by conversion to their hydrides and aspiration into an argon-hydrogen or nitrogen-hydrogen flame.

F. Determination of mercury by the cold vapor technic.

1. General Discussion

a. Principle: Atomic absorption spectrophotometry resembles emission flame photometry in that a sample is aspirated into a flame and atomized. The major difference is that in flame photometry the amount of light emitted is measured, whereas in atomic absorption spectrophotometry a light beam is directed through the flame, into a monochromator, and on to a detector that measures the amount of light absorbed by the atomized element in the flame. For many metals difficult to analyze by flame emission, atomic absorption exhibits superior sensitivity. Because each metal has its own characteristic absorption wavelength, a source lamp composed of that element is used; this makes the method relatively free from spectral or radiation interferences. The amount of energy of the characteristic wavelength absorbed in the flame is proportional to the concentration of the element in the sample. The flame emission method produces good results for analysis of sodium, potassium, and strontium. Most atomic absorption instruments also are equipped for operation in an emission mode.

b. Interference: Many metals can be determined by direct aspiration of sample into an air-acetylene flame. The most troublesome type of interference is termed "chemical" and results from the lack of absorption by atoms bound in molecular combination in the flame. This can occur when the flame is not hot enough to dissociate the molecules or when the dissociated atom is oxidized immediately to a compound that will not dissociate further at the flame temperature. The interference of phosphate in the magnesium determination can be overcome by adding lanthanum. Similarly, introduction of calcium eliminates silica interference in the determination of manganese. However, silicon and metals such as aluminum, barium, beryllium, and vanadium require the higher-temperature, nitrous oxide-acetylene flame to dissociate their molecules.

MIBK extractions with APDC (see 303B) are particularly useful where the salt matrix interferes, for example, in seawater. This procedure also concentrates the sample so that the detection limits are extended.

Barium and other metals ionize in the flame, thereby reducing the ground state (potentially absorbing) population. The addition of an excess of a cation (sodium, potassium, or lithium) having a similar or lower ionization potential will overcome this problem. The wavelength of maximum absorption for arsenic is 193.7 nm and for selenium 196.0 nm—wavelengths at which the air-acetylene flame absorbs intensely. The sensitivity of the method for these metals can be improved by using the nitrogen-hydrogen flame.

Nonatomic absorption caused by molecular absorption and/or light scattering by solid particles in the flame also may lead to erroneous results when trace amounts of an element are determined. This interference can be corrected by using a "continuum source corrector". Two types of continuum sources commonly are used with atomic absorption spectrophotometers: a hydrogen-filled hollow cathode lamp with a metal cathode or a deuterium arc lamp. Subtracting absorption measurements obtained with the continuum source from those obtained with the hollow cathode line source yields the true atomic absorption of the sample.

c. Preparation of standards: Prepare standard solutions of known metal concentrations in water with a matrix similar to the sample. Use standards that bracket expected sample concentration and are within the method's working range. If very dilute standards are needed for a period longer than 1 day, prepare stock solutions in concentrations greater than 500 mg/L. Store stock solutions in a refrigerator and dilute as needed. Stock standard solutions can be obtained from several commercial sources. They also can be prepared from National Bureau of Standards (NBS) reference materials or by procedures outlined in the following sections.

For samples containing high and variable concentrations of matrix materials, make the major ions in the sample and the dilute standard similar. If the sample matrix is complex and components cannot be matched accurately with standards, use the method of standard additions to correct for matrix effects. Ensure that the calibration curve is linear over the concentration range of interest and that no nonatomic absorption is present (see ¶ *b,* above). Do this by preparing several portions of sample solution to each of which is added a different amount of analyte, making sure that the total volumes of all solutions are identical. Prepare a blank solution of similar matrix, measure response of each sample, and plot a calibration curve. Extrapolate to zero addition to give concentration in the sample. If digestion is used, carry standards through the same digestion procedure used for samples.

d. Sensitivity, detection limits, and optimum concentration ranges for determination: The sensitivity of flame atomic absorption spectrophotometry is defined as the metal concentration that produces an absorption of 1% (approximately an absorbance of 0.0044). The detection limit is defined as the concentration that produces absorption equivalent to twice the magnitude of the background fluctuation. Sensitivity and detection limits vary with the instrument, the element analyzed, and the technic selected. The optimum concentration range usually starts from the concentration of several times the sensitivity and extends to the concentration at which the calibration curve starts to flatten. To achieve best results, use concentrations of samples and standards within the optimum concentration range of the spectrophotometer. See Table 303:I for indication of concentration ranges measurable with conventional atomization. In many instances the concentration range shown in Table 303:I may be extended downward with scale expansion and upward by using a less sensitive wavelength or by rotating the burner.

To determine low concentrations of easily volatilized elements, use a microsampling cup or boat system to improve detection limits. Place sample in cup or boat-shaped vessel and dry. Insert vessel into an air-acetylene flame burning on a three-slot burner head. Only a small sample (from several hundred microliters to 1 mL) is required. Insure that sample is completely dry before introducing it into the flame. For dry samples prepared in an organic solvent, prevent solvent ignition by keeping them at a distance of about 5 to 10 cm from the flame. Because the vessels never reach the flame temperature, interferences may be more common; therefore use the standard addition method or standards with a matrix similar to the samples.

2. Apparatus

a. Atomic absorption spectrophotometer, consisting of a light source emitting the line spectrum of an element (hollow cathode lamp or electrodeless discharge lamp), a device for vaporizing the sample (usually a flame), a means of isolating an absorption line (monochromator or filter and adjustable slit), and a photoelectric detector with its associated electronic amplifying and measuring equipment. Both direct current (DC) and alternating current (AC) systems are used in atomic absorption instruments. The AC or chopped-beam system is preferred because with this system flame emission can be distinguished from lamp emission. For waters high in salt, or for the 200- to 300-nm range, preferably use either a continuum source background corrector or a double-

TABLE 303:I. ATOMIC ABSORPTION CONCENTRATION RANGES WITH CONVENTIONAL FLAME ATOMIZER

Element	Wavelength *nm*	Flame Gases*	Detection Limit *mg/L*	Sensitivity *mg/L*	Optimum Concentration Range *mg/L*
Ag	328.1	A–Ac	0.01	0.06	0.1–4
Al	309.3	N–Ac	0.1	1	5–100
As†	193.7	N–H	0.002	—	0.002–0.02
Au	242.8	A–Ac	0.01	0.25	0.5–20
Ba	553.6	N–Ac	0.03	0.4	1–20
Be	234.9	N–Ac	0.005	0.03	0.05–2
Bi	223.1	A–Ac	0.06	0.4	1–50
Ca	422.7	A–Ac	0.003	0.08	0.2–20
Cd	228.8	A–Ac	0.002	0.025	0.05–2
Co	240.7	A–Ac	0.03	0.2	0.5–10
Cr	357.9	A–Ac	0.02	0.1	0.2–10
Cs	852.1	A–Ac	0.02	0.3	0.5–15
Cu	324.7	A–Ac	0.01	0.1	0.2–10
Fe	248.3	A–Ac	0.02	0.12	0.3–10
Hg	253.6	A–Ac	0.2	7.5	10–300
Ir	264.0	A–Ac	0.6	8	—
K	766.5	A–Ac	0.005	0.04	0.1–2
Li	670.8	A–Ac	0.002	0.04	0.1–2
Mg	285.2	A–Ac	0.0005	0.007	0.02–2
Mn	279.5	A–Ac	0.01	0.05	0.1–10
Mo	313.3	N–Ac	0.1	0.5	1–20
Na	589.0	A–Ac	0.002	0.015	0.03–1
Ni	232.0	A–Ac	0.02	0.15	0.3–10
Os	290.9	N–Ac	0.08	1	—
Pb	283.3	A–Ac	0.05	0.5	1–20
Pt	265.9	A–Ac	0.1	2	5–75
Rh	343.5	A–Ac	0.5	0.3	—
Ru	349.9	A–Ac	0.07	0.5	—
Sb	217.6	A–Ac	0.07	0.5	1–40
Se†	196.0	N–H	0.002	—	0.002–0.02
Si	251.6	N–Ac	0.3	2	5–150
Sn	224.6	A–Ac	0.8	4	10–200
Sr	460.7	A–Ac	0.03	0.15	0.3–5
Ti	365.3	N–Ac	0.3	2	5–100
V	318.4	N–Ac	0.2	1.5	2–100
Zn	213.9	A–Ac	0.005	0.02	0.05–2

* A–Ac = air-acetylene; N–Ac = nitrous oxide-acetylene; N–H = nitrogen-hydrogen.
† Gaseous hydride method.

beam instrument that permits simultaneous measurement of absorption at two different wavelengths.

b. Burner: The most common type of burner is a premix, which introduces the spray into a condensing chamber for removal of large droplets. The burner may be fitted with a conventional head containing a single slot, which is most useful for aspiration when organic solvents are used; a three-slot Boling head, which may be preferred for direct aspiration with an air-acetylene flame; or a special head for use with nitrous oxide and acetylene.

c. Recorder: Most instruments are equipped with either a digital or null meter readout mechanism. A good-quality 10-mV recorder with high sensitivity and a fast response time is needed to record the peaks resulting from the determination of As and Se by aspiration of their gaseous hydrides.

d. Lamps: Use either a hollow-cathode lamp or an electrodeless discharge lamp (EDL). Use one lamp for each element being measured. Multi-element hollow-cathode lamps generally provide lower sensitivity than single-element lamps. For determining As or Se by aspiration of their gaseous hydrides, use EDL's for better sensitivity. EDL's take a longer time to warm up and stabilize.

e. Pressure-reducing valves: Maintain supplies of fuel and oxidant at pressures somewhat higher than the controlled operating pressure of the instrument by suitable reducing valves. Use a separate reducing valve for each gas.

f. Vent: Place a vent about 15 to 30 cm above the burner to remove fumes and vapors from the flame. This precaution protects laboratory personnel from toxic vapors, protects the instrument from corrosive vapors, and prevents flame stability from being affected by room drafts. A damper or variable-speed blower is desirable for modulating air flow and preventing flame disturbance. Select blower size to provide the air flow recommended by the instrument manufacturer. In laboratory locations with heavy particulate air pollution, use clean laboratory facilities (¶ 301.3).

3. Precision and Accuracy

Some data typical of the precision and accuracy obtainable with the methods discussed are presented in Table 303:II.

TABLE 303:II. PRECISION AND ACCURACY DATA FOR ATOMIC ABSORPTION METHODS*

Metal	Metal Concentration $\mu g/L$	Relative Standard Deviation %	Relative Standard Deviation %
Direct determination:			
Barium	500	10.0	8.6
Cadmium	50	21.6	8.2
Chromium	50	26.4	2.3
Copper	1,000	11.2	3.4
Iron	300	16.5	0.6
Magnesium	200	10.5	6.3
Manganese	50	13.5	6.0
Silver	550	17.5	10.6
Zinc	500	8.2	0.4
Extracted samples:			
Aluminum	300	22.2	0.7
Beryllium	5	34.0	20.0
Cadmium	50	43.8	13.3
Lead	50	23.5	19.0
Flameless:			
Mercury	0.4	21.2	2.4
As hydride:			
Arsenic	10	6.0	1.0
Selenium	10	11.0	0.0

* Data from: Water Metals No. 4, Study No. 30. 1968. U.S. Public Health Service Publ. No. 999-UIH-8, Dept. Health, Education and Welfare, Cincinnati, Ohio.

4. Bibliography

COON, E., J.E. PETLEY, M.H. MCMULLEN & S.E. WIBERLEY. 1953. Fluorometric determination of aluminum by use of 8 quinolinol. *Anal. Chem.* 25:608.

ALLAN, J.E. 1961. The use of organic solvents in atomic absorption spectrophotometry. *Spectrochim. Acta* 17:467.

WILLIS, J.B. 1962. Determination of lead and other heavy metals in urine by atomic absorption spectrophotometry. *Anal. Chem.* 34:614.

WILLIS, J.B. 1965. Nitrous oxide-acetylene flame in atomic absorption spectroscopy. *Nature* 207:715.

SLAVIN, W. 1968. Atomic Absorption Spectroscopy. John Wiley and Sons, New York, N.Y.

RAMIRIZ-MUNOZ, J. 1968. Atomic Absorption Spectroscopy and Analysis by Atomic Absorption Flame Photometry. American Elsevier Publishing Co., New York, N.Y.

KAHN, H.L. 1968. Principles and Practice of Atomic Absorption. Advan. Chem. Ser. No. 73, Washington, D.C.

HATCH, W.R. & W.L. OTT. 1968. Determination of sub-microgram quantities of mercury by atomic absorption spectrophotometry. *Anal. Chem.* 40:2085.

SACHDEV, S.L. & P.W. WEST. 1970. Concentration of trace metals by solvent extraction and their determination by atomic absorption spectrophotometry. *Environ. Sci. Technol.* 4:749.

UTHE, J.F., F.A.J. ARMSTRONG & M.P. STAINTON. 1970. Mercury determination in fish samples by wet digestion and flameless atomic absorption spectrophotometry. *J. Fish. Res. Board Can.* 27:805.

FERNANDEZ, F.J. & D.C. MANNING. 1971. The determination of arsenic at sub-microgram levels by atomic absorption spectrophotometry. *Atomic Absorption Newsletter* 10:86.

PAUS, P.E. 1971. The application of atomic absorption spectroscopy to the analysis of natural waters. *Atomic Absorption Newsletter* 10:86.

MANNING, D.C. 1971. A high sensitivity arsenic-selenium sampling system for atomic absorption spectroscopy. *Atomic Absorption Newsletter* 10:123.

High Sensitivity Arsenic Determination by Atomic Absorption. 1971. Jarrel-Ash Atomic Absorption Applications Laboratory Bull. No. As-3.

KOPP, J.F., M.C. LONGBOTTOM & L.B. LOBRING. 1972. "Cold vapor" method for determining mercury. *J. Amer. Water Works Ass.* 64:20.

PAUS, P.E. 1973. Determination of some heavy metals in seawater by atomic absorption spectroscopy. *Fresenius Zeitschr. Anal. Chem.* 264:118.

CALDWELL, J.S., R.J. LISHKA & E.F. MCFARREN. 1973. Evaluation of a low cost arsenic and selenium determination of microgram per liter levels. *J. Amer. Water Works Ass.* 65:71.

EDIGER, R.D. 1973. A review of water analysis by atomic absorption. *Atomic Absorption Newsletter* 12:151.

BURRELL, D.C. 1975. Atomic Spectrometric Analysis of Heavy-Metal Pollutants in Water. Ann Arbor Science Publishers, Inc., Ann Arbor, Mich.

303 A. Determination of Antimony, Bismuth, Cadmium*, Calcium, Cesium, Chromium*, Cobalt*, Copper, Gold, Iridium, Iron*, Lead*, Lithium, Magnesium, Manganese*, Nickel*, Platinum, Potassium, Rhodium, Ruthenium, Silver*, Sodium, Strontium, Thallium, Tin, and Zinc* by Direct Aspiration into an Air-Acetylene Flame

1. Apparatus

Atomic absorption spectrophotometer and associated equipment: See Section 303.2. Use burner head recommended by the manufacturer.

*For low concentrations of Cd, Cr, and Pb (<50, 200, and 500 μg/L respectively) and Co, Fe, Mn, Ni, Ag, and Zn, see Section 303B.

2. Reagents

a. Air, cleaned and dried through a suitable filter to remove oil, water, and other foreign substances. The source may be a compressor or commercially bottled gas.

b. Acetylene, standard commercial grade. Acetone, which always is present in acetylene cylinders, can be prevented from entering and damaging the burner head by replacing a cylinder when its pressure has fallen to 689 kPa acetylene.

c. Metal-free water: Use metal-free water for preparing all reagents and calibration standards and as dilution water. Prepare metal-free water by deionizing tap water and/or by using one of the following processes, depending on the metal concentration in the sample: single distillation, redistillation, or sub-boiling. Always check deionized or distilled water to determine whether the element of interest is present in trace amounts. (CAUTION: If the source water contains Hg or other volatile metals, deionized and single- or redistilled water may not be suitable for trace analysis because these metals distill over with the distilled water. In such cases, use sub-boiling to prepare metal-free water).

d. Calcium solution: Dissolve 630 mg calcium carbonate, $CaCO_3$, in 50 mL of 1 + 5 HCl. If necessary, heat and boil gently to obtain complete solution. Cool and dilute to 1,000 mL with water.

e. Hydrochloric acid, HCl, conc.

f. Lanthanum solution: Dissolve 58.65 g lanthanum oxide, La_2O_3, in 250 mL conc HCl. Add acid slowly until the material is dissolved and dilute to 1,000 mL with water.

g. Hydrogen peroxide, 30%.

h. Nitric acid, HNO_3, conc.

i. Aqua regia: Add 3 volumes conc HCl to 1 volume conc HNO_3.

j. Iodine solution, 1*N*: Dissolve 20 g potassium iodide, KI, in 50 mL water, add 12.7 g iodine, and dilute to 100 mL.

k. Cyanogen iodide (CNI) solution: To 50 mL water add 6.5 g potassium cyanide, KCN, 5.0 mL 1*N* iodine solution, and 4.0 mL conc NH_4OH. Mix and dilute to 100 mL with water. Prepare fresh solution every 2 wk.

l. Standard metal solutions: Prepare a series of standard metal solutions in the optimum concentration range by appropriate dilution of the following stock metal solutions with water containing 1.5 mL conc HNO_3/L. Thoroughly dry reagents before use. In general, use reagents of the highest purity. For hydrates, use fresh reagents.

1) *Antimony:* Dissolve 2.7426 g antimony potassium tartrate hemihydrate (analytical reagent grade), $K(SbO)C_4H_4O_6 \cdot {}^1/_2H_2O$, in 1,000 mL water; 1.00 mL = 1.00 mg Sb.

2) *Bismuth:* Dissolve 1.000 g bismuth metal in a minimum volume of 1 + 1 HNO_3. Dilute to 1,000 mL with 2% (v/v) HNO_3; 1.00 mL = 1.00 mg Bi.

3) *Cadmium:* Dissolve 1.000 g cadmium metal in a minimum volume of 1 + 1 HCl. Dilute to 1,000 mL with water; 1.00 mL = 1.00 mg Cd.

4) *Calcium:* To 2.4972 g $CaCO_3$ add 50 mL water and add dropwise a minimum volume of conc HCl (about 10 mL) to complete solution. Dilute to 1,000 mL with water; 1.00 mL = 1.00 mg Ca.

5) *Cesium:* Dissolve 1.267 g cesium chloride, CsCl, in 1,000 mL water; 1.00 mL = 1.00 mg Cs.

6) *Chromium:* Dissolve 2.828 g anhydrous potassium dichromate, $K_2Cr_2O_7$, in about 200 mL water, add 1.5 mL conc HNO_3, and dilute to 1,000 mL with water; 1.00 mL = 1.00 mg Cr.

7) *Cobalt:* Dissolve 1.407 g cobaltic oxide, Co_2O_3, in 20 mL hot conc HCl. Cool and dilute to 1,000 mL with water; 1.00 mL = 1.00 mg Co.

8) *Copper:* Dissolve 1.000 g copper metal in 15 mL of 1 + 1 HNO_3 and dilute to 1,000 mL with water; 1.00 mL = 1.00 mg Cu.

9) *Gold:* Dissolve 0.1000 g gold metal in a minimum volume of aqua regia. Evaporate to dryness, dissolve residue in 5 mL conc HCl, cool, and dilute to 100 mL with water; 1.00 mL = 1.00 mg Au.

10) *Iridium:* Dissolve 1.147 g ammonium chloroiridate, $(NH_4)_2IrCl_6$, in a minimum volume of 1% (v/v) HCl and dilute to 100

mL with 1% (v/v) HCl; 1.00 mL = 5.00 mg Ir.

11) *Iron:* Dissolve 1.000 g iron wire in 50 mL of 1 + 1 HNO_3 and dilute to 1,000 mL with water; 1.00 mL = 1.00 mg Fe.

12) *Lead:* Dissolve 1.598 g lead nitrate, $Pb(NO_3)_2$, in about 200 mL water, add 1.5 mL conc HNO_3, and dilute to 1,000 mL with water; 1.00 mL = 1.00 mg Pb.

13) *Lithium:* Dissolve 5.324 g lithium carbonate, Li_2CO_3, in a minimum volume of 1 +1 HCl and dilute to 1,000 mL with water; 1.00 mL = 1.00 mg Li.

14) *Magnesium:* Dissolve 4.952 g magnesium sulfate, $MgSO_4$, in 200 mL water, add 1.5 mL conc HNO_3, and dilute to 1,000 mL with water; 1.00 mL = 1.00 mg Mg.

15) *Manganese:* Dissolve 3.076 g manganous sulfate, $MnSO_4 \cdot H_2O$, in about 200 mL water, add 1.5 mL conc HNO_3, and dilute to 1,000 mL with water; 1.00 mL = 1.00 mg Mn.

16) *Nickel:* Dissolve 1.273 g nickel oxide, NiO, in a minimum volume of 10% (v/v) HCl and dilute to 1,000 mL with water; 1.00 mL = 1.00 mg Ni.

17) *Platinum:* Dissolve 0.1000 g platinum metal in a minimum volume of aqua regia and evaporate just to dryness. Add 5 mL conc HCl and 0.1 g NaCl and again evaporate just to dryness. Dissolve residue in 20 mL of 1 + 1 HCl and dilute to 100 mL with water; 1.00 mL = 1.00 mg Pt.

18) *Potassium:* Dissolve 1.907 g potassium chloride, KCl, in water and make up to 1,000 mL; 1.00 mL = 1.00 mg K.

19) *Rhodium:* Dissolve 0.412 g ammonium hexachlororhodate, $(NH_4)_3Rh\ Cl_6 \cdot 1.5\ H_2O$, in a minimum volume of 10% (v/v) HCl and dilute to 100 mL with 10% (v/v) HCl; 1.00 mL = 1.00 mg Rh.

20) *Ruthenium:* Dissolve 0.2052 g ruthenium chloride, $RuCl_3$, in a minimum volume of 20% (v/v) HCl and dilute to 100 mL with 20% (v/v) HCl; 1.00 mL = 1.00 mg Ru.

21) *Silver:* Dissolve 1.575 g silver nitrate, $AgNO_3$, in water, add 1.5 mL conc HNO_3, and make up to 1,000 mL; 1.00 mL = 1.00 mg Ag.

22) *Sodium:* Dissolve 2.542 g sodium chloride, NaCl, dried at 140 C, in water and make up to 1,000 mL; 1.00 mL = 1.00 mg Na.

23) *Strontium:* Dissolve 2.415 g strontium nitrate, $Sr(NO_3)_2$, in 1,000 mL of 1% (v/v) HNO_3; 1.00 mL = 1.00 mg Sr.

24) *Thallium:* Dissolve 1.303 g thallium nitrate, $TlNO_3$, in water. Add 10 mL conc HNO_3 and dilute to 1,000 mL with water; 1.00 mL = 1.00 mg Tl.

25) *Tin:* Dissolve 1.000 g tin metal in 100 mL conc HCl and dilute to 1,000 mL with water; 1.00 mL = 1.00 mg Sn.

26) *Zinc:* Dissolve 1.000 g zinc metal in 20 mL 1 + 1 HCl and dilute to 1,000 mL with water; 1.00 mL = 1.00 mg Zn.

3. Procedure

a. Instrument operation: Because of differences between makes and models of atomic absorption spectrophotometers, it is not possible to formulate instructions applicable to every instrument. See manufacturer's operating manual. In general, proceed according to the following: Install a hollow cathode lamp for the desired metal in the instrument and roughly set the wavelength dial according to Table 303:I. Set slit width according to manufacturer's suggested setting for the element being measured. Turn on instrument, apply to the hollow cathode lamp the current suggested by the manufacturer, and let instrument warm up until energy source stabilizes, generally about 10 to 20 min. Readjust current as necessary after warmup. Optimize wavelength by adjusting wavelength dial until optimum energy gain is obtained. Align lamp in accordance with manufacturer's instructions.

Install suitable burner head and adjust burner head position. Turn on air and adjust flow rate to that specified by manufacturer to give maximum sensitivity for the metal being measured. Turn on acetylene, adjust flow rate to value specified, and ignite flame. Aspirate a standard solution and adjust aspiration rate of the nebulizer to obtain maximum sensitivity. Atomize a standard (usually one near the middle of the linear working range) and adjust burner both up and down and sideways to obtain maximum response. Record absorbance of this standard when freshly prepared and with a new hollow cathode lamp. Refer to these data on subsequent determinations of the same element to check consistency of instrument setup and aging of hollow cathode lamp and standard.

The instrument now is ready to operate. When analyses are finished, extinguish flame by turning off first acetylene and then air.

b. Standardization: Select at least three concentrations of each standard metal solution (prepared as in ¶2*l* above) to bracket the expected metal concentration of a sample. Aspirate each in turn into flame and record absorbance. For calcium and magnesium calibration, mix 100 mL of standard with 10 mL lanthanum solution (see ¶2*f* above) before aspirating. For chromium calibration mix 1 mL 30% H_2O_2 with each 100 mL chromium solution before aspirating. For iron and manganese calibration, mix 100 mL of standard with 25 mL calcium solution (¶ 2*d*) before aspirating.

Prepare a calibration curve by plotting on linear graph paper absorbance of standards versus their concentrations. For instruments equipped with direct concentration readout, this step is unnecessary. With some instruments it may be necessary to convert percent absorption to absorbance by using a table generally provided by the manufacturer. Plot calibration curves for calcium and magnesium based on original concentration of standards before dilution with lanthanum solution. Plot calibration curves for iron and manganese based on original concentration of standards before dilution with calcium solution. Plot calibration curve for chromium based on original concentration of standard before addition of H_2O_2.

Check standards periodically during a run. Recheck calibration curve by aspirating at least one standard after completing analysis of a group of samples. For instruments with built-in memory, enter one to three standards to register a calibration curve for use in subsequent sample analysis.

c. Analysis of samples: Rinse nebulizer by aspirating water containing 1.5 mL conc HNO_3/L. Atomize blank and zero instrument. Atomize sample and determine its absorbance.

When determining calcium or magnesium, dilute and mix 100 mL sample with 10 mL lanthanum solution (¶ 2*f*) before atomization. When determining iron or manganese, mix 100 mL with 25 mL of calcium solution (¶ 2*d*) before aspirating. When determining chromium, mix 1 mL 30% H_2O_2 with each 100 mL sample before aspirating.

Analyze standards at the beginning and end of a run and at intervals during longer runs. Run a blank or solvent between each sample or standard to verify baseline stability. Determine metal concentration from calibration curve.

4. Calculations

Calculate concentration of each metal ion, in micrograms per liter, by referring to the appropriate calibration curve prepared according to ¶3*b*.

303 B. Determination of Low Concentrations of Cadmium, Chromium, Cobalt, Copper, Iron, Lead, Manganese, Nickel, Silver, and Zinc by Chelation with Ammonium Pyrrolidine Dithiocarbamate (APDC) and Extraction into Methyl Isobutyl Ketone (MIBK)

1. Apparatus

a. Atomic absorption spectrophotometer and associated equipment: See Section 303.2.

b. Burner head, conventional. Consult manufacturer's operating manual for suggested burner head.

c. Separatory funnels, 250 mL, with TFE stopcocks.

2. Reagents

a. Air: See 303A.2*a*.

b. Acetylene: See 303A.2*b*.

c. Metal-free water: See 303A.2*c*.

d. Methyl isobutyl ketone (MIBK), reagent grade. For trace analysis, purify MIBK by redistillation or by sub-boiling distillation.

e. Ammonium pyrrolidine dithiocarbamate (APDC) solution: Dissolve 4 g APDC in 100 mL water. If necessary, purify APDC with an equal volume of MIBK. Shake 30 sec in a separatory funnel, let separate, and withdraw lower portion. Discard MIBK layer.

f. Nitric acid, HNO_3, conc, ultrapure.

g. Standard metal solutions: See 303A.2*l*.

h. Potassium permanganate solution, $KMnO_4$, 5% aqueous.

i. Sodium sulfate, Na_2SO_4, anhydrous.

j. Water-saturated MIBK: Mix one part purified MIBK with one part water in a separatory funnel. Shake 30 sec and let separate. Discard aqueous layer. Save MIBK layer.

3. Procedure

a. Instrument operation: See Section 303A.3*a*. After final adjusting of burner position, aspirate water-saturated MIBK into flame and gradually reduce fuel flow until flame is similar to that before aspiration of solvent.

b. Standardization: Select at least three standard metal solutions (prepared as in 303A.2*l*) to bracket expected sample metal concentration and to be, after extraction, in the optimum concentration range of the instrument. Adjust 100 mL of each standard and 100 mL of a metal-free water blank to pH 3 by adding 1*N* HNO_3 or 1*N* NaOH. For individual element extraction, use the following pH ranges to obtain optimum extraction efficiency:

Element	pH Range for Optimum Extraction
Ag	3–5 (complex unstable)
Cd	1–6
Co	2–10
Cr	3–9
Cu	0.1–8
Fe	2–5
Mn	2–4 (complex unstable)
Ni	2–4
Pb	0.1–6
Zn	2–6

Transfer each standard solution and blank to an individual 250-mL separatory funnel, add 1 mL APDC solution, and shake to mix. Add 10 mL MIBK and shake vigorously for 30 sec. (The maximum volume ratio of sample to MIBK is 40.) Let contents of each separatory funnel separate into aqueous and organic layers, drain off aqueous layer, and discard. Make sure that none of the aqueous layer remains in funnel stem. Drain organic layer into a 10-mL glass-stoppered graduated cylinder or glass-stoppered centrifuge tube. If a centrifuge tube is used, the extract can be centrifuged to remove entrained water.

Aspirate organic extracts directly into the flame (zeroing instrument on a water-saturated MIBK blank) and record absorbance. With some instruments it may be necessary to convert percent absorption to absorbance by using a table generally provided by the manufacturer.

Prepare a calibration curve by plotting on linear graph paper absorbances of extracted standards against their concentrations before extraction.

c. Analysis of samples: Rinse atomizer by aspirating water-saturated MIBK. Aspirate organic extracts treated as above directly into the flame and record absorbances.

With the above extraction procedure only hexavalent chromium is measured. To determine total chromium, oxidize trivalent chromium to hexavalent chromium by bringing sample to a boil and adding sufficient $KMnO_4$ solution dropwise to give a persistent pink color while the solution is boiled for 10 min. Cool extract and aspirate.

During extraction, if an emulsion forms at the water-MIBK interface, add anhydrous Na_2SO_4 to obtain a homogeneous organic phase.

The extraction period (partitioning time) for silver is critical; keep constant to within 30 sec.

To avoid problems associated with instability of extracted metal complexes, determine metals immediately after extraction.

4. Calculations

Calculate the concentration of each metal ion in micrograms per liter by referring to the appropriate calibration curve.

303 C. Determination of Aluminum*, Barium, Beryllium*, Molybdenum, Osmium, Silicon, Thorium, Titanium, and Vanadium, by Direct Aspiration into a Nitrous Oxide-Acetylene Flame

1. Apparatus

a. Atomic absorption spectrophotometer and associated equipment: See Section 303.2.

b. Nitrous oxide burner head: Use special burner head as suggested in manufacturer's manual. For a carbon rod, about every 20 min of operation, dislodge the carbon crust that forms along the slit surface.

c. T-junction valve or other switching valve for rapidly changing from nitrous oxide to air, so that flame can be turned on or off with air as oxidant to prevent flashbacks.

2. Reagents

a. Air: See 303A.2*a*.

b. Acetylene: See 303A.2*b*.

c. Metal-free water: See 303A.2*c*.

d. Hydrochloric acid, HCl, conc.

e. Nitric acid, HNO_3, conc.

f. Nitrous oxide, commercially available cylinders. Fit nitrous oxide cylinder with a special nonfreezable regulator or wrap a heating coil around an ordinary regulator to prevent flashback at the burner caused by reduction in nitrous oxide flow through a frozen regulator. (Some

*For low concentrations of Al and Be (<900 and 30 $\mu g/L$, respectively) see Section 303D.

atomic absorption instruments have automatic gas control systems that will safely shut down a nitrous oxide-acetylene flame in the event of a reduction in nitrous oxide flow rate.)

g. Potassium chloride solution: Dissolve 250 g KCl in water and dilute to 1,000 mL.

h. Aluminum nitrate solution: Dissolve 139 g $Al(NO_3)_3 \cdot 9H_2O$ in 150 mL water. Warm to dissolve completely. Cool and dilute to 200 mL.

i. Standard metal solutions: Prepare a series of standard metal solutions in the optimum concentration ranges by appropriate dilution of the following stock metal solutions with water containing 1.5 mL conc HNO_3/L:

1) *Aluminum:* Dissolve 1.000 g aluminum metal in 20 mL conc HCl by heating gently and diluting to 1,000 mL with water; alternatively dissolve 17.584 g aluminum potassium sulfate (also called potassium alum), $AlK(SO_4)_2 \cdot 12H_2O$ in 200 mL water, add 1.5 mL conc HNO_3, and dilute to 1,000 mL with water; 1.00 mL = 1.00 mg Al.

2) *Barium:* Dissolve 1.779 g barium chloride, $BaCl_2 \cdot 2H_2O$, in about 200 mL water, add 1.5 mL conc HNO_3, and dilute to 1,000 mL with water; 1.00 mL = 1.00 mg Ba.

3) *Beryllium:* Dissolve 20.76 g beryllium nitrate, $Be(NO_3)_2 \cdot 3H_2O$, in about 200 mL water, add 1.5 mL conc HNO_3, and dilute to 1,000 mL with water; 1.00 mL = 1.00 mg Be. CAUTION: *Beryllium nitrate is an extremely toxic and hazardous material. Preferably handle in a glove box using disposable gloves. If not available, handle in a hood with a face velocity of 45 m/sec. Use disposable gloves and laboratory clothing. Avoid inhalation and contact with skin.*

4) *Molybdenum:* Dissolve 1.500 g molybdenum oxide, MoO_3, in a minimum volume of 10% (v/v) HCl and dilute to 1,000 mL with water; 1.00 mL = 1.00 mg Mo.

5) *Osmium:* Obtain standard 0.1 *M* osmium tetroxide solution† and store in glass bottle; 1.00 mL = 1.90 mg Os. Make dilutions daily as needed using 1% (v/v) H_2SO_4. (CAUTION: *OsO_4 is extremely toxic and highly volatile.)*

6) *Silicon:* Dissolve 10.12 g sodium metasilicate, $Na_2SiO_3 \cdot 9H_2O$, in about 600 mL water. Neutralize to a litmus endpoint (or pH 5.0) with 1 + 1 HCl and make up to 1,000 mL with water; 1.00 mL = 0.10 mg Si.

7) *Thorium:* Dissolve 0.2380 g thorium nitrate, $Th(NO_3)_4 \cdot 4H_2O$ in 100 mL water; 1.00 mL = 1.00 mg Th.

8) *Titanium:* Dissolve 3.960 g pure (99.8 or 99.9%) titanium chloride, $TiCl_4$‡, in a mixture of equal volumes of 1*N* HCl and 1*N* HF. Make up to 1,000 mL with this acid mixture; 1.00 mL = 1.00 mg Ti.

9) *Vanadium:* Dissolve 2.296 g ammonium metavanadate, NH_4VO_3, in about 800 mL water, add 10 mL conc HNO_3, and dilute to 1,000 mL with water; 1.00 mL = 1.00 mg V.

3. Procedure

a. Instrument operation: See Section 303A.3*a*. After adjusting wavelength, install a nitrous oxide burner head. Turn on acetylene (without igniting flame) and adjust flow rate to value specified by manufacturer for a nitrous oxide-acetylene flame. Turn off acetylene. With both air and nitrous oxide supplies turned on, set T-junction valve to nitrous oxide and adjust flow rate according to manufacturer's specifications. Turn switching valve to the air position and verify that flow rate is the same. Turn acetylene on and ignite to a bright yellow flame. With a rapid motion,

†G. Frederick Smith Chemical Co., P.O. Box 23214, Columbus, Ohio 43223, Cat. No. 64, or equivalent.

‡Alpha Ventron, P.O. Box 299, 152 Andover St., Danvers, Mass. 01923, or equivalent.

turn switching valve to nitrous oxide. The flame should have a red cone above the burner. If it does not, adjust fuel flow to obtain red cone. After nitrous oxide flame has been ignited, let burner come to thermal equilibrium before beginning analysis.

Atomize water containing 1.5 mL conc HNO_3/L and check aspiration rate. Adjust if necessary to a rate between 3 and 5 mL/min. Atomize a standard of the desired metal with a concentration near the midpoint of the optimum concentration range and adjust burner (both sideways and vertically) in the light path to obtain maximum response. The instrument now is ready to run standards and samples.

To extinguish flame, turn switching valve from nitrous oxide to air and turn off acetylene. This procedure eliminates the danger of flashback that may occur on direct ignition or shutdown of nitrous oxide and acetylene.

b. Standardization: Select at least three standard metal solutions (prepared as in ¶ 2*i*) to bracket the expected metal concentration of a sample. Aspirate each in turn into the flame. Record absorbances. For aluminum, barium, and titanium, add 2 mL KCl solution to 100 mL standard before aspiration. For molybdenum and vanadium add 2 mL $Al(NO_3)_3 \cdot 9H_2O$ solution to 100 mL standard before aspiration.

With some instruments, it may be necessary to convert percent absorption to absorbance using a table generally provided by the manufacturer. Prepare a calibration curve by plotting on linear graph paper absorbance of standards versus concentration. Plot calibration curves for aluminum, barium, and titanium based on original concentration of standard before adding KCl solution. Plot calibration curves for molybdenum and vanadium based on original concentration of standard before adding $Al(NO_3)_3$ solution.

c. Analysis of samples: Rinse atomizer by aspirating water containing 1.5 mL conc HNO_3/L and zero instrument. Atomize a sample and determine its absorbance.

When determining aluminum, barium, and titanium, add 2 mL KCl solution to 100 mL sample before atomization. For molybdenum and vanadium, add 2 mL $Al(NO_3)_3 \cdot 9H_2O$ solution to 100 mL sample before atomization.

4. Calculations

Calculate concentration of each metal ion in micrograms per liter by referring to the appropriate calibration curve prepared according to ¶ 3*b*.

303 D. Determination of Low Concentrations of Aluminum and Beryllium by Chelation with 8-Hydroxyquinoline and Extraction into Methyl Isobutyl Ketone

1. Apparatus

a. Atomic absorption spectrophotometer and associated equipment: See Section 303.2.

b. Separatory funnels, 250 mL, with TFE stopcocks.

2. Reagents

a. Air: See 303A.2*a*.

b. Acetylene: See 303C.2*b*.

c. Buffer: Dissolve 300 g ammonium acetate, $NH_4C_2H_3O_2$ in water, add 105 mL conc NH_4OH, and dilute to 1 L.

d. Metal-free water: See 303A.2*c*.

e. Hydrochloric acid, HCl, conc.

f. 8-Hydroxyquinoline solution: Dissolve 20 g 8-hydroxyquinoline in about 200 mL water, add 60 mL glacial acetic acid, and dilute to 1 L with water.

g. Methyl isobutyl ketone: See 303B.2*d*.

h. Nitric acid, HNO_3, conc.

i. Nitrous oxide: See 303C.2*f*.

j. Standard metal solutions: Prepare a series of standard metal solutions containing 5 to 1,000 μg/L by appropriate dilution of the stock metal solutions prepared according to 303C.2*i*.

3. Procedure

a. Instrumentation operation: See Section 303A.3*a*. After final adjusting of burner position, aspirate MIBK into flame and gradually reduce fuel flow until flame is similar to that before aspiration of solvent. Adjust wavelength setting according to Table 303:I.

b. Standardization: Select at least three standard metal solutions (prepared as in ¶ 2*j* to bracket the expected metal concentration of a sample and transfer 100 mL of each (and 100 mL water blank) to four different 250-mL separatory funnels. Add 2 mL 8-hydroxyquinoline solution to each and shake. Add 10 mL buffer solution to each and shake again. Add 10.0 mL MIBK to each and shake vigorously. Continue as in Section 303B.3*b*.

c. Analysis of samples: Rinse atomizer by aspirating MIBK. Aspirate extracts of treated samples, and record absorbances.

4. Calculations

Calculate concentration of each metal in micrograms per liter by referring to the appropriate calibration curve prepared according to ¶ 3*b*.

303 E. Determination of Arsenic and Selenium by Conversion to Their Hydrides and Aspiration of the Gas into an Argon-Hydrogen or Nitrogen-Hydrogen Flame

General Discussion

Arsenic and selenium are reduced in acid solution and converted to their hydrides. The hydrides are swept directly by an argon or nitrogen stream into the atomizer of an atomic absorption spectrophotometer and burned in a hydrogen flame. Two reduction methods are given based on sodium borohydride reduction and zinc-stannous chloride reduction. CAUTION: *Both arsenic and selenium are toxic. Handle with care.*

I. SODIUM BOROHYDRIDE REDUCTION METHOD

1. Interferences

Interferences are minimized because the volatile hydrides are removed from solution and leave behind many potentially interfering substances. Slight reductions in peak height occur with H_2SO_4, HNO_3, and $HClO_4$ concentrations exceeding 1*M* and in the presence of large amounts of iron, antimony, copper, and tin. No interference in selenium measurements is obtained with arsenic concentrations as high as 250 μg As/L.

2. Apparatus

a. Atomic absorption spectrophotometer and associated equipment: See Section 303.2.

b. Flow meter, capable of measuring 1 L/min, such as that used for auxiliary nitrogen or argon*.

*Gilmont No. 12 or equivalent.

c. Syringe and needle, syringe with 10-mL capacity and needle with 22-gauge bore.

d. Reaction flask, a 50-mL pear-shaped vessel with side arms, both arms having ST 14/20 joint† or a 125-mL erlenmeyer flask with ST 14/20 ground-glass joint. Any commercially available system using sodium borohydride as a reductant is acceptable.

e. Special gas inlet-outlet tube: Construct from a micro cold finger condenser‡ by cutting off the portion below the ST 14/20 ground-glass joint.

f. Drying tube, 100-mm-long polyethylene tube filled with glass wool to keep particulate matter out of burner.

g. Apparatus setup: Connect apparatus with burner of spectrophotometer as shown in Figure 303:1. Connect outlet of reaction vessel to auxiliary oxidant input of burner with clear vinyl plastic§ tubing. Connect inlet of reaction vessel to outlet side of auxiliary oxidant (nitrogen supply) control valve of instrument.

3. Reagents

a. Sodium borohydride reducing solution: Dissolve 4 g $NaBH_4$ in 100 mL 10% (v/v) NaOH solution; this is sufficient for 20 determinations.

b. Sodium iodide prereductant solution: Dissolve 10 g NaI in 100 mL water; this is sufficient for 200 determinations.

c. Argon gas (or nitrogen gas), commercial grade.

d. Hydrogen gas, commercial grade.

e. Arsenic solutions:

1) *Stock arsenic solution:* Dissolve 1.321 g arsenous oxide, As_2O_3, in 100 mL water containing 4 g NaOH and dilute to 1,000 mL with water; 1.00 mL = 1.00 mg As.

2) *Intermediate arsenic solution:* Pipet 1 mL stock arsenic solution into a 100-mL volumetric flask and bring to volume with water containing 1.5 mL conc HNO_3/L; 1.00 mL = 10.0 μg As.

3) *Standard arsenic solution:* Pipet 10 mL intermediate arsenic solution into a 100-mL volumetric flask and bring to volume with water containing 1.5 mL conc HNO_3/L; 1.00 mL = 1.00 μg As.

f. Selenium solutions:

1) *Stock selenium solution:* Dissolve 1.000 g selenium in 5 mL conc HNO_3. Warm until reaction is complete and cautiously evaporate just to dryness. Dilute to 1,000 mL with water; 1.00 mL = 1.00 mg Se.

2) *Intermediate selenium solution:* Pipet 1 mL stock selenium solution into a 100-mL volumetric flask and bring to volume with water containing 1.5 mL conc HNO_3/L; 1.00 mL = 10.0 μg Se.

3) *Standard selenium solution:* Pipet 10 mL intermediate selenium solution into a 100-mL volumetric flask and bring to volume with water containing 1.5 mL conc HNO_3/L; 1.00 mL = 1.00 μg Se.

g. Perchloric acid, 70 to 72% $HClO_4$.

h. Hydrochloric acid, HCl, conc.

i. Sodium hydroxide, NaOH, analytical grade with lowest possible arsenic and selenium content.

j. Sulfuric acid: H_2SO_4, 1 + 1.

k. Nitric acid, HNO_3, conc.

l. Hydrogen peroxide, H_2O_2, 30%.

4. Procedure

a. Instrument operation: See Section 303A.3*a*. After adjusting wavelength, install appropriate burner head; e.g., Boling or air-propane. Turn on nitrogen (or argon) and adjust to a flow rate of about 8 L/min, with auxiliary nitrogen (or argon) flow at 1 L/min. Turn on hydrogen, adjust to a flow rate of about 7 L/min and ignite flame. The flame is essentially colorless; to determine whether it is ignited, carefully pass the hand about 30 cm above burner to detect heat emitted.

Atomize a standard solution (1.00 mL = 1.00 μg) of desired metal and adjust burner

†Scientific Glass JM-5835 or equivalent.
‡Scientific Glass JM-3325 or equivalent.
§Tygon or equivalent.

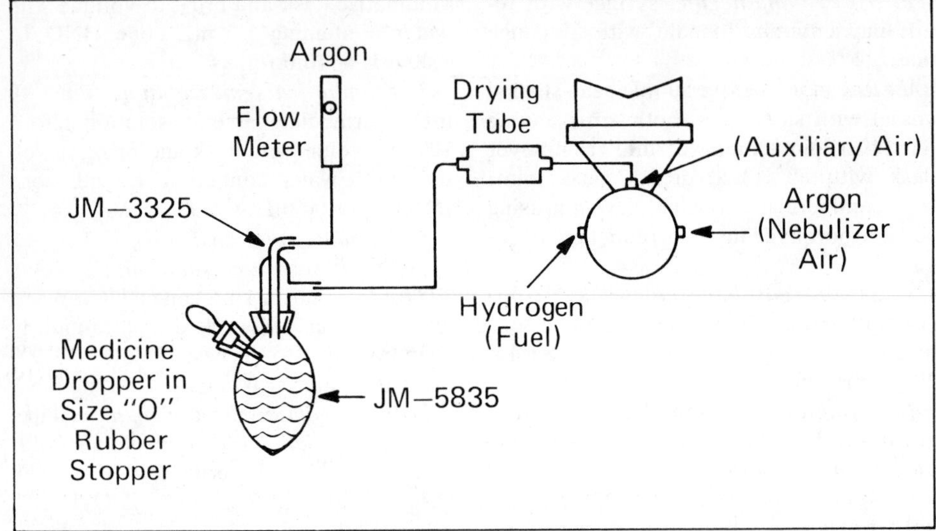

Figure 303:1. Schematic arrangement of equipment for determination of arsenic and selenium.

both sideways and vertically in the light path to obtain maximum response. The instrument is now ready to run standards and samples by the arrangement of Figure 303:1.

b. Sample preparation: For inorganic arsenic or selenium: To a 50-mL volumetric flask, add 25 mL sample, 20 mL conc HCl, and 5 mL 1 + 1 H_2SO_4. CAUTION: When high-chloride samples are digested, low As and Se recoveries may result.

For total (inorganic and organic) arsenic or selenium: To 50 mL sample in a 150-mL beaker, add 10 mL conc HNO_3 and 12 mL 1 + 1 H_2SO_4. Evaporate to SO_3 fumes (a volume of about 20 mL). To avoid loss of arsenic, maintain oxidizing conditions at all times by adding small amounts of HNO_3 whenever red-brown NO_2 fumes disappear. Cool slightly, add 25 mL water and 1 mL $HClO_4$, and again evaporate to SO_3 fumes. Cool, add 40 mL conc HCl, and bring to a volume of 100 mL with water.

c. Preparation of standards: Transfer 0, 0.5, 1.0, 1.5, and 2.0 mL standard arsenic or selenium solution to 100-mL volumetric flasks and bring to volume with water to obtain concentrations of 0, 5, 10, 15, and 20 μg/L arsenic or selenium.

d. Treatment of samples and standards: Transfer 20-mL portion of sample prepared as in ¶ *b* or standard prepared as in ¶ *c* to reaction vessel. For arsenic samples and standards add 0.5 mL NaI pre-reductant solution. Omit NaI for selenium determinations. Allow at least 1 min for the metal to be reduced to its lowest oxidation state. Attach reaction vessel to the special gas inlet-outlet glassware and fit a septum to sidearm. Flush system for at least 10 sec, then inject rapidly, but with uniform pressure, 5 mL $NaBH_4$ solution through the septum using a syringe and 22-gauge needle. Almost instantaneously the generated AsH_3 or SeH_2 is swept into the burner and an absorption signal recorded. Repeat for each standard and sample. Determine concentration from peak heights.

5. Calculations

Draw a standard curve by plotting peak heights of standards versus concentration of standards. Measure peak heights of samples and read concentration from curve. If sample was diluted (or concentrated), apply an appropriate factor.

II. ZINC—STANNOUS CHLORIDE REDUCTION METHOD

1. Apparatus

a. Atomic absorption spectrophotometer and associated equipment: See Section 303.2.

b. Flow meter: See 303E.I.2*b*.

c. Medicine dropper, capable of delivering 1.5 mL, fitted into a size "0" rubber stopper.

d. Reaction flask, a pear-shaped vessel with side arm and 50 mL capacity, both arms having ST 14/20 joint.‖

e. Special gas inlet-outlet tube: See 303E.I.2*e*.

f. Magnetic stirrer, strong enough to homogenize the zinc slurry described in ¶ 2*c* below.

g. Drying tube: See 303E.I.2*f*.

h. Apparatus setup: See 303E.I.2*g*.

2. Reagents

a. Potassium iodide solution: Dissolve 20 g KI in 100 mL water.

b. Stannous chloride solution: Dissolve 100 g $SnCl_2$ in 100 mL conc HCl.

c. Zinc slurry: Add 50 g zinc metal dust (200 mesh) to 100 mL water.

d. Diluent: Add 100 mL 18*N* H_2SO_4 and 400 mL conc HCl to 400 mL water in a 1-L volumetric flask and bring to volume with water.

e. Arsenic solutions: See 303E.I.3*e*.

f. Selenium solutions: See 303E.I.3*f*.

g. Perchloric acid, 70 to 72% $HClO_4$.

3. Procedure

a. Instrument operation: See 303E.I.4*a*.

b. Sample preparation: See 303E.I.4*b*.

c. Preparation of standards: See 303E.I.4*c*.

d. Treatment of samples and standards: Transfer a 25-mL portion of sample prepared in ¶ *b* or standard prepared as in ¶ *c* to the reaction vessel. Add 1 mL KI solution to arsenic samples and standards only. Omit KI for selenium determinations. Add 0.5 mL $SnCl_2$ solution. Allow at least 10 min for the metal to be reduced to its lowest oxidation state. Attach reaction vessel to special gas inlet-outlet glassware. Fill medicine dropper with 1.50 mL zinc slurry that has been kept in suspension with a magnetic stirrer. Firmly insert stopper containing medicine dropper into side neck of reaction vessel. Squeeze bulb to introduce zinc slurry into sample or standard. The metal hydride will produce a peak almost immediately. When recorder pen returns part way to base line, remove reaction vessel.

4. Calculations

See 303E.I.5.

5. Bibliography

BRAMAN, R.S., L.L. JUSTEN & C.C. FOREBACK. 1972. Direct volatilization—Spectral emission type detection system for nanogram amounts of arsenic and antimony. *Anal. Chem.* 44:2195.

FERNANDEZ, F.J. 1973. Atomic absorption of gaseous hydrides utilizing sodium borohydride reduction. *Atomic Absorption Newsletter* 12:93.

CORBIN, D.R. & C.C. BARNARD. 1976. Atomic absorption spectrophotometric determination of arsenic and selenium in water by hydride generation. *Atomic Absorption Newsletter* 15:116.

‖Scientific Glass JM-5835 or equivalent.

303 F. Determination of Mercury by the Cold Vapor Technic

1. Apparatus

a. Atomic absorption spectrophotometer and associated equipment: See Section 303.2. Instruments specifically designed for measurement of mercury by the cold vapor technic are available commercially and may be substituted.

b. Absorption cell, a glass or plastic tube approximately 2.5 cm in diameter. An 11.4-cm-long tube has been found satisfactory but a 15-cm-long tube is preferred. Grind tube ends perpendicular to the longitudinal axis and cement quartz windows in place. Attach gas inlet and outlet ports (6.4 mm diam) 1.3 cm from each end.

c. Cell support: Strap cell to the flat nitrous-oxide burner head or other suitable support and align in light beam to give maximum transmittance.

d. Air pumps: Use any peristaltic pump with electronic speed control capable of delivering 2 L air/min. Any other regulated compressed air system or air cylinder also is satisfactory.

e. Flowmeter, capable of measuring an air flow of 2 L/min.

f. Aeration tubing, a straight glass frit having a coarse porosity for use in reaction flask.

g. Reaction flask, 250-mL erlenmeyer flask fitted with a rubber stopper to hold aeration tube.

h. Drying tube, 150 mm × 18 mm diam containing 20 g $MgClO_4$ or conc sulfuric acid.

i. Connecting tubing, glass tubing to pass mercury vapor from reaction flask to absorption cell and to interconnect all other components. Clear vinyl plastic* tubing may be substituted for glass.

*Tygon or equivalent.

2. Reagents†

a. Metal-free water: See 303A.2*c*.

b. Stock mercury solution: Dissolve 1.3540 g mercuric chloride, $HgCl_2$, in about 700 mL water, add 1.5 mL conc HNO_3, and dilute to 1,000 mL with water; 1.00 mL = 1.00 mg Hg.

c. Standard mercury solutions: Prepare a series of standard mercury solutions containing 0 to 5 μg/L by appropriate dilution of stock mercury solution with water containing 1.5 mL conc HNO_3/L. Prepare standards daily.

d. Nitric acid, HNO_3, conc.

e. Potassium permanganate solution: Dissolve 50 g $KMnO_4$ in water and dilute to 1 L.

f. Potassium persulfate solution: Dissolve 50 g $K_2S_2O_8$ in water and dilute to 1 L.

g. Sodium chloride-hydroxylamine sulfate solution: Dissolve 120 g NaCl and 120 g $(NH_2OH)_2 \cdot H_2SO_4$ in water and dilute to 1 L.

h. Stannous chloride solution: Dissolve 100 g $SnCl_2$ in water containing 12.5 mL conc HCl and dilute to 1 L. On aging, this solution decomposes. If a suspension forms, stir reagent continuously during use.

i. Sulfuric acid, H_2SO_4, conc.

3. Procedure

a. Instrument operation: See Section 303A.3*a*. Install absorption cell and align in light path to give maximum transmission. Connect associated equipment to absorption cell with glass or vinyl plastic tubing as indicated in Figure 303:2. Turn on air and adjust flow rate to 2 L/min. Allow air to flow continuously.

b. Standardization: Transfer 100 mL of each of the 1.0, 2.0, and 5.0 μg/L mercury

†Use specially prepared reagents low in mercury.

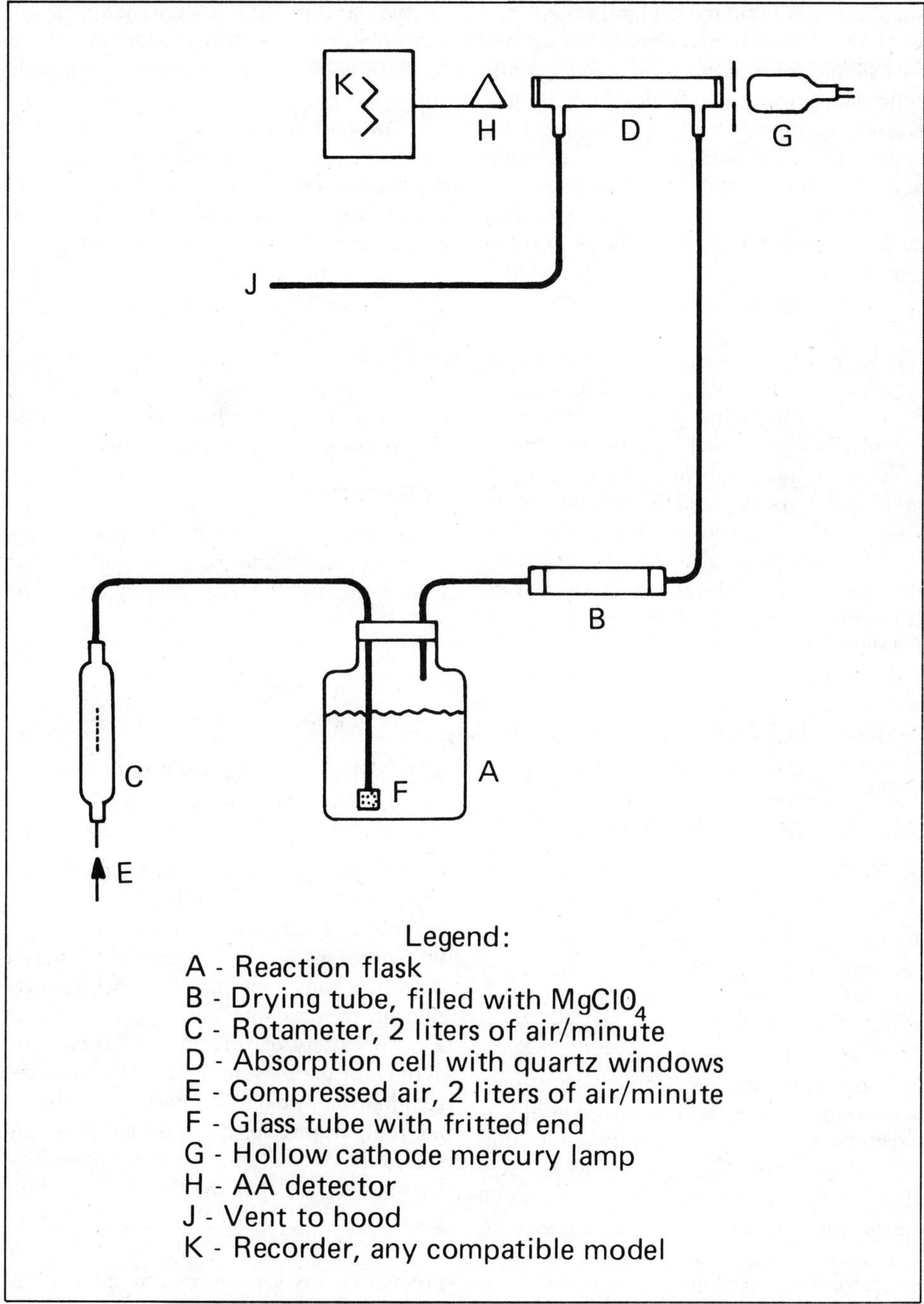

Figure 303:2. Schematic arrangement of equipment for measurement of mercury by cold vapor atomic absorption technic.

standard solutions and a blank of 100 mL water to 250-mL erlenmeyer reaction flasks. Add 5 mL conc H_2SO_4 and 2.5 mL conc HNO_3 to each flask. Add 15 mL $KMnO_4$ solution to each flask and let stand at least 15 min. Add 8 mL $K_2S_2O_8$ solution to each flask and heat for 2 hr in a water-bath at 95 C. For samples such as fish flesh, heat for 4 hr at 70 C. Cool to room temperature.

Treating each flask individually, add enough NaCl-hydroxylamine sulfate solution to reduce excess $KMnO_4$, then immediately add 5 mL $SnCl_2$ solution and attach flask to aeration apparatus. As mercury is volatilized and carried into the absorption cell, absorbance will increase to a maximum within a few seconds. As soon as recorder returns approximately to the base line, remove stopper holding the frit from reaction flask, and replace with a flask containing water. Flush system for a few seconds and run the next standard in the same manner. Construct a standard curve by plotting peak height versus micrograms Hg.

c. Analysis of samples: Transfer 100 mL sample or portion diluted to 100 mL containing not more than 5.0 μg Hg/L to a reaction flask. Treat as in ¶ 3*b*. Seawaters, brines, and effluents high in chlorides require as much as an additional 25 mL $KMnO_4$ solution. During oxidation step, chlorides are converted to free chlorine, which absorbs at 253 nm. Remove all free chlorine before the mercury is reduced and swept into the cell by using an excess (25 mL) of hydroxylamine sulfate reagent.

4. Calculation

Determine peak height of sample from recorder chart and read mercury value from standard curve prepared according to ¶ 3*b*.

304 DETERMINATION OF MICRO AMOUNTS OF METALS BY ELECTROTHERMAL ATOMIC ABSORPTION SPECTROPHOTOMETRY (TENTATIVE)

1. General Discussion

Because a wide variety of instrumentation is available and because technics are developing rapidly, this section does not attempt to present specific methodology. Rather, general information on equipment and procedures is set forth.

The technic of electrothermal atomic absorption, using the heated graphite (or carbon rod) atomizer, permits determination of most metallic elements with sensitivities and detection limits 20 to 1,000 times better than those obtainable with conventional flame atomization. This allows for the determination of metals at concentrations of less than 1 μg/L without sample preconcentration. Additional advantages of using a flameless atomizer are that it simplifies sample handling, permits using a very small sample (5 to 100 μL), reduces interference from suspended matter, increases sensitivity and gives better detection limits by sample preconcentration on the graphite tube, permits use of a wide variety of solvents and frequently avoids an extraction step, and allows determination of elements with high atomizing temperature in the presence of a considerable excess of matrix components. Drawbacks are that the flameless technic generally increases analysis time, requires additional replication to insure adequate precision, has greater matrix effects, requires thorough sample homogenization, and is more subject to contamination errors because of its sensitivity.

a. Principle: With an electrically heated atomizer, a graphite furnace replaces the standard burner head. Samples in solution or colloidal suspension are pipetted into the graphite tube (or cup) through which the light path of the spectrophotometer passes. The sample tube (or cup) is heated in three stages by passing an electrical current through its walls. First, a low current dries the sample (drying stage). Second, an intermediate current ashes or chars the sample (ashing or charring stage). Finally, a high current heats the tube (or cup) to incandescence and atomizes the sample (atomization stage).

b. Interferences: Significant interferences are molecular absorption interference and chemical interference. These have been discussed in Section 303.1*b*. Additionally, light may be scattered or blocked by particulate matter.

c. Absolute sensitivity and detection limit: Absolute sensitivity is the weight of metal that produces an absorbance of 0.0044 (1% absorption). This may vary with the chemical form of the element, spectrophotometer settings, drying, charring, and atomization conditions of the graphite furnace, purging gas, sample matrix, etc. Absolute sensitivities for most metals are between 10^{-10} to 10^{-12} g.

The practical detection limit is that weight of element producing an average absorbance equivalent to twice the magnitude of the baseline fluctuation.

2. Apparatus

a. Atomic absorption spectrophotometer and associated equipment: See Section 303.2. Burner and supplies of fuel and oxidant are not required.

b. Electric furnace mount: The electric furnace mount replaces the burner mount of a conventional atomic absorption spectrophotometer. Adjustments provide for precise alignment of the flameless atomizer in the optical path.

c. Electrothermal atomizer, consisting of a housing constructed to make electrical contact with an atomizer by means of electrodes. Generally the atomizer and electrodes are made of pure, spectrographic graphite although electrodes of inert metals may be used to minimize carbide formation. The atomizer may be designed as a tube or a cup. The tube atomizer is used mainly in the analysis of solutions; it has an average operational life of 50 to 200 uses. The graphite cup is useful for analysis of both solutions and solid samples. An integral part of the housing is a water jacket for cooling and an inert gas purging device.

d. Gas control unit: The gas control unit meters inert gas (usually nitrogen or argon) to prevent combustion of the graphite itself and to minimize formation of oxides of atomized elements. It also removes sample vapors and fumes formed during the drying, charring, and atomization states.

e. Power unit: This supplies variable electrical power to the graphite furnace to provide and control time and temperature programming of the drying, charring, and atomization cycles.

f. Sample dispensers: Use micropipets to introduce a discrete volume of sample into atomizer. Both fixed-volume and variable-volume micropipets are available. Automated sample dispensers are available that increase precision by eliminating manual injections and speeding up analysis.

3. Procedure

For an unknown sample matrix optimize operation of the graphite furnace according to the manufacturer's operational manual. In general, the settings for the drying stage are dictated by a furnace temperature only high enough to evaporate the solvent without vigorous boiling and a time interval long enough to ensure complete evaporation of the solvent.

In optimizing the charring stage provide

a sufficiently long charring time and a high enough charring temperature to volatilize as completely as possible any interference matrix and a charring time short enough and a temperature low enough to insure no loss of the element of interest. To optimize atomization temperature select minimum temperature required for complete atomization of the element as indicated by the maximum absorption peak. Use the shortest possible atomization time while still providing for complete atomization in order to eliminate "memory."

a. Determination of necessity for standard additions: Prepare an analytical curve by replicate analysis of standard solutions in the concentration range of interest, e.g., 0, 10, 20, 30, 50 μg metal ion/L. Prepare standard solution in 0.15N HNO_3. Compute slope of linear portion of curve.

Prepare a standard addition analysis curve by replicate analysis of suitable mixtures of sample and standard solutions, e.g., 1 part of sample + 1 part of 0.15N HNO_3; 1 part of sample + 1 part of 20 μg metal/L; 1 part of sample + 1 part of 40 μg metal/L; 1 part of sample + 1 part of 60 μg metal/L. Compute slope of linear portion of curve. Compare slopes of curves from standard solutions and standards with additions; select a criterion for equality of slopes, such as, "Slopes should not differ by more than 10%." Determine if slopes are equal by this criterion. If the slopes of the analytical and standard addition curves are not equal by the definition chosen, make all analyses of the sample matrix for the metal investigated in the test by the method of standard additions. The analysis of samples with complex matrices by heated furnace atomization atomic absorption frequently requires that the method of standard additions be used. This holds true even if samples are first digested by wet procedures.

4. Bibliography

FERNANDEZ, F.J. & D.C. MANNING. 1971. Atomic absorption analyses of metal pollutants in water using a heated graphite atomizer. *Atomic Absorption Newsletter* 10:65.

SEGAR, D.A. & J.G. GONZALEZ. 1972. Evaluation of atomic absorption with a heated graphite atomizer for the direct determination of trace transition metals in sea water. *Anal. Chim. Acta* 58:7.

BARNARD, W.M. & M.J. FISHMAN. 1973. Evaluation of the use of heated graphite atomizer for the routine determination of trace metals in water. *Atomic Absorption Newsletter* 12:118.

KAHN, H.L. 1973. The detection of metallic elements in wastes and waters with the graphite furnace. *Int. J. Environ. Anal. Chem.* 3:121.

PERKIN-ELMER CO. 1973. Analytical methods for atomic absorption spectroscopy using the HGA graphite furnace. Norwalk, Conn.

RATTONETTI, A. 1974. Determination of soluble cadium, lead, silver, and indium in rainwater and stream water with the use of flameless atomic absorption. *Anal. Chem.* 46:73.

CRUZ, R.B. & J.C. VAN LOON. 1974. A critical study of the application of graphite furnace non-flame atomic absorption spectrophotometry to the determination of trace base metals in complex heavy matrix sample solutions. *Anal. Chim. Acta* 72:231.

MORGENTHALER, L. 1975. A primer for flameless atomization. *Amer. Lab.* 7:41.

VARIAN TECHTRON. 1975. Analytical methods for carbon rod atomizers. Springvale, Vic., Australia.

305 POLAROGRAPHIC METHODS

The 14th edition of this book contained a section on polarographic methods for the analysis of cadmium, copper, lead, nickel, and zinc (301B). It was not reviewed by the Joint Task Groups for the 15th edition. In the opinion of several Standard Methods Committee members the material in this section was antiquated but, unfortunately, time was unavailable to generate the extensive revisions necessary for update. Therefore, the material has been deleted from this 15th edition because it was judged unworthy of standard status. An active Joint Task Group has been formed for the 16th edition to revise and bring this method up to date.

306 ALUMINUM

Aluminum is the third most abundant element of the earth's crust, occurring in minerals, rocks, and clays. This wide distribution accounts for the presence of aluminum in nearly all natural water as a soluble salt, a colloid, or an insoluble compound. Soluble, colloidal, and insoluble aluminum also may appear in treated water or wastewater as a residual from alum coagulation. Recent work indicates that filtered water from a modern rapid sand filtration plant should have an aluminum concentration no greater than 50 μg/L.

Selection of method: The atomic absorption spectrophotometric method is free from such common interferences as fluoride and phosphate, and is preferred. The Eriochrome cyanine R colorimetric method provides a means for estimating aluminum with simpler instrumentation.

306 A. Atomic Absorption Spectrophotometric Method

See Sections 303C and D.

306 B. Eriochrome Cyanine R Method

1. General Discussion

a. Principle: With Eriochrome cyanine R dye, dilute aluminum solutions buffered to a pH of 6.0 produce a red to pink complex that exhibits maximum absorption at 535 nm. The intensity of the developed color is influenced by the aluminum concentration, reaction time, temperature, pH, alkalinity, and concentration of other ions in the sample. To compensate for color and turbidity, the aluminum in one portion of sample is complexed with EDTA to provide a blank. The interference of iron and manganese, two elements often found in water, is eliminated by adding ascorbic acid. The optimum aluminum range lies between 20 and 300 μg/L but can be extended upward by sample dilution.

b. Interference: Negative errors are caused by both fluoride and polyphosphates. When the fluoride concentration is constant, the percentage error decreases with increasing amounts of aluminum. Because the fluoride concentration often is known or can be determined readily, fairly accurate results can be obtained by adding the known amount of fluoride to a set of standards. A simpler correction can be determined from the family of curves in Figure 306:1. A procedure is given for the removal of complex phosphate interference. Orthophosphate in concentrations under 10 mg/L does not interfere. The interference caused by even small amounts of alkalinity is removed by acidifying the sample just beyond the neutralization point of methyl orange. Sulfate does not interfere up to a concentration of 2,000 mg/L.

c. Minimum detectable concentration: The minimum aluminum concentration detectable by this method in the absence of fluorides and complex phosphates is approximately 6 μg/L.

d. Sample handling: Collect samples in clean, acid-rinsed bottles, preferably plastic, and examine them as soon as possible after collection. If only soluble aluminum is to be determined, filter a portion of sample through a 0.45-μm membrane filter; discard first 50 mL of filtrate and use succeeding filtrate for the determination. Do not use filter paper, absorbent cotton, or glass wool for filtering any solution that is to be tested for aluminum, because they will remove most of the soluble aluminum.

2. Apparatus

a. Colorimetric equipment: One of the following is required:

1) *Spectrophotometer,* for use at 535 nm, with a light path of 1 cm or longer.

2) *Filter photometer,* providing a light path of 1 cm or longer and equipped with a green filter with maximum transmittance between 525 and 535 nm.

3) *Nessler tubes,* 50-mL, tall form, matched.

b. Glassware: Treat all glassware with warm 1 + 1 HCl and rinse with aluminum-free distilled water to avoid errors due to materials absorbed on the glass. Rinse sufficiently to remove all acid.

3. Reagents

Use reagents low in aluminum, and aluminum-free distilled water.

a. Stock aluminum solution: Use either the metal (1) or the salt (2) for preparing stock solution; 1.00 mL = 500 μg Al:

1) Dissolve 500.0 mg aluminum metal in 10 mL conc HCl by heating gently. Dilute to 1,000 mL with water, or

2) Dissolve 8.792 g aluminum potassium sulfate (also called potassium alum), $AlK(SO_4)_2 \cdot 12H_2O$, in water and dilute to 1,000 mL.

b. Standard aluminum solution: Dilute 10.00 mL stock aluminum solution to 1,000 mL with water; 1.00 mL = 5.00 μg Al. Prepare daily.

c. Sulfuric acid, H_2SO_4, 0.02*N* and 6*N*.

d. Ascorbic acid solution: Dissolve 0.1 g ascorbic acid in water and make up to 100 mL in a volumetric flask. Prepare fresh daily.

e. Buffer reagent: Dissolve 136 g sodium acetate, $NaC_2H_3O_2 \cdot 3H_2O$, in water, add 40 mL 1*N* acetic acid, and dilute to 1 L.

f. Stock dye solution: Use any of the following products:

1) *Solochrome cyanine R-200* or Eriochrome cyanine†:* Dissolve 100 mg in water and dilute to 100 mL in a volumetric flask. This solution should have a pH of about 2.9.

2) *Eriochrome cyanine R‡:* Dissolve 300 mg dye in about 50 mL water. Adjust pH from about 9 to about 2.9 with 1 + 1 acetic acid (approximately 3 mL will be required). Dilute with water to 100 mL.

3) *Eriochrome cyanine R§:* Dissolve 150 mg in about 50 mL water. Adjust pH from about 9 to about 2.9 with 1 + 1 acetic acid (approximately 2 mL will be required). Dilute with water to 100 mL.

Stock solutions have excellent stability and can be kept for at least a year.

g. Working dye solution: Dilute 10.0 mL of selected stock dye solution to 100 mL in a volumetric flask with water.

---

*A product of Arnold Hoffman & Co., Providence, R.I.

†A product of K & K Laboratories, Plainview, N.Y.

‡A product of Pfaltz & Bauer, Inc., Flushing, N.Y.

§A product of Hartman-Leddon Co., Philadelphia, Pa.

Working solutions are stable for at least 6 months.

h. Methyl orange indicator solution, or the mixed bromcresol green-methyl red indicator solution specified in the total alkalinity determination (Section 403.3*d*).

i. EDTA (sodium salt of ethylenediaminetetraacetic acid dihydrate), 0.01*M:* Dissolve 3.7 g in water, and dilute to 1 L.

j. Sodium hydroxide, NaOH, 1*N* and 0.1*N*.

4. Procedure

a. Preparation of calibration curve:

1) Prepare a series of aluminum standards from 0 to 7 μg (0 to 280 μg/L based on a 25-mL sample) by accurately measuring the calculated volumes of standard aluminum solution into 50-mL volumetric flasks or nessler tubes. Add water to a total volume of approximately 25 mL.

2) Add 1 mL 0.02*N* H_2SO_4 to each standard and mix. Add 1 mL ascorbic acid solution and mix. Add 10 mL buffer solution and mix. With a volumetric pipet, add 5 mL working dye reagent and mix. Immediately make up to 50 mL with distilled water. Mix and let stand for 5 to 10 min. The color begins to fade after 15 min.

3) Read transmittance or absorbance on a spectrophotometer, using a wavelength of 535 nm or a green filter providing maximum transmittance between 525 and 535 nm. Adjust instrument to zero absorbance with the standard containing no aluminum.

Plot concentration of Al (micrograms Al in 50 mL final volume) against absorbance.

b. Sample treatment in absence of fluoride and complex phosphates: Place 25.0 mL sample, or a portion diluted to 25 mL, in a porcelain dish or flask, add a few drops of methyl orange indicator, and titrate with 0.02*N* H_2SO_4 to a faint pink color. Record reading and discard sample. To two similar samples at room temperature add the same amount of 0.02*N* H_2SO_4 used in the titration and 1 mL in excess.

To one sample add 1 mL EDTA solution. This will serve as a blank by complexing any aluminum present and compensating for color and turbidity. To both samples add 1 mL ascorbic acid, 10 mL buffer reagent, and 5.00 mL working dye reagent as prescribed in ¶*a*2) above.

Set instrument to zero absorbance or 100% transmittance using the EDTA blank. After 5 to 10 min contact time, read transmittance or absorbance and determine aluminum concentration from the calibration curve previously prepared.

c. Visual comparison: If photometric equipment is not available, prepare and treat standards and a sample, as described above, in 50-mL nessler tubes. Make up to mark with water and compare sample color with the standards after 5 to 10 min contact time. A sample treated with EDTA is not needed when nessler tubes are used. If the sample contains turbidity or color, the use of nessler tubes may result in considerable error.

d. Removal of phosphate interference: Add 1.7 mL 6*N* H_2SO_4 to 100 mL sample in a 200-mL erlenmeyer flask. Heat on a hot plate for at least 90 min, keeping solution temperature just below the boiling point. At the end of the heating period solution volume should be about 25 mL. Add water if necessary to keep it at or above that volume.

After cooling, neutralize to a pH of 4.3 to 4.5 with NaOH, using 1*N* NaOH at the start and 0.1*N* for the final fine adjustment, and a pH meter. Make up to 100 mL with water, mix, and use a 25-mL portion for the aluminum test.

Run a blank in the same manner, using 100 mL distilled water and 1.7 mL 6*N* H_2SO_4. Subtract blank reading from sample reading or use it to set instrument to zero absorbance before reading the sample.

e. Sample treatment in presence of

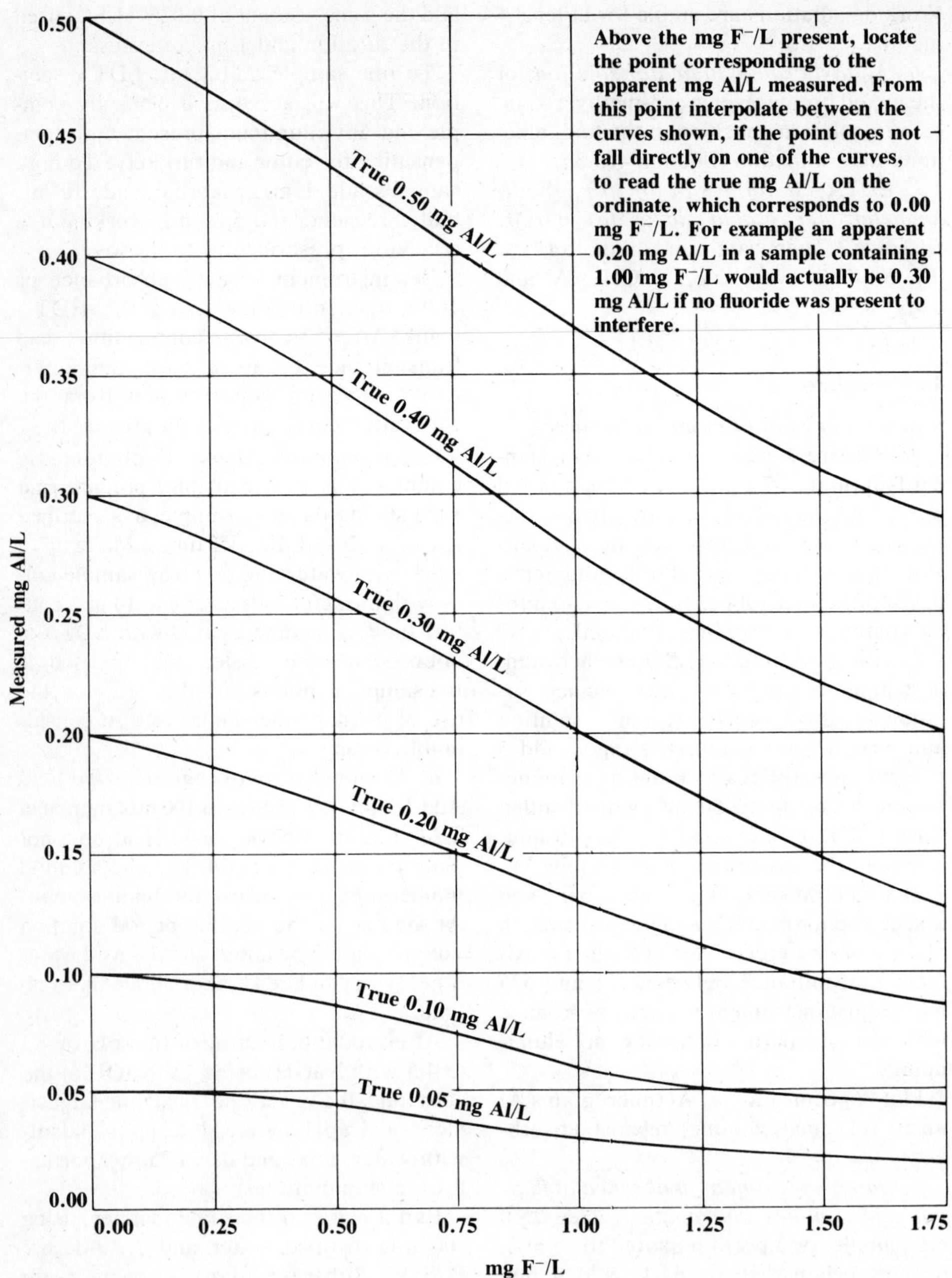
Above the mg F⁻/L present, locate the point corresponding to the apparent mg Al/L measured. From this point interpolate between the curves shown, if the point does not fall directly on one of the curves, to read the true mg Al/L on the ordinate, which corresponds to 0.00 mg F⁻/L. For example an apparent 0.20 mg Al/L in a sample containing 1.00 mg F⁻/L would actually be 0.30 mg Al/L if no fluoride was present to interfere.
True 0.50 mg Al/L
True 0.40 mg Al/L
True 0.30 mg Al/L
True 0.20 mg Al/L
True 0.10 mg Al/L
True 0.05 mg Al/L
Measured mg Al/L
0.50
0.45
0.40
0.35
0.30
0.25
0.20
0.15
0.10
0.05
0.00
0.00
0.25
0.50
0.75
1.00
1.25
1.50
1.75
mg F⁻/L

Figure 306:1. Correction curves for estimation of aluminum in the presence of fluoride.

fluoride: Measure sample fluoride concentration by the SPADNS or electrode method. Either:

1) Add the same amount of fluoride as in the sample to each aluminum standard, or

2) Determine fluoride correction from the set of curves in Figure 306:1.

5. Calculation

$$\text{mg Al/L} = \frac{\mu\text{g Al (in 50 mL final volume)}}{\text{mL sample}}$$

6. Precision and Accuracy

A synthetic sample containing 520 μg Al/L and no interference in distilled water was analyzed by the Eriochrome cyanine R method in 27 laboratories. Relative standard deviation was 34.4% and relative error 1.7%.

A second synthetic sample containing 50 μg Al/L, 500 μg Ba/L, and 5μg Be/L in distilled water was analyzed in 35 laboratories. Relative standard deviation was 38.5% and relative error 22.0%.

A third synthetic sample containing 500 μg Al/L, 50 μg Cd/L, 110μg Cr/L, 1,000 μg Cu/L, 300 μg Fe/L, 70 μg Pb/L, 50 μg Mn/L, 150 μg Ag/L, and 650 μg Zn/L in distilled water was analyzed in 26 laboratories. Relative standard deviation was 28.8% and relative error 6.2%.

A fourth synthetic sample containing 540 μg Al/L was 2.5 mg polyphosphate/L in distilled water was analyzed in 16 laboratories that hydrolyzed the sample in the prescribed manner. Relative standard deviation was 44.3% and relative error 1.3%. In 12 laboratories that applied no corrective measures, the relative standard deviation was 49.2% and the relative error 8.9%.

A fifth synthetic sample containing 480 μg Al/L and 750 μg F/L in distilled water was analyzed in 16 laboratories that relied on the curve to correct for the fluoride content. Relative standard deviation was 25.5% and relative error 2.3%. The 17 laboratories that added fluoride to the aluminum standards showed a relative standard deviation of 22.5% and a relative error of 7.1%.

7. Bibliography

SHULL, K.E. & G.R. GUTHAN. 1967. Rapid modified Eriochrome cyanine R method for determination of aluminum in water. *J. Amer. Water Works Ass.*59:1456.

307 ARSENIC

Severe poisoning can arise from the ingestion of as little as 100 mg arsenic; chronic effects can appear from its accumulation in the body at low intake levels. Carcinogenic properties also have been imputed to arsenic. The arsenic concentration of most potable waters seldom exceeds 10 μg/L, although values as high as 100 μg/L have been reported. Arsenic may occur in water as a result of mineral dissolution, industrial discharges, or the application of insecticides.

Selection of method: The atomic absorption method (A), which converts arsenic to its hydride and uses an argon-hydrogen flame, is the method of choice, although the direct electrothermal method is simpler in the demonstrated absence of interference. The silver diethyldithiocarbamate method (B) is applicable when interferences are absent. The mercuric bromide stain method (C) requires care and experience and is suitable only for qualitative or semiquantitative determinations (±5 μg As).

307 A. Atomic Absorption Spectrophotometric Method

See Section 303E.

307 B. Silver Diethyldithiocarbamate Method

1. General Discussion

a. Principle: Inorganic arsenic is reduced to arsine, AsH_3, by zinc in acid solution in a Gutzeit generator. The arsine is then passed through a scrubber containing glass wool impregnated with lead acetate solution and into an absorber tube containing silver diethyldithiocarbamate dissolved in pyridine or chloroform. In the absorber, arsenic reacts with the silver salt, forming a soluble red complex suitable for photometric measurement.

b. Interference: Although certain metals—chromium, cobalt, copper, mercury, molybdenum, nickel, platinum, and silver—interfere in the generation of arsine, the concentrations of these metals normally present in water do not interfere significantly. Antimony salts in the sample form stibine, which interferes with color development by yielding a red color with maximum absorbance at 510 nm.

c. Minimum detectable quantity: 1 μg As.

2. Apparatus

a. Arsine generator and absorption tube: See Figure 307:1.*

b. Photometric equipment:

1) *Spectrophotometer,* for use at 535 nm with 1-cm cells.

2) *Filter photometer,* with green filter having a maximum transmittance in the range 530 to 540 nm, with 1-cm cells.

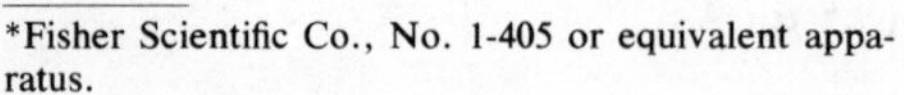

*Fisher Scientific Co., No. 1-405 or equivalent apparatus.

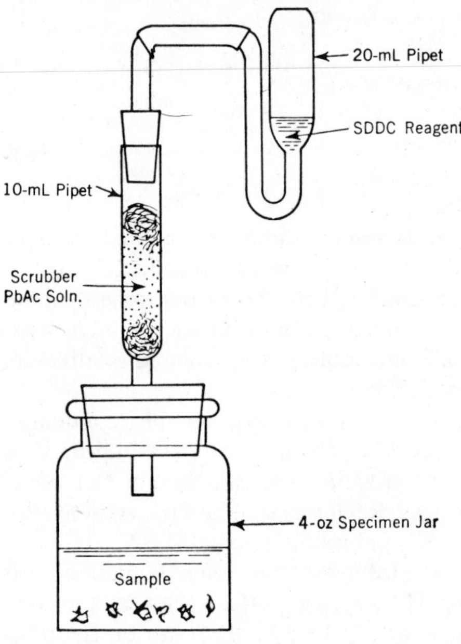

Figure 307:1. Arsine generator and absorber assembly.

3. Reagents

a. Hydrochloric acid, HCl, conc.

b. Potassium iodide solution: Dissolve 15 g KI in 100 mL distilled water. Store in a brown bottle.

c. Stannous chloride reagent: Dissolve 40 g arsenic-free $SnCl_2 \cdot 2H_2O$ in 100 mL conc HCl.

d. Lead acetate solution: Dissolve 10 g $Pb(C_2H_3O_2)_2 \cdot 3H_2O$ in 100 mL distilled water.

e. Silver diethyldithiocarbamate reagent: Prepare this reagent as described in either 1) or 2):

1) Dissolve 410 mg 1-ephedrine in 200 mL chloroform ($CHCl_3$), add 625 mg $AgSCSN(C_2H_5)_2$, and adjust volume to 250 mL with additional $CHCl_3$. Filter and store in brown bottle.

2) Dissolve 1 g $AgSCSN(C_2H_5)_2$ in 200 mL pyridine. Store in brown bottle.

f. Zinc: 20 to 30 mesh, arsenic-free.

g. Stock arsenic solution: Dissolve 1.320 g arsenic trioxide, As_2O_3, in 10 mL distilled water containing 4 g NaOH, and dilute to 1,000 mL with distilled water; 1.00 mL = 1.00 mg As. (CAUTION: *Toxic—take care to avoid ingestion of arsenic solutions.*)

h. Intermediate arsenic solution: Dilute 5.00 mL stock solution to 500 mL with distilled water; 1.00 mL = 10.0 μg As.

i. Standard arsenic solution: Dilute 10.00 mL intermediate solution to 100 mL with distilled water; 1.00 mL = 1.00 μg As.

4. Procedure

For total arsenic digest sample by the procedure in 307C.4*a*. Report if sample has been digested or not.

a. Treatment of sample: Pipet 35.0 mL sample into a clean generator bottle. Add successively, with thorough mixing after each addition, 5 mL conc HCl, 2 mL KI solution, and 8 drops (0.40 mL) $SnCl_2$ reagent. Allow 15 min for reduction of arsenic to the trivalent state.

b. Preparation of scrubber and absorber: Impregnate glass wool in the scrubber with lead acetate solution. Do not make too wet because water will be carried over into the reagent solution. Pipet 4.00 mL silver diethyldithiocarbamate reagent into absorber tube.

c. Arsine generation and measurement: Add 3 g zinc to generator and connect scrubber-absorber assembly immediately. Make certain that all connections are fitted tightly.

Allow 30 min for complete evolution of arsine. Warm the generator slightly to insure that all arsine is released. Pour solution from absorber directly into a 1-cm cell and measure absorbance at 535 nm, using the reagent blank as the reference.

d. Preparation of standard curve: Treat portions of standard solution containing 0, 1.0, 2.0, 5.0, and 10.0 μg As described in this section, ¶s *a* through *d*. Plot absorbance versus concentration of arsenic in the standard.

5. Calculation

$$\text{mg As/L} = \frac{\mu\text{g As (in 4.00 mL final volume)}}{\text{mL sample}}$$

6. Precision and Accuracy

A synthetic sample containing 40 μg As/L, 250 μg Be/L, 240 μg B/L, 20 μg Se/L, and 6 μg V/L in distilled water was analyzed in 46 laboratories by the silver diethyldithiocarbamate method, with a relative standard deviation of 13.8% and a relative error of 0%.

307 C. Mercuric Bromide Stain Method

1. General Discussion

a. Principle: After sample concentration arsenic is liberated as arsine, AsH_3, by zinc in acid solution in a Gutzeit generator. The generated arsine is then passed through a column containing a roll of cotton moistened with lead acetate solution. The generated arsine produces a yellow-brown stain on test paper strips impregnated with mercuric bromide. The length of the stain is roughly proportional to the amount of arsenic present.

b. Interference: Antimony (>0.10 mg) interferes by giving a similar stain.

c. Minimum detectable quantity: 1 μg As.

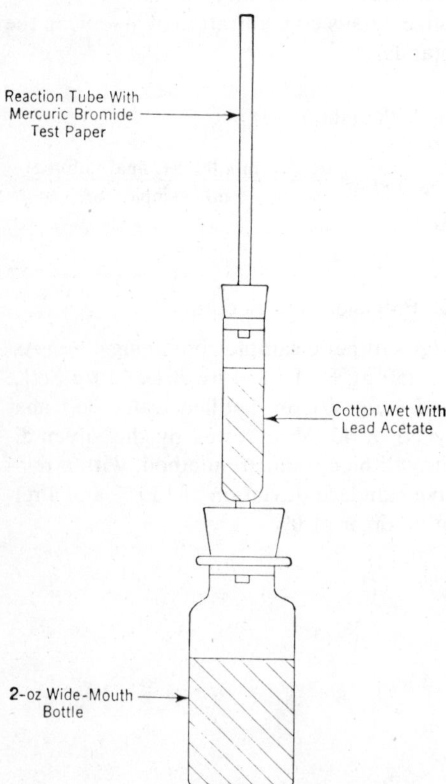

Figure 307:2. Generator used with mercuric bromide stain method.

2. Apparatus

Arsine generator: See Figure 307:2.

3. Reagents

a. Sulfuric acid, H_2SO_4, 1 + 1.

b. Nitric acid, HNO_3, conc.

c. Roll cotton: Cut a roll of dentist's cotton into 25-mm lengths.

d. Lead acetate solution: Prepare as directed in Method B, ¶ 3*d*.

e. Mercuric bromide paper: Use commercial arsenic papers cut uniformly into strips about 12 cm long and 2.5 mm wide (papers can be obtained already cut and sensitized). Soak strips for at least 1 hr in filtered solution prepared by dissolving 3 to 6 g $HgBr_2$ in 95% ethyl or isopropyl alcohol; dry by waving in air. Store in dry, dark place. For best results, make up papers just before use.

f. Potassium iodide solution: Prepare as directed in Method B, ¶ 3*b*.

g. Stannous chloride reagent: Prepare as directed in Method B, ¶ 3*c*.

h. Zinc, 20 to 30 mesh, arsenic-free.

i. Standard arsenic solution: Prepare as directed in Method B, ¶ 3*i*.

4. Procedure

a. Concentration of sample and oxidation of organic matter: Place a suitable sample containing from 2 to 30 μg As in flask or beaker, add 7 mL 1 + 1 H_2SO_4 and 5 mL conc HNO_3. Evaporate to SO_3 fumes. Cool, add about 25 mL distilled water, and again evaporate to SO_3 fumes to expel oxides of nitrogen. Maintain an excess of HNO_3 until the organic matter is destroyed. Do not let solution darken while organic matter is being destroyed because arsenic is likely to be reduced and lost.

Cool, add about 25 mL distilled water, and transfer to generator bottle.

b. Preparation of guard column and re-

action tube: Dip one end of the 2.5-cm length of cotton into lead acetate solution and introduce into glass column. Then put the dried narrow glass tube in place and insert $HgBr_2$ test paper. Make sure paper strip is straight.

c. Treatment of sample concentrate: To the 25-mL sample concentrate in the generator, add 7 mL 1 + 1 H_2SO_4 and cool. Add 5 mL KI solution, 4 drops $SnCl_2$ reagent, and 2 to 5 g zinc. Immediately connect reaction tube to generator. Immerse apparatus to within 2.5 cm of the top of the narrow tube in a water bath kept at 20 to 25 C and allow evolution to proceed for 1.5 hr. Remove strip and compute average length of stains on both sides. Using a calibration curve, the preparation of which is described below, estimate amount of arsenic present.

d. Preparation of calibration curve: Prepare a blank and standards at 3-μg intervals in the 0- to 30-μg As range with 14 mL 1 + 1 H_2SO_4 and bring total volume to 25 mL. Place in generator and treat as described for the sample concentrate. Remove strip and compute average length, in millimeters, of stains on both sides. Plot length in millimeters against micrograms arsenic and use as a standard curve.

5. Precision and Accuracy

A synthetic sample containing 50 μg As/L, 400 μg Be/L, 180 μg B/L, and 50 μg Se/L in distilled water was analyzed in five laboratories by the mercuric bromide stain method with a relative standard deviation of 75.0% and a relative error of 60.0%.

307 D. Bibliography

Silver Diethyldithiocarbamate Method

VASAK, V. & V. SEDIVEC. 1952. Colorimetric determination of arsenic. *Chem. Listy* 46:341.

STRATTON, G. & H.C. WHITEHEAD. 1962. Colorimetric determination of arsenic in water with silver diethyldithiocarbamate. *J. Amer. Water Works Ass.* 54:861.

BALLINGER, D.C., R.J. LISHKA & M.E. GALES. 1962. Application of silver diethyldithiocarbamate method to determination of arsenic. *J. Amer. Water Works Ass.* 54:1424.

Mercuric Bromide Stain Method

FURMAN, N.H., ed. 1962. Standard Methods of Chemical Analysis, 6th ed. Vol. I. D. Van Nostrand Co., Princeton, N.J., pp. 118–124.

308 BARIUM

Barium stimulates the heart muscle. However, a barium dose of 550 to 600 mg is considered fatal to human beings. Afflictions arising from its consumption, inhalation, or absorption involve the heart, blood vessels, and nerves.

Despite a relative abundance in nature (16th in order of rank), barium occurs only in trace amounts in water. The barium concentration of U.S. drinking waters ranges between 0.7 and 900 μg/L, with a mean of 49 μg/L. Higher concentrations in drinking water often signal undesirable industrial waste pollution.

Perform analyses by the Atomic Absorption Spectrophotometric Method, Section 303C.

309 BERYLLIUM

Beryllium and its compounds are very poisonous and in high concentrations can cause death. Inhalation of beryllium dust can cause a serious disease called berylliosis. Beryllium disease also can take the form of dermatitis, conjunctivitis (eye disease), acute pneumonitis (lung disease), and chronic pulmonary berylliosis.

In the form of the element, compounds, or alloys, beryllium is used in atomic reactors, aircraft, rockets, and missile fuels. Entry into water can result from the discharges of such industries. Beryllium has been reported to occur in U.S. drinking waters in the range of 0.01 to 0.7 $\mu g/L$, with a mean of 0.013 $\mu g/L$.

Selection of method: The atomic absorption spectrophotometric method is the method of choice and the flameless methods are convenient for low Be concentrations. The colorimetric method is useful when atomic absorption instrumentation is not available.

309 A. Atomic Absorption Spectrophotometric Method

See Sections 303C and D.

309 B. Aluminon Method

1. General Discussion

a. Principle: The addition of a small amount of an ethylenediaminetetraacetic acid (EDTA) complexing solution prevents interference from moderate quantities of aluminum, cobalt, copper, iron, manganese, nickel, titanium, zinc, and zirconium. An aluminon buffer reagent is added to form a beryllium lake and the color developed is measured at 515 nm.

b. Interference: Under the conditions specified in the method, not more than 10 mg copper can be tolerated. If more is present, increase the amount of EDTA reagent. The complexed copper absorbs slightly at 515 nm; eliminate this interference by adding an equivalent amount of copper to the standards.

c. Minimum detectable concentration: 5 $\mu g/L$.

d. Molar absorptivity: 900 L g^{-1} cm^{-1}.

2. Sample Handling

Acidify all samples at time of collection to keep metals in solution and prevent their plating out on container walls. With relatively clean waters containing no particulate matter, normally 1.5 mL conc nitric acid (HNO_3)/L is sufficient to reduce the pH to 2.0. Surface waters that may contain sediment, such as those from streams, lakes, and wastewater treatment plant effluents, require more acid. If the sample contains particulate matter and only the "dissolved" metal content is desired, filter sample through a 0.45-μm membrane filter and acidify filtrate with 1.5 mL conc HNO_3/L.

3. Apparatus

a. Spectrophotometer, for use at 515 nm, with a light path of 5 cm.

b. Filter photometer, providing a light path of 5 cm and equipped with a green filter, exhibiting a maximum transmittance near 515 nm.

4. Reagents

a. Stock beryllium solution: Dissolve 9.82 g beryllium sulfate tetrahydrate,

$BeSO_4 \cdot 4H_2O$, in 100 mL distilled water, filter if necessary, and dilute to 500 mL; 1.00 mL = 1.00 mg Be.

b. Standard beryllium solution: Dilute 5.00 mL stock beryllium solution with distilled water to 1,000 mL in a volumetric flask; 1.00 mL = 5 μg Be.

c. EDTA reagent: Add 30 mL distilled water and a drop of an alcoholic solution of methyl red (50 mg/100 mL) to 2.5 g ethylenediaminetetraacetic acid. Neutralize with ammonium hydroxide (NH_4OH), cool, and dilute to 100 mL.

d. Aluminon buffer reagent: Transfer 500 g ammonium acetate, $NH_4C_2H_3O_2$, to 1 L distilled water in a 2-L beaker. Add 80 mL conc (glacial) acetic acid and stir until completely dissolved. Filter if necessary. Dissolve 1 g aurintricarboxylic acid triammonium salt (aluminon), in 50 mL distilled water and add to the buffer solution in the 2-L beaker. Dissolve 3.0 g benzoic acid ($C_7H_6O_2$) in 20 mL methyl alcohol and add to buffer solution while stirring. Dilute mixture to 2 L. Transfer 10 g gelatin to 250 mL distilled water in a 400-mL beaker. Place beaker in a boiling water bath and stir occasionally until gelatin has dissolved completely. Pour warm gelatin into a 1,000-mL volumetric flask containing 500 mL distilled water. Cool to room temperature, dilute to mark, and mix. Transfer gelatin solution and buffer solutions to a 4-L chemical-resistant dark glass bottle. Mix and store in a cool dark place. The reagent is stable for at least a month.

5. Procedure

a. Treatment of sample: If organic matter is present and it is desired to determine total beryllium, digest sample with HNO_3 and H_2SO_4 as indicated in Section 302E. If only dissolved beryllium is desired, filter sample through a 0.45-μm membrane filter.

b. Reaction with aluminon: Pipet 0, 0.1, 0.5, 1.00, 2.00, 3.00, and 4.00 mL standard beryllium solution to a series of 100-mL volumetric flasks. Transfer a 50-mL sample or portion containing less than 200 μg Be to a 100-mL volumetric flask. Add 2 mL EDTA reagent to each flask and dilute with distilled water to approximately 75 mL.

Add 15 mL aluminon buffer reagent, dilute to 100 mL with distilled water, and mix thoroughly. Let stand away from light for 20 min after the aluminon buffer is added. Filter if necessary. Read absorbancy of standard and unknown as compared to the blank in a spectrophotometer or filter photometer at a 515-nm wavelength using 5-cm cells. Construct a calibration curve by plotting absorbance of standards versus micrograms beryllium in 100 mL final volume. Determine amount of beryllium in sample by referring to the corresponding absorbance on the calibration curve.

6. Calculation

$$\text{mg Be/L} = \frac{\mu\text{g Be (in 100 mL final volume)}}{\text{mL sample}}$$

7. Precision and Accuracy

In 32 laboratories a synthetic sample of distilled water containing 250 μg Be/L, 40 μg As/L, 240 μg B/L, 20 μg Se/L, and 6 μg V/L, the beryllium was analyzed with a relative standard deviation of 7.13% and a relative error of 12%.

8. Bibliography

LUKE, C.L. & M.E. CAMPBELL. 1952. Photometric determination of beryllium in beryllium-copper alloys. *Anal. Chem.* 24:1056.

LUKE, C.L. & K.C. BROWN. 1952. Photometric determination of aluminum in manganese, bronze, zinc die casting alloys, and magnesium alloys. *Anal. Chem.* 24:1120.

310 CADMIUM

Cadmium is highly toxic and has been implicated in some cases of poisoning through food. Minute quantities of cadmium are suspected of being responsible for adverse changes in arteries of human kidneys. A cadmium concentration of 200 $\mu g/L$ is toxic to certain fish. The cadmium concentration of U.S. drinking waters has been reported to vary between 0.4 and 60 $\mu g/L$, with a mean of 8.2 $\mu g/L$. Cadmium may enter water as a result of industrial discharges or the deterioration of galvanized pipe.

Selection of method: The atomic absorption spectrophotometric method is preferred. The dithizone method is suitable for the determination of cadmium when an atomic absorption spectrophotometer is unavailable.

310 A. Atomic Absorption Spectrophotometric Method

See Sections 303A and B.

310 B. Dithizone Method

1. General Discussion

a. Principle: Cadmium ions under suitable conditions react with dithizone to form a pink to red color that can be extracted with chloroform ($CHCl_3$). $CHCl_3$ extracts are measured photometrically and the cadmium concentration is obtained from a calibration curve prepared from a standard cadmium solution treated in the same manner as the sample.

b. Interference: Under the conditions of this method, concentrations of metal ions normally found in water do not interfere. Lead up to 6 mg, zinc up to 3 mg, and copper up to 1 mg in the portion analyzed do not interfere. Ordinary room lighting does not affect the cadmium dithizonate color.

c. Minimum detectable concentration: 0.5 μg Cd in approximately 15 mL final volume with a 2-cm light path.

2. Apparatus

a. Colorimetric equipment: One of the following is required:

1) *Spectrophotometer,* for use at 518 nm with a minimum light path of 1 cm.

2) *Filter photometer,* equipped with a green filter having a maximum light transmittance near 518 nm, with a minimum light path of 1 cm.

b. Separatory funnels, 125 mL, preferably with TFE stopcocks.

c. Glassware: Clean all glassware, including sample bottles, with 1 + 1 HCl and rinse thoroughly with tap water and distilled water.

3. Reagents

a. Water, cadmium-free: Redistill distilled water in an all-glass still. Use this water to prepare all reagents and solutions.

b. Stock cadmium solution: Weigh 100.0 mg pure cadmium metal and dissolve in a solution composed of 20 mL water plus 5 mL conc HCl. Use heat to assist dissolution of the metal. Transfer quantitatively to a 1-L volumetric flask and dilute to 1,000 mL; 1.00 mL = 100 μg Cd. Store in a polyethylene container.

c. Standard cadmium solution: Pipet 10.00 mL stock cadmium solution into a 1-L volumetric flask, add 10 mL conc HCl, and dilute to 1,000 mL with water. Prepare as needed and use the same day; 1.00 mL = 1.00 μg Cd.

d. Sodium potassium tartrate solution: Dissolve 250 g $NaKC_4H_4O_6 \cdot 4H_2O$ in water and make up to 1 L.

e. Sodium hydroxide-potassium cyanide solutions:

1) *Solution I:* Dissolve 400 g NaOH and 10 g KCN in water and make up to 1 L. Store in a polyethylene bottle. This solution is stable for 1 month.

2) *Solution II:* Dissolve 400 g NaOH and 0.5 g KCN in water and make up to 1 L. Store in a polyethylene bottle. This solution is stable for 1 to 2 months.

CAUTION—*Potassium cyanide is extremely poisonous. Be especially cautious when handling it. Never use mouth pipets to deliver cyanide solutions.*

f. Hydroxylamine hydrochloride solution: Dissolve 20 g $NH_2OH \cdot HCl$ in water and make up to 100 mL.

g. Stock dithizone solution I: See 102.7*b*1).

h. Working dithizone solution: Dilute stock dithizone solution I with $CHCl_3$ to produce a working solution of 10 μg/mL. Prepare daily.

i. Chloroform, ACS grade passed for "suitability for use in dithizone test." Test for a satisfactory $CHCl_3$ by adding a minute amount of dithizone to a portion of the $CHCl_3$ in a stoppered test tube so that a faint green is produced; the green color should be stable for a day.

j. Tartaric acid solution: Dissolve 20 g $H_2C_4H_4O_6$ in water and make up to 1 L. Store in the refrigerator and use while still cold.

k. Hydrochloric acid, HCl, conc.

l. Thymol blue indicator solution: Dissolve 0.4 g thymolsulfonephthalein sodium salt in 100 mL water.

m. Sodium hydroxide, NaOH, 6*N*.

4. Procedure

a. Preparation of standard curve: Prepare a blank and a series of standards from 1 to 10 μg by pipeting the appropriate amounts of standard Cd solution into separatory funnels. Dilute to 25 mL and proceed as in ¶ 4*c*. Plot a calibration curve.

b. Treatment of samples: Digest sample as directed in Section 302. Pipet a volume of digested sample containing 1 to 10 μg Cd to a separatory funnel and dilute to 25 mL as necessary. Add 3 drops thymol blue and adjust with 6 *N* NaOH to the first permanent yellow color, pH 2.8.

c. Color development, extraction, and measurement: Add reagents in the following order, mixing after each addition: 1 mL sodium potassium tartrate solution, 5 mL NaOH-KCN solution I, 1 mL $NH_2OH \cdot HCl$ solution, and 15 mL stock dithizone solution I. Stopper funnels and shake for 1 min, relieving vapor pressure in the funnels through the stopper rather than the stopcock. Drain $CHCl_3$ layer into a second funnel containing 25 mL cold tartaric acid solution. Add 10 mL $CHCl_3$ to first funnel; shake for 1 min and drain into second funnel. Do not permit aqueous layer to enter second funnel. Because time of contact of $CHCl_3$ with the strong alkali must be kept to a minimum, make the two extractions immediately after adding dithizone (cadmium dithizonate decomposes on prolonged contact with strong alkali saturated with $CHCl_3$).

Shake second funnel for 2 min and discard $CHCl_3$ layer. Add 5 mL $CHCl_3$, shake 1 min, and discard $CHCl_3$ layer, making as close a separation as possible. In the following order, add 0.25 mL $NH_2OH \cdot HCl$ solution and 15.0 mL working dithizone solution. Add 5 mL NaOH-KCN solution II; *immediately* shake for 1 min and transfer $CHCl_3$ layer into a dry photometer tube. Read absorbance at 518 nm against the blank. Obtain Cd concentration from calibration curve.

5. Calculation

$$\text{mg Cd/L} = \frac{\mu\text{g Cd (in approx. 15 mL final volume)}}{\text{mL sample}}$$

6. Precision and Accuracy

A synthetic sample containing 50 μg Cd/L, 500 μg Al/L, 110 μg Cr/L, 470 μg Cu/L, 300 μg Fe/L, 70 μg Pb/L, 150 μg Ag/L, and 650 μg Zn/L was analyzed in 44 laboratories by the dithizone method with a relative standard deviation of 24.6% and a relative error of 6.0%.

7. Bibliography

SALTZMAN, B.E. 1953. Colorimetric microdetermination of cadmium with dithizone. *Anal. Chem.* 25:493.

GANOTES, J., E. LARSON & R. NAVONE. 1962. Suggested dithizone method for cadmium determination. *J. Amer. Water Works Ass.* 54:852.

310 C. Polarographic Method

See Section 305.

311 CALCIUM

The presence of calcium (fifth among the elements in order of abundance) in water supplies results from passage through or over deposits of limestone, dolomite, gypsum, and gypsiferous shale. The calcium content may range from zero to several hundred milligrams per liter, depending on the source and treatment of the water. Small concentrations of calcium carbonate combat corrosion of metal pipes by laying down a protective coating. Appreciable calcium salts, on the other hand, break down on heating to form harmful scale in boilers, pipes, and cooking utensils. Calcium carbonate saturation is discussed in Section 203.

Calcium contributes to the total hardness of water. Chemical softening treatment, reverse osmosis, electrodialysis, or ion exchange is used to reduce calcium and the associated hardness.

1. Selection of Method

The atomic absorption method is an accurate means of determining calcium. The permanganate and EDTA titration methods give good results for control and routine applications. The simplicity and rapidity of the EDTA titration procedure make it the method of choice for general use.

2. Storage of Samples

The customary precautions are sufficient if care is taken to redissolve any calcium carbonate that may precipitate on standing.

311 A. Atomic Absorption Spectrophotometric Method

See Section 303A.

311 B. Permanganate Titrimetric Method

1. General Discussion

a. Principle: Ammonium oxalate precipitates calcium quantitatively as calcium oxalate. An excess of oxalate overcomes the adverse effects of magnesium. Optimum crystal formation and minimum occlusion are obtained only when the pH is brought slowly to the desired value. This is accomplished in two stages, with intervening digestion to promote seed crystal formation. The precipitated calcium oxalate is dissolved in acid and titrated with permanganate. The amount of permanganate required to oxidize the oxalate is proportional to the amount of calcium.

b. Interference: The sample should be free of interfering amounts of strontium, silica, aluminum, iron, manganese, phosphate, and suspended matter. Strontium may precipitate as the oxalate and cause high results. In such a case, determine strontium by flame photometry and apply the appropriate correction. Eliminate silica interference by the classical dehydration procedure. Precipitate aluminum, iron, and manganese by ammonium hydroxide after treatment with persulfate. Precipitate phosphate as the ferric salt. Remove suspended matter by centrifuging or by filtration through paper, sintered glass, or a cellulose acetate membrane (see Residue, Section 209).

2. Apparatus

a. Vacuum pump, or other source of vaccum.

b. Filter flasks.

c. Filter crucibles: Medium-porosity, 30-mL crucibles are recommended. Use either glass or porcelain crucibles. Crucibles of all-porous construction are difficult to wash quantitatively.

3. Reagents

a. Methyl red indicator solution: Dissolve 0.1 g methyl red sodium salt and dilute to 100 mL with distilled water.

b. Hydrochloric acid, HCl, 1 + 1.

c. Ammonium oxalate solution: Dissolve 10 g $(NH_4)_2C_2O_4 \cdot H_2O$ in 250 mL distilled water. Filter if necessary.

d. Ammonium hydroxide, NH_4OH, *3N:* Add 240 mL conc NH_4OH to about 700 mL distilled water and dilute to 1 L. Filter before use to remove suspended silica flakes.

e. Ammonium hydroxide, 1 + 99.

f. Special reagents for removal of aluminum, iron, and manganese interference:

1) *Ammonium persulfate,* solid.

2) *Ammonium chloride solution:* Dissolve 20 g NH_4Cl in 1 L distilled water. Filter if necessary.

g. Sodium oxalate: Use primary standard grade $Na_2C_2O_4$. Dry at 105 C overnight and store in a desiccator.

h. Sulfuric acid, H_2SO_4, 1 + 1.

i. Standard potassium permanganate titrant, 0.05*N:* Dissolve 1.6 g $KMnO_4$ in 1 L distilled water. Keep in a brown glass-stoppered bottle and age for at least 1 wk. Carefully decant or pipet supernate without stirring up any sediment. Standardize this solution frequently by the following procedure (standard $KMnO_4$ solution, exactly 0.0500*N*, is equivalent to 1.002 mg Ca/1.00 mL):

Weigh to the nearest 0.1 mg several 100- to 200-mg samples of anhydrous $Na_2C_2O_4$ into 400-mL beakers. To each beaker, in turn, add 100 mL distilled water and stir to dissolve. Add 10 mL 1+ 1 H_2SO_4 and heat rapidly to 90 to 95 C. Titrate rapidly with permanganate solution to be standardized, while stirring, to a slight pink endpoint color that persists for at least 1 min. Do not let temperature fall below 85 C. If necessary, warm beaker contents during titration; 100 mg will consume about 30 mL solution. Run a blank on distilled water and H_2SO_4.

$$\text{Normality of } KMnO_4 = \frac{\text{g } Na_2C_2O_4}{(A - B) \times 0.06701}$$

where:

A = mL titrant for sample and
B = mL titrant for blank.

. Average the results of several titrations.

4. Procedure

a. Pretreatment of polluted water and wastewater samples: Follow procedure described in Section 302D or G.

b. Treatment of sample: Use 200 mL sample, containing not more than 50 mg Ca (or a smaller portion diluted to 200 mL). If interfering substances are present, proceed as follows:

1) Removal of silica interference—Remove interfering amounts of silica by the gravimetric procedure described in Silica, Section 425. Discard silica precipitate and save filtrate for removal of interfering amounts of combined oxides described in ¶*c* below. If combined oxides are absent, proceed to ¶*d*.

2) Removal of combined oxides interference—Remove interfering amounts of aluminum, iron, and manganese by concentrating filtrate from gravimetric silica determination to 120 to 150 mL. Add enough HCl so that filtrate from silica removal contains at least 10 mL conc HCl. Add 2 to 3 drops methyl red indicator and 3*N* NH_4OH until indicator color turns yellow.

Add 1 g ammonium persulfate; when boiling begins, carefully add 3*N* NH_4OH until solution becomes slightly alkaline and the steam bears a distinct but not strong odor of ammonia. Test with litmus paper. Boil for 1 to 2 min and let stand 10 min, until the hydroxides coagulate, but no longer. Filter precipitate and wash three or four times with NH_4Cl solution. Treat filtrate as described in the following.

c. pH adjustment of sample: To 200 mL sample, containing not more than 50 mg Ca, or to a smaller portion diluted to 200 mL, add 2 or 3 drops methyl red indicator solution. Neutralize with 1 + 1 HCl and boil for 1 min. Add 50 mL $(NH_4)_2C_2O_4 \cdot H_2O$ solution and if any precipitate forms, add just enough 1 + 1 HCl to redissolve.

d. Precipitation of calcium oxalate: Keeping solution just below boiling point, add 3*N* NH_4OH dropwise from a buret, stirring constantly. Continue addition until the solution is quite turbid (about 5 mL are required.) Digest for 90 min at 90 C. Filter, preferably through a filter crucible, using suction. Wash at once with 1 + 99 NH_4OH. Although it is not necessary to transfer all of the precipitate to the filter crucible, remove all excess $(NH_4)_2C_2O_4 \cdot H_2O$ from the beaker. (If magnesium is to be determined gravimetrically, set aside combined filtrate and washings for this purpose.)

e. Titration of calcium oxalate: Place filter crucible on its side in beaker and cover with distilled water. Add 10 mL H_2SO_4 and, while stirring, heat rapidly to 90 to 95 C. Titrate rapidly with $KMnO_4$ titrant to a slightly pink end point that persists for at least 1 min. Do not let temperature fall below 85 C; if necessary, warm beaker contents during titration. Agitate crucible sufficiently to insure reaction of all the oxalate. Run a blank, using a clean beaker and crucible, 10 mL H_2SO_4, and about the same volume of distilled water used in titrating sample.

5. Calculation

$$\text{mg Ca/L} = \frac{(A - B) \times N \times 20{,}040}{\text{mL sample}}$$

where:

A = mL titrant for sample,
B = mL titrant for blank, and
N = normality of $KMnO_4$.

6. Precision and Accuracy

A synthetic sample containing 108 mg Ca/L, 82 mg Mg/L, 3.1 mg K/L, 19.9 mg Na/L, 241 mg chloride/L, 1.1 mg NO_3^--N/L, 0.25 mg NO_2^-−N/L, 259 mg sulfate/L, and 42.5 mg total alkalinity/L (contributed by $NaHCO_3$) in distilled water was analyzed in six laboratories by the $KMnO_4$ titrimetric method, with a relative standard deviation of 3.5% and a relative error of 2.8%.

311 C. EDTA Titrimetric Method

1. General Discussion

a. Principle: When EDTA (ethylenediaminetetraacetic acid or its salts) is added to water containing both calcium and magnesium, it combines first with the calcium. Calcium can be determined directly, with EDTA, when the pH is made sufficiently high that the magnesium is largely precipitated as the hydroxide and an indicator is used that combines with calcium only. Several indicators give a color change when all of the calcium has been complexed by the EDTA at a pH of 12 to 13.

b. Interference: Under conditions of this test, the following concentrations of ions cause no interference with the calcium hardness determination: copper, 2 mg/L; ferrous iron, 20 mg/L; ferric iron, 20 mg/L; manganese, 10 mg/L; zinc, 5 mg/L; lead, 5 mg/L; aluminum, 5 mg/L; and tin, 5 mg/L. Orthophosphate precipitates calcium at the pH of the test. Strontium and barium interfere with the calcium determination and alkalinity in excess of 300 mg/L may cause an indistinct end point in hard waters.

2. Reagents

a. Sodium hydroxide, NaOH, 1*N*.

b. Indicators: Many indicators are available for the calcium titration. Some are described in the literature (see Bibliography, Section 311D); others are commercial preparations and also may be used. Murexide (ammonium purpurate) was the first indicator available for detecting the calcium end point, and directions for its use are presented in this procedure. Individuals who have difficulty recognizing the murexide end point may find the indicator Eriochrome Blue Black R (color index number 202) or Solochrome Dark Blue an improvement because of the color change from red to pure blue. Eriochrome Blue Black R is sodium-1-(2-hydroxy-1-naphthylazo) - 2 - naphthol - 4 - sulfonic acid. Other indicators specifically designed for use as end-point detectors in EDTA titration of calcium may be used.

1) *Murexide (ammonium purpurate) indicator:* This indicator changes from pink to purple at the end point. Prepare by dissolving 150 mg dye in 100 g absolute ethylene glycol. Water solutions of the dye are not stable for longer than 1 day. A ground mixture of dye powder and sodium chloride provides a stable form of the indicator. Prepare by mixing 200 mg murexide with 100 g solid NaCl and grinding the mixture to 40 to 50 mesh. Titrate immediately after adding indicator because it is unstable under alkaline conditions. Facilitate end-point recognition

by preparing a color comparison blank containing 2.0 mL NaOH solution, 0.2 g solid indicator mixture (or 1 to 2 drops if a solution is used), and sufficient standard EDTA titrant (0.05 to 0.10 mL) to produce an unchanging color.

2) *Eriochrome Blue Black R indicator;* Prepare a stable form of the indicator by grinding together in a mortar 200 mg powdered dye and 100 g solid NaCl to 40 to 50 mesh. Store in a tightly stoppered bottle. Use 0.2 g of ground mixture for the titration in the same manner as murexide indicator. During titration the color changes from red through purple to bluish purple to a pure blue with no trace of reddish or purple tint. The pH of some (not all) waters must be raised to 14 (rather than 12 to 13) by the use of 8*N* NaOH to get a good color change.

c. Standard EDTA titrant, 0.01*M:* Prepare standard EDTA titrant as described for the EDTA total-hardness method (Section 314). Standard EDTA titrant, 0.0100*M*, is equivalent to 400.8 μg Ca/1.00 mL.

3. Procedure

a. Pretreatment of polluted water and wastewater samples: Follow the procedure described in Section 302D or G.

b. Sample preparation: Because of the high pH used in this procedure, titrate immediately after adding alkali and indicator. Use 50.0 mL sample, or a smaller portion diluted to 50 mL so that the calcium content is about 5 to 10 mg. Analyze hard waters with alkalinity higher than 300 mg $CaCO_3$/L by taking a smaller portion and diluting to 50 mL, or by neutralizing the alkalinity with acid, boiling 1 min, and cooling before beginning the titration.

c. Titration: Add 2.0 mL NaOH solution or a volume sufficient to produce a pH of 12 to 13. Stir. Add 0.1 to 0.2 g indicator mixture selected (or 1 to 2 drops if a solution is used). Add EDTA titrant slowly, with continuous stirring to the proper end point. When using murexide, check end point by adding 1 to 2 drops of titrant in excess to make certain that no further color change occurs.

4. Calculation

$$\text{mg Ca/L} = \frac{A \times B \times 400.8}{\text{mL sample}}$$

Calcium hardness as mg $CaCO_3$/L

$$= \frac{A \times B \times 1{,}000}{\text{mL sample}}$$

where:

A = mL titrant for sample and
B = mg $CaCO_3$ equivalent to 1.00 mL EDTA titrant at the calcium indicator end point.

5. Precision and Accuracy

A synthetic sample containing 108 mg Ca/L, 82 mg Mg/L, 3.1 mg K/L, 19.9 mg Na/L, 241 mg chloride/L, 1.1 mg NO_3^--N/L, 0.25 mg NO_2^--N/L, 259 mg sulfate/L, and 42.5 mg total alkalinity/L (contributed by $NaHCO_3$) in distilled water was analyzed in 44 laboratories by the EDTA titrimetric method, with a relative standard deviation of 9.2% and a relative error of 1.9%.

311 D. Bibliography

Permanganate Titrimetric Methods

KOLTHOFF, I.M., E.J. MEEHAN, E.B. SANDELL & S. BRUCKENSTEIN. 1969. Quantitative Chemical Analysis, 4th ed. Macmillan Co., New York, N.Y.

EDTA Titrimetric Method

DIEHL, H. & J.L. ELLINGBOE. 1956. Indicator for titration of calcium in the presence of magnesium using disodium dihydrogen ethylenediamine tetraacetate. *Anal. Chem.* 28:882.

PATTON, J. & W. REEDER. 1956. New indicator

for titration of calcium with (ethylenedinitrilo) tetraacetate. *Anal. Chem.* 28:1026.

HILDEBRAND, G.P. & C.N. REILLEY. 1957. New indicator for complexometric titration of calcium in the presence of magnesium. *Anal. Chem.* 29:258.

SCHWARZENBACH, G. 1957. Complexometric Titrations. Interscience Publishers, New York, N.Y.

FURMAN, N.H. 1962. Scotts Standard Methods of Chemical Analysis, 6th ed. D. Van Nostrand Co., Inc., Princeton, N.J.

KATZ, H. & R. NAVONE. 1964. Method for simultaneous determination of calcium and magnesium. *J. Amer. Water Works Ass.* 56:121.

312 CHROMIUM

The hexavalent chromium concentration of U.S. drinking waters has been reported to vary between 3 and 40 μg/L with a mean of 3.2 μg/L. Chromium salts are used extensively in industrial processes and may enter a water supply through the discharge of wastes. Chromate compounds frequently are added to cooling water for corrosion control. Chromium may exist in water supplies in both the hexavalent and the trivalent state although the trivalent form rarely occurs in potable water.

1. Selection of Method

Use the colorimetric method for the determination of hexavalent chromium in a natural or treated water intended to be potable. Use the atomic absorption spectrophotometric and colorimetric methods for the determination of total chromium in water and wastewater.

2. Sample Handling

If only the dissolved metal content is desired, filter sample through a 0.45-μm membrane filter at the time of collection. After filtration acidify filtrate with conc nitric acid (HNO_3) to pH <2. If the total chromium content is desired, acidify unfiltered sample at time of collection with conc HNO_3 to pH <2.

312 A. Atomic Absorption Method for Total Chromium

See Section 303A and B.

312 B. Colorimetric Method

1. General Discussion

a. Principle: This procedure measures only hexavalent chromium. Therefore, to determine total chromium, convert all the chromium to the hexavalent state by oxidation with potassium permanganate. The hexavalent chromium is determined colorimetrically by reaction with diphenylcarbazide in acid solution. A red-violet color of unknown composition is produced. The reaction is very sensitive, the molar absorptivity based on chromium being about 40,000 L g^{-1} cm^{-1} at 540 nm. To determine total chronium, digest the sample with a sulfuric-nitric acid mixture and then oxidize with potassium permanganate before reacting with the diphenylcarbazide.

b. Interferences: The reaction with diphenylcarbazide is nearly specific for chromium. Hexavalent molybdenum and mercury salts will react to form color with the reagent but the intensities are much lower than that for chromium at the specified pH. Concentrations of molybdenum and mercury up to 200 mg/L can be tolerated. Vanadium interferes strongly but concentrations up to 10 times that of chromium will not cause trouble. Potential interference from permanganate is eliminated by prior reduction with azide. Iron in concentrations greater than 1 mg/L may produce a yellow color but the ferric ion color is not strong and no difficulty is encountered normally if the absorbance is measured photometrically at the appropriate wavelength. Interfering amounts of molydenum, vanadium, iron, and copper can be removed by extraction of the cupferrates of these metals into chloroform ($CHCl_3$). A procedure for this extraction is provided but should not be used unless necessary, because residual cupferron and $CHCl_3$ in the aqueous solution complicate the later oxidation. Therefore, the extraction is followed by additional treatment with acid fuming in order to decompose these compounds.

2. Apparatus

a. Colorimetric equipment: One of the following is required:

1) *Spectrophotometer,* for use at 540 nm, with a light path of 1 cm or longer.

2) *Filter photometer,* providing a light path of 1 cm or longer and equipped with a greenish yellow filter having maximum transmittance near 540 nm.

b. Separatory funnels, 125-mL, Squibb form, with glass or TFE stopcock and stopper.

3. Reagents

Use redistilled water to prepare reagents.

a. Stock chromium solution: Dissolve 141.4 mg $K_2Cr_2O_7$ in water and dilute to 1,000 mL; 1.00 mL = 50.0 μg Cr.

b. Standard chromium solution: Dilute 10.0 mL stock chromium solution to 100 mL; 1.00 mL = 5.00 μg Cr.

c. Nitric acid, HNO_3, conc.

d. Sulfuric acid, H_2SO_4, 1 + 1.

e. Methyl orange indicator solution.

f. Hydrogen peroxide, H_2O_2, 30%.

g. Redistilled water: Distilled water redistilled in all-glass apparatus.

h. Ammonium hydroxide, NH_4OH, conc.

i. Potassium permanganate solution: Dissolve 4 g $KMnO_4$ in 100 mL water.

j. Sodium azide solution: Dissolve 0.5 g NaN_3 in 100 mL water.

k. Diphenylcarbazide solution: Dissolve 250 mg 1,5-diphenylcarbazide (1,5-diphenylcarbohydrazide) in 50 mL acetone. Store in a brown bottle. Discard when solution becomes discolored.

l. Chloroform, $CHCl_3$: Avoid or redistill material that comes in containers with metal or metal-lined caps.

m. Cupferron solution: Dissolve 5 g $C_6H_5N(NO)ONH_4$ in 95 mL water.

n. Phosphoric acid, H_3PO_4, conc.

o. Sulfuric acid, H_2SO_4, 0.2*N*: Dilute 17 mL 6*N* H_2SO_4 to 500 mL with water.

4. Procedure

a. Preparation of calibration curve: To compensate for possible slight losses of chromium during digestion or other analytical operations, treat chromium standards by the same procedure as the sample. Accordingly, pipet measured volumes of standard chromium solution (5 μg/mL) ranging from 2.00 to 20.0 mL, to give standards for 10 to 100 μg Cr, into 250-mL beakers or conical flasks. Depending on pretreatment used in ¶ *b* below, proceed with subsequent treatment of standards as if they were samples, also carrying out cupferron treatment of standards if this is required for samples.

Develop color as for samples, transfer a suitable portion of each colored solution to a 1-cm absorption cell, and measure absorbance at 540 nm. As reference, use distilled water. Correct absorbance readings of standards by subtracting absorbance of a reagent blank carried through the method.

Construct a calibration curve by plotting corrected absorbance values against micrograms chromium in 102 mL final volume.

b. Treatment of sample: If sample has been filtered and acidified and only hexavalent chromium is desired, proceed to ¶ 4*e*. If total dissolved chromium is desired and there are interfering amounts of molybdenum, vanadium, copper, or iron present, proceed to ¶ 4*c*. If interferences are not present, proceed to ¶ 4*d*. If sample is unfiltered and total chromium is desired, digest with HNO_3 and H_2SO_4 as in Section 302E. If interferences are present, proceed to ¶s 4*c*, 4*d*, and 4*e*. If there are no interferences, proceed to ¶s 4*d* and 4*e*.

c. Removal of molybdenum, vanadium, iron, and copper with cupferron: Pipet a portion of digested sample containing 10 to 100 μg chromium into a 125-mL separatory funnel. Dilute to about 40 mL with distilled water and chill in an ice bath. Add 5 mL ice-cold cupferron solution, shake well, and let stand in ice bath for 1 min. Extract in separatory funnel with three successive 5-mL portions of $CHCl_3$; shake each portion thoroughly with aqueous solution, let layers separate, and withdraw and discard $CHCl_3$ extract. Transfer extracted aqueous solution to a 125-mL conical flask. Wash separatory funnel with a small amount of distilled water and add wash water to flask. Boil for about 5 min to volatilize $CHCl_3$ and cool. Add 5 mL HNO_3 and sufficient H_2SO_4 to have about 3 mL present. Boil samples to the appearance of SO_3 fumes. Cool slightly, carefully add 5 mL HNO_3, and again boil to fumes to complete decomposition of organic matter. Cool, wash sides of flask, and boil once more to SO_3 fumes to eliminate all HNO_3. Cool and add 25 mL water.

d. Oxidation of trivalent chromium: Pipet a portion of digested sample with or without interferences removed, and containing 10 to 100 μg chromium, into a 125-mL conical flask. Using methyl orange as indicator, add conc NH_4OH until solution is just basic to methyl orange. Add 1 + 1 H_2SO_4 dropwise until it is acidic, plus 1 mL (20 drops) in excess. Adjust volume to about 40 mL, add a boiling chip, and heat to boiling. Add 2 drops $KMnO_4$ solution to give a dark red color. If fading occurs, add $KMnO_4$ dropwise to maintain an excess of about 2 drops. Boil for 2 min longer. Add 1 mL NaN_3 solution and continue boiling gently. If red color does not fade completely after boiling for approximately 30 sec, add another 1 mL NaN_3 solution. Continue boiling for 1 min after color has faded completely, then cool. Add 0.25 mL (5 drops) H_3PO_4.

e. Color development and measurement: Use 0.2*N* H_2SO_4 and a pH meter to adjust solution to pH 1.0 ± 0.3. Transfer solution to a 100-mL volumetric flask, dilute to 100 mL, and mix. Add 2.0 mL diphenylcarbazide solution, mix, and let stand 5 to 10 min for full color development. Transfer an appropriate portion to a 1-cm absorption cell and measure its absorbance at 540 nm. Use distilled water as reference. Correct absorbance reading of sample by subtracting absorbance of a blank carried through the method (see also note below). From the corrected absorbance, determine micrograms chromium present by reference to the calibration curve.

NOTE: If the solution is turbid after dilution to 100 mL in ¶ *e* above, take an absorbance reading before adding carbazide reagent and correct absorbance reading of final colored solution by subtracting the absorbance measured previously.

5. Calculation

$$\text{mg Cr/L} = \frac{\mu\text{g Cr (in 102 mL final volume)}}{A \times B} \times 100$$

where:

A = mL original sample, and
B = mL portion from 100 mL digested sample.

6. Precision and Accuracy

The dissolved (trivalent plus hexavalent) chromium was determined in 31 laboratories in a synthetic sample containing 110 μg Cr/L, 500 μg Al/L, 50 μg Cd/L, 470 μg Ca/L, 300 μg Fe/L, 70 μg Pb/L, 120 μg Mn/L, 150 μg Ag/L, and 650 μg Zn/L in distilled water. The relative standard deviation was 47.8% and the relative error 16.3%.

7. Bibliography

ROWLAND, G.P., JR. 1939. Photoelectric colorimetry—Optical study of permanganate ion and of chromium-diphenylcarbazide system. *Anal. Chem.* 11:442.

SALTZMAN, B.E. 1952. Microdetermination of chromium with diphenylcarbazide by permanganate oxidation. *Anal. Chem.* 24:1016.

URONE, P.F. 1955. Stability of colorimetric reagent for chromium, 5-diphenylcarbazide, in various solvents. *Anal. Chem.* 27:1354.

ALLEN, T.L. 1958. Microdetermination of chromium with 1,5-diphenylcarbohydrazide. *Anal. Chem.* 30:447.

SANDELL, E.B. 1959. Colorimetric Determination of Traces of Metals, 3rd ed. Interscience Publishers, New York, N.Y.

313 COPPER

Copper salts are used in water supply systems to control biological growths in reservoirs and distribution pipes and to catalyze the oxidation of manganese. Corrosion of copper-containing alloys in pipe fittings may introduce measurable amounts of copper into the water in a pipe system.

Copper is essential to humans and the adult daily requirement has been estimated at 2.0 mg.

1. Selection of Method

The atomic absorption spectrophotometric and the neocuproine methods are recommended because of their freedom from interferences. The bathocuproine method may be used for potable waters.

2. Sampling and Storage

Copper ion tends to be adsorbed on the surface of sample containers. Therefore, analyze samples as soon as possible after collection. If storage is necessary, use 0.5 mL 1 + 1 HCl/100 mL sample to prevent this adsorption.

313 A. Atomic Absorption Spectrophotometric Method

See Section 303A and B.

313 B. Neocuproine Method

1. General Discussion

a. Principle: Cuprous ion in neutral or slightly acidic solution reacts with 2,9-dimethyl-1,10-phenanthroline (neocuproine) to form a complex in which 2 moles of neocuproine are bound by 1 mole of Cu^+ ion. The complex can be extracted by a number of organic solvents, including a chloroform-methanol ($CHCl_3$-CH_3OH) mixture, to give a yellow solution with a molar absorptivity of about 8,000 at 457 nm. The reaction is virtually specific for copper; the color follows Beer's law up to a concentration of 0.2 mg Cu/25 mL solvent; full color development is obtained when the pH of the aqueous solution is between 3 and 9; the color is stable in $CHCl_3$-CH_3OH for several days.

The sample is treated with hydroxylamine-hydrochloride to reduce cupric ions to cuprous ions. Sodium citrate is used to complex metallic ions that might precipitate when the pH is raised. The pH is adjusted to 4 to 6 with NH_4OH, a solution of neocuproine in methanol is added, and the resultant complex is extracted into $CHCl_3$. After dilution of the $CHCl_3$ to an exact volume with CH_3OH, the absorbance of the solution is measured at 457 nm.

b. Interference: Large amounts of chromium and tin may interfere. Avoid interference from chromium by adding sulfurous acid to reduce chromate and complex chromic ion. In the presence of much tin or excessive amounts of other oxidizing ions, use up to 20 mL additional hydroxylamine-hydrochloride solution.

Cyanide, sulfide, and organic matter interfere but can be removed by a digestion procedure.

c. Minimum detectable concentration: The minimum detectable concentration, corresponding to 0.01 absorbance or 98% transmittance, is 3 μg Cu when a 1-cm cell is used and 0.6 μg Cu when a 5-cm cell is used.

2. Apparatus

a. Colorimetric equipment—One of the following is required:

1) *Spectrophotometer,* for use at 457 nm, providing a light path of 1 cm or longer.

2) *Filter photometer,* providing a light path of 1 cm or longer and equipped with a narrow-band violet filter having maximum transmittance in the range 450 to 460 nm.

b. Separatory funnels, 125-mL, Squibb form, with glass or TFE stopcock and stopper.

3. Reagents

a. Redistilled water, copper-free: Because most ordinary distilled water contains detectable amounts of copper, use redistilled water, prepared by distillation of singly distilled water in a resistant-glass still, or distilled water passed through an ion-exchange unit, to prepare all reagents and dilutions.

b. Stock copper solution: To 200.0 mg polished electrolytic copper wire or foil in a 250-mL conical flask, add 10 mL water and 5 mL conc HNO_3. After the reaction has slowed, warm gently to complete dissolution of the copper and boil to expel oxides of nitrogen, using precautions to avoid loss of copper. Cool, add about 50 mL water, transfer quantitatively to a 1-L volumetric flask, and dilute to the mark with water; 1 mL = 200 μg Cu.

c. Standard copper solution: Dilute 50.00 mL stock copper solution to 500 mL with water; 1.00 mL = 20.0 μg Cu.

d. Sulfuric acid, H_2SO_4, conc.

e. Hydroxylamine-hydrochloride solution: Dissolve 50 g $NH_2OH{\cdot}HCl$ in 450 mL water.

f. Sodium citrate solution: Dissolve 150 g $Na_3C_6H_5O_7{\cdot}2H_2O$ in 400 mL water. Add 5 mL $NH_2OH{\cdot}HCl$ solution and 10 mL neocuproine reagent. Extract with 50 mL

$CHCl_3$ to remove copper impurities and discard $CHCl_3$ layer.

g. Ammonium hydroxide, NH_4OH, 5*N*: Dilute 330 mL conc NH_4OH (28-29%) to 1,000 mL with water. Store in a polyethylene bottle.

h. Congo red paper, or other pH test paper showing a color change in the pH range of 4 to 6.

i. Neocuproine reagent: Dissolve 100 mg 2,9-dimethyl-1,10-phenanthroline hemihydrate* in 100 mL methanol. This solution is stable under ordinary storage conditions for a month or more.

j. Chloroform, $CHCl_3$: Avoid or redistill material that comes in containers with metal-lined caps.

k. Methanol, CH_3OH, reagent grade.

l. Nitric acid, HNO_3, conc.

m. Hydrochloric acid, HCl, conc.

4. Procedure

a. Preparation of calibration curve: Pipet 50 mL water into a 125-mL separatory funnel for use as a reagent blank. Prepare standards by pipetting 1.00 to 10.00 mL (20.0 to 200 μg Cu) standard copper solution into a series of 125-mL separatory funnels, and dilute to 50 mL with water. Add 1 mL conc H_2SO_4 and use the extraction procedure given in ¶ 4*b* below.

Construct a calibration curve by plotting absorbance versus micrograms of copper.

To prepare a calibration curve for smaller amounts of copper, dilute 10.0 mL standard copper solution to 100 mL. Carry 1.00- to 10.00-mL volumes of this diluted standard through the previously described procedure, but use 5-cm cells to measure absorbance.

b. Treatment of sample: Transfer 100 mL sample to a 250-mL beaker, add 1 mL conc H_2SO_4 and 5 mL conc HNO_3. Add a few boiling chips and cautiously evaporate to dense white SO_3 fumes on a hot plate. If solution remains colored, cool, add another 5 mL conc HNO_3, and again evaporate to dense white fumes. Repeat, if necessary, until solution becomes colorless.

Cool, add about 80 mL water, and bring to a boil. Cool and filter into a 100-mL volumetric flask. Make up to 100 mL with water using mostly beaker and filter washings.

Pipet 50.0 mL or other suitable portion containing 4 to 200 μg Cu, from the solution obtained from preliminary treatment, into a 125-mL separatory funnel. Dilute, if necessary, to 50 mL with water. Add 5 mL $NH_2OH{\cdot}HCl$ solution and 10 mL sodium citrate solution, and mix thoroughly. Adjust pH to approximately 4 by adding 1-mL increments of NH_4OH until Congo Red paper is just definitely red (or other suitable pH test paper indicates a value between 4 and 6).

Add 10 mL neocuproine reagent and 10 mL $CHCl_3$. Stopper and shake vigorously for 30 sec or more to extract the copper-neocuproine complex into the $CHCl_3$. Let mixture separate into two layers and withdraw lower $CHCl_3$ layer into a 25-mL volumetric flask, taking care not to transfer any of the aqueous layer. Repeat extraction of the water layer with an additional 10 mL $CHCl_3$ and combine extracts. Dilute combined extracts to 25 mL with CH_3OH, stopper, and mix thoroughly.

Transfer an appropriate portion of extract to a suitable absorption cell (1 cm for 40 to 200 μg Cu; 5 cm for lesser amounts) and measure absorbance at 457 nm or with a 450- to 460-nm filter. Use a sample blank prepared by carrying 50 mL water through the complete digestion and analytical procedure.

Determine micrograms copper in final solution by reference to the appropriate calibration curve.

5. Calculation

$$\text{mg Cu/L} = \frac{\mu\text{g Cu (in 25 mL final volume)}}{\text{mL portion taken for extraction}}$$

*G. F. Smith Chemical Company, Columbus, Ohio, or equivalent.

313 C. Bathocuproine Method

1. General Discussion

a. Principle: Cuprous ion forms a water-soluble orange-colored chelate with bathocuproine disulfonate (2,9-dimethyl-4,7-diphenyl-1,10-phenanthrolinedisulfonic acid, disodium salt). While the color forms over the pH range 3.5 to 11.0, the recommended pH range is between 4 and 5.

The sample is buffered at a pH of about 4.3 and reduced with hydroxylamine hydrochloride. The absorbance is measured at 484 nm. The method can be applied to copper concentrations up to at least 5 mg/L with a sensitivity of 20 μg/L.

b. Interference: The following substances can be tolerated with an error of less than ±2%:

Substance	Concentration *mg/L*
Cations	
Aluminum	100
Beryllium	10
Cadmium	100
Calcium	1,000
Chromium (III)	10
Cobalt (II)	5
Iron (II)	100
Iron (III)	100
Lithium	500
Magnesium	100
Manganese (II)	500
Nickel (II)	500
Sodium	1,000
Strontium	200
Thorium (IV)	100
Zinc	200
Anions	
Chlorate	1,000
Chloride	1,000
Fluoride	500
Nitrate	200
Nitrite	200
Orthophosphate	1,000
Perchlorate	1,000
Sulfate	1,000
Compounds	
Residual chlorine	1
Linear alkylate sulfonate (LAS)	40

Cyanide, thiocyanate, persulfate, and EDTA also can interfere.

c. Minimum detectable concentration: 20 μg/L with a 5-cm cell.

2. Apparatus

a. Colorimetric equipment: One of the following, with a light path of 1 to 5 cm (unless nessler tubes are used):

1) *Spectrophotometer,* for use at 484 nm.

2) *Filter photometer,* equipped with a blue-green filter exhibiting maximum light transmission near 484 nm.

3) *Nessler tubes,* matched, 100-mL, tall form.

b. Acid-washed glassware: Rinse all glassware with conc HCl and then with copper-free water.

3. Reagents

a. Copper-free water: See Method B, ¶ 3*a*.

b. Stock copper solution: Prepare as directed in Method B, ¶ 3*b*, but use 20.00 mg copper wire or foil; 1.00 mL = 20.00 μg Cu.

c. Standard copper solution: Dilute 250 mL stock copper solution to 1,000 mL with water; 1.00 mL = 5.00 μg Cu. Prepare daily.

d. Hydrochloric acid, HCl, 1 + 1.

e. Hydroxylamine hydrochloride solution: See Method B, ¶3*e*.

f. Sodium citrate solution: Dissolve 300 g $Na_3C_6H_5O_7 \cdot 2H_2O$ in water and make up to 1,000 mL.

g. Disodium bathocuproine disulfonate solution: Dissolve 1.000 g $C_{12}H_4N_2(CH_3)_2(C_6H_4)_2(SO_3Na)_2$ in water and make up to 1,000 mL.

4. Procedure

Pipet 50.0 mL sample, or a suitable portion diluted to 50.0 mL, into a 250-mL erlenmeyer flask. In separate 250-mL erlenmeyer flasks, prepare a 50.0-mL water blank and a series of 50.0-mL copper standards containing 5.0, 10.0, 15.0, 20.0, and 25.0 μg Cu. To sample, blank, and standards add, mixing after each addition, 1.00 mL 1 + 1 HCl, 5.00 mL $NH_2OH{\cdot}HCl$ solution, 5.00 mL sodium citrate solution, and 5.00 mL disodium bathocuproine disulfonate solution. Transfer to cells and read sample absorbance against the blank at 484 nm. Plot absorbance against micrograms Cu in standards for the calibration curve. Estimate concentration from the calibration curve.

5. Calculation

$$\text{mg Cu/L} = \frac{\mu\text{g Cu (in 66 mL final volume)}}{\text{mL sample}}$$

6. Precision and Accuracy

A synthetic sample containing 1,000 μg Cu/L, 500 μg Al/L, 50 μg Cd/L, 110 μg Cr/L, 300 μg Fe/L, 70 μg Pb/L, 50 μg Mn/L. 150 μg Ag/L, and 650 μg Zn/L was analyzed in 33 laboratories by the bathocuproine method, with a relative standard deviation of 4.1% and a relative error of 0.3%.

313 D. Bibliography

Neocuproine Method

SMITH, G.F. & W.H. MCCURDY. 1952. 2,9-Dimethyl-1,10-phenanthroline: New specific in spectrophotometric determination of copper. *Anal. Chem.* 24:371.

LUKE, C.L. & M.E. CAMPBELL. 1953. Determination of impurities in germanium and silicon. *Anal. Chem.* 25:1586.

GAHLER, A.R. 1954. Colorimetric determination of copper with neocuproine. *Anal. Chem.* 26:577.

FULTON, J.W. & J. HASTINGS. 1956. Photometric determinations of copper in aluminum and lead-tin solder with neocuproine. *Anal. Chem.* 28:174.

FRANK, A.J., A.B. GOULSTON & A.A. DEACUTIS. 1957. Spectrophotometric determination of copper in titanium. *Anal. Chem.* 29:750.

Bathocuproine Method

SMITH, G.F. & D.H. WILKINS. 1953. New colorimetric reagent specific for copper. *Anal. Chem.* 25:510.

BORCHARDT, L.G. & J.P. BUTLER. 1957. Determination of trace amounts of copper. *Anal. Chem.* 29:414.

ZAK, B. 1958. Simple procedure for the single sample determination of serum copper and iron. *Clinica Chim. Acta* 3:328.

BLAIR, D. & H. DIEHL. 1961. Bathophenanthrolinedisulfonic acid and bathocuproinedisulfonic acid, water soluble reagents for iron and copper. *Talanta* 7:163.

314 HARDNESS

Originally, water hardness was understood to be a measure of the capacity of water to precipitate soap. Soap is precipitated chiefly by the calcium and magnesium ions present. Other polyvalent cations also may precipitate soap, but they often are in complex forms, frequently with organic constituents, and their role in water hardness may be minimal and difficult to define. In conformity with current practice, total hardness is defined as the sum of the calcium and magnesium concentrations, both expressed as calcium carbonate, in milligrams per liter.

When hardness numerically is greater than the sum of carbonate and bicarbonate alkalinity, that amount of hardness equivalent to the total alkalinity is called "carbonate hardness''; the amount of hardness in excess of this is called ''noncarbonate hardness.'' When the hardness numerically is equal to or less than the sum of carbonate and bicarbonate alkalinity, all hardness is carbonate hardness and noncarbonate hardness is absent. The hardness may range from zero to hundreds of milligrams per liter in terms of calcium carbonate, depending on the source and treatment to which the water has been subjected.

1. Selection of Method

Two methods are presented for the determination of hardness. Method A, hardness by calculation, is applicable to all waters and yields the higher accuracy. If a mineral analysis is performed, hardness by calculation can be reported. Method B, the EDTA titration method, measures the calcium and magnesium ions and may be applied with appropriate modification to any kind of water. The procedure described affords a means of rapid analysis.

2. Reporting Results

When reporting hardness, state the method used, for example, ''hardness (calc.)'' or ''hardness (EDTA)''.

314 A. Hardness by Calculation

1. Discussion

The preferred method for determining hardness is to compute it from the results of separate determinations of calcium and magnesium.

2. Calculation

$$\text{Hardness, mg equivalent } CaCO_3/L = 2.497\,[\text{Ca, mg/L}] + 4.118\,[\text{Mg, mg/L}]$$

314 B. EDTA Titrimetric Method

1. General Discussion

a. Principle: Ethylenediaminetetraacetic acid and its sodium salts (abbreviated EDTA) form a chelated soluble complex when added to a solution of certain metal cations. If a small amount of a dye such as Eriochrome Black T or Calmagite is added to an aqueous solution containing calcium and magnesium ions at a pH of 10.0 ± 0.1, the solution becomes wine red. If EDTA is added as a titrant, the calcium and magnesium will be complexed, and when all of the magnesium and calcium has been complexed the solution turns from wine red to blue, marking the end point of the titration. Magnesium ion must be present to yield a satisfactory end point. To insure this, a small amount of complexometrically neutral magnesium salt of EDTA is added to the buffer; this automatically introduces sufficient magnesium and obviates the need for a blank correction.

The sharpness of the end point increases with increasing pH. However, the pH cannot be increased indefinitely because of the danger of precipitating calcium carbonate, $CaCO_3$, or magnesium hydroxide, $Mg(OH)_2$, and because the dye changes color at high pH values. The specified pH of 10.0 ± 0.1 is a satisfactory compromise. A limit of 5 min is set for the duration of the titration to minimize the tendency toward $CaCO_3$ precipitation.

b. Interference: Some metal ions interfere by causing fading or indistinct end points or by stoichiometric consumption of EDTA. Reduce this interference by adding certain inhibitors before titration. Adding MgCDTA [see 2*b*3)], which is *not* an inhibitor, permits titrating all polyvalent cations listed in Table 314:I but yields erroneously high hardness values in proportion to the concentration of such cations. Because it is nontoxic it is the complexing agent of choice if the interferences are known to be so low that no significant addition to the hardness will result from their titration. When the indicated cations are present in significant concentrations, use the inhibitors listed in Table 314:I. The figures in Table 314:I are intended as a rough guide only and are based on using a 25-mL sample diluted to 50 mL.

TABLE 314:I. MAXIMUM CONCENTRATIONS OF INTERFERENCES PERMISSIBLE WITH VARIOUS INHIBITORS*

Interfering Substance	Max. Interference Concentration *mg/L*	
	Inhibitor I	Inhibitor II
Aluminum	20	20
Barium	†	†
Cadmium	†	20
Cobalt	over 20	0.3
Copper	over 30	20
Iron	over 30	5
Lead	†	20
Manganese (Mn^{2+})	†	1
Nickel	over 20	0.3
Strontium	†	†
Zinc	†	200
Polyphosphate		10

*Based on 25-mL sample diluted to 50 mL.
†Titrates as hardness.

Suspended or colloidal organic matter also may interfere with the end point. Eliminate this interference by evaporating the sample to dryness on a steam bath and heating in a muffle furnace at 550 C until the organic matter is completely oxidized. Dissolve the residue in 20 mL 1*N* hydrochloric acid (HCl), neutralize to pH 7 with 1*N* sodium hydroxide (NaOH), and make up to 50 mL with distilled water; cool to room temperature and continue according to the general procedure.

c. Titration precautions: Conduct titrations at or near normal room temperature. The color change becomes impractically slow as the sample approaches freezing temperature. Indicator decomposition becomes a problem in hot water.

The specified pH may produce an environment conducive to $CaCO_3$ precipitation. Although the titrant slowly redissolves such precipitates, a drifting end point often yields low results. Completion of the titration within 5 min minimizes the tendency for $CaCO_3$ to precipitate. The following three methods also reduce precipitation loss:

1) Dilute the sample with distilled water to reduce the $CaCO_3$ concentration. This simple expedient has been incorporated in the procedure. If precipitation occurs at this dilution of 1 + 1 use modification 2) or 3). Using too small a sample contributes a systematic error due to the buret-reading error.

2) If the approximate hardness is known or is determined by a preliminary titration, add 90% or more of titrant to sample *before* adjusting the pH with buffer.

3) Acidify the sample and stir for 2 min to expel CO_2 *before* pH adjustment. Determine alkalinity to indicate the amount of acid to be added.

2. Reagents

a. Buffer solution:

1) Dissolve 16.9 g ammonium chloride (NH_4Cl) in 143 mL conc ammonium hydroxide (NH_4OH). Add 1.25 g magnesium salt of EDTA (available commercially) and dilute to 250 mL with distilled water.

2) If the magnesium salt of EDTA is unavailable, dissolve 1.179 g disodium salt of ethylenediaminetetraacetic acid dihydrate (analytical reagent grade) and 780 mg magnesium sulfate ($MgSO_4 \cdot 7H_2O$) or 644 mg magnesium chloride ($MgCl_2 \cdot 6H_2O$) in 50 mL distilled water. Add this solution to 16.9 g NH_4Cl and 143 mL conc NH_4OH with mixing and dilute to 250 mL with distilled water. To attain the highest accuracy, adjust to exact equivalence through appropriate addition of a small amount of EDTA or $MgSO_4$ or $MgCl_2$.

Store Solution 1) or 2) in a plastic or resistant-glass container for no longer than 1 month. Stopper tightly to prevent loss of ammonia (NH_3) or pickup of carbon dioxide (CO_2). Dispense buffer solution by means of a bulb-operated pipet. Discard buffer when 1 or 2 mL added to the sample fails to produce a pH of 10.0 ± 0.1 at the titration end point.

3) Satisfactory alternate "odorless buffers" also are available commercially. They contain the magnesium salt of EDTA and have the advantage of being relatively odorless and more stable than the NH_4Cl-NH_4OH buffer. They usually do not provide as good an endpoint as NH_4Cl-NH_4OH because of slower reactions and they may be unsuitable when this method is automated. Prepare one of these buffers by mixing 55 mL conc HCl with 400 mL distilled water and then, slowly and with stirring, adding 300 mL 2-aminoethanol (free of aluminum and heavier metals). Add 5.0 g magnesium salt of EDTA and dilute to 1 L with distilled water.

b. Complexing agents: For most waters no complexing agent is needed. Occasionally water containing interfering ions requires the addition of an appropriate complexing agent to give a clear, sharp change in color at the end point. The following are satisfactory:

1) *Inhibitor I:* Adjust acid samples to pH 6 or higher with buffer or 0.1*N* NaOH. Add 250 mg sodium cyanide (NaCN) in powder form to the sample. Add sufficient buffer to adjust to pH 10.0 ± 0.1 (CAUTION: *NaCN is extremely poisonous. Take extra precautions in its use.* Flush solutions containing this inhibitor down the drain with large quantities of water after insuring that no acid is present to liberate volatile poisonous hydrogen cyanide.)

2) *Inhibitor II:* Dissolve 5.0 g sodium sulfide nonahydrate ($Na_2S \cdot 9H_2O$) or 3.7 g $Na_2S \cdot 5H_2O$ in 100 mL distilled water. Exclude air with a tightly fitting rubber stopper. This inhibitor deteriorates through air oxidation. It produces a sulfide precipitate that obscures the end point when appreciable concentrations of heavy metals are present. Use 1 mL in ¶ 3*b* below.

3) *MgCDTA:* Magnesium salt of 1, 2-cyclohexanediaminetetraacetic acid. Add 250 mg per 100 mL sample and dissolve completely before adding buffer solution. Use this complexing agent to avoid using toxic or odorous inhibitors when interfering substances are present in concentrations that affect the end point but will not contribute significantly to the hardness value.

Commercial preparations incorporating a buffer and a complexing agent are available. Such mixtures must maintain pH 10.0 ± 0.1 during the titration and give a clear, sharp end point when the sample is titrated.

c. Indicators: Many types of indicator solutions have been advocated and may be used if the analyst demonstrates that they yield accurate values. The prime difficulty with indicator solutions is deterioration with aging, giving indistinct end points. For example, alkaline solutions of Eriochrome Black T are sensitive to oxidants and aqueous or alcoholic solutions are unstable. In general, use the least amount of indicator providing a sharp end point. It is the analyst's responsibility to determine individually the optimal indicator concentration.

1) *Eriochrome Black T:* Sodium salt of 1-(1-hydroxy-2-naphthylazo)-5-nitro-2-naphthol-4-sulfonic acid; No. 203 in the Color Index. Dissolve 0.5 g dye in 100 g 2,2′,2″-nitrilotriethanol (also called triethanolamine) or 2-methoxymethanol (also called ethylene glycol monomethyl ether). Add 2 drops per 50 mL solution to be titrated. Adjust volume if necessary.

2) *Calmagite:* 1-(1-hydroxy-4-methyl-2-phenylazo)-2-naphthol-4-sulfonic acid. This is stable in aqueous solution and produces the same color change as Eriochrome Black T, with a sharper end point. Dissolve 0.10 g Calmagite in 100 mL distilled water. Use 1 mL per 50 mL solution to be titrated. Adjust volume if necessary.

3) Indicators 1 and 2 can be used in dry powder form if care is taken to avoid excess indicator. Prepared dry mixtures of these indicators and an inert salt are available commercially.

If the end point color change of these indicators is not clear and sharp, it usually means that an appropriate complexing agent is required. If NaCN inhibitor does not sharpen the end point, the indicator probably is at fault.

d. Standard EDTA titrant, 0.01*M:* Weigh 3.723 g analytical reagent-grade disodium ethylenediaminetetraacetate dihydrate, also called (ethylenedinitrilo)-tetraacetic acid disodium salt (EDTA), dissolve in distilled water, and dilute to 1,000 mL. Standardize against standard-calcium solution (¶ 2*e*) as described in ¶ 3*b* below.

Because the titrant extracts hardness-producing cations from soft-glass containers, store in polyethylene (preferable) or borosilicate glass bottles. Compensate for gradual deterioration by periodic restandardization and by using a suitable correction factor.

e. Standard calcium solution: Weigh 1.000 g anhydrous $CaCO_3$ powder (primary standard or special reagent low in heavy metals, alkalis, and magnesium) into a 500-mL erlenmeyer flask. Place a funnel in the flask neck and add, a little at a time, 1 + 1 HCl until all $CaCO_3$ has dissolved. Add 200 mL distilled water and boil for a few minutes to expel CO_2. Cool, add a few drops of methyl red indicator, and adjust to the intermediate orange color by adding 3*N* NH_4OH or 1 + 1 HCl, as required. Transfer quantitatively and dilute to 1,000 mL with distilled water; 1 mL = 1.00 mg $CaCO_3$.

f. Sodium hydroxide, NaOH, 0.1 *N*.

3. Procedure

a. Pretreatment of polluted water and wastewater samples: Follow the procedure described in Section 302E or F.

b. Titration of sample: Select a sample volume that requires less than 15 mL EDTA titrant and complete titration within 5 min, measured from the time of buffer addition.

Dilute 25.0 mL sample to about 50 mL with distilled water in a porcelain casserole or other suitable vessel. Add 1 to 2 mL buffer solution. Usually 1 mL will be sufficient to give a pH of 10.0 to 10.1. The absence of a sharp end-point color change in the titration usually means that an inhibitor must be added at this point in the procedure (¶ 2*b* et seq.) or that the indicator has deteriorated.

Add 1 to 2 drops indicator solution or an appropriate amount of dry-powder indicator formulation [¶ 2*c*3)]. Add standard EDTA titrant slowly, with continuous stirring, until the last reddish tinge disappears from the solution. Add the last few drops at 3- to 5-sec intervals. At the end point the solution normally is blue. Daylight or a daylight fluorescent lamp is highly recommended because ordinary incandescent lights tend to produce a reddish tinge in the blue at the end point.

If sufficient sample is available and interference is absent, improve accuracy by increasing sample size, as described in ¶ 3*c* below.

c. Low-hardness sample: For ion-exchanger effluent or other softened water and for natural waters of low hardness (less than 5 mg/L), take a larger sample, 100 to 1,000 mL, for titration and add proportionately larger amounts of buffer, inhibitor, and indicator. Add standard EDTA titrant slowly from a microburet and run a blank, using redistilled, distilled, or deionized water of the same volume as the sample, to which identical amounts of buffer, inhibitor, and indicator have been added. Subtract volume of EDTA used for blank from volume of EDTA used for sample.

4. Calculation

$$\text{Hardness (EDTA) as mg } CaCO_3\text{/L} = \frac{A \times B \times 1{,}000}{\text{mL sample}}$$

where:

A = mL titration for sample and
B = mg $CaCO_3$ equivalent to 1.00 mL EDTA titrant.

5. Precision and Accuracy

A synthetic sample containing 610 mg/L total hardness as $CaCO_3$ contributed by 108 mg Ca/L and 82 mg Mg/L, and the following supplementary substances: 3.1 mg K/L, 19.9 mg Na/L, 241 mg Cl/L, 0.25 mg NO_2^--N/L, 1.1 mg NO_3^--N/L, 259 mg sulfate/L, and 42.5 mg total alkalinity/L (contributed by $NaHCO_3$) in distilled water was analyzed in 56 laboratories by the EDTA titrimetric method with a relative standard deviation of 2.9% and a relative error of 0.8%.

314 C. Bibliography

CONNORS, J.J. 1950. Advances in chemical and colorimetric methods. *J. Amer. Water Works Ass.* 42:33.

DIEHL, H., C.A. GOETZ & C.C. HACH. 1950. The versenate titration for total hardness. *J. Amer. Water Works Ass.* 42:40.

BETZ, J.D. & C.A. NOLL. 1950. Total hardness determination by direct colorimetric titration. *J. Amer. Water Works Ass.* 42:49

GOETZ, C.A., T.C. LOOMIS & H. DIEHL. 1950. Total hardness in water: The stability of standard disodium dihydrogen ethylenediaminetetraacetate solutions. *Anal. Chem.* 22:798.

DISKANT, E.M. 1952. Stable indicator solutions for complexometric determination of total hardness in water. *Anal. Chem.* 24:1856.

BARNARD, A.J., JR., W.C. BROAD & H. FLASCHKA. 1956 & 1957. The EDTA titration. *Chemist Analyst* 45:86 & 46:46.

GOETZ, C.A. & R.C. SMITH. 1959. Evaluation of various methods and reagents for total hardness and calcium hardness in water. *Iowa State J. Sci.* 34:81 (Aug. 15).

SCHWARZENBACH, G. & H. FLASCHKA. 1969. Complexometric Titrations, 2nd ed. Barnes & Noble, Inc., New York, N.Y.

315 IRON

In filtered samples of oxygenated surface waters iron concentrations seldom reach 1 mg/L. Some groundwaters and acid surface drainage may contain considerably more iron. Iron in water can cause staining of laundry and porcelain. A bittersweet astringent taste is detectable by some persons at levels above 1 or 2 mg/L.

Under reducing conditions, iron exists in the ferrous state. In the absence of complex-forming ions, ferric iron is not significantly soluble unless the pH is very low. On exposure to air or addition of oxidants, ferrous iron is oxidized to the ferric state and may hydrolyze to form insoluble hydrated ferric oxide. This is the predominant form found in most laboratory samples unless the samples are collected and maintained under anoxic conditions to avoid oxidation.

The growth of bacteria in the sample during storage or shipment also may alter the form of iron present (see Section 918A). In acid wastes at pH less than 3.5, ferric iron also may be soluble.

Iron may be in true solution, in a colloidal state that may be peptized by organic matter, in inorganic or organic iron complexes, or in relatively coarse suspended particles. It may be either ferrous or ferric, suspended or filtrable.

Silt and clay in suspension may contain acid-soluble iron. Iron oxide particles sometimes are collected with a water sample as a result of flaking of rust from pipes. Iron may come from a metal cap used to close the sample bottle.

1. Selection of Method

For natural and treated waters, the orthophenanthroline method has attained the greatest acceptance for simplicity and reliability. The atomic absorption spectrophotometric method is relatively easy and accurate. The precision and accuracy data developed by a collaborative study of the atomic absorption method for iron showed results that were superior to those of the colorimetric methods.

It is difficult to distinguish between dissolved and suspended iron because, on exposure to air, soluble ferrous iron can be oxidized rapidly by dissolved oxygen and hydrolyzed at neutral pH values to insoluble ferric oxides. Dissolved iron can be determined (¶ 315B.4*b*) by subsequent analysis of a sample portion that has been filtered and acidified immediately after collection at the sampling site. This procedure may yield low results because of possible oxidation of ferrous iron and hydrolysis during filtration.

A rigorous quantitative distinction between ferrous and ferric iron may be obtained with a special procedure using bathophenanthroline.[1–2] The orthophenanthroline reagent tends to shift the soluble ferric-ferrous equilibrium to ferrous iron. The suggested procedure (¶ 315B.4*d*) has limited application and requires a large excess of orthophenanthroline (mole ratio to ferrous plus ferric greater than 30). The sample is stabilized with hydrochloric rather than acetic acid because the latter does not provide a pH low enough to stabilize the ferrous iron. However, take care while using an acid for stabilizing ferrous iron in the presence of ferric iron. Ferric iron may interfere through photochemical reduction if an acidified sample is exposed to light before adding bathophenanthroline and/or after the extraction of the colored complex.[1–2] Of the colorimetric methods, only the phenanthroline procedure and its extraction modification are outlined here.

Methods using bathophenanthroline[1–6] or ferrozine [3-(2-pryridyl)-5,6-bis-(4 phenylsulfonic acid)-1,2,4 triazine disodium salt][7] may be used to determine iron concentrations as low as 1 μg/L. A chemiluminescence procedure has a detection limit of 5 ng/L.[8] Other methods are described elsewhere.[9–12]

2. Sampling and Storage

Plan methods of collecting, storing, and pretreating samples in advance. Clean sample container with acid and rinse with distilled water. Equipment for membrane filtration of samples in the field may be required to determine iron in solution (filtrable iron). The value of the determination depends greatly on the care taken to obtain a representative sample. Iron in well water or tap samples may vary in concentration and form with duration and degree of flushing before and during sampling. When taking a sample portion for determining iron in suspension, shake the sample bottle often and vigorously to obtain a uniform suspension of precipitated iron. Use particular care when colloidal iron adheres to the sample bottle. This problem can be acute with plastic bottles.

For a precise determination of total iron, use a separate container for sample collection. Treat with acid at the time of collection to place the iron in solution and prevent adsorption or deposition on the walls of the sample container. Take account of the added acid in measuring portions for analysis. The addition of acid to the sample may eliminate the need for adding acid before digestion (¶ 315B.4*b*).

315 A. Atomic Absorption Spectrophotometric Method

See Sections 303A and B.

315 B. Phenanthroline Method

1. General Discussion

a. Principle: Iron is brought into solution, reduced to the ferrous state by boiling with acid and hydroxylamine, and treated with 1,10-phenanthroline at pH 3.2 to 3.3. Three molecules of phenanthroline chelate each atom of ferrous iron to form an orange-red complex. The colored solution obeys Beer's law; its intensity is independent of pH from 3 to 9. A pH between 2.9 and 3.5 insures rapid color development in the presence of an excess of phenanthroline. Color standards are stable for at least 6 months.

b. Interference: Among the interfering substances are strong oxidizing agents, cyanide, nitrite, and phosphates (polyphosphates more so than orthophosphate), chromium, zinc in concentrations exceeding 10 times that of iron, cobalt and copper in excess of 5 mg/L, and nickel in excess of 2 mg/L. Bismuth, cadmium, mercury, molybdate, and silver precipitate phenanthroline. The initial boiling with acid converts polyphosphates to orthophosphate and removes cyanide and nitrite that otherwise would interfere. Adding excess hydroxylamine eliminates errors caused by excessive concentrations of strong oxidizing reagents. In the presence of interfering metal ions, use a larger excess of phenanthroline to replace that complexed by the interfering metals. Where excessive concentrations of interfering metal ions are present, the extraction method may be used.

If noticeable amounts of color or organic matter are present, it may be necessary to evaporate the sample, gently ash the residue, and redissolve in acid. The ashing may be carried out in silica, porcelain, or platinum crucibles that have been boiled for several hours in 1 + 1 HCl. The presence of excessive amounts of organic matter may necessitate digestion before use of the extraction procedure.

c. Minimum detectable concentration: Total, dissolved, or ferrous iron concentrations between 0.02 and 4.0 mg/L can be determined directly, and higher concentrations can be determined by using smaller samples or dilutions. The minimum detectable quantity is 50 μg with a spectrophotometer (510 nm) with a 1-cm cell.

2. Apparatus

a. Colorimetric equipment: One of the following is required:

1) *Spectrophotometer,* for use at 510 nm, providing a light path of 1 cm or longer.

2) *Filter photometer,* providing a light path of 1 cm or longer and equipped with a green filter having maximum transmittance near 510 nm.

3) *Nessler tubes,* matched, 100 mL tall form.

b. Acid-washed glassware: Wash all glassware with conc hydrochloric acid (HCl) and rinse with distilled water before use to remove deposits of iron oxide.

c. Separatory funnels: 125 mL, Squibb form, with ground-glass or TFE stopcocks and stoppers.

3. Reagents

Use reagents low in iron. Use iron-free distilled water. Store reagents in glass-stoppered bottles. The hydrochloric acid and ammonium acetate solutions are stable indefinitely if tightly stoppered. The hydroxylamine, phenanthroline, and stock iron solutions are stable for several months. The standard iron solutions are not stable; prepare daily as needed by diluting the stock solution. Visual standards in nessler tubes are stable for several months if sealed and protected from light.

a. Hydrochloric acid, HCl, conc, containing less than 0.00005% iron.

b. Hydroxylamine solution: Dissolve 10 g $NH_2OH \cdot HCl$ in 100 mL distilled water.

c. Ammonium acetate buffer solution: Dissolve 250 g $NH_4C_2H_3O_2$ in 150 mL distilled water. Add 700 mL conc (glacial) acetic acid. Because even a good grade of $NH_4C_2H_3O_2$ contains a significant amount of iron, prepare new reference standards with each buffer preparation.

d. Sodium acetate solution: Dissolve 200 g $NaC_2H_3O_2 \cdot 3H_2O$ in 800 mL distilled water.

e. Phenanthroline solution: Dissolve 100 mg 1,10-phenanthroline monohydrate, $C_{12}H_8N_2 \cdot H_2O$, in 100 mL distilled water by stirring and heating to 80 C. Do not boil. Discard the solution if it darkens. Heating is unnecessary if 2 drops conc HCl are added to the distilled water. (NOTE: One milliliter of this reagent is sufficient for no more than 100 μg Fe.)

f. Stock iron solution: Use metal (1) or salt (2) for preparing the stock solution.

1) Use electrolytic iron wire, or "iron wire for standardizing", to prepare the solution. If necessary, clean wire with fine sandpaper to remove any oxide coating and to produce a bright surface. Weigh 200.0 mg wire and place in a 1-L volumetric flask. Dissolve in 20 mL 6*N* sulfuric acid (H_2SO_4) and dilute to mark with iron-free distilled water; 1.00 mL = 200 μg Fe.

2) If ferrous ammonium sulfate is preferred, slowly add 20 mL conc H_2SO_4 to 50 mL distilled water and dissolve 1.404 g $Fe(NH_4)_2(SO_4)_2 \cdot 6H_2O$. Add 0.1*N* potassium permanganate ($KMnO_4$) dropwise until a faint pink color persists. Dilute to 1,000 mL with iron-free distilled water and mix; 1.00 mL = 200 μg Fe.

g. Standard iron solutions: Prepare daily for use.

1) Pipet 50.00 mL stock solution into a 1-L volumetric flask and dilute to mark with iron-free distilled water; 1.00 mL = 10.0 μg Fe.

2) Pipet 5.00 mL stock solution into a 1-L volumetric flask and dilute to mark with iron-free distilled water; 1.00 mL = 1.00 μg Fe.

h. Diisopropyl or isopropyl ether.

4. Procedure

a. Preparation of calibration curves for samples analyzed in accordance with ¶ 4f below:

1) Range 0 to 100 μg Fe/100 mL final so-

lution—Pipet 2.0, 4.0, 6.0, 8.0, and 10.0 mL standard iron solution [¶ 3*g*1)] into 100-mL volumetric flasks. Add 1.0 mL $NH_2OH \cdot HCl$ solution and 1 mL sodium acetate solution to each flask. Dilute each to about 75 mL with distilled water, add 10 mL phenanthroline solution, dilute to volume, mix thoroughly, and let stand for 10 min. Measure absorbance in a 5-cm cell at 510 nm against a reference blank prepared by treating distilled water with the specified amounts of all reagents except the standard iron solution. Alternatively, if distilled water is used as a reference, correct absorbance values for standard concentrations of iron by subtracting absorbance for a reagent blank against that for distilled water. Construct a calibration curve for absorbance against milligrams iron.

2) Range 50 to 500 μg Fe/100 mL final solution—Follow the procedure specified in the preceding paragraph, but use 10.0, 20.0, 30.0, 40.0, and 50.0 mL standard iron solution and measure absorbance in 1-cm cells.

b. Total iron: Mix sample thoroughly and measure 50.0 mL into a 125-mL erlenmeyer flask. (If the sample contains more than 2 mg Fe/L, dilute an accurately measured portion containing not more than 100 μg to 50 mL, or use more phenanthroline and a 1- or 2-cm light path.) Add 2 mL conc HCl and 1 mL $NH_2OH \cdot HCl$ solution. Add a few glass beads and heat to boiling. To insure dissolution of all the iron, continue boiling until volume is reduced to 15 to 20 mL. (If the sample is ashed as described in ¶ 1*b*, above, take up residue in 2 mL conc HCl and 5 mL distilled water.) Cool to room temperature and transfer to a 50- or 100-mL volumetric flask or nessler tube. Add 10 mL $NH_4C_2O_3H_2$ buffer solution and 2 mL phenanthroline solution, and dilute to mark with distilled water. Mix thoroughly and allow at least 10 to 15 min for maximum color development.

c. Filtrable iron: Immediately after collection filter sample through a 0.45-μm membrane filter into a vacuum flask containing 1 mL conc HCl/100 mL sample. Analyze filtrate for total filtrable iron (¶ 4*b*) and/or filtrable ferrous iron (¶ 4*d*). (This procedure also can be used in the laboratory with the understanding that normal sample exposure to air during shipment may result in precipitation of iron.)

Calculate suspended iron by subtracting filtrable from total iron.

d. Ferrous iron: To determine ferrous iron, acidify a separate sample with 2 mL conc HCl/100 mL sample at the time of collection. Fill bottle directly from sampling source and stopper. Immediately before analysis, withdraw a 50-mL portion of acidified sample and add 20 mL phenanthroline solution and 10 mL $NH_4C_2O_3H_2$ solution with vigorous stirring. Dilute to 100 mL and measure color intensity within 5 to 10 min. Do not expose to sunlight. (Color development is rapid in the presence of excess phenanthroline. The phenanthroline volume given is suitable for less than 50 μg total iron; if larger amounts are present, use a correspondingly larger volume of phenanthroline or a more concentrated reagent.)

Calculate ferric iron by subtracting ferrous from total iron.

e. Color measurement: Prepare a series of standards by accurately pipetting calculated volumes of standard iron solutions (use weaker solution to measure 1- to 10-μg portions) into 125-mL erlenmeyer flasks, diluting to 50 mL, and carrying out the steps in ¶ 4*b*.

For visual comparison, prepare a set of at least 10 standards, ranging from 1 to 100 μg Fe in the final 100-mL volume. Compare colors in 100-mL tall-form nessler tubes.

For photometric measurement, use Table 315:I as a rough guide for selection of proper light path. Read standards against distilled water set at zero absorbance and plot a calibration curve, includ-

TABLE 315:I. SELECTION OF LIGHT PATH LENGTH FOR VARIOUS IRON CONCENTRATIONS

Fe µg		
50-mL Final Volume	100-mL Final Volume	Light Path *cm*
50–200	100–400	1
25–100	50–200	2
10–40	20–80	5
5–20	10–40	10

ing a blank (see ¶ 3*c* and General Introduction.)

If samples are colored or turbid, carry a second set of samples through all steps of the procedure without adding phenanthroline. Instead of distilled water, use the prepared blanks to set photometer to zero absorbance and read each developed sample with phenanthroline against the corresponding blank without phenanthroline. Translate observed photometer readings into iron values by means of the calibration curve. This procedure does *not* compensate for interfering ions.

f. Samples containing organic interferences: Digest samples containing organic substances in substantial amounts according to the directions given in Sections 302E and F.

1) If a digested sample has been prepared according to the directions given in Section 302E or F, pipet 10.0 mL or other suitable portion containing 20 to 500 µg Fe into a 125-mL separatory funnel. If the volume taken is less than 10 mL, add distilled water to make up to 10 mL. To the separatory funnel add 15 mL conc HCl for a 10-mL aqueous volume; or, if the portion taken was greater than 10.0 mL, add 1.5 mL conc HCl/mL of sample. Mix, cool, and proceed with 4*f*3) below.

2) To prepare a sample solely for determining iron, measure a suitable volume containing 20 to 500 µg Fe and carry it through either of the digestion procedures described in Sections 302E and F. However, use only 5 mL H_2SO_4 or $HClO_4$ and omit H_2O_2. When digestion is complete, cool, dilute with 10 mL distilled water, heat almost to boiling to dissolve slowly soluble salts, and, if the sample is still cloudy, filter through a glass-fiber, sintered-glass, or porcelain filter, washing with 2 to 3 mL distilled water. Quantitatively transfer filtrate or clear solution to a 25-mL volumetric flask or graduate and make up to 25 mL with distilled water. Empty flask or graduate into a 125-mL separatory funnel, rinse with 5 mL conc HCl that is added to the funnel, and add 25 mL conc HCl measured with the same graduate or flask. Mix and cool to room temperature.

3) Extract the iron from the HCl solution in the separatory funnel by shaking for 30 sec with 25 mL isopropyl ether. Draw off lower acid layer into a second separatory funnel. Extract acid solution again with 25 mL isopropyl ether, drain acid layer into a suitable clean vessel, and combine the two portions of isopropyl ether. Pour acid layer back into second separatory funnel and re-extract with 25 mL isopropyl ether. Withdraw and discard acid layer and add ether layer to original funnel. Persistence of a yellow color in the HCl solution after three extractions does not signify incomplete separation of iron because copper, which is not extracted, gives a similar yellow color.

Shake combined ether extracts with 25 mL distilled water to return iron to aqueous phase and transfer lower aqueous layer to a 100-mL volumetric flask. Repeat extraction with a second 25-mL portion of distilled water, adding this to the first aqueous extract. Discard ether layer.

4) Add 1 mL $NH_2OH \cdot HCl$ solution, 10 mL phenanthroline solution, and 10 mL $Na_2C_2O_3H_2$ solution. Dilute to 100 mL with distilled water, mix thoroughly, and let stand for 10 min. Measure absorbance

at 510 nm using a 5-cm absorption cell for amounts of iron less than 100 μg or 1-cm cell for quantities from 100 to 500 μg. As reference, use either distilled water or a sample blank prepared by carrying the specified quantities of acids through the entire analytical procedure. If distilled water is used as reference, correct sample absorbance by subtracting absorbance of a sample blank.

Determine micrograms of iron in the sample from the absorbance (corrected, if necessary) by reference to the calibration curve prepared according to ¶ 4*a* above.

5. Calculation

When the sample has been treated according to 4*b, c, d, e,* or 4*f*2):

$$\text{mg Fe/L} = \frac{\mu\text{g Fe (in 100 mL final volume)}}{\text{mL sample}}$$

When the sample has been treated according to 4*f*1):

$$\text{mg Fe/L} = \frac{\mu\text{g Fe (in 100 mL final volume)}}{\text{mL sample}} \times \frac{100}{\text{mL portion}}$$

Report details of sample collection, storage, and pretreatment if they are pertinent to interpretation of results.

6. Precision and Accuracy

a. Precision and accuracy depend on the method of sample collection and storage, the method of color measurement, the iron concentration, and the presence of interfering color, turbidity, and foreign ions. In general, optimum reliability of visual comparison in nessler tubes is not better than 5% and often only 10%, whereas, under optimum conditions, photometric measurement may be reliable to 3% or 3 μg, whichever is greater. The sensitivity limit for visual observation in nessler tubes is approximately 1 μg Fe. Sample variability and instability may limit precision and accuracy of this determination more than will the errors of analysis itself. Serious divergences have been found in reports of different laboratories because of variations in methods of collecting and treating samples.

b. A synthetic sample containing 300 μg Fe/L, 500 μg Al/L, 50 μg Cd/L, 110 μg Cr/L, 470 μg Cu/L, 70 μg Pb/L, 120 μg Mn/L, 150 μg Ag/L, and 650 μg Zn/L in distilled water was analyzed in 44 laboratories by the phenanthroline method, with a relative standard deviation of 25.5% and a relative error of 13.3%.

315 C. References

1. LEE, G.F. & W. STUMM. 1960. Determination of ferrous iron in the presence of ferric iron using bathophenanthroline. *J. Amer. Water Works Ass.* 52:1567.
2. GHOSH, M.M., J.T. O'CONNOR & R.S. ENGELBRECHT. 1967. Bathophenanthroline method for the determination of ferrous iron. *J. Amer. Water Works Ass.* 59:897.
3. BLAIR, D. & H. DIEHL. 1961. Bathophenanthroline-disulfonic acid and bathocuproine-disulfonic acid, water soluble reagents for iron and copper. *Talanta* 7:163.
4. SHAPIRO, J. 1966. On the measurement of ferrous iron in natural waters. *Limnol. Oceanogr.* 11:293.
5. MCMAHON, J.W. 1967. The influence of light and acid on the measurement of ferrous iron in lake water. *Limnol. Oceanogr.* 12:437.
6. MCMAHON, J.W. 1969. An acid-free bathophenanthroline method for measuring dissolved ferrous iron in lake water. *Water Res.* 3:743.
7. STOOKEY, L.L. 1970. Ferrozine—a new re-

agent for iron. *Anal. Chem.* 42:779.
8. SEITZ, W.R. & D.M. HERCULES. 1972. Determination of trace amounts of iron (II) using chemiluminescence analysis. *Anal. Chem.* 44:2143.
9. MOSS, M.L. & M.G. MELLON. 1942. Colorimetric determination of iron with 2,2′-bipyridine and with 2,2′,2″-tripyridine. *Ind. Eng. Chem.*, Anal. Ed. 14:862.
10. WELCHER, F.J. 1947. Organic Analytical Reagents. D. Van Nostrand Co., Princeton, N.J., Vol. 3, pp. 100–104.
11. MORRIS, R.L. 1952. Determination of iron in water in the presence of heavy metals. *Anal. Chem.* 24:1376.
12. DOIG, M.T., III, & D.F. MARTIN. 1971. Effect of humic acids on iron analyses in natural water. *Water Res.* 5:689.

315 D. Bibliography

CHRONHEIM, G. & W. WINK. 1942. Determination of divalent iron (by o-nitrosophenol). *Ind. Eng. Chem.*, Anal. Ed. 14:447.

MEHLIG, R.P. & R.H. HULETT. 1942. Spectrophotometric determination of iron with o-phenanthroline and with nitro-o-phenanthroline. *Ind. Eng. Chem.*, Anal. Ed. 14:869.

CALDWELL, D.H. & R.B. ADAMS. 1946. Colorimetric determination of iron in water with o-phenanthroline. *J. Amer. Water Works Ass.* 38:727.

WELCHER, F.J. 1947. Organic Analytical Reagents. D. Van Nostrand Co., Princeton, N.J., Vol. 3, pp. 85–93.

KOLTHOFF, I.M., T.S. LEE & D.L. LEUSSING. 1948. Equilibrium and kinetic studies on the formation and dissociation of ferroin and ferrin. *Anal. Chem.* 20:985.

RYAN, J.A. & G.H. BOTHAM. 1949. Iron in aluminum alloys: Colorimetric determination using 1,10-phenanthroline. *Anal. Chem.* 21:1521.

REITZ, L.K., A.S. O'BRIEN & T.L. DAVIS. 1950. Evaluation of three iron methods using a factorial experiment. *Anal. Chem.* 22:1470.

SANDELL, E.B. 1959. Colorimetric Determination of Traces of Metals, 3rd ed. Interscience Publishers, New York, N.Y. Chapter 22.

BROWN, E., M.W. SKOUGSTAD & M.J. FISHMAN. 1970. Methods for Collection and Analysis of Water Samples for Dissolved Minerals and Gases. Chapter A1 in Book 5, Techniques of Water Resources Investigations of the United States Geological Survey. U.S. Geol. Surv., Washington, D.C.

316 LEAD

Lead is a serious cumulative body poison. Natural waters seldom contain more than 20 μg/L, although values as high as 400 μg/L have been reported. Lead in a water supply may come from industrial, mine, and smelter discharges or from the dissolution of old lead plumbing. Tap waters that are soft, acid, and not suitably treated may contain lead resulting from an attack on lead service pipes.

Selection of method: The atomic absorption spectrophotometric method is subject to interference in the flame mode and requires an extraction procedure for the low concentrations common in potable water. The dithizone method is more sensitive and is preferred by some analysts for low concentrations.

316 A. Atomic Absorption Spectrophotometric Method

See Section 303A and B.

316 B. Dithizone Method

1. General Discussion

a. Principle: An acidified sample containing microgram quantities of lead is mixed with ammoniacal citrate-cyanide reducing solution and extracted with dithizone in chloroform ($CHCl_3$) to form a cherry-red lead dithizonate. The color of the mixed color solution is measured photometrically.[1,2] Sample volume taken for analysis may be 2 L when digestion is used.

b. Interference: In a weakly ammoniacal cyanide solution (pH 8.5 to 9.5) dithizone forms colored complexes with bismuth, stannous tin, and monovalent thallium. In strongly ammoniacal citrate-cyanide solution (pH 10 to 11.5) the dithizonates of these ions are unstable and are extracted only partially.[3] This method uses a high pH, mixed color, single dithizone extraction. Interference from stannous tin and monovalent thallium is reduced further when these ions are oxidized during preliminary digestion. A modification of the method allows detection and elimination of bismuth interference. Excessive quantities of bismuth, thallium, and tin may be removed.[4]

Dithizone in $CHCl_3$ absorbs at 510 nm; control its interference by using nearly equal concentrations of excess dithizone in samples, standards, and blank.

The method is without interference for the determination of 0.0 to 30.0 μg Pb in the presence of 20 μg monovalent thallium, 100 μg stannous tin, 200 μg trivalent indium, and 1,000 μg each of divalent Ba, Cd, Co, Cu, Mg, Mn, Hg, Sr, and Zn, trivalent Al, Sb, As, Cr, Fe, and V, and phosphate and sulfate. Gram quantities of alkali metals do not interfere. A modification is provided to avoid interference from excessive quantities of bismuth or tin.

c. Preliminary sample treatment: At time of collection acidify with conc HNO_3 to pH < 2 but avoid excess HNO_3. Add 5 mL 0.1*N* iodine solution to avoid losses of volatile organo-lead compounds during handling and digesting of samples. Prepare a blank of lead-free distilled water and carry through the procedure.

d. Digestion of samples: Unless digestion is shown to be unnecessary, digest all samples for dissolved or total lead as described in 302E or F.

e. Minimum detectable concentration: 1.0 μg Pb/10 mL dithizone solution.

2. Apparatus

a. Spectrophotometer for use at 510 nm, providing a light path of 1 cm or longer.

b. pH meter.

c. Separatory funnels: 250-mL Squibb type. Clean all glassware, including sample bottles, with 1 + 1 HNO_3. Rinse thoroughly with distilled or deionized water.

d. Automatic dispensing burets: Use for all reagents to minimize indeterminate contamination errors.

3. Reagents

Prepare all reagents in lead-free distilled water.

a. Stock lead solution: Dissolve 0.1599 g lead nitrate, $Pb(NO_3)_2$ (minimum purity 99.5%), in approximately 200 mL water. Add 10 mL conc HNO_3 and dilute to 1,000 mL with water. Alternatively, dissolve 0.1000 g pure Pb metal in 20 mL 1 + 1 HNO_3 and dilute to 1,000 mL with water; 1.00 mL = 100 μg Pb.

b. Working lead solution: Dilute 20.0 mL stock solution to 1,000 mL with water; 1 mL = 2.00 μg Pb.

c. Nitric acid, HNO_3, 1 + 4: Dilute 200 mL conc HNO_3 to 1,000 mL with water.

d. Ammonium hydroxide, NH_4OH, 1 + 9: Dilute 10 mL conc NH_4OH to 100 mL with water.

e. Citrate-cyanide reducing solution: Dissolve 400 g dibasic ammonium citrate, $(NH_4)_2HC_6H_5O_7$, 20 g anhydrous sodium sulfite, Na_2SO_3, 10 g hydroxylamine hydrochloride, $NH_2OH{\cdot}HCl$, and 40 g potassium cyanide, KCN (CAUTION: *Poison*) in water and dilute to 1,000 mL. Mix this solution with 2,000 mL conc NH_4OH. *Do not pipet by mouth.*

f. Stock dithizone solution: See 102.7*b*, stock dithizone solution I.

g. Dithizone working solution: Dilute 100 mL stock dithizone solution to 250 mL with $CHCl_3$; 1 mL = 40 μg dithizone.

h. Special dithizone solution: Dissolve 250 mg dithizone in 250 mL $CHCl_3$. This solution may be prepared without purification because all extracts using it are discarded.

i. Sodium sulfite solution: Dissolve 5 g anhydrous Na_2SO_3 in 100 mL lead-free distilled water.

j. Iodine solution, 0.1*N*: Dissolve 40 g KI in 25 mL distilled water, add 12.7 g resublimed iodine, and dilute to 1,000 mL.

4. Procedure

a. With sample digestion: To a digested sample containing not more than 1 mL conc acid add 20 mL 1 + 4 HNO_3 and filter through lead-free filter paper* and filter funnel directly into a 250-mL separatory funnel. Rinse digestion beaker with 50 mL lead-free water and add to filter. Add 50 mL ammoniacal citrate-cyanide solution, mix, and cool to room temperature. Add 10 mL dithizone working solution, shake stoppered funnel vigorously for 30 sec, and let layers separate. Insert lead-free cotton in stem of separatory funnel and draw off lower layer. Discard 1 to 2 mL $CHCl_3$ layer, then fill absorption cell. Measure absorbance of extract at 510 nm, using dithizone solution 3*g* to zero spectrophotometer.

b. Without sample digestion: To 100 mL acidified sample (pH 2) in a 250-mL separatory funnel add 20 mL 1 + 4 HNO_3 and 50 mL citrate-cyanide reducing solution; mix. Add 10 mL dithizone working solution and proceed as in ¶ 4*a*.

c. Calibration curve: Plot concentration of at least five standards and a blank against absorbance. Determine concentration of Pb in extract from curve. All concentrations are μg Pb/10 mL final extract.

d. Removal of excess interferences: The dithizonates of bismuth, tin, and thallium differ from lead dithizonate in maximum absorbance. Detect their presence by measuring sample absorbance at 510 nm and at 465 nm. Calculate corrected absorbance of sample at each wavelength by subtracting absorbance of blank at same wavelength. Calculate ratio of corrected absorbance at 510 nm to corrected absorbance at 465 nm. The ratio of corrected absorbances for lead dithizonate is 2.08 and for bismuth dithizonate it is 1.07. If the ratio for the sample indicates interference, i.e., is markedly less than 2.08, proceed as follows with a new 100-mL sample: If the sample has not been digested, add 5 mL Na_2SO_3 solution to reduce iodine preservative. Adjust sample to pH 2.5 using a pH meter and 1 + 4 HNO_3 or 1 + 9 NH_4OH as required. Transfer sample to 250-mL separatory funnel, extract with a minimum of three 10-mL portions special dithizone solution, or until the $CHCl_3$ layer is dis-

*Whatman No. 541 or equivalent.

tinctly green. Extract with 20-mL portions $CHCl_3$ to remove dithizone (absence of green). Add 20 mL 1 +4 HNO_3, 50 mL citrate-cyanide reducing solution, and 10 mL dithizone working solution. Extract as in ¶ 4*a* and measure absorbance.

5. Calculation

$$\text{mg Pb/L} = \frac{\mu\text{g Pb (in 10 mL, from calibration curve)}}{\text{mL sample}}$$

6. Precision and Accuracy

Single operator precision in recovering 0.0104 mg Pb/L from Mississippi River water was 6.8% relative standard deviation and -1.4% relative error. At the level of 0.026 mg Pb/L, recovery was made with 4.8% relative standard deviation and 15% relative error.

316 C. References

1. SNYDER, L.J. 1947. Improved dithizone method for determination of lead—mixed color method at high pH. *Anal. Chem.* 19:684.
2. SANDELL, E.B. 1959. Colorimetric Determination of Traces of Metals, 3rd ed. Interscience, New York, N.Y.
3. WICHMANN, H.J. 1939. Isolation and determination of trace metals—the dithizone system. *Ind. Eng. Chem.*, Anal. Ed. 11:66.
4. AMERICAN SOCIETY FOR TESTING AND MATERIALS. 1977. Annual Book of ASTM Standards. Part 26, Method D3112-77, ASTM, Philadelphia, Pa.

317 LITHIUM

A minor constituent of minerals, lithium is present in fresh waters in concentrations below 10 mg/L. Brines and thermal waters may contain higher lithium levels. The use of lithium or its salts in dehumidifying units, medicinal waters, metallurgical processes, and the manufacture of some types of glass and storage batteries may contribute to its presence in wastes. Lithium hypochlorite is available commercially as a source of chlorine and may be used in swimming pools.

317 A. Atomic Absorption Spectrophotometric Method

See Section 303A.

317 B. Flame Emission Photometric Method

1. General Discussion

a. Principle: Like the other low-atomic-weight alkali metals, sodium and potassium, lithium can be determined in trace amounts by flame photometric methods. The measurement can be made at a wavelength of 670.8 nm.

b. Interference: Barium, strontium, and calcium interfere in the flame photometric determination of lithium and can be removed by adding a sodium sulfate-sodium carbonate (Na_2SO_4-Na_2CO_3) solution that precipitates barium sulfate ($BaSO_4$), strontium carbonate ($SrCO_3$), and calcium carbonate ($CaCO_3$). The content of magnesium must not exceed 10 mg in the portion taken for analysis.

c. Minimum detectable concentration: The minimum lithium concentration detectable is about 0.1 mg/L.

d. Sampling and storage: Collect sample in a borosilicate glass bottle. At time of collection adjust sample to pH <2 with nitric acid (HNO_3).

2. Apparatus

Flame photometer: An instrument demonstrated by the analyst to be suitable for the Li concentrations to be measured (currently available flame photometers usually are designed for clinical medical purposes) or an atomic absorption spectrophotometer operating in the emission mode using a lean air-acetylene flame.

3. Reagents

a. Sodium sulfate and sodium carbonate reagent: Dissolve 5 g Na_2SO_4 and 10 g Na_2CO_3 in distilled water and dilute to 1 L.

b. Stock lithium solution: Dissolve 152.7 mg anhydrous lithium chloride, LiCl, in distilled water and dilute to 250 mL; 1.00 mL = 100 μg Li. Dry salt overnight in an oven at 105 C. Weigh LiCl very rapidly because it is highly deliquescent.

c. Standard lithium solution: Dilute 10.00 mL stock LiCl solution to 500 mL with distilled water; 1.00 mL = 2.0 μg Li.

4. Procedure

a. Pretreatment of polluted water and wastewater samples: See Section 302D or G.

b. Removal of interference by barium and strontium: Take a sample of 50.0 mL if required or less, containing not more than 10 mg Mg. Add 5.0 mL Na_2SO_4-Na_2CO_3 reagent. Bring to a boil to coagulate precipitate of $BaSO_4$, $SrCO_3$, $CaCO_3$, and possibly $MgCO_3$. Allow approximately 30 min for complete precipitation; otherwise a feathery precipitate of $BaSO_4$ will appear after filtration. Filter*, wash with distilled water, and dilute to 50.0 mL for the flame photometric measurement.

c. Treatment of standard solutions: Add 5 mL Na_2SO_4-Na_2CO_3 reagent to 50.00 mL standard Li solution. Mix to prepare a standard of 2 μg Li/55 mL. Similarly, prepare dilutions of the Li standard solution to bracket sample concentration or to establish at least three points on a calibration curve of absorbance against μg Li/55 mL.

d. Flame photometric measurement: Determine lithium concentration by direct intensity measurements at a wavelength of 670.8 nm. (The bracketing method can be used with some photometric instruments, while the construction of a calibration curve is necessary with others.) Run sample, distilled water, and lithium standard as nearly simultaneously as possible. For best results, average several readings on each solution.

Follow the manufacturer's instructions for instrument operation.

5. Calculation

a. When Ba and Sr are not removed:

$$\text{mg Li/L} = \frac{\mu\text{g Li/55mL}}{\text{mL sample}}$$

b. When Ba and Sr are removed:

$$\text{mg Li/L} = \frac{\mu\text{g Li/55 mL}}{\text{mL sample}} \times 0.9$$

Actual final volume is 50 mL; calibration curve assumed based on 55 mL final volume as in ¶ 4*c*.

*Whatman No. 42 or equivalent.

6. Accuracy

The lithium concentration can be determined with an accuracy of ±0.1 to 0.2 mg/L in the lithium range of 0.7 to 1.2 mg/L.

7. Bibliography

KUEMMEL, D.F. & H.L. KARL. 1954. Flame photometric determination of alkali and alkaline earth elements in cast iron. *Anal. Chem.* 26:386.

BRUMBAUGH, R.J. & W.E. FANUS. 1954. Determination of lithium in spodumene by flame photometry. *Anal. Chem.* 26:463.

ELLESTAD, R.B. & E.L. HORSTMAN. 1955. Flame photometric determination of lithium in silicate rocks. *Anal. Chem.* 27:1229.

WHISMAN, M. & B.H. ECCLESTON. 1955. Flame spectra of twenty metals using a recording flame spectrophotometer. *Anal. Chem.* 27:1861.

HORSTMAN, E.L. 1956. Flame photometric determination of lithium, rubidium, and cesium in silicate rocks. *Anal. Chem.* 28:1417.

318 MAGNESIUM

Magnesium ranks eighth among the elements in order of abundance and is a common constituent of natural water. Important contributors to the hardness of a water, magnesium salts break down when heated, forming scale in boilers. Concentrations greater than 125 mg/L can also exert a cathartic and diuretic action. Chemical softening, reverse osmosis, electrodialysis, or ion exchange reduces the magnesium and associated hardness to acceptable levels. The magnesium concentration may vary from zero to several hundred milligrams per liter, depending on the source and treatment of the water.

Selection of method: The three methods presented for the determination of magnesium are applicable to all natural waters. Magnesium can be determined by the gravimetric method only after removal of calcium salts. This method is applied to the filtrate and washings from the permanganate calcium determination (see Section 311). Direct determinations can be made with the atomic absorption spectrophotometric method.

These methods can be applied to all concentrations by the selection of suitable sample portions. Choice of method is largely a matter of personal preference.

318 A. Atomic Absorption Spectrophotometric Method

See Section 303A.

318 B. Gravimetric Method

1. General Discussion

a. Principle: Diammonium hydrogen phosphate quantitatively precipitates magnesium in ammoniacal solution as magnesium ammonium phosphate. The precipitate is ignited to, and weighed as, magnesium pyrophosphate. A choice is presented between: (*a*) destruction of ammonium salts and oxalate, followed by single precipitation of magnesium ammonium phosphate; and (*b*) double precipitation without pretreatment. Where time is not a factor, double precipitation is preferable because, while pretreatment is faster, it requires close attention to avoid mechanical loss.

b. Interference: The solution should be reasonably free from aluminum, calcium, iron, manganese, silica, strontium, and suspended matter. It should not contain more than about 3.5 g NH_4Cl.

2. Reagents

a. Nitric acid, HNO_3, conc.

b. Hydrochloric acid, HCl, conc; also 1 + 1; 1 + 9; and 1 + 99.

c. Methyl red indicator solution: Dissolve 100 mg methyl red sodium salt in distilled water and dilute to 100 mL.

d. Diammonium hydrogen phosphate solution: In distilled water, dissolve 30 g $(NH_4)_2HPO_4$ and make up to 100 mL.

e. Ammonium hydroxide, NH_4OH, conc; also 1 + 19.

3. Procedure

a. By removal of oxalate and ammonium salts: To the combined filtrate and washings from the calcium determination, containing not more than 60 mg Mg, or to a portion containing less than this amount in a 600- or 800-mL beaker, add 50 mL conc HNO_3 and evaporate carefully to dryness on a hot plate. Do not let reaction become too violent during the latter part of the evaporation; stay in constant attendance to avoid losses through spattering. Moisten residue with 2 to 3 mL conc HCl; add 20 mL distilled water, warm, filter, and wash. To the filtrate add 3 mL conc HCl, 2 to 3 drops methyl red solution, and 10 mL $(NH_4)_2HPO_4$ solution. Cool and add conc NH_4OH, drop by drop, stirring constantly, until the color changes to yellow. Stir for 5 min, add 5 mL conc NH_4OH, and stir vigorously for 10 min more. Let stand overnight and filter through filter paper.* Wash with 1 + 19 NH_4OH. Transfer to an ignited, cooled, and weighed crucible. Dry precipitate thoroughly and burn paper off *slowly,* allowing circulation of air. Heat at about 500 C until residue is white. Ignite for 30-min periods at 1,100 C to constant weight.

b. By double precipitation: To the combined filtrate and washings from the calcium determination, containing not more than 60 mg Mg, or to a portion containing less than this amount, add 2 to 3 drops methyl red solution; adjust volume to 150 mL and acidify with 1 + 1 HCl. Add 10 mL $(NH_4)_2HPO_4$ solution. Cool. Add conc NH_4OH, drop by drop, stirring constantly, until the color changes to yellow. Stir for 5 min, add 5 mL conc NH_4OH, and stir vigorously for 10 min more. Let stand overnight and then filter through filter paper.* Wash with 1 + 19 NH_4OH. Discard filtrate and washings. Dissolve precipitate with 50 mL warm 1 + 9 HCl and wash paper well with hot 1 + 99 HCl. Add 2 to 3 drops methyl red solution, adjust volume to 100 to 150 mL, add 1 to 2 mL $(NH_4)_2HPO_4$ solution, and precipitate as before. Let stand in a cool place for at least 4 hr or preferably overnight. Filter through filter paper* and wash with 1 + 19 NH_4OH. Transfer to an ignited, cooled, and weighed crucible. Dry precipitate thoroughly and burn paper off *slowly,* allowing circulation of air. Heat at about 500 C until residue is white. Ignite for 30-min periods at 1,100 C to constant weight.

4. Calculation

$$\text{mg Mg/L} = \frac{\text{mg } Mg_2P_2O_7 \times 218.5}{\text{mL sample}}$$

5. Precision and Accuracy

A synthetic sample containing 82 mg Mg/L, 108 mg Ca/L, 3.1 mg K/L, 19.9 mg

*Carl Schleicher and Schuell Co., S & S No. 589 White Ribbon, or equivalent.

Na/L, 241 mg chloride/L, 1.1 mg NO_3^--N/L, 0.250 mg NO_2^--N/L, 259 mg sulfate/L, and 42.5 mg total alkalinity/L (contributed by $NaHCO_3$) was analyzed in eight laboratories by the gravimetric method, with a relative standard deviation of 6.3% and a relative error of 4.9%.

318 C. Magnesium By Calculation

Magnesium may be estimated as the difference between hardness and calcium as $CaCO_3$ if interfering metals are present in noninterfering concentrations in the calcium titration (Section 311C) and suitable inhibitors are used in the hardness titration (Section 314B).

$$\text{mg Mg/L} = [\text{total hardness (as mg } CaCO_3\text{/L)} - \text{calcium hardness (as mg } CaCO_3\text{/L)}] \times 0.244$$

318 D. Bibliography

EPPERSON, A.W. 1928. The pyrophosphate method for the determination of magnesium and phosphoric anhydride. *J. Amer. Chem. Soc.* 50:321.

KOLTHOFF, I.M. & E.B. SANDELL. 1952. Textbook of Quantitative Inorganic Analysis, 3rd ed. Macmillan Co., New York, N.Y., Chapter 22.

HILLEBRAND, W.F. et al. 1953. Applied Inorganic Analysis, 2nd ed. John Wiley & Sons, New York, Chapter 41 and pp. 133–134.

319 MANGANESE (TOTAL)

Although manganese in groundwater generally is present in the soluble divalent ionic form because of the absence of oxygen, part or all of the manganese in a water treatment plant may be in a higher valence state. Determination of total manganese does not differentiate the various valence states. The heptavalent permanganate ion is used to oxidize manganese and/or organic matter causing taste. Excess permanganate, complexed trivalent manganese, or a suspension of quadrivalent manganese must be detected with great sensitivity to control treatment and to prevent their discharge into a distribution system. There is evidence that manganese occurs in surface waters both in suspension in the quadrivalent state and in the trivalent state in a relatively stable, soluble complex. Although rarely present in excess of 1 mg/L, manganese imparts objectionable and tenacious stains to laundry and plumbing fixtures. The low manganese limits imposed on an acceptable water stem from these, rather than toxicological, considerations. Special means of removal often are necessary, such as chemical precipitation, pH adjustment, aeration, and use of special ion-exchange materials. Manganese occurs in domestic wastewater, industrial effluents, and receiving streams.

1. Selection of Method

The atomic absorption spectrophotometric method permits direct determination with acceptable sensitivity and is the method of choice. Of the various colorimetric methods, the persulfate method is preferred because the use of mercuric ion can control interference from a limited chloride ion concentration.

2. Sampling and Storage

Manganese may exist in a soluble form in a neutral water when first collected, but it oxidizes to a higher oxidation state and precipitates or becomes adsorbed on the container walls. Determine manganese very soon after sample collection. When delay is unavoidable, total manganese can be determined if the sample is acidified at the time of collection with HNO_3 to pH <2. See Section 301.2.

319 A. Atomic Absorption Spectrophotometric Method

See Section 303A and B.

319 B. Persulfate Method

1. General Discussion

a. Principle: Persulfate oxidation of soluble manganous compounds to form permanganate is carried out in the presence of silver nitrate. The resulting color is stable for at least 24 hr if excess persulfate is present and organic matter is absent.

b. Interference: As much as 0.1 g chloride in a 50-mL sample can be prevented from interfering by adding 1 g mercuric sulfate ($HgSO_4$) to form slightly dissociated complexes. Bromide and iodide still will interfere and only trace amounts may be present. The persulfate procedure can be used for potable water with trace to small amounts of organic matter if the period of heating is increased after more persulfate has been added.

For wastewaters containing organic matter, use preliminary digestion with nitric and sulfuric acids (HNO_3 and H_2SO_4) (see 302E). If large amounts of chloride also are present, boiling with HNO_3 helps remove the chloride ion. Interfering traces of chloride are eliminated by $HgSO_4$ in the special reagent.

Colored solutions from other inorganic ions are compensated for in the final colorimetric step.

Samples that have been exposed to air may give low results due to precipitation of manganese dioxide (MnO_2). Add 1 drop 30% hydrogen peroxide (H_2O_2) to the sample, after adding the special reagent, to redissolve precipitated manganese.

c. Minimum detectable concentration: The molar absorptivity of permanganate ion is about 2,300 L g^{-1} cm^{-1}. This corresponds to a minimum detectable concentration (98% transmittance) of 210 μg Mn/L when a 1-cm cell is used or 42 μg Mn/L when a 5-cm cell is used.

2. Apparatus

Colorimetric equipment: One of the following is required:

a. Spectrophotometer, for use at 525 nm, providing a light path of 1 cm or longer.

b. Filter photometer, providing a light path of 1 cm or longer and equipped with a green filter having maximum transmittance near 525 nm.

c. Nessler tubes, matched, 100-mL, tall form.

3. Reagents

a. Special reagent: Dissolve 75 g $HgSO_4$ in 400 mL conc HNO_3 and 200 mL distilled water. Add 200 mL 85% phosphoric acid (H_3PO_4), and 35 mg silver nitrate ($AgNO_3$). Dilute the cooled solution to 1 L.

b. Ammonium persulfate, $(NH_4)_2S_2O_8$, solid.

c. Standard manganese solution: Prepare a 0.1N potassium permanganate ($KMnO_4$) solution by dissolving 3.2 g $KMnO_4$ in distilled water and making up to 1 L. Age for several weeks in sunlight or heat for several hours near the boiling point, then filter through a fine fritted-glass filter crucible and standardize against sodium oxalate as follows:

Weigh several 100- to 200-mg samples of $Na_2C_2O_4$ to 0.1 mg and transfer to 400-mL beakers. To each beaker, add 100 mL distilled water and stir to dissolve. Add 10 mL 1 + 1 H_2SO_4 and heat rapidly to 90 to 95 C. Titrate rapidly with the $KMnO_4$ solution to be standardized, while stirring, to a slight pink end-point color that persists for at least 1 min. Do not let temperature fall below 85 C. If necessary, warm beaker contents during titration; 100 mg $Na_2C_2O_4$ will consume about 15 mL permanganate solution. Run a blank on distilled water and H_2SO_4.

$$\text{Normality of } KMnO_4 = \frac{\text{g } Na_2C_2O_4}{(A - B) \times 0.06701}$$

where:

A = mL titrant for sample and
B = mL titrant for blank.

Average results of several titrations. Calculate volume of this solution necessary to prepare 1 L of solution so that 1.00 mL = 50.0 μg Mn, as follows:

$$\text{mL } KMnO_4 = \frac{4.55}{\text{normality } KMnO_4}$$

To this volume add 2 to 3 mL conc H_2SO_4 and $NaHSO_3$ solution dropwise, with stirring, until the permanganate color disappears. Boil to remove excess SO_2, cool, and dilute to 1,000 mL with distilled water. Dilute this solution further to measure small amounts of manganese.

d. Standard manganese solution (alternate): Dissolve 1.000 g manganese metal (99.8% min.) in 10 mL redistilled HNO_3. Dilute to 1,000 mL with 1% (v/v) HCl; 1 mL = 1.000 mg Mn. Dilute 10 mL to 200 mL with distilled water; 1 mL = 0.05 mg Mn. Prepare dilute solution daily.

e. Hydrogen peroxide, H_2O_2, 30%.

f. Nitric acid, HNO_3, conc.

g. Sulfuric acid, H_2SO_4, conc.

h. Sodium nitrite solution: Dissolve 5.0 g $NaNO_2$ in 95 mL distilled water.

i. Sodium oxalate, $Na_2C_2O_4$, primary standard.

j. Sodium bisulfite: Dissolve 10 g $NaHSO_3$ in 100 mL distilled water.

4. Procedure

a. Treatment of sample: If a digested sample has been prepared according to directions for reducing organic matter and/or excessive chlorides in Section 302E, pipet a portion containing 0.05 to 2.0 mg Mn into a 250-mL conical flask. Add distilled water, if necessary, to 90 mL and proceed as in *b*.

b. To a suitable sample portion add 5 mL special reagent and 1 drop H_2O_2. Concentrate to 90 mL by boiling or dilute to 90 mL. Add 1 g $(NH_4)_2S_2O_8$, bring to a boil, and boil for 1 min. Do not heat on a water bath. Remove from heat source, let stand 1 min, then cool under the tap. (Boiling too long results in decomposition of excess persulfate and subsequent loss of permanganate color; cooling too slowly has the same effect.) Dilute to 100 mL with distilled water free from reducing substances and mix. Prepare standards containing 0, 5.00, . . . 1,500 μg Mn by treating various amounts of standard Mn solution in the same way.

c. Nessler tube comparison: Use standards prepared as in 4*b* and containing 5 to 100 μg Mn/100 mL final volume. Compare samples and standards visually.

d. Photometric determination: Use a series of standards from 0 to 1,500 μg Mn/100 mL final volume. Make photometric measurements against a distilled water blank. The following table shows light path length appropriate for various amounts of manganese in 100 mL final volume:

Mn Range *μg*	Light Path *cm*
5-200	15
20-400	5
50-1,000	2
100-1,500	1

Prepare a calibration curve of manganese concentration vs. absorbance from the standards and determine Mn in the samples from the curve. If turbidity or interfering color is present, make corrections as in 4*e*.

e. Correction for turbidity or interfering color: Avoid filtration because of possible retention of some permanganate on the filter paper. If visual comparison is used, the effect of turbidity only can be estimated and no correction can be made for interfering colored ions. When photometric measurements are made, use the following "bleaching" method, which also corrects for interfering color: As soon as the photometer reading has been made, add 0.05 mL H_2O_2 solution directly to the sample in the optical cell. Mix and, as soon as the permanganate color has faded completely and no bubbles remain, read again. Deduct absorbance of bleached solution from initial absorbance to obtain absorbance due to Mn.

5. Calculation

a. When all of the original sample is taken for analysis:

$$\text{mg Mn/L} = \frac{\mu\text{g Mn (in 100 mL final volume)}}{\text{mL sample}}$$

b. When a portion of the digested sample (100 mL final volume) is taken for analysis:

$$\text{mg Mn/L} = \frac{\mu\text{g Mn/100 mL}}{\text{mL sample}} \times \frac{100}{\text{mL portion}}$$

6. Precision and Accuracy

A synthetic sample containing 120 μg Mn/L, 500 μg Al/L, 50 μg Cd/L, 110 μg Cr/L, 470 μg Cu/L, 300 μg Fe/L, 70 μg Pb/L, 150 μg Ag/L, and 650 μg Zn/L in distilled water was analyzed in 33 laboratories by the persulfate method, with a relative standard deviation of 26.3% and a relative error of 0%.

A second synthetic sample, similar in all respects except for 50 μg Mn/L and 1,000 μg Cu/L, was analyzed in 17 laboratories by the persulfate method, with a relative standard deviation of 50.3% and a relative error of 7.2%.

319 C. Bibliography

RICHARDS, M.D. 1930. Colorimetric determination of manganese in biological material. *Analyst* 55:554.

NYDAHL, F. 1949. Determination of manganese by the persulfate method. *Anal. Chem. Acta.* 3:144.

MILLS, S.M. 1950. Elusive manganese. *Water Sewage Works* 97:92.

SANDELL, E.B. 1959. Colorimetric Determination of Traces of Metals, 3rd ed. Interscience Publishers, New York, N.Y., Chapter 26.

320 MERCURY

Organic and inorganic mercury salts are very toxic and their presence in the environment, especially in water, should be monitored.

Selection of method: The flameless atomic absorption method is the method of choice for all samples, while the dithizone method can be used for determining high levels of mercury in potable waters.

320 A. Flameless Atomic Absorption Method

See Section 304.

320 B. Dithizone Method

1. General Discussion

a. Principle: Mercury ions react with a dithizone solution in chloroform to form an orange color. The various shades of orange are measured in a spectrophotometer and unknown concentrations are estimated from a standard curve.

b. Interference: Copper, gold, palladium, divalent platinum, and silver react with dithizone in acid solution. Copper is separated during the procedure by remaining in the organic phase while the mercury is left in the aqueous phase. The other contaminants usually are not present.

The mercury dithizonate must be measured quickly because it is photosensitive.

c. Minimum detectable concentration: 1 μg Hg/10 mL final volume, corresponding to 2 μg Hg/L when a 500-mL sample is used. Acceptable precision is obtained when this concentration is exceeded.

2. Apparatus

a. Spectrophotometer, for measurements at 492 nm, providing a light path of 1 cm or longer.

b. Separatory funnels: 250 and 1,000 mL, with TFE stopcocks.

c. Glassware: Clean all glassware with potassium dichromate-sulfuric acid cleaning solution.

3. Reagents

a. Mercury-free water: Use redistilled or deionized distilled water for preparing all reagents and dilutions.

b. Stock mercury solution: Dissolve 135.4 mg mercuric chloride, $HgCl_2$, in about 700 mL mercury-free water, add 1.5 mL conc HNO_3, and make up to 1,000 mL with mercury-free water; 1.00 mL = 100 μg Hg.

c. Standard mercury solution: Dilute 10.00 mL stock solution to 1,000 mL with mercury-free water; 1.00 mL = 1.00 μg Hg. Prepare immediately before use.

d. Potassium permanganate solution: Dissolve 5 g $KMnO_4$ in 100 mL mercury-free water.

e. Sulfuric acid, H_2SO_4, conc.

f. Potassium persulfate solution. Dissolve 5 g $K_2S_2O_8$ in 100 mL mercury-free water.

g. Hydroxylamine hydrochloride solution: Dissolve 50 g $NH_2OH{\cdot}HCl$ in 100 mL mercury-free water.

h. Dithizone solution: See 102.7*b*1). Dilute 60 mL stock dithizone solution I with $CHCl_3$ to 1,000 mL; 1 mL = 6 μg dithizone.

i. Sulfuric acid, 0.25*N:* Dilute 250 mL 1*N* H_2SO_4 to 1L with mercury-free water.

j. Potassium bromide solution: Dissolve 40 g KBr in 100 mL mercury-free water.

k. Chloroform, $CHCl_3$.

l. Phosphate-carbonate buffer solution: Dissolve 150 g $Na_2HPO_4{\cdot}12H_2O$ and 38 g anhydrous K_2CO_3 in 1 L mercury-free water. Extract with 10-mL portions of dithizone until the last portion remains blue. Wash with $CHCl_3$ to remove excess dithizone.

m. Sodium sulfate, Na_2SO_4, anhydrous.

4. Procedure

a. Preparation of calibration curve: Pipet 0 (blank), 2.00, 4.00, 6.00, 8.00, and 10.00 μg mercury into separate beakers. To each beaker, add 500 mL mercury-free water (or any other volume selected for sample), 1 mL $KMnO_4$ solution, and 10 mL conc H_2SO_4. Stir and bring to a boil. If necessary, add more $KMnO_4$ until a pink color persists. After boiling has ceased, cautiously add 5 mL $K_2S_2O_8$ solution and let cool for 0.5 hr. Add one or more drops $NH_2OH{\cdot}HCl$ solution to discharge the pink color. When cool, transfer each solution to individual 1-L separatory funnels. Add about 25 mL dithizone solution. Shake funnel vigorously and transfer each organic layer to a 250-mL funnel. Repeat this extraction at least three times, making sure that the color in the last dithizone layer is as intense a blue as that of the original dithizone solution. Wash accumulated dithizone extracts in the 250-mL separatory funnel by shaking with 50 mL 0.25*N* H_2SO_4. Transfer washed dithizonate extract to another 250-mL funnel. Add 50 mL 0.25*N* H_2SO_4 and 10 mL KBr solution and shake vigorously to transfer mercury dithizonate from organic layer to aqueous layer. Discard lower dithizone layer. Wash aqueous layer with a small volume of $CHCl_3$ and discard the $CHCl_3$. Transfer 20 mL phosphate-carbonate buffer solution to each separatory funnel and add 10 mL standard dithizone solution. Shake thoroughly, and after separation, transfer the mercury dithizone to beakers. The final dithizone extract should be slightly blue. Dry contents with anhydrous Na_2SO_4. Transfer mercury dithizonate solution to a cuvette and record absorbance at 492 nm. On linear graph paper, plot absorbance against micrograms mercury in 10 mL final volume.

b. Treatment of samples: Samples containing 1.5 mL conc HNO_3/L usually do not affect dithizone, although strong solutions of HNO_3 will oxidize it. Use a 500-mL sample to increase absorbance readings, and prepare an absorbance blank consisting of all reagents. When necessary, filter sample through glass wool into the separatory funnel after oxidation step. Complete procedure as described under ¶ 4*a* above. Read mercury content from calibration curve.

5. Calculation

$$\text{mg Hg/L} = \frac{\mu\text{g Hg (in 10 mL final volume)}}{\text{mL sample}}$$

6. Precision and Accuracy

Five portions of inorganic mercury and five portions of organic mercury as methyl mercuric chloride each yielded a 95% recovery. Two of the ten samples were spiked with bayou water.

7. Bibliography

SANDELL, E.B. 1959. Colorimetric Determination of Traces of Metals, 3rd. ed. Interscience Publishers. New York, N.Y., pp. 637-638.

321 NICKEL

Selection of method: The atomic absorption spectrophotometric method is the method of choice for all samples. The heptoxime or dimethylglyoxime method can be used if atomic absorption equipment is not available.

321 A. Atomic Absorption Spectrophotometric Method

See Section 303A and B.

321 B. Heptoxime Method (TENTATIVE)

1. Principle

After preliminary digestion with nitric acid-sulfuric acid (HNO_3-H_2SO_4) mixture, iron and copper are removed by extraction of the cupferrates with chloroform ($CHCl_3$). Nickel is separated from other ions by extraction of the nickel heptoxime complex with $CHCl_3$, reextracted into the aqueous phase with hydrochloric acid (HCl), and determined colorimetrically in the acidic solution with heptoxime in the presence of an oxidant.

2. Apparatus

a. Colorimetric equipment: One of the following is required:

1) *Spectrophotometer,* for use at 445 nm, providing a light path of 1 cm or longer.

2) *Filter photometer,* providing a light path of 1 cm or longer and equipped with a violet filter with maximum transmittance near 445 nm.

b. Separatory funnels, 125-mL, Squibb form, with ground-glass stoppers.

3. Reagents

a. Standard nickel sulfate solution: Dissolve 447.9 mg $NiSO_4 \cdot 6H_2O$ in 1,000 mL distilled water; 1.00 mL = 100 μg Ni.

b. Hydrochloric acid, HCl, 1.0*N*.

c. Bromine water: Saturate distilled water with bromine.

d. Ammonium hydroxide, NH_4OH, conc.

e. Heptoxime reagent: Dissolve 0.1 g 1,2-cycloheptanedionedioxime* (heptoxime) in 100 mL 95% ethyl alcohol.

f. Ethyl alcohol, 95%.

*Hach Chemical Company, Ames, Iowa, or equivalent.

g. Sodium tartrate solution: Dissolve 10 g $Na_2C_4H_4O_6 \cdot 2H_2O$ in 90 mL distilled water.

h. Methyl orange indicator solution.

i. Sodium hydroxide, NaOH, 6*N*.

j. Acetic acid, conc.

k. Cupferron solution: Dissolve 1 g cupferron in 100 mL distilled water. Store in refrigerator or prepare fresh for each series of determinations.

l. Chloroform, $CHCl_3$.

m. Hydroxylamine-hydrochloride solution: Dissolve 10 g $NH_2OH \cdot HCl$ in 90 mL distilled water. Prepare fresh daily.

4. Procedure

a. Preparation of calibration curve: Pipet portions of standard $NiSO_4$ solution into 100-mL volumetric flasks. Use a series from 50 to 250 μg Ni if 1-cm cells are used. Add 25 mL 1.0*N* HCl and 5 mL bromine water. Cool with cold running tap water and add 10 mL conc NH_4OH. Immediately add 20 mL heptoxime reagent and 20 mL ethyl alcohol. Dilute to volume with distilled water and mix.

Measure absorbance at 445 nm, 20 min after adding reagent, using a reagent blank as reference.

b. Treatment of sample:

1) Separation of copper and iron—Take a portion of original sample, prepared by digesting with HNO_3-H_2SO_4 mixture as directed in Section 302E and containing from 50 to 250 μg Ni, place in a separatory funnel, and add 10 mL sodium tartrate solution, 2 drops (0.1 mL) methyl orange indicator, and enough 6*N* NaOH to make the solution basic.

Add 1 mL acetic acid and cool funnel under tap water. Add 4 mL cupferron reagent, add 10 mL $CHCl_3$, and shake. Let layers separate and if necessary add more cupferron until a white precipitate forms, indicating excess cupferron. Shake mixture again, let separate, and discard $CHCl_3$ layer. Reextract with 10 mL $CHCl_3$ and discard $CHCl_3$ layer. Add 1 mL fresh $NH_2OH \cdot HCl$ solution, mix, and let stand for 10 min.

2) Separation of nickel—Add 10 mL heptoxime reagent and extract nickel complex with at least three 10-mL portions of $CHCl_3$. Continue repetitive extractions until final $CHCl_3$ layer is colorless. Collect $CHCl_3$ layers in a separatory funnel and extract with 10 mL 1*N* HCl. Draw off $CHCl_3$ layer into another separatory funnel and reextract $CHCl_3$ with 10 mL 1*N* HCl. Combine HCl layers and determine absorbance as directed in ¶ 4*a*. Plot absorbance against micrograms Ni in 20 mL final volume.

5. Calculation

$$\text{mg Ni/L} = \frac{\mu\text{g Ni (in 20 mL final volume)}}{\text{mL sample}} \times \frac{100}{\text{mL portion}}$$

321 C. Dimethylglyoxime Method

Dimethylglyoxime may be used instead of heptoxime to develop the color with nickel. The conditions of color formation are identical, but prepare separate calibration curves. The rate of color development is slightly different for the two reagents; therefore, with dimethylglyoxime, make readings exactly 10 min after adding reagent, whereas, with heptoxime, make readings exactly 20 min after adding reagent. In both systems make measurements at 445 nm. The heptoxime system is more stable. Dimethylglyoxime cannot be substituted for heptoxime in the extraction process, Section 321B.4*b*2), under the conditions prescribed.

Calculate nickel concentration as in Section 321B.5.

321 D. Bibliography

Butts, P.G., A.R. Gahler & M.G. Mellon. 1950. Colorimetric determination of metals in sewage and industrial wastes. *Sewage Ind. Wastes* 22:1543.

Ferguson, R.C. & C.V. Banks. 1951. Spectrophotometric determination of nickel using 1,2-cycloheptanedionedioxime (heptoxime). *Anal. Chem.* 23:448, 1486.

Serfass, E.J. & R.F. Muraca. 1954. Procedures for Analyzing Metal Finishing Wastes. Ohio River Valley Water Sanitation Commission, Cincinnati, Ohio.

American Society for Testing and Materials. 1977. Book of ASTM Standards, Part 31. Water. ASTM, Philadelphia, Pa.

322 POTASSIUM

Potassium ranks seventh among the elements in order of abundance, yet its concentration in most drinking waters seldom reaches 20 mg/L. However, occasional brines may contain more than 100 mg/L potassium.

1. Selection of Method

Two methods for the determination of potassium are given. Both the atomic absorption spectrophotometric method (A) and the flame photometric method (B) are rapid, sensitive, and accurate.

2. Storage of Sample

Do not store samples in soft-glass bottles because of the possibility of contamination from leaching of the glass. Use polyethylene or borosilicate glass bottles. Adjust sample to pH <2 with nitric acid. This will dissolve insoluble potassium and reduce adsorption on vessel walls.

322 A. Atomic Absorption Spectrophotometric Method

See Section 303A.

322 B. Flame Photometric Method

1. General Discussion

a. Principle: Trace amounts of potassium can be determined in either a direct-reading or internal-standard type of flame photometer at a wavelength of 766.5 nm. Because much of the information pertaining to sodium applies equally to the potassium determination, carefully study the entire discussion dealing with the flame photometric determination of sodium (Section 325B) before making a potassium determination.

b. Interference: Interference in the internal-standard method may occur at sodium-to-potassium ratios of 5:1 or greater. Calcium may interfere if the calcium-to-potassium ratio is 10:1 or more. Magnesium begins to interfere when the magnesium-to-potassium ratio exceeds 100:1.

c. Minimum detectable concentration:

Potassium levels of approximately 0.1 mg/L can be determined.

2. Apparatus

See Sodium, Section 325B.2.

3. Reagents

To minimize potassium pickup, store all solutions in plastic bottles. Use small containers to reduce amount of dry element that may be picked up from bottle walls when the solution is poured. Shake each container thoroughly to wash accumulated salts from walls before pouring.

a. Deionized distilled water: Use this water for preparing all reagents and calibration standards, and as dilution water.

b. Stock potassium solution: Dissolve 1.907 g KCl dried at 110 C and dilute to 1,000 mL with deionized distilled water; 1 mL = 1.00 mg K.

c. Intermediate potassium solution: Dilute 10.0 mL stock potassium solution with deionized distilled water to 100 mL; 1.00 mL = 100 μg K. Use this solution to prepare calibration curve in potassium range of 1 to 10 mg/L.

d. Standard potassium solution: Dilute 10.0 mL intermediate potassium solution with deionized distilled water to 100 mL; 1.00 mL = 10.0 μg K. Use this solution to prepare calibration curve in potassium range of 0.1 to 1.0 mg/L.

e. Standard lithium solution: See Sodium, Section 325B.3*e*.

4. Procedure

Make determination as described under Sodium, Section 325B.4, but measure emission intensity at 768 nm.

5. Calculation

See Sodium, Section 325B.5 et seq.

6. Precision and Accuracy

A synthetic sample containing 3.1 mg K/L, 108 mg Ca/L, 82 mg Mg/L, 19.9 mg Na/L, 241 mg chloride/L, 0.25 mg NO_2^--N/L, 1.1 mg NO_3^--N/L, 259 mg sulfate/L, and 42.5 mg total alkalinity/L (contributed by $NaHCO_3$) was analyzed in 33 laboratories by the flame photometric method, with a relative standard deviation of 15.5% and a relative error of 2.3%.

322 C. Bibliography

Flame Photometric Method

All references cited in bibliography for the flame photometric method of sodium (Section 325) apply equally to potassium.

See also:

MEHLICH, A. & R.J. MONROE. 1952. Report on potassium analyses by means of flame photometer methods. *J. Ass. Offic. Agr. Chem.* 35:588.

323 SELENIUM

Selenium-deficiency diseases occur in animals but this element's role in human nutrition is not understood. Above trace levels of ingested selenium it is toxic to animals and may be toxic to humans.

The selenium concentration of most drinking waters is less than 10 μg/L. Concentrations exceeding 500 μg/L are rare and limited to seepage from seleniferous soils. The sudden appearance of selenium in a water supply might indicate industrial pollution. Little is known about the valence state of selenium in natural waters, but because both selenate and selenite are found in soils, it is reasonable to expect that both may be present in seleniferous water. Water contaminated with wastes may contain selenium in any of its four valence states. Many organic compounds of selenium are known.

Selection of method: An atomic absorption method is referenced in 323A. A colorimetric method using diaminobenzidine is described in two procedures. Method 323B requires less time and is preferred in the absence of large amounts of iodide and bromide. Use 323C, which incorporates a predistillation step, when interferences such as iodide and bromide are expected.

323 A. Atomic Absorption Spectrophotometric Method

See Section 303E.

323 B. Diaminobenzidine Method

1. General Discussion

a. Principle: Oxidation by acid permanganate converts all selenium compounds to selenate. Many carbon compounds are not oxidized completely by acid permanganate, but it is improbable that the selenium-carbon bond will remain intact through this treatment. Inorganic selenium is oxidized by acid permanganate in the presence of much greater concentrations of organic matter than would be anticipated in water supplies.

There is substantial loss of selenium when solutions of sodium selenate are evaporated to complete dryness, but in the presence of calcium all the selenium is recovered. An excess of calcium over the selenate is not necessary.

Selenate is reduced to selenite in warm 4*N* HCl. Temperature, time, and acid concentrations are specified to obtain quantitative reduction without loss of selenium. The optimum pH for formation of piazselenol is approximately 1.5. Above pH 2, the rate of formation of the colored compound is critically dependent on pH. When indicators are used to adjust pH, the results frequently are erratic. Extraction of piazselenol is not quantitative, but equilibrium is attained rapidly. Above pH 6, the partition ratio of piazselenol between water and toluene is almost independent of hydrogen ion concentration.

b. Interference: No inorganic compounds give a positive interference. Colored organic compounds extractable by toluene may exist, but it seems improbable that interference of this nature will resist the initial acid permanganate oxidation. Negative interference results from compounds that reduce the concentration of diaminobenzidine by oxidizing it. Addition of EDTA eliminates negative interference from at least 2.5 mg ferric iron. Manganese has no effect in any reasonable concentration, probably because it is reduced with the selenate. Iodide, and to a lesser extent bromide, cause low results. Recovery of selenium from a standard containing 25 μg Se in the presence of varying amounts of iodide and bromide is shown in Table 323:I. The percentage recovery improves slightly as the amount of selenium is decreased.

TABLE 323:I. SELENIUM RECOVERY IN THE PRESENCE OF BROMIDE AND IODIDE INTERFERENCE

Iodide *mg*	Selenium Recovered—%		
	Br^- 0 mg	Br^- 1.25 mg	Br^- 2.50 mg
0	100	100	96
0.5	95	94	95
1.25	84	80	
2.50	75		70

c. Minimum detectable quantity: 1 μg Se with a 4-cm light path.

2. Apparatus

a. Colorimetric equipment: One of the following is required:

1) *Spectrophotometer,* for use at 420 nm, providing a light path of 1 cm or longer.

2) *Filter photometer,* providing a light path of 1 cm or longer and equipped with a violet filter having a maximum transmittance near 420 nm.

b. Separatory funnel, 250 mL.

c. Centrifuge for 12- or 15-mL tubes (optional).

3. Reagents

a. Stock selenium solution: Place an accurately weighed pellet of ACS-grade metallic selenium in a small beaker. (CAUTION—*Handle selenium metal and solutions with extreme care.*) Add 5 mL conc HNO_3. Warm until the reaction is complete and cautiously evaporate just to dryness. Dilute to 1,000 mL with distilled water.

b. Standard selenium solution: Dilute an appropriate volume of stock selenium solution with distilled water; 1.00 mL = 1.00 μg Se.

c. Hydrochloric acid, HCl, conc and 0.1*N*.

d. Calcium chloride solution: Dissolve 30 g $CaCl_2 \cdot 2H_2O$ in distilled water and dilute to 1 L.

e. Potassium permanganate, 0.1*N*: Dissolve 3.2 g $KMnO_4$ in 1,000 mL distilled water.

f. Sodium hydroxide, NaOH, 0.1*N*.

g. Ammonium chloride solution: Dissolve 250 g NH_4Cl in 1 L distilled water.

h. EDTA-sulfate reagent: Dissolve 100 g disodium ethylenediamine tetraacetate dihydrate and 200 g Na_2SO_4 in 1 L distilled water. Add conc NH_4OH dropwise while stirring until dissolution is complete.

i. Ammonium hydroxide, NH_4OH, 5*N*.

j. Diaminobenzidine solution: Dissolve 100 mg 3,3′-diaminobenzidine hydrochloride in 10 mL distilled water. Prepare no more than 8 hr before use because this solution is unstable. (CAUTION–*Handle this reagent with extreme care.*)

k. Toluene, reagent grade.

l. Sodium sulfate, Na_2SO_4, anhydrous. Required if no centrifuge is available.

4. Procedure

a. Oxidation to selenate: Prepare standards containing 0, 10.0, 25.0, and 50.0 μg Se in 500-mL erlenmeyer flasks. Dilute to approximately 250 mL, add 2 mL 0.1*N* HCl, 5 mL $CaCl_2$ solution, 3 drops 0.1*N* $KMnO_4$, and a 5-mL measure of glass beads to prevent bumping. Boil vigorously for approximately 5 min.

Using a pH meter, adjust a 1,000-mL sample in a 2-L beaker to pH 2.3 to 2.7 with 0.1*N* HCl. Add 3 drops $KMnO_4$, 5 mL $CaCl_2$ solution, and a 5-mL measure of glass beads to prevent bumping. Heat to boiling, adding $KMnO_4$ as required to maintain a purple tint. Ignore a precipitate of MnO_2. Reduce volume to approximately 250 mL and transfer quantitatively to a 500-mL erlenmeyer flask.

b. Evaporation: Add 5 mL 0.1*N* NaOH to each flask and carefully evaporate to dryness using a salt water bath. Avoid prolonged heating of the residue.

c. Reduction to selenite: Cool and add 5 mL conc HCl and 10 mL NH_4Cl solution. Heat in a boiling water bath or steam bath for 10 ± 0.5 min.

d. Formation of piazselenol: Transfer warm solution and NH_4Cl precipitate, if present, to a beaker and wash flask with 5 mL EDTA-sulfate reagent and 5 mL 5*N* NH_4OH. Adjust to pH 1.5 ± 0.3 with NH_4OH, using a pH meter. The precipitate of EDTA will not interfere. Add 1 mL diaminobenzidine solution and heat in a boiling water bath or steam bath for approximately 5 min.

e. Extraction of piazselenol: Cool and add NH_4OH to adjust pH to 8 ± 1; the EDTA precipitate will dissolve. Pour into a 50-mL graduate and adjust volume to 50 ± 1 mL with washings from the beaker. Pour into a 250-mL separatory funnel. Add 10 mL toluene and shake for 30 ± 5 sec. Discard aqueous layer and transfer organic phase to a 12- or 15-mL centrifuge tube. Centrifuge briefly to clear toluene of water droplets. If a centrifuge is not available, filter through a dry filter paper to which approximately 0.1 g anhydrous Na_2SO_4 has been added.

f. Determination of absorbance: Read absorbance at approximately 420 nm, using toluene to establish zero absorbance. The piazselenol color is stable but evaporation of toluene concentrates the color to a marked degree in a few hours. Beer's law is obeyed up to 50 μg.

5. Calculation

$$\text{mg Se/L} = \frac{\mu\text{g Se (in 10 mL final volume)}}{\text{mL sample}}$$

6. Precision and Accuracy

A synthetic sample containing 20 μg Se/L, 40 μg As/L, 250 μg Be/L, 240 μg B/L, and 6 μg V/L in distilled water was analyzed in 35 laboratories by the diaminobenzidine method, with a relative standard deviation of 21.2% and a relative error of 5.0%.

323 C. Distillation and Diaminobenzidine Method

1. General Discussion

a. Principle: Selenium is quantitatively separated from most other elements by distillation of the volatile tetrabromide from an acid solution containing bromine. Bromine is generated by the reaction of bromide with hydrogen peroxide to avoid the inconvenience of handling the element. Selenium tetrabromide, with a minimum of excess bromine, is absorbed in the water. Excess bromine is removed by precipitation as tribromophenol and quadrivalent selenium is determined with diaminobenzidine as in Method B.

b. Interference: No substances are known to interfere.

c. Minimum detectable quantity: 1 μg Se with a 4-cm light path.

2. Apparatus

All of the apparatus in Section 323B.2*a*-*c* plus:

Distillation assembly, all borosilicate glass, for use with 500-mL erlenmeyer flasks with interchangeable ground-glass necks. Figure 323:1 shows another suitable apparatus.

3. Reagents

All the reagents described in Section 323A.3 are needed except Reagents *g* through *i*. The following also are required:

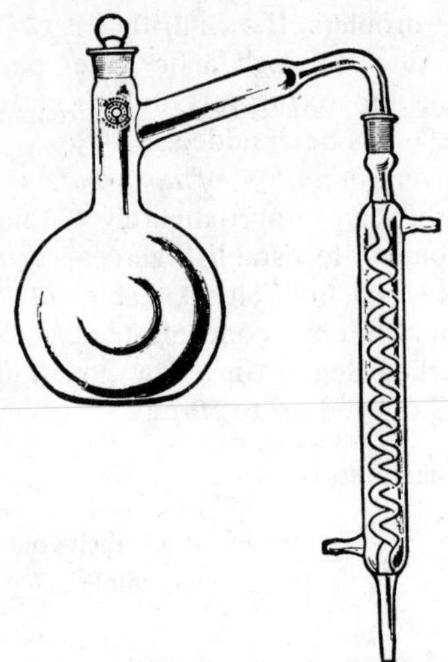

Figure 323:1. Distillation apparatus for ammonia, phenol, selenium, and fluoride determinations.

a. Potassium bromide-acid reagent: Dissolve 10 g KBr in 25 mL distilled water. Cautiously add 25 mL conc H_2SO_4, mixing and cooling under tap water as each acid increment is added. Prepare immediately before use, because $KHSO_4$ precipitates on cooling. Reheating to dissolve this salt drives off some HBr.

b. Hydrogen peroxide, H_2O_2, 30%.

c. Phenol solution: Dissolve 5 g phenol in 100 mL distilled water.

d. Ammonium hydroxide, NH_4OH, conc.

e. Hydrochloric acid, HCl, 1 + 1.

4. Procedure

a. Oxidation to selenate: Proceed as in Section 323B.4*a*, but use glass-stoppered erlenmeyer flasks.

b. Evaporation: Proceed as in Section 323B.4*b*.

c. Distillation: Add 50 mL KBr-H_2SO_4 reagent to cool flask (Figure 323:1). Add 1 mL 30% H_2O_2 and immediately fit flask to condenser. Distill under a fume hood until bromine color is gone. Use a beaker suitable for subsequent pH adjustment as a receiver. Add just enough distilled water to immerse tip of condenser. Wash the small amount of distillate remaining in condenser into the beaker with 5 mL distilled water. If a globule of undissolved Br_2 is present, decant aqueous phase to a similar beaker. Wash first beaker with 5 mL distilled water and decant to second beaker. Discard undissolved Br_2.

d. Formation of piazselenol: Add phenol solution dropwise until Br_2 color is discharged. A white precipitate of tribromophenol forms, but a small proportion of the yellow tetrabromophenol causes no trouble. Using a pH meter, adjust pH to 1.5 ± 0.3 using conc NH_4OH and 1 + 1 HCl. Add 1 mL diaminobenzidine solution and heat in a boiling water bath or steam bath for approximately 5 min.

e. Extraction of piazselenol: Proceed as in Section 323B.4*e*.

f. Determination of absorbance: Read absorbance at approximately 420 nm, using toluene to establish zero absorbance. Make absorbance readings within 2 hr after extraction because phenol used in ¶ 4*d* above causes yellow piazselenol slowly to acquire a greenish tint.

5. Calculation

See Section 323B.5.

6. Precision and Accuracy

See Section 323B.6.

323 D. Bibliography

HOSTE, J. & J. GILLIS. 1955. Spectrophotometric determination of traces of selenium with 3,3′-diaminobenzidine. *Anal. Chem. Acta* 12:158.

CHENG, K. 1956. Determination of traces of selenium. *Anal. Chem.*28:1738.

MAGIN, G.B. et al. 1960. Suggested modified method for colorimetric determination of selenium in natural water. *J. Amer. Water Works Ass.*52:1199.

ROSSUM, J.R. & P.A. VILLARRUZ. 1962. Suggested methods for determining selenium in water. *J. Amer. Water Works Ass.* 54:746.

324 SILVER

Silver can cause argyria, a permanent, blue-gray discoloration of the skin and eyes that imparts a ghostly appearance. Concentrations in the range of 0.4 to 1 mg/L have caused pathologic changes in the kidneys, liver, and spleen of rats. The silver concentration of U.S. drinking waters has been reported to vary between 0 and 2 μg/L with a mean of 0.13 μg/L. Relatively small quantities of silver are bactericidal or bacteriostatic and find limited use for the disinfection of swimming pool waters.

1. Selection of Method

The atomic absorption spectrophotometric method is preferred. The dithizone method is useful in the absence of an atomic absorption spectrophotometer.

2. Sampling and Storage

If total silver is to be determined, acidify sample with conc nitric acid (HNO_3) to pH <2 at time of collection. If sample contains particulate matter and only the "dissolved" metal content is to be determined, filter through a 0.45-μm membrane filter at time of collection. After filtration, acidify filtrate with HNO_3 to pH <2.

324 A. Atomic Absorption Spectrophotometric Method

See Section 303A or B.

324 B. Dithizone Method

1. General Discussion

a. Principle: Many metals can react with dithizone to produce colored coordination compounds. Under proper conditions or on removal of all interferences, the reaction can be made selective. In this mixed-color method, separation of the two colors is not attempted; either the green color of dithizone or the yellow color of silver dithizonate can be measured. In view of the sensitivity of the reaction and numerous interferences among the common metals, the method is empirical and demands careful adherence to the procedure. Final color can be evaluated visually or photometrically. The visual finish has been found as accurate as, and more efficient than, the photometric measurement because it circumvents extra handling involved with a photometer. The use of cells having a volume greater than 1 mL requires final dilution with carbon tetrachloride (CCl_4) or selection of a larger sample.

b. Interference: Ferric ion, residual chlorine, and other oxidizing agents convert dithizone to a yellow-brown color. However, extraction of silver along with other metals into a CCl_4 solution of dithizone overcomes such oxidation interference. Silver is removed selectively from other carry-over metals by using an ammonium thiocyanate solution. The extreme sensitivity of the method, as well as silver's affinity for being adsorbed, makes it desirable to prepare and segregate glassware for this determination and to take unusual precautions at every step. The necessity for checking and preventing contamination cannot be overemphasized. Dithizone and silver dithizonate decompose rapidly in strong light; therefore, do not leave them in the photometer light beam longer than is necessary. Avoid direct sunlight.

c. Minimum detectable quantity: 0.2 μg Ag in 1.0 mL final volume.

2. Apparatus

a. Colorimetric equipment: One of the following is required:

1) *Spectrophotometer* for measurements at either 620 nm or 462 nm, with a light path of 1 cm.

2) *Filter photometer* providing a light path of 1 cm and equipped with a red filter having maximum transmittance at or near 620 nm or a blue filter having maximum transmittance at or near 460 nm.

3) *Micro test tubes,* 10-mL capacity, 1 × 7.5-cm size.

b. Separatory funnels, with a capacity of 500 mL or larger, and also funnels with a capacity of 60 mL, preferably with inert TFE stopcocks.

c. Glassware: Treat all glassware, dishes, and crucibles with a sulfuric-chromic acid mixture and wash in 1 + 1 HNO_3 to dissolve any trace of chromium or silver adsorbed on the glassware. Thoroughly rinse with silver-free water and apply silicone coating fluid* to establish a repellent surface. Omitting these steps will result in serious errors. Oven-dry glassware and do not use acetone rinses because this solvent frequently contains interferences.

d. High-silica-glass dishes† or silica crucibles.

3. Reagents

a. Silver-free water: Use silver-free redistilled or deionized distilled water for preparing all reagents and dilutions.

b. Sulfuric acid, H_2SO_4, 1*N* and conc.

c. Carbon tetrachloride, CCl_4: Store in a glass container and do not allow contact with any metals before use. If this reagent contains traces of an interfering metal, as evidenced by a dithizone solution that is not pure green, redistill in an all-borosilicate-glass apparatus. CAUTION: *CCl_4 is very toxic. Perform all operations with CCl_4 in a fume hood.*

d. Working dithizone solution A: See Section 102.7*b*3). Dilute 50 mL stock dithizone solution III with CCl_4 to 250 mL; 1 mL = 50 μg dithizone. Store in a brown glass bottle in the dark.

e. Working dithizone solution B: Dilute 2.0 mL working dithizone solution A with CCl_4 to 250 mL; 1 mL = 0.4 μg dithizone. Prepare daily.

f. Ammonium thiocyanate reagent: Dissolve 10 g NH_4CNS in water to which 5 mL conc H_2SO_4 have been added and dilute to 500 mL with water. Store in a bottle containing 25 mL dithizone solution A.

g. Nitric acid, HNO_3, 1*N*.

---

*Desicote, manufactured by Beckman Instruments, Inc., or equivalent.

†Vycor, manufactured by Corning Glass Works, or equivalent.

h. Urea solution: Dissolve 10 g $(NH_2)_2CO$ in water and dilute to 100 mL. Store in a bottle containing 25 mL dithizone solution A. Discard if a red film forms.

i. Hydroxylamine sulfate solution: Dissolve 20 g $(NH_2OH)_2 \cdot H_2SO_4$ in water and dilute to 100 mL. Store in a bottle containing 25 mL dithizone solution A.

j. Stock silver solution: Dissolve 157.5 mg anhydrous silver nitrate, $AgNO_3$, in water to which 14 mL conc H_2SO_4 have been added and dilute to 1,000 mL; 1.00 mL = 100 μg Ag.

k. Standard silver solution: Immediately before use, dilute 10.00 mL stock solution to 1,000 mL; 1.00 mL = 1.00 μg Ag.

4. Procedure

a. Pretreatment of sample: If organic matter is present and total silver is to be determined, digest sample with HNO_3 and H_2SO_4 as directed in Section 302E. If only dissolved silver is to be determined, filter sample through a 0.45-μm membrane filter.

b. Preliminary extraction: To 100 mL sample in a 500-mL separatory funnel add 11 mL conc H_2SO_4. Extract silver by adding 5 mL dithizone solution A and shaking for 1 min. Collect organic phase and any scum in a 25-mL centrifuge tube. Transfer scum formed to centrifuge tube because it may contain an appreciable amount of silver. Repeat extraction twice more with 5-mL portions dithizone solution A and add extracts and scum to centrifuge tube. Reject aqueous phase. If a larger original sample is required, use two centrifuge tubes to collect dithizone extracts. For a 500-mL sample, use at least four 5-mL portions dithizone solution A. For either sample size centrifuge, discard aqueous phase and add 2 mL water. Recentrifuge and discard aqueous phase. Transfer CCl_4 layer to a 60-mL separatory funnel. Add 4 mL NH_4CNS reagent to centrifuge tube to collect any remaining extract, gently agitate, and transfer quantitatively to the separatory funnel. Shake for 1 min and with a suction pipet transfer as much aqueous phase as possible to a high-silica-glass dish. Repeat addition of 4 mL NH_4CNS reagent, gently agitate, and transfer aqueous phase two more times. Run off organic layer and add last few drops of aqueous phase to dish. Add 1.5 mL conc H_2SO_4 and evaporate to dryness by first evaporating to fumes by heating from above with an infrared lamp and then by heating from below with a hot plate and above with an infrared lamp. Keep heating temperature low enough to prevent bumping. Add 0.6 mL 1*N* HNO_3 and warm to dissolve. Add 1 mL each of urea and hydroxylamine sulfate solutions and digest for 5 min near the boiling point, adding water dropwise to prevent caking. Let cool to room temperature. Transfer to a 10-mL micro test tube, rinsing dish twice with 2-mL portions 1*N* H_2SO_4.

c. Final extraction of silver: Add 1 mL dithizone solution B and extract silver by mixing for 2 min with the aid of a thin glass rod flattened at the bottom. If organic phase has a greenish hue, the amount of silver is less than 1.5 μg. If organic phase is clear yellow (showing that there is no excess of dithizone), add further 1-mL portions dithizone solution B and repeat extraction until a mixed color is obtained. Record total volume, *B*, of dithizone solution B used.

d. Visual colorimetric estimation: Prepare standards by placing in each of nine micro test tubes 1 mL dithizone solution B and then 0, 0.20. . .1.60 mL standard silver solution, and 3.0, 2.8. . .1.4 mL 1*N* H_2SO_4.

Extract as described in ¶ 4*c* above, starting with the solution of lowest concentration. Compare colors of organic phases of sample and standards in the micro test tubes.

e. Photometric measurement: Prepare a standard curve by adding known amounts of silver in the range 0.20 to 1.50 μg to 0.3 mL 1*N* HNO_3, and 1 mL each of urea and hydroxylamine sulfate solutions. Add 1 mL dithizone solution B and extract the silver, using the same method as with the samples. Measure absorbance at or near 620 nm, using special cells of 1-cm light path but of reduced width so as to contain, when full, no more than about 1 mL. Zero spectrophotometer on a cell containing dithizone solution B at an absorbance reading of 1.000 and read samples and standards against this setting. Determine absorbance of blank by carrying 100 mL (500 mL if 500-mL samples are used) silver-free water through the entire process.

Because samples and standards will give lesser absorbance readings than the dithizone solution, correct all absorbances by *adding* absorbance of the blank. Because final volumes of dithizone solution B necessary to produce a mixed color may not be equal, convert concentrations of standards to silver concentration in final dithizone extract:

$$\mu\text{g Ag/mL} = \frac{\mu\text{g Ag in standard}}{\text{mL final extract}}$$

Obtain a positive-sloping standard curve by subtracting corrected absorbances of standards from 1.000 and plotting differences against concentration of standards in micrograms Ag per milliliter.

Determine absorbance of sample and add absorbance of blank. Subtract corrected absorbance from 1.000 and from standard curve read sample concentration in micrograms Ag per milliliter final volume dithizone. Calculate sample concentration.

5. Calculation

$$\text{mg Ag/L} = \frac{\mu\text{g Ag/mL (from standard curve)}}{\text{mL sample}} \times C$$

where:

C = total volume of dithizone solution B used to extract silver for final colorimetric measurement.

324 C. Bibliography

PIERCE, T.B. 1960. Determination of trace quantities of silver in trade effluents. *Analyst* 85:166.

WEST, F.K., P.W. WEST & F.A. IDDINGS. 1966. Adsorption of traces of silver on container surfaces. *Anal. Chem.* 38:1566.

DYCK, W. 1968. Adsorption of silver on borosilicate glass. Effect of pH and time. *Anal. Chem.* 40:454.

325 SODIUM

Sodium ranks sixth among the elements in order of abundance and is present in most natural waters. The levels may vary from less than 1 mg Na/L to more than 500 mg Na/L. Relatively high concentrations may be found in brines and hard waters softened by the sodium exchange process. The ratio of sodium to total cations is important in agriculture and human pathology. Soil permeability has been harmed by a high sodium ratio. Persons afflicted with certain diseases require water with low sodium concentration. A limiting concentration of 2 to 3 mg/L is recommended in feedwaters destined for high-pressure boilers. When necessary, sodium is removed by the hydrogen-exchange process or by distillation.

1. Selection of Method

Method A uses an atomic absorption spectrophotometer in the flame absorption mode whereas a flame emission photometer is used in Method B. When both instruments are available the analyst's choice will depend on factors including relative quality of the instruments, precision required, number of samples, matrix effects, and relative ease of operation of the instruments. Generally, atomic absorption instruments operating in the emission mode will be preferred.

2. Storage of Sample

Store alkaline samples or samples containing low sodium concentrations in polyethylene bottles to eliminate the possibility of sample contamination due to leaching of the glass container.

325 A. Atomic Absorption Spectrophotometric Method

See Section 303A.

325 B. Flame Emission Photometric Method

1. General Discussion

a. Principle: Trace amounts of sodium can be determined in either a direct-reading or an internal-standard-type flame photometer at a wavelength of 589 nm. The sample is sprayed into a gas flame and excitation is carried out under carefully controlled and reproducible conditions. The desired spectral line is isolated by the use of interference filters or by a suitable slit arrangement in light-dispersing devices such as prisms or gratings. The intensity of light is measured by a phototube potentiometer or other appropriate circuit. The intensity of light at 589.0 nm is approximately proportional to the concentration of the element. The calibration curve may be linear but has a tendency to level off at higher concentrations. The optimum lithium concentration may vary among individual flame photometers operating on the internal-standard principle and therefore must be ascertained for the instrument used. If alignment of the wavelength dial with the prism is not precise in the available photometer, the exact wavelength setting, which may be slightly more or less than 589.0 nm, can be determined from the maximum needle deflection and then used for the emission measurements.

b. Interference: Remove burner-clogging particulate matter from the sample by filtration through a quantitative filter paper of medium retentiveness. Incorporate a nonionic detergent in the lithium standard to assure proper aspirator function.

Minimize interference by the following:

1) Operate in the lowest practical sodium range.

2) Use the internal-standard or standard-addition technic.

3) Add radiation buffers.

4) Introduce identical amounts of the same interfering substances present in the sample into the calibration standards.

5) Prepare a family of calibration curves embodying added concentrations of a common interference.

6) Apply an experimentally determined correction in those instances where the sample contains a single important interference.

7) Remove interfering ions.

The standard-addition approach is described in the flame photometric method for strontium. Its use involves adding an identical portion of sample to each standard and determining the sample concentration by mathematical or graphical evaluation of the calibration data.

Potassium and calcium interfere with the sodium determination by the internal-standard method if the potassium-to-sodium ratio is ≥5:1 and the calcium-to-sodium ratio is ≥10:1. When these ratios are exceeded, measure calcium and potassium first so that the approximate concentration of interfering ions may be added, if necessary, to the sodium calibration standards. Magnesium interference does not appear until the magnesium-to-sodium ratio exceeds 100, a rare occurrence. Among the common anions capable of causing radiation interference are chloride, sulfate, and bicarbonate in relatively large amounts.

c. Minimum detectable concentration: The better flame photometers can be used to determine sodium levels approximating 100 μg/L. With proper modifications in technic the sodium level can be extended to 10 μg/L or lower.

2. Apparatus

a. Flame photometer, either direct-reading or internal-standard type.

b. Glassware: Rinse all glassware with 1 + 15 HNO_3 followed by several portions of deionized distilled water.

3. Reagents

To minimize sodium pickup, store all solutions in plastic bottles. Use small containers to reduce the amount of dry element that may be picked up from the bottle walls when the solution is poured. Shake each container thoroughly to wash accumulated salts from walls before pouring solution.

a. Deionized distilled water: Use deionized distilled water to prepare all reagents and calibration standards, and as dilution water.

b. Stock sodium solution: Dissolve 2.542 g NaCl dried at 140 C and dilute to 1,000 mL with water; 1.00 mL = 1.00 mg Na.

c. Intermediate sodium solution: Dilute 10.00 mL stock sodium solution with water to 100.0 mL; 1.00 mL = 100 μg Na. Use this intermediate solution to prepare calibration curve in sodium range of 1 to 10 mg/L.

d. Standard sodium solution: Dilute 10.00 mL intermediate sodium solution with water to 100 mL; 1.00 mL = 10.0 μg Na. Use this solution to prepare calibration curve in sodium range of 0.1 to 1.0 mg/L.

e. Standard lithium solution: Use either lithium chloride (1) or lithium nitrate (2) to prepare standard lithium solution containing 1.00 mg Li/1.00 mL.

1) Dry LiCl overnight in an oven at 105 C. Weigh rapidly 6.108 g, dissolve in water, and dilute to 1,000 mL.

2) Dry $LiNO_3$ overnight in an oven at 105 C. Weigh rapidly 9.933 g, dissolve in water, and dilute to 1,000 mL.

Prepare a new calibration curve whenever the standard lithium solution is changed. Where circumstances warrant, alternatively prepare a standard lithium solution containing 2.00 mg or even 5.00 mg Li/1.00 mL.

4. Procedure

a. Pretreatment of polluted water and wastewater samples: Follow the procedure described in Section 302.

b. Precautions: Locate flame photometer in an area away from direct sunlight or constant light emitted by an overhead fixture and free of drafts, dust, and tobacco smoke. Guard against contamination from corks, filter paper, perspiration, soap, cleansers, cleaning mixtures, and inadequately rinsed apparatus.

c. Instrument operation: Because of differences between makes and models of flame photometers, it is impossible to formulate detailed instructions applicable to every instrument. Follow manufacturer's recommendation for selecting the proper photocell and wavelength, adjustment of slit width and sensitivity, appropriate fuel and air or oxygen pressures, and the steps for warm-up, correcting for flame background, rinsing of burner, igniting sample, and measuring emission intensity.

d. Direct-intensity measurement: Prepare a blank and sodium calibration standards in stepped amounts in any of the following applicable ranges: 0 to 1.0, 0 to 10, or 0 to 100 mg/L. Starting with the highest calibration standard and working toward the most dilute, measure emission at 589.0 nm. Repeat the operation with both calibration standards and samples enough times to secure a reliable average reading for each solution. Construct a calibration curve from the sodium standards. Determine sodium concentration of sample from the calibration curve. Where a large number of samples must be run routinely, the calibration curve provides sufficient accuracy. If greater precision and accuracy are desired and time is available, use the bracketing approach described in ¶ 4*f* below.

e. Internal-standard measurement: To a carefully measured volume of sample (or diluted portion), each sodium calibration standard, and a blank, add, with a volumetric pipet, an appropriate volume of standard lithium solution. Then follow all steps prescribed in ¶ 4*d* above for direct-intensity measurement.

f. Bracketing approach: From the calibration curve, select and prepare sodium standards that immediately bracket the emission intensity of the sample. Determine emission intensities of the bracketing standards (one sodium standard slightly less and the other slightly greater than the sample) and the sample as nearly simultaneously as possible. Repeat the determination on bracketing standards and sample. Calculate the sodium concentration by the equation in ¶ 5*b* and average the findings.

5. Calculation

a. For direct reference to the calibration curve:

$$\text{mg Na/L} = (\text{mg Na/L in portion}) \times D$$

b. For the bracketing approach:

$$\text{mg Na/L} = \left[\frac{(B - A)(s - a)}{(b - a)} + A\right] D$$

where:

B = mg Na/L in upper bracketing standard,
A = mg Na/L in lower bracketing standard,
b = emission intensity of upper bracketing standard,
a = emission intensity of lower bracketing standard,
s = emission intensity of sample, and
D = dilution ratio

$$= \frac{\text{mL sample} + \text{mL distilled water}}{\text{mL sample}}$$

6. Precision and Accuracy

A synthetic sample containing 19.9 mg Na/L, 108 mg Ca/L, 82 mg Mg/L, 3.1 mg K/L, 241 mg chloride/L, 0.25 mg NO_2^--N/L, 1.1 mg NO_3^--N/L, 259 mg sulfate/L, and 42.5 mg total alkalinity/L was analyzed in 35 laboratories by the flame photometric method, with a relative standard deviation of 17.3% and a relative error of 4.0%.

7. Bibliography

BARNES, R.B. et al. 1945. Flame photometry: A rapid analytical method. *Ind. Eng. Chem.*, Anal. Ed. 17:605.

BERRY, J.W., D.G. CHAPPELL & R.B. BARNES. 1946. Improved method of flame photometry. *Ind. Eng. Chem.*, Anal. Ed. 18:19.

PARKS, T.D., H.O. JOHNSON & L. LYKKEN. 1948. Errors in the use of a model 18 Perkin-Elmer flame photometer for the determination of alkali metals. *Anal. Chem.* 20:822.

BILLS, C.E. et al. 1949. Reduction of error in flame photometry. *Anal. Chem.* 21:1076.

GILBERT, P.T., R.C. HAWES & A.O. BECKMAN. 1950. Beckman flame spectrophotometer. *Anal. Chem* 22:772.

WEST, P.W., P. FOLSE & D. MONTGOMERY. 1950. Application of flame spectrophotometry to water analysis. *Anal. Chem.* 22:667.

FOX, C.L. 1951. Stable internal-standard flame photometer for potassium and sodium analyses. *Anal. Chem.* 23:137.

AMERICAN SOCIETY FOR TESTING AND MATERIALS. 1952. Symposium on flame photometry. Spec. Tech. Publ. 116, ASTM, Philadelphia, Pa.

COLLINS, C.G. & H. POLKINHORNE. 1952. An investigation of anionic interference in the determination of small quantities of potassium and sodium with a new flame photometer. *Analyst* 77:430.

WHITE, J.U. 1952. Precision of a simple flame photometer. *Anal. Chem.* 24:394.

MAVRODINEANU, R. 1956. Bibliography on analytical flame spectroscopy. *Appl. Spectrosc.* 10:51.

MELOCHE, V.W. 1956. Flame photometry. *Anal. Chem.* 28:1844.

BURRIEL-MARTI, F. & J. RAMIREZ-MUNOZ. 1957. Flame Photometry: A Manual of Methods and Applications. D. Van Nostrand Co., Princeton, N.J.

DEAN, J.A. 1960. Flame Photometry. McGraw-Hill Publishing Co., New York, N.Y.

326 STRONTIUM

A typical alkaline-earth element, strontium chemically resembles calcium and causes a positive error in gravimetric and titrimetric methods for the determination of calcium. Because strontium has a tendency to accumulate in bone, radioactive strontium 90, with a half-life of 28 yr, presents a well-recognized peril to health. Naturally occurring strontium is not radioactive. For this reason, the determination of strontium in a water supply should be supplemented by a radiological measurement to exclude the possibility that the strontium content may originate from radioactive contamination (see Section 704).

Although most potable supplies contain little strontium, some well waters in the midwestern United States have levels as high as 39 mg/L.

326 A. Atomic Absorption Spectrophotometric Method

See Section 303A.

326 B. Flame Emission Photometric Method

1. General Discussion

a. Principle: The flame photometric method makes possible the determination of strontium in the low concentrations prevalent in natural water supplies. The strontium emission is measured at a wavelength of 460.7 nm. Because the background intensity at a wavelength of 466 nm equals that at 460.7 nm and is unaffected by the strontium concentration, the difference in readings obtained at these two wavelengths allows an estimate of the light intensity emitted by strontium.

b. Interference: Emission intensity is a linear function of strontium concentration and concentration of other constituents. The standard addition technic distributes the same ions throughout the standards and the sample, thereby equalizing the radiation effect of possible interfering substances.

c. Minimum detectable concentration: Strontium levels of about 0.2 mg/L can be detected by the flame photometric method without prior concentration of sample.

d. Sampling and storage: Polyethylene bottles are preferable for sample storage, although borosilicate glass containers also may be used. At time of collection adjust sample to pH <2 with nitric acid (HNO_3).

2. Apparatus

Spectrophotometer, equipped with photomultiplier tube and flame accessories; or an atomic absorption spectrophotometer capable of operation in flame emission mode.

3. Reagents

a. Stock strontium solution: Weigh into a 500-mL erlenmeyer flask 1.685 g anhydrous strontium carbonate ($SrCO_3$) powder dried at 140 C. Place a funnel in flask neck and gradually add 1 + 1 HCl until all $SrCO_3$ has dissolved. Add 200 mL distilled water and boil for a few minutes to expel CO_2. Cool, add a few drops of methyl red indicator, and adjust to the intermediate orange color by adding 3*N* ammonium hydroxide (NH_4OH) or 1 + 1 hydrochloric acid (HCl) as required. Transfer quantitatively to a 1-L volumetric flask and dilute to 1,000 mL with distilled water; 1.00 mL = 1.00 mg Sr.

b. Standard strontium solution: Dilute 25.00 mL stock strontium solution to 1,000 mL with distilled water; 1.00 mL = 25.0 μg Sr. Use this solution for preparing Sr standards in the 1- to 25-mg/L range.

c. Nitric acid, HNO_3, conc.

4. Procedure

a. Pretreatment of polluted water and wastewater samples: Follow the procedure described in Section 302D or G.

b. Preparation of strontium standards: Add 25.0 mL sample containing less than 10 mg calcium or barium and less than 1 mg strontium to 25.0 mL of each of a series of strontium standards. Use a minimum of four strontium standards from 0 mg/L to a concentration exceeding that of the sample. For most natural waters 0, 2.0, 5.0, and 10.0 mg Sr/L standards are sufficient. Brines may require a strontium series containing 0, 25, 50, and 75 mg/L. Dilute the brine sufficiently to eliminate burner splatter and clogging. Best results are obtained when the strontium concentration of the sample is less than 100 mg/L.

c. Concentration of low-level strontium samples: Concentrate samples containing less than 2 mg Sr/L. Add 3 to 5 drops conc HNO_3 to 250 mL sample and evaporate to about 25 mL. Cool and make up to 50.0 mL with distilled water. Proceed as in ¶ *b*. The HNO_3 concentration in the sample prepared for atomization can approach 0.2 mL/25 mL without producing interference.

d. Flame photometric measurement: Measure emission intensity of prepared samples (standards plus sample) at wavelengths of 460.7 and 466 nm. Follow manufacturer's instructions for correct instrument operation. Use a fuel-rich nitrous oxide-acetylene flame, if possible.

5. Calculation

a. Plot net intensity (reading at 460.7 nm minus reading at 466 nm) against strontium concentration added to the sample. Because the plot forms a straight calibration line that intersects the ordinate, compute sample strontium concentration from the equation:

$$\text{mg Sr/L} = \frac{A - B}{C} \times \frac{D}{E}$$

where:

A = sample emission-intensity reading at 460.7 nm,
B = background radiation reading at 466 nm, and
C = slope of calibration line.

Use the ratio D/E only when E mL of sample are evaporated to form a concentrate of 25.0 mL, the value for D.

b. Graphical method: Strontium concentration also can be evaluated by the graphical method illustrated in Figure 326:1. Plot net intensity against strontium concentration added to sample. The calibration line in the example intersects the ordinate at 12. Thus, Y = 12 and $2Y$ = 24. Find strontium concentration of sample by locating abscissa value of the point on the calibration line having an ordinate value of 24. In the example, the strontium concentration is 9.0 mg/L.

c. Report a strontium concentration below 10 mg/L to the nearest 0.1 mg/L and one above 10 mg/L to the nearest whole number.

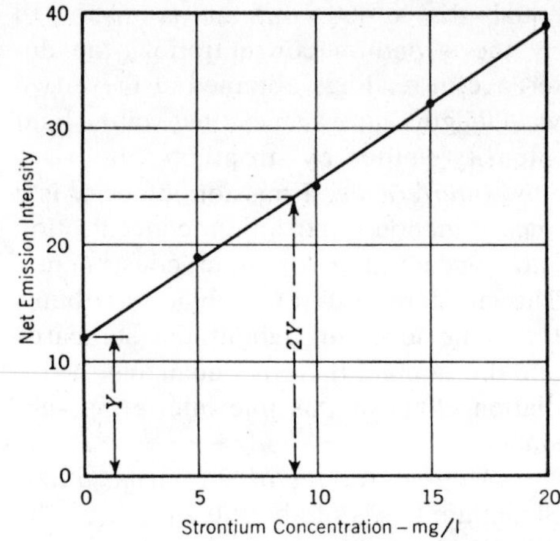

Figure 326:1. Graphical method of computing strontium concentration.

6. Accuracy

Strontium concentrations in the range 12.0 to 16.0 mg/L can be determined with an accuracy within ±1 to 2 mg/L.

326 C. Bibliography

CHOW, T.J. & T.G. THOMPSON. 1955. Flame photometric determination of strontium in sea water. *Anal. Chem.* 27:18.

NICHOLS, M.S. & D.R. MCNALL. 1957. Strontium content of Wisconsin municipal waters. *J. Amer. Water Works Ass.* 49:1493.

HORR, C.A. 1959. A survey of analytical methods for the determination of strontium in natural water. U.S. Geol. Surv. Water Supply Pap. No. 1496A.

327 VANADIUM

Laboratory and epidemiological evidence suggests that vanadium may play a beneficial role in the prevention of heart disease. In New Mexico, which has a low incidence of heart disease, vanadium has been found in concentrations of 20 to 150 μg/L. In a state where incidence of heart disease is high, vanadium was not found in water supplies. However, vanadium pentoxide dust causes gastrointestinal and respiratory disturbances. The mean concentration found in U.S. drinking waters is 6

μg/L. Industrial applications of vanadium include dyeing, ceramics, ink, and catalyst manufacture. Discharges from such sources can contribute to its presence in a water supply.

Selection of method: Both the atomic absorption spectrophotometric method and the gallic acid method are suitable for potable water samples. The atomic absorption spectrophotometric method is preferable for polluted samples.

327 A. Atomic Absorption Spectrophotometric Method

See Section 303C.

327 B. Gallic Acid Method

1. General Discussion

a. Principle: The concentration of trace amounts of vanadium in water is determined by measuring the catalytic effect it exerts on the rate of oxidation of gallic acid by persulfate in acid solution. Under the given conditions of concentrations of reactants, temperature, and reaction time, the extent of oxidation of gallic acid is proportional to the concentration of vanadium. Vanadium is determined by measuring the absorbance of the sample at 415 nm and comparing it with that of standard solutions treated identically.

b. Interference: The substances listed in Table 327:I will interfere in the determination of vanadium if the specified concentrations are exceeded. This is not a serious problem for chromium, cobalt, molybdenum, nickel, silver, and uranium because the tolerable concentration is greater than that commonly encountered in fresh water. However, in some samples the tolerable concentration of copper and iron may be exceeded. Because of the high sensitivity of the method, interfering substances in concentrations only slightly above tolerance limits can be rendered harmless by dilution.

Traces of bromide and iodide interfere seriously and dilution alone will not always reduce the concentration below tolerance limits. Mercuric ion may be added to complex these halides and minimize their interference; however, mercuric ion itself interferes if in excess. Adding 350 μg mercuric nitrate, $Hg(NO_3)_2$, per sample permits determination of vanadium in the presence of up to 100 mg Cl/L, 250 μg Br/L, and 250 μg I/L. Dilute samples containing high concentrations of these ions to concentrations below the values given above and add $Hg(NO_3)_2$.

TABLE 327:I. CONCENTRATION AT WHICH VARIOUS IONS INTERFERE IN THE DETERMINATION OF VANADIUM

Ion	Concentration *mg/L*
Chromium (VI)	1.0
Cobalt (II)	1.0
Copper (II)	0.05
Iron (II)	0.3
Iron (III)	0.5
Molybdenum (VI)	0.1
Nickel (II)	3.0
Silver	2.0
Uranium (VI)	3.0
Bromide	0.1
Chloride	100.0
Iodide	0.001

c. Minimum detectable concentration: 0.025 μg in approximately 13 mL final volume.

2. Apparatus

a. Water bath, capable of being operated at 25 ± 0.5 C.

b. Colorimetric equipment: One of the following is required:

1) *Spectrophotometer,* for measurements at 415 nm, with a light path of 1 to 5 cm.

2) *Filter photometer,* providing a light path of 1 to 5 cm and equipped with a violet filter with maximum transmittance near 415 nm.

3. Reagents

a. Stock vanadium solution: Dissolve 229.6 mg ammonium metavanadate, NH_4VO_3, in a volumetric flask containing approximately 800 mL distilled water and 15 mL 1 + 1 nitric acid (HNO_3). Dilute to 1,000 mL; 1.00 mL = 100 μg V.

b. Intermediate vanadium solution: Dilute 10.00 mL stock vanadium solution with distilled water to 1,000 mL; 1.00 mL = 1.00 μg V.

c. Standard vanadium solution: Dilute 10.00 mL intermediate vanadium solution with distilled water to 1,000 mL; 1.00 mL = 0.010 μg V.

d. Mercuric nitrate solution: Dissolve 350 mg $Hg(NO_3)_2 \cdot H_2O$ in 1,000 mL distilled water.

e. Ammonium persulfate-phosphoric acid reagent: Dissolve 2.5 g $(NH_4)_2S_2O_8$ in 25 mL distilled or demineralized water. Bring just to a boil, remove from heat, and add 25 mL conc H_3PO_4. Let stand approximately 24 hr before use. Discard after 48 hr.

f. Gallic acid solution: Dissolve 2 g $H_6C_7O_5$ in 100 mL warm distilled water, heat to a temperature just below boiling, and filter through filter paper.* Prepare a fresh solution for each set of samples.

*Whatman No. 42 or equivalent.

4. Procedure

a. Preparation of standards and sample: Prepare both blank and sufficient standards by diluting 0- to 8.0-mL portions (0 to 0.08 μg V) of standard vanadium solution to 10 mL with distilled or demineralized water. Pipet sample (10.00 mL maximum) containing less than 0.08 μg V into a suitable container and adjust volume to 10.0 mL with distilled or demineralized water. Filter colored or turbid samples. Add 1.0 mL $Hg(NO_3)_2$ solution to each blank, standard, and sample. Place containers in a water bath regulated to 25 ± 0.5 C and allow 30 to 45 min for samples to come to the bath temperature.

b. Color development and measurement: Add 1.0 mL ammonium persulfate-phosphoric acid reagent (temperature equilibrated), swirl to mix thoroughly, and return to water bath. Add 1.0 mL gallic acid solution (temperature equilibrated), swirl to mix thoroughly, and return to water bath. Add gallic acid to successive samples at intervals of 30 sec or longer to permit accurate control of reaction time. Exactly 60 min after adding gallic acid, remove sample from water bath and measure its absorbance at 415 nm, using distilled water as a reference. Subtract absorbance of blank from absorbance of each standard and sample. Construct a calibration curve by plotting absorbance values of standards versus micrograms vanadium. Determine amount of vanadium in a sample by referring to the corresponding absorbance on the calibration curve. Prepare a calibration curve with each set of samples.

5. Calculation

$$\text{mg V/L} = \frac{\mu\text{g V (in 13 mL final volume)}}{\text{mL sample}}$$

6. Precision and Accuracy

In a synthetic sample containing 6 μg V/L, 40 μg As/L, 250 μg Be/L, 240 μg

B/L, and 20 μg Se/L in distilled water, vanadium was measured in 22 laboratories with a relative standard deviation of 20% and no relative error.

7. Bibliography

Fishman, M.J. & M.V. Skougstad. 1964. Catalytic determination of vanadium in water. *Anal. Chem.* 36:1643.

328 ZINC

Zinc is an essential and beneficial element in body growth. Concentrations above 5 mg/L can cause a bitter astringent taste and an opalescence in alkaline waters. The zinc concentration of U.S. drinking waters varies between 0.06 and 7.0 mg/L with a mean of 1.33 mg/L. Zinc most commonly enters the domestic water supply from deterioration of galvanized iron and dezincification of brass. In such cases lead and cadmium also may be present because they are impurities of zinc used in galvanizing. Zinc in water also may result from industrial waste pollution.

1. Selection of Method

The atomic absorption spectrophotometric method is preferred. Dithizone method I is intended for unpolluted water and II for polluted water.

2. Sampling and Storage

Analyze samples within 6 hr after collection. Addition of HCl preserves the metallic ion content but requires that: (*a*) the acid be zinc-free; (*b*) the sample bottles be rinsed with acid before use; and (*c*) the samples be evaporated to dryness in silica dishes to remove excess HCl before analysis.

328 A. Atomic Absorption Spectrophotometric Method

See Sections 303A and B.

328 B. Dithizone Method I

1. General Discussion

a. Principle: Nearly 20 metals are capable of reacting with dithizone to produce colored coordination compounds. These dithizonates are extractable into organic solvents such as carbon tetrachloride. Most interferences in the zinc-dithizone reaction can be overcome by adjusting the pH to 4.0 to 5.5 and by adding sufficient sodium thiosulfate. Zinc also forms a weak thiosulfate complex that tends to retard the slow and incomplete reaction between zinc and dithizone. For this reason, the determination is empirical and demands the use of an identical technic in standard and sample analysis. The duration and vigor of shaking, the volumes of sample, sodium thiosulfate, and dithizone, and the pH should be kept constant.

b. Interference: Interference from bismuth, cadmium, cobalt, copper, gold, lead, mercury, nickel, palladium, silver, and stannous tin in the small quantities found in potable waters is eliminated by complexing with sodium thiosulfate and by pH adjustment. Ferric iron, residual chlorine, and other oxidizing agents convert dithizone to a yellow-brown color. The zinc-dithizone reaction is extremely sensitive and unusual precautions must be taken to avoid contamination. High and erratic blanks often are traceable to glass containing zinc oxide, surface-contaminated glassware, rubber products, stopcock greases, reagent-grade chemicals, and distilled water. Because of the extreme sensitivity of the reaction, prepare and segregate glassware especially for this determination and extract reagents with dithizone solution to remove all traces of zinc and contaminating metals. Dithizone and dithizonates decompose rapidly in strong light. Perform analyses in subdued light and do not expose solutions to the light of the photometer longer than is necessary. Avoid direct sunlight.

c. Minimum detectable quantity: 1 μg Zn.

2. Apparatus

a. Colorimetric equipment: Use one of the following:

1) *Spectrophotometer,* for use at either 535 or 620 nm, providing a light path of 1 cm or longer.

2) *Filter photometer,* providing a light path of 2 cm or longer and equipped with either a green filter having maximum transmittance near 535 nm or a red filter having maximum transmittance near 620 nm.

3) *Nessler tubes,* matched.

b. Separatory funnels, capacity 125 to 150 mL, Squibb form, preferably with inert TFE stopcocks.

c. Glassware: Rinse all glassware with 1 + 1 HNO_3 and zinc-free water.

d. pH meter.

3. Reagents

a. Zinc-free water: Use redistilled or deionized distilled water for rinsing apparatus and preparing solutions and dilutions.

b. Stock zinc solution: Dissolve 100.0 mg 30-mesh zinc metal in a slight excess of 1 + 1 HCl; about 1 mL is required. Dilute to 1,000 mL with zinc-free water; 1.00 mL = 100 μg Zn.

c. Standard zinc solution: Dilute 10.00 mL zinc stock solution to 1,000 mL with zinc-free water; 1.00 mL = 1.00 μg Zn.

d. Hydrochloric acid, HCl, 0.02*N:* Dilute 1.0 mL conc HCl to 600 mL with zinc-free water. If high blanks are traced to this reagent, dilute conc HCl with an equal volume of distilled water and redistill in an all-borosilicate-glass still.

e. Sodium acetate, 2*N:* Dissolve 68 g $NaC_2H_3O_2 \cdot 3H_2O$ and dilute to 250 mL with zinc-free water.

f. Acetic acid, 1 + 7: Use zinc-free water.

g. Acetate buffer solution: Mix equal volumes of 2*N* sodium acetate solution and 1 + 7 acetic acid solution. Extract with 10-mL portions of dithizone solution I until the last extract remains green; then extract with CCl_4 to remove excess dithizone.

h. Sodium thiosulfate solution: Dissolve 25 g $Na_2S_2O_3 \cdot 5H_2O$ in 100 mL zinc-free water. Purify by dithizone extraction as in ¶ 3*g* above.

i. Stock dithizone solution (CCl_4): See Section 102.7*b*3). (CAUTION: *CCl_4 is toxic. Avoid inhalation, ingestion, and contact with the skin.)*

j. Dithizone solution I: Dilute 40 mL stock dithizone solution (CCl_4) to 100 mL with CCl_4. Prepare daily.

k. Dithizone solution II: Dilute 10 mL dithizone solution I to 100 mL with CCl_4. Prepare daily.

l. Carbon tetrachloride, CCl_4, ACS grade.

m. Sodium citrate solution: Dissolve 10 g $Na_3C_6H_5O_7 \cdot 2H_2O$ in 90 mL zinc-free water. Purify by dithizone extraction as in ¶ 3*g* preceding. Use this reagent in final cleansing of glassware.

4. Procedure

a. Preparation of colorimetric standards: To a series of thoroughly cleansed (see ¶ 2*c* above) 125-mL Squibb separatory funnels, add 0, 1.00, 2.00, 3.00, 4.00, and 5.00 mL standard zinc solution to provide standards containing 0, 1.00, 2.00, 3.00, 4.00, and 5.00 μg Zn, respectively. Bring each volume up to 10.0 mL by adding zinc-free water. To each funnel add 5.0 mL acetate buffer and 1.0 mL $Na_2S_2O_3$ solution, and mix. The pH should be between 4 and 5.5. To each funnel add 10.0 mL dithizone solution II, stopper, and shake vigorously for 4.0 min. Let layers separate, dry inside of stem below stopcock of funnel with strips of filter paper, and run lower (CCl_4) layer into a clean, *dry* absorption cell.

b. Photometric measurement: Measure either the red color of zinc dithizonate at 535 nm or the green color of unreacted dithizone at 620 nm.

Set photometer at 100% transmittance with the blank if the 535-nm wavelength is selected. If 620 nm is used, set blank at 10.0% transmittance. Plot a calibration curve. Run a new calibration curve with each set of samples.

c. Treatment of samples: If the zinc content is not within the working range, dilute sample with zinc-free water or concentrate it in a silica dish. If the sample has been preserved with acid, evaporate a portion to dryness in a silica dish to remove excess acid. Do not neutralize with hydroxides because these usually contain excessive amounts of zinc. Using a pH meter, adjust sample to pH 2 to 3 with HCl. Transfer 10.0 mL to a separatory funnel. Complete analysis as in ¶ 4*a*, beginning with "To each funnel add 5.0 mL acetate buffer."

d. Visual comparison: If a photometric instrument is not available, run samples and standards at the same time and transfer to matched test tubes or nessler tubes. The range of colors obtained with various amounts of zinc are roughly:

Zinc μg	Color
0 (blank)	green
1	blue
2	blue-violet
3	violet
4	red-violet
5	red-violet

5. Calculation

$$\text{mg Zn/L} = \frac{\mu\text{g Zn}}{\text{mL sample}}$$

6. Precision and Accuracy

A synthetic sample containing 650 μg Zn/L, 500 μg Al/L, 50 μg Cd/L, 110 μg Cr/L, 470 μg Cu/L, 300 μg Fe/L, 70 μg Pb/L, 120 μg Mn/L, and 150 μg Ag/L in distilled water was analyzed in 46 laboratories by the dithizone method with a relative standard deviation of 18.2% and a relative error of 25.9%.

328 C. Dithizone Method II

1. Principle

Zinc is separated from other metals by extraction with dithizone and is determined by measuring the color of the zinc-dithizone complex in carbon tetrachloride. Specificity in the separation is achieved by extracting from a nearly neutral solution containing bis(2-hydroxyethyl)dithiocarbamyl ion and cyanide ion, which prevent moderate concentrations of cadmium, copper, lead, and nickel from reacting with dithizone. If excessive amounts of these metals are present, follow the special procedure given in ¶ 4*b*2) below.

The color reaction is extremely sensitive; avoid introducing extraneous zinc during analysis. Contamination may arise from water, reagents, and glassware on which zinc has been adsorbed during previous use. Appreciable blanks generally are found and the analyst must be satisfied that these blanks are representative and reproducible.

2. Apparatus

a. Colorimetric equipment: One of the following is required:

1) *Spectrophotometer,* for use at 535 nm, providing a light path of 1 cm or longer.

2) *Filter photometer,* providing a light path of 1 cm or longer and equipped with a greenish yellow filter with maximum transmittance near 535 nm.

b. Separatory funnels, 125-mL, Squibb form, with ground-glass stoppers.

3. Reagents

a. Standard zinc solution: Dissolve 1,000 g zinc metal in 10 mL 1 + 1 HNO_3. Dilute and boil to expel oxides of nitrogen. Dilute to 1,000 mL; 1.00 mL = 1.00 mg Zn.

b. Redistilled water: Redistill distilled water in all-glass apparatus.

c. Methyl red indicator: Dissolve 0.1 g methyl red sodium salt and dilute to 100 mL with distilled water.

d. Sodium citrate solution: Dissolve 10 g $Na_3C_6H_5O_7 \cdot 2H_2O$ in 90 mL water. Shake with 10 mL dithizone solution to remove zinc, then filter.

e. Ammonium hydroxide, NH_4OH, conc: Place 660 mL redistilled water in a 1-L polyethylene bottle and chill by immersion in an ice bath. Pass ammonia gas from a cylinder through a glass-wool trap into chilled bottle until volume of liquid has increased to 900 mL. Alternatively, place 900 mL conc reagent-grade NH_4OH in a 1,500-mL distillation flask and distill into a chilled 1-L polyethylene bottle initially containing 250 mL redistilled water. Continue distilling until volume of liquid in bottle has increased to 900 mL, keeping condenser tip below surface of liquid.

f. Potassium cyanide solution: Dissolve 5 g KCN in 95 mL redistilled water. (CAUTION: *Potassium cyanide is a deadly poison. Avoid skin contact or inhalation of vapors. Do not mouth pipet or bring in contact with acids.*)

g. Acetic acid, conc.

h. Carbon tetrachloride, CCl_4: See Section 328B.3*l*.

i. Bis (2-hydroxyethyl) dithiocarbamate solution: Dissolve 4.0 g diethanolamine and 1 mL CS_2 in 40 mL methyl alcohol. Prepare every 3 or 4 days.

j. Dithizone solution: Dilute 50 mL stock dithizone solution (CCl_4), prepared in accordance with Section 102.7*b*3), to 250 mL with CCl_4. Prepare fresh daily.

k. Sodium sulfide solution I: Dissolve 3.0 g $Na_2S \cdot 9H_2O$ or 1.65 g $Na_2S \cdot 3H_2O$ in 100 mL zinc-free water.

l. Sodium sulfide solution II: Prepare just before use by diluting 4 mL Na_2S solution I to 100 mL.

m. Nitric acid, HNO_3, 6*N*.

n. Hydrogen sulfide, H_2S.

4. Procedure

a. Preparation of calibration curve:

1) Prepare, just before use, a zinc solution containing 2.0 μg Zn/mL by diluting 5 mL standard zinc solution to 250 mL, then diluting 10 mL of the latter solution to 100 mL with redistilled water. Pipet 5.00, 10.00, 15.00, and 20.00 mL, containing 10 to 40 μg Zn, into separate 125-mL separatory funnels and adjust volume to about 20 mL. Set up another funnel containing 20 mL zinc-free water as a blank.

2) Add 2 drops methyl red indicator and 2.0 mL sodium citrate solution to each funnel. If the indicator is not yellow, add conc NH_4OH a drop at a time until it just turns yellow. Add 1.0 mL KCN solution and acetic acid, a drop at a time, until the indicator just turns peach color.

3) Extract methyl red by shaking with 5 mL CCl_4. Discard yellow CCl_4 layer. Add 1 mL dithiocarbamate solution. Extract with 10 mL dithizone solution, shaking for 1 min.

Draw CCl_4 layer into another separatory funnel and repeat the extraction with successive 5-mL portions of dithizone solution until the last one shows no change from the green dithizone color. Discard aqueous layer.

4) Shake combined dithizone extracts with a 10-mL portion of Na_2S solution II, separate layers, and repeat the washing with further 10-mL portions of Na_2S solution until the unreacted dithizone solution has been removed completely, as shown by the color of the aqueous layer, which remains colorless or very pale yellow; usually three washings are sufficient.

Remove water adhering to stem of funnel with a cotton swab and drain pink CCl_4 solution into a dry 50-mL volumetric flask. Use a few milliliters of CCl_4 to rinse the last droplets from funnel and dilute to mark with CCl_4.

5) Determine absorbance of the zinc dithizonate solutions at 535 nm, using CCl_4 as a reference. Plot an absorbance-concentration curve after subtracting the blank absorbance. The calibration curve is linear if monochromatic light is used.

6) Clean separatory funnels by shaking several minutes successively with HNO_3, distilled water, and finally a mixture of 5 mL sodium citrate and 5 mL dithizone, to minimize large or erratic blanks that result from adsorption of zinc on the glass surface. If possible, reserve separatory funnels exclusively for the zinc determination.

b. Treatment of sample:

1) Digest sample as directed under Preliminary Treatment, Section 302C. Transfer a portion containing 10 to 40 μg Zn to a clean 125-mL separatory funnel and adjust volume to about 20 mL. Determine zinc in this solution as described in ¶ 4*a*.

If more than 30 mL dithizone solution is needed to extract zinc completely, the portion taken contains too much zinc or the quantity of other metals that react with dithizone exceeds the amount that can be withheld by the complexing agent. If this occurs, follow the procedure in ¶ 4*b*2) below.

2) Separation of excessive amounts of cadmium, copper, and lead—When the quantity of these metals, separately or jointly, exceeds 2 mg in the portion taken, adjust volume to about 20 mL in a 100-mL beaker. Adjust acidity to 0.4 to 0.5*N** by adding dilute HNO_3 or NH_4OH as necessary. Pass H_2S into cold solution for 5 min. Filter off the precipitated sulfides through a sintered-glass filter and wash precipitate with two small portions of hot water. Boil filtrate 3 to 4 min to remove

*The normalities of the solutions obtained in the preliminary treatment are approximately 3*N* for the HNO_3-H_2SO_4 digestion and approximately 0.8*N* for the HNO_3-$HClO_4$ digestion.

H_2S, cool, transfer to a separatory funnel, and determine zinc as described in ¶ 4*b*1) et seq.

5. Calculation

$$\text{mg Zn/L} = \frac{\mu\text{g Zn}}{\text{mL sample}} \times \frac{100}{\text{mL portion}}$$

328 D. Zincon Method

1. General Discussion

a. Principle: Zinc forms a blue complex with 2-carboxy-2′-hydroxy-5′-sulfoformazyl benzene (zincon) in a solution buffered to pH 9.0. Other heavy metals likewise form colored complexes with zincon. Cyanide is added to complex zinc and heavy metals. Cyclohexanone is added to free zinc selectively from its cyanide complex so that it can be complexed with zincon to form a blue color. Sodium ascorbate reduces manganese interference. The developed color is stable except in the presence of copper.

b. Interferences: The following ions interfere in concentrations exceeding those listed:

Ion	*mg/L*	Ion	*mg/L*
Cd (II)	1	Cr (III)	10
Al (III)	5	Ni (II)	20
Mn (II)	5	Cu (II)	30
Fe (III)	7	Co (II)	30
Fe (II)	9	CrO_4 (II)	50

c. Minimum detectable concentration: 0.02 mg Zn/L.

2. Apparatus

Colorimetric equipment: One of the following is required:

a. Spectrophotometer, for measurements at 620 nm, providing a light path of 1 cm or longer.

b. Filter photometer, providing a light path of 1 cm or longer and equipped with a red filter having maximum transmittance near 620 nm. Deviation from Beer's Law occurs when the filter band pass exceeds 20 nm.

3. Reagents

a. Zinc-free water: Use redistilled or deionized distilled water for rinsing glassware and preparing reagents and standards.

b. Stock zinc solution: Dissolve 1.000 g zinc metal in 10 mL 1 + 1 HNO_3 in a clean erlenmeyer flask by warming the solution. After zinc has dissolved, continue heating until all yellow vapors are driven off and white vapor appears. It is not necessary to boil. Do *not* heat to dryness. Cool to room temperature. Quantitatively transfer to a 1,000-mL volumetric flask and dilute to volume; 1.00 mL = 1.00 mg Zn.

c. Standard zinc solution; Dilute 10.00 mL stock zinc solution to 1,000 mL; 1.00 mL = 10.00 μg Zn.

d. Sodium ascorbate, fine granular powder, USP.

e. Potassium cyanide solution: Dissolve 1.00 g KCN in approximately 50 mL water and dilute to 100 mL. CAUTION: *Potassium cyanide is a deadly poison. Avoid skin contact or inhalation of vapors. Do not mouth pipet or bring in contact with acids.*

f. Buffer solution, pH 9.0: Dissolve 8.4 g NaOH pellets in about 500 mL water. Add 31.0 g H_3BO_3 and swirl or stir to dissolve. Dilute to 1,000 mL with water and mix thoroughly.

g. Zincon reagent: Dissolve 100 mg zincon (2-carboxy-2′-hydroxy-5′-sulfoformazyl benzene) in 100 mL methanol. Because zincon dissolves slowly, stir and/or let stand overnight.

h. Cyclohexanone, purified.

i. Hydrochloric acid, HCl, conc and 6*N*.

j. Sodium hydroxide, NaOH, 6*N*.

4. Procedure

a. Preparation of colorimetric standards: Add 0, 0.5, 1.0, 3.0, 5.0, 10.0, and 14.0 mL standard zinc solution to a series of clean 50-mL graduated mixing cylinders or erlenmeyer flasks. Dilute each to 20.0 mL to yield solutions containing 0, 0.25, 0.5, 1.5, 2.5, 5.0, and 7.0 mg Zn/L, respectively. Add the following to each solution in sequence, mixing thoroughly after each addition: 0.5 g sodium ascorbate, 5.0 mL buffer solution, 2.0 mL KCN solution, and 3.0 mL zincon solution. Pipet 20.0 mL of the solution into a clean 50-mL erlenmeyer flask and add 1.0 mL cyclohexanone. Swirl for 10 sec and note time. Transfer portions of both solutions to clean sample cells. Use solution without cyclohexanone to zero colorimeter. Read and record absorbance for solution with cyclohexanone after 1 min. The calibration curve does not pass through zero because of the color enhancement effect of cyclohexanone on zincon.

b. Treatment of samples: To determine dissolved zinc, filter sample through a 0.45-μm membrane filter. Adjust to pH 7 with 6*N* NaOH or 6*N* HCl if necessary after filtering. For total zinc add 1 mL conc HCl to 50 mL sample and mix thoroughly. Filter and adjust to pH 7. Before analysis cool samples to less than 30 C if necessary. Analyze 20.0 mL of prepared sample as described in ¶ 4*a* above, beginning with "Add the following to each solution . . ." If the zinc concentration exceeds 7 mg/L prepare a sample dilution and analyze a 20.0-mL portion.

5. Calculation

Read zinc concentration (in milligrams per liter) directly from the calibration curve.

6. Precision and Accuracy

A synthetic sample containing 650 μg Zn/L, 500 μg Al/L, 50 μg Cd/L, 110 μg Cr/L, 470 μg Cu/L, 300 μg Fe/L, 70 μg Pb/L, 120 μg Mn/L, and 150 μg Ag/L in doubly demineralized water was analyzed in a single laboratory. A series of 10 replicates gave a relative standard deviation of 0.96% and a relative error of 0.15%.

A wastewater sample from an industry in Standard Industrial Classification (SIC) No. 3333, primary smelting and refining of zinc, was analyzed by 10 different persons. The mean zinc concentration was 3.36 mg Zn/L and the relative standard deviation was 1.7%. The relative error compared to results from an atomic absorption analysis of the same sample was −1.0%.

328 E. Bibliography

Dithizone Methods

HIBBARD, P.L. 1937. A dithizone method for measurement of small amounts of zinc. *Ind. Eng. Chem.*, Anal. Ed. 9:127.

SANDELL, E.B. 1937. Determination of copper, zinc, and lead in silicate rocks. *Ind. Eng. Chem.*, Anal. Ed. 9:464.

HIBBARD, P.L. 1938. Estimation of copper, zinc, and cobalt (with nickel) in soil extracts. *Ind. Eng. Chem.*, Anal. Ed. 10:615.

WICHMAN, H.J. 1939. Isolation and determination of traces of metals: The dithizone system. *Ind. Eng. Chem.*, Anal. Ed. 11:66.

COWLING, H. & E.J. MILLER. 1941. Determination of small amounts of zinc in plant materials: A photometric dithizone method. *Ind. Eng. Chem.*, Anal. Ed. 13:145.

ALEXANDER, O.R. & L.V. TAYLOR. 1944. Improved dithizone procedure for determination of zinc in foods. *J. Ass. Offic. Agr. Chem.* 27:325.

SERFASS, E.J. et al. 1947. *Chem. Anal.* 35:55.

Serfass, E.J. 1947. Research Report Serial No. 3, American Electroplaters Society, Newark, N.J., p. 22.

Serfass, E.J. et al. 1949. Determination of impurities in electroplating solutions. *Plating* 36:254, 818.

Snell, F.D. & C.T. Snell. 1949. Colorimetric Methods of Analysis, 3rd ed. D. Van Nostrand Co., Princeton, N.J., Vol. 2, pp. 1-7, 412-419.

Butts, P.G., A.R. Gahler & M.G. Mellon. 1950. Colorimetric determination of metals in sewage and industrial wastes. *Sewage Ind. Wastes* 22:1543.

Barnes, H. 1951. The determination of zinc by dithizone. *Analyst* 76:220.

Cooper, S.S. & M.L. Sullivan. 1951. Spectrophotometric studies of dithizone and some dithizonates. *Anal. Chem.* 23:613.

Serfass, E.J. & R.F. Muraca. 1954. Procedures for Analyzing Metal Finishing Wastes. Ohio River Valley Water Sanitation Commission, Cincinnati, Ohio.

Sandell, E.B. 1959. Colorimetric Determination of Traces of Metals, 3rd ed. Interscience Publishers, New York, N.Y.

Zincon Method

Platte, J.A. & V.M. Marcy. 1959. Photometric determination of zinc with zincon. *Anal. Chem.* 31:1226.

Rush, R.M. & J.H. Yoe. 1954. Colorimetric determination of zinc and copper with 2-carboxy-2′hydroxy-5′-sulfoformazyl-benzene. *Anal. Chem.* 26:1345.

PART 400

DETERMINATION OF INORGANIC NONMETALLIC CONSTITUENTS

401 INORGANIC NON-METALS—INTRODUCTION

The measurements included in this part range from collective measurements such as acidity and alkalinity to specific analyses for individual components such as the various forms of chlorine, nitrogen, and phosphorus. The measurements are conducted for the assessment and control of potable and receiving water quality and for determining process efficiency in waste treatment. Each test procedure contains in its introduction reference to any special field sampling conditions, desirable sample containers, and preservation and storage methods.

402 ACIDITY

Acidity of a water is its quantitative capacity to react with a strong base to a designated pH. The measured value may vary significantly with the end-point pH used in the determination. Acidity is a measure of an aggregate property of water and can be interpreted in terms of specific substances only when the chemical composition of the sample is known. Strong mineral acids, weak acids such as carbonic and acetic, and hydrolyzing salts such as ferrous or aluminum sulfates may contribute to the measured acidity according to the method of determination.

Acids contribute to corrosiveness and influence chemical reaction rates, chemical speciation, and biological processes. The measurement also reflects a change in the quality of the source water.

1. General Discussion

a. Principle: Hydrogen ions present in a sample as a result of dissociation or hydrolysis of solutes react with additions of standard alkali. Acidity thus depends on the end-point pH or indicator used. The construction of a titration curve by recording sample pH after successive small measured additions of titrant permits identification of inflection points and buffering capacity, if any, and allows the acidity to be determined with respect to any pH of interest.

In the titration of a single acidic species, as in the standardization of reagents, the most accurate end point is obtained from the inflection point of a titration curve. The inflection point is the point at which a differential plot (pH change per milliliter added alkali versus volume of alkali added) has a maximum. Figure 407:1 illustrates an inflection point even though millivolts per milliliter is plotted versus milliliters.

Because accurate identification of inflection points may be difficult or impossible in buffered or complex mixtures, the titration in such cases is carried to an arbitrary end-point pH based on practical considerations. For routine control titrations or rapid preliminary estimates of acidity, the color change of an indicator may be used for the end point. Samples of industrial wastes, acid mine drainage, or other solutions that contain appreciable amounts of hydrolyzable metal ions such as iron, aluminum, or manganese are treated with hydrogen peroxide to ensure oxidation of any reduced forms of polyvalent cations, and boiled to hasten hydrolysis.

b. End points: Ideally the end point of the acidity titration should correspond to the stoichiometric equivalence point for neutralization of acids present. The pH at the equivalence point will depend on the sample, the choice among multiple inflection points, and the intended use of the data.

Dissolved carbon dioxide (CO_2) usually is the major acidic component of unpolluted surface waters; handle samples from such sources carefully to minimize the loss of dissolved gases. In a sample containing only carbon dioxide-bicarbonates-carbonates, titration to pH 8.3 at 25 C corresponds to stoichiometric neutralization of carbonic acid to bicarbonate. Because the color change of phenolphthalein indicator is close to pH 8.3, this value generally is accepted as a standard end point for titration of total acidity, including CO_2 and most weak acids.

For more complex mixtures or buffered solutions selection of an inflection point may be subjective. Consequently, use fixed end points of pH 3.7 and pH 8.3 for standard acidity determinations in wastewaters and natural waters where the simple carbonate equilibria discussed above cannot be assumed. The resulting titrations are identified as "methyl orange acidity" (pH 3.7) and "phenolphthalein acidity" (pH 8.3) whether or not colored indicators are used.

c. Interferences: Dissolved gases contributing to acidity or alkalinity, such as CO_2, hydrogen sulfide, or ammonia, may be lost or gained during sampling, storage, or titration. Minimize such effects by titrating to the end point promptly after opening sample container, avoiding vigorous shaking or mixing, and protecting sample from the atmosphere during titration.

In the potentiometric titration, oily matter, suspended solids, precipitates, or other waste matter may coat the glass electrode and cause a sluggish response. Difficulty from this source is likely to be revealed in an erratic titration curve. Do *not* remove interferences from sample because they may contribute to its acidity. Pause between titrant additions to let electrode come to equilibrium.

In samples containing oxidizable or hydrolyzable ions such as ferrous or ferric iron, aluminum, and manganese, the rates of these reactions may be slow enough at room temperature to cause drifting end points.

Do not use indicator titrations with colored or turbid samples that may obscure the color change at the end point. Residual free available chlorine in the sample may bleach the indicator. Eliminate this source of interference by adding 1 drop of $0.1N$ sodium thiosulfate ($Na_2S_2O_3$).

d. Selection of method: Determine sample acidity from the volume of standard alkali required to titrate a portion to a pH of 8.3 (phenolphthalein acidity) or pH 3.7 (methyl orange acidity of wastewaters and grossly polluted waters). Titrate at room temperature using a properly calibrated pH meter, electrically operated titrator, or color indicators.

Construct a titration curve for standardization of reagents.

Use the hot peroxide procedure to pretreat samples known or suspected to contain hydrolyzable metal ions or reduced forms of polyvalent cation, such as iron pickle liquors, acid mine drainage, and other industrial wastes. Cool to room temperature before titration.

Color indicators may be used for routine and control titrations in the absence of interfering color and turbidity and for preliminary titrations to select sample size and strength of titrant (see below).

e. Sample size: The range of acidities found in wastewaters is so large that a single sample size and normality of base used as titrant cannot be specified. Use a sufficiently large volume of titrant (20 mL or more from a 50-mL buret) to obtain relatively good volumetric precision while keeping sample volume sufficiently small to permit sharp end points. For samples having acidities less than about 1,000 mg as calcium carbonate ($CaCO_3$)/L, select a volume with less than 50 mg $CaCO_3$ equivalent acidity and titrate with 0.02*N* sodium hydroxide (NaOH). For acidities greater than about 1,000 mg as $CaCO_3$/L, use a portion containing acidity equivalent to less than 250 mg $CaCO_3$ and titrate with 0.1*N* NaOH. If necessary, make a preliminary titration to determine optimum sample size and/or normality of titrant.

f. Sampling and storage: Collect samples in polyethylene or borosilicate glass bottles and store at a low temperature. Fill bottles completely and cap tightly. Because waste samples may be subject to microbial action and to loss or gain of carbon dioxide (CO_2) or other gases when exposed to air, analyze samples without delay, preferably within 1 day. If biological activity is evident analyze within 6 hr. Avoid sample agitation and prolonged exposure to air.

2. Apparatus

a. Electrometric titrator: Use any commercial pH meter or electrically operated titrator that uses a glass electrode and can be read to 0.05 pH unit. Standardize and calibrate according to the manufacturer's instructions. Pay special attention to temperature compensation and electrode care. If automatic temperature compensation is not provided, titrate at 25 ± 2 C.

b. Titration vessel: The size and form will depend on the electrodes and the sample size. Keep the free space above the sample as small as practicable, but allow room for titrant and full immersion of the indicating portions of electrodes. For conventional-sized electrodes, use a 200-mL, tall-form Berzelius beaker without a spout. Fit beaker with a stopper having three holes, to accommodate the two electrodes and the buret. With a miniature combination glass-reference electrode use a 125-mL or 250-mL erlenmeyer flask with a two-hole stopper.

c. Magnetic stirrer.

d. Pipets, volumetric.

e. Flasks, volumetric, 1,000-, 200- 100-mL.

f. Burets, borosilicate glass, 50-, 25-, 10-mL.

g. Polyolefin bottle.

3. Reagents

a. Carbon dioxide-free water: Prepare all stock and standard solutions and dilution water for the standardization procedure with distilled or deionized water that has been freshly boiled for 15 min and cooled to room temperature. The final pH of the water should be ≥6.0 and its conductivity should be <2 μmhos/cm.

b. Potassium hydrogen phthalate solution, approximately 0.05*N*: Crush 15 to 20 g primary standard $KHC_8H_4O_4$ to about 100 mesh and dry at 120 C for 2 hr. Cool in a desiccator. Weigh 10.0 ± 0.5 g (to the nearest mg), transfer to a 1-L volumetric flask, and dilute to 1,000 mL.

c. Standard sodium hydroxide titrant, 0.1*N*: Dissolve 11 g NaOH in 10 mL distilled water, cool, and filter through a Gooch crucible or hardened filter paper. Dilute 5.45 mL clear filtrate to 1 L with water and store in a polyolefin bottle protected from atmospheric CO_2 by a soda lime tube or tight cap. Standardize by titrating 40.00 mL $KHC_8H_4O_4$ solution (3*b*), using a 25-mL buret. Titrate to the inflection point, which should be close to pH 8.7. Calculate normality of NaOH:

$$\text{Normality} = \frac{A \times B}{204.2 \times C}$$

where:

A = g $KHC_8H_4O_4$ weighed into 1-L flask,
B = mL $KHC_8H_4O_4$ solution taken for titration, and
C = mL NaOH solution used.

Use the measured normality in further calculations or adjust to 0.1000*N*; 1 mL = 5.00 mg $CaCO_3$.

d. Standard sodium hydroxide titrant, 0.02*N*: Dilute 200 mL 0.1*N* NaOH to 1,000 mL and store in a polyolefin bottle protected from atmospheric CO_2 by a soda lime tube or tight cap. Standardize against $KHC_8H_4O_4$ as directed in ¶ 3*c*, using 15.00 mL $KHC_8H_4O_4$ solution and a 50-mL buret. Calculate normality as above (¶ 3*c*); 1 mL = 1.00 mg $CaCO_3$.

e. Hydrogen peroxide, H_2O_2, 30%.

f. Methyl orange indicator solution.

g. Phenolphthalein indicator solution, alcoholic.

h. Sodium thiosulfate, 0.1*N*: Dissolve 25 g $Na_2S_2O_3 \cdot 5H_2O$ and dilute to 1,000 mL with distilled water.

4. Procedure

a. Color change: Select sample size and normality of titrant according to criteria of ¶ 1*e*. Adjust sample to room temperature, if necessary, and with a pipet discharge sample into an erlenmeyer flask, while keeping pipet tip near flask bottom. If free residual chlorine is present add 0.05 mL (1 drop) 0.1*N* $Na_2S_2O_3$ solution, or destroy with ultraviolet radiation. Add 0.1 mL (2 drops) indicator solution and titrate over a white surface to a persistent color change characteristic of the equivalence point.

b. Potentiometric titration curve: Rinse electrodes and titration vessel with distilled water and drain. Select sample size and normality of titrant according to the criteria of ¶ 1*e*. Adjust sample to room temperature, if necessary, and with a pipet discharge sample while keeping pipet tip near the titration vessel bottom.

Measure sample pH. Add standard alkali in increments of 0.5 mL or less. After each addition, mix thoroughly but gently with a magnetic stirrer. Avoid splashing. Record pH when a constant reading is obtained. Continue adding titrant and measure pH until pH 9 is reached. Construct the titration curve by plotting observed pH values versus cumulative milliliters titrant added. A smooth curve showing one or more inflections should be obtained. A ragged or erratic curve may indicate that equilibrium was not reached between successive alkali additions. Determine acidity relative to a particular pH from the curve.

c. Potentiometric titration to pH 3.7 or 8.3: Prepare sample and titration assembly as specified in ¶ 4*b*. Titrate to preselected end point pH (¶ 1*d*) without recording intermediate pH values. As the end point is approached make smaller additions of alkali and be sure that pH equilibrium is reached before making the next addition.

d. Hot peroxide treatment:* Pipet a suitable sample (see ¶ 1*e*) into the titration flask. Measure pH. If pH is above 4.0, add 5-mL increments of 0.02*N* sulfuric acid (H_2SO_4) (Section 403.3*c*) to reduce pH to 4 or less. Remove electrodes. Add 5 drops 30% H_2O_2 and boil for 2 to 5 min. Cool to room temperature and titrate with standard alkali to pH 8.3 according to the procedure of ¶ 4*c*.

5. Calculation

Acidity, as mg $CaCO_3$/L

$$= \frac{[(A \times B) - (C \times D)] \times 50{,}000}{\text{mL sample}}$$

where:

A = mL NaOH titrant used,
B = normality of NaOH,
C = mL H_2SO_4 used (¶ 4*d*), and
D = normality of H_2SO_4.

---

*This procedure is intended to be equivalent to ASTM D1067, Method E.

Report pH of the end point used, as follows: "The acidity to pH ____ = ____ mg $CaCO_3$/L." A negative value signifies alkalinity.

6. Precision

No general statement can be made about precision because of the great variation in sample characteristics. The precision of the titration is likely to be much greater than the uncertainties involved in sampling and sample handling before analysis.

Forty analysts in 17 laboratories analyzed synthetic water samples containing increments of bicarbonate equivalent to 20 mg $CaCO_3$/L. Titration according to the procedure of ¶ 4*c* gave a standard deviation of 1.8 mg $CaCO_3$/L, with negligible bias.

6. Bibliography

WINTER, J.A. & M.R. MIDGETT. 1969. FWPCA Method Study 1. Mineral and Physical Analyses. FWPCA, Washington, D.C.

BROWN, E., M.W. SKOUGSTAD & M.J. FISHMAN. 1970. Methods for collection and analysis of water samples for dissolved minerals and gases. Chapter A1 in Book 5, Techniques of Water-Resources Investigations of United States Geological Survey. U.S. Geol. Surv., Washington, D.C.

403 ALKALINITY

Alkalinity of a water is its quantitative capacity to react with a strong acid to a designated pH. The measured value may vary significantly with the end-point pH used. Alkalinity is a measure of an aggregate property of water and can be interpreted in terms of specific substances only when the chemical composition of the sample is known.

Alkalinity is significant in many uses and treatments of natural and wastewaters. Because the alkalinity of many surface waters is primarily a function of carbonate, bicarbonate, and hydroxide content, it is taken as an indication of the concentration of these constitutents. The measured values may include contributions from borates, phosphates, or silicates if these are present. Alkalinity in excess of alkaline earth metal concentrations is significant in determining the suitability of a water for irrigation. Alkalinity measurements are used in the interpretation and control of water and wastewater treatment processes. Raw domestic wastewater has an alkalinity less than or only slightly greater than that of the water supply. Properly operating anaerobic digesters typically have supernatant alkalinities in the range of 2,000 to 4,000 mg calcium carbonate ($CaCO_3$)/L.[1]

1. General Discussion

a. Principle: Hydroxyl ions present in a sample as a result of dissociation or hydrolysis of solutes react with additions of standard acid. Alkalinity thus depends on the end-point pH used. For methods of determining inflection points from titration curves and the rationale for titrating to fixed pH end points, see Section 402.1*a*.

For samples of low alkalinity (less than 20 mg $CaCO_3$/L) use an extrapolation technic based on the near proportionality of concentration of hydrogen ions to excess of titrant beyond the equivalence point. The amount of standard acid required to reduce pH exactly 0.30 pH unit is measured carefully. Because this change in pH corresponds to an exact doubling of the hydrogen ion concentration, a simple extrapolation can be made to the equivalence point.[2,3]

b. End points: When alkalinity is due entirely to hydroxide, carbonate, or bicarbonate content, the pH at the equivalence point of the titration is determined by the concentration of carbon dioxide (CO_2) at that stage. CO_2 concentration depends, in turn, on the total carbonate species originally present and any losses that may have occurred during titration. The following pH values are suggested as the equivalence points for the corresponding alkalinity concentrations as milligrams $CaCO_3$ per liter:

	End point pH Total	Phenolphthalein
Alkalinity, mg $CaCO_3$/L:		
30	5.1	8.3
150	4.8	8.3
500	4.5	8.3
Silicates, phosphates known or suspected	4.5	8.3
Routine or automated analyses	4.5	8.3
Industrial waste or complex system	3.7	8.3

c. Interferences: Soaps, oily matter, suspended solids, or precipitates may coat the glass electrode and cause a sluggish response. Allow additional time between titrant additions to let electrode come to equilibrium. Do not filter, dilute, concentrate, or alter sample.

d. Selection of method: Determine sample alkalinity from volume of standard acid required to titrate a portion to a designated pH taken from ¶ 1*b*. Titrate at room temperature with a properly calibrated pH meter or electrically operated titrator, or use color indicators.

Report alkalinity less than 20 mg $CaCO_3$/L only if it has been determined by the low-alkalinity method of ¶ 4*d*.

Construct a titration curve for standardization of reagents.

Color indicators may be used for routine and control titrations in the absence of interfering color and turbidity and for preliminary titrations to select sample size and strength of titrant (see below).

e. Sample size: See Section 402.1*e* for selection of size sample to be titrated and normality of titrant, substituting 0.02 *N* or 0.1*N* sulfuric (H_2SO_4) or hydrochloric (HCl) acid for the standard alkali of that method. For the low-alkalinity method, titrate a 200-mL sample with 0.02*N* H_2SO_4 from a 10-mL buret.

f. Sampling and storage: See Section 402.1*f*.

2. Apparatus

See Section 402.2.

3. Reagents

a. Sodium carbonate solution, approximately 0.05*N*: Dry 3 to 5 g primary standard Na_2CO_3 at 250 C for 4 hr and cool in a desiccator. Weigh 2.5 ± 0.2 g (to the nearest mg), transfer to a 1-L volumetric flask, fill flask to the mark with distilled water, and dissolve and mix reagent. Do not keep longer than 1 wk.

b. Standard sulfuric acid or hydrochloric acid, 0.1 *N:* Dilute 3.0 mL conc H_2SO_4 or 8.3 mL conc HCl to 1 L with distilled or deionized water. Standardize against 40.00 mL 0.05 *N* Na_2CO_3 solution, with about 60 mL water, in a beaker by titrating potentiometrically to pH of about 5. Lift out electrodes, rinse into the same beaker, and boil gently for 3 to 5 min under a watch glass cover. Cool to room temperature, rinse cover glass into beaker, and finish titrating to the pH inflection point. Calculate normality:

$$\text{Normality, } N = \frac{A \times B}{53.00 \times C}$$

where:

A = g Na_2CO_3 weighed into 1 L flask,

B = mL Na_2CO_3 solution taken for titration, and
C = mL acid used.

Use measured normality in calculations or adjust to 0.1000*N*; 1 mL 0.1000 *N* solution = 5.00 mg $CaCO_3$.

c. Standard sulfuric acid or hydrochloric acid, 0.02*N:* Dilute 200.00 mL 0.1000*N* standard acid to 1,000 mL with distilled or deionized water. Standardize by potentiometric titration of 15.00 mL 0.05*N* Na_2CO_3 according to the procedure of ¶ 3*b*; 1 mL = 1.00 mg $CaCO_3$.

d. Mixed bromcresol green-methyl red indicator solution: Use either the aqueous or the alcoholic solution:

1) Dissolve 100 mg bromcresol green sodium salt and 20 mg methyl red sodium salt in 100 mL distilled water.

2) Dissolve 100 mg bromcresol green and 20 mg methyl red in 100 mL 95% ethyl alcohol or isopropyl alcohol.

e. Methyl orange solution.

f. Phenolphthalein solution, alcoholic.

g. Sodium thiosulfate, 0.1*N:* See Section 402.3*b*.

4. Procedure

a. Color change: See Section 402.4*a*. The color response of the mixed bromcresol green-methyl red indicator is approximately as follows: above pH 5.2, greenish blue; pH 5.0, light blue with lavender gray; pH 4.8, light pink-gray with bluish cast; and pH 4.6, light pink. Check color changes against reading of a pH meter under the conditions of the titration. Because colors are difficult to distinguish, the method is subject to relatively large operator error.

b. Potentiometric titration curve: Follow the procedure for determining acidity (Section 402.4*b*), substituting the appropriate normality of standard acid solution for standard NaOH, and continue titration to pH 3.7 or lower. Do not filter, dilute, concentrate, or alter the sample.

c. Potentiometric titration to preselected pH: Determine the appropriate end-point pH according to ¶ 1*b*. Prepare sample and titration assembly (Section 402.4*b*). Titrate to the end-point pH without recording intermediate pH values and without undue delay. As the end point is approached make smaller additions of acid and be sure that pH equilibrium is reached before adding more titrant.

d. Potentiometric titration of low alkalinity: For alkalinities less than 20 mg/L titrate 100 to 200 mL according to the procedure of ¶ 4*c*, above, using a 10-mL microburet and 0.02*N* standard acid solution. Stop the titration at a pH in the range 4.3 to 4.7 and record volume and exact pH. Carefully add additional titrant to reduce the pH exactly 0.30 pH unit and again record volume.

5. Calculations

a. Potentiometric titration to end-point pH:

$$\text{Alkalinity, mg } CaCO_3/L = \frac{A \times N \times 50{,}000}{\text{mL sample}}$$

where:

A = mL standard acid used and
N = normality of standard acid

or

$$\text{Alkalinity, mg } CaCO_3/L = \frac{A \times t \times 1{,}000}{\text{mL sample}}$$

where:

t = titer of standard acid, mg $CaCO_3$/mL.

Report pH of end point used as follows: "The alkalinity to pH _____ = _____ mg $CaCO_3$/L" and indicate clearly if this pH corresponds to an inflection point of the titration curve.

b. Potentiometric titration of low alkalinity:

Total alkalinity, mg $CaCO_3$/L

$$= \frac{(2\,B - C) \times N \times 50{,}000}{\text{mL sample}}$$

where:

B = mL titrant to first recorded pH,
C = total mL titrant to reach pH 0.3 unit lower, and
N = normality of acid.

c. Calculation of alkalinity relationships: The results obtained from the phenolphthalein and total alkalinity determinations offer a means for stoichiometric classification of the three principal forms of alkalinity present in many waters. The classification ascribes the entire alkalinity to bicarbonate, carbonate, and hydroxide, and assumes the absence of other (weak) inorganic or organic acids, such as silicic, phosphoric, and boric acids. It further presupposes the incompatibility of hydroxide and bicarbonate alkalinities. Because the calculations are made on a stoichiometric basis, ion concentrations in the strictest sense are not represented in the results, which may differ significantly from actual concentrations especially at pH >10. According to this scheme:

1) Carbonate (CO_3^{2-}) alkalinity is present when phenolphthalein alkalinity is not zero but is less than total alkalinity.

2) Hydroxide (OH^-) alkalinity is present if phenolphthalein alkalinity is more than half the total alkalinity.

3) Bicarbonate (HCO_3^-) ions are present if phenolphthalein alkalinity is less than half the total alkalinity. These relationships may be calculated by the following scheme, where P is phenolphthalein alkalinity and T is total alkalinity (¶ 1*b*):

Select the smaller value of P or (T-P). Then, carbonate alkalinity equals twice the smaller value. When the smaller value is P, the balance (T-2P) is bicarbonate. When the smaller value is (T-P), the balance (2P-T) is hydroxide. All results are expressed as $CaCO_3$. The mathematical conversion of the results is shown in Table 403:I.

TABLE 403.I. ALKALINITY RELATIONSHIPS*

Result of Titration	Hydroxide Alkalinity as $CaCO_3$	Carbonate Alkalinity as $CaCO_3$	Bicarbonate Concentration as $CaCO_3$
P = 0	0	0	T
P < ½T	0	2P	T − 2P
P = ½T	0	2P	0
P > ½T	2P − T	2(T − P)	0
P = T	T	0	0

*Key: P—phenolphthalein alkalinity; T—total alkalinity.

Alkalinity relationships also may be computed nomographically (see Carbon Dioxide, Section 406). Accurately measure pH, calculate OH^- concentration as milligrams $CaCO_3$ per liter, and calculate concentrations of CO_3^{2-} and HCO_3^- as milligrams $CaCO_3$ per liter from the OH^- concentration, and the phenolphthalein and total alkalinities by the following equations:

$$CO_3^{2-} = 2P - 2[OH^-]$$

$$HCO_3^- = T - 2P + [OH^-]$$

Similarly, if difficulty is experienced with the phenolphthalein end point, or if a check on the phenolphthalein titration is desired, calculate phenolphthalein alkalinity as $CaCO_3$ from the results of the nomographic determinations of carbonate and hydroxide ion concentrations:

$$P = 1/2\,[CO_3^{2-}] + [OH^-]$$

6. Precision and Accuracy

No general statement can be made about precision because of the great variation in sample characteristics. The precision of the titration is likely to be much greater than the uncertainties involved in sampling and sample handling before the analysis.

In the range of 10 to 500 mg/L, when the alkalinity is due entirely to carbonates or bicarbonates, a standard deviation of 1 mg $CaCO_3$/L can be achieved. Forty analysts in 17 laboratories analyzed synthetic samples containing increments of bicarbonate equivalent to 120 mg $CaCO_3$/L. The titration procedure of ¶ 4*b* was used, with an end point pH of 4.5. The standard deviation was 5 mg/L and the average bias (lower than the true value) was 9 mg/L.[4]

7. References

1. POHLAND, F.G. & D.E. BLOODGOOD. 1963. Laboratory studies on mesophilic and thermophilic anaerobic sludge digestion. *J. Water Pollut. Control Fed.* 35:11.
2. LARSON, T.E. & L.M. HENLEY. 1955. Determination of low alkalinity or acidity in water. *Anal. Chem.* 27:851.
3. THOMAS, J.F.J. & J.J. LYNCH. 1960. Determination of carbonate alkalinity in natural waters. *J. Amer. Water Works Ass.* 52:259.
4. WINTER, J.A. & M.R. MIDGETT. 1969. FWPCA Method Study 1. Mineral and Physical Analyses. FWPCA, Washington, D.C.

8. Bibliography

AMERICAN SOCIETY FOR TESTING & MATERIALS. 1970. Standard Methods for Acidity or Alkalinity of Water. ASTM Publ. D1067-70, Philadelphia, Pa.

BROWN, E., M.W. SKOUGSTAD & M.J. FISHMAN. 1970. Methods of collection and analysis of water sample for dissolved minerals and gases. Chapter A1 *in* Book 5, Techniques of Water-Resources Investigation of the United States Geological Survey. U.S. Geol. Surv., Washington, D.C.

404 BORON

Although it is an element essential for plant growth, boron in excess of 2.0 mg/L in irrigation water is deleterious to certain plants and some plants may be affected adversely by concentrations as low as 1.0 mg/L (or even less in commercial greenhouses). Drinking waters rarely contain more than 1 mg B/L and generally less than 0.1 mg/L, concentrations considered innocuous for human consumption. Boron may occur naturally in some waters or may find its way into a watercourse through cleaning compounds and industrial waste effluents. Seawater contains approximately 5 mg B/L and this element is found in saline estuaries in association with other seawater salts.

The ingestion of large amounts of boron can affect the central nervous system. Protracted ingestion may result in a clinical syndrome known as borism.

1. Selection of Method

The curcumin method (A) is applicable in the 0.10- to 1.0-mg/L range, while the carmine method (B) is suitable for the determination of boron concentrations in the 1- to 10-mg/L range. The range of these methods can be extended by dilution or concentration of the sample.

2. Sampling and Storage

Store samples in polyethylene bottles or alkali-resistant, boron-free glassware.

404 A. Curcumin Method

1. General Discussion

a. Principle: When a sample of water containing boron is acidified and evaporated in the presence of curcumin, a red-colored product called rosocyanine is formed. The rosocyanine is taken up in a suitable solvent and the red color is compared with standards visually or photometrically.

b. Interference: NO_3^--N concentrations above 20 mg/L interfere. Significantly high results are possible when the total of calcium and magnesium hardness exceeds 100 mg/L as calcium carbonate ($CaCO_3$). Moderate hardness levels also can cause a considerable percentage error in the low boron range. This interference springs from the insolubility of the hardness salts in 95% ethanol and consequent turbidity in the final solution. Filter the final solution or pass the original sample through a column of strongly acidic cation-exchange resin in the hydrogen form to remove interfering cations. The latter procedure permits application of the method to samples of high hardness or solids content. Phosphate does not interfere.

c. Minimum detectable quantity: 0.2 μg B.

2. Apparatus

a. Colorimetric equipment: One of the following is required:

1) *Spectrophotometer,* for use at 540 nm, with a minimum light path of 1 cm.

2) *Filter photometer,* equipped with a green filter having a maximum transmittance near 540 nm, with a minimum light path of 1 cm.

b. Evaporating dishes, 100- to 150-mL capacity, of high-silica glass,* platinum, or other suitable material.

c. Water bath, set at 55 ± 2 C.

d. Glass-stoppered volumetric flasks, 25- and 50-mL capacity.

e. Ion-exchange column, 50 cm long by 1.3 cm in diameter.

*Vycor, manufactured by Corning Glass Works, or equivalent.

3. Reagents

Store all reagents in polyethylene or boron-free containers.

a. Stock boron solution: Dissolve 571.6 mg anhydrous boric acid, H_3BO_3, in distilled water and dilute to 1,000 mL; 1.00 mL = 100 μg B. Because H_3BO_3 loses weight on drying at 105 C, use a reagent meeting ACS specifications and keep the bottle tightly stoppered to prevent entrance of atmospheric moisture.

b. Standard boron solution: Dilute 10.00 mL stock boron solution to 1,000 mL with distilled water; 1.00 mL = 1.00 μg B.

c. Curcumin reagent: Dissolve 40 mg finely ground curcumin† and 5.0 g oxalic acid in 80 mL 95% ethyl alcohol. Add 4.2 mL conc HCl, make up to 100 mL with ethyl alcohol in a 100-mL volumetric flask, and filter if reagent is turbid (isopropyl alcohol, 95%, may be used in place of ethyl alcohol). This reagent is stable for several days if stored in a refrigerator.

d. Ethyl or isopropyl alcohol, 95%.

e. Reagents for removal of high hardness and cation interference:

1) *Strongly acidic cation exchange resin.*

2) *Hydrochloric acid,* HCl, 1 + 5.

4. Procedure

a. Precautions: Closely control such variables as volumes and concentrations of reagents, as well as time and temperature of drying. Use evaporating dishes identical in shape, size, and composition to insure equal evaporation time because increasing the time results in intensification of the resulting color.

b. Preparation of calibration curve: Pipet 0 (blank), 0.25, 0.50, 0.75, and 1.00 μg boron into evaporating dishes of the same type, shape, and size. Add distilled water

†Eastman No. 1179 or equivalent.

to each standard to bring total volume to 1.0 mL. Add 4.0 mL curcumin reagent to each and swirl gently to mix contents thoroughly. Float dishes on a water bath set at 55 ± 2 C and let them remain for 80 min, which is usually sufficient for complete drying and removal of HCl. Keep drying time constant for standards and samples. After dishes cool to room temperature, add 10 mL 95% ethyl alcohol to each dish and stir gently with a polyethylene rod to insure complete dissolution of the red-colored product.

Wash contents of dish into a 25-mL volumetric flask, using 95% ethyl alcohol. Make up to mark with 95% ethyl alcohol and mix thoroughly by inverting. Read absorbance of standards and samples at a wavelength of 540 nm after setting reagent blank at zero absorbance. The calibration curve is linear from 0 to 1.00 μg boron. Make photometric readings within 1 hr of drying samples.

c. Sample treatment: For waters containing 0.10 to 1.00 mg B/L, use 1.00 mL sample. For waters containing more than 1.00 mg B/L, make an appropriate dilution with boron-free distilled water, so that a 1.00-mL portion contains approximately 0.50 μg boron.

Pipet 1.00 mL sample or dilution into an evaporating dish. Unless the calibration curve is being determined at the same time, prepare a blank and a standard containing 0.50 μg boron and run in conjunction with the sample. Proceed as in ¶ *b* preceding, beginning with "Add 4.0 mL curcumin reagent. . . ." If the final solution is turbid, filter through filter paper‡ before reading absorbance. Calculate boron content from calibration curve.

d. Visual comparison: The photometric method may be adapted to visual estimation of low boron concentrations, from 50 to 200 μg/L, as follows: Dilute the standard boron solution 1 + 3 with distilled water; 1.00 mL = 0.20 μg B. Pipet 0, 0.05, 0.10, 0.15, and 0.20 μg boron into evaporating dishes as indicated in ¶ 4*b*. At the same time add an appropriate volume of sample (1.00 mL or portion diluted to 1.00 mL) to an identical evaporating dish. The total boron should be between 0.05 and 0.20 μg. Proceed as in ¶ 4*b*, beginning with "Add 4.0 mL curcumin reagent. . . ." Compare color of samples with standards within 1 hr of drying samples.

e. Removal of high hardness and cation interference: Prepare an ion-exchange column of the type illustrated in Figure 106.1 and described in the Introduction, Section 106.2*b*. Charge column with a strongly acidic cation exchange resin. Backwash column with distilled water to remove entrained air bubbles. Keep the resin covered with liquid at all times. Pass 50 mL 1 + 5 HCl through column at a rate of 0.2 mL acid/mL resin in column/min and wash column free of acid with distilled water.

Pipet 25 mL sample, or a smaller sample of known high boron content diluted to 25 mL, onto the resin column. Adjust rate of flow to about 2 drops/sec and collect effluent in a 50-mL volumetric flask. Wash column with small portions of distilled water until flask is filled to mark. Mix and transfer 2.00 mL into evaporating dish. Add 4.0 mL curcumin reagent and complete the analysis as described in ¶ 4*b* preceding.

5. Calculation

Use the following equation to calculate boron concentration from absorbance readings:

$$\text{mg B/L} = \frac{A_2 \times C}{A_1 \times S}$$

where:

A_1 = absorbance of standard,
A_2 = absorbance of sample,
C = μg B in standard taken, and
S = mL sample.

‡Whatman No. 30 or equivalent.

6. Precision and Accuracy

A synthetic sample containing 240 μg B/L, 40 μg As/L, 250 μg Be/L, 20 μg Se/L, and 6 μg V/L in distilled water was analyzed in 30 laboratories by the curcumin method with a relative standard deviation of 22.8% and a relative error of 0%.

404 B. Carmine Method

1. General Discussion

a. Principle: In the presence of boron, a solution of carmine or carminic acid in concentrated sulfuric acid changes from a bright red to a bluish red or blue, depending on the concentration of boron present.

b. Interference: The ions commonly found in water and wastewater do not interfere.

c. Minimum detectable quantity: 2 μg B.

2. Apparatus

Colorimetric equipment: One of the following is required:

a. Spectrophotometer, for use at 585 nm, with a minimum light path of 1 cm.

b. Filter photometer, equipped with an orange filter having a maximum transmittance near 585 nm, with a minimum light path of 1 cm.

3. Reagents

Store all reagents in polyethylene or boron-free containers.

a. Standard boron solution: Prepare as directed in Method A, ¶ 3*b*.

b. Hydrochloric acid, HCl, conc and 1 + 11.

c. Sulfuric acid, H_2SO_4, conc.

d. Carmine reagent: Dissolve 920 mg carmine N.F. 40, or carminic acid, in 1 L conc H_2SO_4. (If unable to zero spectrophotometer, dilute carmine 1 + 1 with conc H_2SO_4 to replace above reagent.)

4. Procedure

a. Preliminary sample treatment: If sample contains less than 1 mg B/L, pipet a portion containing 2 to 20 μg B into a platinum dish, make alkaline with 1*N* NaOH plus a slight excess, and evaporate to dryness on a steam or hot water bath. If necessary, destroy any organic material by ignition at 500 to 550 C. Acidify cooled residue (ignited or not) with 2.5 mL 1 + 11 HCl and triturate with a rubber policeman to dissolve. Centrifuge if necessary to obtain a clear solution. Pipet 2.00 mL clear concentrate into a small flask or 30-mL test tube. Treat reagent blank identically.

b. Color development: Prepare a series of boron standard solutions (100, 250, 500, 750, and 1,000 μg) in 100 mL with distilled water. Pipet 2.00 mL of each standard solution into a small flask or 30-mL test tube.

Treat blank and calibration standards exactly as the sample. Add 2 drops (0.1 mL) conc HCl, carefully introduce 10.0 mL conc H_2SO_4, mix, and let cool to room temperature. Add 10.0 mL carmine reagent, mix well, and after 45 to 60 min measure absorbance at 585 nm in a cell of 1-cm or longer light path, using the blank as reference.

To avoid error, make sure that no bubbles are present in the optical cell while making photometric readings. Bubbles may appear as a result of incomplete mixing of reagents. Because carmine reagent deteriorates, check calibration curve daily.

5. Calculation

$$\text{mg B/L} = \frac{\mu\text{g B (in approx. 22 mL final volume)}}{\text{mL sample}}$$

6. Precision and Accuracy

A synthetic sample containing 180 μg B/L, 50 μg As/L, 400 μg Be/L, and 50 μg Se/L in distilled water was analyzed in nine laboratories by the carmine method with a relative standard deviation of 35.5% and a relative error of 0.6%.

404 C. Bibliography

Curcumin Colorimetric Method

SILVERMAN, L. & K. TREGO. 1953. Colorimetric microdetermination of boron by the curcumin-acetone solution method. *Anal. Chem.* 25:1264.

DIRLE, W.T., E. TRUOG & K.C. BERGER. 1954. Boron determination in soils and plants—Simplified curcumin procedure. *Anal. Chem.* 26:418.

LUKE, C.L. 1955. Determination of traces of boron in silicon, germanium, and germanium dioxide. *Anal. Chem.* 27:1150.

LISHKA, R.J. 1961. Comparison of analytical procedures for boron. *J. Amer. Water Works Ass.* 53:1517.

BUNTON, N.G. & B.H. TAIT. 1969. Determination of boron in waters and effluents using curcumin. *J. Amer. Water Works Ass.* 61:357.

Carmine Colorimetric Method

HATCHER, J.T. & L.V. WILCOX. 1950. Colorimetric determination of boron using carmine. *Anal. Chem.* 22:567.

405 BROMIDE

Bromide may occur in varying amounts in well supplies in coastal areas as a result of seawater intrusion. The bromide content of some groundwater supplies has been ascribed to connate water. Industrial discharges may contribute the bromide found in some freshwater streams. Under normal circumstances, the bromide content of most drinking waters is negligible, seldom exceeding 1 mg/L.

1. General Discussion

a. Principle: Phenol red undergoes a color change from yellow to red over the pH range 6.4 to 8.0. In a bromine solution, phenol red forms an indicator of the bromphenol blue type, which changes from yellow to blue-purple over the pH range from 3.2 to 4.6. The oxidation of the bromide and the bromination of the phenol red take place readily in the presence of chloramine-T (sodium p-toluenesulfonchloramide). The brominated compound will be reddish to violet, depending upon its concentration. Thus, a sharp differentiation can be made among various quantities of bromide. The concentration of chloramine-T and the timing of the reaction before dechlorination are critical.

b. Interference: Materials present in ordinary tap water do not interfere. Dilution of saline and polluted waters into the range of the method may not be sufficient to eliminate interferences. In such cases, unless comparable values for the bromide concentration are obtained from two dilutions differing by a factor of at least five, the method is inapplicable.

c. Minimum detectable concentration: 100 μg Br/L.

2. Apparatus

a. Colorimetric equipment: One of the following is required:

1) *Spectrophotometer,* for use at 590 nm, providing a light path of at least 2 cm.

2) *Filter photometer,* providing a light path of at least 2 cm and equipped with an orange filter having a maximum transmittance near 590 nm.

3) *Nessler tubes,* matched, 100 mL, tall form.

b. Acid-washed glassware: Wash all glassware with 1 + 6 HNO_3 and rinse with distilled water to remove all trace of adsorbed bromide.

3. Reagents

a. Acetate buffer solution: Dissolve 68 g sodium acetate trihydrate, $NaC_2H_3O_2 \cdot 3H_2O$, in distilled water. Add 30 mL conc (glacial) acetic acid and make up to 1 L. The pH should be 4.6 to 4.7.

b. Phenol red indicator solution: Dissolve 21 mg phenolsulfonephthalein sodium salt and dilute to 100 mL with distilled water.

c. Chloramine-T solution: Dissolve 500 mg chloramine-T and dilute to 100 mL with distilled water. Store in a dark bottle and refrigerate.

d. Sodium thiosulfate, 2N: Dissolve 49.6 g $Na_2S_2O_3 \cdot 5H_2O$ or 31.6 g $Na_2S_2O_3$ and dilute to 100 mL with distilled water.

e. Stock bromide solution: Dissolve 744.6 mg anhydrous KBr in distilled water and make up to 1,000 mL; 1.00 mL = 500 μg Br.

f. Standard bromide solution: Dilute 10.00 mL stock bromide solution to 1,000 mL with distilled water; 1.00 mL = 5.00 μg Br.

4. Procedure

a. Preparation of bromide standards: Prepare at least six standards by diluting 0, 2, 4, 6, 8, and 10 mL standard bromide solution to 50 mL with distilled water. Treat standards the same as samples in ¶ 4*b*,

b. Treatment of sample: To 50.0 mL sample containing 0.1 to 1.0 mg Br/L, add 2 mL buffer solution, 2 mL phenol red solution, and 0.5 mL choramine-T solution. Mix thoroughly. Exactly 20 min after adding chloramine-T, dechlorinate by adding, with mixing, 0.5 mL $Na_2S_2O_3$ solution. Compare visually in nessler tubes against bromide standards prepared simultaneously, or preferably read in a photometer at 590 nm against a reagent blank. Determine the bromide values from a calibration curve of μg Br/55 mL final volume against absorbance. A 2.54-cm light path yields a 0.36 absorbance value at 1 mg Br/L.

5. Calculation

$$\text{mg Br/L} = \frac{\mu\text{g Br (in 55 mL final volume)}}{\text{mL sample}}$$

6. Bibliography

STENGER, V.A. & I.M. KOLTHOFF. 1935. Detection and colorimetric estimation of microquantities of bromide. *J. Amer. Chem. Soc.* 57:831.

HOUGHTON, G.U. 1946. The bromide content of underground waters. *J. Soc. Chem. Ind.* (London) 65:227.

GOLDMAN, E. & D. BYLES. 1959. Suggested revision of phenol red method for bromide. *J. Amer. Water Works Ass.* 51:1051.

406 CARBON DIOXIDE

Surface waters normally contain less than 10 mg free carbon dioxide (CO_2) per liter while some groundwaters may easily exceed that concentration. The carbon dioxide content of a water may contribute significantly to corrosion. Recarbonation of a supply during the last stages of water softening is a recognized treatment process. The subject of saturation with respect to calcium carbonate is discussed in Section 203.

Selection of method: A nomographic and a titrimetric method are described for the estimation of free CO_2 in drinking water. The titration may be performed potentiometrically or with phenolphthalein indicator. Properly conducted, the more rapid, simple indicator method is satisfactory for field tests and for control and routine applications if it is understood that the method gives, at best, only an approximation.

The nomographic method (A) usually gives a closer estimation of the total free CO_2 when the pH and alkalinity determinations are made immediately and correctly at the time of sampling. The pH measurement preferably should be made with an electrometric pH meter, properly calibrated with standard buffer solutions in the pH range of 7 to 8. The error resulting from inaccurate pH measurements grows with an increase in total alkalinity. For example, an inaccuracy of 0.1 in the pH determination causes a CO_2 error of 2 to 4 mg/L in the pH range of 7.0 to 7.3 and a total alkalinity of 100 mg $CaCO_3$/L. In the same pH range, the error approaches 10 to 15 mg/L when the total alkalinity is 400 mg as $CaCO_3$/L.

Under favorable conditions, agreement between the titrimetric and nomographic methods is reasonably good. When agreement is not precise and the CO_2 determination is of particular importance, state the method used.

The calculation of the total CO_2, free and combined, is given in Method C.

406 A. Nomographic Determination of Free Carbon Dioxide and the Three Forms of Alkalinity*

1. General Discussion

Diagrams and nomographs enable the rapid calculation of the CO_2, bicarbonate, carbonate, and hydroxide content of natural and treated waters. These graphical presentations are based on equations relating the ionization equilibria of the carbonates and water. If pH, total alkalinity, temperature, and total mineral content are known, any or all of the alkalinity forms and CO_2 can be determined nomographically.

A set of charts, Figures 406:1 through 4, is presented for use where their accuracy for the individual water supply is confirmed. The nomographs and the equations on which they are based are valid only when the salts of weak acids other than carbonic acid are absent or present in extremely small amounts.

Some treatment processes, such as superchlorination and coagulation, can affect significantly pH and total-alkalinity values of a poorly buffered water of low alkalinity and low total-dissolved-mineral content. In such instances the nomographs may not be applicable.

2. Precision and Accuracy

The precision possible with the nomographs depends on the size and range of the scales. With practice, the recommended nomographs can be read with a preci-

*See also Alkalinity, Section 403 preceding.

sion of 1%. However, the overall accuracy of the results is limited by the accuracy of the analytical data applied to the nomographs and by the validity of the theoretical equations and the numerical constants on which the nomographs are based. An approximate check of the accuracy of the calculations can be made by summing the three forms of alkalinity. Their sum should equal the total alkalinity.

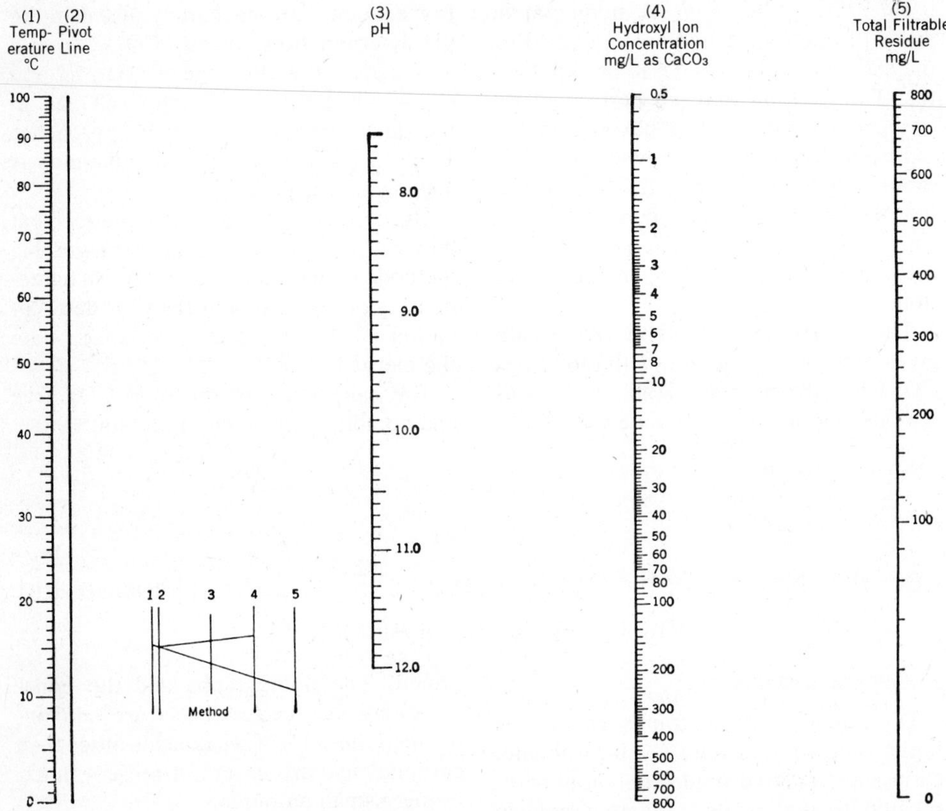

Figure 406:1. Nomograph for evaluation of hydroxide ion concentration.† To use: align temperature (Scale 1) and total filtrable residue (Scale 5); pivot on Line 2 to proper pH (Scale 3); read hydroxide ion concentration, as mg $CaCO_3$/L, on Scale 4. (Example: For 13 C temperature, 240 mg total filtrable residue/L, pH 9.8, the hydroxide ion concentration is found to be 1.4 mg as $CaCO_3$/L.)

†Copies of the nomographs in Figures 406: 1–4, enlarged to 2.5 times the size shown here, may be obtained from The American Water Works Association, 6666 West Quincy Ave., Denver, Colorado 80235, for $3.00 per set of four.

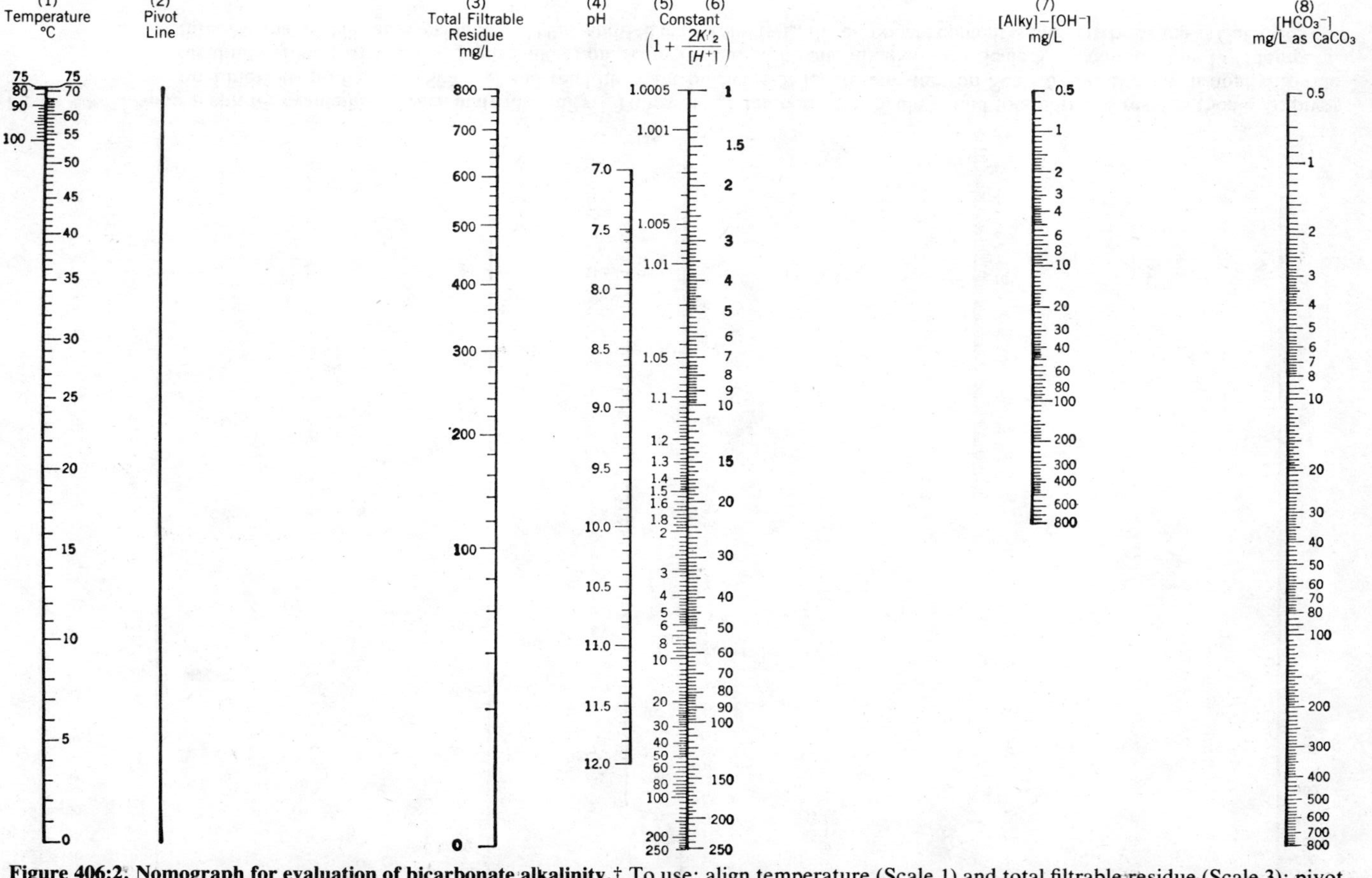

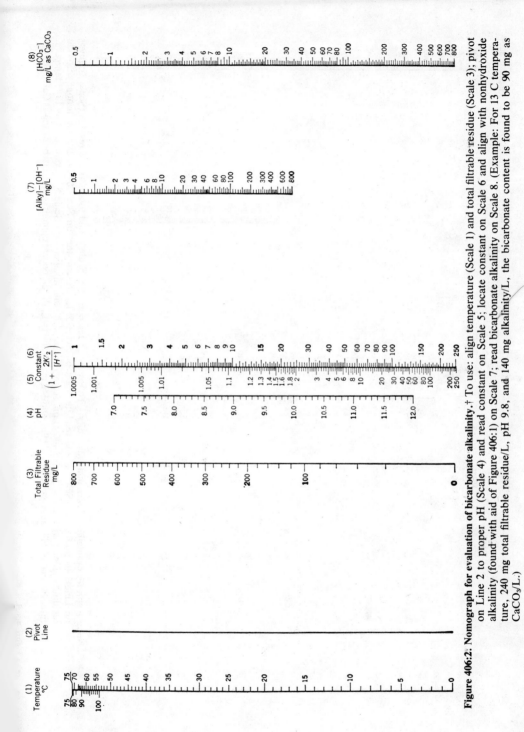

Figure 406:2. Nomograph for evaluation of bicarbonate alkalinity.† To use: align temperature (Scale 1) and total filtrable residue (Scale 3); pivot on Line 2 to proper pH (Scale 4) and read constant on Scale 5; locate constant on Scale 6 and align with nonhydroxide alkalinity (found with aid of Figure 406:1) on Scale 7; read bicarbonate alkalinity on Scale 8. (Example: For 13 C temperature, 240 mg total filtrable residue/L, pH 9.8, and 140 mg alkalinity/L, the bicarbonate content is found to be 90 mg as $CaCO_3$/L.)

†See note to Figure 406:1.

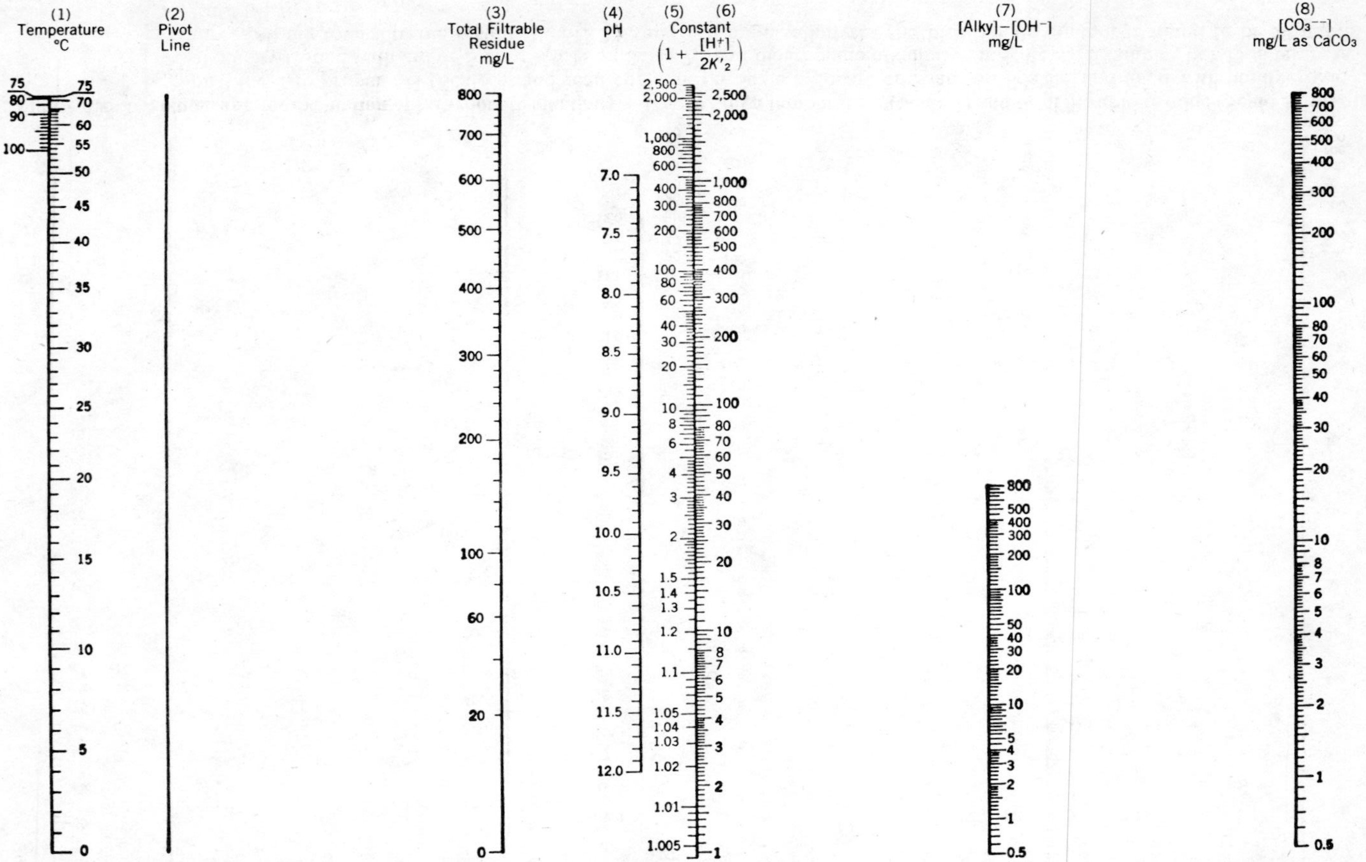

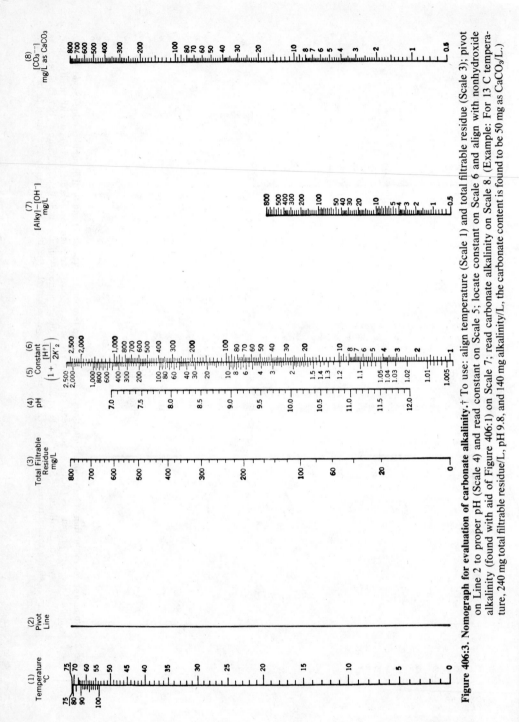

Figure 406:3. Nomograph for evaluation of carbonate alkalinity.† To use: align temperature (Scale 1) and total filtrable residue (Scale 3); pivot on Line 2 to proper pH (Scale 4) and read constant on Scale 5; locate constant on Scale 6 and align with nonhydroxide alkalinity (found with aid of Figure 406:1) on Scale 7; read carbonate alkalinity on Scale 8. (Example: For 13 C temperature, 240 mg total filtrable residue/L, pH 9.8, and 140 mg alkalinity/L, the carbonate content is found to be 50 mg as $CaCO_3$/L.)

†See note to Figure 406:1.

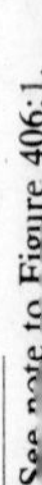

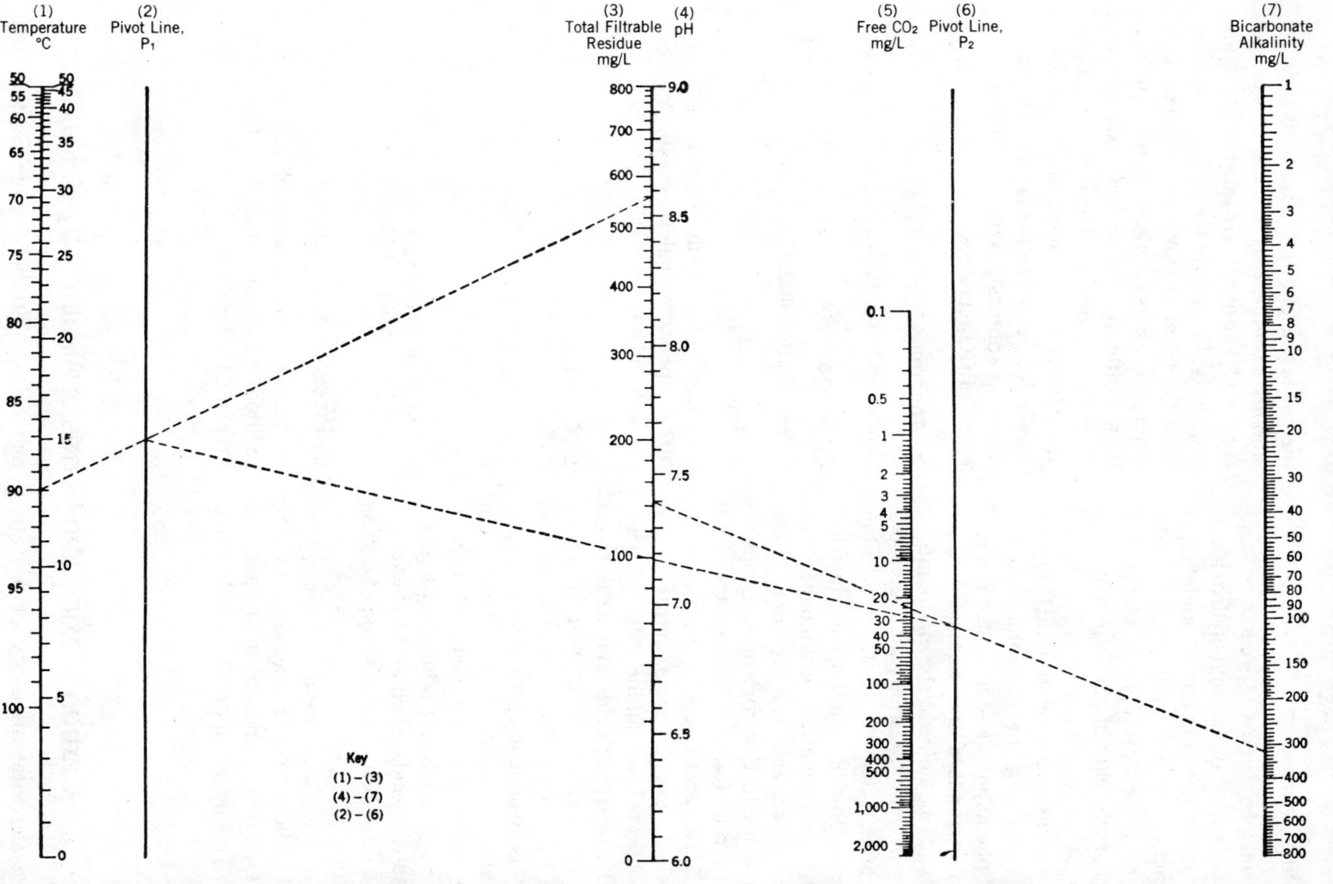

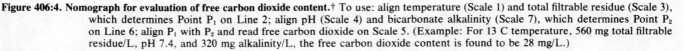

Figure 406:4. Nomograph for evaluation of free carbon dioxide content.† To use: align temperature (Scale 1) and total filtrable residue (Scale 3), which determines Point P_1 on Line 2; align pH (Scale 4) and bicarbonate alkalinity (Scale 7), which determines Point P_2 on Line 6; align P_1 with P_2 and read free carbon dioxide on Scale 5. (Example: For 13 C temperature, 560 mg total filtrable residue/L, pH 7.4, and 320 mg alkalinity/L, the free carbon dioxide content is found to be 28 mg/L.)

†See note to Figure 406:1.

406 B. Titrimetric Method for Free Carbon Dioxide

1. General Discussion

a. Principle: Free CO_2 reacts with sodium carbonate or sodium hydroxide to form sodium bicarbonate. Completion of the reaction is indicated potentiometrically or by the development of the pink color characteristic of phenolphthalein indicator at the equivalence pH of 8.3. A 0.01*N* sodium bicarbonate ($NaHCO_3$) solution containing the recommended volume of phenolphthalein indicator is a suitable color standard until familiarity is obtained with the color at the end point.

b. Interference: Cations and anions that quantitatively disturb the normal carbon dioxide-carbonate equilibrium interfere with the determination. Aluminum, chromium, copper, and iron are some of the metals with salts that contribute to high results. Ferrous ion should not exceed 1.0 mg/L. Positive errors also are caused by amines, ammonia, borate, nitrite, phosphate, silicate, and sulfide. Mineral acids and salts of strong acids and weak bases affect the determination and therefore should be absent. The titrimetric method for CO_2 is inapplicable to samples containing acid mine wastes and effluent from acid-regenerated cation exchangers. Negative errors may be introduced by high total dissolved solids, such as those encountered in seawater, or by addition of excess indicator.

c. Sampling and storage: Even with a careful collection technic, some loss in free CO_2 can be expected in storage and transit. This occurs more frequently when the gas is present in large amounts. Occasionally a sample may show an increase in free CO_2 content on standing. Consequently, determine free CO_2 immediately at the point of sampling. Where a field determination is impractical, fill completely a bottle for laboratory examination. Keep the sample, until tested, at a temperature lower than that at which the water was collected. Make the laboratory examination as soon as possible to minimize the effect of CO_2 changes.

2. Apparatus

See Section 402.2.

3. Reagents

See Section 402.3.

4. Procedure

Follow the procedure given in Section 402.4*a*, phenolphthalein, or 402.4*c*, using endpoint pH 8.3.

5. Calculation

$$\text{mg } CO_2\text{/L} = \frac{A \times N \times 44{,}000}{\text{mL sample}}$$

where:

A = mL titrant and

N = normality of NaOH.

6. Precision and Accuracy

Precision and accuracy of the titrimetric method are on the order of ±10% of the known CO_2 concentration.

406 C. Carbon Dioxide and Forms of Alkalinity by Calculation

When the total alkalinity of a water (Section 403) is due almost entirely to hydroxides, carbonates, or bicarbonates, and the total filtrable residue (Section 209) is not greater than 500 mg/L, the alkalinity forms and free CO_2 can be calculated from the sample pH and total alkalinity. The calculation is subject to the same limitations as the nomographic procedure given above and the additional restriction of using a single temperature, 25 C. The calculations are based on the ionization constants:

$$K_1 = \frac{[H^+][HCO_3^-]}{[H_2CO_3^*]} \qquad (K_1 = 10^{-6.36})$$

and

$$K_2 = \frac{[H^+][CO_3^{2-}]}{[HCO_3^-]} \qquad (K_2 = 10^{-10.33})$$

where:

$$[H_2CO_3^*] = [H_2CO_3] + [CO_2(aq)]$$

Activity coefficients are assumed equal to unity.

Compute the forms of alkalinity and sample pH and total alkalinity using the following equations:

a. Bicarbonate alkalinity:

$$HCO_3^- \text{ as mg } CaCO_3/L = \frac{T - 5.0 \times 10^{(pH-10)}}{1 + 0.94 \times 10^{(pH-10)}}$$

where:

T = total alkalinity, mg $CaCO_3$/L.

b. Carbonate alkalinity:

$$CO_3^{2-} \text{ as mg } CaCO_3/L = 0.94 \times B \times 10^{(pH-10)}$$

where:

B = bicarbonate alkalinity, from *a*.

c. Hydroxide alkalinity:

$$OH^- \text{ as mg } CaCO_3/L = 5.0 \times 10^{(pH-10)}$$

d. Free carbon dioxide:

$$\text{mg } CO_2/L = 2.0 \times B \times 10^{(6-pH)}$$

where:

B = bicarbonate alkalinity, from *a*.

e. Total carbon dioxide:

$$\text{mg total } CO_2/L = A + 0.44\,(2B + C)$$

where:

A = mg free CO_2/L,
B = bicarbonate alkalinity from *a*, and
C = carbonate alkalinity from *b*.

406 D. Bibliography

MOORE, E.W. 1939. Graphic determination of carbon dioxide and the three forms of alkalinity. *J. Amer. Water Works Ass.* 31:51.

DYE, J.F. 1958. Correlation of the two principal methods of calculating the three kinds of alkalinity. *J. Amer. Water Works Ass.* 50:812.

407 CHLORIDE

Chloride, in the form of Cl^- ion, is one of the major inorganic anions in water and wastewater. In potable water, the salty taste produced by chloride concentrations is variable and dependent on the chemical composition of water. Some waters containing 250 mg chloride/L may have a detectable salty taste if the cation is sodium. On the other hand, the typical salty taste may be absent in waters containing as much as 1,000 mg/L when the predominant cations are calcium and magnesium.

The chloride concentration is higher in wastewater than in raw water because sodium chloride (NaCl) is a common article of diet and passes unchanged through the digestive system. Along the sea coast, chloride may be present in high concentrations because of leakage of salt water into the sewerage system. It also may be increased by industrial processes.

A high chloride content may harm metallic pipes and structures, as well as growing plants.

1. Selection of Method

Four methods are presented for the determination of chloride. Because the first two are similar in most respects, selection is largely a matter of personal preference. The argentometric method (A) is suitable for use in relatively clear waters when 0.15 to 10 mg Cl are present in the portion titrated. The end point of the mercuric nitrate method (B) is easier to detect. The potentiometric method (C) is suitable for colored or turbid samples in which color-indicated end points might be difficult to observe. The potentiometric method can be used without a pretreatment step for samples containing ferric ions (if not present in an amount greater than the chloride concentration), chromic, phosphate, and ferrous and other heavy metal ions. The ferricyanide method (D) is an automated technic.

2. Sampling and Storage

Collect representative samples in clean, chemically resistant glass or plastic bottles. The maximum sample portion required is 100 mL. No special preservative is necessary if the sample is to be stored.

407 A. Argentometric Method

1. General Discussion

a. Principle: In a neutral or slightly alkaline solution, potassium chromate can indicate the end point of the silver nitrate titration of chloride. Silver chloride is precipitated quantitatively before red silver chromate is formed.

b. Interference: Substances in amounts normally found in potable waters will not interfere. Bromide, iodide, and cyanide register as equivalent chloride concentrations. Sulfide, thiosulfate, and sulfite ions interfere but can be removed by treatment with hydrogen peroxide. Orthophosphate in excess of 25 mg/L interferes by precipitating as silver phosphate. Iron in excess of 10 mg/L interferes by masking the end point.

2. Apparatus

a. Erlenmeyer flask, 250 mL.

b. Buret, 50 mL.

3. Reagents

a. Potassium chromate indicator solution: Dissolve 50 g K_2CrO_4 in a little distilled water. Add $AgNO_3$ solution until a definite red precipitate is formed. Let stand 12 hr, filter, and dilute to 1 Lwith distilled water.

b. Standard silver nitrate titrant, 0.0141*N:* Dissolve 2.395 g $AgNO_3$ in distilled water and dilute to 1,000 mL. Standardize against 0.0141*N* NaCl by the procedure described in ¶ 4*b* below; 1.00 mL = 500 μg Cl. Store in a brown bottle.

c. Standard sodium chloride, 0.0141*N*: Dissolve 824.0 mg NaCl (dried at 140 C) in distilled water and dilute to 1,000 mL; 1.00 mL = 500 μg Cl.

d. Special reagents for removal of interference:

1) *Aluminum hydroxide suspension:* Dissolve 125 g aluminum potassium sulfate or aluminum ammonium sulfate, $AlK(SO_4)_2 \cdot 12H_2O$ or $AlNH_4(SO_4)_2 \cdot 12H_2O$, in 1 L distilled water. Warm to 60 C and add 55 mL conc ammonium hydroxide (NH_4OH) slowly with stirring. Let stand about 1 hr, transfer to a large bottle, and wash precipitate by successive additions, with thorough mixing and decanting with distilled water, until free from chloride. When freshly prepared, the suspension occupies a volume of approximately 1 L.

2) *Phenolphthalein indicator solution.*

3) *Sodium hydroxide,* NaOH, 1*N*.

4) *Sulfuric acid,* H_2SO_4, 1*N*.

5) *Hydrogen peroxide,* H_2O_2, 30%.

4. Procedure

a. Sample preparation: Use a 100-mL sample or a suitable portion diluted to 100 mL. If the sample is highly colored, add 3 mL $Al(OH)_3$ suspension, mix, let settle, and filter.

If sulfide, sulfite, or thiosulfate is present, add 1 mL H_2O_2 and stir for 1 min.

b. Titration: Directly titrate samples in the pH range 7 to 10. Adjust sample pH to 7 to 10 with H_2SO_4 or NaOH if it is not in this range. Add 1.0 mL K_2CrO_4 indicator solution. Titrate with standard $AgNO_3$ titrant to a pinkish yellow end point. Be consistent in end-point recognition.

Standardize $AgNO_3$ titrant and establish reagent blank value by the titration method outlined above. A blank of 0.2 to 0.3 mL is usual.

5. Calculation

$$\text{mg Cl/L} = \frac{(A - B) \times N \times 35{,}450}{\text{mL sample}}$$

where:

A = mL titration for sample,
B = mL titration for blank, and
N = normality of $AgNO_3$.

$$\text{mg NaCl/L} = (\text{mg Cl/L}) \times 1.65$$

6. Precision and Accuracy

A synthetic sample containing 241 mg chloride/L, 108 mg Ca/L, 82 mg Mg/L; 3.1 mg K/L, 19.9 mg Na/L, 1.1 mg NO_3^--N/L, 0.25 mg NO_2^--N/L, 259 mg sulfate/L, and 42.5 mg total alkalinity/L (contributed by $NaHCO_3$) in distilled water was analyzed in 41 laboratories by the argentometric method, with a relative standard deviation of 4.2% and a relative error of 1.7%.

407 B. Mercuric Nitrate Method

1. General Discussion

a. Principle: Chloride can be titrated with mercuric nitrate, $Hg(NO_3)_2$, because of the formation of soluble, slightly dissociated mercuric chloride. In the pH range 2.3 to 2.8, diphenylcarbazone indicates the titration end point by formation of a purple complex with the excess mercuric ions. Xylene cyanol FF serves as a pH indicator and end-point enhancer. Increasing the strength of the titrant and modifying the indicator mixtures extend the range of measurable chloride concentrations.

b. Interference: Bromide and iodide are titrated with $Hg(NO_3)_2$ in the same manner as chloride. Chromate, ferric, and sulfite ions interfere when present in excess of 10 mg/L.

2. Apparatus

a. Erlenmeyer flask, 250 mL.

b. Microburet, 5 mL with 0.01 mL graduation intervals.

3. Reagents

a. Standard sodium chloride, 0.0141*N*. See Method A, ¶ 3*c* above.

b. Nitric acid, HNO_3, 0.1*N*.

c. Sodium hydroxide, NaOH, 0.1*N*.

d. Reagents for chloride concentrations below 100 mg/L:

1) *Indicator-acidifier reagent:* The HNO_3 concentration of this reagent is an important factor in the success of the determination and can be varied as indicated in a) or b) to suit the alkalinity range of the sample. Reagent a) contains sufficient HNO_3 to neutralize a total alkalinity of 150 mg as $CaCO_3$/L to the proper pH in a 100-mL sample. Adjust amount of HNO_3 to accommodate samples of alkalinity different from 150 mg/L.

a) Dissolve, in the order named, 250 mg s-diphenylcarbazone, 4.0 mL conc HNO_3, and 30 mg xylene cyanol FF in 100 mL 95% ethyl alcohol or isopropyl alcohol. Store in a dark bottle in a refrigerator. This reagent is not stable indefinitely. Deterioration causes a slow end point and high results.

b) Because pH control is critical, adjust pH of highly alkaline or acid samples to 2.5 ± 0.1 with 0.1*N* HNO_3 or NaOH, not with sodium carbonate (Na_2CO_3). Use a pH meter with a nonchloride type of reference electrode for pH adjustment. If only the usual chloride-type reference electrode is available for pH adjustment, determine amount of acid or alkali required to obtain a pH of 2.5 ± 0.1 and discard this sample portion. Treat a separate sample portion with the determined amount of acid or alkali and continue analysis. Under these circumstances, omit HNO_3 from indicator reagent.

2) *Standard mercuric nitrate titrant,* 0.0141*N*: Dissolve 2.3 g $Hg(NO_3)_2$ or 2.5 g $Hg(NO_3)_2 \cdot H_2O$ in 100 mL distilled water containing 0.25 mL conc HNO_3. Dilute to just under 1 L. Make a preliminary standardization by following the procedure described in ¶ 4*a*. Use replicates containing 5.00 mL standard NaCl solution and 10 mg sodium bicarbonate ($NaHCO_3$) diluted to 100 mL with distilled water. Adjust titrant to 0.0141*N* and make a final standardization; 1.00 mL = 500 μg Cl. Store away from light in a dark bottle.

e. Reagent for chloride concentrations greater than 100 mg/L:

1) *Mixed indicator reagent:* Dissolve 0.50 g diphenylcarbazone powder and 0.05 g bromphenol blue powder in 75 mL 95% ethyl or isopropyl alcohol and dilute to 100 mL with 95% ethyl or isopropyl alcohol.

2) *Strong standard mercuric nitrate titrant,* 0.141*N:* Dissolve 25 g $Hg(NO_3)_2 \cdot H_2O$ in 900 mL distilled water containing 5.0 mL conc HNO_3. Dilute to just under 1 L and standardize by following the procedure described in ¶ 4*b*. Use replicates containing 25.00 mL standard NaCl solution and 25 mL distilled water. Adjust titrant to 0.141*N* and make a final standardization; 1.00 mL = 5.00 mg Cl.

4. Procedure

a. Titration of chloride concentrations less than 100 mg/L: Use a 100-mL sample or smaller portion so that the chloride content is less than 10 mg.

Add 1.0 mL indicator-acidifier reagent. (The color of the solution should be green-blue at this point. A light green indicates pH less than 2.0; a pure blue indicates pH more than 3.8.) For most potable waters, the pH after this addition will be 2.5 ± 0.1. For highly alkaline or acid waters, adjust pH to about 8 before adding indicator-acidifier reagent.

Titrate with 0.0141*N* $Hg(NO_3)_2$ titrant to a definite purple end point. The solution turns from green-blue to blue a few drops before the end point.

Determine blank by titrating 100 mL distilled water containing 10 mg $NaHCO_3$.

b. Titration of chloride concentrations greater than 100 mg/L: Use a sample portion (5 to 50 mL) requiring less than 5 mL titrant to reach the end point. Measure into a 150-mL beaker. Add approximately 0.5 mL mixed indicator reagent and mix well. The color should be purple. Add 0.1*N* HNO_3 dropwise until the color just turns yellow. Titrate with 0.141*N* $Hg(NO_3)_2$ titrant to first permanent dark purple. Titrate a distilled water blank using the same procedure.

5. Calculation

$$\text{mg Cl/L} = \frac{(A - B) \times N \times 35{,}450}{\text{mL sample}}$$

where:

A = mL titration for sample,
B = mL titration for blank, and
N = normality of $Hg(NO_3)_2$.

$$\text{mg NaCl/L} = (\text{mg Cl/L}) \times 1.65$$

6. Precision and Accuracy

A synthetic sample containing 241 mg chloride/L, 108 mg Ca/L, 82 mg Mg/L, 3.1 mg K/L, 19.9 mg Na/L, 1.1 mg NO_3^--N/L, 0.25 mg NO_2^--N/L, 259 mg sulfate/L, and 42.5 mg total alkalinity/L (contributed by $NaHCO_3$) in distilled water was analyzed in 10 laboratories by the mercurimetric method, with a relative standard deviation of 3.3% and a relative error of 2.9%.

407 C. Potentiometric Method

1. General Discussion

a. Principle: Chloride is determined by potentiometric titration with silver nitrate solution with a glass and silver-silver chloride electrode system. During titration an electronic voltmeter is used to detect the change in potential between the two electrodes. The end point of the titration is that instrument reading at which the greatest change in voltage has occurred for a small and constant increment of silver nitrate added.

b. Interference: Iodide and bromide also are titrated as chloride. Ferricyanide causes high results and must be removed. Chromate and dichromate interfere and should be reduced to the chromic state or removed. Ferric iron interferes if present in an amount substantially higher than the amount of chloride. Chromic ion, ferrous ion, and phosphate do not interfere.

Grossly contaminated samples usually require pretreatment. Where contamination is minor, some contaminants can be destroyed simply by adding nitric acid.

2. Apparatus

a. Glass and silver-silver chloride electrodes: Prepare in the laboratory or purchase a silver electrode coated with AgCl for use with specified instruments. Instructions on use and care of electrodes are supplied by the manufacturer.

b. Electronic voltmeter, to measure potential difference between electrodes: A pH meter may be converted to this use by substituting the appropriate electrode.

c. Mechanical stirrer, with plastic-coated or glass impeller.

3. Reagents

a. Standard sodium chloride solution, 0.0141*N:* See ¶ 407A.3*c*.

b. Nitric acid, HNO_3, conc.

c. Standard silver nitrate titrant, 0.0141*N:* See ¶ 407A.3*b*.

d. Pretreatment reagents:

1) *Sulfuric acid,* H_2SO_4, 1 + 1.

2) *Hydrogen peroxide,* H_2O_2, 30%.

3) *Sodium hydroxide,* NaOH, 1*N*.

4. Procedure

a. Standardization: The various instruments that can be used in this determination differ in operating details; follow the manufacturer's instructions. Make necessary mechanical adjustments. Then, after allowing sufficient time for warmup (10 min), balance internal electrical components to give an instrument setting of 0 mV or, if a pH meter is used, a pH reading of 7.0.

1) Place 10.0 mL standard NaCl solution in a 250-mL beaker, dilute to about 100 mL, and add 2.0 mL conc HNO_3. Immerse stirrer and electrodes.

2) Set instrument to desired range of millivolts or pH units. Start stirrer.

3) Add standard $AgNO_3$ titrant, recording scale reading after each addition. At the start, large increments of $AgNO_3$ may be added; then, as the end point is approached, add smaller and equal increments (0.1 or 0.2 mL) at longer intervals, so that the exact end point can be determined. Determine volume of $AgNO_3$ used at the point at which there is the greatest change in instrument reading per unit addition of $AgNO_3$.

4) Plot a differential titration curve if the exact end point cannot be determined by inspecting the data. Plot change in instrument reading for equal increments of

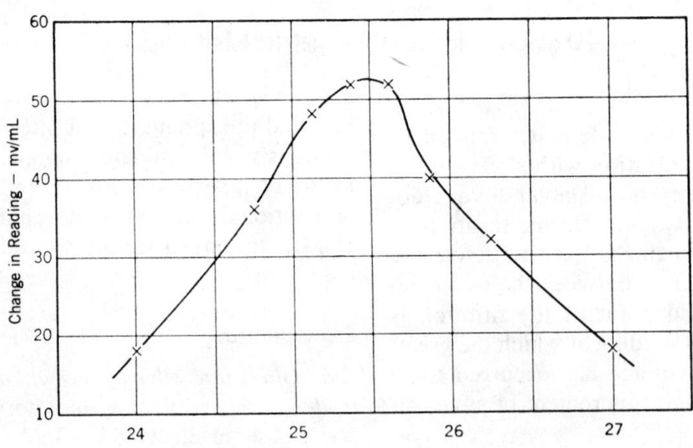

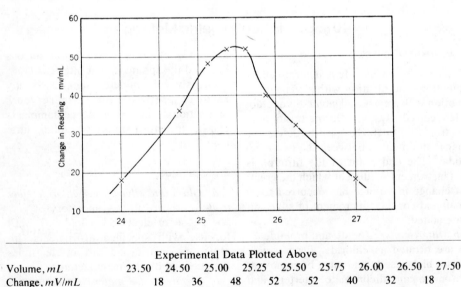

Experimental Data Plotted Above

Volume, *mL*	23.50		24.50		25.00		25.25		25.50		25.75		26.00		26.50		27.50
Change, *mV/mL*		18		36		48		52		52		40		32		18	

Figure 407:1. Example of differential titration curve (end point is 25.5 mL).

$AgNO_3$ against volume of $AgNO_3$ added, using average of buret readings before and after each addition. The procedure is illustrated in Figure 407:1.

b. Sample analysis:

1) Pipet 100.0 mL sample, or a portion containing not more than 10 mg chloride, into a 250-mL beaker. In the absence of interfering substances, proceed with ¶ 3) below.

2) In the presence of organic compounds, sulfite, or other interferences (such as large amounts of ferric iron, cyanide, or sulfide), acidify sample with H_2SO_4, using litmus paper. Boil for 5 min to remove volatile compounds. Add more H_2SO_4, if necessary, to keep solution acidic. Add 3 mL H_2O_2 and boil for 15 min, adding chloride-free distilled water to keep the volume above 50 mL. Dilute to 100 mL, add NaOH solution dropwise until alkaline to litmus, then 10 drops in excess. Boil for 5 min, filter into a 250-mL beaker, and wash precipitate and paper several times with hot distilled water.

3) Add conc HNO_3 dropwise until acidic to litmus paper, then 2.0 mL in excess. Cool and dilute to 100 mL if necessary. Immerse stirrer and electrodes and start stirrer. Make any necessary adjustments according to the manufacturer's instructions and set selector switch to appropriate setting for measuring the difference of potential between electrodes.

4) Complete determination by titrating according to ¶ 4*a*4). If an end-point reading has been established from previous determinations for similar samples and conditions, use this predetermined end point. For the most accurate work, make a blank titration by carrying chloride-free distilled water through the procedure.

5. Calculation

$$\text{mg Cl/L} = \frac{(A - B) \times N \times 35.45 \times 1{,}000}{\text{mL sample}}$$

where:

A = mL $AgNO_3$,
B = mL blank, and
N = normality of titrant.

6. Precision and Accuracy

In the absence of interfering substances, the precision and accuracy are estimated to be about 0.12 mg for 5 mg Cl, or 2.5% of the amount present. When pretreatment is required to remove interfering substances, the precision and accuracy are reduced to about 0.25 mg for 5 mg Cl, or 5% of amount present.

407 D. Automated Ferricyanide Method (TENTATIVE)

1. General Discussion

a. Principle: Thiocyanate ion is liberated from mercuric thiocyanate by the formation of soluble mercuric chloride. In the presence of ferric ion, free thiocyanate ion forms a highly colored ferric thiocyanate, of which the intensity is proportional to the chloride concentration.

b. Interferences: None of significance. However, use a continuous filter on turbid samples.

c. Application: The method is applicable to potable, surface, and saline waters, and domestic and industrial wastewaters. The concentration range can be varied by using the colorimeter controls.

2. Apparatus

a. Automated analytical equipment, consisting of the components (except the heating bath) listed in Section 602.1.

b. Filters, 480 nm.

3. Reagents

a. Stock mercuric thiocyanate solution: Dissolve 4.17 g $Hg(SCN)_2$ in about 500 mL methanol and dilute to 1,000 mL with methanol, mix, and filter through filter paper.

b. Stock ferric nitrate solution: Dissolve 202 g $Fe(NO_3)_3 \cdot 9H_2O$ in about 500 mL distilled water, then carefully add 21 mL conc HNO_3. Dilute to 1,000 mL with distilled water and mix. Filter through paper and store in an amber bottle.

c. Color reagent: Add 150 mL stock $Hg(SCN)_2$ solution to 150 mL stock $Fe(NO_3)_3$ solution. Mix and dilute to 1,000 mL with distilled water. Add 0.5 mL polyoxyethylene 23 lauryl ether.*

d. Stock chloride solution: Dissolve 1.6482 g NaCl, dried at 140 C, in distilled water and dilute to 1,000 mL; 1.00 mL = 1.00 mg Cl.

e. Standard chloride solutions: Prepare chloride standards in the desired concentration range, such as 1 to 200 mg/L, using stock chloride solution.

*Brij 35, available from ICI United States, Chicago, Ill., Technicon Instruments Corporation, Tarrytown, N.Y. 10591, or equivalent.

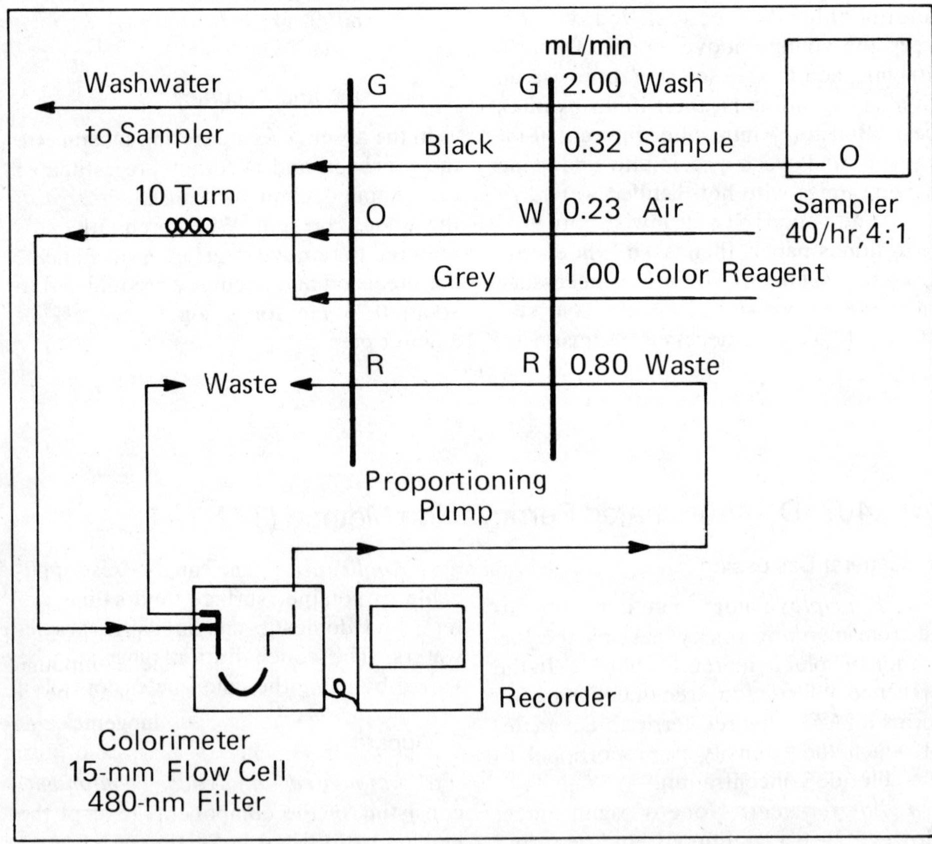

Figure 407:2. Flow scheme for automated chloride analysis.

4. Procedure

Set up manifold as shown in Figure 407:2 and follow general procedure described in Section 602.2.

5. Calculation

See Section 602.3.

407 E. Bibliography

Argentometric Method

HAZEN, A. 1889. On the determination of chlorine in water. *Amer. Chem. J.* 11:409.

KOLTHOFF, L.M. & V.A. STENGER. 1947. Volumetric Analysis, 2nd ed. Vol. 2. Interscience Publishers, New York, pp. 242–245, 256–258.

Mercuric Nitrate Method

KOLTHOFF & STENGER (above), pp. 334–335.

DOMASK, W.C. & K.A. KOBE. 1952. Mercurimetric determination of chlorides and water-soluble chlorohydrins. *Anal. Chem.* 24:989.

GOLDMAN, E. 1959. New indicator for the mercurimetric chloride determination in potable water. *Anal. Chem.* 31:1127.

Potentiometric Method

KOLTHOFF, L.M. & N.H. FURMAN. 1931. Potentiometric Titrations, 2nd ed. John Wiley & Sons, New York.

REFFENBURG, H.B. 1935. Colorimetric determination of small quantities of chlorides in water. *Ind. Eng. Chem.*, Anal. Ed. 7:14.

CALDWELL, J.R. & H.V. MEYER. 1935. Chloride determination. *Ind. Eng. Chem.*, Anal. Ed. 7:38.

WALTON, H.F. 1952. Principles and Methods of Chemical Analysis. Prentice-Hall, Inc., Englewood Cliffs, N.J.

SERFASS, E.J. & R.F. MURACA. 1954. Procedures for Analyzing Metal-Finishing Wastes. Ohio River Valley Water Sanitation Commission, Cincinnati, p. 80.

WILLARD, H.H., L.L. MERRITT & J.A. DEAN. 1958. Instrumental Methods of Analysis, 3rd ed. D. Van Nostrand Co., Princeton, N.J.

FURMAN, N.H., ed. 1962. Standard Methods of Chemical Analysis, 6th ed. D. Van Nostrand Co., Princeton, N.J., Vol. I.

Automated Ferricyanide Method

ZALL, D.M., D. FISHER & M.D. GARNER. 1956. Photometric determination of chlorides in water. *Anal. Chem.* 28:1665.

O'BRIEN, J.E. 1962. Automatic analysis of chlorides in sewage. *Wastes Eng.* 33:670.

408 CHLORINE (RESIDUAL)

The chlorination of water supplies and polluted waters serves primarily to destroy or deactivate disease-producing microorganisms. A secondary benefit is the overall improvement in water quality resulting from the reaction of chlorine with ammonia, iron, manganese, sulfide, and some organic substances.

Chlorination may produce adverse effects. Taste and odor characteristics of phenols and other organic compounds present in a water supply may be intensified. Potentially carcinogenic chloroorganic compounds such as chloroform may be formed. Combined chlorine formed on chlorination of ammonia- or amine-bearing waters adversely affects some aquatic life. To fulfill the primary purpose of chlorination and to minimize any adverse effects, it is essential that proper testing procedures be used with a foreknowledge of the limitations of the analytical determination.

Chlorine applied to water in its elemental or hypochlorite form initially undergoes hydrolysis to form free available chlorine consisting of aqueous molecular chlorine, hypochlorous acid, and hypochlorite ion. The relative proportion of these free chlorine forms is pH- and temperature-dependent. At the pH of most waters, hypochlorous acid and hypochlorite ion will predominate.

Free chlorine reacts readily with ammonia and certain nitrogenous compounds to form combined available chlorine. With ammonia, chlorine reacts to form the chloramines: monochloramine, dichloramine, and nitrogen trichloride. The presence and concentrations of these combined forms depend chiefly on pH, temperature, and initial chlorine-to-nitrogen ratio. Both free and combined chlorine may be present simultaneously. Combined chlorine in water supplies may be formed in the treatment of raw waters containing ammonia or by the addition of ammonia or ammonium salts. Chlorinated wastewater effluents, as well as certain chlorinated industrial effluents, normally contain only combined chlorine. Historically, the principal analytical problem has been to distinguish between free and combined forms of residual chlorine.[1,2]

In two separate but related studies, samples were prepared and distributed to participating laboratories to evaluate residual chlorine methods. Because of poor accuracy and precision and a high overall (average) total error in these studies, all orthotolidine procedures except one were dropped in the 14th edition of this work. The useful Stabilized Neutral Orthotolidine Method has been deleted from this edition because of the toxic nature of orthotolidine.

1. Selection of Method

a. Natural and treated waters: The iodometric methods (A and B) are suitable for measuring total chlorine concentrations greater than 1 mg/L, but the amperometric end point of Methods B and C gives greater sensitivity. All acidic iodometric methods suffer from interferences, generally in proportion to the quantity of potassium iodide (KI) and H^+ added.

The amperometric titration method (C) is a standard of comparison for the determination of free or combined chlorine. It is affected little by common oxidizing agents, temperature variations, turbidity, and color. The method is not as simple as the colorimetric methods and requires greater operator skill to obtain the best reliability. Loss of chlorine can occur because of rapid stirring in some commercial equipment. Electrode cleanliness and conditioning are necessary for sharp end points.

The DPD methods (Methods D and E) are operationally simpler for determining free available chlorine than the amperometric titration. Procedures are given for estimating the separate mono- and dichloramine and combined fractions. High concentrations of monochloramine interfere with the free chlorine determination unless the reaction is stopped with arsenite or thioacetamide. In addition, the DPD methods are subject to interference by oxidized forms of manganese unless compensated for by a blank.

The leuco crystal violet (LCV) method (F) makes possible the determination of free and combined available chlorine but not individual chloramine fractions. The LCV method exhibits less monochloramine interference than DPD in the determination of free available chlorine; however, nitrite and monochloramine in combination, as well as oxidized forms of manganese, interfere in determining free available chlorine.

The amperometric, LCV, and DPD methods are unaffected by dichloramine concentrations in the range of 0 to 9 mg Cl as Cl_2/L in the determination of free chlorine. Nitrogen trichloride, if present, may react partially as free available chlorine in the amperometric, DPD, FACTS, and LCV methods. The extent of this interference in the DPD and LCV free chlorine methods is undetermined but appears not to be significant.

The free available chlorine test, syringaldazine (Method G, Tentative) was developed as a procedure specific for free available chlorine. It is unaffected by significant concentrations of monochloramine, dichloramine, nitrate, nitrite, and oxidized forms of manganese.

Sample color and turbidity may interfere in all colorimetric procedures unless compensated for.

Organic contaminants may produce a false free chlorine reading in most colorimetric methods (see ¶ 1*b* below). Many strong oxidizing agents interfere in the measurement of free available chlorine in all methods. Such interferences include bromine, chlorine dioxide, iodine, and ozone. However, the reduced forms of these compounds—bromide, chloride, iodide, manganous ion, and oxygen—do not interfere. Reducing agents such as ferrous compounds, hydrogen sulfide, and oxidizable organic matter generally do not interfere.

b. Polluted waters: The determination of total residual chlorine in samples containing organic matter presents special problems. Because of the presence of ammonia, amines, and organic compounds, particularly organic nitrogen, residual chlorine exists in a combined state. A considerable residual may exist in this form, but at the same time there may be appreciable unsatisfied chlorine demand. Addition of reagents in the determination may change these relationships so that residual chlorine is lost during the analysis. In wastewater, the differentiation between free available chlorine and combined available chlorine ordinarily is not made because wastewater chlorination is seldom carried far enough to produce free available chlorine.

The determination of residual chlorine in industrial wastes is similar to that in domestic wastewater when the waste contains organic matter, but may be similar to the determination in water when the waste is low in organic matter.

None of these methods is applicable to estuarine or marine waters because of the presence of bromide, which changes the species of residual oxidants.

Although the methods given below are useful for the determination of residual chlorine in wastewaters and treated effluents, select the method in accordance with sample composition. Some industrial wastes, or mixtures of wastes with domestic wastewater, may require special precautions and modifications to obtain satisfactory results.

Determine free chlorine in wastewater by any of the methods (A through G), provided that known interfering substances are absent or compensated for. The amperometric method is the method of choice because it is not subject to interference from color, turbidity, iron, manganese, or nitrite nitrogen. The DPD and LCV methods are subject to interference from high concentrations of monochloramine unless these are compensated for by adding thioacetamide or arsenite immediately after reagent addition. Interference in the LCV and DPD methods is equivalent to approximately 2% of the monochloramine reacting per minute at 25 C. Oxidized forms of manganese at all levels encountered in water will interfere in all methods except in the free chlorine measurement of amperometric titrations and FACTS. In addition, the combination of monochloramine and nitrite interfere seriously with the LCV procedure.

The tentative FACTS method is unaffected by concentrations of monochloramine, dichloramine, nitrite, iron, manganese, and other interfering compounds normally found in domestic wastewaters.

For total available chlorine in samples containing significant amounts of organic matter, use the iodometric back titration method (B) to prevent contact between the full concentration of liberated iodine and the sample. Either the amperometric or the starch-iodide end point may be used if the concentration is greater than 1 mg/L. In the absence of interference, the two modifications give concordant results. The amperometric end point is inherently more sensitive and is free of interference from color and turbidity, which can cause difficulty with the starch-iodide end point. On the other hand, certain metals, surface-active agents, and complex anions in some industrial wastes interfere in the amperometric titration and indicate the need for another method for such wastewaters. Silver in the form of soluble silver cyanide complex, in concentrations as low as 1.0 mg silver/L, poisons the cell at pH 4.0 but not at 7.0. The silver ion, in the absence of the cyanide complex, gives extensive response in the current at pH 4.0 and gradually poisons the cell at all pH levels. Cuprous copper in the soluble copper cyanide ion, in concentrations of 5 mg copper/L or less, poisons the cell at pH 4.0 and 7.0. Although iron and nitrite may interfere with this method, minimize the interference by buffering to pH 4.0 before adding KI. Oxidized forms of manganese interfere in all methods for total available chlorine including amperometric titration. An unusually high content of organic matter may cause uncertainty in the end point.

Regardless of method of end-point detection, either phenylarsine oxide or thiosulfate may be used as the standard reducing reagent at pH 4. The former is more stable and is preferred.

The DPD, titrimetric and colorimetric, and the LCV methods (D, E, and F, respectively) are applicable to determining total available chlorine in polluted waters. In addition, both DPD procedures and the amperometric titration method allow for estimating monochloramine and dichloramine fractions. Because all methods for total chlorine depend on the stoichiometric production of iodine, polluted waters containing iodine-reducing substances may not be analyzed accurately by these methods, especially where iodine remains in the solution for a significant time. This is a problem in Methods A and C. The back titration procedure (B) and the colorimetric methods (E and F) cause immediate reaction of the iodine generated so that it has little chance to react with other iodine-reducing substances.

In all colorimetric procedures, compensate for color and turbidity by using color and turbidity blanks.

2. Sampling and Storage

Chlorine in aqueous solution is not stable, and the chlorine content of samples or solutions, particularly weak solutions, will decrease rapidly. Exposure to sunlight or other strong light or agitation will accelerate the reduction of chlorine. Therefore, start chlorine determinations immediately after sampling, avoiding excessive light and agitation. Do not store samples to be analyzed for chlorine.

408 A. Iodometric Method I

1. General Discussion

a. Principle: Chlorine will liberate free iodine from potassium iodide (KI) solutions at pH 8 or less. The liberated iodine is titrated with a standard solution of sodium thiosulfate ($Na_2S_2O_3$) with starch as the indicator. Preferably carry the reaction out at pH 3 to 4 because at neutral pH the reaction is not stoichiometric because of oxidation of some thiosulfate to sulfate.

b. Interference: Oxidized forms of manganese and other oxidizing agents interfere. Reducing agents such as organic sulfides also interfere. Although the neutral titration minimizes the interfering effect of ferric and nitrite ions, the acid titration is preferred because some forms of combined available residual chlorine do not react at pH 7. Use only acetic acid for the acid titration; sulfuric acid (H_2SO_4) will increase interferences; *never use hydrochloric acid (HCl)*. See Section 408.1 for discussion of other interferences.

c. Minimum detectable concentration: The minimum detectable concentration approximates 40 μg Cl as Cl_2/L if 0.01*N* $Na_2S_2O_3$ is used with a 1,000-mL sample. Concentrations below 1 mg/L can not be determined accurately by the starch-iodide end point used in this method. Lower concentrations can be measured with the amperometric end point in Methods B and C.

2. Reagents

a. Acetic acid, conc (glacial).

b. Potassium iodide, KI, crystals.

c. Standard sodium thiosulfate, 0.1*N*: Dissolve 25 g $Na_2S_2O_3 \cdot 5H_2O$ in 1 L freshly boiled distilled water and standardize against potassium bi-iodate or potassium dichromate after at least 2 wk storage. This initial storage is necessary to allow oxidation of any bisulfite ion present. Use boiled distilled water and add a few milliliters chloroform ($CHCl_3$) to minimize bacterial decomposition.

Standardize the 0.1*N* $Na_2S_2O_3$ by one of the following:

1) Iodate method—Dissolve 3.249 g anhydrous potassium bi-iodate, $KH(IO_3)_2$, primary standard quality, or 3.567 g KIO_3 dried at 103 C ± 2 C for 1 hr, in distilled water and dilute to 1,000 mL to yield a 0.1000*N* solution. Store in a glass-stoppered bottle.

To 80 mL distilled water, add, with constant stirring, 1 mL conc H_2SO_4, 10.00 mL 0.1000*N* $KH(IO_3)_2$, and 1 g KI. Titrate immediately with 0.1*N* $Na_2S_2O_3$ titrant until the yellow color of the liberated iodine almost is discharged. Add 1 mL starch indicator solution and continue titrating until the blue color disappears.

2) Dichromate method—Dissolve 4.904 g anhydrous potassium dichromate, $K_2Cr_2O_7$, of primary standard quality, in distilled water and dilute to 1,000 mL to yield a 0.1000*N* solution. Store in a glass-stoppered bottle.

Proceed as in the iodate method, with the following exceptions: Substitute 10.00 mL 0.1000*N* $K_2Cr_2O_7$ for iodate and let reaction mixture stand 6 min in the dark before titrating with 0.1*N* $Na_2S_2O_3$ titrant.

$$\text{Normality } Na_2S_2O_3 = \frac{1}{\text{mL } Na_2S_2O_3 \text{ consumed}}$$

d. Standard sodium thiosulfate titrant, 0.01*N* or 0.025*N*: Improve the stability of 0.01*N* or 0.025*N* $Na_2S_2O_3$ by diluting an aged 0.1*N* solution, made as directed above, with freshly boiled distilled water. Add 4 g sodium borate and 10 mg mercuric iodide/L solution. For accurate work, standardize this solution daily in accordance with the directions given above, using 0.01*N* or 0.025*N* iodate or $K_2Cr_2O_7$. Use sufficient volumes of these standard solutions so that their final dilution is not greater than 1 + 4. To speed up operations where many samples must be titrated use an automatic buret of a type in which rubber does not come in contact with the solution. Standard titrants, 0.0100*N* and 0.0250*N*, are equivalent, respectively, to 354.5 μg and 886.3 μg available Cl as Cl_2/1.00 mL.

e. Starch indicator solution: To 5 g starch (potato, arrowroot, or soluble), add a little cold water and grind in a mortar to a thin paste. Pour into 1 L of boiling distilled water, stir, and let settle overnight. Use clear supernate. Preserve with 1.25 g salicylic acid, 4 g zinc chloride, or a combination of 4 g sodium propionate and 2 g sodium azide/L starch solution. Some commercial starch substitutes are satisfactory.

f. Standard iodine, 0.1N: See ¶ B.3*g*.

g. Dilute standard iodine, 0.0282N: See ¶ B.3*h*.

3. Procedure

a. Volume of sample: Select a sample volume that will require no more than 20 mL 0.01*N* $Na_2S_2O_3$ and no less than 0.2 mL for the starch-iodide end point. For a chlorine range of 1 to 10 mg/L, use a 500-mL sample; above 10 mg/L, proportionately less sample is required. Use smaller samples and volumes of titrant with the amperometric end point.

b. Preparation for titration: Place 5 mL acetic acid, or enough to reduce the pH to between 3.0 and 4.0, in a flask or white porcelain casserole. Add about 1 g KI estimated on a spatula. Pour in the sample and mix with a stirring rod.

c. Titration: Titrate away from direct sunlight. Add 0.025*N* or 0.01*N* $Na_2S_2O_3$ from a buret until the yellow color of the liberated iodine almost is discharged. Add 1 mL starch solution and titrate until blue color is discharged.

If the titration is made with 0.025*N* $Na_2S_2O_3$ instead of 0.01*N*, then, with a 1-L sample, 1 drop is equivalent to about 50 μg/L. It is not possible to discern the end point with greater accuracy.

d. Blank titration: Correct result of sample titration by determining blank contributed by such reagent impurities as free iodine or iodate in KI that liberates extra iodine or traces of reducing agents that might reduce some liberated iodine.

Take a volume of distilled water corresponding to the sample used for titration in ¶s 3*a–c*, add 5 mL acetic acid, 1 g KI, and 1 mL starch solution. Perform blank titration as in 1) or 2) below, whichever applies.

1) If a blue color develops, titrate with 0.01*N* or 0.025*N* $Na_2S_2O_3$ to disappearance of blue color and record result.

2) If no blue color occurs, titrate with 0.0282*N* iodine solution until a blue color appears. Back-titrate with 0.01*N* or 0.025*N* $Na_2S_2O_3$ and record the difference.

Before calculating the chlorine concentration, subtract the blank titration of ¶ 1) from the sample titration; or, if necessary, add the net equivalent value of the blank titration of ¶ 2).

4. Calculation

For standardizing chlorine solution for temporary standards:

$$\text{mg Cl as Cl}_2\text{/mL} = \frac{(A \pm B) \times N \times 35.45}{\text{mL sample}}$$

For determining total available residual chlorine in a water sample:

$$\text{mg Cl as Cl}_2\text{/L} = \frac{(A \pm B) \times N \times 35{,}450}{\text{mL sample}}$$

where:

A = mL titration for sample,
B = mL titration for blank (positive or negative), and
N = normality of $Na_2S_2O_3$.

5. Precision and Accuracy

References 1 and 2 give the results of nine methods used to analyze synthetic water samples without interferences; variations of five of the methods appear in this edition. It is intended to replace the tables in the 14th Edition with determinations of residual chlorine on real samples to validate the modified methods for which adequate precision and accuracy data are not now available.

408 B. Iodometric Method II

1. General Discussion

a. Principle: In this method, used for wastewater analysis, the end-point signal is reversed because the unreacted standard reducing agent remaining in the sample is titrated with standard iodine or standard iodate, rather than the iodine released being titrated directly. This indirect procedure is necessary regardless of the method of end-point detection, in order to avoid contact between the full concentration of liberated iodine and the wastewater.

When iodate is used as a back titrant, use only phosphoric acid. Do not use acetate buffer.

b. Interference: Oxidized forms of manganese and other oxidizing agents give positive interferences. Reducing agents such as organic sulfides do not interfere as much as in Method A. Minimize iron and nitriteinterference by buffering to pH 4.0 before adding potassium iodide (KI). An unusually high content of organic matter may cause some uncertainty in the end point. Whenever manganese, iron, and other interferences definitely are absent, reduce this uncertainty and improve precision by acidifying to pH 1.0. Control interference from more than 0.2 mg nitrite/L with phosphoric acid-sulfamic acid reagent. A larger fraction of organic chloramines will react at lower pH along with interfering substances. See Section 408.1 for a discussion of other interferences.

2. Apparatus

For a description of the amperometric end-point detection apparatus and a discussion of its use, see Section 408C.

3. Reagents

a. Standard phenylarsine oxide solution, 0.00564*N:* Dissolve approximately 0.8 g phenylarsine oxide powder in 150 mL 0.3*N* NaOH solution. After settling, decant 110 mL into 800 mL distilled water and mix thoroughly. Bring to pH 6 to 7 with 6*N* HCl and dilute to 950 mL with distilled water.

Standardization—Accurately measure 5 to 10 mL freshly standardized 0.0282*N* iodine solution into a flask and add 1 mL KI solution. Titrate with phenylarsine oxide solution, using starch solution as an indicator. See 408A.2*e*. Adjust to 0.00564*N* and recheck against the standard iodine solution; 1.00 mL = 200 μg available chlorine. (CAUTION: *Toxic—take care to avoid ingestion.*)

b. Standard sodium thiosulfate solution, 0.1*N:* See 408A.2*c*.

c. Standard sodium thiosulfate solution, 0.00564*N:* Prepare by diluting 0.1*N* $Na_2S_2O_3$. For maximum stability of the dilute solution, prepare by diluting an aged 0.1*N* solution with freshly boiled distilled water (to minimize bacterial action) and add 10 mg HgI_2 and 4 g $Na_4B_4O_7$/L. Standardize daily as directed in 408A.2*c* using 0.00564*N* $K_2Cr_2O_7$ or iodate solution. Use sufficient volumes of sample so that the final dilution does not exceed 1 + 2. Use an automatic buret of a type in which rubber does not come in contact with the solution. 1.00 mL = 200 μg available chlorine.

d. Potassium iodide, KI, crystals.

e. Acetate buffer solution, pH 4.0: Dissolve 146 g anhydrous $NaC_2H_3O_2$, or 243 g $NaC_2H_3O_2 \cdot 3H_2O$, in 400 mL distilled water, add 480 g conc acetic acid, and dilute to 1 L with chlorine-demand-free water.

f. Standard arsenite solution, 0.1*N:* Accurately weigh a stoppered weighing bottle containing approximately 4.95 g arsenic trioxide, As_2O_3. Transfer without loss to a 1-L volumetric flask and again weigh bottle. Do not attempt to brush out adhering oxide. Moisten As_2O_3 with water and add 15 g NaOH and 100 mL distilled water. Swirl flask contents gently to dissolve. Dilute to 250 mL with distilled water and saturate with CO_2, thus converting all NaOH to $NaHCO_3$. Dilute to mark, stopper, and mix thoroughly. This solution will preserve its titer almost indefinitely. (CAUTION: *Toxic—take care to avoid ingestion.*)

$$\text{Normality} = \frac{\text{g } As_2O_3}{49.455}$$

g. Standard iodine solution, 0.1*N:* Dissolve 40 g KI in 25 mL chlorine-demand-free water, add 13 g resublimed iodine, and stir until dissolved. Transfer to a 1-L volumetric flask and dilute to mark.

Standardization—Accurately measure 40 to 50 mL 0.1*N* arsenite solution into a flask and titrate with 0.1*N* iodine solution, using starch solution as indicator. To obtain accurate results, insure that the solution is saturated with CO_2 at end of titration by passing current of CO_2 through solution for a few minutes just before end point is reached, or add a few drops of HCl to liberate sufficient CO_2 to saturate solution. Iodine also can be standardized against $Na_2S_2O_3$. Refer to ¶ A2.*c*1).

Alternatively, prepare 0.1000*N* iodine solution directly as a standard solution by weighing 12.69 g primary standard resublimed iodine. Because loss of I_2 by volatility can occur from both the solid and the solution, transfer the solid immediately to KI as specified above. Never let the solution stand in open containers for extended periods.

h. Standard iodine titrant, 0.0282*N:* Dissolve 25 g KI in a little distilled water in a 1-L volumetric flask, add correct amount of 0.1*N* iodine solution exactly standardized to yield a 0.0282*N* solution, and dilute to 1 L with chlorine-demand-free water. For accurate work, standardize this solution daily in accordance with directions given in ¶ 3*g* above, using 5 to 10 mL of arsenite or $Na_2S_2O_3$ solution. Store in amber bottles or in the dark; protect solution from direct sunlight at all times and keep it from all contact with rubber.

i. Starch indicator: See 408A.2*e*.

j. Standard iodate titrant, 0.00564*N:* Dissolve 201.2 mg primary standard grade KIO_3, dried for 1 hr at 103 C, or 183.3 mg primary standard anhydrous potassium biiodate in distilled water and dilute to 1 L.

k. Phosphoric acid solution, H_3PO_4, 1 + 9.

l. Phosphoric acid-sulfamic acid solution: Dissolve 20 g NH_2SO_3H in 1 L 1 + 9 phosphoric acid.

m. Chlorine-demand-free water: Prepare chlorine-demand-free water from good-quality distilled or deionized water by adding sufficient chlorine to give 5 mg/L free chlorine residual. After standing 2 days this solution should contain at least 2 mg/L free chlorine; if not, discard and obtain better-quality water. Remove remaining free chlorine by placing container in sunlight or irradiating with an ultraviolet lamp. After several hours take a sample, add KI, and measure total chlorine with a colorimetric method using a nessler tube to increase sensitivity. Do not use before last trace of free and combined chlorine has been removed.

Distilled water commonly contains ammonia and also may contain reducing agents. Collect good-quality distilled or deionized water in a sealed container from which water can be drawn by gravity. To the air inlet of the container add an H_2SO_4 trap consisting of a large test tube half filled with 1 + 1 H_2SO_4 connected in series with a similar but empty test tube. Fit both test tubes with stoppers and inlet tubes terminating near the bottom of the tubes and outlet tubes terminating near the top of the tubes. Connect outlet tube of trap containing H_2SO_4 to the distilled water container, connect inlet tube to outlet of empty test tube. The empty test tube will prevent discharge to the atmosphere of

H_2SO_4 due to temperature-induced pressure changes. Stored in such a container chlorine-demand-free water should be stable for several weeks unless bacterial growth occurs.

4. Procedure

a. Preparation for titration:

1) Volume of sample—For residual chlorine concentrations of 10 mg/L or less, titrate 200 mL. For greater residual chlorine concentrations, use proportionately less sample and dilute to 200 mL with chlorine-demand-free water. Use a sample of such size that not more than 10 mL phenylarsine oxide solution is required.

2) Preparation for titration—Measure 5 mL 0.00564*N* reductant for residual chlorine concentrations from 2 to 5 mg/L, and 10 mL for concentrations of 5 to 10 mg/L, into a flask or casserole for titration with standard iodine or iodate. Start stirring. For titration by amperometry or standard iodine, also add excess KI (approximately 1 g) and 4 mL acetate buffer solution or enough to reduce the pH to between 3.5 and 4.2.

b. Titration: Use one of the following:

1) Amperometric titration—Add 0.0282*N* iodine titrant in small increments from a 1-mL buret or pipet. Observe meter needle response as iodine is added: the pointer remains practically stationary until the end point is approached, whereupon each iodine increment causes a temporary deflection of the microammeter, with the pointer dropping back to its original position. Stop titration at end point when a small increment of iodine titrant gives a definite pointer deflection upscale and the pointer does not return promptly to its original position. Record volume of iodine titrant used to reach end point.

2) Colorimetric (iodine) titration—Add 1 mL starch solution and titrate with 0.0282*N* iodine to the first appearance of blue color that persists after complete mixing.

3) Colorimetric (iodate) titration—To suitable flask or casserole add 200 mL chlorine-demand-free water and add, with agitation, the required volume of reductant, an excess of KI (approximately 0.5 g), 2 mL 10% H_3PO_4 solution, and 1 mL starch solution in the order given, and titrate immediately* with 0.00564*N* iodate solution to the first appearance of a blue color that persists after complete mixing. Designate volume of iodate solution used as *A*. Repeat procedure, substituting 200 mL sample for the 200 mL chlorine-demand-free water. If sample is colored or turbid, titrate to the first change in color, using for comparison another portion of sample with H_3PO_4 added. Designate this volume of iodate solution as *B*.

5. Calculation

a. Titration with standard iodine:

$$\text{mg Cl as Cl}_2\text{/L} = \frac{(A - 5B) \times 200}{C}$$

where:

A = mL 0.00564*N* reductant,
B = mL 0.00282*N* I_2, and
C = mL sample.

b. Titration with standard iodate:

$$\text{mg Cl as Cl}_2\text{/L} = \frac{(A - B) \times 200}{C}$$

where:

A = mL $Na_2S_2O_3$,
B = mL iodate required to titrate $Na_2S_2O_3$, and
C = mL sample.

---

*Titration may be delayed up to 10 min without appreciable error if H_3PO_4 is not added until immediately before titration.

408 C. Amperometric Titration Method

1. General Discussion

Amperometric titration requires a higher degree of skill and care than the colorimetric methods. Chlorine residuals over 2 mg/L are measured best by means of smaller samples or by dilution with water that neither has residual chlorine nor a chlorine demand. The method can be used to determine total residual chlorine and can differentiate between free and combined available chlorine. A further differentiation into monochloramine and dichloramine fractions is possible by control of KI concentration and pH.

a. Principle: The amperometric method is a special adaptation of the polarographic principle. Free available chlorine is titrated at a pH between 6.5 and 7.5, a range in which the combined chlorine reacts slowly. The combined chlorine, in turn, is titrated in the presence of the proper amount of KI in the pH range 3.5 to 4.5. When free chlorine is determined, the pH must not be greater than 7.5 because the reaction becomes sluggish at higher pH values, nor less than 6.5 because at lower pH values some combined chlorine may react even in the absence of iodide. When combined chlorine is determined, the pH must not be less than 3.5 because substances such as oxidized manganese interfere at lower pH values, nor greater than 4.5 because the iodide reaction is not quantitative at higher pH values. The tendency of monochloramine to react more readily with iodide than does dichloramine provides a means for further differentiation. The addition of a small amount of KI in the neutral pH range enables estimation of monochloramine content. Lowering the pH into the acid range and increasing the KI concentration allows the separate determination of dichloramine.

Organic chloramines can be measured as free chlorine, monochloramine, or dichloramine, depending on the activity of the chlorine in the organic compound.

Phenylarsine oxide is stable even in dilute solution and each mole reacts with two equivalents of halogen. A special amperometric cell is used to detect the end point of the residual chlorine-phenylarsine oxide titration. The cell consists of a nonpolarizable reference electrode that is immersed in a salt solution and a readily polarizable noble-metal electrode that is in contact both with the salt solution and the sample being titrated. In some applications end point selectivity is improved by adding +200 mV to the platinum electrode versus silver, silver chloride. Another approach to end point detection uses dual platinum electrodes, a mercury cell with voltage divider to impress a potential across the electrodes, and a microammeter. If there is no chlorine residual in the sample, the microammeter reading will be comparatively low because of cell polarization. The greater the residual, the greater the microammeter reading. The meter acts merely as a null-point indicator—that is, the actual meter reading is not important, but rather the relative readings as the titration proceeds. The gradual addition of phenylarsine oxide causes the cell to become more and more polarized because of the decrease in available chlorine. The end point is recognized when no further decrease in meter reading can be obtained by adding more phenylarsine oxide.

b. Interference: Accurate determinations of free chlorine cannot be made in the presence of nitrogen trichloride, NCl_3, or chlorine dioxide, which titrate partly as free chlorine. When present, NCl_3 can titrate partly as free available chlorine and partly as dichloramine, contributing a positive error in both fractions. Some organic chloramines also can be titrated in each step. Monochloramine can intrude into the free chlorine fraction and dichloramine can interfere in the monochloramine fraction, especially at high temperatures and prolonged titration times. Free halogens other than chlorine also will titrate as free chlorine. Combined chlorine reacts with iodide ions to produce iodine. When titration for free available chlorine follows a combined chlorine titration, which requires addition of KI, erroneous results may occur unless the measuring cell is rinsed thoroughly with distilled water between titrations.

Interference from copper has been noted in samples taken from copper pipe or after heavy copper sulfate treatment of reservoirs, with metallic copper plating out on the electrode. Silver ions also poison the electrode. Interference occurs in some highly colored waters and in waters containing surface-active agents. Very low temperatures slow response of measuring cell and longer time is required for the titration, but precision is not affected. A reduction in reaction rate is caused by pH values above 7.5; overcome this by buffering all samples to pH 7.0 or less. On the other hand, some substances, such as manganese, nitrite, and iron, do not interfere. The violent stirring of some commercial titrators can lower chlorine values by volatilization. When dilution is used for samples containing high chlorine content, take care that the dilution water is free of residual chlorine and ammonia and possesses no chlorine demand.

See 408.1 for a discussion of other interferences.

2. Apparatus

a. End-point detection apparatus, consisting of a cell unit connected to a microammeter, with necessary electrical accessories. The cell unit includes a noble-metal electrode of sufficient surface area, a salt bridge to provide an electrical connection without diffusion of electrolyte, and a reference electrode of silver-silver chloride in a saturated sodium chloride solution connected into the circuit by means of the salt bridge.

Keep noble-metal electrode free of deposits and foreign matter. Vigorous chemical cleaning generally is unnecessary. Occasional mechanical cleaning with a suitable abrasive usually is sufficient. Keep salt bridge in good operating condition; do not allow it to become plugged nor permit appreciable flow of electrolyte through it. Keep solution surrounding reference electrode free of contamination and maintain it at constant composition by insuring an adequate supply of undissolved salt at all times. A cell with two metal electrodes polarized by a small DC potential also may be used. (See Bibliography.)

b. Agitator, designed to give adequate agitation at the noble-metal electrode surface to insure proper sensitivity. Thoroughly clean agitator and exposed electrode system to remove all chlorine-consuming contaminants by immersing them in water containing 1 to 2 mg/L free available residual chlorine for a few minutes. Add KI to the same water and let agitator and electrodes remain immersed for 5 min. After thorough rinsing with chlorine-demand-free water or the sample to be tested, sensitized electrodes and agitator are ready for use. Remove iodide reagent completely from cell.

c. Buret: Commercial titrators usually are equipped with suitable burets (1 mL). Manual burets are available.*

d. Glassware, exposed to water containing at least 10 mg/L residual chlorine for 3 hr or more before use and rinsed with chlorine-demand-free water.

3. Reagents

a. Standard phenylarsine oxide titrant: See Method B, ¶ 3*a*.

b. Phosphate buffer solution, pH 7:

*Kimax 17110-F, 5 mL, Kimble Products, Box 1035, Toledo, Ohio, or equivalent.

Dissolve 25.4 g anhydrous KH_2PO_4 and 34.1 g anhydrous Na_2HPO_4 in 800 mL distilled water. Add 2 mL sodium hypochlorite solution containing 1% available chlorine and mix thoroughly. Protect from sunlight for 2 days. Determine that free chlorine still remains in the solution. Then expose to sunlight until no residual chlorine remains. If necessary, carry out the final dechlorination with an ultraviolet lamp. Determine that no total chlorine remains by adding KI and measuring with one of the colorimetric tests. Dilute to 1 L with distilled water and filter if any precipitate is present.

c. Potassium iodide solution: Dissolve 50 g KI and dilute to 1 L with freshly boiled and cooled distilled water. Store in a brown glass-stoppered bottle, preferably in the refrigerator. Discard when solution becomes yellow.

d. Acetate buffer solution, pH 4: See Method B, ¶ 3*e*.

4. Procedure

a. Sample volume: Select a sample volume requiring no more than 2 mL phenylarsine oxide titrant. Thus, for residual chlorine concentrations of 2 mg/L or less, take a 200-mL sample; for chlorine levels in excess of 2 mg/L, use 100 mL or proportionately less.

b. Free available chlorine: Unless sample pH is known to be between 6.5 and 7.5, add 1 mL pH 7 phosphate buffer solution to produce a pH of 6.5 to 7.5. Titrate with standard phenylarsine oxide titrant, observing current changes on microammeter. Add titrant in progressively smaller increments until all needle movement ceases. Make successive buret readings when needle action becomes sluggish, signaling approach of end point. Subtract last very small increment that causes no needle response because of overtitration.

Continue titrating for combined available chlorine as described in ¶ 4*c* below or for the separate monochloramine and dichloramine fractions as detailed in ¶s 4*e* and 4*f*.

c. Combined available chlorine: To sample remaining from free-chlorine titration add 1.00 mL KI solution and 1 mL acetate buffer solution, in that order. Titrate with phenylarsine oxide titrant to the end point, as above. Do not refill buret but simply continue titration after recording figure for free available chlorine. Again subtract last increment to give amount of titrant actually used in reaction with chlorine. (If titration was continued without refilling buret, this figure represents total residual chlorine. Subtracting free available chlorine from total gives combined residual chlorine.) Wash apparatus and sample cell thoroughly to remove iodide ion to avoid inaccuracies when the titrator is used subsequently for a free available chlorine determination.

d. Separate samples: If desired, determine total residual chlorine and free available chlorine on separate samples. If sample pH is between 3.5 and 9.5 and total available chlorine alone is required, treat sample immediately with 1 mL KI solution followed by 1 mL acetate buffer solution, and titrate with phenylarsine oxide titrant as described in ¶ 4*c* preceding.

e. Monochloramine: After titrating for free available chlorine, add 0.2 mL KI solution to same sample and, without refilling buret, continue titration with phenylarsine oxide titrant to end point. Subtract last increment to obtain net volume of titrant consumed by monochloramine.

f. Dichloramine: Add 1 mL acetate buffer solution and 1 mL KI solution to same sample and titrate final dichloramine fraction as described above.

5. Calculation

Convert individual titrations for free available chlorine, combined available chlorine, total available chlorine, monochloramine, and dichloramine by the following equation:

$$\text{mg Cl as Cl}_2\text{/L} = \frac{A \times 200}{\text{mL sample}}$$

where:

A = mL phenylarsine oxide titration.

6. Precision and Accuracy

See 408A.5.

408 D. DPD Ferrous Titrimetric Method

1. General Discussion

a. Principle: N,N-diethyl-p-phenylenediamine (DPD) is used as an indicator in the titrimetric procedure with ferrous ammonium sulfate (FAS). Where complete differentiation of chlorine species is not required, the procedure may be simplified to give only free and combined available chlorine or total residual available chlorine.

In the absence of iodide ion, free available chlorine reacts instantly with DPD indicator to produce a red color. Subsequent addition of a small amount of iodide ion acts catalytically to cause monochloramine to produce color. Addition of iodide ion to excess evokes a rapid response from dichloramine. Nitrogen trichloride (NCl_3) is included with dichloramine; however, if iodide ion is added before DPD, a proportion of the NCl_3 appears with free available chlorine. A supplementary procedure based on this altered order of adding reagents permits estimating NCl_3.

Chlorine dioxide (ClO_2) appears, to the extent of one-fifth of its total available chlorine content, with free available chlorine. A full response from ClO_2, corresponding to its total available chlorine content, may be obtained if the sample first is acidified in the presence of iodide ion and subsequently is brought back to an approximately neutral pH by adding bicarbonate ion. Bromine, bromamine, and iodine react with DPD indicator and appear with free available chlorine.

b. pH control: For accurate results careful pH control is essential. At the proper pH of 6.2 to 6.5, the red colors produced may be titrated to sharp colorless end points. *Titrate as soon as the red color is formed in each step.* Too low a pH in the first step tends to make the monochloramine show in the free-chlorine step and the dichloramine in the monochloramine step. Too high a pH causes dissolved oxygen to give a color.

c. Temperature control: In all methods for differentiating free chlorine from chloramines, higher temperatures increase the tendency for chloramines to react and leads to increased apparent free-chlorine results. Higher temperatures also increase color fading. Complete measurements rapidly, especially at higher temperature.

d. Interference: The most significant interfering substance likely to be encountered in water is oxidized manganese. To correct for this, place 5 mL buffer solution and 0.5 mL sodium arsenite solution in the titration flask. Add 100 mL sample and mix. Add 5 mL DPD indicator solution, mix, and titrate with standard FAS titrant until red color is discharged. Subtract reading from Reading *A* obtained by the normal procedure as described in ¶ 3*a*1) of this method or from the total available chlorine reading obtained in the simplified procedure given in ¶ 3*a*4). If the combined reagent in powder form (see below) is used, first add KI and arsenite to the sample and mix, then add combined buffer-indicator reagent.

As an alternative to sodium arsenite use

a 0.25% solution of thioacetamide, adding 0.5 mL to 100 mL sample.

Interference by copper up to approximately 10 mg copper/L is overcome by the EDTA incorporated in the reagents. EDTA enhances stability of DPD indicator solution by retarding deterioration due to oxidation, and in the test itself, provides virtually complete suppression of dissolved oxygen errors by preventing trace metal catalysis.

High concentrations of combined chlorine can break through into the free chlorine fraction. At 10 C this amounts to 1% and at 25 C to 2% of the monochloramine present that reacts after standing 1 min. Adding thioacetamide (0.5 mL 0.25% solution to 100 mL) immediately after mixing DPD reagent with sample completely stops further reaction with combined chlorine in the free chlorine measurement. Continue immediately with FAS titration to obtain free chlorine. Obtain total available chlorine from the normal procedure, i.e., without using thioacetamide.

See 408.1 for a discussion of other interferences.

e. Minimum detectable concentration: Approximately 18 μg Cl as Cl_2/L.

2. Reagents

a. Phosphate buffer solution: Dissolve 24 g anhydrous Na_2HPO_4 and 46 g anhydrous KH_2PO_4 in distilled water. Combine with 100 mL distilled water in which 800 mg disodium ethylenediamine tetraacetate dihydrate (EDTA) have been dissolved. Dilute to 1 L with distilled water and add 20 mg $HgCl_2$ to prevent mold growth and interference in the free available chlorine test caused by any trace amounts of iodide in the reagents.

b. N,N-Diethyl-p-phenylenediamine (DPD) indicator solution: Dissolve 1 g DPD oxalate,* or 1.5 g DPD sulfate pentahydrate,† or 1.1 g anhydrous DPD sulfate in chlorine-free distilled water containing 8 mL 1 + 3 H_2SO_4 and 200 mg EDTA. Make up to 1 L, store in a brown glass-stoppered bottle, and discard when discolored. (The buffer and indicator sulfate are commercially available as a combined reagent in stable powder form.) CAUTION: *The oxalate is toxic—take care to avoid ingestion.*

c. Standard ferrous ammonium sulfate (FAS) titrant: Dissolve 1.106 g $Fe(NH_4)_2(SO_4)_2 \cdot 6H_2O$ in distilled water containing 1 mL of 1 + 3 H_2SO_4 and make up to 1 L with freshly boiled and cooled distilled water. This primary standard may be used for 1 month, and the titer checked by potassium dichromate. For this purpose add 10 mL 1 + 5 H_2SO_4, 5 mL conc H_3PO_4, and 2 mL 0.1% barium diphenylamine sulfonate indicator to a 100-mL sample of FAS and titrate with 0.1000*N* primary standard potassium dichromate to a violet end point that persists for 30 sec. The FAS titrant is equivalent to 100 μg Cl as Cl_2/1.00 mL.

d. Potassium iodide, KI, crystals.

e. Potassium iodide solution: Dissolve 500 mg KI and dilute to 100 mL, using freshly boiled and cooled distilled water. Store in a brown glass-stoppered bottle, preferably in a refrigerator. Discard when solution becomes yellow.

f. Potassium dichromate solution: See 408A.2*c*2).

g. Barium diphenylaminesulfonate, 1%: Dissolve 0.1 g $(C_6H_5NHC_6H_4\text{-}4\text{-}SO_3)_2Ba$ in 100 mL distilled water.

h. Sodium arsenite solution: Dissolve 5.0 g $NaAsO_2$ in distilled water and dilute to 1 L (CAUTION: *Toxic—take care to avoid ingestion.*)

i. Thioacetamide solution: Dissolve 250

*Eastman chemical No. 7102 or equivalent.

†Available from Gallard-Schlesinger Chemical Mfg. Corp., 584 Mineola Avenue, Carle Place, N.Y., 11514, or equivalent.

mg CH_3CSNH_2 in 100 mL distilled water.

j. Chlorine-demand-free water: See 408B.3*m*.

3. Procedure

The quantities given below are suitable for concentrations of total available chlorine up to 5 mg/L. If total chlorine exceeds 5 mg/L, use a smaller sample and dilute to a total volume of 100 mL. Mix usual volumes of buffer reagent and DPD indicator solution, or usual amount of DPD powder, with distilled water before adding sufficient sample to bring total volume to 100 mL.

a. Free available chlorine or chloramine: Place 5 mL each of buffer reagent and DPD indicator solution in titration flask and mix (or use about 500 mg of DPD powder). Add 100 mL sample, or diluted sample, and mix.

1) Free available chlorine—Titrate rapidly with standard FAS titrant until red color is discharged (Reading *A*).

2) Monochloramine—Add one very small crystal of KI (about 0.5 mg) or 0.1 mL (2 drops) KI solution and mix. Continue titrating until red color is discharged again (Reading *B*).

3) Dichloramine—Add several crystals KI (about 1 g) and mix to dissolve. Let stand for 2 min and continue titrating until red color is discharged (Reading *C*). For dichloramine concentrations greater than 1 mg/L, let stand 2 min more if color driftback indicates slightly incomplete reaction. When dichloramine concentrations are not expected to be high, use half the specified amount of KI.

4) Simplified procedure for free and combined available chlorine or total availble chlorine—Omit 2) above to obtain monochloramine and dichloramine together as combined available chlorine. To obtain total available chlorine in one reading, add full amount of KI at the start, with the specified amounts of buffer reagent and DPD indicator, and titrate after 2 min standing.

b. Nitrogen trichloride: Place one very small crystal of KI (about 0.5 mg) or 0.1 mL KI solution in a titration flask. Add 100 mL sample and mix. Add contents to a second flask containing 5 mL each of buffer reagent and DPD indicator solution (or add about 500 mg DPD powder direct to the first flask). Titrate rapidly with standard FAS titrant until red color is discharged (Reading *N*).

4. Calculation

For a 100-mL sample, 1.00 mL standard FAS titrant = 1.00 mg/L available residual chlorine.

Reading	NCl_3 Absent	NCl_3 Present
A	Free Cl	Free Cl
B-A	NH_2Cl	NH_2Cl
C-B	$NHCl_2$	$NHCl_2$ + $\frac{1}{2}NCl_3$
N	—	Free Cl + $\frac{1}{2}NCl_3$
2(N-A)	—	NCl_3
C-N	—	$NHCl_2$

In the unlikely event that monochloramine is present with NCl_3, it will be included in *N*, in which case obtain NCl_3 from 2(*N*-B).

Chlorine dioxide, if present, is included in *A* to the extent of one-fifth of its total available chlorine content.

In the simplified procedure for free and combined available chlorine, only *A* (free Cl) and *C* (total Cl) are required. Obtain combined available chlorine from *C*-*A*.

The result obtained in the simplified total available chlorine procedure corresponds to *C*.

5. Precision and Accuracy

See 408A.5.

408 E. DPD Colorimetric Method

1. General Discussion

a. Principle: This is a colorimetric version of the DPD method and is based on the same principles. Instead of titration with standard ferrous ammonium sulfate (FAS) solution as in the titrimetric method, a colorimetric procedure is used.

b. Interference: See 408.1 and 408D.1*d*. Compensate for color and turbidity by using sample to zero photometer.

c. Minimum detectable concentration: Approximately 10 μg Cl as Cl_2/L.

2. Apparatus

Colorimetric equipment: One of the following is required:

a. Spectrophotometer, for use at a wavelength of 515 nm and providing a light path of 1 cm or longer.

b. Filter photometer, equipped with a filter having maximum transmission in the wavelength range of 490 to 530 nm and providing a light path of 1 cm or longer.

3. Reagents

See Section 408D.2*a, b, d, e, h, i,* and *j*.

4. Procedure

a. Calibration of photometer or colorimeter: Calibrate instrument with chlorine (1) or potassium permanganate (2) solutions.

1) Chlorine solutions—Prepare chlorine standards in the range of 0.05 to 4 mg/L from about 100 mg/L chlorine water standardized as directed in Section 409A.3*g*. Use chlorine-demand-free water and glassware to prepare these standards. Develop color by first placing 5 mL phosphate buffer solution and 5 mL DPD indicator reagent in flask and then adding 100 mL chlorine standard with thorough mixing as described in *b* and *c* below. Fill photometer or colorimeter cell from flask and read color at 515 nm. Return cell contents to flask and titrate with standard FAS titrant as a check on chlorine concentration.

2) Potassium permanganate solutions—Prepare a stock solution containing 891 mg $KMnO_4$/1,000 mL. Dilute 10.00 mL stock solution to 100 mL with distilled water in a volumetric flask. When 1 mL of this solution is diluted to 100 mL with distilled water, a chlorine equivalent of 1.00 mg/L will be produced in the DPD reaction. Prepare a series of $KMnO_4$ standards covering the chlorine equivalent range of 0.05 to 4 mg/L. Develop color by first placing 5 mL phosphate buffer and 5 mL DPD indicator reagent in flask and adding 100 mL standard with thorough mixing as described in *b* and *c* below. Fill photometer or colorimeter cell from flask and read color at 515 nm. Return cell contents to flask and titrate with FAS titrant as a check on any absorption of permanganate by distilled water.

b. Volume of sample: Use a sample volume appropriate to the photometer or colorimeter. The following procedure is based on using 10-mL volumes; adjust reagent quantities proportionately for other sample volumes. Dilute sample with chlorine-demand-free water when total available chlorine exceeds 4 mg/L.

c. Free chlorine: Place 0.5 mL each of buffer reagent and DPD indicator reagent in a test tube or photometer cell. Add 10 mL sample and mix. Read color immediately (Reading *A*).

d. Monochloramine: Continue by adding one very small crystal of KI (about 0.1 mg) and mix. If dichloramine concentration is expected to be high, instead of small crystal add 0.1 mL (2 drops) freshly prepared KI solution (0.1 g/100 mL). Read color immediately (Reading *B*).

e. Dichloramine: Continue by adding a few crystals of KI (about 0.1 g) and mix to dissolve. Let stand about 2 min and read color (Reading *C*).

f. Nitrogen trichloride: Place a very small crystal of KI (about 0.1 mg) in a clean test tube or photometer cell. Add 10 mL sample and mix. To a second tube or cell add 0.5 mL each of buffer and indicator reagents; mix. Add contents to first tube or cell and mix. Read color immediately (Reading *N*).

5. Calculation

Reading	NCl_3 Absent	NCl_3 Present
A	Free Cl	Free Cl
B-A	NH_2Cl	NH_2Cl
C-B	$NHCl_2$	$NHCl_2$+ $\frac{1}{2}NCl_3$
N	—	Free Cl+ $\frac{1}{2}NCl_3$
2(N-A)	—	NCl_3
C-N	—	$NHCl_2$

In the unlikely event that monochloramine is present with NCl_3, it will be included in Reading *N*, in which case obtain NCl_3 from 2(*N*–*B*).

408 F. Leuco Crystal Violet Method

1. General Discussion

The leuco crystal violet method measures separately the free and the total available chlorine. The combined available chlorine may be determined by difference. Residual chlorine may be determined by visual comparison with chlorine standards or by reference to a standard calibration curve.

a. Principle: The compound 4,4′,4″-methylidynetris (N,N-dimethylaniline), also known by the common name of leuco crystal violet, reacts instantaneously with free chlorine to form a bluish color. Interference from combined available chlorine can be avoided by completing the test within a 5-min interval.

The total chlorine determination involves the reaction of free and combined chlorine with iodide ion to produce hypoiodous acid, which in turn reacts instantaneously with leuco crystal violet to form the dye crystal violet. The color is stable for days and follows Beer's law over a wide range of total chlorine.

Improved accuracy in the determination of residual chlorine with leuco crystal violet is possible through photometric measurements.

b. Interference: No significant interference from combined available chlorine occurs when free chlorine is determined within 5 min after indicator addition. Fifteen minutes after adding indicator the apparent error in the free chlorine determination is about 0.04 mg/L at 25 C in a sample containing 5.0 mg/L combined residual chlorine.

For combined chlorine concentrations above 5.0 mg/L use the arsenite addition procedure to minimize interference. The major interference in the determination of free residual chlorine is manganic ion, which increases the apparent residual chlorine reading. When manganic ion is known to be present, use the photometric procedure in which absorbance due to manganic ion is determined separately and subtracted from total absorbance to yield that produced by free chlorine alone.

Ferric and nitrate compounds do not interfere and nitrite ion does not interfere in the absence of monochloramine. Where nitrite ion and monochloramine are present together, as in certain wastewaters, serious interference will occur in the determination of free chlorine. Adding arsenite will minimize but not entirely eliminate this interference.

If suspended matter or organic color is present, compensate by incorporating appropriate turbidity or color blanks into the visual or photometric procedures.

c. Minimum detectable concentration: 10 μg free available chlorine/L; 5 μg total available chlorine/L.

2. Apparatus

a. Illumination: Make all readings by looking through samples against an illuminated white surface. This surface may be opaque and illuminated by reflection, or it may be an opal diffusing glass illuminated from behind. Make all comparisons with a standard artificial light. Permanent standards give greater accuracy when used with either of the two artificial light sources specified, both of which are close approximations of average "north" daylight.

b. Colorimetric equipment: One of the following is required:

1) *Nessler tubes,* matched, 50- and 100-mL, tall form.

2) *Test tubes,* matched, with a capacity of at least 10 mL sample when the sample surface is near the top of the test tube.

3) *Volumetric flasks,* 100-mL, with plastic caps or ground-glass stoppers.

4) *Filter photometer,* providing a light path of 1 cm or longer and equipped with an orange filter having maximum transmittance near 592 nm.

5) *Spectrophotometer,* for use at 592 nm, providing a light path of 1 cm or longer.

c. Glassware: Use glassware or plastic containers, including containers for storing reagent solutions, that are free of organic matter. Use either chlorination (1) or chromic acid (2) method after glassware has been cleaned thoroughly with suitable detergent and rinsed with distilled water. The chromic acid method requires less total time, but care is necessary to protect laboratory personnel from contact with the cleaning mixture.

1) Chlorination—Expose all glassware or plastic containers to water containing at least 10 mg/L chlorine for 3 hr or more before use and rinse with chlorine-demand-free water. After rinsing, oven- or air-dry in an atmosphere free from organic fumes.

2) Chromic acid—Add 1 L conc H_2SO_4 to 35 mL saturated sodium dichromate solution in a 2-L beaker. Stir carefully to dissolve. When cleaning glassware, carefully heat a suitable volume of chromic acid solution to approximately 50 C (CAUTION: *Use rubber gloves, safety goggles, and protective clothing in handling this cleaning agent*) and carefully pour it into glassware to be cleaned so that contact is made with entire inside surface of container. Leave in glassware for 2 to 3 min or longer. Drain chromic acid solution and rinse thoroughly with chlorine-demand-free water. Oven- or air-dry glassware away from organic or other chlorine-consuming fumes.

3. Reagents

a. Chlorine-demand-free water: See Method B, ¶ 3*m*.

Prepare reagent solutions and dilute samples with chlorine-demand-free water.

b. Stock chlorine solution: Prepare 100 mg/L stock chlorine solution as described in Section 409A.3*g*.

c. Chlorine solutions for temporary total chlorine standards: For measuring combined residual chlorine, mix an ammonium sulfate solution with chlorine solution in an ammonia-to-chlorine ratio of at least 20 to 1. In distilled water, dissolve 3.89 g $(NH_4)_2SO_4$ and dilute to 1,000 mL; 1.0 mL = 1.0 mg NH_3. To approximately 800 mL chlorine-demand-free water, add 2.0 mL $(NH_4)_2SO_4$ solution for each 1.0 mL stock chlorine solution that contains 100 μg Cl/1.0 mL and dilute to 1,000 mL. Standardize combined chlorine solution and express concentration as milligrams total Cl per liter. Use immediately for calibration.

d. Buffer solution for free chlorine determination, pH 4.0:

1) *Potassium hydroxide, 4N:* Dissolve 224.4 g KOH and dilute to 1 L with water.

2) *Citric acid, 2M:* Dissolve 384.3 g $C_6H_8O_7$, or 420.3 g $C_6H_8O_7 \cdot H_2O$, and dilute to 1 L with water.

3) *Potassium citrate solution:* To 350 mL 4*N* KOH add, with stirring, 700 mL 2*M* citric acid. If desired, prepare smaller volumes in the ratio of 1 volume KOH to 2 volumes citric acid. Use immediately to prepare final buffer solution (5) and discard remainder.

4) *Acetate solution:* Dissolve 161.2 g conc (glacial) acetic acid and 49.5 g sodium acetate, $NaC_2H_3O_2$ or 82.1 g $NaC_2H_3O_2 \cdot 3H_2O$ and dilute to 1 L with water.

5) *Final buffer solution:* Mix equal volumes of potassium citrate solution 3) with acetate solution 4) to make final pH 4.0 buffer solution. Add 20 mg $HgCl_2$/L solution to prevent mold growth.

e. Stock leuco crystal violet reagent: Measure 500 mL chlorine-demand-free water and 14.0 mL 85% orthophosphoric acid into a brown glass container of at least 1-L capacity. Introduce a magnetic stirring bar and mix at moderate speed. Add 3.0 g 4,4′,4″-methylidynetris-(N,N-dimethylaniline)* and with a small amount of water wash down any reagent adhering to neck or sides of container.

Continue agitation until solution is complete. Finally, add 500 mL water. Store in brown bottle at room temperature away from direct sunlight. Discard after 6 months. If a rubber stopper must be used, wrap with plastic wrapping material to protect from contact with reagent.

f. Saturated mercuric chloride solution: To 20 g $HgCl_2$ contained in a 300-mL glass-stoppered flask, add 200 mL water. Gently agitate for a few minutes and let stand for 24 hr. (CAUTION: *Label container with warning that $HgCl_2$ is poisonous and corrosive.*)

g. Mixed indicator: To 600 mL stock leuco crystal violet reagent in a brown bottle, add 50 mL saturated $HgCl_2$ solution and swirl to insure complete mixing. If desired, prepare smaller volumes of mixed indicator in the ratio of 12 volumes stock leuco crystal violet reagent to 1 volume saturated $HgCl_2$ solution. Follow storage directions prescribed in ¶ 3*e* above.

h. Buffer solution for total chlorine determination, pH 4.0: Dissolve 480 g glacial acetic acid and 146 g sodium acetate, $NaC_2H_3O_2$, or 243 g $NaC_2H_3O_2 \cdot 3H_2O$ in 400 mL water and dilute to 1 L. Transfer to a brown bottle. Add 3.0 g KI and mix to dissolve. Store in brown bottle and avoid undue exposure to air.

i. Solutions for preparation of semi-permanent total chlorine standards:

1) *Buffer solution,* pH 4.0: Use solution *h* above.

2) *Crystal violet solution:* Dissolve 40.0 mg crystal violet in 500 mL distilled water containing 20 mL pH 4.0 buffer solution [¶3*i*1) preceding]. Stir for 30 min or more to dissolve completely and dilute to 1,000 mL with distilled water.

j. Sodium hydroxide, 1*N*.

k. Dilute sodium arsenite solution: Dissolve 26 mg $NaAsO_2$ in water and dilute to 100 mL.

l. Sodium arsenite solution: Dissolve 5.0 g $NaAsO_2$ in water and dilute to 1 L. (CAUTION: *Toxic—take care to avoid ingestion.*)

m. Potassium peroxymonosulfate solution: Obtain this reagent, $KHSO_5$, as the commercial product,† a stable powdered mixture containing 42.8% $KHSO_5$ by weight and a mixture of $KHSO_4$ and K_2SO_4. Dissolve 1.0 g powder in water and dilute to 1 L.

*Eastman chemical No. 3651 or equivalent.

†Oxone, a product of E.I. du Pont de Nemours and Co., Inc., Wilmington, Del., or equivalent.

4. Procedure

a. Temporary chlorine standards: Use temporary standards for photometric calibration as well as for visual comparison. Two separate sets of temporary chlorine standards are mandatory because of the divergent colors developed by free and combined residual chlorine. Prepare semipermanent color standards for the total chlorine determination from crystal violet dye; they are suitable for about 3 months.

The color system produced with free residual chlorine differs from the normal crystal violet shade and is stable for only a few days.

Commercially prepared standards are available in test kits for free chlorine determination.

1) Preparation of temporary free chlorine standards—Thoroughly clean all glassware as described in ¶ 2*c* et seq. Prepare temporary chlorine standards from a suitable volume of stock chlorine solution added to 2 L water contained in a brown glass bottle. For visual comparison studies, set up a chlorine series in the range of 0.1 to 2.0 mg/L at increments of 0.1 or 0.2 mg/L. Standardize dilute chlorine solutions by the $Na_2S_2O_3$ or amperometric titration methods.

After standardization measure 50.0 mL dilute chlorine solution with minimum agitation into a 100-mL glass-stoppered volumetric flask.

Add 1.0 mL pH 4.0 buffer solution, ¶ 3*d*5), and gently swirl to mix. With another pipet, add 1.0 mL mixed indicator, ¶ 3*g*. Standardize mixed indicator addition in the following manner: After filling pipet to the mark, position pipet tip inside neck of volumetric flask so that tip makes contact with inside glass surface and let mixed indicator flow down inside glass surface to sample, with a minimum of initial agitation. Remove pipet and swirl flask with a quick firm motion to mix. *Do not dilute sample to 100 mL after adding mixed indicator.*

Transfer colored temporary standards to 50-mL nessler tubes for visual comparison or prepare a photometric calibration curve.

Visual comparison: If temporary standards are prepared directly in 50-mL nessler tubes, stopper tube after adding mixed indicator and mix quickly by inverting tube several times. If a smaller sample volume is taken, for example, 10 mL in a test tube, reduce quantity of pH 4.0 buffer and mixed indicator to 0.2 mL each.

Photometric calibration: Transfer colored temporary standards to cells of 1-cm light path or longer, and read absorbance in a photometer at a wavelength of 592 nm against a distilled water reference. Plot absorbance versus chlorine concentration. Beer's law is followed in the lower free chlorine range but there is a slight curvature with higher free chlorine concentrations.

2) Preparation of temporary total chlorine standards—Prepare standardized total chlorine solutions as prescribed in ¶ 3*c* above in total chlorine range of 0.1 to 2.0 mg/L. Pipet 50-mL sample into a 100-mL volumetric flask or 100-mL nessler tube. Add 0.5 mL total chlorine buffer, ¶ 3*h*, mix, and let stand at least 60 sec. Add 1.0 mL mixed indicator, mix, and dilute to 100 mL. No special precautions are necessary in adding and mixing these solutions.

Photometric calibration: Construct a calibration curve by measuring absorbance of temporary total chlorine standards at 592 nm, preferably in 1-cm cells.

b. Semipermanent total chlorine standards: The variable composition of commercially available crystal violet dye necessitates reconciling absorbance of semipermanent standards with the photometric calibration curve. Adjust final semipermanent standards to agree with calibration absorbance values obtained on temporary total chlorine standards.

Add specified volume of crystal violet solution to a 200-mL volumetric flask containing 100 mL distilled water and 4.0 mL pH 4.0 buffer solution, ¶ 3*i*1). Dilute to volume with distilled water and compare absorbance at 592 nm with suggested values given in Table 408:I or with the photometric calibration curve. Protect standards from direct sunlight and exposure to air to maintain stability for approximately 3 months. Seal standards in glass ampuls for maximum protection.

TABLE 408:I. PREPARATION OF SEMIPERMANENT CRYSTAL VIOLET STANDARDS FOR VISUAL DETERMINATION OF RESIDUAL CHLORINE

Total Chlorine Standard *mg/L*	Crystal Violet Solution *mL*	Absorbance of Final 200-mL Standard at 592 nm in 1-cm cell
0.1	2.84	0.131
0.2	5.80	0.268
0.3	8.60	0.396
0.4	11.60	0.530
0.5	14.40	0.660
0.6	17.30	0.790
0.7	20.00	0.925
0.8	23.20	1.060
0.9	26.60	1.192
1.0	28.80	1.320

c. Color development of free chlorine sample: Measure 50-mL sample into the same type of flask or tube used to prepare temporary standards in ¶ 4*a*1). Add 1.0 mL pH 4.0 buffer, ¶ 3*d*5), and 1.0 mL mixed indicator, ¶ 3*g*. Add mixed indicator in same uniform manner prescribed in ¶ 4*a*1). Match sample visually with temporary standards or read absorbance photometrically and refer to standard calibration curve for free chlorine equivalent. Complete determination within 5 min of adding mixed indicator. For free chlorine concentrations greater than 2.0 mg/L, dilute sample with water to contain 2.0 mg/L or less before analysis.

Color development due to combined chlorine residuals is negligible at sample temperatures as high as 40 C and is slightly accelerated at higher temperatures. Color with free chlorine develops instantaneously, and, in the absence of combined residual chlorine, is stable for several days.

d. Color development of total chlorine samples:

1) Concentrations below 2.0 mg/L—Measure 50-mL sample into suitable flask or tube and add 0.5 mL total chlorine buffer. Mix and wait at least 60 sec. Add 1.0 mL mixed indicator, mix, and dilute to 100 mL. Visually match with standards or read absorbance photometrically and compare with calibration curve.

2) Concentrations above 2.0 mg/L—Place approximately 30 mL water in a flask or tube calibrated to contain at least 100 mL. Add 0.5 mL total chlorine buffer and a measured volume of 20 mL or less of sample. After mixing, let stand for at least 60 sec. Add 1.0 mL mixed indicator, mix, and dilute to mark with water. Match visually with standards or read absorbance photometrically and compare with calibration curve. Select one of the following sample volumes to remain within optimum chlorine range:

Total Chlorine *mg/L*	Sample Volume Required *mL*
2.0–4.0	20.0
4.0–8.0	10.0
8.0–10.0	5.0

Total chlorine color develops instantaneously and remains stable for days. Dilute final color, if it is too intense for visual matching, with water buffered at pH 4.0, then match with standards and estimate initial total chlorine by applying the dilution factor.

e. Elimination of interference from high concentrations of combined chlorine: To determine free chlorine in the presence of high concentrations of combined chlorine, immediately add 5.0 mL $NaAsO_2$ solution after adding and mixing 1.0 mL mixed indicator. Compare visually or photometrically with standards prepared by adding 5.0 mL distilled water after adding mixed indicator to compensate for dilution by $NaAsO_2$ solution.

f. Compensation for manganic [Mn(IV)] manganese: For free chlorine determination, follow the procedure as in ¶ 4*c* and record absorbance as A_1. To a second 50-mL sample, add 0.4 mL 1*N* NaOH. Add 1.0 mL dilute $NaAsO_2$ solution and let react for 2 min. Add 2.0 mL potassium peroxymonosulfate solution, and wait 1.0 min. Add 2.0 mL pH 4.0 buffer, and 1.0 mL mixed indicator, mix, and record absorbance as A_2. Calculate absorbance, A_3, due to free chlorine as follows:

$$A_3 = A_1 - 1.084\,A_2$$

Refer absorbance A_3 to the free chlorine standard curve to obtain free chlorine concentration.

For total chlorine determination, make the total chlorine test as in ¶ 4*d*1) and record absorbance as B_1. To a second 50-mL sample, add 1*N* NaOH (approximately 0.4 mL) to adjust to pH 11.0. Add dilute $NaAsO_2$, potassium peroxymonosulfate, pH 4.0 buffer, and mixed indicator solutions as in ¶ 4*f*. Dilute to 100 mL and record absorbance as B_2. Calculate absorbance, B_3, due to total chlorine as follows:

$$B_3 = B_1 - B_2$$

Refer absorbance B_3 to total chlorine standard curve to obtain total chlorine concentration.

g. Compensation for turbidity and color: Compensate for interference by natural color or turbidity as follows:

1) Visually—View sample and standard horizontally after placing an untreated sample of the same thickness behind standard and the same thickness of distilled water behind the sample.

2) Photometrically—Measure sample absorbance at 592 nm and subtract this reading from absorbance of treated free or total chlorine sample.

5. Calculation

$$\text{mg/L total Cl as Cl}_2 = \frac{A \times 50}{\text{mL sample}}$$

$$\text{mg/L combined Cl as Cl}_2 = B - C$$

where:

A = total chlorine in mg/L measured in diluted sample,
B = total chlorine in mg/L in sample, and
C = free chlorine in mg/L.

6. Precision and Accuracy

See 408A.5.

408 G. Syringaldazine (FACTS) Method (TENTATIVE)

1. General Discussion

a. Principle: The free available chlorine test, syringaldazine (FACTS) measures free available chlorine over the range of 0.1 to 10 mg/L. A saturated solution of syringaldazine (3,5-dimethoxy-4-hydroxybenzaldazine) is used. Syringaldazine is stable when stored as a solid or as a solution in 2-propanol. It is oxidized by free available chlorine on a 1:1 molar basis to produce a colored product with an absorption maximum of 530 nm. The color product is only slightly soluble in water; therefore, at chlorine concentrations greater than 1 mg/L, the final reaction mixture must contain 2-propanol to prevent product precipitation and color fading.

The optimum color and solubility (minimum fading) are obtained in a solution having a pH between 6.5 and 6.8. At a pH less than 6, color development is slow and reproducibility is poor. At a pH greater than 7, the color develops rapidly but fades quickly. A buffer is required to maintain the reaction mixture pH at approximately 6.7. Care should be taken with waters of high acidity or alkalinity to assure that the added buffer maintains the proper pH.

Temperature has a minimal effect on the color reaction. The maximum error observed at temperature extremes of 5 and 35 C is $\pm 10\%$.

b. Interferences: Interferences common to other methods for determining free available chlorine do not affect the FACTS procedure. Monochloramine concentrations up to 18 mg/L, dichloramine concentrations up to 10 mg/L, and manganese concentrations (oxidized forms) up to 1 mg/L do not interfere. Very high concentrations of monochloramine ($\geq$35 mg/L) and oxidized manganese ($\geq$2.6 mg/L) produce a color with syringaldazine slowly. Ferric iron can react with syringaldazine; however, concentrations up to 10 mg/L do not interfere. Nitrite ($\leq$250 mg/L), nitrate ($\leq$100 mg/L), sulfate ($\leq$1,000 mg/L), and chloride ($\leq$1,000 mg/L) do not interfere. Waters with high hardness ($\geq$500 mg/L) will produce a cloudy solution but hardness does not interfere.

Other strong oxidizing agents, such as iodine, bromine, and ozone, will produce a color.

c. Minimum detectable concentration: The FACTS procedure is sensitive to free available chlorine concentrations of 0.1 mg/L or less.

2. Apparatus

Colorimetric equipment: One of the following is required:

a. Filter photometer, providing a light path of 1 cm for chlorine concentrations $\leq$1 mg/L or a light path from 1 to 10 mm for chlorine concentration above 1 mg/L; also equipped with a filter having a band pass of 500 to 560 nm.

b. Spectrophotometer, for use at 530 nm, providing the light paths noted above.

3. Reagents

a. Chlorine-demand-free water: See Method B.3*m*.

b. Syringaldazine indicator: Dissolve 115 mg 3,5-dimethoxy-4-hydroxybenzaldazine* in 1 L 2-propanol.

c. Buffer: Dissolve 17.01 g KH_2PO_4 in 250 mL chlorine-demand-free water. The pH of this solution should be 4.4. Dissolve 17.75 g Na_2HPO_4 in 250 mL chlorine-demand-free water. The pH should be 9.9. Mix equal volumes of these solutions to obtain FACTS buffer, pH 6.6. Verify pH with pH meter.

4. Procedure

a. Calibration of photometer: Prepare a calibration curve by making dilutions of a standardized hypochlorite solution prepared as directed in Section 409A.3*g*. Develop and measure colors as described in ¶ 4*b*, below.

b. Free available chlorine analysis: Add 3 mL sample and 0.1 mL buffer to a 5-mL-capacity test tube. Add 1 mL syringaldazine indicator, cap tube, and invert twice to mix. Transfer to a photometer tube or spectrophotometer cell and measure absorbance. Compare absorbance value obtained with calibration curve and report corresponding value as milligrams per liter free available chlorine.

*Aldrich No. 17, 753-9, Aldrich Chemical Company, Inc., Cedar Knolls, N.J. 07927, or equivalent.

5. Precision and Accuracy

This method has not received extensive testing; however, it has been shown to be the most specific colorimetric test for measuring free available chlorine. The accuracy and precision are comparable to those of the DPD and leuco crystal violet methods. See 408A.5.

408 H. References

1. Water Chlorine (Residual) No. 1. 1969. Analytical Reference Service Rep. No. 35, EPA, Cincinnati, Ohio.

2. Water Chlorine (Residual) No. 2. 1971. Analytical Reference Service Rep. No. 40, EPA, Cincinnati, Ohio.

408 I. Bibliography

General

GUTER, W.J., W.J. COOPER & C.A. SORBER. 1974. Evaluation of existing field test kits for determining free chlorine residuals in aqueous solutions. *J. Amer. Water Works Ass.* 66:38.

WHITTLE, G.P. & A. LAPTEFF, JR. 1973. New analytical techniques for the study of water disinfection. *In* Chemistry of Water Supply, Treatment, and Distribution. p. 63. Ann Arbor Science Publishers, Ann Arbor, Mich.

Iodometric Method

LEA, C. 1933. Chemical control of sewage chlorination: The use and value of orthotolidine test. *J. Soc. Chem. Ind.* (London) 52:245T.

AMERICAN WATER WORKS ASSOCIATION. 1943. Committee report. Control of chlorination. *J. Amer. Water Works Ass.* 35:1315.

MARKS, H.C., R. JOINER & F.B. STRANDSKOV. 1948. Amperometric titration of residual chlorine in sewage. *Water Sewage Works* 95:175.

STRANDSKOV, F.B., H.C. MARKS & D.H. HORCHIER. 1949. Application of a new residual chlorine method to effluent chlorination. *Sewage Works J.* 21:23.

NUSBAUM, I. & L.A. MEYERSON. 1951. Determination of chlorine demands and chlorine residuals in sewage. *Sewage Ind. Wastes* 23:968.

MARKS, H.C., D.B. WILLIAMS & G.U. GLASGOW. 1951. Determination of residual chlorine compounds. *J. Amer. Water Works Ass.* 43:201.

MARKS, H.C. & N.S. CHAMBERLIN. 1953. Determination of residual chlorine in metal finishing wastes. *Anal. Chem.* 24:1885.

Amperometric Titration

FOULK, C.W. & A.T. BAWDEN. 1926. A new type of endpoint in electrometric titration and its application to iodimetry. *J. Amer. Chem. Soc.* 48:2045.

MARKS, H.C. & J.R. GLASS. 1942. A new method of determining residual chlorine. *J. Amer. Water Works Ass.* 34:1227.

HALLER, J.F. & S.S. LISTEK. 1948. Determination of chlorine dioxide and other active chlorine compounds in water. *Anal. Chem.* 20:639.

MAHAN, W.A. 1949. Simplified amperometric titration apparatus for determining residual chlorine in water. *Water Sewage Works* 96:171.

MARKS, H.C., D.B. WILLIAMS & G.U. GLASGOW. 1951. Determination of residual chlorine compounds. *J. Amer. Water Works Ass.* 43:201.

KOLTHOFF, I.M. & J.J. LINGANE. 1952. Polarography, 2nd ed. Interscience Publishers. New York, N.Y.

MORROW, J.J. 1966. Residual chlorine determination with dual polarizable electrodes. *J. Amer. Water Works Ass.* 58:363.

DPD Methods

PALIN, A.T. 1957. The determination of free and combined chlorine in water by the use of diethyl-p-phenylene diamine. *J. Amer. Water Works Ass.* 49:873.

PALIN, A.T. 1960. Colorimetric determination of chlorine dioxide in water. *Water Sewage Works* 107:457.

PALIN, A.T. 1961. The determination of free residual bromine in water. *Water Sewage Works* 108:461.

NICOLSON, N.J. 1963, 1965, 1966. Determination of chlorine in water, Parts 1, 2 and 3. Water Res. Ass. Tech. Pap. Nos. 29, 47, and 53.

NICOLSON, N.J. 1965. An evaluation of the methods for determining residual chlorine in water. Part I. Free chlorine. *Analyst* 90:187.

PALIN, A.T. 1967. Methods for the determination, in water, of free and combined available chlorine, chlorine dioxide and chlorite, bromine, iodine, and ozone using diethyl-p-phenylenediamine (DPD). *J. Inst. Water Eng.* 21:537.

PALIN, A.T. 1968. Determination of nitrogen trichloride in water. *J. Amer. Water Works Ass.* 60:847.

PALIN, A.T. 1975. Current DPD methods for residual halogen compounds and ozone in water. *J. Amer. Water Works Ass.* 67:32.

Leuco Crystal Violet Method

BLACK, A.P. & G.P. WHITTLE. 1967. New methods for the colorimetric determination of halogen residuals. Part II. Free and total chlorine. *J. Amer. Water Works Ass.* 59:607.

FACTS Method

BAUER, R. & C. RUPE. 1971. Use of syringaldazine in a photometric method for estimating "free" chlorine in water. *Anal. Chem.* 43:421.

GUTER, K.J., W.J. COOPER & C.A. SORBER. 1974. Evaluation of existing field test kits for determining free chlorine residuals in aqueous solution. *J. Amer. Water Works Ass.* 66:38.

COOPER, W.J., C.A. SORBER & E.P. MEIER. 1975. A rapid, free, available chlorine test with syringaldazine (FACTS). *J. Amer. Water Works Ass.* 67:34.

409 CHLORINE DEMAND

Chlorine demand is the quantity of chlorine that is reduced or converted to inert or less active forms of chlorine by substances in the water. In most cases, chlorine demand implies complete reaction with all chlorine-reactable materials and is defined as the difference between the amount of chlorine applied and the amount of free available chlorine (hypochlorous acid, HOCl, or hypochlorite ion, OCl^-), remaining at the end of the contact period. The term "breakpoint chlorination" frequently is applied where only free available chlorine remains after the contact period.

Chlorine-consuming substances include ammonia, organics, cyanide, and such inorganic reductants as ferrous, manganous, nitrite, sulfide, and sulfite ions. Chlorine reacts with organics and cyanide in a complex manner involving substitution, addition, or oxidation. Chlorine is reduced to chloride ion by most inorganic reductants.

The reaction of chlorine with ammonia, or possibly amino compounds, to yield combined available forms of chlorine (chloramines) is common. Chloramines can be destroyed by breakpoint or excess chlorination until a free available chlorine residual, equal to the total available chlorine, is attained. Under some test conditions, particularly at low pH values, an apparent breakpoint chlorination is obtained in which free available and combined available chlorine coexist. See Section 408 for methods to differentiate these forms of chlorine.

If the test objective does not involve complete satisfaction of the chlorine demand, then the procedures of Section 411, Chlorine Requirement, may be more appropriate.

The chlorine demand varies with the amount of chlorine applied, time of contact, pH, and temperature. For comparative purposes, *state all test condi-*

tions, including the method of determining residual chlorine. The smallest amount of residual chlorine considered significant is 0.1 mg Cl/L. Presented here are procedures for laboratory and field use.

409 A. Laboratory Method

1. Discussion

The laboratory method standardizes the procedures for determining chlorine demand; however, because of the dependency of chlorine demand on several factors the analyst must define the experimental conditions such as time of contact, pH, and temperature to achieve the desired objectives.

2. Apparatus

a. See Section 408C.2, 408E.2, 408F.2, or 408G.2 for the method used.

b. Glassware: Expose glassware to water containing at least 10 mg residual chlorine/L for 3 hr or more and rinse with chlorine-demand-free water before use.

3. Reagents

a. Chlorine-demand-free water: Prepare as described in 408B.3*m*.

b. Acetic acid, conc (glacial).

c. Potassium iodide, KI, crystals.

d. Standard sodium thiosulfate titrant, 0.025*N:* Prepare as directed in Residual Chlorine, Section 408A.2*d*.

e. Starch indicator solution: Prepare as directed in Section 408A.2*e*.

f. Appropriate reagents for determining residual chlorine by one of the methods described in Section 408.

g. Standard chlorine solution: Obtain a suitable solution from a chlorinator solution hose or by bubbling chlorine gas through distilled water. Improve solution stability by storing in the dark or in a brown, glass-stoppered bottle. Standardize each day of use. Alternatively, dilute household hypochlorite solution, which contains about 30,000 to 50,000 mg chlorine equivalent/L. This is more stable than a chlorine solution, but do not use it for more than 1 wk without restandardizing. Use the same chlorine concentration actually applied in plant treatment to determine chlorine demand. Depending on intended use, a suitable strength of chlorine solution usually will be between 100 and 1,000 mg/L. Use a solution of sufficient concentration so that adding the chlorine solution will not increase the volume of treated portions by more than 5%.

Standardization: Place 2 mL acetic acid and 10 to 25 mL chlorine-demand-free water in a flask. Add about 1 g KI. Measure into the flask a suitable volume of chlorine solution. In choosing a convenient volume, note that 1 mL 0.025*N* thiosulfate titrant is equivalent to about 0.9 mg chlorine.

Titrate with standardized 0.025*N* $Na_2S_2O_3$ titrant until the yellow iodine color almost disappears. Add 1 to 2 mL starch indicator solution and continue titrating to disappearance of blue color.

Determine the blank by adding identical quantities of acid, KI, and starch indicator to a volume of chlorine-demand-free water corresponding to the sample used for titration. Perform blank titration A or B, whichever applies, according to 408A.3*d*.

$$\text{mg Cl as } Cl_2/\text{mL} = \frac{(A + B) \times N \times 35.45}{\text{mL sample}}$$

where:

N = normality of $Na_2S_2O_3$,
A = mL titrant for sample,
B = mL titrant for blank (to be added or subtracted according to required blank titration. See Section 408A.3*d*).

4. Procedure

a. Volume of sample: Measure at least 10 equal sample portions, preferably into brown, glass-stoppered bottles or erlenmeyer flasks of ample capacity to permit mixing. If the test object is to determine chlorine demand, measure 200-mL portions; if it is to relate chlorine demand to bacterial removal, effect on taste and odor, or sample chemical constituents, use portions of 500 mL or more. Properly sterilize all glassware for bacteriological use.

b. Addition of chlorine water: Add an amount of chlorine to the first portion that leaves no chlorine residual at end of contact period. Add increasing amounts of chlorine to successive portions in the series. Increase dosage between portions in increments of 0.1 mg/L for determining low demands and up to 1.0 mg/L or more for higher demands. Mix while adding. Dose sample portions according to a staggered schedule that will permit determining chlorine residual at predetermined contact times. An approximation to the ultimate chlorine demand can be made by dosing (1 mg available chlorine/L minimum) so that the residual is one-half the dosage. Confirm this demand by doubling the dosage; the second demand should be within 10% of the first.

c. Contact time: Conduct test over desired contact period. If test objective is to duplicate in the laboratory the temperature and plant contact time, match contact time and temperature in the plant as closely as possible. For contact times greater than 15 min keep sample temperature constant. Record contact time. Protect chlorinated samples from strong sunlight throughout test.

d. Examination of samples: At end of contact period, determine free and/or combined available chlorine residual by a method described in Section 408 and record method used. Plot residual chlorine or amount consumed versus dosage. If necessary, remove samples for bacteriological examination at desired intervals and process them immediately.

e. Taste and odor: Measure taste and odor of treated samples at ordinary temperatures with or without dechlorination. CAUTION: *The procedures of this test do not insure disinfection. Do not make taste tests unless water is known to be safe to taste.* To determine odor at elevated temperatures, dechlorinate samples before heating. Choose a dechlorinating agent that will not affect sample odor. Generally, sodium thiosulfate, $Na_2S_2O_3$, is satisfactory; however, the contact times and amount of $Na_2S_2O_3$ needed may depend on whether the residual is free or combined. $Na_2S_2O_3$ in excess of the stoichiometric amount may affect taste and odor tests.

409 B. Field Method

1. General Discussion

This procedure may be used in the plant or field when facilities or personnel are not adequate to use the more exact method. Results are approximations only.

2. Apparatus

a. Chlorine comparator, color- and turbidity-compensating.

b. Medicine dropper to deliver 20 drops/mL. Clean end of dropper so that water adheres all around the periphery. Hold dropper vertically and let drops form slowly.

c. Ten flasks of approximately 1 L capacity, or 1-qt wide-mouth bottles marked at the 500-mL level.

d. Ten 60-mL (2-oz) bottles, marked at the 20-mL level.

e. Glass stirring rod.

f. Glass-stemmed thermometer, readable to at least 0.2 C.

3. Reagents

a. Standard chlorine solution: Dilute a 5% household bleaching solution 1 + 4. Standardize as directed in Method A, ¶ 3*g*, but take 20 drops diluted hypochlorite solution as sample to be titrated; use same dropper that will be used in procedure. For each drop:

$$\text{mg available Cl as } Cl_2 = \frac{A \times N \times 35}{20}$$

where:

A = mL titration for sample and
N = normality of $Na_2S_2O_3$.

Adjust solution to 10 mg chlorine/mL (0.5 mg Cl as Cl_2/drop), so that 1 drop added to a 500-mL sample gives a dosage of 1.0 mg/L.

b. Test reagent: Use appropriate reagents for determining residual chlorine. See Section 408.

4. Procedure

a. Sample measurement: Fill each container to the 500-mL mark with sample. Record temperature.

b. Addition of chlorine: While stirring constantly, add 1 drop of chlorine solution to first container, 2 drops to second container, 3 drops to third container, etc.

c. Contact time: Follow directions given in Method A, ¶ 4*c* above.

d. Examination of samples: At end of contact period, remove a portion from each sample and determine residual chlorine by a method described in Section 408. Record method used.

5. Calculation

$$\text{mg Cl demand/L} = \text{mg Cl as } Cl_2\text{/L added} - \text{mg Cl as } Cl_2\text{/L residual}$$

6. Interpretation of Results

The chlorine demand refers only to the particular dosage, contact time, and temperature used in this test. Plot residual chlorine or amount consumed versus chlorine added to aid in studying results.

409 C. Bibliography

Griffin, A.E. & N.S. Chamberlin. 1941. Relation of ammonia-nitrogen to breakpoint chlorination. *Amer. J. Pub. Health* 31:803.

American Water Works Association. 1943. Committee report. Control of chlorination. *J. Amer. Water Works Ass.* 35:1315.

Palin, A.T. 1950. Chemical aspects of chlorination. *J. Inst. Water Eng.* 4:565.

Taras, M.J. 1953. Effect of free residual chlorination on nitrogen compounds in water. *J. Amer. Water Works Ass.* 45:47.

410 CHLORINE DIOXIDE

Because the physical and chemical properties of chlorine dioxide resemble those of chlorine in many respects, read the entire discussion of Residual Chlorine (Section 408) before attempting a chlorine dioxide determination.

Chlorine dioxide, ClO_2, has been widely used as a bleaching agent in the paper and pulp industry. It has received attention as an alternative to chlorine for disinfection of potable water.

Chlorine dioxide has been applied to water supplies to combat tastes and odors due to phenolic-type wastes, actinomycetes, and algae, as well as to oxidize soluble iron and manganese to a more easily removable form. It acts as a disinfectant, and some results suggest that it may be stronger than free chlorine or hypochlorite. The difficulties of generation, handling, and storage have limited both application and experimentation.

Chlorine dioxide is a deep yellow, volatile, unpleasant-smelling gas that is toxic and under certain conditions may react explosively. It should be handled with care in a vented area. There are several methods of generating ClO_2; for laboratory purposes the acidification of a solution of sodium chlorite followed by suitable scrubbing is the most practical.

1. Selection of Method

The iodometric method (A) gives a very precise measure of total available strength of a solution in terms of its ability to liberate iodine from iodide. However, ClO_2, chlorine, chlorite, and hypochlorite are not distinguished easily by this technic. It is designed primarily, and best used, for standardizing ClO_2 solutions needed for preparation of temporary standards. It often is inapplicable to industrial wastes.

The amperometric method (B) is useful when a knowledge of the various chlorine fractions in a water sample is desired. It distinguishes various chlorine compounds of interest with good accuracy and precision, but requires specialized equipment and considerable analytical skill.

The tentative N,N-diethyl-p-phenylenediamine (DPD) method (C) has the advantages of a relatively easy-to-perform colorimetric test with the ability to distinguish between ClO_2 and various forms of chlorine. This technic is not as accurate as the amperometric method, but should yield results adequate for most common applications.

2. Sampling and Storage

Determine ClO_2 promptly after collecting the sample. Do not expose sample to sunlight or strong artificial light and do not aerate to mix. Minimum ClO_2 losses occur when the determination is completed immediately at the site of sample collection.

410 A. Iodometric Method

1. General Discussion

a. Principle: A pure solution of ClO_2 is prepared by slowly adding dilute H_2SO_4 to a sodium chlorite ($NaClO_2$) solution. Contaminants such as chlorine are removed by a $NaClO_2$ scrubber and passing the gas into distilled water in a steady stream of air.

ClO_2 releases free iodine from a KI solution acidified with acetic acid or H_2SO_4. The liberated iodine is titrated with a standard solution of sodium thiosulfate ($Na_2S_2O_3$), with starch as the indicator.

b. Interference: There is little interference in this method, but temperature and strong light affect solution stability. Minimize ClO_2 losses by storing stock ClO_2 solution in a dark refrigerator and by preparing and titrating dilute ClO_2 solutions for standardization purposes at the lowest practicable temperature and in subdued light.

c. Minimum detectable concentration: One drop (0.05 mL) of 0.01*N* $Na_2S_2O_3$ is equivalent to 20 μg ClO_2/L (or 40 μg/L in terms of available chlorine) when a 500-mL sample is titrated.

2. Reagents

All reagents listed for the determination of residual chlorine in Section 408A.2*a-g* are required. Also needed are the following:

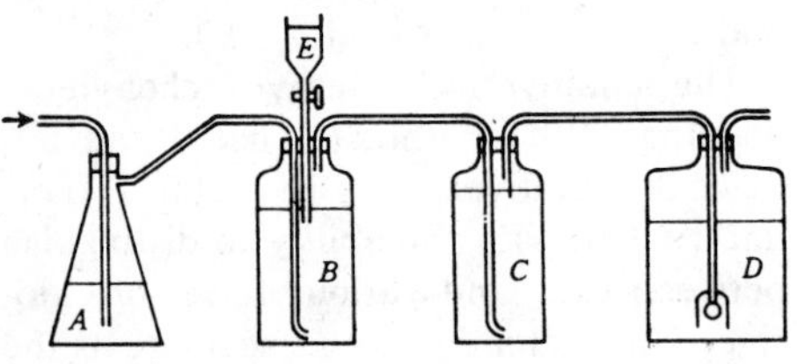

Figure 410:1. Chlorine dioxide generation and absorption system.

a. Stock chlorine dioxide solution: Prepare a gas generating and absorbing system as illustrated in Figure 410:1. Connect aspirator flask (A), 500-mL capacity, with rubber tubing to a source of compressed air. Let air bubble through a layer of 300 mL distilled water in flask and then pass through a glass tube ending within 5 mm of the bottom of the 1-L gas-generating bottle (B). Conduct evolved gas via glass tubing through a scrubber bottle (C) containing saturated $NaClO_2$ solution or a tower packed with flaked $NaClO_2$, and finally, via glass tubing, into a 2-L borosilicate glass collecting bottle (D) where the gas is absorbed in 1,500 mL distilled water. Provide an air outlet tube on collecting bottle (D) for escape of air. Select for gas generation a bottle constructed of strong borosilicate glass and having a mouth wide enough to permit insertion of three separate glass tubes: the first leading almost to the bottom for admitting air, the second reaching below the liquid surface for gradual introduction of H_2SO_4, and the third near the top for exit of evolved gas and air. Fit to second tube a graduated cylindrical separatory funnel (E) to contain H_2SO_4. Locate this system in a fume hood with an adequate shield.

Dissolve 10 g $NaClO_2$ in 750 mL distilled water and place in generating bottle (B).

Carefully add 2 mL conc H_2SO_4 to 18 mL distilled water and mix. Transfer to funnel.

Connect flask to generating bottle, generating bottle to scrubber, and the latter to collecting bottle. Pass a smooth current of air through the system, as evidenced by the bubbling rate in all bottles.

Introduce 5-mL increments of H_2SO_4 from funnel into generating bottle at 5-min intervals.

Continue air flow for 30 min after last portion of acid has been added.

Store yellow stock solution in glass-stoppered dark-colored bottle in a dark refrigerator. The concentration of ClO_2 thus prepared varies between 250 and 600 mg/L corresponding to approximately 500 to 1,200 mg/L available chlorine.

b. Standard chlorine dioxide solution: Use this solution for preparing temporary ClO_2 standards. Dilute required volume of stock ClO_2 solution to desired strength with chlorine-demand-free water (see Section 408B.3*m*). Standardize solution by titrating with standard 0.01*N* or 0.025*N* $Na_2S_2O_3$ titrant in the presence of KI, acid, and starch indicator by following the procedure given in ¶ 3 below. A full or nearly full bottle of chlorine or ClO_2 solution retains its titer longer than a partially full one. When repeated withdrawals reduce volume to a critical level, standardize diluted solution at the beginning, midway in the series of withdrawals, and at the end of the series. Shake contents thoroughly before drawing off needed solution from middle of the glass-stoppered dark-colored bottle. Prepare this solution frequently.

3. Procedure

Select volume of sample, prepare for titration, and titrate sample and blank as described in Section 408A.3. The only exception is the following: *Let ClO_2 react in the dark with acid and KI for 5 min before starting titration.*

4. Calculations

Express ClO_2 concentrations in terms of ClO_2 or as available chlorine content. Available chlorine is defined as the total oxidizing power of ClO_2 measured by titrating iodine released by ClO_2 from an acidic solution of KI. Calculate result in terms of chlorine itself.

For standardizing ClO_2 solution:

$$\text{mg } ClO_2/\text{mL} = \frac{(A \pm B) \times N \times 13.49}{\text{mL sample titrated}}$$

$$\text{mg } ClO_2 \text{ as } Cl_2/\text{mL} = \frac{(A \pm B) \times N \times 35.45}{\text{mL sample titrated}}$$

For determining ClO_2 temporary standards:

$$\text{mg } ClO_2/\text{L} = \frac{(A \pm B) \times N \times 13{,}490}{\text{mL sample}}$$

$$\text{mg } ClO_2 \text{ as } Cl_2/\text{L} = \frac{(A \pm B) \times N \times 35{,}450}{\text{mL sample}}$$

where:

A = mL titration for sample,
B = mL titration for blank (positive or negative, see 408A.3*d*), and
N = normality of $Na_2S_2O_3$.

410 B. Amperometric Method

1. General Discussion

a. Principle: The amperometric titration of ClO_2 is an extension of the amperometric method for residual chlorine. By performing four titrations with phenylarsine oxide, free chlorine (including hypochlorite and hypochlorous acid), chloramines, chlorite, and ClO_2 may be determined separately. In the first titration ClO_2 is converted to chlorite and chlorate through addition of sufficient NaOH to produce a pH of 12, followed by neutralization to a pH of 7 and titration of free chlorine. In the second titration KI is added to a sample that has been treated similarly with alkali and the pH readjusted to 7; titration yields free chlorine and monochloramine. The third titration involves addition of KI and pH adjustment to 7, followed by titration of free chlorine, monochloramine, and one-fifth of the available ClO_2. In the fourth titration, addition of sufficient H_2SO_4 to lower the pH to 2 enables all available ClO_2 and chlorite, as well as the total available chlorine, to liberate an equivalent amount of iodine from the added KI and thus be titrated.

b. Interference: The interferences described in Section 408C.1*b* apply also to determination of ClO_2.

2. Apparatus

The apparatus required is given in Sections 408C.2*a* through *d*.

3. Reagents

All reagents listed for the determination of residual chlorine in Section 408C.3 are required. Also needed are the following:

a. Sodium hydroxide, NaOH, 6*N*.

b. Sulfuric acid, H_2SO_4, 6*N*, 1 + 5.

4. Procedure

Minimize effects of pH, time, and temperature of reaction by standardizing all conditions.

a. Titration of free available chlorine (hypochlorite and hypochlorous acid): Add sufficient 6*N* NaOH to raise sample pH to 12. After 10 min, add 6*N* H_2SO_4 to lower pH to 7. Titrate with standard phenylarsine oxide titrant to the amperometric end point as given in Section 408C. Record result as *A*.

b. Titration of free available chlorine and chloramine: Add 6*N* NaOH to raise sample pH to 12. After 10 min, add 6*N* H_2SO_4 to reduce pH to 7. Add 1 mL KI solution. Titrate with standard phenylarsine oxide titrant to the amperometric end point. Record result as *B*.

c. Titration of free available chlorine, chloramine, and one-fifth of available ClO_2: Adjust sample pH to 7 with pH 7 phosphate buffer solution. Add 1 mL KI solution. Titrate with standard phenylarsine oxide titrant to the amperometric end point. Record result as *C*.

d. Titration of free available chlorine, chloramines, ClO_2, and chlorite: Add 1 mL KI solution to sample. Add sufficient 6*N* H_2SO_4 to lower sample pH to 2. After 10 min, add sufficient 6*N* NaOH to raise pH to 7. Titrate with standard phenylarsine oxide titrant to the amperometric end point. Record result as *D*.

5. Calculation

Convert individual titrations (*A*, *B*, *C*, and *D*) into chlorine concentration by the following equation:

$$\text{mg Cl as Cl}_2\text{/L} = \frac{E \times 200}{\text{mL sample}}$$

where:

E = mL phenylarsine oxide titration for each individual sample *A*, *B*, *C*, or *D*.

Calculate ClO_2 and individual chlorine fractions as follows:

$$\text{mg ClO}_2\text{/L as ClO}_2 = 1.9\ (C - B)$$
$$\text{mg ClO}_2 \text{ as Cl}_2\text{/L} = 5\ (C - B)$$
$$\text{mg free available residual chlorine/L} = A$$
$$\text{mg chloramine/L as chlorine} = B - A$$
$$\text{mg chlorite/L as chlorine} = 4B - 5C + D$$

410 C. DPD Method (TENTATIVE)

1. General Discussion

a. Principle: This method is an extension of the N,N-diethyl-p-phenylenediamine (DPD) method for determining free chlorine and chloramines in water. ClO_2 appears in the first step of this procedure but only to the extent of one-fifth of its total available chlorine content corresponding to reduction of ClO_2 to chlorite ion. If the sample is then acidified in the presence of iodide the chlorite also reacts. When neutralized by subsequent addition of bicarbonate, the color thus produced corresponds to the total available chlorine content of the ClO_2. If chlorite is present in the sample, this will be included in the step involving acidification and neutralization. Chlorite that did not result from ClO_2 reduction by the procedure will cause a positive error equal to twice this chlorite concentration. In evaluating mixtures of these various chloro-compounds, it is necessary to suppress free chlorine by adding glycine before reacting the sample with DPD reagent. Differentiation is based on the fact that glycine converts free chlorine instantaneously into chloroaminoacetic acid but has no effect on ClO_2.

b. Interference: The interference by oxidized manganese described in Section 408D.1*d* applies also to ClO_2 determination. Manganese interference appears as an increase in the first titrations after addition of DPD, with or without KI, and irrespective of whether there has been prior addition of glycine. Titration readings must be corrected suitably. Interference by chromate in wastewaters may be corrected similarly.

Iron contributed to the sample by adding ferrous ammonium sulfate (FAS) titrant may activate chlorite so as to interfere with the first end point of the titration. Suppress this effect with additional EDTA, disodium salt.

2. Reagents

Reagents required in addition to those for the DPD free-combined chlorine method as listed in Section 408D.2 are as follows:

a. Glycine solution: Dissolve 10 g NH_2CH_2COOH in 100 mL distilled water.

b. Sulfuric acid solution: Dilute 5 mL conc H_2SO_4 to 100 mL with distilled water.

c. Sodium bicarbonate solution: Dissolve 27.5 g $NaHCO_3$ in 500 mL distilled water.

d. EDTA: Disodium salt of ethylenediamine tetraacetic acid, solid.

3. Procedure

For samples containing more than 5 mg/L total available chlorine follow the dilution procedure given in Section 408D.3.

a. Chlorine dioxide: Add 2 mL glycine solution to 100 mL sample and mix. Place 5 mL each of buffer reagent and DPD indicator solution in a separate titration flask and mix (or use about 500 mg DPD powder). Add about 200 mg EDTA, disodium salt. Then add glycine-treated sample and mix. Titrate rapidly with standard FAS titrant until red color is discharged (reading *G*).

b. Free available chlorine and chloramine: Using a second 100-mL sample follow the procedures of Section 408D.3*a* adding about 200 mg EDTA, disodium salt, initially with the DPD reagents (readings *A*, *B*, and *C*).

c. Total available chlorine including chlorite: After obtaining reading *C* add 1 mL H_2SO_4 solution to the same sample in titration flask, mix, and let stand about 2 min. Add 5 mL $NaHCO_3$ solution, mix, and titrate (reading *D*).

d. Colorimetric procedure: Instead of titration with standard FAS solution, the readings at each stage may be obtained by colorimetric procedures. Calibrate colorimeters with standard permanganate solution as given in Section 408E.4*a*. Use of additional EDTA, disodium salt, with the DPD reagents is not required in colorimetric procedures.

4. Calculations

For 100 mL sample, 1 mL FAS solution = 1 mg available chlorine/L.

In the absence of chlorite:

$$\text{Chlorine dioxide} = 5G \text{ (or } 1.9G \text{ expressed as } ClO_2)$$

$$\text{Free available chlorine} = A - G$$

$$\text{Monochloramine} = B - A$$

$$\text{Dichloramine} = C - B$$

$$\text{Total available chlorine} = C + 4G$$

If the step leading to reading *B* is omitted, monochloramine and dichloramine are obtained together when:

$$\text{Combined available chlorine} = C - A$$

If it is desired to check for presence of chlorite in sample, obtain reading *D*. Chlorite is indicated if *D* is greater than *C* + 4*G*.

In the presence of chlorite:

$$\text{Chlorine dioxide} = 5G \text{ (or } 1.9G \text{ expressed as } ClO_2)$$

$$\text{Chlorite} = D - (C + 4G)$$

$$\text{Free available chlorine} = A - G$$

$$\text{Monochloramine} = B - A$$

$$\text{Dichloramine} = C - B$$

$$\text{Total available chlorine} = D$$

If *B* is omitted,

$$\text{Combined available chlorine} = C - A$$

410 D. Bibliography

General

INGOLS, R.S. & G.M. RIDENOUR. 1948. Chemical properties of chlorine dioxide in water treatment. *J. Amer. Water Works Ass.* 40:1207.

PALIN, A.T. 1948. Chlorine dioxide in water treatment. *J. Inst. Water Eng.* 11:61.

HODGDEN, H.W. & R.S. INGOLS. 1954. Direct colorimetric method for determination of chlorine dioxide in water. *Anal. Chem.* 26:1224.

FEUSS, J.V. 1964. Problems in determination of chlorine dioxide residuals. *J. Amer. Water Works Ass.* 56:607.

MASSCHELEIN, W. 1966. Spectrophotometric determination of chlorine dioxide with acid chrome violet K. *Anal. Chem.* 38:1839.

MASSCHELEIN, W. 1969. Les Oxydes de Chlore et le Chlorite de Sodium. Dunod, Paris, Chapter XI.

Iodometric Method

POST, M.A. & W.A. MOORE. 1959. Determination of chlorine dioxide in treated surface waters. *Anal. Chem.* 31:1872.

Amperometric Method

HALLER, J.F. & S.S. LISTEK. 1948. Determination of chlorine dioxide and other active chlorine compounds in water. *Anal. Chem.* 20:639.

DPD Method

PALIN, A.T. 1960. Colorimetric determination of chlorine dioxide in water. *Water Sewage Works* 107:457.

PALIN, A.T. 1967. Methods for the determination, in water, of free and combined available chlorine, chlorine dioxide and chlorite, bromine, iodine, and ozone using diethyl-p-phenylenediamine (DPD). *J. Inst. Water Eng.* 21:537.

PALIN, A.T. 1974. Analytical control of water disinfection with special reference to differential DPD methods for chlorine, chlorine dioxide, bromine, iodine and ozone. *J. Inst. Water Eng.* 28:139.

PALIN, A.T. 1975. Current DPD methods for residual halogen compounds and ozone in water. *J. Amer. Water Works Ass.* 67:32.

411 CHLORINE REQUIREMENT

Chlorine demand is the quantity of chlorine that must be added to a unit volume of water to react with all the chlorine-reactable materials. Because some chlorination processes do not require complete satisfaction of the chlorine demand, chlorine requirement is the more applicable term.

Chlorine requirement is the quantity of chlorine that must be added to a unit volume of water under specified conditions (pH, contact time, temperature) to achieve a specified result. Examples include the quantity of chlorine required to limit the maximum bacterial count in wastewater or to oxidize iron and manganese in a potable water for subsequent removal to specified levels.

Where applicable, chlorine residuals may be determined by any of the methods of Section 408; the same method must be used for both laboratory testing and operational control.

In reporting results, include all conditions of testing such as pH, contact time, temperature, and the analytical procedures used.

Chlorine requirement is not an absolute test and the results of laboratory studies must be applied with caution to plant operations. The primary purpose of the test is to provide guidance in the control of chlorination for disinfection or other purposes.

411 A. Method for Control of Disinfection

Chlorine requirement can be determined on a plant or laboratory scale but, in most cases, it is better determined in the plant, under plant conditions, and with plant equipment.

In the plant test, the flow of wastewater, quantity of chlorine used, contact time, residual chlorine concentration, and bacteriological results are determined. The tests are conducted at minimum and average contact times corresponding to different flow conditions, to establish the average and variations from the average of the number of organisms in the effluent.

Sufficient replication may establish a correlation between bacteriological results and residual chlorination. Make bacteriological tests periodically to verify the correlation.

If laboratory studies are made, use more than one contact time to establish minimum and average chlorine requirements to attain the stipulated microbial densities and permissible variations.

2. Reagents

All reagents necessary for determining residual chlorine by the selected method are required, and in addition:

a. Standard chlorine solution: Pass chlorine gas through distilled water or tap water until the solution contains approximately 1.0 mg Cl as Cl_2/mL. Because this solution is not stable, prepare fresh daily or standardize it each time it is used (see Section 409A.3*g*) by using 5 mL chlorine water and 0.025 *N* sodium thiosulfate solution.

Perform blank titration A or B, whichever applies, according to the procedures given in 408A.3*d*. Calculate strength as follows:

$$\text{mg Cl as } Cl_2/\text{mL} = \frac{(A + B) \times N \times 35.45}{\text{mL sample}}$$

where:

N = normality of $Na_2S_2O_3$,

A = mL titrant for sample, and

B = mL titrant for blank, either added or subtracted according to the required blank titration (see 408A.3*d*).

b. Sodium sulfite solution: Dissolve 10 g anhydrous Na_2SO_3 in 100 mL distilled water and heat to boiling to sterilize. Prepare daily.

3. Procedure

a. Sample measurement: In each of a series of 1-L beakers, jars, or flasks, place 500 mL sample.

b. Addition of chlorine: Select dosages and increments of chlorine suited to the sample and to the purpose of chlorination. Use a range of dosages that includes at least one believed certain to produce the desired result. With gentle and constant stirring, add the selected quantities of chlorine to samples.

c. Determination of residual chlorine: At end of stipulated contact times, determine residual chlorine by one or more of the procedures given in Section 408.

d. Determination of degree of disinfection: Immediately after removing portions for determining residual chlorine, add 0.5 mL Na_2SO_3 solution to each portion and determine the number of organisms surviving by the appropriate procedure(s) given in Part 900.

4. Calculation

The chlorine requirement is the amount of chlorine added per unit volume of sample to produce the desired result.

5. Precision

The precision of this test on a single sample is poor because of inaccuracies in enumerating surviving organisms. To establish the chlorine requirement for a given coliform density with suitable precision, repeat test at least ten times on different samples using otherwise identical conditions.

411 B. Methods for Purposes Other Than Disinfection Control

In potable water treatment, chlorine can oxidize constituents such as iron and manganese for subsequent removal. In some applications, chlorine improves coagulation, flocculation, and sedimentation processes. The laboratory test involves adding chlorine to water samples and determining both the residual chlorine and the degree of iron or magnanese oxidation as a function of dosage, contact time, pH, and other controllable conditions. Alternatively, add chlorine to water samples during jar tests to determine the chlorine required for specific objectives such as improved flocculation and sedimentation.

In wastewater treatment, chlorine can be used for odor control, slime and insect control on trickling filters, and control of activated sludge bulking. In these cases, determine the chlorine requirement in the field.

When wastewaters are chlorinated for reduction of such compounds as phenols and cyanides, use laboratory procedures similar to those used for control of disinfection and potable water constituents. For these applications, accompany the determination of chlorine requirement by determination of the constituent or property to be controlled by chlorination. In all cases, contact time, pH, temperature, and other conditions are important controlling factors.

412 CYANIDE

1. General Discussion

"Cyanide" refers to all of the CN groups in cyanide compounds that can be determined as the cyanide ion, CN^-, by the methods used. The cyanide compounds in which cyanide can be obtained as CN^- are classed as simple and complex cyanides.

The simple cyanides are represented by the formula $A(CN)_x$, where A is an alkali (sodium, potassium, ammonium) or a metal, and x, the valence of A, is the number of CN groups. In aqueous solutions of the simple alkali cyanides, the CN group is present as CN^- and molecular HCN, the ratio depending on pH. In most natural waters HCN greatly predominates.[1] In solutions of simple metal cyanides, the CN group may occur also in the form of complex metal-cyanide anions of varying stability. Many of the simple metal cyanides are sparingly soluble or almost insoluble,[2] but they form a variety of highly soluble, complex metal cyanides in the presence of alkali cyanides.

The complex cyanides have a variety of formulae, but the alkali-metallic cyanides normally can be represented by $A_yM(CN)_x$. In this formula, A represents the alkali present y times, M the heavy metal (ferrous and ferric iron, cadmium, copper, nickel, silver, zinc, or others), and x the number of CN groups; x is equal to the valence of A taken y times plus that of the heavy metal. The initial dissociation of each of these soluble, alkali-metallic, complex cyanides yields an anion that is the radical $M(CN)_x^{y-}$. This may dissociate to some extent, depending on several factors, with the liberation of CN^- ion and consequent formation of HCN.

The great toxicity to aquatic life of molecular HCN, formed in solutions of cyanides by hydrolytic reaction of CN^- with water, is well known.[3–6] The toxicity of

CN⁻ is less than that of molecular HCN; it usually is unimportant because most of the free cyanide (CN group present as CN^- or as HCN) exists as HCN.[3,6] The toxicity to fish of most tested solutions of complex cyanides is attributable mainly to the HCN resulting from dissociation of the complexes.[3,5,6] Analytical distinction between HCN and other cyanide species in solutions of complex cyanides now is possible.[3,6–10]

The degree of dissociation of the various metallocyanide complexes at equilibrium, which may not be attained for a long time, increases with decreased concentration and decreased pH, and is inversely related to their highly variable stability.[3,5,6] The zinc- and cadmium-cyanide complexes are almost totally dissociated in very dilute solutions that are acutely toxic to fish at any ordinary pH of natural waters. In equally dilute solutions the dissociation of the nickel-cyanide complex and the more stable cyanide complexes formed with copper (I) and silver is much less. The acute toxicity to fish of dilute solutions containing copper-cyanide or silver-cyanide complex anions can be due mainly or entirely to the toxicity of the undissociated ions, although the complex ions are much less toxic than HCN.[3,6]

The iron-cyanide complex ions are very stable and not materially toxic; in the dark, acutely toxic levels of HCN are attained only in solutions that are not very dilute and have been aged for a long time. However, these complexes are subject to extensive and rapid photolysis, yielding toxic HCN, on exposure of dilute solutions to direct sunlight.[3,11] The photodecomposition is slow in deep, turbid, and shaded receiving waters. Loss of HCN to the atmosphere and its bacterial and chemical destruction concurrent with its production tend to prevent increases of HCN concentrations to harmful levels. Regulatory distinction between cyanide complexed with iron and that bound in less stable complexes, as well as between the complexed cyanide and free cyanide or HCN, can, therefore, be justified.

Historically, the generally accepted industrial waste treatment of cyanide compounds is alkaline chlorination:

$$NaCN + Cl_2 \rightarrow CNCl + NaCl \quad (1)$$

The first reaction product on chlorination is cyanogen chloride (CNCl), a highly toxic gas of limited solubility. The toxicity of CNCl may exceed that of equal concentrations of cyanide.[3,4,12] At an alkaline pH, CNCl hydrolyzes to the cyanate ion (CNO^-), which has only limited toxicity.

There is no known natural reduction reaction that may convert CNO^- to CN^-.[13] On the other hand, breakdown of toxic CNCl is pH- and time-dependent. At pH 9, with no excess chlorine present, CNCl may persist for 24 hr.[14,15]

$$CNCl + 2NaOH \rightarrow NaCNO + NaCl + H_2O \quad (2)$$

CNO^- can be oxidized further with chlorine at a nearly neutral pH to CO_2 and N_2:

$$2NaCNO + 4NaOH + 3Cl_2 \rightarrow 6NaCl + 2CO_2 + N_2 + 2H_2O \quad (3)$$

CNO^- also will be converted on acidification to NH_4^+:

$$2NaCNO + H_2SO_4 + 4H_2O \rightarrow (NH_4)_2SO_4 + 2NaHCO_3 \quad (4)$$

The alkaline chlorination of cyanide compounds is relatively fast, but depends equally on the dissociation constant, which also governs the toxicity. Metal cyanide complexes, such as nickel, cobalt, silver, and gold, do not dissociate readily. The chlorination reaction therefore requires more time and a significant chlorine excess.[16] Iron cyanides, because they do not dissociate to any degree, are not oxidized by chlorination. The refractory properties of the noted complexes, in their resistance to chlorination and lack of toxicity, overlap.

Thus, it is advantageous to differentiate between *total cyanide* and *cyanides amenable to chlorination.* When total cyanide is determined, the almost nondissociable cyanides, as well as cyanide bound in complexes that are readily dissociable and complexes of intermediate stability, are measured. Cyanides amenable to chlorination are free or are potentially dissociable, almost wholly or in large degree, and therefore, potentially toxic at low concentrations, even in the dark. The chlorination test procedure is carried out under the most rigorous conditions appropriate for measurement of the more dissociable forms of cyanide.

The *cyanogen chloride* procedure is common with the colorimetric test for cyanides amenable to chlorination. This test is based on the addition of chloramine-T and subsequent color complex formation with barbituric acid. Without the addition of chloramine-T, only the existing CNCl is measured. CNCl is a gas that hydrolyzes to CNO^-; sample preservation is not possible. Because of this, spot testing of CNCl levels may be best. This procedure can be adapted and used when the sample is collected.

There may be analytical requirements for the determination of CNO^-, even though the reported toxicity level is low. On acidification, CNO^- decomposes to ammonia (NH_3).[4] Molecular ammonia and metal-ammonia complexes are highly toxic.[17]

Thiocyanate (CNS^-) itself is not very toxic to aquatic life.[3] However, upon chlorination, toxic CNCl is formed, as discussed above.[3,4,12] At least where subsequent chlorination is anticipated, the determination of CNS^- is desirable.

2. Cyanide in Solid Waste

a. Soluble cyanide: Determination of soluble cyanide requires sample leaching with distilled water until solubility equilibrium is established. One hour of stirring in distilled water should be satisfactory. Low cyanide concentration in the leachate (<5 mg/L) will indicate the presence of sparingly soluble metal cyanides. The cyanide content of the leachate is indicative of the residual solubility of the insoluble metal cyanides in the waste.

High levels of cyanide in the leachate indicate soluble cyanide in the solid waste. When 500 mL distilled water are stirred into a 500-mg solid waste sample, the cyanide concentration (mg/L) of the leachate multiplied by 1,000 will give the solubility level of the cyanide in the solid waste in milligrams per kilogram. Cyanide determination on the leachate may be for total cyanide and/or cyanide amenable to chlorination.

b. Insoluble cyanide: The insoluble cyanide of the solid waste can be determined with the total cyanide method by placing a 500-mg sample with 500 mL distilled water into the distillation flask and in general following the procedure from the distillation step (Section 412B). In calculating include multiplication by 1,000 to give the cyanide content of the solid waste sample in milligrams per kilogram. Insoluble iron cyanides in the solid waste can be leached out earlier by stirring a weighed sample for 12 to 16 hr in a 10% NaOH solution. The leachate and wash waters of the solid waste will give the iron cyanide content of the sample with the distillation procedure. Prechlorination will have eliminated all cyanide amenable to chlorination from the sample. Do not expose sample to sunlight.

3. Selection of Method

a. Total cyanide after distillation: After removal of interfering substances, the metal cyanide is converted to HCN gas, which is distilled and absorbed in sodium hydroxide (NaOH) solution.[18] Only the cobalticyanide complex is not recovered completely. This is due to the catalytic decomposition of cyanide in the presence of cobalt at high temperature in a strong acid solution.[19,20] The distillation also separates cyanide from other color-producing and possibly interfering organic or inorganic contaminants. Subsequent analysis is for the simple salt, sodium cyanide (NaCN). Some organic cyanide compounds, such as nitriles, are decomposed under the distillation conditions. Aldehydes convert cyanide to nitrile. The absorption liquid is analyzed by either a titrimetric, colorimetric, or cyanide-ion-selective electrode procedure:

1) The titration method (C) is suitable for cyanide concentrations above 1 mg/L.

2) The colorimetric method (D) is suitable for cyanide concentration to a lower limit of 20 μg/L. Analyze higher concentrations by taking a portion and diluting the sample.

3) The ion-selective electrode method (E) using the cyanide ion electrode is applicable in the concentration range of 0.05 to 10 mg/L.

b. Cyanide amenable to chlorination:

1) Distillation of two samples is required, one that has been chlorinated to destroy all amenable cyanide present and the other unchlorinated. Analyze absorption liquids from both tests for total cyanide. The observed difference equals cyanides amenable to chlorination.

2) The colorimetric method, by conversion of amenable cyanide and CNS^- to CNCl and developing the color complex with barbituric acid, is used for the determination of the total of these cyanides.

c. Cyanogen chloride:

1) The colorimetric method for measuring cyanide amenable to chlorination may be used, but omit the chloramine-T addition.

2) The spot test also may be used.

d. Spot test for sample screening: This procedure allows a quick sample screening to establish if more than 50 μg/L cyanide amenable to chlorination is present. The test also may be used to estimate the CNCl content at the time of sampling.

e. Cyanate: CNO^- is converted to ammonium carbonate, $(NH_4)_2CO_3$, by acid hydrolysis at elevated temperature. Ammonia (NH_3) is determined before the conversion of the CNO^- and again afterwards. The CNO^- is estimated from the difference of NH_3 found in the two tests.[21–23] Measure the NH_3 by either:

1) The selective electrode method, using the NH_3 gas electrode.

2) The colorimetric method, using direct nesslerization or the phenate method for NH_3 (Section 417B or C).

f. Thiocyanate: Use the colorimetric determination with ferric nitrate as a color-producing compound.

412 A. Preliminary Treatment of Samples

CAUTION—*Use care in manipulating cyanide samples because of toxicity. Process in a hood or other well-ventilated area. Avoid contact, inhalation, or ingestion.*

1. General Discussion

The nature of the preliminary treatment will vary according to the interfering substance present. Sulfides, fatty acids, and oxidizing agents are removed by special procedures. Most other interfering substances are removed by distillation. The importance of the distillation procedure cannot be overemphasized.

2. Preservation of Samples

Oxidizing agents, such as chlorine, decompose most cyanides. Test by placing a drop of sample on a strip of potassium iodide (KI)-starch paper. If a bluish discoloration is noted, add a few crystals of sodium thiosulfate, $Na_2S_2O_3$, to the sample, stir, and retest. If necessary, repeat procedure until no bluish discoloration of test paper occurs; then add 0.1 g $Na_2S_2O_3$/L to the sample in excess. Manganese dioxide, nitrosyl chloride, etc., if present, also may cause discoloration of the test paper. If possible, carry out this procedure before preserving sample as described below. When the following test indicates presence of sulfide, oxidizing compounds would not be expected.

Sulfide will convert CN^- to CNS^- rapidly, especially at high pH. Test for S^{2-} by placing a drop of sample on lead acetate test paper previously moistened with acetic acid buffer solution, pH 4 (Section 408B.3*e*). Darkening of the paper indicates presence of S^{2-}. Add powdered cadmium nitrate, $Cd(NO_3)_2 \cdot 4H_2O$, to precipitate yellow cadmium sulfide (CdS). If S^{2-} content is high, add instead powdered cadmium carbonate ($CdCO_3$), lead carbonate ($PbCO_3$), lead acetate, or bismuth citrate. Repeat this operation until a drop of treated sample no longer darkens the acidified lead acetate test paper. Filter sample before raising pH for stabilization. When particulate, metal cyanide complexes are suspected in the sample, filter solution before S^{2-} is removed. Reconstitute sample by returning filtered particulates to the sample bottle after S^{2-} removal. Homogenize particulates before analyses.

Because most cyanides are very reactive and unstable, analyze samples as soon as possible. If sample cannot be analyzed immediately, add NaOH pellets or a strong NaOH solution to raise sample pH to 12 to 12.5 and store in a closed, dark bottle in a cool place.

To analyze for CNCl collect a separate sample and omit NaOH addition because CNCl is converted rapidly to CNO^- at high pH. Make colorimetric estimation immediately after sampling.

3. Interferences

a. Oxidizing agents may destroy most of the cyanide during storage and manipulation. Add $Na_2S_2O_3$ as directed in Section 412A.2 above.

b. Sulfide will distill over with cyanide and, therefore, adversely affect colorimetric, titrimetric, and electrode procedures. Test for and remove S^{2-} as directed above. Treat 25 mL more than required for the distillation to provide sufficient filtrate volume.

c. Fatty acids that distill and form soaps under alkaline titration conditions make the end point almost impossible to detect. Remove fatty acids by extraction.[24] Acidify sample with acetic acid (1 + 9) to pH 6.0 to 7.0 (CAUTION—*Perform this operation in a hood as quickly as possible*). Immediately extract with iso-octane, hexane, or $CHCl_3$ (preference in order named). Use a solvent volume equal to 20% of sample volume. One extraction usually is adequate to reduce the fatty acid concentration below the interference level. Avoid multiple extractions or a long contact time at low pH to minimize loss of HCN. When extraction is completed, immediately raise pH to >12 with NaOH solution.

d. Carbonate in high concentration may affect distillation by causing excessive gasing when acid is added. The carbon dioxide (CO_2) released also may significantly reduce the NaOH content in the absorber.

When sampling effluents such as coal

gasification wastes, atmospheric emission scrub waters, and other high-carbonate wastes, use hydrated lime to stabilize the sample; slowly add with stirring to raise pH to 12 to 12.5. Decant sample into sample bottle after precipitate has settled.

e. Other possible interferences include substances that might contribute color or turbidity. In most cases, distillation will remove these.

f. Aldehydes convert cyanide to cyanohydrin, which forms nitrile under the distillation conditions. Only direct titration without distillation can be used, which reveals only non-complex cyanides. Formaldehyde interference is noticeable in concentrations exceeding 0.5 mg/L; eliminate by adding silver nitrate ($AgNO_3$) to the sample. Use the following spot test to establish absence or presence of aldehydes (detection limit 0.05 mg/L):[25–27]

1) Reagents—

a) *MBTH indicator solution*—Dissolve 0.05 g 3-methyl, 2-benzothiazolone hydrazone hydrochloride in 100 mL water. Filter if turbid.

b) *Ferric chloride oxidizing solution*—Dissolve 1.6 g sulfamic acid and 1 g $FeCl_3 \cdot 6H_2O$ in 100 mL water.

c) *Silver nitrate solution*, 0.1*N*—Dissolve 17.0 g $AgNO_3$ crystals in water and dilute to 1 L.

d) *EDTA solution*, 0.1*M*—Dissolve 37.2 g disodium ethylenediaminetetraacetate in water and dilute to 1 L.

2) Procedure—Place 1 drop of sample and 1 drop distilled water for a blank in separate cavities of a white spot plate. Add 1 drop MBTH solution and then 1 drop $FeCl_3$ oxidizing solution to each spot. Allow 10 min for color development. The color change will be from a faint green to a deeper color, tending to blue-green at higher concentrations. Add 0.1*N* $AgNO_3$ solution dropwise and retest on the spot plate. For each drop of $AgNO_3$, add 2 drops EDTA solution. A formaldehyde concentration of 1 mg/L in a 100-mL sample will require approximately 2 drops $AgNO_3$ solution and 4 drops EDTA solution.

g. Glucose and other sugars, especially at the pH of preservation, lead to cyanohydrin formation by reaction of cyanide with aldose.[28] Cyanohydrin can be reduced to cyanide with $AgNO_3$ but the MBTH is not applicable.

412 B. Total Cyanide after Distillation

1. General Discussion

Hydrogen cyanide (HCN) is liberated from an acidified sample by distillation and purging with air. The HCN gas is collected by passing it through an NaOH scrubbing solution. Cyanide concentration in the scrubbing solution is determined by either titrimetric, colorimetric, or potentiometric procedures.

2. Apparatus

The apparatus is shown in Figure 412:1. It includes:

a. Boiling flask, 1 L, with inlet tube and provision for water-cooled condenser.

b. Gas absorber, with gas dispersion tube equipped with medium-porosity fritted outlet.

c. Heating element, adjustable.

d. Ground glass ST joints, TFE-sleeved or with an appropriate lubricant for the boiling flask and condenser. Rubber stopper joints also may be used.

3. Reagents

a. Sodium hydroxide solution: Dissolve

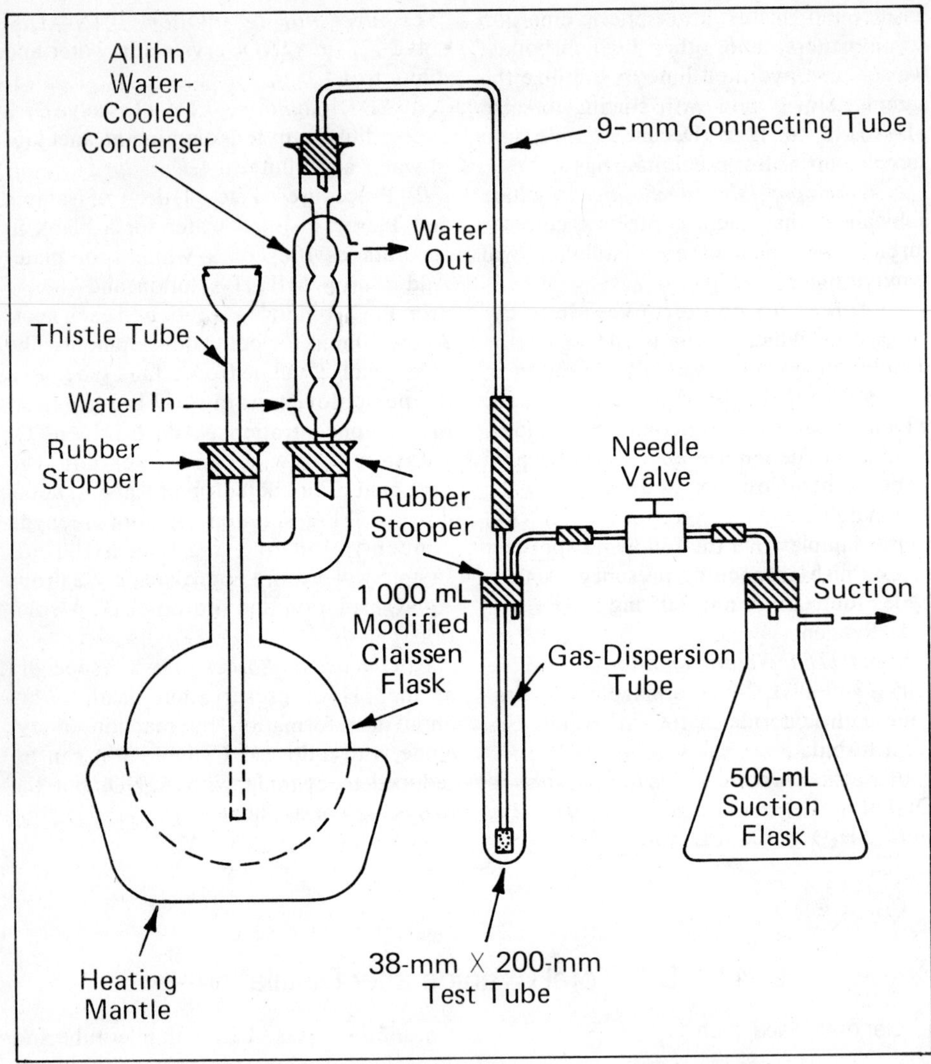

Figure 412:1. Cyanide distillation apparatus.

10 g NaOH in water and dilute to 1 L.

b. Magnesium chloride reagent: Dissolve 510 g $MgCl_2 \cdot 6H_2O$ in water and dilute to 1 L.

c. Sulfuric acid, H_2SO_4, 1 + 1.

4. Procedure

a. Add 500 mL sample, containing not more than 100 mg CN/L (diluted if necessary with distilled water) to the boiling flask. Add 50 mL NaOH solution to the gas washer and dilute, if necessary, with distilled water to obtain an adequate depth of liquid in the absorber. Connect the train, consisting of boiling flask air inlet, flask, condenser, gas washer, suction flask trap, and aspirator. Adjust suction so that approximately 1 air bubble/sec enters the boiling flask. This air rate will carry HCN gas from flask to absorber and usually will prevent a reverse flow of HCN through the air inlet. If this air rate does not prevent sample backup in the delivery tube, increase air-flow rate to 2 air bubbles/sec. Observe air purge rate in the absorber where the liquid level should be raised not more than 6.5 to 10 mm. Maintain air flow throughout the reaction.

b. Add 50 mL 1 + 1 H_2SO_4 through the air inlet tube. Rinse tube with distilled water and let air mix flask contents for 3 min. Add 20 mL $MgCl_2$ reagent through air inlet and wash down with stream of water. A precipitate that may form redissolves on heating.

c. Heat with rapid boiling, but do not flood condenser inlet or permit vapors to rise more than halfway into condenser. Adequate refluxing is indicated by a reflux rate of 40 to 50 drops/min from the condenser lip. Reflux for at least 1 hr. Discontinue heating but continue air flow. Cool for 15 min and drain gas washer contents into a separate container. Rinse connecting tube between condenser and gas washer with distilled water, add rinse water to drained liquid, and dilute to 250 mL in a volumetric flask.

d. Determine cyanide content by titration method (C) if cyanide concentration exceeds 1 mg/L or by the colorimetric method (D) if the cyanide concentration is less. Use titration, the electrode probe method, or the spot test to approximate CN content. Alternatively, use the cyanide-selective electrode in the concentration range 0.05 to 10 mg CN/L (Method E).

e. Distillation gives quantitative recovery of even refractory cyanides such as iron complexes. To obtain complete recovery of cobalticyanide use ultraviolet radiation pretreatment.[29,30] If incomplete recovery is suspected, distill again by refilling the gas washer with a fresh charge of NaOH solution and refluxing 1 hr more. The cyanide from the second reflux, if any, will indicate completeness of recovery.

f. As a quality control measure, periodically test apparatus, reagents, and other potential variables in the concentration range of interest. As an example a minimum 98% recovery from 1 mg CN/L standard should be obtained.

412 C. Titrimetric Method

1. General Discussion

a. Principle: CN^- in the alkaline distillate from the preliminary treatment procedure is titrated with standard silver nitrate ($AgNO_3$) to form the soluble cyanide complex, $Ag(CN)_2^-$. As soon as all CN^- has been complexed and a small excess of Ag^+ has been added, the excess Ag^+ is detected by the silver-sensitive indicator, p-dimethylaminobenzalrhodanine, which immediately turns from a yellow to a salmon color.[31] The distillation has provided a 2:1 concentration. The indicator is sensitive to about 0.1 mg Ag/L. If titration shows that CN^- is below 1 mg/L, examine another portion colorimetrically.

2. Apparatus

Koch microburet, 5-mL capacity.

3. Reagents

a. Sodium hydroxide solution, NaOH, 1*N*.

b. Sulfuric acid solution, H_2SO_4, 1*N*.

c. Indicator solution: Dissolve 20 mg p-dimethylaminobenzalrhodanine in 100 mL acetone.

d. Standard silver nitrate titrant, 0.0192*N*: Dissolve 3.27 g $AgNO_3$ in 1 L distilled water. Standardize against standard NaCl solution, using the argentometric method with K_2CrO_4 indicator, as directed in Chloride, Section 407A.

Dilute 500 mL $AgNO_3$ solution according to the titer found so that 1.00 mL is equivalent to 1.00 mg CN.

4. Procedure

a. From the absorption solution take a measured volume of sample so that the titration will require approximately 1 to 10 mL $AgNO_3$ titrant. Dilute to 250 mL or some other convenient volume to be used for all titrations. For samples with low cyanide concentration (<5 mg/L) do not dilute. Add 0.5 mL indicator solution.

b. Titrate with standard $AgNO_3$ titrant to the first change in color from a canary yellow to a salmon hue. Titrate a blank containing the same amount of alkali and water. As the analyst becomes accustomed to the end point, blank titrations decrease from the high values usually experienced in the first few trials to 1 drop or less, with a corresponding improvement in precision.

5. Calculation

$$\text{mg CN/L} = \frac{(A - B) \times 1000}{\text{mL original sample}} \times \frac{250}{\text{mL portion used}}$$

where:

A = mL standard $AgNO_3$ for sample and
B = mL standard $AgNO_3$ for blank.

6. Precision and Accuracy

For samples containing more than 1 mg CN/L that have been distilled or for relatively clear samples without significant interference, the coefficient of variation is 2%. Extraction and removal of S^{2-} or oxidizing agents tend to increase the coefficient of variation to a degree determined by the amount of manipulation and the type of sample. The limit of sensitivity is approximately 0.1 mg CN/L, but at this concentration the end point is indistinct. At 0.4 mg/L the coefficient of variation is four times that at CN concentration levels>1.0 mg/L.

412 D. Colorimetric Method

1. General Discussion

a. Principle: CN^- in the alkaline distillate from preliminary treatment is converted to CNCl by reaction with chloramine-T at pH <8 without hydrolyzing to CNO^-.[32] (CAUTION: *CNCl is a toxic gas; avoid inhalation.*) After the reaction is complete, CNCl forms a red-blue dye on addition of a pyridine-barbituric acid reagent. If the dye is kept in an aqueous solution, the absorbance is read at 578 nm. To obtain colors of comparable intensity, have the same salt content in sample and standards.

b. Interference: All known interferences are eliminated or reduced to a minimum by distillation.

2. Apparatus

Colorimetric equipment: One of the following is required:

a. Spectrophotometer, for use at 578

nm, providing a light path of 10 mm or longer.

b. Filter photometer, providing a light path of at least 10 mm and equipped with a red filter having maximum transmittance at 570 to 580 nm.

3. Reagents

a. Chloramine-T solution: Dissolve 1.0 g white, water-soluble powder in 100 mL water. Prepare weekly and store in refrigerator.

b. Stock cyanide solution: Dissolve approximately 2 g KOH and 2.51 g KCN in 1 L distilled water. (CAUTION—*KCN is highly toxic; avoid contact or inhalation.)* Standardize against standard silver nitrate ($AgNO_3$) titrant as described in Section 412C.4, using 25 mL KCN solution. Check titer weekly because the solution gradually loses strength; 1 mL = 1 mg CN.

c. Standard cyanide solution: Based on the concentration determined for the KCN stock solution (¶ 3*b*) calculate volume required (approximately 10 mL) to prepare 1 L of a 10 μg CN/mL solution. Dilute with the 2 g NaOH/L solution (¶ 3*f*). Dilute 10 mL of the 10 μg CN/mL solution to 100 mL with the 2 g NaOH/L solution; 1.0 mL = 1.0 μg CN. Prepare fresh daily and keep in a glass-stoppered bottle. (CAUTION: *Toxic–take care to avoid ingestion.)*

d. Pyridine-barbituric acid reagent: Place 15 g barbituric acid in a 250-mL volumetric flask and add just enough water to wash sides of flask and wet barbituric acid. Add 75 mL pyridine and mix. Add 15 mL conc hydrochloric acid (HCl), mix, and cool to room temperature. Dilute to mark with water and mix. This reagent is stable for up to 1 month; discard if a precipitate develops.

e. Sodium dihydrogen phosphate. 1M: Dissolve 138 g $NaH_2PO_4 \cdot H_2O$ in 1 L distilled water. Refrigerate.

f. Sodium hydroxide solution: Dissolve 2 g NaOH in 1 L distilled water.

4. Procedure

a. Preparation of calibration curve: Prepare a blank of NaOH solution. From the standard KCN solution prepare a series of standards containing from 0.2 to 6 μg CN in 20 mL solution using the 2 g NaOH/L solution for all dilutions. Treat standards in accordance with ¶ *b* below. Plot absorbance of standards against CN concentration (micrograms).

Recheck calibration curve periodically and each time a new reagent is prepared.

On the basis of the first calibration curve, prepare additional standards containing less than 0.2 and more than 6 μg CN to determine the limits measurable with the photometer being used.

b. Color development: Adjust photometer to zero absorbance each time using a blank consisting of the NaOH dilution solution and all reagents. Take a portion of absorption liquid obtained in Method B, such that the CN concentration falls in the measurable range, and dilute to 20 mL with NaOH solution. Place the portion in a 50-mL volumetric flask. Add 4 mL phosphate buffer and mix thoroughly. Add 2.0 mL chloramine-T solution and swirl to mix. *Immediately* add 5 mL pyridine-barbituric acid solution and swirl gently to mix. Dilute to mark with water; mix well by inversion.

Measure absorbance with the photometer at 578 nm after 8 min but within 15 min from the time of adding the pyridine-barbituric acid reagent. Even with the specified time of 8 to 15 min there is a slight change in absorbance. To minimize this, standardize time for all readings. Using the calibration curve and the formula in ¶ 5 below, determine CN concentration in original sample.

5. Calculations

$$\text{CN, mg/L} = \frac{A \times B}{C \times D}$$

where:

A = μg CN read from calibration curve

(50 mL final volume),
B = total volume of absorbing solution from the distillation, mL,
C = volume of original sample used in the distillation, mL, and
D = volume of absorbing solution used in colorimetric test, mL.

6. Precision

The analysis of a mixed cyanide solution containing sodium, zinc, copper, and silver cyanides in tap water gave a precision within the designated range as follows:

$$S_T = 0.115X + 0.031$$

where:
S_T = overall precision and
X = CN concentration, mg/L.

412 E. Cyanide-Selective Electrode Method

1. General Discussion

CN^- in the alkaline distillate from the preliminary treatment procedures can be determined potentiometrically by the known addition technic using a CN^--selective electrode in combination with a double-junction reference electrode and a pH meter having an expanded millivolt scale, or a specific ion meter. This method can be used to determine CN concentration in place of either the colorimetric or titrimetric procedures in the concentration range of 0.05 to 10 mg CN/L.[33–35] If the CN^--selective electrode method is used, the previously described titration screening step can be omitted.

2. Apparatus

a. Expanded-scale pH meter or specific-ion meter.

*b. Cyanide-ion-selective electrode.**

c. Reference electrode, double-junction.

d. Magnetic mixer with TFE-coated stirring bar.

3. Reagents

a. Stock standard cyanide solution: See Section 412D.3*b*.

b. Sodium hydroxide diluent: Dissolve 2 g NaOH in water and dilute to 1 L.

c. Intermediate standard cyanide solution: Dilute a calculated volume (approximately 100 mL) of stock KCN solution, based on the determined concentration, to 1,000 mL with NaOH diluent. Mix thoroughly; 1 mL= 100 μg CN.

d. Dilute standard cyanide solution: Dilute 100.0 mL intermediate standard CN solution to 1,000 mL with NaOH diluent; 1.00 mL =10.0 μg CN. Prepare daily and keep in a dark, glass-stoppered bottle.

e. Potassium nitrate solution: Dissolve 100 g KNO_3 in water and dilute to 1 L. Adjust to pH 12 with KOH. This is the outer filling solution for the double-junction reference electrode.

4. Procedure

a. Calibration: Use the dilute and intermediate standard CN solutions and the NaOH diluent to prepare a series of three standards, 0.1, 1.0, and 10.0 mg CN/L. Transfer approximately 100 mL of each of these standard solutions into a 250-mL beaker prerinsed with a small portion of standard being tested. Immerse CN^- and double-junction reference electrodes in the solution. Mix well on a magnetic stirrer at 25 C and maintain as closely as pos-

*Orion Model 94-06A or equivalent.

sible the same stirring rate for all solutions.

Always progress from the lowest to the highest concentration of standard because otherwise equilibrium is reached only slowly. The electrode membrane dissolves in solutions of high CN concentration; do not use with a concentration above 10 mg/L. After making measurements remove electrode and soak in water.

After equilibrium is reached (at least 5 min and not more than 10 min), record potential (millivolt) readings and plot CN concentrations versus readings on semilogarithmic graph paper. A straight line with a slope of approximately 59 mV per decade indicates that the instrument and electrodes are operating properly. Record slope of line obtained (millivolts/decade of concentration). The slope may vary somewhat from the theoretical value of 59.2 mV/decade because of manufacturing variation and reference electrode (liquid-junction) potentials. The slope is necessary for calculating sample concentration. Obtain tables of concentration ratios (used in the calculation) for several different electrode slopes from electrode manufacturers, or correct for slope variation at the time of calculation, using an electronic calculator with antilog capability.

b. Measurement of sample: Place 100 mL of absorption liquid obtained in Section 412B.4*c* into a 250-mL beaker. When measuring low CN concentrations, first rinse beaker and electrodes with a small volume of sample. Immerse CN^- and double-junction reference electrodes and mix on a magnetic stirrer at the same stirring rate used for calibration. After equilibrium is reached (at least 5 min and not more than 10 min), record the potential reading (A).

Choose one of the CN^- standards (3*a*, 3*c*, or 3*d* above) that is 70 to 700 times as concentrated as the expected sample concentration. Obtain the approximate concentration by comparing the potential reading with the calibration curve drawn as directed in 4*a*.

Sample CN Concentration mg/L	Standard Used
0.05–0.1	10 mg/L
0.1–1.0	100 mg/L
1.0–10.0	stock standard

Pipet 1 mL of chosen CN^- standard solution into sample. Allow sufficient time for equilibration (at least 5 min and not more than 10 min) and record the new potential reading (B).

Periodically test electrode slope and instrument performance by repeating calibration procedure.

5. Calculations

$$CN^-,\ mg/L = \frac{C_o \times B}{D}$$

and

$$C_o = \frac{0.01}{1.01\ \text{antilog}\,[(A - B)/S] - 1}\,C_s$$

where:

C_o = CN concentration in sample used for potentiometric measurement, mg/L,
B = volume of absorbing solution, mL,
D = volume of sample distilled, mL,
$(A - B)$ = absolute value of potential change (positive number), mV,
C_s = CN concentration of standard, mg/L, and
S = electrode slope, mV/decade (without regard to sign).

Tables of the ratio C_o/C_s (for various values of S) are available from electrode manufacturers.

6. Precision

The precision of the CN^- ion selective electrode method was determined from four levels of CN concentration by five laboratories and seven analysts. The precision of the method within its designated range may be expressed as follows:

$$S_T = 0.113X + 0.024$$

where:

S_T = overall precision and
X = CN concentration, mg/L.

412 F. Cyanides Amenable to Chlorination after Distillation

1. General Discussion

This method is applicable to the determination of cyanides amenable to chlorination, to determine the dissociable CN content of the sample.

After part of the sample is chlorinated to decompose the cyanides, both the chlorinated and the untreated sample are subjected to distillation as described in Section 412B. The difference between the CN concentrations found in the two samples is expressed as cyanides amenable to chlorination.

Use the titration procedure when it is known that the concentration of cyanides not amenable to chlorination is more than 1 but less than 10 mg/L. With higher concentrations, use a smaller portion as described in 412B.4*a*. Use a colorimetric determination when the cyanides not amenable to chlorination are known to be 1 mg/L or less. The selective-ion electrode method is useful in the concentration range of 0.05 to 10 mg CN/L. For estimation of the concentration of cyanides amenable to chlorination use the spot test procedure (412I).

Some unidentified organic chemicals may oxidize or form breakdown products during chlorination, giving higher results for cyanide after chlorination than before chlorination. This may lead to a negative value for cyanides amenable to chlorination after distillation for wastes from, for example, the steel industry, petroleum refining, and pulp and paper processing. Where such interferences are known or suspected, use other methods for determining dissociable cyanide. When this substitution is made, other procedures, such as 412G, may not give comparable results.

2. Apparatus

a. Distillation apparatus: See Section 412B.2.

b. Apparatus for determining cyanide by either the titrimetric method, Section 412C.2, the colorimetric method, Section 412D.2, or the electrode method, Section 412E.2.

3. Reagents

a. All reagents listed in Section 412B.3.

b. All reagents listed in Section 412C.3, 412D.3 or 412E.3, depending on method of estimation.

c. Calcium hypochlorite solution: Dissolve 5 g $Ca(OCl)_2$ in 100 mL distilled water. Store in an amber-colored glass bottle in the dark. Prepare monthly.

d. Potassium iodide(KI)-starch test paper.

4. Procedure

a. Divide sample into two equal parts and chlorinate one as in ¶ *b*. Analyze both portions for CN. The difference in determined concentrations is the cyanide amenable to chlorination.

b. Add $Ca(OCl)_2$ solution dropwise to sample while agitating and maintaining pH between 11 and 12 by adding NaOH solution. Test for chlorine by placing a drop of treated sample on a strip of KI-starch paper. A distinct blue color indicates sufficient chlorine (approximately 50 to 100 mg Cl_2/L. Maintain excess residual chlorine for 1 hr while agitating. If necessary, add more $Ca(OCl)_2$.

c. Add approximately 500 mg/L sodium thiosulfate ($Na_2S_2O_3$) as crystals to reduce residual chlorine. Test with KI-starch paper; there should be no color change. Add approximately 0.1 g/L more $Na_2S_2O_3$ to ensure a slight excess.

d. Minimize sample exposure to ultraviolet radiation before distillation.

e. Distill both chlorinated and unchlorinated samples as in Section 412B. Test according to Methods C, D, or E.

5. Calculation

$$\text{mg CN amenable to chlorination/L} = G - H$$

where:

G = mg CN/L found in unchlorinated portion of sample and
H = mg CN/L found in chlorinated portion of sample.

For samples containing significant quantities of iron cyanides, it is possible that the second distillation will give a higher value for CN than the test for total cyanide, leading to a negative result. In this case and when a negative result could be due to interferences from unknown organic compounds (See Section 412F.1) report "no detectable quantities of cyanide amenable to chlorination."

6. Precision

The precision, with the titrimetric finish, for cyanides amenable to chlorination was determined from a mixed cyanide solution containing sodium, zinc, copper, and silver cyanides and sodium ferrocyanide. The precision of the method within its designated range may be expressed as follows:

$$S_T = 0.049X + 0.162$$

where:

S_T = overall precision and
X = CN concentration, mg/L.

412 G. Thiocyanates and Cyanides Amenable to Chlorination without Distillation (Short-Cut Method)

1. General Discussion

This method covers the determination of HCN and of CN complexes that are amenable to chlorination and also thiocyanates (CNS^-). The procedure does not measure cyanates (CNO^-) or iron cyanide complexes, but does determine cyanogen chloride (CNCl). This test requires neither lengthy distillation nor the chlorination of one sample before distillation. The recovery of CN from metal cyanide complexes will be comparable to that in Method F.

The cyanides are converted to CNCl by chloramine-T after the sample has been heated. In the absence of nickel, copper, silver, and gold cyanide complexes or CNS^-, the CNCl may be developed at room temperature. The pyridine-barbituric acid reagent produces a red-blue color in the sample. The color can be estimated visually against standards or photometrically at 578 nm. The limits of the determination are 0.2 μg to 6 μg CN, representing 0.01 to 0.30 mg/L in a 20-mL sample. Higher CN concentrations may be determined by dilution. The dissolved salt content in the standards used for the development of the calibration curve should be near the salt content of the sample, including the added NaOH and phosphate buffer. See 412D.1*a*.

The test sensitivity may be extended to the 5- to 150-μg/L level if a fresh, unstabilized sample is used. In these circumstances (pH <9), add phosphate buffer dropwise to a pH of 6.5 (pH 6.0 to 6.6) and use a 40-mL sample, minimizing dilution before color development. Add 1 g sodium chloride (NaCl) to the 40-mL sample to make up for the salt content that would have been added if 2 g sodium hydroxide (NaOH)/L and the required amount of phosphate buffer had been added.

2. Interferences

Remove interfering agents as described in Section 412A. The CNS^- ion, which also reacts with chloramine-T, will give a positive error equivalent to its concentration. When CNS^- is present and its determination is not required, or other color- or turbidity-producing interferences are encountered, use Method F.

3. Apparatus

a. Apparatus listed in 412D.2.

b. Hot water bath.

4. Reagents

a. Reagents listed in Sections 412A *and* D.3.

b. Sodium chloride, NaCl, crystals.

c. Sodium carbonate, Na_2CO_3, crystals.

d. Sulfuric acid solution, H_2SO_4, 1*N*.

5. Procedure

a. Calibrate as directed in Section 412D.1*a* and 4*a*. Adjust absorbance to zero, using the 2 g NaOH/L solution for preparation of the blank. For samples with more than 3,000 mg total filtrable residue/L, prepare a calibration curve from standards and blank NaOH solutions containing 6 g NaCl/L. Samples containing total filtrable residue exceeding 10,000 mg/L require appropriate standards and a new calibration curve.

b. To 20 mL sample add 0.2 to 0.4 g Na_2CO_3 and stir to dissolve. Add slowly, with stirring, 1*N* H_2SO_4 solution to adjust pH to 11.4 to 11.6. Transfer to a 50-mL volumetric flask. If more than 300 mg CN/L is known to be present, use a smaller sample diluted to 20 mL with 2 g NaOH/L dilution solution.

c. Add 4 mL phosphate buffer and swirl to mix. Add 1 drop 0.1*M* EDTA solution. Heat in a water bath kept between 48 and 51 C for 1 min while swirling.

d. While sample is still hot, add 2 mL chloramine-T solution and swirl to mix. Place a drop of solution on a strip of acidified starch-iodide test paper. The test paper should show the presence of chlorine. If reducing agents in the sample consume all the chloramine-T, add more and recheck. After 1 min, add 5 mL pyridine-barbituric acid and swirl in water bath for 1 min.

e. Remove from water bath, dilute to 50.0 mL, and allow 7 min more for color development. Cool to room temperature, if necessary, and read the absorbance at 578 nm within a total of 15 min from the time the pyridine-barbituric acid solution was added.

f. Standardize instrument with an appropriate blank each time it is used. Recheck calibration curve periodically using prepared standards and each time a new reagent is prepared.

6. Calculation

$$\text{Cyanide amenable to chlorination plus thiocyanate, mg CN/L} = A/B$$

where:

A = μg CN read from calibration curve (50 mL final volume) and

B = mL sample used.

7. Precision

The analysis of a mixed cyanide solution in tap water, containing sodium, zinc, copper, and silver cyanides, gave a precision within the designated range as follows:

$$S_T = 0.097X + 0.004$$

where:

S = overall precision and

X = CN concentration, mg/L.

412 H. Cyanogen Chloride

1. General Discussion

Cyanogen chloride (CNCl) is the first reaction product when cyanide compounds are chlorinated. It is a volatile gas, only slightly soluble in water, but highly toxic even in low concentrations (CAUTION: *Avoid inhalation or contact.*) A mixed pyridine-barbituric acid reagent produces a red-blue color with CNCl.

Because CNCl hydrolyzes to cyanate (CNO^-) at a pH of 12 or more, collect a separate sample for CNCl analysis (See Section 412A.2) in a closed container without sodium hydroxide (NaOH). A quick test with a spot plate or comparator as soon as the sample is collected may be the only procedure for avoiding hydrolysis of CNCl due to the lapse of time between sampling and analysis.

If starch-iodide (KI) test paper indicates presence of chlorine or other oxidizing agents, add sodium thiosulfate ($Na_2S_2O_3$) immediately as directed in Section 412A.2.

2. Apparatus

See Section 412D.2.

3. Reagents

See Sections 412D.3 and 412G.4.

4. Procedure

Calibrate as directed in Sections 412D.4 and 412G.1, third paragraph. Add a portion of unstabilized sample, diluted if necessary, to contain 0.2 to 6 μg of CN/40 mL to a beaker. Add phosphate buffer to a pH of 6.5 (pH 6.0 to 6.6). Record exact volume of phosphate buffer required. Prepare a second sample as before and add to a 50-mL volumetric flask. Add phosphate buffer in the previously established volume. Add 5 mL pyridine-barbituric acid solution. Dilute to mark with distilled water and mix well by inversion. Allow 8 min for color development. Measure absorbance at 578 nm within 8 to 15 min from the addition of the pyridine-barbituric acid reagent. Using the calibration curve, determine the CNCl as CN.

5. Calculation

$$\text{mg CNCl (as CN)/L} = \frac{A}{B}$$

where:

A = μg CN read from calibration curve (50 mL final volume) and
B = mL original unstabilized sample.

6. Precision

The instability of CNCl precludes round-robin testing procedures and a precision statement is not possible.

412 I. Spot Test for Sample Screening

1. General Discussion

The spot test procedure allows a quick screening of the sample to establish if more than 50 μg/L of cyanide amenable to chlorination is present. The test also establishes the presence or absence of cyanogen chloride (CNCl). With practice and dilution, the test reveals the approximate concentration range of these compounds by the color development compared with similarly treated standards.

When chloramine-T is added to cyanides amenable to chlorination, CNCl is formed. CNCl forms a red-blue color with the mixed reagent pyridine-barbituric acid. When testing for CNCl omit the chloramine-T addition. (CAUTION: *CNCl is a toxic gas; avoid inhalation.)*

The presence of formaldehyde in excess of 0.5 mg/L interferes with the test. A spot test for the presence of aldehydes and a method for removal of this interference are given in Section 412A.3.

Thiocyanate (CNS^-) reacts with chloramine-T, thereby creating a positive interference. The CN^- can be masked with formaldehyde and the sample retested. This makes the spot test specific for CNS^-. In this manner it can be established if the spot discoloration is due to the presence of CN^-, CNS^-, or both.

2. Apparatus

a. Porcelain spot plate with 6 to 12 cavities.

b. Dropping pipets.

c. Glass stirring rods.

3. Reagents

a. Chloramine-T solution: See Section 412D.3*a*.

b. Stock cyanide solution: See Section 412D.3*b*.

c. Pyridine-barbituric acid reagent: See Section 412D.3*d*.

d. Hydrochloric acid, HCl, 1 + 9.

e. Phenolphthalein indicator aqueous solution.

f. Sodium carbonate, Na_2CO_3, anhydrous.

g. Formaldehyde, 37%, pharmaceutical grade.

4. Procedure

If the solution to be tested has a pH value greater than 10, neutralize a 20- to 25-mL portion. Add about 250 mg Na_2CO_3 and swirl to dissolve. Add 1 drop phenolphthalein indicator. Add 1 + 9 HCl dropwise with constant swirling until the solution becomes colorless. Place 3 drops sample and 3 drops distilled water (for blanks) in separate cavities of the spot plate. To each cavity, add 1 drop chloramine-T solution and mix with a clean stirring rod. Add 1 drop pyridine-barbituric acid solution to each cavity and again mix. After 1 min, the sample spot will turn pink to red if 50 μg/L or more of CN are present. The blank spot will be faint yellow because of the color of the reagents. Until familiarity with the spot test is gained, use, in place of the water blank, a standard solution containing 50 μg CN/L for color comparison. This standard can be made by diluting the stock cyanide solution (¶3*b*).

If CNS^- is suspected, test a second sample pretreated as follows: Heat a 20- to 25-mL sample in a water bath at 50 C; add 0.6 mL formaldehyde and hold for 10 min. This treatment will mask up to 5 mg CN/L, if present. Repeat spot testing procedure. Color development indicates presence of CNS^-. Comparing color intensity in the two spot tests is useful in judging relative concentration of CN^- and CNS^-.

412 J. Cyanates

1. General Discussion

Cyanate (CNO^-) may be of interest in analysis of industrial waste samples because the alkaline chlorination process used for the oxidation of cyanide yields cyanate in the second reaction.

Cyanate is unstable at neutral or low pH; therefore, stabilize the sample as soon as collected by adding sodium hydroxide

(NaOH) to pH >12. Remove residual chlorine by adding sodium thiosulfate ($Na_2S_2O_3$) (see Section 412A.2).

a. Principle: Cyanate hydrolyzes to ammonia when heated at low pH.

$$2NaCNO + H_2SO_4 + 4H_2O \rightarrow (NH_4)_2SO_4 + 2NaHCO_3$$

The ammonia concentration must be determined on one sample portion before acidification. The ammonia content before and after hydrolysis of cyanate may be measured by the direct nesslerization, (417B), phenate (417C), or ammonia-selective electrode (417E) method.[36] The test is applicable to cyanate compounds in natural waters and industrial waste.

b. Interferences:

1) Organic nitrogenous compounds may hydrolyze to ammonia (NH_3) upon acidification. To minimize this interference, control acidification and heating closely.

2) Metal compounds may precipitate or form colored complexes with nessler reagent. Adding Rochelle salt or EDTA in the determination of ammonia overcomes these interferences. Metal precipitates do not interfere with the ion-selective electrode method.

3) Reduce oxidants that oxidize cyanate to carbon dioxide and nitrogen with $Na_2S_2O_3$ (see Section 412F).

4) Industrial waste containing organic material may contain unknown interferences.

c. Detection limit: 1 to 2 mg CNO^-/L.

2. Apparatus

a. Expanded-scale pH meter or selective-ion meter.

*b. Ammonia-selective electrode.**

c. Magnetic mixer, with TFE-coated stirring bar.

d. Heat barrier: Use a 3-mm-thick asbestos board under beaker to insulate against heat produced by stirrer motor.

3. Reagents

a. Stock ammonium chloride solution: See Section 417B.3*d*.

b. Standard ammonium chloride solution: From the stock NH_4Cl solution prepare standard solutions containing 1.0, 10.0, and 100.0 mg NH_3-N/L by diluting with ammonia-free water.

c. Sodium hydroxide, 10*N*: Dissolve 400 g NaOH in water and dilute to 1 L.

d. Sulfuric acid solution, H_2SO_4, 1 + 1.

e. Ammonium chloride solution: Dissolve 5.4 g NH_4Cl in distilled water and dilute to 1 L. (Use only for soaking electrodes.)

4. Procedure

a. Calibration: Daily, calibrate the ammonia electrode as in 417E.4*b* and *c* using standard NH_4Cl solutions.

b. Treatment of sample: Dilute sample, if necessary, so that the CNO^- concentration is 1 to 200 mg/L or NH_3-N is 0.5 to 100 mg/L. Take or prepare at least 200 mL. From this 200 mL, take a 100-mL portion and, following the calibration procedure, establish the potential (millivolts) developed from the sample. Check electrode reading with prepared standards and adjust instrument calibration setting daily. Record NH_3-N content of untreated sample (*B*).

Acidify 100 mL of prepared sample by adding 0.5 mL 1 + 1 H_2SO_4 to a pH of 2.0 to 2.5. Heat sample to 90 to 95 C and maintain temperature for 30 min. Cool to room temperature and restore to original volume by adding ammonia-free water. Pour into a 150-mL beaker, immerse electrode, start magnetic stirrer, then add 1 mL 10*N* NaOH solution. With pH paper check that pH is greater than 11. If neces-

*Orion Model 95-10, EIL Model 8002-2, Beckman Model 39565, or equivalent.

sary, add more NaOH until pH 11 is reached.

After equilibrium has been reached (30 sec) record the potential reading. Estimate NH_3-N content from calibration curve.

5. Calculations

$$\text{mg } NH_3\text{-N derived from } CNO^-/L = A - B$$

where:

A = mg NH_3-N/L found in the acidified and heated sample portion and

B = mg NH_3-N/L found in untreated portion.

$$\text{mg } CNO^-/L = 3.0 \times (A - B)$$

6. Precision

No data on precision of this method are available. See 417E.6 for precision of ammonia-selective electrode method.

412 K. Thiocyanate

1. General Discussion

When wastewater containing thiocyanate (CNS^-) is chlorinated, highly toxic cyanogen chloride (CNCl) is formed. At an acidic pH, ferric ion (Fe^{3+}) forms an intense red color with CNS^-, which is suitable for colorimetric determination.

a. Interference:

1) Hexavalent chromium (Cr^{6+}) interferes and is removed by adding ferrous sulfate ($FeSO_4$) after adjusting to pH 1 to 2, with nitric acid (HNO_3). Raising the pH to 9 with 1*N* sodium hydroxide (NaOH) precipitates Fe(III) and Cr(III), which are than filtered out.

2) Reducing agents that reduce Fe(III) to Fe(II), thus preventing formation of ferric thiocyanate complex, are destroyed by a few drops of hydrogen peroxide (H_2O_2).

3) Industrial wastes may be highly colored or contain various interfering organic compounds. Test method applicability with samples to which CNS^- has been added.

b. Application: 1 to 10 mg CNS^-/L in natural waters. Colored samples may reduce the sensitivity.

2. Apparatus

Spectrophotometer or filter photometer, for use at 480 nm, providing a light path of 1 cm.

3. Reagents

a. Ferric nitrate solution: Dissolve 50 g $Fe(NO_3)_3$ in 500 mL distilled water. Add 25 mL conc HNO_3 and dilute to 1 L.

b. Nitric acid, HNO_3, 1 + 1.

c. Standard thiocyanate solution: Dissolve 1.673 g potassium thiocyanate (KCNS) in distilled water and dilute to 1,000 mL; 1.00 mL = 1.00 mg CNS^-.

4. Procedure

a. Preparation of calibration curve: From the standard KCNS solution, prepare a series of 50-mL standards containing 50 to 500 μg CNS^-. Develop color in accordance with ¶ *b* below. Plot absorbance against CNS^- concentration in μg/50-mL volume on arithmetic paper.

b. Color development: Use a filtered sample containing 50 to 500 μg CNS^-. Adjust to pH 5 to 7 by adding 1 + 1 HNO_3 dropwise. Transfer sample to a 50-mL volumetric flask and add 5 mL $Fe(NO_3)_3$ solution. The pH should be between 1 and 2. Using a glass stirring rod, place a drop of the solution on a narrow-range pH paper. If necessary, adjust pH with one or more drops of 1 + 1 HNO_3. Dilute to volume with distilled water and shake well. Measure sample absorbance at 480 nm using distilled water as a blank.

5. Calculation

$$\text{mg CNS}^-/\text{L} = \frac{\mu\text{g CNS}^-}{\text{mL sample}}$$

6. Precision

No data on the precision of this method are available.

412 L. References

1. MILNE, D. 1950. Equilibria in dilute cyanide waste solutions. *Sewage Ind. Wastes* 23:904.
2. MILNE, D. 1950. Disposal of cyanides by complexation. *Sewage Ind. Wastes* 22:1192.
3. DOUDOROFF, P. 1976. Toxicity to fish of cyanides and related compounds. A review. EPA 600/3-76-038. U.S. Environmental Protection Agency, Duluth, Minn.
4. DOUDOROFF, P. & M. KATZ. 1950. Critical review of literature on the toxicity of industrial wastes and their components to fish. *Sewage Ind. Wastes* 22:1432.
5. DOUDOROFF, P. 1956. Some experiments on the toxicity of complex cyanides to fish. *Sewage Ind. Wastes* 28:1020.
6. DOUDOROFF, P., G. LEDUC & C.R. SCHNEIDER. 1966. Acute toxicity to fish of solutions containing complex metal cyanides, in relation to concentrations of molecular hydrocyanic acid. *Trans. Amer. Fish. Soc.* 95:116.
7. SCHNEIDER, C.R. & H. FREUND. 1962. Determination of low level hydrocyanic acid. *Anal. Chem.* 34:69.
8. CLAEYS, R. & H. FREUND. 1968. Gas chromatographic separation of HCN. *Environ. Sci. Technol.* 2:458.
9. MONTGOMERY H.A.C., D.K. GARDINER & J.G. GREGORY. 1969. Determination of free hydrogen cyanide in river water by a solvent-extraction method. *Analyst* 94:284.
10. NELSON, K.H. & L. LYSYJ. 1971. Analysis of water for molecular hydrogen cyanide. *J. Water Pollut. Control Fed.* 43:799.
11. BURDICK, G.E. & M. LIPSCHUETZ. 1948. Toxicity of ferro and ferricyanide solutions to fish. *Trans. Amer. Fish Soc.* 78:192.
12. ZILLICH, J.A. 1972. Toxicity of combined chlorine residuals to freshwater fish. *J. Water Pollut. Control Fed.* 44:212.
13. RESNICK, J.D., W. MOORE & M.E. ETTINGER. 1958. The behavior of cyanates in polluted waters. *Ind. Eng. Chem.* 50:71.
14. PETTET, A.E.J. & G.C. WARE. 1955. Disposal of cyanide wastes. *Chem. Ind.* 1232.
15. BAILEY, P.L. & E. BISHOP. 1972. Hydrolysis of cyanogen chloride. *Analyst* 97:691.
16. LANCY, L. & W. ZABBAN. 1962. Analytical methods and instrumentation for determining cyanogen compounds. Amer. Soc. Testing & Materials STP No. 337.
17. CALAMARI, D. & R. MARCHETTI. 1975. Predicted and observed acute toxicity of copper and ammonia to rainbow trout. *Progr. Water Technol.* 7, 3-4:569.
18. SERFASS, E.J. & R.B. FREEMAN. 1952. Analytical method for the determination of cyanides in plating wastes and in effluents from treatment processes. *Plating* 39:267.
19. LESCHBER, R. & H. SCHLICHTING. 1969. Uber die Zersezlichkeit Komplexer Metallcyanide bei der Cyanidbestimmung in Abwasser. *Z. Anal. Chem. ZANCA* 245:300.
20. BASSETT, H., JR. & A.S. CORBET. 1924. The hydrolysis of potassium ferricyanide and potassium cobalticyanide by sulfuric acid. *J. Chem. Soc.* 125:1358.
21. DODGE, B.F. & W. ZABBAN. 1952. Analytical methods for the determination of cyanates in plating wastes. *Plating* 39:381.
22. GARDNER, D.C. 1956. The colorimetric determination of cyanates in effuents. *Plating* 43:743.
23. Procedures for Analyzing Metal Finishing Wastes. 1954. Ohio River Valley Sanitation Commission, Cincinnati, Ohio.
24. KRUSE, J.M. & M.G. MELLON. 1951. Colorimetric determination of cyanides. *Sewage Ind. Wastes* 23:1402.
25. SAWICKI, E., T.W. STANLEY, T. R. HAUSER & W. ELBERT. 1961. The 3-methyl-2-benzothiazolone hydrazone test. Sensitive

new methods for the detection, rapid estimation, and determination of aliphatic aldehydes. *Anal. Chem.* 33:93.

26. HAUSER, T.R. & R.L. CUMMINS. 1964. Increasing sensitivity of 3-methyl-2-benzothiazone hydrazone test for analysis of aliphatic aldehydes in air. *Anal. Chem.* 36:679.
27. Methods of Air Sampling and Analysis, 1st ed. 1972. Inter Society Committee, Air Pollution Control Association, pp. 199–204.
28. RAAF, S.F., W.G. CHARACKLIS, M.A. KESSICK & C.H. WARD. 1977. Fate of cyanide and related compounds in aerobic microbial systems. *Water Res.* 11:477.
29. CASAPIERI, P., R. SCOTT & E.A. SIMPSON. 1970. The determination of cyanide ions in waters and effluents by an Auto Analyzer procedure. *Anal. Chim. Acta.* 49:188.
30. GOULDEN, P.D., K.A. BADAR & P. BROOKSBANK. 1972. Determination of nanogram quantities of simple and complex cyanides in water. *Anal. Chem.* 44:1845.
31. RYAN, J.A. & G.W. CULSHAW. 1944. The use of p-dimethylaminobenzylidene rhodanine as an indicator for the volumetric determination of cyanides. *Analyst* 69:370.
32. ASMUS, E. & H. GARSCHAGEN. 1953. Über die Verwendung der Barbitsäure für die photometrische Bestimmung von Cyanid und Rhodanid. *Z. Anal. Chem.* 138:414.
33. ORION RESEARCH, INC. 1975. Cyanide Ion Electrode Instruction Manual.
34. FRANT, M.S., J.W. ROSS & J.H. RISEMAN. 1972. An electrode indicator technique for measuring low levels of cyanide. *Anal. Chem.* 44:2227.
35. SEKERKA, J. & J.F. LECHNER. 1976. Potentiometric determination of low levels of simple and total cyanides. *Water Res.* 10:479.
36. THOMAS, R.F. & R.L. BOOTH. 1973. Selective electrode determination of ammonia in water and wastes. *Environ. Sci. Technol.* 7:523.

413 FLUORIDE

A fluoride concentration of approximately 1.0 mg/L in drinking water effectively reduces dental caries without harmful effects on health. Fluoride may occur naturally in water or it may be added in controlled amounts. Some fluorosis may occur when the fluoride level exceeds the recommended limits. The natural fluoride concentration rarely may approach 10 mg/L. Waters with a fluoride content exceeding health recommendations may require defluoridation.

Accurate determination of fluoride has increased in importance with the spread of fluoridation of water supplies as a public health measure. Maintenance of an optimum fluoride concentration is essential for effectiveness and safety of the fluoridation procedure.

The electrode and colorimetric methods are the most satisfactory. The colorimetric methods are based on the reaction between fluoride and a zirconium-dye lake. Fluoride reacts with the dye lake, dissociating a portion of it into a colorless complex anion (ZrF_6^{2-}) and the dye. As the amount of fluoride increases, the color produced becomes progressively lighter or different in hue, depending on the reagent used.

Because the colorimetric methods are subject to errors due to interfering ions, it may be necessary to distill the sample as directed in Section 413A.5. When interfering ions are not present in excess of the tolerances of the method, the fluoride determination may be made directly without distillation. The analysis is completed by using one of the two colorimetric methods (C and D).

1. Selection of Method

The electrode method usually is preferred because of its convenience. Adding the prescribed buffer frees the electrode method from the interference caused by such relatively common ions as aluminum, hexametaphosphate, iron, and orthophosphate, which adversely affect the colorimetric methods and necessitate preliminary distillation. Distill samples containing fluoroborate ion (BF_4) to convert the fluoroborate to free fluoride before measurement by the electrode or colorimetric method. Although a special electrode selective for fluoroborate is available commercially for estimating fluoroborate ion, preferably distill before electrode measurement of the released fluoride. Both colorimetric methods are applicable directly to samples in the fluoride range 0.05 to 1.4 mg/L, while the electrode method can be applied to fluoride concentrations from 0.1 mg/L to 5 mg/L or more. The SPADNS and electrode methods allow measurements to be made at any time after reagent addition. Although the alizarin visual method does not require accurate time control because sample and standards are treated simultaneously under the same conditions, preferably wait 1 hr after reagent addition for color development. The visual colorimetric method requires inexpensive laboratory glassware while Methods B and C are instrumental.

Permanent colored standards, commercially or otherwise prepared, may be used if appropriate precautions are taken. These include strict adherence to the manufacturer's directions and careful calibration of the permanent standards against standards prepared by the analyst. (See General Introduction, Section 102.7.)

Fluoride also may be determined by the automated complexone method, Method E.

2. Interference in Colorimetric Methods

In general, the different colorimetric methods are susceptible to the same interfering substances, but to different degrees. Table 413:I lists common interferences. Because these are neither linear in effect nor algebraically additive, mathematical compensation is unsatisfactory. Whenever any one substance is present in sufficient quantity to produce an error of 0.1 mg/L or whenever the total interfering effect is in doubt, distill the sample. (Also

TABLE 413:I. CONCENTRATION OF INTERFERING SUBSTANCES CAUSING 0.1-MG/L ERROR AT 1.0 MG F/L IN COLORIMETRIC METHODS

Substance	Method C (SPADNS)		Method D (Alizarin Visual)	
	Conc *mg/L*	Type of Error	Conc *mg/L*	Type of Error
Alkalinity ($CaCO_3$)	5.000	−	400	−
Aluminum (Al^{3+})	0.1†	−	0.25	−
Chloride (Cl^-)	7.000	+	2.000	−
Iron (Fe^{3+})	10	−	2	+
Hexametaphosphate ($[NaPO_3]_6$)	1.0	+	1.0	+
Phosphate (PO_4^{3-})	16	+	5	+
Sulfate (SO_4^{2-})	200	+	300	+

*Completely remove residual chlorine with arsenite reagent. Remove color and turbidity or compensate for their presence.

†On immediate reading. Tolerance increases with time: after 2 hr, 3.0; after 4 hr, 30.

distill colored or turbid samples.) In some instances, sample dilution or adding appropriate amounts of interfering substances to the standards may be used to eliminate the interference effect. If alkalinity is the only significant interference, neutralize it with either HCl or HNO_3.

Chlorine interferes in all colorimetric methods and provision for its removal is made.

The measurement of sample volume and reagent volume are extremely important. Use samples and standards at the same temperature or at least within 2 C. Maintain constant temperature throughout the color development period. For the SPADNS method, different calibration curves may be prepared for different temperatures.

3. Sampling and Storage

Preferably use polyethylene bottles for collecting and storing samples for fluoride analysis. Glass bottles are satisfactory if they have not previously contained high-fluoride solutions. Always rinse bottle with a portion of sample.

Never use an excess of dechlorinating agent. Sodium thiosulfate in excess of 100 mg/L will interfere by producing a precipitate.

413 A. Preliminary Distillation Step

1. Discussion

Fluoride can be separated from other constituents in water by distillation as fluosilicic (or hydrofluoric) acid from an acid solution with a boiling point higher than that of water. Quantitative fluoride recovery is approached by using a relatively large sample. Sulfate carryover from H_2SO_4 is minimized by distilling over a broad temperature range with strict control of maximum temperature.

2. Apparatus

Distillation apparatus consists of a 1-L round-bottom long-neck borosilicate glass boiling flask, a connecting tube, an efficient condenser, a thermometer adapter, and a thermometer reading to 200 C (see Figure 413:1). Use any comparable apparatus with the essential design features. Figure 323:1 shows the general type of distillation apparatus* satisfactory for fluoride, ammonia, phenol, and selenium distillations. Avoid obstruction in the vapor path and trapping liquid in adapter and condenser to minimize sulfate carryover. Use an asbestos shield or similar device to protect upper part of distilling flask from burner flame. If desired, modify apparatus so that the heat is automatically shut off when distillation is completed.

3. Reagents

a. Sulfuric acid, H_2SO_4, conc.

b. Silver sulfate, Ag_2SO_4, crystals.

4. Procedure

a. Place 400 mL distilled water in the distilling flask and carefully add 200 mL conc H_2SO_4. Swirl until flask contents are homogeneous, or use a magnetic stirrer to insure complete mixing that is essential for safety. Add 25 to 35 glass beads and connect the apparatus as shown in Figure 413:1, making sure all joints are tight. Begin heating slowly at first, then as rapidly as condenser efficiency will permit (dis-

*Corning No. 3360 or equivalent.

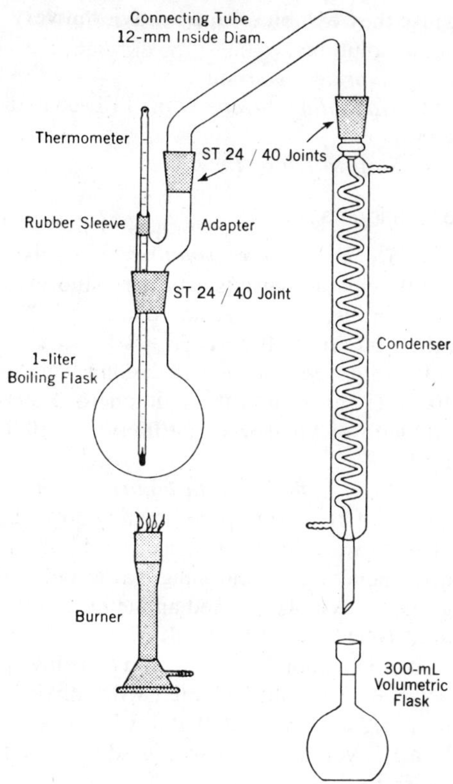

Figure 413:1. Direct distillation apparatus for fluoride.

tillate must be cool) until the temperature of flask contents reaches exactly 180 C. Discard distillate. This process removes fluoride contamination and adjusts the acid-water ratio for subsequent distillations.

b. Cool acid mixture remaining in the steps outlined in ¶ 4*a*, or previous distillations, to 120 C or below. Add 300 mL of sample, mix thoroughly, and distill until the temperature reaches 180 C. To prevent sulfate carryover, do not heat above 180 C.

c. Add Ag_2SO_4 at the rate of 5 mg/mg Cl when high-chloride samples are distilled.

d. Use H_2SO_4 solution in the flask repeatedly until contaminants from samples accumulate to such an extent that recovery is affected or interferences appear in the distillate. Check acid suitability periodically by distilling standard fluoride samples. After distilling high-fluoride samples, flush still with 300 mL distilled water and combine the two fluoride distillates. If necessary, repeat flushing until the fluoride content of the distillates is at a minimum. Include additional fluoride recovered with that of the first distillation. After periods of inactivity, similarly flush still and discard distillate.

5. Interpretation of Results

The recovery of fluoride is quantitative within the accuracy of the methods used for its measurement.

413 B. Electrode Method

1. General Discussion

a. Principle: The fluoride electrode is a selective ion sensor. It is designed to be used with a standard calomel reference electrode and any modern pH meter having an expanded millivolt scale. The key element in the fluoride ion-activity electrode is the laser-type doped single lanthanum fluoride crystal across which a potential is established by the presence of fluoride ions. The crystal contacts the sample solution at one face and an internal reference solution at the other. The cell may be represented by:

$$Ag|AgCl,\ Cl^-(0.3M),\ F^-(0.001M)\ |LaF_3|\ \text{test solution}|\text{reference electrode}$$

The fluoride ion-selective electrode can be used to measure the activity or concentration of fluoride in aqueous samples by using an appropriate calibration curve. However, the fluoride activity depends on the total ionic strength of the sample. The electrode does not respond to bound or complexed fluoride. These difficulties largely are overcome by adding a buffer solution of high total ionic strength to swamp variations in sample ionic strength and containing a chelate to complex aluminum preferentially.

b. Interference: Polyvalent cations such as Al(III), Fe(III), and Si(IV) will complex fluoride ion. The extent to which complexation takes place depends on solution pH and relative levels of fluoride and complexing species. However, adding CDTA (cyclohexylenediaminetetraacetic acid) or sodium citrate preferentially will complex concentrations of aluminum up to 5.0 mg/L and release fluoride as the free ion. Likewise, in acid solution, hydrogen ions form complexes with fluoride ion but the complexing is negligible if the pH is above 5. In alkaline solution hydroxide ion also interferes with electrode response whenever the hydroxide ion concentration is greater than one-tenth the level of fluoride ion. At pH ≤ 8 the hydroxide concentration is $\leq 10^{-6}$ molar and no interference occurs.

The fluoride electrode does not respond to the fluoroborate ion (BF_4). If a sample is suspected of containing fluoroborates, distill it to achieve hydrolysis of the fluoroborate to free fluoride.

2. Apparatus

a. Expanded-scale or digital pH meter or ion-selective meter.

b. Sleeve-type reference electrode:* Do not use fiber-tip reference electrodes because they exhibit erratic behavior in very dilute solutions.

c. Fluoride electrode.

d. Magnetic stirrer, with TFE-coated stirring bar.

e. Stop watch or timer.

3. Reagents

a. Stock fluoride solution: Dissolve 221.0 mg anhydrous sodium fluoride, NaF, in distilled water and dilute to 1,000 mL; 1.00 mL = 100 μg F.

b. Standard fluoride solution: Dilute 100 mL stock fluoride solution to 1,000 mL with distilled water; 1.00 mL = 10.0 μg F.

c. Total ionic strength adjustment buffer (TISAB): Place approximately 500 mL distilled water in a 1-L beaker and add 57 mL glacial acetic acid, 58 g NaCl, and 4.0 g 1,2 cyclohexylenediaminetetraacetic acid (CDTA).† Stir to dissolve. Place beaker in a cool water bath and add slowly 6*N* NaOH (about 125 mL) with stirring, until pH is between 5.0 and 5.5. Transfer to a 1-L volumetric flask and add distilled water to the mark.

4. Procedure

a. Instrument calibration: No major adjustment of any instrument is normally required to use electrodes in the fluoride range of 0.2 to 2.0 mg/L. For those instruments with zero at center scale adjust calibration control so that the 1.0 mg F/L standard reads at the center zero (100 mV) when the meter is in the expanded-scale position. This cannot be done on some meters that do not have a millivolt calibration control. To use a selective-ion meter follow the manufacturer's instructions.

b. Preparation of fluoride standards:

*Orion 90-01-00, Beckman 43462, Corning 476012, or equivalent.

†1,2 cyclohexylenedinitrilotetraacetic acid, J.T. Baker 5-G083, Eastman 15411, MCB CX2390, or equivalent. Alternatively, use 12 g sodium citrate dihydrate, $Na_3C_6H_5 \cdot 2H_2O$ in place of CDTA, but there may be some loss of sensitivity.

Prepare a series of standards by adding, respectively, 2.5, 5.0, and 10.0 mL standard fluoride solution to each of three 100-mL volumetric flasks. To each flask, add by pipet 50 mL of TISAB solution and dilute to 100 mL with distilled water; mix well. These standards are equivalent to 0.5, 1.0, and 2.0 mg F/L. (Because the concentration of the sample is reduced by half by adding TISAB solution, doubling the standards' true concentration enables the analyst to read the samples' original concentration directly.)

c. Treatment of sample: To a 100-mL volumetric flask, add by pipet 50 mL sample, dilute to mark with TISAB, and mix well. Bring standards and sample to the same temperature, preferably room temperature.

d. Measurement with electrode: Transfer each standard and sample to a series of 150-mL beakers. Immerse electrodes and measure developed potential while stirring on a magnetic stirrer. Avoid stirring before immersing electrodes because entrapped air around the crystal can produce erroneous readings or needle fluctuations. Let electrodes remain in the solution 3 min before taking a final positive millivolt reading. Rinse electrodes with distilled water and blot dry between readings. In some cases, extend measurement period to 5 min to achieve equilibrium. A layer of insulating material, such as cork, between the stirrer and sample beaker is helpful in minimizing temperature changes.

When using an expanded-scale pH meter or selective-ion meter, frequently recalibrate the electrode by checking potential reading of the 1.00-mg F/L standard and adjusting the calibration control, if necessary, until meter reads as before. Confirm calibration after each unknown and also after reading each standard when preparing the standard curve.

Plot potential measurement of fluoride standards against concentration on two-cycle semilogarithmic graph paper. Plot milligrams F per liter on the logarithmic axis, with the lowest concentration at the bottom of the page. Using the potential measurement for each sample, read the corresponding fluoride concentration from the standard curve.

413 C. SPADNS Method

1. Discussion

The reaction rate between fluoride and zirconium ions is influenced greatly by the acidity of the reaction mixture. By increasing the proportion of acid in the reagent, the reaction can be made practically instantaneous. Under such conditions, however, the effect of various ions differs from that in the conventional alizarin methods. The selection of dye for this rapid fluoride method is governed largely by the resulting tolerance to these ions.

2. Apparatus

Colorimetric equipment: One of the following is required:

a. Spectrophotometer, for use at 570 nm, providing a light path of at least 1 cm.

b. Filter photometer, providing a light path of at least 1 cm and equipped with a greenish yellow filter having maximum transmittance at 550 to 580 nm.

3. Reagents

a. Standard fluoride solution: Prepare as directed in the electrode method, Section 413B.3*b*.

b. SPADNS solution: Dissolve 958 mg SPADNS, sodium 2-(parasulfophenylazo)-1,8-dihydroxy-3,6-naphthalene disulfonate, also called 4,5-dihydroxy-3-(parasulfophenylazo)-2,7-naphthalenedisulfonic acid trisodium salt, in distilled water and dilute to 500 mL. This solution is stable indefinitely if protected from direct sunlight.

c. Zirconyl-acid reagent: Dissolve 133 mg zirconyl chloride octahydrate, $ZrOCl_2 \cdot 8H_2O$, in about 25 mL distilled water. Add 350 mL conc HCl and dilute to 500 mL with distilled water.

d. Acid zirconyl-SPADNS reagent: Mix equal volumes of SPADNS solution and zirconyl-acid reagent. The combined reagent is stable for at least 2 yr.

e. Reference solution: Add 10 mL SPADNS solution to 100 mL distilled water. Dilute 7 mL conc HCl to 10 mL and add to the diluted SPADNS solution. The resulting solution, used for setting the instrument reference point (zero), is stable and may be reused indefinitely. Alternatively, use a prepared standard as a reference.

f. Sodium arsenite solution: Dissolve 5.0 g $NaAsO_2$ and dilute to 1 L with distilled water. (CAUTION: *Toxic—avoid ingestion.*)

4. Procedure

a. Preparation of standard curve: Prepare fluoride standards in the range of 0 to 1.40 mg/L by diluting appropriate quantities of standard fluoride solution to 50 mL with distilled water. Pipet 5.00 mL each of SPADNS solution and zirconyl-acid reagent, or 10.00 mL mixed acid-zirconyl-SPADNS reagent, to each standard and mix well. Set photometer to zero absorbance with the reference solution and obtain absorbance readings of standards immediately. Plot a curve of the fluoride-absorbance relationship. Prepare a new standard curve whenever a fresh reagent is made or a different standard temperature is desired. If no reference solution is used, set photometer at some convenient point established with a prepared fluoride standard.

b. Sample pretreatment: If the sample contains residual chlorine, add 1 drop (0.05 mL) $NaAsO_2$ solution/0.1 mg Cl and mix. (Sodium arsenite concentrations of 1,300 mg/L produce an error of 0.1 mg/L at 1.0 mg F/L.)

c. Color development: Use a 50.0-mL sample or a portion diluted to 50 mL. Adjust sample temperature to that used for the standard curve. Add 5.00 mL each of SPADNS solution and zirconyl-acid reagent, or 10.00 mL acid-zirconyl-SPADNS reagent; mix well and read absorbance immediately or at any subsequent time, first setting the reference point as above. If the absorbance falls beyond the range of the standard curve, repeat using a smaller sample.

5. Calculation

$$\text{mg F/L} = \frac{A}{\text{mL sample}} \times \frac{B}{C}$$

where:

A = μg F determined photometrically.

The ratio B/C applies only when a sample is diluted to a volume B, and a portion C taken from it for color development.

6. Precision and Accuracy

A synthetic sample containing 830 μg F/L and no interference in distilled water was analyzed in 53 laboratories by the SPADNS method, with a relative standard deviation of 8.0% and a relative error of 1.2%. After direct distillation of the sample, the relative standard deviation was 11.0% and the relative error 2.4%.

A synthetic sample containing 570 μg F/L, 10 mg Al/L, 200 mg sulfate/L, and 300 mg total alkalinity/L was analyzed in 53 laboratories by the SPADNS method without distillation, with a relative standard deviation of 16.2% and a relative error of 7.0%. After direct distillation of the sample, the relative standard deviation was 17.2% and the relative error 5.3%.

A synthetic sample containing 680 μg F/L, 2 mg Al/L, 2.5 mg sodium hexametaphosphate/L, 200 mg sulfate/L, and 300 mg total alkalinity/L was analyzed in 53 laboratories by direct distillation and SPADNS methods with a relative standard deviation of 2.8% and a relative error of 5.9%.

413 D. Alizarin Visual Method

1. Apparatus

Color comparison equipment: One of the following is required:

a. Nessler tubes, matched, 100 mL tall form.

b. Comparator, visual.

2. Reagents

a. Standard fluoride solution: Prepare as directed in Section 413B.3*b*; 1.00 mL = 10.0 μg F.

b. Zirconyl-alizarin reagent: Dissolve 300 mg zirconyl chloride octahydrate, $ZrOCl_2 \cdot 8H_2O$, in 50 mL distilled water contained in a 1-L glass-stoppered volumetric flask. Dissolve 70 mg 3-alizarinsulfonic acid sodium salt (also called alizarin red S) in 50 mL distilled water and pour slowly into the zirconyl solution while stirring. The resulting solution clears on standing for a few minutes.

c. Mixed acid solution: Dilute 101 mL conc HCl to approximately 400 mL with distilled water. Add carefully 33.3 mL conc H_2SO_4 to approximately 400 mL distilled water. After cooling, mix the two acids.

d. Acid-zirconyl-alizarin reagent: To the clear zirconyl-alizarin reagent in the 1-L volumetric flask, add mixed acid solution. Add distilled water to the mark and mix. The reagent changes from red to yellow within an hour and is then ready for use. Store away from direct sunlight to extend reagent stability to 6 months.

e. Sodium arsenite solution: Prepare as directed in Section 413C.3*f*.

3. Procedure

a. Sample pretreatment: If the sample contains residual chlorine, remove by adding 1 drop (0.05 mL) of arsenite/0.1 mg Cl and mix.

b. Preparation of standards: Prepare a series of standards by diluting various volumes of standard fluoride solution to 100 mL in nessler tubes. Choose standards so that there is at least one with lower and one with higher fluoride concentration than that of sample. The interval between standards determines the accuracy of the determination. An interval of 50 μg/L usually is sufficient.

c. Color development: Adjust temperature of samples and standards so that the deviation between them is no more than 2 C. A temperature near that of the room is satisfactory. To 100 mL of clear sample, or a portion diluted to 100 mL, and to the standards in nessler tubes, add 5.00 mL acid-zirconyl-alizarin reagent from a volumetric pipet. Mix thoroughly, avoiding contamination, and compare samples and standards after 1 hr.

4. Calculation

$$\text{mg F/L} = \frac{A}{\text{mL sample}} \times \frac{B}{C}$$

where:

A = μg F determined visually.

The ratio B/C applies only when a sample is diluted to a volume B, and a portion C is taken from it for color development.

5. Precision and Accuracy

A synthetic sample containing 830 μg F/L and no interference in distilled water was analyzed in 20 laboratories by the alizarin visual method, with a relative standard deviation of 4.9% and a relative error of 3.6%. After direct distillation of the sample, the relative standard deviation was 6.4% and the relative error 2.4%.

A synthetic sample containing 570 μg F/L, 10 mg Al/L, 200 mg sulfate/L, and 300 mg total alkalinity/L was analyzed in 20 laboratories by the alizarin visual method without distillation, with a relative standard deviation of 51.8% and a relative error of 29.8%. After direct distillation of the sample, the relative standard deviation was 11.1% and the relative error 0%.

A synthetic sample containing 680 μg F/L, 2 mg Al/L, 2.5 mg sodium hexametaphosphate/L, 200 mg sulfate/L, and 300 mg total alkalinity/L was analyzed in 20 laboratories by the direct distillation and alizarin visual methods, with a relative standard deviation of 10.6% and a relative error of 1.5%.

413 E. Complexone Method (TENTATIVE)

1. General Discussion

a. Principle: The sample is distilled and the distillate is reacted with alizarin fluorine blue-lanthanum reagent to form a blue complex that is measured colorimetrically at 620 nm.

b. Interferences: Interferences normally associated with the determination of fluoride are removed by distillation.

c. Application: This method is applicable to potable, surface, and saline waters as well as domestic and industrial wastewaters. The range of the method, which can be modified by using the adjustable colorimeter, is 0.1 to 2.0 mg F/L.

2. Apparatus

*Automated analytical equipment,** consisting of the components listed in Section 602 with the following additions: heating bath with distillation head, 15-mm tubular flow cell, and 620-nm filters.

3. Reagents

a. Stock fluoride solution: Dissolve 2.210 g anhydrous NaF in about 600 mL distilled water and dilute to 1,000 mL; 1.00 mL = 1.00 mg F.

b. Standard fluoride solution: Prepare fluoride standards in concentrations of 0.1 to 2.0 mg/L, using stock fluoride solution.

c. Distillation reagent: Add 50 mL conc H_2SO_4 to about 600 mL distilled water. Add 1.00 mL stock fluoride solution and dilute to 1,000 mL.

d. Acetate buffer solution: Dissolve 60 g anhydrous $NaC_2H_3O_2$ in about 600 mL distilled water. Add 100 mL conc (glacial) acetic acid and dilute to 1 L.

e. Alizarin fluorine blue stock solution: Add 960 mg alizarin fluorine†, $C_{14}H_7O_4{\cdot}CH_2N(CH_2{\cdot}COOH)_2$, to 100 mL distilled water. Add 2 mL conc NH_4OH and mix until dye is dissolved. Add 2 mL conc (glacial) acetic acid, dilute to 250 mL and store in an amber bottle in the refrigerator.

f. Lanthanum nitrate stock solution: Dissolve 1.08 g $La(NO_3)_3$ in about 100 mL distilled water, dilute to 250 mL, and store in refrigerator.

*Technicon™ AutoAnalyzer™ II, or equivalent.

†J.T. Baker Catalog number J-112 or equivalent.

g. Working color reagent: Mix in the following order: 300 mL acetate buffer solution, 150 mL acetone, 50 mL tertiary butanol, 36 mL alizarin fluorine blue stock solution, 40 mL lanthanum nitrate stock solution, and 2 mL polyoxyethylene 23 lauryl ether‡. Dilute to 1 L with distilled water. This reagent is stable for 2 to 4 days.

‡Brij-35, available from ICI United States, Chicago, Ill., or Technicon Instruments Corp., Tarrytown, N.Y., or equivalent.

4. Procedure

No special handling or preparation of sample is required.

Set up manifold as shown in Figure 413:2 and follow the general procedure described in Section 602.

5. Calculation

See Section 602.3.

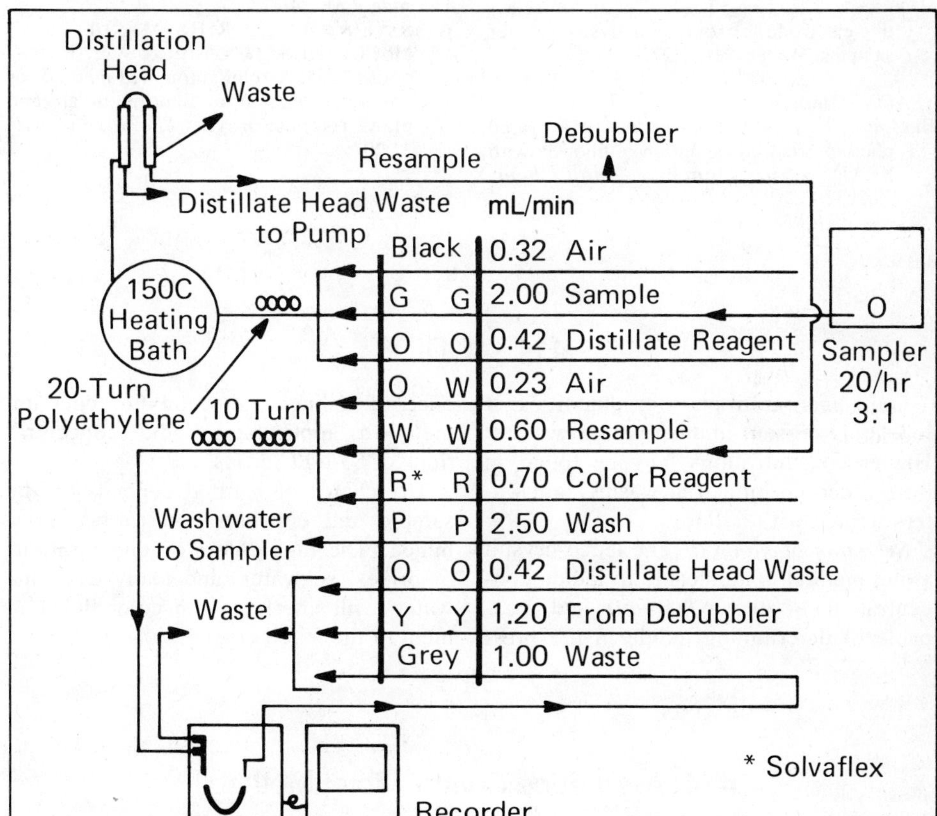

Figure 413:2. Fluoride manifold.

413 F. Bibliography

Direct Distillation Step

BELLACK, E. 1958. Simplified fluoride distillation method. *J. Amer. Water Works Ass.* 50:530.

BELLACK, E. 1961. Automatic fluoride distillation. *J. Amer. Water Works Ass.* 53:98.

Electrode Method

FRANT, M.S. & J.W. ROSS, JR. 1968. Use of total ionic strength adjustment buffer for electrode determination of fluoride in water supplies. *Anal. Chem.* 40:1169.

HARWOOD, J.E. 1969. The use of an ion-selective electrode for routine analysis of water samples. *Water Res.* 3:273.

SPADNS Method

BELLACK, E. & P.J. SCHOUBOF. 1968. Rapid photometric determination of fluoride with SPADNS-zirconium lake. *Anal. Chem.* 30:2032.

Alizarin Visual Method

SANCHIS, J. M. 1934. Determination of fluorides in natural waters. *Ind. Eng. Chem.*, Anal. Ed. 6:134.

SCOTT, R.D. 1941. Modification of fluoride determination. *J. Amer. Water Works Ass.* 33:2018.

TARAS, M.J., H.D. CISCO & M. GARNELL. 1950. Interferences in alizarin method of fluoride determination. *J. Amer. Water Works Ass.* 42:583.

Automated Method

WEINSTEIN, L.H., R.H. MANDL, D.C. MCCUNE, J.S. JACOBSON & A.E. HITCHCOCK. 1963. A semi-automated method for the determination of fluorine in air and plant tissues. *Boyce Thompson Inst.* 22:207.

414 IODIDE

Only microgram-per-liter quantities of iodide are present in most natural waters. Higher concentrations may be found in brines, certain industrial wastes, and waters treated with iodine.

Selection of method: The leuco crystal violet method is applicable to iodide concentrations of 50 to 6,000 μg/L and is capable of determining iodide in the presence of iodine. The catalytic reduction method is applicable to iodide concentrations of 80 μg/L or less.

The choice of method depends on the sample and concentration to be determined. The high chloride concentrations of brines, seawater, and many estuarine waters will interfere with color development in the leuco crystal violet method.

414 A. Leuco Crystal Violet Method

1. General Discussion

a. Principle: Iodide is selectively oxidized to iodine by the addition of potassium peroxymonosulfate, $KHSO_5$. The iodine produced reacts instantaneously with the colorless indicator reagent containing 4,4′4″-methylidynetris (N,N-dimethylaniline), also known as leuco crystal violet, to produce the highly colored crystal violet dye. The developed color is sufficiently stable for the determination of an absorbance value and adheres to Beer's law over a wide range of iodine concentrations.

b. Interference: Chloride concentrations greater than 200 mg/L may interfere with color development. Reduce these interferences by diluting sample to contain less than 200 mg Cl/L.

2. Apparatus

a. Colorimetric equipment: One of the following is required:

1) *Filter photometer,* providing a light path of 1 cm or longer, equipped with an orange filter having maximum transmittance near 592 nm.

2) *Spectrophotometer,* for use at 592 nm, providing a light path of 1 cm or longer.

b. Volumetric flasks: 100-mL with plastic caps or ground-glass stoppers.

c. Glassware: Completely remove any reducing substances from all glassware or plastic containers, including containers for storing reagent solutions using either the chlorination or the chromic acid method (Section 408).

3. Reagents

a. Iodine-demand-free water: Prepare iodine-demand-free water by the ion-exchange method presented in Section 408. Use ordinary distilled water if reducing substances are known to be absent. Do not chlorinate as in the preparation of chlorine-demand-free water, because excess chloride ion concentrations interfere.

Prepare all stock iodide and reagent solutions with iodine-demand-free water.

b. Stock iodide solution: Dissolve 1.3081 g KI in water and dilute to 1,000 mL; 1 mL = 1 mg I.

c. Citric buffer solution, pH 3.8:

1) *Citric acid,* 0.3333*N:* Dissolve 192.2 g $C_6H_8O_7$ or 210.2 g $C_6H_8O_7 \cdot H_2O$ and dilute to 1 L with water.

2) *Ammonium hydroxide, 2N:* Add 131 mL conc NH_4OH to about 700 mL water and dilute to 1 L. Store in a polyethylene bottle.

3) *Final buffer solution:* Slowly add, with mixing, 350 mL 2*N* NH_4OH solution to 670 mL 0.3333*N* citric acid. Add 80 g ammonium dihydrogen phosphate ($NH_4H_2PO_4$) and stir to dissolve.

d. Leuco crystal violet indicator: Measure 200 mL water and 3.2 mL conc sulfuric acid (H_2SO_4) into a brown glass container of at least 1-L capacity. Introduce a magnetic stirring bar and mix at moderate speed. Add 1.5 g 4,4′,4″-methylidynetris (N,N-dimethylaniline)* and with a small amount of water wash down any reagent adhering to neck or sides of container. Mix until dissolved.

To 800 mL water, add 2.5 g mercuric chloride ($HgCl_2$) and stir to dissolve. With mixing, add $HgCl_2$ solution to leuco crystal violet solution. For maximum stability, adjust pH of final solution to 1.5 or less, adding, if necessary, conc H_2SO_4 dropwise. Store in a brown glass bottle away from direct sunlight. Discard after 6 months. Do not use a rubber stopper.

e. Potassium peroxymonosulfate solution: Obtain $KHSO_5$ as a commercial product†, which is a stable powdered mixture containing 42.8% $KHSO_5$ by weight and a mixture of $KHSO_4$ and K_2SO_4. Dissolve 1.5 g powder in water and dilute to 1 L.

f. Sodium thiosulfate solution: Dissolve 5.0 g $Na_2S_2O_3 \cdot H_2O$ in distilled water and dilute to 1L.

4. Procedure

a. Preparation of temporary iodine standards: Add suitable portions of stock iodide solution, or of dilutions of stock iodide solution, to water to prepare an iodide series of 0.1 to 6.0 mg/L in increments of 0.1 mg/L or larger.

*Eastman chemical No. 3651 or equivalent.

†Oxone, E. I. du Pont de Nemours and Co., Inc., Wilmington, Del., or equivalent.

Measure 50.0 mL dilute KI standard solution into a 100-mL glass-stoppered volumetric flask. Add 1.0 mL citric buffer and 0.5 mL $KHSO_5$ solution. Swirl to mix and let stand approximately 1 min. Add 1.0 mL leuco crystal violet indicator, mix, and dilute to 100 mL. For best results, read absorbance as described below within 5 min after adding indicator solution.

b. Photometric calibration: Transfer colored temporary standards of known iodide concentrations to cells of 1-cm light path and read absorbance in a photometer or spectrophotometer at a wavelength of 592 nm against a distilled water reference. Plot absorbance values against iodide concentrations to construct a curve that follows Beer's law.

c. Color development of sample: Measure a 50.0-mL sample into a 100-mL volumetric flask and treat as described for preparation of temporary iodide standards, ¶ 4*a*. Read absorbance photometrically and refer to standard calibration curve for iodide equivalent.

d. Samples containing >6.0 mg/L iodide: Place approximately 25 mL water in a 100-mL volumetric flask. Add 1.0 mL citric buffer and a measured volume of 25 mL or less of sample. Add 0.5 mL $KHSO_5$ solution. Swirl to mix and let stand approximately 1 min. Add 1.0 mL leuco crystal violet indicator, mix, and dilute to 100 mL.

Read absorbance photometrically and compare with calibration curve from which the initial iodide is obtained by applying the dilution factor. Select one of the following sample volumes to remain within the optimum iodide range.

Iodide *mg/L*	Sample Volume Required *mL*
6.0–12	25.0
12–30	10.0
30–60	5.0

e. Determination of iodide in the presence of iodine: On separate samples determine (1) total iodide and iodine, and (2) iodine. The iodide concentration is the difference between the iodine determined and the total iodine-iodide obtained. Determine iodine by not adding $KHSO_5$ solution in the iodide method and by comparing the absorbance value to the calibration curve developed for iodide.

f. Compensation for turbidity and color: Compensate for natural color or turbidity by adding 5 mL $Na_2S_2O_3$ solution to a 50-mL sample. Add reagents to sample as described previously and use as the blank to set zero absorbance on photometer. Measure all samples in relation to this blank and, from the calibration curve, determine concentrations of iodide or total iodine-iodide.

414 B. Catalytic Reduction Method

1. General Discussion

a. Principle: Iodide can be determined by using its ability to catalyze the reduction of ceric ions by arsenious acid. The effect is nonlinearly proportional to the amount of iodide present. The reaction is stopped after a specific time interval by the addition of ferrous ammonium sulfate. The resulting ferric ions are directly proportional to the remaining ceric ions and develop a color complex with potassium thiocyanate that is relatively stable.

Digestion with chromic acid and distillation are necessary to estimate the nonsusceptible bound forms of iodine in addition to the usual iodide ion. Procedures for these special applications are available.[1]

b. Interferences: The formation of non-catalytic forms of iodine and the inhibitory effects of silver and mercury are reduced by adding an excess of sodium chloride (NaCl) that sensitizes the reaction.

2. Apparatus

a. Water bath, capable of temperature control to 30 ± 0.5 C.

b. Colorimetric equipment: One of the following is required:

1) *Spectrophotometer,* for use at wavelengths of 510 or 525 nm and providing a light path of 1 cm.

2) *Filter photometer,* providing a light path of 1 cm and equipped with a green filter having maximum transmittance near 525 nm.

c. Test tubes, 2 × 15 cm.

d. Stopwatch.

3. Reagents

Store all stock solutions in tightly stoppered containers in the dark.

a. Distilled water, containing less than 0.3 μg/L total iodine.

b. Sodium chloride solution: Dissolve 200.0 g NaCl in distilled water and dilute to 1 L. Recrystallize the NaCl if an interfering amount of iodine is present, using a water-ethanol mixture.

c. Arsenious acid, 0.1*N*: Dissolve 4.946 g As_2O_3 in distilled water, add 0.20 mL conc H_2SO_4, and dilute to 1,000 mL.

d. Sulfuric acid, H_2SO_4, conc.

e. Ceric ammonium sulfate, 0.02*N*: Dissolve 13.38 g $Ce(NH_4)_4(SO_4)_4 \cdot 4H_2O$ in distilled water, add 44 mL conc H_2SO_4, and make up to 1 L.

f. Ferrous ammonium sulfate reagent: Dissolve 1.50 g $Fe(NH_4)_2(SO_4)_2 \cdot 6H_2O$ in 100 mL distilled water containing 0.6 mL conc H_2SO_4. Prepare daily.

g. Potassium thiocyanate solution: Dissolve 4.00 g KSCN in 100 mL distilled water.

h. Stock iodide solution: Dissolve 261.6 mg anhydrous KI in distilled water and dilute to 1,000 mL; 1.00 mL = 200 μg I.

i. Intermediate iodide solution: Dilute 20.00 mL stock iodide solution to 1,000 mL with distilled water; 1.00 mL = 4.00 μg I.

j. Standard iodide solution: Dilute 25.00 mL intermediate iodide solution to 1,000 mL with distilled water; 1.00 mL = 0.100 μg I.

4. Procedure

a. Sample size: Add 10.00 mL sample, or a portion made up to 10.00 mL with iodine-free distilled water, to a 2 × 15 cm test tube. If possible, keep iodide content in the range 0.2 to 0.6 μg. Use thoroughly clean glassware and apparatus.

b. Color measurement: Add reagents in the following order: 1.00 mL NaCl solution, 0.50 mL As_2O_3 solution, and 0.50 mL conc H_2SO_4.

Place reaction mixture and ceric ammonium sulfate solution in 30 C water bath and let come to temperature equilibrium. Add 1.0 mL ceric ammonium sulfate solution, mix by inversion, and start stopwatch to time reaction. Use an inert clean test tube stopper when mixing. After 15 ± 0.1 min remove sample from water bath and add immediately 1.00 mL ferrous ammonium sulfate reagent with mixing, whereupon the yellow ceric ion color should disappear. Then add, with mixing, 1.00 mL KSCN solution. Replace sample in water bath. Within 1 hr after adding thiocyanate read absorbance in a photometric instrument. Maintain temperature of solution and cell compartment at 30 ± 0.5 C until absorbance is determined. If several samples are run, start reactions at 1-min intervals to allow time for additions of ferrous ammonium sulfate and thiocyanate. (If temperature control of cell compartment is not possible, let final solution come to room temperature and measure absorbance with cell compartment at room temperature.)

c. Calibration standards: Treat standards containing 0, 0.2, 0.4, 0.6, and 0.8 μg I/10.00 mL of solution as in ¶ 4*b* above. Run with each set of samples to establish a calibration curve.

5. Calculation

$$\text{mg I/L} = \frac{\mu\text{g I (in 15 mL final volume)}}{\text{mL sample}}$$

6. Precision and Accuracy

Results obtained by this method are reproducible on samples of Los Angeles source waters, and have been reported to be accurate to ±0.3 μg I/L on samples of Yugoslavian water containing from 0 to 14.0 μg I/L.

414 C. References

1. Standard Methods for the Examination of Water, Sewage and Industrial Wastes, 10th ed. 1955. APHA, AWWA & FSIWA, New York, pp. 120–124.

414 D. Bibliography

Leuco Crystal Violet Method

BLACK, A.P. & G.P. WHITTLE. 1967. New methods for the colorimetric determination of halogen residuals. Part I. Iodine, iodide, and iodate. *J. Amer. Water Works Ass.* 59:471.

Catalytic Reduction Method

ROGINA, B. & M. DUBRAVCIC. 1953. Microdetermination of iodides by arresting the catalytic reduction of ceric ions. *Analyst* 78:594.

DUBRAVCIC, M. 1955. Determination of iodine in natural waters (sodium chloride as a reagent in the catalytic reduction of ceric ions). *Analyst* 80:295.

415 IODINE

Iodine is used to disinfect certain potable and swimming pool waters. For wastewaters, iodine has had limited application. Use of iodine generally is restricted to personal or remote water supplies where ease of application, storage stability, and an inertness toward organic matter are important considerations. Some swimming pool waters are treated with iodine to lessen eye burn among swimmers and to provide a stable disinfectant residual less affected by adverse environmental conditions.

Iodine is applied in the elemental form or produced in situ by the simultaneous addition of an iodide salt and a suitable oxidant. In the latter case, an excess of iodide may be maintained to serve as a reservoir for iodine production; the determination of iodide is desirable for disinfectant control (see Iodide, Section 414).

Because of hydrolysis, active iodine exists in the forms of elemental I_2, hypoiodous acid (HOI) or a form thereof, hypoiodite ion (OI^-), and, in the presence of excess iodide, the triiodide ion (I^{3-}). Most analytical methods use the oxidizing power of all forms of active iodine for its determination and the results usually are expressed as an equivalent concentration of elemental iodine.

Selection of method: For potable and swimming pool waters treated with elemental iodine, both the amperometric titration and leuco crystal violet colorimetric methods give acceptable results. However, oxidized forms of manganese interfere with the leuco crystal violet method. Where the iodide and chloride ion concentrations are above 50 mg/L and 200 mg/L, respectively, interference in color production may occur in the leuco crystal violet method and the amperometric method is preferred. However, because of the extreme sensitivity of the leuco crystal violet method, this interference may be eliminated by sample dilution to obtain halogen ion concentrations less than 50 mg/L.

For wastewaters or highly polluted waters, organic constituents normally do not interfere with either the amperometric or leuco crystal violet procedures. Determine which of the methods yields the more acceptable results, because specific substances present in these waters may interfere in one method but not in the other. Certain metallic cations such as copper and silver interfere in the amperometric titration procedure. The leuco crystal violet method is relatively free of interference from these and other cations and anions with the exceptions noted previously.

For waters containing iodine coexisting with free chlorine, combined chlorine, or other excess oxidants, of the methods described only the leuco crystal violet method can determine iodine specifically. This condition occurs in the in-situ production of iodine by the reaction of iodide and excess oxidant. Under these conditions, the amperometric method would continue to titrate the iodine produced in a cyclic reaction until exhaustion of the oxidant.

415 A. Leuco Crystal Violet Method (TENTATIVE)

1. General Discussion

The leuco crystal violet method determines aqueous iodine present as elemental iodine and hypoiodous acid. Excess common oxidants do not interfere. While the method utilizes the sum of the oxidative power of all forms of active iodine residuals, the results are expressed as the equivalent concentration of iodine. The method also is capable of determining the sum of iodine and free iodide concentrations; the free iodide concentration can be determined by difference (see Iodide, Section 414).

a. Principle: Mercuric chloride added to aqueous elemental iodine solutions causes essentially complete hydrolysis of iodine and the stoichiometric production of hypoiodous acid. The compound 4,4′,4″-methylidynetris (N,N-dimethylaniline), also known by the common name of leuco crystal violet, reacts instantaneously with the hypoiodous acid to form crystal violet dye. The maximum absorbance of the developed crystal violet dye solution is produced in the pH range of 3.5 to 4.0 and measured at a wavelength of 592 nm. The absorbance follows Beer's law over a wide range of iodine concentrations and the developed color is stable for several hours.

In the presence of certain excess oxidants such as free chlorine or chloramines, the iodine residual will exist exclusively in the form of hypoiodous acid. The leuco crystal violet is relatively insensitive to the combined forms of chlorine while any free chlorine is converted to chloramine by reaction with an ammonium salt incorporated in the test reagents. All the hypoiodous acid is determined and, when expressed as an equivalent elemental I_2 concentration, will yield a weight concentration value twice that found in an elemental I_2 solution of the same weight concentration.

b. Interference: Oxidized forms of manganese interfere by oxidizing the indicator to crystal violet dye and yield apparent high iodine concentrations.

Iodide and chloride ion concentrations above 50 mg/L and 200 mg/L, respectively, interfere by inhibiting full color production. Dilute the sample to eliminate this interference.

Combined chlorine residuals normally do not interfere provided that the test is completed within 5 min after adding of the indicator solution. Eliminate interference from free chlorine by adding an ammonium salt buffer to form combined chlorine.

c. Minimum detectable concentration: 10 μg I as I_2/L.

2. Apparatus

a. Colorimetric equipment: One of the following is required:

1) *Filter photometer,* with a light path of 1 cm or longer, equipped with an orange filter having maximum transmittance near 592 nm.

2) *Spectrophotometer,* for use at 592 nm, with a light path of 1 cm or longer.

b. Volumetric flasks, 100-mL, with plastic caps or ground-glass stoppers.

c. Glassware: Completely remove reducing substances from glassware or plastic containers, including containers for storage of reagent solutions, by using either the chlorination or the chromic acid method (Section 408F.2*c*).

3. Reagents

a. Iodine-demand-free water: See Section 414A.3*a*.

Prepare all stock iodine and reagent solutions with iodine-demand-free water.

b. Stock iodine solution: Prepare a saturated iodine solution by dissolving 20 g elemental iodine in 300 mL water. Let stand several hours. Decant iodine solution and dilute 170 mL to 2,000 mL. Standardize solution by titrating with standard sodium thiosulfate ($Na_2S_2O_3$) titrant as described in Section 408A.3.*b* and *c* or amperometrically as in Section 415B.

Calculate iodine concentration:

$$\text{mg I as } I_2/\text{mL} = \text{normality of iodine solution} \times 126.9$$

Prepare a working solution of 10 μg I as I_2/mL by appropriate dilution of the standardized stock solution.

c. Citric buffer solution, pH 3.8: See Section 414A.3*c*.

d. Leuco crystal violet indicator: See Section 414A.3*d*.

e. Sodium thiosulfate solution: See Section 414A.3*f*.

4. Procedure

a. Preparation of temporary iodine standards: For greater accuracy, standardize working solution immediately before use by the amperometric titration method, Section 415B. Prepare standards in the range of 0.1 to 6.0 mg I as I_2/L by adding 1 mL to 60 mL working solution to 100 mL glass-stoppered volumetric flasks, in increments of 1 mL or larger. Adjust these volumes if the measured iodine concentration of working solution varies by 5% or more from 10 μg I as I_2/mL.

Measure 50.0 mL of each diluted iodine

working solution into a 100-mL glass-stoppered volumetric flask. Add 1.0 mL citric buffer solution, gently swirl to mix, and let stand for at least 30 sec. Add 1.0 mL leuco crystal violet indicator and swirl to develop color. Dilute to 100 mL and mix.

b. Photometric calibration: Transfer colored temporary standards of known iodine concentrations to cells of 1-cm light path and read absorbance in a photometer or spectrophotometer at a wavelength of 592 nm against a distilled water reference. Plot absorbance values against iodine concentrations to construct a curve that follows Beer's law.

c. Color development of iodine sample: Measure 50.0 mL sample into a 100-mL volumetric flask and treat as described for preparation of temporary iodine standards, ¶ 4*a*. Match test sample visually with temporary standards or read absorbance photometrically and refer to standard calibration curve for the iodine equivalent.

d. Samples containing >6.0 mg/L iodine: Place approximately 25 mL water in a 100-mL volumetric flask. Add 1.0 mL citric buffer solution and a measured volume of 25 mL or less of sample. Mix and let stand for at least 30 sec. Add 1.0 mL leuco crystal violet indicator, mix, and dilute to mark. Match visually with standards or read absorbance photometrically and compare with calibration curve from which the initial iodine is obtained by applying the dilution factor. Select one of the following sample volumes to remain within optimum iodine range:

Iodine *mg/L*	Sample Volume Required *mL*
6.0-12.0	25.0
12.0-30	10.0
30-60	5.0

e. Samples containing both chlorine and iodine: For samples containing free or combined chlorine and iodine, follow procedure given in ¶ 4*c* or *d* above but read absorbance within 5 min after adding leuco crystal violet indicator.

f. Compensation for turbidity and color: Compensate for natural color or turbidity by adding 5 mL $Na_2S_2O_3$ solution to a 50-mL sample. Add reagents to sample as described previously and use as blank to set zero absorbance on the photometer. Measure all samples in relation to this blank and, from calibration curve, determine concentrations of iodine.

415 B. Amperometric Titration Method

1. General Discussion

The amperometric titration method for iodine is a modification of the amperometric method for residual chlorine (see Section 408C). Iodine residuals over 7 mg/L are best measured with smaller samples or by dilution. In most cases the titration results represent free available iodine because combined iodine rarely is encountered.

a. Principle: The principle of the amperometric method as described for the determination of total available residual chlorine is applicable to the determination of residual iodine. Iodine is determined on the titrator using buffer solution, pH 4.0, and potassium iodide (KI) solution. Maintain pH at 4.0 because at pH values less than 3.5 substances such as oxidized forms of manganese interfere, while at pH values greater than 4.5, the reaction is not quantitative. Adding KI improves the sharpness of the endpoint.

b. Interference: Free available chlorine and the interferences described in Section

408C.1*b* also interfere in the iodine determination.

2. Apparatus

See Section 408C.2*a* through *d*.

3. Reagents

With the exception of phosphate buffer solution, pH 7.0, all reagents listed for the determination of residual chlorine in Section 408C.3 are required. Standardized phenylarsine oxide solution (1 mL = 1 mg chlorine/L for a 200-mL sample) is equivalent to 3.58 mg I as I_2/mL for a 200-mL sample.

4. Procedure

a. Sample volume: Select a sample volume that will require no more than 2 mL phenylarsine oxide titrant. For iodine concentrations of 7 mg/L or less, take a 200-mL volume; for iodine levels above 7 mg/L, use 100 mL or proportionately less diluted to 200 mL with water.

b. Free available iodine: To the sample add 1 mL KI solution and 1 mL acetate buffer, pH 4.0 solution. Titrate with phenylarsine oxide titrant to the end point described in Section 408C.4.

5. Calculation

Calculate the iodine concentration by the following equation:

$$\text{mg I as } I_2\text{/L} = \frac{A \times 3.58 \times 200}{\text{mL sample}}$$

where:

A = mL phenylarsine oxide titration to the end point.

415 C. Bibliography

Leuco Crystal Violet Method

BLACK, A.P. & G.P. WHITTLE. 1967. New methods for the colorimetric determination of halogen residuals. Part I. Iodine, iodide, and iodate. *J. Amer. Water Works Ass.* 59:471.

Amperometric Titration Method

MARKS, H.C. & J.R. GLASS. 1942. A new method of determining residual chlorine. *J. Amer. Water Works Ass.* 34:1227.

416 NITROGEN

In waters and wastewaters the forms of nitrogen of greatest interest are, in order of decreasing oxidation state, nitrate, nitrite, ammonia, and organic nitrogen. All these forms of nitrogen, as well as nitrogen gas (N_2), are biochemically interconvertible and are components of the nitrogen cycle. They are of interest for many reasons.

Organic nitrogen is defined functionally as organically bound nitrogen in the trinegative oxidation state. It does not include all organic nitrogen compounds. Analytically, organic nitrogen and ammonia can be determined together and have been referred to as "kjeldahl nitrogen," a term that reflects the technic used in their determination. Organic nitrogen includes such natural materials as proteins and peptides, nucleic acids and urea, and numerous synthetic organic materials. Typical organic nitrogen concentrations vary from a few hundred micrograms per liter in some lakes to more than 20 mg/L in raw sewage.

Total oxidized nitrogen is the sum of nitrate and nitrite nitrogen. Nitrate generally occurs in trace quantities in surface water but may attain high levels in some groundwater. In excessive amounts, it contributes to the illness known as methemoglobinemia in infants. A limit of 10 mg nitrate as nitrogen/L has been imposed on drinking water to prevent this disorder. Nitrate is found only in small amounts in fresh domestic wastewater but in the effluent of nitrifying biological treatment plants nitrate may be found in concentrations of up to 30 mg nitrate as nitrogen/L. It is an essential nutrient for many photosynthetic autotrophs and in some cases has been identified as the growth-limiting nutrient.

Nitrite is an intermediate oxidation state of nitrogen, both in the oxidation of ammonia to nitrate and in the reduction of nitrate. Such oxidation and reduction may occur in wastewater treatment plants, water distribution systems, and natural waters. Nitrite can enter a water supply system through its use as a corrosion inhibitor in industrial process water. Nitrite is the actual etiologic agent of methemoglobinemia. Nitrous acid, which is formed from nitrite in acidic solution, can react with secondary amines (RR'NH) to form nitrosamines (RR'N-NO), many of which are known to be potent carcinogens. The toxicologic significance of nitrosation reactions in vivo and in the natural environment is the subject of much current concern and research.

Ammonia naturally is present in surface and wastewaters. Its concentration generally is low in groundwaters because it adsorbs to soil particles and clays and is not readily leached from soils. It is produced largely by the deamination of organic nitrogen-containing compounds and by the hydrolysis of urea. At some water treatment plants ammonia is added to react with chlorine to form a combined chlorine residual.

In the chlorination of wastewater effluents containing ammonia, virtually no free residual chlorine is obtained until the ammonia has been oxidized. Rather, the chlorine reacts with ammonia to form mono- and dichloramines. Ammonia concentrations encountered in water vary from less than 10 μg ammonia nitrogen/L in some natural surface and groundwaters to more than 30 mg/L in some wastewaters.

In this manual, organic nitrogen is referred to as organic N, nitrate nitrogen as NO_3^--N, nitrite nitrogen as NO_2^--N, and ammonia nitrogen as NH_3-N.

417 NITROGEN (AMMONIA)

1. Selection of Method

The two major factors that influence selection of the method to determine ammonia are concentration and presence of interferences. In general, direct manual determination of low concentrations of ammonia is confined to drinking waters, clean surface water, and good-quality nitrified wastewater effluent. In other instances, and where interferences are present and greater precision is necessary, a preliminary distillation step (A) is required. For high ammonia concentrations a distillation and titration technic is preferred. The data presented in 417.4 below and Table 417:I should be helpful in selecting the appropriate method of analysis.

Two manual colorimetric technics—the

nesslerization (B) and phenate (C) methods—and one titration method (D) are presented. An ammonia-selective electrode method (E), which may be used either with or without prior sample distillation, and an automated version of the phenate method (F) also are included. While the stated maximum concentration ranges for the manual methods are not rigorous limits, titration is preferred at concentrations higher than the stated maximum levels for the photometric procedure.

The nessler method is sensitive to 20 μg NH_3-N/L under optimum conditions and may be used for up to 5 mg NH_3-N/L. Turbidity, color, and substances precipitated by hydroxyl ion, such as magnesium and calcium, interfere and may be removed by preliminary distillation or, less satisfactorily, by precipitation with zinc sulfate and alkali.

The manual phenate method has a sensitivity of 10 μg NH_3-N/L and is useful for up to 500 μg NH_3-N/L. Preliminary distillation is required if the alkalinity exceeds 500 mg $CaCO_3$/L or if color or turbidity is present. This step also must precede the phenate technic if the sample has been preserved with acid.

The distillation and titration procedure is used especially for NH_3-N concentrations greater than 5 mg/L.

Distillation into sulfuric acid (H_2SO_4) absorbent is mandatory for the phenate method when interferences are present. Boric acid must be the absorbent following distillation if the distillate is to be nesslerized or titrated.

The ammonia-selective electrode method is applicable over the range from 0.03 to 1,400 mg NH_3-N/L.

2. Interferences

Glycine, urea, glutamic acid, cyanates, and acetamide hydrolyze very slowly in solution on standing but, of these, only urea and cyanates will hydrolyze on distillation at pH of 9.5. Hydrolysis amounts to about 7% at this pH for urea and about 5% for cyanates. Glycine, hydrazine, and some amines will react with nessler reagent to give the characteristic yellow color in the time required for the test. Similarly, volatile alkaline compounds such as hydrazine and amines will influence titrimetric results. Some organic compounds such as ketones, aldehydes, alcohols, and some amines may cause a yellowish or greenish off-color or a turbidity on nesslerization following distillation. Some of these, such as formaldehyde, may be eliminated by boiling off at a low pH before nesslerization. Remove residual chlorine by sample pretreatment.

3. Storage of Samples

Most reliable results are obtained on fresh samples. Destroy residual chlorine immediately after sample collection to prevent its reaction with ammonia. If prompt analysis is impossible, preserve samples with 0.8 mL conc H_2SO_4/L sample and store at 4 C. The pH of the acid-preserved samples should be between 1.5 and 2. Some wastewaters may require more conc H_2SO_4 to achieve this pH. If acid preservation is used, neutralize samples with NaOH or KOH immediately before making the determination.

4. Precision and Accuracy

Six synthetic samples containing ammonia and other constituents dissolved in distilled water were analyzed by five procedures. The first three samples were subjected to direct nesslerization alone, distillation followed by nesslerization, and distillation followed by titration. Samples 4 through 6 were analyzed by direct nesslerization, by distillation followed by nesslerization, by the phenate method alone, and by distillation followed by the phenate method. Results obtained by the participating laboratories are summarized in Table 417:I.

Sample 1 contained the following addi-

TABLE 417:I. PRECISION AND ACCURACY DATA FOR AMMONIA METHODS

Number of Labora-tories	Ammonia Nitrogen Concen-tration μg/L	Relative Standard Deviation					Relative Error				
				Distillation Plus					Distillation Plus		
		Direct Nessleri-zation %	Direct Manual Phenate Method %	Nessler Method %	Manual Phenate Method %	Titri-metric Method %	Direct Nessleri-zation %	Direct Manual Phenate Method %	Nessler Method %	Manual Phenate Method %	Titri-metric Method %
20	200	38.1	—	—	—	—	0	—	—	—	—
44	200	—	—	46.3	—	—	—	—	10.0	—	—
21	200	—	—	—	—	69.8	—	—	—	—	20.0
20	800	11.2	—	—	—	—	0	—	—	—	—
42	800	—	—	21.2	—	—	—	—	8.7	—	—
20	800	—	—	—	—	28.6	—	—	—	—	5.0
21	1,500	11.6	—	—	—	—	0.6	—	—	—	—
42	1,500	—	—	18.0	—	—	—	—	4.0	—	—
21	1,500	—	—	—	—	21.6	—	—	—	—	2.6
70	200	—	39.2	—	—	—	—	2.4	—	—	—
3	200	—	—	—	15.1	—	—	—	—	16.7	—
9	200	22.0	—	—	—	—	8.3	—	—	—	—
5	200	—	—	15.7	—	—	—	—	2.0	—	—
66	800	—	15.8	—	—	—	—	1.5	—	—	—
3	800	—	—	—	16.6	—	—	—	—	1.7	—
9	800	16.1	—	—	—	—	0.3	—	—	—	—
6	800	—	—	16.3	—	—	—	—	3.1	—	—
71	1,500	—	26.0	—	—	—	—	10.0	—	—	—
3	1,500	—	—	—	7.3	—	—	—	—	0.4	—
8	1,500	5.3	—	—	—	—	1.2	—	—	—	—
6	1,500	—	—	7.5	—	—	—	—	3.6	—	—

tional constituents: 10 mg chloride/L, 1.0 mg NO_3^--N/L, 1.5 mg organic N/L, 10.0 mg phosphate/L, and 5.0 mg silica/L.

Sample 2 contained the following additional constituents: 200 mg chloride/L, 1.0 mg NO_3^--N/L, 0.8 mg organic N/L, 5.0 mg phosphate/L, and 15.0 mg silica/L.

Sample 3 contained the following additional constituents: 400 mg chloride/L, 1.0 mg NO_3^--N/L, 0.2 mg organic N/L, 0.5 mg phosphate/L, and 30.0 mg silica/L.

Sample 4 contained the following additional constituents: 400 mg chloride/L, 0.05 mg NO_3^--N/L, 0.23 mg organic phosphorus/L added in the form of adenylic acid, 7.00 mg orthophosphate phosphorus/L, and 3.00 mg polyphosphate phosphorus/L added as sodium hexametaphosphate.

Sample 5 contained the following additional constituents: 400 mg chloride/L, 5.00 mg NO_3^--N/L, 0.09 mg organic phosphorus/L added in the form of adenylic acid, 0.6 mg orthophosphate phosphorus/L, and 0.3 mg polyphosphate phosphorus/L added as sodium hexametaphosphate.

Sample 6 contained the following additional constituents: 400 mg chloride/L, 0.4 mg NO_3^--N/L, 0.03 mg organic phosphorus/L added in the form of adenylic acid, 0.1 mg orthophosphate phosphorus/L, and 0.08 mg polyphosphate phosphorus/L added as sodium hexametaphosphate.

For the ammonia-selective electrode in a single laboratory using surface water samples at concentrations of 1.00, 0.77, 0.19, and 0.13 mg NH_3-N/L, standard deviations were +0.038, +0.017, +0.007, and +0.003, respectively. In a single laboratory using surface water samples at concentrations of 0.10 and 0.13 NH_3-N/L, recoveries were 96% and 91%, respectively. The results of an interlaboratory study involving 12 laboratories using the ammonia-selective electrode* on distilled water and effluents are summarized in Table 417:II.

For the automated phenate system, with Automated System I† in a single laboratory using surface water samples at concentrations of 1.41, 0.77, 0.59, and 0.43 mg NH_3-N/L, the standard deviation was +0.005, and at concentrations of 0.16 and 1.44 mg NH_3-N/L, recoveries were 107 and 99%, respectively.

*American Society For Testing and Materials. ASTM Method 1426-79.

†AutoAnalyzer I, Technicon Instrument Corporation, Tarrytown, N.Y. 10591.

TABLE 417:II. PRECISION AND BIAS OF AMMONIA-SELECTIVE ELECTRODE

Level *mg/L*	Matrix	Mean Recovery %	Precision	
			Overall, S_T	Operator, S_O
0.04	Distilled water	200	0.05	0.01
	Effluent water	100	0.03	0.00
0.10	Distilled water	180	0.05	0.01
	Effluent water	470	0.61	0.01
0.80	Distilled water	105	0.11	0.04
	Effluent water	105	0.30	0.06
20	Distilled water	95	2	1
	Effluent water	95	3	2
100	Distilled water	98	5	2
	Effluent water	97	—	—
750	Distilled water	97	78	12
	Effluent water	99	106	10

417 A. Preliminary Distillation Step

1. General Discussion

The sample is buffered at pH 9.5 with a borate buffer to decrease hydrolysis of cyanates and organic nitrogen compounds. It is distilled into a solution of boric acid when nesslerization or titration is to be used or into H_2SO_4 when the phenate method is used. The ammonia in the distillate can be determined either colorimetrically by nesslerization or the phenate method or titrimetrically with standard H_2SO_4 and a mixed indicator or a pH meter. The choice between the colorimetric or acidimetric method depends on the concentration of ammonia. Ammonia in the distillate also can be determined by the ammonia-selective electrode method, using 0.04*N* H_2SO_4 to trap the distillate.

2. Apparatus

a. Distillation apparatus: Arrange a borosilicate glass flask of 800 to 2,000-mL capacity attached to a vertical condenser so that the outlet tip may be submerged to at least 2 cm below the surface of the receiving acid solution. Use an all-borosilicate-glass apparatus or one with condensing units constructed of block tin or aluminum tubes.

b. pH meter.

3. Reagents

a. Ammonia-free water: Prepare by ion-exchange or distillation methods:

1) Ion exchange—Prepare ammonia-free water by passing distilled water through an ion-exchange column containing a strongly acidic cation-exchange resin mixed with a strongly basic anion-exchange resin. Select resins that will remove organic compounds that interfere with the ammonia determination. Some anion-exchange resins tend to release ammonia. If this occurs, prepare ammonia-free water with a strongly acidic cation-exchange resin. Regenerate the column according to the manufacturer's instructions. Check ammonia-free water for the possibility of a high blank value.

2) Distillation—Eliminate traces of ammonia in distilled water by adding 0.1 mL conc H_2SO_4 to 1 L distilled water and redistilling. Alternatively, treat distilled water with sufficient bromine or chlorine water to produce a free halogen residual of 2 to 5 mg/L and redistill after standing at least 1 hr. Discard the first 100 mL distillate. Check redistilled water for the possibility of a high blank.

It is very difficult to store ammonia-free water in the laboratory without contamination from ammonia fumes. However, if storage is necessary, store in a tightly stoppered glass container to which is added about 10 g ion-exchange resin (preferably a strongly acidic cation-exchange resin) per liter ammonia-free water. For use, let resin settle and decant ammonia-free water. If a high blank value is produced, replace the resin or prepare fresh ammonia-free water.

Prepare all reagents with ammonia-free distilled water.

b. Borate buffer solution: Add 88 mL 0.1*N* NaOH solution to 500 mL approximately 0.025*M* sodium tetraborate ($Na_2B_4O_7$) solution (9.5 g $Na_2B_4O_7 \cdot 10H_2O$/L) and dilute to 1 L.

c. Sodium hydroxide, 6N: Dissolve 240 g NaOH in ammonia-free distilled water and dilute to 1 L.

d. Dechlorinating agent, N/70: Use 1 mL of any of the following reagents to remove 1 mg/L residual chlorine in 500 mL sample.

1) *Phenylarsine oxide:* Dissolve 1.2 g C_6H_5AsO in 200 mL 0.3*N* NaOH solution, filter if necessary, and dilute to 1 L with ammonia-free water. (CAUTION: *Toxic—take care to avoid ingestion.*)

2) *Sodium arsenite:* Dissolve 1.0 g $NaAsO_2$ in ammonia-free water and dilute to 1 L. (CAUTION: *Toxic—take care to avoid ingestion.*) Prepare fresh weekly.

3) *Sodium sulfite:* Dissolve 0.9 g Na_2SO_3 in ammonia-free water and dilute to 1 L. Prepare fresh daily.

4) *Sodium thiosulfate:* Dissolve 3.5 g $Na_2S_2O_3 \cdot 5H_2O$ in ammonia-free water and dilute to 1 L. Prepare fresh weekly.

e. Neutralization agent: Prepare with ammonia-free water.

1) *Sodium hydroxide,* NaOH, 1*N*.

2) *Sulfuric acid,* H_2SO_4, 1*N*.

f. Absorbent solution, plain boric acid: Dissolve 20 g H_3BO_3 in ammonia-free water and dilute to 1 L.

g. Indicating boric acid solution: See Section 417D.3*a* and *b*.

h. Sulfuric acid, 0.04*N*: Dilute 1.0 mL conc H_2SO_4 to 1 L.

4. Procedure

a. Preparation of equipment: Add 500 mL ammonia-free water and 20 mL borate buffer to a distillation flask and adjust pH to 9.5 with 6*N* NaOH solution. Add a few glass beads or boiling chips and use this mixture to steam out the distillation apparatus until distillate shows no traces of ammonia.

b. Sample preparation: Use 500 mL of dechlorinated sample or a portion diluted to 500 mL with ammonia-free water. When NH_3-N concentration is less than 100 μg/L, use a sample volume of 1,000 mL. Remove residual chlorine by adding, at the time of collection, dechlorinating agent equivalent to the chlorine residual. If necessary, neutralize to approximately pH 7 with dilute acid or base, using a pH meter.

Add 25 mL borate buffer solution and adjust to pH 9.5 with 6*N* NaOH using a pH meter.

c. Distillation: To minimize contamination, leave distillation apparatus assembled after steaming out and until just before starting sample distillation. Disconnect steaming-out flask and immediately transfer sample flask to distillation apparatus. Distill at a rate of 6 to 10 mL/min with the tip of the delivery tube submerged to at least 2 cm below the surface of acid receiving solution. Collect distillate in a 500-mL erlenmeyer flask containing 50 mL plain boric acid solution for nesslerization method. Use 50 mL indicating boric acid solution for titrimetric method. Distill ammonia into 50 mL 0.04*N* H_2SO_4 for the phenate method and for the ammonia-selective electrode method. Collect at least 200 mL distillate. Lower collected distillate free of contact with delivery tube and continue distillation during the last minute or two to cleanse condenser and delivery tube. Dilute to 500 mL with ammonia-free water.

When the phenate method is used for determining NH_3-N, neutralize distillate with 1*N* NaOH solution.

d. Ammonia determination: Determine ammonia by the nesslerization method (Section 417B), the phenate method (Section 417C), the titrimetric method (Section 417D) or the ammonia-selective electrode method (Section 417E).

417 B. Nesslerization Method (Direct and Following Distillation)

1. General Discussion

Use direct nesslerization only for purified drinking waters, natural water, and highly purified wastewater effluents, all of which should be low in color and have NH_3-N concentrations exceeding 20 μg/L. Apply the direct nesslerization method to domestic wastewaters only when errors of 1 to 2 mg/L are acceptable. Use this method only after it has been established that it yields results comparable to those obtained after distillation. Check validity of direct nesslerization measurements periodically.

Pretreatment before direct nesslerization with zinc sulfate and alkali precipitates calcium, iron, magnesium, and sulfide, which form turbidity when treated with nessler reagent. The floc also removes suspended matter and sometimes colored matter. Addition of EDTA or Rochelle salt solution inhibits precipitation of residual calcium and magnesium ions in the presence of the alkaline nessler reagent. However, use of EDTA demands an extra amount of nessler reagent to insure a sufficient nessler reagent excess for reaction with the ammonia.

The graduated yellow to brown colors produced by the nessler-ammonia reaction absorb strongly over a wide wavelength range. The yellow color characteristic of low ammonia nitrogen concentration (0.4 to 5 mg/L) can be measured with acceptable sensitivity in the wavelength region from 400 to 425 nm when a 1-cm light path is available. A light path of 5 cm extends measurements into the nitrogen concentration range of 5 to 60 μg/L. The reddish brown hues typical of ammonia nitrogen levels approaching 10 mg/L may be measured in the wavelength region of 450 to 500 nm. A judicious selection of light path and wavelength thus permits the photometric determination of ammonia nitrogen concentrations over a considerable range.

Departures from Beer's law may be evident when photometers equipped with broad-band color filters are used. For this reason, prepare the calibration curve under conditions identical with those adopted for the samples.

A carefully prepared nessler reagent may respond under optimum conditions to as little as 1 μg NH_3-N/50 mL. In direct nesslerization, this represents 20 μg/L. However, reproducibility below 100 μg/L may be erratic.

2. Apparatus

a. Colorimetric equipment: One of the following is required:

1) *Spectrophotometer,* for use at 400 to 500 nm and providing a light path of 1 cm or longer.

2) *Filter photometer,* providing a light path of 1 cm or longer and equipped with a violet filter having maximum transmittance at 400 to 425 nm. A blue filter can be used for higher NH_3-N concentrations.

3) *Nessler tubes,* matched, 50-mL, tall form.

b. pH meter, equipped with a high-pH electrode.

3. Reagents

Prepare all reagents with ammonia-free water. All the reagents listed in Preliminary Distillation, Section 417A, except the borate buffer and absorbent solution, are required, plus the following:

a. Zinc sulfate solution: Dissolve 100 g $ZnSO_4 \cdot 7H_2O$ and dilute to 1 L.

b. Stabilizer reagent: Use either EDTA or Rochelle salt to prevent calcium or magnesium precipitation in undistilled samples after addition of alkaline nessler reagent.

1) *EDTA reagent:* Dissolve 50 g disodium ethylenediamine tetraacetate dihydrate in 60 mL water containing 10 g NaOH. If necessary, apply gentle heat to complete dissolution. Cool to room temperature and dilute to 100 mL.

2) *Rochelle salt solution:* Dissolve 50 g potassium sodium tartrate tetrahydrate, $KNaC_4H_4O_6 \cdot 4H_2O$, in 100 mL water. Remove ammonia usually present in the salt by boiling off 30 mL of solution. After cooling, dilute to 100 mL.

c. Nessler reagent: Dissolve 100 g HgI_2 and 70 g KI in a small quantity of water and add this mixture slowly, with stirring, to a cool solution of 160 g NaOH in 500 mL water. Dilute to 1 L. Store in rubber-stoppered borosilicate glassware and out of sunlight to maintain reagent stability for up to a year under normal laboratory conditions. Check reagent to make sure that it yields the characteristic color with 0.1 mg NH_3-N/L within 10 min after addition and does not produce a precipitate with small amounts of ammonia within 2 hr. (CAUTION: *Toxic—take care to avoid ingestion.*)

d. Stock ammonium solution: Dissolve 3.819 g anhydrous NH_4Cl, dried at 100 C, in water, and dilute to 1,000 mL; 1.00 mL = 1.00 mg N = 1.22 mg NH_3.

e. Standard ammonium solution: Dilute 10.00 mL stock ammonium solution to 1,000 mL with water; 1.00 mL = 10.00 μg N = 12.2 μg NH_3.

f. Permanent color solutions:

1) *Potassium chloroplatinate solution:* Dissolve 2.0 g K_2PtCl_6 in 300 to 400 mL distilled water; add 100 mL conc HCl and dilute to 1 L.

2) *Cobaltous chloride solution:* Dissolve 12.0 g $CoCl_2 \cdot 6H_2O$ in 200 mL distilled water. Add 100 mL conc HCl and dilute to 1 L.

4. Procedure

a. Treatment of undistilled samples: If necessary, remove residual chlorine from the freshly collected sample by adding an equivalent amount of *N*/70 dechlorinating agent. (Do not store chlorinated samples without prior dechlorination.) Add 1 mL $ZnSO_4$ solution to 100 mL sample and mix thoroughly. Add 0.4 to 0.5 mL 6*N* NaOH solution to obtain a pH of 10.5, as determined with a pH meter and a high-pH glass electrode, and mix gently. Let treated sample stand for a few minutes, whereupon a heavy flocculent precipitate should fall, leaving a clear and colorless supernate. Clarify by centrifuging or filtering. Pretest any filter paper used to be sure no ammonia is present as a contaminant. Do this by running ammonia-free water through the filter and testing the filtrate by nesslerization. Filter sample, discarding first 25 mL filtrate. (CAUTION: Samples containing more than about 10 mg NH_3-N/L may lose ammonia during this treatment of undistilled samples because of the high pH. Dilute such samples to the sensitive range for nesslerization before pretreatment.)

b. Color development:

1) Undistilled samples—Use 50.0 mL sample or a portion diluted to 50.0 mL with ammonia-free water. If the undistilled portion contains sufficient concentrations of calcium, magnesium, or other ions that produce turbidity or precipitate with nessler reagent, add 1 drop (0.05 mL) EDTA reagent or 1 to 2 drops (0.05 to 0.1 mL) Rochelle salt solution. Mix well. Add 2.0 mL nessler reagent if EDTA reagent is used or 1.0 mL nessler reagent if Rochelle salt is used.

2) Distilled samples—Neutralize the boric acid used for absorbing the ammonia distillate by either adding 2 mL nessler reagent, an excess that raises the pH to the desired high level, or alternatively, neutralizing the boric acid with NaOH before adding 1 mL nessler reagent.

3) Mix samples by capping nessler tubes with clean rubber stoppers (washed thoroughly with ammonia-free water) and then inverting tubes at least six times. Keep such conditions as temperature and reaction time the same in blank, samples, and standards. Let reaction proceed for at least 10 min after adding nessler reagent. Measure color in sample and standards. If NH_3-N is very low, use a 30-min contact time for sample, blank, and standards. Measure color either photometrically or visually as directed in ¶s *c* or *d* below.

c. Photometric measurement: Measure absorbance or transmittance with a spectrophotometer or filter photometer. Prepare calibration curve at the same temperature and reaction time used for samples.

Measure transmittance readings against a reagent blank and run parallel checks frequently against standards in the nitrogen range of the samples. Redetermine complete calibration curve for each new batch of nessler reagent.

For distilled samples, prepare standard curve under the same conditions as the samples. Distill reagent blank and appropriate standards, each diluted to 500 mL, in the same manner as the samples. Dilute 300 mL distillate plus 50 mL boric acid absorbent to 500 mL using ammonia-free water and take a 50-mL portion for nesslerization.

d. Visual comparison: Compare colors produced in sample against those of ammonia standards. Prepare temporary or permanent standards as follows:

1) Temporary standards—Prepare a series of visual standards in nessler tubes by adding the following volumes of standard NH_4Cl solution and diluting to 50 mL with ammonia-free water: 0, 0.2, 0.4, 0.7, 1.0, 1.4, 1.7, 2.0, 2.5, 3.0, 3.5, 4.0, 4.5, 5.0, and 6.0 mL. Nesslerize standards and portions of distillate by adding 1.0 mL nessler reagent to each tube and mixing well.

2) Permanent standards—Measure into 50-mL nessler tubes the volumes of K_2PtCl_6 and $CoCl_2$ solutions indicated in Table 417:III, dilute to mark, and mix thoroughly. The values given in the table are *approximate;* actual equivalents of the ammonium standards will differ with the quality of nessler reagent, the kind of illumination used, and the color sensitivity of the analyst's eye. Therefore, compare color standards with nesslerized temporary ammonia standards and modify the tints as necessary. Make such comparisons for each newly prepared nessler reagent and satisfy each analyst as to the aptness of the color match. Protect standards from dust to extend their usefulness for several months. Compare either 10 or 30 min after nesslerization, depending on reaction time used in preparing nesslerized ammonium standards against which they were matched.

5. Calculation

a. Deduct amount of NH_3-N in ammonia-free water used for diluting original sample before computing final nitrogen value.

b. Deduct also reagent blank for volume of borate buffer and 6*N* NaOH solutions used with sample.

c. Compute total NH_3-N by the following equation:

$$\text{mg } NH_3\text{-N/L (51 mL final volume)} = \frac{A}{\text{mL sample}} \times \frac{B}{C}$$

where:

A = µg NH_3-N (51 mL final volume),
B = total volume distillate collected, mL, including acid absorbent, and

TABLE 417:III. PREPARATION OF PERMANENT COLOR STANDARDS FOR VISUAL DETERMINATION OF AMMONIA NITROGEN

Value in Ammonia Nitrogen *µg*	Approximate Volume of Platinum Solution *mL**	Approximate Volume of Cobalt Solution *mL**
0	1.2	0.0
2	2.8	0.0
4	4.7	0.1
7	5.9	0.2
10	7.7	0.5
14	9.9	1.1
17	11.4	1.7
20	12.7	2.2
25	15.0	3.3
30	17.3	4.5
35	19.0	5.7
40	19.7	7.1
45	19.9	8.7
50	20.0	10.4
60	20.0	15.0

* In matched 50-mL nessler tubes.

C = volume distillate taken for nesslerization, mL.

The ratio B/C applies only to distilled samples; ignore in direct nesslerization.

6. Precision and Accuracy

See Section 417A and Table 417:I.

417 C. Phenate Method

1. General Discussion

a. Principle: An intensely blue compound, indophenol, is formed by the reaction of ammonia, hypochlorite, and phenol catalyzed by a manganous salt.

b. Interference: Alkalinity over 500 mg as $CaCO_3$/L, acidity over 100 mg as $CaCO_3$/L, color, and turbidity interfere. Remove these interferences by preliminary distillation.

2. Apparatus

a. Colorimetric equipment: One of the following is required:

1) *Spectrophotometer,* for use at 630 nm with a light path of approximately 1 cm.

2) *Filter photometer,* equipped with a red-orange filter having a maximum transmittance near 630 nm and providing a light path of approximately 1 cm.

b. Magnetic stirrer.

3. Reagents

a. Ammonia-free water: Prepare as directed in Section 417A.3*a*.

b. Hypochlorous acid reagent: To 40 mL distilled water add 10 mL 5% NaOCl solution prepared from commercial bleach. Adjust pH to 6.5 to 7.0 with HCl. Prepare this unstable reagent weekly.

c. Manganous sulfate solution, 0.003*M*: Dissolve 50 mg $MnSO_4 \cdot H_2O$ in 100 mL distilled water.

d. Phenate reagent: Dissolve 2.5 g NaOH and 10 g phenol, C_6H_5OH, in 100 mL ammonia-free water. Because this reagent darkens on standing, prepare weekly. (CAUTION: *Handle phenol with care.*)

e. Stock ammonium solution: Dissolve 381.9 mg anhydrous NH_4Cl, dried at 100 C, in ammonia-free water, and dilute to 1,000 mL; 1.00 mL = 100 μg N = 122 μg NH_3.

f. Standard ammonium solution: Dilute 5.00 mL stock ammonium solution to 1,000 mL with ammonia-free water; 1.00 mL = 0.500 μg N = 0.607 μg NH_3.

4. Procedure

a. Treatment of sample: To a 10.0-mL sample in a 50-mL beaker, add 1 drop (0.05 mL) $MnSO_4$ solution. Place on a magnetic stirrer and add 0.5 mL hypochlorous acid reagent. Immediately add, a drop at a time, 0.6 mL phenate reagent. Add reagent without delay using a bulb pipet or a buret for convenient delivery. Mark pipet for hypochlorous acid at the 0.5-mL level and deliver the phenate reagent from a pipet or a buret that has been calibrated by counting the number of drops previously found to be equivalent to 0.6 mL. Stir vigorously during addition of reagents. Because color intensity is affected by age of reagents, carry a blank and a standard through the procedure with each batch of samples. Measure absorbance using reagent blank to zero the spectrophotometer. Color formation is complete in 10 min and is stable for at least 24 hr. Although the blue color has a maximum absorbance at 630 nm, satisfactory measurements can be made in the 600- to 660-nm region.

b. Preparation of standards: Prepare a calibration curve in the NH_3-N range of 0.1 to 5 μg, treating standards exactly as the sample. Beer's Law governs.

5. Calculation

Calculate ammonia concentration as follows:

$$\text{mg } NH_3\text{-N/L (11.1 mL final volume)} = \frac{A \times B}{C \times S} \times \frac{D}{E}$$

where:

A = absorbance of sample,
B = NH_3-N in standard, μg,
C = absorbance of standard,
S = volume of sample used, mL,
D = volume of total distillate collected, mL, including acid absorbent, neutralizing agent, and ammonia-free water added, and
E = volume of distillate used for color development, mL.

The ratio D/E applies only to distilled samples.

6. Precision and Accuracy

See Section 417A.4 and Table 417:I.

417 D. Titrimetric Method

1. General Discussion

The titrimetric method is used only on samples that have been carried through preliminary distillation (see Section 417A). The following table is useful in selecting sample volume for the distillation and titration method.

Ammonia Nitrogen in Sample *mg/L*	Sample Volume *mL*
5–10	250
10–20	100
20–50	50.0
50–100	25.0

2. Apparatus

Distillation apparatus: See Section 417A.2*a* and *b*.

3. Reagents

a. Mixed indicator solution: Dissolve 200 mg methyl red indicator in 100 mL 95% ethyl or isopropyl alcohol. Dissolve 100 mg methylene blue in 50 mL 95% ethyl or isopropyl alcohol. Combine solutions. Prepare monthly.

b. Indicating boric acid solution: Dissolve 20 g H_3BO_3 in ammonia-free distilled water, add 10 mL mixed indicator solution, and dilute to 1 L. Prepare monthly.

c. Standard sulfuric acid titrant, 0.02*N:* Prepare and standardize as directed in Alkalinity, Section 403.3*c*. For greatest accuracy, standardize titrant against an amount of Na_2CO_3 that has been incorporated in the indicating boric acid solution to reproduce the actual conditions of sample titration; 1.00 mL = 280 μg N.

4. Procedure

a. Proceed as described in Section 417A using indicating boric acid solution as absorbent for the distillate.

b. Sludge or sediment samples: Rapidly weigh to within ±1% an amount of wet sample, equivalent to approximately 1 g dry weight, in a weighing bottle or crucible. Wash sample into a 500-mL kjeldahl flask with ammonia-free distilled water and dilute to 250 mL. Proceed as in ¶ 4*a* but add a piece of paraffin wax to distillation flask and collect only 100 mL distillate.

c. Titrate ammonia in distillate with

standard 0.02*N* H_2SO_4 titrant until indicator turns a pale lavender.

d. Blank: Carry a blank through all steps of the procedure and apply the necessary correction to the results.

5. Calculation

a. Liquid samples:

$$\text{mg } NH_3\text{-N/L} = \frac{(A - B) \times 280}{\text{mL sample}}$$

b. Sludge or sediment samples:

$$\text{mg } NH_3\text{-N/kg} = \frac{(A - B) \times 280}{\text{g dry wt sample}}$$

where:

A = volume of H_2SO_4 titrated for sample, mL, and

B = volume of H_2SO_4 titrated for blank, mL.

6. Precision and Accuracy

See Section 417A.4 and Table 417:I.

417 E. Ammonia-Selective Electrode Method (TENTATIVE)

1. General Discussion

a. Principle: The ammonia-selective electrode uses a hydrophobic gas-permeable membrane to separate the sample solution from an electrode internal solution of ammonium chloride. Dissolved ammonia ($NH_{3(aq)}$ and NH_4^+) is converted to $NH_{3(aq)}$ by raising the sample pH to above 11 with a strong base. $NH_{3(aq)}$ diffuses through the membrane and changes the internal solution pH that is sensed by a pH electrode. The fixed level of chloride in the internal solution is sensed by a chloride ion-selective electrode that serves as the reference electrode. Potentiometric measurements are made with a pH meter having an expanded millivolt scale or with a specific ion meter.

b. Scope and application: This method is applicable to the measurement of 0.03 to 1,400 mg NH_3-N/L in potable and surface waters and domestic and industrial wastes. Color and turbidity have no effect on the measurement. The measurement is affected by high concentrations of dissolved ions. Sample distillation is unnecessary. Use standard solutions and samples that have the same temperature and contain about the same total level of dissolved species. The ammonia-selective electrode responds slowly below 1 mg NH_3-N/L; hence, use longer times of electrode immersion (5 to 10 min) to obtain stable readings.

c. Interference: Amines are a positive interference. Mercury and silver interfere by complexing with ammonia.

d. Sample preservation: Do not use $HgCl_2$ as a sample preservative. Refrigerate at 4 C for samples to be analyzed within 24 hr. Preserve samples high in organic and nitrogenous matter, and any other samples for a prolonged period, by lowering pH to 2 or less with conc H_2SO_4.

2. Apparatus

a. Electrometer: A pH meter with expanded millivolt scale capable of 0.1 mV resolution between −700 mV and +700 mV or a specific ion meter.

*b. Ammonia-selective electrode**.

c. Magnetic stirrer, thermally insulated, with TFE-coated stirring bar.

*Orion Model 95-10, EIL Model 8002-2, Beckman Model 39565, or equivalent.

3. Reagents

a. Ammonia-free water: See Section 417A.3*a*.

b. Sodium hydroxide, 10*N*: Dissolve 400 g NaOH in 800 mL ammonia-free water. Cool and dilute to 1,000 mL with ammonia-free water.

c. Stock ammonium chloride solution: See Section 417B.3*d*.

d. Standard ammonium chloride solutions: See 417E.4*a*.

4. Procedure

a. Preparation of standards: Prepare a series of standard solutions covering the concentrations of 1,000, 100, 10, 1, and 0.1 mg NH_3-N/L by making decimal dilutions of stock NH_4Cl solution with ammonia-free water.

b. Electrometer calibration: Place 100 mL of each standard solution in a 150-mL beaker. Immerse electrode in standard of lowest concentration and mix with a magnetic stirrer. Do not stir so rapidly that air bubbles are sucked into the solution because they will become trapped on the electrode membrane. Maintain the same stirring rate and a temperature of about 25 C throughout calibration and testing procedures. Add a sufficient volume of 10*N* NaOH solution (1 mL is usually sufficient) to raise pH above 11. Keep electrode in solution until a stable millivolt reading is obtained. CAUTION: Check electrode sensing element performance according to manufacturer's instructions to make sure that electrode is operating properly. Do not add NaOH solution before immersing electrode, because ammonia may be lost from a basic solution. Repeat procedure with remaining standards, proceeding from lowest to highest concentration. Wait for at least 5 min before recording millivolts for standards and samples containing $\leqq$1 mg NH_3-N/L.

c. Preparation of standard curve: Using semilogarithmic graph paper, plot ammonia concentration in milligrams NH_3-N per liter on the log axis vs. potential in millivolts on the linear axis starting with the lowest concentration at the bottom of the scale. If the electrode is functioning properly a tenfold change of NH_3-N concentration produces a potential change of 59 mV.

d. Calibration of specific ion meter: Refer to manufacturer's instructions and proceed as in ¶s 4*a* and *b*.

e. Measurement of samples: Dilute if necessary to bring NH_3-N concentration to within calibration curve range. Place 100 mL sample in 150-mL beaker and follow procedure in ¶ 4*b* above. Record volume of 10*N* NaOH added in excess of 1 mL. Read NH_3-N concentration from standard curve.

5. Calculation

$$\text{mg } NH_3\text{-N/L} = A \times B \times \left[\frac{101 + C}{101}\right]$$

where:

A = dilution factor,
B = concentration of NH_3-N/L, mg/L, from calibration curve, and
C = volume of added 10*N* NaOH in excess of 1 mL, mL.

6. Precision and Accuracy

See Section 417.4.

417 F. Automated Phenate Method (TENTATIVE)

1. General Discussion

a. Principle: Alkaline phenol and hypochlorite react with ammonia to form indophenol blue that is proportional to the ammonia concentration. The blue color formed is intensified with sodium nitroprusside.

b. Interferences: Seawater contains calcium and magnesium ions in sufficient concentrations to cause precipitation problems during analysis. Adding EDTA and sodium potassium tartrate reduces the problem. Eliminate any marked variation in acidity or alkalinity among samples because intensity of the measured color is pH-dependent. Likewise, insure that the pH of the wash water and standard ammonia solutions approximates that of sample. For example, if sample has been preserved with 0.8 mL conc H_2SO_4/L, include 0.8 mL conc H_2SO_4/L in wash water and standards. Mercuric chloride used as a preservative gives a negative interference by complexing with the ammonia. Overcome this effect by adding a comparable amount of $HgCl_2$ to the ammonia standards. Remove interfering turbidity by filtration before analysis. Color in the samples that absorbs in the photometric range used for analysis interferes.

c. Application: Ammonia nitrogen can be determined in potable, surface, and saline waters as well as domestic and industrial wastewaters over a range of 0.02 to 2.0 mg/L when photometric measurement

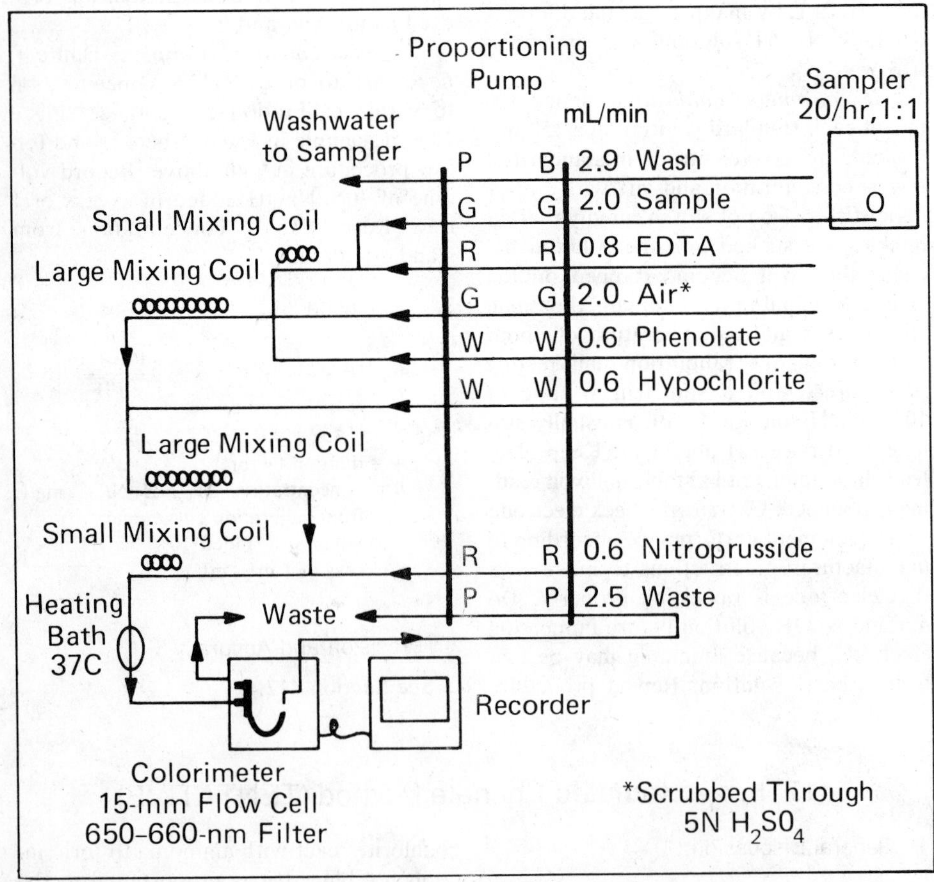

Figure 417:1. Ammonia manifold for Automated System I.

is made at 630 to 660 nm in a 15- or 50-mm tubular flow cell. Determine higher concentrations by diluting the sample.

2. Apparatus

Automated analytical equipment, consisting of the components listed in Part 600 with the following additions: heating bath with double delay coil, 15- or 50-mm tubular flow cell, and 630- or 650-nm filters.

3. Reagents

a. Ammonia-free distilled water: See Section 417A.3*a*. Use ammonia-free water for preparing all reagents and dilutions.

b. Sulfuric acid, H_2SO_4, 5*N*, air scrubber solution: Carefully add 139 mL conc H_2SO_4 to approximately 500 mL ammonia-free water, cool to room temperature, and dilute to 1 L.

c. Sodium phenate solution: In a 1-L erlenmeyer flask, dissolve 83 g phenol in 500 mL ammonia-free water. In small increments and with agitation, cautiously add 32 g NaOH. Cool flask under running water and dilute to 1 L.

d. Sodium hypochlorite solution: Dilute 250 mL bleach solution containing 5.25% NaOCl to 500 mL with ammonia-free water.

e. EDTA reagent: Dissolve 50 g disodium ethylenediamine tetraacetate and approximately six pellets NaOH in 1 L ammonia-free water. For salt-water samples where EDTA reagent does not prevent precipitation of cations, use sodium potassium tartrate solution prepared as follows:

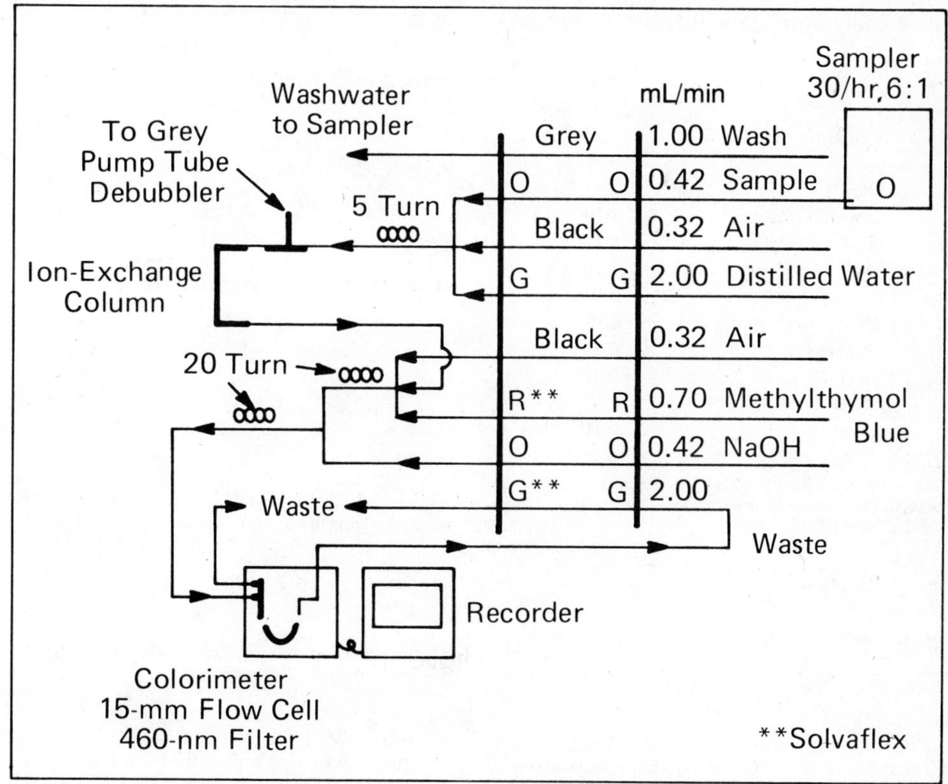

Figure 417:2. Ammonia manifold for Automated System II.

Sodium potassium tartrate solution: To 900 mL ammonia-free water add 100 g $NaKC_4H_4O_6 \cdot 4H_2O$, two pellets NaOH, and a few boiling chips, and boil gently for 45 min. Cover, cool, and dilute to 1 L. Adjust pH to 5.2 ± 0.05 with H_2SO_4. Let settle overnight in a cool place and filter to remove precipitate. Add 0.5 mL polyoxyethylene 23 lauryl ether* solution and store in stoppered bottle.

f. Sodium nitroprusside solution: Dissolve 0.5 g $Na_2(NO)Fe(CN)_5 \cdot 2H_2O$ in 1 L ammonia-free water.

g. Ammonia standard solutions: See Sections 417C.3*e* and *f*. Use standard ammonia solution and ammonia-free water to prepare the calibration curve in the appropriate ammonia concentration range. To analyze saline waters use substitute ocean water of the following composition to prepare calibration standards:

Constituent	Concentration *g/L*
NaCl	24.53
$MgCl_2$	5.20
$CaCl_2$	1.16
KCl	0.70
$SrCl_2$	0.03
Na_2SO_4	4.09
$NaHCO_3$	0.20
KBr	0.10
H_3BO_3	0.03
NaF	0.003

*Brij-35, available from IG United States, Chicago, Ill. 60631 or Technicon Instrument Corporation, Tarrytown, N.Y. 10591.

Subtract blank background response of substitute seawater from standards before preparing standard curve.

4. Procedure

a. Eliminate marked variation in acidity or alkalinity among samples. Adjust pH of wash water and standard ammonia solutions to approximately that of the sample.

b. Set up manifold and complete system as shown in Figure 417:1. (Automated System I*) or Figure 417:2 (Automated System II*).

c. Obtain a stable baseline with all reagents, feeding ammonia-free water through sample line.

d. For Automated System I, sample at a rate of 20/hr, 1:1; for Automated System II, use a 60/hr, 6:1 cam with a common wash.

e. Follow the general procedure described in Part 600.

5. Calculation

See Part 600.

6. Precision and Accuracy

See Section 417.4.

*Autoanalyzers I and II, Technicon Instrument Corporation, Tarrytown, N.Y. 10591.

417 G. Bibliography

Distillation & Nesslerization Methods

JACKSON, D.D. 1900. Permanent standards for use in the analysis of water. *Mass. Inst. Technol. Quart.* 13:314.

NICHOLS, M.S. & M.E. FOOTE. 1931. Distillation of free ammonia from buffered solutions. *Ind. Eng. Chem.*, Anal. Ed. 3:311.

GRIFFIN, A.E. & N.S. CHAMBERLIN, 1941. Relation of ammonia nitrogen to breakpoint chlorination. *Amer. J. Pub. Health* 31:803.

PALIN, A.T. 1950. Symposium on the sterilization of water. Chemical aspects of chlorination. *J. Inst. Water Eng.* 4:565.

SAWYER, C.N. 1953. pH adjustment for determination of ammonia nitrogen. *Anal. Chem.* 25:816.

TARAS, M.J. 1953. Effect of free residual chlorination of nitrogen compounds in water. *J. Amer. Water Works Ass.* 45:47.

BOLTZ, D.F., ed. 1958. Colorimetric Determination of Nonmetals. Interscience Publishers, New York, N.Y.

JENKINS, D. 1967. The differentiation, analysis and preservation of nitrogen and phosphorus forms in natural waters. *In* Trace Inorganics in Water. American Chemical Society, Washington, D.C.

Phenate Method

ROSSUM, J.R. & P.A. VILLARRUZ. 1963. Determination of ammonia by the indophenol method. *J. Amer. Water Works Ass.* 55:657.

WEATHERBURN, M.W. 1967. Phenolhypochlorite reaction for determination of ammonia. *Anal. Chem.* 39:971.

Distillation and Titration Method

MEEKER, E.W. & E.C. WAGNER. 1933. Titration of ammonia in the presence of boric acid. *Ind. Eng. Chem.*, Anal. Ed. 5:396.

WAGNER, E.C. 1940. Titration of ammonia in the presence of boric acid. *Ind. Eng. Chem.* Anal. Ed. 12:711.

Ammonia-Selective Electrode Method

BANWART, W.L., J.M. BREMNER & M.A. TABATABAI. 1972. Determination of ammonium in soil extracts and water samples by an ammonia electrode. *Comm. Soil Sci. Plant Anal.* 3:449.

MIDGLEY, C. & K. TERRANCE. 1972. The determination of ammonia in condensed steam and boiler feed-water with a potentiometric ammonia probe. *Analyst* 97:626.

BOOTH, R.L. & R.F. THOMAS. 1973. Selective electrode determination of ammonia in water and wastes. *Environ. Sci. Technol.* 7:523.

U.S. ENVIRONMENTAL PROTECTION AGENCY. 1974. Methods for Chemical Analysis of Water and Wastes. EPA-625-/6-74-003.

Automated Phenate Method

HILLER, A. & D. VAN SLYKE. 1933. Determination of ammonia in blood. *J. Biol. Chem.* 102:499.

FIORE, J. & J.E. O'BRIEN. 1962. Ammonia determination by automatic analysis. *Wastes Eng.* 33:352.

AMERICAN SOCIETY FOR TESTING AND MATERIALS. 1966. Manual on Industrial Water and Industrial Waste Water. 2nd ed. ASTM, Philadelphia, Pa.

O'CONNOR, B., R. DOBBS, B. VILLIERS & R. DEAN. 1967. Laboratory distillation of municipal waste effluents. *J. Water Pollut. Control Fed.* 39:25.

BOOTH, R.L. & L.B. LOBRING. 1973. Evaluation of the AutoAnalyzer II: A progress report. *In* Advances in Automated Analysis: 1972 Technicon International Congress, Vol. 8, p. 7. Mediad Inc., Tarrytown, N.Y.

418 NITROGEN (NITRATE)

1. Selection of Method

Determination of nitrate (NO_3^-) is difficult because of the relatively complex procedures required, the high probability that interfering constituents will be present, and the limited concentration ranges of the various technics. This section includes two screening technics for determining approximate NO_3^- concentration. A choice of technics, depending on concentration range and/or presence of interferences, is presented.

Screening methods are: (*a*) an ultraviolet (UV) technic that measures the absorbance of NO_3^- at 220 nm and is suitable for uncontaminated waters (low in organic matter) and (*b*) an NO_3^- electrode method that may be used in either unpolluted water or wastewater.

Once a sample is screened, if necessary, select a method suitable for its concentration range. For concentrations below 0.1

mg NO_3^--N/L, use the cadmium reduction method (C). An automated version of this method also is presented (F). For concentrations from 0.1 to 1.0 mg NO_3^--N/L, the cadmium reduction technic (C) may be used. The chromotropic acid procedure (D) is applicable to concentrations from 0.1 to 5.0 mg NO_3^--N/L. For higher NO_3^--N concentrations, dilute into the range of the other methods or use the Devarda's alloy reduction method (E) for "total oxidized nitrogen" (NO_3^--N + NO_2^--N).

2. Storage of Samples

Start NO_3^- determinations promptly after sampling. If storage is necessary, preserve samples with 40 mg $HgCl_2$/L (1 mL saturated $HgCl_2$/L) and store at 4 C or freeze at −20 C. Note that mercuric ion accelerates the degradation of the cadmium column used in the cadmium reduction method.

418 A. Ultraviolet Spectrophotometric Screening Method

1. General Discussion

a. Principle: Use this technic only for screening samples that have low organic matter contents, i.e., uncontaminated natural waters and potable water supplies.

Measurement of UV absorption at 220 nm enables rapid determination of NO_3^-. The NO_3^- calibration curve follows Beer's law up to 11 mg N/L. Because dissolved organic matter also may absorb at 220 nm and NO_3^- does not absorb at 275 nm, a second measurement made at 275 nm may be used to correct the NO_3^- value. The extent of this empirical correction is related to the nature and concentration of organic matter and may vary from one water to another. Consequently, this method is not recommended for waters requiring a significant correction for organic matter absorbance although it may be useful in monitoring NO_3^- levels within a water body in which the nature of the dissolved organic matter remains constant. Correction factors for organic matter absorbance can be established by the method of additions in combination with analysis of the original NO_3^- content by another method. Filtration of the sample is intended to remove possible interference from suspended particles. Acidification with 1 *N* HCl is designed to prevent interference from hydroxide or carbonate concentrations up to 1,000 mg $CaCO_3$/L. Chloride has no effect on the determination.

b. Interference: Dissolved organic matter, surfactants, nitrite, and hexavalent chromium interfere. Various inorganic ions not normally found in natural waters, such as chlorite and chlorate, may interfere. Inorganic substances can be compensated for by independent analysis of their concentrations and preparation of individual correction curves.

2. Apparatus

a. Spectrophotometer, for use at 220 nm and 275 nm with matched silica cells of 1-cm or longer light path.

b. Filter: One of the following is required:

1) *Membrane filter:* 0.45-μm pore diameter membrane filter and appropriate filter assembly.

2) *Paper:* Acid-washed, ashless hard-finish filter paper sufficiently retentive for fine precipitates.

c. Nessler tubes, 50 mL, short form.

3. Reagents

a. Nitrate-free water: Use redistilled or distilled, deionized water of highest purity to prepare all solutions and dilutions.

b. Stock nitrate solution: Dry potassium nitrate (KNO_3), in an oven at 105 C for 24 hr. Dissolve 0.7218 g in water and dilute to 1,000 mL. Preserve with 2 mL $CHCl_3$/L; 1.00 mL = 100 μg NO_3^--N. This solution is stable for at least 6 months.

c. Standard nitrate solution: Dilute 50.0 mL stock nitrate solution to 500 mL with water; 1.00 mL = 10.0 μg NO_3^--N.

d. Hydrochloric acid solution, HCl, 1*N*.

4. Procedure

a. Treatment of sample: To 50 mL clear sample, filtered if necessary, add 1 mL HCl solution and mix thoroughly.

b. Preparation of standard curve: Prepare NO_3^- calibration standards in the range 0 to 7 mg NO_3^--N/L (0 to 350 μg NO_3^--N/50 mL) by diluting to 50 mL the following volumes of standard nitrate solution: 0, 1.00, 2.00, 4.00, 7.00 . . . 35.0 mL. Treat NO_3^- standards in same manner as samples.

c. Spectrophotometric measurement: Read absorbance or transmittance against redistilled water set at zero absorbance or 100% transmittance. Use a wavelength of 220 nm to obtain NO_3^- reading and a wavelength of 275 nm to determine interference due to dissolved organic matter.

5. Calculation

a. Correction for dissolved organic matter: Subtract two times absorbance reading at 275 nm from reading at 220 nm to obtain absorbance due to NO_3^-. Convert this absorbance value to NO_3^--N from standard curve. Note: If correction value is more than 10% of the reading at 220 nm, do not use this method.

b. Calculate as follows:

$$\text{mg } NO_3^-\text{-N/L} = \frac{\text{net } \mu\text{g } NO_3^-\text{-N (in approx. 50 mL final volume)}}{\text{mL sample}}$$

418 B. Nitrate Electrode Screening Method

1. General Discussion

a. Principle: The NO_3^- ion electrode is a selective sensor that develops a potential across a thin, porous, inert membrane that holds in place a water-immiscible liquid ion exchanger. The electrode responds only to NO_3^- ion activity between about 10^{-5} and 10^{-1} *M* (0.14 to 1,400 mg NO_3^--N/L). The lower limit of detection is determined by the small but finite solubility of the liquid ion exchanger.

b. Interferences: Chloride and bicarbonate ions interfere when their ratios to NO_3^--N are >10 or >5, respectively. Ions that are potential interferences but do not normally occur at significant levels in potable waters are NO_2^-, CN^-, S^{2-}, Br^-, I^-, ClO_3^-, and ClO_4^-. Although the electrodes function satisfactorily in buffers over the range pH 3 to 9, erratic responses have been noted where pH is not held constant. Because the electrode responds to NO_3^- activity rather than concentration, ionic strength must be constant in all samples and standards. Minimize these problems by using a buffer solution containing Ag_2SO_4, to remove Cl^-, Br^-, I^-, S^{2-}, and CN^-, sulfamic acid to remove NO_2^-, a buffer at pH 3 to eliminate HCO_3^- and to maintain a constant pH and ionic strength, and $Al_2(SO_4)_3$ to complex organic acids.

2. Apparatus

a. pH meter, expanded-scale or digital, capable of reading 0.1 mV.

*b. Single-junction reference electrode.** Fill chamber with saturated K_2SO_4.

c. Nitrate ion electrode:† (NOTE: Carefully follow manufacturer's instructions regarding care, storage, and recharge of electrode.)

d. Magnetic stirrer: TFE-coated stirring bar.

3. Reagents

a. Nitrate-free water: Prepare as described in Section 418A.3*a*. Use for all solutions and dilutions.

b. Stock nitrate solution: Prepare as described in Section 418A.3*b*.

c. Standard nitrate solutions: Dilute 1.0, 10, and 50 mL stock nitrate solution to 100 mL with water to obtain standards solutions of 1.0, 10, and 50 mg NO_3^--N/L, respectively.

d. Buffer solution: Dissolve 6.66 g $Al_2(SO_4)_3 \cdot 18H_2O$, 3.12 g Ag_2SO_4, 1.24 g H_3BO_3, and 1.94 g sulfamic acid (H_2NSO_3H), in about 400 mL water. Adjust to pH 3.0 by slowly adding 0.10 *N* NaOH. Dilute to 1,000 mL.

e. Sodium hydroxide, NaOH, 0.1 *N*.

*Orion Model 90-01, Radiometer Model 601, or equivalent.

†Orion Model 92-07, Corning Model 476134, or equivalent.

4. Procedure

a. No major adjustment of any instrument normally is required to use the electrodes in the concentration range of 1.0 to 50 mg NO_3^--N/L.

b. Preparation of calibration curve: Transfer 10 mL of 1 mg NO_3^--N/L standard to a 50-mL beaker, add 10 mL buffer, and stir for a constant time (2 or 3 min) with a magnetic stirrer. Immerse tips of electrodes and record millivolt reading after 1 min. Remove electrodes, rinse, and blot dry. Repeat for 10 mg NO_3^--N/L and 50 mg NO_3^--N/L standards. Plot potential measurements against NO_3^--N concentration on two-cycle semilogarithmic graph paper, with NO_3^--N concentration on the logarithmic axis (abscissa) and potential (in millivolts) on the linear axis (ordinate). A straight line with a slope of +59 (+58 to +59 for solutions at 24 to 26 C) mV/decade should result. Recalibrate electrodes several times daily by checking potential reading of the 10 mg NO_3^--N standard and adjusting the calibration control on the meter until the reading plotted on the calibration curve is displayed again.

c. Measurement of sample: Transfer 10 mL sample to a 50-mL beaker, add 10 mL buffer solution, and stir with a magnetic stirrer for a constant time (2 or 3 min). Immerse electrode tips in sample and record potential reading after 1 min. Read concentration from calibration curve.

418 C. Cadmium Reduction Method

1. General Discussion

a. Principle: NO_3^- is reduced almost quantitatively to nitrite (NO_2^-) in the presence of cadmium (Cd). The method recommended here uses commercially available Cd granules treated with copper sulfate ($CuSO_4$), to form a Cu coating.

The NO_2^- produced thus is determined by diazotizing with sulfanilamide and coupling with N-(1 naphthyl)-ethylenediamine to form a highly colored azo dye that is measured colorimetrically. A correction may be made for any NO_2^- present in the sample by analyzing without the reduction step. The applicable range of this method is 0.01 to 1.0 mg NO_3^--N/L. The method especially is recommended for NO_3^- levels below 0.1 mg N/L where other methods lack adequate sensitivity.

b. Interferences: Suspended matter in the reduction column will restrict sample flow. Prefilter turbid samples through glass fiber or 0.45-μm pore size membrane filters. Concentrations of iron, copper, or other metals above several milligrams per liter lower reduction efficiency. Add EDTA to samples to eliminate this interference. Oil and grease will coat the Cd surface. Remove by pre-extraction with an organic solvent. Residual chlorine can interfere by oxidizing the Cd column, reducing its efficiency. Check samples for residual chlorine (see DPD method, Section 408). Remove residual chlorine by adding sodium thiosulfate ($Na_2S_2O_3$) solution [see Section 417A.3*d*4)].

2. Apparatus

*a. Reduction column:** Purchase or construct the column (Figure 418:1) from a 100-mL volumetric pipet by removing the top portion. The column also can be constructed from two pieces of tubing joined end to end: join a 10-cm length of 3-cm-ID tubing to a 25-cm length of 3.5-mm-ID tubing.

b. Colorimetric equipment: One of the following is required:

1) *Spectrophotometer,* for use at 543 nm, providing a light path of 1 cm or longer.

2) *Filter photometer,* with light path of 1 cm or longer and equipped with a yellow-green filter having maximum transmittance near 540 nm.

3. Reagents

a. Nitrate-free water: See 418A.3*a*. The absorbance of a reagent blank prepared with this water should not exceed 0.01. Use for all solutions and dilutions.

b. Copper-cadmium granules: Wash 25 g 40- to 60-mesh Cd granules† with 6*N* HCl and rinse with water. Swirl Cd with 100 mL 2% $CuSO_4$ solution for 5 min or until blue color partially fades. Decant and repeat with fresh $CuSO_4$ until a brown colloidal precipitate develops. Wash Cu-Cd copiously with water (at least 10 times) to remove all precipitated Cu.

c. Sulfanilamide reagent: Dissolve 5 g sulfanilamide in a mixture of 50 mL conc HCl and 300 mL water. Dilute to 500 mL with water. The solution is stable for many months.

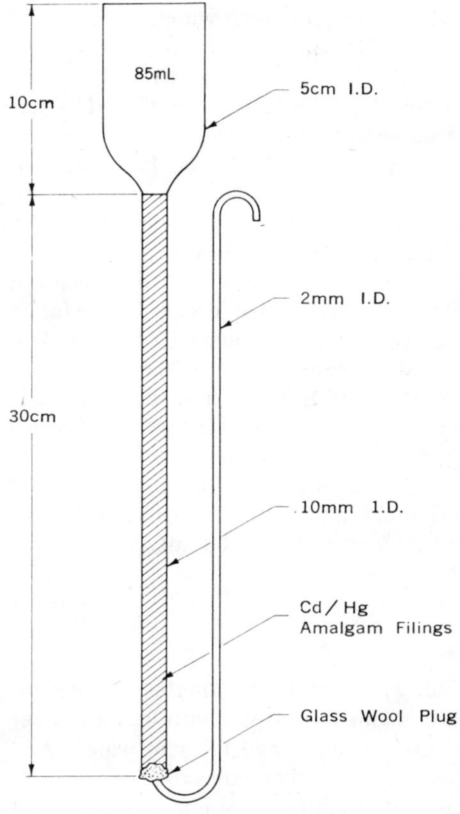

Figure 418:1: Reduction column.

*Tudor Scientific Glass Co., 555 Edgefield Road, Belvedere, S.C. 29841, Cat. TP-1730, or equivalent.

†EM Laboratories, Inc., 500 Exec. Blvd., Elmsford, N.Y. 10523, Cat. 2001, or equivalent.

d. N-(1-naphthyl)-ethylenediamine dihydrochloride (NED dihydrochloride) solution: Dissolve 500 mg NED dihydrochloride in 500 mL water. Store in a dark bottle. Replace monthly or as soon as a brown color develops. Prepare new calibration curve for each new batch of NED-dihydrochloride.

e. Ammonium chloride—EDTA solution: Dissolve 13 g NH_4Cl and 1.7 g disodium ethylenediamine tetraacetate in 900 mL water. Adjust pH to 8.5 with conc NH_4OH and dilute to 1 L.

f. Dilute ammonium chloride—EDTA solution: Dilute 300 mL NH_4Cl-EDTA solution to 500 mL with water.

g. Hydrochloric acid, HCl, 6*N*.

h. Copper sulfate solution, 2%: Dissolve 20 g $CuSO_4 \cdot 5H_2O$ in 500 mL water and dilute to 1 L.

i. Stock nitrate solution: Prepare as directed in Section 418A.3*b*.

j. Standard nitrate solution: Prepare as directed in Section 418A.3*c*.

k. Stock nitrite solution: Dissolve 0.6072 g KNO_2 (dried in a desiccator for 24 hr) in nitrite-free water (see Section 419) and dilute to 1,000 mL; 1.00 mL = 100 μg NO_2^--N. Preserve with 2 mL $CHCl_3$ and refrigerate. This is stable for approximately 3 months.

l. Standard nitrite solution: Dilute 50.0 mL stock nitrite solution to 500 mL with nitrite-free water; 1.00 mL = 10.0 μg NO_2^--N.

4. Procedure

a. Preparation of reduction column: Insert a glass wool plug into bottom of reduction column and fill with water. Add sufficient Cu-Cd granules to produce a column 18.5 cm long. Maintain water level above Cu-Cd granules to prevent entrapment of air. Wash column with 200 mL dilute NH_4Cl-EDTA solution. Activate column by passing through it, at 7 to 10 mL/min, 100 mL of a solution composed of 25 mL of a 1.0 mg NO_3^--N/L standard and 75 mL NH_4Cl-EDTA solution.

b. Treatment of sample:

1) Turbidity removal—If turbidity or suspended solids are present, remove by filtering through a 0.45-μm pore diameter membrane or glass fiber filter.

2) pH adjustment—If sample pH is below 5 or above 9, adjust to between 7 and 9 using a pH meter and dilute HCl or NaOH. This insures a pH of 8.5 after adding NH_4Cl-EDTA solution.

3) Sample reduction—To 25.0 mL sample or a portion diluted to 25.0 mL, add 75 mL NH_4Cl-EDTA solution and mix. Pour mixed sample into column and collect at a rate of 7 to 10 mL/min. Discard first 25 mL. Collect the rest in original sample flask. There is no need to wash columns between samples, but if columns are not to be reused for several hours or longer, pour 50 mL dilute NH_4Cl-EDTA solution on to the top and let it pass through the system. Store Cu-Cd column in this solution and never allow it to dry.

4) Color development and measurement—As soon as possible, and not more than 15 min after reduction, add 2.0 mL sulfanilamide reagent to 50 mL sample. Let reagent react for more than 2 min but less than 8 min. Add 2.0 mL NED-dihydrochloride solution and mix immediately. Between 10 min and 2 hr afterward, measure absorbance at 540 nm against a distilled water-reagent blank. NOTE: If NO_3^- concentration exceeds the standard curve range (about 1 mg N/L), use remainder of reduced sample to make an appropriate dilution and analyze again.

c. Standards: Using the standard NO_3^--N solution, prepare standards in the range 0.05 to 1.0 mg NO_3^--N/L by diluting the following volumes of standard to 100 mL in volumetric flasks: 0.5, 1.0, 2.0, 5.0, and 10.0 mL. Carry out reduction of standards exactly as described for samples. Compare at least one NO_2^- standard to a reduced NO_3^- standard at the same concentration to verify reduction column efficiency. Reactivate Cu-Cd granules as described in ¶ 3*b* above when efficiency of reduction falls below about 75%.

5. Calculation

Obtain a standard curve by plotting absorbance of standards against NO_3^--N concentration. Compute sample concentrations directly from standard curve. Report as milligrams oxidized N per liter (the sum of NO_3^--N plus NO_2^--N) unless the concentration of NO_2^--N is separately determined and corrected for.

418 D. Chromotropic Acid Method

1. General Discussion

a. Principle: Two moles of NO_3^- react with one mole of chromotropic acid to form a yellow reaction product with maximum absorbance at 410 nm. The maximum color develops within 10 min and is stable for 24 hr. A cooling bath dissipates sufficient heat to prevent boiling of the solutions; for this reason, the temperature of the cooling bath may vary from 10 to 20 C without critically affecting results. The method is recommended for the concentration range 0.1 to 5 mg NO_3^--N/L.

b. Interferences: Residual chlorine, certain oxidants, and NO_2^- yield yellow colors with chromotropic acid. Addition of sulfite eliminates interference from residual chlorine and oxidizing agents. Addition of urea converts NO_2^- to N_2 gas. Antimony effectively masks up to 2,000 mg Cl^-/L, a tolerance level that can be raised to 4,000 mg/L by doubling the strength of the specified antimony reagent. The yellow color of the chloroferrate (III) complex in amounts up to 40 mg Fe (III)/L is discharged by adding antimony. Ba^{2+}, Pb^{2+}, Sr^{2+}, I^-, IO_3^-, selenite, and selenate ions are incompatible with the system and form precipitates. However, their occurrence in significant amounts is unlikely in most samples. Concentrations of Cr^{3+} ion exceeding 20 mg/L contribute interfering color.

c. Minimum detectable concentration: 50 μg NO_3^--N/L.

2. Apparatus

Colorimetric equipment: One of the following is required:

a. Spectrophotometer, for use at 410 nm, providing a light path of 1 cm or longer.

b. Filter photometer, providing a light path of 1 cm or longer and equipped with a violet filter having maximum transmittance near 410 nm.

3. Reagents

a. Nitrate-free water: Prepare as directed in Section 418A.3*a*. Use for all solutions and dilutions.

b. Stock nitrate solution: Prepare as directed in Section 418A.3*b*.

c. Standard nitrate solution: Prepared as directed in Section 418A.3*c*.

d. Sulfite-urea reagent: Dissolve 5 g urea and 4 g anhydrous Na_2SO_3 in water and dilute to 100 mL.

e. Antimony reagent: Heat 500 mg Sb metal in 80 mL conc H_2SO_4 until all the metal has dissolved. Cool and cautiously add to 20 mL iced water. If crystals form upon standing overnight, redissolve by heating.

f. Chromotropic acid reagent: Purify chromotropic acid (4, 5-dihydroxy 2, 7-naphthalene disulfonic acid) as follows:

Boil 125 mL distilled water in a beaker and gradually add 15 g 4, 5-dihydroxy 2, 7-naphthalene disulfonic acid disodium salt, with constant stirring. Add 5 g decolorizing activated charcoal. Boil mixture for about 10 min. Add distilled water to make up loss due to evaporation. Filter hot solution through cotton wool. Add 5 g activated charcoal to filtrate and boil for 10 min more. Filter, first through cotton wool and then through filter paper to remove charcoal completely. Cool solution and slowly add 10 mL nitrate-free conc H_2SO_4. Boil solution until about 100 mL are left in beaker. Let stand overnight. Transfer crystals of chromotropic acid to a Buchner funnel and wash thoroughly with 95% ethyl alcohol until crystals are white. Dry crystals at 80 C.

Dissolve 100 mg purified chromotropic acid in 100 mL conc H_2SO_4 and store in a brown bottle. Prepare every 2 wk. A colorless reagent solution signifies the absence of NO_3^- contamination from H_2SO_4.

g. Sulfuric acid, H_2SO_4, conc, nitrate-free.

4. Procedure

a. Preparation of nitrate standards: Prepare NO_3^- standards in the range 0.10 to 5.0 mg N/L by diluting 0, 1.0, 5.0, 10, 25, 40, and 50 mL standard NO_3^- solution to 100 mL with water.

b. Color development: If appreciable amounts of suspended matter are present in sample, remove by centrifuging or filtering. Pipet 2.0-mL portions of standards, samples, and a water blank into dry 10-mL volumetic flasks. Use dilutions of standards and samples in the range 0.1 to 5.0 mg NO_3^--N/L. To each flask add 1 drop sulfite-urea reagent. Place flasks in a tray of cold water (10 to 20 C) and add 2 mL Sb reagent. Swirl flasks during addition of each reagent. After about 4 min in the bath, add 1 mL chromotropic acid reagent, swirl, and let stand in cooling bath for 3 min. Add conc H_2SO_4 to bring volume near the 10-mL mark. Stopper flasks and mix by inverting each flask four times. Let stand for 45 min at room temperature and adjust volume to 10 mL with conc H_2SO_4. Perform final mixing very gently to avoid introducing gas bubbles. Read absorbance at 410 nm between 15 min and 24 hr after last volume adjustment. Use nitrate-free water in the reference cell of the spectrophotometer. Rinse sample cell with sample solution, then fill carefully, to avoid trapping bubbles, by holding cell on a slant and pouring solution slowly down side of cell.

5. Calculation

$$\text{mg } NO_3^-\text{-N/L} = \frac{\mu\text{g } NO_3^-\text{-N (in 10 mL final volume)}}{\text{mL sample}}$$

6. Precision and Accuracy

Synthetic unknown samples containing nitrate and other constituents dissolved in distilled water were analyzed by various laboratories with the following results:

Nitrate Concentration *mg N/L*	No. of Laboratories	Relative Standard Deviation %	Relative Error %
0.050	32	61	13
0.50	34	16	4
1.00	5	8	3
5.00	36	11	3

418 E. Devarda's Alloy Reduction Method

1. General Discussion

a. Principle: This method is recommended for oxidized N (NO_3^--N + NO_2^--N) concentrations greater than 2 mg/L. It is especially convenient if an ammonia determination using the preliminary distillation step (Section 417A) has been made. In this technic NO_3^- and NO_2^- are reduced to NH_3 under hot alkaline conditions in the presence of the reducing agent, Devarda's alloy (an alloy of 50% Cu, 45% Al, and 5% Zn). The reduction is carried out in a kjeldahl distillation apparatus. Under hot alkaline conditions the NH_3 formed distills and is trapped in a receiving flask containing boric acid (Section 417A). The NH_3 can be determined either by direct nesslerization (Section 417B.4*b*) or acidimetrically (Section 417D.4).

b. Interferences: Remove ammonia from sample by preliminary distillation as described in Section 417A.4*b*-*d* if NH_3 is not determined on the same sample portion. Nitrite also is reduced to NH_3 under the conditions of the test. Make a separate determination of NO_2^- (Section 419) and subtract the result. If this is not done, report result as "total oxidized nitrogen". The method is not recommended for levels of NO_3^- below 2 mg/L. At this level, especially in the presence of high concentrations of amino nitrogen, use at least 50-mL and preferably 100-mL samples because of the possibility of positive interference by decomposition of this nitrogenous matter.

2. Apparatus

a. Distillation apparatus: Glass kjeldahl flask, 800 mL, with a condenser and adapter so arranged that the distillate can be collected in boric acid solution. For smaller samples, a steam-distillation microkjeldahl unit also can be used.

b. Measuring scoop: To contain 1 g Devarda's alloy.

c. Colorimetric equipment: See Section 417.2*a* 1), 2), and 3).

3. Reagents

a. Ammonia-free water: See Section 417A.3*a*.

b. Borate buffer solution: See Section 417A.3*b*.

c. Sodium hydroxide, NaOH, 6*N*.

d. Devarda's alloy: 200 mesh or smaller containing less than 0.0005% N.

e. For acidimetric titration: All reagents listed in Section 417D.3*a*, *b*, and *c*.

f. For nesslerization: All reagents listed in Section 417B.3*c*, *d*, and *e*.

1) *Nessler reagent:* See Section 417B.3*d*.

2) *Stock ammonia solution:* See Section 417B.3*e*.

3) *Standard ammonia solution:* See Section 417B.3*f*.

4. Procedure

a. If NH_3 has not been determined by a method involving preliminary distillation, dilute a portion of sample to 500 mL with ammonia-free water. Add 25 mL borate buffer and adjust to pH 9.5 with 6*N* NaOH using a pH meter or short-range pH paper. Distill 250 to 300 mL into a dry receiving flask and discard. Make sure that the last part of the distillation is conducted with condenser tip out of the liquid in receiving flask.

b. To the residue after removing NH_3 (¶ 4*a* above), add 1 g Devarda's alloy and sufficient ammonia-free distilled water to bring total volume to about 350 mL.

Place in a receiving flask 50 mL H_3BO_3 absorbent for each mg NO_3^--N in sample. Immerse end of condenser in absorbent.

Heat distillation flask until boiling or vigorous bubbling occurs. Reduce heat and distill at a rate of 5 to 10 mL/min until at least 150 mL distillate have been collected. Lower receiver so that liquid is below end of condenser and continue distillation for 1 to 2 min to cleanse condenser.

Determine NH_3-N either by nesslerization (Section 417B) or titration with standard strong acid (Section 417D).

5. Calculation

Calculate NH_3-N as in Section 417B or 417D. This represents the NH_3 produced from the reduction of NO_2^- and NO_3^- (total oxidized nitrogen). To obtain NO_3^-, separately determine NO_2^- (Section 419) and subtract.

6. Precision

The recovery of 200 to 400 μg NO_3^--N from partially treated wastewater effluents was found to average 96% with a coefficient of variation of 7.7%.

418 F. Automated Cadmium Reduction Method (TENTATIVE)

1. General Discussion

a. Principle: See Section 418C.7*a*.

b. Interferences: Sample turbidity may interfere with this method. Remove by filtration before analysis. Sample color that absorbs in the photometric range used for analysis also will interfere.

c. Application: Nitrate and nitrite, singly or together in potable, surface, and saline waters and domestic and industrial wastewaters, can be determined over a range of 0.5 to 10 mg N/L.

2. Apparatus

a. Automated analytical equipment: Two automated systems*, which use the same principle but differ in detail and rate of analysis, are in common use. They consist of the components (except a heating bath) listed in Section 602.1.

b. Tubular flow cell, 15 or 50 mm length.

c. Filters, 540 nm.

3. Reagents

a. Deionized distilled water: Use distilled water of highest purity, preferably prepared by mixed-bed ion-exchange deionization of distilled water. Regenerate ion-exchange column according to manufacturer's instructions. Use deionized distilled water to prepare all reagents and dilutions.

b. Copper sulfate solution: Dissolve 20 g $CuSO_4 \cdot 5H_2O$ in 500 mL water and dilute to 1 L.

c. Wash solution: Use deionized distilled water for unpreserved samples. For samples preserved with H_2SO_4, add 2 mL conc H_2SO_4/L wash water.

d. Copper-cadmium granules: Clean Cd granules†, new or used, with 1 + 1 HCl and treat with copper solution in the following manner: Wash Cd with 1 + 1 HCl and rinse with water. Swirl 10 g Cd in 100-mL portions of $CuSO_4$ solution for 5 min or until blue color partially fades, decant, and repeat with fresh $CuSO_4$ solution until a brown colloidal precipitate forms. Wash Cd-Cu with wash solution at least 10 times to remove all precipitated copper.

e. Hydrochloric acid, HCl, conc.

f. Ammonium hydroxide, NH_4OH, conc.

*AutoAnalyzer I and AutoAnalyzer II: Technicon Instrument Corporation, Tarrytown, N.Y. 10591.

†E.M. Laboratories, Inc., 500 Exec. Blvd., Elmsford, N.Y. 10523. Cat. No. 2001, Cadmium, Coarse Powder, 40-60 mesh, or equivalent.

g. Color reagent: To approximately 800 mL water, add, while stirring, 100 mL conc H_3PO_4, 40 g sulfanilamide, and 2 g N-(1-naphthyl) ethylenediamine dihydrochloride. Stir until dissolved and dilute to 1 L. Store in brown bottle and keep in the dark when not in use. This solution is stable for several months.

h. Ammonium chloride solution: Dissolve 85 g NH_4Cl in water and dilute to 1 L. Add 0.5 mL polyoxyethylene 23 lauryl ether.‡

i. Stock nitrate solution: Dissolve 7.218 g anhydrous KNO_3 and dilute to 1,000 mL with water; 1.00 mL = 1.00 mg N. Preserve with 2 mL $CHCl_3$/L. Solution is stable for 6 months.

‡Brij 35, available from ICI United States, Chicago, Ill., or Technicon Instruments Corporation, Tarrytown, N.Y. 10591, or equivalent.

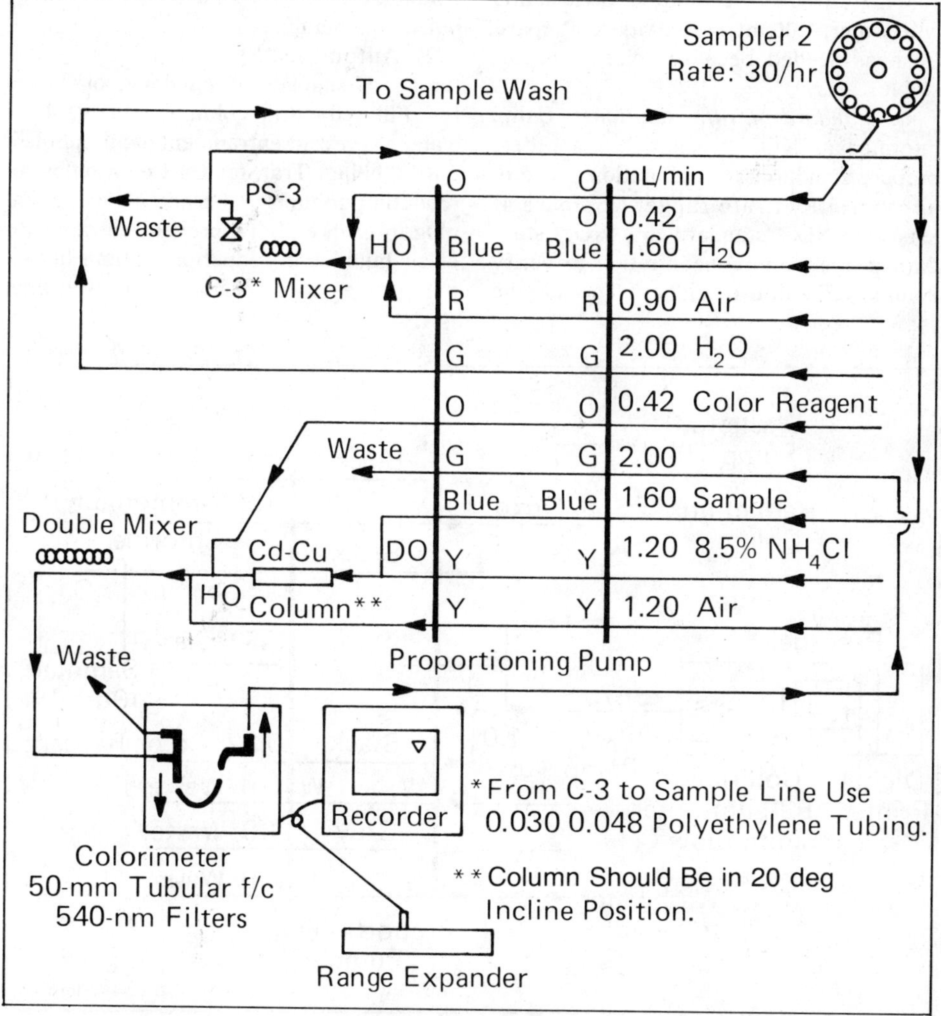

Figure 418:2. Nitrate-nitrite manifold for Automated System I.

j. Stock nitrite solution: Dissolve 6.072 g anhydrous KNO_2 in 500 mL water and dilute to 1,000 mL; 1.00 mL = 1.00 mg N. Preserve with 2 mL $CHCl_3$/L and refrigerate.

k. Intermediate nitrate solution: Dilute 10.0 mL stock nitrate solution to 1,000 mL with water; 1.00 mL = 10.0 μg N. Preserve with 2 mL $CHCl_3$/L. Solution is stable for 6 months.

l. Intermediate nitrite solution: Dilute 10.0 mL stock nitrite solution to 1,000 mL with water; 1.00 mL = 10.0 μg N. Prepare fresh as needed because solution is unstable.

m. Standard nitrate solutions: Using intermediate NO_3^--N solution and water, prepare standards for calibration curve in appropriate nitrate range. Compare at least one NO_2^- standard to a NO_3^- standard at the same concentration to verify column reduction efficiency. To examine saline waters prepare standard solutions with the substitute ocean water described in Section 417F.3*g*.

4. Procedure

a. Preparation of reduction column:

1) Automated System I—Use an 8- by 50-mm glass tube with the ends reduced in diameter to permit insertion into the system. Place granules in column between glass wool plugs. Set packed reduction column in an upflow 20-deg incline to minimize channeling.

2) Automated System II—Use glass tubing, U-shaped, 35 cm long, of 2-mm ID. Fill reduction column with distilled water to prevent entrapment of air bubbles during filling. Transfer Cu-Cd granules to reduction column and place a glass wool plug in each end. To prevent entrapment of air bubbles in reduction column fill all pump tubes with reagents before putting

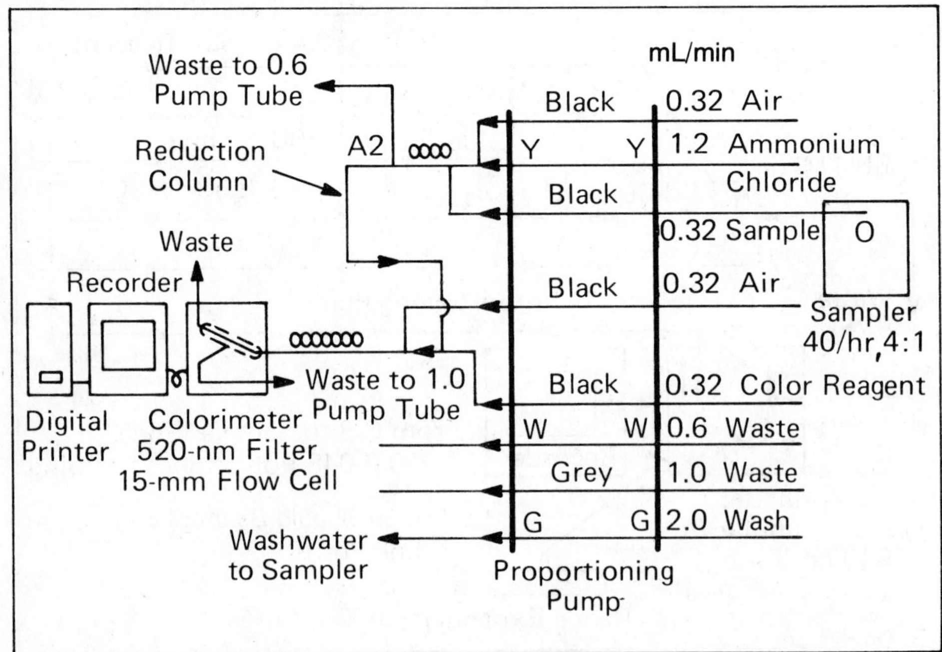

Figure 418:3: Nitrate-nitrite manifold for Automated System II.

column into analytical system. A 0.2-cm-ID pump tube may be used in place of the 2-mm glass tube. When apparatus is not in use, cover metal in column with NH_4Cl solution (3*h*).

b. If sample pH is below 5 or above 9, adjust to between 5 and 9 with either conc HCl or conc NH_4OH.

c. Set up manifold as shown in Figure 418:2 or Figure 418:3.

d. For the Automated System I, sample at a rate of 30/hr, 1:1 cam; for for Automated System II, use a 40/hr, 4:1 cam and a common wash.

e. Follow the general procedure in Section 602.2.

5. Precision and Accuracy

With Automated System I three laboratories analyzed four natural water samples containing exact increments of inorganic nitrate with the following results:

Increment as NO_3^--N *μg/L*	Standard Deviation *μg N/L*	Bias *%*	Bias *μg N/L*
290	12	+5.75	+17
350	92	+18.10	+63
2,310	318	+4.47	+103
2,480	176	−2.69	−67

In a single laboratory using surface water samples at concentrations of 100, 200, 800, and 2,100 μg N/L, the standard deviations were 0, ±40, ±50, and ±50 μg/L, respectively, and at concentrations of 200 and 2,200 μg N/L, recoveries were 100 and 96%, respectively.

Single-laboratory data on Automated System II show comparable precision and accuracy.

418 G. Bibliography

Ultraviolet Screening Method

HOATHER, R.C. & R.F. RACKMAN. 1959. Oxidized nitrogen and sewage effluents observed by ultraviolet spectrophotometry. *Analyst* 84:549.

GOLDMAN, E. & R. JACOBS. 1961. Determination of nitrates by ultraviolet absorption. *J. Amer. Water Works Ass.* 53:187.

ARMSTRONG, F.A.J. 1963. Determination of nitrate in water by ultraviolet spectrophotometry. *Anal. Chem.* 35:1292.

NAVONE, R. 1964. Proposed method for nitrate in potable waters. *J. Amer. Water Works Ass.* 56:781.

Nitrate Electrode Screening Method

LANGMUIR, D. & R.I. JACOBSON. 1970. Specific ion electrode determination of nitrate in some freshwaters and sewage effluents. *Environ. Sci. Technol.* 4:835.

KEENEY, D.R., B.H. BYRNES & J.J. GENSON. 1970. Determination of nitrate in waters with nitrate-selective ion electrode. *Analyst* 95:383.

MILHAM, P.J., A.S. AWAD, R.E. PAUL & J.H. BULL. 1970. Analysis of plants, soils and waters for nitrate by using an ion-selective electrode. *Analyst* 95:751.

SOMERFELDT, T.C., R.A. MILNE & G.C. KOZUB. 1971. Use of the nitrate specific ion electrode for the determination of nitrate nitrogen in surface and groundwater. *Commun. Soil Sci. Plant Anal.* 2:415.

Cadmium Reduction Method

HENRIKSON, A. & A.R. SELMER-OLSEN. 1970. Automatic methods for determining nitrate and nitrite in water and soil extracts. *Analyst* 95:514.

STRICKLAND, J.D.H. & T.R. PARSONS. 1972. A Practical Handbook of Sea Water Analysis, 2nd ed. Fish. Res. Board Can., Ottawa, Bull. No. 167.

ENVIRONMENTAL PROTECTION AGENCY. 1974.

Methods for Chemical Analysis of Water and Wastes. Off. Technol. Transfer, EPA, Washington, D.C.

STANTON, M.P. 1974. Simple efficient column for use in the automated determination of nitrate in water. *Anal. Chem.* 46:1616.

NYDAHL, F. 1976. On the optimum conditions for the reduction of nitrate by cadmium. *Talanta* 23:349.

CONNORS, J.J. & J. BELAND. 1976. Analytical notes. *J. Amer. Water Works Ass.* 68:55.

Chromotropic Acid Method

WEST, P.W. & T.P. RAMACHANDRAN. 1966. Spectrophotometric determination of nitrate using chromotropic acid. *Anal. Chem. Acta.* 35:317.

Devarda's Alloy Method

MINISTRY OF HOUSING AND LOCAL GOVERNMENT. 1956. Methods of Chemical Analysis as Applied to Sewage and Sewage Effluents, 2nd ed. H.M. Stationery Office, London.

ASSOCIATION OF BRITISH CHEMICAL MANUFACTURERS & SOCIETY FOR ANALYTICAL CHEMISTRY. 1958. Recommended Methods for the Analysis of Trade Effluents. W. Heffer & Sons, Ltd., Cambridge, England.

BREMMER, J.M. & D.R. KEENEY. 1965. Steam distillation methods for determination of ammonium, nitrate and nitrite. *Anal. Chem. Acta.* 32:485.

EVANS, W.H. & J.G. STEVENS. 1972. An investigation into the determination of nitrate in potable waters and effluents by the reduction method employing Devarda's alloy. *J. Inst. Water Pollut. Control* 71:98.

Automated Cadmium Reduction Method

FIORE, J. & J.E. O'BRIEN. 1962. Automation in sanitary chemistry—Parts 1 and 2. Determination of nitrates and nitrites. *Wastes Eng.* 33:128 & 238.

FEDERAL WATER POLLUTION CONTROL ADMINISTRATION. 1966. Chemical Analyses for Water, Quality Manual. Dep. Interior, R.A. Taft Sanitary Engineering Center Training Program, Cinicnnati, Ohio.

AMERICAN SOCIETY FOR TESTING AND MATERIALS. 1966. Manual of Industrial Water and Industrial Waste Water. ASTM, Philadelphia, Pa., pp. 418 & 465.

ARMSTRONG, F.A., C.R. STEARNS & J.D. STRICKLAND. 1967. The measurement of upwelling and subsequent biological processes by means of the Technicon AutoAnalyzer and associated equipment. *Deep Sea Res.* 14:381.

U.S. ENVIRONMENTAL PROTECTION AGENCY. FWQA Method Study 4, Automated Methods. National Environmental Research Center, Cincinnati, Ohio (in preparation).

419 NITROGEN (NITRITE)

1. General Discussion

a. Principle: Nitrite (NO_2^-) is determined through formation of a reddish purple azo dye produced at pH 2.0 to 2.5 by coupling diazotized sulfanilic acid with N-(1-naphthyl)-ethylenediamine dihydrochloride (NED dihydrochloride). The method is suitable for determination of NO_2^- down to 1 μg NO_2^--N/L. Photometric measurements can be made in the range 5 to 50 μg N/L if a 5-cm light path and a green color filter are used. The color system obeys Beer's law up to 180 μg N/L with a 1-cm light path at 543 nm. High NO_2^- concentrations can be determined by diluting a sample to 50 mL in the nessler tube used to conduct the reaction.

b. Interferences: Chemical incompatibility makes it unlikely that NO_2^-, free available chlorine, and nitrogen trichloride (NCl_3) will co-exist in a sample. NCl_3 imparts a false red color when the normal order of reagent addition is followed. Although this effect may be minimized by adding the NED dihydrochloride reagent first and then the sulfanilic acid reagent, an orange color still may result when a substantial NCl_3 concentration is present. Under such circumstances, check for free available chlorine and NCl_3 residuals. The following ions interfere because of precipitation under test conditions and should be absent: Sb^{3+}, Au^{3+}, Bi^{3+}, Fe^{3+},

Pb^{2+}, Hg^{2+}, Ag^{+}, chloroplatinate ($PtCl_6^{2-}$), and metavanadate (VO_3^{2-}). Cupric ion may cause low results by catalyzing decomposition of the diazonium salt. Colored ions that alter the color system also should be absent. Remove suspended solids by filtration through a 0.45-μm pore diameter membrane filter before color development.

c. Storage of sample: Make the determination promptly on fresh samples to prevent bacterial conversion of NO_2^- to NO_3^- or NH_3.

Never use acid preservation for samples to be analyzed for NO_2^-. For short-term preservation for 1 to 2 days, freeze at −20 C or add 40 mg $HgCl_2$/L sample and store at 4 C.

2. Apparatus

Colorimetric equipment: One of the following is required:

a. Spectrophotometer, for use at 543 nm, providing a light path of 1 cm or longer.

b. Filter photometer, providing a light path of 1 cm or longer and equipped with a green filter having maximum transmittance near 540 nm.

c. Nessler tubes, matched, 50-mL, tall form.

3. Reagents

a. Nitrite-free water: If it is not known that the distilled or demineralized water is free from NO_2^-, use either of the following procedures to prepare nitrite-free water:

1) Add to 1 L distilled water one small crystal each of $KMnO_4$ and either $Ba(OH)_2$ or $Ca(OH)_2$. Redistill in an all-borosilicate glass apparatus and discard the initial 50 mL of distillate. Collect the distillate fraction that is free of permanganate. (A red color with DPD reagent (Section D.2*b*), indicates the presence of permanganate).

2) Add 1 mL conc H_2SO_4 and 0.2 mL $MnSO_4$ solution (36.4 g $MnSO_4 \cdot H_2O$/100 mL distilled water) to each 1 L distilled water, and make pink with 1 to 3 mL $KMnO_4$ solution (400 mg $KMnO_4$/L distilled water). Redistill as described in the preceding paragraph.

b. Sulfanilamide reagent: Dissolve 5 g sulfanilamide in a mixture of 50 mL conc HCl and about 300 mL nitrite-free water. Dilute to 500 mL with nitrite-free water. The solution is stable for many months.

c. N-(1-naphthyl)-ethylenediamene-dihydrochloride solution: Dissolve 500 mg N-(1-naphthyl)-ethylenediamine dihydrochloride in 500 mL nitrite-free distilled water. Store in a dark bottle. Replace monthly or immediately when it develops a strong brown color.

d. Hydrochloric acid, HCl, 1 + 3: Use nitrite-free distilled water for dilution.

e. Sodium oxalate, 0.05*N*: Dissolve 3.350 g $Na_2C_2O_4$, primary standard grade in 1,000 mL nitrite-free water.

f. Ferrous ammonium sulfate, 0.05*N*: Dissolve 19.607 g $Fe(NH_4)_2(SO_4)_2 \cdot 6H_2O$ plus 20 mL conc H_2SO_4 in nitrite-free water and dilute to 1,000 mL. Standardize as in Section 508.3.

g. Stock nitrite solution: Commercial reagent-grade $NaNO_2$ assays at less than 99%. Because NO_2^- is readily oxidized in the presence of moisture, use a fresh bottle of reagent for preparing the stock solution and keep bottles tightly stoppered against the free access of air when not in use. To determine $NaNO_2$ content, add a known excess of standard 0.05*N* $KMnO_4$ solution prepared and standardized as described in 311B.3*i*, discharge permanganate color with a known quantity of standard reductant such as 0.05*N* $Na_2C_2O_4$ or 0.05*N* $Fe(NH_4)_2(SO_4)_2$, and back-titrate with standard permanganate solution.

1) Preparation of stock solution—Dissolve 1.232 g $NaNO_2$ in nitrite-free distilled water and dilute to 1,000 mL; 1.00 mL = 250 μg N. Preserve with 1 mL $CHCl_3$.

2) Standardization of stock solution—

Pipet, in order, 50.00 mL standard 0.05N $KMnO_4$, 5 mL conc H_2SO_4, and 50.00 mL stock NO_2^- solution into a glass-stoppered flask or bottle. Submerge pipet tip well below surface of permanganate-acid solution while adding stock NO_2^- solution. Shake gently and warm to 70 to 80 C on a hot plate. Discharge permanganate color by adding sufficient 10-mL portions of standard 0.050N $Na_2C_2O_4$. Titrate excess $Na_2C_2O_4$ with 0.050N $KMnO_4$ to the faint pink end point. Carry a nitrite-free distilled water blank through the entire procedure and make the necessary corrections in the final calculation.

If standard 0.050N ferrous ammonium sulfate solution is substituted for $Na_2C_2O_4$, omit heating to 70 to 80 C and extend reaction period between $KMnO_4$ and Fe^{2+} to 5 min before making final $KMnO_4$ titration.

Calculate NO_2^--N content of stock solution by the following equation:

$$A = \frac{[(B \times C) - (D \times E)] \times 7}{F}$$

where:

A = mg NO_2^--N/mL in stock $NaNO_2$ solution,
B = total mL standard $KMnO_4$ used,
C = normality of standard $KMnO_4$,
D = total mL standard reductant added,
E = normality of standard reductant, and
F = mL stock $NaNO_2$ solution taken for titration.

Each 1.00 mL 0.050N $KMnO_4$ consumed by the $NaNO_2$ solution corresponds to 350 μg NO_2^--N.

h. Intermediate nitrite solution: Calculate the volume, G, of stock NO_2^- solution required for the intermediate NO_2^- solution from $G = 12.5/A$. Dilute the volume G (approximately 50 mL) to 250 mL with nitrite-free water; 1.00 mL = 50.0 μg N. Prepare daily.

i. Standard nitrite solution: Dilute 10.00 mL intermediate NO_2^- solution to 1,000 mL with nitrite-free water; 1.00 mL = 0.500 μg N. Prepare daily.

4. Procedure

a. Removal of turbidity: If sample contains suspended solids, filter through a 0.45-μm pore diam membrane filter.

b. Color development: To 50.0 mL clear sample neutralized to pH 7, or to a portion diluted to 50.0 mL, add 1 mL sulfanilamide solution. Let reagent react for more than 2 min but not longer than 8 min. Add 1.0 mL NED-dihydrochloride solution and mix immediately. Let stand at least 10 min but not more than 2 hr.

c. Photometric measurement: Measure absorbance at 543 nm. As a guide use the following light paths for the indicated NO_2^--N concentrations:

Light Path Length *cm*	NO_2^--N *μg/L*
1	2–25
5	2–6
10	<2

Run parallel checks frequently against NO_2^- standards, preferably in the concentration range of the sample. Redetermine complete calibration curves after preparing new reagents.

d. Color standards for visual comparison: Prepare a suitable series of visual color standards in nessler tubes by adding the following volumes of standard NO_2^- solution and diluting to 50 mL with nitrite-free water: 0, 0.1, 0.2, 0.4, 0.7, 1.0, 1.4, 1.7, 2.0, and 2.5 mL, corresponding, respectively, to 0, 1.0, 2.0, 4.0, 7.0, 10, 14, 17, 20, and 25 μg NO_2^--N/L. Develop color as described in ¶ 4*b*. Compare samples to visual standards in matched nessler tubes between 10 and 120 min after adding NED-dihydrochloride reagent. Select the concentration where the sample tube color matches the standard tube color.

5. Calculation

$$\text{mg } NO_2^-\text{-N/L} = \frac{\mu g\ NO_2^-\text{-N (in 52 mL final volume)}}{\text{mL sample}}$$

6. Bibliography

RIDER, B.F. & M.G. MELLON. 1946. Colorimetric determination of nitrites. *Ind. Eng. Chem.*, Anal. Ed. 18:96.

BARNES, H. & A.R. FOLKARD. 1951. The determination of nitrites. *Analyst* 76:599.

BOLTZ, D.F., ed. 1958. Colorimetric Determination of Nonmetals. Intersciences Publishers, New York, N.Y.

420 NITROGEN (ORGANIC)

1. Selection of Method

The major factor that influences the selection of a macro- or semi-micro-kjeldahl method to determine organic nitrogen is the concentration of organic nitrogen. The macro-kjeldahl method is applicable for samples containing low concentrations of organic nitrogen and requires a relatively larger sample volume than the semi-micro-kjeldahl method, which is applicable to samples containing high concentrations of organic nitrogen. In the tentative semi-micro-kjeldahl method, the sample volume should be chosen such that it contains organic plus ammonia nitrogen (kjeldahl nitrogen) in the range of 0.2 to 2 mg.

2. Storage of Samples

The most reliable results are obtained on fresh samples. If an immediate analysis is not possible, preserve samples by adding 0.8 mL conc H_2SO_4/L and store at 4 C. The pH of acidified samples should be in the range of 1.5 to 2. Samples of some wastewater may require more conc H_2SO_4 than the indicated volume.

420 A. Macro-Kjeldahl Method

1. General Discussion

The kjeldahl method determines nitrogen in the trinegative state. It fails to account for nitrogen in the form of azide, azine, azo, hydrazone, nitrate, nitrite, nitrile, nitro, nitroso, oxime, and semi-carbazone. If ammonia nitrogen is not removed in the initial phase (¶ 4*b* below) of the procedure, the term "kjeldahl nitrogen" is applied to the result. Should kjeldahl nitrogen and ammonia nitrogen be determined individually, "organic nitrogen" can be obtained by difference.

a. Principle: In the presence of H_2SO_4, potassium sulfate (K_2SO_4), and mercuric sulfate ($HgSO_4$) catalyst, amino nitrogen of many organic materials is converted to ammonium sulfate [$(NH_4)_2SO_4$]. Free ammonia and ammonium-nitrogen also are converted to $(NH_4)_2SO_4$. During sample digestion, a mercury ammonium complex is formed. After the mercury ammonium complex in the digestate has been decomposed by sodium thiosulfate ($Na_2S_2O_3$), the ammonia is distilled from an alkaline medium and absorbed in boric acid. The ammonia is determined colorimetrically or by titration with a standard mineral acid.

b. Selection of modification: The sensitivity of colorimetric methods makes them useful for determining organic nitrogen levels below 5 mg/L. The titrimetric method of measuring ammonia in the distillate is suitable for determining a wide range of organic nitrogen concentrations, depending on volume of boric acid absorbent used and concentration of standard acid titrant. The ammonia selective electrode likewise can be used over a wide concentration range.

2. Apparatus

a. Digestion apparatus: Kjeldahl flasks with a total capacity of 800 mL yield the best results. Digest over a heating device adjusted so that 250 mL water at an initial temperature of 25 C can be heated to a rolling boil in approximately 5 min. A heating device meeting this specification should provide the temperature range of 365 to 370 C for effective digestion.

b. Distillation apparatus: See Section 417A.2*a*.

c. Apparatus for ammonia determination: See Sections 417B.2, 417C.2, 417D.2, or 417E.2.

3. Reagents

Prepare all reagents in ammonia-free water.

All of the reagents listed for the determination of Nitrogen (Ammonia), Sections 417B.3, 417C.3, 417D.3 or 417E.3, are required, plus the following:

a. Digestion reagent: Dissolve 134 g K_2SO_4 in 650 mL water and 200 mL conc H_2SO_4. Add, with stirring, a solution prepared by dissolving 2 g red mercuric oxide, HgO, in 25 mL 6*N* H_2SO_4. Dilute the combined solution to 1 L with water. Keep at a temperature close to 20 C to prevent crystallization.

b. Phenolphthalein indicator solution.

c. Sodium hydroxide-sodium thiosulfate reagent: Dissolve 500 g NaOH and 25 g $Na_2S_2O_3 \cdot 5H_2O$ in water and dilute to 1 L.

d. Borate buffer solution: See Section 417A.3*b*.

e. Sodium hydroxide, NaOH, 6*N*.

4. Procedure

a. Selection of sample volume: Place a measured volume of sample in an 800-mL kjeldahl flask. Select sample size from the following tabulation:

Organic Nitrogen in Sample *mg/L*	Sample Size *mL*
0–1	500
1–10	250
10–20	100
20–50	50.0
50–100	25.0

If necessary, dilute sample to 300 mL and neutralize to pH 7.

b. Ammonia removal: Add 25 mL borate buffer and then 6*N* NaOH until pH 9.5 is reached. Add a few glass beads or boiling chips and boil off 300 mL. If desired, distill this fraction and determine ammonia nitrogen. Alternately, if ammonia has been determined by the distillation method, use residue in distilling flask for organic nitrogen determination.

For sludge and sediment samples, weigh wet sample in a crucible or weighing bottle, transfer contents to a kjeldahl flask, and determine kjeldahl nitrogen. Follow a similar procedure for ammonia nitrogen and organic nitrogen determined by difference. Determinations of organic and kjeldahl nitrogen on dried sludge and sediment samples are not accurate because drying results in loss of ammonium salts. Measure dry weight of sample on a separate portion.

c. Digestion: Cool and add carefully 50 mL digestion reagent (or substitute 10 mL conc H_2SO_4, 6.7 g K_2SO_4, and 1.5 mL $HgSO_4$ solution) to distillation flask. If large quantities of nitrogen-free organic matter are present, add an additional 50 mL digestion reagent for each gram of solid matter in sample. Add a few glass beads and after mixing, heat under a hood or with suitable ejection equipment to remove SO_3 fumes. Continue to boil briskly until solution clears (becomes colorless or a pale straw color). Then digest for an additional 30 min. Let flask and contents cool, dilute to 300 mL with ammonia-free water, add 0.5 mL phenolphthalein indicator solution, and mix. Tilt flask and carefully add sufficient hydroxide-thiosulfate reagent (approximately 50 mL/50 mL digestion reagent used) to form an alkaline layer at flask bottom. Connect flask to steamed-out distillation apparatus and shake flask to insure complete mixing. Add more hydroxide-thiosulfate reagent in the prescribed manner if red phenolphthalein color fails to appear at this stage.

d. Distillation: Distill and collect 200 mL distillate below surface of 50 mL boric acid solution. Use plain boric acid solution when ammonia is to be determined by nesslerization and use indicating boric acid for a titrimetric finish. Use 50 mL 0.04*N* H_2SO_4 solution for collecting distillate for manual phenate and electrode methods. Extend tip of condenser well below level of boric acid solution and do not let temperature in condenser rise above 29 C. Lower collected distillate free of contact with delivery tube and continue distillation during last 1 or 2 min to cleanse condenser.

e. Final ammonia measurement: Use the nesslerization (417B), titration (417D), manual phenate (417C), or ammonia selective electrode (417E) method.

f. Blank: Carry a blank through all steps of the procedure and apply necessary corrections to the results.

5. Calculation

See Sections 417B.5, 417C.5, 417D.5, or 417E.5.

6. Precision and Accuracy

Three synthetic samples containing various organic nitrogen concentrations and other constituents were analyzed by three procedural modifications of the macro-kjeldahl method: kjeldahl-nessler finish, kjeldahl-titrimetric finish, and calculation of the difference between kjeldahl nitrogen and ammonia nitrogen, both determined by a nessler finish. The results obtained by participating laboratories are summarized in Table 420:I.

No data on the precision of the macro-kjeldahl-phenate method are available.

Sample 1 contained the following additional constituents: 400 mg chloride (Cl^-)/

TABLE 420:I. PRECISION AND ACCURACY DATA FOR ORGANIC NITROGEN, MACRO-KJELDAHL PROCEDURE

			Relative Standard Deviation			Relative Error		
Sample	No. of Laboratories	Organic Nitrogen Concentration $\mu g/L$	Nessler Finish %	Titrimetric Finish %	Calculation of Total Kjeldahl N Minus NH_3-N %	Nessler Finish %	Titrimetric Finish %	Calculation of Total Kjeldahl N Minus NH_3-N %
1	26	200	94.8	—	—	55.0	—	—
	29	200	—	104.4	—	—	70.0	—
	15	200	—	—	68.8	—	—	70.0
2	26	800	52.1	—	—	12.5	—	—
	31	800	—	44.8	—	—	3.7	—
	16	800	—	—	52.6	—	—	8.7
3	26	1,500	43.1	—	—	9.3	—	—
	30	1,500	—	54.7	—	—	22.6	—
	16	1,500	—	—	45.9	—	—	4.0

L, 1.50 mg ammonia nitrogen (NH_3-N)/L, 1.0 mg nitrate nitrogen (NO_3^--N)/L, 0.5 mg phosphate (PO_4^{3-})/L, and 30.0 mg silica (SiO_2)/L.

Sample 2 contained the following additional constituents: 200 mg Cl^-/L, 0.8 mg NH_3-N/L, 1.0 mg NO_3^--N/L, 5.0 mg PO_4^{3-}/L, and 15.0 mg SiO_2/L.

Sample 3 contained the following additional constituents: 10 mg Cl^-/L, 0.2 mg NH_3-N/L, 1.0 mg NO_3^--N/L, 10.0 mg PO_4^{3-}/L, and 5.0 mg SiO_2/L.

7. Bibliography

KJELDAHL, J. 1883. A new method for the determination of nitrogen in organic matter. *Z. Anal. Chem.* 22:366.

PHELPS, E.B. 1905. The determination of organic nitrogen in sewage by the Kjeldahl process. *J. Infect. Dis.* (Suppl.) 1:225.

MEEKER, E.W. & E.C. WAGNER. 1933. Titration of ammonia in the presence of boric acid. *Ind. Eng. Chem.* Anal. Ed. 5:396.

WAGNER, E.C. 1940. Titration of ammonia in the presence of boric acid. *Ind. Eng. Chem.* Anal. Ed. 12:771.

MCKENZIE, H.A. & H.S. WALLACE. 1954. The Kjeldahl determination of nitrogen: A critical study of digestion conditions. *Aust. J. Chem.* 7:55.

MORGAN, G.B., J.B. LACKEY & F.W. GILCREAS. 1957. Quantitative determination of organic nitrogen in water, sewage, and industrial wastes. *Anal. Chem.* 29:833.

BOLTZ, D.F., ed. 1958. Colorimetric Determination of Nonmetals. Interscience Publishers, New York, N.Y.

420 B. Semi-Micro-Kjeldahl Method (TENTATIVE)

1. General Discussion

See Section 420A.1.

2. Apparatus

a. Digestion apparatus: Use kjeldahl flasks with a capacity of 100 mL in a semi-micro-kjeldahl digestion apparatus* equipped with heating elements to accomodate kjeldahl flasks and a suction outlet to vent fumes. The heating elements should provide the temperature range of 365 to 380 C for effective digestion.

b. Distillation apparatus: Use an all-glass unit equipped with a steam-generating vessel containing an immersion heater† (Figure 420:1).

c. pH meter.

3. Reagents

All of the reagents listed for the determination of Nitrogen (Ammonia) (417A.3) and Nitrogen (Organic) macro-kjeldahl (420A.3) are required. Prepare all reagents with ammonia-free water.

4. Procedure

a. Selection of sample volume: Determine the sample size from the following tabulation:

Organic Nitrogen in Sample *mg/L*	Sample Size *mL*
4–40	50
8–80	25
20–200	10
40–400	5

For sludge and sediment samples weigh a portion of wet sample containing between 0.2 and 2 mg organic nitrogen in a crucible or weighing bottle. Transfer

*Rotary kjeldahl digestion unit, Kontes, Model K551100, or equivalent.

†ASTM E-147 or equivalent.

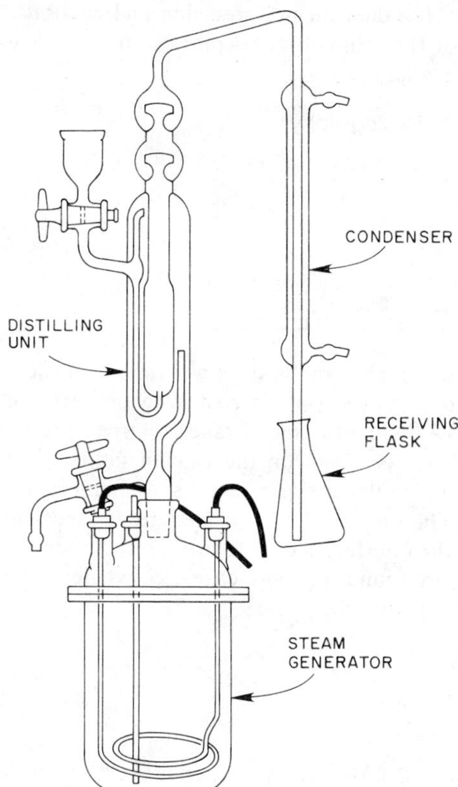

Figure 420:1. Micro kjeldahl distillation apparatus.

sample quantitatively to a 100-mL beaker by diluting it and rinsing the weighing dish several times with small quantities of water. Make the transfer using as small a quantity of water as possible and do not exceed a total volume of 50 mL. Measure dry weight of sample on a separate portion.

b. Ammonia removal: Pipet 50 mL sample or an appropriate volume diluted to 50 mL with water into a 100-mL beaker. Add 3 mL borate buffer and adjust to pH 9.5 with 6*N* NaOH, using a pH meter. Quantitatively transfer sample to a 100-mL kjeldahl flask and boil off 30 mL.

Alternatively, if ammonia removal is not required, digest samples directly as described in ¶ *c* below. Distillation following this direct digestion yields kjeldahl nitrogen concentration rather than organic nitrogen.

c. Digestion: Carefully add 10 mL digestion reagent to kjeldahl flask containing sample. Add 5 or 6 glass beads (3- to 4-mm size) to prevent bumping during digestion. Set each heating unit on the micro-kjeldahl digestion apparatus to its medium setting and heat flasks under a hood or with suitable ejection equipment to remove fumes of SO_3. Continue to boil briskly until solution clears and all water is driven off (becomes colorless or a pale straw color). Then turn each heating unit up to its maximum setting and digest for an additional 30 min. Cool. Quantitatively transfer digested sample by diluting and rinsing several times into micro-kjeldahl distillation apparatus so that total volume in distillation apparatus does not exceed 30 mL. Add 10 mL hydroxide-thiosulfate reagent and turn on steam.

d. Distillation: Control rate of steam generation to boil contents in distillation unit so that neither escape of steam from tip of condenser nor bubbling of contents in receiving flask occurs. Distill and collect 30 to 40 mL distillate below surface of 10 mL boric acid solution contained in a 125-mL erlenmeyer flask. Use plain boric acid solution when ammonia is to be determined by nesslerization and use indicating boric acid for a titrimetric finish. Use 10 mL 0.04*N* H_2SO_4 solution for collecting distillate for the phenate and electrode methods. Extend tip of condenser well below level of boric acid solution and do not let temperature in condenser rise above 29 C. Lower collected distillate free of contact with delivery tube and continue distillation during last 1 or 2 min to cleanse condenser.

e. Blank: Carry a blank through all steps of procedure and apply necessary correction to results.

f. Final ammonia measurement: Determine ammonia by nesslerization, titration, manual phenate, or ammonia selective electrode method.

5. Calculation

See Section 420A.5.

6. Precision and Accuracy

No data on the precision and accuracy of the semi-micro-kjeldahl method are available.

7. Bibliography

See Section 420A.7.

421 OXYGEN (DISSOLVED)

Dissolved oxygen (DO) levels in natural and wastewaters depend on the physical, chemical, and biochemical activities in the water body. The analysis for DO is a key test in water pollution and waste treatment process control.

Two methods for DO analysis are described: the Winkler or iodometric method and its modifications and the electrometric method using membrane electrodes. The iodometric method[1] is a titrimetric procedure based on the oxidizing property of DO while the membrane electrode procedure is based on the rate of diffusion of molecular oxygen across a membrane.[2] The choice of test procedure depends on the interferences present, the accuracy desired, and, in some cases, convenience or expedience.

421 A. Iodometric Methods

1. Principle

Improved by variations in technic and equipment and aided by instrumentation, the iodometric test remains the most precise and reliable titrimetric procedure for DO analysis. The test is based on the addition of divalent manganese solution, followed by strong alkali, to the sample in a glass-stoppered bottle. DO rapidly oxidizes an equivalent amount of the dispersed divalent manganous hydroxide precipitate to hydroxides of higher valency states. In the presence of iodide ions and acidification, the oxidized manganese reverts to the divalent state, with the liberation of iodine equivalent to the original DO content. The iodine is then titrated with a standard solution of thiosulfate.

The titration end point can be detected visually, with a starch indicator, or electrometrically, with potentiometric or dead-stop technics.[3] Experienced analysts can maintain a precision of ± 50 μg/L with visual end-point detection and a precision of ± 5 μg/L with electrometric end-point detection.[2,3]

The liberated iodine also can be determined directly by simple absorption spectrophotometers.[4] This method can be used on a routine basis to provide very accurate estimates for DO in the microgram-per-liter range provided that interfering particulate matter, color, and chemical interferences are absent.

2. Selection of Method

Before selecting a method consider the effect of interferences, oxidizing or reducing materials that may be present in the sample. Certain oxidizing agents liberate iodine from iodides (positive interference) and some reducing agents reduce iodine to iodide (negative interference). Most organic matter is oxidized partially when the oxidized manganese precipitate is acidified, thus causing negative errors.

Several modifications of the iodometric method are given to minimize the effect of interfering materials.[2] Among the more commonly used procedures are the azide modification,[5] the permanganate modification,[6] the alum flocculation modification,[7] and the copper sulfate-sulfamic acid flocculation modification.[8,9] The azide modification (B) effectively removes interference caused by nitrite, which is the most common interference in biologically treated effluents and incubated BOD samples. Use the permanganate modification (C) in the presence of ferrous iron. When the sample contains 5 or more mg ferric iron salts/L, add potassium fluoride (KF) as the first reagent in the azide modification or after the permanganate treatment for ferrous iron. Alternately, eliminate Fe(III) interference by using 85–87% phosphoric acid (H_3PO_4) instead of sulfuric acid (H_2SO_4) for acidification. This pro-

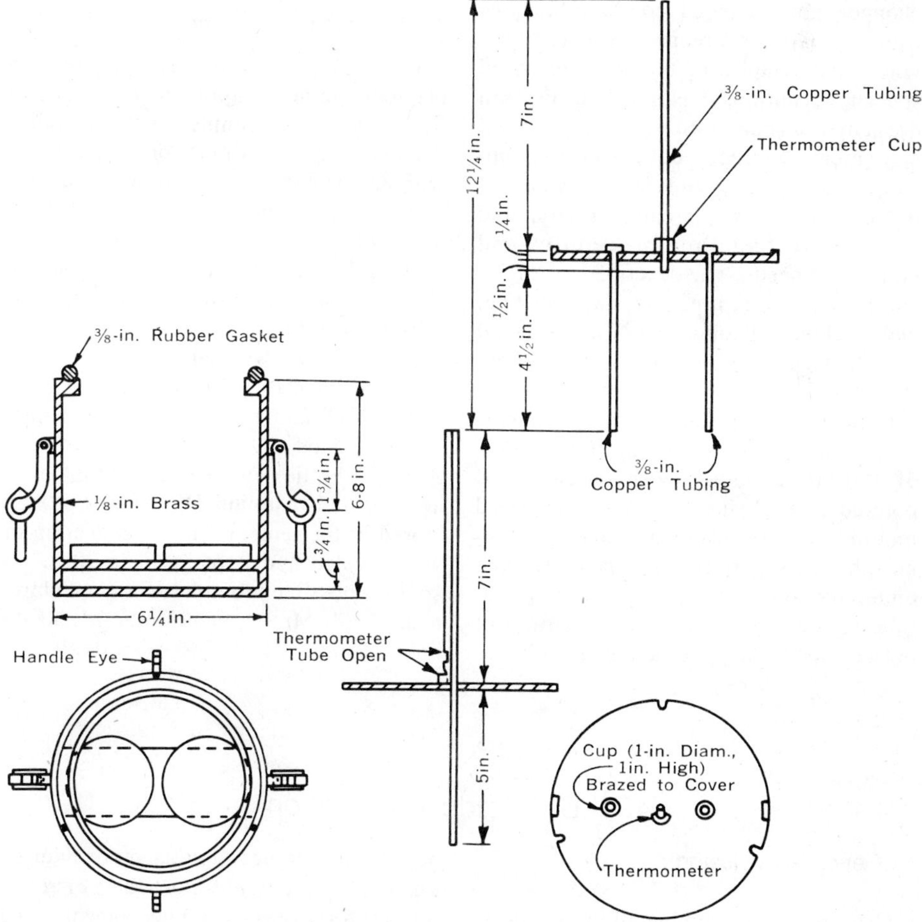

Figure 421:1. DO and BOD sampler assembly.

cedure has not been tested for Fe(III) concentrations above 20 mg/L.

Use the alum flocculation modification (D) in the presence of suspended solids that cause interference and the copper sulfate-sulfamic acid flocculation modification (E) on activated-sludge mixed liquor.

3. Collection of Samples

Collect samples very carefully. Methods of sampling are highly dependent on source to be sampled and, to a certain extent, on method of analysis. Do not let sample remain in contact with air or be agitated, because either condition causes a change in its gaseous content. Samples from any depth in streams, lakes, or reservoirs, and samples of boiler water, need special precautions to eliminate changes in pressure and temperature. Procedures and equipment have been developed for sampling waters under pressure and unconfined waters (e.g., streams, rivers, and reservoirs). Sampling procedures and equipment needed are described in American Society for Testing and Materials Special Technical Publication No. 148-1 and in U.S. Geological Survey Water Supply Paper No. 1454.

Collect surface water samples in narrow-mouth glass-stoppered BOD bottles of 300-mL capacity with tapered and pointed ground-glass stoppers and flared mouths. Avoid entraining or dissolving atmospheric oxygen. In sampling from a line under pressure, attach a glass or rubber tube to the tap and extend to bottom of bottle. Let bottle overflow two or three times its volume and replace stopper so that no air bubbles are entrained.

Suitable samplers for streams, ponds, or tanks of moderate depth are of the APHA type shown in Figure 421:1. Use a Kemmerer-type sampler for samples collected from depths greater than 2 m. Bleed sample from bottom of sampler through a tube extending to bottom of a 250- to 300-mL BOD bottle. Fill bottle to overflowing (overflow for approximately 10 sec), and prevent turbulence and formation of bubbles while filling. Record sample temperature to nearest degree Celsius or more precisely.

4. Preservation of Samples

Determine DO immediately on all samples containing an appreciable oxygen or iodine demand. Samples with no iodine demand may be stored for a few hours without change after addition of manganous sulfate ($MnSO_4$) solution, alkali-iodide solution, and H_2SO_4, followed by shaking in the usual way. Protect stored samples from strong sunlight and titrate as soon as possible.

For samples with an iodine demand, preserve for 4 to 8 hr by adding 0.7 mL conc H_2SO_4 and 1 mL sodium azide solution (2 g NaN_3/100 mL distilled water) to the BOD bottle. This will arrest biological activity and maintain DO if the bottle is stored at the temperature of collection or water-sealed and kept at 10 to 20 C. As soon as possible, complete the procedure, using 2 mL $MnSO_4$ solution, 3 mL alkali-iodide solution, and 2 mL conc H_2SO_4.

421 B. Azide Modification

1. General Discussion

Use the azide modification for most sewage, effluent, and stream samples, especially if samples contain more than 50 μg NO_2^--N/L and not more than 1 mg ferrous iron/L. Other reducing or oxidizing materials should be absent. If 1 mL KF solution is added before the sample is acidified and there is no delay in titration, the method is applicable in the presence of 100 to 200 mg ferric iron/L.

2. Reagents

a. Manganous sulfate solution: Dissolve 480 g $MnSO_4 \cdot 4H_2O$, 400 g $MnSO_4 \cdot 2H_2O$, or 364 g $MnSO_4 \cdot H_2O$ in distilled water, filter, and dilute to 1 L. The $MnSO_4$ solution should not give a color with starch when added to an acidified potassium iodide (KI) solution.

b. Alkali-iodide-azide reagent: Dissolve 10 g NaN_3 in 500 mL distilled water. Add 480 g sodium hydroxide (NaOH) and 750 g sodium iodide (NaI), and stir until dissolved. There will be a white turbidity due to sodium carbonate (Na_2CO_3), but this will do no harm. CAUTION—*Do not acidify this solution because toxic hydrazoic acid fumes may be produced.*

c. Sulfuric acid, H_2SO_4, conc; One milliliter is equivalent to about 3 mL alkali-iodide-azide reagent.

d. Starch: Use either an aqueous solution or soluble starch powder mixtures.

To prepare an aqueous solution, dissolve 2 g laboratory-grade soluble starch and 0.2 g salicylic acid, as a preservative, in 100 mL hot distilled water.

e. Standard sodium thiosulfate titrant: Dissolve 6.205 g $Na_2S_2O_3 \cdot 5H_2O$ in distilled water. Add 1.5 mL 6*N* NaOH or 0.4 g solid NaOH and dilute to 1,000 mL. Standardize with bi-iodate solution.

f. Standard potassium bi-iodate solution, 0.0250*N:* Dissolve 812.4 mg $KH(IO_3)_2$ in distilled water and dilute to 1,000 mL.

Standardization: Dissolve approximately 2 g KI, free from iodate, in an erlenmeyer flask with 100 to 150 mL distilled water. Add 1 mL 6*N* H_2SO_4 or a few drops of conc H_2SO_4 and 20.00 mL standard bi-iodate solution. Dilute to 200 mL and titrate liberated iodine with thiosulfate titrant, adding starch toward end of titration, when a pale straw color is reached. When the solutions are of equal strength, 20.00 mL 0.0250*N* $Na_2S_2O_3$ should be required. If not, adjust the $Na_2S_2O_3$ solution to 0.0250*N*.

g. Potassium fluoride solution: Dissolve 40 g $KF \cdot 2H_2O$ in distilled water and dilute to 100 mL.

3. Procedure

a. To the sample collected in a 250- to 300-mL bottle, add 1 mL $MnSO_4$ solution, followed by 1 mL alkali-iodide-azide reagent. If pipets are dipped into sample, rinse them before returning them to reagent bottles. Alternatively, hold pipet tips just above liquid surface when adding reagents. Stopper carefully to exclude air bubbles and mix by inverting bottle a few times. When precipitate has settled sufficiently (to approximately half the bottle volume) to leave clear supernate above the manganese hydroxide floc, add 1.0 mL conc H_2SO_4. Restopper and mix by inverting several times until dissolution is complete. Titrate a volume corresponding to 200 mL original sample after correction for sample loss by displacement with reagents. Thus, for a total of 2 mL (1 mL each) of $MnSO_4$ and alkali-iodide-azide reagents in a 300-mL bottle, titrate $200 \times 300/(300\text{-}2) = 201$ mL.

b. Titrate with 0.0250*N* $Na_2S_2O_3$ solution to a pale straw color. Add a few drops of starch solution and continue titration to first disappearance of blue color. If end point is overrun, back-titrate with 0.0250*N* bi-iodate solution added dropwise, or by adding a measured volume of treated sample. Correct for amount of bi-iodate solution or sample. Disregard subsequent recolorations due to the catalytic effect of nitrite or to traces of ferric salts that have not been complexed with fluoride.

4. Calculation

a. For titration of 200 mL sample, 1 mL 0.0250*N* $Na_2S_2O_3$ = 1 mg DO/L.

TABLE 421:I. SOLUBILITY OF OXYGEN IN WATER EXPOSED TO WATER-SATURATED AIR*

Temperature *C*	Chloride Concentration in Water *mg/L*				
	0	5,000	10,000	15,000	20,000
0	14.60	13.72	12.90	12.13	11.41
1	14.19	13.35	12.56	11.81	11.11
2	13.81	12.99	12.23	11.51	10.83
3	13.44	12.65	11.91	11.22	10.56
4	13.09	12.33	11.61	10.94	10.30
5	12.75	12.02	11.32	10.67	10.05
6	12.43	11.72	11.05	10.41	9.82
7	12.12	11.43	10.78	10.17	9.59
8	11.83	11.16	10.53	9.93	9.37
9	11.55	10.90	10.29	9.71	9.16
10	11.27	10.65	10.05	9.49	8.96
11	11.01	10.40	9.83	9.28	8.77
12	10.76	10.17	9.61	9.08	8.58
13	10.52	9.95	9.41	8.89	8.41
14	10.29	9.73	9.21	8.71	8.24
15	10.07	9.53	9.01	8.53	8.07
16	9.85	9.33	8.83	8.36	7.91
17	9.65	9.14	8.65	8.19	7.78
18	9.45	8.95	8.48	8.03	7.61
19	9.26	8.77	8.32	7.88	7.47
20	9.07	8.60	8.16	7.73	7.33
21	8.90	8.44	8.00	7.59	7.20
22	8.72	8.28	7.85	7.45	7.07
23	8.56	8.12	7.71	7.32	6.95
24	8.40	7.97	7.57	7.19	6.83
25	8.24	7.83	7.44	7.06	6.71
26	8.09	7.69	7.31	6.94	6.60
27	7.95	7.55	7.18	6.83	6.49
28	7.81	7.42	7.06	6.71	6.38
29	7.67	7.30	6.94	6.60	6.28
30	7.54	7.17	6.83	6.49	6.18
31	7.41	7.05	6.71	6.39	6.08
32	7.28	6.94	6.61	6.29	5.99
33	7.16	6.82	6.50	6.19	5.90
34	7.05	6.71	6.40	6.10	5.81
35	6.93	6.61	6.30	6.01	5.72
36	6.82	6.51	6.20	5.92	5.64

* At a total pressure of 101.3 kPa. Under any other barometric pressure, P, obtain the solubility, S (mg/L) from the corresponding value in the table by the equation:

$$S' = S \frac{P-p}{760-p}$$

in which S is the solubility at 101.3 kPa and p is the pressure (mm) of saturated water vapor at the water temperature. For elevations less than 1,000 m and temperatures below 25 C, ignore p. The equation then becomes:

$$S' = S \frac{P}{760} = S \frac{P'}{29.92}$$

Dry air is assumed to contain 20.90% oxygen. (Calculations made by Whipple and Whipple, 1911. *J. Amer. Chem. Soc.* 33:362.)

TABLE 421:I. SOLUBILITY OF OXYGEN IN WATER EXPOSED TO WATER-SATURATED AIR*

Temperature *C*	Chloride Concentration in Water *mg/L*				
	0	5,000	10,000	15,000	20,000
37	6.71	6.40	6.11	5.83	5.56
38	6.61	6.31	6.02	5.74	5.48
39	6.51	6.21	5.93	5.66	5.40
40	6.41	6.12	5.84	5.58	5.33
41	6.31	6.03	5.76	5.50	5.25
42	6.22	5.94	5.68	5.42	5.18
43	6.13	5.85	5.60	5.35	5.11
44	6.04	5.77	5.52	5.27	5.04
45	5.95	5.69	5.44	5.20	4.98
46	5.86	5.61	5.37	5.13	4.91
47	5.78	5.53	5.29	5.06	4.85
48	5.70	5.45	5.22	5.00	4.78
49	5.62	5.38	5.15	4.93	4.72
50	5.54	5.31	5.08	4.87	4.66

b. To obtain results in milliliters oxygen gas per liter, corrected to 0 C and 101.3 kPa, multiply mg DO/L by 0.70.

c. To express results as percent saturation at 101.3 kPa, use the solubility data in Table 421:I. Equations for correcting solubilities to barometric pressures other than mean sea level are given below the table.

5. Precision and Accuracy

DO can be determined with a precision, expressed as a standard deviation, of about 20 μg/L in distilled water and about 60 μg/L in wastewater and secondary effluents. In the presence of appreciable interference, even with proper modifications, the standard deviation may be as high as 100 μg/L. Still greater errors may occur in testing waters having organic suspended solids or heavy pollution. Avoid errors due to carelessness in collecting samples, prolonging the completion of test, or selecting an unsuitable modification.

421 C. Permanganate Modification

1. General Discussion

Use the permanganate modification only on samples containing ferrous iron. Interference from high concentrations of ferric iron (up to several hundred milligrams per liter), as in acid mine water, may be overcome by the addition of 1 mL potassium fluoride (KF) and azide, provided that the final titration is made immediately after acidification.

This procedure is ineffective for oxidation of sulfite, thiosulfate, polythionate, or the organic matter in wastewater. The error with samples containing 0.25% by volume of digester waste from the manufacture of sulfite pulp may amount to 7 to 8 mg DO/L. With such samples, use the alkali-hypochlorite modification.[10] At best, however, the latter procedure gives low results, the deviation amounting to 1 mg/L for samples containing 0.25% digester wastes.

2. Reagents

All the reagents required for Method B, and in addition:

a. Potassium permanganate solution: Dissolve 6.3 g $KMnO_4$ in distilled water and dilute to 1 L.

b. Potassium oxalate solution: Dissolve 2 g $K_2C_2O_4 \cdot H_2O$ in 100 mL distilled water; 1 mL will reduce about 1.1 mL permanganate solution.

3. Procedure

a. To a sample collected in a 250- to 300-mL bottle add, below the surface, 0.70 mL conc H_2SO_4, 1 mL $KMnO_4$ solution, and 1 mL KF solution. Stopper and mix by inversion. Never add more than 0.7 mL conc H_2SO_4 as the first step of pretreatment. Add acid with a 1-mL pipet graduated to 0.1 mL. Add sufficient $KMnO_4$ solution to obtain a violet tinge that persists for 5 min. If the permanganate color is destroyed in a shorter time, add additional $KMnO_4$ solution, but avoid large excesses.

b. Remove permanganate color completely by adding 0.5 to 1.0 mL $K_2C_2O_4$ solution. Mix well and let stand in the dark to facilitate the reaction. Excess oxalate causes low results; add only an amount of $K_2C_2O_4$ that completely decolorizes the $KMnO_4$ without having an excess of more than 0.5 mL. Complete decolorization in 2 to 10 min. If it is impossible to decolorize the sample without adding a large excess of oxalate, the DO result will be inaccurate.

c. From this point the procedure closely parallels that in Section 421B.3. Add 1 mL $MnSO_4$ solution and 3 mL alkali-iodide-azide reagent. Stopper, mix, and let precipitate settle a short time; acidify with 2 mL conc H_2SO_4. When 0.7 mL acid, 1 mL $KMnO_4$ solution, 1 mL $K_2C_2O_4$ solution, 1 mL $MnSO_4$ solution, and 3 mL alkali-iodide-azide (or a total of 6.7 mL reagents) are used in a 300-mL bottle, take $200 \times 300/(300 - 6.7) = 205$ mL for titration.

This correction is slightly in error because the $KMnO_4$ solution is nearly saturated with DO and 1 mL would add about 0.008 mg oxygen to the DO bottle. However, because precision of the method (standard deviation, 0.06 mL thiosulfate titration, or 0.012 mg DO) is 50% greater than this error, a correction is unnecessary. When substantially more $KMnO_4$ solution is used routinely, use a solution several times more concentrated so that 1 mL will satisfy the permanganate demand.

421 D. Alum Flocculation Modification

1. General Discussion

Samples high in suspended solids may consume appreciable quantities of iodine in acid solution. The interference due to solids may be removed by alum flocculation.

2. Reagents

All the reagents required for the azide modification (Section 421B.2) and in addition:

a. Alum solution: Dissolve 10 g aluminum potassium sulfate, $AlK(SO_4)_2 \cdot 12H_2O$, in distilled water and dilute to 100 mL.

b. Ammonium hydroxide, NH_4OH, conc.

3. Procedure

Collect sample in a glass-stoppered bottle of 500 to 1,000 mL capacity, using the same precautions as for regular DO samples. Add 10 mL alum solution and 1 to 2 mL conc NH_4OH. Stopper and invert gently for about 1 min. Let sample settle for about 10 min and siphon clear supernate into a 250- to 300-mL DO bottle until it overflows. Avoid sample aeration and keep siphon submerged at all times. Continue sample treatment as in Section 421B.3 or an appropriate modification.

421 E. Copper Sulfate-Sulfamic Acid Flocculation Modification

1. General Discussion

This modification is used for biological flocs such as activated sludge mixtures, which have high oxygen utilization rates.

2. Reagents

All the reagents required for the azide modification (Section 421B.2) and, in addition:

Copper sulfate-sulfamic acid inhibitor solution: Dissolve 32 g technical-grade NH_2SO_2OH without heat in 475 mL distilled water. Dissolve 50 g $CuSO_4 \cdot 5H_2O$ in 500 mL distilled water. Mix the two solutions and add 25 mL conc acetic acid.

3. Procedure

Add 10 mL $CuSO_4$-NH_2SO_2OH inhibitor to a 1-L glass-stoppered bottle. Insert bottle in a special sampler designed so that bottle fills from a tube near bottom and overflows only 25 to 50% of bottle capacity. Collect sample, stopper, and mix by inverting. Let suspended solids settle and siphon relatively clear supernatant liquor into a 250- to 300-mL DO bottle. Continue sample treatment as rapidly as possible by the azide (Section 421B.3) or other appropriate modification.

421 F. Membrane Electrode Method

1. General Discussion

Various modifications of the iodometric method have been developed to eliminate or minimize effects of interferences; nevertheless, the method still is inapplicable to a variety of industrial and domestic wastewaters.[11] Moreover, the iodometric method is not suited for field testing and cannot be adapted easily for continuous monitoring or for DO determinations in situ.

Polarographic methods using the dropping mercury electrode or the rotating platinum electrode have not been reliable always for the DO analysis in domestic and industrial wastewaters because impurities in the test solution can cause electrode poisoning or other interferences.[12,13] With membrane-covered electrode systems these problems are minimized, because the sensing element is protected by an oxygen-permeable plastic membrane that serves as a diffusion barrier against impurities.[14–16] Under steady-state conditions the current is directly proportional to the DO concentration.*

Membrane electrodes of the polarographic[14] as well as the galvanic[15] type have been used for DO measurements in lakes and reservoirs,[17] for stream survey and control of industrial effluents,[18,19] for continuous monitoring of DO in activated sludge units,[20] and for estuarine and oceanographic studies.[21] Being completely submersible, membrane electrodes are suited for analysis in situ. Their portability and ease of operation and maintenance make them particularly convenient for field applications. In laboratory investigations, membrane electrodes have been used for continuous DO analysis in bacterial cultures, including the BOD test.[15,22]

Membrane electrodes provide an excellent method for DO analysis in polluted waters, highly colored waters, and strong waste effluents. They are recommended for use especially under conditions that are unfavorable for use of the iodometric method, or when that test and its modifications are subject to serious errors caused by interferences.

a. Principle: Oxygen-sensitive membrane electrodes of the polarographic or galvanic type are composed of two solid metal electrodes in contact with supporting electrolyte separated from the test solution by a selective membrane. The basic difference between the galvanic and the polarographic systems is that in the former the electrode reaction is spontaneous (similar to that in a fuel cell), while in the latter an external source of applied voltage is needed to polarize the indicator electrode. Polyethylene and fluorocarbon membranes are used commonly because they are permeable to molecular oxygen and are relatively rugged.

Membrane electrodes are commercially available in some variety. In all these instruments the "diffusion current" is linearly proportional to the concentration of molecular oxygen. The current can be converted easily to concentration units (e.g., milligrams per liter) by a number of calibration procedures.

Membrane electrodes exhibit a relatively high temperature coefficient largely due to changes in the membrane permeability.[16] The effect of temperature on the electrode sensitivity, ϕ (microamperes per milligram per liter), can be expressed by the following simplified relationship:[16]

$$\log \phi = 0.43\, mt + b$$

where:

t = temperature, degrees C,
m = constant that depends on the membrane material, and
b = constant that largely depends on membrane thickness.

If values of ϕ and m are determined for one temperature (ϕ_0 and t_0), it is possible to calculate the sensitivity at any desired temperature (ϕ and t) as follows:

$$\log \phi = \log \phi_0 + 0.43\, m\, (t - t_0)$$

Nomographic charts for temperature correction can be constructed easily[2] and are available from some manufacturers. An example is shown in Figure 421:2, in which, for simplicity, sensitivity is plotted versus temperature in degrees Celsius on semilogarithmic coordinates. Check one or two points frequently to confirm original calibration. If calibration changes, the new calibration should be parallel to the original, provided that the same membrane material is used.

Temperature compensation also can be made automatically by using thermistors in the electrode circuit.[14] However, thermistors may not compensate fully over a wide temperature range. For certain applications where high accuracy is required,

*Fundamentally the current is directly proportional to the activity of molecular oxygen.[2]

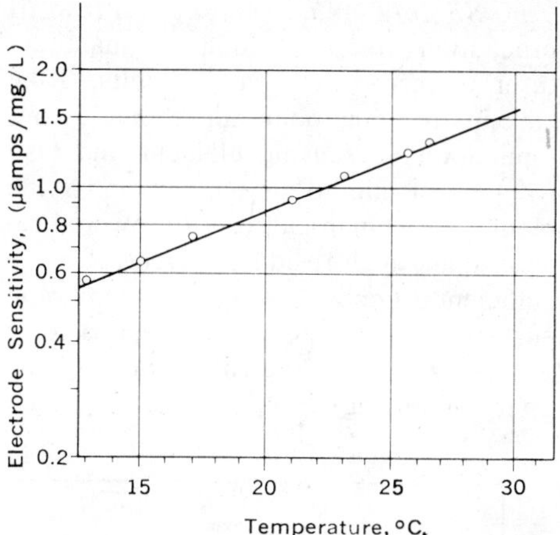

Figure 421:2. Effect of temperature on electrode sensitivity.

use calibrated nomographic charts to correct for temperature effect.

To use the DO membrane electrode in estuarine waters or in wastewaters with varying ionic strength, correct for effect of salting-out on electrode sensitivity.[2,16] This effect is particularly significant for large changes in salt content. Electrode sensitivity varies with salt concentration according to the following relationship:

$$\log \phi_S = 0.43\ m_S C_S + \log \phi_0$$

where:

ϕ_S, ϕ_0 = sensitivities in salt solution and distilled water, respectively,

C_S = salt concentration (preferably ionic strength), and

m_S = constant (salting-out coefficient).

If ϕ_0 and m_S are determined, it is possible to calculate sensitivity for any value of C_S. Conductivity measurements can be used to approximate salt concentration (C_S). This is particularly applicable to estuarine waters. Figure 421:3 shows calibration curves for sensitivity of varying salt solutions at different temperatures.

b. Interference: Plastic films used with membrane electrode systems are permeable to a variety of gases besides oxygen, although none is depolarized easily at the indicator electrode. Prolonged use of membrane electrodes in waters containing such gases as hydrogen sulfide (H_2S) tends to lower cell sensitivity. Eliminate this interference by frequently changing and calibrating the membrane electrode.

c. Sampling: Because membrane electrodes offer the advantage of analysis in situ they eliminate errors caused by sample handling and storage. If sampling is required, use the same precautions suggested for the iodometric method.

2. Apparatus

Oxygen-sensitive membrane electrode, polarographic or galvanic, with appropriate meter.

3. Procedure

a. Calibration: Follow manufacturer's calibration procedure exactly to obtain guaranteed precision and accuracy. Generally, calibrate membrane electrodes by reading against air or a sample of known DO concentration (determined by iodometric method) as well as in a sample with

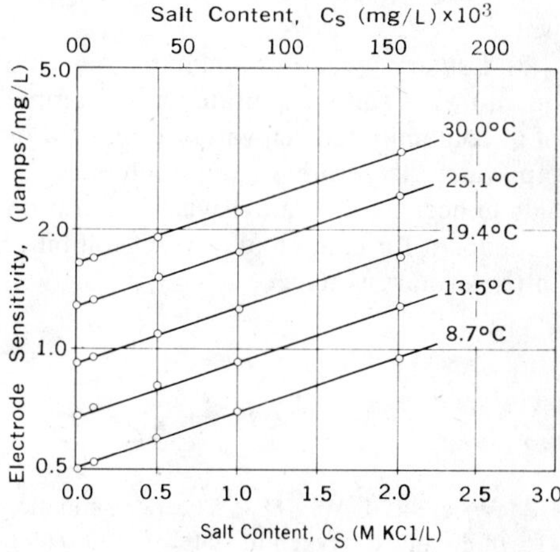

Figure 421:3. The salting-out effect at different temperatures.

zero DO. (Add excess sodium sulfite, Na_2SO_3, and a trace of cobalt chloride, $CoCl_2$, to bring DO to zero.) Preferably calibrate with samples of water under test. Avoid an iodometric calibration where interfering substances are suspected. The following illustrate the recommended procedures:

1) Fresh water—For unpolluted samples where interfering substances are absent, calibrate in the test solution or distilled water, whichever is more convenient.

2) Salt water—Calibrate directly with samples of seawater or waters having a constant salt concentration in excess of 1,000 mg/L.

3) Fresh water containing pollutants or interfering substances—Calibrate with distilled water because erroneous results occur with the sample.

4) Salt water containing pollutants or interfering substances—Calibrate with a sample of clean water containing the same salt content as the sample. Add a concentrated potassium chloride (KCl) solution (see Conductivity, Section 205 and Table 205:I) to distilled water to produce the same specific conductance as that in the sample. For polluted ocean waters, calibrate with a sample of unpolluted seawater.

5) Estuary water containing varying quantities of salt—Calibrate with a sample of uncontaminated seawater or distilled or tap water. Determine sample chloride or salt concentration and revise calibration to account for change of oxygen solubility in the estuary water.[2]

b. Sample measurement: Follow all precautions recommended by manufacturer to insure acceptable results. Take care in changing membrane to avoid contamination of sensing element and also trapping of minute air bubbles under the membrane, which can lead to lowered response and high residual current. Provide sufficient sample flow across membrane surface to overcome erratic response (see Figure 421:4 for a typical example of the effect of stirring).

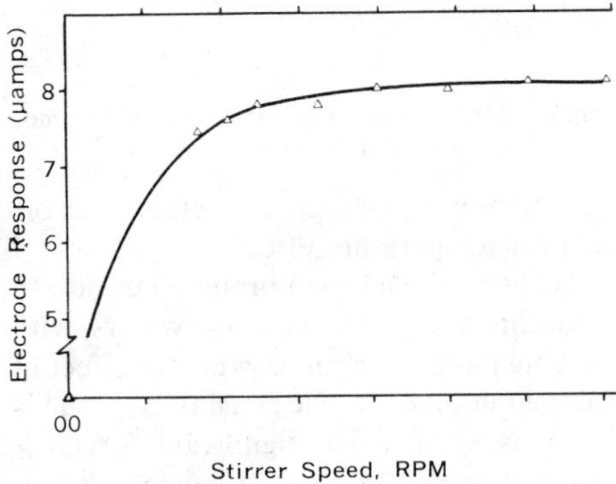

Figure 421:4. Typical trend of effect of stirring on electrode response.

c. Validation of temperature effect: Check frequently one or two points to verify temperature correction data.

4. Precision and Accuracy

With most commercially available membrane electrode systems an accuracy of ±0.1 mg DO/L and a precision of ±0.05 mg DO/L can be obtained.

421 G. References

1. WINKLER, L.W. 1888. The determination of dissolved oxygen in water. *Berlin. Deut. Chem. Ges.* 21:2843.
2. MANCY, K.H. & T. JAFFE. 1966. Analysis of dissolved oxygen in natural and waste waters. USPHS Publ. No. 999-WP-37. Washington, D.C.
3. POTTER, E.C. & G.E. EVERITT. 1957. Ad-

vances in dissolved oxygen microanalysis. *J. Appl. Chem.* 9:642.

4. OULMAN, C.S. & E.R. BAUMANN. 1956. A colorimetric method for determining dissolved oxygen. *Sewage Ind. Wastes* 28:1461.
5. ALSTERBERG, G. 1925. Methods for the determination of elementary oxygen dissolved in water in the presence of nitrite. *Biochem. Z.* 159:36.
6. RIDEAL, S. & G.G. STEWART. 1901. The determination of dissolved oxygen in waters in the presence of nitrites and of organic matter. *Analyst* 26:141.
7. RUCHHOFT, C.C. & W.A. MOORE. 1940. The determination of biochemical oxygen demand and dissolved oxygen of river mud suspensions. *Ind. Eng. Chem.* Anal. Ed. 12:711.
8. PLACAK, O.R. & C.C. RUCHHOFT. 1941. Comparative study of the azide and Rideal-Stewart modifications of the Winkler method in the determination of biochemical oxygen demand. *Ind. Eng. Chem.*, Anal. Ed. 13:12.
9. RUCHHOFT, C.C. & O.R. PLACAK. 1942. Determination of dissolved oxygen in activated-sludge sewage mixtures. *Sewage Works J.* 14:638.
10. THERIAULT, E.J. & P.D. MCNAMEE. 1932. Dissolved oxygen in the presence of organic matter, hypochlorites, and sulfite wastes. *Ind. Eng. Chem.*, Anal. Ed. 4:59.
11. MCKEOWN, J.J., L.C. BROWN & G.W. GOVE. 1967. Comparative studies of dissolved oxygen analysis methods. *J. Water Pollut. Control Fed.* 39:1323.
12. LYNN, W.R. & D.A. OKUN. 1955. Experience with solid platinum electrodes in the determination of dissolved oxygen. *Sewage Ind. Wastes* 27:4.
13. MANCY, K.H. & D.A. OKUN. 1960. Automatic recording of dissolved oxygen in aqueous systems containing surface active agents. *Anal. Chem.* 32:108.
14. CARRITT, D.E. & J.W. KANWISHER. 1959. An electrode system for measuring dissolved oxygen. *Anal. Chem.* 31:5.
15. MANCY, K.H. & W.C. WESTGARTH. 1962. A galvanic cell oxygen analyzer. *J. Water Pollut. Control Fed.* 34:1037.
16. MANCY, K.H., D.A. OKUN & C.N. REILLEY. 1962. A galvanic cell oxygen analyzer. *J. Electroanal. Chem.* 4:65.
17. WEISS, C.M. & R.T. OGLESBY. 1963. Instrumentation for monitoring water quality in reservoirs. Amer. Water Works Ass. 83rd Annual Conf., New York, N.Y.
18. CLEARY, E.J. 1962. Introducing the ORSANCO robot monitor. *Proc. Water Quality Meas. Instrum.* Publ. No. 108, USPHS, Washington, D.C.
19. MACKERETH, F.J.H. 1964. An improved galvanic cell for determination of oxygen concentrations in fluids. *J. Sci. Instrum.* 41:38.
20. SULZER, F. & W.M. WESTGARTH. 1962. Continuous D.O. recording in activated sludge. *Water Sewage Works* 109:376.
21. DUXBURY, A.C. 1963. Calibration and use of a galvanic type oxygen electrode in field work. *Limnol. Oceanogr.* 8:483.
22. LIPNER, H.J., L.R. WITHERSPOON & V.C. CHAMPEAUS. 1964. Adaptation of a galvanic cell for microanalysis of oxygen. *Anal. Chem.* 36:204.
23. POSTMA, H., A. SYANSSON, H. LABOMBE & K. GRASSHOFF. 1976. Letter to the Editors—The International Oceanographic Tables for the Solubility of Oxygen in Sea Water. *J. Cons. Int. Explor. Mer.* 36:295.

422 OZONE (RESIDUAL)

Ozone, O_3, a potent germicide, also is used as an oxidizing agent for the destruction of organic compounds producing taste and odor in water, for the destruction of organic coloring matter, and for the oxidation of reduced iron or manganese salts to insoluble oxides, which can be precipitated or filtered from the water.

1. General Discussion

The iodometric method described is quantitative, subject to few interferences except for most volatile oxidants, and capable of good precision. The method also can be used for the determination of ozone in air by absorption of the ozone in iodide solution.

If iodometric titration is considered undesirable, any method for measuring residual chlorine (see Section 408) may be applied after ozone stripping into the adsorber has been completed. These methods determine iodine with acceptable accuracy.

The ozone concentration in water, air, or oxygen also can be determined continuously by photometric instruments, which measure the strong absorption ozone exerts at the wavelength of 253.7 nm.

a. Principle: Ozone liberates iodine from a potassium iodide (KI) solution. For accurate results make the solution alkaline during absorption of ozone. In practice, KI solutions quickly become alkaline during the process, so buffering is not required. After immediate acidification, the liberated iodine is titrated with standard $0.005N$ sodium thiosulfate ($Na_2S_2O_3$) with starch indicator.

b. Interference: Because ozonated water may contain manganese dioxide, ferric ion, chlorine, possibly peroxide, and other oxidation products, avoid these interferences by passing the ozone through the gaseous phase into a KI solution for titration. Titrating a sample added directly to the KI solution (without first passing ozone through the gaseous phase), as compared with titrating after ozone stripping, will indicate whether interfering oxidants are present. If such interferences are absent or negligible, ozone transfer through inert gas stripping may be eliminated.

The stability of ozone solutions decreases progressively at each increment in temperature above freezing and with each increment in pH above 3.0.

c. Sampling and storage: Determine ozone immediately; samples cannot be preserved or stored because of instability of the residual. The stability of residual ozone is markedly improved at low temperatures and low pH. Minimize aeration during sample collection.

d. Minimum detectable concentration: Approximately 30 $\mu g\ O_3/L$.

2. Apparatus

The following are required for sample collection:

a. Gas-washing bottles and absorbers. 1-L and 500-mL capacities. To prevent loss of ozone on glass frits found on standard wash bottles, cut off glass diffusers and draw tube into a 1-mm tip extending to approximately 5 mm from bubbler bottom.

b. Pure air or pure nitrogen gas supply, 0.2- to 1.0-L/min capacity.

c. Glass, stainless steel, or aluminum piping, for carrying ozonized air. Good-quality polyvinyl* tubing also may be used for short runs, but do not use rubber.

3. Reagents

a. Potassium iodide solution: Dissolve 20 g KI, free from iodine, iodate, and reducing agents, in 1 L freshly boiled and cooled distilled water. Store in a brown bottle in a refrigerator. Store at least 1 day before using.

b. Sulfuric acid, H_2SO_4, $1N$.

c. Standard sodium thiosulfate, $0.1N$*:* Dissolve 25 g $Na_2S_2O_3 \cdot 5H_2O$ in 1 L freshly boiled distilled water. Standardize against potassium bi-iodate (also called potassium hydrogen iodate) or potassium dichromate according to the procedure described in Section 408A.2*c*.

d. Standard sodium thiosulfate titrant, $0.005N$*:* Dilute proper volume (approxi-

*Tygon, US Stoneware Company, or equivalent.

mately 50 mL) of 0.1N $Na_2S_2O_3$ to 1,000 mL. For accurate work, standardize this solution daily, using either 0.005N potassium bi-iodate or potassium dichromate solution. Standardize as described in ¶ 4*c* below; 1.00 mL standard $Na_2S_2O_3$ titrant, 0.005N = 120 μg O_3.

e. Starch indicator solution: Use either an aqueous solution or soluble starch powder mixtures.

To prepare aqueous solution, add a cold water suspension of 5 g arrowroot or soluble starch to approximately 800 mL boiling water, with stirring. Dilute to 1 L, boil a few minutes, and let settle overnight. Use clear supernate. Preserve with 1.25 g salicylic acid/L or by adding a few drops of toluene.

f. Standard iodine, 0.1N: Dissolve 40 g KI in 25 mL distilled water. Add 13 g resublimed iodine and stir until dissolved. Dilute to 1 L and standardize as described in Section 408B.3*g*.

g. Standard iodine, 0.005N: Dissolve 16 g KI in a little distilled water in a 1-L volumetric flask, add proper volume (approximately 50 mL) of 0.1N iodine solution, and dilute to mark. For accurate work, standardize daily. Store solution in a brown bottle or in the dark. Protect from direct sunlight at all times and keep from all contact with rubber.

4. Procedure

a. Sample collection: Collect an 800-mL sample in a 1-L gas washing bottle.

b. Ozone absorption: Pass a stream of pure air or N_2 through sample and then through an absorber containing 400 mL KI solution. Continue for 5 to 10 min at a rate of 0.2 to 1.0 L/min to insure that all ozone is swept from sample and absorbed in KI solution.

c. Titration: Transfer KI solution to a 1-L beaker, rinse absorber, and add 20 mL 1N H_2SO_4 to reduce pH below 2.0. Titrate with 0.005N $Na_2S_2O_3$ titrant until yellow color of liberated iodine almost is discharged. Add 4 mL starch indicator solution and continue titrating carefully but rapidly to the end point, at which the blue color just disappears. Long contact of iodine and starch develops a blue compound that is difficult to decolorize. The end point may be determined amperometrically as described in Section 408C.4*b* except that $Na_2S_2O_3$ can be used as the titrant. Other procedures given in Section 408 for measuring iodine may be used.

d. Blank test: Correct sample titration result by determining blank contributed by such reagent impurities as free iodine or iodate in KI, or traces of reducing agents that might reduce liberated iodine.

Take 400 mL KI solution, 20 mL 1N H_2SO_4, and 4 mL starch indicator solution. Perform whichever blank titrator below applies:

1) If a blue color appears, titrate with 0.005N $Na_2S_2O_3$ to disappearance of blue and record result.

2) If no blue color appears, titrate with 0.005N iodine solution until a blue color appears. Back-titrate with 0.005N $Na_2S_2O_3$ to disappearance and record difference.

Before calculating ozone concentration subtract blank titration in ¶ 4*d*1) from sample titration, or add result of ¶ 4*d*2) above.

5. Calculation

$$\text{mg } O_3/L = \frac{(A \pm B) \times N \times 24{,}000}{\text{mL sample}}$$

where:

A = mL titrant for sample,
B = mL titrant for blank (positive or negative), and
N = normality of $Na_2S_2O_3$.

6. Interpretation of Results

The precision of the test is within ±1% for concentrations of 3 mg O_3/L or greater. However, rapid decrease of the residual occurs in the time elapsing between sampling and testing. Temperature also is an important factor in the decrease.

7. Bibliography

BIRDSALL, C.M., A.C. JENKINS & E. SPADINGER. 1952. The iodometric determination of ozone. *Anal. Chem.* 24:662.

ZEHENDER, F. & W. STUMM. 1953. Determination of ozone in drinking water. *Mitt. Gebiete Lebensm. Hyg.* 44:206.

INGOLS, R.S., R.H. FETNER & W.H. EBERHARDT. 1956. Determination of ozone in solution. *Proc. Int. Ozone Conf.*, American Chemical Society, Advan. Chem. Ser. No. 21.

BYERS, T.H. & B.E. SALTZMAN. 1958. Determination of ozone and air by neutral and alkaline iodide procedures. *J. Ind. Hyg. Ass.* 19:251.

SALTZMAN, B.E. & N. GILBERT. 1959. Iodometric microdetermination of organic oxidants and ozone. *Anal. Chem.* 31:1914.

ALTSHULLER, A.P., C.M. SCHWAB & M. BARE. 1959. Reactivity of oxidizing agents with potassium iodide reagent. *Anal. Chem.* 31:1987.

U.S. DEPARTMENT HEALTH, EDUCATION AND WELFARE. 1965. Selected Methods for the Measurement of Air Pollutants. PHS Publ. No. 999-AP-11, Washington, D.C.

AMERICAN PUBLIC HEALTH ASSOCIATION. 1972. Method of Air Sampling and Analysis. Intersociety Committee, APHA, Washington, D.C.

U.S. DEPARTMENT HEALTH, EDUCATION AND WELFARE. 1974. NIOSH Manual of Analytical Methods. HEW Publ. No. (NIOSH) 75-121, Washington, D.C.

423 pH VALUE

Measurement of pH is one of the most important and frequently used tests in water chemistry. Practically every phase of water supply and wastewater treatment, e.g., acid-base neutralization, water softening, precipitation, coagulation, disinfection, and corrosion control, is pH-dependent. pH is used in alkalinity and carbon dioxide measurements and many other acid-base equilibria. At a given temperature the *intensity* of the acidic or basic character of a solution is indicated by pH or hydrogen ion activity. Pure water is very slightly ionized and at equilibrium the ion product is

$$[H^+][OH^-] = K_W = 1.01 \times 10^{-14} \text{ at } 25\text{ C} \quad (1)$$

and

$$[H^+] = [OH^-] = 1.005 \times 10^{-7}$$

where:

$[H^+]$ = concentration of hydrogen ions, moles/L,

$[OH^-]$ = concentration of hydroxyl ions, moles/L, and

K_w = ion product of water.

Because of ionic interactions in all but very dilute solutions, it is necessary to use the "activity" of an ion and not its molar concentration. All subsequent references to pH assume that the activity of the hydrogen ion, a_{H^+}, is being considered and that its *approximate* equivalence to molarity, $[H^+]$ can be presumed only in very dilute solutions.

A logarithmic scale is convenient for expressing a wide range of ionic activities. Equation 1 in logarithmic form is:

$$(-\log_{10}[H^+]) + (-\log_{10}[OH^-]) = 14 \quad (2)$$

or

$$pH + pOH = pK_W$$

where:

pH* = $-\log_{10}[H^+]$, and

pOH = $-\log_{10}[OH^-]$.

*p is used as an abbreviation of "power" or exponent, after the usage of Sorensen.

Equation 2 defines the pH scale for aqueous solutions between 0 and 14, the practical limits of the electrometric method described herein. At 25 C, pH 7.0 is neutral, the activities of the hydrogen and hydroxyl ions are equal, and each corresponds to an approximate concentration of 10^{-7} moles/L. The neutral point is temperature-dependent and is pH 7.5 at 0 C and pH 6.5 at 60 C.

The pH value of a highly dilute solution represents hydrogen ion activity. Natural waters usually have pH values in the range of 4 to 9, and most are slightly basic because of the presence of bicarbonates and carbonates of the alkali and alkaline earth metals.

1. General Discussion

a. Principle: The basic principle of electrometric pH is determination of the activity of the hydrogen ions by potentiometric measurement using a glass electrode and a reference electrode. The glass electrode method is widely accepted because the glass electrode is much easier to use and is less likely to be poisoned than the primary standard hydrogen electrode.

Because single ion activities such as a_{H^+} cannot be measured, pH is defined operationally on a potentiometric scale. The pH measuring instrument is calibrated potentiometrically with an indicating (glass) electrode and a reference electrode using National Bureau of Standards buffers having assigned values so that:

$$pH_B = -\log_{10} a_{H^+}$$

where:

pH_B = assigned pH of NBS buffer.

The operational pH scale is used to measure sample pH and is defined† as:

$$pH_S = pH_B \pm \frac{F(E_S - E_B)}{2.303\ RT}$$

where:

pH_S = potentiometrically measured sample pH,
F = Faraday; 23,060 cal·V^{-1}·equivalent^{-1},
E_S = sample electromotive force (emf), V,
E_B = buffer emf, V,
R = gas constant; 1.987 cal·deg^{-1}·mole^{-1}, and
T = absolute temperature, K.

This equation assumes that the emf of the cells containing the sample and buffer is due solely to hydrogen ion activity unaffected by sample composition. In practice, a difference in ionic species causes a change in liquid junction potential and ionic strength affects H^+ activity. This imposes an experimental limitation on pH measurement, so that to obtain meaningful results, the differences between E_S and E_B should be minimal. Samples must be dilute aqueous solutions of simple solutes ($<0.2M$). (Choose buffers to bracket the sample.) Determination of pH cannot be made accurately in nonaqueous media, suspensions, colloids, or high-ionic-strength solutions. Under ideal conditions a tenfold change in hydrogen ion activity shifts the emf by 59.16 mV at 25 C.

b. Interferences: The glass electrode is relatively free from interference from color, turbidity, colloidal matter, oxidants, reductants, or high salinity, except for a sodium error at pH>10. Reduce this error by using special "low sodium error" electrodes.

pH measurements are subject to two different temperature effects: (*a*) the effect on the pH measuring system, and (*b*) changes in the solution pH due to shifts in ionic equilibria. Most pH meters are equipped with temperature compensators that can correct errors of the first type, but the measurement can show only the actual pH at the temperature of measurement.

†Although the equation for pH_s appears in the literature with a plus sign, the sign of emf readings in millivolts for most pH meters manufactured in the U.S. is negative. The choice of negative sign is consistent with the IUPAC Stockholm convention concerning the sign of electrode potential.[1,2]

2. Apparatus

a. pH meter consisting of potentiometer, a glass electrode, a reference electrode, and a temperature compensating device. A balanced circuit is completed through the potentiometer when the electrodes are immersed in the test solution. Many pH meters are capable of reading pH or millivolts and some have scale expansion that permits reading to 0.001 pH unit, but most instruments are not that accurate.

For routine work use a pH meter accurate and reproducible to 0.1 pH unit with a range of 0 to 14 and equipped with a temperature compensation adjustment.

b. Reference electrode consisting of a half cell that provides a standard electrode potential. Commonly used are calomel and silver: silver-chloride electrodes. Either is available with several types of liquid junctions.

The liquid junction of the reference electrode is critical because at this point the electrode forms a salt bridge with the sample or buffer and a liquid junction potential is generated that in turn affects the potential produced by hydrogen ions. Reference electrode junctions may be annular ceramic, quartz, or asbestos fiber, or the sleeve type. The quartz type is most widely used. The asbestos fiber type is not recommended for strongly basic solutions and annular ceramic and sleeve types are not suitable for strongly acidic solutions. Follow the manufacturer's recommendation on use and care of the reference electrode.

Except for sealed electrodes, refill only with the correct electrolyte to proper level and make sure junction is properly wetted.

c. Glass electrode: The sensor electrode is a bulb of special glass containing a fixed concentration of HCl or a buffered chloride solution in contact with an internal reference electrode. Upon immersion of a new electrode in a solution the outer bulb surface becomes hydrated and exchanges sodium ions for hydrogen ions to build up a surface layer of hydrogen ions. This, together with the repulsion of anions by fixed, negatively charged silicate sites, produces at the glass-solution interface a potential that is a function of hydrogen ion activity in solution.

Several types of glass electrodes are available. A "low sodium error" electrode that can operate at high temperatures is recommended for measuring pH over 10. Combination electrodes incorporate the glass and reference electrodes into a single probe.

d. Beakers: Preferably use polyethylene or TFE‡ beakers.

e. Stirrer: Use either a magnetic, TFE-coated stirring bar or a mechanical stirrer with inert plastic-coated impeller.

f. Flow chamber: Use for continuous flow measurements or for poorly buffered solutions.

3. Reagents

a. General preparation: Calibrate the electrode system against standard buffer solutions of known pH. Because buffer solutions may deteriorate as a result of mold growth or contamination, prepare fresh as needed for accurate work by weighing the amounts of chemicals specified in Table 423:I, dissolving in distilled water at 25 C, and diluting to 1,000 mL. This is particularly important for borate and carbonate buffers.

Boil and cool distilled water having a conductivity of less than 2 μmhos/cm. To 50 mL add 1 drop of saturated KCl solution suitable for reference electrode use. If the pH of this test solution is between 6.0 and 7.0, use it to prepare all standard solutions.

Dry KH_2PO_4 at 110 C to 130 C for 2 hr before weighing but do not heat unstable

‡Teflon or equivalent.

hydrated potassium tetroxalate above 60 C nor dry the other specified buffer salts.

Although ACS-grade chemicals generally are satisfactory for preparing buffer solutions, use certified materials available from the National Bureau of Standards when the greatest accuracy is required. For routine analysis, use commercially available buffer tablets, powders, or solutions of tested quality. In preparing buffer solutions from solid salts, insure complete solution.

As a rule, select and prepare buffer solutions classed as primary standards in Table 423:I; reserve secondary standards for extreme situations encountered in wastewater measurements. Consult Table 423:II for accepted pH of standard buffer solutions at temperatures other than 25 C. In routine use, store buffer solutions and samples in polyethylene bottles. Replace buffer solutions every 4 wk.

b. Saturated potassium hydrogen tartrate solution: Shake vigorously an excess (5 to 10 g) of finely crystalline $KHC_4H_4O_6$ with 100 to 300 mL distilled water at 25 C in a glass-stoppered bottle. Separate clear solution from undissolved material by decantation or filtration. Preserve for 2 months or more by adding one thymol crystal (8 mm diam) per 200 mL solution.

c. Saturated calcium hydroxide solution: Calcine a well-washed, low-alkali grade $CaCO_3$ in a platinum dish by igniting for 1 hr at 1,000 C. Cool, hydrate by slowly adding distilled water with stirring, and heat to boiling. Cool, filter, and collect solid $Ca(OH)_2$ on a fritted glass filter of medium porosity. Dry at 110 C, cool, and pul-

TABLE 423:I. PREPARATION OF pH STANDARD SOLUTIONS*

Standard Solution (molality)	pH at 25 C	Weight of Chemicals Needed/1,000 mL Aqueous Solution at 25 C
Primary standards:		
Potassium hydrogen tartrate (saturated at 25 C)	3.557	6.4 g $KHC_4H_4O_6$†
0.05 potassium dihydrogen citrate	3.776	11.41 g $KH_2C_6H_5O_7$
0.05 potassium hydrogen phthalate	4.008	10.12 g $KHC_8H_4O_4$
0.025 potassium dihydrogen phosphate+0.025 disodium hydrogen phosphate	6.865	3.388 g KH_2PO_4 + 3.533 g Na_2HPO_4‡
0.008695 potassium dihydrogen phosphate+0.03043 disodium hydrogen phosphate	7.413	1.179 g KH_2PO_4 + 4.302 g Na_2HPO_4‡
0.01 sodium borate decahydrate (borax)	9.180	3.80 g $Na_2B_4O_7 \cdot 10H_2O$‡
0.025 sodium bicarbonate+0.025 sodium carbonate	10.012	2.092 g $NaHCO_3$ + 2.640 g Na_2CO_3
Secondary standards:		
0.05 potassium tetroxalate dihydrate	1.679	12.61 g $KH_3C_4O_8 \cdot 2H_2O$
Calcium hydroxide (saturated at 25 C)	12.454	1.5 g $Ca(OH)_2$†

* BOWER, V. E. & R. G. BATES. 1957. Standards for pH measurements from 60° to 95° C. *J. Res. Nat. Bur. Standards* 59:261, and BATES, R. G., 1962. Revised values for pH measurements from 0° to 95° C. *J. Res. Nat. Bur. Standards* 66A:179.

† Approximate solubility.

‡Prepare with freshly boiled and cooled distilled water (carbon-dioxide-free).

verize to uniformly fine granules. Vigorously shake an excess of fine granules with distilled water in a stoppered polyethylene bottle. Let temperature come to 25 C after mixing. Filter supernatant under suction through a sintered glass filter of medium porosity and use filtrate as the buffer solution. Discard buffer solution when atmospheric CO_2 causes turbidity to appear.

d. Auxiliary solutions: 0.1*N* NaOH, 0.1*N* HCl, 5*N* HCl (dilute five volumes 6*N* HCl with one volume distilled water), and 20% ammonium bifluoride (NH_4HF_2) (dissolve 20 g NH_4HF_2 in distilled water and dilute to 100 mL).

4. Procedure

a. Instrument calibration: In each case follow manufacturer's instructions for pH meter used and for storage and preparation of electrodes for use. Recommended solutions for short-term storage of electrodes vary with type of electrode and manufacturer, but generally have a conductivity greater than 4,000 μmhos/cm. Tap water is a better substitute than distilled water. Keep electrodes wet by returning them to storage solution whenever pH meter is not in use.

Before use, remove electrodes from storage solution and rinse with distilled or demineralized water. Dry electrodes by gently blotting with a soft tissue. Bring sample and buffer to same temperature, which may be the room temperature, a fixed temperature such as 25 C, or the temperature of a fresh sample. Record temperature of measurement and adjust temperature dial on meter to this temperature.

Standardize instrument with electrodes immersed in a buffer solution within 2 pH units of sample pH. Remove electrodes from buffer, rinse thoroughly, and blot dry. Immerse in a second buffer below pH 10, approximately 3 pH units different from the first; the reading should be within 0.1 unit for the pH of the second buffer. If the meter response shows a difference greater than 0.1 pH unit from expected value, look for trouble with the electrodes or potentiometer (see 5 below).

The purpose of standardization is to compensate for changes in potentiometer or electrodes. When only occasional pH measurements are made, standardize instrument before each measurement. When frequent measurements are made, and the instrument is stable, standardize less frequently. If sample pH values vary widely, standardize for each sample with a buffer having a pH within 1 to 2 pH units of the sample.

b. Sample analysis: Establish equilibrium between electrodes and sample by stirring sample to insure homogeneity. For buffered samples or those of high ionic strength, condition electrodes after cleaning by dipping them into sample for 1 min. Blot dry, immerse in a fresh portion of the same sample, and read pH.

With dilute, poorly buffered solutions, equilibrate electrodes by immersing in three or four successive portions of sample. Take a fresh sample to measure pH.

5. Trouble Shooting

a. Potentiometer: To locate trouble source disconnect electrodes and, using a short-circuit strap, connect reference electrode terminal to glass electrode terminal. Observe change in pH when instrument calibration knob is adjusted. If potentiometer is operating properly, it will respond rapidly and evenly to changes in calibration over a wide scale range. A faulty potentiometer will fail to respond, will react erratically, or will show a drift upon adjustment. Switch to the millivolt scale on which the meter should read zero. If inexperienced, do not attempt potentiometer repair other than maintenance as described in instrument manual.

b. Electrodes: If potentiometer is functioning properly, look for the instrument

TABLE 423:II. STANDARD pH VALUES ASSIGNED BY THE NATIONAL BUREAU OF STANDARDS*

	Primary Standards							Secondary Standards	
Temperature C	Tartrate (Saturated)	Citrate (0.05M)	Phthalate (0.05M)	Phosphate (1 : 1)	Phosphate (1 : 3.5)	Borax (0.01M)	Carbonate (0.025M)	Tetroxalate (0.05M)	Calcium Hydroxide (Saturated)
0		3.863	4.003	6.984	7.534	9.464	10.317	1.666	13.423
5		3.840	3.999	6.951	7.500	9.395	10.245	1.668	13.207
10		3.820	3.998	6.923	7.472	9.332	10.179	1.670	13.003
15		3.802	3.999	6.900	7.448	9.276	10.118	1.672	12.810
20		3.788	4.002	6.881	7.429	9.225	10.062	1.675	12.627
25	3.557	3.776	4.008	6.865	7.413	9.180	10.012	1.679	12.454
30	3.552	3.766	4.015	6.853	7.400	9.139	9.966	1.683	12.289
35	3.549	3.759	4.024	6.844	7.389	9.102	9.925	1.688	12.133
38	3.548		4.030	6.840	7.384	9.081		1.691	12.043
40	3.547	3.753	4.035	6.838	7.380	9.068	9.889	1.694	11.984
45	3.547	3.750	4.047	6.834	7.373	9.038	9.856	1.700	11.841
50	3.549	3.749	4.060	6.833	7.367	9.011	9.828	1.707	11.705
55	3.554		4.075	6.834		8.985		1.715	11.574
60	3.560		4.091	6.836		8.962		1.723	11.449
70	3.580		4.126	6.845		8.921		1.743	
80	3.609		4.164	6.859		8.885		1.766	
90	3.650		4.205	6.877		8.850		1.792	
95	3.674		4.227	6.886		8.833		1.806	

* BATES, R. G. 1962. Revised standard values for pH measurements from 0 to 95 C. *J. Res. Nat. Bur. Standards* 66A:179.

fault in the electrode pair. Substitute one electrode at a time and cross-check with two buffers that are about 4 pH units apart. A deviation greater than 0.1 pH unit indicates a faulty electrode. Glass electrodes fail because of scratches, deterioration, or accumulation of debris on the glass surface. Rejuvenate electrode by alternately immersing it three times each in 0.1*N* HCl and 0.1*N* NaOH. If this fails, immerse tip in 20% NH_4HF_2 for 3 min. After rejuvenation, rinse with water and immerse briefly in 5*N* HCl. Rinse and store in 0.1*N* HCl. Rinse again with distilled water before use.

To check reference electrode, oppose the emf of a questionable reference electrode against another one of the same type that is known to be good. Using an adapter, plug good reference electrode into glass electrode jack of potentiometer; then plug questioned electrode into reference electrode jack. Set meter to read millivolts and take readings with both electrodes immersed in same electrolyte (KCl) solution and then in same buffer solution. The millivolt readings should be 0 ± 5 mV for both solutions. If different electrodes are used, i.e., silver: silver-chloride against calomel or *vice versa*, the reading will be 44 ± 5 mV for a good reference electrode.

Reference electrode troubles generally are traceable to a clogged junction. Interruption of the continuous trickle of electrolyte through the junction causes increase in resistance and drift in reading. Clear a clogged junction by applying suction to the tip or by boiling tip in distilled water until the electrolyte flows freely when suction is applied to tip or pressure is applied to the fill hole.

6. Precision and Accuracy

By careful use of a laboratory pH meter, a precision of ±0.02 pH unit and an accuracy of ±0.05 pH unit can be achieved. However, ±0.1 pH unit represents the limit of accuracy under normal conditions, especially for measurement of water and poorly buffered solutions. For this reason, report pH values to the nearest 0.1 pH unit. A synthetic sample of a Clark and Lubs buffer solution of pH 7.3 was analyzed electrometrically by 30 laboratories with a standard deviation of ±0.13 pH unit.

7. References

1. BATES, R.G. 1978. Concept and determination of pH. *In* I.M. Kolthoff & P.J. Elving, eds. Treatise on Analytical Chemistry. Part 1, Vol. 1, p. 821. Wiley-Interscience, New York, N.Y.
2. LICHT, T.S. & A.J. DE BÉTHUNE. 1957. Recent developments concerning the signs of electrode potentials. *J. Chem. Educ.* 34:433.

8. Bibliography

CLARK, W.M. 1928. The Determination of Hydrogen Ions, 3rd ed. Williams & Wilkins Co., Baltimore, Md.

DOLE, M. 1941. The Glass Electrode. John Wiley & Sons, New York, N.Y.

BATES, R.G. & S.F. ACREE. 1945. pH of aqueous mixtures of potassium dihydrogen phosphate and disodium hydrogen phosphate at 0 to 60 C. *J. Res. Nat. Bur. Standards* 34:373.

LANGELIER, W.F. 1946. Effect of temperature on the pH of natural water. *J. Amer. Water Works Ass.* 38:179.

FELDMAN, I. 1956. Use and abuse of pH measurements. *Anal. Chem.* 28:1859.

BRITTON, H.T.S. 1956. Hydrogen Ions, 4th ed. D. Van Nostrand Co., Princeton, N.J.

KOLTHOFF, I.M. & H.A. LAITINEN. 1958. pH and Electrotitrations. John Wiley & Sons, New York, N.Y.

KOLTHOFF, I.M. & P.J. ELVING. 1959. Treatise on Analytical Chemistry. Part I, Vol. 1, Chapter 10. Wiley-Interscience, New York, N.Y.

BATES, R.G. 1962. Revised standard values for pH measurements from 0 to 95 C. *J. Res. Nat. Bur. Standards* 66A:179.

AMERICAN WATER WORKS ASSOCIATION. 1964. Simplified Procedures for Water Examination. Manual M12, AWWA, New York, N.Y.

WINSTEAD, M. 1967. Reagent Grade Water: How, When and Why? Amer. Soc. Medical Technol., The Steck Company, Austin, Tex.

STAPLES, B.R. & R.G. BATES. 1969. Two new standards for the pH scale. *J. Res. Nat. Bur. Standards* 73A:37.

BATES, R.G. 1973. Determination of pH, Theory and Practice, 2nd ed. John Wiley & Sons, New York, N.Y.

424 PHOSPHORUS

Phosphorus occurs in natural waters and in wastewaters almost solely as phosphates. These are classified as orthophosphates, condensed phosphates (pyro-, meta-, and other polyphosphates), and organically bound phosphates. They occur in solution, in particles or detritus, or in the bodies of aquatic organisms.

These forms of phophate arise from a variety of sources. Small amounts of certain condensed phosphates are added to some water supplies during treatment. Larger quantities of the same compounds may be added when the water is used for laundering or other cleaning, because these materials are major constituents of many commercial cleaning preparations. Phosphates are used extensively in the treatment of boiler waters. Orthophosphates applied to agricultural or residential cultivated land as fertilizers are carried into surface waters with storm runoff and to a lesser extent with melting snow. Organic phosphates are formed primarily by biological processes. They are contributed to sewage by body wastes and food residues and also may be formed from orthophosphates in biological treatment processes or by receiving water biota.

Phosphorus is essential to the growth of organisms and can be the nutrient that limits the primary productivity of a body of water. In instances where phosphate is a growth-limiting nutrient, the discharge of raw or treated wastewater, agricultural drainage, or certain industrial wastes to that water may stimulate the growth of photosynthetic aquatic micro- and macroorganisms in nuisance quantities.

Phosphates also occur in bottom sediments and in biological sludges, both as precipitated inorganic forms and incorporated into organic compounds.

1. Definition of Terms

Phosphorus analyses embody two general procedural steps: *(a)* conversion of the phosphorus form of interest to dissolved orthophosphate, and *(b)* colorimetric determination of dissolved orthophosphate. The separation of phosphorus into its various forms is defined analytically but the analytical differentiations have been selected so that they may be used for interpretive purposes.

Filtration through a 0.45-μm membrane filter separates "filtrable" from "nonfiltrable" forms of phosphorus. No claim is made that filtration through 0.45-μm filters is a true separation of suspended and dissolved forms of phosphorus; it is merely a convenient and replicable analytical technic designed to make a gross separation. This is reflected in the use of the term "filtrable" (rather than dissolved) to describe the phosphorus forms determined in the filtrate that passes the 0.45-μm filter.

Membrane filtration is selected over depth filtration because of the greater likelihood of obtaining a consistent separation of particle sizes. Prefiltration through a glass fiber filter may be used to increase the filtration rate.

Phosphates that respond to colorimetric tests without preliminary hydrolysis or oxidative digestion of the sample are termed "reactive phosphorus." While reactive phosphorus is largely a measure of orthophosphate, a small fraction of any condensed phosphate present usually is hydrolyzed unavoidably in the procedure. Reactive phosphorus occurs in both filtrable and nonfiltrable forms.

Acid hydrolysis at boiling-water temperature converts filtrable and particulate condensed phosphates to filtrable orthophosphate. The hydrolysis unavoidably releases some phosphate from organic compounds, but this may be reduced to a minimum by judicious selection of acid strength and hydrolysis time and temperature. The term "acid-hydrolyzable phosphorus" is preferred over "condensed phosphate" for this fraction.

The phosphate fractions that are converted to orthophosphate only by oxidative destruction of the organic matter present are considered "organic" or "organically bound" phosphorus. The severity of the oxidation required for this conversion depends on the form—and to some extent on the amount—of the organic phosphorus present. Like reactive phosphorus and acid-hydrolyzable phosphorus, organic phosphorus occurs both in the filtrable and nonfiltrable fractions. With minor variations, the filtrable and nonfiltrable fractions of a sample correspond to dissolved and particulate phosphates, respectively.

The total phosphorus as well as the filtrable and nonfiltrable phosphorus fractions each may be divided analytically into the three chemical types that have been described; reactive, acid-hydrolyzable, and organic phosphorus. Figure 424:1 shows the steps for analysis of individual phosphorus fractions. As indicated, determinations usually are conducted only on the unfiltered and filtered samples. Nonfiltrable fractions generally are determined by difference.

2. Selection of Method

a. Digestion methods: Because phosphorus may occur in combination with organic matter, a digestion method to determine total phosphorus must be able to oxidize organic matter effectively to release phosphorus as orthophosphate. Three digestion methods are given. The perchloric acid method, the most drastic and time-consuming method, is recommended only for particularly difficult samples such as sediments. The nitric acid-sulfuric acid method is recommended for most samples. By far the simplest method is the persulfate oxidation technic. It is recommended that this method be checked against one or more of the more drastic digestion technics and be adopted if identical recoveries are obtained.

b. Colorimetric methods: Three methods of orthophosphate determination are described. Selection depends largely on the concentration range of orthophosphate. The vanadomolybdic acid method (D) is most useful for routine analyses in the range of 1 to 20 mg P/L. The stannous chloride method (E) or the ascorbic acid method (F) is more suited for the range of 0.01 to 6 mg P/L. An extraction step is recommended for the lower levels of this range and when interferences must be overcome. An automated version of the ascorbic acid method also is presented.

3. Precision and Accuracy

To aid in method selection, Table 424:I presents the results of various combinations of digestion, hydrolysis, and colorimetric technics for three synthetic samples of the following compositions:

Sample 1: 100 μg orthophosphate phosphorus (PO_4-P)/L, 80 μg condensed phosphate phosphorus/L (sodium hexametaphosphate), 30 μg organic phosphorus/L (adenylic acid), 1.5 mg NH_3-N/L, 0.5 mg NO_3-N/L, and 400 mg chloride/L.

Sample 2: 600 μg PO_4-P/L, 300 μg con-

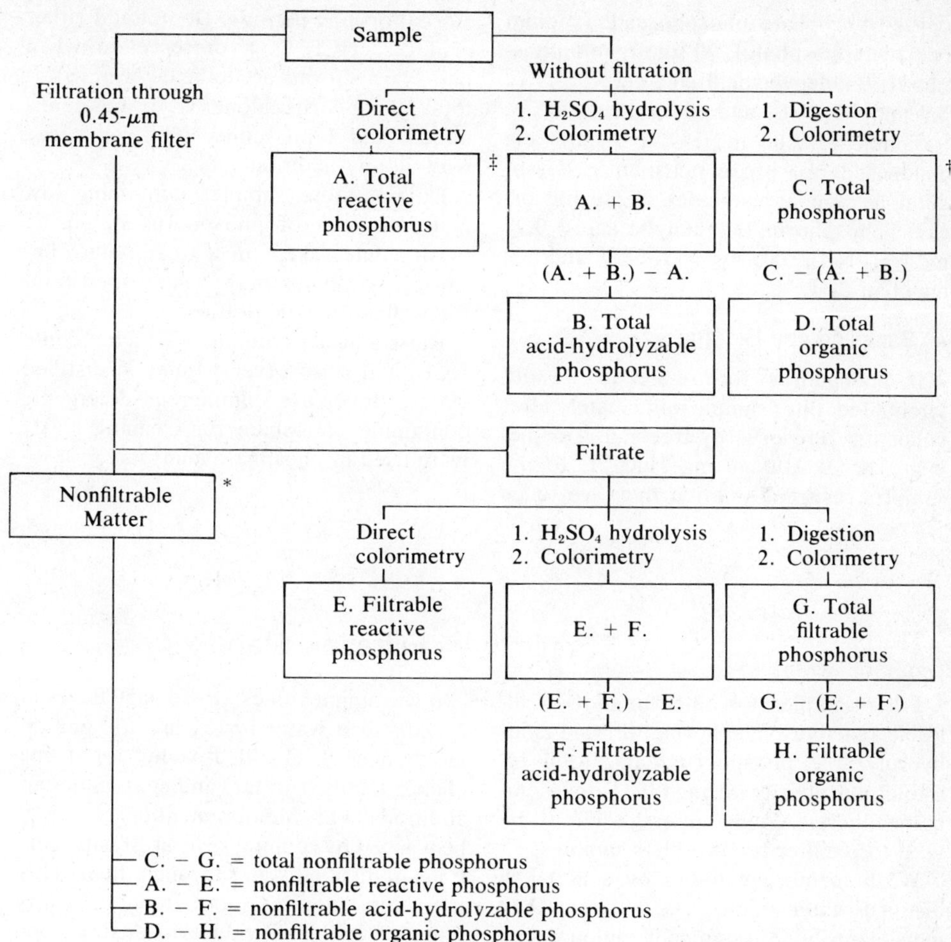

Figure 424:1. Steps for analysis of phosphate fractions.

*Direct determination of phosphorus on the membrane filter containing nonfiltrable matter will be required where greater precision than that obtained by difference is desired. Digest filter with HNO_3 and follow by perchloric acid. Then perform colorimetry.

†Total phosphorus measurements on highly saline samples may be difficult because of precipitation of large quantities of salt as a result of digestion technics that drastically reduce sample volume. For total phosphorus analyses on such samples, directly determine total filtrable phosphorus and total nonfiltrable phosphorus and add the results.

‡When determining total filtrable or total nonfiltrable reactive phosphorus, anomalous results may be obtained on samples containing large amounts of suspended sediments. Very often results depend largely on the degree of agitation and mixing to which samples are subjected during analysis because of a time-dependent desorption of orthophosphate from the suspended particles.

densed phosphate phosphorus/L (sodium hexametaphosphate), 90 μg organic phosphorus/L (adenylic acid), 0.8 mg NH_3-N/L, 5.0 mg NO_3-N/L, and 400 mg chloride/L.

Sample 3: 7.00 mg PO_4-P/L, 3.00 mg condensed phosphate phosphorus/L (sodium hexametaphosphate), 0.230 mg organic phosphorus/L (adenylic acid), 0.20 mg NH_3-N/L, 0.05 mg NO_3-N/L, and 400 mg chloride/L.

4. Sampling and Storage

If phosphorus forms are to be differentiated, filter sample immediately after collection. Preserve by freezing at or below −10 C. Add 40 mg $HgCl_2$/L to the samples, especially when they are to be stored for long periods. Do not add either acid or $CHCl_3$ as a preservative when phosphorus forms are to be determined. If total phosphorus alone is to be determined, add 1 mL conc HCl/L or freeze without any additions.

Do not store samples containing low concentrations of phosphorus in plastic bottles unless kept in a frozen state because phosphates may be adsorbed onto the walls of plastic bottles.

Rinse all glass containers with hot dilute HCl, then rinse several times in distilled water. Never use commercial detergents containing phosphate for cleaning glassware used in phosphate analysis.

424 A. Preliminary Filtration Step

Filter samples for determination of filtrable reactive phosphorus, filtrable acid-hydrolyzable phosphorus, and total filtrable phosphorus through 0.45-μm membrane filters. A glass fiber filter may be used to prefilter hard-to-filter samples.

Wash membrane filters by soaking in distilled water before use because they may contribute significant amounts of phosphorus to samples containing low concentrations of phosphate. Use one of two washing technics: *(a)* soak 50 filters in 2 L distilled water for 24 hr; *(b)* soak 50 filters in 2 L distilled water for 1 hr, change distilled water, and soak filters an additional 3 hr. Membrane filters also may be washed by running several 100-mL portions of distilled water through them. This procedure requires more frequent determination of blank values to ensure consistency in washing and to evaluate different lots of filters.

424 B. Preliminary Acid Hydrolysis Step for Acid-Hydrolyzable Phosphorus

1. Discussion

The acid-hydrolyzable phosphorus content of the sample is defined operationally as the difference between reactive phosphorus as measured in the untreated sample and phosphate found after mild acid hydrolysis. Generally, it includes condensed phosphates such as pyro-, tripoly-, and higher-molecular-weight species such as hexametaphosphate. In addition, some natural waters contain organic phosphate compounds that are hydrolyzed to orthophosphate under the test conditions. Polyphosphates generally do not respond to reactive phosphorus tests but can be hydrolyzed to orthophosphate by boiling with acid.

After hydrolysis, determine reactive phosphorus by a colorimetric method (D, E, or F). Interferences, precision, accuracy, and sensitivity will depend on the colorimetric method used.

2. Apparatus

Autoclave or pressure cooker, capable of operating at 98 to 137 kPa.

3. Reagents

a. Phenolphthalein indicator aqueous solution.

b. Strong acid solution: Slowly add 300 mL conc H_2SO_4 to about 600 mL distilled water. When cool, add 4.0 mL conc HNO_3 and dilute to 1 L.

c. Sodium hydroxide, NaOH, 6*N*.

4. Procedure

To 100-mL sample or a portion diluted to 100 mL, add 0.05 mL (1 drop) phenolphthalein indicator solution. If a red color develops, add strong acid solution dropwise, to just discharge the color. Then add 1 mL more.

Boil gently for at least 90 min, adding distilled water to keep the volume between 25 and 50 mL. Alternatively, heat for 30 min in an autoclave or pressure cooker at 98 to 137 kPa. Cool, neutralize to a faint pink color with NaOH solution, and restore to the original 100-mL volume with distilled water.

Prepare a calibration curve by carrying a series of standards containing orthophosphate (see colorimetric method D, E, or F) through the hydrolysis step. Do not use orthophosphate standards without hydrolysis, because the salts added in hydrolysis cause an increase in the color intensity in some methods.

Determine reactive phosphorus content of treated portions, using Method D, E, or F. This gives the sum of polyphosphate and orthophosphate in the sample. To calculate its content of acid-hydrolyzable phosphorus, determine reactive phosphorus in a sample portion that has not been hydrolyzed, using the same colorimetric method as for treated sample, and subtract.

424 C. Preliminary Digestion Steps for Total Phosphorus

Total phosphorus includes all orthophosphates and condensed phosphates, both dissolved and particulate, organic and inorganic. To release phosphorus from combination with organic matter, digest and oxidize. The rigor of digestion required depends on the type of sample. The three digestion technics presented, in order of decreasing rigor, are perchloric acid digestion, sulfuric acid-nitric acid digestion, and persulfate digestion. Compare phosphorus recovery by each digestion technic for the specific type of sample being tested; if the less tedious persulfate method gives good phosphorus recovery, use this method.

After digestion, determine liberated orthophosphate by Method D, E, or F. The colorimetric method used, rather than the digestion procedure, governs in matters of interference and minimum detectable concentration.

I—PERCHLORIC ACID DIGESTION

1. Apparatus

a. Hot plate: A 30- × 50-cm heating surface is adequate.

b. Safety shield.

c. Safety goggles.

d. Erlenmeyer flasks, 125-mL, acid-washed and rinsed with distilled water.

2. Reagents

a. Nitric acid, HNO_3, conc.

b. Perchloric acid, $HClO_4 \cdot 2H_2O$, purchased as 70 to 72% $HClO_4$, reagent grade.

c. Sodium hydroxide, NaOH, 6*N*.

d. Methyl orange indicator solution.

e. Phenolphthalein indicator aqueous solution.

3. Procedure

CAUTION—*Heated mixtures of $HClO_4$ and organic matter may explode violently. Avoid this hazard by taking the following precautions: (a) Do not add $HClO_4$ to a hot solution that may contain organic matter. (b) Always initiate digestion of samples containing organic matter with HNO_3. Complete digestion using the mixture of HNO_3 and $HClO_4$. (c) Do not fume with $HClO_4$ in ordinary hoods. Use hoods especially constructed for $HClO_4$ fuming or a glass fume eradicator* connected to a water pump. (d) Never let samples being digested with $HClO_4$ evaporate to dryness.*

Measure sample containing the desired amount of phosphorus (this will be determined by whether Method D, E, or F is to be used) into a 125-mL erlenmeyer flask. Acidify to methyl orange with conc HNO_3, add another 5 mL conc HNO_3, and evaporate on a steam bath or hot plate to 15 to 20 mL.

Add 10 mL each of conc HNO_3 and $HClO_4$ to the 125-mL conical flask, cooling the flask between additions. Add a few boiling chips, heat on a hot plate, and evaporate gently until dense white fumes of $HClO_4$ just appear. If solution is not clear, cover neck of flask with a watch glass and keep solution barely boiling until it clears. If necessary, add 10 mL more HNO_3 to aid oxidation.

Cool digested solution and add 1 drop aqueous phenolphthalein solution. Add 6*N* NaOH solution until the solution just turns pink. If necessary, filter neutralized solution and wash filter liberally with distilled water. Make up to 100 mL with distilled water.

Determine the PO_4-P content of the treated sample by Method D, E, or F.

Prepare a calibration curve by carrying a series of standards containing orthophosphate (see Method D, E, or F) through digestion step. Do not use orthophosphate standards without treatment.

*Such as those obtainable from G.F. Smith Chemical Co., Columbus, Ohio.

II—SULFURIC ACID-NITRIC ACID DIGESTION

1. Apparatus

a. Digestion rack: An electrically or gas-heated digestion rack with provision for withdrawal of fumes is recommended. Digestion racks typical of those used for microkjeldahl digestions are suitable.

b. Microkjeldahl flasks.

2. Reagents

a. Sulfuric acid, H_2SO_4, conc.

b. Nitric acid, HNO_3, conc.

c. Phenolphthalein indicator aqueous solution.

d. Sodium hydroxide, NaOH, 1*N*.

3. Procedure

Into a microkjeldahl flask, measure a sample containing the desired amount of phosphorus (this is determined by the colorimetric method used). Add 1 mL conc H_2SO_4 and 5 mL conc HNO_3.

Digest to a volume of 1 mL and then continue until solution becomes colorless to remove HNO_3.

Cool and add approximately 20 mL distilled water, 0.05 mL (1 drop) phenolphthalein indicator, and as much 1*N* NaOH solution as required to produce a faint pink tinge. Transfer neutralized solution, filtering if necessary to remove particulate material or turbidity, into a 100-mL volumetric flask. Add filter washings to flask and adjust sample volume to 100 mL with distilled water.

Determine phosphorus by Method D, E, or F, for which a separate calibration curve has been constructed by carrying standards through the acid digestion procedure.

III—PERSULFATE DIGESTION METHOD

1. Apparatus

a. Hot plate: A 30- × 50-cm heating surface is adequate.

b. Autoclave: An autoclave or pressure cooker capable of developing 98 to 137 kPa may be used in place of a hot plate.

c. Glass scoop, to hold required amounts of persulfate crystals.

2. Reagents

a. Phenolphthalein indicator aqueous solution.

b. Sulfuric acid solution: Carefully add 300 mL conc H_2SO_4 to approximately 600 mL distilled water and dilute to 1 L with distilled water.

c. Ammonium persulfate, $(NH_4)_2S_2O_8$, solid, or potassium persulfate, $K_2S_2O_8$, solid.

d. Sodium hydroxide, NaOH, 1*N*.

3. Procedure

Use 50 mL or a suitable portion of thoroughly mixed sample. Add 0.05 mL (1 drop) phenolphthalein indicator solution. If a red color develops, add H_2SO_4 solution dropwise to just discharge the color. Then add 1 mL H_2SO_4 solution and either 0.4 g solid $(NH_4)_2S_2O_8$ or 0.5 g solid $K_2S_2O_8$.

Boil gently on a preheated hot plate for 30 to 40 min or until a final volume of 10 mL is reached. Cool, dilute to 30 mL with distilled water, add 0.05 mL (1 drop) phenolphthalein indicator solution, and neutralize to a faint pink color with NaOH. Alternatively, heat for 30 min in an autoclave or pressure cooker at 98 to 137 kPa. Cool, add 0.05 mL (1 drop) phenolphthalein indicator solution, and neutralize to a faint pink color with NaOH. Make up to 100 mL with distilled water. In some samples a precipitate may form at this stage, but do not filter. For any subsequent subdividing of the sample, shake well. The precipitate (which is possibly a calcium phosphate) redissolves under the acid conditions of the colorimetric reactive phosphorus test. Determine phosphorus by Method D, E, or F, for which a separate calibration curve has been constructed by carrying standards through the persulfate digestion procedure.

424 D. Vanadomolybdophosphoric Acid Colorimetric Method

1. General Discussion

a. Principle: In a dilute orthophosphate solution, ammonium molybdate reacts under acid conditions to form a heteropoly acid, molybdophosphoric acid. In the presence of vanadium, yellow vanadomolybdophosphoric acid is formed. The intensity of the yellow color is proportional to phosphate concentration.

b. Interference: Positive interference is caused by silica and arsenate only if the sample is heated. Negative interferences are caused by arsenate, fluoride, thorium, bismuth, sulfide, thiosulfate, thiocyanate, or excess molybdate. Blue color is caused by ferrous iron but this does not affect results if ferrous iron concentration is less than 100 mg/L. Sulfide interference may be removed by oxidation with bromine water. Ions that do not interfere in concentrations up to 1,000 mg/L are Al, Fe^{3+}, Mg, Ca, Ba, Sr, Li, Na, K, NH_4^+, Cd, Mn, Pb, Hg^+, Hg^{2+}, Sn^{2+}, Cu, Ni, Ag, U, Zr, AsO_3^-, Br^-, CO_3^{2-}, ClO_4, CN^-, IO_3^-, SiO_4, NO_3^-, NO_2^-, SO_4^{2-}, SO_3^{2-}, pyrophosphate, molybdate, tetraborate, selenate, benzoate, citrate, oxalate, lactate, tartrate, formate, and salicylate. If HNO_3 is used in the test, chloride interferes at 75 mg/L.

c. Minimum detectable concentration: The minimum detectable concentration is 200 μg P/L in 1-cm spectrophotometer cells.

2. Apparatus

a. Colorimetric equipment: One of the following is required:

1) *Spectrophotometer,* for use at 400 to 490 nm.

2) *Filter photometer,* provided with a blue or violet filter exhibiting maximum transmittance between 400 and 470 nm.

The wavelength at which color intensity is measured depends on sensitivity desired, because sensitivity varies tenfold with wavelengths 400 to 490 nm. Ferric iron causes interference at low wavelengths, particularly at 400 nm. A wavelength of 470 nm usually is used. Concentration ranges for different wavelengths are:

P Range *mg/L*	Wavelength *nm*
1.0– 5.0	400
2.0–10	420
4.0–18	470

b. Acid-washed glassware: Use acid-washed glassware for determining low concentrations of phosphorus. Phosphate contamination is common because of its absorption on glass surfaces. Avoid using commercial detergents containing phosphate. Clean all glassware with hot dilute HCl and rinse well with distilled water. Preferably, reserve the glassware only for phosphate determination, and after use, wash and keep filled with water until needed. If this is done, acid treatment is required only occasionally.

*c. Filtration apparatus and filter paper.**

3. Reagents

a. Phenolphthalein indicator aqueous solution.

b. Hydrochloric acid, HCl, 1 + 1. H_2SO_4, $HClO_4$, or HNO_3 may be substituted for HCl. The acid concentration in the determination is not critical but a final sample concentration of 0.5*N* is recommended.

c. Activated carbon.† Remove fine particles by rinsing with distilled water.

d. Vanadate-molybdate reagent:

1) *Solution A:* Dissolve 25 g ammonium molybdate, $(NH_4)_6Mo_7O_{24}\cdot 4H_2O$, in 300 mL distilled water.

2) *Solution B:* Dissolve 1.25 g ammonium metavanadate, NH_4VO_3, by heating to boiling in 300 mL distilled water. Cool and add 330 mL conc HCl. Cool Solution B to room temperature, pour Solution A into Solution B, mix, and dilute to 1 L.

e. Standard phosphate solution: Dis-

*Whatman No. 42 or equivalent.

†Darco G60 or equivalent.

solve in distilled water 219.5 mg anhydrous KH_2PO_4 and dilute to 1,000 mL; 1.00 mL = 50.0 μg PO_4-P.

4. Procedure

a. Sample pH adjustment: If sample pH is greater than 10, add 0.05 mL (1 drop) phenolphthalein indicator to 50.0 mL sample and discharge the red color with 1 + 1 HCl before diluting to 100 mL.

b. Color removal from sample: Remove excessive color in sample by shaking about 50 mL with 200 mg activated carbon in an erlenmeyer flask for 5 min and filter to remove carbon. Check each batch of carbon for phosphate because some batches produce high reagent blanks.

c. Color development in sample: Place 35 mL or less of sample, containing 0.05 to 1.0 mg P, in a 50-mL volumetric flask. Add 10 mL vanadate-molybdate reagent and dilute to the mark with distilled water. Prepare a blank in which 35 mL distilled water is substituted for the sample. After 10 min or more, measure absorbance of sample versus a blank at a wavelength of 400 to 490 nm, depending on sensitivity desired (see ¶ 2*a* above). The color is stable for days and its intensity is unaffected by variation in room temperature.

d. Preparation of calibration curve: Prepare a calibration curve by using suitable volumes of standard phosphate solution and proceeding as in ¶ 4*c*. When ferric ion is low enough not to interfere, plot a family of calibration curves of one series of standard solutions for various wavelengths. This permits a wide latitude of concentrations in one series of determinations. Analyze at least one standard with each set of samples.

5. Calculation

$$\text{mg P/L} = \frac{\text{mg P (in 50 mL final volume)} \times 1{,}000}{\text{mL sample}}$$

6. Precision and Accuracy

See Table 424:I.

424 E. Stannous Chloride Method

1. General Discussion

a. Principle: Molybdophosphoric acid is formed and reduced by stannous chloride to intensely colored molybdenum blue. This method is more sensitive than Method D and makes feasible measurements down to 7 μg P/L by use of increased light path length. Below 100 μg P/L an extraction step may increase reliability and lessen interference.

b. Interference: See Section 424D.1*b*.

c. Minimum detectable concentration: The minimum detectable concentration is about 3 μg P/L. The sensitivity at 0.3010 absorbance is about 10 μg P/L for an absorbance change of 0.009.

2. Apparatus

The same apparatus is required as for Method D, except that a pipetting bulb is required for the extraction step. Set spectrophotometer at 625 nm in the measurement of benzene-isobutanol extracts and at 690 nm for aqueous solutions. If the instrument is not equipped to read at 690 nm, use a wavelength of 650 nm for aqueous solutions, with somewhat reduced sensitivity and precision.

3. Reagents

a. Phenolphthalein indicator aqueous solution.

b. Strong-acid solution: Prepare as directed in Method B, ¶ 3*b* above.

c. Ammonium molybdate reagent I: Dissolve 25 g $(NH_4)_6Mo_7O_{24}\cdot 4H_2O$ in 175 mL distilled water. Cautiously add 280 mL conc H_2SO_4 to 400 mL distilled water. Cool, add molybdate solution, and dilute to 1L.

d. Stannous chloride reagent I: Dissolve 2.5 g fresh $SnCl_2\cdot 2H_2O$ in 100 mL glycerol. Heat in a water bath and stir with a glass rod to hasten dissolution. This reagent is stable and requires neither preservatives nor special storage.

e. Standard phosphate solution: Prepare as directed in Method D, ¶ 3*e*.

f. Reagents for extraction:

1) *Benzene-isobutanol solvent:* Mix equal volumes of benzene and isobutyl alcohol. (CAUTION—*This solvent is highly flammable.*)

2) *Ammonium molybdate reagent II:* Dissolve 40.1 g $(NH_4)_6Mo_7O_{24}\cdot 4H_2O$ in approximately 500 mL distilled water. Slowly add 396 mL ammonium molybdate reagent I. Cool and dilute to 1 L.

3) *Alcoholic sulfuric acid solution:* Cautiously add 20 mL conc H_2SO_4 to 980 mL methyl alcohol with continuous mixing.

4) *Dilute stannous chloride reagent II:* Mix 8 mL stannous chloride reagent I with 50 mL glycerol. This reagent is stable for at least 6 months.

4. Procedure

a. Preliminary sample treatment: To 100 mL sample containing not more than 200 μg P and free from color and turbidity, add 0.05 mL (1 drop) phenolphthalein indicator. If sample turns pink, add strong acid solution dropwise to discharge the color. If more than 0.25 mL (5 drops) are required, take a smaller sample and dilute to 100 mL with distilled water after first discharging the pink color with acid.

b. Color development: Add, with thorough mixing after each addition, 4.0 mL molybdate reagent I and 0.5 mL (10 drops) stannous chloride reagent I. Rate of color development and intensity of color depend on temperature of the final solution, each 1 C increase producing about 1% increase in color. Hence, hold samples, standards, and reagents within 2 C of one another and in the temperature range between 20 and 30 C.

c. Color measurement: After 10 min, but before 12 min, using the same specific interval for all determinations, measure color photometrically at 690 nm and compare with a calibration curve, using a distilled water blank. Light path lengths suitable for various concentration ranges are as follows:

Approximate P Range *mg/L*	Light Path *cm*
0.3–2	0.5
0.1–1	2
0.007–0.2	10

Always run a blank on reagents and distilled water. Because the color at first develops progressively and later fades, maintain equal timing conditions for samples and standards. Prepare at least one standard with each set of samples or once each day that tests are made. The calibration curve may deviate from a straight line at the upper concentrations of the 0.3 to 2.0-mg/L range.

d. Extraction: When increased sensitivity is desired or interferences must be overcome, extract phosphate as follows: Pipet a 40-mL sample, or one diluted to that volume, into a 125-mL separatory funnel. Add 50.0 mL benzene-isobutanol solvent and 15.0 mL molybdate reagent II. Close funnel at once and shake vigorously for exactly 15 sec. If condensed phosphate is present, any delay will increase its conversion to orthophosphate. Remove stopper and withdraw 25.0 mL of separated organic layer, using a pipet with safety bulb.

TABLE 424: I. PRECISION AND ACCURACY DATA FOR MANUAL PHOSPHORUS METHODS

Method	Phosphorus Concentration			No of. Laboratories	Relative Standard Deviation %	Relative Error %
	Ortho-phosphate $\mu g/L$	Poly-phosphate $\mu g/L$	Total $\mu g/L$			
D. Vanadomolybdate	100			45	75.2	21.6
	600			43	19.6	10.8
	7,000			44	8.6	5.4
E. Stannous chloride	100			45	25.5	28.7
	600			44	14.2	8.0
	7,000			45	7.6	4.3
F. Ascorbic acid	100			3	9.1	10.0
	600			3	4.0	4.4
	7,000			3	5.2	4.9
Acid hydrolysis + vanadomolybdate		80		37	106.8	7.4
		300		38	66.5	14.0
		3,000		37	36.1	23.5
Acid hydrolysis + stannous chloride		80		39	60.1	12.5
		300		36	47.6	21.7
		3,000		38	37.4	22.8
Persulfate + vanadomolybdate			210	32	55.8	1.6
			990	32	23.9	2.3
			10,230	31	6.5	0.3
Sulfuric-nitric acids + vanadomolybdate			210	23	65.6	20.9
			990	22	47.3	0.6
			10,230	20	7.0	0.4
Perchloric acid + vanadomolybdate			210	4	33.5	45.2
			990	5	20.3	2.6
			10,230	6	11.7	2.2
Persulfate + stannous chloride			210	29	28.1	9.2
			990	30	14.9	12.3
			10,230	29	11.5	4.3
Sulfuric-nitric acids + stannous chloride			210	20	20.8	1.2
			990	17	8.8	3.2
			10,230	19	7.5	0.4

Transfer to a 50-mL volumetric flask, add 15 to 16 mL alcoholic H_2SO_4 solution, swirl, add 0.50 mL (10 drops) dilute stannous chloride reagent II, swirl, and dilute to the mark with alcoholic H_2SO_4. Mix thoroughly. After 10 min, but before 30 min, read against the blank at 625 nm. Prepare blank by carrying 40 mL distilled water through the same procedure used for the sample. Read phosphate concentration from a calibration curve prepared by taking known phosphate standards through the same procedure used for samples.

5. Calculation

Calculate as follows:

a. Direct procedure:

$$\text{mg P/L} = \frac{\text{mg P (in approximately 104.5 mL final volume)} \times 1{,}000}{\text{mL sample}}$$

b. Extraction procedure:

$$\text{mg P/L} = \frac{\text{mg P (in 50 mL final volume)} \times 1{,}000}{\text{mL sample}}$$

6. Precision and Accuracy

See Table 424:I.

424 F. Ascorbic Acid Method

1. General Discussion

a. Principle: Ammonium molybdate and potassium antimonyl tartrate react in acid medium with orthophosphate to form a heteropoly acid—phosphomolybdic acid—that is reduced to intensely colored molybdenum blue by ascorbic acid.

b. Interference: Arsenates react with the molybdate reagent to produce a blue color similar to that formed with phosphate. Concentrations as low as 0.1 mg arsenic/L interfere with the phosphate determination. Hexavalent chromium and nitrite interfere to give results about 3% low at concentrations of 1 mg/L and 10 to 15% low at 10 mg/L. Sulfide (Na_2S) and silicate do not interfere at concentrations of 1.0 and 10 mg/L.

c. Minimum detectable concentration: Approximately 10 μg P/L. P ranges are as follows:

Approximate P Range *mg/L*	Light Path *cm*
0.30–2.0	0.5
0.15–1.30	1.0
0.01–0.25	5.0

2. Apparatus

a. Colorimetric equipment: One of the following is required:

1) *Spectrophotometer,* with infrared phototube for use at 880 nm, providing a light path of 2.5 cm or longer.

2) *Filter photometer,* equipped with a red color filter and a light path of 0.5 cm or longer.

b. Acid-washed glassware: See Method D, ¶ 2*b* above.

3. Reagents

a. Sulfuric acid, H_2SO_4, 5*N*: Dilute 70 mL conc H_2SO_4 to 500 mL with distilled water.

b. Potassium antimonyl tartrate solution: Dissolve 1.3715 g $K(SbO)C_4H_4O \cdot \frac{1}{2}H_2O$ in 400 mL distilled water in a 500-mL volumetric flask and dilute to volume. Store in a glass-stoppered bottle.

c. Ammonium molybdate solution: Dissolve 20 g $(NH_4)_6Mo_7O_{24} \cdot 4H_2O$ in 500 mL distilled water. Store in a glass-stoppered bottle.

d. Ascorbic acid, 0.01*M*: Dissolve 1.76 g ascorbic acid in 100 mL distilled water. The solution is stable for about 1 week at 4 C.

e. Combined reagent: Mix the above reagents in the following proportions for 100 mL of the combined reagent: 50 mL 5*N* H_2SO_4, 5 mL potassium antimonyl tartrate solution, 15 mL ammonium molybdate solution, and 30 mL ascorbic acid solution. *Mix after addition of each reagent.* Let all reagents reach room temperature before they are mixed and mix in the order given. If turbidity forms in the combined reagent, shake and let stand for a few minutes until turbidity disappears before proceeding. The reagent is stable for 4 hr.

f. Stock phosphate solution: See Method D, ¶ 3*e*.

g. Standard phosphate solution: Dilute 50.0 mL stock phosphate solution to 1,000 mL with distilled water; 1.00 mL = 2.50 μg P.

4. Procedure

a. Treatment of sample: Pipet 50.0 mL sample into a clean, dry test tube or 125-mL erlenmeyer flask. Add 0.05 mL (1 drop) phenolphthalein indicator. If a red color develops add 5*N* H_2SO_4 solution dropwise to just discharge the color. Add 8.0 mL combined reagent and mix thoroughly. After at least 10 min but no more than 30 min, measure absorbance of each sample at 880 nm, using reagent blank as the reference solution.

b. Correction for turbidity or interfering color: Natural color of water generally does not interfere at the high wavelength used. For highly colored or turbid waters, prepare a blank by adding all reagents except ascorbic acid and antimonyl potassium tartrate to the sample. Subtract blank absorbance from absorbance of each sample.

c. Preparation of calibration curve: Prepare individual calibration curves from a series of six standards within the phosphate ranges indicated in Section 424F.1*c*. Use a distilled water blank with the combined reagent to make photometric readings for the calibration curve. Plot absorbance vs. phosphate concentration to give a straight line passing through the origin. Test at least one phosphate standard with each set of samples.

5. Calculation

$$\text{mg P/L} = \frac{\text{mg P (in approximately 58 mL final volume)} \times 1{,}000}{\text{mL sample}}$$

6. Precision and Accuracy

The precision and accuracy values given in Table 424:I are for a single-solution procedure given in the 13th edition. Procedure 424F differs in reagent-to-sample ratios, no addition of solvent, and acidity conditions. It is superior in precision and accuracy to the previous technic in the analysis of both distilled water and river water at the 228 μg P/L level (Table 424:II).

TABLE 424:II. COMPARISON OF PRECISION AND ACCURACY OF ASCORBIC ACID METHODS

Ascorbic Acid Method	Phosphorus Concentration, Filtrable Orthophosphate *μg/L*	No. of Laboratories	Relative Standard Deviation %		Relative Error %	
			Distilled Water	River Water	Distilled Water	River Water
13th Edition (Edwards, Molof, and Schneeman)	228	8	3.87	2.17	4.01	2.08
Current method (Murphy and Riley)	228	8	3.03	1.75	2.38	1.39

424 G. Automated Ascorbic Acid Reduction Method (TENTATIVE)

1. General Discussion

a. Principle: Ammonium molybdate and potassium antimonyl tartrate react with orthophosphate in an acid medium to form an antimony-phosphomolybdate complex, which, on reduction with ascorbic acid, yields an intense blue color suitable for photometric measurement.

b. Interferences: As much as 50 mg ferric ion/L, 10 mg copper/L, and 10 mg silica/L can be tolerated. High silica concentrations cause positive interference.

In terms of phosphorus, the results are high by 0.005, 0.015, and 0.025 mg/L for silica concentrations of 20, 50, and 100 mg/L, respectively. Salt concentrations up to 20% (w/v) cause an error of less than 1%. Arsenate (AsO_4^{3-}) is a positive interference.

Eliminate interference from NO_2^- and S^{2-} by adding an excess of bromine water or a saturated potassium permanganate ($KMnO_4$) solution. Remove interfering turbidity by filtration before analysis. Filter samples for total or total hydrolyzable phosphorus only after digestion. Sample color that absorbs in the photometric range used for analysis also will interfere. See also Section 424F.1*b*.

c. Application: Orthophosphate can be determined in potable, surface, and saline waters as well as domestic and industrial wastewaters over a range of 0.001 to 10.0 mg P/L when photometric measurements are made at 650 to 660 or 880 nm in a 15-mm or 50-mm tubular flow cell. Determine higher concentrations by diluting sample. Although the automated test is designed for orthophosphate only, other phosphorus compounds can be converted to this reactive form by various sample pretreatments described in Methods 424A, B, and C.III.

2. Apparatus

a. Automated analytical equipment: Two automated systems (I and II)* that employ the same principle but differ in detail and rate of analysis, are in common use. They consist of the components listed in Part 600, with the following additions: heating bath adjusted at 50 C for the Automated System I or 37 C for the Automated System II, 15- or 50-mm tubular flow cell, and 650-to 660- or 880-nm filters.

b. Hot plate or autoclave.

c. Acid-washed glassware: Wash all glassware with hot 1 + 1 HCl and rinse with distilled water. Fill the acid-washed glassware with distilled water and treat with all reagents to remove traces of phosphate that might be adsorbed. Preferably, reserve this glassware for determination of phosphate. After use, rinse with distilled water and keep covered until glassware is needed again. If this is done, treatment with 1 + 1 HCl and reagents is required only occasionally. *Never use commercial detergents.*

3. Reagents

a. Potassium antimonyl tartrate solution: Dissolve 0.3 g $K(SbO)\,C_2H_4O_6 \cdot {}^1/_2H_2O$ in approximately 50 mL distilled water and dilute to 100 mL. Store at 4 C in a dark, glass-stoppered bottle.

b. Ammonium molybdate solution: Dissolve 4 g $(NH_4)_6Mo_7O_{24} \cdot 4H_2O$ in 100 mL distilled water. Store in a plastic bottle at 4 C.

c. Ascorbic acid solution: Dissolve 1.8 g ascorbic acid in 100 mL distilled water. The solution is stable for a week if prepared with water containing no more than

*AutoAnalyzer I and AutoAnalyzer II, Technicon Instrument Co., Tarrytown, N.Y. 10591.

a trace amount of heavy metals and if stored at 4 C.

d. Combined reagent: To prepare 100 mL mixed reagent, mix in the following proportions: 50 mL H_2SO_4 (¶ 3*e*), 5.00 mL potassium antimonyl tartrate solution, 15 mL ammonium molybdate solution, and 30 mL ascorbic acid solution. Mix after addition of each reagent. Let all reagents reach room temperature before mixing. Mix in the order given. If turbidity forms in the combined reagent, shake and let stand for several minutes until the turbidity disappears, then continue. This 100 mL of reagent is enough for 4 hr operation. Because stability is limited, prepare fresh for each run. To prepare a stable solution, exclude ascorbic acid from the combined reagent. If this reagent is used, pump the mixed reagent (molybdate, tartrate, and acid) through the distilled water line and the ascorbic acid solution (30 mL of the reagent described in ¶ 3*c* diluted to 100 mL with distilled water) through the original mixed reagent line.

e. Dilute sulfuric acid solution: Slowly add 140 mL conc H_2SO_4 to 600 mL distilled water. When cool, dilute to 1 L.

f. Ammonium persulfate, $(NH_4)_2S_2O_8$, crystalline.

g. Phenolphthalein indicator aqueous solution.

h. Stock phosphate solution: Dissolve 439.3 mg anhydrous KH_2PO_4, dried for 1 hr at 105 C, in distilled water and dilute to 1,000 mL; 1.00 mL = 100 μg P.

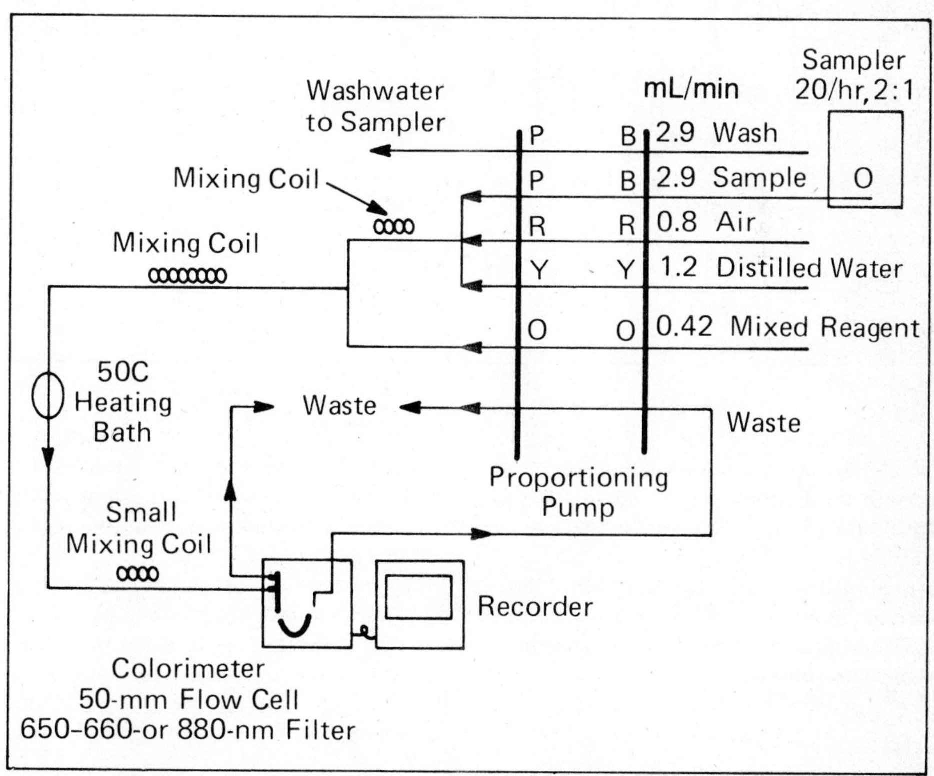

Figure 424:2. Phosphate manifold for Automated System I.

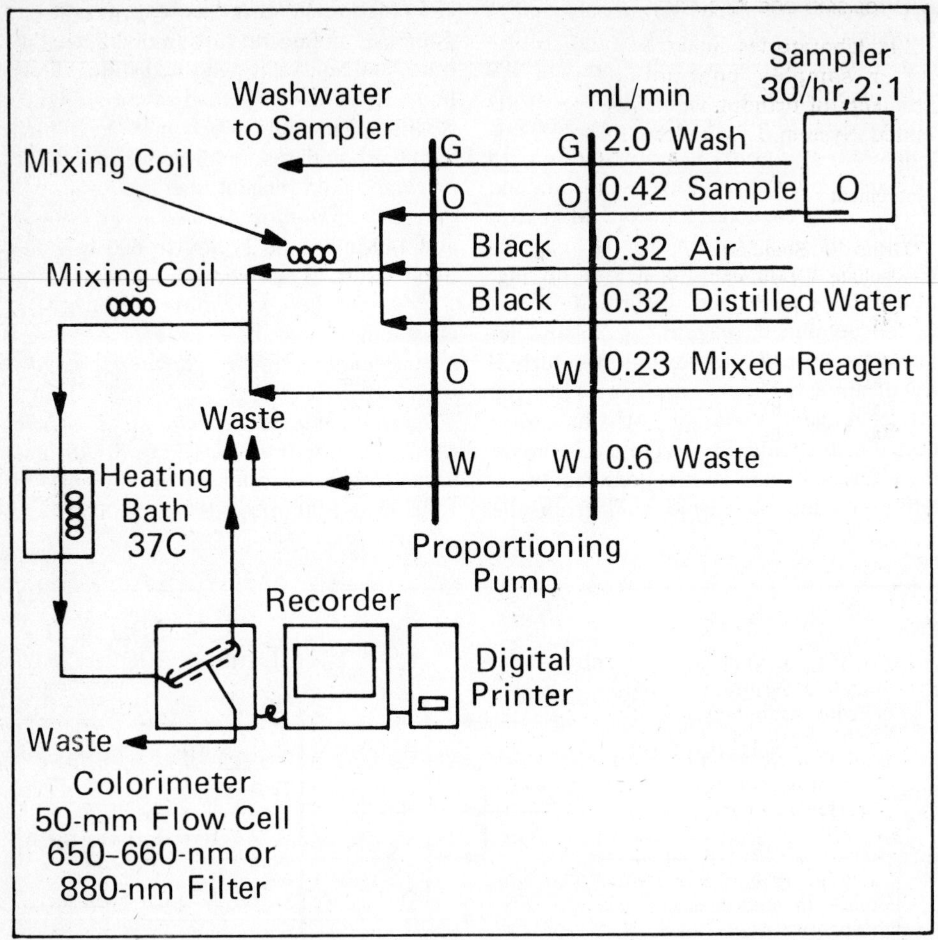

Figure 424:3. Phosphate manifold for Automated System II.

i. Intermediate phosphate solution: Dilute 100.0 mL stock phosphate solution to 1,000 mL with distilled water; 1.00 mL = 10.0 μg P.

j. Standard phosphate solutions: Prepare a suitable series of standards by diluting appropriate volumes of intermediate phosphate solution.

4. Procedure

Set up manifold and complete system as shown in Figure 424:2 or Figure 424:3.

For Automated System I, sample at a rate of 20/hr, 1 min sample, 2 min wash; for Automated System II, use a 30/hr, 2:1 cam and a common wash.

Add 0.05 mL (1 drop) phenolphthalein indicator solution to approximately 50 mL sample. If a red color develops, add H_2SO_4 (¶ 3*e*) dropwise to just discharge the color.

Follow general procedure in Part 600.

5. Calculation

See Part 600.

6. Precision and Accuracy

Six laboratories analyzed four natural water samples containing exact increments of orthophosphate, using Automated System I, with the following results:

Increment as Orthophosphate *µg P/L*	Standard Deviation *µg P/L*	Bias *%*	Bias *µg P/L*
40	19	+16.7	+7
40	14	−8.3	−3
290	87	−15.5	−50
300	66	−12.8	−40

In a single laboratory, using surface water samples at concentrations of 40, 190, 350, and 840 µg P/L, standard deviations were 5, 0, 3, and 0, respectively, and at concentrations of 70 and 760 µg P/L, recoveries were 99 and 100%, respectively.

Single-laboratory data on the Automated System II show comparable precision and accuracy.

424 H. Bibliography

KITSON, R.E. & M.G. MELLON. 1944. Colorimetric determination of phosphorus as molydovanadophosphoric acid. *Ind. Eng. Chem.*, Anal. Ed. 16:379.

BOLTZ, D.F. & M.G. MELLON. 1947. Determination of phosphorus, germanium, silicon, and arsenic by the heteropoly blue method. *Ind. Eng. Chem.*, Anal. Ed. 19:873.

GREENBERG, A.E., L.W. WEINBERGER & C.N. SAWYER. 1950. Control of nitrite interference in colorimetric determination of phosphorus. *Anal. Chem.* 22:499.

YOUNG, R.S. & A. GOLLEDGE. 1950. Determination of hexametaphosphate in water after threshold treatment. *Ind. Chem.* 26:13.

GRISWOLD, B.L., F.L. HUMOLLER & A.R. MCINTYRE. 1951. Inorganic phosphates and phosphate esters in tissue extracts. *Anal. Chem.* 23:192.

BOLTZ, D.F., ed. 1958. Colorimetric Determination of Nonmetals. Interscience Publishers, New York, N.Y.

AMERICAN WATER WORKS ASSOCIATION. 1958. Committee report. Determination of orthophosphate, hydrolyzable phosphate, and total phosphate in surface waters. *J. Amer. Water Works Ass.* 50:1563.

JACKSON, M.L. 1958. Soil Chemical Analysis. Prentice-Hall, Englewood Cliffs, N.J.

SLETTEN, O. & C.M. BACH. 1961. Modified stannous chloride reagent for orthophosphate determination. *J. Amer. Water Works Ass.* 53:1031.

MURPHY, J. & J. RILEY. 1962. A modified single solution method for the determination of phosphate in natural waters. *Anal. Chem. Acta* 27:31.

ABBOT, D.C., G.E. EMSDEN & J.R. HARRIS. 1963. A method for determining orthophosphate in water. *Analyst* 88:814.

GOTTFRIED, P. 1964. Determination of total phosphorus in water and wastewater as molybdovanadophosphoric acid. *Limnologica* 2:407.

STRICKLAND, J.D.H. & T.R. PARSONS. 1965. A Manual of Sea Water Analysis, 2nd ed. Fish. Res. Board, Ottawa, Canada.

BLACK, C.A., D.D. EVANS, J.L. WHITE, L.E. ENSMINGER & F.E. CLARK, eds. 1965. Methods of Soil Analysis, Part 2. Chemical and Microbiological Properties. American Society for Agronomy, Madison, Wisc.

EDWARDS, G.P., A.H. MOLOF & R.W. SCHNEEMAN. 1965. Determination of orthophosphate in fresh and saline waters. *J. Amer. Water Works Ass.* 57:917.

LEE, G.F., N.L. CLESCERI & G.P. FITZGERALD. 1965. Studies on the analysis of phosphates in algal cultures. *J. Air Water Pollut.* 9:715.

JENKINS, D. 1965. A study of methods suitable for the analysis and preservation of phosphorus forms in an estuarine environment. SERL Report No. 65-18, Sanitary Engineering Research Laboratory, Univ. of California, Berkeley.

HENRIKSEN, A. 1966. An automatic method for determining orthophosphate in sewage and highly polluted waters. *Analyst* 91:652.

SHANNON, J.E. & G.F. LEE. 1966. Hydrolysis of condensed phosphates in natural waters. *J. Air Water Pollut.* 10:735.

GALES, M.E., JR., E.C. JULIAN & R.C. KRONER. 1966. Method for quantitative determination of total phosphorus in water. *J. Amer. Water Works Ass.* 58:1363.

LEE, G.F. 1967. Analytical chemistry of plant nutrients. *In* Proc. Int. Conf. Eutrophication, Madison, Wisc.

FITZGERALD, G.P. & S.L. FAUST. 1967. Effect of water sample preservation methods on the release of phosphorus from algae. *Limnol. Oceanogr.* 12:332.

Automated Ascorbic Acid Reduction Method

U.S. ENVIRONMENTAL PROTECTION AGENCY. 1971. Methods for Chemical Analysis of Water and Wastes. National Environmental Research Center, Cincinnati, Ohio.

LOBRING, L.B. & R.L. BOOTH. 1973. Evaluation of the AutoAnalyzer II; A progress report. *In* Advances in Automated Analysis: 1972 Technicon International Congress. Vol. 8, p. 7. Mediad, Inc., Tarrytown, N.Y.

U.S. ENVIRONMENTAL PROTECTION AGENCY. MDQARL Method Study 4, Automated Methods. National Environmental Research Center, Cincinnati, Ohio (in preparation).

425 SILICA

Silicon ranks next to oxygen in abundance in the earth's crust. It appears as the oxide (silica) in quartz and sand and is combined with metals in the form of many complex silicate minerals, particularly igneous rocks. Degradation of silica-containing rocks results in the presence of silica in natural waters as suspended particles, in a colloidal or polymeric state, and as silicic acids or silicate ions. Volcanic and geothermally heated waters often contain an abundance of silica.

A more complete discussion of the occurrence and chemistry of silica in natural waters is available.[1]

The silica (SiO_2) content of natural water most commonly is in the 1- to 30-mg/L range, although concentrations as high as 100 mg/L are not unusual and concentrations exceeding 1,000 mg/L are found in some brackish waters and brines.

Silica in water is undesirable for a number of industrial uses because it forms difficult-to-remove silica and silicate scales in equipment, particularly on high-pressure steam-turbine blades. Silica is removed most often by the use of strongly basic anion-exchange resins in the deionization process, by distillation, or by reverse osmosis. Some plants use precipitation with magnesium oxide in either the hot or cold lime softening process.

1. Selection of Method

Method B determines total silica. Methods C, D, and E determine molybdate-reactive silica. As noted in Section 425C.4, it is possible to convert other forms of silica to the molybdate-reactive form for determination by these methods. Method A, like Method B, determines more than one form of silica. It will determine all dissolved silica and colloidally dispersed silica. The determination of silica present in micrometer and submicrometer particles will depend on the size distribution, composition, and structure of the particles; thus Method A cannot be said to determine total silica.

Use Method B to standardize sodium silicate solutions used as standards for

Methods C, D, and E. It is the preferred method for water samples that contain at least 20 mg SiO_2/L, but is not recommended for determining lower concentrations. Method C is recommended for relatively pure waters containing from 0.4 to 25 mg SiO_2/L. As with most colorimetric methods, the range can be extended, if necessary, by taking smaller portions, by concentrating, or by varying the light path. Interferences due to tannin, color, and turbidity are more severe with this method than with Method D. Moreover, the yellow color produced by Method C has a limited stability and attention to timing is necessary. When applicable, however, it offers greater speed and simplicity than Method D because one less reagent is used; one timing step is eliminated; and many natural waters can be analyzed without dilution, which is not often the case with Method D. Method D is recommended for the low range, from 0.04 to 2 mg SiO_2/L. This range also can be extended if necessary. Such extension may be desirable if interference is expected from tannin, color, or turbidity. A combination of factors renders Methods D and E less susceptible than Method C to those interferences; also, the blue color in Methods D and E is more stable than the yellow color in Method C. However, many samples will require dilution because of high sensitivity. Permanent artificial color standards are not available for the blue color developed in Method D.

The yellow color produced by Method C and the blue color produced by Methods D and E are affected by high concentrations of salts. With sea water the yellow color intensity is decreased by 20 to 25% and the blue color intensity is increased by 10 to 15%. When waters of high ionic strength are analyzed by these methods, use silica standards of approximately the same ionic strengths.[2]

Method E may be used where large numbers of samples are analyzed regularly. Method A is recommended for broad-range use. Although Method A is usable from 1 to 300 mg SiO_2/L, optimal results are obtained from about 20 to 300 mg/L. The range can be extended upward by dilution if necessary. This method is rapid and does not require any timing step.

2. Sampling and Storage

Collect samples in bottles of polyethylene, other plastic, or hard rubber, especially if there will be a delay between collection and analysis. Borosilicate glass is a less desirable choice, particularly with waters of pH above 8 or with seawater, in which cases a significant amount of silica in the glass can dissolve. Freezing to preserve samples for analysis of other constituents can lower soluble silica values by as much as 20 to 40% in waters that have a pH below 6.

425 A. Atomic Absorption Spectrophotometric Method

See Section 303C.

425 B. Gravimetric Method

1. General Discussion

a. Principle: Hydrochloric acid decomposes silicates and dissolved silica, forming silicic acids that are precipitated as partially dehydrated silica during evaporation and baking. Ignition completes dehydration of the silica, which is weighed and then volatilized as silicon tetrafluoride, leaving any impurities behind as nonvolatile residue. The residue is weighed and silica is determined as loss on volatilization. Perchloric acid ($HClO_4$) may be used to dehydrate the silica instead of HCl. A single fuming with $HClO_4$ will recover more silica than one with HCl, although for complete silica recovery two dehydrations with either acid are necessary. The use of $HClO_4$ lessens the tendency to spatter, yields a silica precipitate that is easier to filter, and shortens the time required for the determination.

b. Interference: Because glassware may contribute silica, avoid its use as much as possible. Use agents and distilled water low in silica. Carry out a blank determination to correct for silica introduced by the reagents and apparatus.

2. Apparatus

a. Platinum crucibles, with covers.

b. Platinum evaporating dishes, 200-mL. In dehydration steps, acid-leached glazed porcelain evaporating dishes free from etching may be substituted for platinum.

3. Reagents

For maximum accuracy, set aside batches of chemicals low in silica for this method. Store all reagents in plastic containers and run blanks.

a. Hydrochloric acid, HCl, 1 + 1 and 1 + 50.

b. Sulfuric acid, H_2SO_4, 1 + 1.

c. Hydrofluoric acid, HF, 48%.

d. Perchloric acid, $HClO_4$, 72%.

4. Procedure

Before determining silica, test H_2SO_4 and HF for interfering nonvolatile matter by carrying out the procedure of ¶ 4*a*5) below. Use a clean empty platinum crucible. If any increase in weight is observed, make a correction in the silica determinations.

a. HCl dehydration:

1) Sample evaporation—To a clear sample containing at least 10 mg silica, add 5 mL 1 + 1 HCl. Evaporate to dryness in a 200-mL platinum evaporating dish, in several portions if necessary, on a water bath or suspended on an asbestos ring over a hot plate. Protect against contamination by atmospheric dust. During evaporation, add a total of 15 mL 1 + 1 HCl in several portions. Dry dish and place it in a 110 C oven or over a hot plate to bake for 30 min.

2) First filtration—Add 5 mL 1 + 1 HCl, warm, and add 50 mL hot distilled water. While hot, filter through an ashless medium-texture filter paper, decanting as much liquid as possible. Wash dish and residue with hot 1 + 50 HCl and then with a minimum volume of distilled water until washings are chloride-free. Save all washings. Set aside filter paper with its residue.

3) Second filtration—Evaporate filtrate and washings from the above operation to dryness in the original platinum dish. Bake residue in a 110 C oven or over a hot plate for 30 min. Repeat steps in ¶ 2) above. Use a separate filter paper and a rubber policeman to aid in transferring residue from dish to filter.

4) Ignition—Transfer the two filter papers (one if dehydrated by 4*b*) and residues to a covered platinum crucible, dry at 110 C, and ignite at 1,200 C to constant weight. Avoid mechanical loss of residue when first charring and burning off the paper. Cool in desiccator, weigh, and repeat ignition and weighing until constant weight is attained. Record weight of crucible and contents.

5) Volatilization with HF—Thoroughly moisten weighed residue with distilled water. Add 4 drops 1 + 1 H_2SO_4, followed by 10 mL HF, measuring the latter in a plastic graduated cylinder or pouring an esti-

mated 10 mL directly from the reagent bottle. Slowly evaporate to dryness over an air bath or hot plate in a hood and avoid loss by splattering. Ignite crucible to constant weight at 1,200 C. Record weight of crucible and contents.

b. $HClO_4$ dehydration: Follow procedure in ¶ 4*a*1) above until all but 50 mL of sample has been evaporated. Add 5 mL $HClO_4$ and evaporate until dense white fumes appear. (CAUTION: *Explosive—Place a shield between analyst and fuming dish.*)

Continue dehydration for 10 min. Cool, add 5 mL 1 + 1 HCl, and 50 mL hot distilled water. Bring to a boil and filter through an ashless quantitative filter paper. Wash thoroughly ten times with hot distilled water and proceed as directed in ¶s 4*a*4) and 5) preceding. For most work, the silica precipitate often is sufficiently pure for the purpose intended and may be weighed directly, omitting HF volatilization. Make an initial check against the longer procedure, however, to be sure that the result is within the limits of accuracy required.

5. Calculation

Subtract weight of crucible and contents after HF treatment from the corresponding weigh before HF treatment. The difference, A, in milligrams is "loss on volatilization" and represents silica:

$$\text{mg } SiO_2/L = \frac{A \times 1{,}000}{\text{mL sample}}$$

6. Precision and Accuracy

The accuracy is limited both by the finite solubility of silica in water under the conditions of analysis and by the analytical balance sensitivity. Under optimum conditions, the precision is approximately ± 0.2 mg SiO_2. If a 1-L sample is taken for analysis, this represents a precision of ± 0.2 mg/L.

425 C. Molybdosilicate Method

1. General Discussion

a. Principle: Ammonium molydate at pH approximately 1.2 reacts with silica and any phosphate present to produce heteropoly acids. Oxalic acid is added to destroy the molybdophosphoric acid but not the molybdosilicic acid. Even if phosphate is known to be absent, the addition of oxalic acid is highly desirable and is a mandatory step in both this method and Method D. The intensity of the yellow color is proportional to the concentration of "molybdate-reactive" silica. In at least one of its forms, silica does not react with molybdate even though it is capable of passing through filter paper and is not noticeably turbid. It is not known to what extent such "unreactive" silica occurs in waters. Terms such as "colloidal", "crystalloidal", and "ionic" have been used to distinguish among various forms of silica but such terminology cannot be substantiated. "Molybdate-unreactive" silica can be converted to the "molybdate-reactive" form by heating or fusing with alkali. Molybdate-reactive or unreactive does imply reactivity, or lack of it, toward *other* reagents or processes.

b. Interference: Because both apparatus and reagents may contribute silica, avoid using glassware as much as possible and use reagents low in silica. Also, make a blank determination to correct for silica so introduced. In both this method and Method D, tannin, large amounts of iron, color, turbidity, sulfide, and phosphate interfere. Treatment with oxalic acid eliminates interference from phosphate and decreases interference from tannin. If necessary, use photometric compensation to cancel interference from color or turbidity.

c. Minimum detectable concentration: Approximately 1 mg SiO_2/L can be detected in 50-mL nessler tubes.

2. Apparatus

a. Platinum dishes, 100-mL.

b. Colorimetric equipment: One of the following is required:

1) *Spectrophotometer,* for use at 410 nm, providing a light path of 1 cm or longer.

2) *Filter photometer,* providing a light path of 1 cm or longer and equipped with a violet filter having maximum transmittance near 410 nm.

3) *Nessler tubes,* matched, 50-mL, tall form.

3. Reagents

For best results, set aside and use batches of chemicals low in silica. Store all reagents in plastic containers to guard against high blanks.

a. Sodium bicarbonate, $NaHCO_3$, powder.

b. Sulfuric acid, H_2SO_4, 1*N*.

c. Hydrochloric acid, HCl, 1 + 1.

d. Ammonium molybdate reagent: Dissolve 10 g $(NH_4)_6Mo_7O_{24}\cdot 4H_2O$ in distilled water, with stirring and gentle warming, and dilute to 100 mL. Filter if necessary. Adjust to pH 7 to 8 with silica-free NH_4OH or NaOH and store in a polyethylene bottle to stabilize. (If the pH is not adjusted, a precipitate gradually forms. If the solution is stored in glass, silica may leach out and cause high blanks.) If necessary, prepare silica-free NH_4OH by passing gaseous NH_3 into distilled water contained in a plastic bottle.

e. Oxalic acid solution: Dissolve 7.5 g $H_2C_2O_4\cdot 2H_2O$ in distilled water and dilute to 100 mL.

f. Stock silica solution: Dissolve 4.73 g sodium metasilicate nonahydrate, $Na_2SiO_3\cdot 9H_2O$, in freshly boiled and cooled distilled water and dilute to approximately 900 mL. Analyze 100.0-mL portions by Method A and adjust remainder of solution to contain 1,000 mg SiO_2/L. Store in a tightly stoppered plastic bottle.

g. Standard silica solution: Dilute 10.00 mL stock solution to 1,000 mL with freshly boiled and cooled distilled water; 1.00 mL = 10.0 μg SiO_2. Store in a tightly stoppered plastic bottle.

h. Permanent color solutions:

1) *Potassium chromate solution:* Dissolve 630 mg K_2CrO_4 in distilled water and dilute to 1 L.

2) *Borax solution:* Dissolve 10 g sodium borate decahydrate, $Na_2B_4O_7\cdot 10H_2O$, in distilled water and dilute to 1 L.

4. Procedure

To detect the presence of molybdate-unreactive silica, digest sample with $NaHCO_3$. This digestion is not necessarily sufficient to convert all molybdate-unreactive silica to the molybdate-reactive form. Complex silicates and higher silica polymers may require extended fusion with alkali at high temperatures or digestion under pressure for complete conversion. Omit digestion if all the silica is known to react with molybdate.

a. Digestion with $NaHCO_3$: Prepare a clear sample by filtration if necessary. Place 50.0 mL, or a smaller portion diluted to 50 mL, in a 100-mL platinum dish. Add 200 mg silica-free $NaHCO_3$ and digest on a steam bath for 1 hr. Cool and add slowly, with stirring, 2.4 mL 1*N* H_2SO_4. Do not interrupt analysis but proceed *at once* with remaining steps. Transfer quantitatively to a 50-mL nessler tube and make up to mark with distilled water. (Tall-form 50-mL nessler tubes are convenient for mixing even if the solution subsequently is transferred to an absorption cell for photometric measurement.)

b. Color development: To prepared sample, or to 50.0 mL of an untreated sample if digestion is omitted, add in rapid succession 1.0 mL 1 + 1 HCl and 2.0 mL ammonium molybdate reagent. Mix by inverting at least six times and let stand for 5 to 10 min. Add 2.0 mL oxalic acid solution and mix thoroughly. Read color after 2 min but before 15 min, measuring time from addition of oxalic acid. Because the yellow color obeys Beer's law, measure photometrically or visually.

c. Preparation of standards: If $NaHCO_3$ pretreatment is used, add to the standards (approximately 45 mL total volume) 200 mg $NaHCO_3$ and 2.4 mL 1*N* H_2SO_4, to compensate both for the slight amount of silica introduced by the reagents and for the effect of the salt on color intensity. Dilute to 50.0 mL.

d. Photometric measurement: Prepare a calibration curve from a series of approximately six standards to cover the optimum ranges cited in Table 425:I. Follow directions of ¶ 4*b* above on suitable portions of standard silica solution diluted to 50.0 mL in nessler tubes. Set photometer at zero absorbance with distilled water and read all standards, including a reagent blank, against distilled water. Plot micrograms silica in the final (55 mL) developed solution against photometer readings. Run a reagent blank and at least one standard with each group of samples to confirm that the calibration curve previously established has not shifted.

TABLE 425:I. SELECTION OF LIGHT PATH LENGTH FOR VARIOUS SILICA CONCENTRATIONS

Light Path *cm*	Method C: Silica in 55 mL Final Volume *μg*	Method D: Silica in 55 mL Final Volume *μg*, 650 nm Wavelength	Method D: Silica in 55 mL Final Volume *μg*, 815 nm Wavelength
1	200–1,300	40–300	20–100
2	100–700	20–150	10–50
5	40–250	7–50	4–20
10	20–130	4–30	2–10

TABLE 425:II. PREPARATION OF PERMANENT COLOR STANDARDS FOR VISUAL DETERMINATION OF SILICA

Value in Silica *mg*	Potassium Chromate Solution *mL*	Borax Solution *mL*	Water *mL*
0.00	0.0	25	30
0.10	1.0	25	29
0.20	2.0	25	28
0.40	4.0	25	26
0.50	5.0	25	25
0.75	7.5	25	22
1.0	10.0	25	20

e. Visual comparison: Make a set of permanent artificial color standards, using K_2CrO_4 and borax solutions. Mix liquid volumes specified in Table 425:II and place them in well-stoppered, appropriately labeled 50-mL nessler tubes. Verify correctness of these permanent artificial standards by comparing them visually against standards prepared by analyzing portions of the standard silica solution. Use permanent artificial color standards only for visual comparison.

f. Correction for color or turbidity: Prepare a special blank for every sample that needs such correction. Carry two identical portions of each such sample through the procedure, including $NaHCO_3$ treatment if this is used. To one portion add all reagents as directed in ¶ 4*b* preceding. To the other portion add HCl and oxalic acid but no molybdate. Adjust photometer to zero absorbance with the blank containing no molybdate before reading absorbance of molybdate-treated sample.

5. Calculation

$$\text{mg SiO}_2\text{/L} = \frac{\mu\text{g SiO}_2 \text{ (in 55 mL final volume)}}{\text{mL sample}}$$

Report whether $NaHCO_3$ digestion was used.

6. Precision and Accuracy

A synthetic sample containing 5.0 mg SiO_2/L, 10 mg Cl^-/L, 0.20 mg NH_3-N/L, 1.0 mg NO_3^--N/L, 1.5 mg organic N/L, and 10.0 mg PO_4^{3-}/L in distilled water was analyzed in 19 laboratories by the molybdosilicate method with a relative standard deviation of 14.3% and a relative error of 7.8%.

Another synthetic sample containing 15.0 mg SiO_2/L, 200 mg Cl^-/L, 0.800 mg NH_3-N/L, 1.0 mg NO_3^--N/L, 0.800 mg organic N/L, and 5.0 mg PO_4^{3-}/L in distilled water was analyzed in 19 laboratories by the molybdosilicate method, with a relative standard deviation of 8.4% and a relative error of 4.2%.

A third synthetic sample containing 30.0 mg SiO_2/L, 400 mg Cl^-/L, 1.50 mg NH_3-N/L, 1.0 mg NO_3^--N/L, 0.200 mg organic N/L, and 0.500 mg PO_4^{3-}/L, in distilled water was analyzed in 20 laboratories by the molybdosilicate method, with a relative standard deviation of 7.7% and a relative error of 9.8%.

All results were obtained after sample digestion with $NaHCO_3$.

425 D. Heteropoly Blue Method

1. General Discussion

a. Principle: The principles outlined under Method C, ¶ 1*a*, also apply to this method. The yellow molybdosilicic acid is reduced by means of aminonaphtholsulforonic acid to heteropoly blue. The blue color is more intense than the yellow color of Method C and provides increased sensitivity.

b. Interference: See Section 425C.1*b*.

c. Minimum detectable concentration: Approximately 20 μg SiO_2/L can be detected in 50-mL nessler tubes and 50 μg SiO_2/L spectrophotometrically.

2. Apparatus

a. Platinum dishes, 100-mL.

b. Colorimetric equipment: One of the following is required:

1) *Spectrophotometer,* for use at approximately 815 nm. The color system also obeys Beer's law at 650 nm, with appreciably reduced sensitivity. Use light path of 1 cm or longer.

2) *Filter photometer,* provided with a red filter exhibiting maximum transmittance in the wavelength range of 600 to 815 nm. Sensitivity improves with increasing wavelength. Use light path of 1 cm or longer.

3) *Nessler tubes,* matched, 50 mL, tall form.

3. Reagents

For best results, set aside and use batches of chemicals low in silica. Store all reagents in plastic containers to guard against high blanks. Use distilled water that does not contain detectable silica after storage in glass.

All of the reagents listed in Section 425C.3 are required, and in addition:

Reducing agent: Dissolve 500 mg 1-amino-2-naphthol-4-sulfonic acid and 1 g Na_2SO_3 in 50 mL distilled water, with gentle warming if necessary; add this to a solution of 30 g $NaHSO_3$ in 150 mL distilled water. Filter into a plastic bottle. Discard when solution becomes dark. Prolong reagent life by storing in a refrigerator and away from light. Do not use aminonaphtholsulfonic acid that is incompletely soluble or that produces reagents that are dark even when freshly prepared.*

4. Procedure

a. Color development: Proceed as in ¶ 425C.4*a* and *b* up to and including the words, "Add 2.0 mL oxalic acid solution and mix thoroughly." Measuring time from the moment of adding oxalic acid, wait at least 2 min but not more than 15 min, add 2.0 mL reducing agent, and mix thoroughly. After 5 min, measure blue color photometrically or visually. If $NaHCO_3$ pretreatment is used, follow ¶ 425C.4*c*.

b. Photometric measurement: Prepare a calibration curve from a series of approximately six standards to cover the optimum range indicated in Table 425:I. Carry out the steps described above on suitable portions of standard silica solution diluted to 50.0 mL in nessler tubes; pretreat standards if $NaHCO_3$ digestion is used (see 425C.4*c*). Adjust photometer to zero absorbance with distilled water and read all standards, including a reagent blank, against distilled water. If necessary to correct for color or turbidity in a sample, see ¶ 425C.4*f*. To the special blank add HCl and oxalic acid, but no molybdate or reducing agent. Plot micrograms silica in the final 55 mL developed solution against absorbance. Run a reagent blank and at least one standard with each group of samples to check the calibration curve.

*Eastman No. 360 has been found satisfactory.

c. Visual comparison: Prepare a series of not less than 12 standards, covering the range 0 to 120 μg SiO_2, by placing the calculated volumes of standard silica solution in 50-mL nessler tubes, diluting to mark with distilled water, and developing color as described in ¶ *a* preceding.

5. Calculation

$$\text{mg } SiO_2/L = \frac{\mu g \text{ } SiO_2 \text{ (in 55 mL final volume)}}{\text{mL sample}}$$

Report whether $NaHCO_3$ digestion was used.

6. Precision and Accuracy

A synthetic sample containing 5.0 mg SiO_2/L, 10 mg Cl^-/L, 0.200 mg NH_3-N/L, 1.0 mg NO_3^--N/L, 1.5 mg organic N/L, and 10.0 mg PO_4^{3-}/L in distilled water was analyzed in 11 laboratories by the heteropoly blue method, with a relative standard deviation of 27.2% and a relative error of 3.0%.

A second synthetic sample containing 15 mg SiO_2/L, 200 mg Cl^-/L, 0.800 mg NH_3-N/L, 1.0 mg NO_3^--N/L, 0.800 mg organic N/L, and 5.0 mg PO_4^{3-}/L in distilled water was analyzed in 11 laboratories by the heteropoly blue method, with a relative standard deviation of 18.0% and a relative error of 2.9%.

A third synthetic sample containing 30.0 mg SiO_2/L, 400 mg Cl^-/L, 1.50 mg NH_3-N/L, 1.0 mg NO_3^--N/L, 0.200 mg organic N/L, and 0.500 mg PO_4^{3-}/L in distilled water was analyzed in 10 laboratories by the heteropoly blue method with a relative standard deviation of 4.9% and a relative error of 5.1%.

All results were obtained after sample digestion with $NaHCO_3$.

425 E. Automated Method for Molybdate-Reactive Silica

1. General Discussion

a. Principle: This method is an adaptation of the heteropoly blue method (Method D) utilizing the continuous-flow analytical instrument and general procedure given in Section 602.

b. Interferences: See Section 425C.1*b*. If particulate matter is present, filter sample or use a continuous filter as an integral part of the system.

c. Application: This method is applicable to potable, surface, domestic, and other waters containing 0 to 20 mg/L. The range of concentration can be broadened to 0 to 80 mg/L by substituting a 15-mm flow cell for the 50-mm flow cell shown in Figure 425:1.

2. Apparatus

Use an automated system capable of adding and mixing sample and reagents in the sequence and amounts detailed in Figure 425:1. Determine absorbance at 660

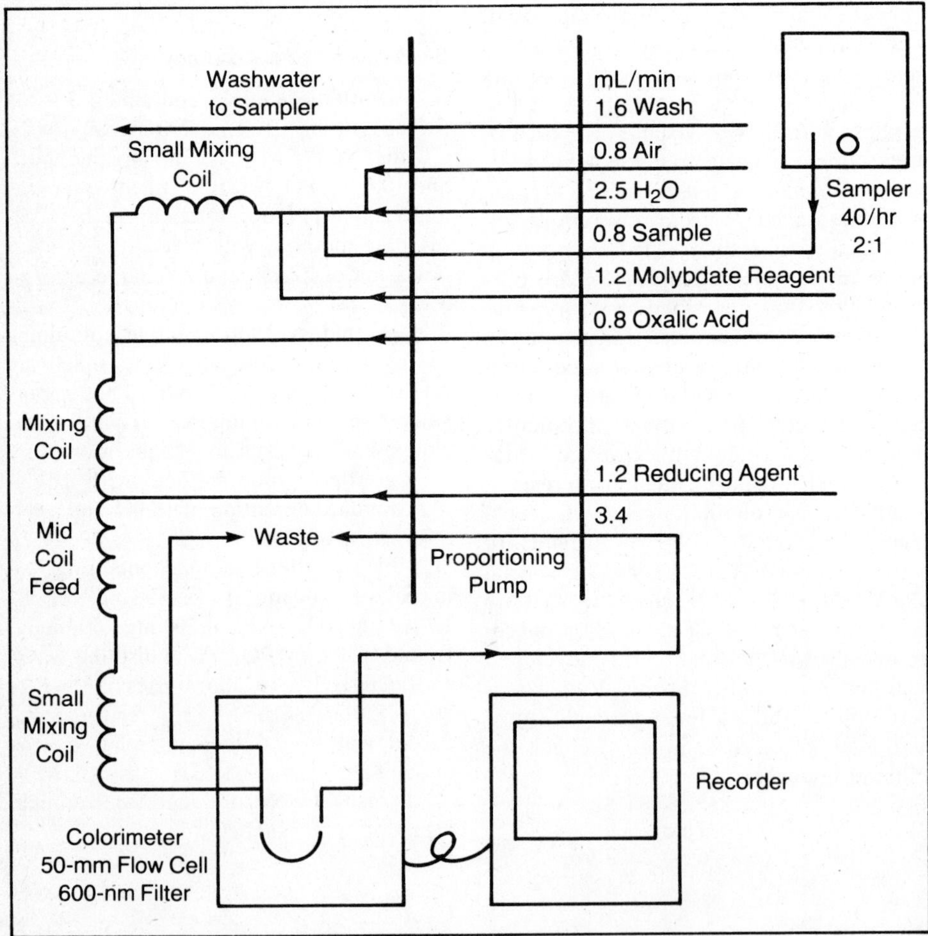

Figure 425:1. Silica manifold.

nm. Assembled modules for this method are available commercially.

3. Reagents

a. Sulfuric acid, H_2SO_4, 0.1*N*.

b. Ammonium molybdate reagent: Dissolve 10 g $(NH_4)_6Mo_7O_{24}\cdot 4H_2O$ in 1 L 0.1*N* H_2SO_4. Filter and store in an amber plastic bottle.

c. Oxalic acid solution: Dissolve 50 g oxalic acid in 900 mL distilled water and dilute to 1 L.

d. Reducing agent: Dissolve 120 g $NaHSO_3$ and 4 g Na_2SO_3 in 800 mL warm distilled water. Add 2 g 1-amino-2-naphthol-4-sulfonic acid, mix well, and dilute to 1 L. Filter into amber plastic bottle for storage.

To prepare working reagent, dilute 100 mL to 1 L with distilled water. Make working reagent daily.

e. Standard silica solution: See 425B.3*f*.

4. Precision and Accuracy

For 0 to 20 mg SiO_2/L, when a 50-mm flow cell was used at 40 samples/hr, the detection limit was 0.1 mg/L, sensitivity (concentration giving 0.398 absorbance) was 7.1 mg/L, and the coefficient of variation (95% confidence level at 7.1 mg/L) was 1.6%. For 0 to 80 mg SiO_2/L, when a 15-mm flow cell was used at 50 samples/hr, detection limit was 0.5 mg/L, sensitivity was 31 mg/L, and coefficient of variation at 31 mg/L was 1.5%.

425 F. References

1. HEM, J.D. 1959. Study and interpretation of the chemical characteristics of natural water. U.S. Geol. Surv. Water Supply Pap. No. 1473.
2. FANNING, K.A. & M.E.Q. PILSON. 1973. On the spectrophotometric determination of dissolved silica in natural waters. *Anal. Chem.* 45:136.

425 G. Bibliography

General

ROY, C.J. 1945. Silica in natural waters. *Amer. J. Sci.* 243:393.

VAIL, J.G. 1952. The Soluble Silicates, Their Properties and Uses. Reinhold Publishing Corp., New York, N.Y. Vol. 1, pp. 95–97, 100–161.

Gravimetric Method

HILLEBRAND, W.F. et al. 1953. Applied Inorganic Analysis, 2nd ed. John Wiley & Sons, New York, N.Y. Chapter 43.

KOLTHOFF, I.M., E.J. MEEHAN, E.B. SANDELL & S. BRUCKENSTEIN. 1969. Quantitative Chemical Analysis, 4th ed. Macmillan Co., New York, N.Y.

Colorimetric Methods

DIENERT, F. & F. WANDENBULCKE. 1923. On the determination of silica in waters. *Bull. Soc. Chim. France* 33:1131, *Compt. Rend.* 176:1478.

DIENERT, F. & F. WANDENBULCKE. 1924. A study of colloidal silica. *Compt. Rend.* 178:564.

SWANK, H.W. & M.G. MELLON. 1934. Colorimetric standards for silica. *Ind. Eng. Chem.*, Anal. Ed. 6:348.

TOURKY, A.R. & D.H. BANGHAM. 1936. Colloidal silica in natural waters and the "silicomolybdate" colour test. *Nature* 138:587.

BIRNBAUM, N. & G.H. WALDEN. 1938. Coprecipitation of ammonium silicomolybdate and ammonium phosphomolybdate. *J. Amer. Chem. Soc.* 60:66.

KAHLER, H.L. 1941. Determination of soluble silica in water: A photometric method. *Ind. Eng. Chem.*, Anal. Ed. 13:536.

NOLL, C.A. & J.J. MAGUIRE. 1942. Effect of

container on soluble silica content of water samples. *Ind. Eng. Chem.*, Anal. Ed. 14:569.

SCHWARTZ, M.C. 1942. Photometric determination of silica in the presence of phosphates. *Ind. Eng. Chem.*, Anal. Ed. 14:893.

BUNTING, W.E. 1944. Determination of soluble silica in very low concentrations. *Ind. Eng. Chem.*, Anal. Ed. 16:612.

STRAUB, F.G. & H. GRABOWSKI. 1944. Photometric determination of silica in condensed steam in the presence of phosphates. *Ind. Eng. Chem.*, Anal. Ed. 16:574.

GUTTER, H. 1954. Influence of pH on the composition and physical aspects of the ammonium molybdates. *Compt. Rend.* 200:146.

BOLTZ, D.F. & M.G. MELLON. 1947. Determination of phosphorus, germanium, silicon, and arsenic by the heteropoly blue method. *Ind. Eng. Chem.*, Anal. Ed. 19:873.

MILTON, R.F. 1951. Formation of silicomolybdate. *Analyst* 76:431.

MILTON, R.F. 1951. Estimation of silica in water. *J. Appl. Chem.* (London) 1: (Supplement No. 2) 126.

CARLSON, A.B. & C.V. BANKS. 1952. Spectrophotometric determination of silicon. *Anal. Chem.* 24:472.

KILLEFFER, D.H. & A. LINZ. 1952. Molybdenum Compounds, Their Chemistry and Technology. Interscience Publishers, New York, N.Y. pp. 1-2, 42-45, 67-82, 87-92.

STRICKLAND, J.D.H. 1952. The preparation and properties of silicomolybdic acid. *J. Amer. Chem. Soc.* 74:862, 868, 872.

CHOW, D.T.W. & R.J. ROBINSON. 1953. The forms of silicate available for colorimetric determination. *Anal. Chem.* 25:646.

426 SULFATE

Sulfate is widely distributed in nature and may be present in natural waters in concentrations ranging from a few to several thousand milligrams per liter. Mine drainage wastes may contribute large amounts of sulfate through pyrite oxidation. Sodium and magnesium sulfate exert a cathartic action.

1. Selection of Method

Base the choice of method on the equipment available and the analyst's preference.

2. Sampling and Storage

In the presence of organic matter certain bacteria may reduce sulfate to sulfide. To avoid this, store heavily polluted or contaminated samples at low temperatures.

426 A. Gravimetric Method With Ignition of Residue

1. General Discussion

a. Principle: Sulfate is precipitated in a hydrochloric acid (HCl) solution as barium sulfate ($BaSO_4$) by the addition of barium chloride ($BaCl_2$).

The precipitation is carried out near the boiling temperature, and after a period of digestion the precipitate is filtered, washed with water until free of chlorides, ignited or dried, and weighed as $BaSO_4$.

b. Interference: The gravimetric determination of sulfate is subject to many errors, both positive and negative. In potable waters where the mineral concentration is low, these may be of minor importance.

1) Interferences leading to high results—Suspended matter, silica, $BaCl_2$ precipitant, nitrate, sulfite, and occluded mother liquor in the precipitate are the principal factors in positive errors. Suspended matter may be present in both the sample and the precipitating solution; soluble silicate may be rendered insoluble and sulfite may be oxidized to sulfate during analysis. Barium nitrate [$Ba(NO_3)_2$], $BaCl_2$, and water are occluded to some extent with the $BaSO_4$ although water is driven off if the temperature of ignition is sufficiently high.

2) Interferences leading to low results—Alkali metal sulfates frequently yield low results. This especially is true of alkali hydrogen sulfates. Occlusion of alkali sulfate with $BaSO_4$ causes substitution of an element of lower atomic weight than barium in the precipitate. Hydrogen sulfates of alkali metals act similarly and, in addition, decompose on being heated. Heavy metals, such as chromium and iron, cause low results by interfering with the complete precipitation of sulfate and by formation of heavy metal sulfates. $BaSO_4$ has small but significant solubility, which is increased in the presence of acid. Although an acid medium is necessary to prevent precipitation of barium carbonate and phosphate, it is important to limit its concentration to minimize the solution effect.

2. Apparatus

a. Steam bath.

b. Drying oven, equipped with thermostatic control.

c. Muffle furnace, with heat indicator.

d. Desiccator.

e. Analytical balance, capable of weighing to 0.1 mg.

f. Filter. Use one of the following:

1) *Filter paper,* acid-washed, ashless hard-finish, sufficiently retentive for fine precipitates.

2) *Membrane filter,* with a pore size of about 0.45 μm.

g. Filtering apparatus, appropriate to the type of filter selected. (Coat holder used for the membrane filter with silicone fluid to prevent precipitate from adhering.)

3. Reagents

a. Methyl red indicator solution: Dissolve 100 mg methyl red sodium salt in distilled water and dilute to 100 mL.

b. Hydrochloric acid, HCl, 1 + 1.

c. Barium chloride solution: Dissolve 100 g $BaCl_2 \cdot 2H_2O$ in 1 L distilled water. Filter through a membrane filter or hard-finish filter paper before use; 1 mL is capable of precipitating approximately 40 mg SO_4.

d. Silver nitrate-nitric acid reagent: Dissolve 8.5 g $AgNO_3$ and 0.5 mL conc HNO_3 in 500 mL distilled water.

*e. Silicone fluid.**

4. Procedure

a. Removal of silica: If the silica concentration exceeds 25 mg/L, evaporate sample nearly to dryness in a platinum dish on a steam bath. Add 1 mL HCl, tilt dish, and rotate it until the acid comes in complete contact with the residue. Continue evaporation to dryness. Complete drying in an oven at 180 C and if organic matter is present, char over flame of a burner. Moisten residue with 2 mL distilled water and 1 mL HCl, and evaporate to dryness on a steam bath. Add 2 mL HCl, take up soluble residue in hot water, and filter. Wash insoluble silica with several small portions of hot distilled water. Combine filtrate and washings. Discard residue.

b. Precipitation of barium sulfate: Adjust volume of clarified sample to contain

*"Desicote" (Beckman), or equivalent.

approximately 50 mg sulfate ion in a 250-mL volume. Lower concentrations of sulfate ion may be tolerated if it is impracticable to concentrate sample to the optimum level, but in such cases limit total volume to 150 mL. Adjust pH with HCl to pH 4.5 to 5.0, using a pH meter or the orange color of methyl red indicator. Add 1 to 2 mL HCl. Heat to boiling and, while stirring gently, add warm $BaCl_2$ solution slowly until precipitation appears to be complete; then add about 2 mL in excess. If amount of precipitate is small, add a total of 5 mL $BaCl_2$ solution. Digest precipitate at 80 to 90 C, preferably overnight but for not less than 2 hr.

c. Filtration and weighing: Mix a small amount of ashless filter paper pulp with the $BaSO_4$, quantitatively transfer to a filter, and filter at room temperature. The pulp aids filtration and reduces the tendency of the precipitate to creep. Wash precipitate with small portions of warm distilled water until washings are free of chloride, as indicated by testing with $AgNO_3$-HNO_3 reagent. Dry filter and precipitate and ignite at 800 C for 1 hr. *Do not let filter paper flame.* Cool in desiccator and weigh.

5. Calculation

$$\text{mg } SO_4/L = \frac{\text{mg } BaSO_4 \times 411.6}{\text{mL sample}}$$

6. Precision and Accuracy

A synthetic sample containing 259 mg SO_4/L, 108 mg Ca/L, 82 mg Mg/L, 3.1 mg K/L, 19.9 mg Na/L, 241 mg Cl^-/L, 0.250 mg NO_2^--N/L, 1.1 mg NO_3^--N/L and 42.5 mg total alkalinity/L (contributed by $NaHCO_3$) was analyzed in 32 laboratories by the gravimetric method, with a relative standard deviation of 4.7% and a relative error of 1.9%.

426 B. Gravimetric Method with Drying of Residue

1. General Discussion

See Method A, preceding.

2. Apparatus

With the exception of the filter paper, all of the apparatus cited in Section 426A.2 is required, plus the following:

a. Filters: Use one of the following:

1) *Fritted-glass filter,* fine ("F") porosity, with a maximum pore size of 5 μm.

2) *Membrane filter,* with a pore size of about 0.45 μm.

b. Vacuum oven.

3. Reagents

All the reagents listed in Section 426A.3 are required.

4. Procedure

a. Removal of interference: See Section 426A.4*a*.

b. Precipitation of barium sulfate: See Section 426A.4*b*.

c. Preparation of filters:

1) Fritted glass filter—Dry to constant weight in an oven maintained at 105 C or higher, cool in desiccator, and weigh.

2) Membrane filter—Place filter on a piece of filter paper or a watch glass and dry to constant weight* in a vacuum oven at 80 C, while maintaining a vacuum of at least 85 kPa or in a conventional oven at a temperature of 103 to 105 C. Cool in desiccator and weigh membrane only.

d. Filtration and weighing: Filter $BaSO_4$ at room temperature. Wash precipitate with several small portions of warm distilled water until washings are free of

*Constant weight is defined as a change of not more than 0.5 mg in two successive operations consisting of heating, cooling in desiccator, and weighing.

chloride, as indicated by testing with $AgNO_3$-HNO_3 reagent. If the membrane filter is used add a few drops of anticreep solution to the suspension before filtering, to prevent adherence of precipitate to holder. Dry filter and precipitate by the same procedure used in preparing filter. Cool in a desiccator and weigh.

5. Calculation

$$\text{mg } SO_4/L = \frac{\text{mg } BaSO_4 \times 411.6}{\text{mL sample}}$$

426 C. Turbidimetric Method

1. General Discussion

a. Principle: Sulfate ion is precipitated in a hydrochloric acid (HCl) medium with barium chloride ($BaCl_2$) so as to form barium sulfate ($BaSO_4$) crystals of uniform size. Light absorbance of the $BaSO_4$ suspension is measured by a nephelometer or transmission photometer and the sulfate ion concentration is determined by comparison of the reading with a standard curve.

b. Interference: Color or suspended matter in large amounts will interfere. Some suspended matter may be removed by filtration. If both are small in comparison with the sulfate ion concentration, correct for interference as indicated in ¶ 4*d* below. Silica in excess of 500 mg/L will interfere, and in waters containing large quantities of organic material it may not be possible to precipitate $BaSO_4$ satisfactorily.

Sulfite may be oxidized to sulfate during analysis to give a positive error. Rinse glassware thoroughly after using chromic acid cleaning solution to remove all traces of sulfate.

In potable waters there are no ions other than sulfate that will form insoluble compounds with barium under strongly acid conditions. Sample temperature control is important for reproducibility of results. Prepare calibration curve and analyze samples adjusted to the same temperature in the range 20 to 25 C. For conditioning reagent use reagents low in sulfates.

c. Minimum detectable concentration: Approximately 1 mg SO_4/L.

2. Apparatus

a. Magnetic stirrer: Use a constant stirring speed. It also is convenient to incorporate a fixed resistance in series with the motor operating the magnetic stirrer to regulate speed of stirring. Use magnets of identical shape and size. The exact speed of stirring is not critical, but keep constant for each run of samples and standards and adjust to about the maximum at which no splashing occurs.

b. Photometer: One of the following is required, with preference in the order given:

1) *Nephelometer.*

2) *Spectrophotometer,* for use at 420 nm, providing a light path of 4 to 5 cm.

3) *Filter photometer,* equipped with a violet filter having maximum transmittance near 420 nm and providing a light path of 4 to 5 cm.

c. Stopwatch or electric timer.

d. Measuring spoon, capacity 0.2 to 0.3 mL.

3. Reagents

a. Conditioning reagent: Mix 50 mL glycerol with a solution containing 30 mL conc HCl, 300 mL distilled water, 100 mL 95% ethyl or isopropyl alcohol, and 75 g NaCl.

b. Barium chloride, $BaCl_2$, crystals sized for turbidimetric work.* To ensure uniformity of results, construct a standard curve for each batch of $BaCl_2$ crystals.

*Baker No. 0974 or equivalent.

c. Standard sulfate solution: Prepare a standard sulfate solution as described in 1) or 2) below; 1.00 mL = 100 μg SO_4.

1) Dilute 10.41 mL standard 0.0200*N* H_2SO_4 titrant specified in Alkalinity, Section 403.3*c*, to 100 mL with distilled water.

2) Dissolve 147.9 mg anhydrous Na_2SO_4 in distilled water and dilute to 1,000 mL.

4. Procedure

a. Formation of barium sulfate turbidity: Measure 100 mL sample, or a suitable portion made up to 100 mL, into a 250-mL erlenmeyer flask. Add 5.00 mL conditioning reagent and mix in stirring apparatus. While stirring, add a spoonful of $BaCl_2$ crystals and begin timing immediately. Stir for 1.0 min at constant speed.

b. Measurement of barium sulfate turbidity: Immediately after stirring period has ended, pour solution into absorption cell of photometer and measure turbidity at 30-sec intervals for 4 min. Because maximum turbidity usually occurs within 2 min and readings remain constant thereafter for 3 to 10 min, consider turbidity to be the maximum reading obtained in the 4-min interval.

c. Preparation of calibration curve: Estimate sulfate concentration in sample by comparing turbidity reading with a calibration curve prepared by carrying sulfate standards through the entire procedure. Set photometer or nephelometer at zero sulfate concentration using distilled water. Determine reading on a distilled water control sample treated for sulfate analysis and, by subtraction, use this result to correct readings on standard sulfate and samples. Space standards at 5-mg/L increments in the 0- to 40-mg/L sulfate range. Above 40 mg/L the accuracy of the method decreases and the suspensions of $BaSO_4$ lose stability. Check reliability of calibration curve by running a standard with every three or four samples. Periodically inspect photometer or nephelometer sample cell for $BaSO_4$ deposition and keep cell clean.

d. Correction for sample color and turbidity: Correct for color and turbidity in the sample by running blanks from which the $BaCl_2$ is withheld.

5. Calculation

$$\text{mg } SO_4/\text{L} = \frac{\text{mg } SO_4 \times 1{,}000}{\text{mL sample}}$$

6. Precision and Accuracy

A synthetic sample containing 259 mg SO_4/L, 108 mg Ca/L, 82 mg Mg/L, 3.1 mg K/L, 19.9 mg Na/L, 241 mg Cl^-/L, 0.250 mg NO_2^--N/L, 1.1 mg NO_3^--N/L, and 42.5 mg total alkalinity/L (contributed by $NaHCO_3$) was analyzed in 19 laboratories by the turbidimetric method, with a relative standard deviation of 9.1% and a relative error of 1.2%.

426 D. Automated Methylthymol Blue Method (TENTATIVE)

1. General Discussion

a. Principle: Barium sulfate is formed by the reaction of the sulfate ion with barium chloride ($BaCl_2$) at a low pH. At high pH excess barium reacts with methylthymol blue to produce a blue chelate. The uncomplexed methylthymol blue is gray. The amount of gray uncomplexed methylthymol blue indicates the concentration of sulfate ion.

b. Interferences: Because many cations interfere, use an ion exchange column to remove interferences.

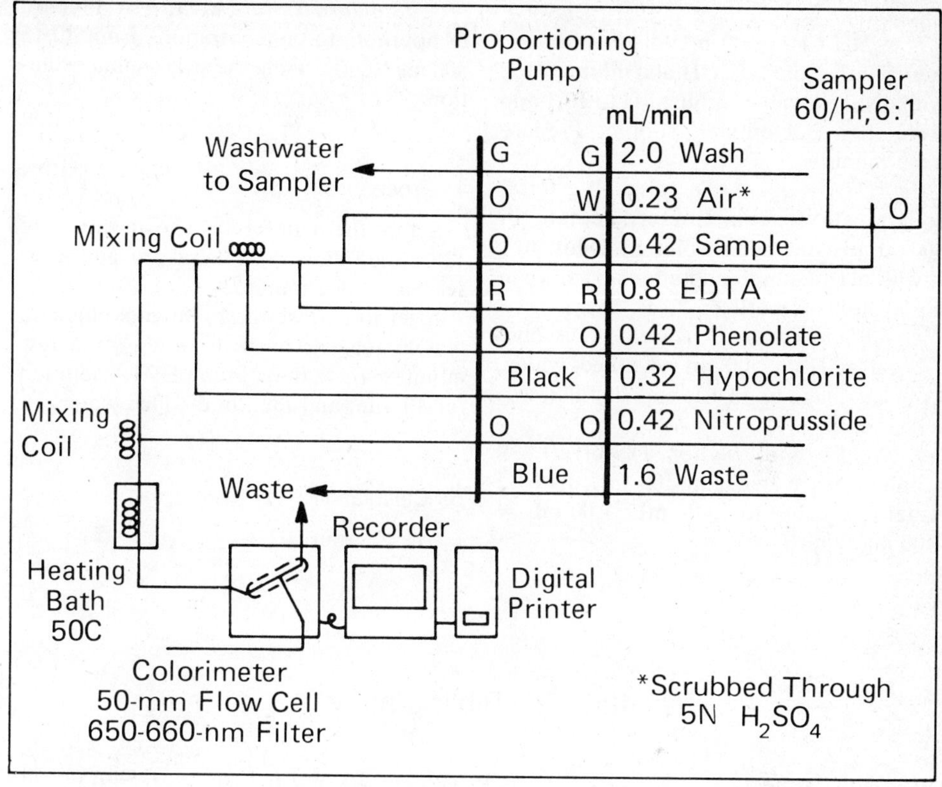

Figure 426:1. Sulfate manifold.

c. Application: This method is applicable to potable, surface, and saline waters as well as domestic and industrial wastewaters over a range from about 10 to 300 mg SO_4/L.

2. Apparatus

a. Automated analytical equipment, consisting of the components listed in Section 602.1 and 460-nm filters.

b. Ion exchange column: Fill a piece of 2-mm-ID glass tubing about 20 cm long with the ion exchange resin.* To simplify filling column put resin in distilled water and aspirate it into the tubing, which contains a glass wool plug. After filling, plug other end of tube with glass wool. Avoid trapped air in the column.

3. Reagents

a. Barium chloride solution: Dissolve 1.526 g $BaCl_2 \cdot 2H_2O$ in 500 mL distilled water and dilute to 1 L. Store in a polyethylene bottle.

b. Methylthymol blue reagent: Dissolve 118.2 mg methylthymol blue† in 25 mL $BaCl_2$ solution. Add 4 mL 1*N* HCl and 71 mL distilled water and dilute to 500 mL with ethanol. Store in a brown glass bottle. Prepare fresh daily.

*Ion exchange resin Bio-Rex 70, 20-50 mesh, sodium form, available from Bio-Rad Laboratories, Richmond, Calif. 94804, or equivalent.

†Eastman Organic Chemicals, Rochester, N.Y. 14615, No. 8068. 3′, 3″ Bis [N,N-bis(carboxymethyl)-aminol-methyl] thymolsulfonphthalein pentasodium salt.

c. Buffer solution, pH 10.1: Dissolve 6.75 g NH_4Cl in 500 mL distilled water. Add 57 mL conc NH_4OH and dilute to 1 L with distilled water. Adjust pH to 10.1 and store in a polyethylene bottle. Prepare fresh monthly.

d. EDTA reagent: Dissolve 40 g tetrasodium ethylenediaminetetraacetate in 500 mL pH 10.1 buffer solution. Dilute to 1 L with pH 10.1 buffer solution and store in a polyethylene bottle.

e. Sodium hydroxide solution: Dissolve 7.2 g NaOH in 500 mL distilled water. Cool and make up to 1 L with distilled water.

f. Stock sulfate solution: Dissolve 1.479 g anhydrous Na_2SO_4 in 500 mL distilled water and dilute to 1,000 mL; 1.00 mL = 1.00 mg SO_4.

g. Standard sulfate solutions: Prepare in appropriate concentrations from 10 to 300 mg SO_4/L, using the stock sulfate solution.

4. Procedure

Set up the manifold as shown in Figure 426:1 and follow the general procedure described in Section 602.2.

After use, rinse methylthymol blue and NaOH reagent lines in water for a few minutes, rinse them in the EDTA solution for 10 min, and then in distilled water.

5. Calculation

See Section 602.3.

426 E. Bibliography

Gravimetric Methods

HILLDBRAND, W.F. et al. 1953. Applied Inorganic Analysis, 2nd ed. John Wiley & Sons, New York, N.Y.

KOLTHOFF, I.M., E.J. MEEHAN, E.B. SANDELL & S. BRUCKENSTEIN. 1969. Quantitative Chemical Analysis, 4th ed. Macmillan Co., New York, N.Y.

Turbidimetric Method

SHEEN, R.T., H.L. KAHLER & E.M. ROSS. 1935. Turbidimetric determination of sulfate in water. *Ind. Eng. Chem.*, Anal. Ed. 7:262.

THOMAS, J.F. & J.E. COTTON. 1954. A turbidimetric sulfate determination. *Water Sewage Works* 101:462.

ROSSUM, J.R. & P. VILLARRUZ. 1961. Suggested methods for turbidimetric determination of sulfate in water. *J. Amer. Water Works Ass.* 53:873.

Automated Methyl Thymol Blue Method

LAZRUS, A.L., K.C. HILL & J. P. LODGE. 1965. A new colorimetric microdetermination of sulfate ion. *Automation Anal. Chem.* p. 291.

427 SULFIDE

1. General Discussion

Sulfide often is present in groundwater, especially in hot springs. Its common presence in wastewaters comes partly from the decomposition of organic matter, sometimes from industrial wastes, but mostly from the bacterial reduction of sulfate. Hydrogen sulfide escaping into the air from sulfide-containing wastewater causes odor nuisances. The threshold odor concentration of H_2S in clean water is between 0.025 and 0.25 μg/L. H_2S is very toxic and has claimed the lives of numerous workmen in sewers. It attacks metals directly and indirectly has caused serious corrosion of concrete sewers because it is oxidized biologically to H_2SO_4 on the pipe wall.

From an analytical standpoint, three categories of sulfide in water and wastewater are distinguished:

a. Total sulfide includes dissolved H_2S and HS^-, as well as acid-soluble metallic sulfides present in suspended matter. The S^{2-} is negligible, amounting to less than 0.5% of the dissolved sulfide at pH 12, less than 0.05% at pH 11, etc. Copper and silver sulfides are so insoluble that they do not respond in ordinary sulfide determinations; they can be ignored for practical purposes.

b. Dissolved sulfide is that remaining after suspended solids have been removed by flocculation and settling.

c. Un-ionized hydrogen sulfide may be calculated from the concentration of dissolved sulfide, the sample pH, and the practical ionization constant of H_2S.

2. Sampling and Storage

Take samples with minimum aeration. To preserve a sample for a total sulfide determination put zinc acetate solution into bottle before filling it with sample. Use 4 drops of 2*N* zinc acetate solution per 100 mL sample. Fill bottle completely and stopper.

3. Qualitative Tests

A qualitative test for sulfide often is useful. It is advisable in the examination of industrial wastes containing interfering substances that may give a false negative result in the methylene blue procedure.

a. Antimony test: To about 200 mL sample, add 0.5 mL saturated solution of potassium antimony tartrate and 0.5 mL 6*N* HCl in excess of phenolphthalein alkalinity.

Yellow antimony sulfide (Sb_2S_3) is discernible at a sulfide concentration of 0.5 mg/L. Comparisons with samples of known sulfide concentration make the technic roughly quantitative. The only known interferences are metallic ions such as lead, which hold the sulfide so firmly that it does not produce Sb_2S_3, and dithionite, which decomposes in acid solution to produce sulfide.

b. Silver-silver sulfide electrode test: The potential of a silver-silver sulfide electrode assembly relative to a reference electrode varies with the activity of the sulfide ion in solution. By correcting for the ion activity coefficient and pH, this potential estimates sulfide concentration. Standardize electrode frequently against a sulfide solution of known strength. An electrode of this type can be used as an endpoint indicator for titrating dissolved sulfide with a standard solution of a silver or lead salt, but slow response always is a problem.

c. Lead acetate paper and silver foil tests: Confirm odors attributed to H_2S with lead acetate paper. On exposure to the vapor of a slightly acidified sample, the paper becomes blackened by formation of PbS. A strip of silver foil is more sensitive than lead acetate paper. Clean the silver by dipping in NaCN solution and rinse. Silver is suitable particularly for long-time exposure in the vicinity of possible H_2S sources because black Ag_2S is permanent whereas PbS slowly oxidizes.

4. Selection of Quantitative Methods

Iodine reacts with sulfide in acid solution, oxidizing it to sulfur. A titration based on this reaction is an accurate method for determining sulfide at concentrations above 1 mg/L if interferences are absent and if loss of H_2S is avoided. The

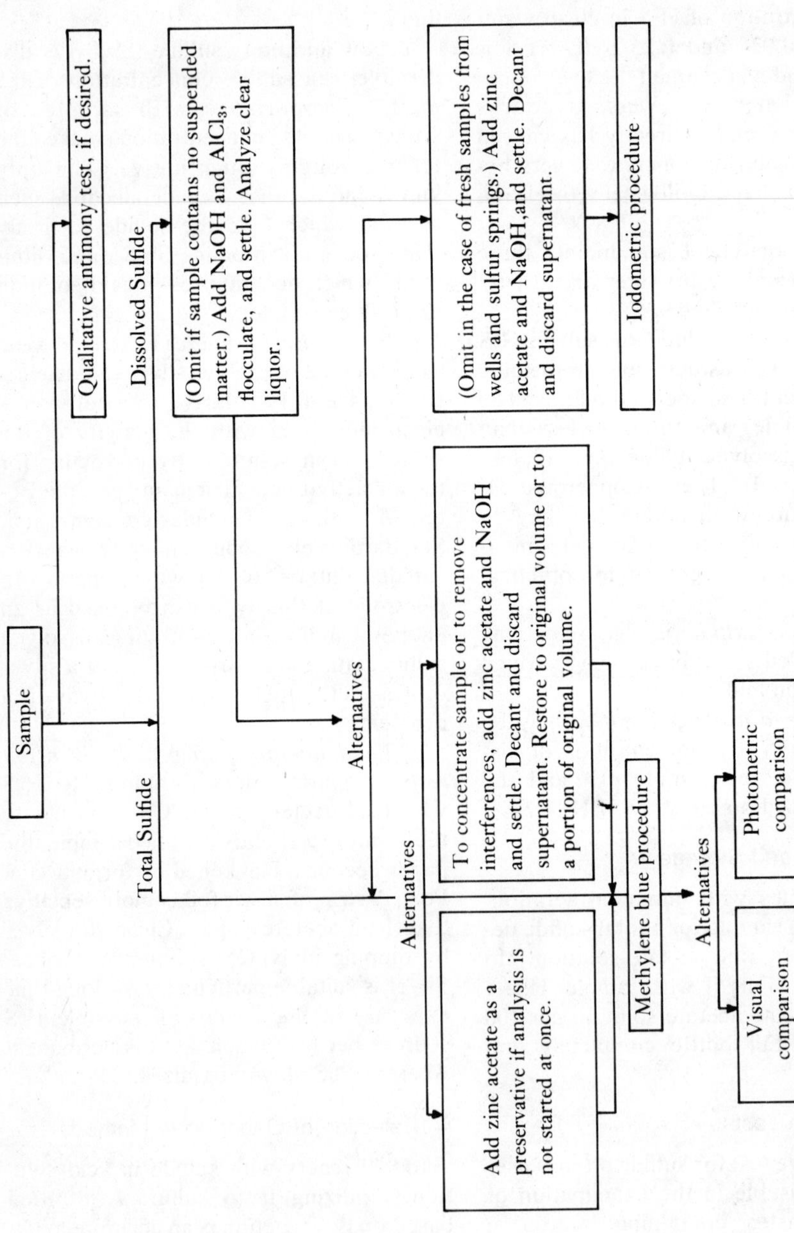

Figure 427:1. Analytical flow paths for sulfide determinations.

iodometric method (D) is useful for standardizing the methylene blue colorimetric method and is suitable for analyzing samples freshly taken from wells or springs. The method can be used for wastewater and partly oxidized water from sulfur springs if interfering substances are removed first.

The methylene blue method (C) is based on the reaction of sulfide, ferric chloride, and dimethyl-p-phenylenediamine to produce methylene blue. Ammonium phosphate is added after color development to remove ferric chloride color. The procedure is applicable at sulfide concentrations up to only 20 mg/L.

Potentiometric methods utilizing a silver electrode are suitable for certain purposes. From the potential of the electrode relative to a reference electrode an estimate can be made of the sulfide concentration, but careful attention to details of procedures and frequent standardizations are needed for securing good results. The electrode is useful particularly as an end-point indicator for titration of dissolved sulfide with silver nitrate.

Figure 427:1 shows analytical flow paths for sulfide determinations under various conditions and options.

427 A. Separation of Soluble and Insoluble Sulfides

Unless the sample is entirely free from suspended solids (dissolved sulfide equals total sulfide), to measure dissolved sulfide first remove insoluble matter. This can be done by producing an aluminum hydroxide floc that is settled, leaving a clear supernatant for analysis.

1. Apparatus

Glass bottles with stoppers. Use 100 mL if sulfide will be determined by the methylene blue method and 500 to 1,000 mL if by the iodometric method.

2. Reagents

a. Sodium hydroxide solution, NaOH, 6*N*.

b. Aluminum chloride solution, 6N: Because of the hygroscopic and caking tendencies of this chemical, purchase 100-g (or 1/4-lb) bottles of $AlCl_3{\cdot}6H_2O$. Dissolve contents of a previously unopened 100-g bottle in 144 mL distilled water (or contents of a 1/4-lb bottle in 164 mL water).

3. Procedure

a. To a 100-mL glass bottle add 0.2 mL (4 drops) 6*N* NaOH. Fill bottle with sample and add 0.2 mL (4 drops) 6*N* $AlCl_3$. Stopper bottle with no air under stopper. Rotate back and forth about a transverse axis vigorously for 1 min or longer to flocculate contents. Vary volumes of these chemicals to get good clarification without using excessively large amounts and to produce a pH of 6 to 9. If a 500- or 1,000-mL bottle is used, add proportionally larger amounts of reagents.

b. Let settle until reasonably clear supernatant can be drawn off. With proper flocculation, this may take 5 to 15 min. Do not wait longer than necessary.

427 B. Sample Pretreatment to Remove Interfering Substances or to Concentrate the Sulfide

The iodometric method suffers interference from reducing substances that react with iodine, including thiosulfate, sulfite, and various organic compounds, both solid and dissolved.

Strong reducing agents also interfere in the methylene blue test by preventing formation of the blue color. Thiosulfate at concentrations above 10 mg/L may retard color formation or completely prevent it. Sulfide itself prevents the reaction if its concentration is very high, in the range of several hundred milligrams per liter. To avoid the possibility of false negative results, use the antimony method to obtain a qualitative result in industrial wastes likely to contain sulfide but showing no color by the methylene blue method. Iodide, which is likely to be present in oilfield wastewaters, may diminish color formation if its concentration exceeds 2 mg/L. Ferrocyanide produces a blue color.

Eliminate interferences due to sulfite, thiosulfate, iodide, and many other soluble substances, but not ferrocyanide, by first precipitating ZnS, then removing the supernatant and replacing with distilled water. Use the same procedure, even when not needed for removal of interferences, to concentrate sulfide.

1. Apparatus

Glass bottles with stoppers (see Section 427A).

2. Reagents

a. Zinc acetate, 2N: Dissolve 220 g $Zn(C_2H_3O_2)_2 \cdot 2H_2O$ in 870 mL water; this makes 1 L solution.

b. Sodium hydroxide solution, NaOH, 6*N*.

3. Procedure

a. Put 0.15 mL (3 drops) 2*N* zinc acetate solution into a 100-mL glass bottle, fill with sample, and add 0.10 mL (2 drops) 6*N* NaOH solution. Stopper with no air bubbles under stopper and mix by rotating back and forth vigorously about a transverse axis. For the iodometric procedure, use a 500-mL bottle or other convenient size, with proportionally larger volumes of reagents. Vary volume of reagents added according to sample so that the resulting precipitate is not excessively bulky and settles readily. Add enough NaOH to produce a pH above 9. Let precipitate settle for 30 min. The treated sample is relatively stable and can be held for several hours. However, if much iron is present, oxidation may be fairly rapid.

b. If the iodometric method is to be used, filter precipitate through glass fiber filter paper and continue at once with titration according to the procedure of Section 427D. If the methylene blue method is used, let precipitate settle for 30 min and decant as much supernatant as possible without loss of precipitate. Refill bottle with distilled water, resuspend precipitate, and withdraw a sample. If interfering substances are present in high concentration, settle, decant, and refill a second time. If sulfide concentration is known to be low, add only enough water to bring volume to one-half or one-fifth of original volume. Use this technic for analyzing samples of very low sulfide concentrations. After determining the sulfide concentration colorimetrically, multiply the result by the ratio of final to initial volume.

Cadmium salts sometimes are used instead of zinc, but CdS is more susceptible to oxidation than ZnS.

427 C. Methylene Blue Method

1. Apparatus

a. Matched test tubes, approximately 125 mm long and 15 mm OD.

b. Droppers, delivering 20 drops/mL methylene blue solution. To obtain uniform drops hold dropper in a vertical position and let drops form slowly.

c. If photometric rather than visual color determination will be used, either;

1) *Spectrophotometer,* for use at a wavelength of 664 nm with cells providing light paths of 1 cm and 1 mm, or

2) *Filter photometer,* with a filter providing maximum transmittance near 600 nm.

2. Reagents

a. Amine-sulfuric acid stock solution: Dissolve 27 g N,N-dimethyl-*p*-phenylenediamine oxalate* in a cold mixture of 50 mL conc H_2SO_4 and 20 mL distilled water. Cool and dilute to 100 mL with distilled water. Use fresh oxalate because an old supply may be oxidized and discolored to a degree that results in interfering colors in the test. Store in a dark glass bottle. When this stock solution is diluted and used in the procedure with a sulfide-free sample, it first will be pink but then should become colorless within 3 min.

b. Amine-sulfuric acid reagent: Dilute 25 mL amine-sulfuric acid stock solution with 975 mL 1 + 1 H_2SO_4. Store in a dark glass bottle.

c. Ferric chloride solution: Dissolve 100 g $FeCl_3 \cdot 6H_2O$ in 40 mL water.

d. Sulfuric acid solution, H_2SO_4, 1 + 1.

e. Diammonium hydrogen phosphate solution: Dissolve 400 g $(NH_4)_2HPO_4$ in 800 mL distilled water.

f. Methylene blue solution I: Use USP grade dye or one certified by the Biological Stain Commission. The dye content should be reported on the label and should be 84% or more. Dissolve 1.0 g in distilled water and make up to 1 L. This solution will be approximately the correct strength, but because of variation between different lots of dye, standardize against sulfide solutions of known strength and adjust its concentration so that 0.05 mL (1 drop) = 1.0 mg sulfide/L.

Standardization—Put several grams of clean, washed crystals of $Na_2S \cdot 9H_2O$ into a small beaker. Add somewhat less than enough water to cover crystals. Stir occasionally for a few minutes, then pour solution into another vessel. This solution reacts slowly with oxygen but the change is unimportant within a few hours. Make solution daily. To 1 L distilled water add 1 drop of solution and mix. Immediately determine sulfide concentration by the methylene blue procedure and by the iodometric procedure. Repeat, using more than 1 drop Na_2S solution or smaller volumes of water, until at least five tests have been made, with a range of sulfide concentrations between 1 and 8 mg/L. Calculate average percent error of the methylene blue result as compared to the iodometric result. If the average error is negative, that is, methylene blue results are lower than iodometric results, dilute methylene blue solution by the same percentage, so that a greater volume will be used in matching colors. If methylene blue results are high, increase solution strength by adding more dye.

g. Methylene blue solution II: Dilute 10.00 mL of adjusted methylene blue solution I to 100 mL.

3. Procedure

a. Color development: Transfer 7.5 mL sample to each of two matched test tubes,

*Eastman catalog No. 5672 has been found satisfactory for this purpose.

using a special wide-tip pipet or filling to marks on test tubes. Add to Tube A 0.5 mL amine-sulfuric acid reagent and 0.15 mL (3 drops) $FeCl_3$solution. Mix immediately by inverting slowly, only once. (Excessive mixing causes low results by loss of H_2S as a gas before it has had time to react). To Tube B add 0.5 mL 1 + 1 H_2SO_4 and 0.15 mL (3 drops) $FeCl_3$ solution and mix. The presence of sulfide ion will be indicated by the appearance of blue color in Tube A. Color development usually is complete in about 1 min, but a longer time often is required for fading out of the initial pink color. Wait 3 to 5 min and add 1.6 mL $(NH_4)_2HPO_4$ solution to each tube. Wait 3 to 15 min and make color comparisons. If zinc acetate was used, wait at least 10 min before making a visual color comparison.

b. Color determination:

1) Visual color estimation—Add methylene blue solution I or II, depending on sulfide concentration and desired accuracy, dropwise, to the second tube, until color matches that developed in first tube. If the concentration exceeds 20 mg/L, repeat test with a portion of sample diluted to one tenth.

With methylene blue solution I, adjusted so that 0.05 mL (1 drop) = 1.0 mg sulfide/L when 7.5 mL of sample are used:

$$\text{mg sulfide/L} = \text{no. drops solution I} + 0.1\ (\text{no. drops solution II})$$

2) Photometric color measurement—A cell with a light path of 1 cm is suitable for measuring sulfide concentrations from 0.1 to 2.0 mg/L. Use shorter or longer light paths for higher or lower concentrations. The upper limit of the method is 20 mg/L. Zero instrument with a portion of treated sample from Tube B. Prepare calibration curves on basis of colorimetric tests made on Na_2S solutions simultaneously analyzed by the iodometric method, plotting concentration vs. absorbance. A straight-line relationship between concentration and absorbance can be assumed from 0 to 1.0 mg/L.

Read sulfide concentration from calibration curve.

4. Precision and Accuracy

The accuracy is about ±10%. The standard deviation has not been determined.

427 D. Iodometric Method

1. Reagents

a. Hydrochloric acid, HCl, 6*N*.

b. Standard iodine solution, 0.0250*N:* Dissolve 20 to 25 g KI in a little water and add 3.2 g iodine. After iodine has dissolved, dilute to 1,000 mL and standardize against 0.0250*N* $Na_2S_2O_3$, using starch solution as indicator.

c. Standard sodium thiosulfate solution, 0.0250*N:* See Section 421B.2*e*.

d. Starch solution: See Section 421B.2*d*.

2. Procedure

a. Measure from a buret into a 500-mL flask an amount of iodine solution estimated to be an excess over the amount of sulfide present. Add distilled water, if necessary, to bring volume to about 20 mL. Add 2 mL 6*N* HCl. Pipet 200 mL sample into flask, discharging sample under solution surface. If iodine color disappears, add more iodine so that color remains. Back-titrate with $Na_2S_2O_3$ solution, adding a few drops of starch solution as end point is approached, and continuing until blue color disappears.

b. If sulfide was precipitated with zinc and ZnS filtered out, return filter with precipitate to original bottle and add about 100 mL water. Add iodine solution and HCl and titrate as in ¶ 2*a* above.

3. Calculation

One milliliter 0.0250*N* iodine solution reacts with 0.4 mg sulfide:

$$\text{mg S/L} = \frac{[(A \times B) - (C \times D)] \times 16{,}000}{\text{mL sample}}$$

where:

A = mL iodine solution,
B = normality of iodine solution,
C = mL $Na_2S_2O_3$ solution, and
D = normality of $Na_2S_2O_3$ solution.

4. Precision

The precision of the end point varies with the sample. In clean waters it should be determinable within 1 drop, which is equivalent to 0.1 mg/L in a 200-mL sample.

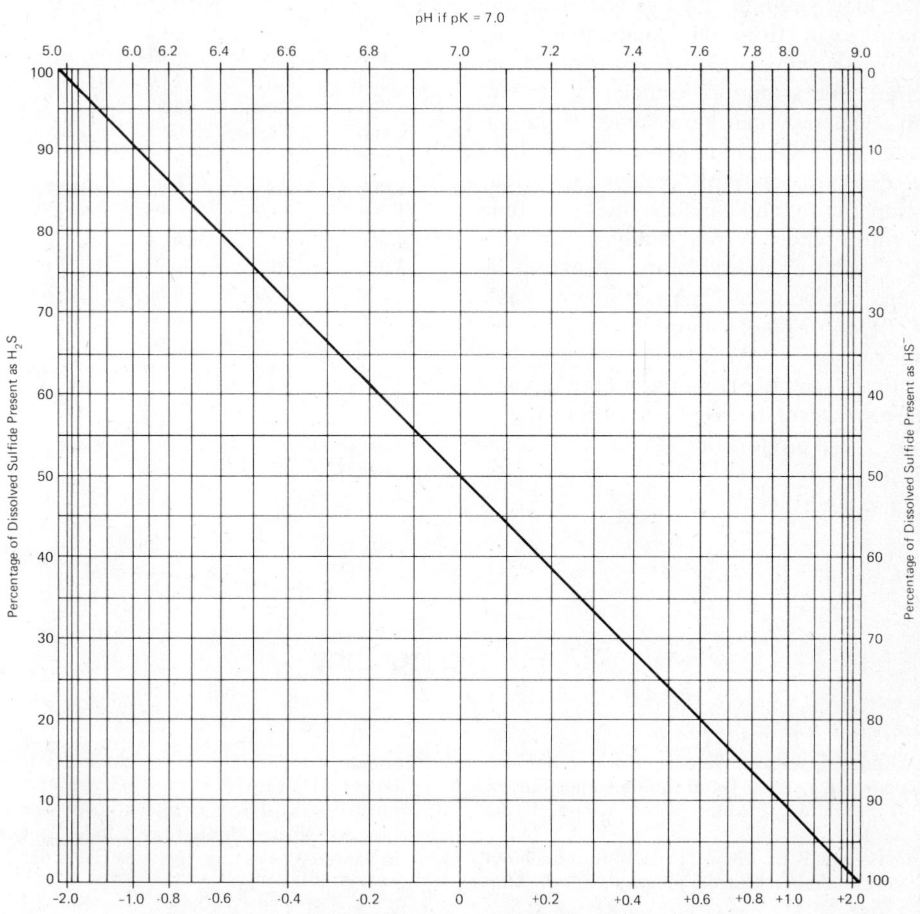

Figure 427:2. Proportions of H_2S and HS^- in dissolved sulfide.

427 E. Calculation of Un-ionized Hydrogen Sulfide

Hydrogen sulfide and HS^-, which together constitute dissolved sulfide, are in equilibrium with hydrogen ions:

$$H_2S \rightleftharpoons H^+ + HS^-$$

The ionization constant of H_2S is used to calculate the distribution of dissolved sulfide between the two forms. The practical constant written in logarithmic form, pK′, is used. The constant varies with temperature and ionic strength of the solution. The ionic strength effect can be estimated most easily from the conductivity. Because the effect of ionic strength is not large, values that are sufficiently dependable generally can be assumed if the nature of the sample is known. Table 427:I gives approximate pK′ values for various temperatures and conductivities. The temperature effect is practically linear from 15 C to 35 C; interpolations or extrapolations can be used. The last line of Table 427:I corresponds approximately to seawater.

From sample pH and appropriate value of pK′, calculate pH − pK′. From Figure 427:2 read proportions of dissolved sulfide present as H_2S (left-side scale of Figure 427:2). Let this proportion equal J,

$$J \times \text{(dissolved sulfide)} = \text{un-ionized } H_2S \text{ expressed as S}$$

TABLE 427:I. VALUES OF pK′, LOGARITHM OF PRACTICAL IONIZATION CONSTANT FOR HYDROGEN SULFIDE

Conductivity at 25 C *μmhos/cm*	pK′ at Given Temperature 20 C	25 C	30 C
0	—	7.03*	
100	7.08	7.01	6.94
200	7.07	7.00	6.93
400	7.06	6.99	6.92
700	7.05	6.98	6.91
1,200	7.04	6.97	6.90
2,000	7.03	6.96	6.89
3,000	7.02	6.95	6.88
4,000	7.01	6.94	6.87
5,200	7.00	6.93	6.86
7,200	6.99	6.92	6.85
10,000	6.98	6.91	6.84
14,000	6.97	6.90	6.83
22,000	6.96	6.89	6.82
50,000	6.95	6.88	6.81

* Theoretical.

427 F. Bibliography

POMEROY, R.D. 1936. The determination of sulfides in sewage. *Sewage Works J.* 8:572.

NUSBAUM, I. 1965. Determining sulfides in water and waste water. *Water Sewage Works* 112:113.

PLATFORD, R.F. 1965. The activity coefficient of sodium chloride in seawater. *J. Mar. Res.* 23:55.

CRUSE, H. & R.D. POMEROY. 1969. Hydrogen sulfide odor threshold. *J. Amer. Water Works Ass.* 61:677.

ENVIRONMENTAL PROTECTION AGENCY. 1974. Process Design Manual for Sulfide Control in Sanitary Sewerage Systems. EPA 625/1-74-005.

428 SULFITE

Sulfite ions may occur in boilers and boiler feedwaters treated with sulfite for dissolved oxygen control, in natural waters or wastewaters as a result of industrial pollution, and in treatment plant effluents dechlorinated with sulfur dioxide. Excess sulfite ion in boiler waters is deleterious because it lowers the pH and promotes corrosion. Control of sulfite ion in wastewater treatment and discharge may be important environmentally, principally because of its toxicity to fish and other aquatic life and its rapid oxygen demand.

1. General Discussion

a. Principle: An acidified water sample containing sulfite is titrated with a standardized potassium iodide-iodate titrant. Free iodine, liberated by the iodide-iodate reagent, reacts with the sulfite ion. The titration endpoint is signalled by the blue color resulting from the first excess of iodine reacting with a starch indicator.

b. Interferences: The presence of other oxidizable materials, such as sulfide, thiosulfate, and Fe(II) ions, can cause apparently high results for sulfite. Some metal ions, such as Cu(II), may catalyze the oxidation of sulfite to sulfate when the sample is exposed to air, thus leading to low results. Nitrite ion will react with sulfite ion in the acidic reaction medium and lead to low sulfite results unless sulfamic acid is added to destroy nitrite. Addition of EDTA as a complexing agent at the time of sample collection inhibits Cu(II) catalysis and promotes oxidation of Fe(II) to Fe(III) before analysis. Sulfide and thiosulfate ions normally would be expected only in samples containing certain industrial discharges, but must be accounted for if present. Sulfide may be removed by adding about 0.5 g zinc acetate and analysis of the supernatant of the settled sample. However, thiosulfate may have to be determined by an independent method (e.g. the formaldehyde/iodometric method[1]), and then the sulfite determined by difference.

c. Minimum detectable concentration: 2 mg SO_3^{2-}/L.

2. Reagents

a. Sulfuric acid: H_2SO_4, 1 + 1.

b. Standard potassium iodide-iodate titrant, 0.0125*N:* Dissolve 445.8 mg primary grade anhydrous KIO_3 (dried for 4 hr at 120 C), 4.35 g KI, and 310 mg sodium bicarbonate ($NaHCO_3$) in distilled water and dilute to 1,000 mL; 1.00 mL = 500 μg SO_3^{2-}.

c. Sulfamic acid, NH_2SO_3H, crystalline.

d. EDTA reagent: Dissolve 2.5 g EDTA in 100 mL distilled water.

e. Starch indicator: Use either of the starch indicators described below:

1) To 5 g starch (potato, arrowroot, or soluble) in a mortar, add a little cold distilled water and grind to a paste. Add mixture to 1 L boiling distilled water, stir, and let settle overnight. Use clear supernatant. Preserve by adding either 1.3 g salicylic acid, 4 g $ZnCl_2$, or a combination of 4 g sodium propionate and 2 g sodium azide to 1 L starch solution.

2) A dual-purpose sulfite indicator powder* composed of cold-water-soluble starch in a sulfamic acid medium is used extensively in control work. When this proprietary formulation is used, omit H_2SO_4 and add 3 to 4 drops phenolphthalein indicator solution to sample followed by sufficient dipperfuls (about 1 g) indicator powder to discharge the alkaline red color. Add a final dipperful in excess

*The use of this product, known as Dual-Purpose Sulfite Indicator Powder, is covered by U.S. Patent No. 2,963,443 issued to E.T. Erickson.

and titrate sample with standard KI-KIO_3 titrant to appearance of a permanent blue color.

3. Procedure

a. Sample collection: Collect a fresh water sample, taking care to minimize contact with air. Fix cooled samples (<50 C) immediately by adding 1 mL EDTA solution for each 100 mL of sample. Cool hot samples to 50 C or below in cooling apparatus depicted in Figure 107.1. Do not filter.

b. Titration: Add 1 mL H_2SO_4 and 0.1 g NH_2SO_3H crystals (or 1 g dual-purpose sulfite indicator) to a 250-mL erlenmeyer flask or other suitable titration vessel. Accurately measure 50 to 100 mL EDTA-stabilized sample into flask, keeping pipet tip below surface of liquid. If dual-purpose sulfite indicator is not used, add 1 mL starch indicator solution or 0.1 g starch powder. Titrate immediately with standard KI-KIO_3 titrant, while swirling flask, until a faint permanent blue color develops. Analyze a reagent blank using distilled water instead of sample.

4. Calculation

$$\text{mg } SO_3^{2-}/\text{L} = \frac{(A - B) \times N \times 40{,}000}{\text{mL sample}}$$

where:

A = mL titrant for sample,
B = mL titrant for blank, and
N = normality of KI-KIO_3 titrant.

5. Reference

1. KURTENACKER, A. 1924. The aldehyde-bisulfite reaction in mass analysis. *Z. Anal. Chem.*, 64:56.

PART 500

DETERMINATION OF ORGANIC CONSTITUENTS

501 INTRODUCTION

The analysis of organic matter in water and wastewater can be classified into two general types of measurements: those that seek to express either the total amount of organic matter or some fraction of the total in general terms and those that are specific for individual organic compounds. Both types of analyses are presented below.

Methods for total organic carbon, chemical oxygen demand, and carbon chloroform extract are used to assess the total amount of organics present. Gross fractions of the organic matter can be identified analytically, as in the measurement of BOD, which is an index of the biodegradable organics present, or oil and grease, which represents material extractable from a sample by a nonpolar solvent. Examples of specific tests for individual organic compounds or groups of compounds are measurements of pesticides, phenols, and synthetic detergents.

Analyses of organics are made to evaluate possible effects on health of water consumers, measure the efficiency of waste treatment processes, and assess the quality of receiving waters.

Determination of some organic compounds, including pesticides and trihalomethanes in drinking waters, is mandatory in the U.S. The list of controlled organics is lengthening, both for drinking water and waste discharges. Many of these analyses are beyond the instrumental capabilities of the average laboratory but there is a need for understanding the analytical procedures and their limitations by those water purveyors and waste dischargers who rely on outside analytical support.

In addition to correct sampling (see Section 105), take special precautions when analyses are performed by independent laboratories. Identify samples only by numbers that are not sequential with the sampling program. For example, do not always number a treated water sample one higher than the corresponding raw water sample. Random numbering, with accurate records, serves as a necessary check on the independent laboratory's competence in sample handling.

Reliable use of independent laboratories deserves the same quality assurance procedures observed for in-house analyses: replicate samples, samples with known additions, and blanks. Preparation of samples with known additions may be technically impossible. In such cases, consider using a mixture, in varying ratios, of several samples. Use the reported concentrations in the samples and the proportions in which they were mixed to calculate the expected concentration in the mixture. Apply this method to volatile compounds with extreme care.

The sampling, field treatment, preservation, and storage of samples taken for organic matter analysis are covered in detail in the individual introductions to the methods. If possible, analyze immediately because preservatives often interfere with the tests. Otherwise, store at a low temperature (4 C) to preserve most samples until the day following collection. Use chemical preservatives only when they are shown not to interfere with the examinations to be made. Never use preservatives for samples to be analyzed for BOD. When preservatives are used, add them to the sample bottle initially so that all portions are preserved as soon as collected.

No single method of preservation is entirely satisfactory; choose the preservative with due regard to the determinations that are to be made. All methods of preservation may be inadequate when applied to samples containing significant amounts of suspended matter.

502 METHANE

Methane (CH_4) is a colorless, odorless, tasteless combustible gas occasionally found in groundwaters. Escape of this gas from water may cause an explosive atmosphere not only in a utility's tanks, pumphouses, and other facilities, but also on the consumer's property, particularly where water is sprayed through poorly ventilated spaces such as public showers.

The explosive limits of CH_4 in air are 5 to 15% by volume. At sea level, a 5% CH_4 concentration in air theoretically could be reached in a poorly ventilated space sprayed with hot (68 C) water having a CH_4 concentration of only 0.7 mg/L. At higher water temperatures, the vapor pressure of water is so great that no explosive mixture can form. At lower barometric pressures, the theoretical hazardous concentration of methane in water will be reduced proportionately. In an atmosphere of N_2 or other inert gas, at least 12.8% O_2 must be present for there to be an explosion hazard.

Selection of method: The combustible-gas indicator method (A) offers the advantages of simplicity, speed, and great sensitivity. The volumetric method (B) can be made more accurate for concentrations of 4 to 5 mg/L and higher, but will not be satisfactory for very low concentrations. The volumetric method also can be applied to differentiate between CH_4 and other gases, as when a water supply is contaminated by liquid petroleum gas or other volatile combustible materials.

Methane also may be determined with the gas chromatograph as described in Sludge Digester Gas, Section 511. The gas chromatograph permits differentiation between H_2, CH_4, and/or its higher homologs.

502 A. Combustible-Gas Indicator Method

1. General Discussion

a. Principle: An equilibrium according to Henry's law is established between CH_4 in solution and the partial pressure of CH_4 in the gas phase above the solution. The partial pressure of CH_4 can be determined with a combustible-gas indicator. The operation of the instrument is based on the catalytic oxidation of a combustible gas on a heated platinum filament that is made a part of a Wheatstone bridge. The heat generated by the oxidation of the gas increases the electrical resistance of the filament. The resulting imbalance of the electrical circuit causes deflection of a milliammeter that may be calibrated in terms of percentage of CH_4 or percentage of the lower explosive limit of the gas sampled.

b. Interference: Small amounts of ethane usually are associated with CH_4 in natural gas and presumably would be present in water that contains methane. Hydrogen gas has been observed in well waters and would behave similarly to CH_4 in this procedure. Hydrogen sulfide may interfere if the pH of the water is low enough for an appreciable fraction of the total sulfide to exist in the un-ionized form. The vapors of combustible oils also may interfere. In general, these interferences are of no practical importance because primary interest is in calculating the explosion hazard to which all combustible gases and vapors contribute.

Interference due to H_2S can be reduced by the addition of solid NaOH to the container before sampling.

c. Minimum detectable concentration: The limit of sensitivity of the test is approximately 0.2 mg/L.

d. Sampling: If the water is supersaturated with CH_4, a representative sample cannot be obtained unless the water is under sufficient pressure to keep all of the gas dissolved. Operate wells for a period sufficient to insure sampling water coming directly from the aquifer. Representative samples can be expected only when the well is equipped with a pump operating at sufficient submergence to assure that no gas escapes from the water.

2. Apparatus

*a. Combustible-gas indicator:** Connect a three-way stopcock to the inlet to zero instrument on atmospheric air immediately before obtaining sample reading. For laboratory use, replace the suction bulb with a filter pump throttled to draw gas through the instrument at a rate of approximately 600 mL/min. See Figure 502:1.

b. Laboratory filter pump.

c. Glass bottle, 4-L (1 gal) fitted with a two-hole rubber stopper. Extend inlet tube to within 1 cm of bottom. End outlet tube approximately 1 cm below stopper. Use metal or glass tubes, each fitted with stopcocks or with short (approximately 5-

*Marketed under the following trade names: "Explosimeter," "Methane Gas Detector," and "Methane Tester," all manufactured by Mine Safety Appliance Company, Pittsburgh, Pa. 15235, and "J-W Combustible Gas Indicator," manufactured by Bacharach Instrument Co., Mountain View, Calif. 94043.

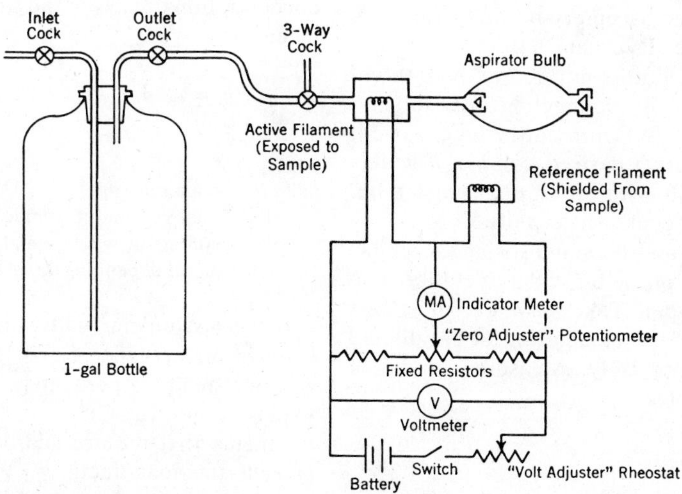

Figure 502:1. Combustible-gas indicator circuit and flow diagram.

cm) lengths of rubber tubing and pinchcocks. The entire assembly should be capable of holding a low vacuum for several hours. Determine volume of assembly by filling with water and measuring volume, or weight, of water contained.

3. Reagent

Sodium hydroxide, NaOH, pellets.

4. Procedure

a. Rough estimation of CH_4 concentration: Fill bottle about half full of water, using a rubber tube connecting sampling tap and inlet tube, with outlet tube open. With both inlet and outlet tubes closed, shake bottle vigorously for approximately 15 sec and let stand for approximately 1 min. Sample gas phase by withdrawing gas from the outlet, leaving inlet open to admit air. If the needle swings rapidly to a high level on the meter and then drops to zero, the CH_4-air mixture is too rich to burn; take a smaller sample for the final test. If needle deflection is too small to be read accurately, take a larger volume of water.

b. Accurate determination: If the water contains H_2S, add approximately 0.5 g NaOH pellets to empty bottle to suppress interference. Evacuate bottle, using filter pump. Fill bottle not more than three-quarters full by connecting inlet tube to sampling cock, with outlet tube closed. After collecting desired volume of water, let bottle fill with air through inlet tube. Close inlet cock, shake bottle vigorously for 60 sec, and let stand for at least 2 hr. Sample gas phase through outlet tube with inlet cock open. Take reading as rapidly as possible before the entering air has diluted sample appreciably. Measure volume of water sampled.

5. Calculation

The weight of CH_4 (w), in mg, in the sample is given by the equation:

$$w = P\left(\frac{0.257\ V_g}{T + 273} + \frac{890\ V_1}{H}\right)$$

where:

P = partial pressure of CH_4, torr,
T = temperature, C,
V_g = volume of gas phase, mL,
V_1 = volume of liquid phase, mL, and
H = Henry's law constant, torr/mole CH_4/mole of water.

Values for Henry's constant are as follows:

Temperature C	Henry's Constant 10^6H	Temperature C	Henry's Constant 10^6H
0	16.99	40	39.46
5	19.69	45	41.83
10	22.58	50	43.85
15	25.60	60	47.57
20	28.53	70	50.62
25	31.36	80	51.84
30	34.08	90	52.60
35	36.95	100	53.30

For most determinations, it may be assumed that atmospheric pressure is 760 torr and that the temperature is 20 C. The concentration of CH_4 in the sample is then given by:

$$\text{mg } CH_4/L = Rf\left(6.7\,\frac{V_0 - V_1}{V_1} + 0.24\right)$$

where:

R = scale reading,
V_0 = total volume of sample bottle, mL,
V_1 = volume of water sampled, mL, and
f = factor depending on instrument used.

If the instrument reads directly in percentage of methane, f = 1.00. If the instrument reads in percentage of the lower explosive limit of CH_4, f = 0.05. For instruments that require additional factors, consult the manufacturer. For example, one commercial instrument with a scale that reads in percentage of the lower explosive limit of combustible gases requires an additional factor of 0.77 for CH_4. Hence, the value of f in the above equation would be 0.77 × 0.05, or 0.0385.

For more accurate work, or in locations where normal barometric pressure is significantly lower than 760 torr use the equation:

$$\text{mg CH}_4\text{/L} = RBf\left(2.57\ \frac{V_0 - V_1}{TV_1} + \frac{8{,}900}{H}\right)$$

where:

B = barometric pressure, torr,

and other symbols are as above.

6. Accuracy

The accuracy of the determination is limited by the accuracy of the instrument used. Errors of approximately 10% may be expected. Calibration of instrument on known CH_4-air mixtures will improve accuracy.

502 B. Volumetric Method

1. General Discussion

a. Principle: If CH_4 is slowly mixed with an excess of O_2 in the presence of a platinum coil heated to yellow incandescence, most of the CH_4 will be converted to CO_2 and H_2O in a smooth reaction. Several passes of the mixed gases may be needed to burn substantially all the CH_4. An excess of O_2 is mixed with the sample before passage through the assembly.

b. Interference: Low-boiling hydrocarbons other than ethane and vapors from combustible oils interfere. These substances, however, are not likely to be present in water in sufficiently high concentration to affect the results significantly.

c. Minimum detectable concentration: This method is not satisfactory for determining CH_4 in water where the concentration is less than 2 mg/L.

d. Sampling: Collect sample as directed in Method A and observe the same precautions to obtain representative samples (Section 502A.1*d*). Omit NaOH pellets and fill sample bottle with water up to 90% of capacity.

Consult Section 511A for a description of apparatus, reagents, procedure, calculation, and precision and accuracy pertaining to the volumetric method.

Use percentage of CH_4 found by this method with Henry's law to obtain the CH_4 concentration in original sample. Substitute CH_4 percentage for R (scale reading) and $f = 1$ in the calculation given under Section 502A.5 preceding.

502 C. Bibliography

Combustible-Gas Indicator Method

ROSSUM, J.R., P.A. VILLARRUZ & J.A. WADE. 1950. A new method for determining methane in water. *J. Amer. Water Works Ass.* 42:413.

Volumetric Method

DENNIS, L.M. & M.L. NICHOLS. 1929. Gas Analysis. Macmillan Co., New York, N.Y.

HALDANE, J.S. & J.I. GRAHAM. 1935. Methods of Air Analysis. Charles Griffin & Co., London.

BUSWELL, A.M. & T.E. LARSON. 1937. Methane in ground waters. *J. Amer. Water Works Ass.* 29:1978.

BERGER, L.B. & H.H. SCHRENK. 1938. Bureau of Mines Haldane gas analysis apparatus. U.S. Bur. Mines Information Circ. No. 7017.

LARSON, T.E. 1938. Properties and determination of methane in ground waters. *J. Amer. Water Works Ass.* 30:1828.

503 OIL AND GREASE

In the determination of oil and grease, an absolute quantity of a specific substance is not measured. Rather, groups of substances with similar physical characteristics are determined quantitatively on the basis of their common solubility in trichlorotrifluoroethane. "Oil and grease" is any material recovered as a substance soluble in trichlorotrifluoroethane. It includes other material extracted by the solvent from an acidified sample (such as sulfur compounds, certain organic dyes, and chlorophyll) and not volatilized during the test. It is important that this limitation be understood clearly. Unlike some constituents that represent distinct chemical elements, ions, compounds, or groups of compounds, oils and greases are defined by the method used for their determination.

The methods presented here are suitable for biological lipids and mineral hydrocarbons. They also may be suitable for most industrial wastewaters or treated effluents containing these materials, although sample complexity may result in either low or high results because of lack of analytical specificity. The method is not applicable to measurement of low-boiling fractions that volatilize at temperatures below 70 C.

1. Significance

Certain constituents measured by the oil and grease analysis may influence wastewater treatment systems. If present in excessive amounts, they may interfere with aerobic and anaerobic biological processes and lead to decreased wastewater treatment efficiency. When discharged in wastewater or treated effluents, they may cause surface films and shoreline deposits leading to environmental degradation.

A knowledge of the quantity of oil and grease present is helpful in proper design and operation of wastewater treatment systems and also may call attention to certain treatment difficulties.

2. Selection of Method

For liquid samples, three methods are presented: the partition-gravimetric method (A), the partition-infrared method (B), and the Soxhlet method (C). Method B is designed for samples that might contain volatile hydrocarbons that otherwise would be lost in the solvent removal operations of the gravimetric procedure. Method C is the method of choice when relatively polar, heavy petroleum fractions are present, or when the levels of nonvolatile greases may challenge the solubility limit of the solvent. For low levels of oil and grease (<10 mg/L), Method B is the method of choice because gravimetric methods do not provide the needed precision.

Method D is a modification of the Soxhlet Method and is suitable for sludges and similar materials. Method E can be used in conjunction with Methods A, B, C, or D to obtain a hydrocarbon measurement in addition to, or instead of, the oil and grease measurement. This method separates hydrocarbons from the total oil and grease on the basis of polarity.

3. Sampling and Storage

Collect a representative sample in a wide-mouth glass bottle that has been rinsed with the solvent to remove any detergent film, and acidify in the sample bottle. Collect a separate sample for an oil and grease determination and do not subdivide in the laboratory. When information is required about average grease concentration over an extended period, examine individual portions collected at prescribed time intervals to eliminate losses of grease on sampling equipment during collection of a composite sample.

In sampling sludges, take every possible precaution to obtain a representative sample. When analysis cannot be made immediately, preserve samples with 1 mL conc HCl/80 g sample. Never preserve samples with $CHCl_3$ or sodium benzoate.

503 A. Partition-Gravimetric Method

1. General Discussion

a. Principle: Dissolved or emulsified oil and grease is extracted from water by intimate contact with trichlorotrifluoroethane. Some extractables, especially unsaturated fats and fatty acids, oxidize readily; hence, special precautions regarding temperature and solvent vapor displacement are included to minimize this effect.

b. Interference: Trichlorotrifluoroethane has the ability to dissolve not only oil and grease but also other organic substances. No known solvent will selectively dissolve only oil and grease. Solvent removal results in the loss of short-chain hydrocarbons and simple aromatics by volatilization. Significant portions of petroleum distillates from gasoline through No. 2 fuel oil are lost in this process. In addition, heavier residuals of petroleum may contain a significant portion of materials that are not extractable with the solvent.

2. Apparatus

a. Separatory funnel, 1 L, with TFE* stopcock.

b. Distilling flask, 125 mL.

c. Water bath.

d. Filter paper, 11 cm diam.†

3. Reagents

a. Hydrochloric acid, HCl, 1 + 1.

b. Trichlorotrifluoroethane‡ (1,1,2-trichloro-1,2,2-trifluoroethane), boiling point 47 C. The solvent should leave no measurable residue on evaporation; distill if necessary. Do not use any plastic tubing to transfer solvent between containers.

c. Sodium sulfate, Na_2SO_4, anhydrous crystal.

4. Procedure

Collect about 1 L of sample and mark sample level in bottle for later determination of sample volume. Acidify to pH 2 or lower; generally, 5 mL HCl is sufficient. Transfer to a separatory funnel. Carefully rinse sample bottle with 30 mL trichlorotrifluoroethane and add solvent washings to separatory funnel. Preferably shake vigorously for 2 min. However, if it

*Teflon or equivalent.

†Whatman No. 40 or equivalent.

‡Freon or equivalent.

is suspected that a stable emulsion will form, shake gently for 5 to 10 min. Let layers separate. Drain solvent layer through a funnel containing solvent-moistened filter paper into a clean, tared distilling flask. If a clear solvent layer cannot be obtained, add 1 g Na_2SO_4 to the filter paper cone and slowly drain emulsified solvent onto the crystals. Add more Na_2SO_4 if necessary. Extract twice more with 30 mL solvent each but first rinse sample container with each solvent portion. Combine extracts in tared distilling flask and wash filter paper with an additional 10 to 20 mL solvent. Distill solvent from distilling flask in a water bath at 70 C. Place flask on a water bath at 70 C for 15 min and draw air through it with an applied vacuum for the final 1 min. Cool in a desiccator for 30 min and weigh.

5. Calculation

If the organic solvent is free of residue, the gain in weight of the tared distilling flask is mainly due to oil and grease. Total gain in weight, *A*, of tared flask less calculated residue, *B*, from solvent blank is the amount of oil and grease in the sample:

$$\text{mg oil and grease/L} = \frac{(A - B) \times 1{,}000}{\text{mL sample}}$$

6. Precision and Accuracy

Methods A, B, and C were tested by a single laboratory on a sewage sample. By this method the oil and grease concentration was 12.6 mg/L. When 1-L portions of the sewage were dosed with 14.0 mg of a mixture of No. 2 fuel oil and Wesson oil, recovery of added oils was 93% with a standard deviation of 0.9 mg.

503 B. Partition-Infrared Method (TENTATIVE)

1. General Discussion

a. Principle: Although the extraction procedure for this method is identical to that of Method A, infrared detection permits the measurement of many relatively volatile hydrocarbons. Thus, the lighter petroleum distillates, with the exception of gasoline, may be measured accurately. Adequate instrumentation allows for the measurement of as little as 0.2 mg oil and grease/L.

b. Interference: Some degree of selectivity is offered by this method to overcome some of the coextracted interferences discussed in Method A. Heavier residuals of petroleum may contain a significant portion of materials insoluble in trichlorotrifluoroethane.

c. Definitions: A "known oil" is defined as a sample of oil and/or grease that represents the only material of that type used or manufactured in the processes represented by a wastewater. An "unknown oil" is defined as one for which a representative sample of the oil or grease is not available for preparation of a standard.

2. Apparatus

a. Separatory funnel, 1 L, with TFE* stopcock.

b. Infrared spectrophotometer, double beam, recording.

c. Cells, near-infrared silica.

d. Filter paper, 11 cm diam.†

3. Reagents

a. Hydrochloric acid, HCl, 1 + 1.

b. Trichlorotrifluoroethane. See 503A.3*b*.

c. Sodium sulfate, Na_2SO_4, anhydrous, crystal.

*Teflon or equivalent.
†Whatman No. 40 or equivalent.

d. Reference oil: Prepare a mixture, by volume, of 37.5% iso-octane, 37.5% hexadecane, and 25% benzene. Store in sealed container to prevent evaporation.

4. Procedure

Refer to Method A for sample collection, acidification, and extraction. Collect combined extracts in a 100-mL volumetric flask and adjust final volume to 100 mL with solvent.

Prepare a stock solution of known oil by rapidly transferring about 1 mL (0.5 to 1.0 g) of the oil or grease to a tared 100-mL volumetric flask. Stopper flask and weigh to nearest milligram. Add solvent to dissolve and dilute to mark. If the oil identity is unknown (¶ 1*c*) use the reference oil (¶ 3*d*) as the standard. Using volumetric technics, prepare a series of standards over the range of interest. Select a pair of matched near-infrared silica cells. A 1-cm-path-length cell is appropriate for a working range of about 4 to 40 mg. Scan standards and samples from 3,200 cm^{-1} to 2,700 cm^{-1} with solvent in the reference beam and record results on absorbance paper. Measure absorbances of samples and standards by constructing a straight baseline over the scan range and measuring absorbance of the peak maximum at 2,930 cm^{-1} and subtracting baseline absorbance at that point. If the absorbance exceeds 0.8 for a sample, select a shorter pathlength or dilute as required. Use scans of standards to prepare a calibration curve.

5. Calculation

$$\text{mg oil and grease/L} = \frac{A \times 1{,}000}{\text{mL sample}}$$

where:

A = mg of oil or grease in extract as determined from calibration curve.

6. Precision and Accuracy

See 503A.6. By this method the oil and grease concentration was 17.5 mg/L. When 1-L portions of the sewage were dosed with 14.0 mg of a mixture of No. 2 fuel oil and Wesson oil, the recovery of added oils was 99% with a standard deviation of 1.4 mg.

503 C. Soxhlet Extraction Method

1. General Discussion

a. Principle: Soluble metallic soaps are hydrolyzed by acidification. Any oils and solid or viscous grease present are separated from the liquid samples by filtration. After extraction in a Soxhlet apparatus with trichlorotrifluoroethane, the residue remaining after solvent evaporation is weighed to determine the oil and grease content. Compounds volatilized at or below 103 C will be lost when the filter is dried.

b. Interference: The method is entirely empirical and duplicate results can be obtained only by strict adherence to all details. By definition, any material recovered is oil and grease and any filtrable trichlorotrifluoroethane-soluble substances, such as elemental sulfur and certain organic dyes, will be extracted as oil and grease. The rate and time of extraction in the Soxhlet apparatus must be exactly as directed because of varying solubilities of different greases. In addition, the length of time required for drying and cooling extracted material cannot be varied. There may be a gradual increase in weight, presumably due to the absorption of oxygen, and/or a gradual loss of weight due to volatilization.

2. Apparatus

a. Extraction apparatus. Soxhlet.

b. Vacuum pump or other source of vacuum.

c. Buchner funnel, 12 cm.

d. Electric heating mantle.

e. Extraction thimble, paper.

f. Filter paper, 11 cm diam.*

g. Muslin cloth disks, 11 cm diam.

3. Reagents

a. Hydrochloric acid, HCl, 1 + 1.

b. Trichlorotrifluoroethane: See 503A.3*b*.

c. Diatomaceous-silica filter aid suspension,† 10 g/L distilled water.

4. Procedure

Collect about 1 L of sample in a wide-mouth glass bottle and mark sample level in bottle for later determination of sample volume. Acidify to pH 2 or lower; generally, 5 mL HCl is sufficient. Prepare a filter consisting of a muslin cloth disk overlaid with filter paper. Wet paper and muslin and press down edges of paper. Using a vacuum, pass 100 mL filter aid suspension through prepared filter and wash with 1 L distilled water. Apply vacuum until no more water passes filter. Filter acidified sample. Apply vacuum until no more water passes through filter. Using forceps, transfer filter paper to a watch glass. Add material adhering to edges of muslin cloth disk. Wipe sides and bottom of collecting vessel and Buchner funnel with pieces of filter paper soaked in solvent, taking care to remove all films caused by grease and to collect all solid material. Add pieces of filter paper to filter paper on watch glass. Roll all filter paper containing sample and fit into a paper extraction thimble. Add any pieces of material remaining on watch glass. Wipe watch glass with a filter paper soaked in solvent and place in paper extraction thimble. Dry filled thimble in a hot-air oven at 103 C for 30 min. Fill thimble with glass wool or small glass beads. Weigh extraction flask. Extract oil and grease in a Soxhlet apparatus, using trichlorotrifluoroethane at a rate of 20 cycles/hr for 4 hr. Time from first cycle. Distill solvent from extraction flask in a water bath at 70 C. Place flask on a water bath at 70 C for 15 min and draw air through it using an applied vacuum for the final 1 min. Cool in a desiccator for 30 min and weigh.

*Whatman No. 40 or equivalent.

†Hyflo Super-Cel, Johns-Manville Corp., or equivalent.

5. Calculation

See Section 503A.5.

6. Precision and Accuracy

See Section 503A.6. By this method the oil and grease concentration was 14.8 mg/L. When 1-L portions of the sewage were dosed with 14.0 mg of a mixture of No. 2 fuel oil and Wesson oil, the recovery of added oils was 88% with a standard deviation of 1.1 mg.

503 D. Extraction Method for Sludge Samples

1. General Discussion

a. Principle: Drying acidified sludge by heating leads to low results. Magnesium sulfate monohydrate is capable of combining with 75% of its own weight in water in forming $MgSO_4 \cdot 7H_2O$ and is used to dry sludge. After drying, the oil and grease can be extracted with trichlorotrifluoroethane.

b. Interference: See 503C.1*b*.

2. Apparatus

a. Extraction apparatus, Soxhlet.

b. Vacuum pump or other source of vacuum.

c. Extraction thimble, paper.

d. Grease-free cotton: Extract nonabsorbent cotton with solvent.

3. Reagents

a. Hydrochloric acid, HCl, conc.

b. Magnesium sulfate monohydrate: Prepare $MgSO_4 \cdot H_2O$ by overnight drying of a thin layer in an oven at 150 C.

c. Trichlorotrifluoroethane: See 503A.3*b*.

4. Procedure

In a 150-mL beaker weigh a sample of wet sludge, 20 ± 0.5 g, of which the dry-solids content is known. Acidify to pH 2.0 (generally, 0.3 mL conc HCl is sufficient). Add 25 g $MgSO_4 \cdot H_2O$. Stir to a smooth paste and spread on sides of beaker to facilitate subsequent removal. Let stand until solidified, 15 to 30 min. Remove solids and grind in a porcelain mortar. Add the powder to a paper extraction thimble. Wipe beaker and mortar with small pieces of filter paper moistened with solvent and add to thimble. Fill thimble with glass wool or small glass beads. Extract in a Soxhlet apparatus, using trichlorotrifluoroethane, at a rate of 20 cycles/hr for 4 hr. If any turbidity or suspended matter is present in the extraction flask, remove by filtering through grease-free cotton into another weighed flask. Rinse flask and cotton with solvent. Distill solvent from extraction flask in water at 70 C. Place flask on a water bath at 70 C for 15 min and draw air through it using an applied vacuum for the final 1 min. Cool in a desiccator for 30 min and weigh.

5. Calculation

$$\text{Oil and grease as \% of dry solids} = \frac{\text{gain in weight of flask, g} \times 100}{\text{weight of wet solids, g} \times \text{dry solids fraction}}$$

6. Precision

The examination of six replicate samples of sludge yielded a standard deviation of 4.6%.

503 E. Hydrocarbons

1. Significance

In the absence of specially modified industrial products, oil and grease is composed primarily of fatty matter from animal and vegetable sources and hydrocarbons of petroleum origin. A knowledge of the percentage of each of these constituents in the total oil and grease minimizes the difficulty in determining the major source of the material and simplifies the correction of oil and grease problems in wastewater treatment plant operation and stream pollution abatement.

2. General Discussion

a. Principle: Silica gel has the ability to absorb polar materials. If a solution of hydrocarbons and fatty materials in trichlorotrifluoroethane is mixed with silica gel, the fatty acids are selectively removed from solution. The materials not eliminated by silica gel adsorption are designated hydrocarbons by this test.

b. Interference: The more polar hydrocarbons, such as complex aromatic compounds and hydrocarbon derivatives of chlorine, sulfur, and nitrogen, may be adsorbed by the silica gel. Any compounds other than hydrocarbons and fatty matter recovered by the procedures for the determination of oil and grease also interfere.

3. Reagents

a. Trichlorotrifluoroethane: See 503A.3*b*.

b. Silica gel, 60 to 200 mesh.* Dry at 110 C for 24 hr and store in a tightly sealed container.

4. Procedure

Use the oil and grease extracted by Method A, B, C, or D for this test. When only hydrocarbons are of interest, introduce this procedure in any of the previous methods before final measurement. When hydrocarbons are to be determined after total oil and grease has been measured, redissolve, if necessary, the extracted oil and grease in trichlorotrifluoroethane. To 100 mL solvent containing less than 100 mg fatty material, add 3.0 g silica gel. Stopper container and stir on a magnetic stirrer for 5 min. For infrared measurement of hydrocarbons no further treatment is required before measurement as described in Method B. For gravimetric determinations, filter solution through filter paper and wash silica gel and filter paper with 10 mL solvent and combine with filtrate before performing the solvent stripping steps outlined in Methods A, C, or D.

*Davidson Grade 950 or equivalent.

5. Calculation

Calculate hydrocarbon concentration, in milligrams per liter, as in oil and grease (Method A, B, C, or D).

6. Precision and Accuracy

The accuracy of this determination cannot be measured directly in wastewaters. The following data, obtained on synthetic samples, are indicative for natural animal, vegetable, and mineral products, but cannot be applied to the specialized industrial products previously discussed.

For hydrocarbon determinations on 10 synthetic solvent extracts containing known amounts of a wide variety of petroleum products, average recovery was 97.2%. Similar synthetic extracts of Wesson oil, olive oil, Crisco, and butter gave 0.0% recovery as hydrocarbons measured by infrared analysis.

503 F. Bibliography

HATFIELD, W.D. & G.E. SYMONS. 1945. The determination of grease in sewage. *Sewage Works J.* 17:16.

KIRSCHMAN, H.D. & R. POMEROY. 1949. Determination of oil in oil field waste waters. *Anal. Chem.* 21:793.

GILCREAS, F.W., W.W. SANDERSON & R.P. ELMER. 1953. Two new methods for the determination of grease in sewage. *Sewage Ind. Wastes* 25:1379.

ULLMANN, W.W. & W.W. SANDERSON. 1959. A further study of methods for the determination of grease in sewage. *Sewage Ind. Wastes* 31:8.

CHANIN, G., E.H. CHOW, R.B. ALEXANDER & J.F. POWERS. 1967. A safe solvent for oil and grease analyses. *J. Water Pollut. Control Fed.* 39:1892.

GRUENFELD, M. 1973. Extraction of dispersed oils from water for quantitative analysis by infrared spectrophotometry. *Environ. Sci. Technol.* 7:636.

504 ORGANIC ACIDS AND VOLATILE ACIDS

The measurement of organic acids either by adsorption and elution from a chromatographic column or by distillation (with or without steam) can be used as a control test for anaerobic digestion. The chromatographic separation method is presented for organic acids (Section 504A), while methods using distillation are presented for volatile acids.

Volatile fatty acids are classified as water-soluble fatty acids that can be distilled at atmospheric pressure. These volatile acids can be removed from aqueous solution by steam distillation, in spite of their high boiling points, because of their high vapor pressure. This group includes water-soluble fatty acids with up to six carbon atoms.

Selection of method: The methods are straight distillation and steam distillation. The straight distillation method is empirical and gives incomplete and somewhat variable recovery. It is suitable for routine control purposes. The steam distillation method is more tedious, taking some 4 hr to complete, but gives 92 to 98% recovery of the volatile acids from sewage sludge. The steam distillation is conducted on a magnesium sulfate- ($MgSO_4$-) saturated, acidified, sludge-free liquor produced by chemical treatment with ferric chloride ($FeCl_3$) and filter aid followed by vacuum filtration. By this technic, the effects of varying concentrations of dissolved solids and mineral acid during distillation are avoided. Sludge separation before distillation also reduces the possibility of hydrolysis of complex materials to volatile acids.

504 A. Chromatographic Separation Method for Organic Acids

1. General Discussion

a. Principle: An acidified aqueous sample containing organic acids is adsorbed on a column of silicic acid and the acids are eluted with *n*-butanol in chloroform ($CHCl_3$). The eluate is collected and titrated with standard base. All short-chain (C_1 to C_6) organic acids are eluted by this solvent system and are reported collectively as total organic acids.

b. Interference: The $CHCl_3$-butanol solvent system is capable of eluting organic acids other than the volatile acids and also some synthetic detergents. Besides the so-called volatile acids, crotonic, adipic, pyruvic, phthalic, fumaric, lactic, succinic, malonic, gallic, aconitic, and oxalic acids; alkyl sulfates; and alkyl-aryl sulfonates are adsorbed by silicic acid and eluted.

c. Precautions: Basic alcohol solutions decrease in strength with time, particularly when exposed repeatedly to the atmosphere. These decreases usually are accompanied by the appearance of a white precipitate. The magnitude of such changes normally is not significant in ordinary process control if tests are made within a few days of standardization. To minimize this effect, store standard sodium hydroxide (NaOH) titrant in a tightly stoppered borosilicate glass bottle and protect from atmospheric carbon dioxide (CO_2) by attaching a tube of CO_2-absorbing material, as described in the inside front cover. For more precise analyses, standardize titrant or prepare before each analysis.

Although the procedure is adequate for routine analysis of most sludge samples, volatile-acids concentrations above 5,000 mg/L may require an increased amount of organic solvent for quantitative recovery. Elute with a second portion of solvent and titrate to reveal possible incomplete recoveries.

2. Apparatus

a. Centrifuge or filtering assembly.

b. Crucibles, Gooch or medium-porosity fritted-glass, with filtering flask and vacuum source. Use crucibles of sufficient size (30 to 35 mL) to hold 12 g silicic acid.

c. Separatory funnel, 1,000 mL.

3. Reagents

a. Silicic acid, specially prepared for chromatography, 50 to 200 mesh: Remove fines by slurrying in distilled water and decanting supernatant after settling for 15 min. Repeat several times. Dry washed acid in an oven at 103 C until *absolutely dry,* then store in a desiccator.

b. Chloroform-butanol reagent: Mix 300 mL reagent-grade $CHCl_3$, 100 mL *n*-butanol, and 80 mL 0.5*N* H_2SO_4 in a separatory funnel. Let water and organic layers separate. Drain off lower organic layer through a fluted filter paper into a dry bottle.

c. Thymol blue indicator solution: Dissolve 80 mg thymol blue in 100 mL absolute methanol.

d. Phenolphthalein indicator solution: Dissolve 80 mg phenolphthalein in 100 mL absolute methanol.

e. Sulfuric acid, H_2SO_4, conc.

f. Standard sodium hydroxide, NaOH, 0.02*N*: Dilute 20 mL 1.0*N* NaOH stock solution to 1 L with absolute methanol. Prepare stock in water and standardize in accordance with the methods outlined in Section 402.3*d*.

4. Procedure

a. Pretreatment of sample: Centrifuge or vacuum-filter enough sludge to obtain 10 to 15 mL clear sample in a small test tube or beaker. Add a few drops of thymol blue indicator solution, then conc H_2SO_4 dropwise, until definitely red to thymol blue (pH = 1.0 to 1.2).

b. Column chromatography: Place 12 g silicic acid in a Gooch or fritted-glass crucible and apply suction to pack column. Tamp column while applying suction to reduce channeling when the sample is applied. With a pipet, distribute 5.0 mL of acidified sample as uniformly as possible over column surface. Apply suction momentarily to draw sample into silicic acid. Release vacuum as soon as last portion of sample has entered column. Quickly add 65 mL $CHCl_3$-butanol reagent and apply suction. Discontinue suction just before the last of reagent enters column. Use a new column for each sample.

c. Titration: Remove filter flask and purge eluted sample with N_2 gas or CO_2-free air immediately before titrating. (Obtain CO_2-free air by passing air through a CO_2 absorbant.*)

Titrate sample with standard 0.02*N* NaOH to phenolphthalein end point, using a fine-tip buret and taking care to avoid aeration. The fine-tip buret aids in improving accuracy and precision of the titration. Use N_2 gas or CO_2-free air delivered through a small glass tube to purge and mix sample and to prevent contact with atmospheric CO_2 during titration.

d. Blank: Carry a distilled water blank through steps ¶s 4*a* through 4*c*.

5. Calculation

$$\text{Total organic acids (mg as acetic acid/L)} = \frac{(a - b) \times N \times 60{,}000}{\text{mL sample}}$$

where:

a = mL NaOH used for sample,
b = mL NaOH used for blank, and
N = normality of NaOH.

*Ascarite or equivalent.

6. Precision

Average recoveries of about 95% are obtained for organic acid concentrations above 200 mg as acetic acid/L. Individual tests generally vary from the average by approximately 3%. A greater variation results when lower concentrations of organic acids are present. Titration precision expressed as the standard deviation is about ±0.1 mL (approximately ±24 mg as acetic acid/L).

504 B. Steam Distillation Method for Volatile Acids

1. Apparatus

The required apparatus listed below is shown in Figure 504:1.

a. Steam generator: 1-L capacity, with a safety tube and a steam bypass valve.

b. Distillation flask, 1-L.

c. Condenser, at least 76 cm long.

d. Receiving flask, provided with a soda-lime tube to protect distillate from air.

e. Buchner funnel, 14 cm diam with suction.

2. Reagents

a. Sulfuric acid, H_2SO_4, 1 + 1.

b. Ferric chloride solution: Dissolve 82.5 g $FeCl_3 \cdot 6H_2O$ in 1 L distilled water.

*c. Diatomaceous-silica filter aid.**

*Hyflo Super-Cell (Johns-Manville Corp.).

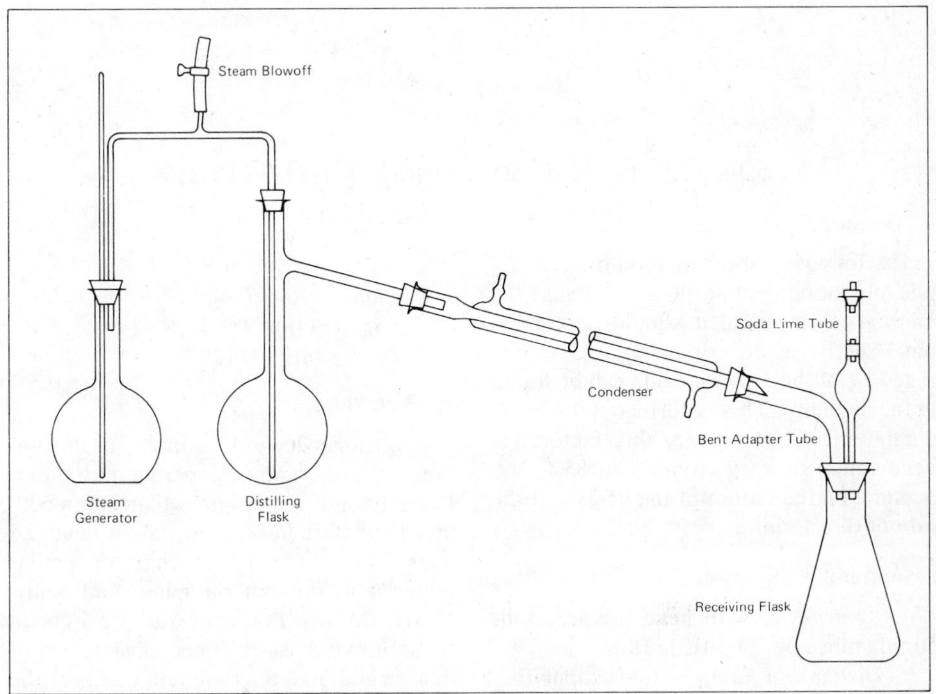

Figure 504:1. Volatile-acid distillation apparatus.

d. Magnesium sulfate, $MgSO_4 \cdot 7H_2O$.

e. Standard sodium hydroxide titrant, 0.1*N:* See Section 402.3*c*.

f. Phenolphthalein indicator solution.

3. Procedure

Adjust a sludge sample, 200 to 1,000 mL, to pH 3.5 with 1 + 1 H_2SO_4. Add 6 mL $FeCl_3$ solution/L sample (equivalent to 500 mg/L); add 50 g filter aid/L. Mix well. Filter by suction, using a Buchner funnel containing a filter paper freshly coated with a thin layer of filter aid. Wash residue thoroughly three or four times with water and adjust filtrate to pH 11 with NaOH solution. Concentrate by evaporation on a steam bath to 150 mL and cool in a refrigerator.

Adjust cooled filtrate to pH 4 with 1 + 1 H_2SO_4 and add it quantitatively and quickly to distilling flask. Add $MgSO_4$ to slight excess of saturation. Apply heat with a small flame to steam generator until rapid evolution of volatile acids commences. This will prevent excessive increase in volume of mixture. Steam-distill slowly so that about 200 mL distillate will be collected in 25 min. Increase rate of distillation and continue until a total of 600 mL is collected. Carefully control steam rate to prevent excessive foaming.

Titrate distillate with 0.1*N* NaOH, using phenolphthalein indicator.

4. Calculation

$$\text{mg volatile acids as acetic acid/L} = \frac{\text{mL NaOH} \times N \times 60{,}000}{\text{mL sample}}$$

where:

N = normality of NaOH.

504 C. Distillation Method (TENTATIVE)

1. General Discussion

The following short method often is applicable for control purposes. Because the method is empirical it should be carried out exactly as described. It is assumed that 70% of the volatile acids will be found in the distillate. This is corrected for in the computations. However, this factor has been found to vary from 68 to 85%, depending on the nature of the acids and the rate of distillation.

2. Apparatus

a. Centrifuge, with head to carry four 50-mL tubes or 250-mL bottles.

b. Distillation flask, 500-mL capacity.

c. Condenser, about 76 cm long.

d. Adapter tube.

3. Reagents

a. Sulfuric acid, H_2SO_4, 1 + 1.

b. Standard sodium hydroxide titrant, 0.1*N:* See Section 402.3*c*.

c. Phenolphthalein indicator solution.

4. Procedure

Centrifuge 200 mL sample for 5 min. Pour off and combine supernatant liquors. Place 100 mL supernatant liquor in a 500-mL distillation flask. Add 100 mL distilled water, four to five clay chips or similar material to prevent bumping, and 5 mL H_2SO_4. Mix so that acid does not remain on bottom of flask. Connect flask to a condenser and adapter tube and distill at the rate of about 5 mL/min. Collect 150 mL distillate in a 250-mL conical flask and titrate with 0.1*N* NaOH, using phenolphthalein as an indicator. The endpoint is the first pink coloration that persists on standing a short time. Titration at 95 C produces a stable endpoint.

5. Calculation

$$\text{mg volatile acids as acetic acid/L} = \frac{\text{mL NaOH} \times N \times 60{,}000}{\text{mL sample} \times 0.7}$$

where:

N = normality of NaOH.

504 D. Bibliography

OLMSTEAD, W.H., W.M. WHITAKER & C.W. DUDEN. 1929–1930. Steam distillation of the lower volatile fatty acids from a saturated salt solution. *J. Biol. Chem.* 85:109.

OLMSTEAD, W.H., C.W. DUDEN, W.M. WHITAKER & R.F. PARKER. 1929–1930. A method for the rapid distillation of the lower volatile fatty acids from stools. *J. Biol. Chem.* 85:115.

BUSWELL, A.M. & S.L. NEAVE. 1930. Laboratory studies of sludge digestion. *Ill. State Water Surv. Bull.* 30:76.

HEUKELEKIAN, H. & A.J. KAPLOVSKY. 1949. Improved method of volatile-acid recovery from sewage sludges. *Sewage Works J.* 21:974.

KAPLOVSKY, A.J. 1951. Volatile-acid production during the digestion of seeded, unseeded, and limed fresh solids. *Sewage Ind. Wastes* 23:713.

MUELLER, H.F., A.M. BUSWELL & T.E. LARSON. 1956. Chromatographic determination of volatile acids. *Sewage Ind. Wastes* 28:255.

MUELLER, H.F., T.E. LARSON & M. FERRETTI. 1960. Chromatographic separation and identification of organic acids. *Anal. Chem.* 32:687.

WESTERHOLD, A.F. 1963. Organic acids in digester liquor by chromatography. *J. Water Pollut. Control Fed.* 35:1431.

HATTINGH, W.H.J. & F.V. HAYWARD. 1964. An improved chromatographic method for the determination of total volatile fatty acid content in anaerobic digester liquors. *Int. J. Air Water Pollut.* 8:411.

POHLAND, F.G. & B.H. DICKSON, JR. 1964. Organic acids by column chromatography. *Water Works Wastes Eng.* 1:54.

505 ORGANIC CARBON (TOTAL)

Combustion-Infrared Method

The organic carbon in water and wastewater represents many different compounds and oxidation states. Some of these carbon compounds can be oxidized further by biological or chemical processes, and the biochemical oxygen demand (BOD) and chemical oxygen demand (COD) are used to determine these fractions. Because of the presence of organic carbon that does not respond to either BOD or COD, these tests may not be a satisfactory measure of total organic carbon. Total organic carbon (TOC) is therefore a more convenient and direct expression of the total organic content than either BOD or COD, but does not provide the same kind of information. If a sufficiently constant empirical relationship is established between TOC and BOD or COD, TOC can be used to estimate the accompanying BOD or COD. This relationship must be established for each separate condition, such as various points in a treatment process. Unlike BOD or COD, TOC is independent of the oxidation state of the organic matter and does not measure other organically bound elements, such as nitrogen and hydrogen, that can contribute to the oxygen demand measured by BOD and COD.

A distinction must be made and reported in regard to purgeable and nonpurgeable TOC. In the combustion infrared method, as well as others, the removal of inorganic carbon by purging will cause loss of purgeable organics, and the result of the determination is reported as nonpurgeable TOC. Instrumentation is available for the determination of both purgeable and nonpurgeable organic carbon and the sum can be reported as true TOC.

TOC analyzers offer a means of measuring total organic carbon in the range normally found in water and wastewater. Organic carbon is oxidized to carbon dioxide (CO_2) by heat and oxygen, ultraviolet irradiation, chemical oxidants, or by various combinations of these. The CO_2 may be measured directly by a nondispersive infrared analyzer or it may be reduced to methane and measured by a flame ionization detector in a gas chromatograph or in a TOC analyzer so equipped. The CO_2 may be titrated chemically.

Sample inorganic carbon must be eliminated or compensated for, because it usually is a very large part of the total carbon. The determination of total carbon and total inorganic carbon, with the estimation of total organic carbon by difference, is common. These determinations are dependent on the amount of organic carbon present and the precision of the two values the difference of which is sought. The uncertainties in the values can make their difference meaningless if the organic carbon is low. Because volatile organic carbon will be lost during purging of an acidified solution, the organic carbon is reported as total nonpurgeable organic carbon. Instrumentation is available for samples low in organic carbon.

Selection of an analyzer must include consideration not only of the carbon content of the samples to be analyzed but also of the need for presentation to the instrument of portions that are representative of the bulk sample. This requires particle size reduction of the suspended material, which is usually difficult and will therefore establish the precision obtainable.

1. General Discussion

The combustion-infrared method has been used for a wide variety of samples, but it is very dependent on particle size reduction because it uses small-orifice syringes.

a. Principle: The sample is homogenized and diluted as necessary and a microportion is injected into a heated, packed tube in a stream of oxygen or purified air. The water is vaporized and the organic carbon is oxidized to CO_2 and H_2O. The CO_2 is measured by means of a nondispersive type of infrared analyzer. Because the carbon analyzer measures all carbon in a sample, procedural modifications are needed to limit the determination to organic carbon. Inorganic carbonates may be decomposed with acid and volatilized in the form of CO_2 before the organic carbon is determined.

It may be necessary to homogenize the sample again after acidification and removal of the carbonates because of coagulation of suspended material. Alternatively, the total carbon and the inorganic carbon can be determined and the organic carbon calculated by difference.

b. Interference: Removal of carbonate and bicarbonate by acidification and purging with nitrogen or other inert gas can result in the loss of volatile organic substances. The volatiles also can be lost during sample blending, particularly if the sample is allowed to heat up. Another important loss can occur if large carbon-containing particles fail to enter the hypodermic needle used for injection. These particles also may cause interference by clogging the pump tubes in instruments. Filtration, although desirable to limit insoluble organic matter when only dissolved TOC is to be determined, can result in loss or gain of TOC, depending on the physical properties of the carbon-containing compounds and the adsorption of carbonaceous material on the filter, or its desorption from it. Check filters for their contribution to TOC when exposed to the type of sample being analyzed. Any sample treatment may alter the measurable carbon. Record such treatment and consider it in interpreting results.

c. Minimum detectable concentration: 1 mg carbon/L. This can be achieved with most combustion-infrared analyzers. The minimum detectable concentration may be reduced by concentrating the sample, if feasible, or by increasing the portion taken for analysis.

d. Sampling and storage: Collect and store samples in glass bottles, preferably brown. Plastic containers are acceptable after tests have demonstrated the absence of extractable carbonaceous substances. Use a Kemmerer or similar type sampler for collecting samples from a depth exceeding 2 m. Preserve samples that cannot be examined promptly by holding at 4 C with minimal exposure to light and atmosphere. Acidification with HCl to a pH not over 2 may be used only if inorganic carbon is subsequently purged. Under any conditions, minimize storage time.

2. Apparatus

a. Sample blender or homogenizer, Waring type, ultrasonic, or other type shown to be effective.

b. Magnetic stirrer.

c. Hypodermic syringe, 0 to 50 or 0 to 500 μL capacity.*

d. Total organic carbon analyzer.†

3. Reagents

a. Redistilled water: Prepare blank and standard solutions with redistilled, CO_2-free water.

b. Hydrochloric acid, HCl, conc.

c. Standard carbon solution: Dissolve 2.1254 g dried potassium biphthalate, $C_8H_5KO_4$, in redistilled CO_2-free water and dilute to 1,000 mL; 1.00 mL = 1.00 mg carbon. Alternatively, use any other organic carbon-containing compound of adequate purity, stability, and water solubility. Preserve standard solution by acidifying with HCl to pH ≤ 2.

d. Standard carbonate solution: Dissolve 8.824 g dried sodium carbonate, Na_2CO_3, in water and dilute to 1,000 mL; 1.00 mL = 1.00 mg carbon. Alternatively, use any other inorganic carbonate compound of adequate purity, stability, and water solubility. Keep tightly stoppered.

e. Packing for oxidation tube: Follow the directions supplied with the total organic carbon analyzer.

f. Oxygen gas, CO_2-free.

g. Air, dry, CO_2-free (alternative to oxygen).

h. Nitrogen gas, CO_2-free.

4. Procedure

a. Instrument operation: Follow manufacturer's instructions for analyzer assembly, testing, calibration, and operation. Vary injected sample size from that recommended by the manufacturer if a different response is desired. If an enlarged

*Hamilton No. 705 N or 750 N; CR-700-20 or CR 700-200 with needle point style No. 3. Spring-loaded syringes may improve reproducibility.

†Beckman Instruments, Inc., or equivalent.

combustion tube is available increase sample size to obtain greater sensitivity. Do not increase sample volume significantly beyond the manufacturer's specifications.

b. Sample treatment: If a sample contains gross solids or insoluble matter, homogenize until satisfactory replication is obtained.

If inorganic carbon must be removed before analysis, transfer a representative portion of 10 to 15 mL to a 30-mL beaker, add 2 drops (0.1 mL) conc HCl to reduce pH to 2 or less, and purge with CO_2-free nitrogen gas for 10 min. Do not use plastic tubing. Inorganic carbon also may be removed by stirring the acidified sample in a beaker while directing a stream of CO_2-free nitrogen into the beaker. Because volatile organic carbon will be lost during purging of the acidified solution, report organic carbon as total nonpurgeable organic carbon.

If the available instrument provides for a separate determination of inorganic carbon (carbonate, bicarbonate, free CO_2), omit decarbonation and proceed according to the manufacturer's directions.

c. Sample injection: While stirring with a magnetic stirrer, withdraw a portion from the beaker using a syringe fitted with a hypodermic needle. Select needle size so as to obtain the most reproducible results. Inject portion into analyzer and obtain peak-height reading. Repeat injection twice or until three consecutive peaks are obtained that are reproducible to within ±10%.

d. Preparation of standard curve: Prepare a standard carbon series of 10, 20, 30, 40, 50, 60, 80, and 100 mg/L with water by diluting 10, 20, 30, 40, and 50 mL standard carbon solution to 1,000 mL and 30, 40, and 50 mL standard carbon solution to 500 mL. Inject and record peak height of these standards and a dilution water blank. Plot carbon concentration in milligrams per liter against corrected peak height in millimeters on rectangular coordinate paper. This is unnecessary for instruments provided with a digital readout of concentration. If desirable, prepare a standard curve having concentrations of 1 to 10 mg/L by making appropriate dilutions of the above standards. Because even redistilled water is not carbon-free and the HCl may contain carbon, determine blanks for both and apply appropriate corrections.

Inject sample portions and blanks and determine sample concentrations from corrected peak heights by reference to the calibration curve.

5. Calculation

a. Calculate corrected peak height in millimeters by deducting the blank correction in the standards and samples as follows:

$$\text{Corrected peak height, mm} = A - B$$

where:

A = peak height, mm, of standards or sample, and
B = peak height, mm, of blank.

b. Apply appropriate dilution factor when necessary.

6. Precision

The difficulty of sampling particulate matter on unfiltered samples limits the precision of the method to approximately 5 to 10%. On clear samples or on those that have been filtered before analysis, precision approaches 1 to 2% or 1 to 2 mg carbon/L, whichever is greater.

7. Bibliography

VAN HALL, C.E., J. SAFRANKO & V.A. STENGER. 1963. Rapid combustion method for the determination of organic substances in aqueous solutions. *Anal. Chem.* 35:315.

VAN HALL, C.E., D. BARTH & V.A. STENGER. 1965. Elimination of carbonates from aqueous solutions prior to organic carbon determinations. *Anal. Chem.* 37:769.

SCHAFFER, R.B. et al. 1965. Application of a carbon analyzer in waste treatment. *J. Water Pollut. Control Fed.* 37:1545.
BUSCH, A.W. 1966. Energy, total carbon, and oxygen demand. *Water Resour. Res.* 2:59.
WILLIAMS, R.T. 1967. Water-pollution instrumentation—Analyzer looks for organic carbon. *Instrum. Technol.* 14:63.
BLACKMORE, R.H. & D. VOSHEL. 1967. Rapid determination of total organic carbon (TOC) in sewage. *Water Sewage Works* 114:398.

506 CARBON-CHLOROFORM EXTRACT (CCE-m)

Organic contaminants—natural substances, plus insecticides, herbicides, and other agricultural chemicals—enter water in precipitation runoff. Domestic and industrial wastewaters, depending on their degree of treatment, also contribute contaminants in various amounts. As a result of accidental spills and leaks, industrial organic wastes may enter bodies of water. Some contaminants, extremely persistent and only partially removed by treatment, reach the consumer in drinking water.

Both natural and man-made contaminants can have undesirable effects on health.[1,2] Some of these materials degrade water quality by causing tastes and odors and by killing fish. The isolation and recovery of insecticides, nitriles, ortho-nitrochlorobenzene, aromatic ethers, waste hydrocarbons, and many other synthetic chemicals suggest that a method for assessing these materials in water is desirable.

The first group of methods for total organic contaminants includes direct determination of contaminating components by Carbon-Chloroform Extract (CCE-m) or Total Organic Carbon (TOC) analysis. In principle, the CCE-m is most appropriate because it is based on the direct recovery and weighing of organics. However, the method is time-consuming. The TOC method can be carried out in a few minutes and gives a quantitative measure of carbon (see Section 505). The results are related to the weight of organic contaminants. Special effort is needed to make this method useful for drinking water analysis.

The second group of methods for measuring organic contaminants depends on determining the equivalence of oxidizing agents reacting with the organic substances. The two common procedures are Oxygen Demand (Biochemical) (BOD) (Section 507) and Oxygen Demand (Chemical) (COD) (Section 508). These methods, while not directly measuring organic contaminants, are widely used and a rationale for interpreting the data has been developed. Both methods have relatively low sensitivity.

Carbon-Chloroform Extract (CCE-m) has an operational definition. CCE-m is a mixture of organic compounds that can be adsorbed on activated carbon under prescribed conditions and then desorbed with chloroform. The lower-case letter "m" denotes the use of a miniaturized sampler and extraction technic and distinguishes it from the high-flow (hf) and low-flow (lf) carbon adsorption method (CAM) technic recommended in the 13th edition of this manual.[3–6]

1. General Discussion

In this gravimetric adsorption-extraction method organic materials are adsorbed on activated carbon and removed by extraction with an organic solvent. The extract is processed by volume reduction through distillation and drying. The quantity of organic materials is determined gravimetrically.

a. Inadequacies: Some organic compounds may not be adsorbed on activated carbon or recovered from the activated carbon by the solvent used, causing a negative error. Inorganic substances may contribute to the weight of extract obtained, causing a positive error.

b. Application: This method is useful for sampling water that contains organic matter mostly in the dissolved form. Because the adsorption capacity of activated carbon is limited, this technic is not suited to wastewaters containing high concentrations of organics.

The method is used primarily for monitoring the general organic concentration and not as a collector of organics for further identification, although with certain precautions, the resulting extract can be reprocessed.* All adsorption-desorption technics have inherent limitations; however, this method does not require expensive instrumentation and should be within the capabilities of most laboratories. Toxicological studies on the components of CCE-m are not complete but control of the concentration of CCE-m in drinking water will provide some consumer protection.

Typical values are shown in Table 506:I.

c. Sampling and storage: In addition to the normal precautions to obtain a representative sample, use the special sampler to give about 60 L. If necessary, store dried, exposed activated carbon in a sealed glass container at about 4 C. Dry as soon as possible to prevent biodegradation of the adsorbed organics.

TABLE 506:I. CONCENTRATIONS OF CCE-m IN VARIOUS CLASSES OF WATER

Class of Sample	CCE-m *mg/L*	Number of Samples Averaged
Well water	0.1	2
Spring water stored in small pond	0.1	3
Finished water from lightly polluted surface water:		
Autumn	0.3	10
Finished water from moderately polluted surface water:		
Summer	0.4	11
Winter	0.5	5
Finished water from heavily polluted surface water:		
Summer and autumn	0.9	9
Winter	1.2	9

2. Apparatus

a. Boiling flask, 300-mL round or flat bottom, borosilicate glass, Standard Taper (ST) 24/40 joint (two required).

b. Condenser, Graham, outer ST 24/40 joint.

c. Distilling column packing: Berl saddles, porcelain, 6 mm.

d. Drying tray, stainless steel, 22 cm long, 13 cm wide, and 5 cm deep. Do not use aluminum or galvanized metals for trays, because they react with wet activated carbon.†

e. Extraction apparatus, Soxhlet, with Allihn condenser, borosilicate glass, 50-mm-ID extractor, ST 55/50 top joint, ST 24/40 bottom joint.

f. Extraction thimbles, paper, seamless, fat-free, single thickness, 43 mm ID, 123 mm length.

g. Filter paper, unwashed, 125 mm diam.‡ Determine blank correction on fil-

*Reference 7 contains details of a technic for re-extracting the activated carbon with 95% ethyl alcohol. Although the resulting extract is too heavily contaminated with inorganic salts to produce useful gravimetric data, it can be used as a source of additional organics from the original water sample that can be reprocessed for identification.

†Matheson Scientific No. 63127-10 trays, instrument, stainless steel, round corners, 1-L capacity have been found satisfactory for this purpose.

‡Whatman No. 1 or equivalent.

ter paper and protect supply from contamination.

h. Filtering screen, 40-mesh, stainless steel.

i. Forceps, aquarium, metal.§

j. Forceps, small, metal, to handle Berl saddles and vials.

k. Funnel, filling, powder, polypropylene, 80 mm diam.

l. Funnel, short stem, fluted, borosilicate glass, 75 mm diam.

m. Heating mantle for extraction step.‖

n. Heating mantle for distillation step.#

o. Mechanical convection oven regulated to a temperature of 40 ± 1C.**

p. Miniature CAM sampler:†† See Figure 506:1.

q. Miniature CAM sample columns:[8] See Figure 506:2.

r. Stopper, polyethylene, hollow, ST 24/40 cone size.

s. Stopwatch.

t. Transformer, variable output, 10 A, 120 V.

u. Tube, connecting, distilling, borosilicate glass, ST 24/40 joints.

v. Vial, 18.5 mL, flint glass, with polyethylene stopper.

3. Reagents

a. Activated carbon, granular, coal base, 14 × 40 mesh grain size.‡‡

b. Chloroform, $CHCl_3$, spectroanalyzed.§§

c. Compressed air, dry, oil-free.‖‖

4. Procedure

a. A schematic drawing of the miniature CAM sampler apparatus is shown in Figure 506:1. Operate sampler with a low constant head of water of less than 1 m. Supply water to be sampled by gravity from a very low constant-head tank (less than 13 cm) and let it flow upward through 70.0 g activated carbon at a rate of about 20 mL/min (4.5-min contact time) for a 48-hr period, thereby sampling about 60 L. Once each 30 min the activated carbon is flushed automatically for 7 to 8 sec (in the same direction as sample flow) at a flow rate of about 400 mL/min. This increased upward flow rate (flushing) expands carbon in column slightly and washes out air and fine particles that would impede flow.

The sampler can be operated on high- or low-pressure water supplies and requires 115 V to operate flushing-cycle timers, solenoid valves, and volume-measuring ap-

§Turtox No. 205A40, or any forceps with a large-tip surface area to avoid tearing the thimbles, is satisfactory.

‖Glas-Col No. STM-800 or equivalent. When purchasing this item, specify if it is to be used for 300-mL round- or flat-bottom flasks according to the choice made for the boiling flask. Do not use flat-bottom flasks in mantles designed for round-bottom flasks.

#Glas-Col No. M-104 or equivalent. See note for extraction-heating mantle.

**The Blue M, Model OV-490 A-C, is satisfactory. If a gravity convection incubator operating at 35 ± 0.5 C (as used for incubation of total coliform tubes or plates) is available, it can be used for activated carbon drying, but determine the minimum acceptable drying time for the activated carbon, because excess residual moisture will interfere with extraction.

††C.F.H. Research Laboratories, Box 269, Springfield, Mass. 01101, and Belcan Corp., 9546 Montgomery Rd., Cincinnati, Ohio 45242, or equivalent. A parts list, construction drawings, and a manual on the installation, operation, and maintenance of the miniaturized CAM sampler[8] are available from the Director, Water Supply Research Division, Municipal Environmental Research Laboratory, USEPA, 26 W. St. Clair St., Cincinnati, Ohio 45268.

‡‡Filtrasorb 200, manufactured by the Calgon Corp., Pittsburgh, Pa., is a satisfactory adsorbant. Calgon manufactures a special sampling activated carbon with the CCE-m blank value per 70.0 g activated carbon predetermined.

§§$CHCl_3$ distilled during reduction of solvent volume may be reused. Gas chromatographic (GC) analysis of this distillate shows that some impurities are present, but less than 1%. Prevent buildup of impurities to significant concentrations in the reused $CHCl_3$ by diluting with fresh $CHCl_3$ (required to offset solvent losses that are up to 60%).

‖‖If a central source of compressed air is not available, use a small air pump such as Fisher Scientific Co. No. 1-092-5, Dyna-Pump, or a small tank of compressed air.

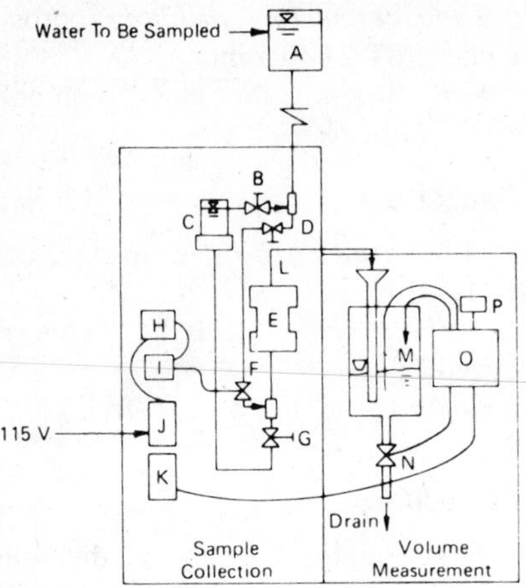

Figure 506:1. Schematic of miniature CAM sampler, Model A. Model B does not have parts M-P and replaces them with a calibrated collection vessel. A—constant-head main tank, B—sample column feed-tank valve, C—sample column feed tank, D—flushing valve, E—sample column, F—flushing solenoid valve, G—sample column flow-regulating valve, H—30-min timer, I—delay timer set for 7-8 sec, J—power on-off switch, K—duplex outlet, L—sample column outlet tube, M—volume-measuring tank, N—volume-measuring solenoid valve, O—volume-measurement control, P—digital-counter volume recorder. SOURCE: BUELOW, R.W., J.K. CARSWELL & J.M. SYMONS.[7]

paratus if Model A is used. If sufficient water pressure is not available or cannot be created by using a siphon, use a small pump constructed of materials that will not introduce organics into the water or allow lubricants to contaminate sample. Do not use rubber or plastic tubing to connect sampler to sampling point because plasticizers in the tubing will contaminate the sample.

b. Preparation of sample column: Place 70.0 g activated carbon in miniature CAM sample column (Figure 506:2). Apply a double layer of TFE tape on column threads for the first use and a single layer for repeat uses. When tape builds up on threads so that it interferes with assembling column, strip it off completely and retape as for first use. Assemble column and prevent leakage by hand-tightening polyvinylchloride (PVC) end cap. Disassembly by hand is possible but a vise and strap wrench may be used. Care in wrapping the TFE tape on the threads is the best leak preventive. Avoid excessive tightening because it shortens effective length of sample column, reduces volume for activated carbon, and restricts cleansing of activated carbon during flushing.

c. Start of sampling: Attach sample column to column inlet connection, clamp column to board, and connect sample column outlet tube. Turn power switch on and open sample column flow-regulating valve fully until a stream of water flows from sample column outlet tube. This should take less than 1 min. Close sample column regulating valve, then reopen 2.5 turns. Measure sampling flow in graduated cylinder and adjust to about 20 mL/min.

If flushing valve was not previously set, open about three-fourths of a turn. Activate flushing timer and check flushing volume—50 to 55 mL during the 7 to 8 sec flushing. Adjust flushing valve as necessary. If desired, set the 30-min timer so that one flush occurs at any fixed minute during an hour. Recheck sampling flow with graduated cylinder about 1 to 2 hr after startup. Readjust if necessary.

d. Daily, during sampling: Once the flushing volume has been set, make a daily visual check of flushing flow. A decrease in flushing volume can result in column blocking. Check sample flow rate daily. If it is outside the range 18 to 22 mL/min, adjust flow rate to about 20 mL/min with column flow-regulating valve. The sample flow rate usually does not require adjustment during sampling.

Record volume counter reading and date and time of reading if Model A is

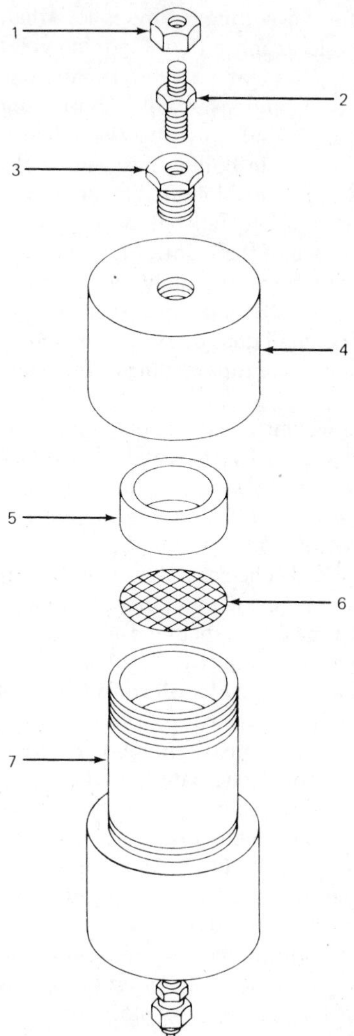

Figure 506:2. Miniature CAM sample column assembly. 1—fitting cap; 2—adapter, 1/4-in. pipe to 1/4-in. OD tubing; 3—bushing, pipe, brass, 1/2 × 1/4-in.; 4—PVC end cap, pipe thread, PVC schedule 80; 5—spacer-screen support, 1 1/2 in., PVC schedule 80, 5/8-in. high; 6—screen, 2 1/4-in. diameter stainless steel wire cloth 40 × 40 mesh, wire 0.012-in. diameter; 7—nipple, pipe, PVC schedule 80, 2-in. pipe threaded, 3-in. long (note: both ends are identical). SOURCE: BUELOW, R.W., J.K. CARSWELL & J.M. SYMONS.[7]

used. If Model B is used, mark expected daily levels on collection container and check. About 31 L should pass through sampler each 24 hr.

e. End of sampling: Shut off sample column flow-regulating valve and power switch. Record final volume counter reading (Model A) or final volume in collection container (Model B). Remove sample column outlet tubing from top of column. Remove column clamp and tilt column forward slightly. Place fitting cap on top fitting of column. Loosen sample column inlet connection and remove column. Drain water from column by removing fitting cap from top fitting. Place fitting caps on fittings at both ends of column. If screen in main tank is coated with floc or particulates, clean it. If main tank or sample column feed tank shows a buildup of sediment, clean by brushing and flushing. Between runs store sampler full of water with sample column flow-regulating valve and power switch off.

f. Emptying sample column: Remove activated carbon sample from column as soon as possible after collection. Loosen one PVC end cap and remove fitting caps to permit air to enter sample column. Remove loosened PVC end cap while holding column over and close to drying tray. Remove bulk of activated carbon from column and PVC end cap by tapping together and catching the activated carbon in the drying tray. Do not spill activated carbon outside of tray. Rinse activated carbon adhering to column surfaces directly into drying tray by a stream of distilled water from a squeeze bottle. Ignore the few granules of activated carbon remaining after thorough rinsing but take care to keep activated carbon loss to a minimum.

Shake drying tray lightly to distribute activated carbon evenly. In the recommended tray the activated carbon layer will be about 0.5 cm deep. Without shifting the carbon distribution, remove excess water by slowly pouring from a corner of the tray. Drain this excess water through a screening device## that will retain any activated carbon particles that flow off with the water. Return to tray any particles retained by the screening device.

g. Drying of activated carbon: Dry wet activated carbon sample in a mechanical convection drying oven regulated to a temperature of 40 ± 1 C for at least 24 but not more than 72 hr. Alternatively, use a gravity convection drying oven or incubator. Locate drying oven in a dust- and organic-vapor-free area.

If an alternate drying device is used, check its drying capabilities as follows: Weigh 70.0 g unexposed activated carbon and soak in distilled water in a covered glass container for at least 24 hr, drain in standard manner, and place in drying device. Weigh at 24-hr intervals until weight of dried activated carbon returns to 70.0 g. The time required to reach this weight will be the minimum allowable drying time for the device.

h. Storage of dried activated carbon samples: If extraction is delayed, store dried activated carbon samples at about 4 C in clean 500-mL glass container with screw-on lid.

i. Extraction: Add 250 mL $CHCl_3$ to the 300-mL boiling flask. Add 10 Berl saddles, prerinsed in $CHCl_3$, as boiling stones. Stopper flask to prevent solvent evaporation and place in extraction-type heating mantle. Place a sample of dried activated carbon in extraction thimble and with aquarium forceps transfer thimble to Soxhlet extractor. Put extractor on boiling flask and connect condenser.

Extract for 44 hr after adjusting transformer to attain an initial extraction cycle time to 6 min ± 15 sec/cycle. As a minimum, check cycle times after about 2 hr operation and three or four times during the second day of extraction. Readjust transformer setting if time is less than 5.5 min/cycle or more than 6.5 min/cycle.

Some loss of solvent through glass joints and condenser will occur during extraction. Periodically observe solvent level remaining in boiling flask when extractor is ready to siphon. Never allow this solvent level to fall below the aluminum top surface of the mantle. If extractor is to run unattended overnight, adjust this carefully. Add $CHCl_3$ as necessary by slowly pouring small quantities (approximately 25 mL) down the top opening of the condenser.

The weight of extract obtained is directly influenced by the total number of extraction cycles to which the activated carbon is subjected. To insure that a total number of cycles within the desired range is obtained, check cycle times throughout extraction period and compute an average cycle time. The specified cycle time will produce 405 to 480 total cycles in 44 hr and is easily obtainable. Because ambient air temperature will have an effect on cycle time, continuously monitor air temperature in the immediate vicinity of the extraction equipment.

After completing extraction, turn off heating mantle and let apparatus cool 15 min before removing condenser. Transfer remaining extraction solvent in extractor to the 300-mL flask by draining and removing thimble with forceps. Tip extractor-flask assembly to start siphon. Remove flask from heating mantle. Drain additional solvent from thimble into a beaker and add to the 300-mL flask. Stopper flask to prevent evaporation.

j. Reduction of solvent volume: Before beginning solvent-volume reduction, filter through filter paper the contents of 300-mL flask to remove carbon particles. In a hood, using distillation-type heating mantle, connecting tube, and Graham condenser, distill off excess $CHCl_3$ until volume remaining is less than 20 mL. Use the same transformer as in the extraction at a

##A 40-mesh circular stainless steel screen pressed into an 80-mm polypropylene powder funnel has been found satisfactory for this purpose.

setting of about 60. Occasionally swirl contents during distillation to dissolve extract from sidewalls. Until boiling is well established, swirl contents continuously and vigorously for 3 to 5 min. This will avoid abrupt, massive boilover of flask contents, which will invalidate the determination. If the solvent is not to be collected, boil off excess in a hood. To recover solvent, condense in an apparatus such as a Kuderna-Danish evaporative concentrator.

Remove boiling flask from still while hot, using a towel or gloves. Swirl contents vigorously to dissolve extract from sidewalls and let cool. Transfer residue to a clean, tared 18.5-mL vial. Avoid tipping Berl saddles into vial. Rinse flask and Berl saddles with 2 mL $CHCl_3$ and add to vial. Repeat with 1 mL $CHCl_3$ and add to vial.

k. Drying of extract: Evaporate contents of tared vial to dryness using a gentle stream of dry, oil-free air directed into the vial with a tube or in an unheated mechanical convection oven in which no other samples are being processed. After about 24 hr of drying, tilt vial, using small forceps, to determine if contents will still

TABLE 506:II. REPLICATE CCE-m ANALYSES FROM DIFFERENTLY EXPOSED ACTIVATED CARBON SAMPLES

Type of Carbon Sample	Analyst No.	No. of Replicates	Mean Weight of Extract /70.0 g Adsorbant *mg*	Standard Deviation *mg*	Coefficient Variation, Mean *%*
Unexposed activated carbon (from shipping container)	1	5	2.2*	0.1	4.5
	1	5	3.6*	1.0	28
Activated carbon exposed in large container, dried, mixed, and divided in 70.0-g portions.	1	6	87.0†	1.7	2.0
	1	6	86.6†	2.8	3.2
	2	6	66.3†	5.2	7.8
	2	4	63.8†	2.0	3.1
			Calculated mean, CCE-m conc *mg/L*‡	Standard Deviation *mg/L*	
Activated carbon exposed to same water in separate minisamplers (collected by Analyst No. 3)	1	4	0.330§	0.010	3.0
	2	6	0.503§	0.013	2.6

* Extract weighed daily until Δ weight/day ≦ 0.1 mg.

† Extract weighed daily until Δ weight/day ≦ 1% previous day's weight.

‡ Data expressed as calculated CCE-m concentration because total volume of water sampled varied for each minisampler.

§ Extracts weighed daily until dry to constant concentration. See Procedure (drying of extract).

"flow." If "flow" is evident, continue drying and re-examine at 24-hr intervals.

If "flow" is no longer evident, weigh vial and calculate CCE-m concentration. Place vial in a desiccator with $CaSO_4$ desiccant for 24 hr, reweigh, and recalculate concentration. If it is unchanged report the calculated concentration. If concentration has decreased, continue desiccator drying and daily reweighing until concentration is the same on two successive days. If it is necessary to delay the start of extract drying, cover vial with aluminum foil and store at room temperature for not more than 2 or 3 days.

If extract is to be reprocessed for identification of components, test also an extract from blank activated carbon to insure that any compound identified did not originate from the activated carbon, the solvent, or the thimble.

5. Calculation

$$\text{mg CCE-m/L} = \frac{[(A - B) \times 1000] - C}{D}$$

where:

A = weight of vial plus extract, g,
B = tare weight of vial, g,
C = weight of CCE-m average blank, mg,*** and
D = volume of water sampled, L.

Express result to nearest 0.1 mg/L.

6. Precision and Accuracy

Very limited data are available to calculate precision. See Table 506:II. The concentrations in these data were calculated to three decimals and analyzed statistically.

Although percentage-recovery determinations with known compounds have not been made, the increase in extract yield[7] over that obtained by previous methods[5] indicates improved accuracy.

7. References

1. HEUPER, W.C. & W.W. PAYNE. 1963. Carcinogenic effects of raw and finished water supplies. *Amer. J. Clin. Pathol.* 39:475.
2. MCCABE, L.J. 1964. Life table analysis of Heuper's data. Interoffice memo, Robert A. Taft Sanitary Engineering Center, Cincinnati, Ohio (unpublished).
3. Standard Methods for the Examination of Water and Wastewater, 13th ed. 1971. American Public Health Ass., New York, N.Y.
4. BOOTH, R.L., J.N. ENGLISH & G.N. MCDERMOTT. 1965. Evaluation of sampling conditions in the carbon adsorption methods. *J. Amer. Water Works Ass.* 57:215.
5. REID, B.H., H. STIERLI, C. HENKE & A.W. BREIDENBACH. 1965. Field Evaluation of Low-Flow-Rate Carbon Adsorption Equipment and Methods for Organics Sampling of Surface Waters. PHS Water Pollution Surveillance System Applications and Development Rep. No. 14, Div. Water Supply and Pollution Control, U.S. Dep. HEW, Cincinnati, Ohio (mimeo).
6. Installation, Operation, and Maintenance of Models No. LF-1 and LF-2, Organics Samplers for Water. 1967. Div. Pollution Surveillance, FWPCA. U.S. Dep. Interior, Cincinnati, Ohio.
7. BUELOW, R.W., J.K. CARSWELL & J.M. SYMONS. 1973. An improved method for determining organics in water by activated carbon adsorption and solvent extraction. *J. Amer. Water Works Ass.* 65:57 and 65:195.
8. CARSWELL, J.K., R.W. BUELOW & J.M. SYMONS. 1973. The Determination of Organics—Carbon Adsorbable in Water. Water Supply Research Lab., National Environmental Research Center, USEPA, Cincinnati, Ohio (mimeo).

***CCE-m blanks are dried by the standard procedure of air followed by desiccator drying until their weight changes at successive 24-hr intervals are $\leq$0.1 mg.

507 OXYGEN DEMAND (BIOCHEMICAL)

1. Discussion

The biochemical oxygen demand (BOD) determination is an empirical test in which standardized laboratory procedures are used to determine the relative oxygen requirements of wastewaters, effluents, and polluted waters. The test measures the oxygen required for the biochemical degradation of organic material (carbonaceous demand) and the oxygen used to oxidize inorganic material such as sulfides and ferrous iron. It also may measure the oxygen used to oxidize reduced forms of nitrogen (nitrogenous demand) unless their oxidation is prevented by an inhibitor.

The method consists of placing a sample in a full, airtight bottle and incubating the bottle under specified conditions for a specific time. Dissolved oxygen (DO) is measured initially and after incubation. The BOD is computed from the difference between initial and final DO.

The bottle size, incubation temperature, and incubation period are all specified. Most wastewaters contain more oxygen-demanding materials than the amount of DO available in air-saturated water. Therefore, it is necessary to dilute the sample before incubation to bring the oxygen demand and supply into appropriate balance. Because bacterial growth requires nutrients such as nitrogen, phosphorus, and trace metals, these are added to the dilution water, which is buffered to ensure that the pH of the incubated sample remains in a range suitable for bacterial growth. Complete stabilization of a sample may require a period of incubation too long for practical purposes; therefore, 5 days has been accepted as the standard incubation period.

Measurements of BOD that include both carbonaceous oxygen demand and nitrogenous oxygen demand generally are not useful; therefore, where appropriate, an inhibiting chemical may be used to prevent ammonia oxidation. With this technic carbonaceous and nitrogenous demands can be measured separately. The inclusion of ammonia in the dilution water demonstrates that there is no intent to include the oxygen demand of reduced nitrogen forms in the BOD test. If this ammonia were oxidized, errors would result because the oxygen use would not be due exclusively to pollutants in the sample.

The extent of oxidation of nitrogenous compounds during the 5-day incubation period depends on the presence of microorganisms capable of carrying out this oxidation. Such organisms usually are not present in raw sewage or primary effluent in sufficient numbers to oxidize significant quantities of reduced nitrogen forms in the 5-day BOD test. Currently, many biological treatment plant effluents contain significant numbers of nitrifying organisms. Because oxidation of nitrogenous compounds can occur in such samples, inhibition of nitrification is recommended for samples of secondary effluent, for samples seeded with secondary effluent, and for samples of polluted waters.

The method included here contains both a dilution water check (5*b*) and a dilution water blank (5*h*). The dilution water check is to determine the acceptability of a particular batch of dilution water before it is used for BOD analysis. Seeded dilution waters are checked further for acceptable quality by measuring their consumption of oxygen from a known organic mixture, usually glucose and glutamic acid (5*c*).

The dilution water blank, made at the same time that samples are analyzed, provides a further quality control on dilution water at the time of analysis as well as on the cleanliness of apparatus such as BOD bottles.

The procedure for determining immediate oxygen demand (IDOD) has been eliminated because: (*a*) it was not clear whether IDOD should be reported in 5-day BOD data; (*b*) the measurement was inaccurate because of the small differences between initial DO and DO after 15 min; (*c*) arbitrary selection of 15 min for measuring IDOD did not necessarily include all short-term oxygen-consuming reactions; and (*d*) the IDOD is, in some instances, an iodine demand (during the DO determination) rather than a true DO demand. The methods outlined here require determining initial DO immediately after making the dilution. In this fashion all oxygen uptake (including that occurring during the first 15 min) is included in the BOD measurement.

Although only the 5-day BOD is described here, many variations of oxygen demand measurements exist. These include using shorter and longer incubation periods, tests to determine rates of oxygen uptake, continuous oxygen uptake measurements by respirometric technics, etc.

2. Sampling and Storage

Samples for BOD analysis may degrade significantly during storage between collection and analysis, resulting in low BOD values. Minimize reduction of BOD by analyzing the sample promptly or by cooling it to near-freezing temperature during storage. However, even at low temperature, keep the holding time to a minimum. Warm the chilled samples to 20 C before analysis; some storage time can be used to accomplish this conveniently.

a. Grab samples: If analysis is initiated within 2 hr of collection, cooling is unnecessary. If analysis is not started within 2 hr of sample collection, keep sample at or below 4 C from the time of collection. Begin analysis within 6 hr of collection; when this is not possible because the sampling site is distant from the laboratory, store at or below 4 C and report length and temperature of storage with the results. In no case start analysis more than 24 hr after grab sample collection. When samples are to be used for regulatory purposes make every effort to deliver samples for analysis within 6 hr of collection.

b. Composite samples: Keep samples at or below 4 C during compositing. Limit compositing period to 24 hr. Use the same criteria as for storage of grab samples, starting the measurement of holding time from the end of the compositing period. State storage time and conditions as part of the results.

3. Apparatus

a. Incubation bottles: 250 to 300 mL capacity, with ground-glass stoppers. Clean bottles with a detergent, rinse thoroughly, and drain before use. As a precaution against drawing air into the dilution bottle during incubation, use a water-seal. Obtain satisfactory water seals by inverting bottles in a water bath or adding water to the flared mouth of special BOD bottles. Place a paper or plastic cup or foil cap over the flared mouth of the bottle to reduce evaporation of the water seal during incubation.

b. Air incubator or water bath: Thermostatically controlled at 20 ±1 C. Exclude all light to prevent possibility of photosynthetic production of DO.

4. Reagents

a. Phosphate buffer solution: Dissolve 8.5 g KH_2PO_4, 21.75 g K_2HPO_4, 33.4 g $Na_2HPO_4 \cdot 7H_2O$, and 1.7 g NH_4Cl in about 500 mL distilled water and dilute to 1 L. The pH should be 7.2 without further adjustment. Discard reagent (or any of the following reagents) if there is any sign of biological growth in the stock bottle.

b. Magnesium sulfate solution: Dissolve 22.5 g $MgSO_4 \cdot 7H_2O$ in distilled water and dilute to 1 L.

c. Calcium chloride solution: Dissolve 27.5 g $CaCl_2$ in distilled water and dilute to 1 L.

d. Ferric chloride solution: Dissolve

0.25 g $FeCl_3 \cdot 6H_2O$ in distilled water and dilute to 1 L.

e. Acid and alkali solutions, 1N: For neutralization of caustic or acidic waste samples.

f. Sodium sulfite solution, 0.025*N:* Dissolve 1.575 g Na_2SO_3 in 1,000 mL distilled water. This solution is not stable; prepare daily.

g. Nitrification inhibitor: Reagent-grade 2-chloro-6-(trichloro methyl) pyridine.*

h. Glucose-glutamic acid solution: Dry reagent-grade glucose and reagent-grade glutamic acid at 103 C for 1 hr. Add 150 mg glucose and 150 mg glutamic acid to distilled water and dilute to 1 L. Prepare fresh immediately before use.

5. Procedure

a. Preparation of dilution water: Place desired volume of water in a suitable bottle and add 1 mL each of phosphate buffer, $MgSO_4$, $CaCl_2$, and $FeCl_3$ solutions/L of water. Seed dilution water, if desired, as described in 5*d*. Test and store dilution water as described in 5*b* and 5*c* so that water of assured quality always is on hand.

b. Dilution water check: Use this procedure as a rough check on quality of dilution water. If dilution water has not been stored for quality improvement, add sufficient seeding material to produce a DO uptake of 0.05 to 0.1 mg/L in 5 days at 20 C. Do not seed dilution water that has been stored for quality improvement. Incubate a BOD bottle full of dilution water for 5 days at 20 C. Determine initial and final DO as in 5*g* and 5*j*. The DO uptake in 5 days at 20 C should not be more than 0.2 mg/L and preferably not more than 0.1 mg/L.

If the oxygen depletion of a candidate water exceeds 0.2 mg/L obtain a satisfactory water by improving purification or from another source. Alternatively, if nitrification inhibition is used, store the seeded dilution water at 20 C until the oxygen uptake is sufficiently reduced to meet the dilution water check criteria. Storage is not recommended when BOD's are to be determined without nitrification inhibition because nitrifying organisms may develop during storage. Check stored dilution water to determine whether sufficient ammonia remains after storage.

Before use bring dilution water temperature to 20 C. Saturate with DO by shaking in a partially filled bottle or by aerating with filtered air. Alternatively, store in cotton-plugged bottles long enough for water to become saturated with DO. Protect water quality by using clean glassware, tubing, and bottles.

c. Glucose-glutamic acid check: Because the BOD test is a bioassay the results can be influenced greatly by the presence of toxicants or by use of a poor seeding material. Distilled waters frequently are contaminated with copper; some sewage seeds are relatively inactive. Low results always are obtained with such seeds and waters. Periodically check dilution water quality, seed effectiveness, and analytical technic by making BOD measurements on pure organic compounds. In general, for BOD determinations not requiring an adapted seed, use a mixture of 150 mg glucose/L and 150 mg glutamic acid/L as a "standard" check solution. Glucose has an exceptionally high and variable oxidation rate but when it is used with glutamic acid, the oxidation rate is stabilized and is similar to that obtained with many municipal wastes. Alternatively, if a particular wastewater contains an identifiable major constituent that contributes to the BOD, use this compound in place of the glucose-glutamic acid.

Determine the 5-day 20 C BOD of a 2% dilution of the glucose-glutamic acid standard check solution using the technics out-

*N-Serve, Dow Chemical Co., Nitrification Inhibitor 2533, Hach Chemical Co., or equivalent.

lined in 5*d–j*. If the 5-day 20 C BOD value of the check is outside the range of 200 ± 37 mg/L, reject any BOD determinations made with the seed and dilution water and seek the cause of the problem.

d. Seeding: It is necessary to have present a population of microorganisms capable of oxidizing the biodegradable organic matter in the sample. Domestic wastewater, unchlorinated or otherwise-undisinfected effluents from biological waste treatment plants, and surface waters receiving wastewater discharges contain satisfactory microbial populations. Some samples do not contain a sufficient microbial population (for example some untreated industrial wastes, disinfected wastes, high-temperature wastes, or wastes with extreme pH values). For such wastes seed the dilution water by adding a population of microorganisms. The preferred seed is effluent from a biological treatment system processing the waste. Where this is not available, use supernatant from domestic wastewater after settling at 20 C for at least 1 hr but no longer than 36 hr.

Some samples may contain materials not degraded at normal rates by the microorganisms in settled domestic wastewater. Seed such samples with an adapted microbial population obtained from the undisinfected effluent of a biological process treating the waste. In the absence of such a facility, obtain seed from the receiving water below (preferably 3 to 8 km) the point of discharge. When such seed sources also are not available, develop an adapted seed in the laboratory by continuously aerating a sample of settled domestic wastewater and adding small daily increments of waste. Optionally use a soil suspension or activated sludge to obtain the initial microbial population. Determine the existence of a satisfactory population by testing the performance of the seed in BOD tests on the sample. BOD values that increase with time of adaptation to a steady high value indicate successful seed adaptation. In making tests, use enough seed to assure satisfactory numbers of microorganisms but not so much that the oxygen demand of the seed itself is a major part of the oxygen used during incubation.

Determine BOD of the seeding material as for any other sample. This is the seed control. From the value of the seed control and a knowledge of the seeding material dilution (in the dilution water) determine seed DO uptake. To determine a sample DO uptake subtract the seed DO uptake from the total DO uptake. The DO uptake of the seeded dilution water should be between 0.6 and 1.0 mg/L.

Technics for adding seeding material to dilution water are described for two sample dilution methods (¶ 5*f*).

e. Sample pretreatment:

1) Samples containing caustic alkalinity or acidity—Neutralize samples to pH 6.5 to 7.5 with a solution of sulfuric acid (H_2SO_4) or sodium hydroxide (NaOH) of such strength that the quantity of reagent does not dilute the sample by more than 0.5%. The pH of seeded dilution water should not be affected by the lowest sample dilution.

2) Samples containing residual chlorine compounds—If possible, avoid samples containing residual chlorine by sampling ahead of chlorination processes. If the sample has been chlorinated but no detectable chlorine residual is present, seed the dilution water. If residual chlorine is present, dechlorinate and seed the dilution water (5*f*). Do not test chlorinated/dechlorinated samples without seeding the dilution water. In some samples chlorine will dissipate within 1 to 2 hr of standing in the light. This often occurs during sample transport and handling. For samples in which chlorine residual does not dissipate in a reasonably short time, destroy chlorine residual by adding Na_2SO_3 solution. Determine required volume of Na_2SO_3 solution on a 100- to 1,000-mL portion of

neutralized sample by adding 10 mL of 1 + 1 acetic acid or 1 + 50 H_2SO_4, 10 mL potassium iodide (KI) solution (10 g/100 mL), and titrating with 0.025*N* Na_2SO_3 solution to the starch-iodine end point. Add to the neutralized sample the volume of Na_2SO_3 solution determined by the above test, mix, and after 10 to 20 min check sample for residual chlorine.

3) Samples containing other toxic substances—Certain industrial wastes, for example, plating wastes, contain toxic metals. Such samples often require special study and treatment.

4) Samples supersaturated with DO—Samples containing more than 9 mg DO/L at 20 C may be encountered in cold waters or in water where photosynthesis occurs. To prevent loss of oxygen during incubation of such samples, reduce DO to saturation at 20 C by bringing sample to about 20 C in a partially filled bottle while agitating by vigorous shaking or by aerating with compressed air.

5) Sample temperature adjustment—Bring samples to 20 ± 1 C before making dilutions.

6) Nitrification inhibition—If nitrification inhibition is desired add 10 mg 2-chloro-6 (trichloro methyl) pyridine/L of dilution water at the same time as adding nutrient and buffer solutions. Samples that may require nitrification inhibition include, but are not limited to, biologically treated effluents, samples seeded with biologically treated effluents, and river waters. Note the use of nitrogen inhibition in reporting results.

f. Dilution technic: Dilutions that result in a residual DO of at least 1 mg/L and a DO uptake of at least 2 mg/L after 5 days incubation produce the most reliable results. Make several dilutions of prepared sample to obtain DO uptake in this range. Experience with a particular sample will permit use of a smaller number of dilutions. A more rapid analysis, such as COD, may be correlated approximately with BOD and serve as a guide in selecting dilutions. In the absence of prior knowledge, use the following dilutions: 0.0 to 1.0% for strong industrial wastes, 1 to 5% for raw and settled wastewater, 5 to 25% for biologically treated effluent, and 25 to 100% for polluted river waters.

Prepare dilutions either in graduated cylinders and then transfer to BOD bottles or prepare directly in BOD bottles. Either dilution method can be combined with any DO measurement technic. The number of bottles to be prepared for each dilution depends on the DO technic and the number of replicates desired.

When using graduated cylinders to prepare dilutions, and when seeding is necessary, either add seed directly to dilution water or to individual cylinders before dilution. Seeding of individual cylinders avoids a declining ratio of seed to sample as increasing dilutions are made. When dilutions are prepared directly in BOD bottles and when seeding is necessary, add seed directly to dilution water.

1) Dilutions prepared in graduated cylinders—If the azide modification of the titrimetric iodometric method (Section 421B) is used, carefully siphon dilution water, seeded if necessary, into a 1- to 2-L-capacity graduated cylinder. Fill cylinder half full without entraining air. Add desired quantity of carefully mixed sample and dilute to appropriate level with dilution water. Mix well with a plunger-type mixing rod; avoiding entraining air. Siphon mixed dilution into two BOD bottles. Determine initial DO on one of these bottles. Stopper the second bottle tightly, water-seal, and incubate for 5 days at 20 C. If the membrane electrode method is used for DO measurement, siphon dilution mixture into one BOD bottle. Determine initial DO on this bottle and replace any displaced contents with sample dilution to fill the bottle. Stopper tightly, water-seal, and incubate for 5 days at 20 C.

2) Dilutions prepared directly in BOD

bottles—Using a wide-tip volumetric pipet, add the desired sample volume to individual BOD bottles of known capacity. Fill bottles with enough dilution water, seeded if necessary, so that insertion of stopper will displace all air, leaving no bubbles. For dilutions greater than 1:100 make a primary dilution in a graduated cylinder before making final dilution in the bottle. When using titrimetric iodometric methods for DO measurement, prepare two bottles at each dilution. Determine initial DO on one bottle. Stopper second bottle tightly, water-seal, and incubate for 5 days at 20 C. If the membrane electrode method is used for DO measurement, prepare only one BOD bottle for each dilution. Determine initial DO on this bottle and replace any displaced contents with dilution water to fill the bottle. Stopper tightly, water-seal, and incubate for 5 days at 20 C.

g. Determination of initial DO: If materials are present in the sample that react rapidly with DO, determine initial DO immediately after filling BOD bottle with diluted sample. If rapid initial DO uptake is insignificant, the time period between preparing dilution and measuring initial DO is not critical.

Use the azide modification of the iodometric method (Section 421B) or the membrane electrode method (Section 421F) to determine initial DO on all sample dilutions, dilution water blanks, and where appropriate, seed controls.

For activated sludge samples use either the membrane electrode method or the $CuSO_4$-sulfamic acid modification of the iodometric method (Section 421E). For muds use either the membane electrode method or the alum flocculation modification of the iodometric method (Section 421D).

h. Dilution water blank: Use a dilution water blank as a rough check on the quality of unseeded dilution water and cleanliness of incubation bottles. Together with each batch of samples incubate a bottle of unseeded dilution water. Determine initial and final DO as in 5*g* and 5*j*. The DO uptake should not be more than 0.2 mg/L and preferably not more than 0.1 mg/L.

i. Incubation: Incubate at 20 C ± 1 C BOD bottles containing desired dilutions, seed controls, dilution water blanks, and glucose-glutamic acid checks. Water-seal bottles as described in 5*f*.

j. Determination of final DO: After 5 days incubation determine DO in sample dilutions, blanks, and checks as in 5*g*.

6. Calculation

When dilution water is not seeded:

$$\text{BOD, mg/L} = \frac{D_1 - D_2}{P}$$

When dilution water is seeded:

$$\text{BOD, mg/L} = \frac{(D_1 - D_2) - (B_1 - B_2)f}{P}$$

where:

D_1 = DO of diluted sample immediately after preparation, mg/L,
D_2 = DO of diluted sample after 5 days incubation at 20 C, mg/L,
P = decimal volumetric fraction of sample used,
B_1 = DO of seed control before incubation, mg/L,
B_2 = DO of seed control after incubation, mg/L, and
f = ratio of seed in sample to seed in control = (% seed in D_1)/(% seed in B_1).

If more than one sample dilution meets the criteria of a residual DO of at least 1 mg/L and a DO depletion of at least 2 mg/L and there is no evidence of toxicity at higher sample concentrations or the existence of an obvious anomaly, average results in the acceptable range.

In these calculations, corrections are not made for DO uptake by the dilution water blank during incubation. This correction is unnecessary if dilution water meets the blank criteria stipulated above. If the dilution water does not meet these criteria, proper corrections are difficult and results become questionable.

7. Precision and Accuracy

In a series of interlaboratory studies, each involving 86 to 102 laboratories (and as many river water and wastewater seeds), 5-day BOD measurements were made on synthetic water samples containing a 1:1 mixture of glucose and glutamic acid in the total concentration range of 5 to 340 mg/L. The regression equations for mean value, $\overline{X}$, and standard deviation, S, from these studies were:

$$\overline{X} = 0.665 \text{ (added level, mg/L)} - 0.149$$

$$S = 0.120 \text{ (added level, mg/L)} + 1.04$$

For the 300-mg/L mixed primary standard, the average 5-day BOD was 199.4 mg/L with a standard deviation of 37.0 mg/L.

8. References

1. YOUNG, J.C. 1979. Chemical methods for nitrification control. *J. Water Pollut. Control Fed.* 45:637.
2. U.S. ENVIRONMENTAL PROTECTION AGENCY, OFFICE OF RESEARCH & DEVELOPMENT, ENVIRONMENTAL MONITORING & SUPPORT LABORATORY, CINCINNATI, OHIO. 1978. Personal communication, D.W. Ballinger to G.N. McDermott.

9. Bibliography

THERIAULT, E.J., P.D. MCNAMEE & C.T. BUTTERFIELD. 1931. Selection of dilution water for use in oxygen demand tests. *Pub. Health Rep.* 48:1084.

LEA, W.L. & M.S. NICHOLS. 1937. Influence of phosphorus and nitrogen on biochemical oxygen demand. *Sewage Works J.* 9:34.

RUCHHOFT, C.C. 1941. Report on the cooperative study of dilution waters made for the Standard Methods Committee of the Federation of Sewage Works Associations. *Sewage Works J.* 13:669.

SAWYER, C.N. & L. BRADNEY. 1946. Modernization of the BOD test for determining the efficiency of the sewage treatment process. *Sewage Works J.* 18:1113.

RUCHHOFT, C.C., O.R. PLACAK, J. KACHMAR & C.E. CALBERT. 1948. Variations in BOD velocity constant of sewage dilutions. *Ind. Eng. Chem.* 40:1290.

ABBOTT, W.E. 1948. The bacteriostatic effects of methylene blue on the BOD test. *Water Sewage Works* 95:424.

MOHLMAN, F.W., E. HURWITZ, G.R. BARNETT & H.R. RAMER. 1950. Experience with modified methods for BOD. *Sewage Ind. Wastes* 22:31.

SAWYER, C.N., P. CALLEJAS, M. MOORE & A.Q.Y. TOM. 1950. Primary standards for BOD work. *Sewage Ind. Wastes* 22:26.

508 OXYGEN DEMAND (CHEMICAL)

The chemical oxygen demand (COD) is a measure of the oxygen equivalent of the organic matter content of a sample that is susceptible to oxidation by a strong chemical oxidant. For samples from a specific source, COD can be related empirically to BOD, organic carbon, or organic matter content.

1. Selection of Method

The dichromate reflux method is preferred over other methods using oxidants because of superior oxidizability, applicability to a wide variety of samples, and ease of manipulation. The test is most useful for monitoring and control, especially after correlations with constituents[1,2] such as BOD and organic carbon have been developed. For most organic compounds oxidation is 95 to 100% of the theoretical value.[2,3] Pyridine is not oxidized.[2] Benzene and other volatile organics are oxidized if they have sufficient contact with the oxidants.[2] While the carbonaceous portion of nitrogen-containing organic matter is oxidized, no oxidation of ammonia, either present in a waste or liberated from the nitrogen-containing organic matter, takes place in the absence of significant chloride concentrations.

2. Sampling and Storage

Test unstable samples without delay. Homogenize samples containing settleable solids in a blender to permit representative sampling. If there is to be a delay before analysis, preserve the sample by acidification to pH 2 or lower with conc sulfuric acid (H_2SO_4). Make preliminary dilutions for wastes containing a high COD to reduce the error inherent in measuring small volumes of sample.

508 A. Dichromate Reflux Method

1. General Discussion

a. Principle: Most types of organic matter are oxidized by a boiling mixture of chromic and sulfuric acids. A sample is refluxed in strongly acid solution with a known excess of potassium dichromate ($K_2Cr_2O_7$). After digestion the remaining unreduced $K_2Cr_2O_7$ is titrated with ferrous ammonium sulfate (FAS), the amount of $K_2Cr_2O_7$ consumed is determined, and the amount of oxidizable organic matter is calculated in terms of oxygen equivalent.

b. Interferences and limitations: Volatile straight-chain aliphatic compounds are not oxidized to any appreciable extent. This failure occurs partly because volatile organics are present in the vapor space and do not come in contact with the oxidizing liquid. Straight-chain aliphatic compounds are oxidized more effectively when silver sulfate (Ag_2SO_4) is added as a catalyst. However, Ag_2SO_4 reacts with chloride, bromide, and iodide to produce precipitates that are oxidized only partially. The difficulties caused by the presence of halides can be largely, though not completely, overcome by complexing with mercuric sulfate ($HgSO_4$) before the refluxing procedure.[4] Do not use the test for samples containing more than 2,000 mg chloride/L.

Nitrite (NO_2^-) exerts a COD of 1.1 mg O_2/mg NO_2^--N. Because concentrations of NO_2^- in polluted waters rarely exceed 1 or 2 mg NO_2^--N/L the interference is considered insignificant and usually is ignored. To eliminate a significant interference due to NO_2^-, add 10 mg sulfamic acid/mg NO_2^--N present in the refluxing flask. Also add the same amount of sulfamic acid to the reflux flask containing the distilled water blank.

Reduced inorganic species such as ferrous iron, sulfide, manganous manganese, etc., are oxidized quantitatively under the test conditions. For samples containing significant levels of these species, stoichiometric oxidation can be assumed from known initial concentration of the interfering species and corrections can be made to the COD value obtained.

c. Minimum detectable concentration: Determine COD values of >50 mg/L using 0.250*N* $K_2Cr_2O_7$. With 0.025*N* $K_2Cr_2O_7$, COD values from 5 to 50 mg/L can be determined but with lesser accuracy.[5]

2. Apparatus

Reflux apparatus, consisting of 500-mL

or 250-mL erlenmeyer flasks with ground-glass 24/40 neck* and 300-mm jacket Liebig, West, or equivalent condensers,† with 24/40 ground-glass joint, and a hot plate having sufficient power to produce at least 1.4 W/cm^2 of heating surface, or equivalent.

3. Reagents

a. Standard potassium dichromate solution, 0.250*N*: Dissolve 12.259 g $K_2Cr_2O_7$, primary standard grade, previously dried at 103 C for 2 hr, in distilled water and dilute to 1,000 mL.

b. Silver sulfate, Ag_2SO_4, reagent or technical grade, crystals or powder.

c. Sulfuric acid reagent: Add Ag_2SO_4 to conc H_2SO_4 at the rate of 22 g Ag_2SO_4/4 kg bottle. Let stand 1 to 2 days to dissolve Ag_2SO_4.

d. Sulfuric acid, H_2SO_4, conc.

e. Ferroin indicator solution: Dissolve 1.485 g 1,10-phenanthroline monohydrate and 695 mg $FeSO_4{\cdot}7H_2O$ in distilled water and dilute to 100 mL. This indicator solution may be purchased already prepared.‡

f. Standard ferrous ammonium sulfate titrant, approximately 0.25*N*: Dissolve 98 g $Fe(NH_4)_2(SO_4)_2{\cdot}6H_2O$ (FAS) in distilled water. Add 20 mL conc H_2SO_4, cool, and dilute to 1,000 mL. Standardize this solution daily against standard $K_2Cr_2O_7$ solution, as follows:

Dilute 10.0 mL standard $K_2Cr_2O_7$ solution to about 100 mL. Add 30 mL conc H_2SO_4 and cool. Titrate with FAS titrant, using 0.10 to 0.15 mL (2 to 3 drops) ferroin indicator.

Normality of FAS solution

$$= \frac{\text{Volume } 0.25N\ K_2Cr_2O_7 \text{ solution titrated, mL}}{\text{Volume FAS used in titration, mL}} \times 0.25$$

g. Mercuric sulfate: $HgSO_4$, crystals or powder.

h. Sulfamic acid: Required only if the interference of nitrites is to be eliminated (see ¶ 1*b* above).

i. Potassium hydrogen phthalate standard: Lightly crush and then dry potassium acid phthalate ($HOOCC_6H_4COOK$) to constant weight at 120 C, dissolve 425 mg in distilled water, and dilute to 1,000 mL. Potassium hydrogen phthalate has a theoretical COD of 1.176 g O_2/g and this solution has a theoretical COD of 500 mg O_2/L. Prepare fresh for each use.

4. Procedure

a. Treatment of samples with $\geqq$*50 mg COD/L:* Place 50.0 mL sample (for samples with COD >900 mg COD/L, use a smaller sample portion diluted to 50.0 mL) in the 500-mL refluxing flask. Add 1 g $HgSO_4$, several glass beads, and very slowly add 5.0 mL sulfuric acid reagent, with mixing to dissolve $HgSO_4$. Cool while mixing to avoid possible loss of volatile materials. Add 25.0 mL 0.250*N* $K_2Cr_2O_7$ solution and mix. Attach flask to condenser and turn on cooling water. Add remaining sulfuric acid reagent (70 mL) through open end of condenser. Continue swirling and mixing while adding sulfuric acid reagent. CAUTION: *Mix reflux mixture thoroughly before applying heat to prevent local heating of flask bottom and a possible blowout of flask contents.* If sample volumes other than 50 mL are used, keep ratios of reagent weights, volumes, and strengths constant. See Table 508:I for examples of applicable ratios. Maintain these ratios and follow the procedure as outlined above.

Use 1 g $HgSO_4$ with a 50.0-mL sample to complex up to a maximum of 100 mg chloride (2,000 mg/L) For smaller samples use less $HgSO_4$, according to the chloride concentration; maintain a 10:1 ratio of $HgSO_4$:Cl. A slight precipitate does not affect the determination adversely. Gener-

*Corning 5000 or equivalent.

†Corning 2360, 91548, or equivalent.

‡G. F. Smith Chemical Co., Columbus, Ohio.

TABLE 508:I. REAGENT QUANTITIES AND NORMALITIES FOR VARIOUS SAMPLE SIZES

Sample Size *mL*	0.25*N* Standard Dichromate *mL*	Sulfuric Acid Reagent *mL*	$HgSO_4$ *g*	Normality of FAS	Final Volume before Titration *mL*
10.0	5.0	15	0.2	0.05	70
20.0	10.0	30	0.4	0.10	140
30.0	15.0	45	0.6	0.15	210
40.0	20.0	60	0.8	0.20	280
50.0	25.0	75	1.0	0.25	350

ally, COD cannot be measured accurately in samples containing more than 2,000 mg chloride/L.

Reflux mixture for 2 hr. Use a shorter period for particular wastes if it has been shown that the shorter period yields the same COD as that found by 2-hr refluxing. Cover open end of condenser with a small beaker to prevent foreign material from entering refluxing mixture. Cool and wash down condenser with distilled water.

Disconnect reflux condenser and dilute mixture to about twice its volume with distilled water. Cool to room temperature and titrate excess $K_2Cr_2O_7$ with FAS, using 0.10 to 0.15 mL (2 to 3 drops) ferroin indicator. Although the quantity of ferroin indicator is not critical, use the same volume for all titrations. Take as the end point of the titration the first sharp color change from blue-green to reddish brown. The blue-green may reappear.

Reflux and titrate in the same manner a blank containing the reagents and a volume of distilled water equal to that of sample.

b. Alternate procedure for low-COD samples: Follow the above procedure, ¶ 4*a*, with two exceptions: (*i*) Use standard 0.025*N* $K_2Cr_2O_7$, and (*ii*) titrate with 0.025*N* FAS. Exercise extreme care with this procedure because even a trace of organic matter on glassware or from the atmosphere may cause gross errors.

If a further increase in sensitivity is required, concentrate a larger volume of sample before digesting under reflux as follows: Add all reagents to a sample larger than 50 mL and reduce total volume to 150 mL by boiling in the refluxing flask open to the atmosphere without the condenser attached. Compute amount of $HgSO_4$ to be added (before concentration) on the basis of a weight ratio of 10:1, $HgSO_4$:Cl, using the amount of chloride present in the original volume of sample. Carry a blank reagent through the same procedure.

This technic has the advantage of concentrating the sample without significant losses of easily digested volatile materials. Hard-to-digest volatile materials such as volatile acids are lost, but an improvement is gained over ordinary evaporative concentration methods.

c. Determination of standard solution: Evaluate the technic and quality of reagents by testing a standard potassium hydrogen phthalate solution.

5. Calculation

$$\text{mg COD/L} = \frac{(A - B) \times N \times 8{,}000}{\text{mL sample}}$$

where:

A = volume FAS used for blank, mL,
B = volume FAS used for sample, mL, and
N = normality of FAS.

6. Precision and Accuracy

A set of synthetic samples containing potassium hydrogen phthalate and NaCl was tested by 74 laboratories.[5] At 200 mg COD/L in the absence of chloride, the standard deviation was ± 13 mg/L (coefficient of variation, 6.5%). At 160 mg COD/L and 100 mg chloride/L the standard deviation was ± 14 mg/L (coefficient of variation, 10.8%).

508 B. References

1. MOORE, W.A., R. C. KRONER & C.C. RUCHHOFT. 1949. Dichromate reflux method for determination of oxygen consumed. *Anal. Chem.* 21:953.
2. MOORE, W.A., F. J. LUDZACK & C.C. RUCHHOFT. 1951. Determination of oxygen-consumed values of organic wastes. *Anal. Chem.* 23:1297.
3. MEDALIA, A.I. 1951. Test for traces of organic matter in water. *Anal. Chem.* 23:1318.
4. DOBBS, R.A. & R.T. WILLIAMS. 1963. Elimination of chloride interference in the chemical oxygen demand test. *Anal. Chem.* 35:1064.
5. ANALYTICAL REFERENCE SERVICE, USHEW-PHS. 1965. Oxygen Demand No. 2. Study No. 21, Environmental Health Ser., Water. PHS Publ. No. 999-WP-26.

509 PESTICIDES (ORGANIC)

Large-scale application of pesticides in agricultural and forest areas can contribute to the presence of these toxic materials in surface and groundwaters and ultimately in water supplies. Contamination can occur through drainage from surrounding terrain, precipitation from the atmosphere, accidental spills of pesticides in the watershed area, or a cross-connection on a distribution system.

Gas chromatographic methods for the determination of organochlorine pesticides and chlorinated phenoxy acid herbicides in water are presented here.

509 A. Organochlorine Pesticides (TENTATIVE)

1. General Discussion

a. Principle: This gas chromatographic procedure is suitable for quantitative determination of the following specific compounds: BHC, lindane, heptachlor, aldrin, heptachlor epoxide, dieldrin, endrin, Captan, DDE, DDD, DDT, methoxychlor, endosulfan, dichloran, mirex, and pentachloronitrobenzene. Under favorable circumstances, Strobane, toxaphene, chlordane (tech.), and others also may be determined. Certain organophosphorus pesticides, such as parathion, methylparathion, and malathion, which respond to the electron-capture detector, also may be measured. However, the usefulness of the method for organophosphorus or other specific pesticides must be demonstrated before it is applied to sample analysis.

In this procedure the pesticides are extracted with a mixed solvent, diethyl ether/hexane or methylene chloride/hexane. The extract is concentrated by evaporation and, if necessary, is cleaned up by column absorption-chromatography. The individual pesticides then are determined by gas chromatography.

In gas chromatography a mobile phase (a carrier gas) and a stationary phase (column packing) are used to separate individual compounds. The carrier gas is nitrogen, argon-methane, helium, or hydrogen. The stationary phase is a liquid that has been coated on an inert granular solid, called the column packing, that is held in borosilicate glass tubing. The column is installed in an oven so that the inlet is attached to a heated injector block and the outlet is attached to a detector. Precise and constant temperature control of the injector block, oven, and detector is maintained. Stationary-phase material and concentration, column length and diameter, oven temperature, carrier-gas flow, and detector type are the controlled variables.

The sample solution is injected through a silicone rubber septum onto the column with a microsyringe. The pesticides are vaporized and moved through the column by the carrier gas. They travel through the column at different rates, depending on differences in partition coefficients between the mobile and stationary phases. As each component passes through the detector a quantitatively proportional change in electrical signal is measured on a strip-chart recorder. Each component is observed as a peak on the recorder chart. The retention time is indicative of the particular pesticide and peak height/peak area is proportional to its quantity.

Variables may be manipulated to obtain important confirmatory identification data. For example, the detector system may be selected on the basis of the specificity and sensitivity needed. The detector used in this method is an electron-capture detector that is very sensitive to chlorinated compounds. Additional confirmatory identification can be made from retention data on two or more columns where the stationary phases are of different polarities. A two-column procedure that has been found particularly useful is specified.

b. Interference: Some substances other than chlorinated compounds respond to the electron-capture detector. Among these are oxygenated and unsaturated compounds. Sometimes plant or animal extractives obscure pesticide peaks. These interfering substances often can be removed by ancillary cleanup technics. A magnesia-silica gel column cleanup and separation procedure is used for this purpose. Such cleanup usually is not required for potable waters.

1) Polychlorinated biphenyls (PCB's)—Industrial plasticizers and hydraulic fluids such as PCB's are a potential source of interference in pesticide analysis. The presence of PCB's is suggested by a large number of partially resolved or unresolved peaks that may occur throughout the entire chromatogram. Particularly severe PCB interference will require special separation procedures.[1]

2) Phthalate esters—These compounds, widely used as plasticizers, respond to the electron-capture detector and are a source of interferences. Water leaches these materials from plastics, such as polyethylene bottles and plastic tubing. Phthalates can be separated from many important pesticides by the magnesia-silica gel column cleanup. They do not respond to halogen-specific detectors such as microcoulometric or electrolytic conductivity detectors.

c. Detection limits: The ultimate detection limit of a substance is affected by many factors, for example, detector sensitivity, extraction and cleanup efficiency, concentrations, and detector signal-to-noise level. Lindane usually can be determined at 10 ng/L in a sample of relatively unpolluted water; the DDT detection limit is somewhat higher, 20 to 25 ng/L. Increased sensitivity is likely to increase interference with all pesticides.

d. Sample preservation: Some pesticides are unstable. Transport under iced conditions, store at 4 C until extraction, and do not hold more than 7 days. When possible, extract upon receipt in the laboratory and store extracts at 4 C until analyzed.

2. Apparatus

Clean thoroughly all glassware used in sample collection and pesticide residue analyses. Clean glassware as soon as possible after use. Rinse with water or the solvent that was last used in it, wash with soapy water, rinse with tap water, distilled water, redistilled acetone, and finally with pesticide-quality hexane. As a precaution, glassware may be rinsed with the extracting solvent just before use. Heat heavily contaminated glassware in a muffle furnace at 400 C for 15 to 30 min. High-boiling-point materials, such as PCB's, may require overnight heating at 500 C, but no borosilicate glassware can exceed this temperature without risk. Do not heat volumetric ware. Clean volumetric glassware with special reagents.* Rinse with water and pesticide-quality hexane. After drying, store glassware to prevent accumulation of dust or other contaminants. Store inverted or cover mouth with foil.

a. Sample bottles: 1 L capacity, glass, with TFE-lined screw cap.

b. Evaporative concentrator, Kuderna-Danish, 500-mL flask and 10-mL graduated lower tube, or equivalent.

c. Separatory funnels, 2 L capacity, with TFE stopcock.

d. Graduated cylinders, 1 L capacity.

e. Funnels, 125 mL.

f. Glass wool, filter grade.

g. Chromatographic column, 20 mm diam by 400 mm long, with coarse fritted disk at bottom.

h. Microsyringes, 10 and 25 μL capacity.

i. Hot water bath.

j. Gas chromatograph, equipped with:

1) *Glass-lined injection port.*

2) *Electron-capture detector.*

3) *Recorder:* Potentiometric strip chart, 25 cm (10 in.), compatible with detector and associated electronics.

4) *Borosilicate glass column,* 1.8 m × 4 mm ID or 2 mm ID.

Variations in available gas chromatographic instrumentation necessitate different operating procedures for each. Therefore, refer to the manufacturer's operating manual as well as gas chromatography catalogs and other references (see Bibliography). In general, use equipment with the following features:

- Carrier gas line with a molecular sieve drying cartridge and a trap for removal of oxygen from the carrier gas. A special purifier† may be used. Use only dry carrier gas and insure that there are no gas leaks.
- Oven temperature stable to ±0.5 C or better at desired setting.
- Chromatographic columns—A well-prepared column is essential to an acceptable gas chromatographic analysis. Obtain column packings and pre-packed columns from commercial sources or prepare column packing in the laboratory.

It is inappropriate to give rigid specifications on size or composition to be used because some instruments perform better with certain columns than do others. Columns with 4-mm ID are used most commonly. The carrier gas flow is approxi-

*No Chromix, Godax, 6 Varick Place, New York, N.Y., or equivalent.

†Hydrox, Matheson Gas Products, P.O. Box E, Lyndhurst, N.J., or equivalent.

mately 60 mL/min. When 2-mm-ID columns are used, reduce carrier gas flow to about 25 mL/min. Adequate separations have been obtained using 5% OV-210 on 100/120 mesh dimethyl-dichlorosilane-treated diatomaceous earth‡ in a 2-m column. The 1.5% OV-17 and 1.95% QF-1 column is recommended for confirmatory analysis. Two additional column options are included: 3% OV-1 and mixed phase 6% OV-210 + 4% SE-30, each on dimethyl-dichlorosilane-treated diatomaceous earth, 100–120 mesh. OV-210, which is a refined form of QF-1, may be substituted for QF-1. A column is suitable when it effects adequate and reproducible resolution.

3. Reagents§

Use solvents, reagents, and other materials for pesticide analysis that are free from interferences under the condition of the analysis. Specific selection of reagents and distillation of solvents in an all-glass system may be required. "Pesticide quality" solvents usually do not require redistillation; however, determine a blank before use.

a. Hexane.

b. Petroleum ether, boiling range 30 to 60 C.

c. Diethyl ether.

d. Ethyl acetate.

e. Methylene chloride.

f. Magnesia-silica gel‖ (Florisil)™ PR (60 to 100 mesh). Purchase activated at 676 C and store in the dark in glass container with glass stopper or foil-lined screw cap; do not accept in plastic container. Before use, activate each batch overnight at 130 C in foil-covered glass container.

g. Sodium sulfate, Na_2SO_4, anhydrous, granular; do not accept in plastic container.

h. Silanized glass wool.

i. Column packing:

1) Solid support—Dimethyl dichlorosilane-treated diatomaceous earth (Gas-Chrom Q™) (100 to 120 mesh).

2) Liquid phases—OV-1, OV-210, 1.5% OV-17 + 1.95% QF-1, and 6% QF-1 +4% SE-30.

j. Carrier gas: One of the following is required:

1) *Nitrogen gas,* purified grade, moisture- and oxygen-free.

2) *Argon-methane* (95 + 5%) for use in pulse mode.

k. Pesticide reference standards: Obtain purest standards available (95 to 98%) from gas chromatographic and chemical supply houses.

l. Stock pesticide solutions: Dissolve 100 mg of each pesticide in ethyl acetate and dilute to 100 mL in a volumetric flask; 1.00 mL = 1.00 mg.

m. Intermediate pesticide solutions: Dilute 1.0 mL stock solution to 100 mL with ethyl acetate; 1.0 mL = 10 μg.

n. Working standard solutions for gas chromatography: Prepare final concentration of standards in hexane solution as required by detector sensitivty and linearity.

4. Procedure

a. Preparation of chromatograph:

1) Packing the column—Use a column constructed of borosilicate glass because other tubing materials may catalyze sample component decomposition. Pack column to a uniform density not so compact as to cause unnecessary back pressure and not so loose as to create voids during use. Do not crush packing. Rinse and dry column tubing with solvent, e.g., methylene chloride, then methanol, before

‡Gas Chrom Q, Applied Science Labs., Inc., P.O. Box 440, State College, Pa., or equivalent.

§Gas chromatographic methods are extremely sensitive to the materials used. Mention of trade names by "Standard Methods" does not preclude the use of other existing or as-yet-undeveloped products that give *demonstrably* equal results.

‖Floridin Co., 3 Penn Center, Pittsburgh, Pa., or equivalent.

packing. Fill column through a funnel connected by flexible tubing to one end. Plug other end of column with about 1.3 cm silanized glass wool and fill with aid of *gentle* vibration or tapping but do not use an electric vibrator because it tends to fracture packing. Optionally, apply a vacuum to plugged end. Plug open end with silanized glass wool. In a similar manner, fill one-half of a "U"-shaped column and then the other, and plug ends with silanized glass wool.

2) Conditioning—Proper thermal and pesticide conditioning are essential to eliminate column bleed and to provide acceptable gas chromatographic analysis. The following procedure provides excellent results: Connect packed column to the injection port. *Do not* connect column to detector; however, maintain gas flow through detector by using the purge gas line, or in dual-column ovens, by connecting an unpacked column to the detector. Adjust carrier gas flow to about 50 mL/min and slowly (over a 1-hr period) raise oven temperature to 230 C. After 24 to 48 hr at this temperature the column is ready for pesticide conditioning.

Adjust oven temperature and carrier gas flow rate to approximate operating levels. Make six consecutive 10-μL injections of a concentrated pesticide mixture through column at about 15-min intervals. Prepare this injection mixture from lindane, heptachlor, aldrin, heptachlor epoxide, dieldrin, endrin, and *p,p'*-DDT, each compound at a concentration of 200 ng/μL. After pesticide conditioning, connect column to detector and let equilibrate for at least 1 hr, preferably overnight. Column is then ready for use.

3) Injection technic—

a) One acceptable technic of loading the syringe and measuring volume injected is as follows: Wet syringe needle and barrel with solvent solution of the standard or sample to be injected and expel all air bubbles. Draw entire quantity of solution into calibrated barrel and note volume. Inject into chromatograph rapidly and withdraw syringe immediately. Then partially withdraw plunger and note volume remaining in syringe. Determine volume injected by subtracting remaining volume from original volume. Clean syringe thoroughly after each injection with several solvent rinses.

Develop an injection technic with constant rhythm and timing. The "solvent flush" technic described below has been used successfully and is recommended to prevent sample blowback or distillation within the syringe needle. Flush syringe with solvent, then draw a small volume of clean solvent into syringe barrel (e.g., 1 μL in a 10-μL syringe). Remove needle from solvent and draw 1 μL of air into barrel. Draw 3 to 4 μL of sample extract into barrel. Remove needle from sample extract and draw approximately 1 μL air into barrel. Record volume of sample extract between air pockets. Rapidly insert needle through inlet septum, depress plunger, and withdraw syringe. After each injection thoroughly clean syringe by rinsing several times with solvent.

b) Inject standard solutions of such concentration that the injection volume and peak height of the standard are approximately the same as those of the sample.

b. Treatment of samples:

1) Sample collection—Fill sample bottle to neck. Collect samples in duplicate.

2) Extraction of samples—Shake sample well and accurately measure all the sample in a 1-L graduated cylinder in two measuring operations if necessary. Pour sample into a 2-L separatory funnel. Rinse sample bottle and cylinder with 60 mL 15% diethyl ether or methylene chloride in hexane, pour this solvent into separatory funnel, and shake vigorously for 2 min. Let phases separate for at least 10 min.

Drain water phase from separatory funnel into sample bottle and carefully pour organic phase through a 2-cm-OD column containing 8 to 10 cm of Na_2SO_4 into a Kuderna-Danish apparatus fitted with a 10-mL concentrator tube.

Rinse sample bottle with 60 mL mixed solvent, use solvent to repeat sample extraction, and pass organic phase through Na_2SO_4. Complete a third extraction with 60 mL of mixed solvent that was used to rinse sample bottle again, and pass organic phase through Na_2SO_4. Wash Na_2SO_4 with several portions of hexane and drain well. Fit Kuderna-Danish apparatus with a three-ball Snyder column and reduce volume to about 7 mL in a hot water bath (90 to 95 C). At this point all methylene chloride present in the initial extracting solvent has been distilled off. Cool, remove concentrator tube from Kuderna-Danish apparatus, rinse ground-glass joint, and dilute to 10 mL with hexane. Make initial gas chromatographic analysis at this dilution.

3) Gas chromatography—Inject 3 to 4 μL of extract solution into a column. Always inject the same volume. Inspect resulting chromatogram for peaks corresponding to pesticides of concern and for presence of interferences.

a) If there are presumptive pesticide peaks and no significant interference, rechromatograph the extract solution on an alternate column.

b) Inject standards frequently to insure optimum operating conditions. If neces-

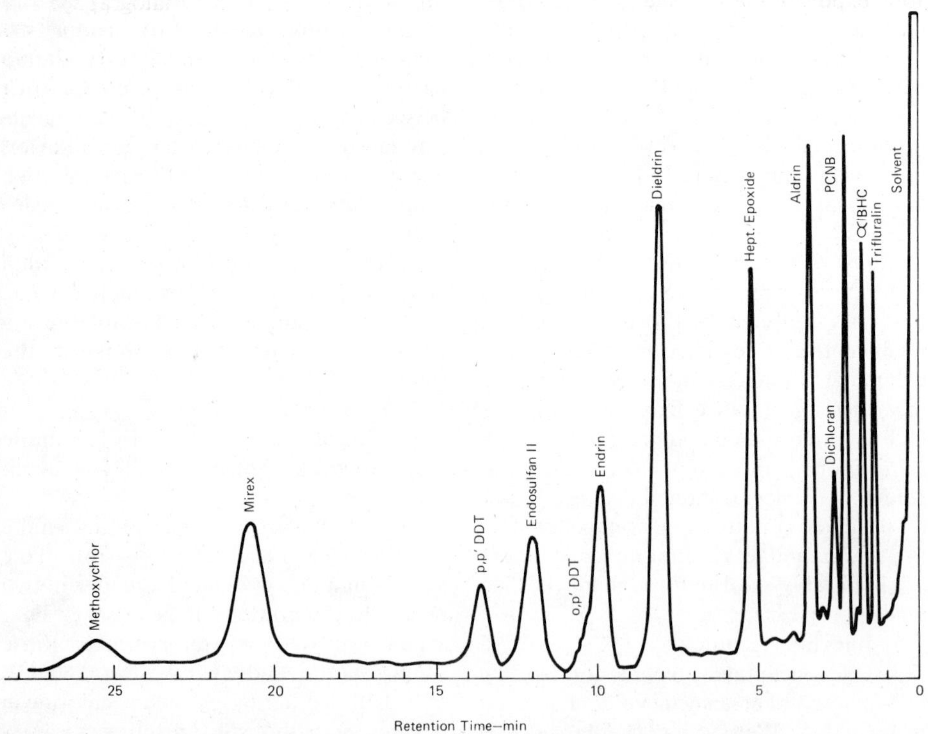

Figure 509:1. Results of gas chromatographic procedure for organochlorine pesticides. Column packing: 1.5% OV-17 + 1.95% QF-1; carrier gas: argon/methane at 60 mL/min; column temperature: 200 C; detector: electron capture in pulse mode.

sary, concentrate or dilute (*do not use methylene chloride*) extract so that peak height of sample is very close to height of corresponding peaks in standard. (See dilution factor, ¶ 5*a*).

c) If significant interference is present, separate interfering substances from pesticide materials by using cleanup procedure described in the following paragraph.

4) Magnesia-silica gel cleanup—Adjust sample extract volume to 10 mL with hexane. Place a charge of activated magnesia-silica gel (weight determined by lauric-acid value, see Appendix) in a chromatographic column. After settling gel by tapping column, add about 1.3 cm anhydrous granular Na_2SO_4 to the top. Pre-elute column, after cooling, with 50 to 60 mL petroleum ether. Discard eluate and just before exposing sulfate layer to air, quantitatively transfer sample extract into column by careful decantation and with subsequent petroleum ether washings (5 mL maximum). Adjust elution rate to about 5 mL/min and, separately, collect up to four eluates in 500-mL Kuderna-Danish flasks equipped with 10-mL receivers.

Make first elution with 200 mL 6% ethyl ether in petroleum ether, and the second with 200 mL 15% ethyl ether in petroleum ether. Make third elution with 200 mL 50% ethyl ether-petroleum ether and the fourth with 200 mL 100% ethyl ether. Add 50 to 100 mL petroleum ether to the fourth eluate to insure removal of all ethyl ether.

Concentrate eluates in Kuderna-Danish evaporator in a hot water bath as in ¶ 4*b*2) preceding, dilute to appropriate volume, and analyze by gas chromatography.

Eluate composition—By use of an equivalent quantity of any batch of magnesia-silica gel as determined by its lauric acid value (see Appendix) the pesticides will be separated into the eluates indicated below:

6% Ethyl Ether Eluate

Aldrin	Heptachlor	Pentachloro-nitrobenzene
BHC	Heptachlor epoxide	Strobane
Chlordane	Lindane	Toxaphene
DDD	Methoxychlor	Trifluralin
DDE	Mirex	PCB's
DDT		

15% Ethyl Ether Eluate	*50% Ethyl Ether Eluate*
Endosulfan I	Endosulfan II
Endrin	Captan
Dieldrin	
Dichloran	
Phthalate esters	

If present, certain thiophosphate pesticides will occur in each of the above fractions as well as in the 100% ether fraction. For additional information regarding eluate composition and the procedure for determining the lauric acid value, refer to

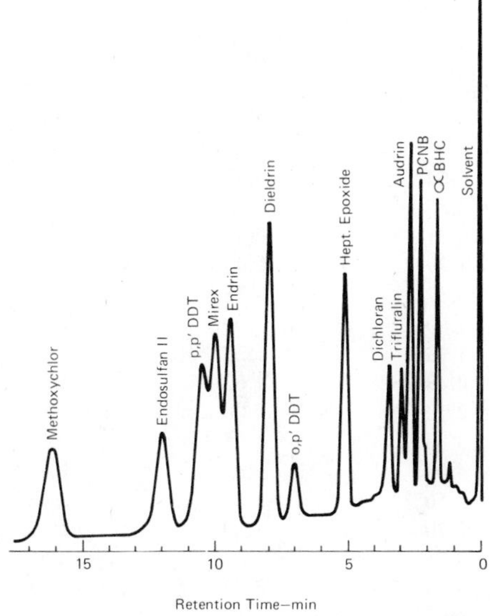

Figure 509:2. Results of gas chromatographic procedure for organochlorine pesticides. Column packing: 5% OV-210; carrier gas: argon/methane at 70 mL/min; column temperature: 180 C; detector: electron capture.

the FDA Pesticide Analytical Manual (see Bibliography).

5) Determination of extraction efficiency—Add known amounts of pesticides in ethyl acetate solution to 1 L water sample and carry through the same procedure as for samples. Dilute an equal amount of intermediate pesticide solution (¶3*m* above) to the same final volume. Call peak height from standard "*a*" and peak height from sample to which pesticide was added "*b*", whereupon the extraction efficiency equals *b*/*a*. Periodically determine extraction efficiency and a control blank to test the procedure. Also analyze one set of duplicates with each series of samples as a quality control check.

5. Calculation

a. Dilution factor: If a portion of the extract solution was concentrated, the dilution factor, *D*, is a decimal; if it was diluted, the dilution factor exceeds 1.

b. Determine pesticide concentrations by direct comparison to a single standard when the injection volume and response

TABLE 509:I. RETENTION RATIOS OF VARIOUS ORGANOCHLORINE PESTICIDES RELATIVE TO ALDRIN

Liquid phase*	1.5% OV-17 + 1.95% QF-1	5% OV-210	3% OV-1	6% QF-1 + 4% SE-30
Column temperature	200 C	180 C	180 C	200 C
Argon/methane carrier flow	60 mL/min	70 mL/min	70 mL/min	60 mL/min
Pesticide	RR	RR	RR	RR
∝-BHC	0.54	0.64	0.35	0.49
PCNB	0.68	0.85	0.49	0.63
Lindane	0.69	0.81	0.44	0.60
Dichloran	0.77	1.29	0.49	0.70
Heptachlor	0.82	0.87	0.78	0.83
Aldrin	1.00	1.00	1.00	1.00
Heptachlor epoxide	1.54	1.93	1.28	1.43
Endosulfan I	1.95	2.48	1.62	1.79
p,p'-DDE	2.23	2.10	2.00	1.82
Dieldrin	2.40	3.00	1.93	2.12
Captan	2.59	4.09	1.22	1.94
Endrin	2.93	3.56	2.18	2.42
o,p'-DDT	3.16	2.70	2.69	2.39
p,p'-DDD	3.48	3.75	2.61	2.55
Endosulfan II	3.59	4.59	2.25	2.72
p,p'-DDT	4.18	4.07	3.50	3.12
Mirex	6.1	3.78	6.6	4.79
Methoxychlor	7.6	6.5	5.7	4.60
Aldrin (Min absolute)	3.5	2.6	4.0	5.6

*All columns glass, 180 cm × 4 mm ID, solid support Gas-Chrom Q (100/120 mesh).

are close to that of the sample (Table 509:I, Figures 509:1 and 2). Calculate concentration of pesticide:

$$\mu g/L = \frac{A \times B \times C \times D}{E \times F \times G}$$

where:

A = ng standard pesticide,
B = peak height of sample, mm,
C = extract volume, μL,
D = dilution factor,
E = peak height of standard, mm,
F = volume of extract injected, μL, and
G = volume of sample extracted, mL.

Typical chromatograms of certain pesticide mixtures are shown in Figures 509:3 and 509:4.

Report results in micrograms per liter without correction for efficiency.

6. Precision and Accuracy

Precision and accuracy data are given in Table 509:II.

TABLE 509:II. PRECISION AND ACCURACY DATA FOR SELECTED ORGANOCHLORINE PESTICIDES

Pesticide	Level Added *ng/L*	Pre-treatment	Mean Recovery *ng/L*	Recovery %	Precision* *ng/L*	
					S_T	S_O
Aldrin	15	No cleanup	10.42	69	4.86	2.59
	110		79.00	72	32.06	20.19
	25	Cleanup†	17.00	68	9.13	3.48‡
	100		64.54	65	27.16	8.02‡
Lindane	10	No cleanup	9.67	97	5.28	3.47
	100		72.91	73	26.23	11.49‡
	15	Cleanup†	14.04	94	8.73	5.20
	85		59.08	70	27.49	7.75‡
Dieldrin	20	No cleanup	21.54	108	18.16	17.92
	125		105.83	85	30.41	21.84
	25	Cleanup	17.52	70	10.44	5.10‡
	130		84.29	65	34.45	16.79‡
DDT	40	No cleanup	40.30	101	15.96	13.42
	200		154.87	77	38.80	24.02
	30	Cleanup†	35.54	118	22.62	22.50
	185		132.08	71	49.83	25.31

*S_T = overall precision and S_O = single operator precision.
†Use of magnesia-silica gel column cleanup before analysis.
‡$S_o < S_T/2$.

Appendix—Standardization of Magnesia-Silica Gel* Column by Weight Adjustment Based on Adsorption of Lauric Acid

A rapid method for determining adsorptive capacity of magnesia-silica gel is based on adsorption of lauric acid from hexane solution. An excess of lauric acid is used and the amount not adsorbed is measured by alkali titration. The weight of

*Florisil or equivalent.

lauric acid adsorbed is used to calculate, by simple proportion, equivalent quantities of gel for batches having different adsorptive capacities.

1. Reagents

a. Ethyl alcohol, USP or absolute, neutralized to phenolphthalein.

b. Hexane, distilled from all-glass apparatus.

c. Lauric acid solution: Transfer 10.000 g lauric acid to a 500-mL volumetric flask, dissolve in hexane, and dilute to 500 mL; 1.00 mL= 20 mg.

d. Phenolphthalein indicator: Dissolve 1 g in alcohol and dilute to 100 mL.

e. Sodium hydroxide, 0.05*N:* Dilute 25 mL 1*N* NaOH to 500 mL with distilled water. Standardize as follows: Weigh 100 to 200 mg lauric acid into 125-mL erlenmeyer flask; add 50 mL neutralized ethyl alcohol and 3 drops phenolphthalein indicator; titrate to permanent end point; and calculate milligrams lauric acid per milliliter NaOH (about 10 mg/mL).

2. Procedure

Transfer 2.000 g magnesia-silica gel to a 25-mL glass-stoppered erlenmeyer flask. Cover loosely with aluminum foil and heat overnight at 130 C. Stopper, cool to room temperature, add 20.0 mL lauric acid solu-

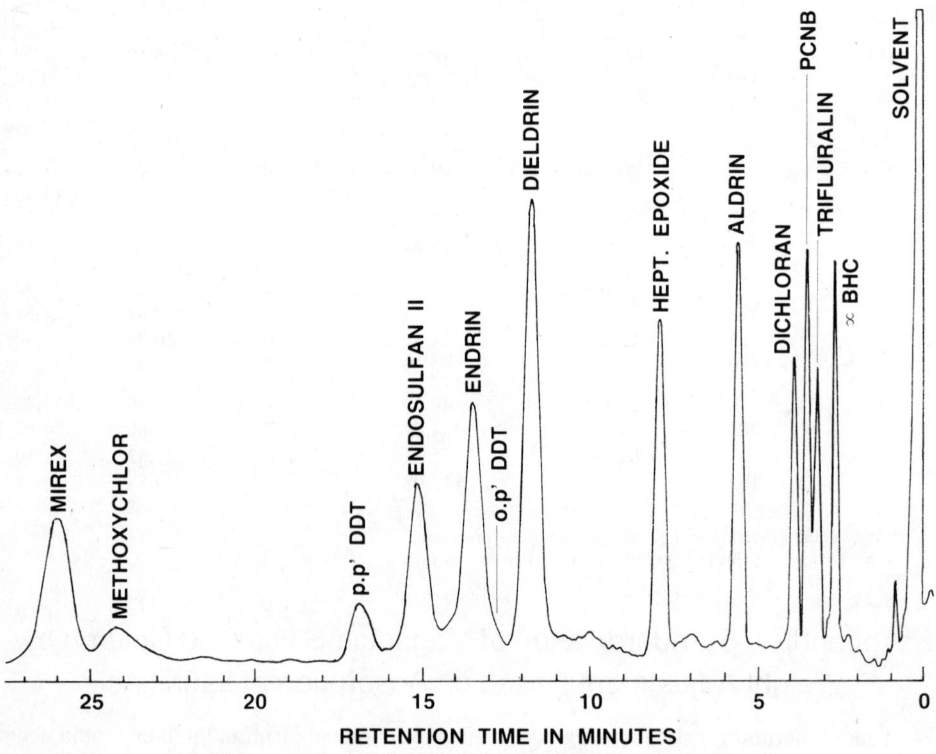

Figure 509:3. Chromatogram of pesticide mixture. Column packing: 6% QF-1 + 4% SE-30; carrier gas: argon/methane at 60 mL/min; column temperature: 200 C; detector: electron capture.

tion (400 mg), stopper, and shake occasionally during 15 min. Let adsorbent settle and pipet 10.0 mL supernatant into a 125-mL erlenmeyer flask. Avoid including any gel. Add 50 mL neutral alcohol and 3 drops phenolphthalein indicator solution; titrate with 0.05*N* NaOH to a permanent end point.

3. Calculation of Lauric Acid Value and Adjustment of Column Weight

Calculate amount of lauric acid adsorbed on gel as follows:

Lauric acid value = mg lauric acid/g gel = 200 − (mL required for titration × mg lauric acid/mL 0.05*N* NaOH).

To obtain an equivalent quantity of any batch of gel, divide 110 by lauric acid value for that batch and multiply by 20 g. Verify proper elution of pesticides by the procedure given below.

4. Test for Proper Elution Pattern and Recovery of Pesticides

Prepare a test mixture containing aldrin, heptachlor epoxide, p,p′-DDE, dieldrin, parathion, and malathion. Dieldrin and parathion should elute in the 15% eluate; all but a trace of malathion in the 50% eluate, and the others in the 6% eluate.

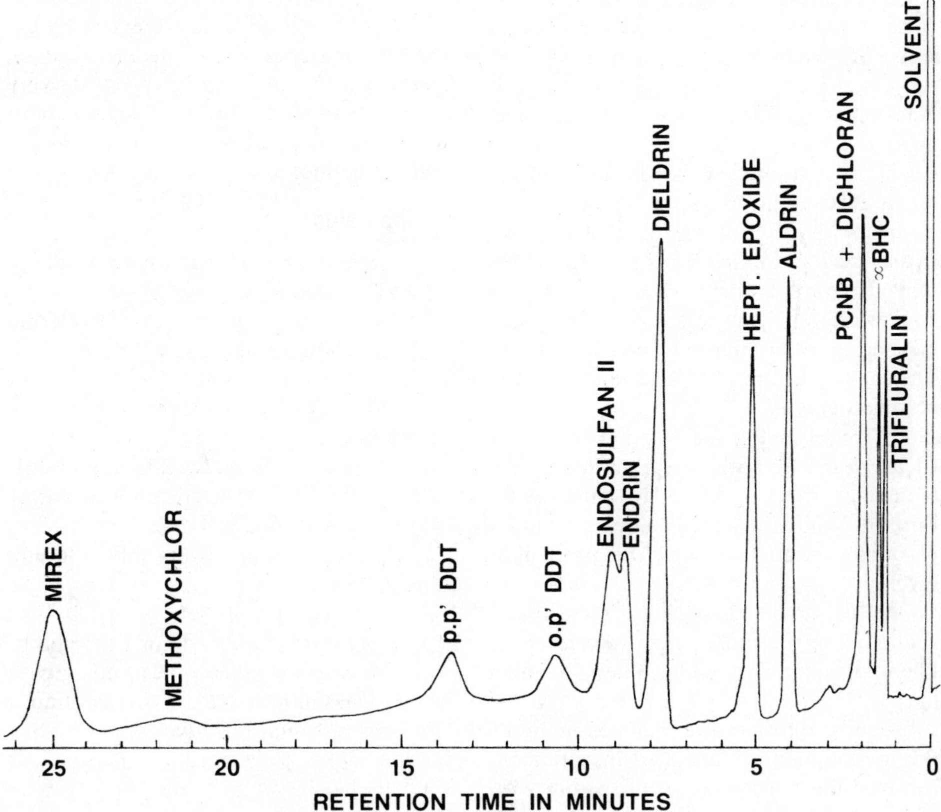

Figure 509:4. Chromatogram of pesticide mixture. Column packing: 3% OV-1; carrier gas: argon/methane at 70 mL/min; column temperature: 180 C; detector: electron capture.

509 B. Chlorinated Phenoxy Acid Herbicides (TENTATIVE)

Phenoxy acid herbicides are used extensively for weed control. Esters and salts of 2.4-D and silvex have been used as aquatic herbicides in lakes, streams, and irrigation canals. Phenoxy acid herbicides are very potent even at low concentrations.

1. General Discussion

a. Principle: Chlorinated phenoxy acid herbicides such as 2,4-D [2,4-dichlorophenoxyacetic acid], silvex [2-(2,4,5-trichlorophenoxy) propionic acid], 2,4,5-T [2,4,5-trichlorophenoxyacetic acid], and similar chemicals may be determined by a gas chromatographic procedure.

Because these compounds may occur in water in various forms (e.g., acid, salt, ester) a hydrolysis step is included to permit determination of the active part of the herbicide.

Chlorinated phenoxy acids and their esters are extracted from the acidified water sample with ethyl ether. The extracts are hydrolyzed and extraneous material is removed by a solvent wash. The acids are converted to methyl esters and are further cleaned up on a microadsorption column. The methyl esters are determined by gas chromatography.

b. Interference: See Section 509A.1*b*. Clean glassware with detergent in the usual manner, rinse in dilute HCl, and finally rinse in distilled water. To assure removal of organic matter, follow the procedure given in Section 509A.2.

Organic acids, especially chlorinated acids, cause the most direct interference. Phenols, including chlorophenols, also may interfere. Alkaline hydrolysis and subsequent extraction eliminates many of the predominant chlorinated insecticides. Because the herbicides react readily with alkaline substances, loss may occur if there is alkaline contact at any time except in the controlled alkaline hydrolysis step. Acid-rinse glassware and glass wool and acidify sodium sulfate (Na_2SO_4) to avoid this possibility.

c. Detection limits: The practical lower limits for measurement of phenoxy acid herbicides depend primarily on sample size and instrumentation used. If the extract from a 1-L sample is concentrated to 2.00 mL and 5.0 μL of concentrate is injected into the electron-capture gas chromatograph, reliable measurement of 50 ng 2,4-D/L, 10 ng silvex/L, and 10 ng 2,4,5-T/L is feasible. Concentrating extract to 0.50 mL permits detection of approximately 10 ng 2,4-D/L, 2 ng silvex/L, and 2 ng 2,4,5-T/L. The sensitivity of the electron-capture detector often is affected adversely by extraneous material in sample or reagents. Concentrating the extract progressively amplifies this complication. Thus, the practical lower limits of measurement are difficult to define.

2. Apparatus

a. Sample bottles: 1-L capacity, glass, with TFE-lined screw cap.

b. Evaporative concentrator, Kuderna-Danish, 250-mL flask and 5-mL volumetric receiver.*

c. Snyder columns, three-ball macro, one-ball micro.

d. Separatory funnels, 2-L and 60-mL sizes with TFE stopcocks and taper ground-glass stoppers.*

e. Pipets, Pasteur, disposable, 140 mm long and 5 mm ID, glass.

f. Microsyringes, 10 μL.

g. Sand bath, fluidized† or water bath.

h. Erlenmeyer flask, 250-mL with ground-glass mouth to fit Snyder columns.

i. Gas chromatographic system: See Section 509A.2*j*. Operating parameters

*Kontes or equivalent.

†TeCam or equivalent.

that produce satisfactory chromatograms for herbicide analyses are: injector temperature, 215 C; oven temperature, 185 C; and carrier gas flow 70 mL/min in a 4-mm-ID column.

3. Reagents‡

Check all reagents for purity by the gas chromatographic procedure. Save time and effort by selecting high-quality reagents that do not require further preparation. Some purification of reagents may be necessary as outlined below. If more rigorous treatment is indicated, obtain reagent from an alternate source.

a. Ethyl ether, reagent grade. Test for peroxide and if necessary redistill in glass after refluxing over granulated sodium-lead alloy for 4 hr.

b. Toluene, pesticide quality, distilled in glass, or equivalent.

c. Sodium sulfate, Na_2SO_4, anhydrous, granular. Store at 130 C.

d. Sodium sulfate solution: Dissolve 50 mg anhydrous Na_2SO_4 in distilled water and dilute to 1 L.

e. Sodium sulfate, acidified: Add 0.1 mL conc H_2SO_4 to 100 g Na_2SO_4 slurried with enough ethyl ether to just cover the solid. Remove ether by vacuum drying. Mix 1 g of resulting solid with 5 mL distilled water and confirm that mixture pH is below 4. Store at 130 C.

f. Sulfuric acid, H_2SO_4, conc.

g. Sulfuric acid, H_2SO_4, 1 + 3. Store in refrigerator.

h. Potassium hydroxide solution: Dissolve 37 g KOH pellets in distilled water and dilute to 100 mL.

i. Boron trifluoride-methanol, 14% boron trifluoride by weight.

j. Magnesia-silica gel (Florisil™), PR grade (60 to 100 mesh). Purchase activated at 676 C and store at 130 C.

k. Glass wool, filtering grade, acid-washed.

l. Herbicide standards, acids, and methyl esters, analytical reference grade or highest purity available.

m. Stock herbicide solutions: Dissolve 100 mg herbicide or methyl ester in 60 mL ethyl ether; dilute to 100 mL in a volumetric flask with hexane; 1.00 mL = 1.00 mg.

n. Intermediate herbicide solution: Dilute 1.0 mL stock solution to 100 mL in a volumetric flask with a mixture of equal volumes of ethyl ether and benzene; 1.00 mL = 10.0 μg.

o. Standard solution for chromatography: Prepare final concentration of methyl ester standards in benzene solution according to the detector sensitivity and linearity.

4. Procedure

a. Sample extraction: Accurately measure sample (850 to 1,000 mL) in a 1-L graduated cylinder. Acidify to pH 2 with conc H_2SO_4 and pour into a 2-L separatory funnel. Add 150 mL ethyl ether and shake vigorously for 1 min. Let phases separate for at least 10 min. Occasionally, emulsions prevent adequate separation. If emulsion forms, drain off separated aqueous layer, invert separatory funnel, and shake rapidly. CAUTION: *Vent funnel frequently to prevent excessive pressure buildup.* Collect extract in a 250-mL ground-glass-stoppered erlenmeyer flask containing 2 mL KOH solution. Extract sample twice more, using 50 mL ethyl ether each time, and combine extracts in erlenmeyer flask.

b. Hydrolysis: Add 15 mL distilled water and a small boiling stone and fit flask with a three-ball Snyder column. Remove ether on a steam bath and continue heating for a total of 60 min. Transfer concentrate to a 60-mL separatory funnel. Extract

‡Chromatographic methods are extremely sensitive to the materials used. Mention of trade names by "Standard Methods" does not preclude the use of other existing or as-yet-undeveloped products that give *demonstrably* equal results.

twice, with 20 mL ethyl ether each time, and discard ether layers. The herbicides remain in the aqueous phase.

Acidify by adding 2 mL cold (4 C) 1 + 3 H_2SO_4. Extract once with 20 mL and twice with 10 mL ethyl ether each. Collect extracts in a 125-mL erlenmeyer flask containing about 0.5 g acidified anhydrous Na_2SO_4. Let extract remain in contact with Na_2SO_4 for at least 2 hr.

c. Esterification: Fit a Kuderna-Danish apparatus with a 5-mL volumetric receiver. Transfer ether extract to Kuderna-Danish apparatus through a funnel plugged with glass wool. Use liberal washing of ether. Crush any hardened Na_2SO_4 with a glass rod. Before concentrating, add 0.5 mL toluene. Reduce volume to less than 1 mL on a sand or steam bath heated to 60 to 70 C. Attach a Snyder micro-column to Kuderna-Danish receiver and concentrate to less than 0.5 mL.

Cool and add 0.5 mL boron trifluoride-methanol reagent. Use the small one-ball Snyder column as an air-cooled condenser and hold contents of receiver at 50 C for 30 min in the sand bath. Cool and add enough Na_2SO_4 solution (¶3*d* above) so that the toluene-water interface is in the neck of the Kuderna-Danish volumetric receiver flask (about 4.5 mL). Stopper flask with a ground-glass stopper and shake vigorously for about 1 min. Let stand for 3 min for phase separation.

Pipet solvent layer from receiver to top of a small column prepared by plugging a

TABLE 509:III. RETENTION TIMES FOR METHYL ESTERS OF SOME CHLORINATED PHENOXY ACID HERBICIDES RELATIVE TO 2,4-D METHYL ESTER

Herbicide	Relative Retention Time for Given Liquid Phase*	
	1.5% OV-17 + 1.95% QF-1	5% OV-210
2,4-D	1.00	1.00
Silvex	1.34	1.22
2,4,5-T	1.72	1.51
2,4-D (min. absolute)	2.00	1.62

*All columns glass, 180 cm × 4 mm ID, solid support Gas Chrom Q (100/120 mesh); column temperature 185 C; argon/methane carrier flow, operated in pulse mode, 70 mL/min.

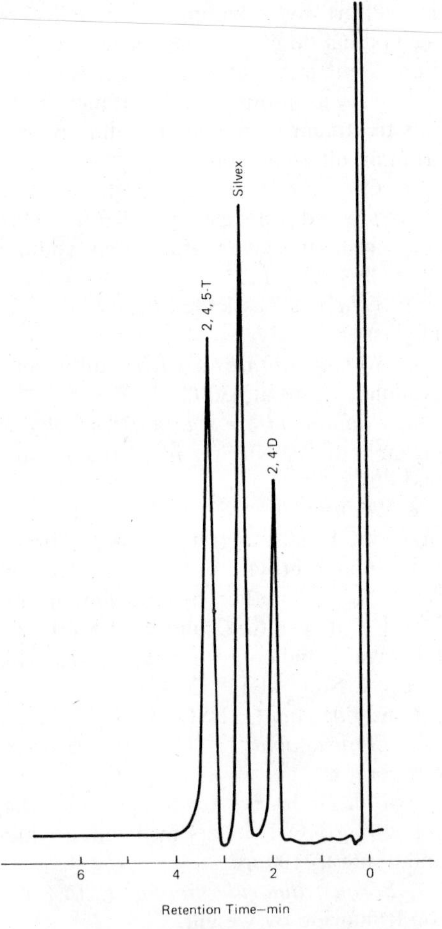

Figure 509:5. Results of gas chromatographic procedure for chlorinated phenoxy acid herbicides. Column: 1.5% OV-17; + 1.95% QF-1; carrier gas: argon (5%)/methane at 70 mL/min; column temperature: 185 C; detector: electron capture.

disposable Pasteur pipet with glass wool and packing with 2.0 cm Na_2SO_4 over 1.5 cm magnesia-silica gel adsorbent. Collect eluate in a 2.5-mL graduated centrifuge tube. Complete transfer by repeatedly rinsing volumetric receiver with small quantities of toluene until a final eluate volume of 2.0 mL is obtained. Check calibration of centrifuge tubes to insure that graduations are correct.

d. Gas chromatography: Analyze a suitable portion, 5 to 10 μL, by gas chromatography, using at least two columns for identification and quantification. Inject standard herbicide methyl esters frequently to insure optimum operating conditions. Always inject the same volume. Adjust sample volume extract with toluene, if necessary, so that the heights of the peaks obtained are close to those of the standards (see ¶ 5*a* below). For a sample chromatogram, see Figure 509:5.

e. Determination of recovery efficiency: Add known amounts of herbicides to 1-L water sample, carry through the same procedure as the samples, and determine recovery efficiency. Periodically determine recovery efficiency and a control blank to test the procedure. Analyze one set of duplicates with each series of samples as a quality control check.

5. Calculation

a. Dilution factor: If a portion of the extract solution was concentrated, the dilution factor, *D*, is less than 1; if it was diluted, the dilution factor exceeds 1.

Compare peak height of a standard to peak height of sample to determine amount of herbicide injected.

Calculate concentration of herbicide:

$$\mu g/L = \frac{A \times B \times C \times D}{E \times F \times G}$$

where:

A = weight of herbicide standard injected, ng,
B = peak height of sample, mm,
C = extract volume, μL,
D = dilution factor,
E = peak height of standard, mm,
F = volume injected, μL, and
G = volume of sample extracted, mL.

b. Molecular weight of herbicides: Molecular weights of herbicides are as follows:

Compound	*Molecular Weight*
2,4-D	222.0
2,4-D methyl ester	236.0
Silvex	269.5
Silvex methyl ester	283.5
2,4,5-T	255.5
2,4,5-T methyl ester	269.5

Report results in micrograms per liter without correction for recovery efficiency.

7. Precision and Accuracy

No collaborative study has been conducted.

509 C. Bibliography

Organochlorine Pesticides

MILL, P.A. 1968. Variation of Florisil activity: simple method for measuring absorbent capacity and its use in standardizing Florisil columns. *J. Ass. Off. Anal. Chem.* 51:29.

FOOD AND DRUG ADMINISTRATION. 1968 (revised 1978). Pesticide Analytical Manual, 2nd ed. U.S. Dep. Health, Education and Welfare. Washington, D.C.

MONSANTO CHEMICAL COMPANY. 1970. Monsanto Methodology for Arochlors—Analysis of Environmental Materials for Biphenyls, Analytical Chemistry Method 71-35. St. Louis, Mo.

U.S. ENVIRONMENTAL PROTECTION AGENCY. 1971. Method for Organic Pesticides in Water and Wastewater. National Environmental Research Center, Cincinnati, Ohio.

STEERE, N.V., ed. 1971. Handbook of Laboratory Safety. Chemical Rubber Company, Cleveland, Ohio.

GOERLITZ, D.F. & E. BROWN. 1972. Methods for analysis of organic substances in water. *In* Techniques of Water Resources Investigations of the United States Geological Survey, Book 5, Chapter A3, p. 24, U.S. Dept. of Interior, Geological Survey, Washington, D.C.

U.S. ENVIRONMENTAL PROTECTION AGENCY. 1973. Method for Organophosphorus Pesticides in Industrial Effluents. National Environmental Research Center, Cincinnati, Ohio.

U.S. ENVIRONMENTAL PROTECTION AGENCY. 1973. Method for Polychlorinated Biphenyls in Industrial Effluents. National Environmental Research Center, Cincinnati, Ohio.

U.S. ENVIRONMENTAL PROTECTION AGENCY. 1973. Handbook for Analytical Quality Control in Water and Wastewater Laboratories. National Environmental Research Center, Analytical Quality Control Laboratory, Cincinnati, Ohio.

U.S. ENVIRONMENTAL PROTECTION AGENCY. 1976. Analysis of Pesticide Residues in Human and Environmental Samples. Environmental Toxicology Div., Health Effects Laboratory, Research Triangle Park, N.C.

Chlorinated Phenoxy Acid Herbicides

METCALF, L.D. & A.A. SCHMITZ. 1961. The rapid preparation of fatty acid esters for gas chromatographic analysis. *Anal. Chem.* 33:363.

GOERLITZ, D.F. & W.L. LAMAR. 1967. Determination of phenoxy acid herbicides in water by electron-capture and microcoulometric gas chromatography. U.S. Geol. Surv. Water-Supply Paper 1817-C.

510 PHENOLS

Phenols, defined as hydroxy derivatives of benzene and its condensed nuclei, may occur in domestic and industrial wastewaters, natural waters, and potable water supplies. Chlorination of such waters may produce odorous and objectionable-tasting chlorophenols. Phenol removal processes in water treatment include superchlorination, chlorine dioxide or chloramine treatment, ozonation, and activated carbon adsorption.

Of the three analytical procedures offered here, two use the 4-aminoantipyrine colorimetric method that determines phenol, ortho- and meta-substituted phenols, and, under proper pH conditions, those para-substituted phenols in which the substitution is a carboxyl, halogen, methoxyl, or sulfonic acid group. The 4-aminoantipyrine method does not determine those para-substituted phenols where the substitution is an alkyl, aryl, nitro, benzoyl, nitroso, or aldehyde group. A typical example of these latter groups is paracresol, which may be present in certain industrial wastewaters and in polluted surface waters. The third procedure is a direct aqueous gas-liquid chromatographic technic.

1. Selection of Method

The 4-aminoantipyrine method is given in two forms: Method B, for extreme sensitivity, is adaptable for use in water samples containing less than 1 mg phenol/L. It concentrates the color in a nonaqueous solution. Method C retains the color in the aqueous solution. Because the relative amounts of various phenolic compounds in a given sample are unpredictable, it is not possible to provide a universal standard containing a mixture of phenols. For this reason, phenol (C_6H_5OH) itself has been selected as a standard for colorimetric procedures and any color produced by the reaction of other phenolic compounds is reported as phenol. Because substitution generally reduces response, this value represents the minimum concentration of phenolic compounds. Method D, a gas-liquid chromatographic procedure, may be applied to samples or concentrates that contain more than 1 mg phenolic compounds/L.

2. Interferences

a. Interferences such as phenol-decomposing bacteria, oxidizing and reducing substances, and alkaline pH values are dealt with by acidification with phosphoric acid (H_3PO_4). Some highly contaminated wastewaters may require specialized technics for eliminating interferences and for quantitative recovery of phenolic compounds.

b. Eliminate major interferences as follows (see Section 510A for reagents):

1) Oxidizing agents, such as chlorine and those detected by the liberation of iodine on acidification in the presence of potassium iodide (KI)—Remove immediately after sampling by adding excess ferrous sulfate ($FeSO_4$). If oxidizing agents are not removed, the phenolic compounds will be oxidized partially.

2) Sulfur compounds—Remove by acidifying to pH 4.0 with H_3PO_4 and aerating briefly by stirring. This eliminates the interference of hydrogen sulfide (H_2S) and sulfur dioxide (SO_2).

3) Oils and tars—Make an alkaline extraction by adjusting to pH 12 to 12.5 with NaOH pellets. Extract oil and tar from aqueous solution with 50 mL chloroform ($CHCl_3$). Discard oil- or tar-containing layer. Remove excess $CHCl_3$ in aqueous layer by warming on a water bath before proceeding with the distillation step.

3. Sampling

Sample in accordance with the instructions of Section 105.

4. Preservation and Storage of Samples

a. Phenols in concentrations usually encountered in wastewaters are subject to biological and chemical oxidation. Preserve and store samples at 4 C or lower unless analyzed within 4 hr after collection.

b. Acidify to pH 4.0, or slightly below, with H_3PO_4. If H_2S or SO_2 is known to be present, briefly aerate or stir sample with caution. Store at 4 C or lower.

c. Analyze preserved and stored samples within 24 hr after collection.

510 A. Cleanup Procedure

1. Principle

Phenols are distilled from nonvolatile impurities. Because the volatilization of phenols is gradual, the distillate volume must ultimately equal that of the original sample.

2. Apparatus

a. Distillation apparatus, all-glass, consisting of a 1-L borosilicate glass distilling apparatus with Graham condenser* (see Figure 323:1.)

b. pH meter.

3. Reagents

Prepare all reagents with distilled water free of phenols and chlorine.

*Corning No. 3360 or equivalent.

a. Phosphoric acid solution, H_3PO_4, 1 + 9: Dilute 10 mL 85% H_3PO_4 to 100 mL with water.

b. Methyl orange indicator solution.

c. Special reagents for turbid distillates:

1) *Sulfuric acid,* H_2SO_4, 1*N*.

2) *Sodium chloride,* NaCl.

3) *Chloroform,* $CHCl_3$, *or methylene chloride,* CH_2Cl_2.

4) *Sodium hydroxide,* NaOH, 2.5*N*: Dilute 41.7 mL 6*N* NaOH to 100 mL or dissolve 10 g NaOH pellets in 100 mL water.

4. Procedure

a. Measure 500 mL sample into a beaker, adjust pH to approximately 4.0 with H_3PO_4 solution using methyl orange indicator or a pH meter, and transfer to distillation apparatus. Use a 500-mL graduated cylinder as a receiver. Omit adding H_3PO_4 if sample was preserved as described in 510.4*b*.

b. Distill 450 mL, stop distillation and, when boiling ceases, add 50 mL warm water to distilling flask. Continue distillation until a total of 500 mL has been collected.

c. One distillation should purify the sample adequately. Occasionally, however, the distillate is turbid. If so, acidify with H_3PO_4 solution and distill as described in ¶ 4*b*. If second distillate is still turbid, use extraction process described in ¶ 4*d* before distilling sample.

d. Treatment when second distillate is turbid: Extract a 500-mL portion of original sample as follows: Add 4 drops methyl orange indicator and make acidic to methyl orange with 1*N* H_2SO_4. Transfer to a separatory funnel and add 150 g NaCl. Shake with five successive portions of $CHCl_3$, using 40 mL in the first portion and 25 mL in each successive portion. Transfer $CHCl_3$ layer to a second separatory funnel and shake with three successive portions of 2.5*N* NaOH solution, using 4.0 mL in the first portion and 3.0 mL in each of the next two portions. Combine alkaline extracts, heat on a water bath until $CHCl_3$ has been removed, cool, and dilute to 500 mL with distilled water. Proceed with distillation as described in ¶s 4*a* and *b*.

NOTE: CH_2Cl_2 may be used instead of $CHCl_3$, especially if an emulsion forms when the $CHCl_3$ solution is extracted with NaOH.

510 B. Chloroform Extraction Method

1. General Discussion

a. Principle: Steam-distillable phenols react with 4-aminoantipyrine at pH 7.9 ± 0.1 in the presence of potassium ferricyanide to form a colored antipyrine dye. This dye is extracted from aqueous solution with $CHCl_3$ and the absorbance is measured at 460 nm. This method covers the phenol concentration range from 1.0 μg/L with a sensitivity of 1 μg/L.

b. Interference: All interferences are eliminated or reduced to a minimum if the sample is preserved, stored, and distilled in accordance with the foregoing instructions.

c. Minimum detectable concentration: The minimum detectable quantity is 0.5 μg phenol when a 25-mL $CHCl_3$ extraction with a 5-cm cell or a 50-mL $CHCl_3$ extraction with a 10-cm cell is used in the photometric measurement. The minimum detectable quantity is 1 μg phenol/L in 500 mL distillate.

2. Apparatus

a. Photometric equipment: A spectrophotometer for use at 460 nm equipped with absorption cells providing light paths of 1 to 10 cm depending on the absorbances of the colored solutions and the individual characteristics of the photometer.

b. Filter funnels: Buchner type with fritted disk.*

c. Filter paper: Alternatively use an appropriate 11-cm filter paper for filtration of $CHCl_3$ extracts instead of the Buchner-type funnels and anhydrous Na_2SO_4.

d. pH meter.

e. Separatory funnels, 1,000-mL, Squibb form, with ground-glass stoppers and TFE stopcocks. At least eight are required.

3. Reagents

Prepare all reagents with distilled water free of phenols and chlorine.

a. Stock phenol solution: Dissolve 1.00 g phenol in freshly boiled and cooled distilled water and dilute to 1,000 mL. Ordinarily this direct weighing yields a standard solution; if extreme accuracy is required, standardize as follows:

1) To 100 mL water in a 500-mL glass-stoppered conical flask, add 50.0 mL stock phenol solution and 10.0 mL 0.1*N* bromate-bromide solution. Immediately add 5 mL conc HCl and swirl gently. If brown color of free bromine does not persist, add 10.0-mL portions of bromate-bromide solution until it does. Keep flask stoppered and let stand for 10 min; then add approximately 1 g KI. Usually four 10-mL portions of bromate-bromide solution are required if the stock phenol solution contains 1,000 mg phenol/L.

2) Prepare a blank in exactly the same manner, using distilled water and 10.0 mL 0.1*N* bromate-bromide solution. Titrate blank and sample with 0.025*N* sodium thiosulfate, using starch solution indicator.

3) Calculate the concentration of phenol solution as follows:

$$\text{mg phenol/L} = 7.842\,[(A \times B) - C]$$

where:

A = mL thiosulfate for blank,
B = mL bromate-bromide solution used for sample divided by 10, and
C = mL thiosulfate used for sample.

b. Intermediate phenol solution: Dilute 10.0 mL stock phenol solution in freshly boiled and cooled distilled water to 1,000 mL; 1 mL = 10.0 μg phenol. Prepare daily.

c. Standard phenol solution: Dilute 50.0 mL intermediate phenol solution to 500 mL with freshly boiled and cooled distilled water; 1 mL = 1.0 μg phenol. Prepare within 2 hr of use.

d. Bromate-bromide solution, 0.10*N*: Dissolve 2.784 g anhydrous $KBrO_3$ in water, add 10 g KBr crystals, dissolve, and dilute to 1,000 mL.

e. Hydrochloric acid, HCl, conc.

f. Standard sodium thiosulfate titrant, 0.025*N*: See Section 421B.2*e*.

g. Starch solution: See Section 421B.2*d*.

h. Ammonium hydroxide, NH_4OH, 0.5 *N*: Dilute 35 mL fresh, conc NH_4OH to 1 L with water.

i. Phosphate buffer solution: Dissolve 104.5 g K_2HPO_4 and 72.3 g KH_2PO_4 in water and dilute to 1 L. The pH should be 6.8.

j. 4-Aminoantipyrine solution: Dissolve 2.0 g 4-aminoantipyrine in water and dilute to 100 mL. Prepare daily.

k. Potassium ferricyanide solution: Dissolve 8.0 g $K_3Fe(CN)_6$ in water and dilute to 100 mL. Filter if necessary. Store

*15-mL Corning No. 36060 or equivalent.

in a brown glass bottle. Prepare fresh weekly.

l. Chloroform, $CHCl_3$.

m. Sodium sulfate, anhydrous Na_2SO_4, granular.

n. Potassium iodide, KI, crystals.

4. Procedure

Ordinarily, use Procedure *a*; however, Procedure *b* may be used for infrequent analyses.

a. Place 500 mL distillate, or a suitable portion containing not more than 50 μg phenol, diluted to 500 mL, in a 1-L beaker. Prepare a 500-mL distilled water blank and a series of 500-mL phenol standards containing 5, 10, 20, 30, 40, and 50 μg phenol.

Treat sample, blank, and standards as follows: Add 12.0 mL 0.5*N* NH_4OH and adjust pH to 7.9 ± 0.1 with phosphate buffer. About 10 mL phosphate buffer are required. Transfer to a 1-L separatory funnel, add 3.0 mL aminoantipyrine solution, mix well, add 3.0 mL $K_3Fe(CN)_6$ solution, mix well, and let color develop for 3 min. The solution should be clear and light yellow.

Extract immediately with $CHCl_3$ using 25 mL for 1- to 5-cm cells and 50 mL for a 10-cm cell. Shake separatory funnel at least 10 times, let $CHCl_3$ settle, shake again 10 times, and let $CHCl_3$ settle again. Filter each $CHCl_3$ extract through filter paper or fritted glass funnels containing a 5-g layer of anhydrous Na_2SO_4. Collect dried extracts in clean cells for absorbance measurements; do not add more $CHCl_3$.

Read absorbance of sample and standards against the blank at 460 nm. Plot absorbance against micrograms phenol concentration. Construct a separate calibration curve for each photometer and check each curve periodically to insure reproducibility.

b. For infrequent analyses prepare only one standard phenol solution. Prepare 500 mL standard phenol solution of a strength approximately equal to the phenolic content of that portion of original sample used for final analysis. Also prepare a 500-mL distilled water blank.

Continue as described in ¶ *a*, above, but measure absorbances of sample and standard phenol solution against the blank at 460 nm.

5. Calculation:

a. For Procedure a:

$$\mu\text{g phenol/L} = \frac{A}{B} \times 1{,}000$$

where:

A = μg phenol in sample, from calibration curve, and
B = mL original sample.

b. For Procedure b, calculate the phenol content of the original sample:

$$\mu\text{g phenol/L} = \frac{C \times D \times 1000}{E \times B}$$

where:

C = μg standard phenol solution,
D = absorbance reading of sample,
E = absorbance of standard phenol solution, and
B = mL original sample.

6. Precision and Accuracy

Because the "phenol" value is based on C_6H_5OH, this method yields only an approximation and represents the minimum amount of phenols present. This is true because the phenolic reactivity to 4-aminoantipyrine varies with the types of phenols. Using the extraction procedure for concentration of color, six laboratories analyzed samples at concentrations of 9.6, 48.3, and 93.5 μg C_6H_5OH/L with standard deviations of, respectively, ±0.99, ±3.1, and ±4.2 μg/L.

510 C. Direct Photometric Method

1. General Discussion

a. Principle: Steam-distillable phenolic compounds react with 4-aminoantipyrine at pH 7.9 ± 0.1 in the presence of potassium ferricyanide to form a colored antipyrine dye. This dye is kept in aqueous solution and the absorbance is measured at 500 nm. Because extreme sensitivity is not required in this method, smaller distillate volumes may be used.

b. Interference: Interferences are eliminated or reduced to a minimum by using the distillate from the preliminary distillation procedure.

c. Minimum detectable concentration: This method has less sensitivity than Method B. The minimum detectable quantity is 10 μg phenol when a 5-cm cell and 100 mL distillate are used.

2. Apparatus

a. Photometric equipment: Spectrophotometer equipped with absorption cells providing light paths of 1 to 5 cm for use at 500 nm.

b. pH meter.

3. Reagents

See Section 510B.3.

4. Procedure

Place 100 mL distillate, or a portion containing not more than 0.5 mg phenol diluted to 100 mL, in a 250-mL beaker. Prepare a 100-mL distilled water blank and a series of 100-mL phenol standards containing 0.1, 0.2, 0.3, 0.4, and 0.5 mg phenol. Treat sample, blank, and standards as follows: Add 2.5 mL 0.5 *N* NH_4OH solution and adjust to pH 7.9 ± 0.1 with phosphate buffer. Add 1.0 mL 4-aminoantipyrine solution, mix well, add 1.0 mL $K_3Fe(CN)_6$ solution, and mix well.

After 15 min, transfer to cells and read absorbance of sample and standards against the blank at 500 nm.

5. Calculation

a. Use of calibration curve: Estimate sample phenol content from photometric readings by using a calibration curve as directed in Section 510B.5*a*.

$$\text{mg phenol/L} = \frac{A}{B} \times 1000$$

where:

A = mg phenol in sample, from calibration curve, and
B = mL original sample.

b. Use of single phenol standard:

$$\text{mg phenol/L} = \frac{C \times D \times 1000}{E \times B}$$

where:

C = mg standard phenol solution,
D = absorbance of sample, and
E = absorbance of standard phenol solution.

6. Precision and Accuracy

Using the direct photometric procedure, six laboratories analyzed samples at concentrations of 4.7, 48.2, and 97.0 mg C_6H_5OH/L with standard deviations of ±0.18, ±0.48, and ±1.58 mg/L, respectively.

510 D. Gas-Liquid Chromatographic Method*

1. General Discussion

This method is a direct aqueous-injection procedure for the gas-liquid chromatographic determination of phenols, cresols, and mono- and dichlorophenols in water using a flame ionization detector. The method may be applied to samples containing more than 1 mg phenolic compounds/L.

Three tables (Tables 510:III-V) are appended (see Section 510D.7) that provide general analytical guidance but are not intended to define standard methodology.

a. Principle: This method specifies a single gas-liquid chromatographic column for the separation of phenolic compounds and a flame-ionization detector for their measurement. The peak area of each component is measured and compared with that of a known standard to obtain quantitative results. Elution of characteristic phenols occurs in the following order: (1) *o*-chlorophenol, (2) phenol and *o*-cresol, (3) *m*- and *p*-cresol, (4) 2,3- 2,4- 2,5- and 2,6-dichlorophenols, (5) *m*- and *p*-chlorophenols, and (6) 3,4-dichlorophenol.

b. Interferences: Particulate or suspended matter, unless very finely subdivided, may plug the microsyringe needle used for sample injection. Remove particulates by centrifugation or filtration, provided that compounds of interest are not removed also. Use a colloid mill, if necessary, to prepare a colloidal suspension suitable for injection. Particulate matter may serve as condensation nuclei for samples; acid treatment often may dissolve such interfering solids.

Nonphenolic organic compounds that have the same retention time as phenolic compounds interfere with the test. Eliminate these by distillation (Section 510A).

Under strongly alkaline conditions, some chlorophenols may form salts that reduce their volatility. Also, some nonphenolic organics—for example, tar bases—may be more volatile in basic solutions. Adjusting pH to near neutral or slightly acid will eliminate these interferences.

A "ghost" is an interference, showing as a peak, that appears at the same elution time as an organic component of a previous analysis. To minimize or eliminate this effect, inject 3 μL water between samples. The water wash usually clears the injection port, column, and detector of artifacts; however, repeated wash injections may be necessary to clear the system. Set electrometer at maximum sensitivity during wash injections to facilitate detection of ghosts. Use glass injector inserts that are easy to clean or replace.

2. Apparatus†

a. Gas chromatograph, equipped with a hydrogen flame-ionization detector: A commercial or custom-designed gas chromatograph with a column oven capable of isothermal temperature control to at least 210 ± 0.2 C. A unit for temperature programming will facilitate elution of a mixture of phenolics of wide boiling-point range. Temperature programming is optional.

b. Recorder, to measure chromatographic output at a full-scale range of 1 mV with a response time of 1 sec.

c. Chromatographic columns: Purchase columns or prepare in the laboratory. Variations in column loading, length, diameter, support size, treatment, etc.,

*Adapted from ASTM D-2580-68-Standard, as published in ASTM Book of Standards, Part 23 (1968).

†Gas chromatographic methods are extremely sensitive to the materials used. Mention of trade names by "Standard Methods" does not preclude the use of other existing or as-yet-undeveloped products that give demonstrably equivalent results.

are possible. Three columns are cited in this procedure. Modifications may alter elution time and sensitivity.

1) *Polyethylene glycol* (Carbowax 20M™): A 3.2-mm by 3-m stainless steel column packed with 60/80 mesh white diatomaceous earth (Chromosorb W™) (acid-washed and hexamethyldisilazane-treated) coated with 20% by weight of polyethylene glycol TPA (terephthalic acid).

2) *Free Fatty Acid Phase*, 1.5 m: A 3.2-mm by 1.5-m stainless steel column packed with 70/80 mesh white diatomaceous earth (Chromosorb W) (acid-washed) coated with 5% by weight Free Fatty Acid Phase (FFAP).‡

3) *Free Fatty Acid Phase*, 3m: A 3.2-mm by 3-m stainless steel column packed with 60/80 mesh screened TFE (Chromosorb T™) coated with 10% Free Fatty Acid Phase. The screened TFE is a fluorocarbon-6 product; it melts at 327 C and may begin to fuse above 250 C. It is available from suppliers of gas-chromatographic materials.

d. Syringe, 10 μL.

3. Reagents

a. Carrier gases: Research-grade N_2 or He of highest purity.

b. Hydrogen, for use with the flame-ionization detector; obtain by using a hydrogen generator, or from a high-purity tank supply.

c. Water, redistilled deionized: Treat ordinary distilled water by *both* redistillation and deionization as described in Section 417A.3*a*. Test treated water in the chromatograph to assure freedom from "ghost" peaks.

d. Phenolic compounds: Use research grades of high purity. Prepare highest-purity compounds by redistillation or recrystallization, or by using a preparatory gas-chromatographic instrument. The following phenolic compounds are suggested: *o*-chlorophenol, *m*-chlorophenol, *p*-chlorophenol, *o*-cresol, *m*-cresol, *p*-cresol, 2,3-dichlorophenol, 2,4-dichlorophenol, 2,5-dichlorophenol, 2,6-dichlorophenol, 3,4-dichlorophenol, and phenol. Prepare 100 mg/L solutions in redistilled deionized water.

4. Procedure

a. Preparation of chromatograph: Install packed column in chromatograph, using suitable fittings. Use antigalling thread lubricant.

b. Check for leaks: Test for leaks at approximately 103 kPa (15 psig) above operating pressure by shutting off the downstream end of the system and pressurizing from the carrier gas supply. Shut off cylinder valve and observe pressure gauge. If no drop occurs in 10 to 15 min, consider system to be tight. Locate minor leaks with aqueous soap solution but do so with caution, because soap solution entering the system may cause extraneous peaks or affect system stability. Do not use soap for leak-testing near the ionization detector.

c. Column conditioning: Before use, condition columns for at least 24 hr at temperatures 30 to 50 C above expected operating temperature. Do not exceed maximum allowable temperature for either packing or substrate. Disconnect column at end near detector base to avoid depositing volatiles on detector during conditioning. Adjust carrier gas flow to 20 to 40 mL/min for a 3.2-mm-diam column. Occasionally inject 3 to 5 μL water during conditioning to aid elution of impurities.

After conditioning, connect column to flame-ionization detector. Adjust H_2 flow to the detector to about 25 mL/min for a 3.2-mm-diam column. Adjust air-flow as specified for the instrument being used and ignite H_2 flame at detector. Adjust column temperature to desired level and car-

‡Available from Varian Aerograph, Walnut Creek, Calif. 94598.

rier gas flow rate to 20 to 40 mL/min. Observe recorder base line. When a baseline drift is no longer apparent, column is ready for use.

When the series of analyses is completed and column is to be removed and stored, seal or cap ends.

d. Operating conditions for analysis: Typical operating conditions are summarized in Table 510:I. If these operating parameters are varied, reconcile analytical and calibration test variations in calculating results. For example, either N_2 or He may be used as carrier gas; recorder chart speeds of approximately 76 cm/hr (30 in./hr) commonly are used; sample sizes of 3 to 5 μL usually are injected.

e. Method of compound identification: Compound identification is based on retention time—the time that elapses from introduction of sample until peak maximum is reached. Compare retention times of sample peaks with those of known standards obtained under the same operating conditions. When several related materials are eluted at the same time, reanalyze sample with a different column that produces a better separation or supplement the chromatographic procedure with spectrographic analyses. If necessary, trap the various sample components as they emerge from the system and analyze them by other appropriate methods.

To determine retention time of phenol, cresols, and mono- and dichlorophenols, take the following steps:

With the column at operating condi-

TABLE 510:I. TYPICAL OPERATING CONDITIONS FOR CHROMATOGRAPHIC COLUMN

Column and Packing	Column No. (Chromatographic Columns)		
	1 ¶2c 1)	2 ¶2c 2)	3 ¶2c 3)
Carrier gas	He	He	N_2
Carrier gas flow, mL/min	25	35	60
Temperature, C			
Injection port	250	205	250
Column	210	147	188
H_2 flow to detector, mL/min	25	25	30
Chart speed, cm/hr	30	30	30
Sensitivity, mV	1	1	1
Electrometer range	1	0.1	1
Attenuation	1	1	1
Sample volume, μL	1	1	1
Results	*	†	‡

*See Figures 510:1 and 510:2.
†See Figure 510:5.
‡See Figures 510:3 and 510:4.

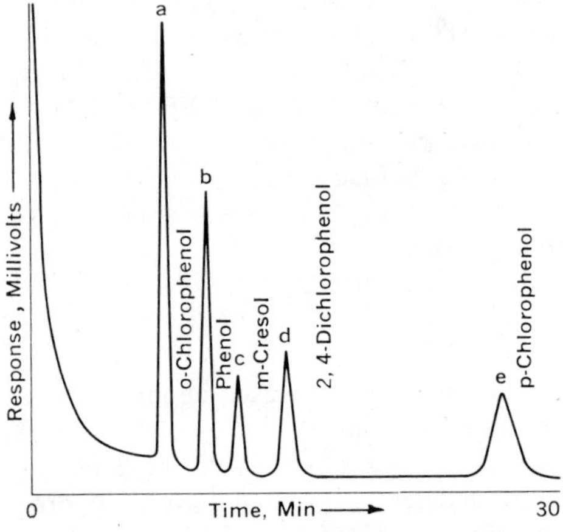

Figure 510:1. Analyses were made with 3 m by 3.2 mm stainless steel column coated with 20% Carbowax-terephthalic acid on 60/80-mesh diatomite, HMDS treated. Column temperature was 210 C, and injection temperature, 250 C. H_2 and He flow rates were each 20 mL/min at electrometer range 1 and attenuation 1, with chart speed at 30 cm/hr, 1-mV full-scale response, and 1-μL sample of approximately 100-mg/L solutions of each phenolic. Peak A is for *o*-chlorophenol; B, phenol; C, *m*-cresol; D, 2,4-dichlorophenol; E, *p*-chlorophenol.

tions, and using a 10-μL syringe, inject in turn a 1-μL sample of a 100 mg/L aqueous solution of each phenolic compound. Adjust instrument attenuation so that peak height is near 50% of full scale. Mark injection point on recorder chart. Measure retention time in minutes to at least two significant figures.

To eliminate errors induced by ghosting, inject distilled water after each sample. Set electrometer range and attenuation at maximum sensitivity during water washing. Inject same volume of water as that used for sample. Repeat water injections until a steady base line free of ghosts has been attained, then inject the next sample.

Make triplicate determinations for each phenolic compound and record average retention value. (Mixtures of phenols having different retention intervals may be injected simultaneously to expedite standardizations.)

CAUTION—Flush injection syringe used for calibration with each new sample at least three times before injecting into the chromatograph.

Calculate retention of all compounds relative to phenol. The relative retention times with approximate chromatograph calibration factors are shown in Table 510:II for two columns. These calibration values are presented for information only. Determine and regularly recheck calibra-

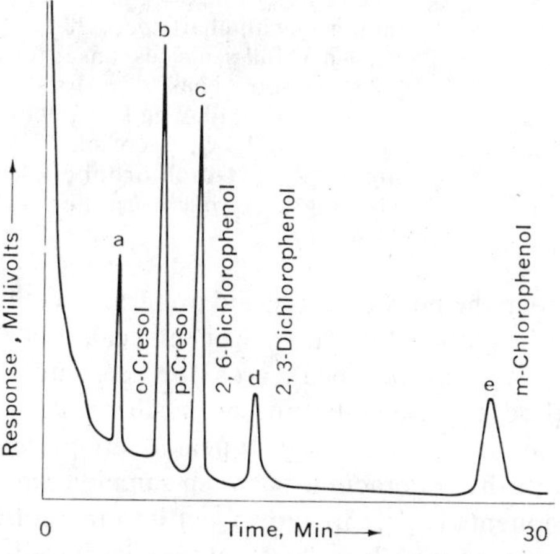

Figure 510:2. Analyses were made with 3 m by 3.2 mm stainless steel column coated with 20% Carbowax-terephthalic acid on 60/80-mesh diatomite, HMDS treated. Column temperature was 210 C, and injection temperature, 250 C. H_2 and He flow rates were each 20 mL/min, at electrometer range 1 and attenuation 1, with chart speed at 30 cm/hr, 1-mV full-scale response, and 1-μL sample of approximately 100 mg/L solutions of each phenolic. Peak A is for *o*-cresol; B, *p*-cresol; C, 2-6-dichlorophenol; D, 2,3-dichlorophenol; E, *m*-chlorophenol.

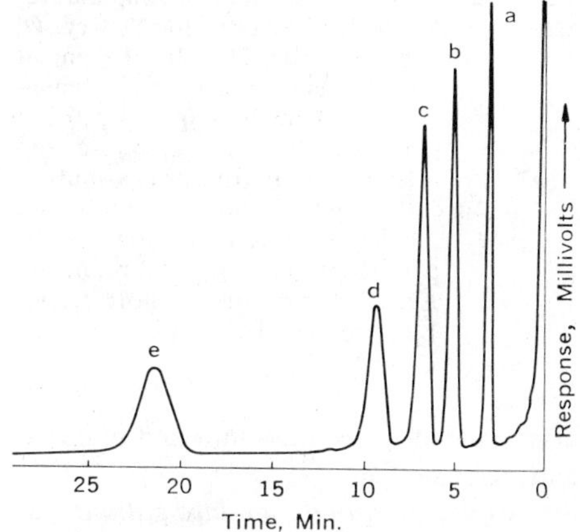

Figure 510:3. Analyses were made with 3 m by 3.2 mm stainless steel column coated with 10% polyester on 60/80-mesh fluorocarbon resin medium. Column temperature was 188 C, and injection temperature, 250 C. Flow rates for N_2 were 60 mL/min, for H_2, 30 mL/min, at electrometer range 1 and attenuation 1, with chart speed at 30 cm/hr, 1-mV full-scale response, and 1-μL sample of approximately 100-mg/L solutions of each phenolic. Peak A is for *o*-chlorophenol; B, phenol; C, *m*-cresol; D, 2,4-dichlorophenol; E, *p*-chlorophenol.

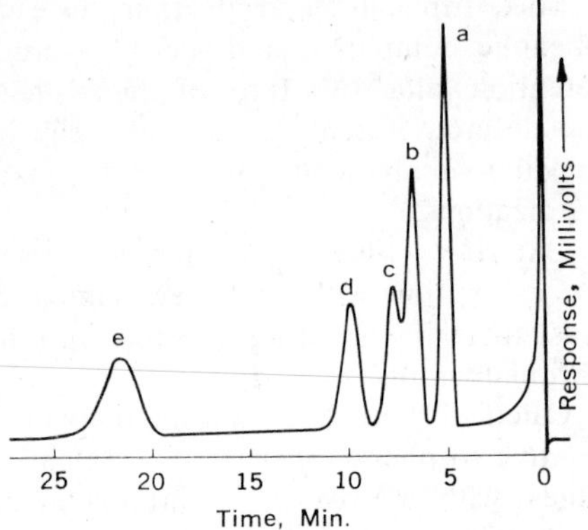

Figure 510:4. Analyses were made with 3 m by 3.2 mm stainless steel column coated with 10% polyester on 60/80-mesh fluorocarbon resin medium. Column temperature was 188 C, and injection temperature, 250 C. Flow rates for N_2 were 60 mL/min, for H_2, 30 mL/min, at electrometer range 1 and attenuation 1, with chart speed at 30 cm/hr, 1-mV full-scale response, and 1-μL sample of approximately 100-mg/L solutions of each phenolic. Peak A is for *o*-cresol; B, *p*-cresol, C, 2,6-dichlorophenol; D, 2,3-dichlorophenol; E, *m*-chlorophenol.

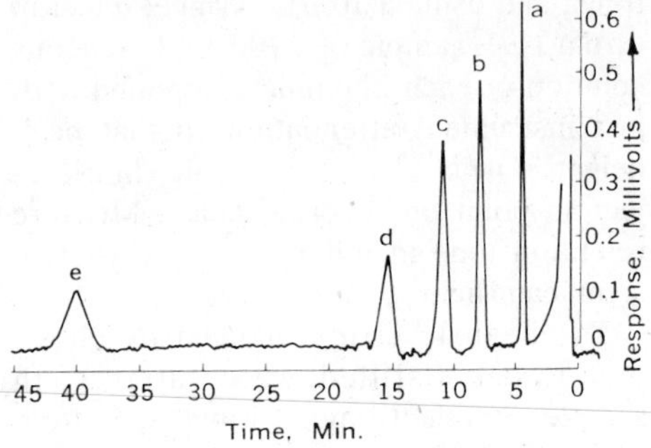

Figure 510:5. Analyses were made with 1.5 m by 3.2 mm stainless steel column coated with 5% polyester on acid-washed 70/80-mesh diatomite. Column temperature was 147 C. Flow rates for He were 35 mL/min, for H_2, 25 mL/min, at electrometer range 0.1 and attenuation 1, with chart speed 30 cm/hr, 1-mV full-scale response, for 1-μL sample. Peak A is for *o*-chlorophenol, 10.4 mg/L; B, phenol, 9.3 mg/L; C, *m*-cresol, 10.7 mg/L; D, 2,4-dichlorophenol, 10.7 mg/L; E, *p*-chlorophenol, 11.2 mg/L.

tion values for each column and phenolic material used.

f. Calibration and standardization: The area under the peak of the chromatogram is a quantitative measure of the amount of corresponding compound. To calibrate, select phenolic concentration range desired (for example, 1 mg/L or 10 mg/L) and prepare fresh solutions of each compound in redistilled deionized water.

With the column at equilibrium operating conditions, inject measured volumes (e.g., 3 μL) of standard solutions, using the previously described procedure and observing all precautions. Continue until at least three peaks are obtained that deviate by no more than $\pm 1\%$ in area at the same attenuation. Adjust attenuation to keep the peak on scale and preferably with a height of close to 50% of full-scale.

Measure the peak area for each standard by triangulation, or mechanical or electronic integration. If peak overlapping or other interaction between sample components is anticipated, inject into the column prepared mixtures of standards representing the phenolic compounds expected. Measure the peak areas as for the pure compounds. Express results as nanograms per unit area.

g. Sample treatment: With the column at operating conditions, inject 1 to 3 μL sample into the injection port. Determine retention times of phenols in the sample. If necessary, adjust attenuation to keep the highest peak on scale for the major phenolic component. Make triplicate determinations at identical column and instrument conditions; flush as required to eliminate artifacts. Characterize and measure peak

TABLE 510:II. CHROMATOGRAPHIC RETENTION TIMES

Phenolic Compound	Boiling Point, C	Relative Retention	Calibration Factor*† *ng/in.*^{2}	Relative Retention	Calibration Factor†‡ *ng/in.*^{2}
Phenol	182	1.0	45.2	1.0	28.8
o-Cresol	192	1.0	51.0	1.0	27.0
m-Cresol	203	1.3	51.9	1.3	30.5
p-Cresol	202	1.3	51.3	1.3	31.3
o-Chlorophenol	176	0.8	86.7	0.6	44.8
m-Chlorophenol	214	3.6	106	3.6	45.4*
p-Chlorophenol	217	3.6	125	3.6	46.5*
2,3-Dichlorophenol	—	1.8	76.5	1.9	46.8
2,4-Dichlorophenol	210	1.8	117	1.9	55.3
2,5-Dichlorophenol	210	1.8	113	1.9	50.4
2,6-Dichlorophenol	220	1.6	71.5	1.5	50.0
3,4-Dichlorophenol	254	—	—	11.5	43.0*

*Column 1 [¶2*c* 1)] at 210 C.
†Calibration in ng (10^{-9}g)/in.^{2} at chart speed of 90 in./hr, 1-mV response, range 1, attenuation 1.
‡Column 2 [¶2*c*3)] at 188 C.

areas obtained. Average results of triplicate determinations.

5. Calculation

a. Characterize each peak by a retention time. Report supplementary tests, such as infrared and ultraviolet used in characterizing trapped fractions.

b. For those peak areas representing two or more phenolic compounds, use an average value of the calibration obtained with the standards in the calculation or measure the area as a given material, with the notation in the results that other components have comparable elution intervals and may be represented.

c. Calculate concentration of each component by the equation:

$$\text{Phenolic compound(s), mg/L} = \frac{A \times B}{C}$$

where:

A = area of sample peak, cm^2,
B = calibration factor, ng/cm^2, and
C = sample volume, μL.

6. Precision

The precision of this method has been tested with four master solutions, each containing four phenolic compounds. The compositions of the master solutions were as follows:

Phenolic Compounds	Master Solution *mg/L* 1	2	3	4
Phenol	20	80	40	10
m-Cresol	40	10	20	80
o-Chlorophenol	10	20	80	40
2,4-Dichlorophenol	80	40	10	20

The respective precisions may be expressed as follows:

Phenol: $S_T = 0.048X + 0.6$
$S_O = 0.017X + 0.3$
m-Cresol: $S_T = 0.029X + 2.2$
$S_O = 0.014X + 0.7$
o-Chlorophenol: $S_T = 0.083X + 1.2$
$S_O = 0.031X + 0.2$
2,4-Dichlorophenol: $S_T = 0.172X + 0.1$
$S_O = 0.036X + 1.2$

where:

S_T = overall precision, mg/L,
S_O = single-operator precision, mg/L, and
X = concentration of phenolic determined, mg/L.

TABLE 510: III. COMPARISON OF PHENOLIC ANALYTICAL PROCEDURES

Phenolic Compound	Concentration mg/L		
	By Weight	4-Amino-anti-pyrine*	GLC
Phenol	1.06	0.97	0.97
o-Cresol	1.04	0.64	1.03
m-Cresol	1.02	0.38	1.03
p-Cresol	1.00	0.00	1.00
			3.94†
Composite	4.12‡	2.40	4.11§

*Reported as phenol; ASTM Method D 1783, Test for Phenolic Compounds in *Industrial Water and Industrial Wastewater*; average of two analyses.

†Based on *m*- and *p*-cresol of 2.01, plus *o*-cresol and phenol as phenol of 1.93; average of four analyses.

‡Composite of the four phenolics.

§Based on *m*- and *p*-cresols of 2.01, plus *o*-cresol and phenol, equal concentrations, with calibration factors averaged.

TABLE 510:IV. EFFECT OF pH IN GLC ANALYSES OF 2.4-DICHLOROPHENOL*

pH	4.5	7.7	9.7	10.8	11.7
Peak area in.^{2}	2.88	2.90	2.90	2.93	1.08

* Column and operating conditions: column, 5 ft by 1/8 in., stainless steel; 5% FFAP, 60/80 Chromosorb W. Flow rates and temperature: H_2, 25 mL/min; N_2, 25 mL/min; column temperature, 176 C; injector temperature, 205 C; range, 0.1; attenuation, 1; 3 samples in distilled water; chart, 90 in./hr; pH adjusted by NaOH; 1 mV full scale. Initial dichlorophenolic concentration, 5 mg/L.

7. Additional Information

The information in Tables 510:III through 510:V is presented to aid the analyst.

8. References

1. RUMP, O. 1974. Phenolic acids as indicators of pollution with liquid manure. A method for their detection. *Water Res.* 8:889.
2. CHRISWELL, C.D., R.C. CHANG & J.S. FRITZ. 1975. Chromatographic determination of phenols in water. *Anal. Chem.* 47:1325.
3. COOPER, R.L. & K.C. WHEATSTONE. 1973. The determination of phenols in aqueous effluents. *Water Res.* 7:1375.
4. KAWAHARA, F.K. 1971. Gas chromatographic analysis of mercaptans, phenols and organic acids in surface waters with use of pentafluorobenzyl derivatives. *Environ. Sci. Technol.* 5:235.
5. SAUNDERS, B.J. & A.R. PHILLIPS. 1968. A note on the detection of pentachlorophenol in water. *J. Ass. Pub. Analysts* 6:29.
6. EICHELBERGER, J.W., R.C. DRESSMAN & J.E. LONGBOTTOM. 1970. Separation of phenolic compounds from carbon chloroform extract for individual chromatographic identification and measurement. *Environ. Sci. Technol.* 4:576.
7. HUNT, G.T. 1976. The concentration, separation, and identification of trace quantities of phenolic compounds from natural water systems. M.S. thesis, Rutgers, The State University, New Brunswick, N.J.
8. CHAU, A.S.Y. & J.A. COBURN. Determination of pentachlorophenol in natural and waste waters. *J. Ass. Off. Anal. Chem.* 57:389.
9. BAIRD, R.B., C.L. KUO, J.S. SHAPIRO & W.A. YANKO. 1974. The fate of phenolics in wastewater—determination by direct injection GLC & washing respirometry. *Arch. Environ. Contam. & Toxicol.* 2:165.

TABLE 510:V. GAS-LIQUID CHROMATOGRAPHIC PROCEDURES FOR PHENOLS IN AQUEOUS SYSTEMS

Reference	Type Phenol	Extractive Technic	Column Construction	Column Temp. *C*	Detector*	Recovery or Sensitivity Claimed
1	*m*-OH benzoic acid, *m*-OH phenylacetic acid, *m*-OH phenylpropionic acid	Ethyl-acetate	10% SE 52 on Chromosorb W 80/100	180	F.I.D.	—
2	Phenol, *o*-cresol, *p*-cresol, 3,5-dimethyl-phenol	XAD−26 resin	5% OV-17 on Chromosorb W-AW-DMCS, 60/80	115-230	F.I.D.	97.7%
3	Phenol, cresols, xylenols, ethylphenols	Methyl isobutyl ketone—then silylestered	5% Tri-2,4 xylenyl phosphate on Chromosorb W	75–125	F.I.D.	0.1 mg/L
4	Various	CAM-CCE, pentafluorobenzyl derivatives	10% FFAP on Chromosorb T&W, 60/80	195	F.I.D.	0.1 μg
5	Pentachlorophenol	Petroleum ether	5% Carbowax 15,000 & 0.5% terephthalic acid on Supasorb BDH 100/120	185	E.C.	95%
6	Various	CAM-CCE Florisil	10% Carbowax 20 M terephthalic acid on Chromosorb W, 60/80	210	F.I.D.	0–111%
7	Phenol, *o*-cresol, *m*-cresol, 2,6 xylenol, 2,5 xylenol, 2,4,6 trimethylphenol, 2,4,5 trimethylphenol	CH_2Cl_2	3% OV-1 on Supelcoport, 100/120	113	F.I.D.	—

TABLE 510:V. (*Continued*)

Reference	Type Phenol	Extractive Technic	Column Construction	Column Temp. *C*	Detector*	Recovery or Sensitivity Claimed
	Phenol, *o*-cresol, *m*-cresol, 2,6 xylenol, 2,5 xylenol, 2,4,6 trimethylphenol, 2,4,5 trimethylphenol	CH_2Cl_2	9% SP-2100 on Supelcoport	115	F.I.D.	—
	o,*m*,*p*-Cresol, *p*-ethylphenol, 2,6 xylenol, 3,5 xylenol, 3,4 xylenol	CH_2Cl_2	0.2% SP-1000 on Carbopack A	223	F.I.D.	—
	o-Chlorophenol, *o*-cresol, 6 chloro-*o*-cresol, *o*-nitrophenol, 3,5 xylenol	XAD-2 0.4% NaOH, H_3PO_4, CH_2Cl_2	3% OV-1 on Supelcoport 100/120	113	F.I.D.	—
8	Pentachlorophenol	Benzene, acetate derivative	*a*. 11% OV-17-QF-1 on Chromosorb Q, 80/100 *b*. 3.6% OV-101, 5.5% OV-210, Chromosorb W, DMCS, 80/100 *c*. 3.0% OV-225 on Chromosorb Q-(HP) 80/100	200	E.C.	0.01 μg/L
9	Various	None: direct aqueous injection	4% Dinonylphthalate on Chromosorb G, 80/100	125	F.I.D.	0.2–0.5 mg/L

* F.I.D. = flame ionization detector; E.C. = electron capture detector.

510 E. Bibliography

SCOTT, R.D. 1931. Application of a bromine method in the determination of phenols and cresols. *Ind. Eng. Chem.*, Anal. Ed. 3:67.

EMERSON, E., H.H. BEACHAM & L.C. BEEGLE. 1943. The condensation of aminoantipyrine. II. A new color test for phenolic compounds. *J. Org. Chem.* 8:417.

ETTINGER, M.B., S. SCHOTT & C.C. RUCHHOFT. 1943. Preservation of phenol content in polluted river water samples previous to analysis. *J. Amer. Water Works Ass.* 35:299.

ETTINGER, M.B. & R.C. KRONER. 1949. The determination of phenolic materials in industrial wastes. *Proc. 5th Ind. Waste Conf.* (Purdue Univ.), p. 345.

ETTINGER, M.B., C.C. RUCHHOFT & R.J. LISHKA. 1951. Sensitive 4-aminoantipyrine method for phenolic compounds. *Anal. Chem.* 23:1783.

DANNIS, M. 1951. Determination of phenols by the aminoantipyrine method. *Sewage Ind. Wastes* 23:1516.

MOHLER, E.F., JR. & L.N. JACOB. 1957. Determination of phenolic-type compounds in water and industrial waste waters: Comparison of analytical methods. *Anal. Chem.* 29:1369.

BURTSCHELL, R.H., A.A. ROSEN, F. M. MIDDLETON & M.B. ETTINGER. 1959. Chlorine derivatives of phenol causing taste and odor. *J. Amer. Water Works Ass.* 51:205.

GORDON, G.E. 1960. Colorimetric determination of phenolic materials in refinery waste waters. *Anal. Chem.* 32:1325.

OCHYNSKI, F.W. 1960. The absorptiometric determination of phenol. *Analyst* 85:278.

FAUST, S.D. & O.M. ALY. 1962. The determination of 2,4-dichlorophenol in water. *J. Amer. Water Works Ass.* 54:235.

BAKER, R.A. 1966. Phenolic analyses by direct aqueous injection gas chromatography. *J. Amer. Water Works Ass.* 58:751.

BAKER, R.A. 1966. Trace organic analyses by aqueous gas-liquid chromatography. *J. Air Water Pollut.* 10:591.

FAUST, S.D. & E.W. MIKULEWICZ. 1967. Factors influencing the condensation of 4-aminoantipyrine with derivatives of hydroxybenzene. II. Influence of hydronium ion concentration on absorptivity. *Water Res.* 1:509.

BAKER, R.A. & B.A. MALO. 1967. Phenolics by aqueous injection gas chromatography. *Environ. Sci. Technol.* 1:997.

SMITH, L.S. 1976. Evaluation of Instrument for the (Ultraviolet) Determination of Phenol in Water. EPA-600/4-76-048, U.S. EPA, Cincinnati, Ohio.

511 SLUDGE DIGESTER GAS

Gas produced during the anaerobic decomposition of wastes contains methane (CH_4) and carbon dioxide (CO_2) as the major components with minor quantities of hydrogen (H_2), hydrogen sulfide (H_2S), nitrogen (N_2), and oxygen (O_2). It is saturated with water vapor. Common practice is to analyze the gases produced to estimate their fuel value and to check on the treatment process. The relative proportions of CO_2, CH_4, and N_2 are normally of most concern and the easiest to determine because of the relatively high percentages of these gases.

1. Selection of Method

Two procedures are described for gas analysis, the Volumetric Method (A), and the Gas Chromatographic Method (B). The volumetric analysis is suitable for the determination of CO_2, H_2, CH_4, and O_2. Nitrogen is estimated indirectly by difference. Although the method is time-consuming, the equipment is relatively simple. Because no calibration is needed before use, the procedure is particularly appropriate when analyses are conducted infrequently.

The principal advantage of gas chromatography is speed. Commercial equipment is designed specifically for ambient-temperature gas analysis and permits the routine separation and measurement of CO_2, N_2, O_2, and CH_4 in less than 5 min. The requirement for a recorder, pressure-regulated bottles of carrier gas, and certified standard gas mixtures for calibration raise costs to the point where infrequent analyses by this method may be uneconomical. The advantages of this system are freedom from the cumulative errors found in sequential volumetric measurements, adaptability to other gas component analyses, adaptability to intermittent on-line sampling and analysis, and the use of samples of 1 mL or less.[1]

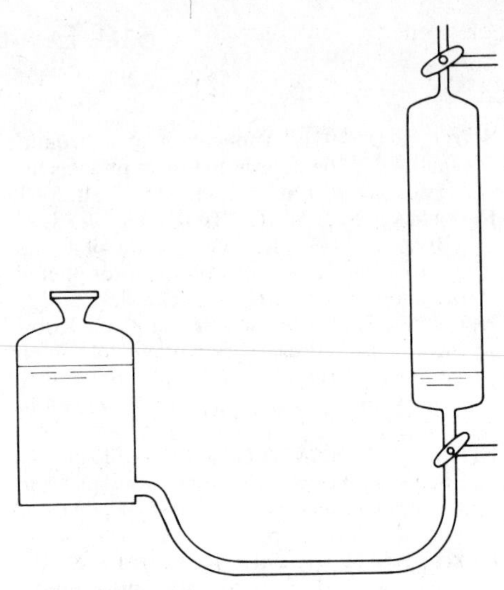

Figure 511:1. Gas collection apparatus.

2. Sample Collection

When the source of gas is some distance from the apparatus used for analysis, collect samples in sealed containers and bring to the instrument. Displacement collectors are the most suitable containers. Long glass tubes with three-way glass stopcocks at each end, as indicated in Figure 511:1, are particularly useful. These also are available with centrally located ports provided with septa for syringe transfer of samples. Connect one end of collector to gas source and vent three-way stopcock to the atmosphere. Clear line of air by passing 10 to 15 volumes of gas through vent and open stopcock to admit sample. If large quantities of gas are available, sweep air away by passing 10 to 15 volumes of gas through tube. If the gas supply is limited, fill tube with a liquid that is displaced by gas. Use either mercury or an acidified salt solution. The latter solution is easier and less expensive to use, but dissolves gases to some extent. Therefore, fill collection tube completely with the gas and seal off from any contact with displacement fluid during temporary storage. When transferring gas to the gas-analyzing apparatus, do not transfer any fluid.

511 A. Volumetric Method

1. General Discussion

a. Principle: This method may be used for the analysis either of digester gas or of methane in water (see Section 502, Methane). A measured volume of gas is passed first through a solution of potassium hydroxide (KOH) to remove CO_2, next through a solution of alkaline pyrogallol to remove O_2, and then over heated cupric oxide, which removes H_2 by oxidation to water. After each of the above steps, the volume of gas remaining is measured; the decrease that results is a measure of the relative percentage of volume of each component in the mixture. Finally, CH_4 is determined by conversion to CO_2 and H_2O in a slow-combustion pipet or a catalytic oxidation assembly. The volume of CO_2 formed during combustion is measured to determine the fraction of methane originally present. Nitrogen is estimated by assuming that it represents the only gas remaining and equals the difference between 100% and the sum of the measured percentages of the other components.

When only CO_2 is measured, report only CO_2. No valid assumptions may be made about the remaining gases present without making a complete analysis.

Follow the equipment manufacturers' recommendations with respect to oxidation procedures.

CAUTION: *Do not attempt any slow-combustion procedure on digester gas because of the high probability of exceeding the explosive 5% by volume concentration of CH_4.*

2. Apparatus

Orsat-type gas-analysis apparatus, consisting of at least: (1) a water-jacketed gas buret with leveling bulb; (2) a CO_2-absorption pipet; (3) an O_2-absorption pipet; (4) a cupric oxide-hydrogen oxidation assembly; (5) a shielded catalytic CH_4-oxidation assembly or slow-combustion pipet assembly; and (6) a leveling bulb. With the slow-combustion pipet use a controlled source of current to heat the platinum filament electrically. Mercury is recommended as the displacement fluid; aqueous Na_2SO_4-H_2SO_4 solution also has been used successfully for sample collection. Use any commercially available gas analyzer having these units.

3. Reagents

a. Potassium hydroxide solution: Dissolve 500 g KOH in distilled water and dilute to 1 L.

b. Alkaline pyrogallol reagent: Dissolve 30 g pyrogallol (also called pyrogallic acid) in distilled water and make up to 100 mL. Add 500 mL KOH solution.

c. Oxygen gas: Use approximately 100 mL for each gas sample analyzed.

d. Displacement liquid: Use either

1) *Mercury* or

2) *Sodium sulfate-sulfuric acid solution:* Dissolve 200 g Na_2SO_4 in 800 mL distilled water; add 30 mL conc H_2SO_4.

4. Procedure

a. Sample introduction: Transfer 5 to 10 mL gas sample into gas buret through a capillary-tube connection to the collector. Expel this sample to the atmosphere to purge the system. Transfer up to 100 mL gas sample to buret. Bring sample in buret to atmospheric or reference pressure by adjusting leveling bulb. Measure volume accurately and record as V_1.

b. Carbon dioxide absorption: Remove CO_2 from sample by passing it through the CO_2-absorption pipet charged with the KOH solution. Pass gas back and forth until sample volume remains constant. Before opening stopcocks between buret and any absorption pipet, make sure that the gas in the buret is under a slight positive pressure to prevent reagent in the pipet from contaminating stopcock or manifold. After absorption of CO_2, transfer sample to buret and measure its volume. Record as V_2.

c. Oxygen absorption: Remove O_2 by passing sample through O_2-absorption pipet charged with alkaline pyrogallol reagent until sample volume remains constant. Measure volume and record as V_3. For digester gas samples, continue as directed in ¶ 4*d*. For CH_4 in water, store gas in CO_2 pipet and proceed to ¶ 4*e* below.

d. Hydrogen oxidation: Remove H_2 by passing sample through CuO assembly maintained at a temperature in the range 290 to 300 C. When a constant volume has been obtained, transfer sample back to buret, cool, and measure its volume. Record as V_4.

Waste to the atmosphere all but 20 to 25

mL of remaining gas. Measure volume and record as V_5. Store temporarily in CO_2-absorption pipet.

e. Methane oxidation: Purge inlet connections to buret with O_2 by drawing 5 to 10 mL into buret and expelling to the atmosphere. Oxidize CH_4 either by the catalytic oxidation process for digester gas and gas phase of water samples or by the slow-combustion process for gas phase of water samples.

1) Catalytic oxidation process—For catalytic oxidation of digester gas and gas phase of water samples, transfer 65 to 70 mL O_2 to buret and record this volume as V_6. Pass O_2 into CO_2-absorption pipet so that it will mix with sample stored there. Return this mixture to buret and measure its volume. Record as V_7. This volume should equal closely V_5 plus V_6. Pass O_2-sample mixture through catalytic oxidation assembly, which should be heated in accordance with directions from the manufacturer. Keep rate of gas passage less than 30 mL/min. After first pass, transfer mixture back and forth through the assembly between buret and reservoir at a rate not faster than 60 mL/min until a constant volume is obtained. Record as V_8.

2) Slow-combustion process—For slow combustion of the gas phase of water samples, transfer 35 to 40 mL O_2 to buret and record this volume as V_6. Transfer O_2 to slow-combustion pipet and then transfer sample from CO_2-absorption pipet to buret. Heat platinum coil in combustion pipet to yellow heat while controlling temperature by adjusting current. Reduce pressure of O_2 in pipet to somewhat less than atmospheric pressure by means of the leveling bulb attached to the pipet. Pass sample into slow-combustion pipet at rate of approximately 10 mL/min. After the first pass, transfer sample and O_2 mixture back and forth between pipet and buret several times at a faster rate, allowing mercury in pipet to rise to a point just below heated coil. Collect sample in combustion pipet, turn off coil, and cool pipet and sample to room temperature with a jet of compressed air. Transfer sample to buret and measure its volume. Record as V_8.

f. Measurement of carbon dioxide produced: Determine amount of CO_2 formed in the reaction by passing sample through CO_2-absorption pipet until volume remains constant. Record volume as V_9.

Check accuracy of determination by absorbing residual O_2 from sample. After this absorption, record final volume as V_{10}.

5. Calculation

a. CH_4 and H_2 usually are the only combustible gases present in sludge digester gas. When this is the case, determine percentage by volume of each gas as follows:

$$\% \ CO_2 = \frac{(V_1 - V_2) \times 100}{V_1}$$

$$\% \ O_2 = \frac{(V_2 - V_3) \times 100}{V_1}$$

$$\% \ H_2 = \frac{(V_3 - V_4) \times 100}{V_1}$$

$$\% \ CH_4 = \frac{V_4 \times (V_8 - V_9) \times 100}{V_1 \times V_5}$$

$$\% \ N_2 = 100 - (\%CO_2 + \%O_2 + \%H_2 + \%CH_4)$$

b. Alternatively, calculate CH_4 by either of the two following equations:

$$\% \ CH_4 = \frac{V_4 \times (V_6 + V_{10} - V_9) \times 100}{2 \times V_1 \times V_5}$$

$$\% \ CH_4 = \frac{V_4 \times (V_7 - V_8) \times 100}{2 \times V_1 \times V_5}$$

Results from the calculations for CH_4 by the three equations should be in reasonable agreement. If not, repeat analysis after checking apparatus for sources of error, such as leaking stopcocks or connections. Other combustible gases, such as ethane, butane, or pentane, will cause a lack of agreement among the calculations; however, the possibility that digester gas contains a significant amount of any of these is remote.

6. Precision and Accuracy

A gas buret measures gas volume with a precision of 0.05 mL and a probable accuracy of 0.1 mL. With the large fractions of CO_2 and CH_4 normally present in digester gas, the overall error for their determination can be made less than ±1%. The error in the determination of O_2 and H_2, however, can be considerable because of the small concentrations normally present. For a concentration as low as 1%, an error as large as ±20% can be expected. When N_2 is present in a similar low-volume percentage, the error in its determination would be even greater, since errors in each of the other determinations would be reflected in the calculation for N_2.

511 B. Gas Chromatographic Method

1. General Discussion

a. Principle: See Section 509A for a discussion of gas chromatography.

b. Equipment selection: Many columns have been proposed for gas mixture analyses. Any that is capable of the desired separation is acceptable, provided that all of the exact conditions of analysis are reported with the calibration standards. The following directions are necessarily general. Follow the manufacturer's recommendations for the specific instrumentation.

2. Apparatus

a. Gas chromatograph: Use any commercially available instrument equipped with a thermal conductivity detector. With some column packings, ovens and temperature controls are necessary. Preferably use a unit with a gas sample valve.

b. Recorder: Use a 10-mV full-span strip chart recorder with the gas chromatograph. When minor components such as H_2 and H_2S are to be detected, a 1-mV full-span recorder is preferable.

*c. Column packing:** Some commercially available column packings useful for separating sludge gas components are listed below along with the routine separations possible at room temperature:[2,3]

1) Silica gel at room temperature: H_2, air (O_2 + N_2), CH_4, (CO_2-slow);

2) Molecular Sieve 13X: H_2, O_2, N_2, CH_4;

3) HMPA (hexamethylphosphoramide) 30% on Chromosorb P: CO_2 from (O_2, N_2, H_2, CH_4);

4) DEHS (di-2-ethylhexylsefacate) 30% on Chromosorb P: CO_2 from (O_2, N_2, H_2, CH_4).

Combinations of Columns 1 and 2, 3 and 2, or 4 and 2 when properly sized and used in the sequence: 1st column, detector, 2nd column, detector, readily will separate H_2, O_2, N_2, CH_4, and CO_2. Commercial equipment specifically designed for such operations is available.[3]

d. Sample introduction apparatus: An instrument equipped with gas-sampling valves is designed to permit automatic injection of a specific sample volume into the chromatograph. If such an instrument is not available, introduce samples with a 2-mL syringe fitted with a 27-gauge hypodermic needle. Reduce escape of gas by

*Gas chromatographic methods are extremely sensitive to the materials used. Use of trade names in "Standard Methods" does not preclude the use of other existing or as-yet-undeveloped products that give demonstrably equivalent results.

greasing plunger lightly with mineral oil or by using a special gas-tight syringe.

3. Reagents

a. Carrier gases: Use helium for separating digester gases. If H_2 is to be determined, use argon as a carrier gas to increase the sensitivity greatly.

b. Calibration gases: Use samples of CH_4, CO_2, and N_2 of known purity, or mixtures of known composition, for calibration. Also use samples of O_2, H_2, and H_2S of known purity if these gases are to be measured.

4. Procedure

a. Preparation of gas chromatograph: Adjust carrier gas flow rate to 60 to 80 mL/min. Turn on oven heaters, if used, and detector current and adjust to desired values. The instrument is ready for use when the recorder yields a stable base line. Silica gel and molecular sieve columns gradually lose activity because of adsorbed moisture or materials permanently adsorbed at room temperature. If insufficient separations occur, reactivate by heating or repacking.

b. Calibration: For accurate results, prepare a calibration curve for each gas to be measured because different gas components do not give equivalent detector responses on either a weight or a molar basis. Calibrate with synthetic mixtures or with pure gases.

1) Synthetic mixtures—Use purchased gas mixtures of known composition or prepare in the laboratory. Inject a standard volume of each mixture into the gas chromatograph and note response for each gas. Compute detector response, either as area under a peak or as height of peak, after correcting for attenuation. Read peak heights accurately and correlate with concentration of component in sample. Reproduce operating parameters exactly from one analysis to the next. If sufficient reproducibility cannot be obtained by this procedure, use peak areas for calibration. Prepare calibration curve by plotting either peak area or peak height against volume percent for each component.

2) Pure gases—Introduce pure gases into chromatograph individually with a syringe. Inject sample volumes of 0.25, 0.5, 1.0 mL, etc., and plot detector response, corrected for attenuation, against gas volume.

When the analysis system yields a linear detector response with increasing gas component concentration from zero to the range of interest, run standard mixtures along with samples. If the same sample size is used, calculate gas concentration by direct proportions.

c. Sample analysis: If samples are to be injected with a syringe, equip sample collection container with a port closed by a rubber or silicone septum. To take a sample for analysis, expel air from barrel of syringe by depressing plunger and force needle through the septum. Withdraw plunger to take gas volume desired, pull needle from collection container, and inject sample rapidly into chromatograph.

When samples are to be injected through a gas-sampling valve, connect sample collection container to inlet tube. Permit gas to flow from collection tube through the valve to purge dead air space and fill sample tube. About 15 mL normally are sufficient to clear the lines and to provide a sample of 1 to 2 mL. Transfer sample from loop into carrier gas stream by following manufacturer's instructions. Bring samples to atmospheric pressure before injection.

When calibration curves have been prepared with synthetic mixtures, use the same sample volume as that used during calibration. When calibration curves are prepared by the procedure using varying volumes of pure gases, inject any convenient gas sample volume up to about 2 mL.

5. Calculation

a. When calibration curves have been prepared with synthetic mixtures and the volume of the sample analyzed is the same as that used in calibration, read volume percent of each component directly from calibration curve after detector response for that component is computed.

b. When calibration curves are prepared with varying volumes of pure gases, calculate the percentage of each gas in the mixture as follows:

$$\text{Volume \%} = \frac{A}{B} \times 100$$

where:

A = partial volume of component (read from calibration curve) and
B = volume sample injected.

c. Where standard mixtures are run with samples and instrument response is linear from zero to the concentration range of interest:

$$\text{Volume \%} = \text{Volume \% (std)} \times \frac{C}{D}$$

where:

C = recorder value of sample and
D = recorder value of standard.

6. Precision and Accuracy

Precision and accuracy depend on the instrument used and the technics of operation. With proper care, a precision of 2% generally can be achieved. With digester gas the sum of the percent CH_4, CO_2, and N_2 should approximate 100%. If it does not, suspect errors in collection, handling, storage, and injection of gas, or in instrumental operation or calibration.

511 C. References

1. GRUNE, W.N. & C.F. CHUEH. 1962–63. Sludge gas analysis using gas chromatograph. *Water Sewage Works* 109:468; 110:43, 77, 102, 127, 171, 220, and 254.
2. ANDREWS, J.F. 1968. Chromatographic analysis of gaseous products and reactants for biological processes. *Water Sewage Works* 115:54.
3. Column Systems for the Fisher Gas Partitioner. Tech. Bull. TB-154, Fisher Scientific Co., Atlanta, Ga. Catalog 77, Fisher Scientific Co.

511 D. Bibliography

Volumetric Method

YANT, W.F. & L.B. BERGER. 1936. Sampling of Mine Gases and the Use of the Bureau of Mines Portable Orsat Apparatus in Their Analysis. Miner's Circ. No. 34, U.S. Bur. Mines, Washington, D.C.

MULLEN, P.W. 1955. Modern Gas Analysis. Interscience Publishers, New York, N.Y.

512 SURFACTANTS (ANIONIC)

The most widely used anionic surfactant is linear alkylbenzene sulfonate (LAS); therefore, it is used to standardize the following analytical methods. Neither of these methods is specific for LAS and other means must be used if unequivocal proof of its presence is desired.

LAS is not a single compound but may comprise any or all of 26 related compounds with various chain lengths and isomers. The composition of a commercial product will depend partly on the properties desired by the manufacturer or user. The composition of the LAS found in environmental waters will reflect the amount of each LAS component that has entered the environment minus the amount of each that has been destroyed or removed by biological or physicochemical processes.

The methylene blue method is relatively simple and precise, giving the net amount of methylene blue active substances (MBAS). MBAS includes LAS in particular, other anionic surfactants in general, and many other materials.

The carbon adsorption-infrared method involves a multistep, time-consuming prepurification followed by infrared spectroscopy. It responds to LAS in particular, alkylbenzene sulfonates in general, and to any other amphiphilic sulfonates that absorb in the same infrared (IR) region. Prepurification may be followed by some non-IR finish, such as the MBAS method if IR equipment is not available, or by desulfonation-gas chromatography if the LAS content and composition are to be determined unequivocally.

512 A. Methylene Blue Method for Methylene-Blue-Active Substances (MBAS)

1. General Discussion

a. Principle: This method depends on the formation of a blue salt or ion pair when methylene blue, a cationic dye, reacts with anionic surfactants (including LAS, other sulfonates, and sulfate esters) as well as with other strongly amphiphilic anions, natural or man-made. Because of this lack of specificity, the materials determined are designated simply as MBAS. The methylene blue complex (in contrast with methylene blue itself) is extractable into chloroform ($CHCl_3$), where the intensity of color is proportional to the MBAS concentration. After an aqueous backwash the color remaining in the $CHCl_3$ is measured by spectrophotometry at 652 nm. The method is sensitive to 0.025 mg MBAS/L (calculated as LAS).

b. Interferences: If a direct determination of LAS (or of any other individual MBAS species) is sought, all other MBAS species interfere. Cationic amphiphilic materials will give negative interference by competing with the cationic methylene blue. Some of the proven interferences can be predicted on the basis of chemical properties. Organic sulfates, sulfonates, carboxylates, and phenols that complex methylene blue, and inorganic cyanates, chlorides, nitrates, and thiocyanates that form ion pairs with methylene blue are among the positive interferences. Organic materials, especially amines that compete with methylene blue in the reaction, can cause low results. Colored materials present in the sample and extractable into $CHCl_3$ will interfere to the extent of their absorbance at the wavelength used.

c. Molecular weight: Test responses will appear to differ if expressed in terms of weight rather than in molar quantities. Equimolar amounts of two anionic surfactants with different molecular weights should give substantially equal colors in the $CHCl_3$ layer, although the amount by weight may differ significantly. If results are to be expressed by weight, as is generally desirable, the average molecular weight of the surfactant measured must be known or a calibration curve must be made using that particular compound. Because neither of these is possible with environmental samples, report the result in terms of a suitable standard calibration curve, for example "0.65 mg MBAS/L (calculated as LAS, mol wt 318)."

d. Minimum detectable quantity: 10 μg MBAS (calculated as LAS).

e. Application: The methylene blue method has been applied successfully to the examination of the anionic surfactant content in drinking water supplies. Unfortunately, the numerous materials normally present in wastewater, industrial wastes, and sludge can interfere seriously with the determination and lead to incorrect results and conclusions, if these interferences are not recognized and taken into account.

2. Apparatus

a. Colorimetric equipment: One of the following is required:

1) *Spectrophotometer,* for use at 652 nm, providing a light path of 1 cm or longer.

2) *Filter photometer,* providing a light path of 1 cm or longer and equipped with a red color filter exhibiting maximum transmittance near 652 nm.

b. Separatory funnels, 500-mL, preferably with inert TFE stopcocks.

3. Reagents

a. Stock LAS solution: Weigh an amount of the reference material* equal to 1.00 g LAS on a 100% active basis. Dissolve in distilled water and dilute to 1,000 mL; 1.00 mL = 1.00 mg LAS. Store in a refrigerator to minimize biodegradation. If necessary, prepare weekly.

b. Standard LAS solution: Dilute 10.00 mL stock LAS solution to 1,000 mL with distilled water; 1.00 mL = 10.0 μg LAS. Prepare daily.

c. Phenolphthalein indicator solution, alcoholic.

d. Sodium hydroxide, NaOH, 1*N*.

e. Sulfuric acid, H_2SO_4, 1*N*.

f. Chloroform, $CHCl_3$: CAUTION: *Chloroform vapors are toxic. Take appropriate precautions against inhalation.*

g. Methylene blue reagent: Dissolve 100 mg methylene blue† in 100 mL distilled water. Transfer 30 mL to a 1,000-mL flask. Add 500 mL distilled water, 6.8 mL conc H_2SO_4, and 50 g monosodium dihydrogen phosphate monohydrate, $NaH_2PO_4 \cdot H_2O$. Shake until dissolution is complete. Dilute to 1,000-mL.

h. Wash solution: Add 6.8 mL conc H_2SO_4 to 500 mL distilled water in a 1,000-mL flask. Add 50 g $NaH_2PO_4 \cdot H_2O$ and shake until dissolution is complete. Dilute to 1,000-mL.

4. Procedure

a. Preparation of calibration curve: Prepare a series of 10 separatory funnels with 0, 1.00, 3.00, 5.00, 7.00, 9.00, 11.00, 13.00, 15.00, and 20.00 mL standard LAS solution. Add sufficient water to make the total volume 100 mL in each separatory funnel. Treat each standard as described in ¶s 4*c* and 4*d* following, and plot a calibration curve of micrograms LAS versus absorbance.

---

*A suitable reference material can be obtained from U.S. Environmental Protection Agency, Environmental Monitoring and Support Laboratory, Cincinnati, Ohio 45268.

†Eastman No. P573 or equivalent.

b. Volume of sample: Select sample volume on the basis of expected MBAS concentration:

Expected MBAS Concentration *mg/L*	Sample Taken *mL*
0.025–0.080	400
0.08 –0.40	250
0.4 –2.0	100

If expected MBAS concentration is over 2 mg/L, dilute a sample containing 40 to 200 μg MBAS to 100 mL with distilled water.

c. Extraction and color development:

1) Add sample to a separatory funnel. Make alkaline by dropwise addition of 1*N* NaOH, using phenolphthalein indicator. Discharge pink color by dropwise addition of 1*N* H_2SO_4.

2) Add 10 mL $CHCl_3$ and 25 mL methylene blue reagent. Rock funnel vigorously for 30 sec and let phases separate. Alternatively, place a magnetic stirring bar in the separatory funnel; lay funnel on its side on a magnetic mixer and adjust speed of stirring to produce a rocking motion. Excessive agitation may cause difficulties due to emulsion formation. To break persistent emulsions add a small volume of isopropyl alcohol (<10 mL); add same volume of isopropyl alcohol to all standards. Some samples require a longer period of phase separation than others. Before draining $CHCl_3$ layer, swirl gently, then let settle.

3) Draw off $CHCl_3$ layer into a second separatory funnel. Rinse delivery tube of first separatory funnel with a small amount of $CHCl_3$. Repeat the extraction three times, using 10 mL $CHCl_3$ each time. If the blue color in the water phase becomes faint and disappears, discard sample and repeat, using a smaller sample.

4) Combine all $CHCl_3$ extracts in the second separatory funnel. Add 50 mL wash solution and shake vigorously for 30 sec. Emulsions do not form at this stage. Let settle, swirl, and draw off $CHCl_3$ layer through glass wool that has been pre-extracted with $CHCl_3$ into a 100-mL volumetric flask. Extract wash solution twice with 10 mL $CHCl_3$ and add to flask. Rinse glass wool and funnel with $CHCl_3$. Collect washings in volumetric flask, dilute to mark with $CHCl_3$, and mix well.

d. Measurement: Determine the absorbance of the solution at 652 nm against a blank of $CHCl_3$.

5. Calculation

Read micrograms apparent LAS from calibration curve.

$$\text{mg total apparent LAS/L} = \frac{\mu\text{g LAS (in 100 mL final volume)}}{\text{mL sample}}$$

Report as methylene-blue-active substances (MBAS).

6. Precision and Accuracy

A synthetic sample containing 270 μg LAS/L in distilled water was analyzed in 110 laboratories with a relative standard deviation of 14.8% and a relative error of 10.6%.

A tap water sample to which were added 480 μg LAS/L was analyzed in 110 laboratories with a relative standard deviation of 9.9% and a relative error of 1.3%.

A river water sample to which were added 2.94 mg LAS/L was analyzed in 110 laboratories with a relative standard deviation of 9.1% and a relative error of 1.4%.[1]

512 B. Carbon Adsorption-Infrared Method* (TENTATIVE)

1. General Discussion

a. Principle: The method uses carbon adsorption, desorption, acid hydrolysis, and conversion to an amine salt for purification and isolation. Quantitation is by measurement of the infrared absorption.

b. Interferences: Most natural interferences and organic sulfate surfactants are removed in the purification, leaving mainly sulfonate-type surfactants. The infrared measurements at 9.6 μm (sulfonate) and 9.9 μm (substituted benzene ring) will distinguish the alkylbenzene sulfonates from the other sulfonates, but cannot distinguish between LAS and other alkylbenzene sulfonates. Accordingly, the materials measured by this procedure are best termed CIAS (carbon adsorption-infrared method active substances) in the absence of any other identification.

c. Application: This method is applicable only to raw and potable water samples.

d. Precaution: Surfactants are markedly attracted to interfaces and thusconcentrate in particulates and in the bubble surfaces of foams. For accurate results it is essential that particulates be representatively sampled or excluded (whichever meets the purposes of the analysis) and that foams be allowed to subside before sampling.

2. Apparatus

a. Carbon adsorption tube: Charge the glass column, about 5 × 60 cm, with 100 g carbon. Screens of stainless steel or brass, about 30-mesh, divide the carbon into sections of 20, 30, 40, and 10 g (Figure 512:1).

b. Buchner funnel, 500-mL, medium-porosity, sintered-glass.

c. pH meter.

d. Volumetric flasks, either 2 or 5 mL.

e. Separatory funnels, 500 mL.

f. Infrared spectrophotometer, for use at 2 to 15 μm.

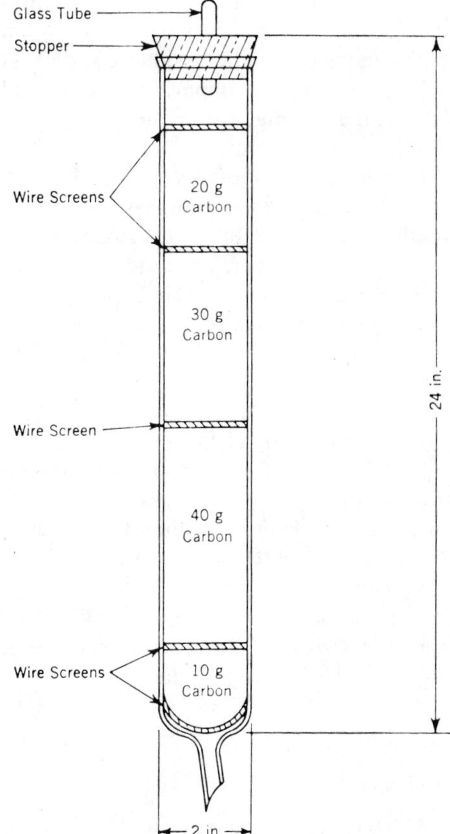

Figure 512:1. Carbon adsorption tube.

g. Acid-washed glassware: Keep all glassware used in the infrared method free of contamination. Rinse thoroughly with a solution composed of equal volumes of water and conc HCl to remove adsorbed LAS.

*This method is identical in source and substance to that developed by the Subcommittee on Analytical Methods, Technical Advisory Committee, The Soap and Detergent Association.[2]

3. Reagents

a. Standard LAS, for calibration. (See Section 512A.3*a*.)

b. Activated carbon, unground, 20-mesh, for the carbon adsorption tube.†

Test for impurities in carbon—Extract 100 g carbon by boiling 1 hr with 1 L benzene-alcohol solution (see ¶ 3*c* following). Filter, wash with 100 mL methyl alcohol, add washings to remainder of solvent mixture, evaporate to dryness on a steam bath, and weigh. The residue consists of extractable organic impurities and should be less than 10 mg, not including any residue from the solvent.

c. Benzene-alcohol solution: Mix 500 mL thiophene-free benzene, 420 mL methyl alcohol, and 80 mL 0.5*N* KOH.

d. Methyl alcohol, absolute.

e. Hydrochloric acid, HCl, conc.

f. Sodium hydroxide, NaOH, 1*N*.

g. Petroleum ether, boiling range 35 to 60 C.

h. Ethyl alcohol, 95%.

i. Sulfuric acid, H_2SO_4, 1*N*.

j. Buffer solution: Dissolve 6.8 g monopotassium dihydrogen phosphate, KH_2PO_4, in 1 L distilled water. Adjust to pH 6.8 to 6.9 with 6*N* NaOH.

k. 1-methylheptylamine.‡

l. Solution for extracting LAS: Dissolve 400 mg (20 drops) 1-methylheptylamine in 400 mL $CHCl_3$. Prepare daily.

m. Chloroform, $CHCl_3$.

n. Carbon disulfide, CS_2.

4. Procedure

CAUTION: *Benzene, methanol, and chloroform vapors are toxic; carry out operations in an efficient fume hood and take care to avoid vapor inhalation.*

a. Preparation of calibration curve: Place 25 mg standard LAS in a 20-L glass vessel and dilute with about 15 L distilled water. Mix thoroughly and, using synthetic rubberlike tubing, siphon the entire solution through the carbon column. Treat as described in ¶s 4*c* 1) through 7) below. Repeat with 20, 15, 10, 5, and 0 mg LAS. Make two calibration curves by plotting LAS added as abscissa and absorbances of the maxima at 9.6 and 9.9 μm as ordinate. The baseline technic is best used in determining absorbance of the maxima.

b. Volume of sample: Estimate concentration of sulfonates (CIAS) in sample. Calculate volume of sample required to supply 10 to 25 mg CIAS. If 2 L or less, measure about 10 g granular activated carbon into a 2-L glass-stoppered graduated cylinder, add sample, and shake well for 2 min. Filter on a medium-porosity, sintered-glass Buchner funnel. If samples larger than 2 L are required, pass through the carbon column at the rate of 630 mL/min or less.

c. Extraction and measurement of CIAS:

1) Transfer carbon from Buchner funnel or column, treating the sections separately, to porcelain evaporating dishes and dry at 105 to 110 C. Brush dried carbon from each dish into separate 2-L bottles or flasks with standard-taper necks and add 1 L benzene-alcohol solution. Add boiling chips and reflux under an air condenser for 1 hr. Filter with vacuum through a Buchner funnel, draw off all liquid, release vacuum, and add 100 mL methyl alcohol. Stir with a glass rod and draw off the wash with vacuum. Wash a second time with another 100-mL portion of methyl alcohol. Return carbon to flask, add solvent as before, and reflux for 1 hr. While making this second extraction, evaporate solvent from first extract and washes. Carry out this evaporation in a 2-L beaker on a steam bath. (A gentle stream of nitrogen or air on the surface will hasten evaporation.)

2) Filter off second extract and wash carbon as before. Add extract and washes to beaker containing first extract. Discard carbon. Evaporate sufficiently to combine

†Nuchar C190 (Westvaco), or equivalent.

‡Eastman No. 2439 or equivalent.

in one beaker extracts from the 20-, 30-, and 40-g sections of the column. Treat extracts of the 10-g section separately throughout entire procedure. After removing solvent, take up residue in 50 mL warm distilled water. Transfer to a 250-mL standard-taper erlenmeyer flask. Rinse beaker with 30 mL conc HCl and add slowly to flask. Carbon dioxide is evolved. Rinse beaker with 50 mL distilled water and combine with other washings in the flask. Reflux under an air condenser for 1 hr.

3) Remove condenser and continue boiling until volume is reduced to 20 to 30 mL, transfer to a steam bath, and evaporate to near dryness. (A jet of air directed on the surface of the liquid will greatly aid evaporation.) Take solids up in 100 mL distilled water and neutralize with NaOH solution to a pH of 8 or 9. Extract once with 50 mL petroleum ether. Add up to 70% ethyl alcohol, if necessary, to break emulsions. Wash petroleum ether twice with 25-mL portions of distilled water, discard petroleum ether layer, and add washes to aqueous solution. Boil off any alcohol that was added.

4) Cool and transfer quantitatively to a 500-mL separatory funnel. Neutralize by adding H_2SO_4 until just acidic to litmus. Add 50 mL buffer solution (¶ 3*j*) and 2 drops methylheptylamine (¶ 3*k*), and shake vigorously. Add 50 mL LAS extracting solution and 25 mL $CHCl_3$. Shake for 3 min and let phases separate. If an emulsion forms, draw off lower $CHCl_3$ phase, including any emulsion, and filter through a plug of glass wool wet with $CHCl_3$, using suction if necessary, into a 500-mL separatory funnel. Draw off $CHCl_3$ phase into a 400-mL beaker and return any aqueous solution to first separatory funnel. Wash glass wool plug with 10 mL $CHCl_3$ and add to $CHCl_3$ extract.

5) Make an additional extraction with 50 mL LAS extracting solution and 25 mL $CHCl_3$. Shake 2 min and separate phases as in ¶ 4*c*4) if necessary. Extract a third time with 5 mL amine solution and 45 mL $CHCl_3$. Evaporate combined $CHCl_3$ extracts on a steam bath. With 10 mL $CHCl_3$, quantitatively transfer residue to a 50-mL beaker, using three 5-mL portions of $CHCl_3$ as rinses. Evaporate to dryness and continue heating on steam bath for 30 min to remove excess amine. Take up residue in about 1 mL CS_2 and filter through a plug of glass wool in a funnel stem (2-mm bore) into a 2- or 5-mL volumetric flask. Dilute to volume through the filter with several rinsings from beaker.

6) Transfer a sample portion to an infrared cell without further dilution. Run infrared absorption curve from 9.0 to 10.5 μm against a solvent blank. Measure absorbance of the 9.6- and 9.9-μm peaks, using baselines from 9.5 to 9.8 and from 9.8 to 10.1 μm. From appropriate calibration curves calculate CIAS as apparent LAS in original sample. Report values based on each wavelength separately. If infrared equipment is unavailable, use colorimetric method 512A. Break sulfonate-amine complex by boiling with aqueous alkali. Boil off the amine (as indicated by a lack of amine odor) and make suitable dilutions. Colorimetric results should check well with infrared values.

7) Evaporate a 0.5- to 1.0-mL portion of CIAS solution on a sodium chloride (NaCl) flat. Record absorption spectrum from 2 to 15 μm for positive qualitative identification of LAS or use desulfonation-gas chromatography for unequivocal identification and quantitation of LAS.

PRECAUTION: Use carbon adsorption on all samples. It separates CIAS from many of the other substances present and reduces emulsion difficulties.

NOTE: From 10 to 50 mL of water may be lost through a 60 × 1-cm air condenser during acid hydrolysis. This loss, while not affecting hydrolysis, reduces amount of water that needs to be boiled off after removal of the condenser.

512 C. References

1. LISHKA, R.J. & J.H. PARKER. 1968. *Water Surfactant* No. 3, Number 32, USPHS, Cincinnati.
2. SALLEE, E.M., J.D. FAIRING, R.W. HESS, R. HOUS & P.M. MAXWELL. 1956. Determination of trace amounts of alkyl benzenesulfonates in water. *Anal. Chem.* 28:1822.

512 D. Bibliography

BARR, T., J. OLIVER & W.V. STUBBINGS. 1948. The determination of surface-active agents in solution. *J. Soc. Chem. Ind.* (London) 67:45.

EPTON, S.R. 1948. New method for the rapid titrimetric analysis of sodium alkyl sulfates and related compounds. *Trans. Faraday Soc.* 44:226.

EVANS, H.C. 1950. Determination of anionic synthetic detergents in sewage. *J. Soc. Chem. Ind.* (London) 69:Suppl. 2:576.

DEGENS, P.N., JR., H.C. EVANS, J.D. KOMMER & P.A. WINSOR. 1953. Determination of sulfate and sulfonate anion-active detergents in sewage. *J. Appl. Chem.* (London) 3:54.

AMERICAN WATER WORKS ASSOCIATION. 1954. Task group report. Characteristics and effects of synthetic detergents. *J. Amer. Water Works Ass.* 46:751.

EDWARDS, G.P. & M.E. GINN. 1954. Determination of synthetic detergents in sewage. *Sewage Ind. Wastes* 26:945.

LONGWELL, J. & W.D. MANIECE. 1955. Determination of anionic detergents in sewage, sewage effluents, and river water. *Analyst* 80:167.

MOORE, W.A. & R.A. KOLBESON. 1956. Determination of anionic detergents in surface waters and sewage with methyl green. *Anal. Chem.* 28:161.

ROSEN, A.A., F.M. MIDDLETON & N. TAYLOR. 1956. Identification of synthetic detergents in foams and surface waters. *J. Amer. Water Works Ass.* 48:1321.

AMERICAN WATER WORKS ASSOCIATION. 1958. Task group report. Determination of synthetic detergent content of raw water supplies. *J. Amer. Water Works Ass.* 50:1343.

OGDEN, C.P., H.L. WEBSTER & J. HALLIDAY. 1961. Determination of biologically soft and hard alkylbenzenesulfonate in detergents and sewage. *Analyst* 86:22.

MAGUIRE, O.E., F. KENT, L-R.L. MILLER & G.J. PAPENMEIER. 1962. Field test for analysis of anionic detergents in well waters. *J. Amer. Water Works Ass.* 54:665.

ABBOTT, D.C. 1962. The determination of traces of anionic surface-active materials in water. *Analyst* 87:286.

ABBOTT, D.C. 1963. A rapid test for anionic detergents in drinking water. *Analyst* 88:240.

REID, V.W., G.F. LONGMAN & E. HEINERTH. 1967. Determination of anionic-active detergents by two-phase titration. *Tenside* 4:292.

SWISHER, R.D. 1970. Surfactant Biodegradation. Marcel Dekker, N.Y., pp. 47–54 and 62–63.

WANG, L.K., P.J. PANZARDI, W.W. SCHUSTE & D. AULENBACH. 1975. Direct two-phase titration method for analyzing anionic nonsoap surfactants in fresh and saline waters. *J. Environ. Health* 38:159.

513 TANNIN AND LIGNIN

Lignin is a plant constituent that often is discharged as a waste during the manufacture of paper pulp. Another plant constituent, tannin, may enter the water supply through the process of vegetable matter degradation or through the wastes of the tanning industry. Tannin also is applied in the so-called internal treatment of boiler waters, where it reduces scale formation by causing the production of a more easily handled sludge.

Both lignin and tannin contain aromatic hydroxyl groups that react with tungstophosphoric and molybdophosphoric acids to form a blue color. However, the reaction is not specific for lignin or tannin, inasmuch as other reducing materials respond similarly.

The nature of the substance suspected in the water sample will dictate the choice of tannic acid or lignin for use in the preparation of the standard solution. This is necessary because it is impossible to distinguish among hydroxylated aromatic compounds. Unless tannin or lignin is definitely known to be present, the results of this determination logically may be reported in the more general terms of "tannin-like," "lignin-like," or simply as "hydroxylated aromatic" compounds.

1. General Discussion

a. Principle: Tannins and lignins reduce tungstophosphoric and molybdophosphoric acids to produce a blue color suitable for the estimation of concentrations up to at least 9 mg/L for tannic acid or lignin.

b. Interference: Such reducing substances as 2 mg ferrous iron/L and 125 mg sodium sulfite/L individually produce a color equivalent to 1 mg tannic acid/L.

c. Minimum detectable concentration: Approximately 0.1 mg/L for tannic acid and 0.3 mg/L for lignin.

2. Apparatus

Colorimetric equipment: One of the following is required:

a. Spectrophotometer, for use at 700 nm. A light path of 1 cm or longer yields satisfactory results.

b. Filter photometer, provided with a red filter exhibiting maximum transmittance in the wavelength range of 600 to 700 nm. Sensitivity improves with increasing wavelength. A light path of 1 cm or longer yields satisfactory results.

c. Nessler tubes, matched, 100-mL, tall form, marked at 50-mL volume.

3. Reagents

a. Tannin-lignin reagent: Transfer 100 g sodium tungstate, $Na_2WO_4 \cdot 2H_2O$, and 25 g sodium molybdate, $Na_2MoO_4 \cdot 2H_2O$, together with 700 mL distilled water, to a 2,000-mL flat-bottom boiling flask. Add 50 mL 85% H_3PO_4 and 100 mL conc HCl. Connect to a reflux condenser and boil gently for 10 hr. Add 150 g Li_2SO_4, 50 mL distilled water, and a few drops of liquid bromine. Boil without condenser for 15 min to remove excess bromine. Cool to 25 C, dilute to 1 L, and filter. Store finished reagent, which should have no greenish tint, in a tightly stoppered bottle to protect against reduction by dust and organic materials.

b. Carbonate-tartrate reagent: Dissolve 200 g Na_2CO_3 and 12 g sodium tartrate, $Na_2C_4H_4O_6 \cdot 2H_2O$, and 750 mL hot distilled water, cool to 25 C, and dilute to 1 L.

c. Stock solution: Weigh 1.000 g tannic acid or tannin, or lignin compound being used for boiler water treatment or known to be a contaminant of the water sample. Dissolve in distilled water and dilute to 1,000 mL.

d. Standard solution: Dilute 10.00 mL or 50.00 mL stock solution to 1,000 mL with distilled water; 1.00 mL = 10.0 or 50.0 μg active ingredient.

4. Procedure

Bring 50 mL clear sample and standards

to a temperature above 20 C and maintain within a ±2 C range. Add in rapid succession 1 mL tannin-lignin reagent and 10 mL carbonate-tartrate reagent. Allow 30 min for color development. Compare visually against simultaneously prepared standards or make photometric readings against a reagent blank prepared at the same time. Because different tannin and lignin compounds react with variable sensitivity, use the appropriate tannin or lignin material for preparation of the calibration curve and visual standards. Use the following guide for the instrumental measurements in the wavelength region of 600 to 700 nm:

Tannic Acid in 61-mL Final Volume *μg*	Lignin in 61-mL Final Volume *μg*	Light Path *cm*
50–600	100–1,500	1
10–150	30– 400	5

5. Bibliography

BERK, A.A. & W.C. SCHROEDER. 1942. Determination of tannin substances in boiler water. *Ind. Eng. Chem.*, Anal. Ed. 14:456.

KLOSTER, M.B. 1973. Determination of tannin and lignin. *J. Amer. Water Works Ass.* 66:44.

514 HALOGENATED METHANES AND ETHANES BY PURGE AND TRAP (TENTATIVE)

Organohalides, particularly the trihalomethanes, have been reported in nearly every chlorinated water supply tested in the United States.[1] Common organohalide solvents, traceable to industrial effluents, have been detected in many raw source waters and in the corresponding finished drinking waters, but they are most often the result of chlorination. Toxicological studies suggest that chloroform ($CHCl_3$) and other organohalides have had detrimental effects on human health. Their presence in water supplies should be monitored closely so that measures may be taken to minimize or eliminate them whenever concentrations approach levels of concern.

1. General Discussion

a. Principle: This method is applicable to the determination of the following organohalides contained in water; all chloromethanes, chloroethanes, and chloroethylenes, bromomethane, dibromomethane, bromochloromethane, bromodichloromethane, chlorodibromomethane, bromoform, and 1, 2-dibromoethane. Other organohalides that boil at less than 150 C and that are less than 2% soluble in water are determined also. However, the analyst must demonstrate the usefulness of the method by collecting accuracy and precision data using dosed and actual samples.

An inert gas is bubbled through the sample to transfer the volatile constituents from the aqueous phase into the purge gas. The organics contained in the purge gas are then trapped in a short column (trap) containing a suitable sorbant (Figure 514:1). After purging is complete, the compounds are desorbed thermally from the trap and backflushed (Figure 514:2) into a temperature-programmed gas chromatographic column for separation and analysis (Figure 514:3). Detection is by use of a halogen-specific detector, and for this analysis, either a microcoulometric or an electrolytic conductivity detector operated in the specific halogen mode is acceptable.

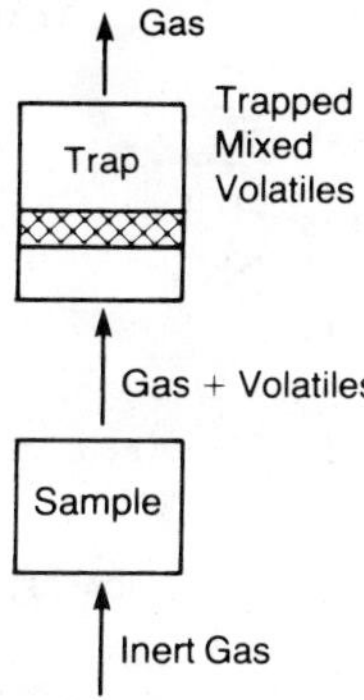

Figure 514:1. Removal of volatiles from the sample.

b. Interferences: Impurities contained in the purge gas and organic compounds outgassing from the system ahead of the trap usually account for most contamination problems. Such compounds will be concentrated in the trap and ultimately analyzed with the sample components. The analysis of blank samples (organic-free water) is a convenient means of monitoring for this problem. Whenever potential interferences are noted in blank samples, change the purge gas source and replace the molecular sieve gas filter (see ¶ 2*h*). Outgassing from the system generally cures itself with time. Do not subtract blank values.

c. Detection limits: The detection limits are determined by the halogen-specific gas chromatographic detector selected for the analysis, the degree of halogen substitution, and the specific halogen species contained in the compound. Most of the above organohalides can be analyzed over a concentration range of approximately 0.5 to 1,500 μg/L.

2. Apparatus*

Wash sample bottles and seals in detergent solution, rinse with distilled water, let

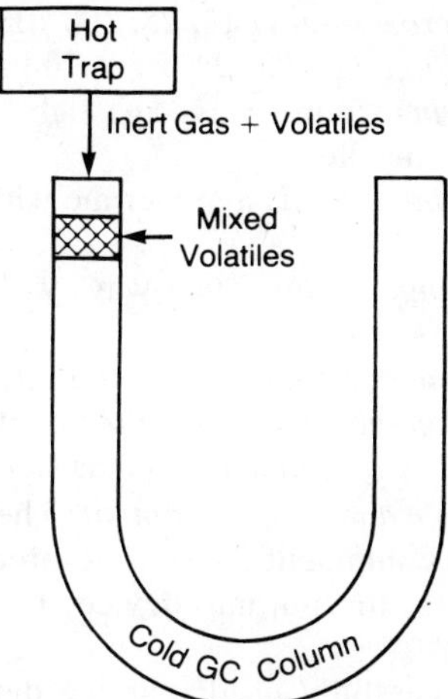

Figure 514:2. Transfer of trapped volatiles to chromatographic column.

air-dry, then heat to 105 C for 1 hr. Once cool, seal sample bottles using same septa intended for sample. Use more vigorous heat treatment of glassware (such as heating to 400 C in a muffle furnace) when high-boiling organic materials present contamination problems.

a. Sample bottles and seals, 40-mL screw-cap vials† sealed with TFE‡-faced compressible silicone rubber septa.§

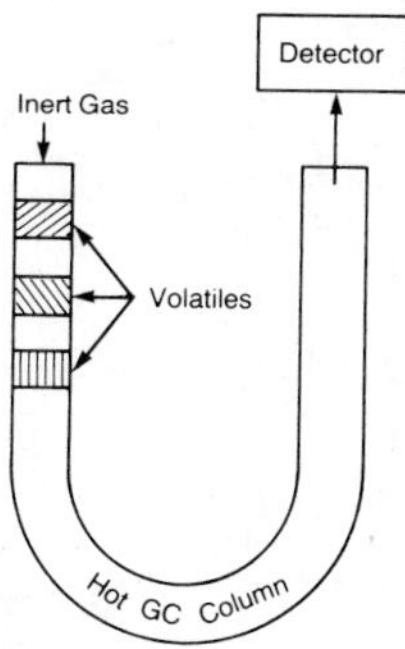

Figure 514:3. Separation of the volatiles.

*Gas chromatographic methods are extremely sensitive to differences in materials used. Mention of specific products by "Standard Methods" does not preclude the use of other existing or as-yet-undeveloped products that give *demonstrably* equivalent results.

†Pierce No. 13075 or equivalent. Crimp-top serum vials also are acceptable.

‡Teflon or equivalent.

§Pierce No. 12722 or equivalent.

b. Microsyringes, 10, 25, 100 μL.

c. Syringe needle, 20 cm × 20 gauge.

d. Syringe, 5 mL valved gastight with 2-in (5-cm) needle.

e. Syringe, 5 mL hypodermic with luer-lok tip.

f. Syringe valve, two-way with luer ends.

g. Volumetric flasks, 10, 100 mL.

h. Purge-gas scrubber filter containing about 225 g granular molecular sieve.‖

i. Purge and trap equipment: The purge and trap equipment consists of three separate parts, the purging device, trap, and desorber.

A successfully used[2] purging device is shown in Figure 514:4. The glass frit installed at base of sample chamber allows finely divided gas bubbles to pass through sample while it is retained above the frit. Gas volumes above the sample are kept to a minimum to eliminate dead-volume effects, yet allow sufficient space for most foams to disperse. Inlet and exit ports are constructed of heavy-walled 6.4-mm (1/4-in.) OD glass tubing so that leak-free removable connections can be made using

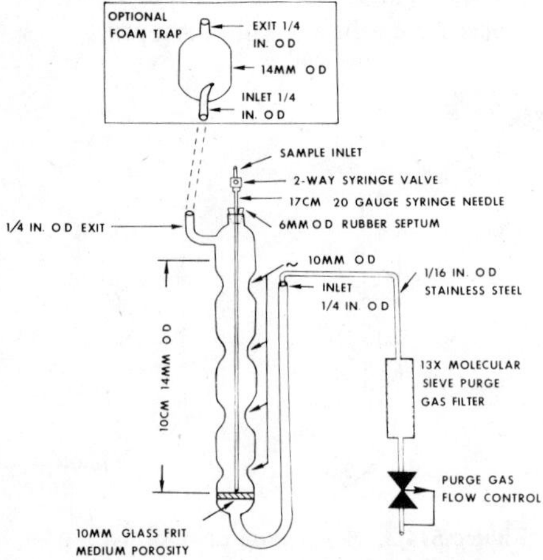

Figure 514:4. Purging device.

‖Molecular Sieve 5-A, Applied Science Laboratories, P.O. Box 440, State College, Pa., or equivalent.

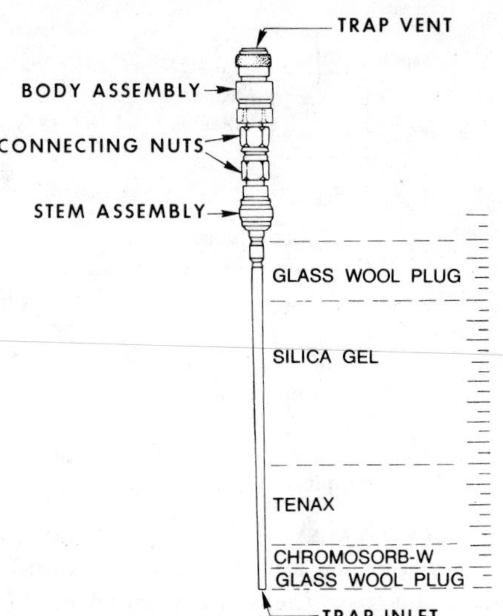

Figure 514:5. Trap.

"finger-tight" compression fittings containing TFE‡ ferrules. The removable foam trap controls excessive foaming. Any similar device with a configuration such that a purge gas flow of 20 mL/min quantitatively (>90%) will strip $CHCl_3$ from solution in approximately 10 min should be acceptable.

Examples of traps and desorbers are shown in Figures 514:5 to 10. A trap is a short gas chromatographic column that at 22 C retards flow of compounds of interest while venting purge gas and, depending on sorbent used, much of the water vapor. Use a trap with a low thermal mass so that it can be heated rapidly for efficient desorption, then cooled rapidly to room temperature for reuse. The Tenax™#-silica gel trap utilizes the sorptive properties of two sorbents, providing a trap that effectively sorbs and desorbs a wide variety of organic compounds.

A desorber is essentially a heated tube into which the trap is placed and maintained at the desorb temperature (Figures

‡Teflon or equivalent

#A source of this material is Applied Science Division, Milton Roy Co. Laboratory Group, P.O. Box 440, State College, Pa. 16801.

514:6 and 7) or is heated rapidly in the desorb mode (Figure 514:9), depending on the type of purge-trap-desorb apparatus. Using this device, the trap is heated rapidly to the desorb temperature (180 C, <45 sec) and maintained at that temperature with minimal temperature overshoot. Trapped compounds are released as a plug to the gas chromatograph by this heat and backflush step.

The trap illustrated in Figure 514:5 and two associated alternative desorbers in Figures 514:6 and 7 are easy to assemble and are relatively inexpensive, but they require considerable manipulation during analysis.

A more complex and expensive-to-build purge-trap-desorb system is shown in the purge-sorb mode in Figure 514:8 and the desorb mode in Figure 514:9. This system is easier to operate reproducibly because all manipulations requiring manual transfers of trap to purge device, trap to desorber, etc., in the first system (Figures 514:5 to 7) are accomplished by switching a single six-port valve (Figures 514:8 and 9).

The purge-trap-desorb apparatus used need not conform necessarily to the exact configurations shown so long as it functions to purge and trap the compounds of interest quantitatively or in a way linearly related to concentration. The apparatus must quantitatively desorb sorbed organics to the chromatographic column sufficiently quickly that they are deposited as a narrow band. Rapid desorption is necessary to produce good chromatographic resolution.

To prepare trap, pack a 25-cm-long, 2.7-mm-(0.105-in.-) ID × 3.2-mm-(0.125-in.-) OD stainless steel tube as follows: place a 5-mm plug of glass wool in inlet end of trap. Add 1 cm of 3% OV-1 on Chromosorb-W™# (60/80 mesh), 15 cm of Tenax™# GC (60/80 mesh), 8 cm of Grade 15 silica gel# (35/60 mesh), and a 5-mm plug of glass wool at exit end of trap. Variations in critical trap parameters (ID, length, sorbant type, amount of sorbant, and sorbant packing order) will affect adversely trap/desorb efficiencies.

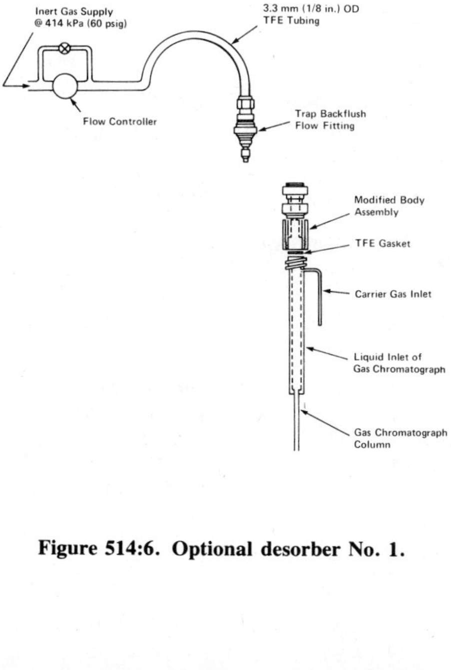

Figure 514:6. Optional desorber No. 1.

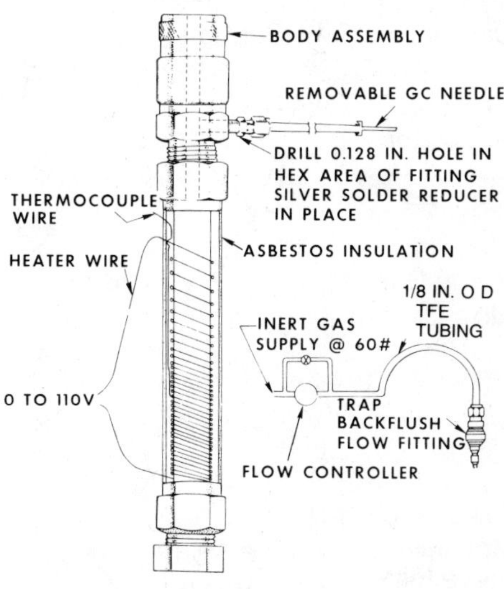

Figure 514:7. Optional desorber No. 2.

#A source of this material is Applied Science Division, Milton Roy Co. Laboratory Group, P.O. Box 440, State College, Pa. 16801.

j. Gas chromatograph, equipped with:

1) *Temperature-programmable oven,*

2) *Injection port* converted into a desorbing device (Figure 514:6[3] for use with trap if an external desorbing device (Figure 514:7, 8, or 9) is not used.

3) *Fittings* for 3.2-mm (1/8-in.) columns.

4) *Halogen-specific detector,* electrolytic conductivity (halide mode) or microcoulometric titration (halide mode). The electron capture detector usually is too sensitive but is acceptable if satisfactory results can be demonstrated.

k. Gas chromatographic columns:

1) Analytical column: 8 ft (2.44 m) long, 0.1 in. ID by 0.125 in. OD (or closest available metric equivalents) stainless steel or equivalent glass, packed as in ¶ 3*a*. Use helium as carrier gas at flow rate of 40 mL/min. Program temperature as follows: With column at room temperature desorb sample into column for 4 min, rapidly heat column to 60 C for 3 min, then raise temperature 8 C/min to 160 C. Hold at 160 C until all compounds have been eluted. A sample chromatogram obtained with this column is shown in Figure 514:11.

2) Confirmatory column: 6 ft (1.83 m) long, 0.1 in. ID by 0.125 in. OD (or closest available metric equivalents) stainless steel packed as in ¶ 3*b*. Use helium at flow rate of 40 mL/min. Program temperature as follows: With column at room temperature desorb sample for 4 min, heat column to 50 C for 3 min, then program at 6 C/min to 170 C. Hold at 170 C until all compounds have been eluted. A sample chromatogram obtained with this column is shown in Figure 514:12.

3. Reagents**

a. 0.2% Carbowax™ 1500 on Carbopack™-C (80/100 mesh).††

**Gas chromatographic methods are extremely sensitive to differences in materials used. Mention of specific products by "Standard Methods" does not preclude the use of other existing or as-yet-undeveloped products that give *demonstrably* equivalent results.

††Available from Supelco: Request Batch No. R-1579.

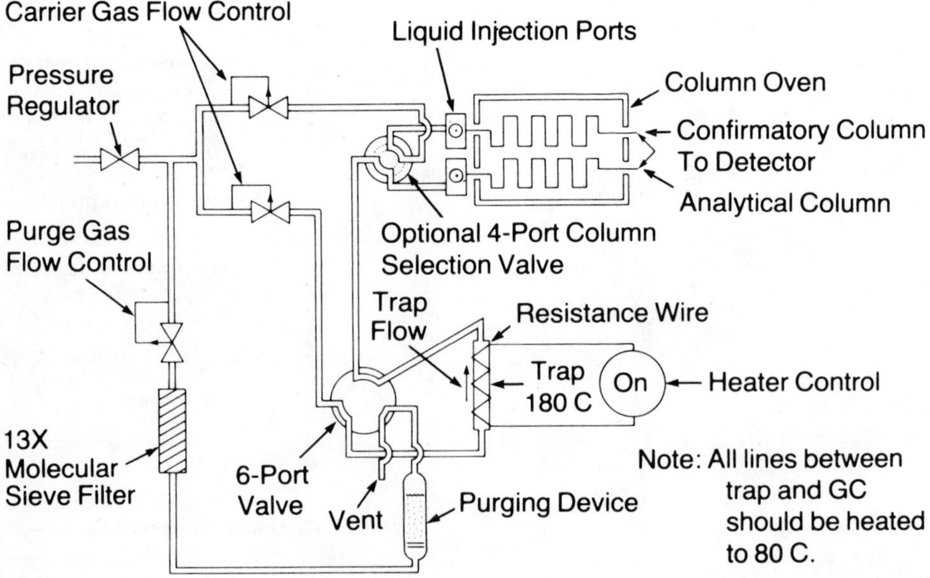

Figure 514:8. A complete purge-trap system (purge-sorb mode).

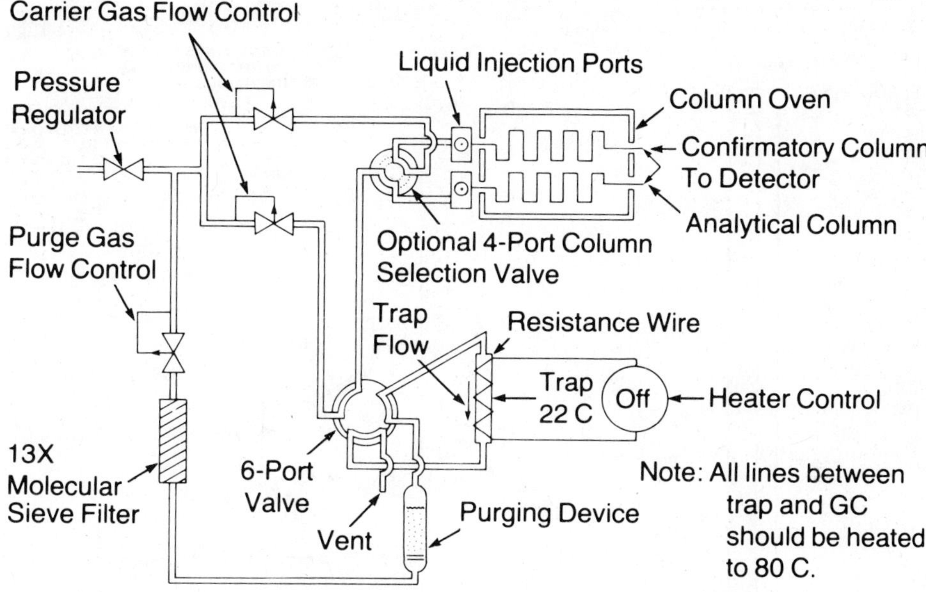

Figure 514:9. A complete purge-trap system (desorb mode).

b. n-Octane/Porasil™*-C* (100/120 mesh).‡‡

c. Three percent OV-1 on Chromosorb™*-W* (60/80 mesh).

d. Tenax™ *GC* (60/80 mesh).§§

e. Silica gel Grade 15 (35/60 mesh).

f. Reference standards: Whenever possible obtain pre-analyzed reagents with guaranteed purities higher than 95%.

g. Sodium thiosulfate, $Na_2S_2O_3 \cdot 5H_2O$.

h. Methyl alcohol.

i. Purgeable-organic-free water: Boil distilled or deionized water for 15 min, then maintain at 90 C while bubbling contaminant-free inert gas through water for 1 hr. While still hot, transfer to narrow-mouth screw-cap bottle with TFE seal. Test purgeable-organic-free water daily by analyzing according to this method. Do not use if significant interfering peaks are observed.

‡‡Available from Waters Associates.

§§ENKA, N.V. Holland. Available from Applied Science Laboratories, P.O. Box 440, State College, Pa.

j. Standard stock solutions of compounds boiling above room temperature: Place about 9.8 mL methyl alcohol in a ground-glass-stoppered 10.0-mL volumetric flask. Let flask stand unstoppered about 10 min or until all alcohol-wetted surfaces have dried. Weigh to nearest 0.1 mg. Using a 100-μL syringe, add 4 drops of reference standard to flask and reweigh. Be sure that the 4 drops fall directly into the alcohol without contacting flask neck. Dilute to volume with methyl alcohol, restopper, and mix by inverting flask several times. Calculate concentration in micrograms per microliter from the net gain in weight. Store at 4 C. Such standards are stable up to 4 wk.

k. Standard stock solutions of compounds boiling below room temperature: Place about 9.8 mL methyl alcohol in a ground-glass-stoppered 10.0-mL volumetric flask. Let flask stand unstoppered about 10 min or until all alcohol-wetted surfaces have dried. Weigh to nearest 0.1

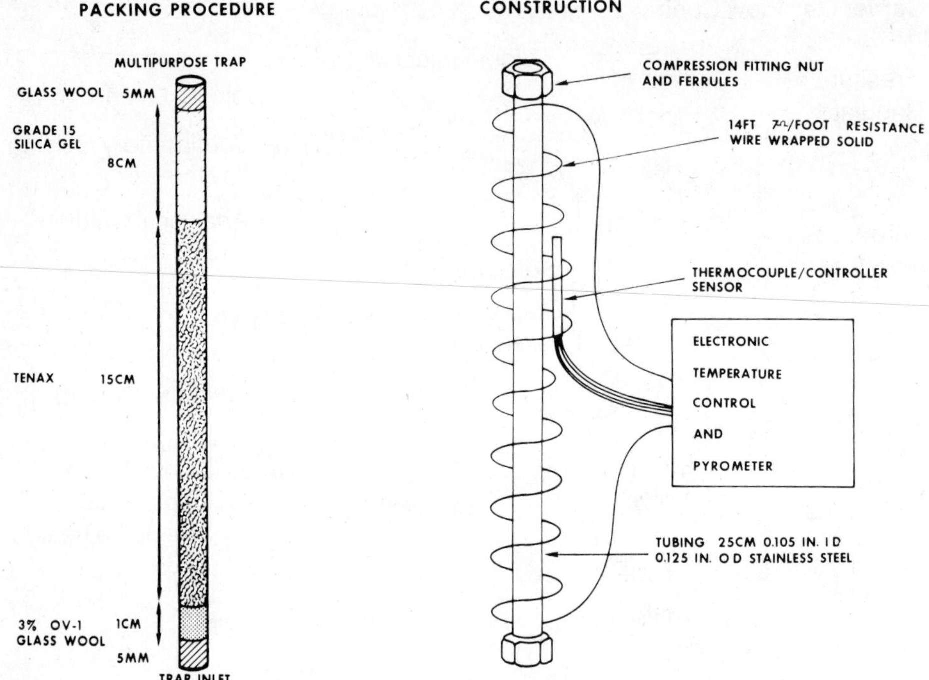

Figure 514:10. Trap and desorber for complete purge-trap system.

mg. Fill a valved gastight syringe with reference standard to 5.0-mL mark. Lower needle to 5 mm above methyl alcohol meniscus. Slowly inject gaseous reference standard into flask neck (the gas will dissolve rapidly in methyl alcohol). Immediately, reweigh to nearest 0.1 mg, dilute to volume, stopper, and mix by inverting flask several times. Store at 0 C or below. Calculate concentration in micrograms per microliter from net gain in weight. Stock standards prepared in methyl alcohol from gaseous reference standards stored with ground-glass seals generally are not stable for more than 1 wk even when stored at <0 C. Preferably store such standards in glass vials with a TFE-lined screw cap. Fill vials at least 90% full.

CAUTION: *Because of toxicity of most organohalides, prepare primary dilutions in a hood. Preferably use a NIOSH/MESA-approved toxic gas respirator when handling high concentrations of such materials.*

l. Calibration standards: From standard stock solutions prepare secondary dilutions in methyl alcohol such that a 20-μL injection into 100 mL organic-free water will generate a calibration standard that produces a response close ($\pm$10%) to that of the samples (see ¶ 7). Purge and analyze aqueous calibration standards in the same manner as samples. Aqueous standards are not stable; discard after 1 hr. Inject standard stock solutions below the water surface, ideally near the flask bottom.

m. Quality check standard (2.0 μg/L): From standard stock solutions prepare a secondary dilution in methyl alcohol containing 10 ng/μL of each organohalide to be determined. Daily, inject 20.0 μL of this mixture into 100.0 mL organic-free water and analyze.

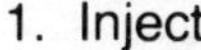

Column: 0.2% Carbowax 1500 on Carbopack-C
Program: 60 C-3 minutes 8/minute to 160 C
Detector: Electrolytic conductivity

1. Inject
2. Chloromethane
3. Bromomethane
4. Vinyl Chloride
5. Chloroethane
6. Methylene Chloride
7. 1.1-Dichloroethylene
8. Bromochloromethane
9. 1.1-Dichloroethane
10. trans-1,2-Dichloroethylene
11. Chloroform
12. 1.2-Dichloroethane
13. 1.1.1-Trichloroethane
14. Carbon Tetrachloride
15. Bromodichloromethane
16. 1.2-Dichloropropane + 2.3-Dichloropropene
17. trans-1.3-Dichloropropene
18. 1.1.2-Trichloroethane
19. 1.1.2-Trichloroethane + Chlorodieromomethane + cis-1.3-Dichloropropene
20. 1.2-Dibromoethane
21. 2-Bromo-1-Chloropropane
22. Bromoform + 1.1.1.2-Tetrachloroethane
23. 1.1.2.2-Tetrachloroethylene
24. 1.1.2.2-Tetrachloroethane
25. 1.4-Dichlorobutane
26. Chlorobenzene

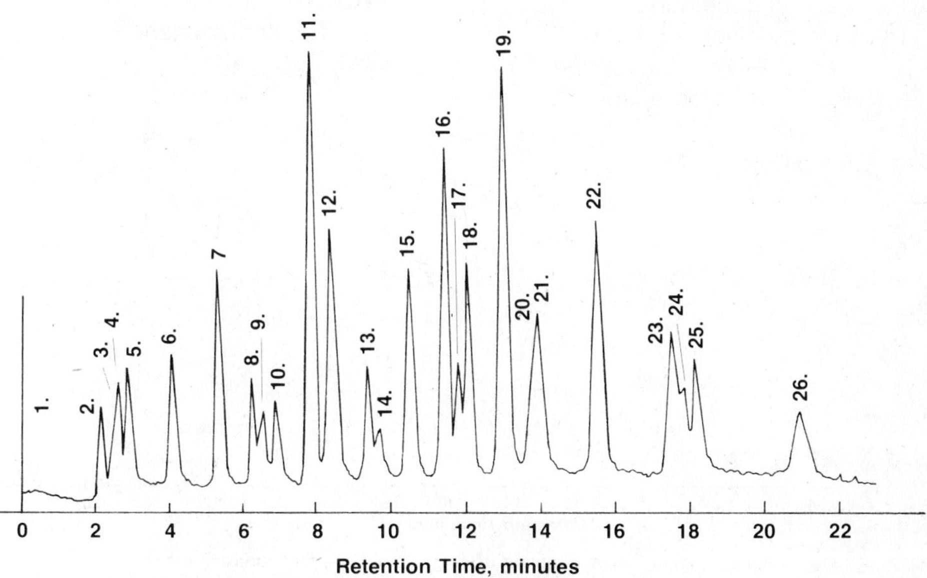

Figure 514:11. Chromatogram of organohalides—analytical column.

4. Sampling

Collect all samples in duplicate. Fill sample bottles so that no air bubbles pass through sample as bottle is filled. Carefully seal bottles so that no air bubbles are entrapped. Maintain hermetic seal on sample bottle until analysis.

Column: n-Octane on Porasil-C
Program: 50 C-3 minutes 6/minute to 170 C
Detector: Electronic conductivity

1. Vinyl Chloride & Chloromethane
2. Bromomethane
3. 1.1-Dichloroethylene
4. Chloroethane
5. trans-1.2-Dichloroethylene
6. Methylene Chloride
7. Carbon Tetrachloride
8. Chloroform + cis-1.2-Dichloroethylene
9. 1.1-Dichloroethane + Bromochloromethane
10. 1.1.1-Trichloroethane + 1.1.2-Trichloroethylene
11. Bromodichloromethane
12. Dibromomethane + Tetrachloroethylene
13. 1.2-Dichloroethane
14. Dibromochloromethane + trans-1.3-Dichloropropene + 1.2-Dichloropropane
15. cis-1.3-Dichloropropene + 1.1.2-Trichloroethane
16. 2-Bromo-1-Chloropropane Chlorobenzene 1.2-Dibromoethane
17. Bromoform
18. Chlorohexene
19. Chlorohexane
20. 1.1.2.2-Tetrachloroethane + Pentachloroethane o-Chlorotoluene
21. m-Dichlorobenzene + Hexachloroethane + p-Dichlorobenzene
22. 1.4-Dichlorobutane + o-Dichlorobenzene
23. Hexachlorobutadiene
24. 1.2 4-Trichlorobenzene

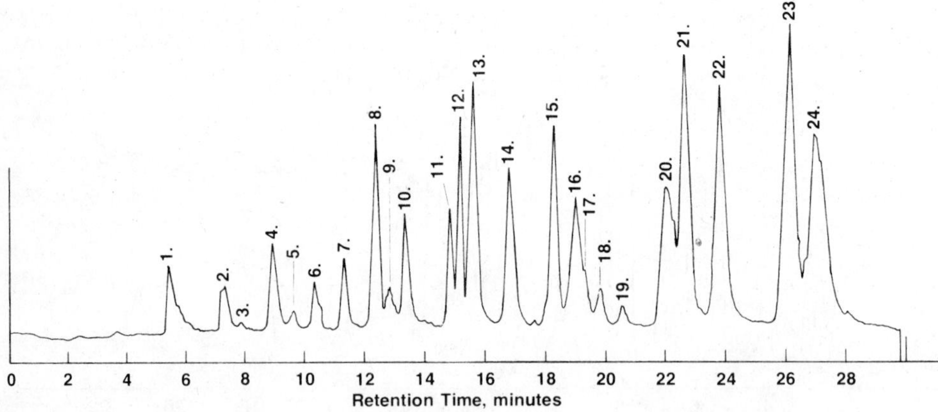

Figure 514:12. Chromatogram of organohalides—confirmatory column.

If free chlorine is present in sample add an inorganic chemical dechlorinating agent such as $NaSO_3$ or $Na_2S_2O_3$ to prevent continued formation of trihalomethanes. Add a stoichiometric excess of dechlorinating agent based on free chlorine residual measurements at time of sample collection. Store samples at 4 C until analysis. Analyze samples as soon as possible after collection because little is known about possible sample matrix effects.

5. Conditioning Traps

Condition newly packed traps in desorption device overnight at 200 C with an inert gas flow of at least 20 mL/min. Vent trap effluent to the room; do not connect to analytical column. Before daily use, condition traps for 30 min while backflushing at 180 C.

6. Procedure

a. Sample extraction: Warm sealed sample to room temperature (22 C) in a water bath. Adjust purge gas (nitrogen or helium) flow rate of 40 mL/min. Attach trap inlet to purging device. (On a valved system turn valve to purge-sorb position.) Remove plungers from two clean 5-mL syringes and attach a closed syringe valve to each syringe. Open sample bottle and carefully pour sample into one of the syringe barrels until it overflows. Replace syringe plunger and compress sample. Open syringe valve and vent any residual air while adjusting sample volume to 5.0 mL. Close valve. Fill and seal second syringe in an identical manner from same sample bottle. Reserve second syringe for a duplicate analysis, if needed. Attach 20-cm needle to first syringe. Open syringe valve and inject sample into purge chamber. Seal chamber by closing valve and purge sample for 11.0 ± 0.05 min.

b. Analysis: As sample is being purged, cool chromatograph column oven to room temperature by opening oven door and turning off oven heater. After purging sample for 11 min, desorb trapped compounds into gas chromatograph by placing trap in desorber and attaching trap backflush flow fitting. On a valved system, turn valve to desorb position. Desorb for 4 min while heating trap to 180 C.

Empty and clean syringe, syringe valve, needle, and purging device as sample is being desorbed. Rinse purging device and sample introduction syringe twice with organic-free water between analyses. Whenever high concentrations are encountered (>100 μg/L) or when suspended solids are contained in the sample, remove purging device from instrument, wash with detergent solution, rinse in distilled water, and dry in an oven at 105 C.

After 4 min of desorption, disconnect trap backflush flow fitting from trap. On valved system turn valve back to purge-sorb position. Immediately, close oven door of gas chromatograph and heat column oven to initial operating temperature: 60 C for analytical column and 50 C for confirmatory column. Start collecting retention data and start stripchart recorder as soon as column over heater is turned on. Program column according to ¶ 2*l* above.

c. Calibration: Prepare calibration standards from standard stock solutions in organic-free water that are close to the unknown in composition and concentration. Use calibration standards of such concentration that 20 μL or less of the secondary dilution need be added to 100 mL of organic-free water to produce a standard at the same level as the unknown.

7. Analytical Quality Control

Daily analyze the 2-μg/L quality check standard before analyzing any samples. Calculate instrument status checks and lower limit of detection estimations from these data. In addition, use the 2-μg/L quality check standard to estimate the concentration in samples. From this information determine appropriate standard dilutions to be made. Analyze at least one blank sample containing only organic-free water daily to monitor for potential interferences as described in ¶ 1*b*.

Qualitative misidentifications are potential problems in gas chromatography. Whenever samples of unknown nature are analyzed, make duplicate analyses using the two recommended columns and whenever possible use mass spectrometric detection to provide unequivocal identification.

8. Calculations

Compare sample peak height to standard peak height to calculate concentration:

$$\mu g/L = \frac{S}{P} \times C$$

where:

S = sample peak height,
P = standard peak height, and
C = concentration of standard, µg/L.

Calculate and report limit of detection (LOD) for each normally reported sample component not detected for each analysis using the following criterion:

$$LOD\ (\mu g/L) = \frac{2\ (A \times ATT)}{(B \times ATT)}$$

where:

B = peak height (mm for 2-µg/L quality check standard),
A = twice the noise level in millimeters at the exact retention time of the compound or the baseline displacement in millimeters from theoretical zero at the exact retention time for the compound. (Tracing these compounds through treatment processes is simplified by expressing concentrations in nanomoles per liter or smaller

TABLE 514:I. SINGLE OPERATOR PRECISION AND ACCURACY FOR SELECTED ORGANOHALIDES

Compound	Dose µg/L	Number of Samples	Mean µg/L	Standard Deviation
Vinyl chloride	2.0	14	1.90	0.10
1,1-dichloroethylene	2.0	13	2.03	0.13
Trans 1,2-dichloroethylene	2.0	14	1.96	0.10
1,1,2-trichloroethylene	2.0	14	1.82	0.08
1,1,2,2-tetrachloroethylene	2.0	14	1.90	0.10
Methylene chloride	1.07	12	1.11	0.23
Chloroform	1.19	12	1.21	0.14
Carbon tetrachloride	1.28	12	1.21	0.08
Bromodichloromethane	1.60	12	1.52	0.05
Chlorodibromomethane	1.96	12	1.91	0.09
Bromoform	2.31	12	2.33	0.16
1,2-dichloroethane	1.00	12	1.00	0.04
Methylene chloride	10.7	8	10.0	0.89
Chloroform	11.9	8	11.3	0.16
Carbon tetrachloride	12.8	8	12.3	0.51
Bromodichloromethane	16.0	8	15.1	0.39
Chlorodibromomethane	19.6	8	19.1	0.70
Bromoform	23.1	8	22.5	1.38
1,2-dichloroethane	10.0	8	9.57	0.24
Methylene chloride	107	11	96.3	10.9
Chloroform	119	11	105	7.9
Carbon tetrachloride	128	11	98.5	6.3
Bromodichloromethane	160	11	145	10.2
Chlorodibromomethane	196	11	185	10.6
Bromoform	231	11	223	16.3
1,2-dichloroethane	100	11	92.5	3.9

units. Regulatory reporting is in micrograms per liter.), and

ATT = attenuation factor.

9. Precision and Accuracy

Single-laboratory accuracy and precision data were obtained by dosing 1-L volumes of organic-free water with the organohalides listed in Table 514:I. The dosed water was used to fill vials that were sealed and stored under ambient conditions. Dosed samples were analyzed randomly over a period of 2 wk. The data in Table 514:I reflect errors due to the analytical procedure and storage.

10. References

1. SYMONS, J.M., T.A. BELLAR, J.K. CARSWELL, J. DEMARCO, K.L. KROPP, G.C. ROBECK, D.R. SEEGER, C.J. SLOCUM, B.I. SMITH & A.A. STEVENS. 1975. National organics reconnaissance survey for halogenated organics. *J. Amer. Water Works Ass.* 67:634.
2. BELLAR, T.A. & J.J. LICHTENBERG. 1974. The determination of volatile organic compounds in water at the μg/L level in water by gas chromatography. *J. Amer. Water Works Ass.* 66:739.
3. DRESSMAN, R.C. & E.F. MCFARREN. 1977. A sample bottle purging method for the determination of vinyl chloride in water at submicrogram per liter levels. *J. Chromat. Sci.* 15:69.

PART 600

AUTOMATED LABORATORY ANALYSES

601 INTRODUCTION

This book is basically a compilation of chemical, physical, and biological methods that have been selected to yield reliable results when applied to specified types of samples. For chemical methods emphasis is placed on the analytical steps that must be followed to achieve such results. These include such preliminary steps as homogenization, sample volume measurement, concentration, digestion, dilution, distillation, filtration, and refluxing. Analytical steps include reagent addition, mixing, incubation, absorbance measurement, titration, and end-point detection by potentiometric or colorimetric means. All these steps potentially are automatable and indeed, most of them have been automated. The availability of instruments to perform the various steps in a specific analysis depends more on the market for the instruments than on the technology required. Instruments and instrumental technics are available for many of the methods presented as manual methods in this book. Among the chemical methods that have been automated are: alkalinity, ammonia, calcium, chemical oxygen demand, chloride, fluoride, hardness, nitrite, nitrate, pH, phosphate, silica, sulfate, and various metals by colorimetry. Some of these constituents can be determined automatically either by measurement of the color developed in a treated sample or by colorimetric determination of a titration end point.

The increasing availability and variety of automated analytical instrumentation makes it impossible to give detailed descriptions and operating instructions for all of the instrumentation applicable to water and wastewater analysis. It is important for the analyst to recognize that the chemical principles on which the automated methods are based are the same as, or comparable to, the chemical principles governing the manual methods. In some instances, a method differing from the manual method is chosen for automation because of simplicity or stability; sometimes a loss of sensitivity results. In no case can an automated instrument improve a method that is analytically unsuitable for the measurement required. However, it is possible to use in automated versions, methods that would be operationally difficult to perform manually. For example, where the time period between reagent additions and colorimetric measurement must be precisely the same from sample to sample, an automated method may be suitable for a technic that would be unmanageable manually.

Automation also has been applied to many of the physical-chemical measurements presented in this book, including the determination of pH, conductivity, and metals by either flame or flameless atomic absorption spectrophotometry.

The advantages of automated instrumentation, which include unattended operation and reduced operator bias, make analytical quality control of even greater importance in automated methods. An instrument does not have analytical judgement; it will not recognize when "something is wrong" and may produce a false reading. This disadvantage is largely balanced by the ease of including significant numbers of quality-control samples to determine recovery and efficiency and to eliminate possible interferences. Sample replication also is less burdensome than for manual methods.

The treatment of automation herein is based on the principle that the automation of one or more parts of an accepted manual method does not change the validity of the method. If there is significant deviation from the nature and sequence of the steps detailed in the manual method, it is the responsibility of the analyst to acquire sufficient quality-control information to demonstrate unequivocally that the automated method gives results equivalent or superior to those of the manual method.

In previous editions of this book this section contained several automated chemical methods. These methods have been recognized by regulatory agencies and are relied on by many users of this book to meet regulatory requirements. In this edition they appear under the constituent to be determined, in keeping with the principle that the automation of an accepted manual method does not change its validity. There is no exclusivity implied in the presentation of one type of automated system in these methods. The analyst may substitute equivalent instrumentation with the proviso that the substitute instrumentation gives results equivalent or superior to those of the methods presented.

602 MODULAR INSTRUMENT SYSTEMS USING FLEXIBLE TUBING

Automated analytical instruments for both chemical and physical measurements are available and in use to analyze individual samples at rates of 10 to 60 samples/hr. The instruments consist of a group of interchangeable modules joined in series by a tubing system. Each module performs one operation such as filtering, heating, digesting, time delay, color sensing, etc.

Readout includes sensing elements with indicators, alarms, and/or recorders. For monitoring applications, automatic electrical and chemical standardization compensation is done by a self-adjusting recorder when known chemical standards are sent periodically through the same analytical train.

Appropriate methodology is supplied by the manufacturer for many common water and wastewater constituents. Some methods are based on procedures described in this manual, while others originate from the manufacturer's adaptation of published research, Because a number of methods of varying reliability may be available for a single constituent, a critical appraisal of the method adopted is mandatory.

Automated methodology is susceptible to the same interferences as the original method from which it derives. For this reason, new methods developed for automated analysis must be subjected to exacting tests for accuracy and freedom from adverse response already met by the accepted standard methods.

Abnormal color and turbidity produced during an analysis will be visible to an analyst manually performing a determination and the result properly will be discarded. Such effects caused by unsuspected interferences may escape notice in an automated analysis.

When such interferences are not known to be absent, analyze replicate samples to which known additions of the constituent being determined have been made. The percentage recovery of the increment is useful, but not absolute, in detecting the presence of interferences.

Practice is to check instrument action routinely and to guard against questionable results by the insertion of standards and blanks at regular intervals, at least after every 10 samples in the train. Proper sample identification is essential.

A fair degree of operator skill and knowledge, together with adequately detailed instructions, is required for successful automated analysis.

1. Apparatus

The required continuous-flow analytical instrument consists of the following interchangeable components assembled in the number and manner indicated in the figures appearing with each method:

a. Sampler.

b. Manifold or analytical cartridge.

c. Proportioning pump.

d. Heating bath operable at the temperature specified.

e. Colorimeter equipped with tubular flow cell of specified length.

f. Filters of specified transmittance.

g. Recorder.

h. Digital printer (optional).

2. General Procedure

The following general procedure applies to each automated method described:

Set up the manifold and complete system as shown in the figure(s) for each method.

Let colorimeter and recorder warm up for 30 min.

Run a baseline with all reagents, feeding distilled water through the sample line. Adjust colorimeter to obtain a stable baseline.

Sample at rate indicated under each method. Arrange standards in sampler in order of decreasing concentration, then load sampler tray with samples.

Switch sample line from distilled water to sampler and begin analysis.

3. Calculation

Prepare standard curves by plotting peak heights of standards processed through the manifold against constituent concentrations in standards. Compute sample concentrations by comparing sample peak height with standard curve.

PART 700
EXAMINATION OF WATER AND WASTEWATER FOR RADIOACTIVITY

701 INTRODUCTION

Radioactivity in water and wastewater originates from natural and artificial sources. Natural or background radioactivity generally contributes picocurie quantities or less of alpha activity and tens of picocuries of beta activity per liter of surface water. Gamma activity also can be associated with alpha and beta emissions. Artificial sources of radioactivity include fission, fusion, or particle acceleration, giving rise largely to alpha, beta, and gamma radioactivity. The development of nuclear science and its application to power development, industrial operations, and medical uses require that attention be given to formulating technics to assess the resultant environmental radioactive contamination. Adequate warnings of unsafe conditions are necessary so that proper precautions can be taken. It is of nearly equal importance to know that conditions are indeed safe when they are, in fact, safe.

It is necessary to establish baselines for the kinds and amounts of radionuclides present naturally and to measure man-made additions to this background. In this way, information is provided for sound judgments regarding the hazardous or nonhazardous nature of increased concentrations.

Measurement technics are not difficult to devise because radiation counting equipment of high sensitivity, selectivity, and stability is fairly commonplace. Furthermore, the guides provided by the Federal Radiation Council[1] on permissible daily intake of some radionuclides, the recommendations on radionuclide concentrations in water made by the National Council on Radiation Protection and Measurements (NCRP),[2] those made by the International Commission on Radiation Protection (ICRP),[3] and the Environmental Protection Agency National Interim Primary Drinking Water Regulations[4] are, with few exceptions, at concentrations that are readily measured by current methods and instruments.

Meaningful measurements require careful application of good scientific technics. Gross alpha and gross beta measurements are relatively inexpensive and serve a useful purpose for screening samples and for checking long-term trends. Samples sufficiently low in radioactivity may not require further analyses. Samples at intermediate concentrations may be composited for more complete and expensive analysis of specific radionuclides. To be effective, however, a gross screening technic should be based on knowledge of the relationship between gross measurements and the radionuclides of greatest concern.

Both natural and artificial sources of radiation from samples emitting alpha, beta, or gamma activity are included in the examination (to the exclusion of radiation external to the sample, i.e., cosmic, gamma, X-ray, and hard beta radiation in the environment). Because the rate of decay and the energy of radiation are unique characteristics of each radioelement, strict adherence to a standard procedure is essential to proper interpretation of a radioactivity examination. Frequently the procedure may have rigid timing requirements to discriminate among radioelements. For example, the rapid alpha analysis of airborne particulates usually consists of the measurement of radon daughter products (largely ^{218}Po and ^{214}Po)

from which the equilibrium parent radon concentration and each of its descendent products may be estimated. Subsequent alpha-counting of the same sample could be designed to measure thoron daughter activity or long-lived alpha emitters. The beta activity of fresh rain, a few minutes to several hours after collection, includes significant contamination by radon daughter products. If the analysis is postponed for 6 hr, the radon daughters will disappear along with some short-lived artificial radionuclides. The loss of activity resulting from delayed counting can be estimated by the extrapolation of decay data. During concentration of water samples by evaporation, radionuclides such as elemental iodine or hydrogen iodide (in acid solution) may be lost by volatilization at temperatures below 105 C. If the sample is ignited, the chance of volatilization is even greater. Radioactive substances such as carbon 14 and tritium may be present as volatile chemicals. Groundwater generally contains nuclides of the uranium and thorium series. Use special care in sampling and analyzing because members of these series often are not in secular equilibrium. This is particularly true of gaseous ^{222}Rn, ^{220}Rn, and their daughter products, which may be present far in excess of the equilibrium concentration from radium in solution.

701 A. Collection of Samples

The principles of representative sampling of water and wastewater apply to sampling for radioactivity examinations.

Because a radioactive element often is present in submicrogram quantities, a significant fraction may be lost by adsorption on the surface of containers used in the examination. Similarly, a radionuclide may be largely or wholly adsorbed on the surface of suspended particles.

When radioactive industrial wastes or comparable materials are sampled, consider the possibility of deposition of radioactivity on surfaces of glassware, plastic containers, and equipment that may cause a loss of radioactivity and possible contamination of subsequent samples collected in inadequately cleansed containers. Sample containers vary in size from 0.5 L to 18 L, depending on required analyses. Use containers of plastic (polyethylene or equivalent) or glass, except for tritium samples (use glass only).

See Section 105 for general information on sample collection and preservation. Add preservative at time of collection unless sample is to be separated into suspended and dissolved fractions but do not delay acid addition beyond 5 days. Use conc hydrochloric (HCl) or nitric (HNO_3) acid to obtain a pH <2, except for radiocesium (use only conc HCl) and radioiodine and tritium (use no preservative). Hold acidified sample at least 16 hr before analysis. For further details see references.[5,6]

Test preservatives and reagents for radioactive content.

701 B. Counting Room

Design and construction of a counting room may vary widely. Provide a room free of dust and fumes that may affect electrical stability of instruments. Stabilize and reduce background radiation by making the walls, floor, and ceiling out of at least 5 cm of concrete but avoid using shales, granites, and sands containing sufficient natural activity to affect instrument background.

A modern chemical laboratory can be used to process routine environmental samples but preferably segregate monitoring work from other laboratory operations.

Provide air-conditioning and humidity control depending on the number of instruments to be used and the prevailing climatic conditions. Generally, electronic instruments perform best when the temperature remains constant within 3 C and does not exceed 30 C. Keep the temperature inside the instrument chassis below the maximum specified by the manufacturer.

Humidity affects instrument performance even more than extremes of temperature because of moisture buildup on critical components. This causes leakage and arcing, and shortens the life of these components. A humidity between 30 and 80% usually is satisfactory.

Most scalers are supplied with constant-voltage regulators suitable for controlling minor fluctuations in line voltage. For unusual fluctuations use an auxiliary voltage regulator transformer. Use a manually reset voltage-sensitive device in series with a voltage regulator placed in the main power line to instruments to protect them in case of power failure or fluctuating line voltage.

Store samples containing appreciable activity at a distance so as not to affect instrument background counting rate.

Cover floors and desk tops with a material that can be cleaned easily or replaced as necessary.

701 C. Counting Instruments

The operating principle of Geiger-Mueller and proportional counters is that the expenditure of energy by a radiation event causes ionization of counter gas and electron collection at the anode of the counting chamber. Through gas or electronic amplification, or both, the ion-collection event triggers an electronic scaler recorder.

The principle of scintillation counters is similar in that quanta of light produced by the interaction of a radiation event and the detection phosphor are seen by a photomultiplier tube. The tube converts the light pulse into an amplified electrical pulse that is recorded by an electronic scaler. Thallium-activated sodium iodide crystals and silver-activated zinc sulfide screens form useful scintillation detectors for counting gamma and alpha radioactivity, respectively. Semiconductor detectors are new and are undergoing rapid changes. Alpha and gamma spectrometers using, respectively, silicon surface-barrier and lithium-drifted germanium detectors are available.

Characteristic of most counters is a background or instrument counting rate usually due to cosmic radiation, to radioactive contaminants of instrument parts and counting room construction material, and to the nearness of radioactive sources. The background is roughly proportional to the size or mass of the counting chamber or detector, but it can be reduced by metal shielding.

Instrument "noise", or the false recording of radiation events, may be caused by faulty circuitry, too sensitive a gain setting, high humidity, and variable line voltage or transients. Control these problems by using constant-voltage transformers with transient filters, properly adjusting gain setting as specified by the manufacturer, and air-conditioning the counting room.

The internal proportional counter accepts counting pans within the counting chamber and at the beta operating voltage, records all alpha, all beta, and some gamma radiation emitted into the counting gas. Theoretically, half the radiation is emitted in the direction of the counting pan. Some of the beta radiation, but only 1 to 2% of the alpha radiation, is back-scattered into the counting gas by sample solids, the counting pan, or the walls of the counting chamber. For substantially weightless samples, considerably more than 60% of the beta radiation and about 50% of the alpha radiation is counted. However, take considerable care in sample preparation to prevent the sample or counting pan from distorting the electrical field of the counter and depressing the counting rate. Avoid nonconducting surfaces, airborne dusts, and vapor from moisture or solvents.

The end-window Geiger-Mueller counting tubes are rugged and stable counting detectors. Samples usually are mounted 5 to 15 mm from the window. Under these conditions, most alpha and weak beta radiations are stopped completely by the air gap and mica window and are not counted. Counting efficiencies for mixed fission products frequently are less than 10% for substantially weightless samples having an area less than that of the window. Because most Geiger-Mueller tubes have diameters of about 2.5 cm, the pan size—and, as a consequence, the sample volume—is restricted. Under these conditions, the detectability is low and uncertain, particularly for unknown sources of radiation. On the other hand, Geiger-Mueller tubes are excellent for counting samples of tracers or purified radionuclides. Prepare standard sample mounts that yield reproducible counting efficiencies for which counting is not affected by the electrical conductance of sample pans.

Thin-window (polyester plastic film* less than 250 $\mu g/cm^2$) tubes approximately 5 cm in diameter provide counting efficiencies intermediate between those of conventional Geiger-Mueller tubes and internal counters. Counting alpha activity in these thin-window counters is satisfactory.

Thin-window counters with chamber diameter greater than 60 mm and sample mount diameter greater than 50 mm are preferable to Geiger-Mueller tubes. These counting chambers have good operational stability and less interference from nonconducting surfaces and moisture vapors than internal proportional counters.

1. Internal Proportional Counters

a. Uses: Internal proportional counters are suitable for determining alpha activity at the alpha operating plateau and alpha-plus-beta activity at the beta operating plateau. The alpha or beta activity, or both, can refer to a single or to several radionuclides.

Use instruments consisting of a counting chamber, a preamplifier, and a scaler with high-voltage power supply, timer, and register. Use the specified counting gas and accessories, make adjustments for sensitivity, and use in accordance with manufacturer's operating instructions.

b. Plateau (alpha or beta): Find the operating voltage where the counting rate is constant, i.e., varies less than 5% over a 150-V change in anode voltage:

1) With the instrument in operating order, place the alpha or beta standard (see Section 701D.3) in the chamber, close,

*Mylar or equivalent.

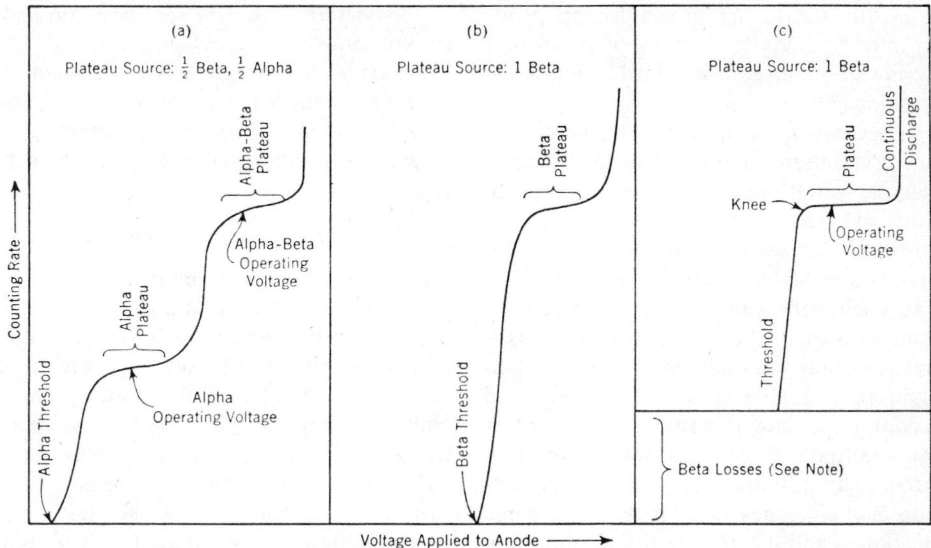

Figure 701:1. Shape of counting rate—anode voltage curves. Key: (a) and (b) are for internal proportional counter with P-10 gas; (c) is for end-window Geiger-Mueller counter with Geiger gas. (Note: Beta losses are dependent on energy of radiation and thickness of window and air path.)

and flush with counter gas for 2 to 5 min.

2) Use the manufacturer's recommended operating voltage and count for a convenient time giving an acceptable coefficient of variation, preferably 2%. Repeat at voltages higher and lower than the suggested operating voltage in increments of 50 V. (CAUTION: *Instrument damage will result from prolonged continuous discharge at too high a voltage.*)

3) Plot relative counting rate (ordinate) against anode voltage (abscissa). A plateau at least 150 V long with a slope of 5% or less should result (see Figure 701:1). Select an anode voltage near the center of this plateau as the operating voltage.

c. Counter stability: Check instrument stability at the operating voltage by counting the plateau source daily (see Section 701D.3). If the source count is within two standard deviations of the count rate, proceed as in *d* following. If the source count is not so reproduced, repeat the test. If stability is not attained, service the instrument.

d. Background: Determine the background (with an empty counting pan in the counting chamber). Use a background counting time as long as the longest sample-counting time. Make control charts as an aid in stability testing.

e. Sample counting: Place dry sample on a counting pan in the counting chamber and ground the pan to the chamber piston. Flush with counter gas and count for a preset time, or preset count, to give the desired counting precision (see Section 701F).

f. Calibration of overall counter efficiency: Correct observed counting rate for geometry, back-scatter, and self-absorption (sample absorption).

Although it is useful to know the variation in these individual factors, determine overall efficiency by preparing standard sample sources and unknowns.

1) For measuring mixed fission products or beta radioactivity of unknown composition, use a standard solution of cesium 137.

Prepare a standard (known disintegration rate) in an aqueous solution of sample solids similar in composition to that present in samples. Dispense increments of solution in tared pans and evaporate. Make a series of samples having a solids thickness of 1 to 10 mg/cm^2 of bottom area in the counting pan. Evaporate carefully to obtain uniform solids deposition. Dry (103 to 105 C), weigh, and count. Calculate the ratio of counts per minute to disintegrations per minute (efficiency) for different weights of sample solids. Plot efficiency as a function of sample thickness and use the resulting calibration curve to convert counts per minute (cpm) to disintegrations per minute (dpm).

2) If other radionuclides are to be tested, repeat the above procedure, using certified solutions of each radionuclide. Avoid unequal distribution of sample solids, particularly in the 0- to 3-mg/cm^2 range, in both calibration and sample preparation.

3) For alpha calibration, proceed as above, using a standard solution of natural uranium salt (not depleted uranium), plutonium 239, or americium 241. Report calibration standard used with results.

2. End-Window Counters

End-window counters may be used for beta-gamma and absorption examinations. Most alpha and soft beta radiations are stopped by the air gap and window. Use a sample pan with a diameter less than that of the window and, for maximum efficiency, place it as close to the window as possible. House the detector inside a 5-cm-thick lead shield to improve counting sensitivity by decreasing the background by about 50%. Use associated equipment consisting of a scaler having a timer, a register, and a high-voltage power supply.

See Section 701C.1*a-f* for operation and calibration.

Coincidence correction: Geiger-Mueller counters commonly have resolving times of 100 to 400 μsec; therefore, correct data on samples of high counting rate for loss in counts.

3. Thin-Window Proportional Counter

The thin-window proportional counter has application for counting moderate to high levels and for counting residues that adhere poorly to the counting pan. The counters detect alpha and low-energy beta emitters. They are about one-half as sensitive as internal proportional counters because the geometry of counting is not as favorable and absorption losses (air path and window) are greater. Because the sample is outside, this counter is less affected than the internal proportional counter by contamination from loose residues, losses due to residue moisture, and poor electrical conductance. See Section 701C.1*a-f* for operation and calibration.

4. Low-Background Beta Counter

The low-background counter is useful for measurements as low as 0.1 and as high as 50 pCi/sample. Higher activity levels, to about 1,000 pCi, can be counted if other beta detectors are not available. The counters are designed primarily for beta emitters having a maximum beta energy above 0.3 MeV.

The detector window thickness usually is less than 1.0 mg/cm^2 and attenuation of high-energy beta rays is relatively minor. Use sample pans with a diameter less than that of the window. The counting efficiencies for weightless samples vary from 30 to 55% for beta radiation of moderate energy.

To obtain a background counting rate of 1 cpm or less use an instrument with a lead or steel shield and an anticoincidence device of one or more guard detectors with the electronics needed to prevent counting in the sample detectors when a count is recorded in the guard. An instrument with an automatic sample changer is desirable. Most counters of this type use helium-isobutane or similar gases that operate in the Geiger region. Instruments using proportional counting gas also are available. Geiger-Mueller counters commonly have resolving times of 100 to 400 μsec. Correct data on samples of high counting rate for loss of counts.

5. Gamma Spectrometer

Gamma spectrum analysis may be made with a minimum of sample preparation. Unless a very complex spectrum with overlapping photopeaks is obtained, chemical separation followed by gamma analysis for quantitative measurements on each sample fraction is unnecessary. This method normally excludes nongamma-emitting radionuclides, and those having photon emission energies less than 0.01 MeV usually are measured with considerable uncertainty. Two types of gamma spectrometers are currently in use: the sodium iodide, thallium-activated [NaI(Tl)] crystal system using a scintillation phenomenon and germanium diodes, some with lithium activation [Ge(Li)]. For details on operation and calibration of gamma spectrum analysis see manufacturer's instructions.

a. Principle: Gamma photons from a sample enter the sensitive volume of the detector and interact with detector atoms. The interactions are converted to an electrical voltage pulse proportional to the energy of the photon. The pulses are stored in sequence in finite energy-equivalent increments (such as 0.02 MeV/channel for NaI systems and 0.001 MeV/channel for Ge systems) over the entire spectrum range (such as 0.1 to 2 MeV) depending on instrument capabilities and operator choice.

After counting of the sample, the accumulated counts in each energy increment of an entire spectrum are analyzed for the number and energy of photopeaks (a qualitative test) or for the number of pulses associated with each photopeak, corrected for background count and interference from gamma emissions (a quantitative test). Because each gamma-emitting radionuclide usually has several photopeaks, one of which yields the greatest abundance of pulses, the number of radionuclides in the sample to be analyzed is limited by the probability that overlapping or interfering photopeaks will cause errors in a quantitative estimation. For NaI systems, analysis of four to eight components is practical. With more complex mixtures, use chemical separation followed by gamma spectrum analysis of each fraction, or use a Ge system.

1) In NaI crystal systems, the interaction of gamma photons with the detector gives rise to pulses of light that are proportional in intensity to the gamma photon energy. The light pulses enter a photomultiplier tube (PMT) and are converted to electrical voltage pulses proportional to the light (scintillation) intensity. Because of the components involved, the resolving time is about 10^{-9} sec.

2) In Ge diode detector systems, the interaction of gamma photons with the detector ionizes detector atoms. A bias voltage applied to the detector allows collection of electrons proportional to the deposited photon energy with a resolving time of 10^{-9} to 10^{-13} sec. Such a system has exceptionally high resolution.

b. Components: A gamma spectrometer consists of a detector, pulse-height analyzer system, data readout capability, and a shielded enclosure. Connect the detector to a preamplifier and a high-voltage power supply. Place the detector and sample in a metal shield (10 to 20 cm steel or equivalent) to reduce external radiation background level. The pulse-height analyzer system consists of a linear amplifier, a biased linear amplifier, an analog-to-digital converter (ADC), memory storage, and a logic control mechanism. The logic control capability permits data storage in various modes and display or recall of data. The data readout system contains one or more of the following: an oscilloscope for visual display, a readout indicator, an electric typewriter or digital printer, a paper-tape perforator, a magnetic-tape recorder, a strip-chart recorder, a keypunch card unit, an x-y recorder or plotter, and a computer terminal with associated capabilities. The oscilloscope is useful in aligning the instrument with standards such as ^{60}Co, ^{137}Cs, and ^{207}Bi. Computer capability is essential in data reduction and in complex spectrum stripping procedures.

A common scintillation (NaI) detector is a crystal 10 cm in diameter by 10 cm thick hermetically sealed in a container that is optically coupled to a photomultiplier tube.

A common high-resolution germanium detector consists of a diode of over 30 cm^3 sensitive volume encased in a 7.6-cm-diam vacuum-sealed cylinder with a dip-stick immersed in liquid nitrogen in a large cryostat, a preamplifier, and a detector bias voltage supply. Effectively cool the detector with liquid nitrogen to protect the intrinsic qualities of the diode and to reduce electronic noise generation. Use a linear amplifier that will maintain the pulse resolution provided by the detector.

A single-channel gamma spectrum analyzer is similar to a multichannel analyzer but is limited to the examination of a single energy range at a time. The instrument is best used for continuous monitoring of a waste having fixed radionuclide composition, making gross gamma measurements, or measuring a single gamma-emitting radionuclide. It is similar to the multichannel analyzer except that the design is comparatively inexpensive.

6. Alpha Spectrometer

Semiconductor particle detectors, that is, silicon surface-barrier detectors, are used for alpha spectrometry. Detector performance is affected primarily by resolution, active area, and depletion depth. Chemical separation of the sample followed by monomolecular electrodeposition on counting planchets is required. Count under a high vacuum. For details on operation and calibration of alpha spectrum analysis see manufacturer's instructions.

a. Principle: Alpha emissions from a sample enter a sensitive detector volume and interact with detector atoms. Ionization of the detector and collection of the electrons by use of a bias voltage applied to the detector allows the generation of an electrical voltage pulse proportional to the deposited alpha particle energy.

b. Components: An alpha spectrometer consists of a detector, a vacuum chamber for the detector and sample, a preamplifier, a detector bias voltage supply, a linear amplifier, a biased amplifier, a mechanical vacuum pump, a multichannel analyzer (with ADC and memory storage) or single-channel analyzer, and data readout capability similar to that discussed under gamma spectrometry.

7. Alpha Scintillation Counter

When an alpha particle bombards an impure crystal of zinc sulfide, a portion of the kinetic energy is transformed into visible light. The sulfide scintillates more efficiently when it contains silver impurities and when the duration of the light pulse is shortened by the presence of nickel ions.

The alpha scintillation counter consists of a phosphor detector coupled to a photomultiplier, a high-voltage supply, an amplifier-discriminator, and a scaler. Generally, the photomultiplier tube has a window diameter greater than that of the sample unless the phosphor is coupled to a light-focusing optical system.

Place the silver-activated and nickel-quenched zinc sulfide phosphor near, or in contact with, the alpha-emitting sample and arrange it so that a photomultiplier tube observes the light pulses, which are amplified and recorded on the scaler.

a. Mount solid samples in a thin layer (less than 3 mg/cm^2) on a planchet. Place the phosphor between the sample and the photomultiplier tube. Enclose the sample and detector in a light-tight chamber 3 to 5 mm from the phototube window. Under these conditions the counting efficiency is from 35 to 40%.

b. Gaseous samples contained in a dome-shaped cell coated with zinc sulfide "paint" are observed more efficiently than solid samples. See Section 706 for a description of such a system.

8. Liquid Beta Scintillation Counter

When a sample having radionuclides is mixed with an organic liquid scintillator, light is produced. The flashes of light are detected and amplified by one or more photomultiplier tubes.

Liquid scintillation counters are well suited for counting low-energy beta emitters such as tritium or carbon 14 because self-absorption losses are eliminated. Counting efficiencies approach 100% for high-energy betas, but for tritium the efficiency is much lower because the beta pulses, lowest in energy, are at the level of the "dark current" pulses from the photomultiplier tube and are discriminated against to reduce background. Most liquid scintillation instruments use two photomultipliers in coincidence to reduce background from "dark current". Most liquid scintillation counter systems incorporate at least a two-channel analyzer, which permits more than one beta emitter to be counted at the same time if the respective maximum beta energies differ by a factor of at least three.

Dissolve or suspend the sample in a scintillator solvent such as toluene, xylene, or 1-4-dioxane. Place the sample in a transparent bottle to enable the light flashes to be transmitted to the phototube. Use a calibration standard containing the same radionuclide prepared in the same medium. Determine background by counting a bottle containing both solvent and scintillator. Measure background at least once daily.

701 D. Laboratory Reagents and Apparatus

See Section 102 for basic standards applying to laboratory reagents and apparatus. The following special instructions are pertinent:

1. Reagents and Distilled Water

Periodically check the background radioactivity of all solutions and reagents used. Discard those having a radioactivity level that significantly interferes with the test.

2. Apparatus

Before reuse, thoroughly decontaminate apparatus and glassware with detergents and complexing agents, followed, if necessary, by acid and distilled-water rinses. Segregate equipment and glassware for storage and reuse on samples of comparable activity—i.e., keep apparatus for background and low-level counting separate from that for higher levels by using distinctive markings and different storage cabinets or laboratories. Preferably use single-use counting pans, planchets, and auxiliary supplies. Glassware that is slightly radiocontaminated may be entirely satisfactory for use in chemical tests but is unsatisfactory for radioanalysis.

3. Radioactivity Sources

a. Solutions: Use standard solutions having calibrations traceable to sources of radioactivity certified by the National Bureau of Standards.

b. Plateau or check sources:

1) Alpha—Uranium oxide (U_3O_8) plated, not less than 45 mm in diameter, having an alpha activity of about 10,000 cpm, or plutonium or americium plated as a weightless alpha standard source.

2) Beta—Cover uranium oxide (U_3O_8), plated as described above, with 8 to 10 mg/cm² aluminum foil, or use cesium 137.

701 E. Expression of Results

Preferably report results of radioactivity analyses in terms of picocuries per liter (pCi/L) at 20 C or, for samples of specific gravity significantly different from 1.00, picocuries per gram where 1 pCi = 10^{-12} Ci = 2.22 dpm. For samples normally containing 1,000 to 1,000,000 pCi/unit volume or weight, use the nanocurie (nCi) unit (1 nCi = 10^{-9} Ci = 1,000 pCi). For values higher than 1,000 nCi, use the microcurie (μCi) unit.

Report results in such a way that they do not imply greater or lesser accuracy than can be obtained by the method used. See Part 100.

"Gross alpha" implies unknown alpha sources in which natural uranium, ^{239}Pu, or ^{241}Am has been used to determine self-absorption and efficiency factors.

"Gross beta" implies unknown sources of beta, including some gamma radiation, and calibration with ^{137}Cs as in Section 701C.1*f* above.

701 F. Statistics

Section 104 discusses the statistics of analytical problems as applied to chemical constituents. These remarks also are generally applicable to radioactivity examinations.

1. Standard Deviation and Counting Error

The variability of any measurement is described by the standard deviation, which can be obtained from replicate determinations. There is an inherent variability in radioactivity measurements because disintegrations occur in a random manner described by the Poisson distribution. This distribution is characterized by the standard deviation of a large number of events, N, that equals its square root, or:

$$\sigma(N) = N^{1/2}$$

For ease in mathematical application, the normal (Gaussian) approximation to the Poisson distribution ordinarily is used. This approximation, which generally is valid at $N \geq 20$, is the particular normal distribution with a mean of N and standard deviation of $N^{1/2}$.

More often, the concern is not with the standard deviation of the number of counts but rather with the deviation in rate (number of counts per unit time):

$$R' = \frac{N}{t}$$

where:

t = time of observation.

The standard deviation in the counting rate, $\sigma(R')$, is calculated by usual methods for propagation of error:

$$\sigma(R') = \frac{N^{1/2}}{t} = \left(\frac{R'^{1/2}}{t}\right)$$

In practice, all counting instruments have a background counting rate, B, when no sample is present. With a sample, the counting rate increases to R_O. The counting rate R due to the sample is:

$$R - R_O - B$$

By propagation-of-error methods, calculate the standard deviation of R as follows:

$$\sigma(R) = \left(\frac{R_O}{t_1} + \frac{B}{t_2}\right)^{1/2}$$

where:

t_1, t_2 = times at which the gross sample and background counting rates were measured, respectively.

Practical counting times often are 30 min, or 2,500 total counts above background, whichever takes less time. It is desirable to divide the counting time into equal periods to check constancy of the observed counting rate. For low-level counting, use $t_1 = t_2$. The error thus calculated includes only the error caused by inherent variability of the radioactive disintegration process. Report it as the "counting error."

Use a confidence level of 95%, or 1.96 standard deviations, as the counting error.

2. Limit of Detection

Different conventions with differing terminology and mathematics have been used to estimate the lower limit of detection (LLD) or the minimum detectable activity (MDA).[6–8] To eliminate confusion and the production of noncomparable data, it is proposed that the Health and Safety Laboratory procedure[6] be used exclusively. The basis of this procedure is hypothesis testing. LLD is defined as the smallest quantity of sample radioactivity that will yield a net count for which there is a predetermined level of confidence that radioactivity is present. Two errors may occur: Type I, in which a false conclusion is reached that radioactivity is present and Type II, with a false conclusion that radioactivity is absent from the sample.

The LLD may be approximated as

$$LLD \cong (K_\alpha + K_\beta)\, S_O$$

where:

K_α = value for the upper percentile of the standardized normal variate corresponding to the preselected risk of concluding falsely that activity is present (α),

K_β = the corresponding value for the predetermined degree of confidence for detecting presence of activity ($1 - \beta$), and

S_O = estimated standard error of the net sample counting rate.

For sample and background counting rates that are similar (as is expected at or near the LLD) and for α and β equal to 0.05, the smallest amount of radioactivity that has a 95% probability,

$$LLD_{95} = 4.66\, S_b$$

where:

S_b = standard deviation of the instrument background counting rate.

The LLD thus calculated is in units of counts per minute; to convert to concentration use the appropriate factors of sample volume, counting efficiency, etc.

701 G. Quality Assurance

The continuous application of quality assurance principles to radiological measurements results in consistent data, but does not guarantee accuracy. Accurate data are both precise and unbiased. To obtain them requires day-to-day control over instrumentation, chemical processing, and associated factors. Use of three allied but independent procedures is most useful: (*a*) "blind" replicate analysis of real samples to evaluate within laboratory precision or reproducibility; (*b*) cross-check analysis of natural samples among several laboratories to determine agreement with other laboratories; and (*c*) analysis of standard samples to determine accuracy, i.e., the agreement of results with a known value. Process real samples as part of the normal laboratory workload but submit replicates to analysis without the knowledge of the analyst. Obtain natural samples for interlaboratory checking by collecting larger-than-normal samples and subdividing them for distribution. Standard samples may be natural samples or carefully prepared simulated environmental samples to which have been added known or accurately determined concentrations of appropriate radionuclides.

To conduct cross-check and standard analysis evaluations most effectively, use an independent referee laboratory. Standard samples may be used within a single laboratory, but exercise extreme care to prepare the samples by an accurate means divorced from the analytical method being studied. Statistically analyze all quality-assurance data to permit making firm conclusions about validity. Use control charts to simplify measuring the amounts by which the mean value of several individual determinations is biased from the true value. If a mean value fails to fall within the limits established by the mean chart, the analytical results are not from the expected normal distribution.

The overriding factor in a successful quality-control program is proper selection of criteria for acceptable and attainable accuracy. Through experimentation and experience, select criteria that reflect both the capabilities of the laboratory and the requirements of the users of the laboratory output. Secondly, select a sufficient number of samples to test properly whether the analytical data meet the established criteria.

At present standard samples of environmental materials and standardized radioisotope solutions are available from the International Atomic Energy Agency (IAEA),* the U.S. National Bureau of Standards,† and the U.S. Environmental Protection Agency.‡ Participation is possible in intercomparison programs sponsored by IAEA, EPA, or the World Health Organization.§

*International Atomic Energy Agency, Div. of Research and Laboratories, Kaerntner Ring, 1010 Vienna, Austria.

†National Bureau of Standards, Center for Radiation Research, Radiochemistry Section, Washington, D.C. 20234.

‡Environmental Monitoring and Support Laboratory, Quality Assurance Branch, P.O. Box 15027, Las Vegas, Nev. 89114.

§World Health Organization, Geneva, Switzerland.

701 H. References

1. FEDERAL RADIATION COUNCIL. 1961. Background Material for the Development of Radiation Protection Standards. Rep. No. 2 (Sept.), U.S. Govt. Printing Off., Washington, D.C.
2. NATIONAL COMMITTEE ON RADIATION PROTECTION AND MEASUREMENTS. 1959. Maximum Permissible Body Burdens and Maximum Permissible Concentrations of Radionuclides in Air and Water for Occupational Exposure. NBS Handbook No. 69, pp. 1, 17, 37, 38, and 93.
3. Recommendation of the International Commission on Radiological Protection (rev. Dec. 1, 1954). 1960. *Health Phys.* 3:1.
4. U.S. ENVIRONMENTAL PROTECTION AGENCY, Office of Water Supply. 1977. National Interim Primary Drinking Water Regulations, EPA-570/9-76. U.S. Govt. Printing Off., Washington, D.C.
5. U.S. GEOLOGICAL SURVEY. 1977. Methods for determination of radioactive substances in water and fluvial sediments. U.S. Govt. Printing Off., Washington, D.C.
6. HARLEY, J.H., ed. 1972. Health and Safety Laboratory Procedures Manual. HASL-300. U.S. Dep. Energy, New York, N.Y.
7. U.S. DEPARTMENT OF COMMERCE, NATIONAL BUREAU OF STANDARDS. 1961. Handbook 80, A Manual of Radioactivity Procedures. Superintendent of Documents, Washington, D.C.
8. AMERICAN NATIONAL STANDARDS INSTITUTE. 1974. American National Standard Specifications and Performance of On-Site Instrumentation for Continuously Monitoring Radioactivity in Effluents. ANSI N13 10-1974, IEEE, Inc., New York, N.Y.

701 I. Bibliography

JARRETT, A.A. 1946. Statistical Methods Used in the Measurement of Radioactivity (Some Useful Graphs). U.S. Atomic Energy Comm. Document No. AECU-262 (June 17). Washington, D.C.

CORYELL, C.D. & N. SUGARMAN, eds. 1951. Radiochemical Studies: The Fission Products. McGraw-Hill Book Co., New York, N.Y.

NADER, J.S., G.R. HAGEE & L.R. SETTER. 1954. Evaluating the performance of the internal counter. *Nucleonics* 12(6):29.

COMAR, C.I. 1955. Radioisotopes in Biology and Agriculture. McGraw-Hill Book Co., New York, N.Y.

CROUTHAMEL, C.E., ed. 1960. Applied Gamma-Ray Spectrometry. Pergamon Press, New York, N.Y., Vol. II.

TAYLOR, J.M. 1963. Semiconductor Particle Detectors. Butterworths, Washington, D.C.

FRIEDLANDER, G., J.W. KENNEDY & J.M. MILLER. 1964. Nuclear and Radiochemistry, 2nd ed. John Wiley & Sons, New York, N.Y.

GOULDING, F.S. 1964. A survey of the applications and limitations of various types of detectors in radiation energy measurement. *IEEE Trans. Nucl. Sci.* NS-11:177.

GOULDING, F.S. 1964. Semiconductor detectors—their properties and applications. *Nucleonics* 22(5):54.

HEATH, R.L. 1964. Scintillation Spectrometry, Gamma Ray Spectrum. IDO-16880, Technical Information Div., U.S. Atomic Energy Comm., Washington, D.C., Vols. 1 and 2.

SIEGBAHN, K., ed. 1965. Alpha, Beta, and Gamma Ray Spectroscopy. Vol. 1. North Holland Publishing Co., Amsterdam.

DEARNALEY, G. & D.C. NORTHROP. 1966. Semiconductor Counters for Nuclear Radiations, 2nd ed. John Wiley & Sons, New York, N.Y.

DEARNALEY, G. 1966. Nuclear detection by solid state devices. *J. Sci. Instrum.* 43:869.

LEDERER, C.M., J.M. HOLLANDER & I. PERLMANN. 1967. Table of Isotopes, 6th ed. John Wiley & Sons, New York, N.Y.

LOS ALAMOS SCIENTIFIC LABORATORY, RADIOCHEMISTRY GROUP J-11. 1967. Collected Radiochemical Procedures. U.S. Atomic Energy Comm. Rep. No. LA-1721, 3rd ed., Washington, D.C.

BOLOGNA, J.A. & S.B. HELMICH. 1969. An Atlas of Gamma Ray Spectra, LA 4312.

FRENCH, W.R., JR., R.L. LASHURE & J.L. CURRAN. 1969. Lithium drifted germanium detectors. *Amer. J. Phys.* 37:11.

FRENCH, W.R., JR., W.M. WEHRBEIN & S.E. MOORE. 1969. Measurement of photoelectric, Compton, and pair-production cross sections in germanium. *Amer. J. Phys.* 37:391.

MCKENZIE, J.M. 1969. Index to the Literature of Semiconductor Detectors. U.S. Govt. Printing Off., Washington, D.C.

GOULDING, F.S. & Y. STONE. 1970. Semiconductor radiation detectors. *Science* 170:281.

HEATH, R.L. 1974. Gamma Ray Spectrum Catalogue, Ge(Li) and Si(Li) Spectrometry, ANCR-1000-2. National Technical Information Serv., Springfield, Va.

702 RADIOACTIVITY IN WASTEWATER

1. Discussion

Factors considered in sampling and sample preservation and the behavior of radioactive species are of great significance in the analysis of wastewater. Usually wastewater contains larger amounts of nonradioactive suspended and dissolved solids than does water and often most of the radioactivity is in the solid phase. Generally, the use of carriers in the analysis is ineffective without prior conversion of the solid phase to the soluble phase; even then the high fixed solids may interfere with radioanalytical procedures. Table 702:I shows the usual solubility characteristics of common radioelements in wastewater.

Moreover, the radioelements may exhibit unusual chemical characteristics because of the presence of complexing agents or the method of waste production. For example, tritium may be combined in an organic compound when used in the manufacture of luminous articles; radioiodine from hospitals may occur as complex organic compounds, compared to elemental and iodide forms found in fission products from the processing of spent nuclear fuels; uranium and thorium daughter products often exist as inorganic complexes other than oxides after processing in uranium mills; the strontium 90 titanate waste from a radioisotope heat source would be quite insoluble compared to most other strontium wastes.

Valuable information on the chemical composition of wastes, the behavior of radioelements, and the quantity of radioisotopes in use appears in the literature.[1,2] Radionuclides having or likely to have a public health significance are emphasized

TABLE 702:I. THE USUAL DISTRIBUTION OF COMMON RADIOELEMENTS BETWEEN THE SOLID AND LIQUID PHASES OF WASTEWATER

In Solution	In Suspension
HCO_3	Ce
Co	Cs
Cr	Mn
Cs	Nb
H	P
I	Pm
K	Pu
Ra	Ra
Rn	Sc
Ru	Th
Sb	U
Sr	Y
	Zn
	Zr

here. Methods are provided for radionuclides of high radiotoxicity. Some of these are beta emitters. The levels of most gamma emitters that have public health significance can be measured by gamma spectrometry without chemical separation. This is usually true for ^{106}Ru, ^{137}Cs, ^{131}I, and ^{60}Co.

Information on the determination of radioactivity in wastes, as well as other environmental samples, may be found, for example, in manuals on Radioassay Procedures of the Environmental Protection Agency,[3,4] the AEC manual,[5] and the American Society for Testing and Materials' Book of Standards.[6] General information on the behavior of radioelements and on analytical methods can be found in the monographs of the National Research Council.[7]

Radionuclide standards for elements commonly encountered in wastewater are available from one or more of the following: The National Bureau of Standards and Amersham/Searle Corporation in the United States; the Radiochemical Centre, Amersham, England; the International Atomic Energy Agency, Vienna, Austria; and CEA-Saclay, France. General information on radionuclide standards is available in publications issued in the United States.[8,9]

Data on half-lives and decay schemes are available.[10,11] The Environmental Protection Agency, through the Quality Assurance Branch, Environmental Monitoring and Support Laboratory* assists laboratories in achieving radioanalytical proficiency.

Generally it is not feasible to perform collaborative (interlaboratory) analyses of wastewater samples because of the variable composition of elements and solids from one facility to the next, but the methods that follow have been evaluated by use of homogeneous samples and are useful for nonhomogeneous samples after sample preparation (wet or dry oxidation and/or fusion and solution) resulting in homogeneity. A potential problem or characteristic of reference samples used for collaborative testing is that they may be deficient in radioelements exhibiting interferences because of decay during shipment of short-half-life radionuclides. Generally, however, analytical steps have been incorporated into the methods to eliminate these interferences, even though they may not be necessary for the reference samples under study.

*Quality Assurance Branch, Environmental Monitoring and Support Laboratory, P.O. Box 15027, Las Vegas, Nev. 89114.

2. References

1. INTERNATIONAL ATOMIC ENERGY AGENCY. 1960. Disposal of Radioactive Wastes. IAEA, Vienna, Austria.
2. NEMEROW, N.L. 1963. Industrial Waste Treatment. Addison-Wesley, Reading, Mass.
3. JOHNS, F.B., ed. 1975. Handbook of Radiochemical Analytical Methods. EPA—680/4-75-001, National Environmental Research Center, Office of Research and Development, USEPA, Las Vegas, Nev.
4. KRIEGER, H.L. 1976. Interim Radiochemical Methodology for Drinking Water. EPA-600/4-75-008 (revised), Environmental Monitoring and Support Laboratory, Office of Research and Development, USEPA, Cincinnati, Ohio.
5. HARLEY, J.H., ed. 1972. Health and Safety Laboratory Procedures Manual. HASL-300. U.S. Atomic Energy Comm., New York, N.Y.
6. AMERICAN SOCIETY FOR TESTING AND MATERIALS. *1974 Book of ASTM Standards.* ASTM. Philadelphia, Pa.
7. NAS-NRC. 1960 to 1971. Radiochemistry of the Elements. Rep. Nos. NAS-NS-3001 through 3058 and Radiochemical Techniques. Rep. Nos. 3101 through 3113. National Technical Information Service, U.S. Dep. Commerce, Springfield, Va.
8. NAS-NRC. 1974. Users' Guides for Radioactivity Standards. Rep. No. NAS-NS-3115.

9. INTERNATIONAL COMMISSION ON RADIATION UNITS AND MEASUREMENTS. 1968. Certification of Standardized Radioactive Sources. ICRU Rep. 12, Washington, D.C.
10. BOWMAN, W.W. & K.W. MACMURDO. 1974. Atomic Data and Nuclear Data Tables, Radioactive-Decay, Gammas Ordered by Energy and Nuclide. Academic Press, New York, N.Y.
11. LEDERER, C.M., J.M. HOLLANDER & I. PERLMAN. 1967. Table of Isotopes. John Wiley & Sons, New York, N.Y.

703 GROSS ALPHA AND GROSS BETA RADIOACTIVITY (TOTAL, SUSPENDED, AND DISSOLVED)

1. General Discussion

a. Natural radioactivity: Uranium, thorium, and radium are naturally occurring radioactive elements that have a long series of radioactive daughters that emit alpha or beta and gamma radiations until a stable end-element is produced. These naturally occurring elements, through their radioactive daughter gases, radon and thoron, cause an appreciable airborne particulate activity and contribute to the radioactivity of rain and groundwaters. Additional naturally radioactive elements include potassium 40, rubidium 87, samarium 147, lutetium 176, and rhenium 187.

b. Artificial radioactivity: With the development and operation of nuclear reactors and other atom-smashing machines, large quantities of radioactive elements are being produced. These include almost all the elements in the periodic table.

c. Significance of gross alpha and gross beta concentrations in water: The Environmental Protection Agency has established maximum contaminant levels for radium 226, radium 228, and gross alpha as follows: combined radium 226 and 228, 5 pCi/L; gross alpha (including radium 226 but excluding radon and uranium), 15 pCi/L. For beta particles and photon radioactivity from man-made radionuclides in community water systems, the maximum contaminant levels are intended to produce an annual dose equivalent to the total body or any internal organ of less than 4 millirem/yr. Specifically, if the average annual concentration of gross beta activity is less than 50 pCi/L and if the average annual concentrations of tritium and strontium 90 are less than 20,000 pCi/L and 8 pCi/L, respectively, no further analyses are required. If the gross beta activity exceeds 50 pCi/L, the major radioactive contaminants must be identified and organ and whole body doses calculated; the doses shall not exceed 4 millirem/yr.

With the simpler technics for routine measurement of gross beta activity, the presence of contamination may be determined in a matter of minutes, whereas hours or even days may be required to make the radiochemical analyses necessary to identify radionuclides present.

Regular measurements of gross alpha and gross beta activity in water may be invaluable for early detection of radioactive contamination and indicate the need for supplemental data on concentrations of more hazardous radionuclides.

d. Preferred counting instrument and calibration standard: The internal proportional counter is the recommended instrument for counting gross beta radioactivity because of its superior operating characteristics. These include a high sensitivity to detect and count a wide range of low- to high-energy beta radiation and a high geometry (2π) due to the introduction of the sample into the counting chamber. In this case the system of assay is calibrated by adding standard nuclide portions to media comparable to the samples and preparing, mounting, and counting the standards exactly as the samples.

Thin-window proportional or Geiger counters may be used although they have lower counting efficiencies than the internal proportional counter. When a Geiger counter is used, alpha activity cannot be determined separately. Alpha counting efficiency in end-window counters may be very low because of absorption in the air and in the window.

When gross beta activity is assayed in samples containing mixtures of naturally radioactive elements and fission products, the choice of a calibration standard may influence the beta results significantly because self-absorption factors and counting chamber characteristics are beta-energy-dependent.

A standard solution of cesium 137, which is certified by the National Bureau of Standards or is traceable to a certified source, is recommended for calibration of counter efficiency and self-absorption for gross beta determinations. The half-life of cesium 137 is about 30 yr. The daughter products after beta decay of cesium 137 are stable barium 137 and metastable barium 137, which in turn disintegrates by gamma emission. For this reason, the standardization of cesium 137 solutions may be stated in terms of the gamma emission rate per milliliter or per gram. To convert gamma rate to equivalent beta disintegration rate, multiply the calibrated gamma emission rate by 1.29.

e. Radiation lost by self-absorption: The radiation from alpha emitters having an energy of 8 MeV and from beta emitters having an energy of 60 KeV will not escape from the sample if the emitters are covered by a sample thickness of 5.5 mg/cm^2. The radiation from a weak alpha emitter will be stopped if covered by only 4 mg/cm^2 of sample solids. Consequently, for low-level counting it is imperative to evaporate all moisture and preferable to destroy organic matter before depositing a thin film of sample solids from which radiation may enter the counter. In counting water samples for gross beta radioactivity, a solids thickness of 10 mg/cm^2 or less on the bottom area of the counting pan is recommended. For the most accurate results, determine the self-absorption factor as outlined in Section 701C.1*f*.

2. Apparatus

a. Counting pans, of metal resistant to corrosion from sample solids or reagents, about 50 mm diam, 6 to 10 mm in height, and thick enough to be serviceable for one-time use. Stainless steel or aluminum pans are satisfactory, depending on the kind of sample and reagents added.

b. Internal proportional counting chambers, capable of receiving and maintaining good electrical contact with counting pans, complete with preamplifier, scaler, timer, register, constant-voltage supply, counting gas equipment, and counting gas.

c. Alternate counters: Other beta counters are thin end-window proportional and Geiger counters.

*d. Membrane filter,** 0.45-μm pore diam.

e. Gooch crucibles.

3. Reagents

a. Methyl orange indicator solution.

b. Hydrochloric acid, HCl, 1*N* (1 + 11).

c. Nitric acid, HNO_3, 1*N*.

d. Clear acrylic solution: Dissolve 50 mg clear acrylic† in 100 mL acetone.

e. Ethyl alcohol, 95%.

f. Conducting fluid:‡ Prepare according to manufacturer's directions.

*Type HA, Millipore Filter Corp., Bedford, Mass., or equivalent.

†Lucite or equivalent.

‡Anstac 2M, Chemical Development Corporation, Danvers, Mass, or equivalent.

g. Standard certified cesium 137 solution.§

h. Standard certified americium 241 solution.§

i. Reagents for wet-combustion procedure:

1) *Nitric acid,* HNO_3, 6*N*.

2) *Hydrogen peroxide solution:* Dilute 30% H_2O_2 with an equal volume of water.

4. Procedure

a. Total sample activity:

1) For each 20 cm^2 of counting pan area, take a volume of sample containing not more than 200 mg residue for beta examination and not more than 100 mg residue for alpha examination. The specific conductance test helps to select the appropriate sample volume.

2) Evaporate by either of the following technics:

a) Add sample directly to a tared counting pan in small increments, with evaporation just below boiling temperature.

b) Place sample in a borosilicate glass beaker or evaporating dish, add a few drops of methyl orange indicator solution, add 1*N* HCl or 1*N* HNO_3 dropwise to pH 4 to 6, and evaporate on a hot plate or steam bath to near dryness. Avoid baking solids on evaporation vessel. Transfer to a tared counting pan with the aid of a rubber policeman and distilled water from a wash bottle. Using a rubber policeman, thoroughly wet walls of evaporating vessel with a few drops of acid and transfer washings to counting pan. (Excess alkalinity or mineral acidity is corrosive to aluminum counting pans.)

3) Complete drying in an oven at 103 to 105 C, cool in a desiccator, weigh, and keep dry until counted.

§Quality Assurance Branch, Environmental Monitoring and Support Laboratory, P.O. Box 15027, Las Vegas, Nev. 89114.

4) Treat sample residues having particles that tend to be airborne, which are to be counted in internal counters, with a few drops of clear acrylic solution, then air- and oven-dry and weigh.

5) With an internal counter, count alpha activity at the alpha plateau and beta-gamma activity at the beta plateau.

6) Store sample in a desiccator and count for decay if necessary. Avoid heat treatment because it will increase the escape rate of gaseous daughter products.

b. Activity of dissolved matter: Proceed as in ¶ 4*a*1) above, using a sample filtered through a 0.45-μm membrane filter.

c. Activity of suspended matter:

1) For each 10 cm^2 of membrane filter area, take a volume of sample not to exceed 50 mg suspended matter for alpha assay and not to exceed 100 mg for beta assay.

2) Filter sample through membrane filter with suction; then wash sides of filter funnel with a few milliliters of distilled water.

3) Transfer filter to a tared counting pan and oven-dry.

4) If sample is to be counted in an internal counter, saturate membrane with alcohol and ignite. (When beta or alpha activity is counted with another type counter, ignition is not necessary provided that the sample is dry and flat.) When burning has stopped, direct flame of a Meker burner down on the partially ignited sample to fix sample to pan.

5) Cool, weigh, and count at the alpha and the beta plateaus.

6) If sample particles tend to be airborne, treat sample with a few drops of clear acrylic solution, air-dry, and count.

7) Alternatively, prepare membrane filters for counting in internal counters by wetting filters with conducting fluid, drying, weighing, and counting. (Include weight of membrane filter in the tare.)

d. Activity of suspended matter (alternate): If it is impossible to filter sewage, highly polluted waters, or industrial wastes through membrane filters in a reasonable time, proceed as follows:

1) Determine total and dissolved activity by the procedures given in ¶s 4*a* and 4*b* and estimate suspended activity by difference.

2) Filter sample through an ashless mat or filter paper of stated porosity. Dry, ignite, and weigh suspended fixed residue. Transfer and fix a thin uniform layer of sample residue to a tared counting pan with a few drops of clear acrylic solution. Dry, weigh, and count in an internal counter for alpha and beta, or beta count with a thin end-window counter and alpha count with an alpha scintillation counter.

e. Activity of nonfatty semisolid samples: Use the following procedure for samples of sludge, vegetation, soil, etc.:

1) Determine total residue and fixed residue of representative samples according to Sections 209A and B.

2) Reduce fixed residue of a granular nature to a fine powder with pestle and mortar.

3) Transfer a maximum of 100 mg fixed residue for alpha assay and 200 mg fixed residue for beta assay for each 20 cm^2 of counting pan area (see NOTE below).

4) Distribute residue to uniform thickness in a tared counting pan by (*a*) spreading a thick aqueous cream of residue that is weighed after oven-drying, or (*b*) dispensing dry residue of known weight and spread with acetone and a few drops of clear acrylic solution.

5) Oven-dry at 103 to 105 C, weigh, and count.

NOTE: The fixed residue of vegetation and similar samples usually is corrosive to aluminum counting pans. To avoid difficulty, use stainless steel pans or treat a weighed amount of fixed residue with HCl or HNO_3 in the presence of methyl orange indicator to pH 4 to 6, transfer to an aluminum counting pan, dry at 103 to 105 C, reweigh, and count.

f. Alternate wet-combustion procedure for biological samples: Some samples, such as fatty animal tissues, are difficult to process according to ¶ 4*e* above. An alternate procedure consists of acid digestion. Because a highly acid and oxidizing state is created, volatile radionuclides may be lost under these conditions.

1) To a 2- to 10-g sample in a tared silica dish or equivalent, add 20 to 50 mL 6*N* HNO_3 and 1 mL 15% H_2O_2 and digest at room temperature for a few hours or overnight. Heat gently and, when frothing subsides, heat more vigorously but without spattering, until nearly dry. Add two more 6*N* HNO_3 portions of 10 to 20 mL each, heat to near boiling, and continue gentle treatment until dry.

2) Ignite in a muffle furnace for 30 min at 600 C, cool in a desiccator, and weigh.

3) Continue the test as described in ¶s 4*e*3)-5) above.

5. Calculation and Reporting

a. Counting error: Determine the counting error, *E* (in picocuries per sample), at the 95% confidence level from:

$$E = \frac{1.96\ \sigma(R)}{2.22e}$$

where $\sigma(R)$ is calculated as shown in Section 701F, using $t_1 = t_2$ (in minutes); and *e*, the counter efficiency, is defined and calculated as in Section 701C.1*f*.

b. Alpha activity: Calculate alpha activity, in picocuries per liter, by the equation

$$\text{Alpha} = \frac{\text{net cpm} \times 1{,}000}{2.22e\ v}$$

where:

e = calibrated overall counter efficiency (see Section 701C.1*f*), and
v = volume of sample counted, mL.

Express the counting error in picocuries per liter by dividing picocuries per sample by sample volume in liters. Similarly, calculate and report alpha activity in picocuries or nanocuries per kilogram of moist biological material or per kilogram of moist and per kilogram of dry silt.

c. Gross beta activity when alpha activity is insignificant: For samples having an alpha activity less than one-half the beta counting error, calculate and report gross beta activity and counting error in picocuries or nanocuries per liter of fluid, per kilogram of moist (live weight) biological material, or per kilogram of moist and per kilogram of dry silt, according to ¶s *a* and *b* above. Disregard the slight amount of alpha activity.

For calculation of picocuries of beta activity per liter, determine the value of *e* in the above equation as described in Section 701C.1*f*.

d. Beta activity when alpha activity is significant: For samples containing an alpha activity that exceeds one-half the beta error, deduct net alpha counts per minute from net beta counts per minute to give net corrected beta counts per minute. Proceed as in ¶ *c* above to calculate and report beta radioactivity. When the count of alpha activity at the beta plateau represents a small fraction of the activity, a rough approximation of the beta counting error consists of the gross beta counting error. Where greater precision is desired—for example, when the count of alpha activity at the beta plateau is a substantial fraction of the net counts per minute of gross beta activity—the beta counting error equals $(E_a^2 + E_b^2)^{1/2}$, where E_a is the alpha counting error and E_b the gross beta counting error.

e. Miscellaneous information to be reported: In reporting radioactivity data, identify adequately the sample, sampling station, date of collection, volume of sample, type of test, type of activity, type of counting equipment, standard calibration solutions used (particularly when standards other than americium 241 for alpha or cesium 137 for beta are used), time of counting (particularly if short-lived isotopes are involved), weight of sample solids, and kind and amount of radioactivity. So far as possible, tabulate the data for ease of interpretation and incorporate repetitious items in the table heading or in footnotes. Unless especially inconvenient, do not change quantity units within a given table. For low-level assays, optimally report the counting error to assist in the interpretation of results.

6. Precision and Accuracy

In a collaborative study of two sets of paired water samples containing known additions of radionuclides, 15 laboratories determined the gross alpha activity and 16 analyzed the gross beta activity. The water samples contained simulated water minerals of approximately 350 mg fixed solids/L. The alpha results of one laboratory were rejected as outliers.

The average recoveries of added gross alpha activity were 86, 87, 84, and 82%. The precision (random error) at the 95% confidence level was 20 and 24% for the two sets of paired samples. The method was biased low, but not seriously.

The average recoveries of added gross beta activity were 99, 100, 100, and 100%. The precision (random error) at the 95% confidence level was 12 and 18% for the two sets of paired samples. The method showed no bias.

7. Bibliography

BURTT, B.P. 1949. Absolute beta counting. *Nucleonics* 5:8, 28.

GOLDIN, A.S., J.S. NADER & L.R. SETTER. 1953. The detectability of low-level radioactivity in water. *J. Amer. Water Works Ass.* 45:73.

SETTER, L.R., A.S. GOLDIN & J.S. NADER. 1954. Radioactivity assay of water and industrial wastes with internal proportional counter. *Anal. Chem.* 26:1304.

SETTER, L.R. 1964. Reliability of measurements of gross beta radioactivity in water. *J. Amer. Water Works Ass.* 56:228.

JOHNS, F.B., ed. 1975. Handbook of Radiochemical Analytical Methods. EPA-680/4-

75-001, National Environmental Research Center, Office of Research and Development, USEPA, Las Vegas, Nev.

ENVIRONMENTAL PROTECTION AGENCY. 1976. *Fed. Reg.* 41:28402.

THATCHER, L.L., V.J. JANZER & K.W. EDWARDS. 1977. Techniques of water resources investigations of the US Geological Survey. Chap. A5 *in* Methods for Determination of Radioactive Substances in Water and Fluvial Sediments. USGPO, Stock No. 024-001-02928-6, Washington, D.C.

704 TOTAL RADIOACTIVE STRONTIUM AND STRONTIUM 90 IN WATER

The important radioactive nuclides of strontium produced in nuclear fission are ^{89}Sr and ^{90}Sr. Strontium 90 is one of the most hazardous of all fission products. It decays slowly, with a half-life of 28 yr. Upon ingestion, strontium is concentrated in the bone. The total occupational permissible exposure to ^{90}Sr allows a concentration in water of 100 pCi/L, as compared to 10,000 pCi/L for ^{89}Sr, which has a half-life of only 50.5 days. The 1976 Environmental Protection Agency Interim Primary Drinking Water regulations limit the concentration of ^{90}Sr in water to 8 pCi/L when other sources of intake are not considered.

1. General Discussion

a. Principle: The following method is designed to measure total radioactive strontium (^{89}Sr and ^{90}Sr) or ^{90}Sr alone in drinking water or in filtered raw water. It is applicable to sewage and industrial wastes provided that steps are taken to destroy organic matter and eliminate other interfering ions. A known amount of inactive strontium ions, in the form of strontium nitrate, $Sr(NO_3)_2$, is added as a "carrier." The carrier, alkaline earths, and rare earths are precipitated as the carbonate to concentrate the radiostrontium. The carrier, along with the radionuclides of strontium, is separated from other radioactive elements and inactive sample solids by precipitation as $Sr(NO_3)_2$ from fuming nitric acid solution. The strontium carrier, together with the radionuclides of strontium, is finally precipitated as strontium carbonate, $SrCO_3$, which is dried, weighed to determine recovery of carrier, and measured for radioactivity. The activity in the final precipitate is due to radioactive strontium only, because all other radioactive elements have been removed. A correction is applied to compensate for losses of carrier and activity during the various purification steps. A delay in the count will give an increased counting rate due to the ingrowth of ^{90}Y.

b. Concentration technics: Because of the very low amount of radioactivity, a large sample must be taken and the activity concentrated by precipitation. $Sr(NO_3)_2$ and barium nitrate, $Ba(NO_3)_2$, carriers are added to the sample. Sodium carbonate is then added to concentrate radiostrontium by precipitation of alkaline earth carbonates along with other radioactive elements. The supernate is discarded. The precipitate is dissolved and reprecipitated to remove interfering radionuclides.

c. Interference: Radioactive barium (^{140}Ba, ^{140}La) interferes in the determination of radioactive strontium inasmuch as it precipitates with the radioactive strontium. Eliminate this interference by adding inactive $Ba(NO_3)_2$ carrier and separating this from the strontium by precipitating barium chromate in acetate buffer solution. Radium isotopes also are eliminated by this treatment.

In hard water, some calcium nitrate may be coprecipitated with $Sr(NO_3)_2$ and can cause errors in recovery of the final precipitate and in measuring its activity. Eliminate this interference by repeated precipitations of strontium as the nitrate followed by leaching the $Sr(NO_3)_2$ with acetone (CAUTION).

For total radiostrontium, count the precipitate within 3 to 4 hr after the final separation and before ingrowth of ^{90}Y.

d. Determination of ^{90}Sr: Because it is impossible to separate the isotopes ^{89}Sr and ^{90}Sr by any chemical procedure, the amount of ^{90}Sr is determined by separating and measuring the activity of ^{90}Y, its daughter. After equilibrium is reached, the activity of ^{90}Y is exactly equal to the activity of ^{90}Sr. Two alternate procedures are given for the separation of ^{90}Y. In the first method, ^{90}Y is separated by extraction into tributyl phosphate from concentrated nitric acid (HNO_3) solution. It is back-extracted into dilute HNO_3 and evaporated to dryness for beta counting. The second method consists of adding yttrium carrier, separating by precipitation as yttrium hydroxide, $Y(OH)_3$, and finally precipitating yttrium oxalate for counting.

2. Apparatus

a. Counting instruments: Use either an internal proportional counter, gas-flow, with scaler, timer, and register; or a thin end-window (polyester plastic film*) proportional or G-M counting chamber with scaler, timer, register amplifier, and preferably having an anticoincident system (low background).

b. Filter paper,† 2.4 cm diam; or glass fiber filters, 2.4 cm diam.

c. Two-piece filtering apparatus for 2.4-cm filters such as TFE filter holder,‡ stainless steel filter holder, or equivalent.

d. Stainless steel pans, about 50 mm diam and 7 mm deep, for counting solids deposited on pan bottom. For counting precipitates on 2.4-cm filters, use nylon disk with ring§ on which the filter samples are mounted and covered by 0.25 mil film.*

3. Reagents

a. Strontium carrier (10 mg Sr^{2+}/mL) standardized: Carefully add 24.16 g $Sr(NO_3)_2$ to a 1-L volumetric flask and dilute with distilled water to the mark. For standardization, pipet three 10.0-mL portions of strontium carrier solution into 40-mL centrifuge tubes and add 15 mL 2*N* Na_2CO_3 solution. Stir, heat in a boiling water bath for 15 min, and cool. Filter $SrCO_3$ precipitate through a tared fine-porosity sintered-glass crucible of 15-mL size. Wash precipitate with three 5-mL portions of water and then with three 5-mL portions of absolute ethanol (or acetone). Wipe crucible with absorbent tissue and dry to constant weight in an oven at 110 C (20 min). Cool in a desiccator and weigh.

$$\text{Sr, mg/mL} = \frac{(\text{mg SrCO}_3)\ (0.5935)}{10}$$

b. Barium carrier (10 mg Ba^{2+}/mL): Dissolve 19.0 g $Ba(NO_3)_2$ in distilled water and dilute to 1 L.

c. Rare earth carrier, mixed: Dissolve 12.8 g cerous nitrate hexahydrate, $Ce(NO_3)_3 \cdot 6H_2O$, 14 g zirconyl chloride octahydrate, $ZrOCl_2 \cdot 8H_2O$, and 25 g ferric chloride hexahydrate, $FeCl_3 \cdot 6H_2O$, in 600

*Mylar, E.I. du Pont de Nemours, Wilmington, Del., or equivalent.

†Whatman No. 42 or equivalent.

‡Flurolon Laboratory, Box 305, Caldwell, N.J.

§Control Molding Corp., Staten Island, N.Y., or equivalent.

mL distilled water containing 10 mL conc HCl, and dilute to 1 L.

d. Yttrium carrier: Dissolve 12.7 g yttrium oxide,‖ Y_2O_3, in 30 mL conc HNO_3 by stirring and warming. Add an additional 20 mL conc HNO_3 and dilute to 1 L with distilled water; 1 mL is equivalent to 10 mg Y, or approximately 34 mg $Y_2(C_2O_4)_3 \cdot 9H_2O$. Determine exact equivalence by precipitating yttrium carrier in acid solution according to Section 704.4*c*2)-8) or by extracting yttrium carrier in acid solution according to Section 704.4*b*3)-11), following.

e. Acetate buffer solution: Dissolve 154 g $NH_4C_2H_3O_2$ in 700 mL distilled water, add 57 mL conc acetic acid, adjust pH to 5.5 by dropwise addition of conc acetic acid or 6*N* NH_4OH as necessary, and dilute to 1 L.

f. Acetic acid, 6*N*.

g. Acetone, anhydrous.

h. Ammonium hydroxide, NH_4OH, 6*N*.

i. Hydrochloric acid, HCl, 6*N*.

j. Methyl red indicator, 0.1%: Dissolve 0.1 g methyl red in 100 mL distilled water.

k. Nitric acid, HNO_3, fuming (90%), conc, 14*N*, 6*N*, and 0.1*N*.

l. Oxalic acid, saturated solution: Dissolve approximately 11 g $H_2C_2O_4 \cdot 2H_2O$ in 100 mL distilled water.

m. Sodium carbonate solution, 2*N*: Dissolve 124 g $Na_2CO_3 \cdot H_2O$ in distilled water and dilute to 1 L.

n. Sodium chromate solution, 0.5*M*: Dissolve 117 g $Na_2CrO_4 \cdot 4H_2O$ in distilled water and dilute to 1 L.

o. Sodium hydroxide, 6*N*: Dissolve 240 g NaOH in distilled water and dilute to 1 L.

p. Tributyl phosphate, reagent grade: Shake with an equal volume of 14*N* HNO_3 to equilibrate. Separate and discard the HNO_3 washings.

4. Procedure

a. Total radiostrontium:

1) To 1 L of drinking water, or a filtered sample of raw water in a beaker, add 2.0 mL conc HNO_3 and mix. Add 2.0 mL each of strontium and barium carriers and mix well. (A precipitate of $BaSO_4$ may form if the water is high in sulfate ion, but this will cause no difficulties.) A smaller sample may be used if it contains at least 25 pCi strontium. The suspended matter that has been filtered off may be digested [see Gross Alpha and Gross Beta Radioactivity, 703.4*f*1)], diluted, and analyzed separately.

2) Heat to boiling, then add 20 mL 6*N* NaOH and 20 mL 2*N* Na_2CO_3. Stir and let simmer at 90 to 95 C for about 1 hr.

3) Set beaker aside until precipitate has settled (about 1 to 3 hr).

4) Decant and discard clear supernate. Transfer precipitate to a 40-mL centrifuge tube and centrifuge. Discard supernate.

5) Add, dropwise (Caution—*effervescence*), 4 mL conc HNO_3. Heat to boiling, stir, then cool under running water.

6) Add 20 mL fuming HNO_3, cool 5 to 10 min in ice bath, stir, and centrifuge. Discard supernate.

7) Add 4 mL distilled water, stir, and heat to boiling to dissolve the strontium. Centrifuge while hot to remove remaining insolubles and decant supernate to a clean centrifuge tube. Add 2 mL 6*N* HNO_3, heat to boiling, centrifuge while hot, and combine supernate with aqueous supernate. Discard insoluble residue of SiO_2, $BaSO_4$, etc.

8) Cool combined supernates, then add 20 mL fuming HNO_3, cool 5 to 10 min in ice bath, stir, centrifuge, and discard supernate.

9) Add 4 mL distilled water and dissolve by heating. Repeat Step 8) preceding.

‖Yttrium oxide, Code 1118, American Potash and Chemical Corp., West Chicago, Ill., or equivalent. Yttrium oxide of purity less than Code 1118 may require purification because of radioactivity contamination.

10) Repeat Step 9) preceding if more than 200 mg Ca were present in the sample.

11) After last HNO_3 precipitation, invert tube in a beaker for about 10 min to drain off most excess HNO_3. Add 20 mL anhydrous acetone, stir thoroughly, cool, and centrifuge. Discard supernate (CAUTION).

12) Dissolve precipitate of $Sr(NO_3)_2+Ba(NO_3)_2$ in 10 mL distilled water and boil for 30 sec to remove any remaining acetone.

13) Add 0.25 mL (5 drops) mixed rare earth carrier and precipitate rare earth hydroxides by making solution basic with 6*N* NH_4OH. Digest in a boiling water bath for 10 min. Cool, centrifuge, and decant supernate to a clean tube. Discard precipitate.

14) Repeat Step 13) preceding.

Note the time of rare earth precipitation, which marks the beginning of the ^{90}Y ingrowth period. Do not delay procedure more than a few hours after the separation; otherwise, false results will be obtained because of ingrowth of ^{90}Y.

15) Add 2 drops methyl red indicator and then add 6*N* acetic acid dropwise with stirring until indicator changes from yellow to red.

16) Add 5 mL acetate buffer solution, heat to boiling, and add dropwise, with stirring, 2 mL Na_2CrO_4 solution. Digest in a boiling water bath for 5 min. Cool, centrifuge, and decant supernate to a clean tube. Discard residue.

17) Add 2 mL 6*N* NaOH, add 5 mL 2*N* Na_2CO_3 solution, and heat to boiling. Cool in an ice bath (about 5 min) and centrifuge. Discard supernate.

18) Add 15 mL distilled water, stir, centrifuge, and discard wash water.

19) Repeat Step 18), and proceed either as in Step 20)a) or 20)b), below. *Save this precipitate if a determination of* ^{90}Sr *is required.*

20) Either

a) Slurry precipitate with a small volume of distilled water and transfer to a tared stainless steel pan; dry under an infrared lamp, cool, weigh, and count# the precipitate of $SrCO_3$;** or

b) Transfer precipitate to a tared paper or glass filter mounted in a two-piece funnel. Allow gravity settling for uniform deposition and then apply suction. Wash precipitate with three 5-mL portions of water, three 5-mL portions of 95% alcohol, and three 5-mL portions of ethyl ether or acetone. Dry in an oven at 110 to 125 C for 15 to 30 min, cool, weigh,** mount on a nylon disk and ring with polyester plastic film cover, and count.

21) Calculation:

$$\text{Total Sr activity in pCi/L} = \frac{b}{adf \times 2.22}$$

where:

a = beta counter efficiency [see Step 22) below],

$$d = \frac{\text{mg final } SrCO_3 \text{ precipitate}}{\text{mg } SrCO_3 \text{ in 2 mL of carrier}}$$

= correction for carrier recovery [see Step 23) below],

f = sample volume, L,
b = beta activity, net cpm = $(i/t)-k$,
i = total counts accumulated,
t = time of counting, min, and
k = background, cpm.

22) Counting efficiency: As a first estimate, when mounting sample according to Step 20)a), convert counts per minute to

#Strontium 90 in thick samples is counted with low efficiency; hence, a first count within hours favors ^{89}Sr counting, and a recount after 3 to 6 days that exceeds the first count provides a rough estimate of the ^{90}Y ingrowth—see Figure 704:1 and R.J. Velten (1966) below.

**When a determination of total strontium is not required, weigh precipitate [Step 20)a) or 20)b)] for carrier recovery but do not count. Then proceed with ^{90}Sr determination according to Section 704.4*b* following.

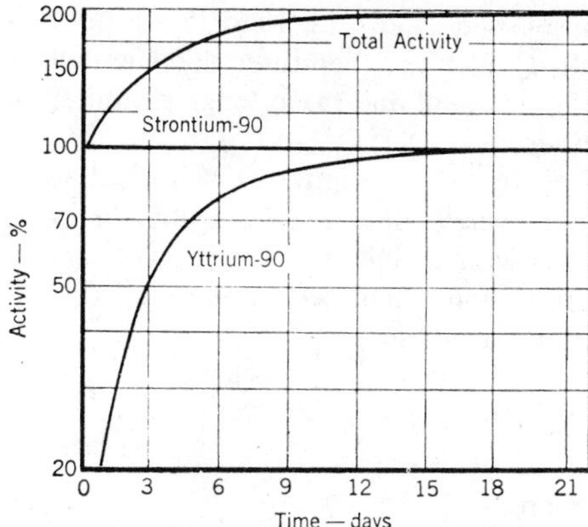

Figure 704:1. Yttrium 90 vs. strontium 90 activity as a function of time.

disintegrations per minute, based on the beta activity of cesium 137 standard solutions having a sample thickness equivalent to that of the $SrCO_3$ precipitate. More precise measurements may follow a second count after substantial ingrowth of ^{90}Y from ^{90}Sr, but this precision is not warranted for the usual total radiostrontium determination. When mounting samples according to Step 20)b), determine self-absorption curves by separately precipitating standard solutions of ^{89}Sr and ^{90}Sr as the carbonate (see gross beta in Section 703).

23) Correction for carrier recovery: 20 mg Sr are equivalent to 33.7 mg $SrCO_3$. Should more than traces of stable strontium be present in the sample, it would act as carrier; hence its determination by flame photometric or atomic absorption spectrophotometric method would be required.

b. Strontium 90†† by extraction of yttrium 90: Store $SrCO_3$ precipitate, as in ¶ 4*a*20), for at least 2 wk to allow ingrowth of ^{90}Y and then proceed as directed here or in an alternate procedure in Section 704.4*c* following.

††See footnote to Step 20a) when a determination for only ^{90}Sr is required.

1) Transfer of precipitate to separatory funnel—Either

a) Place a small funnel upright into mouth of a 60-mL separatory funnel; then place pan with precipitate, as in Step 20)a), in funnel and add, dropwise, 1 mL 6*N* HNO_3 (CAUTION—*effervescence*); tilt pan to empty into funnel and rinse pan twice with 2-mL portions of 6*N* HNO_3; or

b) Uncover precipitate from filter, as in Step 20)b), and transfer filter with forceps to upright funnel in mouth of 60-mL separatory funnel as in ¶ a) above. Dislodge bulk of precipitate into funnel stem. Dropwise, add with caution 1 mL 6*N* HNO_3 to filter, removing residual precipitate and dissolving bulk precipitate. Rinse filter and funnel twice with 2-mL portions 6*N* HNO_3.

2) Remove filter or pan and add 10 mL fuming HNO_3 to separatory funnel through upright funnel.

3) Remove upright funnel and add 1 mL yttrium carrier in a separatory funnel.

4) Add 5.0 mL tributyl phosphate reagent, shake thoroughly for 3 to 5 min, allow phases to separate, and transfer aqueous layer to a second 60-mL separatory funnel.

5) Add 5.0 mL tributyl phosphate reagent, shake 5 min, allow phases to separate, and transfer aqueous layer to a third 60-mL separatory funnel.

6) Combine organic extractants in the first and second funnels into one funnel and wash organic phase twice with 5-mL portions 14*N* HNO_3. Record time as the beginning of ^{90}Y decay (combine acid washings with aqueous phase in third funnel if a second ingrowth of ^{90}Y is desired).

7) Back-extract ^{90}Y from combined organic phases with 10 mL 0.1*N* HNO_3 for 5 min.

8) Continue as in Section 704.4*c*, Steps 6)-8) below or transfer aqueous phase from Step 7) immediately above into a 50-mL beaker and evaporate on a hotplate to 5 to 10 mL.

9) Repeat Step 7) above and transfer aqueous phase to beaker in Step 8) preceding; evaporate to 5 to 10 mL.

10) Transfer residual solution in beaker to a tared stainless steel counting pan and evaporate.

11) Rinse beaker twice with 2-mL portions of 0.1*N* HNO_3; add rinsings to counting pan, evaporate to dryness, and weigh.

12) Count in an internal proportional or end-window counter and calculate ^{90}Sr as given in Section 4*c*9) following.

c. Strontium 90 by oxalate precipitation of yttrium 90:††

1) Quantitatively transfer $SrCO_3$ precipitate to a 40-mL centrifuge tube with 2 mL 6*N* HNO_3. Add acid dropwise during dissolution. (CAUTION—*effervescence.*) Use 0.1*N* HNO_3 for rinsing.

2) Add 1 mL yttrium carrier, 2 drops methyl red indicator and, *dropwise,* add conc NH_4OH to the methyl red end point.

3) Add 5 mL more conc NH_4OH and *record the time,* which is the end of ^{90}Y ingrowth and the beginning of decay; centrifuge and decant supernate to a beaker (save supernate and washings for a second ingrowth if desired).

4) Wash precipitate twice with 20-mL portions hot distilled water.

5) Add 5 to 10 drops of 6*N* HNO_3, stir to dissolve precipitate, add 25 mL distilled water, and heat in a water bath at 90 C.

6) Gradually add 15 to 20 drops saturated oxalic acid reagent with stirring and adjust to pH 1.5 to 2.0 (pH meter or indicator paper) by adding conc NH_4OH dropwise. Digest precipitate for 5 min and cool in an ice bath with occasional stirring.

7) Transfer precipitate to a tared glass fiber filter in a two-piece funnel. Let precipitate settle by gravity (for uniform deposition) and apply suction. Wash precipitate in sequence with 10 to 15 mL hot distilled water and then three times with 95% ethyl alcohol and three times with diethyl ether.

8) Air-dry precipitate with suction for 2 min, weigh, mount on a nylon disk and ring with polyester plastic film cover, count, and calculate ^{90}Sr as follows.

9) Calculation:

$$^{90}\text{Sr pCi/L} = \frac{\text{net cpm}}{a\ b\ c\ d\ f\ g \times 2.22}$$

where:

a = counting efficiency for ^{90}Y,
b = chemical yield of extracting or precipitating ^{90}Y,
c = ingrowth correction factor if not in secular equilibrium,
d = chemical yield of strontium determined gravimetrically or by flame photometry,
f = volume of original sample, L,
g = ^{90}Y decay factor, $e^{-\lambda t}$, and
e = base of natural logarithms,
λ = $0.693/T_{1/2}$, where $T_{1/2}$ for ^{90}Y is 64.2 hr, and
t = time between separation and counting, hr.

5. Precision and Accuracy

In a collaborative study of two sets of paired, moderately hard water samples containing known additions of radionuclides, 12 laboratories determined the total radiostrontium and 10 laboratories determined ^{90}Sr. The results of one sample from one laboratory were rejected as outliers.

The average recoveries of added total radiostrontium from the four samples were 99, 99, 96, and 93%. The precision (random error) at the 95% confidence level was 10 and 12% for the two sets of paired samples. The method was slightly biased on the low side.

6. Bibliography

HAHN, R.B. & C.P. STRAUB. 1955. Determination of radioactive strontium and barium in

††See footnote to Step 20a) when a determination for only ^{90}Sr is required.

water. *J. Amer. Water Works Ass.* 47:335.
GOLDIN, A.S., R.J. VELTEN & G.W. FRISHKORN. 1959. Determination of radioactive strontium. *Anal. Chem.* 31:1490.
GOLDIN, A.S. & R.J. VELTEN. 1961. Application of tributyl phosphate extraction to the determination of strontium 90. *Anal. Chem.* 33:149.
VELTEN, R.J. 1966. Resolution of Sr-89 and Sr-90 in environmental media by an instrumental technique. *Nucl. Instrum. Methods* 42:169.

705 RADIUM IN WATER BY PRECIPITATION

The determination of radium by precipitation is a screening technic particularly applicable to drinking water. As long as the concentration of radium is less than the ^{226}Ra plus ^{228}Ra drinking water standard, examination by a more specific method is seldom needed.

There are four naturally occurring radium isotopes—11.6-day radium 223, 3.6-day radium 224, 1,600-yr radium 226, and 5.75-yr radium 228. Radium 223 is a member of the uranium 235 series, radium 224 and radium 228 are members of the thorium series, and radium 226 is a member of the uranium 238 series. The contribution of radium 228 (a beta emitter) to the total radium alpha activity is negligible because of the 1.9-yr half-life of its first alpha-emitting daughter product, thorium 228. The other three radium isotopes are alpha emitters; each gives rise to a series of relatively short-lived daughter products, including three more alpha emitters. Because of the difference in half-lives of the nuclides in these series, the isotopes of radium can be identified by the rate of ingrowth and decay of their daughters in a barium sulfate precipitate.[1-3] The ingrowth of alpha activity from radium 226 increases at a rate governed primarily by the 3.8-day half-life radon 222. The ingrowth of alpha activity in radium 223 is complete by the time a radium-barium precipitate can be prepared for counting. The ingrowth of the first two alpha-emitting daughters of radium 224 are complete within a few minutes and the third alpha daughter activity increases at a rate governed by the 10.6-hr half-life of lead 212. The activity of the radium 224 itself, with a 3.6-day half-life, also is decreasing, leading to a rather complicated ingrowth and decay curve.

The 1976 Environmental Protection Agency Interim Primary Drinking Water regulations established a maximum contaminant level for combined radium 226 and 228 of 5 pCi/L and for radium 226 of 3 pCi/L.

The principles of the two common methods for measuring radium are (*a*) alpha-counting a barium-radium sulfate precipitate that has been purified, and (*b*) measurement of radon 222 produced from the radium 226 in a sample or in a soluble concentrate isolated from the sample. The former technic includes all alpha-emitting radium isotopes present, whereas the latter (emanation) technic is nearly, but not absolutely, specific for radium 226.

1. General Discussion

Principle: This method is designed to measure radium in clear water. It is applicable to sewage and industrial wastes, provided that steps are taken to destroy organic matter and eliminate other interfering ions (see Gross Alpha and Gross Beta Radioactivity, 703.4*f*). However, ignition of sample ash should be avoided or a fusion will be necessary. Radium carried by barium sulfate is determined by alpha-counting. Lead and barium carriers are added to the sample containing alkaline citrate, then sulfuric acid (H_2SO_4) is added to precipitate radium, barium, and lead as sulfates. The precipitate is purified by washing with nitric acid (HNO_3), dissolving in alkaline EDTA, and reprecipitating as radium-barium sulfate after pH adjustment to 4.5. This slightly acidic EDTA keeps other naturally occurring alpha emitters and the lead carrier in solution.

2. Apparatus

a. Counting instruments: One of the following is required:

1) *Internal proportional counter,* gas-flow, with scaler and register.

2) *Alpha scintillation counter,* silver-activated zinc sulfide phosphor deposited on thin polyester plastic, with photomultiplier tube, scaler, timer, and register; or

3) *Proportional counter,* thin end-window, gas-flow, with scaler and register.

b. Membrane filter holder, or stainless steel or TFE filter funnels, with vacuum source.*

c. Membrane filters† or *glass fiber filters.*‡

3. Reagents

a. Citric acid, 1*M*: Dissolve 210 g $H_3C_6H_5O_7 \cdot H_2O$ in distilled water and dilute to 1 L.

b. Ammonium hydroxide, conc and 5*N*: Verify strength of old 5*N* NH_4OH solution before use.

c. Lead nitrate carrier: Dissolve 160 g $Pb(NO_3)_2$ in distilled water and dilute to 1 L; 1 mL = 100 mg Pb.

d. Stock barium chloride solution: Dissolve 17.79 g $BaCl_2 \cdot 2H_2O$ in distilled water and dilute to 1 L in a volumetric flask; 1 mL = 10 mg Ba.

e. Barium chloride carrier: To a 100-mL volumetric flask, add 20.00 mL stock $BaCl_2$ solution using a transfer pipet, dilute to 100 mL with distilled water, and mix; 1 mL = 2.00 mg Ba.

f. Methyl orange indicator solution.

g. Phenolphthalein indicator solution.

h. Bromcresol green indicator solution: Dissolve 0.1 g bromcresol green sodium salt in 100 mL distilled water.

i. Sulfuric acid, H_2SO_4, 18*N*.

j. Nitric acid, HNO_3, conc.

k. EDTA reagent, 0.25*M*: Add 93 g disodium ethylenediaminetetraacetate dihydrate to distilled water, dilute to 1 L, and mix.

l. Acetic acid, conc.

m. Ethyl alcohol, 95%.

n. Acetone.

o. Clear acrylic solution:§ Dissolve 50 mg clear acrylic in 100 mL acetone.

p. Standard radium 226 solution: Prepare as directed in method for radium 226 by radon 222, Section 706.3*d*-*f*, except that in ¶ *f* (standard radium 226 solution), add 0.50 mL $BaCl_2$ stock solution (Section 706.3*d*, method for total radium) before adding the ^{226}Ra solution; 1 mL final standard radium solution so prepared contains 2.00 mg Ba/mL and approximately 3 pCi ^{226}Ra/mL after the necessary correcting factors are applied.

4. Procedure for Radium in Drinking Water and for Dissolved Radium

a. To 1 L sample in a 1,500-mL beaker, add 5 mL 1*M* citric acid, 2.5 mL conc NH_4OH, 2 mL $Pb(NO_3)_2$ carrier, and 3.00 mL $BaCl_2$ carrier. In each batch of samples include a distilled water blank.

b. Heat to boiling and add 10 drops methyl orange indicator.

c. While stirring, slowly add 18*N* H_2SO_4 to obtain a permanent pink color; then add 0.25 mL acid in excess.

*Fisher Filtrator or equivalent.

†Millipore Type HAWP or equivalent.

‡No. 934-AH, diameter 2.4 cm, H. Reeve Angel and Co., or equivalent.

§Lucite or equivalent.

d. Boil gently 5 to 10 min.

e. Set beaker aside and let stand until precipitate has settled (3 to 5 hr or more).‖

f. Decant and discard clear supernate. Transfer precipitate to a 40-mL or larger centrifuge tube, centrifuge, decant, and discard supernate.

g. Rinse wall of centrifuge tube with a 10-mL portion of conc HNO_3, stir precipitate with a glass rod, centrifuge, and discard supernate. Repeat rinsing and washing two more times.

h. To precipitate, add 10 mL distilled water and 1 to 2 drops phenolphthalein indicator solution. Stir and loosen precipitate from bottom of tube (using a glass rod if necessary) and add 5*N* NH_4OH, dropwise, until solution is definitely alkaline (red). Add 10 mL EDTA reagent and 3 mL 5*N* NH_4OH. Stir occasionally for 2 min. Most of the precipitate should dissolve, but a slight turbidity may remain.

i. Warm in a steam bath to clear solution (about 10 min), but do not heat for an unnecessarily long period.# Add conc acetic acid dropwise until red color disappears; add 2 or 3 drops bromcresol green indicator solution and continue to add conc acetic acid dropwise, while stirring with a glass rod, until indicator turns green (aqua).** $BaSO_4$ will precipitate. Note date and time of precipitation as zero time for ingrowth of alpha activity. Digest in a steam bath for 5 to 10 min, cool, and centrifuge. Discard supernate. The final pH should be about 4.5, which is sufficiently low to destroy the Ba-EDTA complex, but not Pb-EDTA. A pH much below 4.5 will precipitate $PbSO_4$.

j. Wash Ba-Ra sulfate precipitate with distilled water and mount in a manner suitable for counting as given in ¶s *k,l*, or *m* following.

k. Transfer Ba-Ra sulfate precipitate to a tared stainless steel planchet with a minimum of 95% ethyl alcohol and evaporate under an infrared lamp. Add 2 mL acetone, 2 drops clear acrylic solution, disperse precipitate evenly, and evaporate under an infrared lamp. Dry in oven at 110 C, weigh, and determine alpha activity, preferably with an internal proportional counter. Calculate net counts per minute and weight of precipitate.

l. Weigh a membrane filter, a counting dish, and a weight (glass ring) as a unit. Transfer precipitate to tared membrane filter in a holder and wash with 15 to 25 mL distilled water. Place membrane filter in dish, add glass ring, and dry at 110 C. Weigh and count in one of the counters mentioned under ¶ 2*a* above. Calculate net counts per minute and weight of precipitate.

m. Add 20 mL distilled water to the Ba-Ra sulfate precipitate, let settle in a steam bath, cool, and filter through a special funnel with a tared glass fiber filter. Dry precipitate at 110 C to constant weight, cool, and weigh. Mount precipitate on a nylon disk and ring with an alpha phosphor on polyester plastic film,[4] and count in an alpha scintillation counter. Calculate net counts per minute and weight of precipitate.

n. If the isotopic composition of the precipitate is to be estimated, perform additional counting as mentioned in the calculation below.

o. Determination of combined efficien-

‖If original concentrations of isotopes of radium other than ^{226}Ra are of interest, note date and time of this original precipitation as the separation of the isotopes from their parents; use a minimal settling time and complete procedure through ¶ *j* without delay. Assuming the presence of and separation of parents, decay of ^{223}Ra and ^{224}Ra begins at the time of the first precipitation, but ingrowth of decay products is timed from the second precipitation (¶ *i*). The time of the first precipitation is not needed if the objective is to check the final precipitate for its ^{226}Ra content only.

#If solution does not clear in 10 min, cool, add another mL 5*N* NH_4OH, let stand 2 min, and heat for another 10-min period.

**The end point is most easily determined by comparison with a solution of similar composition that has been adjusted to pH 4.5 using a pH meter.

cy and self-absorption factor: Prepare standards from 1 L distilled water and the standard radium 226 solution (¶ 3*p* preceding). Include at least one blank. The barium content will impose an upper limit of 3.0 mL on the volume of the standard radium 226 solution that can be used. If x is volume of standard radium 226 solution added, then add $(3.00 - x)$ mL $BaCl_2$ carrier (¶ 3*e* above). Analyze standards as samples, beginning with ¶ 4*a*, but omit 3.00-mL $BaCl_2$ carrier.

From the observed net count rate, calculate the combined factor, bc, from the formula:

$$bc = \frac{\text{net cpm}}{ad \times 2.22 \times \text{pCi radium 226}} \quad \dagger\dagger$$

where:

ad = ingrowth factor (see below) multiplied by chemical yield.

If all chemical yields on samples and standards are not essentially equal, the factor bc will not be a constant. In this event, construct a curve relating the factor bc to varying weights of recovered $BaSO_4$.

5. Calculation

$$\text{Radium, pCi/L} = \frac{\text{net cpm}}{a\ b\ c\ d\ e \times 2.22}$$

where:

a = ingrowth factor (as shown in the following tabulation):

Ingrowth (hr)	Alpha Activity from ^{226}Ra
0	1.000
1	1.016
2	1.036
3	1.058
4	1.080
5	1.102
6	1.124
24	1.489
48	1.905
72	2.253

b = efficiency factor for alpha counting,
c = self-absorption factor,
d = chemical yield, and
e = sample volume, L.

The calculations are based on the assumption that the radium is radium 226. If the observed concentration approaches 3 pCi/L, it may be desirable to follow the rate of ingrowth and estimate the isotopic content[2,3] or, preferably, to determine radium 226 by radon 222.

The optimum ingrowth periods can be selected only if the ratios and identities of the radium isotopes are known. The number of observed count rates at different ages must be equal to or greater than the number of radium isotopes present in a mixture. In the general case, suitable ages for counting are 3 to 18 hr for the first count; for isotopic analysis, additional counting at 7, 14, or 28 days is suggested, depending on the number of isotopes in mixture. The amounts of the various radium isotopes can be determined by solving a set of simultaneous equations.[4] This approach is most satisfactory when radium 226 is the predominant isotope; in other situations, the approach suffers from statistical counting errors.

6. Precision and Accuracy

In a collaborative study, 20 laboratories analyzed four water samples for total (dissolved) radium. The radionuclide composition of these reference samples is shown in Table 705:I. Note that Samples C and D had a ^{224}Ra concentration equal to that of ^{226}Ra.

The four results from each of two laboratories and two results from a third laboratory were rejected as outliers. The average recoveries of radium 226 from the remaining A, B, C, and D samples were 97.5, 98.7, 94.9, and 99.4%, respectively. At the 95% confidence level, the precision (random error) was 28% and 30% for the two sets of paired samples. The method is

††See calculation that follows.

TABLE 705:I. CHEMICAL AND RADIOCHEMICAL COMPOSITION OF SAMPLES USED TO DETERMINE ACCURACY AND PRECISION OF RADIUM 226 METHOD

Radionuclide Composition	Samples			
	Pair 1		Pair 2	
	A	B	C	D
Radium 226,* pCi/L	12.12	8.96	25.53	18.84
Thorium 228,* pCi/L	none	none	25.90	19.12
Uranium, natural, pCi/L	105	77.9	27.7	20.5
Lead 210,* pCi/L	11.5	8.5	23.7	17.5
Strontium 90,* pCi/L	49.1	36.3	13.9	10.2
Cesium 137, pCi/L	50.3	37.2	12.7	9.5
NaCl, mg/L	60	60	300	300
$CaSO_4$, mg/L	30	30	150	150
$MgCl_2 \cdot 6H_2O$, mg/L	30	30	150	150
KCl, mg/L	5	5	10	10

* Daughter products were in substantial secular equilibrium.

biased low for radium 226, but not seriously. The method appears satisfactory for radium 226 alone or in the presence of an equal activity of radium 224 when correction for radium 224 interference is made from a second count.

For the determination of ^{224}Ra in Samples C and D, the results of two laboratories were excluded. Hence the average recoveries were 51 and 45% for Samples C and D, respectively. At the 95% confidence level, the precision was 46% for this pair of samples. The results indicated that the method for ^{224}Ra is seriously biased low. When the recoveries for radium 224 did not agree with those for radium 226, this may have been due, in part, to incomplete instructions given in the method to account for the transitory nature of ^{224}Ra activity. The method as given here contains footnotes calling attention to the importance of the time of counting. Still uncertain is the degree of separation of radium 224 from its parent, thorium 228, in ¶s 4*a* through *g* above.

Radium 223 and radium 224 analysis by this method may be satisfactory, but special refinements and further investigations are required.

7. References

1. KIRBY, H.W. 1954. Decay and growth tables for naturally occurring radioactive series. *Anal. Chem.* 26:1063.
2. SILL, C. 1960. Determination of radium-226, thorium-230, and thorium-232, U.S. Atomic Energy Comm. Rep. No. TID 7616 (Oct.). USAEC, Washington, D.C.
3. GOLDIN, A.S. 1961. Determination of dissolved radium. *Anal. Chem.* 33:406.
4. HALLDEN, N.A. & J.H. HARLEY. 1960. An improved alpha-counting technique. *Anal. Chem.* 32:1961.

706 RADIUM 226 BY RADON IN WATER (SOLUBLE, SUSPENDED, AND TOTAL)

1. General Discussion

a. Introduction: The discussion of radium, particularly in drinking water, presented in Section 705 preceding, also is pertinent to the determination of radium 226 by radon 222. In this method, *total* radium 226 means the sum of suspended and dissolved radium 226. Radon means radon 222 unless otherwise specified.

b. Principle: Radium in water is concentrated and separated from sample solids by coprecipitation with a relatively large amount of barium as the sulfate. The precipitate is treated to remove silicates, if present, and to decompose insoluble radium compounds, fumed with phosphoric acid to remove sulfite (SO_3), and dissolved in hydrochloric acid (HCl). The completely dissolved radium is placed in a bubbler, which is then closed and stored for a period of several days to 4 wk for ingrowth of radon. The bubbler is connected to an evacuated system and the radon gas is removed from the liquid by aeration, dried with a desiccant, and collected in a counting chamber. The counting chamber consists of a dome-topped scintillation cell coated inside with silver-activated zinc sulfide phosphor; a transparent window forms the bottom (Figure 706:1). The chamber rests on a photomultiplier tube during counting. About 4 hr after radon collection, the alpha-counting rate of radon and decay products is at equilibrium, and a count is obtained and related to radium 226 standards similarly treated.

The counting gas used to purge radon from the liquid to the counting chamber may be helium, nitrogen, or aged air.

Some radon (emanation) technics employ a minimum of chemistry but require high dilution of the sample and large chambers for counting the radon 222.[1] Others involve more chemical separation, concentration, and purification of radium 226 before de-emanation into counting cells of either the ionization or alpha scintillation types. The method[2] given here requires a moderate amount of chemistry coupled with a sensitive alpha scintillation count of radon 222 plus daughter products in a small chamber.[3]

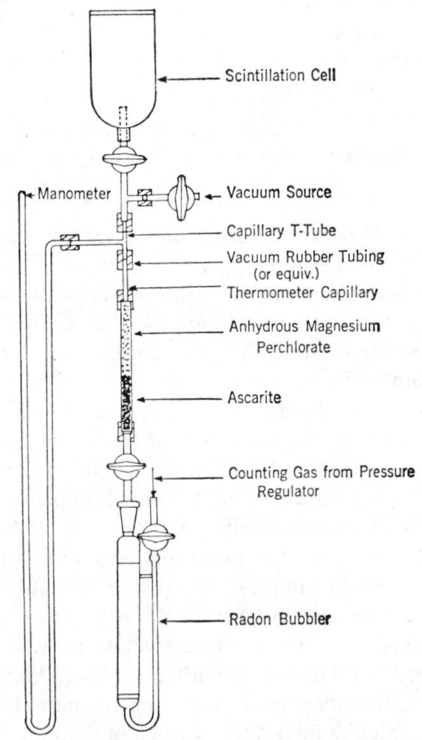

Figure 706:1. De-emanation assembly.

c. Concentration technics: The chemical properties of barium and radium are similar; therefore, because barium does not interfere with de-emanation, as much as 100 mg may be used to aid in coprecipitating radium from a sample to be placed in a single radon bubbler. However, because some radium 226 is present in barium salts, reagent tests are necessary to account for radium 226 introduced in this way.

d. Interferences: Only the gaseous alpha-emitting radionuclides, radon 219 (actinon) and radon 220 (thoron), can interfere. Interference from these radionuclides would be expected to be very rare in water not contaminated by such industrial wastes as uranium mill elements.[2] The half-lives of these nuclides are only 3.92 and 54.5 sec, respectively, so only their alpha-emitting decay products interfere.

Interference from stable chemicals is limited. Small amounts of lead, calcium, and strontium, collected by the barium sulfate, do not interfere. However, lead may cause deterioration of platinum ware. Calcium at a concentration of 300 mg/L and other dissolved solids (in brines) at 269,000 mg/L cause no difficulty.[4]

The formation of precipitates in excess of a few milligrams during the radon 222 ingrowth period is a warning that modifications[2] may be necessary because radon 222 recovery may be impaired.

e. Minimum detectable concentration: The minimum detectable concentration depends on counter characteristics, background-counting rate of scintillation cell, length of counting period, and contamination of apparatus and environment by radium 226. Without reagent purification, the overall reagent blank (excluding background) should be between 0.03 and 0.05 pCi radium 226, which may be considered the minimum detectable amount under routine conditions.

2. Apparatus

a. Scintillation counter assembly with a photomultiplier (PM) tube 5 cm or more in diameter, normally mounted, face up, in a light-tight housing. The photomultiplier tube, preamplifier, high-voltage supply, and scaler may be contained in one chassis; or the PM tube and preamplifier may be used as an accessory with a proportional counter or a separate scaler. A high-voltage safety switch should open automatically when the light cover is removed, to avoid damage to the photomultiplier tube.

Use a preamplifier with a variable gain adjustment. Equip counter with a flexible ground wire attached to the chassis and to the neck of the scintillation cell by an alligator clip or similar device. Ascertain operating voltage by determining a plateau using ^{222}Rn in the scintillation cell as the alpha source; the slope should not exceed 2%/100 V. Calibrate and use counter and scintillation cell as a unit when more than one counter is available. The background-counting rate for the counter assembly without the scintillation cell in place should be 0.00 to 0.03 cpm.

b. Scintillation cells,[2,3] Lucas-type, preferably having a volume of 95 to 140 mL, made in the laboratory, or commercially available.*

c. Radon bubblers, capacity 18 to 25 mL, as shown in Figure 706:1.† Use gas-tight glass stopcocks and a fritted glass disk of medium porosity.‡ Use one bubbler for a standard ^{226}Ra solution and one for each sample and blank in a batch.[2]

d. Manometer, open-end capillary tube or vacuum gauge having volume that is small compared to volume of scintillation cell, 0 to 760 mm Hg.

e. Gas purification tube, 7 to 8 mm OD standard-wall glass tubing, 100 to 120 mm long, constricted at lower end to hold glass wool plug (Figure 706:1); thermometer capillary tubing.

---

*William H. Johnston Laboratories, 3617 Woodland Ave., Baltimore, Md. 21215.

†Available from Corning Glass Works, Special Sales Section, Corning, N.Y. 14830.

‡Corning or equivalent.

f. Sample bottles, polyethylene, 2- to 4-L capacity.

g. Membrane filters.§

h. Gas supply: Helium, nitrogen, or air aged in high-pressure cylinder with two-stage pressure regulator and needle valve. Helium is preferred.

i. Silicone grease, high-vacuum.

j. Sealing wax, low-melting.||

k. Laboratory glassware: Excepting bubblers, decontaminate all glassware before and between uses by heating for 1 hr in EDTA decontaminating solution at 90 to 100 C, then rinse in water, 1*N* HCl, and again in distilled water to dissolve barium (radium) sulfate, $Ba(Ra)SO_4$.

Removal of previous samples from bubblers and rinsing is described in Section 706.4*b*17). More extensive cleaning of bubblers requires removal of wax from joints, silicone grease from stopcocks, and the last traces of barium-radium compounds.

l. Platinum ware: Crucibles (20 to 30 mL) or dishes (50 to 75 mL), large dish (for flux preparation), and platinum-tipped tongs (preferably Blair type). Clean platinum ware by immersion and rotation in a molten bath of potassium pyrosulfate, remove, cool, rinse in hot tap water, digest in hot 6*N* HCl, rinse in distilled water, and finally flame over a burner.

3. Reagents

a. Stock barium chloride solution: Dissolve 17.79 g $BaCl_2 \cdot 2H_2O$ in distilled water and dilute to 1 L; 1 mL = 10 mg Ba.

b. Dilute barium chloride solution: Dilute 200.0 mL stock barium chloride solution to 1,000 mL as needed; 1 mL = 2.00 mg Ba. Let stand 24 hr and filter through a membrane filter.

Optionally, add approximately 40,000 dpm of ^{133}Ba to this solution before dilution. Take account of the stable barium carrier added with the ^{133}Ba and with the diluting solution, so that the final barium concentration is near 2 mg/L. (The use of ^{133}Ba provides a convenient means of checking on the recovery of ^{226}Ra from the sample (Section 706.6). Use the $BaCl_2$ solution containing ^{133}Ba in Sections 706.4*b*3), 4*c*8), and 4*d*3). Do *not* use in *d* below; instead, use a separate dilution of stock $BaCl_2$ solution for preparing ^{226}Ra standard solutions.

c. Acid barium chloride solution: To 20 mL conc HCl in a 1-L volumetric flask, add dilute $BaCl_2$ solution to the mark and mix.

d. Stock radium 226 solution: Take every precaution to avoid unnecessary contamination of working area, equipment, and glassware, preferably by preparing ^{226}Ra standards in a separate area or room reserved for this purpose. Obtain a National Bureau of Standards gamma ray standard containing 0.1 μg ^{226}Ra as of date of standardization. Using a heavy glass rod, cautiously break neck of ampul, which is submerged in 300 mL acid $BaCl_2$ solution in a 600-mL beaker. Chip ampul unit until it is thoroughly broken or until hole is large enough to give complete mixing. Transfer solution to a 1-L volumetric flask, rinse beaker with acid $BaCl_2$ solution, dilute to mark with same solution, and mix; 1 mL = approximately 100 pg ^{226}Ra.

Determine the time in years, *t*, since the NBS standardization of the original ^{226}Ra solution. Calculate pCi ^{226}Ra/mL as:

$$\text{pCi } ^{226}\text{Ra} = [1 - (4.3 \times 10^{-4})(t)]\,[100]\,[0.990]$$

e. Intermediate radium 226 solution: Dilute 100 mL stock radium 226 solution to 1,000 mL with acid $BaCl_2$ solution; 1 mL = approximately 10 pCi ^{226}Ra.

§Type HAWP (Millipore Filter Corp., Bedford, Mass.), or equivalent.

||Pyseal, Fisher Scientific Co., Pittsburgh, Pa., or equivalent.

f. Standard radium 226 solution: Add 30.0 mL intermediate radium 226 solution to a 100-mL volumetric flask and dilute to mark with acid $BaCl_2$ solution; 1 mL = approximately 3 pCi ^{226}Ra and contains about 2 mg Ba. See ¶ *d* et seq above for correction factors.

g. Hydrochloric acid, HCl, conc, 6*N*, 1*N*, and 0.1*N*.

h. Sulfuric acid, H_2SO_4, conc and 0.1*N*.

i. Hydrofluoric acid, HF, 48%, in a plastic dropping bottle. (CAUTION.)

j. Ammonium sulfate solution: Dissolve 10 g $(NH_4)_2SO_4$ in distilled water and dilute to 100 mL in a graduated cylinder.

k. Phosphoric acid, H_3PO_4, 85%.

l. Ascarite, 8 to 20 mesh.

m. Magnesium perchlorate, anhydrous desiccant.

n. EDTA decontaminating solution: Dissolve 10 g disodium ethylenediaminetetraacetate dihydrate and 10 g Na_2CO_3 in distilled water and dilute to 1 L in a graduated cylinder.

o. Special reagents for total and suspended radium:

1) *Flux*: Add 30 mg $BaSO_4$, 65.8 g K_2CO_3, 50.5 g Na_2CO_3, and 33.7 g $Na_2B_4O_7 \cdot 10H_2O$, to a large platinum dish (500-mL capacity). Mix thoroughly and heat cautiously to expel water, then fuse and mix thoroughly by swirling. Cool flux, grind in a porcelain mortar to pass a 10- to 12-mesh (or finer) screen, and store in an airtight bottle.

2) *Dilute hydrogen peroxide solution*: Dilute 10 mL 30% H_2O_2 to 100 mL in a graduated cylinder. Prepare daily.

4. Calibration of Scintillation Counter Assembly

a. Test bubblers by adding about 10 mL distilled water and passing air through them at the rate of 3 to 5 mL (free volume)/min. Air should form many fine bubbles rather than a few large ones; the latter condition indicates nonuniform pores. Do not use bubblers requiring excessive pressure to initiate bubbling. Fritted-glass disks of medium porosity (¶ 2*c*) usually are satisfactory.

b. Apply silicone grease to stopcocks of a bubbler and, with gas inlet stopcock closed, add 1 mL stock $BaCl_2$ solution and 10 mL (30 pCi) standard radium 226 solution, and fill bubbler two-thirds to three-fourths full with additional acid $BaCl_2$ solution.

c. With bubbler in a clamp or rack, dry joint with lint-free paper or cloth, warm separate parts of the joint, apply sealing wax sparingly to the male part, and make the connection with a twisting motion to spread the wax uniformly in the ground joint. Let cool. Establish zero ingrowth time by purging liquid with counting gas for 15 to 20 min according to ¶ 4*j* below and adjust inlet pressure to produce a froth a few millimeters thick. Close stopcocks, record date and time, and store bubbler, preferably for 3 wk or more (with most samples) before collecting and counting ^{222}Rn. A much shorter ingrowth period of 16 to 24 hr is convenient for a standard bubbler. Obtain an estimate of ^{222}Rn present at any time from the B columns in Table 706:I.

d. Attach scintillation cell as shown in Figure 706:1; # substitute a glass tube with a stopcock for bubbler so that the compressed gas can be turned on or off conveniently. Open stopcock on scintillation cell, close stopcock to gas, and gradually open stopcock to vacuum source to evacuate cell. Close stopcock to vacuum source and check manometer reading for 2 min to test system, especially the scintillation cell, for leaks.

#The system as described and shown in Figure 706:1 is considered minimal. In routine work, use manifold systems and additional, more precise needle valves. An occasional drop of solution will escape from the bubbler; provide enough free space beyond the outlet stopcock to accommodate this liquid, preventing its entrance into the gas-purifying train.

TABLE 706:I. FACTORS FOR DECAY OF RADON 222, GROWTH OF RADON 222 FROM RADIUM 226, AND CORRECTION OF RADON 222 ACTIVITY FOR DECAY DURING COUNTING

Time	Factor for Decay of Radon 222 $A = e^{-\lambda t}$		Factor for Growth of Radon 222 from Radium 226 $B = 1 - e^{-\lambda t}$		Factor for Correction of Radon 222 Activity for Decay during Counting $C = \lambda t/(1 - e^{-\lambda t})$
	Hours	Days	Hours	Days	Hours
0.0	1.0000		0.000 00		1.000
0.2	0.9985		0.001 51		1.001
0.4	0.9970		0.003 01		1.001
0.6	0.9955		0.004 52		1.002
0.8	0.9940		0.006 02		1.003
1	0.9925	0.8343	0.007 52	0.1657	1.004
2	0.9850	0.6960	0.014 99	0.3040	1.008
3	0.9776	0.5807	0.022 40	0.4193	1.011
4	0.9703	0.4844	0.029 75	0.5156	1.015
5	0.9630	0.4041	0.037 05	0.5959	1.019
6	0.9557	0.3372	0.044 29	0.6628	1.023
7	0.9485	0.2813	0.051 48	0.7187	1.027
8	0.9414	0.2347	0.058 61	0.7653	1.031
9	0.9343	0.1958	0.065 69	0.8042	1.034
10	0.9273	0.1633	0.072 72	0.8367	1.038
11	0.9203	0.1363	0.079 69	0.8637	1.042
12	0.9134	0.1137	0.086 62	0.8863	1.046
13	0.9065	0.0948	0.093 49	0.9052	1.050
14	0.8997	0.0791	0.100 31	0.9209	1.054
15	0.8929	0.0660	0.107 07	0.9340	1.058
16	0.8862	0.0551	0.1138	0.9449	1.062
17	0.8795	0.0459	0.1205	0.9541	1.066
18	0.8729	0.0383	0.1271	0.9617	1.069
19	0.8664	0.0320	0.1336	0.9680	1.073
20	0.8598	0.0267	0.1402	0.9733	1.077
21	0.8534	0.0223	0.1466	0.9777	1.081
22	0.8470	0.0186	0.1530	0.9814	1.085
23	0.8406	0.0155	0.1594	0.9845	1.089
24	0.8343	0.0129	0.1657	0.9871	1.093
25	0.8280	0.0108	0.1720	0.9892	1.097

TABLE 706:I. FACTORS FOR DECAY OF RADON 222, GROWTH OF RADON 222 FROM RADIUM 226, AND CORRECTION OF RADON 222 ACTIVITY FOR DECAY DURING COUNTING

Time	Factor for Decay of Radon 222		Factor for Growth of Radon 222 from Radium 226		Factor for Correction of Radon 222 Activity for Decay during Counting
	$A = e^{-\lambda t}$		$B = 1 - e^{-\lambda t}$		$C = \lambda t/(1 - e^{-\lambda t})$
	Hours	Days	Hours	Days	Hours
26	0.8218	0.0090	0.1782	0.9910	1.101
27	0.8156	0.0075	0.1844	0.9925	1.105
28	0.8095	0.0063	0.1905	0.9937	1.109
29	0.8034	0.0052	0.1966	0.9948	1.113
30	0.7973	0.0044	0.2027	0.9956	1.118
31	0.7913	0.0036	0.2087	0.9964	1.122
32	0.7854	0.0030	0.2146	0.9970	1.126
33	0.7795	0.0025	0.2205	0.9975	1.130
34	0.7736	0.0021	0.2264	0.9979	1.134
35	0.7678	0.0018	0.2322	0.9982	1.138
36	0.7620	0.0015	0.2380	0.9985	1.142
37	0.7563	0.0012	0.2437	0.9988	1.146
38	0.7506	0.0010	0.2494	0.9990	1.150
39	0.7449	0.0009	0.2551	0.9991	1.154
40	0.7393	0.0007	0.2607	0.9993	1.159
41	0.7338	0.0006	0.2662	0.9994	1.163
42	0.7283	0.0005	0.2717	0.9995	1.167
43	0.7228	0.0004	0.2772	0.9996	1.171
44	0.7173	0.0003	0.2827	0.9997	1.175
45	0.7120	0.0003	0.2880	0.9997	1.179
46	0.7066	0.0002	0.2934	0.9998	1.184
47	0.7013	0.0002	0.2987	0.9998	1.188
48	0.6960	0.0002	0.3040	0.9998	1.192
49	0.6908	0.0001	0.3092	0.9999	1.196
50	0.6856	0.0001	0.3144	0.9999	1.201
51	0.6804	0.0001	0.3196	0.9999	1.205
52	0.6753	0.0001	0.3247	0.9999	1.209
53	0.6702	0.0001	0.3298	0.9999	1.213
54	0.6652	0.0001	0.3348	0.9999	1.218
55	0.6602	0.0000	0.3398	1.0000	1.222
56	0.6552	0.0000	0.3448	1.0000	1.226
57	0.6503	0.0000	0.3497	1.0000	1.231
58	0.6454	0.0000	0.3546	1.0000	1.235
59	0.6405	0.0000	0.3595	1.0000	1.239
60	0.6357	0.0000	0.3643	1.0000	1.244

e. Open stopcock to counting gas and cautiously admit gas to scintillation cell until atmospheric pressure is reached.

f. Center scintillation cell on photomultiplier tube, cover with light-tight hood and, after 10 min, obtain a background counting rate (preferably over a 100- to 1,000-min period, depending on concentration of ^{226}Ra in samples). *Do not expose phototube to external light with the high voltage applied.*

g. Repeat Steps *d* through *f* above for each scintillation cell.

h. If the leakage test and background are satisfactory, continue calibration.

i. With scintillation cell and standard bubbler (¶ 4*c*) on vacuum train, open stopcock on scintillation cell and evacuate scintillation cell and purification system (Figure 706:1) by opening stopcock to vacuum source. Close stopcock to vacuum source. Check system for leaks as in Step *d* above.

j. Adjust gas regulator (diaphragm) valve so that a very slow stream of gas will flow with the needle valve open. Attach gas supply to inlet of bubbler.

k. Note time as beginning of an approximately 20-min de-emanation period. Very cautiously open bubbler outlet stopcock to equalize pressure and transfer all or most of the fluid in the inlet side arm to bubbler chamber.

l. Close outlet stopcock and very cautiously open inlet stopcock to flush remaining fluid from side arm and fritted disk. Close inlet stopcock.

m. Repeat Steps *h* and *l* above, four or five times, to obtain more nearly equal pressures on the two sides of bubbler.

n. With outlet stopcock fully open, cautiously open inlet stopcock so that gas flow produces a froth a few millimeters thick at surface of bubbler solution. Maintain flow rate by gradually increasing pressure with regulator valve and continue de-emanation until pressure in cell reaches atmospheric pressure. Total elapsed time for the de-emanation should be 15 to 25 min.

o. Close stopcocks to scintillation cell, close bubbler inlet and outlet, shut off and disconnect gas supply, and record date and time as the ends of the ^{222}Rn ingrowth and de-emanation periods and as the beginnings of decay of ^{222}Rn and ingrowth of decay products.

p. Store bubbler for another ^{222}Rn ingrowth in the event a subsequent de-emanation is desired (Table 706:I). The standard bubbler may be kept indefinitely.

q. Four hours after de-emanation, when daughter products are in virtual transient equilibrium with ^{222}Rn, place scintillation cell on photomultiplier tube, cover with light-tight hood, let stand for at least 10 min, then begin counting. Record date and time counting was started and finished.

r. Correct net counting rate for ^{222}Rn decay (Table 706:I) and relate it to picocuries ^{226}Ra in standard bubbler (see Section 706.6*a*). Unless the scintillation cell is physically damaged, the calibration will remain essentially unchanged for years. Occasional calibration is recommended.

s. Repeat Steps *h* through *r* above on each scintillation cell.

t. To remove ^{222}Rn and prepare scintillation cell for reuse, evacuate and cautiously refill with counting gas. Routinely, repeat evacuation and refilling twice, and repeat process more times if the cells have contained a high ^{222}Rn activity. (Decay products with a half-life of approximately 30 min will remain in the cell. Do not check background on cells until activity of decay products has had time to decay to insignificance.)

5. Procedure

a. Soluble radium 226:

1) Using a membrane filter, filter at least 1 L sample or a volume containing up to 30 pCi ^{226}Ra and transfer to a polyethylene bottle as soon after sampling as pos-

sible. Save the suspended matter for determination by the procedure described in Section 706.5*b*. Record sample volume filtered if suspended solids are to be analyzed as in the procedure for ^{226}Ra in suspended matter.

2) Add 20 mL conc HCl/L of filtrate and continue analysis when convenient.

3) Add 50 mL dilute $BaCl_2$ solution, with vigorous stirring, to 1,020 mL acidified filtrate [Section 706.5*a*2) preceding] in a 1,500-mL beaker. In each batch of samples include a reagent blank consisting of distilled water plus 20 mL conc HCl.

4) Cautiously, with vigorous stirring, add 20 mL conc H_2SO_4. Cover beaker and let precipitate overnight.

5) Filter supernate through a membrane filter, using 0.1*N* H_2SO_4 to transfer Ba-Ra precipitate to filter, and wash precipitate twice with 0.1*N* H_2SO_4.

6) Place filter in a platinum crucible or dish, add 0.5 mL HF and 3 drops (0.15 mL) $(NH_4)_2SO_4$ solution, and evaporate to dryness.

7) Carefully ignite over a small flame until carbon is burned off; cool. (After filter is charred a Meker burner may be used.)

8) Add 1 mL H_3PO_4 with a calibrated dropper and heat on hot plate at about 200 C. Gradually raise temperature and maintain at about 300 to 400 C for 30 min.

9) Swirl vessel over a low Bunsen flame, adjusted to avoid spattering, while covering the walls with hot H_3PO_4. Continue to heat for a minute after precipitate fuses into a clear melt (just below redness) to insure complete removal of SO_3.

10) Fill cooled vessel one-half full with 6*N* HCl, heat on steam bath, then gradually add distilled water to within 2 mm of top of vessel.

11) Evaporate on boiling steam bath until there are no more vapors of HCl.

12) Add 6 mL 1*N* HCl, swirl, and warm to dissolve $BaCl_2$ crystals.

13) Close gas inlet stopcock, add a drop of water to the fritted disk of the fully greased and tested radon bubbler, and transfer sample from platinum vessel to bubbler with a medicine dropper. Use dropper to rinse vessel with at least three 2-mL portions of distilled water. Add distilled water until bubbler is two-thirds to three-fourths full.

14) Dry, wax if necessary, and seal joint. Establish zero ingrowth time as instructed in Section 706.4*c* preceding.

15) Close stopcocks, record date and time, and store bubbler for ^{222}Rn ingrowth, preferably for 3 wk for low concentrations of radium 226.

16) De-emanate and count ^{222}Rn as instructed for calibrations in ¶s 4*i* through *r*, with sample replacing standard bubbler.

17) The sample in the bubbler may be stored for a second ingrowth or it may be discarded and the bubbler cleaned for reuse. (A bubbler is readily cleaned while in an inverted position by attaching a tube from a beaker containing 100 mL 0.1*N* HCl to the inlet and attaching another tube from outlet to a suction flask. Alternately open and close outlet and inlet stopcocks to pass the acid rinse water sequentially through the fritted disk, accumulate in the bubbler, and flush into the suction flask. Drain bubbler with the aid of vacuum, heat ground joint gently to melt wax, and separate joint. More extensive cleaning, as indicated in Section 706.2*k* above, may be necessary if the bubbler contained more than 10 pCi ^{226}Ra.)

b. Radium 226 in suspended matter:

1) Suspended matter in water usually contains siliceous materials that require fusion with an alkaline flux to insure recovery of radium. Dry suspended matter (up to 1,000 mg inorganic material) retained on the membrane filter specified in ¶ 5*a*1) above in a tared platinum crucible and ignite as in ¶ 5*a*7).

2) Weigh crucible to estimate residue.

3) Add 8 g flux/g residue, but not less than 2 g flux, and mix with a glass rod.

4) Heat over a Meker burner until melting begins, being careful to prevent spattering. Continue heating for 20 min after bubbling stops, with an occasional swirl of the crucible to mix contents and achieve a uniform melt. A clear melt usually is obtained only when the suspended solids are present in small amount or have a high silica content.

5) Remove crucible from burner and rotate as melt cools to distribute it in a thin layer on crucible wall.

6) When cool, place crucible in a covered beaker containing 120 mL distilled water, 20 mL conc H_2SO_4, and 5 mL dilute H_2O_2 solution for each 8 g flux. (Reduce acid and H_2O_2 in proportion to flux used.) Rotate crucible to dissolve melt if necessary.

7) When melt is dissolved, remove and rinse crucible into beaker. Save crucible for Step 10) below.

8) Heat solution and slowly add 50 mL dilute $BaCl_2$ solution with vigorous stirring. Cover beaker and let stand overnight for precipitation. (Precipitation with cool sample solution also is satisfactory.)

9) Add about 1 mL dilute H_2O_2 and, if yellow color (from titanium) deepens, add more H_2O_2 until there is no further color change.

10) Continue analysis according to Section 706.5*a*5) through 16).

11) Calculate result as directed in Section 706.6*a* and *b*, taking into account that the suspended solids possibly were contained in a sample volume other than 1 L [see Section 706.5*a*1)].

c. Total radium 226:

1) Total ^{226}Ra in water is the sum of soluble and suspended ^{226}Ra as determined in Sections 5*a* and *b* preceding, or it may be determined directly by examining the original water sample that has been acidified with 20 mL conc HCl/L sample and stored in a polyethylene bottle.

2) Thoroughly mix acidified sample and take 1,020 mL or a measured volume containing not more than 1,000 mg inorganic suspended solids.

3) Add 50 mL dilute $BaCl_2$ solution and slowly, with vigorous stirring, add 20 mL conc H_2SO_4/L sample. Cover and let precipitate overnight.

4) Filter supernate through membrane filter and transfer solids to filter as in ¶ 5*a*5) preceding.

5) Place filter and precipitate in tared platinum crucible and proceed as in ¶s 5*b*2) through 10) above but with the following changes in the procedure given in ¶ 5*b*8): Omit adding dilute $BaCl_2$ solution, digest for 1 hr on a steam bath, and filter immediately after digestion without stirring up $BaSO_4$. (If these changes are not made, filtration will be very slow.)

6) Calculate total radium 226 concentration as directed in Sections 706.6*a* and *b*.

6. Calculations

a. Calculate the ^{226}Ra in a bubbler, including reagent blank, as follows:

$$^{226}\text{Ra in pCi} = \frac{R_s - R_b}{R_e} \times \frac{1}{1 - e^{-\lambda t_1}} \times \frac{1}{e^{-\lambda t_2}} \times \frac{\lambda t_3}{1 - e^{-\lambda t_3}}$$

where:

λ = decay constant for ^{222}Rn, 0.0755/hr,

t_1 = time interval allowed for ingrowth of ^{222}Rn, hr,

t_2 = time interval between de-emanation and counting, hr,

t_3 = time interval of counting, hr,

R_s = observed counting rate of sample in scintillation cell, cph,

R_b = (previously) observed background counting rate of scintillation cell with counting gas, cph,

R_e = calibration constant for scintillation cell (i.e., observed net counts per hour, corrected by use of ingrowth and decay factors (C/AB from below) per picocurie of Ra in standard), or:

$$^{226}\text{Ra in pCi} = \frac{(R_s - R_b)}{R_e} \times \frac{C}{AB}$$

where:

A = factor for decay of ^{222}Rn (see Table 706:I),

B = factor for growth of ^{222}Rn from ^{226}Ra (see Table 706:I), and

C = factor for correction of ^{222}Rn activity for decay during counting (see Table 706:I).

For nontabulated times, obtain decay factors for ^{222}Rn by multiplying together the appropriate tabulated ''day'' and ''hour'' decay factors, interpolating for less than 0.2 hr if indicated by the precision desired. Obtain radon 222 growth factors for nontabulated times most accurately, especially for short periods (e.g., in calibrations), by calculation from ^{222}Rn decay factors given in Column A and using formula given in heading for Column B (of Table 706:I). Linear interpolations are satisfactory for routine samples. Obtain the decay-during-counting factors by linear interpolation for all nontabulated times.

In calculating cell calibration constants, use the same equation, but picocuries of ^{226}Ra is known and R_c is unknown.

b. Convert the activity into picocuries per liter of soluble, suspended, or total ^{226}Ra by the following equation:

$$^{226}\text{Ra, pCi/L} = \frac{(D - E) \times 1{,}000}{\text{mL sample}}$$

where:

D = pCi ^{226}Ra found in sample, and

E = pCi ^{226}Ra found in reagent blank.

7. Recovery of Barium (Radium 226) (Optional)

If ^{133}Ba was added in reagent *b*, check recovery of Ba by removing sample from bubbler, adjusting its volume appropriately, gamma-counting it under standardized conditions, and comparing the result with the count obtained from a 50-mL portion (evaporated if necessary to reduce volume) of dilute barium solution also counted under standardized conditions; add 1 mL H_3PO_4 to the latter portion before counting. The assumption that the Ba and ^{226}Ra are recovered to the same extent is valid in the method described.

Note that ^{226}Ra and its decay products interfere slightly even if a gamma spectrometer is used. The technic works best when the ratio of ^{133}Ba to ^{226}Ra is high.

Determinations of recovery are particularly helpful with irreplaceable samples, both in gaining experience with the method and in applying the general method to unfamiliar media.

8. Precision and Accuracy

In a collaborative study, seven laboratories analyzed four water samples for dissolved radium 226 by this method. No result was rejected as an outlier. The average recoveries of added radium 226 from Samples A, B, C, and D (below) were 97.1, 97.3, 97.6, and 98.0%, respectively. At the 95% confidence level, the precision (random error) was 6% and 8% for the two sets of paired samples. Because of the small number of participating laboratories and the low values for random and total errors, there was no evidence of laboratory systematic errors. Neither radium 224 at an activity equal to that of the radium 226 nor dissolved solids up to 610 mg/L produced a detectable error in the results.

Test samples consisted of two pairs of simulated moderately hard and hard water samples containing known amounts of added radium 226 and other radionuclides. The composition of the samples with respect to nonradioactive substances was the same for a pair of samples but varied for the two pairs. The radiochemical composition of the samples is given in Table 705:I.

9. References

1. HURSH, J.B. 1954. Radium-226 in water supplies of the U.S. *J. Amer. Water Works Ass.* 46:43.

2. RUSHING, D.E., W.J. GARCIA & D.A. CLARK. 1964. The analysis of effluents and environmental samples from uranium mills and of biological samples for radium, polonium, and uranium. *In* Radiological Health and Safety in Mining and Milling of Nuclear Materials. International Atomic Energy Agency, Vienna, Austria, Vol. 11, p. 187.
3. LUCAS, H.F. 1957. Improved low-level alpha scintillation counter for radon. *Rev. Sci. Instrum.* 28:680.
4. RUSHING, D.E. 1967. Determination of dissolved radium-226 in water. *J. Amer. Water Works Ass.* 59:593.

707 RADIUM 228 (SOLUBLE) (TENTATIVE)

1. General Discussion

a. Introduction: The discussion of radium, particularly in drinking water, presented in Section 705 also is pertinent here. In this sequential method either radium 228 alone or radium 228 and radium 226 may be determined. The EPA recommended detection limits are satisfied by this method.

b. Principle: Radium 228 and radium 226 in water are concentrated and separated by coprecipitation with barium and lead as sulfates and purified by EDTA chelation. After 36-hr ingrowth of actinium 228 from radium 228, actinium 228 is carried on yttrium oxalate, purified, and beta-counted. Radium 226 in the supernatant is precipitated as the sulfate, purified, and alpha-counted (Section 705) or it is transferred to a radon bubbler and determined by the emanation procedure (Section 706), which is the preferred method.

If analysis of radium 226 is not required, the procedure for radium 228 may be terminated by beta-counting the yttrium oxalate precipitate with a follow-up precipitation of barium sulfate for yield determination. If it is determined that radium 228 is absent, the radium 226 fraction may be alpha-counted directly. If radium 228 is present, radium 226 must be determined by radon emanation.

c. Sampling and storage: To drinking water or a filtered sample of turbid water, add 2 mL conc nitric acid (HNO_3)/L sample at the time of collection or immediately after filtration.

2. Apparatus

a. Counting instruments: One of the following is required:

1) *Internal proportional counter,* gas flow, with scaler, timer, and register; or a thin end-window (polyester plastic)* proportional counting chamber with scaler, timer, register amplifier, and preferably having an anticoincident system (low background).

2) *Scintillation counter assembly.* See Section 706. This equipment is necessary only if radium 226 is determined sequentially with radium 228 and is analyzed by emanation of radon.

b. Centrifuge, bench-size clinical, with polypropylene tubes.

c. Filter funnels, for 2.4-cm filter paper.

d. Stainless steel pans, 5.1 cm.

e. Infrared drying lamp assembly.

f. Magnetic stirrer hot plate.

g. Membrane filters, 47 mm diam, 0.45 μm pore diam.†

3. Reagents

a. Acetic acid, conc.

b. Acetone, anhydrous.

c. Ammonium hydroxide, NH_4OH, conc.

*Mylar or equivalent.

†Gelman Ga-6 or equivalent.

d. Ammonium oxalate solution: Dissolve 25 g $(NH_4)_2C_2O_4$ in distilled water and dilute to 500 mL.

e. Ammonium sulfate solution: Dissolve 20 g $(NH_4)_2SO_4$ in a minimum of distilled water and dilute to 100 mL.

f. Ammonium sulfide solution: Dilute 10 mL $(NH_4)_2S$ (20 to 24%) to 100 mL with distilled water.

g. Barium carrier standardized: Dissolve 2.846 g $BaCl_2{\cdot}2H_2O$ in distilled water, add 0.5 mL conc HNO_3, and dilute to 100 mL; 1 mL = 16 mg Ba.

h. Citric acid, 1 *M*: See Section 705.3*a*.

i. EDTA reagent, 0.25 *M*: See Section 705.3*k*.

j. Ethanol, 95%.

k. Lead carrier, Solution A: Dissolve 2.397 g $Pb(NO_3)_2$ in distilled water, add 0.5 mL conc HNO_3, and dilute to 100 mL; 1 mL = 15 mg Pb. *Solution B:* Dilute 10 mL Solution A to 100 mL with distilled water; 1 mL = 1.5 mg Pb.

l. Methyl orange indicator solution: Dissolve 0.1 g methyl orange powder in 100 mL distilled water.

m. Nitric acid, HNO_3, conc, 6*N*, and 1*N*.

n. Sodium hydroxide, 18*N*: Dissolve 720 g NaOH in 500 mL distilled water and dilute to 1 L.

o. Sodium hydroxide, 10*N*: Dissolve 400 g NaOH in 500 mL distilled water and dilute to 1 L.

p. Sodium hydroxide, NaOH, 1*N*.

q. Strontium-yttrium mixed carrier: Solution A: Dilute 10.0 mL yttrium carrier to 100 mL. *Solution B:* Dissolve 0.4348 g $Sr(NO_3)_2$ in distilled water and dilute to 100 mL. Combine equal volumes of Solutions A and B; 1 mL = 0.9 mg Sr and 0.9 mg Y.

s. Sulfuric acid, H_2SO_4, 18*N*.

t. Yttrium carrier: Add 12.7 g Y_2O_3 (Section 704.3*d*) to an erlenmeyer flask containing 20 mL distilled water. Heat to boiling and, while stirring with a magnetic stirring hot plate, add small portions of conc HNO_3. (About 30 mL is necessary to dissolve the Y_2O_3. Small additions of distilled water also may be needed to replace water lost by evaporation.) After total dissolution, add 70 mL conc HNO_3 and dilute to 1 L with distilled water; 1 mL = 10 mg Y.

4. Procedure

a. Radium 228

1) For 1L sample add 5 mL 1*M* citric acid and a few drops methyl orange indicator. The solution should be red. Add 10 mL lead carrier (Solution A), 2.0 mL barium carrier, and 2 mL yttrium carrier; stir well. Heat to incipient boiling and maintain at this temperature for 30 min.

2) Add conc NH_4OH until a definite yellow color is obtained; add a few drops excess. Precipitate lead and barium sulfates by adding 18*N* H_2SO_4 until the red color reappears; add 0.25 mL excess. Add 5 mL $(NH_4)_2SO_4$ solution/L sample. Stir frequently and hold at about 90 C for 30 min.

3) Cool and filter with suction through a membrane filter. Quantitatively transfer precipitate to filter. Carefully place filter in a 250-mL beaker. Add about 10 mL conc HNO_3 and heat gently until the filter dissolves completely. Using conc HNO_3 transfer precipitate to a centrifuge tube. Centrifuge and discard supernatant.

4) Wash precipitate with 15 mL conc HNO_3, centrifuge, and discard supernatant. Repeat wash and centrifuge again. Add 25 mL EDTA reagent, heat in a hot water bath, and stir well. Add a few drops 10*N* NaOH if the precipitate does not dissolve readily.

5) Add 1 mL strontium-yttrium mixed carrier and stir thoroughly. Add a few drops 10*N* NaOH if any precipitate forms. Add 1 mL $(NH_4)_2SO_4$ solution and stir thoroughly. Add conc acetic acid until $BaSO_4$ precipitates; add 2 mL excess. The pH should be about 4.5. Digest in a hot water bath (80 C) until precipitate settles.

Centrifuge and discard supernatant.

6) Add 20 mL EDTA reagent, heat in a hot water bath, and stir until precipitate dissolves. Repeat Step 5. Note time of last $BaSO_4$ precipitation as zero time for in-growth of ^{228}Ac. Dissolve precipitate in 20 mL EDTA reagent, add 0.5 mL yttrium carrier and 1 mL lead carrier (Solution B). If any precipitate forms, dissolve by adding a few drops 10*N* NaOH. Mix well, cap tube, and age at least 36 hr.

7) Add 0.3 mL $(NH_4)_2S$ solution and mix well. Add 10*N* NaOH dropwise with vigorous stirring until PbS precipitates; add 10 drops excess. Stir intermittently for about 10 min. Centrifuge and decant supernatant into a clean tube.

8) Add 1 mL lead carrier (Solution B), 0.1 mL $(NH_4)_2S$ solution, and a few drops 10*N* NaOH. Repeat precipitation of PbS. Centrifuge and filter supernatant through Whatman No. 42 filter paper into a clean tube. Wash filter with a few milliliters of distilled water. Discard residue.

9) Add 5 mL 18*N* NaOH (make at least 2*N* in OH^-). Because of the short half-life of ^{228}Ac (6.13 hr) complete the following procedure without delay. Mix well and digest in a hot water bath until $Y(OH)_3$ coagulates. Centrifuge and decant supernatant into a beaker. Cover beaker and save supernatant for ^{226}Ra analysis, ¶s *b* or *c* below. Note time of $Y(OH)_3$ precipitation; this is the end of ^{228}Ac ingrowth and beginning of ^{228}Ac decay. (t_3 = time in minutes between last $BaSO_4$ and first $Y(OH)_3$ precipitations.) Dissolve precipitate in 2 mL 6*N* HNO_3. Heat and stir in a hot water bath about 5 min. Add 5 mL distilled water and reprecipitate $Y(OH)_3$ with 3 mL 10*N* NaOH. Heat and stir in a hot water bath until precipitate coagulates. Centrifuge and discard supernatant.

10) Dissolve precipitate with 1 mL 1*N* HNO_3 and heat in hot water bath for several minutes. Dilute to 5 mL with distilled water and add 2 mL ammonium oxalate solution. Heat to coagulate, centrifuge, and discard supernatant. Add 10 mL distilled water, 6 drops 1*N* HNO_3, and 6 drops ammonium oxalate solution. Heat and stir in a hot water bath for several minutes. Centrifuge and discard supernatant. Transfer quantitatively to a tared stainless-steel planchet using a minimum quantity of distilled water. Dry under an infrared lamp to constant weight and count in a low-background beta counter. (t_1 = time in minutes between first $Y(OH)_3$ precipitation and counting.)

If analysis of radium 226 is not required, complete Steps *b*1) and 3) below to obtain the fractional barium yield to be used in calculating ^{228}Ra activity.

b. Radium by precipitation

1) To the supernatant saved in ¶ *a*9) above add 4 mL conc HNO_3 and 2 mL $(NH_4)_2SO_4$ solution, mixing well after each addition. Add conc acetic acid until $BaSO_4$ precipitates; add 2 mL excess. Digest on a hot plate until precipitate settles. Centrifuge and discard supernatant.

2) Add 20 mL EDTA reagent, heat in a hot water bath, and stir until precipitate dissolves. Add a few drops 10*N* NaOH if precipitate does not dissolve readily. Add 1 mL strontium-yttrium mixed carrier and 1 mL lead carrier (Solution B), and stir thoroughly. Add a few drops 10*N* NaOH if any precipitate forms. Add 1 mL $(NH_4)_2SO_4$ solution and stir thoroughly. Add conc acetic acid until $BaSO_4$ precipitates; add 2 mL excess. Digest in a hot water bath until precipitate settles. Centrifuge, discard supernatant, and note time.

3) Wash precipitate with 10 mL distilled water. Centrifuge and discard supernatant. Transfer quantitatively to a tared stainless-steel planchet using a minimum quantity distilled water. Dry under an infrared lamp to constant weight. If after sufficient beta decay of the actinium fraction ^{228}Ra is found to be absent, make a direct alpha count for ^{226}Ra. If ^{228}Ra is present, determine ^{226}Ra by radon emanation, ¶ *c* below.

4) Count immediately in an alpha proportional counter.

c. Radium 226 by radon: Transfer the final precipitate obtained in *b* above to a small beaker using a rubber policeman and 14 mL EDTA reagent. Add a few drops 10*N* NaOH and heat on a hot plate to dissolve. Cool and transfer to a radon bubbler (Figure 706:1) rinsing beaker with 1 mL EDTA reagent. Proceed as in Section 706 beginning with 5*a* 14).

5. Calculation

a. Calculation of ^{228}Ra concentration:

$$^{228}\text{Ra, pCi/L} = \frac{C}{2.22 \times EVR} \times \frac{\lambda t_2}{(1 - e^{-\lambda t_2})} \times \frac{1}{(1 - e^{-\lambda t_3})} \times \frac{1}{e^{-\lambda t_1}}$$

where:

C = average net count rate, cpm,
E = counter efficiency, for ^{228}Ac,
V = sample volume, L,
R = fractional chemical yield of yttrium carrier, ¶4*a* 10), multiplied by fractional chemical yield of barium carrier, ¶ *b*3),
λ = decay constant of ^{228}Ac, 0.001884/min,
t_1 = time between first $Y(OH)_3$ precipitation and start of counting, min,
t_2 = counting time, min, and
t_3 = ingrowth time of ^{228}Ac between last $BaSO_4$ precipitation and first $Y(OH)_3$ precipitation, min.

The factor $\lambda t_2/(1 - e^{-\lambda t_2})$ corrects average count rate to count rate at beginning of counting time.

b. Calculation of ^{226}Ra (plus any ^{224}Ra and ^{223}Ra) concentration: See Section 705.5.

c. Calculation of ^{226}Ra (emanation) concentration: See Section 706.6.

6. Bibliography

JOHNSON, J.O. 1971. Determination of radium 228 in natural waters. Radiochemical Analysis of Water. U.S. Geol. Surv., Water Supply Paper 1696-G, U.S. Govt. Printing Office, Washington, D.C.

KRIEGER, H.L. 1976. Interim Radiochemical Methodology for Drinking Water. EPA-600/4-75-008 (revised).

708 TRITIUM

Tritium exists fairly uniformly in the environment as a result of natural production by cosmic radiation and residual fallout from nuclear weapons tests. This background level gradually is being increased by the use of nuclear reactors to generate electricity, although tritium from this source comprises only a small proportion of environmental tritium. Nuclear reactors and fuel-processing plants are localized sources of tritium because of discharges during normal operation. This industry is expected to become the major source of environmental tritium contamination in the future. Tritium is produced in light-water nuclear reactors by ternary fission, neutron capture in coolant additives, control rods and plates, and activation of deuterium. About 1% of the tritium in the primary coolant is released in gaseous form to the atmosphere; the remainder eventually is released in liquid waste discharges. Most tritium produced in reactors remains in the fuel and is released when fuel is reprocessed.

Naturally occurring tritium is most abundant in precipitation and lowest in aged water because of its physical decay by beta emission to helium. The maximum beta energy of tritium is 0.018 MeV and its half-life is 12.26 yr.

The Environmental Protection Agency

Interim Primary Drinking Water Regulations set a maximum contaminant level of 20,000 pCi/L.

1. General Discussion

a. Principle: A sample is treated by alkaline permanganate distillation to hold back most quenching materials, as well as radioiodine and radiocarbon. Complete transfer of tritiated water is assured by distillation to near dryness. A subsample of distillate is mixed with scintillation solution and the beta activity is counted on a coincidence-type liquid scintillation spectrometer. The scintillation solution consists of 1,4-dioxane, naphthalene, POPOP and PPO.* The spectrometer is calibrated with standard solutions of tritiated water; then background and unknown samples are prepared and counted alternately, thus nullifying errors that could result from instrument drift or from aging of the scintillation solution.

b. Interferences: Sample distillation effectively removes nonvolatile radioactivity and the usual quenching materials. For waters containing volatile organic or radioactive materials, use wet oxidation (Section 420) to remove interference from quenching due to volatile organic material. Distillation at about pH 8.5 holds back volatile radionuclides such as iodides and bicarbonates. Double distillation with an appropriate delay (10 half-lives) between distillations may be required to eliminate interference from volatile daughters of radium isotopes. Some clear-water samples collected near nuclear facilities may be monitored satisfactorily without distillation, especially when the monitoring instrument is capable of discriminating against beta radiation energies higher than those in the tritium range.

*POPOP = 1,4-di-2-(5-phenyloxazolyl) benzene; PPO = (2,5-diphenyloxazole).

2. Apparatus

a. Liquid scintillation spectrometer, concidence-type.

b. Liquid scintillation vial: 20-mL; polyethylene, low-K glass, or equivalent bottles.

c. Distillation apparatus: 250-mL round-bottom distillation flask, connecting side-arm adapter, condenser, and heating mantle.

3. Reagents

a. Scintillation solution: Thoroughly mix 4 g PPO, 0.05 g POPOP, and 120 g solid naphthalene in 1 L spectroquality 1,4-dioxane. Store in dark bottle. Solution is stable for 2 months. Alternatively, use a commercially prepared scintillation solution available from suppliers of liquid scintillation materials.

b. Low-background water: Use water with no detectable tritium activity (most deep well waters are low in tritium).

c. Standard tritium solution: Dilute available tritium standard solution to approximately 1,000 dpm/mL with low-background water.

d. Sodium hydroxide, NaOH, pellets.

e. Potassium permanganate, $KMnO_4$.

4. Procedure

Add three pellets NaOH and 0.1 g $KMnO_4$ to 100 mL sample in 250-mL distillation flask. Distill at 100 to 105 C, discard first 10 mL distillate, and collect next 50 mL. Thoroughly mix 4 mL distillate with 16 mL scintillation solution in tightly capped vial.

Prepare low-background water and standard tritium solution in same manner as samples.

Hold samples, background, and standards in the dark for 3 hr. Count samples containing less than 200 pCi/mL for 100 min and samples containing more than 200 pCi/mL for 50 min.

5. Calculations and Reporting

a. Calculate and report tritium, ^3H, in picocuries per milliliter (pCi/mL) or its equivalent, nanocuries per liter (nCi/L) as follows:

$$^3H = \frac{(C - B)}{(E \times 4 \times 2.22)}$$

where:

C = gross counting rate for sample, cpm,
B = background counting rate, cpm,
E = counting efficiency, $(S - B)/D$,
S = gross counting rate for standard solution, cpm, and
D = tritium activity in standard sample, dpm, corrected for decay to time of counting.

b. Calculate the counting error at the 95% confidence level based on the equation for $\sigma(R)$ given in Section 701F. A total count of 40,000 within 1 hr for a background count rate of about 50 cpm gives a counting error slightly in excess of 1% at the 95% confidence level.

6. Precision and Accuracy

Samples with tritium activity above 200 pCi/mL can be analyzed with precision of less than ±6% at the 95% confidence level and those with 1 pCi/mL can be analyzed with a precision of less than ±10%.

7. Bibliography

LIBBY, W.F. 1946. Atmospheric helium-3 and radiocarbon from cosmic radiation. *Phys. Rev.* 69:671.

NATIONAL COUNCIL ON RADIATION PROTECTION, SUBCOMMITTEE ON PERMISSIBLE INTERNAL DOSE. 1959. Maximum Permissible Body Burdens and Maximum Permissible Concentrations of Radionuclides in Air and in Water for Occupational Exposure. NBS Handbook 69 (June). National Bureau of Standards, Washington, D.C.

INTERNATIONAL COMMISSION ON RADIATION PROTECTION. 1960. Report of Committee II on permissible dose for internal radiation, 1959. *Health Phys.* 3:41.

BUTLER, F.E. 1961. Determination of tritium in water and urine. *Anal. Chem.* 33:409.

FAO, IAEA & WHO. 1966. Methods of Radiochemical Analysis. World Health Organization, Geneva.

SMITH, J.M. 1967. The Significance of Tritium in Water Reactors. General Electric Co., San Jose, Calif.

YOUDEN, W.J. 1967. Statistical Techniques for Collaborative Tests. Ass. Official Analytical Chemists, Washington, D.C.

PETERSON, H.T.J., J.E. MARTIN, C.L. WEAVER & E.D. HARWARD. 1969. Environmental tritium contamination from increasing utilization of nuclear energy sources. Seminar on Agricultural and Public Health Aspects of Environmental Contamination by Radioactive Materials, International Atomic Energy Ass., Vienna, pp. 35–60.

SODD, V.J. & K.L. SCHOLZ. 1969. Analysis of tritium in water; a collaborative study. *J. Ass. Offic. Anal. Chem.* 52:1.

WEAVER, C.L., E.D. HARWARD & H.T. PETERSON. 1969. Tritium in the environment from nuclear power plants. *Pub. Health Rep.* 84, 363.

ENVIRONMENTAL PROTECTION AGENCY. 1975. Tentative Reference Method for Measurement of Tritium in Environmental Waters. EPA 600/4-75-013, Environmental Monitoring and Support Laboratory, Las Vegas, Nev.

709 RADIOACTIVE CESIUM

Radioactive cesium has been considered one of the more hazardous radioactive nuclides produced in nuclear fission. Upon ingestion, like potassium, cesium distributes itself throughout the soft tissue and has a relatively short residence time in the body. Half-lives of ^{134}Cs and ^{137}Cs are 2 and 30 yr, respectively, both being beta- and gamma-emitters. The EPA Interim Primary Drinking Water Regulations' limit of 4 mrem/yr is equivalent to 80 and 200 pCi/L, respectively; the recommended detection limit for ^{134}Cs is 10 pCi/L.

1. General Discussion

a. Principle: If the activity of cesium is high, radioactive cesium can be determined directly by gamma-counting a large liquid sample (4 L) or the sample can be evaporated to dryness and counted. For lower-level environmental samples, add cesium carrier to an acidified sample and collect the cesium as phosphomolybdate. This is purified and precipitated as Cs_2PtCl_6 for counting. If total radiocesium determined by beta-counting exceeds 30 pCi/L, ^{134}Cs and ^{137}Cs must be determined by gamma spectrometry.

2. Apparatus

a. Magnetic stirrer with TFE-coated magnet bar.

b. Centrifuge: bench size clinical, and centrifuge tubes.

c. Filter papers and glass fiber filter* (2.4 cm diam).

d. pH paper: Wide range, 1 to 11 pH.

e. Filtering apparatus: See Section 704.2*c*.

f. Counting instruments: Use either a low-background beta counter (see Section 701C.4) or a gamma spectrometer (see Section 701C.5).

3. Reagents

a. Ammonium phosphomolybdate reagent, $H_{12}Mo_{12}N_3O_{40}P$: Dissolve 100 g molybdic acid (85% MoO_3) in a mixture of 240 mL distilled water and 140 mL conc ammonium hydroxide (NH_4OH). When solution is complete, filter and add 60 mL conc nitric acid (HNO_3). Separately mix 400 mL conc HNO_3 and 960 mL distilled water. After both solutions cool to room temperature, add, with constant stirring, the $(NH_4)_6Mo_7O_{24}$ solution to the HNO_3 solution. Let stand for 24 hr. Filter† and discard insoluble material.

Collect filtrate in a 3-L beaker and heat to 50 to 55 C (never above 55 C). Remove from heating unit. Add 25 g sodium dihydrogen phosphate (NaH_2PO_4) dissolved in 100 mL distilled water, stir occasionally for 15 min, and let settle (approximately 30 min). Filter† and wash precipitate with 1% potassium nitrate (KNO_3) and finally with distilled water. Dry precipitate and paper at 100 C for 3 to 4 hr. Transfer solid $(NH_4)_3PMo_{12}O_{40}$ to a weighing bottle and store in a desiccator.

b. Chloroplatinic acid, 0.1*M*: Dissolve 51.8 g $H_2PtCl_6 \cdot 6H_2O$ in distilled water and dilute to 1,000 mL.

c. Cesium carrier: Dissolve 1.267 g cesium chloride (CsCl) in distilled water and dilute to 100 mL; 1 mL = 10 mg Cs.

d. Calcium chloride, 3M: Dissolve 330 g $CaCl_2$ in distilled water and dilute to 1,000 mL.

e. Ethanol, 95%.

f. Hydrochloric acid, HCl, conc, 6*N*, 1*N*.

g. Sodium hydroxide, NaOH, 6*N*.

4. Procedure

a. To a 1-L sample, add 1.0 mL cesium carrier and enough 12*N* HCl to make the solution about 0.1*N* HCl (about 8.6 mL). Slowly add 1 g $(NH_4)_3PMo_{12}O_{40}$ and stir for 30 min using a magnetic stirrer at 800 rpm. Let precipitate settle for at least 4 hr and discard supernatant by decanting or using suction (provided by an inverted glass funnel connected to a vacuum source). Using a stream of 1*N* HCl, quantitatively transfer precipitate to a centrifuge tube. Centrifuge and discard super-

*Whatman No. 41, 9 cm diam; Whatman No. 42, 2.4 cm diam; or equivalents.

†Whatman No. 42 filter paper or equivalent.

natant. Wash precipitate with 20 mL 1*N* HCl and discard wash solution.

b. Dissolve precipitate by dropwise addition of 3 to 5 mL 6*N* NaOH. Heat over a flame for several minutes to remove ammonium ions. (Moist pH paper turns green as long as NH_3 vapors are evolved.) Dilute to 20 mL with distilled water. Add 10 mL 3*M* $CaCl_2$ and adjust to pH 7 with 6*N* HCl to precipitate $CaMoO_4$. Stir, centrifuge, and filter‡ supernatant into a 50-mL centrifuge tube. Wash precipitate remaining in the original centrifuge tube with 10 mL distilled water, filter through the same filter paper, and combine the wash with filtrate. Discard precipitate and filter paper.

c. Add 2 mL 0.1*M* H_2PtCl_6 and 5 mL ethanol. Cool and stir in ice bath for 10 min. Using distilled water transfer to a tared glass-fiber filter. Wash with successive portions of distilled water, 1*N* HCl, and ethanol.

d. Dry at 110 C for 30 min, cool, weigh, mount on a nylon disk and ring with polyester plastic§ cover and beta-count or gamma-scan for ^{134}Cs and ^{137}Cs.

‡Whatman No. 41 filter paper or equivalent.
§Mylar or equivalent.

5. Calculation

Calculate the concentration of radiocesium as follows:

$$\text{Cs, pCi/L} = \frac{C}{2.22 \times E\ V\ R}$$

where:

C = net count rate, cpm,
E = counter efficiency,
V = volume of sample, L, and
R = fractional chemical yield

$$= \frac{\text{recovered } Cs_2PtCl_6\text{, mg} \times 0.3945}{\text{added Cs carrier, mg}}$$

6. Bibliography

FINSTON, H.L. & M.T. KINSLEY. 1961. The Radiochemistry of Cesium. AEC Rep. NAS-NS-3035.

ENVIRONMENTAL PROTECTION AGENCY. 1976. Interim Radiochemical Methodology for Drinking Water. Environmental Monitoring and Support Laboratory, Cincinnati, Ohio. EPA-600/4-75-008 (revised).

710 RADIOACTIVE IODINE

Radioiodine that results from testing nuclear devices or is released during use and processing of reactor fuels is a major concern in radioactivity monitoring. Fission products may contain iodine 129 through iodine 135. Iodine 129 has a half-life of 1.6×10^7 yr but a relatively low specific activity (1.73×10^{-4} pCi/g for ^{129}I as compared to 1.24×10^5 pCi/g for ^{131}I). The half-life of ^{131}I is 8 days while for the other isotopes it is shorter (35 min to 21 hr). At present, only ^{131}I is likely to be found in water. When ingested or inhaled, it concentrates in the thyroid gland and may cause thyroid cancer.

The EPA drinking water maximum contaminant level for ^{131}I is 3 pCi/L.

Selection of method: Of the three methods, the precipitation method (A) is preferred because it is simple and involves the least time. Method B, in which iodide is concentrated by absorption on an anion resin, purified, and counted in a beta-gamma coincidence system, is sensitive and accurate. Method C uses distillation. With each method it is possible to reach the EPA recommended detection limit of 1 pCi^{131}I/L.

710 A. Precipitation Method

1. General Discussion

Principle: Iodate carrier is added to an acidified sample and, after reduction with Na_2SO_3 to iodide, the ^{131}I is precipitated with $AgNO_3$. The precipitate is dissolved and purified with zinc powder and H_2SO_4 and the solution is reprecipitated as PdI_2 for counting.

2. Apparatus

a. Counting instrument: Low-background beta counter (see Section 701C.4) or beta-gamma coincidence counter (see Section 701C.9).

b. Fine-fritted glass funnel.

c. Filter apparatus: Two-piece filter funnel with filtering equipment.*

d. Filter materials: Filter paper†; glass-fiber filter, 2.4 cm diam; or 0.8-μm pore-diam membrane filter, 4.7 cm diam.

3. Reagents

a. Ammonium hydroxide, NH_4OH, 6*N*.

b. Ethanol, 95%.

c. Hydrochloric acid, HCl, 6*N*.

d. Iodate carrier: Dissolve 1.685 g KIO_3 in distilled water and dilute to 100 mL. Store in dark flask; 1 mL = 10 mg I.

e. Nitric acid, HNO_3, conc.

f. Palladium chloride, $PdCl_2$: Dissolve 3.3 g $PdCl_2$ in 100 mL 6*N* HCl; 1 mL = 20 mg Pd.

g. Silver nitrate, $AgNO_3$, 0.1*N*: Dissolve 17 g $AgNO_3$ in distilled water and dilute to 1,000 mL. Store in dark flask.

h. Sodium sulfite, Na_2SO_3, 1*M* (freshly prepared): Dissolve 6.3 g Na_2SO_3 in distilled water and dilute to 50 mL.

i. Sulfuric acid, H_2SO_4, 2*N*.

j. Zinc, powder, reagent grade.

*Fisher Filtrator or equivalent.

†Whatman No. 42 or equivalent.

4. Procedure

a. To a 2,000-mL sample, add 15 mL conc HNO_3 and 1.0 mL iodate carrier. Mix well. Add 4 mL freshly prepared 1*M* Na_2SO_3 and stir for 30 min. Add 20 mL 0.1*M* $AgNO_3$, stir for 1 hr, and let settle for 1 hr. Decant and discard as much of the supernatant as possible. Filter remainder through a glass-fiber filter and discard filtrate.

b. Transfer filter to a centrifuge tube and slurry with 10 mL distilled water. Add 1 g zinc powder and 2 mL 2*N* H_2SO_4 and stir frequently for at least 30 min. Filter, with vacuum, through a fine-fritted glass funnel and collect filtrate in an erlenmeyer flask. Wash both residue and filter with a minimum quantity of distilled water and add wash water to filtrate. Discard residue.

c. Add 2 mL 6*N* HCl and heat in water bath at 80 C for 10 min. Add 1 mL 0.2*M* $PdCl_2$ and digest for at least 5 min. Centrifuge and discard supernatant.

d. Dissolve precipitate in 5 mL 6*N* NH_4OH and heat in boiling water bath for 5 min. Filter through a glass-fiber filter and collect filtrate in a centrifuge tube. Discard filter and residue.

e. Neutralize filtrate with 6*N* HCl, add 2 mL in excess, and heat in a water bath. Add 1 mL 0.2*M* $PdCl_2$ to reprecipitate PdI_2 and digest for 10 min. Cool slightly and transfer to a tared filter with distilled water. Wash successively with 5-mL portions of distilled water and 95% ethanol. Dry in a vacuum oven at 60 C for 1 hr, weigh precipitate, mount, and beta-count.

f. If final PdI_2 precipitate on a glass-fiber filter is counted in a low-background beta counter, the background counting rate is relatively high (about 1.3 cpm). If precipitate is collected on a 0.8-μm membrane filter and dried for 30 min at 70 C it may be counted in a beta-gamma coincidence scintillation system with a background rate of less than 0.1 cpm.

If a low-background counter is used, confirm identity of ^{131}I by recounting precipitate after about 1 wk to check the half-life.

5. Calculation

Calculate concentration of radioiodine as follows:

$$^{131}\text{I, pCi/L} = \frac{C}{2.22 \times EVR \times A}$$

where:

C = net count rate, cpm,
E = counting efficiency of ^{131}I as function of mass of PdI_2 precipitate,
V = volume of sample, L,
R = fractional chemical yield
$= \frac{\text{recovered } PdI_2 \times 0.0704}{\text{added iodine carrier}}$, and
A = ^{131}I decay factor for the time interval between sample collection and measurement.

710 B. Ion Exchange Method

1. General Discussion

Principle: A known amount of inactive iodine in the form of KI is added as a carrier and the sample is taken through an oxidation-reduction step using hydroxylamine and sodium bisulfite to convert all iodine to iodide. Iodine, as the iodide, is concentrated by absorption on an anion exchange column. Following an NaCl wash, iodine is eluted with sodium hypochlorite. Iodine in the iodate form is reduced to I_2, extracted into CCl_4, and back-extracted as iodide into water. The iodine finally is precipitated as PdI_2.

2. Apparatus

a. Counting instrument: Low-background beta counter (see Section 701C.4) or beta-gamma coincidence counter (see Section 701C.9).

b. Chromatographic column, 2 cm × 15 cm.

c. Vacuum filter holder, 2.5 cm^2 filter area.

*d. Filter paper,** 2.4 cm diam.

e. Vacuum oven.

3. Reagents

a. Iodine carrier: Weigh approximately 13 g dried KI to the nearest 0.1 mg. Dissolve in a 1,000-mL volumetric flask containing 100 mL distilled water. Add 10 mL 1*M* $NaHSO_3$ and dilute to mark with distilled water. Concentration of carrier I, mg/L = g KI × 0.7644.

b. Ethanol, absolute.

c. Hydroxylamine hydrochloride, 1*M*: Dissolve 6.95 g $NH_2OH{\cdot}HCl$ in distilled water and dilute to 100 mL.

d. Nitric acid, HNO_3, conc, 8*N*, 1.6*N*.

e. Sodium bisulfite, 1*M*: Dissolve 1.04 g $NaHSO_3$ in distilled water and dilute to 10 mL.

f. Sodium hydroxide, 12*N*: Dissolve 480 g NaOH in distilled water and dilute to 1 L.

g. Sodium hypochlorite, NaOCl, 5%: Use available household bleach.

h. Anion exchange resin.†

i. Carbon tetrachloride, CCl_4, reagent grade.

j. Hydrochloric acid, HCl, 3*N*, 1*N*.

k. Palladium chloride: Dissolve 3.3 g

*Whatman No. 42 or equivalent.

†Dowex 1 × 8, 50-100 mesh, chloride form, or equivalent.

$PdCl_2$ in 100 mL 6*N* HCl; 1 mL = 20 mg Pd.

l. Sodium chloride, NaCl, 2*N*. Dissolve 117 g NaCl in distilled water and dilute to 1 L.

m. Hydroxylamine hydrochloride wash solution: Add 20 mL conc HNO_3 and 20 mL 1*M* $NH_2OH{\cdot}HCl$ to 100 mL distilled water.

4. Procedure

a. To 1 L sample in a beaker add, while stirring, 2.0 mL iodine carrier and 5 mL 5% NaOCl, and heat for 2 to 3 min to complete oxidation. After the interchange reaction (2 to 3 min), slowly add 5 mL conc HNO_3. Add 25 mL 1*M* $NH_2OH{\cdot}HCl$ and stir. Let reaction go on for a few seconds, add 10 mL 1*M* $NaHSO_3$, and adjust pH to 6.5 with 12*N* NaOH or 1.6*N* HNO_3. Stir thoroughly for a few minutes. (Stir samples containing a large amount of organic material, such as muddy water, for 45 min.) Filter through a glass-fiber filter to remove suspended matter. Discard residue.

b. Pour 20 mL anion exchange resin into a column and wash sides down with distilled water. Pass sample through ion exchange column at a flow rate of 20 mL/min. Discard effluent. Wash column with 200 mL distilled water and then with 100 mL 2*N* NaCl at a flow rate of 4 mL/min. Discard wash solutions.

c. Add 50 mL 5% NaOCl in 10- to 20-mL increments, stirring the resin as needed to eliminate gas bubbles, and maintain a flow rate of 2 mL/min. To the eluted volume of 50 to 60 mL, collected in a beaker, carefully add 10 mL conc HNO_3 to make sample 2 to 3*N* in HNO_3 and transfer to a separatory funnel. (Add acid slowly with stirring until vigorous reaction subsides.)

d. Add 50 mL CCl_4 and 10 mL 1*M* $NH_2OH{\cdot}HCl$. Extract iodine into organic phase by shaking for about 2 min. Let phases separate and transfer organic phase to another separatory funnel. Add 25 mL CCl_4 and 5 mL 1*M* $NH_2OH{\cdot}HCl$ to the first separatory funnel and shake for 2 min. Combine organic phase with the one obtained from the first extraction. Discard aqueous phase. Add 20 mL $NH_2OH{\cdot}HCl$ wash solution to the organic phase and shake for 2 min. Let phases separate and transfer organic phase to a clean separatory funnel. Discard wash solution.

e. Add 25 mL distilled water and 10 drops 1*M* $NaHSO_3$ to organic phase. Shake for 2 min, let phases separate, and discard organic phase. Transfer aqueous phase to a beaker. Add 10 mL 3*N* HCl. Using a stirrer-hot plate, boil and stir the sample until it evaporates to 10 to 15 mL or begins to turn yellow.

f. Add 1.0 mL $PdCl_2$ solution dropwise. Rinse sides of beaker with 1*N* HCl and add sufficient 1*N* HCl to make a volume of 30 mL. Continue stirring until cool. Place beaker in a stainless steel tray and store at about 4 C overnight.

g. Filter through a tared filter mounted in a filter holder. Wash residue with 1*N* HCl and then with absolute alcohol. Dry in a vacuum oven at 60 C for 1 hr. Cool in a desiccator, weigh precipitate, then seal it between polyester tape and polyester plastic film,‡ with the film over the precipitate. Count with a beta-gamma coincidence system.

5. Calculation

Calculate ^{131}I, pCi/L, as in Section 710A.5.

‡Mylar or equivalent.

710 C. Distillation Method

1. General Discussion

Principle: Iodine carrier is added to an acidified sample and iodine is distilled into a caustic solution. The distillate is acidified and the iodine is extracted into CCl_4. After back-extraction as iodide, the iodine is purified as PdI_2 for counting.

2. Apparatus

a. Distillation apparatus and 3-L round-bottom flask.

b. Separatory funnel, 60 mL.

c. Filter apparatus: Two-piece filter funnel with filtering equipment.*

d. Filter paper: See Section 710A.2*d*.

3. Reagents

a. Ammonium hydroxide, NH_4OH, conc.

b. Carbon tetrachloride, CCl_4.

c. Ethanol, 95%.

d. Hydrochloric acid, HCl, 6*N*, 1*N*.

e. Iodide carrier: Dissolve 2.616 g KI in distilled water, add 2 drops $NaHSO_3$, and dilute to 100 mL. Store in dark flask. 1 mL = 20 mg I.

f. Nitric acid, HNO_3, conc.

g. Palladium chloride: Dissolve 3.3 g $PdCl_2$ in 100 mL 6*N* HCl; 1 mL = 20 mg Pd.

h. Sodium bisulfite, $NaHSO_3$, 1*M*: Dissolve 5.2 g $NaHSO_3$ in distilled water and dilute to 50 mL. Prepare only in small quantities.

i. Sodium hydroxide, NaOH, 0.5*N*.

j. Sodium nitrite, $NaNO_2$, 1*M*: Dissolve 69 g $NaNO_2$ in distilled water and dilute to 1 L.

k. Sulfuric acid, H_2SO_4, 12*N*.

l. Tartaric acid, $C_4H_6O_6$, 50%: Dissolve 50 g $C_4H_6O_6$ in distilled water and dilute to 100 mL.

*Fisher Filtrator or equivalent.

4. Procedure

a. To a 2,000-mL sample in a 3-L round-bottom flask, add 15 mL 50% $C_4H_6O_6$ and 1.0 mL iodide carrier. Mix well, cautiously add 25 mL cold conc HNO_3, and close distillation apparatus (Figure 710:1).

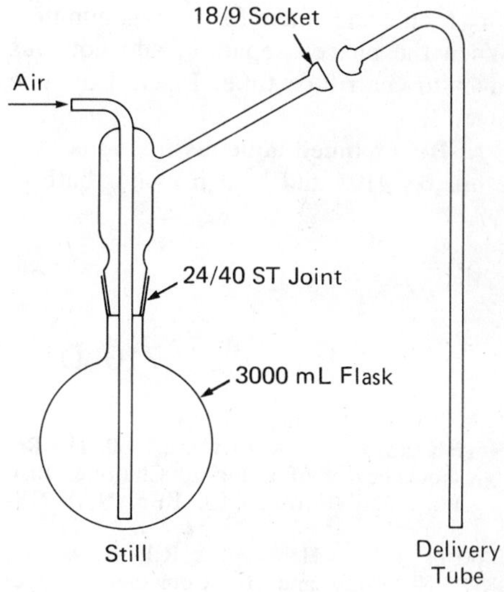

Figure 710:1. Distillation apparatus for iodine analysis (not to scale).

b. Connect an air line to still inlet, adjust flow rate to about 2 bubbles/sec, and distill for at least 15 min into 15 mL 0.5*N* NaOH. Cool and transfer NaOH solution to a 60-mL separatory funnel. Discard still residue.

c. Adjust distillate to slightly acid with 1 mL 12*N* H_2SO_4 and oxidize with 1 mL 1*M* $NaNO_2$. Add 10 mL CCl_4 and shake for 1 to 2 min. Transfer organic layer to a clean 60-mL separatory funnel containing 2 mL 1*M* $NaHSO_3$.

d. Add 5 mL CCl_4 and 1 mL 1*M* $NaNO_2$ to original separatory funnel containing the aqueous layer and shake for 2 min.

Combine organic fractions. Repeat and discard aqueous layer.

e. Shake separatory funnel thoroughly until CCl_4 layer is decolorized; let phases separate and transfer aqueous layer to a centrifuge tube. Add 2 mL 1*M* $NaHSO_3$ to the separatory funnel containing CCl_4 and shake for several minutes. When phases separate, add aqueous layer to centrifuge tube. Add 1 mL distilled water to separatory funnel and shake for several minutes. When the phases separate, add aqueous layer to centrifuge tube. Discard organic layer.

f. To combined aqueous fractions, add 2 mL 6*N* HCl and heat in water bath at 80 C for 10 min. Add 1.0 mL $PdCl_2$ solution dropwise, with stirring, and digest for 15 min.

g. Cool, stir precipitate, and transfer to a tared filter mounted in a two-piece funnel. Let precipitate settle by gravity for uniform deposition, then apply suction. Wash residue with 10 mL 1*N* HCl, 10 mL distilled water, and then with 10 mL 95% ethanol. Dry in a vacuum oven at 60 C for 1 hr. Cool in desiccator, weigh, mount, and make beta count.

5. Calculation

Calculate the concentration of radioiodine as given in Section 710A.5.

710 D. Bibliography

KLEINBERG, J. & G.A. COWAN. 1960. The Radiochemistry of Fluorine, Chlorine, Bromine and Iodine. AEC Rep. NAS-NS-3005.

BRAUER, F.P., J.H. KAYE & R.E. CONNALY. 1970. X-ray and β-γ Coincidence Spectrometry Applied to Radiochemical Analysis of Environmental Samples. Advances in Chemistry Ser., No. 93, Radionuclides in the Environment, pp. 231–253. American Chemical Society.

AMERICAN SOCIETY FOR TESTING AND MATERIALS. 1972 Book of ASTM Standards. Part 23. ASTM, Philadelphia, Pa., D 2334-68.

GABAY, J.J., C.J. PAPERIELLO, S. GOODYEAR, J.C. DALY & J.M. MATUSZEK. 1974. A method of determining ^{129}I in milk and water. *Health Phys.* 26:89.

ENVIRONMENTAL PROTECTION AGENCY. 1975. Interim Radiochemical Methodology for Drinking Water. EPA-600/4-75-008, Environmental Monitoring and Support Laboratory, Cincinnati, Ohio.

PART 800

BIOASSAY METHODS FOR AQUATIC ORGANISMS

801 INTRODUCTION

Bioassays are necessary in water pollution evaluations because chemical and physical tests alone are not sufficient to assess potential effects on aquatic biota.[1,2] For example, the interaction of chemical factors and the toxic effects of complex matrices cannot be determined. Different kinds of aquatic organisms are not equally susceptible to the same toxic substances nor are organisms equally susceptible throughout the life cycle. Even previous exposure to toxicants can alter susceptibility.

The procedures given below measure biological response to known and unknown concentrations of materials in both fresh and saline waters. These bioassay tests are applicable to routine monitoring requirements as well as research needs. Refer to Part 900 for microbiological methods and Part 1000 for field and other types of biological laboratory methods for water quality evaluations. Refer to Section 1007 for identification aids for aquatic organisms.

The bioassay is useful for a variety of purposes that include determining: (*a*) suitability of environmental conditions for aquatic life, (*b*) favorable and unfavorable concentrations of environmental factors, such as DO, pH, temperature, salinity, or turbidity, for aquatic life, (*c*) effect of environmental factors on waste toxicity, (*d*) toxicity of wastes to a test species, (*e*) relative sensitivity of aquatic organisms to an effluent or toxicant, (*f*) amount of waste treatment needed to meet water pollution control requirements, (*g*) effectiveness of waste treatment methods, (*h*) permissible effluent discharge rates, and (*i*) compliance with water quality standards, effluent requirements, and discharge permits.

Reasonable uniformity of procedures and of data presentation is essential. The use of standardized methods described below will ensure adequate uniformity, reproducibility, and general usefulness of results without interfering unduly with the adaptability of the tests to local circumstances.

There is still much "art" in determining biological response. Thus there is a need to use correct and uniform tests and terminology (see Section 801A, Terminology), to apply relevant tests for meeting legal, effluent testing, monitoring, and research requirements, and to meet the uniqueness of a given environment.[3–14]

801 A. Terminology

An aquatic bioassay is a procedure in which the responses of aquatic organisms are used to detect or measure the presence or effect of one or more substances, wastes, or environmental factors, alone or in combination. This section explains bioassay conditions and analysis and use of test results.

1. General Terms

Acclimate—to accustom test organisms to different environmental conditions, such as temperature, light, and water quality.[13]

Response—the measured biological effect of the material tested. In acute toxicity tests this usually is death. In biostimulation tests it is biomass increase.

Control—test organisms in test chamber under test conditions exposed to dilution water alone and/or the natural water to which they are normally exposed.

2. Toxicity Terms

Dose—amount of toxicant that enters the organism.[12]

Toxicity—adverse effect to a test organism caused by "pollutants", generally a poison or mixture of poisons. Toxicity is a resultant of concentration and time, modified by variables such as temperature and chemical form and availability.

Exposure time—time of exposure of test organism to test solution.

Acute toxicity—a relatively short-term lethal or other effect, usually defined as occurring within 4 days for fish and macroinvertebrates and shorter times for smaller organisms.

Chronic toxicity—long-term effects that may be related to changes in appetite, growth, metabolism, reproduction, and even death or mutations.[13]

Lethal concentration (LC)—toxicant concentration producing death of test organism. Usually defined as median (50%) lethal concentration, LC50, i.e., concentration killing 50% of exposed organisms at a specific time of observation, for example, 96-hr LC50.[12]

Effective concentration (EC)—toxicant concentration affecting a specific response, e.g., respiration rate, loss of equilibrium, in a given time, for example, 96-hr EC50.[12]

Asymptotic LC50—toxicant concentration at which LC50 becomes a constant for a prolonged exposure time.

Median tolerance limit (TL_m)—test material concentration at which 50% of test organisms survive for a specified exposure time. This term has been superseded by median lethal concentration (LC50).[11]

3. Biostimulation Terms

Limiting nutrient—the nutrient most needed for growth in relation to the quantities of other nutrients.

Nutrient—a specific substance required for organism growth.

Specific growth rate—unit rate of mass change for a population of organisms.

Maximum standing crop—maximum dry weight of organisms during a test.

4. Flow Terms

Static bioassay—test in which solutions and test organisms are placed in test chambers and kept there for the duration of the test.[12]

Recirculation bioassay—static test with circulation of test solution through test chamber. Test solution may be treated by aeration, filtration, sterilization, etc., to maintain water quality.[12]

Renewal bioassay—static test with periodic exposure (usually at 24-hr intervals) of test organisms to fresh test solution of the same composition. This is accomplished by transferring test organisms or replacing test solution.[12]

Flow-through bioassay—test in which solution is replaced continuously in test chambers for the test duration.[12]

Mixing—agitation performed by mechanical stirrers, pumps (air, water), or inflow currents. Note that aeration or vigorous mixing may increase losses of volatile substances.

5. Evaluation of Results

Maximum allowable toxicant concentration (MATC)—toxicant concentration that may be present in a receiving water without causing significant harm to productivity or other uses. MATC is determined by long-term tests of either partial life cycle with sensitive life stages or a full life cycle of the test organism.

Application factor—a factor applied to acute toxicity tests to estimate toxicant concentration that is safe for chronic or lifetime exposure of test organism.[11] (However, a recent review[15] questions this concept and considers it invalid.)

$$\text{Application factor (AF)} = \frac{\text{MATC}}{\text{96-hr LC50}}$$

801 B. Basic Requirements for Bioassays

The basic requirements and desirable conditions for bioassay tests are: (*a*) an abundant supply of water of desired quality, (*b*) an adequate and effective flowing water system constructed of nonpolluting materials, (*c*) adequate space and well-planned holding, culturing, and testing equipment and facilities, and (*d*) an adequate source of healthy experimental organisms. Much valuable information and advice are available for planning and constructing water supply systems.[16–20] Make a complete chemical analysis of water used in bioassay testing (see 801D.4*b*).

The facilities, equipment, and water supplies needed for effective tests depend on the type of tests and their objectives. For effluent and monitoring compliance tests, take dilution water outside the zone of influence of the waste. Use a water supply free from pollution and choose facilities and equipment as indicated above for the following tests: (*a*) determination of most sensitive species and life stage, (*b*) effect of different toxicants, (*c*) effects of water quality and environmental factors alone and in combination with toxicants, and (*d*) maximum concentration of waste that does not taint the flesh of edible organisms.

801 C. Conducting Bioassays

1. Types of Bioassays: Their Uses, Advantages, and Disadvantages

Bioassays are classified according to (*a*) duration—short-term, intermediate, and/or long-term, (*b*) method of adding test solutions—static, recirculation, renewal, or flow-through, and (*c*) purpose—effluent quality monitoring, relative toxicity, relative sensitivity, taste or odor, or growth rate, etc. Use short-term tests for routine monitoring suitable for effluent discharge permit requirements and for exploratory tests.

Short-term definitive tests determine LC50 or EC50. These tests also indicate toxicant concentrations to be used in intermediate and long-term tests.

Use intermediate tests when LC50 determination requires additional time, for studies of life stages of long-life-cycle organisms, and to indicate toxicant concentrations for life-cycle tests.

Use long-term tests for estimating MATC. Long-term tests almost always are flow-through tests.

Short-term tests are valuable for quickly

supplying an estimate of toxicity, for assessing relative toxicity of different toxicants or wastes to a selected test organism, or relative sensitivity of different organisms to different conditions of such variables as temperature and pH. They also indicate the maximum allowable concentrations for very short exposures, such as those that might occur to organisms passing through a thermal electric power plant or a zone of heated water.

Do not use static tests for high-BOD wastes because DO depletion may stress test organisms. Volatile or unstable toxicants decrease in concentration during the test, so the exposure of test organisms becomes progressively less. Metabolic products may build up and undesirably high concentrations of CO_2 or NH_3 may occur. Toxicant concentration may be reduced by sorption on sediments and test chamber walls or by combination with the mucus or metabolic products of the test organisms and in their bodies.

Flow-through tests are desirable for high-BOD samples and for those that contain unstable or volatile substances. Organisms with high metabolic rates are difficult to maintain in standing water, whereas flow-through tests provide well-oxygenated test solutions, nonfluctuating toxicant concentrations, and continuous removal of metabolic wastes. Use flow-through tests whenever there is evidence or expectation of rapid toxicity changes of the test solution. Such a change is indicated when the survival time of test animals in a fresh solution is significantly shorter than the survival time in a corresponding 2-day-old solution, provided that adequate DO is present throughout both tests. Flow-through tests of industrial effluents and chemicals that are removed appreciably from solution by precipitation, by test organisms, or by other means are preferable to static tests and their modifications.

The LC50 values may be useful measures of acute toxicity but they *do not* represent concentrations that are safe or harmless in aquatic habitats subject to pollution. Concentrations of wastes that are not demonstrably toxic in 96 hr may be very toxic under conditions of continuous exposure in a receiving water. Thus the 96-hr LC50 may represent only a small fraction of long-term toxicity. When estimating safe discharge rates or dilution ratios for effluents or other pollutants on the basis of acute toxicity evaluations, use AF's determined by life-cycle tests. Even the provision of an apparently ample margin of safety can fail to accomplish its purpose when there is cumulative toxicity that cannot be predicted from acute toxicity results.

No single, simple AF is valid for all wastes or toxicants. The constituents of a complex waste responsible for acute toxicity may be, but are not necessarily, the constituents responsible for chronic or cumulative toxicity demonstrable in diluted waste that is no longer acutely toxic. The chronic toxicity may be lethal after a long time or it may cause only nonlethal impairment of function. Knowledge of acute toxicity of a waste often can be very helpful in predicting and preventing acute damage to aquatic life in receiving waters as well as in regulating toxic waste discharges.

2. Short-Term Bioassays

a. Range-finding tests: For effluents or materials of unknown toxicity conduct short-term (usually 24-hr), small-scale range-finding or exploratory tests to determine approximate concentration range to be covered in full-scale short-term tests. For effluents with low or slow-acting toxicity, 48- or 96-hr tests may be necessary. Expose test organisms to a wide range of concentrations, usually in a logarithmic ratio, such as 0.01, 0.1, 1, 10, and 100%. Attempt to include concentrations that kill all organisms and others that kill very few or no organisms. For short-term, definitive tests, select a geometrically-spaced series of concentrations between the highest concentration that killed no, or only a few, test organisms and the lowest concentration that killed most or all test organisms.

Prepare test concentrations as described in Section 801E.2*b*.

b. Short-term definitive tests: Because death is an important, easily detected adverse effect, the most commonly used tests are for acute lethality. These tests are most appropriate for routine monitoring and checking conformity with NPDES requirements.[13]

Short-term tests may be static, renewal, recirculation, or flow-through. Static tests often are used when the test organisms are phyto- or zooplankton because these organisms are easily washed out in flow-through tests. With these small organisms, do not renew the toxic solution before 96 hr unless there is an oxygen demand, in which case renew the test solution every 24 hr. Renewal tests are required most often with macroinvertebrates and fish. If the test material has high BOD, is volatile, or is relatively unstable, use the renewal or flow-through technic.

Test duration is determined by the toxicant and the test objectives and usually is the same for different groups of organisms. For short-life-cycle organisms such as phytoplankton, the usual exposure time can cover many generations. Determine test duration in part by the length of the life cycle. Generally, expose fish and large invertebrates in static tests for 4 days and in flow-through tests for 8 days. Expose *Daphnia* for 48 hr. Short-term tests have been limited arbitrarily to 96 hr, but longer tests sometimes are desirable because death does not occur always within the 96-hr period. When some test animals, though still alive, are dying or evidently affected after 96-hr exposure, prolong the test. If tests are continued for longer periods, feed the test organisms.

For feeding requirements see Sections 804 through 810. Record feeding and ensure that it is equivalent in each container.

Special tests may be conducted on altered or treated samples of effluent to obtain additional toxicity information. For example, effluent dilution water mixtures may be aged 24 to 48 hr before adding the test organisms, to determine changes in toxicity. When special tests are conducted, describe methodology in detail.

3. Intermediate-Term Bioassays

No sharp time separation exists between short- and intermediate- or between intermediate- and long-term tests. Usually tests lasting 8 days or less are considered short-term while intermediate tests last from 8 to 90 days.

Intermediate-length tests may be static, renewal, or flow-through, but flow-through tests are recommended for most situations. For conduct of tests see Section 801E.3*a*.

4. Long-Term, Partial- or Complete-Life-Cycle Bioassays

With few exceptions, mostly the short-life-cycle forms (phyto- and zooplankton), use flow-through bioassays with exposure extending over as much of the life cycle as possible. Continue tests from egg to egg or beyond, or for several life cycles for smaller forms. Determine the maximum concentrations of toxicant not producing harmful effects with continuous exposure. The overall objective of this type of test is to determine MATC's of effluents, toxicants, or wastes. Use life cycle tests to determine AF and the effects on growth, reproduction, development of sex products, maturation, spawning, success of spawning and hatching, survival of larvae or fry, growth and survival of different life stages, deformities, behavior, and bioaccumulation.

In life-cycle or partial-life-cycle tests, ensure that water quality factors such as temperature, pH, salinity, and DO follow the natural seasonal cycle unless the test objective is to study one of these factors. It may be essential that the natural annual cycle be duplicated if the development of sex products, spawning, and development of eggs and larvae are to be normal. Do not allow toxicant concentration to vary by more than ±10% from the selected concentration because of uptake by test organisms, absorption, precipitation, or other factors.

In these tests, select five or more concentrations on the basis of short- or intermediate-term tests and set up at least in duplicate. Vary exposure chambers, spawning chambers, hatching containers, growth chambers, and other equipment to meet the needs of the different organisms. (See Sections 802 through 810.) Other apparatus, water supplies, and analytical determinations are listed in Section 801D.

5. Special-Purpose Bioassays

a. Relative sensitivity to a toxicant: To rank the sensitivity of different species to a toxicant, use a standard water and standard exposure conditions. Select exposure conditions (e.g., temperature, DO, pH, CO_2, and salinity) in a favorable range for each species and keep conditions constant throughout the test.

b. Relative toxicity of various toxicants to selected species: These tests resemble sensitivity tests because the selected test conditions, dilution waters, and test species are kept constant and standard. Prevent any change in sensitivity of test organisms during the tests. If possible, select species from several different groups, an alga, microcrustacean, macrocrustacean, insect, mollusk, or fish sensitive to the toxicants being rated.

c. Flesh tainting tests: Use these tests to determine the maximum concentrations of wastes and materials that do not taint the flesh of edible aquatic organisms. Expose organisms that are large enough to supply portions for a taste panel. Set up exposure tanks as for other flow-through tests. Perform range-finding tests over a wide concentration range to determine the concentrations for a more definitive series of tests.

After exposure, prepare test organisms for taste testing. Clean, prepare for cooking, wrap in aluminum foil, and bake in an oven. When organisms are cooked, divide them into portions, wrap in aluminum foil, assign a code number, and distribute to a taste panel while still warm, along with samples of unexposed organisms similarly cooked, wrapped, and coded. Record the observations of the panel on a prepared form and determine the highest concentration of test material not causing detectable tainting. Several tests may be necessary.[21–24]

d. Growth-rate determinations: Growth rate is an important response of both algae and fish to toxicants and environmental factors. This section discusses the topic with respect to fish.[25] For a discussion related to algae see Section 802G.3*c*. Always report details of the method of feeding fish in growth studies. Three technics are available:

Unrestricted food supply—Provide attractive and palatable food (usually live food such as *Daphnia* or tubificid worms) uninterruptedly in greater quantities than fish can consume. Make a mass balance on food consumed by weighing food introduced and uneaten food removed.

Intermittent satiated food supply—Provide all the attractive food that fish can consume at time of feeding once or twice daily. After fish cease to feed, remove all uneaten food.

Uniformly restricted food supply—Once per day provide all fish with an amount of food that they will consume completely and without exception. Ideally, hold fish separately in individual aquariums or compartments. For fish held together feed so that all fish have an equal opportunity to consume food. Uniformity of temperature and DO helps to ensure equal feeding of a group of fish.

While growth studies usually have been conducted with unrestricted and intermittent satiated feeding technics, it is recommended that each study include at least one test series using uniformly restricted food supply. Only this technic can reveal whether growth rate differences are not due solely to the effect of toxicant on appetite or food consumption rate. The presence of an abundant food supply can obscure toxic effects. For example, fish exposed to toxicants such as cyanide or pentachlorophenol increase food consumption rate to compensate partially for loss of efficiency of food utilization caused by the toxicant. This may not be possible in natural conditions where food supply may be limited.

Ideally, include a series of tests with different, uniformly restricted food rations with the lowest ration near that which results in no growth (or loss of weight) in the controls. This is the maintenance level. Determine the effect of the variable under study at any level of food availability and consumption by relating observed growth rates to, for example, toxicant concentration, at each feeding level.[26]

Juvenile fish may gain enough weight in 2 to 3 wk to determine growth rate satisfactorily. Longer exposures with weighings at intervals of approximately 10 days are needed to determine long-term effects such as acclimation or accumulative toxicity.

Report results as growth rates computed as follows:

$$\text{Growth rate} = \frac{\text{weight gain, g}}{\text{time interval, day}} \times \frac{1}{\text{mean weight, g}}$$

where:

$$\text{mean weight} = (\text{weight at start of time interval, g} + \text{weight at end of time interval, g}) \div 2$$

Determine dry weight, wet weight, and fat content of fish at the beginning and end of a test. Weight gain due to increased fat content is not universally considered true growth; some investigators consider that true growth occurs only when there is an increase of protein. However, fat storage is important ecologically and bioenergetically because fat can be used as an energy source during periods of malnutrition and weight loss.[26]

801 D. Preparing Organisms for Bioassays

1. Selecting Test Organisms

The prime considerations in selecting test organisms are: their sensitivity to the factors under consideration; their geographical distribution, abundance, and availability within a practical size range throughout the year; their recreational, economic, and ecological importance; the availability of culture methods for rearing them in the laboratory and a knowledge of their requirements; and their general physical condition and freedom from parasites and disease. Few studies have been made to determine the important species most sensitive to a potential toxicant. To select a best species consider available information on sensitivity or determine sensitivity with short-term tests. Select the test species based on the considerations listed above as well as organism size and life-cycle length. Generally, organisms not over 5 to 8 cm long and having a short life cycle are most desirable, but some tests require larger organisms with long life cycles.

For studies to determine effluent requirements, use the most sensitive locally important species and the most sensitive life stage. When using another species, make comparative tests to relate sensitivity of the selected test species to the most sensitive locally important species. For each series of tests, collect organisms at one time from the same source. Use organisms that are nearly uniform in size, with the largest individual not more than 50% longer than the shortest. Use organisms of the same age group or life stage. Report time, place, and method of collection, transportation, and handling.

In designing a test, consider any unusual past conditions to which the organisms have been exposed (pesticides, effluents from industries, waste treatment plants, return flows, etc.). Synergistic effects of a new toxicant mixed with those presently being discharged to the receiving water may be important. Do not collect test organisms from polluted areas where they are in poor condition, diseased, parasitized, or deformed, or where they have unusually high body burdens of potential toxicants.

Knowledge of the environmental requirements and food habits is important in selecting test organisms. Because methods for laboratory holding and culturing throughout the life cycle are available for only a few organisms, it often is necessary to collect certain life stages of selected organisms from the field for testing.

2. Collecting Test Organisms

If laboratory-reared specimens are used, report original source and strain. Cultures of many algae species are maintained and can be obtained readily (see Section 802F.1).

Many smaller invertebrates and fish can be collected along the shore in dip nets, in coarse plankton nets, or by hand. Catch larger species that occur near shore in seines. Traps and fyke nets are good for freshwater areas but are selective for some species. Use various types of trawls to sample in the sea. Otter trawls are effective for collecting benthic species and midwater trawls for pelagic species. Various dredges are available to collect benthic species from different types of bottoms or to collect different sizes of organisms. Catch commercially important species such as lobster, blue crab, and dungeness crab in traps or take by deep-water trawls. Harvest species that colonize surfaces, such as barnacles, from plates of wood, plastic, or glass suspended in the water. Insure that organisms are not damaged during collection, transfer, and transport. When seining or using trawls, make short hauls. Do not collect significant amounts of plant materials, debris, mud, sand, or gravel in net or in bag of seine because these will injure the animals. Always leave seine bag in the water at end of haul, stretch out wings of seine, open bag entrance, dip out organisms with a bucket, and transfer directly to prepared holding tanks. Do not expose delicate, easily-damaged species to air. Take out larger, more hardy species with soft mesh dip nets. Do not collect too many animals at one time. After bringing a trawl up to the boat, very quickly bring it over the side without letting catch hit side of the boat. Immerse that portion of net containing collection in a tank of water. Open trawl and remove desired animals by dipping with a bucket or a hand net with small soft mesh. Have adequate quantities of clean water available in tanks before beginning a haul. Transfer organisms to tanks as rapidly and carefully as possible.

If organisms are to be transported any distance by boat, hold in aerated live boxes. If they are transported by truck, put them in large baffled and insulated tanks filled with water from area in which they were collected. Aerate the water and cool in summer or warm in winter. Determine water temperature, salinity, DO, and pH at the collecting site. Do not handle organisms more than necessary. Make transfers with suitable containers or hand nets, or for small organisms, by large-bore pipets. Use hand nets made of soft material with several layers around the net rim and free from sharp points or projections. Clean and sterilize all equipment before use. Do not crowd organisms during transport. Watch them carefully for signs of distress. Oxygenation, water exchange, and cooling may reduce distress.

Observe collected animals for possible injury resulting from transport to the laboratory. Examine smaller forms under a dissecting microscope. Criteria for assessing injury depend on the species and are more difficult for sluggish ones. Useful criteria include loss of appendages, inability to maintain a normal body posture, e.g., dorsal side uppermost, abnormal locomotion, refusal to feed, and uncoordinated movements of the mouth or other body parts.

For additional information on collecting aquatic organisms, see Part 1000.

3. Handling, Holding, and Conditioning Test Organisms

During transport to the laboratory, organisms often are crowded, bruised, and otherwise stressed, thereby increasing their susceptibility to disease. To avoid introducing disease into stock tanks, treat organisms during transit or on arrival in accordance with procedures in ¶ 5 below and as suggested for each of the different groups (Sections 804 through 810). Hold field-collected animals in quarantine for at least 7 days to observe for parasites and disease, and to recover from collection and transport stress. If more than 10% of the collected animals die after the second day or if they are parasitized or diseased beyond control, do not use them. Clean and sterilize all contacted containers and equipment and collect another supply from a different area if possible. Combat disease or parasites brought in with test organisms or water supply by disinfection with UV light, or use a rapid sand filter to remove unwanted organisms.

Because it is not always possible to collect from unpolluted areas and the collector cannot always be sure that a particular organism has not been exposed to a toxicant, sample collected individuals to determine if they have accumulated one or more potential toxicants. Check animals or materials collected as food for test organisms for disease and content of pesticides or other toxic materials. Feed test organisms daily during quarantine.

After quarantine period transfer disease-free animals to regular stock tanks. Discard organisms that touch dry surfaces, are dropped, or are injured during handling. To avoid unnecessary stress, do not subject organisms to rapid temperature or water-quality changes. In general, change water temperature less than 3 C in any 24-hr period. For stenothermal, deep-water species use an even smaller rate of temperature change. Keep DO levels preferably at or near saturation but never at less than 60% saturation. After transfer to stock holding tanks, begin a slow acclimation to laboratory conditions such as temperature, salinity, and hardness. The period of acclimation will be governed by type of organism and extent of changes in water quality. For forms with a life cycle of several months or more, use an acclimation period of at least 2 to 3 wk. Inspect organisms closely and frequently to determine stress, unusual behavior, parasites or disease, changes in color, or failure to eat. Avoid crowding. Provide adequate flow-through water so that characteristics such as DO, pH, CO_2, salinity, hardness, and NH_3 are favorable. Check temperature and DO frequently. Do not let metabolic products accumulate. Generally, use a flow-through rate of 6 to 10 tank volumes/day. Usually, greater amounts of flow-through water are required for smaller organisms on a weight-volume basis. For small organisms, use a water flow of at least 3 L/day/g. When brood stock are being held, periodic or continuous treatment for parasite and disease control may be required.[27,28] Clean tanks and equipment thoroughly and often, removing or flushing out all growths and wastes, preferably daily but at least twice per week. Remove all uneaten food within 24 hr. Use different sets of nets and other equipment for different groups of organisms and clean and sterilize them between use. Cover tanks and containers to prevent organisms from jumping out. Shield tanks with curtains or by some other means to protect organisms from nearby movements and noise. Provide photoperiods and light intensities favorable to the organisms (see Section 801E.3*f*). Begin acclimation to test conditions at a suitable interval in advance of testing. Use only those groups of organisms that are free from gross parasitic infection and disease and in which mortality is less than 10% during laboratory holding period. When handling is necessary, clean hands and nets before touching organisms.

It is of utmost importance that animals be kept in excellent condition before the tests. Make no abrupt changes in environmental conditions; preferably follow natural seasonal variations in environmental conditions such as temperature and daylight patterns. Do not supersaturate with gases, especially in winter when very cold water is brought into the laboratory and warmed. If there is danger of gas bubble disease, keep incoming water in an open system and let it cascade over baffles or otherwise aerate it to bring dissolved gases into equilibrium with the air.[29]

Acclimate freshwater arthropods by rearing them in the dilution water at the test temperatures, unless temperature is one of the factors being studied. Acclimate other organisms to the dilution water and test temperatures by gradually changing the water from 100% holding water to 100% dilution water over a period of several days. Keep all organisms in 100% dilution water for at least 2 days before use. Do not use a group of organisms if more than 3 to 5% die during the 48 hr immediately before the beginning of the test.[12] If a group fails to meet these criteria discard or re-treat, hold an additional 10 days, and reacclimate if necessary.

Make necessary provisions for organisms that require a special substrate, cover, or materials to use for clinging, support, the building of cases, or hiding.

During acclimation, hold test organisms below the optimum temperature to reduce metabolic rate and number and severity of disease outbreaks. Hold cold-water, freshwater organisms between 5 C and 15 C. Hold warm-water organisms at between 10 C and 25 C depending on season. Hold aquatic invertebrates within the temperature range of the water from which they were obtained unless they are being acclimated for special temperature or other tests. If possible, follow natural seasonal variation in temperatures.

4. Culturing Test Organisms

a. Facilities, construction materials, and equipment: Do not use construction materials in contact with dilution water that contain leachable substances or adsorb significant amounts of substances from the water. Stainless steel probably is the best construction material for freshwater systems. Glass significantly adsorbs some trace organics. Do not use rubber or plastics containing fillers, additives, stabilizers, plasticizers, etc. Fluorocarbon plastic, nylon, and their equivalents usually are acceptable. Glass-fiber-reinforced polyester, polyester resin, and epoxy resins also may be used. Test the toxicity of all materials before purchasing large quantities. Clean and flush all new tanks, troughs, and similar equipment with dilution water for several days before use. Use a glass or titanium interface between the water and heating elements for marine waters and glass or stainless steel for fresh waters. If concrete tanks are used, leach with frequent water changes over a period of weeks before use. Some wooden tanks are acceptable for fresh water after leaching.

Provide adequate space for test organisms, holding facilities, water storage reservoirs, and water supply systems. Provide distribution of hot and cold water and mixing facilities to obtain any desired temperature. Aerate or vigorously mix to prevent gas supersaturation caused by heating dilution water. Use air compressors with water seals to prevent oil from entering air lines and contaminating tanks. When large volumes of air are needed, use low-pressure blowers. Do not locate air intakes in shops or furnace rooms or near outlets from hoods, chemical laboratories, or vehicle exhausts. Provide acclimation and culturing tanks with temperature control and aeration. Design holding facilities for ease of cleaning and prevention of growths. For holding and culturing fish and many macroinvertebrates, preferably use round tanks of at least 1 to 3 m diam. Provide a standpipe drain in the center, threaded below the tank floor so that, when the standpipe is removed, the opening is flush with the tank bottom. Slope tank bottom gently to center. Use tanks with smooth surfaces to facilitate cleaning, to prevent injuries to organisms, and to insure that no material will collect in corners, cracks, and crevices. Use square or rectangular tanks for special purposes or when space is scarce. Provide standpipes at one end for draining, with threads for securing the pipe on the underside. Ensure that tank corners are rounded and smooth. Introduce water into a circular tank as a jet along the edge and above the surface to create a circular movement of water around the central standpipe. Fit another pipe, with half-moon cutouts at its base, over standpipe and screen, so that outflowing water passes out at the bottom, goes up through the outside pipe, then down the standpipe. This results in a circular current and a certain amount of self-cleaning (Figure 801:1).

b. Water supply: Provide a flowing water system for holding, spawning, and rearing a variety of aquatic organisms. Reconstituted fresh water or artificial seawater are not cost-effective for large-scale rearing or for flow-through tests. Use natural, unpolluted water supplies that have low turbidity, high DO, low BOD, and an annual temperature cycle that approximates that of the test organisms.

1) Freshwater supplies—A good freshwater supply is constant in quality and does not contain more than the designated amounts of the following: suspended solids, 20 mg/L; total organic carbon or chemical oxygen demand (TOC or COD), 10 mg/L; un-ionized NH_3, 20 μg/L; total residual chlorine, 0.5 μg/L; total organophosphorus pesticides, 50 ng/L;* total organochloride pesticides plus PCB's, 50 ng/L.* Consider water to be of constant quality if the monthly ranges of hardness, alkalinity, conductivity, TOC or COD, and salinity are less than 10% of the average values and the pH range is less than 0.4 units.

Check municipal water supplies to determine their acceptability from the standpoint of, for example, copper, lead, zinc,

*No individual pesticide should exceed the allowable concentration limit set in the National Water Quality Guidelines, EPA, as set in accordance with the Federal Water Pollution Control Act 92-500 as amended 1972.

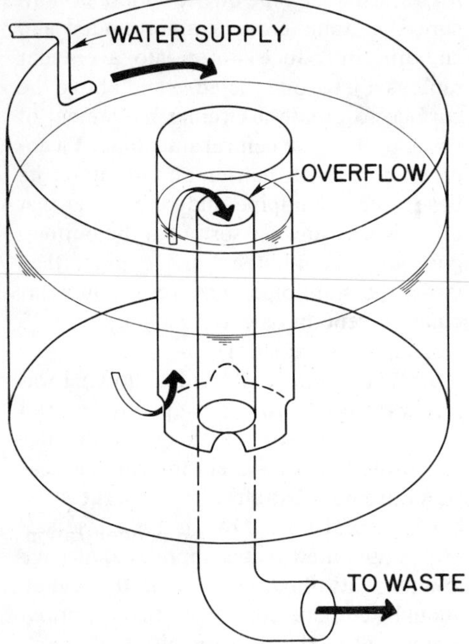

Figure 801:1. Holding tank design for fish and macroinvertebrates.

fluoride, and free or combined chlorine concentrations. If a satisfactory fresh-water supply is not available or if a standard water is required for comparative toxicity tests, relative sensitivity tests, or tests to determine the effects of hardness, pH, or total alkalinity on the toxicity of various materials, use a reconstituted standard water.

Prepare standard fresh water (Tables 801:I and 801:II) by adding reagent-grade chemicals to glass-distilled and/or deionized water. For special studies, determine that the distilled and/or deionized water contains less than the indicated constituents:

Conductivity	1 μmho/cm
Total organic carbon (TOC) or chemical oxygen demand (COD)	1 mg/l 2 mg/l
Boron, fluoride	100 μg/L each
Un-ionized ammonia	20 μg/L
Aluminum, arsenic, chromium, cobalt, copper, iron, lead, nickel, zinc	1 μg/L each
Total residual chlorine	3 μg/L
Cadmium, mercury, silver	100 ng/L each
Total organophosphorous pesticides	50 ng/L*
Total organochlorine pesticides plus polychlorinated biphenyls (PCB's)	50 ng/L*

Carbon-filtered deionized water usually is acceptable. Determine conductivity of distilled and/or deionized water for each batch of reconstituted water. Check other constituents periodically. If the water is prepared from a dechlorinated water (use sodium bisulfite), test the reconstituted water to determine that first instar daphnids survive for 48 hr (804B).[12]

The pH, alkalinity, and hardness of a receiving water influence toxicity of some materials, especially metals. Therefore, it is desirable to have a supply of both hard and soft waters.

It is advantageous to have water with temperatures between 3 and 7 C during the winter and between 20 and 25 C at peak summer temperatures. For general use, the pH should be in the range of 7 to 8.2 and dissolved CO_2 should be 1 mg/L or less.

2) Marine water supplies—Use unpolluted marine water with low turbidity and settleable solids and a pH and salinity favorable for the test organism. Ensure that annual salinity variations are not so wide as to be harmful to the organisms. In general, it is preferable to have a source of higher-salinity water (e.g., ocean water) from which brackish water can be prepared by dilution.

If a suitable marine water supply is not available use artificial seawater for limited culturing and toxicity testing. Prepare artificial seawater as in Table 801:III. When

*No individual pesticide should exceed the allowable concentration limit set in the National Water Quality Guidelines, EPA, as set in accordance with the Federal Water Pollution Control Act 92-500 as amended 1972.

TABLE 801:I. RECOMMENDED COMPOSITION FOR RECONSTITUTED FRESH WATER

Water Type	Salts Required mg/L				Water Quality		
	$NaHCO_3$	$CaSO_4 \cdot 2H_2O$	$MgSO_4$	KCl	pH*	Hardness mg $CaCO_3$/L	Alkalinity mg $CaCO_3$/L
Very soft	12	7.5	7.5	0.5	6.4–6.8	10–13	10–13
Soft	48	30	30	2.0	7.2–7.6	40–48	30–35
Hard	192	120	120	8.0	7.6–8.0	160–180	110–120
Very hard	384	240	240	16.0	8.0–8.4	280–320	225–245

*Approximate equilibrium pH after aeration with fish in water.

all chemicals are dissolved, add distilled, deionized water to make 1 L. The salinity should be 34 ± 0.5‰ and pH 8.0 ± 0.2. Obtain desired salinity at time of use by dilution with deionized water.

To increase salinity of a natural water, use a strong natural brine prepared by freezing and then partially thawing seawater. This is satisfactory if only limited amounts of water are needed; for larger volumes, use commercial sea salts or a stronger solution of artificial seawater.

In the preparation of artificial seawater, be sure that an undesirable concentration of metals does not occur. Even reagent chemicals contain traces of several metals and their extensive use can result in a buildup of metals. If large volumes of artificial seawater are not required, remove the metals by passing the seawater through a column containing a cation-exchange resin in the sodium form. Equipment has been devised for supplying flow-through water of constant salinity.[35]

TABLE 801:II. QUANTITIES OF REAGENT-GRADE CHEMICALS TO BE ADDED TO AERATED SOFT RECONSTITUTED FRESH WATER FOR BUFFERING PH[12,30]

Desired pH*	Quantity of Chemical to Be Added mL/L water		
	1.0*N* NaOH	1.0*M* KH_2PO_4	0.5*M* H_3BO_3
6.0	1.3	80.0	—
6.5	5.0	30.0	—
7.0	19.0	30.0	—
7.5	—	—	—
8.0	19.0	20.0	—
8.5	6.5	—	40.0
9.0	8.8	—	30.0
9.5	11.0	—	20.0
10.0	16.0	—	18.0

*Approximate equilibrium pH with fish in water. Do not aerate after adding these chemicals.

c. Food and feeding:

1) Culture of microorganisms—Culture phytoplankton and zooplankton for testing biostimulation, environmental requirements, and toxicity. Culture also as food for other organisms such as copepods, daphnia and other microcrustaceans, the larvae and adults of mollusks, and young and adult fish.

a) Culture medium for freshwater algae—Prepare reconstituted fresh water by adding reagent-grade macro- and micronutrients to glass-distilled and/or deionized water in the concentrations given in Tables 801:IV. A and B.

Prepare a separate stock solution of each macronutrient salt in 1,000 times the specified final concentration in glass-distilled or deionized water. Combine trace metals and EDTA in a single micronutrient stock solution in glass-distilled or deionized water at 1,000 times the final concentration of each.

To prepare algal culture medium, add 1 mL of each macronutrient stock solution to 900 mL glass-distilled or deionized water, then add 1 mL trace metal EDTA mixture and make up to 1 L with glass-distilled or deionized water.

TABLE 801:III. PROCEDURE FOR PREPARING RECONSTITUTED SEAWATER*[31,32]

Compound in Order of Addition	Final Concentration *mg/L*
NaF	3
$SrCl_2 \cdot 6H_2O$	20
H_3BO_3	30
KBr	100
KCl	700
$CaCl_2 \cdot 2H_2O$	1,470
Na_2SO_4	4,000
$MgCl_2 \cdot 6H_2O$	10,780
NaCl	23,500
$Na_2SiO_3 \cdot 9H_2O$	20
Na_4EDTA*	1
$NaHCO_3$	200

*Tetrasodium ethylenediaminetetraacetate. Omit when toxicity tests are conducted with metals. Omit when tests are conducted with plankton or larvae. Strip the medium of trace metals.[33,34]

Use sterile technics for tests with algae or other organisms using axenic cultures if bacterial growth interferes.

Prepare freshwater algal culture medium as follows: to 800 mL glass-distilled or deionized water add 1 mL of each macronutrient stock solution in the order listed, mixing after each addition. Filter-sterilize by passing through a sterile 0.22-μm-porosity membrane filter (pre-rinsed with 100 mL double-distilled water) into an autoclave-sterilized container. Add 1 mL filter-sterilized micronutrient solution, and make up to 1 L with sterile distilled or deionized water.

Sterilization may not be required for some tests using freshly-prepared culture medium. However, maintain stock cultures in previously sterilized culture medium. Store uninoculated sterile reference medium in the dark to avoid photochemical changes.

When sterility is desired in algal tests, check sterility periodically by adding 1 mL inoculated test culture to tubes of sterile nutrient test medium and incubate in the dark at the test temperature for 2 wk. The appearance of opalescence in the test medium indicates contamination.

Prepare sterile nutrient test medium by adding the following quantities of chemicals to 1 L glass-distilled water:

Sodium glutamate	250 mg
Sodium acetate	250 mg
Glycine	250 mg
Sucrose	250 mg
Sodium lactate	250 mg
DL alanine	250 mg
Nutrient agar	50 mg

b) Culture medium for marine algae—To artificial seawater (Table 801:III) add nutrients listed in Table 801:V to give the indicated concentrations in the algal culture medium.

When an unpolluted seawater is available, prepare the medium by enriching fil-

TABLE 801:IV.A. MACRONUTRIENT STOCK SOLUTION

Compound	Concentration *mg/L*	Element	Resulting Concentration *mg/L*
$NaNO_3$	25.5	N	4.20
		Na	11.0
$NaHCO_3$	15.0	C	2.14
K_2HPO_4	1.04	K	0.469
		P	0.186
$MgSO_4 \cdot 7H_2O$	14.7	S	1.91
		Mg	2.90
$MgCl_2$	5.70		
$CaCl_2 \cdot 2H_2O$	4.41	Ca	1.20

TABLE 801:IV.B. MICRONUTRIENT STOCK SOLUTION

Compound	Concentration *μg/L*	Element	Resulting Concentration *μg/L*
H_3BO_3	186	B	32.5
$MnCl_2$	264	Mn	115
$ZnCl_2$	3.27	Zn	1.57
$CoCl_2$	0.780	Co	0.354
$CuCl_2$	0.009	Cu	0.004
$Na_2MoO_4 \cdot 2H_2O$	7.26	Mo	2.88
$FeCl_3$	96.0	Fe	33.0
$Na_2EDTA \cdot 2H_2O$	300	—	—

ter-sterilized seawater with micronutrients at one-half the indicated concentrations. If sterile technics are required, follow procedures for fresh water. When sterilization is performed by autoclaving for 4 hr at 60 C, add vitamins after autoclaving. When filter-sterilization is used, use a positive pressure of 72 kPa.

2) Mass production of algae as food for other organisms—Rearing zooplankton, various filter-feeders, the larvae of crustaceans, and fish requires large quantities of phytoplankters. Supply these needs with an apparatus capable of producing continuous amounts of desired organisms at high densities. Such an apparatus (Figure 801:2) permits easy assembly, cleaning, and sterilization and efficient utilization of light energy, and is constructed for continuous use.†

The main body of the unit is a 60-cm section of 15-cm-diam borosilicate glass drainage pipe. The top section is a 15-cm to 5-cm concentric reducer and the bottom section is a 15-cm to 5-cm ell. Each section accommodates a No. 12 silicone rubber stopper held in place by a carboy or similar type clamp. Hold the three sections together by aluminum ring clamps and seal adjoining surfaces by silicone "O" rings made from small-diameter tub-

†For details of construction and use, contact Dr. Richard Steele, National Marine Water Quality Laboratory, Narragansett, R.I. 02882.

TABLE 801:V. NUTRIENTS FOR ALGAL CULTURE MEDIUM IN SEAWATER

Compound	Concentration	Concentration of Nutrient
$NaNO_3$	25.0 mg/L	4.2 mg N/L
K_2HPO_4	1.05 mg/L	0.19 mg P/L
$FeCl_3$	72.6 μg/L	
$MnCl_2$	2.30 μg/L	
$ZnCl_2$	2.10 μg/L	
$Na_2MoO_4 \cdot 2H_2O$	2.50 μg/L	
$CuCl_2$	0.20 μg/L	
Na_2EDTA	300 μg/L	
Vitamins:		
Thiamine	0.100 mg/L	
Biotin	0.50 μg/L	
B_{12}	0.50 μg/L	

ing. Use material that is autoclavable and nontoxic. Set up assembly, light source, culture medium supply bottle, air filter, pumps, and stands as shown in Figure 801:3.

Use this device to supply cells on a periodic or continuous basis. As the cells are withdrawn, add more medium. The following species have been grown at the indicated concentrations: *Skeletonema costatum,* 4.3 × 10^6 cells/mL; *Dunaliella tertiolecta,* 4.4 × 10^6 cells/mL; *Isochrysis galbana,* 7.0 × 10^6 cells/mL; *Monochrysis lutheri,* 5.0 × 10^6 cells/mL.

3) Food for macroinvertebrates and fish—A suitable food is essential for rearing various macroinvertebrate and fish life stages. Distinguish carnivores from herbivores to supply the correct type of food. Organisms taken as food differ for different life stages of a species, but related species may take the same food. As organisms grow they require progressively larger food organisms. Many feed on pelagic organisms whose movements should be sufficient to attract the predator but slow enough so they can be caught readily. Use food organisms that are nutritious, easily digested, and readily obtainable. Distribute zooplankton food in rearing tanks to match distribution of organisms using it. Provide an adequate amount of food with a ratio of number of prey to predator varying from 50:1 to 200:1. If a small number of organisms is reared in a large tank, provide more food organisms to insure that enough are captured. Some

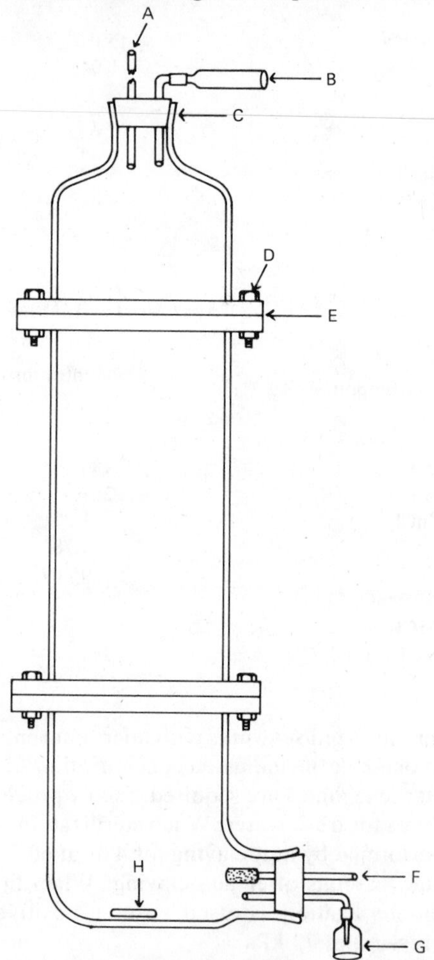

Figure 801:2. Algal culture units. A—medium inlet tube; B—air exhaust tube; C—No. 12 silicone stopper with two holes (may be held in place with carboy clamp); D—aluminum flange clamp ring; E—silicone tube, 0.24 cm OD, in a groove between adjoining glass surfaces; F—air inlet tube; G—aseptic filling bell for withdrawing cells from culture; H—magnetic stirring bar.

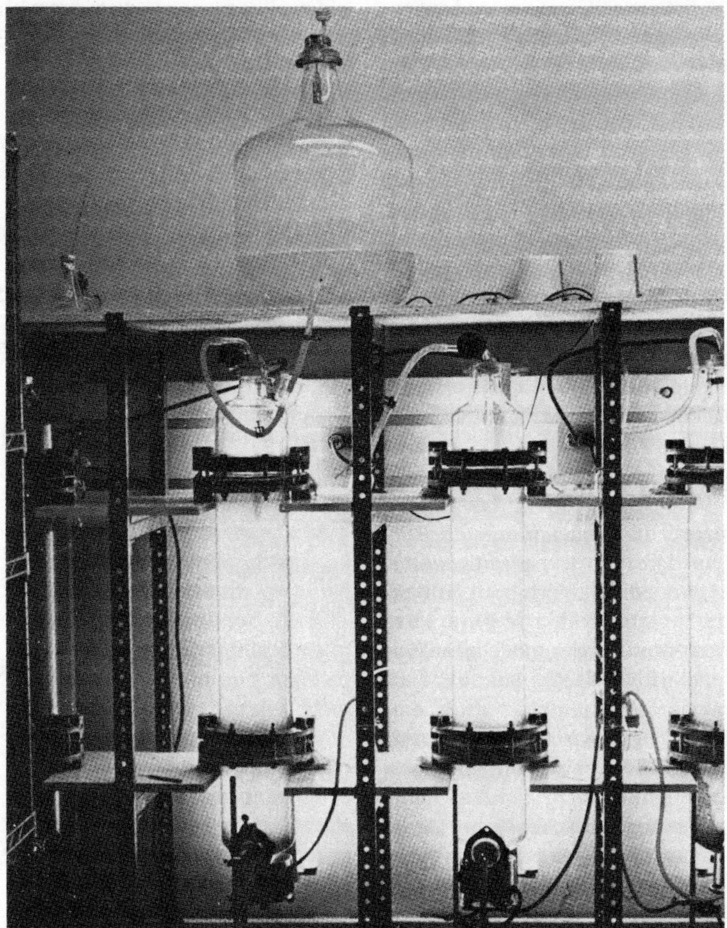

Figure 801:3. Method of lighting, free-standing frames, and placement of medium source, air pumps, and other apparatus for mass algal culture device.

food algae and diatoms have a tendency to settle to the bottom. Circular movement of the water in rearing tanks, provided as described in Section 801D.4*a*, keeps food materials in suspension.

Methods for rearing freshwater organisms have been described by Needham et al.[36] Methods for rearing larvae of marine animals with special reference to their food organisms have been summarized by Hirano and Oshima.[37] May[38] has reviewed the literature on laboratory feeding larvae of marine fish. Bruhn and Schoettger[39] have reviewed needs of freshwater fish.

When using natural and cultured plankton, provide facilities for passing the water supply directly from the source to the holding, acclimation, and testing tanks. Provide test organisms with uncontaminated, palatable, and readily-taken food. Use cold rooms and deep freezers for food preparation and storage.

When using cultured microorganisms as food, be aware of possible environmental changes they may cause. In addition to the possible presence of toxic metabolites, algal blooms may occur that produce excess oxygen and result in supersaturation and gas bubble disease.[29] Use live food whenever possible. Analyze food for toxicants, especially pesticides and heavy metals. Supplement natural foods with commercially available dried and pelleted foods.‡ (See 809A.3 for fish feeding.) These foods should be attractive to the organisms, supply necessary nutrients and trace elements, and contain binders to insure pellet stability.[40]

d. Cleaning containers and equipment:

1) Cleaning holding, acclimation, testing, and dilution water tanks—Clean test containers and toxicant delivery systems before use. Wash new containers with laboratory detergent and rinse with 100% acetone, water, acid (such as 5% HNO_3), and twice with tap water. At the end of each test empty the toxicant delivery system or test chambers, rinse with water, clean by a procedure appropriate for removing the toxicant tested (e.g., acid to remove metals and bases; detergent, sodium hypochlorite (NaOCl) solution (200 mg/L), organic solvent, or activated carbon to remove organic compounds). Do not use acid and hypochlorite together. Immediately before testing, rinse again with dilution water.[12] Hesselberg and Burres[41] have developed labor-saving devices useful when many tests are performed.

2) Removal of unused food and wastes—Do not let unused food or fecal material accumulate. Whenever possible, build holding and testing containers with sloping bottoms so food and feces can be removed easily with a siphon. The amount and frequency of cleaning depends on the organism, ratio of dilution water to weight and volume of organisms, and feeding schedule. Clean holding containers at least once every other day. If growths occur on sides of containers, dislodge with a rubber spatula and let settle for removal.

‡Some foods that have been widely used include Glencoe Trout Food, Glencoe, Minn. 53336; Biorell and Tetramine, available from local pet shops; Oregon Moist, Warrenton, Ore. 97146; and Cerophyll, Cerophyll Laboratories, Inc., Kansas City, Mo. 64111. The latter has been used for the smaller forms and as a food for organisms providing food for the higher species.

5. Parasites and Disease

a. Stress in relation to parasites and disease: Unexpected and often unexplained mortalities in experimental and control animals interfere with acute or chronic test results. While many factors may be responsible for the death of an animal, diseases due to specific pathogens are among the most significant. In general obtain fish and other animals from pathogen-free stocks (specific hatcheries, etc.) rather than stressing populations by parasite and disease controls. Also optimize laboratory conditions for the individual species to prevent the fostering of disease conditions.

When large numbers of organisms are retained in a relatively small space, undesirable growths, diseases, and parasites become a problem. If the water is unpolluted and poor in nutrients, problems often can be controlled by strict sanitation. However, if the water is somewhat enriched by organic materials and potential toxicants are present, problems increase greatly. Pathogens and parasites that might be very rare in natural waters become potential and ever-present dangers in intensive culture. Bacteria grow on uneaten food, fecal material, and other wastes and compete for DO. They provide a potential source of unwanted growths, disease, and toxic products. Even with good flow-through, each corner, crevice, and dead area of a tank may become a trouble area.

Use filtration and/or sterilization of water, regular cleaning of holding vessels, strict sanitation practices, and sterilization of equipment as the first lines of defense. Uniform food distribution and limiting the amount of unused food, removal of unused food and waste materials are important.

Organisms exposed to toxicants become stressed, weakened, and much more susceptible to parasites and disease. Because other environmental factors contribute to reduced resistance, pay careful attention to nutrition, oxygen supply, and water quality.

b. Control methods: UV light and ozonation have been used successfully to control disease and parasites. Antibiotics used in holding tanks reduce bacterial populations. To reduce mortality and to avoid introduction of disease into stock tanks, treat with a wide-spectrum antibiotic immediately after collection, during transport, or on arrival at the laboratory. Holding in a tetracycline-based antibiotic,§ 15 mg/L for 24 to 48 hr, can be very helpful. Other chemotherapeutic agents are available, but use care in their application because some are toxic at low concentrations.[42] Do not use treated organisms for tests for at least 10 days after treatment. If contamination is suspected, disinfect tanks and containers with 200 mg NaOCl/L for 1 hr.

Adequate nutrition and good water quality minimize disease, but even with good sanitation disease outbreaks occur. No manual for diagnosis and control of disease in marine fish and shellfish exists.

For larval tests, use strict sanitary measures including sterilization of utensils and containers, filtration and UV sterilization of water, and removal of metabolic products. If disease signs appear in larval cultures, discard the entire culture.

For tests using adult fish and shellfish, early diagnosis and prompt treatment, when available, can prevent losses.

§Terramycin or equivalent.

801 E. Bioassay Systems, Test Materials, and Procedures

1. Water Supply Systems and Testing Equipment

Construct all components of a test system, including water heating and cooling units, pipes, constant-level troughs and head boxes, valves and fittings, diluters, pumps, mixing equipment, tanks, and exposure chambers, from nontoxic materials. Obtain dilution water of the desired temperature by mixing hot and cold water of constant temperatures in the correct proportions, by heat exchangers, or by heaters or coolers in constant-level troughs and head boxes. A heated room with thermostatic controls usually is suitable for static tests on warm-water organisms. Hold dilution water in tanks until it reaches ambient temperature for conducting static tests. For cold water species use a specially insulated constant-temperature room or large water bath equipped with temperature controls and adequate circulating water. A satisfactory design for a small laboratory to conduct short-term static tests has been described.[43,44] Special facilities required for different groups of organisms are described in Sections 802 through 810.

In flow-through tests use metering pumps or other devices for accurate delivery of toxicant or test material into the dilution water. Most toxicant delivery systems have been designed for fresh water and may not be applicable to all wastes. If necessary, stir test chamber contents to maintain suspended solids and nonhomogeneous wastes in flow-through and static tests. Deliver dilution water from constant-head troughs or head boxes by siphons, constricted tubing, nozzles, or pumps. Deliver toxicants by siphons from constant-head reservoirs, pumps, cali-

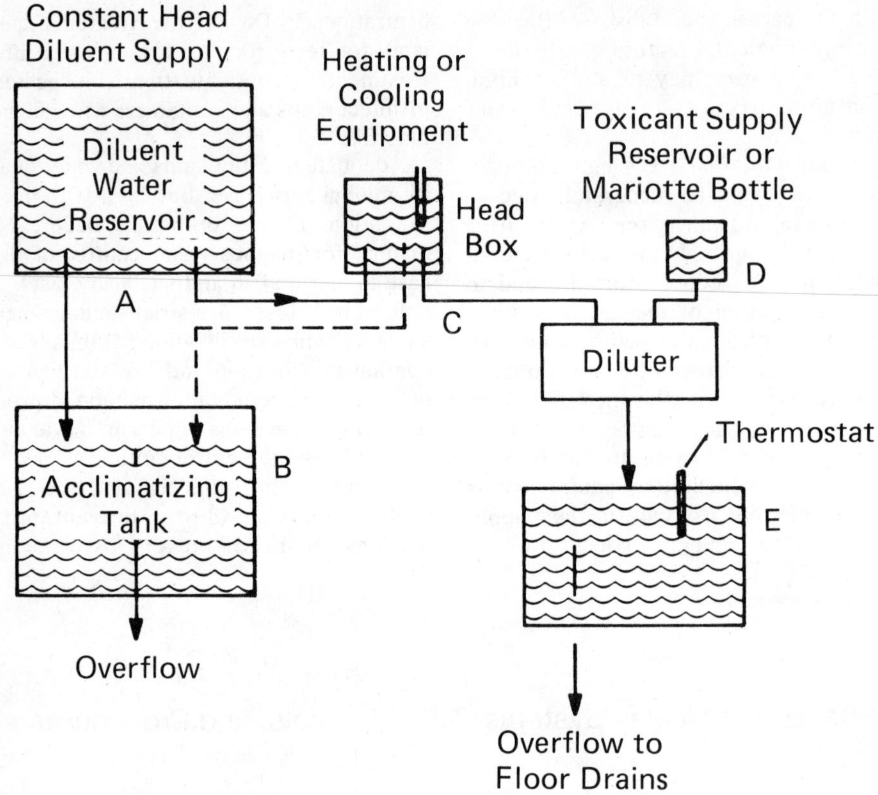

Figure 801:4. Basic components of flow-through system.

brated glass nozzles, or Mariotte bottles.[45] Mix dilution water and toxicant in tanks with baffles or stirrers or in mixing troughs.[46] Since the introduction of the serial diluter,[47] various methods and types of diluters have been described[14,48–55] but the proportional diluter[56] and its modifications are used most commonly.

The basic components of a flow-through system are shown in Figure 801:4. The diluent water reservoir is large enough to provide water for at least 5 days. If dilution water is added to this reservoir continuously, a smaller capacity is preferred. Dilution water flows at a constant rate by gravity from this reservoir to a constant-head diluent-supply head box through a nonmetallic float-controlled valve or other device and then to the diluter. Provide head box with heating or cooling equipment and a thermostat to maintain constant temperature. Equip test containers with an overflow system designed to prevent organisms from entering outlets. Clean test containers as described in 801D.4*d*.[57]

The constant-head toxicant supply is a constant-level tank, a Mariotte bottle, or other device. If the toxicant stock solution is unstable, renew it before it degrades. If metering pumps are used, the toxicant supply system need not maintain a constant pressure.

A simple valve control system for regulation of flow rates of dilution water and toxicant solution has been described.[58] For more toxic materials, less toxicant is required and a Mariotte bottle that delivers a very slow but constant flow is useful.[59,60]

A diluter meters dilution water from the constant-head box and toxicant from the constant-head tank or other containers and mixes them in the proper proportions for each of the test chambers. After proper calibration of the diluter, make toxicant stock solutions to the proper concentration. When a Mariotte bottle is used, insulate it against rapid temperature fluctuations to reduce flow variation resulting from pressure changes.

The proportional diluter generally is more useful and easier to construct, calibrate, and operate, but the serial dilution apparatus is more applicable when the desired dilution factor (i.e., the value by which a concentration is multiplied to obtain the next lower concentration) is less than 0.50. The serial diluter has been modified to provide a very narrow range of concentrations with a dilution factor of 0.90.[61]

The proportional diluter[47] can deliver toxicant dilutions at flow rates of up to 400 mL/min for each concentration, with dilution factors ranging from 0.75 to 0.50. Metering cells can be exchanged to provide dilution factors outside this range.[62,63]

Provide a mixing chamber between diluter and test container for each concentration. If duplicate test containers are used, run separate delivery tubes from the mixing chamber to each duplicate. Use flow rates through test containers of at least 6 tank water volumes/24 hr. Do not let rates through test containers vary temporally or between containers by more than $\pm 10\%$. Calibrate the toxicant delivery system before, during, and after each test. Determine stock solution and dilution water volumes used in each portion of the toxicant delivery system and the flow rate through each test container. Check operation of toxicant delivery system daily during test.

The diluter is one of the best methods for mixing and delivering dilution water and toxicants because, if the dilution water is shut off, the system stops and no more toxicant is added. Many relatively inexpensive but accurate metering pumps are available for supplying toxicant.

A special diluter may be required for effluents containing large amounts of suspended materials and when there is a need for large volumes of toxicant.

2. Preparing Test Materials

a. Dilution water: Whenever possible, test toxicity of effluents on site where ample supplies of toxicant and dilution water are available. On-site testing permits temperature, DO, pH, hardness, salinity, turbidity, and other qualities of the dilution water to vary with those of the receiving water. Convey the effluent sample to testing chambers with as little modification as possible. Do not aerate, heat, cool, or unnecessarily agitate. If the diluted effluent is low in DO, adjust flow-through and loading in the test chambers so that DO is not reduced significantly; hold temperature at or near that of the receiving water.

If the receiving water is deficient in DO and has temperatures above the locally pertinent water quality standards, bring these into compliance so that allowable levels of a specific waste can be assessed meaningfully. Determine toxicity of test waste in conjunction with other contaminants present in receiving water by taking dilution water from receiving water just outside the area of effect of test waste. This is especially necessary when effluents contain metal salts, cyanide complexes, ammonium compounds, or other materials, the toxicity of which is greatly influenced by changes in pH, hardness, etc. If there are wide variations in receiving water quality characteristics, determine waste toxicity at the upper and lower limits of the range.

Evaluate receiving water stress on aquatic biota by using two controls, one with the receiving water and another (either natural or synthetic) with an unpolluted water of similar quality. Do not let calcium, magnesium, sulfate, and DO content for fresh-water controls differ by more than 10% from the natural content of the water receiving the test waste. Adjust pH, alkalinity, and hardness to those of the receiving water before adding wastes to determine if dilution water itself is unfavorable for the more sensitive aquatic species in the area. In addition, evaluate existing information on composition of the aquatic biota and study aquatic populations above and below the discharge and in other areas to compare qualitative and quantitative population makeup.

When the purpose of the test is other than to determine toxicity of an effluent, use for dilution water only a nonpolluted natural or synthetic water of constant and reproducible quality that is favorable for aquatic life. Warm or cool dilution water to test temperature and bring to equilibrium with atmospheric gases before use.

Use standard water conditions and organisms for comparative toxicity and sensitivity tests. Use reconstituted fresh or marine water (Sections 801D.4*b*1) and 2) if natural supply is not suitable. Because of the effects of water quality on toxicity, use both hard and soft water for tests on freshwater organisms.

Many marine organisms spend a portion of their life cycles in estuaries. In life-cycle tests, change dilution water in accordance with their requirements at different life stages. If the effects of temperature are not being studied, keep it within a favorable range.

For warm-water species keep DO at the highest of the following: 4 mg/L, 60% of saturation, or level specified by state or federal standard. For cold-freshwater species, keep minimum DO above 5 mg/L unless local standards differ. Some larval forms, such as those of marine crustaceans, require higher DO.

Determine pH requirements for test organisms. In long-term studies, other than effluent and monitoring tests, keep pH within 0.4 units of the desired value. Avoid rapid changes in pH or CO_2 content. A rapid increase in the CO_2 content of marine waters indicates that some significant change has occurred that should be investigated at once. Freshwater organisms are more tolerant of pH changes and accommodate to much wider variations than strictly marine forms. Changes in pH drastically alter toxicity of some materials such as cyanide and NH_3.

In working with estuarine and marine organisms and different life stages that may be marine or estuarine, salinity is of prime importance. Use the natural salinity for each test species and its different developmental stages.[34]

Dilution water turbidity is an important environmental factor in determining harmful concentrations of potential toxicants because some toxicants are sorbed on particles. Turbid dilution water may limit visual inspection and photosynthesis of algae, form deposits, and clog water systems. Where large amounts of settleable solids significantly remove toxicants from the water, determine concentrations of toxicants in bottom sediments and their toxicity by appropriate tests with benthic organisms.

Keep acidity, total alkalinity, and hardness of dilution water constant. Alkalinity and hardness influence toxicity of some metals and total alkalinity is an important factor in photosynthesis and algal growth.

b. Toxicant solution: Prepare toxicant solution in advance and add immediately to the dilution water for static tests. If a toxicant is unstable, determine its stability and replace as necessary. If possible, measure toxicant concentrations during the test. Prepare all solutions for each series of tests from the same source sample. Disperse undissolved material uniformly by shaking.

Some effluents, especially oily wastes, are difficult to distribute evenly. Agitate test tank contents to maintain solids in suspension using a magnetic stirring device or other mixers. The nature of the test material governs the preparation of test concentrations and frequency of test medium replacement. Common problems include insolubility, adsorption on exposed surfaces, decomposition, photolysis, loss of volatiles, high BOD, and bacterial growth. These can change the apparent test concentration and lead to erroneous results.

Do not composite effluent samples that vary in composition with time because maximum, rather than average, toxicity data are required.

Store effluent samples in completely filled, stoppered containers at between 0 and 4 C. Do not store samples longer than absolutely necessary.

Shake wastes thoroughly before use. Use waste directly as a stock solution of toxicant or prepare a stock solution using filtered dilution water. Make stock solutions with dilution water on a volume-to-volume basis. If the effluent is liquid, designate the percentage waste in each test concentration. If the waste is a solid, dilute on a weight-to-volume basis, e.g., milligrams per liter.

If the waste contains both solids and liquids, shake thoroughly to disperse before using as a stock toxicant. Provide agitation in the stock reservoir and test containers. If small organisms are tested, use a magnetic stirrer in the test chamber. If larger organisms are tested, use a propeller placed under a screen or perforated false bottom. If the solids settle out rapidly and do not contact pelagic organisms, test only the liquid portion. After thorough mixing, let settle and decant or drain off the liquid for use as test toxicant. If the solid waste portion is toxic, set up test chambers having a certain weight-to-volume ratio of bottom material and expose benthic and burrowing organisms of the receiving water area. Mix wastes and let settle before adding organisms. If waste contains sparingly water-soluble materials check solubility and if below or at very low toxic levels, use solvents, emulsifying agents, or water-miscible solvents to disperse.

Use acetone, dimethylformamide (DMF), ethanol, methanol, isopropanol, acetonitrile, dimethylacetamide, ethylene glycol, or triethylene glycol to prepare stock solutions. Certain surfactants|| may be useful. Use only the minimal amount of solvent necessary to disperse the toxicant. Do not exceed 0.5 mg/L in static and 0.1 mg/L in flow-through test solutions.

If an additive is used, make two sets of controls, one containing no additive and the other containing the highest concentration of additive used.

c. Test organisms: Select test organisms as described in Section 801D.1 and handle as indicated in Sections 801D.3 and 4. For long-term tests, use only test organisms in excellent condition. At the end of the test, control organisms still should be in good condition.

3. Test Procedures

a. Experimental design: Expose test organisms in at least duplicate containers. Using more organisms and replicate test containers for each toxicant concentration is desirable to evaluate variability, but use only true replicates with no water connection between test containers. Each test consists of a minimum of five test concentrations and a control, with an additional control if additives are used. Arrange test containers at random in the testing area. If replicates are used, randomize each series of test containers separately. Distribute organisms randomly to test containers either adding one at a time to each container if there are to be less than 11 organisms

||Triton X-100, Rohm and Haas Co., or equivalent.

per container or two at a time if there are to be more. In short-term static tests, add organisms to interim containers and then add them to test chambers containing the toxicant all at the same time.

Repeat all short-term tests to determine reproducibility. In long-term flow-through tests, results are not acceptable if 10% of the test organisms in the control die, show the effect under study, fail to spawn, develop abnormally, or are in apparent poor condition. Do not include cannibalism in these percentages when it takes place equally in the control and test chambers. When larvae of some marine crustaceans are tested it often is difficult to keep losses in the controls below 15%.

In short-term static or renewal tests with fish, use 10 or more test organisms in each toxicant concentration. Use larger numbers of organisms per test concentration for smaller organisms. The number of organisms exposed in each test concentration is governed by size of the organism; expected normal mortality; extent of cannibalism; availability of dilution water, toxicant, and test organisms; and desired test precision. Test precision depends on variability of organism response, number of organisms exposed to each concentration, number of replications, toxicant and its variability, and relationship of toxicant concentration to LC50 concentration.

Arbitrarily setting the number of test organisms will not assure a certain precision. For a given test under similar conditions, increasing the number of test organisms increases precision. It is recommended that the 95% confidence interval be less than ±30% of the mean. The number of test organisms and test concentrations required for each test is the number required to attain this precision. With test organisms for which culture methods have not been worked out well, this precision may be difficult or impossible to attain. As a general rule, use no less than 10 test organisms at each concentration; preferably use 20 or more, especially with small organisms and in tests with larvae where cannibalism and natural mortality are high. Because industrial effluents vary in their volume and toxicity, determine the LC50 periodically and especially when there is reason to believe that there is some change in the effluent.

b. Selecting test concentrations: Express liquid waste concentrations as percent on a volume-to-volume basis. Express concentrations of nonaqueous wastes and of individual chemicals as milligrams or micrograms per liter. Indicate inclusion of water of hydration as part of the weight of the solute (e.g., $CuSO_4 \cdot 5H_2O$). Likewise, when an impure chemical is tested, especially in a formulation containing added inert ingredients, indicate the chemical composition by weight and whether the LC value is based on concentration of total material or active ingredient.

Although LC may be determined by using any appropriate series of test concentrations, the geometric series of concentration values is simplest to use. Multiply the highest and succeeding concentrations by a constant factor (0.5 to 0.6) to obtain concentrations that are evenly spaced on a logarithmic scale.

The magnitude of concentration intervals to establish an LC50 or EC50 by interpolation depends on the required degree of precision and on the experimental data.

c. Loading: For static tests do not exceed an organism loading of 0.8 g/L in the test container. In tests with small organisms and tropical forms, decrease loading to as low as 0.1 g/L and accommodate larger test organisms by using larger or duplicate test containers. Limit the number of test organisms per volume of test solution so that during the test (*a*) DO is greater than 60% saturation; (*b*) toxicant concentration is not lowered significantly; (*c*) concentrations of metabolic products

TABLE 801:VI. PERCENTAGE OF AMMONIA UN-IONIZED IN DISTILLED WATER*

Temperature *C*	Percentage Un-ionized at Given pH								
	6.0	6.5	7.0	7.5	8.0	8.5	9.0	9.5	10.0
5	0.01	0.04	0.11	0.40	1.1	3.6	10	27	54
10	0.02	0.06	0.18	0.57	1.8	5.4	15	36	64
15	0.03	0.08	0.26	0.83	2.6	7.7	21	45	72
20	0.04	0.12	0.37	1.2	3.7	11	28	55	80
25	0.05	0.17	0.51	1.7	5.1	14	35	63	84
30	0.07	0.23	0.70	2.3	7.0	19	43	70	88

*Prepared from data given in Sillen and Martell.[64]

(e.g., NH_3, CO_2) do not become too high; and (*d*) organisms are not stressed by crowding. Do not let the concentration of un-ionized ammonia exceed 20 μg/L (Table 801.VI). For flow-through studies, use a flow rate of at least 6 tank volumes/ 24 hr to maintain desirable temperature and DO and safe concentrations of metabolites.

d. Physical and chemical determinations:

1) Dilution water analysis—For fresh water measure hardness, alkalinity, pH, TOC, COD, and suspended solids at the beginning of the test and at least once every 30 days thereafter. If water quality is variable, test more frequently. If a treated water is used, measure residual chlorine.[65]

Analyze controls weekly for pH, alkalinity, and hardness to define test-water variability. If characteristics are affected by the toxicant, test samples from each toxic concentration at least once every other week. For brackish or marine dilution water, measure salinity, pH, suspended solids, and TOC or COD at least once every 30 days and at the beginning and end of each test.

2) Toxicant analysis—For flow-through life-cycle tests do not make routine detailed analyses but make periodic tests to insure that the correct ratio of effluent to dilution water is maintained in exposure tanks.

For studies to determine water quality criteria measure concentration of toxicants in each container at the beginning and at least once during the test, in at least one container at the next-to-lowest toxicant concentration at least weekly, and in at least one container whenever a malfunction is detected in any part of the toxicant delivery system. For replicate test containers use a ratio of the highest measured concentration to the lowest measured concentration of less than 1.15; if this is exceeded, check the toxicant delivery system and analyze additional samples from test containers to determine if the sampling or analytical method is sufficiently precise. Do not accept measured toxicant concentrations differing by more than 10 to 15% from the calculated concentration.

Record temperature at least hourly throughout the test (24 hr/day) in at least one test container and make additional measurements on dilution water and other test solutions. Measure DO, pH, and salinity at the beginning of the test and daily thereafter in the control, high, medium, and low toxicant concentrations. Generally, variation of ±0.5 C is allowable, but do not exceed ±1 C. Correct any change in DO beyond the limits specified as soon as noted.

Take water samples for chemical analysis at the center of the exposure tank; do not include surface scum or material from tank bottom or sides. If analytical results are not affected by storage, collect daily, equal-volume, grab samples and composite for a week. Analyze sufficient samples throughout the test to determine whether the concentration of toxicant is reasonably constant. If it is not, analyze enough samples weekly to show the variability of toxicant concentration. If methods are available, determine on the next-to-lowest concentration the loss of toxicant. If the loss is more than 10%, attempt to alleviate by using either a faster flow rate or a lower loading.

When possible and necessary, analyze mature and immature test organisms for toxicant residues. For larger organisms analyze muscle and liver and possibly gills, blood, brain, bone, kidney, GI tract, gonads, and skin. For higher organisms, analysis of whole specimens does not replace analysis of individual tissues, especially muscle.

e. Biological data and observations: In short-term tests with macroinvertebrates and fish, count the number of dead or affected organisms in each container at least daily throughout the test. During the first day count the number of dead or affected organisms in each container at 1.5, 3, 6, 12, and 24 hr after the beginning of the test. Remove dead organisms as soon as observed. It is more important to obtain data that will define the shape of the toxicity curve than to obtain data at prespecified times.

Death is the adverse effect most often used to reflect acute toxicity. The usual criterion for death is no movement, especially no gill movement in fish, and no reaction to gentle prodding. Death is not easily determined for some invertebrates. Cessation of movement of antennae, mouth parts, or other organs may be used. When death cannot be determined, use EC50 rather than LC50. The effect usually used for determining EC50 with daphnia, midge larvae, copepods, and other organisms is immobilization, defined as inability to move, except for minor activity of appendages. Other effects can be used to determine EC50, but always report the effect and its definition. Also report such effects as erratic swimming, loss of reflex, discoloration, changes in behavior, excessive mucus production, hyperventilation, opaque eyes, curved spine, hemorrhaging, molting, and cannibalism.

In short-term tests, organism reactions during the first few hours may indicate the nature of the toxicant and serve as a guide for further tests. Determine length and weight of representative organisms before the test and of all live organisms after the test. After acclimation has begun, do not weigh or handle test organisms. However, to determine increases in growth rate or in weight, add more organisms initially so that some can be removed periodically to make the necessary determinations.

In long-term partial- or full-life-cycle tests use a photographic method for counting and measuring small test organisms.[66] This is rapid and accurate, and does not entail handling the organism. With this method, use exposure tanks with glass bottoms and drains that allow the water level to be drawn down. To count and measure test organisms, draw water down to a depth of 2 to 3 cm and transfer tank to a light box having fluorescent lights under a square millimeter grid of adequate size. Photograph aquarium bottom; this shows the organisms over the grid. On an enlargement of the picture, count and measure organisms.

f. Photoperiod and artificial light: In long-term studies to determine water quality requirements for those species requiring annual-light-cycle photoperiods, simulate the natural seasonal daylight and darkness periods at the locality or some central location.[67] (See Section 810C for more information on light cycles for fish.)

Use wide-spectrum fluorescent tubes as a light source similar to daylight.# Some organisms require subdued light, others need a place to hide, and some, such as lake trout eggs, require darkness during certain life stages. Base exposure to light on what is normal to, and required by, the species. Measure light intensity at the water surface. In short-term tests a standard photoperiod of 16 hr light, 8 hr dark is suggested but often usual laboratory lighting is adequate.

g. Exposure chambers: For organisms weighing more than 0.5 g, use a test solution between 15 and 30 cm deep. In short-term tests, these organisms often are exposed in about 15 L solution in 20-L wide-mouth, soft-glass bottles. Fabricate test containers of other sizes by welding (not soldering) stainless steel, by gluing double-strength or stronger window glass with clear silicone adhesive, or by modifying glass bottles, battery jars, or beakers to provide screened overflow holes or V-notches. Because silicone adhesives absorb some organic chlorine and organophosphorous pesticides, expose as little of the adhesive as possible to the water. Place extra beads of adhesive for added strength only on the outside of containers. Expose smaller organisms in 4-L wide-mouth, soft-glass bottles or battery jars that contain 2 to 3 L solution. Expose daphnids, midge larvae, copepods, and other small organisms in loosely-covered beakers or other containers.

Keep surface areas small in relation to volume to limit sorption on vessel walls. With flow-through tests keep liquid surface area/volume ratio small to reduce loss of volatiles. For various exposure chamber designs see Sections 804 through 810.

#For example, OPTIMA 50, Duro-Test Corp., North Bergen, N.J. 07047, or equivalent.

801 F. Calculating, Analyzing, and Reporting Results of Bioassays

The precision of a biological test is limited by a number of factors including the normal biological variation among individuals of a species. Toxicity studies with a randomly selected species do not give accurate information on toxicity to other species and life stages or to an entire biota. A test with one species yields an accurate estimate of the toxicity only to others of that species of similar size, age, and physiological condition, in water with the same or similar characteristics, and under similar test conditions. It is important to use a test organism that is the most sensitive important species to the toxicant in the area under consideration when the objective is the protection of the biota.

1. Analyzing Results of Quantal Bioassays[68]

Responses are of two kinds: quantal and quantitative. In a *quantal* test, an organism either shows the response under study or does not show it, for example, it dies or does not die. Thus, at any concentration greater than that tolerated without effect, a certain percentage of test organisms will show the response within some stated time. In a *quantitative* or graded test, each organism yields a response that is variable in degree, e.g., amount of growth.

Methods for quantal tests are designed to estimate the concentrations of a test material that affects 50% of the test organisms. Determine either death or a sublethal effect, such as immobilization, turnover, fatigue in a swimming test,[69] avoidance reaction,[70] or a significant effect on such factors as growth, fertility, or tissue structure. For all responses except death, substitute "EC50" for "LC50" in the text below and substitute the word or phrase describing the sublethal response for "mortality." Always report 95% confidence limits for LC50 and EC50 values. The most widely used methods for calculating an LC50 and confidence limits are: probit,[71] logit,[72] moving average,[73] and Litchfield-Wilcoxon.[74]

a. Estimating LC50 and EC50 by probit method using graphical analysis: Tabulate observations of mortality as in Table 801:VII for at least one selected exposure time; this time ordinarily is the longest one used in the test, often 96 hr. Use only one successive 0% and one successive 100% mortality value, plus the ones nearest the center of the range of concentrations.

To construct the graph, plot percentage mortality as the ordinate against concentration as the abscissa on probit paper. Death is on a probit or probability scale and concentration on a logarithmic scale (Figure 801:5). Because the probit scale never reaches 0 or 100%, plot any such points with an arrow indicating their true position.

Next, fit a line to the points by eye. Give most consideration to points between 16 and 84% mortality and try to minimize total vertical deviations of the line from the points. If there is doubt about placing the line, draw it as horizontally as possible because this acknowledges more variability in the data.

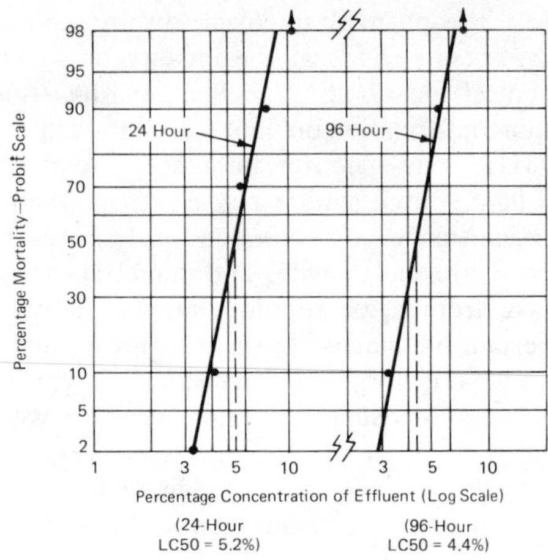

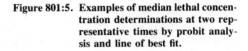

Figure 801:5. Examples of median lethal concentration determinations at two representative times by probit analysis and line of best fit.

Read the concentration causing 50% mortality from the fitted line; this is the estimated LC50 for the selected exposure time. Report this as the result of the test. In the example, in Figure 801:5, the estimated 96-hr LC50 is approximately 4.4%, that is, the estimated concentration that would kill the average or typical test organism in 96 hr. The LC50 estimated by graphical procedures is usually of sufficient accuracy. For example, the first graphical estimates shown in Table 801:VII are very close to those obtained by formal probit analysis with a computer.[75]

b. Confidence limits of the LC50: Calculate confidence limits of the LC50 by probit analysis. Estimate confidence limits for the LC50 of the longest exposure time. Use simplified nomographic methods for field work or when a computer is not available.[74] For more precise calculations computer programs are available.[75] Even when a formal probit analysis is carried out with a computer, make a graph such as Figure

801:5 to check the reasonableness of the computed LC50.

Test for significant difference between two LC50's by examining confidence limits for overlap. If there is no overlap the LC50's are significantly different. However, the LC50's still may be different if the confidence limits overlap. Test for significant differences more exactly by the formula:

$$f_{1,2} = \frac{1.96\ SE_{diff}}{LC50_{1,2}}$$

$$= \text{antilog } \sqrt{(\log f_1)^2 + (\log f_2)^2}$$

where:

f = factor for 95% confidence limits of LC50, i.e., the confidence limits are LC50 × f and LC50 ÷ f (f = antilog of two standard deviations of the log LC50).

This formula has been adapted to simple nomogram use.[74] If the ratio (greater LC50)/(smaller LC50) exceeds the value for $f_{1,2}$, the LC50's are significantly different.

The confidence limits about the LC50 do *not* describe variability of the LC50 under conditions other than those tested. The limits indicate the accuracy of the estimate of replicate tests at the same time under the same conditions. A precision of about 10% is sometimes attainable, but better than that is not to be expected unless more than 10 organisms are exposed at each concentration.

c. Other methods of analyzing results: The graph for estimating the LC50 (Figure 801:5) can be constructed with an arithmetic scale for percentage mortality. However, the probit scale is preferred because it usually gives a straight line.

Logits have been used instead of probits with equivalent results.[72] Reciprocal transformations and angle (arcsine) transformations with estimation of LC50 by a moving average also have been used.[73] These methods have limitations but the last one is recommended for estimating

TABLE 801.VII. EXPERIMENTAL DATA FROM HYPOTHETICAL TOXICITY TEST SUBJECTED TO PROBIT ANALYSIS

Concentration of Waste *% by volume*	No. of Test Organisms	Number of Test Organisms Dead at							
		2 hr	4 hr	6 hr	8 hr	24 hr	48 hr	72 hr	96 hr
10	10	1	4	7	9	10	10	10	10
7.5	10	0	1	2	6	9	9	10	10
5.6	10	0	0	0	2	7	7	8	9
4.2	10	0	0	0	0	1	4	4	4
3.2	10	0	0	0	0	0	1	1	1
0	10	0	0	0	0	0	0	0	0
LC50, %, estimated from graph		10	10	9	7.1	5.2	4.7	4.5	4.4
LC50, estimated by probit analysis		—	—	8.96	7.02	5.27	4.70	4.46	4.34
95% confidence limits		—	—	7.60	5.82	4.53	3.95	3.87	3.49
		—	—	10.5	8.42	6.12	5.59	5.14	5.40
Slope of probit line		—	—	10.9	8.42	10.1	7.03	9.54	11.3

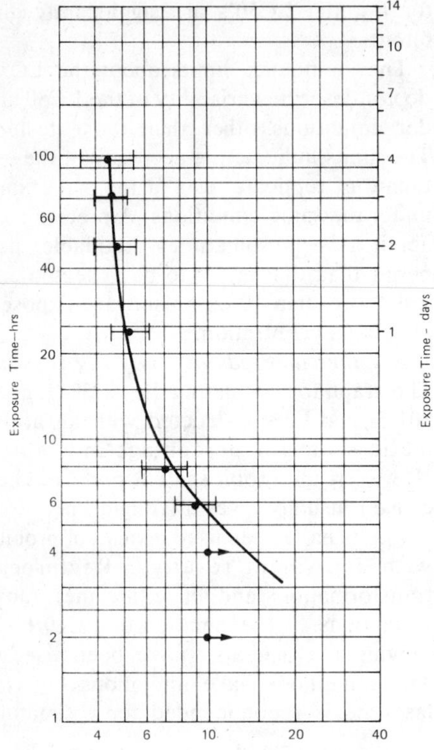

Figure 801:6. Toxicity curve, drawn from LC50's determined in Figure 801:5. Curve almost asymptotic. The 95% confidence limits are shown for each LC50.

LC50 and confidence limits when fewer than two partial lethalities have been obtained. Highly sophisticated computerized approaches describing multivariate response surfaces are available.[76,77] Median effective time (ET50) for mortality at each concentration is estimated by plotting percentage mortality on a probit scale against time on a logarithmic scale and then using probit analysis technics similar to those given above.[71,78–80] Procedures are the same, although more frequent observations of mortality may be required.

d. Plotting toxicity curves: Most tests provide information on mortality at times before the final selected time. Use such information to plot a toxicity curve. Estimate LC50 from a graph plotted as in Figure 801:5 for each observation time. Use the series of LC50's to construct a toxicity curve as the experiment proceeds (Figure 801:6).

A toxicity curve gives an overall picture of test progress and indicates when acute lethality has ceased. This is indicated by the curve becoming asymptotic to the time axis.[68] In Figure 801:6 the toxicity curve closely approaches an asymptote. The LC50 for an exposure time in the asymptotic part of the curve (asymptotic LC50) also can be termed the "threshold" or "incipient" LC50. The threshold of acute lethality for the median organism has greater theoretical significance than an LC50 for some arbitrary time.

Asymptotic LC50's usually can be determined for most macroinvertebrates and fish within a 96- to 168-hr exposure.[68] If no threshold is found, report this fact.

e. Mortality in controls: Control mortality should not be greater than 10% and preferably not more than 5%, representing an occasional weak organism in a group. More than this is unsatisfactory and requires repetition of the test. Sometimes with long tests or with some invertebrates that have considerable mortality under the best possible conditions, it is necessary to use Abbot's formula:[81]

$$P = \frac{P^* - C}{1 - C}$$

where:

P and P^* = corrected and observed proportions responding to the experimental stimulus and

C = proportion responding in the control

This approach does not solve the problem of probable interaction of stress from the toxicant with whatever stress is causing mortality in the control.

2. Analyzing Results of Graded or Quantitative Bioassays

In quantitative or graded tests, each organism gives a response that is measurable on a continuous graded scale. For example, each organism might show a measurable percentage increase in body weight. Because there usually are many test organisms, a series of several graded measurements is generated for each test concentration. A one-way analysis of variance may be used initially to assess significance of differences.

Use one of several technics such as the Student-Newman-Keuls test,[82] Duncan's new multiple-range test,[83,84] or Dunnett's test[85] to determine whether responses for a given concentration are significantly different from responses for the control.

These technics are applicable to most graded responses described in Sections 802 through 809.

3. Reporting Results

Report the LC50 with specified exposure time and the confidence limits of the LC50. Graphically show the toxicity curve or a list of the LC50's for different exposure times. State mortality in controls.

Provide descriptions of: (*a*) test organisms, their species, number, source, weight, and condition, acclimation to test conditions, treatment for disease and parasites before use, and observations on behavior during the test; (*b*) tested material, its source, storage, and physical and chemical properties; (*c*) dilution water, its source, physical and chemical properties and their variations during the test, pretreatment, additives, unusual constituents, or known contaminants; (*d*) test solution, its physical and chemical properties, especially the concentration of toxic component, if measurable, and the test temperature; (*e*) test method, a brief mention if standard, or its description if different, plus the specific experimental design; (*f*) test conditions, including type of container with volume and depth of solution, number of organisms and loading rate, toxicant delivery system, flow rate, or frequency of renewal; and (*g*) the criterion of response.

801 G. Interpreting and Applying Results of Bioassays

The 48- and 96-hr LC50 values are useful measures of relative acute lethal toxicity to organisms under specified conditions. These values do not represent safe concentrations in natural habitats. Long-term exposure to much lower concentrations may be lethal to fish and other organisms and may cause nonlethal impairment of their function. The possibility of synergism or antagonism and other multiple effects of multiple toxicants must be considered.

A review[25] of water quality requirements of fish serves as a general introduction to physiological, toxicological, and ecological fundamentals. Other reviews[10,11,68] also are of value in interpreting results. Several discussions of the uses and value of biological tests in the establishment of effluent requirements and water quality standards for meeting water pollution problems have appeared.[4,25,68,86–88]

801 H. References

1. TARZWELL, C.M. 1958. The use of bioassays in the safe disposal of electroplating wastes. *Amer. Electroplaters Soc. 44th Annu. Tech. Proc.*:60.
2. TARZWELL, C.M. 1971. Bioassays to determine allowable waste concentrations in the aquatic environment. I. Measurement of pollution effects on living organisms. *Proc. Royal Soc. London B.* 177:279.
3. BROWN, L.M. 1975. Concepts and outlook in testing the toxicity of substances to fish. *In* G.E. Glass, Bioassay Techniques and Environmental Chemistry. Ann Arbor Science Publ., Inc. pp. 73–79.
4. GREENE, J.C., W.E. MILLER, T. SHIROYAMA & T.E. MALONEY. 1975. Utilization of algal assays to assess the effects of municipal, industrial, and agricultural wastewater effluents upon phytoplankton production in the Snake River System. *Water, Air, Soil Pollut.* 4:415.
5. FRY, F.E.J. 1947. Effects of the environment on animal activity. Univ. Toronto Stud. Biol. Ser. 55., *Publ. Ont. Fish. Res. Lab.* 68:1.
6. DOUDOROFF, P. et al. 1951. Bioassay methods for the evaluation of acute toxicity of industrial wastes to fish. *Sewage Ind. Wastes* 23:1380.
7. LLOYD, R. & D.H.M. JORDAN. 1963. Predicted and observed toxicities of several sewage effluents to rainbow trout. *J. Proc. Inst. Sewage Purif.* Pt.2:167.
8. BALL, I.R. 1967. The relative susceptibilities of some species of freshwater fish to poisons—I. Ammonia. *Water Res.* 1:767.
9. SPRAGUE, J.B. 1969. Measurement of pollutant toxicity to fish. I. Bioassay methods for acute toxicity. *Water Res.* 3:793.
10. Water Quality Criteria. Report of the National Technical Advisory Committee to the Secretary of the Interior. 1968. FWPCA, U.S. Dep. Interior, U.S. Government Printing Office I-X.
11. NATIONAL ACADEMY OF SCIENCE AND NATIONAL ACADEMY OF ENGINEERING. 1972. Water Quality Criteria. U.S. Government Printing Office, Washington, D.C.
12. COMMITTEE ON METHODS FOR TOXICITY TESTS WITH AQUATIC ORGANISMS. 1975. Methods for Acute Toxicity Tests with Fish, Macroinvertebrates, and Amphibians. Ecol. Res. Ser., EPA-660/3-75-009.
13. KOPPERDAHL, F.R. 1976. Guidelines for Performing Static Acute Toxicity Fish Bioassays in Municipal and Industrial Waste Waters. Rep. to Calif. State Water Resources Control Board, Sacramento.
14. MOUNT, D.I. & C. STEPHAN. 1967. A method for establishing acceptable toxicant limits for fish—Malathion and the butoxyethanol ester of 2,4-D. *Trans. Amer. Fish. Soc.* 96:185.
15. MOUNT, D.I. 1977. An Assessment of Application Factors in Aquatic Toxicology. U.S. EPA-600/3-77-085, Corvallis, Ore.
16. U.S. DEPARTMENT OF COMMERCE. 1970. Aquarium Design Criteria, Special ed. National Fish. Center Aquarium.
17. CLARK, J.R. & R.L. CLARK, eds. 1964. Sea water systems for experimental aquariums. U.S. Fish & Wildl. Serv. Bur. Sports Fish & Wildl., Res. Rep. 63:1.
18. SPOTTE, S. 1973. Marine Aquarium Keeping-The Science, the Animals, the Art. Wiley Interscience Publ., New York, N.Y.
19. LASKER, R. & L.L. VLYMER. 1969. Experimental seawater aquarium. U.S. Fish & Wildl. Serv. Bur. Commercial Fish., Circ. 334:1.
20. TARZWELL, C.M. 1962. Development of water quality criteria for aquatic life. *J. Water Pollut. Control Fed.* 34:1178.
21. SURBER, E.W., J.N. ENGLISH & G.N. MCDERMOTT. 1962. Tainting of fish by outboard motor exhaust wastes as related to gas and oil consumption. PHS Publ. No. 999-WP-25, Environmental Health Ser.:170. Washington, D.C.
22. AMAN, C.W. 1955. The relation of taste and odor to flavor. *Taste Odor Control J.* 21(10):1.
23. BALDWIN, R.E., D.H. STRONG & J.H. TORRIE. 1961. Flavor and aroma of fish taken from four fresh-water sources. *Trans. Amer. Fish. Soc.* 90:176.

24. DAWSON, E.H. & B.L. HARRIS. 1951. Sensory methods for measuring differences in food quality. *U.S. Dep. Agr. Agr. Inform. Bull.* 34:1.

25. BROWN, M.E., ed. 1957. The Physiology of Fishes. Vol. 1, Metabolism. Academic Press Inc., New York, N.Y.

26. WARREN, C.E. & P. DOUDOROFF. 1971. Biology and Water Pollution Control. W.B. Saunders Co., Philadelphia, Pa.

27. MOUNT, D.I. & W.A. BRUNGS. 1967. A device for continuous treatment of fish in holding chambers. *Trans. Amer. Fish. Soc.* 96:55.

28. CLINE, T.F. & G. POST. 1972. Therapy for trout eggs infected with *Saprolegnia*. *Progr. Fish-Cult.* 34:148.

29. RUCKER, R.R. & K. HODGEBOOM. 1953. Observations on gas-bubble disease of fish. *Progr. Fish-Cult.* 15:24.

30. MARKING, L.L. & V.K. DAWSON. 1973. Toxicity of Quinaldine Sulfate to Fish. Invest. Fish Control No. 48, U.S. Fish & Wildl. Serv., Washington, D.C.

31. KESTER, E., I. DREDALL, D. CONNERS & R. PYTOWICZ. 1967. Preparation of artificial seawater. *Limnol. Oceanogr.* 12:176.

32. ZAROOGIAN, G.E., G. PESCH & G. MORRISON. 1969. Formulation of an artificial sea water media suitable for oyster larvae development. *Amer. Zool.* 9:1141.

33. ZILLIOUS, E.J., H.R. FOUCK, J.C. PRAGER & J.A. CARDIN. 1973. Using *Artemia* to assay oil dispersement toxicities. *J. Water Pollut. Control Fed.* 45:2389.

34. DAVEY, E.W., J.H. GENTILE, S.J. ERICKSON & P. BETZER. 1970. Removal of trace metals from marine culture medium. *Limnol. Oceanogr.* 15:486.

35. BAHNER, L.H. & D.R. NIMMO. 1975. A salinity controller for flow-through bioassays. *Trans. Amer. Fish. Soc.* 104:388.

36. NEEDHAM, J.G., P.S. GALTSOFF, F.E. LUTZ & P.S. WELSH. 1937. Culture Methods for Invertebrate Animals. Comstock Publ. Co., Inc. Ithaca, N.Y. XXXII.

37. HIRANO, R. & Y. OSHIMA. 1963. Rearing of larvae of marine animals with special reference to their food organisms. *Bull. Jap. Soc. Sci. Fish.* 29:282.

38. MAY, R.C. 1970. Feeding larval marine fishes in the laboratory, A review. Calif. Mar. Res. Comm., CalCOFI Rep. 14:76.

39. BRAUHN, J.L. & R.A. SCHOETTGER. 1975. Acquisition and Culture of Research Fish: Rainbow Trout, Fathead Minnows, Channel Catfish, and Bluegills. Ecol. Res. Ser. EPA-660/3-75-011.

40. MEYERS, S.P. & Z.P. ZEIN-ELDIN. 1972. Binders and pellet stability in development of crustacean diets. *Proc. 3rd Annu. Workshop World Mariculture Soc.*:351.

41. HESSELBERG, R.J. & R.M. BURRESS. 1967. Labor saving devices for bioassay laboratories. U.S. Bur. Sport Fish Wildl. Invest. Fish Control 21:1.

42. WILLFORD, W.A. 1967. Toxicity of 22 therapeutic compounds to six fishes. U.S. Bur. Sport Fish Wildl. Invest. Fish. Control 18:1

43. HENDERSON, D. & C.M. TARZWELL. 1957. Bioassays for the control of industrial effluents. *Sewage Ind. Waste* 29:1002.

44. LENNON, R.E. & C.R. WALKER. 1964. Investigations in fish control I. Laboratories and methods for screening fish-control chemicals. U.S. Bur. Sport Fish Wildl. Circ. 185:1.

45. MCALLISTER, W.A., JR., W.L. MAUCH & F.L. MAYER, JR. 1972. A simplified device for metering chemicals in intermittent-flow bioassays. *Trans. Amer. Fish. Soc.* 101:555.

46. LOWE, J.I. 1964. Chronic exposure of spot, *Leiostomus xanthurus*, to sublethal concentrations of toxaphene in seawater. *Trans. Amer. Fish. Soc.* 93:396.

47. MOUNT, D.I. & R.E. WARNER. 1965. A Serial Dilution Apparatus for Continuous Delivery of Various Concentrations of Material in Water. PHS Publ. No. 999-WP-23, Environ. Health Ser., U.S. Dep. HEW, Washington, D.C.

48. CHANDLER, J.H., H.O. SANDERS & D.F. WALSH. 1974. An improved chemical delivery apparatus for use in intermittent-flow bioassays. *Bull. Environ. Contam. Toxicol.* 12:123.

49. SCHIMMEL, S.C., D.J. HANSEN & J. FORESTER. 1974. Effects of aroclor® 1254 on laboratory-reared embryos and fry of sheepshead minnows (*Cyprinodon variegatus*). *Trans. Amer. Fish Soc.* 103:582.

50. FREEMAN, R.A. 1971. A constant flow delivery device for chronic bioassay. *Trans. Amer. Fish. Soc.* 100:135.

51. Bengtsson, B.E. 1972. A simple principle for dosing apparatus in aquatic systems. *Arch. Hydrobiol.* 70:413.
52. Granmo, A. & S.C. Kollberg. 1972. A new simple water flow system for accurate continuous flow tests. *Water Res.* 6:1597.
53. Benoit, D.A. & F.A. Puglisi. 1973. A simplified flow-splitting chamber and siphon for proportional diluters. *Water Res.* 7:1915.
54. Lichatowich, J.A., P.W. O'Keefe, J.A. Strand & W.L. Templeton. 1973. Development of methodology and apparatus for the bioassay of oil. *In* Proc. Joint Conf. Prevention and Control of Oil Spills, p. 659. American Petroleum Inst., EPA, & U.S. Coast Guard, Washington, D.C.
55. Abram, F.S.H. 1973. Apparatus for control of poison concentration in toxicity studies with fish. *Water Res.* 7:1875.
56. Mount, D.I. & W.A. Brungs. 1967. A simplified dosing apparatus for fish toxicology studies. *Water Res.* 1:21.
57. Lemke, A.E. 1964. A new device for constant-flow test chambers. *Progr. Fish-Cult.* 26:136.
58. Jackson, H.W. & W.A. Brungs. 1966. Biomonitoring of industrial effluents. *Proc. 21st. Ind. Waste Conf.*, Purdue Univ., Eng. Ext. Bull. 121:117.
59. Surber, E.W. & T.O. Thatcher. 1963. Laboratory studies of the effects of alkyl benzene sulfonate (ABS) on aquatic invertebrates. *Trans. Amer. Fish. Soc.* 92:152.
60. Burrows, R.E. 1949. Prophylactic treatment for control of fungus, *Saprolegnia parasitica*. *Prog. Fish-Cult.* 11:97.
61. Thatcher, T.O. & J.F. Santner. 1966. Acute toxicity of LAS to various fish species. *Proc. 21st Ind. Waste Conf.*, Purdue Univ., Eng. Ext. Bull. No. 121:996.
62. Brungs, W.A. & G.W. Bailey. 1966. Influence of suspended solids on the acute toxicity of endrin to fathead minnows. *Proc. 21st Ind. Waste Conf.*, Purdue Univ., Eng. Ext. Bull. No. 121:4.
63. Esvelt, L.A. & J.D. Conners. 1971. Continuous-flow fish bioassay apparatus for municipal and industrial effluents. *In* L.A. Esvelt, W.J. Kaufman & R.E. Selleck, eds. Toxicity Removal from Municipal Wastewaters. Vol. IV of A Study of Toxicity and Biostimulation in San Francisco Bay-Delta Waters. p. 155. Sanitary Engineering Research Lab., Univ. Calif., Berkeley.
64. Sillen, L.C. & A.E. Martell. 1964. Stability constants of metal ion complexes. Spec. Publ. 17, Chemical Soc., London, England.
65. Andrew, R.W. & G.E. Glass. 1974. Amperometric methods for determining residual chlorine, ozone and sulfite. U.S. EPA, National Water Quality Lab., Duluth, Minn.
66. McKim, J.M. & D.A. Benoit. 1971. Effect of long-term exposures to copper on survival, reproduction and growth of brook trout *Salvelinus fontinalis* (Mitchill). *J. Fish. Res. Board Can.* 28:655.
67. Drummond, R.A. & W.F. Dawson. 1970. An inexpensive method for simulating a diel pattern of lighting in the laboratory. *Trans. Amer. Fish. Soc.* 99:434.
68. Sprague, J.B. 1970. Review paper, Measurement of pollutant toxicity to fish—II. Utilizing and applying bioassay results. *Water Res.* 4:3.
69. Brett, J.R. 1967. Swimming performance of sockeye salmon (*Oncorhynchus nerka*) in relation to fatigue time and temperature. *J. Fish. Res. Board Can.* 24:1731.
70. Sprague, J.B. 1968. Avoidance reactions of rainbow trout to zinc sulphate solutions. *Water Res.* 2:367.
71. Finney, D.J. 1971. Probit Analysis, 3rd ed. Cambridge Univ. Press, London and New York.
72. Berkson, J. 1953. A statistically precise and relatively simple method of estimating the bioassay with quantal response based on the logistic function. *J. Amer. Statist. Ass.* 48:565.
73. Pickering, O.H. & W.N. Vigor. 1965. The acute toxicity of zinc to eggs and fry of the fathead minnow. *Progr. Fish-Cult.* 27:153.
74. Litchfield, J.T. & F. Wilcoxon. 1949. A simple method of evaluating dose-effect experiments. *Pharmacol. Exp. Ther.* 96:99.
75. Dixon, W.J., ed. 1970. BMD biomedical computer programs. *In* Automatic Computation Ser. No. 2, 2nd ed., Univ. of Calif. Press, Los Angeles.

76. Alderdice, D.F. 1972. Factor combinations. Responses of marine poikilotherms to environmental factors acting in concert. *In* O. Kinne, ed. Marine Ecology. Vol. 1, Part 3:1659. Wiley-Interscience, London and New York.
77. Vernberg, W.B., P. De Coursey & W.J. Padgett. 1973. Synergistic effects of environmental variables on larvae of *Uca pugilator*. *Mar. Biol.* 22:307.
78. Litchfield, J.T. 1949. A method for rapid graphic solution of time-percent effect curves. *Pharmacol. Exp. Ther.* 97:399.
79. Shepard, M.P. 1955. Resistance and tolerance of young speckled trout (*Salvelinus fontinalis*) to oxygen lack, with special reference to low oxygen acclimation. *J. Fish. Res. Board Can.* 12:387.
80. Sprague, J.B. 1973. The ABC's of pollutant bioassay using fish. *In* J. Cairns and K.L. Dickson, eds. Biological Methods for the Assessment of Water Quality. p. 6. American Soc. for Testing and Materials, Philadelphia, Pa. ASTM STP 528.
81. Tattersfield, F. & H.M. Morris. 1924. An apparatus for testing the toxic values of contact insecticides under controlled conditions. *Bull. Entomol. Res.* 14:223.
82. Keuls, M. 1952. The use of the "studentized range" in connection with an analysis of variance. *Euphytica* 1:112.
83. Duncan, D.B. 1955. Multiple range and multiple F tests. *Biometrics* 11:1.
84. Middlebrooks, E.J., D.B. Porcella, E.A. Pearson, P.H. McGauhey & G.A. Rohlich. 1971. Biostimulation and algal growth kinetics of wastewater. *J. Water Pollut. Control Fed.* 43:454.
85. Dunnett, C.W. 1955. A multiple comparison procedure for comparing several treatments with a control. *J. Amer. Statist. Ass.* 50:1096.
86. Sprague, J.B. 1971. Review paper, Measurement of pollutant toxicity to fish—III. Sublethal effects and "safe" concentrations. *Water Res.* 5:245.
87. Mount, D.I. 1968. Chronic toxicity of copper to fathead minnows, *Pimephales promelas*, (Rafinesque). *Water Res.* 2:215.
88. Cardwell, R.D., D.G. Poreman, T.R. Payne & D.J. Wilbur. 1976. Acute Toxicity of Selected Toxicants to Six Species of Fish. EPA-600/3-76-0011. NTIS, Springfield, Va.

802 BIOSTIMULATION (ALGAL PRODUCTIVITY)

802 A. General Principles

The algal test procedure for determining primary productivity of a water sample is based on Liebig's "Law of the Minimum," which states that "growth is limited by the substance that is present in minimal quantity in respect to the need of the organism." Biostimulants are substances that increase algal growth or the potential for algal growth.

Algal species used in biostimulation tests are selected to allow for a standardized test of growth response using a well-characterized organism under standard laboratory conditions. See Sections 1001,

1002, and 1003 for methods appropriate to field studies.

Effects of various substances on maximum specific growth rate and maximum crop of selected algal species cultured under specified conditions are measured. Results are assessed by comparing growth in the presence of selected substances to growth in controls. Experimental designs must incorporate sufficient replication to permit statistical evaluation of results.

The algal assays consist of three steps: (*a*) selection and measurement of appropriate factors or conditions during the assay (for example, biomass indicators such as total cell carbon); (*b*) presentation and statistical evaluation of measurements; and (*c*) interpretation of results.

Interpretation of results involves assessment of receiving water to determine its nutritional status and its sensitivity to change, potential effects of materials on algal growth in receiving waters, effects of changes in waste treatment processes on algal growth in receiving waters, impact of nutrients in tributary waters on algal growth in lakes and receiving waters, and effects of measures such as those used in lake restoration.

The maximum specific growth rate and the maximum standing crop are responses that can be estimated from growth measurements. The maximum specific growth rate is related to concentration of rate-limiting nutrient present. The maximum standing crop is proportional to the initial amount of limiting nutrient available.

802 B. Planning and Evaluating Algal Assays

Because water quality may vary greatly with time and point of collection, establish sampling programs to obtain representative and comparable data. It may be valuable to sample both epilimnion and hypolimnion in stratified water bodies. Use transection lines to secure adequate samples and locate sampling stations. In rivers and streams, take samples upstream and downstream from suspected nutrient sources and from tributary streams. When general conditions are evaluated, include samples from a number of natural waters having a range of representative water qualities.

Because the nutrient contents of natural waters and wastewaters often vary daily and seasonally, use composite or frequent grab sampling.

Deficiency of any essential nutrient may limit algal growth, but tests usually are made for those few nutrients most likely to be growth-limiting (nitrogen, phosphorus, trace elements).

To evaluate the potential effect of a substance on receiving waters, consider the following factors: amount and distribution, chemical and/or physical nature, fate and persistence, pathways by which it will reach the receiving water, dilution by the receiving body, and selection of appropriate test water.[1,2]

When the algal assay is used to measure stimulation of growth by a given effluent, include the following in the overall evaluation: effluent conditions, growth measurements and test organisms, concentration of growth-limiting nutrient, and potential nutrient concentration and changes in availability.

802 C. Apparatus

1. Sampling and Sample Preparation

a. Sampler, non-metallic.

b. Sample bottles, borosilicate glass, linear polyethylene, polycarbonate, or polypropylene, capable of being autoclaved.

c. Membrane filter apparatus, for use with 47-mm petri prefilter pads and 0.45-μm-porosity filters.

d. Autoclave or pressure cooker, capable of producing 108 kPa at 121 C.

2. Culturing and Incubation

a. Culture vessels: Use permanently numbered erlenmeyer flasks of good-quality borosilicate glass. Use the same brand of glass throughout the laboratory. When trace nutrients are being studied, use special glassware such as high-silica glass, polycarbonate, or coated (e.g., silicone) glassware. While flask size is not critical, the surface-to-volume ratios of the growth medium are, because of CO_2 limitation. Use the following:

40 mL sample in 125-mL flask
60 mL sample in 250-mL flask
100 mL sample in 500-mL flask

Eliminate those flasks having anomalous growth from future tests.

b. Culture closures: Use demonstrably nontoxic foam plugs,* loose-fitting aluminum foil, or inverted beakers to permit some gas exchange and prevent contamination. Determine for each batch of closures whether that batch has any significant effect on maximum specific growth rate and/or maximum standing crop.

c. Constant-temperature room: Provide constant-temperature room, or equivalent incubator, capable of maintaining temperature of 18 ± 2 C (marine) to 24 ± 2 C (freshwater).

d. Illumination: Use "cool-white" fluorescent lighting to provide 4,304 lux ± 10% or 2,152 lux ± 10% measured adjacent to the flask at the liquid level with closure in place.

An illumination meter can be used in place of a spectroradiometer to obtain the required light energy level. For example, the energy level output of a bank of six 144-cm "cool white" fluorescent lamps (GE 40-W, @ 60 Hz) was approximately 1,300 $\mu W/cm^2$ (range, 380 to 760 nm) at a distance of 67 cm, as measured with an ISCO Model SRC spectroradiometer. With the same measurement geometry, a Weston Model 756 Illumination Meter reading was 4,304 lux. All reflecting surfaces were matte white. Therefore, utilizing a calibrated illumination meter, one may, by adjusting the height of the lights, achieve a known energy level output of 1,300 $\mu W/cm^2$. For further discussion of problems of differences in absorption of light by photosynthesizing organisms and by the human eye and their measurement, see Tyler.[3]

e. Light meter: Calibrate meter against a standard light source or light meter.

3. General Equipment

a. Analytical balance capable of weighing 100 g with a precision of ± 0.1 mg.

b. Microscope and illuminator, good quality, general purpose.

c. Hemocytometer or plankton counting slide.

d. pH meter to measure to ± 0.1 pH unit.

e. Dry-heat oven capable of operating at up to 120 C.

f. Centrifuge capable of a relative centrifugal force of at least 1,000 × *g*.

g. Spectrophotometer or colorimeter for use at 600 to 750 nm.

*Gaymar white, polyurethane foam plugs, VWR Scientific or Gaymar Industries, Inc., 701 Seneca St., Buffalo, N.Y. 14210, or *demonstrably nontoxic* equivalent.

4. Optional Equipment

a. Electronic particle (cell) counter.

b. Fluorometer, suitable for chlorophyll *a*.

c. Shaker table, capable of 100 oscillations/min.

802 D. Sample Handling

Use a nonmetallic water sampler and autoclavable storage container. Do not reuse containers when toxic or nutrient contamination is suspected. Leave a minimum of air space in the transport container and keep it in the dark at 0 to 4 C.

To use unialgal test species, "remove" indigenous algae before assay by either separation or destruction. Filter, autoclave, or autoclave and filter, according to information sought.[4] In either case, prepare sample as soon as possible (within 24 hr) after collection.

Use membrane filtration to remove indigenous algae before determining deficiencies in soluble nutrients that have not been taken up by filtrable organisms or to predict effect of added nutrients to a test water at a specific time. Pretreat 0.45-μm-porosity membrane filters by filtering at least 50 mL distilled or deionized water. Discard filtrate and use another collecting vessel. Filter the quantity of sample as needed under reduced pressure of 51 kPa or more. If the sample contains a large amount of suspended material, filter through an appropriate prefilter (for example, glass fiber), also washed with water, before filtration through the 0.45-μm-porosity filter.

Use autoclaving to determine amount of algal biomass that can be grown from all nutrients in the water, including those contained in filtrable organisms. Autoclave fresh-water samples at 108 kPa and 121 C for 30 min or 10 min/L of sample, whichever is longer. Pasteurize marine or estuarine samples for 4 hr at 60 C. After autoclaving, cool and let sample equilibrate either in an air or CO_2-rich atmosphere in order to restore lost CO_2. If an electronic particle counter is to be used for cell counting, pass the CO_2-equilibrated sample through a 0.45-μm membrane filter.

Changes occur in water samples during storage regardless of storage conditions. The extent and nature of these changes is not well known; therefore, keep storage duration to a minimum after sample preparation. Store samples in full containers with no air space. Before sample preparation, store samples in the dark at 0 to 4 C. If prolonged storage is anticipated, prepare sample first and then store in the dark at 0 to 4 C.

802 E. Synthetic Algal Culture Medium

See Section 801D.4*c* 1).

802 F. Inoculum

1. Recommended Test Algae

a. Fresh water:

1) *Selenastrum capricornutum* Printz.

2) *Microcystis aeruginosa* Kutz (*Anacystis cyanae* Drouet and Daily).

3) *Anabaena flos-aquae* (Lyngb.) de Brebisson.

4) *Cyclotella* sp.

5) *Nitzschia* sp.

6) *Synedra* sp.

For any of the latter three algae, add 20 mg Si/L (101.214 mg $Na_2SiO_3 \cdot 9H_2O$/L) to the culture medium as noted in Section 801D.4*c*.

b. Marine:

1) *Dunaliella tertiolecta* Butcher (DUN Clone) Woods Hole Oceanographic Institution.

2) *Thalassiosira pseudonana* (Hasle and Heimdal) (CN Clone) (old *Cyclotella nana*) Univ. Rhode Island. Do not shake.

3) *Skeletonema costatum* (Greville) Clevel.

2. Sources of Test Algae

Obtain algal cultures from: Dr. Richard C. Starr, Culture Collection of Algae, Department of Botany, University of Texas, Austin, Tex. 78712 or Special Studies Branch, Corvallis Environmental Research Laboratory, Environmental Protection Agency, 200 SW 35th Street, Corvallis, Ore. 97330.

Other sources of cultures are: Virginia Institute of Marine Science, Gloucester Point, Va. 23062; Chesapeake Biological Laboratory, Box 38, Solomons, Md. 20688; Dr. Robert Guillard, Woods Hole Oceanographic Institution, Woods Hole, Mass. 024543; or Graduate School of Oceanography, University of Rhode Island, Narragansett, R.I. 02881.

After receipt of cultures, check authenticity and purity.

3. Maintaining Stock Cultures

a. Medium: See Section 801D.4*c*1).

b. Incubation conditions:

1) Freshwater species—Temperature 24 ± 2 C under continuous cool-white fluorescent lighting at 4,304 lux ± 10% for *S. capricornutum* and the diatoms; 2,152 lux ± 10% for *M. aeruginosa* and *A. flos-aquae*; shake at 110 oscillations/min. If other species are used, always relate growth of those species to *S. capricornutum* to insure comparability.

2) Marine species—Temperature 18 ± 2 C under continuous cool-white fluorescent lighting at 4,304 lux ± 10% for *D. tertiolecta* (shake at 110 oscillations/min) and for *T. pseudonana* (do *not* shake but swirl daily). Higher temperatures (up to 24 C) may be justified for appropriate test species used in the Gulf of Mexico and other warm-water marine systems. If other species are used, always relate growth of those species to *D. tertiolecta* or *S. costatum* to insure comparability.

c. First stock transfer: Upon receipt of inoculum species, transfer a portion to the algal culture medium. (Example: 1 mL of inoculum in 100 mL in a 500-mL erlenmeyer flask).

d. Subsequent stock transfers: Make a new stock transfer, using aseptic technic, as the first operation on opening a stock culture. The volume transfered is not critical so long as enough cells are included to overcome significant growth lag. Make weekly stock transfers to provide a continuing supply of "healthy" cells. Check algal cultures microscopically to insure that the stock cultures remain unialgal.

e. Age of inoculum: Use cultures 1 to 3 wk old as a source of inoculum. For *Selenastrum, Dunaliella*, and the diatoms, a 1-wk incubation often is sufficient to provide enough cells. The blue-green species require a longer time to achieve maximum crop and 2 to 3 wk may be required to provide test inocula.

4. Preparing Inoculum

Centrifuge stock culture and discard supernatant. Resuspend sediment cells in an appropriate volume of glass-distilled water containing 15 mg $NaHCO_3$/L for freshwater species and artificial seawater minus nutrients for marine species (Section 801D.4*c*1); Table 801:II) diluted to appropriate salinity, and again centrifuge. Resuspend sedimented algae again in the proper solution and use as the inoculum.

5. Amount of Inoculum

Count cells suspended in the prepared inoculum and pipet into the test water to give a starting cell concentration as follows:

S. capricornutum	10^3 cells/mL
M. aeruginosa	50×10^3 cells/mL
A. flos-aquae	50×10^3 cells/mL
Diatoms	10^3 cells/mL
D. tertiolecta	10^2 cells/mL

Calculate volume of transfer to result in the above concentrations in the test flasks (Example: for *S. capricornutum*, if there are 5×10^5 cells/mL in stock culture, transfer 0.2 mL/100 mL test water).

802 G. Test Conditions and Procedures

1. Temperature

Keep temperature at 18 ± 2 C for marine species and 24 ± 2 C for freshwater species.

2. Illumination

See Section 802F.3*b*. Measure light intensity adjacent to the flask at the liquid level. A convenient and acceptable way to achieve these two light intensities is to set up the illumination for 4,304 lux and then place cheesecloth over the blue-green algal culture flask to reduce the light intensity at the liquid level to 2,052 lux.

3. Procedure

a. Preparation of glassware: Wash all glassware with detergent (nonphosphate or sodium carbonate) and rinse thoroughly with tap water. Never use chromic acid or similar cleaning solutions. Then rinse with a warm 10% (v/v) solution of reagent-grade HCl. Fill vials and centrifuge tubes with 10% HCl solution for a few minutes. Fill all large containers to about one-tenth capacity with HCl solution and swirl to bathe entire inner surface. After HCl rinse, rinse glassware five times with tap water, then five times with deionized water.

Place pipets in 10% HCl solution for 12 hr or longer and rinse at least 10 times with tap water in an automatic pipet washer, then three times with deionized water. Preferably use disposable pipets to eliminate the need for pipet washing and to minimize possibility of contamination.

Dry cleaned glassware at 105 C in an oven and store either in closed cabinets or on open shelves with tops covered with aluminum foil.

Before use, autoclave culture flasks

covered with aluminum foil at 108 kPa for 15 min. After autoclaving, prerinse flasks with culture medium and invert on absorbent paper for 20 to 30 min to drain.

b. pH control: To insure the availability of CO_2, keep the pH below 8.5 by using optimum surface-to-volume ratios (Section 802C.2), continuously shaking the flask (approximately 100 oscillations/min), and/or ventilating with air or air/CO_2 mixture.

c. Growth measurements: Describe the growth of a test alga in the bottle test[1] by maximum specific growth rate and maximum standing crop.[4] Generally, these measurements are made at different times, the former early in the test and the latter near the end.

1) Maximum specific growth rate—The maximum specific growth rate (μ_{max}) for an individual flask is the largest specific growth rate (μ) occurring at any time during incubation. Determine μ_{max} for a set of replicate flasks by averaging μ_{max} of the individual flasks.

The specific growth rate, μ, is defined by:

$$\mu = \frac{\ln (X_2/X_1)}{t_2 - t_1} \text{ days}^{-1}$$

where:

X_2 = biomass concentration at end of selected time interval,

X_1 = biomass concentration at beginning of selected time interval, and

$t_2 - t_1$ = elapsed time between selected intervals, days.

If biomass (dry weight) is determined indirectly, e.g., by cell counts, compute specific growth rate directly without converting to biomass, if the factor relating the direct determination to biomass remains constant for the time period. Cell counts, chlorophyll content, and other measures of biomass do not necessarily remain constant in relation to one another or to dry weight.

a) Laboratory measurements—The maximum specific growth rate occurs during the logarithmic phase of growth, usually between Day 0 and Day 5; therefore, measure biomass at least daily during the first 5 days of incubation. Indirect measurements of biomass, such as cell counts, normally will be required because of the difficulty in making accurate gravimetric measurements at low cell densities.

b) Computation of maximum specific growth rate—Calculate maximum specific growth rate (μ_{max}) by using the equation in ¶ 1) to determine the daily specific growth rate (μ) for each replicate flask. Average all replicates, using the largest value for each flask. Alternatively, prepare a semi-log plot of biomass concentration versus time for each flask. Ideally, the exponential growth phase is drawn on the plot. If it appears that the data describe two straight lines, use the line of steepest slope. A linear regression analysis of the data also may be used to determine the best-fit straight line. Select two data points that most closely fit the line and determine the specific growth rate (μ). Average the largest specific growth rates for the replicate flasks to obtain μ_{max}.

2) Maximum standing crop—Maximum standing crop in any flask is defined as the maximum algal biomass achieved during incubation. For practical purposes, it may be assumed that the maximum standing crop has been achieved when the increase in biomass is less than 5%/day.

After attaining the maximum standing crop, determine the dry weight of algal biomass gravimetrically. If biomass is determined indirectly, convert results to an equivalent dry weight using appropriate conversion factors.

4. Biomass Monitoring

Several methods may be used, but relate each to dry weight.

a. Dry weight: Use either the aluminum dish or membrane filter method. To use the first, centrifuge a suitable portion of algal suspension, wash sedimented cells three times in distilled water containing 15 mg $NaHCO_3$/L, transfer to tared crucibles or aluminum cups, dry overnight in a hot-air oven at 105 C, and weigh. This method is more sensitive than the second method but cells may be lost during washing.

For the second method, filter a measured portion of algal suspension through a tared 0.45-μm-pore-diam membrane filter. Dry filters for several hours at 60 C in an oven. Place filters in folded sheets of paper or aluminum weighing dishes on which the weights or codes are written. Cool in a desiccator and weigh. Filter a suitable portion of culture under a vacuum of 51 kPa (50 mL or less as the volume or cell density dictates). Rinse filter funnel with 50 mL distilled water containing appropriate salts (802F.4) using a wash bottle and let the rinsings pass through the filter. Dry at 60 C, cool in desiccator, and weigh. To correct for loss of weight of filters during washing, wash two blank filters with 50 mL distilled water, pouring it through slowly under reduced vacuum. Dry and weigh filters and record weight loss. This correction is not large, but is essential for meaningful results on dilute cultures.

b. Cell counting:

1) Electronic particle counting—Suspend *S. capricornutum* cells in a 1% NaCl electrolyte solution in a ratio of 1.0 mL cell suspension to 9 mL of 0.22-μm-filtered saline (10:1 dilution). Pass the resulting suspension through a 100-μm-diam aperture. Each cell that passes through the aperture causes a voltage drop proportional to its displaced electrolyte volume, which is recorded as a count. A knowledge of both the number of particles (cells) counted per unit volume of sample (usually 0.5 mL) and the mean particle (cell) volume displaced allows changes in cell biomass (in microliters per liter) to be calculated. Equations have been developed that accurately relate volume to dry weight.

2) Direct microscopic counting—Use a hemocytometer or plankton counting cell (Part 900). For filamentous algae break up the algal filaments by using a syringe, an ultrasonic bath, a high-speed blender, or vigorous stirring with glass beads. Each of these technics has drawbacks, but expelling the sample forcefully through a syringe against the inside of the flask is most satisfactory. Other methods of biomass measurement such as dry weight, absorbance, or chlorophyll fluorescence are more precise than cell counts for growth assessment of filamentous algae.

c. Absorbance: Measure absorbance with a spectrophotometer or colorimeter at a wavelength of 750 nm. Report instrument make and model, geometry and path length of the cuvette, wavelength used, and the equivalence to biomass (absorbance units per milligram dry weight per liter).

Limit photometric measurement of absorbance to a range of $0.05<D<1.0$, where D represents optical density.

d. Chlorophyll: All algae contain chlorophyll and measuring this pigment can yield some insight into the relative amount of algal biomass present. Measure chlorophyll either in vivo or in vitro by fluorescence.

1) In vivo procedure—Swirl flasks to homogenize. Pipet a portion of cell suspension (5 mL minimum) into a small beaker or vial. Zero fluorometer with a distilled water blank before each sample reading. Pour well-mixed sample into cuvette and read fluorescence.

2) In vitro procedure—Filter a measured sample under vacuum (51 kPa) through a glass fiber filter. Add approximately 1.0 mL $MgCO_3$ suspension (1.0 g finely powdered $MgCO_3$ in 100 mL distilled water) and drain filter thoroughly un-

der suction. Place filter in bottom of a tissue-grinding tube. Add 2 mL 90% acetone to the grinding tube and insert pestle. Grind sample 1 to 2 min (in subdued light) and wash pestle and grinding tube with 5 mL 90% acetone into a 15-mL screw-cap centrifuge tube. Centrifuge (2,000 × g) for 1 to 5 min and let stand in the dark 1 to 2 hr to extract pigment. Measure fluorescence as outlined above. If pheophytin is to be measured, acidify with 2 drops 1N HCl, and reread fluorescence. Record fluorescence values as relative chlorophyll values (fluorescence units) or as chlorophyll a as calculated from the equation:

$$\text{chlorophyll } a \text{ (mg/m}^3\text{)} = \frac{\dfrac{F_o/F_{a_{max}}}{(F_o/F_{a_{max}}) - 1}(kx)(F_o - F_a)}{\text{L filtered}}$$

$$\text{pheophytin } a \text{ (mg/m}^3\text{)} = \frac{\dfrac{F_o/F_{a_{max}}}{(F_o/F_{a_{max}}) - 1}(kx)[(F_o/F_{a_{max}})(F_a) - F_o]}{\text{L filtered}}$$

where:

F_o = fluorescence before acidification,
F_a = fluorescence after acidification,
$F_o/F_{a_{max}}$ = maximum acid factor that can be expected in the absence of pheophytin, and
kx = calibration constant for a specific sensitivity scale.

Note that kx, $F_o/F_{a_{max}}$, and acid factors are functions of the combination of photomultiplier and color filters.

e. Total cell carbon: Determine by carbon analyzer. Report equivalence between total cell carbon and dry weight in milligrams per liter.

802 H. Effect of Additions

The quantity of cells produced in a given medium is limited by the concentration of nutrient present in the lowest relative quantity with respect to the needs of the organism. If a quantity of the limiting nutrient is added to the medium, cell production increases until this additional supply is depleted or until some other nutrient becomes limiting. Additions of substances other than that which is limiting would yield no increase in cell production. Nutrient additions may be made singly or in combination and the growth response compared to that of untreated controls to identify those substances that limit growth rate or cell production. The selection of additives, e.g., nitrogen, phosphorus, iron, wastewater effluents, will depend on the requirements of the test.

In all cases, keep the volume of added nutrient solution as small as possible, but make it large enough to yield a potentially measurable response. Relate the concentrations of additions to nutrient levels in the sample. To assess the effect of nutrient additions, compare treated sample to an untreated control. For highly productive controls, flask-to-flask variations may be high and might mask the effect of small additions of the limiting nutrient.

It is sometimes necessary to check the test water for the presence of material. To do this, treat the sample with an appropriate dilution of the complete synthetic medium. If no growth, or less than expected growth, occurs, toxic materials are suspected. In some situations, sample dilution or addition of a chelating agent will eliminate toxic effects.

802 I. Data Analysis and Interpretation

The fundamental measure used in the algal assay to determine biostimulation is the amount of suspended solids (dry weight) produced and determined gravimetrically. Other biomass indicators may be used, but all results must include experimentally determined conversion factors and the dry weight of suspended solids. Use several biomass indicators whenever possible, because biomass indicators respond differently to any given nutrient-limiting condition.

Report results of addition assays with results from two types of reference samples: the assay reference medium and untreated water samples. Present entire growth curves for each of the two types of reference sample. Report results of individual assays as maximum specific growth rate (with time of occurrence) and maximum standing crop (with time at which it was reached).

To determine the nutrients that limit growth rate by single-nutrient additions, treat a number of replicate flasks with single nutrients, determine the maximum specific growth rate for each flask, and compare the averages by Student's t-test or other appropriate statistical tests.

To identify growth-rate-limiting nutrients by multiple-nutrient tests, make analysis of variance calculations. Account for possible interaction between different nutrients by using factorial analysis.[2]

The methods for finding growth-rate-limiting nutrients are used also to determine the nutrient that limits maximum standing crop. Determine the available concentration of growth-limiting nutrient by comparing maximum standing crop in an untreated sample with a maximum standing crop in reference medium.

Report both maximum specific growth rate and maximum standing crop with their confidence intervals. Base the calculation of confidence interval for the average values on at least five samples. Consequently, make a minimum of five replications when an unfamiliar source water is first analyzed. Use these results to calculate the standard deviation. For subsequent samples from the same source use only three replicates and report with the confidence interval established for that source water.

In algal assays, occasionally one flask among replicates shows a considerable growth difference from the others. Eliminate such outliers from the results (and flasks from apparatus) if they fall outside the 95% confidence limits.

The overall evaluation of assay results consists of first determining whether a result is significant when considered as a laboratory measurement. Several methods are available, such as the Student's t-test and analysis of variance technics.

The second part of the evaluation is the correlation of laboratory assay results to effects observed or predicted in the field. Specific guidelines are not yet available, but note the general considerations in Section 802B.

802 J. References

1. NATIONAL EUTROPHICATION RESEARCH PROGRAM. 1971. Algal Assay Procedure: Bottle Test. EPA, Pacific Northwest Environmental Research Lab., Corvallis, Ore.

2. EUTROPHICATION AND LAKE RESTORATION BRANCH. 1974. Marine Algal Assay Procedure: Bottle Test. EPA, Pacific Northwest Environmental Research Lab., Corvallis, Ore.
3. TYLER, J.E. 1973. Applied radiometry. *Oceanogr. Mar. Biol. Annu. Rev.* 11:11.
4. JOINT INDUSTRY—GOVERNMENT TASK FORCE ON EUTROPHICATION. 1969. Provisional Algal Assay Procedure. EPA, Pacific Northwest Environmental Research Lab., Corvallis, Ore.
5. STRICKLAND, J.D.H. & T.R. PARSONS. 1965. A Manual of Sea Water Analysis, 2nd rev. ed. Fisheries Research Board Canada Bull. 125.

802 K. Bibliography

MCGAUHEY, P.H., D.B. PORCELLA & G.L. DUGAN. 1970. Eutrophication of surface waters—Indian Creek reservoir. First Progress Rep., FWQA Grant No. 16010 DNY. EPA, Pacific Northwest Environmental Research Lab., Corvallis, Ore.

MALONEY, T.E. W.E. MILLER & T. SHIROYAMA. 1971. Algal responses to nutrient additions in natural waters. Spec. Symp., American Society Limnology & Oceanography. Special Symposium on Nutrients and Eutrophication: Limiting-Nutrient Controversy 1:134.

MILLER, W.E. & T.E. MALONEY. 1971. Effects of secondary and tertiary wastewater effluents on algal growth in a lake-river system. *J. Water Pollut. Control Fed.* 43:2361.

WEISS, C.M. & R.W. HELMS. 1971. Interlaboratory Precision Test—An Eight Laboratory Evaluation of the Provisional Algal Assay Procedure: Bottle Test. Dep. Environmental Science & Engineering, School of Public Health, Univ. North Carolina, Chapel Hill.

MALONEY, T.E., W.E. MILLER & N.L. BLIND. 1972. Use of algal assays in studying eutrophication problems. *Proc. Int. Conf. Water Pollut. Res. 6th.* p. 205. Pergamon Press, Oxford, England and New York, N.Y.

SCHERFIG, J., P.S. DIXON, R. APPLEMAN & C.A. JUSTICE. 1973. Effect of Phosphorus Removal on Algal Growth, EPA Ecology Research Ser. 660/3-75-015.

MILLER, W.E., T.E. MALONEY & J.C. GREENE. 1974. Algal productivity in 49 lakes as determined by algal assays. *Water Res.* 8:667.

SPECHT, D.T. 1975. Seasonal variation of algal biomass production potential and nutrient limitation in Yaquina Bay, Oregon. *In* E.J. Middlebrooks, D.H. Falkenborg, and T.E. Maloney, eds. Proceedings Workshop on Biostimulation and Nutrient Assessment, Utah State Univ., Logan, Sept. 10-12, 1975. PRWG 168-1; also published as Biostimulation and Nutrient Assessment. Ann Arbor Science Publ., Ann Arbor, Mich.

803 TOXICITY TESTING WITH PHYTOPLANKTON (TENTATIVE)

1. Test Technics

The phytoplankton are primary producers in the aquatic community and, as such, are at the base of aquatic food chains. Because of this, they must be tested in bioassays that predict and determine the potential effects of a substance on the aquatic environment. The same general principles and technics used in determining biostimulation (Section 802) are used to determine toxicity to phytoplankton. The procedure applies to freshwater, estuarine, and marine phytoplankton.

In addition to the marine or estuarine algae listed in 802B.1*b*, *Monochrysis lutheri* Droop may be used. Maintain the test species in full-strength media (Section 802 and Tables 801:III, 801:IV. A and B). Test species must be in the logarithmic growth phase; therefore, transfer them to fresh culture medium every 4 to 5 days. Add test material to test vessels to give desired concentrations. Prepare triplicate vessels for each concentration. Use dilutions of culture medium to simulate chemical conditions of specific receiving waters. For optimum surface-to-volume ratios, see Section 802C.2*a*.

The maximum specific growth rate (μ_{max}) occurs during the logarithmic phase of growth, usually between Day 0 and Day 5. Therefore, measure biomass at least daily during the first 5 days of incubation. Indirect measurements of biomass, such as chlorophyll *a* or cell numbers, usually will be required because accurate gravimetric measurements at low cell densities are difficult. See Section 802G.3*c* and 802G.4 for methods.

Test a geometric series of concentrations initially (see Section 801E). After this preliminary test, progressively bisect intervals on a logarithmic scale. Narrow the range of test concentrations to determine the concentration that reduces the maximum specific growth rate (μ_{max}) to 50% of the control. This requires that two of the concentrations tested fall on each side of the concentration that inhibited (μ_{max}) to 50% (see Section 801F).

Compare the maximum specific growth rate (μ_{max}) to that obtained in the synthetic freshwater or artificial seawater culture medium. Regional and seasonal variations in quality make natural waters unsuitable as standard test media for comparative toxicity tests. Therefore, use a synthetic freshwater medium and/or artificial seawater. Add various concentrations of toxicants to the culture medium in triplicate and inoculate with test species.

For other types of tests, such as those to determine effluent requirements or compliance with water quality standards, take dilution water from the receiving body near the outfall but outside its influence. Remove undesirable organisms before making growth rate tests with selected sensitive species (Section 802D). Determine maximum specific growth rates (μ_{max}) in test vessels and compare with controls and EC50's based on percent of growth reduction. An alternative approach that provides a number that may relate to natural conditions should be reviewed.[1]

2. Reference

1. Miller, W.E., J.C. Greene & T. Shiroyama. 1978. The *Selenastrum capricornutum* Printz Algal Assay Bottle Test. U.S. EPA Rep. EPA-600/9-78-018. NTIS, U.S. Dep. of Commerce, Springfield, Va.

3. Bibliography

Erickson, S.J., N. Lackie & T.E. Maloney. 1970. A screening technique for estimating copper toxicity to estuarine phytoplankton. *J. Water Pollut. Control Fed.* 42:R270.

Walsh, G.E. 1972. Effects of herbicides on photosynthesis and growth of marine unicellular algae. *Hyacinth Control J.* 10:45.

Green, J.C., W.E. Miller, T. Shiroyama & E. Merwin. 1975. Toxicity of Zinc to the Green Alga *Selenastrum capricornutum* as a Function of Phosphorus or Ionic Strength. U.S. EPA Rep. EPA-660/3-75-034. NTIS, U.S. Dep. of Commerce, Springfield, Va.

Green, J.C., W.E. Miller, T. Shiroyama, R.A. Soltero & K. Putnam. 1976. Use of algal assays to assess the effects of municipal and smelter wastes upon phytoplankton production. *In* Terrestrial and Aquatic Ecological Studies of the Northwest. Eastern Washington State College Press, Cheney, Wash.

804 PROCEDURES FOR TESTING ZOOPLANKTON

804 A. Procedures for Ciliated Protozoa (TENTATIVE)

Protozoa, algae, and bacteria form the broad base of aquatic food chains. Ciliated protozoa are the most numerous animals of the estuarine benthos[1,2] and may be more important than bacteria as regenerators of nutrients, particularly nitrogen and phosphorus.[3,4] Further, some ciliates can concentrate certain persistent pesticides and related chemicals[5–7] and thereby aid in their translocation to higher trophic levels. Thus it is possible that the effects of such toxicants could be exerted at higher trophic levels either through disruption of nutrient cycles or through biological concentration of the toxicants higher in the food chain.[8,9]

The procedures described herein have been used successfully to test toxicity and bioaccumulation of toxic materials.[5,6] Responses measured are the effects on population growth rate and maximum population density and degree of accumulation and concentration of toxicants. *Tetrahymena pyriformis* is the recommended test organism because it occurs in freshwater and salt marshes, has worldwide distribution, and is grown readily in axenic culture. Its physiology has been studied extensively.[10,11] Strain W is used in this test; however, several other strains can be used. These may be obtained from the American Type Culture Collection (ATCC), 12301 Parklawn Drive, Rockville, Md. 20852. Instructions for cultivation will be furnished when requested. If other strains are used, note their source.

1. Holding and Culturing Test Organisms

a. Culture of test organisms: Use standard bacteriological technics to prepare and autoclave culture media and to transfer axenic cultures of *T. pyriformis*. Maintain stock cultures at 26 ± 0.5 C. Slant culture tubes containing 10 mL of medium at 60 deg from the vertical to enhance aeration. Keep duplicate stock cultures at room temperature and hold them upright to retard growth by decreasing aeration. Grow cultures axenically in 18- × 150-mm bacteriological culture tubes capped with polypropylene or stainless steel closures and supported in vinyl-coated 40-tube racks.

b. Culture medium: Use the same medium for stock and test cultures: 2% (w/v) proteose peptone, 0.2% (w/v) yeast extract, 0.5% glucose, and 90μM Fe:EDTA chelate/L. Adding Fe:EDTA chelate eliminates population growth variation due to differences in iron content of different lots of proteose peptone. Prepare chelate by the method of Conner and Cline.[12] Omit Fe:EDTA chelate when testing metal toxicity.

Prepare medium in distilled water or artificial seawater, Section 801D.4*c*1)b), diluted to a lower salinity, dispense among culture tubes, and autoclave for 15 min at 108kPa. Before inoculation let stand for a time long enough to restore DO. Best growth occurs in the distilled water formulation, although there is no significant difference between the growth rates of populations grown in the distilled water medium and the 5‰ salinity medium.

2. Testing Procedures

a. Preparation of toxicant stock solutions: Prepare a sterile stock solution for each concentration as described in Section

801E.3*b*. For water-insoluble toxicants, use acetone and polyethylene glycol 200 as solvents at a final concentration of 0.1%. During range-finding tests always test for interaction of toxicant and solvent at the solvent concentrations used.

b. Conducting tests:

1) Test procedures—Use test procedures described in Section 801E. Select and prepare test concentrations of toxicants as described in Sections 801E.2*b* and 3*b*. Prepare 11 or 12 optically matched culture tubes for each concentration, 10 tubes for test solutions and a control and special controls for solvent and emulsifier if used. Inoculate each tube with 1 mL of a 65-hr culture diluted by adding 3 mL culture to 7 mL sterile medium. Adjust volume so that each tube contains 10 mL of mixed toxicant and inoculum. Run tests for 96 hr in an incubator in which temperature is either kept constant or varied daily and seasonally in accordance with the natural temperature regime of the area. Concurrently use five concentrations of toxicant (see Section 801E.3*b*). Because large variations usually are observed among replicates, use six replicates to insure statistical validity at $p = 0.05$. Analyze effects of toxicant on growth rate and population density by analysis of variance (see Section 801F).

Run tests to determine bioaccumulation for 7 days in replicate using 10 cultures per concentration plus controls.

2) Analytical procedures—Chemically analyze toxicant solutions at beginning and end of test (see Section 801E.3*d*).

At end of test, pool all cultures from each concentration and concentrate cells by centrifugation at 1,000 *g*. Wash cells twice in toxicant-free medium and store, if necessary, before analysis, in sealed tubes at −20 C. Typically 100 mL of a 96-hr control culture yields about 1 g packed cells. Analyze medium and cells separately.[5,13] Calculate residues on a wet-weight basis. To obtain dry weight values prepare replicate cultures and dry cells at 60 C for 65 hr.

3. Reporting Results

Measure population density in a spectrophotometer* as absorbance at 540 nm. Use blanks of sterile proteose-peptone medium with the appropriate concentration of toxicant and carrier. Absorbance values cannot be converted to absolute numbers of ciliates;[14] therefore, use mean absorbance to estimate differences in population density between cultures. Estimate effects of toxicants on growth rate and population density from absorbance measurement of control and test cultures at 0, 4, 8, 16, 24, 36, 48, 60, 72, 84, and 96 hr. Plot these data to determine the period of exponential growth for each population. Because growth in this period is logarithmic, estimate growth rate by the quantity b of the least squares estimate of $y = a + bx$. The calculated line gives a close fit to the data, generally $r \geq 0.9$. Determine effect of toxicant on growth rate and population density as described in Sections 801F.1 and 2.

4. References

1. Borror, A.C. 1963. Morphology and ecology of the benthic ciliated protozoa of Alligator Harbor, Florida. *Arch. Protistenk.* 106:465.
2. Fenchel, T. 1967. The ecology of marine microbenthos. I. The quantitative importance of ciliates as compared with metazoans in various types of sediments. *Ophelia* 4:121.
3. Johannes, R.E. 1965. Influence of marine protozoa on nutrient regeneration. *Limnol. Oceanogr.* 10:434.

*The Bausch and Lomb Spectronic 20 has been used successfully in this type of study because it accepts large culture tubes. Other available instruments that can be adapted to large tubes and should be satisfactory are Coleman Junior and Junior II and Turner Model 330 Spectrophotometers.

4. JOHANNES, R.E. 1968. Nutrient regeneration in lakes and oceans. *In* M.R. Droop & E.J. Ferguson Wood, eds. Advances in Microbiology of the Sea 1:203. Academic Press, New York, N.Y.
5. COOLEY, N.R., J.M. KELTNER, JR. & J. FORESTER. 1972. Mirex and Aroclor® 1254: Effect on and accumulation by *Tetrahymena pyriformis* strain W. *J. Protozool.* 19:636.
6. COOLEY, N.R., J.M. KELTNER, JR. & J. FORESTER. 1973. The polychlorinated biphenyls, Aroclors® 1248 and 1260: Effect on and accumulation by *Tetrahymena pyriformis*. *J. Protozool.* 20:443.
7. GREGORY, W.W., JR., J.K. REED & L.E. PRIESTER JR. 1969. Accumulation of parathion and DDT by some algae and protozoa. *J. Protozool.* 16:69.
8. BURDICK, G.E., E.J. HARRIS, H.J. DEAN, T.M. WALKER, J. SKEA & D. COLBY. 1964. The accumulation of DDT in lake trout and the effect on reproduction. *Trans. Amer. Fish. Soc.* 93:127.
9. BUTLER, P.A. 1969. Monitoring pesticide pollution. *Bio-Science* 19:889.
10. CORLISS, J.O. 1970. The comparative systematics of species comprising the hymenostome ciliate genus *Tetrahymena*. *J. Protozool.* 17:198.
11. ELLIOTT, A.M., ed. 1973. Biology of *Tetrahymena*. Dowden, Hutchinson & Ross, Inc., Stroudsburg, Pa.
12. CONNER, R.L. & S.G. CLINE. 1964. Iron deficiency and the metabolism of *Tetrahymena pyriformis*. *J. Protozool.* 11:486.
13. HATCH, W.R. & W.L. OTT. 1968. Determination of sub-microgram quantities of mercury by atomic absorption spectrophotometry. *Anal. Chem.* 40:2085.
14. SLATER, J.V. & A.M. ELLIOTT. 1951. Volume change in *Tetrahymena* in relation to age of the culture. *Proc. Amer. Soc. Protozool.* 2:20.

804 B. Bioassay Procedures for *Daphnia* (TENTATIVE)

Daphnia have been used in tolerance studies for over a century. *Daphnia magna* is the largest of the *Daphnia*, reaching a maximum size of over 5 mm. Large numbers can be reared in a relatively small space. Neonates (first-instar young) are 0.8 and 1.0 mm long and can be observed without optical aids. This stage is the one most commonly used for tolerance studies. Species smaller than *D. magna* and *D. pulex* may require special handling.

1. General Considerations

Individual female *D. magna* have been known to live for as long as 4 months at 20 C. They have been cultured in natural waters and in dechlorinated tap water. Foods include bacteria, algae, and yeast, together with soil extracts and organic materials such as cotton-seed meal, herring meal, powdered dried grass, and enriched trout fry granules. *Daphnia* have been reared individually in small vessels and in mass culture in large aquariums.[1–5]

Reproduction can be restricted to the production of females by diploid parthenogenesis when suitable culture conditions are maintained, thereby insuring a supply of experimental animals whose genetic variability is limited to the heterozygosity of the parent.[6] Genetic uniformity is expected within parthenogenetic clones derived from a single female.[7] When culture conditions are optimum *D. magna* females release their first broods of young within 10 days at 20 C and 7 days at 25 C. Thereafter young are released every 3 to 4 days at 20 C and 2 to 3 days at 25 C. Twenty or more young per brood are produced as long as culture conditions remain adequate. A single female may release over 400 young during her lifetime.

Obtain *Daphnia* from established cultures or by field collection. While *Daphnia* of any age can be available at any time, it is more convenient to use neonates than older animals for testing. When reared at 20 C, they undergo their first ecdysis about 30 hr after release from the brood chamber and at 25 C after about 20 hr. Neonates can be segregated from stock animals at 24-hr intervals when reared at 20 C and at 12-hr intervals at 25 C. Neonates are less tolerant of many substances than older animals. *Daphnia* are more susceptible to most substances at ecdysis than between molts.

Results obtained by different investigators using *Daphnia* are comparable and in general, *Daphnia* are less tolerant of toxic substances than are fish.[8]

2. Culturing

One of the simplest media is the manure-soil medium[3,9] supplemented periodically with yeast. Make this medium by mixing 5 g dried sheep manure, 25 g garden soil or sandy muck, and 1 L pond, spring, or tap water. Let stand for 2 days at room temperature, then strain through bolting cloth with mesh openings approximately 0.15 mm. During straining, work some of the finer soil particles through the cloth. Set filtrate aside for 1 wk or more and discard sediment. To make final medium, mix 1 part filtrate with 6 to 8 parts pond, spring, or dechlorinated tap water. The original filtrate may be kept indefinitely before final medium is made.

Use final medium for individual or mass culturing. For individual rearing, dispense 100 mL medium into 125-mL wide-mouth flint-glass bottles or equivalent vessels and inoculate with one daphnid per bottle. Beginning 1 day after inoculation, add 1 mL of a suspension containing 1 mg active dry yeast in water to each bottle on alternate days. For mass rearing use 3.8-L wide-mouth glass jars filled with 3 L medium to which a suspension of 30 mg active dry yeast is added on alternate days. Once the cultures are initiated, do not change the culture medium, but occasionally add water to replace that lost by evaporation and in removal of young. To retard evaporation, use a perforated cover that permits air diffusion. Aeration is unnecessary because the critical DO for *Daphnia* is less than 15% of saturation at 20 C.

With a stock of 100 individually reared stock females, over 300 neonates can be available daily. When stock females begin reproducing, periodically remove young, preferably every 24 hr at 20 C or 12 hr at 25 C. When stock animals reach old age and their reproductive rate drops, replace with young females in fresh medium.

Other methods of culturing, such as those using algae,[4] have been used successfully. Biesinger and Christensen[5] developed a medium consisting of a suspension of 0.5 g powdered dried grass† and 10 g enriched trout fry granules in 250 mL unpolluted lake, well, or river water mixed vigorously in a blender for 5 min. Strain suspension through a stainless steel screen with about 0.066-cm opening and add an additional 50 mL water to rinse blender. Refrigerate suspension. Before a portion is withdrawn, mix thoroughly. Feed it at the rate of 1 mL/wk/L.

3. Testing Procedures

a. Static tests:

1) Setting up the tests—Prepare test materials, dilution water, and toxicant solutions as described in Sections 801E.2 and 3. Select test concentrations as described in Section 801E.3*b*. Make up test solutions and controls in 100-mL quantities in 125-mL wide-mouth flint-glass bottles or equivalent vessels.

†Cerophyll, Cerophyll Laboratories, Inc., 4722 Broadway, Kansas City, Mo. 64111, or equivalent.

2) Performing the tests—After preparing test solutions, segregate neonate *Daphnia* that have been released from the mothers' brood chambers during the preceding 24 hr at 20 C or 12 hr at 25 C and collect in one vessel. Wash in three changes of diluent, allowing 5 min in each wash solution to reduce carryover of materials. Introduce 10 neonates into each test vessel and the control. Use a piece of 8-mm glass tubing 20 cm long, with one end drawn to a diameter of 2 mm and the other end fire-polished and fitted with a 2-mL-capacity rubber bulb, for collecting and transferring neonates. Use identical tubing with both ends fire-polished, and fitted with a rubber bulb, for handling adults.

After introducing neonates to test solutions, observe them regularly, usually after 1, 2, 4, 8, and 16 hr and daily thereafter. Record number of motile animals in each test vessel. Consider an animal nonmotile if it shows no independent movement even after a test vessel is rotated. Nonmotile animals are not necessarily dead. At threshold concentrations of such substances as ethanol, acetone, and chlorobutanol, animals may show no movement and the heart may have ceased to beat but on transfer to dilution water they will recover. However, if such animals are maintained in the test medium they will die. Continue observations for 5 days or as long as most of the animals in some of the test solutions remain motile. Run tests in triplicate.

Do not feed animals during tests. *D. magna* will live for as long as 1 wk without food in well-balanced salt solutions.[10] Different species and longer-term tests will require modifications of standard conditions.

b. Long-term tests:

1) Determination of reproductive impairment—Reproduction may be impaired at much lower toxicant concentrations than those causing acute toxicity, sometimes as low as 0.00003 of that producing acute toxicity. Precede these tests by acute tolerance tests to establish the maximum concentration to be used.

2) Preparation of test medium—Prepare test medium in the same way as regular culture medium, except using water representative of that receiving the effluent discharge. Prepare a series of 6 to 10 1-L quantities of medium to which graded amounts of toxicant have been added. Use as the highest concentration of toxicant the equivalent to the LC50 or EC50 at 96 hr. Reduce each successive concentration in a geometric progression by a factor of three or more in preliminary experiments. Use undiluted culture medium as control. Dispense each liter of test medium in 100-mL quantities to each of 10 bottles.

3) Performing tests—Segregate and collect neonate *Daphnia* as for static tests. Do not wash animals. Introduce one neonate into each bottle. On the following day and on alternate days thereafter, add a 1-mL suspension containing 1 mg active dry yeast to each bottle. Make daily observations and note dead or immobilized animals. As the animals grow and reproduce, remove young and record their number. Replace lost water. Continue observations until control animals have released at least six broods of young, which will take about 21 days at 25 C and 30 days at 20 C. Handle animals as individual cultures of stock animals.

At end of observation period analyze results and test for significance the differences in the number of young produced. Note time of appearance of first broods and number of broods.

If the number of young produced in the lowest concentration of toxicant differs significantly from that in the controls, repeat the test with reduced concentrations of toxicant until no significant differences are obtained.

After reaching toxicant concentrations where no differences are observed between tests and controls in number of young and broods and time of appearance of first broods, make a final set of tests with 30 or more animals in each of four concentrations of toxicant bracketed about the lowest concentration for which significant differences in production of young were found. Use an equal number of animals for the control.

Assemble, analyze, evaluate, and report data as described in Section 801F.

4. References

1. NEEDHAM, J.G., P.S. GALTSOFF, F.E. LUTZ & P.S. WELCH. 1937. Culture Methods for Invertebrate Animals. Comstock Publishing Company. Reprinted by Dover Publ., Inc., New York, N.Y.
2. NAUMANN, E. 1933. *Daphnia magna* Straus als Versuchstier. *Kgl. Fysiogr. Sallsk. Lund Forhandl.* 3:15.
3. ANDERSON, B.G. 1944. The toxicity thresholds of various substances found in industrial wastes as determined by the use of *Daphnia magna*. *Sewage Works J.* 16:1156.
4. PARKER, B.L. & J.E. DEWEY. 1969. Further improvements on the mass rearing of *Daphnia magna*. *J. Econ. Entomol.* 62:725.
5. BIESINGER, K.E. & G.M. CHRISTENSEN. 1972. Effects of various metals on survival, growth, reproduction and metabolism of *Daphnia magna*. *J. Fish. Res. Board Can.* 29:1691.
6. BACCI, G., G. COGNETTI & A.M. VACCARI. 1961. Endomeiosis and sex determination in *Daphnia pulex*. *Experientia* 17:505.
7. HEBERT, P.P. & R.D. WARD. 1972. Inheritance during parthenogenesis in *Daphnia magna*. *Genetics* 71:639.
8. KEMP, H.T., J.P. ABRAMS & R.C. OVERBECK. 1971. Water Quality Criteria Data Book Vol. 3. Effects of Chemicals on Aquatic Life, Selected Data from the Literature through 1968. EPA Water Pollution Center Research Ser. 18050GWV05/71, U.S. Government Printing Office, Washington, D.C.
9. ANDERSON, B.G. 1950. The apparent thresholds of toxicity to *Daphnia magna* for chlorides of various metals when added to Lake Erie water. *Trans. Amer. Fish. Soc.* 78:96.
10. STAMPER, W.R. 1969. The determination of the optimal combination of concentrations of sodium, potassium, calcium and magnesium chlorides for the survival of *Daphnia pulex*. Ph.D. thesis, The Pennsylvania State Univ.

804 C. Bioassay Procedures for the Calanoid Copepod, *Acartia tonsa* (Dana) (TENTATIVE)

This method provides data on the effects of toxicants on a marine copepod, *Acartia tonsa*. Use general procedures described in Sections 801C, D, and E. *A. tonsa* is sensitive; therefore, minimize handling or handle gently.

1. Collecting and Holding Test Organisms

a. Collection: Collect *A. tonsa* by towing (≥4 km/hr) a plankton net (aperture 150 to 250 μm) at a depth of 1 to 3 m. Transfer animals carefully by pouring gently into insulated containers three-fourths filled with ambient seawater. Remove predators immediately, especially jellyfish and ctenophores. Do not exceed a copepod density of 250/L to assure that the DO concentration remains adequate. Measure and record temperature, salinity, DO, and pH at time of collection and maintain these conditions during initial holding stages.

b. Holding: Immediately on return to the laboratory, transfer samples to 2.3 L

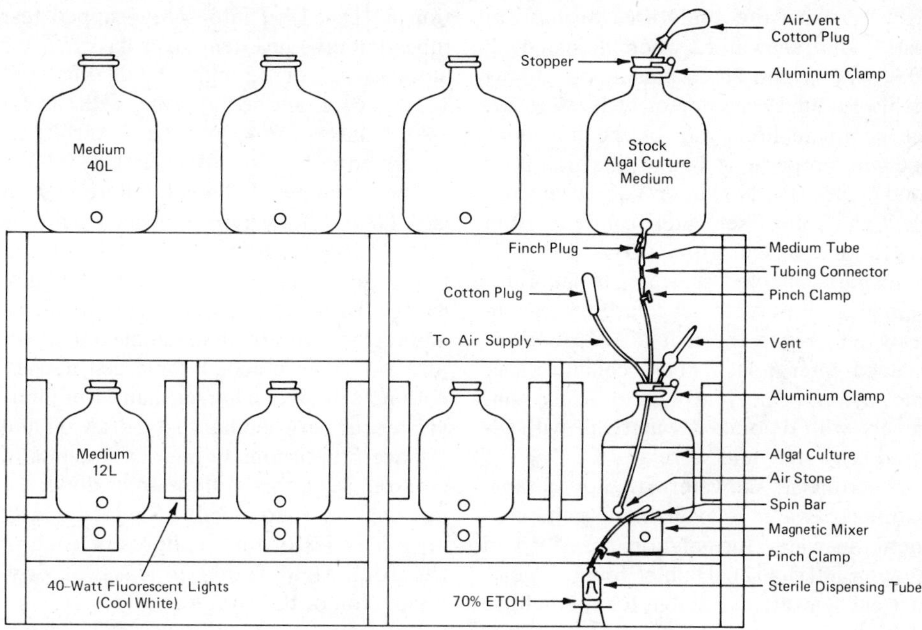

Figure 804:1. Algal culture system.

(190 × 100 mm) borosilicate crystallizing dishes. Adjust volume of water in each dish to 2 L with filtered seawater at ambient temperature and salinity and add the algal diet for adult copepods, described in ¶ 2*b* below, in the quantity necessary to attain the density of organisms listed in Table 804:I. Incubate cultures at collection temperature and illuminate as described in Section 801E.3*f*. After 24 hr, begin to acclimate cultures to 20 C and 20‰ salinity. Change salinity and temperature in increments of 3‰ and 3 C/day. If organisms are held in the original vessel during acclimation, alternately siphon off and add seawater of a different salinity. Transfer organisms either by gently pouring, pipetting, or siphoning (with 1-cm-ID tubing) to new vessels. During acclimation, maintain a daily feeding schedule to provide the density of algae listed in Table 804:I.

c. Sorting and identification: For basic information on taxonomy and biology of *Acartia* and other coastal calanoids consult other sources.[2–7] Separate *A. tonsa* from other organisms collected in the tow. To facilitate copepod capture, reduce culture volume from 2.0 L to 0.5 L by slowly siphoning the seawater, with a 150-μm plankton netting screen over the siphon intake. Carefully draw individual adults up into a wide-bore (≥2 mm) transfer pipet and place in depression slides. Identify and transfer to food-enriched filtered seawater of 20‰ salinity at 20 C [see ¶ 2*a* 1) and Section 801D.4*c* 1)b)]. Exclude all nauplii and juvenile forms.

d. Water supply:

1) Natural water—Collect sufficient

seawater from the study area to perform tests. This seawater, when adjusted to 20‰ salinity and 20 C, must support survival of adult *A. tonsa* for at least 96 hr, the complete life cycle of the copepod, and with proper enrichment, the growth of food algae. Use Niskin or Van Dorn samplers and collect seawater from 3 to 10 m depth. Record salinity, DO, and pH. Transport collected seawater to laboratory in glass or polyethylene carboys aged in seawater. Filter through a 1.0-μm acid-washed filter (glass fiber, cellulose-acetate, nylon, or polycarbonate). Rinse containers with deionized water, fill with filtered seawater, and store at 4 C.

2) Artificial seawater—If no suitable natural seawater is available use the synthetic seawater formulation described in Section 801D.4*b*2). Heinle[7] found a commercial seawater* suitable for the culture of both *A. tonsa* and *Eurytemora affinis*. Algal assays indicate that this product is unsuitable for tests involving heavy metals; use it only for culture until more extensive comparative data are available.

2. Culturing Test Organisms

a. Production of food for test organisms:

1) Culture medium for algae—See Section 801D.4*c* 1)b).

2) Growth chambers—Grow algal cultures in standard screw-capped test tubes, flasks, or in the fill-and-draw semi-continuous system described below.

3) Culturing and harvesting algae for food—Grow algae for feeding copepods and other zooplankton in filtered natural or synthetic seawater at 20‰ salinity and 20 C with 2,500 to 5,000 lux continuous illumination for 14 hr daylight and 10 hr darkness (14L:10D). Use nutrient enrichments of Guillard and Ryther.[8]

Dispense enriched seawater [see Section 801D.4*c* 1)b)] into screw-capped test tubes (50 mL) or erlenmeyer flasks. After autoclaving for 15 min at 108 kPa and 121 C, cool and equilibrate with atmospheric gases for 48 hr. Check sterility as described in Section 801D.4*c* 1)a).

The recommended algal culture system is a fill-and-draw type in which cultures are maintained near their maximum log-phase cell density and growth rate (Figure 804:1). Draw off medium and organisms and replace with fresh medium so that culture will have reached same cell density within 24 hr. When longer than 24-hr intervals occur between harvests, draw off and replace proportionally greater amounts of culture. Scale this system up or down depending on food needs. Simultaneously keep a series of tube cultures of each of the four algal foods in case of contamination of the large cultures.

Determine algal cell densities by direct microscopic counts using a hemacytometer, Palmer-Maloney chamber, or Utermohl chamber (inverted microscope).[9,10] An electronic particle counter is an accurate and rapid method for determining unialgal densities. If visual counts are necessary, relate these to specific absorbance at 750 nm with a spectrophotometer (see Section 802G.4). A curve relating cells per milliliter to absorbance can be used to replace the cell count.

b. Mass culture of test organisms: This system provides large quantities of *A. tonsa* of standard age for tests.

1) Growth chambers—Use mass culture units of Mullin and Brooks[11] and Frost.[12] The culture vessel in Figure 804:2 is a borosilicate glass aspirator bottle ranging from 13 to 45 L in capacity, depending on the number of copepods needed. Gently mix the contents with a low-speed motor (≤25 rpm) with a stirring rod mounted above the culture vessel. Slow mixing maintains algal food in suspension where planktonic copepods normally feed. Keep water movement gentle and free of vortices such

*Instant Ocean®, available from Aquarium Systems, Inc., 1450 East 298th St., Wickliffe, Ohio 44092.

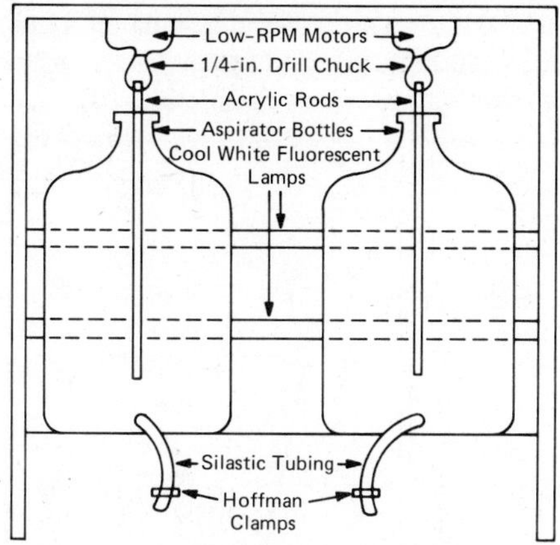

Figure 804:2. Apparatus for mass copepod culture (static).

as those produced by magnetic stirrers. Use cool white fluorescent lights to provide 1,000 lux illumination incident to the culture surface on a 14L:10D cycle.

Alternatively use a continuous-flow system (Figure 804:3). Fit a cylindrical vessel (15 to 40 L) with a standpipe and drain. Collar the standpipe with a 200-μm plastic† screen to prevent loss of eggs and nauplii. Mount a low-speed motor (≤25 rpm) with stirrer above the culture vessel. Add filtered seawater by a peristaltic pump or siphon from a constant-head tank at a rate of 1 to 3 tank volumes/24 hr. Feed by mixing the four algal species (Table 804:I) in a single container and introduce them into the culture chambers with a pump (Figure 804:3) at a rate necessary to maintain a food cell density of 2 to 5 × 10^7 cells/L. Place the stirring rod in the culture vessel so that the lower 25% of the vessel receives only gentle mixing. This permits a quiet area for mating and egg laying and keeps sediment from being stirred up and clogging the screen. Drain and clean the system every 3 to 4 wk. Brush the screen daily with a fine brush to avoid clogging.

†Nitex or equivalent.

2) *Algal diet for copepods*—Although various algal diets have been used for copepod cultures[10,13–15] use the algal diet given in Table 804:I.[1] *Skeletonema costatum* has been added because it is naturally occurring food for *A. tonsa*.

3) *Setting up cultures of test organisms*—*A. tonsa* females are each capable of producing more than 30 eggs/day when fed the recommended ration. If 250 or more gravid females are kept as breeders, theoretically over 4,000 eggs will be produced within 24 hr. For this potential number of adults, use a 40-L culture vessel. Generally, the relationship between culture volume and organism density is 10:1.

For production and collection of copepod eggs, use generation cages (Figure 804:4), consisting of a clear acrylic plastic cylinder 125 mm in diam and 90 mm high, with one end covered by plankton netting having an aperture size of 250 μm. Place this cylinder in a 2.3-L (190 × 100 mm)

TABLE 804:I. COMPOSITION OF ALGAL DIET AND RECOMMENDED CONCENTRATIONS FOR ADULT AND NAUPLIAR FEEDING AND EGG-LAYING[1]

Species	Adult & Copepodite	Naupliar	Egg-Laying
Skeletonema costatum	5.0×10^6	5.0×10^5	1.5×10^7
Thalassiosira pseudonana	7.0×10^6	7.0×10^5	2.1×10^7
Isochrysis galbana	5.0×10^6	5.0×10^5	1.5×10^7
Rhodomonas baltica	3.0×10^6	3.0×10^5	9.0×10^6
Total cells/L of copepod culture	2.0×10^7	2.0×10^6	6.0×10^7

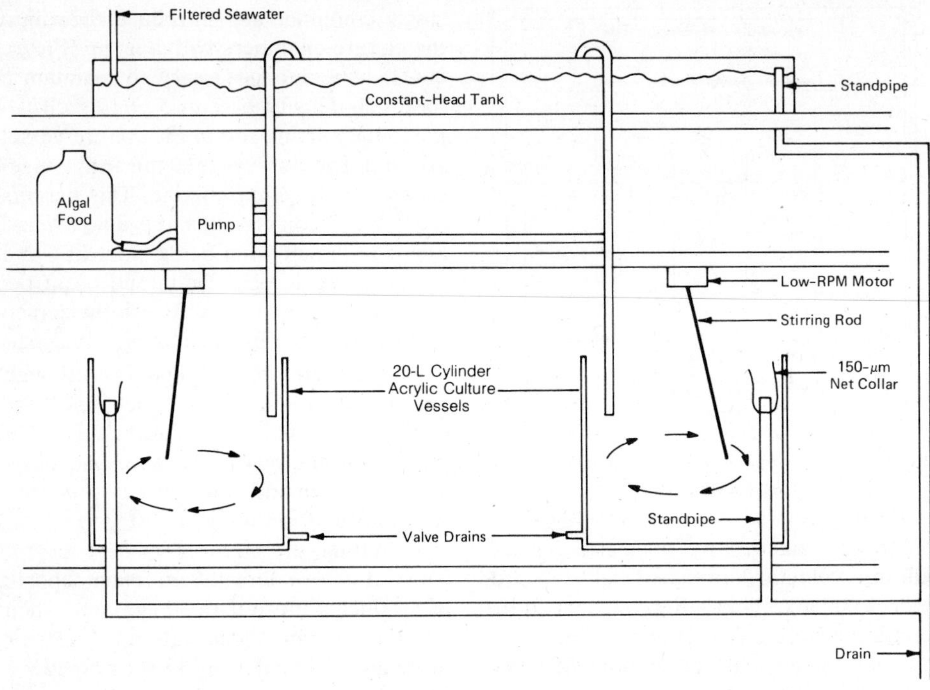

Figure 804:3. Apparatus for mass copepod culture (flowing).

crystallizing dish with covered end 25 mm from bottom. Add 2 L filtered seawater or artificial seawater and place 50 to 100 gravid females in each screen-bottom cylinder. Feed three times the adult algal diet (Table 804:I). The plankton netting allows eggs to pass through and hatch in crystallizing dish where they are protected from adult predation. After 24 hr remove adults by gently lifting each generation cage from crystallizing dish and quickly immersing it in another dish with three times the usual food density. Carefully siphon remaining seawater from all dishes containing eggs and nauplii into a glass aspirator bottle containing filtered seawater. Adjust final volume and feed nauplioid culture as indicated in Table 804:I. If a second mass culture is desired, repeat after 24 hr, using required number of generation cages to produce needed organisms.

The average length of each developmental stage in the life cycle of *A. tonsa* at 20 C and 20‰ is as follows:

Stage	Length days
Egg (newly oviposited)	1
Nauplius (6 instars)	7
Copepodite (6 instars)	6
Adult (until gravid)	3
Total life cycle	17

During the first 6 days of mass culture only naupliar stages are present. Feed daily with 2×10^6 cells/L. On the third and seventh days, slowly siphon off 50% of culture medium and replace with clean medi-

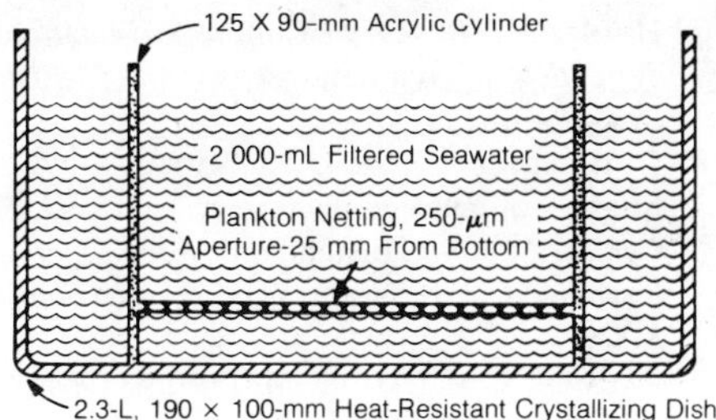

Figure 804:4. Generation cage (after Heinle).

um. Cover intake end of siphon with 60-μm netting to prevent loss of nauplii.

After the seventh day, copepodites should be present. From this time on, feed 2×10^7 cells/L/day, replacing 50% of the culture volume with filtered seawater every third day. Within 16 to 17 days the population will have reached maturity and can be used for tests or to start new cultures. Average adult life span at 20 C is approximately 30 days.

Maintain a non-age-standardized mass culture in reserve. Use gravid females from original generation cages to start a 12-L system. Feed adult food ration and replace 50% of culture water every third day. In addition, harvest approximately one-third of culture, including organisms, periodically (10 to 14 days) to keep population at about 50 adults and copepodites.

4) Harvesting test organisms—Harvest mass cultures of copepods that have reached the adult stage as follows: Reduce culture volume by 75% using a slow siphon with its intake covered with 60-μm plankton netting. Transfer carefully the remaining 25%, including organisms, to 2.3-L borosilicate glass crystallizing dishes to provide a total volume of 2 L/dish. Because of organism fragility do not constrict discharge tube to reduce flow; instead, control flow through ventral tubulation on aspirator bottle (Figure 804:2) by minimizing head between culture vessel and crystallizing dish. A slow flow minimizes turbulence and opportunity for organisms to collide with vessel walls.

Further concentrate harvested animals in crystallizing dishes by siphoning off culture medium. Facilitate capture by using the positive phototactic response of the animals.

5) Other organisms—The culture system designed for *A. tonsa* has worked well for *Eurytemora affinis* and *Pseudodiaptomus coronatus*. However, the generation cages are suitable only for *A. tonsa* because it releases eggs individually. Both *E. affinis* and *P. coronatus* produce egg sacs.

3. Test Procedures

With adult *A. tonsa* and the previously described culture method, make short-term tests as follows:

a. Range-finding tests: Use 10 adult *Acartia* per replicate with three replicates per test concentration and control. For a test container, use a suitable flat-bottom borosilicate glass dish containing 100 mL seawater ≥2.0 cm deep. Generally use a broad range of concentrations (see Sections 801E.3*b* and 801C.2*a*). Prepare toxicant solutions as described in Section 801E.2*b*. Capture 10 adult *Acartia* for each test chamber from stock cultures with a wide-bore transfer pipet and transfer to a 20-mL beaker containing 5 mL filtered seawater. Adjust final volume to 15 mL. Add animals and 15 mL medium to 85 mL of toxicant-dosed medium in test chamber by immersing beaker and rinsing gently. Expose for 96 hr. Observe and record number of dead, moribund, and living copepods after 1.5, 3, 6, 12, 24, 48, 72, and 96 hr exposure. To ascertain if a motionless animal is dead, touch it gently with a sealed glass capillary probe. If control mortalities exceed 15%, reject results. Analyze results as directed in Section 801F.

b. Definitive short-term tests: Follow

general culture conditions and handling as described in Sections 801D and 801E. Specifically test 15 adults in each of four replicate test vessels per toxicant concentration and control. Select test concentrations of toxicants as described in Section 801E.3*b*. Expose animals and collect data as described in Sections 801C and E.

Calculations, data presentation, and expression of results are as described in Section 801F.

4. References

1. WILSON, D.F. & K.K. PARRISH. 1971. Remating in a planktonic marine calanoid copepod. *Mar. Biol.* 9:202.
2. CONOVER, R.J. 1956. Oceanography of Long Island Sound, 1952–1954. VI. Biology of *Acartia clausi* and *A. tonsa*. *Bull. Bingham Oceanogr. Coll.* 15:156.
3. HEINLE, D.R. 1966. Production of a calanoid copepod *Acartia tonsa*, in the Patuxent River Estuary. *Chesapeake Sci.* 7:59.
4. HEINLE, D.R. 1969. Effects of Temperature on the Population Dynamics of Estuarine Copepods. Ph.D. thesis, Univ. of Maryland, College Park.
5. WILSON, C.B. 1932. The Copepods of the Woods Hole Region. Massachusetts. Smithsonian Inst., U.S. National Museum Bull.
6. ROSE, M. 1933. Faune de France. No. 26. Copepodes Pelagiques. Librarie de la Faculte des Sci. Reprinted 1970 by Kraus Reprint, Nendeln, Lichtenstein.
7. HEINLE, D.R. 1969. Culture of calanoid copepods in synthetic seawater. *J. Fish Res. Board Can.* 26:150.
8. GUILLARD, R.R. & J.H. RYTHER. 1962. Studies of marine planktonic diatoms. I. *Cyclotella nana* Hustedt and *Detonula confervacia* (Cleve) Grant. *Can. J. Microbiol.* 8:229.
9. PALMER, C.M. & T.E. MALONEY. 1954. A new counting slide for nannoplankton. *J. Amer. Soc. Limnol. Oceanogr.*, Spec. Publ. 21:1.
10. SCHWOERBEL, J. 1970. Methods of Hydrobiology. Pergamon Press, New York, N.Y.
11. MULLIN, M.M. & E.R. BROOKS. 1967. Laboratory culture, growth rate and feeding behavior of a planktonic marine copepod. *J. Amer. Soc. Limnol. Oceanogr.* 12:657.
12. FROST, B.W. 1972. Effects of size and concentration of food particles on the feeding behavior of the marine planktonic copepod *Calanus pacificus*. *J. Amer. Soc. Limnol. Oceanogr.* 17:805.
13. ZILLIOUX, E.J. & D.F. WILSON. 1966. Culture of a planktonic calanoid copepod through multiple generations. *Science* 151:996.
14. KATONA, S.K. 1970. Growth characteristics of the copepods *Eurytemora affinis* and *E. herdmani* in laboratory cultures. *Helgolander wiss. Meeresunters* 20:373.
15. NASSOGNE, A. 1970. Influence of food organisms on the development and culture of pelagic copepods. *Helgolander wiss. Meeresunters* 20:333.

805 BIOASSAY PROCEDURES FOR SCLERACTINIAN CORAL (TENTATIVE)

The scleractinian corals[1–7] comprise one of the most conspicuous and important components of many tropical reef ecosystems. They are especially sensitive to environmental perturbations and are valuable indicator organisms of water quality in shallow tropical marine environments.

The reef-forming corals flourish only within a narrow range of chemical and physical conditions including clear water, low inorganic and organic nutrients, low sedimentation rate, and tropical open-ocean temperature and salinity.

Corals are valued for their beauty. They also form protecting reefs about islands and lagoons.[8] These massive structures provide an essential environment for many organisms and are the basis for an entire ecosystem. Reef corals are slow to reestablish themselves; the process often requires decades. Once the corals are killed, the imbalance may lead to erosion and substantial environmental modification such that the area becomes permanently unsuited for coral recovery.

Corals lack specialized circulatory and excretory systems and are highly sensitive to changes in their physical-chemical environment. Unlike most reef animals, corals cannot migrate or burrow to protect themselves from localized short-term environmental stress, nor can they isolate themselves temporarily from harsh conditions by withdrawing into a shell or tube.

805 A. Selecting and Preparing Test Organisms

1. Suggested Test Species

a. Tropical, Indo-Pacific area (species listed in order of their importance):

1) Branching *Acropora*
   Acropora formosa (Dana)
   Acropora is not present in Hawaii. Substitute the branching form of *Montipora verrucosa* Lamarck.
2) Finely branched *Pocillopora*
   Pocillopora damicornis (Linnaeus)
   Synonyms: *Pocillopora caespitosa* Dana, *Pocillopora bulbosa*
   Alternate: *Pocillopora brevicornis* Lamarck.
3) Branching *Porites*
   Porites compressa Dana or small *Porites lobata* Dana
   Alternate: *Porites andrewsi* Vaughan.
4) Representative of solitary hermatypic corals: *Fungia scutaria* Lamarck.
5) Representative of the ahermatypic (lacking zooxanthellae) corals: *Tubastrea aurea* (Quoy and Giamard).

b. Tropical Atlantic area (test groups listed in order of their importance):

1) Branching *Acropora*
   Acropora cervicornis (Lamarck).
2) Branching *Porites*
   Porites porites (Pallas)
   Alternate: *Porites furcata* (Lamarck).
   Pocillopora is not present in the Atlantic.
3) Other widely used corals:
   Meandrina meandrites (Linnaeus) forma *meandrites*
   Montastrea annularis (Ellis & Solander)
   Montastrea cavernosa (Linnaeus) forma *areolata*.
4) Representative of solitary hermatypic corals:
   Scolymia lacera (Pallas) formerly *Mussa lacera* (Pallas).
5) Representative of ahermatypic (lacking zooxanthellae) corals:
   Tubastrea aurea (Quoy and Giamard).

This list of recommended species is not intended to restrict selection of test species, but rather to act as a guide in choosing test organisms.

2. Selecting Test Organisms

Because basic differences exist between Caribbean and Indo-Pacific coral fauna, it can be difficult to find comparable species from the two areas. Wells[2] estimated that 36 species occur in the Caribbean as opposed to at least 500 for the Indo-Pacific area. Few coral species are common to both oceans. Genera studied intensively in the Indo-Pacific include *Acropora, Pocillopora, Porites,* and *Fungia. Meandrina, Montastrea,* and *Manicina* often are used as laboratory subjects in the Caribbean.

Except for *Acropora cervicornis*, all suggested species have been used successfully. *A. cervicornis* is listed as the primary test species for the Atlantic because it is widely distributed, very sensitive, and similar to the important Indo-Pacific species *A. formosa*. Although *A. cervicornis* could eventually prove to be too delicate to maintain in some experimental systems, it has been used in short-term laboratory studies[9] and in field transplants.[10] With proper care it should be possible to culture this species in the laboratory.

The ahermatypic corals do not contain symbiotic algae and long-term maintenance requires that *Tubastrea* be kept in flowing seawater with a source of plankton for food. Although this organism is not an important reef-former, it has been listed because it is common to both oceans and lacks symbiotic algae. It has been used as a comparison species in tests designed to assess the role of zooxanthellae in hermatypic corals.[11]

Whenever possible, use the suggested species because of their previous use as laboratory subjects, widespread occurrence, and relative importance as reef-builders. Preferably use small-polyp forms.[12]

In determining adverse effects of toxicants, the high species diversity of coral reefs dictates the use of at least several genera. Include members of other locally dominant species as well as species that may be the most sensitive. Ideally, make comparative tests to determine tolerance of all locally common species so that the most sensitive species can be used in long-term studies.

3. Collecting, Handling, and Holding Test Organisms

a. Collecting corals: Guidelines for the operation of collecting boats in tropical reef areas are given by Domm.[13] The field procurement program should produce large numbers of small, unattached specimens that are similar in size, morphology, genetic makeup, and environmental history. Take extreme care in selecting specimens because corals display a wide variety of growth forms within a given species.[3,14–16]

For a branching species, use a single large colony as the source of all specimens for a test. Clip branches of approximately 10 g wet weight from the colony with pruning shears. Selection of specimens of solitary and massive corals will depend on local availability. Unattached species, such as *Fungia,* are easily collected, whereas the massive and encrusting forms generally grow firmly attached to the reef and must be pried or broken away with a pry bar or chisel. Because corals vary widely in size, morphology, and availability, it is difficult to give firm rules on collection for all species.

b. Field preparation: If broken fragments are left in a protected reef area where there is no heavy wave action to roll them about, coral tissue will cover the surface within 1 month and a small colony completely covered with living tissue will be produced. Use a substrate in the holding area consisting of gravel-size reef rubble and not subject to abnormally high temperatures or land runoff. On unprotected coasts use a deep offshore area for holding. Corals may be allowed to heal over in a laboratory seawater system.

Larger corals are tagged easily by tying commercially available vinyl "spaghetti" tags to the specimen. Use monofilament tags on small specimens. Small coral colonies readily overgrow a monofilament or vinyl strand and produce a solidly tagged colony.

c. Handling and holding corals: Keep animals submerged in seawater at all times and handle lightly and only when necessary. When transferring corals, submerge transfer container, place coral in container, and transport submerged coral to its new location; then again submerge container and remove coral. If coral is to be transported over a distance and must remain in a transfer container for more than 5 min, insure that the DO remains within ±10% saturation and that the temperature does not change by more than 1 C. A boat supplied with a live well is desirable for transportation. If a large volume of water is used to contain a small volume of coral, there is less danger of these limits being reached. Preferably renew water in containers at a rate sufficient to keep DO and temperature within safe levels, rather than aerating. It can be assumed that when DO has been altered adversely by coral metabolic activity, other biologically important characteristics also have been affected adversely. Aeration will not correct these other changes.

d. Long-term holding: For long-term holding in the laboratory, use a substrate of natural reef rubble with its associated organisms covering the container bottom to a depth of 5 cm. In addition to invertebrate herbivores included with coral rubble, maintain small herbivorous fish (such as *Acanthurus triostegus*) in the tanks to control algal growth. Use holding tanks of several hundred liters capacity and maintain by pumping a continuous flow of seawater at a rate sufficient to flush tanks at least hourly. If DO concentrations or temperatures between inlet and outlet change by more than 10% or 1 C, respectively, increase flow rate.

4. Culturing Test Organisms

a. Water supply: Take seawater from an area that supports good, natural coral growths (Section 801D). For static studies, an artificial seawater can be used [Section 801D.4*b*2)].

Successful laboratory maintenance of corals requires continuous flow of large amounts of uncontaminated open-ocean water. Use the seawater system, pumps, pipe, cleanouts, and delivery systems described in Sections 801D and E.

b. Food and feeding: Specific nutritional requirements of hermatypic corals are more complex than those of most other organisms because of the symbiotic relationship between the coral and the zooxanthellar algae it contains. It is due also to the rapid calcification process that produces the massive coral skeleton.

Symbiotic algae utilize sunlight to produce food, a portion of which is transferred to the coral.[17] In addition, corals capture plankton[18] and are capable of digesting bacteria[19,20] and absorbing dissolved organics from the water.[20,21] Coral also takes up various inorganic nutrients (e.g., HCO_3) for the calcification process and certain inorganic plant nutrients.[22–25] Corals kept in full sunlight continue to grow in filtered, continuous-flow seawater systems but cease to grow and eventually die if deprived of light but still supplied with unfiltered seawater.[11,26]

Observe feeding behavior during tests. The carnivorous feeding habits of 15 Pacific species of corals have been described by Abe.[27] In the Atlantic, *Manicina areolata, Montastrea cavernosa,* and *Porites porites* have been used in feeding studies[28,29] and readily will ingest freshly hatched nauplii of the brine shrimp, *Artemia.* Zooplankton captured with a fine mesh net also can be used.

5. Parasites and Predators

For a general discussion, see Section

801D.5. The known coral parasites and predators have been reviewed.[30] Many of these can enter laboratory tanks in larval form and grow rapidly to maturity while feeding on coral tissue. White patches appear, especially on the undersides of the coral, when predators are present and feeding. Generally, predators will be attached to the coral near damaged areas. Corals of the genus *Montipora*, and possibly other Acroporids, are attacked by the Polyclad flatworm *Prosthiostomum*.[31] *Porites* is eaten by the acolid nudibranch *Phestilla sibogae*, while *Tubastrea* is eaten by a similar species, *Phestilla melanobranchia*.[32] *Fungia* is attacked by the wentletrap *Epitomium ulu*.[33] If predators of these types appear, remove them by drawing them up into an ordinary household basting pipet. Damaged specimens of larval corals, as well as adult corals, may be destroyed rarely by holotrich ciliate protozoa.[34]

805 B. Test Procedures

1. General Considerations

Many coral reef communities are "biologically accommodated," rather than "physically controlled."[35] Because corals rapidly modify water chemistry of closed systems, give special attention to loading (Section 801E.3*c*). Closed recirculating systems using carbonate and charcoal filters have been used to hold corals for several months.[36–38] Lighting systems needed to simulate reef environments tend to raise water temperatures. Because of these factors, use only continuous-flow systems for long-term tests. For facilities, equipment, and construction materials, see Sections 801D.3 and 4 and 801E.1.

2. Preparing Test Materials

a. Dilution water: Use tropical open-ocean surface water as the dilution medium. Take water from an offshore area, away from strong terrestrial influence (see Sections 801B through E).

b. Toxicant solutions: See Section 801E.2*b*.

c. Test organisms: See Sections 801D and E for general procedures.

3. Test Procedures and Conditions

For general procedures, see Section 801E. Use at least 20 coral colonies in each test concentration and in controls. The number and size of exposure chambers will be governed by size of colonies used. Load as indicated in Section 801E.3*c*. Select toxic concentrations as described in Section 801E.3*b*. Replace tank volume hourly at a flow rate that varies no more than ±5%. Use toxicant delivery systems described in Section 801E.1. Keep test chambers clean. Maintain salinity at 33 to 35.0‰ or at that of the area from which the coral were collected. Hold temperatures at 27.0 ± 1 C or at those of the natural habitat. Keep DO within ±10% of saturation. The pH may vary from 8.1 to 8.4. Use light source described in Section 801E.3*f* with a midday intensity at the water surface. Cover top of coral colonies with at least 2 cm water. Illuminate for tropical conditions, 12 hr light and 12 hr dark with appropriate twilight periods (Section 801E.3*f*). Supply nutrients in long-term studies. The need for animal food varies among species and probably with environmental conditions within a single species. When animal food is needed, feed as described in Section 805A.4*b*.

4. Conducting Tests

Conduct flow-through tests of effluent or toxicants under local conditions of light, temperature, and water quality. Carry out tests in large outdoor aquariums exposed to full sunlight. Supply tanks with a continuous flow of seawater pumped from the receiving water.

Measure and report frequently physical, chemical, and biological characteristics of the water. The light intensity (especially ultraviolet) in shallow tanks can be excessive. Simulate light levels at the intended depth by screening part of the natural light with a filter such as glass-fiber window screening.[39,40]

Treat coral colonies as described in Sections 805A.3*b* and *c*. Acclimate coral fragments that have developed polyps on all sides in holding facilities for 30 days before use. If more than 10% of colonies exhibit damage, discard and use a new group. Because it will not be known if corals used in a test are functionally autotrophic or heterotrophic[24] under stress conditions, run duplicate series for each concentration with and without a source of zooplankton. Use field-collected zooplankton or rear in the laboratory as described in Section 804C.2.

Use uniform-size colonies weighing approximately 10 g wet weight or more depending on available space.

The various life stages of corals have been photographed[34] and described.[41,42] Coral planulae have been used successfully to test relative wastewater toxicity.[43] The advantage of using larvae is their small size, permitting use of small containers and large numbers of organisms. However, free planulae are quite unlike adult colonies: they are noncalcifying, nonfeeding, solitary, planktonic organisms with some mobility. Planulae may not reflect accurately the effects of environmental alterations on adult colonies. In some species, planulae may be the most resistant life stage.[42]

Planulae are not always available.[42,44–49] They usually can be obtained from *Pocillopora damicornis* in the Pacific and from *Agaricia agaricites* in the Atlantic. Gather planulae by collecting mature coral heads and placing them in an aquarium with a continuous flow of seawater. Pass the outlet water from the aquarium into a submerged upright cylinder closed at the ends with plankton netting. This retains planulae expelled by the coral polyps. Hold them for up to several days until needed.[50] Release of planulae has been stimulated by warming the water to 35 C for a few minutes;[41,42] however, this is not recommended because it tends to produce damaged and immature specimens.

Conduct static tests with planulae in 150-mL borosilicate glass beakers containing 100 mL test solution held in a water bath or air-conditioned room. Use flow-through systems for medium- and long-term studies. Conduct these tests in 100-mL or larger glass containers, tightly covered with plankton mesh and containing 20 organisms. Replace container volume hourly.

Because some planulae will settle in the containers, also measure setting success. The extremely high mortality involved in setting and early colony growth[50,51] indicates that this life stage is most sensitive to environmental disruption.

Coral larvae can be settled on glass slides[4,27] to produce new colonies for testing. Newly settled corals can be substituted for planulae as described previously.

Follow procedures described in Section 801F for data analysis and reporting results.

805 C. Evaluating and Reporting Results

1. Lethal Response

Most corals are colonial; therefore, they present the problem of what should be counted as one individual. Although coral fragments differ in total number of polyps, a lethal concentration generally will kill all polyps and thus allow treatment of the colony as an individual. Interdependence between polyps has been suggested.[50,52] Exceptions result from strong gradients affecting only one portion of the colony. For example, accumulation of sediments around the base of the colony may kill only those polyps contacted. High light intensity may kill only the exposed polyps. These problems usually can be eliminated by proper test design.

Death of the colony becomes obvious when polyps become insensitive to stimulation and opaque in appearance. Within a few hours after death, corals begin to disintegrate. At the end of a test, return corals to optimal conditions and observe for recovery.

2. Visible Sublethal Effects

Healthy corals generally contain dense concentrations of zooxanthellae, containing plant pigments that give the polyps a characteristic color. In response to sublethal toxicant concentrations, the zooxanthellae are extruded over a period of hours or days, leaving the polyps transparent. This may be a response to high temperature, low light intensity, and starvation,[26] as well as to runoff of fresh water and silt[53] and estuarine effluents.[54]

Under some conditions, the first sign of damage is destruction of the thin tissue covering the septae or cenosarc; the skeleton then becomes visible. Some corals will extrude mesenterial filaments when subjected to stress.

Irritated or damaged corals may produce large amounts of mucus. Often distressed corals will contract their normally expanded polyps. Edmondson[55] and Mayer[56] reported that corals cease feeding long before lethal temperatures are reached. Samoan corals exposed to pesticides show abnormal feeding behavior long before death occurs.

Abe[27] described a method of using carmine granules to observe ciliary currents on the surfaces of corals. This may be a useful technic for observing signs of distress.

3. Other Tests

In situ field measurement of response can be used in support of laboratory tests. Transplanting corals into a different environment frequently has been used to evaluate coral response to various environmental differences.[10,14,16,57–59]

Locate transplant stations to give a span of extreme gradients while being similar in all other respects (depth, wave action, salinity, substrate type, etc.). Use as a platform for attachment of transplanted coral a section of vinyl-coated steel wire fencing material firmly anchored to the reef with iron stakes or concrete blocks. Tie the corals to the platform with plastic-coated electrical wire. In areas of fine sediment, support the frame off the bottom, unless sedimentation is the factor being investigated.

Select, handle, tag, and field-prepare specimens in the same manner as for laboratory tests. Use larger heads weighing up to several hundred grams because space is not limited.

Continue such studies for a minimum of 1 yr and check corals for mortality, growth, or other responses monthly. Monitor the environment for significant water quality factors such as salinity, temperature, DO, light, sediment load, dissolved nutrients, particulate matter, and dissolved organic carbon.

In situ measurements of coral calcification and oxygen metabolism have been carried out in submerged enclosures.[57,60] This approach can be used in connection with field transplants.

805 D. References

1. WELLS, J.W. 1954. Recent corals of the Marshall Island, Bikini and nearby atolls. Part 2, Oceanogr. (Biologic). *Geol. Survey Paper* 260:385. Pl. 94–185. Washington, D.C.
2. WELLS, J.W. 1956. Scleractinia. *In* R.C. Moore, ed. Treatise on Invertebrate Paleontology. Geol. Soc. Amer. F328–F444.
3. VAUGHAN, T.W. & J.W. WELLS. 1943. Revision of the suborders, families and genera of the Scleractinia. *Geol. Soc. Amer. Spec. Papers* 44:1.
4. CROSSLAND, C. 1952. Madreporaria, Hydrocorallinae, Heliopora and Tubipora. Great Barrier Reef Expedition. VI:86. British Museum (Natural History).
5. HICKSON, S.J. 1924. An Introduction to the Study of Corals. Manchester Univ. Press, London & New York.
6. VAUGHAN, T.W. 1907. Recent Madreporaria of the Hawaiian Islands and Laysan. Smithsonian Inst., U.S. National Museum, Bull. LIX, Washington, D.C.
7. SMITH, F.G. 1971. Atlantic Reef Corals. Univ. of Miami Press, Miami, Fla.
8. YONGE, C.M. 1963. The biology of coral reefs. *Advan. Mar. Biol.* I(4):209.
9. PEARSE, V.B. & L. MUSCATINE. 1971. Role of symbiotic algae (zooxanthellae) in coral calcification. *Biol. Bull.* 141:350.
10. SHINN, E.A. 1966. Coral growth rate, an environmental indicator. *J. Paleontol.* 40:233.
11. FRANZISKET, L. 1969. Riffkorallen konnen autotroph leben. *Naturwissenschaften* 56:144.
12. FISHELSON, L. 1973. Ecological and biological phenomena influencing coral-species composition on the reef tables at Eilat (Gulf of Aqaba, Red Sea). *Marine Biol.* 19:183.
13. DOMM, S.B. 1971. The safe use of open boats in the coral reef environment. *Atoll Res. Bull.* 143:1.
14. WOOD-JONES, F. 1907. On the growth forms and supposed species in corals. *Zool. Soc. London, Proc.*, 518.
15. WOOD-JONES, F. 1910. Corals and Atolls. L. Reeve, London.
16. MARAGOS, J.E. 1972. A Study of the Ecology of Hawaiian Reef Corals. Ph.D. thesis, Univ. of Hawaii, Honolulu.
17. MUSCATINE, L. & E. CERNICHIARI. 1969. Assimilation of photosynthetic products of zooxanthellae by a reef coral. *Biol. Bull.* 137:506.
18. YONGE, C.M. 1973. The nature of reef-building (hermatypic) corals. *Bull. Mar. Sci.* 23:1.
19. DISALVO, L.H. 1971. Ingestion and assimilation of bacteria by two scleractinian coral species. *In* H.M. Lenhoff, L. Muscatine & L.V. Davis, eds. Experimental Coelenterate Biology. p. 129. Univ. Hawaii Press, Honolulu.
20. SOROKIN, Y.I. 1973. On the feeding of some scleractinian corals with bacteria and dissolved organic matter. *J. Amer. Soc. Limnol. Oceanogr.* 18:380.
21. STEPHENS, G.C. 1962. Uptake of organic material by aquatic invertebrates. I. Uptake of glucose by the solitary coral *Fungia scutaria*. *Biol. Bull.* 123:648.
22. FRANZISKET, L. 1973. Uptake and accumulation of nitrate and nitrite by reef corals. *Naturwissenschaften* 60:552.
23. FRANZISKET, L. 1974. Nitrate uptake by reef corals. *Int. Rev. Gesamten Hydrobiol.* 59:1.
24. GOREAU, T.F., N.I. GOREAU & C.M. YONGE. 1971. Reef corals; autotrophs or heterotrophs? *Biol. Bull.* 141:247.
25. KAWAGUTI, S. 1953. Ammonia metabolism of the reef corals. *Biol. J. Okayama Univ.* 1:171.
26. YONGE, C.M. & A.G. NICHOLLS. 1931. Studies on the physiology of corals. IV. The structure, distribution and physiology of the zooxanthellae. *Sci. Rep. Great Bar-*

rier Reef Exped. 1:135.

27. ABE, N. 1938. Feeding behavior and the nematocyst of *Fungia* and 15 other species of corals. *Palau Trop. Biol. Sta. Stud.* 1:469.
28. COLES, S.L. 1969. Quantitative estimates of feeding and respiration for three scleractinian corals. *J. Amer. Soc. Limnol. Oceanogr.* 14:949.
29. LEHMAN, J.T. & J.E. PORTER. 1973. Chemical activation and feeding in the Caribbean reef-building coral *Montastrea cavernosa. Biol. Bull.* 145:140.
30. ROBERTSON, R. 1970. Review of the predators and parasites of stony corals with special reference to symbiotic prosobranch gastropods. *Pac. Sci.* 24:43.
31. JOKIEL, P.L. & S.J. TOWNSLEY. 1974. Biology of the polyclad *Prosthiostomum* sp., a new coral parasite from Hawaii. *Pac. Sci.* 28:361.
32. HARRIS, L.G. 1971. Nudibranch associations as symbioses. *In* T.C. Cheng, ed. Aspects of the Biology of Symbiosis. p. 77. Univ. Park Press, Baltimore, Md.
33. BOSCH, H.F. 1965. A gastropod parasite of solitary corals in Hawaii. *Pac. Sci.* 19:267.
34. SISSON, R.F. 1973. Life cycle of a coral captured in color. *Nat. Geogr. Mag.* 143:780.
35. SANDERS, H.L. 1968. Marine benthic diversity: A comparative study. *Amer. Natur.* 102:243.
36. ABE, N. 1937. Post-larval development of the coral *Fungia actiniformis* var. *palawensis* Doderlein. *Palao Trop. Biol. Sta. Studies* 1:73.
37. YONGE, C.M., M.J. YONGE & A.G. NICHOLLS. 1932. Studies on the physiology of corals. VI. The relationship between respiration and the production of oxygen by their zooxanthellae. Rep. Great Barrier Reef Exped., British Museum (Natural History) p. 213.
38. CATALA, R. 1964. Carnival Under the Sea. R. Sicard, Paris, France.
39. COLES, S.L. 1973. Some Effects of Temperature and Related Physical Factors on Hawaiian Reef Corals. Ph.D. dissertation. Dep. Zool., Univ. of Hawaii, Honolulu (unpublished).
40. JONES, R.S. & R.H. RANDALL. 1973. A study of biological impact caused by natural and man-induced changes on a tropical reef. Univ. Guam Tech. Rep. No. 7:1.
41. EDMONDSON, C.H. 1929. Growth of Hawaiian Corals. B.P. Bishop Museum, Bull. No. 58:1. Honolulu, Hawaii.
42. EDMONDSON, C.H. 1946. Behavior of coral planulae under altered saline and thermal conditions. *Occas. Pap. Bernice Pauahi Bishop Mus.* 18:283.
43. ENGINEERING SCIENCE, INC. 1971. Water Quality Program for Oahu with Special Emphasis on Waste Disposal. Final Report of Work Area 5. Chapter III, Toxicity Bioassays. City and County of Honolulu, Dep. Pub. Works.
44. WILSON, H.V. 1888. On the development of *Manicina areolata. J. Morphol.* 2:191.
45. DUERDEN, J.E. 1904. The coral *Siderastrea radians* and its postlarval development. *Carnegie Inst. Wash. Publ.* 20:1. Washington, D.C.
46. YABE, H. & EGUCHI. 1939. Ecological studies on *Rhizopsammia minuta* var. *mutsuensis*. Jubilee Publ. Prof. H. Yabe 60th Birthday Vol. 1:175. (Japanese with English summary).
47. YONGE, C.M. 1935. Studies on the biology of Tortugas corals. I. Observations on *Meandrina areolata* Linn. *Carnegie Inst. Wash. Publ.* 452:185.
48. MARSHALL, S.M. & T.A. STEPHENSON. 1933. The breeding of reef animals. Part I. The corals. *Sci. Rep. Great Barrier Reef Exped.* 3:219.
49. KAWAGUTI, S. 1941. On the physiology of reef corals. V. Tropisms of coral planulae, considered as a factor of distribution of the reefs. *Palao Trop. Biol. Sta. Stud.* 2:319.
50. HARRIGAN, J.F. 1972. The Planulae Larvae of *Pocillopora damicornis:* Lunar Periodicity of Swarming and Substratum Selection Behavior. Ph.D. dissertation, Dep. of Zoology, Univ. of Hawaii, Honolulu.
51. FISHELSON, L. 1973. Ecology of coral reefs in the Gulf of Aqaba (Red Sea) influenced by pollution. *Oecologia* (Berlin) 12:55.
52. GOREAU, T.F. 1963. Calcium carbonate deposition by coralline algae and corals in relation to their roles as reef builders. *Ann. N.Y. Acad. Sci.* 109:127.
53. GOREAU, T.F. 1964. Mass expulsion of zooxanthellae from Jamaican reef communities after Hurricane Flora. *Science* 145:383.

54. WELLS, J.M., A.H. WELLS & J.G. VANDERWALKER. 1973. *In situ* studies in benthic reef communities. *Helgolander wiss. Meeresunters* 24:78.
55. EDMONDSON, G.H. 1928. The ecology of a Hawaiian coral reef. B.P. Bishop Museum, Bull. No. 45:1.
56. MAYER, J.W. 1915. On the development of the coral *Agaricia fragilis* Dana. *Amer. Acad. Arts Sci. Proc.* 51:483.
57. GLYNN, P.W. & R.H. STEWART. 1973. Distribution of coral reefs in the Pearl Islands (Gulf of Panama) in relation to thermal conditions. *Limnol. Oceanogr.* 18:367.
58. MARSHALL, S.M. & A.P. ORR. 1931. Sedimentation on Low Isles Reef and its relation to coral growth. *Sci. Rep. Great Barrier Reef Exped.* 1:93.
59. JOKIEL, P.L. & S.L. COLES. 1974. Effects of heated effluent on hermatypic corals at Kahe Point, Oahu. *Pac. Sci.* 28:1.
60. BARNES, D.J. & D.L. TAYLOR. 1973. *In situ* studies of calcification in the coral *Montastrea annularis. Helgolander wiss. Meeresunters* 24:284.

806 BIOASSAY PROCEDURES FOR MARINE POLYCHAETE ANNELIDS (TENTATIVE)

Polychaete annelids are an important component of marine and estuarine biota. In subtidal benthic environments, they compose about 30 to 75% of the macroinvertebrate species and individuals. They include a variety of feeding types with the majority being either filter or detritus feeders. They are important food for snails, larger crustaceans, fishes, and birds. Many species have short life cycles.

806 A. Selecting and Preparing Test Organisms

1. Selecting Test Organisms

In accordance with the criteria listed in Section 801D.1, the following organisms are recommended for tests:

a. Family Nereidae:

1) *Neanthes arenaceodentata* (New England, Florida, California coast).

2) *Neanthes succinea* (all U.S. coastlines).

3) *Neanthes virens* (East coast).

b. Family Capitellidae:

Capitella capitata (all U.S. coasts).

c. Family Dorvilleidae:

Ophryotrocha sp.

d. Ctenodrilidae:

Ctenodrilus serratus (cosmopolitan).

2. Collecting and Culturing Test Organisms

a. Collection technics: Neanthes arenaceodentata, N. succinea, and *Capitella capitata* inhabit intertidal mud flats in estuarine areas and the fouling communities on pilings, boat floats, or submerged objects. To obtain worms, bring substrate or fouling material into the laboratory, place in white enameled pans, and cover with seawater. After a period of time, the worms come to the surface; remove them with a fine brush and transfer to petri dishes containing seawater. Examine each specimen under a dissecting microscope and discard all injured worms. Transfer uninjured specimens to 4-L aquariums or other suitable containers.

Ophryotrocha and *Ctenodrilus* occur on fouling organisms attached to floats and pilings. Because these species are minute, look for them under a dissecting microscope. Because only a small number can be collected at one time, establish a laboratory colony [Section 806A.2*b*1)].

b. Culturing:

1) Condition of animals—Because polychaetes may be physiologically altered during gamete maturation, discard specimens containing gametes in the coelom. Also discard animals injured during collection.

2) Food and feeding—Feed worms in all long-term experiments; the green alga, *Enteromorpha* sp., is most convenient. It grows abundantly in nearly all estuarine areas of the United States. Collect in quantity, wash with seawater, dry, and store indefinitely. Before use, soak the alga in seawater and knead to separate individual filaments. For each worm weighing less than 100 mg wet weight, use about 0.15 g dry weight (0.25 to 0.3 g wet weight)/wk. Feed this amount of food twice per week to polychaetes weighing more than 100 mg. Alternatively prepare a fine powder from *Enteromorpha* or different commercially available fish food by grinding the dry material in a blender and passing it through a fine sieve with 0.061-mm openings. This powder is particularly suitable for small species such as *Ophryotrocha* and *Ctenodrilus*. Feed these species at the rate of 0.1 mL of a mixture of 1.0 g powder/100 mL seawater/specimen/wk or for *Ophryotrocha* sp. feed frozen spinach at the rate of 1.0 mg dry weight/worm/wk.[1]

Because polychaetes vary greatly in size, use worms large enough (at least 10 to 15 mm) to permit determination of feeding rate. Feed adult nereids and *C. capitata* 0.15 g dry weight (0.25 to 0.3 g wet weight)/wk of *Enteromorpha*.

Feed living *Dunaliella* sp. to larvae or recently settled juveniles of *Neanthes arenaceodentata, N. succinea,* and *C. capitata*. Use *Dunaliella* as the initial food source until the larvae settle. For culturing instructions see Section 801D.4*c*1)b). Feed *Dunaliella* at the rate of 10 mL culture (having a minimum cell count of 20,000 cells/mL)/L of worm culture or at a rate great enough to maintain a green color in the test solution. After settling, feed *Enteromorpha* to the worms until the larvae mature to the swimming reproductive epitoke stage.

Because many polychaetes fail to feed in the presence of a toxicant, examine each container daily to ensure that the worms are feeding and food is not spoiling. Accumulation of fecal pellets is an indication of feeding.

3) Producing test organisms

a) *Ophryotrocha*—Several species of the genus have been cultured. They reproduce rapidly under laboratory conditions and because of their minute size can be transported easily.[1]

b) *Capitella capitata*—Laboratory-cultured specimens begin to mature in about 15 to 25 days, as indicated by the appearance of maturing white masses of eggs from about Segment 10 posteriorly in the female and the appearance of specialized setae on the dorsal surface of Segments 8 and 9 in the male. The female lays fertilized eggs along the inside lining of her tube where larval development continues until the trochophore larvae emerge. Obtain free-swimming trochophore larvae from the tubes of females. Examine these tube masses under a dissecting microscope to detect those containing eggs or larvae. Recently fertilized eggs appear white, but as they mature they become grey-green and can be seen moving about. Place females containing larvae in a petri dish. Under a dissecting microscope open the tubes to free the trochophores. One female provides 200 to 300 trochophores.

Remove and discard the parent and the tube containing the larvae that did not swim free. Use the free-swimming larvae in tests or let them develop for later use.

c) *Neanthes succinea*—Take nearly mature epitokes from the field or laboratory colony and hold until they complete sexual metamorphosis. Mature epitokes swim to the water surface and release gametes. If fertilization is successful, resulting in the production of zygotes, separate the zygotes into several 4-L jars containing aerated water and allow them to develop to the three-setiger stage. This requires about 1 wk. These larvae are ready for use. One fertilization provides more than 2,000 larvae.

d) *Neanthes arenaceodentata*—Before spawning, either the male or female enters the tube or burrow of another worm. If the worms are of different sex, they remain together and spawn within the tube. The female dies within a day after spawning and the male incubates the eggs for about 3 wk, at which time they have 18 to 21 setigerous segments. At that time the young worms leave the tube, begin feeding, and construct their own tubes. Feed them *Enteromorpha* as indicated in Section 806A.2*b*2). Under laboratory conditions, sexual maturity is reached in 3 to 4 months. It is impossible to distinguish the sex of immature forms morphologically. Distinguish by observing whether or not they fight when placed together. Use a female with maturing eggs in her coelom as a known individual. The most convenient time to obtain larvae is shortly after they have left the parent's tube and have begun to feed.

e) *Ctenodrilus serratus*—This species reproduces asexually about every 14 days at 20 C by transverse division. Each individual produces about five to eight new specimens. Large colonies can be maintained with minimum care.

c. Parasites and diseases: Microbial growth can result from overfeeding, improper conditioning of food, or insufficient DO. Prevent fungal growths by proper sanitation and periodic care. To minimize the possibilities of overfeeding, examine each aquarium before feeding to make certain food is required. Generally there is adequate DO in 4-L aquariums; however, aeration can be increased to correct for any deficiency.

The internal protozoan parasitic gregarines have been observed to reduce the vitality of *Ophryotrocha* laboratory populations.[1] Gregarines are widespread in polychaetes but it is not known if they cause similar problems in other species.

806 B. Bioassay Procedures

1. General Procedures

Use exploratory tests (see Section 801C) to determine toxicant concentrations for short-term tests (acute static and renewal) and intermediate and long-term tests. Prepare dilution water and toxicant solutions and introduce them into test containers as described in Section 801E.

2. Water Supply

a. Artificial seawater: See Section 801D.4*b*2). Use a salinity of approximately 35.5 ‰ and a pH of about 7.8 for marine populations; use lower salinity for estuarine worms.

b. Natural seawater: Determine and report quality routinely. Maintain dilution water salinity at or near selected or normal concentration. During a test, do not allow salinity to vary by more than ±3 ‰. In all except effluent tests, filter seawater through a 0.45-μm membrane filter.

3. Exposure Chambers

Use 4-L aquariums or glass jars for short-term and intermediate static and renewal tests and for long-term tests where flow-through facilities are not available. Cover aquariums to prevent entrance of foreign materials. Do not add more than 2.5 L test solution to each 4-L aquarium. Use 500-mL erlenmeyer flasks, containing 100 mL seawater, for either short-term or long-term experiments when only one organism is placed in each flask. Close the flask with a No. 7 TFE stopper fitted with a glass tube for aeration. Use small stender dishes (30 mL) for larval tests. For life-cycle, flow-through tests, use exposure chambers described in Section 807A.3*b*3).

4. Conducting the Bioassays

a. Setting up test chambers: For static and renewal tests, set up as described in Section 801C. When adult worms are used, place a minimum of two worms in each chamber and use 10 replicate test chambers to supply at least 20 worms for each test concentration. If larvae are used, use a minimum of four exposure chambers, each containing at least five larvae, for each test concentration and for the controls. If a single adult organism is used per container, make tests in erlenmeyer flasks. If aeration is required, see Section 801E.3*c*.

In short-term tests do not clean exposure containers. In long-term tests in which the organisms are fed, remove unused food and other materials as described in Section 801D.4*d*.

It is unnecessary to provide a bottom substrate or certain light intensities or photoperiods. Keep temperature within ±2 C of the natural habitat unless the effect of temperature is being tested.

b. Duration and type of test:

1) Short-term tests—The length of short-term or acute tests depends on the length of the organism's life cycle (see Section 801E.3*a*) which varies from 1 to 14 days. Short-term tests may be static if they last for 4 days or less or they may be renewal tests.

2) Intermediate-length tests—Use tests of intermediate length for determining adult survival. For most species conduct these renewal or flow-through tests for 28 days.

3) Long-term tests—Long-term tests are either lifetime tests beginning with the trochophore larval stage and continuing through sexual maturity or life-cycle tests beginning with the newly settled larval stage and continuing through reproduction and subsequent larval settlement of the offspring. Use only flow-through-type tests.

Select and prepare test concentrations as described in Section 801E.2*b*. Measure and mix dilution water and stock toxicant solutions by proportional diluters and deliver to exposure chambers as described in Section 801E.1. Make tests in flow-through exposure chambers similar to those used for the dungeness crab [Sections 807A.3*b*3) and 807B.4*c*2)]. Renewal tests using 4-L exposure chambers may be necessary if flowing seawater is unavailable.

The duration of long-term tests depends on length of life cycle of the organism and varies from about 1 wk with *Ophryotrocha* to 1 month with *C. capitata,* to 3 or more months with *N. succinea, N. virens,* and *N. arenaceodentata.*

c. Test organisms: See Section 806A.

d. Performing tests:

1) Short-term tests—Set up and conduct renewal tests as described in Section 801C.2. Determine survival of adults by checking exposure chambers at 1.5, 3, 6,

12, and 24 hr, then once or twice daily thereafter. Dead specimens generally are pale and swollen and lie on the bottom; live specimens usually move when the container is rotated. If the tests are more than 4 days long, renew solutions, preferably daily but at least every fourth day. In short-term tests with larvae, determine survival after 96 or 168 hr by microscopic examination. The absence of larvae generally indicates death because decomposition of small larvae is rapid.

2) Intermediate-length tests—Set up test chambers described in ¶s 4*a* and *c* to determine adult lethality (LC50 or incipient LC50). Examine test containers daily to determine survival during each of the first 5 days; thereafter make an examination 2 to 3 times/wk. Report results as 28-day LC50. If no organisms are killed after a certain length of exposure, report the period beyond which there is no further kill and the percentage killed in each test concentration. Calculate asymptotic LC50.

3) Life-cycle tests beginning with the trochophore larval stages—Set up as described previously. Carry organisms through sexual maturity with test periods varying from 3 to 4 wk with *C. capitata,* and from 2 to 3 months or longer for *N. succinea* and *N. virens*. Feed larvae as described in Section 806A.2*b*2). Determine survival at least twice per week for *C. capitata* and once per week for *Neanthes*. During the early part of the study, count organisms on bottom of exposure chambers. For 4-L chambers, decant the supernatant fluid and examine under a dissecting microscope. For flow-through chambers remove chambers from the glass tray, count organisms, and return test chamber to the glass tray. After counting 4-L chambers, replace fluid with fresh test solution. If no organisms are observed by the third examination, terminate that test chamber. When *C. capitata* is the test organism, remove test chambers after about 15 to 16 days and every 2 days thereafter to check with a dissecting microscope for presence of eggs in the coelom and later for the presence of zygotes along the sides of the tube. Remove females when developing eggs are in the trochophore stage and count the larvae. Discard females and larvae after counting larvae and recording other data such as length and number dead and deformed. Continue to examine each exposure chamber every 2 days for detection of females incubating larvae until all females have been removed and the total number of larvae recorded.

For *N. succinea,* set up exposure chambers as described in ¶s 4*a* and *c* with 25 larvae in each 1-L exposure chamber or 10 larvae in each flow-through exposure chamber. Use 10 chambers per concentration tested. Because these worms fight and are cannibalistic when crowded, after the first month set up additional exposure chambers or reduce numbers in each test chamber to five individuals. Continue tests until animals reach the epitoke stage, then determine individual lengths and total weights and compare with those in the control.

4) Life-cycle tests beginning with the newly settled larval stage—These tests will vary in duration from about 1 wk with *Ophryotrocha,* to 1 month for *C. capitata,* and 3 months or more for *N. arenaceodentata*. Set up tests as described previously with newly settled larvae. Use a minimum of two specimens per flask and 10 flasks per concentration. As tests progress, count organisms as above. Because *Ophryotrocha* is very small, examine chambers for survival only at completion of the test period in 4-L chambers. Periodically count the larvae in flow-through chambers by removing chambers from glass trays and setting them on a grid under a microscope. For other species, examine for survival once or twice per week as in ¶ *d*3).

For *N. arenaceodentata,* use recently

emerged larvae having approximately 18 to 21 setigerous segments. Place four worms in each 4-L exposure chamber with 2.5 L test solution. Set up five jars for each test concentration. Prepare appropriate controls. For flow-through tests, place two larvae in each of 10 exposure chambers for each test concentration and the controls. At 25 days, examine worms by viewing from outside the jar for the presence of eggs in the coelom. Mature eggs reach 450 μm diam and are yellowish-orange. Examine at 5-day intervals until eggs are noted and then at 2- to 3-day intervals to determine whether eggs are being laid. The females die within a day after laying eggs and the males incubate them for about 3 wk. The life cycle is complete when the worm leaves the male's tube. Remove males and count larvae.

806 C. Data Evaluation

1. Short-term and Intermediate Adult Survival Studies

Determine the LC50 values for each exposure period as described in Section 801F.

2. Life-Cycle Studies Beginning With the Trochophore and Settled Larval Stages

The number of females forming and laying eggs and the number of offspring produced are inversely related to sublethal toxicant concentrations at levels below the LC50. They provide a more sophisticated measure of effects than the LC50. Record life cycle data for each concentration of toxicant as follows: number of females forming eggs, number of females laying eggs, and number of eggs and live offspring produced. Compare these data, expressed on a percentage basis, for all test concentrations with those obtained from the controls, using statistical and reporting technics described in Sections 801F and G.

806 D. Reference

1. Akesson, B. 1970. *Orphryotrocha labronia* as a test animal for the study of marine pollution. *Helgolander wiss. Meeresunters* 20:293.

806 E. Bibliography

Eisig, H. Zur Entwickelungsgeschichte der Capitelliden. *Zool. Stat. Neapel, Mitt.* 13:1.

Korschelt, E. 1931. Art und Daver der ungeschlechtlichen Fortpflanzungen bei *Ctenodrilus*. *Zool. Anz.* 93:227.

THORSON, G. 1946. Reproduction and larval development of Danish marine bottom invertebrates with special reference to the planktonic larvae in the sound (Øresund). *Komm. Dan. Fisk. -Havundergelser. Ser. Plankton. Meddel.* 4:1.

REISH, D.J. 1957. The life history of the polychaetous annelid *Neanthes caudata* (delle Chiaje), including a summary of development in the family Nereidae. *Pac. Sci.* 11:216.

REISH, D.J. & J.L. BARNARD. 1960. Field toxicity tests in marine waters utilizing the polychaetous annelid *Capitella capitata* (Fabricius). *Pac. Natur.* 1(21):1.

REISH, D.J. & T.L. RICHARDS. 1966. A technique for studying the effect of varying concentrations of dissolved oxygen on aquatic organisms. *Air Water Pollut. Inst.* 10:69.

RICHARDS, T.L. 1969. Physiological Ecology of Selected Polychaetous Annelids Exposed to Different Temperature, Salinity, and Dissolved Oxygen Combination. Ph.D. dissertation, Univ. of Maine.

REISH, D.J. 1970. The effects of varying concentrations of nutrients, chlorinity and dissolved oxygen on polychaetous annelids. *Water Res.* 4:721.

BELLAN, G., D.J. REISH & J.P. FORET. 1972. The sublethal effects of a detergent on the reproduction, development and settlement in the polychaetous annelid *Capitella capitata. Mar. Biol.* 14:183.

REISH, D.J. 1974. The establishment of laboratory colonies of polychaetous annelids. *Thallassia Jugoslav.* 10:181.

REISH, D.J., F.M. PILTZ, J.M. MARTIN & J.Q. WORD. 1974. The induction of abnormal polychaete larvae by heavy metals. *Mar. Pollut. Bull.* 5:125.

OSHIDA, P.S., A.J. MEARNS, D.J. REISH & C.S. WORD. 1976. The Effects of Hexavalent and Trivalent Chromium on *Neanthes arenaceodentata* (Polychaeta: Annelids). So. Calif. Coastal Water Res. Project, TM 225.

807 BIOASSAY PROCEDURES FOR CRUSTACEANS (TENTATIVE)

Crustaceans are a large, mostly marine group of organisms. The class, Crustacea, contains more than 25,000 species grouped into the two subclasses, Entomostraca and Malacostraca. Many of the smaller Entomostraca, the copepods and cladocerans, commonly are referred to as microcrustaceans and are important macroplanktonic organisms in both marine and fresh waters. Bioassay methods for this group are given in Section 804. The subclass Malacostraca contains the larger and economically important crustaceans. Although most are marine there are some important freshwater forms. In the family Mysidae, *Mysis relicta* of the Great Lakes is an important food of salmonids. The order Amphipoda, containing over 3,000 species, is almost exclusively marine but has important freshwater species in the genera, *Hyalella, Gammarus, Crangonyx,* and *Pontoporeia*. The order Isopoda also is almost exclusively marine but has a few freshwater forms in the genus *Asellus*. Of greater economic importance is the order Decapoda, containing the lobsters, spiny lobsters, crabs, shrimp, prawns, and crayfish. Crustaceans are especially relevant for determining the toxicity of pesticides in the aquatic environment because of their phylogenic relationship to the insects for whose control many pesticides have been developed.

807 A. Selecting and Preparing Test Species

1. Selecting Test Organisms

The general principles governing the selection of test organisms are described in 801D.1. The following are suggested as test organisms:

Freshwater species:
- *Gammarus lacustris*
- *Gammarus pseudolimnaeus*
- *Gammarus fasciatus*
- *Hyalella azteca*
- *Pontoporeia affinis*
- *Mysis relicta*
- *Palaemonetes cummingi*
- *Palaemonetes paludosus*
- *Palaemonetes kadiakensis*
- Crayfish—*Cambarus*
  - *Potamobius*
  - *Orconectes rusticus*

Marine and brackish water species:
- *Palaemonetes pugio*—grass shrimp
- *Palaemonetes vulgaris*
- *Palaemonetes intermedius*
- *Crangon septemspinosa*—sand shrimp
- *Penaeus duorarum*—pink shrimp
- *Penaeus aztecus*—brown shrimp
- *Penaeus setiferus*—white shrimp
- *Homarus americanus*—American lobster
- *Callinectes sapidus*—blue crab
- *Cancer irroratus*—rock crab
- *Cancer borealis*—jonah crab
- *Cancer magister*—dungeness crab
- *Panopeus herbstii*—mud crab
- *Rhithropanopeus harrisii*—mud crab
- *Menippe mercenaria*—stone crab

Accurately identify test animals before use. Most inshore macrocrustaceans in the United States have been classified and regional keys are available.[1–5]

2. Collecting and Handling Test Organisms

Collect smaller forms in dip nets or coarse plankton nets; collect larger near-shore forms in small-mesh seines. Trawl for most marine forms. For collecting and transporting methods see Sections 1005B and 801D.2.

3. Holding, Acclimating, and Culturing Test Organisms

a. Water supply: See Section 801D.4*b*.

b. Acclimating, holding, and maintaining stock cultures: See Section 801B.3 and 4. Risks in handling most adult crustaceans usually are not great because of their rigid exoskeleton and general durability. Both larval and adult forms of many species are cannibalistic and readily attack their fellows in the soft-shell stage. Hold juveniles and adults in individual compartments in long troughs or divided tanks. Form the compartments with perforated separators that slide into slots on the sides. Use stainless steel for freshwater forms and glass, acrylic, plastic, or plywood covered with fiberglass for marine forms. Provide rigid, transparent covers to prevent loss of the highly motile specimens. Use perforated separators to ensure a flow of water through each compartment to remove metabolic products and provide DO. The crustacean growth process, which involves a periodic ecdysis or sloughing of the rigid exoskeleton, imposes a lack of uniformity in test animals that is not readily detectable in advance. In the pre-ecdysis stage and during ecdysis animals are heavily stressed and more sensitive to unsatisfactory environmental conditions and toxicants.[6]

1) Amphipods—Collect freshwater amphipods from their natural habitat or rear in the laboratory. A few species have been reared through three or more generations. Maintain several stock cultures and supplement by collections from the natural habitat. In the laboratory, a new generation can be produced in about 8 to 20 wk, depending on species and water temper-

ature.[7–9] For holding and for tests, use a known favorable temperature and a light intensity of 540 to 1,600 lux at the water surface.

Keep amphipods collected from the field in flow-through systems with water temperatures held initially near the temperature of the water from which they were collected. Gradually change temperature to that at which tests are to be conducted. Handle test animals carefully and as little as possible. Use small dip nets having soft flexible netting and no projections.

Feed young and adults on aspen, maple, and birch leaves that have been soaked for several weeks in flowing water. Test leaves to insure that they are free from pesticides or other toxicants. Supply leaves in sufficient quantities to support a stable population but not to the extent that they cause DO depletion or excessive fungal growth. Supply some leaves at all times to provide cover. Use commercial fish food to supplement the leaf diet.

2) Crayfish—Collect specimens from their natural habitat by trapping, seining, or by hand (Section 801D.2). General procedures for holding and acclimating are as described in Sections 801D.3 and 4.

Because crayfish are cannibalistic, hold all but the young stages in separate compartments. Suitable holding, acclimating, and culturing chambers are stainless steel, glass, fiberglass-covered wood, or plastic troughs, 180 cm long, 30 cm wide, and 20 cm deep, with a divider down the center to make two long troughs. Make shallow channels on the sides and central divider every 15 cm into which separators can be slipped to make 12 compartments on each side, each approximately 15 × 15 cm square and 20 cm deep. This size is suitable for crayfish. The number and size of compartments depends on the size and number of organisms to be tested. To hold a large number of small crayfish, remove the separators to make a tank of the desired length. Provide separators with a large number of perforations so they operate as screens. Control water depth in test chambers by a standpipe in the last compartment of the trough.[9] When cleaning the separators, temporarily raise them a short distance from the bottom to allow excess food and wastes to be washed out and remove the standpipe in the last compartment to insure strong flows. Clean routinely with a siphon and a brush to loosen materials from compartments, screens, walls, and bottoms. Supply water adjusted to the desired temperature and DO to the two head compartments by a siphon from a constant-head box. Use a minimum flow of 10 trough volumes/day. Adjust volume to maintain favorable water quality in each compartment. Water depth required depends on size of organisms but 15 cm is preferred. Provide each set of troughs with a transparent lid.

For life-cycle studies beginning with eggs or newly hatched young, collect ovigerous females and place in flow-through troughs under natural water conditions. Begin acclimation to different conditions after 2 days. Hold animals in troughs until young hatch. Remove compartment dividers to provide freedom of movement of young. Clean as described in Section 801D.4*d*.

Use macerated fish food for juveniles and adults. Alternatively use prepared dry fish food. Use very finely divided pieces of fish and commercial fish food pellets as food for the newly hatched.

3) Crabs—Successful static culture of brachyuran crab larvae has been made for several species of Atlantic coast crabs.[10–13] Long-term static or renewal bioassays with these species have been performed.[14–16] Culture of dungeness crab, *Cancer magister*, larvae has been reported.[17–19] Culturing crab larvae requires a favorable water supply and control of competitors, predators, and disease. Filter water and disinfect by UV light treatment.

For unpolluted open ocean water little or no treatment is required. If the supply is from an estuary receiving organic wastes, purify before use. Filter seawater for the flow-through system by gravity flow through a coarse, quartz sand filter and adjust to the desired salinity, approximately 25 to 30 ‰, by adding fresh water. To remove other organisms, refilter under pressure through sequential layers of 40/60-mesh garnet, 20/30-mesh silica sand, and 0.3 cm hard coal.* Follow by filtration through a polishing filter† and treat with UV light. Use constant-level head boxes equipped with heating, cooling, and stirring devices, to deliver constant measured flows by siphons, selected nozzles, or constant and accurate delivery pumps.

Collect ovigerous females or purchase from fishermen and place in holding tanks or in flow-through troughs similar to, but larger than, those described for crayfish. Acclimate and condition as described in Section 801C.3. When eggs are ready to hatch, transfer females to static tanks provided with aerated and UV-disinfected water at 30 ‰ and 13 C. As eggs hatch, dip out swimming first-stage larvae with beakers and transfer to culture beakers with large-bore pipets.

Dungeness crab larvae have long delicate spines that make their culture in flowing systems difficult. Culture larvae to the fourth and fifth stage in 250-mL beakers that have a hole 15 mm in diameter blown through their sides near the bottom. Using silicone cement, fasten a plastic‡ screen having 360-μm openings over this hole on the inside of the beaker and plastic screen with 210-μm openings over the hole on the outside.[19] Because of the lip created by blowing the glass, the two screens are 3 to 4 mm apart. The larger-mesh screen on the inside is less likely to catch and damage spines of larval crabs while the smaller-mesh screen on the outside does not come in contact with larvae but does retain food organisms, brine shrimp nauplii.

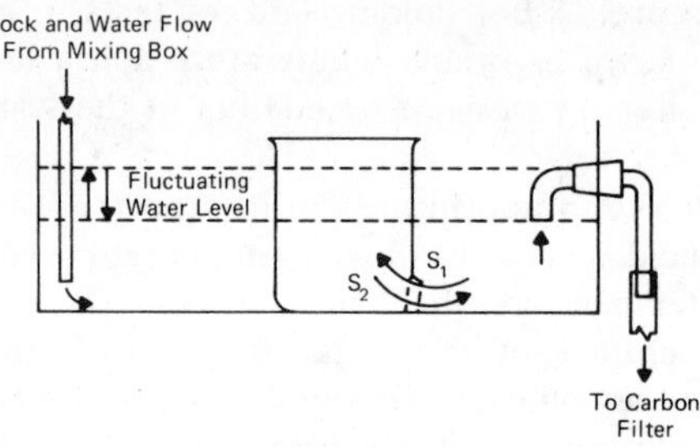

Figure 807:1. Rearing and exposure beaker and automatic siphon for dungeness crab larvae.[19]

Set the 250-mL beakers in glass trays or aquariums large enough to accomodate 10 beakers and provide a depth of at least 10 cm.[19] Supply trays with a constant flow of water by a tube that discharges near the tray bottom. Provide an automatic siphon at the outlet so there is continual fill and drawdown (Figure 807:1). Construct the automatic siphon so that when the water reaches the high point and the siphon is activated the beakers contain approximately 200 mL and when the siphon is broken the beakers contain about 150 mL.

The automatic siphon consists of a silicone rubber stopper drilled to receive an 8-mm-ID, right-angle glass tube on one end and a 5-mm-ID, right-angle glass tube on the other end. In a 1.3-cm-diam hole blown through side of tray, insert stopper with 8-mm hole on inside of tray. Insert tubes into stopper as shown in Figure 807:1. Placement of hole inside of tray controls water level in beakers at 200 mL. The distance between the top of the inside hole in the stopper and the bottom of the inside siphon leg is equal to the difference

*Filter design patented by Microfloc Corp., Corvallis, Ore. 97330.

†The Commercial Filter Corp., Lebanon, Ind. 46052, Fulflo, Model F15-10, or equivalent.

‡Nitex or equivalent.

in depth between 200 mL and 150 mL. Make the siphon intake perfectly flat and smooth to prevent air from being drawn[19] into siphon. Adjust tube diameters to give a 15-min cycle—10 min filling and 5 min drawdown.

When culture chambers are set up and functioning, place 10 first-stage larvae in each beaker with a smooth large-bore pipet. The larvae can be fed nonliving food but preferably feed first-stage brine shrimp nauplii at the rate of 70 for each crab larva 3 times/wk through the third stage, then 100 brine shrimp for each crab larva. Keep density of crab larvae low and that of food organisms high, to minimize crab larvae contacts that may result in cannibalism. Before feeding, transfer larvae to clean chlorine-disinfected and rinsed beakers. Use a temperature of 12 to 13 C, pH 8, and a salinity of 25 to 30 ‰. Adjust the photoperiod to correspond with natural conditions or, if the cycle is off-season, to correspond to the normal annual cycle of light and dark. Exclude natural light and use fluorescent light (Section 801D.3*f*). Under these conditions survival of 80 to 90% through the fourth zoeal stage has been attained. Larvae usually begin molting into the fifth zoeal stage by the 45th day. Mortalities then increase.

Juvenile and adult dungeness crabs are much less susceptible to disease than larvae. With strict sanitation and unpolluted open-sea water sand filtration alone provides sufficient water quality control. Hold juvenile and adult crabs in trough compartments similar to, but larger than, those used for crayfish. To allow sufficient space for each juvenile crab use compartments 15 × 15 cm and 15 to 20 cm deep. For adult crabs use 30- × 30-cm or 40-× 40-cm compartments with a depth of about 30 cm. For large specimens use deeper water. For ease of supplying water, arrange troughs on stands having three shelves with space on each for two troughs. Feed cut-up or macerated fresh fish, clams, or mussels, or commercial dried fish foods to juveniles and adults. Remove unused food within 24 hr to reduce fouling.

Routinely clean sides and bottoms of compartments and remove wastes with vacuum or siphon cleaners. Raise screen separators a few millimeters and flush as suggested for crayfish troughs.

4) American lobster, *Homarus americanus*—Obtain adult lobsters by trapping or purchase from lobster fishermen. Ovigerous females can be obtained most readily in the early spring from lobster fishermen who have permits. Select females with brownish eggs because these eggs will hatch within a few weeks to a few months, depending in part on water temperature.[20] Immediately place ovigerous females in holding tanks 300 × 100 × 30 cm. Use 25 to 30 lobsters per tank at a ratio of 2 females to 1 male. Fasten the claws of these lobsters with elastic bands but not wooden pegs. Pass uncontaminated seawater continually through the tank at a rate that maintains the DO at or above 80% of saturation. Maintain salinity between 30 ‰ and that normal to seawater. Maintain temperature at or above 15 C.[20] Feed clams, quahogs, bay scallop viscera, fish (such as alewives), crabs, abalone scraps, squid, or commercial dry pelleted foods.

To provide an egg-hatching tank, place a partition across the lower end of the holding tank 30 cm from the end. Locate standpipe to maintain desired water levels at one side of this 100- × 30-cm area. Remove a piece 5 cm deep and 30 cm long from the top central portion of the partition for the outflow from the hatching tank. Fit screen box, 25 cm wide, 15 cm deep, and 30 cm long with a notch 5 × 30 cm in the top of one side of the frame, against the notch in the partition. Use plastic window screen with 2-mm-square[4] openings for screen box. Place female lobsters with eggs about to hatch in this

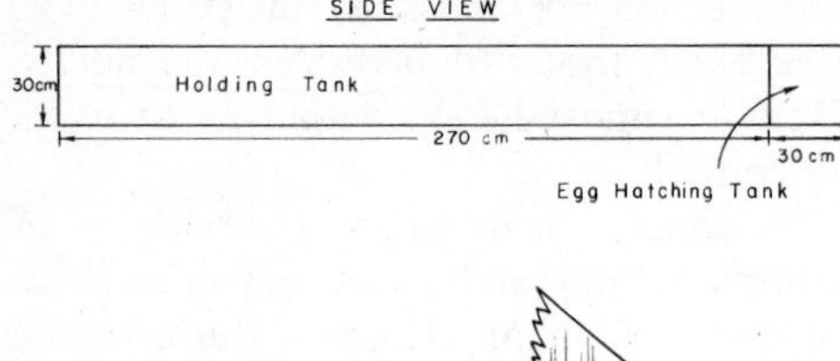

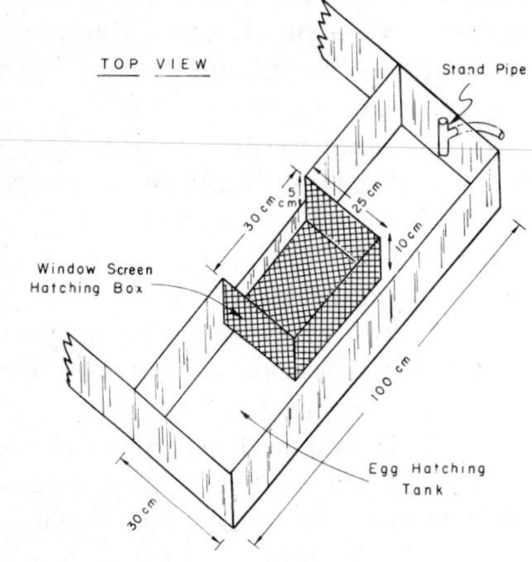

Figure 807:2. Egg-hatching tank for lobsters.

hatching box (Figure 807:2). Supply with flowing seawater at a rate sufficient to maintain DO above 80% of saturation. Maintain temperature at 19 to 20 C.

The larval development stages of *H. americanus* have been fully described.[21,22] The larval period extends from the time of hatching to the fourth molt or attainment of the fifth stage. Duration of the larval period depends somewhat on water temperature.[23] During the first three stages larvae are free-swimming, move toward light, and remain near the water surface. After the fourth stage they become bottom-crawlers and seek dark places, lead a nocturnal life, and acquire the defensive instincts of the adult lobster.

Construct from molded fiberglass a special 40-L culture tank for rearing larval stages[24] with a combined water circulator and overflow device at the center (Figure 807:3).[24] For tests use a 5-L tank. Maintain water flow for this modified tank between 0.5 and 1 L/min through the standpipe at the top. Discharge into the tank at the bottom through small slits in the manifold at the bottom of the overflow circulator.

Maintain temperature variations to within ± 1 C and regulate flow rate using ball valves.§ If water supply contains silt or suspended matter pass through a sand filter. If supply is unpolluted open-ocean water, use without treatment. If water is contaminated, use fine filtration and disinfection by ozone or UV light. Use a 10-μm wound polypropylene filter and a 40-W UV unit‖ for fine filtration.[24] For more information on filtration refer to Spotte.[25] For flows of >40 L/min use a diatomaceous earth filter. In some instances recirculating systems may be advantageous, especially when treatment with streptomycin is required to to prevent growth of filamentous bacteria or when large amounts of water are required. The maximum concentration of lobster larvae is about 45/L. At higher concentrations there are more lobster-to-lobster contacts and cannibalism.

As eggs begin to hatch, wash first-stage larvae over the 5- × 30-cm notch in the partition into the screen box. Dip larvae from this box with a small beaker and place in modified Hughes larval rearing chamber by submerging the beaker and gently removing it. Stock with <225 larvae.

Feed newly hatched larvae foods such as clams, mussels, liver, or other meat, chopped in a blender at a ratio of 1 part meat to 2 parts seawater. Continually drip this mixture into the rearing tanks from a stock bottle, the contents of which are mixed continually by a magnetic stirrer. To insure that food is well dispersed and

§Chemcock ball valve, Celanese Piping Systems, Aquafin Corp., Burbank, Calif. or equivalent.

‖PVCL-1, Aquafine Corp., Burbank, Calif., or equivalent.

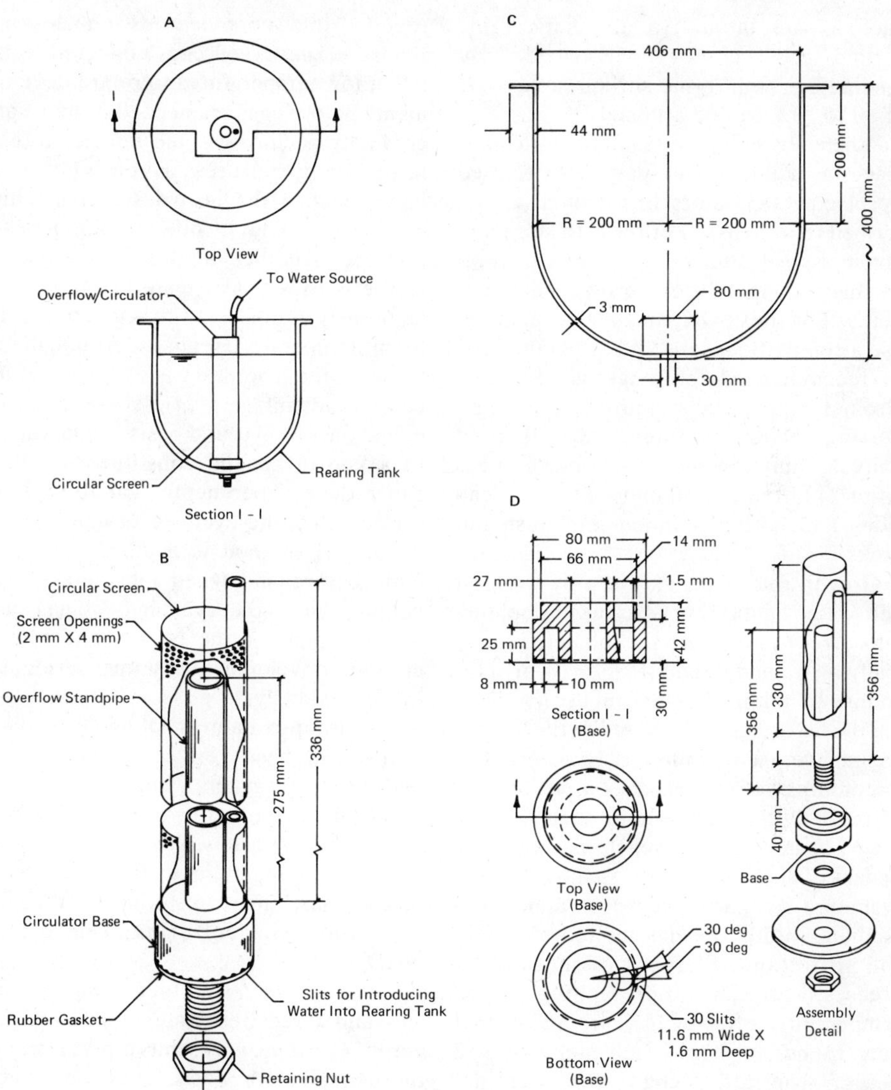

Figure 807:3. Hughes lobster-rearing tank.[24] **A**—general views; **B**—views of overflow/circulator; **C**—details of rearing tank construction; **D**—construction and assembly details for rearing tank and overflow/circulator. This is a 40-L tank. For bioassays, scale to 5-L volume.

held in suspension for as long as possible in the rearing chamber, introduce water under pressure through openings in overflow circulator. Because lobster larvae are cannibalistic, keep dispersed by currents and by providing many food particles for each larva. Best results are obtained by feeding newly hatched nauplii of brine shrimp, *Artemia salina*. Maintain a density of 50 to 200 brine shrimp nauplii for each lobster larva. An automatic brine shrimp feeder has been described.[26] By feeding live adult brine shrimp, a survival of 80 to 90% can be attained.[24]

When lobster larvae reach the fifth stage place in individual compartments formed by placing separators in a trough as described for crayfish. For the fifth stage and juveniles use 15- × 15- × 15-cm compartments. As lobsters grow use larger tanks. For those weighing 460 g or more, use a compartment 60 × 45 × 30 cm. Feed as recommended for ovigerous females. Ground, whole crabs improve coloration. During spring, summer, and fall, feed daily but during winter, feed once a week. After 24 hr remove all unused food. Clean sides and bottom; siphon and flush out tanks.

Growth rate depends not only on food and water quality but also on the holding tank size. Long before the lobster is physically restrained it reduces its growth in response to holding compartment size. During the first calendar year of life the lobster has an average of 10 molts. In nature, larval molting actively reaches a peak in the 15 to 20 C range; it seldom occurs below 5 C. Lobsters usually reach maturity at a weight of about 460 g. For mating, place a male in a compartment with a female immediately after she has molted and is in the soft stage. Success of mating decreases with time after molting. When temperatures of 22 to 24 C are maintained year round, lobsters reach maturity in 2 years.[27] The rate of egg development depends in part on temperature. For extrusion of eggs, place females in a deeper tank because they need at least 45 cm of water over them. Provide the egg-laying tank with a rough or nonslip bottom that allows female to assume and remain in the egg-laying position until all eggs are laid and attached to the nonplumose hairs of the swimmerets.

With stable temperatures, it should be possible to maintain larval cultures year round. Hold nonovigerous females and those bearing green eggs collected in the fall at low temperatures to retard development. Before eggs are needed, remove and gradually acclimate some females to egg-laying temperatures. Even when eggs have reached the brown stage, hatching can be spread out by different temperature regimes. Another method of producing larvae is to rear and mate lobsters in the laboratory at different times and under different temperature regimes. Although culturing in the laboratory is expensive it has certain advantages: (*a*) larvae are produced on a year-round basis; (*b*) larvae are of a known genetic constituency, which can reduce experimental variability; and (*c*) complete-life-cycle tests can be conducted. A method is available to determine beforehand when lobster eggs will hatch.[28] Once the eye pigment has been formed, monitor the course and rate of embryo development by measuring the eye periodically.

5) Shrimp-Natantia—Obtain by collection or purchase from bait dealers. Seine shrimp of the genera *Penaeus, Palaemonetes*, and *Crangon* from estuaries. Check animals for parasites, disease, and general condition. For general instructions on collecting, handling, transfering, holding, acclimating, and culturing, see Section 801D.

a) *Palaemonetes*—Three marine species and three freshwater species of the genus *Palaemonetes* have been reared through metamorphosis. They are suitable for life-cycle studies and can be brought from the field for direct use or for laboratory rearing. Place field-collected adult shrimp in suitable flow-through aquarium water. Feed freshwater species macerated parts of local fishes;[29] feed marine forms macerated mollusks or fish.[30] Examine shrimp periodically to detect ovigerous females. When eggs are nearly ready to hatch, remove desired number of females from tank and put into individual containers. Keep females in these containers, preferably with flow-through water, until eggs begin to hatch. During this period feed macerated fish or other suitable food. After eggs hatch, remove female and feed prelarvae or prezoeae on day-old *Artemia salina* nauplii. The rearing procedure for larvae is similar for all six species.[31] Use equipment and procedures similar to those for the dungeness crab, but with rearing chambers with a capacity of 1 L set in a deeper tray. Place 10 larvae in each beaker and feed with newly hatched brine shrimp nauplii and ideally maintain at 25 C. Filter and disinfect water. During the larval period, provide 14 hr light and 10 hr dark cycle[32] (see Section 801E.3*f*). Inspect larvae and feed daily. If sediments or wastes tend to collect, remove them daily with a siphon. At 25 C the larval period lasts 16 to 24 days. The average length of larval life is between 19 and 20 days. To rear through entire life cycle, immediately place females that have laid and hatched their eggs in an aquarium with males. Mating takes place, producing a second batch of fertilized eggs. The egg incubation period depends on temperature; usually 24 to 28 days are required. The number of eggs laid varies. There are six larval stages, the first being the protozoea. The seventh stage is a post-larval or juvenile shrimp, which marks the end of metamorphosis.

Keep larvae of marine species in seawater adjusted to 25 ‰ salinity. Feed with newly hatched *Artemia salina* nauplii. Rear at between 23 and 27 C using procedures similar to those used for freshwater species. Larval development has been described.[33–35] Remove chelipeds of ovigerous females with fine surgical scissors to prevent removal of eggs. When rearing larvae to a particular age, maintain a 10 to 15% surplus to compensate for mortality and to provide for other uses. The larvae are relatively hardy to temperature and salinity and can be reared at 25 C and a salinity between 15 and 25 ‰.

b) *Penaeus*—Hold shrimp in glass tanks of at least 30 L capacity. Provide each tank with flow-through water, 2 to 3 cm of sand over the bottom, and a screen over the top to prevent the shrimp from jumping out. Avoid overloading; keep no more than 22 to 24 animals in a 30-L tank. For *Penaeus* spp. use a minimum flow of 7.5 L/g/day; flows up to 22 L/g/day may be desirable to insure DO above 60% of saturation and the removal of metabolic products. Acclimate to laboratory test conditions for about 2 wk. For short-term or medium-length tests with adults and juveniles, shrimp can be field-collected. Cut-up fish is a satisfactory food. Cut a fillet from mullet, grouper, or other abundant species into 1-cm^2 pieces; feed one piece for each shrimp once each 2 to 3 days, depending on the size of the shrimp. Remove uneaten food daily. For larval tests or life-cycle studies collect gravid females offshore, let them spawn, and rear larvae at least to the postlarval stage. Penaeid shrimp can be reared from egg to postlarvae in the laboratory.[36–38] Mock and Murphy[39] have described improvements on procedures for spawning females, rearing larvae, and culturing the diatom *Skeletonema* as food for protozoeal stages. If diatoms are used as food, use air-lift pumps to prevent accumulation of the diatoms.[40] Feed freshly hatched brine shrimp, *Artemia*, to the mysis and postlarval stages. See Mock[41] for additional data on culture of algae and their feeding to shrimp. Equipment and procedures for continuous mass culture of algae as food are described in Section 801D.4*c*2).

The protozoeal stages, 1 through 3, of the penaeid shrimp require algae as food. Because larval shrimp are pelagic and unable to search for food during the early part of their life cycle, maintain the required density of the phytoplankton *Skeletonema costatum* and *Tetraselmis* sp. Add these to larval culture chambers according to stage of development, number of larvae present, and volume of water:

Protozoeal I	*Skeletonema*	50,000 cells/mL
Protozoeal II	*Skeletonema*	150,000 cells/mL
Protozoeal III	*Tetraselmis*	20,000 cells/mL
Mysis I	*Artemia* nauplii	3/mL
Mysis II	*Artemia* nauplii	3/mL
Mysis III	*Artemia* nauplii	3/mL
Postlarvae I-IV	*Artemia* nauplii	3/mL

Maintain phytoplankton in continuous culture or harvest and freeze to use later. Algae culture production units shown in Figures 801:1 and 801:2, Section 801D.4c2) will produce daily 7.5 L of culture containing 4.3×10^6 *Skeletonema costatum*/mL or 7.0×10^6 *Isochrysis galbana*/mL and several other species of algae at similar concentrations.

Add algae as food for the larval shrimp as either a fresh or frozen concentrate. Concentrate algae by centrifuging and discard the growth medium.

Use a temperature of 28 to 30 C and a salinity of 27 to 35 ‰. Omit antibiotics from the larval culture medium if the EDTA (disodium salt) is substituted at a concentration of 10 mg/L seawater.[42] Feed juvenile shrimp with fresh pieces of fish, clams, or mussels[42] or with prepared dried foods.[43] Knowledge of environmental requirements is essential for correct feeding and rearing of shrimp.[44,45]

4. Parasites, Diseases, and Harmful Growths

Much remains to be learned concerning harmful growths, parasites, and diseases of crustaceans.[46] For general problems and control procedures see Section 801D.5.

Adult shellfish in recirculated or flow-through systems are susceptible to biotoxins and pathogens. Remove metabolites and dead individuals from recirculating systems.

Juvenile and adult lobsters, crabs, and shrimps are subject to bacterial and fungal infections. *Gaffkya*, a bacterial pathogen, is particularly prevalent in tank-held lobsters, while *Vibrio* disease occurs in tank-held postlarval adult shrimp. Most captive crustaceans are subject to "shell disease," produced by chitin-destroying bacteria. A systemic fungal disease has been described in European prawns and several fungal infections occur in wild shrimp populations.

The larval stages of the lobster and several other crustaceans are prone to infections of the ubiquitous marine bacterium *Leucothrix mucor*, which has produced mortalities of over 90% in larval cultures.[47] The exuvia and the new exoskeleton after molting become entangled in the long dense filaments of the bacteria and the larvae are unable to swim or feed adequately. This organism also can produce high mortalities by causing pelagic eggs to sink and by interfering with the filtering apparatus of larval forms and the functioning of gills.[48]

In some instances it may be necessary to culture larvae in artificial seawater to avoid *L. mucor* infection. Place ovigerous females in a bath of malachite green (5 mg/L) for 1 min *only* or rinse them several times in artificial seawater of the correct salinity that contains streptomycin, 2 mL/L, from a stock solution containing 2 g/L of antibiotic. Maintaining a 1-mg/L concentration of antibiotic throughout the larval culture period prevents infections. Twice daily cleaning also is a good preventive method. Seawater, filtered and exposed to UV radiations, should be nearly bacteria-free.

A disease of lobster larvae tentatively has been associated with the phycomycete *Haliphtorus*. It appears as a scab on the first segment of the thoracic appendages up to and surrounding the first row of gills. Thorough cleaning and UV treatment of the water supply is the only known treatment. In most cases these scabs adhere to both old and new carapaces and thus cause a mechanical impediment to molting. Mortality appears to be restricted to larvae and young juveniles. No deaths of specimens with a carapace length over 27 mm have been observed.

The fungus *Lagenidium* sp. causes serious problems in rearing larval shrimp.[49] The disease first becomes apparent in the second protozoeal stage and disappears as the shrimp reach the first mysis stage. Shrimp become immobilized by replacement of muscle tissue by fungal mycelium.

Parasites have been found in many species of crustaceans and their presence can influence results. In *Uca*, an ectoparasitic isopod is found on the gills, nematodes in the gut, and metacercaria in the green glands. Species of *Lagenidium* similar to the one that occurs in shrimp occur in other marine crustaceans. *L. callinectes* occurs in eggs and larvae of the blue crab.[50,51] The blue crab has a barnacle *(Octolasmus lowei)* living in association with its gills and gill chamber, metacercariae in various organs, and the sacculinid *Loxothylacus taxanas* beneath its abdomen.

Saprolegia parasitica attacks larvae of the shrimp *Palaemonetes kadiakensis*.[52] Amphipods sometimes are parasitized by larval stages of acanthocephalan worms.[53]

807 B. Conducting the Bioassays

1. General Considerations

Procedures for crustaceans in general follow the methods outlined in Sections 801C and E. However, many adult crustaceans and some larvae must be segregated into individual compartments because of their aggressive and cannibalistic tendencies. As with most other groups, larval and juvenile crustaceans usually are significantly more sensitive than adults and therefore are preferred for short-term tests.

Determining toxic effects on decapods is complicated by three factors: initial paralysis, delayed response, and much greater sensitivity at molting periods. A true sublethal effect may be shown either by increased irritability or by inactivity. Penaeid shrimp may lie motionless for days without dying and become covered by silt, or they may be so irritable that they damage themselves by hitting aquarium walls when lights are turned on or when someone walks by. These effects may be reversible, disappearing when clean water is restored. A more definitive indication of toxicity is partial or complete paralysis of adults or lack of swimming by larvae. Crabs may lie paralyzed for days before dying. In general, paralysis is not a reversible effect. In nature, a paralyzed animal would be easy prey for predators and would not survive for long. The second factor that may complicate interpretation of crustacean tests is the "delayed response." After short exposure to toxic materials, test animals appear unharmed. If they are held in clean water, however, they may begin to die, the mortality eventually reaching 100%, even though no deaths occurred in the first few days after the exposure period.

Because of delayed mortality, place all surviving individuals from each test exposure in compartmented tanks receiving clean dilution water and hold under favorable environmental conditions for 2 wk to detect delayed effects.

For tests to assess effects of periodic spills, discharges, or dumping, base results on several series of short-term tests at the indicated exposures made over a period of weeks to determine the cumulative effects of intermittent exposure. This is essential because at least some decapods show cumulative effects and three 8-hr exposures at a given concentration, even though they are as much as 3 wk apart, are equivalent to a 24-hr exposure to the same concentration.[9]

Tests with embryonic and larval stages may last from a few hours to as much as 30 days, depending on the species and larval stages used. Toxicity criteria include egg hatchability, rate and success of molting, swimming ability, tendency to lose appendages, and metamorphosis of the larvae as well as death. Tests with selected larvae, juveniles, and adults may measure acute toxicity with paralysis or death as the end point. Long-term and life-cycle tests usually measure effects on respiration, behavior, osmo-regulation, growth, molting rates, reproduction, and general well-being.

2. Water Supply—Dilution Water and Water Distribution System

See Section 801E.1.

3. Equipment and Materials

See Section 801E.

4. Test Procedures

a. Amphipods:

1) Short-term tests—Make exploratory or range-finding tests as described in Section 801C.2 either under flow-through or static conditions. For static exposure chambers, use 5- to 6-L-capacity wide-mouth glass containers containing 4 L test solution. For long-term tests use flow-through test chambers. Prepare toxicant solutions and place in exposure chambers as described in Section 801C. In exploratory and short-term definitive tests keep water temperature constant using water baths or constant-temperature rooms. Preferably use a test temperature of 16 to 18 C. For effluent tests match test temperature to that of receiving water as described in Section 801C. Use juveniles for these tests. Use five organisms in each of the widely spaced concentrations for exploratory tests. Use five or more concentrations of toxicant in definitive tests with duplicate test chambers for each concentration and the controls. Place 10 organisms in each test container, thus obtaining 20 test organisms for each control and concentration tested. Select organisms randomly and immediately place in exposure chambers as described in Section 801E.3*a*. Do not feed. Use a photoperiod of 16 hr light and 8 hr dark with changes from L to D and D to L made during a 0.5-hr twilight period. Supply light as described in Section 801E.3*f* and at an intensity of 540 to 1,600 lux at the water surface. Maintain DO as described in Section 801E.3*a*. Limit static tests, in which the organisms are not fed, to 4 days.

2) Partial- or complete-life-cycle tests—Use long-term tests to determine waste toxicity to various species of amphipods.[54–59] Use equipment for the water system described in the literature[54–59] and in Section 801E.1. Prepare toxic solutions as described in Section 801E.2*b* and select test concentrations on the basis of short-term test results (Section 801E.3*b*). When methods of analysis are available, sample test concentrations immediately after tests are begun and at monthly or more frequent intervals thereafter. Analyze as described in Section 801E.3*d* to determine actual exposure concentrations.

Use glass aquariums with volumes of 8 to 20 L. To enhance mixing, deliver toxicant near one end of exposure chamber and position screened standpipe at other end. Maintain a minimum solution depth of 15 cm. Hold test solution flow rate to each chamber sufficient to maintain DO and remove metabolic products. Stabilize test solution flow to chambers before introducing test organisms. For effluent tests use DO and temperature of receiving water. For other tests, maintain DO above 60% saturation but do not supersaturate.

Use naturally occurring photoperiods during the stages of development under investigation (Section 801E.3*f*).

Start partial- or full-life-cycle tests with newly hatched young, juveniles, or adults. To test newly hatched young, collect ovigerous females about to shed young from the holding tanks of field-collected specimens or from culture tanks. Place each one in a clean 400- to 600-mL beaker containing dilution water to which it has been acclimated to test temperatures, feed as described in Section 807A.3*b*1), and hold until young are shed. Record number produced by each female.

When sufficient newly hatched young are available, begin test by selecting at random 30 young. Place them in each of duplicate growth aquariums for each test concentration and the controls. Feed test organisms as described in Section 807A.3*b*1) and clean aquariums as described in Section 801D.4*d*. Observe test chamber daily; remove dead organisms and preserve for future study. At the end of 60 days siphon the contents of each test chamber into pans and record all living young. Randomly select 15 from each exposure chamber and return them to their respective exposure chambers. Use the methods described in Clemens[6] for measuring the remaining organisms to determine growth. Preserve all organisms for toxicant accumulation and histopathological analyses. Observe test chambers daily and record behavior and number dead. As the amphipods approach maturity and mate, look for gravid females about to shed young. Place one in each beaker and handle as for the F_1 generation. Record number of young produced by each female and randomly place 30 in each duplicate exposure chamber for each test concentration and control. Terminate adult exposure, record number and sex of each animal, measure animals, and preserve for additional analyses. Continue exposure of young and record survival in each concentration after 1 month. Continue exposure to 60 days and then terminate. Count, measure length and weight, record condition of all survivors, and preserve for other studies.

b. Crayfish:

1) Short-term tests—Follow general procedures in Sections 801C and E for range-finding and definitive tests. Make special modifications to apparatus for crayfish. Preferably, make short-term tests with juveniles and young adults. For a test series use animals of similar size, with the largest individual being no more than 50% longer than the shortest, and in approximately the same intermolt or molt stage. Expose all organisms in individual compartments. Duration of exposure depends on study objectives and may be 96 hr, 1 wk, or 2 wk. Make exposures in the compartmented troughs described in 807A.3*b*2). Tests may be static, renewal, or flow-through. If the test lasts more than 96 hr, use flow-through procedures with some feeding and thorough cleaning of compartments.

2) Partial-life-cycle tests—Make exposures in the compartmented troughs described in Section 807A.3*b*2) with one trough used for each concentration tested and each control. Run tests in duplicate. Prepare dilution water and stock solutions of toxicants and arrange test troughs as described for other crustaceans. Use modified proportional diluters (Section 801E.1) to deliver toxicant to exposure chambers. Place 100 newly hatched young in each of the large chambers formed by removing the screens and converting 10 compartments into one large chamber. Adjust water quality to follow the normal seasonal cycle in local favorable waters. Regulate water flow, feed, and clean as described in Section 807A.3*b*2) and in the general procedures. Before the young become cannibalistic, remove them from the

common exposure chamber and count; randomly select 20 individuals and place one in each of the 20 compartments of their respective exposure troughs. Count, measure, weigh, and preserve. Make daily observations on mortality and condition and record results. Continue exposure until animals become adults. Record numbers, weight, condition, and sex. Preserve organisms for additional studies.

c. *Brachyura-Crabs:* Numerous investigators have used short- and long-term tests.[11,14–16,18,60–62] Basic procedures are those for fish and other organisms (see Sections 801C and E). Methods in current use for the dungeness crab are presented as a guide for conducting tests with this and other species of the group.[63,64]

1) Short-term tests—Use zoeal stages because juveniles and adults are more resistant to toxicants. Conduct short-term studies to determine relative sensitivity, relative toxicity, or test concentrations for long-term tests under conditions of temperature, pH, salinity, etc., described for rearing the dungeness crab, Section 807A.3*b*3). Condition and prepare dilution water as described for crab rearing.

For short-term tests of egg hatchability and early larval development to the first zoeal stage, treat ovigerous females with eggs hatching to control bacterial growths, parasites, and diseases (Section 807A.4). Then place in tanks with flowing seawater at 12 to 13 C and a salinity of 25 to 30 ‰. When initial hatching occurs, carefully remove unhatched eggs from the egg mass and place 30 directly into each of the 250-mL test chambers described in Section 807A.3*b*3), which have been set up 10 in each tray for each test concentration and the controls. Measure, then mix, dilution water and toxicant as described in Section 801E.1. Operate the automatic siphon on a 15-min or longer cycle [Section 807A.3*b*3)]. Development from egg to prezoeal stage to the first zoeal stages usually requires less than 24 hr. Determine hatching success, molting from prezoeal to first zoeal stage, and percentage of motile first-stage zoeae. To make short-term tests with first or later zoeal stages, treat female crabs as in Section 807A.3*b*3), place directly into filtered, UV-disinfected water at 13 C and 30 ‰ salinity, and hold until hatching occurs. When large numbers of first-stage zoeae come near the surface or collect in corners, dip them out with a beaker and place in beaker exposure chambers, 10 larvae in each beaker, and 10 beakers for each concentration and for controls. Expose for 96 to 168 hr. Use larvae of a comparable stage of development.

To conduct tests with later zoeal stages, rear the larvae in seawater to the selected stage as described. Before beginning the bioassay, collect megalops stages for testing from the field and hold for short periods until they metamorphose into the early juvenile stage so that this stage can be used in the bioassay. Determine DO and pH during the test. When possible, make chemical analyses to determine concentrations to which organisms actually are exposed. Use photoperiods normal for the season during which the larval stages develop. Because of the cannibalistic tendencies of crabs beyond the larval stages, expose each individual in a separate exposure chamber. The compartmented troughs described in Section 807A.3*b*3) provide the necessary exposure chambers and form a compact unit for testing of each effluent or toxicant concentration. With other than effluent bioassays, which are carried out as described in Section 801E, recommended test conditions for these stages are 12 to 13 C, 25 to 33 ‰ salinity, and DO above 60% saturation. Feed test organisms macerated fish or mollusks.

2) Partial-life-cycle tests—Conduct flow-through tests in the beaker test chambers described for rearing. Supply dilution water and toxicants by pumps or siphons

from constant-head tanks and stock solution containers. Use procedures described in Section 801E.1. Obtain eggs or first-zoeal-stage larvae from females, as described for short-term tests. To begin the long-term study with eggs, place 30 eggs in each of 10 beaker test chambers in the tray, making 300 eggs for each concentration. After hatching, determine number of dead, deformed, and active first-zoeal-stage larvae and randomly reduce number of active larvae to 20 in each test beaker. Continue exposure through the desired number of zoeal stages. Transfer test organisms to clean beakers three times per week. At that time, feed 70 first-stage brine shrimp nauplii to each crab larva. After the third zoeal stage, increase feeding to 100 nauplii per crab larva. Examine for bacterial growths or disease. If needed, treat larvae continuously with streptomycin as described in the rearing procedures. If necessary, pass effluents through a filter to remove coarse materials to prevent clogging of proportional diluter and screens. In partial-life-cycle tests continue test from the egg stage through the fourth zoeal stage. Usually, zoeae molt into the fourth zoeal stage by the 45th day. If mortality in controls is not excessive, carry on until the early juvenile stage. In other effluent tests, condition dilution water and keep water quality the same as that used to rear crab larvae.

When crabs reach the juvenile stage, transfer to compartmented trough chambers, one to each compartment; continue exposure. Enumerate, measure, and weigh all crabs to be discarded; preserve for additional tests.

Feed juvenile crabs pieces of fish or some other marine animal every 2 or 3 days and remove excess food after 24 hr. Clean exposure chambers by scraping down sides to remove algae; brush crabs to remove growth twice per week. In tests with small juvenile crabs, provide a sand substrate if it is found that they have difficulty in shedding the old exoskeleton during ecdysis. The criteria for death of zoeae are cessation of heartbeat and swimming, failure to recover after transfer to pure water, and opaqueness. The criterion of death for juveniles is absence of movement after stimulation. Terminate test at 4 months and record data.

d. American lobster, Homarus americanus: By laboratory rearing it is possible to produce any live stage for short-term tests at most times of year.

1) Short-term tests—Conduct short-term static, renewal, or flow-through tests with juvenile or adult lobsters in compartmented rearing troughs [Section 807A.3*b*4)]. Use dilution water as described in Section 801E.2*a*. Use procedures outlined in Sections 801C and E for dungeness crab and crayfish. Preferably make flow-through tests. For tests longer than 96 hr, feed lobsters, clean, and carry out other functions described for rearing [Section 807A.3*b*4)]. Use the adaptation of the Hughes rearing chamber for larval stages (Figure 807:3). Filter, disinfect, and treat for parasites and disease as for dungeness crab larvae. Obtain eggs and larvae, handle, and transfer as described in Section 807A.3*b*4). When conducting larval tests with effluents, filter receiving water and other effluent to remove other organisms and clogging materials. Because water movement is necessary to keep larvae separated and off the bottom, use flow-through tests with flow rates of 0.5 to 1 L/min. Meter stock toxic solutions and dilution water by a proportional diluter; discharge into a mixing chamber and then pump to larval test chamber to provide necessary currents. Feed with newly hatched nauplii of brine shrimp at a rate of at least 50 per animal [Section 807A.3*b*4)]. Use five concentrations and a control. Stock each 10-L exposure chamber with a minimum of 100 first-stage larvae. Con-

tinue tests for 96 to 168 hr. To stock each duplicate exposure chamber and the control, dip larvae from screen box of hatching tank with a small beaker and place in a glass tray for counting by the photographic method (Section 801E.3*e*). At termination, count survivors photographically or visually. Use the care, feeding, and cleaning procedures given for rearing. Controls give an indication of cannibalism; see Abbott's formula (Section 801F.1*e*).

2) Partial-life-cycle and life-cycle tests—Determine test concentrations from short-term test results. Handle test animals as in rearing [Section 807A.3*b*4)]. To speed up the tests and complete various life stages within a shorter time, expose at 22 to 24 C. To obtain test concentrations, use proportional diluters (Section 801E). Use duplicate larval rearing chambers for each test concentration and control and place in operation before introducing test organisms.

Dip larvae from screen box of hatching tank with a small beaker and place in a glass tray to count by the photographic method, Section 801E.3*e*. Place 150 first-stage larvae in each chamber. Immediately after photographing, dip larvae from tray and gently float them from beaker into larval exposure chamber. Concentrate larvae that cannot be dipped in one corner of tray by tipping and immerse that corner in exposure chamber to transfer all larvae. Maintain food density (brine shrimp) above 750/L. Check density periodically by dipping out a measured amount of test solution and counting. Observe all exposure chambers daily. At completion of larval stages and attainment of fifth stage, shut off water flow, reduce volume by a screened siphon, dip out lobsters, and place in a glass tray. To insure recovery of all lobsters from larval rearing chamber, disconnect it, remove overflow circulator, and wash those not dipped out into a tray. Randomly select 20 lobsters and place one the tray, photograph to measure length, in each of the 20 compartments of compartmented troughs and continue exposure. Anesthetize remaining lobsters in remove, count, and preserve.

Set up a compartmented trough for each larval exposure chamber, i.e., 40 lobsters for each exposure concentration and the controls. Feed and care for lobsters and clean troughs as described previously. Transfer lobsters to larger compartmented troughs as they grow, randomly reducing the number to 20 for each test concentration when they reach about 250 g. Weigh, measure and preserve those discarded. Continue exposure in each test concentration and control until test objectives are obtained or until maturity. When animals are mature and ready to mate, randomly select two pairs, one from each side of the compartmented trough, and place in mating chambers immediately after female has molted and is in the soft stage. After mating, remove males. Measure, weigh, and preserve all lobsters from each exposure concentration and controls. Place two females in compartmented troughs for each concentration. Remove separators so each has the entire length of one side of the trough. Continue exposure until they are nearly ready to extrude eggs. Place them in deep egg-laying chambers with special rough bottom until eggs are extruded and fastened. Then place them in egg-laying tanks, one for each toxicant concentration and controls. Continue exposure through egg hatching. Collect and place larvae in larval exposure chambers, two for each test concentration and controls, and expose as for F_1 generation. Count and record all live larvae produced, all dead first-stage larvae, and all unhatched eggs. Measure, weigh, and preserve females. Continue F_2 larvae only through fourth stage. Count and measure all larvae. Record number of active larvae, number deformed, and number dead. Measure, weigh, and preserve. Also record and report results of other tests such as histology, disease, parasites, and toxicant accumulation.

e. Shrimp:

1) *Palaemonetes*—By manipulating temperature and photoperiods it has been possible to induce spawning in the laboratory[65] and so make various life stages available throughout the year.

a) Short-term tests—For adults and the six larval stages use glass jars or aquariums as described in Section 801C. Use static, renewal, or flow-through tests.

Rear adults and larvae as described in Section 807A.3*b*5)a). A large number of ovigerous females is needed to produce larvae required for tests. Feed adults macerated fish or mollusks and feed the larvae brine shrimp daily before and during tests. Rear a 10 to 15% surplus of larvae to insure an adequate number for the test. Ideally, hatch and rear larvae for a series of tests at one time and in mass culture. Conduct short-term tests with larvae of specified ages. Most first-stage larvae metamorphose to postlarvae in 18 to 21 days. Therefore, set up two or more series of tests to cover the larval and postlarval stages. Because of variability of each age group of larvae and increased sensitivity before and during molting, run two to three replicates simultaneously for each test concentration in each series of tests.

In static tests, monitor DO in each test container and control to indicate if and when to renew test solutions. Do not let DO concentrations fall more than 0.5 mg/L below the minimum levels occurring at that period in the receiving water. For other types of tests, use the same DO and temperature at which test organisms are reared, about 25 C and 60% of saturation or above. For marine species use salinities between 15 and 25 ‰ but do not vary salinity more than ±1 ‰ in a test series.

b) Life-cycle tests—If feasible, begin tests with ovigerous females. Place 25 females with well-developed eggs in a compartmented trough from which the separators have been removed. Stock six or seven of these double-compartmented troughs for five test concentrations and controls. Feed with macerated fish, mollusks, or other suitable organisms. Observe daily to detect females with eggs nearly ready to hatch. Remove these females, place in separate glass jars supplied with appropriate test concentration, and hold until eggs hatch. Renew test solutions as required. Supply brine shrimp nauplii as food for those larvae hatching first. When hatching is complete, remove females and transfer larvae by pipet or gentle pouring into test chambers receiving toxicant. These chambers are the same as the rearing chambers for dungeness crabs (Figure 807:1). Alternatively use rearing chambers like those for lobster larvae. For dungeness crab chambers, use 1-L capacity and set up in duplicate 20 exposure chambers for each concentration. Stock with five larvae each. If the exposure chamber for lobster larvae is used (Figure 807:3), stock with 50 larvae each in duplicate. After transfering active larvae, check hatching chamber to determine percentage of eggs that did not hatch, first-stage larvae that died, and total number of live larvae. Count, measure, and preserve. Continue exposure with daily feeding of brine shrimp until the postlarval stage, which should occur in 18 to 21 days. Randomly select 25 postlarvae from each duplicate group for each exposure concentration. Place groups of 25 postlarvae on each side of a compartmented trough similar to the troughs in which the ovigerous females were exposed. There will be two compartments in the trough, one for each of the two groups of 25 postlarvae for each concentration and the controls. Preserve those not needed for further exposure to determine lengths, weights, histopathological effects, and toxicant accumulation. Feed with macerated fish and/or mollusks and continue exposure until eggs develop and are about to hatch. Select at random and transfer five ovigerous females from each group to hatching chambers; continue exposure. Determine hatching success and number and percentage survival of active first-stage larvae. Count, measure, weigh, and preserve those not taken for further exposure.

2) *Penaeus*

a) Short-term tests—Use flow-through short-term tests for penaeid shrimp. Hold, acclimate, and culture shrimp as described in Section 807A.3*b*5). For delivery of toxicants see Section 801E.1. For flow-through tests use aquariums 60 × 30 × 30 cm deep. Place 2 to 3 cm clean sand in the bottom of aquariums and cover tops with screens. After exposure chambers are operating under equilibrium conditions place 20 to 22 juvenile shrimp having a total weight of not more than 50 g in each aquarium. Use duplicate aquariums for each concentration and control. Use uniform-sized shrimp. Conduct tests at 25 C and salinity between 25 and 34 ‰, depending on the water supply. Maintain salinity to within ±1 ‰.[67] Use a flow rate of not less than 7 L/day/g of organism.

If tests are for effluents, use procedures described in Section 801C.

Use short-term test with larval stages to determine toxicant concentrations for long-term studies. Obtain larval stages of shrimp from gravid females brought in from offshore. While this is a time-consuming and expensive process, once secured, the first protozoeal shrimp are obtained readily. Handle and feed as described in Section 807A.3*b*5)b). Set up desired test toxicants in duplicate for five test concentrations and controls with 100 mL each in 250-mL beakers covered with a watch glass. For range-finding tests use concentrations of 0.01, 0.1, 1, 10, and 100%. Set up definitive tests at the concentrations indicated by the range-finding tests and as described in Section 801C. Using a smooth-bore pipet, introduce 10 larvae into each test beaker. Place beakers in a temperature-controlled shaker bath. Add food (algae for protozoeal stages and brine shrimp for mysis stages) as indicated in Section 807A.3*b*5)b). Test duration is governed by length of stage tested. At conclusion of test, examine beaker contents in petri dishes under a microscope and determine number of deformed, dead, and live larvae.

b) Partial-life-cycle tests—Begin with egg and continue to postlarval or juvenile stage (Section 801C). Maintain temperature at 28 to 28.5 C and salinity at 28 to 30 ‰. As soon as possible acclimate ovigerous females brought in from offshore to 28 to 28.5 C. While on the collecting boat, control temperature in insulated chests. Females usually spawn the first night after collection. As soon as spawning is complete, remove them from the spawning chamber and take samples to estimate the number of eggs in the chamber.[39] Set up a series of exposure beakers as described for the dungeness crab [Section 807A.3*b*3)]. Use screens that will retain shrimp larvae and brine shrimp but allow algae to pass through. Set up 10 test beakers in a glass tray for each test concentration and controls. Larvae usually pass through five nauplii stages to the protozoea stage in about 35 hr. On the basis of egg count, take portions from the mixed spawning tank containing about 10 nauplii and put in each exposure beaker; this will give 100 larvae for each test concentration and control. Just before the nauplii metamorphose to the protozoeal stage add algal food to each tank. Maintain toxicant flow at 50 mL/min. Add sufficient algae to each exposure tray to supply the quantities indicated in Section 807A.3*b*5)b). Add only enough algal concentrate to maintain the desired concentration. Culture algae in units shown in Section 801D.4*c*2). Monitor algal cell concentrations in the exposure trays daily.

Feed algae throughout the protozoeal

stages. At transformation to the mysis stage, phase out algal feed and add newly hatched brine shrimp. Because algae tend to remove metabolites, increase the flow-through in the mysis stage. Feed brine shrimp nauplii to excess. Do not let their concentration drop below 3/mL. Sample by pipet daily and count to insure adequate concentrations.[39] Feed brine shrimp to larvae through the fourth day of the postlarval stages. At that time remove them from beakers, place in glass dishes, and count. Count active, deformed, and dead larvae. Record percentage living and dead. Transfer living post-larvae to compartmented troughs, one trough for each concentration. Change diet to macerated fish, mollusks, or dry food.[41,42] Remove or flush out sediment and organic material as described previously. Examine for presence of disease or parasites. For bacterial or fungal growth, treat as in Section 801D.5 and 807A.4. If test chambers become too crowded, set up additional duplicate chambers. If desired, continue into the juvenile stage. At the end of the study, collect all organisms from each tank, weigh, measure, and make histological and toxicant accumulation studies as appropriate.

807 C. Reporting Results

Analyze and evaluate data and report results as recommended in Section 801F and G.

807 D. References

1. SMITH, R.I., ed. 1964. Keys to Marine Invertebrates of the Woods Hole Region, 1st ed. Systematics-Ecology Program, Marine Biology Lab., Woods Hole, Mass.
2. WILLIAM, A.B. 1965. Marine decapod crustaceans of the Carolinas. *Fish Bull.* 65:10298.
3. WASS, M.L. 1955. The decapod crustaceans of Alligator Harbor and adjacent inshore areas of northwestern Florida. *Quart. J. Fla. Acad. Sci.* 18:129.
4. LIGHT, S.F., R.I. SMITH, F.A. PIDELKA, E.P. ABBOTT & F.M. WEESNER. 1957. Intertidal Invertebrates of the Central California Coast, 2nd ed. Univ. of Calif. Press, Berkeley.
5. HALSINGER, J.R. 1972. The Freshwater Amphipod Crustaceans (Gammaridae) of North America. Biota of Freshwater Ecosystems, Identification Manual No. 5, Water Pollution Research Ser. No. 5, 18050 E.L.D.O. 4/72.
6. HUBSCHMAN, J.H. 1967. Effects of copper on the crayfish *Orconectes rusticus* (Girard). I. Acute toxicity. *Crustaceana* 12:33.
7. CLEMENS, H.P. 1950. Life cycle and ecology of *Gammarus fasciatus* Say. Ohio State Univ., *Franz T. Stone Inst. Hydrobiol. Contrib.* 12:1.
8. COOPER, W.E. 1965. Dynamics and productivity of a natural population of a freshwater amphipod *Hyalella azteca*. *Ecol. Monogr.* 35:377.
9. SMITH, W.E. 1973. Thermal tolerances of

two species of *Gammarus. Trans. Amer. Fish. Soc.* 102:431.

10. COSTLOW, J.D. & C.G. BOOKHOUT. 1960. A method of developing brachyuran crab eggs in vitro. *Limnol. Oceanogr.* 5:212.
11. COSTLOW, J.D., C.G. BOOKHOUT & R. MONROE. 1962. Salinity-temperature effects on the larval development of the crab *Panopeus berbstii* Milne-Edwards reared in the laboratory. *Physiol. Zool.* 35:78.
12. COSTLOW, J.D. & C.G. BOOKHOUT. 1962. The larval development of *Sesarma recticulatum* Say reared in the laboratory. *Crustaceana* 4:281.
13. COSTLOW, J.D. & C.G. BOOKHOUT. 1971. The effect of cyclic temperature on larval development in the mud-crab *Rhithropanopeus barrisii.* D.J. Crisp. 4th European Marine Biology Symp.:211. Cambridge Press, London.
14. BOOKHOUT, C.G., A.J. WILSON, JR., T.W. DUKE & J.I. LOWE. 1972. Effects of mirex on the larval development of two crabs. *Water, Air Soil Pollut.* 1:165.
15. EPIFANIO, C.E. 1971. Effects of dieldrin in seawater on the development of two species of crab larvae, *Leptodius floridanus* and *Panopeus berbstii. Mar. Biol.* 11:356.
16. LOWE, J.I. 1965. Chronic exposure of blue crabs, *Callinectes sapidus* to sublethal concentrations of DDT. *Ecology* 46:899.
17. REED, P.H. 1969. Culture methods and effects of temperature and salinity on survival and growth of dungeness crab, *Cancer magister* larvae in the laboratory. *J. Fish. Res. Board Can.* 26:389.
18. BUCHANAN, D.V., R.E. MILLEMANN & N.E. STEWART. 1970. Effects of the insecticide sevin on various stages of the dungeness crab, *Cancer magister. J. Fish. Res. Board Can.* 27:93.
19. BUCHANAN, D.V., M.J. MYERS & R.S. CALDWELL. 1975. An improved flowing water apparatus for culture of brachyuran crab larvae. (unpublished).
20. HUGHES, J.T. & G.C. MATTHIESSEN. 1962. Observations on the biology of the American lobster, *Homarus americanus. Limnol. Oceanogr.* 7:414.
21. HERRICK, F.H. 1896. The American lobster: A study of its habits and development. *Bull. U.S. Fish Comm.* 15:1.
22. HERRICK, F.H. 1911. Natural history of the American lobster. *Bull. U.S. Bur. Fish.* 29:147.
23. TEMPLEMAN, W. 1948. Growth per molt in the American lobster. *Bull. Newfoundland Govt. Lab.* 18:26.
24. HUGHES, J.T., R.A. SHLESER & G. TCHOBANOGLOUS. 1974. A rearing tank for lobster larvae and other aquatic species. *Progr. Fish-Cult.* 36:129.
25. SPOTTE, S.H. 1970. Fish and Invertebrate Culture, Water Management in Closed Systems. John Wiley and Sons, Inc., New York, N.Y.
26. SMITH, R.A., J.A. HOLMAN & R.H. KRAMER. 1974. Automatic brine shrimp feeder. *Progr. Fish-Cult.* 36:133.
27. HUGHES, J.T., J.J. SULLIVAN & R. SHLESER. 1972. Enhancement of lobster growth. *Science* 177:1110.
28. PERKINS, H.C. 1972. Developmental rates at various temperatures of embryos of the northern lobster (*Homarus americanus* Milne-Edwards). *Fish. Bull.* 70:96.
29. BROAD, A.C. & J.H. HUBSCHMAN. 1963. The larval development of *Palaemonetes kadiakensis,* M.J. Rathbun, in the laboratory. *Trans. Amer. Microsc. Soc.* 82:185.
30. HUBSCHMAN, J.H. & A.C. BROAD. 1974. The larval development of *Palaemonetes intermedius* Holthuis 1949 (Decapoda, Palaemonidae) reared in the laboratory. *Crustaceana* 26:89.
31. DOBKIN, S. 1963. The larval development of *Palaemonetes paludosus* (Gibbes 1850) (Decapoda Palaemonidae) reared in the laboratory. *Crustaceana* 6:41.
32. HUBSCHMAN, J.H. & J.A. ROSE. 1969. *Palaemonetes kadiakensis* Rathbun: Post embryonic growth in the laboratory (Decapoda, Palaemonidae). *Crustaceana* 16:81.
33. FAXON, W. 1879. On the development of *Palaemonetes vulgaris. Bull. Mus. Comp. Zool.* (Harvard) 5:303.
34. BROAD, A.C. 1957. Larval development of *Palaemonetes pugio* Holthuis. *Biol. Bull.* 112:144.
35. BROAD, A.C. 1957. The relationship between diet and larval development of *Palaemonetes. Biol. Bull.* 112:162.
36. COOK, H.L. & M.A. MURPHY. 1966. Rearing penaeid shrimp from eggs to postlarvae. *Proc. 19th Annu. Conf. S.E. Ass. Game Fish. Comm.* 19:283.

37. Cook, H.L. & M.A. Murphy. 1969. The culture of larval penaeid shrimp. *Trans. Amer. Fish. Soc.* 98:751.
38. Cook, H.L. 1967. A method of rearing penaeid shrimp larva for experimental studies. *FAO (Food Agr. Organ. U.N.) Fish. Rep.* 3:709.
39. Mock, C.R. & M.A. Murphy. 1970. Techniques for raising penaeid shrimp from egg to postlarvae. *Proc. 1st Annu. Workshop World Maricult. Soc.* 1:143.
40. Salser, B.R. & C.R. Mock. 1973. An air-lift circulator for algal culture tanks. *Proc. 4th Annu. Workshop World Maricult. Soc.* 4:295.
41. Mock, C.R. 1974. Larval Culture of Penaeid Shrimp at the Galveston Biological Laboratory. NOAA Tech. Rep. NMFS Circ. 388:33.
42. Mock, C.R. 1973. Shrimp culture in Japan. *Mar. Fish. Rev.* 35(3–4):71.
43. Meyers, S.P. & Z.P. Zein-Eldin. 1972. Binders and pellet stability in development of crustacean diets. *Proc. 3rd Annu. Workshop World Maricult. Soc.* 3:351.
44. Mock, C.R., R.A. Neal & B.R. Salser. 1973. A closed raceway for the culture of shrimp. *Proc. 4th Annu. Workshop World Maricult. Soc.* 4:247.
45. Zein-Eldin, Z.P. & G.W. Griffith. 1969. An appraisal of the effects of salinity and temperature on growth and survival of postlarval penaeids. *FAO (Food Agr. Organ. U.N.) Fish. Rep.* 3:1015.
46. Anderson, J.I.W. & D.A. Conroy. 1968. The significance of disease in preliminary attempts to raise crustacea in sea water. *Bull. Off. Inform. Epizoot.* 69:1239.
47. Brock, T.D. 1966. The habitat of *Leucothrix mucor,* a widespread marine organism. *Limnol. Oceanogr.* 11:303.
48. Johnson, P.W., J.M. Sieburth, A. Sastry, C.R. Arnold & M.S. Doty. 1971. *Leucothrix mucor* infestation of benthic crustacea, fish eggs and tropical algae. *Limnol. Oceanogr.* 16:962.
49. Lightner, D.V. & C.T. Fontain. 1973. A new fungus disease of the white shrimp *Penaeus setiferus. J. Invertebr. Pathol.* 22:94.
50. Couch, J.H. 1942. A new fungus on crab eggs. *J. Elisha Mitchell Sci. Soc.* 58:158.
51. Rogers-Talbert, R. 1948. The fungus *Lagenidium callinectes* Couch on eggs of the blue crab in Chesapeake Bay. *Biol. Bull.* 95:214.
52. Hubschman, J.H. & J.A. Schmitt. 1969. Primary mycosis in shrimp larvae. *J. Invertebr. Pathol.* 13:351.
53. Spencer, L.T. 1974. Parasitism of *Gammarus lacustris* (Crustacea; Amphipoda) by *Polymorphus minutus* (Acanthocephala) in Colorado. *Amer. Midland Natur.* 91:505.
54. Arthur, J.W. & E.N. Leonard. 1970. Effects of copper on *Gammarus pseudolimnaeus, Physa integra* and *Campeloma decisum* in soft water. *J. Fish. Res. Board Can.* 27:1277.
55. Arthur, J.W. & J.G. Eaton. 1971. Chloramine toxicity to the amphipod, *Gammarus pseudolimnaeus* and the fathead minnow, *Pimephales promelas. J. Fish. Res. Board Can.* 28:1841.
56. Nebeker, A.V. & F.A. Puglisi. 1974. Effect of polychlorinated biphenyls (PCB's) in survival and reproduction of *Daphnia, Gammarus* and *Tanytarsus. Trans. Amer. Fish. Soc.* 103:722.
57. Oseid, D.M. & L.L. Smith. 1974. Chronic toxicity of hydrogen sulfide to *Gammarus pseudolimnaeus. Trans. Amer. Fish. Soc.* 103:819.
58. Arthur, J.W., R.W. Andrew, V.R. Mattson, D.T. Olson, G.E. Glass, B.J. Halligan & C.T. Walbridge. 1975. Comparative toxicity of sewage-effluent disinfection to freshwater aquatic life. Ecol. Res. Ser., EPA-600/3-75-012.
59. Arthur, J.W., A.E. Lemke, V.R. Mattson & J.B. Halligan. 1974. Toxicity of sodium nitrilotriacetate (NTA) to the fathead minnow and an amphipod in soft water. *Water Res.* 8:187.
60. Collier, R.S., J.E. Miller, M.A. Dawson & F.P. Thurberg. 1973. Physiological response of the mud crab *Eurypanopeus depressus* to cadmium. *Bull. Environ. Contam. Toxicol.* 10:378.
61. Thurberg, F.P., M.A. Dawson & R.S. Collier. 1973. Effects of copper and cadmium on osmoregulation and oxygen consumption in two species of estuarine crabs. *Mar. Biol.* 23:171.
62. Vernberg, W.B. & J. Vernberg. 1972. The synergistic effects of temperature, salinity and mercury on survival and metabo-

lism of the adult fiddler crab, *Uca pugilator*. *Fish. Bull.* 70:415.

63. ARMSTRONG, D.A., D.V. BUCHANAN, M.H. MALLON, R.S. CALDWELL & R.E. MILLIMAN. 1975. Toxicity of the insecticide methoxychlor to the dungeness crab, *Cancer magister* Dana. Oregon State Univ. Marine Science Center, Newport (unpublished).
64. CALDWELL, R.S., D.V. BUCHANAN, D.A. ARMSTRONG, M.H. MALLON & R.E. MILLIMAN. 1975. Toxicity of pesticides to the dungeness crab *Cancer magister* Dana. I. the fungicide captan. Oregon State Univ. Marine Science Center, Newport (unpublished).
65. LITTLE, G. 1968. Induced winter breeding and larval development in the shrimp. *Palaemonetes pugio* Holthuis (Caridea Palaemonidae) studies in decapod larval development. *Crustaceana Suppl.* 2:19.
66. SANDIFER, P.A. 1973. Effects of temperature and salinity on larval development of grass shrimp, *Palaemonetes vulgaris* (Decapoda Caridea). *Fish. Bull.* 71:115.
67. BAHNER, L.H., C.D. CRAFT & D.R. NIMMO. 1975. A salt water flow-through bioassay method with controlled temperature and salinity. *Progr. Fish-Cult.* 37:126.

807 E. Bibliography

HADLEY, P.B. 1906. Regarding the rate of growth of the American lobster. *Homarus americanus. 36th Annu. Rep., Comm. Inland Fish. R.I.*:153.

TEMPLEMAN, W. 1934. Mating in the American lobster. *Contrib. Can. Biol. Fish.* 8:423.

TEMPLEMAN, W. 1936. Further contributions to mating in the American lobster. *J. Biol. Board Can.* 2:223.

MACKAY, D.C.G. 1943. Temperature and world distribution of the genus *Cancer*. *Ecology* 24:113.

GRUNBAUM, B.W., B.V. SIEGEL, A.R. SCHULZ & P.L. KIRK. 1955. Determination of oxygen uptake by tissue growth in all glass differential microrespirometer. *Mikrochim. Acta.* 1955:1069.

ROBERTS, J.L. 1957. Thermal acclimation of metabolism of *Pachygrapsus crassipes* Randall. II. Mechanisms and the influence of season and latitude. *Physiol. Zool.* 30:242.

BOUSFIELD, E.L. 1958. Fresh water amphipod crustaceans of glaciated North America. *Can. Field Natur.* 72:55.

VERNBERG, F.J. & R.E. TASHIAN. 1959. Studies on the physiological variation between tropical and temperate zone fiddler crabs of the genus *Uca*. I. Thermal death limits. *Ecology* 40:589.

COSTLOW, J., C.G. BOOKHOUT & R. MONROE. 1960. The effect of salinity and temperature on larval development of *Sesarma cinerium* (Boxc) reared in the laboratory. *Biol. Bull.* 118:183.

KINNE, O. 1963. The effect of temperature and salinity on marine and brackish water animals: *Oceanogr. Mar. Biol. Annu. Rev.* 1:301.

COSTLOW, J. & C.G. BOOKHOUT. 1964. An approach to the ecology of marine invertebrate larvae. *Symp. Exp. Mar. Ecol., Occas. Publ.* 2:69. Grad. School Oceanog., Univ. R.I.

VERNBERG, F.J. & J.D. COSTLOW. 1966. Studies on the physiological variation between tropical and temperate zone fiddler crabs of the genus *Uca*. IV. Oxygen consumption of larvae and young crabs reared in the laboratory. *Physiol. Zool.* 39:36.

HUGHES, J.T. 1968. Grow your own lobsters commercially. *Ocean Ind.* 3(12):46.

SAILA, S., J. FLOWERS & J.T. HUGHES. 1968. Fecundity of the American lobster *Homarus americanus*. *Trans. Amer. Fish. Soc.* 98:537.

BALLARD, B.S. & R.E. TASHIAN. 1969. Osmotic accomodation in *Callinectes supidus* Rathbun. *Comp. Biochem. Physiol.* 29:671.

EISLER, R. 1969. Acute toxicities of insecticides to marine decapod crustaceans. *Crustaceana* 16:302.

VERNBERG, F.J. 1969. Acclimation of intertidal crabs. *Amer. Zool.* 9:333.

HARGRAVE, G.T. 1970. The utilization of benthic microflora by *Hyalella azteca* (Amphipoda). *J. Anim. Ecol.* 39:427.

NIMMO, D.R., A.J. WILSON, JR. & R.R. BLACKMAN. 1970. Localization of DDT in the body organs of pink and white shrimp. *Bull. Environ. Contam. Toxicol.* 5:333.

RICE, A.L. & D.I. WILLIAMSON. 1970. Methods for rearing larval decapod crustacea. *Helgolander wiss. Meeresunters* 20.

SASTRY, A.N. 1970. Culture of brachyuran crab larvae using a recirculating sea water system in the laboratory. *Helgolander wiss. Meeresunters* 20:406.

SASTRY, A.N. 1971. Culture of brachyuran crab larvae under controlled conditions. *In* M. Uda, ed. The Ocean World, p. 475. Joint Oceanographic Assembly. Japan Soc. Promotion Science, Tokyo.

HYNES, H.B.N. & F. HARPER. 1972. The life histories of *Gammarus lacustris* and *Gammarus pseudolimnaeus* in southern Ontario. *Crustaceana,* Suppl. 3:329.

JENIO, F., JR. 1972. The *Gammarus* of Elm Spring, Union County, Illinois (Amphipoda: Gammaridae). Ph.D. dissertation, Southern Ill. Univ.

PORTMAN, J.E. 1972. Results of acute toxicity tests with marine organisms, using a standard method. *In* M. Ruivo, ed. Marine Pollution and Sea Life. Fishing News (Brooks) Ltd., London.

REES, C.P. 1972. The distribution of the amphipod *Gammarus pseudolimnaeus* Bousfield as influenced by oxygen concentration, substratum and current velocity. *Trans. Amer. Microsc. Soc.* 19:514.

STRONG, D.R., JR. 1972. Life history variation among populations of an amphipod *(Hyalella azteca). Ecology* 53:1103.

BARLOCHER, F. & B. KENDRICK. 1973. Fungi and food preferences of *Gammarus pseudolimnaeus. Arch. Hydrobiol.* 72:501.

LOCKWOOD, A.P.M. & C.B.E. INMAN. 1973. Changes in the apparent permeability to water at moult in the amphipod *Gammarus duebeni* and the isopod *Idotea linearis. Comp. Biochem. Physiol.* 44A:943.

VERNBERG, W.B., P. DECOURSEY & W.J. PADGETT. 1973. Synergistic effects of environmental variables on larvae of *Uca pugilator. Mar. Biol.* 22:307.

HYNES, H.B.N., N.K. KAUSHIK, M.A. LOCK, D.L. LUSH, Z.S.J. STOCKER, R.R. WALLACE & D.P. WILLIAMS. 1974. Benthos and allochthonous organic matter in streams. *J. Fish. Res. Board Can.* 31:545.

NILSSON, L.M. 1974. Energy budget of a laboratory population of *Gammarus pulex* (Amphipoda). *Oikos* 25:35.

808 BIOASSAY PROCEDURES FOR AQUATIC INSECTS (TENTATIVE)

Aquatic insects are important components of lake and stream biota. In trout streams, they comprise 50 to 90% of the macroinvertebrate species. Such groups as mayflies, stoneflies, caddisflies, and midges are major food items for many species of fish.[1] Many aquatic insects are more sensitive to organic pesticides than are fish.

The wide variety of aquatic insects, their abundance in unpolluted streams, their sensitivity to low concentrations of pollutants, and the ease of maintenance of many species under laboratory conditions make them useful test animals. Procedures using aquatic insects have been developed for determining acceptable environmental conditions or concentrations of toxicants.[2] Most studies have been short-term tests, but the procedures can be used for long-term tests.

Toxicants may interfere with survival, growth, reproduction, emergence, and metabolism of aquatic insects. Because effects of long-term exposure to sublethal concentrations of toxicants may be more important than effects of infrequent short-term exposure to higher concentrations, flow-through, long-term tests are recommended.

808 A. Selecting and Preparing Test Organisms

1. Suggested Test Organisms

Use insects that are important food for fishes, readily available and abundant, relatively easy to keep and culture in the laboratory, and most sensitive to the materials under investigation. The following organisms are suggested:

a. Stoneflies: *Pteronarcys dorsata*, *Pteronarcys californica*, *Acroneuria lycorias*, *Acroneuria pacifica*

b. Mayflies: *Hexagenia limbata*, *Ephemerella subvaria*

c. Caddisflies: *Brachycentrus americanus*, *Brachycentrus occidentalis*

d. Other species that have been used are:

1) Stoneflies— *Isogenus frontalis*, *Perlesta placida*, *Paragnetina media*, *Phasganophora capitata*, *Acroneuria californica*

2) Mayflies— *Ephemerella grandis*, *Ephemerella doddsi*, *Ephemerella needhami*, *Ephemerella tuberculata*, *Stenonema ithaca*

3) Caddisflies— *Hydropsyche betteni*, *Macronemum zebratum*, *Arctopsyche grandis*, *Hydropsyche bifida*

4) Diptera— *Chironomus plumosus*, *Chironomus attenuatus*, *Chironomus tentans*, *Chironomus californicus*, *Glyptochironomus labiferus*, *Goeldichironomus holoprasinus*, *Tanypus grodhausi*, *Tanytarsus (Paratanytarsus) dissimilis*

For each test, use insects of the same year, class, and, as nearly as possible, the same size. Use early instar (first-year) larvae or nymphs when possible, especially for growth studies. Many of the listed species complete a generation in one summer. Use late instars for adult emergence tests.

2. Collecting Test Animals

Collect all test specimens from clean, natural waters rich in aquatic insects (see Section 1005, Benthic Macroinvertebrates). Collect larger stream species from riffle areas of clean, well-aerated gravel rubble streams with hand screens or bottom samplers such as the Surber sampler. Stir bottom and let current carry dislodged insects downstream into net.

Immediately after collection, gently place net contents in a 15- to 20-L insulated container partly filled with stream water. Transport to laboratory. Remove and discard larger rocks after it has been determined that they are free of insects. If transportation time exceeds 30 min, provide for aeration and temperature control. In laboratory, swirl water in containers and dip it out. Pour through a screen-bottom container (of a mesh that will retain insects required), held partly submerged in a tank of water. Wash screenings into a holding tank. If it is desired to separate insects, wash into a large white enamel pan containing 3 to 5 cm of water. Remove desired species with a large-bore pipet or small spoon-shaped screen and place in holding tanks. For riffle insects use oval annular flow-through tanks[2] provided with rocks for cover and paddle wheels to provide a current in dilution water.[2] Alternatively collect insects by gently picking up rocks, rubble, or gravel, and carefully washing or picking, then placing desired insects in insulated containers for transport to laboratory.

To obtain benthic insects, sample bottom materials with Eckman, Petersen, or Ponar dredges. Empty dredge into a large pail, add water, and swirl by hand. Partly submerge an appropriate mesh washing screen, pour a portion of swirling sample into it, and wash by moving up and down in the water. Place washed insects in an insulated container and continue until enough insects have been collected.

Chironomids probably will be the dominant insect species in silt bottom material. However, other important insects such as dragonflies, damsel-flies, several species of Diptera, beetle larvae, and mayflies may be found in and on silt bottoms. The mayfly, *Hexagenia limbata,* is a large species often occurring in great abundance in soft, unpolluted muds rich in organic matter that occur in deep pools, ponds, lakes, and reservoirs. Obtain these by collecting top 8 cm of mud and washing as described previously.

3. Holding, Acclimating, and Culturing

a. General considerations: As soon after collection as possible, examine insects for injury. Place all uninjured specimens in holding chambers, supply them with food, and hold for at least 1 wk for observation and acclimation to desired temperature. Acclimate stream species in flowing water. Keep in oval troughs that have a current of water or in stainless steel wire cages in running water.[2] In these troughs include flat stones covered with attached algae as cover and food for herbivorous species. Supply insects with materials to build larval and pupal cases. For caddisflies, use sand grains, small pieces of wood, and plant materials retained by a 16-mesh screen. Permit insects that construct tubes or cases to do so. Hold benthic species in aquariums provided with a 3- to 5-cm layer of unsterilized mud from the site where they were collected. *Hexagenia* require a substrate in which to burrow.[3] For chironomids use the highly organic ooze that overlays the bottom where they were collected.

Provide water, DO, and other conditions as described in Section 801D and E. Maintain final holding temperature within 3 C of temperature at which organisms were collected. For long holding periods, maintain natural seasonal temperatures. When aquatic insects are collected in winter at water temperatures of 1 C or lower, acclimate them to higher temperatures if they are to be used in short-term tests (Section 801D.3).

Different species require different light intensities. Stoneflies require stones under which they can hide from direct light. Fix light cycle at a certain day length, or vary it seasonally to correspond with natural annual photoperiod. For *Chironomus plumosus,* use a 16-hr photoperiod. Lamps and fixtures are described in Section 801E.3*f*.

b. Food and feeding: Acroneuria, Brachycentrus, Isogenus, and *Paragnetina* are predators requiring live food. Feed to excess with small midges, blackfly larvae, mosquitoes, or small caddisfly larvae from an unpolluted environment.[4] Feed *Pteronarcys* and *Ephemerella* to excess with coarse, chopped maple, birch, or aspen leaves that have fallen naturally and have been dried and then soaked in test water for at least 2 wk before feeding. Feed *Hexagenia, Hydropsyche,* and *Arctopsyche* finely ground leaves and fish-food pellets. If the substrate is rich in organic matter, additional food may not be required for *Hexagenia*. Avoid overfeeding with fish food because it causes DO depletion. The larvae of some Hydropsychidae are highly carnivorous and cannibalistic; keep them well-fed with plankton, microcrustacea, blackfly larvae, and other organisms, collected from fish hatcheries, ponds, lakes, and streams with a net of No. 20 bolting silk.

Feed chironomids twice per week. Keep in jars supplied with algal culture medium [Section 801D.4*c*1)a)] inoculated with algae and diatoms. Alternatively use a mixture of 5 g fish food plus 1 g powdered dried grass* shaken in 1 L of water. Add about 100 mL of this suspension to each culture per feeding. If there is no flow-through, remove 100 mL of test solution before feeding. Use 10-L culture jars containing 8 L or less of medium with a screen cover to retain adults.[5,6] Keep in a constant-temperature room at 21 to 24 C. For long-term studies follow natural temperature cycle of water from which chironomids were taken. Because the jars have a mud substrate, do not clean, but do not overfeed. Collect emerging adults for breeding in wire screen cylinders placed over the culture jars.[7,8]

808 B. Bioassay Procedures

1. Procedures

Conduct tests as described in Section 801C. Use a minimum of 20 specimens for each toxicant concentration with an additional 40 animals for growth studies. Two species may be tested in the same tank if precautions are taken to avoid predation.

Do not use static testing with stream insects. Use static tests with certain lake or reservoir species if required DO levels are maintained. For long-term tests, see Section 801C.

a. Test tanks: Use glass and stainless steel aquariums of either 8-L or 20-L size for quiet-water species. For stream species, use oval, stainless steel annular troughs[2,5,8] (90 cm long, 15 cm wide, and 15 cm deep) in which natural stream flow is simulated. Set tanks side by side so paddle wheels on one long shaft can be used to circulate water in them all.[2]

b. Flow rate: Use flows to each tank of no less than 6 to 10 tank volumes/24 hr. In aquariums without water-circulating devices use much higher flows for stream species to simulate stream flow. In oval test tanks use velocities near 0.5 cm/sec. For quiet-water forms, such as *Hexagenia* and *Chironomus,* do not disturb mud substrate with water flow.

c. Aeration: Aeration is unnecessary; however, use if desired with nonvolatile toxicants to increase or control water movement, especially for tank tests with lake and reservoir species.

d. Cleaning: See Section 801D.4*d*. Siphon out detritus on tank bottom weekly during long-term testing. If a mud substrate is used, no cleaning is necessary. Avoid overfeeding.

e. Substrate: For all stream riffle species use fine-mesh stainless steel screens formed into cylinders or cubes, which provide 10 to 15 cm²/insect. Place cages in oval, annular troughs or in glass cylinders.[2] For 30- to 90-day adult emergence tests, obtain clean rocks, 5 to 10 cm in diameter (one for every three insects) from collection site for a substrate. Provide fine screen or sticks that protrude above water surface for adult emergence tests.

f. Light and photoperiod: See Section 801E.3*f*. Use natural photoperiod at time

*Cerophyll, Cerophyll Laboratories, Inc., Kansas City, Mo. 64111, or equivalent.

of testing for locality in which test is conducted. Increase day length during adult emergence tests by 0.5 hr every 2 wk.

g. Temperature: See Section 801E.3*a*. Use 10 C as a winter temperature. For trout stream insects, use summer temperatures near 15 C. Increase temperature during adult emergence tests by 1 C each week up to a maximum of 5 C above initial temperature. When using warm-water stream or lake insects, follow natural temperature cycle.

h. Time of year: Under natural conditions, most species emerge as adults in spring. Therefore start adult emergence tests no later than March 1st. *Hexagenia limbata* is an exception, emerging throughout the summer in most localities.

2. Toxicant Preparation

See Sections 801E.1 and 2*b*.

3. Test Procedures for *Hexagenia*

Use *Hexagenia* for short-term survival (96 to 168 hr), survival for 5 to 60 days, adult emergence, or full-life-cycle tests (90 to 120 days). Use a minimum of 20 organisms per aquarium of not less than 8 L capacity. Use a water depth of 8 to 20 cm. Provide a fine organic ooze substrate 4 to 5 cm deep and as similar as possible to that where naiads occur naturally. When using newly hatched *Hexagenia* to start a test, use 50/tank. When *Hexagenia* eggs are used as a source of larvae, pipet them into petri dishes (about 200/dish) with 200 mL test water at about 20 C and let hatch.

When substrate is mud, determine survival by counting number of dead animals that have left their burrows and/or by counting number of new burrows formed after disturbing mud surface sufficiently to destroy entrances to old burrows. If counts do not agree, use the latter. For acute toxicity tests, alternatively use an artificial substrate of epoxy resin to facilitate observation and monitoring of test animals.[3] For growth or emergence tests, set up an additional set of containers so that naiads can be removed periodically for measurement. Remove 10 naiads from their burrows after 20 to 60 days to determine growth. Do not remove more than 50% of surviving animals before conclusion of these tests. Keep a record of total number removed. Use these animals to provide additional data on growth and emergence. Record body length, head capsule width, and live weight.

In acute toxicity tests, determine survival after 1.5, 3, 6, and 12 hr and twice daily thereafter. As a sign of death, use failure of specimens to respond by movement to gentle probing or flashlight illumination. In longer-term studies, check tanks daily to remove and record dead animals and cast naiad skins, which indicate successful molting.

For growth studies, determine initial range and mean of total length, head capsule width, and weight from specimens in holding tank. Kill all animals in warm water (40 to 50 C) before measurement. Take measurements twice during testing, using animals that are to be discarded. Obtain final measurements for all survivors. Make two counts: number of adults and cast skins; if different, use cast skins because some adults may have escaped.

Determine and record percentage of adults that emerge, their sex, incidence of incomplete emergence (i.e., half-out of nymphal skin, wings unsuccessfully unfolded, etc.), adult length, weight, and head capsule width, and number of mature eggs.

4. Test Procedures for *Chironomus*

Follow procedures described in Section 801E. For each concentration, use duplicate 20-L aquariums with mud substrate and screen covers. Maintain flow to each test container at about 2 L/hr. Use a mud substrate similar to that for *Hexagenia*. Use lighting and photoperiod as described in Section 801E.3*f*. Do not feed animals during short-term tests. Feed during 30-day and emergence tests as in Section 808A.3*b*. If prepared food is used, add about 100 mL food suspension to each container twice per week.

For long-term tests, place 50 first-instar larvae (about 1.5 mm long and less than 24 hr old) in each test aquarium. Transfer larvae with an eyedropper. Determine number of emerging adult males and females. Count both adults and pupal cases. If counts differ, use pupal case count. At 25 ±1 C emergence takes about 1 month. To determine success of fertilization of eggs, take 50 eggs and determine percent hatchability. If it is impossible to separate and count eggs, hatch fertilized egg masses in beakers with same test water from which adults emerged. Count 60 larvae into a petri dish and examine for injured larvae. Transfer often will injure early instar larvae; if a correction for this is not made in the count, errors in percent survival result. After examination, count 50 larvae back to test chamber and rear them to adult stage. End points for taking and analyzing data are emergence of adults, egg production, and hatching of young. Repeat complete test at least once.

808 C. Data Evaluation

Analyze, evaluate, and report data from various tests as described in Section 801F.

808 D. References

1. Hynes, H.B. 1970. Ecology of Running Waters. Univ. of Toronto Press, Buffalo, N.Y. and Toronto, Ont., Canada.
2. Surber, E.W. & T.O. Thatcher. 1963. Laboratory studies of the effects of alkyl benzene sulfonate on aquatic invertebrates. *Trans. Amer. Fish. Soc.* 92:152.
3. Fremling, C.R. & G.L. Schoening. 1973. Artificial substrates for *Hexagenia* may-fly nymphs. *In* Proc. 1st Int. Conf. Ephemeroptera: 209.
4. Nebeker, A.V. & F.A. Puglisi. 1974. Effect of polychlorinated biphenyls on survival and reproduction of *Daphnia, Gammarus* and the midge *Tantarus*. *Trans. Amer. Fish Soc.* 103:722.
5. Nebeker, A.V. 1972. Effect of high winter water temperatures on adult emergence of aquatic insects. *Water Res.* 5:777.
6. Bay, E.C. 1967. An inexpensive filter-aquarium for rearing and experimenting with aquatic invertebrates. *Turtox News* 45:146.
7. Brever, K.D. 1965. A rearing technique for the colonization of chironomid midges. *Ann. Entomol. Soc. Amer.* 58:135.
8. Nebeker, A.V. & A.E. Lemke. 1968. Preliminary studies on the tolerance of aquatic insects to heated waters. *J. Kans. Entomol. Soc.* 41:413.

808 E. Bibliography

Pennak, R. 1953. Fresh Water Invertebrates of the United States. Ronald Press, New York, N.Y.

Usinger, R.L., ed. 1956. Aquatic Insects of California—With Keys to North American Genera and California Species. Univ. of

Calif. Press, Berkeley.
ROBACK, S.S. 1957. The Immature Tendipedids of the Philadelphia Area. Monographs Acad. Natural Science Philadelphia, No. 9.
EDMUNDSON, W.T. 1959. Freshwater Biology, 2nd ed. John Wiley & Sons, Wiley Interscience, New York, N.Y.
MACAN, T.T. 1963. Freshwater Ecology. John Wiley & Sons, Wiley Interscience, New York, N.Y.
FREMLING, C.R. 1967. Methods for mass-rearing *Hexagenia* (Ephemeroptera: Ephemeridae). *Trans. Amer. Fish. Soc.* 96:407.
SANDERS, H.O. & O.B. COPE. 1968. The relative toxicities of several pesticides to naiads of three species of stone-flies. *Limnol. Oceanogr.* 13:112.
HYNES, H.B. 1970. Biology of Polluted Waters. Univ. of Toronto Press, Buffalo, N.Y. and Toronto, Ont., Canada.
GAUFIN, A.R. & S. HERN. 1971. Laboratory studies on tolerance of aquatic insects to heated waters. *J. Kans. Entomol. Soc.* 44:240.
Standard methods for detection of insecticide resistance of *Diabrotica* and *Hypera* beetles. 1972. *Bull. Entomol. Soc. Amer.* 18:179.
GAUFIN, A.R. 1972. Water Quality Requirements of Aquatic Insects. Final Rep. Contract 14-12-438, Water Quality Office, Environmental Protection Agency.
NEBEKER, A.V. 1972. Effect of low oxygen concentration on survival and emergence of aquatic insects. *Trans. Amer. Fish. Soc.* 101:675.
GAUFIN, A.R., R. CLUBB & R. NEWELL. 1974. Studies on the tolerance of aquatic insects to low oxygen concentrations. *Great Basin Natur.* 31:45.

809 BIOASSAY PROCEDURES FOR MOLLUSKS (TENTATIVE)

Oysters, clams, scallops, and mussels are widely distributed and are of great value as human food. These mollusks and others are suitable for toxicity evaluation in short- and long-term tests. Oyster and clam embryos have been used to measure effects of chemical and environmental variables in estuarine and marine environments.

Methods for using freshwater mollusks in standard tests are being developed; methods for laboratory culture and maintenance of some hydrobiid snails are available.[1]

Toxicants affect bivalves by interfering with fertilization, normal embryonic development, growth (shell deposition), byssal thread secretion, reproduction, and normal tissue histology. These toxic effects are the bases for short- and long-term toxicity tests. Adult bivalves generally are not suitable for determination of acute lethal concentrations of toxicants because of their ability to close their shells and protect themselves from toxicant.

809 A. Selecting and Preparing Test Organisms

1. Selecting Test Organisms

For oil and heavy metals, the bay scallop is the most sensitive species. Some adult mollusks are resistant to many materials and accumulate them to high concentrations. In comparison to fish and to other invertebrates, oyster larvae may be more or less sensitive, especially to pesticides. Species recommended for testing include:

Crassostrea gigas	Pacific oyster
Crassostrea virginica	Eastern oyster
Ostrea lurida	Olympia oyster
Argopecten irradians irradians	Bay scallop
Mytilus edulis	Mussel
Mercenaria mercenaria	Quahog
Spisula solidissima	Surf clam
Mulinia lateralis	Coot clam
Macoma balthica	
Rangia cuneata	

2. Water Supply and Water System

Tests require a marine laboratory with a supply of clean, unfiltered estuarine or open-ocean water of the desired temperature and salinity range. (Section 801D.4*b* and 801E.1 and 2*a*). If necessary, use artificial seawater.

For filter-feeding mollusks use a natural seawater containing planktonic organisms. In long-term growth tests, supply adult oysters with a minimum of 5 L unfiltered seawater/hr/oyster. Supply clams and scallops with comparable amounts of natural seawater rich in plankton. In flow-through tests, distribute water from a constant-head tank (Section 801E.1).

3. Collecting, Conditioning, and Culturing Test Organisms

Collect test organisms from the field, purchase from commercial dealers, or rear them.

Use natural seawater rich in plankton for growing adults or spawners. Clean intake pipes and entire water system frequently to insure that growth in pipes does not remove plankton. If there is not sufficient food, or if a continuous flow-through of unfiltered seawater results in problems of competitors, parasites, and disease, produce planktonic food in rearing chambers [Section 801D.4*c*2)].

a. Oysters: For tests with adults, or for producing embryos, use adults of the Pacific or Eastern oyster, 7.5 to 15 cm in height. Cull oysters to singles and condition at 2- to 4-wk intervals, depending on need. In the laboratory, hold each lot of oysters as a separate population in conditioning trays. Oysters collected from December to April need longer conditioning than those collected between April and July. After August, oysters begin resorbing their unspawned gametes and are unsatisfactory spawners. To have spawning oysters after August, collect during spring months and keep in year-round cool water, or in refrigerated, flowing seawater in laboratory at less than 12 C.[2]

Clean oysters of fouling organisms and other extraneous materials. Use 15 oysters per conditioning tray (58 × 46 × 8 cm). Provide each tray with a minimum of 7 L/hr of flowing seawater at 20 ± 1 C. Oysters require 2 to 6 wk of thermal conditioning before they are ready to spawn and can be held 2 to 4 wk for use as spawners before discarding. Conditioned mature mollusks will spawn in either natural or artificial seawater. When they are induced to spawn in natural seawater, transfer immediately to receiving water or artificial medium for collection of gametes.

Check oysters daily and remove moribund individuals. If any die, empty tray and clean with detergent and warm water. Scrub remaining oysters with clean seawater, rinse several times, and replace in tray.

Clean accumulated feces and silt from trays at least once and preferably twice a week. Should an unplanned spawning occur, discard all oysters in that tray. When fertilized eggs are required, rinse conditioned oysters with clean seawater and place individually in spawning dishes. Raise water temperature 5 to 10 C to induce spawning and add a sperm suspension as a further inducement. Prepare sperm suspension by opening a male oyster, rupturing gonad, and gently washing sperm free from gonadal tissue into a 1-L beaker with a fine jet of 20 C seawater. Take care not to rupture other body organs during the process and use enough water so that DO remains near saturation. After a spawning attempt, return oysters that have not spawned to conditioning trays. Discard females once they have spawned. Place any surplus males that spawn in a separate tray and use for making sperm suspensions.

b. Clams: Collect clams from the field or secure from commercial fishermen. Keep adults in live boxes or in containers with adequate flow-through water during transit; then place in suitable holding trays. Use specimens in long-term tests or for production of larvae for short-term tests. If adults are not taken during normal spawning season, thermally condition for spawning by placing in trays with flow-through seawater at 18 to 22 C, depending on species, for 3 to 4 wk. Stimulate adult clams with ripe gonads to spawn naturally by increasing water temperature to 24 to 28 C, depending on species, and add a sperm suspension as described for oysters. Produce larvae only from naturally liberated eggs. A stock of 30 to 40 thermally conditioned adults is adequate to assure spawning at any time. Place eggs in test solutions as soon as possible, and not more than 2 hr after fertilization.

To use larvae in longer-term tests in the absence of natural food, use a mixture of *Isochrysis galbana* and *Monochrysis lutheri* or some other cultured algae[3] grown in standard enriched seawater described in Section 801D.4*c* 1)b).

c. Mussels: Collect *Mytilus edulis* from pilings, rocks, floating boat docks, jetties, and other suitable habitats. Those from floating docks or platforms are preferred because they are easiest to collect and clean. Because larger specimens may contain maturing gametes that may be freed during tests, use only small specimens in intermediate-term tests with adults. Use specimens of uniform size, 15 to 20 mm wide.[4] Separate this size class by passing mussels through a 20-mm-diam hole drilled in a piece of sheet metal; mussels of the correct size class will pass through this hole but not through one 15 mm in diameter. Discard all other specimens. Clean surface and trim byssal threads, but do not remove byssal thread stalk.

Acclimate cleaned specimens in aquariums for 1 wk at not more than 4 C above field water temperature but below 26 C and at a density of about 10 specimens/4 L seawater. Provide aeration and filtration in static aquariums. Observe holding aquariums daily for deaths. Discard all specimens if more than 10% die during acclimation. Occasionally some 15- to 20-mm animals have mature gametes and will spawn. Change water to prevent fouling and remove the mature specimen if it can be identified. If spawning is widespread discard all specimens.

d. Scallops: Collect bay scallops from the field or purchase from fishermen taking special precautions to insure proper handling. Use adults as described for the oyster. Culture methods and conditioning technics are available.[2,5–8] Condition at 20 to 22 C for 3 to 8 wk in same type of tray used for oysters. Check gonads periodically to determine development. Place a finger in a gaping scallop to hold shells open so the gonads can be seen. The bay scallop is a functional hermaphrodite. The testis comprises the anterior border of the gonad and the ovary the posterior portion. When ripe, the ovarian portion is reddish orange and the testis cream-colored. After gonads have ripened, induce spawning by raising water temperature to 27 to 30 C. Procedures for spawning, handling ova, fertilization, and rearing larvae have been described.[6] Feed larvae with marine phytoplankton cultured as described in Section 801D.4*c* 1)b).

4. Parasites and Diseases

Cyclopoid copepods may be present in the mantle cavity or digestive tract of mollusks.[9] Species inhabiting the mantle cavity do not cause known pathological damage. However, cyclopoid copepods, such as *Mytilicola intestinalis*, inhabiting the digestive tract, may damage cellular linings and cause a higher incidence of mortality. Examine digestive tracts of a randomly selected sample of 20 mollusks for cyclopoid copepods before using a group taken in one collection, especially in areas where *M. intestinalis* is known to occur. Do not use mollusks if incidence of infestation exceeds 10%.

809 B. Conducting the Bioassays

1. Short-Term Tests

a. Oyster embryo tests: Produce oyster eggs in the laboratory and fertilize to initiate development. They usually become shelled, straight-hinge larvae in 48 hr. Use normality of development in receiving water samples to determine its quality. When embryos are cultured in artificial seawater use normality of development as a criterion of relative toxicity of added toxicants.

Use artificial seawater[10] (Section 801C, Table 801:II) for spawning adults and culturing embryos.[11] Artificial seawater has been proposed as a standard testing medium.[12,13] Methodology for oyster embryo culture was developed in studies of the eastern oyster, *Crassostrea virginica*,[2] and adapted for the Pacific oyster, *C. gigas*,[14] and the blue mussel, *Mytilus edulis*.[15] Use this method for other bivalves that can spawn under controlled laboratory conditions.

Use the following general steps for tests with fertilized oyster eggs on a year-round basis:

1) Two hours before spawning is desired, place 15 thermally conditioned ripe female oysters in an equal number of borosilicate glass baking dishes about 22 × 12 × 8 cm filled with filtered, UV-light-treated seawater or artificial seawater at 20 C.

2) Raise water temperature by placing dishes in a water bath at 28 to 30 C.

3) About 30 min before spawning is desired, add to each dish 20 mL sperm suspension, prepared as in Section 809A.3*a*. The combination of increased temperature and sperm usually induces accelerated pumping by the oysters and one or more ripe females to spawn. The sperm fertilizes the eggs as they are discharged.

4) If spawning is not achieved in about 1 hr or oysters stop vigorous pumping activity, replace water in spawning dishes with fresh 20 C water and repeat the process. Pipet additional sperm suspension into water being drawn in by the oyster to initiate spawning.

5) About 30 to 45 min after spawning, pour eggs from a single female (usually 6 to 40 × 10^6) into a 2-L beaker. Determine egg density from two counts, in a Sedgwick-Rafter cell, of number of eggs in 1-mL samples of a 1:99 dilution of homogeneous egg suspension.

6) Bring temperature of control and test water to 20 C ± 0.5 C, (25 C ±0.5 C in southern areas), before inoculation with oyster embryos. Add enough embryo sus-

pension to each test container to give a population density of 20,000 to 30,000/L. Use at least 10% of the cultures as controls and at least two replicates of each experimental condition.

7) Incubate cultures in a 20 ± 1 C water bath for 48 hr (25 ± 1 C in southern areas) and then pour through a 37-μm sieve to retain and concentrate larvae.

8) Wash larvae into a 100-mL graduated cylinder. Take a 2-mL sample containing 150 to 250 larvae with an automatic pipet and preserve in vials with 3% neutral formalin for microscopic examination.

9) Count preserved larvae in a Sedgwick-Rafter cell and record number of normal and abnormal larvae. Normal larvae are fully shelled, even though they may be misshapen or undersized. This criterion avoids the need to make an excessive number of value judgments to classify a larva as normal or abnormal. The percentage of normal larvae is the basic measure of biological response. Further details, test variability, and computer methods for data processing and analysis are available.[14]

b. Scallop, clam, and mussel embryo tests: Make tests with bay scallop, clam, and mussel embryos using procedures described for oyster embryos.

c. Oyster shell deposition test:

1) General considerations—This 96-hr test demonstrates the comparative toxicity of pollutants to young oysters. Conduct tests in flowing, unfiltered seawater at a temperature between 15 and 30 C. Actively feeding oysters extend their mantle edges to the periphery of the shell or valves. However, the body can contract to occupy a much smaller area. If peripheral valve edges are mechanically ground away, oysters respond by depositing new shell to replace this loss.[16]

New shell growth is primarily linear during the first week and deposition rate is an index of the animals' reaction to ambient water quality. With acceptable water

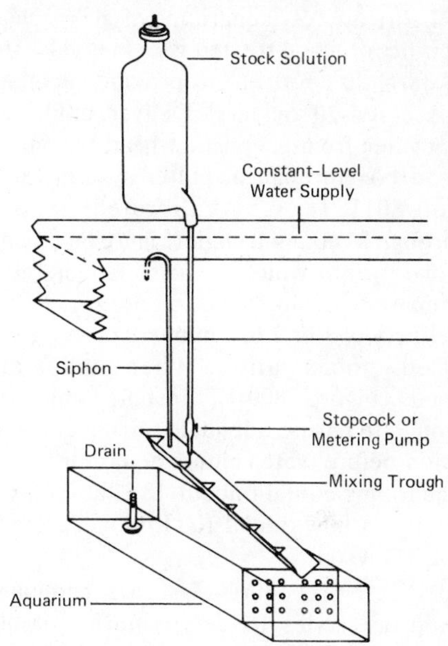

Figure 809:1. Diagram of constant-flow apparatus.[17]

conditions, 25 mm and larger oysters deposit as much as 1 mm/day of peripheral new shell. Small oysters are more suitable than large ones because typically they form new shell deposits within a broader temperature range than mature oysters. Interpretation of test data is independent of minor fluctuations in temperature and salinity during the 96-hr exposure because the simultaneous shell deposition in control oysters is considered to be the norm or 100%.

2) Procurement and preparation of oysters—Cull to singles oysters about 25 to 50 mm in height (i.e., the long axis) with reasonably flat, rounded shape, brush clean, and maintain in trays in natural environment. To test, reclean oysters and remove 3 to 5 mm of shell periphery by hand-holding oyster against an electric disc grinder. Insure uniform removal from shell rim to produce a smoothly-rounded blunt profile. Discard oysters damaged by removing too wide a rim of shell.

Fabricate test aquariums of glass, clear acrylic, or wood treated with fiberglass (64 × 38 × 10 cm deep) to provide adequate space for 20 oysters. Deliver unfiltered seawater from a constant-head trough or head box through a diluter system (Section 801E.1) or by calibrated siphons through a mixing trough (Figure 809:1 and Lowe[17]) into which toxicant is metered.* Prepare toxicant stock solutions so that a delivery of 1 or 2 mL/min will produce desired concentration. When pumps are used (Figure 809:1), install baffles in trough to ensure adequate mixing and aeration before water enters aquariums. The aquariums contain about 18 L at 75% capacity. A flow rate of 100 L/hr will provide 5 L/hr/oyster.

3) Test procedure—Use a preliminary exposure series to determine a suitable range of toxicant concentrations (see Section 801C.2*a*). Expose five oysters for 48 hr to concentrations of 100, 10, 1.0, 0.1, and 0.01 mg/L to bracket range of toxicant concentrations required to determine 96-hr EC50. Lower concentrations may need to be checked.

Prepare and distribute oysters randomly so that each control and test aquarium contains 20 individuals. Place oysters with left, cupped valve downward and anterior hinged ends all oriented in one direction. Establish one control aquarium, another receiving only toxicant solvent, and one aquarium for each toxicant concentration. After 96 hr, remove all oysters and measure shell increments. Because shell deposition is not uniform on periphery, record length of longest "finger" of new shell, measured to nearest 0.5 mm.

4) Calculation—Calculate ratio of mean growth of a group of test oysters to mean growth of control oysters to provide a percentage index of response. For 96-hr exposure, calculate concentrations allowing 50% relative shell growth and 5% relative shell growth.

*Sage syringe pump, Sage Instruments, Inc., Cambridge, Mass. 02139, MilRoyal® controlled pump, Milton Roy Co., St. Petersburg, Fla. 33733, or equivalent.

2. Long-term Tests

a. Oyster growth test:

1) General considerations—Oysters will grow from setting size to sexual maturity in about 3 to 4 months under optimum conditions. Use 50 to 100 individuals to determine chronic effects. Toxic effects may be manifested by accumulation of chemical residues, changes in resistance to disease, or interference with reproduction.

Evaluate oyster growth by recording weight increases. If oysters are weighed under water, day-to-day changes in shell deposition can be detected.[18,19]

For meaningful weight data, carefully clean oysters and remove all fouling organisms and debris. Do not destroy shell integrity before each weighing. Insure that oyster's valves are closed when it is exposed to air and that shell is free from air bubbles when weighed. Weighings can be replicated to 0.05 g and, under satisfactory conditions, a 15-g oyster (underwater weight) will gain 1 g or more/wk.

Sublethal effects may become apparent only slowly; extended exposure may be required.[20]

2) Test procedure—Place small (2- to 3-cm) single oysters, randomly selected from a population of approximately known age, in test aquariums large enough to accommodate their anticipated growth. Supply water and toxicant as described for 96-hr shell deposition test. Use a minimum of 5 L seawater/hr/oyster. Select toxicant concentration on basis of 96-hr tolerance data (Section 801C.2*a*).

Position groups of 50 oysters on shallow compartmented racks to facilitate handling and identification of individual animals.[20] At weekly intervals remove oysters, clean, and keep immersed in water until weighed. Clean containers at this time.

Weigh individual immersed oysters with a top-loading balance with suspension attachment. Estimate weights to nearest 0.01 g. Record changes in shell length at each weighing. Expose a suitable number of extra oysters in test and control groups to serve as periodic subsamples to determine residue accumulations and histological changes.

b. Scallop, clam, and mussel growth tests: Bay scallops, clams, and mussels can be used in growth tests similar to those described for oysters if provisions are made to meet their special requirements.

809 C. Reporting and Analyzing Results

Except for special studies, analyze data, calculate results, and report results as described in Section 801F.

809 D. References

1. VAN DER SCHALIE, H. & G.M. DAVIS. 1968. Culturing *Oncomelania* snails (Prosobranchia: *Hydrobiidae*) for studies of oriental Schistosomiasis. *Malacologia* 6:321.
2. LOOSANOFF, V.L. & H.C. DAVIS. 1963. Rearing of bivalve mollusks. *Advan. Mar. Biol.* 1:1.
3. CHANLEY, P. & M. CASTAGNA. 1966. Larval development of the pelecypod *Lyonsia hyalina*. *Nautilus* 79(4):123.
4. REISH, D.J. & J.C. AYRES, JR. 1968. Studies on the *Mytilus edulis* community in Alamitos Bay, California. III. The effects of reduced dissolved oxygen and chlorinity concentrations on survival and byssal thread production. *Veliger* 10:384.
5. BELDING, D.L. 1910. A Report upon the Scallop Fishery of Massachusetts, Including the Habits, Life History of *Pecten irradians*, Its Rate of Growth and Other Factors of Economic Value. Spec. Rep., Comm. Fish and Game, Boston, Mass.
6. CASTAGNA, M. & W. DUGGAN. 1971. Rearing the bay scallop, *Argopecten irradians*. *Proc. Nat. Shellfish. Ass.* 61:80.
7. TURNER, H. & J.E. HANKS. 1960. Experimental stimulation of gametogenesis in *Hydroides dianthus* and *Pecten irradians* during the winter. *Biol. Bull.* 119:145.
8. SASTRY, A.N. 1966. Temperature effects in reproduction of the bay scallop, *Argopecten irradians* Lamarck. *Biol. Bull.* 130:118.
9. CHENG, T.C. 1967. Marine mollusks as hosts for symbiosis with a critical review of known parasites of commercially important species. *Advan. Mar. Biol.* 5.
10. ZAROOGIAN, G.E., G. PESCH & G. MORRISON. 1969. Formulation of an artificial sea water medium suitable for oyster larvae development. *Amer. Zool.* 9:1144.
11. CALABRESE, A., R.S. COLLIER, D.A. NELSON & J.R. MACINNES. 1973. The toxicity of heavy metals to embryos of the American oyster *Crassostrea virginica*. *Mar. Biol.* 18:162.
12. TARZWELL, C.M. 1969. Standard methods

for determination of relative toxicity of oil dispersants and mixtures of dispersants and various oils to aquatic life. *In* Proc. Joint Conf. on Prevention and Control of Oil Spills. p. 179. American Petroleum Inst.

13. LaRoche, G., R. Eisler & C.M. Tarzwell. 1970. Bioassay procedures for oil and oil dispersant toxicity evaluation. *J. Water Pollut. Control Fed.* 42:1982.
14. Woelke, C.E. 1972. Development of a Receiving Water Quality Bioassay Criterion Based on the 48-hr Pacific Oyster (*Crassostrea gigas*) Embryo. Tech. Rep. 9, Wash. Dep. Fisheries, Olympia.
15. Dimick, R.E. & W.P. Breese. 1965. Bay mussel embryo bioassay. *In* Proc. 12th Pacific Northwest Industrial Waste Conf., Univ. of Wash., Seattle. p. 165.
16. Butler, P.A. 1965. Reaction of some estuarine mollusks to environmental factors. *In* C.M. Tarzwell, ed. Biological Problems in Water Pollution. U.S. PHS Publ. 999-WP-25, p. 92.
17. Lowe, J.I. 1964. Chronic exposure of spot, *Leiostomus xanthurus,* to sub-lethal concentrations of toxaphene in seawater. *Trans. Amer. Fish. Soc.* 93:396.
18. Havinga, B. 1928. The daily rate of growth of oysters during summer. *J. Cons. Perma. Int. Explor. Mer.* 3:231.
19. Andrews, J.D. 1961. Measurement of shell growth in oysters by weighing in water. *Proc. Nat. Shellfish. Ass.* 52:1.
20. Lowe, J.I., P.D. Wilson, A.J. Rick & A.J. Wilson, Jr. 1971. Chronic exposure of oysters to DDT, toxaphene and parathion. *Proc. Nat. Shellfish. Ass.* 61:231.

809 E. Bibliography

Gutsell, J.S. 1930. Natural history of the bay scallop. *Bull. U.S. Bur. Fish.* 46:569.

Loosanoff, V.L. & H.C. Davis. 1951. Delayed spawning of lamellibranchs by low temperature. *J. Mar. Res.* 10:197.

Davis, H.C. 1953. On food and feeding of larvae of the American oyster, *C virginica. Biol. Bull.* 104:334.

Okubo, K. & T. Okubo. 1962. Study of the bioassay method for the evaluation of water pollution. II. Use of fertilized eggs of sea urchins and bivalves. *Bull. Tokai Reg. Fish. Res. Lab.* No. 32.

Tubiash, H.S. & P.E. Chanley. 1963. Bacterial necrosis of bivalve larvae. *Bacteriol. Proc.* 1963.

Davis, H.C. & A. Calabrese. 1964. Combined effects of temperature and salinity on development of eggs and growth of larvae of *M. mercenaria* and *C. virginica. U.S. Bur. Commer. Fish. Bull.* 63:643.

Galtsoff, P.S. 1964. The American oyster, *Crassostrea virginica. U.S. Bur. Commer. Fish. Bull.* 64:1.

Woelke, C.E. 1965. Bioassays of pulp mill wastes with oysters. *In* C.M. Tarzwell, ed. Biological Problems in Water Pollution. U.S. PHS Publ. 999-WP-25, p. 67.

Woelke, C.E. 1965. Development of a Bioassay Method Using the Marine Alga, *Monochrysis lutheri*. Wash. Dep. Fish. Shellfish Prog. Rep., Olympia.

Matthiessen, G.C. & R.C. Toner. 1966. Possible Methods of Improving the Shellfish Industry of Martha's Vineyard, Duke's County, Massachusetts. Marine Research Foundation Inc., Edgartown, Mass.

Woelke, C.E. 1967. Measurement of water quality with the Pacific oyster embryo bioassay. *Spec. Tech. Pub.* 416:112. American Society for Testing and Materials, Philadelphia, Pa.

Woelke, C.E. 1968. Application of shellfish bioassay results to the Puget Sound pulp mill problem. *Northwest Sci.* 42(4):125.

Woelke, C.E., T.D. Schink & E.W. Sanborn. 1970. Development of an *in situ* marine bioassay with clams. Annu. Rep. Oct. 1, 1969–Sept. 30, 1970. July 6, 1970. Wash. Dep. Fish., Olympia.

Brown, B.E. 1972. The effect of copper and zinc on the metabolism of the mussel *Mytilus edulis. Mar. Biol.* 16:108.

Favretto, L. & F. Tunis. 1974. Typical level of lead in *Mytilus galloprovincialis* Lmk from the Gulf of Trieste. *Rev. Int. Oceanogr. Med.* 33:67.

810 BIOASSAY PROCEDURES FOR FISH (TENTATIVE)

Fish have been used more widely for bioassays than any other group of aquatic organisms. However, only about 15 species have been used extensively, and of these, only a few freshwater species[1] have been used in life-cycle tests. Until recently, few marine fish were cultured, reared in the laboratory, or used for testing.

810 A. Selecting and Preparing Fish

1. Selection of Test Species

For general guidelines for selecting test organisms see Section 801D.1. A prime consideration in the selection of fish is their individual sensitivity to the effluent, material, or environmental factor under consideration and their sensitivity relative to that of other organisms in the ecosystem. Because some fish have been reared in hatcheries, it is possible to conduct life-cycle studies with a number of freshwater fish. For those that have not been reared, collect various life stages from the field for partial-life-cycle tests.

The following is a partial list of the freshwater, estuarine, and marine fish that have been used. Some species will survive under wide variations of temperature, DO, pH, and salinity and are not well suited for determining favorable environmental factors for the survival of a biota. Further, because certain levels of environmental factors and certain toxicant concentrations under which adults may live indefinitely may not be suitable for the survival of the species, preferably use the most sensitive life stages. If data are not available on relative sensitivity of different species to the waste, use short-term tests with local species to determine the one most sensitive.

a. Freshwater fishes:

Clupeidae:	
Alosa pseudoharengus	Alewife
Dorosoma petenense	Threadfin shad
Salmonidae:	
Coregonus artedii	Lake herring
Coregonus clupeaformis	Lake whitefish
Prosopium williamsoni	Mountain whitefish
Oncorhynchus gorbuscha	Pink salmon
Oncorhynchus keta	Chum salmon
Oncorhynchus kisutch	Coho salmon
Oncorhynchus nerka	Sockeye salmon
Oncorhynchus tschawytscha	Chinook salmon
Salmo clarki	Cutthroat trout
Salmo gairdneri	Rainbow trout
Salmo salar	Atlantic salmon
Salmo trutta	Brown trout
Salvelinus fontinalis	Brook trout
Salvelinus namaycush	Lake trout
Osmeridae:	
Osmerus mordax	Rainbow smelt
Esocidae:	
Esox lucius	Northern pike
Cyprinidae:	
Carassius auratus	Goldfish
Cyprinus carpio	Carp
Notropis atherinoides	Emerald shiner
Notemigonus crysoleucas	Golden shiner
Pimephales notatus	Bluntnose minnow

Pimephales promelas	Fathead minnow
Catostomidae:	
Catostomus commersoni	White sucker
Ictaluridae:	
Ictalurus melas	Black bullhead
Ictalurus natalis	Yellow bullhead
Ictalurus nebulosus	Brown bullhead
Ictalurus punctatus	Channel catfish
Cyprinodontidae:	
Jordanella floridae	Flagfish
Poeciliidae:	
Gambusia affinis	Mosquitofish
Poecilia reticulata	Guppy
Percichthyidae:	
Morone chrysops	White bass
Centrarchidae:	
Lepomis macrochirus	Bluegill
Micropterus dolomieui	Smallmouth bass
Micropterus salmoides	Largemouth bass
Pomoxis annularis	White crappie
Pomoxis nigromaculatus	Black crappie
Percidae:	
Perca flavescens	Yellow perch
Stizostedion canadense	Sauger
Stizostedion v canadense	Walleye pike

b. Marine and estuarine fishes:

Clupeidae:	
Brevoortia patronus	Gulf menhaden
Brevoortia tyrannus	Atlantic menhaden
Clupea harengus	Atlantic herring
Harengula pensacolae	Scaled sardine
Sardinops sagax	Pacific sardine
Engraulidae:	
Anchoa mitchilli	Bay anchovy
Cyprinodontidae:	
Cyprinodon variegatus	Sheepshead minnow
Fundulus heteroclitus	Mummichog
Fundulus similis	Longnose killifish
Fundulus parvipinnis	California killifish
Atherinidae:	
Menidia beryllina	Tidewater silverside
Menidia menidia	Atlantic silverside
Gasterosteidae:	
Gasterosteus aculeatus	Threespine stickleback
Percichthyidae:	
Morone saxatilis	Striped bass
Serranidae:	
Centropristis striata	Black sea bass
Sparidae:	
Lagodon rhomboides	Pinfish
Sciaenidae:	
Leiostomus xanthurus	Spot
Micropogon undulatus	Atlantic croaker
Mugilidae:	
Mugil cephalus	Striped mullet
Mugil curema	White mullet
Pleuronectidae:	
Pseudopleuronectes americanus	Winter flounder

2. Collecting, Handling, and Treating Test Fish

Collecting equipment and methods are described in Sections 801D.2 and 1006A. Handling, holding, and treatment are discussed in Section 801D.3.

a. Freshwater fish: Collect fish from the field, or preferably obtain species routinely raised in hatcheries. Some minnows can be obtained from bait dealers; larger species can be obtained from fishermen. Salmonid fish usually are available from private, state, and federal hatcheries. For trout, obtain fish certified free from infectious pancreatic necrosis, furunculosis, kidney disease, and whirling disease. Collecting permits usually are required by state agencies.[2]

b. Marine and estuarine fish: Collect

various life stages of marine fish in the field for laboratory tests. Vertical movement of early larval stages may necessitate night-time collection. Because most marine fish larvae are extremely fragile, handle carefully during collecting, sorting, and transferring. To transfer, use large-bore pipets or dip or pour gently. Fine-mesh dip nets are suitable if transfers are made quickly.

3. Holding and Acclimating

See Sections 801D and E.

Keep fish stocks in aquariums, small ponds, live boxes, screen pens, or tanks, depending on size and number. Use good-quality water. Feed fish natural or prepared foods* daily during acclimation. Detailed information on handling, holding, care, and feeding of fish is available.[2–6] The food required varies with the fish; use care in selecting the diet. Feed fish obtained from a hatchery with food to which they are accustomed. Many fish can be maintained for long periods on dried food but live food supplements are desirable. Do not overfeed. Analyze food and reject if contaminated with PCB's, pesticides, or heavy metals.

To maintain fish in good condition during holding and acclimation, watch carefully for signs of disease, stress, physical damage, and mortality. Remove dead and abnormal individuals immediately.

Handle gently, carefully, and as quickly as possible.[2] For extensive handling such as weighing, measuring, or taking other data, anesthetize† fish.[5] Before using fish from a given area, determine body burdens of pesticides, PCB's, heavy metals, and other toxicants. For short-term tests, use fish weighing between 0.5 and 5 g. In any one series of tests, use fish from the same year class. The length of the largest fish should not be more than 50% greater than that of the shortest fish.[2]

Use temperature and salinity regimes for larval marine fish that approximate field conditions. Do not acclimate larvae to temperature or salinity combinations that differ from those normally encountered because many marine larvae migrate from saline to brackish water[7,8] or from fresh to tidal water.[9]

Most larvae of marine fish require special holding tanks. Prevent accumulation of toxic metabolites by using properly designed holding tanks[10] with correct lighting intensity.[11–14] Pay particular attention to the diet for test fish. Field-collected zooplankton,[13–17] dense cultures of phytoplankton[12,18,19] [see Section 801D.4*c*1)], and live brine shrimp, *Artemia salina,* have been used successfully for feeding larval fish.

4. Culturing Test Fish

a. Freshwater fish: More than 30 species of freshwater fish have been reared for stocking fresh waters. Cultural methods can be adapted to laboratory conditions to produce different life stages of fish.[20–33]

b. Marine and estuarine fish: Reviews on the feeding of larval marine fish[34] and other larvae are available.[35] Use laboratory spawning, either natural or artificially induced by hormones, to provide early life stages. Technics are available for rearing herring, bay anchovy, scaled sardine, and Pacific sardine from eggs.[14,18,36–38] The sheepshead minnow and the common mummichog have been reared through their life cycle and the silverside through parts of its live cycle.[39–41] The mullet, croaker, black sea bass, and spot have been cultured to or past yolk-sac absorp-

*Examples are Oregon Moist Fish Food, Warrenton, Ore. 97146, and Glencoe Trout Food, Glencoe Mills, Glencoe, Minn. 53336.

†MS222, Sandoz (tricaine methane-sulfonate), now marketed as Finquel by Ayerst, or equivalent, may be used as an anaesthetic.

tion.[15,42–46] Difficulty in rearing many marine species past yolk-sac absorption may limit test duration. If hormones are used to ripen mature adults artificially, avoid atrophy (over-ripeness), which will cause poor fertilization of eggs and emergence of larvae.[42,47] Carefully handle developing eggs and pro-larvae. Developing zygotes of some marine fish are especially susceptible to mechanical damage[48] and even gentle handling can result in significant reduction of percentage that emerge as larvae.[44] Hold adult brood stocks, eggs, and larvae under conditions approximating those encountered by adults spawning naturally.

1) Sheepshead minnow—The sheepshead minnow thrives over a wide range of salinities and temperatures. To establish cultures, collect adult fish over 30 mm long and acclimate to laboratory temperatures. They should begin to spawn after 2 wk at 30 C. During this period feed liberally on fresh or frozen adult brine shrimp. Alternatively obtain sheepshead minnow eggs by injecting hormones to cause ripening of sex products.[49] To induce egg production artificially, inject each female intraperitoneally with 50 IU human chorionic gonadotrophic hormone. Repeat after 2 days. On the third day most females produce eggs that are readily stripped. Strip or dissect eggs into seawater in a beaker and add macerated testes.[48] Check eggs for cleavage approximately 1 to 1.5 hr later. Usually 90% or more of the eggs are fertile. Collect eggs from natural spawning by placing a pair of adult fish in a spawning chamber about 12 × 18 × 10 cm high into which is fitted a 2-cm-deep tray formed by an 0.5-mm nylon screen over a frame and covered with nylon screen having openings 2 mm square. As eggs are deposited they fall through the screen into the tray, so they escape predation and can be removed readily. A pair may spawn 100 eggs/day but average production is about 8 eggs/pair/day. Place eggs in a hatching chamber formed by placing a 9-cm collar of 0.5-mm mesh plastic‡ screen around a petri dish. Place chambers in flow-through aquariums receiving test solutions.[50] Sheepshead minnow fry hatch after 5 days at 30 C and a salinity between 15 and 20 ‰. As fry hatch, transfer to a rearing aquarium and immediately feed newly hatched brine shrimp nauplii. Feed adults and juveniles adult brine shrimp, live or frozen, and a dry food.§ Supplement a dry food diet occasionally with live organisms. Survival from fertile eggs to 28 days is approximately 85%.[50] Juveniles become sexually distinguishable when about 24 mm long and females produce eggs within 3 months after hatching.

2) Atlantic silverside, *Menidia menidia*—The silverside that occurs in east-coast estuarine areas throughout the year is ecologically significant as food for predacious fish such as sea bass, mackerel, bluefish, striped bass, and sea trout.[51] Collect sexually mature and ripe individuals from April to July at water temperatures above 15 C. Collect mature adults for spawning with a 5-mm-mesh bag seine. Maintain in the laboratory on a diet of chopped fish, mollusks, or grass shrimp, and live brine shrimp. The Atlantic silverside spawns from March to August and produces approximately 500 eggs. Hold mature adults brought in from the field in circular fiberglass tanks constructed as in Section 801D.4*d*, 1 m diam and 60 cm deep and containing a 45-cm depth of seawater. Make interior tank surfaces almost glass-smooth. Place adults in rearing tank in early spring and subject them to a temperature and daily light regime that follows natural conditions. Supply light as recom-

‡Nitex or equivalent.

§Biorell® or equivalent.

mended in Section 801E.3*f* with an illumination of 2,000 lux at the water surface. Raise temperatures gradually to 22 C and keep salinities between 24 and 26 ‰. Feed live and dry food to excess. Clean tanks and siphon unused food and wastes from bottom of tank daily. Information on rearing the silverside is given elsewhere.[52,53]

When adults are ready to spawn, either place them in spawning chambers or strip them to have better control of the eggs. Do not handle more than absolutely necessary.

Strip females into 200-mm-diam culture dishes containing a 1-cm depth of seawater. After four to six have been stripped, strip a few males and gently stir the milt and eggs.[54] Immediately after fertilization the eggs form gelatinous strands that bind them together.[20] Remove fertilized eggs by rolling a 25-cm length of nylon or polyethylene string around the gelatinous mass and suspend in a 20- to 40-L, aerated, all-glass, flow-through aquarium. Hold water temperature between 18 and 22 C and salinity between 25 and 30 ‰. Hatching of larvae will begin 7 to 9 days after fertilization. Feed copepod nauplii or *Artemia* nauplii less than 8 hr old to newly hatched larvae. Silversides are omnivorous feeders. Provide 200 to 800 nauplii/L. As fish grow, increase feed and vary the diet using dry food, nauplii of *Balanus,* sea urchin eggs, copepods, rotifers, annelids, small bivalve larvae, and mysid shrimp. Occasional live food is desirable. When larval fish begin to be crowded, transfer to the circular tanks described earlier and continue feeding to adult stage.

5. Parasites and Diseases

Fish held in confined quarters at unusually high densities are prey to parasites and disease. These must be controlled.

Larval fish are particularly vulnerable to fungi and parasites, such as copepods, that have direct life cycles and may quickly increase to overwhelming numbers. Parasites also may be transmitted to fish larvae by ingestion of phytoplankton. Treat with antibiotics to reduce bacterial populations (15-mg tetracycline solution/L for 24 to 48 hr). Carefully select and use chemotherapeutic agents.[54] Also see Sections 801C.5 and 810A.5.

a. Control methods for freshwater fish: Treat freshwater fish to cure or prevent disease by the methods in Section 801D.5 and Table 810:I. These have been found dependable, but their efficacy may be altered by temperature or water quality. *Test treatments on small lots of fish before making large-scale applications.* Treat newly acquired fish with the formalin-malachite green combination on 3 alternate days if possible; do not treat fish on the first day in the laboratory.[2] If fish are severely diseased, destroy the entire lot. A number of good reviews of fish diseases and parasites and methods for their control have been published.[56–64]

b. Control methods for marine and estuarine fish: Species such as the clupeoids are particularly susceptible to bacterial (usually *Vibrio*) infections. Tetracycline-based antibiotics in the food provide limited protection.

Ectoparasitic protozoa (particularly the ciliate *Cryptocaryon* and the dinoflagellate *Oodinium*) may multiply quickly to epizootic proportions in marine aquariums. Treat with formalin and cupric acetate.[64]

Treat monogenetic trematodes on gills and body surfaces by brine and sodium pyrophosphate dips.

To treat other microbial and parasitic diseases (such as lymphocystis, fungus, and parasitic copepods) see Sindermann.[65]

For a summary of advances and unsolved problems in marine fish larval culture and a description of a successful larval culture system, see Houde.[12,66] For disinfection of water supplies see Hoffman.[67]

TABLE 810:I. RECOMMENDED PROPHYLACTIC AND THERAPEUTIC TREATMENTS FOR FRESHWATER FISH

Disease	Chemical	Concentration* mg/L	Application
External bacteria	Benzalkonium chloride†	1-2 AI	30-60 min in flow-through system‡
	Nitrofurazone (water mix)	3-5 AI	30-60 min in flow-through system‡
	Neomycin sulfate	25	30-60 min in flow-through system‡
	Oxytetracycline hydrochloride (water-soluble)	25 AI	30-60 min in flow-through system‡
Monogenetic trematodes, fungi, and external protozoa§	Formalin *plus* zinc-free malachite green oxalate	25 0.1	1-2 hr in static system
	Formalin	150-250	30-60 min in flow-through system‡
	$KMnO_4$	2-6	30-60 min in flow-through system‡
	NaCl	15,000-30,000 2,000-4,000	5-10 min dip 24 hr minimum, but may be continued indefinitely
	Para-dimethylaminobenzene-diazo sodium sulfonate (35% AI)‖	20	30-60 min in flow-through system‡
Parasitic copepods	Trichlorfon#	0.25 AI	Weekly for up to 4 wk if necessary in static or flow-through systems. Do not use at >27 C.

*AI = active ingredient.
†Hyamine 1622® or equivalent.
‡Add concentrated stock solution to the inflowing water by a drip system or by the technic of Brungs and Mount.[57]
§One treatment is usually sufficient except for *Ichthyophthirius,* which must be treated daily or every other day until no sign of the protozoans remains. This may take 4 to 5 wk at 5 to 10 C and 11 to 13 days at 15 to 21 C. A temperature of 32 C is lethal to *Ichthyophthirius* in 1 wk.
‖Dexon® or equivalent.
#Masoten® or equivalent.

810 B. General Procedures for Fish

1. Short-Term Tests

Conduct range-finding tests as described in Sections 801C through F and 810A.

2. Long-Term Partial- and Complete-Life-Cycle Testing

Use juvenile or sexually immature adults, newly spawned eggs, or newly hatched larval fish to initiate these tests. The life stage selected depends on the species, its sensitivity in relation to other life stages, laboratory space and facilities, availability of different stages, and the test purpose. When space and water supply permit, expose additional fish for special histological, residue, or other examinations. When holding mature adults for spawning, establish the sex ratio as soon as possible. Record total length and weight of all fish at the beginning of a test, at selected intervals during the test, after mortality, and at end of the test. To prevent injury, anesthetize large fish before handling. When large numbers of small fish are used in the test, measure them by the photographic method.[68] Stop treatment of parental fish after stripping, at a fixed number of days after the last spawning during the test. Record sex, condition of gonads, and total length and weight at end of test.

For egg viability and hatchability tests, incubate eggs from each spawning at the optimum temperature. If effluent tests are performed, use temperature of receiving water. Count live and dead eggs and remove dead ones daily. Evaluate egg viability for all spawnings by incubating eggs until development can be determined. Determine egg hatchability for all spawnings in all exposure chambers or from a predetermined number of spawnings when the species tested is one that spawns continuously or many times per female. Count number of dead, deformed, and normal larvae hatched daily, using a dissecting microscope if necessary.

For larval growth and survival, collect a uniform number of normal larvae (usually 20 to 50) at random from two or more successful hatches and place in chambers for each toxicant concentration. If there is a prolonged hatching period use the median hatching date as the test starting date. Determine length and number of larvae upon transfer of growth chambers, preferably by the photographic method.[68] Determine total length of larvae at selected intervals and at end of test. Record deaths daily.

Tests that begin with eggs require approximately equal numbers of eggs pooled from at least three females. For tests beginning with juvenile fish of species that have a long life cycle and spawn only once a year, (e.g., brook trout) use sexually immature fish that will spawn for the first time at the upcoming spawning season. Acclimate for at least 1 month before the test (approximately 8 months before time of spawning). For fish maturing and spawning more often, use shorter periods.

For methods of toxicant mixing and delivery see Section 801E.1. Install an automatically triggered emergency alarm system to indicate failure of diluter, temperature controller, or water supply.

Vary site of spawning tanks, exposure tanks, and growth chambers according to the species. Generally use a water depth of at least 15 to 30 cm. In growth chambers for larvae, use lesser depths. Design each growth chamber so that test solutions can be drained down to 2.5 to 3 cm and the chamber transferred to a fluorescent light box provided with a millimeter grid for photographing fish.

Construct incubation cups or chambers from 8-cm sections of 5-cm-OD plastic or glass tubing by cementing nylon screen to retain eggs and larvae over one end of tube. Oscillate incubation cups in test water by a rocker arm driven by a 2-rpm electric motor.[69]

Retain fish and eggs produced for physiological, biochemical, and histological tests. Report all pertinent data for each test container at the beginning, about a third of the way through the test, and at the end. Include number and weight of individuals, number of spawnings, number of eggs, and total lengths of normal, deformed, and injured mature and immature males and females. Record mortality. Calculate mean incubation time using date of spawning and median hatch dates. Record hatchability of eggs and fry survival, growth, and deformities.

810 C. Conducting Tests

Examples of partial-life-cycle and life-cycle tests are given for two freshwater and two marine fishes.

1. Conducting Short-Term Tests

Conduct tests as described in Sections 801C through E. A stepwise procedure has been described.[70]

2. Long-Term Partial- and Complete-Life-Cycle Tests

Use the following rearing technics to provide stock fish for short-term tests.

a. Partial-life-cycle tests with the brook trout, Salvelinas fontinalis:

1) Equipment and physical conditions—For mixing toxicants and measuring and delivering dilution water, see Section 801E. Set up duplicate tanks for each concentration and control. For proportional diluters use a mixing tank or flow-splitting chamber[71] to mix each concentration before delivery to duplicate spawning tanks and growth chambers.

Construct alevin-to-juvenile growth chambers of glass or stainless steel with a glass bottom with approximate dimensions of 38 × 15 × 18 cm. Maintain water depth at about 13 cm. Design each chamber so that the water can be drained down to a depth of 2 to 3 cm to allow the chamber to be placed over a millimeter grid on a fluorescent light box for photographing fish (Section 801E.3*e*).

Use flow rates of 6 to 10 tank volumes/day, depending on amount needed to keep DO above 60% of saturation. Siphon unused food and wastes from growth chambers daily and brush and clean interior surfaces and remove attached growths at least weekly.

Construct spawning tanks, preferably of No. 316 stainless steel, with dimensions of 90 × 30 × 40 cm. Use a 30-cm water depth. Place a spawning substrate or nest[72] in spawning tanks at appropriate time. Use a spawning nest 28 × 33 × 7.5 cm made of double-strength glass or stainless steel. Larger fish require a larger nest. Drill three 2.5-cm holes in each end 2.5 cm from the bottom and cover with 10 mesh stainless steel wire to allow water in box to drain down to 2.5 cm when box is removed from spawning chamber. Place in spawning box a bottomless screen egg retainer (27 × 32 × 1.3 cm with 2.5-cm-square compartments, constructed from 1.3-cm-wide strips of 7 mesh stainless steel screen). Place a 2.0 mesh stainless steel screen, 27 × 32 cm, to which 1.3 to 2.5 cm gravel is attached with silicone adhesive, on top of the screen egg retainer. Use smooth gravel to prevent injury to active, spawning fish. This spawning box is readily removed from the spawning tanks to collect eggs for transfer to the incubation cups described in Section 810B.2. Select yearling fish that will not grow too large in the confinement of the spawning box. Use fish weighing not more than 50 to 70 g at time of selection and 150 g at spawning. If they weigh more than 150 g, use a larger box.

For brook trout simulate the photoperiod, and dawn to dusk times, at Evansville, Ind., or at local area. Photoperiod

TABLE 810:II. TEST (EVANSVILLE, IND.) PHOTOPERIOD FOR BROOK TROUT, PARTIAL LIFE CYCLE

Dawn to Dusk Time	Date	Day Length *hr & min*	
6:00-6:15	Mar. 1	12:15	Juvenile-adult exposure
6:00-7:00	15	13:00	
6:00-7:30	Apr. 1	13:30	
6:00-8:15	15	14:15	
6:00-8:45	May 1	14:45	
6:00-9:15	15	15:15	
6:00-9:30	June 1	15:30	
6:00-9:45	15	15:45	
6:00-9:45	July 1	15:45	
6:00-9:30	15	15:30	
6:00-9:00	Aug. 1	15:00	
6:00-8:30	15	14:30	
6:00-8:00	Sept. 1	14:00	
6:00-7:30	15	13:30	
6:00-6:45	Oct. 1	12:45	Spawning and egg incubation
6:00-6:15	15	12:15	
6:00-5:30	Nov. 1	11:30	
6:00-5:00	15	11:00	
6:00-4:45	Dec. 1	10:45	Alevin-juvenile exposure
6:00-4:30	15	10:30	
6:00-4:30	Jan. 1	10:30	
6:00-4:45	15	10:45	
6:00-5:15	Feb. 1	11:15	
6:00-5:45	15	11:45	

adjustments for brook trout are shown in Table 810:II, which is arranged so that adjustments need be made only at dusk. Adapt this schedule for establishing photoperiods for other test organisms. The dawn and dusk times listed in Table 810:II need not correspond to the actual test times where the test is being conducted. For example, a test started on March 1 would require use of the photoperiod for the Evansville test date of March 1. The lights could go on at any time on that day just as long as they remained on for 12 hr and 15 min; 15 days later, the photoperiod would be changed to 13 hr. Gradual changes in light intensity at dawn and dusk[52] may be included within the photoperiod shown but they should not last for more than 0.5 hr from "full on" to "full off" and vice versa (see Section 801E.3*f*).

The temperature regime is indicated in Table 810:III. Do not vary temperature by more than ±2 C from that specified and do not permit a variation of ±1 C for more than 48 hr at a time.

Cover spawning tanks and growth chambers with a screen to confine the fish and place behind curtains so the fish will not be distrubed. Shield tanks and chambers from extraneous light.

Ideally, take water for other than effluent studies from a well or spring or, alternatively, from an unpolluted surface source. If water is contaminated with fish pathogens, disinfect with UV immediately before the test.

TABLE 810:III. TEMPERATURE REGIME FOR BROOK TROUT LIFE CYCLE TESTS

Months	Stage	Temperature *C*	Comment
Mar.		9	
Apr.		12	
May		14	
June	Juvenile-adult exposure	15	
July		15	
Aug.		15	
Sept.		12	
Oct.		9	
Nov.	Spawning and egg incubation	9	Establish constant temperature just prior to spawning and egg incubation, and maintain throughout the 3-month alevin-juvenile exposure.
Dec.		9	
Jan.		9	
Feb.	Alevin-juvenile exposure	9	
Mar.		9	

2) Exposure procedures—When the test is begun with juveniles, collect them no later than March 1 and acclimate for at least 1 month. Judge suitability of fish for testing on the basis of their acceptance of food, apparent lack of disease, and occurrence of less than 2% mortality during acclimation and no mortality during the 2 wk prior to the test. Set aside enough fish to supply an adequate number for use in short-term tests to determine asymptotic LC50's.

Begin exposure no later than April 1 by placing 12 acclimated yearling brook trout in each duplicate tank for each test concentration and the controls using a stratified random assignment (Section 801E.3*a*). This allows about 4-month exposure to toxicant before the onset of secondary or rapid-growth phase of the gonads. Add extra test animals at the beginning so that fish can be removed periodically for special examinations or for chemical analysis.

Use a good particulate or pelleted trout food. Feed fish the largest particle or pellet they will take at least twice daily. Base amount on a reliable hatchery feeding schedule. Analyze each batch of prepared food for pesticides (Section 810A.3).

Record mortalities daily and measure total length and weight of fish directly at initiation of tests, after 3 months, and when number of test fish are reduced. Do not feed for 24 hr before weighing. Lightly anesthetize them to facilitate measuring.

When secondary sexual characteristics are well developed (approximately 2 wk before spawning), separate males, females, and undeveloped fish in each tank and randomly reduce number of sexually mature fish to two males and four females per tank. Record number of mature, immature, deformed, and injured males and females in each tank and number from each category to be discarded. After they have been thoroughly cleaned, sterilized and rinsed, place one spawning substrate in each spawning tank. As soon as spawning begins, set up incubation cups to receive eggs for hatching. Remove eggs from the substrate at a fixed time each day (preferably after 1:00 PM, Evansville time, so fish are not disturbed during the morning).

Randomly select 50 eggs from the first eight spawnings of 50 eggs or more in each duplicate spawning chamber and place in an egg incubator cup. Count remaining eggs from the first eight spawnings and all eggs from subsequent spawnings and place them in separate egg incubator cups for determining viability (formation of neural keel after 11 to 12 days at 9 C). Remove and record number of dead eggs from each spawn. Never place more than

250 eggs in one incubator cup. Incubate all eggs to determine viability and discard after 12 days. Use discards for chemical analysis and other measurements.

Obtain additional information on hatchability and alevin survival by transferring eggs from control tanks immediately after spawning (*a*) to tanks having test concentrations where spawning is reduced or absent and (*b*) to tanks where an effect is seen on survival of eggs oralevin, and by transferring eggs from those test concentrations to control tanks. Always reserve two growth chambers for each duplicate spawning tank for eggs produced in that tank.

Remove dead eggs daily from hatchability cups. When hatching begins, record number of alevins hatching daily in each cup. On completion of hatching in any cup, transfer fish to a culture dish and randomly select 25 alevins. Count dead or deformed alevins. Transfer 25 selected alevins to a growth chamber and place it over the light box to measure by the photographic method (Section 801E.3*e*). After photographing, return alevins to incubation cup. Never net alevins, but transfer by gentle pouring and by large-bore pipets. Transport in growth chambers containing a 2.5-cm depth of test solution. Preserve unused alevins in formalin for further tests. Record length (and weight) of discarded alevins separate from those of fish kept for subsequent exposure.

For 90-day growth and survival exposures, randomly select 25 alevins from each duplicate incubation cup for each test concentration and control. Because hatching from one spawn may be spread out over a 3- to 6-day period, use the median hatch date to establish the 90-day growth and survival period. If it is determined that the median-hatch dates for the eight groups will be more than 3 wk apart, select the two groups of 25 alevins from those that are less than 3 wk old. Use the remaining groups only for hatchability results in the duplicate tests for those test concentrations where hatching does not occur during the 3-wk period. After photographing to determine lengths, preserve for weight determination. To equalize effects of incubation cups on growth, keep all groups selected for 90-day exposure in the incubation cups 3 wk after the median hatch date, then release into growth chambers. Begin feeding immediately. Keep separate the two groups selected from the duplicate exposure chambers for each test concentration. Record mortalities daily, total lengths 30 and 60 days after hatching by the photographic method, and total length and weight at 90 days after hatching. At the end of the test cease feeding juveniles for 24 hr, then weigh. End survival and growth studies after 3 months and use fish for chemical analysis of tissue and physiological measurements of toxicant-related effects.

End exposure of all parental fish when 3 wk pass in which no spawning occurs in any spawning tank. Record mortality and weight, measure total length of parental fish, and check sex and conditions of gonads (e.g., reabsorption, degree of maturation, spent ovaries).

Report, for each tank of a partial-life-cycle test, number and individual weights and total lengths of immature males and females at initiation of test, after 3 months, at reduction in numbers, and at end of the test. Report individual weights and total lengths of normal, deformed, and injured fish, number maturing, mortality during test, number of spawnings and eggs, hatchability and fry survival, growth, and deformities. Calculate a mean incubation time using date of spawning and median hatch dates.

3) Measurement of toxicant concentrations—If possible, measure toxicant concentration in one tank at each concentration weekly. Use composites of equal-volume daily grab samples for 1 wk if it has

been shown that results are unaffected by storage.

Analyze enough pooled and grouped grab samples periodically throughout the test to determine if toxicant concentration is constant. If it is not, analyze enough samples weekly to establish variability of toxicant concentration (see Section 801E.3*d*).

Record temperature continuously. Measure DO daily at least 5 days/wk on an alternating basis so that each tank is analyzed once per week. If the toxicant depresses DO, also analyze toxicant concentration with lowest DO daily.

Analyze control and one test concentration at least weekly for pH, alkalinity, hardness, acidity, and conductivity. If any characteristic is affected by the toxicant, analyze for that characteristic at least 5 days/wk, rotating between tanks so that each tank is analyzed once every other week. At a minimum, analyze test water at the beginning, middle, and end of exposure period for Ca^{2+}, Mg^{2+}, Na^{+}, K^{+}, Cl^{-}, SO_4^{2-}, conductivity, total residue, and filtrable residue.

When possible and necessary, analyze mature fish and/or eggs, larvae, and juveniles for toxicant residues. For fish, analyze muscles and gills; also consider analyzing blood, brain, liver, bone, kidney, GI tract, gonads, and skin. Do not analyze whole organisms in place of individual tissues, especially muscle.

For additional information on cycles of brook trout consult other sources.[33,68,73–81]

b. Life-cycle with the fathead minnow Pimephales promelas:

1) Physical system—See Section 801E and 810B. The physical systems are similar to those for the brook trout.

Use one of two arrangements of test tanks (glass or stainless steel with glass ends): duplicate spawning tanks for each of the five or more test concentrations and controls, measuring 30 × 30 × 90 cm with a 30-cm-square portion at one end, screened off and divided in half to form two larval chambers for the progeny. Deliver test water separately to larval and spawning chambers of each tank, with about one-third of water volume going to larval chamber.

Alternatively, arrange duplicate spawning tanks measuring 30 × 30 × 60 cm with duplicate progeny tanks for each spawning tank. Use a larval tank with minimum dimensions of 30 × 30 × 30 cm, divided to form two separate larval chambers with separate standpipes; or use separate 30- × 15- × 30-cm tanks. Supply test solutions and water for controls as in Section 801E.1. Maintain a water depth of 15 cm in all tanks.

Flow rate, DO requirements, aeration, cleaning, and operation are as described for the brook trout.

Fathead minnows deposit eggs on the underside of submerged objects. For spawning substrates use inverted halves of tile 7.5 cm ID and 7 to 10 cm long, or equivalent. Place tiles parallel to the long axis of the spawning tank so each end is readily accessible to the fish.

Fasten egg incubation cups, such as those described for the brook trout, to a rocker arm with a vertical travel distance of 3 to 4 cm.[69] Lamps and illumination are described in Section 801E.3*f*.

Photoperiods (Table 810:IV) simulate dawn-to-dusk times at Evansville, Ind. Make adjustments in day length on the 1st and 15th day of every Evansville test month. Regardless of actual starting date, adjust the Evansville test photoperiod so that the mean or estimated hatching date of the fish used to start the experiment corresponds to the Evansville test day length for December 1. This gives a consistent prespawning exposure. The dawn and dusk times listed in Table 810:IV need not correspond to the actual times at the testing location.

Control temperature to 25 C ± 2 C. Record temperature continuously.

TABLE 810:IV. TEST PHOTOPERIOD FOR FATHEAD MINNOW LIFE CYCLE TESTS*

Dawn to Dusk Time	Date	Day Length *hr & min*	
6:00–4:45	Dec. 1	10:45	5-month prespawning growth period
6:00–4:30	15	10:30	
6:00–4:30	Jan. 1	10:30	
6:00–4:45	15	10:45	
6:00–5:15	Feb. 1	11:15	
6:00–5:45	15	11:45	
6:00–6:15	Mar. 1	12:15	
6:00–7:00	15	13:00	
6:00–7:30	Apr. 1	13:30	
6:00–8:15	15	14:15	
6:00–8:45	May 1	14:45	4-month spawning period
6:00–9:15	15	15:15	
6:00–9:30	June 1	15:30	
6:00–9:45	15	15:45	
6:00–9:45	July 1	15:45	
6:00–9:30	15	15:30	
6:00–9:00	Aug. 1	15:00	
6:00–8:30	15	14:30	post-spawning period
6:00–8:00	Sept. 1	14:00	
6:00–7:30	15	13:30	
6:00–6:45	Oct. 1	12:45	
6:00–6:15	15	12:15	
6:00–5:30	Nov. 1	11:30	
6:00–5:00	15	11:00	

*Based on Evansville, Ind., times.

2) Biological systems—Obtain test fish from laboratory cultures, from the field, or from bait dealers. Treat fish for control of parasites and disease as described in Sections 801D.5 and 810A.5. For starting tests use a mixture of approximately equal numbers of eggs or larvae from at least three different females.

Set aside enough eggs or larvae at the start of the test to supply an adequate number of fish for acute mortality tests used in determining application factors. Conduct all acute mortality tests to determine concentrations to be used in life-cycle tests with 2- to 3-months-old fish.

Begin life-cycle test by distributing 50 1- to 5-day-old larvae in each duplicate spawning tank for each test concentration by a stratified random assignment. If 1- to 5-day-old larvae are not available, use fish up to 30 days old. Add extra fish at the beginning so that some can be removed periodically for special examinations.

During the test, feed fish once or twice a day with live brine shrimp nauplii for 30 to 60 days and then frozen adult brine shrimp as the main diet, supplemented by pelleted trout food, *Daphnia,* and chopped earthworms. Feed once daily live young zooplankton from mixed cultures of small copepods, rotifers, and protozoans. Check each batch of commercial feed for pesticides.

When test fish are 60 ±1 or 2 days old, discard injured or crippled individuals and randomly reduce the number in each tank to 15. Record number, length, and weight of discarded and deformed fish. If necessary to obtain 15 fish per tank, select one or two fish for transfer from one duplicate to the other. Continue routine feeding and cleaning until fish mature and are almost ready to spawn. Place five spawning tiles in each duplicate spawning tank, separated fairly widely to reduce fighting between the guarding male fish. Place tiles so their undersides and the guard males can be seen from the tank end. During spawning, remove sexually maturing males so that there are no more than four per tank. Reserve the fifth tile as cover for females. Do not remove those males having well-established territories under tiles where recent spawnings have occurred.

Daily, remove eggs from spawning tiles starting at 12:00 noon Evansville test time (Table 810:IV). Loosen eggs from spawning tiles and at the same time separate them from one another by lightly placing a finger on the egg mass and moving it in a circular pattern with increasing pressure until the eggs begin to roll. Wash groups of eggs into separate containers and return tiles to spawning aquariums. Count eggs, select those needed for incubation, and discard remainder after counting. Check all eggs for different stages of development. If more than one distinct development stage is present, consider each stage as one spawning and handle separately as described below.

Randomly select 50 unbroken eggs from a single spawning and place in an egg incubator cup to determine viability and hatchability. Count and discard remaining eggs. Determine viability and hatchability on each spawning of more than 49 eggs until the number of spawnings (>49 eggs) in each tank equals the number of females in that tank. Subsequently, test for hatchability only eggs from every third spawning of more than 49 eggs. Remove weekend spawns from tiles and count eggs but discard.

If no spawning occurs for a week, cease parental fish testing. Record total length and weight of parental fish, their sex, and gonad conditions. Do not freeze fish before making these sexing examinations.

Daily, record live and dead eggs in incubator cups, remove dead ones, and clean cup screens. The total number of eggs accounted for always should add up to within two of 50; if not, discard the entire batch. After 4 to 6 days, when larvae begin to hatch, cease handling eggs or removing them from egg cups until all have hatched. At that time, if enough larvae are still alive, select 40 at random and transfer immediately to a larval growth chamber to determine survival and growth of second generation. Count and discard entire egg-cup groups not used for survival and growth studies.

From early spawned eggs in each duplicate tank, use larvae for 30- and 60-day growth and survival exposures. Plan the distribution of eggs for hatchability tests so that a new group of larvae is ready to be tested as soon as possible after the previously-tested group leaves the larval chambers. Record mortality and larval length at 30 and 60 days after hatching. Weigh them when the larval test is ended. Do not feed fish (larvae, juveniles, or adults) for 24 hr before weighing.

Transfer 50 of the 60-day post-hatching fish from each growth chamber to the corresponding spawning chamber and adjust the photoperiod to December 1. Follow procedures used for the F_1 generation to determine survival of eggs, larvae, and juveniles of the F_2 generation. Cease testing adult fish on completion of spawning. Continue post-hatching study to 60 days.

Use fish and eggs obtained from the test for physiological, biochemical, histological, and other tests that may indicate certain toxicant-related effects.

Record the following data for each test tank and the controls: total number and length of normal and deformed individuals at the end of 30 and 60 days for each generation; total length, weight, and number of each sex, both normal and deformed, at the end of the tests; mortality during tests; number of spawnings and eggs produced in each and total egg production by each generation; percentage of eggs hatching; number and percentage of larvae surviving and growth of fry as well as deformities produced. For additional information on the life cycle of the fathead minnow and flow-through life-cycle tests, consult other sources.[28–30,82–88]

3) Chemical analyses—Analyze test concentrations periodically to determine exposure as described in Section 801E.3*d* and for brook trout. Measure other variables and make chemical analyses as described for brook trout bioassays.

Partial-life-cycle bioassays have been made with the bluegill, *Lepomis macrochirus,* and the flagfish, *Jordanella floridae.*[31,32,89–93]

c. Bioassay for sheepshead minnow, Cyprinodon variegatus:

1) Short-term tests—Handle, hold, acclimate, and culture test fish as described in Section 810A. Make tests as described in Sections 801C and E. Use 10 or 20 fish in each test concentration, depending on the variability of test fish and precision desired. Conduct range-finding tests and carry out definitive acute studies at indicated concentrations. Where possible, test two different age classes, for example, larvae and juveniles, to obtain best possible data.

2) Life-cycle tests—Begin with adult fish or preferably with eggs. Secure eggs either by natural spawning or by hormone-induced spawning as described in Section 810A.4*b*1). For natural spawning, place spawning pairs in individual aquariums at least 30 × 18 × 20 cm. Keep water temperature above 22 C, preferably at 30 C, with salinities above 15 ‰. Set up five or six spawning aquariums for each test concentration and controls. Use breeding fish, all from the same stock, that have been kept in holding tanks for at least 2 wk, during which deaths were less than 2%. Feed spawners a combination of frozen adult brine shrimp and dry trout food. Maintain water flow through spawning aquariums at 6 to 10 tank volumes/day. Use natural seawater filtered to remove planktonic larvae 15 μm and larger.

Use a test apparatus similar to that used for other fish. Use spawning chambers that have a coarse screen with 2-mm-square openings in the bottom over a net tray, as described in Section 810A.4*b*1).

When eggs become available, remove them from spawning chambers and place in each of eight hatching chambers for each concentration being tested and the controls. Start toxicant dosing in exposure chambers before hatching chambers are placed inside them. Construct hatching chambers by cementing a 9-cm-wide strip of 500-μm plastic* screen around a petri dish. Place them in the 90- × 30- × 30-cm exposure chambers in 7 cm of water with flow-through of the toxicant. As eggs hatch, feed fry with newly hatched brine shrimp nauplii. Clean screens on incubation cups or chambers daily. Check and record daily survival of embryos and fry, which constitute the parental stock, F_1. On the first day after hatching, remove each chamber and photograph. Count and measure fry as described in Section 801E.3*e*. During the first 2 wk feed them newly hatched brine shrimp nauplii. During the following 2 wk supplement this diet with dry trout pellets or dry mollie flakes. At 4 wk count and measure fish by the photographic method (Section 801E.3*e*) and reduce the number to 50 for each test concentration and controls. Record length, weight, condition, and number of living, deformed, and dead fish remaining. Determine total number dying in each test concentration and controls. Preserve

*Nitex or equivalent.

specimens for future tests or discard. Place the 50 chosen fish, 25 each, in growth chambers having a glass bottom and provisions for drawing the water level down to 1 to 2 cm. Feed a mixed diet of brine shrimp and dry trout food twice daily and examine daily for dead specimens. At 8 wk measure again by the photographic method. Twice daily, feed dry food supplemented with frozen adult brine shrimp until maturity. Check each batch of food for pesticides, PCB's, and other toxicants. Clean all exposure aquariums and spawning and hatching chambers as previously described two to three times per week. Siphon out all wastes.

As fish mature to adults and approach spawning, place separate pairs in spawning chambers, five from each duplicate exposure chamber, i.e., 10 pair for each test concentration and controls, and continue exposure. Count, measure, and weigh all unused fish from each duplicate exposure chamber. Record number deformed and dying in each test concentration and controls, condition of fish, and other pertinent data. Preserve some for whole-body analyses. As fertilized eggs are produced, remove at a specified time daily and place 25 each in hatching chamber as for the F_1 generation. Keep a record of all eggs produced in each chamber, time required to hatch, hatching success, and survival of embryos. Test those not placed in hatching chambers for fertility and record percent fertile and percent from which fry did not emerge.

If no spawning occurs in higher toxicant concentrations, transfer eggs from controls and incubate in the higher concentrations. Further, place eggs from the high test concentrations in control aquariums.

Keep pairs in each spawning chamber until all needed eggs have been obtained. At termination, measure and weigh spawning pairs and record all data. Preserve for toxicant analyses.

Expose eggs in hatching chambers in their respective duplicate exposure chambers for each test concentration and controls as before. Count and measure by photographic method as for the F_1 generation. Feed fish and record results. At the end of 4 wk terminate the test. Weigh and measure all fish; record number of deformed fish and determine number that died. Preserve for histological and accumulation studies. Determine different effects of each test concentration and calculate indicated safe levels. Analyze, handle, and report data as in Section 801F. During tests, periodically record temperature, DO, pH, and salinity. If possible, make chemical analyses at the beginning, at various times during exposure, and on completion of tests. Analyze lots of 10 fish from highest and lowest exposure concentrations and from controls for toxicant accumulation. Analyze dilution water and toxicant at beginning and end of test.

d. Bioassay procedures for Atlantic silverside, Menidia menidia:

1) Short-term tests—Perform range-finding and short-term acute tests with the silverside as described in Sections 801C through E. Handle, hold, transfer, and feed the silverside during the tests as described in Section 810A. Perform short-term tests as described in Sections 801C through F.

2) Partial-life-cycle tests—Because of their extreme sensitivity, collect adult silverside carefully to prevent injury. Bring adult silversides into the laboratory in early spring. Dip them from the transfer tank with a beaker or other suitable container, then submerge it in the circular holding tanks to free the fish. Use holding tanks described in Section 801D.4*a*. Feed dry food, adult brine shrimp, copepods, and minced clams or mussels. Siphon all unused food and waste from tank daily. When it appears that fish are ready to spawn, strip sufficient females into a 200-mm-diam culture dish containing a 1-cm depth of seawater to provide an estimated

400 eggs. Fertilize with milt from ripe males.

Allow 15 min for fertilization and suspend the egg mass from a nylon or polyethylene string. After eggs have been attached to string, suspend in flow-through treatment aquariums with dimensions 30 × 45 × 30 cm. Maintain flow rate sufficient to keep eggs well aerated. Prepare additional aquariums for the other four concentrations of toxicant and for controls. Insure that eggs are suspended, not touching sides or bottom. Place one 500-μm plastic† screen over each drain tube.

Maintain hatching jars at between 20 and 22 C and a salinity between 25 and 30 ‰ or higher.

With a pipet, transfer larvae that will be exposed continuously to toxicants to rearing aquariums measuring 30 × 45 × 30 cm. Do not exceed a stocking density of 100 larvae per aquarium. Examine larvae not used in the rearing aquariums for deformities and then preserve. Shortly after larvae have been added to rearing tanks, determine number and length by the photographic method.

Begin feeding newly hatched brine shrimp nauplii immediately and feed to excess. Observe aquarium daily for dead or deformed larvae. Remove with a large-bore pipet and record number and apparent condition. Clean tank three times per week or daily if necessary, using a siphon. Cover outlet of each rearing aquarium with a screen having a mesh that will retain the larval fish and brine shrimp nauplii. Arrange standpipes so that the water levels in the aquariums can be drawn down for a photographic count of the larvae.

Larvae in hatching jars usually emerge before the 12th day. At the 12th day, examine egg clumps and if it appears that no more will hatch, discontinue the procedure, count and record live active larvae, deformed larvae, and unhatched eggs. Examine egg clumps thoroughly and record number of viable eggs from which larvae failed to emerge. Note developmental stage.

Because of difficulty in maintaining silversides in the laboratory for long intervals, conduct tests within a certain segment of the life cycle. Partial-life-cycle tests beginning with ripening adults or recently fertilized eggs and extending through the juvenile stage are preferred.

†Nitex or equivalent.

810 D. References

1. KEMP, H.T., J.P. ABRAMS & R.C. OVERBECK. 1971. Water Quality Criteria Data Book Vol. 3. Effects of Chemicals on Aquatic Life, Selected Data from the Literature Through 1968. EPA Water Pollution Center, Research Ser. 18050GWV05/71, U.S. Government Printing Office, Washington, D.C.
2. COMMITTEE ON METHODS FOR TOXICITY TESTS WITH AQUATIC ORGANISMS. 1975. Methods for Acute Toxicity Tests with Fish, Macroinvertebrates, and Amphibians, Ecol. Res. Ser. EPA-660/3-75-009.
3. HUNN, J.B., R.A. SCHOETTGER & E.W. WHEALDON. 1968. Observations on the handling and maintenance of bioassay fish. *Progr. Fish-Cult.* 30:164.
4. INNES, W.T. 1966. Exotic Aquarium Fishes, 19th ed. Metaframe Corp., Maywood, N.J.

5. Lewis, W.M. 1962. Maintaining Fishes for Experimental and Instructional Purposes. Southern Ill. Univ. Press., Carbondale.
6. Hessleberg, R.J. & R.M. Burress. 1967. Investigations in Fish Control. No. 21. Labor-Saving Devices for Bioassay Laboratories. Bur. Sport Fish. Wildl. U.S. Dep. Interior.
7. Massman, W.H. 1954. Marine fishes in fresh and brackish waters of Virginia rivers. *Ecology* 35:75.
8. Raney, E.C. & W.H. Massman. 1953. The fishes of the tidewater section of the Pamunkey River, Virginia. *J. Wash. Acad. Sci.* 43:424.
9. Raney, E.C. 1952. The life history of the striped bass. *Roccus saxatilis,* Walbaum. *Bull. Bingham Oceanogr. Coll.* 14:5.
10. Spotte, S.H. 1970. Fish and Invertebrate Culture. John Wiley and Sons, Inc., New York, N.Y.
11. Garstang, W. 1900. Preliminary experiments on the rearing of seafish larvae. *J. Mar. Biol. Ass., U.K.* 6:76.
12. Houde, E.D. 1973. Some recent advances and unresolved problems in the culture of marine fish larvae. *Proc. World Maricult. Soc.* 3:83.
13. Shelbourne, J.E. 1964. The artificial propagation of marine fish. *Advan. Mar. Biol.* 2:1.
14. Blaxter, J.H.S. 1968. Rearing herring larvae to metamorphosis and beyond. *J. Mar. Biol. Ass. U.K.* 48:17.
15. Kuo, C., Z.H. Shehadeh & K.K. Milisen. 1973. A preliminary report on the development, growth and survival of laboratory reared larvae of the grey mullet, *Mugil cephalus* L. *J. Fish. Biol.* 5:459.
16. Kramer, D. & J.R. Zweifel. 1970. Growth of anchovy larvae, *Engraulis mordax,* Girard in the laboratory as influenced by temperature. *Rep. Calif. Coop. Oceanic Fish. Invest.* 14:84.
17. O'Connell, C.P. & L.P. Raymond. 1970. The effect of food density on survival and growth of early post yolk sac larvae of the northern anchovy, *Engraulis mordax,* Girard, in the laboratory. *J. Exp. Mar. Biol. Ecol.* 5:187.
18. Saskena, V.P. & E.D. Houde. 1972. Effect of food level on the growth and survival of laboratory reared larvae of the bay anchovy, *Anchoa mitchelli,* Valenciennes and scaled sardine, *Harengula pensacolae,* Goode and Bean. *J. Exp. Mar. Biol. Ecol.* 8:249.
19. Qasim, S.Z. 1959. Laboratory experiments on some factors affecting the survival of marine teleost larvae. *J. Mar. Biol. Ass. India* 1:13.
20. Hildebrand, S.F. 1923. Notes on habits and development of eggs and larvae of the silversides *Menidia menidia* and *Menidia beryllina. U.S. Bur. Fish. Bull.* 38:113.
21. National Academy of Sciences. 1973. Nutrient requirements of trout, salmon and catfish. 11:1. Publ. Off. NAS, Washington, D.C.
22. Stalnaker, C.B. & R.E. Gresswell. 1974. Early life history and feeding of young mountain white fish. EPA 660/3-73-019, Off. Research and Development EPA, U.S. Government Printing Office, Washington, D.C.
23. Carlson, A.R. & J.G. Hale. 1972. Successful spawning of largemouth bas *Micropterus salmoides* (Lacepede) under laboratory conditions. *Trans. Amer. Fish. Soc.* 101:539.
24. Hokanson, K.E.F., J.H. McCormick, B.R. Jones & J.H. Tucker. 1973. Thermal requirements for maturation, spawning, and embryo survival of the brook trout. *Salvelinus fontinalis. J. Fish. Res. Board Can.* 30:975.
25. McCormick, J.H., K.E.F. Hokanson & B.R. Jones. 1972. Effects of temperature on growth and survival of young brook trout. *Salvelinus fontinalis. J. Fish. Res. Board Can.* 29:1107.
26. Hokanson, K.E.F., J.H. McCormick & B.R. Jones. 1973. Temperature requirements for embryos and larvae of the northern pike, *Esox lucius* (Linnaeus). *Trans. Amer. Fish. Soc.* 102:89.
27. Siefert, R.E. 1972. First food of larval yellow perch, white sucker, bluegill, emerald shiner and rainbow smelt. *Trans. Amer. Fish. Soc.* 101:219.
28. Pickering, Q.H. 1974. Chronic toxicity of nickel to the fathead minnow. *J. Water Pollut. Control Fed.* 46:760.
29. Mount, D.I. & C.E. Stephan. 1969. Chronic toxicity of copper to the fathead minnow *(Pimephales promelas)* in soft wa-

ter. *J. Fish. Res. Board Can.* 26:2449.

30. EATON, J.G. 1973. Chronic toxicity of a copper, cadmium and zinc mixture to the fathead minnow (*Pimephales promelas*, Rafinesque). *Water Res.* 7:1723.
31. EATON, J.G. 1974. Chronic cadmium toxicity to the bluegill (*Lepomis macrochirus* Rafinesque). *Trans. Amer. Fish. Soc.* 103:729.
32. SMITH, W.E. 1973. A cyprinodontid fish. *Jordanella floridae*, as a laboratory animal for rapid chronic bioassays. *J. Fish. Res. Board Can.* 30:329.
33. MCKIM, J.M. & D.H. BENOIT. 1974. Duration of toxicity tests for establishing "no effect" concentrations for copper with brook trout *(Salvelinus fontinalis). J. Fish. Res. Board Can.* 31:449.
34. MAY, R.C. 1970. Feeding larval marine fishes in the laboratory: A review. Calif. Mar. Res. Comm., CalCOFI Rep. 14:76.
35. HIRANO, R. & Y. OSHIMA. 1963. Rearing of larvae of marine animals with special reference to their food organisms. *Bull. Jap. Soc. Sci. Fish.* 29:282.
36. HOUDE, E.D. & B.J. PALKO. 1970. Laboratory rearing of the clupeid fish, *Harengula pensacolae*, from fertilized eggs. *Mar. Biol.* 5:354.
37. SAKSENA, V.P., C. STEINMETZ & E.D. HOUDE. 1972. Effect of temperature on growth and survival of laboratory reared larvae of the scaled sardine *Harengula pensacolae*, Goode and Bean. *Trans. Amer. Fish. Soc.* 101:691.
38. LASKER, R. 1964. An experimental study of the effect of temperature on the incubation time, development and growth of Pacific sardine embryos and larvae. *Copeia* (2):399.
39. BOYD, J.F. & R.C. SIMMONS. 1974. Continuous laboratory production of fertile *Fundulus heteroclitus*, Walbaum eggs lacking chorionic fibrils. *J. Fish. Biol.* 6:389.
40. MIDDAUGH, D.P. & J.M. DEAN. 1974. The Toxicity of Cadmium to the Eggs, Larvae and Adults of the Mummichog, *Fundulus heteroclitus* and the Silverside, *Menidia menidia*. Bears Bluff Field Station, EPA (unpublished).
41. BAYLIFF, W.H. 1950. The life history of the silverside (*Menidia menidia* Linnaeus). *Chesapeake Biol. Lab. Publ.* 50:1.
42. MIDDAUGH, D.P. & R.L. YOAKUM. 1974. The use of chorionic gonadotropin to induce laboratory spawning of the Atlantic Croaker, *Micropogon undulatus*, with notes on subsequent embryonic development. *Chesapeake Sci.* 15:110.
43. HOFF, F.H. 1972. Artificial spawning of black seabass, *Centropristis striata*, aided by chorionic gonadotropin hormones. Florida Dep. Natural Resources Marine Research Lab (mimeograph).
44. MIDDAUGH, D.P. & A.C. BADGER. 1974. Laboratory Spawning, Development and Rearing of the Spot, *Leiostomus xanthurus*, Lacépède. Bears Bluff Field Station, EPA (unpublished).
45. HILDEBRAND, S.F. & L. GABLE. 1931. Development and life history of fourteen teleostean fishes at Beaufort, N.C. *Bull. U.S. Bur. Fish.* 46:383.
46. DAWSON, C.E. 1959. A study of the biology and life history of the spot *Leiostomus xanthurus* Lacépède with special reference to South Carolina. *Bears Bluff Lab. Contrib.* 28:1.
47. STEVENS, R.E. 1966. Hormone-induced spawning of striped bass for reservoir stocking. *Progr. Fish-Cult.* 28:19.
48. BLAXTER, J.H.S. 1969. Development: Eggs and larvae. *In* Fish Physiology, 3, Reproduction and Growth, Bioluminescence, Pigments and Poisons. Academic Press, New York, N.Y.
49. SCHIMMEL, S.C., D.J. HANSEN & J. FORESTER. 1974. Effects of Aroclor 1254 on laboratory-reared embryos and fry of sheepshead minnows *(Cyprinodon variegatus). Trans. Amer. Fish. Soc.* 103:582.
50. HANSEN, D.J., S.C. SCHIMMEL & J. FORESTER. 1973. Aroclor 1254 in eggs of sheepshead minnows: Effect on fertilization success and survival of embryos and fry. *Proc. 27th Annu. Conf. S.E. Ass. Game Fish. Comm.*:420.
51. BIGELOW, H.B. & W.C. SCHROEDER. 1953. Fishes of the Gulf of Maine. *U.S. Bur. Fish. Bull.* 53.
52. RUBINOFF, I. 1958. Raising the atherinid fish *Menidia menidia* in the laboratory. *Copeia* (2):146.
53. RUBINOFF, I. & E. SHAW. 1960. Hybridization in two sympatric species of atherinid fishes, *Menidia menidia* Linnaeus and

Menidia beryllina, Cope. *Amer. Mus. Natur. Hist.* No. 1999:1.

54. WILLFORD, W.A. 1967. Investigations in Fish Control, Toxicity of 22 Therapeutic Compounds to Six Fishes. U.S. Bur. Sport Fish Wildl. Pap. No. 18.
55. RUCKER, R.R. & K. HODGEBOOM. 1953. Observations on gas-bubble disease of fish. *Progr. Fish-Cult.* 15:24.
56. MARKING, L.L. & V.K. DAWSON. 1973. Toxicity of quinaldine sulfate to fish. *U.S. Bur. Sport Fish. Wildl. Invest. Fish. Control* 48:1.
57. BRUNGS, W.A. & D.I. MOUNT. 1967. A device for continuous treatment of fish in holding chambers. *Trans. Amer. Fish. Soc.* 96:55.
58. SNIESZKO, S.F. 1970. A Symposium on Diseases of Fishes and Shellfishes. Spec. Publ. 5, American Fisheries Soc.
59. HOFFMAN, G.L. 1967. Parasites of North American Freshwater Fishes. Univ. of Calif. Press, Berkeley and Los Angeles.
60. VAN DUIJN, C., JR. 1973. Diseases of Fishes, 3rd ed. Charles C. Thomas Co., Springfield. Ill.
61. REICHENBACK-KLINKE, H. & E. ELKAN. 1965. The Principal Diseases of Lower Vertebrates. Academic Press, London and New York.
62. DAVIS, H.S. 1953. Culture and Diseases of Game Fishes. Univ. of Calif. Press, Berkeley and Los Angeles.
63. HOFFMAN, G.L. & F.P. MEYER. 1974. Parasites of Freshwater Fishes: A Review of Their Control and Treatment. T.F.H. Publications Inc., Ltd., Neptune City, N.J.
64. NIGRELLI, R.F. & G.D. RUGGIERI. 1966. Enzootics in the New York aquarium caused by *Cryptocaryon irritans* Brown, 1951 (=*Ichthyophthirius marinus* 1961) a histophagous ciliate in the skin, eyes and gills of marine fishes. *Zoologica* 51:97.
65. SINDERMANN, C.J. 1970. Principal Diseases of Marine Fish and Shellfish. Academic Press, New York, N.Y.
66. HOUDE, E.D. & A.J. RAMSEY. 1971. A culture system for marine fish larvae. *Progr. Fish-Cult.* 33:156.
67. HOFFMAN, G.L. 1974. Disinfection of contaminated water by ultraviolet irradiation with emphasis on whirling disease *(Myxosoma cerebralis),* and its effects on fish. *Trans. Amer. Fish. Soc.* 103:541.
68. MCKIM, J.M. & D.A. BENOIT. 1971. Effect of long-term exposures to copper on survival, reproduction and growth of brook trout, *Salvelinus fontinalis* (Mitchell). *J. Fish. Res. Board Can.* 28:655.
69. MOUNT, D.I. 1968. Chronic toxicity of copper to fathead minnows, *Pimephales promelas* Rafinesque. *Water Res.* 2:215.
70. KOPPERDAHL, F.R. 1976. Guidelines for Performing Static Acute Toxicity Fish Bioassays in Municipal and Industrial Waste Waters. Report to Calif. State Resource Control Board, Sacramento.
71. BENOIT, D.A. & F.A. PUGLISI. 1973. A simplified flow-splitting chamber and siphon for proportional diluters. *Water Res.* 7:1915.
72. BENOIT, D.A. 1974. Artificial laboratory spawning substrate for brook trout (*Salvelinus fontinalis* Mitchell). *Trans. Amer. Fish. Soc.* 103:144.
73. ALLISON, L.N. 1951. Delay of spawning in eastern brook trout by means of artificially prolonged light intervals. *Progr. Fish-Cult.* 13:111.
74. CARSON, B.W. 1955. Four years progress in the use of artificially controlled light to induce early spawning of brook trout. *Progr. Fish-Cult.* 17:99.
75. FABRICIUS, E. 1953. Aquarium observations on the spawning behavior of the char, *Salmo alpinus. Rep. Inst. Freshwater Res. Drottingholm* 34:14.
76. HALE, J.G. 1968. Observations on brook trout, *Salvelinus fontinalis* spawning in 10-gallon aquaria. *Trans. Amer. Fish. Soc.* 97:299.
77. HENDERSON, N.E. 1962. The annual cycle in the testes of the eastern brook trout. *Salvelinus fontinalis* (Mitchell). *Can. J. Zool.* 40:631.
78. HENDERSON, N.E. 1963. Influence of light and temperature on the reproductive cycle of the eastern brook trout *Salvelinus fontinalis* (Mitchell). *J. Fish. Res. Board Can.* 20:859.
79. HOOVER, E.E. & H.E. HUBBARD. 1937. Modification of the sexual cycle in trout by control of light. *Copeia* (4):206.
80. PYLE, E.A. 1969. The effect of constant light or constant darkness on the growth and sexual maturity of brook trout. Fish.

Res. Bull. No. 31. The nutrition of trout, Cortland Hatchery Rep. No. 36:13.

81. WYDOSKI, R.S. & E.L. COOPER. 1966. Maturation and fecundity of brook trout from infertile streams. *J. Fish. Res. Board Can.* 23:623.
82. BRUNGS, W.A. 1969. Chronic toxicity of zinc to the fathead minnow, *Pimephales promelas* Rafinesque. *Trans. Amer. Fish. Soc.* 98:272.
83. BRUNGS, W.A. 1971. Chronic effects of low dissolved oxygen concentrations on the fathead minnow *(Pimephales promelas). J. Fish. Res. Board Can.* 28:1119.
84. CARLSON, D.R. 1967. Fathead minnow, *Pimephales promelas* Rafinesque, in the Des Moines River, Boone County, Iowa and the Skunk River drainage, Hamilton and Story Counties, Iowa. *Iowa State J. Sci.* 41:363.
85. MARKUS, H.C. 1934. Life history of the fathead minnow *(Pimephales promelas). Copeia* (3):116.
86. MOUNT, D.I. & C.E. STEPHAN. 1967. A method for establishing acceptable toxicant limits for fish—malathion and the butoxyethanol ester of 2,4-D. *Trans. Amer. Fish. Soc.* 96:185.
87. PICKERING, Q.H. & T.O. THATCHER. 1970. The chronic toxicity of linear alkylate sulfonate (LAS) to *Pimephales promelas,* Rafinesque. *J. Water Pollut. Control Fed.* 42:243.
88. PICKERING, Q.H. & W.N. VIGOR. 1965. The acute toxicity of zinc to eggs and fry of the fathead minnow. *Progr. Fish-Cult.* 27:153.
89. BREDER, C.M. 1936. The reproduction behavior of North American sunfish. *Zoologica* 21:1.
90. EATON, J.G. 1970. Chronic malathion toxicity to the bluegill, *Lepomis marcrochirus. Water Res.* 4:673.
91. MCCOMISH, T.S. 1968. Sexual differentiation of bluegills by the urogenital opening. *Progr. Fish-Cult.* 30:28.
92. FOSTER, N.R., J. CAIRNS, JR. & R.L. KAESLER. 1969. The flagfish *Jordanella floridae,* as a laboratory animal for behavioral bioassay studies. *Proc. Acad. Natur. Sci. Philadelphia* 121(5):129.
93. BENOIT, D.A. 1975. Chronic effects of copper on survival, growth and reproduction of the bluegill *(Lepomis macrochirus). Trans. Amer. Fish. Soc.* 104:353.

PART 900
MICROBIOLOGICAL EXAMINATION OF WATER

901 INTRODUCTION

901 A. General Discussion

The following sections describe procedures for making microbiological examinations of water samples to determine sanitary quality. The methods are intended to indicate the degree of contamination with wastes. They are the best technics currently available; however, their limitations must be understood thoroughly.

Tests for detection and enumeration of indicator organisms, rather than of pathogens, are used. The coliform group of bacteria, as herein defined, is the principal indicator of suitability of a water for domestic, dietetic, or other uses. The cultural reactions and characteristics of this group of bacteria have been studied extensively.

Experience has established the significance of coliform group density as a criterion of the degree of pollution and thus of the sanitary quality of the sample. The significance of the tests and the interpretation of results are well authenticated and have been used as a basis for standards of bacteriological quality of water supplies.

The membrane filter technic, which involves a direct plating for detection and estimation of coliform densities, is as effective a method as the multiple-tube fermentation test for detection of bacteria of the coliform group. Modification of details of this method, particularly of the culture medium, has made the results comparable with those given by the multiple-tube fermentation procedure. Although there are limitations in the application of the membrane filter technic for the examination of all types of water, it is equivalent when used with strict adherence to these limitations and to the specified technical details. Thus, two standard methods are presented for the detection of bacteria of the coliform group.

It is customary to report results of the coliform test by the multiple-tube fermentation procedure as a Most Probable Number (MPN) index. This is merely an index of the number of coliform bacteria that, more probably than any other number, would give the results shown by the laboratory examination. It is not an actual enumeration of coliform bacteria. By contrast, direct plating methods such as the membrane filter procedure permit a direct count of coliform colonies. In both procedures coliform density is reported conventionally as the MPN or membrane filter count per 100 mL. Either procedure is a valuable tool for appraising the sanitary quality of water and the effectiveness of treatment processes.

Increasing attention to the potential value of fecal streptococci as indicators of significant fecal pollution of water has led to inclusion of methods for detection and enumeration of such microorganisms. A multiple-tube dilution and a membrane filter procedure are included as standard as well as a tentative pour plate method.

Since the 13th edition, additional standard methods for the differentiation of that segment of the coliform group designated as fecal coliforms have been included. Such differentiation generally is considered of limited value in assessing drinking water quality because the presence of either type of coliform bacteria renders the water potentially unsatisfactory and unsafe. Investigations strongly indicate that the portion of the coliform group present in the gut and feces of warm-blooded animals generally includes organisms capable of producing gas from lactose in a suitable culture medium at 44.5 ± 0.2 C. Inasmuch as coliform organisms from other sources generally cannot produce gas under these conditions, this criterion is used to define the fecal component of the coliform group. Both the multiple-tube dilution technic and the membrane filter procedure have been modified to incorporate incubation in confirmatory tests at 44.5 C to provide estimates of the density of fecal organisms, as defined. This differentiation will yield valuable information concerning the possible source of pollution in water, and especially the remoteness of this pollution, because the *nonfecal* members of the coliform group may be expected to survive longer than the *fecal* members in the unfavorable environment provided by the water.

The Standard Plate Count provides an approximate enumeration of total numbers of bacteria multiplying at 35 C that may yield useful information about water quality and may provide supporting data on the significance of coliform test results. The Standard Plate Count is useful in judging the efficiency in operation of various water treatment processes and may have significant application as an in-plant control test. It also is valuable for checking quality of finished water in a distribution system as an indicator of microbial regrowth and sediment buildup in slow-flow sections and dead ends.

Experience in the shipment of un-iced samples by mail indicates that changes in type or numbers of bacteria during such shipment for even limited periods of time are not negligible. Therefore, refrigeration during transportation is recommended, particularly when ambient air temperature exceeds 13 C.

Procedures for the isolation of certain pathogenic bacteria are presented. These procedures are tedious and complicated and are not recommended for routine use. Likewise, tentative procedures for enteric viruses are included but their routine use is not advocated now.

Examination of routine bacteriological samples of water cannot be regarded as providing complete or final information concerning water quality. Bacteriological results must be considered in the light of information available concerning the sanitary conditions surrounding the source of any particular sample. Precise evaluation of the quality of a water supply can be made only when the results of laboratory examinations are interpreted in the light of sanitary survey data. Therefore, the results of the examination of a single sample from a given source must be considered inadequate. When possible, evaluation of the quality of a water supply must be based on examination of a series of samples collected over a known and protracted period of time.

Pollution problems of tidal estuaries and other bodies of saline water has focused attention on necessary modification of existing bacteriological technics so that they may be used effectively in the examination of samples from such sources. In the following section, application of specific technics to saline water has not been discussed because the methods used for fresh waters also can be used satisfactorily with saline waters.

Methods for examination of the waters of swimming pools and other bathing places are included. The standard procedures for the plate count, fecal coliforms, and fecal streptococci are identical with those used for other waters. Procedures for *Staphylococcus* and *Pseudomonas aeruginosa,* organisms commonly associated with the upper respiratory tract or the skin, are included on a tentative basis.

Procedures for aquatic fungi, actinomycetes, and nematodes are included.

Sections have been added dealing with rapid methods for coliform testing and for the recovery of stressed organisms. Because of increased interest and concern with analytical quality control, this section has been completely rewritten and significantly expanded.

The bacteriological methods in Part 900, developed primarily to permit prompt and rapid examination of water samples, have been considered frequently to apply only to routine examinations. These same methods, however, are the basic technics required for research investigations in sanitary bacteriology and water treatment. Their value in routine studies must not be allowed to overshadow or limit their value in research studies. Similarly, all technics should be the subject of investigations to establish their specificity, improve their procedural details, and expand their application to the measurement of the sanitary quality of water supplies or polluted waters.

901 B. U.S. EPA Regulations for Drinking Water Quality

In the United States the quality of public water supplies is judged in terms of the 1975 U.S. Environmental Protection Agency (EPA) Interim Primary Drinking Water Regulations.[1] These regulations provide for a minimum number of samples to be examined per month and establish the maximum number of coliform organisms (Maximum Contaminant Level, MCL) allowable per 100 mL of finished water. They also require that analyses be made in a certified laboratory.[2]

1. Sampling

Make bacteriological examinations on samples collected at representative points throughout the distribution system. Select the frequency of sampling and the location of sampling points to insure accurate determination of the bacteriological quality of the treated water supply, which may be controlled in part by the known quality of the untreated water and thus by the need for treatment. Base the minimum number of samples to be collected and examined each month on the population served by the supply. It is important to examine repetitive samples from a designated point, as well as samples from a number of widely distributed sampling points. Take samples at reasonably evenly spaced time intervals. Consider daily samples collected after an unsatisfactory sample has been taken as special samples and do not count in the total number of samples examined monthly.

2. Application

For the multiple-tube fermentation technic the maximum number of allowable coliform organisms is prescribed in terms of standard portion volume (10 mL or 100 mL) and the number of portions examined. The absence of gas in all tubes, when five 10-mL portions are examined by the fermentation tube method (equivalent to an MPN of less than 2.2 coliforms/100 mL), is generally interpreted to indicate that the single sample meets the standards. A positive Confirmed Test for coliform organisms in three or more tubes (10-mL portions) or five portions (100-mL portions) indicates the need for immediate remedial action and additional examinations. Analyze repeat samples of finished water from the same location that consistently show 3 or more positive 10-mL portions by the Completed Test. Similarly, for the membrane filter technic, the standard portion volume is 100 mL, the quality limit is 1 coliform colony/100 mL, and the action limit is more than 4 coliform colonies/100 mL. Analyze repeat samples of finished water from the same location that consistently give positive results by the verification procedure. For either technic collect daily samples from the sampling point and examine promptly until the results obtained from at least two consecutive samples show the water to be of satisfactory quality.

Although the Standard Plate Count is not required in the EPA Interim Primary Drinking Water Regulations, its use may be required in conjunction with modification of the turbidity limit.

These standards also specify limiting concentrations of chemical and physical constituents of water as related to its safety and potability.

The World Health Organization has established International Standards of Drinking Water Quality.[3] These are similar to the U.S. drinking water standards, but have been modified and liberalized to apply to water supply conditions in all parts of the world.

3. References

1. U.S. ENVIRONMENTAL PROTECTION AGENCY. 1976. National Interim Primary Drinking Water Regulations. EPA-570/9-76-003, Washington, D.C.
2. U.S. ENVIRONMENTAL PROTECTION AGENCY. 1978. Manual for the Interim Certification of Laboratories Involved in Analyzing Public Drinking Water Supplies. EPA-600/8-78-008, Washington, D.C.
3. WORLD HEALTH ORGANIZATION. 1971. International Standards for Drinking-Water, 3rd ed. WHO, Geneva.

902 LABORATORY QUALITY ASSURANCE

The growing emphasis on water quality standards, enforcement, and monitoring requires the establishment of a quality assurance program to substantiate the validity of analytical data.

A laboratory quality assurance program is the orderly application of the practices necessary to remove or reduce errors that may occur in any laboratory operation, caused by personnel, equipment, supplies, sampling procedures, and analytical methodology.

The program must be practical and integrated. It should require only a reasonable amount of time or it will be bypassed. When properly administered, a balanced, conscientiously applied quality assurance program will yield uniformly high-quality data without interfering with the primary analytical functions of the laboratory. Detailed descriptions of quality control practices are available.[1,2] Generally, 15% of total analyst time should be spent on different aspects of a quality assurance program. All intralaboratory and interlaboratory quality control practices should be documented and the records should be available for inspection. The quality assurance guidelines discussed below are recommended strongly as a minimal program for a microbiology laboratory.

902 A. Intralaboratory Quality Control

All laboratories have some intralaboratory quality control practices that have evolved from common sense and the principles of controlled experimentation. Special problems exist in microbiology because analytical standards, known additions, and reference samples usually are not available. Personal judgment is required more frequently. An effective program must control all factors, from sample collection through data reporting, that can influence the results. The factors include sampling technics, facilities, personnel, equipment, supplies, media, and analytical test procedures. It is especially important that laboratories performing only a limited amount of microbiological testing exercise strict quality control.

1. Laboratory Operations

a. Ventilation: Plan well-ventilated laboratories that can be maintained free of dust, drafts, and extreme temperature changes. Central air conditioning is recommended to reduce contamination, permit more stable operation of incubators, and decrease moisture problems with media and analytical balances.

b. Space utilization: Design and operate the laboratory to minimize through traffic and visitors. Provide a separate area for preparing and sterilizing media, glassware, and equipment. Use a special work area such as a vented laminar-flow hood for dispensing and preparing sterile media, transferring microbial cultures, or working with pathogenic materials. In smaller laboratories it may be necessary, although undesirable, to carry out these activities in the same room.

c. Laboratory bench areas: Provide a minimum of 2 m linear bench space per analyst and additional areas for preparation and support activities. For stand-up work, use bench tops 90 to 97 cm high and 70 to 76 cm deep. For sit-down activities such as microscopy and plate counting, provide benches 75 to 80 cm high. Specify bench tops of stainless steel, epoxy plastic, or other smooth, impervious surface that is inert, corrosion-resistant, and has a minimum number of seams. Install even, glare-free lighting with about 1,000 lux intensity at the working surface.

d. Walls and floors: Cover walls with a smooth finish that is easily cleaned and disinfected. Specify floors of sealed, smooth concrete, vinyl, asphalt tile, or other impervious, washable surface.

e. Air monitoring: Maintain high standards of cleanliness in work areas. Monitor air and bench tops with RODAC plates, air density plates, or the swab method.[2]

f. Laboratory cleanliness: Regularly clean laboratory rooms and wash benches, shelves, floors, and windows. Wet-mop floors and treat with a disinfectant solution; do not sweep or dry-mop. Wipe bench tops and treat with a disinfectant before and after use. Do not permit laboratory to become cluttered.

g. Personnel: Ideally, bacteriological testing should be done by a professional microbiologist. If that is not possible, have a professional microbiologist or trained analyst available for guidance and assistance.

Clearly define work assignments. Train the analyst in basic laboratory procedures. The supervisor periodically should review procedures of sample collecting and handling, media and glassware preparation, sterilization, routine testing procedures, counting, data handling, and quality control technics to identify and eliminate problems. Effective personnel training produces greater technical competence and improved laboratory results.

h. Monitoring laboratory equipment and instrumentation: Verify that each item of equipment meets the user's needs for accuracy and precision. Perform equipment maintenance on a regular basis as recommended by the manufacturer or obtain preventive maintenance contracts on autoclave and balances whenever economically feasible. Directly record all quality control checks in a permanent log book.

Use the following equipment control procedures:

1) Thermometer/temperature-recording instruments—Check accuracy of thermometers or recording instruments semiannually against a certified National Bureau of Standards (NBS) thermometer or one traceable to NBS and conforming to NBS specifications. For general purposes use thermometers graduated in increments of 0.5 C or less. For a 44.5 C water bath, use a submersible thermometer graduated to at least 0.2 C. Record temperature check data in a quality control log. Mark NBS calibration corrections on each thermometer used with an incubator, refrigerator, or freezer. When possible, equip incubators and water baths with temperature-recording instruments providing a continuous record of operating temperature.

2) Balance—Wipe balance before and after each use with a camel's hair or polonium brush. Clean balance pans after each use and wipe spills up immediately with a damp towel. Inspect weights with each use and discard if corroded. Check weights monthly against certified weights. For weighing 2 g or less, use an analytical balance with a sensitivity less than 1 mg at a 10-g load. For larger quantities, use a pan balance with a sensitivity of 0.1 g at a 150-g load. Perform preventive maintenance or obtain on contract on an annual basis.

3) pH meter—Standardize pH meter with at least two standard buffers (pH 4.0, 7.0, or 10.0) and compensate for temperature before each series of tests. Date buffer solutions when opened and check monthly against another pH meter if available. See Section 423.

4) Water deionization unit—A deionization column produces a good grade of pure water and when combined with filtration and activated carbon in a recirculating system can produce a water of excellent quality (ultra-pure water). Deionization systems tend to produce the same quality water until resins or activated carbon are near exhaustion and quality abruptly becomes unacceptable.

Monitor deionized water continuously or daily with a conductivity meter and analyze at least annually for trace metals. Replace cartridges at intervals recommended by the manufacturer or as indicated by analytical results. Filter product water through a 0.22-μm-pore-diam membrane filter to remove bacterial contamination. Monitor at least monthly and replace filter when the Standard Plate Count exceeds 1,000/mL.

5) Water still—Stills produce water of a good grade that characteristically deteriorates slowly over time as corrosion, leaching, and fouling occur. These conditions can be controlled with proper maintenance and cleaning. Stills efficiently remove dissolved substances but not dissolved gases or volatile organic chemicals. Freshly distilled water may contain chlorine and ammonia (NH_3). On storage, additional NH_3 and CO_2 are absorbed from the air. Use softened water as the source water to reduce frequency of cleaning the still. Drain and clean still and reservoir according to manufacturer's instructions and usage. Check product water continuously or daily with a conductivity meter. Perform other checks as described in Table 902:I.

6) Reverse osmosis units—Commercial reverse osmosis (RO) units consisting of prefilters and reverse osmosis cartridges are sold as water purification systems. These RO units remove only about 90% of impurities; consequently, do not use the

TABLE 902:I. QUALITY OF PURIFIED WATER USED IN BACTERIOLOGICAL TESTING

Test	Monitoring Frequency	Limit
Chemical tests:		
Conductivity	Continuously or with each use	< 5 μmhos/cm at 25 C
pH	With each use	5.5–7.5
Total organic carbon	Monthly	< 1.0 mg/L
Heavy metals, single (Cd, Cr, Cu, Ni, Pb, and Zn)	Monthly	< 0.05 mg/L
Heavy metals, total	Monthly	< 1.0 mg/L
Ammonia/organic nitrogen	Monthly	< 0.1 mg/L
Free chlorine	With each use	< 0.1 mg/L
Biological tests:		
Standard plate count:		
Freshly distilled or ultra-pure water	Monthly	< 1,000 colonies/mL
Stored or deionized water (See Section 907)	Monthly	< 10,000 colonies/mL
Water suitability test	Annually and for a new source	0.8–3.0 ratio
Use test	Annually and for a new source	Student's $t \leq 2.78$

water for microbiological or other analyses sensitive to water quality. The major use of RO units should be initial cleanup of water before deionization or distillation. Commercial units combining prefiltration, reverse osmosis, and mixed-bed ion-exchange resins are available.

7) Media dispensing apparatus—Check accuracy of dispensing with a graduated cylinder at start of each volume change and periodically throughout extended runs. If the unit is used more than once per day, pump a large volume of hot distilled water through the dispenser to rinse. Correct leaks, loose connections, or malfunctions immediately. At the end of the work day, break apparatus down into parts, wash, rinse with distilled water, and dry. Lubricate parts according to manufacturer's instructions or at least once per month.

8) Hot-air oven—Test performance with commercially available spore strips or spore suspensions quarterly. Monitor temperature with a thermometer accurate in the 160 to 180 C range and record results. Use heat-indicating tape to identify supplies and materials that have been exposed to sterilization temperatures.

9) Autoclave—Record temperatures, pressure, and time for each run. Optimally use a recording thermometer. Check operating temperature weekly with a minimum/maximum thermometer. Test performance with spore strips or suspensions monthly. Use heat-sensitive tape to identify supplies and materials that have been sterilized.

10) Refrigerator—Check and record temperature daily and clean unit monthly. Identify and date materials stored. Defrost as required and discard outdated materials quarterly.

11) Freezer—Check and record temperature daily. A recording thermometer and alarm system are highly desirable. Identify and date materials stored. Defrost and clean semiannually; discard outdated materials.

12) Membrane filter equipment—Before use, assemble filtration units and check for leaks. Coat units with silicone to improve drainage. Discard units if inside surfaces are scratched. Clean filtration assemblies thoroughly after use, wrap, and autoclave. See Section 909A.1*f*.

13) Ultraviolet sterilization lamps—Disconnect unit monthly and clean lamps by wiping with a soft cloth moistened with ethanol. Test lamps quarterly with UV light meter* and replace if they emit less than 70% of initial output or if agar spread plates containing 200 to 250 microorganisms, exposed to the light for 2 min, do not show a count reduction of 99%.

14) Safety cabinet (hood)—Check filters monthly for plugging or dirt accumulation and clean or replace as needed. Once per month expose blood agar plates to air flow for 1 hr. Incubate plates at 35 C for 24 hr and examine for contamination. Disconnect UV lamps and clean monthly by wiping with a soft cloth moistened with ethanol. Check lamps' efficiency as specified above. Inspect cabinet for leaks and rate of air flow quarterly. Use a pressure monitoring device to measure efficiency of hood performance. Maintain as directed by the manufacturer.

15) Water bath—Keep an accurate thermometer immersed in the water bath; monitor and record temperature daily unless a recording thermometer and alarm system are used. Use only stainless steel, plastic-coated, or other corrosion-proof racks. Clean bath as needed.

16) Incubator (air or water jacketed)—Check and record temperature twice daily (morning and afternoon) on the shelf areas in use. For walk-in incubators adjust to a reasonable number of test points. If a glass thermometer is used, submerge bulb and stem in water or glycerine to the stem mark. For best results use a recording thermometer and alarm system. Locate incubator where room temperature is in the range of 16 to 27 C.

17) Microscopes—Use lens paper to clean optics and stage after each use. Cover microscope when not in use.

18) Microscope, fluorescence—Allow only trained technicians to use microscope and light source. Log lamp operation time, lamp efficiency, and alignment. Monitor fluorescence lamp with a light meter and replace when a significant loss in fluorescence is observed. Periodically check lamp alignment, particularly when the bulb has been changed; realign if necessary. Use known positive 4+ fluorescence slides as controls.

i. Maintenance of laboratory supplies:

1) Glassware—With each use, examine glassware and discard items with chipped edges or etched surfaces. Particularly examine screw-capped dilution bottles and flasks for chipped edges that can leak and contaminate the work area or create aerosols. Inspect glassware after washing; if water beads excessively on cleaned surfaces, rewash. Before using a new supply of detergent, test for inhibitory residual. Because some cleaning solutions are difficult to remove completely, spot check batches of clean glassware for pH reaction with bromthymol blue or other indicator, especially if soaked in alkali or acid.

Test for inhibitory residues on glassware: Certain wetting agents or detergents used in washing glassware may contain bacteriostatic or inhibiting substances re-

*Shortwave UV meter, UV Products, Inc., San Gabriel, Calif., or equivalent.

quiring 6 to 12 successive rinsings to remove all traces and insure freedom from residual bacteriostatic action. Use this test for biological examination of glassware where bacteriostatic or inhibitory residues may be present. If prewashed, presterilized plasticware is used, test it for inhibitory residues.

a) Procedure for test—Wash six petri dishes according to usual laboratory practice and designate as Group A.

Wash six petri dishes as above, rinse 12 times with successive portions of distilled water, and designate as Group B.

Rinse six petri dishes with detergent wash water (in use concentration), dry without further rinsing, and designate as Group C.

Sterilize dishes in Groups A, B, and C by the usual procedure. To test presterilized plasticware, set up Group D, consisting of six sterile petri dishes, and proceed.

Add not more than 1 mL of a water sample yielding 50 to 150 colonies and proceed according to the procedure described for the Standard Plate Count. If there is difficulty in obtaining a suitable sample, inoculate three plates of each group with 0.1 mL and the other three plates of each group with 1 mL.

b) Interpretation of results—Difference in average number of colonies of less than 15% on plates of Groups A, B, C, and D indicates that the detergent has no toxicity or inhibitory characteristics or that the presterilized dishes are acceptable.

Difference in colony count of 15% or more between Groups A and B or D and B demonstrates inhibitory residue.

Disagreement in averages of less than 15% between Groups A and B and greater than 15% between Groups A and C indicates that the cleaning detergent has inhibitory properties that are eliminated during routine washing.

2) Utensils and containers for media preparation—Use utensils and containers of borosilicate glass, stainless steel, aluminum, or other noncorrosive and noncontaminating material (see Section 903).

3) Monitoring pure water quality—The quality of water obtainable from a pure water system differs with the system used and its maintenance. Acceptable limits of water quality are given in Table 902:I.

Test for bacteriological quality of distilled water: The test is based on the growth of *Enterobacter aerogenes* in a chemically defined minimal growth medium. The presence of a toxic agent or a growth-promoting substance will alter the 24-hr population by an increase or decrease of 20% or more when compared to a control.

a) Apparatus and material—Use borosilicate glassware and rinse in water freshly redistilled from a glass still before sterilizing it with dry heat; steam sterilization will recontaminate these specially cleaned items. Test sensitivity and reproducibility depend in part on cleanliness of sample containers, flasks, tubes, and pipets. It often is convenient to set aside new glassware for exclusive use in this test. Use any strain of coliform IMViC type − − + + *(E. aerogenes)* obtained from a polluted river or wastewater sample.

b) Reagents—Use only reagents of highest purity. Some brands of potassium dihydrogen phosphate, KH_2PO_4, contain large amounts of impurities. Test sensitivity is controlled in part by the reagent purity. Prepare reagents in water freshly distilled from a glass still.

Sodium citrate solution: Dissolve 0.29 g sodium citrate, $Na_3C_6H_6O_7 \cdot 2H_2O$, in 500 mL water.

Ammonium sulfate solution: Dissolve 0.60 g $(NH_4)_2SO_4$ in 500 mL water.

Salt-mixture solution: Dissolve 0.26 g magnesium sulfate, $MgSO_4 \cdot 7H_2O$; 0.17 g calcium chloride, $CaCl_2 \cdot 2H_2O$; 0.23 g ferrous sulfate, $FeSO_4 \cdot 7H_2O$; and 2.50 g sodium chloride, NaCl, in 500 mL water.

Phosphate buffer solution: Stock phosphate buffer solution (see Media Specifications, Section 905C) diluted 1:25 in water.

Boil all reagent solutions 1 to 2 min to kill vegetative cells. Store solutions in sterilized glass-stoppered bottles in the dark at 5 C for up to several months provided that they are tested for sterility before each use. Because the salt-mixture solution will develop a slight turbidity within 3 to 5 days as the ferrous salt converts to the ferric state, prepare the salt-mixture solution without $FeSO_4$ for long-term storage. To use the mixture, add an appropriate amount of freshly prepared and freshly boiled iron salt. Discard solutions with a heavy turbidity and prepare a new solution. Discard phosphate buffer solution if it becomes turbid.

To prepare a test sample collect 150 to 200 mL water in a sterile borosilicate glass flask and boil for 1 to 2 min. Avoid longer boiling to prevent chemical changes.

c) Procedure—Label five flasks or tubes, A, B, C, D, and E. Add water samples, media reagents, and redistilled water to each flask as indicated in Table 902:II.

Add a suspension of *E. aerogenes* (IMViC type − − + +) of such density that each flask will contain 30 to 80 cells/mL, prepared as directed below. Cell densities below this range result in inconsistent ratios while densities above 100 cells/mL result in decreased sensitivity to nutrients in the test water. Make an initial bacterial count by plating triplicate 1-mL portions from each culture flask in plate count agar. Incubate Flasks A through E at 35 C for 24 ± 2 hr. Prepare final plate counts from each flask, using dilutions of 1, 0.1, 0.01, 0.001, and 0.0001 mL.

d) Preparation of bacterial suspension: Bacterial growth—On the day before making the distilled-water suitability test, inoculate a strain of *E. aerogenes* onto a nutrient agar slant with a slope approximately 6.3 cm long contained in a 125- × 16-mm screw-cap tube. Streak entire agar surface to develop a continuous-growth film and incubate 18 to 24 hr at 35 C.

Harvesting of viable cells—Pipet 1 to 2 mL sterile dilution water from a 99-mL water blank onto the 18- to 24-hr culture.

TABLE 902:II. REAGENT ADDITIONS FOR WATER QUALITY TEST

Media Reagents	Control Test *mL*		Optional Tests *mL*		
	Control A	Unknown Distilled Water B	Food Available C	Nitrogen Source D	Carbon Source E
Sodium citrate solution	2.5	2.5	—	2.5	—
Ammonium sulfate solution	2.5	2.5	—	—	2.5
Salt-mixture solution	2.5	2.5	2.5	2.5	2.5
Phosphate buffer (7.3 ± 0.1)	1.5	1.5	1.5	1.5	1.5
Unknown water	—	21.0	21.0	21.0	21.0
Redistilled water	21.0	—	5.0	2.5	2.5
Total volume	30.0	30.0	30.0	30.0	30.0

Emulsify growth on slant by gently rubbing bacterial film with pipet, being careful not to tear agar; pipet suspension back into original 99-mL water blank.

Dilution of bacterial suspension—Make a 1:100 dilution of original bottle into a second water blank, a further 1:100 dilution of second bottle into a third water blank, then 10 mL of third bottle into a fourth water blank, shaking vigorously after each transfer. Pipet 1.0 mL of the fourth dilution (1:10^6) into each of Flasks A, B, C, D, and E. This procedure should result in a final dilution of the organisms to a range of 30 to 80 viable cells for each milliliter of test solution.

Verification of bacterial density—Variations among strains of the same organism, different organisms, media, and surface area of agar slopes possibly will necessitate adjustment of the dilution procedure to arrive at a specific density range between 30 to 80 viable cells. To establish the growth range numerically for a specific organism and medium, make a series of plate counts from the third dilution to determine bacterial density. Choose proper volume from this third dilution, which, when diluted by the 30 mL in Flasks A, B, C, D, and E, will contain 30 to 80 viable cells/mL. If the procedures are standardized as to slant surface area and laboratory technic, it is possible to reproduce results on repeated experiments with the same strain of microorganism.

Procedural difficulties—Problems in this method may be due to: storage of unknown distilled water sample in soft-glass containers or in glass containers without liners for metal caps; use of chemicals in reagent preparation not of analytical-reagent grade or not of recent manufacture; contamination of reagent by distilled water with a bacterial background (to avoid this, prepare a Standard Plate Count on all media reagents before starting the suitability test, as a check on stock solution contamination); failure to obtain desired initial bacterial concentration or incorrect choice of dilution used to obtain 24-hr plate count; delay in pouring plates; and prolongation of incubation time beyond 26-hr limit, resulting in desensitized growth response.

Calculation—For growth-inhibiting substances:

$$\text{Ratio} = \frac{\text{colony count/mL, Flask B}}{\text{colony count/mL, Flask A}}$$

A ratio of 0.8 to 1.2 (inclusive) shows no toxic substances; a ratio of less than 0.8 shows growth-inhibiting substances in the water sample. For nitrogen and carbon sources that promote growth:

$$\text{Ratio} = \frac{\text{colony count/mL, Flask C}}{\text{colony count/mL, Flask A}}$$

For nitrogen sources that promote growth:

$$\text{Ratio} = \frac{\text{colony count/mL, Flask D}}{\text{colony count/mL, Flask A}}$$

For carbon sources that promote bacterial growth:

$$\text{Ratio} = \frac{\text{colony count/mL, Flask E}}{\text{colony count/mL, Flask A}}$$

Do not calculate the last three ratios when the first ratio indicates a toxic reaction. For these ratios a value above 1.2 indicates an available source for bacterial growth.

e) Interpretation of results—The colony count from Flask A after 20 to 24 hr at 35 C will depend on number of organisms initially planted in Flask A and strain of *E. aerogenes* used. For this reason, run the control, Flask A, for each individual series of tests. However, for a given strain of *E. aerogenes* under identical environmental conditions, the terminal count should be reasonably constant when the initial plant is the same. The difference in initial plant of 30 to 80 will be about threefold larger for the 80 organisms initially planted in

Flask A, provided that the growth rate remains constant. Thus, it is essential that initial colony counts on Flasks A and B be approximately equal.

When the ratio exceeds 1.2, assume that growth-stimulating substances are present. However, this procedure is extremely sensitive and ratios up to 3.0 have little significance in actual practice. Therefore, if the ratio is between 1.2 and 3.0, do not make Tests C, D, and E, except in special circumstances.

Usually Flask C will be very low and Flasks D and E will have a ratio of less than 1.2 when the ratio of Flask B to Flask A is between 0.8 and 1.2. Limiting factors of growth in Flask A are nitrogen and organic carbon. An extremely large amount of ammonia nitrogen with no organic carbon could increase the ratio in Flask D above 1.2, or absence of nitrogen with high carbon concentration could give ratios above 1.2 in Flask E, with a B:A ratio between 0.8 and 1.2.

A ratio below 0.8 indicates that the water contains toxic substances, and this ratio includes all allowable tolerances. As indicated in the preceding paragraph, the ratio could go as high as 3.0 from 1.2 without any undesirable consequences.

Specific corrective measures cannot be recommended in specific instances of defective distillation apparatus. However, make a careful inspection of the distillation equipment and review production and handling of distilled water to help locate and correct the cause of difficulty.

Feedwater to a still often is passed through a deionizing column and a carbon filter. If these columns are well maintained, most inorganic and organic contaminants will be removed. If maintenance is poor, input water may be degraded to a quality lower than that of raw tap water.

The best distillation system is made of quartz or high-silica-content borosilicate glass with special thermal endurance. Tin-lined stills are not recommended. For connecting plumbing, use stainless steel, borosilicate glass, or special plastic pipes made of polyvinyl chloride (PVC). Use stainless steel storage reservoirs and protect them from dust.

f) Test sensitivity—Taking copper as one relative measurement of distilled water toxicity, maximum test sensitivity is 0.05 mg copper/L in a distilled water sample.

4) Reagents—Because reagents are an integral part of microbiological analyses, their quality must be assured. Use only chemicals of ACS or equivalent grade because impurities can inhibit bacterial growth, provide nutrients, or fail to produce the desired reaction. Date chemicals and reagents when received and when first opened for use. Make reagents to volume in volumetric flasks and transfer for storage to good-quality inert plastic or borosilicate glass bottles with borosilicate or polyethylene or other plastic stoppers or caps. Label prepared reagents with name and concentration, date prepared, and initials of preparer. Include positive and negative control cultures with each series of cultural or biochemical tests.

5) Dyes and stains—In microbiological analyses, organic chemicals are used as selective agents (i.e., brilliant green), as indicators (i.e., phenol red lactose), and as microbiologic stains (i.e., Gram stain). Dyes from commercial suppliers vary from lot to lot in percent dye, dye complex, insolubles, and inert materials present. Because dyes for microbiology must be of proper strength and stability to produce correct reactions, use only dyes certified for biological use by the Biological Stain Commission. Check bacteriological stains before use with at least one positive and one negative control culture and record results.

6) Membrane filters and pads—The quality and performance of membrane filters vary with the manufacturer, type, brand, lot number, and storage conditions

(see Section 909A.1) resulting from differences in manufacturing methods, materials, and quality control. Pores in the filters must be distributed uniformly and have a diameter of 0.45 ± 0.02 μm. Filters should retain bacteria quantitatively, be free of bacterial-growth-inhibiting or stimulating substances, and be free of materials that directly or indirectly interfere with bacterial indicator systems in the medium. The filter and absorbent pad should not be degraded by sterilization at 121 C for 10 min. The ink used to delineate the surface grid should be nontoxic. Filter uniformity should be sufficient so that the variation in five filter culture replicates should be no more than 10% of the average.

According to the Defense Personnel Support Center, DSA,[3,4] 0.45-μm-pore-diam, 47-mm-diam membrane filters and pads shall meet the following specifications:

a) Porosity—At least 77% of the filter disk shall be pores.

b) Bubble point—Bubble point shall be 32 ± 2 psi (221 ± 14 kPa) when tested by the mercury intrusion method.

c) Diffusibility—Laboratory pure water shall diffuse uniformly through the filter disk in 15 sec and there shall be no dry spots when floated on water surface.

d) Flow rate—Flow rates shall be at least 55 mL/min/cm^2 at 25 C and a differential pressure of 70 cm mercury (93 kPa).

e) Organism retention—A 100-mL suspension of *Serratia marcescens* containing 1 × 10^3 cells/mL shall be retained 100%.

f) Toxicity—Filter disk shall not inhibit growth of test organism. The filters shall be considered nontoxic if the arithmetic mean of five counts is at least 90% of the arithmetic mean of the counts on five agar spread plates using the same sample volumes and agar media.

g) Extractables—Weight loss of a membrane filter shall not exceed 2.5% after boiling in 100 mL laboratory pure water for 20 min, drying, cooling, and bringing to constant weight.

h) Sterility—No growth shall result when a membrane filter is placed on a pad saturated with tryptone glucose extract broth or tryptone glucose extract agar and incubated at 35 C for 24 hr.

i) Absorbent pad dimensions (nominal)—Diameter shall be 47 mm and thickness shall be 0.8 mm.

j) Absorbent pad absorptivity—Pads shall absorb 1.5 ± 0.2 mL Endo broth.

k) Absorbent pad acidity—Pads shall release less than 1 mg total acidity calculated as $CaCO_3$ when titrated to the phenolphthalein endpoint with 0.02*N* NaOH.

In addition to the above specifications a test for recovery of fecal coliforms has been proposed.[5] This test compares the fecal coliform counts on test membranes to the counts on spread plates using M-FC agar for both procedures. Analyze four polluted water and one raw wastewater sample.

a) To determine sample test volumes, prepare serial dilutions of each sample in 0.1% peptone water to produce a suspension containing 20 to 60 fecal coliforms/0.1 mL. Hold original samples at 4 C. Determine fecal coliform density of each sample or dilution by membrane filter or spread plate test. Read the results after 22 to 24 hr incubation at 44.5 ± 0.2 C. If the fecal coliform density in a water sample is less than 10/0.1 mL, seed sample with raw wastewater, stabilize at 4 C for 24 hr, and recheck density.

b) Using the selected dilutions, test membranes and set up the spread plate control. Make membrane filter tests according to the procedure described in Section 909. Before beginning the spread plate tests, air-dry the surface of the M-FC agar contained in 100 mL petri dishes. Aseptically deliver 0.1 mL of selected sample dilution to the agar surface and spread with a sterile bent glass rod. Do not invert dish until sample liquid is completely absorbed. Alternate membrane filter tests with spread plate controls to randomize systematic errors.

Insert petri dishes into waterproof bags or seal with waterproof tape, and submerge in water bath incubator. Incubate for 22 to 24 hr at 44.5 ± 0.2 C.

c) Count blue colonies. If more than one dilution was prepared, select plates with between 10 and 100 colonies, but preferably with 20 to 60 colonies. Convert counts to logarithms and calculate the mean of the logarithms of the five replicate membrane filter counts ($\bar{x}_{mf}$) and the mean of the logarithms of the five replicate spread plates ($\bar{x}_{sp}$). Determine percent recovery:

$$\% \text{ recovery} = \frac{\bar{x}_{mf}}{\bar{x}_{sp}} \times 100$$

d) Pick 20 blue colonies from each of two randomly-selected filters and two spread plates. If plates contain less than 20 colonies, pick all blue colonies. Verify the colonies as described below in Section 902A.2*d*3). For membrane filters to be acceptable, 80% of the colonies must verify. If the samples are known or suspected to have a high concentration of pseudomonads or aeromonads, perform the cytochrome oxidase test. As in the fecal coliform verification procedure, a minimum of 80% of blue colonies should have negative oxidase tests for the membranes to be considered acceptable.

Some manufacturers provide information beyond that required by specifications and certify that their membranes are satisfactory for water analysis. They report retention, pore size, flow rate, sterility, pH, percent recovery, and limits for specific inorganic and organic chemical extractables.

To maintain quality control inspect each lot of membranes before use and during testing to insure they are round and pliable, with undistorted gridlines after autoclaving. After incubation, colonies should be well-developed with well-defined color and shape as defined by the test procedure. The gridline ink should not channel growth along the ink line nor restrict colony development. Colonies should be distributed evenly across the membrane surface.

7) Culture media—Because cultural methods depend on properly prepared media, use the best available materials and technics in media preparation, storage, and application. For control of quality, use commercially prepared media whenever available but note that such media may vary in quality among manufacturers and even from lot to lot from the same manufacturer. A reference Standard Plate Count agar is available from APHA for use by manufacturers who, after appropriate testing, may certify that their medium meets the specifications and standards of APHA. Such standardization is not available for other media listed in Section 905C and the use test described below should be used.

Order media in quantities to last no longer than 1 yr. Use media on a first-in, first-out basis. When practical, order media in quarter pound (114 g) multiples rather than one pound (454 g) bottles, to keep the supply sealed as long as possible. Record kind, amount, and appearance of media received, lot number, and dates received and opened. Check inventory quarterly for reordering. Discard media that are caked, discolored, or show other deterioration.

Because temperature, light, and moisture conditions differ among laboratories, it is impossible to establish absolute shelf-life limits for unopened bottles of media. A conservative limit for unopened bottles is 2 yr at room temperature. If bottles of media are older than 1 yr, compare recovery of recent pure culture isolates and natural samples using the old medium and another proven lot.

Use open bottles of media within 6

months after opening. Once bottles are opened, store them in a large hinged-door desiccator immediately after use if humidity is a problem.

Use test for media, membranes, and laboratory pure water—When a new lot of culture medium, membrane filters, or a new source of laboratory pure water is to be used, or at annual testing of water, make comparison tests of the current lot in use (reference lot) against the new lot (test lot) as follows:

a) Use a single batch of pure water, glassware, membrane filters, or other needed materials as specified to control all other variables except the one under study. Make parallel pour plate or membrane filter plate tests on reference lot and test lot, according to procedures in Sections 907 and 909. As a minimum, make single analyses on five positive water samples. Replicate analyses and additional samples can be tested to increase the sensitivity of detecting differences between reference and test lots.

When comparing sources of water, make the tests in parallel using reference water and test water separately for all water used in the tests (dilution, rinse, media preparation, etc.).

b) After incubation, compare bacterial colonies from the two lots for size and appearance. If colonies on the test lot are atypical or noticeably smaller than colonies on the reference lot plates, record the evidence of inhibition or other problem, regardless of count differences. Count plates and calculate the individual count per 1 mL or per 100 mL. Transform the count to logarithms and compile the log-transformed results for the two lots in parallel columns. Calculate the difference, d, between the two transformed results for each sample, including the algebraic sign, the mean, $\bar{d}$, and the standard deviation S_d of these differences.

Calculate Student's t statistic, using the number of samples as n:

$$t = \frac{\bar{d}}{S_d/\sqrt{n}}$$

c) Use the critical t value, 2.78, from a Student's t table[5] for comparison against the calculated value. At the 0.05 significance level this value is 2.78 for five samples (four degrees of freedom). If the calculated t value does not exceed 2.78, the lots do not produce significantly different results and the test lot is acceptable. If the calculated t value exceeds 2.78, the lots produce significantly different results. If the test lot t exceeds the reference lot result, the test lot is more stimulatory. If the test lot result is less than that of the reference lot, the test lot is less stimulatory.

If the colonies are atypical or noticeably smaller on the test lot and the Student's t exceeds 2.78, review test conditions, repeat the test, and/or reject the test lot and obtain another one.

8) Preparation of media—Prepare media in containers that are at least twice the volume of the medium being prepared. Stir media, particularly agars, while heating. Avoid scorching or boil-over by using a boiling water bath for small batches of media and by continually attending to larger volumes heated on a hot plate or gas burner. Preferably use hot plate-magnetic stirrer combinations. Identify and date prepared media. Prepare all media in deionized or distilled water of proven quality. Measure water volumes and media with approved graduates or pipets conforming to NBS and APHA standards, respectively. For potentially polluted samples, do not use blow-out pipets.

Check medium pH after solution and again after sterilization. Record results. Make minor adjustments in pH (<0.5 pH units) with NaOH or HCl solution to the pH specified. If the pH difference is larger, discard the batch. Incorrect pH values may indicate a problem with distilled water quality, medium deterioration, or improper preparation. Review instructions for makeup and check water pH. If water pH is unsatisfactory, remake batch of medium using water from a new source [see 902A.1*h*4) and 5)]. If water is satisfactory, remake the medium and check. If pH of second batch is incorrect, prepare medium from a second bottle.

Record pH problems in the media record book and report to the manufacturer if the medium is indicated as the source of error. Examine prepared media for unusual color, darkening, or precipitation and record observations. Consider variations of sterilization time and temperature as possible causes for problems. If any of the above occur, discard the medium.

9) Sterilization—Expose media to sterilization temperatures for the minimal time specified. A double-walled autoclave permits maintenance of full pressure and temperature in the jacket between loads and reduces chance for heat damage. Follow manufacturer's directions for sterilization of specific media. The required exposure time varies with form and type of material, type of medium, presence of carbohydrates, and volume. Table 902:III gives guidelines for typical items.

Remove sterilized media from autoclave as soon as chamber pressure reaches zero. Never reautoclave media.

Check effectiveness of sterilization with each run by using spore suspensions or strips (commercially available). Sterilization at 121 C for 15 min kills the spores; if growth of the autoclaved spores occurs after incubation at 55 C, sterilization was inadequate.

Sterilize nonautoclavable solutions or media by membrane filtration through a 0.22-μm-pore-diam filter in a sterile filtration and receiving apparatus. Filter and dispense medium in a safety cabinet or biohazard hood if available. Sterilize glassware (pipets, petri dishes, sample bottles) in an oven at 170 C for 2 hr. Sterilize equipment, supplies, and other solid or dry materials that are heat-sensitive, by exposing to ethylene oxide in a gas sterilizer. Use commercially available spore strips or suspensions to check dry heat and ethylene oxide sterilization.

TABLE 902:III. TIME AND TEMPERATURE FOR AUTOCLAVE STERILIZATION

Material	Temperature/Time
Membrane filters and pads	121 C for 10 min
Carbohydrate-containing media (lauryl tryptose, BGB broth, etc.)	121 C for 12 to 15 min
Contaminated materials and discarded cultures	121 C for 30 min
Membrane filter assemblies (wrapped), sample collection bottles (empty)	121 C for 15 min
Dilution water, 99 mL in screw-cap bottles	121 C for 15 min
Rinse water volumes of 500 mL to 1000 mL	121 C for 30 min
Rinse water in excess of 1 L	121 C with time adjusted for volume; check for sterility

10) Use of agars and broths—Temper melted agars in a water bath at 44 to 46 C until used but do not hold longer than 3 hr. To monitor agar temperature, expose a bottle of water or medium to the same heating and cooling conditions as the agar. Insert a thermometer in the monitoring bottle to determine when the temperature is 44 to 46 C and safe for use in pour plates. After pouring agar plates for streaking, keep covers open slightly for at least 15 min to dry agar surface.

Handle sterile fermentation tubes carefully to avoid entrapping air in inner tubes thereby producing false positive reactions. Examine freshly prepared tubes to determine that gas bubbles are absent.

11) Storage of media—Prepare sterile media in amounts that will be used within holding time limits given in Table 902:IV. If fermentation tube media are refrigerated, incubate overnight before use and check for false positive gas bubbles. Prepare media that are to be stored for more than 1 wk in screw-capped tubes and flasks to prevent loss of moisture. Seal prepoured agar plates in plastic bags and refrigerate to retain moisture.

To check for loss of moisture in broth tubes, mark the original liquid level in several tubes of each batch and monitor loss of moisture. If estimated loss exceeds 10%, discard tubes. Protect media containing dyes from light; if color changes are observed, discard medium.

Prepared sterile broths and agars available from commercial sources may offer advantages when analyses are done intermittently, when staff is not available for preparation work, or when cost can be balanced against other factors of laboratory operation. Check performance of these media as described in ¶ 12) below. Time limits for holding prepared media are given in Table 902:IV.

12) Quality control of prepared media—Maintain in a bound book a complete record of each batch of medium prepared with name of preparer and date, name and lot number of medium, amount of medium weighed, volume of medium prepared, sterilization time and temperature, pH measurements and adjustments, and preparations of labile components. Include sterility and positive and negative control culture checks on all media as described below.

TABLE 902:IV. HOLDING TIMES FOR PREPARED MEDIA

Medium	Holding Time
Membrane filter (MF) broth in screw-cap flasks at 4 C	96 hr
MF agar in plates with tight-fitting covers at 4 C	2 wk
Agar or broth in loose-cap tubes at 4 C	1 wk
Agar or broth in tightly closed screw-cap tubes at 4 C	3 months
Poured agar plates with loose-fitting covers in sealed plastic bags at 4 C	2 wk
Large volume of agar in tightly closed screw-cap flask or bottle at 4 C	3 months

2. Analytical Quality Control Procedures

Quality control over microbiological procedures includes checks on sterility, confirmation of test results, and measurements of precision.

a. General quality control procedures:

1) For membrane filter tests, check sterility once each run of media, membrane filters, dilution and rinse water, and glassware and equipment, using sterile water as the sample. For multiple-tube procedures, check sterility of media, dilution water, and glassware. If contamination is indicated, reject analytical data from samples tested with these materials and request immediate resampling.

2) For each lot of medium check analytical procedures with known positive and negative control cultures for the organism(s) under test. See Table 902:V for examples of test cultures.

3) Make duplicate analyses on 10% of samples and on at least one sample per test run.

4) In laboratories with more than one analyst, have each make parallel analyses

TABLE 902:V. CONTROL CULTURES FOR MICROBIOLOGICAL TESTS

Group	Control Culture	
	Positive	Negative
Total coliforms	*Escherichia coli* *Enterobacter aerogenes*	*Staphylococcus aureus* *Pseudomonas* sp.
Fecal coliforms	*E. coli*	*E. aerogenes* *Streptococcus faecalis*
Fecal streptococci	*Streptococcus faecalis*	*Staphylococcus aureus* *E. coli*

on at least one positive sample monthly to compare performance.

b. Measurement of method precision: Calculate precision of duplicate analyses for each type of sample examined, according to the following procedure and using a table similar to Table 902:VI:

1) Make duplicate analyses on the first 15 positive samples of a specific type. Have each set of duplicates analyzed by the same analyst, but include all analysts within the laboratory. Note duplicate analyses as D_1 and D_2 in Column 2 of the table.

2) Calculate the logarithm of each result (Column 3). If either of a set of duplicate results is zero, add 1 to both values before calculating the logarithms.

3) Calculate the range (R) for each pair of transformed duplicates as the mean ($\overline{R}$) of these ranges (Column 4 and bottom of table).

4) Thereafter, analyze 10% of routine samples in duplicate. Transform the duplicates as in ¶ 2) and calculate their range. If

TABLE 902:VI. CALCULATION OF PRECISION CRITERION

Sample No.	Duplicate Analyses		Logarithms of Counts		Range of Logarithms (R_{log}) ($L_1 - L_2$)
	D_1	D_2	L_1	L_2	
1	89	71	1.9494	1.8513	0.0981
2	38	34	1.5798	1.5315	0.0483
3	58	67	1.7634	1.8261	0.0627
•	•	•	•	•	•
•	•	•	•	•	•
•	•	•	•	•	•
14	7	6	0.8451	0.7782	0.0669
15	110	121	2.0414	2.0828	0.0414

Calculations:

1) Σ of $R_{log} = 0.0981 + 0.0483 + 0.0627 + \cdots + 0.0669 + 0.0414$
$= 0.71889$

2) $\overline{R} = \frac{\Sigma R_{log}}{n} = \frac{0.71889}{15} = 0.0479$

3) Precision criterion $= 3.27\ \overline{R} = 3.27\ (0.0479) = 0.1566$

the range is greater than 3.27 $\overline{R}$, analyst precision is out of control. Discard all analytical results since the last precision check (see Table 902:VII). Identify and resolve the analytical problem before making further analyses.

5) Update the criterion used in ¶ 4) by periodically repeating the procedures of ¶ 1) through ¶ 3) using the most recent sets of 15 duplicate results.

c. Quality control on multiple-tube dilution tests:

1) For routine analyses, do the Completed Test on 10% of positive samples. If no positives result from potable water samples, complete at least one positive source water quarterly.

2) For public water supply samples with a history of heavy growth without gas in presumptive tubes, submit tubes to the Confirmed Test to check for coliform bacteria.

d. Quality control on membrane filter procedures:

1) Verify the colonies monthly for each type of test from a known positive sample. If laboratory has two or more analysts, require each to count typical colonies on the same membrane from one positive sample per month. Verify colonies on the membrane and compare the analysts' counts to the verified count.

2) Verify total coliform analyses by picking at least five colonies counted as coliforms and inoculating into lauryl tryptose broth. Incubate at 35 C for 24 to 48 hr and read for gas production. Transfer growth from positive tubes into brilliant green bile broth, incubate at 35 C and examine at 24 and 48 hr. Gas production verifies total coliform organisms. It also is desirable to pick at least 10 non-sheen-producing colonies and conduct the same verification tests to determine that false negatives do not occur. If no positives result from analyzing potable water samples, verify at least one positive source water quarterly.

3) Verify fecal coliform analyses by picking at least 10 isolated colonies from membranes containing typical blue colonies and transferring to lauryl tryptose broth. Incubate at 35 C for 24 and 48 hr and examine for gas production. Transfer growth from positive tubes to EC broth and incubate at 44.5 C for 24 hr. Gas production in EC broth verifies presence of fecal coliform organisms.

4) Verify analyses for fecal streptococci by picking at least 10 isolated pink to red colonies and transferring to brain heart infusion (BHI) agar and broth. Make catalase test on 24-hr-old cultures. Transfer catalase-negative cultures (possible fecal streptococci) to 40% bile BHI broth and incubate at 35 C. Transfer also to BHI broth and incubate at 45 C. Growth in 40% bile and at 45 C verify fecal streptococci.

TABLE 902:VII. DAILY CHECKS ON PRECISION OF DUPLICATE COUNTS*

Date of Analysis	Duplicate Analyses		Logarithms of Counts		Range of Logarithms	Acceptance of Range†
	D_1	D_2	L_1	L_2		
8/29	71	65	1.8513	1.8129	0.0383	A
8/30	110	121	2.0414	2.0828	0.0414	A
8/31	73	50	1.8633	1.6990	0.1643	U

*Precision criterion = $(3.27\overline{R})$ = 0.1566.
†A = acceptable; U = unacceptable.

3. Data Handling

a. Distribution of bacterial populations: In most chemical analyses the distribution of analytical results follows the Gaussian curve, which has symmetrical distribution of values about the mean (see Section 104A).

In microbiology, distributions are not symmetrical. Bacterial counts often are characterized as having a skewed distribution because of many extremely low values and a few extremely high values. These characteristics lead to an arithmetic mean that is considerably larger than the median. The frequency curve of this distribution has a long right tail, as shown in Figure 902:1 and is said to display positive skewness.

Application of the most rigorous statistical technics requires the assumption of symmetric distributions such as the normal curve. Therefore it usually is necessary to convert skewed data so that a symmetric distribution resembling the normal distribution results. An approximately normal distribution can be obtained from positively skewed data by converting numbers to their logarithms, as shown in Table 902:VIII. Comparison of the frequency tables for the original data (Table 902:IX) and their logarithms (Table 902:X) shows that the logarithms closely approximate a symmetrical distribution.

TABLE 902:VIII. COLIFORM COUNTS AND THEIR LOGARITHMS

MPN Coliform Count *no./100 mL*	log MPN
11	1.041
27	1.431
36	1.556
48	1.681
80	1.903
85	1.929
120	2.079
130	2.114
136	2.134
161	2.207
317	2.501
601	2.779
760	2.881
1,020	3.009
3,100	3.491
$\bar{x}$ = 442	$\bar{x}_g$ = antilog 2.1825 = 152

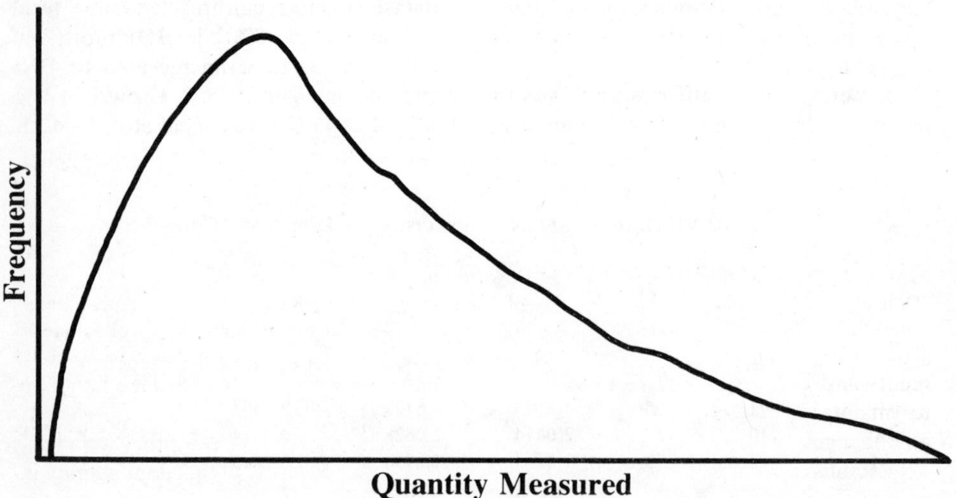

Figure 902:1. Frequency curve (positively skewed distribution).

TABLE 902:IX. COMPARISON OF FREQUENCY OF MPN DATA

Class Interval	Frequency (MPN)
0 to 400	11
400 to 800	2
800 to 1,200	1
1,200 to 1,600	0
1,600 to 2,000	0
2,000 to 2,400	0
2,400 to 2,800	0
2,800 to 3,200	1

b. Central tendency measures of skewed distribution: If the logarithms of numbers from a positively skewed distribution are approximately normally distributed, the original data have a log-normal distribution. The best estimate of central tendency of log-normal data is the geometric mean, defined as:

TABLE 902:X. COMPARISON OF FREQUENCY OF LOG MPN DATA

Class Interval	Frequency (log MPN)
1.000 to 1.300	1
1.300 to 1.600	2
1.600 to 1.900	1
1.900 to 2.200	5
2.200 to 2.500	1
2.500 to 2.800	2
2.800 to 3.100	2
3.100 to 3.400	0
3.400 to 3.700	1

$$\bar{x}_g = n\sqrt{(x_1)\ (x_2) \cdots (x_n)}$$

and

$$\log \bar{x}_g = \frac{\Sigma\ (\log x_i)}{n}$$

that is, the geometric mean is equal to the antilog of the arithmetic mean of the logarithms. For example, the following means calculated from the data in Table 902:VIII are drastically different.

$$\log \bar{x}_g = \frac{\Sigma\ (\log x_i)}{n} = \frac{32.737}{15} = 2.1825$$

geometric mean

$$\bar{x}_g = \text{antilog}\ (2.1825) = 152$$

and arithmetic mean

$$\bar{x} = \frac{\Sigma x_i}{n} = \frac{6632}{15} = 442$$

Therefore, although regulations or tradition may require or cause microbiological data to be reported as the arithmetic mean or median, the preferred statistic for summarizing microbiological data is the geometric mean.

902 B. Interlaboratory Quality Control

An interlaboratory quality control program is a system of agreed-upon requirements and laboratory practices necessary to maintain minimal quality standards among a group of participant laboratories. To establish such a program the participants first approve uniform sampling procedures and standardized analytical methodology. Minimal standards are set for laboratory operations (personnel, facilities, equipment, supplies, data handling, and quality control).

After methods and laboratory standards are established, an independent agency inspects laboratory facilities and conducts methods validation and performance evaluation studies to measure acceptable levels of performance. Both method validation and performance evaluation studies specify the analytical methodology but the former is done to establish precision and accuracy of selected methods while the latter is to establish acceptable performance by each laboratory. An important part of interlaboratory quality control is follow-up with technical assistance on problems observed.

To safeguard drinking water quality and assure a level of data reliability, most states have approval, registration, or certification programs for commercial and noncommercial laboratories that test water. These state programs provided the base for a new federal/state program for the certification of water supply laboratories, developed under the Safe Drinking Water Act.[6]

In the certification program for public water supply laboratories, all laboratories that analyze public water supplies must be certified according to minimal criteria and procedures described in the EPA manual on certification[7]: criteria are established for laboratory operations and methodology; on-site inspections are required by the certifying State agency or its surrogate to verify minimal standards; annually, laboratories are required to perform acceptably on unknown samples in formal studies, as samples are available; the responsible authority follows up on problems identified in the on-site inspection or performance evaluation and requires corrections within a set period of time. Individual state programs may exceed the federal criteria.

On-site inspections of laboratories in the present certification program and the earlier USPHS Interstate Carrier Program showed that primary causes for discrepancies have been inadequate equipment, improperly prepared media, incorrect analytical procedures, and insufficiently trained personnel.[8] The EPA manual of microbiological methods[2] provides added guidance on sampling, analytical methodology, and quality assurance practices. A handbook developed under the superseded interstate carrier program[9] gives useful background information on aquatic microbiology.

Interlaboratory studies also may compare laboratories, methods, and media by the analysis of randomly distributed split samples.[10] Such studies evaluate laboratory results and indicate general problem areas. Acceptability of laboratory performance compared to referee laboratories and other participating laboratories can be established.

902 C. References

1. INHORN, S.L., ed. 1977. Quality Assurance Practices for Health Laboratories. Amer. Pub. Health Ass., Washington, D.C.
2. BORDNER, R.H., J.A. WINTER & P.V. SCARPINO, eds. 1978. Microbiological Methods for Monitoring the Environment, Water and Wastes. EPA 600/8-78-017, Environmental Monitoring and Support Laboratory, U.S. EPA, Cincinnati, Ohio.
3. Interim Federal Specification, Disk Filter, Membrane, Bacteriological/Particulates. 1965. NNN-D-00370 (DSA-DM), Defense Personnel Support Center, Defense Supply Agency, Brooklyn, N.Y.
4. Military Specification, Disk, Filtering, Microporous. 1973. MIL-D-37005 (DSA-OM), Defense Personnel Support Center, DPSC-ATT, Philadelphia, Pa.
5. AMERICAN SOCIETY FOR TESTING AND MATERIALS. 1977. Annual Book of ASTM Standards, Part 31, Water. ASTM, Philadelphia, Pa.

6. Safe Drinking Water Act, Public Law 93-523. Dec. 16, 1974. 88 Stat. 1660, 42 U.S. Code (USC) 300 f.
7. Manual for the Interim Certification of Laboratories Involved in Analyzing Public Drinking Water Supplies, Criteria and Procedures. 1977. EPA 600/8-78-008. U.S. EPA, Washington, D.C.
8. GELDREICH, E.E. 1971. Application of bacteriological data in potable water surveillance. *J. Amer. Water Works Ass.* 63:225.
9. GELDREICH, E.E. 1975. Handbook for Evaluating Water Laboratories, 2nd ed. EPA-670/9-75-006, Municipal Environmental Research Laboratory, U.S. EPA, Cincinnati, Ohio.
10. GREENBERG, A.E., J.S. THOMAS, T.W. LEE & W.R. GAFFEY. 1967. Interlaboratory comparisons in water bacteriology. *J. Amer. Water Works Ass.* 59:237.

903 LABORATORY APPARATUS

In addition to the specifications given below, see Section 902.

1. Incubators

Incubators must maintain a uniform and constant temperature at all times in all areas, that is, they must not vary more than ±0.5 C in the areas used. Obtain such accuracy by using a water-jacketed or anhydric-type incubator with thermostatically controlled low-temperature electric heating units properly insulated and located in or adjacent to the walls or floor of the chamber and preferably equipped with mechanical means of circulating air.

Incubators equipped with high-temperature heating units are unsatisfactory, because such sources of heat, when improperly placed, frequently cause localized overheating and excessive drying of media, with consequent inhibition of bacterial growth. Incubators so heated may be operated satisfactorily by replacing high-temperature units with suitable wiring arranged to operate at a lower temperature and by installing mechanical air-circulation devices. It is desirable, where ordinary room temperatures vary excessively, to keep laboratory incubators in special rooms maintained at a few degrees below the recommended incubator temperature.

Alternatively, use special incubating rooms well insulated and equipped with properly distributed heating units and with forced air circulation, provided that they conform to desired temperature limits. When such rooms are used, record the daily temperature range in areas where plates or tubes are incubated. Provide incubators with open metal wire or sheet shelves so spaced as to assure temperature uniformity throughout the chamber. Leave a 2.5-cm space between walls and stacks of dishes or baskets of tubes.

Maintain an accurate thermometer with the bulb immersed in liquid (glycerine, water, or mineral oil) on each shelf in use within the incubator and record daily temperature readings (preferably morning and afternoon). It is desirable, in addition, to maintain a maximum and minimum registering thermometer within the incubator on the middle shelf to record the gross temperature range over a 24-hr period. At intervals, determine temperature variations within the incubator when filled to maximum capacity. Install a recording thermometer whenever possible, to maintain a continuous and permanent record of temperature.

Ordinarily, a water bath with a gabled cover to reduce water and heat loss, or a solid heat sink incubator, is required to maintain a temperature of 44.5 ± 0.2 C. If satisfactory temperature control is not achieved, provide water recirculation.

Keep water depth in the incubator sufficient to immerse tubes to upper level of media.

2. Hot-Air Sterilizing Ovens

Use hot-air sterilizing ovens of sufficient size to prevent internal crowding; constructed to give uniform and adequate sterilizing temperatures of 170 ± 10 C; and equipped with suitable thermometers. Optionally use a temperature-recording instrument.

3. Autoclaves

Use autoclaves of sufficient size to prevent internal crowding; constructed to provide uniform temperatures within the chambers (up to and including the sterilizing temperature of 121 C); equipped with an accurate thermometer the bulb of which is located properly on the exhaust line so as to register minimum temperature within the sterilizing chambers (temperature-recording instrument is optional); equipped with pressure gauge and properly adjusted safety valves connected directly with saturated-steam power lines or directly to a suitable special steam generator (Do not use steam from a boiler treated with amines for corrosion control.); and capable of reaching the desired temperature within 30 min.

Use of a vertical autoclave or pressure cooker is not recommended because of difficulty in adjusting and maintaining sterilization temperature and the potential hazard. If a pressure cooker is used in emergency or special circumstances, equip it with an efficient pressure gauge and a thermometer the bulb of which is 2.5 cm above the water level.

4. Gas Sterilizers

Use a sterilizer equipped with automatic controls capable of carrying out a complete sterilization cycle. As a sterilizing gas use ethylene oxide diluted to 10 to 12% with an inert gas. Provide an automatic control cycle to evacuate sterilizing chamber to at least 0.06 kPa, to hold the vacuum for 30 min, to adjust humidity and temperature, to charge with the ethylene oxide mixture to a pressure dependent on mixture used, to hold such pressure for at least 4 hr, to vent gas, to evacuate to 0.06 kPa, and finally, to bring to atmospheric pressure with sterile air. The humidity, temperature, pressure, and time of sterilizing cycle depend on the gas mixture used.

Store overnight sample bottles with loosened caps that were sterilized by gas, to allow last traces of gas mixture to dissipate. Incubate overnight media sterilized by gas, to insure dissipation of gas.

In general, mixtures of ethylene oxide with chlorinated hydrocarbons such as freon are harmful to plastics, although with temperatures below 55 C, gas pressure not over 35 kPa, and time of sterilization less than 6 hr, the effect is minimal. If carbon dioxide is used as a diluent of ethylene oxide, increase exposure time and pressure, depending on temperature and humidity that can be used.

Determine proper cycle and gas mixture for objects to be sterilized and confirm by sterility tests.

5. Colony Counters

Use Quebec-type colony counter, dark-field model preferred, or one providing equivalent magnification (1.5 diameters) and satisfactory visibility.

6. pH Equipment

Use electrometric pH meters, accurate to at least 0.1 pH units, for determining pH values of media.

7. Balances

Use balances providing a sensitivity of at least 0.1 g at a load of 150 g, with appropriate weights. Use an analytical balance having a sensitivity of 1 mg under a load of 10 g for weighing small quantities (less than 2 g) of materials. Single-pan rapid-weigh balances are most convenient.

8. Media Preparation Utensils

Use borosilicate glass or other suitable noncorrosive equipment such as stainless steel. Use glassware that is clean and free of residues, dried agar, or other foreign materials that may contaminate media.

9. Pipets and Graduated Cylinders

Use pipets of any convenient size, provided that they deliver the required volume accurately and quickly. The error of calibration for a given manufacturer's lot must not exceed 2.5%. Use pipets having graduations distinctly marked and with unbroken tips. Bacteriological transfer pipets or pipets conforming to the APHA standards given in the latest edition of "Standard Methods for the Examination of Dairy Products" may be used. Optimally, protect the mouth end of all pipets by a cotton plug to eliminate hazards to the worker or possible sample contamination by saliva.

Use graduated cylinders meeting ASTM Standards (D-86 and D-216) and with accuracy limits established by the National Bureau of Standards where appropriate.

10. Pipet Containers

Use boxes of aluminum or stainless steel, end measurement 5 to 7.5 cm, cylindrical or rectangular, and length about 40 cm. When these are not available, paper wrappings may be substituted. To avoid excessive charring during sterilization, use best-quality sulfate pulp (kraft) paper. *Do not use copper or copper alloy cans or boxes as pipet containers.*

11. Dilution Bottles or Tubes

Use bottles or tubes of resistant glass, preferably borosilicate glass, closed with glass stoppers or screw caps equipped with liners that do not produce toxic or bacteriostatic compounds on sterilization. Do not use cotton plugs as closures. Mark graduation levels indelibly on side of dilution bottle or tube. Plastic bottles of nontoxic material and acceptable size may be substituted for glass provided that they can be sterilized properly.

12. Petri Dishes

For the Standard Plate Count, use glass or plastic petri dishes about 100 × 15 mm. Use dishes the bottoms of which are free from bubbles and scratches and flat so that the medium will be of uniform thickness throughout the plate. For the membrane filter technic use loose-lid glass or plastic dishes, 60 × 15 mm, or tight-lid dishes, 50 × 12 mm. Sterilize petri dishes and store in metal cans (aluminum or stainless steel, but not copper), or wrap in paper—preferably best-quality sulfate pulp (kraft)—before sterilizing.

13. Fermentation Tubes and Vials

Use fermentation tubes of any type, if their design permits conforming to medium and volume requirements for concentration of nutritive ingredients as described subsequently. Where tubes are used for a test of gas production, enclose a shell vial, inverted. Use tube and vial of such size that the vial will be filled completely with medium and at least partly submerged in the tube.

14. Inoculating Equipment

Use wire loops made of 22- or 24-gauge nickel alloy* or platinum-iridium for flame sterilization. Single-service transfer loops of aluminum or stainless steel are satisfactory. Use loops at least 3 mm in diameter. Sterilize by dry heat or steam. Single-service hardwood applicators also may be used. Make these 0.2 to 0.3 cm in diameter and at least 2.5 cm longer than the fermentation tube; sterilize by dry heat and store in glass or other nontoxic containers.

*Chromel, nichrome, or equivalent.

15. Sample Bottles

For bacteriological samples, use sterilizable bottles of glass or plastic of any suitable size and shape. Use bottles holding a sufficient volume of sample for all required tests, permitting proper washing, and maintaining samples uncontaminated until examinations are completed. Ground-glass-stoppered bottles, preferably wide-mouthed and of resistant glass, are recommended. Plastic bottles of suitable size, wide-mouthed, and made of nontoxic materials such as polypropylene that can be sterilized repeatedly are satisfactory as sample containers. These eliminate the possibility of breakage during shipment.

Metal or plastic screw-cap closures with liners may be used on sample bottles provided that no toxic compounds are produced on sterilization.

Before sterilization, cover tops and necks of sample bottles having glass closures with aluminium foil or heavy kraft paper.

16. Bibliography

COLLINS, W.D. & H.B. RIFFENBURG. 1923. Contamination of water samples with material dissolved from glass containers. *Ind. Eng. Chem.* 15:48.

CLARK, W.M. 1928. The Determination of Hydrogen Ion Concentration, 3rd ed. Williams & Wilkins, Baltimore, Md.

ARCHAMBAULT, J., J. CUROT & M.H. MCCRADY. 1937. The need of uniformity of conditions for counting plates (with suggestions for a standard colony counter). *Amer. J. Pub. Health* 27:809.

RICHARDS, O.W. & P.C. HEIJN. 1945. An improved dark-field Quebec colony counter. *J. Milk Technol.* 8:253.

COHEN, B. 1957. The measurement of pH, titratable acidity, and oxidation-reduction potentials. *In* Manual of Microbiological Methods. Society of American Bacteriologists. McGraw-Hill Book Co., New York, N.Y.

MCGUIRE, O.E. 1964. Wood applicators for the confirmatory test in the bacteriological analysis of water. *Pub. Health Rep.* 79:812.

AMERICAN PUBLIC HEALTH ASSOCIATION. 1978. Standard Methods for the Examination of Dairy Products, 14th ed. APHA, Washington, D.C.

904 WASHING AND STERILIZATION

Cleanse all glassware thoroughly with a suitable detergent and hot water, rinse with hot water to remove all traces of residual washing compound, and finally rinse with distilled water. If mechanical glassware washers are used, equip them with influent plumbing of stainless steel or other nontoxic material. Do not use copper piping to distribute distilled water. Use stainless steel or other nontoxic material for the rinse water system.

Sterilize glassware, except when in metal containers, for not less than 60 min at a temperature of 170 C, unless it is known from recording thermometers that oven temperatures are uniform, under which exceptional condition use 160 C. Heat glassware in metal containers to 170 C for not less than 2 hr.

Sterilize sample bottles not made of plastic as above or in an autoclave at 121 C for 15 min.

For plastic bottles that distort on autoclaving, use low-temperature ethylene oxide gas sterilization.

905 PREPARATION OF CULTURE MEDIA

905 A. General Procedures

1. Storage of Culture Media

Store dehydrated media (powders) in tightly closed bottles in the dark at less than 30 C in an atmosphere of low humidity. Do not use them if they discolor or become caked so as to lose their free-flowing power. Purchase dehydrated media in small quantities that will be used within 6 months after opening. Additionally, use stocks of dehydrated media containing selective agents such as sodium azide, bile salts or derivatives, antibiotics, sulfur-containing amino acids, etc., of relatively current lot number (within a year of purchase) so as to maintain optimum selectivity. See also Section 902.

Prepare culture media in batches of such size that the entire batch will be used in less than 1 wk. However, if the media are contained in screw-capped tubes they may be stored for up to 3 months. Store media out of direct sun and avoid contamination and excessive evaporation.

Liquid media in fermentation tubes, if stored at refrigeration or even moderately low temperatures, may dissolve sufficient air to produce, upon incubation at 35 C, a bubble of air in the tube. Incubate fermentation tubes that have been stored at a low temperature overnight before use and discard tubes containing air.

Fermentation tubes may be stored at approximately 25 C; but because evaporation may proceed rapidly under these conditions—resulting in marked changes in concentration of the ingredients—do not store at this temperature for more than 1 wk. Discard tubes with an evaporation loss exceeding 1 mL.

2. Adjustment of Reaction

State reaction of culture media in terms of hydrogen ion concentration, expressed as pH.

The increase in hydrogen ion concentration (decrease in pH) during sterilization will vary slightly with the individual sterilizer in use, and the initial reaction required to obtain the correct final reaction will have to be determined. The decrease in pH usually will be 0.1 to 0.2 but occasionally may be as great as 0.4. When buffering salts such as phosphates are present in the media, the decrease in pH value will be negligible.

Make tests to control adjustment to required hydrogen ion concentration with a pH meter. Measure pH of prepared medium as directed in pH Value, Glass Electrode Method (Section 423). Titrate a known volume of medium with a solution of NaOH to the desired pH. Calculate amount of NaOH solution that must be added to the bulk medium to reach this reaction. After adding and mixing thoroughly, check reaction and adjust if necessary. The required final pH is given in the directions for preparing each medium. If a specific pH is not prescribed, adjustment is unnecessary.

The pH of reconstituted dehydrated media seldom will require adjustment if made according to directions. Such factors as errors in weighing dehydrated medium or overheating reconstituted medium may produce an unacceptable final pH. Measure pH, especially of rehydrated selective media, regularly to insure quality control and media specifications.

3. Sterilization

After rehydrating a medium, dispense promptly to culture vessels and sterilize within 2 hr. Do not refrigerate or otherwise store nonsterile media.

Sterilize all media, except sugar broths or broths with other specifications, in an autoclave at 121 C for 15 min after the temperature has reached 121 C. When the pressure reaches zero, remove medium from autoclave and cool quickly to avoid decomposition of sugars by prolonged exposure to heat. To permit uniform heating and rapid cooling, pack materials loosely and in small containers. Sterilize sugar broths at 121 C for 12 to 15 min. The maximum elapsed time for exposure of sugar broths to any heat (from time of closing loaded autoclave to unloading) is 45 min. Preheat autoclave before loading to reduce total needed heating time to within the 45-min limit.

905 B. Water

To prepare culture media and reagents, use only distilled or demineralized water that has been tested and found free from traces of dissolved metals and bactericidal or inhibitory compounds. Toxicity in distilled water may be derived from fluoridated water high in silica. Other sources of toxicity are silver, lead, and various unidentified organic complexes. Where condensate return is used as feed for a still, toxic amines or other boiler compounds may be present in distilled water. Residual chlorine or chloramines also may be found in distilled water prepared from chlorinated water supplies. If chlorine compounds are found in distilled water, neutralize them by adding an equivalent amount of sodium thiosulfate or sodium sulfite.

Distilled water also should be free of contaminating nutrients. Such contamination may be derived from flashover of organics during distillation, continued use of exhausted carbon filter beds, deionizing columns in need of recharging, solder flux residues in new piping, dust and chemical fumes, and storage of water in unclean bottles. Store distilled water out of direct sunlight to prevent growth of algae. Good housekeeping practices usually will eliminate nutrient contamination.

See Section 902 for distilled water suitability test.

905 C. Media Specifications

The need for uniformity dictates the use of dehydrated media. Never prepare media from basic ingredients when suitable dehydrated media are available. Follow manufacturer's directions for rehydration and sterilization. Commercially prepared media in liquid form (sterile ampoule or other) also may be used if known to give equivalent results. See Section 902 for quality control specifications.

The terms used for protein source in most media, for example, peptone, tryptone, tryptose, were coined by the developers of the medium and may reflect commercial products rather than clearly defined entities. It is not intended to preclude the use of alternative materials provided they produce equivalent results.

NOTE—The term "percent solution" as used in these directions is to be understood to mean "grams of solute per 100 mL solution."

1. Dilution Water

a. Buffered water: To prepare stock phosphate buffer solution, dissolve 34.0 g potassium dihydrogen phosphate (KH_2PO_4), in 500 mL distilled water, adjust to pH 7.2 ± 0.5 with 1*N* sodium hydroxide (NaOH), and dilute to 1 L with distilled water.

Add 1.25 mL stock phosphate buffer solution and 5.0 mL magnesium chloride solution (38 g $MgCl_2$/L distilled water) to 1 L distilled water. Dispense in amounts that will provide 99 ± 2.0 mL or 9 ± 0.2 mL after autoclaving for 15 min.

b. Peptone water: Prepare a 10% solution of peptone in distilled water. Dilute a measured volume to provide a final 0.1% solution. Final pH should be 6.8.

Dispense in amounts to provide 99 ± 2.0 mL or 9 ± 0.2 mL after autoclaving for 15 min.

Do not suspend bacteria in any dilution water for more than 30 min at room temperature because death or multiplication may occur.

2. Tryptone Glucose Extract Agar

Beef extract	3.0	g
Tryptone	5.0	g
Glucose	1.0	g
Agar	15.0	g
Distilled water	1	L

pH should be between 6.8 and 7.0 after sterilization.

3. Plate Count Agar (Tryptone Glucose Yeast Agar)

Tryptone	5.0	g
Yeast extract	2.5	g
Glucose	1.0	g
Agar	15.0	g
Distilled water	1	L

pH should be 7.0 ± 0.1 after sterilization.

4. Lauryl Tryptose Broth

Tryptose	20.0	g
Lactose	5.0	g
Dipotassium hydrogen phosphate, K_2HPO_4	2.75	g
Potassium dihydrogen phosphate, KH_2PO_4	2.75	g
Sodium chloride, NaCl	5.0	g
Sodium lauryl sulfate	0.1	g
Distilled water	1	L

pH should be approximately 6.8 after sterilization. Before sterilization, dispense in fermentation tubes with an inverted vial sufficient medium to cover inverted vial at least partially after sterilization.

Make lauryl tryptose broth of such strength that adding 100-mL or 10-mL portions of sample to medium will not reduce ingredient concentrations below those of the standard medium. Prepare in accordance with Table 905:I.

5. Brilliant Green Lactose Bile Broth

Peptone	10.0	g
Lactose	10.0	g
Oxgall	20.0	g
Brilliant green	0.0133	g
Distilled water	1	L

pH should be 7.2 after sterilization. Before sterilization, dispense in fermentation tubes with an inverted vial sufficient medium to cover inverted vial at least partially after sterilization.

TABLE 905:I. PREPARATION OF LAURYL TRYPTOSE BROTH

Inoculum *mL*	Amount of Medium in Tube *mL*	Volume of Medium + Inoculum *mL*	Dehydrated Lauryl Tryptose Broth Required *g/L*
1	10 or more	11 or more	35.6
10	10	20	71.2
10	20	30	53.4
100	50	150	106.8
100	35	135	137.1
100	20	120	213.6

6. Eosin Methylene Blue (EMB) Agar (Levine's Modification)

Peptone	10.0	g
Lactose	10.0	g
Dipotassium hydrogen phosphate, K_2HPO_4	2.0	g
Agar	15.0	g
Eosin Y	0.4	g
Methylene blue	0.065	g
Distilled water	1	L

pH should be 7.1 after sterilization. Decolorization of the medium occurs during sterilization, but color returns after cooling.

7. Nutrient Agar

Peptone	5.0	g
Beef extract	3.0	g
Agar	15.0	g
Distilled water	1	L

pH should be approximately 6.8 after sterilization.

8. EC Medium

Tryptose or trypticase	20.0	g
Lactose	5.0	g
Bile salts mixture or bile salts No. 3	1.5	g
Dipotassium hydrogen phosphate, K_2HPO_4	4.0	g
Potassium dihydrogen phosphate, KH_2PO_4	1.5	g
Sodium chloride, NaCl	5.0	g
Distilled water	1	L

pH should be 6.9 after sterilization. Before sterilization, dispense in fermentation tubes with an inverted vial sufficient medium to cover the inverted vial at least partially after sterilization.

9. A-1 Broth

This medium may not be available in dehydrated form and may require preparation from the basic ingredients.

Lactose	5.0	g
Tryptone	20.0	g
Sodium chloride, NaCl	5.0	g
Salicin	0.5	g
Polyethylene glycol *p* isooctylphenyl ether*	1.0	mL
Distilled water	1	L

Heat to dissolve solid ingredients, add polyethylene glycol *p* isooctylphenyl ether, and adjust to pH 6.9 ± 0.1. Before sterilization dispense in fermentation tubes with an inverted vial sufficient medium to cover the inverted vial at least partially after sterilization. Sterilize by autoclaving at 121 C for 10 min. Store in dark at room temperature for not longer than 7 days. Ignore formation of precipitate.

Make A-1 broth of such strength that adding 10-mL sample portions to medium will not reduce ingredient concentrations below those of the standard medium. For 10-mL samples prepare double-strength medium.

10. LES Endo Agar

Yeast extract	1.2	g
Casitone or trypticase	3.7	g
Thiopeptone or thiotone	3.7	g
Tryptose	7.5	g
Lactose	9.4	g
Dipotassium hydrogen phosphate, K_2HPO_4	3.3	g
Potassium dihydrogen phosphate, KH_2PO_4	1.0	g
Sodium chloride, NaCl	3.7	g
Sodium desoxycholate	0.1	g
Sodium lauryl sulfate	0.05	g
Sodium sulfite, Na_2SO_3	1.6	g
Basic fuchsin	0.8	g
Agar	15.0	g
Distilled water	1	L

Rehydrate in distilled water containing 20 mL 95% ethanol. Bring to a boil, cool to

*Triton X-100, Rohm and Haas Co., or equivalent.

45 to 50 C and dispense in 4-mL quantities into lower section of 60-mm glass or plastic petri dishes. If dishes of any other size are used, adjust quantity to give an equivalent depth. Store plates in the dark at 2 to 10 C and discard unused medium after 2 wk. Do not expose to direct sunlight.

11. M-Endo Medium†

Tryptose or polypeptone	10.0	g
Thiopeptone or thiotone	5.0	g
Casitone or trypticase	5.0	g
Yeast extract	1.5	g
Lactose	12.5	g
Sodium chloride, NaCl	5.0	g
Dipotassium hydrogen phosphate, K_2HPO_4	4.375	g
Potassium dihydrogen phosphate, KH_2PO_4	1.375	g
Sodium lauryl sulfate	0.050	g
Sodium desoxycholate	0.10	g
Sodium sulfite, Na_2SO_3	2.10	g
Basic fuchsin	1.05	g
Distilled water	1	L

Rehydrate in 1 L distilled water containing 20 mL 95% ethanol. Heat to boiling, promptly remove from heat, and cool to below 45 C. Do not sterilize by autoclaving. Final pH should be between 7.1 and 7.3.

Store finished medium in the dark at 2 to 10 C and discard any unused medium after 96 hr.

NOTE—This medium may be solidified by adding 1.2 to 1.5% agar before boiling.

12. LES MF Holding Medium

Tryptone	3.0	g
M-Endo broth MF	3.0	g
Dipotassium hydrogen phosphate, K_2HPO_4	3.0	g
Sodium benzoate	1.0	g
Sulfanilamide	1.0	g
Paraaminobenzoic acid	1.2	g
Cycloheximide	0.5	g
Distilled water	1	L

†Dehydrated Difco M-Endo Broth MF (No. 0749), dehydrated BBL m-Coliform Broth (No. 11119), or equivalent may be used.

Rehydrate in distilled water without heating. Final pH should be 7.1 ± 0.1.

13. M-FC Broth

Tryptose or biosate	10.0	g
Proteose peptone No. 3 or polypeptone	5.0	g
Yeast extract	3.0	g
Sodium chloride, NaCl	5.0	g
Lactose	12.5	g
Bile salts No. 3 or bile salts mixture	1.5	g
Aniline blue	0.1	g
Distilled water	1	L

Rehydrate in distilled water containing 10 mL 1% rosolic acid in 0.2*N* NaOH.‡ Heat to boiling, promptly remove from heat, and cool to below 45 C. Do not sterilize by autoclaving. Final pH should be 7.4.

Store finished medium at 2 to 10 C and discard unused medium after 96 hr.

NOTE—This medium may be solidified by adding 1.2 to 1.5 percent agar before boiling.

14. M-VFC Holding Medium

This medium may not be available in dehydrated form and may require preparation from the basic ingredients.

Casitone, vitamin-free	0.2	g
Sodium benzoate	4.0	g
Sulfanilamide	0.5	g
Ethanol (95%)	10.0	mL
Distilled water	1	L

Heat to dissolve medium and sterilize by filtration through a membrane filter (pore diam, 0.22 μm). Final pH should be 6.7.

Store finished medium at 2 to 10 C and discard unused medium after 1 month. To prepare 100 mL medium, make a 1:100

‡Rosolic acid reagent will decompose if sterilized by autoclaving. Store stock solution in the dark at 2 to 10 C and discard after 2 wk or sooner if its color changes from dark red to muddy brown. Rosolic acid may be omitted from the medium if minimal background colony counts occur and equivalent results are obtained without it.

aqueous solution of casitone and add 2 mL.

15. Azide Dextrose Broth

Beef extract	4.5 g
Tryptone or polypeptone	15.0 g
Glucose	7.5 g
Sodium chloride, NaCl	7.5 g
Sodium azide, NaN_3	0.2 g
Distilled water	1 L

pH should be about 7.2 after sterilization.

16. KF Streptococcus Agar

Proteose peptone No. 3 or polypeptone	10.0 g
Yeast extract	10.0 g
Sodium chloride, NaCl	5.0 g
Sodium glycerophosphate	10.0 g
Maltose	20.0 g
Lactose	1.0 g
Sodium azide, NaN_3	0.4 g
Agar	20.0 g
Distilled water	1 L

Mix 7.64 g dehydrated medium with 100 mL distilled water in a flask. Heat in a boiling water bath to dissolve the agar. After solution is complete, heat for an additional 5 min. Cool to 50 or 60 C and add 1 mL sterile aqueous 1% solution of 2,3,5-triphenyltetrazolium chloride/100 mL. Adjust pH to 7.2 with 10% Na_2CO_3 if necessary. Hold medium for not more than 4 hr at 45 to 50 C before plates are poured. Store poured plates in the dark at 2 to 10 C. Discard after 30 days.

17. Brain-Heart Infusion Agar

Brain-heart infusion agar contains the same ingredients as brain-heart infusion (18) except that 15.0 g agar are added. The pH should be 7.4 after sterilization. Tube for slants.

18. Brain-Heart Infusion

Infusion of calf brains	200 g
Infusion of beef heart	250 g
Proteose peptone	10.0 g
Glucose	2.0 g
Sodium chloride, NaCl	5.0 g
Disodium hydrogen phosphate, Na_2HPO_4	2.5 g
Distilled water	1 L

pH should be 7.4 after sterilization.

19. Pfizer Selective Enterococcus (PSE) Agar

Peptone C	17.0 g
Peptone B	3.0 g
Yeast extract	5.0 g
Bacteriological bile	10.0 g
Sodium chloride, NaCl	5.0 g
Sodium citrate	1.0 g
Esculin	1.0 g
Ferric ammonium citrate	0.5 g
Sodium azide, NaN_3	0.25 g
Agar	15.0 g
Distilled water	1 L

pH should be 7.1 after sterilization. Hold medium for not more than 4 hr at 45 to 50 C before plates are poured.

20. Tryptophane Broth

Tryptophane broth contains 10.0 g tryptone or trypticase/L distilled water. Dispense in 5-mL portions in test tubes.

21. Buffered Glucose Broth

Proteose peptone or equivalent peptone	5.0 g
Glucose	5.0 g
Dipotassium hydrogen phosphate, K_2HPO_4	5.0 g
Distilled water	1 L

Dispense in 5-mL portions in test tubes and sterilize by autoclaving at 121 C for 12 to 15 min, making sure that total time of exposure to heat is not longer than 30 min.

22. Salt Peptone Glucose Broth

Polypeptone or proteose peptone	10.0 g
Sodium chloride, NaCl	5.0 g
Glucose	10.0 g
Distilled water	1 L

pH should be 7.0 to 7.2 before sterilization. Dispense in 5-mL portions in test tubes and sterilize by autoclaving at 121 C for 12 to 15 min, making sure that total time of exposure to heat is not longer than 30 min.

23. Koser's Citrate Broth

Sodium ammonium hydrogen phosphate, $NaNH_4HPO_4 \cdot 4H_2O$	1.5 g
Dipotassium hydrogen phosphate, K_2HPO_4	1.0 g
Magnesium sulfate heptahydrate, $MgSO_4 \cdot 7H_2O$	0.2 g
Sodium citrate dihydrate, crystals	3.0 g
Distilled water	1 L

Dispense in 5-mL portions in test tubes.

24. Simmons' Citrate Agar

Magnesium sulfate heptahydrate, $MgSO_4 \cdot 7H_2O$	0.2 g
Ammonium dihydrogen phosphate, $NH_4H_2PO_4$	1.0 g
Dipotassium hydrogen phosphate, K_2HPO_4	1.0 g
Sodium citrate dihydrate	2.0 g
Sodium chloride, NaCl	5.0 g
Agar	15.0 g
Bromthymol blue	0.08 g
Distilled water	1 L

Tube for long slants.

25. M-Staphylococcus Broth

Tryptone	10.0 g
Yeast extract	2.5 g
Lactose	2.0 g
Mannitol	10.0 g
Dipotassium hydrogen phosphate, K_2HPO_4	5.0 g
Sodium chloride, NaCl	75.0 g
Sodium azide, NaN_3	0.049 g
Distilled water	1 L

pH should be 7.0 after sterilization by boiling for 5 min.

26. Lipovitellin-Salt-Mannitol Agar

This agar medium may not be available in dehydrated form and may require preparation from the basic ingredients or addition of egg yolk to a dehydrated base.

Beef extract	1.0 g
Polypeptone	10.0 g
Sodium chloride, NaCl	75.0 g
d-Mannitol	10.0 g
Agar	15.0 g
Phenol red	0.025 g
Egg yolk	20.0 g
Distilled water	1 L

pH should be 7.4 after sterilization.

27. M-PA Agar

This agar medium may not be available in dehydrated form and may require preparation from the basic ingredients.

L-lysine HCl	5.0 g
Sodium chloride, NaCl	5.0 g
Yeast extract	2.0 g
Xylose	2.5 g
Sucrose	1.25 g
Lactose	1.25 g
Phenol red	0.08 g
Ferric ammonium citrate	0.8 g
Sodium thiosulfate, $Na_2S_2O_3$	6.8 g
Agar	15.0 g
Distilled water	1 L

Adjust pH to 6.5 and sterilize. Cool to 55 to 60 C; carefully readjust pH to 7.1 ± 0.1 and add the following dry antibiotics per liter of agar base: sulfapyridine,§ 176 mg; kanamycin,‖ 8.5 mg; nalidixic acid,# 37.0 mg; and cycloheximide,** 150 mg. After mixing dispense in 3-mL quantities in 50- by 12-mm petri plates. Store poured plates at 2 to 10 C. Discard unused medium after 1 month.

28. Milk Agar (Brown and Scott Foster Modification)

Mixture A:		
Instant non-fat milk††	100	g
Distilled water	500	mL
Mixture B:		
Nutrient broth	12.5	g
Sodium chloride, NaCl	2.5	g
Agar	15.0	g
Distilled water	500	mL

Separately sterilize Mixtures A and B; cool rapidly to 55 C; aseptically combine mixtures and pour into 100- by 15-mm petri plates, about 20 mL/plate.

29. Asparagine Broth

This medium may not be available in dehydrated form and may require preparation from the basic ingredients.

Asparagine, DL	3.0	g
Anhydrous dipotassium hydrogen phosphate, K_2HPO_4	1.0	g
Magnesium sulfate, $MgSO_4{\cdot}7H_2O$	0.5	g
Distilled water	1	L

Adjust pH to 6.9 to 7.2 before sterilization.

30. Acetamide Broth

This medium may not be available in dehydrated form and may require preparation from the basic ingredients.

Acetamide	10.0	g
Sodium chloride, NaCl	5.0	g
Anhydrous dipotassium hydrogen phosphate, K_2HPO_4	1.39	g
Anhydrous potassium dihydrogen phosphate, KH_2PO_4	0.73	g
Magnesium sulfate, $MgSO_4{\cdot}7H_2O$	0.5	g
Phenol red	0.012	g
Distilled water	1	L

Adjust pH to 6.9 to 7.2 before sterilization.

Prepare acetamide agar slants as above, except add 15 g agar, boil to dissolve agar, and dispense in 8-mL quantities to 16-mm tubes. After autoclaving, incline tubes while cooling to provide a large slant surface.

31. Neopeptone-Glucose-Rose Bengal-Aureomycin Agar

This medium may not be available in dehydrated form and may require preparation from the basic ingredients.

Neopeptone	5.0	g
Glucose	10.0	g
Rose bengal	0.035	g
Agar	20.0	g
Chlortetracycline (Aureomycin) or tetracycline	35.0	μg
Distilled water	1	L

pH should be about 6.5 after sterilization. Prepare rose bengal in advance by dissolving 1 g rose bengal, aqueous, in 100 mL distilled water. Add 3.5 mL/L of medium before autoclaving.

Prepare the antibiotic chlortetracycline (or tetracycline) separately and add after autoclaving but just before plates are poured. Add 1 g of water-soluble antibiotic to 150 mL distilled water to prepare stock solution. Refrigerate. Sterilize by filtration before each use. Add 0.05 mL sterile solution/10 mL agar medium.

§Available from Nutritional Biochemicals, Cleveland, Ohio.

‖Available from Bristol-Myers, Syracuse, N.Y.

#Available from Calbiochem, La Jolla, Calif.

**Actidione, Upjohn Company, Kalamazoo, Mich., or equivalent.

††Carnation or equivalent.

Because this medium is used for preparing pour plates, prepare and store basal agar either in bulk, or more conveniently, in tubes in 10-mL amounts. After melting stored medium, cool to about 45 C, add 0.05 mL antibiotic solution/10 mL, and pour plate.

Dehydrated Cooke's rose bengal agar may be used in place of neopeptone-glucose-rose bengal agar base.

32. Neopeptone-Glucose Agar

Neopeptone (or equivalent)	5.0 g
Glucose	10.0 g
Agar	20.0 g
Distilled water	1 L

pH should be about 6.5 after sterilization.

This medium is known also as Emmons' Sabouraud Agar or Emmons' Sabouraud Dextrose Agar.

33. Czapek (or Czapek Dox) Agar

Sucrose	30.0 g
Sodium nitrate, $NaNO_3$	3.0 g
Dipotassium hydrogen phosphate, K_2HPO_4	1.0 g
Magnesium sulfate, $MgSO_4$	0.5 g
Potassium chloride, KCl	0.5 g
Ferrous sulfate, $FeSO_4$	0.01 g
Agar	15.0 g
Distilled water	1 L

pH should be 7.3 after sterilization.

34. Yeast Nitrogen Base-Glucose Broth

Yeast nitrogen base	13.4 g
Distilled water	1 L

Sterilize by filtration. Prepare 500 mL each of 2% and 40% aqueous glucose solutions. Sterilize each separately by filtration. To use final medium, aseptically add to a sterile 250-mL erlenmeyer flask 25 mL yeast nitrogen base and 25 mL of either 2% or 40% glucose solutions to make 1% or 20% final glucose concentrations. Stopper flask with a gauze-wrapped cotton stopper and store until used.

35. Yeast Extract - Malt Extract - Glucose Agar

Yeast extract	3.0 g
Malt extract	3.0 g
Neopeptone (or equivalent)	5.0 g
Glucose	10.0 g
Agar	20.0 g
Distilled water	1 L

No pH adjustment is required.

36. Diamalt Agar

Diamalt	150.0 g
Agar	20.0 g
Distilled water	1 L

No pH adjustment is required. The medium will be turbid but filtration is not required.

37. Starch-Casein Agar

Soluble starch	10.0 g
Casein	0.3 g
Potassium nitrate, KNO_3	2.0 g
Sodium chloride, NaCl	2.0 g
Dipotassium hydrogen phosphate, K_2HPO_4	2.0 g
Magnesium sulfate, hydrate, $MgSO_4{\cdot}7H_2O$	0.05 g
Calcium carbonate, $CaCO_3$	0.02 g
Ferrous sulfate, hydrate, $FeSO_4{\cdot}7H_2O$	0.01 g
Agar	15.0 g
Distilled water	1 L

No pH adjustment is required. Because medium is used to prepare double-layer plates, store medium for bottom layer in bulk or in tubes in about 15-mL amounts. Store medium for surface layer in tubes containing 17.0 mL. Add 1 mL cycloheximide‡‡ (1 mg/mL, sterilized at 121 C for 15 min) to the liquefied surface medium at time of inoculation.

‡‡Actidione, Upjohn Company, Kalamazoo, Mich., or equivalent.

38. Casitone-Glycerol-Yeast Autolysate Broth (CGY)

This medium may not be available in dehydrated form and may require preparation from the basic ingredients. It may be solidified by adding 1.5% agar.

Casitone	5.0	g
Glycerol	10.0	g
Yeast autolysate	1.0	g
Distilled water	1	L

39. Isolation Medium (Iron Bacteria)

This medium may not be available in dehydrated form and may require preparation from the basic ingredients.

Glucose	0.15	g
Ammonium sulfate, $(NH_4)_2SO_4$	0.5	g
Calcium nitrate, $Ca(NO_3)_2$	0.01	g
Dipotassium hydrogen phosphate, K_2HPO_4	0.05	g
Magnesium sulfate, $MgSO_4 \cdot 7H_2O$	0.05	g
Potassium chloride, KCl	0.05	g
Calcium carbonate, $CaCO_3$	0.1	g
Agar	10.0	g
Vitamin B_{12}	0.01	mg
Thiamine	0.4	mg
Distilled water	1	L

40. Maintenance (SCY) Medium (Iron Bacteria)

This medium may not be available in dehydrated form and may require preparation from the basic ingredients.

Sucrose	1.0	g
Casitone	0.75	g
Yeast extract	0.25	g
Trypticase soy broth without dextrose	0.25	g
Agar	10.0	g
Vitamin B_{12}	0.01	mg
Thiamine	0.4	mg
Distilled water	1	L

41. Mn-Agar

This medium may not be available in dehydrated form and may require preparation from the basic ingredients.

Manganous carbonate, $MnCO_3$	2.0	g
Beef extract	1.0	g
Ferrous ammonium sulfate, $Fe(NH_4)_2(SO_4)_2$	150	mg
Sodium citrate	150	mg
Yeast extract	75	mg
Cyanocobalamin	0.005	mg
Agar	10.0	g
Distilled water	1	L

Prepare and sterilize the medium without cyanocobalamin. Separately sterilize cobalamin by filtration and add aseptically just before medium solidifies.

42. Iron Oxidizing Medium (*Thiobacillus ferrooxidans*)

This medium may not be available in dehydrated form and may require preparation from the basic ingredients.

Basal salts:		
Ammonium sulfate, $(NH_4)_2SO_4$	3.0	g
Potassium chloride, KCl	0.10	g
Dipotassium hydrogen phosphate, K_2HPO_4	0.50	g
Magnesium sulfate, $MgSO_4 \cdot 7H_2O$	0.50	g
Calcium nitrate, $Ca(NO_3)_2$	0.01	g
H_2SO_4, 10*N*	1.0	mL
Distilled water	700	mL
Energy source:		
Ferrous sulfate, $FeSO_4 \cdot 7H_2O$, 14.74% solution (w/v)	300	mL

Separately sterilize basal salts and energy source and combine when cool. Store in the refrigerator and discard after 2 wk. A precipitate will form and the medium will be opalescent and green. The pH should be 3.0 to 3.6.

43. Ferrous Sulfide Agar (*Gallionella ferruginea*)

This medium may not be available in dehydrated form and may require preparation from the basic ingredients.

Agar layer:		
Ferrous sulfide, FeS (washed precipitate and liquid)	500	mL
Sodium sulfide, Na_2S	15.6	g
Ferrous ammonium sulfate, $Fe(NH_4)_2(SO_4)_2 \cdot 6H_2O$	78.4	g
Boiling distilled water	1	L
Agar (liquid) (30 g/L)	500	mL
Liquid overlay:		
Ammonium chloride, NH_4Cl	1.0	g
Dipotassium phosphate, K_2HPO_4	0.5	g
Magnesium sulfate, $MgSO_4 \cdot 7H_2O$	0.2	g
Calcium chloride, $CaCl_2$	0.1	g
Distilled water	1	L

Prepare FeS by reacting equal molar quantities of Na_2S and $Fe(NH_4)_2(SO_4)_2$ in boiling distilled water. Let precipitate settle from the hot solution in a completely filled and stoppered bottle. Wash precipitate four times by decanting supernatant and replacing with boiling water. Store FeS in a glass stoppered bottle completely filled with additional boiling distilled water.

Add equal volumes of FeS and 3% agar at 45 C. Prepare slants in screw-capped tubes. Prepare liquid overlay, bubble CO_2 through it for 10 to 15 sec, and add several milliliters to agar slant.

A variation of the basic medium requires adding 0.5 mL formalin (40% formaldehyde solution) to a screw-capped dilution bottle containing 10 mL FeS agar and 100 mL of liquid overlay. Add 0.001% bromthymol blue and 0.004% bromcresol purple to liquid overlay.

44. Sulfate-Reducing Medium

This medium may not be available in dehydrated form and may require preparation from the basic ingredients.

Sodium lactate	3.5	g
Beef extract	1.0	g
Peptone	2.0	g
Magnesium sulfate, $MgSO_4 \cdot 7H_2O$	2.0	g
Sodium sulfate, Na_2SO_4	1.5	g
Dipotassium phosphate, K_2HPO_4	0.5	g
Ferrous ammonium sulfate, $Fe(NH_4)_2(SO_4)_2 \cdot 6H_2O$	0.392	g
Calcium chloride, $CaCl_2$	0.10	g
Sodium ascorbate	0.10	g
Distilled water	1	L

pH should be 7.5 ± 0.3 after sterilization. Prepare medium excluding ferrous ammonium sulfate and sodium ascorbate, dispense in screw-capped test tubes, and sterilize. For use, the tubes must be completely filled; therefore, in a flask sterilize extra medium to be added to tubes for filling. On day of use, prepare separate solutions of ferrous ammonium sulfate (3.92 g/100 mL) and sodium ascorbate (1.00 g/100 mL), sterilize by filtration through a 0.45-μm membrane filter, and aseptically add 0.1 mL each solution/10 mL basal medium.

45. Sulfate-Reducing Medium (*Thiobacillus thioparus*)

This medium may not be available in dehydrated form and may require preparation from the basic ingredients.

Sodium thiosulfate, $Na_2S_2O_3 \cdot 5H_2O$	10.0	g
Dipotassium hydrogen phosphate, K_2HPO_4	2.0	g
Magnesium sulfate, $MgSO_4 \cdot 7H_2O$	0.1	g
Calcium chloride, $CaCl_2 \cdot 2H_2O$	0.1	g
Ammonium sulfate, $(NH_4)_2SO_4$	0.1	g
Ferric chloride, $FeCl_3 \cdot 6H_2O$	0.02	g
Distilled water	1	L

pH should be 7.8 after sterilization. Separately sterilize $Na_2S_2O_3$ and $(NH_4)_2SO_4$ and add before use.

46. Sulfur Medium (*Thiobacillus thiooxidans*)

This medium may not be available in dehydrated form and may require preparation from the basic ingredients.

Sulfur, elemental	10.0	g
Potassium dihydrogen phosphate, KH_2PO_4	3.0	g
Magnesium sulfate, $MgSO_4 \cdot 7H_2O$	0.5	g
Ammonium sulfate, $(NH_4)_2SO_4$	0.3	g
Calcium chloride, $CaCl_2 \cdot 2H_2O$	0.25	g
Ferric chloride, $FeCl_3 \cdot 6H_2O$	0.02	g
Distilled water	1	L

pH should be 4.8 after sterilization. Weigh sulfur into 250-mL flasks using 1 g/flask. Add 100 mL medium to each flask and sterilize with intermittent steam (30 min for each of 3 consecutive days).

47. M-7 Hr FC Agar

This medium may not be available in dehydrated form and may require preparation from the basic ingredients.

Proteose peptone No. 3 or polypeptone	5.0	g
Yeast extract	3.0	g
Lactose	10.0	g
d-Mannitol	5.0	g
Sodium chloride, NaCl	7.5	g
Sodium lauryl sulfate	0.2	g
Sodium desoxycholate	0.1	g
Bromcresol purple	0.35	g
Phenol red	0.3	g
Agar	15.0	g
Distilled water	1	L

Heat in boiling water bath. After ingredients are dissolved heat additional 5 min. Cool to 55 to 60 C and adjust pH to 7.3 ± 0.1 with 0.1*N* NaOH (0.35 mL/L usually required). Cool to about 45 C and dispense in 4- to 5-mL quantities to petri plates with tight-fitting covers. Store at 2 to 10 C. Discard after 30 days.

905 D. Bibliography

LEVINE, M. 1918. Differentiation of *B. coli* and *B. aerogenes* on a simplified eosine methylene blue agar. *J. Infect. Dis.* 23:43.

LEVINE, M. 1918. A simplified fuchsin sulphite (Endo) agar. *Amer. J. Pub. Health* 8:864.

LEVINE, M. 1921. Further observations on the eosine methylene blue agar. *J. Amer. Water Works Ass.* 8:151.

LEVINE, M. 1921. Bacteria fermenting lactose and their significance in water analysis. *Iowa State Coll. Agr. Mech. Arts Bull.* 62:117.

BUNKER, G.C. & H. SCHUBER. 1922. The reaction of culture media. *J. Amer. Water Works Ass.* 9:63.

JORDAN, H.E. 1932. Brilliant green bile for *Coli-Aerogenes* group determinations. *J. Amer. Water Works Ass.* 24:1027.

RUCHHOFT, C.C. 1935. Comparative studies of media for the determination of the *Coli-Aerogenes* group in water analysis. *J. Amer. Water Works Ass.* 27:1732.

RUCHHOFT, C.C. & J.F. NORTON. 1935. Study of selective media for *Coli-Aerogenes* isolations. *J. Amer. Water Works Ass.* 27:1134.

MCCRADY, M.H. 1937. A practical study of procedures for the detection of the presence of coliform organisms in water. *Amer. J. Pub. Health* 27:1243.

DARBY, C.W. & W.L. MALLMANN. 1939. Studies on media for coliform organisms. *J. Amer. Water Works Ass.* 31:689.

KELLY, C.B. 1940. Brilliant green lactose bile and the *Standard Methods* completed test in isolation of coliform organisms. *Amer. J. Pub. Health* 30:1034.

RICHEY, D. 1941. Relative value of 2 per cent and 5 per cent brilliant green bile confirmatory media. *J. Amer. Water Works Ass.* 33:649.

HOWARD, N.J., A.G. LOCHHEAD & M.H. McCRADY. 1941. A study of methods for the detection of the presence of coliform organisms in water. *Can. J. Pub. Health* 32:29.

MALLMANN, W.L. & C.W. DARBY. 1941. Uses of a lauryl sulphate tryptose broth for the detection of coliform organisms. *Amer. J. Pub. Health.* 31:127.

MALLMANN, W.L. & R.S. BREED. 1941. A comparative study of standard agars for determining bacterial counts in water. *Amer. J. Pub. Health* 31:341.

HOWARD, N.J., A.G. LOCHHEAD & M.H. McCRADY. 1942. Report of the committee on bacteriological examination of water and sewage. *Can. J. Pub. Health* 33:49.

ARCHAMBAULT, J. & M.H. McCRADY. 1942. Dissolved air as a source of error in fermentation tube results. *Amer. J. Pub. Health* 32:1164.

WATTIE, E. 1943. Coliform confirmation from raw and chlorinated waters with brilliant green bile lactose broth. *Pub. Health Rep.* 58:377.

McCRADY, M.H. 1943. A practical study of lauryl sulfate tryptose broth for detection of the presence of coliform organisms in water. *Amer. J. Pub. Health* 33:1199.

LEVINE, M. 1944. The effect of concentration of dyes on differentiation of enteric bacteria on eosin methylene blue agar. *J. Bacteriol.* 45:471.

MALLMANN, W.L. & E.B. SELIGMANN. 1950. A comparative study of media for the detection of streptococci in water and sewage. *Amer. J. Pub. Health* 40:286.

LITSKY, W., W.L. MALLMANN & C.W. FIFIELD. 1955. Comparison of the most probable numbers of *Escherichia coli* and enterococci in river waters. *Amer. J. Pub. Health* 45:1049.

STRAKA, R.P. & J.L. STOKES. 1957. Rapid destruction of bacteria in commonly used diluents and its elimination. *Appl. Microbiol.* 5:21.

SLANETZ, L.W. & C.H. BARTLEY. 1957. Numbers of enterococci in water, sewage, and feces determined by the membrane filter technique, with an improved medium. *J. Bacteriol.* 74:591.

FIFIELD, C.W. & C.P. SCHAUFUS. 1958. Improved membrane filter medium for the detection of coliform organisms. *J. Amer. Water Works Ass.* 50:193.

KENNER, B.A., H.F. CLARK & P.W. KABLER. 1961. Fecal streptococci. I. Cultivation and enumeration of streptococci in surface waters. *Appl. Microbiol.* 9:15.

McCARTHY, J.A., J.E. DELANEY & R.J. GRASSO. 1961. Measuring coliforms in water. *Water Sewage Works* 108:238.

DELANEY, J.E., J.A. McCARTHY & R.J. GRASSO. 1962. Measurement of *E. coli* Type I by the membrane filter. *Water Sewage Works* 109:289.

GELDREICH, E.E., H.F. CLARK, C.B. HUFF & L.C. BEST. 1965. Fecal coliform-organism medium for the membrane filter technique. *J. Amer. Water Works Ass.* 57:208.

GELDREICH, E.E. & H.F. CLARK. 1965. Distilled water suitability for microbiological applications. *J. Milk Food Technol.* 28:351.

AMERICAN PUBLIC HEALTH ASSOCIATION. 1972. Standard Methods for the Examination of Dairy Products, 13th ed. APHA, New York, N.Y.

MACLEOD, R.A., S.C. KUO & R. GELINAS. 1967. Metabolic injury to bacteria. II. Metabolic injury induced by distilled water or Cu^{++} in the plating diluent. *J. Bacteriol.* 93:961.

ISENBERG, H.D., D. GOLDBERG & J. SAMPSON. 1970. Laboratory studies with a selective enterococcus medium. *Appl. Microbiol.* 20:433.

GUNN, B.A., W.E. DUNKELBERG, JR. & J.R. CREITZ. 1972. Clinical evaluation of 2%-LSM medium for primary isolation and identification of staphylococci. *Amer. J. Clin. Pathol.* 57:236.

LEVIN, M.A. & V.J. CABELLI. 1972. Membrane filter technique for enumeration of *Pseudomonas aeruginosa*. *Appl. Microbiol.* 24:864.

TAYLOR, R.H., R.H. BORDNER & P.V. SCARPINO. 1973. Delayed incubation membrane filter test for fecal coliforms. *Appl. Microbiol.* 25:363.

LENNETTE, E.H., E.H. SPAULDING & J.P. TRUANT, eds. 1974. Manual of Clinical Microbiology. Amer. Soc. Microbiol., Washington, D.C.

OLIVIERI, V.P., C.W. KRUSE & K. KAWATA. 1977. Microorganisms in urban stormwater. EPA-600/2-77-087, Cincinnati, Ohio.

906 SAMPLES

906 A. Collection

1. Containers

Collect samples for bacteriological examination in bottles that have been cleansed and rinsed carefully, given a final rinse with distilled water, and sterilized as directed in Sections 903 and 904.

2. Dechlorination

Add reducing agent to bottles intended for the collection of water containing residual chlorine or other halogens unless they contain broth for direct planting of sample. Sodium thiosulfate ($Na_2S_2O_3$) is a satisfactory dechlorinating agent. Its presence at the instant of collection of a sample from a halogenated supply will neutralize any residual disinfectant and prevent a continuation of bactericidal action during sample transit. The bacteriological examination then will indicate more accurately the true bacterial content of the water at the time of sampling.

Add sufficient $Na_2S_2O_3$ to clean sample bottle before sterilization to give a concentration about 100 mg/L in the sample. To a 120-mL bottle add 0.1 mL 10% solution of $Na_2S_2O_3$ (this will neutralize a sample containing about 15 mg/L residual chlorine). Stopper bottle, cap, and sterilize by either dry or moist heat, as directed previously.

Collect water samples high in copper or zinc and wastewater samples high in heavy metals in sample bottles containing a chelating agent that will reduce metal toxicity. This is particularly significant when such samples are in transit for 4 hr or more. Use 372 mg/L of the tetrasodium salt of ethylenediaminetetraacetic acid (EDTA). Adjust EDTA solution to pH 6.5 before use. Add EDTA separately to sample bottle before bottle sterilization (0.3 mL 15% solution in a 120-mL bottle) or combine it with the $Na_2S_2O_3$ solution before addition.

3. Sampling Procedures

When the sample is collected, leave ample air space in the bottle (at least 2.5 cm) to facilitate mixing by shaking, preparatory to examination. Exercise care to take samples that will be representative of the water being tested and avoid sample contamination at time of collection or in period before examination.

Keep sampling bottle closed until the moment it is to be filled. Remove stopper and hood or cap as a unit, taking care to avoid soiling. During sampling, do not handle stopper or cap and neck of bottle, and protect them from contamination. Hold bottle near base, fill it without rinsing, replace stopper or cap immediately, and secure hood around neck of bottle.

If the water sample is to be taken from a distribution-system tap without attachments, select a tap that is supplying water from a service pipe directly connected with the main, and is not, for example, served from a cistern or storage tank. Open tap fully and let water run to waste for 2 or 3 min, or for a time sufficient to permit clearing the service line. Restrict water flow to permit filling bottle without splashing. Do not use as sampling points leaking taps that allow water to flow over the outside of the tap. In sampling from a mixing faucet remove faucet attachments such as screen or splash guard, run hot water for 2 min, then cold water for 2 to 3 min. Collect sample as indicated above.

In collecting samples directly from a river, stream, lake, reservoir, spring, or shallow well, the aim is to obtain a sample representative of the water that will be the source of supply to consumers. It is undesirable to take samples too near the bank or too far from the point of drawoff, or at a depth above or below the point of drawoff.

Location of sampling sites and frequency of sampling are critical factors in obtaining reliable information about bacterial pollution in any body of water. Single or unscheduled grab samples from a river, stream, or lake often can be collected for control data or to satisfy regulatory requirements. A grab sample can be taken near the surface.

For extensive stream studies whereby source and extent of pollution are to be determined, take more representative samples with consideration of site, method, and time of sampling. The number of sampling sites usually represents a compromise based on physical limitations of the laboratory, detection of pollution peaks, and frequency of sample collection. The number of samples to be processed depends on whether the survey objective is to measure cycles of immediate pollution, duration of peak pollution, or probable average pollution. Sites for measuring cyclic pollution and its duration are immediately below the pollution source. Sample as frequently as possible.

Choose a site designated to measure estimated average pollution conditions far enough downstream to insure complete mixing of pollutant and water. Sampling at such points does not eliminate all possible variations but will minimize any sharp fluctuations in quality. Downstream site sampling need not be done as frequently as cyclic pollution sampling.

Collect samples one-quarter, one-half, or three-quarters the stream width at each site or at other distances, depending on survey objectives. Avoid areas of relative stagnation. Often only one sample, taken near the surface, is collected in the stream channel.

Collect samples of bathing-beach water at locations and time of greatest bather load, and, in natural bathing places, periods of stormwater runoff during the bathing season.

Take samples from a river, stream, lake, or reservoir by holding the bottle near its base in the hand and plunging it, neck downward, below the surface. Turn bottle until neck points slightly upward and mouth is directed toward the current. If there is no current, as in the case of a reservoir, create a current artificially by pushing bottle forward horizontally in a direction away from the hand. When sampling from a boat, obtain samples from upstream side of boat. If it is not possible to collect samples from these situations in this way, attach a weight to base of bottle and lower it into the water. In any case, take care to avoid damage to bank or stream bed; otherwise, water fouling may occur.

Special apparatus that permits mechanical removal of bottle stopper below water surface is required to collect samples from depths of a lake or reservoir. Various types of deep sampling devices are available. The most common is the ZoBell J-Z sampler, which uses a sterile 350-mL bottle and a rubber stopper through which a piece of glass tubing has been passed. This tubing is connected to another piece of glass tubing by a rubber connecting hose. The unit is mounted on a metal frame containing a cable and a messenger. When the messenger is released, it strikes the glass tubing at a point that has been slightly weakened by a file mark. The glass tube is broken by the messenger and the tension set up by the rubber connecting hose is released and the tubing swings to the side. Water is sucked into the bottle as a consequence of the partial vacuum created by sealing the unit at time of autoclaving. Commercial adaptations of this sampler and others are available.

Bottom sediment sampling also requires special apparatus. The sampler described by Van Donsel and Geldreich has been found effective for a variety of bottom materials for remote (deep water) or hand (shallow water) sampling. Construct this sampler preferably of stainless steel and fit with a sterile plastic bag. A nylon cord closes the bag after the sampler penetrates the sediment. A slide bar keeps the bag closed during descent and is opened, thereby opening the bag, during sediment sampling.

If the sample is to be taken from a well fitted with a hand pump, pump water to waste for about 5 min before collecting sample. If the well is equipped with a mechanical pump, collect sample from a tap on the discharge. If there is no pumping machinery, collect a sample directly from the well by means of a sterilized bottle fitted with a weight at the base; take care to avoid contaminating samples by any surface scum.

For sampling wastewaters or effluents the technics described above generally are adequate; in addition see Section 105.

4. Size of Sample

The volume of sample should be sufficient to carry out all tests required, preferably not less than 100 mL.

5. Identifying Data

Accompany samples by complete and accurate identifying and descriptive data. Do not accept for examination inadequately identified samples.

906 B. Preservation and Storage

Start bacteriological examination of a water sample promptly after collection to avoid unpredictable changes. If samples cannot be processed within 1 hr after collection, use an iced cooler for storage during transport to the laboratory.

Hold temperature of all stream pollution samples below 10 C during a maximum transport time of 6 hr. Refrigerate these samples upon receipt in the laboratory and process within 2 hr. When local conditions necessitate delays in delivery of samples longer than 6 hr, consider either making field examinations using field laboratory facilities located at the site of collection or the use of delayed-incubation procedures.

If it is known that the results will be used in legal action, employ a special messenger to deliver samples to the laboratory within 6 hr.

Unfortunately, these requirements are seldom realistic in the case of individual potable water samples sent to the laboratory by mail service, but in no case may the time elapsing between collection and examination exceed 30 hr. Where refrigeration of individual water samples sent by mail is not possible, a thermos-type insulated sample bottle that can be sterilized may be used. Record time and temperature of storage of all samples and consider this information in the interpretation of data.

906 C. Bibliography

CALDWELL, E.L. & L.W. PARR. 1933. Present status of handling water samples—Comparison of bacteriological analyses under varying temperatures and holding conditions, with special reference to the direct method. *Amer. J. Pub. Health* 23:467.

ZOBELL, C.E. 1941. Apparatus for collecting water samples from different depths for bacteriological analysis. *J. Mar. Res.* 4:173.

COX, K.E. & F.B. CLAIBORNE. 1949. Effect of age and storage temperature on bacteriological water samples. *J. Amer. Water Works Ass.* 41:948.

PUBLIC HEALTH LABORATORY SERVICE WATER SUB-COMMITTEE. 1952. The effect of storage on the coliform and *Bacterium coli* counts of water samples. Overnight storage at room and refrigerator temperatures. *J. Hyg.* 50:107.

PUBLIC HEALTH LABORATORY SERVICE WATER SUB-COMMITTEE. 1953. The effect of storage on the coliform and *Bacterium coli* counts of water samples. Storage for six hours at room and refrigerator temperatures. *J. Hyg.* 51:559.

PUBLIC HEALTH LABORATORY SERVICE WATER SUB-COMMITTEE. 1953. The effect of sodium thiosulphate on the coliform and *Bacterium coli* counts of non-chlorinated water samples. *J. Hyg.* 51:572.

SHIPE, E.L. & A. FIELDS. 1956. Chelation as a method for maintaining the coliform index in water samples. *Pub. Health Rep.* 71:974.

MCCARTHY, J.A. 1957. Storage of water sample for bacteriological examinations. *Amer. J. Pub. Health* 47:971.

HOATHER, R.C. 1961. The bacteriological examination of water. *J. Inst. Water Eng.* 61:426.

COLES, H.G. 1964. Ethylenediamine tetra-acetic acid and sodium thiosulphate as protective agents for coliform organisms in water samples stored for one day at atmospheric temperature. *Proc. Soc. Water Treat. Exam.* 13:350.

LONSANE, B.K., N.M. PARHAD & N.U. RAO. 1967. Effect of storage temperature and time on the coliform in water samples. *Water Res.* 1:309.

LUCKING, H.E. 1967. Death rate of coliform bacteria in stored Montana water samples. *J. Environ. Health* 29:576.

VAN DONSEL, D.J. & E.E. GELDREICH. 1971. Relationships of Salmonellae to fecal coliforms in bottom sediments. *Water Res.* 5:1079.

907 STANDARD PLATE COUNT

1. Introduction

The Standard Plate Count procedure provides a standardized means of determining the density of aerobic and facultative anaerobic heterotrophic bacteria in water. This is an empirical measurement because bacteria occur singly, in pairs, chains, clusters, or packets, and no single growth medium or set of physical and chemical conditions can satisfy the physiological requirements of all bacteria in a water sample. Consequently, the number of colonies may be lower substantially than the actual number of viable bacteria present. To facilitate the collection of reliable data for water quality control measurements, especially for comparative and legal purposes, a standardized plate count procedure is useful.

2. Work Area

Provide a level table or bench top with ample area in a clean, draft-free, well-lighted room. Use table and bench tops having a nonporous surface and disinfect before any analysis is made.

3. Samples

Collect water as directed in Section 906A. Initiate analysis as soon as possible to minimize changes in bacterial population. The recommended maximum elapsed time between collection and examination of unrefrigerated samples is 8 hr (maximum transit time 6 hr, maximum processing time 2 hr). When analysis cannot begin within 8 hr, maintain sample at a temperature below 10 C. Do not let maximum elapsed time between collection and analysis exceed 30 hr.

Hold or transport bottled water samples obtained from retail outlets unrefrigerated provided the temperature does not exceed 20 to 25 C. Examine freshly bottled samples (less than 48 hr old) within 6 hr of collection if unrefrigerated and within 30 hr if refrigerated.

4. Sample Preparation

Mark each plate with sample number, dilution, date, and any other necessary information before examination. Prepare at least duplicate plates for each volume of sample or dilution examined.

Thoroughly mix all samples or dilutions by making 25 complete up-and-down (or back-and-forth) movements of about 0.3 m in 7 sec. Optionally, use a mechanical shaker to shake dilution blanks for 15 sec.

5. Sample Dilution

Prepare water used for dilution blanks as directed in Media Specifications, Section 905C.

a. Selecting dilutions: Select the dilution(s) so that the total number of colonies on a plate will be between 30 and 300 (Figure 907:1). For example, where a Standard Plate Count as high as 3,000 is suspected, prepare plates containing 1:100 dilution.

For most potable water samples, plates suitable for counting will be obtained by planting 1 mL and 0.1 mL undiluted sample and 1 mL sample diluted 1:100.

b. Measuring sample portions: Use a sterile pipet for initial and subsequent transfers from each container. If pipet becomes contaminated before transfers are completed, replace with a sterile pipet. Use a separate sterile pipet for transfers from each different dilution. Do not prepare dilutions and pour plates in direct sunlight. Use caution when removing sterile pipets from the container; to avoid contamination, do not drag pipet tip across exposed ends of pipets or across lips and necks of dilution bottles. When removing sample, do not insert pipets more than 2.5 cm below the surface of sample or dilution.

c. Measuring dilutions: When discharging sample portions, hold pipet at an angle of about 45° with tip touching bot-

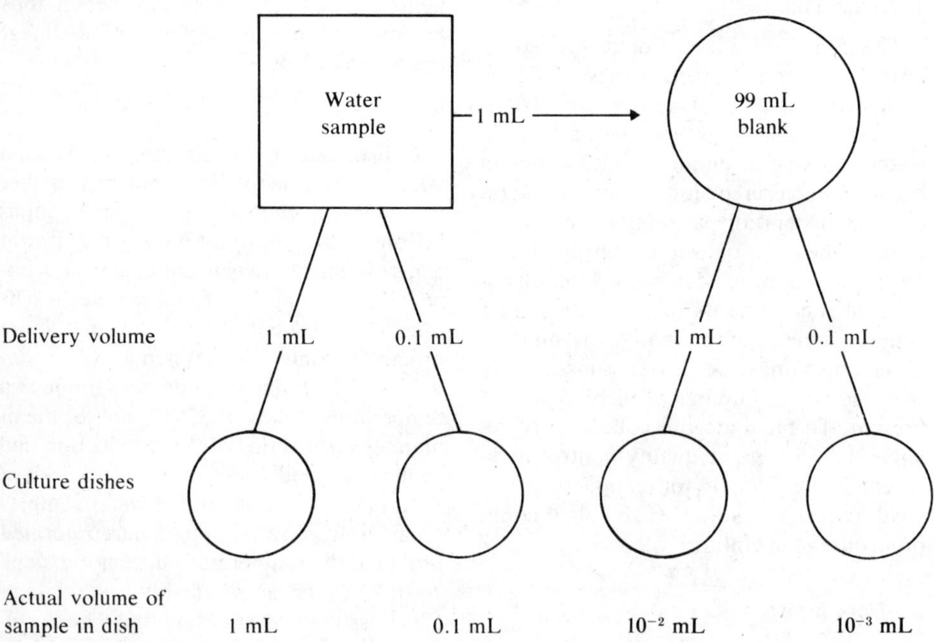

Figure 907:1. Preparation of dilutions.

tom of petri dish or inside neck of dilution bottle. Lift cover of petri dish just high enough to insert pipet. Allow 2 to 4 sec for liquid to drain from 1-mL graduation mark to tip of pipet. If pipet is not a blow-out type, touch tip of pipet *once* against a dry spot on petri dish bottom. Less preferably, use a blow-out-type pipet and insert a cotton plug in the mouthpiece before sterilizing it; gently blow out remaining volume of sample dilution. When 0.1-mL quantities are measured, let diluted sample drain from chosen reference graduation until 0.1 mL has been delivered. Remove pipet without retouching it to dish. Pipet 1 mL, 0.1 mL, or other suitable volume in sterile petri dish before adding melted culture medium. Use decimal dilutions in preparing sample volumes of less than 1 mL; in examining sewage or turbid water, do not measure a 0.1-mL inoculum of original sample, but prepare an appropriate dilution. Prepare at least two replicate plates for each sample dilution used. After depositing test portions for each series of plates, pour culture medium.

6. Plating

a. Melting medium: Melt sterile solid agar medium in boiling water or by exposure to flowing steam in a partially closed container, but avoid prolonged exposure to unnecessarily high temperatures during and after melting. Do not resterilize plating medium. If the medium is melted in two or more batches, use all of each batch in order of melting, provided that the contents remain fully melted. Discard melted agar that contains precipitate.

Maintain melted medium in a water bath between 44 C and 46 C until used. In a separate container place a thermometer in water or medium that has been exposed to the same heating and cooling as the plating medium. Do not depend on the sense of touch to indicate proper temperature when pouring agar.

Use tryptone glucose extract agar or plate count agar, as specified in Section 905C. Before using a new lot of medium test its suitability as directed in 905A.4.

b. Pouring plates: Limit the number of samples to be plated in any one series so that no more than 20 min (preferably 10 min) elapse between dilution of the first sample and pouring of the last plate in the series. Pour at least 10 to 12 mL liquefied medium at 44 to 46 C into each dish by gently lifting cover just high enough to pour. Carefully avoid spilling medium on outside of container or on inside of dish lid when pouring. As each plate is poured mix melted medium thoroughly with test portions in petri dish, taking care not to splash mixture over the edge, by rotating the dish first in one direction and then in the opposite direction, or by rotating and tilting. Let plates solidify (within 10 min) on a level surface. After medium solidifies, invert plates and place in the incubator.

c. Sterility controls: Check sterility of medium and dilution water blanks by pouring control plates for each series of samples. Prepare additional controls also to determine contamination of plates, pipets, and room air.

7. Incubation

Incubate for the Standard Plate Count for all water samples except bottled water at 35 ± 0.5 C for 48 ± 3 hr.

For the Standard Plate Count of bottled water, incubate at 35 ± 0.5 C for 72 ± 4 hr. Because bacteria found in bottled water demonstrate a prolonged lag phase during adaptation to growth on tryptone glucose extract agar or plate count agar, such bacteria do not form colonies that can be counted after 48 hr incubation; thus an additional 24 hr incubation is required to obtain a reliable Standard Plate Count.

Pack plates as directed under Laboratory Apparatus, Section 903, without crowding in the incubator. State any deviation from this method in the examination report.

8. Counting and Recording

Count all colonies on selected plates promptly after incubation. If counting must be delayed temporarily, store plates at 5 to 10 C for no more than 24 hr, but avoid this as routine practice. Record results of sterility controls on the report for each lot of samples.

Use an approved counting aid, such as the Quebec colony counter, for manual counting. If such equipment is not available, count with any other counter provided that it gives equivalent magnification and illumination. Automatic plate counting instruments now are available. These generally use a television scanner coupled to a magnifying lens and an electronics package. Their use is acceptable if evaluation in parallel with manual counting gives comparable results.

In preparing plates, plant sample volumes that will give from 30 to 300 colonies/plate. The aim is to have at least one dilution for which the plates give colony counts between these limits, except as provided below.

Ordinarily, do not plant more than 1.0 mL water in a plate; therefore, when the total number of colonies developing from 1.0 mL is less than 30, disregard the rule above and record result observed. With this exception, consider only plates showing 30 to 300 colonies in determining the Standard Plate Count. Compute bacterial count per milliliter by multiplying average number of colonies per plate by the dilution used. Report as the "Standard Plate Count" per milliliter.

If there is no plate with 30 to 300 colonies, and one or more plates have more than 300 colonies, use the plate(s) having a count nearest 300 colonies. Compute the count by multiplying average count per plate by the reciprocal of the dilution used and report as the "Estimated Standard Plate Count" per milliliter.

If plates from all dilutions of any sample have no colonies, report the count as less than one (<1) times the reciprocal of the corresponding lowest dilution.

If the number of colonies per plate far exceeds 300, do not report result as "too numerous to count" (TNTC). If there are fewer than 10 colonies/cm^2, count colonies in 13 squares (of the colony counter) having representative colony distribution. If possible, select seven consecutive squares horizontally across the plate and six consecutive squares at right angles, being careful not to count a square more than once. Multiply sum of the number of colonies in 13 representative cm^2 by 5 to compute estimated colonies per plate when the plate area is 65 cm^2. When there are more than 10 colonies/cm^2, count four representative squares, take average count per square centimeter, and multiply by the appropriate factor to estimate colonies per dish (usually about 65 for glass petri dishes). When bacterial counts on crowded plates are greater than 100 colonies/cm^2, report result as greater than (>) 6,500 times the reciprocal of the highest dilution plated.

If spreading colonies (spreaders) are encountered on the plate(s) selected, count colonies on representative portions only when (*a*) colonies are well distributed in spreader-free areas, and (*b*) the area covered by the spreader(s) does not exceed one-half the plate area.

When spreading colonies must be counted, count each unit of the following types as one: (*a*) a chain of colonies that appears to be caused by disintegration of a bacterial clump as agar and sample were mixed. Count each such chain as a single colony, do not count each individual colony in the chain; (*b*) a spreader that develops as a film of growth between the agar and bottom of petri dish; (*c*) a colony that forms in a film of water at the edge or over agar surface. Types *b* and *c* largely develop because of an accumulation of moisture at the point from which the spreader originates. They frequently cover more than half the plate and interfere with obtaining a reliable plate count.

Count as individual colonies similar appearing colonies growing in close proximity but not touching provided that the distance between them is at least equal to the diameter of the smallest colony. Count impinging colonies that differ in appearance, such as morphology or color, as individual colonies.

If plates have excessive spreader growth, report as "Spreaders" (Spr). When plates are uncountable because of missed dilution, accidental dropping, and contamination, or the control plates indicate that the medium or other material or labware was contaminated, report as "Laboratory Accident" (LA).

9. Computing and Reporting Counts

To compute the Standard Plate Count, multiply total number of colonies or average number (if duplicate plates of the same dilution) per plate by the reciprocal of the dilution used. Record dilutions used and number of colonies on each plate counted or estimated.

When colonies on duplicate plates and/or consecutive dilutions are counted and results are averaged before being recorded, round off counts to two significant figures only when converting to the Standard Plate Count.

Avoid creating fictitious precision and accuracy when computing Standard Plate Counts, by recording only the first two left-hand digits. Raise the second digit to the next highest number when the third digit from the left is 5, 6, 7, 8, or 9; use zeros for each successive digit toward the right from the second digit. For example, report a count of 142 as 140 and a count of 155 as 160, but report a count of 35 as 35.

Report counts as "Standard Plate Count" or "Estimated Standard Plate Count" per milliliter.

10. Personal Errors

Avoid inaccuracies in counting due to carelessness, damaged or dirty optics that impair vision, or failure to recognize colonies. Laboratory workers who cannot duplicate their own counts on the same plate within 5% and the counts of other analysts within 10% should discover the cause and correct such disagreements.

11. Bibliography

BREED, R.S. & W.D. DOTTERER. 1916. The number of colonies allowable on satisfactory agar plates. Tech. Bull. 53, N.Y. Agr. Exp. Sta.

BUTTERFIELD, C.T. 1933. The selection of a dilution water for bacteriological examinations. *J. Bacteriol.* 23:355; *Pub. Health Rep.* 48:681.

ARCHAMBAULT, J., J. CUROT & M.H. MCCRADY. 1937. The need of uniformity of conditions for counting plates (with suggestions for a standard colony counter). *Amer. J. Pub. Health* 27:809.

RICHARDS, O.W. & P.C. HEIJN. 1945. An improved darkfield Quebec colony counter. *J. Milk Technol.* 8:253.

BERRY, J.M., D.A. MCNEILL & L.D. WITTER. 1969. Effect of delays in pour plating on bacterial counts. *J. Dairy Sci.* 52:1456.

AMERICAN PUBLIC HEALTH ASSOCIATION. 1972. Standard Methods for the Examination of Dairy Products, 13th ed. APHA, New York, N.Y.

GELDREICH, E.E., H.D. NASH, D.J. REASONER & R.H. TAYLOR. 1972. The necessity of controlling bacterial populations in potable waters: Community water supply. *J. Amer. Water Works Ass.* 64:596.

GELDREICH, E.E., H.D. NASH, D.J. REASONER & R.H. TAYLOR. 1975. The necessity for controlling bacterial populations in potable waters: Bottled water and emergency water supplies. *J. Amer. Water Works Ass.* 67:117.

908 MULTIPLE-TUBE FERMENTATION TECHNIC FOR MEMBERS OF THE COLIFORM GROUP

The coliform group comprises all aerobic and facultative anaerobic, gram-negative, nonspore-forming, rod-shaped bacteria that ferment lactose with gas formation within 48 hr at 35 C.*

The standard test for the coliform group may be carried out either by the multiple-tube fermentation technic (presumptive test, confirmed test, or completed test) described herein or by the membrane filter technic (Section 909). Each technic is applicable within the limitations specified and with due consideration of the purpose of the examination.

As applied to the membrane filter technic, the coliform group may be redefined as comprising all aerobic and facultative anaerobic, gram-negative, nonspore-forming, rod-shaped bacteria that produce a dark colony with a metallic sheen within 24 hr at 35 C on an Endo-type medium containing lactose.

Even after the prescribed shaking, the distribution of bacteria in water is irregular. It is entirely possible to divide a water sample into portions and, after testing, find that the number of organisms in any portion may be none, or at least less than the arithmetic average based on examination of the total volume. It also is quite probable that the growth in a fermentation tube may result not from one organism but from many organisms. It is reasonable, however, to assume that growth develops from a single individual.

Results of the examination of replicate tubes and dilutions are reported in terms of the Most Probable Number (MPN). This term is actually an estimate based on certain probability formulas. Theoretical considerations and large-scale replicate determinations indicate that this estimate tends to be greater than the actual number and that the disparity tends to diminish with increasing numbers of tubes in each dilution examined.

The accuracy of any single test depends on the number of tubes used. The most satisfactory information will be obtained when the largest portion examined shows gas in some or all of the tubes and the smallest portion shows no gas in all or a majority of the tubes. The numerical value of the estimation of the bacterial content is determined largely by the dilution that shows both positive and negative results. The number of portions scheduled, especially in the critical dilution, will be governed by the desired accuracy of the result. The increased interest in the multiple-tube technic, the numerous investigations into its precision, and the expression of test results as MPN's should not lead the analyst to regard this method as a statistical exercise rather than a means of estimating the coliform density of a water and its sanitary quality. The best assessment of the sanitary quality of a water still must depend on the interpretation of results of the multiple-tube technic—or of other methods, possibly more precise—and of all other information regarding a water that may be obtained by surveys or otherwise.

1. Water of Drinking Water Quality

When water is examined for evidence of drinking water quality that meets the standards of the U.S. Environmental Protection Agency (EPA), use five fermentation tubes, each containing 10 mL or 100 mL of sample. It is generally impractical to use portions larger than 100 mL. Use the Confirmed Test (908A.2) when ex-

*The "coliform group" as defined above is equivalent to the "*B. coli* group" as used in the third, fourth and fifth editions of this manual, and to the "*coli-aerogenes* group" as used through the eighth edition.

amining drinking water and use the Completed Test (908A.3) as a reference standard on selected samples.

For water examined frequently, or even daily, inoculating five 10-mL or five 100-mL portions generally provides sufficient definite information. In examining other waters presumed to be of drinking-water quality, use at least three dilutions with five tubes per dilution to provide acceptable precision and a reasonably accurate coliform estimate.

For the routine examination of most potable water supplies, particularly those that are disinfected, the object of the test is to determine compliance with EPA standards as a measure of either the efficiency of operation or the presence of bacterial contamination. The assessment of potability generally is based on knowledge of the sanitary condition of the supply as determined by bacteriological monitoring. It is expected that more than 95% of all samples examined will yield negative results. An occasional positive result, unless repeated from the same sampling point (resampling), or unless it is one yielding three or more positive tubes when five tubes are inoculated, usually is of limited significance but should not be ignored. An increase in the number of positive samples over a period of time or an abrupt increase in a short period of time indicates a change in the quality of the water, the significance of which should be studied, with correction made as necessary.

2. Water of Other than Drinking Water Quality

In the examination of nonpotable waters inoculate a series of tubes with decimal quantities of the water; the selection of portion sizes depends on the probable coliform density as indicated by the experience of the analyst and how much is known about the character of the water. The object of the examination of nonpotable water generally is to estimate the density of bacterial contamination, determine a source of pollution, enforce water quality standards, or trace the survival of microorganisms. Each objective requires a numerical value for reporting results. The multiple-tube fermentation technic may be used; however, to obtain statistically valid MPN values, run a series of five tubes for each sample volume, each inoculated with decimal quantities of sample. Examine a sufficient number of samples to yield representative results for the sampling station. Generally, the log average or median value of the results of a number of samples will yield a value in which the effect of individual extreme values is minimized. The membrane filter technic may prove the better procedure to accomplish this objective.

3. Other Samples

The multiple-tube fermentation technic is applicable to the analysis of salt or brackish waters as well as muds, sediments, or sludges. Follow the precautions given above on portion sizes and numbers of tubes per dilution.

To prepare solid or semisolid samples weigh sample and add diluent to make a 10^{-1} dilution. For example, place 50 g sample in sterile blender jar, add 450 mL phosphate buffer or 0.1% peptone dilution water, and blend for 1 to 2 min at low speed (8,000 rpm).

908 A. Standard Total Coliform MPN Tests

1. Presumptive Test

Use lauryl tryptose broth in the Presumptive Test, but do not use positive results without confirmation.

a. Procedure:

1) Inoculate a series of fermentation tubes ("primary" fermentation tubes) with appropriate graduated quantities (multiples and submultiples of 1 mL) of sample. If 100-mL sample portions are used prewarm bottles at 35 C. After adding sample mix thoroughly. Be sure that the concentration of nutritive ingredients in the mixture of medium and added sample conforms to the requirements given in Section 905C, Media Specifications. The portions of sample used for inoculating lauryl tryptose broth fermentation tubes will vary in size and number with the character of the water under examination, but in general use decimal multiples and submultiples of 1 mL. Select these in accordance with the discussion of the multiple-tube test above.

In making dilutions and measuring diluted sample volumes, follow the precautions given in Section 907.5. Use Figure 907:1 as a guide to preparing dilutions.

2) Incubate inoculated fermentation tubes at 35 ± 0.5 C. After 24 ± 2 hr shake each tube gently and examine it and, if no gas has formed and been trapped in the inverted vial, reincubate and reexamine at the end of 48 ± 3 hr. Record presence or absence of gas formation regardless of amount at each examination of the tubes.

b. Interpretation: Formation of gas in any amount in the inner fermentation tubes or vials within 48 ± 3 hr constitutes a positive Presumptive Test.

Do not confuse the appearance of an air bubble in a clear tube with actual gas production. If gas is formed as a result of fermentation, the broth medium will become cloudy. Active fermentation may be shown by the continued appearance of small bubbles of gas throughout the medium outside the inner vial when the fermentation tube is shaken gently.

The absence of gas formation at the end of 48 ± 3 hr of incubation constitutes a negative test. An arbitrary limit of 48 hr for observation doubtless excludes from consideration occasional members of the coliform group that form gas very slowly and generally are of limited sanitary significance.

2. Confirmed Test

Use lauryl tryptose broth for the primary fermentation. Use brilliant green lactose bile broth fermentation tubes for the Confirmed Test.

a. Procedure: Submit all primary fermentation tubes showing any amount of gas within 24 hr of incubation to the confirmed test. If active fermentation appears in the primary fermentation tube earlier than 24 hr preferably transfer to the confirmatory medium without waiting for the full 24-hr period to elapse. If additional primary fermentation tubes show gas production at the end of 48-hr incubation, submit these to the Confirmed Test.

b. Procedure with brilliant green lactose bile broth:

Gently shake or rotate primary fermentation tube showing gas and, either with a sterile metal loop, 3 mm in diameter, transfer one loopful of culture to a fermentation tube containing brilliant green lactose bile broth, or insert a sterile wooden applicator at least 2.5 cm into the culture, promptly remove, and plunge applicator to bottom of fermentation tube containing brilliant green lactose bile broth. Remove and discard applicator.

Incubate the inoculated brilliant green lactose bile broth tube for 48 ± 3 hr at 35 ± 0.5 C.

Formation of gas in any amount in the inverted vial of the brilliant green lactose bile broth fermentation tube at any time within 48 ± 3 hr constitutes a positive Confirmed Test.

c. Alternative procedure: Use this alternative only for polluted water or wastewater known to produce positive results consistently.

If all presumptive tubes are positive in two or more consecutive dilutions within 24 hr, submit to the Confirmed Test only the tubes of the highest dilution (smallest volume) in which all tubes are positive and any positive tubes in still higher dilutions. Submit to the Confirmed Test all tubes in which gas is produced only after 48 hr.

3. Completed Test

Use the Completed Test on positive confirmed tubes to establish definitively the presence of coliform bacteria and to provide quality control data. Double confirmation into brilliant green lactose bile broth for total coliforms and EC broth for fecal coliforms (see Section 908C below) may be used. Consider positive EC broth results as a positive Completed Test response. Submit all other confirmation positive tubes, not doubly confirmed, to the Completed Test procedure.

a. Procedure:

1) Streak one or more eosin methylene blue plates from each tube of brilliant green lactose bile broth showing gas, as soon as possible after the appearance of gas. Streak plates to insure presence of some discrete colonies separated by at least 0.5 cm. Observe the following precautions when streaking plates to obtain a high proportion of successful isolations if coliform organisms are present: *(a)* Use an inoculating needle slightly curved at the tip; *(b)* tap and incline the fermentation tube to avoid picking up any membrane or scum on the needle; *(c)* insert end of needle into the liquid in the tube to a depth of approximately 5.0 mm; and *(d)* streak plate with curved section of the needle in contact with the agar to avoid a scratched or torn surface.

Incubate plates (inverted) at 35 ± 0.5 C for 24 ± 2 hr.

2) The colonies developing on eosin methylene blue agar are called *typical* (nucleated, with or without metallic sheen); *atypical* (opaque, unnucleated, mucoid, pink after 24 hr incubation), or *negative* (all others). From each of these plates pick one or more typical well-isolated coliform colonies or, if no typical colonies are present, pick two or more colonies considered most likely to consist of organisms of the coliform group, and transfer growth from each isolate to a lauryl tryptose broth fermentation tube and to a nutrient agar slant.

Use a colony magnifying device to provide optimum magnification when colonies are picked from the plates of selective medium.

If possible, when transferring colonies, choose well-isolated colonies and barely touch the surface of the colony with a flame-sterilized, air-cooled transfer needle to minimize the danger of transferring a mixed culture.

Incubate secondary broth tubes at 35 ± 0.5 C for 24 ± 2 hr; if gas is not produced within 24 ± 2 hr reincubate and examine again at 48 ± 3 hr. Microscopically examine Gram-stained preparations (see Section 908A.4 below) from those 24-hr agar slant cultures corresponding to the secondary tubes that show gas.

The Gram stain may be omitted from the Completed Test for potable water samples only.

b. Interpretation: Formation of gas in the secondary tube of lauryl tryptose broth within 48 ± 3 hr and demonstration of gram-negative, nonspore-forming, rod-shaped bacteria in the agar culture constitute a satisfactory Completed Test, demonstrating the presence of a member of the coliform group.

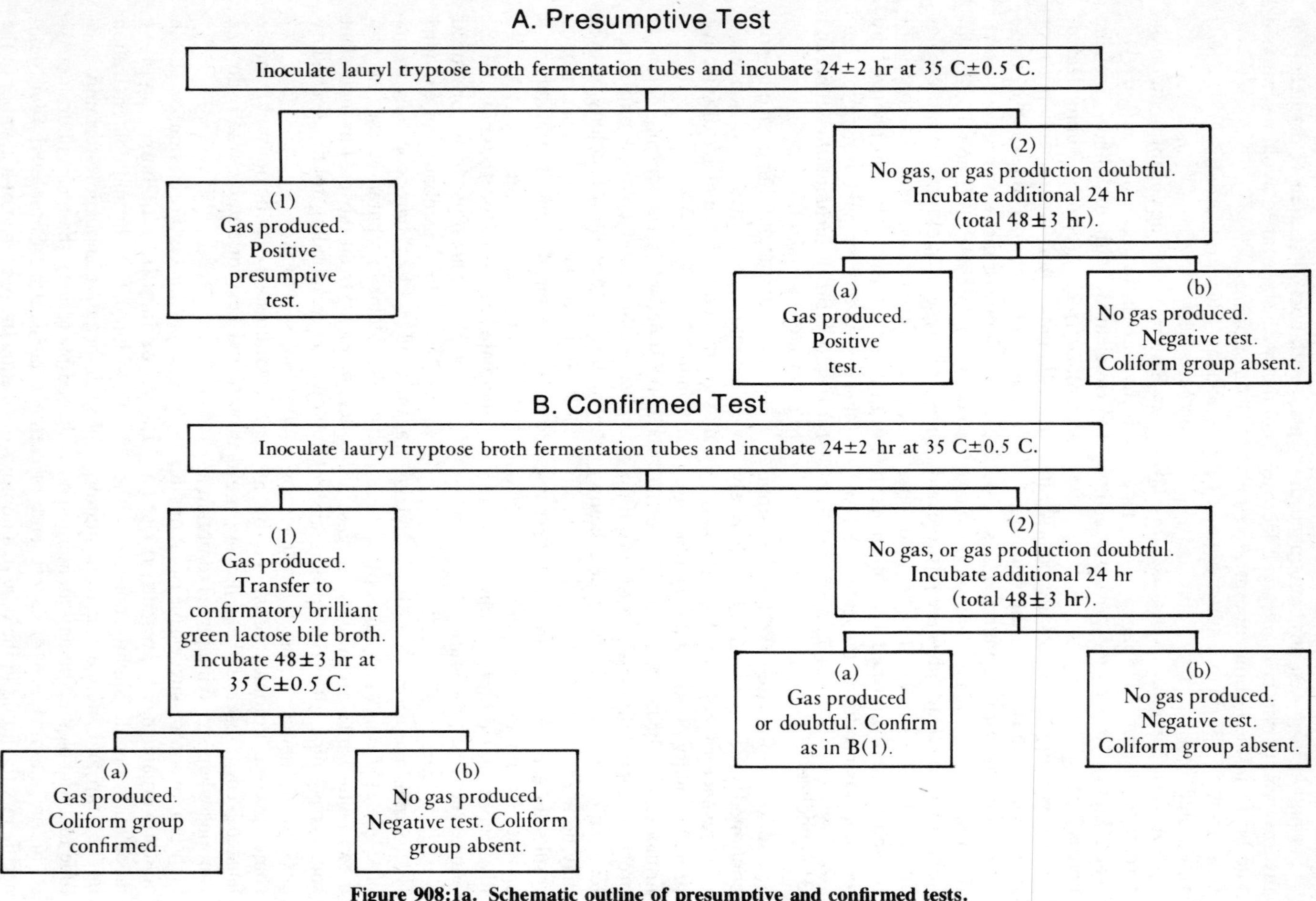

Figure 908:1a. Schematic outline of presumptive and confirmed tests.

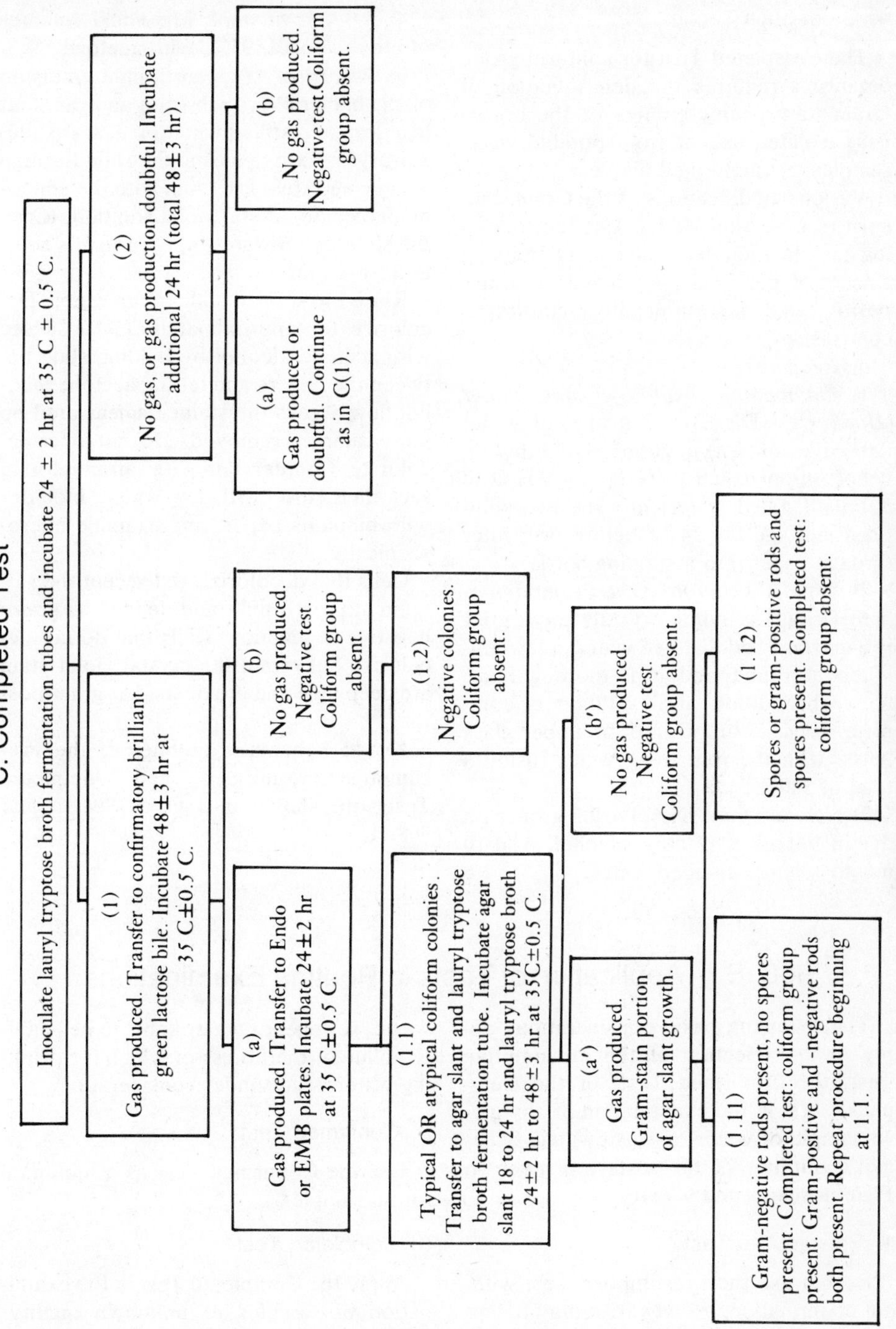

Figure 908:1b. Schematic outline of completed test.

4. Gram-Stain Technic

The Completed Test for coliform-group organisms requires the determination of Gram-stain characteristics of the organisms isolated except from potable water samples, as mentioned above.

Various modifications of the Gram stain exist (see Section 908E). Use the following modification by Hucker for staining smears of pure culture; include a gram-positive and a gram-negative culture as controls.

a. Reagents:

1) *Ammonium oxalate-crystal violet (Hucker's):* Dissolve 2 g crystal violet (90% dye content) in 20 mL 95% ethyl alcohol; dissolve 0.8 g $(NH_4)_2C_2O_4 \cdot H_2O$ in 80 mL distilled water; mix the two solutions and age for 24 hr before use; filter through paper into a staining bottle.

2) *Lugol's solution, Gram's modification:* Grind 1 g iodine crystals and 2 g KI in a mortar. Add distilled water, a few milliliters at a time, and grind thoroughly after each addition until solution is complete. Rinse solution into an amber glass bottle with the remaining water (using a total of 300 mL).

3) *Counterstain:* Dissolve 2.5 g safranin dye in 100 mL 95% ethyl alcohol. Add 10 mL to 100 mL distilled water.

4) *Acetone alcohol:* Mix equal volumes of ethyl alcohol, 95%, with acetone.

b. Procedure: Prepare a light emulsion of the bacterial growth from an agar slant in a drop of distilled water on a glass slide. Air-dry or fix by passing the slide through a flame and stain for 1 min with the ammonium oxalate-crystal violet solution. Rinse the slide in tap water; apply Lugol's solution for 1 min.

Rinse the stained slide in tap water. Decolorize for approximately 15 to 30 sec with acetone alcohol by holding slide between the fingers and letting acetone alcohol flow across the stained smear until no more stain is removed. Do not over-decolorize. Counterstain with safranin for 15 sec, then rinse with tap water, blot dry with bibulous paper, and examine microscopically.

Cells that decolorize and accept the safranin stain are pink and defined as gram-negative in reaction. Cells that do not decolorize but retain the crystal violet stain are deep blue and are defined as gram-positive.

NOTE: Schematic outlines of the Presumptive, Confirmed, and Completed Tests are shown in Figures 908:1a and 908:1b.

908 B. Application of Tests to Routine Examinations

The following basic considerations apply to the selection of the Presumptive Test, the Confirmed Test, or the Completed Test in the examination of any given sample of water or wastewater. Schematic outlines of the tests are given in Figures 908:1a and 908:1b.

1. Presumptive Test

Do not use the Presumptive Test without confirmation. However, it may be applied to the examination of any sample of waste, sewage, or water know to be heavily polluted, the fitness of which for drinking water is not under consideration.

2. Confirmed Test

Use the Confirmed Test as a minimum for all samples.

3. Completed Test

Apply the Completed Test in the examination of samples of unknown quality, samples yielding unexpected results, samples with high non-coliform counts, and for quality control purposes (see Section 902). If the Completed Test is not applied to all samples, apply it to at least 10% of positive samples to establish beyond reasonable doubt the value of the Confirmed Test in determining sanitary quality of water supplies. Analyze by the Completed Test repeat samples of finished water from the same location that consistently show three or more positive 10-mL portions by the Confirmed Test.

908 C. Fecal Coliform MPN Procedure

Elevated-temperature tests for the separation of organisms of the coliform group into those of fecal origin and those derived from nonfecal sources are available. Recent modifications in technical procedures, standardization of methods, and detailed studies of members of the coliform group found in the feces of various warm-blooded animals compared with those from other environmental sources have established the value of a fecal coliform determination. Make test by one of the multiple-tube procedures described here or by membrane filter methods as described in the section on the membrane filter technic. The procedure using EC medium yields adequate information about the source of the coliform group (fecal or nonfecal) when used as a *confirmatory test*. Do not use it for direct isolation of coliforms from water because prior enrichment in a Presumptive Test medium for optimum recovery of fecal coliforms is required. The tentative procedure using A-1 broth is a single-step method not requiring confirmation.

The fecal coliform test (EC medium) is applicable to investigations of stream pollution, raw water sources, wastewater treatment systems, bathing waters, seawaters, and general water-quality monitoring. The procedure is not recommended as a substitute for the coliform test in the examination of potable waters, because no coliform bacteria of any kind should be tolerated in a treated water. The test using A-1 medium is applicable to seawaters.

1. Fecal Coliform Test (EC Medium)

The fecal coliform test may be expected to differentiate between coliforms of fecal origin (intestines of warm-blooded animals) and coliforms from other sources. Use EC medium as described in Section 905C, Media Specifications.

a. Procedure: Make transfers from all positive presumptive tubes from the total coliform MPN test to EC medium. Make this examination simultaneously with the confirmatory procedure using brilliant green lactose bile broth. Use a sterile metal loop with a minimum 3-mm diam or a sterile wooden applicator to transfer from the positive fermentation tube to EC medium. When making such transfers, first gently shake the presumptive tube or mix by rotating. Incubate inoculated tubes in a water bath at 44.5 ± 0.2 C for 24 ± 2 hr. Place all EC tubes in the water bath within 30 min after inoculation. Maintain the water depth in the incubator sufficient to immerse tubes to the upper level of the medium.

b. Interpretation: Gas production in a fermentation tube within 24 hr or less is considered a positive reaction indicating fecal origin. Failure to produce gas (growth sometimes occurs) constitutes a negative reaction indicating a source other than the intestinal tract of warm-blooded animals. Calculate fecal coliform densities as described under Estimation of Bacterial Density (Section 908D below).

2. Fecal Coliform Test (A-1 Medium) (TENTATIVE)

a. Procedure: Inoculate tubes of A-1 broth (see Section 905C) as directed in Section 908A.1*a*1). Incubate for 3 hr at 35 ± 0.5 C. Transfer tubes to a water bath at 44.5 ± 0.2 C and incubate for an additional 21 ± 2 hr.

b. Interpretation: Gas production in a fermentation tube within 24 hr or less is a positive reaction indicating coliforms of fecal origin. Calculate fecal coliform densities as described in Section 908D.

908 D. Estimation of Bacterial Density

1. Precision of Fermentation Tube Test

Unless a large number of sample portions is examined, the precision of the fermentation tube test is rather low. For example, even when the sample contains 1 coliform organism/mL, about 37% of 1-mL tubes may be expected to yield negative results because of irregular distribution of the bacteria in the sample. When five tubes, each with 1 mL sample, are used under these conditions, a completely negative result may be expected less than 1% of the time.

Even when five fermentation tubes are used, the precision of the results obtained is not of a high order. Consequently, exercise great caution when interpreting the sanitary significance of coliform results obtained from the use of a few tubes with each sample dilution, especially when the number of samples from a given sampling point is limited.

2. Computing and Recording of MPN

Record the number of positive findings of coliform group organisms (either presumptive, confirmed, or completed) resulting from multiple-portion decimal-dilution plantings as the combination of positives and compute in terms of the Most Probable Number (MPN). The MPN, for a variety of planting series and results, is given in Tables 908:I and 908:II. Included in these tables are the 95% confidence limits for each MPN value determined.

TABLE 908:I. MPN INDEX AND 95% CONFIDENCE LIMITS FOR VARIOUS COMBINATIONS OF POSITIVE AND NEGATIVE RESULTS WHEN FIVE 10-ML PORTIONS ARE USED

No. of Tubes Giving Positive Reaction out of 5 of 10 mL Each	MPN Index /100 mL	95% Confidence Limits	
		Lower	Upper
0	< 2.2	0	6.0
1	2.2	0.1	12.6
2	5.1	0.5	19.2
3	9.2	1.6	29.4
4	16.	3.3	52.9
5	>16.	8.0	Infinite

The sample volumes indicated in Table 908:II relate more specifically to finished waters. Use the values in computing the MPN in larger or smaller portion plantings in the following manner: If, instead of portions of 10, 1.0, and 0.1 mL, a combination of portions of 100, 10, and 1 mL is used, record the MPN as 0.1 times the value given in the applicable table. If, on the other hand, a combination of corresponding portions at 1.0, 0.1, and 0.01 mL is planted, record 10 times the value shown in the table; if a combination of portions of 0.1, 0.01, and 0.001 mL is planted, record 100 times the value shown

TABLE 908.II. MPN INDEX AND 95% CONFIDENCE LIMITS FOR VARIOUS COMBINATIONS OF POSITIVE RESULTS WHEN FIVE TUBES ARE USED PER DILUTION (10 mL, 1.0 mL, 0.1 mL)

Combination of Positives	MPN Index /100 mL	95% Confidence Limits		Combination of Positives	MPN Index /100 mL	95% Confidence Limits	
		Lower	Upper			Lower	Upper
0-0-0	<0	—	—	4-2-0	22	7	67
0-0-1	2	<0.5	7	4-2-1	26	9	78
0-1-0	2	<0.5	7	4-3-0	27	9	80
0-2-0	4	<0.5	11	4-3-1	33	11	93
				4-4-0	34	12	93
1-0-0	2	<0.5	7	5-0-0	23	7	70
1-0-1	4	<0.5	11	5-0-1	31	11	89
1-1-0	4	<0.5	11	5-0-2	43	15	110
1-1-1	6	<0.5	15	5-1-0	33	11	93
1-2-0	6	<0.5	15	5-1-1	46	16	120
				5-1-2	63	21	150
2-0-0	5	<0.5	13	5-2-0	49	17	130
2-0-1	7	1	17	5-2-1	70	23	170
2-1-0	7	1	17	5-2-2	94	28	220
2-1-1	9	2	21	5-3-0	79	25	190
2-2-0	9	2	21	5-3-1	110	31	250
2-3-0	12	3	28	5-3-2	140	37	340
3-0-0	8	1	19	5-3-3	180	44	500
3-0-1	11	2	25	5-4-0	130	35	300
3-1-0	11	2	25	5-4-1	170	43	490
3-1-1	14	4	34	5-4-2	220	57	700
3-2-0	14	4	34	5-4-3	280	90	850
3-2-1	17	5	46	5-4-4	350	120	1,000
4-0-0	13	3	31	5-5-0	240	68	750
4-0-1	17	5	46	5-5-1	350	120	1,000
4-1-0	17	5	46	5-5-2	540	180	1,400
4-1-1	21	7	63	5-5-3	920	300	3,200
4-1-2	26	9	78	5-5-4	1,600	640	5,800
				5-5-5	≥2,400	—	—

in the table; and so on for other combinations.

When more than three dilutions are used in a decimal series of dilutions, use the results from only three of these in computing the MPN. To select the three dilutions to be used in determining the MPN index, choose the highest dilution that gives positive results in all five portions tested (no lower dilution giving any negative results) and the two next succeeding higher dilutions. Use the results at these three volumes in computing the MPN index. In the examples given below, the significant dilution results are shown in boldface. The number in the numerator represents positive tubes; that in the denominator, the total tubes planted; the combination of positives simply represents the total number of positive tubes per dilution:

Example	1 mL	0.1 mL	0.01 mL	0.001 mL	Combination of positives
(a)	5/5	**5/5**	**2/5**	**0/5**	5-2-0
(b)	**5/5**	**4/5**	**2/5**	0/5	5-4-2
(c)	**0/5**	**1/5**	**0/5**	0/5	0-1-0

In *c*, take the first three dilutions so as to throw the positive result in the middle dilution.

When a case such as that shown below in line *d* arises, where a positive occurs in a dilution higher than the three chosen according to the rule, incorporate it in the result for the highest chosen dilution, as in *e*:

When it is desired to summarize with a single MPN value the results from a series of samples, use the geometric mean, the arithmetic mean, or the median.

Example	1 mL	0.1 mL	0.01 mL	0.001 mL	Combination of positives
(d)	**5/5**	**3/5**	**1/5**	1/5	5-3-2
(e)	**5/5**	**3/5**	**2/5**	0/5	

Table 908:II does not include all positive combinations; however, the most likely ones are shown. If unlikely combinations occur with a frequency greater than 1% it is an indication that the technic is faulty or that the statistical assumptions underlying the MPN estimate are not being fulfilled. The MPN for combinations not appearing in the table, or for other combinations of tubes or dilutions, may be estimated by Thomas' simple formula:

$$\text{MPN/100 mL} = \frac{\text{no. of positive tubes} \times 100}{\sqrt{\left(\begin{matrix}\text{mL sample in}\\ \text{negative tubes}\end{matrix}\right) \times \left(\begin{matrix}\text{mL sample in}\\ \text{all tubes}\end{matrix}\right)}}$$

While the MPN tables and calculations are described for use in the coliform test, they are equally applicable to determination of the MPN of any organisms provided a suitable test is available.

908 E. Bibliography

Standard Tests

MEYER, E.M. 1918. An aerobic spore-forming bacillus giving gas in lactose broth isolated in routine water examination. *J. Bacteriol.* 3:9.

HUCKER, G.J. & H.J. CONN. 1923. Methods of Gram staining. N.Y. State Agr. Exp. Sta. Tech. Bull. No. 93.

NORTON, J.F. & J.J. WEIGHT. 1924. Aerobic spore-forming lactose fermenting organisms and their significance in water analysis. *Amer. J. Pub. Health* 14:1019.

HUCKER, G.J. & H.J. CONN. 1927. Further studies on the methods of gram staining. N.Y. State Agr. Exp. Sta. Tech. Bull. No. 128.

PORTER, R., C.S. MCCLESKEY & M. LEVINE. 1937. The facultative sporulating bacteria producing gas from lactose. *J. Bacteriol.* 33:163.

COWLES, P.B. 1939. A modified fermentation tube. *J. Bacteriol.* 38:677.

BREED, R.S., E.G.D. MURRAY & N.R. SMITH. 1957. Bergey's Manual of Determinative

Bacteriology, 7th ed. Williams & Wilkins, Baltimore, Md.

AMERICAN SOCIETY FOR MICROBIOLOGY. 1957. Manual of Microbiological Methods. McGraw Hill, New York, N.Y.

SHERMAN, V.B.D. 1967. A Guide to the Identification of the Genera of Bacteria. Williams & Wilkins, Baltimore, Md.

AMERICAN PUBLIC HEALTH ASSOCIATION. 1970. Recommended Procedures for the Examination of Sea Water and Shellfish, 4th ed. APHA, New York, N.Y.

GELDREICH, E.E. 1975. Handbook for Evaluating Water Bacteriological Laboratories, 2nd ed. EPA-670/9-75-006.

Fecal Coliform Tests

PERRY, C.A. & A.A. HAJNA. 1933. A modified Eijkman medium. *J. Bacteriol.* 25:419.

PERRY, C.A. & A.A. HAJNA. 1944. Further evaluation of EC medium for the isolation of coliform bacteria and *Escherichia coli. Amer. J. Pub. Health* 34:735.

VAUGHN, R.H. et al. 1951. A buffered boric acid lactose medium for enrichment and presumptive identification of *Escherichia coli. Food Res.* 16:10.

LEVINE, M., R.H. TANIMOTO, H. MINETTE, J. ARAKAKI & G. FERNANDES. 1955. Simultaneous determination of coliform and *Escherichia coli* indices. *Appl. Microbiol.* 3:310.

CLARK, H.F., E.E. GELDREICH, P.W. KABLER, R.H. BORDNER & C.B. HUFF. 1957. The coliform group. I. The boric acid lactose broth reaction of coliform IMViC types. *Appl. Microbiol.* 5:396.

GELDREICH, E.E., H.F. CLARK, P.W. KABLER, C.B. HUFF & R.H. BORDNER. 1958. The coliform group. II. Reactions in EC medium at 45 C. *Appl. Microbiol.* 6:347.

GELDREICH, E.E., R.H. BORDNER, C.B. HUFF, H.F. CLARK & P.W. KABLER. 1962. Type distribution of coliform bacteria in the feces of warm-blooded animals. *J. Water Pollut. Control Fed.* 34:295.

GELDREICH, E.E. 1966. Sanitary significance of fecal coliforms in the environment. FWPCA Publ. WP-20-3 (Nov.). U.S. Dep. Interior, Washington, D.C.

ANDREWS, W.H. & M.W. PRESNELL. 1972. Rapid recovery of *Escherichia coli* from estuarine water. *Appl. Microbiol.* 23:521.

Numerical Interpretation

MCCRADY, M.N. 1915. The numerical interpretation of fermentation tube results. *J. Infect. Dis.* 12:183.

GREENWOOD, M. & G.U. YULE. 1917. On the statistical interpretation of some bacteriological methods employed in water analysis. *J. Hyg.* 16:36.

WOLMAN, A. & H.L. WEAVER. 1917. A modification of the McCrady method of the numerical interpretation of fermentation tubes results. *J. Infect. Dis.* 21:287.

MCCRADY, M.H. 1918. Tables for rapid interpretation of fermentation tube results. *Can. J. Pub. Health* 9:201.

REED, L.J. 1925. *B. coli* densities as determined from various types of samples. *Pub. Health Rep.* 40:704 (Reprint 1029).

HOSKINS, J.K. 1933. The most probable number of *B. coli* in water analysis. *J. Amer. Water Works Ass.* 25:867.

HOSKINS, J.K. 1934. Most Probable Numbers for evaluation of *Coli-Aerogenes* tests by fermentation tube method. *Pub. Health Rep.* 49:393 (Reprint 1621).

HOSKINS, J.K. & C.T. BUTTERFIELD. 1935. Determining the bacteriological quality of drinking water. *J. Amer. Water Works Ass.* 27:1101.

HALVORSON, H.O. & N.R. ZIEGLER. 1933-35. Application of statistics to problems in bacteriology. *J. Bacteriol.* 25:101; 26:331, 559; 29:609.

SWAROOP, S. 1938. Numerical estimation of *B. coli* by dilution method. *Indian J. Med. Res.* 26:353.

DALLA VALLE, J.M. 1941. Notes on the most probable number index as used in bacteriology. *Pub. Health Rep.* 56:229.

THOMAS, H.A., JR. 1942. Bacterial densities from fermentation tube tests. *J. Amer. Water Works Ass.* 34:572.

AMERICAN PUBLIC HEALTH ASSOCIATION, AMERICAN WATER WORKS ASSOCIATION & FEDERATION OF SEWAGE AND INDUSTRIAL WASTES ASSOCIATIONS. 1955. Standard Methods for the Examination of Water, Sewage, and Industrial Wastes, 10th ed. APHA, New York, N.Y.

WOODWARD, R.L. 1957. How probable is the Most Probable Number? *J. Amer. Water Works Ass.* 49:1060.

MCCARTHY, J.A., H.A. THOMAS & J.E. DELANEY. 1958. Evaluation of reliability of coliform density tests. *Amer. J. Pub. Health* 48:12.

U.S. ENVIRONMENTAL PROTECTION AGENCY. 1975. Interim primary drinking water standards. *Fed. Reg.* 40(51):11990 (Mar. 14, 1975).

909 MEMBRANE FILTER TECHNIC FOR MEMBERS OF THE COLIFORM GROUP

The membrane filter technic is highly reproducible, can be used to test relatively large volumes of sample, and yields definite results more rapidly than the multiple-tube procedure, although it has limitations in testing waters high in turbidity and in noncoliform bacteria. Its use has been approved in U.S. EPA standards. The membrane filter technic is extremely useful in emergencies and in the examination of a variety of natural waters. However, when the membrane filter technic has not been used previously, it is desirable to conduct parallel tests with the multiple-tube fermentation technic (Section 908) to demonstrate applicability.

Turbidity caused by the presence of algae or other interfering material may not permit testing of a sample volume sufficient to yield significant results. Low coliform estimates may be caused by the presence of high numbers of noncoliforms or of toxic substances. The membrane filter technic is applicable to examination of saline waters, but not wastewaters that have received only primary treatment followed by chlorination or wastewaters containing toxic metals or phenols. For total coliforms analyze chlorinated secondary or tertiary effluents by the two-step procedure (Section 909A.5*c*). For fecal coliforms use the multiple-tube fermentation technic (Section 908C) if the results are to be used in enforcement actions. A modified membrane filter technic for fecal coliforms (Section 920) may be used if parallel testing with the multiple-tube fermentation technic shows significant agreement.

The standard volume to be filtered for drinking water samples is 100 mL (see Section 901B). This may be distributed among multiple membranes if necessary. Smaller or larger samples may be used for other waters or special analyses.

Statistical comparisons of results obtained by the multiple-tube method and the membrane filter technic show that the membrane filter is more precise (compare Tables 908:I and II with Table 909:II). Although data from each test yield approximately the same water quality information, numerical results are not identical (see Section 901B for drinking water). For raw water sources it would be expected that 80% of the membrane filter test results would be within the 95% confidence limits of the multiple-tube completed test results. Results from the multiple-tube test would be expected to be higher than membrane filter results because of a built-in positive statistical bias.

909 A. Standard Total Coliform Membrane Filter Procedure

1. Laboratory Apparatus

For membrane filter analyses use glassware and other apparatus composed of material free from agents that may have unfavorable effects on bacterial growth. Carefully note any deviations from the recommendations presented below and make quantitative tests to demonstrate that such deviations have not introduced agents or factors resulting in conditions less favorable for growth.

Sterilize glassware as described in Washing and Sterilization, Section 904.

a. Sample bottles: See Laboratory Apparatus, Section 903.15.

b. Dilution bottles: See Laboratory Apparatus, Section 903.11.

c. Pipets and graduated cylinders: See Laboratory Apparatus, Section 903.9. Before sterilization, cover opening of graduated cylinders with metal foil or a suitable paper substitute.

d. Containers for culture medium: Use clean borosilicate glass flasks presterilized to reduce bacterial contamination. Any size or shape of flask may be used, but erlenmeyer flasks with metal caps, metal foil covers, or screw caps provide for adequate mixing of the medium contained and are convenient for storage.

e. Culture dishes: Petri-dish type, 60 by 15 mm, 50 × 12 mm, or other appropriate size. The bottom of the dish should be flat and large enough so that the absorbent pad for the culture nutrient will lie flat. Wrap clean culture dishes before sterilization, singly or in convenient numbers, in metal foil if sterilized by dry heat, or suitable paper substitute when autoclaved. If glass petri dishes are used, use borosilicate or equivalent glass. Take precautions to prevent possible loss of medium by evaporation with resultant change in medium concentration, because covers for such dishes are loose-fitting, and to maintain a humid environment for optimum colony development.

Disposable plastic dishes that are tight-fitting and meet the specifications noted above also may be used. Suitable sterile plastic dishes are available commercially. To reuse these culture dishes, treat by immersing cleaned, opened dishes in 70% ethanol for 30 min and air-drying on a sterile towel, protected from dust. Reassemble when dry. Alternatively, use ultraviolet radiation or other appropriate chemical or physical agents. To use alternate sanitation procedures, test to demonstrate the effectiveness of such methods, including freedom of the culture containers from residual growth-suppressive effects. After sanitizing and removing the sanitizing agent, close containers, using aseptic technics, and store in a dustproof container until needed.

f. Filtration units: The filter-holding assembly (constructed of glass, autoclavable plastic, porcelain, or any noncorrosive bacteriologically inert metal) consists of a seamless funnel fastened by a locking device or held in place by magnetic force or gravity. The design should be such that the membrane filter will be held securely on the porous plate of the receptacle without mechanical damage and all fluid will pass through the membrane during filtration.

Separately wrap the two parts of the assembly in heavy wrapping paper for sterilization by autoclaving and storage until use. Alternatively treat unwrapped parts by ultraviolet radiation before using them. Field units may be sanitized by igniting methyl alcohol or immersing in boiling water for 5 min. Do not ignite plastic parts.

For filtration, mount receptacle of filter-holding assembly in a 1-L filtering flask with a side tube or other suitable device such that a pressure differential can be exerted on the filter membrane. Connect flask to an electric vacuum pump, a filter pump operating on water pressure, a hand aspirator, or other means of securing a pressure differential. Connect an additional flask between filtering flask and vacuum source to trap carry-over water.

g. Filter membranes: Use membrane filters (for additional specifications, see Section 902) with a rated pore diameter such that there is complete retention of coliform bacteria. Use only those filter membranes that have been found, through adequate quality control testing and *certification by the manufacturer*, to exhibit full retention of the organisms to be cultivated, stability in use, freedom from chemical extractables inimical to the growth and development of bacteria, a satisfactory speed of filtration, no significant influence on medium pH, and no increase in number of confluent colonies or spreaders. Preferably use membranes grid-marked in such a manner that bacterial growth is neither inhibited nor stimulated along the grid lines. Store membrane filters held in stock in an environment without extremes of temperature and humidity. Obtain no more than a year's supply at any one time.

If presterilized membrane filters are to be used, use those for which the manufacturer has certified that the sterilization technic has neither induced toxicity nor altered the chemical or physical properties of the membrane. If membranes are sterilized in the laboratory, remove paper separators—but not the absorbent paper pads—from the packaged filters. Divide filters into groups of 10 to 12, or other convenient units, and place in 10-cm petri dishes or wrap in heavy wrapping paper. Autoclave for 10 min at 121 C. At the end of the sterilization period, let the steam escape rapidly to minimize accumulation of water of condensation on filters.

h. Absorbent pads consist of disks of filter paper or other material known to be of high quality and free of sulfites or other substances that could inhibit bacterial growth. Use pads approximately 48 mm in diameter and of sufficient thickness to absorb 1.8 to 2.2 mL of medium. Presterilized absorbent pads or pads subsequently sterilized in the laboratory should release less than 1 mg total acidity (calculated as $CaCO_3$) when titrated to the phenolphthalein end point, pH 8.3, using 0.02*N* NaOH. Where there is evidence of absorbent pad toxicity, pre-soak pads in distilled water at 121 C (in an autoclave) for 15 min, decant the water, and repackage pads in a large petri dish for sterilization and subsequent use. Sterilize pads simultaneously with membrane filters available in resealable kraft envelopes, or separately in other suitable containers. Dry pads so they are free of visible moisture before use. See sterilization procedure described for membrane filters above and Section 902 for additional pad specifications.

As a substrate substitution for nutrient-saturated absorbent pads, 1.5% agar may be added to any MF broth medium.

i. Forceps: Round-tipped, without corrugations on the inner sides of the tips. Sterilize before use by dipping in 95% ethyl or absolute methyl alcohol and flaming.

j. Incubators: Use incubators to provide a temperature of 35 ± 0.5 C and to maintain a high level of humidity (approximately 90% relative humidity).

k. Microscope and light source: Count membrane filter colonies with a magnification of 10 to 15 diameters and a light source adjusted to give maximum sheen discernment. Optimally use a binocular wide-field dissecting microscope. However, a small fluorescent lamp with magnifier is acceptable. Use cool white fluorescent lamps. Do not use a microscope illuminator with optical system for light concentration from an incandescent light source for coliform colony identification on Endo-type media.

2. Materials and Culture Media

Refer to Preparation of Culture Media, Sections 905A, B, and C and Section 902. Test each new medium lot for satisfactory productivity by preparing dilutions of a culture of *Enterobacter aerogenes* (Section 902) and filtering appropriate volumes to give 20 to 80 colonies per filter. With each new lot of Endo-type medium, verify enough colonies, obtained from natural samples, to establish the differential accuracy of the medium lot.

3. Samples

Collect samples as directed under Samples, Sections 906A and B.

4. Coliform Definition

All organisms that produce a colony with a golden-green metallic sheen within

24 hr of incubation are considered members of the coliform group. The sheen may cover the entire colony or may appear only in a central area or on the periphery. The coliform group thus defined is based on the production of aldehydes from fermentation of lactose. While this biochemical characteristic is part of the metabolic pathway of gas production in the multiple-tube test, some variations may be observed among coliform strains. However, this slight difference in indicator definition is not considered to change its sanitary significance, particularly if suitable studies have been conducted to establish the relationship between results obtained by the filter and those obtained by the standard tube dilution procedure.

Coliform organisms occasionally may produce atypical colonies. If only atypical forms are found, verify their identity as coliform bacteria by transfer of growth from doubtful colonies to tubes of lauryl tryptose broth, followed by transfer of growth from positive cultures to brilliant green lactose bile broth. Gas formation in the confirmatory medium within 48 hr of incubation at 35 ± 0.5 C is evidence of coliform colonies. (See also Section 902.)

To obtain more rapid colony verification, simultaneously inoculate a tube of lauryl tryptose broth and a tube of brilliant green lactose bile broth from each doubtful colony; incubate at 35 ± 0.5 C for 24 hr. If gas is produced in both tubes, coliforms are verified. If one or both tubes show no gas, reincubate the lauryl tryptose broth tube for an additional 24 hr. If, after a total of 48 hr, no gas has been produced, the colony was not a coliform colony; if gas is produced, transfer to brilliant green lactose bile broth to complete verification.

5. Procedures

Generally, an enrichment procedure will give the best assessment of the quality of drinking water. However, this step may be eliminated in the routine examination of drinking water where repeated determinations have shown that adequate results are obtained by a single-step technic. Enrichment usually is not necessary in the examination of nonpotable water or sewage. Verify all samples of finished water giving positive results as described above.

In the following sections, methods are offered with and without enrichment that provide for use of agar-based medium or M-Endo medium without agar. When reporting results, state method followed.

a. Selection of sample size: Size of sample will be governed by expected bacterial density, which in finished-water samples will be limited only by the degree of turbidity (Table 909:I).

An ideal sample volume will yield growth of about 50 coliform colonies and not more than 200 colonies of all types. Examine finished waters by filtering duplicate portions of the same volume, such as 100 to 500 mL or more, or by filtering two diluted volumes. Examine other waters by filtering three different volumes, depending on the expected bacterial density. See Section 907.5 for preparation of dilutions. When less than 20 mL of sample (diluted or undiluted) is filtered, add a small amount of sterile dilution water to the funnel before filtration. This increase in water volume aids in uniform dispersion of the bacterial suspension over the entire effective filtering surface.

b. Filtration of sample: Using sterile forceps, place a sterile filter over porous plate of receptacle, grid side up. Carefully place matched funnel unit over receptacle and lock it in place. Filter sample under partial vacuum. With filter still in place, rinse funnel by filtering three 20- to 30-mL portions of sterile dilution water. Unlock and remove funnel, immediately remove filter with sterile forceps, and place it on sterile pad or agar with a rolling motion to avoid entrapment of air.

TABLE 909:I. SUGGESTED SAMPLE VOLUMES FOR MEMBRANE FILTER TOTAL COLIFORM TEST

Water Source	Volume to be Filtered *mL*							
	100	50	10	1	0.1	0.01	0.001	0.0001
Drinking water	X							
Swimming pools	X							
Wells, springs	X	X	X					
Lakes, reservoirs	X	X	X					
Water supply intake			X	X	X			
Bathing beaches			X	X	X			
River water				X	X	X	X	
Chlorinated sewage				X	X	X		
Raw sewage					X	X	X	X

Use sterile filtration units at the beginning of each filtration series as a minimum precaution to avoid accidental contamination. A filtration series is considered to be interrupted when an interval of 30 min or longer elapses between sample filtrations. After such interruption, treat any further sample filtration as a new filtration series and sterilize all membrane filter holders in use. Decontaminate this equipment between successive filtrations by use of an ultraviolet (UV) sterilizer, flowing steam, or boiling water. In the UV sterilization procedure, a 2-min exposure to UV radiation is sufficient. Do not expose membrane-filter culture preparations to random UV radiation leaks that might emanate from the sterilization cabinet. Eye protection is recommended; either safety glasses or prescription-ground glasses afford adequate eye protection against stray radiation from a UV sterilization cabinet that is not light-tight during the exposure interval. Clean UV tube regularly and check it periodically for effectiveness to insure that it will produce a 99.9% bacterial kill in a 2-min exposure. See also Section 902.

c. Enrichment technic: Place a sterile absorbent pad in the upper half of a sterile culture dish and pipet enough enrichment medium (1.8 to 2.0 mL lauryl tryptose broth) to saturate pad. Carefully remove any surplus liquid. Aseptically place filter through which the sample has been passed on pad. Incubate filter, without inverting dish, for 1.5 to 2 hr at 35 ± 0.5 C in an atmosphere of at least 90% relative humidity.

If the agar-based medium is used, prepare final culture dish as directed under Preparation of Culture Media, Section 905C. Remove enrichment culture from incubator, lift filter from enrichment pad, and roll it onto the agar surface. Incorrect filter placement is at once obvious, because patches of unstained membrane indicate entrapment of air. Where such patches occur, carefully reseat filter on agar surface. If the liquid medium is used, prepare final culture by removing enrichment culture from incubator and separating the dish halves. Place a fresh sterile pad in bottom half of dish and saturate it with 1.8 to 2.0 mL of final M-Endo medium (Section 905C). Transfer filter, with same precautions as above, to new pad. Discard used pad.

With either the agar or the liquid medium, invert dish and incubate for 20 to 22 hr at 35 ± 0.5 C. Proceed to Counting (*e* below).

d. Alternative single-step direct technic: If the agar-based medium is used, place prepared filter directly on agar as described in preceding section and incubate for 22 to 24 hr at 35 ± 0.5 C.

If liquid medium is used, place a pad in the culture dish and saturate with 1.8 to 2.0 mL M-Endo medium. Place prepared filter directly on pad, invert dish, and incubate for 22 to 24 hr at 35 ± 0.5 C.

e. Counting: The typical coliform colony has a pink to dark-red color with a metallic surface sheen. The sheen area may vary in size from a small pinhead to complete coverage of the colony surface. Count sheen colonies with the aid of a low-power (10 to 15 magnifications) binocular wide-field dissecting microscope or other optical device, with a cool white fluorescent light source directed above, and as nearly perpendicular as possible to, the plane of the filter. The total count of colonies (coliform and noncoliform) on Endo-type medium has no relation to the total number of bacteria present in the original sample and, so far as is known, no significance can be inferred or correlation made with the quality of the water sample.

6. Calculation of Coliform Density

Report coliform density as (total) coniforms/100 mL. Compute the count, using membrane filters with 20 to 80 coliform colonies and not more than 200 colonies of all types per membrane, by the following equation:

(Total) coliform colonies/100 mL

$$= \frac{\text{coliform colonies counted} \times 100}{\text{mL sample filtered}}$$

a. Water of drinking water quality: With water of good quality, the number of coliform colonies will be less than 20 per membrane. In this event, count all coliform colonies and use the formula given above to obtain coliform density.

If confluent growth occurs, that is, growth covering either the entire filtration area of the membrane or a portion thereof, and colonies are not discrete, report results as "confluent growth with or without coliforms". If the total number of bacterial colonies, coliforms plus noncoliforms, exceeds 200 per membrane, or if the colonies are too indistinct for accurate counting, report results as "too numerous to count" (TNTC). In either case, request a new sample and select more appropriate volumes to be filtered per membrane, remembering that the standard drinking water portion is 100 mL. Thus, instead of filtering 100 mL per membrane, 50-mL portions may be filtered through each of two membranes, 25-mL portions may be filtered through each of four membranes, etc. Total the coliform counts observed on the membranes and report as number per 100 mL.

b. Water of other than drinking water quality: As with potable water samples, if no filter has a coliform count falling in the ideal range, total the coliform counts on all filters and report as number per 100 mL. For example, if duplicate 50-mL portions were examined and the two membranes had five and three coliform colonies, respectively, report the count as eight coliform colonies per 100 mL, i.e.,

$$\frac{[(5 + 3) \times 100]}{(50 + 50)}$$

Similarly, if 50-, 25-, and 10-mL portions were examined and the counts were 15, 6, and <1 coliform colonies, respectively, report the count as 25/100 mL, i.e.,

$$\frac{[(15 + 6) \times 100]}{(50 + 25 + 10)}$$

On the other hand, if 10-, 1.0-, and 0.1-mL portions were examined with counts of 40, 9, and <1 coliform colonies, respectively, select the 10-mL portion only for calculating the coliform density because this filter had a coliform count falling in the ideal range. The result is 400/100 mL, i.e.,

$$\frac{(40 \times 100)}{10}$$

In this last example, if the membrane with 40 coliform colonies also had a total bacterial colony count greater than 200, report the coliform count as >400/100 mL.

Report confluent growth or membranes with colonies too numerous to count as described in *a* above. Request a new sample and select more appropriate volumes for filtration.

c. Statistical reliability of membrane filter results: Although the statistical reliability of the membrane filter technic is greater than that of the MPN procedure, membrane counts really are not absolute numbers. Table 909:II illustrates some 95% confidence limits.

TABLE 909:II. 95% CONFIDENCE LIMITS FOR MEMBRANE FILTER RESULTS USING 100-ML SAMPLE

Number of Coliform Colonies Counted	95% Confidence Limits	
	Lower	Upper
1	0.05	3.0
2	0.35	4.7
3	0.81	6.3
4	1.4	7.7
5	2.0	9.2

909 B. Delayed-Incubation Total Coliform Procedure

Modification of the standard membrane filter technic permits membrane shipment or transport after filtration to a distant laboratory for incubation and completion of the test. This delayed-incubation test may be used where it is impractical to apply conventional procedures. It also may be used where it is not possible to maintain the desired sample temperature during transport; when the elapsed time between sample collection and analysis would exceed the approved time limit; where the sampling location is remote from laboratory services; when it is necessary to monitor streams for water quality or pollution control activities by a standardized procedure; or for other reasons that prevent analysis of the sample at or near the sample site.

Data secured by the delayed-incubation test have yielded results consistent with those from the immediate standard test in independent studies of samples from both fresh and salt waters. Determine the applicability of the delayed-incubation test for a specific water source by comparing with results of test procedures using conventional methods.

To conduct the delayed-incubation test, filter sample in the field immediately after collection, place filter on the transport medium, and ship to the laboratory. Complete the coliform determination in the laboratory by transferring the membrane to standard M-Endo or LES Endo medium, incubating at 35 ± 0.5 C for 20 to 22 hr, and counting typical coliform colonies that developed. Transport media are designed to keep coliform organisms viable and generally they do not permit visible growth during the time of transit. Bacteriostatic agents suppress growth of microorganisms en route but allow normal coliform growth after transfer to a fresh medium.

The delayed-incubation test follows the methods outlined for the total coliform membrane filter procedure, except as indicated below. Two alternative methods are given, one using the M-Endo preservative medium and the other the LES MF holding medium.

1. Apparatus

a. Culture dishes: Disposable, sterile, moisture-tight plastic petri dishes (50 by 12 mm). Such containers are light in weight and are less likely to break in transit. In an emergency or when plastic dishes are unavailable, use sterile glass petri dishes wrapped in plastic film or similar material. See Section 909A.1*e* for specifications.

b. Field filtration units: See Section 909A.1*f* for specifications. Disinfect by adding methyl alcohol to the filtering chamber, igniting the alcohol, and covering unit to produce formaldehyde. Glass or metal filtration units may be sterilized by immersing in boiling water for 5 min. Use a hand aspirator to obtain necessary vacuum.

2. Materials and Transport Media

a. M-Endo methods:

1) *M-Endo preservative medium:* Prepare as described in Section 905C, but add 3.84 g sodium benzoate (USP grade)/L or 3.2 mL 12% sodium benzoate solution/100 mL of medium.

2) *Sodium benzoate solution:* Dissolve 12 g $NaC_7H_5O_2$ in sufficient distilled water to make 100 mL. Sterilize by autoclaving or filtration. Discard after 6 months.

3) *Cycloheximide*:* Addition of cycloheximide to M-Endo preservative medium is optional. It may be used for samples that previously have shown overgrowth by molds or fungi. Prepare modification by adding 50 mg/100 mL M-Endo preservative medium. Store cycloheximide solution at 5 to 10 C and discard after 6 months. Cycloheximide is a powerful skin irritant; handle with caution according to the manufacturer's directions.

b. LES method:

LES MF holding medium, coliform: Prepare as in Section 905C.

3. Procedure

a. Sample preservation and shipment: Place an absorbent pad in the bottom of a sterile petri dish and saturate with selected coliform holding medium (see Section 909A.5*c* above). Remove membrane filter from filtration unit with sterile forceps and roll it, grid side up, onto the surface of the absorbent pad that has been saturated with the transport medium. Protect membrane from moisture loss by tight closure of the plastic petri dish. While it is important that the membrane does not become dehydrated during transit, an excess of liquid in the dish also is undesirable. Place culture dish containing membrane in an appropriate shipping container and send to the laboratory for completion of the examination. The sample can be held without visible growth for a maximum of 72 hr on the transport medium. This usually allows use of the mail or a common carrier. Visible growth occasionally begins on the transport medium when high temperatures are encountered.

b. Transfer: At the laboratory, transfer membrane from plastic dish in which it was shipped to a second sterile petri dish containing M-Endo or LES Endo medium.

c. Incubation:

1) M-Endo method—Transfer membrane from M-Endo preservative medium to a pad and petri dish containing M-Endo medium without the growth-suppressing reagents and incubate at 35 ± 0.5 C for 20 to 22 hr.

2) LES method—Transfer membrane from LES MF holding medium to LES Endo agar (see Section 905C) and incubate at 35 ± 0.5 C for 20 to 22 hr. If distinct colonies are observable without magnification at time of transfer, store petri dish containing transferred membrane at 5 to

*Actidione, manufactured by the Upjohn Company, Kalamazoo, Mich., or equivalent.

10 C until it can be incubated at 35 ± 0.5 C for 16 to 18 hr. This manipulation of incubation time permits the analyst a measure of control over the problems of overgrowth and sheen dissipation that interfere with the coliform colony count.

4. Estimation of Coliform Density

Proceed as in Section 909A.6 above. Record times of collection, filtration, and laboratory examination, and calculate elapsed time.

909 C. Fecal Coliform Membrane Filter Procedure

Fecal coliform bacterial densities may be determined either by the multiple-tube procedure or by a membrane filter technic. If the membrane filter procedure is used for chlorinated effluents, demonstrate that it gives comparable information to that obtainable by the multiple-tube test before accepting it as an alternative. The following procedure gives 93% accuracy for differentiating between coliforms from warm-blooded animals and coliforms from other sources. The membrane filter procedure calls for using an enriched lactose medium and depends on an incubation temperature of 44.5 ± 0.2 C for its selectivity. Because incubation temperature is critical, submerge membrane filter cultures in a water bath for incubation at the elevated temperature or use an appropriate, accurate solid heat sink incubator. Areas of application for this method are stated in the introduction to the multiple-tube fecal coliform procedures, Section 908C.

1. Materials and Culture Medium

a. M-FC medium: See Media Specifications, Section 905C.

Test each medium lot for satisfactory productivity by preparing dilutions of a culture of *Escherichia coli* (compare Section 905B.2*e* and Section 902) and filtering appropriate volumes to give 20 to 80 colonies per filter. With each new lot of medium verify enough colonies, obtained from natural samples, to establish the absence of false positives.

b. Culture dishes: Use tight-fitting plastic dishes because the membrane-filter cultures are submerged in a water bath during incubation. Enclose groups of fecal coliform cultures in plastic bags or seal individual dishes with waterproof (freezer) tape to prevent leakage during submersion. Specifications for plastic culture dishes are given in Section 909A.1*e* above.

c. Incubator: The specificity of the fecal coliform test is related directly to the incubation temperature. Air incubation is undesirable because of heat layering within the chamber and the slow recovery of temperature each time the incubator is opened during daily operations. To meet the need for greater temperature control use a water bath or a heat-sink incubator. A temperature tolerance of 44.5 ± 0.2 C can be obtained with most types of water baths that also are equipped with a gable top for the reduction of water and heat losses. A circulating water bath is excellent but may not be essential to this test if the maximum permissible variation of ±0.2 C in temperature can be maintained with other equipment.

2. Procedure

a. Selection of sample size: Select volume of water sample to be examined in accordance with the information in Table 909:III.

When the bacterial density of the sample is unknown, filter several decimal volumes to establish fecal coliform den-

TABLE 909:III. SUGGESTED SAMPLE VOLUMES FOR MEMBRANE FILTER FECAL COLIFORM TEST

Water Source	Volume to be Filtered *mL*						
	100	50	10	1	0.1	0.01	0.001
Lakes, reservoirs	X	X					
Wells, springs	X	X					
Water supply intake		X	X	X			
Natural bathing waters		X	X	X			
Sewage treatment plant, secondary effluent			X	X	X		
Farm ponds, rivers				X	X	X	
Stormwater runoff				X	X	X	
Raw municipal sewage					X	X	X
Feedlot runoff					X	X	X

sity. Estimate volume expected to yield a countable membrane and select two additional quantities representing one-tenth and ten times this volume, respectively. Use sample volumes that will yield counts between 20 and 60 fecal coliform colonies per membrane.

b. Filtration of sample: Follow the same procedure and precautions as prescribed under Section 909A.5*b* above.

c. Preparation of culture dish: Place a sterile absorbent pad in each culture dish and pipet approximately 2 mL M-FC medium, prepared as directed under Media Specifications (Section 905C) to saturate pad. Carefully remove any surplus liquid from culture dish. Place prepared filter on medium-impregnated pad as described in Section 909A above.

As a substrate substitution for the nutrient saturated absorbent pad, 1.5 percent agar may be added to M-FC broth.

d. Incubation: Place parpared cultures in waterproof plastic bags or seal petri dishes, submerge in water bath, and incubate for 24 ± 2 hr at 44.5 ± 0.2 C. Anchor dishes below water surface to maintain critical temperature requirements. Place all prepared cultures in the water bath within 30 min after filtration. Alternatively, use an appropriate, accurate solid heat sink incubator.

e. Counting: Colonies produced by fecal coliform bacteria are blue. Nonfecal coliform colonies are gray to cream-colored. Background color on the membrane filter will vary from a yellowish cream to faint blue, depending on age of the rosolic acid salt reagent. Normally, few nonfecal coliform colonies will be observed on M-FC medium because of the selective action of the elevated temperature and addition of the rosolic acid salt reagent. Count colonies with the aid of a low-power (10 to 15 magnifications) binocular wide-field dissecting microscope or other optical device.

3. Calculation of Fecal Coliform Density

Compute the density from the sample quantities that produced membrane filter counts within the desired range of 20 to 60 fecal coliform colonies. This colony density range is more restrictive than the 20 to 80 total coliform range because of larger colony size on M-FC medium. Calculate fecal coliform density as directed in Section 909A.6 above. Record densities as fecal coliforms per 100 mL.

909 D. Delayed-Incubation Fecal Coliform Procedure

This delayed-incubation procedure is comparable to the delayed-incubation total coliform procedure (Section 909B). It eliminates the need for a field water bath incubator and frees the field investigator from the time-consuming task of counting colonies. Examination at a central laboratory, rather than in the field, permits colony confirmation and complete biochemical identification of the organisms, as necessary.

Results obtained by this delayed method have been consistent with results from the immediate standard test under various laboratory and field use conditions. However, determine the applicability of this test for a specific water source by comparison with the standard membrane filter test, especially for saline waters, chlorinated wastewaters, and waters containing toxic substances. Use the delayed incubation test only when the standard immediate fecal coliform test cannot be performed.

To conduct the delayed-incubation test filter the sample in the field immediately after collection, place filter on transport medium, and ship to the laboratory. Complete fecal coliform test in the laboratory by transferring filter to M-FC medium, incubating at 44.5 C for 24 ± 2 hr, and counting the fecal coliform colonies.

The transport medium keeps fecal coliform organisms viable but prevents visible growth during transit. Membrane filters can be held for up to 3 days on VFC holding medium with little effect on the fecal coliform counts.

1. Apparatus

a. Culture dishes: See Section 909B.1*a* for specifications.

b. Field filtration units: See Section 909B.1*b*.

2. Materials and Transport Medium

a. VFC holding medium: Prepare as directed under Media Specifications, Section 905C.

b. M-FC medium: Prepare as in Media Specifications, Section 905C.

3. Procedure

a. Membrane filter transport: Place an absorbent pad in a tight-lid plastic petri dish and saturate with VFC holding medium. After filtering sample remove filter from filtration unit and place it on medium-saturated pad. It is important to use tight-lid dishes to prevent moisture loss from pad and filter, but it is undesirable to have excess liquid in the dish. Place culture dish containing membrane in an appropriate shipping container and send to the examining laboratory. Membranes can be held on the transport medium at ambient temperature for a maximum of 72 hr with little effect on fecal coliform counts.

b. Transfer: At the laboratory remove membrane from holding medium and place it in another dish containing M-FC medium (broth-saturated pad or agar).

c. Incubation: After transfer of filter to M-FC medium, place the tight-lid dishes in waterproof plastic bags and submerge in a water bath at 44.5 C ± 0.2 C for 24 ± 2 hr or use a solid heat sink incubator.

d. Counting: Colonies produced by fecal coliform bacteria are blue. Nonfecal coliform colonies are gray to cream-colored. Make the colony count with the aid of a binocular wide-field dissecting microscope at 10 to 15 magnifications.

4. Estimation of Coliform Density

Count as directed in Section 909C.2*e* above and calculate as described in Section 909C.3. Record time of collection, filtration, and laboratory examination, and calculate the elapsed time.

909 E. Bibliography

CLARK, H.F., E.E. GELDREICH, H.L. JETER & P.W. KABLER. 1951. The membrane filter in sanitary bacteriology. *Pub. Health Rep.* 66:951.

GOETZ, A. & N. TSUNEISHI. 1951. Application of molecular filter membranes to bacteriological analysis of water. *J. Amer. Water Works Ass.* 43:943.

VELS, C.J. 1951. Graphical approach to statistics. IV. Evaluation of bacterial density. *Water Sewage Works* 98:66.

TASK GROUP, AMERICAN WATER WORKS ASSOCIATION. 1953. Technic of bacterial examination of water. *J. Amer. Water Works Ass.* 45:1196.

KABLER, P.W. 1954. Water examinations by membrane filter and MPN procedures. *Amer. J. Pub. Health* 44:379.

GELDREICH, E.E., P.W. KABLER, H.L. JETER & H.F. CLARK. 1955. A delayed incubation membrane filter test for coliform bacteria in water. *Amer. J. Pub. Health* 45:1462.

THOMAS, H.A. & R.L. WOODWARD. 1956. Use of molecular filter membranes for water potability control. *J. Amer. Water Works Ass.* 48:1391.

CLARK, H.F., P.W. KABLER & E.E. GELDREICH. 1957. Advantages and limitations of the membrane filter procedure. *Water Sewage Works* 104:385.

FIFIELD, C.W. & C.P. SCHAUFUS. 1958. Improved membrane filter medium for the detection of coliform organisms. *J. Amer. Water Works Ass.* 50:193.

MCCARTHY, J.A. & J.E. DELANEY. 1958. Membrane filter media studies. *Water Sewage Works* 105:292.

MCKEE, J.E., R.T. MCLAUGHLIN & P. LESGOURGUES. 1958. Application of molecular filter technics to the bacterial assay of sewage. III. Effects of physical and chemical disinfection. *Sewage Ind. Wastes* 30:245.

MCCARTHY, J.A., J.E. DELANEY & R.J. GRASSO. 1961. Measuring coliforms in water. *Water Sewage Works* 108:238.

JUDIS, J. 1962, 1963, 1964, 1965. Studies on the mechanism of action of phenolic disinfectants, I-V. *J. Pharm. Sci.* 51:261; 52:126; 53:196; 54:417, 541.

RHINES, C.E. & W.P. CHEEVERS. 1965. Decontamination of membrane filter holders by ultraviolet light. *J. Amer. Water Works Ass.* 57:500.

GELDREICH, E.E., H.F. CLARK, C.B. HUFF & L.C. BEST. 1965. Fecal-coliform-organism medium for the membrane filter technic. *J. Amer. Water Works Ass.* 57:208.

PANEZAI, A.K., T.J. MACKLIN & H.G. COLES. 1965. *Coli-aerogenes* and *Escherichia coli* counts on water samples by means of transported membranes. *Proc. Soc. Water Treat. Exam.* 14:179.

MCCARTHY, J.A. & J.E. DELANEY. 1965. Methods for measuring the coliform content of water. Sec. III. Delayed holding procedure for coliform bacteria. PHS Res. Grant WP 00202 NIH Rep.

GELDREICH, E.E., H.L. JETER & J.A. WINTER. 1967. Technical considerations in applying the membrane filter procedure. *Health Lab. Sci.* 4:113.

DEPARTMENT HEALTH & SOCIAL SECURITY, WELSH OFFICE. 1969. The Bacteriological Examination of Water Supplies. Rep. 71, H. M. Stationery Office, London.

BREZENSKI, F.T. & J.A. WINTER. 1969. Use of the delayed incubation membrane filter test for determining coliform bacteria in sea water. *Water Res.* 3:583.

LIN, S. 1973. Evaluation of coliform test for chlorinated secondary effluents. *J. Water Pollut. Control Fed.* 45:498.

TAYLOR, R.H., R.H. BORDNER & P.V. SCARPINO. 1973. Delayed incubation membrane-filter test for fecal coliforms. *Appl. Microbiol.* 25:363.

910 TESTS FOR THE FECAL STREPTOCOCCUS GROUP

The terms "fecal streptococcus" and "Lancefield's Group D Streptococcus" have been used synonymously. When this group is used as an indicator of fecal contamination, the following species and subspecies are implied: *S. faecalis*, *S. faecalis* subsp. *liquefaciens*, *S. faecalis* subsp. *zymogenes*, *S. faecium*, *S. bovis*, and *S. equinus*. Other biochemical types, formerly considered as biotypes of *S. faecalis* and *S. faecium*, have been isolated from warm-blooded animal wastes. Studies indicate that these streptococci belong to Lancefield's serological Group Q and occur in the feces of humans and other warm-blooded animals, especially chickens. *S. avium* is the characteristic species. It has been isolated frequently from the feces of chickens, and occasionally from the feces of humans, dogs, and pigs. The Group Q antigen occurs in the cell wall of these organisms, and in addition, the Group D antigen is present between the cell wall and the cytoplasmic membrane where it occurs naturally in the established Group D species. The common antigens indicate a relationship between Group D and Group Q organisms. On the basis of antigenic and biochemical similarities and occurrence in warm-blooded animal feces, the Group Q organisms should be considered in the fecal streptococcus group.

The term "enterococcus group" has been substituted erroneously for the fecal streptococcus group. The former group excludes *S. bovis, S. equinus,* and Group Q organisms. Consequently, media that are specific for the enterococcus group are restrictive and may not indicate the full extent of contamination by streptococci derived from fecal sources. The following diagram illustrates the subgroups within the fecal streptococcus group.

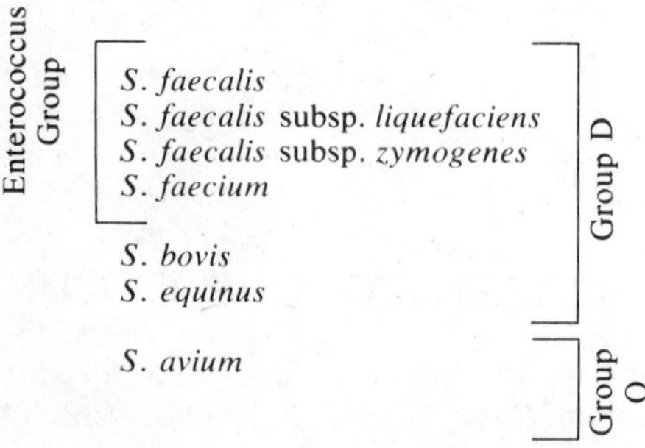

The normal habitat of fecal streptococci is the intestines of man and animals; thus these organisms are indicators of fecal pollution. Assays for fecal streptococci may provide valuable supplementary data on the bacteriological quality of lakes, streams, and estuaries. Because of limited survival in the environment, it is not recommended to use *only* fecal streptococci when determining water quality. In combination with fecal coliform data, data on fecal streptococci may provide more specific information about pollution sources because certain fecal streptococci are host-specific. Biochemical characterization or speciation may provide source information. For example, a predominance of *S. bovis* and *S. equinus* would indicate pollution due to the excrement of nonhuman, warm-blooded animals. Investigations have demonstrated high numbers of these species associated with pollution involving meat-processing plants, dairy wastes, and feedlot and farmland runoff. *S. bovis* and *S. equinus* have limited or short survival times outside of their natural habitat; thus, their presence in water indicates very recent contamination.

S. faecalis subsp. *liquefaciens* is not restricted to the intestines of humans and animals. It has been found associated with vegetation, insects, and certain types of soils. This may be detrimental, especially for indicating low-density fecal contamination, because when the count is below 100 fecal streptococci/100 mL this organism generally predominates. Because media currently in use do not exclude such strains selectively, use of fecal streptococcus limits for recreational water based on densities below 100 organisms/100 mL must be assumed unreliable unless confirmed by concurrent fecal coliform testing. Biochemical characterization of fecal streptococci also may eliminate the possibility of predominance of *S. faecalis* subsp. *liquefaciens*.

Fecal coliform/fecal streptococcus ratios may provide information on possible sources of pollution. Estimated per capita contributions of fecal coliforms and fecal streptococci for animals were used to develop the following FC/FS ratios:

Human	4.4
Duck	0.6
Sheep	0.4
Chicken	0.4
Pig	0.4
Cow	0.2
Turkey	0.1

A ratio greater than 4.1 is considered indicative of pollution derived from domestic wastes composed of human excrement whereas ratios less than 0.7 suggest that pollution was due to nonhuman sources. Ratios between 0.7 and 4.4 usually indicate wastes of mixed human and animal sources. Speciation will provide more definitive information on potential sources. To minimize misinterpretation of ratios, take the following precautions: (*a*) measure sample pH because streptococcal densities can be altered significantly if water pH is above 9.0 or below 4.0; (*b*) sample as close as possible to the pollution source because fecal streptococci have relatively short lives outside the animal host. Points downstream, where travel time from pollution sources exceeds 24 hr, will provide erroneous ratios; (*c*) inspect source(s) of pollution when various pollution sources are present because ratios may yield deceptive assessments; (*d*) carefully use ratios for samples from marine waters, bays, and estuaries, because ratios may be of little value in differentiating between human and nonhuman sources; and (*e*) do not use ratios when fecal streptococcus counts are below 100/100 mL.

The fecal streptococci are valuable pollution indicators in the study of rivers, streams, lakes, and marine systems, especially when used with the fecal coliform bacteria. Periodic verification of isolates is necessary to ensure adequate medium performance. Further identification of species will eliminate organisms of little sanitary significance while verifying human and animal sources of pollution.

910 A. Multiple-Tube Technic

1. Presumptive Test Procedure

a. Inoculate a series of tubes of azide dextrose broth (see Section 905C) with appropriate graduated quantities of sample. Use 10 mL single-strength broth for inocula of 1 mL or less and 10 mL double-strength broth for 10-mL inocula. The portions used will vary in size and number with the sample character. Use only decimal multiples of 1 mL (see Section 908 for suggested sample sizes).

b. Incubate inoculated tubes at 35 ± 0.5 C. Examine each tube for turbidity at the end of 24 ± 2 hr. If no definite turbidity is present, reincubate, and read again at the end of 48 ± 3 hr.

2. Confirmed Test Procedure

Subject all azide dextrose broth tubes showing turbidity after 24- or 48-hr incubation to the Confirmed Test.

Streak a portion of growth from each positive azide dextrose broth tube on a petri dish containing PSE agar (see Section 905C) or equivalent esculin-azide agar. Incubate the inverted dish at 35 ± 0.5 C for 24 ± 2 hr. Brownish-black colonies with brown halos confirm the presence of fecal streptococci.

3. Computing and Recording of MPN

Refer to Tables 908:I and II and to Section 908D, Estimation of Bacterial Density.

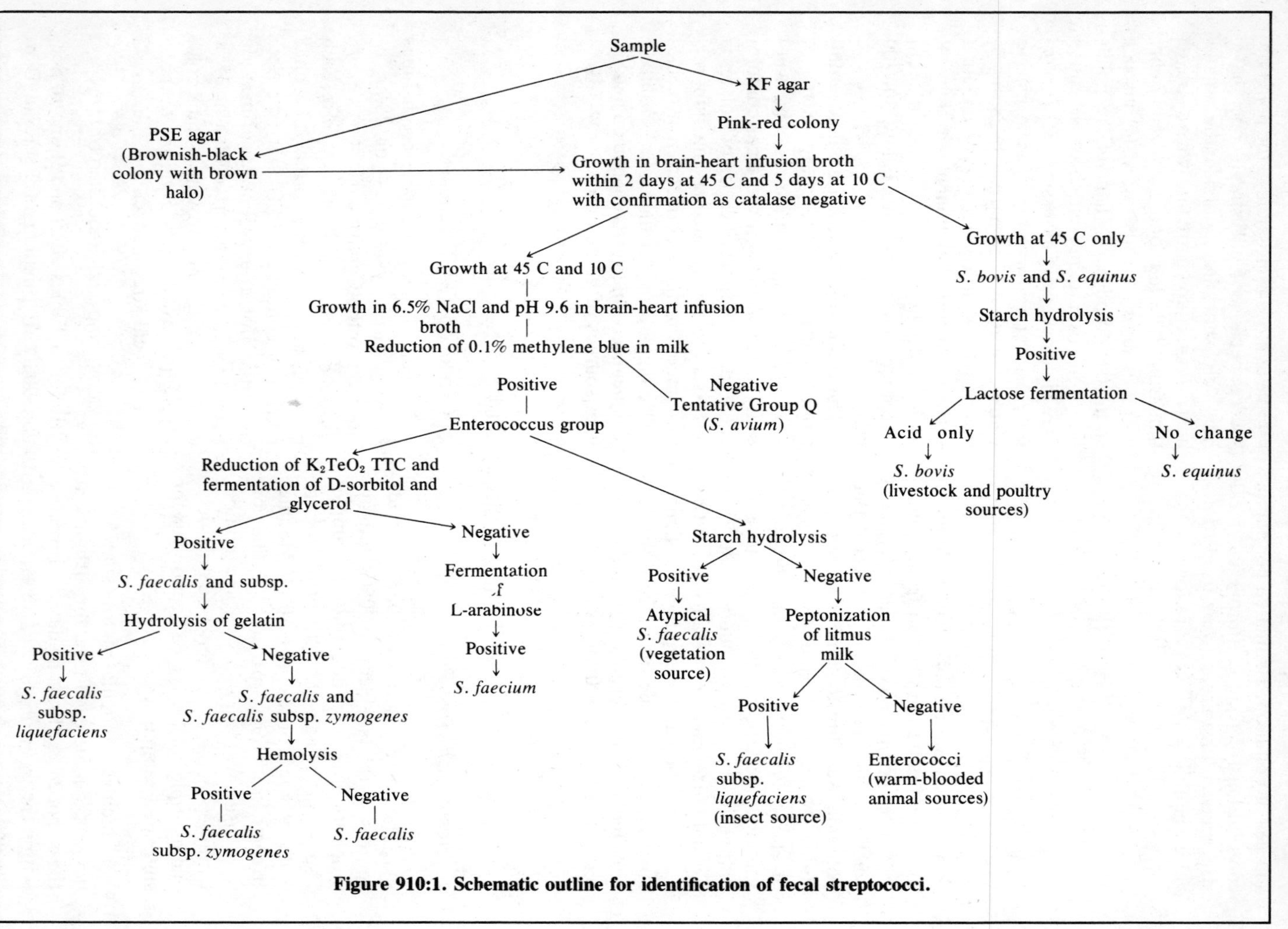

Figure 910:1. Schematic outline for identification of fecal streptococci.

4. Test Application

This test is designed primarily for raw wastewater and chlorinated wastewater effluent samples. It is applicable to other fresh, but not saline, waters.

910 B. Membrane Filter Technic

1. Laboratory Apparatus

See Section 909A.1.

2. Materials and Culture Media

a. Culture medium: KF Streptococcus agar; see Section 905C.

b. Culture dish preparation: Pour or pipet 4 to 5 mL liquefied medium into culture dishes (60 by 15 mm); flame surface if necessary to eliminate bubbles. If tight-fitting plastic dishes (50 × 12 mm) are used, make a stock of prepared dishes in advance and store at 4 to 10 C for use within a 4-wk period.

3. Procedure

a. Selection of sample size and filtration: Filter samples of water through a sterile membrane to give 20 to 100 colonies on the membrane surface. Use volumes of from 100 to 10, 1, 0.1, or 0.01 mL, depending on pollution present (see Section 909A.5*a*). Transfer filter directly to agar medium in petri dish, avoiding air bubbles.

b. Incubation: Invert culture plates and incubate at 35 ± 0.5 C for 48 hr.

c. Counting: Colonies produced by fecal streptococci are dark red to pink. Count with the aid of a low-power (10 to 15 magnifications) binocular wide-field dissecting microscope or equivalent optical device.

4. Calculation of Fecal Streptococcal Density

Compute the density from sample quantities producing membrane filter counts within the desired 20- to 100-fecal streptococcus colony range. This colony density range is greater than the 20- to 80-total coliform range because of increased selectivity of the fecal streptococcus medium. Calculate as in Section 909A.6. Record densities as fecal streptococci per 100 mL.

5. Confirmed Test

In examining samples from sources other than swimming pools, results reported to date indicate that practically 100% of the red and pink colonies growing on filters placed on KF agar are fecal streptococci. If further confirmation is indicated, use the following procedures:

a. Pick selected typical colonies from membrane and inoculate onto a brain-heart infusion agar slant (see Section 905C). Incubate at 35 ± 0.5 C for 24 to 48 hr. When growth is detected, continue as in ¶s *b* and *c*.

b. Transfer a loopful of growth from the brain-heart infusion agar slant to a clean glass slide and add a few drops of freshly tested 3% hydrogen peroxide to the smear. The absence of bubbles constitutes a negative catalase test, indicating a probable streptococcus culture. Continue confirmation as indicated below. The presence of bubbles constitutes a positive catalase test, which indicates the presence of nonstreptococcal species. Discontinue confirmation.

c. Transfer a loopful of growth from the brain-heart infusion agar into brain-heart infusion broth and incubate at 45 C for 48 hr. Also transfer a loopful of growth into bile broth medium and incubate at 35 C for 3 days. Prepare this latter medium by adding 40 mL sterile 10% oxgall solution to 60 mL sterile brain-heart infusion broth.

d. Growth in the above media constitutes a positive test for fecal streptococci.

6. Differentiation of Fecal Streptococcal Organisms

Further identification of fecal streptococcal types requires additional biochemical tests. See Figure 910:1.

Initial separation of Group Q streptococci is based on failure to reduce 0.1% methylene blue in milk. Other prominent characteristics of Group Q organisms include fermentation of sorbose, growth in sorbose medium at pH 10, no growth in 0.04% tellurite medium, and no hydrolysis of starch and gelatin.

7. Test Application

This test may be used for the examination of fresh and saline water samples. Like other membrane filter technics, it is not appropriate for highly turbid waters and chlorinated wastewater.

910 C. Fecal Streptococcal Plate Count (TENTATIVE)

Fecal streptococci may be enumerated by the pour-plate technic using KF Streptococcus Agar, Pfizer Selective Enterococcus Agar (PSE), or equivalent esculin-azide agar. This direct-count procedure may be considered an alternative to the membrane filter procedure and should be used preferentially for those samples containing few fecal streptococci and significant turbidity.

1. Preparation and Dilution

See Section 907.5.

2. Plating

Prepare KF Streptococcus or PSE agar as directed in Section 905C and hold in a water bath at 45 C before making pour plates. Discard any liquid agar medium held over 4 hr.

Place 1 mL, 0.1 mL, or other suitable volume of sample or dilution in a 100 × 15-mm petri dish. Make dilutions in preparing volumes less than 1 mL. For wastewater or turbid water samples do not measure a 0.1-mL inoculum of the original sample but prepare an appropriate dilution. Do not let more than 20 min elapse between making the dilution and pouring the plate.

Pour 12 to 15 mL liquefied agar medium into each culture dish containing the measured sample. Lift dish cover just enough to permit introduction of pipet or culture medium. Thoroughly mix agar and sample together for a uniform dispersion of organisms and medium over dish bottom by gently tilting and rotating dish, taking care not to splash upper portion. Solidify agar plates as rapidly as possible after pouring and place immediately, in inverted position, in an incubator.

3. Incubation

Incubate KF fecal streptococcus pour plates at 35 ± 0.5 C for 48 ± 3 hr. Incubate PSE fecal streptococcus pour plates for 18 to 24 hr at 35 to 37 C.

4. Counting

Surface and subsurface colonies produced by fecal streptococci on KF agar are dark red to pink with entire edges. Subsurface colonies frequently are ellipsoidal or lens-shaped. Normally, few nonfecal streptococcus colonies will be observed on KF streptococcus agar because of medium selectivity. However, occasional stream samples may contain gram-positive soil organisms, such as *Corynebacterium* species, that develop yellow or orange colonies on this medium. Infrequently, *Bacillus* species may produce fuzzy-white colonies with or without minute red-dot centers.

Fecal streptococci on PSE agar give brownish-black colonies with brown halos. PSE agar is a selective and differential medium that inhibits proliferation of gram-negative bacteria. The only gram-positive cocci that will grow and exhibit esculin hydrolysis are the Group D streptococci. *Listeria monocytogenes* shows only pinpoint colonies at 24 hr and small (0.5-mm) brown-black colonies with brownish halos at the end of 48 hr. Coloration of colonies due to esculin hydrolysis is less marked with *Listeria* than with fecal streptococci.

Count with either a Quebec colony counter or a low-power (10 to 15 magnifications) binocular wide-field dissecting microscope and suitable light source, preferably cool white flourescent lamps.

5. Calculation of Fecal Streptococcal Density

See Section 907.8. Report results as fecal streptococci per 100 mL.

910 D. Bibliography

SHERMAN, J.M. 1937. The streptococci. *Bacteriol. Rev.* 1:3.

SKAUDHAUGE, K. 1950. Studies on Enterococci, with Special Reference to the Serological Properties. Einar Munksgaards, Copenhagen.

MALLMANN, W.L. & E.B. SELIGMANN. 1950. A comparative study of media for the detection of streptococci in water and sewage. *Amer. J. Pub. Health* 40:286.

LITSKY, W., W.L. MALLMANN & C.W. FIFIELD. 1953. A new medium for the detection of enterococci in water. *Amer. J. Pub. Health* 43:873.

LITSKY, W. 1955. Comparison of the Most Probable Number of *Escherichia coli* and enterococci in river waters. *Amer. J. Pub. Health* 45:1049.

SHATTOCK, P.M.F. 1955. The identification and classification of *Streptococcus faecalis* and some associated streptococci. *Ann. Inst. Pasteur* (Lille) 7:95.

COOPER, K.E. & F.M. RAMADAN. 1955. Studies in the differentiation between human and animal pollution by means of fecal streptococci, *J. Gen. Microbiol.* 12:180.

LAKE, D.E., R.H. DEIBEL & C.F. NIVEN, JR. 1957. The identity of *Streptococcus faecium. Bacteriol. Proc.*, p. 13.

BARNES, E.M. 1957. Reduction as a means of differentiating *Streptococcus faecalis* from *Streptococcus faecium. J. Gen. Microbiol.* 14:57.

BUCHANAN, R.E. & N.E. GIBBONS, eds. 1974. Bergey's Manual of Determinative Bacteriology, 8th ed. Williams & Wilkins, Baltimore, Md.

MORELIS, P. & L. COLOBERT. 1958. Un milieu selectif permettant l'identification et le denombrement rapides de *Streptococcus faecalis. Ann. Inst. Pasteur* 95:667.

SUREAU, P. 1958. Isolation and enumeration of faecal streptococci in waters by means of filtering membranes. *Ann. Inst. Pasteur* 95:6.

MEDREK, T.F. & W. LITSKY. 1959. Comparative incidence of coliform bacteria and enterococci in undisturbed soil. *Appl. Microbiol.* 8:60.

CROFT, C.C. 1959. A comparative study of media for detection of enterococci in water. *Amer. J. Pub. Health* 49:1379.

BARTLEY, C.H. & L.W. SLANETZ. 1969. Types and sanitary significance of fecal streptococci isolated from feces, sewage and water. *Amer. J. Pub. Health* 50:1545.

KENNER, B.A., H.F. CLARK & P.W. KABLER. 1960. Fecal streptococci. II. Quantification of streptococci in feces. *Amer. J. Pub. Health* 50:1553.

KENNER, B.A., H.F. CLARK & P.W. KABLER. 1961. Fecal streptococci. I. Cultivation and enumeration of streptococci in surface waters. *Appl. Microbiol.* 9:15.

NIVEN, C.F., JR. 1963. Microbial indices of food quality: Fecal streptococci. *In* Proc. Conf.

Microbiol. Qual. Foods. Academic Press, New York, N.Y.

SHATTOCK, P.M.F. 1963. Enterococci: Chemical and Biological Hazards in Food. Iowa Univ. Press, Des Moines.

MUNDT, J.C. 1963. Occurrence of enterococci on plants in a wild environment. *Appl. Microbiol.* 11:141.

GELDREICH, E.E., H.F. CLARK & C.B. HUFF. 1964. A study of pollution indicators in a waste stabilization pond. *J. Water Pollut. Control Fed.* 36:1372.

NOWLAN, S.S. & R.H. DEIBEL. 1967. Group Q streptococci. I. Ecology, serology, physiology, and relationship to established enterococci. *J. Bacteriol.* 94:291.

GELDREICH, E.E. & B.A. KENNER. 1969. Concepts of fecal streptococci in stream pollution. *J. Water Pollut. Control Fed.* 41:R336.

PAVLOVA, M. T., W. LITSKY & F.J. FRANCIS. 1971. A comparative study of starch hydrolysis by fecal streptococci employing plate and tube techniques. *Health Lab. Sci.* 8:67.

PAVLOVA, M.T., F.T. BREZENSKI & W. LITSKY. 1972. Evaluation of various media for isolation, enumeration and identification of fecal streptococci from natural sources. *Health Lab. Sci.* 9:289.

PAVLOVA, M.T., E. BEAUVAIS, F.T. BREZENSKI & W. LITSKY. 1972. Fluorescent-antibody techniques for the identification of Group D streptococci: Direct staining method. *Appl. Microbiol.* 23:571.

PAVLOVA, M.T., E. BEAUVAIS, F.T. BREZENSKI & W. LITSKY. 1973. Rapid assessment of water quality by fluorescent antibody identification of fecal streptococci. *In* Int. Conf. Water Pollut. Res., 6th, Jerusalem, Proc., Pergamon Press, New York, N.Y.

GELDREICH, E.E., 1976. Fecal coliform and fecal streptococcus density relationships in waste discharges and receiving waters. *CRC Critical Rev. Environ. Control* 6:349.

DROOP, M.R. & H.W. JANNASCH, eds. 1977. Advances in Aquatic Microbiology. Vol. 1, Academic Press, pp. 273–378.

PFIZER DIAGNOSTICS DIVISION. Pfizer Selective Enterococcus (PSE) Agar. Pfizer Diagnostics Tech. Bull., New York, N.Y.

911 DIFFERENTIATION OF COLIFORM GROUP OF ORGANISMS (TENTATIVE)

The methods given previously for the detection and estimation of the coliform group of bacteria provide information on pollution and sanitary quality of water supplies. Differentiation of fecal coliform organisms as a group has been described in test procedures given in preceding sections.

Occasionally, it is of value to differentiate and identify coliform strains for special study. Tentative methods for such differentiation are presented below. Commercial differential test systems are becoming widely accepted for use in more comprehensive identification of gram-negative bacteria. These systems give 95 to 98% agreement with conventional tests, provide results in 24 to 48 hr, and have a shelf life of 12 months. The IMViC tests are a minimal differential system giving limited separation of coliform bacteria and, although they are the least expensive differential test system, require up to 5 days for completion.

It should be noted that some lactose-positive bacteria have been found to be associated with regrowth problems due to high temperature and excessive nutrients in receiving waters. These organisms include *Aeromonas* strains as well as coliforms of the genus *Klebsiella*. *Klebsiella* are oxidase-negative organisms, while the *Aeromonas* are oxidase-positive; thus the oxidase reaction can be used as a differential characteristic.

911 A. Culture Purification

The accuracy of the Completed Test and of differential tests may be impaired by failure to purify cultures. To help insure that a pure culture is obtained, restreak to plate count agar from a well-isolated colony on EMB agar (from Completed Test) or from a membrane filter and incubate at 35 ± 0.5 C for 24 hr. Pick a single well-isolated colony and transfer to an agar slant, incubate at 35 ± 0.5 C for 24 hr, and make a Gram stain to confirm the presence of gram-negative, nonspore-forming rods.

Variation in organisms of the coliform group, particularly the "unstable" variation characteristic of the *mutabile* type, occurs occasionally. When purifying cultures, be alert for apparent mixtures of organisms that really may consist of a single strain showing variation. Persistent plus-minus reactions may indicate inadequate purification of culture.

911 B. IMViC Tests

Differentiation of coliforms into subgroups can be carried out on the basis of the results of four tests (indole, methyl red, Voges-Proskauer, and sodium citrate) often referred to collectively as the "IMViC tests." A simplified grouping of the organism types is given in Table 911:I. Additional tests are required for the separation of *Klebsiella* and *Enterobacter*.

The significance of various coliform organisms in water is a subject of considerable study (see discussion under Fecal Coliform MPN Procedure, Section 908C). All types of coliform organisms may occur in feces. Although *E. coli* nearly always will be found in fresh pollution derived from warm-blooded animals, some other type or types of coliform organisms, not accompanied by *E. coli*, occasionally may be found in fresh pollution from a particular source.

Multiplication of coliform organisms occurs on various materials in contact with water including leather washers, wood, liners or walls of storage reservoirs, swimming pool ropes, valve packing material, and slimes and sediments inside pipes. One of the most practical applications of coliform differentiation may be in the study of unexpected coliform densities due to multiplication on or in organic materials. The presence of a large number of coliform organisms of the same type in water from a well or spring—or from a

TABLE 911:I. INTERPRETATION OF IMViC REACTIONS

Organism	Indole	Methyl Red	Voges Proskauer	Citrate
Escherichia coli	+	+	–	–
Shigella sp	+ or –	+	–	–
Citrobacter freundii	–	+	–	+
Citrobacter diversus	+	+	–	+
Klebsiella-Enterobacter group*	+ or –	–	+	+

* See HOMoC series (Section 911C) for further separation of this group.

single tap on a distribution system, for example—is suggestive of such multiplication.

1. Indole Test

a. Reagents:

1) *Medium*—Use tryptophane broth as described under Media Specifications, Section 905C.

2) *Test reagent*—Dissolve 5 g paradimethylaminobenzaldehyde in 75 mL isoamyl (or normal amyl) alcohol, ACS grade, and add 25 mL conc HCl. The reagent should be yellow. Some brands of paradimethylaminobenzaldehyde are not satisfactory and some good brands become unsatisfactory on aging.

The amyl alcohol solution should have a pH value of less than 6.0. Purchase both amyl alcohol and benzaldehyde compound in as small amounts as will be consistent with the volume of work to be done.

b. Procedure: Inoculate 5-mL portions of medium from a pure culture and incubate at 35 ± 0.5 C for 24 ± 2 hr. Add 0.2 to 0.3 mL test reagent and shake. Let stand for about 10 min and observe results.

A dark red color in the amyl alcohol surface layer constitutes a positive indole test; the original color of the reagent, a negative test. An orange color probably indicates the presence of skatole.

2. Methyl Red Test

a. Reagents:

1) *Medium:* Use buffered glucose broth as described under Media Specifications, Section 905C.

2) *Indicator solution:* Dissolve 0.1 g methyl red in 300 mL 95% ethyl alcohol and dilute to 500 mL with distilled water.

b. Procedure: Inoculate 10-mL portions of medium from a pure culture. Incubate at 35 C for 5 days. To 5 mL of the culture add 5 drops methyl red indicator solution.

Record a distinct red color as methyl red-positive and a distinct yellow color as methyl red-negative. Record a mixed shade as questionable and possibly indicative of incomplete culture purification.

3. Voges-Proskauer Test

a. Reagents:

1) *Media:* Use the medium described for the methyl red differential test or, alternatively, the salt peptone glucose medium described under Media Specifications, Section 905C.

2) *Naphthol solution:* Dissolve 5 g purified α-naphthol (melting point 92.5 C or higher) in 100 mL absolute ethyl alcohol. When stored at 5 to 10 C, this solution is stable for 2 wk.

3) *Potassium hydroxide, 7N:* Dissolve 40 g KOH in 100 mL distilled water.

b. Procedure: Separate 5 mL of the culture inoculated for the methyl red test after 48 hr or inoculate 5 mL of salt peptone glucose medium from a pure culture and incubate for 48 hr at 35 ± 0.5 C. To 5 mL of culture add 3 mL naphthol solution and 1 mL KOH solution, and shake vigorously. Development of a pink to crimson color within 15 min to 1 hr constitutes a positive test.

4. Sodium Citrate Test

a. Alternate media: Use either Koser's citrate broth or Simmons' citrate agar (see Media Specifications, Section 905C).

b. Procedure:

1) Lightly inoculate liquid medium with a straight needle, never with a pipet. Incubate at 35 ± 0.5 C for 72 to 96 hr. Record visible growth as positive, no growth as negative.

2) Inoculate agar medium with straight needle, using both a stab and a streak. Incubate 48 hr at 35 ± 0.5 C. Record growth on the medium with (usually) a blue color as a positive reaction; record absence of growth as negative.

911 C. *Klebsiella* Differentiation

The coliform group as defined herein includes the genus *Klebsiella. Klebsiella* strains exhibit seven different IMViC patterns with −−++, ++++, and −+++ being the most common.[1] The confusion between *E. aerogenes* and *Klebsiella pneumoniae* has resulted from their identical −−++ IMViC pattern. Obviously, more biochemical tests are necessary to separate these two organisms.

Klebsiella pneumoniae is an important species because of its association with human respiratory and genitourinary infections.[2–9] In addition, *K. pneumoniae* infects approximately 30% of the human population, being found in human excretions (feces, urine, and sputum) and, consequently, in sewage.[10–13] *Klebsiella* occurrences, in both healthy livestock and infected animals, have been estimated to range from 30 to 40%.[14,15]

Large concentrations of *Klebsiella* in certain industrial wastes such as pulp and paper, sugar beet, and food-processor effluents are significant because they reflect the high bacterial nutrient levels in these wastes. Such wastewaters presumably contain large amounts of carbohydrates and are capable of supporting significant "aftergrowth" of these organisms in the effluents and receiving stream.[12] Aftergrowth also may occur in potable water.[16]

1. Procedure

a. Using a pure culture (Section 911A above), streak on a nitrogen-deficient medium,[17] on which *Klebsiella* colonies are larger, more convex, and gummier in appearance than *Enterobacter*. More detailed and definitive differentiation requires biochemical testing for oxidase and the HOMoC series (hydrogen sulfide and ornithine decarboxylase production, motility, and citrate utilization).[18] Commercially available differential test systems may be used in preliminary screening before serological confirmation.[16] These systems give 95 to 98% agreement with conventional tests, although more significant differences occasionally occur. In some instances supplementary tests will be necessary to differentiate further among strains of Enterobacteriaceae. *Klebsiella* is oxidase-negative, does not produce H_2S, does not decarboxylate ornithine, is nonmotile, and utilizes citrate as the sole source of carbon.[18–20]

b. Oxidase test: The oxidase test can be made by spreading the reagent over colonies of a 24-hr culture on a noncarbohydrate medium agar plate or, preferably, by smearing a colony on a reagent-impregnated filter paper.

1) Reagents—Prepare a 1% aqueous solution of dimethylparaphenylenediamine oxalate and saturate filter paper* with it. Air-dry paper by hanging strips on strings or glass rods; do not use metal clips or tacks. Store dried paper strips in a screw-capped jar containing a desiccant.

Alternatively, prepare a 1% aqueous solution of diethylparaphenylenediamine hydrochloride. Freeze in 1- to 2-mL portions and store frozen for up to 12 months.

2) Procedure—Use a clean nichrome or platinum loop and smear part of a colony growing on an agar plate on to test paper strip; do not use iron wire because this gives false positive results. Record the development of a red color within 10 sec as oxidase-positive.

If the liquid reagent is used, flood a portion of recently thawed reagent (3 to 4 hr) over the agar surface. Oxidase-positive colonies develop a red color within 10 sec and become dark red to black within 60 sec.

2. References

1. Brown, C. & R.J. Seidler. 1973. Potential pathogens in the environment: *Kleb-*

*Whatman No. 1 or equivalent.

siella pneumoniae, a taxonomic and ecological enigma. *Appl. Microbiol.* 25:900.

2. TALERMAN, A. 1968. Multiple liver abscesses caused by *Klebsiella aerogenes*. *J. Med. Microbiol.* 1, 164.
3. CRUICKSHANK, R. 1965. Medical Microbiology, A Guide to Laboratory Diagnosis and Control of Infection. Williams and Wilkins, Baltimore, Md.
4. EICKHOFF, T.C., B.W. STEINHAUER & M. FINLAND. 1966. The *Klebsiella-Enterobacter-Serratia* division. Biochemical and serologic characteristics and susceptibility to antibiotics. *Ann. Intern. Med.* 65:1163.
5. HENSHAW, V., J. PUNCH, M.J. ALLISON & H.P. DALTON. 1969. Frequency of R factor-mediated multiple drug resistance in *Klebsiella* and *Aerobacter*. *Appl. Microbiol.* 17:214.
6. KAYYALI, M.Z., D.P. NICHOLSON & I.M. SMITH. 1972. A *Klebsiella* outbreak in a pediatric nursery: emergency action and preventive surveillance. *Clin. Pediat.* 11:422.
7. ROONEY, J.C., D.J. HILL & D.M. DANKS. 1971. Jaundice associated with bacterial infection in the newborn. *Amer. J. Dis. Child.* 122:39.
8. PARKKULAINEN, K.V. & T.V. KOSUNEN. 1971. A follow-up study of bacteriuria in female patients treated for recurrent urinary tract infections using dip-slides. *Ann. Clin. Res.* 3:163.
9. GAVRILLA, I. 1969. Current clinico-bacteriological and therapeutic aspects of enterocolitis in children. *Pediatria* (Bucur) 18:63.
10. THRONE, B.T. 1970. *Klebsiella* in faeces. *Lancet* 2:1033.
11. SELDEN, R., S. LEE, W.L. LOW, J.V. BENNETT & T.C. EICKHOFF. 1971. Nosocomial *Klebsiella* infections: intestinal colonization as a reservoir. *Ann. Intern. Med.* 74:657.
12. BORDNER, R.H. & B.J. CARROLL, eds. 1972. Proc. Seminar on the Significance of Fecal Coliforms in Industrial Wastes. EPA, Office of Enforcement. National Field Investigations Center, Denver, Colo.
13. BERGERSEN, F.J. & E.H. HIPSLEY. 1970. The presence of N_2-fixing bacteria in the intestines of man and animals. *J. Gen. Microbiol.* 60:61.
14. CARROLL, E.J. 1970. Bactericidal activity of bovine serum against coliform organisms isolated from milk of mastitic udders, udder skin, and environment. *Amer. J. Vet. Res.* 32:689.
15. SHIMAKURA, S., H. IWAMORI & K. HIRAI. 1970. Studies on the omphalitis in baby chicks. I. On the organisms isolated from baby chicks with omphalitis. *Jap. J. Poult. Sci.* 7:57.
16. PTAK, D.J., W. GINSBURG & B.F. WILLEY. 1973. Identification and incidence of *Klebsiella* in chlorinated water supplies. *J. Amer. Water Works Ass.* 65:604.
17. ELLER, C. & F.F. EDWARDS. 1968. Nitrogen-deficient medium in the differential isolation of *Klebsiella* and *Enterobacter* from feces. *Appl. Microbiol.* 16:896.
18. CLOSS, O. & A. DIOGRANES. 1971. Rapid identification of prompt lactose-fermenting genera within the family *Enterobacteriaceae*. *Acta Pathol. Microbiol. Scand.* Sec. B 79:673.
19. COWAN, S.T. & K.J. STEEL. 1965. Manual for the Identification of Medical Bacteria. Cambridge Univ. Press, New York, N.Y.
20. TRAUB, W.H., E.A. RAYMOND & J. LINEHAN. 1970. Identification of *Enterobacteriaceae* in the clinical microbiological laboratory. *Appl. Microbiol.* 20:303.

911 D. Bibliography

CLARK, W.M. 1915. The final hydrogen ion concentrations of cultures of *Bacillus coli*. *Science* 42:71.

CLARK, W.M. & W.A. LUBS. 1915. The differentiation of bacteria of the colon-aerogenes family by the use of indicators. *J. Infect. Dis.* 17:160.

LEVINE, M. 1916. On the significance of the Voges-Proskauer reaction. *J. Bacteriol.* 1:153.

LEVINE, M. 1921. Notes on *Bact. coli* and *Bact. aerogenes*. *Amer. J. Pub. Health* 11:21.

KOSER, S.A. 1924. Correlation of citrate utilization by members of the colon-aerogenes group with other differential characteristics and with habitat. *J. Bacteriol.* 9:59.

SIMMONS, J.S. 1926. A culture medium for differentiating organisms of typhoid-colon-aerogenes groups and for isolation of certain fungi. *J. Infect. Dis.* 39:309.

KOVACS, N. 1928. A simplified method for detecting indol formation by bacteria. *Z. Immunitatsforsch.* 56:311; *Chem. Abstr.* 22:3425.

RUCHHOFT, C.C., J.G. KALLAS, B. CHINN & E.W. COULTER. 1930 and 1931. Coli-aerogenes differentiation in water analysis. *J. Bacteriol.* 21:407; 22:125.

EPSTEIN, S.S. & R.H. VAUGHN. 1934. Differential reactions in the coli group of bacteria. *Amer. J. Pub. Health* 24:505.

BARRITT, M.W. 1936. The intensification of the Voges-Proskauer reaction by the addition of alpha-naphthol. *J. Pathol. Bacteriol.* 42:441.

VAUGHN, R., N.B. MITCHELL & M. LEVINE. 1939. The Voges-Proskauer and methyl red reactions in the coli-aerogenes group. *J. Amer. Water Works Ass.* 31:993.

BORMAN, E.K., C.A. STUART & K.M. WHEELER. 1944. Taxonomy of the family Enterobacteriaceae. *J. Bacteriol.* 48:351.

EWING, W.H. 1966. Enterobacteriaceae: Taxonomy and Nomenclature. U.S. Dep. HEW, Nat. Center for Disease Control, Atlanta, Ga.

WOLFE, M.W. & D. AMSTERDAM. 1968. New diagnostic system for the identification of lactose-fermenting gram-negative rods. *Appl. Microbiol.* 16:1528.

BARRY, A.L. & K.L. BERSOHN. 1969. Methods for storing oxidase test reagents. *Appl. Microbiol.* 17:933.

LENNETTE, E.H., E.H. SPAULDING & J.P. TRUANT, eds. 1974. Manual of Clinical Microbiology, 2nd ed. American Society for Microbiology, Washington, D.C.

912 DETECTION OF PATHOGENIC MICROORGANISMS

Among pathogenic microorganisms that can be demonstrated in wastewater and, under certain conditions, surface and groundwaters of the United States, the most common and important are *Salmonella, Shigella,* enteropathogenic *Escherichia coli, Leptospira,* and the enteric viruses. Organisms such as hookworm larvae, the cysts of *Entamoeba histolytica,* and *Giardia lamblia,* and other animals occasionally may find their way into poorly constructed wells, particularly in areas where the infections they cause are endemic. Most cysts will pass through water treatment systems that use only disinfection or operate poorly designed filtration beds. Other organisms not normally associated with the climate in the U.S., such as *Vibrio cholerae*, also may be found in water because of extensive and rapid world travel.

Routine examination of water and wastewater for pathogenic microorganisms is not recommended. No single procedure is available that can be used to isolate and identify all pathogens. Therefore, negative findings for specific pathogens are provisional because state-of-the-art methodology is not sufficiently sensitive to detect a level of 1 pathogen/100 mL. Salmonellae are extremely common in the environment. Unfortunately, isolation technics even for these ubiquitous organisms involve relatively complicated procedures that exceed the capabilities of many water laboratories. Monitoring water or wastewater for enteric viruses can be carried out only in very well-equipped laboratories and then usually as a special research study. Thus, a combination of factors—among them lack of facilities, lack of trained personnel, lack of laboratory time, high costs, and lack of adequate methods—makes routine examination of water for pathogens impossible. In view of the foregoing, it is apparent that there is a strong need for intensive research in this area and that such research should be encouraged at every opportunity.

Some suspicion has been cast on the validity of the coliform test as an indicator of the biological safety of water.[1-3] These reports suggest that under unusual circumstances pathogenic bacteria can be isolated from waters containing few if any coliform bacteria. *The circumstances surrounding these isolations are not at all clear and it should not be concluded that the coliform test is unreliable or even needs to be supplemented by routine examinations for pathogens at this time.* The coliform test has, over the years, clearly proven its value. The pathogen isolation procedures that follow are offered for the specialist who may wish to initiate a research study, for example, to obtain background data on the numbers, types, and frequency of occurrence of pathogens in water as related to the coliform or fecal coliform index.

912 A. General Qualitative Isolation and Identification Procedures for *Salmonella*

The methods presented below for the isolation of *Salmonella* from water or wastewater are not standardized; they must be considered research procedures that may need modification to fit a particular set of circumstances. Check recovery efficiency of given lots of media using several recently isolated strains of *Salmonella*.

Rather than a specific protocol for *Salmonella* detection in water, a brief summary of methods suitable for recovery of these organisms is given. Methods currently available have been used in numerous field investigations to demonstrate *Salmonella* in both fresh and estuarine water environments. The technics available must be evaluated carefully for the development of a protocol that will yield optimum isolation of these organisms during a specific investigation. Finally, the occurrence of *Salmonella* in water is highly variable; there are limitations and variations in both the sensitivity and selectivity of accepted *Salmonella* isolation procedures for the detection of the more than 1,700 *Salmonella* serotypes currently recognized. Thus, a negative result by any of these methods does not imply the absence of all salmonellae, nor does it imply the absence of other pathogens.

1. Concentration Technics

Generally, it is necessary to examine a relatively large sample in order to isolate pathogenic organisms. These organisms usually are present in small numbers compared to coliforms, because their sporadic occurrence is related to the incidence of disease or infection at a given period.

a. Swab technic: Prepare swabs from cheesecloth 23 cm wide, folded five times at 36-cm lengths and cut lengthwise to within 10 cm from the head into strips approximately 4.5 cm wide. Securely wrap the uncut or folded end of each swab with 16-gauge wire for use in suspending the swab in water. Place the swabs in kraft-type bags and sterilize at 121 C for 15 min. Place swab just below the surface of the stream, lake, or estuary sampling location for 1 to 3 days. (Longer swab exposure will not increase entrapment of pathogens.) Gauze pads of similar thickness—for example, maternity pads—may be substituted. During the period of sampling, particulate matter and microorganisms are concentrated from the water passing through or over the swab. After exposure, retrieve the swab, place it in a sterile plastic bag, ice, and send to the laboratory. Maximum storage-transit time allowable is

6 hr. Do not transport swabs in enrichment media; ambient transport temperature may cause sufficient proliferation of competitive organisms to mask the salmonellae. In the laboratory, place pad or portions of it in enrichment media. When flasks of enrichment medium containing iced swabs are to be incubated at 40 to 41 C, place flasks in a 44.5 C water bath for 5 min before incubation in an air incubator.

b. Diatomaceous earth technic: The filtering ability of diatomaceous earth may be used to concentrate relatively large proportions of microorganisms present in a sample. Place an absorbent pad (not a membrane filter) on a membrane filter funnel receptacle, assemble funnel, and add sufficient sterile diatomaceous earth* to pack the funnel neck loosely. Apply vacuum and slowly filter 2 L of sample. After filtration, disassemble the funnel, divide resulting "plug" of diatomaceous earth and adsorbent pad in half aseptically with a knife-edged, sterile spatula, and add to suitable enrichment media. Alternatively, place entire plug in the enrichment medium.

c. Large-volume sampler: Use a filter composed of borosilicate glass microfibers bonded with epoxy resin to examine several liters or more of sample, provided that sample turbidity does not limit filtration.[4] The filter apparatus consists of a 2.5- × 6.4-cm cartridge filter and a filter holder.† Sterilize by autoclaving at 121 C for 15 min. Place sterile filter apparatus (connected in series with tubing to a 20-L water bottle reservoir and vacuum pump) into the 20-L sample container appropriately calibrated to measure the volume of sample filtered. Apply vacuum and filter an appropriate volume. When filtration is complete, remove filter and place it in a selective enrichment medium.

d. Membrane filter technic: To examine low-turbidity water, filter several liters through a sterile 142-mm-diam membrane of 0.45-μm pore size.[5] For turbid waters, precoat the filter: make 1 L of sterile diatomaceous earth suspension (5 g/L distilled water) and filter about 500 mL. Without interrupting filtration, quickly add sample (1 L or more) to remaining suspension and filter. After filtration place membrane in a sterile blender jar containing 100 mL sterile 0.1% peptone water and homogenize at high speed for 1 min. Add entire homogenate to 100 mL double-strength selective enrichment medium.

Qualitative detection of *Salmonella* in suspect potable water also may be achieved successfully by further analysis of selected M-Endo MF cultures (from 100 mL sample volume) that contain significant background growth and total coliforms.[6] After completing routine coliform count, place entire filter with mixed growth into 10 mL tetrathionate broth (containing 1:50,000 brilliant green dye) for *Salmonella* enrichment before differential colony isolation on brilliant green agar. This unique approach requires no special large sample collections and can be an extension of the routine total coliform analysis.

2. Enrichment

Selectively enrich the concentrated sample in a growth medium that suppresses growth of coliform bacteria. Sample enrichment is essential, because the pathogens usually are present in low numbers and solid selective media for colony isolation are somewhat toxic, even to the pathogens. No single enrichment medium can be recommended that allows optimum growth of all *Salmonella* serotypes. Use two selective enrichment media in parallel for optimum detection.

a. Dulcitol selenite broth inhibits many

*Celite, Johns-Manville Co., Denver, Colo., or equivalent.
†Balston Type AA filter with Type 90 holder, or equivalent.

nonpathogenic enterobacteria during the early hours of incubation following inoculation and allows *Salmonella* strains to multiply rather rapidly. Optimum incubation time for maximum recovery of *Salmonella* is 24 hr. However, recovery of relatively slow-growing organisms such as variants of *S. enteritidis* (Montevideo, Enteritidis, and Worthington) require longer incubation periods. Make repeated streakings from the same inoculated medium after each 24-hr period. Streak from broth cultures that develop turbidity and any orange-red color due to selenite reduction onto suitable selective solid media.

b. Tetrathionate broth may yield more salmonellae than selenite broth. However, extend incubation beyond 24 hr, with repeat streaking from the same tube several times during the first day and daily up to 5 days to increase potential recovery of all serotypes that may be present. Transfer 1 mL tetrathionate broth culture to a fresh tube of the same medium for continued incubation to enrich further for *Salmonella* growth and enhance recovery on streak plates. Improve suppression of nonpathogenic organisms by adding 1:50,000 brilliant green dye. Improve sensitivity by adding 3 mg *l*-cystine/L tetrathionate broth.

3. Selective Growth

Further separation of pathogens from the remaining nonpathogenic bacterial population is facilitated by proper choice of incubation temperature for primary enrichment followed by secondary differentiation on selective solid media. These two factors, incubation temperature and choice of media, are interrelated. More *Salmonella* may be recovered at either 35 or 37 C with bismuth sulfite agar than at higher temperatures. Great skill at screening for these pathogens is necessary because of the competing growth of various nonpathogens. Use of an incubation temperature of 41.5 C for both primary enrichment broth and differential media enhances the isolation of many salmonellae while reducing the number of interfering organisms. Some *Salmonella* serotypes, among them *S. typhi,* will not grow at this elevated temperature.

Solid media commonly used for enteric pathogen detection may be classed into three groups: (*a*) differential media with little or no inhibition toward nonpathogenic bacteria, such as EMB (containing sucrose); (*b*) selective media containing bile salts or sodium desoxycholate as inhibitors, such as MacConkey's agar, desoxycholate agar, or xylose lysine desoxycholate (XLD) agar; and (*c*) selective media containing brilliant green dye, such as brilliant green agar or bismuth sulfite agar. Any medium selected must provide optimum suppression of coliforms while permitting good recovery of the pathogenic group. Streaking duplicate plates, one heavily and one lightly, often aids in recognition of enteric pathogens in the presence of large numbers of interfering organisms.

a. Brilliant green agar: Typical well-isolated *Salmonella* colonies grown on this medium are pinkish white with a red background. *S. typhi* and a few other species of *Salmonella* grow poorly because of the brilliant green dye content. Lactose-fermenters not subject to growth suppression will form greenish colonies or may produce other colorations. Occasionally, slow lactose-fermenters (*Proteus, Citrobacter,* and *Pseudomonas*) will produce colonies resembling those of a pathogen. Suppress the spreading effect of pseudomonads by increasing the agar concentration to 2%. In some instances, *Proteus* has been observed to "swarm"; this tendency can be reduced by using agar plates dried to remove surface moisture. If suspect *Salmonella* colonies are not observed after 24 hr incubation, reincubate for an additional 24 hr to permit slow-growing or partially inhibited organisms to develop visible colonies. If typical colonies are not observed or if the streak plate is crowded, it may be necessary to isolate in pure culture a few colonies for biochemical characterization tests. Non-lactose-fermenting colonies in close proximity to lactose-fermenting colonies may be masked.

b. Bismuth sulfite agar: Luxuriant growth of many *Salmonella* species (including *S. typhi*) can be expected on this medium. Examine bismuth sulfite plates after 24 hr incubation for suspect colonies; reincubate for 24 hr to detect slow-growing strains. Typical colonies usually develop a black color, with or without a metallic sheen, and frequently this blackening extends beyond the colony to give a "halo" effect. A few species of *Salmonella* develop a green coloration; therefore, it may be necessary to isolate some of these colony types when typical colonies are absent. As with brilliant green agar, typical colony coloration may be masked by the presence of numerous bordering colonies after 48 hr incubation. A black color also is developed by other H_2S-producing colonies, for example, *Proteus* and certain coliforms.

c. Xylose lysine desoxycholate agar: Compared to brilliant green dye, sodium desoxycholate is only slightly toxic to fastidious *Salmonella*. *Salmonella* and *Arizona* organisms produce black-centered red colonies. Coliform bacteria, *Proteus*, and many *Enterobacter* produce yellow colonies. Optimum incubation time is 24 hr. If plates are incubated longer, an alkaline reversion and subsequent blackening occurs with H_2S-positive nonpathogens (*Citrobacter, Proteus vulgaris*, and *P. mirabilis*).

4. Biochemical Reactions

Many enteric organisms of little or no pathogenicity share certain major biochemical characteristics with *Salmonella*. The identification of pathogens by colony characteristics on selective solid media has limitations inherent in the biological variations of certain organisms. Suspected colonies grown on selective solid media must be purified and further characterized by biochemical reactions; final verification is based on serological identification. Usually a large number of cultures will be obtained from the screening procedure.

Commercially available differential media kits may be used as an alternative to Phases 1, 2, and 3 described below, before serological confirmation. These kits give 95 to 98% agreement with conventional tests, although more significant tests will be necessary to achieve further differentiation among strains of *Enterobacteriaceae*.

When such kits are not used, follow a sequential pattern of biochemical testing that will result in a greater saving of media and time for laboratory personnel.[7]

Phase 1—Preliminary screening, phenylalanine deaminase activity: Discard phenylalanine deaminase-positive cultures immediately as indicative of the *Proteus* group. Subject phenylalanine deaminase-negative cultures to the biochemical tests of Phase 2. In this test, spot isolates on phenylalanine agar and incubate for 24 hr at either 35 or 37 C. Phenylalanine deaminase activity is indicated by a green zone that develops around the colony after flooding the plate with a 0.5*M* $FeCl_3$ solution.

Simultaneously, test for inability to ferment lactose on a selective agar, such as MacConkey agar.

Phase 2—Biochemical tests: The tests used are:

Medium	Purpose of Test
TSI	Fermentation pattern: H_2S production; ONPG for β-D-galactosidase
SIM	Production of indole and H_2S, motility

Conformance to the typical biochemical patterns of the *Salmonella* will determine whether to process the cultures further (Phase 3). Aberrant cultures may be encountered that do not conform to all the classical reactions attributed to each pathogenic group. In all cases, therefore, review the reactions as a whole and do not discard cultures on the basis of a small number of apparent anomalies.

Phase 3—Fermentation reactions: Test fermentation reactions in dextrose, mannitol, maltose, dulcitol, xylose, rhamnose, and inositol broths to characterize further the biochemical capabilities of the isolates. This additional sorting of isolates reduces the possible number of positive cultures to be processed for serological confirmation. If the testing laboratory is equipped for serological identification, this series of biochemical tests may be eliminated.

5. Genus Identification by Serological Technics

Upon completion of the recommended biochemical tests, inoculate the suspected *Salmonella* pure culture onto a brain heart infusion agar slant and incubate for 18 to 24 hr at 35 C. With wax pencil (china marker) divide an alcohol-cleaned glass slide into four sections. Prepare a dense suspension of the test organism by suspending growth from an 18- to 24-hr agar slant in 0.5 mL 0.85% NaCl solution. Place a drop of *Salmonella* "O" polyvalent antiserum in the first section and antiserum plus 0.85% NaCl in the second section. Using a clean inoculating loop, transfer a loopful of bacterial suspension to the third section containing 0.85% NaCl solution and to the fourth section containing 0.85% NaCl solution plus antiserum. Gently rock slide back and forth. If agglutination is not apparent in the fourth section at the end of 1 min, the test is negative. All other sections should remain clear.

If the biochemical reactions are characteristic of *S. typhi* and the culture reacts with "O" polyvalent antiserum, check other colonies from the same plate for Vi antigen reaction. If there is no agglutination with *Salmonella* Vi antiserum, the culture is not *S. typhi*. Identification of *Salmonella* serotypes requires determination of H antigens and phase of the organism as described by Edwards and Ewing.[7]

912 B. Immunofluorescence Technic for Detection of *Salmonella*

The direct fluorescent antibody (FA) technic is a rapid and effective means of detecting salmonellae in fresh and seawater samples. It may be used as a screening technic to provide rapid results for large numbers of samples, such as recreational or shellfish-harvesting waters. Sample volumes used depend on the degree of contamination. Where gross pollution is present, use smaller samples. When background information is absent, analyze a 2-L sample.

1. Apparatus for Fluorescence Microscopy

Standard fluorescent antibody microscopy equipment may be obtained separately or in a package containing the essential instrumentation:

a. Light microscope with microscope stand.

b. Very bright light source, providing energy in the short-wavelength region of the spectrum. A high-pressure mercury arc enclosed in a quartz envelope satisfies this requirement. A significant portion of the energy should be emitted in the ultraviolet and blue region of the spectrum.

c. Power pack to provide constant voltage and wattage output for the high-pressure mercury bulb. Include a starter button that ignites the bulb when released.

d. Basic filters including heat-absorbing filter (KG-1 or KG-2, or equivalent): red-absorbing filter (BG-38, or equivalent); exciter filter (BG-12, or equivalent, BG-12 being also a blue filter); and barrier filter (OG-1 or blue-absorbing filter).

e. Cardioid dark-field condenser for illuminating the specimen. A 95 × oil immersion objective with built-in iris diaphragm is desirable. True dark-field illumination can be achieved only if the numerical aperture of the objective is smaller than the numerical aperture of the condenser, i.e., of the illuminating cone of light. (Difference in numerical aperture between objective and condenser should be at least 0.05). Reduce the numerical aperture of an oil immersion objective by using the built-in diaphragm or by putting a funnel stop onto the objective.

2. Reagents for Fluorescent Microscopy

a. Non-drying immersion oil, Type A (low fluorescence).*

b. Fluorescent antibody pre-cleaned micro slides, 7.6 by 2.5 cm, 0.8 to 1.0 mm thickness.

c. Cover glass for FA slides, No. 1-1/2, 0.16 to 0.19 mm thickness.

d. FA Kirkpatrick fixative.†

e. Phosphate buffered saline (PBS): Add 10 g buffer‡ to 1,000 mL freshly prepared distilled water. Stir until the powder dissolves completely. Adjust with NaOH to pH 8.0.

f. FA mounting fluid: Use standardized reagent-grade glycerine adjusted to pH 9.0 to 9.6 and intended for mounting slides to be viewed with the FA microscope.

g. Distilled water: Use double-distilled water from an all-glass still or other high-quality analytic-grade laboratory water.

h. Staining assembly consisting of dish, cover, and slide rack with handle. Five dishes are required: for Kirkpatrick's fixative; 95% ethanol; first PBS rinse; second PBS rinse; and distilled water.

i. FA Salmonella panvalent conjugate is a fluorescein-conjugated anti-*Salmonella* globulin.§ To rehydrate, add 5 mL distilled water to a vial of conjugate. Determine working dilution (see ¶5*e*). Store unused rehydrated conjugate in a freezer, preferably at −60 C. Avoid repeated freezing and thawing.

j. Moist chamber used to incubate slides containing smears with added conjugate. A simple chamber consists of water-saturated toweling with a culture dish bottom (150 by 20 mm) placed over the wet toweling.

3. Concentration Technic

Place an absorbent pad on a membrane filter funnel and add sufficient sterile diatomaceous earth ‖ to pack the funnel neck loosely. Filter 2 L of sample. Rinse funnel with 50 to 100 mL sterile phosphate-buffered dilution water or 0.1% peptone water. Disassemble funnel and remove resulting "plug" of diatomaceous earth and the absorbent pad. Repeat with a second 2-L sample.

4. Enrichment

Immerse one plug and absorbent pad in a flask containing 300 mL dulcitol selenite broth. Immerse second plug and absorb-

*R.P. Cargille Laboratories, Inc., Cedar Grove, N.J., or equivalent.
†Difco No. 3188 or equivalent.
‡Difco Bacto-FA Buffer, dried, or equivalent.
§Difco or equivalent.
‖Celite, Johns Manville Co., Denver, Colo., or equivalent.

ent pad in a flask containing 300 mL tetrathionate broth supplemented with 3 mL 1:1,000 aqueous solution of brilliant green dye and 3 mg *l*-cystine. Incubate at either 35 or 37 C for 24 hr.

5. Fluorescent Antibody Reaction and Analysis

a. Prepare spot plates of brilliant green agar (BGA) and xylose lysine brilliant green agar (XLBG) by placing 1 drop (about 0.01 mL, delivered with a wire loop) of the enrichment medium (dulcitol selenite or tetrathionate broth) at each of four separate points on the agar surface. Space drops on agar plate so that FA microscope slide will cover two inoculation points. This is essential because glass slide impression smears of the inoculated points will be made after incubation of plates.

b. Incubate BGA and XLBG plates at either 35 or 37 C for 3 hr. After incubation make impression smears by taking a *clean* FA microscope glass slide and placing it over two inoculated points on the medium. Press down lightly, being careful not to move the glass slide horizontally. Do not apply too much pressure, because it will cause movement of the slide and collection of additional agar. Repeat this process for the other two inoculation points and for inoculation points on second agar medium. Prepare a total of four FA slides in this manner.

c. Air-dry smears and fix for 2 min in Kirkpatrick's fixative. Rinse slides briefly in 95% ethanol and let air dry. *Do not blot.*

d. Cover fixed smears with 1 drop of *Salmonella* panvalent conjugate. Before use, dilute commercial conjugate and determine appropriate working dilution. Most batches are effective at a 1:4 dilution but this will vary with the type of fluorescence equipment used, light source, alignment, magnification, cultures, etc. Determine working dilution (titer) of each lot of conjugate.

e. To determine conjugate titer use a known 18- to 24-hr *Salmonella* culture grown in veal infusion broth and make smears on FA glass slide. Dilute conjugate and treat as outlined in *c* and *d* above. For example, if the following results are obtained:

Dilution of Conjugate	Fluorescence
1:2	4+
1:4	4+
1:6	4+
1:8	2+
1:10	1+

use a 1:4 dilution of conjugate. Diluting conjugate insures minimum cross-reactivity. Prepare fresh diluted conjugate daily.

f. After covering each smear with 1 drop of the appropriate dilution of conjugate, place slides in a moist chamber to prevent evaporation of staining reagent. After 30 min wash away excess reagent by dipping slides into phosphate-buffered saline (pH 8.0). Place slides in second bath of buffered saline for 10 min. Remove, rinse in distilled water, and drain dry. *Do not blot.*

g. Place a small drop of mounting fluid (pH 9.0 to 9.6) on the smear and cover with a No. 1-1/2 cover slip. Seal edges of cover slip with clear fingernail polish. Examine sealed slides within a few days while fluorescence is of optimum intensity.

h. Examine under a fluorescence microscope unit fitted with appropriate filters.

6. Recording and Interpreting Results

The intensity of organisms fluorescing in any given field is important in assessing positive *Salmonella* smears. If the majority of cells present fluoresce (4+ or 3+) the smear is positive. Carefully scrutinize smears showing only a few scattered fluorescing cells. Critical examination of cellu-

Reaction	Description	Fluorescence Intensity
Positive	Brilliant yellow-green fluorescence, cells sharply outlined.	4+
Positive	Bright yellow-green fluorescence, cells sharply outlined with dark center.	3+
Negative	Dull yellow-green fluorescence, cells not sharply outlined.	2+
Negative	Faint green fluorescence discernible in dense areas, cells not outlined.	1+
Negative	No fluorescence.	0

lar morphology may distinguish between these cells and salmonellae. The degree of fluorescence is the criterion on which positivity is based. Consider weakly fluorescing cells (2+ and 1+) negative. Confirm all positive FA results by cultural technics (see Section 912A).

912 C. Quantitative *Salmonella* Procedures

1. Multiple-Tube Fermentation Technic

Because of the high ratio of coliform bacteria to pathogens, large samples (1 L or more) are required. Preferably concentrate the sample by the membrane filter technic (Section 912A.1*d*). After blending the membrane with 100 mL sterile 0.1% peptone water, use a quantitative MPN procedure by proportioning homogenate into a five-tube, three-dilution multiple-tube procedure using either dulcitol selenite or tetrathionate broth as the selective enrichment medium. Incubate for 24 hr at 35 C and streak from each tube to plates of brilliant green and xylose lysine desoxycholate agars. Incubate for 24 hr at 35 C. Select from each plate at least one colony suspected of being *Salmonella*, inoculate a slant each of TSI and lysine iron agars, and incubate for 24 hr at 35 C. Test cultures giving a positive reaction for *Salmonella* by serological technics (see Section 912A.5). From the combination of *Salmonella* negative and positive tubes, calculate the MPN/10 L of original sample (see Section 908D).

2. Membrane Filter Procedure for *S. typhi*

A quantitative procedure for *S. typhi* is available. The method utilizes M-bismuth sulfite broth and the membrane filter procedure for bacterial concentration. It is applicable only with samples low in organic and particulate materials, because quantities of 100 mL or more generally are filtered. After filtration (see Section 909A.5), incubate filter on a pad saturated with M-bismuth sulfite broth for 18 to 20 hr at 35 C. Transfer to a fresh pad saturated with M-bismuth sulfite broth and continue incubation to give a total of 30 hr. Transfer suspect colonies (smooth glistening colonies with jet-black centers surrounded by a thin clear white border) to triple sugar iron agar (TSI) and incubate at 35 C for 18 hr. Proceed with additional biochemical and serological procedures as described under qualitative methods (Section 912A.3 and 4).

912 D. *Shigella*

While most shigellosis epidemics are food-borne or spread by person-to-person contact, they also may be due to contaminated drinking water. Outbreaks of waterborne shigellosis frequently result from accidental interruption of water treatment, flood-borne excreta contamination of well water supplies, cross-connections between contaminated water pipes and potable water supply lines, untreated water supplies, or wastewater seepage into water supply lines.

Shigellae have been found in various polluted waters, but methodology is qualitative and low in sensitivity. Instability of some biochemical characteristics can occur in *Shigella* strains introduced into the water environment. Coliform bacteria and most strains of *Proteus vulgaris* are antagonistic to *Shigella*.

1. Concentration Technics

See Section 912A.1. Use a sample of 1 to 10 L.

2. Enrichment

Choose a selective enrichment medium to minimize accumulation of volatile acid by-products derived from growth of coliforms and other antagonistic organisms. Use nutrient broth adjusted to pH 8.0 (a less favorable growth pH for coliforms) and incubate for 6 to 18 hr at 35 C. Streak cultures at 6 and 18 hr to selective differential agar plates to optimize *Shigella* recovery. Alternatively, use an autocytotoxic enrichment medium prepared by adding 4-chloro-2-cyclopentylphenyl β-D-galactopyranoside (CPPG) to a final concentration of 1m*M* in lactose broth buffered to pH 6.5 with citrate buffer (at a final concentration of 0.05*M*). Solubilize in lactose broth by use of a magentic stirrer and careful heating at 45 C. Sterilize lactose-CPPG broth by membrane filtration. Add sample concentrates to lactose-CPPG broth and incubate for 24 hr at 35 C.

3. Selective Growth

Use xylose lysine desoxycholate (XLD) agar for primary isolation of *Shigella* strains. Suspected *Shigella* colonies are red. Incubate for 24 hr at 35 C. The major interfering organisms on this differential agar include strains of *Proteus*, *Providencia*, and *Pseudomonas*.

4. Biochemical Reactions and Serological Identification

Follow procedures described in Section 912A.4. For serological identification use the slide agglutination technic (see Section 912A.4) with *Shigella* antisera (polyvalent and type-specific sera).

912 E. Enteropathogenic *Escherichia coli*

Enteropathogenic *Escherichia coli* has been isolated from tap water,[8] drinking water sources,[9] and mountain streams.[10] It is unlikely that *E. coli* organisms could initiate disease by transmission through a properly treated potable water. Historically, at least in the United States, these organisms reportedly caused disease almost exclusively in infants, but because infants normally are given boiled or sterilized water, such waterborne infections probably are infrequent. Diarrhea of travelers may be caused by enteropathogenic *E. coli*.[11]

Examination of potable water supplies for enteropathogenic *E. coli* can be made by use of the membrane filter technic (Section 909A), preferably with M-FC broth.[12] Pick characteristic blue colonies, purify, and determine IMViC reactions (Section 911B). Test IMViC reactive strains ++−−, producing gas from lactose, by the serological technics of Edwards and Ewing.[7] There is, however, no biochemical marker that will separate pathogenic from nonpathogenic strains and the relationship between serotype and pathogenicity is questionable.[13]

Three classes of antigens are important in serological grouping of *E. coli*: the heat-stable and major grouping factor, the "O" antigen associated with the cell; the envelope or capsular "K" antigen; and the "H" flagellar antigen. Use slide agglutination to determine O and K antigens and the microscopic tube test for confirmation of O and H antigens.

912 F. Pathogenic Leptospires

The occurrence of pathogenic leptospires in natural waters is extremely variable. Many factors make interpretation of results difficult, for example, intermittent leptospire discharge from infected wildlife or farm animals and the effects of stormwater runoff and flooding of contaminated land.[14] Leptospire persistence in warm, slow-moving waters of pH 6.0 to 8.0[15–18] and moderate levels of bacterial nutrients[19] also complicate interpretation. Even when pathogenic leptospires are present, their detection is difficult because of the competitive growth of other organisms[20] and the need to differentiate between pathogenic and saprophytic strains.[16,20–23] Failure to isolate pathogenic leptospires from natural waters does not necessarily indicate their absence.

These factors explain why qualitative methodology evolved to detect leptospires in polluted water. Long-term incubation on various media is necessary because of the relatively slow growth of the organisms. During incubation, check inoculated media weekly for the appearance of leptospires and for culture contamination using dark field microscopy. Upon detection, characterize the leptospire isolates further by various biochemical and serological tests to separate pathogenic and saprophytic strains. Use animal tests for pathogenic leptospires but only on primary pure-culture isolates because pathogenic strains may become avirulent through subsequent culture passages.

1. Preliminary Concentration Technic

Pathogenic leptospires tend to concentrate in near-shore bottom sediments of streams and farm ponds. Gently agitate bottom sediment before sampling to insure collection of bacteria-laden material from the sediment-water interface. Use the bacteriological bottom sampler or standard sample bottles (see Section 906A) to collect this finely suspended material. Upon return to the laboratory, or preferably at a field site, shake sample vigorously to release entrapped bacteria from the sediment and prefilter immediately through either filter paper* or a membrane filter absorbent pad. Pass prefiltered sample through a Swinney hypodermic adapter containing a fiberglass prefilter and a membrane filter of 0.45-μm pore size to separate leptospires (which can pass through the pores into the filtrate) from other organisms. Rinse with an equal volume of sterile dilution water.

*Whatman No. 1 or equivalent.

2. Enrichment

Inoculate portions of sample filtrate (1 mL and 0.1 mL) into Fletcher's semisolid medium containing 10% rabbit serum.[23] Incubate inoculated medium at 30 C for 6 wk. Examine each tube at least weekly for leptospiral growth and culture contamination: use dark-field illumination and 250 × magnification.[24] Strains of *Vibrio, Spirillum,* or *Paraspirillum* are the most common contaminants observed, particularly when filtrate volumes greater than 0.1 mL are examined.[24]

Leptospires are helicoidal, usually 6 to 20 μm long with each coil about 0.2 to 0.3 μm in diameter. The coils of leptospires are more compact than those of other spirochaetes.[25] If leptospires are not observed microscopically within a 6-wk incubation period, the test is negative.

As an alternate enrichment procedure, inoculate spread plates of SM agar[26] or bovine albumin polysorbate 80 medium[27,28] with 0.1 to 1.0 mL sample filtrate. Incubate at 30 C for 7 to 9 days. When bovine albumin polysorbate 80 medium is used, an agar overlay of 0.7% distilled water agar is recommended. Regardless of agar medium used, prepare it 1 to 2 days before inoculation to condition the agar and promote even spreading of the inoculum over the surface. Identify all colonies morphologically by dark-field microscopy before making biochemical and serological tests or animal inoculations.

3. Differentiation of Leptospires

Detection of pathogenic leptospires in lakes and streams indicates leptospirosis in domestic or wild animals that frequent these waters and signals a health risk to bathers. It is critically important to differentiate pathogenic from saprophytic leptospire strains.

a. Culture reactions: Saprophytic leptospires grow well in Stuart's medium containing 10% rabbit serum supplemented with 10 μg copper sulfate ($CuSO_4$)/mL[17,18] or 100 μg 8-azaguanine/mL.[15,18] Only saprophytic leptospires grow in a 10% rabbit serum medium at 13 C.[15] Saprophytic strains demonstrate higher oxidase response[29] and higher egg yolk decomposition activity[30] than pathogenic leptospires. Optimum incubation temperature for pathogenic leptospires is 30 C. Incubate all tests for 5 days. Use no single test to differentiate saprophytic from pathogenic leptospires.[31]

b. Verification of pathogenicity: Commercial antisera are available that permit tentative identification of pathogenic leptospires. Make final verification for pathogenicity by intraperitoneal injection of the suspect strain into guinea pigs. After 4 weeks, sacrifice animals, test blood for serum antibody titers above 100, and demonstrate presence of pathogenic leptospires by cultivation of aseptically removed kidney tissue in Stuart's medium.

912 G. *Vibrio Cholerae* and NAG Vibrios

While waterborne outbreaks of cholera often are a serious problem in the Near East and Orient, the potential risk of similar epidemics in other parts of the world cannot be ignored. The spread of cholera in 1970 may reflect a lack of international quarantine enforcement by some countries having primitive public water supplies and inadequate sanitary regulations, the international mobility of convalescent carriers in the world population, and the quick transport of contaminated food and water.

Various noncholera *Vibrio* strains are found in both fresh water and estuarine environments. Some strains of *V. cho-*

lerae isolated under endemic conditions have been known to yield nonpathogenic mutants. Whether some noncholera vibrios revert back to pathogenicity is unknown. The viability of *V. cholerae* in surface waters may range from 1 hr to 13 days, with greater persistence in alkaline waters of pH 8.2 to 8.7 that contain chlorides and organic nutrients. Nonagglutinating (NAG) vibrios are vibrios that morphologically and biochemically resemble *V. cholerae* but usually can be distinguished serologically because they do not agglutinate in group 0:1 antiserum.

1. Concentration Technics

See Section 912A.1. Use a 1- to 10-L sample.

2. Enrichment Procedures

Choose selective enrichment medium to minimize the competitive, antagonistic action of many organisms in the polluted water sample. Use alkaline peptone water (pH 9.0). Add sample concentrates to alkaline peptone water and incubate for 6 to 18 hr at 35 C. Streak cultures at 6 and 18 hr to selective differential agar plate to optimize *V. cholerae* recovery.

3. Selective Growth

Use TCBS agar (thiosulfate-citrate-bile-salt-sucrose-agar) for primary isolation of *V. cholerae*. Incubate for 24 hr at 35 C. Suspected *V. cholerae* colonies are distinguished easily by their yellow color on TCBS agar. The predominant interfering organisms are sucrose-positive bacteria and include *V. anguillarum, V. alginolyticus,* and some strains of *V. parahaemolyticus* and *Proteus*.

4. Biochemical Reaction

Check suspect cultures for purity before making biochemical tests. Do not use colonies from selective media to determine the oxidase reaction nor use growth from sugar-containing media in the agglutination test.

Phase 1—Preliminary screening for urease and oxidase activity: Discard urea-positive cultures immediately as indicative of the *Proteus* group or other nonpathogenic forms. Discard oxidase-negative cultures. Subject urea-negative, oxidase-positive cultures to the biochemical tests of Phase 2.

Phase 2—Biochemical tests: The tests described below are minimal requirements for screening suspect strains:

Test	Purpose
No. of flagella	Single, polar flagellum
Ornithine decarboxylase	+
Lysine decarboxylase	+
Arginine	−
Indole 22 C	+
35 C	Delayed (+)
Sugar fermentation, acid from:	
Arabinose	+ or d (more than 10% strains positive or negative)
Mannose	+ or d (more than 10% of strains positive or negative)
Sucrose	+
Growth at 5 C	−
Growth in:	
1% tryptone + 4% NaCl	+
1% tryptone + 8% NaCl	−
1% tryptose, pH 10.0	+

Use conformance to the typical biochemical patterns for *V. cholerae* and its biotypes to determine whether to process cultures further. Aberrant cultures may be encountered that do not satisfy all the classical reactions attributed to *V. cholerae*. Do not discard cultures on the basis of a small number of apparent anomalies.

5. Serological Identification

Serological identification of *V. cholerae* involves the slide agglutination technic (see Section 912A.4) but utilizes *V. cholerae* antisera (polyvalent and type-specific sera). Final verification of strains to distinguish between classical type, El-Tor biotypes, and NAG vibrio must be done in a specialized national laboratory.

912 H. Pathogenic Protozoa

1. *Giardia lamblia*

The causative agent of giardiasis is a flagellate protozoan shed in the feces of man and animals, most often in the cyst stage, although with cases of severe watery diarrhea, the fragile trophozoite reproductive stage may be shed. As few as 10 cysts ingested in capsules have been found to be infective for man[32] and it is possible that fewer viable cysts ingested in drinking water may be sufficient to initiate infection. An infected individual may shed more than 10^6 cysts/g of stool.[33] Cysts may survive up to 16 days in drinking water at 8 C.[34] Fecal contamination of water supplies by infected persons could, therefore, lead to waterborne transmission of giardiasis.

Most documented waterborne outbreaks[35–37] associated with municipal supplies and recreational areas have occurred as the result of drinking surface water where the only treatment was chlorination. There are no reliable data on the inactivation of *Giardia* cysts by chlorine, but they are believed to be resistant to chlorination at conventional dosages and contact times. *Giardia* cysts, when present in water supplies, would occur most likely at low concentrations.[38–40]

The method described below has been used to detect cysts in raw and treated supplies.[41–42] It should be regarded as tentative and experimental with limited recovery efficiency data. A negative result may reflect intermittent shedding patterns, poor recovery efficiency with a given water, and insufficient sample volume or sampling frequency. The significance of a positive finding is difficult to assess unless suitable infectivity determinations and/or epidemiological studies are performed.

a. Concentration technics: Collect a sample using the apparatus shown in Figures 912:1 and 912:2, consisting of an inlet hose, plastic filter holder with 25-cm-(10-in.-) long yarn-wound orlon filter (7-μm porosity)* outlet hose, water meter, and a limiting orifice flow control device with a flow rate of 3.15×10^{-5} m^3/sec.† Components of the apparatus need not be sterile, but thoroughly drain and rinse the equipment between samples. Use aseptic technic during sample collection to protect the sample collector and to prevent sample cross-contamination. A line pres-

*Commercial Filters Division, Carborundum Company, Lebanon, Ind. 46052, or equivalent.
†Eaton Corporation, Carol Stream, Ill. 60187, or equivalent.

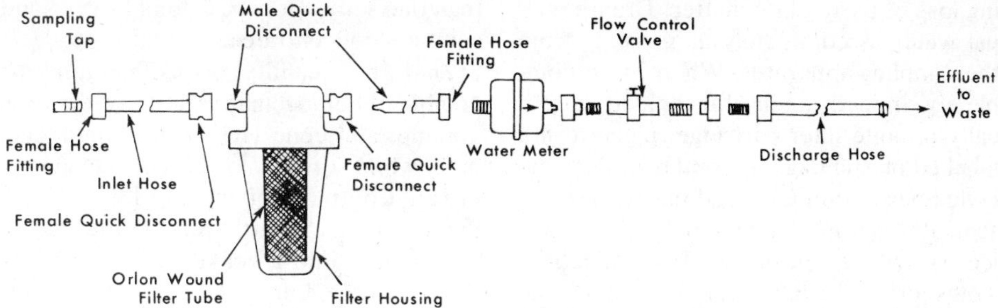

Figure 912:1. ***Giardia*** **sampling device schematic.**

sure of 100 to 500 kPa is satisfactory; if water under pressure is unavailable, use a pump.

The volume of water to be sampled depends on the intent of the investigation; 1,900 L collected over 18 to 24 hr is suggested.[41]

To collect a sample, connect inlet hose to an appropriate sampling tap. The direction of water flow is from the outside to the inside of the yarn-wound filter cartridge. Record time and meter reading and open sampling tap. Collect sample, turn off sampling tap, record time and meter reading, and disconnect sampling apparatus. Take care to maintain opening on the inlet hose above the level of opening on outlet hose to prevent filter backwash-

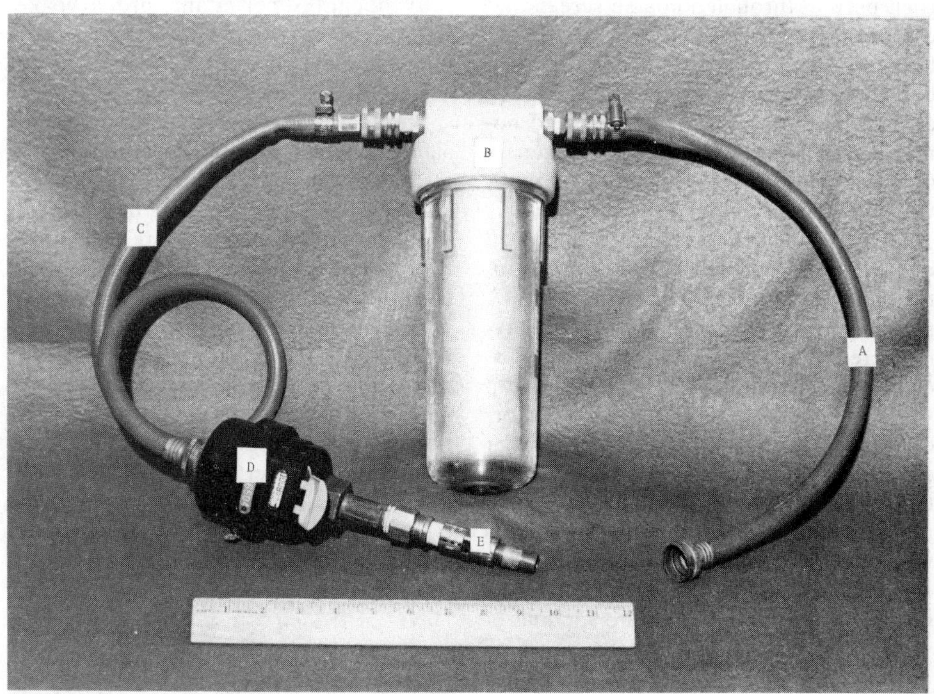

Figure 912:2. ***Giardia*** **sampling device. A—inlet hose; B—filter housing; C—outlet hose; D—water meter; E—limiting orifice flow controller.**

ing loss of particulate matter. Drain residual water as completely as possible from the sampling apparatus. When unit is completely drained, open filter holder, aseptically remove filter cartridge, place it in a labelled plastic bag, and seal bag. Place labelled bag within a second plastic bag and seal. Refrigerate samples or place on wet ice as soon as possible after collection. Transport to the laboratory on wet ice and process as soon as practical but within 48 hr. Do not freeze.

b. Sample processing: Using aseptic technic and wearing rubber gloves, remove filter cartridge from plastic bag and transfer to an appropriate tray or pan. Cut orlon fibers from stainless steel support core of filter with a razor knife and divide fiber mat into four approximately equal portions. Blend each portion at low speed for 10 sec with 250 mL isotonic saline or distilled water. Pass homogenate from each portion through a coarse screen and combine. Express residual fluid from fibers by placing them in a sealable plastic bag with one corner removed. Knead bag and add expressed fluid to combined homogenate. Pour combined homogenate (approximately 1 L) into a 1- to 1.5-L graduated cylinder and let settle for 15 min.

1) If the quantity of sediment obtained in the graduated cylinder is <5% of the total volume, filter homogenate through 47-mm-diam 45-, 30-, and 7-μm porosity nylon screens‡ on a membrane holder. Discard material retained by 45- and 30-μm screens. Transfer 7-μm screen with forceps to the inside wall of a 50-mL beaker containing 1 to 2 mL distilled water or isotonic saline. Using a capillary pipet and rubber bulb, flush screen 10 to 15 times with water or saline, holding tip of pipet as close as possible to the screen. Discard screen and retain concentrate for microscopic examination. If needed, use more than one screen for the 7-μm filtration and combine concentrates.

2) If the quantity of sediment in the combined homogenate is >5% of the total volume, add conc HCl (2 mL/L homogenate) while stirring slowly with a magnetic mixer. Continue mixing for 15 min. If a floc forms, let it settle for 10 min, decant supernatant into a beaker, and strain floc through three 8-in. × 4-in. "post-op" surgical sponges§ into beaker containing supernatant. Filter mixture through 30- and 7-μm porosity nylon screens. Process the 7-μm screen as described above in 1) and retain concentrate for microscopic examination.

If no floc forms on adding HCl, slowly add formalin solution (40% w/v) to acidified homogenate with slow mixing. Add formalin until a floc forms or until a 5% (v/v) final concentration of formalin is attained. Let floc settle for 10 min and carefully decant supernatant into a beaker without disturbing settled floc. Strain, filter, and complete preparation of floc as described in preceding paragraph.

Sterilize all materials used in processing samples by autoclaving or other appropriate means before reuse or discard.

c. Microscopic examination: Pipet 2 drops of 7-μm screen washing concentrate onto a clean microscope slide, mix with 1 drop of Lugol's iodine solution, and cover with a cover slip. Scan under a 10× objective for *Giardia* cysts over entire area of cover slip. Confirm suspect cysts by examining with a 43× objective. Estimate total number of cysts per sample by multiplying the number of cysts counted by one-half the number of drops in the screen washing concentrate. If no cysts are found, examine two additional preparations from the same concentrate. Consider

‡Tetko, Elmsford, N.Y. 10523, or equivalent.

§Absorbent cellulose cotton covered with gauze, Absorbent Cotton Company, Valley Park, Mo. 63088, or equivalent.

sample negative if all three preparations yield negative findings.

d. Cultivation: No in vitro method for cultivating *Giardia* from the cyst stage is available currently. Trophozoites may be cultured;[43] however, it is less likely that trophozoites would be present in water supplies in detectable numbers or that they would survive long.

e. Animal testing: In the absence of a cultivation technic, the feeding of cyst suspensions to specific pathogen-free beagle puppies[40] is the best available method to determine infectivity of *Giardia* cysts. This procedure should be performed only in a specialized laboratory equipped with the necessary isolation and handling facilities. Divide the 1-L orlon filter homogenate using one-half for microscopic examination and the other half for animal feeding studies. Alternatively, collect and process duplicate water samples.

2. *Entamoeba histolytica*

Fecal contamination of drinking water is a major source of amebiasis although the oral-fecal route and consumption of uncooked vegetables are important. Plumbing defects involving cross-connections between sewer and water lines, back-siphonage from toilets, drainage from defective sewer lines over an open water cooler, and leaking low-pressure water lines submerged in wastewater have caused disease outbreaks. *E. histolytica* occurs in wastewater in low densities and 1 to 5 cysts/L have been found in the effluent of a wastewater treatment plant. Cysts may remain viable for many days but their initial densities are low and these numbers are reduced drastically in the receiving stream through dilution, water temperature changes, and settling.

a. Concentration technics: Use a sample of 4 L or more and membrane filters of 7 to 10 μm pore size if turbidity is not limiting. Filter sample but avoid drying the filter; discontinue suction, transfer filter to side wall of a 100-mL beaker, and repeatedly flush filter surface with several milliliters sterile distilled water.

b. Direct microscopic examination: Place washings in a Sedgewick-Rafter counting cell and examine under low-power magnification for cysts and/or trophozoites.

c. Cultivation: Inoculate sample concentrate into modified liver infusion medium,[39] incubate at 37 C for at least 3 but not more than 6 days, and examine microscopically for trophozoites. By concentrating and culturing replicate sample portions an analysis for the most probable number of *E. histolytica* can be made.

912 I. References

1. AHMED, J., I.A. POSHNI & M.A. SIDIQUI. 1964. Bacteriological examination of drinking water of Karachi and isolation of enteric pathogens. *Pakistan J. Sci. Ind. Res.* 7:103.
2. SELIGMANN, R. & R. REITLER. 1965. Enteropathogens in water with low *Esch. coli* titers. *J. Amer. Water Works Ass.* 57:1572.
3. GREENBERG, A.E. & H.J. ONGERTH. 1966. Salmonellosis in Riverside, California. *J. Amer. Water Works Ass.* 58:1145.
4. LEVIN, M.A., J.R. FISCHER & V.J. CABELLI. 1974. Quantitative large-volume sampling technique. *Appl. Microbiol.* 28:515.
5. PRESNELL, M.W. & W.H. ANDREWS. 1976. Use of the membrane filter and a filter aid for concentrating and enumerating indicator bacteria and *Salmonella* from estuarine waters. *Water Res.* 10:549.
6. CANLAS, L. Feb. 5, 1975. Personal communication. Guam Environmental Protection

Agency, Agana, Guam.

7. EDWARDS, P.R. & W.H. EWING. 1972. Identification of Enterobacteriaeceae, 3rd ed. Burgess Publ. Co., Minneapolis, Minn.
8. EWING, W.H. 1962. Sources of *Escherichia coli* cultures that belong to O-antigen groups associated with infantile diarrheal disease. *J. Infect. Dis.* 110:114.
9. SEIGNEURIN, R., R. MAGNIN & M.L. ACHARD. 1951. Types d'*Escherichia coli* isolés des eaux d'alimentation. *Ann. Inst. Pasteur* 89:473.
10. PETERSEN, N. & J.R. BORING. 1960. A study of coliform densities and *Escherichia coli* serotypes in two mountain streams. *Amer. J. Hyg.* 71:134.
11. MERSON, M.H., G.K. MORRIS, D.A. SACK, J.G. WELLS, J.C. FEELEY, B. SACK, W.B. CREECH, A.Z. KAPIKAN & E.J. GANGAROSA. 1976. Travellers diarrhea in Mexico: A prospective study of physicians and family members attending a congress. *N. England J. Med.* 294:1299.
12. GLANTZ, P.J. & T.M. JACKS. 1968. An evaluation of the use of *Escherichia coli* serogroups as a means of tracing microbial pollution of water. *Water Resour. Res.* 4:625.
13. SACK, R.B. 1975. Human diarrheal disease caused by enterotoxigenic *Escherichia coli*. *Ann. Rev. Microbiol.* 29:333.
14. CRAWFORD, R.P., J.M. HEINEMANN, W.F. MCCULLOCH & S.L. DIESCH. 1971. Human infections associated with waterborne leptospires, and survival studies on serotype pomona. *J. Amer. Vet. Med. Ass.* 159:1477.
15. GALTON, M.M., R.W. MENGES & J.H. STEELE. 1958. Epidemiological patterns of leptospirosis. *Ann. N.Y. Acad. Sci.* 70:427.
16. JOHNSON, R.C. & V.G. HARRIS. 1967. Differentiation of pathogenic and saprophytic leptospires. I. Growth at low temperatures. *J. Bacteriol.* 94:27.
17. OKAZAKI, W. & L.M. RINGEN. 1957. Some effects of various environmental conditions on the survival of *Leptospira pomona*. *Amer. J. Vet. Res.* 18:219.
18. RYU, E. & C.K. LIU. 1966. The viability of leptospires in the summer paddy water. *Jap. J. Microbiol.* 10:51.
19. DIESCH, S.L., W.F. MCCULLOCH, J.L. BRAUN & R.P. CRAWFORD, JR. 1969. Environmental studies on the survival of leptospires in a farm creek following a human leptospirosis outbreak in Iowa. Proc. Annu. Conf., *Bull. Wildlife Dis. Ass.* 5:166.
20. CHANG, S.L., M. BUCKINGHAM & M.P. TAYLOR. 1948. Studies of *Leptospira icterohemorrhagiae*. IV. Survival in water and sewage: Destruction in water by halogen compounds, synthetic detergents and heat. *J. Infect. Dis.* 82:256.
21. JOHNSON, R.C. & P. ROGERS. 1964. Differentiation of pathogenic and saprophytic leptospires with 8-azaguanine. *J. Bacteriol.* 88:1618.
22. FUZI, M. & R. CSOKA. 1960. Differentiation of pathogenic and saprophytic leptospire by means of a copper sulfate test. *Zentralbl. Bakteriol. Parasitenk. Infektionskr. Hyg. Abt. Orig.*, I. 179:231.
23. CRAWFORD, R.P., J.L. BRAUN, W.F. MCCULLOCH & S.L. DIESCH. 1969. Characterization of leptospires isolated from surface waters in Iowa. *Bull. Wildlife Dis. Ass.* 5:157.
24. BRAUN, J.L., S.L. DIESCH & W.F. MCCULLOCH. 1968. A method for isolating leptospires from natural surface waters. *Can. J. Microbiol.* 14:1011.
25. TURNER, L.H. 1970. Leptospirosis III. Maintenance, isolation and demonstration of leptospires. *Trans. Roy. Soc. Trop. Med. Hyg.* 64:623.
26. BASEMAN, J.B., R.C. HENNEBERRY & C.D. COX. 1966. Isolation and growth of leptospira on artificial media. *J. Bacteriol.* 91:1374.
27. ELLINGHAUSEN, H.C., JR. & W.G. MCCULLOUGH. 1965. Nutrition of *Leptospira pomona* and growth of 13 other serotypes. Fractionation of oleic albumin complex and a medium of bovine albumin and polysorbate 80. *Amer. J. Vet. Res.* 26:45.
28. TRIPATHY, D.N. & L.E. HANSON. 1971. Agar overlay method for growth of leptospires in solid medium. *Amer. J. Vet. Res.* 32:1125.
29. FUZI, M. & R. CSOKA. 1961. Rapid method for differentiation of parasitic and saprophytic leptospirae. *J. Bacteriol.* 81:1008.
30. FUZI, M. & R. CSOKA. 1961. An egg-yolk reaction test for the differentiation of leptospira. *J. Pathol. Bacteriol.* 82:208.
31. KMETY, E., I. OKESJI, P. BAKASS & B.

CHORVATH. 1966. Evaluation of methods for differentiating pathogenic and saprophytic leptospira strains. *Ann. Soc. Belge. Med. Trop.* 46:111.

32. RENDTORFF, R.C. 1954. The experimental transmission of human intestinal protozoan parasites. II. *Giardia lamblia* cysts given in capsules. *Amer. J. Hyg.* 59:209.
33. DANCIGER, M. & M. LOPEZ. 1975. Numbers of *Giardia* in the feces of infected children. *Amer. J. Trop. Med. Hyg.* 24:237.
34. RENDTORFF, R.C. & C.J. HOLT. 1954. The experimental transmission of human intestinal protozoan parasites. IV. Attempts to transmit *Endamoeba coli* and *Giardia lamblia* cysts by water. *Amer. J. Hyg.* 60:327.
35. CRAUN, G.F. 1977. Waterborne outbreaks. *J. Water Pollut. Control Fed.* 49:1268.
36. KIRNER, J.C., J.D. LITTLER & L.A. ANGELO. 1978. A waterborne outbreak of giardiasis in Camas, Washington. *J. Amer. Water Works Ass.* 70:35.
37. NATIONAL CENTER FOR DISEASE CONTROL. 1977. Waterborne giardiasis outbreaks—Washington, New Hampshire. *Morbidity & Mortality Weekly Rept.* NCDC, Atlanta, Ga. 26:169.
38. MOORE, G.T., W.M. CROSS, D. MCGUIRE, C.S. MOLLOHAN, N.N. GLEASON, G.P. HEALY & L.H. NEWTON. 1969. Epidemic giardiasis at a ski resort. *N. England J. Med.* 281:402.
39. CHANG, S.L. & P.W. KABLER. 1956. Detection of cysts of *Entamoeba histolytica* in tap water by the use of membrane filter. *Amer. J. Hyg.* 64:170.
40. SHAW, P.K., R.E. BRODOKY, D.O. LYMAN, B.T. WOOD, C.P. HIBLER, G.R. HEALY, K.I.E. MACLEOD, W. STAHL & M.G. SCHULTZ. 1977. A communitywide outbreak of giardiasis with evidence of transmission by a municipal water supply. *Ann. Intern. Med.* 87:426.
41. JAKUBOWSKI, W., T.H. ERICKSEN & S.L. CHANG. 1977. Detection, identification and enumeration of *Giardia* cysts in water supplies. *In* Proceedings AWWA Water Quality Technology Conf., Kansas City, Mo., Dec. 4-7, 1977.
42. JAKUBOWSKI, W., S.L. CHANG, T.H. ERICKSEN, E.C. LIPPY & E.W. AKIN. 1978. Large volume sampling of water supplies for microorganisms. *J. Amer. Water Works Ass.* 70:702.
43. MEYER, E.A. 1976. *Giardia lamblia:* isolation and axenic cultivation. *Exp. Parasitol.* 30:101.

912 J. Bibliography

LEIFSON, E. 1935. New culture media based on sodium desoxycholate for the isolation of intestinal pathogens and for enumeration of colon bacilli in milk and water. *J. Pathol. Bacteriol.* 40:581.

CHANG, S.L. & G.M. FAIR. 1941. Viability and destruction of the cysts of *Entamoeba histolytica. J. Amer. Water Works Ass.* 33:1705.

MÜLLER, G. 1947. Der Nachweis von Keimen der Typhus-Paratyphusgruppe im Wasser. H.H. Nolke Verlag, Hamburg, Germany.

CLARK, H.F., E.E. GELDREICH, H.L. JETER & P.W. KABLER. 1951. The membrane filter in sanitary bacteriology—Culture of *Salmonella typhosa* from water samples on a membrane filter. *Pub. Health Rep.* 66:951.

RUDOLFS, W., L.L. FALK & R.A. RAGOTZKIE. 1951. Contamination of vegetables grown in polluted soil. II. Field and laboratory studies on entamoeba cysts. *Sewage Ind. Wastes* 23:478.

GREENBERG, A.E., R.W. WICKENDEN & T.W. LEE. 1957. Tracing typhoid carriers by means of sewage. *Sewage Ind. Wastes* 29:1237.

KABLER, P. 1959. Removal of pathogenic microorganisms by sewage treatment processes. *Sewage Ind. Wastes* 31:1373.

MCCOY, J.H. 1964. Salmonella in crude sewage, sewage effluent, and sewage polluted natural waters. *In* Int. Conf. Water. Pollut. Res., 1st, London, 1962. Vol. 1:205, Macmillan, New York, N.Y.

BREZENSKI, F.T., R. RUSSOMANNO & P. DEFALCO, JR. 1965. The occurrence of *Salmonella* and *Shigella* in post-chlorinated and nonchlorinated sewage effluents and receiving waters.. *Health Lab. Sci.* 2:40.

TAYLOR, W.I. & B. HARRIS. 1965. Isolation of

Shigellae, II. Comparison of plating media and enrichment broths. *Amer. J. Clin. Pathol.* 44:476.

U.S. DEPT. HEALTH, EDUCATION, & WELFARE. 1965. Proceedings Cholera Research Symposium, PHS Pub. No. 1328, Washington, D.C.

RAJ, H. 1966. Enrichment medium for selection of *Salmonella* from fish homogenate. *Appl. Microbiol.* 14:12.

SPINO, D.E. 1966. Elevated temperature technique for the isolation of *Salmonella* from streams. *Appl. Microbiol.* 14:591.

MEYER, E.A. & J.A. CHADD. 1967. Preservation of *Giardia* trophozoites by freezing. *J. Parasitol.* 53:1108.

GALTON, M.M., G.K. MORRIS & W.T. MARTIN. 1968. Salmonella in foods and feeds. Review of isolation methods and recommended procedures. PHS Bureau of Disease Prevention & Environmental Control, NCDC, Atlanta, Ga.

SCHULTE, S.J., J.S. WITZEMAN & W.M. HALL. 1968. Immunofluorescent screening for *Salmonella* in foods: comparison with culture methods. *J. Amer. Org. Agr. Chem.* 51:1334.

TAYLOR, W.I. & D. SCHELHART. 1968. Isolation of *Shigella*. V. Comparison of enrichment broths and stools. *Appl. Microbiol.* 16:1383.

TAYLOR, W.I. & D. SCHELHART. 1968. Isolation of *Shigella*. VI. Performance of media with stool specimens. *Appl. Microbiol.* 16:1387.

BREZENSKI, F.T. & R. RUSSOMANNO. 1969. The detection and use of Salmonella in studying polluted tidal estuaries. *J. Water Pollut. Control Fed.* 41:725.

HENTGES, D.J. 1969. Inhibitions of *Shigella flexneri* by the normal intestinal flora. II. Mechanism of inhibition by coliform organisms. *J. Bacteriol.* 97:513.

GORBACK, S.L., J.B. BANWELL, N.P. PIERCE, B.O. CHATTERJEE & R.C. MITRA. 1970. Intestinal microflora in a chronic carrier of *Vibrio cholerae*. *J. Infect. Dis.* 121:383.

CHRISTIE, D.W., R.S. ANDERSON, E.T. BELL & G.L. GALLAGHER. 1971. Ulceration of the ileum and giardiasis in a beagle. *Vet. Rec.* 88:214.

THOMASON, B.M. & J.G. WALLS. 1971. Preparation and testing of polyvalent conjugates for F.A. detection of Salmonellae. *Appl. Microbiol.* 22:876.

THOMASON, B.M. 1971. Rapid detection of *Salmonella* microcolonies by fluorescent antibody. *Appl. Microbiol.* 22:1064.

CHERRY, W.B., J.B. HANKS, B.M. THOMASON, A.M. MURLIN, J.W. BIDDLE & J.M. GROOM. 1972. Salmonellae as an index of pollution of surface waters. *Appl. Microbiol.* 24:334.

FRANKOWSKI, I. 1975. Detectability of parasite eggs and cysts of lamblia in the feces as a function of the time of viewing the preparation under the microscope. *Wiad Lek* 28:1841.

MEYER, E.A. 1976. *Giardia lamblia:* isolation and axenic cultivation. *Exp. Parasitol.* 39:101.

PARK, C.E., M.K. RAYMAN, R. SZABO & Z.K. STANKSEWICZ. 1976. Selective enrichment of *Shigella* in the presence of *Escherichia coli* by use of 4-chloro-2-cyclopentylphenyl B-D-galactopyranoside. *Can. J. Microbiol.* 22:654.

SHEFFIELD, H.G. & B. JORVATN. 1977. Ultrastructure of the cyst of *Giardia lamblia*. *Amer. J. Trop. Med. Hyg.* 26:23.

913 DETECTION OF ENTERIC VIRUSES

Viruses excreted with feces or urine from any species of animal may pollute water. Especially numerous, and of particular importance to health, are the viruses that infect the gastrointestinal tract of man and are excreted with the feces of infected individuals. These viruses are transmitted most frequently from person to person by the fecal-oral route; however, they also are present in domestic sewage which, after various degrees of treatment, enters waterways to become a part of the rivers and streams that are the sources of drinking water for most large communities. The viruses known to be excreted in relatively large numbers with feces include polioviruses, coxsackieviruses, echoviruses, and other enteroviruses, adenoviruses, reoviruses, rotaviruses, the hepatitis A (infectious hepatitis) virus(es), and the parvovirus-like agents, such as the Norwalk agent, that can cause acute infectious nonbacterial gastroenteritis. With the possible exception of hepatitis A, each group or subgroup consists of a number of different serological types; thus more than 100 different human enteric viruses are recognized. Other viruses may be present in domestic sewage, but not usually in large numbers.[1–7]

In temperate climates enteric viruses occur at peak levels in sewage during the late summer and early fall. Infectious hepatitis virus(es) may be an important exception because the incidence of the disease it produces increases in the colder months. The etiological agent of this disease has been isolated only recently and is not yet fully characterized because it is not easily propagated in the laboratory.

Viruses are not normal flora in the intestinal tract; they are excreted only by infected individuals, mostly infants and young children. Infection rates vary considerably from area to area, depending on sanitary and socioeconomic conditions. Viruses usually are excreted in numbers several orders of magnitude lower than those of coliform bacteria. Because enteric viruses multiply only within living, susceptible cells, their numbers cannot increase in sewage. Sewage treatment, dilution, natural inactivation, and water treatment further reduce viral numbers by the time water is drunk. Thus, although large outbreaks of waterborne viral disease may occur when massive sewage contamination of a water supply takes place,[8,9] transmission of viral infection and disease in technologically advanced nations depends on whether minimal quantities of viruses are capable of producing infections. It has been demonstrated that infection can be produced by a very few virus units.[10] However, the risk of infection incurred by an individual in a community with a water supply containing a very few virus units has not been determined.[2]

It has been argued that transmission of small numbers of viruses through water supplies may produce inapparent infections. However, the subsequent transmission of viruses from these cases of inapparent infections to susceptible contacts probably involves large quantities of viruses. This may result in a considerable amount of disease transmission in a community, epidemiologically consistent with contact and not with transmission from a common source such as water.

Most recognized waterborne virus disease outbreaks in the U.S. have been caused by obvious sewage contamination of untreated or inadequately treated private and semipublic water supplies. Virus disease outbreaks in community water supply systems usually are caused by contamination through the distribution system.[11]

The routine examination of water and wastewater for enteric viruses is not recommended now. However, in special circumstances such as wastewater reclamation, disease outbreaks, or special research studies, it may be prudent or essential to conduct virus testing. Such testing should be done only by competent and specially trained water virologists having adequate facilities.

Laboratories planning to concentrate viruses from water and wastewater should do so with the clear understanding that the available methodology has important limitations. Even the most current methods for concentrating viruses from water still are being researched and continue to be modified and improved. The efficiency of a virus concentration method may vary widely depending on water quality. Furthermore, none of the available virus detection methods have been tested adequately with representatives from all of the virus groups of public health importance. Most virus concentration methods have achieved adequate virus recoveries with water or wastewater samples that have been contaminated experimentally with known quantities of a few specific enteric viruses. Although method effectiveness in field trials is difficult to evaluate, some virus concentration methods have been used successfully to recover naturally occurring enteric viruses. Some of these methods require expensive equipment and materials for sample processing and all virus assay and identification procedures require expensive cell culture and related virology laboratory facilities.

Detecting viruses in water through recovery of infectious virus requires three general steps: (*a*) collecting a representative sample, (*b*) concentrating the viruses in the sample, and (*c*) identifying and estimating quantities of the concentrated viruses. Particular problems associated with virus detection are: (*a*) the small size of virus particles (about 20 to 100 nm in diameter for those of public health interest in the water environment), (*b*) the low virus concentrations in water and the variability in amounts and types that may be present, (*c*) the inherent instability of viruses as biological entities, (*d*) the various dissolved and suspended impurities in water and wastewater that interfere with virus concentration and estimation procedures, and (*e*) the present limitations of virus estimation and identification methods.

The densities of enteric viruses in water and wastewater usually are so low that virus concentration is necessary, except possibly for raw sewage at certain areas or seasons.[12,13] Numerous methods for concentrating waterborne enteric viruses have been proposed, tested under laboratory conditions with experimentally contaminated samples, and in some cases used to detect viruses under field conditions.[14,15]

Virus concentration methods often are capable of processing only limited volumes of water of a given quality. In selecting a virus concentration method consider both the probable virus density and the volume limitations of the concentration method for that type of water. A sample volume less than 1 L and possibly as small as a few milliliters may suffice for recovery of viruses from raw or primary treated sewage. For drinking water and other relatively nonpolluted waters, the virus levels are likely to be so low that hundreds or perhaps thousands of liters must be sampled to increase the probability of virus detection.

Three technics used to concentrate viruses from water are described herein: (*a*) absorption to and elution from microporous filters, (*b*) aluminum hydroxide adsorption-precipitation, and (*c*) polyethylene glycol (PEG) hydroextraction-dialysis.[15] Virus concentration by adsorption to and elution from microporous filters can be used for both small volumes of wastewater and large volumes of drinking or other highly finished waters. The aluminum hydroxide adsorption-precipitation and PEG hydroextraction-dialysis methods are impractical for processing large fluid volumes and therefore are suitable only for concentrating viruses from wastewater or other waters having relatively high virus densities.

In examining a particular water with one of these methods include a preliminary evaluation of virus recovery efficiency. To do this add a known quantity of one or more test virus types to the required volume of water sample, process the sample by the concentration method, and assay the sample concentrate for test viruses to determine the virus recovery efficiency.

913 A. Virus Concentration by Adsorption to and Elution from Microporous Filters (TENTATIVE)

Viruses can be concentrated from aqueous samples by reversibly adsorbing them to microporous filters and then eluting them off the filters in a small liquid volume.[14–17] The virus-containing water sample is pressure-filtered through microporous filters having large surface areas to which viruses adsorb, presumably by electrostatic interactions. Several filter materials, including cellulose nitrate, cellulose acetate, fiberglass, and acrylonitrile-polyvinylchloride, have been used as virus adsorbents. To maximize virus adsorption, the water usually is acidified and polyvalent cations such as Al^{3+} and Mg^{2+} often are added in the form of chloride salts.

Adsorbed viruses usually are eluted from the surfaces of microporous filters by pressure-filtering a small volume of eluent fluid through the filters in situ. The eluent is either a slightly alkaline proteinaceous fluid such as serum, beef extract, or nutrient broth or a highly alkaline buffer consisting of 0.05*M* glycine adjusted to pH 11.5 with NaOH. A distinct advantage in using the latter eluent is that large eluent volumes obtained from the processing of large volumes of clean water with adsorbent filters can be concentrated further (reconcentrated) by a second adsorption-elution step using a small microporous filter as a secondary adsorbent. The eluate from the initial or primary virus adsorbent is adjusted to pH and ionic conditions for optimum virus adsorption, filtered through the secondary adsorbent, and the adsorbed viruses are eluted with a small volume of eluent. Alternatively, if primary elution is done with a proteinaceous fluid such as beef extract, the viruses can be precipitated from the primary eluate by acidifying to pH 3.5.[18] The precipitate is recovered by centrifugation and resuspended in a small volume of alkaline buffer.

Microporous filter methods suffer from three main limitations. Sample suspended matter tends to clog the adsorbent filter, thereby limiting the volume that can be processed and possibly interfering with the elution process.[19] Dissolved and colloidal organic matter in some waters can interfere with virus adsorption to filters, presumably by competing with viruses for adsorption sites.[20–22] As a sample is passed through an adsorbent filter, these interfering components can accumulate to such an extent that viruses will no longer be adsorbed. Finally, viruses adsorbed to suspendedmatter may be removed in any clarification procedure applied before virus adsorption. These solids-associated viruses are lost from the sample unless special efforts are made to recover the solids and process them for viruses.[19]

Despite these limitations, virus concentration by adsorption to and elution from microporous filters is a most promising technic for detecting viruses.[15] Modifications of the single-step microporous filter method have been used to concentrate viruses from raw and treated wastewater.[23–25] One such modification is described below.

Modifications of a two-step microporous filter procedure have been used to concentrate viruses from large volumes of potable water and other finished waters[21,26–30] and a tentative method was introduced in the 14th edition of this work.[16] Modified two-step procedures are given here. Although it is effective for drinking water and other highly finished waters, the two-step microporous filter method has not been applied successfully to large volumes of natural water or wastewater for the reasons already given. Such difficulties can occur even during processing of large volumes of certain tap waters. To improve recovery, attempts have been made to reconcentrate viruses in primary eluates by alternative procedures such as polymer two-phase separation,[27,31] aluminum hydroxide,[17,32,33] iron,[33,34] and beef extract[18] adsorption-precipitation, and polyethylene glycol (PEG) hydroextraction-dialysis.[19,35] These alternative reconcentration procedures have yet to be developed and tested adequately with a variety of viruses and water quality conditions. The beef extract adsorption-precipitation procedure for the reconcentration of primary eluates from large volumes of finished waters is included below as an alternative reconcentration procedure.

I. SINGLE-STAGE MICROPOROUS FILTER ADSORPTION-ELUTION METHOD FOR CONCENTRATING VIRUSES FROM SMALL VOLUMES OF WATER AND WASTEWATER (TENTATIVE)

1. Equipment and Apparatus

a. Adsorbent filter holder, 47, 90, or 142 mm diam, equipped with pressure relief valve.

b. Pressure vessel, 12 or 20 L capacity.

c. Positive pressure source up to about 400 kPa with regulator: laboratory air line, air pump, or cylinder of compressed air or nitrogen gas.

d. Autoclavable vinyl plastic tubing with plastic or metal connectors (quick-disconnect type), for connecting positive pressure source, pressure vessel, and filter holder in series.

e. pH meter.

f. Beakers, 50 to 500 mL.

g. Laboratory balance.

h. Graduated cylinders, 25 to 100 mL.

i. Pipets, 1, 5, and 10 mL.

2. Materials

a. Virus adsorbent filter. Use either:

1) *Cellulose nitrate filter,* 0.45 μm porosity.*

2) *Fiberglass-asbestos-epoxy filter,* 0.45 μm porosity.†

b. Prefilter: Use one or more cellulose nitrate, fiberglass-asbestos-epoxy, or fiberglass-acrylic resin filters or equivalent, with porosities greater than 0.45 μm to prevent clogging of the virus adsorbent filter by suspended matter. Place prefilters on top of the 0.45-μm-porosity virus adsorbent filter in the same filter holder.

3. Reagents

a. Hydrochloric acid, HCl, 0.1, 1.0, and 10*N*.

b. Sodium hydroxide, NaOH, 0.1, 1.0, and 10*N*.

c. Aluminum chloride, $AlCl_3 \cdot 6H_2O$, 0.05*M*, or magnesium chloride, $MgCl_2 \cdot 6H_2O$, 2.5*M*.

d. Sodium thiosulfate, $Na_2S_2O_3 \cdot 5H_2O$, 0.5% (w/v).

e. Sodium chloride, 0.14*N*, pH 3.5. Dissolve 8.18 g in 1 L distilled water and adjust to pH 3.5 with HCl.

f. Virus eluent. Use either:

1) *Glycine-NaOH,* pH 11.5: Prepare 0.05*M* glycine solution, autoclave, and adjust to pH 11.5 with 1 to 10*N* NaOH. Add phenol red to a concentration of 0.0005% as a pH indicator.

2) *Beef extract,* 3%, pH 9.0: Dissolve 30 g beef extract paste or 24 g beef extract powder in 1,000 mL distilled water, adjust to pH 9.0 with 1 to 10*N* NaOH, and sterilize by autoclaving.

g. Glycine-HCl, pH 1.5: Prepare 0.05*M* glycine solution, autoclave, and adjust to pH 1.5 with 1 to 10*N* HCl. Add phenol red, 0.0005% as a pH indicator.

h. Nutrient broth, 10X, pH 7.5: Dissolve 8.0 g nutrient broth in 90 mL distilled water, adjust to pH 7.5, dilute to 100 mL with distilled water, and sterilize by autoclaving.

i. Penicillin-streptomycin, 10X: Contains 5,000 I.U. penicillin/mL and 5,000 μg

*Type HA, Millipore Corp., Bedford, Mass., or equivalent.

†Type M-780, Series AA, Cox Instrument Div., Lynch Corp., Detroit, Mich., or equivalent.

streptomycin/mL. Available commercially or prepare by dissolving powdered sodium or potassium penicillin-G and streptomycin sulfate in distilled water and sterilizing by filtration. Store frozen.

j. Hanks balanced salt solution, 10X: Available commercially or prepared following a standard protocol.[36,37]

k. Sodium hypochlorite, 5.25% available chlorine (household bleach).

4. Procedures

a. Sterilization of apparatus, materials, and reagents: Most reagents, virus adsorbent filters, filter holders, tubing, and labware can be sterilized by autoclaving. To sterilize filters load into their holders; if several filters are to be placed in one holder, place filter with smallest porosity on the bottom with progressively larger filters on top. Do not use an automatic drying cycle when autoclaving virus adsorbent filters. Sterilize apparatus and material that cannot be autoclaved by treating with 10-mg/L free chlorine solution, pH 7.0, for 30 min and rinse or flush with 50-mg/L sterile $Na_2S_2O_3$ solution. Use aseptic technic during all virus concentration operations to prevent extraneous microbial contamination.

b. Sample size and choice of filter size: Sample size and, hence, filter diameter depends partly on water quality and the probable virus concentration. Single-stage microporous filter adsorption-elution methods have been used to recover viruses from 100 mL raw sewage on 47-mm-diam filters[24] and from 3.8 to 4.6 L secondary and tertiary sewage effluent on 90- or 142-mm-diam filters.[20,22,24]

c. Sample collection and storage: Collect samples aseptically in sterile containers. If they contain residual chlorine, immediately add $Na_2S_2O_3$ solution to give a final concentration of 50 mg/L. Process samples as soon as possible after collection; do not hold samples for more than 2 hr at up to 25 C or 48 hr at 2 to 10 C. Do not freeze samples unless they cannot be processed within 48 hr; then freeze and store at −70 C or less.

d. Sample processing: Adjust sample to pH 3.5 and 0.0005*M* $AlCl_3$ or to pH 6.0 to 3.5 and 0.05*M* $MgCl_2$. Make sample adjustments either in a pressure vessel or in another appropriate container. Mix sample vigorously during addition of 1.0 or 0.1*N* HCl and stock 0.05*M* $AlCl_3$ solution (1 part stock solution to 100 parts sample) or 2.5*M* $MgCl_2$ solution (1 part stock solution to 50 parts sample). Because $AlCl_3$ is an acid salt, it may decrease sample pH slightly. Do not let sample pH fall below 3.0.

Place sample in a pressure vessel connected to a source of positive pressure and connect pressure vessel outlet to inlet of virus adsorbent filter holder. With pressure relief valve on filter holder opened, apply a slight positive pressure to purge air from filter holder. When sample just begins to flow from pressure relief valve, quickly close valve and continue filtration at a rate not exceeding 28 mL/min per cm^2 of filter area (about 130, 250, and 4,000 mL/min for 47-, 90-, and 142-mm-diam filters, respectively). After filtering entire sample let positive pressure source purge excess fluid from filter holder.

Wash filters with 0.14*N* NaCl to remove excess Al^{3+} or Mg^{2+} ions from virus adsorbent filter. Use about 1.5 mL NaCl solution/cm^2 filter area (25, 100, and 240 mL for 47-, 90-, and 142-mm-diam filters, respectively). Place wash solution in a pressure vessel connected to filter holder inlet, use positive pressure to filter solution through virus adsorbent filter, discard filtrate, and let positive pressure purge virus adsorbent filter of excess wash solution.

Elute viruses from filters with either recommended eluent. Use about 0.45 mL eluent/cm^2 filter surface area (about 7.5, 28, and 71 mL for 47-, 90-, and 142-mm-diam filters, respectively). With pressure relief valve on filter holder open, add eluent to filter holder so that it completely covers filter surface. When eluent begins to discharge from pressure relief valve, quickly close valve. If pH 11.5 glycine-NaOH is the eluent, place a sterile beaker under filter outlet and apply positive pressure so that filtrate flows slowly from filter holder outlet. Collect filtrate in sterile beaker and, when filtrate no longer flows, slowly increase pressure to force retained fluid from filters. Quickly check eluate (filtrate) pH. If it is less than 11.0, elute with additional pH 11.5 glycine-NaOH until an eluate with a pH $\geq$11.0 is obtained. Immediately after checking pH, adjust eluate to a pH between 9.5 and 7.5 with pH 1.5 glycine-HCl or 0.1*N* HCl while mixing vigorously. Complete elution and eluate pH adjustment to 7.5 to 9.5 in 5 min or less to avoid the possibility of appreciable virus inactivation.

If 3% beef extract, pH 9.0, is the eluent, place a sterile beaker under filter outlet and apply a slight positive pressure to eluent-containing filter holder so that eluent is just forced into the void volume of the adsorbent filter material; release pressure. Allow eluent to remain in contact with the filter for 30 min, apply positive pressure, and collect filtrate. Slowly increase pressure to force additional retained fluid from filters.

Measure eluate volume and add 1/10 of the measured volume each of penicillin-streptomycin, Hanks balanced salt solution, and 10X nutrient broth (add last item to glycine eluates only). Adjust sample to pH 7.4 with glycine-HCl or 0.1*N* HCl while mixing vigorously. Store at either 4 or −70 C, depending on the time until virus assay.

II. TWO-STAGE MICROPOROUS FILTER ADSORPTION-ELUTION METHOD FOR CONCENTRATING VIRUSES FROM LARGE VOLUMES OF FINISHED WATERS (TENTATIVE)

1. Equipment and Apparatus

a. Apparatus for first-stage concentration (Figure 913:1):

1) *First-stage virus adsorbent filter holder.*

2) *Fluid proportioner* with four feed pumps (quadraplex) and a mixing chamber.‡

3) *Water flow meter.*

4) *Pressure gauge,* 0 to 400 kPa.

5) *Vinyl plastic tubing,* autoclavable, with plastic or metal connectors (quick-disconnect type).

6) *Pressure relief valve* (optional).

7) *Carboys,* 20 to 50 L, or similar containers.

8) *Positive pressure source* up to 400 kPa with regulator: laboratory air line, positive pressure pump, or cylinder of compressed air or nitrogen gas.

9) *Pump* (if source water is not under pressure).

b. pH meter.

c. Laboratory balance.

d. Beakers, 2 or 4 L.

e. Pressure vessel, 4 L.

f. Graduated cylinders, 1 and 2 L.

g. Pipets, 1, 5, and 10 mL.

h. Centrifuge with rotor and buckets for 250- to 500-mL-capacity bottles.§

i. Centrifuge bottles, 250 to 500 mL.

2. Materials

a. First-stage virus adsorbent filters: Use one of the following:

1) 293-mm-diam, 8.0- and 1.2-μm-porosity cellulose nitrate filter series.‖

2) 267-mm-diam, 5.0- and 1.0-μm-porosity fiberglass-asbestos-epoxy filter series.#

‡Johanson and Son Machine Corp., Clifton, N.J., or equivalent.

§Required for alternative reconcentration procedure using 3% beef extract.

‖Millipore Corp., Bedford, Mass., or equivalent.

#Type M-780, Series AA, Cox Instrument Div., Lynch Corp., Detroit, Mich., or equivalent.

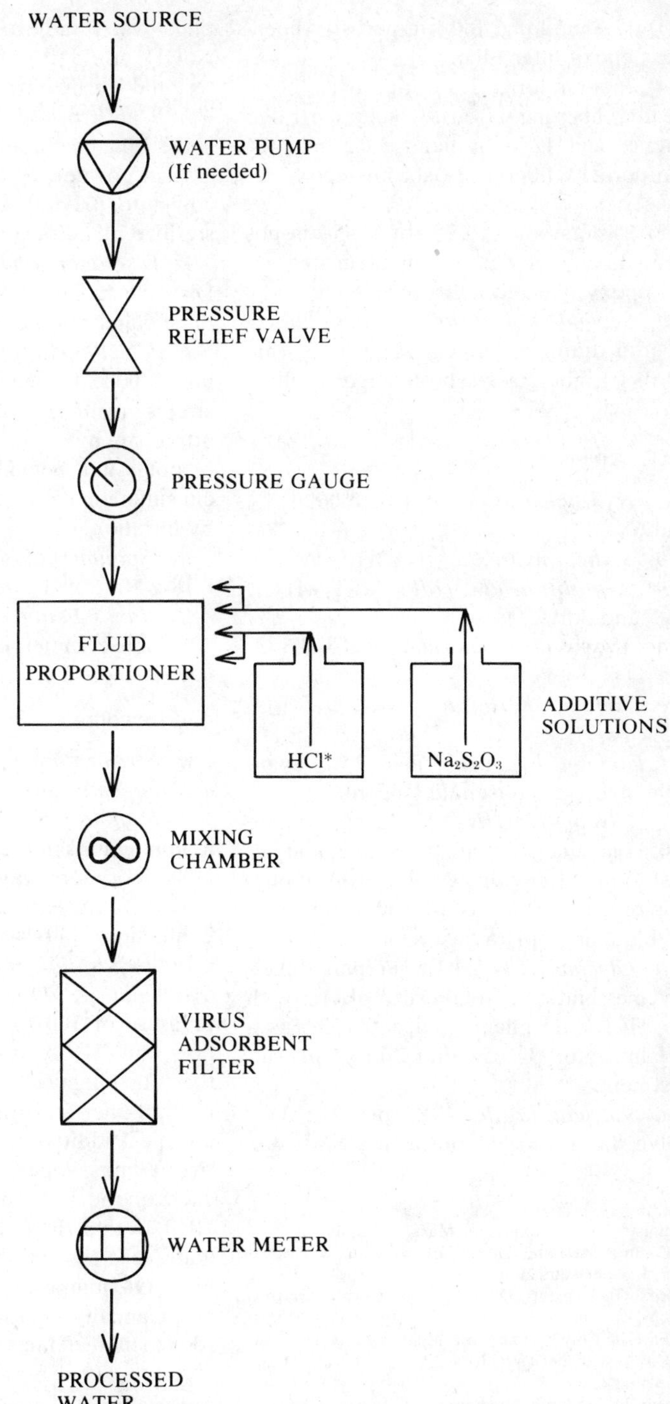

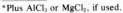

*Plus $AlCl_3$ or $MgCl_2$, if used.

Figure 913:1. Schematic of apparatus for first-stage concentration.

3) 17.8-cm-long, 8.0-μm-porosity fiberglass-epoxy filter tube.**

4) 25.4-cm-long, 3.0-μm-porosity heat-treated fiberglass, yarn-wound cartridge filter†† and 127-mm-diam, 5.0- and 1.0-μm-porosity fiberglass-asbestos-epoxy filter series.‡‡

5) 25.4-cm-long, 0.25- or 0.45-μm-porosity fiberglass-melamine-impregnated paper-epoxy, pleated filter tube.§§

b. Second-stage virus adsorbent filters: 47-mm-diam, 5.0-, 1.0-, and 0.45-μm-porosity fiberglass-asbestos-epoxy filter series.‡‡

3. Reagents

a. Hydrochloric acid, HCl, 0.06, 1, ‖ ‖ and 6*N*.

b. Sodium hydroxide, NaOH, 10*N*.

c. Aluminum chloride, $AlCl_3 \cdot 6H_2O$, 0.05 and 2.0*M*.‖ ‖

d. Magnesium chloride, $MgCl_2 \cdot 6H_2O$, 2.5*M*‖ ‖

e. Sodium thiosulfate, $Na_2S_2O_3 \cdot 5H_2O$, 0.5% (w/v).

f. Sodium hypochlorite, 5.25% available chlorine (household bleach).

g. Glycine-NaOH, pH 11.5: Prepare 0.05*M* glycine solution, autoclave, and adjust to pH 11.5 with 10*N* NaOH. Add phenol red, 0.0005%, as pH indicator. Use within 2 hr of pH adjustment.

h. Glycine-HCl, pH 1.5: Prepare 0.05*M* glycine solution and adjust to pH 1.5 with 6*N* HCl. Add phenol red, 0.0005%, as a pH indicator. Use within 2 hr of pH adjustment.

i. Nutrient broth, 10X, pH 7.5: Dissolve 8.0 g nutrient broth in 90 mL distilled water, adjust to pH 7.5 with 10*N* NaOH, dilute to 100 mL with distilled water, and sterilize by autoclaving.

j. Beef extract, 3%, pH 9.0:## Dissolve 3.0 g beef extract paste or 2.4 g beef extract powder in 100 mL distilled water, adjust to pH 9.0 with 10*N* NaOH, and sterilize by autoclaving.

k. Disodium phosphate, 0.15*M*: ## Dissolve 40.2 g $Na_2HPO_4 \cdot 7H_2O$ in 1 L distilled water and sterilize by autoclaving.

l. Penicillin-streptomycin, 10X: Contains 5,000 I.U. penicillin/mL and 5,000 μg streptomycin/mL. Available commercially or prepare by dissolving powdered sodium or potassium penicillin-G and streptomycin sulfate in distilled water and sterilizing by filtration.

m. Sodium chloride, 0.14*M*. Dissolve 8.18 g NaCl in 1 L distilled water.

n. Hanks balanced salt solution, 10X. Available commercially or prepare following a standard protocol.[36,37]

4. Procedures

a. Sterilization of apparatus, materials, and reagents: See Section 913A.I.4*a*.

b. Sample size: For drinking water use a minimum sample of 400 L, although 2,000 L or more may have to be processed to detect viruses at a concentration of 1 to 2 infectious units/400 L.

c. Preparation of fluid proportioner and feed solutions: Operate proportioner at a pressure of 100 to 700 kPa and a water flow rate of 4 to 40 L/min. Adjust each of the four chemical additive pumps of the proportioner for a ratio of 1 to 100 (1 part chemical additive to 100 parts water). Use two pumps, operating reciprocally, for each of the two additives ($Na_2S_2O_3$ and HCl) so that the overall dilution for each additive is 1 to 50. Place lines from the two $Na_2S_2O_3$ pumps in the $Na_2S_2O_3$ solution and manually operate the pump metering rods to fill feed lines and purge them of air.

**Balston, Inc., Lexington, Mass., or equivalent.

††Commercial Filter Div., Carborundum Co., Lebanon, Ind., or equivalent.

‡‡Cox Instrument Div., Lynch Corp., Detroit, Mich., or equivalent.

§§Filterite Corp., Timonium, Md., or equivalent.

‡‡Cox Instrument Div., Lynch Corp., Detroit, Mich., or equivalent.

‖ ‖May be needed for first-stage virus adsorption.

##For alternative reconcentration procedure.

Connect virus concentrator assembly to source water and operate for several minutes without a virus adsorbent in place to purge the unit of chlorine solution. Collect a water sample from outlet of meter to insure absence of chlorine.

Use an HCl additive solution to adjust sample pH to 3.5 for virus adsorption to filters. For some waters, however, acidification to pH 3.5 may be inadequate for obtaining efficient virus adsorption; therefore, add either $AlCl_3$ or $MgCl_2$ solution.

When only HCl is used, prepare additive solution as follows: Determine concentration of HCl additive solution by titrating a 1-L sample of dechlorinated water to pH 3.5 with 0.06*N* HCl and noting volume required. Divide volume, in milliliters, by 2 to determine milliliters 6*N* HCl needed/L distilled water for making the additive solution. Make at least 10 L additive solution for 400 L of sample. Place lines from the two HCl additive pumps in the HCl solution and manually operate pump metering rods to fill feed lines and purge them of air. Connect virus concentrator assembly to source water and operate briefly without a virus adsorbent in place. Sample conditioned water from meter outlet and check pH. The pH should be 3.5 ± 0.3.

When $AlCl_3$ is used to enhance virus adsorption use pH 3.5 and a final concentration of added $AlCl_3$ of 0.0005*M*. Because $AlCl_3$ is an acid salt, titrate a 1-L sample to about pH 4.0 with 0.06*N* HCl, add $AlCl_3$ to a concentration of 0.005*M*, and continue titration to pH 3.5. When preparing additive solution, add 12.5 mL 2.0*M* $AlCl_3$/L HCl additive solution.

When $MgCl_2$ is used to enhance virus adsorption, use a pH between 3.5 and 6.0 and a final concentration of added $MgCl_2$ of 0.05*M*. To prepare additive solution titrate a 1-L sample to desired pH with 0.06 *N* HCl as previously described. Add one half the titrant volume of 6.0*N* HCl/L 2.5*M* $MgCl_2$ to make the additive solution.

d. First-stage concentration: After preparing concentration apparatus and additive solutions and checking conditioned water for proper pH and absence of chlorine, attach a virus adsorbent filter to mixing chamber outlet. Attach water meter and effluent hose to virus adsorbent outlet. Record initial meter reading and add to this value the desired volume to be processed plus an additional 4% (to account for volume of additive solutions). This gives meter reading at which sampling is to be stopped. Turn on water and start a timer (or record starting time). Shortly after filtration begins collect a sample from filter outlet and check for absence of chlorine and for appropriate pH value. Also check flow rate. Do not use a flow rate above 40 L/min. Recheck pH and chlorine residual several times during sample processing. When desired volume has been processed, turn water off. Purge filter holder of excess water with positive pressure from an air or nitrogen gas source.

e. Washing and virus elution: If $AlCl_3$ or $MgCl_2$ has been used, wash excess Al^{3+} or Mg^{2+} ions from filter with 4 L 0.14*N* NaCl. Omit washing if only HCl was used. Place wash solution in a 4-L pressure vessel and pass through filter with positive pressure. Purge filter of excess wash solution with positive pressure and discard entire filtrate.

Using aseptic technic, elute virus from filter as soon as possible in the field or after returning to the laboratory. If filter holders with adsorbed viruses must be returned to the laboratory, seal filter holder openings, place filter holder in a sterile plastic bag, and chill.

Use either pH 11.5 glycine-NaOH or 3% beef extract, pH 9.0, to elute viruses from first-stage adsorbent filters. When using pH 11.5 glycine-NaOH for elution, readjust eluate to pH and ionic conditions for virus adsorption and further concentrate (reconcentrate) viruses by adsorption to and elution from a second, smaller filter series. Because some viruses are inactivated or inefficiently eluted when pH 11.5 glycine-NaOH is used, alternatively elute with 3% beef extract, pH 9.0. Concentrate (reconcentrate) the viruses thus eluted by acid precipitation at pH 3.5.

To elute with pH 11.5 glycine-NaOH, place 1 L eluent in a pressure vessel or to elute with 3% beef extract, pH 9.0, place 300 to 1,000 mL eluent in a pressure vessel. Connect pressure vessel to inlet of filter holder and with pressure relief valve on filter holder open, apply a small positive pressure to the system so that eluent fills void volume of filter holder. When eluent begins to discharge from pressure relief valve, quickly close it. Filter remaining eluent slowly through filter within 1 to 2 min and collect filtrate (eluate) in a sterile 2- or 4-L beaker. When filtrate no longer appears, slowly increase pressure to force additional fluid from filter. If using pH 11.5 glycine-NaOH eluent, immediately check filtrate pH and if it is less than 11.0, elute with 1 L more of pH 11.5 glycine-NaOH. Immediately after checking pH, adjust filtrate to a pH between 7.5 and 9.5 with pH 1.5 glycine-HCl while mixing vigorously. Complete elution and pH adjustment to 7.5 to 9.5 in 5 min or less to avoid possibility of appreciable virus inactivation.

f. Reconcentration of primary eluates: Concentrate (reconcentrate) viruses in glycine eluates by adsorption to and elution from filters. While mixing vigorously, adjust to pH 3.5 with pH 1.5 glycine-HCl and add 0.05M $AlCl_3$ to a final concentration of 0.0005M. Transfer sample to a 4-L pressure vessel. Filter through a 47-mm-diam 5-, 1-, and 0.45-μm-porosity fiberglass-asbestos-epoxy filter series at a flow rate of no more than 130 mL/min and discard filtrate. Rinse filters with 25 mL 0.14M NaCl to remove excess Al^{3+} ions. Pipet NaCl solution directly into filter inlet or place in a small pressure vessel connected to the inlet. Use positive pressure to pass NaCl solution through filter and discard filtrate. Elute adsorbed viruses from filter with two 7-mL portions of pH 11.5 glycine-NaOH. Pipet 7 mL eluent directly into filter holder inlet or into a small pressure vessel connected to filter inlet and connect to a positive pressure source. Carefully apply positive pressure so that eluate flows slowly from filter outlet into a sterile container. When filtrate no longer flows from outlet, increase pressure to force retained fluid from filters. Measure eluate pH and immediately adjust to pH between 7.5 and 9.5 with pH 1.5 glycine-HCl. Repeat this elution procedure with another 7-mL portion of pH 11.5 glycine-NaOH. Complete reconcentration within 5 min. If neither eluate portion had an initial pH of 11.0 or more, repeat elution procedure with additional 7-mL portions of pH 11.5 glycine-NaOH until an eluate portion has a pH of at least 11.0. Combine all eluates and measure total volume. Add 1/10 of the measured volume each of penicillin-streptomycin, Hanks balanced salt solution, and 10X nutrient broth and adjust to pH 7.4 with 1.5 glycine-HCl. Store at 4 C or −70 C, depending on time until virus assay.

Further concentrate viruses in beef extract eluates by precipitation at pH 3.5. While mixing vigorously, adjust eluate to pH 3.5 by adding 1N HCl dropwise. Continue to mix at slow speed for 30 min and centrifuge at 3,000 × g for 10 min. Decant and discard supernatant. With vigorous mixing, resuspend sediment in 1/20 the initial sample volume of 0.15M Na_2HPO_4. Add antibiotics (1/10 final sample volume) and while mixing vigorously adjust to pH 7.4 with 1.0 or 0.1N NaOH. Store at 4 or −70 C, depending on time until virus assay.

5. Diagram of Overall Concentration Procedures

See Figure 913:2 for the overall two-stage microporous filter adsorption-elution method.

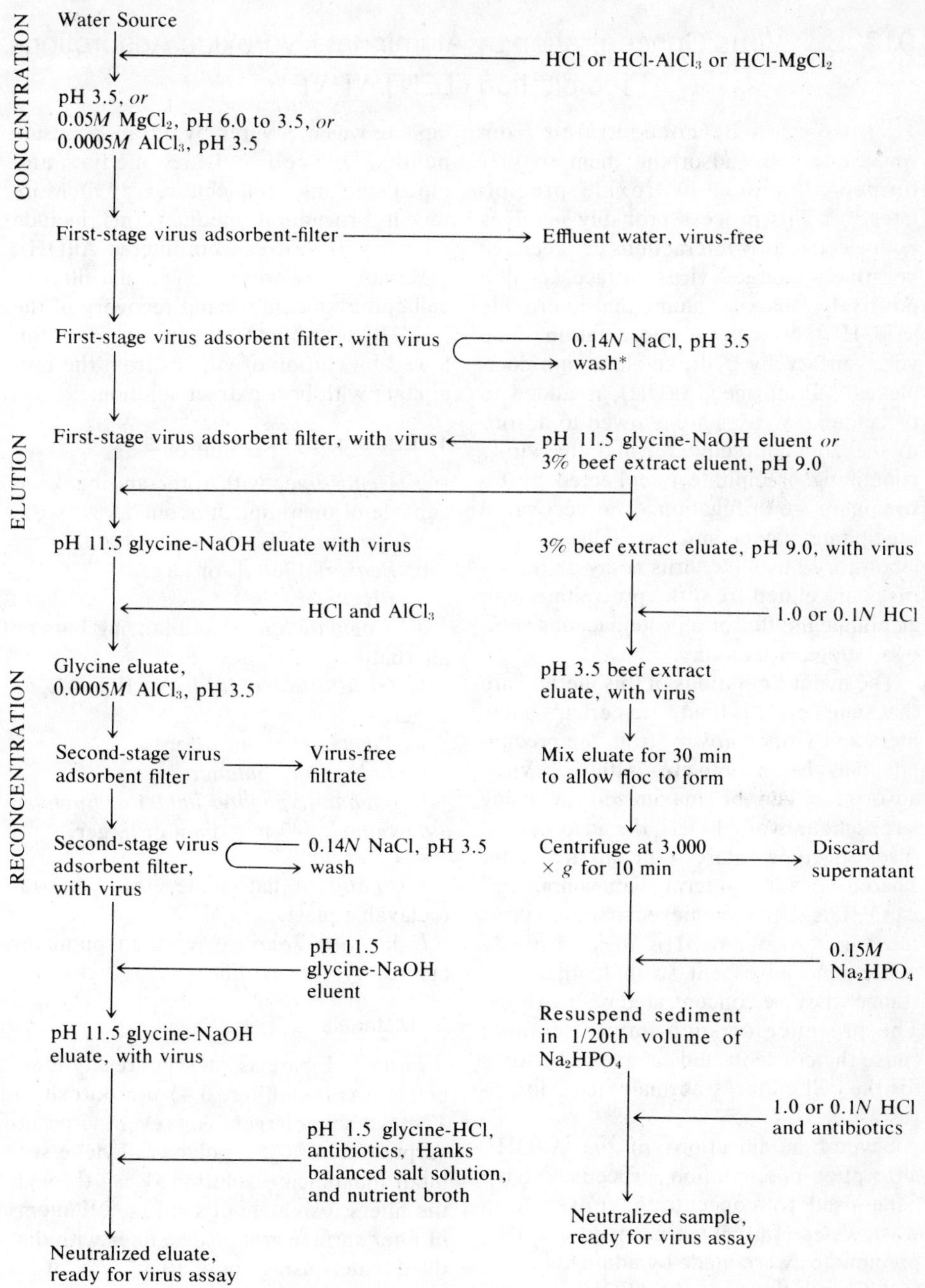

*When using $MgCl_2$ or $AlCl_3$ to enhance virus adsorption.

Figure 913:2. Two-stage microporous filter adsorption-elution method for concentrating viruses from large volumes of water.

913 B. Virus Concentration by Aluminum Hydroxide Adsorption-Precipitation (TENTATIVE)

Viruses can be concentrated from wastewaters by adsorbing them to preformed aluminum hydroxide precipitates.[38,39] This process probably involves both electrostatic interactions between the negatively charged virus surface and the positively charged aluminum hydroxide [$Al(OH)_3$] surfaces and coordination of the virus surface by hydroxo-aluminum complexes.[40] Preformed $Al(OH)_3$ is added to the sample, viruses are allowed to adsorb to the added precipitate, and the virus-containing precipitate is collected by filtration or centrifugation. The recovered precipitate may be inoculated directly into laboratory hosts for virus assay or the viruses are eluted from the precipitate with an alkaline buffer or a proteinaceous solution before virus assay.

The major limitations of this method are that sample size is limited to perhaps a few liters and virus recovery from the precipitate may be incomplete. Although virus adsorption can be maximized by using large amounts of $Al(OH)_3$ the adsorbed viruses become more difficult to elute. Therefore, some intermediate amount of $Al(OH)_3$ is used to achieve maximum virus recovery. Also, $Al(OH)_3$ is a relatively nonspecific adsorbent so that other substances may be concentrated with viruses. The presence of such impurities may cause the concentrated sample to be toxic for the cell cultures normally used for virus assay.

Several modifications of the $Al(OH)_3$ adsorption-precipitation procedure have been used to concentrate viruses from wastewater. Initially preformed $Al(OH)_3$ precipitates were made by adding Na_2CO_3 to $AlCl_3$ solutions and the $Al(OH)_3$ precipitate was resuspended in 0.15*N* NaCl. This was added to the wastewater and the mixture was stirred gently for 1 hr to allow viruses to adsorb to the precipitate. The precipitate was recovered by filtration, resuspended in cell culture media, and inoculated into cell cultures.[23,38,39] More recent procedural modifications include recovery of viruses from filtered $Al(OH)_3$ precipitates by trituration of the filter in cell culture medium[23] and recovery of the $Al(OH)_3$ precipitate by centrifugation followed by elution of viruses from the precipitate with beef extract solution.[41]

1. Equipment and Apparatus

a. Centrifuge, with rotor and buckets, capable of operating at about 1,900 × *g*.

b. Centrifuge bottles and tubes.

c. Beaker, 100 mL or larger.

d. pH meter.

e. Magnetic stirrer and stirring bars or alternative mixing device.

f. Graduated cylinders, 100 mL or larger.

g. Pipets, 1, 5, and 10 mL.

h. Laboratory balance.

*i. Vacuum-type filter holder or Buchner filter funnel,** 47 mm diam or larger.

*j. Filter flask.**

*k. Spatula,** flat blade, metal or autoclavable plastic.

*l. Vacuum source,** vacuum pump or laboratory vacuum line.

2. Materials

*Filter.** Fiberglass-acrylic resin filter† or microporous filter, 0.45 μm porosity,‡ 47 mm diam or larger. To prevent virus adsorption, filter 0.1% polyoxyethylene sorbitan monooleate solution (¶ 3*h*) through the filters, using about 1 mL solution/cm² of filter surface area. Rinse filter with distilled water, using about 10 mL/cm² of fil-

*Required for optional method for collecting $Al(OH)_3$ precipitate from sample.
†Millipore AP20 or equivalent.
‡Millipore HA or equivalent.

ter surface area. Sterilize treated filters by autoclaving.

3. Reagents

a. Hydrochloric acid, HCl, 0.1 and 1.0*N*.

b. Sodium hydroxide, NaOH, 0.1 and 1.0*N*.

c. Sodium carbonate, Na_2CO_3, 4*N*.

d. Aluminum chloride, $AlCl_3$, 0.025*M*.

e. Sodium chloride, NaCl, 0.14*N*.

f. Beef extract, 3%, pH 7.4. Dissolve 3 g beef extract paste or 2.4 g beef extract powder in 90 mL distilled water, adjust to pH 7.4 with 1.0 or 0.1*N* NaOH, dilute to 100 mL with distilled water, and sterilize by autoclaving.

g. Penicillin-streptomycin, 10X, containing 5,000 I.U. penicillin/mL and 5,000 µg streptomycin/mL. Available commercially or prepare by dissolving powdered sodium or potassium penicillin-G and streptomycin sulfate in distilled water and sterilizing by filtration.

h. Polyoxyethylene sorbitan monooleate,§ 0.1% (v/v) in distilled water.

4. Procedures

a. Sterilization of apparatus, materials, and reagents: See Section 913A.I.4*a*.

b. Preparation of $Al(OH)_3$ precipitate: While mixing 100 mL 0.025*M* $AlCl_3$ at room temperature, slowly add 4*N* Na_2CO_3 solution to form precipitate and adjust to pH 7.2. Continue mixing for 15 min and, if necessary, add more Na_2CO_3 to maintain pH 7.2. Centrifuge at 1,100 × *g* for 15 min and discard supernatant. Resuspend sediment in 0.14*N* NaCl and recentrifuge. Discard supernatant, resuspend sediment in 0.14*N* NaCl, and sterilize by autoclaving. Cool, centrifuge again, decant supernatant, and resuspend $Al(OH)_3$ sediment in 50 mL sterile 0.14*N* NaCl. Store at 4 C.

c. Sample size, collection, and storage: Process samples of no more than several liters because the method is too cumbersome and time-consuming for larger volumes. See Section 913A.I.4*c* for sample collection and storage procedures.

d. Sample processing: Do not prefilter sample[21,39] because substantial virus losses can occur. Adjust sample to pH 6.0 with 1.0 or 0.1*N* HCl while mixing vigorously. Add 1 part stock $Al(OH)_3$ suspension/100 parts sample and mix slowly for 2 hr at 4 to 10 C to allow for virus adsorption.

Collect virus-containing $Al(OH)_3$ precipitate by centrifugation or filtration. To collect precipitate by centrifugation, centrifuge at 1,700 × *g* for 15 to 20 min, discard supernatant, and resuspend sediment in 1/1,000 to 1/20 original sample volume of 3% beef extract, pH 7.4.

To collect precipitate by filtration, vacuum filter sample through a treated filter (Section 913B.2*a*) held in a vacuum-type filter holder or Buchner funnel, using additional filters if filter clogs before entire sample is filtered. Carefully scrape precipitate from filter(s) with a sterile spatula and resuspend in 1/1,000 to 1/20 original sample volume of 3% beef extract, pH 7.4.

Regardless of collection method, vigorously mix the $Al(OH)_3$ beef extract suspension and, if necessary, adjust to pH 7.4 with 0.1*N* HCl or NaOH. Continue mixing for a total of 10 min. Centrifuge at 1,900 × *g* for 30 min. Decant supernatant, add 1/10 the volume of the concentrate of penicillin-streptomycin solution and store at 4 or −70 C.

§Tween 80®, ICI United States, Inc., Wilmington, Del., or equivalent. Required for optional method for collecting $Al(OH)_3$ precipitate from sample.

913 C. Hydroextraction-Dialysis with Polyethylene Glycol (TENTATIVE)

Polyethylene glycol (PEG) hydroextraction is an ultrafiltration process in which the sample is placed in a cellulose dialysis bag and exposed to PEG, a hygroscopic material. Water and microsolutes leave the sample by passing across the semipermeable dialysis membrane into the hygroscopic PEG[15] Viruses and other macrosolutes, including PEG, cannot cross the dialysis membrane. The sample volume in the dialysis bag is reduced by water loss to the PEG, thereby concentrating viruses and other macrosolutes. The viruses retained in the dialysis bag are recovered by opening the bag, collecting the remaining sample, and eluting any viruses possibly adsorbed to the inner walls of the bag with a small volume of slightly alkaline proteinaceous solution such as 3% beef extract, pH 9.0. The collected concentrate and eluate are combined and assayed for viruses.

The main limitations of this method are that only small samples (less than 1 L) can be processed conveniently, virus elution from the walls of the dialysis bag may be incomplete unless the elution is done painstakingly, and other macrosolutes in the sample that are concentrated with viruses may interfere with virus assays by being cytotoxic.

Initial investigations of this method reported low and highly variable virus recoveries from wastewater.[42,43] The type of dialysis tubing and eluent solution as well as the thoroughness of the elution step have been found to influence virus recovery efficiency. More recently, with a modified procedure, efficient and consistent virus recoveries have been obtained.[17,19,35]

1. Equipment and Apparatus

a. Beakers, 100 mL or larger.

b. Graduated cylinders, 100 mL or larger.

*c. Dialysis tubing clamps.**

d. Pan, approximately 30 × 30 × 12 cm, autoclavable.

e. Magnetic stirrer and stirring bars or alternative mixing device.

f. Centrifuge, with rotor and buckets, capable of operating at about 1,900 × *g*.

g. pH meter.

h. Pipets, 1, 5, and 10 mL.

i. Tape roller† or similar device to aid in washing the inside walls of dialysis bags with eluting fluid.

j. Ultrasonic disruptor-emulsifier,‡ probe type, capable of generating 100 W of acoustical output.

2. Materials

a. Dialysis tubing, seamless, regenerated cellulose, 4.8 nm average pore diameter.§

b. Polyethylene glycol (PEG),|| dry flakes.

3. Reagents

See Section 913B.3.

4. Procedure

a. Sterilization of apparatus, materials and reagents: See Section 913A.I.4*a*. Do not sterilize PEG.

b. Sample size, collection, and storage: Process samples of no more than a few hundred milliliters because the method is too cumbersome for larger volumes. See Section 913A.I.4*c* for sample collection and storage procedures.

*Fisher Scientific No. 8-670-11A, or equivalent.
†Optional. Fisher Scientific No. 14-245-21 or equivalent.
‡Optional.
§Made by Union Carbide Corp. and available from many scientific supply companies.
||Carbowax® 20,000 or equivalent.

c. Preparation of dialysis tubing: Cut a length of dialysis tubing long enough to accommodate entire sample. Close one end with a clamp, fill tubing bag with distilled water, sterilize by autoclaving, and let cool.

d. Sample processing: Aseptically remove dialysis bag from distilled water and drain. Fill bag with sample and close open end with a second clamp. Place bag in a pan containing a 5-cm layer of PEG, making sure that bag does not touch pan walls. Cover tubing with an additional 5 cm PEG and store at 4 C (for about 18 hr) until sample volume has been reduced to no more than a few milliliters. (If PEG 6000 is used the process time is reduced to 4 to 6 hr.) Although sample may be allowed to dewater completely, do not let it remain in this state.

Remove dialysis bag from PEG and quickly wash PEG from outside of bag with sterile distilled water. Remove clamp from one end of bag and carefully collect sample concentrate. Add about 1/200 to 1/20 the original sample volume of 3% beef extract, pH 9.0, and clamp closed. Thoroughly wash inside walls of bag with beef extract by rubbing fluid from one end to the other several times using either fingers or a roller device. Remove clamp from one end of bag and collect fluid, kneading or squeezing to recover the last traces. Add recovered fluid to previously collected sample concentrate.

Adjust to pH 7.5 with 1.0 or 0.1*N* HCl while mixing vigorously. To disperse solids-associated viruses in sample, stir overnight (about 18 hr) in the cold (about 4 C) or treat with ultrasonics at 100 W for 1 to 2 min. Prevent sample temperature from rising above 37 C during ultrasonic treatment by chilling in an ice bath. Centrifuge at 1,900 $\times g$ for 30 min. Decant supernatant, add 1/10 the volume of the concentrate of penicillin-streptomycin solution, and store at 4 or −70 C.

913 D. Assay and Identification of Viruses in Sample Concentrates (TENTATIVE)

1. Storage of Sample Concentrates

Because it often is impossible to assay sample concentrates immediately, store at room temperature (about 25 C) for up to 2 hr or at refrigerator temperatures (4 to 10 C) for up to 48 hr to minimize virus losses. Freeze samples requiring storage longer than 48 hr at −70 C or less. Do not freeze samples at −10 to −20 C because extensive inactivation of some enteric viruses may occur. Store sample concentrates from finished waters in separate freezers or physically separated from other virus-containing material in common freezers.

2. Decontamination of Sample Concentrates

Sample concentrates, especially those from wastewater, are likely to be contaminated with bacteria and fungi that can overgrow cell cultures and interfere with virus detection and assay. Do not decontaminate by centrifugation or filtration because virus losses are likely to occur. For many samples, especially those from finished waters, contamination is controlled adequately by the penicillin and streptomycin that is added immediately after the sample is obtained. To provide additional protection against fungal contamination, add amphotericin B or nystatin at concentrations of 2.5 and 50 μg/mL, respectively.[44] If penicillin and streptomycin are inadequate, use one or more additional antibiotics such as aureomycin, gentamycin, kanamycin, neomycin, or polymyxin B. To maximize the antibiotic effects, incubate samples for 1 to 3 hr at 25 to 37 C after adding the antibiotics. Bacterial destruction is further enhanced by freezing at −70 C after incubation. Keep samples frozen until assayed for viruses. To determine if antibiotic treatment has been effective, plate a small subsample on a general-purpose medium such as plate count agar by the spread plate technic and incubate at 37 C for 24 to 48 hr.

If extensive bacterial contamination persists after antibiotic treatment, treat with chloroform. Add 1/10 volume of sample of $CHCl_3$ and mix for 30 min at room temperature. Store overnight in a refrigerator. Separate sample (upper layer) from $CHCl_3$ (bottom layer) by aspirating with a pipet and bubble with filter-sterilized air for about 15 min to remove dissolved $CHCl_3$. It may be necessary to expose sample to the atmosphere in a sterile air environment (laminar air flow clean bench or biological safety cabinet) for several hours to remove remaining traces of $CHCl_3$. Although ether also has been used to decontaminate samples, do not use it because of the hazard of explosion or fire.

3. Laboratory Facilities and Host Systems for Virus Assay

Because viruses are obligate, intracellular parasites, they grow (multiply) only in living host cells. This ability to multiply in, and thereby destroy, their host cells is the basis for virus detection and assay. The two major host cell systems for human enteric viruses are whole animals (usually mice) and mammalian cell cultures of primate origin.

A complete description of facilities, equipment, materials, and methods for conducting virus assays is beyond the scope of this book; see standard handbooks on virology and cell culture.[36,37,44–46] Virus assay is beyond the capability of most water and wastewater microbiology laboratories. It should be done only by a trained virologist working in specially equipped virology laboratory facilities. Take particular care to prevent samples or inoculated hosts from becoming contaminated with viruses from other sources and to prevent virus cross-contamination arising from sample concentrates or inoculated hosts. Process and handle samples in a Class II Type I biological safety cabinet[47] or in a "sterile" room or cubicle. The use of such cabinets or facilities is mandatory for testing drinking water or other finished water samples.

There is no single, universal host system for all enteric viruses. Some enteric viruses, notably hepatitis A virus, the human rotaviruses, and parvovirus-like gastroenteritis viruses (Norwalk and related agents), cannot be assayed routinely in any convenient laboratory host systems. However, most enteric viruses can be detected by using one or more cell culture systems and perhaps suckling mice. The latter previously were considered essential for the detection of group A coxsackieviruses, but recent studies indicate that the RD cell line may be nearly as sensitive as suckling mice for the isolation of these viruses.[48] In general, the more different host systems used, the greater the enteric virus recovery rate. However, the number of different host systems used is limited by practical and economic considerations.

There have been several comparative studies on relative sensitivities of various cell culture systems for enteric virus detection,[50–65] but no systematic, comprehensive study has been reported for enteric virus recoveries from water and wastewater. Primary or secondary human embryonic kidney (HEK) cell cultures appear to be the single most sensitive host system for enteric virus isolations,[53,61,65] but they are becoming increasingly more difficult to obtain regularly and, when available from commercial sources, they are expensive. Primary or secondary African green, cynomolgus, or rhesus monkey or baboon kidney cells are sensitive hosts for many enteroviruses and reoviruses, but are not particularly suitable for recovering adenoviruses or group A coxsackieviruses. Recent studies indicate that BGM, a continuous line derived from African green monkey kidney cells, may be comparable in sensitivity to primary monkey kidney cells for enteric virus recovery.[62,63] A number of other continuous cell lines as well as human fetal diploid cell strains have been evaluated for enteric virus recoveries. Some human fetal diploid cell strains give virus isolation rates comparable to primary monkey kidney cells, but plentiful supplies of specific human fetal diploid cell strains are not readily available and many are difficult to maintain. Furthermore, each different cell strain must be characterized for virus susceptibility. Most continuous cell lines generally are less effective than primary cells, but some comparable isolation rates have been obtained with Hep-2[52] and HeLa[58] cells.

Assay the entire sample concentrate for enteric viruses, using at least two different host systems and dividing entire sample equally among the hosts. Preferably use primary (or secondary) HEK cells with either primary (or secondary) monkey kidney or BGM cells for the recovery of most enteroviruses, adenoviruses, and reoviruses. Additional use of either suckling mice or RD cells provides for enhanced recovery of group A coxsackieviruses. Different host systems may be substituted for these if it is demonstrated that they have equivalent sensitivity.

4. Virus Quantitation Procedures for Sample Concentrates

a. Advantages and disadvantages of different quantitation procedures: Virus assays in suckling mice or other animals are quantal assays and in cell cultures they can be done either by quantal (most probable number or 50% endpoint) or enumerative (plaque) methods. Selection between cell culture assay methods depends on the sample and the choice between achieving either maximum virus sensitivity or maximum precision and accuracy in estimating virus concentration. The plaque technic generally is more precise and accurate than the quantal assay because relatively large numbers of individual infectious units can be counted directly as discrete, localized areas of infection (plaques). However, quantal assays are more sensitive.

Because virus plaques are discrete areas of infection arising from a single infectious virus unit, it is relatively easy to recover viruses from individual plaques and then to inoculate them into additional cell cultures to obtain a pure virus culture for identification.[49] However, the use of specific plaque assay conditions for optimizing the recovery of certain enteric virus groups may preclude efficient recovery of other enteric groups requiring different plaque assay conditions. Furthermore, some viruses, such as adenoviruses, do not form plaques efficiently under any conditions. Cytotoxicity due to water or wastewater constituents in sample concentrates is difficult to control in plaque assay systems because the agar overlay medium is difficult to remove and replace.

When a quantal assay cell culture contains a mixture of two or more virus types, it is difficult to obtain pure cultures of each virus because the different types are intimately mixed together in the liquid medium of the culture. When a cell culture is inoculated with a mixture of two or more different virus types, the faster replicating type(s) may mask the presence of other, slower-growing virus types. Cytotoxicity due to constituents of sample concentrates usually can be controlled in quantal assay cell cultures by replacing the culture medium before the cells die.

b. Cell culture procedures for virus isolation and assay: To assay sample concentrates in cell cultures by quantal or plaque methods, drain the medium from newly confluent cultures and inoculate with unit volumes of sample. Use no more than 0.06 mL sample/cm^2 of cell layer surface, e.g., maximum volumes of 1.5, 4.5, and 7.4 mL in cell culture flasks with areas of 25, 74, and 150 cm^2, respectively. If samples are expected to contain such large quantities of viruses that it would be difficult to make reliable estimates of concentration, inoculate cell cultures with dilutions of concentrates. Allow viruses to adsorb to cells for 2 hr at 37 ± 0.5 C, redistributing inoculum over the cell layer every 15 min. Add liquid maintenance medium to cultures for quantal assays or agar-containing medium for plaque assays. Incubate at 37 C and invert plaque assay cultures so that cell (agar) side of culture faces up.

Microscopically examine quantal assay cultures for the appearance of cytopathic effects (CPE) daily during the first 3 days and then periodically for a total of 14 days. Do not change cell culture medium unless cytotoxicity develops. Freeze cultures developing CPE at −70 C when more than 75% of the cells become involved. At 14 days, freeze at −70 C all remaining cultures, including those remaining negative for CPE as well as controls. Thaw cultures and clarify culture fluid-cell lysate by slow-speed centrifugation. Inoculate clarified material from each initial (first-passage) culture into a second (second-passage) culture by transferring 20% of the total initial culture into newly confluent cell cultures of the same type. Microscopically examine second-passage cultures for development of CPE periodically over a 14-day period. Consider second-passage cultures developing CPE as confirmed virus-positive. Freeze and store at −70 C for virus identification. Discard as negative any virus cultures negative for CPE after the second 14-day period.

Periodically examine plaque assay cultures for appearance of plaques over a 14-day period. Mark and tally plaques as they appear. Transfer viruses from each plaque to at least two newly confluent, liquid-medium cell cultures of the same type[49] before plaques become too large and grow together or before the entire cell layer deteriorates. Microscopically examine these second-passage cultures periodically over 14 days for development of CPE. Freeze cultures developing CPE at −70 C for virus identification.

c. Virus isolation and assay in mice: To detect group A and B coxsackieviruses in mice, inoculate samples into animals no older than 24 hr using standard procedures.[45,46,49] Use either the intracerebral or intraperitoneal route, inoculating 0.02 and 0.05 mL, respectively. Observe mice daily over a 14-day period for development of weakness, tremors, and either flaccid (due to group A coxsackieviruses) or spastic (due to group B coxsackieviruses) paralysis. Sacrifice animals developing symptoms, and using sterile technic, prepare 20% tissue suspensions in Hanks' balanced salt solution of the entire skinned, eviscerated torso or just the brain and legs. Store suspensions at −70 C until used for further passage and identification. For second passage in mice, use the same general procedures employed for the initial inoculations. However, making a second passage in cell cultures is preferable to making a second passage in mice because it is easier to do subsequent virus identification by neutralization tests.

d. Estimating virus concentration: Determining the amount of virus in a sample concentrate depends on the assay used. If a sample concentrate is assayed in cell cultures by the plaque technic, count all plaques and calculate the virus concentration, expressed as plaque-forming units (PFU).

If a sample concentrate is assayed by the quantal method, estimate the virus concentration by the most probable number (MPN) method and express as most probable number of infectious units (MPNIU), or by a 50% end point method and express as 50% infectious or lethal dose (ID_{50} or LD_{50}).[37,45,66–68] If the undiluted sample concentrate or a single sample dilution was inoculated into a series of replicate cell cultures (or mice), calculate the MPNIU from the number of confirmed CPE-negative cultures (or mice), q, per total number of cultures (or mice) inoculated, n, according to the formula

$$MPN = -\ln(q/n)$$

If more than one sample dilution was inoculated into cell cultures (or mice), calculate the MPNIU from the formula developed by Thomas:[69]

$$MPN/mL = \frac{P}{\sqrt{(NQ)}}$$

where:

P = total number of positive cultures (or mice) from all dilutions,

N = total mL sample inoculated for all dilutions, and

Q = total mL sample in all negative cultures (or mice).

In using this formula, exclude from the computation all dilutions containing only positive cultures (or mice).

For MPN values obtained from a single sample dilution, the 95% confidence interval is based on the standard error of the binomial distribution when more than 30 cultures (or mice) are inoculated or from the confidence coefficient table of Crow[67,70] when 30 or fewer cultures (or mice) are inoculated.

Make 50% end-point estimates arithmetically by either the Reed-Muench or Karber method.[37,45] These methods require results from several equally spaced sample dilutions, preferably with about the same number of dilutions above and below the 50% end point, and may not be useful for sample concentrates containing relatively low virus levels.

e. Identification of virus isolates: Identify enteric viruses isolated from sample concentrates by standard serological technics, although preliminary identification of genus (enterovirus, reovirus, or adenovirus) sometimes can be made on the basis of information obtained from the isolation procedure. Enteric viruses recovered in suckling mice are likely to be either group A or B coxsackieviruses. For enteric viruses isolated in cell cultures, preliminary identification of genus often can be made from the characteristic appearance of cytopathic effects (CPE) in infected cell cultures.

Confirm preliminary identification of suspected adenovirus and reovirus isolates by detecting their respective group specific antigens by complement fixation tests using clarified, second-passage cell-culture lysate as the antigen. Identify specific reovirus serotypes by hemagglutination-inhibition (HI) or neutralization (Nt) tests. The 31 adenovirus serotypes can be separated into four groups on the basis of their ability (or inability) to hemagglutinate rhesus monkey or rat erythrocytes.[45,46,49] Except for type 18, the first 28 numbered adenoviruses can be identified as to specific serotype by HI. Alternatively, identify all 31 adenovirus serotypes by Nt tests using either individual type-specific antisera or intersecting antisera pools. Also identify specific enterovirus serotypes by neutralization tests in cell cultures using intersecting pools of hyperimmune sera.[45,46,49,71] Use mice for Nt tests for group A and B coxsackieviruses only if the virus isolates fail to propagate in cell cultures.[72] Because polioviruses often are the most prevalent enteroviruses in water and wastewater, test enterovirus isolates for neutralization by an antisera pool against the three types of poliovirus before making neutralization tests with intersecting antisera pools.

913 E. References

1. BERG, G. 1966. Virus transmission by the water vehicle. I. Viruses. *Health Lab. Sci.* 3:86.
2. BERG, G. 1966. Virus transmission by the water vehicle. II. Virus removal by sewage treatment procedures. *Health Lab. Sci.* 3:90.
3. BERG, G. 1966. Virus transmission by the water vehicle. III. Removal of viruses by water treatment procedures. *Health Lab. Sci.* 3:170.
4. CLARKE, N.A. & S.L. CHANG. 1959. Enteric viruses in water. *J. Amer. Water Works Ass.* 51:1299.
5. CLARKE, N.A., G. BERG, P.W. KABLER & S.L. CHANG. 1962. Human enteric viruses in water: Source, survival and removability. *In* Int. Conf. Water Pollut. Res., 1st, London, Proc. Vol. 2:523. Macmillan, New York, N.Y.
6. SOBSEY, M.D. 1975. Enteric viruses and drinking water supplies. *J. Amer. Water Works Ass.* 67:414.
7. MELNICK, J.L. 1976. Enteroviruses. *In* A.S. Evans, ed. Viral Infections of Man: Epidemiology and Control. Plenum, New York, N.Y.
8. VISNAWATHAN, R. 1957. Epidemiology. *Indian J. Med. Res.* 45:1 (supplementary number).
9. MELNICK, J.L. 1957. A water-borne urban epidemic of hepatitis. *In* Hepatitis Frontiers. Little, Brown and Company, Boston, Mass.
10. PLOTKIN, S.A. & M. KATZ. 1967. Minimal infective doses of viruses for man by the oral route. *In* G. Berg, ed. Transmission of Viruses by the Water Route. Interscience Publ., New York, N.Y.
11. CRAUN, G.F. & J.L. MCCABE. 1973. Review of the causes of waterborne disease outbreaks. *J. Amer. Water Works Ass.* 65:74.
12. BURAS, N. 1974. Recovery of viruses from waste-water and effluent by the direct inoculation method. *Water Res.* 8:19.
13. BURAS, N. 1976. Concentration of enteric viruses in wastewater and effluent: A two year survey. *Water Res.* 10:295.
14. HILL, W.F., JR., E.W. AKIN & W.H. BENTON. 1971. Detection of viruses in water: A review of methods and application. *Water Res.* 5:967.
15. SOBSEY, M.D. 1976. Methods for detecting enteric viruses in water and wastewater. *In* G. Berg, H.L. Bodily, E.H. Lennette, J.L. Melnick & T.G. Metcalf, eds. Viruses in Water. APHA, Washington, D.C.
16. AMERICAN PUBLIC HEALTH ASSOCIATION. 1976. Standard Methods for the Examination of Water and Wastewater. 14th ed. APHA, Washington, D.C.
17. FARRAH, S.R., S.M. GOYAL, C.P. GERBA, C. WALLIS & J.L. MELNICK. 1977. Concentration of enteroviruses from estuarine water. *Appl. Environ. Microbiol.* 33:1192.
18. KATZENELSON, E., B. FATTAL & T. HOSTOVESKY. 1976. Organic flocculation: an efficient second-step concentration method for the detection of viruses in tapwater. *Appl. Environ. Microbiol.* 32:638.
19. WELLINGS, F.M., A.L. LEWIS & C.W. MOUNTAIN. 1976. Demonstration of solids-associated virus in wastewater and sludge. *Appl. Environ. Microbiol.* 31:354.
20. WALLIS, C. & J.L. MELNICK. 1967. Concentration of viruses from sewage by adsorption on Millipore membranes. *Bull. World Health Org.* 36:219.
21. SOBSEY, M.D., C. WALLIS, M. HENDERSON & J.L. MELNICK. 1973. Concentration of enteroviruses from large volumes of wa-

ter. *Appl. Microbiol.* 26:529.

22. FARRAH, S.R., S.M. GOYAL, C.P. GERBA, C. WALLIS & P.T.B. SHAFFER. 1976. Characteristics of humic acid and organic compounds concentrated from tapwater using the aquella virus concentrator. *Water Res.* 10:897.
23. MOORE, M., P.P. LUDOVICI & W.S. JETER. 1970. Quantitative methods for the concentration of viruses in wastewater. *J. Water Pollut. Control Fed.* 42:R21.
24. RAO, V.C., U. CHANDORKAR, N.U. RAO, P. KUMARAN & S.B. LAKHE. 1972. A simple method for concentrating and detecting viruses in wastewater. *Water Res.* 6:1565.
25. SOBSEY, M.D., C. WALLIS, M.F. HOBBS, A.C. GREEN & J.L. MELNICK. 1973. Virus removal and inactivation by physical-chemical waste treatment. *J. Environ. Eng. Div., Proc. Amer. Soc. Civil Eng.* 99:245.
26. WALLIS, C., M. HENDERSON & J.L. MELNICK. 1972. Enterovirus concentration on cellulose membranes. *Appl. Microbiol.* 23:476.
27. HILL, W.F., JR., E.W. AKIN, W.H. BENTON & T.G. METCALF. 1972. Virus in water. II. Evaluation of membrane cartridge filters for recovering low multiplicities of poliovirus from water. *Appl. Microbiol.* 23:880.
28. WALLIS, C., A. HOMMA & J.L. MELNICK. 1972. A portable virus concentrator for testing water in the field. *Water Res.* 6:1249.
29. JAKUBOWSKI, W., W.F. HILL, JR. & N.A. CLARKE. 1975. Comparative study of four microporous filters for concentrating viruses from drinking water. *Appl. Microbiol.* 30:58.
30. HILL, W.F., JR., W. JAKUBOWSKI, E.W. AKIN & N.A. CLARKE. 1976. Detection of virus water: Sensitivity of the tentative standard method for drinking water. *Appl. Environ. Microbiol.* 31:254.
31. FIELDS, H.A. & T.G. METCALF. 1975. Concentration of adenovirus from seawater. *Water Res.* 9:357.
32. FARRAH, S.R., C.P. GERBA, C. WALLIS & J.L. MELNICK. 1976. Concentration of viruses from large volumes of tapwater using pleated membrane filters. *Appl. Environ. Microbiol.* 31:221.
33. PAYMENT, P., C.P. GERBA, C. WALLIS & J.L. MELNICK. 1976. Methods for concentrating viruses from large volumes of estuarine water on pleated membranes. *Water Res.* 10:893.
34. SOBSEY, M.D., C.P. GERBA, C. WALLIS & J.L. MELNICK. 1977. Concentration of enteroviruses from large volumes of turbid estuary water. *Can. J. Microbiol.* 23:770.
35. WELLINGS, F.M., A.L. LEWIS, C.W. MOUNTAIN & L.V. PIERCE. 1975. Demonstration of virus in groundwater after effluent discharge onto soil. *Appl. Microbiol.* 29:751.
36. SCHMIDT, N.J. 1969. Tissue culture technics for diagnostic virology. *In* E.H. Lennette & N.J. Schmidt, eds. Diagnostic Procedures for Viral and Rickettsial Infections, 4th ed. APHA, Washington, D.C.
37. ROVOZZO, G.C. & C.N. BURKE. 1973. A Manual of Basic Virological Techniques. Prentice-Hall, Englewood Cliffs, N.J.
38. WALLIS, C. & J.L. MELNICK. 1967. Concentration of viruses on aluminum hydroxide precipitates. *In* G. Berg, ed. Transmission of Viruses by the Water Route. Interscience Publ., New York, N.Y.
39. WALLIS, C. & J.L. MELNICK. 1967. Virus concentration on aluminum and calcium salts. *Amer. J. Epidemiol.* 85:459.
40. COOKSON, J.T., JR. 1974. The chemistry of virus concentration by chemical methods. *Develop. Ind. Microbiol.* 15:160.
41. ENGLAND, B. 1977. Concentration of virus by adsorption to and elution from $Al(OH)_3$. *In* E. Lund, ed. Manual on Analyses for Water Pollution Control. Annex 1, chapter on Virological Examination. World Health Organization Regional Office for Europe, Copenhagen, Denmark (in press).
42. CLIVER, D.O. 1967. Detection of enteric viruses by concentration with polyethylene glycol. *In* G. Berg, ed. Transmission of Viruses by the Water Route. Interscience Publ., New York, N.Y.
43. SHUVAL, H.I., S. CYMBALISTA, B. FATTAL & N. GOLDBLUM. 1967. Concentration of enteric viruses in water by hydro-extraction and two-phase separation. *In* G. Berg, ed. Transmission of Viruses by the Water Route. Interscience Publ., New York, N.Y.
44. PAUL, J. 1975. Cell and Tissue Culture, 5th

ed. Churchill Livingstone, New York, N.Y.

45. LENETTE, E.H. & N.J. SCHMIDT, eds. 1969. Diagnostic Procedures for Viral and Rickettsial Infections, 4th ed. APHA, Washington, D.C.

46. LENNETTE, E.H., E.H. SPAULDING & J.P. TRUANT, eds. 1974. Manual of Clinical Microbiology, 2nd ed. American Society for Microbiology, Washington, D.C.

47. U.S. PUBLIC HEALTH SERVICE. 1976. Guidelines for Research Involving Recombinant DNA Molecules. Appendix D-I, Biological Safety Cabinets. National Inst. of Health, Bethesda, Md.

48. SCHMIDT, N.J., H.H. HO & E.H. LENNETTE. 1975. Propagation and isolation of group A coxsackieviruses in RD cells. *J. Clin. Microbiol.* 2:183.

49. HSIUNG, G.D. 1973. Diagnostic Virology, revised ed. Yale Univ. Press, New Haven, Conn.

50. KELLY, S., J. WINSSER & W. WINKELSTEIN. 1957. Poliomyelitis and other enteric viruses in sewage. *Amer. J. Pub. Health* 47:72.

51. KELLEY, S. & W.W. SANDERSON. 1962. Comparison of various tissue cultures for the isolation of enteroviruses. *Amer. J. Pub. Health* 52:455.

52. PAL, S.R., J. MCQUILLIN & P.S. GARDNER. 1963. A comparative study of susceptibility of primary monkey kidney cells, Hep 2 cells and HeLa cells to a variety of faecal viruses. *J. Hyg.*, Camb. 61:493.

53. LEE, L.H., C.A. PHILLIPS, M.A. SOUTH, J.L. MELNICK & M.D. YOW. 1965. Enteric virus isolations in different cell cultures. *Bull. World Health Org.* 32:657.

54. SCHMIDT, N.J., H.H. HO & E.H. LENNETTE. 1965. Comparative sensitivity of human fetal diploid kidney cell strains and monkey kidney cell cultures for isolation of certain human viruses. *Amer. J. Clin. Pathol.* 43:297.

55. BERQUIST, K.R. & G.J. LOVE. 1966. Relative efficiency of three tissue culture systems for the primary isolation of viruses from feces. *Health Lab. Sci.* 3:195.

56. HERRMANN, E.C. 1967. The usefulness of human fibroblast cell lines for the isolation of viruses. *Amer. J. Epidemiol.* 85:200.

57. FAULKNER, R.S. & C.E. VAN ROOYEN. 1969. Studies on surveillance and survival of viruses in sewage in Nova Scotia. *Can. J. Pub. Health* 60:345.

58. LUND, E. & C.E. HEDSTROM. 1969. A study on sampling and isolation methods for the detection of virus in sewage. *Water Res.* 3:823.

59. SHUVAL, H., B. FATTAL, S. CYMBALISTA & N. GOLDBLUM. 1969. The phase-separation method for the concentration and detection of viruses in water. *Water Res.* 3:225.

60. SCHMIDT, N.J. 1972. Tissue culture in the laboratory diagnosis of virus infections. *Amer. J. Clin. Pathol.* 57:820.

61. COONEY, M.K. 1973. Relative efficiency of cell cultures for detection of viruses. *Health Lab. Sci.* 4:295.

62. DAHLING, D.R., G. BERG & D. BERMAN. 1974. BGM: A continuous cell line more sensitive than primary rhesus and African green kidney cells for the recovery of viruses from water. *Health Lab. Sci.* 11:275.

63. SCHMIDT, N.J., H.H. HO & E.H. LENNETTE. 1976. Comparative sensitivity of the BGM cell line for the isolation of enteric viruses. *Health Lab. Sci.* 13:115.

64. HATCH, M.H. & G.E. MARCHETTI. 1971. Isolation of echoviruses with human embryonic lung fibroblast cells. *Appl. Microbiol.* 22:736.

65. RUTALA, W.A., D.F. SHELTON & D. ARBITER. 1977. Comparative sensitivities of viruses to cell cultures and transport media. *Amer. J. Clin. Pathol.* 67:397.

66. CHANG, S.L., G. BERG, K.A. BUSCH, R.E. STEVENSON, N.A. CLARKE & P.W. KABLER. 1958. application of the 'Most Probable Number' method for estimating concentrations of animal viruses by tissue culture technique. *Virology* 6:27.

67. CHANG, S.L. 1965. Statistics of the infective units of animal viruses. *In* G. Berg, ed. Transmission of Viruses by the Water Route. Interscience Publ., New York, N.Y.

68. SOBSEY, M.D. 1976. Field monitoring techniques and data analysis. *In* L.B. Baldwin, J.M. Davidson & J.F. Gerber, eds. Virus Aspects of Applying Municipal Waste to Land. Univ. of Florida, Gainsville.

69. THOMAS, H.A., JR. 1942. Bacterial densities from fermentation tube tests. *J. Amer. Water Works Ass.* 34:572.
70. CROW, E.L. 1956. Confidence intervals for a proportion. *Biometrika* 43:423.
71. MELNICK, J.L., V. RENNICK, B. HAMPIL, N.J. SCHMIDT & H.H. HO. 1973. Lyophilized combination pools of enterovirus equine antisera: Preparation and test procedures for the identification of field strains of 42 enteroviruses. *Bull. World Health Org.* 48:263.
72. MELNICK, J.L., N.J. SCHMIDT, B. HAMPIL & H.H. HO. 1977. Lyophilized combination pools of enterovirus equine antisera: Preparation and test procedures for the identification of field strains of 19 group A coxsackievirus serotypes. *Intervirology* 8:1720.

914 BACTERIOLOGICAL EXAMINATION OF RECREATIONAL WATERS

Recreational waters can be categorized as "fresh-water swimming pools" and "naturally" occurring fresh and marine surface waters. Historically, they have been examined for coliform bacteria and/or Standard Plate Count. Described below are acceptable, available methods for the microorganisms most frequently suggested as indicators of recreational water quality. The type of water being examined must be considered in selecting the microbiological method to be used.

914 A. Swimming Pools

A swimming pool is a body of water of limited size contained in a holding structure. The water generally is chlorinated potable water but it also may be derived from thermal springs or salt water. The modern pool has a recirculating system so that the water can be filtered and disinfected. Whirlpools used in hospitals, convalescent homes, health spas, and gymnasiums also may be classed as recirculating systems that supply heated water. Microorganisms of concern typically are those from the bather's body and its orifices and include those causing infections of the ear, upper respiratory tract, skin, and intestinal tract. Water quality depends on the efficacy of disinfection, the number of bathers in the pool at any one time, and the total number of bathers per day. Pools may be characterized as disinfected or untreated.

For disinfected indoor pools, the residual levels of disinfectant and turbidity should be determined periodically, frequently during periods of peak bather load, when a disinfectant residual cannot be detected, or when the turbidity is above 1 NTU and the Standard Plate Count exceeds 100 colonies/mL. The Standard Plate Count is the primary indicator of disinfection efficiency. Supporting indicators may include *Streptococcus mitis-salivarius* (common to saliva and sinus drainage but difficult to identify), *Staphylococcus aureus* (skin pathogen), and *Pseudomonas aeruginosa* (organism associated with eye and ear infections). These organisms account for a large percentage of swimming-pool-associated illnesses and both the *Streptococcus* species and *Staphylococcus aureus* are relatively resistant to the effect of chlorine. For disinfected outdoor pools, use the measurement of residual disinfectant and turbidity as described for indoor pools. However, fecal coliform bacteria are the primary indicators to monitor for contamination from animal pets, rodents, and stormwater runoff. Supporting indicators include the Standard Plate Count, *Streptococcus mitis-salivarius*, and *Staphylococcus aureus*.

For untreated pools, the primary indicator should be fecal coliform bacteria that are used to detect fecal contamination derived from source water or from convalescent bathers. Supporting indicators should be those described for disinfected pools.

1. Samples

a. Containers: Collect samples for bacteriological examination of swimming pool water as directed in Section 906A. Use containers of from 120 to 480 mL capacity, depending on analyses to be made. Add sodium thiosulfate, $Na_2S_2O_3$, (for disinfected pool waters) in an amount sufficient to provide an approximate concentration of 100 mg/L in the sample. Do this by adding from 0.1 mL to a 120-mL bottle to 0.4 mL to a 480-mL bottle of 10% solution of $Na_2S_2O_3$ (this will neutralize about 15 mg residual chlorine/L). Stopper, cap, and sterilize the bottle as outlined in Section 904.2.

b. Sampling procedure: Collect samples in the area of, and during the time of, maximum bather density. Information on bathing load also will be helpful in subsequent interpretation of laboratory results.

For pools equipped with a filter, samples may be collected conveniently from sampling cocks provided in the return and discharge lines from the filter. Alternatively, carefully remove the cap of a sterile sample bottle and hold the bottle near its base at an angle of 45 deg. Plunge the bottle vertically into the water approximately 20 cm to fill, while making sure that the dechlorinating agent (necessary only for disinfected pools) is not washed out. Collect samples where water depth is approximately 1 m. To prevent contamination of the sample during collection, wear sterile rubber or plastic gloves or attach the sample bottle to a device such as a modified telescoping golf ball retriever.

Because most bacteria shed by bathers are in body oils, saliva, and mucus discharges that layer near the surface, collect additional samples of the surface microlayer in 1-m-deep water. Collect microlayer samples by plunging a sterile glass plate (approximately 20 cm by 20 cm) vertically through the water surface and withdrawing it upward at a rate of approximately 6 cm/sec. Remove the surface film and water layer adhering to both sides of the plate with a sterile silicone rubber scraper and collect in a glass bottle. Repeat until desired volume is obtained. To minimize microbial contamination, wrap glass plate and scraper in metal foil and sterilize by autoclaving before use. Wear rubber or plastic gloves during sampling or hold glass plate with forceps, clips, or tongs.

Determine residual chlorine or other disinfectant at poolside, at the time of sample collection.

c. Sample storage: Refrigerate samples immediately upon collection and hold at less than 10 C during transport to the laboratory. Test samples within 6 hr of collection.

d. Sample dilutions: If dilutions are required during processing of disinfected pool samples, use 0.1% peptone water as diluent to optimize recovery of stressed organisms. Because peptone water has a tendency to foam, avoid including air bubbles when pipetting to assure accurate measure.

2. Standard Plate Count

Determine the total bacterial count as directed under Standard Plate Count, Section 907. Use at least two plates per dilution.

3. Tests for Fecal Coliforms

Apply this indicator system only to outdoor chlorinated and untreated pools. Test for fecal coliforms as directed under the Multiple Tube Fermentation Technic (Section 908), the Membrane Filter Technic (Section 909), or the Rapid Methods (Section 919).

4. Test for *Staphylococcus aureus* (TENTATIVE)

Use a modified multiple-tube procedure because currently available membrane filter technics are inadequate. As a selective enrichment medium, use M-Staphylococcus broth to which is added 0.75 m*M* sodium azide. If 10-mL portions are included in the multiple-tube technic, use double-strength medium. Incubate cultures at 35 C for 24 hr. Streak cultures containing turbidity after incubation on plates of lipovitellenin-salt-mannitol agar (LSM) and incubate at 35 C for 48 hr. Opaque (24 hr), yellow (48 hr) zones around the colonies are positive evidence of lipovitellenin-lipase activity (opaque) and mannitol fermentation (yellow). Restreak negative plates from the original enrichment tube cultures before discarding tube. Lipovitellenin-lipase activity has a 95% positive correlation with coagulase production. If necessary, confirm positive isolates as catalase negative, coagulase positive, DNase positive, fermenting mannitol, fermenting glucose anaerobically, yielding typical microscopic morphology, and gram positive. These additional tests are equivalent to the Completed Test for coliform bacteria.

5. Tests for *Pseudomonas aeruginosa*

Tests for *P. aeruginosa* are presented below in Sections 914C and D and include a membrane filter procedure and a multiple-tube technic.

6. *Streptococcus mitis-salivarius*

There are no simple reliable procedures for directly detecting *S. mitis-salivarius;* therefore, determine fecal streptococcus as described in Section 910 as an alternative, and if necessary, make additional biochemical tests to identify species.

914 B. Natural Bathing Beaches

A natural bathing beach is any shoreline area of a stream, ocean, inland lake, or impoundment that is used for recreation. A wide variety of pathogenic microorganisms potentially can be transmitted to humans through use of natural fresh and marine recreational waters that may be contaminated by wastewater. These include: (*a*) enteropathogenic agents, such as salmonellae, shigellae, enteroviruses, and multicellular parasites; (*b*) human pathogens or "opportunists", such as *P. aeruginosa, Klebsiella, Vibrio parahemolyticus,* and *Aeromonas hydrophila,* which may multiply in recreational waters in the presence of sufficient nutrients; (*c*) organisms carried into the water from the skin and upper orifices of the recreationists, such as *S. aureus;* and (*d*) other organisms, e.g., pathogenic mycobacteria and leptospira, *Francisella tularensis*, and pathogenic *Naegleria* species (amoebic meningoencephalitis) that also may be found in swimming pools.

Methods suitable for the routine examination of recreational waters are not available now for most of the above organisms. Even with the methods described herein, and particularly with reference to the marine environment, there may be local conditions that compromise the accuracy or selective and differential characteristics of these methods.

The best available bacteriological methods for monitoring fecal contamination of naturally occurring recreational waters is the fecal coliform test. For the special case of thermal waters, because of possible false positive fecal coliform results, verify positives by the IMViC tests and check for pigmentation and cytochrome oxidase activity.

The most valuable application of the fecal streptococcus test in natural waters is in the development of fecal coliform:fecal streptococcus ratios. Fecal coliform:fecal streptococcus ratios of 4.0 or higher typically indicate domestic waste while ratios of 0.6 or lower are common to discharges from farm animals or stormwater runoff. Optimally, determine this ratio near the point of waste discharge. Because several biotypes of *S. faecalis* are known to be ubiquitous in the environment, it may be difficult to interpret the sanitary significance of densities of fecal streptococci below 100 organisms/100 mL.

Methods are available for *P. aeruginosa*, *Salmonella*, and *Klebsiella*. The enumeration of *P. aeruginosa*, *Aeromonas hydrophila*, and *Klebsiella* species in recreational waters can be of considerable value with reference to the discharge of highly nutritive wastes into receiving waters, e.g., pulp mill wastes, effluents from textile finishing plants, etc.

1. Samples

a. Containers: Collect samples as directed in Section 906A. The size of the container varies with the number and variety of tests to be performed. Adding $Na_2S_2O_3$ to the bottle is unnecessary.

b. Sampling procedure: Collect samples just below the water surface in the areas of greatest bather density. Take samples over the range of environmental and climate conditions, especially during times when maximal pollution can be expected, i.e., periods of stormwater runoff, sewage by-passing, tidal, current, and wind influences, etc. See Section 914A.1*b* for methods for sample collection.

c. Sample storage: See Section 914A.1*c*.

2. Tests for Fecal Coliforms

Perform tests for fecal coliforms as directed under the Multiple-Tube Fermentation Technic (Section 908), the Membrane Filter Technic (Section 909), or the Rapid Methods (Section 919).

3. Tests for Fecal Streptococci

Make tests for fecal streptococci as directed under Multiple-Tube Technic (Section 910A) or Membrane Filter Technic (Section 910B).

4. Tests for *Pseudomonas aeruginosa*

Perform tests for *P. aeruginosa* as directed in Sections 914C and D. Use the multiple-tube test with turbid samples but note that the procedures may not be applicable to marine samples.

5. Tests for *Salmonella*

See Section 912.

6. Test for Enterovirus

See Section 913.

7. Tests for *Klebsiella*

See Section 911C.

914 C. Membrane Filter Technic for *Pseudomonas aeruginosa* (TENTATIVE)

1. Laboratory Apparatus

Refer to membrane filter assembly and laboratory apparatus under Standard Total Coliform Membrane Filter Procedure (Section 909A).

2. Culture Media

Refer to Media Specifications, M-PA and milk agar. Modify M-PA agar by adding 1.5 g $MgSO_4 \cdot 7H_2O$/L and reducing concentrations of $Na_2S_2O_3$ to 5 g/L and xylose to 1.25 g/L.

3. Procedure

Filter 200 mL or smaller portions of natural waters or up to 500 mL of swimming pool waters through sterile membrane filters. Place each membrane on a poured plate of modified M-PA agar so that there is no air space between the membrane and the agar surface. Invert plates and incubate at 41.5 ± 0.5 C for 72 hr.

Typically, *P. aeruginosa* colonies are 0.8 to 2.2 mm in diameter and flat in appearance with light outer rims and brownish to greenish-black centers. Count typical colonies, preferably from filters containing 20 to 80 colonies. Use a 10- to 15-power magnifier as an aid in colony counting.

4. Confirmation

Use milk agar to confirm a number of typical and atypical colonies. Make a single streak (2 to 4 cm long) from an isolated colony on a milk agar plate and incubate at 35 C for 24 hr. *P. aeruginosa* hydrolyzes casein and produces a yellowish to green diffusible pigment.

5. Interpretation and Calculation of Density

Except as noted above, confirmation is not required routinely. In the absence of confirmation, report results as "presumptive". Calculate and record as the number of *P. aeruginosa*/100 mL.

914 D. Multiple-Tube Technic for *Pseudomonas aeruginosa* (TENTATIVE)

1. Laboratory Apparatus

Refer to Multiple-Tube Technic for total coliforms (Section 908).

2. Materials and Culture Media

Refer to asparagine broth and acetamide medium in Section 905C.

3. Procedure

a. Presumptive test: Inoculate five 10-mL, five 1-mL, and five 0.1-mL samples into asparagine broth. Use 10 mL single-strength broth for inocula of 1 mL or less and 10 mL double-strength broth for 10-mL inocula. For artificial swimming pools these sample sizes are usually adequate; for natural recreational waters, higher dilutions may be necessary. Incubate inoculated tubes at 35 to 37 C. After 24 and again after 48 hr of incubation, examine tubes under long-wave ultraviolet light (black light) in a darkened room. Production of a greenish fluorescent pigment constitutes a positive presumptive test.

b. Confirmed test: Confirm positive

tubes by inoculating 0.1 mL of culture into acetamide broth or onto the surface of acetamide agar slants. A positive confirmed reaction is the development of a high pH as indicated by a purple color within 24 to 36 hr of incubation at 35 to 37 C.

c. Computing and recording MPN: Refer to Table 908:II and to Section 908D.

914 E. Bibliography

ROBINTON, E.D., E.W. MOOD & L.R. ELLIOT. 1957. A study of bacterial flora in swimming pool water treated with high-free residual chlorine. *Amer. J. Pub. Health* 47:1101

MALLMAN, W.L. 1962. Cocci test for detecting mouth and nose pollution of swimming pool waters. *Amer. J. Pub. Health* 52:2001.

MCLEAN, D.M. 1963. Infection hazards in swimming pools. *Pediatrics* 31:811.

FAVERO, M.S., C.H. DRAKE & G.B. RANDALL. 1964. Use of staphylococci as indicators of swimming pool pollution. *Pub. Health Rep.* 79:61.

BØE, J., C.O. SOLBERG, T.M. VOGELSANG & A. WORMNES. 1964. Perianal carriers of staphylococci. *Brit. Med. J.* 2:280.

FAVERO, M.S. & C.H. DRAKE. 1964. Comparative study of microbial flora of iodated and chlorinated pools. *Pub. Health Rep.* 79:251.

COWAN, S.T. & K.J. STEEL. 1965. Manual for the Identification of Medical Bacteria. Cambridge Univ. Press, New York, N.Y.

WORKING PARTY OF THE PUBLIC HEALTH LABORATORY SERVICE. 1965. A bacteriological survey of swimming baths in primary schools. *Monthly Bull. Min. Health & Pub. Health Lab. Serv.* 24:116.

DRAKE, C.H. 1966. Evaluation of culture media for the isolation and enumeration of *Pseudomonas aeruginosa. Health Lab. Sci.* 3:10.

ROBINTON, E.D. & E.W. MOOD. 1966. A quantitative and qualitative appraisal of microbial pollution of water by swimmers: A preliminary report. *J. Hyg.* 64:489.

ROBINTON, E.D. & E.W. MOOD. 1967. An evaluation of the inhibitory influence of cyanuric acid upon swimming pool disinfection. *Amer. J. Pub. Health* 57:301.

KEIRN, M.A. & H.D. PUTNAM. 1968. Resistance of staphylococci to halogens as related to a swimming pool environment. *Health Lab. Sci.* 3:180.

BROWN, M.R.W. & J.H. SCOTT FOSTER. 1970. A simple diagnostic milk medium for *Pseudomonas aeruginosa. J. Clin. Pathol.* 23:172.

FAVERO, M.S., L.A. CARSON, W.W. BOND & N.J. PETERSEN. 1971. *Pseudomonas aeruginosa:* growth in distilled water from hospitals. *Science* 173:836.

ADAMS, J.C. 1972. Unusual organism which gives a positive elevated temperature test for fecal coliforms. *Appl. Microbiol.* 23:172.

LEVIN, M.A. & V.J. CABELLI. 1972. Membrane filter technique for enumeration of *Pseudomonas aeruginosa. Appl. Microbiol.* 24:864.

GRUN, L. & H. KLEYBRINK. 1972. Staphylokokken-Mikrokokken im Badewasser. *Zentralbl. Bakteriol. Parasitenk. Infektionskr. Hyg. Abt. Orig. B.* 155:384.

GUNN, B.A., W.E. DUNKELBERG, JR. & J.R. CRUTZ. 1972. Clinical evaluation of 2% LSM medium for primary isolation and identification of staphylococci. *Amer. J. Clin. Pathol.* 57:236.

MORTON, H.E. & J. COHN. 1972. Coagulase and desoxyribonuclease activities of staphylococci isolated from clinical sources. *Appl. Microbiol.* 23:725.

HATCHER, R.F. & B.C. PARKER. 1974. Investigations of Freshwater Surface Microlayers. Virginia Polytechnic Inst. and State Univ. Bull. VPI-SRRC-BULL 64.

CABELLI, V.J., H. KENNEDY & M.A. LEVIN. 1976. *Pseudomonas aeruginosa* and fresh recreational waters. *J. Water Pollut. Control Fed.* 48:367.

DUTKA, B.J. & K.K. KWAN. 1977. Confirmation of the single-step membrane filter procedure for estimating *Pseudomonas aeruginosa* densities in water. *Appl. Environ. Microbiol.* 33:240.

OLIVIERI, V.P., C.W. KRUSÉ & K. KAWATA. 1977. Microorganisms in Urban Stormwater. U.S. Environmental Protection Agency, EPA 600/2-77-087, Cincinnati, Ohio.

DUTKA, B.J. & K.K. KWAN. 1978. Health indicator bacteria in water surface microlayers. *Can. J. Microbiol.* 24:187.

915 DETECTION OF FUNGI

Fungi are ubiquitous achlorophyllous, heterotrophic organisms with an organized nucleus. They may be found wherever nonliving organic matter occurs, although some species are pathogenic and others are parasitic. In spring water near the source, the number of fungus spores is usually minimal. Unpolluted river water has relatively large numbers representing the true aquatic fungi (including flagellated zoospores and gametes), aquatic Hyphomycetes, and soil fungi. Moderately polluted water may carry cells or spores of the three types; however, it has fewer true aquatic fungi and aquatic Hyphomycetes, and soil fungi are more numerous. Heavily polluted water has large numbers of soil fungi only. The group designated as soil fungi includes the yeasts, many species of which have been isolated from polluted waters.

The association between fungus numbers and organic loading suggests that fungi may be useful indicators of pollution. Unfortunately, no single species or group of fungi has been identified as important in this role. There are some exceptional special cases; for example, the principal distinction between the yeasts *Candida lambica* and *C. krusei* is the ability to use pentose sugars. Because the former fungus grows well on pentoses, it could be used as an indicator of paper mill wastes, which are high in such sugars. Certain species of yeasts and filamentous fungi are characteristic of warmer waters and may be useful indicators of thermal pollution. The amount of chlorine or other disinfectant required for fungus control is essentially unknown.

In water there are two basic patterns of fungal growth. True aquatic fungi produce zoospores or gametes that are motile by means of flagella, either of the whiplash or tinsel type. Some fungi, particularly the Trichomycetes (fungi that inhabit the hind gut of certain worms, mosquito larvae, etc.), have amoeboid stages. Aquatic fungi typically are collected by exposing suitable baits (solid foodstuffs) in the habitat being examined or in a sample within the laboratory. Relatively little work on these fungi in polluted water has been done in the U.S. They have been studied more extensively in polluted waters in England, Germany, and Japan.

The second fungal growth form is nonmotile in all stages of the life cycle. Growth and reproduction are usually asexual. Three growth processes have been recognized: (*a*) filamentous growth with blastic spores or spores produced in special structures; (*b*) filamentous growth with the filaments breaking up in an arthric (fragmenting) manner to form separate spores called arthrospores; and (*c*) single-celled growth with buds produced on each parent cell. The fungus *Geotrichum* and its relatives belong to the second type, while most yeasts have the budding form.

Identification of fungi, (including yeasts) which are considerably larger than bacteria, is dependent on colonial morphology on a solid medium, growth and reproduction morphology, and, for yeasts, physiological activity in laboratory cultures. Increasing numbers of fungi usually indicate increasing organic loadings in water or soil. Large numbers of similar fungi suggest excessive organic load while a highly diversified mycobiota indicates populations adjusted to the environmental organics.

In a survey of the literature of fungi occurring in water, wastewater, and related organically polluted substrata, of the 984 species listed, 133 species are assigned to the Mastigomycotina including the Chytridiomycetes and Oomycetes, fungi with flagellated zoospores; 79 to the Zygomycotina, mostly mucoraceous fungi; 161 to the Ascomycetina, including perfect states of some of those assigned to the Fungi Imperfecti; 18 to the Basidiomycotina, including perfect states of several yeasts; and 593 to the Deuteromycotina or Fungi Imperfecti. Of the total, 133 species are zoosporic, 131 species are yeast-like, and 718 species are filamentous. Most of the zoosporic species were recovered from weakly or mildly polluted or from unpolluted waters; of the remaining species fewer than half were recovered in numbers large enough to indicate membership in a population for even a brief period of time.

915 A. Technic for Fungi

1. Samples

a. Containers: Collect samples as directed in Section 906A. Alternatively, use cylindrical plastic vials with snap-on caps. These vials usually are sterile as received. Transport them in an upright position to minimize the chance of leakage and discard after use.

b. Sample storage: Hold samples not more than 24 hr. If analysis is not begun within 1 hr after sample collection, refrigerate.

2. Procedure

As many as 40 samples can be analyzed simultaneously by the following procedure; however, the optimum number is about 20 samples.

a. Preparation and dilution: To a sterile 250-mL erlenmeyer flask add 135 mL sterile distilled water and 15 mL sample. Use a sterile measuring device for each sample, or, less preferably, rinse the measure with sterile distilled water between samples. Mix sample well before withdrawing the 15-mL portion. Shake flask on a rotary shaker at about 120 to 150 oscillations/min for about 30 min or transfer flask contents to a blender jar, cover and blend at low speed for 1 min or high speed for 30 sec. (Avoid contamination of laboratory air for public health reasons). Wash jar thoroughly between samples and rinse with sterile distilled water. Further dilutions may be made by adding 45 mL sterile distilled water to 5 mL 1:10 diluted suspension.

For stream water samples a dilution of 1:10 usually is adequate. Dilute samples with large amounts of organic material, such as sediments, to 1:100 or 1:1,000. Dilute stream bank or soil samples to 1:1,000 or 1:10,000.

b. Plating: Prepare five plates for each dilution to be examined. Neopeptone-glucose-rose bengal-aureomycin agar is the usual medium of choice, although experience may indicate that Czapek agar (for *Aspergillus*, *Penicillium*, and related fungi) and yeast extract-malt extract-glucose agar or Diamalt agar (for yeasts) may be preferable. Refer to Media Specifications, Section 905C. To use neopeptone-glucose-rose bengal-aureomycin agar, aseptically transfer 10 mL of medium (containing 0.05 mL aureomycin solution, that has been filter-sterilized) at 45 C to a 9-cm petri dish. Add 1 mL of appropriate sample dilution and mix thoroughly by tilting and rotating the dish (see plating procedure under Standard Plate Count, Section 907). Alternatively add to petri dish 1 mL sample, 0.05 mL aureomycin solution, and 10 mL liquefied agar medium at 43 to 45 C. Solidify agar as rapidly as possible.

c. Incubation: Stack plates but do not invert. Incubate at room conditions of temperature and lighting but avoid direct sunlight. Examine and count plates after 5 to 7 days.

d. Counting and inventory: The fungus plate count will provide the basis for rough quantitative comparisons among samples; an inventory will give relative importance of at least the more readily identifiable species or genera.

In preparing plates, use sample portions that will give about 50 to 60 colonies on a plate. Estimates of up to 300 colonies may be made, but discard more crowded plates. The medium containing rose bengal tends to produce discrete colonies and permits slow-growing organisms to develop.

The inventory includes the direct identification of fungi based on colonial morphology and the counting of colonies assignable to various species or genera. When discrete colonies cannot be identified and identification is important, with a nichrome wire the tip of which is bent in an L-shape pick from each selected colony and streak on a slant of neopeptone-glucose agar. If five plates are used per sample, the average number of colonies on all plates (total number of colonies counted/5), times the reciprocal of the dilution (10/1, 100/1, 1,000/1, etc.) equals the fungus colony count per milliliter of original sample. For solid or semisolid samples, use a correction for the water content to report fungus colonies per gram dry weight. Determine water content by drying paired 15-mL units of original sample at 100 C overnight; the difference between wet and dry weights is the amount of water lost from the sample.

915 B. Technic for Yeasts

Of the total number of fungal colonies obtained from polluted waters, as many as 50% may be yeast colonies. Solid media such as those described above do not permit growth of all yeasts; thus, a quantitative enrichment technic may be useful in addition to the plate count (see also Pathogenic Fungi, 915E).

1. Sample Preparation and Dilution

Prepare sample as directed in Section 915A.

2. Enrichment

In 250-mL erlenmeyer flasks prepare one flask each of yeast nitrogen base medium containing 1% and 20% glucose (Media Specifications, Section 905C). Inoculate with 1 mL of appropriate sample dilution and incubate at room temperature on a rotary shaker operating at 120 to 150 oscillations/min for at least 64 hr. Shaken cultures are necessary to prevent overgrowth by filamentous fungi.

3. Isolation

Remove flasks from shaking machine and let settle 4 to 5 hr. Yeast cells, if present, will settle to the bottom, bacteria will remain in suspension, and filamentous fungi will remain in suspension, will float on the surface, or will be attached to the glass surface. With a nichrome wire loop remove a loopful of sediment at the sediment-supernatant interface and smear-streak on malt extract-yeast extract-glucose agar. Use three plates per flask. Incubate at room temperature but out of direct sunlight for 2 to 3 days. Do not invert dishes. To obtain pure cultures, pick from reasonably isolated colonies and restreak on the same medium or on Diamalt agar plates. Obtain pure cultures of as many different colonies as can be recognized.

4. Counting

It is impossible to obtain a meaningful plate count after enrichment isolation. If one cell in the original sample is assumed to produce one or more colonies on the plates following enrichment, it can be stated that yeasts, or specific types of yeasts, occur at a minimal number dependent on the highest positive dilution. The reciprocal of this dilution is the indicated number of yeasts in the sample.

915 C. Zoosporic Fungi

Most fungi found in lacustrine (lake) and lotic (river) habitats that reproduce asexually by motile spores and have determinate growth of the fungal body belong to the class Chytridiomycetes. Fungi with indeterminate growth, asexual reproduction by motile spores, and oogamous sexual reproduction are members of the class Oomycetes. Recent evidence indicates that a reduction in numbers of species of both classes occurs in polluted areas of rivers. However, more species of Oomycetes can be found in polluted situations than Chytridiomycetes. Species of the Oomycete genus *Saprolegnia*, particularly *S. ferax*, appear to be more tolerant than other forms. Bioassay studies indicate that Oomycetes are more tolerant to zinc, cyanide, and mannitol than are Chytridiomycetes. The latter appear to be more tolerant to treatment with surfactants than do the Oomycetes.

Some Chytridiomycetes may parasitize planktonic and other algae. In the case of epidemic fungal infections of phytoplankton species, the activities of fungi may affect the composition of phytoplankton communities by delaying the time of algal maximum and by reducing the population of certain algae so that other phytoplankters will replace the infected algal populations. In the case of nonepidemic infections, fungi may not influence algal populations; instead, they may infect only phytoplankters during periods of decline and thus only hasten decomposition of the algae.

Filamentous Oomycetes, particularly members of the Saprolegniaceae and Pythiaceae, are found in virtually all types of fresh water habitats and damp-to-wet soils. Most of the nearly 250 species involved occur as saprobes on dead and decaying organic matter such as insect exuviae, algae, and submerged vascular plant remains. A few occur as parasites of algae, aquatic invertebrates, fish, and vascular plants; none are associated with human disease.

Rarely do any of these fungi develop in sufficient numbers to be observed or collected directly. Consequently, various technics have been devised for their collection and isolation.

1. Sampling and Baiting

Collect samples in sterile 35-mL glass or polyethylene bottles, refrigerate, and start analysis within 6 to 8 hr. Place each sample in a sterile plate (20 × 100 mm) and dilute with 10 to 15 mL sterile distilled or deionized water. Add three to four hemp seed halves (*Cannabis sativa*, or use whole seeds of *Brassica* or *Sesamun*) as bait to each culture. Incubate at 18 to 20 C and examine daily for fungus growth on the bait. As isolates appear, usually within 72 hr, remove the infected bait, wash it thoroughly with water from a wash bottle, and transfer to a fresh plate of water containing two to three halves of hemp seed. Genera may be identified from spore arrangement within the sporangium and the manner in which spores are released. Specific determination requires microscopic examination of sexual reproductive structures.

To collect the few naturally occurring parasites or pathogens, place the host organism in a plate containing sterile water and hemp seed.

2. Isolation

Although most filamentous Oomycetes can be cultivated on plain cornmeal agar, selective media for isolating *Saprolegnia* from fresh water have been developed.[1]

Obtain axenic cultures by drawing spores into a micropipet as they emerge from the sporangium. Less preferably, use hyphal tips, but note that several different genera and species frequently occur on a single piece of bait. Transfer the spore suspension or hyphal tip to a plate of cornmeal agar. When growth on the agar has occurred, remove bacteria-free hyphal tips aseptically by cutting out a small block of agar. Transfer to fresh medium or water. If growth is not free from contamination after one transfer, make additional transfers to insure pure cultures. Other methods have been outlined by Seymour.[2]

3. Dilution Plating

Make serial dilutions with sterile distilled water (1:100,000 to 1:700,000) and spread 1 mL over surface of a *freshly* prepared cornmeal agar plate. Remove each developing colony and transfer to water for identification. This method permits numerical estimation and composition of the Oomycete community but requires many plates.

915 D. Aquatic Hyphomycetes

Freshwater Hyphomycetes are a very specialized group of conidial fungi that usually occur on the partially decayed, submerged leaves, and occasionally wood of Angiosperms. The mycelium, which is branched and septate, ramifies through the leaf tissue, especially in petioles and veins. The conidiophores project into the water and conidia that usually develop are liberated under water. Mature conidia also can be found in the surface foam of most rivers, streams, and lakes. The conidia of the majority of these fungi are hyaline, thin-walled, and either tetraradiately branched, that is, with four divergent arms, or sigmoid (worm-like) with the curvature in more than one plane. A special feature of the conidia is that while suspended in water, even for long periods, they do not germinate. However, if they come to rest on a solid surface, germ tubes are produced within a few hours. The size and morphology of these spores make them potentially more prominent in plankton analysis work than the spores of other fungi.

Ecological investigations of freshwater Hyphomycetes have been limited to substrate, habitat, dispersal, and their role in the enhancement of leaf substrates as food for aquatic invertebrates. The most common substrates of these organisms are submerged, decayed leaves of Angiosperms such as *Alnus*, *Quercus*, *Corylus*, *Ulmus*, *Acer*, *Castanea*, *Rubus*, *Fraxinus*, and *Salix*. Submerged Gymnosperm leaves usually are free of aquatic Hyphomycetes. The usual habitat of these fungi is well-oxygenated water, such as alpine brooks, mountain streams, and fast-flowing rivers. However, they also have been found in slow-running, often contaminated, rivers, stagnant or temporary pools, melting snow, and soil. There is often an increase in the numbers of species and individuals of aquatic Hyphomycetes between autumn and spring, with a decline between spring and summer.

1. Sample Collection and Storage

For most freshwater environments, collect foam or partially decayed, submerged, angiosperm leaves in sterile bottles. Refrigerate sample until it is examined.

2. Sample Treatment and Analysis

Wash the leaf samples in sterile distilled water and place one to three leaves in a sterile petri dish about 1 cm deep containing sterile pond, river, or lake water. Incubate at room temperature. Within 1 to 2 days, the mycelium and conidia develop. Conidiophores and conidia can be observed with a dissecting microscope on any portion of a leaf surface, but most frequently are seen on petioles and veins. When released, the conidia either remain suspended in the water or settle to the dish bottom. Using a dissecting microscope, pick up single conidium with a micropipet. Transfer each conidium to a microscope slide in a drop of water for identification. The conidium may be transferred with a sterile needle to a plate of 2% malt extract agar for colony production. Search for conidia in foam samples with a dissecting microscope and isolate single conidia as described above. Submerge mycelial plugs from stock culture isolates of aquatic Hyphomycetes in autoclaved pond water in deep petri dishes; conidiogenesis usually occurs within 2 to 10 days.

Conidia in all stages of development can be preserved on slides with lacto-phenol mounting medium in which either acid fuchsin or cotton blue is dissolved, and sealed with clear fingernail polish.

915 E. Pathogenic Fungi

Routine isolations of fungi from polluted streams and sewage treatment plants usually have yielded relatively few species of fungi pathogenic to human beings and other higher animals. *Geotrichum candidum*, an arthrospore-producing fungus, in which there is presumptive evidence of an association with disease, is isolated almost universally. When its sexual stage (an ascus) develops, it is known as *Endomyces candidus*. *Rhinocladiella mansonii*, a causal agent of one form of chromomycosis, usually in the tropics, is equally widespread. It is known also as *Phialophora jeanselmei* or *Trichosporium heteromorphum*. *Aspergillus fumigatus*, a causal agent for pulmonary aspergillosis, is commonly isolated. *Petriellidium (Allescheria) boydii*, a causal agent of eumycotic mycetoma following a puncture wound with contaminated materials, is sometimes recovered, usually as its asexual state, *Scedosporium apiospermum*. Their presence in stream water probably represents soil runoff because virtually all zoopathogenic fungi exist saprobically in soil as their natural reservoir. Other zoopathogenic fungi occasionally are recovered in low frequencies from streams, polluted or not. Another fungus, the yeast *Candida albicans*, can be recovered in varying numbers from sewage treatment plant effluents and streams receiving effluent. In human beings, this fungus is usually a commensal organism, like *Geotrichum candidum*, coexisting in harmony with its host organism; up to 80% of normal, healthy adults have detectable levels of *C. albicans* in their feces, while about 35% harbor it in their oral cavities in the absence of any overt disease. A very large proportion of the female population has vaginal candidiasis in varying degrees of severity. The presence of *C. albicans* in raw sewage, sewage treatment plant effluent, or contaminated water is not surprising. *Candida albicans* has been isolated from these habitats on routine media heavily supplemented with antibacterial drugs, but not on media or with technics described in Sections 915A or 915B. It also has been isolated from estuarine and marine habitats on a maltose-yeast nitrogen base-chloramphenicol-cycloheximide medium. *C. albicans* can be identified among the white and pink yeasts growing on an 0.8-μm black membrane filter on this medium. From each colony, inoculate a 0.5-mL portion of calf or human blood serum, incubate at 37 C for 2 to 3 hr, transfer a drop or two to a slide, and examine microscopically for the production of germ tubes from a majority of the cells. Of the white yeasts, only *C. albicans* produces these short hyphae from the parent cell within 2 to 3 hr incubation.

915 F. References

1. HO, H.H. 1975. Selective medium for isolation of *Saprolegnia* spp. from freshwater. *Can. J. Microbiol.* 21:1126.
2. SEYMOUR, R.L. 1970. The genus *Saprolegnia*. *Nova Hedwigia* Beiheft 19:1.

915 G. Bibliography

EMERSON, R. 1958. Mycological organization. *Mycologia* 50:589.

COOKE, W.B. 1958. Continuous sampling of trickling filter populations. I. Procedures. *Sewage Ind. Wastes* 30:21.

COOKE, W.B. & A. HIRSCH. 1958. Continuous sampling of trickling filter populations. II. Populations. *Sewage Ind. Wastes* 30:139.

COOKE, W.B. 1959. Trickling filter ecology. *Ecology* 40:273.

SPARROW, F.K. 1959. Fungi. (Ascomycetes, Phycomycetes); including W.W. Scott, Key to genera, Fungi Imperfecti (Aquatic Hyphomycetes only). *In* W.T. Edmondson, ed. Ward & Whipple's Fresh Water Biology, 2nd ed. John Wiley & Sons, Inc., New York.

COOKE, W.B. 1961. Pollution effects on the fungus population of a stream. *Ecology* 42:1.

ALEXOPOULOS, C.J. 1962. Introductory Mycology, 2nd ed. John Wiley & Sons, New York, N.Y.

WILLOUGHBY, L.G. 1962. The occurrence and distribution of reproductive spores of Saprolegniales in fresh water. *J. Ecol.* 50:733.

COOKE, W.B. 1963. A Laboratory Guide to Fungi in Polluted Waters, Sewage, and Sewage Treatment Systems, Their Identification and Culture. USPHS Publ. 999-WP-1, Cincinnati, Ohio.

FULLER, M.S. & R.O. PAYTON. 1964. A new technique for the isolation of aquatic fungi. *BioScience* 14:45.

COOKE, W.B. 1965. The enumeration of yeast populations in a sewage treatment plant. *Mycologia* 57:696.

WILLOUGHBY, L.C. & V.G. COLLINS. 1966. A study of fungal spores and bacteria in Blelham Tarn and its associated streams. *Nova Hedwigia* 12:150.

COOKE, W.B. & G.S. MATSUURA. 1969. Distribution of fungi in a waste stabilization pond system. *Ecology* 50:689.

BROCK, T.D. 1970. Biology of Microorganisms. Prentice-Hall, Englewood Cliffs, N.J.

COOKE, W.B. 1970. Our Mouldy Earth. FWPCA Res. Contract Ser. Publ. No. CWR-, Cincinnati, Ohio.

COOKE, W.B. 1970. Fungi in the Lebanon sewage treatment plant and in Turtle Creek, Warren Co., Ohio. *Mycopathol. Mycol. Appl.* 42:89.

LODDER, J., ed. 1970. The Yeasts, A Taxonomic Study, 2nd ed. North Holland Publ. Co., Amsterdam.

PAGAN, E.F. 1970. Isolation of human pathogenic fungi from river water. Ph.D. dissertation, Botany Dept., The Ohio State University, Columbus.

COOKE, W.B. 1971. The role of fungi in waste treatment. *CRC Critical Rev. Environ. Control* 1:581.

PATERSON, R.A. 1971. Lacustrine fungal communities. *In* J. Cairns, ed. Structure and Function of Microbial Communities. Symposium, Amer. Microscopical Soc., Burlington, Vt. 1969. Virginia Polytechnic Inst. and State Univ. Res. Div. Mono. 3:209.

JONES, E.B.G. 1971. Aquatic Fungi, *in* C. Booth, ed. Methods in Microbiology 4:335, Academic Press, New York, N.Y.

PARK, D. 1972. Methods of detecting fungi in organic detritus in water. *Trans. Brit. Mycol. Soc.* 58:281.

FARR, D.F. & R.A. PATERSON. 1974. Aquatic fungi in rivers: Their distribution and response to pollutants. VPI-WRRC Bull. 68, Virginia Water Resources Research Center, Virginia Polytechnic Inst. and State Univ., Blacksburg, Va.

BUCK, J.D. 1975. Distribution of aquatic yeasts—effect of inoculation temperature and chloramphenicol concentration on isolation. *Mycopathologia* 56:73.

COOKE, W.B. 1976. Fungi in sewage. p. 389 *in* Gareth Jones, E.B., ed. Recent Advances in Aquatic Mycology. Elek Science, London.

GARETH JONES, E.B., ed. 1976. Recent Advances in Aquatic Mycology. Elek Science, London.

916 DETECTION OF ACTINOMYCETES

1. General Discussion

Earthy-musty odors affect the quality and public acceptance of municipal water supplies in many parts of the world. They are among the naturally occurring odors that plant operators find most difficult to remove by conventional treatment. As early as 1929, it was assumed that these odors could be attributed to volatile metabolites formed during normal actinomycete development.[1] Two compounds, geosmin and 2-methyl-isoborneol, have been isolated[2–8] and identified as the agents responsible for earthy-musty odor problems in surface water.[8–10] Geosmin, however, also is produced by filamentous blue-green algae.[11–13] Both compounds have threshold odor concentrations well below the microgram-per-liter level. Thus, traces of these products are sufficient to impart a disagreeable odor to water or a muddy flavor to fish. In areas periodically plagued by this problem, it is prudent to enumerate actinomycetes. Identification of their relative abundance in a drinking water source can provide yet another means to assess water quality. The methods described are well-established technics that have been used with success in the isolation and enumeration of actinomycetes related to public water supplies.[14–15] Actinomycetes also have been recognized as the cause of disruptions in wastewater treatment. Massive growths are capable of producing thick foam in the activated sludge process.[16–17]

Of the general properties of actinomycetes, the most striking is their fungal-type morphology. Although actinomycetes were looked upon initially as fungi, later research revealed that they were filamentous, branching bacteria.[18] The actinomycetes are represented most commonly by saprophytic forms that have an extensive impact on the environment by decomposing and transforming a wide variety of complex organic residues. Widely distributed in nature, actinomycetes constitute a considerable proportion of the population of soil and lake and river muds. Most actinomycetes from which geosmin and 2-methylisoborneol have been identified are members of the genus *Streptomyces*, which is considered the most likely to be significant in water supply problems.

2. Samples

a. Collection: Collect samples as directed in Section 906A.

b. Storage: Analyze samples as promptly after collection as possible. Store water samples below 10 C if they cannot be processed promptly.

3. Actinomycete Plate Count

A plating method using a double-layer agar technic has been adapted for determining actinomycete density. Because only the thin top layer of the medium is inoculated with sample, surface colonies predominate and identification and counting of colonies is facilitated.

a. Preparation and dilution: Prepare and dilute samples as directed in Section 907 or 915A. Dilutions up to 1:1,000 (10^{-3}) usually are suitable for raw water, while treated waters may be examined directly. For soil samples, use dilutions from 1:1,000 (10^{-3}) to 1:1,000,000 (10^{-6}).

b. Plating: Prepare three plates for each dilution to be examined. Aseptically transfer 15 mL of sterile starch-casein agar (see Section 905C) to a petri dish and let agar solidify, thus forming the bottom layer. To a test tube containing 17.0 mL liquefied starch-casein agar at 45 to 48 C, add 2 mL of appropriately diluted sample and 1 mL of the anti-fungal antibiotic, cycloheximide,* prepared in distilled water (1 mg/mL) and sterilized by autoclaving for 15 min at 121 C. Pipet 5 mL of inoculated agar over the hardened bottom layer with gentle swirling to obtain even distribution of the surface layer.

*Actidione, Upjohn and Company, Kalamazoo, Mich., or equivalent.

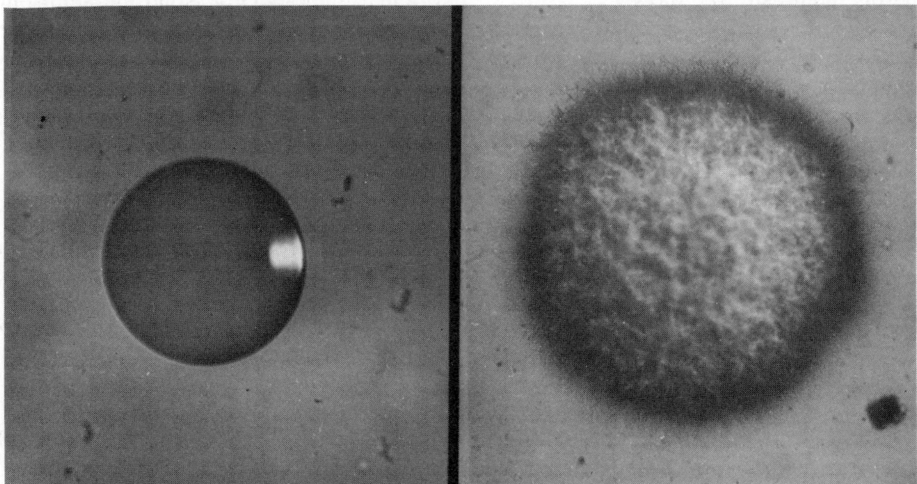

Figure 916:1. Bacterial colonies—typical colony type vs. actinomycete colony type 50X.

Left: A typical bacterial colony characterized by a smooth mucoid appearance and a relatively distinct smooth border.

Right: An actinomycete colony characterized by the mass of branching filaments that result in the fuzzy appearance of its border and by the dull powdery appearance of the spore-laden, aerial hyphae.

TABLE 916:I. GENERAL MACROSCOPIC PROPERTIES OF BACTERIAL COLONIES ON SOLID MEDIUM

Characteristic	Typical Colony Type	Actinomycete Colony Type*
Appearance	Shiny or opalescent	When young it is composed of hyphae, but in some species these may later fragment. Substrate and surface hyphae have no distinctive color. As the colony matures, fluffy aerial hyphae that carry spores form and give to colonies of different species various colors and sometimes a chalky appearance. Soluble pigments, either melanin or brightly colored type that diffuse into the medium, also are common.
Texture	Soft	Strong and leathery
Degree of adherence to solid medium	Weak	Strong
Edge of colony	Regular, continuous, and not different from colony as a whole	Irregular, intermittent, slightly less dense than colony as a whole, and of hyphal appearance.

* Actinomycetes are authentic bacteria by all modern criteria, except for their hyphal character and mode of spore formation.

c. Incubation: Invert and incubate at 28 C until no new colonies appear. Usually this requires 6 to 7 days.

d. Counting: Plates suitable for counting contain 30 to 300 colonies. Identify actinomycetes by gross colony appearance. If necessary, verify by microscopic examination at a magnification of 50 to 100X, as shown in Figure 916:1. Actinomycete colonies, because of filamentous growth, typically have a fuzzy colonial border. Table 916:I lists the distinguishing characteristics commonly used to differentiate actinomycete from other bacterial colonies. Cycloheximide generally suppresses fungal growth; however, fungal colonies, if present, can be recognized by their wooly appearance. Microscopically, fungi reveal a considerably larger cell diameter than actinomycetes.

e. Calculation: Report actinomycetes per milliliter of water or gram (dry weight) of soil. If three plates are used per sample, the average number of colonies on all plates (total number of colonies/3), times 2, times the reciprocal of the dilution (10/1, 100/1, 1,000/1, etc.) equals the actinomycete colony count per milliliter of original sample. For solid or semisolid samples, correct for water content and report actinomycete colonies per gram, dry weight, of sample.

4. References

1. ADAMS, B.A. 1929. *Cladothrix dichotoma* and allied organisms as a cause of an "indeterminate" taste in chlorinated water. *Water & Water Eng.* 31:327.
2. GERBER, N.N. & H.A. LECHEVALIER. 1965. Geosmin, an earthy-smelling substance isolated from actinomycetes. *Appl. Microbiol.* 13:935.
3. GERBER, N.N. 1968. Geosmin, from microorganisms, is trans-1,10-dimethyl-trans-9-decalol. *Tetrahedron Lett.* 25:2971.
4. MARSHALL, J.A. & A.R. HOCHSTETLER. 1968. The synthesis of (±)-geosmin and the other 1,10-dimethyl-9-decalol isomers. *J. Org. Chem.* 33:2593.
5. ROSEN, A.A., R.S. SAFFERMAN, C.I. MASHNI & A.H. ROMANO. 1968. Identity of odorous substances produced by

Streptomyces griseoluteus. *Appl. Microbiol.* 16:178.

6. MEDSKER, L.L., D. JENKINS & J.F. THOMAS. 1969. Odorous compounds in natural waters: 2-exo-hydroxy-2-methylbornane, the major odorous compound produced by several actinomycetes. *Environ. Sci. Technol.* 3:476.
7. GERBER, N.N. 1969. A volatile metabolite of actinomycetes, 2-methylisoborneol. *J. Antibiot.* 22:508.
8. ROSEN, A.A., C.I. MASHNI & R.S. SAFFERMAN. 1970. Recent developments in the chemistry of odour in water: The cause of earthy/musty odour. *Water Treat. Exam.* 19:106.
9. PIET, G.J., B.C.J. ZOETEMAN & A.J.A. KRAAYEVELD. 1972. Earthy-smelling substances in surface waters of the Netherlands. *Water Treat. Exam.* 21:281.
10. YURKOWSKI, M. & J.A.L. TABACHEK. 1974. Identification, analysis, and removal of geosmin from muddy-flavored trout. *J. Fish. Res. Board Can.* 31:1851.
11. SAFFERMAN, R.S., A.A. ROSEN, C.I. MASHNI & M.E. MORRIS. 1967. Earthy-smelling substances from a blue-green alga. *Environ. Sci. Technol.* 1:429.
12. MEDSKER, L.L., D. JENKINS & J.F. THOMAS. 1968. Odorous compounds in natural waters. An earthy-smelling compound associated with blue-green algae and actinomycetes. *Environ. Sci. Technol.* 2:461.
13. KIKUCHI, T., T. MIMURA, K. HARIMAYA, H. YANO, M. ARIMOTO, Y. MASADA & T. INOUE. 1973. Odorous metabolites of blue-green alga *Schizothrix muelleri* Nageli collected in the southern basin of Lake Biwa. Identification of geosmin. *Chem. Pharm. Bull.* 21:2342.
14. SAFFERMAN, R.S. & M.E. MORRIS. 1962. A method for the isolation and enumeration of actinomycetes related to water supplies. USPHS, Robert A. Taft Sanitary Eng. Center Tech. Rep. W62-10, Cincinnati, Ohio.
15. KUSTER, E. & S.T. WILLIAMS. 1964. Selection of media for isolation of *Streptomyces*. *Nature* 202:928.
16. LECHEVALIER, H.A. 1975. Actinomycetes of sewage-treatment plants. U.S. EPA, Environmental Protection Technology Ser., EPA-600/2-75-031, Cincinnati, Ohio.
17. LECHEVALIER, M.P. & H.A. LECHEVALIER. 1974. *Nocardia amarae*, sp. nov., an actinomycete common in foaming activated sludge. *Int. J. Syst. Bacteriol.* 24:278.
18. LECHEVALIER, H.A. & M.P. LECHEVALIER. 1967. Biology of actinomycetes. *Annu. Rev. Microbiol.* 21:71.

917 NEMATOLOGICAL EXAMINATION

Free-living nematodes are usually benthal or wet-soil dwellers, thriving in aerobic habitats plentifully supplied with bacteria and other microbial food. Hence, they propagate in slow sand filters, flourish in aerobic biological sewage treatment plants, and appear in large numbers in secondary effluents. Surface waters receiving such effluents may carry much larger numbers of nematodes than those free of waste discharges.[1–3] Because of their active motility and resistance to chlorination, nematodes are not susceptible to conventional water supply treatment and may enter the distribution system in almost unreduced numbers.

Nematodes of sewage-treatment origin may carry ingested human enteric pathogens. While information on free-living nematodes may supplement bacteriological data on pollution history,[3] these organisms bear only a very remote relation to infection transmission potential.

Plant-parasitic nematodes include aquatic forms that feed on algae and submerged aquatic plants and nonaquatic forms that parasitize fungi and higher plants. The aquatic forms generally are included with the free-living nematodes, while the nonaquatic groups are dealt with as plant parasites. In plant-nematode-infested areas, land runoff and effluents from crop-processing factories may carry eggs, egg masses, swollen females, and/or nematode-infested plant tissue galls. Surface waters receiving such runoff and effluents, when used for irrigation, may spread the infestation.

Nematodes, commonly known as round-, eel- or threadworms, are invertebrates without appendages. Nematodes parasitizing animals are usually macroscopic; the smallest, such as pinworms and hookworms, are about 1 cm long but the microfilariae (larvae) of blood or tissue threadworms are microscopic (about 0.25 mm long). Although freshwater, soil, and plant-parasitic nematodes usually are microscopic but visible under a 6 to 10X hand lens, they can be identified only under a microscope. Most of these are about 0.5 to 1 mm long and 0.03 to 0.05 mm wide, with both ends tapered. Sexes are separate in most genera, but some are hermaphroditic (laying fertilized eggs) or parthenogenetic (laying diploid eggs).

A nematode has a head, a body, and a tail, which are indistinctly separated. The head consists of a six-lipped mouth and a stoma leading to the esophagus. In many forms feeding on zoomicrobes the stoma is enlarged into a buccal cavity equipped with teeth. Nematodes adapted to feed on bacteria, yeast, and minute algae usually have a tubular stoma while those adapted to suck cell juice have a long funnel-shaped stoma modified into a hollow stylet. The tail is posterior to the anus, usually tapered, and may end abruptly or be elongated into a filament. The males of some free-living species have a bursa appearing as a transparent membrane.

Between the head and tail is the lengthy body containing the alimentary canal—

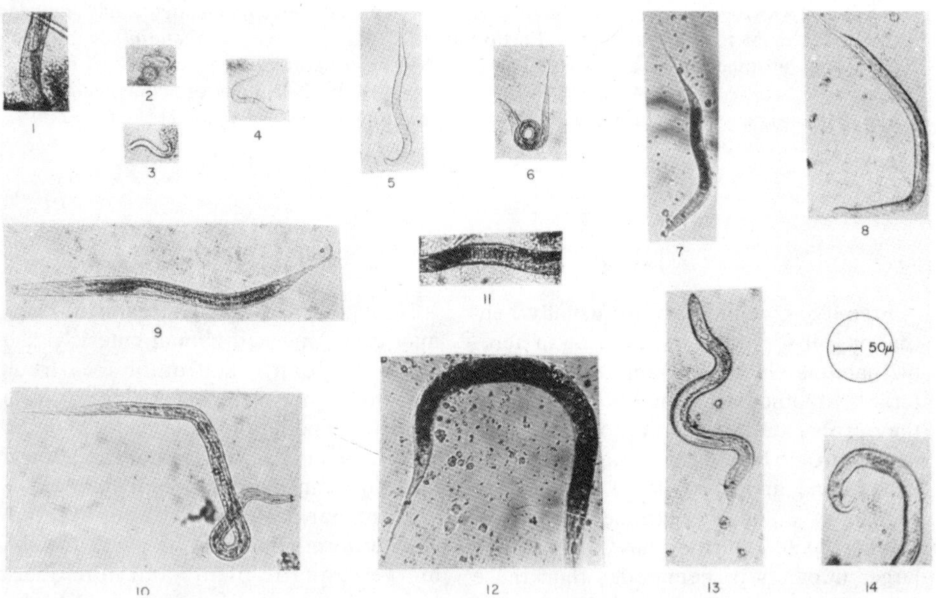

Figure 917:1. Life cycle of nematodes. 1. Section of female worm showing 3 ova. 2. Ovum freed from a female worm with a well formed larva. 3. Larva hatching out from an ovum. 4. 1st stage larva with tail. 5. 2nd stage larva. 6–7. 3rd stage larvae. 8. 4th stage larva. 9–10. Female worms. 11. Section of mature female worm showing striated cuticle. 12. Mature female worm with fully developed uterus and one ovum. 13. Male worm with spicules and gubernaculum. 14. Posterior section of male worm showing spicules on side view.

esophagus, esophageal bulb, and intestine. The body cavity is the space between the body wall and the alimentary canal; it contains the reproductive, nervous, and excretory systems. The latter two are hard to discern without staining and high magnification, although the excretory canal and pore usually are easy to see. Many nematodes have caudal glands in the wide part of the tail while others have papilla-like phasmids on the ventrolateral sides of the tail.

The female reproductive system consists of one or two ovaries, oviducts, and uteri, with the latter open to the outside through the genital pore or vulva. In most free-living nematodes, the genital pore is located posterior to the midpoint of the body; in some parasitic genera it is located in the anterior one-fourth to accommodate the much expanded gonads. The counterparts in males are the testes, seminal vesicles, vas deferens, and ejaculatory duct. A pair of conspicuous spicules (the male copulatory organs) are located near the opening of the ejaculatory duct.

The life cycle of free-living nematodes consists of the egg, four larval stages, and one adult stage. Most eggs are hard to identify in raw water or sewage effluent. In finished water free from most microfaunal forms, eggs laid by live, or freed from dead, females in the distribution system can be recognized readily. Larvae resemble adults but differ from them by a lack of a reproductive system. Newly hatched larvae are about one-fifth adult size. These developmental stages are illustrated in Figure 917:1.

Plant-parasitic nematodes are divided into parasites of the aerial parts and parasites of the underground roots and stems. Some infected plant roots harbor swollen females and/or egg masses. Potatoes, beets, and carrots are the common host-plants. Control of nematodes in infested soils is difficult. Cleaning infested plant parts before marketing can release nematodes, galls, and eggs into wastewater.

917 A. Technic for Nematodes (TENTATIVE)

1. Samples

Collect all samples from sampling stations where bacteriological samples are taken so that findings can be interpreted together. If examination is intended to obtain data on stream pollution, take samples above and below an outfall, with an appropriate number of stations below the outfall to ascertain nematode settling. When samples are taken for plankton analysis, nematodes may be included but report them separately.

In plant-nematode-infested areas, especially in waters polluted with effluents from cleaning processes, also collect samples of effluent.

a. Sample collection: Collect samples in same manner as for bacteriological examination (Section 906) with the following modifications: Use sample size of not less than 2 L, preferably 4 L. Square plastic containers are convenient for collection and shipment of samples. Containers need not be sterilized. Wash them thoroughly with tap water and rinse with distilled water. For examination of potable water, samples of 20 to 100 L may be used.

b. Sample storage: It is preferable to examine material containing live nematodes. Although sample refrigeration is unnecessary, analyze as soon as possible and at least within 2 days of collection. If samples cannot be examined within 2 days, preserve according to instructions for plankton samples (Section 1002B).

2. Nematode Count Procedure

a. Sample concentration: If convenient, concentrate sample at site of collection. Filter sample as directed under Section 909A.5*b* but replace the membrane filter with a woven nylon strainer (pore size 25 to 30 μm). It usually will be possible to filter 4 L of moderately polluted water through a single strainer in a reasonable time. If strainer clogging occurs, use two or more strainers for the entire sample. As last amount of water disappears from the surface, disconnect holder top. Remove strainer with a pair of forceps and place it on wall of a clean 100-mL beaker containing 2 to 4 mL phosphate-buffered dilution water or distilled water. Using a capillary pipet, flush surface of strainer about 8 to 10 times with water in beaker to dislodge nematodes. If more than one strainer is used to concentrate a single sample, pool washings.

After a sample is filtered, wash strainer-holder assembly in running tap water and rinse with distilled water before reuse. Sterilizaton is unnecessary.

b. Counting: Transfer 1 mL thoroughly mixed wash water from the beaker to a Sedgwick-Rafter counting chamber. Using 100X magnification, scan entire chamber and count all nematodes, larvae as well as adults. If the number of nematodes counted per milliliter of concentrate exceeds 10, multiply the number by milliliters concentrate, divide by liters of sample, and report nematodes per liter. If the number of nematodes is between 5 and 10/mL concentrate, count nematodes in a second milliliter of concentrate; if the number is less than 5/mL, count entire concentrate. It may be useful to separate live nematode count from dead; this can be done only in unpreserved samples.

In examining samples taken from waters suspected of carrying plant-parasitic nematodes, look for egg masses and swollen females, and count them, if present.

3. Identification of Common Aquatic Nematodes

Pour scanned portions of sample concentrate into a 10-mL conical centrifuge tube and centrifuge at 1,000 rpm for 5 min. Without disturbing sediment, remove supernate with a capillary pipet until 4 to 5 drops remain, mix, and transfer a drop of the remainder to a microscope slide. Place a ring of vaseline around the drop and cover with a coverslip. Examine essential anatomical features according to the illustrated key for identification of genera.

4. Interpretation

Because presence of free-living nematodes in open waters of lakes and rivers is chiefly, if not entirely, attributable to pollution by secondary effluents from aerobic sewage treatment plants, their number provides a rough, but reliable, indication of level of pollution. Adult nematodes, especially large ones, settle to the bottom depending on water quiescence or flow rate. Hence, their presence in samples taken at specified distances from the point of discharge provides useful information on the history of pollution. The rapidity of obtaining results gives this analysis an obvious advantage.

When nematological results are examined in conjunction with bacteriological findings, interferences can be made about pollution history, presence of toxic substances, anaerobiosis, and related matters. For instance, a combination of high fecal bacterial count and very low nematode count suggests that the water was polluted by a primary effluent, by a secondary effluent from an anaerobic biological treatment plant, or by effluent from a stabilization pond or lagoon. A very low fecal bacterial count and a high nematode count strongly indicate that the water had been used as a carrier of disinfected secondary effluent because nematodes are much more resistant to chlorine and other common water disinfectants than most bacteria. A moderately high fecal bacteria count and a low nematode count, with mostly small or larval forms, indicates that the polluted water is stagnant or very slow-flowing, thus permitting settling of large nematodes. If examination reveals a large percentage of dead or dying nematodes, long-time anaerobiosis or the presence of a nematicide, such as ethylene dibromide or dibromochloropropane, is likely.

If nematodes are included in examination of a raw water source, and if treatment processes include prechlorination, sedimentation, and filtration, a much lower nematode count in the finished water indicates effective chlorination that immobilized nematodes and facilitated their removal by flocculation. In such a treatment practice, use the following as a guide to treatment efficiency:

Nematode Removal %	Efficiency Of Treatment
90	Very good
75–90	Good
50–74	Fair
50	Poor

When repeated examination reveals nematode counts equal to, or greater than, 20/L of raw water, consideration may be given to reducing the nematode load in the finished water by special treatment. When plant-parasitic nematodes are found in an irrigation water, its use on cropland without special treatment may be undesirable.

917 B. Illustrated Key to Fresh Water Nematodes

The following key was devised so that trained biologists, but not necessarily nematologists, could use it. The illustrations include original drawings, photocopies of published drawings, or photocopies on which figures were redrawn. The two most important references were Goodey[4] and Chitwood & Chitwood.[5] Other publications used as references and for illustrative material are listed in 917D.

		Refer to couplet No.
1.	Cephalic setae indistinct or absent	2
	Cephalic setae absent but setae-like head appendages present	64
	Cephalic setae present	69
2.(1)	Stylet present	3
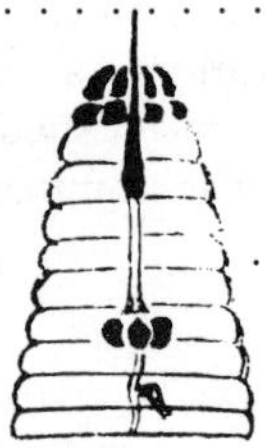	Stylet absent	38
3.(2)	Base of stylet knobbed or flanged	4
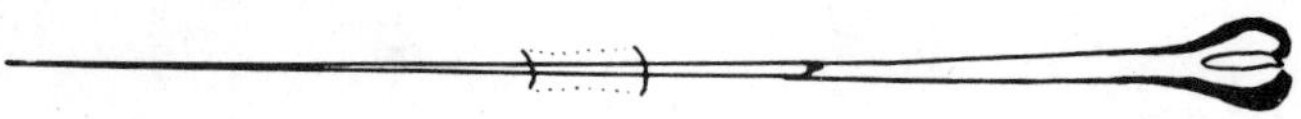	Stylet knobs or flanges absent	29
4.(3)	Valvate median esophageal bulb present	5
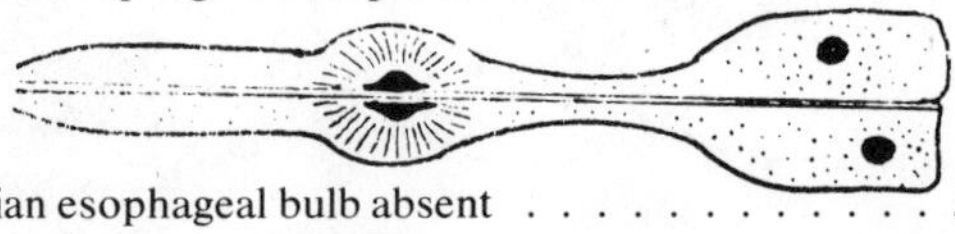	Valvate median esophageal bulb absent	22
5.(4)	Females eel-like	6
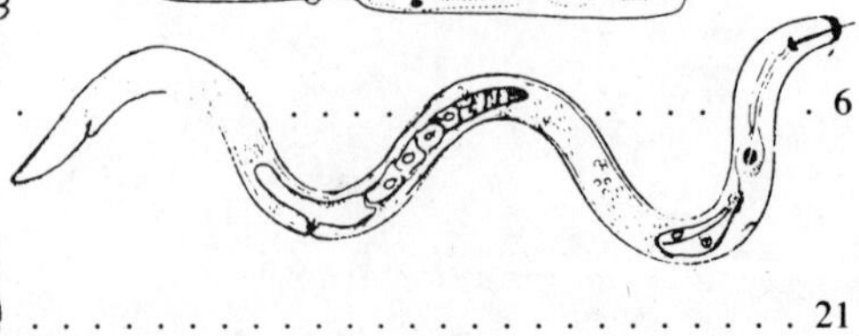	Females swollen	21

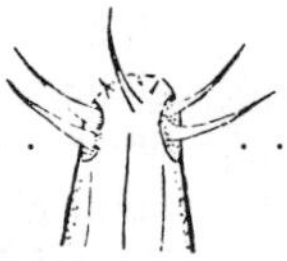

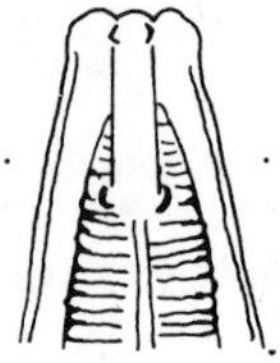

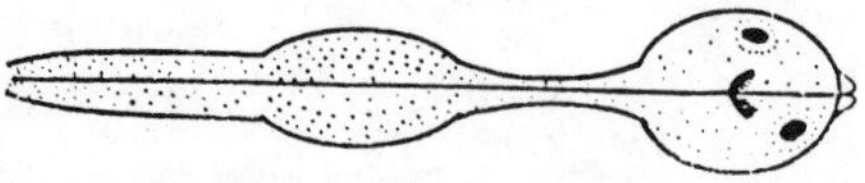

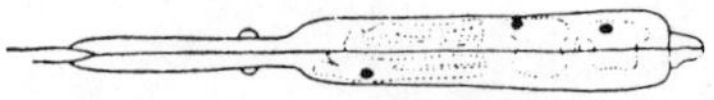

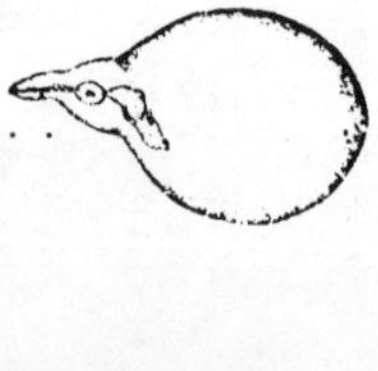

Refer to couplet No.

6.(5) Vulva at mid-body 7

Vulva on lower third of body 14

7.(6) Esophagus not overlapping intestine. . . 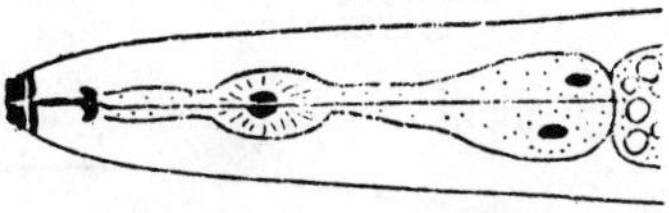 8

Esophagus overlapping intestine 11

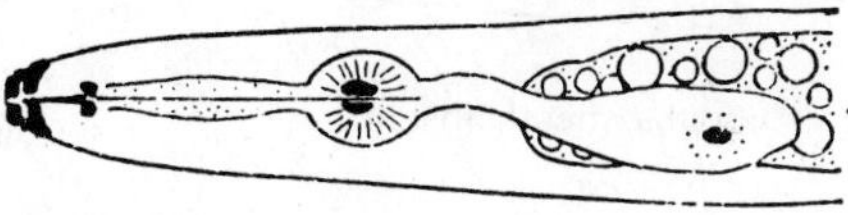

8.(7) Stylet length less than 50 μm. 9

Stylet length greater than 80 μm . . 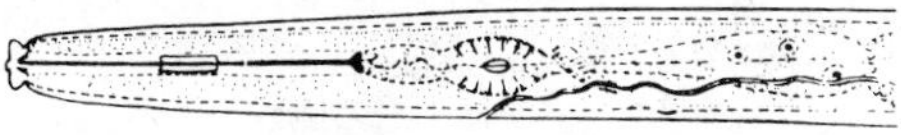 . . *Dolichodorus*

9.(8) Tail terminus pointed. . . . . *Tetylenchus*

Tail terminus not pointed 10

10.(9) Tail terminus knobbed. . . . . *Psilenchus*

Tail terminus never knobbed or pointed . . 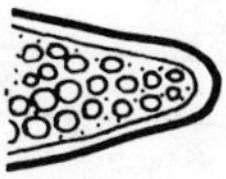 . .  . . *Tylenchorhynchus*

11.(7) Labium offset 12

Labium flattened, amalgamated or nearly so 13

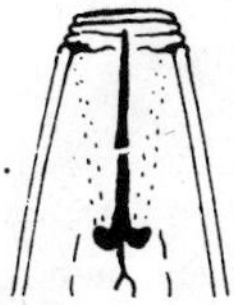

12.(11) Stylet massive, 40-50 μm long . . . . *Hoplolaimus*

Stylet long and thin, longer than 90 μm . . . . *Belonolaimus*

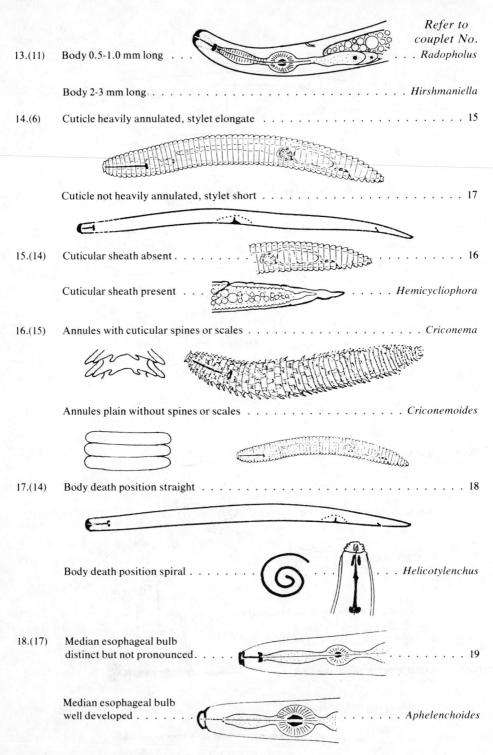

Refer to couplet No.

13.(11) Body 0.5-1.0 mm long . . . 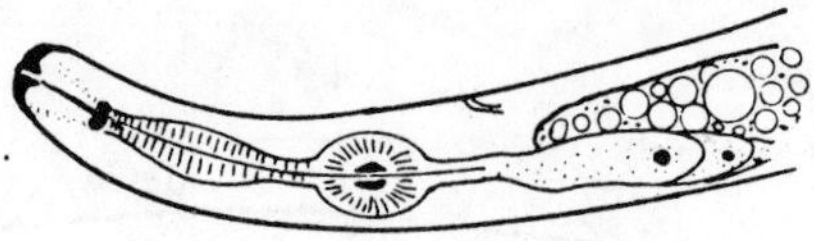 . . . *Radopholus*

Body 2-3 mm long. . . . . *Hirshmaniella*

14.(6) Cuticle heavily annulated, stylet elongate 15

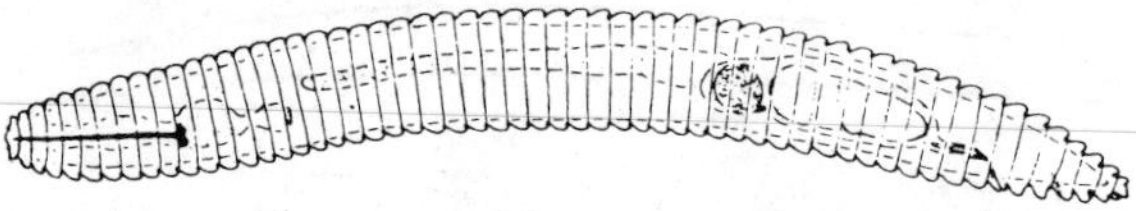

Cuticle not heavily annulated, stylet short 17

15.(14) Cuticular sheath absent 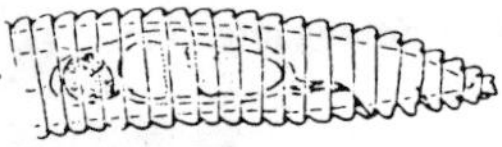 16

Cuticular sheath present . . . . *Hemicycliophora*

16.(15) Annules with cuticular spines or scales . . . . *Criconema*

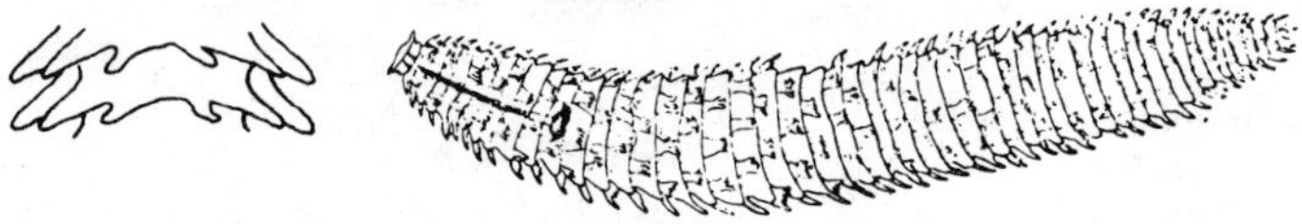

Annules plain without spines or scales . . . . *Criconemoides*

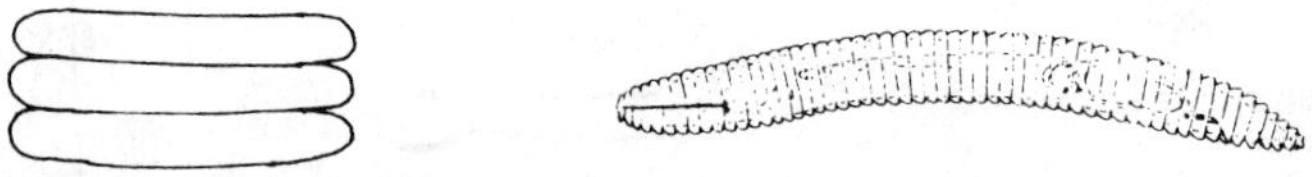

17.(14) Body death position straight 18

Body death position spiral  . . . . *Helicotylenchus*

18.(17) Median esophageal bulb distinct but not pronounced. 19

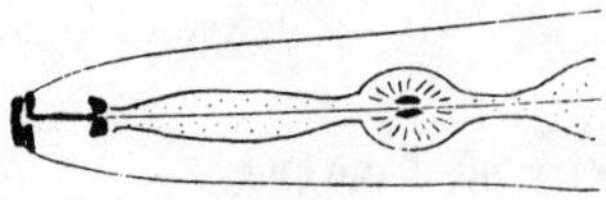

Median esophageal bulb well developed . . . . *Aphelenchoides*

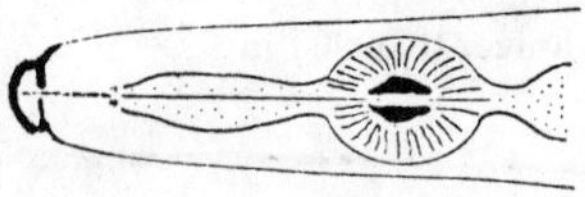

		Refer to couplet No.
19.(18)	Esophagus overlapping intestine	20
	Esophagus not overlapping intestine	*Tylenchus*
20.(19)	Median bulb and valves small, stylet usually weak	*Ditylenchus*
	Median bulb valves and stylet well developed, labium flattened	*Pratylenchus*
21.(5)	Female body white without eggs	*Meloidogyne*
	Female body brown, usually with eggs	*Heterodera*
22.(4)	Stylet short, less than 100 μm	23
	Stylet long, greater than 100 μm	*Xiphinema*
23.(22)	Stylet complex	24
	Stylet simple	25
24.(23)	Stylet with anterior arch-like portion	*Diphtherophora*
	Stylet with dorsal thickening piece	*Tylencholaimellus*
25.(23)	Stylet knobs elongate, flange-like	26
	Stylet knobs round	27

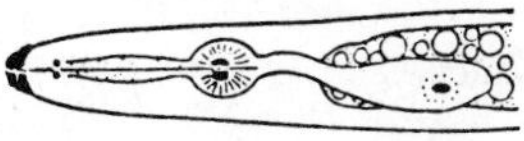

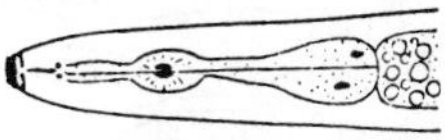

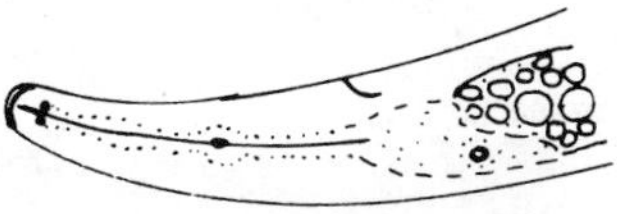

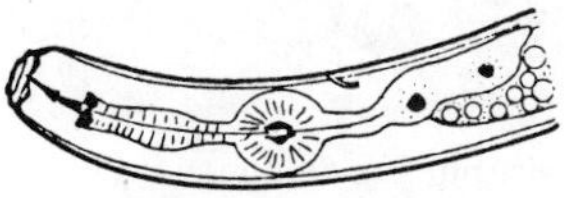

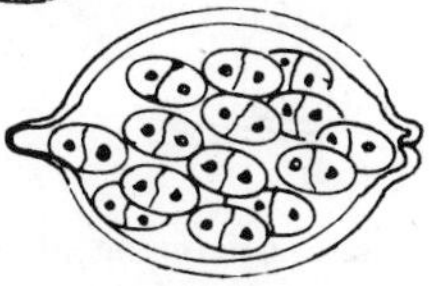

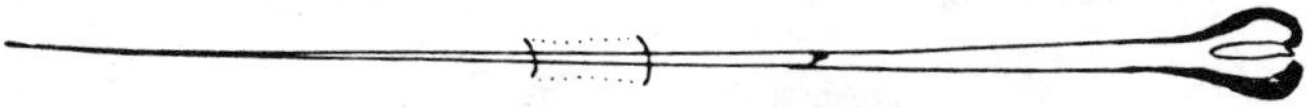

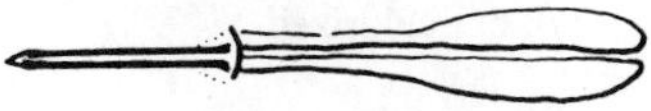

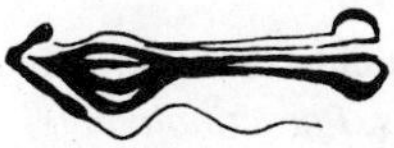

		Refer to couplet No.
26.(25)	Filiform tail	*Aulolaimoides*
	Round tail	*Enchodelus*
27.(25)	Tail rounded	28
	Tail pointed	*Nothotylenchus*
28.(27)	Basal part esophagus elongate	*Tylencholaimus*
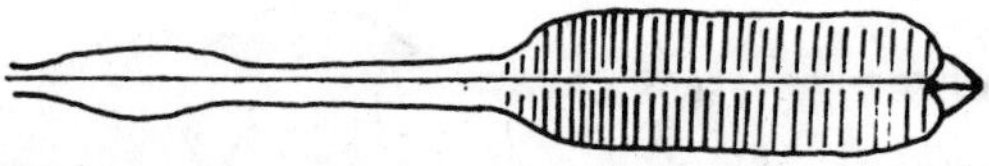	Basal part esophagus oval	*Doryllium*
29.(3)	Valvate median esophageal bulb absent	30
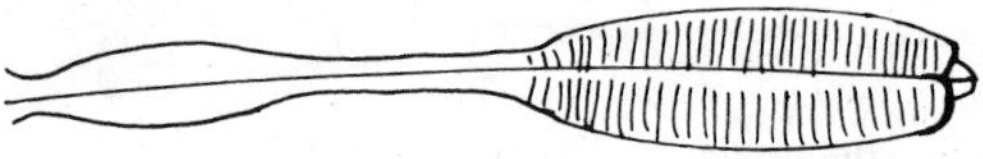	Valvate median esophageal bulb present	37
30.(29)	Stomal walls not cuticularized	31
	Stomal walls cuticularized *(Actinolaimus, Metactinolaimus, Paractinolaimus)*	*Actinolaiminae*
31.(30)	Esophagus with basal expansions	32
	Esophagus expanding uniformly	*Oionchus*

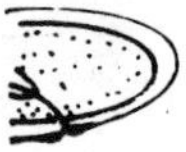

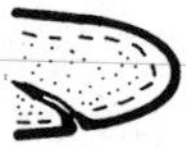

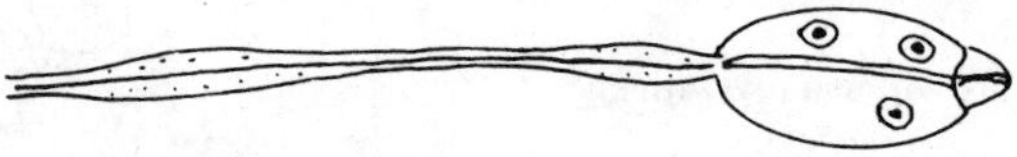

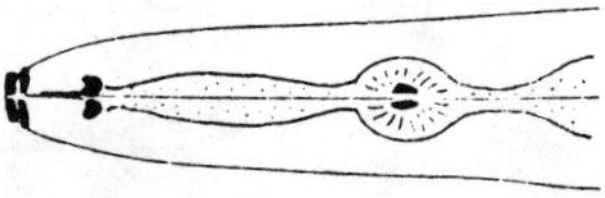

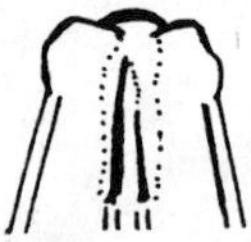

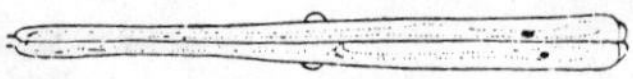

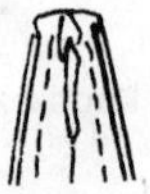

Refer to couplet No.

32.(31) Terminal fifth or sixth of esophagus an ovoid bulb 33

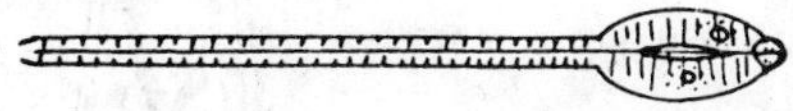

Posterior third of esophagus swollen 36

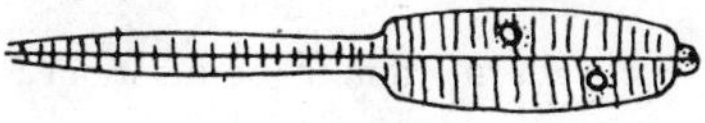

33.(32) Stylet axial, positioned centrally 34

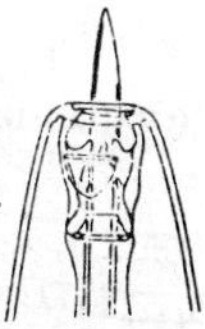

Stylet not axial, originating from tooth in stoma wall. . . . . *Campydora*

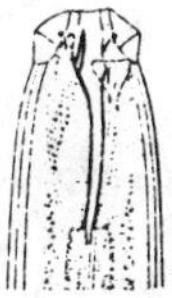

34.(33) Gonads paired; vulva usually near mid-body 35

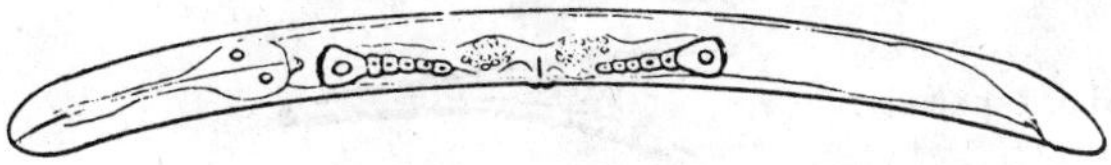

Gonad single, posterior to vulva; vulva usually posterior to mid-body. . . . . . *Tyleptus*

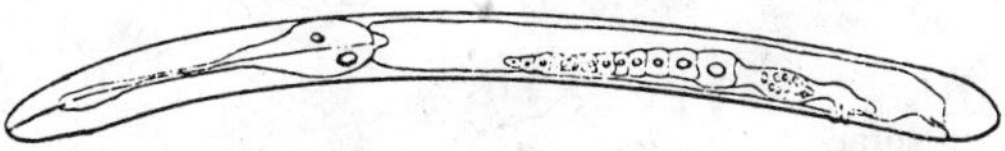

35.(34) Stylet slender . . . . *Leptonchus*

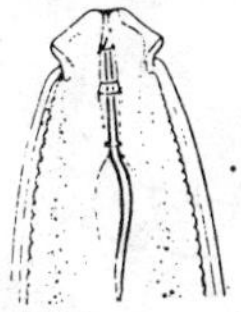

Stylet not slender . . . . *Dorylaimoides*

36.(32) Stylet axial, positioned centrally (*Dorylaimus, Eudorylaimus, Labronema, Mesodorylaimus, Thornia, Laimydorus, Prodorylaimus*). . . . . *Dorylaiminae*

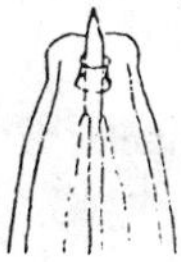

Stylet not axial, originating from tooth in stoma wall. . . . . *Nygolaimus*

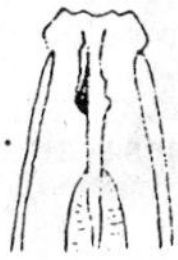

37.(29) Tail pointed . . . . *Seinura*

Tail rounded . . . . *Aphelenchus*

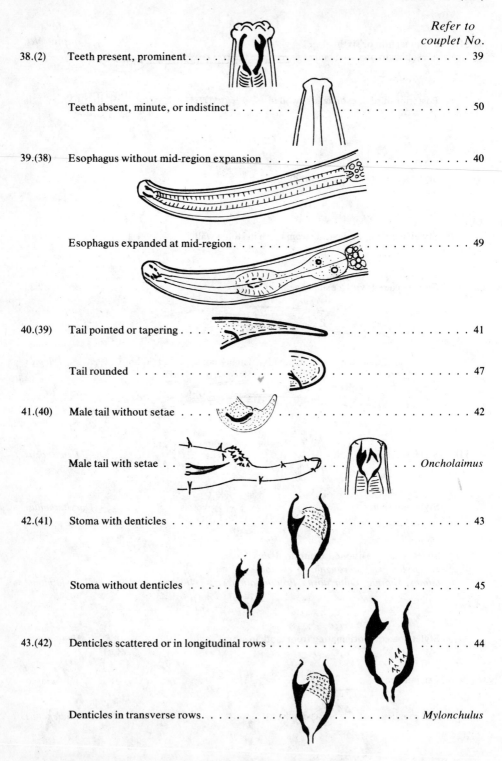

Refer to couplet No.

38.(2) Teeth present, prominent 39

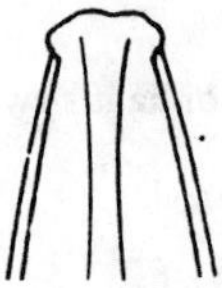

Teeth absent, minute, or indistinct 50

39.(38) Esophagus without mid-region expansion 40

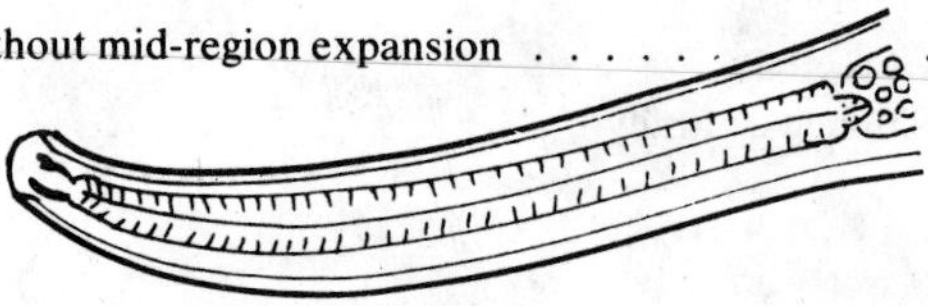

Esophagus expanded at mid-region 49

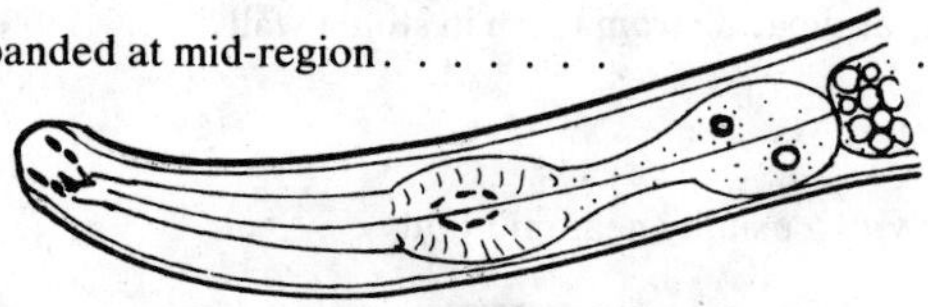

40.(39) Tail pointed or tapering 41

Tail rounded 47

41.(40) Male tail without setae 42

Male tail with setae . . . . . *Oncholaimus*

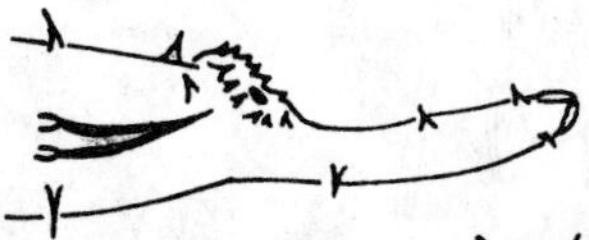

42.(41) Stoma with denticles 43

Stoma without denticles 45

43.(42) Denticles scattered or in longitudinal rows 44

Denticles in transverse rows . . . . . *Mylonchulus*

		Refer to couplet No.
44.(43)	Denticles situated on longitudinal rib of stoma	*Prionchulus*
	Denticles scattered on stoma wall	*Sporonchulus*
45.(42)	Tooth anteriorly directed	46
	Tooth retrorse	*Anatonchus*
46.(45)	Tooth in basal part of stoma	*Iotonchus*
	Tooth in anterior part of stoma	*Mononchus*
47.(40)	Stoma with prominent medial or apical tooth	48
	Stoma with small basal tooth	*Bathyodontus*
48.(47)	Stoma with 3 teeth, without small basal tooth, caudal glands terminal	*Enoplocheilus*
	Stoma with large anterior and small basal tooth, caudal glands ventral	*Mononchulus*
49.(39)	Lip region with rib-like armature	*Mononchoides*
	Lip region without rib-like armature	*Diplogaster*

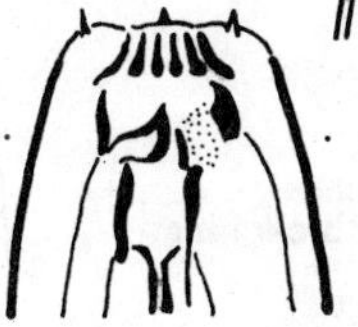

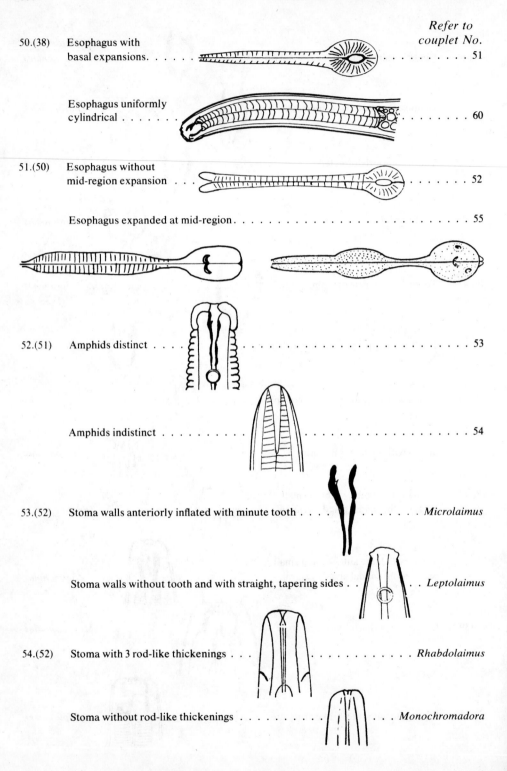

Refer to couplet No.

50.(38) Esophagus with basal expansions. 51

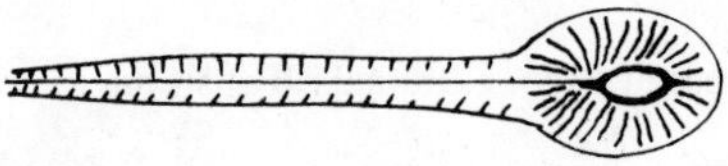

Esophagus uniformly cylindrical 60

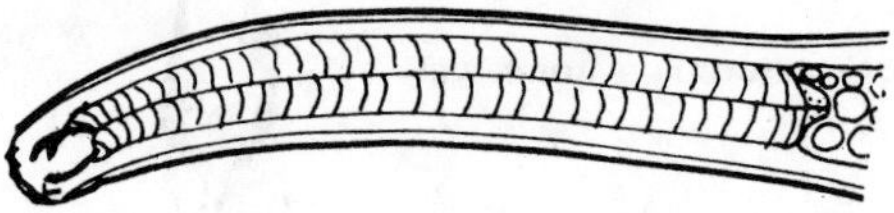

51.(50) Esophagus without mid-region expansion 52

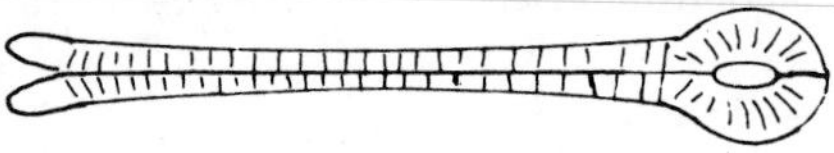

Esophagus expanded at mid-region 55

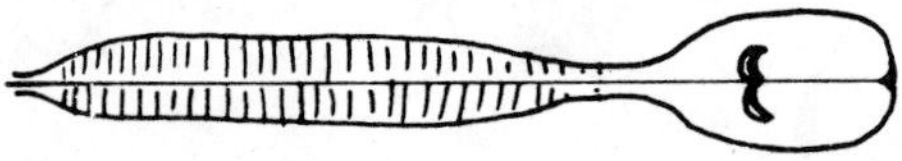

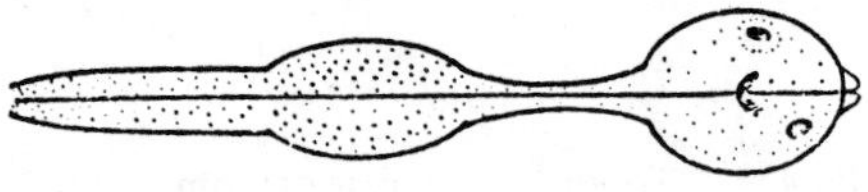

52.(51) Amphids distinct 53

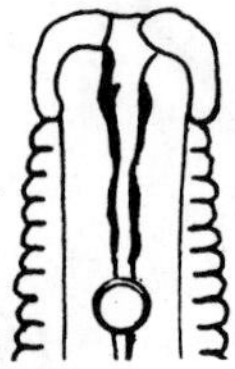

Amphids indistinct 54

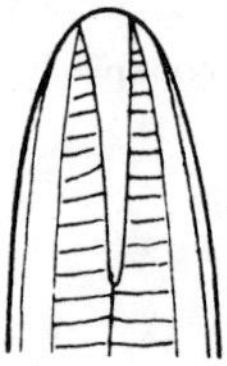

53.(52) Stoma walls anteriorly inflated with minute tooth . . . . *Microlaimus*

Stoma walls without tooth and with straight, tapering sides . . . . *Leptolaimus*

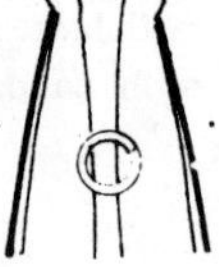

54.(52) Stoma with 3 rod-like thickenings . . . . *Rhabdolaimus*

Stoma without rod-like thickenings . . . . *Monochromadora*

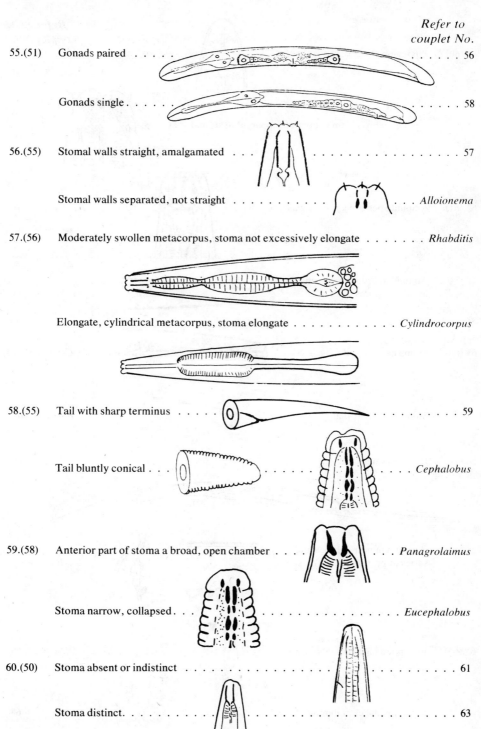

		Refer to couplet No.
55.(51)	Gonads paired	56
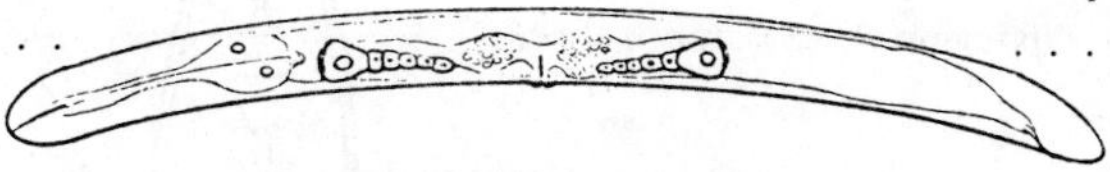	Gonads single	58
56.(55)	Stomal walls straight, amalgamated	57
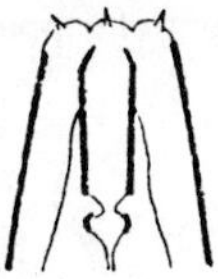	Stomal walls separated, not straight	*Alloionema*
57.(56)	Moderately swollen metacorpus, stoma not excessively elongate	*Rhabditis*
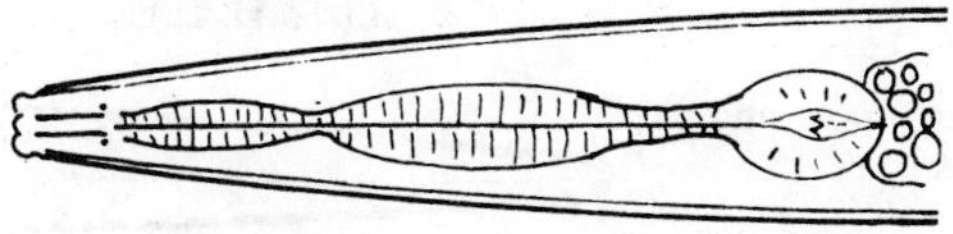	Elongate, cylindrical metacorpus, stoma elongate	*Cylindrocorpus*
58.(55)	Tail with sharp terminus	59
	Tail bluntly conical	*Cephalobus*
59.(58)	Anterior part of stoma a broad, open chamber	*Panagrolaimus*
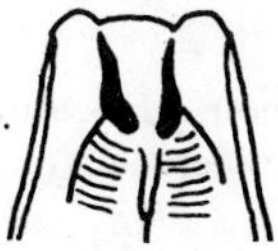	Stoma narrow, collapsed	*Eucephalobus*
60.(50)	Stoma absent or indistinct	61
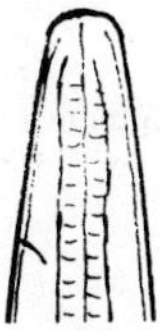	Stoma distinct	63

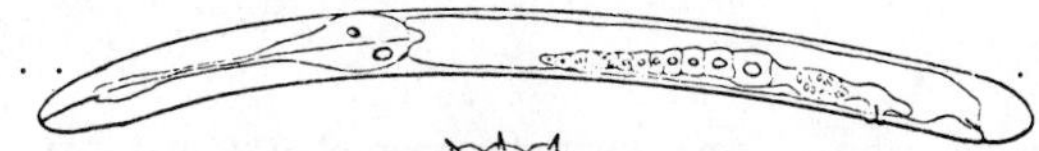

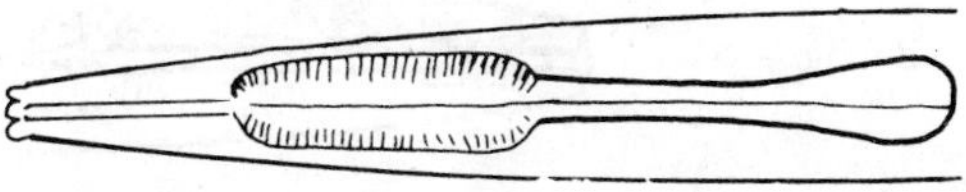

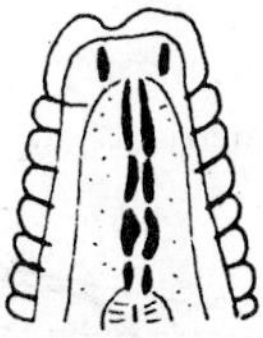

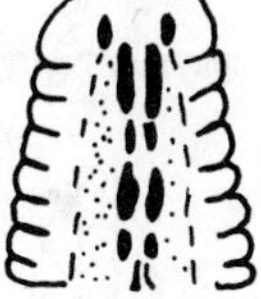

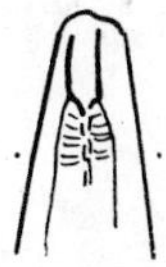

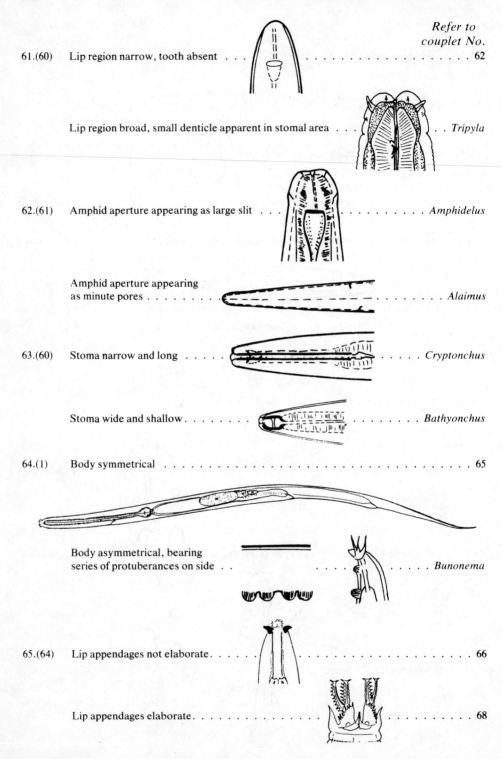

Refer to couplet No.

61.(60) Lip region narrow, tooth absent 62

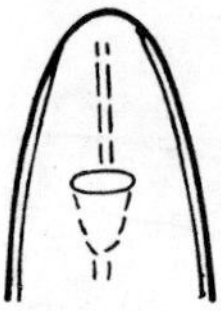

Lip region broad, small denticle apparent in stomal area . . . . . *Tripyla*

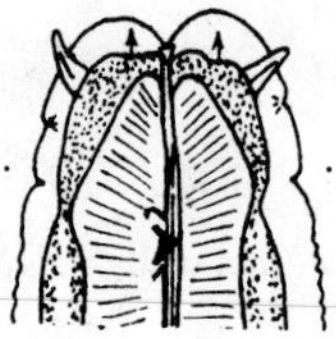

62.(61) Amphid aperture appearing as large slit . . . . *Amphidelus*

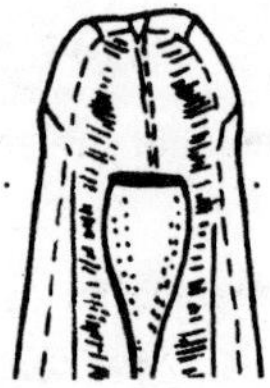

Amphid aperture appearing as minute pores . . . . *Alaimus*

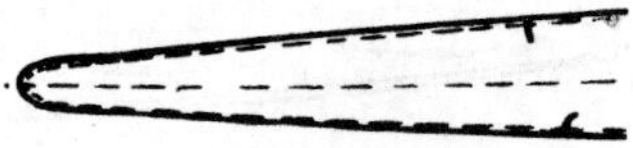

63.(60) Stoma narrow and long . . . . *Cryptonchus*

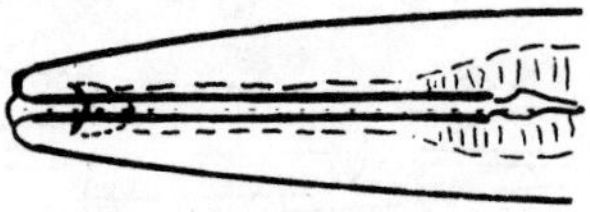

Stoma wide and shallow . . . . *Bathyonchus*

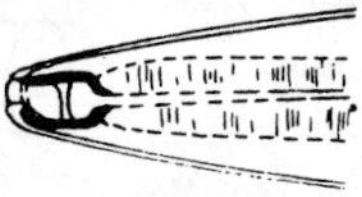

64.(1) Body symmetrical 65

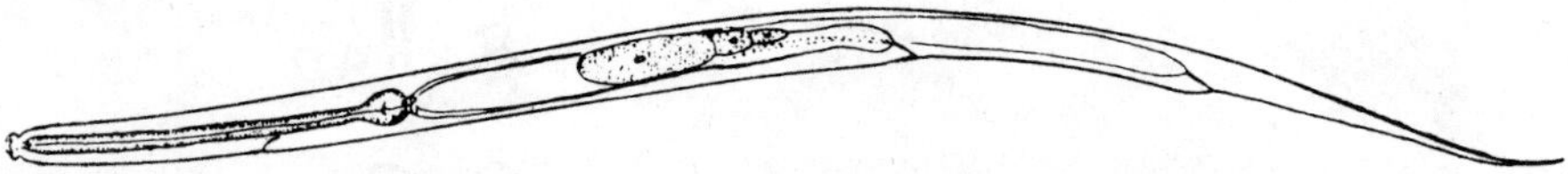

Body asymmetrical, bearing series of protuberances on side . . . . *Bunonema*

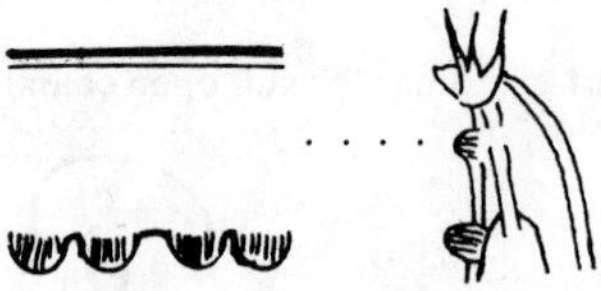

65.(64) Lip appendages not elaborate 66

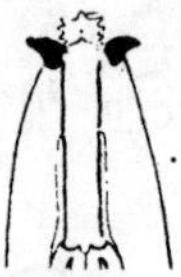

Lip appendages elaborate 68

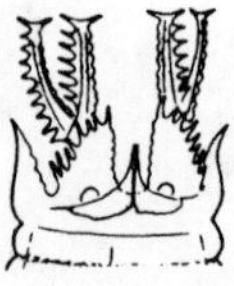

		Refer to couplet No.
66.(65)	Lateral lip appendages thorn-like, directed laterally	*Diploscapter*
	Lateral lip appendages not thorn-like or directed laterally	67
67.(66)	Papillae or setae horn-like.	*Macrolaimus*
	Lips flap-like and pointed anteriorly	*Teratocephalus*
68.(65)	Lip appendages forked and elaborately fringed.	*Acrobeles*
	Lip appendages membranous and wing-like.	*Wilsonema*
69.(1)	Post-cephalic setae absent	70
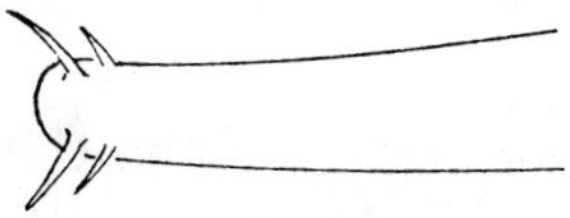	Post-cephalic setae present (may be very faint ex. *Tobrilus*)	92
70.(69)	Stylet absent	71
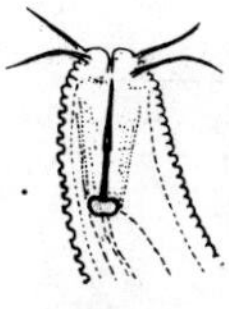	Stylet present	91

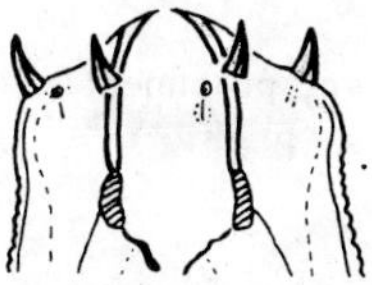

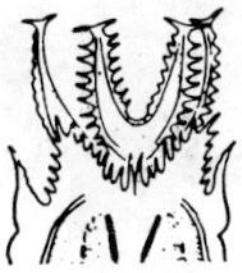

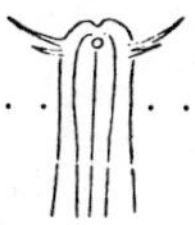

		Refer to couplet No.
71.(70)	Teeth absent, minute or indistinct	72
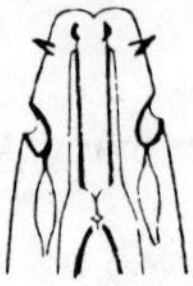	Teeth usually present, prominent	85
72.(71)	Esophagus with basal expansions.	73
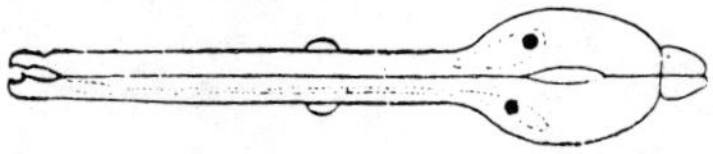	Esophagus uniformly cylindrical	82
73.(72)	Amphids oval, spiral, or stirrup-shaped	74
	Amphids circular	80
74.(73)	Amphids spiral	75
	Amphids not spiral	79
75.(74)	Cuticular punctations absent	76
	Cuticular punctations present.	78
76.(75)	Esophageal bulb without valves	77
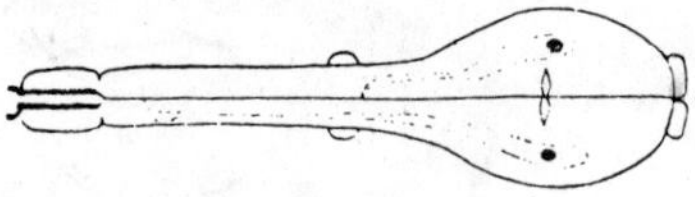	Esophageal bulb valvate	*Plectus & Anaplectus*

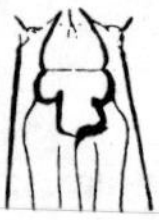

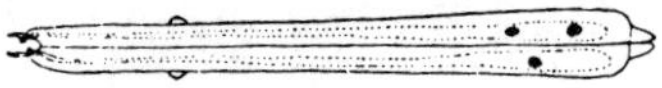

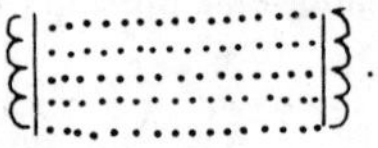

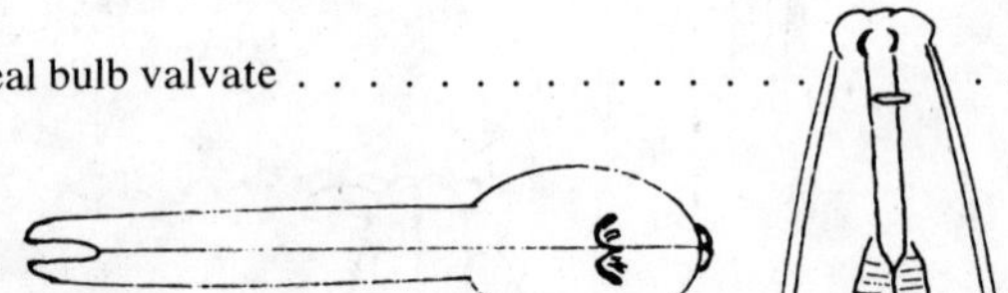

Refer to couplet No.

77.(76) Esophageal-intestinal valve elongate . . . . *Paraplectonema*

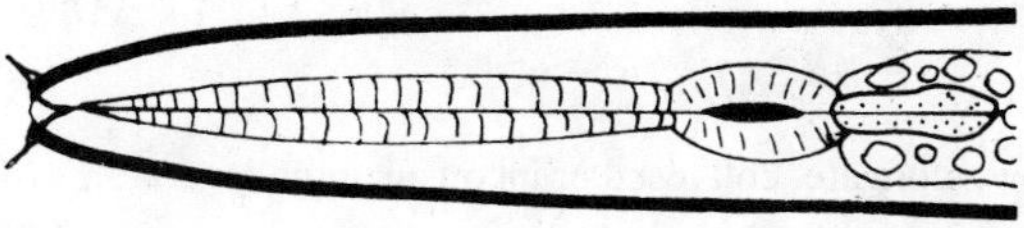

Esophageal-intestinal valve shortened . . . . *Paraphanolaimus*

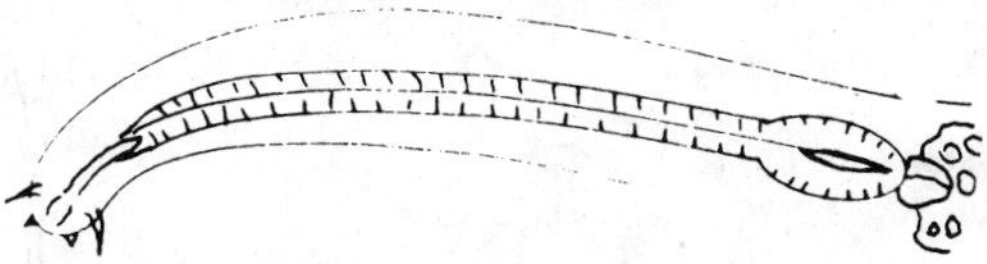

78.(75) Labial region characteristically flap-like . . . 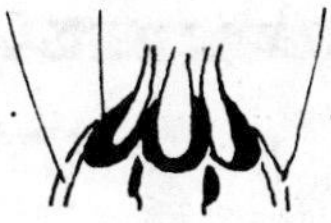 . . . . . *Euteratocephalus*

Labial region not flap-like, lips bluntly rounded. . .  . . *Ethmolaimus*

79.(74) Amphids oval . . . . *Greenenema*

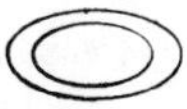

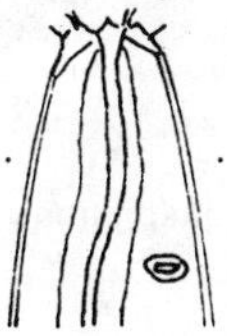

Amphids stirrup-shaped . . . . *Chronogaster*

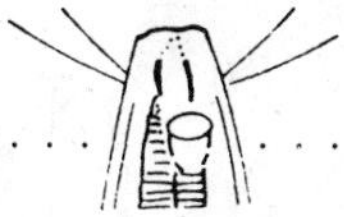

80.(73) Esophageal-intestinal valve shortened 81

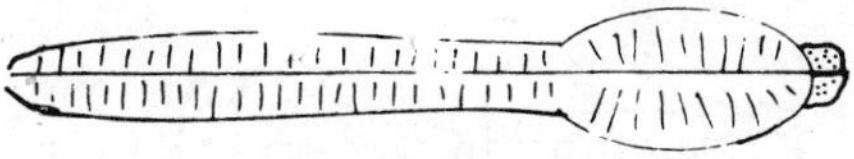

Esophageal-intestinal valve elongate . . . . 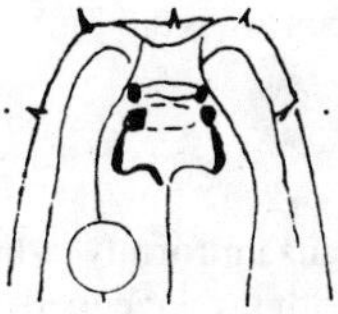 . . . . *Desmolaimus*

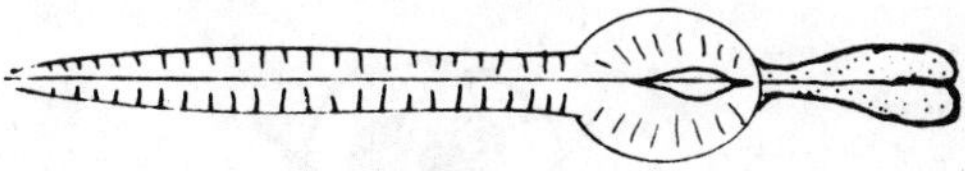

81.(80) Excretory pore and large excretory gland present. . . . . 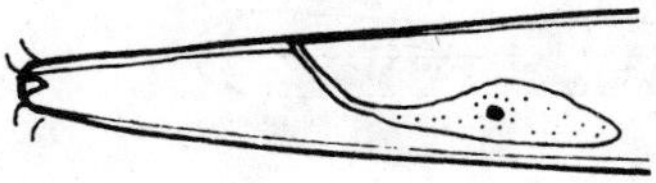 . . . . *Domorganus*

Excretory pore and gland indistinct or absent . . . . . . 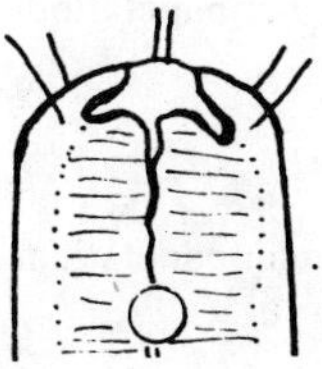 . . *Monhystera*

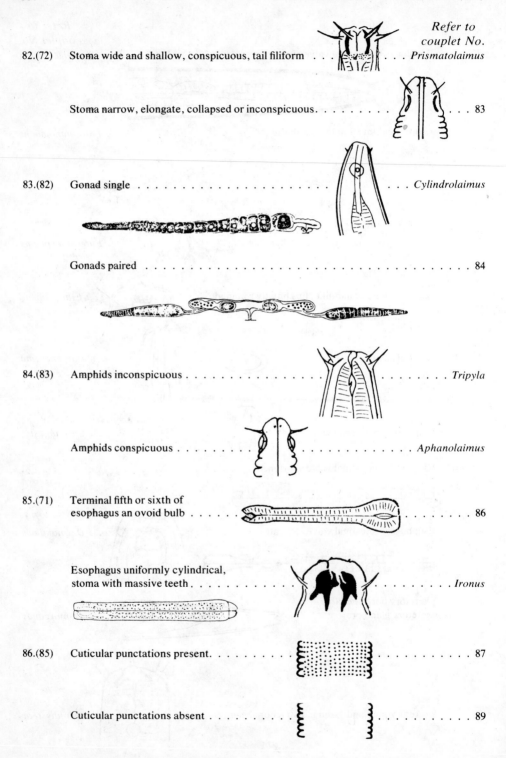

Refer to couplet No.

82.(72) Stoma wide and shallow, conspicuous, tail filiform . . . . . . *Prismatolaimus*

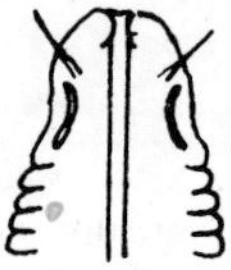

Stoma narrow, elongate, collapsed or inconspicuous 83

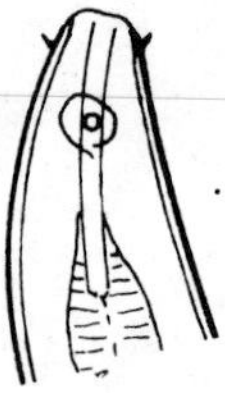

83.(82) Gonad single . . . . *Cylindrolaimus*

Gonads paired 84

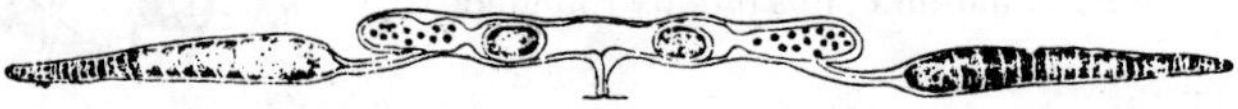

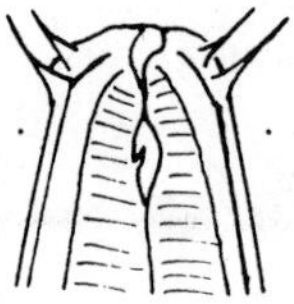

84.(83) Amphids inconspicuous . . . . *Tripyla*

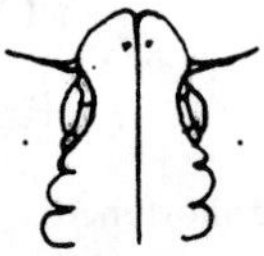

Amphids conspicuous . . . . *Aphanolaimus*

85.(71) Terminal fifth or sixth of esophagus an ovoid bulb 86

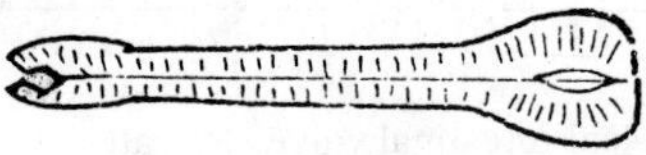

Esophagus uniformly cylindrical, stoma with massive teeth . . . . *Ironus*

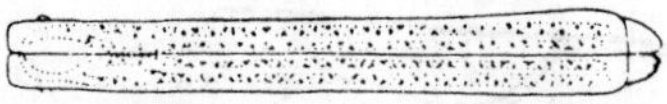

86.(85) Cuticular punctations present. 87

Cuticular punctations absent 89

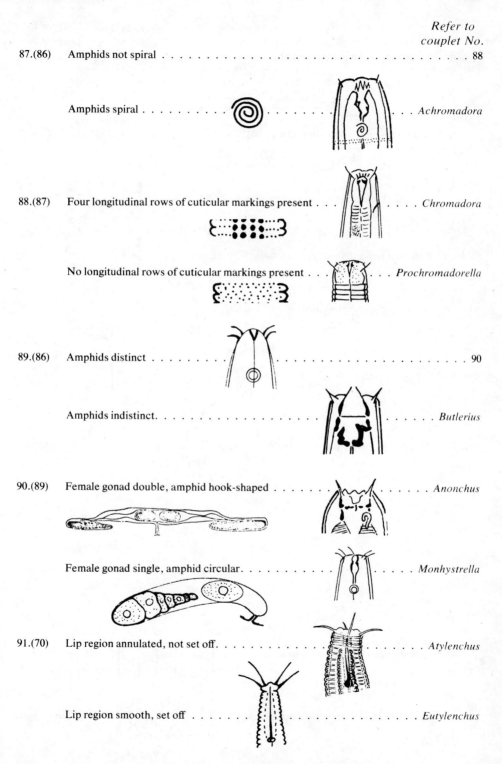

		Refer to couplet No.
87.(86)	Amphids not spiral	88
	Amphids spiral	*Achromadora*
88.(87)	Four longitudinal rows of cuticular markings present	*Chromadora*
	No longitudinal rows of cuticular markings present	*Prochromadorella*
89.(86)	Amphids distinct	90
	Amphids indistinct	*Butlerius*
90.(89)	Female gonad double, amphid hook-shaped	*Anonchus*
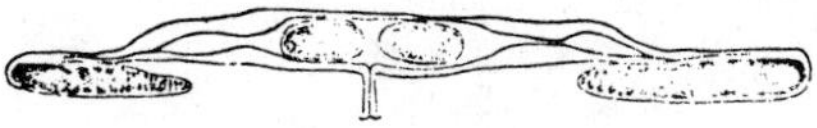	Female gonad single, amphid circular	*Monhystrella*
91.(70)	Lip region annulated, not set off	*Atylenchus*
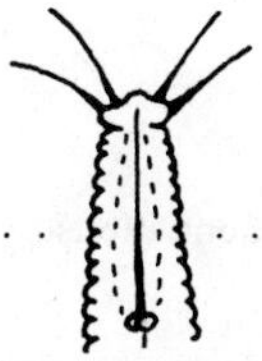	Lip region smooth, set off	*Eutylenchus*

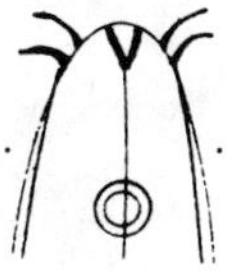

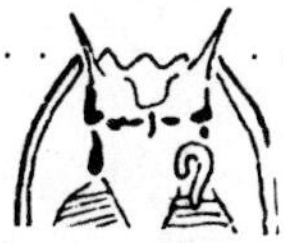

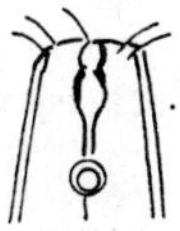

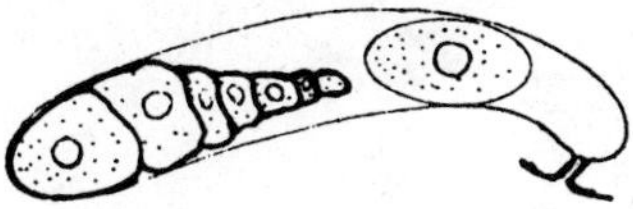

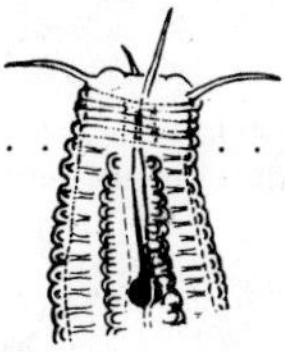

		Refer to couplet No.
92.(69)	Esophagus with basal expansion	93
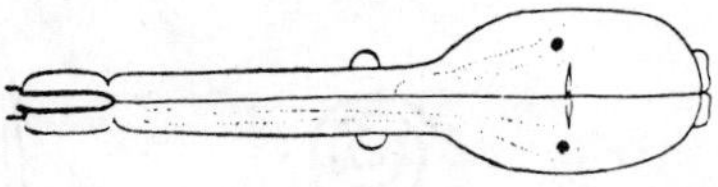	Esophagus uniformly cylindrical	98
93.(92)	Cuticular punctation present, amphids not circular	94
	Cuticular punctation absent, amphids circular	97
94.(93)	Ocelli (eye spots) present	95
	Ocelli absent	96
95.(94)	Stoma with three equal-sized teeth	*Chromadorina*
	Stoma with at least one large tooth	*Punctodora*
96.(94)	Cuticle with lateral longitudinal rows of punctation	*Hypodontolaimus*
	Cuticle without lateral differentiations	*Chromadorita*

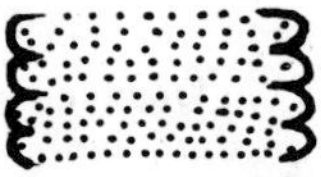

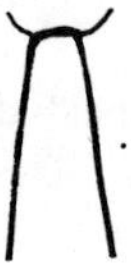

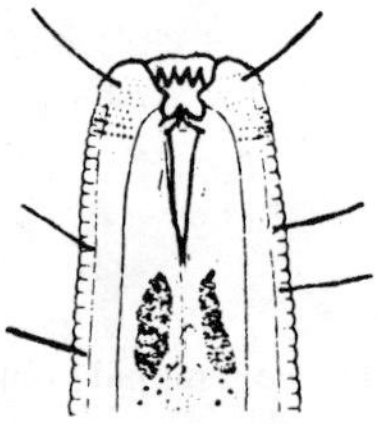

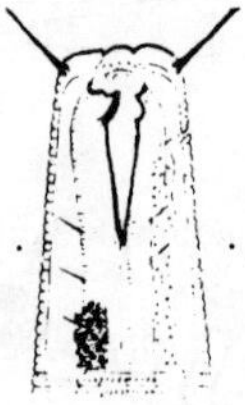

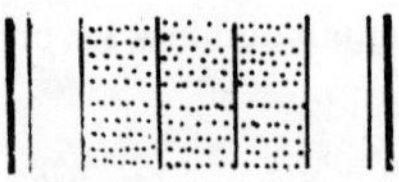

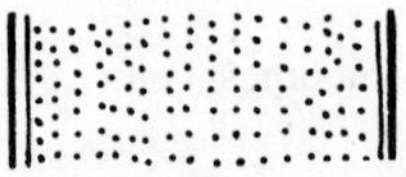

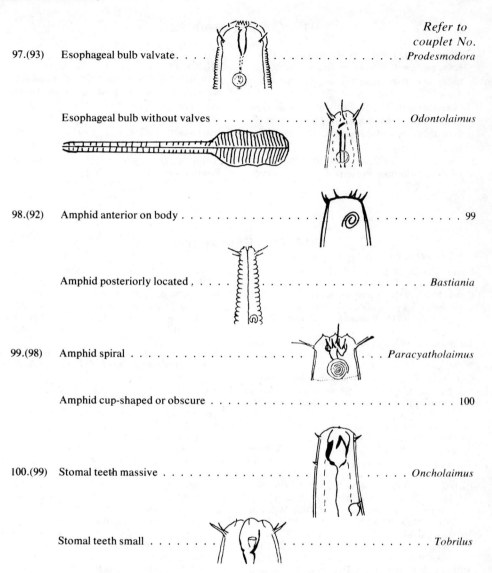

Refer to couplet No.

97.(93) Esophageal bulb valvate . . . . *Prodesmodora*

Esophageal bulb without valves . . . . *Odontolaimus*

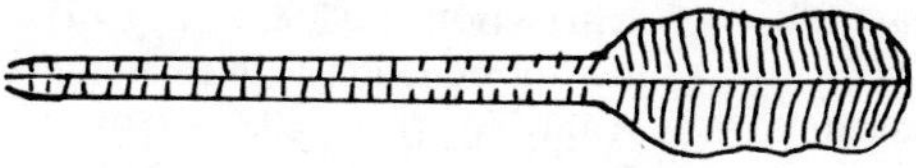

98.(92) Amphid anterior on body 99

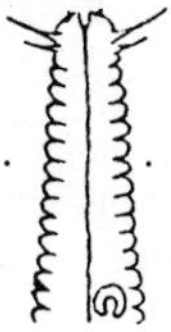

Amphid posteriorly located . . . . *Bastiania*

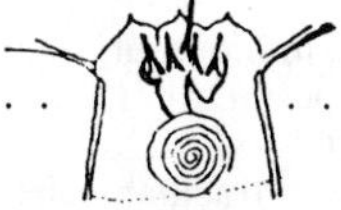

99.(98) Amphid spiral . . . . *Paracyatholaimus*

Amphid cup-shaped or obscure 100

100.(99) Stomal teeth massive . . . . *Oncholaimus*

Stomal teeth small . . . . *Tobrilus*

917 C. References

1. CHANG, S.L., R.L. WOODWARD & P.W. KABLER. 1969. Survey of free-living nematodes and amoebas in municipal supplies. *J. Amer. Water Works Ass.* 52:613.
2. CHAUDHURI, N., R. SIDDIQI & R.S. ENGELBRECHT. 1964. Source and persistence of nematodes in surface waters. *J. Amer. Water Works Ass.* 56:73.
3. CHANG, S.L. 1972. Zoomicrobial indicators of water pollution. Presented at 72nd Annual Meeting of Amer. Soc. Microbiol., Philadelphia, Pa., Apr. 23–28, 1972.
4. GOODEY, T. 1963. Soil and Freshwater Nematodes, 2nd ed. (Revised by J.B. Goodey). The Methuen Co., London & John Wiley, New York, N.Y.
5. CHITWOOD, B.G. & M.B. CHITWOOD. 1937. An Introduction to Nematology, Section I: Anatomy (rev. ed., 1950). Monumental Printing Co., Baltimore, Md.

917 D. Bibliography

PETERS, B.G. 1930. A biological investigation of sewage. *J. Helminth.* 8:133.

COBB, N.A. 1935. Contributions to a Science of Nematology. Collected Papers between 1914 and 1935. Williams & Wilkens, Baltimore, Md.

THORNE, G. 1939. A monograph of the nematodes of the superfamily Dorylaimoidea. *Capita. Zool.* 8:1.

GERLACH, S.A. 1954. Brasilianische Meeres-Nematoden 1. *Bol. Inst. Oceanog.* 5:3.

CHITWOOD, B.G. & A.C. TARJAN. 1957. A redescription of *Atylenchus decalineatus* Cobb, 1913 (Nematoda: Tylenchinae). *Proc. Helminth. Soc. Wash.* 24:48.

ANDRÁSSAY, I. 1959. Nematoden aus dem Psammon des Adige-Flusses, I. *Mem. Mus. Civ. Stor. Nat., Verona* 7:163.

CHITWOOD, B.G. & M.W. ALLEN. 1959. Nemata, Chap. 15 in Freshwater Biology, 2nd ed. (W.T. Edmondson, ed.), John Wiley & Sons, New York, N.Y.

EDMONDSON, W.T., ed. 1959. Ward & Whipple's Fresh Water Biology, 2nd ed. John Wiley & Sons, New York, N.Y.

HOPPER, B.E. & E.J. CAIRNS. 1959. Taxonomic Keys to Plant, Soil and Aquatic Nematodes. Alabama Polytechnic Inst., Auburn.

CHANG, S.L. 1960. Survival and protection against chlorination of human enteric pathogens in free-living nematodes isolated from water supplies. *Amer. J. Trop. Med. Hyg.* 9:136.

CHITWOOD, B.G. 1960. A preliminary contribution on the marine nemas (Adenophorea) of Northern California. *Trans. Amer. Micr. Soc.* 79:347.

LUC, M. 1960. *Dolichodorus profundus* n. sp. (Nematoda-Tylenchida). *Nematologica* 5:1.

SASSER, J.N. & W.R. JENKINS, ed. 1960. Nematology: Fundamental and Recent Advances with Emphasis on Plant Parasitic and Soil Forms. University of North Carolina, Chapel Hill.

LOOF, P.A.A. 1961. The nematode collection of Dr. J.G. de Man. *Meded. Lab. Fytopath.* 190:169.

CHANG, S.L. & P.W. KABLER. 1962. Free-living nematodes in aerobic treatment effluent. *J. Water Pollut. Control Fed.* 34:1356.

CALAWAY, W.T. 1963. Nematodes in wastewater treatment. *J. Water Pollut. Control Fed.* 35:1006.

ENGELBRECHT, R.S. & J.H. AUSTIN. 1964. Nematodes and their detection in public water supplies. Presented to Kentucky-Tennessee Sect., Amer. Water Works Ass., Sept. 15, 1964.

THORNE, G. 1964. Nematodes of Puerto Rico: Belondiroidea new superfamily, Leptonchidae, Thorne, 1935, and Belonenchidae new family (Nemata, Adenophorea, Dorylaimida). Univ. Puerto Rico Agr. Exp. Sta. Tech. Paper 39.

EDWARD, J.C. & S.L. MISRA. 1966. *Criconema vishwanathum* n. sp. and four other hitherto described Criconematinae. *Nematologica* 11:566.

HOPPER, B.E. & S.P. MEYERS. 1967. Folliculous marine nematodes on turtle grass, *Thalassia testudinum* König, in Biscayne Bay, Florida. *Bull. Mar. Sci.* 17:471.

MULVEY, R.H. & H.J. JENSEN. 1967. The Mononchidae of Nigeria. *Can. J. Zool.* 45:667.

WALTERS, J.V. & R.R. HOLCOMB. 1967. Isolation of an enteric pathogen from sewage-borne nematodes. (abstr.) *Nematologica* 13:155.

ALLEN, M.W. & E.M. NOFFSINGER. 1968. Revision of the genus *Anaplectus* (Nematoda: Plectidae). *Proc. Helminth. Soc. Wash.* 35:77.

ANDRÁSSAY, I. 1968. Fauna Paraguayensis 2. Nematoden aus den Galeriewaldern des Acaray-Flusses. *Opusc. Zool. Boest.* 8:167.

DEGRISSE, A. 1968. Bijdrage tot de morfologie en de systematiek van Criconematidae (Taylor, 1936) Thorne, 1949 (Nematoda). Plantenatlas Sleutel, Gent.

CHANG, S.L. 1970. Interactions between animal viruses and higher forms of microbes. *J. San. Eng. Div., Proc. Amer. Soc. Civil Eng.* 96:151.

ANDRÁSSAY, I. 1973. Nematoden aus strand—und höhoenbiotopen von Kuba. *Acta Zool. Hung.* 19(3-4):233.

FERRIS, V.R., J.M. FERRIS, & J.P. TJEPKEMA. 1973. Genera of Freshwater Nematodes (Nematoda) of Eastern North America. Biota of Freshwater Ecosystems. Identification Manual No. 10, U.S. EPA.

TARJAN, A.C. & B.E. HOPPER. 1974. Nomenclatorial Compilation of Plant and Soil Nematodes. Society of Nematologists, O.E. Painter Printing Co., DeLeon Springs, Fla.

918 IDENTIFICATION OF IRON AND SULFUR BACTERIA

The group of nuisance organisms collectively designated "iron and sulfur bacteria" is neither morphologically nor physiologically homogeneous, yet it may be characterized by the ability to transform or deposit significant amounts of iron or sulfur, usually in the form of objectionable slimes. However, iron and sulfur bacteria are not the sole producers of bacterial slimes.

The organisms in this group may be filamentous or single-celled, autotrophic or heterotrophic, aerobic or anaerobic. According to conventional bacterial classification, these organisms are assigned to a variety of orders, families, and genera. They are studied as "iron and sulfur bacteria," because these elements and their transformations may be important in water treatment and distribution systems and may be especially bothersome in waters for industrial use, such as cooling and boiler waters. Iron bacteria may cause fouling and plugging of wells and distribution systems and sulfate-reducing bacteria may cause rusty water and tuberculation of pipes. These organisms also may cause odor, taste, frothing, color, and increases in turbidity in waters.

The nutrient supply of iron and sulfur bacteria may be wholly or partly inorganic and they may extract it, if attached in a gelatinous substrate, from a low concentration in flowing water. This seems quite important in the case of certain sulfur bacteria utilizing small amounts of hydrogen sulfide or in the case of organisms such as *Gallionella*, which obtain their energy from the oxidation of ferrous iron. *Thiobacillus ferrooxidans* and *Ferrobacillus ferrooxidans*, which contribute to the problem of acid mine drainage, can be identified by tests for transformation of ferrous to ferric iron or oxidation of reduced sulfur compounds. Temperature, light, pH, and oxygen supply also affect the growth of these organisms. Under different environmental conditions some bacteria may appear either as iron or as sulfur bacteria.

918 A. Iron Bacteria

1. General Characteristics

"Iron bacteria" are considered to be capable of metabolizing reduced iron present in their aqueous habitat and of depositing it in the form of hydrated ferric oxide on or in their mucilaginous secretions. A somewhat similar mechanism is used by bacteria that utilize manganese. The large amount of brown slime so produced will impart a reddish tinge and an unpleasant odor to drinking water and may render the supply unsuitable for domestic or industrial purposes. Bacteria of this type obtain energy by the oxidation of iron from the ferrous to the ferric state; the ferric form is precipitated as ferric hydroxide. Iron may be obtained from the pipe itself or from the water within it. The amount of ferric hydroxide deposited is very large in comparison with the enclosed cells.

Some bacteria that do not oxidize ferrous iron nevertheless may cause it to be dissolved or deposited indirectly. In their growth, either they liberate iron by utilizing organic radicals to which the iron is attached or they alter environmental conditions to permit the solution or deposition of iron. In consequence, less ferric hydroxide may be produced, but taste, odor, and fouling may be engendered.

2. Collection of Samples and Identification

Identification of nuisance iron bacteria usually has been made on the basis of microscopic examination of the suspected material. Directly examine bulked activated sludge, masses of microbial growth in lakes, rivers, and streams, and slime growths in cooling-tower waters. Suspected development of iron bacteria in water wells or in distribution systems may require special efforts to secure samples useful for identification. Continued heavy deposition of iron caused by the oxidation of ferrous iron by air or by other environ-

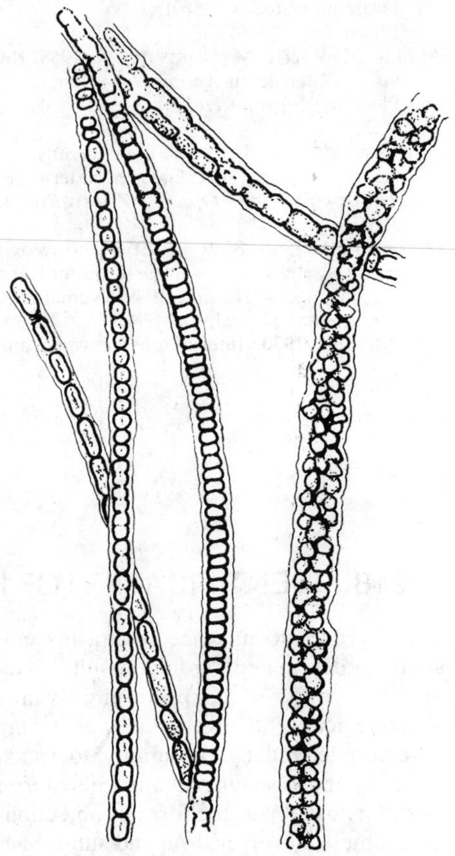

Figure 918:1. Filaments of *Crenothrix polyspora* showing variation of size and shape of cells within the sheath. Note especially the multiple small round cells, or "conidia", found in one of the filaments. This distinctive feature is the reason for the name *polyspora*. Young growing colonies are usually not encrusted with iron or manganese. Older colonies often exhibit empty sheaths that are heavily encrusted. Cells may vary considerably in size: Rod-shaped cells average 1.2 to 2.0 μm in width by 2.4 to 5.6 μm in length; coccoid cells of "conidia" average 0.6 μm in diameter.

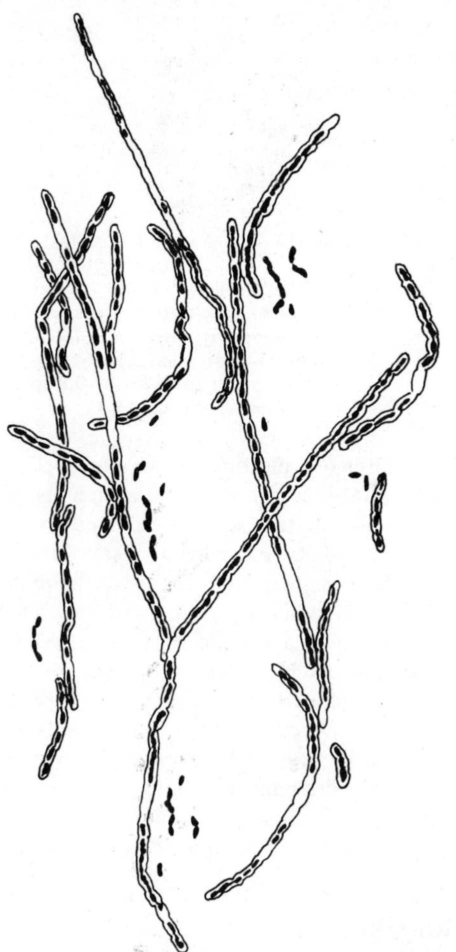

Figure 918:2. Filaments of *Sphaerotilus natans*, showing cells within the filaments and some free "swarmer" cells. Filaments show false branching and areas devoid of cells. Individual cells within the sheath may vary in size, averaging 0.6 to 2.4 μm in width by 1.0 to 12.0 μm in length; most strains are 1.1 to 1.6 μm wide by 2.0 to 4.0 μm long.

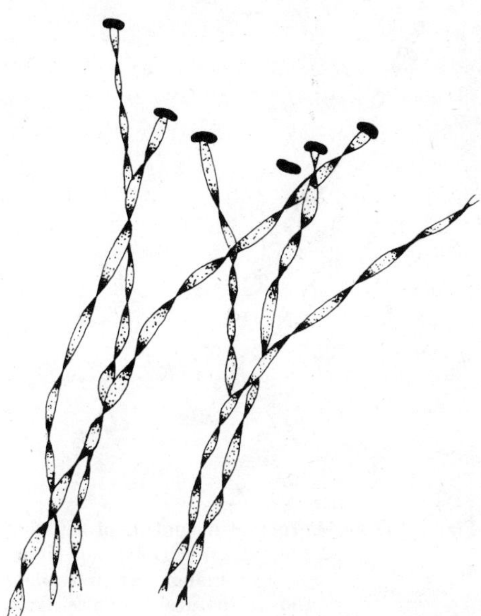

Figure 918:3. Laboratory culture of *Gallionella ferruginea*, showing cells, stalks excreted by cells, and branching of stalks where cells have divided. A precipitate of inorganic iron on and around the stalks often blurs the outlines. Cells at tip of stalk average 0.4 to 0.6 μm in width by 0.7 to 1.1 μm in length.

mental changes often hides the sheaths or stalks of iron bacteria. The cells within the filaments often die and disintegrate and the filaments tend to be fragmented or crushed by the mass of the iron precipitate.

Settle or centrifuge samples of water drawn directly from wells and examine sediment microscopically. Place a portion of sediment on a microscope slide, cover with a cover slip, and examine under a low-power microscope for filaments and iron-encrusted filaments. The material trapped by filters placed in front of back-surge valves often has yielded excellent specimens of iron bacteria. Water pumped from wells may be passed through a 0.45-μm membrane filter and the filter examined microscopically after drying and clearing with immersion oil applied directly to the membrane. Phase-contrast microscopes have made possible the examina-

Figure 918:4. Mixture of fragments of stalks of *Gallionella ferruginea* and inorganic iron-manganese precipitate found in natural samples from wells. Fragmented stalks appear golden yellow to orange when examined under the microscope.

Figure 918:5. Single-celled iron bacterium *Siderocapsa treubii*. Cells are surrounded by a deposit of ferric hydrate. Individual cells average 0.4 to 1.5 μm in width by 0.8 to 2.5 μm in length.

tion of unstained culture material. Use india ink or lactophenol blue for staining when conventional light microscopy is used. To dissolve iron deposits place several drops of 1*N* HCl at one edge of cover slip and draw it under cover slip by applying filter or blotting paper to opposite edge. Reducing compounds such as sodium ascorbate also may be used to dissolve deposits and permit observation of cellular structure. To verify that the material is iron, add a solution of potassium ferrocyanide to a sample on a slide, cover, and draw 1*N* HCl under cover slip. A blue precipitate of Prussian blue will form as iron around cells or filaments is dissolved.

Identify organisms by comparing with available drawings or photographs of iron bacteria.[1–14] Some examples are given in Figures 918:1 through 918:5.

918 B. Sulfur Bacteria

1. General Characteristics

The bacteria that oxidize or reduce significant amounts of organic sulfur compounds exhibit a wide diversity of morphological and biochemical characteristics. One group, the sulfate-reducing bacteria, consists of single-celled forms that grow anaerobically and reduce sulfate to hydrogen sulfide, H_2S. A second group, the photosynthetic green and purple sulfur bacteria, grows anaerobically in the light and uses H_2S as a hydrogen donor for photosynthesis. The sulfide is oxidized to sulfur or sulfate. A third group, the aerobic sulfur-oxidizers, oxidizes reduced sulfur compounds aerobically to obtain energy for chemoautotrophic growth.

The sulfur bacteria of most importance in the water and wastewater field are the sulfate-reducing bacteria, which include *Desulfovibrio*, and the single-celled aerobic sulfur-oxidizers of the genus *Thiobacillus*. The sulfate-reducing bacteria contribute greatly to tuberculations and galvanic corrosion of water mains and to taste and odor problems in water. *Thiobacillus*, by its production of sulfuric acid, has contributed to the destruction of concrete sewers and the acid corrosion of metals.

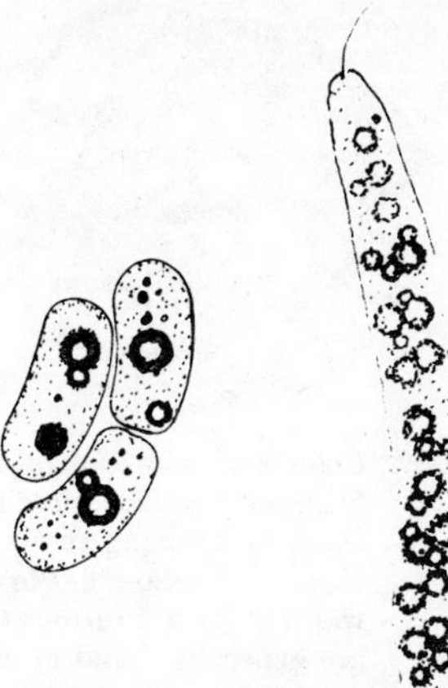

Figure 918:6. Photosynthetic purple sulfur bacteria: Large masses of cells have brown-orange to purple color—may appear chalky if there is a large amount of sulfur within the cells. Left: cells of *Chromatium okenii* (5.0 to 6.5 μm wide by 8 to 15 μm long) containing sulfur granules. Right: *Thiospirillum jenense* (3.5 to 4.5 μm wide by 30 to 40 μm long); cell contains sulfur granules and polar flagellum is visible.

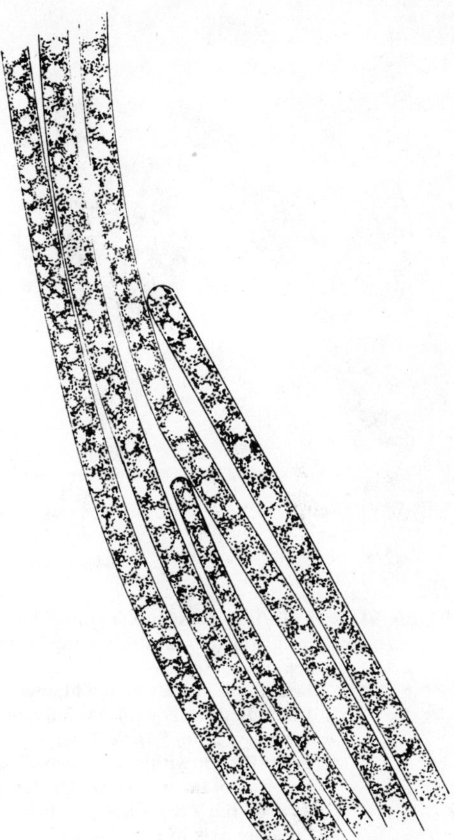

Figure 918:7. Colorless filamentous sulfur bacteria: *Beggiatoa alba* trichomes, containing granules of sulfur. Filaments are composed of a linear series of individual rod-shaped cells that may be visible when not obscured by light reflecting from sulfur granules. Trichomes are 2 to 15 μm in diameter and may be up to 1,500 μm long; individual cells, if visible, are 4.0 to 16.0 μm long.

2. Collection of Samples and Identification

Identification of nuisance sulfur bacteria usually has been made on the basis of microscopic examination of the suspected material. Examine directly samples of slimes suspended in waters, scrapings from exposed surfaces, or sediments.

Three groups of sulfur bacteria may be recognized microscopically; green and purple sulfur bacteria; large, colorless filamentous sulfur bacteria; and large, colorless, nonfilamentous sulfur bacteria. The fourth group, consisting of sulfate-reducing bacteria and sulfur-oxidizing bacteria of the genus *Thiobacillus,* cannot be identified by appearance alone.

a. Green and purple sulfur bacteria:

1) Green sulfur bacteria most frequently occur in waters containing H_2S. They are small, ovoid to rod-shaped, nonmotile organisms, generally less than 1 μm in diameter, and with a yellowish-green color in masses. Sulfur globules are seldom if ever deposited within the cells.

2) Purple sulfur bacteria (Figure 918:6) occur in waters containing H_2S. They are large, generally stuffed with sulfur glob-

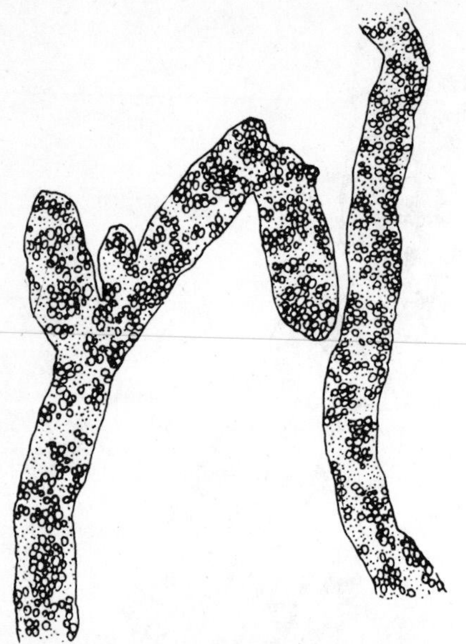

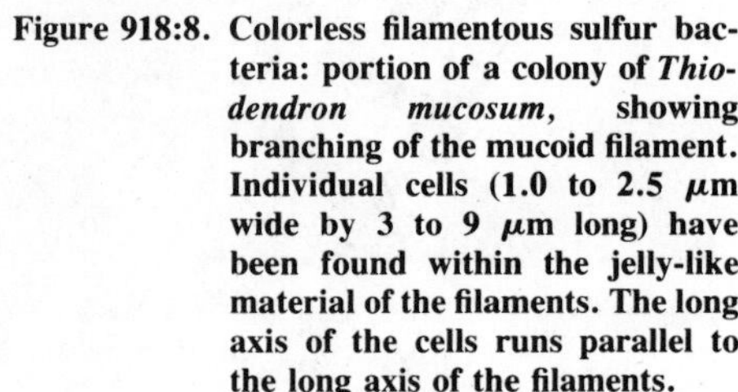

Figure 918:8. Colorless filamentous sulfur bacteria: portion of a colony of *Thiodendron mucosum*, showing branching of the mucoid filament. Individual cells (1.0 to 2.5 μm wide by 3 to 9 μm long) have been found within the jelly-like material of the filaments. The long axis of the cells runs parallel to the long axis of the filaments.

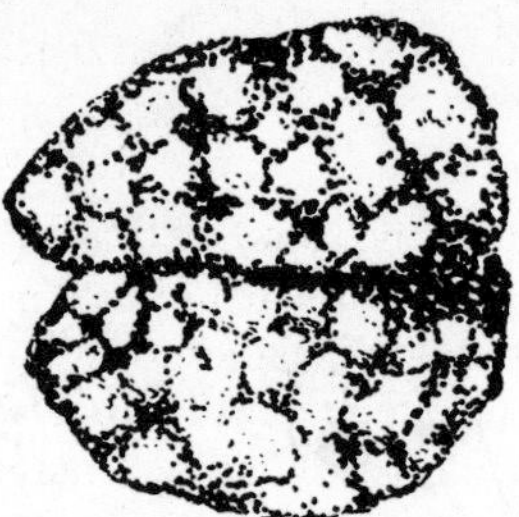

Figure 918:9. Colorless nonfilamentous sulfur bacteria: dividing cell of *Thiovolum majus*, containing sulfur granules. Cells may measure 9 to 17 μm in width by 11 to 18 μm in length and are generally found in nature in a marine littoral zone rich in organic matter and hydrogen sulfide.

ules, and often so intensely pigmented as to make individual cells appear red. Large, dense, highly colored masses easily are detected by the naked eye.

b. Colorless filamentous sulfur bacteria: Colorless filamentous sulfur bacteria (Figures 918:7 and 918:8) occur in waters where both oxygen and H_2S are present. They may form mats with a slightly yellowish-white appearance due to deposition of internal sulfur globules. They generally are large and may be motile with a characteristic gliding movement. Identify by comparing organisms with available photographs.[15–17]

c. Colorless nonfilamentous sulfur bacteria: Colorless, nonfilamentous sulfur bacteria (see Figure 918:9, for example) usually are associated with decaying algae. They are extremely motile, ovoid to rod-shaped with sulfur globules and possible calcium carbonate deposits. They generally are very large.

d. Colorless small sulfur bacteria and sulfate-reducing bacteria: The small single-celled bacteria, *Thiobacillus* spp., and the sulfate-reducing bacteria, such as *Desulfovibrio*, cannot be identified by direct microscopic examination. *Thiobacillus* types are small, colorless, motile, and rod-shaped and are found in an environment containing H_2S. Sulfur globules are absent. Identify *Thiobacillus* types, *Desulfovibrio*, or other sulfate-reducing bacteria physiologically.

918 C. Enumeration, Enrichment, ana Isolation of Iron and Sulfur Bacteria (TENTATIVE)

There are no good means of enumerating iron and sulfur bacteria other than the sulfate-reducing bacteria and the thiobacilli. Laboratory cultivation and isolation of pure cultures is difficult and successful isolation is uncertain. This is especially true of attempts to isolate filamentous bacteria from activated sludge or other sources where many different bacterial types are present.

1. Iron Bacteria

a. Sphaerotilus-Leptothrix:

1) Iron bacteria, especially those belonging to the *Sphaerotilus-Leptothrix* group, thrive in media too dilute to support the proliferation of more rapidly growing organisms. One medium[18] is partially selective for *Sphaerotilus* (BOD dilution water supplemented with 100 mg/L sodium lactate). Dispense 50 mL of this medium in French square bottles and autoclave at 69 kPa for 15 min. To inoculate sample add 25-mL portions of stream water or 1-, 5-, and 10-mL portions of settled wastewater or process liquor to duplicate bottles of medium. Incubate at 22 to 25 C for 5 days and observe for filamentous growth. Isolate pure cultures by picking a filament from the BOD-lactate broth and streaking on 0.05% meat extract agar. After incubating for 24 hr at 25 C, pick typical curling filaments with the aid of a dissecting microscope and transfer to casitone-glycerol-yeast autolysate (CGY) broth (refer to Media Specifications, Section 905C). If a pellicle with no underlying turbidity develops in 2 to 3 days, transfer a filament to a CGY agar slant, incubate at 25 C until growth is visible, and store in a refrigerator.

A detailed key for identifying filamentous microorganisms in complex mixtures such as wastewater and activated sludge is available.[19]

2) Isolation and maintenance media (see Media Specifications, Section 905C) have proven quite successful for identifying various groups of filamentous organisms, including iron bacteria.[20] Prepare agar slants of these media and aseptically pipet 3 mL sterile tap water onto surface of slants. Inoculate tubes and incubate at room temperature until turbid growth has developed in liquid layer. The cells will remain viable for 3 months in the refrigerator.

3) Another good maintenance medium for cultivating the *Sphaerotilus* group is CGY (see Media Specifications, Section 905C).[9]

4) *Leptothrix (Sphaerotilus discophorous)* can be distinguished from *Sphaerotilus natans* by its ability to oxidize manganous ion. Use Mn-agar as the differential medium (see Media Specifications, 905C).[21]

b. Thiobacillus ferrooxidans: Although this organism also is a sulfur-oxidizing bacterium,[22,23] its main importance has been in acid mine drainage. A medium suitable for enumeration of the MPN is available (see Media Specifications, 905C).[24] Some oxidation of iron occurs during sterilization but the loss of ferrous iron is not appreciable. The medium has a precipitate (probably ferrous and ferric phosphates), is opalescent and green, has a pH of 3.0 to 3.6, and contains 9,000 mg/L ferrous iron. Growth of the organism is manifested by a decrease in pH and an increase in concentration of oxidized iron. With practice and use of uninoculated controls, an increase of deep orange-brown color can be seen in positive enrichment tubes or flasks as compared to negative ones. Shake test-tube dilutions daily because these organisms are highly aerobic.

c. Gallionella ferruginea: For cultivation of this organism use ferrous sulfide agar[4,6] (see Media Specifications, Section 905C). Inoculate tubes with a drop of suspension of a suspected *Gallionella* deposit. Growth at room temperature usually occurs in 18 to 36 hr and appears as a white deposit on sides of test tube. The ring of colonies occurring at a certain level reflects a balance between upward diffusion of ferrous ions and downward diffusion of oxygen molecules. Supplement ferrous sulfide agar with formalin[25] to use it for isolation of pure cultures.

d. Other iron bacteria: An acid-tolerant (pH 3.5 to 5.0) filamentous iron-oxidizing *Metallogenium* has been isolated with a medium[26] containing $(NH_4)_2SO_4$, 0.1%; $CaCO_3$, 0.01%; $MgSO_4$, 0.02%; K_2HPO_4, 0.001%; potassium acid phthalate, 0.4%; and 250 mg/L ferrous iron from an acidified $FeSO_4 \cdot 7H_2O$, solution. Add 0.4% formalin to 100 mL of the isolating medium in a 250-mL erlenmeyer flask.

For heterotrophic iron-precipitating bacteria[27] use a ferric ammonium citrate medium consisting of: $(NH_4)_2SO_4$, 0.5 g/L; $NaNO_3$, 0.5 g/L; K_2HPO_4, 0.5 g/L; $MgSO_4 \cdot 7H_2O$, 0.5 g/L; and ferric ammonium citrate, 10.0 g/L. Adjust pH to 6.6 to 6.8 and sterilize. To make the medium solid add 15 g/L agar.

Alternatively grow iron bacteria by combining 20 mL liquid broth medium, 10 mL raw water or inoculum, and 3 g iron oxide. Incubate 48 to 72 hr at 25 C on a wrist-action shaker to produce good, visible growth.

2. Sulfur Bacteria

a. Sulfate-reducing bacteria:

1) To enumerate sulfate-reducing bacteria such as *Desulfovibrio,* use medium described by Lewis[28] (see Media Specifications, Section 905C). Inoculate tubes and fill completely with sterile medium to create anaerobic conditions. For comparative purposes, incubate one or two uninoculated controls with each set of inoculated tubes. To sample volumes greater than 10 mL, pass water sample through a 0.45-μm membrane filter and transfer filter to screw-cap test tube with medium. If sulfate-reducing bacteria are present, tubes will show blackening within 4 to 21 days of incubation at 20 to 30 C.

2) An agar medium suitable for growth and enumeration of sulfate-reducing bacteria also is available.[29] The medium consists of trypticase soy agar, 4.0%, fortified with additional agar, 0.5%, to which is added 60% sodium lactate (0.4% v/v), hydrated magnesium sulfate, 0.2%, and ferrous ammonium sulfate, 0.2%. Adjust pH to 7.2 to 7.4 and sterilize. Medium should be clear and free from precipitate. Inoculate all plates within 1 or at most 4 hr after agar hardens to prevent saturation with oxygen. To prevent moisture condensation on petri dish covers, replace covers with sterile absorbent tops until 10 to 15 min after agar hardens. Place uninverted plates in desiccator or Brewer jars and replace atmosphere with tank hydrogen or nitrogen by successive evacuation and gas replacement. Incubate at room temperature (21 to 24 C) or at 28 to 30 C, the optimum temperature for these organisms. Growth and blackening around the colonies is typical of sulfate-reducing bacteria and may occur at any time between 2 and 21 days, although the usual time is 2 to 7 days.

3) Media suitable for enumeration of various species of sulfate-reducing bacteria have been evaluated.[30,31]

b. Photosynthetic purple and green sulfur bacteria: Because these organisms are so specialized and rarely cause problems in water and wastewater treatment processes, methods for their isolation and enumeration are not included here. An excellent review is available.[32]

c. Thiobacillus spp.: The growth and physiology of different species of the single-celled sulfur-oxidizing bacteria of the genus *Thiobacillus* have been evaluated carefully.[33,34] Media[35] suitable for enumeration of *Thiobacillus thionarus* and *Thiobacillus thiooxidans* by an MPN technic are listed in Media Specifications, Section 905C. Inoculate medium and incubate for 4 to 5 days at 25 to 30 C. Growth of thiobacilli produces elemental sulfur, which sinks to the bottom with a coincident decrease in pH and turbidity of the medium. Chemical tests for formation of sulfate are necessary to confirm presence of *Thiobacillus*.

d. Beggiatoa: Methods for enrichment and isolation of *Beggiatoa* depend on use of a hay extract.[16,36] Extract dried hay at 100 C in large volumes of water, changing water three times during extraction. The final wash water has an amber color. After draining extracted hay, dry it on trays at 37 C. Prepare enrichment medium by suspending 8 g extracted and dried hay in 1 L tap water; dispense 70 mL per 125-mL erlenmeyer flask and sterilize by autoclaving. Inoculate 5-mL portions of mud containing decaying plant material from small ponds, lakes, and streams and incubate at room temperature. In successful enrichments a strong odor of H_2S is noticeable in the flasks and *Beggiatoa* growth appears within 10 days as a white film on surface of medium and submerged upper walls of flask. To isolate organism, wash portions of white surface film several times in sterile tap water, place on surface of agar plates (1% agar and 0.2% beef extract), and incubate at 28 C. From those plates in which filaments have migrated to the periphery of the agar surface and away from contaminants, cut out agar blocks containing single, isolated filaments of *Beggiatoa* and place, filament side down, on fresh plates of same medium. Incubate again at 28 C. After growth has progressed to the extent that isolated single filaments are present, repeat isolation procedure.

An inorganic medium for *Beggiatoa* also has been described.[37]

918 D. References

1. LUESCHOW, L.A. & K.M. MACKENTHUN. 1962. Detection and enumeration of iron bacteria in municipal water supplies. *J. Amer. Water Works Ass.* 54:751.
2. STARKEY, R.L. 1945. Transformations of iron by bacteria in water. *J. Amer. Water Works Ass.* 37:963.
3. STOKES, J.L. 1954. Studies on the filamentous sheathed iron bacterium *Sphaerotilus natans*. *J. Bacteriol.* 67:278.
4. KUCERA, S. & R.S. WOLFE. 1957. A selective enrichment method for *Gallionella ferruginea*. *J. Bacteriol.* 74:344.
5. WAITZ, S. & J.B. LACKEY. 1958. Morphological and biochemical studies on the organism *Sphaerotilus natans*. *Quart. J. Fla. Acad. Sci.* 21:335.
6. WOLFE, R.S. 1958. Cultivation, morphology, and classification of the iron bacteria. *J. Amer. Water Works Ass.* 50:1241.
7. WOLFE, R.S. 1960. Observations and studies of *Crenothrix polyspora*. *J. Amer. Water Works Ass.* 52:915.
8. WOLFE, R.S. 1960. Microbial concentration of iron and manganese in water with low concentrations of these elements. *J. Amer. Water Works Ass.* 52:1335.
9. DONDERO, N.C., R.A. PHILIPS & H. HEUKELEKIAN. 1961. Isolation and preservation of cultures of *Sphaerotilus*. *Appl. Microbiol.* 9:219.
10. MULDER, E.G. 1964. Iron bacteria, particularly those of the *Sphaerotilus-Leptothrix* group, and industrial problems. *J. Appl. Bacteriol.* 27:151.
11. DRAKE, C.H. 1965. Occurrence of *Siderocapsa treubii* in certain waters of the Niederrhein. *Gewässer Abwässer* 39/40:41.
12. BUCHANAN, R.E. & N.E. GIBBONS, eds. 1974. Bergey's Manual of Determinative

Bacteriology, 8th ed. Williams & Wilkens Co., Baltimore, Md.

13. EDMONDSON, W.T., ed. 1959. Ward & Whipple's Fresh Water Biology, 2nd ed. John Wiley & Sons, New York, N.Y.
14. SKERMAN, V.B.D. 1967. A Guide to the Identification of the Genera of Bacteria, 2nd ed. Williams & Wilkens Co., Baltimore, Md.
15. LACKEY, J.B. & E.W. LACKEY. 1961. The habitat and description of a new genus of sulfur bacterium. *J. Gen. Microbiol.* 26:28.
16. FAUST, L. & R.S. WOLFE. 1961. Enrichment and cultivation of *Beggiatoa alba*. *J. Bacteriol.* 81:99.
17. MORGAN, G.B. & J.B. LACKEY. 1965. Ecology of a sulfuretum in a semitropical environment. *Z. Allg. Mikrobiol.* 5:237.
18. ARMBRUSTER, E.H. 1969. Improved technique for isolation and identification of *Sphaerotilus*. *Appl. Microbiol.* 17:320.
19. FARQUHAR, G.J. & W.C. BOYLE. 1971. Identification of filamentous microorganisms in activated sludge. *J. Water Pollut. Control Fed.* 43:604.
20. VANVEEN, W.L. 1973. Bacteriology of activated sludge, in particular the filamentous bacteria. *Antonie van Leeuwenhoek* (Holland) 39:189.
21. MULDER, E.G. & W.L. VANVEEN. 1963. Investigations on the *Sphaerotilus-Leptothrix* group. *Antonie van Leeuwenhoek* (Holland) 29:121.
22. UNZ, R.F. & D.G. LUNDGREN. 1961. A comparative nutritional study of three chemoautotrophic bacteria: *Ferrobacillus ferrooxidans*, *Thiobacillus ferrooxidans*, and *Thiobacillus thiooxidans*. *Soil Sci.* 92:302.
23. MCGORAN, C.J.M., D.W. DUNCAN & C.C. WALDEN. 1969. Growth of *Thiobacillus ferrooxidans* on various substrates. *Can. J. Microbiol.* 15:135.
24. SILVERMAN, M.P. & D.C. LUNDGREN. 1959. Studies on the chemoautotrophic iron bacterium *Ferrobacillus ferrooxidans*. *J. Bacteriol.* 77:642.
25. NUNLEY, J.W. & N.R. KRIEG. 1968. Isolation of *Gallionella ferruginea* by use of formalin. *Can. J. Microbiol.* 14:385.
26. WALSH, F. & R. MITCHELL. 1972. A pH dependent succession of iron bacteria. *Environ. Sci. Technol.* 6:809.
27. CLARK, F.M., R.M. SCOTT & E. BONE. 1967. Heterotrophic, iron-precipitating bacteria. *J. Amer. Water Works Ass.* 59:1036.
28. LEWIS, R.F. 1965. Control of sulfate-reducing bacteria. *J. Amer. Water Works Ass.* 57:1011.
29. IVERSON, W.P. 1966. Growth of *Desulfovibrio* on the surface of agar media. *Appl. Microbiol.* 14:529.
30. LECHEVALIER, H.A. & D. PRAMER. 1970. The Microbes, 1st ed. J.B. Lippincott Co., Philadelphia, Pa.
31. MARA, D.D. & D.J.A. WILLIAMS. 1970. The evaluation of media used to enumerate sulphate reducing bacteria. *J. Appl. Bacteriol.* 33:543.
32. PFENNIG, N. 1967. Photosynthetic bacteria. *Annu. Rev. Microbiol.* 21:285.
33. HUTCHINSON, M., K.I. JOHNSTONE & D. WHITE. 1965. The taxonomy of certain thiobacilli. *J. Gen. Microbiol.* 41:357.
34. HUTCHINSON, M., K.I. JOHNSTONE & D. WHITE. 1966. Taxonomy of the acidophilic thiobacilli. *J. Gen. Microbiol.* 44:373.
35. STARKEY, R.L. 1937. Formation of sulfide by some sulfur bacteria. *J. Bacteriol.* 33:545.
36. SCOTTEN, H.L. & J.L. STOKES. 1962. Isolation and properties of *Beggiatoa*. *Arch. Mikrobiol.* 42:353.
37. KOWALLIK, U. & E.G. PRINGSHEIM. 1966. The oxidation of hydrogen sulfide by beggiatoa. *Amer. J. Bot.* 53:801.

919 RAPID DETECTION METHODS

During emergencies involving water treatment plant failure, line breaks in a distribution network, or other disruptions to water supply caused by natural or man-made disasters, there is urgent need for methods that permit rapid assessment of the sanitary quality of water. Ideally, these procedures would be reliable and have sensitivity levels equal to those of the standard tests routinely used. However, sensitivity of a rapid test may be compromised because the bacterial limit sought may be below the minimal bacterial concentration essential to rapid detection. Rapid tests fall into two categories, those involving modified conventional procedures and those requiring special instrumentation and materials.

919 A. Seven-Hour Fecal Coliform Test (TENTATIVE)

This method[1,2] is similar to the fecal coliform membrane filter procedure (see Section 909C) but uses a different medium and incubation temperature to yield results in 7 hr that generally are comparable to those obtained by the standard fecal coliform method.

To make the test, filter an appropriate sample volume through a membrane filter, place filter on the surface of a plate containing M-7 Hr FC agar medium (see Section 905C), and incubate at 41.5 C for 7 hr. Fecal coliform colonies are yellow (indicative of lactose fermentation).

919 B. Special Technics (TENTATIVE)

Special rapid technics are summarized in Table 919:I. Most are not sensitive enough for potable water quality measurement or are not specific. They may be useful in monitoring wastewater effluents but require reagents not generally available, are tedious, and require special handling or incubation schemes incompatible with most water laboratory schedules. None is suitable for routine use but they may be used as research tools. For these reasons, only the adenosine triphosphate (ATP) procedure (the firefly bioluminescence system) to measure total microbial density and a rapid fecal coliform procedure that uses a ^{14}C-labeled substrate can be recommended.

In using these procedures correlate initial concentration of bacteria with ATP concentration by extracting ATP from decimal dilutions of a bacterial suspension, or for the ^{14}C radiometric method, standardize by determining the $^{14}CO_2$ released by known concentrations of fecal coliform organisms in natural samples, not pure cultures. Determine initial fecal coliform density by the procedures of Sections 908C or 909C.

1. Bioluminescence Test (Total Viable Microbial Measurement)

The firefly luciferase test for ATP in living cells is based on the reaction between the luciferase enzyme, luciferin (enzyme

TABLE 919:I. SPECIAL RAPID TECHNICS

Microbial Group	Rapid Method	Test Time *hr*	Sensitivity *cells/mL*	Reference
Nonspecific microflora	Bioluminescence	1	100,000	3, 4, 5
	Chemiluminescence	1	500,000	12, 13
	Impedance	3–12	100,000	14, 15
Fecal coliforms	Radiometric	4–5	2–20	6, 7, 8, 9, 10, 11
	Glutamate decarboxylase	10.5	500,000	18, 19
	Electrochemical	1–7	1,000,000	16, 17
	Impedance	7	100,000	14, 15
Gram-negative bacteria	Limulus assay	2	—	20, 21
	Fluorescent antibody	2–3	—	22, 23, 24, 25, 26

substrate), magnesium ions, and ATP. Light is emitted during the reaction and can be measured quantitatively and correlated with the quantity of ATP extracted from known numbers of bacteria. When all reactants except ATP are in excess, ATP is the limiting factor. Addition of ATP drives the reactions, producing a pulse of light that is proportional to the ATP concentration.

The assay is completed in less than 1 hr.[5] For monitoring microbial populations in water, the ATP assay is limited primarily by the need to concentrate bacteria from the sample to achieve the minimum ATP sensitivity level, which is 10^5 cells/mL. When combined with membrane filtration of a 1-L sample, ATP assay can provide the sensitivity level needed.

2. Radiometric Detection (Fecal Coliforms)

In this test, $^{14}CO_2$ is released from a ^{14}C-labeled substrate.[11] The technic permits presumptive detection of as few as 2 to 20 fecal coliform bacteria in 4.5 hr. The test uses M-FC broth, uniformly labeled ^{14}C-mannitol, and two-temperature incubation; 2 hr at 35 C followed by 2.5 hr at 44.5 C for fecal coliform specificity. Add labeled substrate at start of 44.5 C incubation. Use membrane filtration to concentrate organisms from sample and place membrane filter on M-FC broth. Except for the use of the ^{14}C-mannitol substrate and liquid scintillation spectrometry to count the activity of the $^{14}CO_2$ released by the fecal coliforms, this procedure is similar to those given in Section 909.

919 C. References

1. VAN DONSEL, D.J., R.M. TWEDT, & E.E. GELDREICH. 1969. Optimum temperature for quantitation of fecal coliforms in seven hours on the membrane filter. *Bacteriol. Proc.* Abs. No. G46, p. 25.
2. REASONER, D.J., J.C. BLANNON & E.E. GELDREICH. 1977. The seven hour fecal coliform test. *In* Proc. Amer. Water Works Ass. Water Quality Technology Conf., Dec. 5–8, 1976, San Diego, Calif.; 1979. Rapid seven hour fecal coliform test. *Appl. Environ. Microbiol.* 38:229.
3. CHAPPELLE, E.W. & G.V. LEVIN. 1968. The use of the firefly bioluminescent assay

for the rapid detection and counting of bacteria. *Biochem. Med.* 2:41.

4. McElroy, W.D., H.H. Seliger & E.H. White. 1969. Mechanism of bioluminescence, chemiluminescence and enzyme function in the oxidation of firefly luciferin. *Photochem. Photobiol.* 10:153.
5. Chappelle, E.W. & G.L. Picciolo. 1975. Laboratory Procedures Manual for the Firefly Luciferase Assay for Adenosine Triphosphate (ATP). NASA GSFC Document X-726-75-1.
6. Levin, G.V., V.R. Harrison & W.C. Hess. 1956. Preliminary report on a one-hour presumptive test for coliform organisms. *J. Amer. Water Works Ass.* 48:75.
7. Levin, G.V., V.L. Strauss & W.C. Hess. 1961. Rapid coliform organisms determination with ^{14}C. *J. Water Pollut. Control Fed.* 33:1021.
8. Scott, R.M., D. Seiz & H.J. Shaughnessy. 1964. I. Rapid carbon^{14} test for coliform bacteria in water. *Amer. J. Pub. Health* 54:827.
9. Scott, R.M., D. Seiz & H.J. Shaughnessy. 1964. II. Rapid carbon^{14} test for sewage bacteria. *Amer. J. Pub. Health* 54:834.
10. Bachrach, U. & Z. Bachrach. 1974. Radiometric method for the detection of coliform organisms in water. *Appl. Microbiol.* 28:169.
11. Reasoner, D.J. & E.E. Geldreich. (In press). Rapid detection of water-borne fecal coliforms by $^{14}CO_2$ release. *In* Proc. Int. Conf. on Mechanized Microbiology, Sept. 10–12, 1975.
12. Seitz, W.R. & M.P. Neary. 1974. Chemiluminescence and bioluminescence. *Anal. Chem.* 46:188A.
13. Oleniaz, W.S., M.A. Pisano, M.H. Rosenfeld & R.L. Elgart. 1968. Chemiluminescent method for detecting microorganisms in water. *Environ. Sci. Technol.* 2:1030.
14. Cady, P. 1973. Rapid Automated Bacterial Identification by Impedance Measurement. Symp. Rapid Methods Autom. Microbiol., Stockholm, John Wiley and Sons, N.Y.
15. Wheeler, T.G. & M.C. Goldschmidt. 1975. Determination of bacterial cell concentrations by electrical measurements. *J. Clin. Microbiol.* 1:25.
16. Wilkins, J.R., G.E. Stoner & E.H. Boykin. 1974. Microbial detection method based on sensing molecular hydrogen. *Appl. Microbiol.* 27:947.
17. Wilkins, J.R. & E.H. Boykin. 1976. Analytical notes—electrochemical method for early detection of monitoring of coliforms. *J. Amer. Water Works Ass.* 68:257.
18. Moran, J.W. & L.D. Witter. 1976. An automated rapid test for *Escherichia coli* in milk. *J. Food Sci.* 41:165.
19. Moran, J.W. & L.D. Witter. 1976. An automated rapid method for measuring fecal pollution. *Water Sewage Works* 123:66.
20. DiLuzio, N.R. & T.J. Friedmann. 1973. Bacterial endotoxins in the environment. *Nature* 244:49.
21. Jorgensen, J.H., J.C. Lee & H.R. Pahren. 1976. Rapid detection of bacterial endotoxins in drinking water and renovated wastewater. *Appl. Environ. Microbiol.* 32:347.
22. Abshire, R.L. 1976. Detection of enteropathogenic *Escherichia coli* strains in wastewater by fluorescent antibody. *Can. J. Microbiol.* 22:365.
23. Cherry, W.B., B.M. Thomason, J.B. Gladden, N. Holsing & A.M. Murlin. 1975. Detection of Salmonella in foodstuffs, feces, and water by immunofluorescence. *Ann. N.Y. Acad. Sci.* 254:350.
24. Abshire, R.L. & R.K. Guthrie. 1973. Fluorescent antibody techniques as a method for the detection of fecal pollution. *Can. J. Microbiol.* 19:201.
25. Cherry, W.B. & B.M. Thomason. 1969. Fluorescent antibody techniques for Salmonella and other enteric pathogens. *Pub. Health Rep.* 84:887.
26. Thomason, B.M. & J.G. Wells. 1971. Preparation and testing of polyvalent conjugates for fluorescent-antibody detection of Salmonellae. *Appl. Microbiol.* 22:876.

920 STRESSED ORGANISMS (TENTATIVE)

1. General Discussion

Stressed organisms are organisms that, as a result of adverse environmental exposures, multiply slowly or not at all under standard test conditions. Because of their physiological deficiencies, their presence may not be detected and a water of poor sanitary quality may appear to be satisfactory. Such false negative bacteriological findings may result in an inaccurate definition of water quality, or even worse, lead to the acceptance of a potentially hazardous condition because stressed pathogens may retain their virulence.

Stressed organisms are present in chlorinated effluents, saline waters, and polluted natural waters, especially those polluted by heavy-metal ions or certain organic industrial wastes. Environmental factors causing stress include extremes of temperature and pH, minimal concentrations of critical nutrients, toxic substances, metabolic products, and exposure to solar radiation and disinfectants. Storage and inappropriate sample handling following collection also may produce stressed organisms. Laboratory conditions causing stress include excessive sample storage time, excessive holding time (more than 30 min) of diluted samples before inoculation into growth media, incorrect media formulations, abrupt changes in temperature, incomplete mixing of sample with concentrated medium, and exposure to untempered liquefied agar media.

2. Recovery Enhancement

Specific methods for recovery of stressed organisms cannot be given but some general procedures are recommended. For chlorinated samples, insure that sufficient dechlorinating agent is present in the sample bottle (see Section 906A.2). Collect samples of waters containing heavy-metal ions in a sample bottle containing a chelating agent (see Section 906A.2). Hold sample storage time to a minimum (see Section 906B). Use peptone dilution water rather than buffered water (see Section 905C.1) when preparing dilutions of samples containing heavy-metal ions. After making dilutions, do not hold for more than 30 min before making inoculations.

Resuscitation, the process of revitalizing stressed or injured organisms, is enhanced by inoculating samples and culturing organisms in an enriched, noninhibitory medium. This suggests that for coliform testing the multiple-tube fermentation technic (Section 908) may be preferable. If the membrane filter technic is used (Section 909), use membrane filters with the largest pore diameter consistent with organism retention and resuscitate as detailed in *b* or *c* below. For total coliforms use the enrichment technic (see Section 909A5.*c*).

To enhance recovery of stressed fecal coliform bacteria by the membrane filter procedure use one or more of the following procedures:

a. Deletion of the medium suppressive agent: Eliminate rosolic acid from M-FC medium and incubate cultures at 44.5 C ± 0.2 C for 24 hr. Fecal coliform colonies are intense blue on the modified medium and are distinguished easily from the cream, gray, and pale-green colonies typically produced by nonfecal coliforms.

b. Temperature acclimation: Modify elevated temperature procedure by preincubation of M-FC cultures for 5 hr at 35 C, followed by 18 ± 1 hr at 44.5 C. Use a temperature-programmed incubator to make the change from 35 to 44.5 C after the 5 hr preincubation period to eliminate inconvenience and provide a practical method of analysis.

c. Enrichment-temperature acclima-

tion: Use two-layer agar (M-FC agar with an overlay of lauryl tryptose agar) with a 2-hr incubation at 35 C followed by 22 hr at 44.5 C. Prepare the M-FC agar plate in advance but do not add the lauryl tryptose agar overlay more than 1 hr before use.

d. Verification of stressed fecal coliform bacteria: Modifications of media and procedures may decrease selectivity and differentiation of fecal coliform colonies. Therefore, before accepting any procedural modifications, verify not less than 10 percent of the blue colonies from a variety of samples. Use lauryl tryptose broth (35 C for 48 hr) with transfer of gas-producing cultures to EC broth (44.5 C for 24 hr). Gas production at 44.5 C confirms the presence of fecal coliforms.

3. Bibliography

CLARK, H.F., E.E. GELDREICH, H.L. JETER & P.W. KABLER. 1951. The membrane filter in sanitary bacteriology. *Pub. Health Rep.* 66:951.

MCKEE, J.E., R.T. MCLAUGHLIN & P. LESGOURGUES. 1958. Application of molecular filter techniques to the bacterial assay of sewage. III. Effects of physical and chemical disinfection. *Sewage Ind. Wastes* 30:245.

MAXCY, R.B. 1970. Non-lethal injury and limitations of recovery of coliform organisms on selective media. *J. Milk Food Technol.* 33:445.

LIN, S.D. 1973. Evaluation of coliform tests for chlorinated secondary effluents. *J. Water Pollut. Control Fed.* 45:498.

BRASWELL, J.R. & A.W. HOADLEY. 1974. Recovery of *Escherichia coli* from chlorinated secondary sewage. *Appl. Microbiol.* 28:328.

LIN, S.D. 1974. Evaluation of fecal streptococci tests for chlorinated secondary effluents. *J. Environ. Eng. Div., Proc. Amer. Soc. Civil. Engr.* 100:253.

STEVENS, A.P., R.J. GRASSO & J.E. DELANEY. 1974. Measurements of fecal coliform in estuarine water. pp. 132–136 *in* D.D. Wilt, ed., Proceedings of the 8th National Shellfish Sanitation Workshop, U.S. Dept. of Health, Education, and Welfare, Washington, D.C.

BISSONNETTE, G.K., J.J. JEZESKI, G.A. MCFETERS & D.G. STUART. 1975. Influence of environmental stress on enumeration of indicator bacteria from natural waters. *Appl. Microbiol.* 29:186.

ROSE, R.E., E.E. GELDREICH & W. LITSKY. 1975. Improved membrane filter method for fecal coliform analysis. *Appl. Microbiol.* 29:532.

LIN, S.D. 1976. Membrane filter method for recovery of fecal coliforms in chlorinated sewage effluents. *Appl. Environ. Microbiol.* 32:547.

BISSONNETTE, G.K., J.J. JEZESKI, G.A. MCFETERS & D.S. STUART. 1977. Evaluation of recovery methods to detect coliforms in water. *Appl. Environ. Microbiol.* 33:590.

GREEN, B.L., E.M. CLAUSEN & W. LITSKY. 1977. Two-temperature membrane filter method for enumerating fecal coliform bacteria from chlorinated effluents. *Appl. Environ. Microbiol.* 33:1259.

PRESSWOOD, W.G. & D. STRONG. 1977. Modification of M-FC medium by eliminating rosolic acid. *Amer. Soc. Microbiol. Abs. Annu. Meeting,* ISSN-0067-2777:272.

STUART, D.S., G.A. MCFETERS & J.E. SCHILLINGER. 1977. Membrane filter technique for quantification of stressed fecal coliforms in the aquatic environment. *Appl. Environ. Microbiol.* 34:42.

PART 1000

BIOLOGICAL EXAMINATION OF WATER

1001 INTRODUCTION

Water quality affects the abundance, species composition and diversity, stability, productivity, and physiological condition of indigenous populations of aquatic organisms. Therefore, an expression of the nature and health of the aquatic communities is an expression of the quality of the water. Biological methods used for measuring water quality include the collection, counting, and identification of aquatic organisms; biomass measurements; measurements of metabolic activity rates; measurements of the toxicity, bioaccumulation, and biomagnification of pollutants; and processing and interpretation of biological data.

Information from these types of measurements may serve one or more of the following purposes:

1. To explain the cause of color and turbidity and the presence of objectionable odors, tastes, and visible particulates in water;

2. To aid in the interpretation of chemical analyses, for example, in relating the presence or absence of certain biological forms to oxygen deficiency or supersaturation in natural waters;

3. To identify the source of a water that is mixing with another water;

4. To explain the clogging of pipes, screens, or filters, and to aid in the design and operation of water and wastewater treatment plants;

5. To determine optimum times for treatment of surface water with algicides and to monitor treatment effectiveness;

6. To determine the effectiveness of drinking water treatment stages and to aid in determining effective chlorine dosage within a water treatment plant as such dosage is related to organic materials in water;

7. To identify the nature, extent, and biological effects of pollution;

8. To indicate the progress of self-purification in bodies of water;

9. To aid in explaining the mechanisms of biological wastewater treatment methods and to serve as an index of the effectiveness of treatment;

10. To aid in determining the condition and effectiveness of unit processes in a wastewater treatment plant;

11. To document short- and long-term variability in water quality as influenced by natural and/or man-induced changes;

12. To provide data on the status of an aquatic system on a regular basis.

The specific nature of a problem and the reasons for collecting samples will dictate which communities of aquatic organisms will be examined and which sampling and analytical technics will be used.

The following communities of aquatic organisms are considered in specific sections that follow:

1. Plankton: A community of plants (phytoplankton) and animals (zooplankton), usually swimming or suspended in water, nonmotile or insufficiently motile to overcome transport by currents. In fresh water they are generally small or microscopic in size; in salt water, larger forms are observed more frequently.

2. Periphyton (Aufwuchs): A community of microscopic plants and animals associated with the surfaces of submersed objects. Some are attached, some move about. Many of the protozoa and other minute invertebrates and algae that are found in the plankton also occur in the periphyton.

3. Macrophyton: The larger plants of all types. They are sometimes attached to the bottom (benthic), sometimes free-

floating, sometimes totally submersed, and sometimes partly emergent. "Higher" types usually have true roots, stems, and leaves; the algae are simpler but may have stem- and leaf-like structures.

4. MACROINVERTEBRATES: The larger invertebrates, defined here as those retained by the US Standard No. 30 sieve. They are generally bottom-dwelling organisms (benthos).

5. FISH: As defined in this publication, only the finned fish.

6. AMPHIBIANS, AQUATIC REPTILES, BIRDS, AND MAMMALS: These vertebrates also may be affected directly or indirectly by spills or other discharges of pollutants and may be useful in monitoring the presence of toxic pollutants or long-term changes in water quality. Discussion of these organisms is not included.

Large numbers of bacteria and fungi are present in the plankton and periphyton and constitute an essential element of the total aquatic ecosystem. Although their interactions with living and dead organic matter profoundly affect the larger aquatic organisms dealt with below, technics for their investigation are not included herein (but see Part 900).

Field observations are indispensable for meaningful biological interpretations, but many biological factors cannot be evaluated directly in the field. These must be examined as field data or field samples within the laboratory. Because the significance of the analytical result depends on the representativeness of the sample taken, attention is given to field methods as well as to associated laboratory procedures. In making biological examinations, field and laboratory personnel often are the same; if they are not, their activities must be coordinated closely.

Before sampling begins, study objectives must be defined clearly. For example, the frequency of a repetitive sampling program may vary from hourly, for a detailed study of diurnal variability, to every third month (quarterly) for a general assessment of seasonal conditions, depending on the objectives. The scope of the study must be adjusted to limitations in personnel, time, and money. An examination of historic data for the study area and a literature search of work by previous investigators should precede the development of a study plan.

Whenever practicable, biologists should collect their own samples. Much of the value of an experienced biologist lies in personal observations of conditions in the field and in the ability to recognize signs of environmental changes as reflected in the various aquatic communities.

Many specialized items of biological collecting equipment may not be available from the usual laboratory supply houses. The American Society of Limnology and Oceanography has compiled a list of manufacturers and distributors of such equipment entitled "Special Publication No. 1, Sources of Limnological and Oceanographic Apparatus and Supplies," which is available on request from the secretary of the society (consult a current issue of *Limnology and Oceanography* for the name and address). The American Association for the Advancement of Science (AAAS) also prepares an equipment/supplier index that is published annually as an issue of *Science*.

The primary orientation of Part 1000 is toward field collection and associated laboratory analyses to aid in determining the status of aquatic communities under existing field conditions and to aid in interpreting the influence of past and present environmental conditions. Many other types of studies may be, and are being, conducted that are oriented more toward laboratory research. Such laboratory studies will develop further basic knowledge of community and/or organism responses under controlled conditions and will aid in predicting effects of future changes in environmental conditions on the aquatic communities. However, such studies are not within the scope of this presentation.

The complex interrelationships existing in an aquatic environment are reflected in the organization of the following sections. Field and laboratory procedures relating to one section also may be appropriate for other sections; consequently, frequent cross-references between sections have been made.

The methods selected are necessary for the appraisal of water quality. Principal emphasis is on methods and equipment, rather than on interpretation or application of results. Preference is given to procedures used in assessing water quality rather than to those used in aquatic resources management.

1002 PLANKTON

1002 A. Introduction

The term "plankton" refers to those microscopic aquatic forms having little or no resistance to currents and living free-floating and suspended in open or pelagic waters. Planktonic plants are referred to as "phytoplankton" and animals as "zooplankton". The phytoplankton (microscopic algae and bacteria) occur as unicellular, colonial, or filamentous forms. Many are photosynthetic and are grazed upon by zooplankton and other aquatic organisms. The zooplankton in fresh water comprise principally protozoans, rotifers, cladocerans, and copepods; in marine waters, a much greater variety of organisms is encountered.

Plankton, particularly phytoplankton, long have been used as indicators of water quality.[1–4] Some species flourish in highly eutrophic waters while others are very sensitive to organic and/or chemical wastes. Phytoplankton reported to be indicators of clean water include *Melosira islandica*,[5] *Cyclotella ocellata*, and species of *Dinobryon*. Species reported to be indicators of polluted water include *Nitzschia palea*, *Microcystis aeruginosa*, and *Aphanizomenon flos-aquae*. The latter two species have been associated with toxic blooms and anoxic conditions. They and other algae have been associated with noxious blooms in polluted waters, creating offensive tastes and odors.[6] As with phytoplankton, the species assemblage of zooplankton in a given area is useful in assessing water quality.

Because of their short life cycles, plankters respond quickly to environmental changes, and hence the standing crop and species composition indicate the quality of water mass in which they are found. Also, because of their small size and often great numbers, they not only influence strongly certain nonbiological aspects of water quality (such as pH, color, taste, and odor), but in a very practical sense, they *are* a part of water quality. Certain taxa often are useful in determining the origin or recent history of a given water mass. Because of their transient nature, and often patchy distribution, however, the reliability or accuracy of plankton as a water quality indicator may be limited. In rivers, their origin can be uncertain and the duration of their exposure to pollutants unknown.

1002 B. Sample Collection

1. General Considerations

Locate sampling stations as near as possible to those selected for chemical and bacteriological sampling to insure maximum correlation of findings. Establish a sufficient number of stations in as many locations as necessary to define adequately the kinds and quantities of plankton in the waters studied. The physical nature of the water (standing, flowing, or tidal) will influence greatly the selection of sampling stations. The use of sampling sites selected by previous investigators usually will assure the availability of historic data that will lead to a better understanding of current results and provide continuity in the study of an area.

In stream and river work, locate stations upstream and downstream from suspected pollution sources and major tributary streams and at appropriate intervals throughout the reach under investigation. If possible, locate stations on both sides of the river because lateral mixing of river water may not occur for great distances downstream. Some effects of pollution on plankton populations may not be apparent for distances downstream as great as several days stream flow. In a similar manner, investigate tributary streams suspected of being polluted but take care in the interpretation of data from a small stream because much of the plankton may be periphytic in origin, arising from scouring of natural substrates by the flowing water. Plankton contributions from adjacent lakes, reservoirs, and backwater areas, as well as soil organisms carried into the stream by runoff, also can influence data interpretation. The depth from which water is discharged from upstream stratified reservoirs also can affect the nature of the plankton.

Because water of rivers and streams usually is well mixed vertically, subsurface sampling, i.e., upper meter or a composite of two or more strata, often is adequate for collection of a representative sample at a given point. There may be problems caused by stratification due to thermal discharges or mixing of warmer or colder waters from tributaries and reservoirs. Always sample in the main channel of a river and avoid sloughs, inlets, or backwater areas that reflect local habitats rather than river conditions. In rivers that are mixed vertically and horizontally, measure plankton populations by examining periodic samples collected at midstream 0.5 to 1 m below the surface.

If it can be determined or correctly assumed that the plankton distribution is uniform and normal, use a scheme of random sampling to accomodate statistical testing. Include both random selection of sampling sites and transects as well as the random collection of samples at each selected site. On the other hand, if it is known or assumed that plankton distribution is variable or patchy, include additional sampling sites, collect composite samples, and increase sample replication. Use appropriate statistical tests to determine population variability.

In sampling a lake or reservoir use a grid network or transect lines in combination with random procedures. Take a sufficient number of samples to make the data meaningful. Sample a circular lake basin at strategic points along a minimum of two perpendicular transects extending from shore to shore; include the deepest point in the basin. Sample a long, narrow basin at several points along a minimum of three regularly spaced parallel transects that are perpendicular to the long axis of the basin, with the first near the inlet and the last near the outlet. Sample a large bay along several parallel transects originating near shore and extending to the lake proper.

Because many samples are required to appraise completely the plankton assemblage, it may be necessary to restrict sampling to strategic points, such as the vicinity of water intakes and discharges, constrictions within the water body, and major bays that may influence the main basin.

In lakes, reservoirs, and estuaries where plankton populations can vary with depth, collect samples from all major depth zones or water masses. The sampling depths will be determined by the water depth at the station, the depth of the thermocline or an isohaline, or other factors. In shallow areas of 2 to 3 m depth, subsurface samples collected at 0.5 to 1 m usually are adequate. In deeper areas, collect samples at regular depth intervals. Sample estuarine plankton at regular intervals from the surface to the bottom. Collect offshore marine samples at intervals of 3 to 6 m or more throughout the euphotic zone, and to the bottom if zooplankton are to be included. The compromise sampling depths for marine waters may be an arbitrary depth between the thermocline and the euphotic zone, above and below an isohaline, or other boundary.

Samples usually are referred to as "surface" or "depth" (subsurface) samples. The latter are samples taken from some stated depth, whereas surface samples may be interpreted as samples collected as near the water surface as possible. A "skimmed" sample of the surface film can be revealing, but ordinarily do not include a disproportionate quantity of surface film in a surface sample because plankton often are trapped on top or at the surface film together with pollen, dust, and other detritus. Various methods have been used for sampling surface organisms. This microlayer has been defined and sampling methods have been compared.[7]

The frequency of sampling is dictated by the purpose and scope of the study as well as the range of seasonal fluctuations, the immediate meteorological conditions, adequacy of equipment, and availability of personnel. Select a sampling frequency at some interval shorter than community turnover time. This requires consideration of life cycle length, competition, predation, flushing, and current displacement. Frequent plankton sampling is desirable because of normal temporal variability and migratory character of the plankton community. Daily vertical migrations occur in response to sunlight and random horizontal migrations or drifts are produced by winds, shifting currents, and tides. Ideally, collect daily samples and, when possible, sample at different times during the day and at different depths. When this is not possible, weekly, biweekly, monthly, or even quarterly sampling still may be useful for determining major population changes.

In tidal areas of fresh-water streams, in saline reaches of tributaries and estuaries, and in marine waters, collect plankton in accordance with tidal oscillations at each preselected depth during the flood and ebb tides for more than one tidal cycle. Data derived from these waters frequently are most meaningful when the samples have been collected near the end or beginning of both the flood and the ebb tides, but carry out some sampling at all tidal stages.

2. Sampling Procedures

Once sampling locations, depths, and frequency have been determined, prepare for field sampling. Label sample containers with sufficient information to avoid confusion or error. On the label indicate date, sampling station, study area (river, lake, reservoir), type of sample, and depth. Use waterproof labels. When possible, enclose collection vessels in a protective container to avoid breakage. If samples are to be preserved immediately after collection, add preservative to container before sampling. Sample size depends on type and number of determinations to be made; the number of replicates depends on the statistical design of the study and the statistical analyses selected for data interpretation. Always design a study around an objective with a statistical approach rather than fit statistical analyses to data already collected.

In a field record book note sample location, depth, type, time, meteorological conditions, turbidity, water temperature, and other significant observations. These field data are invaluable when analytical data are interpreted and often help to explain unusual changes caused by the variable character of the aquatic environment. Collect coincident samples for chemical analyses to help define environmental variations having a potential effect on plankton.

3. Phytoplankton

In oligotrophic waters or where phytoplankton densities are expected to be low collect a 6-L sample. For richer, eutrophic waters collect a smaller sample of 1 to 2 L.

For qualitative and quantitative evaluations collect whole (unfiltered and unstrained) water samples with a device consisting of a cylindrical tube with stoppers at each end and a closing device. Lower the open sampler to the desired depth and close by dropping a weight, called a messenger, which slides down the supporting wire or cord and trips the closing mechanism. The most commonly used samplers that operate on this principle are the Kemmerer,[8] Van Dorn[9] (Figure 1002:1), Niskin, and Nansen samplers. The Kemmerer and Van Dorn samplers are similar except for the closure mechanisms. The Nansen reversing water bottle[10] is used for sampling at greater depths.

It may be useful to divide plankton on the basis of size regardless of the type of collection equipment used. "Net plankton" usually is defined as plankton with a diameter of 60 μm or more. Organisms

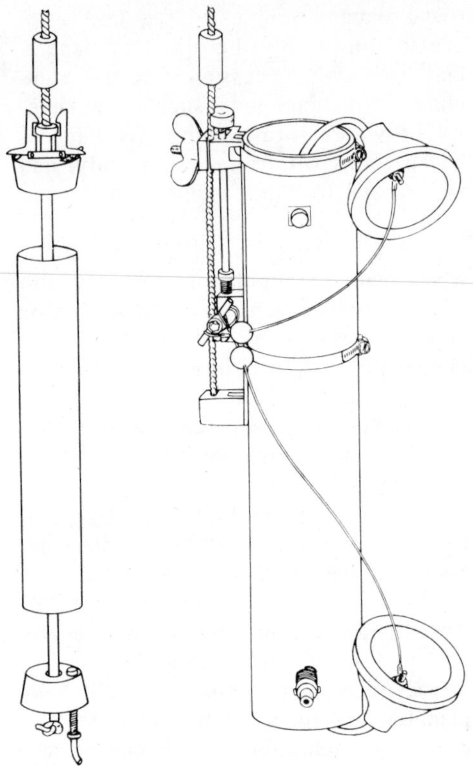

Figure 1002:1. Structural features of common water samplers, Kemmerer (left) and Van Dorn (right).

smaller than 60 μm in diameter are termed "nannoplankton" and those smaller than 10 μm in diameter are called "micro-" or "ultraplankton". In using such a classification system describe it carefully or reference it in a report. Because the Kemmerer and Van Dorn samplers collect whole water samples, both net and nannoplankton are collected. Typically, for net plankton use a No. 20 net with mesh openings of 76 to 80 μm (silk bolting cloth or plastic*) and for nannoplankton a No. 25 net.

The Van Dorn usually is the preferred sampler because its design offers no inhibition to free flow of water through the cylin-

*Nitex or equivalent.

der. Also, the samplers can be cast in a series on a single line for simultaneous sampling at multiple depths with the use of auxiliary messengers. Because the triggering devices of these samplers are very sensitive, avoid rough handling. Always lower the sampler into the water; do not drop. The capacity of Kemmerer and Van Dorn samplers varies from 0.5 to 5 L or more. Polyethylene or polyvinyl chloride sampling devices are preferred to metal samplers because the latter liberate metallic ions that may contaminate the sample. Use polyethylene or glass sample storage bottles. Metallic ion contamination can lead to significant errors when algal assays or productivity measurements are made.

Sampling phytoplankton with nets provides reliable data on forms that cannot pass through the net mesh. Because of selectivity of the mesh size, the smaller plankton (nannoplankton), which may contribute as much as 60% of the total biomass, may not be collected. Qualitative data typically are collected with plankton nets. If the net is equipped with an accurate flow meter, quantitative data for total cell count, biovolume, biomass, and chlorophyll can be collected. As net openings become clogged with large-bodied forms there is increased retention of nanno- and ultraplankton, less water is sampled, and the overall sampling error is increased. Under these conditions plankton cell damage may occur. Various types of plankton nets are shown in Figures 1002:2 and 1002:3.

For primary productivity and other more quantitative determinations, use bottle samplers to collect all organisms present. For greater speed of collection and for obtaining large quantities of organisms, use a pump. Lower a long, weighted hose, attached to a suction pump, to the desired depth, and pump water to the surface. The water may be pumped through a net to collect net plankton or it may be taken whole for nanno- or ultraplankton analyses. The pump is advantageous because it can supply a homogeneous sample from a given depth or an integrated sample from the surface to a particular depth, but consider possible damage to organisms by the pump impeller. Drawing sample from the line before the impeller will protect against organism damage.

For shallow waters use the Jenkins surface mud sampler[11] or one of the bottle samplers modified so that it is held horizontally (Figure 1002:1).[12] Like many other samplers mentioned, the water core sampler[13] is applicable to phytoplankton and zooplankton sampling.

If live samples are to be examined, fill the containers only partially to reduce inhibition of metabolic activities. Store in a portable refrigerator or ice chest in the dark, or preferably, hold at ambient temperature to avoid shocking the organisms. Examine specimens as soon after collection as possible.

If it is impossible to examine living material, preserve the sample. For a sample that will be preserved, fill the container completely. The most universally used phytoplankton preservative is formalin. For delicate forms such as naked flagellates, Lugol's solution is more suitable.

Formalin: To preserve samples with formalin, add 40 mL buffered formalin to 1 L of sample immediately after collection.

Lugol's solution: To preserve samples with Lugol's solution add 1 mL Lugol's solution to 100 mL sample and store in the dark. Prepare Lugol's solution by dissolving 60 g potassium iodide (KI) and 40 g iodine crystals in 1 L distilled water.

Merthiolate: To preserve samples with merthiolate add 36 mL merthiolate solution to 1 L of sample and store in the dark. Prepare merthiolate solution by dissolving 1.0 g merthiolate, 1.5 g sodium borate ($Na_2B_2O_4$), and 1.0 mL Lugol's solution in 1 L distilled water. Merthiolate-preserved samples are not sterile, but can be

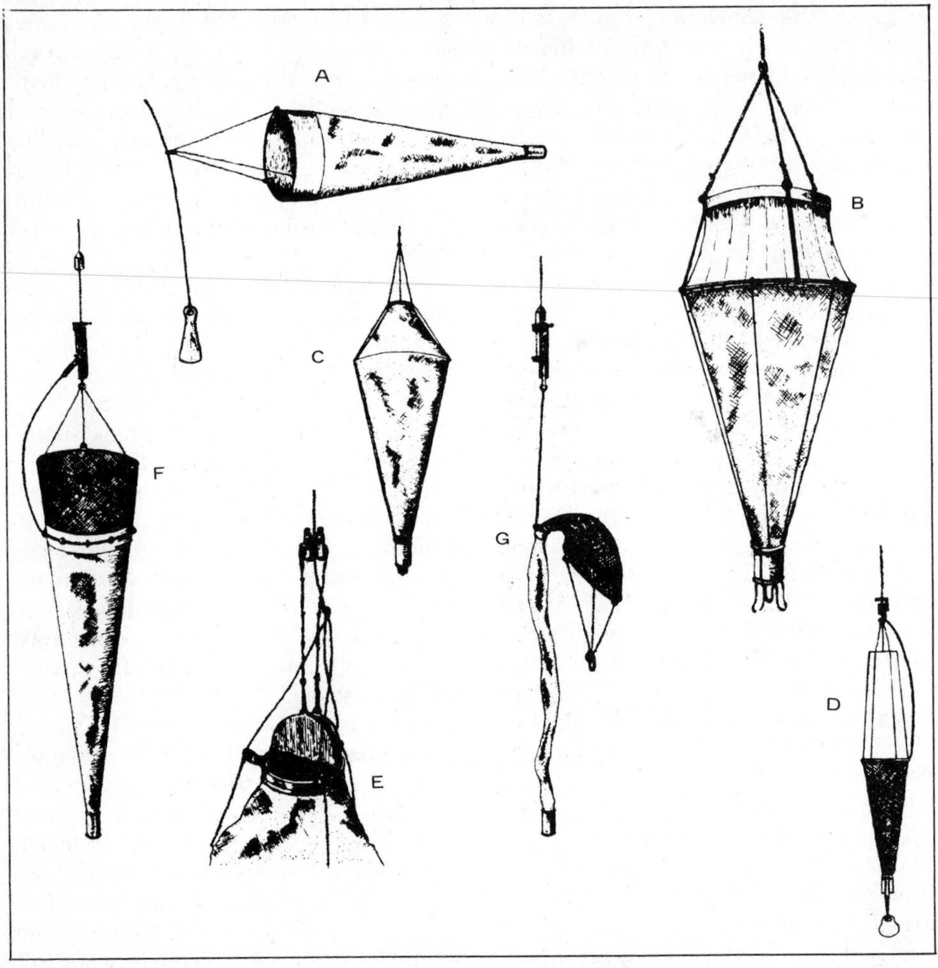

Figure 1002:2. Plankton sampling nets. (A) Simple conical tow-net; (B) Hensen net; (C) Apstein net; (D) Juday net; (E) Apstein net with semicircular closing lids; (F) Nansen closing net, open; (G) Nansen closing net, closed. Source: TRANTER, D.J., ed. 1968. Reviews on Zooplankton Sampling Methods. UNESCO, Switzerland.

kept effectively for 1 yr, after which time formalin must be added.[14]

"M³" fixative: To preserve samples with M^3 fixative add 20 mL fixative to 1 L of sample and store in the dark. Prepare M^3 by dissolving 5 g KI, 10 g iodine, 50 mL glacial acetic acid, and 250 mL formalin in 1 L distilled water (dissolve the iodide in a small quantity of water to aid in solution of the iodine).

Other commonly used preservatives include 95% alcohol, and 6-3-1 preservative, which is composed of 6 parts water, 3 parts 95% alcohol, and 1 part formalin. A relatively large volume of these solutions is necessary for adequate preservation. Usually use equal volumes of preservative and sample.

To retain color in preserved plankton, store samples in the dark or add 1 mL sat-

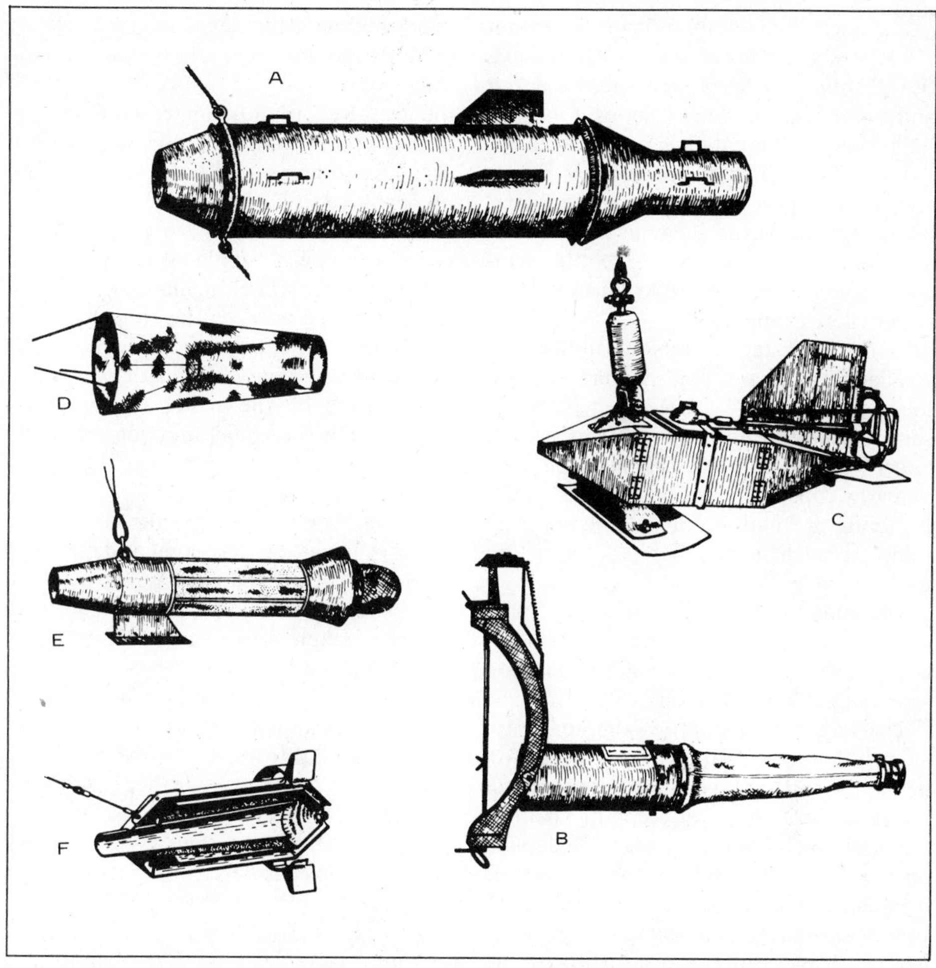

Figure 1002:3. High-speed, oceanographic and/or quantitative zooplankton samplers. (A) Gulf III sampler; (B) Clarke-Bumpus sampler; (C) Hardy continuous plankton recorder; (D) Shread high-speed sampler; (E) Icelandic plankton sampler; (F) Clarke jet net (section). Source: TRANTER, D.J., ed. 1968. Reviews on Zooplankton Sampling Methods. UNESCO, Switzerland.

urated copper sulfate ($CuSO_4$) solution/L.

Most preservatives distort and disrupt certain cells, especially those of delicate forms such as *Euglena, Synura, Chromulina,* and *Mallamonas*. Lugol's iodine solution is recommended for these forms. Become familiar with live specimens and the distortions associated with preservation. A reference collection supplied by a biological supply house or compiled by an experienced co-worker is useful in identifying preserved phytoplankton.

4. Zooplankton

The choice of sampler depends on the kind of study (distribution, productivity) and the body of water being investigated. The spatial distribution of zooplankton in a lake normally is non-uniform. In certain situations, a species can be distributed in a thin, continuous layer at a specific depth and may occur nowhere else, or it may be limited to occasional dense patches.

For collecting small (nanno) zooplankters such as protozoa, small rotifers, and immature microcrustacea, use the bottle samplers described above. Bottle samplers usually are unsuitable for collecting larger (net) zooplankton (such as the mature microcrustacea) which, unlike the smaller forms, can avoid capture and are much less numerous. Usually it is necessary to concentrate the organisms from a larger volume of water. Zooplankton, particularly copepods, are capable of avoiding both net and bottle samplers. Trap samplers such as the Schindler trap are more suitable.

The Juday trap[15] operates on the same principle as the water bottle samplers but is especially designed for net zooplankton sampling. It has a capacity of 10 L.

The larger zooplankton commonly are sampled with a No. 8 mesh plankton net.[16,17] However, the mesh size, type of material, orifice size, length, hauling methods, and volume sampled will depend on the particular needs of the study. The mesh size and net material determine filtration efficiency, clogging tendencies, velocity, drag, and the condition of the sample after collection. Most nets are made of nylon or silk. These materials are stiff yet flexible, durable, and resistant to swelling. Use long nets with wide mouths because they permit maximum filtering surface area and cause minimal population disruption.

Three types of tows are used: vertical, horizontal, and oblique. To make a vertical tow, lower the weighted net to a given depth, then raise vertically at an even speed of 0.5 to 1.0 m/sec. Estimate the volume (cubic meters) of water filtered through the net as $V = \pi r^2 d$ where r is the radius of the net's orifice and d is the depth to which the net is lowered. Determine the volume sampled by using a calibrated flow meter mounted in the mouth of the net. Record flow-meter readings before and after collecting the sample. For oblique and horizontal tows, use the samplers shown in Figure 1002:3.

To collect a sample, tow the plankton net from a boat. Use a boat equipped with a davit, meter wheel, angle indicator, and winch. Attach a 3- to 5-kg weight to hold the net down. Determine the depth of the net by multiplying the length of the extended wire by the cosine of the wire's angle with the vertical direction. Maintain the wire angle by controlling the boat's speed.

For oblique tows, lower the net or sampler to some predetermined depth and then raise at a constant rate as the boat moves forward. Oblique tows do not necessarily sample a true angle from the bottom to the surface. Under best conditions the pattern is somewhat sigmoid due to boat acceleration and slack in the tow line. For horizontal tows, lower the net to the preselected depth, tow at that depth for 5 to 8 min, then close, and raise the net. Vertical and oblique tows collect a composite sample, whereas horizontal tows collect a sample at a particular depth.

Use the Clarke-Bumpus sampler[18] for quantitative collection of zooplankton because it is more versatile than other devices and can be used in various sampling patterns. In standing waters, collect tow samples by filtering 1 to 5 m^3 of water.

For an estimate of net zooplankton populations in flowing waters, collect 20 L of surface water by bucket and filter through a mesh of appropriate size. To prevent the organisms from being forced through the net mesh or otherwise being damaged, do not hold its rim more than a few centimeters above the water surface.

Preserve zooplankton samples with 70% ethanol, 5% buffered formalin, or Lugol's solution. Formalin preservative is better than ethanol for samples held for biomass analysis because there is no loss of soluble cell components. Ethanol preservative is preferred for materials to be stained in permanent mounts or stored. Formalin may be used for the first 48 hr of preservation with subsequent transfer to 70% ethanol. Formalin preservative may cause distortion of plastic forms such as protozoans and rotifers. Use a narcotizing agent such as carbonated water, menthol-saturated water, or neosynephrine to prevent or reduce contraction or distortion of organisms, especially rotifers. Adding a few drops of detergent prevents clumping of preserved organisms. Preserve samples within a few hours of narcotizing. To prevent evaporation, add 5% glycerin to the concentrated sample. In turbid samples, differentiate animal and vegetative material by adding 0.04% rose bengal stain, which is chemically specific for the carapace (shells) of zooplankters.

1002 C. Concentration Technics

The organisms contained in water samples sometimes must be concentrated in the laboratory before analysis. Three technics for concentrating, namely, sedimentation, membrane filtration, and centrifugation, are described below.

1. Sedimentation

Sedimentation is the preferred method of concentration because it is nonselective (unlike filtration) and nondestructive (unlike centrifugation), although blue-green algae with pseudo-vacuoles may not settle. The volume concentrated varies inversely with the abundance of organisms and is affected by sample turbidity. It may be as small as 10 mL for use with an inverted microscope to as much as 1 L for general phytoplankton and zooplankton enumeration.

Settling and the settling rate may be enhanced by adding a liquid household detergent. For an untreated sample allow 1 hr settling/mm of column depth. For a treated sample (10 mL detergent/L) allow about 0.5 hr settling/mm depth. The sample may be concentrated in a series of steps by quantitatively transferring the sediment from the initial container to sequentially smaller ones. Fill settling chambers without forming a vortex, keep them vibration-free, and move them carefully to avoid nonrandom distribution of settled matter. Siphon or decant the supernatants to obtain the desired final volume (5 mL for diatom mounts). Store the concentrated sample in a closed, labeled glass vial.

2. Membrane Filtration

The filtration method is used primarily for phytoplankton. Do not use it when populations are dense and the content of detritus is high because the filter clogs quickly and silt may crush the organisms or obscure them from view. Pour a measured volume of well-mixed sample into a funnel equipped with a nongridded membrane filter having a pore diameter of 0.45 μm. Apply a vacuum of less than 50 kPa to the filter.

For samples with a low phytoplankton and silt content the method permits use of high magnification for enumerating small plankters, does not require counting of individual plankters to assemble enumeration data, and increases the probability of observing less abundant forms.[19]

3. Centrifugation

Plankton can be concentrated by batch or continuous centrifugation. Centrifuge batch samples at 1,000 *g* for 20 min. The Foerst continuous centrifuge* is no longer generally recommended as a quantitative device but it may be desirable to continue its use in existing programs in order to assure continuity with previously collected data. Although centrifugation accelerates sedimentation, it may damage fragile organisms.

*Foerst Mechanical Specialities Co., 2407 N. St. Louis Ave., Chicago, Ill. 60647 or Limnological Apparatus Co., 2406 N. Bernard St., Chicago, Ill. 60647.

1002 D. Mounting and Preparing for Examination

1. Phytoplankton Semi-Permanent Wet Mounts

Agitate the settled sample concentrate and withdraw a subsample with an accurately calibrated automatic pipet. Clean and calibrate the pipet regularly. Prepare wet mounts by transferring 0.1 mL to a glass slide, placing a cover slip over the sample, and ringing the cover slip with an adhesive such as clear nail polish to prevent evaporation. For semipermanent slides, mix glycerin with the sample; as the sample ages the water evaporates, leaving the organisms imbedded in the glycerin. If the cover slip is ringed with adhesive, the slide can be retained for a few years if stored in the dark.

2. Phytoplankton Membrane Filter Mounts

Place two drops of immersion oil on a labeled slide. Immediately after filtering place the filter on top of the oil with a pair of forceps and add two drops of oil on top of the filter. The oil impregnates the filter and makes it transparent. Impregnation time is 24 to 48 hr. This procedure can be completed in 1 to 2 hr by applying heat (70 C). Once the filter has cleared, place a few additional drops of oil on the cleared filter and cover with a cover slip. The mounted filter is now ready for microscopic examination. Alternatively, mount membrane filters in mounting medium.* Immerse filters in 1-propanol to displace residual water and transfer to xylol for several minutes to clear filters. Place a section of filter or entire filter on a microscope slide with the mounting medium, cover with a cover glass, and dry at low temperature.[20]

*Permount, Fisher Scientific Co., or equivalent.

3. Diatom Mounts

Samples concentrated for diatom analysis by settling or centrifugation may contain dissolved materials, such as marine salts, formalin, and detergents, that will leave interfering residues. Wash well with distilled water before slide preparation. Transfer several drops of washed concentrate by means of a large-bore disposable pipet or large-bore dropper to a cover slip on a hot plate warmed enough to increase the evaporation rate but not enough to boil (use a large-bore pipet or dropper to prevent possible selective filtration, thus exclusion, of larger forms or those forming colonies or chains). Evaporate to dryness. Repeat addition and evaporation until a sufficient quantity of sample has been transferred to the cover slip but avoid producing a residue so dense that organisms cannot be recognized. If in doubt about the density, examine under a compound microscope. After evaporation, incinerate the residue on the cover slip on the hot plate at 300 to 500 C. This usually requires 20 to 45 min. Mount† as described below.

†Hyrax, Custom Research and Development, Inc., 8500 Mt. Vernon Road, Auburn, Calif., or equivalent.

Treat samples concentrated for diatom analysis by membrane filtration as described by Patrick and Reimer.[21] Mix equal volumes of conc nitric acid (HNO_3) and sample. Add a few grains of potassium dichromate ($K_2Cr_2O_7$)[22] to facilitate digestion of the filter and cellular organic matter. Add more dichromate if solution color changes from yellow to green. Place sample on a hot plate and boil down to approximately one-third the original volume. Alternatively, let the treated sample stand overnight. This cleaning process destroys organic matter and leaves only the diatom shells (frustules). Cool, wash with distilled water, and mount as described above. CAUTION: *When working with conc HNO_3 wear safety goggles and an acid-resistant apron and gloves, and work under a hood*. Transfer the cleaned frustules to a cover glass and dry as described above.

Place a drop of mounting medium in the center of a labeled slide. Use 25-mm by 75-mm slides with frosted ends. Using a suitable† microscopic mounting medium assures permanent, easily handled mounts for examination under oil immersion. Heat the slide to near 90 C for 1 to 2 min before applying the heated cover slip with its sample residue to hasten evaporation of solvent in the mounting medium. Remove the slide to a cool surface and, during cooling (5 to 10 sec), apply firm but gentle pressure to the cover glass with a broad, flat instrument.

†Hyrax, Custom Research and Development, Inc., 8500 Mt. Vernon Road, Auburn, Calif., or equivalent.

4. Zooplankton Mounts

For zooplankton analyses, withdraw a 5-mL subsample from the concentrate and dilute or concentrate further as necessary. Transfer the sample to a counting cell or chamber (see below) for analysis as a wet mount. Use polyvinyl lactyl phenol‡ for preparing semipermanent zooplankton mounts. The mounts are good for about a year, after which time the clearing agent causes deterioration of organisms. For permanent mounting, other mountants are available.§

‡Biomedical Specialists, Box 1687, Santa Monica, Calif.

§Turtox CMC-10, Turtox/Cambosco, 8200 South Hoyne Avenue, Chicago, Ill.; Hydramount, Biomedical Specialists, Box 1687, Santa Monica, Calif.; or equivalent.

1002 E. Microscopes and Calibrations

1. Compound Microscope

Although most workers prefer the binocular compound microscope, the monocular type can be used. Equip either type with a mechanical stage capable of moving all parts of a counting cell past the aperture of the objective. Standard equipment includes 10× oculars (paired when a binocular microscope is used) and objectives in the following ranges (manufacturers differ slightly in exact specification):

Type of Objective	Approximate Overall Magnification with 10× Ocular
16 mm (low power, 10×)	100×
8 mm (medium power, 20×)	200×
4 mm (high power, dry 43×)	430×
1.8 mm (oil immersion, 90×)	900×

The 8-mm (20×) objective with a working distance of approximately 1.6 mm commonly is used with a standard plankton-counting cell 1 mm deep.

2. Stereoscopic Microscope

The stereoscopic microscope is essentially two complete microscopes assembled into a binocular instrument to give a stereoscopic view and an erect rather than an inverted image. Use this microscope for the study and counting of large plankters such as mature microcrustacea. Include 10× to 15× paired oculars in combination with 1× to 8× objectives in this microscope. This combination of lenses bridges the gap between the hand lens and the compound microscope and provides magnification ranging from 10× to 120×. Alternatively, use a good-quality zoom-type instrument with comparable magnification.

3. Inverted Compound Microscope

The inverted compound microscope is used routinely for plankton counting in many laboratories.[23] This instrument is unique in that the objectives are below a movable stage and the illumination comes from above. Place samples in a cylindrical settling chamber having a thin, clear glass bottom. Chambers of various capacities are available; the appropriate size depends on the density of organisms. After a suitable period of settling (see Section 1002C.1), count organisms in the settling chambers.

The major advantage of the inverted microscope is that by a simple rotation of the nosepiece a specimen can be examined (or counted) directly in the settling chamber at any desired magnification. No other preparation or manipulation is required. An important disadvantage is that floating organisms, such as buoyant blue-green algae, are not included in the count. Using a detergent (see Section 1002C.1) or Lugol's solution may eliminate this problem.

4. Microscope Calibration

Microscope calibration is essential. The usual equipment for calibration is a

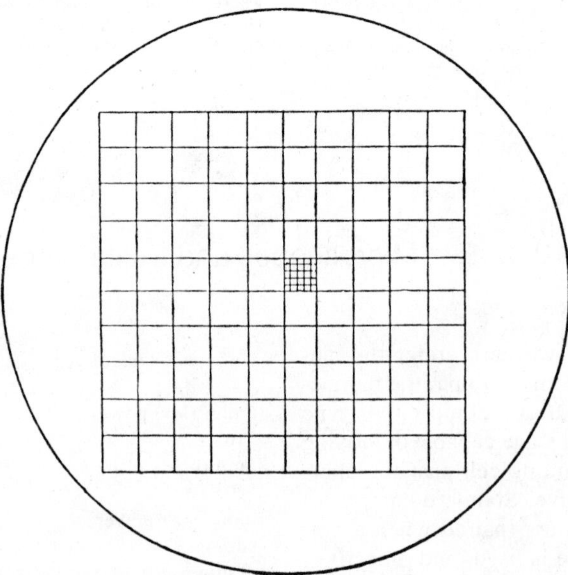

Figure 1002:4. Ocular micrometer ruling. A Whipple micrometer reticule is illustrated.

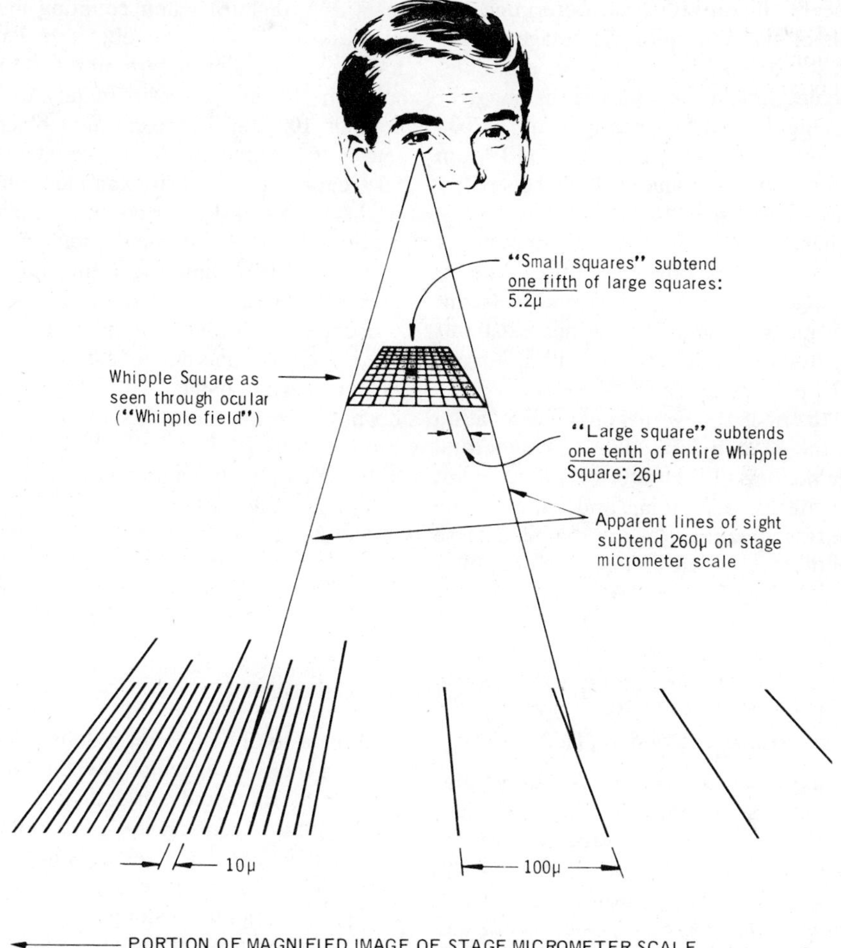

Figure 1002:5. Calibration of Whipple Square, as seen with 10× ocular and 43× objective (approximately 430× total magnification).

Whipple grid (ocular micrometer, reticle, or reticule) placed in an eyepiece of the microscope and a stage micrometer that has a standardized, accurately ruled scale on a glass slide. Purchase a microscope for plankton counting that will accept a standard Whipple disc. The Whipple disc (Figure 1002:4) has an accurately ruled grid subdivided into 100 squares. One square near the center is subdivided further into 25 smaller squares. The outer dimensions of the grid are such that with a 10× (16-mm) objective and a 10× ocular, it delimits an area of approximately 1 mm^2 on the microscope stage. Because this area may differ from one microscope to another, carefully calibrate the Whipple grid for each microscope.

With the ocular and stage micrometers parallel and in part superimposed, match the line at the left edge of the Whipple grid with the zero mark on the stage micrometer scale (Figure 1002:5). Determine the width of the Whipple grid image to the nearest 0.01 mm from the stage micrometer scale. Should the width of the image of the Whipple grid be exactly 1 mm (1,000 μm), the larger squares will be 1/10 mm (100 μm) on a side and each of the smaller squares 1/50 mm (20 μm).

When the microscope is calibrated at higher magnifications, the entire scale on the stage micrometer will not be seen; make measurements to the nearest 0.001 mm. Additional details for calibration are available.[8,24]

With a 10× eyepiece, magnification with the 10× objective lens (giving a total magnification of 10 × 10, or 100×) is not adequate for examining and enumerating plankters 10 μm or less in diameter. Use objective lenses with a magnification of at least 20× (8 mm) when counting nannoplankton. When a 20× objective lens is used, each smaller square on the Whipple micrometer field is approximately 10 × 10 μm, or 100 μm^2. Conventional objective lenses with magnifications greater than 20× cannot be used to examine plankton in the Sedgwick-Rafter counting chamber discussed below because their working distance is less than 1 mm and they will break the cover slip on the chamber. Microscopes with the "zoom" mechanism may give the same magnification as a 20× objective when using a 10× objective, but do not give the same resolution. Special "long working distance" objectives providing magnifications greater than 20× are available. The inverted microscope, as noted above, obviates these problems.

1002 F. Counting Technic

1. Phytoplankton Counting Units

Some phytoplankton are unicellular while others are multicellular (colonial). The variety of configurations poses a problem in enumeration. For example, should a four-celled colony of *Scenedesmus* (Plate 3, A) be reported as one colony or four individual cells? Listed below are suggestions for reporting:

Enumeration Method	Counting Unit	Reporting Unit
Total cell count	one cell	cells/mL
Natural unit count[25] (clump count)	one organism (any unicellular organism or natural colony)	units/mL
Areal standard unit count*	400 μm^2	units/mL

*Areal standard unit equals area of four small squares in Whipple grid at a magnification of 200×.

Making a total cell count is time-consuming and tedious, especially when colonies consist of thousands of individual cells. The natural unit or clump is the most easily used system; however, it is not necessarily accurate because sample handling and preserving may dislodge cells from the colony. The unit method also may not be quantitatively accurate nor reflect abundance of biomass or biovolume. Whatever method is chosen, identify it in reporting results.

2. Counting Chambers

To enumerate plankton use a counting cell or chamber that limits the volume and area for ready calculation of population densities. Do not count dead cells or diatoms with broken frustules. Tally empty centric and pennate diatoms separately as "dead centric diatoms" or "dead pennate diatoms" for use in converting the diatom species proportional count to a count per milliliter.

When counting with a Whipple grid, establish a convention for tallying organisms lying on an outer boundary line. For example, in counting a "field" (entire Whipple square), designate the top and left boundaries as "no-count" sides, and the bottom and right boundaries as "count" sides. Thus, tally every plankter touching a "count" side from the inside or outside but ignore any touching a "no-count" side. If there are present significant numbers of filamentous or other large forms that cross two or more boundaries of the grid, count them separately at a lower magnification and include their number in the total count.

In strip counting (see below), the top and bottom of the grid are the "count" and "no-count" boundaries, respectively, and plankters are counted as they move across the center vertical line.

To identify organisms use standard bench references (see section on Selected Taxonomic References below).

a. Sedgwick-Rafter (S-R) counting cell: The S-R cell is a device commonly used for plankton counting because it is easily manipulated and provides reasonably reproducible data when used with a calibrated microscope equipped with an eyepiece measuring device such as the Whipple grid. The S-R cell is approximately 50 mm long by 20 mm wide by 1 mm deep. The total area of the bottom is approximately 1,000 mm^2 and the total volume is approximately 1,000 mm^3 or 1 mL. Carefully check the exact length and depth of the cell with a micrometer and calipers before use.

The greatest disadvantage associated with the cell is that objectives providing high magnification cannot be used.

1) Filling the cell—Before filling the S-R cell with sample, place the cover glass diagonally across the cell and transfer the sample with a large-bore pipet (Figure 1002:6). Placing the cover slip in this manner will help prevent formation of air bubbles in corners of the cell. The cover slip often will rotate slowly and cover the in-

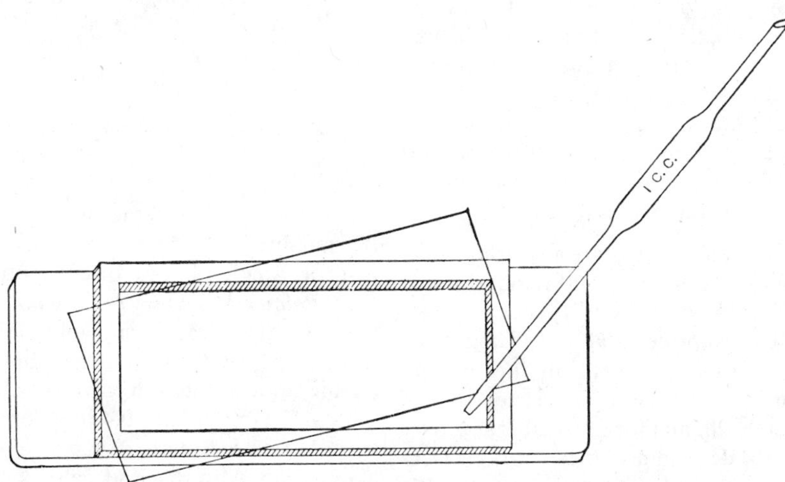

Figure 1002:6. Counting cell (Sedgwick-Rafter), showing method of filling. Source: WHIPPLE, G.C., G.M. FAIR and M.C. WHIPPLE. 1927. The Microscopy of Drinking Water, John Wiley & Sons, New York, N.Y.

ner portion of the S-R cell during filling. Do not overfill because this would yield a depth greater than 1 mm and produce an invalid count. Do not permit large air spaces caused by evaporation to develop in the chamber during a lengthy examination. To prevent formation of air spaces, occasionally place a small drop of distilled water on the edge of the cover glass.

Before proceeding with the count, let the S-R cell stand for at least 15 min to settle plankton. Count the plankton on the bottom of the S-R cell. Some phytoplankton, notably some blue-green algae, may not settle but rise to the underside of the cover slip. When this occurs, count these organisms and add to the total of those counted on the cell bottom to derive the total number of organisms in the sample.

2) Strip counting—A "strip" the length of the cell constitutes a volume approximately 50 mm long, 1 mm deep, and the width of the total Whipple grid. When 10× eyepieces and 20× or greater objectives are used and the width of the total Whipple grid is calibrated to be 0.5 mm (500 μm), the volume of one strip is 25 mm^3, or 1/40 (2.5%) the total cell volume. For making a strip count, use a magnification of at least 200×. Calculate total number of plankters in the S-R cell by multiplying actual count in the "strip" by the number (enumeration factor) representing the portion of the S-R cell counted.

Usually count two or four strips, depending on plankton density; count more strips as plankton density decreases. When 10× eyepieces and a 20× objective are used, the enumeration factor for plankton counted along two strips is approximately 20, and for four strips approximately 10, depending on calibration. Derive number of plankton in the S-R cell from the following:

$$\text{No./mL} = \frac{C \times 1{,}000 \text{ mm}^3}{L \times D \times W \times S}$$

where:

C = number of organisms counted,
L = length of each strip (S-R cell length), mm,
D = depth of a strip (S-R cell depth), mm,
W = width of a strip (Whipple grid image width), mm, and
S = number of strips counted.

Multiply or divide number of cells per milliliter by a correction factor to adjust for sample dilution or concentration.

3) Field counting—On samples containing many plankton (10 or more plankters per field), make field counts rather than strip counts. Count plankters in 10 or more random fields each consisting of one Whipple grid. The number of fields counted will depend on plankton density and variety and statistical accuracy desired. Calculate the numbers per milliliter from field counts:

$$\text{No./mL} = \frac{C \times 1{,}000 \text{ mm}^3}{A \times D \times F}$$

where:

C = number of organisms counted,
A = area of a field (Whipple grid image area), mm^2,
D = depth of a field (S-R cell depth), mm, and
F = number of fields counted.

Multiply or divide the number of cells per milliliter by a correction factor to adjust for sample dilution or concentration.

b. Palmer-Maloney (P-M) nannoplankton cell: The P-M nannoplankton cell[26] is designed specifically for nannoplankton enumeration. It has a circular chamber with a 17.9-mm diam, 0.4-mm depth, and 0.1-mL volume. Use a magnification of about 450× with the P-M cell. Because a relatively small sample portion is examined in the P-M cell do not use it unless the sample contains a dense population (10 or more plankters per field). Examining such a small sample portion from a less dense population results in seriously underestimating it.

Introduce sample with a pipet into one of the 2-mm by 5-mm channels on the side of the chamber with the cover slip in place. After a 10-min settling period examine 10 to 20 Whipple fields, depending on density and variety of plankton and the statistical accuracy desired. Strips may be counted in this or any other circular cell by measuring the effective diameter and counting two perpendicular strips that cross at the center. Calculate the number of plankters per milliliter as follows:

$$\text{No./mL} = \frac{C \times 1{,}000 \text{ mm}^3}{A \times D \times F}$$

where:

C = number of organisms counted,
A = area of a field (Whipple grid image), mm^2,
D = depth of a field (P-M cell depth), mm, and
F = number of fields counted.

Multiply or divide the number of cells per milliliter by a correction factor to adjust for sample dilution or concentration.

c. Other counting cells: The most readily available chamber is the standard medical hemacytometer used for enumerating blood cells. It has a ruled grid machined into a counting plate and is fitted with a ground glass cover slip. The grid is divided into square millimeter divisions; the chamber is 0.1 mm deep. Introduce sample by pipet and view under 450× magnification. Count all cells within the grid. A similar chamber, the Petroff-Hausser, designed for bacterial counts, also can be used. Each of these chambers comes from the manufacturer with a detailed instruction sheet containing directions on calculations and proper usage. The disadvantage to these counting cells is that the sample must have a very high plankton density to yield statistically reliable data.

d. Inverted microscope counts: Prepare a sample for examination by filling the settling chamber. After the desired settling time (see Section 1002C.1), transfer the chamber to the microscope stage. Count perpendicular strips across the center of the bottom cover glass. Strip counts may be made by using a Whipple grid or special counting oculars† that have a pair of adjustable parallel hairs and a single cross hair. Determine the width of the strip with a stage micrometer and tally organisms as they pass the single cross hair that functions as a reference point. Hold strip width constant for any series of samples. Alternatively examine random nonoverlapping fields until at least 100 units of the dominant species are counted. For highest accuracy, particularly because algae distribution may be nonuniform, count the entire chamber floor.

$$\text{Strip count (no./mL)} = \frac{C \times A_t}{L \times W \times S \times V}$$

where:

C = number of organisms counted,
A_t = total area of bottom of settling chamber, mm^2,
L = length of a strip, mm,
W = width of a strip (Whipple grid image width), mm,
S = number of strips counted, and
V = volume of sample settled, mL.

$$\text{Field count (no./mL)} = \frac{C \times A_t}{A_f \times F \times V}$$

where:

A_f = area of a field (Whipple grid image area), mm^2,
F = number of fields counted,

and other terms are as defined above.

†Measuring eyepieces containing graticules for linear and angular measurements and counts, Wild M40 Inverted Biological Microscope, Wild Heerbrugg, Switzerland, or equivalent.

3. Lackey Drop (Microtransect) Counting Method

The Lackey drop (microtransect) method[27] is a simple method of obtaining counts of considerable accuracy with samples containing a dense plankton population. It is similar to the S-R strip count.

Pipet 0.1 mL to a glass slide and cover with a 22- by 22-mm glass cover slip. Count organisms in three or four strips the width of the cover slip. Calculate number of organisms per milliliter as follows:

$$\text{No./mL} = \frac{C \times A_t}{A_s \times S \times V}$$

where:

C = number of organisms counted,
A_t = area of cover slip, mm^2,
A_s = area of one strip, mm^2,
S = number of strips counted, and
V = volume of sample under the cover slip, mL.

4. Membrane Filter Counts

Examine samples, concentrated on unlined membrane filters and mounted in oil as described above, at a magnification of 200× to 450×. Select magnification level and size of microscope field (quadrat) such that the most abundant species appear in at least 70% but not more than 90% of microscopic fields examined (80% is optimum). Adjust microscope field size by using part or all of the Whipple grid. Examine 30 random microscope fields and record number of fields in which each species occurred. Determine percent occurrence (F %) and density per field (N) from Table 1002:I. Report results as organisms per milliliter, calculated as follows:

$$\text{No./mL} = \frac{N \times Q}{V \times D}$$

where:

N = density (organisms/field) from Table 1002:I,
Q = number of fields per filter,
V = milliliters filtered, and
D = dilution factor (0.96 for 4% formalin preservative).

TABLE 1002:I. CONVERSION TABLE FOR MEMBRANE FILTER TECHNIC (Based on 30 Scored Fields)

Total Occurrence	$F\%$*	N†
1	3.3	0.03
2	6.7	0.07
3	10.0	0.10
4	13.3	0.14
5	16.7	0.18
6	20.0	0.22
7	23.3	0.26
8	26.7	0.31
9	30.0	0.35
10	33.3	0.40
11	36.7	0.45
12	40.0	0.51
13	43.3	0.57
14	46.7	0.63
15	50.0	0.69
16	53.3	0.76
17	56.7	0.83
18	60.0	0.91
19	63.3	1.00
20	66.7	1.10
21	70.0	1.20
22	73.3	1.32
23	76.7	1.47
24	80.0	1.61
25	83.3	1.79
26	86.7	2.02
27	90.0	2.30
28	93.3	2.71
29	96.7	3.42
30	100.0	?

$${}^*F = \frac{\text{total number of species occurrences} \times 100}{\text{total number of fields examined}}$$

†N = number of organisms per field.

5. Particle Counters

Several particle counters have been marketed.[28] These can be used effectively for counting pure cultures but are not suited for enumerating natural plankton communities because they do not discriminate between plankton and particles of silt or organic detritus. These systems detect particles only in a specific size range; to accommodate the wide range of sizes of plankton numerous orifice settings are required.

6. Diatom Species Proportional Count

Examine diatom samples prepared as directed above under oil immersion at a magnification of at least 900×. Scan lateral strips the width of the Whipple grid until at least 250 cells are counted. Available time and accuracy required dictate the number of cells to be counted. Determine percentage abundance of each species from tallied counts and calculate counts per milliliter of each species by multiplying percent abundance by total live and dead diatom count obtained from the plankton counting chamber. For greater accuracy distinguish between living and dead diatoms at the species level.

7. Zooplankton

Enumerate small (nanno) zooplankton in a counting chamber during routine phytoplankton count. Report as organisms per milliliter. Count the larger (net) zooplankton, such as mature cladocera and copepoda, from concentrates and report as organisms per cubic meter. For larger forms, use a counting chamber 80 mm by 50 mm and 2 mm deep. Use the chamber with or without a cover. An open chamber is difficult to move because jarring disrupts the count; however, the exposed plankton is accessible. To reduce effects of movement in open chambers use a specially designed (partitioned) Bogorov chamber.[29] When using a larger counting chamber, adjust volume of concentrate to 8 mL and transfer to the chamber. Count and identify rotifers and nauplii in 10 strips scanned at 100× magnification using a compound microscope equipped with an ocular Whipple grid. Count large organisms, such as mature microcrustacea, by scanning the entire chamber under a binocular dissecting microscope at 20 to 40× magnification. Where necessary, identify larger organisms after dissection and subsequent examination under a compound microscope. Calculate the number of organisms per cubic meter:

$$\text{No./m}^3 = \frac{C \times V'}{V'' \times V'''}$$

where:

C = number of organisms counted,
V' = volume of the concentrated sample, mL,
V'' = volume counted, mL, and
V''' = volume of the grab sample, m^3.

8. Phytoplankton Staining and Preparation Technic (TENTATIVE)

Staining algae permits differentiation between "live" and "dead" diatoms.[30] This permits enumerating total phytoplankton in a single sample without sacrificing detailed diatom taxonomy. It also results in permanent reference slides. The procedure is most useful when diatoms are major components of phytoplankton and it is important to distinguish between living and dead diatoms.

Preferably preserve samples in Lugol's solution or alternatively in formalin (see 1002B.3). For analysis thoroughly mix the sample and filter a portion through a 47-mm-diam membrane filter (pore diam 0.45 or 0.65 μm). Use a vacuum of 16 to 20 kPa and never let sample dry. Add 2 to 5 mL aqueous acid fuchsin solution (dissolve 1 g acid fuchsin in 100 mL distilled water to which 2 mL glacial acetic acid have been added; filter) to the filter and let stand for 20 min. After staining, filter sample, wash briefly with distilled water, and filter again. Administer successive rinses of 50%, 90%, and 100% propanol to the sample while filtering. Soak for 2 min in a second 100% propanol wash, filter, and add xylene. At least two washes are required; let the final one soak 10 min before filtering. Trim the xylene-soaked filter and place on a microscope slide on which there are several drops of mounting medium.‡ Apply several more drops of medium to the top of the filter and install a coverglass. Carefully squeeze out excess mounting medium. Make the final mount permanent by lacquering the edges of the coverglass.

Count organisms using the most appropriate magnification. Oil immersion is necessary for species identifications of diatoms and many other algae. Count either strips or random fields and calculate plankton densities per milliliter:

$$\text{No./mL} = \frac{C \times A_t}{A_c \times V}$$

where:

C = number of organisms counted,
A_t = total area of effective filter before trimming and mounting,
A_c = area counted (strips or fields), and
V = volume of sample filtered, mL.

‡Permount, Fisher Scientific Co., or equivalent.

1002 G. Chlorophyll

The characteristic algal pigments are chlorophylls, xanthophylls, and carotenes. The three chlorophylls commonly found in planktonic algae are chlorophylls *a*, *b*, and *c*. Chlorophyll *a* constitutes approximately 1 to 2% of the dry weight of organic material in all planktonic algae and is the preferred indicator for algal biomass estimates. Chlorophyll content of cells varies with species or taxonomic groups and is affected by age, growth rate, light, and nutrient conditions.[31]

Two methods for determining chlorophyll *a* in phytoplankton are available, the spectrophotometric[9,32,33] and fluorometric.[9,34,35,36] The latter is more sensitive, requires less sample, and has been adapted for in vivo measurements.[37] A specific method for chlorophyll *c*, more sensitive than the trichromatic method described below, especially for samples of low pigment content, is available[38] but is not included here.

Pheophytin *a*, a common degradation product of chlorophyll *a*, can interfere with the determination of chlorophyll *a* because it absorbs light and fluoresces in the same region of the spectrum as chlorophyll *a* and, if present, may cause errors in chlorophyll *a* values.[39,40] When measuring chlorophyll *a* measure also the concentration of pheophytin *a*. The ratio of chlorophyll *a* to pheophytin *a* serves as a good indicator of physiological condition of phytoplankton. Another useful water quality indicator is the ratio of biomass to chlorophyll *a* (Autotrophic Index). In unpolluted waters the plankton population is composed largely of autotrophic (food-producing), chlorophyllous algae. As waters become organically enriched, the proportion of heterotrophic (consuming), nonchlorophyllous organisms, such as the filamentous bacteria and stalked protozoa, increases. The Autotrophic Index (AI) is a means of relating changes in plankton species composition to changes in water quality.[41] Calculate as:

$$\text{AI} = \frac{\text{Biomass (ash-free wt of organic matter), mg/m}^3}{\text{Chlorophyll } a\text{, mg/m}^3}$$

Normal AI values range from 50 to 200. Larger AI values (above 200) indicate poor water quality.

1. Spectrophotometric Determination of Chlorophyll *a*, *b*, and *c* (Trichromatic Method)

The pigments are extracted from the plankton concentrate with aqueous acetone and the optical density (absorbance) of the extract is determined with a spectrophotometer. When immediate pigment extraction is not possible (as described below), the samples may be stored frozen for as long as 30 days if kept in the dark. The ease with which the chlorophylls are removed from the cells varies considerably with different algae. To achieve complete extraction of the pigments, it is necessary usually to disrupt the cells mechanically with a tissue grinder.

a. Equipment and reagents:

1) *Spectrophotometer,* with a narrow band (0.5 to 2.0 nm) because the chlorophyll absorption peak is relatively narrow. At a spectral band width of 20 nm the chlorophyll *a* concentration may be underestimated by as much as 40%.

2) *Cuvettes* with 1 cm, 4 cm, and 10 cm path length.

3) *Clinical centrifuge.*

4) *Tissue grinder.** Successfully macerating glass fiber filters in tissue grinders with grinding tube and pestle of conical design may be difficult. Preferably use grinding tubes and pestles with rounded bottoms.

5) *Centrifuge tubes,* 15 mL, graduated, screw-cap.

6) *Filtration equipment,* filters, membrane (0.45 μm porosity, 47-mm diam) or glass fiber (GF/C or GF/A, 4.5-cm diam); vacuum pump.

7) *Magnesium carbonate suspension:* Add 1.0 g finely powdered $MgCO_3$ to 100 mL distilled water.

8) *Aqueous acetone solution:* Mix 90 parts acetone (reagent grade BP 56 C) with 10 parts water (v/v).

b. Procedure:

1) Concentrate the sample by centrifuging or filtering (membrane or glass fiber filter). Add 0.2 mL $MgCO_3$ suspension before centrifuging or during the final phase of filtering. Store concentrated samples frozen in a desiccator in the dark if extraction is delayed. Use glassware and cuvettes that are clean and acid-free.

2) Place sample in a tissue grinder, cover with 2 to 3 mL 90% aqueous acetone solution, and macerate. Use TFE/glass grinder for a glass-fiber filter and glass/glass grinder for a membrane filter.

3) Transfer sample to a screw-cap centrifuge tube, rinse grinder with a few milliliters 90% aqueous acetone, and add the rinse to the extraction slurry. Adjust total volume to a constant level, 5 to 10 mL with 90% aqueous acetone. Use solvent sparingly and avoid excessive dilution of pigments. Steep samples overnight at 4 C in the dark.

4) Clarify extract by centrifuging in closed tubes for 20 min at 500 *g*. Decant the clarified extract into a clean, calibrated, 15-mL, screw-cap centrifuge tube and measure the total volume of extract.

5) Transfer extract to a 1-cm cuvette and measure optical density (OD) at 750, 663, 645, and 630 nm. Choose a cell path length or dilution to provide an OD663 greater than 0.2 and less than 1.0.

c. Calculations: Use the optical density readings at 663, 645, and 630 nm for the determination of chlorophyll *a, b,* and *c,* respectively. The OD reading at 750 nm serves as a correction for turbidity. Subtract this reading from each of the pigment OD values of the other wavelengths before using them in the equations below. Because the OD of the extract at 750 nm is very sensitive to changes in the acetone-to-water proportions, adhere rigidly to the 90 parts acetone: 10 parts water (v/v) formula for pigment extraction.

To avoid using the 750-nm reading, clear the pigment solution by centrifuging for 20 min at 1,000 *g* and use a light path limited to 1 cm. However, when the possibility of resuspending sediment exists, make the 750-nm reading. This is commonly a problem when using glass fiber filters

*Kontes Glass Company, Vineland, N.J. 08360: Glass/glass grinder, Model No. 885500; Glass/teflon grinder, Model No. 886000; or equivalent.

and a centrifuge with a slant head. To reduce this difficulty use a swing-out centrifuge head and additional amounts of $MgCO_3$ added immediately before centrifuging.

1) Calculate the concentrations of chlorophyll *a*, *b*, and *c* in the extract by inserting the corrected optical densities in following equations:

a) Chl *a*, mg/L = 11.64 (OD663) − 2.16 (OD645) + 0.10 (OD630)

b) Chl *b*, mg/L = 20.97 (OD645) − 3.94 (OD663) − 3.66 (OD630)

c) Chl *c*, mg/L = 54.22 (OD630) − 14.81 (OD645) − 5.53 (OD663)

where:

OD663, OD645, and OD630 = corrected optical densities (with a 1 cm light path) at the respective wavelengths.

2) After determining the concentration of pigment in the extract, calculate the amount of pigment per unit volume as follows:

$$\text{Chlorophyll } a\text{, mg/m}^3 = \frac{\text{Chl } a \times \text{extract volume, L}}{\text{Volume of sample, m}^3}$$

where:

Chl *a* = chlorophyll concentration in the extract determined by Equation a) above.

2. Fluorometric Method for Chlorophyll *a*

The fluorometric method for chlorophyll *a* is more sensitive than the spectrophotometric method, requires a smaller sample, and does not require the wavelength resolution needed for the spectrophotometric method. Optimum sensitivity for in vitro chlorophyll *a* measurements is obtained at an excitation wavelength of 430 nm and an emission wavelength of 663 nm. A method for continuous measurement of chlorophyll *a* in vivo is available,[37] but is reported to be less efficient than the in vitro method given here, yielding about one-tenth as much fluorescence per unit weight as the same amount in solution. Pheophytin *a* also can be determined fluorometrically.[9]

a. Equipment and reagents:

1) *Fluorometer*, equipped with a high-intensity F4T.5 blue lamp, photomultiplier tube R-136 (red sensitive), sliding window orifices 1×, 3×, 10×, and 30×, and filters for light emission (CS-2-64) and excitation (CS-5-60), and a high-sensitivity door.†

2) Other equipment and reagents as specified for the Spectrophotometric Determination of Chlorophyll, above.

b. Procedure:

1) Calibrate fluorometer with a chlorophyll solution of known concentration as follows:

a) Prepare chlorophyll extract and analyze spectrophotometrically.

b) Prepare serial dilutions of the extract to provide concentrations of approximately 2, 6, 20, and 60 µg chlorophyll *a*/L.

c) Make readings for each solution at each sensitivity setting (sliding window orifice): 1×, 3×, 10×, and 30×.

d) Using the values obtained above, derive calibration factors to convert fluorometric readings in each sensitivity level to concentrations chlorophyll *a*, as follows:

$$F_s = \frac{C_a}{R_s}$$

where:

F_s = calibration factor for sensitivity setting S,

R_s = reading of the fluorometer for sensitivity setting S, and,

C_a = concentration of chlorophyll *a* determined spectrophotometrically, µg/L.

2) Measure sample fluorescence at sensitivity settings that will provide a mid-

†Model 111, Turner Assoc., 2524 Pulgas Ave., Palo Alto, Calif., or equivalent.

scale reading. Convert fluorescence readings to concentrations of chlorophyll *a* by multiplying the readings by the appropriate calibration factor.

3) Avoid using the 1× window because of quenching effects.

3. Spectrophotometric Determination of Chlorophyll *a* in the Presence of Pheophytin *a*

Chlorophyll *a* may be overestimated by including pheopigments that absorb near the same wavelength as chlorophyll *a*. Chlorophyll *a*, acidified with dilute acid, degrades to pheophytin *a*, which has maximum absorption at wavelengths of 410 and 665 (667) nm. Additional acidification with more concentrated acid results in further degradation to pheophorbide-like compounds.[31] Addition of acid to chlorophyll *a* results in loss of the magnesium atom, converting it to pheophytin *a*. When a solution of pure chlorophyll *a* is converted to pheophytin *a* by acidification, the absorption peak is reduced to approximately 60% of its original value and shifts from 663 nm to 665 nm. This results in a before-to-after acidification absorption-peak-ratio (OD663/OD665) of 1.70 and is used in correcting the apparent chlorophyll *a* concentration for pheophytin *a*.

Samples with an OD663 before/OD665 after acidification ratio ($663_b/665_a$) of 1.70 are considered to contain little if any pheophytin *a* and to be in excellent physiological condition. Solutions of pure pheophytin show no reduction in OD665 upon acidification and have a $663_b/665_a$ ratio of 1.0. Thus, mixtures of chlorophyll *a* and pheophytin *a* have absorption peak ratios ranging between 1.0 and 1.7. These ratios are based on the use of 90% acetone as solvent. Using 100% acetone as solvent results in a chlorophyll *a* before-to-after acidification ratio of about 2.0.[31,32]

a. Equipment and reagents:

1) See Section 1002G.1*a*.

2) *Hydrochloric acid,* HCl, 1*N*.

b. Procedure:

1) Extract the pigment with 90% acetone (v/v), clarify by centrifuging (see Section 1002G.1*b*), and read OD at 750 nm and 663 nm.

2) Acidify extract in a 1-cm cuvette with 2 drops 1*N* HCl. If a larger cell is used add a proportionately larger volume of acid. Gently agitate the acidified extract and read OD at 750 nm and at 665 nm not sooner than 1 min or later than 2 min after acidification. Treat all samples identically.

3) Subtract the 750-nm OD value from the readings before (OD663 nm) and after acidification (OD665 nm).

c. Calculations: Using the corrected values calculate chlorophyll *a* (*C*) and pheophytin *a* (*P*) per cubic meter as follows:

$$1)\ C,\ \text{mg/m}^3 = \frac{26.73\,(663_b - 665_a) \times V_1}{V_2 \times L}$$

$$2)\ P,\ \text{mg/m}^3 = \frac{26.73\,[1.7\,(665_a) - 663_b] \times V_1}{V_2 \times L}$$

where:

V_1 = volume of extract, L,
V_2 = volume of sample, m³,
L = light path length or width of cuvette, cm, and
663_b,
665_a = optical densities of 90% acetone extract before and after acidification, respectively.

The value 26.73 is the absorbance correction and equals $A \times K$
where:

A = absorbance coefficient for chlorophyll *a* at 663 nm = 11.0, and
K = ratio expressing correction for acidification,

$$= \frac{\left(\dfrac{663_b}{665_a}\right)_{\text{pure chlorophyll } a}}{\left(\dfrac{663_b}{665_a}\right)_{\text{pure chlorophyll } a} - \left(\dfrac{663_b}{665_a}\right)_{\text{pure pheophytin } a}}$$

$$= \frac{1.7}{1.7 - 1.0} = 2.43$$

4. Fluorometric Determination of Chlorophyll *a* in the Presence of Pheophytin *a*

To determine fluorometrically the concentration of pheophytin *a* requires the measurement of the fluorescence of acetone extracts before and after acidification. Acidification of acetone extracts of chlorophyll *a* and the resultant conversion of chlorophyll *a* to pheophytin *a* causes a reduction in fluorescence, which can be used to determine the concentration of pheophytin *a* in the extract.

a. Equipment and reagents:

1) See Section 1002G.2*a*.

2) *Hydrochloric acid,* HCl, 1*N*.

3) *Pure chlorophyll a*‡ (or a plankton chlorophyll extract with before-and-after acidification ratio of 1.70).

b. Procedure: Calibrate fluorometer as in Section 1002G.2*b*. Determine extract fluorescence at each sensitivity setting before and after acidification. Calculate calibration factors (F_s) and before-and-after acidification fluorescence ratio by dividing the fluorescence reading obtained before acidification by the reading obtained after acidification. Avoid readings on the 1 × scale and those outside the range of 20 to 80 fluorometric units.

c. Calculations: Determine the "corrected" chlorophyll *a* and pheophytin *a* in extracts of plankton samples, using the following equations:[9,36]

$$\text{Chlorophyll } a, \text{ mg/m}^3 = F_s \frac{r}{r-1} (R_b - R_a)$$

$$\text{Pheophytin } a, \text{ mg/m}^3 = F_s \frac{r}{r-1} (rR_a - R_b)$$

where:

F_s = conversion factor for sensitivity setting "S" (see 1002G.2*b*),
R_b = fluorescence of extract before acidification,
R_a = fluorescence of extract after acidification, and
r = R_b/R_a, as determined with pure chlorophyll *a* for the instrument. Redetermine *r* if filters or light source are changed.

‡Purified chlorophyll *a*, Sigma Chemical Company, St. Louis, Mo., or equivalent.

1002 H. Determination of Biomass (Standing Crop)

The standing crop of plankton can be expressed as numbers of organisms per unit volume. However, because plankton populations vary greatly in their size distribution, numbers alone do not give an adequate picture of population dynamics nor of the diversity and structure of the ecosystem. Methods available to provide more complete information on biomass include determination of total carbon, nitrogen, oxygen, hydrogen, lipids, carbohydrates, phosphorus, silica (diatoms), chitin (zooplankton), and chlorophyll (algae). The only practical methods for assessment of zooplankton biomass are volume and dry weight determinations. Recently, ATP (adenosine triphosphate)[42] and DNA (deoxyribonucleic acid)[43,44] contents of plankton have been evaluated as an estimate of viable biomass. Biomass estimates based on ATP appear to be in excellent agreement with estimates based on measurements such as chlorophyll *a* and cell volume. The DNA determination, however, is not recommended as an accurate indicator of biomass because of the occurrence of large amounts of detrital DNA in surface water, which can cause errors in biomass estimates.

1. Chlorophyll *a*

Chlorophyll *a* is an algal biomass indicator.[45] Assuming that chlorophyll *a* constitutes, on the average, 1.5% of the dry weight of organic matter (ash-free weight) of algae, estimate the algal biomass by multiplying the chlorophyll *a* content by a factor of 67.

2. Biovolume (Cell Volume)

Plankton data derived on a volume-per-volume basis often are more useful than numbers per milliliter.[46] Determine cell volume by using the simplest geometric configuration that best fits the shape of the cell being measured (such as sphere, cone, cylinder).[12] Cell sizes of an organism can differ substantially in different waters and from the same waters at different times during the year; therefore, average measurements from 20 individuals of each species for each sampling period. Calculate the total biovolume of any species by multiplying the average cell volume in cubic micrometers by the number per milliliter.

Compute total wet algal volume as:

$$V_t = \sum_{i=1}^{n} (N_i \times V_i)$$

where:

V_t = total plankton cell volume, mm^3/L,
N_i = number of organisms of the *i*th species/L, and
V_i = average volume of cells of *i*th species, μm^3.

3. Cell Surface Area

An estimation of cell surface area is valuable in analyzing interactions between the cell and surrounding waters. Compute average surface area in square micrometers and multiply by the number per milliliter of the species being considered.

4. Gravimetric Methods

The biomass of the plankton community can be estimated from gravimetric determinations, although silt and organic detritus interfere. Determine dry weight by placing 100 mg wet concentrated sample in a clean, ignited, and tared porcelain crucible and dry at 105 C for 24 hr. Alternatively, filter a known volume of sample through 0.45-μm-pore-diam membrane or a prerinsed, dried, and preweighed glass-fiber filter. (Note that the small sample used in direct filtration may lead to error if not handled properly.) Cool sample in a desiccator and weigh. Obtain ash-free weight by igniting the dried sample at 500 C for 1 hr. Cool, rewet ash with distilled water, and bring to constant weight at 105 C. The ash is rewetted to restore water of hydration of clays and other minerals; this may amount to as much as 10% of weight lost during incineration.[47] The ash-free weight is preferred to dry weight to compare mixed assemblages. The ash content may constitute 50% or more of the dry weight in phytoplankton having inorganic structures, such as the diatoms. In other forms the ash content is only about 5% of dry weight.

5. Adenosine Triphosphate (ATP)

Methods of measuring adenosine triphosphate (ATP) in plankton provide the only means of determining the total viable plankton biomass. ATP occurs in all plants and animals, but only in living cells; it is not associated with nonliving particulate material. The ratio of ATP to biomass varies from species to species, but appears to be constant enough to permit reliable estimates of biomass from ATP measurements.[41] The method is simple and relatively inexpensive and the instrumentation is stable and reliable. The method also has many potential applications in entrainment and bioassay work, especially plankton mortality studies.

a. Equipment and reagents:

1) *Glassware:* clean, sterile, dry borosilicate glass flasks, beakers, and pipets.

2) *Filters:* 47-mm-diam, 0.45-μm-porosity membrane filters.

3) *Filtration equipment.*

4) *Freezer* (−20 C).

5) *Boiling water bath.*

6) *Detection instruments* designed specifically for measuring ATP.*

7) *Microsyringes:* 10, 25, 50, 100, 250 μL.

8) *Reaction cuvettes and vials.*

9) *Tris buffer* (0.02*M*, pH 7.75): Dissolve 7.5 g trishydroxymethylaminomethane in 3,000 mL distilled water, and adjust to pH 7.75 with 20% HCl. Autoclave 150-mL portions at 115 C for 15 min.

10) *Luciferin-luciferase enzyme preparation:*† Rehydrate frozen (−20 C) lyophilized extracts of firefly lanterns with Tris buffer as directed by the supplier; let stand at room temperature 2 to 3 hr, then centrifuge at 300 *g* for 1 min and decant the supernatant into a clean, dry test tube; let stand at room temperature for 1 hr.

11) *Purified ATP standard* for instrument calibration: Dissolve 12.3 mg disodium ATP in 1 L distilled water and dilute 1.0 mL to 100 mL with Tris buffer; 0.2 mL = 20 ng ATP.

b. Procedure:

1) Calibration. To determine the calibration factor (F), prepare a series of dilutions of ATP standard, record the light emission from several portions of each concentration of standard. Correct mean area of standards by subtracting peak reading or mean area of several blanks using 0.2 mL Tris buffer. Calculate calibration factor F_s as:

$$F_s = \frac{C}{A_s}$$

where:

F_s = calibration factor at sensitivity S,
A_s = peak reading or mean area under standard ATP curve corrected for blank, and
C = concentration of ATP in standard solution, ng/mL.

2) Sample analysis. Collect a 1- to 2-L sample in a clean, sterile sampler. Pass through a 250-μm net to remove large zooplankton[48] and filter through a 47-mm 0.45-μm-porosity filter by applying a vacuum of about 30 kPa. (Important: Break the suction before the last film of water is pulled through the filter.) Quickly place filter in a small beaker. Immediately cover filter with 3 mL boiling Tris buffer, using an automatic pipet. Place beaker in boiling water bath for 5 min and transfer the extract to a clean, dry, calibrated test tube with a Pasteur pipet. Rinse filter and beaker with 2 mL boiling Tris buffer; combine extracts, record volume, bring volume up to 5 mL with Tris buffer, cover tubes with parafilm and, if samples cannot be analyzed immediately, freeze at −25 C. Extracts may be stored for many months in a freezer. Prepare at least triplicate extracts of each sample.

The analytical procedure depends on detection equipment used. If a scintillation counter is used, pipet 0.2 mL enzyme preparation into a glass vial. Measure the light emission of the enzyme preparation (blank) for 2 to 3 min at sensitivity settings near that anticipated for the sample. Add 0.2 mL sample extract to the vial, record the time, and swirl contents of the vial. Start recording light output 10 sec after combining ATP extract and enzyme preparation; record output for 2 to 3 min, using the same time period for all samples. Determine the mean of areas under the curves obtained and correct by subtracting mean of areas under the curves obtained from blanks (prepare blanks according to Strickland and Parsons[9]).

c. Calculations: Calculate concentration of ATP as:

$$ATP,\ \text{ng/L} = \frac{A_c \times V_e \times F_s}{V_s}$$

where:

A_c = mean corrected area under extract curves,
V_e = extract volume, mL,
V_s = volume of sample, L, and
F_s = calibration factor.

*Dupont, Bechman, JRB, or equivalent.
†Dupont, Sigma Chemical, or equivalent.

Total living plankton biomass is given as:

$$B\text{, mg/L}\ddagger = \frac{(\text{ng } ATP/\text{L})}{(2.4)\ (1{,}000)}$$

‡Assuming an ATP content of 2.4 μg ATP/mg dry weight organic matter.[41]

where:

B = plankton biomass as dry weight organic matter.

1002 I. Metabolic Rate Measurements

The physiological condition of the aquatic community and the spectrum of biological interactions must be considered. Earlier, numbers, species composition, and biomass were the prime considerations. Recognition of the limitations of this approach, however, led to the measurement of rates of metabolic processes such as photosynthesis (productivity), nitrogen fixation, respiration, and electron transport. These provide a better understanding of the complex nature of the aquatic ecosystem. An indication of photosynthetic efficiency can be determined by the productivity index (mg C fixed/unit chlorophyll *a*).[49]

1. Nitrogen Fixation

The ability of an organism to fix nitrogen is a great competitive advantage and plays a major role in population dynamics. Two reliable methods for estimating nitrogen fixation rates in the laboratory are the ^{15}N isotope tracer method[50,51] and the acetylene reduction method.[52] Because the rate of nitrogen fixation varies greatly with different organisms and with the concentration of combined nitrogen, nitrogen fixation rates cannot be used to estimate biomass of nitrogen-fixing organisms. However, the acetylene reduction method is useful in measuring nitrogen budgets and in algal assay work.[53]

2. Productivity, Oxygen Method

Productivity is defined as the rate at which inorganic carbon is converted to an organic form. Cholorphyll-bearing plants (phytoplankton, periphyton, macrophytes) serve as primary producers in the aquatic food chain. Photosynthesis results in the formation of a wide range of organic compounds, release of oxygen, and depletion of carbon dioxide (CO_2) in the surrounding waters. Primary productivity[54] can be determined by measuring the changes in oxygen and CO_2 concentrations.[55] In poorly buffered waters, pH can be a sensitive property for detecting variations in the system. As CO_2 is removed during photosynthesis, the pH rises. This shift can be used to estimate both photosynthesis and respiration.[56] The sea and many fresh waters are too highly buffered to make this useful, but it has been applied successfully to productivity studies in some lake waters.

Two methods of measuring the rate of carbon uptake and net photosynthesis in situ are: (*a*) the oxygen method[57] and (*b*) the carbon 14 method.[58] In both methods, clear (light) and darkened (dark) bottles are filled with water samples and suspended at regular depth intervals for an incubation period of several hours or samples are incubated under controlled conditions in environmental growth chambers in the laboratory.

The basic reactions in algal photosynthesis involve uptake of inorganic carbon and release of oxygen, summarized by the relationship:

$$CO_2 + H_2O \rightarrow (CH_2O)_x + O_2$$

The chief advantages of the oxygen method are that it provides estimates of gross and net productivity and respiration and that analyses can be performed with inexpensive laboratory equipment and common reagents. The DO concentration is determined at the beginning and end of the incubation period. Productivity is calculated on the assumption that one atom of carbon is assimilated for each molecule of oxygen released.

a. Equipment:

1) *BOD bottles,* numbered, 300-mL, clear borosilicate glass, with ground-glass stopper and flared mouth, for sample incubation. Acid-clean the bottles, rinse thoroughly with distilled water, and just before use, rinse with the water being tested. Do not use phosphorus-containing detergents.

If suitable opaque bottles are not available, make clear BOD bottles opaque by painting them black and wrapping with black waterproof tape. As a further precaution, wrap the entire bottle in aluminum foil or place in a light-excluding container during incubation.

2) *Supporting line or rack* that does not shade the suspended bottles.

3) *Nonmetallic opaque acrylic Van Dorn sampler* or equivalent, of 3- to 5-L capacity.

4) *Equipment and reagents for dissolved oxygen determinations* (see Section 421).

5) *Pyrheliometer.*

6) *Submarine photometer.*

b. Procedure:

1) Obtain a profile of the input of solar radiation for the photoperiod with a pyrheliometer.

2) Determine depth of euphotic zone (the region that receives 1% or more of surface illumination) with a submarine photometer. Select depth intervals for bottle placement. The photosynthesis-depth curve will be closely approximated by placing samples at intervals equal to one-tenth the depth of the euphotic zone. Estimate productivity in relatively shallow water with fewer depth intervals.

3) Introduce samples taken from each preselected depth into duplicate clear, darkened, and initial-analysis bottles. Insert delivery tube of sampler to bottom of sample bottle and fill so that three volumes of water are allowed to overflow. Remove tube slowly and close bottle. Use water from the same grab sample to fill a "set" (one light, one dark, and one initial bottle).

4) Immediately treat (fix) samples taken for the chemical determination of initial dissolved oxygen (see Dissolved Oxygen, Section 421) with manganous sulfate ($MnSO_4$), alkaline iodide, and sulfuric acid (H_2SO_4) or check with an oxygen probe. Analyses may be delayed several hours if necessary, if samples are fixed or iced and stored in the dark.

5) Suspend duplicate paired clear and darkened bottles at the depth from which the samples were taken and incubate for at least 2 hr, but never longer than it takes for oxygen-gas bubbles to form in the clear bottles or DO to be depleted in the dark bottles.

6) At the end of the exposure period, immediately determine DO as described above.

c. Calculations: The increase in oxygen concentration in the light bottle during incubation is a measure of net production which, because of the concurrent use of oxygen in respiration, is somewhat less than the total (or gross) production. The loss of oxygen in the dark bottle is used as an estimate of respiration. Thus:

Net photosynthesis = light bottle DO − initial DO
Respiration = initial DO − dark bottle DO
Gross photosynthesis = light bottle DO − dark bottle DO

Average results from duplicates.

1) Calculate the gross or net production for each incubation depth and plot:

$$\text{mg carbon fixed/m}^3 = \text{mg oxygen released/L} \times 12/32 \times 1{,}000$$

Use the factor 12/32 to convert oxygen to carbon; 1 mole of O_2 (32 g) is released for each mole of carbon (12 g) fixed.

2) Productivity is defined as the rate of production and generally is reported in grams carbon fixed per square meter per day. Determine the productivity of a vertical column of water 1 m square by plotting productivity for each exposure depth and graphically integrating the area under the curve.

3) Using the solar radiation profile and photosynthesis rate during incubation adjust the data to represent phytoplankton productivity for the entire photoperiod. Because photosynthetic rates vary widely during the daily cycle,[59,60] do not attempt to convert data to other test circumstances.

3. Productivity, Carbon 14 Method

A solution of radioactive carbonate ($^{14}CO_3^{2-}$) is added to light and dark bottles that have been filled with sample as described for the oxygen method. After in situ incubation, the plankton is collected on a membrane filter, treated with hydrochloric acid (HCl) fumes to remove inorganic carbon 14, and assayed for radioactivity. The quantity of carbon fixed is proportional to the fraction of radioactive carbon assimilated.

This procedure differs from the oxygen method in that it affords a direct measurement of carbon uptake and measures only net photosynthesis.[61] It is basically more sensitive than the oxygen method, but fails to account for organic materials that leach from cells[62,63] during incubation.

a. Equipment and reagents:

1) *Pyrheliometer.*

2) *Submarine photometer.*

3) *BOD bottles and supporting apparatus:* See Oxygen Method above.

4) *Membrane-filtering device and 25-mm filters* with pore diameters of 0.22, 0.30, 0.45, 0.80, and 1.2 μm.

5) *Counting equipment for measuring radioactivity:* Scaler with end-window tube, gas flow meter, or liquid scintillation counter (see Part 700). The thin-window tube is the least expensive detector and, when used with a small scaler, provides acceptable data at modest cost.

6) *Fuming chamber:* Use a glass desiccator with a depth of about 1.4 cm conc HCl in desiccant chamber. The fuming chamber is recommended for filter decontamination.[64,65]

7) *2-mL hypodermic syringe* with 15-cm needle.

8) *Chemical reagents:* See Sections 406 (Carbon Dioxide) and 403 (Alkalinity).

9) *Radioactive carbonate solutions:*

a) *Sodium chloride dilution solution,* 5% NaCl (w/v): Add 0.3 g sodium carbonate (Na_2CO_3) and one pellet sodium hydroxide (NaOH) per liter. Use for marine studies only.

b) *Carrier-free radioactive carbonate solution,* commercially available in sealed vials having approximately 5 μCi ^{14}C/mL.

c) *Working solutions* with activities of 1, 5, and 25 μCi ^{14}C/2 mL. For fresh-water studies use carrier-free radioactive carbonate and for marine studies prepare by diluting carrier-free radioactive carbonate solution with NaCl dilution solution.

d) *Stock ampules.* Prepare ampules containing 2 mL of required working solution. Fill ampules using hypodermic needles; autoclave sealed ampules at 121 C for 20 min.[9]

b. Procedure:

1) Obtain a record of incident solar radiation for the photoperiod with a pyrheliometer.

2) Determine depth intervals for sampling and incubation as described above.

3) Use duplicate light and dark bottles at each depth. Fill bottles with sample, add 2 mL radioactive carbonate solution (using the syringe with needle) to the bottom of each bottle, and mix thoroughly by repeated inversion. The concentration of carbon 14 should be approximately 10 μCi/L. To obtain statistical significance, have at least 1,000 cpm in the filtered sample. Take duplicate samples at each depth to determine initial concentration of inorganic carbon (CO_2, HCO_3^-, and CO_3^{2-}) available for photosynthesis (see Carbon Dioxide, Section 406).

4) Incubate samples for up to 4 hr. If measurements are required for the entire photoperiod, overlap 4-hr periods from dawn until dusk. A 4-hr incubation period may be sufficient provided energy input is used as the basis for integrating incubation period to entire photoperiod (incubation procedure, see Oxygen Method).

5) After incubating remove sample bottles and immediately place in dark or preserve by adding 40 mL formalin/L. Filter unpreserved samples without delay.

6) Filter two portions of each sample through a membrane filter, taking care that the largest pore size is consistent with quantitative retention of plankton. Although the 0.45-μm pore filter usually is adequate, determine the efficiency of sample retention immediately before analysis, with a wide range of pore sizes.[66,67] Apply approximately 30 kPa of vacuum during filtration. Excess vacuum may cause extensive cell rupture and loss of radioactivity through the membrane.[68] Use maximum sample volume consistent with rapid filtration (1 to 2 min).

7) Place membranes in HCl fumes for 20 min. Count filters as soon as possible, although extended storage in a desiccator is acceptable.

8) Determine radioactivity by counting with an end-window tube, windowless gas flow detector, or liquid scintillation counter. The efficiency of the counting methods is approximately as follows: thin-window, 3%; windowless gas flow, 50%; liquid scintillation, 40%.

9) Determine counting geometry of thin-window and windowless gas flow detectors.[69] Using three ampules of carbon 14, prepare a series of barium carbonate ($BaCO_3$) precipitates on tared 0.45-μm membrane filters, each precipitate containing the same amount of carbon 14 activity but varying in thickness from 0.5 to 6.0 mg/cm^2. Dilute each ampule to 500 mL with a solution of 1.36 g Na_2CO_3/L CO_2-free distilled water. Pipet 0.5-mL portions into each of seven conical flasks containing 0, 0.5, 1.5, 2.5, 3.5, 4.5, and 5.5 mL, respectively, of a solution of 1.36 g Na_2CO_3/L CO_2-free distilled water. Add, respectively, 0.3, 0.6, 1.2, 1.8, 2.4, 3.0, and 3.6 mL of 1.04% barium chloride ($BaCl_2$) solution. Let $BaCO_3$ precipitate stand 2 hr with gentle swirling every half hour. Collect each precipitate on a filter (using an apparatus with a filtration area comparable to that of the samples). With suction, dry filters without washing; place in a desiccator for 24 hr, weigh, and count. The counting rate increases exponentially with decreasing precipitate thickness. Extrapolate graphically (or mathematically) to zero precipitate thickness and multiply the zero-thickness counting rate by 1,000 to correct for ampule dilution. This represents the amount of activity added to each sample bottle used to determine fraction of carbon 14 taken up in light and dark bottles.

c. Calculations:

1) Subtract the mean dark-bottle sample count from the mean light-bottle counts for each replicate pair.

2) Determine the total dissolved in-

organic carbon available for photosynthesis (carbonate, bicarbonate, and free CO_2) from pH and alkalinity measurements; make direct measurement of total CO_2 according to Section 406 or the methods described in the literature.[70–73]

3) Determine quantity of carbon fixed by using the following relationship:

$$\text{mg carbon fixed/L} = \frac{\text{counting rate of filtered sample}}{\text{total activity added to sample}} \times \frac{300}{\text{volume filtered}} \times \text{mg/L initial inorganic carbon} \times 1.064^{*}$$

*Correction for isotope effect.

4) Integrate productivity for the entire depth of euphotic zone and express as grams carbon fixed per square meter per day (see Oxygen Method, preceding).

5) Using the solar radiation records and photosynthesis rates during incubation, adjust data to represent phytoplankton productivity for the entire photoperiod. If samples were incubated for less than the full photoperiod, apply a correction factor.

1002 J. References

1. PALMER, C.M. 1969. A composite rating of algae tolerating organic pollution. *J. Phycol.* 5:78.
2. PALMER, C.M. 1963. The effect of pollution on river algae. *Bull. N.Y. Acad. Sci.* 108:389.
3. RAWSON, D.S. 1956. Algal indicators of trophic lake types. *Limnol. Oceanogr.* 1:18.
4. STOERMER, E.F. & J.J. YANG. 1969. Plankton Diatom Assemblages in Lake Michigan. Spec. Rep. No. 47, Great Lakes Res. Div., Univ. of Michigan, Ann Arbor.
5. HOLLAND, R.E. 1968. Correlation of *Melosira* species with trophic conditions in Lake Michigan. *Limnol. Oceanogr.* 13:555.
6. PRESCOTT, G.W. 1968. The Algae: A Review. Houghton Mifflin Co., Boston, Mass.
7. PARKER, B.C. & R.F. HATCHER. 1974. Enrichment of surface freshwater microlayers with algae. *J. Phycol.* 10:185.
8. WELCH, P.S. 1948. Limnological Methods. Blakiston Co., Philadelphia, Pa.
9. STRICKLAND, J.D.H. & T.R. PARSONS. 1968. A Practical Manual of Sea Water Analysis. Fish. Res. Board Can. Bull. No. 167. Queens Printer, Ottawa, Ont., Canada.
10. NANSEN, F. 1915. Closing nets for vertical hauls and for horizontal towing. *Publ. Circonst. Cons. Perma. Int. Explor. Mer.* 67:1.
11. MORTIMER, C.H. 1942. The exchange of dissolved substances between mud and water in lakes. *J. Ecol.* 30:147.
12. VOLLENWEIDER, R.A. 1969. A Manual on Methods for Measuring Primary Production in Aquatic Environments. IBP Handbook 12, Blackwell Sci. Publ., England.
13. MEYER, R.L. 1971. A study of phytoplankton dynamics in Lake Fayetteville as a means of assessing water quality. Arkansas Water Res. Center, Publ. 10.
14. WEBER, C.I. 1968. The preservation of phytoplankton grab samples. *Trans. Amer. Microsc. Soc.* 87:70.
15. JUDAY, C. 1916. Limnological apparatus. *Trans. Wis. Acad. Sci.* 18:566.
16. SCHWOERBEL, J. 1970. Methods of Hydrobiology. Pergamon Press, Toronto, Ont., Canada.
17. TRANTER, D.J., ed. 1968. Reviews on Zooplankton Sampling Methods. UNESCO, Switzerland.
18. CLARKE, G.L. & D.F. BUMPUS. 1940. The Plankton Sampler: An Instrument for

Quantitative Plankton Investigations. Spec. Publ. No. 5, Limnol. Soc. Amer. 19. McNabb, C.D. 1960. Enumeration of freshwater phytoplankton concentrated on the membrane filter. *Limnol. Oceanogr.* 5:57.

19. McNabb, C.D. 1960. Enumeration of freshwater phytoplankton concentrated on the membrane filter. *Limnol. Oceanogr.* 5:57.
20. Millipore Filter Corp. 1966. Biological examination of water, sludge and bottom materials. Millipore Techniques, Water Microbiology, p. 25.
21. Patrick, R. & C.W. Reimer. 1967. The Diatoms of the United States. Vol. 1. Monogr. 13, Philadelphia Acad. Natur. Sci.
22. Hohn, M.H. & J. Hellerman. 1963. The taxonomy and structure of diatom populations for three eastern North American Rivers using three sampling methods. *Trans. Amer. Microsc. Soc.* 62:250.
23. Lund, J.W.G., C. Kipling & E.D. LeCren. 1958. The inverted microscope method of estimating algal numbers and the statistical basis of estimations by counting. *Hydrobiologia* 11:143.
24. Jackson, H.W. & L.G. Williams. 1962. Calibration and use of certain plankton counting equipment. *Trans. Amer. Microsc. Soc.* 81:96.
25. Ingram, W.M. & C.M. Palmer. 1952. Simplified procedures for collecting, examining, and recording plankton in water. *J. Amer. Water Works Ass.* 44:617.
26. Palmer, C.M. & T.E. Maloney. 1954. A New Counting Slide for Nannoplankton. Spec. Publ. No. 21, Amer. Soc. Limnol. & Oceanogr.
27. Lackey, J.B. 1938. The manipulation and counting of river plankton and changes in some organisms due to formalin preservation. *Pub. Health Rep.*53:2080.
28. Maddux, W.S. & J.W. Kanwischer. 1965. An *in situ* particle counter. *Limnol. Oceanogr.* 10 (Suppl):R162.
29. Gannon, J.E. 1971. Two counting cells for the enumeration of zooplankton microcrustacea. *Trans. Amer. Microsc. Soc.* 90:486.
30. Owen, B.B., Jr., M. Afzal & W.R. Cody. 1978. Staining preparations for phytoplankton and periphyton. *Br. Phycol. J.* 13:155.
31. Hallegraeff, G.M. 1976. Pigment Diversity, Biomass and Species Diversity of Three Dutch Lakes. Bronder-Offset B.V.—Rotterdam.
32. Lorenzen, C.J. 1967. Determination of chlorophyll and pheo-pigments: spectrophotometric equations. *Limnol. Oceanogr.* 12:343.
33. Fitzgerald, G.P. & S.L. Faust. 1967. A spectrophotometric method for the estimation of percentage degradation of chlorophylls to pheo-pigments in extracts of algae. *Limnol. Oceanogr.* 12:335.
34. Yentsch, C.S. & D.W. Menzel. 1963. A method for the determination of phytoplankton chlorophyll and phaeophytin by fluorescence. *Deep Sea Res.*10:221.
35. Loftus, M.E. & J.H. Carpenter. 1971. A fluorometric method for determining chlorophylls *a, b,* and *c*. *J. Mar. Res.* 29:319.
36. Holm-Hansen, O., C.J. Lorenzen, R.W. Holmes & J.D.H. Strickland. 1965. Fluorometric determination of chlorophyll. *J. Cons. Cons. Perma. Int. Explor. Mer* 30:3.
37. Lorenzen, C.J. 1966. A method for the continuous measurement of *in vivo* chlorophyll concentration. *Deep Sea Res.* 13:223.
38. Parsons, T.R. 1963. A new method for the microdetermination of chlorophyll "c" in seawater. *J. Mar. Res.* 21:164.
39. Patterson, J. & T.R. Parsons. 1963. Distribution of chlorophyll *a* and degradation products in various marine materials. *Limnol. Oceanogr.* 8:355.
40. Vernon, L.P. 1960. Spectrophotometric determination of chlorophyll and pheophytins in plant extracts. *Anal. Chem.* 32:1144.
41. Weber, C.I. 1973. Recent developments in the measurement of the response of plankton and periphyton to changes in their environment. *In* G. Glass, ed. Bioassay Techniques and Environmental Chemistry. Ann Arbor Sci. Publ. Inc., Ann Arbor, Mich.
42. Holm-Hansen, O. & C.R. Booth. 1966. The measurement of adenosine triphosphate in the ocean and its ecological significance. *Limnol. Oceanogr.*11:510.
43. Holm-Hansen, O., W.H. Sutcliffe, Jr. & J. Sharp. 1968. Measurement of deoxyribonucleic acid in the ocean and its ecological significance. *Limnol. Oceanogr.* 13:507.

44. HOLM-HANSEN, O. 1969. Determination of microbial biomass in ocean profiles. *Limnol. Oceanogr.* 14:740.

45. CREITZ, G.I. & F.A. RICHARDS. 1955. The estimation and characterization of plankton populations by pigment analysis. *J. Mar. Res.* 14:211.

46. KUTKUHN, J.H. 1958. Notes on the precision of numerical and volumetric plankton estimates from small sample concentrations. *Limnol. Oceanogr.* 3:69.

47. NELSON, D.J. & D.C. SCOTT. 1962. Role of detritus in the productivity of a rock-outcrop community in a Piedmont stream. *Limnol. Oceanogr.* 7:396.

48. RUDD, J.W.M. & R.D. HAMILTON. 1973. Measurement of adenosine triphosphate (ATP) in two precambrian shield lakes of northwestern Ontario. *J. Fish. Res. Board Can.* 30:1537.

49. GUNDERSEN, K. 1973. *In-situ* determination of primary production by means of the new incubator, ISIS. *Helgolander wiss. Meeresunters.*

50. BURRIS, R.H., F.J. EPPLING, H.B. WAHLIN & P.W. WILSON. 1942. Studies of biological nitrogen fixation with isotopic nitrogen. *Proc. Soil Sci. Soc. Amer.* 7:258.

51. NEESS, J.C., R.C. DUGDALE, V.A. DUGDALE & J.J. GOERING. 1962. Nitrogen metabolism in lakes. I. Measurement of nitrogen fixation with N^{15}. *Limnol. Oceanogr.* 7:163.

52. STEWART, W.D.P., G.P. FITZGERALD & R.H. BURRIS. 1967. *In situ* studies on N_2 fixation using the acetylene reduction technique. *Proc. Nat. Acad. Sci.* 58:2071.

53. STEWART, W.D.P., G.P. FITZGERALD & R.H. BURRIS. 1970. Acetylene reduction assay for determination of phosphorus availability in Wisconsin lakes. *Proc. Nat. Acad. Sci.* 66:1104.

54. GOLDMAN, C.R. 1968. Aquatic primary production. *Amer. Zoologist* 8:31.

55. ODUM, H.T. 1957. Primary production measurements in eleven Florida springs and a marine turtle-grass community. *Limnol. Oceanogr.* 2:85.

56. BEYERS, R.J. & H.T. ODUM. 1959. The use of carbon dioxide to construct pH curves for the measurement of productivity. *Limnol. Oceanogr.* 4:499.

57. GAARDER, T. & H.H. GRAN. 1927. Investigations of the production of plankton in Oslo Fjord. *Rapp. Proces-Verbaux. Reunions Cons. Perma. Int. Explor. Mer* 42:1.

58. STEEMAN-NIELSEN, E. 1952. The use of radioactive carbon (C-14) for measuring organic production in the sea. *J. Cons. Perma. Int. Explor. Mer* 18:117.

59. RYTHER, J.H. 1956. Photosynthesis in the ocean as a function of light intensity. *Limnol. Oceanogr.* 1:61.

60. FEE, E.J. 1969. A numerical model for the estimation of photosynthetic production, integrated over time and depth, in natural waters. *Limnol. Oceanogr.* 14:906.

61. STEEMAN-NIELSEN, E. 1964. Recent advances in measuring and understanding marine primary production. *J. Ecol.* 52(Suppl.):119.

62. ALLEN, M.B. 1956. Excretion of organic compounds by *Chlamydomonas*. *Arch. Mikrobiol.* 24:163.

63. FOGG, G.E. & W.D. WATT. 1965. The kinetics of release of extracellular products of photosynthesis by phytoplankton. *In* C. R. Goldman, ed. Primary Productivity in Aquatic Environments. Suppl. 18, Univ. California Press, Berkeley.

64. WETZEL, R.G. 1965. Necessity for decontamination of filters in C^{14} measured rates of photosynthesis in fresh waters. *Ecology* 46:540.

65. MCALLISTER, C.D. 1961. Decontamination of filters in the C^{14} method of measuring marine photosynthesis. *Limnol. Oceanogr.* 6:447.

66. LASKER, R. & R.W. HOLMES. 1957. Variability in retention of marine phytoplankton by membrane filters. *Nature* 180:1295.

67. HOLMES, R.W. & C.G. ANDERSON. 1963. Size fractionation of C^{14}-labelled natural phytoplankton communities. *In* C.H. Oppenheimer, ed. Symp. on Marine Microbiology. Charles C. Thomas, Springfield, Ill.

68. ARTHUR, C.R. & F.H. RIGLER. 1967. A possible source of error in the C^{14} method of measuring primary productivity. *Limnol. Oceanogr.* 12:121.

69. JITTS, H.R. & B.D. SCOTT. 1961. The determination of zero-thickness activity in Geiger counting of C^{14} solutions used in marine productivity studies. *Limnol. Oceanogr.* 6:116.

70. SAUNDERS, G.W., F.B. TRAMA & R.W. BACHMANN. 1962. Inst. of Science and Technology, Univ. of Michigan. Great Lakes Res. Div. Publ. No. 8.
71. DYE, J.F. 1944. The calculation of alkalinities and free carbon dioxide in water by use of nomographs. *J. Amer. Water Works Ass.*36:859.
72. MOORE, E.W. 1939. Graphic determination of carbon dioxide and the three forms of alkalinity. *J. Amer. Water Works Ass.* 31:51.
73. PARK, K., D.W. HOOD & H.T. ODUM. 1958. Diurnal pH variation in Texas bays and its application to primary production estimations. *Publ. Inst. Mar. Sci. Univ. Tex.* 5:47.

1003 PERIPHYTON

1003 A. Introduction

Communities of microorganisms growing on stones, sticks, aquatic macrophytes, and other submerged surfaces are useful in assessing the effects of pollutants on lakes and streams. Included in this group of organisms, here designated periphyton,[1,2] are the zoogleal and filamentous bacteria, attached protozoa, rotifers, and algae, and also the free-living microorganisms found swimming, creeping, or lodged among the attached forms.

Unlike the plankton, which often do not respond fully to the influence of pollution in rivers for a considerable distance downstream, the periphyton show dramatic responses immediately below pollution sources. Examples are the beds of *Sphaerotilus* and other "slime organisms" commonly observed in streams below discharges of organic wastes. Because the abundance and composition of the periphyton at a given location are governed by the water quality at that point, observations of their condition generally are useful in evaluating conditions in streams.

The use of periphyton in assessing water quality often is hindered by the lack of suitable natural substrates at the desired sampling station. Furthermore, it often is difficult to collect quantitative samples from these surfaces. To circumvent these problems artificial substrates have been used to provide a uniform surface type, area, and orientation.[3]

1003 B. Sample Collection

1. Station Selection

In rivers, locate stations a short distance upstream and at one or more points downstream from the suspected pollution source or intended study area. In large rivers, sample both sides of the stream. Because the effects of a pollutant depend on the assimilative capacity of the stream and on the nature of the pollutant, progressive changes in water quality downstream from the pollution source may be caused entirely by dilution and cooling—as in the case of nutrients, toxic industrial wastes, and thermal pollution—or to gradual mineralization of degradable organics. Cursory examination of shoreline and bottom periphyton growths downstream from an outfall may disclose conspicuous zones of biological response to water quality that will be useful in determining appropriate sites for sampling stations. When an intensive sampling program is not feasible, a minimum of two sampling stations, one in a reference area upstream from a pollution source and the other in the community downstream from the source, where complete mixing with the receiving water has occurred, will provide data on the periphyton community.

In lakes, reservoirs, lentic waters, and other standing-water bodies where zones of pollution may be arranged concentrically, locate stations in areas adjacent to a waste outfall and in unaffected areas.

2. Sample Collection

a. Natural substrates: Collect qualitative samples by scraping submerged stones, sticks, pilings, and other available substrates. Many devices have been developed to collect quantitative samples from irregular surfaces, but success rarely is achieved.

b. Artificial substrates: The most widely used artificial substrate is the standard, plain, 25- by 75-mm glass microscope slide, but other materials such as clear vinyl plastic and asbestos also are suitable. Do not change substrate type during a study because colonization varies with substrate. In small, shallow streams and in the littoral regions of lakes where light is transmitted to the bottom, place slides or other substrates in frames anchored to the bottom. In large, deep streams or standing-water bodies where turbidity varies widely, place slides vertically with the slide face at right angles to the prevailing current. A floating rack, as shown in Figure 1003:1, is suitable. Expose several slides for each type of analysis to assure collecting sufficient material and to determine variability in results caused by normal differences in colonization of individual slides. In addition to effects of pollutants, length of substrate exposure and seasonal changes in temperature and other natural environmental conditions may have a profound effect on sample composition. No community on an artificial substrate is completely representative of the natural community.

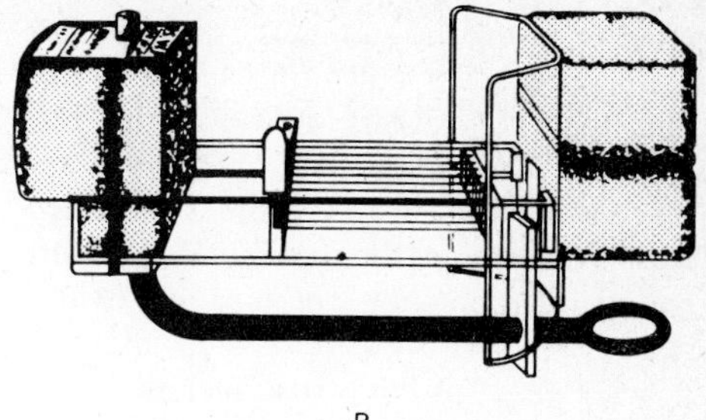

B

Figure 1003:1. Periphyton sampler. Floating slide rack constructed of clear vinyl plastic and styrofoam, used in streams and lakes. Source: PATRICK, R., M.H. HOHN and J.H. WALLACE. 1954. A new method of determining the pattern of the diatom flora. *Bull. Philadelphia Acad. Natur. Sci.* 259:1.

Place, expose, and handle all artificial substrate samplers in as near identical conditions as possible irrespective of their being replicate samplers at a particular sampling location or samplers at different locations. Sampler type and/or construction cause changes in surrounding physical conditions that in turn affect periphyton growth. Variations of 10 to 25% between sample replicates are not uncommon. Therefore, to reduce sampling error and increase interpretive power, reduce the magnitude of all possible test variables and use maximum replication.

c. Exposure period: Colonization on clean slides proceeds at an exponential rate for the first 1 or 2 wk and then slows. Because exposures of less than 2 wk may result in very sparse collections, and exposures of more than 2 wk may result in loss of material due to sloughing, 2 wk usually constitutes the optimum sampling interval during the summer. This exposure period precludes collecting sexually mature thalli of larger, slow-growing filamentous algae such as *Cladophora* and *Stigeoclonium*. For the most exacting work, determine the optimum exposure period by pretesting colonization over a period of about 6 wk.

Secondary problems associated with macroinvertebrate infestation and grazing may occur, often within 7 to 14 days. To reduce the confounding influence of grazing, increase substrate sampling area and expose for 7 to 10 days.

3. Sample Preservation

Preserve samples that are taken for counting and identification in 5% neutralized formalin or merthiolate (see Section 1002B.4).

Preserve slides intact in bottles of suitable size or scrape into containers in the field. Air-dry slides for dry and ash-free dry weight in the field and store in a 3.0- × 7.7-cm glass bottle. Place slides for chlorophyll analyses in acetone in the field or collect and freeze with trichlorotrifluoroethane* or CO_2 and hold on dry ice until returned to the laboratory. Store all samples in the dark.

*Freon or equivalent.

1003 C. Sample Analysis

1. Sedgwick-Rafter Counts

Remove periphyton from slides with a razor blade and rubber policeman. Do not include periphyton on the edges of the slide. Disperse scrapings in 100 mL or other suitable volume of preservative with vigorous shaking, or use a blender. Transfer a 1-mL portion to a Sedgwick-Rafter cell, and make a strip count as described in Section 1002F.2*a*. If material in the Sedgwick-Rafter cell is too dense to count directly, discard and replace with a diluted sample.

Express the counts as cells or filaments per square millimeter of substrate area, calculated as follows:

1) Cells/mL suspended scrapings

$$= \frac{\text{actual count/strip}}{\text{volume of 1 strip, mL}}$$

2) Cells/mm² slide surface

$$= \text{cells/mL suspended scrapings} \times \frac{\text{total volume of scrapings}}{\text{area of slide or slides, mm}^2}$$

2. Diatom Species Proportional Counts

Preparation of permanent diatom mounts from periphyton samples differs from preparation of mounts from plankton samples because of the need to remove extracellular organic matter (such as gelatinous materials). If this is not removed it will produce a thick brown or black carbonaceous deposit on the cover glass when the sample is incinerated. Decompose organic substances by oxidation with ammonium persulfate or HNO_3 (or H_2O_2, 30%) and $K_2Cr_2O_7$ (see Section 1002D.3) before mounting sample. To oxidize with persulfate place approximately 5 mL sample in a disposable 10-mL vial. Let stand 24 hr, withdraw supernatant liquid by aspiration, replace with a 5% solution of $(NH_4)_2S_2O_8$, and mix thoroughly. Do not exceed a total volume of 8 mL. Heat vial to approximately 90 C for 30 min. Let stand 24 hr, withdraw supernatant liquid, and replace with distilled water. After three changes of distilled water, with a disposable pipet transfer a drop of the diatom suspension to a cover glass, evaporate to dryness, and prepare and count a mount as described for plankton (Section 1002). Count as least 500 frustules and express results as organisms per square millimeter.

3. Stained Sample Preparation and Counting

Staining periphyton samples permits distinguishing algae from detritus and "live" from "dead" diatoms. This distinction is especially important because periphyton often acts as a graveyard for dead diatoms of planktonic as well as periphytic origin. In this method all algal components of periphyton may be studied in one preparation, without sacrificing detailed diatom taxonomy.[4] It yields permanent slides for reference collections.

Thoroughly mix preserved samples in the preservative solution by using a blender for a few seconds. Prepare acid fuchsin stain by dissolving 1 g acid fuchsin in 100 mL distilled water, add 2 mL glacial acetic acid, and filter. Place a measured sample in a centrifuge tube with 10 to 15 mL acid fuchsin stain. Mix sample and stain several times during a 20-min staining period; centrifuge at 1,000 *g* for 20 min.

Decant stain, being careful not to disturb sediment or siphon off supernatant. Add 10 to 15 mL 90% propanol, mix, centrifuge for 20 min, and decant supernatant. Repeat using two washes of 100% propanol and one wash of xylene. Centrifuge, decant xylene, and add fresh xylene. At this stage, store sample in well-sealed vials or prepare slides.

Slides for quantitative periphyton examinations require random dispersion of a known amount of xylene suspension. Use a microstirrer to break up clumps of algae before removing sample portion from xylene suspension. Count a number of drops of suspended sample into a thin ring of mounting medium* on a slide. Mix the xylene suspension and medium with a spatula until the xylene has evaporated. Warm the slide on a hot plate at 45 C and cover sample with a cover slip.

*Hyrax, Custom Research and Development, Inc. 8500 Mt. Vernon Rd, Auburn, Calif.

Count organisms on the prepared slides using the magnification most appropriate to the desired level of taxonomic identification. Count strips or random fields. Calculate algal density per unit area of substrate:

$$\text{Organisms/area sampled} = \frac{N \times A_t \times V_t}{A_c \times V_s \times A_s}$$

where:

N = number of organisms counted,
A_t = total area of slide preparation, mm^2,
V_t = total volume of original sample suspension, mL,
A_c = area counted (strips or fields), mm^2,
V_s = sample volume used to prepare slide, mL, and
A_s = surface area of slide or substrate, mm^2.

4. Dry and Ash-Free Weight

Collect at least three replicate slides for weight determinations.[5] Slides air-dried in the field can be stored indefinitely if protected from abrasion, moisture, and dust. While weights can be obtained from the material used for chlorophyll determinations, preferably use slides expressly designated for dry and ash-free weight analysis.

a. Equipment:

1) *Analytical balance,* with a sensitivity of 0.1 mg.

2) *Drying oven,* double-wall, thermostatically controlled to within ±1 C.

3) *Electric muffle furnace* with automatic temperature control.

4) *Crucibles,* porcelain, 30-mL capacity.

5) *Single-edge razor blades or rubber policeman.*

b. Procedure:

1) If dry and ash-free weights are to be obtained from the material used for

chlorophyll determinations, combine the particulate matter and the acetone extract from each slide, evaporate acetone in a hood on a steam bath or in an explosion-proof oven, dry to constant weight at 105 C, and ignite for 1 hr at 500 C. If weights are to be obtained from field-dried material, re-wet dried material with distilled water and remove from slides with a razor blade or rubber policeman. Place scrapings from each slide in a separate prewashed, prefired, tared crucible; dry to constant weight at 105 C; cool in a desiccator and weigh; and ignite for 1 hr at 500 C.

2) Re-wet ash with distilled water and dry to constant weight at 105 C. This reintroduces water of hydration of clay and other minerals, which is not driven off at 105 C but is lost during ashing. If not corrected for, this water loss will be recorded as volatile organic matter.[6]

c. Calculations: Calculate the mean weight from the slides and report as dry weight and ash-free weight per square meter of exposed surface. If 25- by 75-mm slides are used, then

$$\text{g/m}^2 = \frac{\text{g/slide (average)}}{0.00375}$$

5. Chlorophyll and Pheophytin

The chlorophyll content of attached communities is a useful index of the phytoperiphyton biomass. Because quantitative chlorophyll determinations require the collection of periphyton from a known surface area, use artificial substrates. Extract the pigments with aqueous acetone (see Section 1002G) and complete analysis using a spectrophotometer or fluorometer. If immediate pigment extraction is not possible, samples may be stored frozen for as long as 30 days if kept in the dark.[7] The ease with which chlorophylls are removed from cells varies considerably with different algae: to achieve complete pigment extraction disrupt the cells mechanically with a grinder, blender, or sonic disintegrator, or freeze them. Grinding is the most rigorous and effective of these methods.

The Autotrophic Index (AI) is a means of determining the trophic nature of the periphyton community (see Section 1002G). It is calculated as follows:

$$\text{AI} = \frac{\text{Biomass (ash-free weight of organic matter), mg/m}^2}{\text{Chlorophyll } a\text{, mg/m}^2}$$

Normal AI values range from 50 to 200; larger values indicate heterotrophic associations or poor water quality. Nonviable organic material affects this index. Depending on the community, its location and growth habit, and method of sample collection, there may be large amounts of nonliving organic material that may inflate the numerator and produce disproportionately high AI values. Nonetheless, the AI is a useful means of describing changes in periphyton communities between sampling locations.

a. Equipment and reagents: See Section 1002G.

b. Procedure: In the field, place individual glass microscope slides used as substrates directly into 100 mL of a mixture of 90% aqueous acetone and 10% saturated $MgCO_3$ solution. Immediately store on dry ice in the dark. (NOTE: Vinyl plastic is soluble in acetone. If vinyl plastic is used as the substrate, scrape periphyton from it before solvent extraction.) If extraction cannot be carried out immediately, freeze samples in the field and keep frozen until processed.

Rupture cells by grinding in a tissue homogenizer and steep in acetone for 24 hr in the dark at or near 4 C.

To determine pigment concentration, follow the procedures given in Section 1002G.

c. Calculation: After determining pigment concentration in the extract, calculate amount of pigment per unit surface area of sample as follows:

$$\text{mg chlorophyll } a/\text{m}^2 = \frac{C_a \times \text{volume of extract, L}}{\text{area of substrate, m}^2}$$

where:

C_a is as defined in Section 1002G.

1003 D. Productivity

The productivity of periphyton communities is a function of water quality, substrate, and seasonal patterns in temperature and solar illumination. It may be estimated from temporal changes in standing crop (biomass) or from the rate of oxygen evolution or carbon uptake.[8]

1. Biomass Accumulation

a. Ash-free dry weight: The accumulation rate of organic matter on artificial substrates by attachment, growth, and reproduction of colonizing organisms has been used widely to estimate the productivity of streams and reservoirs.[9,10] To use this method, expose several replicate clean substrates for a predetermined period, scrape the accumulated material from the slides, and ash as described previously.

$$P = \frac{\text{mg ash-free weight/slide}}{tA}$$

where:

P = net productivity, mg ash-free weight/ m^2/day,
t = exposure time, days, and
A = area of a slide, m^2.

Obtain estimates of seasonal changes in standing crop of established communities by placing many replicate substrates at a sampling point and then retrieving a few at a time at regular intervals. The recommended collection interval ranges from 2 to 4 wk for a year or longer.[9] Gain in ash-free weight per unit area from one collection period to the next is a measure of net production.

b. ATP estimates: Measurement of adenosine triphosphate (ATP) has been used in recent years to estimate microbial biomass in water. This technic is applicable to periphyton.[11] It provides an additional tool for assessing the magnitude and rate of biomass accumulation on substrates in natural waters. At present, the procedure should be limited to communities colonizing artificial substrates.

1) *Equipment and reagents:* See Section 1002H.5*a*.

2) *Procedure:* Either scrape periphyton from an exposed artificial substrate or, if standard glass microscope slides are used, place them in polyethylene slide mailers containing preheated (99 C) tris buffer. Immerse in a boiling water bath for 10 min to extract ATP. If samples are not assayed immediately, freeze at −25 C; they may be stored in a freezer for up to several months. Complete analysis as directed in Section 1002H.5*b*. Slides exposed in waters containing high turbidity may collect substantial amounts of particulates including clays. ATP sorbs to these materials; the sorption results in a quenching effect.

3) *Calculations:* See Section 1002H.5*c*.

2. Standing Water Productivity Measured by Oxygen Method

Hourly and daily rates of oxygen evolution and carbon uptake by periphyton growing in standing water can be studied by confining this community briefly in bottles, bell jars, or other chambers. In contrast, the metabolism of organisms in flowing water is highly dependent on current velocity and cannot be determined with precision under static conditions. Productivity estimates for flowing waters and those for standing waters present different problems; therefore, separate procedures are given.

Productivity and respiration of epilithic and epipelic periphyton in littoral regions of lakes and ponds can be determined by inserting transparent and opaque bell jars or open-ended plastic chambers into substrata along transects perpendicular to the shoreline.[12,13] Chambers are left in place for one-half the photoperiod. The DO concentration in a chamber is determined at the beginning and end of the exposure period. Gross productivity is the sum of the net gain in DO in the transparent chamber and the oxygen used in respiration. Values obtained are doubled to determine productivity for the entire photoperiod.

Failure to account for changes in DO in chambers caused by plankton photosynthesis and respiration may cause serious errors in the estimates of periphyton metabolism. It is essential that these values be obtained at the time the periphyton is studied by using the light- and dark-bottle method (see Section 1002I).

a. Equipment and reagents:

1) *Clear and darkened glass or plastic* chambers,* approximately 20 cm in diameter and 30 cm high, with a median lateral port, sealed with a serum bottle stopper for removal of small water samples for DO analyses or for the insertion of an oxygen probe. Fit the chamber with a small, manually operated, propeller-shaped stirring paddle.

2) *Dissolved oxygen probe, or equipment and reagents required for Winkler dissolved oxygen determinations:* See Section 421.

b. Procedure: At each station place both a transparent and an opaque chamber over the substrate at sunrise or noon and leave in place for one-half the photoperiod. In extremely productive environments or to define the hourly primary productivity changes throughout the day, use incubation periods shorter than one-half the photoperiod. The minimum incubation period giving reliable results is 2 hr. Determine DO concentration at the beginning of the incubation period.

Include a set of Gaarder-Gran light- and dark-bottle productivity and respiration measurements with each set of chambers to obtain a correction for phytoplankton metabolism. Incubate for the same time period as the chambers. See Section 1002I.

At end of exposure period, carefully mix the water in the chambers and determine DO concentration.

c. Calculations: When the exposure period is one-half of the photoperiod, calculate gross primary productivity of the periphyton community as:

$$P_G = \frac{2\,[V_c(C'_{fc} - C'_{ic}) + V_o(C'_{io} - C'_{fo})]}{A}$$

where:

P_G = gross production, mg $O_2/m^2/day_{12hr}$,
V_c = volume of clear chamber, L,
C'_{fc} and
C'_{ic} = final and initial concentrations, respectively, of DO in the clear chamber, mg/L, corrected for phytoplankton metabolism,
V_o = volume of opaque chamber, L,
C'_{io} and
C'_{fo} = initial and final concentrations, respectively, of DO in the opaque chamber, mg/L, corrected for phyto-

*Plexiglas or equivalent.

plankton metabolism, and
A = substrate area, m^2.

Correct for the effects of phytoplankton metabolism in the overall oxygen change in the clear chamber by the following equations:

$$C'_{fc} = C_{fc} - C_{flb}$$
$$C'_{ic} = C_{ic} - C_{ilb}$$
$$C'_{fo} = C_{fo} - C_{fdb}$$
$$C'_{io} = C_{io} - C_{idb}$$

where:

C_{fc} = final DO concentrations in clear chamber, mg/L,
C_{flb} = final DO concentrations in light bottle, mg/L,
C_{ic} = initial DO concentration in clear chamber, mg/L,
C_{ilb} = initial DO concentration in light bottle, mg/L,
C_{fo} = final DO concentration in opaque chamber, mg/L,
C_{fdb} = final DO concentration in dark bottle, mg/L,
C_{io} = initial DO concentration in opaque chamber, mg/L, and
C_{idb} = initial DO concentration in dark bottle, mg/L.

Calculate periphyton community respiration by:

$$R = \frac{24\ V_0(C'_{io} - C'_{fo})}{tA}$$

where:

R = community respiration, mg $O_2/m^2/day_{24hr}$, and
t = length of exposure, hr.

Determine the net periphyton community (P_N) as the difference:

$$P_N = P_G - R$$

If the incubation time is different from one-half the photoperiod, modify the daily gross production calculation as follows:

$$P_G = \frac{t_p[V_c(C'_{fc} - C'_{ic}) + V_o(C'_{io} - C'_{fo})]}{tA}$$

where:

t_p = length of the daily photoperiod, hr.

Community respiration and net production calculations for incubation periods other than one-half the photoperiod are not changed.

3. Standing Water Productivity Measured by Carbon 14 Method

The approach is similar to that described above for the oxygen method. Transparent and opaque chambers are placed over the substrate, carbon 14-labeled Na_2CO_3 is injected into the chamber by syringe, mixed well, and allowed to incubate with the periphyton for one-half the photoperiod. The concentration of dissolved inorganic carbon available for photosynthesis is determined by titration. At the end of the incubation period, the periphyton is removed from the substrate and assayed for carbon 14.[12,14]

a. Equipment and reagents:

1) *Incubation chamber:* See Section 1003D.2*a*.

2) *Special equipment and reagents:* See Section 1002I.

3) *Carbon 14-labeled solution of sodium carbonate,* having a specific activity of approximately 10 μCi/mL.

4) *Other equipment and reagents:* See Section 406.

b. Procedure: At each station place a transparent and opaque chamber over the substrate and add approximately 10 μCi carbon 14/L of chamber volume. Mix water in the chambers well, taking care to avoid disturbing the periphyton. Determine concentration of dissolved inorganic carbon as described in Section 403. At end of exposure period, remove surface centimeter of periphyton enclosed in the chamber, freeze, and store frozen in a vacuum desiccator.

Immediately before analysis, expose sample to fumes of HCl for 10 to 15 min to drive off all inorganic carbon 14 retained in the periphyton. Combust sample (or portion) by the Van Slyke method[14] and assay radioactivity by one of the following methods: (*a*) flush CO_2 produced by combustion into a gas-flow counter or electrometer; (*b*) take it up in a 0.1*N* solution of Na_2CO_3, precipitate as $BaCO_3$ on a membrane filter, and count with an end-window tube; or (*c*) assay as the Na_2CO_3 solution by the liquid scintillation technic.

c. *Calculations:*

$$P_N = \frac{\text{activity in sample}}{\text{activity added}} \times \frac{\text{dissolved inorganic carbon}}{\text{area of substrate}} \times 1.064$$

where:

P_N = net productivity for exposure period, and

1.064 = correction for isotope effect.

4. Flowing Water Productivity Measured by Oxygen Method

Primary productivity of the periphyton community in a stream or river ecosystem can be related to demand changes in DO. These changes are the integrated effects of photosynthesis, affected by light levels and turbidity, that is carried out during the photoperiod by stream plankton, periphyton, and the submerged portions of macrophytes. Respiration results from metabolism of plant communities, aquatic animals, and attached and free-floating microbial heterotrophs. Water depth, turbulence, and water temperature all influence the process of reaeration. Oxygen also can enter by accrual of groundwater and surface change. Daily fluctuations in photosynthetic production of oxygen are imposed on the relatively steady demand of respiratory activity. However, this latter process may fluctuate greatly in streams receiving a significant load of organic wastes, particularly under intermittent loads such as oxygen demand from urban stormwater runoff. Respiration rates also may vary diurnally under certain conditions, but the factors involved are not well understood.

The rate of change in stream DO (q) in grams per cubic meter per hour is represented by the following function of the photosynthetic rate (p), respiration (r), reaeration (d), and accrual from groundwater inflow and surface runoff (a):[15]

$$q = p - r + d + a$$

If the equation is multiplied through by depth in meters (z), the resulting values are in terms of grams oxygen per square meter per hour. Figure 1003:2 illustrates this conceptual relationship between q, primary productivity, and respiration of the stream community.

The procedure measures the time-variable oxygen concentrations in a stream over a 24-hr period. Compensations are made for oxygen changes due to physical factors (accrual and reaeration) and the rate of oxygen change due to biological activity that is separated into components due to respiration and primary production. The metabolic rates are the sum of the activity of the entire stream community. Planktonic productivity and respiration can be separated easily from overall community activity by the use of the light- and dark-bottle oxygen technic (see Section 1002I). However, in most small streams planktonic production is insignificant. The component of production and respiration due to macrophytes is very difficult to separate from periphytic metabolic activity in systems where vascular plants are common.

Because periphyton attach to plant surfaces as well as nonliving substrates, radiotracer technics are required to separate the component of production due to macrophytes from that due to attached algae.[16] When vascular plants are present use technics discussed in Section 1004 to esti-

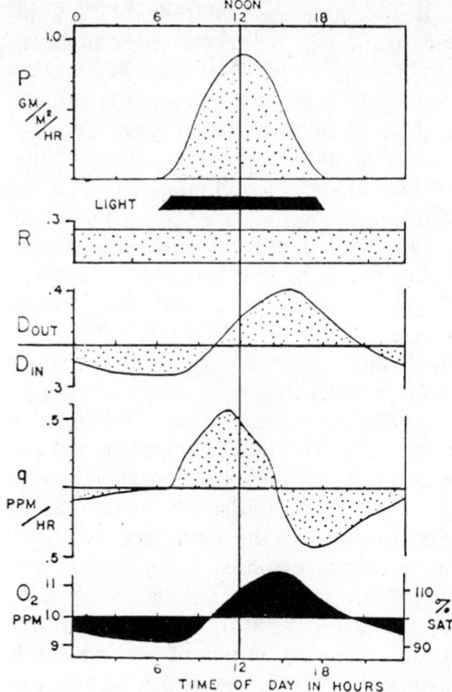

Figure 1003:2. Component processes in the oxygen metabolism of a section of a hypothetical stream during the course of a cloudless day. Production (P), respiration (R), and diffusion (D) are given on an areal basis. The combined effect of these rate processes for a stream 1 m deep is given in mg/L/hr (q). The actual oxygen values that would result in a stream with a long homogeneous community are given in the lowermost curve. Source: ODUM, H.T. 1956. Primary production in flowing waters. *Limnol. Oceanogr.* 1:102.

mate their contribution to net primary productivity.

Respiration by fish and benthic fauna also is difficult to quantitate directly and usually is not separated from periphyton respiration. If compartmentalized animal metabolism is required, calculate this contribution from laboratory respiration rates extrapolated to the field situation based on animal population sizes.[17,18]

Estimate primary productivity in flowing water by either the free water demand method or the Thomas-O'Connell[19] chamber method. The first does not introduce artificiality to the system; however, it is difficult to separate the components of metabolic activity except for the contribution due to plankton. The chamber method measures periphyton activity alone.[20–23]

Depending on the hydrologic characteristics of the stream system, accrual and reaeration may be significant. Accrual can be accounted for by simple mixing equations if estimates of the accrued flow and its oxygen concentration are known. In practice, select for study reaches that do not incur significant accrual. Measure reaeration rates either directly[20–23] or estimate them from physical and hydrodynamic features of the stream itself.[22,23]

a. Equipment:

1) *BOD bottles,* for light- and dark-bottle measurements. See Section 1002I.

2) *DO meter and probe* for measurement of DO.

3) *Bottom chamber,* 60 × 20 × 10 cm, with 32-cm lengthwise dividing baffle, rheostat-controlled submersible pump, temperature thermistor, and DO probe.[19] Use clear and opaque plastic sleeves for covering chamber and petri dishes or other means of placing periphyton within chambers.

4) *Current meter,* capable of detecting water current velocities ranging from 0.03 to 3 m/sec in water depths as shallow as 0.3 m.

5) *Tape measure (30 m) and depth staff,* or similar equipment, as required to measure stream cross sections.

6) *Fluorometer,* capable of detecting fluorescent dye concentration at 0.5 to 100 μg/L (required only if direct measurement of reaeration is made).

7) *Liquid scintillation counter,* capable of sensitive detection of ^{85}Kr and ^{3}H (re-

quired only if direct measurement of reaeration is made).

b. Procedure:

1) Light- and dark-chamber method—Grow samples of typical periphyton communities on artificial substrate or collect natural material. Transfer identical portions to both clear and opaque chambers, taking care to use sufficient periphyton to make the ratio of chamber volume to periphyton area equivalent to the ratio of stream volume to periphyton substrate area. Measure current in the stream and match the circulation rate in the clear and opaque chambers to the current. Measure DO concentrations initially in both clear and opaque chambers and after 1 to 3 hr to estimate the rate of oxygen increase or decrease. Make concurrent measurements of phytoplankton activity using light- and dark-bottle technics as described in Section 1002I.2. Incubate light and dark bottles for the same time interval as the chambers.

Make several measurements during the photoperiod to define daily primary productivity. In addition, collect sufficient natural substrate samples of the study reach to estimate periphyton biomass (see Section 1003B). At end of incubation period harvest enclosed periphyton and determine ash-free biomass (see Section 1003B).

2) Free-water diurnal curve methods—Measure, hourly or continuously, DO concentration and water temperature for a 24-hr period at one or two stations, depending on stream conditions, precision desired, and availability of equipment. If similar conditions exist for some distance upstream from the reach being studied, diurnal measurements of DO at a single station are sufficient to determine productivity. Where upstream conditions are significantly different from those in the reach being studied, make measurements at the upstream and downstream limits of the reach.

If the single-station method is used, measure depth at several points along the study reach to define average depth. Map and/or make physical surveys to estimate magnitude of possible sources of accrual via effluents or tributary streams and springs. If the two-station method is used, measure the wetted cross-sectional stream area as well as current velocity at several points to define flow (in cubic meters per second) and average cross-sectional area. Correct for phytoplankton activity by light- and dark-bottle measurements (see Section 1002I.2).

3) Direct measurement of reaeration[21] (TENTATIVE)—Under special circumstances it may be desirable to estimate reaeration directly although the results may not be more accurate than those of the empirical formulations usually used. The tracer gas technic is satisfactory, but is difficult and requires sophisticated equipment not routinely available. Use this method with care and with full recognition of its restrictions. Depending on stream flow, release 10 to 250 μCi ^{85}Kr with 5 to 125 μCi ^{3}H at the upstream end of the reach together with sufficient fluorescent dye to produce a concentration of 10 μg/L when completely mixed across the river cross section. Make fluorometric measurements at the downstream end of the reach until the dye peak appears, then collect water samples to measure the ^{85}Kr/^{3}H ratio by liquid scintillation technics. Record time of travel for the dye peak from the injection point.

c. Calculations:

1) Chamber method—Calculation is analogous to that used for the bell jar technic discussed in Section 1003D.2.

$$P_n = \frac{V_c(C'_{fc} - C'_{ic})\, B}{tW_c}$$

where:

P_n = hourly rate of net primary production, mg $O_2/m^2/hr$,

V_c = volume of clear chamber, L,

B = average periphyton biomass estimated for the study reach, mg/m^2,
t = incubation period, hr,
W_c = total biomass of periphyton contained in clear chamber, mg,
C'_{fc} = final oxygen concentration in clear chamber, corrected for phytoplankton metabolism, mg/L:
$C'_f = C_{fc} - C_{flb}$
C_{fc} = final DO in clear chamber,
C_{flb} = final DO in light bottle, and
C'_{ic} = initial oxygen concentration in clear chamber corrected for light-bottle measurement, mg/L:
$C'_{ic} = C_{ic} - C_{ilb}$
C_{ic} = initial DO in clear chamber, and
C_{ilb} = initial DO in light bottle.

$$r = \frac{V_o (C'_{io} - C'_{fo})\, B}{t W_o}$$

where:
r = hourly periphyton respiration rate, mg O_2/m^2/hr,
V_o = volume of opaque chamber, L,
B = average periphyton biomass for the study reach, mg/m^2,
W_o = total biomass of periphyton contained in opaque chamber, mg,
C'_{io} = initial oxygen concentration in opaque chamber, corrected for phytoplankton respiration, mg/L:
$C'_{io} = C_{io} - C_{idb}$
C_{io} = initial DO in opaque chamber, mg/L,
C_{idb} = initial DO in dark bottle, mg/L, and
C'_{fo} = final oxygen concentration in opaque chamber, mg/L:
$C'_{fo} = C_{fo} - C_{fdb}$
C_{fo} = final DO in opaque chamber, mg/L, and
C_{fdb} = final DO in dark bottle, mg/L.

For each pair of chamber measurements,

$$P_g = P_n + r$$

where:
P_g = hourly gross periphytic primary production, mg O_2/m^2/hr.

P_G is the area under the curve of primary production per hour through the photoperiod, mg O_2/m^2/day (see Figure 1003:3).

Also,

$$R = \left(\frac{\sum_{1}^{n} r_n}{n} \right) \times 24$$

where:
R = total periphyton community respiration, mg O_2/m^2/day, and
n = number of observations.

Thus,

$$P_N = P_G - R$$

where:
P_N = net periphytic production, mg O_2/m^2/day.

2) Free water methods

a) Calculation of reaeration or diffusion for both the single and upstream-downstream methods—Calculate k_2 from radiotracer data as follows:

$$K_{Kr} = \frac{-1}{t} \ln \frac{(C_{Kr}/C_H)_d}{(C_{Kr}/C_H)_u}$$

and

$$k_2 = \frac{K_{Kr}}{0.83}$$

where:
k_2 = reaeration coefficient (base e), days^{-1},
K_{Kr} = base e transfer coefficient for ^{85}Kr, days^{-1},
t = time of travel, days,
$(C_{Kr}/C_H)_u$ = ratio of released radioactivities (μCi/mL) ^{85}Kr to ^{3}H at the upstream station, and
$(C_{Kr}/C_H)_d$ = ratio of radioactivities (μCi/mL) ^{85}Kr to ^{3}H at the downstream station.

The reaeration coefficient also can be calculated from an equation relating the rate of energy dissipation in a stream to k_2[21,22]:

$$k_2 = K \frac{\Delta h}{t}$$

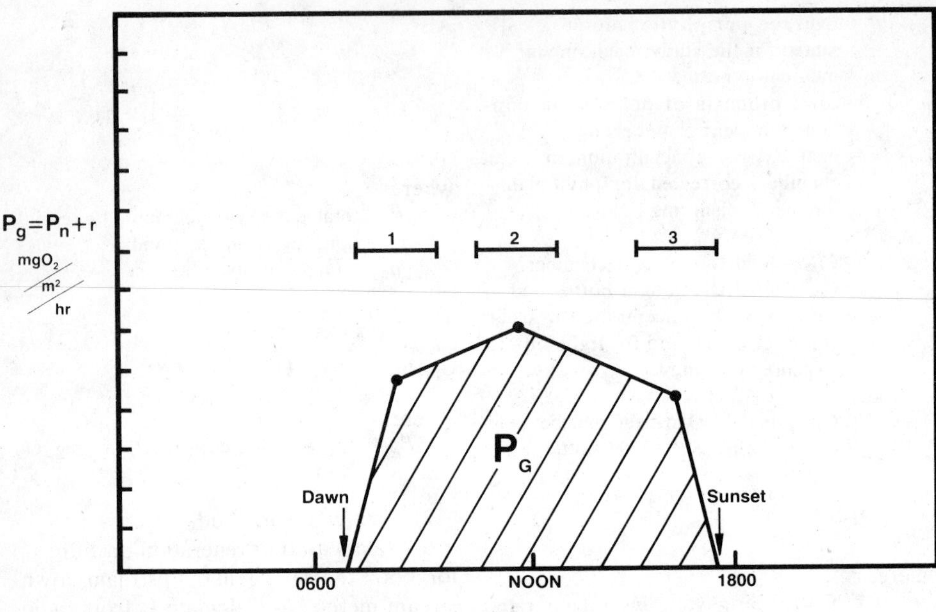

Figure 1003:3. Gross periphytic primary production (P_G) determined by the O'Connell-Thomas Chamber. P_G is the area under the curve obtained by graphical integration planimetry. Each point is the run $P_g = P_n + r$ for incubation periods 1, 2, and 3, which are denoted by the indicated lines.

where:

K = escape coefficient,
Δh = change in water surface elevation in a stream reach, and
t = time of flow through a stream reach.

This can be expressed in terms of hydrodynamic and physical data:

$$k_{2_{20}} = K' \frac{\Delta H}{\Delta X} \times V$$

where:

K' = 28.3×10^3 sec/m·day for stream flows between 0.028 and 0.28 m^3/sec; 21.3×10^3 sec/m·day for stream flows between 0.28 and 0.56 m^3/sec; and 15.3×10^3 sec/m·day for stream flows above 0.56 m^3/sec,
$k_{2_{20}}$ = reaeration coefficient, days^{-1}, at 20 C,
$\frac{\Delta H}{\Delta X}$ = slope, m/km, and
V = velocity, m/sec.

Convert $k_{2_{20}}$ to the temperature of the stream by the following equation:

$$k_{2_T} = k_{2_{20}} (1.024)^{(T-20)}$$

where:

k_{2_T} = k_2 at ambient water temperature, days^{-1}, and
T = ambient water temperature, C.

Convert to D in mg/L/hr:

$$D = \frac{k_{2_T} C_s}{24}$$

where:

C_s = oxygen concentration at saturation at ambient stream temperatures, mg/L.

b) Single-station method—Calculation of primary productivity and respiration from diurnal oxygen and temperature measurements at a single station is summarized in Figure 1003:4 and Table 1003:I.

Tabulate hourly DO measurements and temperatures. Determine C_s (DO of air-saturated H_2O at each temperature from Table 421:I) and compute uncorrected DO consumption, milligrams per liter per hour, for each period:

$$\Delta DO_{\substack{\text{hours} \\ 1 \text{ to } 2}} = DO_{\text{hour } 2} - DO_{\text{hour } 1}$$

Plot on the half hour, as shown in Figure 1003:4*b*.

Calculate the net primary production and respiration of phytoplankton as shown in Section 1002:I. Determine the 24-hr average hourly plankton respiration, $\frac{\sum_{1}^{n} r_p}{n}$, in milligrams per liter per hour every half hour. Calculate the hourly net phytoplankton production and tabulate for the approximate hours during the photoperiod. Plot as shown on Figure 1003:4*c*.

Calculate and tabulate k_{2_T} and substitute D for each C_s, as outlined in ¶ a), above. Plot as shown in Figure 1003.4*c*.

Correct each ΔDO for diffusion and phytoplankton metabolism:

$$\Delta DO_{\text{corrected}}, \text{mg/L/hr} = \Delta DO_{\text{uncorrected}} - D - P_p - R_p$$

Plot each point as shown in Figure 1003.4*d*.

Gross primary productivity of the benthic and attached populations are computed as the area under the curve in Figure 1003.4*d* from sunrise to sunset. This is primary production in grams per cubic meter per day. Multiply by the average depth for a reach, z meters, to obtain P_G in grams

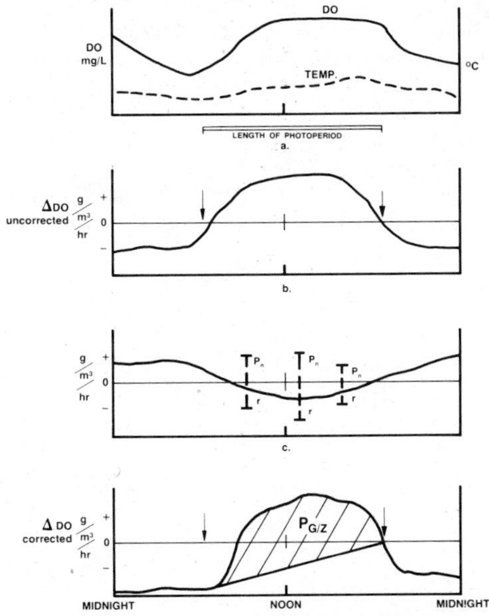

Figure 1003:4. Calculation of gross primary production at a single station. P_g, g O_2/m^2/hr = area of corrected rate of change curve integrated for the length of the photoperiod times average water depth, z, for the reach in meters.

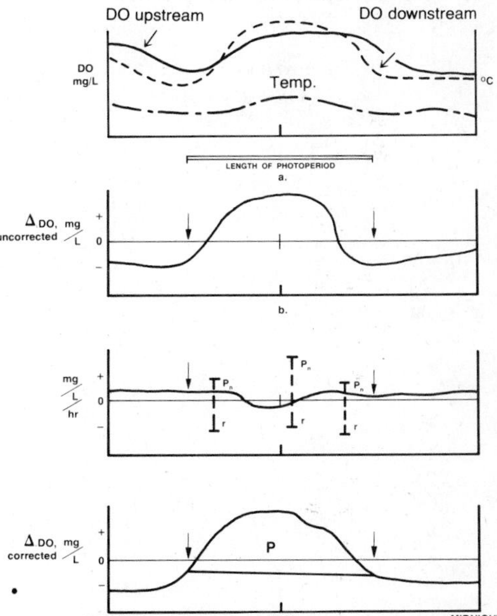

Figure 1003:5. Calculation of gross periphytic primary productivity from upstream-downstream diurnal curves. P is the area under the corrected rate of change graph.

per square meter per day. Calculate community respiration:

$$R = 24\,z\,F$$

where:

R = community respiration, g/m²/day,
z = depth, m, and
F = average hourly ΔDO for the dark period (without regard to sign), mg/L/hr.

Calculate net primary productivity P_N as[38]:

$$P_N = P_G - R$$

c) Upstream-downstream method—Calculation of primary productivity and respiration for a stream reach from upstream and downstream pairs of diurnal curves of oxygen and water temperature is summarized in Figure 1003:5 and Table 1003:II. Alternatively, calculate as below, with oxygen change expressed as the difference between stations rather than as change per hour. The calculations are analogous. Multiply the area under a curve of oxygen change between two stations, corrected for diffusion and plankton metabolism and expressed in milligrams per liter, by the discharge in cubic meters per hour, and divide by the water surface area between the two stations. This, multiplied by 24, yields gross primary productivity in grams per square meter per day.

To compute gross primary productivity by this method, tabulate upstream and downstream DO and average water temperature for the reach at each hour. Calculate ΔDO between upstream and downstream stations for each hour as

$$\Delta DO = DO_{downstream} - DO_{upstream}$$

Tabulate C_s and determine the planktonic activity. Correct for planktonic respiration by relating average hourly dark bottle DO change to the time of travel in the

TABLE 1003:I. SAMPLE CALCULATION LEDGER FOR COMPUTATION OF CORRECTED RATE OF OXYGEN CHANGE FROM A SINGLE-STATION DIURNAL CURVE

Time *hr*	DO *mg/L*	Water Temp. *C*	C_s* *mg/L*	Uncorrected ΔDO† *mg/L/hr*	P_p‡ *mg/L/hr*	R_p§ *mg/L/hr*	k_2 *days*$^{-1}$	D *mg/L/hr*	Corrected ΔDO‖ *mg/L/hr*
Midnight									
0030									
0100									
0130									
•									
•									
•									
Noon									
1230									
1300									
•									
•									
•									
Midnight									

* DO concentration at 100% saturation for a given water temperature, from Table 421:I.
† Hourly rate of change of DO. For example, for noon to 1300, $\Delta DO_{1200-1300} = DO_{1300} - DO_{1200}$; plot at 1230.
‡ Phytoplankton net production.
§ Phytoplankton respiration rate.
‖ $\Delta DO_{corrected} = \Delta DO_{uncorrected} - D - P_p - R_p$

TABLE 1003:II. SAMPLE CALCULATION LEDGER FOR COMPUTATION OF CORRECTED RATES OF OXYGEN CHANGE FROM THE UPSTREAM-DOWNSTREAM DIURNAL CURVES OF OXYGEN CONCENTRATION AND TEMPERATURE

Time *hr*	DO *mg/L*		Uncorrected Δ DO *mg/L*	Water Temp. *C*	C_s* *mg/L*	P_p† *mg/L*	R_p‡ *mg/L*	k_2 *days*$^{-1}$	Correct ΔDO‖ *(mg/L)*
	Upstream	Downstream							
Midnight									
0100									
0200									
•									
•									
•									
Noon									
1300									
•									
•									
•									
Midnight									

* DO concentration at 100% saturation for a given water temperature, from Table 421:I.
† Change in oxygen concentration in the light bottle per hour multiplied by travel time between the upstream and downstream station.
‡ Change in oxygen concentration in the dark bottle multiplied by travel time between the upstream and downstream station.
‖ $\Delta DO_{corrected} = \Delta DO_{uncorrected} - D - P_p - R_p$

stream reach; correct for planktonic production by the hourly change in DO in the light bottle times the time of travel (see Table 1003:II).

Calculate or tabulate k_2 and convert it to the total oxygen diffusion for the reach. Because diffusion, D, is expressed as milligrams per liter per hour, multiply it by the travel time to obtain the diffusion correction.

Correct each hourly upstream-downstream ΔDO as shown in Table 1003:II. Integrate the area under this ΔDO curve from sunrise to sunset to give P as in Figure 1003:5*d*.

$$P_G, \text{ g/m}^2\text{/day} = \frac{Q}{A} P$$

where:

Q = flow, m³/hr, and
A = reach area, m² (average reach width × reach length).

Respiration, R, g O_2/m²/day

$$= \frac{\Delta DO_{dark} \times Q \times 24}{A}$$

and

$$\text{Net production, } P_N = P_G - R$$

1003 E. Interpreting and Reporting Results

Although several systems have been developed to organize and interpret periphyton data, no single method is universally accepted. The methods may be qualitative or quantitative. Qualitative methods deal with the taxonomic composition of the communities in zones of pollution, whereas quantitative methods deal with community structure using diversity indices, similarity indices, and numerical indices of saprobity.

1. Qualitative Methods (Indicator Species and Communities)

The saprobity system developed by Kolkwitz and Marsson is a widely used method of interpreting periphyton data. This scheme divides polluted stream reaches into polysaprobic, α and β mesosaprobic, and oligosaprobic zones, and lists the characteristics of each. The system has been refined[24,25] and enlarged by Fjerdingstad[26,27] and Sladecek.[28,29]

2. Quantitative Methods

These methods use cell counts per unit area of substrate and numerical indices of pollution or water quality. Considerable data on cell densities and species composition of periphyton collected on glass slides in polluted rivers in England are available.[30]

Other indices include the Shannon-Weaver, Simpson's, and Pinkham-Pearson. The saprobity system[31] also may be used where code numbers assigned for the saprobial value and the abundance of individual species are used to calculate a Mean Saprobial Index. Results also may be expressed using the truncated-log normal distribution of diatom species[32,33] as well as the Autotrophic Index (AI).[34]

1003 F. References

1. ROLL, H. 1939. Zur terminologie des periphytons. *Arch. Hydrobiol.* 35:39.
2. YOUNG, O.W. 1945. A limnological investigation of periphyton in Douglas Lake, Michigan. *Trans. Amer. Microsc. Soc.* 64:1.
3. SLADECKOVA, A. 1962. Limnological investigation methods for the periphyton community. *Bot. Rev.* 28:286.
4. OWEN, B.B., JR. 1977. The effect of increased temperatures on algal communities of artificial stream channels. Ph.D. dissertation, Univ. of Alberta, Edmonton.
5. NEWCOMBE, C.L. 1950. A quantitative study of attachment materials in Sodon Lake, Michigan. *Ecology* 31:204.
6. NELSON, D.J. & D.C. SCOTT. 1962. Role of detritus in the productivity of a rock outcrop community in a piedmont stream. *Limnol. Oceanogr.* 7:396.
7. GRZENDA, A.R. & M.L. BREHMER. 1960. A quantitative method for the collection and measurement of stream periphyton. *Limnol. Oceanogr.* 5:190.
8. VOLLENWEIDER, R.A., ed. 1969. A Manual on Methods for Measuring Primary Production in Aquatic Environments. IBP Handbook No. 12, F.A. Davis Co., Philadelphia, Pa.
9. SLADECEK, V. & A. SLADECKOVA. 1964. Determination of periphyton production by means of the glass slide method. *Hydrobiologia* 23:125.
10. KING, D.L. & R.C. BALL. 1966. A qualitative and quantitative measure of aufwuchs production. *Trans. Amer. Microsc. Soc.* 82:232.
11. CLARK, J.R., D.I. MESSENGER, K.L. DICKSON & J. CAIRNS, JR. 1978. Extraction of ATP from aufwuchs communities. *Limnol. Oceanogr..* 23:1055.
12. WETZEL, R.G. 1963. Primary productivity of periphyton. *Nature* 197:1026.
13. WETZEL, R.G. 1964. A comparative study of the primary production of higher aquatic plants, periphyton, and phytoplankton in a large shallow lake. *Int. Rev. Ges. Hydrobiol.* 49:1.
14. ARONOFF, S. 1956. Techniques in Radiochemistry. Iowa State College Press, Ames.
15. ODUM, H.T. 1956. Primary production in

flowing waters. *Limnol. Oceanogr.* 1:102.
16. ALLEN, H.L. 1971. Primary productivity, chemo-organotrophy, and nutritional interactions of epiphytic algae and bacteria or macrophytes in the littoral of a lake. *Ecol. Monogr.* 41:97.
17. HALL, C.A.S. 1972. Migration and metabolism in a temperate stream ecosystem. *Ecology* 53:585.
18. NIXON, S.W. & C.A. OVIATT. 1974. Ecology of a New England salt marsh. *Ecol. Monogr.* 43:463.
19. THOMAS, N.A. & R.L. O'CONNELL. 1966. A method for measuring primary production by stream benthos. *Limnol. Oceanogr.* 11:386.
20. COPELAND, B.J. & W.R. DUFFER. 1964. Use of a clear plastic dome to measure gaseous diffusion rates in natural waters. *Limnol. Oceanogr.* 9:494.
21. TSIVOGLOU, E.C. & L.A. NEAL. 1976. Tracer measurement of reaeration. III. Predicting the capacity of inland streams. *J. Water Pollut. Control Fed.* 48:2669.
22. GRANT, R.S. 1976. Reaeration-coefficient measurements of 10 small streams in Wisconsin. U.S. Geol. Surv. Water Resources Publ. 76-96.
23. ODUM, H.T. & C.M. HOSKIN. 1958. Comparative studies of the metabolism of marine water. *Publ. Inst. Mar. Sci. Univ. Tex.* 4:115.
24. KOLKWITZ, R. 1950. Oekologie der saprobien. *Ver Wasser-, Boden, Lufthyg. Schriftenreihe* (Berlin) 4:1.
25. LIEBMANN, H. 1951. Handbuch der Frischwasser und Abwasserbiologie. Bd. I. Oldenbourg, Munchen, Germany.
26. FJERDINGSTAD, E. 1964. Pollution of streams estimated by benthal phytomicroorganisms. I. A saprobic system based on communities of organisms and ecological factors. *Int. Rev. Ges. Hydrobiol.* 49:63.
27. FJERDINGSTAD, E. 1965. Taxonomy and saprobic valency of benthic phytomicroorganisms. *Int. Rev. Ges. Hydrobiol.* 50:475.
28. SLADECEK, V. 1966. Water quality system. *Verh. Int. Ver. Limnol.* 16:809.
29. SLADECEK, V. 1973. System of water quality from the biological point of view. *Arch. Hydrobiol.* 7:1.
30. BUTCHER, R.W. 1946. Studies in the ecology of rivers. VI. The algal growth in certain highly calcareous streams. *J. Ecol.* 33:268.
31. PANTLE, R. & H. BUCK. 1955. Die biologische uberwachung der Gewasser und der Darstellung der Ergebnisse. *Gas-Wasserfach* 96:604.
32. PATRICK, R., M.H. HOHN & J.H. WALLACE. 1954. A new method for determining the pattern of the diatom flora. *Bull. Philadelphia Acad. Natur. Sci.* 259:1.
33. PATRICK, R. 1973. Use of algae, especially diatoms, in the assessment of water quality. *In* J. Cairns, Jr., ed. Biological Methods for the Assessment of Water Quality. ASTM STP 528, Amer. Soc. for Testing & Materials.
34. WEBER, C. 1973. Recent developments in the measurement of the response of plankton and periphyton to changes in their environment. *In* G. Glass, ed. Bioassay Techniques and Environmental Chemistry. Ann Arbor Sci. Publ., Inc., Ann Arbor, Mich.

1003 G. Bibliography

KOLKWITZ, R. & M. MARSSON. 1908. Okologie der pflanzlichen Saprobien. *Berl. Deut. Bot. Ges.* 26(a):118.

FRITSCH, F.E. 1929. The encrusting algal communities of certain fast-flowing streams. *New Phytol.* 28:166.

BUTCHER, R.W. 1931. An apparatus for studying the growth of epiphytic algae with special relation to the river Tees. *Trans. N. Natur. Union.* 1:1.

BUTCHER, R.W. 1932. Studies in the ecology of rivers. II. The microflora of rivers, with special reference to the algae on the river bed. *Ann. Bot.* 46:813.

BUTCHER, R.W. 1947. Studies in the ecology of rivers. VII. The algae of organically enriched waters, *J. Ecol.* 35:186.

MARGALEF, R. 1948. A new limnological method for the investigation of thin-layered epilithic communities. *Trans. Amer. Mi-*

crosc. Soc. 67:153.

FJERDINGSTAD, F. 1950. The microflora of the River Mølleaa, with special reference to the relation of the benthal algae to pollution. *Folia Limnol. Scand.* 5:1.

BLUM, J.L. 1956. The ecology of river algae. *Bot. Rev.* 22:291.

COOKE, W.B. 1956. Colonization of artificial bare areas by microorganisms. *Bot. Rev.* 22:613.

YOUNT, J.L. 1956. Factors that control species numbers in Silver Springs, Florida. *Limnol. Oceanogr.* 1:286.

BUTCHER, R.W. 1959. Biological assessment of river pollution. *Proc. Linnean Soc. London* 170:159.

HOHN, M.H. 1959. The use of diatom populations as a measure of water quality in selected areas of Galveston and Chocolate Bay, Texas. *Publ. Inst. Mar. Sci. Univ. Tex.* 5:206.

POMEROY, L.R. 1959. Algal productivity in salt marshes. *Limnol. Oceanogr.* 4:386.

CASTENHOLZ, R.W. 1961. An evaluation of a submerged glass method of estimating production of attached algae. *Verb. Int. Ver. Limnol.* 14:155.

HOHN, M.H. 1961. Determining the pattern of the diatom flora. *J. Water Pollut. Control Fed.* 33:48.

PATRICK, R. 1963. The structure of diatom communities under varying ecological conditions. *Ann. N.Y. Acad. Sci.* 108:359.

WHITFORD, L.A. & G.J. SCHUMACHER. 1964. Effect of a current on respiration and mineral uptake in *Spirogyra* and *Oedogonium*. *Ecology* 45:168.

WILLIAMS, L.G. & D.I. MOUNT. 1965. Influence of zinc on periphytic communities. *Amer. J. Bot.* 52:26.

DUFFER, W.R. & T.C. DORRIS. 1966. Primary productivity in a southern Great Plains stream. *Limnol. Oceanogr.* 11:143.

EATON, J.W. & B. MOSS. 1966. The estimation of numbers and pigment content in epipelic algal populations. *Limnol. Oceanogr.* 11:584.

HOHN, M.H. 1966. Artificial substrate for benthic diatoms—collection, analysis, and interpretation. *In* K.W. Cummings, C.A. Tryon, Jr., & R.T. Hartman, eds. Organism-Substrate Relationships in Streams. Pymatuning Lab. of Ecology, Spec. Publ. No. 4, pp. 87–97, Univ. of Pittsburgh.

KEVERN, N.R., J.L. WILHM & G.M. VAN DYNE. 1966. Use of artificial substrata to estimate the productivity of periphyton communities. *Limnol. Oceanogr.* 11:499.

KING, D.L. & R.C. BALL. 1966. A qualitative and quantitative measure of *aufwuchs* production. *Trans. Amer. Microsc. Soc.* 82:232.

MCINTIRE, C.D. 1966. Some factors affecting respiration of periphyton communities in lotic environments. *Ecology* 47:918.

SCHLICHTING, H.E., JR. & R.A. GEARHEART. 1966. Some effects of sewage effluent upon phyco-periphyton in Lake Murray, Oklahoma. *Proc. Okla. Acad. Sci.* 46:19.

SLADECKOVA, A. & V. SLADECEK. 1966. Periphyton as indicator of reservoir water quality. *Technol. Water* (Czech). 7:507.

THOMAS, N.A. & R.L. O'CONNELL. 1966. A method for measuring primary production by stream benthos. *Limnol. Oceanogr.* 11:386.

CUSHING, C.E. 1967. Periphyton productivity and radionuclide accumulation in the Columbia River, Washington, USA. *Hydrobiologia* 29:125.

PHAUP, J.D. & J. GANNON. 1967. Ecology of *Sphaerotilus* in an experimental outdoor channel. *Water Res.* 1:523.

TAYLOR, M.P. 1967. Thermal effects on the periphyton community in the Green River. TVA, Div. Health & Safety, Water Qual. Br., Biol. Sect., Chattanooga, Tenn.

MOSS, B. 1968. The chlorophyll *a* content of some benthic algal communities. *Arch. Hydrobiol.* 65:51.

PATRICK, R. 1968. The structure of diatom communities in similar ecological conditions. *Amer. Natur.* 102:173.

ARTHUR, J.W. & W.B. HORNING. 1969. The use of artificial substrates in pollution surveys. *Amer. Midland Natur.* 82:83.

DICKMAN, M. 1969. A quantitative method for assessing the toxic effects of some water soluble substances, based on changes in periphyton community structure. *Water Res.* 3:963.

BESCH, W.K., M. RICARD & R. CANTIN. 1970. Use of benthic diatoms as indicators of mining pollution in the N.W. Miramichi River. *Tech. Rep. Fish. Res. Board Can.* 202:1.

NUSCH, E.A. 1970. Ecological and systematic studies of the Peritricha (Protozoa, Ciliata) in the periphyton community of reservoirs and dammed rivers with different degrees of saprobity. *Arch. Hydrobiol.* (Suppl.)37:243.

ROSE, F.L. & C.D. MCINTIRE. 1970. Accumulation of dieldrin by benthic algae in laboratory streams. *Hydrobiologia* 35:481.

TIPPETT, R. 1970. Artificial surfaces as a method of studying populations of benthic micro-algae in fresh water. *Brit. Phycol. J.*

5:187.

WHITTON, B.A. 1970. Toxicity of zinc, copper and lead to Chlorophyta from flowing waters. *Arch. Mikrobiol.* 72:353.

BURROWS, E.M. 1971. Assessment of pollution effects by the use of algae. *Proc. Roy. Soc. Lond. Ser. B.* 177:295.

CURTIS, E.J.C. & C.R. CURDS. 1971. Sewage fungus in rivers in the United Kingdom: The slime community and its constituent organisms. *Water Res.* 5:1147.

ERTL, M. 1971. A quantitative method of sampling periphyton from rough substrates. *Limnol. Oceanogr.* 16:576.

HANSMANN, E.W., C.B. LANE & J.D. HALL. 1971. A direct method of measuring benthic primary production in streams. *Limnol. Oceanogr.* 16:822.

PATRICK, R. 1971. The effects of increasing light and temperature on the structure of diatom communities. *Limnol. Oceanogr.* 16:405.

ANDERSON, M.A. & S.L. PAULSON. 1972. A simple and inexpensive woodfloat periphyton sampler. *Progr. Fish-Cult.* 34:225.

ARCHIBALD, R.E.M. 1972. Diversity of some South African diatom associations and its relation to water quality. *Water Res.* 6:1229.

CAIRNS, J., JR., B.R. LANZA & B.C. PARKER. 1972. Pollution-related structural and functional changes in aquatic communities with emphasis on freshwater algae and protozoa. *Proc. Acad. Natur. Sci. Philadelphia* 124:79.

OLSON, T.A. & T.O. ODLAUG. 1972. Lake Superior Periphyton in Relation to Water Quality. Univ. Minn. Sch. Pub. Health, Minneapolis, Water Pollut. Control Res. Ser., 18080 DEM 02/72.

HANSMANN, E.W. 1973. Effects of logging on periphyton in coastal streams of Oregon. *Ecology* 54:194.

RUTHVEN, J.A. & J. CAIRNS, JR. 1973. Response of fresh-water protozoan artificial communities to metals. *J. Protozool.* 20:127.

SCHINDLER, D.W., V.E. FROST & R.V. SCHMIDT. 1973. Production of epilithiphyton in two lakes of the experimental lakes area, northwestern Ontario. *J. Fish. Res. Board Can.* 30:1511.

MIDWEST BENTHOLOGICAL SOCIETY. 1964–1973. (Annual) Current and Select Bibliographies on Benthic Biology, Springfield, Ill.

1004 MACROPHYTON

1004 A. Introduction

The macrophyton consists principally of aquatic vascular flowering plants, but it also includes the aquatic mosses, liverworts, ferns, and the larger marine algae. Like other primary producers, these plants respond to the quality of the water they grow in; thus their analysis is important for water quality evaluation. They often constitute a dominant factor in the habitat of other aquatic organisms.

Freshwater forms range in size from the tiny water meal (*Wolffia* spp), about the size of a pinhead, to plants such as the cattail (*Typha* spp), up to 4 m high, and finally to cypress trees (*Taxodium* spp), up to 50 m high. Higher aquatic plants often are clustered in large numbers, many in pure stands, and covering extensive areas of shallow lakes, marshes, and canals. A few of the larger freshwater algae (*Chara* spp, *Nitella* spp, and *Cladophora* spp) resemble higher plants in size, form, and habit, but generally are not included in the macrophyton. In marine water, the intertidal rockweeds (*Fucus* spp and *Ascophyllum* spp) and offshore kelps (*Fucus* spp and *Macrocystis* spp) are conspicuous. Vascular marine or estuarine plants, such as the eelgrass (*Zostera* spp) and the marshgrass (*Spartina* spp), are essential to the aquatic ecosystem.

Three types of macrophyton are recognized: floating, submersed, and emersed. Floating plants are not rooted; their principal foliage or crown floats on the water surface. All or most of the foliage of submersed plants grows beneath the water surface; the plants may or may not have roots. Growing tips of submersed plants may emerge to flower. Emersed plants may be subdivided into erect-leaved plants and floating-leaved plants. Their principal foliage is in the air at or above the water surface; they are attached by roots to the bottom mud. Floating-leaved emergents lack the supportive tissue found in the erect-leaved emersed plants. In some cases the same species may grow as a floating or emersed type, or as a submersed or emersed type. Submersed and emersed vascular plants typically are rooted to the bottom but they may be found detached and floating.

The distribution and abundance of higher plants is subject to considerable spatial and temporal variation. Among the many factors that determine their presence and density are sediment type, turbidity, nutrients, water depth, shoreline disturbance, herbivores, and humans. Zonation in the littoral zone of lakes and shallow, slow-moving streams is common. Emergent macrophytes generally are found in the most shallow region of the littoral zone. During periods of low water level they may occupy the terrestrial as well as the aquatic habitat. The depth of inhabitation seldom exceeds 1 m. Floating-leaved plants commonly are found in the lower littoral areas at depths between 1 and 3 m. Submersed plants may occur from the edge of the shore to the interface of the littoral and profundal zones, but rarely extend beyond a depth of 10 m because of hydrostatic pressure and limitation of underwater light.

1004 B. Preliminary Survey

Conduct a preliminary survey of the study area before undertaking vegetation studies. Use plant collections to make before-survey identifications. Such preliminary identifications within the study area will increase sampling efficiency greatly during the first portion of any vegetation study. Make onshore collections by hand or with a rake from shallow water. In deep water, collect samples using a grapnel drag, bottom dredge, or grab. Use an inexpensive, yet durable, grapnel drag of about 2 kg constructed of light, pliable steel that can be retrieved easily, even if caught on submerged objects.

Survey the study area to determine its accessibility by boat (e.g., outboard, airboat), its size, its high- and low-water marks, the types of macrophyton, and the distribution of vegetation according to depth and breadth. Note location marks within the study area to aid in constructing vegetation maps, markers for transect lines, or coordinate maps. Using a fathometer, determine depth contours of the study area as well as distribution of submersed vegetation.

If a population estimate of various species is to be made, conduct a pilot study to determine horizontal and vertical homogeneity. Use the same collection method for the pilot study and the definitive study. Select a sampling design to estimate population after completing the pilot study.

1004 C. Vegetation Mapping Methods

1. Baseline

Vegetation maps constructed using the baseline method or the basepoint-stadia rod-alidade method generally are limited to pure stands of floating or emersed littoral macrophyton in small bodies of water. In clear water, determine the outline of pure stands of submersed vegetation with a viewing box (usually a wooden box with a watertight glass lens), snorkel, or SCUBA (self-contained underwater breathing apparatus). The baseline method and the basepoint-stadia rod-alidade method provide accurate maps of vegetation in areas up to 1×10^5 m^2 where most of the vegetation outline is visible. The baseline method uses intercepting lines from each end of a predetermined baseline to closely spaced markers (i.e., chaining pins) around the stand. Preset the map scale by determining the ratio between the length of the baseline and its reduction on the map (drawn on a plane table). The basepoint-stadia rod-alidade method is a modification of the baseline method in that the distance between the vegetational outline and the basepoint is determined with an alidade. One unit on the stadia rod, as viewed between the cross-hairs of the alidade, is equivalent to a distance of 3.05 m between the stadia rod and the alidade. Chaining pins are not required when this method is used. In practice, more readings closely spaced along the vegetational outline usually are taken using the basepoint-stadia rod-alidade method.

2. Line Intercept

Determine vegetation maps of mixed stands and/or large areas with the line intercept method. Select sampling points at equal intervals along the shoreline. Choose the length of the intervals by the degree of accuracy desired: the closer the sampling points, the more accurate the map. Run transects perpendicular from shore to the boundary of the plant stand. Construct the intercept line (transect line) of plastic-coated wire rope to prevent stretching. Weight the line by wrapping it with a lead filament; the lead weights can act as interval markers on the rope to designate sampling units. Use segments of 0.5 m. Use snorkel or SCUBA to determine the plant species that vertically intercept the line at each 0.5 m segment. In sparse vegetation, place a ring of brass or PVC pipe, 0.25 m^2, over the sample point on the intercept line and record plant species within the ring. Use a writing board, suitable for underwater recording, constructed of several sheets of a matte-finish white plastic, notched at one end to fit securely to a board by wing nuts and washers. Use these sheets to record data underwater with a soft lead pencil. Clean them for reuse by erasing with a kitchen cleanser. Construct the vegetation map by placing points where plant species are found within an outline map (or aerial photograph) of the sample area. Encircle the points on the map represented by single species to enclose the area where the species occurred. Obtain the area of this outline with planimetry.

Determine relative frequency at sample points along an intercept line with a set sampler consisting of a 2-cm steel tube, 2 m long, to which five 0.75-cm by 25-cm steel rods are attached on 40-cm centers. Make a "set" of the sampler at sample segments along an intercept line. Record vegetation touching each of the five points within 2.54 cm of the distal tip. If more than one plant species is touching, record the one nearest the tip. If no plant is touched, record bare ground. Use field data to derive frequency of plant species at sample points, as well as by contour elevation and intercept line segment.

Determine percent cover of vegetation from vegetation maps by planimetry; digital compensating polar planimeters are the most accurate.

3. Remote Sensing

Aerial photography from an aircraft with accompanying ground truth observations provides accurate vegetation maps for determining vegetation distribution, its cover, and its growth through time. Adequate photographs, but with low resolution, can be obtained with a good-quality 35-mm camera, vertically mounted on an aircraft. Preferably use a 70-mm camera (80-mm focal length, f/2.8 lens) to provide adequate light-gathering for color film and to accommodate the higher shutter speeds needed for aerial photography. Mount the 70-mm camera under the hull of the aircraft, but do not gyro-stabilize it because photogrammetric measurements are not needed. Aerographic film* is fast enough for aerial photography. It has a fine grain that allows adequate enlargement. Positive film may be viewed directly on a light table but stereo-viewing equipment will enable better distinction between submersed, floating, and emergent species. Use an 8 × 10-in. enlargement for reconnaissance photointerpretation purposes; however, prepare an internegative from the frame before making color prints. Use color prints with a table-top stereo viewer to help distinguish the vegetation types. Do not use a haze filter with color film in aerial photography because it will eliminate portions of the blue spectrum that are very useful in separating some aquatic vegetation types. However, at altitudes above 1,000 m a haze-penetration filter† may be required.

The following climatic conditions provide the best photographs:

Sun angle: minimum = 25°; maximum = 45°; surface wind: less than 7.7 m/sec (15 knots); and clouds and cloud shadows: less than 15%; high cirrus clouds absent.

Outlines of the vegetation types and/or species are made on film or print. Determine the area of vegetation by planimetry and use copies of the outlines as records. Confirm vegetation on aerial photographs by ground observation.

4. Fathometry

Obtain fathometer plots and visual observations by grapnel drag along regularly spaced transect lines to survey submersed plants. The narrower the interval between transect lines, the greater the accuracy in the population reconnaissance. Record fathometer plots while traveling along the transect line at a constant speed. Use a fathometer accurate to the nearest 0.1 m to provide the necessary resolution to distinguish between sediment and plant. Fathometer recorders provide a surface line, a sediment line, and a vegetation line. Darker, more compact lines indicate harder sediments or denser vegetation. The vertical and horizontal distribution of submersed vegetation along the transect line is determined accurately by fathometry.

*Kodak SO-397, Ektachrome EF, or equivalent.

†HF-3/HF-4 or HF-3/HF-5.

1004 D. Population Estimates

1. Sampling Design

The design of a sampling program depends on study aims, collection methods, variation and distribution of vegetation, personnel availability, and accuracy expected. The overall goal of any sampling to obtain population estimates of higher aquatic plants is to remove vegetation from enough known areas to obtain a mean sufficiently accurate to show significant differences between sample periods and sample areas. Variation in space usually is not random; distribution is determined by water depth, shoreline activity, sediment type, or other factors. The parametric statistic for estimating the true population mean assumes that the population being sampled has a normal distribution and that all sample units have the same probability of being selected. Avoid fixed sampling stations in sampling programs to determine population means, unless they are chosen at random at the beginning of the study. Because normally distributed plant populations may not be a characteristic of contiguous plant communities, use parametric statistics with caution.

The simple random sampling design (quadrants within a grid overlay of the sampling area or rectanglar coordinates on intercepting transect lines used to designate sampling sites) is best applied to homogeneous, noncontiguous plant communities. Determine the number of stations required to obtain an estimate of the true population mean with a predetermined level of confidence and permissible error by applying the data from a pilot study to the following equation:

$$N = \left(\frac{t \times S}{d \times \bar{x}}\right)^2$$

where:

N = number of sampling stations,
t = Student's t at a given probability level; because N is unknown, set t = 2.0; t is approximately equal to 2.0 for $N > 30$,
S = standard deviation,
$\bar{x}$ = estimator of true population mean, and
d = permissible error of the final mean; d = 0.1 is recommended for vegetation studies (±10%).

Apply stratified random sampling to populations having many homogeneous stands. This design is best applied to populations with obvious gradients and, in practice, to gain precision by the minimized variance within strata. Determine placement of strata by a pilot study. To maximize precision, place stratum boundaries around homogeneous areas; generally, the fewer strata, the greater precision.

Means for population measurement taken along randomly placed stations on a transect line do not represent large area or lake populations unless the transect line is placed randomly. Arbitrarily placed transect lines within a sampling area may or may not reflect the true variation of the vegetation within.

2. Collection Methods

The choice of collection method depends on vegetation density, vegetation type, height of vegetation in the water column, and nature of the sediment. Sample directly by hand from a known area (quadrat) or by using a sampling apparatus. The amount of plant material taken from a quadrat at a given time is defined as standing crop and may or may not include roots or other inaccessible structures. The biomass of a plant community is, like standing crop, measured in a known area at an instant in time, but it includes all plant parts. Some collection methods may be more useful for standing crop than for biomass. Use a quadrat of 0.25 m^2, but vary it with type of vegetation.

a. Direct collection by hand: Use floating or sinking quadrat frames, 0.25 m^2 (0.5 m × 0.5 m). Construct floating frames of wood, with holes to insert sediment pins. These frames are ideal for use with entangled vegetation because they can be slid under vegetation and fastened into place. Floating frames with fixed corners can be constructed of 1.25-cm-diam PVC pipe. Build sinking frames with stainless steel or brass tubing fastened at the corners with wing bolts that may be loosened for use with entangled, dense vegetation.

Harvest from quadrats in shallow water

by wading or using a snorkel. In deep water use SCUBA, but only if experienced divers are available. Trap vegetation collected by SCUBA with a nylon mesh bag (mesh size 0.25 cm) fastened on the quadrat frame. Use floats placed on the upper edge of the bag to keep it upright in the water. Limit collections in deep water by SCUBA to firm sediments to avoid turbidity. Do not use SCUBA for direct collections of tall, submersed vegetation. Include all vegetation within or overlapping the quadrat in the sample.

b. Ekman grab: Use the Ekman grab (Figure 1005:6) for short, erect plants growing in soft sediment. The standard sampling area of the grab is 0.0225 m^2; thus its use may require a large number of samples to characterize variable plant communities.

c. Cylinder: Construct a cylinder (Figure 1005:10) of light sheet metal, plexiglass, or plastic with an enclosed area of 0.25 m^2. The sampler is ideal for obtaining collections in shallow water while wading. Place the sampler on the selected sample sites and remove vegetation from within by hand.

d. Louisiana box sampler: Construct the Louisiana box sampler of stainless steel or galvanized black-steel sheet metal. The sheet metal box possesses a cutting edge along the bottom while a nylon mesh bag covers the top. The unit is designed to be hoisted above the water by a boom with a quick-release device to lower. Insert a cutter frame, with a brass screen, in a slot near the quadrat bottom by SCUBA or snorkel to cut plants at the top of the sediment. Raise the unit above the water to remove vegetation. Use this sampler to obtain standing crop for short vegetation in deep water.

e. Forsberg frame sampler: The Forsberg frame sampler has a sediment plate or foot attached to a handle that can be extended by adding sections. A movable, U-shaped frame is hinged to one side of the sediment plate and controlled by nylon rope. The movable frame hangs vertically under the sediment plate as it is pressed into the sediment. The frame is moved by rope to describe a 180° arc; the area sampled is equal to the area outlined by the frame. Vegetation taken by the frame is pressed against a rubber pad on the handle. Remove vegetation by retrieving the sampler, using care to avoid loss of stems and roots. Use the Forsberg frame sampler in short stands of vegetation in soft sediments.

f. Osborne aquatic plant sampler: The Osborne aquatic plant sampler (Figure 1004:1) consists of a quadrat constructed from sheets of stainless steel with a thickness of 0.64 cm. The side dimension of 50 cm × 50 cm allows a cutting edge of 200 cm to provide a 0.25 m^2 sample. The cutting teeth on the bottom edge are tempered steel mower blades. The box has a volume of 0.15 m^3. A hollow stainless steel tube (diameter = 2.0 cm; wall thickness = 0.64 cm) is welded around the top edge to provide the guide for attachment of a steel frame and its replaceable wire net (mesh size = 1.25 cm). A hinged slat door of stainless steel, held in an open, upright position by a release clip, lies along the outside of the box in a stainless steel track. A polished stainless steel knife edge on the first slat of the door cuts through the sediment when the door closes. Each slat of the door has a metal wheel on each side to allow for free movement. A steel wheel located in the curve of the door track minimizes friction. The door is closed with a steel cable attached to each corner of the first slat; these are pulled through the track and side pulleys when the door is closed. Stainless steel cables are attached to the top corners of the sampler (load capacity = 1,800 kg). A quick-release latch is fastened through two steel O-rings on the corner cables. Total weight of the sampler is 125 kg.

CAUTION: *For safety in use, keep hands*

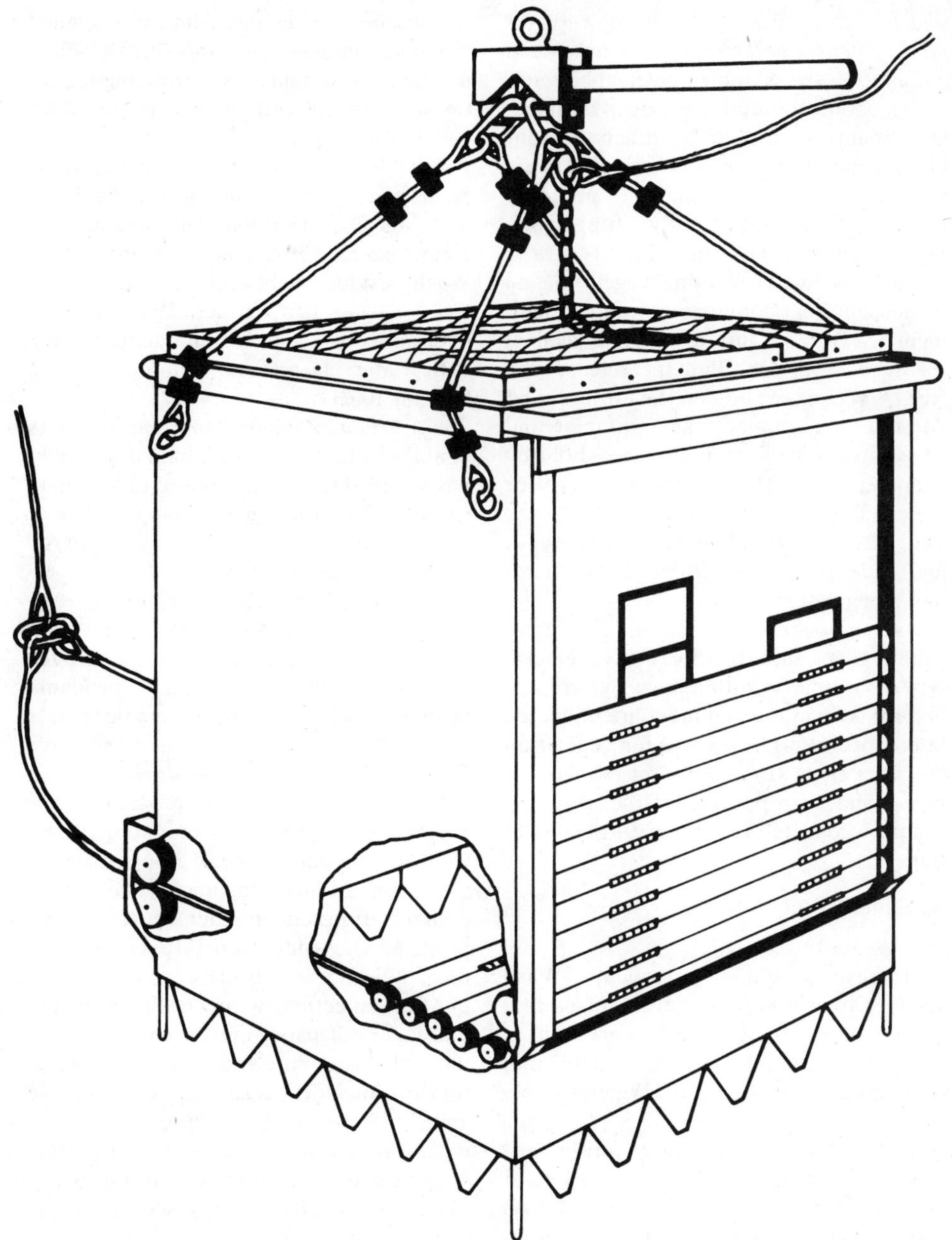

Figure 1004:1. Osborne aquatic plant sampler.

and feet away from mechanical sampler jaws, winches, and triggering devices.

Suspend the sampler over the water with a boom mounted to the deck of a pontoon boat (Figure 1004:2). Attach the door closing and lifting cables to a double drum winch. Attach a chain, leading from the release clip for the door, to the lifting cable. Upon release, the sampler free-falls through the water, cutting vegetation on its descent, and comes to rest in the sediment. Severed plants are caught against the wire mesh top. Pull the release clip free by placing tension on the lifting cable. After door is closed, raise sampler and its contents to water surface, holding sampler away from deck by a roller boom. Remove sampler from the water with the boom, place it on the deck, and slide wire mesh frame aside to remove vegetation.

Use this sampler to obtain emergent, submergent, and free-floating vegetation types. Vegetation within the water column and roots from the sediment are collected in one operation; it is ideal for vegetation biomass from soft sediment (sapropel, peat, sand). Dropped from the side, the sampler collects from undisturbed locations.

3. Sample Preparation and Analysis

a. Biomass:

1) Fresh weight (wet weight)—Wash samples free of silt and debris, place in a nylon bag (mesh size 0.75 cm) and spin in a garment washer at 560 rpm for 6 to 7 min to remove excess moisture. Weigh sample to nearest 0.001 g.

2) Dry weight—Dry subsample (not less than 10%) in a forced-air oven at 105 C for 48 hr or until a constant weight is achieved. The coefficient of variation for a series of subsamples *should not exceed* 10%. Calculate dry weight by dividing dry weight of subsample by fresh weight of subsample times fresh weight of sample.

3) Ash-free dry weight—Transfer dried subsample to a covered and preweighed crucible. Ignite at 550 C for 6 hr. Calculate ash-free dry weight by determining the ratio between ash and dry weight times dry weight of sample.

b. Chlorophyll content: Extract fresh plant material with acetone made basic with $MgCO_3$. Grind the plant material and centrifuge at 2,500 rpm for 10 to 15 min. Wash residue with acetone and add filtered washings to extract. Dry overnight over anhydrous Na_2SO_4 and dilute to 90% with water. Determine chlorophyll (see Section 1002G).

c. Carbon content: Most entire plants contain 46 to 48% carbon on a dry-weight basis. This factor can be used to calculate carbon content and make comparisons.

d. Caloric content: Determine energy content by bomb calorimetry.[1]

e. Species identification:

1) Sample preparation—Use fresh specimens for identification wherever possible. Avoid immature plants or plants lacking flowers. Because aquatic plants contain from 80 to 95% water and have less supportive tissue than terrestrial forms, use a different procedure for drying, preserving, and mounting them. Collect plants during peak growing season at the time of flowering and/or fruit development if practical. Include the entire plant as the specimen (stems, leaves, roots, flowers, and fruit).

After collection, wrap specimen in several layers of paper and submerge in water. Label with date and location of collection on an index card and place sample and card in a plastic bag. Preferably use an ice chest containing crushed ice for storage in the field but do not store for longer than 4 to 6 hr. Plants can be kept for several days under refrigeration at 4 C.

Prepare a dry mount, in which the specimen is displayed through a clear, adhesive plastic sheet, by centering the plant on 100% rag herbarium paper. Clean the plant of all silt and residue. Place emergent

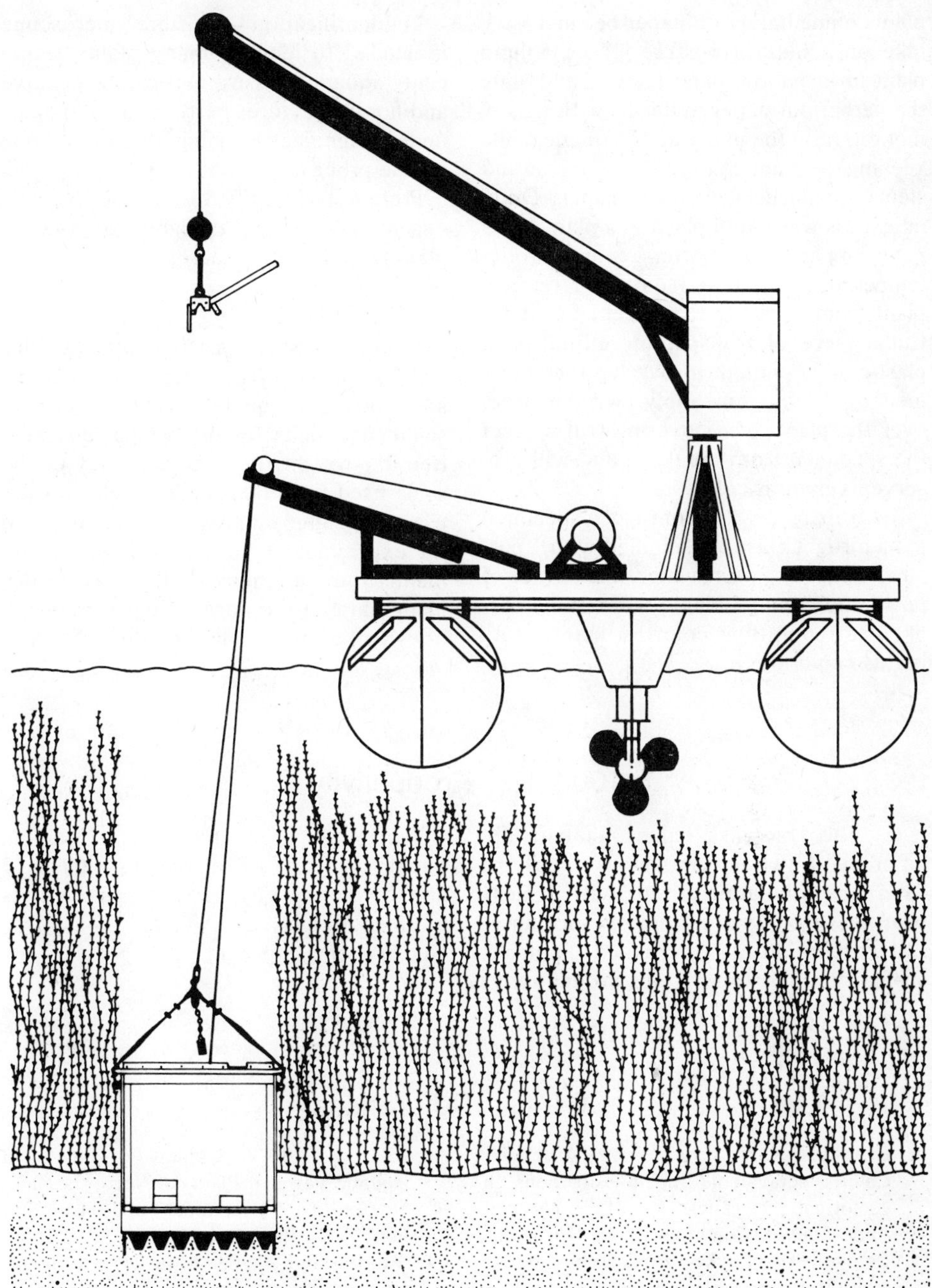

Figure 1004:2. Operation of Osborne aquatic plant sampler.

plants immediately on paper because they take on a natural posture. Place a limp plant in a shallow pan of water and slide the herbarium paper under it; with a slow motion, raise the paper at a 30° angle while keeping the plant centered. The leaves and stems should lie flat on the paper. Drain off excess water and place in a plant press between paper and blotters. Store at room temperature for 3 to 5 days but remove plant from press at the onset of wilting. Cut a piece of transparent self-adhesive plastic large enough to overlap the specimen by 5 cm, and apply, wrinkle free, over the plant. The covering will protect the specimen from breakage and will preserve its natural color.

To prepare a wet mount place specimen in an airtight glass vessel filled with 1 part 10% formalin, 3 parts water, and a trace of powdered copper. Plants will remain lifelike and retain their color for many years in this condition.

2) Identification—A stereomicroscope is needed to identify many plants, especially aquatic grasses. Observe vegetative and floral structures by tearing them apart under magnification using forceps and fine needle probes.

Preferably identify to species. Numerous references are available to assist in identifying aquatic macrophytes.[2–23]

4. Data Presentation

Express fresh weight (wet weight), dry weight, and ash-free weight as grams or kilograms per square meter. Determine significant digits for dry weight and ash-free dry weight from the accuracy of the scale used to obtain fresh weight; do not use more significant digits than those used for expressing fresh weight. Report pigment as grams chlorophyll per gram dry plant matter and caloric value as gram calories per gram dry plant matter.

1004 E. Productivity

Use two biomass values, statistically different at two instantaneous times, to determine population productivity. Unless plant loss from grazing, injury, mortality, and respiration is determined, biomass change through time is net productivity. Calculate net productivity by subtracting biomass at Time 1 from biomass at Time 2 and dividing by days. Express in the units of fresh weight, dry weight, or ash-free weight biomass per unit area per day.

1004 F. References

1. CUMMINS, K.W. & J.C. WUYCHECK. 1971. Caloric equivalents for investigations in ecological energetics. *Int. Ass. Theoret. Appl. Limnol.* Communications No. 18, E. Schweizerbart'scke Verlagesbuchhandlung, Stuttgart.
2. AYENSU, E.S. 1975. Underexploited Tropical Plants with Promising Economic Value. National Academy of Sciences, Washington, D.C.
3. BEAL, E.O. 1977. A Manual of Marsh and Aquatic Vascular Plants of North Carolina. N. Carolina Agricultural Experiment Station, Tech. Bull. No. 247.
4. BREEN, R.S. 1963. Mosses of Florida. Univ. of Florida Press.
5. COBB, B. 1963. A Field Guide to the Ferns. Houghton Mifflin Company, Boston, Mass.
6. CORRELL, D.S. & H.B. CORRELL. 1972. Aquatic and Wetland Plants of South-

western United States. USEPA, 16030 DNL 01/72.

7. EYLES, D.E. & J. ROBERTSON, JR. 1963. A Guide and Key to the Aquatic Plants of the Southeastern United States. U.S. Fish & Wildlife Serv. Circ. 158.
8. FASSETT, N.C. 1960. A Manual of Aquatic Plants. Univ. of Wisconsin Press, Madison.
9. HARRISON, D.S., R.D. BLACKBURN, L.W. WELDON, J.R. ORSENIGO & G.F. RYAN. 1966. Aquatic Weed Control Circular 21913. Univ. of Florida Agr. Extension Service, Gainesville.
10. HITCHCOCK, A.S. 1950. Manual of the Grass of the United States. U.S. Dept. Agr. Misc. Publ. 200.
11. HOTCHKISS, N. 1967. Underwater and Floating-Leaved Plants of the United States and Canada. U.S. Dept. Fish & Wildlife Serv. Res. Publ. 44.
12. HOTCHKISS, N. 1970. Common Marsh Plants of the United States and Canada. Resource Publ., Bur. Sport Fish. & Wildlife, U.S. Dept. Interior, Washington, D.C.
13. LONG, R.W. & O. LAKELA. 1971. A Flora of Tropical South Florida. Univ. of Miami Press, Coral Gables, Fla.
14. LOPINOT, A.C. 1979. Aquatic Weeds. Their Identification and Methods of Control. Fishery Bull. No. 4, Div. of Fisheries, State of Illinois Dept. of Conservation, Springfield.
15. MARTIN, A.C., H.S. ZIM & A.L. NELSON. 1951. American Wildlife and Plants. Dover Publ., New York, N.Y.
16. MASON, H.L. 1969. Flora of the Marshes of California. Univ. of Calif. Press, Berkeley.
17. MUENSCHER, W.C. 1944. Aquatic Plants of the United States. Comstock Publ. Co., Inc., Ithaca, N.Y.
18. POPENOE, H., et al. 1976. Making Aquatic Weeds Useful: Some Perspectives for Developing Countries. Nat. Acad. Sci., Washington, D.C.
19. RADFORD, A.E., H.E. AHLES & C. RITCHIE BELL. 1968. Manual of the Vascular Flora of the Carolinas. Univ. of N. Carolina Press, Chapel Hill.
20. REED, C.F. 1977. Economically Important Foreign Weeds. Potential Problems in the United States. U.S. Dept. of Agr., U.S. Govt. Printing Office, Washington, D.C.
21. SMALL, J.K. 1933. Manual of the Southeastern Flora. Univ. of N. Carolina Press, Chapel Hill.
22. VARSHNEY, C.K. & J. RZOSKA. 1976. Aquatic Weeds in S.E. Asia. Proc. Regional Seminar on Noxious Aquatic Vegetation, New Delhi, 12-17 Dec. 1973. Dr. W. Junk Publ. B.V., The Hague, Netherlands.
23. WELDON, L.W., R.D. BLACKBURN & D.S. HARRISON. 1969. Common Aquatic Weeds. U.S. Dept. Agr. Handbook No. 352.

1004 G. Bibliography

PEARSALL, W.H. 1920. The aquatic vegetation of the English Lakes. *J. Ecol.* 8:163.

BUTCHER, R.W. 1933. Studies on the ecology of rivers. I. On the distribution of macrophytic vegetation in the rivers of Britain. *J. Ecol.* 21:58.

WELCH, P.S. 1948. Limnological Methods. Blakiston Co., Philadelphia, Pa.

MOORE, W.G. 1950. Limnological studies of Louisiana lakes. I. Lake Providence. *Ecology* 31:86.

PENFOUND, W.T. 1953. Plant communities of Oklahoma lakes. *Ecology* 34:561.

PENFOUND, W.T. 1956. Primary production of vascular aquatic plants. *Limnol. Oceanogr.* 1:92.

SWINDALE, D.N. & J.T. CURTIS. 1957. Phytosociology of the larger submersed plants in Wisconsin lakes. *Ecology* 38:397.

FORSBERG, C. 1959. Quantitative sampling of subaquatic vegetation. *Oikos* 10:233.

EDWARDS, R.W. & M.W. BROWN. 1960. An aerial photographic method for studying the distribution of aquatic macrophytes in shallow waters. *J. Ecol.* 48:161.

EDWARDS, R.W. & M. OWENS. 1960. The effects of plants on river conditions. I. Summer crops and estimates of net productivity of macrophytes in a chalk stream. *J. Ecol.* 48:151.

FORSBERG, C. 1960. Subaquatic macrovegetation in Osbysjon, Djursholm. *Oikos* 11:183.

STEEL, R.G.D. & J.H. TORRIE. 1960. Principles and Procedures of Statistics with Special Reference to the Biological Sciences. McGraw-Hill Book Company, Inc. New York, N.Y.

HARROD, J.J. & R.E. HALL. 1962. A method for determining the surface areas of various aquatic plants. *Hydrobiologia* 20:173.

COBB, B. 1963. A Field Guide to the Ferns. Houghton Mifflin Co., Boston, Mass.

WESTLAKE, D.F. 1963. Comparisons of plant productivity. *Biol. Rev.* 38:385.

WOOD, R.D. 1963. Adapting SCUBA to aquatic plant ecology. *Ecology* 44:416.

SCHMID, W.D. 1965. Distribution of aquatic vegetation as measured by line intercept with SCUBA. *Ecology* 46:816.

WESTLAKE, D.F. 1966. The biomass and productivity of *Glyceria maxima*. *J. Ecol.* 54:745.

FAGER, E.W., A.O. FLECHSIG, R.F. FORD, R.I. CLUTTER & R.J. GHELARDI. 1966. Equipment for use in ecological studies using SCUBA. *Limnol. Oceanogr.* 11:503.

OWENS, M., M.A. LEARNER & P.J. MARIS. 1967. Determination of the biomass of aquatic plants using an optical method. *J. Ecol.* 55:671.

WESTLAKE, D.F. 1967. Some effects of low-velocity currents on the metabolism of aquatic macrophytes. *J. Exp. Bot.* 18:187.

WYRICK, G.D. 1967. Common Aquatic Weeds of Kansas Ponds and Lakes. Emporia State Research Studies, Kansas State Teachers College, Emporia.

NEUSHUL, M. 1967. Studies of subtidal marine vegetation in Western Washington. *Ecology* 48:83.

BLACKBURN, R.D., P.F. WHITE & L.W. WELDON. 1968. Ecology of submersed aquatic weeds in South Florida canals. *Weed Sci.* 16:261.

MITCHELL, D.S. 1969. The ecology of vascular hydrophytes on Lake Kariba. *Hydrobiologia* 34:448.

LIND, C.T. & G. COTTAM. 1969. The submerged aquatics of University Bay: A study in eutrophication. *Amer. Midland Natur.* 81:353.

LIVERMORE, D.F. & W.E. WUNDERLICH. 1969. Mechanical removal of organic production from waterways. *In* Eutrophication: Causes, Consequences, Correctives. Nat. Acad. Sci., Washington, D.C.

WESTLAKE, D.F. 1969. Macrophytes. *In* R.A. Vollenweider, ed. A Manual on Methods for Measuring Primary Production in Aquatic Environments. J.B. Lippincott Co., Philadelphia, Pa.

WETZEL, R.G. 1969. Excretion of dissolved organic compounds of aquatic macrophytes. *Bioscience* 19:539.

MULLIGAN, H.F. 1969. Management of aquatic vascular plants and algae. *In* Eutrophication: Causes, Consequences, Correctives. Nat. Acad. Sci., Washington, D.C.

BOYD, C.E. 1970. Vascular aquatic plants for mineral nutrient removal from polluted water. *Econ. Bot.* 24:95.

HANNAN, H.H. & T.C. DORRIS. 1970. Succession of a macrophyte community in a constant temperature river. *Limnol. Oceanogr.* 15:442.

HOLCOMB, D. & W. WEGENER. 1971. Hydrophytic changes related to lake fluctuation as measured by point transects. Proc. 25th Annu. Conf. S.E. Ass. Game & Fish Comm.

BOYD, C.E. 1971. The limnological role of aquatic macrophytes and their relationship to reservoir management. *In* G.E. Hall, ed. Reservoir Fisheries and Limnology. Spec. Publ. Amer. Fish. Soc. No. 8.

WEBER, C.I. 1973. Biological Field and Laboratory Methods for Measuring the Quality of Surface Waters and Effluents. EPA-670/4-73-001, Off. Research and Development, U.S. EPA, Cincinnati, Ohio.

HESTAND, R.S., B.E. MAY, D.P. SCHULTZ & C.R. WALKER. 1973. Ecological implications of water levels on plant growth in a shallow reservoir. *Hyacinth Control J.* 11:54.

DAVIS, T. 1974. Biological control of aquatic vegetation. Phase I: Data analysis report. Applications Projects Branch, Kennedy Space Center, NASA, Florida.

BERRY, C.R., C.B. SCHRECK & R.V. CORNING. 1975. Control of Egeria in a Virginia water supply reservoir. *Hyacinth Control J.* 13:24.

MANNING, J.H. & D.R. SANDERS, SR. 1975. Effects of water fluctuation on vegetation in Black Lake, Louisiana. *Hyacinth Control J.* 13:17.

HUTCHINSON, G.E. 1975. A Treatise on Limnology. John Wiley and Sons, New York, N.Y.

MANNING, J.H. & R.E. JOHNSON. 1975. Water level fluctuation and herbicide application: an integrated control method for hydrilla in a Louisiana reservoir. *Hyacinth Control J.* 13:11.

WETZEL, R.G. 1975. Limnology. W.B. Saunders Company, Inc., Philadelphia, Pa.

WESTLAKE, D.F. 1975. Macrophytes. *In* B.A. Whitton, ed. River Ecology. Univ. of Cali-

fornia Press, Berkeley and Los Angeles.
WOOD, R.D. 1975. Hydrobotanical Methods. University Park Press, Baltimore, Md.
WILE, I. 1975. Lake restoration through mechanical harvesting of aquatic vegetation. *Verh. Int. Ver. Limnol.* 19:660.
NALL, L.E., M.J. MAHLER & J. SCHARDT. 1976. Aquatic macrophyte sampling in Lake Conway. *In* Proc. Research Planning Conference on the Aquatic Plant Control Program, 19-22 October, Atlantic Beach, Fla. Miscellaneous Paper A-77-3, U.S. Army Engineer Waterways Experiment Station, Vicksburg, Miss.
OSBORNE, J.A. 1977. Ground Truth Measurements of *Hydrilla verticillata* Royle and Those Factors Influencing Underwater Light Penetration to Coincide with Remote Sensing and Photographic Analysis. Res. Rep. No. 2, Bureau of Aquatic Plant Research and Control, Florida Department of Natural Resources, Tallahassee.
GOLDSBY, T.L. & D.R. SANDERS, SR. 1977. Effects of consecutive water fluctuations on submersed vegetation of Black Lake, Louisiana. *Hyacinth Control J.* 15:23.
TARVER, D.P. Consultation with Bureau of Aquatic Plant Research and Control, Florida Department of Natural Resources, Tallahassee.
LAZOR, R. Consultation with Bureau of Aquatic Plant Research and Control, Florida Department of Natural Resources, Tallahassee.

1005 BENTHIC MACROINVERTEBRATES

1005 A. Introduction

Benthic macroinvertebrates are animals inhabiting the substratum of lakes and streams. They may construct attached cases or nests in which they live or roam freely over rocks, organic debris, and other substrates. Although immature specimens may be very small, macroinvertebrates are considered by definition to be visible to the unaided eye and are retained on a U.S. Standard No. 30 sieve (0.595-mm openings). (The standard sieve opening for marine benthic fauna is 1.0 mm, U.S. Standard No. 18 Sieve). Included among the macroinvertebrates are aquatic insects, macrocrustaceans, mollusks, annelids, roundworms, flatworms, and other aquatic invertebrates.

The composition and density (number of individuals per unit area) of macroinvertebrate communities in streams and lakes are reasonably stable from year to year in unperturbed environments. However, seasonal fluctuations associated with life-cycle dynamics of individual species may result in extreme variation at specific sites within any calendar year. Most aquatic habitats (particularly free-flowing streams) with acceptable water quality and substrate conditions support diverse macroinvertebrate communities in which there is a reasonably balanced distribution of species among the total number of individuals present. Such communities respond to changing habitat quality by adjustments in community structure. However, many estuaries are dominated by a few species. Small changes in their relative numbers may not be indicative of changes in water quality. Macroinvertebrate community responses to environmental perturbations are useful in assessing the impact of municipal, industrial, oil, and agricultural wastes, as well as impacts from other land uses on natural water bodies. Two situations for which the patterns of macroinvertebrate community structure change have been documented are organic loading and toxic chemical pollution. Severe organic pollution usually results in a restriction in the variety of macroinvertebrates to only the most tolerant ones and a corresponding increase in density of those tolerating the polluted conditions, usually low dissolved oxygen. On the other hand, pollution by toxic chemicals may eliminate the entire macroinvertebrate community from an affected area. Not all cases conform to these described, because conditions may be mediated by other environmental and biological factors.

Assessing the impact of a pollution source generally involves the comparison of macroinvertebrate communities at sites influenced by pollution with those collected from adjacent unaffected sites. The procedure includes sampling and analyzing both communities and subsequently determining whether the presumed pollution-affected community differs from the nonaffected one. The basic information required for most community structure analyses is a count of individuals per species. From the count data, the communities can be characterized and compared according to composition, density, biomass, diversity, or other analysis.

1005 B. Sample Collection

1. General Considerations

Before conducting a survey, determine specific objectives and define clearly the information sought. Ultimate selection of a methodology will depend on whether a stream, lake, or reservoir is to be studied. For example, to determine whether the macroinvertebrate community downstream from a discharge is damaged, only a few sampling stations upstream, usually two, and downstream from the discharge, usually two or more, are needed. However, if the objective is to delimit the extent of impact from a discharge or series of discharges, it is necessary to have reference stations upstream from all discharges, to bracket each discharge with stations, and to establish stations downstream.

After gaining a thorough understanding of the factors involved with a particular body of water, select specific areas to be sampled. There is no set number of sampling stations that will be sufficient to monitor all possible waste discharges. No water quality survey is routine, nor can one be conducted totally on a ''cookbook'' basis. However, if some basic rules such as the following are adhered to carefully, a sound survey can be designed:

1. Always establish a reference station(s) upstream or at a point remote from all possible wastewater discharges. Because most surveys are made to determine the damage that pollution causes to aquatic life, this will be the basis for comparison of the fauna in polluted and unpolluted areas. Preferably have at least two reference stations, one well away from, or upstream from, the discharge and the other directly above, or in the immediate vicinity of, the effluent discharge, but not subject to its influence.

2. Locate a station immediately downstream or in the affected area, in the immediate vicinity of each discharge.

3. If the discharge does not mix completely on entering the receiving water, but channels along one side of the stream or disperses in a specific direction, subdivide stations into left-bank, midchannel, and right-bank sections of the stream or into concentric arcs in a lake.

4. Establish stations at various distances downstream from the last discharge to determine the linear extent of damage.

5. To permit comparison of macroinvertebrate communities, be sure that all sampling stations are ecologically similar. For example, select stations that are simi-

lar with respect to bottom substrate (sand, gravel, rock, or mud), depth, presence of riffles and pools, stream width, flow velocity, and bank cover.

6. Collect samples for physical and chemical analyses close to biological sampling stations to assure correlation of findings.

7. Locate sampling stations for macroinvertebrates in an area not influenced by atypical conditions.

For a long-term biological monitoring program collect macroinvertebrates at each station at least once during each of the annual seasons. More frequent sampling may be necessary if the characteristics of the effluents change or if spills occur. Make allowance for collections at night where "drift" organisms are of special concern. In general, the most critical period for macroinvertebrates in streams is during periods of high temperature and low flow. Therefore, if available time and funds limit sampling frequency, make at least one survey during this time.

2. Sampling Design

A sample usually is considered to be a portion taken from some larger aggregate about which inferences are to be made. The problem in collecting a representative sample arises from the variation usually encountered in successive samples. Without knowledge of sample variation, data cannot be considered correctly to represent the population. Therefore, take replicate samples of a population or population aggregate if correct statistical inferences about the population are to be made.[1]

When developing a standardized sampling design consider the following requirements:

1. Define the set of all samples that can be selected (i.e., separate the population into sampling units). For example, if the area known to contain the population (the sampling universe) to be sampled is 1,000 m^2 and 1 m^2 is to be sampled, there are 1,000 samples in the set of all available sampling sites.

2. Assign to each possible sample a known probability of selection (randomize to give every sample equal probability of being selected). Using the situation cited above, divide the area to be sampled into 1,000 discrete units.

3. Select each sample by a process that assigns the correct probability of selection to all samples. Use a table of random numbers to select sites for sampling or, alternatively, base on proportional allocation of sampling effort according to bottom or substrate type. In the latter case, sampling effort for a given substrate is based on proportion of the total substrate represented by the given substrate.

4. Choose a method for computation of estimates from the sample that provides a unique value for each sample, i.e., mean, standard deviation.

Standardize acquisition and recording of data when practical. Record and report data in metric units.

One of the most difficult decisions to make when collecting aquatic invertebrates is how many replicate samples are required at each station or substation to obtain reliable information. Many taxa are not distributed uniformly over the bottom of a river or lake. Different habitats (sand, mud, gravel, or organic material) support different densities and species of organisms. Even on a relatively homogeneous bottom, animals tend to aggregate. Therefore, take replicate samples to evaluate this variability.

In most cases use at least three replicate samples per station to describe the macroinvertebrate community. More may be necessary to achieve the desired level of precision. If a station must be divided into substations, sample each substation as described above. Ideally, conduct a baseline survey first to determine the number of samples necessary to achieve the desired level of accuracy in each pollution survey.[2]

3. Sampling Devices, Quantitative

Quantitative and qualitative samplers have been designed to collect organisms from stream and lake bottoms. The most common quantitative sampling devices are the Ponar, Petersen, and Ekman grabs and the Surber or square-foot stream bottom sampler, all described below.

a. Grab samplers:

1) *The Ponar grab* (Figure 1005:1) is used increasingly in medium to deep rivers, lakes, and reservoirs.[3] It is similar to the Petersen grab in size, weight, lever system, and sample compartment, but has side plates and a screen on top of the sample compartment to prevent sample loss during closure. With one set of weights, the standard 23- by 23-cm sampler weighs 20 kg. A 15- by 15-cm petite Ponar also is available. The large amount of surface disturbance associated with a Ponar grab can be reduced greatly by installing hinged rather than fixed screen tops, which reduces the pressure wave associated with the sampler's descent. This sampler is best for sand, gravel, or small rocks with mud but it can be used in all substrates except bedrock.

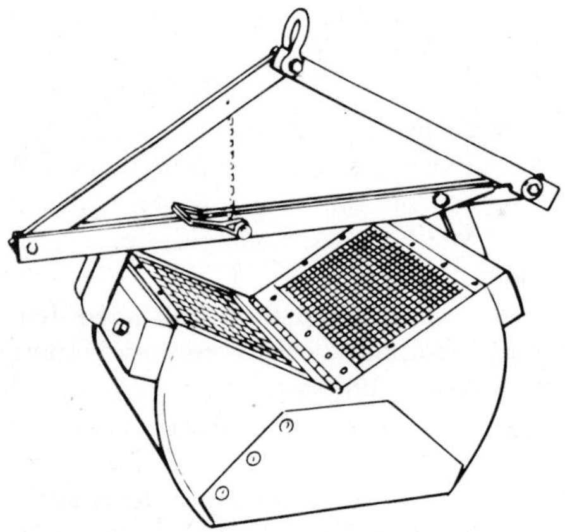

Figure 1005:1. Ponar grab.

2) *The orange-peel grab* (Figure 1005:2) is a multi-jawed round grab with a canvas closure at the top serving as a portion of the sample compartment. The 1,600-cm^3 size generally is used, although larger sizes are available. The area sampled and volume of material collected depend on depth of penetration.[4] This grab is suited for marine environments and deep lakes with sandy substrates.

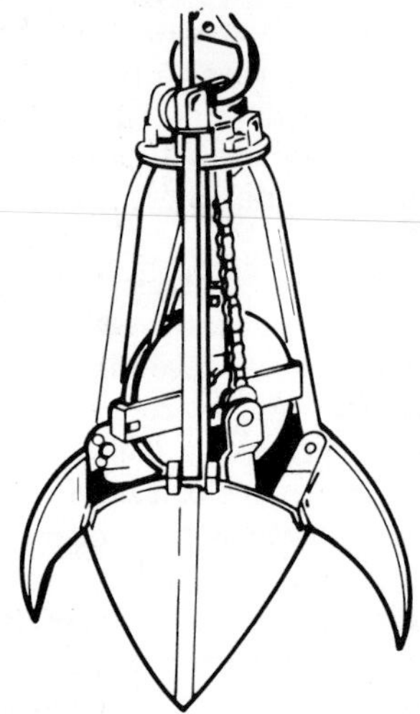

Figure 1005:2. Orange peel sampler.

3) *The Petersen grab* (Figure 1005:3) is used widely for sampling hard bottoms such as sand, gravel, marl, and clay in swift currents and deep water. It is an iron, clam-type grab manufactured in various sizes that will sample an area of from 0.06 to 0.09 m^2. It weighs approximately 13.7 kg, but may weigh as much as 31.8 kg when auxiliary weights are bolted to its sides. The primary advantage of the extra weights is to make the grab stable in swift currents and to give additional cutting force in fibrous or firm bottom materials. Such a sampler may be modified by adding end plates, by cutting large strips out at

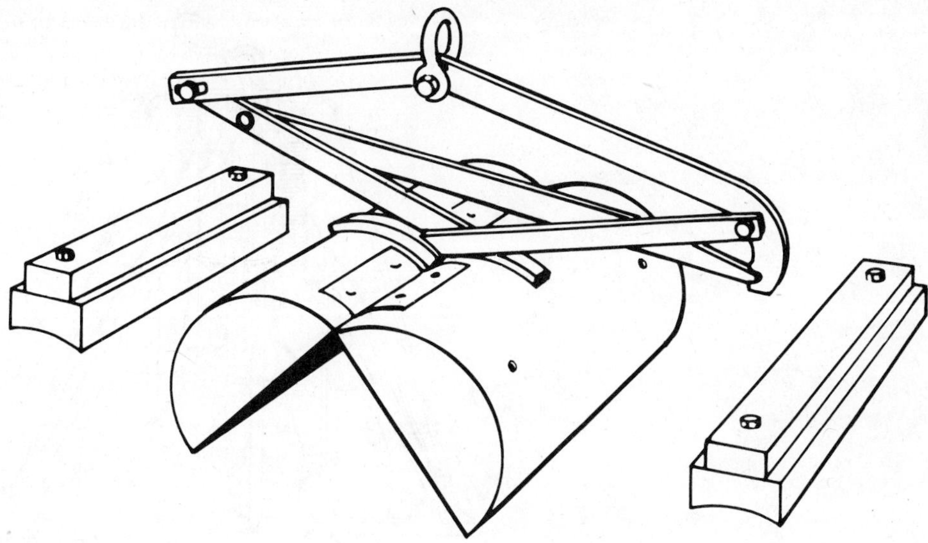

Figure 1005:3. Petersen grab.

the top of each side, and by adding hinged 30-mesh screen as in the Ponar grab.[5]

To use the Petersen grab, set the hinged jaws and slowly lower to the bottom to avoid disturbing lighter bottom materials. Ease rope tension to release the catch. As the grab is raised the lever system closes the jaws.

4) *The Smith-McIntyre grab* (Figure 1005:4) has the heavy steel construction of the Petersen, but its jaws are closed by strong coil springs.[6] Its chief advantages are stability and easier control in rough water. Its bulk and heavy weight require operation from a large boat equipped with a winch. The 45.4-kg grab can sample an area of 0.1 m^2.[7,8]

5) *The Shipek grab* (Figure 1005:5) is designed to take a sample 0.04 m^2 in surface area and approximately 10 cm deep at the center. The sample compartment is composed of two concentric half cylinders. When the grab touches bottom, inertia from a self-contained weight releases a catch and helical springs rotate the inner half cylinder by 180°. The sample bucket may be disengaged from the upper semicylinder by release of two retaining latches. This grab is used primarily in marine waters and large inland bodies of water.

6) *The Ekman grab* (Figure 1005:6) is useful for sampling silt, muck, and sludge in water with little current. It is difficult to use when rocky or sandy bottoms are present because small pebbles or grit prevent proper jaw closure. The grab is made of 12- to 20-gauge brass or stainless steel and weighs approximately 3.2 kg. The box-like compartment holding the sample has spring-operated jaws on the bottom that must be cocked manually (exercise caution in cocking and handling the grab because of possible injuries if jaws are tripped accidentally). At the top of the grab are two hinged overlapping lids that are held partially open during descent by water passing through the sample compartment. These lids are held shut by water pressure when the sampler is being retrieved. The grab is made in three sizes: 15 by 15 cm, 23 by 23 cm, and 30 by 30 cm. The smallest size usually is adequate for

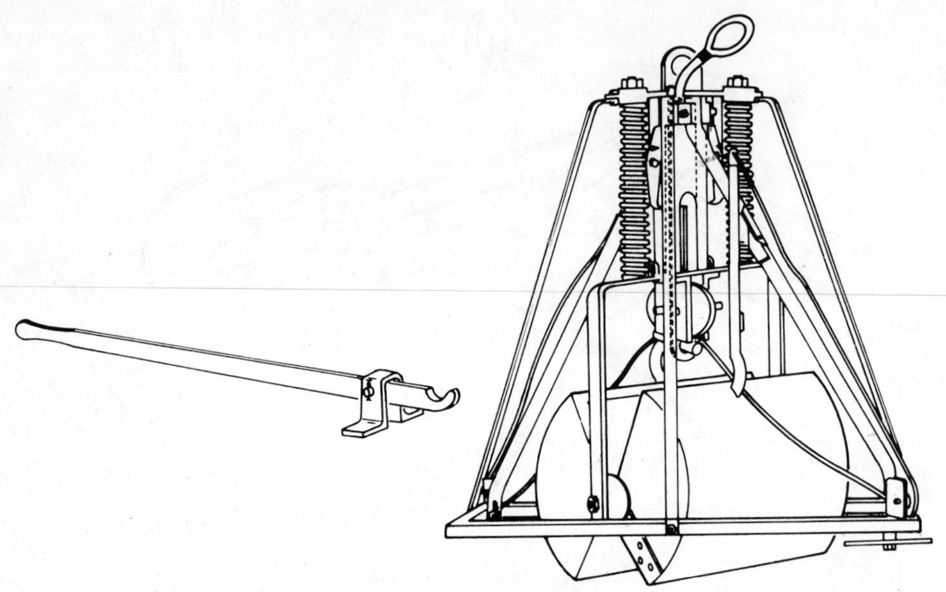

Figure 1005:4. Smith-McIntyre grab.

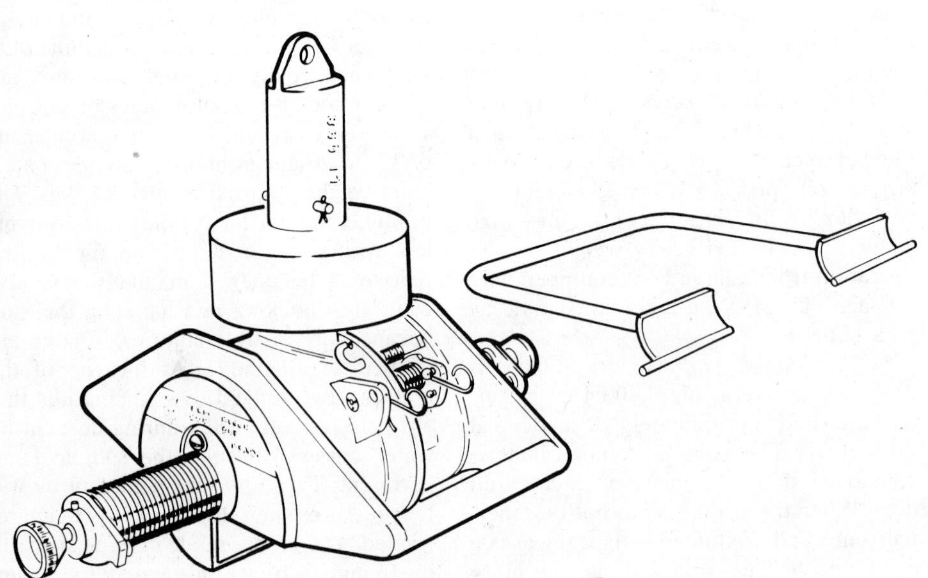

Figure 1005:5. Shipek grab.

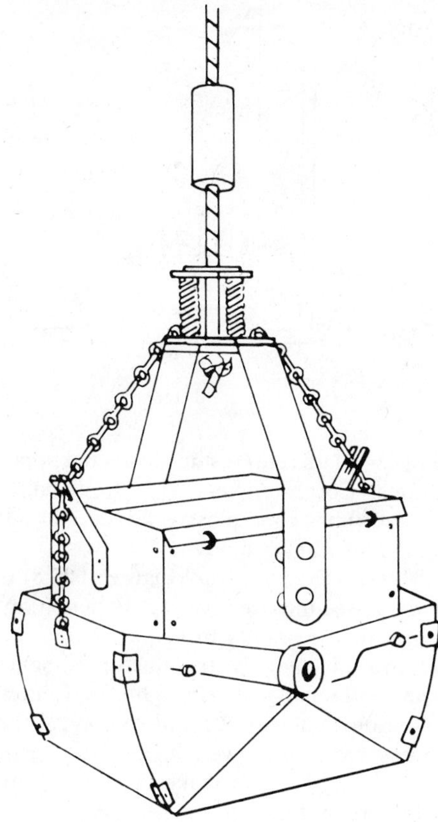

Figure 1005:6. Ekman grab.

general sampling. However, a taller model of this sampler, either 23 cm or 30 cm tall, is better than the 15-cm version. Place a Standard U.S. No. 30 sieve insert in the top for deep sediments.

b. Riffle samplers:

1) *The Surber or square-foot riffle sampler* (Figure 1005:7)[9] consists of two brass frames, each 30.5 cm (1 foot) square, hinged together along one edge. When in use, the two frames are locked at right angles, one frame marking off the area of substrate to be sampled, and the other supporting a net to collect organisms washed into it from the sample area. The net usually is 69 cm long with the first few centimeters and the wings constructed of heavier material (canvas, taffeta) to increase durability. Standard mesh size is 9 threads/cm (23 threads/in.). While a smaller mesh size might increase the number of smaller invertebrates and young instars collected, it also would clog more easily and exert more resistance to the current than a larger mesh. This could result in a loss of organisms due to backwashing from the sample net. This sampler is specific for macrobenthos and many microcomponents of the benthos are not collected.

Use this sampler in shallow, flowing water (30 cm or less). If it is used in deeper water some organisms may be carried over the top of the sampler. Position the sampler securely on the stream bottom parallel to water flow with the net portion downstream. Take care not to disturb the substrate upstream from sampler. Leave no gaps under the edges of the frame that would allow water to wash under the net. Fill gaps that may occur along the back edge of the sampler by carefully shifting rocks and gravel along the outside edge. When the sampler is in place (it may be necessary to hold it in place with one hand in a strong current), carefully turn over and rub lightly all rocks and large stones with the hands to dislodge organisms clinging to them. Examine each stone before discarding it for organisms, larval or pupal cases, etc., that may be clinging to it. Scrape attached algae, insect cases, etc., from the stones into the sampler net. Stir remaining gravel and sand with the hands or a stick to a depth of 5 to 10 cm, depending on the substrate, to dislodge bottom-dwelling organisms; repeat 2 to 3 times. It may be necessary to hand-pick some mussels and snails that are not carried into the net by current.

Remove sample by inverting net into sample container. Carefully examine net for small organisms clinging to it. Remove these, preferably with forceps to avoid

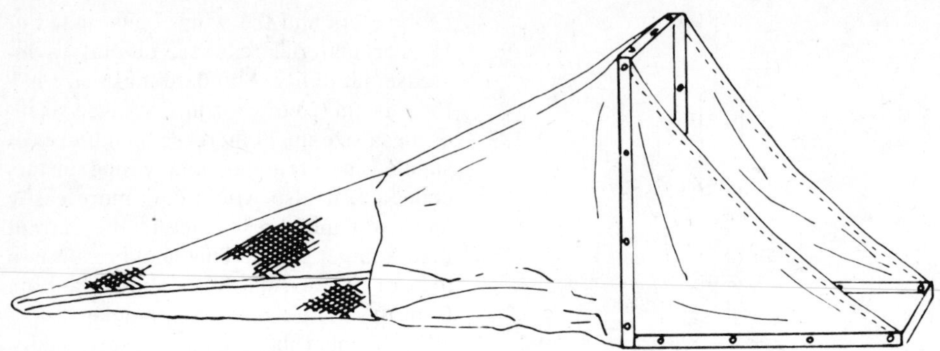

Figure 1005:7. Surber or square-foot sampler.

damage, and include in sample. Rinse sampler net after each use.

A common problem arising during use of the Surber sampler is that organisms wash under the bottom edge of the sampler. The following modifications have been suggested for different substrates:

Method for loose gravel—Extend bottom edge of Surber frame to 5 or more cm allowing for insertion of frame into substrate to a greater depth. This works well in soft substrates such as sand and gravel where the current causes substrate shifting.

Method for coarse gravel and rock—Add serrated extension to back edge of frame to secure it and reduce washing from under this edge. This is helpful in hard gravel and rock substrates where sinking the entire frame is impossible.

Method for gravel and bedrock—Add a 5-cm band of flexible material to bottom edge of sampler to create a seal in rocky, uneven substrates. Make this band from foam rubber or fine-textured synthetic sponge. Remove organisms that stick to foam and include in sample.

2) *Other riffle samplers,* such as the Neils Cylinder, the Hess, the Waters and Knapp, and similar devices, may be used.

c. Core or cylindrical samplers: Use core or cylindrical samplers to sample sediments in depth. Efficient use as surface samplers requires dense animal populations. Core samplers vary from hand-pushed tubes to explosive-driven and automatic-surfacing models.[10]

1) *The Phleger corer* (Figure 1005:8) is widely used in water-quality studies. It operates on the gravity principle. Styles and weights vary among manufacturers, some using interchangeable weights that allow variations between 7.7 and 35.0 kg, while others have fixed weights of 41.0 kg or more. Length of core taken will vary with substrate texture.

2) *The KB core sampler* or a modification thereof known as the Kajak-Brinkhurst corer, may be useful in obtaining estimates of the standing stock of benthic macroinvertebrates inhabiting soft sediments.[11]

3) *The Wilding or stovepipe sampler* (Figure 1005:9) is made in various sizes and with many modifications. It is especially useful for quantitatively sampling a bottom with dense, vascular plant growth. It may be used to sample vegetation, mud-water interface sediment, or most shallow stream substrates. Large volumes of vegetation, when sampled in this way, may require a great deal of time for laboratory processing.

d. Drift samplers:

1) *Drift nets* (Figure 1005:10) are anchored in flowing water for capture of

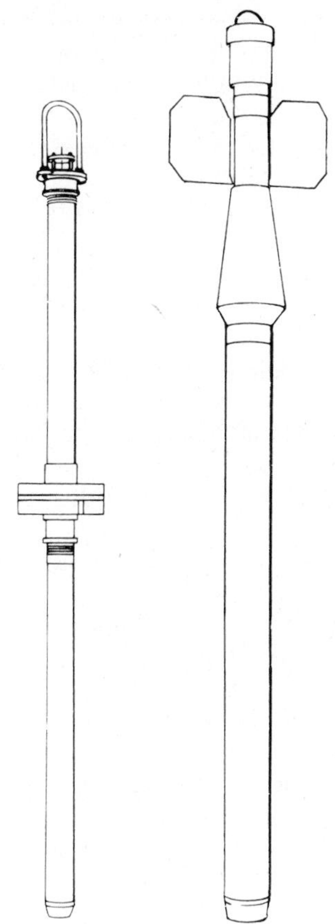

Figure 1005:8. Phleger core sampler.

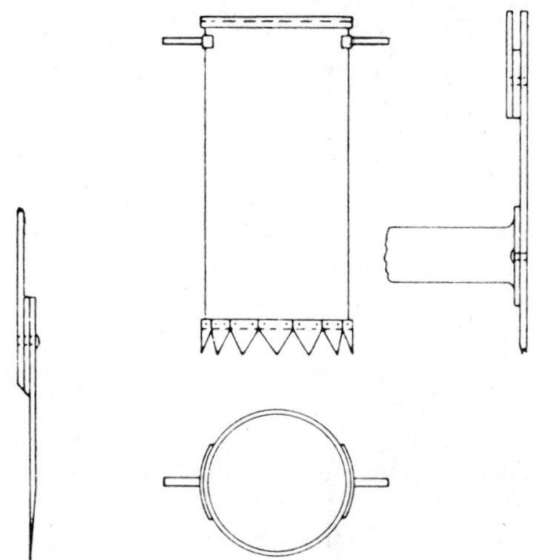

Figure 1005:9. Wilding or stovepipe sampler.

macroinvertebrates that have migrated or have been dislodged from the bottom substrates into the current. Drift organisms are important to the stream ecosystem because they are prey for stream fishes, and thus should be considered in the study of fish populations. Benthic organisms respond to pollutional stresses by drifting from an affected area; thus, drift is important in water-quality investigations, especially those of spills of toxic materials. Drift also is a factor in recolonizing denuded areas and contributes to recovery of disturbed streams.

Use nets having a 929-cm^2 upstream opening and mesh equivalent to U.S. Standard No. 30 screen (0.595-mm pore size). Alternatively, use a plastic net with a 0.471-mm pore size. After placing the net in the water, frequently remove organisms and debris to prevent clogging and subsequent diversion of water at the net opening. Set drift-net samples for any specified period of time (usually 3 hr) but use the same time for each station. Sampling between dusk and 1 AM is optimum.

The total quantity (numbers or biomass) of organisms drifting past a given station per 24 hr divided by the total stream discharge is the best measure of drift intensity. Report data in terms of (number/24 hr)/(m^3/24 hr) or (biomass/24 hr)/(m^3/24 hr).[12–14]

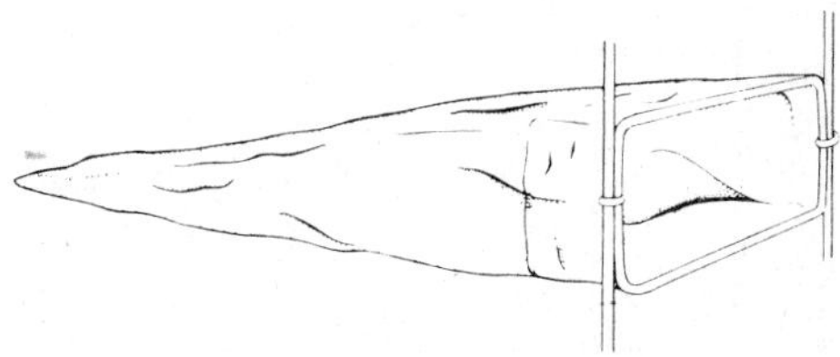

Figure 1005:10. Drift net sampler.

4. Sampling Devices, Qualitative

When sampling qualitatively, search for as many different organisms as possible. Collect samples by any method that will capture representative species.

a. Bottom nets are the most versatile collecting devices for shallow, flowing water, and are useful also for shoreline collecting in lakes. When combined with a standardized kicking technic,[15] bottom nets are appropriate for quantitatively sampling macroinvertebrates.[16]

b. Tow nets or trawls range from simple sled-mounted nets to complicated devices incorporating teeth that dig into the bottom. Some models feature special apparatus to hold the net open during towing and to close it during descent and retrieval. Welch,[17] Barnes,[10] and Usinger[18] discuss different styles available.

c. Miscellaneous devices such as garden rakes, pocket knives, buckets, or sieves are useful for collecting macroinvertebrates. The extent of their use is determined by the type of substrate to be sampled and the collector's ingenuity.

5. Sampling Devices, Artificial Substrate Samplers

Artificial substrate samplers are devices of standard composition and configuration placed in the water for a predetermined period for colonization by macroinvertebrate communities. Because many physical variables encountered in bottom sampling are minimized, e.g., depth, light penetration, temperature differences, and species substrate preferences, artificial substrate sampling complements other types of sampling. Like natural submerged substrates such as logs and pilings, artificial substrates are colonized primarily by larval aquatic insects, crustaceans, coelenterates, bryozoans, and to some extent worms and mollusks. The organisms that colonize artificial substrates are primarily drift organisms carried by water currents. Colonization rates should be similar; thus, the numbers and kinds of organisms reflect capacity to support aquatic life.

Position artificial substrates in the euphotic zone (0.3 m) for maximum abundance and diversity of macroinvertebrates.[19] Optimum time for substrate colonization is 6 wk for most waters in the U.S. For uniformity of depth, suspend sampler from floats on a 3.2-mm steel (preferably stainless) cable. If vandalism is a problem, use subsurface floats or place sampler on the bottom. Regardless of installation technic, use uniform procedures.

At shallow water stations (less than 1.2 m deep), install samplers so that the exposure occurs midway in the water column at low flow. For samplers installed in July when the water depth is about 1.2 m and the August average low flow is 0.6 m, install 0.3 m above the bottom. Take care not to allow samplers to touch the bottom or they may become covered with silt, thereby increasing the sampling error. In shallow streams with sheet rock bottoms, secure artificial substrates to 0.95-cm steel rods that are driven into the substrate or secure to rods that are mounted on low, flat, rectangular blocks.

Before removing samples from the water, enclose them in an oversized plastic bag (double wrapping) that is tightly sealed to prevent possible loss of organisms or use a large dip net (U.S. Standard No. 30 mesh) when the sample is removed. Disassemble sampler and brush in a pan of water in the field or add preservative to the bag containing the intact sampler, and disassemble and brush later in the laboratory.

Many different styles of artificial substrate samplers have been tested.[20] The Fullner[21] modification of the Hester-Dendy[22] multiplate and the basket sampler[19] are used widely.

a. Multiple-plate or modified Hester-Dendy sampler (Figure 1005:11) is con-

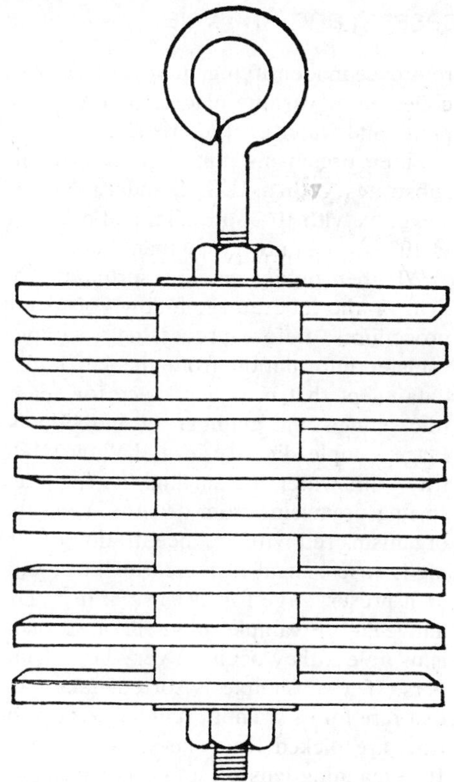

Figure 1005:11. Multiple-plate or Hester-Dendy sampler (unmodified).

structed of 0.3-cm-thick tempered hardboard with 7.6-cm round plates and 2.5-cm round spacers that have center-drilled 1.6-cm holes. The plates are separated by spacers on a 0.63-cm-diam eyebolt, held in place by a nut at the top and bottom. A total of 14 large plates and 24 spacers is used in each sampler. Separate the top nine plates by a single spacer, Plate 10 by two spacers, Plates 11 and 12 by three spacers, and Plates 13 and 14 by four spacers. The sampler is approximately 14 cm long and 7.6 cm in diameter, has an exposed surface area of approximately 1,160 cm^2, and weighs about 0.45 kg. Do not reuse samplers exposed to oils and chemicals that may inhibit colonization. Because of its cylindrical configuration, the sampler fits a wide-mouth container for shipping and storage. The sampler is inexpensive, compact, and lightweight.[19,21,22]

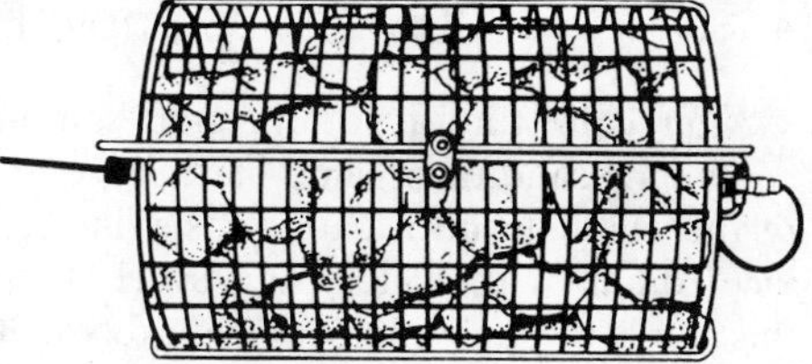

Figure 1005:12. Basket sampler.

b. *The basket sampler*[19] (Figure 1005:12) is a cylindrical "barbecue" basket 28 cm long and 17.8 cm in diameter, filled with approximately 30 5.1-cm-diam rocks or rocklike material weighing 7.7 kg. A hinged side door allows access to the contents. The sampler provides an estimated 0.24 m^2 of surface area for colonization. The factors governing proper installation and collection are the same as those described for the multiplate sampler. Some investigators prefer the basket technic because natural substrate materials are used for colonization.

6. Suction Samplers

"Dome suction" samplers[23] are now widely used for collecting benthic macroinvertebrate samples. This sampler can be placed directly on specific sampling sites but a scuba diver is required to collect samples.[24] Improved accuracy of locating sampling sites and ability to collect a large number of replicate samples outweigh the disadvantage of using a diver. Suction samplers have been used widely in sampling marine environments.

1005 C. Sample Processing and Analysis

After collecting a bottom grab sample containing sand or organic material, empty the contents into a tub or sieving device, dilute with water, and swirl. Pour this slurry through a U.S. Standard No. 30 sieve. Slurries that clog the screen require removal of screened material. A series of one or two coarser screens (e.g., 1-cm and 0.5-cm mesh) will hold back leaves, sticks, etc., while permitting small organisms to pass through to the No. 30 sieve. Carefully check rocks, sticks, and other objects for clinging organisms before discarding.

Use laboratory elutriation devices[25,26] to reduce time required to sort the benthic organisms from samples containing large amounts of silt, mud, or clay. Wash the screened material into a container and fix the contents in a 10% formalin solution or a 70% ethanol solution.[27–30] If ethanol is used, do not fill more than one-half the container with screened material. Preserve animals with calcareous shells or exoskeletons, i.e., mussels, snails, crayfish, and ostracods, in ethanol.[30,31] Label with location, date, type of sampler used, name of collector, and other pertinent information. Some macroinvertebrates, leeches, and turbellarians are identified more easily if they are relaxed to prevent constriction during preservation.[27–29]

For qualitative samples it often is desirable to place rocks, sticks, and other objects in a white pan partially filled with water. Many animals will float free from these objects and can be removed with forceps.

Assign identification numbers either in the field or at the laboratory and transcribe information from the labels to a permanent ledger. The ledger provides a convenient reference in identifying number of samples collected at various places, time of sampling, and water characteristics.

Filter organisms taken from artificial substrates with a U.S. Standard No. 30 sieve, fix with 10% formalin, and preserve in 10% formalin or 70% ethanol.

Whether organisms are sorted in the field or the laboratory, follow consistent procedures. Before processing a sample, transfer information from the label to a data sheet that provides space for scientific names and number of individuals. Place sample directly in a shallow white tray with water for sorting. To facilitate sorting organisms from detritus, stain the organisms red with a concentration of 200 mg/L rose bengal in the formalin or ethanol preservative for at least 24 hr.[32] Examine entire sample and separate organisms unless they occur in very large numbers. If a subsample is sorted, take care that rare forms are not excluded. As organisms are picked from the sample (a 5 to 10× scanning lens or stereoscopic microscope is useful), separate them into different taxonomic categories (e.g., Odonata, Coleoptera, and Ephemeroptera) and identify on the data sheet. Place animals in separate vials according to category and fill vials with 5% formalin or 70% ethanol. Label vials with sample number, date, sampling location, names of organisms, etc.

Identify the animals in each vial using a stereoscopic and compound microscope, according to need, and available experience and resources. See Section 1005F for additional references on laboratory technics and identification of macroinvertebrates.

1005 D. Data Evaluation and Presentation

There are two basic approaches used in evaluating the effects of pollutants on aquatic life. The first is to make a qualitative analysis of fauna and flora "above and below" or "before and after," thereby determining species present or absent. Then, through an understanding of the responses of various species to specific pollutants, determine the significance of damage or change. The second approach involves a quantitative inventory of the number of specimens, species, and structure of the aquatic community affected by the pollutant and comparison with reference information. In most pollution surveys these approaches are integrated because each provides valuable interpretative information.

1. Qualitative Data Evaluation

No two aquatic organisms react identically to a pollutant because of complex interrelationships between genetic factors and environmental conditions. However, certain groups are intolerant of pollution. For example, operculate snails, immature stages of certain mayflies, stoneflies, caddisflies, riffle beetles, and hellgrammites are sensitive to many pollutants. Pollution-tolerant macroinvertebrates such as sludgeworms, certain midge larvae (bloodworms), leeches, and pulmonate snails usually increase in number under organically enriched conditions. Facultative organisms, those that tolerate moderate amounts of pollution, include most snails, sowbugs, scuds, and blackfly larvae. Tolerant organisms may be found in either clean or polluted situations; their presence does not mean that a water body is polluted. However, a population of tolerant organisms combined with an absence of intolerant organisms is a good indication of pollution.

2. Quantitative Data Evaluation

There is a strong trend toward greater use of statistical methods of data evaluation and mathematical expressions of community structure. Statistical analyses of biological data commonly include determination of the mean and confidence interval and use such tests as chi-square, Student's *t*, regression, correlation, one- and two-way analysis of variance, robust analyses, and numerous nonparametric tests. The use of mathematical expressions of community structure to derive numerical indices of diversity of aquatic communities is based on the general, though not invariably true, assumption that the greater the diversity of aquatic life, the greater the structural and functional stability of the system and, therefore, the greater the health of the system.

Diversity indices are useful because they condense considerable biological data into a single numerical value. Diversity indices in current use include $\bar{d}$ (diversity per individual), which follows concepts of information theory, and the SCI (Sequential Comparison Index).[33–36]

To evaluate statistically the data collected in a pollution survey, identify the sources of variability commonly found. Variability in macroinvertebrate data comes from the methods of sampling and the distribution of organisms. Perhaps the major source is sampling error. Organisms generally are clustered in relation to habitat distribution; therefore, random samples often show high variability among replicates. In statistical analyses of quantitative data, large numbers of samples often are required to detect statistically significant differences. Exercise care in using parametric statistical methods because the basic assumption of normal distribution is not always true. Do not assume that a statistically significant difference is ecologically significant.

3. Data Presentation

Presentation of data in reports may take many forms. However, the basic technics include tables, bar graphs (horizontal and vertical), pie diagrams, pictorial charts (ideographs), line graphs, frequency distribution tables and graphs, histograms, frequency polygons, and cumulative frequency polygons. These may be superimposed on maps. Several reports that may be useful in analyzing macroinvertebrate data have been included in the bibliography.

1005 E. References

1. SNEDECOR, G.W. & W.G. COCHRAN. 1967. Statistical Methods. Iowa State Univ. Press, Ames.
2. ELLIOTT, J.M. 1971. Some Methods for the Statistical Analysis of Samples of Benthic Invertebrates. Freshwater Biol. Ass., Sci. Publ. No. 25.
3. POWERS, C.F. & A. ROBERTSON. 1967. Design and Evaluation of an All-Purpose Benthos Sampler. Special Rep. No. 30, Great Lakes Res. Div., Univ. of Michigan, Ann Arbor.
4. MERNA, J.W. 1962. Quantitative sampling with the orange peel dredge. *Limnol. Oceanogr.* 7:432.
5. WEBER, C.I. ed. 1973. Biological Field and Laboratory Methods for Measuring the Quality of Surface Waters and Effluents. EPA-670/4-73-001.
6. SMITH, W. & A.D. MCINTYRE. 1954. A spring-loaded bottom sampler. *J. Mar. Biol. Ass. U.K.* 33:257.
7. MCINTYRE, A.D. 1971. Efficiency of Marine Bottom Samplers. *In* Holme, N.A. & A.D. McIntyre, eds. Methods for the Study of Marine Benthos. IBP Handbook No. 16:140.
8. WIGLEY, R.L. 1967. Comparative efficiency of Van Veen and Smith-McIntyre grab samplers as recorded by motion pictures. *Ecology* 48:168.
9. SURBER, E. 1937. Rainbow trout and bottom fauna production in one mile of stream. *Trans. Amer. Fish. Soc.* 66:193.
10. BARNES, H. 1959. Oceanographic and Marine Biology. George Allen and Unwin, Ltd., London, England.
11. BRINKHURST, R.O., K.E. CHUA & E. BATOOSINGH. 1969. Modifications in sampling procedures as applied to studies on the bacteria and tubificid oligochaetes inhabiting aquatic sediments. *J. Fish Res. Board Can.* 26:2581.
12. WATERS, T.F. 1961. Standing crop and drift of stream bottom organisms. *Ecology* 42:532.
13. DIMOND, J.B. 1967. Pesticides and Stream Insects. Bull. No. 2, Maine Forest Service, Augusta, and Conservation Foundation, Washington, D.C.
14. WATERS, T.F. 1972. The drift of stream insects. *Annu. Rev. Entomol.* 17:253.
15. FROST, S., A. HUN & W. KERSHAW. 1971. Evaluation of a kicking technique for sampling stream bottom fauna. *Can. J. Zool.* 49:167.
16. CROSSMAN, J.S., J. CAIRNS, JR. & R.L. KAESLER. 1973. Aquatic Invertebrate Recovery in the Clinch River Following Hazardous Spills and Floods. Water Resources Res. Center Bull. 63, Virginia Polytechnic Inst. and State Univ., Blacksburg.
17. WELCH, P.S. 1948. Limnological Methods. Blakiston Co., Philadelphia, Pa.
18. USINGER, R.L. 1956. Aquatic Insects of California, with Keys to North American Genera and California Species. Univ. of California Press, Berkeley.
19. MASON, W.T., JR., C.I. WEBER, P.A. LEWIS & E.C. JULIAN. 1973. Factors affecting the performance of basket and multiplate macroinvertebrate samplers. *Freshwater Biol.* 3:409.
20. BEAK, T.W., T.C. GRIFFING & A.G. APPLEBY. 1974. Use of artificial substrates to assess water pollution. *In* Proceedings Bio-

logical Methods for the Assessment of Water Quality. ASTM, Philadelphia, Pa.

21. FULLNER, R.W. 1971. A comparison of macroinvertebrates collected by basket and modified multiple-plate samplers. *J. Water Pollut. Control Fed.* 43:494.
22. HESTER, F.E. & J.B. DENDY. 1962. A multiple-plate sampler for aquatic macroinvertebrates. *Trans. Amer. Fish. Soc.* 91:420.
23. GALE, W. & J. THOMPSON. 1975. A suction sampler for quantitatively sampling benthos on rocky substrates in rivers. *Trans. Amer. Fish. Soc.* 104:398.
24. SIMMONS, G.M., JR. 1977. The Use of Underwater Equipment in Freshwater Research. Virginia Polytechnic Inst. & State Univ., Blacksburg (VPI-SG-77-03).
25. WORSWICK, J.M. & M.T. BARBOUR. 1974. An elutriation apparatus for macroinvertebrates. *Limnol. Oceanogr.* 19:538.
26. LAUFF, G.H., K.W. CUMMINS, C.H. ERIKSON & M. PARKER. 1961. A method for sorting bottom fauna samples by elutriation. *Limnol. Oceanogr.* 6:462.
27. EDMONDSON, W.T., ed. 1959. Ward and Whipple's Freshwater Biology, 2nd ed. John Wiley & Sons, Inc., New York, N.Y.
28. COOK, D.G. & R.O. BRINKHURST. 1973. Marine Flora and Fauna of the Northeastern United States, Annelida: Oligochaeta. NOOA Tech. Rep. NMFS CIRC-374. U.S. Dep. of Commerce, Nat. Oceanic and Atmospheric Admin., Nat. Mar. Fish. Ser.
29. KLEMM, D.J. 1977. Leeches (Hirudinea: Annelida) of North America. Environmental Monitoring and Support Lab. U.S. EPA, Cincinnati, Ohio.
30. PENNAK, R.W. 1953. Freshwater Invertebrates of the United States. The Ronald Press, New York, N.Y.
31. BURCH, J.B. 1972. Freshwater Sphaeriacean Clams (Mollusca: Pelecypoda) of North America. U.S. EPA.
32. MASON, W.T., JR. & P.P. YEVICH. 1967. The use of phloxine B and rose bengal stains to facilitate sorting benthic samples. *Trans. Amer. Microsc. Soc.* 86:221.
33. WILHM, J.L. 1967. Comparison of some diversity indices applied to populations of benthic macroinvertebrates in a stream receiving organic wastes. *J. Water Pollut. Control Fed.* 39:1673.
34. WILHM, J.L. & T.C. DORIS. 1968. Biological parameters for water quality criteria. *Bioscience* 18:477.
35. WILHM, J.L. 1970. Range of diversity index in benthic macroinvertebrate populations. *J. Water Pollut. Control Fed.* 42:R221.
36. WILHM, J.L. 1972. Graphic and mathematical analyses of biotic communities in polluted streams. *Annu. Rev. Entomol.* 17:223.

1005 F. Bibliography

BAKER, F.C. 1928. The Freshwater Mollusca of Wisconsin Part I. Gastropoda and Part II. Pelecypoda. Wisconsin Acad. Science, Madison, Wisc.

FRISON, T.H. 1935. The stoneflies or Plecoptera of Illinois. *Bull. Ill. Natur. Hist. Surv.* 20:281.

NEEDHAM, J.G. & P.R. NEEDHAM. 1941. A Guide to the Study of Freshwater Biology. Comstock Publ. Co., Ithaca, N.Y.

ROSS, H.H. 1944. The caddisflies, or Trichoptera, of Illinois. *Bull. Ill. Natur. Hist. Surv.* 23:1.

CHU, H.F. 1949. How to Know the Immature Insects. William C. Brown Co., Dubuque, Ia.

BERNER, L. 1950. The Mayflies of Florida. Univ. of Florida Studies in Biol. Sci. Ser. 4:1.

PATRICK, R. 1950. Biological measure of stream conditions. *Sewage Ind. Wastes* 22:926.

PRATT, H.W. 1951. A Manual of the Common Invertebrate Animals Exclusive of Insects. The Blakiston Co., Philadelphia, Pa.

WIMMER, G.R. & E.W. SURBER. 1952. Bottom Fauna Studies in Pollution Surveys and Interpretation of the Data. 14th Mid. Wildlife Conf., Des Moines, Ia.

BURKS, B.D. 1953. The mayflies or Ephemeroptera of Illinois. *Bull. Ill. Natur. Hist. Surv.* 26:1.

BECK, W.M. 1954. Studies in stream pollution

biology. I. A simplified ecological classification of organisms. *Quart. J. Fla. Acad. Sci.* 17:211.

NEEDHAM, J.G. & M.J. WESTFALL, JR. 1954. Dragonflies of North America. Univ. of California Press, Berkeley.

BECK, W.M. 1955. Suggested method for reporting biotic data. *Sewage Ind. Wastes* 27:1193.

GAUFIN, A.R. & C.M. TARZWELL. 1956. Aquatic macroinvertebrate communities as indicators of organic pollution in Lytle Creek. *Sewage Ind. Wastes* 28:906.

HUTCHINSON, G.E. 1957. A Treatise on Limnology. John Wiley & Sons, New York, N.Y.

ROBACK, S.S. 1957. The Immature Tendipedids of the Philadelphia Area. Philadelphia Acad. Natur. Sci. Monogr. No. 9.

MACAN, T.T. 1958. Methods of sampling bottom fauna in stony streams. *Mitt. Int. Ver. Limnol.* 8:1.

WALKER, E.M. 1958. The Odonata of Canada and Alaska. Univ. of Toronto Press, Toronto, Vols. 1 and 2.

BOUSFIELD, E.L. 1958. I. Freshwater amphipod crustaceans of glaciated North America. *Can. Field Natur.* 72:55.

INGRAM, W.M. 1960. Effective methods for collecting and recording data from water pollution surveys. *In* C.M. Tarzwell, compiler. Biological Problems in Water Pollution. U.S. Dep. Health, Education & Welfare, Cincinnati, Ohio, pp. 260–263.

INGRAM, W.M. & A.F. BARTSCH. 1960. Graphic expression of biological data in water pollution reports. *J. Water Pollut. Control Fed.* 32:297.

EDDY, S. & A.C. HODSON. 1961. Taxonomic Keys to the Common Animals of the North Central States. 3rd ed. Burgess Publ. Co., Minneapolis, Minn.

WATERS, T.F. & R.J. KNAPP. 1961. An unproved stream bottom fauna sampler. *Trans. Amer. Fish. Soc.* 90:225.

HYNES, H.B.N. 1963. The Biology of Polluted Waters. Liverpool Univ. Press, England.

MACAN, T.T. 1963. Freshwater Ecology. John Wiley & Sons, New York, N.Y.

BRINKHURST, R.O. 1964. Studies on the North American aquatic oligochaeta. Part I. *Proc. Acad. Sci. Philadelphia* 116:195.

KING, D.L. & R.C. BALL. 1964. A quantitative biological measure of stream pollution. *J. Water Pollut. Control Fed.* 36:650.

SINCLAIR, R.M. 1964. Water Quality Requirements for Elmid Beetles. Tenn. Dep. Pub. Health, Nashville, Tenn.

BRINKHURST, R.O. 1965. Studies on the North American aquatic oligochaeta. Part II. *Proc. Acad. Sci. Philadelphia* 117:117.

SUBLETTE, J.E. & M.S. SUBLETTE. 1965. Family Chironomidae (Tendipedidae). A catalog of diptera of America north of Mexico. *Bull. U.S. Dep. Agr.* 276:143.

TACKETT, J.H. 1965. Biological Assessment of Water Quality. Roanoke River-Tinker Creek. (mimeograph) Virginia State Water Control Board.

RUTTNER, F. 1966. Fundamentals of Limnology. Univ. of Toronto Press, Canada.

HEARD, W.H. & J. BURCH. 1966. Keys to the Genera of Freshwater Pelecypods of Michigan. Univ. of Michigan Museum of Zoology Circ. No. 4, Ann Arbor.

BECK, W.M., JR. & E.C. BECK. 1966. Chironomidae (Diptera) of Florida. I. Pentaneurini (Tanypodinae). *Bull. Fla. State Museum* 10:305.

INGRAM, W., K.M. MACKENTHUN & A.F. BARTSCH. 1966. Biological Field Investigative Data for Water Pollution Surveys. FWPCA, U.S. Govt. Printing Off., Washington, D.C.

PIELOU, E.C. 1966. The measurement of diversity in different types of biological collections. *J. Theor. Biol.* 13:131.

MACKENTHUN, K.M. & W.M. INGRAM. 1967. Biological Associated Problems in Freshwater Environments. FWPCA, Washington, D.C.

LLOYD, M., J.H. ZAR & J.R. KARR. 1968. On the calculation of information—Theoretical measures of diversity. *Amer. Midland Natur.* 79:257.

MASON, W.T., JR. 1968. An Introduction to the Identification of Chironomid Larvae. Div. Pollut. Surveillance, FWPCA, Cincinnati, Ohio.

SURBER, E.W. 1969. Procedure in taking stream bottom samples with the stream square foot bottom sampler. *Proc. 23rd Annu. Conf. S.E. Ass. Game Fish Comm.* 587.

CAIRNS, J., JR., K.L. DICKSON, R.E. SPARKS & W.T. WALLER. 1970. A preliminary report on rapid biological information systems for water pollution control. *J. Water Pollut. Control Fed.* 45:685.

HYNES, H.B.N. 1970. The Ecology of Running Waters. Univ. of Toronto Press, Canada.

JACKSON, H.W. 1970. A controlled-depth volumetric bottom sampler. *Progr. Fish-Cult.* 32(2):113.

LARIMORE, R.W. 1970. Two shallow-water bottom samplers. *Progr. Fish-Cult.* 32(2).

CAIRNS, J., JR. 1971. A simple method for the biological assessment of the effects of waste discharges on aquatic bottom-dwelling organisms. *J. Water Pollut. Control*

Fed. 43:755.

ERMAN, D.C. & W.T. HELM. 1971. Comparison of some species importance values and ordination techniques used to analyze benthic invertebrate communities. *Oikos* 22:240.

DICKSON, K.L., J. CAIRNS, JR. & J.C. ARNOLD. 1971. An evaluation of the use of a baket-type artificial substrate for sampling macroinvertebrate organisms. *Trans. Amer. Fish. Soc.* 100:553.

ODUM, E.P. 1971. Fundamentals of Ecology, 3rd ed. Saunders Publ. Co., Philadelphia, Pa.

LEWIS, P.A. 1972. References for the Identification of Freshwater Macroinvertebrates. EPA-R4-F2-006.

MERRITT, R.W. & K.W. CUMMINS. 1978. An Introduction to the Aquatic Insects of North America. Kendall/Hunt Publ. Co., Dubuque, Ia.

PENNAK, R.W. 1978. Freshwater Invertebrates of the United States, 2nd ed. Wiley-Interscience, New York, N.Y.

1006 FISH

Information on the abundance and species composition of fish is useful for assessing the quality of a body of water.[1] Fish occupy the highest trophic level of the aquatic food chain (barring predation by higher vertebrates); hence their condition constitutes a summation of the condition of lower biological forms and is a result of the total quality of the water. Water quality factors that alter the ecological balance of the periphyton, plankton, and macroinvertebrate populations also can alter the fish population. Because fish and invertebrates have differing susceptibilities to certain toxicants, fish might be affected by certain pollutants that do not cause a demonstrable change in the invertebrate and plant communities. It is important to sample fish populations as the final or climax product of the aquatic community. Because fish as aquatic organisms are well known and also have economic value, they are the most intelligible symbol of water quality to the general public and are important for public relations purposes as well as for technical interpretations.

Where commercial fisheries exist locally, hiring of commercial fishermen and equipment should be considered, especially for studies of relatively short duration. Commercial catch data also may be useful in detecting certain long-term changes in water quality.

1006 A. Sample Collection and Preservation

In a fishery survey, secure information on the kinds of fish present and their relative abundance. A one-time study provides information on the species of fish present. This may be sufficient, for example, in investigating a fish kill, but in many rivers and lakes, the changes in fish populations are subtle and should be determined through long-range studies.

Carefully review state regulations on fish collection when field operations are being planned. Scientific collecting permits issued by state fish and game agencies provide authorization for most cases. If use of piscicides is contemplated, make a thorough check of Food and Drug Administration, Environmental Protection Agency, and state regulations, because introduction of toxic substances into any water is stringently controlled.

Fish may be collected by various methods such as seining, trapping, gill or trawl-netting, electrofishing, and use of chemicals.[2–5] To obtain representative data, sample in obscure and unlikely areas as well as at obvious locations. Early life states (eggs and larvae) of many species may be found in the plankton. Trawl waterway bottoms for bottom fish; seine both riffles and pools of streams; fish for free-swimming open-water types with various nets; and take migrating or roving types with traps or gillnets. Sample all depths, not just surface and bottom. Brush, rock, and other types of obstructions sometimes are sampled best by using chemicals or by electrofishing.

Visual observations by a trained individual are very useful. Some methods of fish sampling, such as electrofishing, trapping, and gillnetting, are best undertaken at night because many species of fish are sedentary during daylight hours.

1. Sampling Devices

Devices commonly used in fish sampling are as follows:

a. Haul seines are used to collect fish from shallow water. In small streams, seine with "straight" seines of various lengths, 1.2 m or longer, with square-mesh sizes of 3, 6, or 12 mm.

For shoreline seining of lakes and large rivers use a bag seine. There are two common sizes for bag seines: One is 7.5 m long and 1.8 m deep, with the main portion constructed of 12.6-mm square-mesh netting. The center 1.8-m-long bag is made of 6-mm mesh netting. The second size, often used for larger fish, is 30 m long and 1.8 or 2.4 m deep, with 2.5- to 5-cm square-mesh netting in the main body and 12- to 25-mm netting in the center bag. A waterway free from snags is essential to successful seining. Express results as number of fish per unit area seined, recognizing, however, that quantitative seining is very difficult. The procedure is more useful in determining the variety of fish inhabiting the water than for quantitative sampling.

b. Gillnets (Figure 1006:1A) are used in estuaries, lakes, reservoirs, or large rivers where fish movement can be expected. The most versatile experimental gillnet is 38 m long with five 7.5-m sections of mesh sizes ranging from 19 mm to 62 mm square. Express results as number or weight of fish taken per length of net per day.

c. Trammel nets (Figure 1006:1B) have a layer of large mesh netting on each side of loosely hung smaller gillnetting. Small fish are captured in the gillnetting and larger fish are captured in a "bag" of the gillnetting that is formed in the larger mesh netting. Express results as number or weight of fish captured per length of net per day.

d. Traps range in size from small containers (minnow traps) with inverted cone entrances to semipermanent structures (weirs) (Figure 1006:1C). Use traps and weir nets mainly in rivers and estuaries.

Trap nets, when properly located to intercept fish movements, may be used effectively to sample fish populations in lakes during certain seasons. The trap net most used in fishery studies is the square or round hoop net. This net may have leads or wings attached to the first frame. The second and third frames can each hold tunnel throats that prevent escape of fish entering each section in turn. The opposite (closed) end may be tied with a slip cord to facilitate fish removal. Express results as fish per net-day. Most fish can be sampled when trap nets of various mesh sizes are set in a variety of habitats.

e. Trawls are specialized gear used in large open water areas of reservoirs, lakes, large rivers, estuaries, and offshore marine areas (Figure 1006:1D). Use them to gain information on a particular species of fish rather than on overall fish popu-

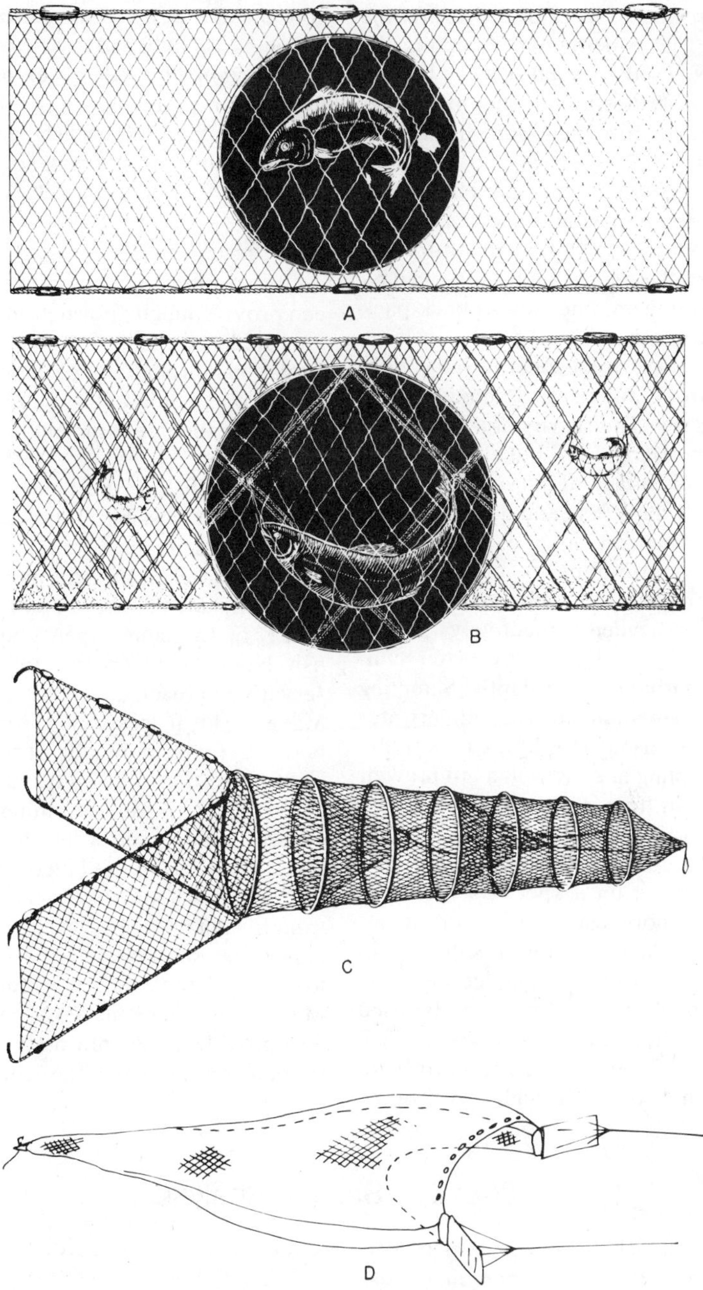

Figure 1006:1. Some types of nets: A—gillnet; **B**—trammel net; **C**—Fyke or hoop net; **D**—otter trawl.

lation. Three basic types of trawls are: the fry trawl, with a permanent opening; the otter trawl, used to capture near-bottom and bottom fishes;[6] and the mid-water trawl, used to collect schooling fish at various depths.

f. Electrofishing devices are effective in collecting most sizes and species of fish. These devices vary in size and complexity. Small streams can be surveyed by a two-person team, one using a back-pack shocker, the other dipping fish. In large rivers and lakes, use a generator and electrical control equipment to achieve satisfactory results. The survey of a large river may require a large crew but electrofishing offers the advantage of surveying a reach of water in a short time and does not entail leaving unattended such expensive and pilferable equipment as nets. Install shockers on boats for sampling waters that cannot be waded effectively. Electrofishing may be limited by water conductance, turbidity, and depth. Sampling is more efficient at night, when underwater lights may be used. Use block nets to delineate sampling area as well as to prevent escape of fish from electrical field.

g. Chemicals, such as rotenone, often are used to collect fish from a restricted area. Use these on a spot basis, for example, in a short reach of river or an embayment of a lake. Express results as fish per unit of surface area sampled.

h. Trotlines (or long lines) can be used to sample a limited segment of the fish population. These are used mainly to sample fish feeding at night. A trotline consists of a series of short drop lines with baited hooks attached to a long main line.

i. Suction tubes or hoses often are effective for removing eggs and fry from bottom nests. Study of fish egg viability is important because the egg stage is a critical one in the life cycle.

2. Procedures

Observations by experienced personnel can provide much information on fish habitats and populations. Enhance such observations through use of underwater television or photographic cameras, underwater breathing equipment, or submersible vessels.

Preserve fish in the field as soon after collection as possible. If they are to be analyzed the same day, icing is sufficient; if not, preserve in 10% formalin, to which has been added 3 g borax and 50 mL glycerin/L of formalin. Open body cavity of fish larger than 7.5 cm to enable preservative to reach internal organs quickly. Make a slit at least one-third as long as body cavity on right side of fish.

After fixation (1 to 7 days, depending on fish size), wash fish in running water or in several changes of water for 24 hr and place in 40% isopropyl alcohol.

Place labels on inside and outside of sample container. Use waterproof ink and paper for inner label. Label with station location, date, collector, gear used, etc. Make field notes that describe the station and provide other information necessary to interpret the data obtained.

1006 B. Sample Analysis

Identify all fish to species. In reports prepared for nonfisheries personnel, use common names adopted by the American Fisheries Society.[7]

Determine weight in grams and total length in centimeters of each fish. Total length is defined as the distance from the anterior end (with mouth closed) to the tip of fins, when drawn together.[2]

Age and growth rates are useful for determining the effects of water quality on fish populations. The methods for these measurements include[2] (*a*) the length-frequency method, (*b*) the otolith and bone methods, and (*c*) the scale method. The lengths of young fish tend to cluster into groups (size classes), permitting separation into age classes. As the fish age, however, the year classes usually become less distinct. The age of fish also can be estimated by counting the annual growth rings in ear stones, vertebrae fin rays, opercles, and scales. The scale method is used most widely.

The coefficient of condition is the length-weight relationship used to express relative plumpness or robustness of fish. In turn it is related to environmental conditions. The equation[8] generally used is:

$$K = \frac{W \times 10^5}{L^3}$$

where:

K = coefficient of condition (or condition factor),
W = weight, g, and
L = length, mm.

The general health and well-being of a fish influences its response to pollution; likewise, sublethal levels of pollution may affect the fish's health and its ability to resist disease. Natural epidemics and debilities of many types occur among fish as among all organisms. Examine fish early in any water quality survey for their general condition, gill lesions, parasites, and other factors that might accentuate or mask effects of pollution.

Parasites, particularly larger attached external parasites such as copepods or leeches, usually can be detected by examining the surface of live or freshly killed fish under a wide-field dissecting microscope. Examination, under greater magnification, of slime scraped from the body surface or from areas of accumulation around the gills may reveal smaller external parasites. Scars from lamprey attacks may appear as round fresh red lesions or older healed areas up to 2 cm or more in diameter. Also report inflamed areas, swellings, frayed or eroded fins, or other gross indications of disability that may be visible externally. Internal parasites and other pathologies may be revealed by appropriate microtechnics. For critical identification of pathological conditions, consult an established laboratory specializing in this field.

When chronic toxicity is suspected or when a general weakness or poor condition of the fish is noted, make histological, physiological, and/or biochemical tests.[9–12]

Fish flesh is often unpalatable because of chemicals present in the water. Many taste-producing substances are from municipal or industrial wastes, although natural sources also are known to contribute.[13] Use taste panels for detecting tainted fish flesh. Test fish may be native fish from the stream, uncontaminated fish held in cages in the test area, or uncontaminated fish held in tanks, with the suspect chemical added. Give the trained taste panel acceptable samples for comparison.[14–17]

1006 C. Production and Productivity

Fish production in a given body of water is indicative of water quality. Consider the species composition of the entire fish community as well as the relative quantities of each species. Productivity in a fishery context includes the following concepts:[3]

PRODUCTIVITY: the rate at which a given body of water is capable of producing fish biomass (usually weight/area/year).

PRODUCTION: the actual amount (weight) produced in a unit of time.

STANDING CROP: The quantity of fish present at any given time (usually weight/unit area). Report this quantity as total for all species or break it down into actual or relative (percentage) weights of each.

HARVEST OR YIELD: the quantity of fish actually removed by human beings (weight/year or weight/unit area/year).

1006 D. Investigation of Fish Kills

Fish kills vary, from the individual fish that dies of old age to the catastrophic kill, from partial to complete, and from natural occurrence to the results of human activity. No single investigative procedure can be appropriate for all situations. The following brief description may serve as an aid in investigating kills. Getting to the scene promptly before the evidence has decomposed or drifted away is vital. If surveillance of a particular body of water or area is involved, have available preset plans and equipment on a standby basis.[18,19]

Fish kills may be caused by such natural events as acute temperature change, storms, ice and snow cover, decomposition of natural materials, salinity change, spawning mortalities, parasites, and bacterial and viral epidemics. Human-caused fish kills may be attributed to municipal or industrial wastes, agricultural activities, and water control activities.

One dead fish in a stream may be called a fish kill; however, in a practical sense, adopt some minimal range in number of dead fish observed, plus additional qualifications, in reporting and classifying fish kills. Any fish kill is significant if it affects fish of sport or commercial value, results from a suspected negligent discharge or malfunctioning waste treatment facility, or causes widespread environmental damage. The following definitions, based on a stream about 60 m wide and 2 m deep, are suggested as guidelines. For other size streams, make proportional adjustments.

1. MINOR KILL: 1 to 100 dead or dying fish confined to a small area or stream stretch. If recurrent, it could be significant; investigate.

2. MODERATE KILL: 100 to 1,000 dead or dying fish of various species in 1 to 2 km of stream or equivalent area of a lake or estuary.

3. MAJOR KILL: 1,000 or more dead or dying fish of many species in a reach of stream up to 16 km or greater, or equivalent area of a lake or estuary.

In preparing for a field investigation, study area maps[1,20,21] and determine kill area and access to it. Identify waste dischargers. Contact participating laboratories to discuss number and size of samples that will be submitted, types of analyses required, dates of sample receipt, method of sample shipment, date by which results are needed, and to whom results are to be reported. Use two information record forms for fish kill investigations, an initial contact form and a field investigation form.

On all fish kill investigations take a thermometer, dissolved oxygen test kit, conductivity and pH meters; or a general chemical kit, biological sampling gear, sample bottles, and other specimen containers. Include in the investigating team at least one person who is experienced in investigating fish kills.

The field investigation consists of visual observations, sampling of fish, water, and other biota, and physical measurements of the environment. The first local observer of the kill is a useful guide to the area, which should be reconnoitered initially to establish that a fish kill actually has occurred.

If a fish kill has taken place, immediately start fish sampling because collection of dying or recently dead fish is critical. For purposes of comparison, if possible, collect healthy fish from an unaffected area.

Place individual fish in well-labeled plastic bags and preserve by freezing until the fish can be examined.

Bleed dying or dead fish at collection time to obtain at least 1 g blood. Collect a blood sample in a chemically clean, solvent-washed glass bottle with a TFE-lined screw cap.

Identify and count dead fish. In a large river count dead fish from a fixed station such as a bridge during a fixed period of time. Extrapolate to the total time involved. Alternatively, in a large river or lake, make a shore count and project to entire area of kill. In smaller water bodies traverse entire area to enumerate dead fish.

Collect water samples representative of unpolluted and polluted areas in accordance with the instruction given in Section 1002. As a minimum, measure temperature, pH, dissolved oxygen, and conductivity. Make additional tests depending on suspected causes of the fish kill. Take samples for examination of plankton, periphyton, macrophyton, and macroinvertebrates.

Record observations on water appearance, streamflow, and weather conditions. Color photographs are valuable in recording conditions.

1006 E. References

1. MACKENTHUN, K.M. 1973. Toward a Cleaner Environment. USEPA. Office of Air & Water Programs, Washington, D.C.
2. LAGLER, K.F., J.E. BARDACH & R.R. MILLER. 1962. Ichthyology. The Study of Fishes. John Wiley & Sons, New York, N.Y.
3. RICKER, W.E. ed. 1971. Methods for Assessment of Fish Production in Fresh Waters. IBP Handbook No. 3, 2nd ed., Blackwell Sci. Publ., Oxford, England.
4. DUMONT, W.H. & G.T. SUNDSTROM. 1961. Commercial Fishing Gear of the United States. U.S. Fish & Wildlife Circ. No. 109, U.S. Govt. Printing Off., Washington, D.C.
5. ROUNSEFELL, G.A. & W.H. EVERHART. 1953. Fishery Science—Its Methods and Applications. John Wiley & Sons, New York, N.Y.
6. STANSBY, M.E. 1963. Industrial Fishery Technology. Reinhold Publ. Co., New York, N.Y.
7. BAILEY, R.M., J.E. FITCH, E.S. HERALD, E.A. LACHNER, C.C. LINDSEY, C.R. ROBINS & W.B. SCOTT. 1970. A List of Common and Scientific Names of Fishes from the United States and Canada, 3rd ed. Spec. Publ. No. 6, Amer. Fish. Soc., Washington, D.C.
8. CARLANDER, K.D. 1969. Handbook of Freshwater Fishery Biology. Vol. I, Iowa State Univ. Press, Ames.
9. CAIRNS, J., JR. & A. SCHEIER. 1963. The acute and chronic effects of standard sodium alkyl benzene sulfonate on the pumpkin seed sunfish, *Lepomis gibbosus* (Linn.) and the bluegill sunfish. *L. macrochirus* Raf. *Proc. 17th Ind. Waste Conf.*, Eng. Ext. Serv. Ser. No. 112, Purdue Univ., Lafayette, Ind. pp. 14–28.
10. JACKIM, E., J.M. HAMLIN & S. SONIS. 1970. Effects of metal poisoning on five liver enzymes in the killfish (*Fundulus heteroclitus*). *J. Fish. Res. Board Can.* 27:383.
11. HINTON, D.E., R.L. SNIPES & M.W. KENDALL. 1972. Morphology and enzyme histochemistry in the liver of largemouth bass (*Micropterus salmoides*). *J. Fish. Res.*

Board Can. 29:531.

12. OLSON, K.R. & P.O. FROMM. 1973. Mercury uptake and ion distribution in gills of rainbow trout (*Salmo gairdneri*): tissue scans with an electron microprobe. *J. Fish. Res. Board Can*. 30:1575.
13. NATIONAL TECHNICAL ADVISORY COMMITTEE. 1968. Water Quality Criteria. FWPCA, Washington, D.C.
14. WINSTON, A.W. 1959. Test for odor imparted to the flesh of fish. 2nd Seminar on Biological Problems in Water Pollution, Cincinnati, Ohio. USPHS, Div. Water Supply & Pollution Control.
15. BOYLE, H.W. 1967. Taste/odor contamination of fish from the Ohio River near Tell City, Indiana. Cincinnati Water Research Lab, FWPCA, Cincinnati, Ohio.
16. THOMAS, N.A. 1969. Flavor of Ohio River channel catfish (*Ictalarus punctatus* Raf.). USEPA, Cincinnati, Ohio.
17. THOMAS, N.A. & D.B. HICKS. 1971. Effects of waste water discharge on the flavor of fishes in the Missouri River (Sioux Falls, Iowa, to Waverly, Missouri). *In* Everyone Can't Live Upstream. USEPA Off. Water Programs, Kansas City, Mo.
18. BURDICK, G.E. 1965. Some problems in the determination of the cause of fish kills. *In* Biological Problems in Water Pollution. USPHS Publ. No. 999-WP-25.
19. Pollution Caused Fish Kills. 1967. 1968. U.S. Dept. Interior, FWPCA Publ. No. CWA-7.
20. SMITH, L.L., JR., et al. 1956. Procedures for Investigation of Fish Kills. A Guide for Field Reconnaissance and Data Collection. ORSANCO, Cincinnati, Ohio.
21. Investigating Fish Mortalities. 1970. U.S. Dept. Interior, FWPCA Publ. No. CWT-5 (also available from U.S. Govt. Printing Off. as No. 0-380-257).

1007 IDENTIFICATION OF AQUATIC ORGANISMS

Experienced aquatic biologists will be familiar with most organisms illustrated in Plates 1 through 38, and will seldom need the assistance of keys to identify organisms to the level illustrated. Because these plates are not intended for critical identification, specific (species) names are not cited. Types most likely to be observed are illustrated. For the convenience of those less familiar with the organisms referred to in preceding sections, a series of short keys is presented to enable them to identify most organisms to the level illustrated by the plates.

In conformity with preceding sections, organisms are arbitrarily divided into microscopic and macroscopic, depending on whether or not they pass through a U.S. Standard No. 30 sieve. For the study of microscopic forms, use a compound microscope. For examination of the smaller macroscopic organisms and to resolve the finer structures of larger forms, use a wide-field stereoscopic microscope.

1007 A. Procedure in Identification

Critical identification of a specimen often is time-consuming, even for an experienced biologist. Before looking at any key or other aid to identification, carefully study the specimen for one to several minutes. If necessary, find other examples and compare them with the unknown.

It is important to know where or under what conditions the organism lived before attempting to identify it. For example, did it come from fresh water—a lake or a stream? Is it marine—from the open ocean, shoreline, or estuary? Was it a free swimmer or floater in the water? Was it a bottom organism, attached, crawling, or burrowing? Finally, turn to the following key to major groups.

Only the more common types of aquatic organisms are illustrated here, with special attention to those most frequently used in water quality evaluation. When specimens do not fit obviously into one of the types listed, consult a professional biologist, a microbiologist for the bacteria and fungi, or some of the references provided. Descriptions of color and movement refer to freshly collected or living specimens, or, in the case of microscopic forms, to those preserved as described in Section 1002.

Sizes of the organisms illustrated in Plates 1 through 38 are shown in parentheses in the legend. These are intended to represent *common* sizes, not absolute maxima or minima. Exceptional individuals and even whole localized populations may be encountered that are considerably larger or smaller than the sizes cited.

1007 B. Key to Major Groups of Aquatic Organisms (Plates 1-38)

Beginning with couplet 1a and 1b of the Keys, compare the descriptions given with the subject specimen. A choice must be made between statement "a" and statement "b." Proceed to the couplet number indicated at the right and repeat the process. Continue until the name of an organism or a plate number is cited instead of another couplet number. Additional information is provided in many of the plate legends.

		Refer to Couplet No.
1a.	Macroscopic: The organism, mass, or colony is visible to the naked eye	13
1b.	Microscopic: Not readily visible to the naked eye	2

1. Key to Microscopic Organisms

2a.	Specimen a single living cell or a mass or colony of relatively independent cells (shapeless, rounded, or threadlike)	3
2b.	Specimen a many-celled, highly organized plant or animal.	7
3a.	Cells contain one or more pigments, including chlorophyll *a* (overall color may range through various shades of green, blue, red, brown, or yellow). ALGAE (for details, see Section 1007D following, "Key for Identification of Freshwater Algae")	4
3b.	Cells typically colorless, lacking chlorophyll *a*	12
4a.	Nuclei present; pigment confined to chloroplasts	5
4b.	Nuclei, plastids, or vacuoles absent (pseudovacuoles may be present in certain filamentous forms). Pigment generally diffused throughout cytoplasm. BLUE-GREEN ALGAE, Plates 1 and 2.	
5a.	Cell wall permanently rigid, composed of SiO_2, geometrical in appearance, and with regular patterns of fine markings; composed of two essentially similar halves, one	

Refer to Couplet No.

placed over the other as a cover. Golden brown to greenish in color. DIATOMS, Plates 5 and 6.

5b. Cell wall, if present, capable of sagging or bending, rigidity depending on internal pressure of cell contents. Cell walls usually of one piece 6

6a. Cells or colonies nonmotile. Usually some shade of green. NONMOTILE GREEN ALGAE, Plates 3 and 4.

6b. Cells or colony move by means of relatively long whiplike flagella. PIGMENTED FLAGELLATES, Plates 11 and 12.

7a. Body with cilia (hairlike structures used for locomotion) 8

7b. Body without cilia 9

8a. Body generally covered with cilia, usually somewhat elongate or wormlike, bilaterally symmetrical. Minute FLATWORMS (Platyhelminthes), relatives of *Planaria*, Plate 19.

8b. Cilia confined to one or two crowns at anterior end, which often present the illusion of rotating wheels. Internal jaws present. ROTIFERS (Rotifera), Plate 17.

9a. Long slender unsegmented worms that move by sinuous crawling or thrashing motion. ROUNDWORMS (Nemathelminthes), Plate 18.

9b. Possess external skeleton and jointed appendages 10

10a. Crawl about or swim by means of jointed appendages thrust out from between two clamlike shells. All appendages can be withdrawn entirely within shells when disturbed. OSTRACODS (Ostracoda), Plate 21.

10b. Swim rapidly by means of a pair of enlarged jointed appendages (antennae) that cannot be withdrawn inside carapace or shell 11

11a. Locomotor appendages (antennae) branched. Microcrustacea, CLADOCERA (Cladocera), Plate 20.

11b. Locomotor appendages (antennae) unbranched; body tapers toward rear. Microcrustacea, COPEPODS (Copepoda), Plate 21.

12a. Ingest and digest food internally (ingested food of various colors may be visible through body wall). Single-celled or colonial, attached or free-living. PROTOZOANS (Protozoa), Plates 13, 14, and 15.

12b. Digest food externally and adsorb products through cell wall. Often secrete masses of slime. BACTERIA and FUNGI, Plate 38.

2. Key to Macroscopic Organisms

13a. Specimen a mass of filaments or a glob of gelatinous or semisolid material containing many tiny units, requiring microscopic examination to determine details of structure . 2

13b. Specimen a well-organized unit or colony 14

14a. Organism plantlike; flowerlike structures, if present, do not respond when touched, generally are colored some shade of green, brown, or red 16

14b. Organism animal-like; usually responds rapidly when touched, whether attached or free-living 15

15a. Internal backbone present (vertebrates) 17

15b. No internal backbone present (macroinvertebrates)* 18

16a. Plant structure relatively simple. Attachment structures may be present, but no true roots or fibrous tissue. Larger ALGAE, Plate 7 and Color Plates A (*Nitella*) and F (*Chara* and *Batrachospermum*).

*Invertebrates retained on a U.S. Standard No. 30 sieve.

Refer to Couplet No.

16b. Plant structure usually includes true roots, stems, and leaves. Fibers or vascular tissue usually present; flowers or seeds may be observed. (One atypical group, "watermeal," consists only of tiny roundish masses, 0.5 to 1 mm in diameter, often misidentified as algae.) HIGHER PLANTS, Plates 8, 9, and 10.

17a. Side appendages, if present, are flat fins. FISHES, Plate 36.

17b. Side appendages, if present, are footlike, with separate digits. AMPHIBIANS, Plate 37.

3. Key to Macroinvertebrates

18a. Body bilaterally symmetrical (with right and left sides, but may be superficially coiled into a spiral); animal usually not attached but may live inside an attached cocoon or case, or crawl about; usually solitary 23

18b. Symmetry not bilateral 19

19a. Body typically radially symmetrical 21

19b. Body or colony nonsymmetrical 20

20a. Body mass generally porous; not a colony, sometimes finger- or antler-like. Freshwater representatives generally are fragile, colored green or brown; marine forms tougher, various colors. SPONGES (Porifera), Plate 16.

20b. Body mass otherwise 22

21a. Animals with soft smooth bodies and tentacles around a mouth; no anus. Solitary or colonial. Larger colonies usually have rigid limy skeleton of massive, branched, or fan-shaped form. HYDRAS, SEA ANEMONES, JELLYFISHES, CORALS, etc. (Coelenterata), Plate 35A,B .

21b. Body covering usually spiny, soft or rigid, flattened or elongate, typically having five radii, with or without spines or arms; anus present. Solitary. Marine only. STARFISHES and relatives (Echinodermata), Plate 34.

22a. Colony a jellylike mass, a network of branching tubes, a plant-like tuft, or a lacy limy crust or mass. MOSS ANIMALS (Bryozoa), Plate 16.

22b. Exclusively marine. Surface of body or colony relatively smooth but tough. Solitary forms, sac-like, with two external openings. Exhibit all degrees of colonialism. Compound forms range from thin slimy masses, with organisms arranged in tiny radial patterns to huge, shapeless masses resembling tough frozen gelatin. SEA SQUIRTS, SEA PORK (Ascidiacea, Urochorda, Chordata), not illustrated.

23a. Animal living within a hard limy shell, soft body (Mollusca) 29

23b. Animal without a limy shell 24

24a. Jointed legs present (may not be functional). Body may be hard or soft 30

24b. Jointed legs absent, body covering mostly soft, animal pliable (a hardened head capsule may be present) 25

25a. Body girded by annulations or creases at regular intervals, dividing it into many small segments much wider than long 26

25b. Segments present or absent; if present, not much wider than they are long 27

26a. Body with suction disk at one or both ends, in length usually less than 10 times its width. LEECHES (Annelida, Hirudinea), Plate 19.

26b. Body without suction disks, in length usually more than 10 times its width; hairs or bristles often evident. SEGMENTED WORMS (Annelida), Plate 19.

27a. Body unsegmented, long and slender, appearing smooth, evenly tapered to a fine point at one end. ROUNDWORMS (Nematoda), Plate 18.

27b. Body otherwise 28

28a. Body flat, elongate, or oblong; unsegmented head is spade-shaped. Pigmented

Refer to Couplet No.

spots on top of head often give the animal a cross-eyed appearance. FLATWORMS (Turbellaria), Plate 19.

28b. Body segmented, cylindrical, oblong, or capsule-like; may or may not have a head capsule and thick fleshy knobs on underside. Larvae of TWO-WINGED FLIES (Diptera), Plate 29 30

29a. Shell consisting of two hinged halves. BIVALES (Pelecypoda), Plate 33.

29b. Shell entire, usually spiral but may be "coolie hat"-shaped. SNAILS (Gastropoda), Plate 32.

30a. Body with functional legs 31

30b. Body without functional legs, mummy- or capsule-like, living in a cocoon. PUPAE (Insecta), Plate 22 38

31a. Body with three pairs of legs. Larvae, nymphs, and some adults (Insecta) 42

31b. Body with more than three pairs of legs 32

32a. Body compact, spider-like, with four conspicuous pairs of legs (two other pairs of appendages present). WATER MITES (Acari), Plate 35.

32b. Body with at least five conspicuous pairs of legs. CRUSTACEANS (Crustacea) . . . 33

4. Key to Crustaceans

33a. Sides of body compressed 34

33b. Body flattened horizontally 36

34a. Eyes on stalks 35

34b. Eyes, if present, only seen as spots on sides of head. SCUDS (Amphipoda), Plate 21.

35a. Pincers on first pair of legs strong and large; other legs stout, cylindrical, and used for walking. CRAYFISH, also marine lobster (Decapoda), Plate 21.

35b. Pincers on first pair of legs weak and small; other legs, thin and flattened, are used for swimming. SHRIMPS (Mysidea and others), Plate 21.

36a. Eyes on stalks, shells generally broad, various shapes (marine and brackish water). CRABS (Decapoda), not illustrated.

36b. Eyes not on stalks 37

37a. Body covering hard; divided into broad head, truncate body, and sharp tail sections (marine). HORSESHOE CRABS (Arthropoda), Plate 35.

37b. Body with three or more joints. SOWBUGS (Isopoda), Plate 21.

5. Key to Insect Pupae

38a. Back of pupa with small, paired, hook-bearing plates. CADDISFLIES (Trichoptera), Plate 27.

38b. Back without paired hook-bearing plates but may have knobs or bristles 39

39a. Developing wings (pads) held free from body. BEETLES (Coleoptera), Plate 30.

39b. Wing pads closely appressed to body, mummy-like, or appendages not evident 40

40a. With one closely appressed pair of wing pads, but not fused to body; or capsule-like, appendages not evident. TWO-WINGED FLIES (Diptera), Plate 28.

40b. Two pairs of wing pads 41

41a. First two or three abdominal segments with spiracles (holes for breathing) on each side; body without numerous projections. AQUATIC MOTHS (Lepidoptera), not illustrated.

Refer to Couplet No.

41b. Body differing from above, may have numerous knobs or other projections on back. HELLGRAMMITES (Neuroptera and Megaloptera), Plate 26.

6. Key to Insect Larvae, Nymphs, and Some Adults

42a. Animal flea-like, with a bifid projecting appendage on the underside. SPRINGTAILS (Collembola), Plate 35.
42b. Animal otherwise 43
43a. Body ending in long segmented filaments 44
43b. Long filaments absent or, if present, not segmented 45
44a. Two tail filaments, legs ending in two claws. STONEFLIES (Plecoptera), Plate 23.
44b. Three tail filaments (with few exceptions); middle filament may be slightly smaller than laterals, legs ending in one claw. MAYFLIES (Ephemeroptera), Plate 24.
45a. Back of body covered with two hard wing covers, a pair of membranous wings underneath the covers. ADULT BEETLES (Coleoptera), Plate 30.
45b. Back without hard wing covers 46
46a. Body with exposed membranous wings or wing pads on back 47
46b. Body without membranous wings or wing pads (larvae) 49
47a. Membranous wings present; held flat and in a V-shape on back. Mouth parts formed into a long, sharply pointed beak folded underneath body. TRUE BUGS (Hemiptera), Plate 31.
47b. Membranous wings absent, wing pads present. Mouth parts formed into an extendable, scoop-like mask that covers face. (Odonata) 48
48a. Body ending in three oblong, fan-like plates. DAMSELFLIES (Zygoptera), Plate 25.
48b. Fan-like plates absent. DRAGONFLIES (Anisoptera), Plate 25.
49a. Mouth parts formed into slender curved rods nearly half as long as body (less than 10 mm). SPONGILLA FLIES (Neuroptera), not illustrated.
49b. Mouth parts adapted for biting or chewing 50
50a. Body with five paired knobs on underside of abdominal segments, legs on first three segments short and stubby. Often found on lily pads. AQUATIC MOTHS (Lepidoptera), not illustrated.
50b. Body without paired knobs on underside of abdomen 51
51a. Sides of each abdominal segment with a slender, tapering process 52
51b. Sides of each abdominal segment without a tapering process, but may have hair-like or tubular processes 53
52a. Body ending in a pair of hook-bearing fleshy legs or in a single tapering filament. HELLGRAMMITES and relatives (Megaloptera), Plate 26.
52b. Body otherwise. BEETLES (Coleoptera), Plate 30.
53a. Body covering mostly hard; knobs, hairlike processes, or other special ornamentation may be present on back, or else body is entirely soft except for a hardened head capsule. BEETLES (Coleoptera), Plate 30.
53b. Most of body soft except for a hardened head capsule and with one to three hard plates on the back of first body segments; tubular processes may be present on sides of the body in various arrangements. Body may end in a pair of hook-bearing legs. Most larvae living in portable cases made of bits of sticks, leaves, or sand or in attached fibrous cases. CADDISFLIES (Trichoptera), Plate 27.

1007 C. List of Common Types of Aquatic Organisms (Plates 1–38), by Trophic Level

1. Producers (plants)

	Plate No.
Blue-green algae (Cyanophyta): coccoid	1
Blue-green algae (Cyanophyta): filamentous	2
Nonmotile green algae (Chlorophyta): coccoid	3
Nonmotile green algae (Chlorophyta): filamentous	4
Diatoms (Bacillariophyceae): pennate	5
Diatoms (Bacillariophyceae): centric	6
Types of larger marine algae	7
Higher plants: floating (Spermatophyta and Pteridophyta)	8
Higher plants: submersed (Spermatophyta)	9
Higher plants: emersed (Spermatophyta)	10
Pigmented flagellates: single cells (various phyla)	11
Pigmented flagellates: colonial types (various phyla)	12

2. Consumers (animals)

	Plate No.
Nonpigmented flagellates (Protozoa)	13
Amoebas (Protozoa)	14
Ciliates (Protozoa)	15
Sponges (Porifera) and bryozoa	16
Rotifers (Rotatoria)	17
Roundworms (Nematoda)	18
Flatworms (Platyhelminthes) and segmented worms (Annelida)	19
Crustaceans: cladoceran types	20
Crustaceans: common types	21
Insects: types of pupae	22
Insects: stoneflies (Plecoptera)	23
Insects: mayflies (Ephemeroptera)	24
Insects: damsel- and dragonflies (Odonata)	25
Insects: hellgrammites and relatives (Megaloptera)	26
Insects: caddisflies (Trichoptera)	27
Insects: two-winged flies (Diptera)	28
Insects: two-winged flies (Diptera)	29
Insects: beetles (Coleoptera)	30
Insects: true bugs (Hemiptera)	31
Mollusks: snails (Gastropoda)	32
Mollusks: bivalves (Pelecypoda)	33
Echinoderm types	34
Miscellaneous invertebrates	35
Fish types	36
Amphibian types	37

3. Reducers (bacteria and fungi)

	Plate No.
Bacteria and fungi	38

ACKNOWLEDGMENTS

Plates 1 through 38, which follow on succeeding pages, present over 200 aquatic organisms commonly found in natural, polluted, and treated waters. These plates were drawn especially for this work by Eugene Schunk of the Cincinnati Art Service, Inc. In a number of instances, it would have been impossible to illustrate a certain organism for the purposes of this manual were it not for the courtesy of other publishers, who permitted illustrations from their publications to be incorporated herein. The following organisms were so reproduced:

Plate

5: B—*Diatoma*,
F—*Achnanthes*,
G—*Gomphonema*,
H—*Cymbella*, and
K—*Surirella*, courtesy of Veb Gustav Fischer Verlag, Jena. Source: Die Susswasser—Flora Mitteleuropas, Heft 10, by F. Hustedt. 1930.

6: C—*Coscinodiscus*, and
D—*Melosira*, courtesy of E. Schweizerbart'sche Verlagsbuchhandlung, Stuttgart. Source: Das Phytoplankton des Susswassers, Die Binnengewasser, Band XVI, Teil II, Halfte II, by G. Huber-Pestalozzi and F. Hustedt, 1942. Plates CVIII-CXVI and CXXIII.
F—*Skeletonema*, courtesy of Academische Verlagsgesellschaft, Leipzig. Source: Die Kieselalgen, by F. Hustedt. In: L. Rabenhorst, Kryptogamen-Flora von Deutschland, Osterreich und der Schweiz, Band VII, 1930.

16: E—*Membranipora monostachys*, reprinted by permission of G.P. Putnam's Sons, Inc., New York. Source: Field Book of Seashore Life, by R.W. Miner. Copyright 1950 by the author. Plate 236, page 817.

17: I—*Notholca* Robert W. Pennak, Fresh-Water Invertebrates of the United States, Copyright © 1953, The Ronald Press Company, New York. Figure 116*N*, page 190, adapted for Figure 171, courtesy of The Ronald Press.

21: A—*Asellus* (sowbug),
C—*Mysis* (shrimp),
D—*Diaptomus* (copepod),
E—*Cypridopsis* (ostracod),
F—*Cyclops* (copepod), and
G—*Cambarus* (crayfish, crawdad), courtesy of Holden-Day, Inc., San Francisco, California. Source: Needham & Needham's Guide to the Study of Freshwater Biology, 1951. Figures 1 and 10, Plate 14, page 37; Figures 16, 18 and 20, Plate 24, page 61; and Figure 9, Plate 14, page 37.

22: Dr. Harold Walters

32: A—*Pomacea* (apple snail),
B—*Marisa*,
E—*Tarebia*,
I—*Lymnaea* (pond snail),
J—*Helisoma* (orb snail), and
M—*Lanx* (limpet) courtesy of John Wiley & Sons, Inc., New York. Source: Ward & Whipple, Fresh Water Biology (2nd ed.), W.T. Edmondson, Editor, 1959. Figures 43.31A(A), 43.31B(B), 43.62B(E), 43.13(I), 43.20(J) and 43.14(M).
C—*Campeloma*,
D—*Bithynia* (faucet snail),
F—*Pleurocera* (river snail),

Plate

G-*Valvata,*
H—*Littorina* (periwinkle),
K—*Nassa* (mud snail),
L—*Ferrissia* (limpet), and
N—*Physa,* courtesy of R.M. Sinclair, Advisor for Biological Sampling and Analysis (American Public Health Association) 13th ed.

34: Connecticut State Geological and Natural History Survey: Echinoderms of Connecticut, by Wesley Roswell Coe, 1912.

35: C—*Limulus* (horseshoe crab) courtesy of Western Publishing Company, Inc., Golden Press Division, Racine, Wisconsin. Source: Seashores, a Golden Nature Guide, 1955. Page 79.

37: C—*Ambystoma* (terrestrial adult), courtesy of Dover Publications, Inc., New York. Source: Biology of the Amphibia, by G.K. Noble, 1931. Figure 147C, page 471.
D—*Ambystoma* (aquatic larva), courtesy of the New York State Museum and Science Service, Albany, New York. Source: The Salamanders of New York, by Sherman C. Bishop, 1941. Figure 33b, page 166. [Bulletin 324, New York State Museum, Albany.]
E—*Necturus,* courtesy of Dover Publications, Inc., New York. Source: Biology of the Amphibia, by G.K. Noble, 1931. Figure 35B, page 99.
G—*Siren intermedia* (siren), reprinted from Sherman Bishop: Handbook of Salamanders. Copyright 1943 by Comstock Publishing Company, Inc. Used by permission of Cornell University Press.

38: A—(a) micrococcus, (b) streptococcus, (c) sarcina, (d) bacillus, (e) vibrio, (f) spirillum, courtesy of John Wiley & Sons, Inc., New York. Source: Ward & Whipple, Fresh Water Biology (2nd ed.), W.T. Edmondson, Editor, 1959. Figure 3.1.
(k) actinomycete growth form, Selman A. Waksman, The Actinomycetes. Copyright © 1957, The Ronald Press Company, New York. Figure 2-6, page 18, adapted for Figure 37A(k), courtesy The Ronald Press.
B—*Tetracladium* and (e) (f), *Achlya,* courtesy of John Wiley & Sons, Inc., New York. Source: Ward & Whipple, Fresh Water Biology (2nd ed.), W.T. Edmondson, Editor, 1959. Figures 4.119 and 4.79.

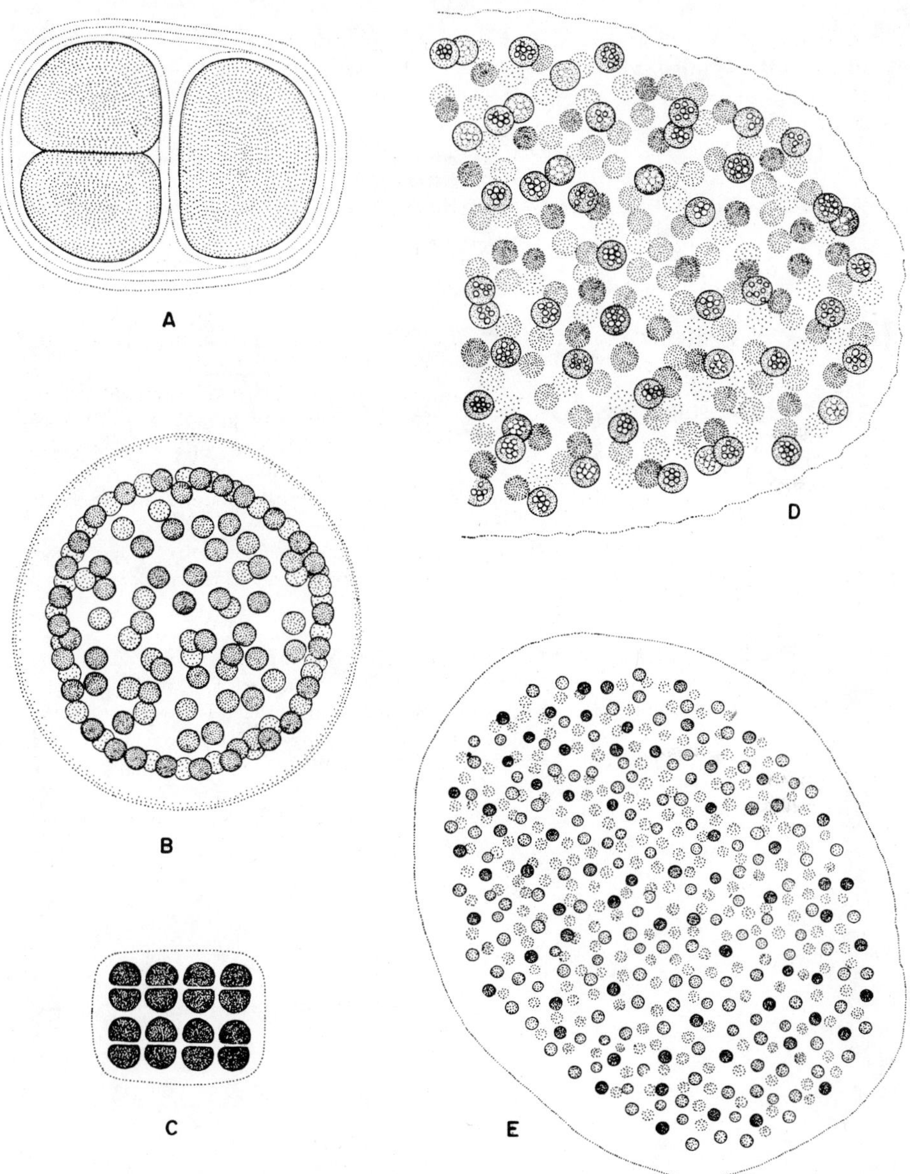

Plate 1. Blue-green algae: Coccoid (Phylum Cyanophyta). Dimensions refer to individual cells.

A—*Anacystis* (4–20 μm)
B—*Gomphosphaeria* (3–6 μm)
C—*Agmenellum* (2–6 μm)
D—*Anacystis* sp. (4–6 μm)
E—*Anacystis* sp. (3–4 μm)

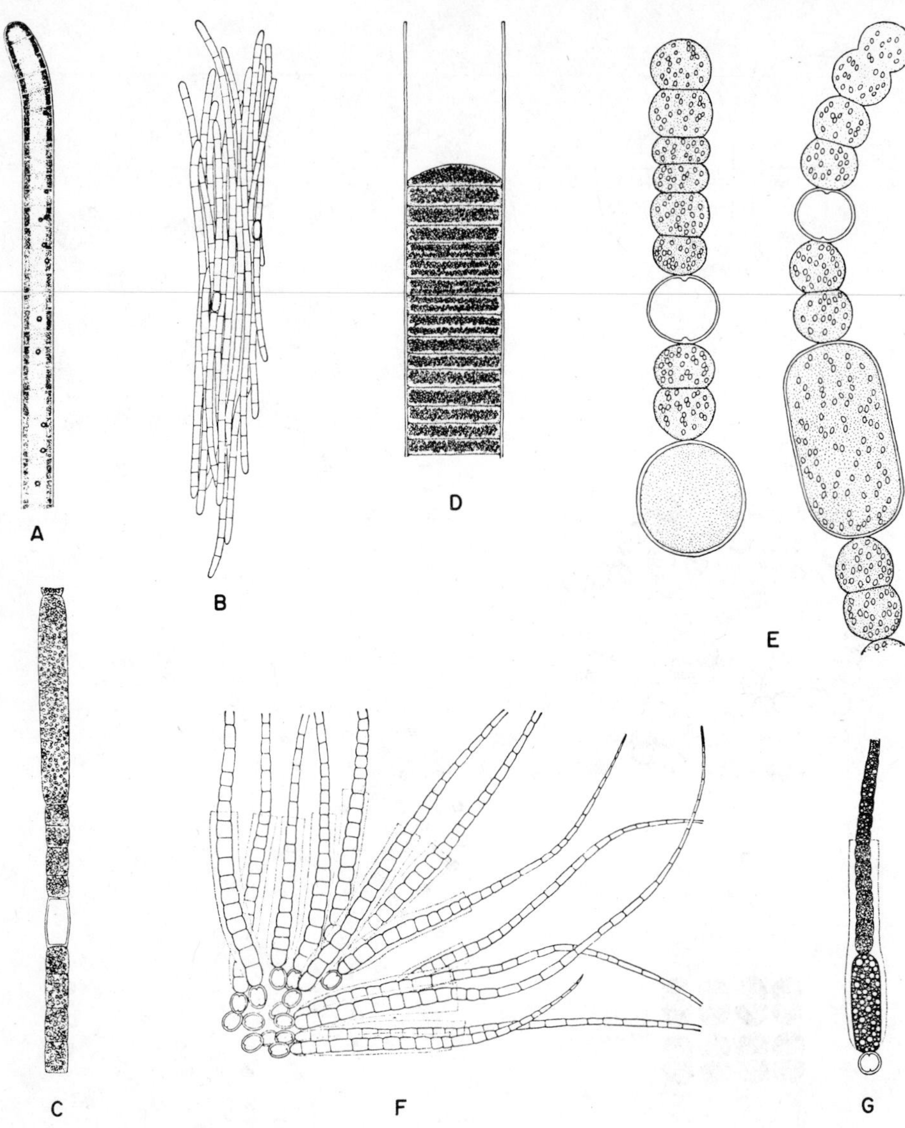

Plate 2. Blue-green algae: Filamentous (Phylum Cyanophyta). Most dimensions refer to diameter of individual filaments.

A—*Oscillatoria*	(4–20 μm)	E—*Anabaena*	(5–12 μm)
B—*Aphanizomenon*, aggregate of filaments	(5–6 μm)	F—*Gleotrichia*, portion of colony	(Cells 7–9 μm diameter near akinete)
C—*Aphanizomenon*, detail		G—*Gleotrichia*, detail	
D—*Lyngbya*	(4–20 μm)		

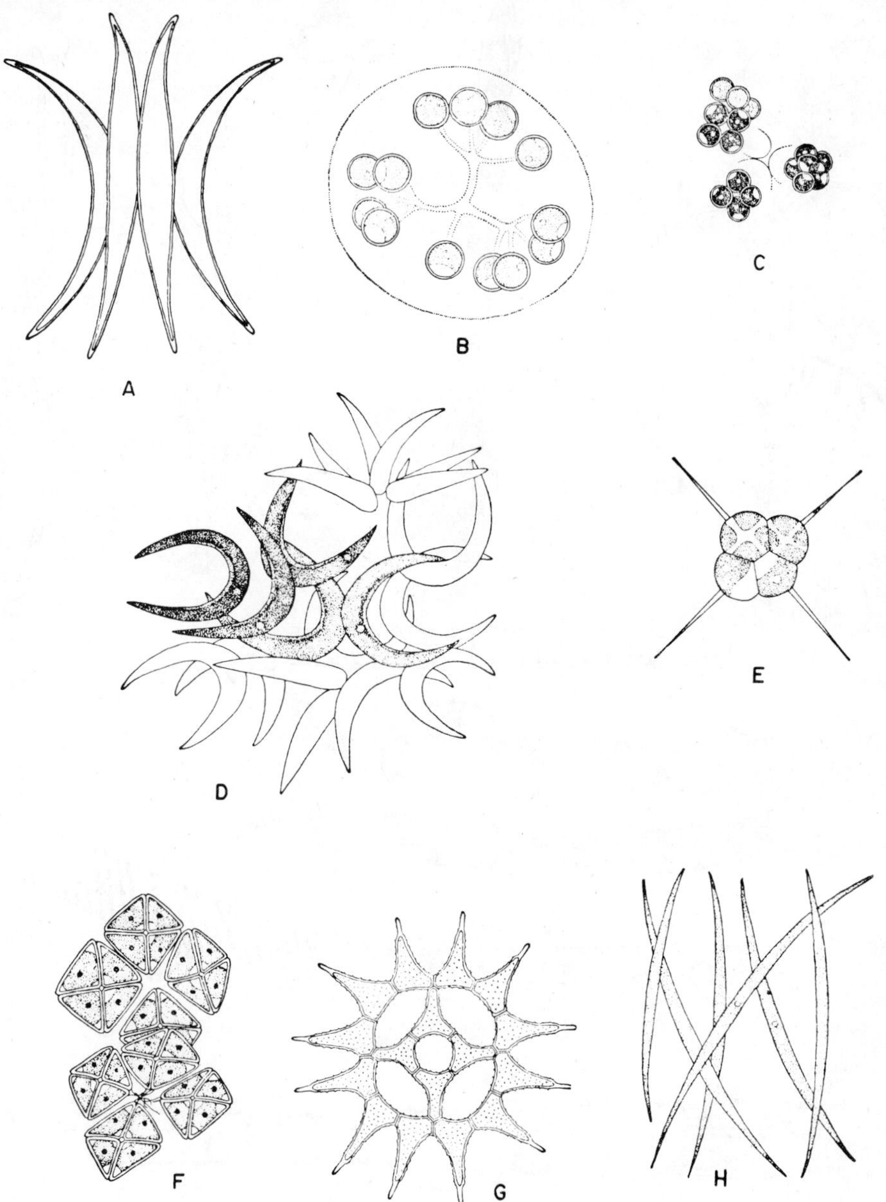

Plate 3. Nonmotile green algae: Coccoid (Phylum Chlorophyta). Dimensions refer to individual cells.

A—*Scenedesmus*	(4-6 μm diameter)	E—*Tetrastrum*	(5-9 μm)
B—*Dictyosphaerium*	(8-14 μm)	F—*Crucigenia*	(5-8 μm)
C—*Westella*	(5-7 μm)	G—*Pediastrum*	(10-20 μm)
D—*Selenastrum*	(6-7 μm)	H—*Ankistrodesmus*	(2-3 μm)

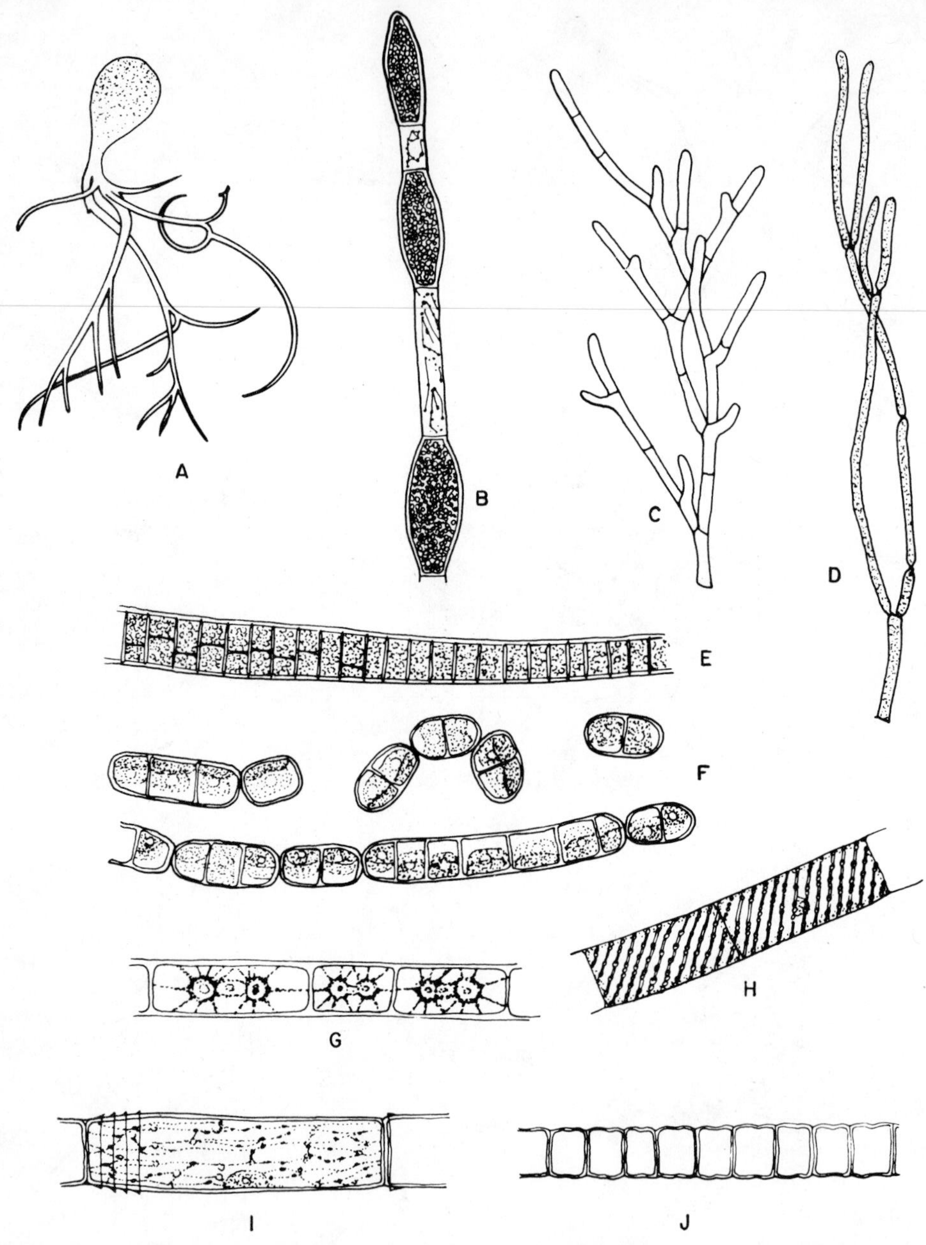

Plate 4. Nonmotile green algae: Filamentous (Phylum Chlorophyta). Dimensions refer to diameters of filaments or to mass.

A—*Botrydium*	(1000–2000 μm)	F—*Stichococcus*	(3 μm)
B—*Pithophora*	(50–100 μm)	G—*Zygnema*	(20–35 μm)
C—*Microthamnion*	(2–4 μm)	H—*Spirogyra*	(15–100 μm)
D—*Dichotomosiphon*	(50–100 μm)	I—*Oedogonium*	(6–40 μm)
E—*Schizomeris*	(12–18 μm)	J—*Hyalotheca*	(12–30 μm)

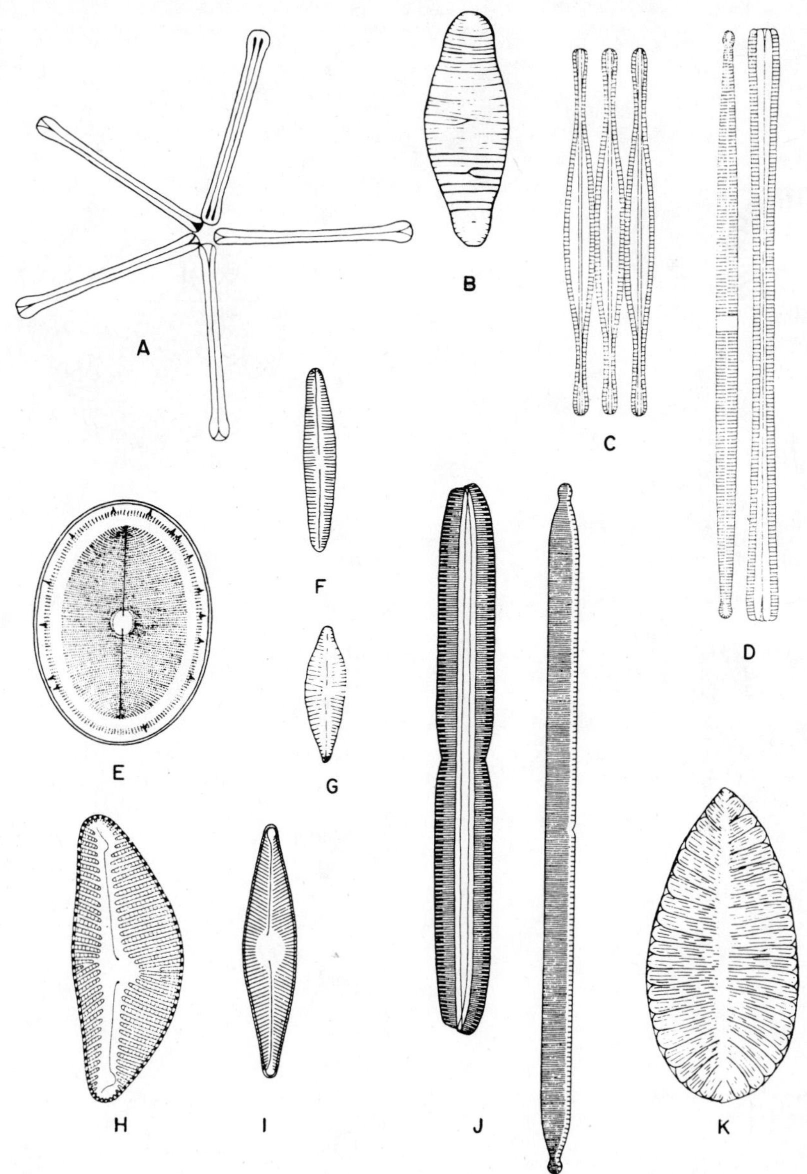

Plate 5. Diatoms: Pennate (Phylum Chrysophyta, Class Bacillariophyceae). Dimensions refer to length of cells unless otherwise specified.

A—*Asterionella*	(300 μm, entire colony)
B—*Diatoma*	(20 μm)
C—*Fragilaria*	(100 μm)
D—*Synedra*	(200 μm)
E—*Cocconeis*	(10 μm)
F—*Achnanthes*	(10 μm)
G—*Gomphonema*	(20 μm)
H—*Cymbella*	(15 μm)
I—*Navicula*	(30 μm)
J—*Nitzschia*	(100 μm)
K—*Surirella*	(20 μm)

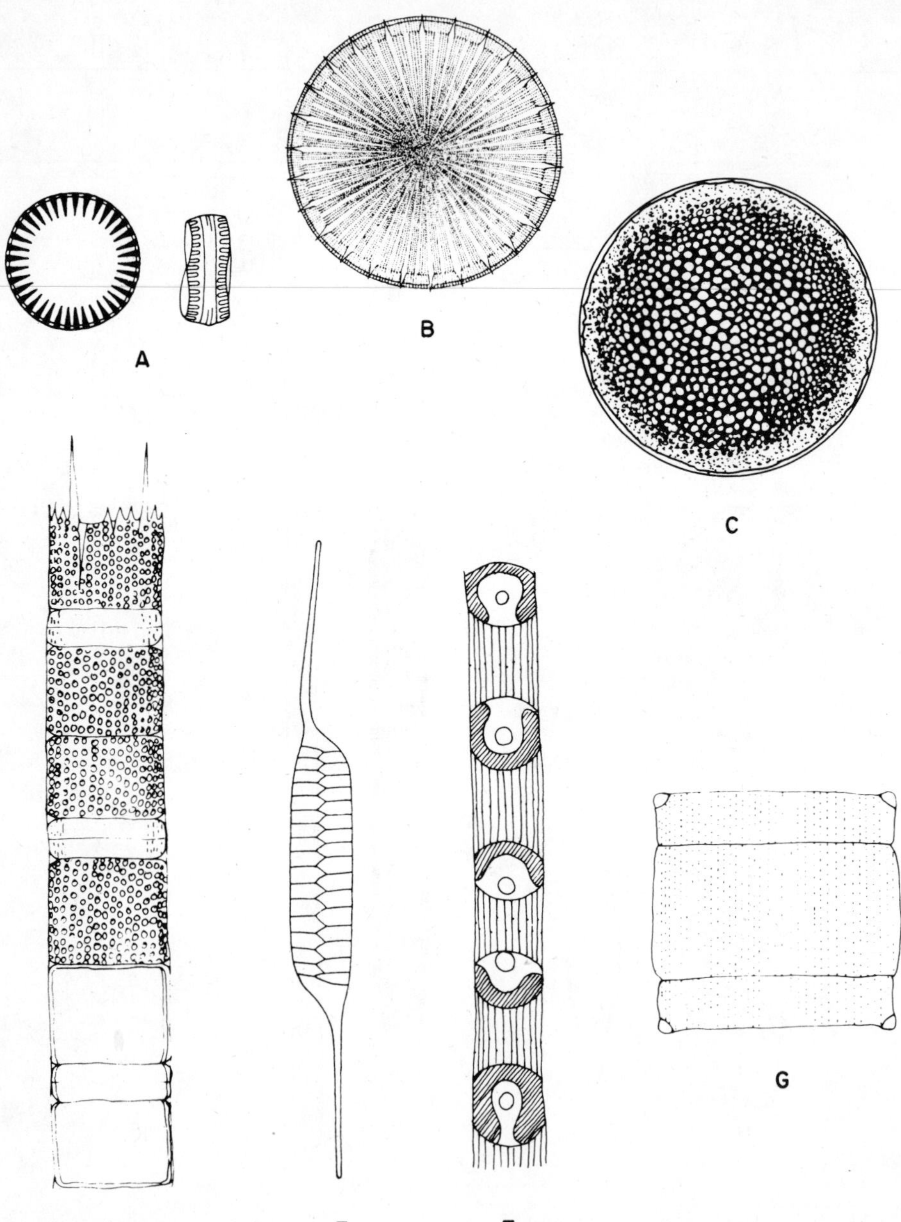

Plate 6. Diatoms: Centric (Phylum Chrysophyta, Class Bacillariophyceae). Dimensions refer to diameter.

A—*Cyclotella*	(10 μm)	E—*Rhizosolenia*	(5–15 μm)
B—*Stephanodiscus*	(30 μm)	F—*Skeletonema*	(3–18 μm)
C—*Coscinodiscus*	(20 μm)	G—*Biddulphia*	(100 μm)
D—*Melosira*	(3–12 μm)		

Plate 7. Types of larger marine algae (green, brown, and red).

Green algae (Phylum Chlorophyta):		Red algae (Phylum Rhodophyta):	
A—*Enteromorpha*	(40 cm)	E—*Gracilaria*	(50 cm)
B—Sea lettuce, *Ulva*	(20 cm)	F—*Corallina*	(4 cm)
Brown algae (Phylum Phaeophyta):			
C—Rockweed, *Fucus*	(75 cm)		
D—Giant kelp, *Nereocystis*	(20 m)		

Plate 8. Higher plants: Floating plants.

A—Great duckweed, *Spirodela* (Phylum Spermatophyta, 8 mm)
B—Water velvet, *Azolla* (Phylum Pteridophyta, 1 cm)
C—Water hyacinth, *Eichhornia* (Phylum Spermatophyta, 22 cm)
D—Lesser duckweed, *Lemna* (Phylum Spermatophyta, 5 mm)
E—Water fern, *Salvinia* (Phylum Pteridophyta, 4 cm)
F—Watermeal, *Wolffia* (Phylum Spermatophyta, 1-1.5 mm)

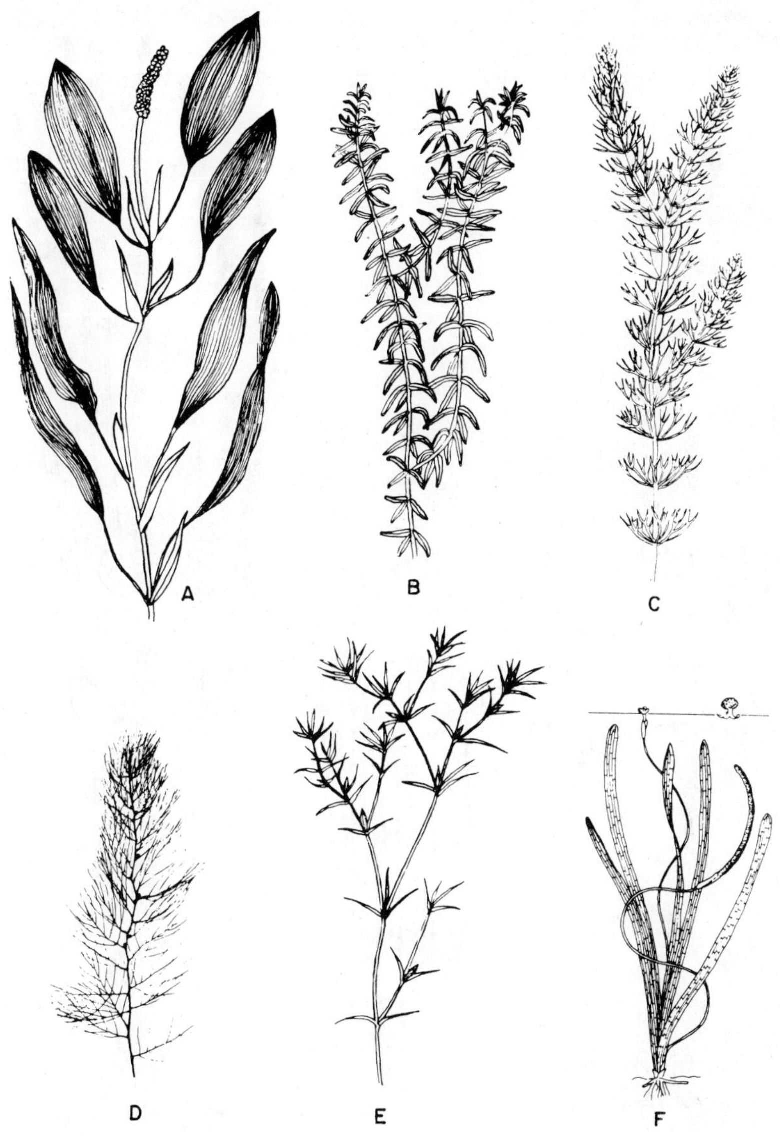

Plate 9. Higher plants: Submersed (all forms illustrated are Spermatophytes).

A—Pondweed, *Potamogeton*	(30–60 cm)	E—Naiad, *Najas*	(60 cm)
B—Waterweed, *Elodea*	(15 cm)	F—Eelgrass, *Vallisneria*	(45 cm)
C—Coontail, *Ceratophyllum*	(30 cm)		
D—Water milfoil, *Myriophyllum*	(30 cm)		

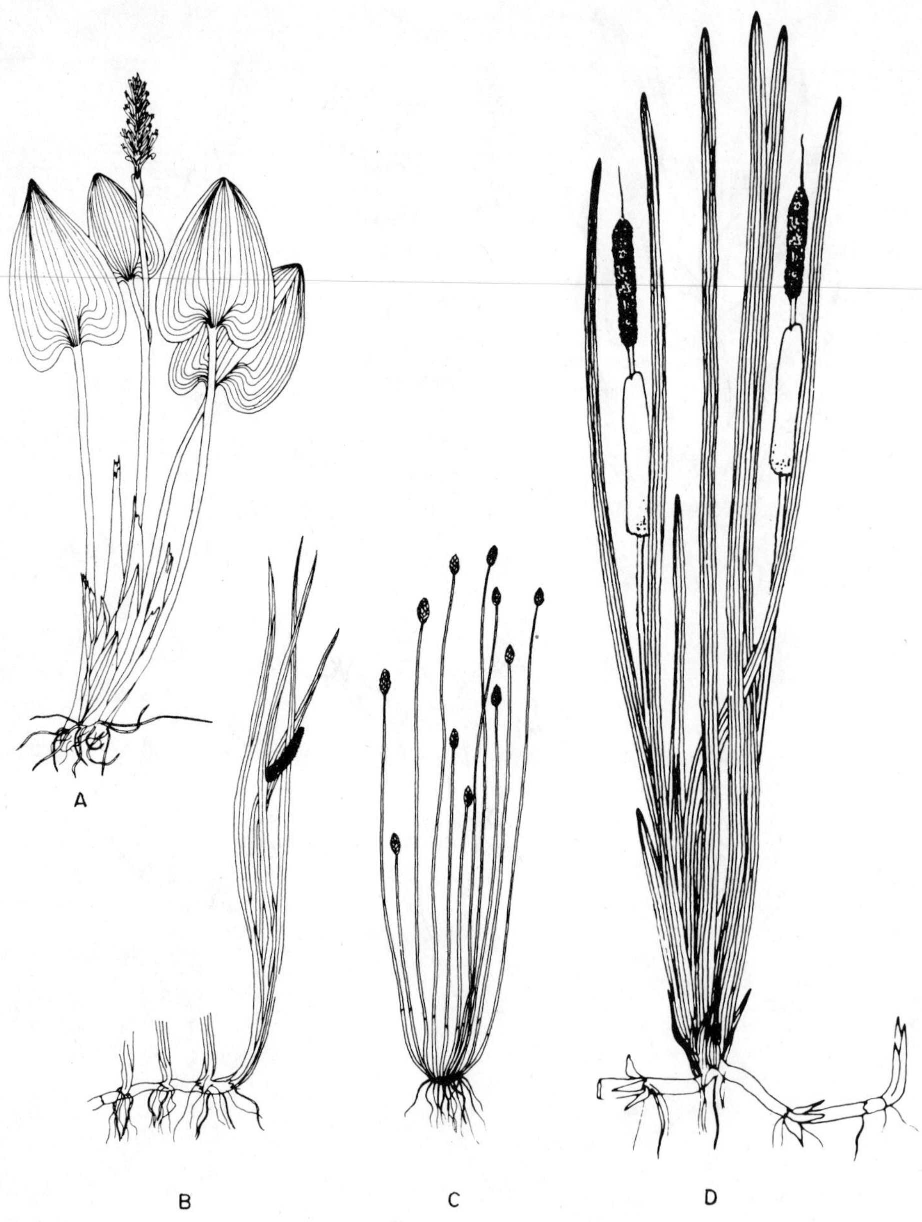

Plate 10. Higher plants: Emersed (all forms illustrated are Spermatophytes).

A—Pickerelweed, *Pontederia* (60 cm)
B—Sweetflag, *Acorus* (30 cm)
C—Spike rush, *Eleocharis* (30 cm)
D—Cattail, *Typha* (1-2 m)

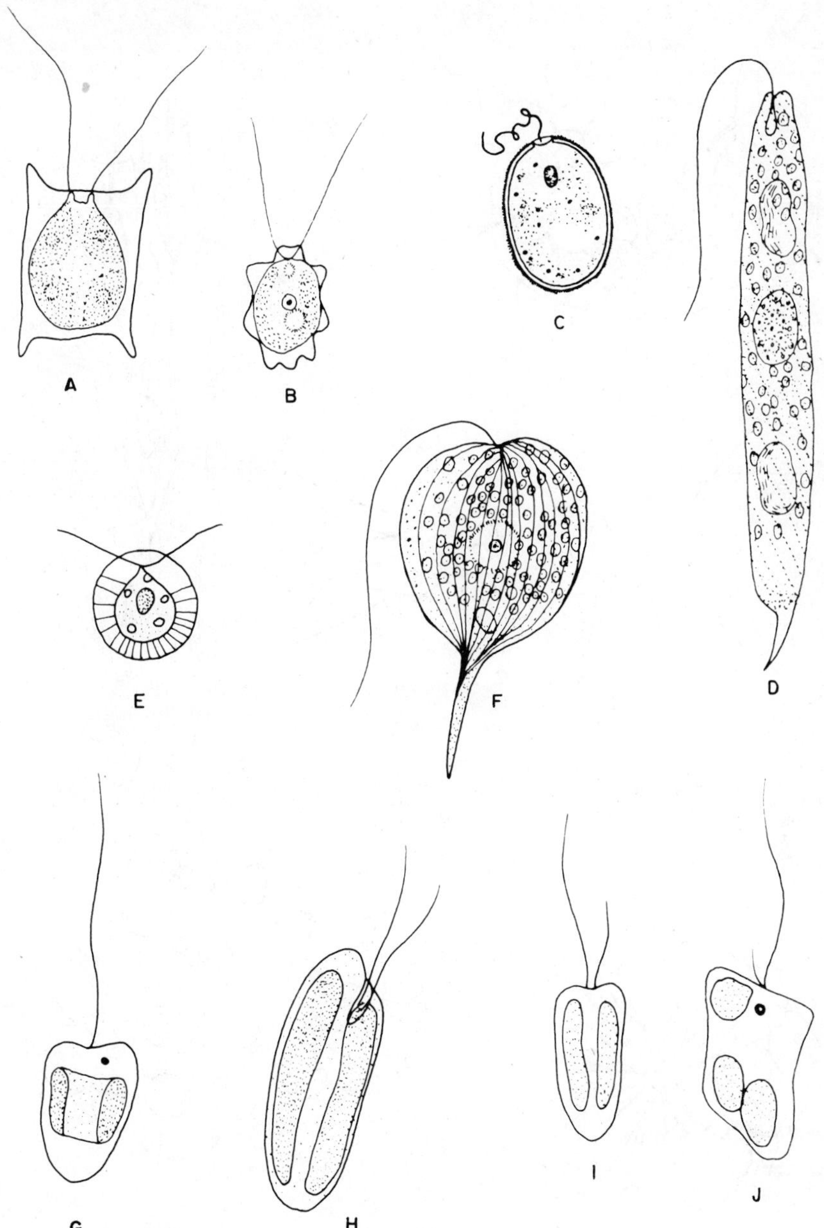

Plate 11. Pigmented flagellates: Single-celled (various phyla).

A—*Pteromonas*	(9–18 μm)	F—*Phacus*	(20–50 μm)
B—*Lobomonas*	(5–14 μm)	G—*Chromulina*	(4–10 μm)
C—*Trachelomonas*	(15–30 μm)	H—*Cryptomonas*	(6–12 μm)
D—*Euglena*	(10–25 μm)	I—*Ochromonas*	(7–14 μm)
E—*Haematococcus*	(40–45 μm)	J—*Chloramoeba*	(10–15 μm)

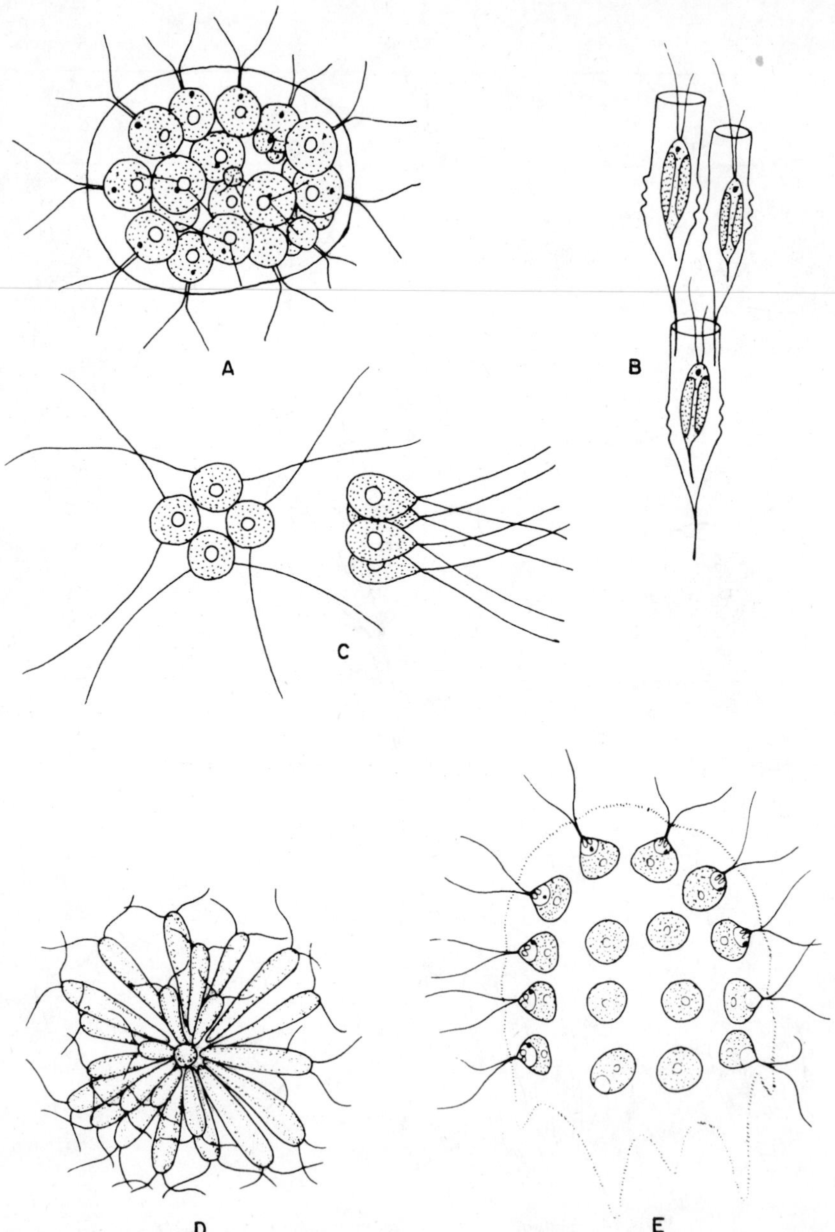

Plate 12. Pigmented flagellates: Colonial types (various phyla). Dimensions refer to individual cells unless otherwise specified.

A—*Pleodorina*	(8–10 μm)	D—*Synura*	(10–15 μm)
B—*Dinobryon*	(7–12 μm)	E—*Platydorina*	(66–70 μm colony)
C—*Gonium*	(7–12 μm)		

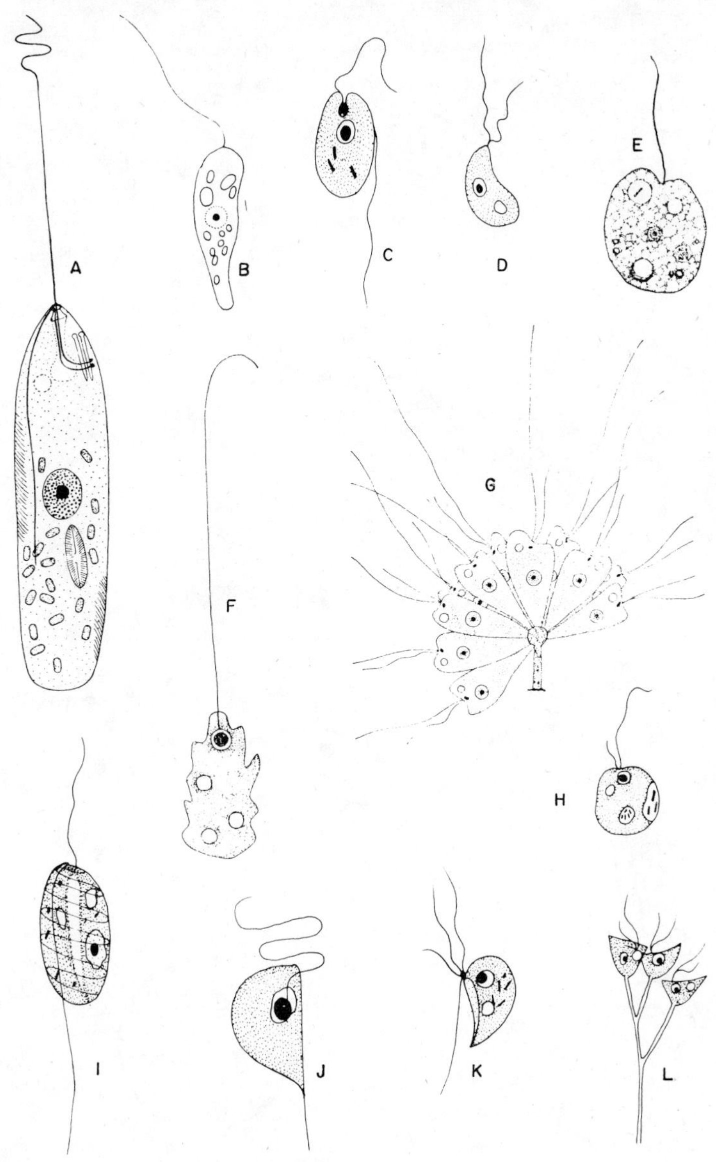

Plate 13. Nonpigmented flagellates (Phylum Protozoa).

A—*Peranema*	(40–70 μm)	G—*Anthophysa*	(5-6 μm)
B—*Astasia*	(40–50 μm)	H—*Monas*	(5-16 μm)
C—*Bodo*	(11-22 μm)	I—*Anisonema*	(14-60 μm)
D—*Dinomonas*	(15-16 μm)	J—*Cercomonas*	(10-36 μm)
E—*Oikomonas*	(5-20 μm)	K—*Tetramitus*	(11-30 μm)
F—*Mastigamoeba*	(28-200 μm)	L—*Dendromonas*	(8 μm)

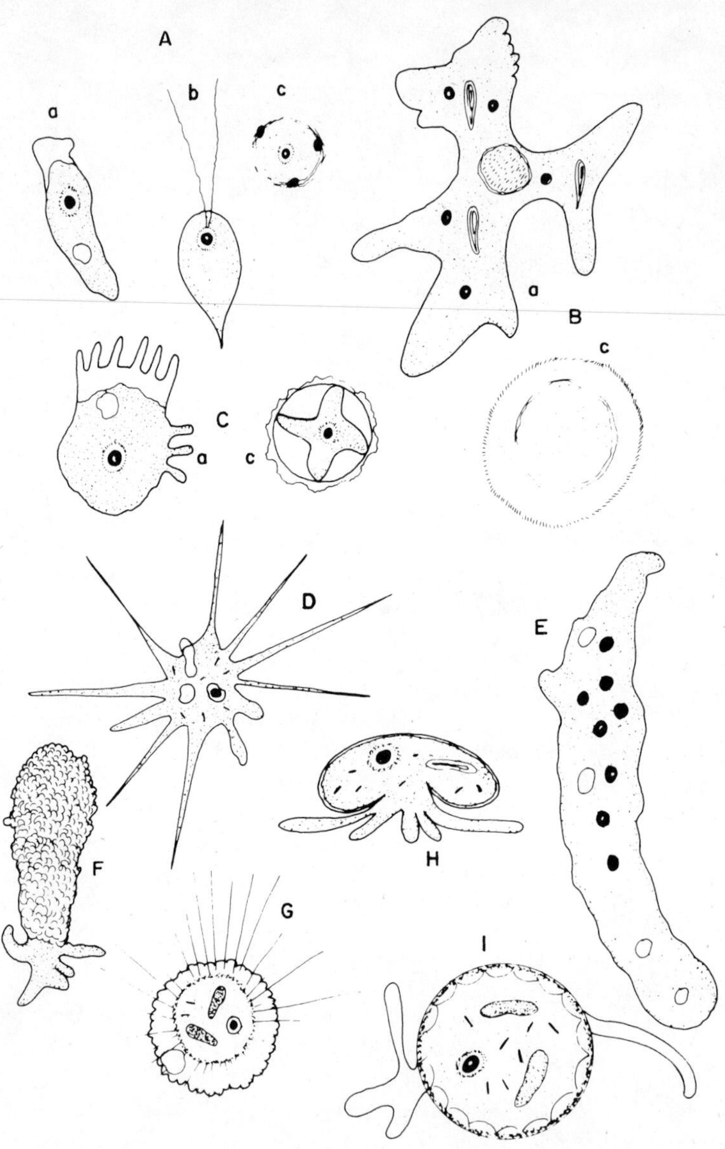

Plate 14. Amoebas (Phylum Protozoa). (a) Amoeboid stages, (b) flagellated stages, (c) cyst stages.

A—*Naegleria*	(10-36 μm)	E—*Pelomyxa*	(0.25-3 mm)
B—*Amoeba* sp.	(30-600 μm)	F—*Difflugia*	(40 μm)
C—*Acanthamoeba* (*Hartmannella*)	(15-25 μm)	G—*Actinophrys*	(25-50 μm)
D—*Amoeba radiosa*	(30-120 μm)	H—*Arcella* (side view)	(30-260 μm)
		I—*Arcella* (top view)	(30-260 μm)

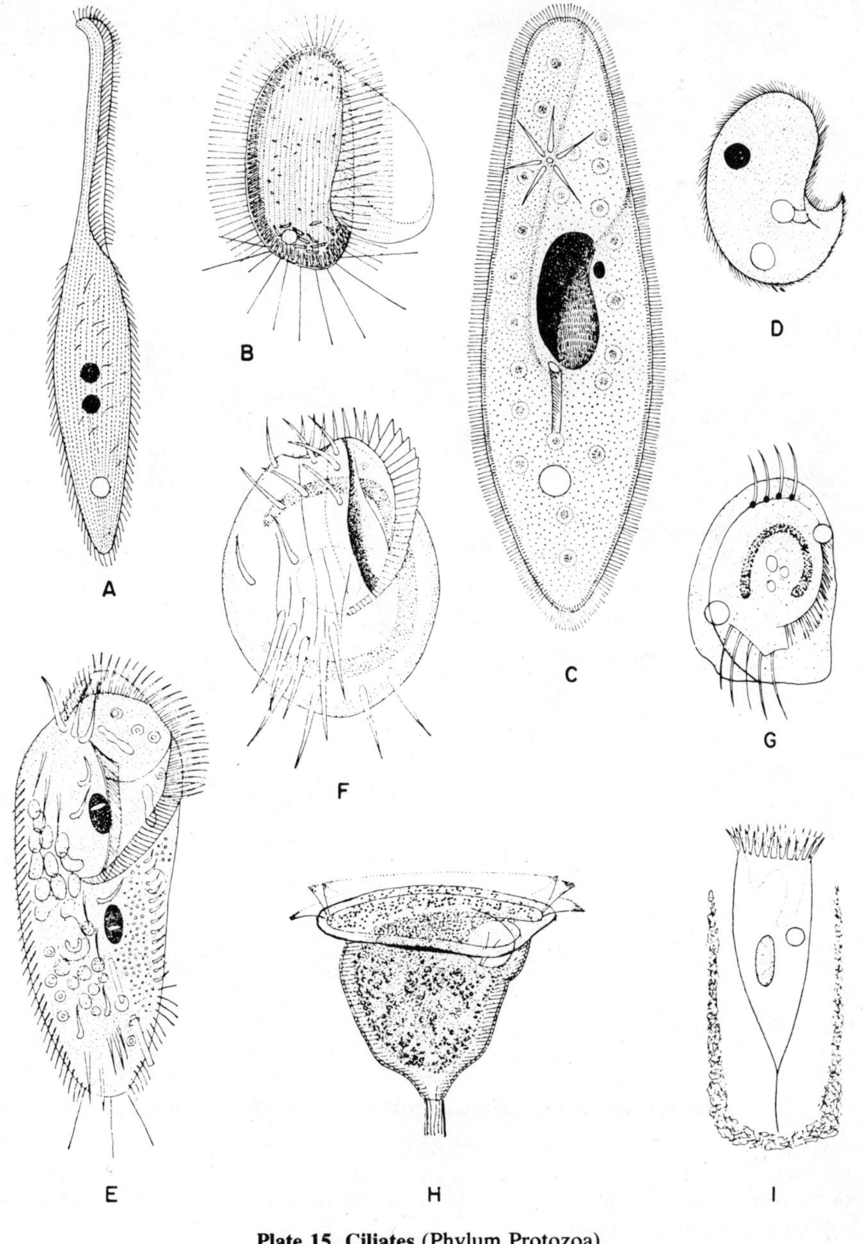

Plate 15. Ciliates (Phylum Protozoa).

A—*Lionotus*	(100 μm)	F—*Euplotes*	(70-195 μm)
B—*Pleuronema*	(38-120 μm)	G—*Aspidisca*	(30-50 μm)
C—*Paramoecium*	(50-330 μm)	H—*Vorticella*	(40-175 μm)
D—*Colpoda*	(12-110 μm)	I—*Tintinnidium*	(40-200 μm)
E—*Stylonychia*	(100-300 μm)		

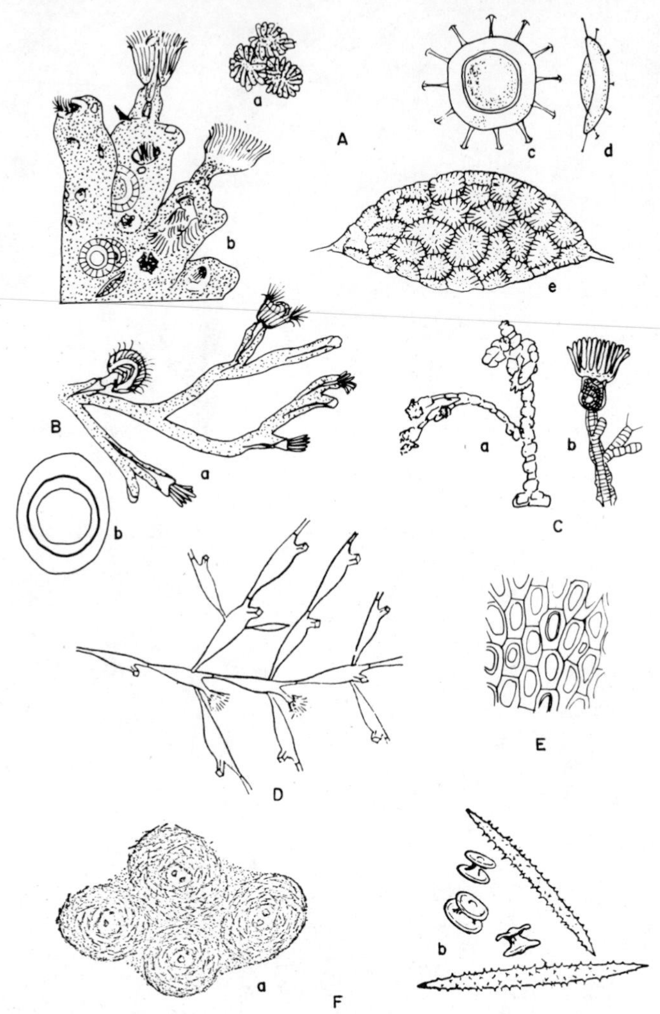

Plate 16. Sponges (Phylum Porifera) and Bryozoans (Phylum Bryozoa).

Bryozoa:

A—Jellyball, *Pectinatella*
- (a) Young colony (15 mm)
- (b) Section (highly magnified)
- (c) Statoblast (1 mm)
- (d) Statoblast (1 mm)
- (e) Colony on a plant stem (10 cm)

B—*Plumatella*
- (a) Colony (4 cm)
- (b) Statoblast (0.5 mm)

C—*Urnatella* (5 mm)
- (a) Colony (7 mm)
- (b) Individual zooid at tip of stalk (0.5 mm)

D—*Paludicella* (6 mm)

E—*Membranipora,* an encrusting marine form (individuals 1 mm, colonies unlimited)

Porifera:

F—*Trochospongilla*
- (a) Gemmules in a colony (1 mm)
- (b) Spicules (0.2 mm)

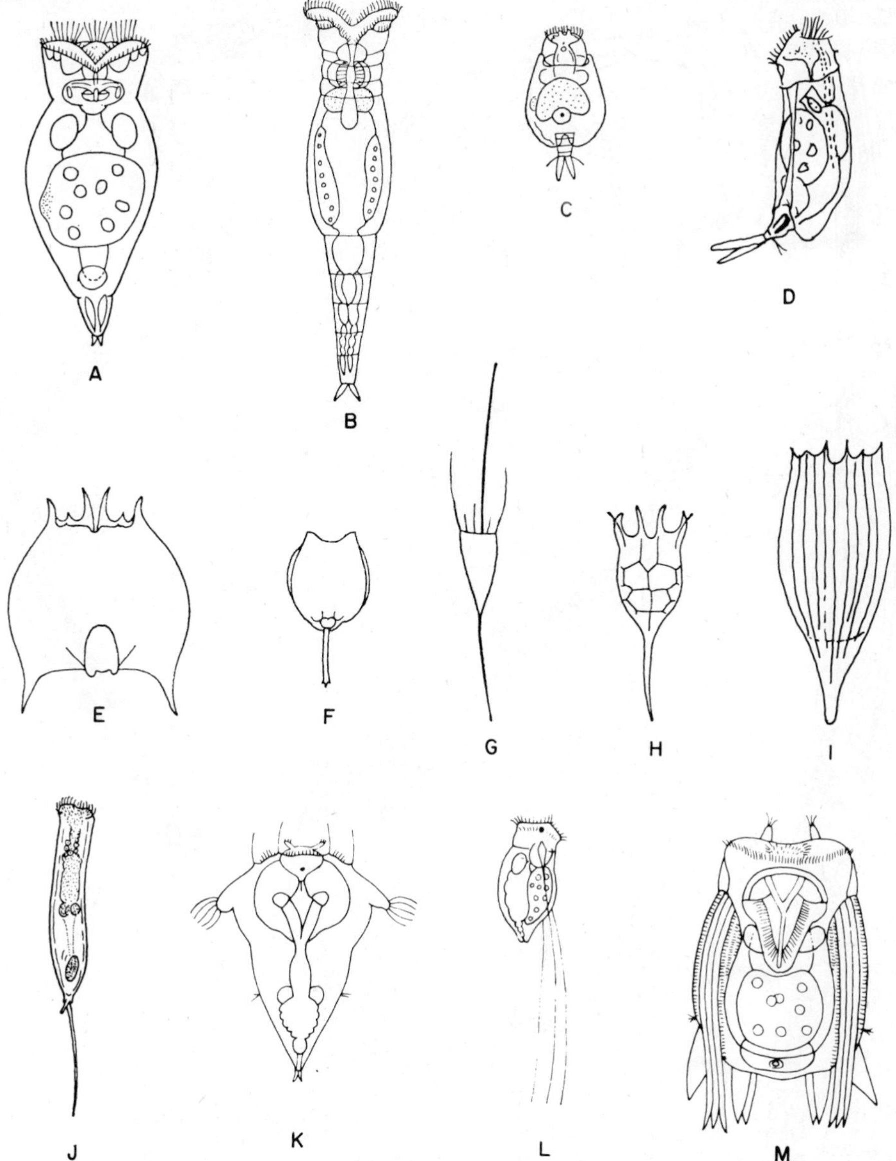

Plate 17. Rotifers (Phylum Rotatoria). Dimensions include spines.

A—*Epiphanes*	(600 μm)	H—*Keratella*	(200 μm)
B—*Philodina*	(400 μm)	I—*Notholca*	(200 μm)
C—*Euchlanis*	(250 μm)	J—*Trichocerca*	(600 μm)
D—*Proales*	(450 μm)	K—*Synchaeta*	(260 μm)
E—*Brachionus*	(200 μm)	L—*Filinia*	(150 μm)
F—*Monostyla*	(150 μm)	M—*Polyarthra*	(175 μm)
G—*Kellicottia*	(1 mm)		

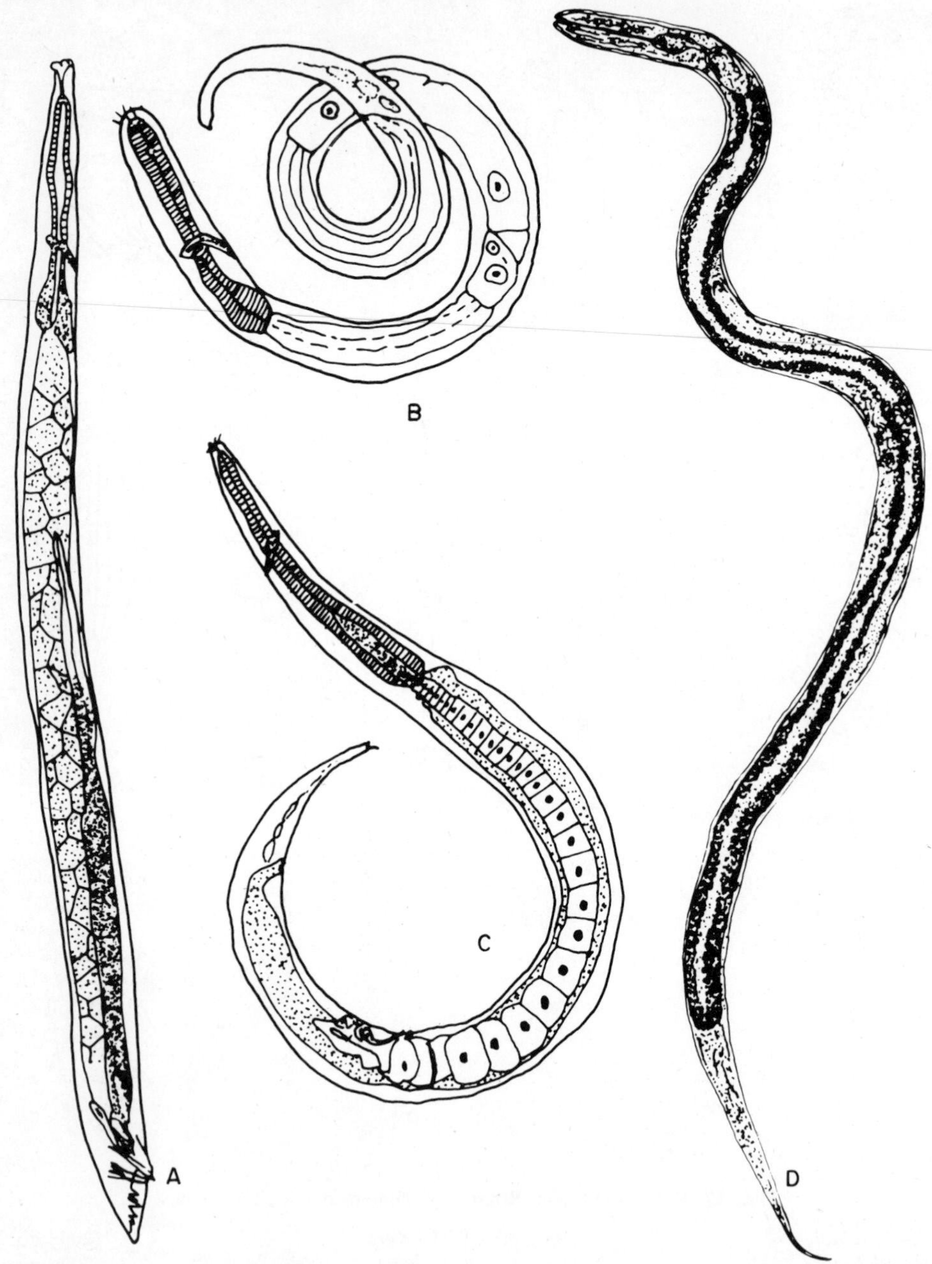

Plate 18. Roundworms (Phylum Nemathelminthes).

A—*Rhabditis* (male)	(1.6–1.9 mm)	D—*Diplogasteroides* (female)	(1.5–1.85 mm)
B—*Achromadora* (female)	(0.3–0.7 mm)		
C—*Monhystera* (female)	(0.8–1.0 mm)		

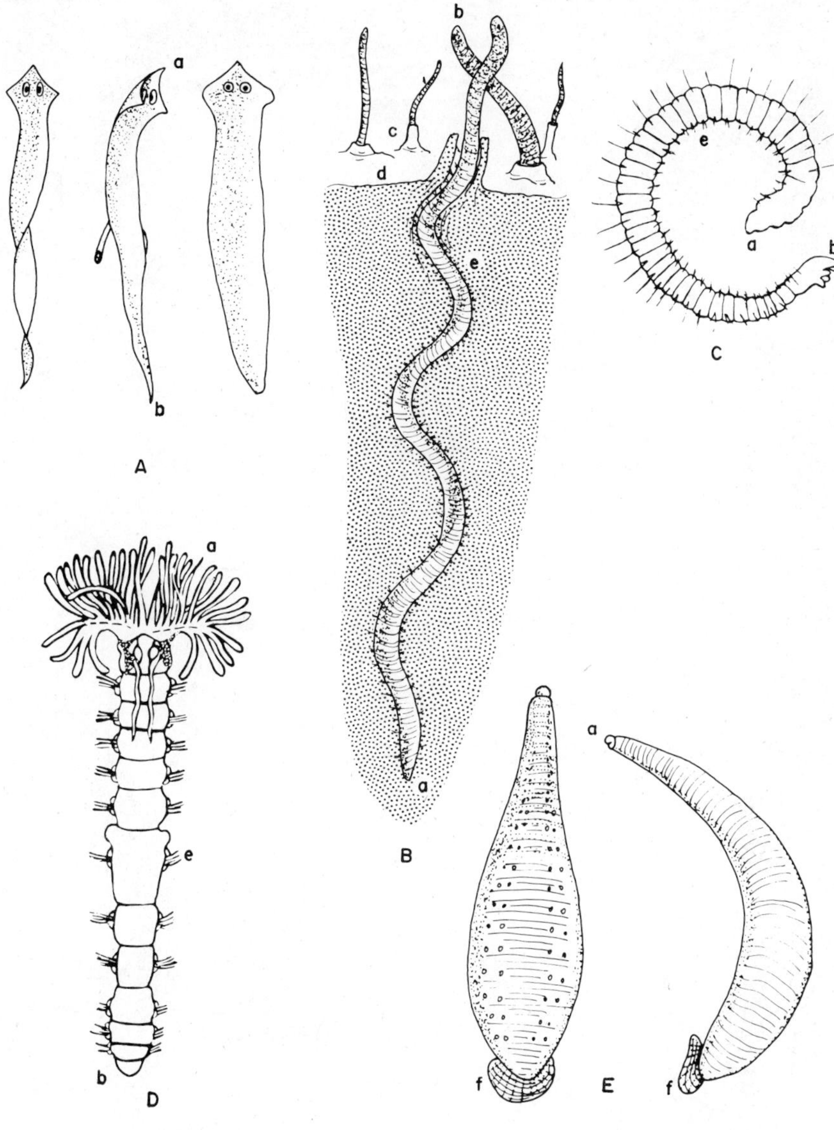

Plate 19. Flatworms (Phylum Platyhelminthes) **and segmented worms** (Phylum Annelida) (a) Anterior end, (b) posterior end. (c) tubes, (d) mud surface, (e) setae, (f) sucker disk.

Platyhelminthes:
A—*Planaria,* a free-living flatworm (5-13 mm)
Annelida:
B—*Tubifex,* a sludgeworm (25-50 mm)
C—*Dero,* a bristle worm (3-7 mm)
D—*Manayunkia,* a freshwater tube-building polychaet worm similar to certain common marine forms (5 mm)
E—Leech (50 mm)

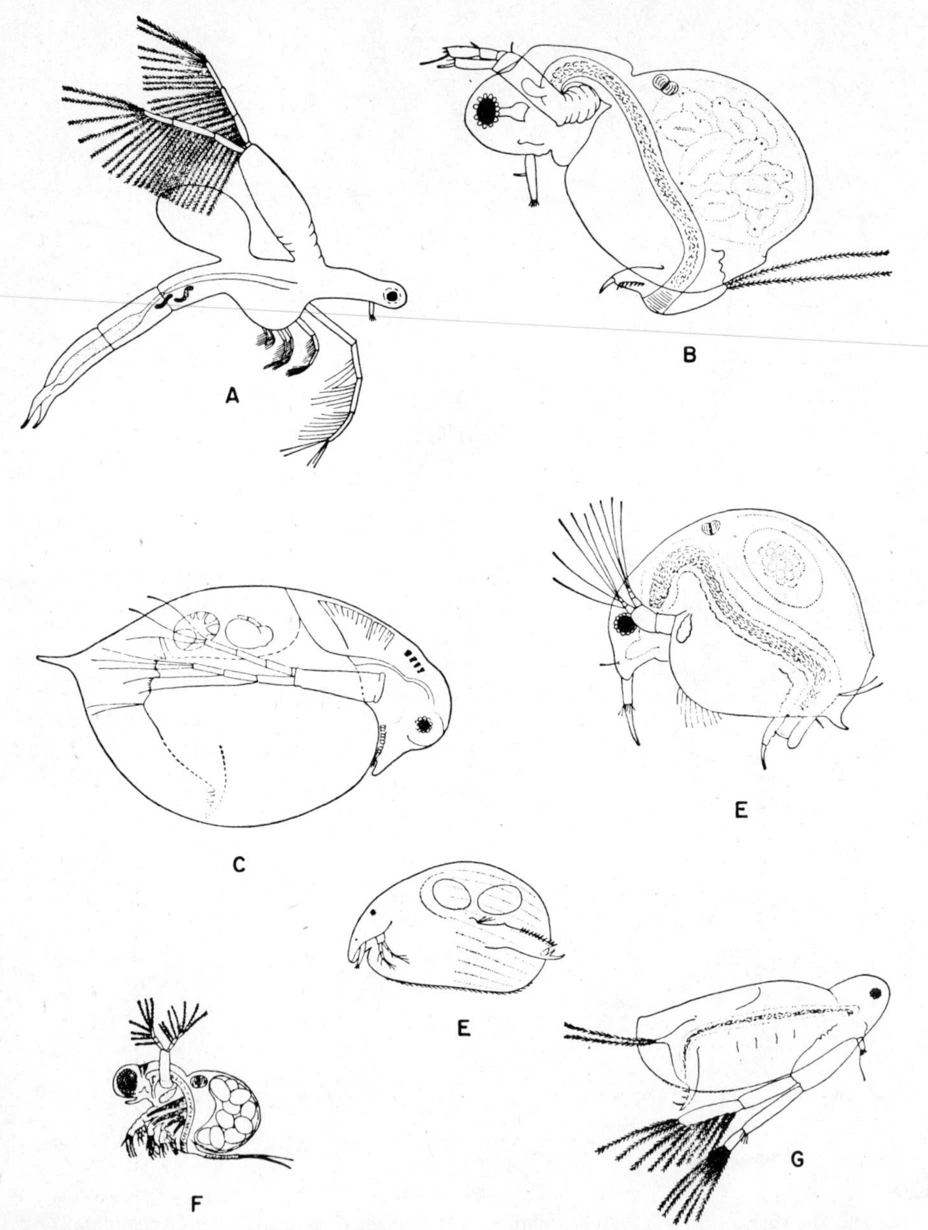

Plate 20. Crustaceans (Phylum Arthropoda, Class Crustacea): Types of cladocerans (Order Cladocera).

A—*Leptodora*	(9 mm)	E—*Bosmina*	(0.4 mm)
B—*Moina*	(1.5 mm)	F—*Polyphemus*	(1.5 mm)
C—*Daphnia*	(2 mm)	G—*Diaphanosoma*	(1.5 mm)
D—*Alona*	(0.4 mm)		

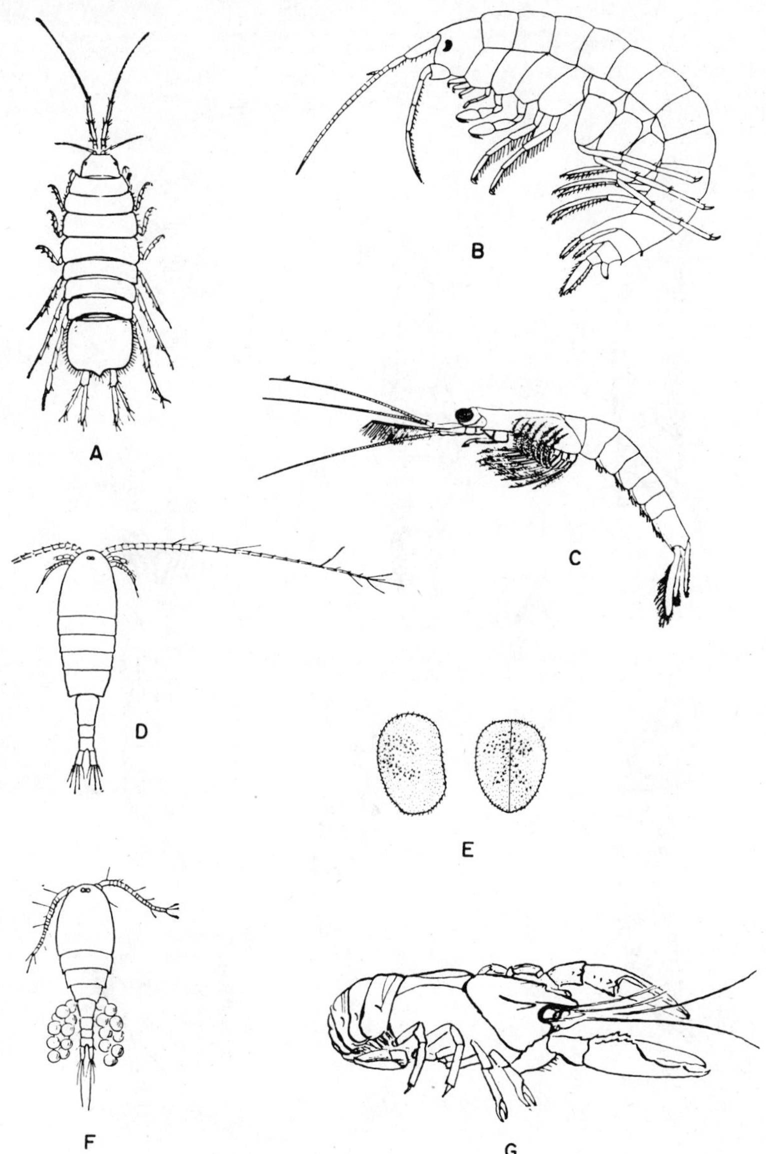

Plate 21. Crustaceans (Phylum Arthropoda, Class Crustacea): Selected common types.

A—Sowbug, *Asellus,* Order Isopoda (20 mm)
B—Scud, *Gammarus,* Order Amphipoda (15 mm)
C—Shrimp, *Mysis,* Order Decapoda (20 mm)
D—Copepod, *Diaptomus,* Order Copepoda (2 mm)
E—Ostracod, *Cypridopsis,* Order Ostracoda (1 mm)
F—Copepod, *Cyclops,* Order Copepoda (1 mm)
G—Crayfish, crawdad, *Cambarus,* Order Decapoda (150 cm)

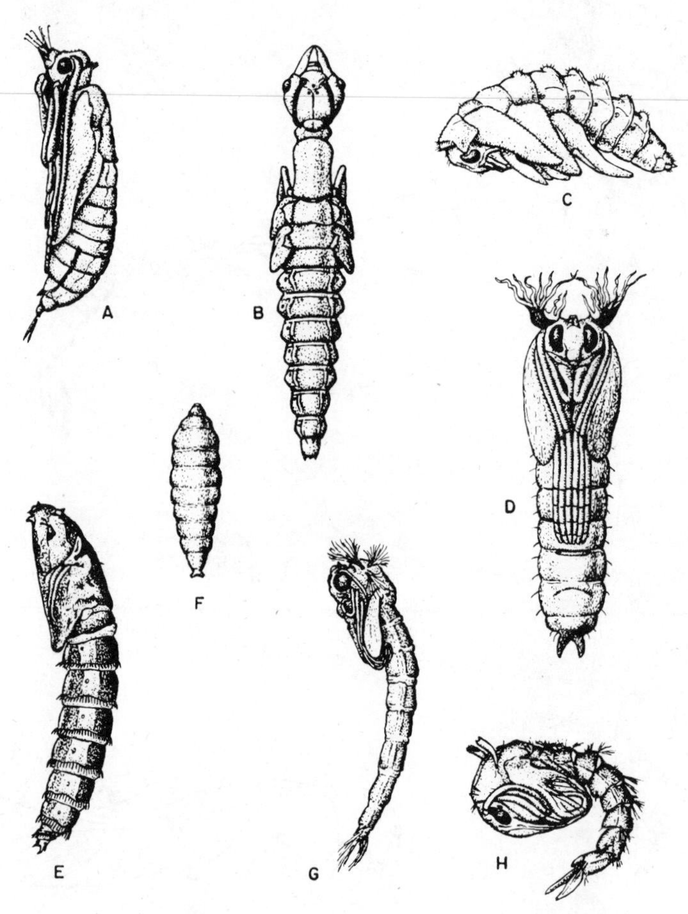

Plate 22. Types of insect pupae.

A—Caddisfly, *Goera,* Order Trichoptera
B—Hellgrammite, *Corydalis,* Order Megaloptera
C—Beetle, *Cybister,* Order Coleoptera
D—Cranefly, *Antocha,* Tipulidae
E—Blowfly, *Tabanus,* Tabinidae
F—Sewage fly, *Limnophorus,* Anthomyidae
G—Midge, *Chironomus,* Chironomidae
H—Mosquito, *Culex,* Culicidae

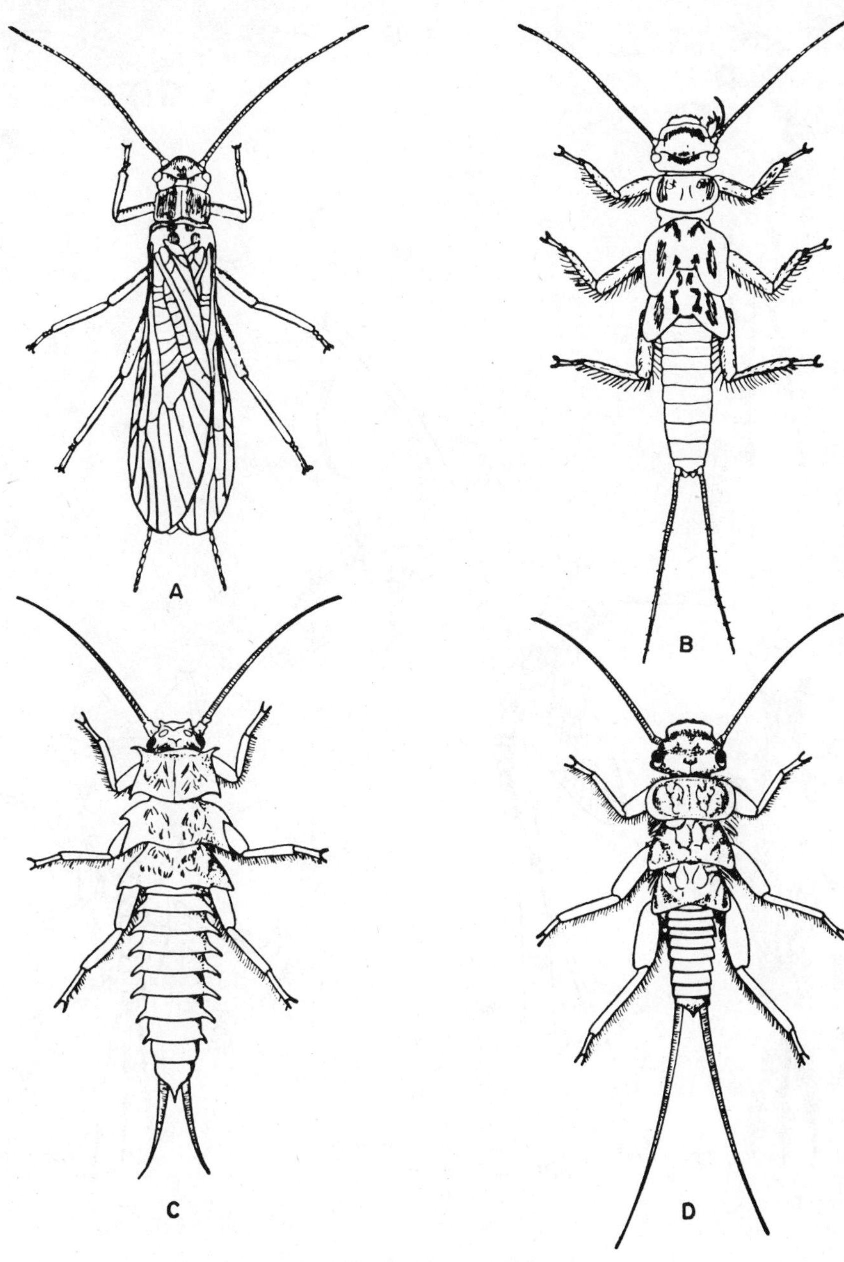

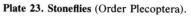

Plate 23. Stoneflies (Order Plecoptera).

A—Adult *Isoperla*, Isoperlidae (14-23 mm)
B—Nymph *Isoperla*, Isoperlidae (10-14 mm)
C—Nymph *Pteronarcys*, Pteronarcidae (10-40 mm)
D—Nymph *Acroneuria*, Perlidae (20-30 mm)

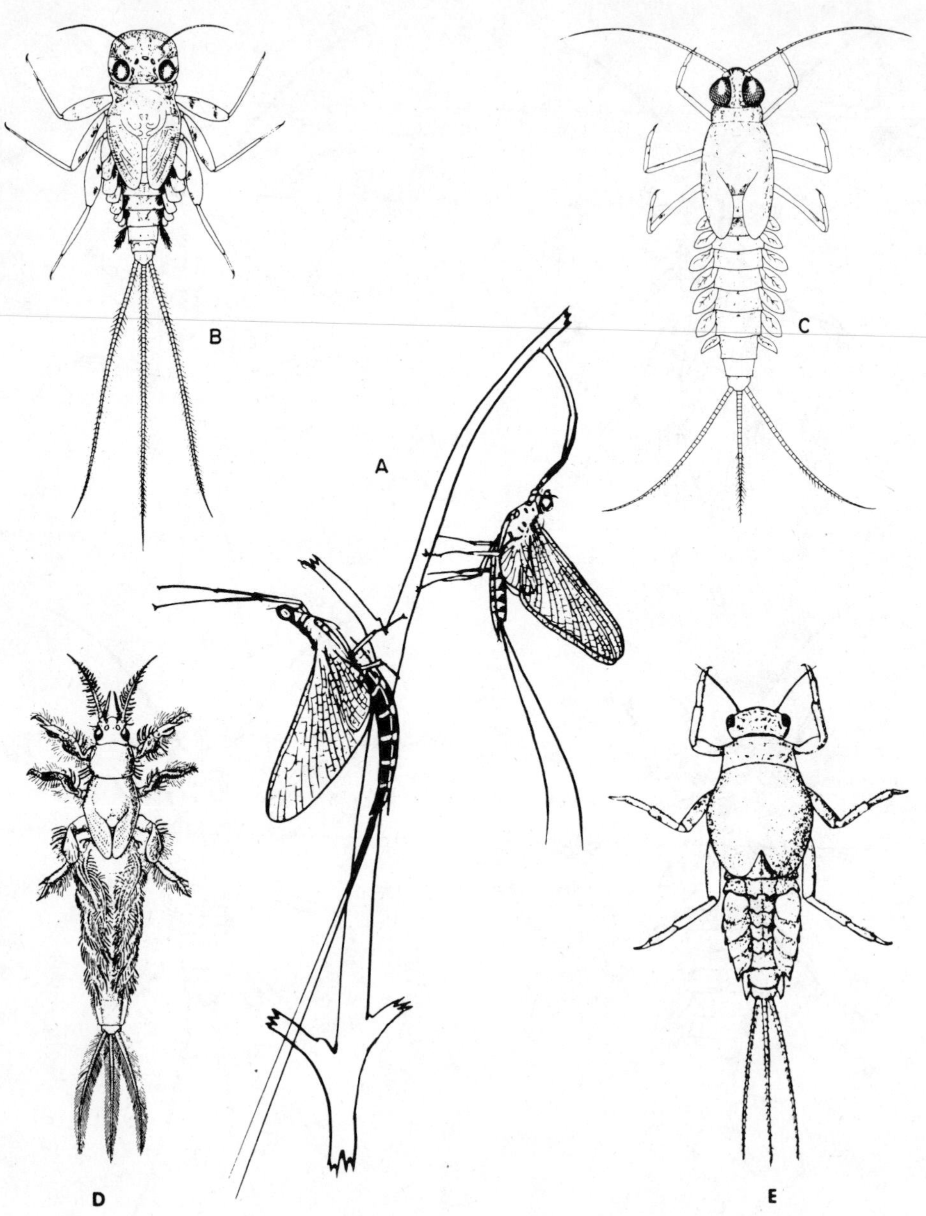

Plate 24. Mayflies (Order Ephemeroptera).

A—Adult mayfly, Heptageniidae (12-18 mm)
B—Nymph *Stenonema*, Heptageniidae (10-14 mm)
C—Nymph *Baetis*, Baetidae (7-14 mm)
D—Nymph *Hexagenia*, Ephemeridae (20-30 mm)
E—Nymph *Ephemerella*, Ephemerellidae (8-15 mm)

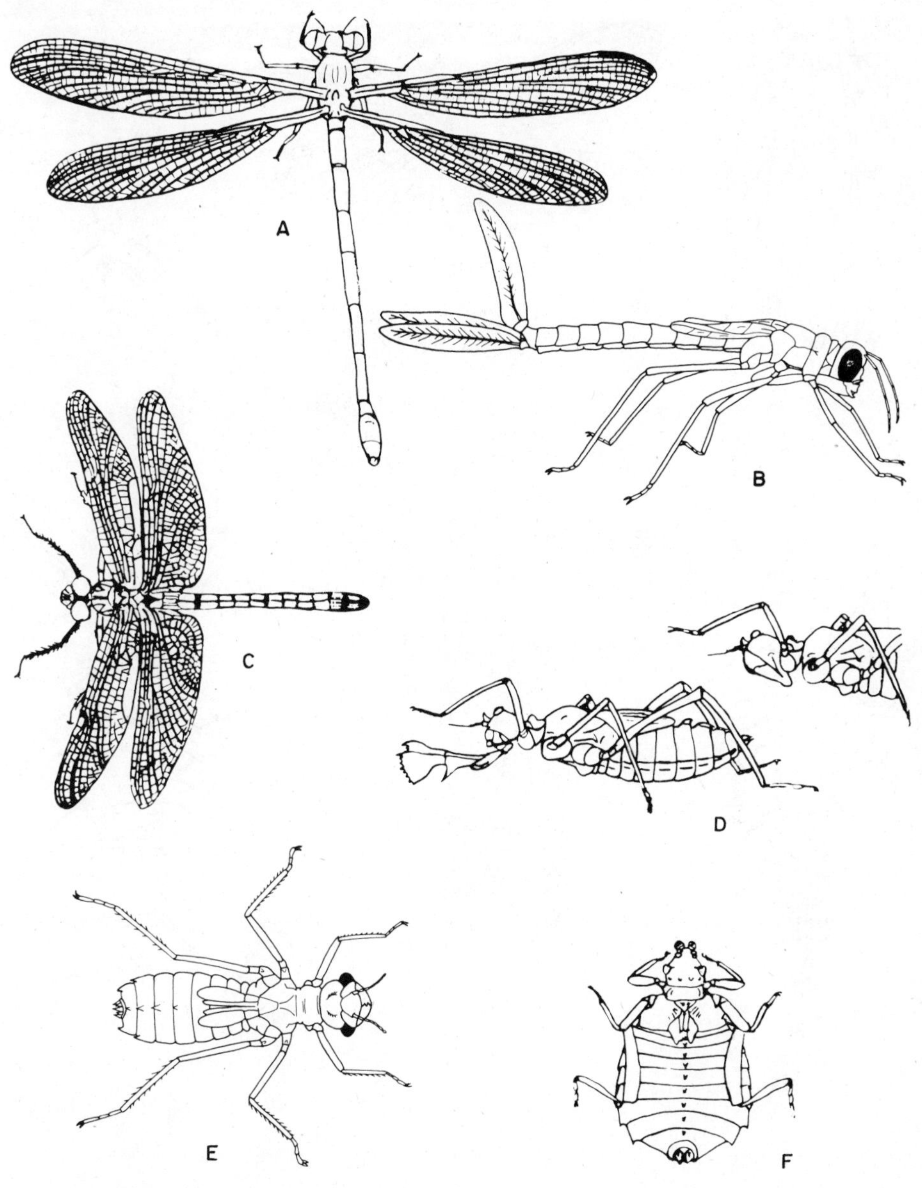

Plate 25. Damselflies, dragonflies (Order Odonata)

A—Adult damselfly (35-55 mm)

B—Damselfly nymph *Lestes,* Coenagrionidae (20-30 mm)

C—Adult dragonfly *Macromia,* Libellulidae (50-70 mm)

D—Dragonfly nymph *Macromia,* showing "mask" both extended and contracted, Libellulidae (15-45 mm)

E—Dragonfly nymph *Helocordulia,* Libellulidae (15-45 mm)

F—Dragonfly nymph *Hagenius,* Gomphidae (15-20 mm)

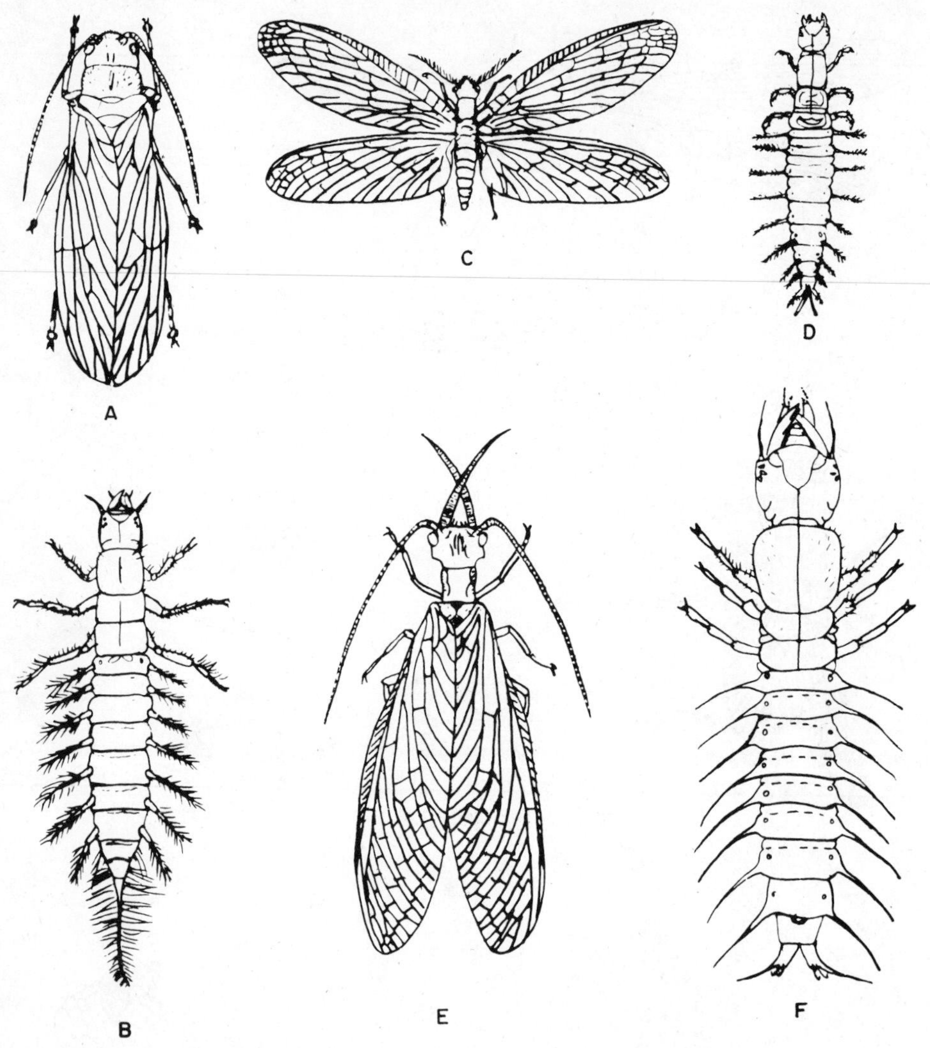

Plate 26. Hellgrammite and relatives.

A—Adult alderfly *Sialis,* Sialidae (9–15 mm)
B—Alderfly larva *Sialis,* Sialidae (15–30 mm)
C—Adult fishfly *Chauliodes,* Corydalidae (15–30 mm)
D—Fishfly larva *Chauliodes,* Corydalidae (20–40 mm)
E—Adult dobsonfly, *Corydalus* (25–70 mm)
F—Dobsonfly larva or hellgrammite, *Corydalus* (25–90 mm)

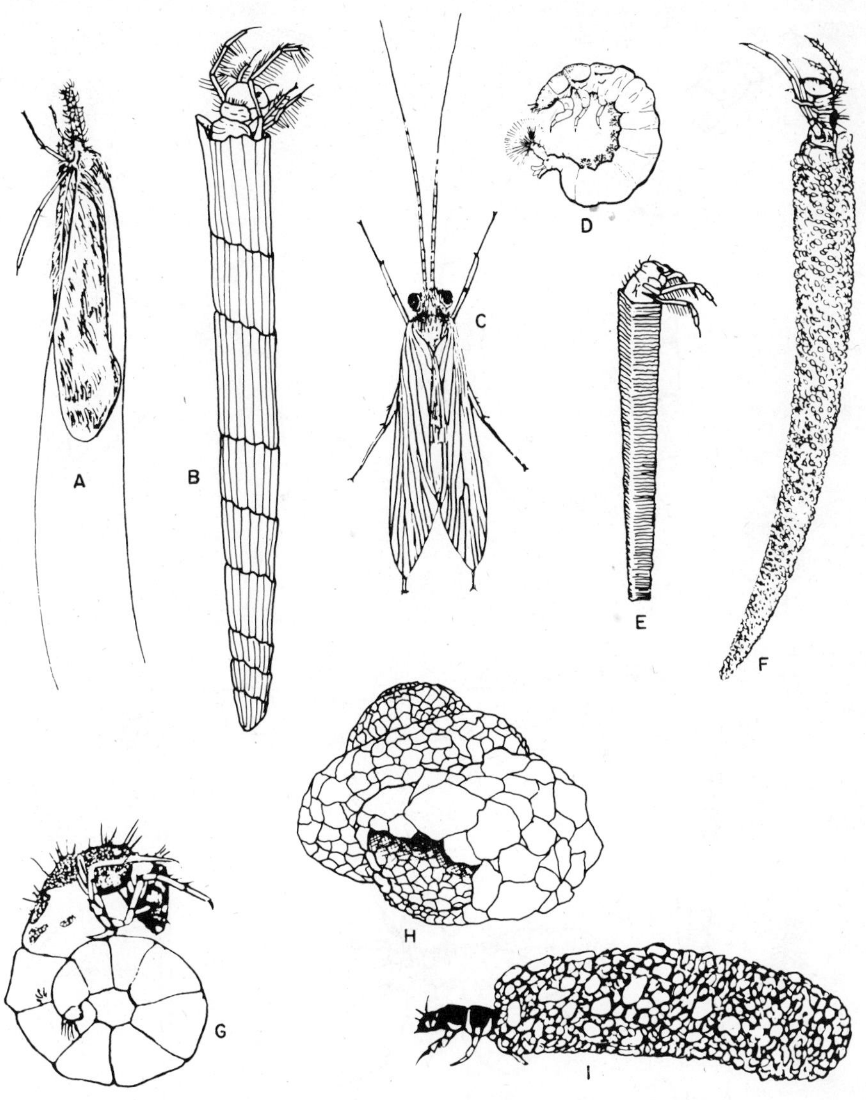

Plate 27. Caddisflies (Order Trichoptera).

A—Adult *Triaenodes,* Leptoceridae (10–20 mm)

B—Larva and case, *Triaenodes,* Leptoceridae (10–14 mm)

C—Adult *Hydropsyche,* Hydropsychidae (20–30 mm)

D—*Hydropsyche* larva, Hydropsychidae (20–30 mm)

E—Larva and case, *Brachycentrus,* Brachycentridae (12–16 mm)

F—Larva and case, *Leptocella,* Leptoceridae (14–18) mm

G—*Helicopsyche* larva, Helicopsychidae (6–10 mm)

H—*Helicopsyche* case, Helicopsychidae (4–6 mm)

I—Larva and case, *Ochrotricha,* Hydroptilidae (4–6 mm)

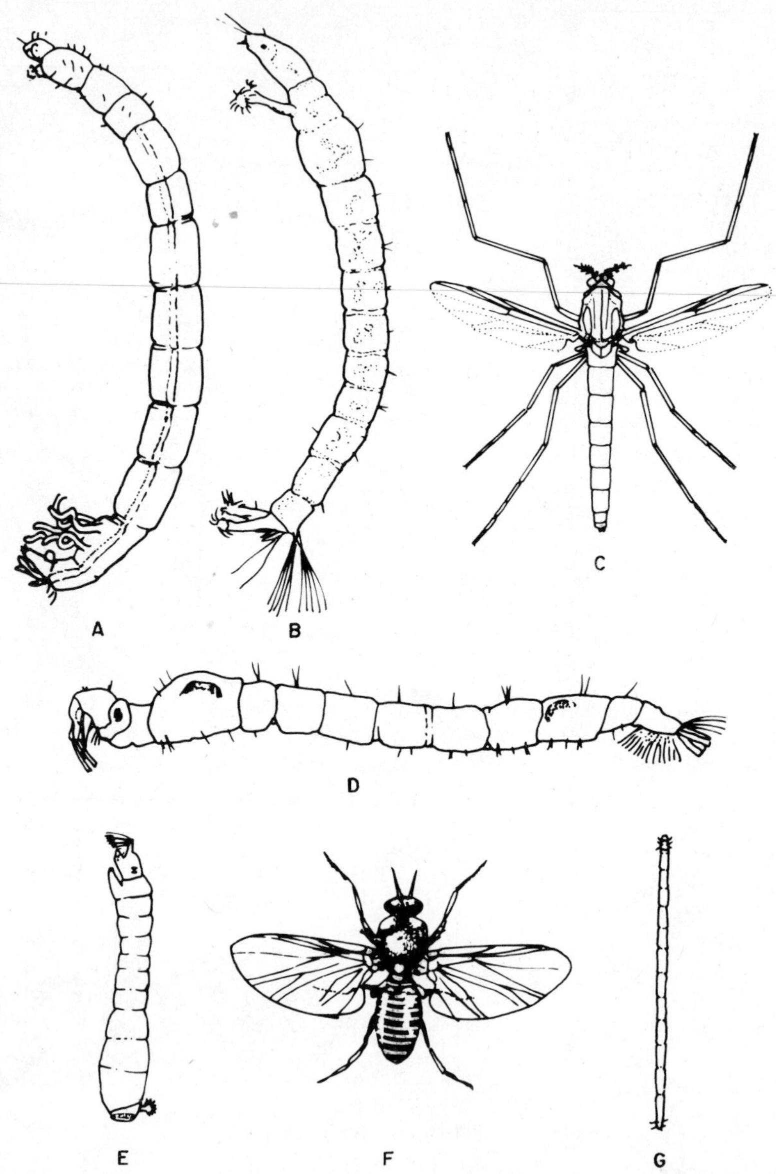

Plate 28. Two-winged flies (Order Diptera).

A—Larva midge *Chironomus*, Chironomidae (5–30 mm)

B—Larva midge *Ablabesmyia*, Chironomidae (5–10 mm)

C—Adult midge, Chironomidae (4–12 mm)

D—Larva phantom midge *Chaoborus*, Culicidae (8–12 mm)

E—Larva black fly *Simulium*, Simuliidae (3–8 mm)

F—Adult black fly *Simulium*, Simuliidae (2–6 mm)

G—Larva biting midge, Ceratopogonidae (3–12 mm)

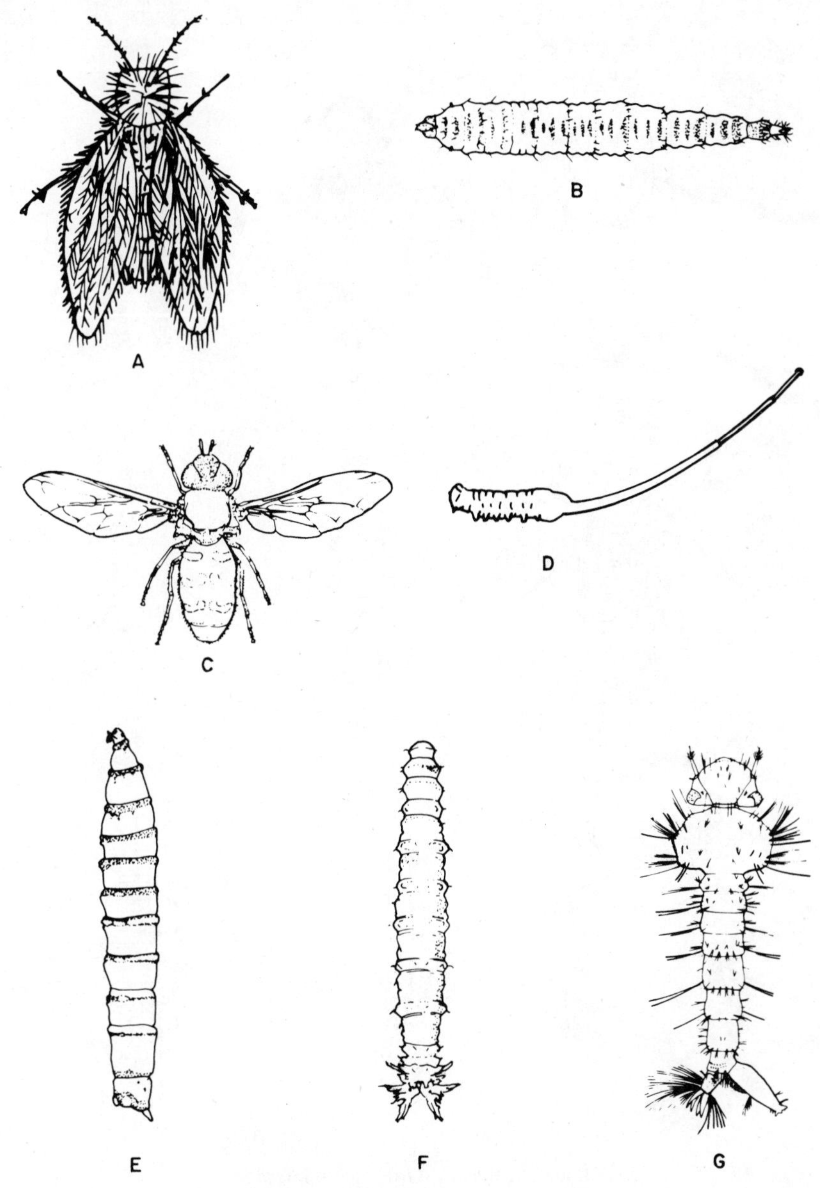

Plate 29. Two-winged flies (Order Diptera)

A—Adult sewage fly *Psychoda,* Psychodidae (2-5 mm)

B—Larva sewage fly *Psychoda,* Psychodidae (4-6 mm)

C—Adult drone fly, Syrphidae (10-15 mm)

D—Rat-tailed maggot *Eristalis,* Syrphidae (15-30 mm)

E—*Tabanus* larva, Tabanidae (30-40 mm)

F—Larva cranefly *Tipula,* Tipulidae (30-40 mm)

G—Larva mosquito *Aedes,* Culicidae (10-15 mm)

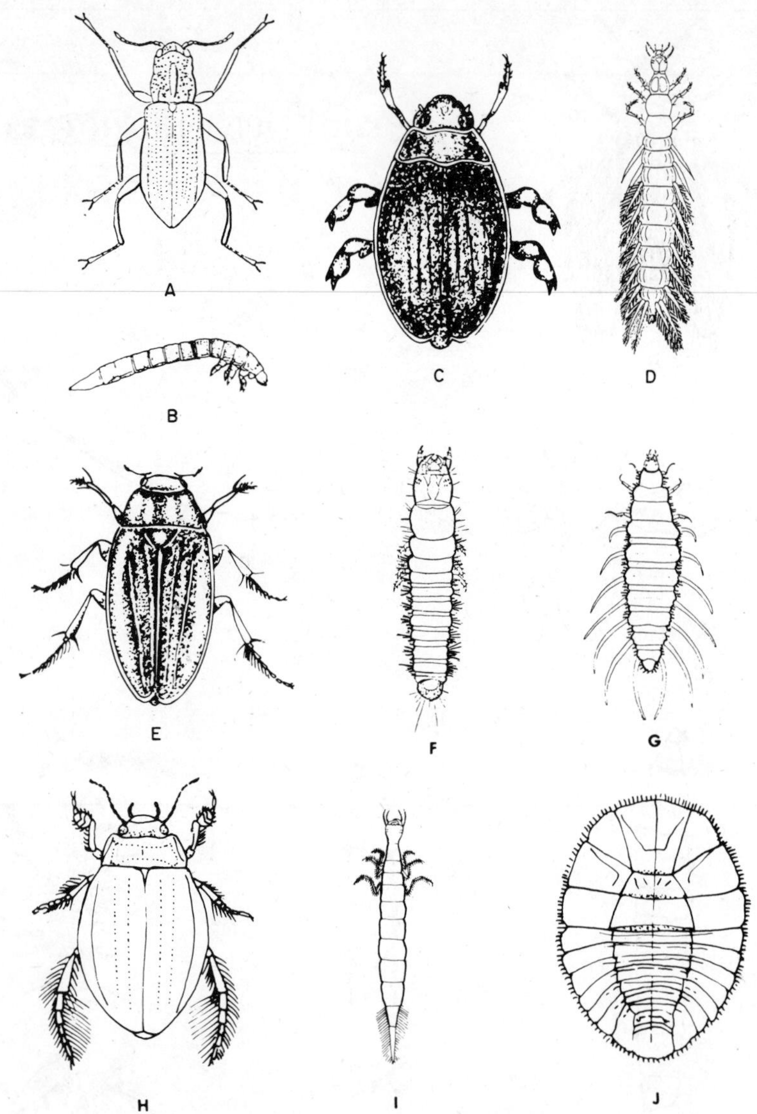

Plate 30. Beetles (Order Coleoptera).

A—Adult riffle beetle *Stenelmis*, Elmidae (2-5 mm)
B—Larva *Narpus*, Elmidae (4-10 mm)
C—Adult whirligig beetle *Dineutus*, Gyrinidae (7-15 mm)
D—Larva *Dineutus*, Gyrinidae (10-30 mm)
E—Adult water scavenger beetle *Hydrophilus*, Hydrophilidae (2-40 mm)
F—Larva *Berosus*, Hydrophilidae (5-20 mm)
G—Larva *Enochrus*, Hydrophilidae (10-25 mm)
H—Adult predacious diving beetle *Dytiscus*, Dytiscidae (2-40 mm)
I—Larva *Cybister*, Dytiscidae (10-25 mm)
J—Larva water penny *Psephenus*, Psephenidae (3-10 mm)

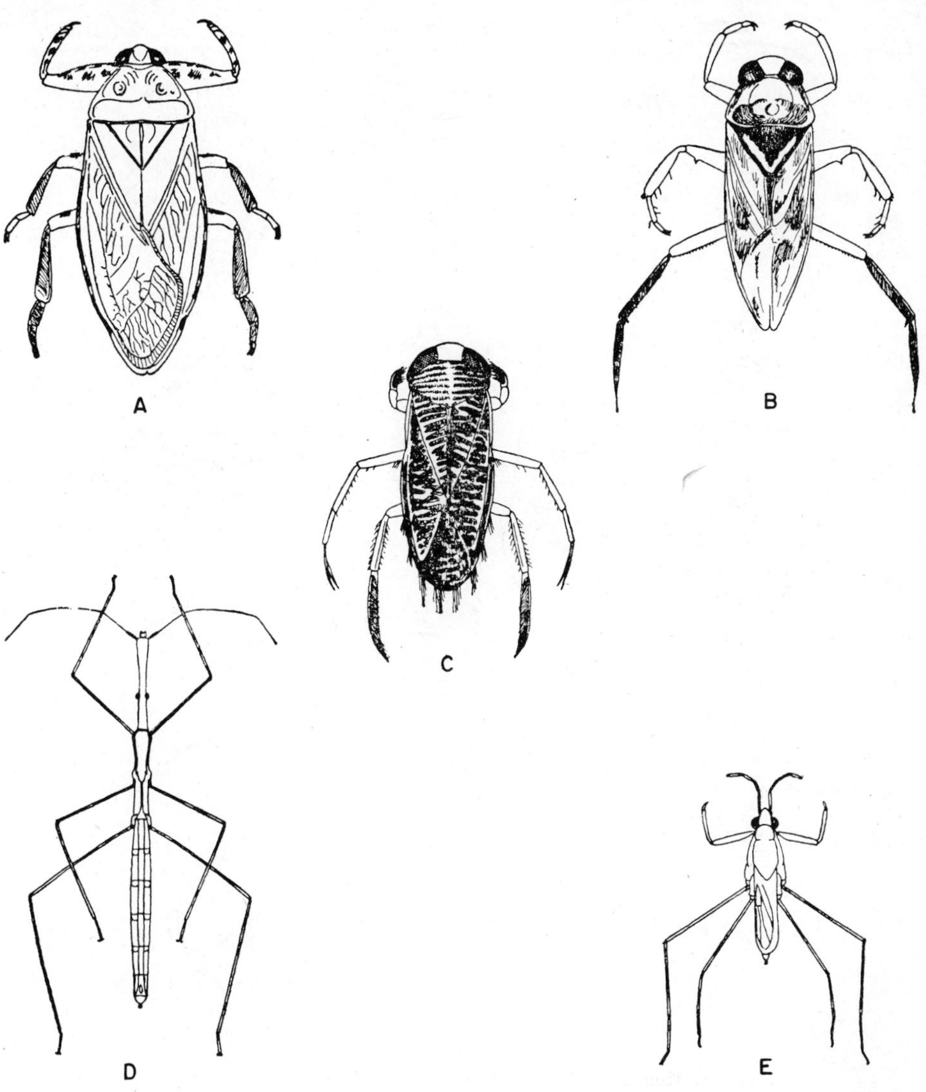

Plate 31. True bugs (Order Hemiptera, all adults).

A—Electric light bug, *Lethocerus,* Belostomidae (20–70 mm)

B—Backswimmer, *Notonecta,* Notonectidae (5–17 mm)

C—Water boatman *Sigara,* Corixidae (3–12 mm)

D—Marsh treader *Hydrometra,* Hydrometridae (8–11 mm)

E—Water strider *Gerris,* Gerridae (2–15 mm)

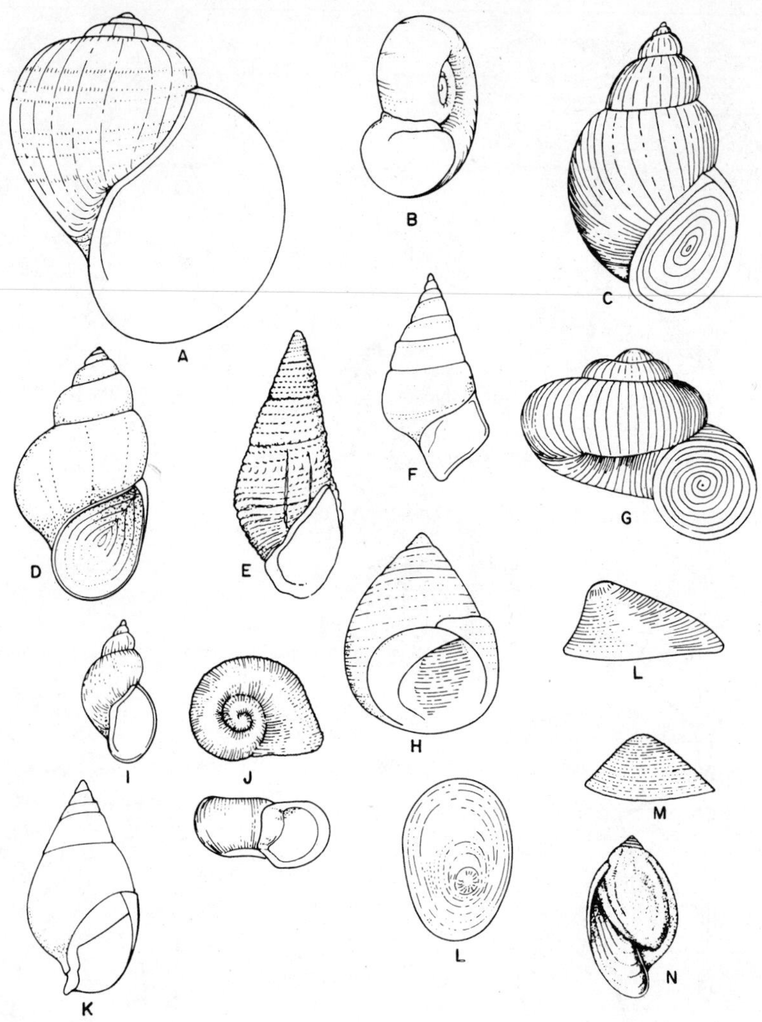

Plate 32. Mollusks (Phylum Mollusca): Snails (Class Gastropoda).

Gill-breathing families:

A—Apple snail *Pomacea*, Pilidae (5 cm)
B—*Marisa*, Pilidae (15 mm)
C—*Campeloma*, Viviparidae (4 cm)
D—Faucet snail *Bithynia*, Amnicolidae (2 cm)
E—*Tarebia*, Thiaridae (15 mm)
F—River snail *Pleurocera*, Pleuroceridae (3 cm)
G—*Valvata*, Valvatidae (1 cm)
H—Periwinkle *Littorina*, Littorinidae (marine, 2 cm)

Lung breathers:

I—Pond snail *Lymnaea*, Lymnaeidae (15 mm)
J—Orb snail *Helisoma*, Planorbidae (1 cm)
K—Mud snail *Nassa*, Nassidae (marine, 2 cm)
L—Limpet *Ferrissia*, Ancylidae (2 mm)
M—Limpet *Lanx*, Lancidae (10 mm)
N—Pouch snail *Physa*, Physidae (5 mm)

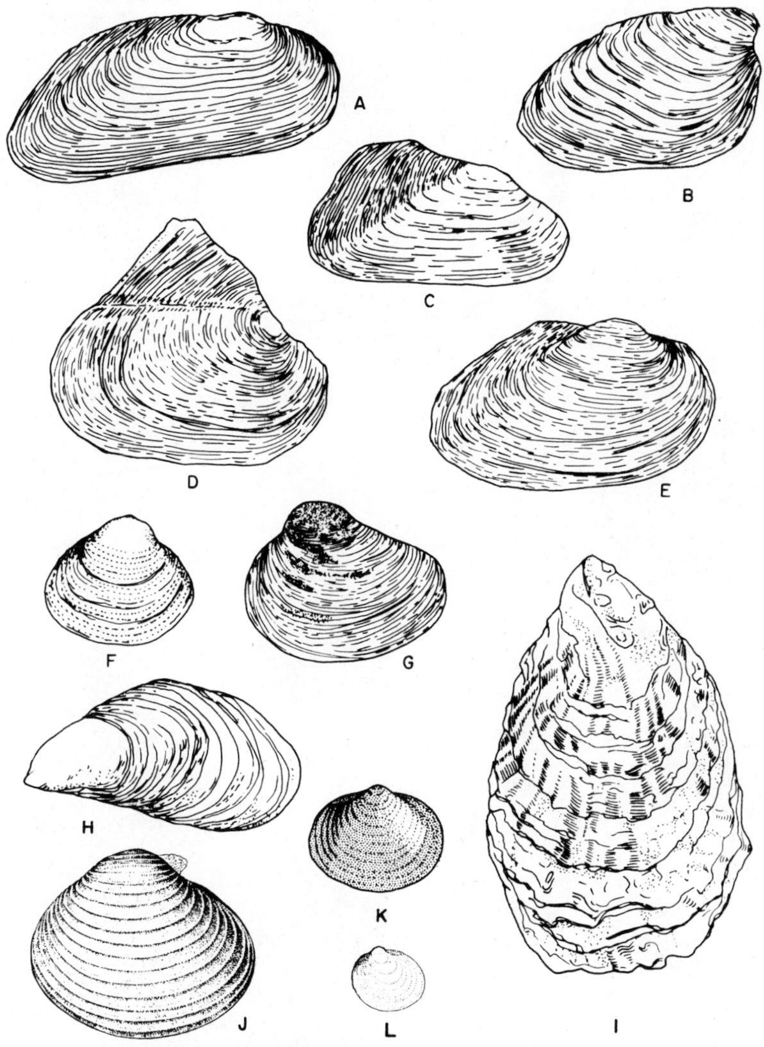

Plate 33. Mollusks (Phylum Mollusca): Bivalves (Class Pelecypoda)

A—Spectacle case *Margaritifera*, Margaritiferidae (10 cm)
B—Pearly mussel *Pleurobema*, Unionidae (10 cm)
C—Pearly mussel *Gonidea*, Unionidae (10 cm)
D—Winged lampshell *Proptera*, Lampsilinae (13 cm)
E—Papershell *Anodonta*, Anodontidae (14 cm)
F—Marsh clam *Polymesoda*, Corbiculidae (marine, 4 cm)
G—Rangia clam *Rangia*, Mactridae (marine, 5 cm)
H—Edible mussel *Mytilus*, Mytilidae (marine, 6 cm)
I—Oyster *Crassostrea*, Ostreidae (marine, 9 cm)
J—Asiatic clam *Corbicula*, Corbiculidae (4 cm)
K—Fingernail clam *Sphaerium*, Sphaeriidae (1 cm)
L—Peashell clam *Pisidium*, Sphaeriidae (5 mm)

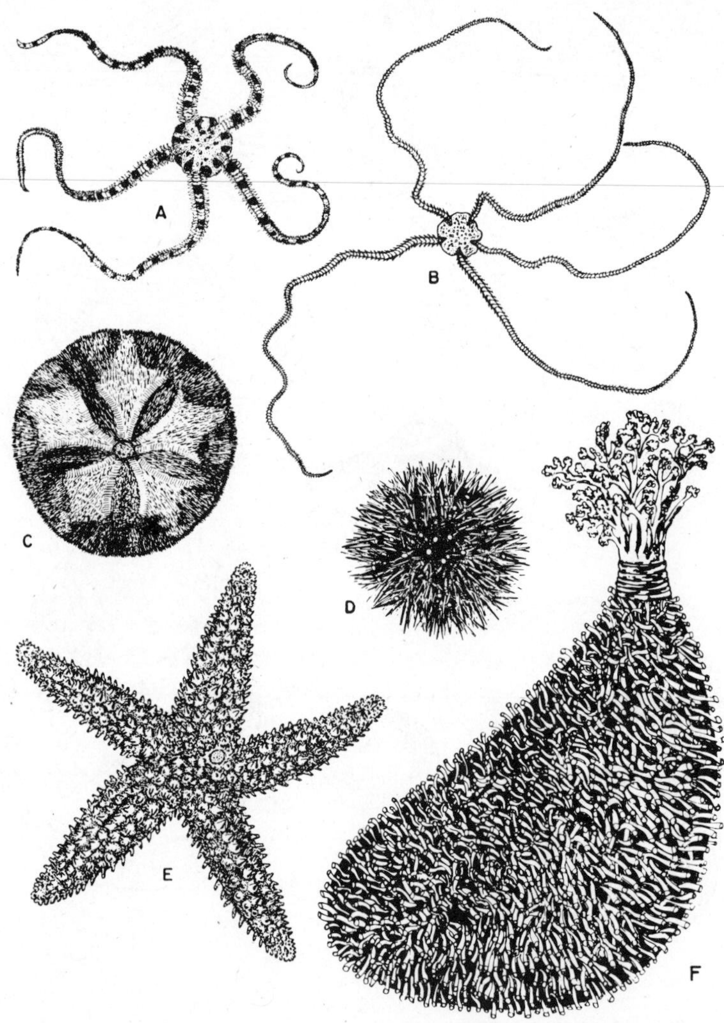

Plate 34. Echinoderm types (Phylum Echinodermata, all marine).

A—Brittle star, class Ophiuroidea: *Ophiopholis* (disc 15 mm)

B—Brittle star, class Ophiuroidea: *Amphioplus* (disc 5 mm)

C—Sand dollar, class Echinoidea: *Echinarachnius* (7 cm)

D—Sea urchin, class Echinoidea: *Strongylocentrotus* (6 cm)

E—Starfish, class Asteroidea: *Asterias* (15 cm)

F—Sea cucumber, class Holothuroidea: *Thyone* (10 cm)

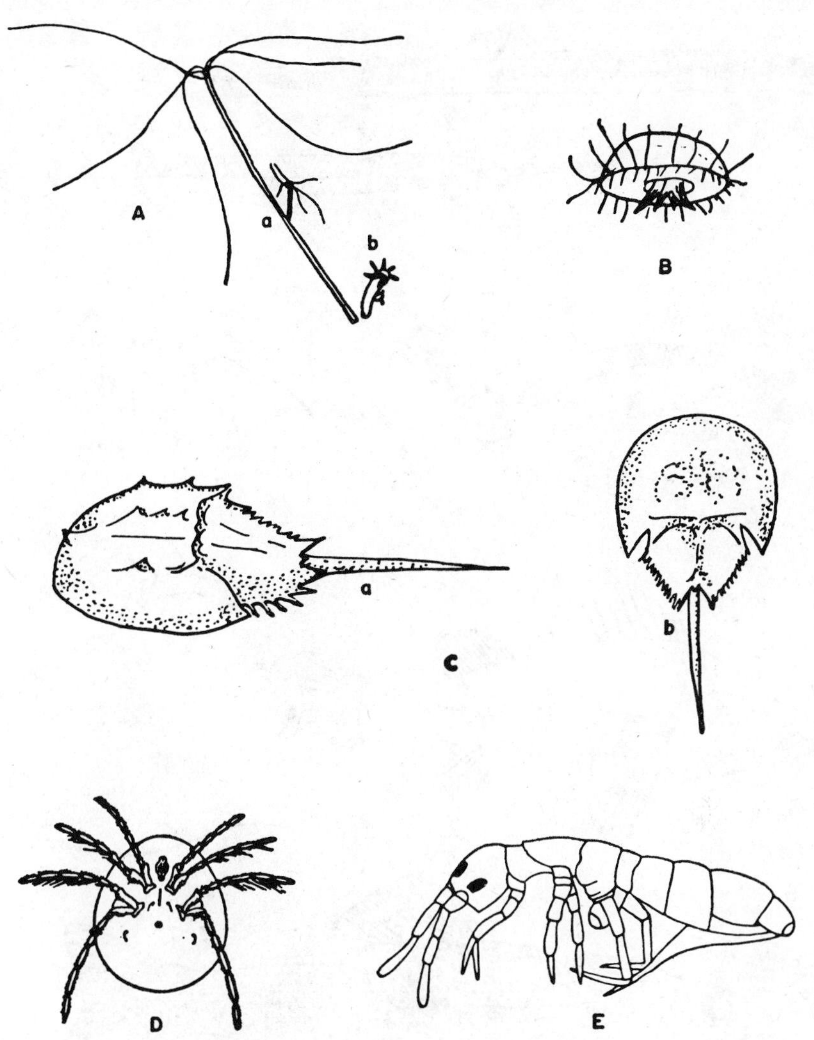

Plate 35. Miscellaneous invertebrates.

Freshwater coelenterates (Phylum Coelenterata):

A—*Hydra*, at (a) extended (2 cm) with bud, and at (b) contracted

B—Jellyfish (Medusa) stage of *Craspedacusta* (2 cm)

Arthropods (Phylum Arthropoda):

C—Horseshoe crab (Class Arachnoidea), marine: *Limulus* (30 cm); (a) shows side view and (b) top view.

D—Water mite (Class Arachnoidea): *Limnochares* (3 mm)

E—Springtail (Class Insecta, Order Collembola): *Orchesella* (2 mm)

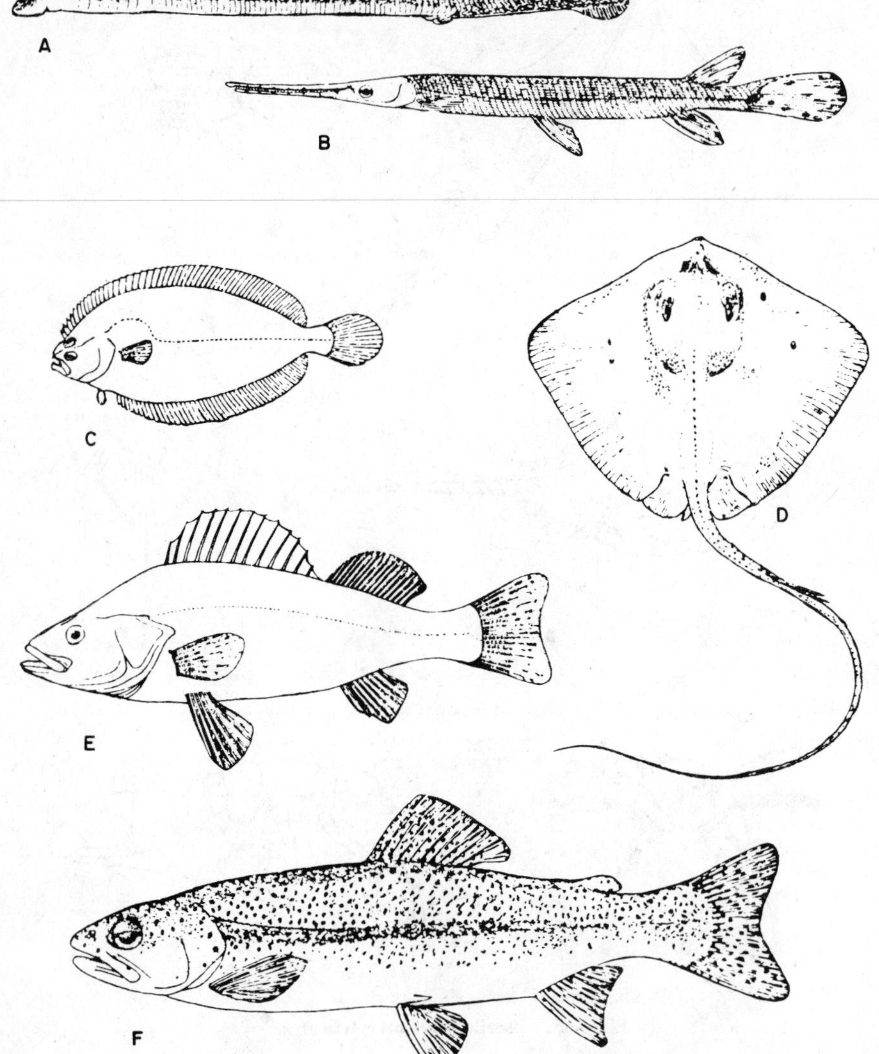

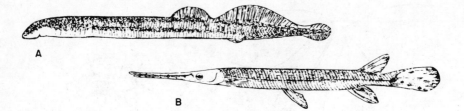

Plate 36. Some types of fishes (Phylum Chordata).

A—Jawless fish (Class Agnatha): lamprey, *Petromyzon* (30–45 cm)

B—Ganoid fish (Class Osteichthys, or Pisces): long-nosed gar, *Lepisosteus* (2.4 m)

C—Flatfish (Class Osteichthys): flounder, *Paralichthys* (30–60 cm)

D—Cartilage fish (Class Chondrichthys): stingray, *Dasyatis* (2 m)

E—Spiny-rayed fish (Class Osteichthys): perch, *Perca* (30 cm)

F—Soft-rayed fish (Class Osteichthys): rainbow trout, *Salmo* (30 cm)

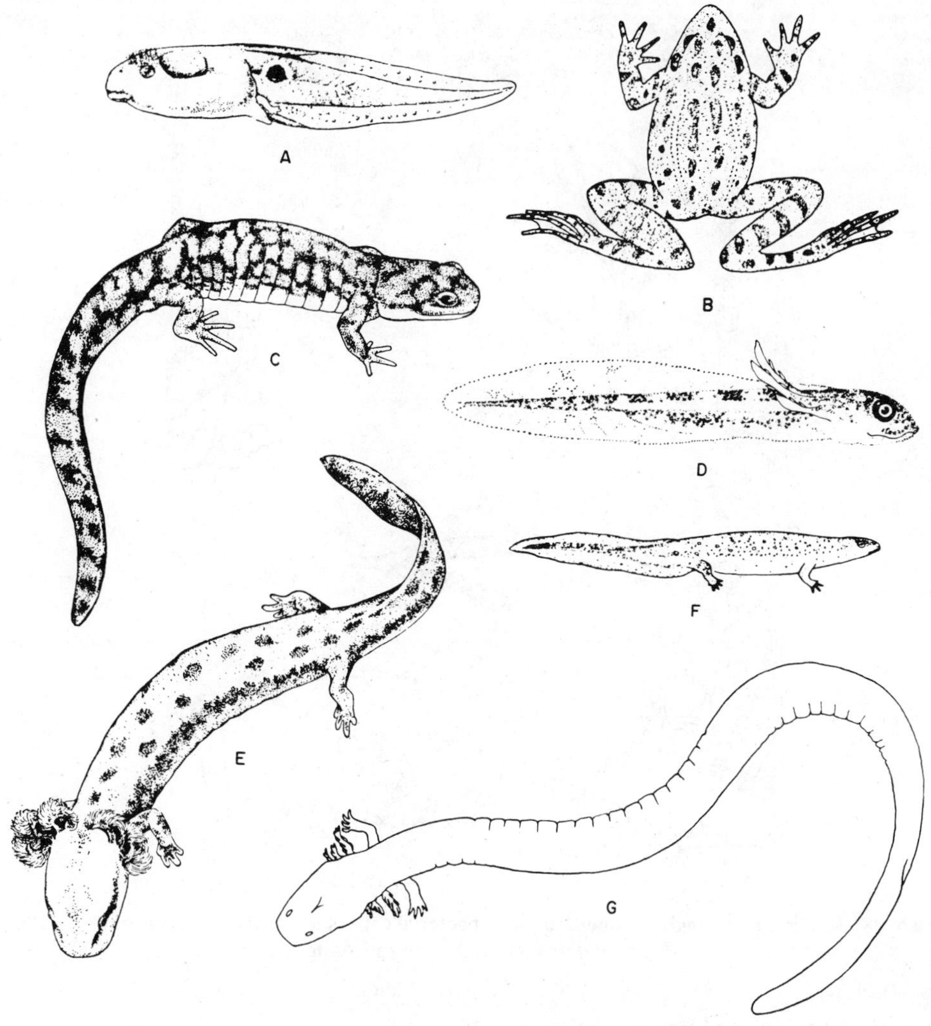

Plate 37. Types of amphibians (Phylum Chordata, Class Amphibia)

Frogs and toads (Order Salientia):

A—The "tadpole" larva (note the developing leg protruding from the body of the tadpole)

B—An adult frog, *Rana* (20 cm). Salientia with dry warty skins are usually called toads.

Salamanders (Order Caudata):

C—*Ambystoma* (20 cm). Adult is typically terrestrial.

D—*Ambystoma* larva is aquatic. Salamander larvae typically have gills.

E—Water dog or mud puppy, *Necturus* (to 30 cm). Larval gills are retained by the adult.

F—An adult aquatic salamander with a flat tail, *Diemictylus* (9 cm).

G—An aquatic salamander. The hind legs have been lost; *Siren* (1 m).

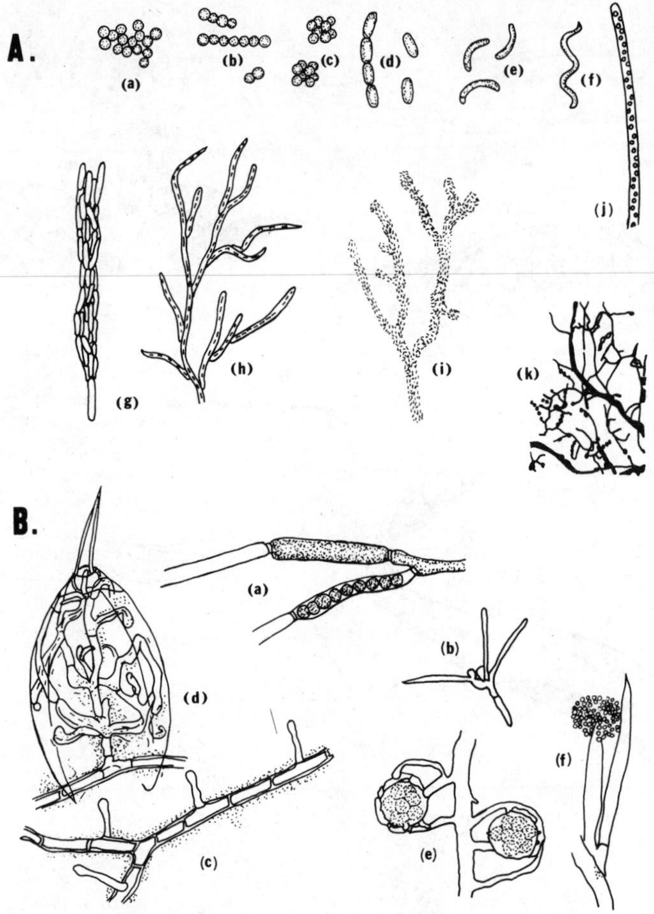

Plate 38. Bacteria and Fungi. (Diameter of most bacterial cells is less than 2 μm, though *Beggiatoa* may range up to 16 μm in diameter, and be of indefinite length.)

A—Bacteria:

(Cellular forms and arrangements)

(a) micrococcus
(b) streptococcus
(c) sarcina
(d) bacillus
(e) vibrio
(f) spirillum

(Sewage organisms)

(g) *Sphaerotilus* ("sewage fungus") cells
(h) A *Sphaerotilus* growth form
(i) A growth form of *Zoogloea*
(j) *Beggiatoa* (sulfur bacterium)
(k) An actinomycete growth form from compost

B—Fungi:

(a) *Leptomitus,* showing zoospores and cellulin plugs (diameter 8.5–16 μm)
(b) *Tetracladium* (diameter 2.5–3.5 μm)
(c) *Zoophagus,* showing mycelial pegs
(d) *Zoophagus* with rotifer impaled on mycelial peg (diameter 3 μm)
(e) *Achlya,* showing oospores
(f) *Achlya,* showing extruded encysted zoospores (Oogonia 50–60 μm, oospores 18.5–22 μm, encysted zoospores 3–5 μm)

1007 D. Key for Identification of Freshwater Algae Common in Water Supplies and Polluted Waters (Color Plates A–F)

By C. Mervin Palmer

Beginning with 1a and 1b, choose one of the two contrasting statements and follow this procedure with the "a" and "b" statements of the number given at the end of the chosen statement. Continue until the name of the alga is given instead of another key number. (Where recent changes in names of algae have been made, the new name is given followed by the old name in parenthesis.)

		Refer to Couplet No.
1a.	Plastid (separate color body) absent; complete protoplast pigmented; generally blue-green; iodine starch test* negative (blue-green algae)	4
1b.	Plastid or plastids present; parts of protoplast free of some or all pigments; generally green, brown, red, etc., but not blue-green; iodine starch test* positive or negative	2
2a.	Cell wall permanently rigid (never showing evidence of collapse), and with regular pattern of fine markings (striations, etc.); plastids brown to green; iodine starch test* negative; flagella absent; wall of two essentially similar halves, one placed over the other as a cover (diatoms)	29
2b.	Cell wall, if present, capable of sagging, wrinkling, bulging, or rigidity, depending on existing turgor pressure of cell protoplast; regular pattern of fine markings on wall generally absent; plastids green, red, brown, etc.; iodine starch test* positive or negative; flagella present or absent; cell wall continuous and generally not of two parts	3
3a.	Cell or colony motile; flagella present (often not readily visible); anterior and posterior ends of cell different from one another in contents and often in shape (flagellate algae)	51
3b.	Nonmotile; true flagella absent; ends of cells often not differentiated (green algae and associated forms)	77

1. Blue-Green Algae

4a.	Cells in filaments (or much elongated to form a thread)	5
4b.	Cells not in (or as) filaments	23
5a.	Heterocysts present	6
5b.	Heterocysts absent	14
6a.	Heterocyst located at one end of filament	7
6b.	Heterocysts at various locations in filament	9
7a.	Filaments radially arranged in a gelatinous bead	*Rivularia*
7b.	Filaments isolated or irregularly grouped	8
8a.	Filament gradually narrowed to one end	*Calothrix*
8b.	Filament not gradually narrowed to one end	*Cylindrospermum*
9a.	Filament unbranched	10
9b.	Filament with occasional (false) branches	13
10a.	Crosswalls in filament much closer together than width of filament	*Nodularia*

*Add 1 drop of Lugol's (iodine) solution, diluted 1:1 with distilled water. In about 1 min, if positive, starch is stained blue and later black. Other structures (such as nucleus, plastids, cell wall) may also stain, but turn brown to yellow.

		Refer to Couplet No.
10b.	Crosswalls in filament at least as far apart as width of filament	11
11a.	Filaments normally in tight parallel clusters; heterocysts and spores cylindric to long oval in shape . . . *Aphanizomenon*	
11b.	Filaments not in tight parallel clusters; heterocysts and spores often round to oval	12
12a.	Filaments in a common gelatinous mass . . . *Nostoc*	
12b.	Filaments not in a common gelatinous mass . . . *Anabaena*	
13a.	False branches in pairs . . . *Scytonema*	
13b.	False branches, single . . . *Tolypothrix*	
14a.	Filament or elongated cell attached at one end, with one or more round cells (spores) at the other . . . *Entophysalis (Chamaesiphon)*	
14b.	Filament generally not attached at one end; no terminal spores present	15
15a.	Filament with regular spiral form throughout	16
15b.	Filament not spiral, or with spiral form limited to a portion of filament	17
16a.	Filament septate . . . *Arthrospira*	
16b.	Filament nonseptate . . . *Spirulina*	
17a.	Filament very narrow, only 0.5 to 2.0 μm wide . . . *Schizothrix*	
17b.	Filament 3 to 95 μm wide	18
18a.	Filaments loosely aggregated or not in clusters	19
18b.	Filaments tightly aggregated and surrounded by a common gelatinous secretion that may be invisible	22
19a.	Filament surrounded by wall-like sheath that frequently extends beyond the ends of the filament of cells; filament generally without movement	20
19b.	Filament not surrounded by a wall-like sheath; filament may show movement	21
20a.	Cells separated from one another by a space . . . *Johannesbaptistia*	
20b.	Cells in contact with adjacent cells . . . *Lyngbya*	
21a.	All filaments short, with less than 20 cells; one or both ends of filament sharply pointed . . . *Raphidiopsis*	
21b.	Filaments long, with more than 20 cells; filaments commonly without sharp-pointed ends . . . *Oscillatoria*	
22a.	Filaments arranged in a tight, essentially parallel bundle . . . *Microcoleus*	
22b.	Filaments arranged in irregular fashion, often forming a mat . . . *Phormidium*	
23a.	Cells in a regular pattern of parallel rows, forming a plate . . . *Agmenellum (Merismopedia)*	
23b.	Cells not regularly arranged to form a plate	24
24a.	Cells regularly arranged near surface of a spherical gelatinous bead	25
24b.	Gelatinous bead, if present, not spherical	26
25a.	Cells ovate to heart-shaped, connected to center of bead by colorless stalks . . . *Gomphosphaeria*	
25b.	Cells round, without gelatinous stalks . . . *Gomphosphaeria (Coelosphaerium* type)	
26a.	Cells cylindric-oval . . . *Coccochloris (Aphanothece)*	
26b.	Cells spherical	27
27a.	Two or more distinct layers of gelatinous sheath around each cell or cell cluster . . . *Anacystis (Gloeocapsa)*	
27b.	Gelatinous sheath around cells not distinctly layered	28
28a.	Cells isolated or in colonies of 2 to 32 cells . . . *Anacystis (Chroococcus)*	
28b.	Cells in colonies composed of many cells . . . *Anacystis (Microcystis, Polycystis)*	

2. Diatoms

29a.	Front (valve) view circular in outline; markings radial in arrangement; cells may form a filament (centric diatoms)	30

Refer to Couplet No.

29b. Front (valve) view elongate, not circular; transverse markings in one or two longitudinal rows; cells, if grouped, not forming a filament (pennate diatoms) 32

30a. Cells in persistent filaments with valve faces in contact; therefore, cells commonly seen in side (girdle) view . . . . *Melosira*

30b. Cells isolated or in fragile filaments, often seen in front (valve) view 31

31a. Radial markings (striations), in valve view, extending from center to margin; short spines often present around margin (valve view) . . . . *Stephanodiscus*

31b. Area of prominent radial markings, in valve view, limited to approximately outer half of circle, marginal spines generally absent . . . . *Cyclotella*

32a. Cell longitudinally symmetrical in valve view 33

32b. Cell longitudinally unsymmetrical (two sides unequal in shape), at least in valve view 49

33a. Raphe at or near the edge of the valve 34

33b. Raphe or pseudoraphe median or submedian 35

34a. Marginal, keeled raphe areas lie opposite one another on the two valves . . . . *Hantzschia*

34b. Marginal, keeled raphe areas lie diagonal to one another on the two valves . . . . *Nitzschia*

35a. Cell transversely symmetrical in valve view 36

35b. Cell transversely unsymmetrical (two ends unequal in shape or size), at least in valve view 44

36a. Cell round-oval in valve view, not more than twice as long as it is wide. *Cocconeis*

36b. Cell elongate, more than twice as long as it is wide 37

37a. Cell flat (girdle face wide, valve face narrow) . . . . *Tabellaria*

37b. Girdle and valve faces about equal in width 38

38a. Cell with several markings (septa) extending without interruption across the valve face; no marginal line of pores present . . . . *Diatoma*

38b. Cross-markings (striations or costae) on valve surface, interrupted by either longitudinal space (pseudoraphe), or line (raphe), or line of pores (carinal dots) . . . 39

39a. Cells attached side by side to form a ribbon of several to many cells . . . . *Fragilaria*

39b. Cells isolated or in pairs 40

40a. Cell narrow, linear, often narrowed to both ends; true raphe absent . . *Synedra*

40b. Cell commonly "boat-shape" in valve view; true raphe present 41

41a. Cell longitudinally unsymmetrical in girdle view; sometimes with attachment stalk . . . . *Achnanthes*

41b. Cell symmetrical in girdle as well as valve view; generally not attached 42

42a. Area without striations extending as a transverse belt around middle of cell . . . . *Stauroneis*

42b. No continuous clear belt around middle of cell 43

43a. Cell with coarse transverse markings (costae), which appear as solid lines even under high magnification . . . . *Pinnularia*

43b. Cell with fine transverse markings (striae), which appear as lines of dots under high magnification . . . . *Navicula*

44a. Cells attached together at one end only to form radiating colony . . *Asterionella*

44b. Cell not forming a loose radiating colony 45

45a. Cells in fan-shaped colonies . . . . *Meridion*

45b. Cells isolated or in pairs 46

46a. Prominent wall markings in addition to striations present just below lateral margins on valve surface of cell . . . . *Surirella*

46b. Wall markings along sides of valve limited to striations 47

47a. Cell elongate, sides almost parallel except for terminal knobs . . . . . . *Asterionella*

Couplet	Description	Refer to Couplet No.
47b.	Sides of cell converging toward one end	48
48a.	Cells bent in girdle view	*Rhoicosphenia*
48b.	Cells straight in girdle view	*Gomphonema*
49a.	Valves with transverse septa or costae	*Epithemia*
49b.	Valves with no transverse septa or costae	50
50a.	Raphe located almost through center of valve	*Cymbella*
50b.	Raphe excentric, near concave edge of valve	*Amphora*

3. Flagellate Algae

Couplet	Description	Refer to Couplet No.
51a.	Cell in a loose, rigid conical sac (lorica); isolated or in a branching colony	*Dinobryon*
51b.	Case or sac, if present, not conical; colony, if present, not branching	52
52a.	Cells isolated or in pairs	53
52b.	Cells in a colony or four or more cells	71
53a.	Prominent transverse groove encircles cell	54
53b.	Cell without transverse groove	56
54a.	Cell with prominent rigid projections, one forward and two or three on posterior end	*Ceratium*
54b.	Cell without several rigid polar projections	55
55a.	Portions above and below transverse groove about equal	*Peridinium*
55b.	Front portion distinctly larger than posterior portion	*Massartia*
56a.	Cell with long bristles extending from surface plates	*Mallomonas*
56b.	Cell without bristles and surface plates	57
57a.	Cell protoplast enclosed in loose rigid covering (lorica)	58
57b.	Cell with tight membrane or wall but no loose rigid covering	60
58a.	Lorica flattened; cell with two flagella	*Phacotus*
58b.	Lorica not flattened; cell with one flagellum	59
59a.	Lorica often opaque, generally dark brown to red; plastid green	*Trachelomonas*
59b.	Lorica often transparent, colorless to light brown; plastid light brown	*Chrysococcus*
60a.	Plastids brown to red to olive or blue-green	61
60b.	Plastids grass green	64
61a.	Plastids blue-green to blue	*Chroomonas*
61b.	Plastids brown to red to olive green	62
62a.	Plastid brown; one or two flagella	63
62b.	Plastids red, red-brown, or olive green; two flagella	*Rhodomonas*
63a.	Anterior end of cell oblique; two flagella	*Cryptomonas*
63b.	Anterior end of cell rounded or pointed; one flagellum	*Chromulina*
64a.	Cell with colorless rectangular wing	*Pteromonas*
64b.	No wing extending from cell	65
65a.	Cells flattened; margin rigid	*Phacus*
65b.	Cell not flattened; margin rigid or flexible	66
66a.	Pyrenoid present in the single plastid; no paramylon; margin not flexible; two or more flagella per cell	67
66b.	Pyrenoid absent; paramylon present; several plastids per cell; margin flexible or rigid; one flagellum per cell	70
67a.	Cells fusiform (tapering at each end)	*Chlorogonium*
67b.	Cells not fusiform, generally almost spherical	68
68a.	Plastids numerous	*Vacuolaria*
68b.	Plastids few, commonly one	69
69a.	Two flagella per cell	*Chlamydomonas*

Refer to Couplet No.

69b. Four flagella per cell . . . *Carteria*
  70a. Cell flexible in form; paramylon a capsule or disk; cell elongate . . . *Euglena*
  70b. Cell rigid in form; paramylon ring-shaped; cell almost spherical . . *Lepocinclis*
71a. Plastids brown . . . 72
71b. Plastids green . . . 73
  72a. Cells in contact with one another . . . *Synura*
  72b. Cells separated from one another by space. . . . *Uroglenopsis*
73a. Colony flat, one cell thick . . . *Gonium*
73b. Colony rounded, more than one cell thick . . . 74
  74a. Cells in contact with one another . . . 75
  74b. Cells separated from one another by space . . . 76
75a. Cells radially arranged . . . *Pandorina*
75b. Cells all facing one direction . . . *Pyrobotrys (Chlamydobotrys)*
  76a. Cells more than 400 per colony . . . *Volvox*
  76b. Cells less than 75 per colony . . . *Eudorina*

4. Green Algae and Associated Forms

77a. Cells jointed together to form a net . . . *Hydrodictyon*
77b. Cells not forming a net . . . 78
  78a. Cells attached side by side to form a plate or ribbon one cell wide and thick; number of cells commonly two, four, or eight . . . *Scenedesmus*
  78b. Cells not attached side by side . . . 79
79a. Cells isolated or in nonfilamentous or nontubular thalli . . . 80
79b. Cells in filaments or other tubular or threadlike thalli . . . 111
  80a. Cells isolated and narrowest at the center because of incomplete fissure (desmids) . . . 81
  80b. Cells isolated or in clusters but without central fissure . . . 84
81a. Each half of cell with three spinelike or pointed knobular extensions . . *Staurastrum*
81b. Cell margin with no such extensions . . . 82
  82a. Semicells with a median incision or depression . . . 83
  82b. Semicells with no median incision or depression . . . *Cosmarium*
83a. Margin with rounded lobes . . . *Euastrum*
83b. Margin with sharp-pointed teeth . . . *Micrasterias*
  84a. Cells elongate . . . 85
  84b. Cells round to oval or angular . . . 92
85a. Cell radiating from a central point . . . *Actinastrum*
85b. Cells isolated or in irregular clusters . . . 86
  86a. Cells with terminal spines . . . *Schroederia*
  86b. Cells without terminal spines . . . 87
87a. Cells with colorless attachment area at one end . . . *Cbaracium*
87b. No attachment area at one end of cell . . . 88
  88a. Plastids two per cell; unpigmented area across center of cell . . . *Closterium*
  88b. Cell with plastid that continues across the center . . . 89
89a. Cell 5 to 10 times as long as it is broad . . . 90
89b. Cell 2 to 4 times as long as it is broad . . . 91
  90a. Pyrenoid absent, or one per cell . . . *Ankistrodesmus*
  90b. Pyrenoids several per cell . . . *Closteriopsis*
91a. Cells semicircular; cell ends pointed but with no terminal spines . . . *Selenastrum*

Refer to Couplet No.

91b. Cells arcuate but less than semicircular; cell ends pointed and each with a short spine . . . *Closteridium*

92a. Cells regularly arranged in a tight, flat colony . . . *Pediastrum*

92b. Cells not in a tight, flat regular colony . . . 93

93a. Cells angular . . . 94

93b. Cells round to oval . . . 95

94a. Two or more spines at each angle . . . *Polyedriopsis*

94b. Spines none or less than two at each angle . . . *Tetraedron*

95a. Cells with long sharp spines . . . 96

95b. Long sharp spines absent . . . 99

96a. Cells round . . . 97

96b. Cells oval . . . 98

97a. Cells isolated . . . *Golenkinia*

97b. Cells in colonies . . . *Micractinium*

98a. Each cell end has one spine . . . *Diacanthos*

98b. Each cell end has more than one spine . . . *Chodatella*

99a. Colony of definite regular form, round to oval . . . 100

99b. Colony, if present, not a definite oval or sphere; or cells may be isolated . . . 104

100a. Colony a tight sphere of cells . . . 101

100b. Colony a loose sphere of cells enclosed by a common membrane . . . 102

101a. Sphere solid, slightly irregular, no connecting processes between cells . . . *Planktosphaeria*

101b. Sphere hollow, regular; short connecting processes between cells . . . *Coelastrum*

102a. Cells round . . . 103

102b. Cells oval . . . *Oocystis*

103a. Cells connected to center of colony by branching stalk . . . *Dictyosphaerium*

103b. No stalk connecting cells . . . *Sphaerocystis*

104a. Oval cells, enclosed in a somewhat spherical, often orange-colored matrix . . . *Botryococcus*

104b. Cells rounds, isolated or in colorless matrix . . . 105

105a. Adjoining cells with straight flat walls between their protoplasts . . . 106

105b. Adjoining cells with rounded walls between their protoplasts . . . 107

106a. Cells embedded in a common gelatinous matrix . . . *Palmella*

106b. No matrix or sheath outside of cell walls . . . *Phytoconis (Protococcus)*

107a. Cells loosely arranged in a large gelatinous matrix . . . *Tetraspora*

107b. Cells isolated or tightly grouped in a small colony . . . 108

108a. Cells located inside of protozoa . . . *Zoochlorella*

108b. Cells not inside of protozoa . . . 109

109a. Plastid filling 2/3 or less of cell . . . 110

109b. Plastid filling 3/4 or more of cell . . . *Chlorococcum*

110a. Cell diameter 2 μm or less; reproduction by cell division . . . *Nannochloris*

110b. Cell diameter 2.5 μm or more; reproduction by internal spores . . . *Chlorella*

111a. Cells attached end to end in an unbranched filament . . . 112

111b. Thallus branched, or more than one cell wide . . . 119

112a. Plastids in form of one or more marginal spiral ribbons . . . *Spirogyra*

112b. Plastids not in form of spiral ribbons . . . 113

113a. Filaments, when breaking, separating through middle of cells . . . 114

113b. Filaments, when breaking, separating irregularly or at ends of cells . . . 115

114a. Starch test positive; cell margin straight; one plastid, granular . . . *Microspora*

114b. Starch test negative; cell margin slightly bulging; several plastids . . . *Tribonema*

	Refer to Couplet No.
115a. Marginal indentations between cells . . . . *Desmidium*	
115b. No marginal indentations between cells	116
116a. Plastids, two per cell . . . . *Zygnema*	
116b. Plastid, one per cell (sometimes appearing numerous)	117
117a. Some cells with walls having transverse wrinkles near one end; plastid an irregular net . . . . *Oedogonium*	
117b. No apical wrinkles in wall; plastid not porous	118
118a. Plastid a flat or twisted axial ribbon . . . . *Mougeotia*	
118b. Plastid an arcuate marginal band . . . . *Ulothrix*	
119a. Thallus a flat plate of cells . . . . *Hildenbrandia*	
119b. Thallus otherwise	120
120a. Thallus a long tube without crosswalls . . . . *Vaucheria*	
120b. Thallus otherwise.	121
121a. Thallus a leathery strand with regularly spaced swellings and a continuous surface membrane of cells . . . . *Lemanea*	
121b. Thallus otherwise	122
122a. Filament unbranched . . . . *Schizomeris*	
122b. Filament branched	123
123a. Branches in whorls (clusters)	124
123b. Branches single or in pairs	126
124a. Thallus embedded in gelatinous matrix . . . . *Batrachospermum*	
124b. Thallus not embedded in gelatinous matrix	125
125a. Main filament one cell thick . . . . *Nitella*	
125b. Main filament three cells thick . . . . *Chara*	
126a. Most of filament surrounded by a layer of cells . . . . *Compsopogon*	
126b. Filament not surrounded by a layer of cells	127
127a. End cell of branches with a rounded or blunt-pointed tip	128
127b. End cell of branches with a sharp-pointed tip	130
128a. Plastids green; starch test positive	129
128b. Plastids red; starch test negative . . . . *Audouinella*	
129a. Some cells dense, swollen, dark green (spores); other cells light green, cylindric . . . . *Pithophora*	
129b. All cells essentially alike, light to medium green, cylindric . . . . *Cladophora*	
130a. Filaments embedded in gelatinous matrix	131
130b. Filaments not embedded in gelatinous matrix	132
131a. Cells of main filament much wider than even the basal cells of branches *Draparnaldia*	
131b. No abrupt change in width of cells from main filament to branches . . . *Chaetophora*	
132a. Branches very short, with no cross-walls . . . . *Rhizoclonium*	
132b. Branches long, with cross-walls	133
133a. Branches ending in an abrupt spine having a bulbous base . . . . *Bulbochaete*	
133b. Branches gradually reduced in width, ending in a long pointed cell, with or without color . . . . *Stigeoclonium*	

1007 E. Recent Changes in Names of Algae

Old Name	New Name
Aphanocapsa	*Anacystis*
Aphanothece	*Coccochloris*
Chamaesiphon	*Entophysalis*
Chantransia	*Audouinella*
Chlamydobotrys	*Pyrobotrys*
Chroococcus	*Anacystis*
Clathrocystis	*Anacystis*
Coelosphaerium	*Gomphosphaeria*
Encyonema	*Cymbella*
Gloeocapsa	*Anacystis*
Gloeothece	*Coccochloris*
Merismopedia	*Agmenellum*
Microcystis	*Anacystis*
Odontidium	*Diatoma*
Polycystis	*Anacystis*
Protococcus	*Phytoconis*
Sphaerella	*Haematococcus*
Synechococcus	*Coccochloris*

ALGAE COLOR PLATES

A through F

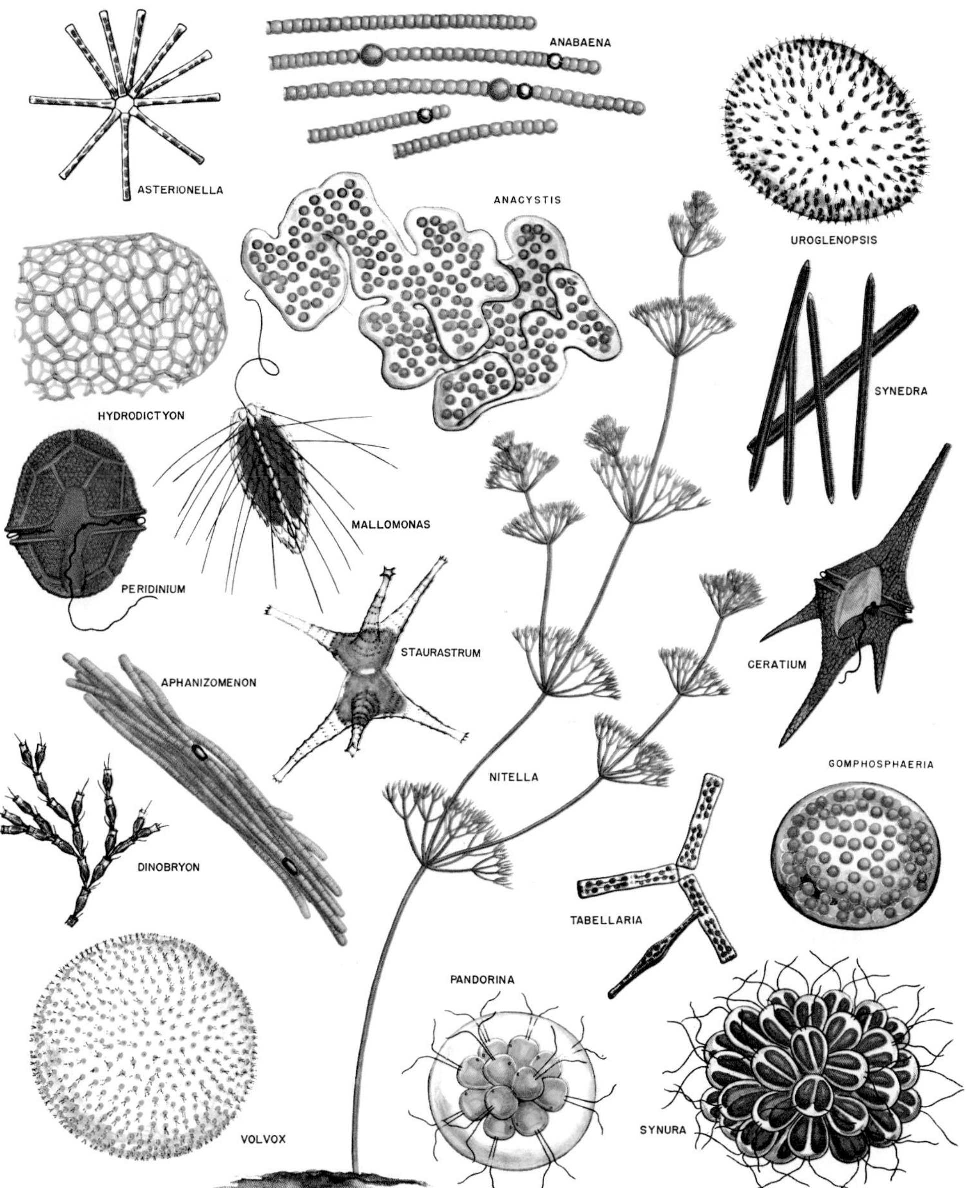

Plate A. Taste and odor algae

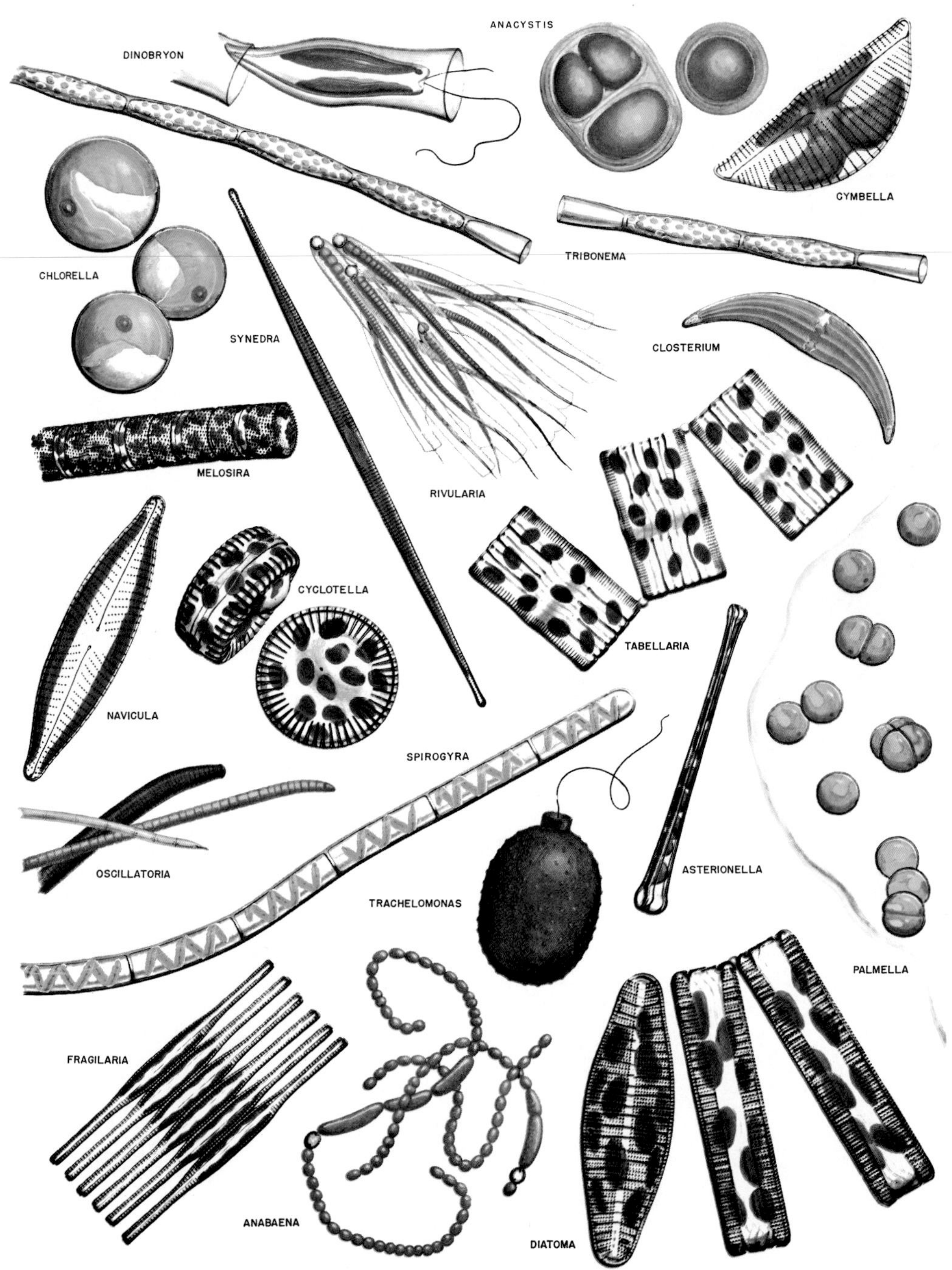

Plate B. Filter clogging algae

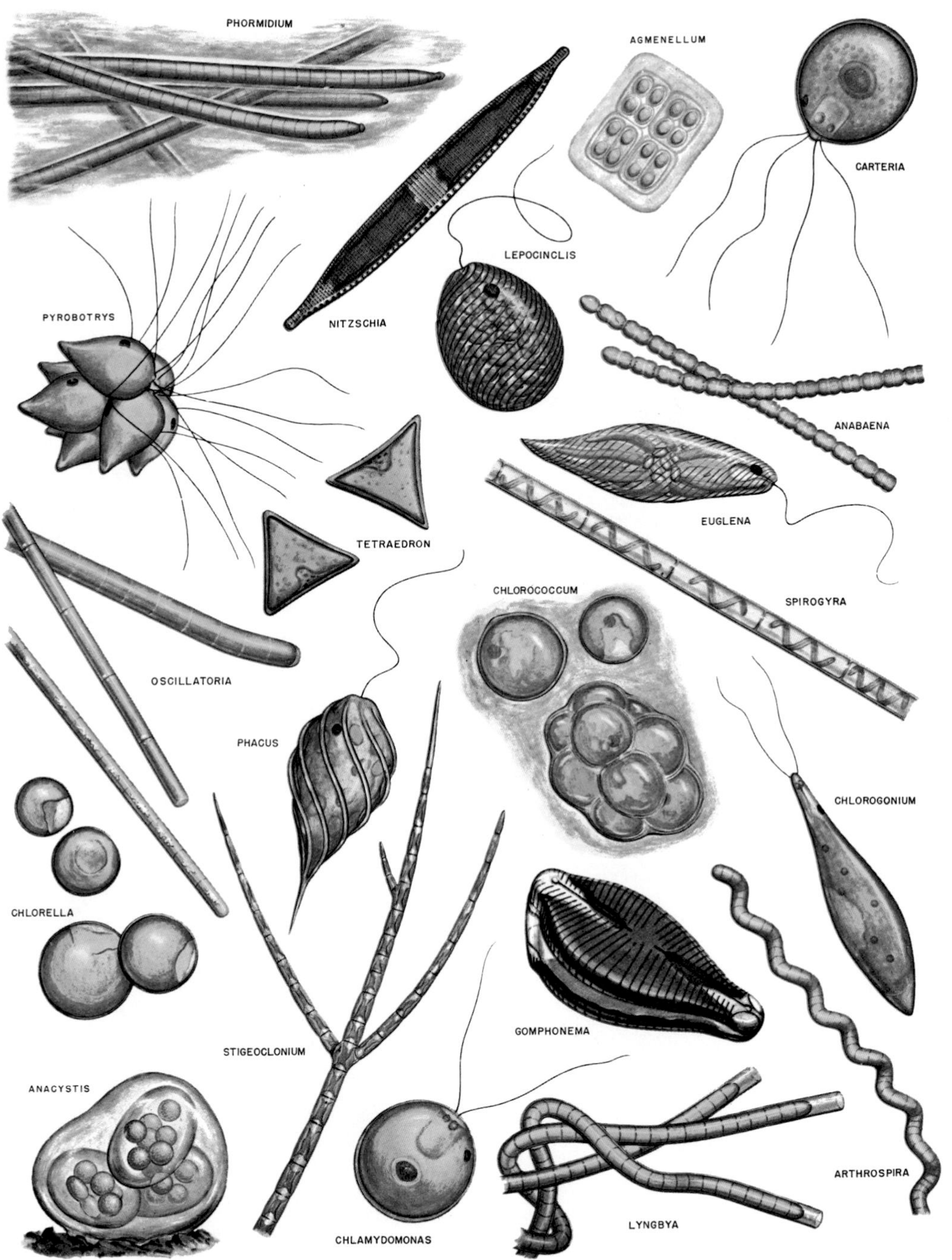

Plate C. Polluted water algae

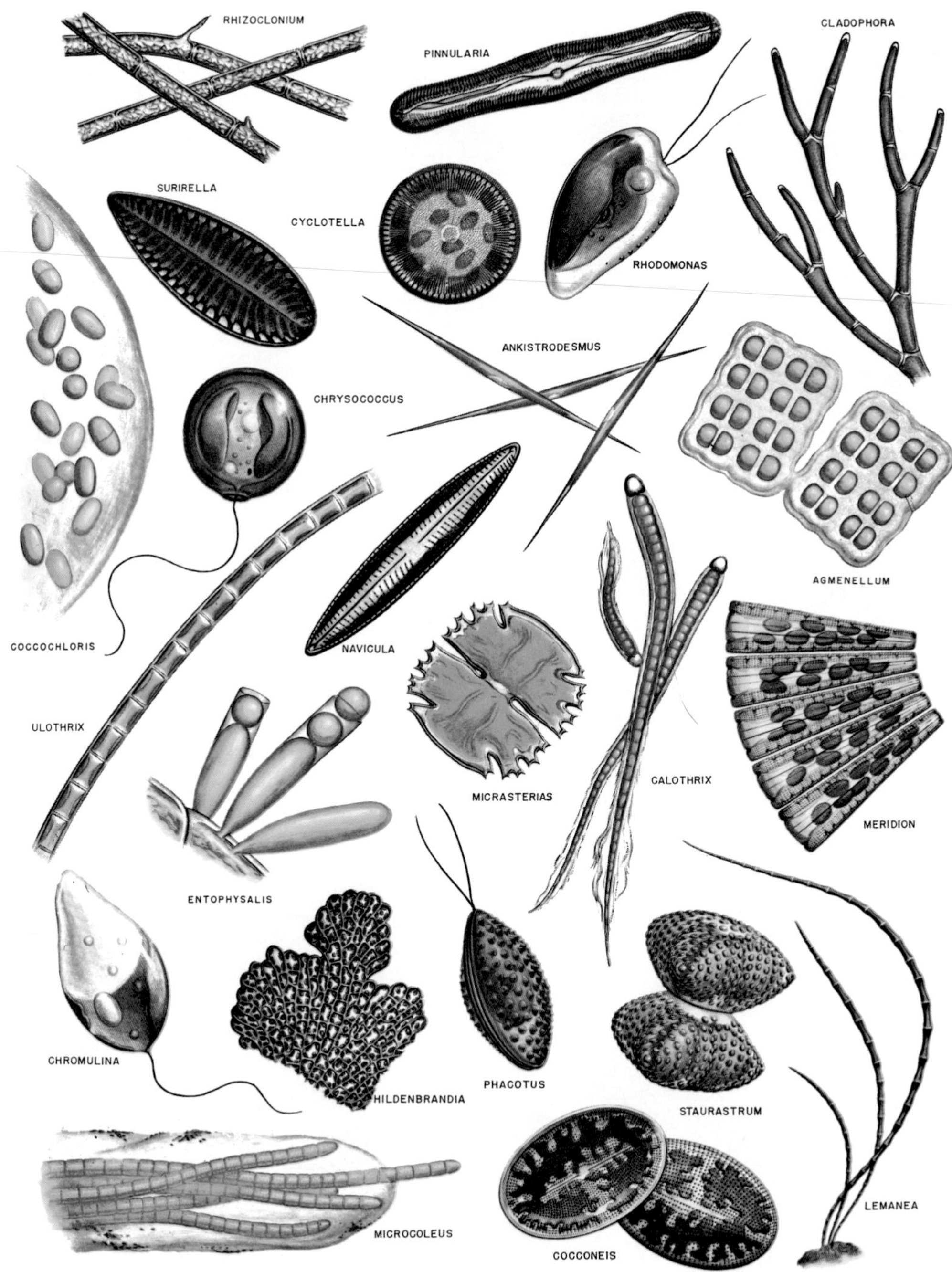
RHIZOCLONIUM
PINNULARIA
CLADOPHORA
SURIRELLA
CYCLOTELLA
RHODOMONAS
ANKISTRODESMUS
CHRYSOCOCCUS
AGMENELLUM
COCCOCHLORIS
NAVICULA
ULOTHRIX
MICRASTERIAS
CALOTHRIX
MERIDION
ENTOPHYSALIS
CHROMULINA
HILDENBRANDIA
PHACOTUS
STAURASTRUM
MICROCOLEUS
COCCONEIS
LEMANEA

Plate D. Clean water algae

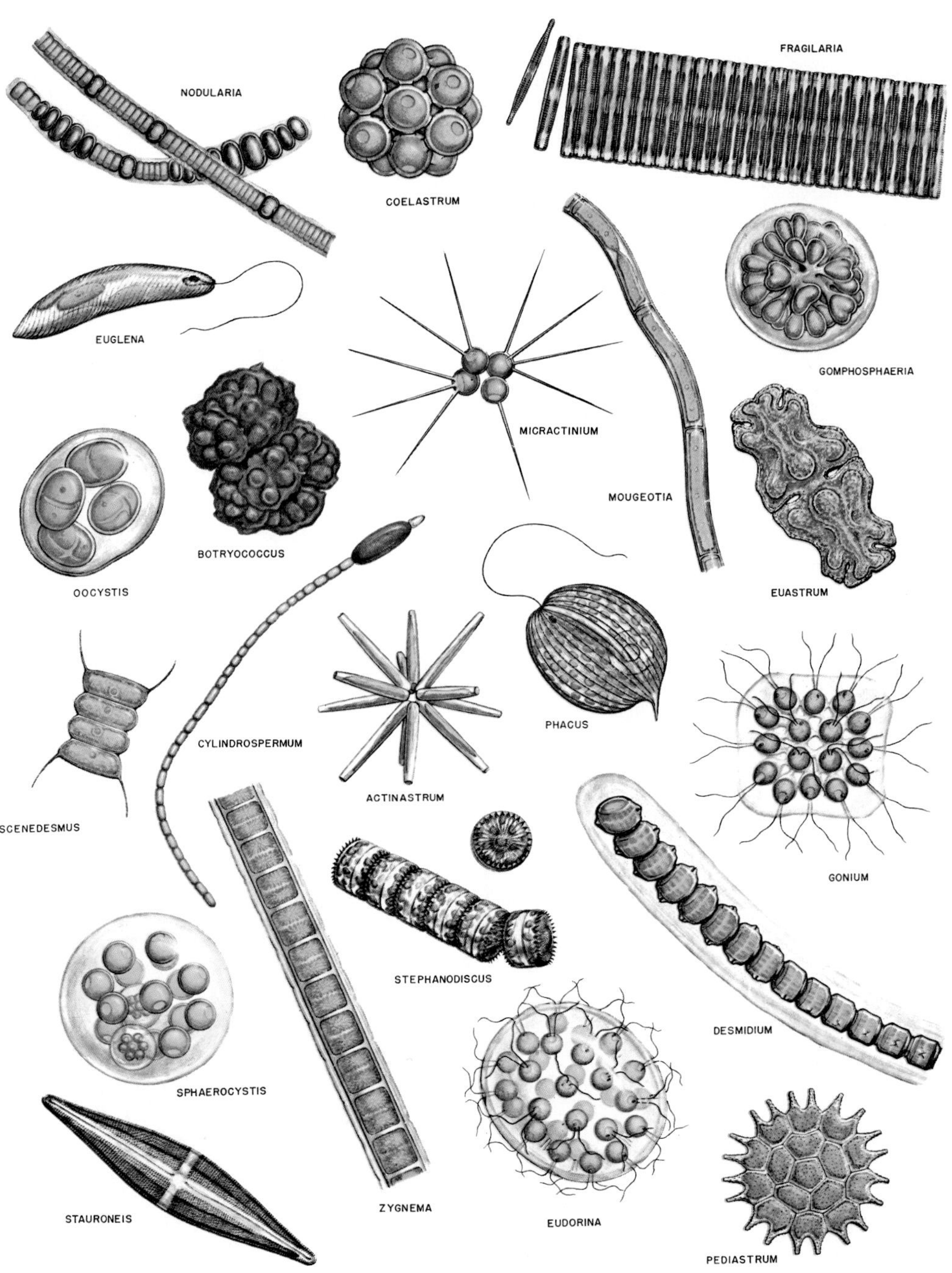

Plate E. Plankton and other surface water algae

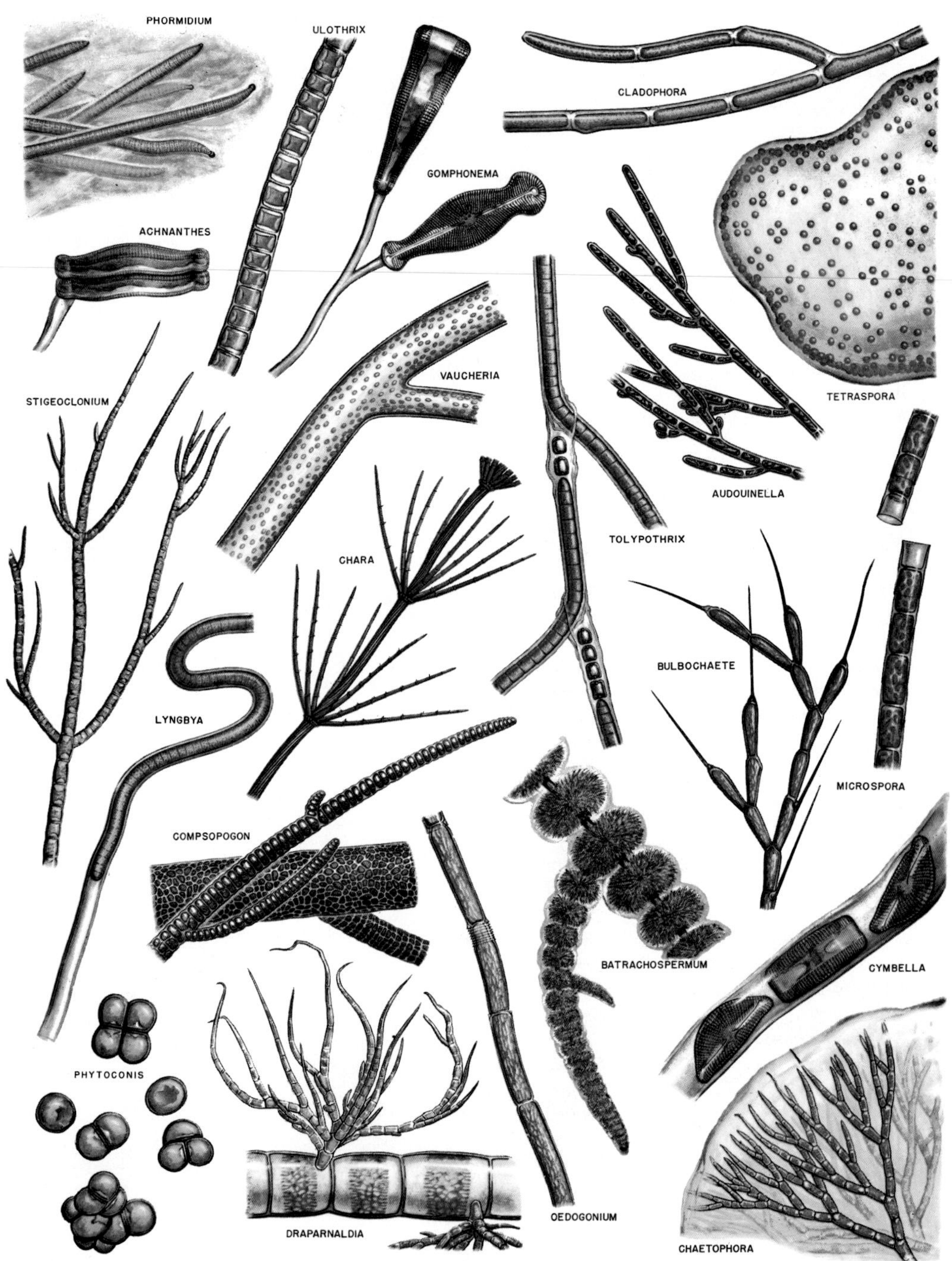

Plate F. Algae growing on reservoir walls

1007 F. Index to Illustrations

NOTE: Arabic numerals refer to black and white plate numbers, capital letters to color plates. See also "Recent Changes in Names of Algae," Section 1007 E, preceding. Family names are not generally included.

1007 G. Selected Taxonomic References

1. General, Introductory

JAQUES, H. E. 1947. Living Things: How To Know Them. William C. Brown Co., Dubuque, Ia.

MINER, R. W. 1950. Field Book of Seashore Life. G. P. Putnam's Sons, N.Y.

DAVIS, C. C. 1955. The Marine and Fresh-Water Plankton. Michigan State Univ. Press, East Lansing.

EDDY, S. & A. C. HODSON. 1961. Taxonomic Keys to the Common Animals of the North Central States, Exclusive of the Parasite Worms, Insects, and Birds. Burgess Publ. Co., Minneapolis, Minn.

HEDGPETH, J. & S. HINTON. 1961. Common Seashore Life of Southern California. Naturegraph Co., Healdsburg, Calif.

NEEDHAM, J. G. & P. R. NEEDHAM. 1962. A Guide to the Study of Fresh-Water Biology, 5th ed. Holden-Day Inc., San Francisco, Calif.

RICKETTS, E. F. & J. CALVIN. 1963. Between Pacific Tides, 3d ed. Revised by Hedgpeth. Stanford Univ. Press, Calif.

KLOTS, E. B. 1966. New Field Book of Freshwater Life. G. P. Putnam's Sons, N.Y.

PIMENTEL, R. A. 1967. Invertebrate Identification Manual. Reinhold Publ. Corp., N.Y.

REID, G. K. 1967. Pond Life. A Guide to Common Plants and Animals of North American Ponds and Lakes. Golden Press, N.Y.

2. General, Advanced

PENNAK, R. W. 1953. Fresh-Water Invertebrates of the United States. The Ronald Press, N.Y.

EDMONDSON, W. T., ed. 1959. Ward and Whipple's Fresh Water Biology, 2nd ed. John Wiley & Sons, N.Y.

BLAIR, W. F. et al. 1968. Vertebrates of the United States. McGraw-Hill, N.Y.

3. Algae, General

BRANDT, K. & C. APSTEIN. 1908. Nordisches Plankton (Botanisher Teil). Asher & Co., Amsterdam, reprinted in 1964.

SETCHELL, W. A. & N. L. GARDNER. 1919–1925. The Marine Algae of the Pacific Coast of North America. Univ. of California Publ. in Botany No. 8 (Parts 1, 2, 3). J. Cramer, Weinheim, Germany. Reprinted in 1967.

DAWSON, E. Y. 1946. Marine algae of the Pacific Coast of North America. *Mem. So. Calif. Acad. Sci.* 3:2.

WHIPPLE, G. C. 1948. The Microscopy of Drinking Water. John Wiley & Sons, N.Y.

SMITH, G. M. 1950. The Fresh-Water Algae of the United States. McGraw-Hill, N.Y.

TIFFANY, L. H. & M. E. BRITTON. 1952. The Algae of Illinois. Univ. of Chicago Press, Chicago. Ill.

PRESCOTT, G. W. 1954. How To Know the Fresh Water Algae. Wm. C. Brown Co., Dubuque, Ia.

FRITSCH, F. E. 1956. The Structure and Reproduction of the Algae. Vol. I: Chlorophyceae, Xanthophyceae, Crysophyceae, Bacillariophyceae, Cryptophyceae, Dinophyceae, Chloromonadineae, Euglenineae, and Colourless Flagellata. Cambridge Univ. Press, Cambridge, England.

TAYLOR, W. R. 1957. Marine Algae of the Northeastern Coast of North America, 2nd ed. Univ. of Michigan Press, Ann Arbor.

PALMER, C. M. 1959. Algae in Water Supplies. USPHS Publ. No. 657, Washington, D.C.

GRIFFITH, R. E. 1961. The Phytoplankton of Chesapeake Bay—An Illustrated Guide to the Genera. Chesapeake Biological Lab., College Park, Md., Contrib. No. 172.

PRESCOTT, G. W. 1962. Algae of the Western Great Lakes Area, rev. ed. Wm. C. Brown Co., Dubuque, Iowa.

FRITSCH, F. E. 1965. The Structure and Reproduction of the Algae. Vol. II: Phaeophyceae, Rhodophyceae, and Myxophyceae. Cambridge Univ. Press, Cambridge, England.

ROUND, F. E. 1965. The Biology of the Algae. Edward Arnold, Ltd., London.

DAWSON, E. Y. 1966. Marine Botany. Holt, Rinehart & Winston, N.Y.

PRESCOTT, G. W. 1968. The Algae: A Review. Houghton Mifflin, Boston, Mass.

WOOD, R. D. & J. LUTES. 1968. Guide to the Phytoplankton of Narragansett Bay, Rhode Island, rev. ed. Kingston Press, Kingston, R.I.

4. Blue-Green Algae

GEITLER, L. 1930. Cyanophyceae. In: Kryptogamenflora von Deutschland, Osterreich, und der Schweiz (L. Rabenhorst, ed.). Akad. Verlags., Leipsig. Reprinted in 1961.

HUBER-PESTALOZZI, G. 1938. Blue-Green Algae, Bacteria and Aquatic Fungi. *In* Die Binnengewasser. Part 1: Das Phytoplankton des Susswassers (A. Thienemann, ed.). E. Schweizerbart'sche Verlagsbuchhandlung, Stuttgart, Germany. Reprinted in 1962.

SMITH, G. W. 1950. The Fresh-Water Algae of the United States. McGraw-Hill, N.Y.

TIFFANY, L. H. & M. E. BRITTON. 1952. The Algae of Illinois. Univ. of Chicago Press, Chicago, Ill.

DROUET, F. & W. A. DAILY. 1956. Revision of the Coccoid Myxophyceae. Butler Univ. Bot. Studies XII. Indianapolis, Ind.

DESIKACHARY, T. V. 1959. Cyanophyta. Indian Counc. Agr. Res., New Delhi.

GEITLER, L. 1960. Schizophyzeen. *In* Encyclopedia of Plant Anatomy (W. Zimmermann and P. Ozeuda, eds.). Gebruder Borntraeger, Berlin, Vol. 6, Part 1.

HUMM, H. J. 1962. Key to the Genera of Marine Bluegreen Algae of Southeastern North America. Virginia Fish. Lab. Spec. Sci. Rep. No. 28.

PRESCOTT, G. W. 1962. Algae of the Western Great Lakes Area, rev. ed. Wm. C. Brown Co., Dubuque, Ia.

WELCH, H. 1964. An introduction to the blue-green algae, with a dichotomous key to all the genera. *Limnol. Soc. S. Afr. News Letter* 1:25.

DROUET, F. 1968. Revision of the classification of the Oscillatoriaceae. Philadelphia Acad. Natur. Sci. Monogr. 15.

5. Green Algae

COLLINS, F. S. 1909. The Green Algae of North America. *Tufts College Studies, Sci. Ser.* 2:79.

TIFFANY, L. H. 1937. Oedogoniales, Oedogoniaceae. *In* North American Flora (New York Botanical Gardens) Hafner Publ. Co., N.Y. 11(1):1.

SMITH, G. M. 1950. The Fresh-Water Algae of the United States. McGraw-Hill, N.Y.

TRANSEAU, E. N. 1951. The Zygnemataceae. Ohio State Univ. Press, Columbus.

TIFFANY, L. H. & M. E. BRITTON. 1952. The Algae of Illinois. Univ. of Chicago Press, Chicago. Ill.

RANDHAWA, M. S. 1959. Zygnemaceae. Indian Counc. Agr. Res., New Delhi.

HIRN, K. E. 1960. Monograph of the Oedogoniaceae. Hafner Publ. Co., N.Y.

PAL, B. P., B. C. KUNDU, U. S. SUNDARALINGAM & G. S. VENKATARAMAN. 1962. Charophyta. Indian Counc. Agr. Res., New Delhi.

PRESCOTT, G. W. 1962. Algae of the Western Great Lakes Area, rev. ed. Wm. C. Brown Co., Dubuque, Ia.

ISLAM, A. K. M. 1963. A revision of the genus *Stigeoclonium. Nova Hedwigia* (Supplement) 10:1.

SODERSTROM, J. 1963. Studies in Cladophora. Almquist Publ. Co., Uppsala, Sweden.

VAN DER HOEK, C. 1963. Revision of the European Species of Cladophora. Brill Publ. Co., Leiden, Netherlands.

RAMANATHAN, K. R. 1964. Ulotrichales. Indian Counc. Agr. Res., New Delhi.

WOOD, R. D. & K. IMAHARI. 1964. A Revision of the Characeae. Vols. I, II. Monograph and Iconograph. J. Cramer Publ. Co., Weinheim, Germany.

6. Flagellates

KOFOID, C. A. & O. SWEZY. 1921. The Free-Living Unarmored Dinoflagellata. Univ. of California Press, Berkeley.

SKVORTZOW, B. V. 1925. The euglenoid genus *Trachelomonas* Ehr. Systematic review. *Proc. Sungari River Sta.* 1:1.

DEFLANDRE, G. 1926. Monographie du genre Trachelomonas. Ehr. Nemours: Impremerie André Lesot.

HUBER-PESTALOZZI, G. 1938. Chrysophyceen. Farblose Flagellaten Heterokonten. *In* Die Binnengewasser. Vol. 16. Das Phytoplankton des Susswassers, Part 2 (A. Thienemann, ed.). E. Schweizerbart'sche Verlagsbuchhandlung, Stuttgart, Germany. Reprinted in 1962.

ALLEGRE, C. F. & T. L. JAHN. 1943. A survey of the Genus *Phacus* Dumardin. *Trans. Amer. Microsc. Soc.* 62:233.

GRAHAM, H. W. & N. BRONIKOVSKY. 1944. The Genus Ceratium in the Pacific and North Atlantic Oceans. Publ. No. 565, Carnegie Inst., Washington, D.C.

HUBER-PESTALOZZI, G. 1950. Chryptophyceen, Chloromonadinen, Peridineen. *In* Die Binnengewasser. Vol. 16. Das Phytoplankton des Susswassers, Part 3 (A. Thienemann, ed.). E. Schweizerbart'sche Verlagsbuchhandlung. Stuttgart, Germany. Reprinted in 1962.

SMITH, G. M. 1950. The Fresh-Water Algae of the United States. McGraw-Hill, N.Y.

TIFFANY, L. H. & M. E. BRITTON. 1952. The Algae of Illinois. Univ. of Chicago Press, Chicago, Ill.

GOJDICS, M. 1953. The Genus Euglena. Univ. of Wisconsin Press, Madison.

HUBER-PESTALOZZI, G. 1955. Euglenophyceen. *In* Die Binnengewasser. Vol. 16. Das Phytoplankton des Susswassers. Part 4 (A. Thienemann, ed.). E. Schweizerbart'sche Verlagsbuchhandlung, Stuttgart, Germany. Reprinted in 1962.

HUBER-PESTALOZZI, G. 1938. Chlorophyceae; Ordnung Volvocales. *In* Die Binnengewasser. Vol. 16. Part 5: Das Phytoplankton des Susswassers (A. Thienemann, ed.). E. Schweizerbart'sche Verlagsbuchhandlung, Stuttgart, Germany. Reprinted in 1962.

PRESCOTT, C. W. 1962. Algae of the Western Great Lakes Area, rev. ed. Wm. C. Brown Co., Dubuque, Ia.

7. Diatoms

CLEVE, P. T. 1894–1896. The Naviculoid Diatoms. Asher & Co., Amsterdam. Reprinted in 1965.

VAN HEURCH, H. 1896. A Treatise on the Diatomaceae. Weldon & Wesley, Ltd., Herts, England. Reprinted in 1962.

BOYER, C. S. 1916. The Diatomaceae of Philadelphia and Vicinity. Reproduced in Xerox by University Microfilms, Ann Arbor, Mich.

ELMORE, C. J. 1922. The Diatoms of Nebraska. Univ. of Nebraska Ser., Lincoln, 21(1–4).

BOYER, C. S. 1927. Synopsis of North American Diatomaceae. *Proc. Acad. Natur. Sci. Philadelphia,* 79 Supplement, Part 2: 229–583.

GRAN, H. H. & E. C. ANGST. 1930. Plankton Diatoms of Puget Sound. Univ. of Washington, Seattle.

HUSTEDT, F. 1930. The Diatoms. *In* Kryptogamenflora von Deutschland, Osterreich und der Schweiz (L. Rabenhorst, ed.). Geest and Partig K-G, Leipzig, Germany, Parts 1, 2, and 3.

HUSTEDT, F. 1930. Bacillariophyta (Diatomeae). *In* Die Susswasserflora Mitteleuropas (A. Pascher, ed.). Vol. 10. Reproduced in Xerox by University Microfilms, Ann Arbor, Mich.

HUBER-PESTALOZZI, G. 1942. The Diatoms. *In* Die Binnengewasser. Part I: Das Phytoplankton des Susswassers (A. Thienemann, ed.). E. Schweizerbart'sche Verlagsbuchhandlung, Stuttgart, Germany. Reprinted in 1962.

CUPP, E. E. 1943. The Marine Plankton Diatoms of the West Coast of North America. Reproduced in Xerox by University Microfilms, Ann Arbor, Mich.

TIFFANY, L. H. & M. E. BRITTON. 1952. The Algae of Illinois. Univ. of Chicago Press, Chicago, Ill.

CLEVE-EULER, A. 1953. The Diatoms of Sweden and Finland. Almquest & Wiksells, Stockholm, Sweden.

HUSTEDT, F. 1955. Marine littoral diatoms, Beaufort, North Carolina. *Duke Univ. Mar. Sta. Bull.* 6:5.

VAN DER WERFF, A. & H. HULS. 1957–1966. Diatomenenflora van Nederlands, Parts 1–8. Published by the author, De Hoef, Netherlands.

MULFORD, R. A. 1962. Diatoms from Virginia Tidal Waters. Virginia Inst. Mar. Sci. Spec. Sci. Rep. No. 30, Gloucester Point, Va.

HENDY, N. I. 1964. An Introductory Account of the Smaller Algae of British Coastal Waters. Part V: Bacillariophyceae (Diatoms). Fishery Invest. Ser. IV, Her Majesty's Stationery Office, London.

CHOLNOKY, B. J. 1966. Diatomaceae, Vol. I. Krebs, Weinheim, Germany.

PATRICK, R. & C. W. REIMER. 1966. The Diatoms of the United States, Vol. I. Philadelphia Acad. Natur. Sci. Monogr. No. 13, Philadelphia, Pa.

WEBER, C. I. 1966. A Guide to the Common Diatoms at Water Pollution Surveillance Stations. U.S. Dept. Interior, FWPCA, Cincinnati, Ohio

8. Higher Plants, Introductory

EYLES, D. E. & J. L. ROBERTSON. 1963. Guide and Key to the Aquatic Plants of the Southeastern United States. U.S. Fish & Wildlife Serv. Circ. 158.

HOTCHKISS, N. 1967. Underwater and Floating-Leaved Plants of the United States and Canada. U.S. Fish & Wildlife Serv. Resour. Publ. No. 44.

WELDON, L. W. 1969. Common Aquatic Weeds. U.S. Dep. Agr., Agr. Handbook No. 352.

9. Higher Plants, Advanced

MUENSCHER, W. C. 1944. Aquatic Plants of the United States. Comstock Publ. Co., Ithaca, N.Y.

OGDEN, E. C. 1953. Key to the North American species of *Potamogeton*. *Circ. N.Y. State Mus.* 31:1.

FASSETT, N. C. 1960. A Manual of Aquatic Plants (with a revision appendix by E. C. Ogden). Univ. of Wisconsin Press, Madison.

SCULTHORPE, C. D. 1967. The Biology of Aquatic Vascular Plants. St. Martin's Press, N.Y.

10. General Invertebrates, Introductory

BUCHSBAUM, R. M. & L. J. MILNE. 1960. The Lower Animals, Living Invertebrates of the World. Chanticleer Press, Garden City, N.Y.

HICKMAN, C. P. 1967. Biology of the Invertebrates. C. V. Mosby, St. Louis, Mo.

11. General Invertebrates, Advanced

PRATT, H. S. 1951. A Manual of the Common Invertebrate Animals Exclusive of Insects. The Blakiston Co., Philadelphia, Pa.

PENNAK, R. W. 1953. Fresh-Water Invertebrates of the United States. The Ronald Press Co., N.Y.

LIGHT, S. F., R. I. SMITH, F. A. PITELKA, D. P. ABBOTT & F. M. WEESNER. 1961. Intertidal Invertebrates of the Central California Coast. Univ. of California Press, Berkeley.

12. Protozoa

JAHN, T. L. & F. F. JAHN. 1949. How To Know the Protozoa. Wm. C. Brown Co., Dubuque, Ia.

KUDO, R. 1950. Protozoology. Charles C. Thomas, Springfield, Ill.

CORLISS, J. O. 1961. Ciliated Protozoa: Characterization, Classification, and Guide to the Literature. Pergamon Press, N.Y.

CALAWAY, W. T. & J. B. LACKEY. 1962. Waste Treatment Protozoa, Flagellata. Univ. of Florida, Fla. Eng. Ser. No. 3.

13. Sponges and Bryozoa

DELAUBENEELS, M. W. 1953. Guide to the Sponges of Eastern North America. Univ. of Miami, Coral Gables, Fla.

ROGICK, M. D. 1960. Ectoprocta. *McGraw-Hill Encyc. Sci. Technol.* 5:7.

ROGICK, M. D. 1960. Bryozoa. *McGraw-Hill Encyc. Sci. Technol.* 2:354.

BUSHNELL, J. H., JR. 1965. On the taxonomy and distribution of the freshwater Ectoprocta in Michigan (Parts I–III). *Trans. Amer. Microsc. Soc.* 84:231; 339; 529.

PENNEY, J. T. & A. A. RACEK. 1968. Comprehensive Revision of a Worldwide Collection of Freshwater Sponges (Porifera: Spongillidae). USNM Bull. 272.

14. Rotifers

VOIGT, M. 1957. Rotataria—Die Radertiere. Mitteleuropas. Borntraeger, Berlin, Vols. I and II.

DONNER, J. 1966. Rotifers. Warne, London & N.Y.

15. Roundworms (Nemathelminthes)

CHITWOOD, B. G. 1951. North American marine nematodes. *Tex. J. Sci.* 3:617.

GOODEY, T. 1963. Soil and Freshwater Nematodes (revised by J. B. Goodey). John Wiley & Sons, N.Y.

WIESER, W. & B. E. HOPPER. 1967. Marine nematodes of the cast coast of North America. 1. *Fla. Bull. Mus. Comp. Zool.* 135:239.

16. Segmented Worms (Annelids)

SPERBER, C. 1948. A Taxonomical Study of the Naididae. Zool. Bidrag Fran Uppsala, Sweden. Band. 28.

BRINKHURST, R. O. 1964–1966. Studies on the North American aquatic Oligochaeta: I. Naididae and Opistocystidae; II. Tubificidae; III. Lumbriculidae and additional notes and records of other families. *Proc. Acad. Natur. Sci. Philadelphia* 116:195; 117:117; 118:1.

BRINKHURST, R. O. 1966. Detection and assessment of water pollution using oligochaete worms. *Water Sewage Works* 113:398, 438. Parts I and II.

17. Flatworms (Platyhelminthes)

EDMONDSON, W. T., ed. 1959. Ward and Whipple's Fresh Water Biology, 2nd ed. John Wiley & Sons, N.Y.

18. Crustaceans

HOBBS, H. H., JR. 1942. The crayfishes of Florida. *Univ. Fla. Publ. Biol. Sci. Ser.* 3:1.

HUBRICHT, L. & J. G. MACKIN. 1949. The freshwater isopods of the genus *Lirceus* (Asellota, Asellidae). *Amer. Midland Natur.* 42:334.

BOUSFIELD, E. L. 1958. Freshwater amphipod crustaceans of glaciated North America. *Can. Field Natur.* 72:55.

WATERMAN, T. H. 1960. The Physiology of Crustacea. Vol. I. Metabolism and Growth. Academic Press, N.Y.

19. Insects, General and Introductory

LUTZ, P. E. 1927. Field Book of Insects. G. P. Putnam's Sons, N.Y.

CHU, H. F. 1949. How To Know the Immature Insects. Wm. C. Brown Co., Dubuque, Ia.

USINGER, R. L. 1956. Aquatic Insects of California, with Keys to North American Genera and California Species. Univ. of California Press, Berkeley.

20. Stoneflies (Plecoptera)

NEEDHAM, J. G. & P. W. CLAASEN. 1925. A Monograph of the Plecoptera or Stoneflies of America North of Mexico, Vol. 2. Thomas Say Foundation, Lafayette, Ind.

FRISON, T. H. 1935. The stoneflies, or Plecoptera, of Illinois. *Bull. Ill. Natur. Hist. Surv.* 20:281.

FRISON, T. H. 1942. Studies of North American Plecoptera, with special reference to the fauna of Illinois. *Bull. Ill. Natur. Hist. Surv.* 22:235.

JEWETT, S. G. 1960. The stoneflies (Plecoptera) of California. *Bull. Calif. Insect Surv.* 6:125.

21. Mayflies (Ephemeroptera)

NEEDHAM, J. G., J. R. TRAVER & Y. HSU. 1935. The Biology of Mayflies. Comstock Publ. Co., Ithaca, N.Y.

BERNER, L. 1950. The Mayflies of Florida. Univ. of Florida Press, Gainesville.

BURKS, B. D. 1953. The mayflies, or Ephemeroptera, of Illinois. *Bull. Ill. Natur. Hist. Surv.* 26:1.

BERNER, L. 1959. A tabular summary of the biology of North American mayfly nymphs (Ephemeroptera). *Bull. Fla. State Mus.* (Gainesville) 4:1.

EDMONDS, G. F., R. K. ALLEN & W. L. PETERS. 1963. An annotated key to the nymphs of the families and subfamilies of mayflies (Ephemeroptera). *Univ. Utah Biol. Ser.* 13:1.

22. Dragonflies and Damselflies (Odonata)

NEEDHAM, J. G. & M. J. WESTFALL, JR. 1955. A Manual of the Dragonflies of North America, Including the Greater Antilles and the Provinces of the Mexican Border. Univ. of California Press, Berkeley.

WALKER, E. M. 1958. The Odonata of Canada and Alaska, Vols. I and II. Univ. of Toronto Press, Toronto, Canada.

23. Hellgrammites and Relatives

PENNAK, R. W. 1953. Fresh-Water Invertebrates of the United States. The Ronald Press, N.Y.

EDMONDSON, W. T., ed. 1959. Ward and Whipple's Fresh Water Biology, 2nd ed. John Wiley & Sons, N.Y.

24. Caddisflies (Trichoptera)

ROSS, H. H. 1944. The caddis flies, or Trichoptera, of Illinois. *Bull. Ill. Natur. Hist. Surv.* 23:1.

FLINT, O. S., JR. 1962. Taxonomy and biology of nearctic limnephilid larvae (Trichoptera), with special reference to species found in eastern United States. *Entomol. Amer.* 40:1.

25. Two-Winged Flies (Diptera)

JOHANNSEN, O. A. 1933, 1935, 1936, 1937. Memoirs of the Cornell University Agricultural Experiment Station. Parts I–IV. (Part V by L. C. Thomsen.) Reproduced in 1969 by Ent. Reprint Specialists, East Lansing, Mich.

ROBACK, S. S. 1957. The Immature Tendipedids of the Philadelphia Area. Philadelphia Acad. Natur. Sci. Monogr. No. 9.

BECK, W. M., JR. & E. C. BECK. 1966. Chironomidae (Diptera) of Florida. I: Pentaneurini (Tanypodinae). *Bull. Fla. State Mus.* 10:305.

MASON, W. T. 1968. An Introduction to the Identification of Chironomid Larvae. FWPCA, Washington, D.C.

SNODDY, E. L. 1969. Simuliidae of Alabama. Ala. Agr. Exp. Sta. Bull. 390.

26. Beetles (Coleoptera)

JAQUES, H. E. 1951. How To Know the Beetles. Wm. C. Brown, Dubuque, Ia.

YOUNG, F. N. 1954. Water Beetles of Florida. Univ. of Florida Press, Gainesville.

DILLON, E. S. & L. S. DILLON. 1961. Manual of Common Beetles of Eastern North America. Harper and Row, N.Y.

27. True Bugs (Hemiptera)

PENNAK, R. W. 1953. Fresh-Water Invertebrates of the United States. The Ronald Press, N.Y.

EDMONDSON, W. T., ed. 1959. Ward and Whipple's Fresh Water Biology, 2nd ed. John Wiley & Sons, N.Y.

28. Mollusks, General and Introductory

PENNAK, R. W. 1953. Fresh-Water Invertebrates of the United States. The Ronald Press, N.Y.

EDMONDSON, W. T., ed. 1959. Ward and Whipple's Fresh Water Biology, 2nd ed. John Wiley & Sons, N.Y.

29. Snails (Gastropoda)

BAKER, F. C. 1928. The Fresh-Water Mollusca of Wisconsin. Part I. Gastropoda. Wisconsin Geol. Natur. Hist. Surv. Bull. No. 70.

KEEN, A. M. & J. C. PEARSON. 1952. Illustrated Key to West North American Gastropod Genera. Stanford Univ. Press, Stanford, Calif.

WALTER, H. J. & J. B. BURCH. 1957. Key to the Genera of Freshwater Gastropods (Snails and Limpets) Occurring in Michigan. Mus. Zool., Univ. Mich. Circ. No. 3.

LEONARD, A. B. 1959. Handbook of Gastropods in Kansas. Univ. Kans. Mus. Natur. Hist. Misc. Publ. No. 20.

LAROCQUE, A. 1968. Pleistocene mollusca of Ohio. *Ohio Geol. Surv. Bull.* 62:357.

30. Bivalves (Pelecypoda)

WALKER, B. 1918. A Synopsis of the Classification of Freshwater Mollusca of North America, North of Mexico, and a Catalogue of the More Recently Described Species, with Notes. Museum of Zoology, Univ. of Michigan Misc. Publ. No. 6.

BAKER, F. C. 1928. The Fresh-Water Mollusca of Wisconsin. Part II. Pelecypoda. Wis. Geol. Natur. Hist. Surv. Bull. No. 70.

KEEN, A. M. & D. FRIZZELL. 1946. Illustrated Key to West North American Pelecepod Genera. Stanford Univ. Press, Stanford, Calif.

MURRAY, H. D. & A. B. LEONARD. 1962. Handbook of Unionid Mussels in Kansas. Univ. Kans. Mus. Natur. Hist. Misc. Publ. No. 28.

HERRINGTON, H. B. 1962. A Revision of the Sphaeriidae of North America (Mollusca: Pelecypoda). Univ. Mich. Mus. Zool. Misc. Publ. No. 118.

NEEL, J. K. & W. R. ALLEN. 1963. The mussel fauna of the Upper Cumberland Basin before its impoundment. *Malacologia* 1:427.

HEARD, W. H. & J. BURCH. 1966. Keys to the Genera of Freshwater Pelecypods of Michigan. Museum of Zoology, Univ. Mich. Circ. No. 4.

LA ROCQUE, A. 1967. Pleistocene mollusca of Ohio. Ohio Geol. Surv. Bull. No. 62.

31. Echinoderms

COE, W. R. 1912. Echinoderms of Connecticut, Conn. State Geol. Natur. Hist. Surv. Bull. No. 19.

MINER, R. W. 1950. Field Book of Seashore Life. G. P. Putnam's Sons, N.Y.

HARVEY, E. B. 1956. The American Arbacia and Other Sea Urchins. Princeton Univ. Press, Princeton, N.J.

32. Fishes

WALFORD, L. A. 1937. Marine Game Fishes of the Pacific Coast from Alaska to the Equator. Univ. Calif. Press, Berkeley.

BREDER, C. M. 1948. Fieldbook of Marine Fishes. G. P. Putnam's Sons, N.Y.

EDDY, S. 1957. How To Know the Freshwater Fishes. Wm. C. Brown Co., Dubuque, Ia.

TRAUTMAN, M. B. 1957. The Fishes of Ohio. Ohio State Univ. Press, Columbus.

BAILEY, R. M., et al. 1960. A List of Common and Scientific Names of Fishes from the United States and Canada. Amer. Fish. Soc. Spec. Publ. No. 2.

PERLMUTTER, A. 1961. Guide to Marine Fishes. N.Y. Univ. Press, N.Y.

HUBBS, C. L. & K. F. LAGLER. 1964. Fishes of the Great Lakes Region. Univ. Mich. Press, Ann Arbor.

CROSS, F. B. 1967. Handbook of Fishes of Kansas. Univ. Kans. Mus. Natur. Hist., Lawrence, Kans.

BLAIR, W. F. & G. A. MOORE. 1968. Fishes. *In* Vertebrates of the United States, 2d ed. McGraw-Hill, N.Y.

33. Amphibians

BISHOP, S. C. 1943. Handbook of Salamanders. Comstock Publ. Co., Ithaca, N.Y.

CONANT, R. 1958. Field Guide to the Reptiles and Amphibians of Eastern North America. Houghton-Mifflin, N.Y.

BRANDON, R. A. 1961. A comparison of the larvae of five northeastern species of *Ambystoma* (Amphibia, Caudata). *Copeia* 4:377.

BRANDON, R. A. 1964. An annotated and illustrated key to multistage larvae of Ohio salamanders. *Ohio J. Sci.* 64:252.

BLAIR, W. F. et al. 1968. Vertebrates of the United States. McGraw-Hill, N.Y.

34. Bacteria and Fungi

BREED, R. S., E. G. D. MURRAY & N. R. SMITH. 1957. Bergey's Manual of Determinative Bacteriology, 7th ed. Williams & Wilkins, Baltimore, Md.

COOKE, W. B. 1963. A Laboratory Guide to Fungi in Polluted Waters. USPHS, Cincinnati, Ohio.

JOHNSON, T. W., JR. 1968. Saprobic Marine Fungi. *In* The Fungi, Vol. III (G. C. Ainsworth & A. S. Sussman, eds.). Academic Press, N.Y.

INDEX

A

B

C

D

I

J

K

L

M

N

O

Q

R

S

U

V

W

X

Y

Z

ABBREVIATIONS

The following symbols and abbreviations are used throughout this book:

Abbreviation	*Referent*
A or amp	ampere(s)
AC	alternating current
ACS	American Chemical Society
APHA	American Public Health Association
ASTM	American Society for Testing and Materials
AWWA	American Water Works Association
BOD	biochemical oxygen demand
C	degree(s) Celsius
c	count(s)
CCE	carbon chloroform extract
Ci	curie(s)
cm, cm^2, cm^3	centimeter(s), square centimeter(s), cubic centimeter(s)
COD	chemical oxygen demand
conc	concentrated
cpm	counts per minute
cps	counts per second
DC	direct current
diam	diameter
DO	dissolved oxygen
dpm	disintegrations per minute
EC	effective concentration
ft	foot (feet)
ft-c	foot-candle(s)
g	gram(s)
g	gravity, unit acceleration of
gal	gallon(s)
gph	gallons per hour
gpm	gallons per minute
hr	hour
ID	inside diameter
in.	inch(es)
JTU	Jackson candle turbidity unit(s)
KeV	kiloelectron volt(s)
kg	kilogram(s)
kPa	kilopascal
L	liter(s)
lb	pound(s)
M	mole or molar
m, m^2, m^3	meter(s), square meter(s), cubic meter(s)
MCL	maximum contaminant level
me	milliequivalent(s)
MeV	megaelectron volt(s)
mg	milligram(s)
min	minute(s)
mL	milliliter(s)
mm, mm^2, mm^3	millimeter(s), square millimeter(s), cubic millimeter(s)
mol wt	molecular weight
MPN	most probable number
mV	millivolt(s)
μA	microampere(s)
μCi	microcurie(s)
μg	microgram(s)
μL	microliter(s)
μm	micrometer(s)
N	normal
NBS	National Bureau of Standards
nCi	nanocurie(s)
ng	nanogram(s)
nm	nanometer(s)
No.	number
NTU	nephelometric turbidity unit(s)
OD	outside diameter
oz	ounce(s)